THE OCCUPATIONAL ERGONOMICS HANDBOOK

THE OCCUPATIONAL ERGONOMICS HANDBOOK

EDITED BY
Waldemar Karwowski
University of Louisville
Louisville, Kentucky
William S. Marras
The Ohio State University
Columbus, Ohio

CRC Press
Boca Raton London New York Washington, D.C.

Library of Congress Cataloging-in-Publication Data

The Occupational ergonomics handbook / edited by Waldemar Karwowski and William S. Marras.
p. cm.
Includes bibliographical references and index.
ISBN 0-8493-2641-9 (alk. paper)
1. Human engineering – Handbooks, manuals, etc. 2. Industrial hygiene – Handbooks, manuals, etc. I. Karwowski, Waldemar, 1953– . II. Marras, William S. (William Steven), 1952– .
TA166.0258 1998
620.8′2–DC21 98-22336
CIP

International Standard Book Number 0-8493-2641-9
Library of Congress Card Number 98-22336
Printed in the United States of America 1 2 3 4 5 6 7 8 9 0
Printed on acid-free paper

The Editors

Waldemar Karwowski, Ph.D., P.E., C.P.E., is Professor of Industrial Engineering and Director of the Center for Industrial Ergonomics at the University of Louisville, Kentucky. He holds an M.S. (1978) in production management from Technical University of Wroclaw, Poland, and a Ph.D. (1982) in Industrial Engineering from Texas Tech University. He is a Board Certified Professional Ergonomist (CPE). His research, teaching, and consulting activities focus on prevention of low back injury and cumulative trauma disorders, human and safety aspects of advanced manufacturing, fuzzy sets and systems, and theoretical aspects of ergonomics.

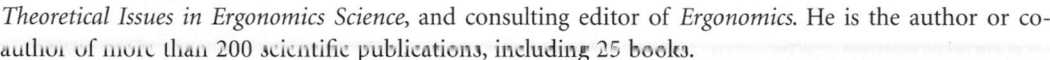

Dr. Karwowski currently serves as Secretary-General of the International Ergonomics Association. He is editor of many international journals, including *Human Factors and Ergonomics in Manufacturing*, and *Theoretical Issues in Ergonomics Science*, and consulting editor of *Ergonomics*. He is the author or co-author of more than 200 scientific publications, including 25 books.

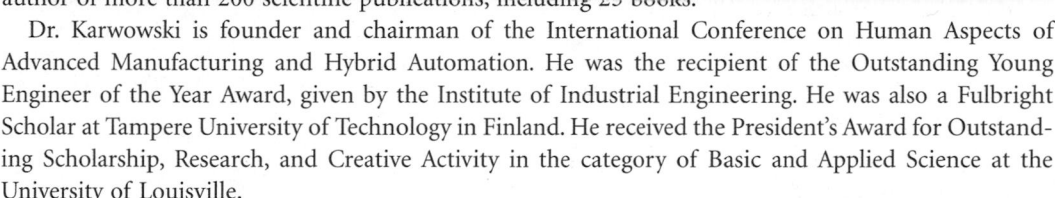

Dr. Karwowski is founder and chairman of the International Conference on Human Aspects of Advanced Manufacturing and Hybrid Automation. He was the recipient of the Outstanding Young Engineer of the Year Award, given by the Institute of Industrial Engineering. He was also a Fulbright Scholar at Tampere University of Technology in Finland. He received the President's Award for Outstanding Scholarship, Research, and Creative Activity in the category of Basic and Applied Science at the University of Louisville.

William S. Marras, Ph.D., C.P.E., holds the Honda Endowed Chair in Transportation in the Department of Industrial, Welding, and Systems Engineering at the Ohio State University, Columbus. He is also the director of the biodynamics laboratory and holds appointments in the departments of physical medicine and biomedical engineering. Professor Marras is also the co-director of the Ohio State University Institute for Ergonomics.

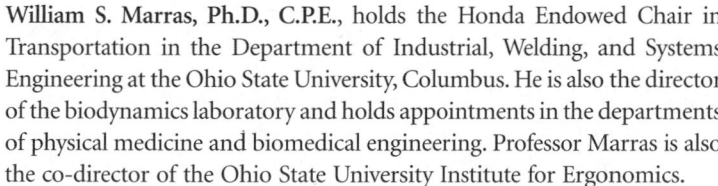

Dr. Marras received his Ph.D. in bioengineering and ergonomics from Wayne State University in Detroit, Michigan. His research centers around biomechanical epidemiologic studies, laboratory biomechanic studies, mathematical modeling, and clinical studies of the back and wrist.

His findings have been published in more than 100 refereed journal articles and 12 book chapters. He also holds two patents, including one for the lumbar motion monitor (LMM). His work has also attracted national and international recognition. He has won the prestigious Swedish Volvo Award for Low Back Pain Research and Austria's Vienna Award for Physical Medicine.

Contributors

Elsayed Abdel-Moty
Department of Industrial Engineering
University of Miami
Coral Gables, FL

Thomas J. Albin
3M Center
St. Paul, MN

David C. Alexander
Auburn Engineers, Inc.
Auburn, AL

W. Gary Allread
The Ohio State University
Columbus, OH

Charles K. Anderson
Advanced Ergonomics, Inc.
Dallas, TX

Gunnar B.J. Andersson
Department of Orthopedic Surgery
St. Luke's Medical Center
Chicago, IL

Michele Battié
Department of Orthopaedics
University of Washington
Seattle, WA

Susan A.H. Benysh
Purdue University
West Lafayette, IN

Patricia Bertsche
Occupational Safety & Health
Directorate of Technical Support
Washington, DC

Ram Bishu
IMSE Department
University of Nebraska
Lincoln, NE

Regina Brauchler
University of Stuttgart
Hohenheim, GERMANY

Heiner Bubb
Technische Universität Munchen
Garching, GERMANY

James R. Buck
Department of Industrial Engineering
University of Iowa
Iowa City, IA

Peter M. Budnick
University of Utah
Salt Lake City, UT

J. Bugajska
Central Institute for Labour Protection
Warsaw, POLAND

A. Kim Burton
Department of Clinical Biomechanics
Huddersfield Polytechnic
Huddersfield, ENGLAND

Ahmet Cakir
Ergonomics Institute for Social and Occupational Sciences
Berlin, GERMANY

Pascale Carayon
École des Mines de Nancy
Nancy, FRANCE

John G. Casali
Industrial/Systems Engineering
Virginia Tech
Blacksburg, VA

Keith Case
Loughborough University
Leicestershire, ENGLAND

Don B. Chaffin
Center for Ergonomics
University of Michigan
Ann Arbor, MI

Klaus Christoffersen
The Ohio State University
Columbus, OH

David R. Clark
IMSE Department
Kettering University
Flint, MI

David J. Cochran
Department of Industrial Engineering
University of Nebraska
Lincoln, NE

E.N. Corlett
Institute for Occupational Ergonomics
University of Nottingham
Nottingham
ENGLAND

Brechtje J. Daams
Delft University of Technology
Delft, THE NETHERLANDS

Marvin J. Dainoff
Psychology Department
Miami University
Oxford, OH

Patrick G. Dempsey
Liberty Mutual Center for Safety and Health
Hopkinton, MA

Marjolein Douwes
TNO Prevention & Health
Leiden, THE NETHERLANDS

Colin G. Drury
State University of New York at Buffalo
Department of Industrial Engineering
Buffalo, NY

J. Dul
TNO Prevention & Health
Leiden, THE NETHERLANDS

Bradley Evanoff
School of Medicine
Washington University
St. Louis, MO

Susan Evans
Sue Evans & Associates, Inc.
Ann Arbor, MI

Chris Forsythe
Statistics and Human Factors
Sandia National Labs
Albuquerque, NM

Martin T. Freer
Loughborough University
Leicestershire, ENGLAND

Andris Freivalds
The Pennsylvania State University
University Park, PA

Paul Gaddie
University of Louisville
Louisville, KY

Sean Gallagher
National Institute for Occupational Safety and Health
Pittsburgh, PA

Robert L. Getty
Lockheed Martin Tactical Aircraft Systems
Ft. Worth, TX

Anand Gramopadhye
Department of Industrial Engineering
Clemson University
Clemson, SC

Katharyn A. Grant
Robert A. Taft Laboratories
National Institute for Safety and Health
Cincinnati, OH

Antonio Grieco
Istituto di Medicina del Lavoro
Universita di Milano
Milan, ITALY

Eric Grose
Statistics and Human Factors
Sandia National Laboratories
Albuquerque, NM

Thomas Hales
National Institute for Occupational Safety and Health

M. Susan Hallbeck
University of Nebraska
Lincoln, NE

Peter A. Hancock
Human Factors Research Laboratory
University of Minnesota
Minneapolis, MN

Martin G. Helander
Department of Mechanical Engineering
Linkoping Institute of Technology
Linkoping, SWEDEN

Hal W. Hendrick
University of Southern California
Los Angeles, CA

Simon M. Hsiang
Liberty Mutual Research Center for Safety and Health
Hopkinton, MA

Sheik N. Imrhan
Industrial Engineering Department
University of Texas
Arlington, TX

Toni Ivergard
Institute of Working Life
Ostersund, SWEDEN

Dieter W. Jahns
SynerTech Associates
Bellingham, WA

Renliu Jang
University of Louisville
Louisville, KY

Jari Järvinen
Motorola
Delray Beach, FL

Iwona Jastrzebska-Fraczek
Technische Universität Munchen
Garching
GERMANY

Benta Rona Jensen
National Institute of Occupational Health
Copenhagen, DENMARK

Roger C. Jensen
Ergonomics Services Group
UES, Inc.
Dayton, OH

Steven L. Johnson
Bell Engineering Center
University of Arkansas
Fayetteville, AR

Bradley S. Joseph
Ford Motor Company
Dearborn, MI

Marilyn Joyce
Joyce Institute
Arthur D. Little, Inc.
Cambridge, MA

Waldemar Karwowski
University of Louisville
Department of Industrial Engineering
Center for Ergonomics
Louisville, KY

Karol Kerns
The MITRE Corporation
McLean, VA

Glenda L. Key
Key Functional Assessments
Minneapolis, MN

Tarek M. Khalil
Department of Industrial Engineering
University of Miami
Coral Gables, FL

Åsa Kilbom
National Institute for Working Life
Solna, SWEDEN

Jung-Yong Kim
Department of Industrial Engineering
Hanyang University
Ansan, KOREA

Barry Kirwan
Air Traffic Management Development Center
National Air Traffic Services Ltd.
Bournemouth Airport
and University of Birmingham
Christchurch, Dorset
ENGLAND

Udo Konradt
Ruhr-Universität Bochum
Bochum, GERMANY

Stephan Konz
Department of IMSE
Kansas State University
Manhattan, KS

Danuta Koradecka
Central Institute for Labour Protection
Warsaw, POLAND

Richard J. Koubek
Wright State University
Dayton, OH

Karl H.E. Kroemer
Industrial Ergonomics Laboratory
Virginia Tech
Blacksburg, VA

Kurt Landau
Institut für Arbeitswissenschaft
Darmstadt, GERMANY

Kenneth A. Laughery
Rice University
Houston, TX

Steven A. Lavender
Department of Orthopedic Surgery
St. Luke's Medical Center
Chicago, IL

Wook Gee Lee
University of Louisville
Louisville, KY

Mark R. Lehto
Purdue University
West Lafayette, IN

Sheau-Farn Liang
Purdue University
West Lafayette, IN

Soo-Yee Lim
National Institute for Occupational Safety and Health
Washington, DC

William E. Lischeid
Travelers Insurance Company
Hartford, CT

Veikko Louhevaara
Institute of Occupational Health
and the University of Kuopio
FINLAND

David R. Lovvoll
Rice University
Houston, TX

Hongzheng Lu
Lucent Technologies
Eatontown, NJ

Holger Luczak
Institute of Industrial Engineering and Ergonomics
Aachen, GERMANY

Barbara Majonica
Ruhr-Universität Bochum
Bochum, GERMANY

Chris J. Main
Department of Behavioural Medicine
Hope Hospital
Salford, ENGLAND

Richard W. Marklin
Department of Mechanical and Industrial Engineering
Marquette University
Milwaukee, WI

William S. Marras
ISE Department
The Ohio State University
Columbus, OH

Markku Mattila
Tampere University of Technology
FINLAND

Stuart M. McGill
University of Waterloo
Waterloo, Ontario
CANADA

C. Elaine McCoy
Ohio University
Athens, OH

Raymond W. McGorry
Liberty Mutual Research Center for Safety and Health
Hopkinton, MA

Donald R. McIntyre
Interlogics
Hillsborough, NC

David Meister
Human Factors Consultant
San Diego, CA

Mathilde C. Miedema
TNO Prevention & Health
Leiden
THE NETHERLANDS

Dwight P. Miller
Sandia National Laboratories
Albuquerque, NM

Giovanni Molteni
Istituto di Medicina del Lavoro
Universita di Milano
Milan, ITALY

J. Moore
Department of Preventive Medicine
Medical College of Wisconsin
Milwaukee, WI

J. Steven Moore
University of Texas
Health Center
Tyler, TX

Stephen J. Morrissey
Portland Field Office
State of Oregon
OSHA
Portland, OR

A. Muralidhar
University of Nebraska
Lincoln, NE

Mitsuo Nagamachi
Kure National Institute of Technology
Hiroshima, JAPAN

Robert W. Norman
Department of Kinesiology
University of Waterloo
Waterloo, Ontario, CANADA

Kageyu Noro
Ergonomics Department
Waseda University
Mikashima, JAPAN

Ewa Nowak
Department of Ergonomics Research
Institute of Industrial Design
Warsaw, POLAND

Judith Orasanu
NASA
Ames Research Center
Palo Alto, CA

Gary B. Orr
US Department of Labor
OSHA
Washington, DC

Mohamad Parnianpour
Department of ISE
The Ohio State University
Columbus, OH

Nils F. Petersson
National Institute for Working Life
Solna, SWEDEN

Malcolm H. Pope
Department of Orthopaedic Surgery
Iowa Spine Research Center
University of Iowa
Iowa City, IA

J. Mark Porter
Department of Design and Technology
Loughborough University
Leicestershire
ENGLAND

Vern Putz-Anderson
Applied Psychology and Ergonomics
National Institute for Occupational Safety and Health
Cincinnati, OH

Peter M. Quesada
Department of Mechanical Engineering
University of Louisville
Louisville, KY

Robert G. Radwin
Department of Industrial Engineering
University of Wisconsin
Madison, WI

Mansour Rahimi
University of Southern California
Industrial and Systems Engineering
Los Angeles, CA

David Rempel
University of California
San Francisco, CA

Valerie J. Rice
U.S. Army Research Institute of Environmental Medicine
Occupational Physiology Division
Natick, MA

Richard G. Ried
The Ohio State University
Columbus, OH

Stephen N. Robinovitch
San Francisco General Hospital
San Francisco, CA

Gary S. Robinson
Industrial/Systems Engineering
Virginia Tech
Blacksburg, VA

Walter Rohmert
Institut für Arbeitswissenschaft
Darmstadt, GERMANY

Hubert L. Rosomoff
Comprehensive Pain and Rehabilitation Center
University of Miami
Coral Gables, FL

Renee Steele-Rosomoff
Comprehensive Pain and Rehabilitation Center
University of Miami
Coral Gables, FL

David L. Roy
Travelers Insurance Company
Hartford, CT

Jorma Saari
University of Waterloo
Waterloo, Ontario, CANADA

Steven L. Sauter
Taft Laboratories
National Institute for Occupational Safety and Health
Cincinnati, OH

Christopher Schlick
Institute of Industrial Engineering and Ergonomics
Aachen, GERMANY

Scott P. Schneider
Ergonomics Program Director
Center to Protect Workers' Rights
Washington, DC

Joseph Sharit
University of Miami
Coral Gables, FL

Aboulfazl Shirazi-Adl
École Polytechnique
Montreal, Quebec
CANADA

Juraj Sinay
Technika Universita Kosice
Kosice, SLOVAKIA

Gisela Sjøgaard
National Institute of Occupational Health
Copenhagen, DENMARK

Philip J. Smith
The Ohio State University
Columbus, OH

Carolyn M. Sommerich
Department of Industrial Engineering
North Carolina State University
Raleigh, NC

Johannes Springer
Institute of Industrial Engineering and Ergonomics
Aachen, GERMANY

Neville A. Stanton
Department of Psychology
University of Southampton
Highfield, Southampton
ENGLAND

Terry L. Stentz
University of Nebraska
Lincoln, NE

Brian L. Stonecipher
University of Nebraska
Lincoln, NE

Leon M. Straker
Curtin University of Technology
Shenton Park, AUSTRALIA

Carol Stuart-Buttle
Stuart-Buttle Ergonomics
Philadelphia, PA

Naomi G. Swanson
Taft Laboratories
National Institute for Occupational Safety and Health
Cincinnati, OH

Donald I. Tepas
Ergonomics Laboratory
Department of Psychology
University of Connecticut
Storrs, CT

Jatin Thaker
Clemson University
Clemson, SC

Steven W. Thompson
United Parcel Service
Atlanta, GA

Ioannis Vasmatzidis
iNautix Technologies, Inc.
Jersey City, NJ

Eira Viikari-Juntura
Department of Physiology
Musculoskeletal Research Unit
Finnish Institute of Occupational Health
Topeliuksenkatu, FINLAND

Mika Vilkki
Tampere University of Technology
Tampere, FINLAND

Donald E. Wasserman
D.E. Wasserman, Inc.
Cincinnati, OH

Thomas R. Waters
National Institute for Occupational Safety and Health
Cincinnati, OH

Richard Wells
Department of Kinesiology
University of Waterloo
Waterloo, Ontario, CANADA

David G. Wilder
University of Iowa
Iowa Spine Research Center
Iowa City, IA

Bengt Å. Willén
Department of Mechanical Engineering
Division of Industrial Ergonomics
Linkoping Institute of Technology
Linkoping, SWEDEN

Michael S. Wogalter
North Carolina State University
Raleigh, NC

David D. Woods
The Ohio State University
ISE Department
Columbus, OH

Thomas Y. Yen
Department of Industrial Engineering
University of Wisconsin
Madison, WI

Mark S. Young
Department of Psychology
University of Southampton
Southampton, ENGLAND

Stephen L. Young
Liberty Mutual Research Center for Safety and Health
Hopkinton, MA

Bernhard Zimolong
Ruhr-Universität Bochum
Bochum, GERMANY

Contents

SECTION II Fundamentals of Work Analysis

SECTION III Cognitive Environment Issues

PART III MUSCULOSKELETAL DISORDERS — ENGINEERING FACTORS

SECTION I Disorders of the Extremities

SECTION II Low Back Disorders

PART IV ADMINISTRATIVE CONTROLS

SECTION I Ergonomics Surveillance

SECTION II Medical Management Prevention

PART V ORGANIZATIONAL DESIGN

SECTION I Ergonomics Quality and Cost-Benefit Issues

SECTION II Ergonomics Processes

PART VI ENVIRONMENTAL ISSUES IN ERGONOMICS

PART VII ERGONOMICS AND THE WORKING ENVIRONMENT

SECTION I The Office Environment

Preface

Although many books, and indeed handbooks, have been published in the area of human factors and ergonomics, there has been no single volume that focuses on the needs of the industrial practitioner and, at the same time, covers a broad scope of occupational problems. With this Handbook, we have set out to provide a comprehensive source of applied ergonomics knowledge useful to a large number of ergonomics practitioners who strive to improve product and process quality, worker health and safety, and productivity in a variety of industries and businesses. Development of this Handbook was also motivated by the quest to facilitate wider acceptance of ergonomics as an effective methodology for work-system design aimed at improving the overall quality of life for millions of workers with a variety of needs and expectations.

These writings will help to properly assess and implement ergonomic solutions in the workplace. It has often been observed that ergonomic solutions are not always as effective as planned. This is not because they do not work, but because they are often improperly implemented. This Handbook compiles the knowledge of more than 150 world-renowned experts in a single volume to help ensure the successful application of ergonomic principles in almost any environment.

As the manufacturing and service industries become more competitive, it is becoming more important for businesses to minimize expenditures and maximize quality. Herein lies the true value of ergonomics. Through the careful inclusion of ergonomic processes, described in this Handbook, not only will employees' health care costs come under control, but the quality of products will improve, as will the quality of life in the workplace for the worker. In this respect, ergonomics, when applied correctly, is a win–win proposition.

Given the rapid expansion of scientific knowledge about human abilities, limitations, and other characteristics relevant to the design, testing, and evaluation of machines, industrial facilities, service systems, tasks, jobs, tools, and working environments, we have invited international experts to present up-to-date information within their domains of expertise. We have also structured the Handbook to include both the generic fundamentals of ergonomic design principles and the specific application of topics such as development of engineering design measures aimed at eliminating or reducing job-related risk factors, methods for identifying worker populations affected by adverse working conditions, early medical intervention efforts, assessment of the economics of ergonomics programs, and methods for optimizing manufacturing processes with respect to the users' perceptual and cognitive abilities, and task reliability. This Handbook is rare in that it has attempted to emphasize the role of both physical ergonomics and cognitive issues in a systems-oriented approach to controlling ergonomic problems.

The above information is presented in the seven main parts of the Handbook: Background, Foundation of Ergonomics Knowledge, Musculoskeletal Disorders: Engineering Factors, Administrative Controls, Organizational Design, Environmental Issues in Ergonomics, and Ergonomics and the Working Environment.

We trust that this Handbook will serve a broad segment of industrial practitioners, including industrial and manufacturing engineers; managers; plant supervisors and ergonomics professionals; students and researchers from academia, business, and government; human factors and safety specialists; physical therapists; cognitive and work psychologists; sociologists; and human–computer communications specialists.

Finally, we would like to express our sincere appreciation to Norm Stanton, Bob Stern, and Gail Renard for their wisdom in carrying this project through each phase of the publishing process, and to Mrs. Laura Abell for her logistical assistance.

Waldemar Karwowski
University of Louisville
Louisville, Kentucky

WIlliam S. Marras
The Ohio State University
Columbus, Ohio

October 1998

Part I
Background

1

Ergonomics: An International Perspective

Hal W. Hendrick
University of Southern California

1.1 Introduction

In comparison with other disciplines, the science and practice of ergonomics, or human factors as it is also known in North America, is very young. Ergonomics does, however, have at least several distinct roots going back to the early 1900s. For example, in the United Kingdom, we can note the classic work of the Industrial Fatigue Research Board between World Wars I and II. These studies greatly contributed to our understanding of environmental effects on human work performance. In the United States, the development of "scientific management" by Frederick W. Taylor and "industrial psychology" by Hugo Munsterberg and others are among the factors that have contributed to the formation of ergonomics as a distinct discipline. Russia has distinguished histories in psychology, mathematics, and engineering that have helped shape the field of ergonomics. Similar precursors of modern ergonomics can be found in a number of other countries (Hendrick, 1993).

Although its roots go back to the early 1900s, ergonomics, as an identifiable profession, is in only its sixth decade of existence. Since World War II, we have seen the field of ergonomics develop from a few

0-8493-2641-9/99/$0.00+$.50

hundred individuals, working on military systems in several industrialized countries, to over 25,000 professional ergonomists working in both industrialized and industrially developing countries throughout the world. These persons, in turn, are supported by thousands of scientists from diverse fields who help develop ergonomically useful data, methods, and related technology. Professional ergonomists now work on a wide variety of systems, ranging from simple hand tools to highly complex equipment, software, and environments. Professor Brian Shackel, the International Ergonomics Association (IEA) historian, has noted the international developmental focus of ergonomics as follows: In the 1950s, military ergonomics; in the 1960s, industrial ergonomics; in the 1970s, ergonomics of consumer goods and services; and in the 1980s, computer ergonomics. The 1990s are proving to be the era of both macro- and cognitive ergonomics, with a major focus on their application to industrial systems. In the 1990s, we also have seen the further maturation of ergonomics into a distinct, stand-alone discipline (Hendrick, 1993).

1.2 The IEA and Its Member Societies

One of the early indications of the development of a clearly definable field of science and practice is the formation of professional societies. The first professional society formed in the field of ergonomics was the Ergonomics Research Society in the United Kingdom in 1949. This society later was renamed The Ergonomics Society, which is how we know it today (IEA, 1996).

The IEA

The idea of forming an international society in ergonomics was first formally put forward at a technical seminar of the European Productivity Agency (EPA) in Leiden, Holland, in 1957. (The EPA, a body working under the auspices of the Organization for European Economic Cooperation, had established a working party on ergonomics in 1955.) The idea was further developed at the meeting of the EPA Steering Committee in Paris, France, in 1958. At a meeting of the Steering Committee in Zurich, Switzerland, in 1959 the name "International Ergonomics Association" was formally adapted. Later in 1959, at an EPA Steering Committee Meeting in Oxford, England, the bylaws drawn up by Etienne Grandjean, Secretary of the Steering Committee, were approved. The first IEA Triennial Congress was held in Stockholm in 1961 (IEA, 1996)

From this beginning, the IEA has steadily grown. Prior to the 1980s, the primary activity of the IEA was to sponsor the triennial congresses. Structurally, the IEA consisted of a Council, comprised of representatives of the federated member societies, and an Executive Committee consisting of the three elected officers: president, secretary general, and treasurer. By 1980, it was clear that ergonomics was entering a phase of rapid growth and expansion internationally. It also was clear that there was a growing need to increase the opportunities for international coordination and exchange of technical and educational information. To meet this need, the IEA began to greatly expand its activities. During the past decade, the IEA developed an expanded committee structure to plan and manage these increased activities; and a series of specialized technical groups which, among other activities, participate in organizing international conferences, symposia, and workshops in their specialty areas.

The IEA now consists of 33 federated and 2 affiliated societies representing over 45 countries. Approximately half of these societies became members between 1991 and 1995. Still other societies, primarily from industrially developing countries, are in the process of forming. Put simply, ergonomics is a young but rapidly expanding science and profession.

The IEA Federated Societies

Most of the 35 IEA federated and affiliated societies now fall into three size groupings: Approximately two thirds have from 30 to 250 members each. Another 20% each have from 450 to 650 members. Three societies fall between 900 and 2000 members; these are the Ergonomics Society (933), the Nordic

Ergonomics Society (a federation of the four Nordic country societies, 1497), and the Japan Ergonomics Research Society (1862). The Human Factors and Ergonomics Society in the United States has approximately 4000 members. All told, the IEA federated and affiliated societies are comprised of approximately 14,500 full and associate members (IEA Treasurer's Report, 1995). Including other membership categories (e.g., student members and affiliates) would increase this number to approximately 18,000.

In a recent survey among 25 IEA federated societies (Brown, Noy, and Robertson, 1995; Helander, 1996), the societies estimated that an average of 61.5% of the eligible ergonomists belong to their society. This would indicate that there are over 20,000 ergonomists within the countries represented by the IEA federated societies. Adding the ergonomists living in countries that do not yet have established IEA federated societies would raise this figure to over 25,000.

1.3 The Technology of Ergonomics

A second, and perhaps the most important factor that defines a profession as a unique, stand-alone discipline is its unique *technology.* To the extent that a given professional area, though empirical science, has developed a technology and, as a practice, applies that technology to some end, it distinguishes itself as a unique discipline. Over the five decades of its formal existence, ergonomics has progressively developed and applied its unique technology. That technology may be thought of as *human–system interface technology* (Hendrick, 1991). As a *science,* ergonomics is concerned with developing knowledge about human performance capabilities, limitations, and other characteristics as they relate to the design of the interfaces between people and other system components. As a *practice,* ergonomics concerns the application of human–system interface technology to the *analysis, design,* and *evaluation* of systems to enhance safety, health, comfort, effectiveness, and quality of life. At present, this unique technology has at least four identifiable major components: Human–machine interface technology or *hardware ergonomics,* human–environment interface technology or *environmental ergonomics,* human–software interface technology or *cognitive ergonomics,* and human–organization interface technology or *macroergonomics.* A brief overview of the historical development, present state, and apparent future of each of these four technical areas follows (Hendrick, 1993).

1.4 Ergonomics Technology and Its Application

Hardware Ergonomics or Human–Machine Interface Technology

Originally known as man–machine interface technology, this technology represents the focus of ergonomics during the first three decades of our profession. It primarily concerns the study of human physical and perceptual characteristics, and the application of that knowledge to the analysis, design, and evaluation of controls, displays, and workspace arrangements. It remains today as the single largest aspect of our profession and is likely to remain so in the future.

In the United States, this aspect of ergonomics began with attempts of engineering psychologists to explain why so many military aviation accidents were being attributed to "pilot error" during the early years of World War II. In general, these psychologists found that the real cause was not "human error" but "engineering error." In short, design engineers had failed to adequately take into account "human factors" (i.e., human performance capabilities and limitations) in designing the aircraft human–machine interfaces, and thus, had inadvertently incorporated error-producing characteristics into their designs. The outcome was the birth of the new field of "human factors" in North America. In Europe, several countries (most notably, Germany) were also concerned with human performance capabilities as applied to design (e.g., design of gun sights; white versus red aircraft cockpit lighting). However, it was after the war that the development and application of human–machine interface technology really began in earnest. Faced with the tremendous task of rebuilding their industries, Europe and Japan became concerned with developing and applying a science of work or "ergonomics."

During the 1950s, and 1960s, the primary applications of this rapidly emerging human–machine interface technology were to factory work systems and to transportation systems, with particular emphasis on aircraft design. The primary funding in the U.S. and other countries for the development of human–machine interface technology came from military research and development budgets. Beginning in the late 1960s, we experienced an expansion of hardware ergonomics to many other types of systems. In part, this expansion was a result of the development of computers and related automation of offices and factories. In the future, world competition is likely to be a prime motivator of greater attention to hardware ergonomics in both workstation design (for greater worker effectiveness and quality of production) and product design (as differences in the adequacy of ergonomic design are likely to be the major discriminators between economically successful and unsuccessful products).

Environmental Ergonomics or Human–Environment Interface Technology

The second ergonomics technology is concerned with human capabilities and limitations with respect to the demands imposed by various environmental modalities (e.g., light, heat, noise, vibration, etc.). It is applied to the design of human environments to minimize environmental stress on human performance — including comfort, health, and safety — and to enhance productivity.

One of the main roots of environmental ergonomics can be traced back to the work of the British Industrial Fatigue Research Board in the early 1900s, which carried through the 1930s. In terms of contemporary environmental ergonomics, significant research was begun in the 1960s, at such places as Aston, Loughborough, Wales, and Birmingham universities in the United Kingdom, and various Department of Defense, NASA, and university (e.g., Cornell) research units in the U.S. During this same period, parallel human–environment technology research, development and application was ongoing in other western European countries, Japan, the U.S.S.R., Australia, and elsewhere.

During the last several decades, the importance of understanding the relation of humans to both their natural and man-made environments has gained increasing focus internationally, and a related ergonomics technology is continuously developing. There has even been the development of an ecological approach to human performance modeling as well as to classic ergonomics methods such as task analysis (Vicente, 1990).

Recently, there has been a growing trend toward more stringent ergonomics-related health and safety legislation in both Europe and North America. A progressively increasing international awareness of the importance of ecological issues to human health and effectiveness will ensure the continued expansion of human–environment technology research and application throughout the world.

Cognitive Ergonomics or Human–Software Interface Technology

The third technology is a relatively new development in our profession, having come into being with the development of the silicon chip in the 1960s, and the modern computer revolution which followed. Because this relatively new technology is primarily concerned with how people conceptualize and process information, it is often referred to as "cognitive ergonomics." The major application of this technology is to the design or modify system software to enhance its usability. In the United States, the development of software/cognitive ergonomics has resulted in an increase of approximately 25% in the number of ergonomists, as reflected by the increase in HFES memberships. For example, over 900 of the approximately 4000 HFES members belong to the Society's Computer Systems Technical Group (HFES, 1996). A similar growth appears to have happened in many other countries.

Because of the continuing growth in software technology and application, and the growing realization of the importance of human–software interface technology to effective software design, cognitive ergonomics will continue to be a strong growth area within our profession. The related interest in the technology of artificial intelligence (AI) and its high potential for expert systems development will further fuel the fire under this part of ergonomics development and growth.

Macroergonomics or Human–Organization Interface Technology

Macroergonomics is the newest part of our profession. The central focus of the first three technologies has been the individual operator and operator teams or subsystems. Thus, the primary application of these technologies has been at the microergonomic level. In contrast, because it deals with the overall structure of the work system as it interfaces with the system's people and technology, the human–organization aspect of the human–system interface tends to be *macro* in its focus; hence, it is referred to as "macroergonomics." *Conceptually,* it is a top-down, sociotechnical systems approach to organizational and work systems design, and the design of related human–machine, human–environment, and human–software interfaces (Hendrick, 1986a,b, 1987, 1991).

For many years, organizational factors occasionally have been considered in ergonomics research and practice; but macroergonomics, as a formally recognized area of ergonomics, grew out of a study of future ergonomics needs by the Human Factors Society which was completed in 1980 (Hendrick, 1991). This study noted such factors as (a) rapid changes in technology, (b) changing value systems, (c) demographic shifts (a greying work force), (d) increasing world competition, and the failure of traditional microergonomics to (e) improve overall *system* administrative productivity, (f) reach potentially achievable production quality, and (g) system safety, health, and related quality of worklife goals as underlying drivers of the need for developing and applying a human–organization interface technology.

Since its formal inception a decade ago, macroergonomics has experienced rapid growth throughout the world. Research over the past decade has shown organizational design and management variables to be critical to effective work system design. Anecdotal reports have shown the potential for effective macroergonomic design and follow-through microergonomic design of systems to greatly improve productivity, safety, health, and quality of worklife. For example, in a series of industrial studies in both Japan and the U.S., Nagamachi and Imada (1992), using macroergonomic approaches, achieved reductions in both industrial injuries and motor vehicle accidents of from 76% to over 90%. A field study at L.L. Bean Corporation in the U.S. used a true macroergonomic approach in a total quality management (TQM) change effort in both their production and distribution units, and achieved over 70% reduction in lost time injuries (Rooney, Morency, and Herrick (1993). Ongoing macroergonomic research, such as that on information systems by HUSAT at Loughborough University in the U.K. and on factory automation by Ann Majchrzak and her colleagues at the University of Southern California, offer equally exciting possibilities. Given its potential, macroergonomics should experience an exponential growth during the next several decades.

The Future

Based on the above, all four major technology areas of ergonomics can look forward to a very bright future throughout the world — one characterized by very healthy expansion and growth. In fact, in terms of potential development and growth, the future for ergonomics appears brighter than for any other design-related field except, possibly, biomedical engineering. In any event, it is brighter than for most other design and engineering fields which, while much larger than ergonomics, are likely to see only modest growth in the foreseeable future. The rapid development of computer-based tools in ergonomics will further enhance the discipline's capability and growth.

1.5 Variations Among Countries

Historically, "human factors," as practiced in North America, and "ergonomics," as practiced in Europe and elsewhere, differed noticeably in both their approach and emphasis. Human factors placed relatively more emphasis on the study of human psychological and perceptual characteristics; and on the laboratory research approach to acquiring human factors knowledge and developing human factors technology. Ergonomics, on the other hand, placed more emphasis on the study and application of knowledge about human physical characteristics, including work physiology and biomechanics; and on the field study approach to acquiring ergonomics knowledge and developing ergonomics technology. Historically, the primary application of "human factors" technology was to moving systems, with particular emphasis on aircraft design. In contrast, ergonomics placed more emphasis on the design of factory jobs and workstations. During the last several decades, both classical "ergonomics" and "human factors" have moved toward one another in terms of emphasis, research methodology, and application. As a result, for well over a decade, those who identify as "human factors professionals" and those who identify as "ergonomists" have been doing essentially the same kinds of things. In short, the two have become synonymous; and for well over a decade the IEA has formally recognized the two as identical.

Having noted the above, it is still important to mention that national differences *do* exist. Most easily recognizable are national differences in the root professions in which the practicing ergonomists were trained (Hendrick, 1989). These differences, in turn, also reflect proportional differences in the kinds of things ergonomists do in different countries. At the risk of over-generalizing, there appear to be two major categories of root training and related applications that can be found internationally.

In the United States, the U.K., Germany, Japan, and many other countries, the root education of the majority of ergonomists is in industrial, systems, or mechanical *engineering* and industrial or applied experimental *psychology*. Often, ergonomists have studied in both fields. For example, in the United States, ergonomists may have a formal major in one and a minor in the other of these two fields. Ergonomics education historically has been offered as a specialty within both engineering and psychology. In comparison with countries in which this is *not* the general training pattern, a proportionately greater emphasis tends to be given to the design of moving vehicles, command and control systems, and consumer products to improve operational effectiveness, usability, comfort, and safety.

In other countries, such as Australia, Denmark, and Italy, the root education of the majority of ergonomists tends to be medical or medical related. In particular, ergonomists have been trained as physical therapists, occupational therapists, industrial hygienists, or as physicians. In comparison with countries having a predominantly engineering and psychology ergonomics training pattern, a proportionately greater emphasis is given to applying ergonomics to the workplace, tending to be more work physiology and biomechanics oriented and having a primary focus on preventing injury in the workplace.

Having noted these differences, it is important to recognize that these differences are *proportional* rather than *absolute*. The full range of educational backgrounds is found in virtually all countries with a developed ergonomics profession. Similarly, the full range of ergonomics applications is found within these countries.

More recently, we are seeing the development internationally of stand-alone professional education programs in ergonomics (Pearson, 1994). Given the maturation of human–system interface technology, this trend toward separate ergonomics professional degree programs should continue. What is common throughout the world is that ergonomics involves the development and application of *human–system interface technology*, and therein lies the identity of our profession internationally. It is this common activity that provides us with our basis for developing internationally harmonized certification criteria and educational guidelines.

1.6 Core Competencies in Ergonomics

One of the major tasks of professional ergonomics societies throughout the world has been, and continues to be, to identify the core competencies required for the practice of ergonomics. This is not surprising, because knowledge of these competencies is essential to determining certification criteria and professional education requirements. For example, a major national study was conducted around 1980 in the U.S. for the Air Force in particular and the Department of Defense (DoD) in general. It is the most extensive study of its kind ever conducted in the field of ergonomics. The purpose of the study was to analyze and evaluate all facets of the development and application of ergonomics (or what then was called human factors engineering) technology to DoD systems, with particular emphasis on how ergonomics is integrated into the system development process (Hendrick, 1981).

One of the major purposes of the study was to identify what human factors practitioners actually *do* in applying their technology to the development of a broad spectrum of systems, tools, jobs, facilities, and environments. These data, obtained for literally hundreds of ergonomists, were analyzed to determine what core competencies were actually required to perform the ergonomics tasks. The competencies identified were as follows:

1. Sufficient background in the behavioral sciences to respond to ergonomics questions and issues having psychological or other behavioral implications. Implies the equivalent of a strong undergraduate behavioral science minor.
2. Sufficient background in the physical and biological sciences to appreciate the interface of these disciplines with ergonomics. Implies the equivalent of an undergraduate minor.
3. Sufficient background in engineering to (a) understand design drawings, electrical schematics, test reports, and similar design tools, (b) appreciate engineering design problems and the general engineering process, and (c) communicate effectively with design engineers. Implies formal knowledge of basic engineering concepts at the familiarization level (i.e., at least two undergraduate level courses, or equivalent, in engineering).
4. Ability to evaluate the adequacy of applied ergonomics research and the generalization of the conclusions to operational settings. Requires formal knowledge of the basic statistical methods and principles of experimental design. Implies the equivalent of two introductory level graduate courses in statistics and research design.
5. Ability to (a) evaluate and (b) conduct the various kinds of traditional ergonomics analyses (e.g., functional task, time-line, link). Requires formal training in these techniques.
6. Ability to (a) evaluate and (b) perform classic human–machine integration including workspace arrangement, controls, displays, and instrumentation. Implies formal knowledge of ergonomics human–machine interface technology at the introductory graduate level or equivalent.
7. Ability to apply knowledge of human performance capabilities and limitations under varying environmental conditions in (a) evaluating environmental design and (b) developing environmental design requirements for new or modified systems. Requires formal knowledge of human performance capabilities and limitations in the various physical environments (e.g., noise, vibration, thermal, visual). Implies formal knowledge of environmental ergonomics at the introductory graduate level, or equivalent.
8. Sufficient knowledge of computer modeling, simulation, and design methodology to appreciate their utility in systems development, including ergonomics utilization in function allocation, task time-line analysis, workload analysis, workstation layout evaluation, and human performance simulation (this does not include the ability to actually design the models and simulations). Requires math through calculus and introductory computer science, and knowledge at the familiarization level of measurement, modeling, and simulation because these are applied to ergonomics.
9. Ability to apply knowledge of learning and training methodology to the evaluation of training programs and to instructional systems development (ISD). Requires knowledge of that portion of learning theory and research applicable to training, and of training methodology at the familiarization level.
10. Ability to assist in the development and evaluation of job aids and related hardware. Requires knowledge of the state of the art at the familiarization level.
11. Ability to apply the organizational behavior and motivational principles of work group dynamics, job enrichment and redesign, and related quality of worklife considerations in (a) developing ergonomic system design requirements and (b) evaluating the design of complex systems. Requires formal knowledge of organizational theory and behavior at the introductory graduate level (i.e., what today, somewhat broadened and deepened in scope, is called macroergonomics).
12. Specialized expertise in at least one area of human–system integration technology that goes beyond the introductory graduate level of understanding and application. Requires additional ergonomics course work and a tutorial research or thesis project at the graduate level, and an M.S. degree or

equivalent in ergonomics or a closely related academic discipline (e.g., engineering, psychology, safety science, physical therapy).

Note that the above analysis was completed a decade and a half ago. Today, we undoubtedly would add knowledge of software or cognitive ergonomics at the familiarization level to this list.

The above core requirements are just that. They represent what the committee concluded a fully qualified professional ergonomist practitioner, *ideally*, should have. In addition, the project team concluded from its study that at least two years of supervised practice experience was necessary to become a fully qualified professional ergonomist. Although not a formal conclusion of the study, project team members noted that additional unsupervised experience should be required for professional certification, if a certification program were to be implemented.

In reality, very few North American certifiable ergonomics practitioners would meet every single one of the above criteria. However, based upon the first several years' experience of the Board of Certification in Professional Ergonomics (BCPE) in evaluating U.S. and Canadian applicants, they would meet at least seven or eight out of the first eleven, at least minimally, and the twelfth criterion fully. In large part, failure of most ergonomists to meet all of the core requirements is a reflection of their education. Put simply, most have not graduated from true ergonomics programs but from degree programs in related fields with a few ergonomics courses thrown in. In time, with clearly established educational guidelines or accreditation programs reinforced by ergonomics certification programs, professional ergonomics education internationally should become better rounded and the number of true ergonomics graduate degree programs should increase. This will result in many more practicing ergonomists meeting all, or most, of the core criteria.

1.7 Professional Standards

A third indication of the development and maturation of a discipline to a unique, stand-alone status is that of establishing formal standards for professional competency (certification or registration) and education programs (either formal guidelines or accreditation). These kinds of standards protect the public, enhance the stature of the profession, and ensure its continuing viability. In ergonomics, major efforts now are well under way to develop these professional standards and guidelines. Because it is a relatively small profession, it is especially important to harmonize these standards internationally if ergonomics is to have a distinct identity; and the IEA is proactively facilitating this process.

Professional Certification/Registration

Professional registration of ergonomists is not new. For example, the Ergonomics Society in the U.K. has had such a program for a number of years. What is new is the recent, greatly increased interest internationally in professional certification or registration programs, and the development of two major international programs: the Board of Certification in Professional Ergonomics (BCPE), based in North America, and the Center for Registration of European Ergonomists (CREE) in Europe. The BCPE, based on a review of one's professional education, work experience, selected work product(s), and the passing of a three-part, comprehensive basic knowledge and practice written examination, certifies persons from any country as either a "Certified Professional Ergonomist" (CPE) or "Certified Human Factors Professional" (CHFP). The CREE, using similar evaluation criteria but without a written examination, certifies persons as a "European Ergonomist" (Eur. Erg.) who reside or work in the "European Ergonomic Space" (EES). Both certification programs were developed following an extensive review of the literature on ergonomics and ergonomists, and independently arrived at a similar conception of the basic requirements for functioning as a professional ergonomist practitioner. Ergonomists in other areas of the world are also exploring the development of professional certification/registration programs.

Professional Educational Standards and Accreditation

The development of guidelines and standards for professional education programs in ergonomics is also being actively explored by many ergonomics societies and the IEA. Because of the broad number of disciplines in which ergonomics specialization is offered internationally, harmonizing educational standards internationally is proving to be a more difficult problem, but progress is being made. The IEA sponsored a major conference on professional standards in conjunction with financial support from, and hosting by, the Italian Ergonomics Society in Palermo in 1992. This was followed by a second conference in conjunction with the IEA Triennial Congress in Toronto, Canada, in 1994. A major aim of both conferences was to work toward international harmonization of ergonomics education standards. To date, the most developed professional education standards program is the accreditation program of the Human Factors and Ergonomics Society in the United States. This program was developed over approximately a ten-year period. Thus far, approximately 15 of the 60+ human factors/ergonomics degree programs, and specialty programs within related degree programs in the United States have been accredited.

1.8 International Trends in Ergonomics

Employment of Ergonomists

Based on a survey of 25 IEA federated societies, 29% of their members are employed in educational institutions as academics, 27% practice in industry, 15% are researchers, 10% are private consultants, 8% work for the government, and 11% are employed in other occupations (Brown, Noy, and Robertson, 1995; Helander, 1996).

A recent all-member survey of the Human Factors and Ergonomics Society in the United States found the following distribution: 35% work in industry, 23% are consultants, 19% are academicians, and 17% are with the government, including the military services. In terms of primary work focus, 41% do design and development, 32% are in research and/or teaching, 14% in safety assurance, and 10% are in management or administration. Only 2% are in medical rehabilitation.

In terms of educational level, 84% hold an advanced degree (46% Ph.D., 38% masters). Of these, 41% are in psychology, 24% in engineering, and 17% in human factors or ergonomics. (Hendrick, 1996)

Major Problem: Industry, Government, and Public Awareness

The recent IEA survey of 25 member societies found the major problem facing the ergonomics profession was a lack of awareness and recognition by industry and government decision makers and the public in general (Brown, Noy, and Robertson, 1995; Helander, 1996). The recent HFES all-member survey further confirms this and indicates that a more proactive effort is needed by the society to raise the consciousness of industry and government decision makers as to the value and cost benefits of ergonomics (Hendrick, 1996).

Most Important Early Versus Current Applications

Based on the survey of 25 IEA federated societies (Brown, Noy, and Robertson, 1995; Helander, 1996), there has been a significant shift in the most important areas of ergonomics since the societies were founded as compared with today. Table 1.1 shows those comparisons for the five most important areas. Note particularly the emergence of safety, workload, and human computer-interaction (HCI) as among the most important application areas internationally. Implicit in this shift is the resurgence of industrial ergonomics and the impact of computers and automation on human performance and work.

Emerging Areas of Ergonomics

Of equal or greater importance, the IEA survey also identified the emerging areas in ergonomics in the countries represented by the 25 societies surveyed. The most frequently cited of these are shown in

TABLE 1.1 The Five Most Important Early and Current Applications of Ergonomics in 25 Ergonomics Societies

Importance	Early applications	Current applications
1	Anthropometry	Safety
2	Work physiology	Industrial engineering
3	Industrial engineering	Biomechanics
4	Biomechanics	Workload
5	Psychology	Human–computer interaction

TABLE 1.2 Important Emerging Areas of Ergonomics in 25 Ergonomics Societies Around the World

Topics	Frequency
Methodology to change work organization and design	7
Work-related musculoskeletal disorders	7
Usability testing for consumer electronic goods	6
Human–computer interface software	6
Organizational design and psychosocial work organization	5
Ergonomic design of physical work environment	4
Control room design of nuclear power plants	3
Training ergonomics	3
Mental workload	3
Workforce cost calculation	3
Product liability	2
Road safety and car design	2
Transfer of technology to industrially developing countries	2

Table 1.2 (Brown, Noy, and Robertson, 1995; Helander, 1996). Note particularly the emergence of organizational design, psychosocial work organization, and related ergonomics methodology as, perhaps, the major emerging new area of ergonomics. Reflected here is the growing recognition that a more macroergonomic, systems approach often is needed to effect major improvements in productivity, health, safety, and quality of worklife in complex sociotechnical systems.

Note also the international concern over preventing work-related musculoskeletal disorders, and the continued growth of HCI software design or cognitive ergonomics. Equally important is the emergence and growth of usability testing (and other methods of employee participation) as a major ergonomics methodology. This also is a reflection of another major trend in ergonomics internationally — the shift from technology-centered ergonomics to a more human-centered ergonomics. All of these emerging areas are highly relevant to the science and practice of industrial ergonomics.

1.9 Managing the Development of Ergonomics

As with any maturing organization or profession, as it develops and expands it will experience growing pains. When this happens, more attention must be paid to actively *managing* that growth and development. With respect to managing the development of the ergonomics discipline internationally, at least four major facets have been identified (Hendrick, 1993): (1) the orientation of ergonomics practice, (2) professional standards, (3) development of a more adequate database for predicting human performance, and (4) supporting ergonomics development in industrially developing countries. More recently, the importance of documenting the cost benefits of ergonomics applications and ensuring that this information is shared with organizational and government decision makers has been highlighted (Brown, Noy, and Robertson, 1995; Helander, 1996; Hendrick, 1996).

Orientation of Professional Practice

Historically, the actual practice of our profession has tended to be technology oriented. New system design or modification often is motivated by the desire to exploit new developments in technology. The ergonomist, if consulted at all, is called in after selection of the technology to "fit the human to the machine." In effect, the method of function and task allocation most frequently employed throughout the world is the leftover approach (i.e., whatever cannot be done by the machine is assigned to humans) (Bailey, 1989). When compared to systems designed without any ergonomic attention, this approach has proven effective in reducing accidents and improving human comfort and effectiveness. Not surprisingly, this success has further encouraged continuation of the technology-oriented approach to employing ergonomics in system design.

There are at least two major problems with a technology-oriented approach to ergonomics. First, it is *reactive* rather than *proactive*. The ergonomist's role becomes one of reacting to the system's hardware once it has already been selected or designed. As a result, the ergonomist's potential for enhancing total system performance is severely limited. Second, history repeatedly has shown this approach to lead to *suboptimization* of work system design, including poor utilization of human capabilities and a failure to incorporate well-known intrinsic motivational factors into jobs (e.g., see Hackman and Oldham, 1975); in short, to a relatively dehumanized work system (Hendrick, 1994).

What has become increasingly evident over the past decade is that we must proactively adopt a more *human-centered* approach to ergonomic practice. Whenever possible, we should begin the function analysis process by *justifying* the use of a human, and then selecting or adopting technology as required to *assist* the human in accomplishing system objectives (Bailey, 1989). This kind of approach to our professional practice transforms our role from one of *reactive advisors* to that of *proactive change agents*. As proactive change agents, the effectiveness of ergonomists can be greatly increased, and the end result will be many more humanized and productive work systems than presently exist in any country.

Professional Standards

The second area where proactive management of ergonomics is required is in the area of professional standards for (1) our academic programs (accreditation), (2) ensuring competency in ergonomics practice (certification), and (3) prescribing how we conduct our business (code of practice). As already noted, these kinds of standards mark the maturation of any profession and are well along in development internationally. Most recently, the IEA developed a draft code of practice for use as guidance by national and regional ergonomics societies in developing their own codes.

Development of a More Adequate Data Base

As Meister (1992) has noted, we need to be able to predict human performance, because this is the essence of ergonomics. Meister further notes that "the ability to predict is power. What design engineer could refuse our advice if he knew that it would cost him heavily in terms of the performance of his system? But to achieve this power, to achieve the credibility that will induce the designer to follow our advice, we need numbers. ...To be able to predict, to develop performance standards, we need a database, and this is much more than what they mean when most ergonomists use the term; it is much more than a series of studies plucked from journals. A database means extracting the raw data formally, combining data from various sources, relating the human performance values to the tasks performed, the system in which the performance occurs, performance-shaping factors, etc. Above all it means transforming the raw data into error probabilities and relating these to design variables" (Meister, 1992, pp. 259-260). Because ergonomics is a numerically small discipline, achieving the kind of database Meister envisions will require a *managed*, collaborative systems effort internationally.

A possible first step in this direction was recently taken by the Human Factors Committee of the National Research Council in the U.S. This group has completed a comprehensive project to identify the research needs in human factors/ergonomics. The report of the study's findings has been published by the National Research Council. Hopefully, this report will help provide a basis for establishing and managing an international ergonomics effort to develop a more adequate ergonomics database. The challenge, of course, will be to identify the necessary funding sources and to develop a management structure for this much needed but highly ambitious research effort.

Supporting Ergonomic Development in Industrially Developing Counties

Perhaps the single greatest weakness in the application of the current ergonomics knowledge base is that well over half of the world's population does not benefit from it, or does so in a very limited way. This is particularly critical in the area of industrial ergonomics for improving health and safety.

The IEA has several projects under way to help support ergonomics growth and awareness in industrially developing countries. These include IEA roving ergonomics seminars and the recently published IEA–ILO developed manual, *Ergonomics Checkpoints,* published by the International Labor Organization (ILO). *Ergonomics Checkpoints* contains over 130 common ergonomic applications, including step-by-step instructions and illustrations on how to apply ergonomics principles, guidelines and specifications. It is designed specifically for use with only limited ergonomics training and, in the IEA roving seminars, has proven to be an excellent "textbook" for providing basic ergonomics instruction — particularly for industrial applications.

The IEA effort is admirable but represents only a limited step in managing and enhancing the development of ergonomics in the industrially developing portions of the world. National and regional ergonomics societies, international and national government agencies, etc., also need to be tapped for assistance.

1.10 Conclusion

In summary, the future of ergonomics internationally has never been brighter. If ergonomists around the world properly and proactively manage their profession's development, the profession not only will greatly expand in numbers, but its contribution to improving the human condition will increase by at least several orders of magnitude.

References

Bailey, W. (1989). *Human Performance Engineering,* Second Edition, Prentice-Hall, Englewood Cliffs, NJ.

Board for Certification in Professional Ergonomics (1992). *BCPE* (brochure). Bellingham, WA: Board for Certification in Professional Ergonomics.

Brown, O. Jr., Noy, I, and Robertson, M. (1995). Special survey of IEA federated societies. International Ergonomics Association, c/o Human Factors and Ergonomics Society, Santa Monica, CA.

Hackman, J. R. and Oldham, G. (1975). Development of the job diagnostic survey, *Journal of Applied Psychology,* 159-170.

Helander, M. G. (1996). The human factors profession, in G. Salvendy (Ed.) *Handbook of Human Factors and Ergonomics,* Second Edition (pp. 3-15). New York: Wiley.

Hendrick, H. W. (1996). All member survey, preliminary results. *Human Factors and Ergonomics Society Bulletin, 39,* 1, 4-6.

Hendrick, H. W.(1993). The IEA and international ergonomics: Past, present and future. In O. Brown, Jr. (Ed.), *Ergonomics in Russia, the other independent states, and around the world,* Volume 2 (pp. 1-12). St Petersburg: Russian Ergonomics Association.

Hendrick, H. W. (1991). Human factors in organizational design and management. *Ergonomics, 34,* 743-756.

Hendrick, H. W. (1989). Human factors/ergonomics societies around the world: characteristics and issues. *Human Factors Society Bulletin, 32,* 8-10.

Hendrick, H. W. (1987). Organizational design, in G. Salvendy (Ed.), *Handbook of Human Factors* (pp. 470-494). New York: Wiley.

Hendrick, H. W. (1986a). Macroergonomics: A conceptual model for integrating human factors with organizational design. In O. Brown, Jr. and H. Hendrick (Eds.), *Human Factors in Organizational Design and Management-II.* Amsterdam: North Holland, pp. 467-478.

Hendrick, H. W., (1986b). Macroergonomics: a concept whose time has come, *Human Factors Society Bulletin, 30,* 1-3.

Hendrick, H. W., (1981). Engineering education's response to the need for human factors engineers. Presented during the Professional Session on Human Factors in Engineering Education, *89th ASSE Annual Conference*, Los Angeles, CA.

Human Factors Society (1996). *Directory of the Human Factors Society.* Santa Monica, CA: Human Factors Society.

IEA (1996). The History of the International Ergonomics Association, in *Basic Documents of the International Ergonomics Association*, Utrecht, Holland: IEA, pp. 1-2.

Meister, D. (1992). Some comments on the future of ergonomics. *International Journal of Industrial Ergonomics, 10,* 257-260.

Nagamachi, M. and Imada, A. S. (1992). A macroergonomic approach for improving safety and work design, in *Proceedings of the Human Factors Society 36th Annual Meeting.* Santa Monica, CA: Human Factors Society, pp. 859-861.

Pearson, R. (Ed.) (1994). *International Directory of Educational Programs in Ergonomics/Human Factors* (Third Edition). International Ergonomics Association, c/o Human Factors and Ergonomics Society, Santa Monica, CA.

Rooney, E. F., Morency, R. R., and Herrick, D. R. (1993). Macroergonomics and total quality Management at L. L. Bean: a case study, in R. Nielsen and K. Jorgensen (Eds.), *Advances in Industrial Ergonomics and Safety V.* London: Taylor & Francis, 493-498.

Vicente, K. J. (1990). A few implications of an ecological approach to human factors. *Human Factors Society Bulletin*, 30, 1-4.

2

The Ergonomics of System Design

David Meister
Human Factors Consultant

2.1 Introduction

It is necessary to define what is meant by *system* and by *design* because these terms may mean different things in various contexts. The term *system* in its most general sense is used to describe the entire spectrum of human-machine equipment in which humans interact with physical objects. Systems vary on a scale of complexity, encompassing very complex systems like aircraft or a manufacturing assembly line. The term also describes somewhat less complex equipment like automobiles or the workstations or subsystems that control higher-level system functioning, such as those in the control room of a process control system like a nuclear power plant (NPP) or the bridge of a steamship. Lower on the system scale is the handheld tool or appliance, like the hammer, shovel, or can opener. In this chapter the term system is used to designate all of these, although the differences among them produce significant differences in personnel requirements and performance. The brevity of this chapter does not permit detailed discussion of this and other topics (but see Meister, 1991, for additional details).

The term *design* is also used very broadly to describe the design of everything from an entire company or facility to the simplest tool. System design is the process in which the developmental ergonomist or *human factors engineer* (*HFE*) performs his/her tasks as part of a design team. System design is not often a concern of the industrial ergonomist or IE, but should be, because system development of some sort always precedes industrial ergonomics and often determines the specific problem the IE must face. If, for example, the designer of a machine tool designs it improperly, the IE may be faced with a high incidence of muscular trauma and/or accidents. Moreover, one of the functions of the IE is to evaluate and redesign a manufacturing facility when one of these preceding conditions arises (Narayan and Rudolph, 1993). The redesign employs some of the same processes as those in the initial design of the primary system. Whatever the specific characteristics of the item being designed or redesigned, the design processes involved are essentially the same, although certain aspects may be emphasized more in one type of system than in another.

The design of the primary system also influences manufacturing processes which are the special purview of the IE (Dockery and Neuman, 1994). These manufacturing processes also involve system

0-8493-2641-9/99/$0.00+$.50

design at two levels: first, of any special machine tools needed to perform assembly, and, second, the design of the total manufacturing facility, e.g., the spatial arrangement of the workstations that comprise it. Many machine tools are quite common, but those developed specifically for the production of the primary system will require design, which should be of interest to the IE.

2.2 System Development

System development is the total life cycle of the system being designed, up to the point at which it is released to the customer (see Meister, 1987a). It has three major functions: *analysis* of the design problem, *solution* of the problem, and *testing* of that solution. System development is *iterative* and progressively more detailed (commonly termed "top-down" design), which means that initial solutions are somewhat general and therefore tentative, and are often refined at more detailed levels. Initial design solutions are often revised after testing, at which point more information is gathered. (this is called "bottom-up" design, see Meister, 1991.)

System design always presents a problem, which begins with the specification of design requirements. The problem is, what is the object to be designed supposed to do, and how? The problem is to create hardware and software that will satisfy design requirements. The problem is made more difficult by the fact that there is more than one possible design solution; there is almost always a number of alternative configuration solutions, from among which the design team will have to select one that is the most desirable. Moreover, the solution has multiple, possibly competing dimensions: performance versus cost versus reliability versus operator ease, etc. For example, a system can be more automated, but automation has a higher cost.

Dockery and Neuman (1994) and Marcotte, Mervin, and Lagemann (1995) have demonstrated that the design process involves more than engineering alone. Moreover, there are several players in the design drama: the design engineer (one or more), who has direct responsibility for the finished product; the manufacturing engineer; the HFE and other specialists, who are part of the design team; engineering management, which makes high-level development decisions based on nonengineering criteria, such as costs; and the user, who may play several roles.

The User

One of the newer developments in system design is the increasing involvement of the *user* in the design process. That comes about in several ways. (1) The user, defined as the customer who has ordered the development of the new system, will be asked at the beginning of the project to provide details of the system mission and its implications for operator performance. From a behavioral standpoint, the design specification is almost always seriously deficient, because operator requirements are lacking. The user (company management and its experts) will be asked to describe typical behaviors and standards of operator performance. (2) The user will be asked to present a profile of those lower-level personnel who will eventually use or operate the system — characteristics such as age, gender, intelligence, available skills, and handicaps, all of which may have implications (such as the effects of handicaps) for system design (see Scerbo, 1995). The preceding applies primarily to the design of new or heavily upgraded systems; if the system is only a minor upgrade, much of this information should already be available. (3) The user can be asked to provide subjects to test alternative configurations (Dolan, Wiklund, Logan, and Augaitis, 1995). (4) As the customer for the new system, the user will ultimately be asked to approve the completed system.

It need hardly be said that user involvement in the design process will help to avoid gross errors in the ultimate design. Nonetheless, the customer cannot be asked to participate in daily design decisions. Moreover, some customers have only a vague notion of how the genuinely new system is to be utilized. The HFE should ask for as much information from the user as is available and make the best of what s/he receives.

2.3 Design Analysis

The starting point for design is the requirement *specification,* almost always in written and graphic form, and providing more or less detail. Usually the requirement is much more specific in terms of the engineering qualities it demands than it is for behavioral requirements. The requirement will specify such items as power usage or output, a target range, a maximum fuel consumption. One almost never finds this for behavioral elements of the system, mostly because the customer does not feel there is enough information to be able to specify personnel qualities. Because of this, the first task of the HFE is to analyze the requirement to determine those aspects that may affect or constrain operator performance.

As part of this analysis, the HFE must collect data about any predecessor system (e.g., test results, task analyses, performance data) before analyzing the requirements of the new system. Among the factors that the behavioral analyst must consider are: the degree of automation, which determines in large part what the operator must do; the primary human functions the system emphasizes — physical, perceptual-motor or cognitive — and in what way these will ultimately influence hardware or software design; the conditions under which the system (including the operator) must perform; the number and sources of information to which the operator must attend while operating the system; how rapidly the system functions and how quickly the operator must respond; how forgiving the system will be if the operator makes errors; and any anthropometric constraints. For systems that are in common use and have a long history, such as automobiles, the preceding considerations may be somewhat insignificant, but they become significant if the system is new or highly advanced over any predecessor system (e.g., the Apollo space module or a new generation fighter aircraft). If the IE is asked to participate in the design of a new or modified manufacturing facility, some of the same questions must be asked. What does the worker have to do? What stresses are likely to be imposed by the manufacturing facility? What requirements in terms of speed, physical activity, communication, decision making, etc. are imposed on the worker?

If the system is being designed for the government, the HFE's design analysis will probably be quite formal. The government will probably insist on a formal human engineering plan. This is part of the various procedures set up by the government in programs such as *Manprint* (Booher, 1990). For a system designed for a commercial customer, or one which is for general use by the public, the formal documentation may be at a minimum.

One factor that impacts the design analysis is the molecularity of the actions performed by the operator. If the operator's actions are largely motoric and repetitive, one can make use of the predetermined time systems (PTS) which describe the time required to perform the action. This is feasible in the design of a manufacturing assembly line, but the design of the primary system usually requires the operator to perform at a much more molar level, involving perceptual processes, analysis of system data, and decision making; for these, PTS are not much good. It would be highly desirable to predict the human error associated with each operator action, but here the available predictive data do not exist, particularly as they relate to cognitive and perceptual-motor processes. It is possible to make use of subject matter experts (SMEs) who can rate the error probability of major actions, and this may be all that one can do.

Depending on the nature of the system, the HFE will be particularly concerned with certain aspects of that system and its operator/user. In the highly automated systems of the upcoming 21st century, such as process control plants, the operator will be a monitor and diagnostician and will work primarily with system status symptomology (see Meister, 1996), and so the HFE will be interested in the nature and amount of information the system provides. The characteristics of the operator are also becoming increasingly important to the HFE. For example, the aging process produces reductions in strength, sight, hearing, and speed of response, and these must be considered in evaluating the adequacy of the design. Use of equipment by the physically and mentally handicapped imposes special demands on design. In the design of a home for paraplegics, for example, anthropometric considerations will be primary. If the user is mentally retarded, mental demands posed by the equipment must be considered (Robertson and Hix, 1994).

The starting point of the design analysis is the determination of what the operator is required to do by the system, to what degree of precision, and how this requirement will affect the probability of successful task accomplishment (e.g., in terms of potential errors and time to accomplish the task). In a manufacturing facility whose outputs are physical, task accomplishment would be measured in terms of quantity and quality of items produced.

The analysis of what the operator must do proceeds in parallel with the development of alternative design concepts. This is because what the operator is required to do is determined by the design concept. The consideration of alternative design configurations is, in its initial stages, almost always free-wheeling discussion ("brainstorming") by the design team. Only after the initial concept is agreed upon, does the engineer develop formal drawings. The HFE has the dual task of contributing to the design discussion and simultaneously analyzing the effect of the alternative design possibilities on operator performance. If, for example, an engineer proposes a design that requires an operator to move 500 pounds without mechanical aids, the HFE will tell him/her that such a requirement is impossible. In actuality, since design engineers ordinarily recognize such disqualifying conditions, most design concepts do not make obvious extraordinary physical or perceptual-motor demands on operators, but the HFE must be alert to less severe demands (such as mental) that, even if not impossible, may cumulatively reduce the effectiveness of personnel responses.

There are formal task analytic (TA) techniques (see Meister, 1985, and Kirwan and Ainsworth, 1992) that decompose required operator actions into their physical, perceptual-motor, and cognitive elements. Whatever the form of the TA, it always breaks required actions down into three elements: the stimulus to which the operator must respond; the operator's internal processes that produce a response; and the physical response. The relationship among these three determines the amount and type of system demand on the operator to which design must respond.

During the initial brainstorming of design concepts the HFE can perform such an analysis only very grossly. It is only when a small number of design configurations is decided upon as viable system candidates that the HFE can engage in a formal analysis and comparison of these alternatives in terms of behavioral factors. The great difficulty is that the design engineers on the team, not being overly concerned with behavioral aspects, are likely to focus precipitously upon a single design configuration without adequate consideration of these aspects. The HFE's job is to slow them down just enough so that behavioral factors can be adequately considered. Since initial design concepts may be relatively gross, the HFE must utilize creative imagination to fill in details.

This chapter cannot describe, even briefly, the available TA procedures because of space limitations. These procedures make it possible to decompose operator actions in minute detail, and to describe these in both verbal and graphic modes in the sequence in which they occur. Initially TA formats were verbal only; since then, verbal/graphic equivalents such as Operational Sequence Diagrams, Decision-Action Diagrams, and cognitive TA have been developed. There are also analytic techniques to determine the workload imposed on the operator, to determine the optimal arrangement of subsystems or machine units in the work area, and to examine the communications or interactions among team members. The length of time it takes the HFE to perform a TA depends on the size and complexity of the system being analyzed, the amount of detail the analysis requires, and the time permitted for the analysis. The TA for a system as complex as the Atlas ICBM, in whose development this author participated, resulted in a series of volumes that almost filled a small room. Engineering, cost and time constraints do not usually permit such an extensive TA, and any TA that can influence design is likely to be rather basic; more detailed ones are more useful in determining training and logistic requirements, as well as operational procedures. These last will not be considered here. TA, like system development as a whole, is an iterative, refining process; as design becomes more detailed, the TA also becomes more detailed.

The culmination of the design analysis process which, depending on system complexity and the novelty of the system, may take days, weeks, or even months, is the selection of a single "best" design configuration and the refinement of the details of that configuration. The selection of the one among several possible configurations the design team will "go with" should ideally be a quantitative decision. There are quantitative techniques for making such a decision (e.g., Sadacca and Root, 1968, Meister, 1985), but these do not consider solely behavioral factors. The latter is only one of the factors the design team and management considers; others are performance output, cost, reliability, manufacturing ease, etc. Moreover, the method of making the design selection may not be formal and quantitative, although it may have quantitative elements. The HFE contributes to the overall decision by providing an analysis of the design alternatives in terms of how each will affect operator performance.

2.4 Design

Some authors divide design into a preliminary design phase, in which analysis is pre-eminent, followed by a detail design phase in which drawings of components and circuitry are produced. These and other categorizations do not mean a great deal in reality, since in every phase there is analysis, design, and test to some degree, and system development as a whole has many feedback loops.

Unless the HFE is also the designer, which is quite rare, s/he acts as the representative of behavioral interests to the design team by evaluating design drawings in terms of their potential impact upon the operator, and by acting as a specialist consultant to the team. The primary task is to review design drawings to ensure that no aspect of the design is likely to lead to inadequate operator performance. One form of review examines the drawing alone by comparing its characteristics with a checklist of characteristics that the design should have. The checklist is likely to be based on such well-known standards as MIL STD 1472 (Department of Defense, 1992) or, in the case of software, Smith and Mosier (1986). This review is performed by the HFE singly and/or with the designer. Another manner of conducting a review is to perform what is called a "walkthrough," which requires both preliminary drawings of the human-machine interface and procedures for system operation. The design team operates the equipment symbolically, examining each step in the procedure with reference to the design drawings to determine whether the operator will be able to perform the step and to note potential problems in that performance. In one form of the walkthrough, someone takes the part of the operator and simulates the action required by motioning at each step to the control on the drawing to be moved and the display to be observed.

The walkthrough is easiest to conduct when system operation is in the form of a step by step procedure but can also be performed even if much of the operator's activity is perceptual (attending to displayed information) and cognitive (analysis of that information). If the operator's role in system operation is to monitor its performance and diagnose its malfunctions, the walkthrough can be couched in terms of the type of malfunction that can arise and how the operator should interpret the symptomology of that malfunction. The focus of such a design review is to determine how difficult or easy it will be to recognize and interpret the symptomology.

As an ergonomics consultant, the HFE is a specialist on the fine points of controls and displays and the human-machine interface generally. The designer *may* ask for recommendations with regard to these; the HFE *should* volunteer his/her opinions.

2.5 Test

Behavioral testing in system development is of two types: that performed to support design (developmental testing) and that performed to verify adequacy of the final design solution (operational testing). Any test may incorporate both aspects, but developmental testing occurs earlier in design, whereas operational testing will always occur toward the close of development (see Meister, 1987b).

It is also necessary to distinguish between research and testing. Research seeks to understand the relationship between system variables and their effect on operator performance. Neither developmental nor operational testing is concerned with such questions, but seeks merely to determine the adequacy of the design and any problems that exist. Only very rarely would system development testing become involved with research on behavioral variables.

Developmental tests are performed to determine (a) which of two or more alternative configurations is best from a behavioral standpoint; (b) if, given a configuration which has already been selected as best, will the operator actually be able to perform effectively with it; (c) what problems, if any, will the operator have with a particular configuration. These questions pertain to both the *operability* and *maintainability* of the system.

It might be supposed that by this time questions of alternative configurations would have been resolved, but these often recur because they are repeated at more detailed levels of design. For example, the individual subsystems to be monitored in a nuclear power plant control room may have been identified early in design, but later the characteristics of the individual workstations in the control room (particularly the information they will present) must be determined; and where alternative configurations of a workstation are possible, testing must decide which is best.

Developmental Testing

The tests that were described previously in the design phase (review of drawings, walkthroughs) were symbolic and conceptual; now actual performance, using subjects, and a physical replication of the system, must be examined. The physical representation of the system is the mock-up or prototype. For some systems, like aircraft, the development of a static or functional mock-up is an accepted means of testing proposed solutions to design problems that arise. This is like a dressmaker's dummy on which changes in dress form can be tried out. In the same manner, if there is a question about which of several design formats or procedures should be accepted, it is possible to test the adequacy of these by requiring subjects to use them under simulated operational conditions. The mock-up can be built of various materials (wood, cardboard, styrofoam) to roughly the same dimensions as the operational equipment will have (machine tolerances are usually not required). The mock-up, although largely static, may have actual controls and displays; these may be instrumented to display operational signals, and the operator may be given the capability of operating controls to produce changes in those displays. Thus, some mock-ups are completely static, others are partially functional, and still others are fully functional (Janousek, 1967). The more functional the mock-up, the more it approaches a simulator, and the more useful are the test results. In attempting to replicate operational performance, procedures used in the mock-up test must also approximate those that will be used operationally. The further along development has proceeded, the closer will the mock-up test replicate operational systems operation. The mock-up test is run as much as possible like the operation of the actual equipment, and data are collected and analyzed just as one would collect operator performance data on actual equipment.

The mock-up test whose purpose is to test the adequacy of software is called "prototyping." This type of test is easier to conduct than that of hardware, because it is easier to write software programs. There are also special software test beds into which the test software can be inserted.

It is easier to conduct a mock-up/prototype test when the system is to be operated in a step-by-step manner. When the system under test has a more flexible, contingent procedure (characteristic of information-driven systems) the instrumentation required to perform the test is more complex and in particular may require the HFE to include various modes of operation and potential malfunctions.

The mock-up or prototype test can be performed in various ways: it can be more or less formal; subjects may be fellow workers or more carefully selected to represent eventual system operators or users; data may be performance measures (e.g., time, errors) or may be judgments made by observers of subject performance. It is presumed that a more formal test with selected subjects, performance data, etc. will be more valid than a less formal test involving only observation and subjective judgment.

What questions can such a test answer? Are required tasks performed successfully (e.g., can a quadriplegic in a motorized wheelchair reach and grasp objects on shelves in a simulated habitat? How long does it take him/her to do so, and with what degree of strain? What significant difficulties are manifested? What changes in design are suggested by the test results? The developmental test enables one to refine the design and also provides assurance that the design selected is the most desirable one.

In attempting to gauge the level of difficulty which the design presents to the operator/user, or to solicit opinions about design adequacy from test subjects, it is necessary to make use of judgmental data, either from the test subject or from an SME observer. This requires the use of subjective test methods, which may be a post-test interview, questionnaire, or rating scale. It is impossible in this chapter to describe how one designs such tools. (Instructions for doing so can be found in Meister, 1985.) Obviously, any questions or judgments regarding difficulties experienced by the test subject or designed to elicit opinions about design adequacy must reflect the system characteristics and operating procedures under test. Examples of interview items, questionnaires, and rating scales can be found in the references provided and can easily be modified to fit the individual test situation. Whether one uses such tools in the developmental test depends on how formal the HFE wishes to be.

Design is iterative because the test results may suggest the need for design revision. However, any performance difficulties found in the test must be sufficiently significant to overcome the inherent resistance of engineers to make potentially expensive modifications in the design.

Operational System Testing (OST)

This has several variations: (1) formal testing in the engineering facility or at a test range, testing which reproduces the operational environment (OE); (2) usability testing conducted in the user's facility during normal operations. The latter is much more formal than the former, because one can have much less control over the OE. The usability test is more often conducted for general commercial systems, such as office furniture and word processors. A third type of OST is what one can call a "redesign test," because it occurs after the system has been accepted and is in operation. This test is performed when difficulties such as excessive muscular trauma or accidents occur (or a need is felt to upgrade performance) in a manufacturing facility.

1. Formal OST

Formal OST immediately precedes the completion of the system and its release to the customer, and verifies that the system complies with the design requirement. To the extent that the customer has specified standards that the OST can demonstrate, it is not difficult to conduct such a test. If the test is to be valid, then the system must be completely operational, that is, it must have not only the same physical characteristics of the system as it will have when fielded operationally, but it must utilize operating and maintenance procedures that would be utilized in routine operations. Subjects who perform in the OST should be either operational personnel themselves or those who have similar characteristics. Note the emphasis on system maintenance as well as operation. In the highly automated systems that are now being developed, the quality of maintenance is as critical as the quality of normal output. Certainly, this is true with regard to operator behavior, since, as was pointed out previously, in such systems the operator is a monitor, diagnostician, and maintenance person when (as is inevitable) the system fails.

If formal operator performance requirements were written into the design specification (although they almost never are), it is possible to compare actual operator performance with the requirement. In the absence of these requirements it may seem as if running an OST would be pointless; but it is in fact quite useful, because, even if formal behavioral standards are not available, SMEs can examine the performance data and comment on whether it represents what should be expected of trained personnel. The performance of operators can be videotaped, and communications (if these are involved) can be recorded.

The same questions that are asked of the mock-up or prototype test are asked also of OST. The criteria for recommending changes as a result of performance discrepancies are much changed, however; only the most serious behavioral inadequacies that threaten system output will warrant consideration. Since the system is about to be released for operational use, proposed changes, except the most minor, are enormously expensive and will be rejected.

2. Usability Testing

In formal OST the system is tested by simulating the OE within the engineering facility; in usability testing the system is transferred to the user and thus becomes part of the OE (Scerbo, 1995). The system must be portable in the sense that it can be taken out of the engineering facility (for example, office equipment). The requirements for usability testing are much the same as those for formal OST. In particular, the user must be willing to use the system for a period of time; where the system requires training of user personnel, these must be given training; and the user must be willing to record the data needed by system developers to evaluate the system.

One does not have the same control over the system in usability testing as one has in formal OST, but whatever performance measures are desired must be collected. If the system is designed to collect performance data automatically, this presents no difficulty; but even a highly automated system may depend on some manual data collection procedures, and the user must be willing to devote a certain amount of time and effort to that collection, e.g., logs. User personnel must be debriefed periodically.

The same questions as those of formal OST are asked in usability testing — Is the system adequate to design requirements? Are any difficulties experienced and changes needed? Is the user satisfied with the system? The degree of user satisfaction with the new system is much more important than in formal OST, because almost all systems in usability testing are commercial systems to be sold generally and in competition with comparable systems. Subjective tools like the interview and the rating scale become especially important.

3. Redesign Testing

IEs are perhaps more familiar with this type of testing, since it is frequently performed in relation to manufacturing but only after performance discrepancies are noted (e.g., high accident rates, muscular trauma incidence). The redesign test (see Schmidt, Petree, and Laughery, 1984, as an example) is essentially a "before and after" test. The situation complained of is analyzed (before) to determine what is causing the problem; then certain changes are introduced (e.g., new office equipment, new machine tools, new procedures); and the system is tested again to determine whether the changes have resolved the problem.

Since redesign testing begins by evaluating the situation, this requires the use of an instrument like task analysis (what is the operator supposed to do?) to determine if the task itself imposes difficulties. Worker operations are observed both directly and by videotaping to aid in diagnosing the situation; the workers themselves are interviewed. The quality of the product produced by the worker is examined, using SMEs to determine whether product quality has been endangered and in what ways.

If the results of this investigation suggest certain design or procedural changes, then the diagnostic phase of the test concludes with redesign and installation of the changes. Following this, performance data are collected (e.g., number of items produced, their quality, worker performance times and errors, number of operator complaints). This is the "after" part of the test. "Before" and "after" data are then systematically compared to determine if the new changes have produced a significant reduction in the problem. Such a comparison should involve statistics of significance of differences (as in Allen and Gerstberger, 1973).

All testing requires substantial periods of time. Management may exert pressure to compress the test period as much as possible, so that the new system can be operational as quickly as possible. Precisely how much time is required depends on the specifics of the new system. The cost in terms of resources, both human and machine, to perform such tests may be high, but without such tests the system will certainly be at risk.

2.6 Conclusions

For reasons of brevity it has been necessary to disregard any consideration of the review of procedures, development of technical manuals, training, and simulator design, any or all of which may become the HFE's responsibility. A detailed list of HFE activities during system development would also include writing analytic and test reports, participating in design reviews, collecting and disseminating data, etc. To be successful, the HFE must be in a continuing relationship with other engineers (e.g., reliability, industrial designers, training) and must be cognizant of difficulties other specialties encounter. Because system development encounters many obstacles involving detours, short cuts, and the repetition of processes, the HFE must be adaptable to accommodate these variations.

The extent to which all the procedures in this chapter are followed depends to a certain extent on the size of the system, equipment, or tool. For example, the design of a new blender may require fewer of these procedures; the design of a control room for a new process control facility will require all of them.

Although industrial ergonomics and HFE logically intersect, there has probably been too little awareness of the other on the part of those who practice each specialty. It would be helpful for each to become more cognizant of the other, because even if the industrial ergonomist does not himself/herself engage in system design, s/he should become more aware of the impact of system design on industrial processes.

Defining Terms

Human factors engineer (HFE): The specialist who, during system development, applies behavioral principles to the analysis, design, and testing of the system.

Specification: The written list of requirements that represent the standards to which the design engineering team performs during system development.

System: Any physical entity operated or used by a human to produce or utilize a desired output.

System design/development: The engineering process by means of which a system is constructed.

User: The management customer who has ordered or will ultimately purchase the system and those employees and the general public who will operate and use the system after it is built.

References

Note: The term "Proceedings" below refers to Proceedings of the annual meetings of the Human Factors and Ergonomics Society.

Allen, T.J. and Gerstberger, P.G. 1973. A field experiment to improve communications in a product engineering department: The non-territorial office, *Human Factors*, 15(5): 487-498.

Booher, H.R. ed. 1990. *Manprint, An Approach to Systems Integration*. Van Nostrand Reinhold, New York.

Casey, S.M., Dick, R.A., and Allen, C.C. 1984. Human factors and performance evaluations of the emergency response information system. *Proceedings*, 225-229.

Clarke, M.M. and Kreifeldt, J.G. 1984. A control room concept for remote maintenance in high radiation areas. *Proceedings*, 230-233.

Department of Defense 1992. MIL-STD-1472. *Human Engineering Design Criteria for Military Systems, Equipment and Facilities*. Washington, D.C.

Dockery, C.A. and Neuman, T. 1994. Ergonomics in product design solves manufacturing problems: Considering users' needs at every stage of the product's life. *Proceedings*, 691-695.

Dolan, W.R., Wiklund, M.E., Logan, R.J., and Augaitis, S. 1995. Participatory design shapes future of telephone handsets. *Proceedings*, 331-335.

Evans, T.E., Jr., Lucaccini, L.F., Hazell, J.W., and Lucas, R.J. 1973. Evaluation of dental hand instruments. *Human Factors*, 15(4): 401-406.

Janousek, J.A. 1970. The use of mock-ups in the design of a deep submergence rescue vehicle. *Human Factors*, 12(1): 63-68.

Kirwan, B. and Ainsworth, L.K., eds. 1992. *A Guide to Task Analysis*. Taylor & Francis, London.

Koch, C.G. and Richardson, R.M.M. 1984. Case study evaluation of color graphics for process control. *Proceedings*, 196-200.

Lowe, B.D., You, H., Bucciaglia, J.D., Gilmore, B.J., and Freivalds, A. 1995. An ergonomic design strategy for the transit bus operator's workspace. *Proceedings*, 1142-1146.

Marcotte, A.J., Mervin, S., and Lagemann, T. 1995. Ergonomics applied to product and process design achieves immediate, measurable cost saving. *Proceedings*, 660-663.

Meister, D. 1985. *Behavioral Analysis and Measurement Methods*. Wiley, New York.

Meister, D. 1987a. Systems design, development, and testing, in *Handbook of Human Factors*, ed., G. Salvendy, pp. 17-42. Wiley, New York.

Meister, D. 1987b. System effectiveness testing, in *Handbook of Human Factors*, ed., G. Salvendy, pp. 1271-1297. Wiley, New York.

Meister, D. 1991. *The Psychology of System Design*. Elsevier, Amsterdam, the Netherlands.

Meister, D. 1996. Human factors test and evaluation in the Twenty-first century, in *Handbook of Test and Evaluation*, eds., T.G. O'Brien and S.G. Charlton, pp. 313-322. Erlbaum, Mahwah, NJ.

Narayan, M. and Rudolph, L. 1993. Ergonomic improvements in a medical device assembly plant: A field study. *Proceedings*, 812-816.

O'Hara, J.M. 1994. Evaluation of complex human-machine systems using HFE guidelines. *Proceedings*, 1008-1012.

Robertson, G.L. and Hix, D. 1994. User interface design guidelines for computer accessibility by mentally retarded adults. *Proceedings*, 300-304.

Sadacca, R. and Root, R.T. 1968. A method of evaluating large numbers of system alternatives. *Human Factors*, 10(1): 5-10.

Scerbo, M.W. 1995. Usability testing, in *Research Techniques in Human Engineering*, ed., J. Weimer, pp. 72-111. Prentice-Hall. Englewood Cliffs, NJ.

Schmidt, J.K., Petree, B.L., and Laughery, K.R., Sr. 1984. The test of a task re-design. *Proceedings*, 829-831.

Smith, S.L. and Mosier, J.N. 1986. *Guidelines for Designing User Interface Software*. Report ESD-TR-86-278. Electronics Systems Division, Hanscom Field, MA.

For Further Information

The Handbook of Human Factors, edited by Gavriel Salvendy, describes every aspect of the discipline. *Research Techniques in Human Engineering*, edited by Jon Weimer, is an excellent applied reference. *Human Factors in Simple and Complex Systems* by Robert W. Proctor and Trisha Van Zandt provides the psychological theory and research which underlies human factors practice.

Further information about various Human Factors specialties can be secured by writing to the Human Factors and Ergonomics Society, Post Office Box 1369, Santa Monica, CA 90406-1369. Telephone 310/394-1811.

3

A Guide to Scientific Sources of Ergonomics Knowledge

Holger Luczak
Institute of Industrial Engineering and Ergonomics
Aachen, Germany

Christopher Schlick
Institute of Industrial Engineering and Ergonomics
Aachen, Germany

Johannes Springer
Institute of Industrial Engineering and Ergonomics
Aachen, Germany

3.1 Generation and Consolidation of Ergonomics Knowledge

Knowledge as a product of scientific discovery grows in the context of people, institutions, organizations, or societies which constitute the special interests of their members. Such a "scientific community" defines a set of rules and procedures to create, publish, discuss, revise, establish, and store information on the one hand, or to reject it on the other. Knowing about these rules can guide the information seeker through the vast amount of scientific information available. Therefore, the ergonomics knowledge generation process can be analyzed separately from the process of knowledge consolidation.

Generation of Ergonomics Knowledge

Analogous to the broad spectrum/scope of ergonomics problems, a similarly broad repertoire of methods is necessary for the generation of knowledge. Beneath specific ergonomics methods, numerous techniques from natural, social, engineering, and human sciences are applied to gather information about "man-at-work." These analytical methods can be assigned to four categories:

- Systematical observations
- Questionnaires, interviews, and surveys
- Physiological measurements
- Physical and chemical measurements

Whereas numerous methods of observing and interviewing are derived from the social sciences, the latter two groups are typical of the approach of natural sciences.

0-8493-2641-9/99/$0.00+$.50

Systematical Observations

Methods of systematical observation can be differentiated according to five criteria (Friedrichs, 1975):

1. Open versus hidden observation: Can the observer (or a technical means, such as a camera or a sensor) be perceived/recognized by the working person?

   If the researcher suspects that the behavior of the working person changes under observation (problem of reactivity of measurement), it is useful to use a hidden method of observation for knowledge acquisition. For ethical reasons, the researcher should inform the working person(s) afterward about the observation, allowing the subject to choose to object to the use of the data gathered. Beneath ethical considerations, numerous legal restrictions have to be taken into account. Thus, hidden observation is a method of minor importance in ergonomics knowledge acquisition.
2. Participatory versus nonparticipatory observation: Does the observer take part in the working situation, or is he or she an external noninvolved person?

   When a researcher doing a field study is employed at a normal workplace in a company, participatory observation is used to avoid disturbance of the working processes and to get more authentic information. In the latter case, one usually combines participatory observation with the hidden approach.
3. Standardized versus nonstandardized observation: Is the observation structure following a standardized scheme or is it more or less explorative and unsystematic?

   The more precisely the hypotheses and goals of the investigation can be formulated and the better the previous knowledge about the object/subject of research available, the more advisable is the use of standardized observation methods due to their better economy in gathering specific information, and in evaluation and interpretation of results.
4. Artificial versus natural situation: Has the situation to be observed been designed solely for the purpose of investigation or does it exist independently of the research project?

   With this differentiation, laboratory and field studies, as well as simulated workplaces (flight simulators, weapon systems simulators, etc.), are taken into account.
5. Self-observation versus observation of the outside world: Is the observer his own subject?

   In ergonomic data acquisition, self-observation is mostly used as one method in conjunction with other methods, for example, an investigated working procedure is performed by the researcher himself to get insights into specific difficulties and problems of execution.

   In the generation of ergonomics knowledge, the open, nonparticipatory observation of the outside world is the predominant method. Open observation does not necessarily mean that the observed persons have to be informed beforehand about the purpose of the investigation. In many cases it seems to be necessary that the real purpose be left unclear to the subjects during the observation in order to avoid results that are influenced voluntarily or involuntarily by the subjects.

Questionnaires, Interviews, and Surveys

These methods can be subdivided into four categories according to the degree of standardization of the question and the respective range of answers:

1. Standardized questions and standardized answers

   The investigation is usually performed in a written manner. Typical representation of this form of knowledge generation is the questionnaire with predetermined answer-categories. The answer-categories can be alternatives (yes/no; right/wrong) or imply a choice on a scale (for example, intensity: no/a bit/somewhat/rather/predominantly/totally; or frequency: never/seldom/sometimes/often).

   A general problem of this type of investigation is the predetermination of all questions and all answer categories beforehand; another problem is the possible misunderstanding of the question by the subject and the coding of possible corrective measures and additional information. Also it is sometimes not obvious, for instance with postal inquiries, who filled in the questionnaire. An advantage is the simple data processing, which can be done automatically by character-recognizing and processing machines.

This type of questionnaire is mostly used in studies of the subjective evaluation of working situations — for instance classifications of stressors and strains by lists of attributes (tiresome, monotonous) to which multistep intensity scales are assigned.

2. Standardized questions and nonstandardized answers

   The investigation is done either as a standardized interview, when a person answers predetermined questions verbally in his/her own words, or in written form with free formulation of answers. The answers can be assigned to categories and classifications afterward. This type of knowledge acquisition seems to be advantageous if the investigator cannot foresee all answer categories. The data processing, however, is more complicated and time consuming.

3. Nonstandardized questions and standardized answers

   This type of investigation is seldom used in the acquisition of ergonomics knowledge. A free-formulated question should be answered by a choice among several figures/drawings/sketches (about alternatives of ergonomic design) or a choice among preformulated statements. Nonstandardized questions are mostly coupled with the verbal presentation/interview.

4. Nonstandardized questions and nonstandardized answers

   This type of investigation, known as free interview or narrative interview is especially used when the prior (*ex ante*) knowledge about the object is scarce and the process of the interview leads to more specific and detailed questions. The evaluation of the information gathered is time consuming and therefore limited to case studies.

In ergonomic knowledge acquisition two other survey techniques are important. *Self-documentation* is used when job items have to be protocolled over a long time. The method can be standardized to different degrees, but it usually pertains to momentary task execution with a time-line protocol. Since the method is very economical for the investigator, it is used mostly in field studies with a broad repertoire of work systems over a long period of time (Frieling and Sonntag, 1987).

The method of *verbal protocols* (thinking aloud) is primarily used in cases in which the mental processes of a person have to be followed in detail (task research, user interface design). Therefore, the structure of cognitive processes is recorded in laboratory settings. The person is asked to vocalize his/her thoughts during task execution. This is usually recorded on tape and categorized afterward according to schemes. Besides the fact that the method is work-intensive, it may also suffer from a person's inability to express appropriately his/her thoughts during a complicated task execution.

Physiological Measurements

A person's state of strain frequently cannot be diagnosed by observation or interview because external signs of strain are difficult to interpret and interviews at short intervals would hinder task execution. Furthermore, persons may want to hide their real status of strain and thus try to give wrong information. Physiological measures (heart rate for example) are considered to be objective, because a person is mostly unable to influence them, and they can be recorded continuously. In addition, physiological measures may demonstrate strain levels which are not perceived by the working person and thus cannot be identified in interviews. The following physiological systems allow measurements for the purpose of ergonomic diagnosis:

- Circulation and respiration system: heart rate, arrhythmia, respiration frequency, blood pressure, etc.
- Motor and limb system: biomechanical variables, electromyogram, tremor, etc.
- Brain and central nervous system: EEG, CNV, evoced potentials, etc.
- Visual system: eye movements, EOG, lid frequency, flicker fusion frequency, etc.
- Dermal system: skin conductance/resistance, EDA, skin con./res. responses.
- Hormonal system: catecholamines, cortisol, etc.
- Metabolic system: respiration volume, O_2/CO_2, energy expenditure, etc.

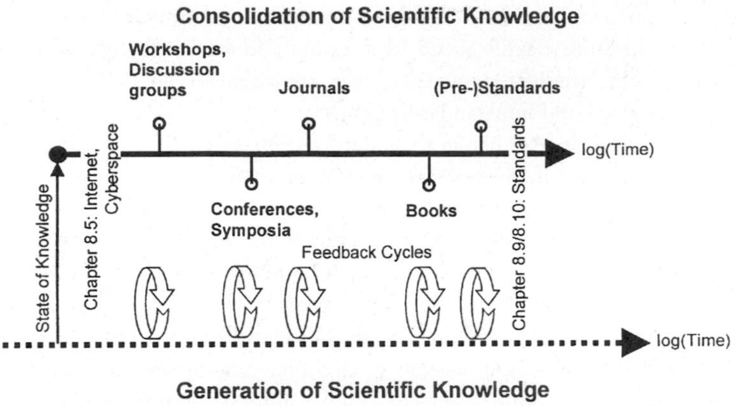

FIGURE 3.1 Time span for the consolidation of scientific knowledge.

In principle, the interpretation of physiological measurements follows two patterns:

- The measured variable directly indicates a bottleneck in the respective organic system. This pattern applies, for example, to heart rate with respect to a heavy muscular work load.
- Variations in organism measurements are interpreted as indicators of a central process. The decrease of flicker fusion frequency, for instance, is seen as an indicator of fatigue.

Physical and Chemical Measures

The physical and chemical methods (beneath the physiological measurements, which are, in principle, physical and chemical measures as well) can be subdivided into those related to the working person and those related to the work environment. The first group consists of methods of time and motion study as well as the analysis of body dimensions and forces. So these measures imply times, distances, forces, and variables derived, like speed, acceleration, and (physical) power. For the description of the work environment techniques of climate, radiation, illumination, vibration, and noise measurement, as well as the quantitative diagnosis of gases, dusts, chemical compositions, etc., are considered.

Consolidation of Ergonomics Knowledge

In order to consolidate the generated ergonomics knowledge there have to be people in a scientific community who discuss, criticize/encourage, transmit, and "store" the information. Knowledge can be differentiated as personal, institutional, organizational, or societal. All these types of knowledge have their own means of distribution and, therefore, need specific guidelines for their use. In this section, scientific knowledge is characterized as consolidated knowledge in the field of ergonomics, which can be acquired by everyone. In principle, that means no barriers hinder peoples' access. The factor "time" structures the consolidation process as illustrated in Figure 3.1.

In the early consolidation phase, different techniques and media are used to discuss ideas and to refine statements and hypotheses. One technique in this phase is the *workshops.* These are often organized in conjunction with conferences and symposia (see next phase) but have a more open and informal character. Another technique is *discussion groups* which are formally or informally organized by special interest groups. (Today these discussion groups are often aided by electronic communication media like the Internet.) The special interest groups are usually established by scientific societies, by scientific projects from a variety of institutions and organizations, or by informal relations among researchers sharing a common interest (for example, the technical groups of the International Ergonomics Association).

After its creation and informal discussion, scientific information is distributed to a broader audience. A characteristic of this phase is that the creator of the information does not exactly know who receives it. In the initial step, a typical way of publishing scientific papers is through *conferences* and *symposia.* Two important examples in the domain of ergonomics are the annual meeting of the Human Factors and Ergonomics Society and the triennial congress of the International Ergonomics Association. Unfortunately, similar events held by scientific societies like the SELF, GfA, NES, etc., not held in the English language do not capture the attention of the worldwide scientific community. In general the whole set of proposed papers has to pass a selection process of a reviewing committee before the researcher can present the results in the conference/symposia "proceedings." Normally, the members of the reviewing committee are well-known experts in a certain domain, which guarantees the scientific quality of the proposals. If the proposal passes the review process and is printed in the proceedings, this presents a good opportunity for discussion and defense of the results of one's research. Today, the presentations — abstracts or long form — are not printed only on paper, but are also often published in an electronic media such as CD-ROM. Due to printing and publication processes, there is a certain time lag between the printed/electronic product and the personal presentation. With reference to the stability of knowledge, this time lag can be either positive or negative: if the material is printed before the conference (preprints), the results of the discussion, the criticism of specific topics, etc., cannot be included in the publication. The results are "older" than the media suggests. If the publication is made after the conference, the results of discussions can be reflected by the "proceedings," including a documentation of the discussion process (protocols, etc.). Thus, the results are more actual than those presented at the conference; but in the worst case the publication is rejected *a posteriori.* Usually the research activities are continued and, depending on the publication time, the scientific information will probably be older than the actual research. This is one reason for using electronic publication provided mainly by the Internet.

The next step of consolidating scientific knowledge is the publication through *national* or *international journals* which are, in some cases, directly linked to a conference. Two examples of primarily basic research-oriented international journals are *Ergonomics* and the *Human Factors. Applied Ergonomics* and the *International Journal of Industrial Ergonomics* are examples of international journals with a stronger focus on application-oriented research. In general, every scientific journal has an established review procedure to guarantee the scientific quality standards of the corresponding community, which are reflected by the journal as well. Therefore, the review process is also a stabilization procedure, because the reviewers' comments and proposals are influencing the discussion of the results and the conclusion. Due to publication processes lasting up to two years, the relevancy compared to contemporary research can be rather low. Nevertheless, the published information in journals can be characterized as "state of the current research."

In addition, *scientific books* are primarily written by researchers having a sound knowledge and experience in a specific domain. But economic criteria must also be taken into account, because a publisher will only be willing to publish a book if the interested community (customer base) is large enough to have market potential. With reference to guides to scientific sources of ergonomics knowledge, the most interesting books are *handbooks.* Some good examples of handbooks with a large circulation are the *Handbook of Human Factors* (Salvendy, 1997), the *Handbook of Industrial Engineering* (Salvendy, 1992), the *Handbook of Perception and Human Performance* (Boff et al., 1986), or the *Handbook of Human-Computer Interaction* (Helander et al., 1997). They normally give an excellent overview of a research field for two reasons. First, the research is broadly represented, and second, the scientific quality of the contributions is high because the editor wants to acquire the "scientific capacities" of a research domain for the authorship.

Concerning the time span between creation and establishment of scientific knowledge, the final step of the whole consolidation process is the definition of *standards.* Because standards have the same status as legislation, they are subject to many discussions from different social perspectives: employers and employees, manufacturers and customers, legislation and politics, etc. Before the establishment of standards, the status of the represented scientific knowledge is a draft or prestandard. Critical reflections of standards are usually part of the whole variety of publications such as conference proceedings, journals, and books.

Design-Oriented Compounds of Ergonomics Knowledge

Beneath the presented generic sources, there are also specific compounds of ergonomics knowledge, which are strongly related to the design of work systems. The process of work systems' design can be characterized as a problem-solving task: based on general ergonomic objectives and technological, economical, psychological, social, etc., constraints, the whole task is divided into meaningful subtasks (analysis). This reduction of general task complexity allows the acquisition of existing knowledge for a solution as well as the development of new solutions. When combining partial solutions to the overall solution (synthesis), conflicting partial solutions must be identified and nonconflicting solutions must be elaborated. Since this iterative design process is costly and time consuming, different *compounds* of ergonomics knowledge are available. A compound of ergonomics knowledge is characterized as a generic, design-oriented piece of knowledge in which the different reasons for specific solutions are hidden. In refining more problem-unspecific to more problem-oriented methods, six types of compounds can be differentiated:

- General principles and rules
- Checklists and data lists
- Cases and best practices
- Databases
- Knowledge bases and expert systems
- Problem-oriented tools

Following are some examples of compounds of ergonomics knowledge.

General Principles and Rules

A frequent question in work systems' design is the allocation of functions between man and machine. MABA–MABA rules (man are better at–machines are better at, according to Fitts, 1951) compare the abilities of humans and technical systems in order to decide how functions should be appropriately allocated. But MABA lists are only of general relevance for the evaluation of functions. Different metrics of human and machine, and therefore the incompatibility of the compared functions, are the major criticism of these rules. Nevertheless, they are a general guideline in systems automation.

Due to the general objective of stress reduction, the maximization of efficiency concerning the relation of the input (physiological costs, attention, etc.) and workers' outcome (assembled devices, identified errors, etc.) can be postulated. This general principle of work design implies two different types of design improvements. First, it is possible to achieve the same result, e.g., assembling the underside of a car, with lower physiological costs by rotating the car instead of assembling over head. Second, the designer can prove that additional tasks can be performed with nearly the same, and tolerable, costs, e.g., the use of machines' process times for the handling of additional machines.

Because the organic system has an exponential characteristic of fatigue and recreation, fatigue should be minimized in order to reduce recreation time. The time organization of work should consider this condition when scheduling breaks: more frequent but shorter breaks are better for efficiency than fewer and longer breaks of the same total duration (short breaks rule).

For purposes of occupational safety, engineering rules can be defined to guarantee a safe function of a machine, equipment, etc. (Rohmert and Becker-Biskaborn, 1974). Normally the principles and rules are formulated in relation to design characteristics, e.g., in the cited safety example, the connection principle (connection of safety and handling equipment) or the economics principle (don't disturb working procedures with safety equipment).

Mainly for human information processing tasks and the respective ergonomic systems' design, general principles of compatibility are helpful: stimulus–response, stimulus–stimulus, and response–response compatibility are useful to reduce the required mental capacities of encoding signals (perception), cognition, and carrying out a response (information output). Especially for those tasks with a high degree of decision making (knowledge-based information processing according to Rasmussen, 1983), mental or conceptual compatibility is required to reduce the costs induced by handling the system, e.g., using a computer program, rather than performing the task itself.

Checklists and Data Lists

The complexity of system characteristics to be designed requires aids to support design decisions. Checklists have been developed for more than 30 years (e.g., *Ergonomic System Analysis Checklist,* N.N., 1964) which compound ergonomics knowledge. Ergonomic objectives, ordered by different ergonomic problems, are related to questions the designer can ask for the evaluation of worksystems (e.g., Rohmert, 1974). Because checklists are merely an analytical tool, the most often cited weaknesses are missing guides to improved alternatives, as well as conflicts between different objectives (e.g., Easterby, 1967). It should be stated that, independent of the type of use, the checklists should be used by ergonomic experts only.

Checklists are also used if the conformity of design parameters of a worksystem has to be compared with regulations of standards or legislation. An example is the design of human–computer interfaces: ISO 9214 comprises a list of design rules. Software-ergonomic checklists (e.g., EVADIS, Oppermann et al., 1992) are formulated to enable the quantitative evaluation of existing or planned systems with respect to the standard. As stated above, the most often cited problem in this case is developing appropriate improvements.

Cases and Best Practices

Improving the workplace is sometimes a time- (and knowledge-) consuming challenge. Especially smaller companies, which have neither the staff nor the time for elaborating the "optimal" ergonomic solution, can best use practice cases that are documented through a variety of publications. For example, there are the "ergonomic checkpoints" (ILO, 1996), which provide an easy-to-understand compound of best practice cases. Mostly, the cases are organized according to different design problems, e.g., the height of desks, the use of colors, work organization, etc.

If the system to be designed is based on a set of design elements that are used for a variety of applications, e.g., the user–interface design in human–computer interaction, best practices are normally elaborated with help from different guidelines (see above) or standards (either company standards or public standards). In the human–computer interface domain each standard (X-Windows, Macintosh, etc.) has its own publication of "best practices" (e.g., OSF/Motif, 1993).

Databases and Information Systems

The designer of a work system needs a variety of data, such as geometrical dimensions, performance measurements (e.g., perception parameters) etc., which are usually stored in conventional or electronic databases. Following the latest developments, the hypertext/hypermedia databases are of great interest, especially concerning remote and online access. For example, the military standard MIL-STD 1472D (1991), "Human Engineering Criteria for Military Systems, Equipment, and Facilities," provides the user with criteria, principles, and practices as well as tables and figures for system design.

The more a database consists of guidance functions, the more it becomes an information system (see, for example, Becker-Biskaborn, 1975). The structuring of data, the rules which should be formulated for the use of data, as well as the relationships between the rules (conflicts, dependencies, etc.) are highly important for an efficient use of the information system.

Knowledge Bases and Expert Systems

The application of artificial intelligence techniques in the ergonomic domain has influenced a variety of system developments. In contrast to information systems, the expert system uses a set of rules (knowledge base) in conjunction with inference mechanisms to elaborate design proposals. Because the defined rules are only valid for a limited application domain, the design recommendations can only be used for partial problems of work systems' design.

An example of an expert system is ErgonExpert (Laurig and Rombach, 1989), which produces design recommendations for manual material handling tasks (anthropometric and biomechanical problems). The rules are generated by a model based on anatomy and biomechanics as well as on different empirical investigations, such as fatigue analysis and epidemiological research results.

Problem-Oriented and Rapid Prototyping Tools

For an increasing number of design problems computer-based tools are used. Because the tool uses information for its design purposes the tool itself compounds ergonomics knowledge. CAD-based man models (see, for example, Kroemer, 1988) are examples of tools used for anthropometrical and biomechanical design purposes. Each model is based on anthropometrical databases, storing the human geometry as basic information. Design-oriented information, and therefore ergonomics knowledge, is implemented if the basic information is combined with higher-level aggregates such as the influence of clothing (shoe heights, limited body angles because of clothing, etc.), or measurements of comfort as defined by models like RAMSIS (Tecmath, 1994). Furthermore, the aggregation of basic information to meta-information, like reaching areas, angles of vision, body angles, etc., refers to a compound of ergonomics knowledge.

Another category of design-oriented tools is user interface management systems (UIMS), which are used for rapid prototyping and programming of human–computer interfaces. The UIMS-tool (e.g., Visual Basic) offers a variety of design components like switches, radio buttons, or common dialog boxes for the graphical user interface. Each component has its predefined function and, therefore, can be used by the designer only in a limited application context (e.g., 1–n choice: take radio buttons; m–n choice: take switches, etc.). Because the graphical controls are predefined, the ergonomics knowledge of different functions and dimensions, as well as the question of color, location, relation to other elements, etc., is also compounded.

3.2 Structure of Ergonomic Problems and Empirical Investigation of Scientific Sources

Ordering Model

With regard to the research domain of ergonomics, scientific information and knowledge is acquired from different disciplines such as medicine, physiology, psychology, engineering, economics, sociology, etc. Hence, the sources of ergonomic knowledge are distributed in a variety of congress proceedings, journals, and books. This fact sometimes makes it costly to find the relevant information and, therefore, the utility of structuring aids is potentially high.

An ordering model of ergonomic problems and knowledge by Luczak and Volpert (1987) is oriented toward different levels for describing structures and processes that result when a working person is followed with different analytical approaches for hours, days, or weeks. Considering procedural aspects of work processes, the following levels can be distinguished (Figure 3.2):

- V1. Activity of sensomotory automatisms of a person, i. e., elementary operations in sequencing and control of movements
- V2. Goal-oriented, consciously controlled action of a person
- V3. Motive-related activity of persons, whose concrete results are produced by the sequential and logical arrangement of action
- V4. Cooperative work in groups, where the working person has to tune his activity to the activities of other persons
- V5. Organization within the company (employers, employees) and between companies to define the roles and orientations to which the working person has to contribute implicitly or explicitly to tasks
- V6. Work-oriented political actions, which shall maintain or modify the frame for the parties within the company and which may have severe consequences for any working person

In a structure-oriented form, the levels can be distinguished from the top in the following manner:

- S7. Political and societal organization of work.
- S6. Forms of industrial relations and organization.
- S5. Forms of cooperation in groups and human relations.

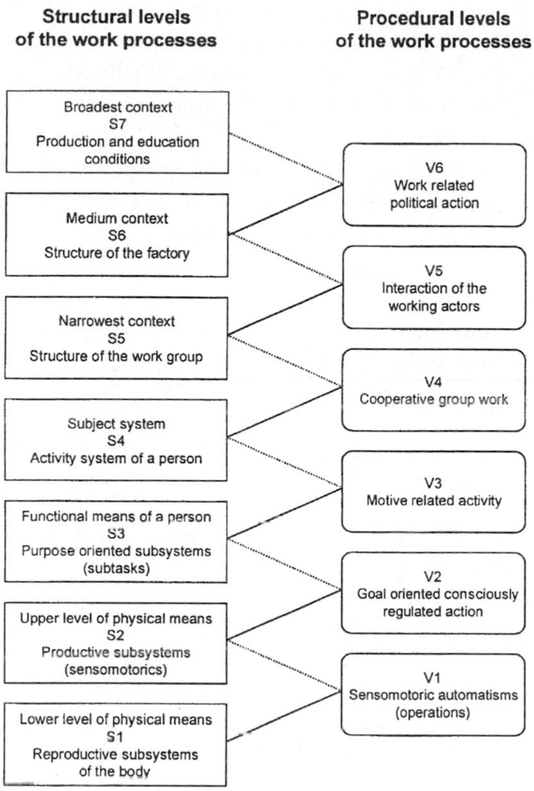

FIGURE 3.2 Structural and procedural levels of the worker process. (From Luczak, H., Volpert, W., 1987, *Arbeitswissenschaft. Kerndefinition–Gegenstandskatalog–Forschungsgebiete,* RKW-Verlag, Eschborn. With permission.)

S4. Forms and types of work and personal activities (individual work).
S3. Subtasks and workplaces.
S2. Operations with tools and working means.
S1. Vegetative systems and environmental factors.

It can easily be seen that these levels have a lot to do with the orientation of work-related scientific disciplines. Thus level S1 is *conditio sine qua non* of occupational physiology and health, but this discipline can widen its approach to levels S2 and S3 as well. Level 7 describes the specialty of political economy and macroeconomics in the sense of labor policy, but excursions to level 6 (industrial relations and organization) and even to level 5 (human relations and group work) may be possible. Levels S2 and S3 may be the focus of industrial engineering (motion study) and ergonomical design, whereas levels S5 and S6 form the core of personnel management and occupational sociology. An integrative approach to work can be assigned to levels S3 to S5.

Empirical Investigation

One of the most helpful sources of ergonomics knowledge is *Ergonomics Abstracts,* edited by The Ergonomics Information Analysis Center at the School of Manufacturing and Mechanical Engineering, University of Birmingham, U.K. The *Ergonomics Abstracts* have been published on paper by Taylor & Francis since 1967. In addition, since 1994 (vol. 26) the scanned abstracts are also available on CD-ROM, which significantly improves the search and retrieval procedure.

Based on the 1996 edition of *Ergonomics Abstracts* an empirical investigation was made with the purpose of an exhausted scanning of relevant sources (abstracts), which are included in conference proceedings, journals, and handbooks. These abstracts were analyzed in relation to different ergonomic aspects as represented by the ordering model introduced in the previous section. The empirical basis stored in the 1996 CD-ROM edition is approximately 44,000 abstracts from the whole variety of ergonomics. Most abstracts are dated no later than 1985. For economic reasons, small congresses and journals with just a small number of scanned abstracts (<10) were excluded from the CD-database, so that approximately 24,000 abstracts were finally investigated.

The classification scheme of the *Ergonomics Abstracts* corresponds with the ordering model. The assignment of classification items of *Ergonomics Abstracts* to the structural levels of the ordering model is shown in Table 3.1 (an additional zero level was added to the ordering model in order to categorize general aspects).

TABLE 3.1 Assignment of the Classification Items of Ergonomics Abstracts (item number in brackets) to the Structural Levels of Work Processes (see Figure 3.2).

0. Basics
- General (1)

S1. Autonomous Organic Systems & Work Environment
- Physiological and Anatomical Aspects (3)
- Environment:
  - *Illumination (29); Noise (30); Vibration (31); Whole Body Movement (32); Climate (33); Atmosphere (34); Altitude, Depth and Space (35); Other Environmental Issues (36)*
- Health and Safety:
  - *General Health and Safety (48); Etiology (49); Injury and Illness Prevention (50)*

S2. Willfully Steered Organic Systems & Tool/Work Means
- Psychological Aspects (2)
- Man-Machine Interface Design:
  - *Choice of Communication Media (10); Input Devices and Controls (21); Visual Displays (22); Auditory Displays (23); Other Modality Displays (24); Display and Control Characteristics (25)*
- Workplace Design:
  - *General Workplace Design and Buildings (26); Workstation Design (27); Equipment Design (28)*

S3. Tasks & Workplace:
- *Visual Communication (8); Auditory and Other Communication Modalities (9); Person-Machine Dialogue Mode (11); System Feedback (12); Error Prevention and Recovery (13); Design of Documentation and Procedures (14); User Control Features (15); Language Design (16); Database Organization and Data Retrieval (17); Programming, Debugging, Editing and Programming Aids (18); Software Performance and Evaluation (19); Software Design, Maintenance and Reliability (20)*

S4. Qualification, Motivation and Work Contents & Types of Work
- Individual and Task Related Organizational Factors:
  - *Individual Differences (5); Psychophysiological State Variables (6); Task Related Factors (7)*
- Job Design:
  - *Job Attitudes and Job Satisfaction (40); Job Design (41)*

S5. Work Group & Cooperative Processes
- Group Factors (4)
- Group Conflicts and Support:
  - *Supervision (45); Use of Support (46)*

S6. Company & Company Wide Organizational Measures
- General System Characteristics:
  - *General System Features (37); Total System Design and Evaluation (38)*
- Hours of Work (39)
- Payment Systems (42)
- Human Resources:
  - *Selection and Screening (43); Training (44)*

S7. Society & Social and Labour Politics
- Technological and Ergonomic Change (47)
- Social and Labour Politics:
  - *Trade Unions (52); Employment, Job Security, and Job Sharing (53); Productivity (54); Women at Work (55); Organizational Design (56); Education (57); Law (58); Privacy (59); Family and Home Life (60); Quality of Working Life (61); Political Comment and Ethical Considerations (62)*

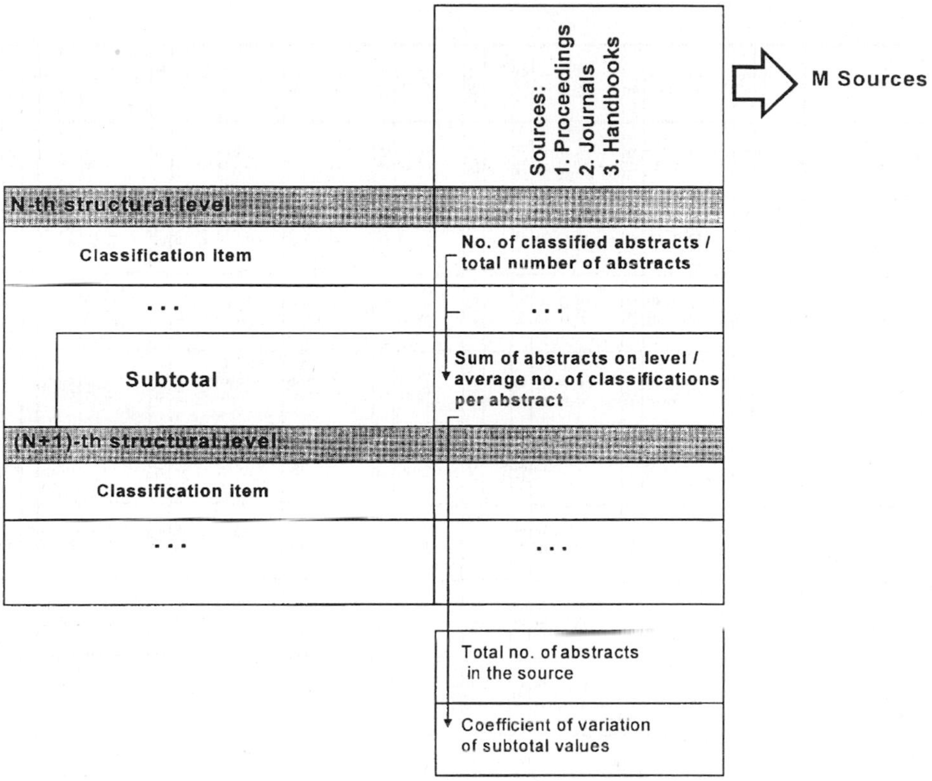

FIGURE 3.3 Calculation scheme of the four "relevance measures."

Based on this systematology, four "relevance measures" were calculated for every classification category (see calculation scheme in Figure 3.3):

1. As a first measure, the total number of scanned abstracts per investigated source was calculated. This was used as a simple indicator of the relevance of the sources in relation to the whole field of ergonomics knowledge. (If the proceedings or journal contained fewer than 10 abstracts, the source was excluded from the calculation.)
2. The number of abstracts which were classified with the corresponding classification item (cf. Table 3.1) divided by the total number of abstracts on the level of the systematology was used as an indicator of the relevance of the source in relation to the specific classification item.
3. A subtotal measure (weighted relevance indicator) was calculated, which equals the sum of the abstracts classified to a whole level of the systematology divided by the average number of classifications per abstract. Thus the subtotal measure is "normalized," because the same abstract might be classified twice or three times with respect to an ordering level, and multiple search results for the same abstract must be taken into account. (Thus, the sum of the subtotal measures over all ordering levels is approximately 100 percent.)
4. The coefficient of variation (CV) of the subtotal measures was calculated. The CV equals the standard deviation of the subtotal measures divided by their average and is therefore an indicator of the degree of specialization of the scientific source as a whole. A large CV indicates a high degree of specialization on few classification items and vice versa.

The results concerning the four measures are shown in Table 3.2 for conference/symposia proceedings, Table 3.3 for journals, and Table 3.4 for handbooks. The titles of the various proceedings which were combined in one heading, as used in Table 3.2, are listed in the appendix.

TABLE 3.2 Results of the Empirical Investigation Concerning Congress/Symposia Proceedings

Ergonomics Abstracts - Classification	Advances in Human-Computer Interaction	Advances in Industrial Ergonomics	Advances in Man-Machine Systems	Analysis, Design and Evaluation of Man-Machine Systems	Computers for Handicapped Persons	Contemporary Ergonomics	Design of Work and Development of Personnel in Advanced Manufacturing	Empirical Studies of Programmers	Environmental Ergonomics	Ergonomics of Hybrid Automated Systems	European Coal and Steel Community	Graphics Interface	HCI
0 Basics													
GENERAL	0,0	10,2	0,0	3,1	0,0	11,5	5,3	0,0	9,2	8,3	19,7	4,2	4,2
Subtotal: [1]	***0,0***	***3,7***	***0,0***	***1,1***	***0,0***	***4,4***	***1,9***	***0,0***	***3,7***	***3,4***	***4,3***	***1,6***	***1,6***
S1 Autonomous Organic Systems & Work Environment													
PHYSIOLOGICAL AND ANATOMICAL ASPECTS	0,0	48,1	9,5	2,6	3,3	21,6	5,3	0,0	64,2	3,4	28,9	0,0	4,0
ENVIRONMENT	0,0	12,7	4,8	0,5	0,0	8,5	0,0	0,0	80,8	3,8	78,6	0,0	2,6
HEALTH AND SAFETY	0,0	72,4	14,3	15,9	0,0	33,0	21,1	0,0	62,5	28,3	75,1	0,0	6,5
Subtotal: [1]	***0,0***	***48,0***	***8,7***	***6,9***	***1,5***	***23,9***	***9,3***	***0,0***	***83,0***	***14,4***	***40,1***	***0,0***	***5,0***
S2 Wilfully Steered Organic Systems & Tool/Work Means													
PSYCHOLOGICAL ASPECTS	23,5	8,1	76,2	46,7	10,0	22,1	42,1	57,4	0,8	19,2	23,1	16,7	21,3
MAN-MACHINE INTERFACE DESIGN	8,8	5,4	38,1	18,5	96,7	18,3	0,0	1,9	0,8	7,9	20,2	50,0	19,0
WORKPLACE DESIGN	0,0	27,1	0,0	6,7	10,0	25,9	0,0	0,0	3,3	8,7	75,7	2,1	4,2
Subtotal: [1]	***11,3***	***14,6***	***34,8***	***26,2***	***53,0***	***25,2***	***14,8***	***19,8***	***2,0***	***14,5***	***26,1***	***26,0***	***17,0***
S3 Tasks & Workplaces													
Subtotal: [1]	***55,7***	***1,6***	***24,6***	***26,2***	***25,8***	***12,1***	***3,7***	***54,3***	***0,0***	***10,2***	***5,1***	***59,1***	***36,3***
S4 Qualification, Motivation and Work Content & Types of Work													
INDIVIDUAL AND TASK RELATED ORGANIZATIONAL FACTORS	14,7	23,2	33,3	16,4	0,0	21,2	21,1	13,0	19,2	9,1	17,9	2,1	21,5
JOB DESIGN	0,0	20,5	4,8	13,3	0,0	11,0	31,6	5,6	0,0	30,2	40,5	2,1	11,9
Subtotal: [1]	***5,2***	***15,8***	***11,6***	***10,8***	***0,0***	***12,2***	***18,5***	***6,2***	***7,7***	***15,9***	***12,8***	***1,6***	***12,8***
S5 Work Group & Cooperative Processes													
GROUP FACTORS	14,7	14,7	0,0	3,1	0,0	8,7	0,0	51,9	7,5	1,9	1,2	2,1	6,9
GROUP CONFLICTS AND SUPPORT	5,9	1,0	9,5	8,2	0,0	2,8	0,0	3,7	0,0	2,3	0,6	2,1	3,7
Subtotal: [1]	***7,2***	***5,7***	***2,9***	***4,1***	***0,0***	***4,4***	***0,0***	***18,5***	***3,0***	***1,7***	***0,4***	***1,6***	***4,0***
S6 Company & Company Wide Organizational Measures													
GENERAL SYSTEM CHARACTERISTICS	55,9	7,2	19,0	50,8	33,3	27,3	31,6	0,0	0,0	34,7	19,7	20,8	35,3
HOURS OF WORK	0,0	4,6	0,0	0,0	0,0	0,8	5,3	0,0	0,8	1,5	1,2	0,0	1,5
PAYMENT SYSTEMS	0,0	0,1	0,0	0,0	0,0	0,0	0,0	0,0	0,0	1,1	0,0	0,0	0,0
HUMAN RESOURCES	2,9	4,4	28,6	5,1	10,0	6,2	31,6	1,9	0,8	5,3	6,4	0,0	3,4
Subtotal: [1]	***20,6***	***5,9***	***14,5***	***20,4***	***19,7***	***13,0***	***24,1***	***0,6***	***0,7***	***17,3***	***6,0***	***7,9***	***15,4***
S7 Society & Social and Labour Politics													
TECHNOLOGICAL AND ERGONOMIC CHANGE	0,0	4,6	0,0	7,7	0,0	5,3	26,3	0,0	0,0	31,3	11,0	2,1	10,5
SOCIAL AND LABOUR POLITICS	0,0	8,6	9,5	4,1	0,0	7,6	52,6	1,9	0,0	24,9	12,7	4,2	9,8
Subtotal: [1]	***0,0***	***4,8***	***2,9***	***4,3***	***0,0***	***4,9***	***27,8***	***0,6***	***0,0***	***22,7***	***5,2***	***2,4***	***7,8***
Average Number of Classifications per Abstract:	2,9	2,8	3,3	2,7	2,2	2,6	2,8	3,0	2,5	2,5	4,6	2,6	2,6
Number of Abstracts:	34	803	21	195	30	792	19	54	120	265	173	48	875
Coefficient of Variation:	1,51	1,22	0,96	0,82	1,55	0,66	0,84	1,50	2,29	0,57	1,1C	1,65	0,89

[1] Weighted Relevance (Subtotal Divided by Average Number of Classifications per Abstract)

Human Decision Making	Human Factors and Industrial Design	Human Factors and Power Plants	Human Factors in Computing Systems	Human Factors in Information Systems	Human Factors in Manufacturing	Human Factors in Organizational Design and Management	Human Factors in Telecommunications	Human Factors Association	Human Factors and Ergonomics Society	Human-Computer Interaction - Interact '87 und '90	IEEE	Industrial Ergonomics and Safety	International Ergonomics Association	Man-Computer Interaction Research	National Board/National Institute of/for Occupational Safety and Health	People and Computers	SAFE Association	The Ergonomics of Manual Work	The Ergonomics of Working Postures	Vision in Vehicles	Visual Search	Work with Display Units	TOTAL:
2,2	10,8	12,8	18,6	0,0	10,0	9,8	9,1	15,1	11,5	2,1	6,3	10,0	13,7	0,0	19,2	4,2	8,4	13,2	2,9	4,9	0,0	8,4	
0,9	4,1	5,4	5,7	0,0	3,8	3,8	3,1	5,2	4,3	0,7	2,5	3,8	6,3	0,0	7,3	1,5	3,6	5,8	1,2	1,6	0,0	2,7	
5,8	20,6	0,9	1,2	0,0	6,9	7,7	0,9	28,8	12,9	0,6	0,0	45,3	28,1	3,9	24,3	0,4	17,6	49,7	97,1	2,5	6,3	11,6	
0,7	1,0	3,2	0,0	0,0	4,6	4,1	0,9	12,1	7,1	0,3	6,3	13,3	10,2	1,3	53,1	0,0	66,1	6,0	2,9	23,3	4,8	19,9	
8,0	21,6	45,7	0,0	0,0	16,2	22,2	1,3	68,2	22,5	0,0	9,4	69,6	36,4	0,0	69,5	1,1	82,0	41,7	41,2	24,5	0,0	56,6	
6,2	16,2	21,0	0,4	0,0	10,6	13,2	1,0	37,7	16,1	0,3	6,2	48,3	34,2	1,6	55,9	0,5	70,3	43,0	59,3	16,5	5,2	28,6	
65,0	10,8	22,8	30,2	29,2	13,8	11,3	15,7	15,6	34,3	22,6	40,6	9,3	14,3	61,0	6,8	24,3	3,3	6,6	0,0	69,3	95,2	22,7	
21,9	12,7	19,2	18,6	8,3	5,4	11,3	73,9	12,8	28,6	17,1	25,0	5,0	9,1	10,4	3,4	8,8	8,4	2,6	2,9	70,6	28,6	32,7	
2,9	73,5	5,9	1,2	0,0	13,1	12,9	12,6	32,5	18,2	0,3	3,1	25,3	18,7	1,3	20,3	1,1	26,8	25,2	32,4	38,7	1,6	10,8	
37,8	36,5	20,2	15,3	14,3	12,4	13,8	30,5	21,0	30,8	14,1	27,2	14,9	19,3	22,5	11,6	12,2	16,3	15,2	14,8	58,7	58,5	21,5	
18,8	14,8	14,0	44,5	23,8	5,6	5,2	32,5	5,9	15,0	52,0	34,6	2,3	4,9	30,9	0,0	54,0	1,8	0,3	0,0	7,3	11,9	8,2	
26,3	4,9	8,2	17,4	33,3	3,8	20,1	8,3	22,2	24,0	16,8	0,0	22,1	20,9	55,8	29,4	10,9	5,0	31,1	14,7	19,0	25,4	45,8	
8,8	2,0	12,3	2,3	12,5	30,8	40,7	1,7	18,8	8,9	6,[illegible]	15,6	19,4	14,3	20,8	11,3	2,8	1,7	25,8	23,5	1,8	4,8	16,3	
14,8	2,6	8,7	6,0	17,5	13,3	23,6	3,4	14,2	12,5	8,[illegible]	6,2	15,6	16,3	23,7	15,5	4,9	2,8	25,1	16,0	6,9	14,1	20,2	
5,8	16,7	1,8	24,4	8,3	4,6	9,8	6,5	13,5	12,9	12,3	0,0	13,3	9,9	18,2	10,2	9,2	2,1	9,3	17,6	14,1	15,9	10,0	
7,3	2,9	15,1	12,8	0,0	2,3	1,0	8,7	0,7	4,0	7,[illegible]	6,2	1,2	0,9	9,1	0,0	6,0	0,0	0,0	0,0	0,6	3,2	1,6	
5,5	7,4	7,1	11,4	3,2	2,7	4,2	5,1	4,9	6,4	7,[illegible]	2,5	5,5	5,0	8,4	3,9	5,4	0,9	4,1	7,4	4,8	8,9	3,8	
27,7	43,1	25,6	45,3	66,7	28,5	26,3	56,1	10,8	19,0	40,2	40,6	4,8	10,7	20,8	1,7	51,1	3,8	4,0	0,0	3,1	0,0	11,2	
0,7	0,0	2,7	0,0	0,0	0,8	1,5	1,3	3,2	1,1	0,0	0,0	4,2	2,5	1,3	1,1	0,0	0,0	2,0	0,0	0,6	0,0	7,2	
0,0	0,0	0,0	0,0	0,0	2,3	0,0	0,0	0,2	0,1	0,3	0,0	0,1	0,1	0,0	0,0	0,0	0,0	0,0	0,0	0,0	0,0	0,4	
5,8	1,0	18,3	0,0	16,7	12,3	8,2	2,2	5,5	10,7	3,7	3,1	4,2	3,5	9,1	1,7	3,5	5,4	0,7	0,0	8,0	3,2	1,6	
14,5	16,6	19,6	13,9	31,7	16,8	14,0	20,1	6,8	11,7	15,[illegible]	17,3	5,0	7,7	9,6	1,7	19,4	3,9	2,9	0,0	3,8	1,5	6,6	
2,2	1,0	1,4	5,8	8,3	42,3	23,7	3,9	4,1	2,5	2,4	0,0	4,9	5,2	7,8	5,6	1,8	0,0	3,3	0,0	0,0	0,0	13,5	
1,5	3,9	8,2	3,5	16,7	48,5	33,5	9,1	8,2	5,9	4,3	9,4	7,6	8,8	2,6	5,1	4,2	0,8	4,6	2,9	1,2	0,0	12,7	
1,5	1,8	4,0	2,8	9,5	34,8	22,2	4,4	4,3	3,2	2,3	3,7	4,7	6,4	3,2	4,1	2,1	0,4	3,5	1,2	0,4	0,0	8,5	
2,4	2,7	2,4	3,3	2,6	2,6	2,6	3,0	2,9	2,6	2,9	2,5	2,7	2,2	3,2	2,6	2,8	2,4	2,3	2,4	3,0	2,1	3,1	
137	102	219	86	24	130	194	230	437	2637	328	32	1078	1438	77	177	284	239	151	34	163	63	251	11965
0,97	0,92	0,57	1,12	0,92	0,82	0,62	1,05	0,94	0,71	1,35	0,99	1,22	0,82	0,93	1,46	1,43	1,91	1,19	1,60	1,54	1,55	0,77	

TABLE 3.3 Results of the Empirical Investigation Concerning International Journals

Ergonomics Abstracts - Classification	Accident Analysis and Prevention	ACM Transactions on Information Systems	Acta Psychologica	American Journal of Physical Medicine & Rehabilitation	American Journal of Psychology	Annals of Occupational Hygiene	Applied Cognitive Psychology	Applied Ergonomics	Applied Occupational and Environmental Hygiene	Australian Journal of Psychology	Australian Safety News	Automatica	Aviation, Space and Environmental Medicine	Behavior and Information Technology
0 Basics														
GENERAL	5,2	2,1	1,4	0,0	0,0	34,9	0,0	11,0	20,0	0,0	16,7	6,3	2,2	6,1
Subtotal: [1]	***1,8***	***0,7***	***0,6***	***0,0***	***0,0***	***11,4***	***0,0***	***3,3***	***6,2***	***0,0***	***6,6***	***2,0***	***0,7***	***2,0***
S1 Autonomous Organic Systems & Work Environment														
PHYSIOLOGICAL AND ANATOMICAL ASPECTS	2,2	2,1	6,8	84,2	0,0	7,0	0,0	46,4	36,7	5,9	8,3	0,0	47,7	1,4
ENVIRONMENT	4,4	0,0	0,7	5,3	0,0	72,1	7,2	16,0	56,7	0,0	31,0	0,0	75,6	2,6
HEALTH AND SAFETY	113,3	0,0	0,7	84,2	0,0	120,9	8,7	49,3	113,3	29,4	119,0	6,3	39,6	4,9
Subtotal: [1]	***41,3***	***0,7***	***3,7***	***70,2***	***0,0***	***65,2***	***6,0***	***33,7***	***63,9***	***12,0***	***63,0***	***2,0***	***53,8***	***2,9***
S2 Wilfully Steered Organic Systems & Tool/Work Means														
PSYCHOLOGICAL ASPECTS	46,7	21,3	93,2	10,5	89,5	2,3	87,0	17,7	0,0	52,9	2,4	50,0	33,7	28,0
MAN-MACHINE INTERFACE DESIGN	12,6	17,0	13,0	5,3	15,8	2,3	15,9	24,2	0,0	23,5	7,1	18,8	10,0	28,8
WORKPLACE DESIGN	12,6	2,1	0,0	0,0	0,0	4,7	0,0	40,2	46,7	5,9	11,9	0,0	2,6	9,8
Subtotal: [1]	***24,7***	***14,1***	***48,3***	***6,4***	***55,6***	***3,0***	***38,8***	***24,7***	***14,4***	***28,0***	***8,5***	***21,6***	***15,3***	***21,8***
S3 Tasks & Workplaces														
Subtotal: [1]	***2,3***	***71,1***	***10,9***	***0,0***	***16,7***	***1,5***	***16,4***	***4,8***	***3,1***	***6,0***	***2,4***	***29,4***	***1,3***	***27,7***
S4 Qualification, Motivation and Work Content & Types of Work														
INDIVIDUAL AND TASK RELATED ORGANIZATIONAL FACTORS	30,4	4,3	37,7	26,3	26,3	14,0	40,6	30,6	3,3	94,1	14,3	18,8	46,1	23,3
JOB DESIGN	4,4	10,6	0,0	0,0	0,0	2,3	2,9	18,4	13,3	29,4	6,0	31,3	5,2	18,2
Subtotal: [1]	***12,0***	***5,2***	***17,1***	***10,6***	***13,9***	***5,3***	***16,4***	***14,8***	***5,2***	***42,0***	***8,1***	***15,7***	***16,9***	***13,6***
S5 Work Group & Cooperative Processes														
GROUP FACTORS	32,6	2,1	17,8	21,1	21,1	14,0	33,3	24,2	10,0	29,4	1,2	0,0	13,3	17,3
GROUP CONFLICTS AND SUPPORT	2,2	4,3	0,0	0,0	0,0	0,0	11,6	4,1	0,0	0,0	1,2	0,0	0,7	8,1
Subtotal: [1]	***12,0***	***0,7***	***8,1***	***8,5***	***11,1***	***4,5***	***16,9***	***8,5***	***3,1***	***10,0***	***0,9***	***0,0***	***4,6***	***8,3***
S6 Company & Company Wide Organizational Measures														
GENERAL SYSTEM CHARACTERISTICS	2,2	10,6	6,2	0,0	0,0	0,0	2,9	12,4	3,3	0,0	1,2	68,8	1,2	35,2
HOURS OF WORK	3,0	0,0	0,7	0,0	0,0	4,7	2,9	3,3	0,0	0,0	6,0	0,0	2,6	0,6
PAYMENT SYSTEMS	0,0	0,0	0,0	0,0	5,3	2,3	0,0	0,7	0,0	0,0	0,0	0,0	0,0	0,3
HUMAN RESOURCES	5,9	0,0	17,1	0,0	0,0	2,3	7,2	6,2	3,3	0,0	2,4	0,0	16,0	7,2
Subtotal: [1]	***3,8***	***3,7***	***10,9***	***0,0***	***2,8***	***3,0***	***4,9***	***6,8***	***2,1***	***0,0***	***3,8***	***21,6***	***6,5***	***14,2***
S7 Society & Social and Labour Politics														
TECHNOLOGICAL AND ERGONOMIC CHANGE	0,0	10,6	0,0	5,3	0,0	9,3	0,0	4,3	6,7	0,0	3,6	12,5	0,3	15,0
SOCIAL AND LABOUR POLITICS	5,9	29,8	0,7	5,3	0,0	9,3	1,4	6,7	0,0	5,9	13,1	12,5	2,2	14,1
Subtotal: [1]	***2,0***	***3,7***	***0,3***	***4,3***	***0,0***	***6,1***	***0,5***	***3,3***	***2,1***	***2,0***	***6,6***	***7,8***	***0,9***	***9,5***
Average Number of Classifications per Abstract:	2,9	2,9	2,2	2,5	1,9	3,1	2,7	3,3	3,2	2,9	2,5	3,2	3,0	3,1
Number of Abstracts:	135	47	146	19	19	43	69	418	30	17	84	16	581	347
Coefficient of Variation:	1,13	1,93	1,24	1,89	1,49	1,72	1,02	0,90	1,69	1,20	1,65	0,89	1,43	0,71

[1] Weighted Relevance (Subtotal Divided by Average Number of Classifications per Abstract)

Behavior Research Methods, Instruments & Computers	British Journal of Psychology	Cognitive Psychology	Cognitive Science	Communications of the ACM	Computer Supported Cooperative Work	Computers & Industrial Engineering	Computers in Human Behavior	CSERIAC Gateway	Ergonomics	Ergonomics in Design	European Journal of Applied Physiology and Occupational Physiology	European Journal of Cognitive Psychology	European Work and Organizational Psychologist	Health and Safety at Work	Human-Computer Interaction	Human Factors	Human Movement Science	Human Performance	Human Relations	IEEE Transactions on Biomedical Engineering	IEEE Transactions on Systems, Man and Cybernetics
1,6	1,2	0,0	0,0	2,6	0,0	13,4	4,2	45,2	9,8	8,7	0,5	0,0	5,6	56,0	4,4	3,7	0,0	2,5	7,7	3,4	2,0
0,9	***0,5***	***0,0***	***0,0***	***1,0***	***0,0***	***6,1***	***1,4***	***23,3***	***5,1***	***3,5***	***0,2***	***0,0***	***1,7***	***21,2***	***1,2***	***1,3***	***0,0***	***1,1***	***2,3***	***2,0***	***0,8***
8,8	3,6	3,7	3,1	1,7	0,0	25,4	3,5	22,6	0,2	10,9	96,0	1102,7	0,0	4,0	0,0	14,6	43,9	11,3	0,0	62,1	5,5
0,8	7,2	0,0	0,0	0,0	0,0	3,0	0,0	12,9	0,0	8,7	25,9	2,7	0,0	48,0	0,0	7,5	1,0	5,0	0,0	13,8	0,0
2,4	7,2	0,0	0,0	3,2	0,0	22,4	2,8	16,1	0,3	45,7	9,2	0,0	16,7	104,0	2,7	18,7	1,0	7,5	17,9	13,8	5,5
6,6	***7,8***	***2,8***	***1,5***	***1,9***	***0,0***	***23,1***	***2,2***	***26,7***	***0,3***	***26,3***	***53,9***	***86,7***	***5,2***	***59,1***	***0,7***	***14,5***	***25,4***	***10,2***	***5,3***	***52,0***	***4,5***
46,4	81,9	92,6	93,8	10,3	36,8	10,4	34,0	22,6	26,9	13,0	3,5	94,6	5,6	8,0	52,2	56,1	84,7	55,0	15,4	13,8	65,4
24,0	4,8	7,4	3,1	14,8	26,3	4,5	22,9	12,9	10,6	17,4	0,2	24,3	5,6	4,0	23,9	41,9	5,1	8,8	0,0	13,8	21,3
0,0	0,0	0,0	0,0	1,3	0,0	9,0	0,0	6,5	16,6	43,5	2,1	0,0	0,0	24,0	0,9	15,2	1,0	0,0	5,1	6,9	2,8
38,9	***37,5***	***75,0***	***47,0***	***10,3***	***21,4***	***10,9***	***19,7***	***21,7***	***28,4***	***29,8***	***2,4***	***9,3***	***3,4***	***13,6***	***21,6***	***40,3***	***50,3***	***27,4***	***6,1***	***20,0***	***36,2***
23,5	***3,6***	***8,3***	***27,3***	***45,1***	***12,5***	***24,5***	***15,9***	***6,7***	***6,0***	***9,6***	***0,0***	***0,8***	***0,0***	***0,0***	***37,5***	***13,2***	***5,1***	***2,7***	***0,0***	***4,0***	***27,8***
27,2	69,9	7,4	25,0	6,5	10,5	11,9	77,8	19,4	44,7	10,9	53,2	21,6	44,4	0,0	18,6	36,0	23,5	65,0	51,3	20,7	13,8
0,8	4,8	0,0	0,0	9,7	57,9	13,4	9,7	0,0	15,0	0,0	6,6	0,0	127,8	4,0	13,3	3,7	0,0	23,8	110,3	0,0	6,3
15,5	***32,3***	***5,6***	***12,1***	***6,3***	***23,2***	***11,6***	***30,3***	***10,0***	***31,3***	***4,4***	***24,6***	***1,7***	***53,4***	***1,5***	***8,9***	***14,2***	***13,0***	***38,2***	***48,1***	***12,0***	***8,1***
6,4	33,7	7,4	18,8	5,8	5,3	9,0	24,3	3,2	21,3	10,9	36,2	16,2	5,6	0,0	27,4	21,3	10,2	15,0	12,8	3,4	5,9
1,6	2,4	0,0	3,1	0,0	0,0	9,0	3,5	0,0	0,8	4,3	0,0	2,7	5,6	0,0	10,6	3,5	0,0	1,3	7,7	0,0	5,9
4,4	***15,6***	***5,6***	***10,6***	***2,3***	***1,8***	***8,2***	***9,6***	***1,7***	***11,6***	***6,1***	***14,9***	***1,5***	***3,4***	***0,0***	***10,7***	***8,8***	***5,6***	***7,0***	***6,1***	***2,0***	***4,8***
5,6	1,2	0,0	3,1	40,6	21,1	14,9	14,6	12,9	8,0	32,6	0,0	0,0	0,0	4,0	39,8	5,5	0,0	1,3	23,1	13,8	30,7
3,2	0,0	0,0	0,0	0,0	0,0	3,0	0,7	0,0	9,8	0,0	3,1	0,0	0,0	0,0	0,0	3,0	0,0	3,8	0,0	0,0	0,0
0,0	0,0	0,0	0,0	0,0	0,0	0,0	0,0	0,0	0,1	0,0	0,0	0,0	0,0	0,0	0,0	0,2	0,0	0,0	0,0	0,0	0,0
8,0	0,0	3,7	0,0	2,6	0,0	9,0	21,5	3,2	4,0	6,5	6,1	0,0	5,6	0,0	8,0	10,2	1,0	18,8	5,1	0,0	6,7
9,3	***0,5***	***2,8***	***1,5***	***16,9***	***7,1***	***12,2***	***12,7***	***8,3***	***11,5***	***15,8***	***3,8***	***0,0***	***1,7***	***1,5***	***13,4***	***6,7***	***0,6***	***10,2***	***8,4***	***8,0***	***15,2***
1,6	1,2	0,0	0,0	13,5	5,3	4.5	12,5	0,0	2,5	2,2	0,0	0,0	27,8	0,0	1,8	1,0	0,0	2,5	7,7	0,0	1,2
0,0	3,6	0,0	0,0	27,7	94,7	3,0	11,1	3,2	8,4	8,7	0,7	0,0	72,2	8,0	19,5	1,6	0,0	5,0	71,8	0,0	5,5
0,9	***2,1***	***0,0***	***0,0***	***16,1***	***33,9***	***3,4***	***8,2***	***1,7***	***5,7***	***4,4***	***0,3***	***0,0***	***31,0***	***3,0***	***6,0***	***0,9***	***0,0***	***3,2***	***23,7***	***0,0***	***2,7***
1,8	2,3	1,3	2,1	2,6	2,9	2,2	2,9	1,9	1,9	2,5	2,4	12,8	3,2	2,6	3,6	2,8	1,8	2,3	3,4	1,7	2,5
125	83	27	32	155	19	67	144	31	1197	46	425	37	18	25	113	508	98	80	39	29	254
1,05	1,18	2,03	1,34	1,17	1,01	0,61	0,76	0,80	0,91	0,84	1,51	2,41	1,55	1,63	0,97	1,00	,40	1,06	1,28	1,38	1,04

TABLE 3.3 (continued) Results of the Empirical Investigation Concerning International Journals

	Industrial Management & Data Systems	Information Processing & Management	Interacting with Computers	International Journal of Aviation Psychology	International Journal of Human-Computer Interaction	International Journal of Human-Computer Studies	International Journal of Human Factors in Manufacturing	International Journal of Industrial Ergonomics	International Journal of Lighting Research and Technology	International Journal of Manpower	International Journal of Production Research	International Journal of Psychology	International Review of Applied Psychology	Japanese Journal of Ergonomics	Journal of Applied Physiology	Journal of Applied Psychology	Journal of Educational Multimedia and Hypermedia	Journal of Experimental Psychology: Human Perception and Performance	Journal of Experimental Psychology: Learning, Memory and Cognition	Journal of Human Ergology	Journal of the Illuminating Engineering Society	Journal of Information Technology
	6,3	4,1	6,1	2,3	7,7	0,0	9,8	20,5	0,0	2,3	6,4	10,0	0,0	7,1	0,0	0,4	0,0	0,0	0,0	3,6	7,9	8,0
	3,2	1,7	2,0	0,8	2,2	0,0	3,0	6,2	0,0	0,8	2,9	4,5	0,0	3,4	0,0	0,2	0,0	0,0	0,0	1,3	3,7	2,7
	0,0	0,0	0,8	3,4	7,7	0,0	3,9	55,3	4,0	0,4	2,1	0,0	15,4	25,3	96,7	3,5	0,0	13,7	4,7	64,3	5,6	0,0
	0,0	0,0	0,8	4,5	3,4	0,0	2,0	22,5	112,0	0,2	0,0	20,0	23,1	10,1	48,4	2,2	0,0	3,4	6,5	16,4	93,3	0,0
	0,0	0,0	0,0	11,4	12,0	1,7	17,6	64,5	0,0	0,8	2,1	0,0	23,1	10,6	9,0	14,3	0,0	1,1	0,0	32,9	10,1	68,0
	0,0	0,0	0,5	7,1	6,7	0,6	7,2	42,8	50,9	0,4	2,0	9,1	17,8	22,2	57,7	7,8	0,0	11,6	6,2	39,9	51,3	23,3
	6,3	16,2	28,0	40,9	40,2	51,7	19,6	8,7	20,0	38,5	31,9	50,0	15,4	37,1	2,5	19,5	42,1	89,7	97,2	15,0	28,1	8,0
	0,0	8,1	22,7	29,5	27,4	17,2	2,9	6,0	32,0	18,1	2,1	10,0	0,0	21,5	0,0	2,6	21,1	11,4	6,5	5,0	40,4	0,0
	0,0	0,0	1,5	0,0	3,4	0,0	6,9	34,5	28,0	1,3	17,0	0,0	0,0	16,1	1,6	1,7	0,0	1,7	0,9	12,9	16,9	0,0
	3,2	10,2	17,0	25,9	20,4	23,7	9,0	14,8	35,1	19,7	23,5	27,3	4,4	36,1	1,5	9,4	26,1	65,5	58,0	11,6	40,2	2,7
	12,9	61,6	42,1	12,6	25,6	49,7	7,5	2,3	0,0	50,1	18,6	4,5	0,0	7,0	0,0	0,7	47,8	9,1	3,6	0,3	1,6	8,2
	0,0	14,9	10,6	50,0	46,2	15,5	17,6	27,5	20,0	20,0	8,5	50,0	84,6	23,7	57,4	51,5	15,8	14,3	29,0	45,7	3,4	0,0
	6,3	2,7	11,4	18,2	23,1	0,0	68,6	29,0	0,0	3,8	17,0	0,0	69,2	6,3	0,8	77,5	5,3	0,0	0,0	17,9	0,0	28,0
	3,2	7,3	7,1	25,1	20,0	5,3	26,4	17,0	8,8	8,1	11,8	22,7	44,4	14,5	21,8	50,7	8,7	9,1	16,1	22,4	1,6	9,6
	6,3	10,8	8,3	8,0	17,1	15,5	2,9	19,3	8,0	17,9	6,4	40,0	23,1	13,6	38,5	23,4	10,5	7,4	23,4	35,7	2,2	0,0
	0,0	6,8	6,1	0,0	7,7	5,2	5,9	1,7	0,0	7,9	2,1	0,0	7,7	0,3	0,0	6,1	0,0	0,0	2,8	10,7	1,1	4,0
	3,2	7,3	4,7	2,9	7,1	7,1	2,7	6,3	3,5	8,8	3,9	18,2	8,9	6,7	14,4	11,6	4,3	4,7	14,5	16,3	1,6	1,4
	25,0	14,9	51,5	11,4	33,3	31,0	33,3	9,2	0,0	25,1	21,3	0,0	23,1	10,6	0,0	0,0	21,1	0,0	2,8	0,7	0,0	28,0
	0,0	0,0	0,0	1,1	3,4	0,0	3,9	6,8	0,0	0,0	0,0	0,0	0,0	2,7	0,0	1,3	0,0	0,0	0,0	7,1	0,0	0,0
	0,0	0,0	0,0	0,0	0,9	0,0	2,9	0,2	0,0	0,0	0,0	0,0	0,0	0,0	0,0	3,5	0,0	0,0	0,0	2,9	0,0	0,0
	0,0	1,4	3,8	42,0	5,1	1,7	5,9	3,6	0,0	4,9	10,6	10,0	0,0	1,9	12,3	13,4	5,3	0,0	0,0	2,9	0,0	4,0
	12,9	6,8	18,0	20,1	12,3	11,2	14,1	6,0	0,0	10,2	14,7	4,5	6,7	7,4	4,6	7,1	10,9	0,0	1,6	4,8	0,0	11,0
	56,3	5,4	8,3	1,1	11,1	3,4	50,0	5,6	0,0	1,1	14,9	0,0	0,0	0,8	0,0	1,3	5,3	0,0	0,0	2,1	0,0	64,0
	62,5	6,8	18,2	13,6	8,5	3,4	48,0	9,7	4,0	4,9	34,0	20,0	61,5	4,9	0,0	30,7	0,0	0,0	0,0	7,9	0,0	56,0
	61,3	5,1	8,6	5,4	5,7	2,4	30,0	4,6	1,8	2,0	22,5	9,1	17,8	2,8	0,0	12,6	2,2	0,0	0,0	3,5	0,0	41,1
	1,9	2,4	3,1	2,7	3,5	2,9	3,3	3,3	2,3	2,9	2,2	2,2	3,5	2,1	2,7	2,5	2,4	1,6	1,8	2,8	2,1	2,9
	16	74	132	88	117	58	102	414	25	530	47	10	13	367	122	231	19	175	107	140	89	25
	1,62	1,61	1,08	0,80	0,68	1,35	0,83	1,06	1,56	1,32	0,70	0,72	1,17	0,92	1,59	1,29	1,33	1,75	1,55	1,08	1,66	1,08

Journal of Managerial Psychology	Journal of Occupational Health and Safety - Australia and New Zealand	Journal of Occupational and Organizational Psychology	Journal of Occupational Medicine	Journal of Organizational Behavior	Journal of Organizational Change Management	Journal of Safety Research	Journal of Science of Labour	Lighting Research and Technology	Military Psychology	New Technology, Work and Employment	Occupational Medicine	Perception	Perception & Psychophysics	Perceptual and Motor Skills	Presence	Professional Safety	Psychophysiology	Quarterly Journal of Experimental Psychology	Reliability Engineering and System Safety	Risk Analysis	Safety Science
5,3	9,5	0,0	6,5	1,1	0,0	6,4	9,0	9,3	0,0	0,0	8,6	0,0	0,6	0,8	2,2	16,4	0,0	2,3	11,1	7,1	23,9
1,6	4,0	0,0	1,8	0,3	0,0	2,0	3,0	3,7	0,0	0,0	2,7	0,0	0,5	0,4	1,0	6,9	0,0	1,3	5,3	2,6	6,9
5,3	11,1	0,0	31,5	2,2	0,0	11,7	23,9	4,0	4,5	0,0	22,9	8,0	4,8	18,9	4 3	4,8	97,7	3,1	0,0	3,6	6,8
0,0	12,7	0,0	24,0	0,0	0,0	8,5	35,8	101,3	4,5	1,4	31,4	11,5	8,9	11,5	0,0	15,8	4,7	3,1	0,0	14,3	14,8
47,4	104 8	42,9	138,0	35,2	15,4	134,0	47,8	13,3	4,5	0,0	120,0	0,9	0,0	5,3	0,0	123,0	2,3	0,0	71,1	78,6	167,0
16,1	53,6	13,3	53,4	9,4	6,1	49,5	36,5	47,1	5,4	0,5	54,5	13,5	10,8	15,6	1,9	60,9	41,3	3,6	34,0	34,6	54,2
5,3	3,2	14,3	6,5	16,5	0,0	23,4	12,7	38,7	40,9	0,0	11,4	74,3	72,5	63,8	28,3	9,1	53,5	93,8	51,1	75,0	23,9
0,0	1,6	0,0	2,5	0,0	0,0	10,6	8,2	29,3	6,8	0,0	0,0	19,5	9,9	15,6	69,6	3,0	7,0	13,2	0,0	0,0	1,1
0,0	3,2	0,0	9,0	2,2	0,0	27,7	12,7	25,3	0,0	0,0	5,7	3,5	1,3	2,9	2,2	16,4	0,0	0,8	0,0	3,6	39,8
1,6	3,3	4,4	5,0	4,7	0,0	19,8	11,4	37,0	18,8	0,0	5,4	65,[illegible]	66,0	35,9	44,2	12,1	23,9	61,8	24,5	28,2	18,6
0,0	0,7	2,2	0,0	0,3	0,0	2,0	2,8	4,2	3,6	0,5	0,0	10,7	7,6	3,9	21,2	2,1	0,0	5,8	13,8	7,7	2,9
84,2	31,7	85,7	54,0	87,9	30,8	31,9	45,5	9,3	56,8	1,4	34,3	8,0	11,8	65,8	30,4	6,7	72,1	31,8	0,0	17,9	14,8
68,4	7,9	103,6	25,0	125,3	38,5	10,6	22,4	0,0	29,5	63,9	20,0	0,0	0,0	2,5	0,0	5,5	0,0	0,0	2,2	3,6	4,5
46,8	16,6	58,9	21,8	53,6	27,3	13,7	23,1	3,7	33,9	21,6	17,0	5,3	9,3	29,8	13,5	5,1	28,4	18,2	1,1	7,7	5,6
15,8	17,5	10,7	24,0	26,4	15,4	21,3	24,6	5,3	9,1	4,2	14,3	7,1	5,8	25,9	0,0	0,6	7,0	12,4	0,0	14,3	12,5
5,3	1,6	17,9	0,0	6,6	0,0	4,3	0,0	1,3	2,3	4,2	0,0	0,0	0,3	1,2	0,0	2,4	0,0	2,3	11,1	0,0	8,0
6,5	7,9	8,9	6,6	8,3	6,1	8,2	8,4	2,6	4,5	2,8	4,5	4,7	4,8	11,8	0,0	1,3	2,8	8,4	5,3	5,1	5,9
0,0	3,2	0,0	0,5	0,0	0,0	2,1	1,5	4,0	0,0	2,8	0,0	0,0	0,0	0,4	30,4	1,2	0,0	0,0	26,7	10,7	2,3
0,0	1,6	0,0	5,0	5,5	0,0	0,0	19,4	0,0	6,8	6,9	8,6	0,0	0,0	0,8	0,0	1,2	2,3	0,0	0,0	0,0	1,1
0,0	1,6	3,6	0,0	2,2	0,0	1,1	0,0	0,0	0,0	2,8	0,0	0,0	0,0	0,0	0,0	0,0	0,0	0,0	0,0	0,0	0,0
21,1	1,6	7,1	8,0	6,6	0,0	3,2	0,7	0,0	72,7	6,9	11,4	0,9	1,3	3,7	6,5	4,8	4,7	1,6	4,4	0,0	2,3
6,5	3,3	3,3	3,7	3,6	0,0	2,0	7,4	1,6	31,3	6,4	6,3	0,6	1,0	2,2	16,3	3,1	2,8	0,9	14,9	3,8	1,6
0,0	1,6	0,0	4,0	7,7	30,8	0,0	5,2	0,0	0,0	87,5	0,0	0,0	0,0	0,0	2,2	3,6	0,0	0,0	0,0	0,0	4,5
68,4	23,8	28,6	24,0	71,4	123,1	8,5	16,4	0,0	6,8	119,4	31,4	0,0	0,0	0,8	2,2	16,4	2,3	0,0	2,2	28,6	10,2
21,0	10,6	8,9	7,7	19,9	60,6	2,7	7,4	0,0	2,7	68,3	9,8	0,0	0,0	0,4	1,9	8,5	0,9	0,0	1,1	10,3	4,2
3,3	2,4	3,2	3,6	4,0	2,5	3,1	2,9	2,5	2,5	3,0	3,2	1,5	1,3	2,3	2,3	2,4	2,5	1,7	2,1	2,8	3,5
19	63	28	200	91	13	94	134	75	44	72	35	113	313	243	46	165	43	129	45	28	88
1,26	1,39	1,54	1,42	1,42	1,72	1,31	0,93	1,48	1,09	1,90	1,42	1,75	1,76	1,10	1,21	1,59	1,30	1,66	0,95	0,96	1,41

TABLE 3.3 (continued) Results of the Empirical Investigation Concerning International Journals

Scandinavian Journal of Work, Environmental & Health	SIGHI Bulletin	Speech Communication	Speech Technology	Travail et Santé	Travail Humain	Vision Research	Work	Work and Stress	Work, Employment & Society	Workplace Ergonomics	
10,3	19,8	0,0	2,1	22,0	3,5	0,0	21,2	4,3	0,0	9,1	
2,9	***7,8***	***0,0***	***1,6***	***9,7***	***1,3***	***0,0***	***7,4***	***1,0***	***0,0***	***5,0***	
42,5	1,0	0,0	0,0	12,0	13,6	15,4	27,3	16,6	0,0	18,2	
57,1	0,0	8,2	2,1	20,0	6,0	6,7	3,0	6,4	0,0	0,0	
115,3	1,5	0,0	2,1	94,0	30,2	0,0	81,8	52,9	3,8	90,9	
59,4	***1,0***	***6,0***	***3,2***	***55,8***	***18,8***	***14,8***	***39,4***	***18,2***	***1,0***	***60,0***	
4,6	12,9	19,7	0,0	10,0	44,2	97,9	3,0	19,8	0,0	0,0	
0,8	20,8	32,8	41,7	2,0	8,5	15,4	3,0	0,5	0,0	0,0	
8,8	2,0	4,9	6,3	14,0	6,0	0,0	27,3	1,6	0,0	45,5	
3,9	***14,1***	***41,7***	***36,5***	***11,5***	***22,2***	***75,9***	***11,7***	***5,2***	***0,0***	***25,0***	
0,0	***42,2***	***41,7***	***25,4***	***0,0***	***10,2***	***3,4***	***0,0***	***0,0***	***0,0***	***0,0***	
47,5	9,9	1,6	0,0	6,0	28,6	2,6	21,2	144,4	15,4	0,0	
32,2	4,0	0,0	0,0	8,0	19,1	0,0	36,4	75,4	57,7	9,1	
22,0	***5,5***	***1,2***	***0,0***	***6,2***	***18,0***	***1,7***	***20,2***	***52,6***	***19,8***	***5,0***	
27,2	7,9	3,3	2,1	10,0	25,1	5,6	18,2	19,3	30,8	0,0	
0,4	7,4	0,0	0,0	2,0	4,0	0,0	0,0	1,1	7,7	0,0	
7,6	***6,1***	***2,4***	***1,6***	***5,3***	***11,0***	***3,8***	***6,4***	***4,9***	***10,4***	***0,0***	
0,0	48,0	8,2	39,6	2,0	8,5	0,0	0,0	0,5	3,8	0,0	
7,7	0,0	0,0	0,0	4,0	11,6	0,0	0,0	19,8	11,5	0,0	
0,4	0,0	0,0	0,0	0,0	0,5	0,0	0,0	0,0	7,7	0,0	
0,4	4,5	1,6	2,1	4,0	5,5	0,5	15,2	1,6	15,4	0,0	
2,3	***20,7***	***7,1***	***31,7***	***4,4***	***9,9***	***0,3***	***5,3***	***5,2***	***10,4***	***0,0***	
1,1	1,0	0,0	0,0	6,0	3,0	0,0	0,0	2,7	42,3	0,0	
5,7	5,9	0,0	0,0	10,0	19,6	0,0	27,3	50,8	173,1	9,1	
1,9	***2,7***	***0,0***	***0,0***	***7,1***	***8,5***	***0,0***	***9,6***	***12,8***	***58,3***	***5,0***	
3,6	2,5	1,4	1,3	2,3	2,6	1,5	2,8	4,2	3,7	1,8	TOTAL:
261	202	61	48	50	199	195	33	187	26	11	11778
1,61	1,09	1,46	1,27	1,43	0,54	2,09	0,98	1,38	1,59	1,67	

With regard to the total number of classified abstracts concerning conferences/symposia (Table 3.2), the annual meeting of the Human Factors and Ergonomics Society (HFES) is the largest source, followed by the triennial congress of the International Ergonomics Association (IEA) second, and the annual International Industrial Ergonomics and Safety Conference (IIESC) third. Although all three conferences have a quite similar profile concerning their weighted relevance (subtotal value according to Table 3.2) of high-order levels (levels S4, S5, S6, and S7), the annual meeting of the HFES has a stronger focus on psychological aspects and man–machine interface design (level S2) as well as on tasks and workplaces (level S3). However, the congress of the IEA and the IIESC emphasize physiological/anatomical aspects and health/safety issues (level S1) as well as general equipment design (level S2). If the CV of the weighted relevance is used as a measure for specialization of the three largest sources, the annual meeting of the HFES covers the broadest range (CV = 0.71), closely followed by the triennial congress of the IEA (CV = 0.82), and finally the IIESC (CV = 1.22). According to the CV, the most specialized periodical conferences are the International Conference on Environmental Ergonomics (CV = 2.29) and the annual symposiums of the SAFE Association (CV = 1.91), which both strongly focus on level S2.

The international journal with the most classified abstracts is the monthly published *Ergonomics,* which holds more than 10% of all abstracts (Table 3.3). According to the subtotal value *Ergonomics* focuses on individual and task related organizational factors (level S4), psychological factors (level S2), and group factors (level S5). The second largest source is *Aviation, Space and Environmental Medicine,* which specializes in physiological/anatomical aspects and environmental and health/safety issues (level S1).

TABLE 3.4 Results of the Empirical Investigation Concerning Handbooks

Ergonomics Abstracts - Classification	Handbook of Human Factors	Handbook of Human Performance	Handbook of Human-Computer Interaction	Handbook of Industrial Engineering	Handbook of Military Psychology	Handbook of Occupational Safety and Health
0 Basics						
GENERAL	4,4	2,9	15,4	10,0	5,6	12,5
Subtotal: [1]	***2,0***	***0,7***	***4,2***	***3,1***	***1,4***	***5,6***
S1 Autonomous Organic Systems & Work Environment						
PHYSIOLOGICAL AND ANATOMICAL ASPECTS	10,3	34,3	1,9	5,0	27,8	6,3
ENVIRONMENT	33,8	60,0	1,9	45,0	66,7	6,3
HEALTH AND SAFETY	17,6	22,9	5,8	30,0	22,2	137,5
Subtotal: [1]	***27,5***	***28,7***	***2,6***	***24,6***	***30,4***	***66,7***
S2 Wilfully Steered Organic Systems & Tool/Work Means						
PSYCHOLOGICAL ASPECTS	27,9	94,3	30,8	25,0	72,2	0,0
MAN-MACHINE INTERFACE DESIGN	20,6	5,7	23,1	10,0	0,0	18,8
WORKPLACE DESIGN	10,3	5,7	9,6	35,0	0,0	6,3
Subtotal: [1]	***26,1***	***25,9***	***17,3***	***21,5***	***18,8***	***11,1***
S3 Tasks & Workplaces						
Subtotal: [1]	***11,8***	***0,7***	***34,0***	***4,6***	***1,4***	***0,0***
S4 Qualification, Motivation and Work Content & Types of Work						
INDIVIDUAL AND TASK RELATED ORGANIZATIONAL FACTORS	13,2	154,3	25,0	25,0	94,4	3,3
JOB DESIGN	11,8	2,9	11,5	30,0	5,6	0,0
Subtotal: [1]	***11,1***	***38,5***	***9,9***	***16,9***	***26,1***	***[illegible],8***
S5 Work Group & Cooperative Processes						
GROUP FACTORS	2,9	17,1	15,4	5,0	5,6	[illegible],0
GROUP CONFLICTS AND SUPPORT	1,5	0,0	11,5	5,0	0,0	[illegible],0
Subtotal: [1]	***2,0***	***4,2***	***7,3***	***3,1***	***1,4***	***[illegible],0***
S6 Company & Company Wide Organizational Measures						
GENERAL SYSTEM CHARACTERISTICS	20,6	0,0	[illegible]2,3	20,0	5,6	[illegible],0
HOURS OF WORK	1,5	2,9	[illegible],0	0,0	5,6	[illegible],0
PAYMENT SYSTEMS	1,5	0,0	[illegible],0	10,0	0,0	[illegible],0
HUMAN RESOURCES	10,3	2,9	[illegible]3,5	15,0	55,6	1[illegible],8
Subtotal: [1]	***15,0***	***1,4***	***[illegible]5,2***	***13,8***	***17,4***	***8,3***
S7 Society & Social and Labour Politics						
TECHNOLOGICAL AND ERGONOMIC CHANGE	4,4	0,0	[illegible],6	20,0	0,0	0,0
SOCIAL AND LABOUR POLITICS	5,9	0,0	25,0	20,0	11,1	12,5
Subtotal: [1]	***4,6***	***0,0***	***[illegible],4***	***12,3***	***2,9***	***5,3***
Average Number of Classifications per Abstract:	2,3	4,1	[illegible],7	3,3	3,8	2,3

							TOTAL:
Number of Abstracts:	68	35	[illegible]2	20	18	1[illegible]	234
Coefficient of Variation:	0,80	1,26	0,30	0,67	0,97	1,[illegible]8	

[1] Weighted Relevance (Subtotal Divided by Average Number of Classifications per Abstract)

Third is the *International Journal of Manpower*, which strongly emphasizes tasks and workplaces on level S3. *Human Factors* is the fourth largest source just a few abstracts behind the third and specializes in psychological aspects of work and man–machine interface design (level S2). Of these four largest sources *Ergonomics* is the least specialized journal with a coefficient of variation of 0.91, followed by *Human Factors* (CV = 1.00), the *International Journal of Manpower* (CV = 1.32), and finally *Aviation, Space and Environmental Medicine* (CV = 1.43). The most specialized sources are the *European Journal of Cognitive Psychology* (CV = 2.41, centered around level S2), *Vision Research* (CV = 2.09, centered around level S3), and *Cognitive Psychology* (CV = 2.03, centered around level S3).

With regard to the investigation of handbooks as sources of ergonomics knowledge, the total number of abstracts is an inappropriate measure because of significant structural differences. Therefore, the coefficients of variation only are discussed. The *Handbook of Industrial Engineering* is the source with the largest diversity (least specialization) of the whole set with a CV of 0.67, followed by the *Handbook of Human Factors* and the *Handbook of Human-Computer Interaction,* both with a CV of 0.80. Although the sources have the same CV, the *Handbook of Human Factors* has a stronger focus on levels S1 and S2 of the systematology and the *Handbook of Human-Computer Interaction* deals primarily with the S2 and S3 levels. The *Handbook of Occupational Safety and Health* is the most specialized handbook (CV = 2.18), as indicated in the title, and is centered around level S1 of the systematology.

3.3 Conclusion and Future Developments

In the preceeding sections, the creation and consolidation process of scientific knowledge was analyzed and different types of scientific sources were discussed. Due to the multidisciplinary character of ergonomics, a systematology considering structural and procedural aspects of work was introduced which served as a framework for an empirical investigation of proceedings, international journals, and handbooks. With regard to future developments of scientific sources of ergonomics knowledge, probably the biggest challenge to the established publication procedures is the development of electronic information and communication media, especially the Internet. The Internet has the potential to accelerate the creation, consolidation, and dissemination process significantly, because of online, worldwide information access and sophisticated content-oriented search methods which overcome the limitations of traditional paper-based publications. Nevertheless, the primary goal of scientific discovery should be quality rather than "volume" or "time to market," independent of the publication media used. Hence, the definition, implementation, and control of appropriate quality-ensuring procedures will become even more important in a world of fast-growing knowledge sources. Within this context, the efforts of the International Ergonomics Association, which advertises an IEA Press label guaranteeing high-quality contents, should be taken into account,.

Acknowledgments

The authors would like to thank Claudia Peters and Ralf Hunecke for the preparation of the empirical data. In addition, we owe special thanks to Taylor & Francis for their valuable contribution of the *Ergonomics Abstracts* on CD-ROM.

References

Becker-Biskaborn, G.U., 1975, Ergonomische Erkenntnissammlung für den Arbeitsschutz mit Informationssystem, BAU Forschungsbericht Nr. 142, Band I und II, Wirtschaftsverlag NW, Bremerhaven.

Boff, K.R., Kaufman, L., Thomas, J.P. (ed.), 1986, *Handbook of Perception and Human Performance*, John Wiley & Sons, New York.

Easterby, R.S., 1967, Ergonomic checklists: an appraisal, *Ergonomics* 10 (5), pp. 549-556.

Fitts, P.M., 1951, Human engineering for an effective air navigation and traffic control system, National Research Council, Washington, D.C.

Friedrichs, J., 1975, *Methoden Empirischer Sozialforschung*, Rowohlt Verlag, Reinbek.
Frieling, E., Sonntag, K., 1987, *Lehrbuch Arbeitspsychologie*, Verlag Hans Huber, Bern.
Helander, M.G., Landaver, T.K. Prabhu, P.V. (eds.), 1997, *Handbook of Human–Computer Interaction*, second edition, North-Holland, Amsterdam.
ILO, 1996, Ergonomic checkpoints — Practical and easy-to-implement solutions for improving safety, health and working conditions, Geneva, Switzerland.
Kroemer, K.H.E. et al., 1988, Ergonomic Models of Anthropometry, Human Biomechanics, and Operator-Equipment Interfaces, Proceedings of a Workshop, National Academy Press, Washington, D.C.
Laurig, W., Rombach, V., 1989, Expert systems in ergonomics: requirements and an approach, *Ergonomics* 32 (7), pp. 795-811.
Luczak, H., Volpert, W., 1987, *Arbeitswissenschaft. Kerndefinition–Gegenstandskatalog–Forschungsgebiete*, RKW-Verlag, Eschborn.
MIL-STD 1472D, 1991, Human engineering criteria for military systems, equipment and facilities, the hypertext version is provided by CSERIAC, Crew System Ergonomics Analysis Center, Wright-Patterson AFB, OH.
N.N., 1964, Ergonomic system analysis checklist, International Ergonomics Association (editor), Proceedings of the 2nd World Congress of the IEA, Dortmund.
Oppermann, R., Murchner, B., Reiterer, H., Koch, M., 1992, *Softwareergonomische Evaluation (EVADIS II)*, 2nd edition, Walter de Gruyter, Berlin.
OSF/Motif, 1993, *Style Guide — Release 1.2*, Prentice Hall, Englewood Cliffs, NJ.
Rasmussen, J., 1983, Skills, rules, knowledge: signals, signs, and symbols and other distinctions in human performance models, *IEEE Transactions on Systems, Man, and Cybernetics* SMC-13 (3), pp. 257-267.
Rohmert, W., Becker-Biskaborn, G.U., 1974, Ergonomische Prüfliste für den Arbeitsschutz mit Literaturanhang, BAU Forschungsbericht Nr. 116, Wirtschaftsverlag NW, Bremerhaven.
Salvendy, G. (ed.), 1997, *Handbook of Human Factors*, 2nd edition, John Wiley & Sons, New York.
Salvendy, G. (ed.), 1992, *Handbook of Industrial Engineering*, 2nd edition, John Wiley & Sons, New York.
Tecmath GmbH, 1994, RAMSIS-Leistungsbeschreibung und Softwarehandbuch, Tecmath, Kaiserslautern.

Appendix

Advances in Human-Computer Interaction	Advances in Human-Computer Interaction Volume 1 - 4
Advances in Industrial Ergonomics	Advances in Industrial Ergonomics and Safety 1, I, II, III, IV, V, VI
Advances in Man-Machine Systems	Advances in Man-Machine Systems Research, Volume 1, 3, 4, 5
Analysis, Design and Evaluation of Man-Machine Systems	Analysis, Design and Evaluation of Man-Machine Systems 1988, 1989, 1992 Analysis, Design and Evaluation of Man-Machine Systems, Proceedings of the 2nd IFAC/IFIP/IFORS/IEA Conference, Varese, Italy, September 10-12 1985
Computers for Handicapped Persons	Proceedings of the 3rd International Congress on Computers for Handicapped Persons, 1992
Contemporary Ergonomics	Contemporary Ergonomics 1986 - 1995
Design of Work and Development of Personnel in Advanced Manufacturing	Design of Work and Development of Personnel in Advanced Manufacturing
Empirical Studies of Programmers	Empirical Studies of Programmers Empirical Studies of Programmers, Second Workshop Empirical Studies of Programmers, Fourth Workshop Empirical Studies of Programmers, Fifth Workshop

Environmental Ergonomics	Environmental Ergonomics, Edited by I.B. Mekjavic, E.W. Banister and J.B. Morrison, Taylor & Francis, London, 1988 Proceedings of the Fifth International Conference on Environmental Ergonomics, Maastricht, the Netherlands, 1992
Ergonomics of Hybrid Automated Systems	Ergonomics of Hybrid Automated Systems I - III
European Coal and Steel Community	European Coal and Steel Community Ergonomics Action Information Bulletin, 1986 European Coal and Steel Community Ergonomics Action, Luxembourg, Final Reports for various Projects
Graphics Interface	Proceedings of Graphics Interface '88, Edmonton, Canada, 6-10th June 1988 Proceedings of Graphics Interface '90, Halifax, Nova Scotia, 14-18 May 1990 Proceedings of Graphics Interface '91, Calgary, Alberta, 3-7 June 1991 Proceedings of Graphics Interface '92, Vancouver, British Columbia, 11-15 May 1992 Proceedings of Graphics Interface '93, Toronto, Ontario, 19-21 May 1993 Proceedings of Graphics Interface '94, Banff, Alberta, 18-20 May 1994
HCI	HCI '87: Social, Ergonomic and Stress Aspects of Work with Computers/Cognitive Engineering in the Design of Human-Computer Interaction HCI '89: Work with Computers: Organizational, Management, Stress and Health Aspects/Designing and Using Human-Computer Interfaces and Knowledge-Based Systems HCI '91: Human Aspects in Computing: Volume 1. Design and Use of Interactive Systems and Work with Terminals/Human Aspects in Computing: Volume 2. Design and Use of Interactive Systems and Information Management HCI '93: Human-Computer Interaction: Applications and Case Studies/Human-Computer Interaction: Software and Hardware Interface
Human Decision Making	Proceedings of the Fourth European Annual Conference on Human Decision Making and Manual Control, Zeist, the Netherlands, 28-30 May 1984 Human Decision Making and Manual Control, 5th EAM, 1985 Proceedings of the 7th Annual Conference on Human Decision Making and Manual Control, Paris, 18-20 October, 1988 Proceedings of the 10th European Annual Conference on Human Decision Making and Manual Control, Liège, Belgium, 11-13 November 1991 11th European Annual Conference on Human Decision Making and Manual Control, November 17-19, 1992 Proceedings of the XII European Annual Conference of Human Decision Making and Manual
Human Factors and Ergonomics Society	Progress for People. Proceedings of the Human Factors Society 29th Annual Meeting, Baltimore, Maryland, September 29 - October 3, 1985, Edited by R.W. Swezey. The Human Factors Society, Santa Monica, California, Human Factors Society Bulletin, 1986 Combining Human and Artificial Intelligence: A New Frontier in Human Factors, Proceedings of a Symposium Sponsored by the Metropolitan Chapter of the Human Factors Society, New York, 15 November 1984, Edited by G. Kohl and S.J. Nassau Ergonomics and Safety in the Workplace: Proceedings of the Europe Chapter of the Human Factors and Ergonomics Society Annual Meeting in Antwerp, November 1992 Proceedings of Interface '93, Raleigh, North Carolina, May 5-8, 1993. Designing for Diversity. Proceedings of the Human Factors and Ergonomics Society 37th Annual Meeting, Seattle, Washington, October 11-15 1993 People and Technology in Harmony. Proceedings of the Human Factors and Ergonomics Society 38th Annual Meeting, Nashville, Tennessee, October 24-28 1994 Human Factors and Ergonomics Society, Santa Monica, California, 1994 Human Factors and Ergonomics Society, Santa Monica, California, 1995
Human Factors and Industrial Design	Interface 87: Human Implications of Product Design, Proceedings of the 5th Symposium on Human Factors and Industrial Design in Consumer Products, Rochester, New York, May 13-15, 1987 Interface 91: Proceedings of the 7th Symposium on Human Factors and Industrial Design in Consumer Products, Dayton
Human Factors and Power Plants	Conference Record for 1985 IEEE Third Conference on Human Factors and Power Plants, Monterey, California, 23-27 June 1985 Conference Record for 1988 IEEE Fourth Conference on Human Factors and Power Plants, Monterey, California, June 5-9, 1988 Conference Record for the 1992 Fifth Conference on Human Factors and Power Plants, Monterey, California, June 7-11, 1992

Human Factors in Computing Systems	Human Factors in Computing Systems III und IV
Human Factors in Information Systems	Human Factors in Information Systems: An Organizational Perspective, New Jersey 1991
Human Factors in Manufacturing	Proceedings of the 2nd International Conference on Human Factors in Manufacturing and 4th IAO Conference, Stuttgart, West Germany, 11-13 June 1985
	Proceedings of the 3rd International Conference on Human Factors in Manufacturing, Stratford-upon Avon, 4-6 November 1986
	Proceedings of the International Ergonomics Association Conference on Human Factors in Design for Manufacturability and Process Planning, Honolulu, Hawaii, 9-11 August, 1990
	Organization and Management of Advanced Manufacturing
Human Factors in Telecommunications	Proceedings of the 12th International Symposium of the Human Factors in Telecommunications, The Hague, the Netherlands, May 24-27, 1988
	Proceedings of the Tenth International Symposium on Human Factors in Telecommunications, Helsinki, Finland, 6-10 June 1983
	Proceedings of the 11th International Symposium on Human Factors in Telecommunications, Cesson Sevigne, France, 9-13 September 1985
	Proceedings of the 13th International Symposium on Human Factors in Telecommunications, Torino, 10-14 September 1990
	Supplement to the Proceedings of the 13th International Symposium on Human Factors in Telecommunications, Torino, 10-14 September 1990
	Proceedings of the 14th International Symposium on Human Factors in Telecommunications, Darmstadt, Germany, May 11-14, 1993
Human Factors in Organizational Design and Management	Human Factors in Organizational Design and Management, II - IV
Human Factors Association	Proceedings of the Human Factors Association of Canada, 18th Annual Meeting, Hull, Quebec, 27-28 September 1985
	Proceedings of the 19th Annual Meeting of the Human Factors Association of Canada, Richmond (Vancouver), British Columbia, August 22-23, 1986
	Proceedings of the 20th Annual Conference of the Human Factors Association of Canada, Montreal, Quebec, October 14-17, 1987
	Proceedings of the Human Factors Association of Canada 21st Annual Conference, Edmonton, Alberta, 14-16 September 1988
	Proceedings of the Human Factors Association of Canada 22nd Annual Conference, Toronto, Ontario, November 26-29, 1989
Human-Computer Interaction - Interact	Human-Computer Interaction - Interact '87 and '90
IEEE	IEEE Proceedings of the international Conference on Cybernetics and Society, Tucson, Arizona, November 1985
	IEEE Virtual Reality Annual Symposium, Seattle, Washington, September 1993
Industrial Ergonomics and Safety	Trends in Ergonomics/Human Factors III, Proceedings of the Annual International Industrial Ergonomics and Safety Conference Held in Louisville, Kentucky, U.S.A., 12-14 June 1986
	Trends in Ergonomics/Human Factors IV, Proceedings of the Annual International Industrial Ergonomics and Safety Conference Held in Miami, Florida, U.S.A., 9-12 June 1987
	Advances in Ergonomics and Safety I-VI, London 1989-1994
International Ergonomics Association	Proceedings of the Ninth Congress of the International Ergonomics Association, Bournemouth, 2-6 September 1985
	Ergonomics International 88, Proceedings of the 10th Congress of the International Ergonomics Association, Sydney, Australia, 1-5 August 1988
	Designing for Everyone: Proceedings of the 11th Congress of the International Ergonomics Association, Paris, 1991, Volume 1
	Proceedings of the 12th Triennial Congress of the International Ergonomics Association, Toronto, Canada, August 15-19, 1994, Volume 1
Man-Computer Interaction Research	Man-Computer Interaction Research - Macinter - I
	Man-Computer Interaction Research - Macinter - II
National Board/National Institute of/for Occupational Safety and Health	National Board of Occupational Safety and Health
	National Institute for Occupational Safety and Health

People and Computers	People and Computers III - IX, Cambridge University Press People and Computers: Designing for Usability People and Computers: Designing the Interface
SAFE Association	Proceedings of the 23rd Annual Symposium of the SAFE Association, Las Vegas, Nevada, December 1-5, 1985 Proceedings of the 24th Annual Symposium of the SAFE Association, San Antonio, Texas, December 11-13, 1986 Proceedings of the 25th Annual Symposium of the SAFE Association, Las Vegas, Nevada, November 16-19, 1987 Proceedings of the Twenty-Sixth Annual Symposium of the SAFE Association, Las Vegas, Nevada, December 5-8, 1988 Proceedings of the 27th Annual Symposium of the SAFE Association, New Orleans, Louisiana, December 5-8, 1989 Proceedings of the Twenty-Ninth Annual Symposium of the SAFE Association, Las Vegas, Nevada, November 11-13, 1991 Proceedings of the 30th Annual Symposium of the SAFE Association, Las Vegas, Nevada, November 2-4, 1992 Proceedings of the 31st Annual Symposium of the SAFE Association, Las Vegas, Nevada, November 8-10, 1993
The Ergonomics of Manual Work	The Ergonomics of Manual Work
The Ergonomics of Working Postures	The Ergonomics of Working Postures
Vision in Vehicles	Vision in Vehicles, Proceedings of the Conference on Vision in Vehicles, Nottingham, 9-13 September 1985 Vision in Vehicles - II, Amsterdam 1988 Vision in Vehicles - III, Amsterdam 1991 Vision in Vehicles - IV, Amsterdam 1993
Visual Search	Visual Search, London, 1990 Visual Search 2, London, 1993
Work with Display Units	Work with Display Units 86, Amsterdam, 1987 Work with Display Units 89, Amsterdam, 1990 Work with Display Units 92, Amsterdam, 1993

4

A Guide to Certification in Professional Ergonomics

Dieter W. Jahns
Board of Certification in Professional Ergonomics

4.1 Introduction

Some form of quality assurance efforts are natural to most professions. These generally involve development of credentialing for educational programs and/or of individuals. Three types of processes are most common: *Accreditation* is established for the regulation of instructional programs. It is voluntary and generally developed and administered by an association of professionals within the field. *Certification* involves a voluntary process of evaluation and measurement of individuals which can then indicate whether they have achieved a professional level of qualifications as judged by professional peers. It is developed and administered by a professional association or a group specifically established for professional development purposes. *Licensure*, while it does credential individuals, is a mandatory process and is administered by a political or governing body. When laws are implemented "to protect the public" from unprofessional practices, it becomes illegal to practice one's profession without a license. Thus, these processes are distinguishable by three aspects: (a) the recipient of the credential, (b) the credentialing body, and (c) the degree of volunteerism involved in obtaining the credential (Jahns, 1991).

Slappendel (1994) reviewed nine ergonomics certification/registration programs in operation around the world. Her findings are summarized in Tables 4.1 and 4.2. The IEA is currently working on policies and procedures to endorse and harmonize credentialing organizations on an international scale. Since IEA (International Ergonomics Association) Federated Societies are more oriented towards information dissemination, and not so much toward control of the profession as a guild structure, there is an increasing trend for cooperative, yet independent credentialing agencies. In "open-market" societies there are also opportunities for sham operators, which makes a supervisory role by IEA Federated Societies desirable. This is happening. For example, the Human Factors Association of Canada (HFAC/ACE) recognizes BCPE (Board of Certification in Professional Ergonomics) as a valid and reliable certification organization operating simultaneously with its own efforts to develop certification processes and criteria for Canadian ergonomists. BCPE has also become a consultant to efforts in South Africa and Japan on an informal basis.

Similarly, in Europe, CREE (Centre for Registration of European Ergonomists) works with the ergonomics societies of member countries in the European Union in evaluating and registering applicants

0-8493-2641-9/99/$0.00+$.50

TABLE 4.1 Certification of the Ergonomist: Programs in Operation as of May 1994*

Certification/Registration Authority		Designation	Acronyms
Non-Society	Board of Certification in Professional Ergonomics (BCPE) U.S.A.	Certified Professional Ergonomist Certified Human Factors Professional Ergonomist in Training: Associate Ergonomics Professional Associate Human Factors Professional	CPE CHFP AEP AHFP
	Centre for Registration of European Ergonomists (CREE) The Netherlands	European Ergonomist	EurErg
	Stichting Registratie ergonomen (SRe) Netherlands	Registered Ergonomist	R.e.
Society	Professional Affairs Board PAB) of The Ergonomics Society, U.K.	Registered Member of the Ergonomics Society (Professional Member) Fellow of the Ergonomics Society Practitioner of the Professional Register	M.Erg.S. F.Erg.S.
	Professional Affairs Board (PAB) of the Ergonomics Society of Australia	Certified Professional Member	C.Erg.
	Membership Subcommittee of the New Zealand Ergonomics Society	Professional Member	M.NZ.Erg.S

* Programs are also in operation in France, Belgium, and Sweden, but information on these was unavailable at the time of writing.

Source: Slappendel, C. 1994. Harmonising the different approaches to the certification of the ergonomist. *Proceedings of the 12th Triennial Congress of the IEA*, Toronto, ON, Canada.

TABLE 4.2 Criteria Applied in Certification Programmes

Designation	Criteria	Recertification
Certified Professional Ergonomist/Certified Human Factors Professional (BCPE)	Masters degree in ergonomics (human factors) or equivalent, *plus* 4 years of full-time professional practice in ergonomics with emphasis on ergonomic design, *plus* submission of a work product, *plus* a passing score on a written certification examination	Not required yet; annual renewal fee
European Ergonomist (CREE)	At least 3 years of academic formation in any field of which the total amount of education in ergonomics is at least 1 year, *plus* at least 1 year of training, plus at least 2 years of experience	Registration is for a 5-year period
Registered Ergonomist (SRe)	Not specified, but are in line with CREE criteria	Every 3 years
Registered Member of The Ergonomics Society (a.k.a. Professional Member)	At least 3 years (or part-time equivalent) in the practice of ergonomics, and/or teaching and/or research of ergonomics relevance since admission to the Society, *plus* evidence of academic achievements	Not required
Fellow of The Ergonomics Society	Registered Member for at least 6 years plus significant contribution to the practice of, teaching of, and/or research in ergonomics for a period of 10 years since becoming an Ordinary Member *plus* substantial contribution to the activities of the Society	Not required
Practitioner on the Professional Register of The Ergonomics Society	Must be a Registered Member of the Society *plus* a minimum of 3 years in active practice during the preceding year	Every 3 years
Certified professional member of the Ergonomics Society of Australia	A suitable qualification *plus* 3 years full-time equivalent experience in the practice of ergonomics	Required
Professional member of the New Zealand Ergonomics Society	A tertiary qualification in ergonomics, or a qualification of which ergonomics made up a substantial portion of the course content, *plus* experience in the practice of ergonomics, or teaching or research of ergonomics relevance	Not required

Source: Slappendel, C. 1994. Harmonising the different approaches to the certification of the ergonomist. *Proceedings of the 12th Triennial Congress of the IEA*, Toronto, ON, Canada.

for the "Eur. Erg." designation. The BCPE and CREE have a reciprocity agreement in place. As CREE President E. N. Corlett (1996) wrote: "Our policy at the moment is to be linked with only one registering body in each country. Because of our constitution, this body has to have certain requirements, as laid out in the European Standard 45013 to which we adhere. We have confirmed that BCPE fulfills these requirements."

As a member of the U.S.-based National Organization for Competency Assurance (NOCA), the BCPE has used the following checklist (provided by the National Commission for Certifying Agencies) for tracking its program relative to accepted criteria:

1. Agency's purpose must be certification of individuals. (Yes)
2. Agency is nongovernmental. (Yes)
3. Agency is national in scope. (Yes — international)
4. Agency is administratively independent. (Yes)
5. Certificants are represented on governing board. (Yes)
6. Governing body is selected from certificants (Yes)
7. Governing body does not select its successors. (No)
8. Governing body has public member. (No)
9. Agency is separate from any accrediting body. (Yes)
10. Agency has completed two national examination administrations. (Yes)
11. Agency has adequate funding. (Yes — barely)
12. Agency has adequately trained staff. (Yes)
13. Certificant evaluation is fair and based on appropriate knowledge and skills. (Yes)
14. Agency reviews evaluation process periodically. (Yes)
15. Agency has adequate examination security. (Yes)
16. Agency sets pass/fail levels appropriately. (Yes)
17. Agency maintains statistical data on the assessment methods. (Yes)
18. Agency publishes certification criteria and descriptive information. (Yes)
19. Agency has nondiscrimination policy. (Yes)
20. Agency has test examination accommodation policy. (Yes)
21. Agency documents uniform policies. (Yes)
22. Agency periodically reviews certification criteria and examinations to show uniform enforcement. (Yes)
23. Agency publicizes process and results nationally. (Yes — internationally)
24. Agency offers examinations in multiple locations. (Yes)
25. Agency has policy on alternate eligibility options. (Yes)
26. Agency notifies examinees promptly of scores, and failures are given specific deficiencies. (Yes)
27. Agency keeps examination scores confidential. (Yes)
28. Examinees may challenge examination results. (No)
29. Agency has policy giving grounds for refusing eligibility. (Yes)
30. Agency maintains a register of certificants. (Yes)
31. Agency maintains an ethics and discipline code. (No)
32. Agency has recertification procedure. (Not yet)

The candidates for certification usually follow the pathways shown in Figure 4.1 (solid lines) by contacting either the certification agency directly or by inquiring of one of the IEA Federated Societies, which then coordinates the certification procedures. Both BCPE and CREE have highly coordinated information exchanges (dashed lines in Figure 4.1) with the IEA and selected, regionally active Federated Societies to harmonize the professional development of ergonomists. Interested readers can contact the organizations listed in "For Further Information" at the end of this chapter. A general overview of BCPE certification criteria and procedures is given next.

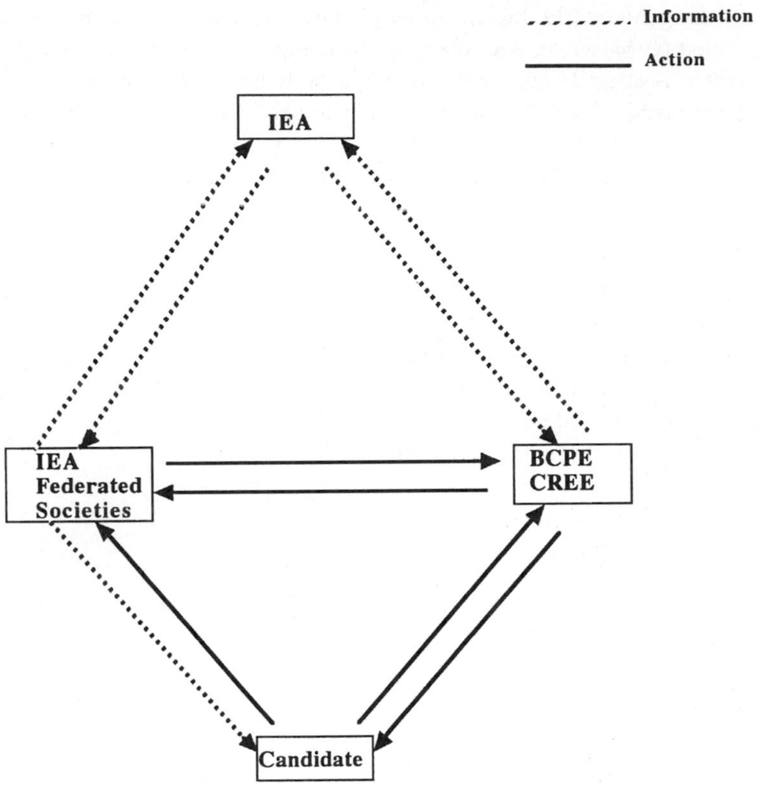

FIGURE 4.1 Communication and actions among ergonomics societies and certification agencies.

4.2 BCPE Certification Requirements

Criteria for Certification

1. CPE/CHFP

   The BCPE job/task analyses led to the following minimum criteria for certification:

   - a master's degree in ergonomics or human factors, or an equivalent educational background in the life sciences, engineering sciences, *and* behavioral sciences to comprise a professional level of ergonomics education, and
   - four years of full-time professional practice as an ergonomist practitioner with emphasis on design involvement (derived from ergonomic analysis and/or ergonomic testing/evaluation), and
   - documentation of education, employment history, and ergonomic project involvement by means of the BCPE "Application for Certification," and
   - a passing score on the BCPE written examination, and
   - payment of all fees levied by the BCPE for processing and maintenance of certification, in accordance with the following schedule:

     Application Fee: U.S. $10.00
     Processing and examination fee: U.S. $290.00
     Annual certificate maintenance fee: U.S. $100.00

2. AEP/AHFP

On March 26, 1995, the BCPE created a new "Associate" category of certification. As of January 1996, a person can be certified by the BCPE as an "Associate Ergonomics Professional" or an "Associate Human Factors Professional" if he or she:

- meets the education requirements for BCPE certification (M.S. in human factors/ergonomics or related field),
- has passed Part-I (on "Basic Knowledge" of human factors/ergonomics) of the BCPE certification examination, and
- is currently working toward fulfilling the BCPE requirement of four years' practical experience as a human factors and ergonomics professional.

Thus, the AEP/AHFP designation is a preprofessional category, while the candidate gains experience. A person can take the Basic Knowledge portion (Part-I) of the BCPE exam immediately after fulfilling the education requirement. Parts-II and III of the exam may be taken after fulfilling the other BCPE requirements. *A person who has graduated from a human factors/ergonomics degree program accredited by an IEA Federated Society (e.g., HFES), will not have to take Part-I of the exam.* BCPE established the "Associate" category to create a path by which individuals could achieve professional certification in progressive steps. The Board also wanted to link this path with accredited educational degree programs by giving some preference (by waiving Part-I of the exam) to applicants who have graduated from programs accredited by IEA Federated Societies. This linkage is an important step in strengthening the human factors and ergonomics profession.

Fees associated with the AEP/AHFP designations are:

Application fee: U.S. $10.00
Processing and examination: U.S. $100.00
Annual AEP/AHFP Maintenance fee: U.S. $60.00

3. CEA

As of July 1998, the new designation CEA (Certified Ergonomics Associate) is available to individuals who:

- have a Bachelor's degree, or equivalent education, in ergonomics (or in one of the scientific disciplines closely related to ergonomics)
- have at least 200 contact hours of ergonomics training commensurate with the "Ergonomist Formation Model (EFM)" adopted by BCPE
- have at least two years of full-time equivalent practice in ergonomics
- obtain a satisfactory score on the BCPE, two-part, multiple-choice answers examination on ergonomics foundations and ergonomics practice methods

These individuals, once obtaining their CEA designation, are expected to be involved in more limited activities of system evaluation and intervention than are CPEs, who have a broader and scientifically deeper involvement with the multidimensionality of work systems development. The CEAs are most likely to use well-established ergonomics principles, methods, and tools in an "intervention strategy" and as an adjunct to reactive workplace issues deriving from worker's compensation insurance, labor negotiations, regulatory compliance audits, and/or standardized ergonomics guidelines.

Fees associated with the CEA designation are:

Application fee: U.S. $10.00
Processing and examination: U.S. $190.00
Annual CEA Maintenance fee: U.S. $75.00

Procedures for Certification

1. The candidate requests application materials by sending a U.S. $10.00 check to: BCPE, P.O. Box 2811, Bellingham, WA 98227-2811. Materials consist of four pages of instructions and seven pages of forms to be filled out by the applicant:
   Section A — Personal data
   Section B — Academic qualifications
   Section C — Employment history
   Section D — Work experience (in ergonomic analysis, design, and testing/evaluation)
   Section E — Work product description (CPE/CHFP only)
   Section F — Signature and payment (US $290.00) record
2. The candidate completes the application and submits it with an official academic degree transcript and a work product (article/technical report/project description/patent application, etc.).
3. The review panel evaluates all submitted materials and makes recommendations to the Board as to whether or not the applicant qualifies to take the written examination.
4. The candidate becomes certified by taking and passing the written examination.

Examination Administration

Applicants who have demonstrated eligibility for the examination will be notified regarding the date and location of the next examination approximately two months before the test date. The examination, requiring a full day, will generally be scheduled for the spring and fall, and will probably be offered as an adjunct to the meetings of ergonomics-related professional societies and associations. Qualified applicants needing accommodations in compliance with the Americans with Disabilities Act (ADA) are asked to specify their accommodation needs to the BCPE prior to signing up for the examination.

Scoring Methods

A panel of prior BCPE certificants with expertise in psychometrics determines the method of establishing passing scores to be used for the examination. Passing scores are established to ensure the applicant's mastery of the knowledge and skills required for professional ergonomics practice. The BCPE will periodically review, evaluate, and, as necessary, revise the examination and scoring to assure that valid and reliable measures of requisite performance capability for ergonomics practice are maintained.

Retaking the Examination

An applicant who does not pass the examination may retake it at the next regularly scheduled examination date and place. A reduced fee will be applied to retakes, and the retake must take place within two years of the original application date.

BCPE Examination: Approximate Weighting of Subject Areas

1. Methods and Techniques (M&T)	30%
2. Design of Human–Machine Interface (DHMI)	25%
3. Humans as Systems Components (Capabilities/Limitations) (HSC)	25%
4. Systems Design and Organization (SDO)	15%
5. Professional Practice (PP)	5%

References

Jahns, D. W. 1991. Certification of professional ergonomists: a status report. *Proceedings 24th Annual Conference of the HFAC/ACE*, Vancouver, B.C., Canada.

Corlett, E. N. 1996. Personal communication, letter dated 12/11/1996.

Slappendel, C. 1994. Harmonising the different approaches to the certification of the ergonomist. *Proceedings of the 12th Triennial Congress of the IEA*, Toronto, ON, Canada.

For Further Information

The Board of Certification in Professional Ergonomics (BCPE)

P. O. Box 2811
Bellingham, WA 98227-2811 U.S. A.
Telephone: 360-671-7601
Fax: 360-671-7681
e-mail: BCPEHQ@aol.com
WWW Home Page: http://bcpe.org

Center for Registration of European Ergonomists (CREE)

Franz Witt, Administrative Director
Aldrinaweg 3
NL-9892 PE Feerwerd
Netherlands
Telephone: 1-31-59-411971
Fax: 1-31-50-134104

National Commission for Certifying Agencies

(A Division of NOCA)
1200 19th Street NW #300
Washington D.C. 20036-2401
Telephone: 202-857-1165
Fax: 202-223-4579

5

Effective Use of the World Wide Web

Peter M. Budnick
ErgoWeb Inc.
University of Utah

5.1 Introduction

The Internet, and the World Wide Web in particular, are revolutionizing the way people seek and access information. For a growing number of people, research begins with the Web, "surfing" around the world in search of the electronic equivalent of a library book — or the electronic equivalent of an art gallery; or, the on-line version of a favorite software program; or, a live or bulletin-board style discussion with others. Or — and this may be the ultimate beauty and utility of the Web — you may find all these elements intertwined into one system through the effective use of "hypermedia," the interactive multimedia capabilities that make up the Web. Using the Web, specialists are communicating and sharing information and ideas faster and more efficiently than ever before, breaking down the traditional barriers of distance and borders, and the information produced by specialists is being distributed to a wider population than ever before.

To traditionalists and hard-copy purists, this may seem a disturbing trend. To this author, it is an exciting human progression, and one that will redefine the level and availability of knowledge around the world. A traditionalist may argue that the speed and ease with which information is distributed in the electronic environment makes it possible for dubious or untested theories to be widely broadcast and accepted as fact by the untrained mind. It is true that this may occur, however, the benefits that improved communications bring far outweigh the potential downsides or misuse of the medium. Further, in disciplines like occupational biomechanics or ergonomics, releasing the knowledge from the world of research to the everyday world of the working person may be the key to practical success. Theorizing and academic debate will continue in the electronic medium, but only a medium like the Web can take the results of such specific bodies of knowledge and bring them to the people who need to know, transcending national, cultural, and corporate borders. It is the understanding and application of ergonomic principles in the everyday working life of individuals that will make work more productive, safer, more fulfilling, more dignified, and ultimately better for both employers and employees.

The Web is constantly evolving, and information, service, and product providers appear and disappear each day. On one hand, it can seem a disorganized environment, bringing equal parts of annoyance and satisfaction. However, as the Internet has proved, this chaotic, even anarchistic environment will self-organize through voluntary persuasion and innovation on behalf of the interested participants. What we know today as the Web was preceded by a number of voluntary Internet protocols that took a set of mere communication links and developed highly organized systems, including e-mail, Telnet, FTP, Gopher,

0-8493-2641-9/99/$0.00+$.50

and Usenet. (See the glossary of terms at the end of this chapter if these terms are new to you.) E-mail formed the basis for personal communication; Telnet provided a common standard for communicating between different computer hardware locations; FTP (File Transfer Protocol) provided a standard method to send or retrieve electronic files, such as data sets, text files, and digitally stored images; Gopher provided an organized and easily accessed indexing method for larger stores of information; and Usenet organized millions of unrelated public e-mail discussions into thousands of topic-specific discussions, allowing you to pick and choose only the information you desire to monitor, or to which you wish to contribute. These advancements were all developed and implemented voluntarily.

The Web was a great leap forward in the organization and accessibility of the Internet. Yet, as captured in the name "Web," this new communication protocol is more of an interface advancement than it is an organizational advancement. Anyone who has "surfed" for any amount of time will attest to the fact that it is easy to rush forward (in terms of the excitement of a new frontier) and get lost (in terms of not finding what you set out to find) in this "Web." It is now up to individuals and organizations to organize the knowledge on the Web so that you can locate what you need, when you need it.

It is the distributed design of the Internet communication network that lends to the distributed nature of the Web. The Internet was intentionally designed to be a system of numerous and redundant links between computer hardware spread across wide geographic locations. This design provided a nearly fail safe communications network no matter how large the local disturbance. In summary, when a message is sent across the Internet, it is disassembled into smaller "packets," which are in turn redundantly sent along a variety of different routes through the network. If a "packet" encounters a break in the system, it gets rerouted around the disturbance. At the message destination, the arriving "packets" are reassembled into the original message. So, there is no central path through which Internet traffic passes. (At least not at this date, in the United States. Conceivably, some will design networking so that all traffic passes through central locations which can be more easily monitored and controlled.) Thus, there is no central authority nor location at which to intervene and organize the vast amounts of knowledge available on the Web.

Individuals and organizations are quickly organizing information on their own, however. Such Web sites form the building blocks for a vast body of distributed knowledge, methodologies, etc., yet they exist independently and are often unknown to each other. The Web provides the protocol necessary for these dispersed sites to easily "link" to each other, allowing the Web user to move easily from one source to another. Thus, at the lowest level, a site is organized locally in some way. The next level of organization involves linking from one site to another. This may be anonymous, or it may be a reciprocal arrangement with mutual links. In this way, a group of sites may form a level of organization among themselves. This is good for the user who happens to locate a site that either contains the information sought, or links to others that do, but it still presents a barrier to the user that does not have a starting place in this Web. "Search engines," or meta-indexing sites are the organizational solution that has risen on the Web to help users locate and focus on the information they seek.

This chapter seeks to provide you with a sense of how the Web and the Internet work, but more important, the author hopes to show how you can successfully utilize this excellent resource. The chapter contains no list of Web sites dealing with occupational biomechanics or ergonomics, since such a list would likely be obsolete by the time this is published. Also, you will find no graphics reproduced from the Web for this chapter, because a black and white book format cannot capture the vivid colors and animation frequently encountered on the Web. After reading this chapter, though, you should know how to locate them on your own by using the Web effectively and responsibly.

To demonstrate the effective use of the Internet and the Web in research, this author has deliberately performed all research for this chapter using only the Web. This brings one of the many issues of protocol and standardization to mind: how should one reference an electronic site? Often, an author's name will not appear on a Web page, nor will a date, nor will the location of the publisher. Further, if it is a large compilation of information, there will rarely be page numbers or any other traditional means to identify a place relative to the larger collection. So, the first task is to settle on a referencing style that will help you find the information referred to here, if it still exists when you read this (see the section titled "Finding

What you Need on the Web"). Walker (1996) recommends the following general style for Web sites (she also addresses referencing styles for other Internet protocols):

Author's Last Name, First Name. "Title of Work." Title of Complete Work.
[protocol and address] [path] (date of message or visit).

Applying this style to the Walker (1996) page referenced here,

Walker, J.R., 1996, "MLA-Style Citations of Electronic Sources,"
http://www.cas.usf.edu/english/walker/mla.html, (19 Jan 1997).

Next, it is instructive to review some of the terms that are commonly used when discussing Internet-based materials. Therefore, the author has compiled a glossary of terms at the end of this chapter that you may want to refer to while reading.

5.2 Copyright Issues

The proliferation of materials on the Internet and World Wide Web has forced many to revisit the issues surrounding intellectual property rights. The Internet makes widespread distribution of multimedia (text, images, video, and sound) easier than ever before. Further, the culture of the Internet, which, until recently, was populated primarily by academics and researchers, is one of sharing information, making it simple to broadcast information of interest to a huge audience at the mere touch of a few keys on a keyboard. This author will not discuss the philosophical aspects related to greater information exchange here, but some mention of copyright issues and the evolving legal concerns is very important for Web users and developers alike. For the interested reader, the author recommends visiting "Copyright and Intellectual Property Rights for Digital Documents" (Berkley Digital Library, 1996), a Web Page with a comprehensive set of links related to this issue. You are also encouraged to consult with the legal authorities representing your organization if you have any questions.

Many believe that in order for a document to be copyrighted, it must contain a copyright statement. However, according to Stanford University (1996), "Currently, the author's rights begin when a work is created. Copyrighted works are not limited to those that bear a copyright notice." So, one should always assume an electronic document is copyrighted material, and that permission must be requested and granted from the copyright owner in order to use it in a publication or to distribute it to others. Copyrighted materials may be written documents, photographs, electronic images, video, software, databases, any digital works or works transformed into digital format. Copyrights are protected regardless of the medium in which they are created or reproduced.

There are many instances in which the definitions of words like "publish" and "distribute," for instance, are vague. In such cases, you should consider what is commonly referred to as "fair use." O'Mahoney (1995-1996) proposes the "Fair Use Test," based on the "Fair Use" provision in U.S. copyright law. Briefly, the fair use test requires consideration of four factors:

Factor 1 — Purpose and Character of Use
Factor 2 — Nature of Copyrighted Work
Factor 3 — Relative Amount
Factor 4 — Effect on the Market

Interpretation of these factors, and thus determination of what fair use of a particular media is, is constantly evolving. Stanford University (1996) provides one current interpretation:

- The purpose and nature of the use — If the copy is used for teaching at a nonprofit institution, distributed without charge, and made by a teacher or students acting individually, then the copy is more likely to be considered as fair use. In addition, an interpretation of fair use is more likely if the copy was made spontaneously, for temporary use, not as part of an "anthology," and not as an institutional requirement or suggestion.

- The nature of the copyrighted work — With multimedia material, there are different standards and permissions for different media: a digitized photo from a *National Geographic*, a video clip from *Jaws*, and an audio selection from Peter Gabriel's CD would be treated differently — the selections are not treated as equivalent chunks of digital data.
- The nature and substantiality of the material used — In general, when other criteria are met, the copying of extracts that are "not substantial in length" when compared to the whole of which they are a part may be considered fair use.
- The effect of use on the potential market for or value of the work — In general, any use that supplants or diminishes the normal market for the original work is considered an infringement, but a use does not have to have an effect on the market to be an infringement.

The authors caution that the last factor is the most important consideration.

There are works that are considered to be in the "public domain" and can be copied freely by anyone. These include works by the U.S. government (and presumably other governments), works for which a copyright has expired, and works which the owner has explicitly identified as public domain.

When in doubt about whether something is copyrighted, or whether a particular use falls within the "fair use" provisions, always seek permission from the author or copyright owner.

5.3 User Interface Issues

There are many issues to consider when designing Web-based materials. These include careful consideration of file size, layout of the page for effective viewing, and browser compatibility with selected page formatting codes (i.e., HTML). User interface design goes far beyond the confines of this chapter, but a few things are worth mentioning if you decide to produce Web pages.

File size can define whether your pages will ever be viewed. Web users quickly become impatient while waiting for large files to download over a modem or other slow Internet connection. Therefore, you should take care to reduce the number of images contained in a page, or minimize the file size of the images you do include. Sometimes there will be a trade-off between download time and the desired visual effect of your pages, since graphic files often account for the majority of the transfer delay. (It is also important to have good Internet connections and fast computer hardware and networks, but that too is well beyond the scope of this chapter.)

The purpose of a page will often define the size of the file. For example, if you intend a particular document to be printed by the user, you will want to include the entire document in one continuous page, even though it may take longer to download the file. Alternatively, you might provide a link to a compressed downloadable file formatted in a common word processor format, allowing users to print the formatted document on their own, rather than viewing and printing directly from the Web. Presumably, the user will wait the extra time in order to obtain the full document.

On the other hand, if a page contains intermediate information, such as an index page that is just one step for the user in a search for more specific information, you should take care to design a smaller file set that will download quickly. Readers interested in designing effective Web pages may refer to, for example, Diehn and Katz (1996). Among other suggestions, these authors recommend that developers minimize bandwidth impacts; test graphics on a variety of computers; make sure graphics add value; and restrict navigation pages to one screen.

The look and feel of a Web page is influenced by the capabilities and limitations of the HTML formatting language and the speed limitations related to the use of graphics, as noted above. An additional barrier to achieving a desired look is the difference in browser capability. Since the Web is evolving fast and there are no universally accepted development standards, different browser software will interpret the same HTML code in different ways. That is, what looks "good" in one browser, may look "bad" in another. For this reason, you may want to minimize the use of advanced HTML features that have not yet been incorporated into the browsers commonly available and in use among the general public.

Navigation is another important point in Web site design. Web users often complain of "getting lost." While some of this is under the control of the user and the particular browser being used, page designers can take steps to assist navigation, at least on the pages under their control. There are a number of methods one can use to assist users, but the important thing to keep in mind is that the user should be provided with some referencing method that defines the relationship between pages within a given site, such as a navigation bar that appears on every page, allowing the user to quickly jump to a primary page when desired.

If you plan to develop Web pages, here are a few places you can go to learn the basics and the latest advancements.

- The Netscape® "Creating Net Sites" (1997) maintains a set of HTML guidelines and pointers to other Web sources for design assistance.
- World Wide Web Consortium is an international organization founded in 1994 to develop common standards for the evolution of the World Wide Web. This is a good starting place for many Web topics, but with respect to HTML and Web page style guides, see "HyperText Markup Language (HTML)" (1996) and "Web Style Sheets" (1996).
- "HTML and Style Manuals," provided by "InfoQuest" (1996) at http://www.fptoday.com/htmlandstyle.htm provides a nice index of style guides.

5.4 Finding What You Need on the Web

The first few times "surfing" the Web can be exhilarating (so can the rest, for that matter, if you learn to utilize it effectively). The vast amounts of information, a connection to people around the world, the innovative pioneering use of multimedia, and all the possibilities have been likened to an addiction. Not to fear, however, the thrill can quickly dissipate if you merely bounce from site to site without ever honing in on the materials you seek. As Digital Corporation notes at the AltaVista® Web searching and indexing site,

> "The Web is immense. If you only spent a minute per page and devoted ten hours a day to it, it would take four and a half years to explore a million Web pages, a lifetime to explore just this index."

Frustration can be minimized, though, by the effective use of search engines and Web indexes.

With the explosive growth of the Web, any list of specific sites compiled today may well be obsolete in a very short time. Therefore, it is best to review the search methodologies you should apply when you turn to the Web to locate and gather information. To illustrate, this author will take you through a series of searches performed while preparing this chapter.

There is a growing number of Web sites dedicated to searching for specific information, sites, people, and places on the Web. Generally called "search engines," they may utilize specialized software programs that "crawl" the Web reading documents and organizing them into a master index at the home site (e.g., AltaVista®, Excite®, Lycos®). Others are simply a large categorical index of links submitted by interested parties (e.g., Yahoo®). The search engines may be organized in a browsable hierarchical index, in which the user narrows a search by point-and-click.

For example, setting out to find "ergonomic" sites, one might choose the broad category of "science and engineering," which then brings up a number of subcategories falling under that topic. Eventually, you might expect, you will find the topic you seek. However, a different user with another viewpoint may begin searching for "ergonomic" by looking at the broad category of "workplace health and safety." Whether "ergonomic" will be found through either or both of these paths depends on the sophistication and breadth of the indexing method. By experience, this author believes few, if any, indexes have successfully developed a robust indexing system that can capture the expectations of users arriving with varying viewpoints on the same topic.

Other search engines rely on keyword searching schemes or a combination of keyword searching and hierarchical indexes. Effectively used, keyword search methods can be an excellent way to quickly narrow a search to Web pages of interest. However, without taking the time to learn the advanced search methods

TABLE 5.1 Keyword Search Results for "Ergonomic" and "Ergonomics" on Three Example Search Engines

Keyword	Search Engine	Number of Pages Found
ergonomic	AltaVista®	about 40,000
	Excite®	17,481
	HOTBOT®	29,216
ergonomics	AltaVista®	about 40,000
	Excite®	21,297
	HOTBOT®	36,564

TABLE 5.2 Keyword Search Results for the String "Occupational Biomechanics" (without using the quotation marks) on Three Example Search Engines

Keywords	Search Engine	Number of Pages Found
occupational biomechanics	AltaVista®	about 30,000
	Excite®	134,685
	HOTBOT®	2,341

TABLE 5.3 Keyword Search Results for the String "Occupational Biomechanics" on Three Example Search Engines

Keyword(s)	Search Engine	Number of Pages Found
occupational biomechanics	AltaVista®	about 200
	Excite®	1,139
	HOTBOT®	353

employed by the particular site, a user may spend hours wandering through unrelated pages. To illustrate, the author performed one-word keyword searches on three different keyword-driven search engines. The results are summarized in Table 5.1.

Without reading any instructions, the author then did a search for the two-word string "occupational biomechanics" (without using quotation marks). The results indicated that each search engine treats this string differently, which helps build the case that to successfully utilize the Web, one must become very familiar with the syntax and capabilities of at least one search engine. The results are compiled in Table 5.2.

After reading the instructions, the author learned that the search engines where not searching for the phrase "occupational ergonomics," but were instead searching for the booleans "occupational" *OR* "ergonomics," or "occupational" AND "ergonomics" (neither site was clear regarding this issue). Using the various syntax required by the different search engines, the author then performed the search for the string "occupational biomechanics," and received the results shown in Table 5.3.

This did narrow the search but still produced a large number of selections from which to choose. So, in order to narrow the search further, the author consulted the "advanced search" or "help" links at each search engine and quickly learned that each system uses different syntax and search methodologies, and the learning curve can be steep. Some search engines may be exclusively keyword based, while others will employ proprietary "intelligent" search methods that attempt to look at related concepts, related words, and so forth. The advanced search methods also provided options such as search location (the Web, Usenet, etc.), date limits, boolean operators, results ranking, case sensitivity, and more. These advanced features make it possible to narrow a search to an exact document, or narrow a topical search to a manageable set of pages to peruse.

Spending the time to learn and use search engines will save you significant amounts of time when doing research on the Web. However, search engine designers have a long way to go before they make searching easy for the casual user, and this is an area where ergonomics and human factors experts could make substantial contributions.

Glossary of Terms

Browser: The software program installed on a client computer through which the user accesses the Web. This might be thought of as the "window" to the Web. Netscape Navigator® and Microsoft Internet Explorer® are two example browsers.

CGI: "Common-Gate Interface," a protocol allowing Web servers to communicate with server-based computer programs. For example, if calculations are to be performed, the server may call on a server-based computer program to perform those calculations. Once the calculations are complete, the server sends the results to the client through the network.

Client: The computer (or computer user) that "requests" a document from a remote server.

Domain: An Internet address conforming to the Internet Protocol (IP), such as fictitiouscompany.com. This evolving naming convention varies from country to country. In the U.S., the last three characters will often represent the nature of the organization, such as "com" for commercial enterprises, "edu" for educational institutions, "mil" for military, and so on. Other countries follow this convention as well, but add a country designation at the end. Some countries have developed their own domain-naming methodology, usually ending with a country designation.

Email: Protocols which allow individuals to send and receive electronic messages. Once restricted to text only, the protocols are evolving to include hypermedia capabilities.

FAQ: "Frequently Asked Questions." FAQs are common documents that summarize and answer frequently asked questions regarding a topic or a specific site. FAQs can provide a quick study on the topic of interest and are often an excellent starting place.

FTP: "File Transfer Protocol," another method, like HTTP, defined above, in which computers may communicate across the Internet. (FTP may also be used via the Web.)

Home Page: A page dedicated to one individual, or the primary "front" page of a collection of Web pages.

HTML: "Hypertext Markup Language," the computer language used to format Web pages.

HTTP: "Hypertext Transfer Protocol," the common scheme through which Web servers communicate. Also see FTP, Telnet, Gopher, etc., in this definition list.

Hypermedia: The interconnected media of text, static graphics, animated graphics, video, and sound that are published on the Web.

Java: An evolving computer language (developed by Sun Microsystems) that is designed to operate across different computer operating systems (e.g., UNIX, Windows®, etc.). A Java "applette" (a mini software application) executes through the browser on the client side, as opposed to executing on the server side, like CGI applications.

Listserv: Listservs are software programs used to manage an e-mail discussion or distribution list. Users join "Listservs" when they want to monitor or participate in discussions (usually dedicated to a particular topic) with other Internet users via e-mail.

Navigating: Linking from page to page, and site to site on the Web. "Navigating" generally refers to successful, planned "surfing."

Netiquette: Commonly accepted or expected behavior while participating in Internet-based communications. Interested readers may refer to Rinaldi (1996) for a compilation of netiquette topics relating to many Internet communication protocols, including the Web.

Search Engine: A computer software system that locates and catalogs Web pages. Most search engines provide a keyword searching method for clients to locate and link to the cataloged pages.

Search Index, or Web Index: An index of Web links organized by topic that may or may not have been gathered by spiders, etc.

Server: The computer hardware that "serves" electronic documents on the World Wide Web.

Spider, Robot, Crawler, Worm, etc.: Terms used to describe the computer software programs that search the Web in conjunction with certain types of search engines.

Surfing: Linking from page to page, and site to site on the Web. "Surfing" often refers to a somewhat random path of linking (getting "lost" is not unusual for new users).

URL: An acronym for "Uniform Resource Locator," a draft standard for specifying an object on the Internet. Think of this as an address, such as http:www.fictitiouscompany.com. (See Connolly, 1990, for an in-depth discussion.)

Usenet: Thousands of discussions organized by topic. One can monitor a usenet group dedicated to discussions and information exchange on some topic by reviewing posted submissions from others, or one can participate by "posting" a message to the group.

Web Page: An electronic document formatted in the HTML language and made accessible to the Web through a server.

Web Site: A collection of HTML formatted documents that make up one primary server or collection of servers managed by one individual or organization.

World Wide Web: Hereafter referred to as the "Web," this is a distributed system of linked hypermedia documents spanning the world.

References

AltaVista®, AltaVista Technology, Inc.,
http://www.altavista.com, (22 Jan 1997).

Connolly, D., 1990 (updated 1996), "Names and Addresses, URIs, URLs, URNs, URCs," World Wide Web Consortium,
http://www.w3.org/pub/WWW/Addressing/Addressing.html, (19 Jan 1997).

Diehn, M. J., and Katz, M., 1996, "WTCS Workshop: Guidelines for Slim Web Pages," Marketing on the Web: Pre-workshop preparation for participants in the WTCS Web Workshop,
http://owl.warren-wilson.edu/~mdiehn/wtcs/guide.htm, (22 Jan 1997).

Digital Corporation, 1996, "AltaVista Search: Surprise CyberSpace Jump," AltaVista ™ Search, (20 Jan 1997).

Excite®, 1996, Excite Inc.,
http://www.excite.com, (22 Jan 1997).

Harnack, A. and G. Kleppinger, "Citing the Sites: MLA-Style Guidelines and Models for Documenting Internet Sources," Version 1.3,
http://falcon.eku.edu/honors/beyond-mla/#citing_sites, (19 Jan 1997).

HOTBOT®, 1996-97, Hotwired Inc.,
http://www.hotbot.com, (22 Jan 1997).

InfoQuest®, 1996, "HTML and Style Manuals," Front Page Today,
http://www.fptoday.com/htmlandstyle.htm, (22 Jan 1997).

Netscape®, 1997, "Creating Net Sites,"
http://www.netscape.com/assist/net_sites/index.html, (30 Jan 1997).

O'Mahoney, B., 1995-1996, "The Fair Use Test," The Copyright Website,
http://www.benedict.com/fairtest.htm, (21 Jan 1997).

Rinaldi, A.H., 1996, "The Net: User Guidelines and Netiquette — Index," Florida Atlantic University,
http://www.fau.edu/rinaldi/net/index.htm, (30 Jan 1997).

Stanford University, 1996, "Main Fair Use Index: Library Copyright Guidelines: Copyright Law: Frequently Asked Questions," Copyright & Fair Use,
http://fairuse.stanford.edu/library/faq.html, (21 Jan 1997).

Walker, J.R., 1996, "MLA-Style Citations of Electronic Sources,"
http://www.cas.usf.edu/english/walker/mla.html, (19 Jan 1997).

World Wide Web Consortium, 1996, "HyperText Markup Language (HTML),"
http://www.w3.org/pub/WWW/MarkUp/

World Wide Web Consortium, 1996, "Web Style Sheets,"
http://www.w3.org/pub/WWW/Style/, (22 Jan 1997).

Yahoo!®, 1994-97, Yahoo! Inc., http://www.yahoo.com, (22 Jan 1997).

6

Professional Ergonomics Issues

Dieter W. Jahns
Board of Certification in Professional Ergonomics

6.1 Introduction

The education, training and career domains of ergonomics in general, and industrial ergonomics in particular, have a long and diverse history. Different countries, cultures, and economies still view the science and technology of work systems from a variety of perspectives. These sometimes divergent perspectives have made it difficult until recently to achieve a consensus on what the scope of ergonomics is, how people should be educated and trained to conduct ergonomics research and practice ergonomics, and how to measure the competencies of those claiming to be ergonomists. It is rather ironic that while ergonomists routinely do job/task analyses, performance/technology assessments, and work-system designs for career fields ranging from astronauts to warehouse workers, they rarely apply these methods to their own career field. A true case of "The cobbler's children have no shoes"! The goal of this chapter is to summarize the current status of professional issues in ergonomics as portrayed in the literature published by the International Ergonomics Association (IEA) and its Federated Societies. This literature also forms the foundation for certification criteria development by the Board of Certification in Professional Ergonomics (BCPE) in the U.S. and by the Center for Registration of European Ergonomists (CREE) in Europe. Readers wanting a broader and deeper survey of the evolutionary development of ergonomics into a transdisciplinary, unique career field should consult the resources listed in "For Further Information." Particularly useful for achieving an awareness of what ergonomics is all about are the books by Chapanis (1996), Booher (1990), Klemmer (1989), MacLeod (1995), Meister (1997), and Salvendy (1987).

6.2 Education

It has long been postulated that ergonomics is akin to engineering because its subject matter involves the design and performance of humans and their technological tools to accomplish work effectively and efficiently. Three root sciences have vied for influence in shaping the quality of working life since the end of the 19th century and since then in an accelerating industrialized society: (a) the biomedical sciences interested in human fatigue, work physiology, biomechanics/kinesiology and anthropometry in occupational settings, (e.g., Amar, 1914), (b) industrial engineering as promulgated by Frederick W. Taylor in his book *The Principles of Scientific Management*, (Taylor, 1911), and (c) industrial/experimental psychology as

0-8493-2641-9/99/$0.00+$.50

exemplified by the work of Hugo Münsterberg in 1912 at Harvard University and later Berlin (e.g., Münsterberg, 1912, 1914). Even today, the graduation certificates of ergonomists may be from one of these "traditional" academic disciplines even though not all graduates from the cognate departments are necessarily educated in ergonomics. The situation is clearer in Europe and some parts of Asia where academic departments with core subjects in ergonomics *per se* exist. Where an ergonomics education program is "housed" can have a significant impact on how and what a student learns about the field. For example, it is unlikely that aviation ergonomists will have been exposed to all of the topics covered in this handbook; similarly, it is unlikely that industrial ergonomists will have in-depth knowledge of the perceptual–motor and cognitive requirements of transportation systems or complex, automated industrial processing plants. Both specialization and diversification have occurred in ergonomics in conjunction with economic, demographic, political, and technological pressures. As jobs have changed with advances in science and technology, so has the influence of those sciences on ergonomics; everything became "work-related." Reich (1992, pp. 174-180) predicts that as human enterprise evolves from an industrial/materialistic base to an information/knowledge base, three job categories will dominate economies: (a) routine production services, (b) in-person services, and (c) symbolic–analytic services. Ergonomics can be viewed as the practice of symbolic-analytic services for the benefit of those people who perform routine production services and in-person services.

In order to provide the most effective services, three levels of education are currently being offered by both public and private vendors of knowledge and skills training (Webb and Stager, 1990):

Awareness education: Provides sufficient education in the structure and application potential of ergonomics to enable students to judge its relevance to their current interests.

Familiarization education: Provides sufficient education in the structure, relevant knowledge, and skills in ergonomics to enable a professional in a different field (such as management, engineering, psychology, human resources, or safety) to interact effectively with an ergonomist.

Professional education: Provides a full understanding of the structure, knowledge, and skills of ergonomics to enable the student, after sufficient practical experience, to practice ergonomics.

Awareness education is most appropriate for workers who are involved with *Participatory Ergonomics* and related organizational and management issues. Normally, a survey course of 40 contact hours specifically oriented toward the job, worker demographics and work system with which attending workers are most familiar will provide a learning experience which is useful. Generalized "one-size-fits-all" pep talks on ergonomics should be avoided, as should building unreasonable expectations of what can be achieved with ergonomic data, methods, and techniques. An excellent resource for awareness education is MacLeod's (1995) book *The Ergonomics Edge.*

Familiarization education should build on awareness education in both breadth and depth. The knowledge base of course participants should be used as a point-of-departure for launching into ergonomics topics and how they differ from (and are related to) their occupational responsibilities. In this regard, the current practice of structuring course topics in accordance with the OSHA (Occupational Safety and Health Administration) agenda and guidelines (as was done for this handbook) is not quite appropriate. For one thing, the goals of ergonomics are

- Reasonable human performance
- Reasonable workload
- Reasonable health maintenance
- Reasonable hazard control and injury-risk management

through optimum human–machine system design, personnel selection, and training. The emphasis is on humane use of technology to achieve good system performance. For another thing, OSHA's agenda is much narrower, reactive, and topically different from ergonomics. "Medical management," for example, is important in work systems, as are safety audits and industrial hygiene surveillance, but none of these falls within the scope of ergonomics as a work-system design profession. Ergonomists and other career

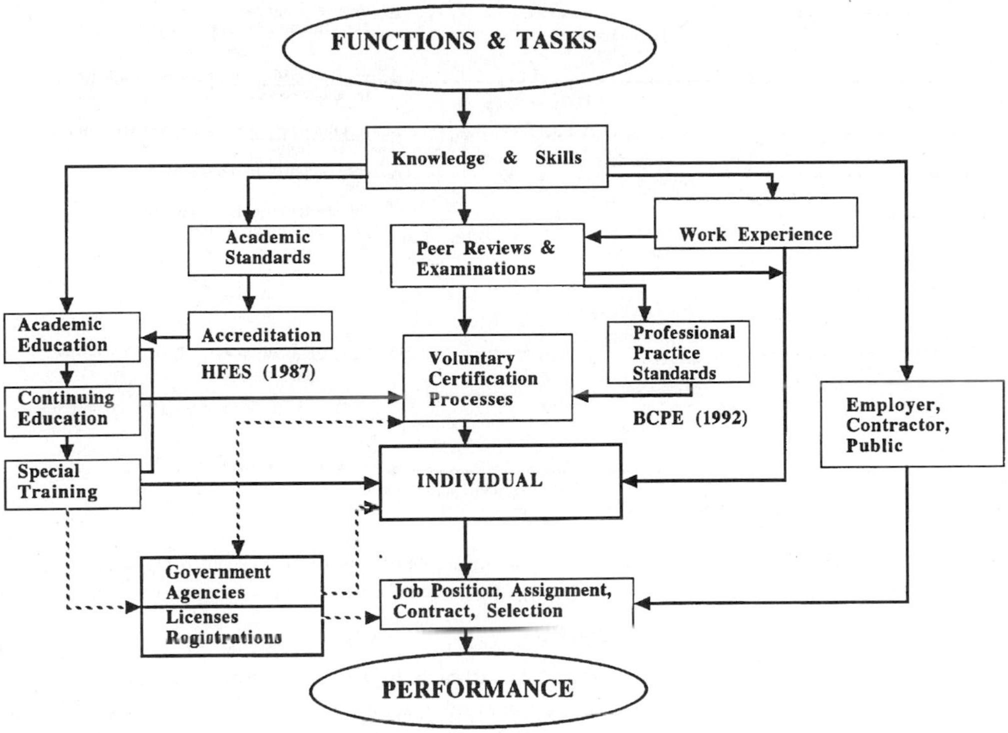

FIGURE 6.1 Professional development and assessment infrastructure for ergonomists.

fields must see their edges clearly (Duncan, 1996) or risk becoming overwhelmed by information overload and/or trivialization of their knowledge and skill base. Being able to drive a car does not mean you can fly an airplane! Try it, and we'll plant daisies on your grave.

Professional education in ergonomics builds on the professional development and assessment model shown in Figure 6.1. The functions and tasks performed by ergonomists have been discussion topics at numerous meetings of related scientific/engineering societies since the 1960s. However, individuals working in the field were often given job titles which reflected the contractual requirements of employers more than academic standards or work experience. "Human Performance Analyst," "Methods Engineer," "Human Factors Specialist," "Human Factors Engineer" and "Engineering Psychologist" have been common titles for ergonomists since the 1920s. Cohesive academic standards for coursework and voluntary certification processes for individuals were slow in coming and still have not been harmonized internationally.

In 1976 a five-day symposium sponsored and financed by the "Special Programme Panel on Human Factors" of the North Atlantic Treaty Organization (NATO) Science Committee brought together 12 experienced university teachers/scientists in ergonomics from seven nations plus four university teachers from other nations who are experts in areas closely related to ergonomics (Bernotat and Hunt, 1977). This group developed recommended curricula structures for six different educational programs in ergonomics:

1. A full program in ergonomics
2. A specialization program in ergonomics for students in engineering
3. An orientation program in ergonomics for students in physiology
4. An orientation program in ergonomics for students in psychology
5. An orientation program in general ergonomics for students in management, engineering, and other disciplines
6. An indoctrination program in general ergonomics for students in physiology, psychology, engineering, and other sciences.

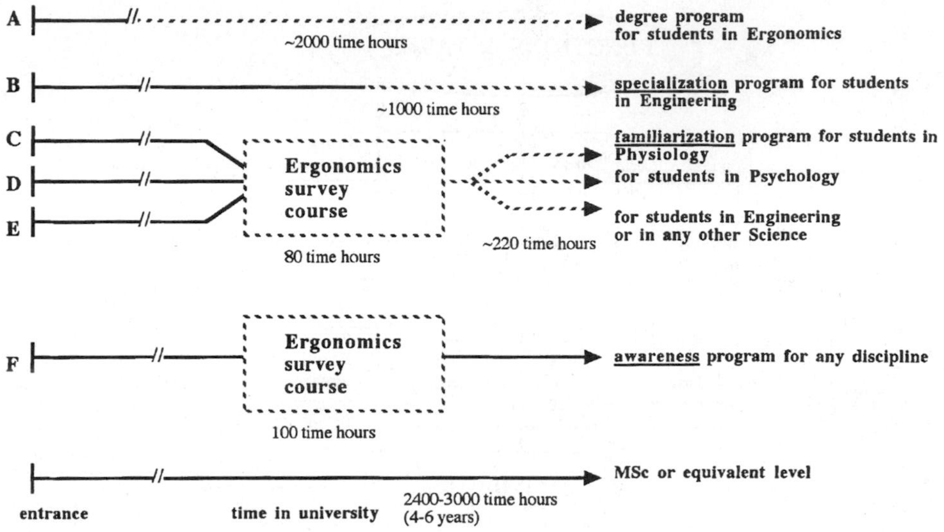

FIGURE 6.2 Recommended curricula structures in ergonomics (see text for details) (From Bernotat, R. and Hunt, D. P. 1977. *University Curricula in Ergonomics.* Wachtberg-Werthhoven, Germany: Forschungsinstitut für Anthropotechnik.)

Figure 6.2 illustrates the academia time history of each of the educational programs. The student/teacher contact hours shown are 60-minute *clock* hours, as opposed to normally shorter *lab/lecture* hours. The solid lines indicate that part of the education which is devoted to the specific discipline listed and general university requirements (GURs); the dashed lines indicate concentration on ergonomics. The bottom line of the figure is meant to show the time line from university entrance up to the granting of a degree at the Master-of-Science level. While there are some programs which are offered to undergraduates in pursuit of a Bachelor's degree (or equivalent diploma), most ergonomics programs are graduate degrees as indicated for programs A, B, C, D, and E of Figure 6.2. The recommended curriculum for a "degreed" ergonomist is shown in Table 6.1. Keep in mind that this was 1976. By 1987, the Human Factors and Ergonomics Society had a voluntary accreditation program in place which used many of the 1976 recommendations for establishing academic standards. The Ergonomics Society (U. K.) also reviews academic programs to establish membership status for its applicants. Table 6.2 lists the HFES and ES (U.K.) accredited programs in the U.S. as of December 1996. Strangely, and most likely for political reasons, these are not titled "ergonomics programs" but follow traditional designators of root sciences or academic departments, even where the campus may house an ergonomics research institute. It should be noted that there are many excellent ergonomics academic programs which for a variety of reasons have not yet applied for accreditation. The interested reader should consult the program directories listed in "For Further Information."

Since "education" is one of the three components upon which professional certification is based, both the BCPE and CREE adopted the Ergonomist Formation Model (EFM) released and published by the HETPEP working group in 1992 (Rookmaaker, et al., 1992). Table 6.3 lists the topics of knowledge, skill, and experience currently used in the evaluative policies, practices, and procedures by BCPE. Notice that certification candidates must be educated, trained, and experienced in at least one "Design" topic of those listed in Table 6.3. It is this tie-in to design knowledge and skills that differentiates ergonomics from other "human factors"-oriented professions like psychology, human resource development, allied healthcare providers, etc., who "help" humans by means of a treatment/behavior modification philosophy.

The "Ergonomics Approach" is based on the premise that tasks, tools, and talents form a system for productive human work. The relationships among humans, their tools, and their talents need to be empirically studied in order to create a "human-centered technology." The other categories and details listed in Table 6.3 form the core foundation upon which this technology is built.

TABLE 6.1 The Curriculum for Degreed Ergonomists

Summary of the Time Hours

	Lecture (hours)	Practicum (hours)	Total (hours)
Part I. Basic Knowledge	**250**	**157**	**407**
Human anatomy and physiology	95	55	150
Human psychology	105	102	207
Developmental psychology	50	—	50
Part II. Methods and Techniques	**35**	**150**	**185**
Methods of measurement	—	150	150
Selection	10	—	10
Training	15	—	10
Instrumentation	10	—	10
Part III. Application	**215**	**165**	**380**
Ergonomics in design	60	85	145
The physical environment	90	65	155
Accidents and safety	20	—	20
Systems ergonomics	45	15	60
Part IV. Other Basic Subjects	**345**	**330**	**675**
Statistics	75	25	100
Mathematics	50	50	100
Physics	30	65	95
Computation	20	45	65
Networks, time series, stochastic processes	45	45	90
Systems theory and optimization	50	50	100
Engineering	50	50	100
Communications and "public relations"	25	—	25
Part V. Background Subjects	**145**	—	**145**
The individual in work organizations	25	—	25
Organizational context of systems design	10	—	10
Project management	10	—	10
Production & process management & control 50	—	50	
Production & process system design	50	—	50

From Bernotat, R. and Hunt, D. P. 1977. *University Curricula in Ergonomics.* Wachtberg-Werthhoven, Germany: Forschungsinstitut für Anthropotechnik.

TABLE 6.2 Accredited Ergonomics/Human Factors Degree Programs in the U.S. as of December 1996

Georgia Institute of Technology (1)
Louisiana State University (2)
New Mexico State University (1)
North Carolina State University (2)
Ohio State University (1)
State University of New York at Buffalo (1,2)
Texas A&M University (1)
University of Central Florida (1)
University of Dayton (1)
University of Illinois at Urbana-Champaign (1)
University of Southern California (1) (Deactivated 1997)
Virginia Polytechnic Institute & State University (1,2)

(1) = Accreditation by Human Factors and Ergonomics Society (U.S. A.)

(2) = "Vetting" by Ergonomics Society (U. K.)

From Ergonomics Society and Human Factors and Ergonomics Society

TABLE 6.3 Ergonomist Formation Model: Topics of Knowledge, Skills and Experience

General Category		Details	
Code	Name	Number	Name
A	Ergonomics Principles	1	Ergonomics Approach
		2	Systems Theory
B	Human Characteristics	1	Anatomy, Demographics, and Physiology
		2	Human Psychology
		3	Social and Organizational Aspects
		4	Physical Environment
C	Work Analysis and Measurement	1	Statistics and Experimental Design
		2	Computation and Information Technology
		3	Instrumentation
		4	Methods of Measurement and Investigation
		5	Work Analysis
D	People and Technology	1	Technology
		2	Human Reliability
		3	Health, Safety and Well-Being
		4	Training and Instruction
		5	Occupational Hygiene
		6	Workplace Design
		7	Information Design
		8	Work Organization Design
E	Applications		Projects pursued by the individual
F	Professional Issues		

From Board of Certification in Professional Ergonomics, June 1995, *Information on Certification Policies, Practices & Procedures,* 3rd Ed., p. 24, BCPE: Bellingham WA.

6.3 Training

One of the most thorough and comprehensive surveys of ergonomics as a career was conducted in 1989 (with results published in 1992) by the U.S. National Research Council's Committee on Human Factors (NRC-HF) (Van Cott and Huey, 1992). This survey revealed that approximately 83% of ergonomic work in the U.S. is centered in six areas: computers (22.3), aerospace (21.6), industrial processes (16.5), health and safety (8.9), communications (8.2), and ground transportation (5.3). The remaining 17% of work is spread over a large variety of other areas, but it is performed by very few ergonomists, often just one or two people. The principal workplaces of ergonomists are shown in Figure 6.3. where the private business category covers both corporate and self-employed ergonomists. It is interesting that this distribution is similar to that of engineering professionals. The nature of the work performed was categorized by the NRC-HF committee into six categories, spanning 52 tasks as derived from unpublished task/job analyses provided by the Human Factors and Ergonomics Society:

1. Systems analysis
2. Risk and error analysis
3. Design support
4. Test and evaluation
5. Instructional systems design
6. Communications

While it is customary in other career fields to receive post-academic training upon entry into the job market from seasoned practitioners and/or supervisors familiar with the career field, such is generally not the case for ergonomists, most likely because there are so few of them to begin with. The NRC-HF committee survey indicated that only 9% of ergonomist supervisors are ergonomists. Thirty-four percent are engineers, with others scattered through business managers, psychologists, industrial designers, etc.

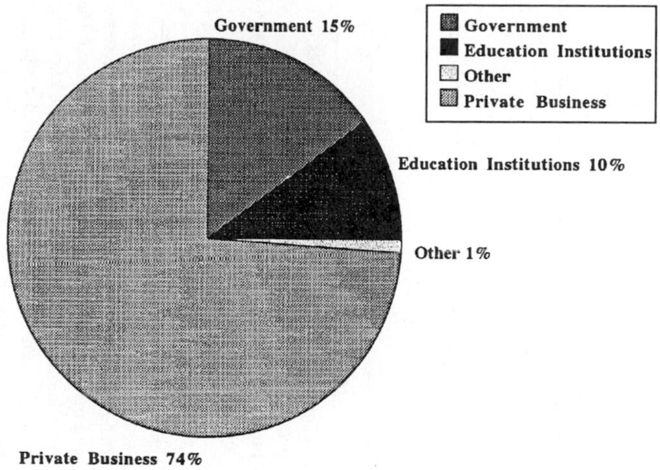

FIGURE 6.3 Principal workplaces of ergonomists. (From Van Cott, H. P. and Huey, B. M. (Eds.), 1992. *Human Factors Specialists' Education and Utilization, Results of a Survey.* Committee on Human Factors, Commission on Behavioral and Social Sciences and Education, National Research Council, Washington, D.C.: National Academy Press.)

This training dilemma in ergonomics may improve as the field grows, but in the short run, more internships and academia–industry coordinated programs are severely needed.

6.4 Experience

As a scientific approach to work systems development, ergonomics often gets caught in the middle between labor and capital; yet to maintain its neutrality, it is frequently reluctant to take sides in business management, political, and/or regulatory issues. Taylorism (and its reliance on time-and-motion studies to establish labor standards) is at times accused of being too "task focused" and not adequately considering individual differences among workers' capabilities and limitations. Occupational safety and health specialists, sociologists, lawyers, workers' compensation insurers, and industrial/organizational psychologists often see all work and technological innovations as risk factors to the quality of working life. Thus, experience in ergonomics is shaped by how "human factors" in systems development and operation are viewed and dealt with. It does make a difference to the derivation of design solutions whether the human operator and/or maintainer of technological devices is considered to be a "mechanical power source," an "information processor and transducer," an "intelligent controller of processes," a "production cost factor," or the most vital "business asset."

Application scenarios will provide different experiences in how to accommodate human factors in work systems. For example, industrial systems dependent on manual materials handling may create more concern about musculoskeletal issues and thus rely on work physiology, biomechanics, and kinesiology, than transportation or computer systems, which require little physical effort but much information search, decision making, and error-free motoric responses and thus depend on the behavioral/cognitive sciences and linguistics. Because ergonomists work in a variety of settings and situations which are influenced by many different parameters and "stakeholders," they must learn through experience how best to work with other career fields (e.g., engineering, hazard/risk management, project management, labor representatives, system-users/operators, health and safety advocates, etc.).

Unpublished studies conducted by BCPE also determined that ergonomics is practiced at four levels (see Table 6.4). In this regard, the situation is similar to other career fields where the technician or draftsman performs routine engineering tasks; the paralegal assists the attorney; the medical assistant, nurse, or therapist aids the physician; the research/teaching assistant helps the professor. The BCPE started defining the competencies and assessment criteria for the "Certified Ergonomics Associate (CEA)" in 1997. Implementation is scheduled for late 1998. This development is particularly relevant for industrial

TABLE 6.4 Levels of Comprehensive Practice in Ergonomics

Purpose:
To define levels of comprehensive practice for the ergonomics profession.

Rationale:
The BCPE needs to define levels of comprehensive practice in order to develop measurement instruments that serve to recognize ergonomists progressing in their professional development. Four levels of comprehensive practice are believed to span the profession.

Level of Practice	What can be expected in terms of the ergonomist's:		
	Background and Experience	Capabilities	Areas of Knowledge
Level-1	Has received at least 40 hours of coursework in ergonomics topics taught by qualified ergonomists. Has no experience in independently applying ergonomics knowledge; works under direct supervision from a Level-3 ergonomist or Level-4 ergonomist.	Functions at an introductory level with supervision; uses evaluation devices (checklists, surveys) contracted by qualified ergonomists to collect data and information for making preliminary system analyses.	Has introductory-level knowledge of ergonomics principles and systems theory, characteristics of humans in systems, and methods of work analysis and measurement.
Level-2 BCPE Certification Level at CEA	Has general knowledge of topics in the broad field of ergonomics as received from a baccalaureate degree program in ergonomics or a related field. Alternatively, may have a baccalaureate degree in a nonergonomics discipline and a minimum of 200 hours of continuing education coursework in ergonomics topics of increasing depth and specificity, or one semester (4, 3-hour courses) in ergonomics from an accredited program. Has at least two years of experience in applying ergonomics knowledge.	Works with minimal supervision and/or consultation; conducts basic system analyses using checklists, surveys, questionnaires, videotaping, human performance measurements, etc.; and makes basic recommendations for addressing ergonomics design criteria and specifications. Analysis and design recommendations follow established practices and guidelines (i.e., refers to standards and ensures implementation).	Has general knowledge of ergonomics principles and systems theory, characteristics of humans in systems, and methods of work analysis and measurement. Has basic knowledge and ability to apply basic knowledge of at least one of the following in system design: the relationship between technology and human performance; human reliability; health and safety; training and instruction; occupational hygiene; workplace design; information design; and work organization design. Example systems include consumer products, manufacturing systems, office work, transport systems, process industry; health care systems; and recreation, arts and leisure activities.

Level-3 BCPE Certification Level at CPE/CHFP	Has a working knowledge of all "core" topics comprising the ergonomics profession based on a master's degree or equivalent in ergonomics and at least 4 years of demonstrated experience in applying that knowledge. Alternatively, may have a master's degree in a nonergonomics discipline and a minimum of 320 hours of continuing education coursework in ergonomics topics of increasing depth and specificity, or two semesters (8, 3-hour courses) in ergonomics from an accredited program. Is able to perform independently and to supervise others in ergonomics work.	Works independently; conducts detailed, specific analyses by collecting, analyzing, and interpreting subjective and objective system and human performance data; consolidates and compares data with relevant current research; develops meaningful alternative solutions to identified problems, and proposes practical design solutions specific to the setting; supervises implementation of solutions; employs valid statistical methods to conduct tests/evaluations to verify that identified problems have been appropriately addressed by control measures; clearly communicates recommendations in writing and in oral presentations to all levels of personnel; conducts mission, function, job, and task analyses; supervises, mentors, and trains others in performing necessary support tasks.	Has broad understanding of ergonomics principles and systems theory, characteristics of humans in systems, methods of work analysis and measurement, and issues of importance to the ergonomics profession. Has working knowledge and ability to apply knowledge of at least three of the following in system design: the relationship between technology and human performance; human reliability; health and safety; training and instruction; occupational hygiene; workplace design; information design; and work organization design. Example systems include consumer products, manufacturing systems, office work, transport systems, process industry; health care systems; and recreation, arts, and leisure activities.
Level-4 BCPE Certification Level at CPE/CHFP	Has extensive working knowledge and experience beyond the Level-3 ergonomist with specialization in selected ergonomics application areas or systems. Has at least a master's degree or professional credentials (for example a doctoral or postdoctoral degree) in ergonomics and has received specialty education or training in at least one area of ergonomics. In lieu of specialty training, may have developed recognized expertise through concentrated experience of at least 8 years.	Works independently; performs beyond the capabilities of the Level-3 ergonomist by providing in-depth consultation in general ergonomics or within area of special expertise.	Has extensive understanding of ergonomics principles and systems theory, characteristics of humans in systems, methods of work analysis and measurement, and issues of importance to the ergonomics profession. Has extensive working knowledge and ability to apply knowledge of at least five of the following in system design: the relationship between technology and human performance; human reliability; health and safety; training and instruction; occupational hygiene; workplace design; information design; and work organization design. Example systems include consumer products, manufacturing systems, office work, transport systems, process industry; health care systems and recreation, arts and leisure activities.

From Board of Certification in Professional Ergonomics, August 1996, *Fact Sheet Levels of Comprehensive Practice*. BCPE: Bellingham, WA.

ergonomics where the demand for practitioners to deal with OSHA issues has outstripped the supply of qualified people at all levels of research, teaching, and practice. The CEA should be able to provide services around Level 2 of Table 6.4. Some people claim that ergonomics is not "rocket science"; they are right! Rocket science is much easier and more orderly than ergonomics. Physics, chemistry, and lots of mathematics are all that rocket scientists need (plus money and a creative vision) to do their job of objective materialism. Ergonomists have to also deal with fuzzy human factors which are subjective in some ways, objective in others, poorly understood, always changing, bounded by ethical and political taboos, and never fully predictable. The job market itself is rapidly changing, as are workforce demographics. The gurus of scientific business management promote new theories faster than they can be implemented and evaluated. Consequently, the budding ergonomist needs about four years of on-the-job experience before he or she can create any sort of stability in systems functioning in unstable equilibrium. Twenty percent of time should be set aside just for tracking changes in technology and their impact on ergonomics. It is up to employers to hire not just experienced ergonomists, but to also provide entry-level experiences. Ergonomics is as vital to business planning as engineering, production, marketing, and financial accounting because ergonomists should build the bridges between human factors and material factors.

6.5 Summary and Conclusions

As was shown in Figure 6.1, a mature career field is established when professional development criteria and assessment standards are in place. In the U.S., accreditation of academic ergonomics programs was started by HFES in 1987. The establishment of professional practice standards and a voluntary certification process for individuals independent of HFES (but in coordination with their resources) was implemented by BCPE in 1992. Hopkin (1994) and Jahns (1996) have predicted that these efforts will impact the infrastructure and further evolutionary development of ergonomics. The IEA is actively involved in harmonizing both professional development and assessment of ergonomists on a global basis, while being sensitive to the unique cultural, demographic, technological, and economic situations of various regions and continents. So long as a core set of criteria and content definitions for ergonomics exists, derivative and suitably modified programs will provide flexibility without weakening the fundamental foundation upon which ergonomics is built.

There are currently more career fields (or occupations) than ergonomists. The U.S. Department of Labor *Dictionary of Occupational Titles* lists over 32,000 occupations; there are an estimated 9,000 to 15,000 practicing ergonomists in the U.S. The challenges ergonomists face are daunting; but then, that is what makes ergonomics an exciting profession.

References

Amar, J. 1914. *Le moteur Humain et les Bases Scientifiques du Travail Professionnel.* Conservatoire Nat. des Arts et Métiers: Paris, France.

Bernotat, R. and Hunt, D. P. 1977. *University Curricula in Ergonomics.* Wachtberg-Werthhoven, Germany: Forschungsinstitut für Anthropotechnik.

Board of Certification in Professional Ergonomics (BCPE). June 1995. *Information on Certification Policies, Practices and Procedures,* 3rd Ed., BCPE: Bellingham WA.

Booher, H.R. (Ed.) 1990. *Manprint: An Approach to Systems Integration,* Van Nostrand Reinhold: New York, NY.

Chapanis, A. 1996. *Human Factors in Systems Engineering,* John Wiley & Sons, Inc.: New York, NY.

Duncan, J. 1996. Do we see our edges clearly?, *The Professional Ergonomist,* Summer 96, BCPE: Bellingham, WA.

Hopkin, C. O. 1994. Accreditation — a mechanism for quality. *Proceedings of 12th Triennial Congress of the IEA,* Vol. 1, 125-127; IEA: Toronto, Canada.

Jahns, D. W. 1996. *A global perspective on ergonomists' education, training and practice assessments for the 21st century.* IEA CybErg Conference: Internet.

Klemmer, E. T., (Ed.) 1989. *Ergonomics: Harness the Power of Human Factors in Your Business.* Ablex: Norwood, NJ.
MacLeod, D. 1995. *The Ergonomics Edge*, Van Nostrand Reinhold: New York, NY.
Meister, D. 1997. *The Practice of Ergonomics.* BCPE: Bellingham, WA.
Münsterberg, H. 1912. *Psychology and Industrial Efficiency*, Harper: New York, NY.
Münsterberg, H. 1914. *Gründzüge der Psychotechnik.* Springer: Leipzig, Germany.
Reich, R. 1992. *The Work of Nations*, Random House: New York, NY.
Rookmaaker, D. P., Hurts, C. M. M., Corlett, E. N., Queinnec, Y., Schwier, W. June 1992. *Towards a European Registration Model for Ergonomists*, Final report of the working group "Harmonising European Training Programs for the Ergonomics Profession" (HETPEP), Leiden, NL.
Salvendy, G. (Ed.) 1987. *Handbook of Human Factors*, John Wiley & Sons: New York, NY.
Taylor, F. W., 1911. *The Principles of Scientific Management*, Harper: New York, NY.
Van Cott, H. P. and Huey, B. M. (Eds.) 1992. *Human Factors Specialists' Education and Utilization, Results of a Survey*, Committee on Human Factors, Commission on Behavioral and Social Sciences and Education, National Research Council, Washington, D.C.: National Academy Press.
Webb, R. and Stager, P. February 1990. *Report of the Human Factors Association/Association Canadienne D'Ergonomie Committee on Professional Education.* HFAC/ACE Unpublished.

For Further Information

Professional issues in ergonomics are most often aired at the technical meetings of related scientific/technical societies and in their meeting proceedings. These are listed below with some auxiliary comments.

International Ergonomics Association (IEA)

c/o HFES
P. O. Box 1369
Santa Monica, CA 90406-1369 U.S.A.
WWW Home Page: http://www.spd.Louisville.edu/~ergonomics/iea/html

The IEA serves as the umbrella agency for 29 Federated Societies in most countries and geographic regions throughout the world. By virtue of membership in a Federated Society, individuals are members of the IEA which only has societies as voting members. The Federated Societies appoint representatives to the IEA.

The official journal of the IEA is *Ergonomics*, published by Taylor & Francis, London, England, which also publishes *Ergonomics International* as a quarterly newsletter (Editor: Dr. Stephan Konz, e-mail: SK@KSU.EDU). The IEA sponsors a triennial congress for dissemination of information on research, methods, and data (2000 San Diego, U.S.). Individuals who want to participate in IEA activities normally do so through technical groups; contact Professor H. Luczak, FIR, Pontdriesch 14/16, 52062 Aachen, GERMANY. e-mail: hluczak@iaw-1.rwth-aachen.de

Human Factors and Ergonomics Society

P.O. Box 1369
Santa Monica, CA 90406-1369 U.S.A.
Telephone: 310-394-1811
Fax: 310-294-2410
e-mail: HFESHQ@aol.com
WWW Home Page: http://hfes.org

HFES was formed in 1957 and currently has over 5,000 members. Its journal is *Human Factors*; it also publishes and distributes many other materials, including *The Directory of Ergonomics/Human Factors Graduate Programs, Consultants Directory, Membership Directory and Yearbook, Meeting Proceedings*, monographs, and brochures. Its annual meeting is usually held in September/October, alternating between east/west U.S. locations.

The Board of Certification in Professional Ergonomics (BCPE)

P. O. Box 2811
Bellingham, WA 98227-2811 U.S. A.
Telephone: 360-671-7601
Fax: 360-671-7681
e-mail: BCPEHQ@aol.com
WWW Home Page: http://bcpe.org

The BCPE is a nonprofit, nonmember certification agency operated by nine directors who establish policies, practices, and procedures administered by a part-time headquarters staff of three people. The BCPE is working with IEA and others to harmonize the teaching and practice of ergonomics. It is a member of the National Organization of Competency Assurance (NOCA) and has a cooperative agreement for "reciprocity" with CREE for the *EurErg* designation operative in the European Union. The BCPE has certified close to 1,000 ergonomists since 1992 and usually offers its certification examinations in the Spring and Fall of each year.

Institute of Industrial Engineers (IIE)

Ergonomics & Work Measurement Division (E & WMD)
25 Technology Park/Atlanta
Norcross, GA 30092 U.S.A.
Contact: Carter J. Kerk
Telephone: 409-862-4149
Fax: 409-847-9005
e-mail: Kerk@zeus.tamu.edu

About 10% of the general IIE membership are active in the E&WM division, which publishes a newsletter and arranges sessions for IIE meetings.

American Industrial Hygiene Association (AIHA)

Ergonomics Committee
2700 Prosperity Avenue, Suite 250
Fairfax, VA 22031 U.S.A.
e-mail: infonet@aiha.org

This is a small committee which addresses environmental and biomechanical factors operating in workplaces from a risk-management perspective.

7

Development of Ergonomics Programs[1]

David C. Alexander
Auburn Engineers, Inc.

Gary B. Orr
USDOL/OSHA

7.1 Introduction

This chapter will describe the development of a "model ergonomics program" for Ergonomics Program Managers. A recent survey[1] of forty ergonomics programs revealed that over half (58%) were floundering, and only 25% were deemed to be successful. "Floundering" was chosen rather than "program failure" and was defined as consuming excessive resources relative to the value provided. In other words, the program could still be operational, but was not expected to be successful in its present form over the long term. Additionally, the survey examined the reasons for floundering and found that, for larger organizations, the managerial aspects of the ergonomics programs were more often the cause than were its technical aspects. The differences are highlighted in Figure 7.1 Clearly, with success rates of less than 50%, there is a critical need for information on the development of effective ergonomics programs.

This chapter provides information that will:

1. Ensure that a comprehensive ergonomics program is developed
2. Ensure that the program is effective and produces the intended results
3. Ensure that the program uses resources effectively
4. Prevent many common program development mistakes
5. Ensure that the program meets both the organization's business needs as well as regulatory/OSHA compliance requirements.

[1]Adapted from *The Model Ergonomics Program* © Auburn Engineers, Inc., 1993. Used with permission.

0-8493-2641-9/99/$0.00+$.50

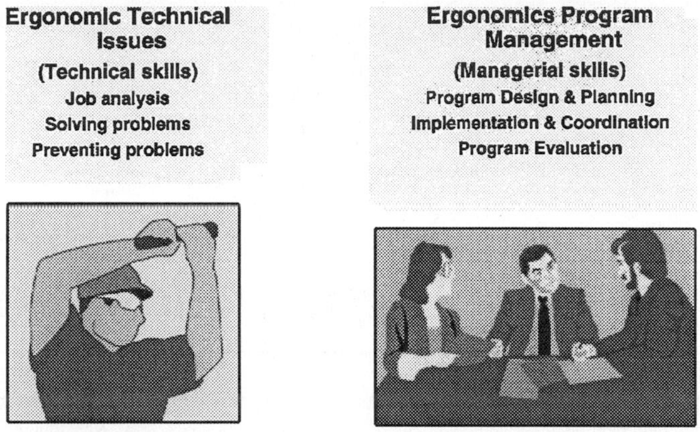

FIGURE 7.1 Ergonomics technical issues vs. ergonomics program management.

This chapter is broken into four sections, each dealing with one major aspect of the ergonomics program. Due to the interaction of the elements, there is some overlap between the sections. The sections are:

- *The Ergonomics Program: An Overview*: This provides an overview of the three major parts of the program: planning, implementing, and evaluating.
- *The Ergonomics Committee*: This covers the structure, responsibilities, logistics, and preparation/training of the ergonomics committee.
- *Activities Necessary to Manage an Ergonomics Program*: These activities include strategic planning, surveillance, tactical planning, ergonomics problem solving, medical management, training, and the prevention of ergonomics problems.
- *Program Management*: This segment provides tools for the management of the program, including the use of measures of success, working the plan, documentation, project evaluation, and audits and assessments.

7.2 The Ergonomics Program: An Overview

This section is based on *An Integrated Plan for an Occupational Ergonomics Program: Your Guide to Developing and Managing an Occupational Ergonomics Program*[2] by Auburn Engineers, Inc. There are three important parts to an effective ergonomics program:

1. Planning
2. Implementation
3. Evaluation

A map of this integrated plan is shown in Figure 7.2.

Planning

Unfortunately, many people fail to plan their ergonomics programs and simply jump into training or project work. While this may seem expedient, it generally leads to floundering and an ineffective program. As a rough rule of thumb, one hour of planning will save forty hours of unproductive and ineffective work. Thus, there are tremendous advantages to planning. Planning consists of two distinct parts: strategic planning and tactical planning.

Strategic planning asks the question, "What do we want our ergonomics program to accomplish?" The answers may be as simple as: reduce our CTD (cumulative trauma disorder) rate by 40%, cut our workers' compensation costs by half, ensure compliance with OSHA regulations, or build a culture which supports

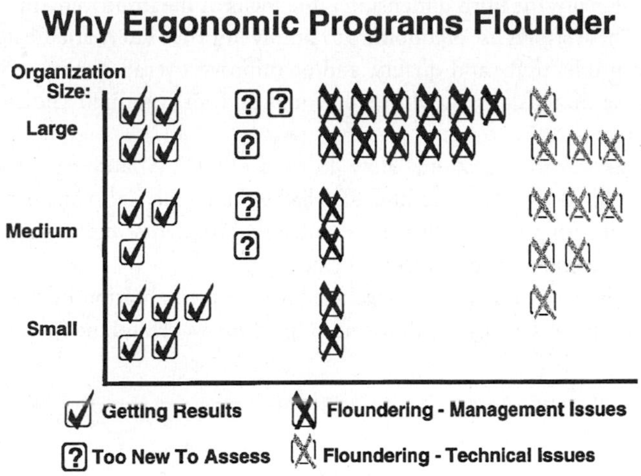

FIGURE 7.2.

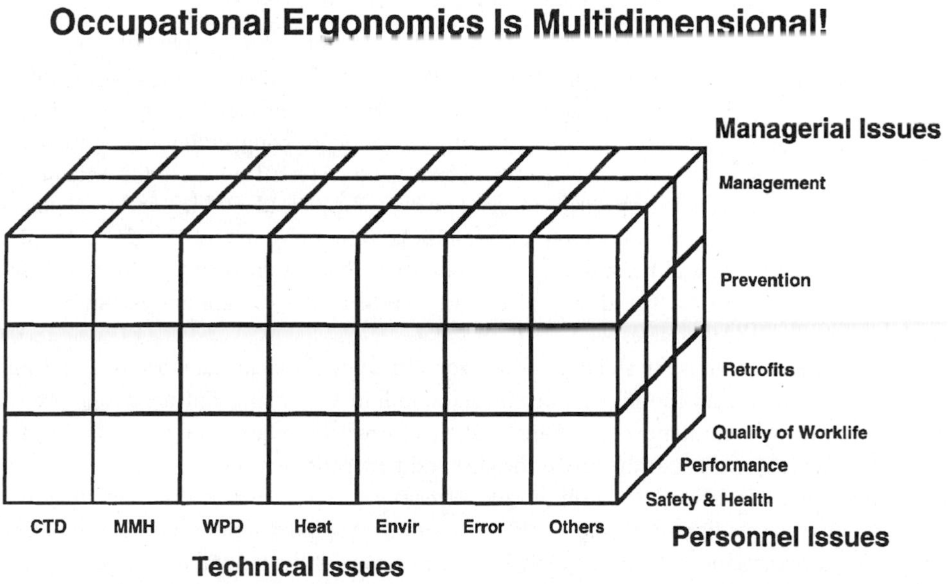

FIGURE 7.3 Possible scope of occupational ergonomics program.

an ergonomics initiative long term. The statement, "If we don't know where we are going, then any path will do," describes the intent of strategic planning.

In addition, strategic planning resolves a number of scope issues as illustrated in Figure 7.3. One helpful way to think about occupational ergonomics is by technical subspecialty, types of problems addressed, and focus of improvement efforts. These are three distinct scales, and as such, should be thought of as three separate dimensions for the ergonomics program. The technical subspecialty dimension determines which aspects of ergonomics will be used. The technical issues include cumulative trauma disorders (CTD), manual material handling (MMH), workplace design (WPD), heat stress, environmental factors such as noise, lighting, thermal comfort, human error issues, and/or other issues. The types of problems addressed are called the managerial issues and include a focus on fixing existing problems (retrofits) and avoiding new problems (prevention), and/or the other managerial and development aspects of the ergonomics program. Finally, the third dimension is the focus of the improvement efforts, which is called the personnel issues. These efforts use ergonomics to simply improve safety and health, improve operating performance such as productivity and quality, and/or improve overall quality of worklife. During the program's design, these dimensions should be explored and an informed choice made regarding the importance of each item.

Tactical planning answers the question, "How do we do that?" Whatever strategic goals are desired, tactical planning is required to determine how to achieve them. Tactical plans provide a list of detailed activities, identify specific financial and human resource requirements, and create a timetable. A written plan is typically one outcome of a sound tactical plan.

Once the strategic and tactical plans are developed, the implementation of the ergonomics program is usually fairly straightforward. Planning allows us to "plan the work and then work the plan." It's simple, and it's effective.

Implementation

Implementation of the ergonomics program usually encompasses four separate, but related, areas:

1. Identifying and resolving current problems (retrofits)
2. Preventing future problems
3. Medical management (prevention and treatment of injured workers)
4. Training

Identifying and resolving ergonomics problems actually consists of three substeps — identifying the ergonomics problems; performing the needed analysis and studies; and developing and implementing solutions. Many organizations use inconsistent methods for identifying ergonomics problems. This substep is necessary for several reasons: it permits problems to be corrected before additional injury/illness occurs, and it permits projects to be prioritized based on risk of injury. Identifying and resolving current problems is often the centerpost of ergonomics activities because they are costly, the injuries place one at risk for OSHA citations, and because injured people (whether assigned to light duty or off work) disrupt production operations. The details of this step are the subject of many technically oriented ergonomics books and short courses.

Preventing future problems is the ability to design equipment, tools, facilities, jobs/tasks, and information so that are ergonomically sound. This step includes both in-house and contract design projects, the purchase of appropriate equipment and tools, and the correct installation of purchased items (e.g., installing purchased equipment at the proper height) and new construction.

Medical management involves the identification and treatment of injured workers. Consistent treatment protocols and appropriate light duty and work conditioning programs are also facets of medical management. Medical management is a particularly important aspect of the treatment of cumulative trauma disorders. Within the medical management area, there are a number of personnel policy issues, such as return to work, light and/or restricted duty, and medical treatment/referral practices.

Training may be required where skills are needed. It is important to realize that training should be performed only to fulfill a need. Training may be required for problem solving, for identification of problems, and for awareness. Some awareness training may be necessary at all levels of the organization. One common error made in implementing ergonomics programs is to do too much training without first determining the need for it. This wastes resources and produces few real results. A second common problem is to perform widespread training too early in the program.

Evaluation

Evaluation should be done at two levels: (1) the project level, and (2) the program level. The project evaluations are done at the completion of each project and evaluate the reduced injury potential, any performance improvements, and costs/benefits.

Program evaluations are time based and are often done semiannually for new programs and annually thereafter. The program evaluations assess progress toward goals and the benefits relative to costs. Following each program evaluation, it is appropriate to review plans and make any necessary adjustments.

7.3 The Ergonomics Committee

An ergonomics steering committee is an effective way to manage the overall implementation of the ergonomics program. The committee should be convened to do this job, and then should be disbanded once the work is complete. This may take from one year to 5 or 6 years, depending on the size of the organization, the exact details of the ergonomics program implemented, and the pace with which the organization chooses to pursue its goals.

Structure

The ergonomics committee for an operating site (a typical site has from 500 to 2,000 people; if it is larger, consider the use of multiple committees) should be comprised of 6 to 10 members. Ideally, there should be representation from loss prevention or safety, medical, human resources, engineering, labor, and production areas. There should be more committee members from production areas than from any other single group. The technical areas may be represented by technical staff or by managers. The production representatives may include labor representatives as well as production management. The production managers on the committee should be at the mid- to senior-management level, and thus will represent significant operating areas of the plant.

The committee can be chaired by the "ergonomist," although strong consideration should be given to having a production manager as chair. Like any committee of this sort, political savvy is important and should take priority over technical skills.

One member of the committee should be the person who will spend the most time with the ergonomics program and its implementation. In general, the more time this person spends with ergonomics, the more that will get done. Realistically, very little will be accomplished if less than 20 to 25% of one person's time is spent with the ergonomics program during its implementation. On the other hand, a dedicated person will do more than twice what a half-time person can accomplish. This person is often called the "site ergonomist," whether or not he or she actually has university training in ergonomics.

Responsibilities

The ergonomics committee is a part-time responsibility for most of its members. It is responsible for planning and managing the ergonomics program, but not necessarily for actually performing all the activities. For example, the committee may identify a need to study and improve a particular job, and then delegate the work to an individual or a problem-solving team. The ergonomics committee is too large and too busy with other activities to really do much detailed problem solving. However, the committee is usually very effective with management functions, such as determining needs, prioritizing activities, obtaining resources, and following up and evaluating the efforts.

Normally, the ergonomics committee will be commissioned by, and will report to, the plant safety committee or a key senior manager. Thus, the role of the committee is similar to that of a middle manager who takes directions from top management, formulates plans, budgets, and schedules, and then oversees the performance of the work. The ergonomics committee, under this scenario, is judged effective when there are plans, ongoing activities, and results. In fact, the "measures of success" are established to measure these aspects of performance.

Logistics

The ergonomics committee meets as needed in order to get the job done. Their time demands are not uniform. The initial time demands are to plan the work and eventually to "work that plan."

Typically, the initial planning (to create the strategic and tactical plans) will take from 8 to 24 hours in a workshop setting, followed by regular meetings of 1 to 2 hours once or twice a month (to implement and manage the plans). The initial tactical plan will provide a list of specific activities, personnel, and resources, on a 6- to 12-month timetable. Subsequent meetings will ensure that the plan is working, and that projects are actually being implemented.

Preparation/Training

The ergonomics committee may need some training to get its job done. There are two types of skill that this committee and its members need: (a) technical skills and (b) program management skills. The technical skills allow the committee to understand ergonomics and ergonomics problem solving. Appropriate technical topics include workplace design, cumulative trauma, manual material handling, heat stress, environmental factors, and human error. Other possible topics include engineering design, maintainability, designing for disabled workers, work shifts and schedules, and job design (macroergonomics).

Program management skills, on the other hand, prepare people for planning, implementation, and evaluation of the ergonomics program. The topics that should be covered include the planning and managing of an ergonomics program, establishing measures, making a program pay off, compliance issues (such as those covered in OSHA's *Ergonomics Program Management Guidelines for Meatpacking Plants*[3]), and building the ergonomics culture.

In closing this section, a simple caveat is necessary: most ergonomics training courses and textbooks deal only with the technical issues and do not provide adequate information on the management of ergonomics programs.

7.4 Activities Necessary to Manage an Ergonomics Program

For some people, the term ergonomics program has come to mean the collection of elements defined by OSHA in the *Ergonomics Program Management Guidelines for Meatpacking Plants:* recognition, evaluation, and control of ergonomics hazards. This is a static model of an ergonomics program, and it fails to recognize all the steps required to initiate an ergonomics program and permit it to mature. This list of activities is broader than the OSHA model and is more business oriented.

The Strategic Plan

The strategic plan will answer the question, "What do we want to do?" Such answers as "We want to do ergonomics" or "We want to implement ergonomics" are *not* appropriate. Ergonomics is a tool that helps to do something, such as reducing workers' compensation costs or reducing the number of lost workday cases. The ergonomics program should be aimed at specific, measurable goals. The strategic plan for ergonomics is best developed by the plant safety committee or a similar group that has a view of overall operations and can see the relative value of different programs (one of their decisions is who should serve on the plant ergonomics committee). However, if strategic planning is not done when the ergonomics committee begins its work, one of its first activities should be to complete the strategic plan and then review it with the plant safety committee or with top management.

To develop the strategic plan, ask the following questions:

1. *What do we want the ergonomics program to do?* This usually becomes the mission, vision, and scope of the ergonomics program.
2. *How do we monitor results?* What data do we measure to demonstrate progress with ergonomics? (These are the "measures of success.")
3. *What are the barriers?* And how can they be overcome?
4. *What policy issues* are likely to be affected?
5. *Who is or should be involved, and what are their roles?* This includes both the ergonomics committee and ergonomics problem-solving groups.

6. *How important is ergonomics* relative to other safety and health issues for our company?
7. What is our general plan?

This information can be documented and shared with the ergonomics committee, the plant safety committee, and management. It should be used to help with program reviews and evaluations.

Surveillance

Some type of surveillance is helpful as the ergonomics program is initiated. With surveillance, one is trying to identify all the ergonomics problems and opportunities — what is wrong and what can be improved? This also will help to determine the magnitude of the ergonomics program and the resources that will be needed. Surveillance does not need to be burdensome or time consuming to be effective. Some things to look for are outlined below:

- *Ergonomics has safety and health implications.* Therefore, one of the first places to look is the OSHA 200 log, workers' compensation, medical records, and restricted work cases. Eventually, as the ergonomics initiative moves from correction to prevention, additional surveillance techniques will be needed to identify ergonomic risk factors and the early warning indicators of potential injuries.
- *Examine traumatic injuries* as well, since their cause is often rooted in an interface problem between the person and some piece of equipment. From these situations, the number and cost of injuries/illness can be determined. Some rough estimates indicate that as many as 25% of traumatic injuries may have an ergonomics root cause.
- *Look for performance problems and cost issues* as well as safety and health concerns. For example, the recruitment, hiring, and training of replacement workers can often cost $2,000 or more. People who work at poorly designed workstations take more breaks, produce less, or turn out lower-quality products. There are a number of ways in which ergonomics can improve job performance, and the ergonomics committee should be alert to these savings.

By determining the overall issues, an organization with an effective ergonomics program may find that ergonomics can easily pay for itself through reduced financial losses.

When looking for ergonomics issues, be sure to look at all aspects of human performance. This is sometimes called the "European Model" of ergonomics because it emphasizes both the physical and cognitive aspects of ergonomics. A useful outline of important ergonomics issues can be found in *The Practice and Management of Industrial Ergonomics*,[4] **Chapter 2.**

The six primary aspects of ergonomics are:

1. The physical size of people and its implications for the fit of the person at the workplace and within the facility.
2. The cardiovascular system and its limitations on work as measured by work physiology.
3. The major musculoskeletal system and its limitations on manual material handling.
4. The minor musculoskeletal systems and their limitations on fine work, manipulation, and dexterity.
5. The environmental factors, such as lighting, noise, and thermal comfort, and their impact on performance.
6. The cognitive capabilities of people and their impact on processing information and "human error."

Once there is an understanding of all the ergonomics issues, including the troublesome jobs, then a list of the highest-priority jobs should be developed. This list may be known by different names — the top five, the worst ten, or the dirty dozen — and it is simply an effective way to prioritize. The ergonomics committee should be aware of these jobs, and the tactical plans should specifically address when and how they will be resolved. When the ergonomics problems with the initial list of jobs are resolved, a revised list can be prepared. Remember that surveillance is an ongoing activity and will be repeated periodically.

Tactical Planning

While strategic planning is the key to deciding what to do, tactical planning is the key to actually getting things done. Tactical planning is best done in a workshop-type setting with the ergonomics committee. Typically, 8 to 16 hours devoted to tactical planning will really help "jump start" the program. Tactical planning is the link between the goals of the ergonomics program and the specific projects undertaken. Tactical planning will answer these questions:

- What should be done?
- When should it be done?
- Who should do it?
- What are the quality standards?

A practical way to lead the tactical planning is to ask what needs to be done in each of the major implementation areas as well as the major management areas of the ergonomics program. For example, there are projects that help correct existing problems, projects that help prevent future problems, training needs, medical management procedures and protocols, project evaluations, and periodic program reviews. By examining each individual area, lists of needed actions can be developed. These individual lists can then be integrated into a tactical plan covering the upcoming 6 to 12 months, as shown in Table 7.1. Managing the plan then becomes as easy as making monthly assignments and following up on actions, and, of course, making periodic updates to the tactical plan so that it remains current.

Solving Ergonomics Problems

There are several ways in which ergonomics problems can be resolved. The ergonomics committee should carefully consider each problem and decide which approach, from among those discussed below, is best for that situation. Ergonomics problems can be easily resolved by using small group problem-solving processes. An excellent six-step process, which has proven itself time and again, follows:

1. Identify the jobs/tasks at risk and select one to improve
2. Analyze the problem to the extent required
3. Develop alternative solutions
4. Select the most appropriate solution
5. Implement the preferred solution or, if that is not possible, an alternate
6. Follow up to ensure that the problem is resolved

The *use of problem-solving teams* can be both highly effective and enjoyable. The teams usually consist of 4 to 8 people. (For example, a team may consist of 2 to 3 operators from different shifts, a production supervisor, the ergonomist, a mechanic, an engineer, and a nurse.) Some training is needed so that the team can do its job well. The team should use the small group problem solving process outlined above. This author's experience shows that a "quick strike team" (dedicated, full-time work for 2 to 5 days) is preferable to the "quality circle approach" (short, 1- to 2-hour meetings spread over several months) for these appointed teams. Quick strike teams function better, and the projects get done much faster. The team should be able to analyze the problem, develop alternatives, and recommend specific, practical solutions within the 2- to 5-day period. This chapter is not a tutorial on problem-solving teams and additional information can be found in "Using a Quality Action Team to Resolve Ergonomics Problems."

Individuals also can be used to resolve ergonomics problems. These projects may take more elapsed time, but fewer overall labor hours, to resolve a problem. They are particularly good for someone who has some depth of experience and skills and who can talk easily with operating personnel and gather information. Individuals are also good for working on projects directed toward the prevention of ergonomics problems. The person can be a project engineer or any other talented performer.

Additionally, *consultants* can help to resolve ergonomics problems. Consultants are helpful when the organization doesn't have adequate human resources, when the problem is particularly challenging, or

TABLE 7.1 Ergonomics Committee Tactical Plan

May

What?	Will Be Done By Whom?	By When?	Check When Completed
Plant strategy developed			
Ergonomics policy written			
Roles/responsibilities developed			

June

What?	Will Be Done By Whom?	By When?	Check When Completed
First problem solving training			
Initial 2 projects started			
Baseline data collected for measurement systems			

July

What?	Will Be Done By Whom?	By When?	Check When Completed
Initial project recommendations implemented			
Next 2 projects initiated			
Illness investigation procedure tested			

August

What?	Will Be Done By Whom?	By When?	Check When Completed
Project recommendations implemented			
Next 2 projects initiated			
Detailed surveillance completed			
Policy issues identified			

September

What?	Will Be Done By Whom?	By When?	Check When Completed
Project recommendations implemented			
Next 2 projects initiated			
Plans for proactive ergonomic program developed			

October

What?	Will Be Done By Whom?	By When?	Check When Completed
Project recommendations implemented			
Next 2 projects initiated			
Engineering training started			

when it is not cost effective to organize and train a team. A consultant should be able to offer fresh insight, novel solutions, and should show personnel how to resolve problems in a quick and cost-effective manner. A note of caution is necessary — there are many newcomers to the field of ergonomics who can speak the lingo but lack the engineering skills to solve in-plant problems. This problem with consultants will get worse before it gets better. So before you employ a consultant, verify his or her ability and skills — do you want to pay people to learn on your time?

Some ways to recognize unacceptable consultants include:

- A consultant who simply reiterates that you have a problem (you already knew that — that's why you hired him/her)
- A consultant who tells you to automate (you can figure that alternative out on your own)
- A consultant who offers common "textbook solutions" (e.g., suggests that you add lift assist devices for your lifting problems)
- A consultant who offers inappropriate solutions (suggests, for example, that you add footrests to workstations where the only problem is hand–wrist cumulative traumas)

At some point, the *natural work group* should be able to identify and resolve ergonomics problems within its own work areas. The "natural work group" is a work group with its supervisor. This is an important point in the evolution of an ergonomics program because it indicates that ergonomics is becoming part of the culture: ergonomics problem solving is seen as "just part of the job — it's nothing special." Long term, this is the best way to deal with ergonomics problems because ergonomic risk factors will be identified and resolved within the work group before injuries or performance problems occur. Natural work groups do not spontaneously develop problem-solving skills, but after some of their members have worked on problem-solving teams, they can carry back some of their skills and experience. The natural work groups can use either a "quality circle approach" or a "quick strike approach" or some other combination, depending on the time available, the urgency of the problem, and the approaches used to resolve quality and production problems.

Projects should be evaluated after their conclusion. This information will help plan future efforts, and it will provide useful feedback to the team or individual involved, to the ergonomics committee, to the plant safety committee, and to management. Some common factors to evaluate are:

1. The time required to work on the problem relative to the quality of the solutions
2. The effectiveness of the solutions
3. The overall costs (time, materials, equipment, etc.)
4. The overall benefits (injury/illness, lost time, productivity, quality, etc.)
5. The enthusiasm for ergonomics that was generated

Medical Management

This section is not intended to replace sound medical judgment from your healthcare providers. However, it may help the ergonomics committee understand some key areas relative to the medical management aspects of ergonomics. Medical management practices affect many plant policies, such as light duty and work restrictions, work hardening and rehabilitation, and work conditioning prior to full performance on the job. There also may be policy issues relative to medical protocols, treatment practices, and/or record keeping.

One important aspect of medical management is that it complements ergonomics problem solving. In general, there are two ways to deal with overuse and cumulative trauma injuries: (1) fix the problem with ergonomics problem solving, or (2) identify injuries in their early stages and treat the injured workers via medical management. Either approach will impact the frequency and severity rate, although aggressive medical treatment will impact severity far more than frequency. Some organizations choose to emphasize medical management initially, followed by the resolution of the ergonomics problems. The only incorrect strategy is to use medical management exclusively as a long-term strategy without the use of ergonomics problem solving.

When developing the tactical plan for ergonomics, some of the "to do" items may be to examine the pros and cons relative to current policies. For example, an organization may have a current policy of "no light duty or restricted work" that the ergonomics committee wishes to examine. However, since the ergonomics committee is not a policy-making committee, it may choose to have a small team evaluate light duty programs, and prepare a "white paper" which explores the advantages and disadvantages of the current policy and of a proposed new policy. At that point, senior management would be asked to review the document, to develop an understanding of the issues, to enter the debate, and then to make a decision regarding the policy. In general, the ergonomics committee should not recommend medical treatment practices and protocols, although they should verify that any treating physicians are aware of the medical protocols found in the *Ergonomics Program Management Guidelines for Meatpacking Plants* and other OSHA documents.

Training

Training is necessary to ensure that people have appropriate skills. However, when training occurs too early in the process, it can be, and often is, counter-productive. Training needs should be carefully assessed to ensure that people get adequate training, and also to ensure that unnecessary training is not conducted. A training matrix can be used by the ergonomics committee to assess training needs and the organization's ability to provide that training. This training review also provides training objectives for the trainer, and serves as a means to assess the quality of the training. Finally, avoid the temptation to make the training overly complex by using too much technical jargon (ergo-babble), especially for the line organization.

While no training is legally mandated, the *Ergonomics Program Management Guidelines for Meatpacking Plants* suggest the training of workers to spot the early signs of cumulative trauma disorders (much like training on early-warning symptoms for hazardous chemical exposure).

Likewise, training for people exposed to CTDs should include on-the-job activities to reduce CTD exposure (adjusting workplaces, using proper tools, etc.) as well as off-the-job awareness (using a keyboard at home can be just as bad as using a keyboard at work).

Some in-plant training courses to consider are:

- In-depth skill training for safety and health professionals and possibly the ergonomics committee
- Ergonomics problem solving for teams or individuals
- Ergonomics design for engineers
- Awareness training for managers and supervisors
- Operator training on ergonomics (awareness of ergonomics plus job-specific Job Safety Analysis type skills for job protection)

The courses should be customized for the specific training objectives. Thus, some courses will be as short as 30 to 60 minutes, while others can last as long as 20 hours.

Preventing Ergonomics Problems

Prevention of ergonomics problems is necessary in order to avoid having to correct problems in the future. When you build or purchase a piece of poorly designed equipment, you have two bad choices:

1. Live with it
2. Pay to change it yourself

Neither is desirable. Thus, the preferred approach is to ensure that all equipment, tools, and facilities are ergonomically designed prior to purchase and installation.

Two things are necessary for good ergonomics design of equipment, tools, and facilities. First is the expectation that all new designs will be ergonomically sound. Engineers and designers must understand that there is a new requirement for their design work. Second, engineers and designers must have access to usable, up-to-date ergonomics design information (such as the **Ergonomics Design Guidelines**[5]) that allows them to design properly.

People other than engineers and designers are involved in this process, however, and they must be included as well. For example, purchasing agents must specify ergonomics as a feature for all new purchases. Contract engineering firms must be required to pay attention to ergonomics. People who install equipment (anything from setting up a new computer workplace, to the alteration of existing equipment, to the building of a new plant) must all understand and apply ergonomics principles when they make the installation. Good ergonomics design should occur at both the central corporate design staff level and at the facility level.

Some policy issues that are likely to surface when the prevention of ergonomics via design is discussed include:

1. Is a policy on ergonomics design necessary?
2. Is ergonomics design required?
3. How is ergonomics design of purchased items evaluated?
4. Will the organization incur increased design and fabrication costs for sound ergonomics design?
5. How are audits of new designs and finished construction projects performed?

7.5 Program Management

Program management is generally the weakest area when it comes to ergonomics programs. As a result, the information in this section may be the most valuable for the ergonomics committee. Program management is usually not covered in academic programs nor in ergonomics short courses, and many ergonomics program managers simply have to learn these skills on the job.

Measures of Success

You can't manage what you don't measure. Measures of success allow you to determine the results of the ergonomics program. These measures have many benefits including:

- Helping you to set goals and establish priorities
- Telling you whether you are making progress
- Allowing you to judge whether you need additional resources
- Letting you know when you're done

There are many different measures of success that can be used. Each organization should carefully identify a number of possible measures and then select the ones that best fit. A handful of measures (from 6 to 12) reflecting short- and long-term changes, with leading and lagging indicators, works best. If your initial measures aren't working for you, then change them. Periodically, over the life of the ergonomics program the measures should be reviewed and modified. (This is especially important as you move through the stages of ergonomics on your way to an ergonomics culture. For example, early in the program you may want to measure how many people call for ergonomics assistance, but as the program matures, you will want to know how many applications occur without your involvement.)

A few notes about measures are important. There are leading and lagging measures. Lagging measures tell you about things you can no longer control (OSHA Log 200 cases; lost workdays), while leading measures predict what will occur. (First aid cases are a predictor of lost workday cases; number of projects resolved predicts ergonomics risk reduction.)

During the maturing process associated with ergonomics program implementation, the measures will change from activity-based measures (committee established, people trained, surveillance completed, etc.) to outcomes/results-based measures (injuries/illnesses, projects completed, etc.) and finally to systems-based measures (medical management system in place and working; engineering design system in place and working).

TABLE 7.2 Possible Measures of Success for Ergonomics Programs

Measure	When Monitored?	Type of Measure
Number of lost workdays for ergonomics reasons	Quarterly	Outcome; Lagging
Compensation costs for ergonomics reasons	Semi-annually	Outcome; Lagging
OSHA Log 200 cases for ergonomics reasons	Monthly	Outcome; Lagging
First aid cases for ergonomics reasons	Monthly	Outcome; Leading
Progress of ergonomics problem solving efforts	Monthly	Activity; Leading
Progress on milestones listed on the ergonomics committee action plan	Monthly	Activity; Lagging
Number of people trained	Annually	Activity; Leading
Number of projects resolved	Quarterly	Outcome; Leading
Number of ergonomic systems in place	Annually	Systems; Leading

Some examples of measures which can be used are shown in Table 7.2. Remember, though, there are many additional measures that can be used. The measures should monitor information about the objectives you hope to accomplish.

Working the Plan

Working the plan involves tracking data to see how things are going, ensuring that adequate resources are available, and overcoming barriers. It is not an easy job yet, this is precisely what the ergonomics committee is being paid to do over the long term.

With the preplanning of the ergonomics activities, the ergonomics committee can focus on the management of these efforts and spend less of its time on specific projects. A typical meeting agenda should contain these elements:

A. Review of the "Measures"
- What do the most recent numbers tell us?
- Are we on target with our plans?

B. Review of the Action Log (assigned projects and activities)
- What was to be accomplished prior to this meeting?
- What was actually accomplished?
- What were the results? Are these results acceptable?
- If there were problems with the solutions, what were they? Why did they occur? How did you overcome them? Is additional help needed?

C. Review the Action Log (planned projects and activities)
- Are the upcoming projects still necessary?
- Who should do them? Do they have time?
- Assign the project; clarify its scope, verify deadlines, verify the person who is responsible for the project; verify other people involved.

The committee should set up a system to track progress on the tactical ergonomics plan. An action plan format is a good tool for monitoring progress because each action item can be checked off when it's finished. Each monthly meeting should have a review of current projects and activities, followed by the assignment of the new activities that will keep the ergonomics effort moving along.

Another specific tool shown in Table 7.2 can be used to track individual projects. Tracking each project as it progresses through the problem-solving process will identify any weak points in that process. For example, one common problem is noted in the Table 7.3. Failure to fully implement solutions to ergonomics problems is clearly shown when projects routinely reach "select the best solution" but then do not move through "implement solution" in the chart. For example, if the following chart is being reviewed in April, then it is clear that the organization is relatively good at developing solutions but poor at implementing them. Tracking problem-solving data in this way brings these weak points to the surface.

TABLE 7.3 Tracking Ergonomics Projects

Project or Sub-Project Name	Identify & Clarify Project	Analyze Appropriate Data	Develop Alternate Solutions	Select the Best Solution	Implement Solution	Follow Up to Ensure Success
Project # 1	Completed Jan 9	Completed Jan 10	Completed Jan 10	Completed Jan 11	Completed Jan 29	Completed Feb 26
Project # 2	Completed Jan 19	Completed Jan 19	Completed Jan 20	Completed Jan 21		
Project # 3	Completed Feb 6	Completed Feb 6	Completed Feb 13	Completed Feb 14		
Project # 4	Completed Feb 8	Completed Feb 9	Completed Feb 19	Completed Feb 19		

Obtaining resources is another major problem. When cost/benefit analysis is used to justify projects, a concerted effort must be made to determine the full costs and benefits. There are other methods by which to obtain funding for ergonomics projects, and the interested reader should review the chapter on the cost and benefits of ergonomics interventions.

Regarding the issue of overcoming barriers, the four most common major obstacles are:

- Lack of time
- Lack of money
- Too few skills
- Lack of management support

There are, however, some effective means for dealing with each of these problems.[6]

Documentation

Documentation by the ergonomics committee is necessary at several stages of the process. First, the committee should record the tactical plan. The written plan will generate a higher degree of commitment and more action than nonwritten plans.

The committee can prioritize the worst jobs (the dirty dozen) and share the list with management as a means to focus action on the "vital few." Since this list is relatively short, action is much easier to obtain.

When the committee meets, it should keep minutes of the topics discussed and of the progress made. Many committees keep a centralized notebook of minutes, plans, projects, and evaluations.

As projects are assigned, they should be noted, and when solutions are implemented (whether successful or not), they should be recorded. A short videotape of the problem job with narrative describing the problem, which is then followed by a short segment showing the changes made, is a very effective way to document progress. This type of video is also useful in the event of OSHA inspections, as well as when the opportunity arises to share ideas among other company operating sites or even with other companies. The teamworking on a problem can be assigned the duties of documentation for that project.

Project Evaluations

There are two methods that can be used to evaluate ergonomics projects. One is based on reductions in the cumulative stress of the job on the people performing the job, and the other is cost based.

Reduction in cumulative stress:

1. Determine which indicators of excessive stress to use. Some typical indicators are lost workdays, medical claims, OSHA log cases, first aid treatment, pain on the job, and discomfort on the job. In order to evaluate changes, leading indicators (discomfort, pain, and first aid treatments) provide information more quickly than lagging (lost workdays, medical claims, OSHA log cases). In addition, the leading indicators will provide information before an on-the-job injury occurs.

2. Use the appropriate indicators both before and after any changes to the job are made. The baseline data taken before changes are made provide a good comparison for improvements.

Cost-based evaluation of a project:

1. Determine the cost of the changes. This will include the study time for the project team, engineering and maintenance time for alterations, equipment costs, work order costs, downtime (if in a sold out condition), training time, time of supervisors in dealing with the changes, and the time of other staff.
2. Determine the benefits from the project. This will include any reduction in injuries/illnesses, lost workdays, etc., resulting in the following dollar benefits: medical costs, workers' comp costs, costs of replacement workers, overtime costs, training costs, changes in productivity or quality for a less experienced worker, equipment downtime, supervisor's time, etc. In addition, the project may result in enhanced productivity or quality improvements on the job. Finally, there may be less waste or rework on this job.
3. Determine net benefits (the costs subtracted from the benefits) and the benefits/cost ratio (the benefits divided by the costs). These may be compiled for periodic summaries by the ergonomics committee.

Audits and Assessment of the Ergonomics Program

Just as there are audits of the safety program, there is a need for audits of the ergonomics program. Audits are often done by an outside party, while assessments may be conducted by the ergonomics committee itself.

An effective audit tool will have two practical benefits. First, it can provide quantitative information about your ergonomics program. It can tell you where you stand with implementation. A second benefit, however, is that the assessment can give some insight into where you may want to take the program, over what time period, and what specific actions are required to meet those ends. More information on audits and assessments can be found in the chapter titled "Evaluation of Ergonomics Programs."

References

1. *Review of Ergonomics Programs, The Top Ten Reasons Why Ergonomics Programs Fail,* Auburn Engineers, Inc., Auburn, AL 1994.
2. *An Integrated Plan for An Occupational Ergonomics Program, Your Guide to Developing and Managing an Occupational Ergonomics Program,* Auburn Engineers, Inc. Auburn, AL, 1991.
3. *Ergonomics Program Management Guidelines for Meatpacking Plants,* U. S. Department of Labor, Occupational Safety and Health Administration, 1990.
4. *The Practice and Management of Industrial Ergonomics,* David C. Alexander, Prentice-Hall, 1986.
5. *Ergonomic Design Guidelines,* Auburn Engineers, Inc., Auburn, AL, 1990, 1992.
6. *Overcoming the Barriers, Managing Your Ergonomics Program,* Auburn Engineers, Inc., Auburn, AL, 1993.

For Further Information

Assessing Your Ergonomics Program, David C. Alexander, PE, CPE, Auburn Engineers Press, 1995.

Overcoming Four Common Barriers, David C. Alexander, Workplace Ergonomics-Stevens Publishing Co., 1996.

Planning a Successful Ergonomics Program, David C. Alexander, PE, CPE, Workplace Ergonomics-Stevens Publishing Co., 1996.

Framework for Assigning Responsibilities, David C. Alexander, PE, CPE, Workplace Ergonomics-Stevens Publishing Co., 1995.

Why Would You Want An Ergonomics Culture? David C. Alexander, PE, CPE, Workplace Ergonomics-Stevens Publishing Co., 1996.
The Ergonomics Program Report Card, David C. Alexander, PE, CPE, Workplace Ergonomics-Stevens Publishing Co., 1996.

Part II
Foundation of Ergonomics Knowledge

Section I
Fundamentals of Ergonomics

8

Design of Information Devices and Controls

Toni Ivergard
Institute of Working Life
Sweden

8.1 Introduction

This chapter covers the design of traditional information devices, and also the design of devices for communication with computers. At the end of the chapter the design of instructions, forms, and tables will be dealt with.

Instructions on how scales and scale markings on visual instruments should be designed, together with the advantages and disadvantages of different types of visual instruments, are included. It appears, for example, that the common round meter with a moving pointer is the best one for most applications. Where more exact quantitative readings are necessary, the direct-reading digital instrument is best.

A relatively detailed specification for the design of VDU screens is included. The main attribute of VDU screens is their flexibility. They can be used for presenting many different forms of information. However, in control rooms access to a large amount of information simultaneously is often required.

0-8493-2641-9/99/$0.00+$.50

Therefore, it may be necessary to have several VDUs, or to have access to other information devices, such as overview displays as a complement to VDUs.

Methods for producing diagrams, codes, and symbols are also described, and various methods for using color symbols are discussed. The use of colors may have some importance in simplifying the reading of process information. However, the use of colors should be limited, bearing in mind that significant proportions of the population are color-blind. Colors should be used to supply additional information so that the VDU can be read correctly even if all colors disappear.

In man/machine communication, man receives information via the various information devices. In the control room, the information is either visual or auditory, and can be either static or dynamic. Dynamic information is constantly changing, such as the information shown on speed meters, height meters, radar, TV, and temperature and pressure meters. Static information does not change over time and includes road markings, maps, notices, manuals, and any printed or written material.

In this chapter we shall deal primarily with the various types of dynamic information. Static information will also be covered to a certain extent, particularly in the section dealing with VDU screens.

Three main types of information device will be covered in this section:

1. Traditional instruments
2. VDU screens
3. Sound signals

8.2 Traditional Information Devices

Traditional instruments are still the most common forms of information device in the control room. In modern control rooms, however, more and more information is being transferred to VDUs. Traditional instruments may be divided into the following subgroups with regard to their areas of use:

1. Instruments with associated control devices
   - *Control regulation instruments.* The instrument is read and, if necessary, the operator adjusts the machine.
   - *Instruments for setting-up.* Instruments used for making changes in the running conditions.
   - *Instruments for following (tracking).* Instruments usually used in different types of vehicles (e.g., cars, airplanes).
   - *Instruments for indication.* They show such information as "entrance," "way ahead closed," "backwards," etc.
2. Instruments without control devices
   - *Instruments for quantitative readings. Instruments for qualitative readings.*
   - *Instruments for check readings.* Instruments used for detecting and reporting a deviation from a normal value.
   - *Instruments for comparison.* For checking that two machines are producing the same value.

Of the instruments above, those for check readings, control regulation, and for setting up are the most common in industry. The introduction of computer systems for process control, however, has meant that qualitative readings are becoming more common.

Different Types of Visual Instruments

The choice from among the many different types of visual instruments available depends on the information to be presented and how it will be used. Figure 8.1 shows some of the more common types of instrument. Figure 8.1a shows a digital instrument, which displays the various numbers directly. The instrument may have mechanically or electronically generated numbers. Figure 8.1b shows instruments with moving pointers and fixed scales, whereas in Figure 8.1c the instruments have fixed pointers and moving scales.

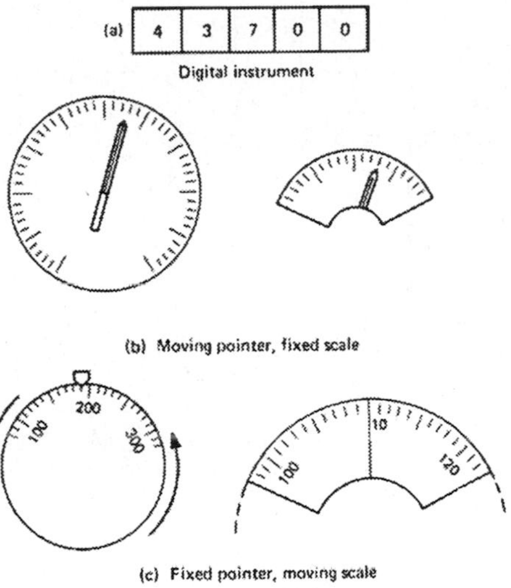

FIGURE 8.1 Various types of instruments.

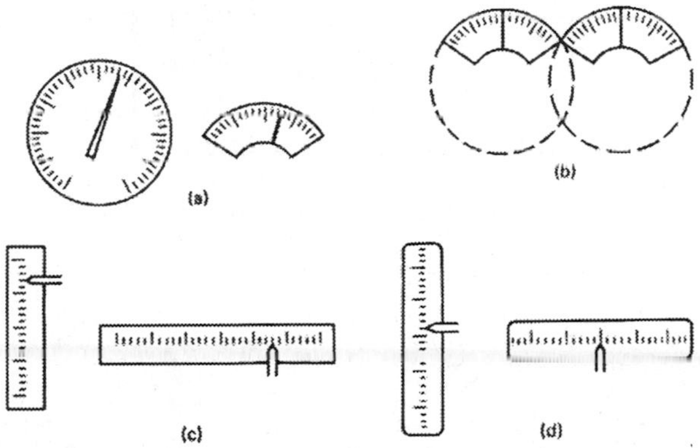

FIGURE 8.2 Varieties of instruments with moving pointers and moving scales.

There are many variants of pointer instruments. Figure 8.2 gives examples of some of these.

1. Round and sector-shaped instruments (Figure 8.2a) are recommended for check and qualitative readings. Round ones are usually better than sector-shaped ones, but take up more space.
2. Vertical and horizontal scales with moving pointers (Figure 8.2b) are also good for check readings. They do not give as much information as that provided by the angle of the pointer in round or sector-shaped instruments. However, this type of instrument takes up little space.
3. Round and sector-shaped instruments with fixed pointers (Figure 8.2c) can be recommended where the whole scale does not need to be seen for quantitative readings which change slowly. Their design allows a relatively long scale to be used without taking up much panel space, but they may need a relatively large area behind the panel.
4. Figure 8.2(d) shows vertical and horizontal instruments with fixed pointers. This type of instrument can also have a very long scale (see Figure 8.3). They cannot be recommended for cases other than where a very long scale is required.

Table 8.1 summarizes the recommended areas of usage for different types of pointer instrument.

In certain cases, it may be desirable to choose other instrument designs. One may wish to use the design to show the function of the instrument. Figure 8.4 gives examples of a round instrument with a

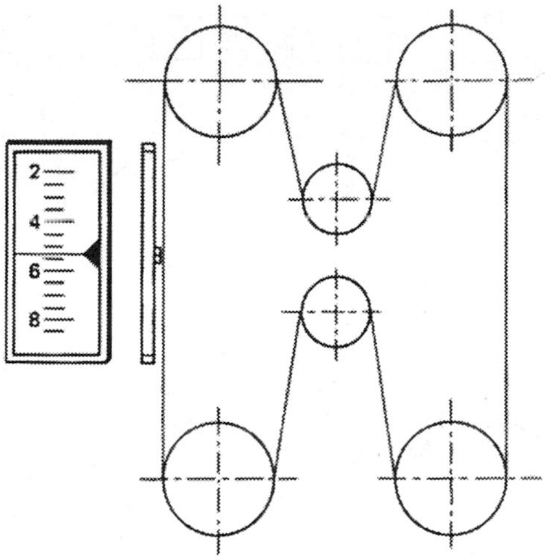

FIGURE 8.3 An instrument that covers a large scale range can be made with a large moving scale and a fixed pointer.

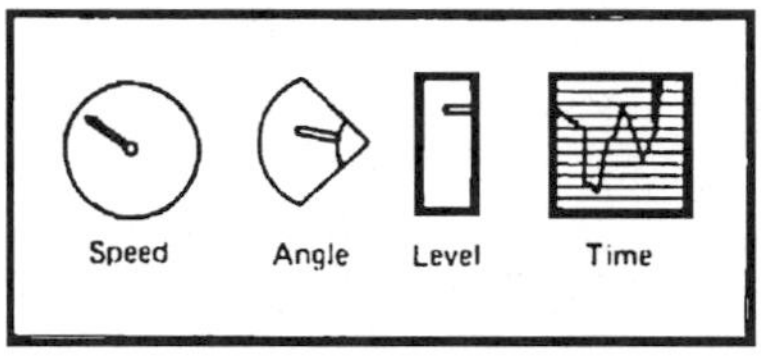

FIGURE 8.4 Form coding through the shape of the instruments.

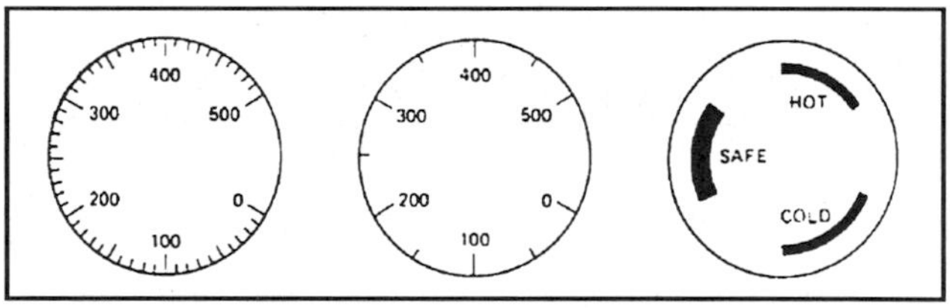

FIGURE 8.5 Different scale complexities.

moving pointer indicating speed. The angle of the material being controlled is shown on a sector-shaped instrument with a moving pointer. Level is shown on a vertical instrument; and time changes are plotted out on paper with a moving pointer; the most recent time is marked at a suitable time interval.

Design of Scales and Markings

The terms to be used in this section are defined thus:

1. *Scales range.* The numerical difference between the lowest and the highest value on the scale.
2. *Numbering interval.* The numerical difference between two successive numbers on the scale.
3. *Scale marking interval.* The numerical difference between two scale markings.

Before choosing a scale for an instrument, one must determine the precision required. Figure 8.5 gives examples of different precisions. It is generally true that the least exact scale, which fulfils system requirements, should be chosen, thus avoiding unnecessary accuracy. If possible, the information should be given in units that require no modification by the operator in order to be used. An example of this

TABLE 8.1 Recommended Areas of Use for Different Types of Pointer Instrument

	A Moving pointer	Linear		B1 Fixed pointer	Linear		B2 Window		C	D	E
	Round	horizontal	vertical	Round	horizontal	vertical	Round	Linear	Counter	Switch	Lamps
Without controls:											
Quantitative reading, slow change	o	o	o	o	o	c	o	o	xx	o	o
Quantitative reading, fast change	x	x	o	o	o	c	(x)	(x)	O	o	o
Qualitative reading, direction	(x)	o	x	o	o	c	o	o	o	o	t
Qualitative reading speed	x	o	o	(x)	o	c	o	o	o	o	o
Control/check reading	xx	(x)	(x)	o	o	c	o	o	o	o	o
Comparison, fast	xx	o	o	o	o	c	o	o	o	o	o
Comparison, slow	(x)	o	o	o	o	c	o	o	xx	o	o
Warning with controls:	(x)	o	o	o	o	c	o	o	o	x	xx
Control/check adjustment	xx	(x)	o	o	o	c	o	o	o	o	o
Setting up	x	x	(x)	o	o	c			o		
(Tracking)	Mostly special instruments										
Indicating	xx	o	o	o	o	c	o	o	o	x	x

xx very good
x good
(x) uncertain
o unsuitable

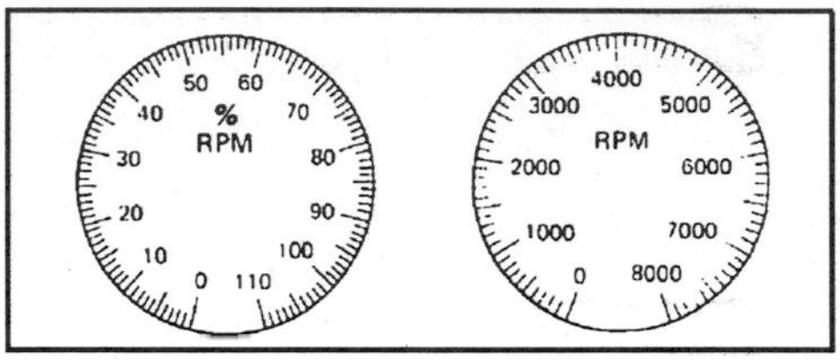

FIGURE 8.6 Comparison of percentage (left) and absolute value (right) scales.

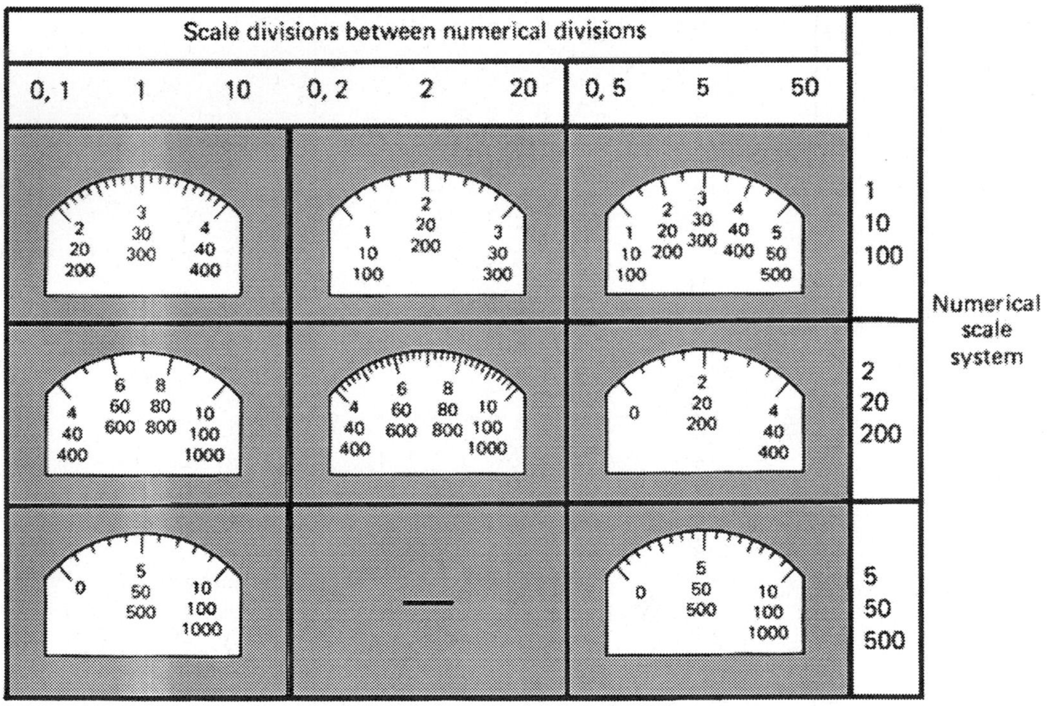

FIGURE 8.7 Examples of recommended scale divisions for different scale sizes.

might be percentage figures being used instead of the actual number of revolutions, so that the operator need not remember different top speeds on different machines (a scale speed is often the same percentage of top speed on all machines). The percentage scale also gives fewer digits to read (see Figure 8.6).

Figure 8.7 shows examples of recommended scale designs for instruments with different scale ranges. On the horizontal axis are the marking intervals, 1, 2, 5 (and 1/10 and 10 times these values). This means that for 2 there are two (or 02 or 20) units between each subsidiary marking. On the vertical axis are the numbering intervals, i.e., the numerical difference between two successive numbers on the scale.

The matrix in Figure 8.7 shows examples of recommended scales and examples of how the markings can be designed. The scale can either be marked with large, medium-sized, or small marks. Certain of the scales use all three sizes. Large markings are always used for the numbers of the scale. The medium-sized and small markings are used for subdivisions, i.e., those divisions that lie between the numbered markings.

There must not be more than nine subdivisions between numerical markings. The scales should be designed so that interpolation between divisions is not necessary. Where there is a lack of space, however, it is better to allow interpolation than to clutter up the instrument with too many markings. If there is room for the scale illustrated in Figure 8.8a, this is the one to choose, otherwise choose the one shown

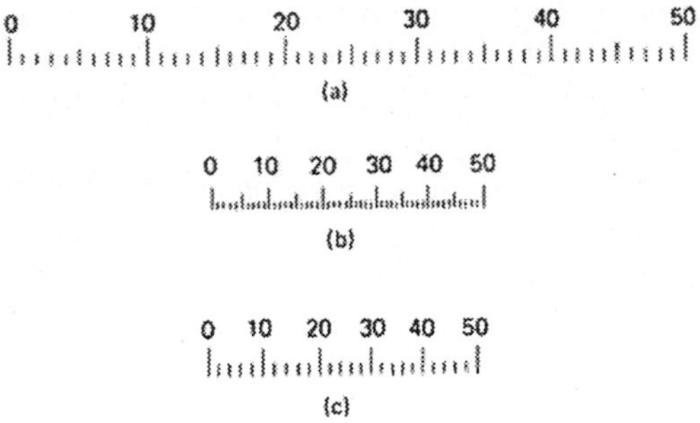

FIGURE 8.8 The physical size of the scale and the number of scale diversions. Scale (b) is not recommended.

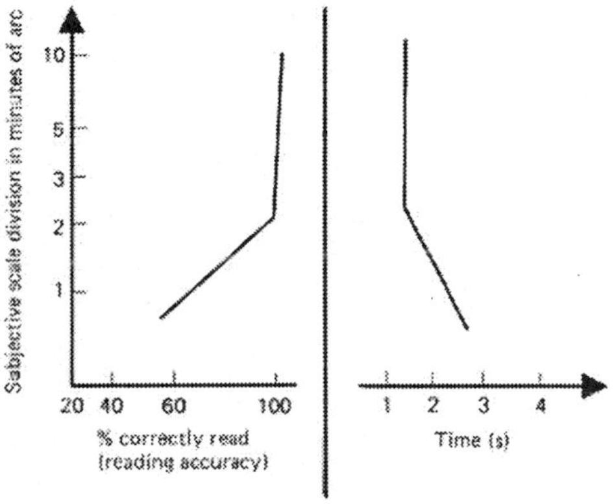

FIGURE 8.9 Effect of subjective scale division on reading accuracy and time. Increasing the subjective scale division to greater than 2 minutes of arc does not increase the reading accuracy or reduce the reading time (From Murrell, K.F.H., 1958, *Fitting the Job to the Worker. A Study of American and European Research into Working Conditions in Industrial Situations,* (Paris: Organisation for European Economic Cooperation–EPA).

in Figure 8.8c. The scale in Figure 8.8b is not suitable. If the alternative in Figure 8.8c is chosen and the scale has to be read off to the nearest whole number, this means having to interpolate one step between markings. The smallest step, which it is necessary to interpolate, is called the *subjective scale division.* There should be zero, two, or five subjective scale divisions for every marked interval.

The basic factor in the reliability of reading an instrument dial is the physical width (measured as an angle of view) of subjective scale divisions (i.e., zero, two, or five subjective scale divisions per marked interval). The scale is subdivided in such a way that the reading tolerance and accuracy one wishes the instrument to have are achieved. Accuracy must not be greater or even less than the allowable pointer error for the instrument in question.

If it is necessary to interpolate fifths in order for the correct degree of accuracy to be obtained, the angle of view must be five times the angle of the subjective scale division, which equals the smallest division which it is necessary to interpolate. If the viewing angle for the subjective scale division is less than a certain critical size, there will be considerable errors in reading it. If, on the other hand, the angle of view of a subjective scale division is greater than this critical size, one cannot expect any further increase in reliability of reading, as shown in Figure 8.9, according to which the width of the critical scale division is two minutes of arc.

The reading reliability over this critical value would be expected to rise to 99% for people with long experience in instrument reading. When designing a scale, therefore, one would start by determining the greatest distance from which the scale is going to have to be read. Based on the distance and the critical angle of two minutes of arc, the smallest size of the subjective scale divisions, which will be needed on a scale, is calculated from the total measuring range the instrument is to have. Knowing that there should be two or five subjective scale divisions between each marked interval, the number of marked intervals can be selected. There should be two, four, or five marked intervals for each numbered interval.

For a scale marked in five subjective scale divisions per marked interval, and which has a total of 100 subjective scale intervals, the total length of the scale is determined according to the reading distance obtained from the following formula:

$$D = 14.4\ L$$

D is the reading distance, and L is the length of the scale. D and L have the same units of length, e.g., cm. If one does not have such a "standard scale" with the above relationship between subjective scale division and marking interval, there is a correction factor, which may be added to the formula:

$$14.4 \times L = \frac{D \times (i \times n)}{100}$$

where i is the number of subjective scale divisions into which each marking interval is to be interpolated, and n is the number of marking intervals.

In practice one is often tied to existing standard instruments. Here one can evaluate the maximum reading distance for the different instruments instead and determine whether this is sufficient for the actual application.

The scale itself should not have long lines joining up the scale markings, but should be open (see Figure 8.10). Figure 8.11 gives the recommended measurements of the scale marking for (a) normal visual conditions and (b) low light levels. The reading distance is approximately 70 cm in this case. The measurements given in Figure 8.11 apply not only to straight instruments, but also to circular and segment-shaped ones.

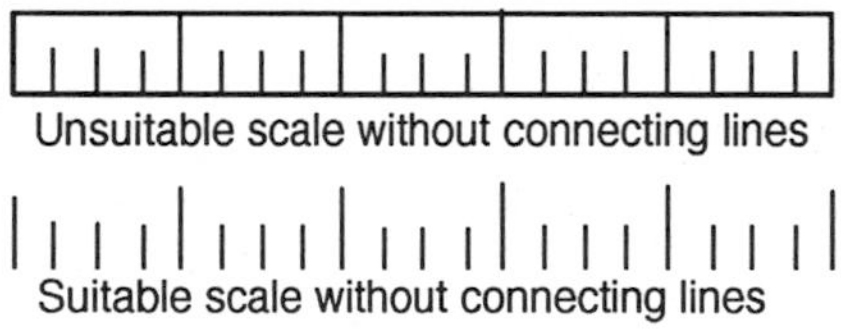

FIGURE 8.10 Examples of suitable and unsuitable scale designs.

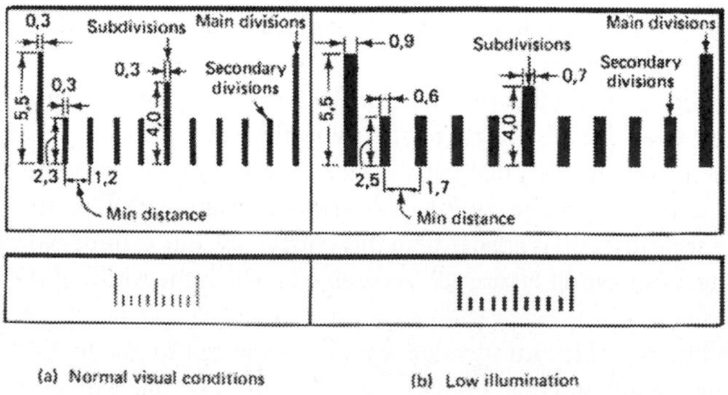

FIGURE 8.11 Recommended design of scales (mm).

The design of the letters and numbers on the scale is important for giving the best possible reading reliability. The following points should be noted especially:

1. The ratio between the thickness of the line and the height of the letter/number for white numbers on a black background should be between 1:10 and 1:20. If the background is dark and the numbers are illuminated from behind, the optimum is between 1:10 and 1:40. The optimum for black letters on a white background lies between 1:6 and 1:8. Black lettering on a white background is normally preferable. For dark adapted and low illumination conditions, white numbers on a black background are preferable (for night viewing in combination with red instrument lighting, the white numbers will be seen as red).
2. The ratio between the height of the numbers/letters and their width should lie between 2:1 and 0:7:1. The optimum is 1.25:1.
3. The appearance of the numbers/letters is important. As a guide, the numbers/letters should be simple in their design. Serifs and too many lines complicate the reading. However, they must not be so simple as to become difficult to understand.
4. Letters/numbers must not be positioned on the scale in such a way that they are shaded by the pointer, and they must be vertical.

The following recommendations are made for the design of the pointer:

1. The pointer must be long enough to reach but not to overlap the scale markings.
2. The pointer should lie as close to the surface of the dial as possible in order to avoid parallax errors.
3. The section of the pointer, which goes from the pivot to the scale markings, should be the same color as the scale markings. The remaining part of the pointer should be the same color as the dial face.
4. On horizontal scales, the pointer should sit on the top of the scale, and for vertical scales it should be on the right-hand side with the numbers on the left.

Sound Signals

Certain types of information are better transmitted as sound signals rather than visual ones. This is the case in the following situations:

1. When the signal is originally acoustic.
2. Warning signals. The advantage is that the operator doesn't need to see the signal in order to detect it; i.e., he or she does not need to look at the instrument constantly.
3. Where the operator lacks training and experience of coded messages.
4. Where two-way communication is required.
5. Where the message concerns something that will occur in the future, e.g., the countdown to the start-up of a process.
6. In stressful situations, where there is a possibility that the operator would forget what a coded message meant.

Tones are preferable in the following situations:

1. For simplicity.
2. Where the operator is trained to understand coded messages.
3. Where rapid action is required.
4. In situations where it is difficult to hear speech (tones can be heard in situations where speech is inaudible).
5. Where it is undesirable or unnecessary for others to understand the message.
6. If the operator's job involves constant talking.
7. In cases where speech could interfere with other speech messages.

Warning signals are without doubt the most common form of sound information.

Sound is an information source which is little used, with the exception of warning signals. Sound signals/information could probably be used to advantage in giving spoken instructions in disaster and other acute situations, for example that there is a fire at a particular location and what action should be taken. In this type of situation one can use a calm voice to give further instructions on necessary actions, such as clear and simple instructions on how to evacuate the building. Various forms of acoustic alarm should be able to be complemented with tones, which would give preliminary information on the type of fault. If there is a hierarchy of alarms, different tones can be used for different groups of alarms. If detailed information is required about the alarms, questions can be entered via a keyboard, or directly by voice. The section on Speech Generation and Recognition later in this chapter includes a description of the design of computer-generated speech, and computers that understand speech.

8.3 Visual Display Units (VDUs)

The information devices traditionally used in control rooms normally have only a single area of application each. The instrument is normally electrically connected directly to some form of measuring device. In certain cases, switches are used that allow the same instrument to be connected directly to several measuring devices. For example, this is usually the case for temperature readings where it is common to have alarms connected to all the measuring points. One then has to turn a special switch in order to be able to read the temperature at all the measuring points.

In the modern computerized process control system, information is collected from many measuring points in the control system. Information is stored and subjected to further processing in the computer, which then sends it on to various receivers, such as the different information devices in the control room. The information devices connected to the computer system are normally the type that can carry out several simultaneous functions, i.e., which can receive different forms of information, such as cathode ray tubes (CRTs).

VDU Design

The collective term for such information devices designed for several different purposes (e.g., for reading temperature, shaft speed, instructions, information on handbook data) is visual display unit (VDU). The most common type of VDU to date is the cathode ray tube (CRT), but there are many other types, such as liquid crystals, plasma displays, and matrices built up of different types of lamp or light-emitting diode (LED).

The important advantage of the VDU is that it can handle most types of presentation although the various technical solutions do have some limitations, e.g., the shape of characters which can be formed on the screen. The pictorial alternative is provided (within the framework of the technical specifications) by programming the computer. It must be stressed, however, that the programming must suit the actual process operator and his or her work and not just reflect how the programmer feels interactive VDUs should be designed.

The latter is unfortunately often the case. This means that many VDU applications in industry have shortcomings or are simply unsuitable and faulty; this does not just mean discomfort and limitations for the operator — it also gives rise to poor performance and an unnecessary number of errors.

There now follow some guidelines on how information on VDU screens should be designed. These recommendations are not final ones that would be applicable to all the different applications and control situations. They must be treated instead as a form of checklist to be used in the design of man–computer communication systems in the control room for process control.

It is also important to stress that it is practically impossible to choose the technical components first, and then expect to get a functional solution. Even if ergonomic requirements are specified for each individual component, there is a risk that the overall solution will be bad. Technical solutions to the design of the interface with the operator can only be produced from detailed descriptions of the job content and analyses of the various work and skill requirements.

Equipment, which can be used for different forms of information presentation, is summarized below, together with some advantages and disadvantages.

Information presentation Technique/Method Equipment

1. Presentation of set value of a variable or deviation from set value
   Pointer or bar with scale (e.g., thermometer).
   Traditional instrument, plasma or LED forming a bar or equivalent on CRT.
   Moving numbers.
   Dot matrix, 7-segment plasma panel, CRT, etc.
2. Trend or time history of a variable
   Graphics.
   Matrix printer or plotter.
   Line diagram:
   Alt. 1: line printer or plotter.
   Alt. 2: CRT.
3. Relationship between set values and several different values
   Line or bar diagram:
   Alt. l: CRT or plasma panel.
   Alt. 2: Multi-pen printer or plotter.
4. The way a circuit works light indicators or displayed mnemonics:
   Alt. l: Lighted buttons, LEDs, rear-projected displays.
   Alt. 2: CRT.
   Alt. 3: Printer.
5. Alarms
   Matrix:
   Alt. l: Signal/number board.
   Alt. 2: Special reserved CRT.
   Tables in chronological, hierarchical, or random order:
   Alt. 1: CRT reserved for the job.
   Alt. 2: Line printer.
6. Text
   Lighted points in certain groupings:
   Alt. l: Plasma grouping (usually a 5 × 7 point matrix per symbol).
   Alt. 2: CRT at 25 Hz with interlacing.
   Alt. 3: CRT at 50 Hz frame frequency with no interlacing.
   Lighted line segments:
   CRT with DC positioning (x/y techniques).
7. Diagrams
   Lighted points in a matrix:
   Alt. l: Plasma grouping.
   Alt. 2: CRT with 25 Hz frame frequency and interlacing.
   Lighted line segments
   CRT with DC positioning.

For the *presentation of the set value,* there are advantages and disadvantages to most types of equipment. The advantage of pointers or thermometer type bars is that they give a certain indication of the size of the changes as they occur, which is more difficult with a digital presentation. The advantage of a digital presentation is, of course, that it is easy to present a large amount of data at the same time. A numerical presentation, on the other hand, gives a poor understanding of size and quantity. It also makes it difficult to compare several values at the same time.

The ordinary printer or plotter is good for *trends and time histories* of values. However, it is difficult to print out several values at the same time, although good documentation (copies) is provided. The bar chart may mean more difficulty with paper handling, for example, and also requires some form of identification of the individual diagrams. CRTs produce a somewhat poorer picture than graphic printers, but they are more flexible. It is also easy to use a CRT to display different variables in turn. No hard copy can be produced directly from the screen, but this can be obtained using cameras or additional equipment.

In order to present the *relationship between variables and set values* CRTs and plasma panels are flexible and suitable in many cases. Resolution and sharpness are, however, relatively poor in comparison with a multi-pen plotter. The disadvantage of the multi-pen plotter is that the format is often limited. Due to the lack of resolution, the same is also true for the CRT.

In order to see how different *circuits are working*, lighted press-buttons are a good system but impractical if many functions have to be presented at the same time. The CRT can be tabulated for a number of different circuits and functions, and can also be used in conjunction with cursors and light pens.

The more traditional type of display for *alarms* has as its greatest disadvantage its lack of flexibility. It also needs a lot of space, but it does give a good overview and continuous accessibility. The system also works well interactively if the diagram has lighted press-buttons on it, where the buttons function for acknowledgment or corrective actions.

The great advantage in using a CRT is that it is easy to update. It is important, however, to have a CRT reserved specifically for this purpose. In addition, the system is compact and easy to use interactively with a light pen, for example.

If the alarms are presented in tabular form with text, chronologically, hierarchically, or in random order (instead of in a matrix), the greatest disadvantage is that the overview becomes lost. On the other hand, it is possible to include an extra line along with the alarm concerning the action which should be taken. Compared with the CRT, the printer also has the advantage that it produces documentation of the alarm. However, printers are often noisy, and it is often difficult to read the actual line being printed. This application would not require any especially reserved CRT.

The plasma display has a number of advantages in the *presentation of text.* For example, it is very thin; it doesn't cause any flickering effect; and the surface is flat. The contrast on the light surfaces can be high, and it is easy to program. The disadvantages are that only one color can be obtained, and the cost is very high. Different types of CRT are often a more realistic alternative. The great advantage of frame frequencies of 50 to 60 Hz is that the picture is clearer and flicker-free (if the brightness is not set too high). CRTs with DC positioning are considerably more costly, but they give much better contrast and sharpness. The color choice, however, is not so great as on other types of CRTs.

Of the visual display units, the CRT is by far the most common type, and will probably be so for some time to come. CRTs are also used for other purposes, e.g., for surveillance and internal TV. However, with today's technology the CRT can give rise to a number of visual problems. A good printer can be a good alternative if the lack of flexibility is acceptable.

If the recommendations given in the next section are followed, the risk of visual difficulties and eye fatigue will be considerably reduced. Problems may remain, however, for those with vision defects. The operator must then obtain glasses which are made and tested specifically for work with CRTs.

Design of Cathode Ray Tubes (CRTs)

Some brief recommendations on the requirements that should be placed on CRTs, currently the most common form of information presentation aid in the computer control system, are presented. The basic design features of a CRT are that the information shall be *visible* and *easy to read.*

It is important to choose a suitable coding and presentation of information for particular applications. These points are dealt with in the next section. The following recommendations assume a screen with a dark background and light text. This form of CRT demands very careful planning of the lighting. There is much to suggest that screens with dark text on a light background give a better result if one can avoid the problems of flicker, which easily occur on light screens (Berns and Herring 1985).

1. Clarity

   Luminance. Minimum 85 cd/m′ for characters (250 ash*), with an optimum of 171 cd/m^2 (500 ash).

   Contrast. The optimal character contrast is 94%, but 90% is acceptable.

   Light/dark characters. If flicker can be avoided, dark characters on a light background are preferable. Otherwise, and if other environmental conditions are suitable, light characters on a dark background can be accepted.

   Flicker. For the optimal luminance level of 171 cd/m^2, a frame regeneration rate of at least 50 Hz is required. Certain types of phosphor (e.g., p-20) may need higher frame rates, up to and even over 60 Hz (frequencies of 60 Hz may interfere with mains electrical frequencies of 50 Hz in Europe, and other places).

2. Legibility

   Character size. The height varies between 16 and 27 minutes of arc (visual angle), a width-to-height ratio of 0:75, and a ratio of height to thickness of the strokes making up the characters of between 10:1 and 6:1.

   Character shape. A dot matrix using 9 × 7 points produces the best symbols. There are many different ways to design letters and numbers, and it is not clear which is best. It is important, however, to choose types which do not cause confusion between letters and numbers as in O and Q, T and Y, 5 and S, I with 2 and K, I with zero 1, O with B and O.

   Resolution. Ten raster lines per character height and more for nonalphanumeric characters.

   Character separation. Fifty percent of character height and 100% between lines.

   Reading distance. For characters of 3.2 mm in height, a reading distance of 69 cm is acceptable. In most practical applications, the reading distance is determined by the height of the characters. The screen should be read from directly in front; angles of more than 30 degrees from the optimum reduce legibility considerably.

   Color. Colors which lie well outside the optimum visual spectrum must be avoided, particularly those in the blue-UV region. Colors in the yellow-green range are good.

   Flashing (e.g., to gain attention). A 3 Hz flashing frequency has no adverse effect on legibility.

   Cursors. A rectangular cursor with a 3 Hz flashing rate is thought to be best.

3. Coding

   Coding improves *performance* considerably, especially for simple information processing tasks.

   Color coding. Best for localization, calculation, and comparison of different information.

   Alphanumeric codes. These are best, and color coding is next best, for recognition and identification of information. Coding is of little or no help in *quantifying* or *size estimation.*

   Components. A code should not comprise more than seven different components (e.g., seven letters or numbers). The fewer components the code uses, the better.

   The *efficiency* of the coding must be seen in relation to the total construction of pictures on the screen.

   Code groups. If possible, different groups of codes, each consisting of components, should not contain the same components.

4. Construction of picture on the screen

   The basic factor in the determination and design of the picture is the function which is to be fulfilled. The following information may be used to provide certain guidelines:

   a. Digital presentation is best for quantitative information.

   b. Pointers on circular scales are best for showing changes.

*1 ash is the luminance of a white surface illuminated by 1 lux.)

c. Vertical and circular scales with moving pointers are good for check readings.
d. Chart recorders are valuable for seeing instrument faults and for obtaining a rapid impression of the system's response.

Diagrams are often preferable to graphs on CRTs. Histograms are especially easy to read. Curves, however, are preferable for reading trends. Several curves can be presented to advantage for comparisons between several variables. Here, semigraphic screens are better, but My screens are to be preferred above all.

Design of tables
a. Whether data should be presented in columns or rows depends on what is more natural for the task in question.
b. Letters are better for identification than numbers.
c. Numbers should be arranged in as few columns as possible.

The design of pictures and the positioning of text within each frame is usually a compromise between several different factors. It is usually very difficult to adhere strictly to the various criteria which are set in designing different frames.

The length of the text also determines the design of frames or pictures because there is only a limited area available, and if the text is very long its length will determine its positioning. In such cases, it may be necessary to abbreviate the text. When doing this, it is of the utmost importance to maintain the comprehensibility of a word as far as possible, even though it is abbreviated.

By placing the same information in the same position in different frames, the possibility of errors by the operators is reduced, together with their search time. This is particularly true for beginners and those who use the terminal infrequently. Consistent positioning of error messages is also very important, because they must always be immediately obvious to the operator, who should not have to search for them.

The most fundamental criteria by which information to be used in a frame should be analyzed are:

1. The importance of the information
2. Its frequency of use
3. The sequence of use

It is clear that even though these criteria have importance on their own, they are very closely connected. One example of this is that the first information to come up on the screen is often both the most important and most frequently used. When the principles for the design of the pictures and the text have been determined, the information should be presented in such a way as to simplify the task of the operator as much as possible.

The CRT-type of display often has poor picture sharpness (focus) at the perimeter caused by the curve on the screen edge, which reduces the clarity of the characters. On many screens, this may mean reducing the amount of information on both the left and the right-hand sides of the screen by 3, 4, or even 5 columns in order to avoid this poor focus area. The effective area of the screen is thus reduced. Another aspect of the design is whether to use every line or every alternate line. No general answer can be given to this; it depends partly on the design of the screen (i.e., the distance between the lines) and partly on the amount of information on the screen.

After determining the above criteria comes the question of how tightly the information should be positioned. The amount of information may be defined as the quantity of information within a defined area. If too much information is displayed, operator performance is decreased (Stewart, 1976; Cakir et al., 1979). Stewart suggests that the search time in seconds is roughly a fifth of the number of alternatives to be searched through. Even if a very experienced operator is able to be selective in searching for information, irrelevant information means that the search time and error frequency will increase, thus reducing the performance of the operator and the system.

Advantages and Disadvantages of VDUs

In older control rooms, each variable was represented by its own instrument, regulator, or control device. These were placed together on a panel. Various ergonomic rules for the design of panels have been produced over the years, with the aim of achieving the best possible operational conditions. The development of VDUs and computers have now provided new forms of information presentation, monitoring, and controlling.

Research by ERGOLAB (Ivergård et al., 1982) has shown that many operators prefer the conventional instrumentation over the computerized versions. They have often described using the computerized alternative as being akin to monitoring the process through a keyhole.

Descriptions have been produced (Stark, 1983) of how people scan a picture for the object they are seeking. It appears that the normal strategy is to fix the gaze on a certain number of points in order to identify the picture; the points chosen on which to set the sight depend on the object. Stark (1983) also found that the experienced viewer will miss some of the points he previously fixed on. Despite jumping over them, he will still be able to make the correct interpretation of the picture. In addition, it will be detected whether there are any changes in the parts of the picture that were skipped over. In other words, people use their peripheral vision for checking whether there are any changes in the picture.

It is likely that a process control room operator will work in a similar fashion in monitoring a VDU or a conventional panel. He fixes on a number of points in order to obtain a picture of the current situation in the process. The operator then updates his mental model of the status of the process. The experienced operator has a considerably higher performance level at monitoring, and probably fixes on considerably fewer points to estimate the status of the whole process. Sometimes the operator only has access to a number of pictures presented on VDU screens. This does not allow parts of the process, other than those currently presented on the screen, to be updated. More conventional instrumentation, where all the information is presented in parallel, gives the operator very different possibilities. By fixing the gaze actively on certain parts and using the peripheral vision for other parts, he or she can continuously update his/her mental model of the status of the whole process.

Given this background information, one can see that there is a natural division of the viewing process into two types: active and passive. When someone fixes their gaze on one or a certain number of points in order to identify an object, this is active vision. In parallel with the active process, passive vision is occurring via the more peripheral parts of the retina.

In active vision, the gaze is turned toward the object and fixed on the central part of the retina where the cones are most dense, i.e., the fovea. In the area around the fovea, the rods dominate. The rods are considerably more sensitive to light and can therefore work under relatively dark conditions. They are also sensitive to movement and changes and are used in pattern recognition. The cones, on the other hand, need more light. The ability to distinguish detail, mainly by the cones, is also thought to increase in proportion to an increase in the light level. There are also cones which have the ability to distinguish colors.

This allows us to draw certain conclusions regarding active and passive vision. Active vision allows the identification of colors and small objects under good lighting conditions. Passive vision permits recognition of patterns and, therefore, especially changes and movements, and passive vision also works if the object is in motion.

In control room work, active vision is used to make detailed readings of a more quantitative nature. It is also used in the identification of color codes and in the detection of small differences in curves or diagrams, for example. Active vision is excellent for VDU viewing. There the operator can call up a particular frame and adjust parameters such as the set value.

Using passive vision, one could tour the control room and identify changes in the process pattern, for example, of lights being lit or extinguished on a panel. Schematic representations of a process with built-in indicators (e.g., signal lamps indicating deviations) are a suitable type of presentation for passive vision. In other words, passive vision is perhaps best applied on a more traditional type of instrument panel.

The choice of presentation method depends very much on the task of the operator. There are three main motives for having an operator in the control room:

1. The operator acts as a supervisor in order to carry out certain standardized routine tasks which, for various reasons, have not been automated. He also has the job of calling for expert help when some unforeseen incident occurs.
2. The operator is himself a qualified expert with the job of carrying out production planning and optimization tasks. The operator deals with simpler, more routine and predictable types of fault. The more serious, unforeseen faults, on the other hand, are passed on to special maintenance experts.
3. The operator is primarily a maintenance-oriented expert who gets information about the process from other sources, especially those which are economic in character. He usually looks after production quality matters himself. The operator is expected to be able to deal with most unforeseen and difficult faults and events in the process.

If the first alternative is chosen, one can determine relatively accurately beforehand what type of information the operator will need in different situations. There are fewer requirements for the more detailed type of overview information. The conventional type of instrumentation therefore provides very little information to this operator and he can largely rely on a number of VDU screens with a predetermined program of frames.

The expert operator who is either production-oriented or maintenance-oriented has a considerably greater need for more detailed, continuous, and parallel presentation of the whole process. If the information is presented on VDUs, most processes would require a large number of VDU screens or extremely large VDU screens. The alternative is to require the operator, even during normal running conditions, to sit down and leaf through all the status frames. He would have to do this in order to update himself actively on the process status and to build up his knowledge of the functioning of the process. Instrumentation of the conventional type offers completely different possibilities for the operator to update his mental picture of the process. Looking at the instrumentation both consciously and unconsciously can do this.

The production-oriented operator needs to be able to see a relatively detailed and dynamic functionally oriented process model. The maintenance-oriented operator, on the other hand, needs to have a more physically oriented model available. The traditional instrument and control panel is a good alternative, but the conventional type of instrumentation is not preferable to a VDU in all processes.

The VDU screen has the great advantage of being flexible. Color monitors with high resolution and detailed pictures also allow the presentation of a large quantity of information in a limited workspace. VDUs are also suitable for presenting different types of information. This may be graphical information (e.g., maps of temperature distributions [isotherms], pressure distributions [isobars], etc.). Even if VDUs cannot always wholly replace conventional instrumentation, they are a necessary complement in modern process control.

Table 8.2 gives a comparison of conventional instrumentation and electronic VDUs. It may be seen that the VDU has many advantageous characteristics, and that the disadvantages are largely of an ergonomic nature. In addition to the characteristics given in Table 8.2, visual problems should be included. These practically always occur when using the CRT-type of visual display. In practice, both VDUs and more conventional instrumentation are required in most cases. When modernizing existing control rooms, it is best to keep the old instrumentation as a reserve and a complement to the VDUs.

When building new control rooms, it is rarely sensible to install both the conventional type of instrumentation and VDUs. On the other hand, it may be a useful complement to the VDUs to produce some form of detailed overview panel. This can schematically and dynamically describe the physical design of the process. It is often desirable, and sometimes necessary, to provide dynamic information on the overview panel. This may, for example, show which valves are open or closed, which pumps are working or not. It may sometimes even be desirable to provide the overview board with quantitative information such as flow rates and levels.

TABLE 8.2 Characteristics of Conventional Electronic Display Devices

Attribute	Conventional	Electronic/Advanced
Nature of presentation	Parallel	Serial
Mode of presentation (digital, analog, etc.)	Fixed	Variable
Availability of information	Continuous	On request
Relationships among items of information	Static	Dynamic
Nature of the interface	Inflexible	Flexible
Incorporation of diagnostic aids	Limited	Readily feasible
Ability to modify or update	Difficult	Easy
Redundancy of displays about control rooms	Costly	Relatively inexpensive
Control room size requirements	Relatively large	Relatively compact
Compatibility between process response and control movement	High	None
Control display feedback	High	Name
Nature of overview information	Detailed	Summarized

Adapted from Seminara, J. L., 1980, *Human Factors Considerations for Advanced Control Board Design,* vol. 4 of *Human Factors Methods for Nuclear Control Room Design* (Palo Alto, CA: Electric Power Research Institute).

The amount of dynamic information provided on the panel depends very much on the type of process. In a continuous, stable process, with relatively few starts and stops, a static overview panel is often sufficient. These may be described as processes with high inertia, where changes in the process variables take place over many minutes, and where a detailed diagram of the principles showing the flow and the interconnections is required. In addition, the most important physical units must be marked on the diagram. A good complement to this form of presentation may be a three-dimensional model of the process.

A dynamic and detailed overview board is required where a complicated network with several alternative connection routes exists. The connections in one part of the network will affect the conditions in another part of the network. A typical example is an electricity supply network, and to a lesser extent a water network. A relatively complex batch process also needs a detailed dynamic overview panel. If the batch process is simple enough to be presented as a picture on a VDU screen, an overview panel is not necessary.

An overview panel is also required in a continuous process with many stops and starts, and where the course of events is relatively rapid. Such processes would have changes occurring within seconds or, at the most, a few minutes. Production-oriented operators will always need a functional overview diagram. This can be suitably presented on a VDU screen. The maintenance-oriented operator has a special need for detail in the overview panel.

8.4 Instructions, Forms, Tables, and Codes

Instructions, notices, and forms give printed information to the operator, often in connection with a product. It is of the utmost importance that this information is presented clearly and concisely and that it is easily read. This is particularly the case when an operator is faced with a new product, one which he rarely uses, or in a situation where time is limited.

The following rules apply:

1. *Letters.* Capitals alone are recommended for general use, but a combination of capitals and lower case (small) letters is allowed; i.e., first letter a capital, followed by lower case letters. The ratio between width and height of the uprights should be between 1:6 and 1:8. The ratio of width to height of letters should be 3:5.
2. *Short messages.* In order to save time and space, the text should be as short as possible, while still maintaining clarity.

3. *Choice of words.* Words and meanings should be kept as simple as possible (see above). Only well-known words should be used. One may use common technical terms for special populations, for example, the use of aeronautical terms when addressing pilots.
4. *Clarity.* Instructions must be short and easily understood. One very simple way to test this is to try out different instructions on different people.
5. *Contrast* in work where dark vision is required, letters should be white or light yellow on a black or other dark (e.g., brown) background. Where no dark vision is required, the letters should be black on a white background. Other colors may also be used for coding purposes, but they should always be chosen with maximum contrast in mind.
6. *Size.* The recommended size of letters depends on the reading distance. At a distance of 70 cm or less, in low lighting conditions, the size should be about 5 mm. In good light, this can be reduced to 2.5 mm.

Diagrams and Tables

The following recommendations apply to diagrams and tables.

1. Diagrams are always better than tables if the shape, variation, or connection between materials is of interest, or if interpolation is necessary. If not, tables are preferable.
2. Simplify the table as much as possible without reducing its accuracy and without the need for interpolations. (An example of this would be where, if the meter marks on the water tank level are not sufficient, several calculations are required to interpret the water level.)
3. Leave at least 4 to 5 mm between columns that are not separated by vertical lines.
4. Where the table columns are long (more than six lines), they should divide into groups of three or four lines. Leave some space between each group.
5. Diagrams should be drawn so that the numbered axis lines are darker than the unnumbered ones. Where only every tenth line is numbered, the fifth line should also be denser than the others, but less dense than the numbered ones.
6. Avoid combining too many parameters in the same diagram — there should not be more than three. If more parameters are shown, more diagrams must be used.

Codes and Symbols

Different sorts of traditional symbols are used to convey a particular meaning. For example the cross represents Christianity for a large proportion of humanity. The alphabet can be used to build up different definitive meanings using various systematic rules. It is important in the construction of an artificial language that all the users agree on the rules, which are to be applied to it. This is the case both in more complex languages (e.g., those which resemble the natural ones) and in the simpler ones, such as those used by machines of different types or as explanations on VDU screens. The assumptions necessary before a language can be built are, first, that everyone agrees what the different characters mean and, second, that there are certain predetermined semantics. There must also be a grammar, which specifies the way in which the different symbols can be combined. Finally, the symbols must be designed in such a way that other people can understand them. As far as visual symbols are concerned, they must be able to be clearly seen and understood in the situation in which they are designed for use.

Languages may be termed natural or artificial. Natural languages are often learned in childhood or after very comprehensive training and/or education. More complex artificial languages, which are to be used in a way similar to the natural languages, also require a very comprehensive education and training. In the design of languages for use in different technical applications in industry, special stress must be put on the need to make the rules simple so that as far as possible they are self-evident.

Even if one can set out such very simple rules for the language, it is still necessary for the rules to be put down in writing so that everyone is agreed on which rules are valid. A semantic, then a grammar, and finally a number of viewpoints on the visual detachability of symbolic languages that can be used on VDU screens are presented.

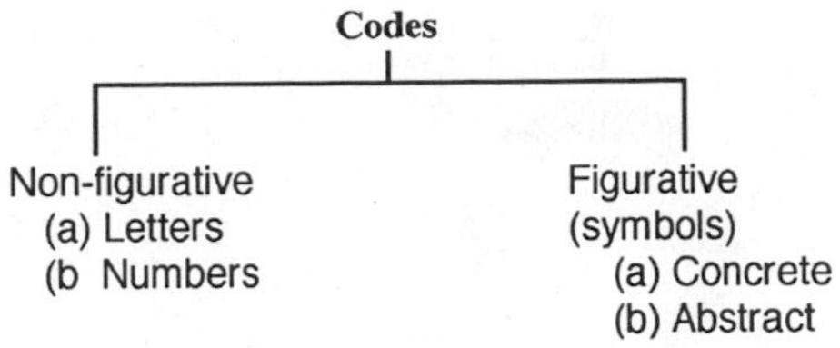

FIGURE 8.12 Categorization of codes.

Semantics

Semantics specify what the different characters mean. In this case, the semantics are summarized in Figure 8.12. The notion of codes is used here as an overall concept covering all types of symbol/character. We are conventionally accustomed to two types of codes:

1. Nonfigurative codes, which consist of several small elements. The individual elements have no meaning in themselves, but when combined they have common unambiguous meanings. The most common form of these nonfigurative codes is numbers and letters.
2. There are also figurative codes (symbols), which are those designed in such a way as to have a meaning themselves without needing to be combined with other symbols. Such figurative codes may be either concrete or abstract. The concrete codes attempt to imitate what they symbolize (e.g., a pedestrian crossing sign is represented by a stylized drawing of a walking person), while the abstract codes symbolize an abstract concept (e.g., Christianity, represented by a cross).

Both figurative and nonfigurative codes are used in process industry applications. The figurative concrete codes should try to resemble the apparatus and machines they represent, and the abstract codes should be used to represent the actual events occurring during the process. Nonfigurative codes, usually in the form of letter and number abbreviations, are also used. Numbers are used both for identification and quantification, although it is preferable to use two different number series for these purposes, for example, Roman numerals for identification and Arabic (ordinary) numbers for quantification.

Grammar

The grammar to be used in this connection must be very simple. It should consist only of:

1. Nouns
2. Adjectives
3. Verbs

Nouns should be used to specify different physical objects such as generators, motors, transformers, and switches. The verb is used to give the condition of the noun, e.g., on, off, running, open, closed. The adjective is used either for specifying which machine/unit is under discussion, or for giving its characteristics, e.g., DC, AC, size. Concrete symbols are suitable for nouns, and abstract symbols are best used for verbs (arrows, colors, etc.). Nonfigurative codes such as abbreviations or numbers are best used for adjectives. Figure 8.13 shows the symbol for a lathe chuck. It has a concrete symbol to specify the noun, arrows are used to denote whether the chuck is opening or closing, and numbers and letters specify which lathe the chuck belongs to.

Comprehensibility of Codes

We shall only look at the ease of understanding of visual codes in this section. It is, however, important to remember that sound codes can be just as useful in many instances as visual codes. One advantage of sound codes is that it is not necessary to be in a particular workplace in order to notice them, and also faster reactions may be expected to sound signals. Hearing can be said to be the dominant sense in terms

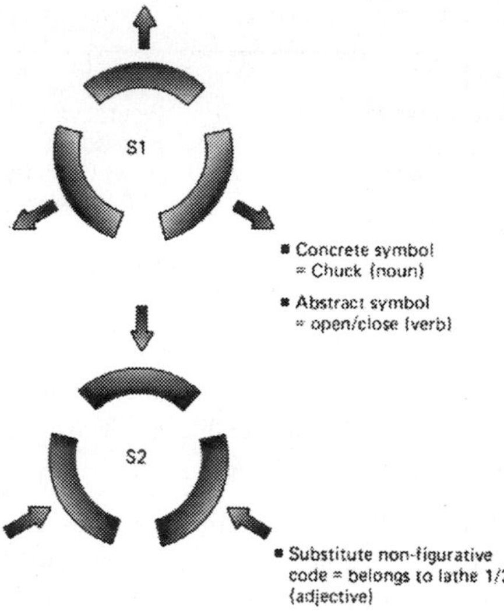

FIGURE 8.13 The symbol for a lath chuck.

of noticing, recognizing, and identifying patterns. The ability of a musician to recognize very small variations in a piece of music (which is a pattern presented in the form of sound) is outstanding. The amount of information and the total number of possible different messages that could be presented in this way are immense. Sound also has the advantage that it is not necessary to focus attention on the equipment supplying the information, and one is free to move around while listening for the information.

Another information channel which gives almost the same degree of freedom is the sense of touch. Experiments have been done on producing a language for the deaf that is transmitted through touch. There has been an attempt, in Ivergård (1982) for example, to produce a "hearing glove"; this has an instrument mounted against each finger which produces pressure at different frequencies. Sound, for example from someone talking, is converted electronically and transferred to this hearing glove. It is known that at least four different frequencies can be sensed, so by this method one can transfer at least 4×4 different symbols (information units). Similar methods have also been used by placing the vibration devices on the chest and other parts of the body.

The sense of touch is not only useful as an information transmission channel for the deaf, but it can also be used by people who are overloaded with visual or sound information. The large number of instruments they have to read, for example, often overloads pilots. If we wish to give the pilot even more information, vibrators could be used to transfer information through the sense of touch instead. Like hearing, touch is better than sight for use as a warning signal; while one can close the eyes to remove the sense of vision, the sense of hearing or feeling cannot be shut off. Four variables can be used in the transferal of information by touch: the positioning, frequency, amplitude, and variation of the stimulus.

The visibility of visual codes will not be discussed in detail here except to say that they must be (a) large enough, (b) bright enough, and (c) have sufficient contrast, in order to be visible.

Experience from Gestalt psychology can be useful regarding the suitability of codes and for general rules on their design. In addition, there are a number of perceptual psychological grounds, which form the basis of the number of elements that can be used in different codes in order to avoid confusion. In particular, there is a body of specialist knowledge on the design of letters and numbers.

Table 8.3 summarizes the number of possible variants that can be obtained with different types of stimuli for designing codes. It is important to remember that the estimations given are only approximations. The maximum number only refers to conditions where the person performing the reading has special education or training for that particular form of stimulus. In most normal applications, e.g., work on VDUs, the recommended number of alternatives must be used. Table 8.3 also gives the number of

TABLE 8.3 Numbers of Recognizable Variations of Different Stimuli

Notes	Stimuli	Maximum	Recommended
	Colors		
Lamps	10	3	Good
Surfaces	50	9	Good
	Design		
Letters and numbers	∞	?	
Geometrical	15	5	
Figures/diagrams	30	10	Good
	Size		
Surfaces	6	3	Satisfactory
Length	6	3	Satisfactory
Lightness	4	2	Poor
Frequency (flashing)	4	2	Poor
	Slope		
Direction of pointer on dial	24	12	Good
	Sound		
Frequency	(Large)	5	Good
Loudness	(Large)	2	Satisfactory

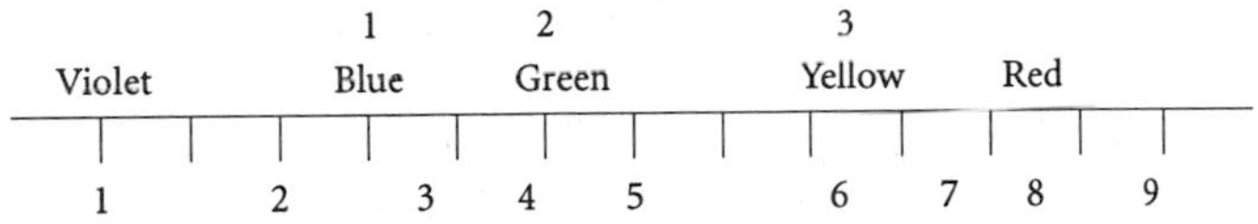

absolutely recognizable units of a particular stimulus. If comparisons are possible, the number of recognizable units becomes considerably greater.

Table 8.4 identifies letters and numbers that are easily confused with each other, and also those that are difficult to read. The risk of confusion on VDU screens is especially great, because the letters and numbers are built in a fairly simple and limited manner.

Figure 8.14 shows the number of seconds it takes to read a particular number compared with the grouping of the digits. It is clear from Figure 8.14 that groups of three digits are the quickest to read. Suggested forms of grouping are given in Table 8.5.

8.5 Using Color

First, it is necessary to define what is meant by color because various color classification methods exist. The CIE system (Commission Internationale de l'Éclairage) (see Figure 8.15) is probably the best-known color classification system, and this has many advantages when it is used for specifying colors for use on VDU screens. This system is described in Wsyzecki and Stiles (1967).

The colors on a CRT are created using three electronic guns, which are aimed at the front of the CRT. This is covered with a phosphor layer. Each gun produces a different color — red, green and blue. The rays travel through a shadow mask which has many small holes in it, and then meet the front of the screen where there are a large number of symmetrically arranged round phosphor dots (see Figure 8.16). This results in different combinations of the colors red, blue, and green. Because the points are very close together, the eye does not see them as individual points, but as mixtures of color.

TABLE 8.4 Interchanging of Letters and Numbers

These:	()* (zero)	8**	B*	D*	1**	Ø** Capital O	Z*	S	G	N	V	Y
Are mistaken for these:	Ø 6 D 9	B		0 (zero) Q P	1	0 (zero) D	2	5 3	6 C	W	U Y	W V 4 T

*Interchanged often
**Interchanged very often
The following difficult to read: 2, 4, 5, 8, 9, N, T, I, Q, X, and K

On VDU Screens

Mutual	One-way	
O and Q	C	read as G
T and Y	D	read as B
5 and 5	H	read as M or N
I and L	J,T	read as I
X and K	K	read as R
1* and 1*	2	read as Z*
	B	read as R. 5 or 8

* These three account for more than 50% of all errors

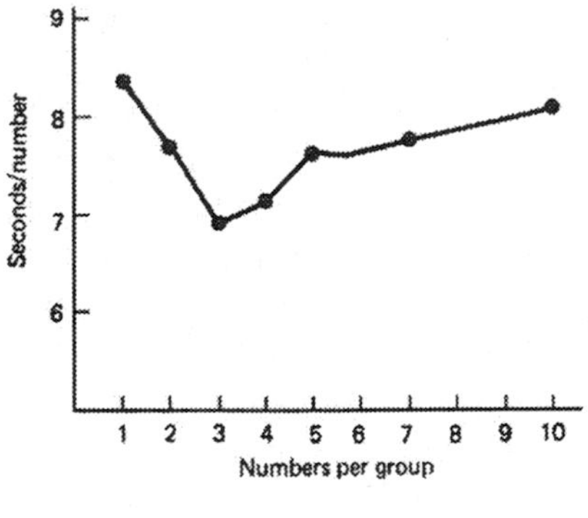

FIGURE 8.14 Average reading time per number for the keying in of 18 to 21 numbers at a time as a function of their grouping.

TABLE 8.5 Examples of How Signs in Codes of Different Lengths Should Be Grouped

Number of digits	Alternative groupings		
7	3, 3, 2	3, 4	3, 2, 2
8	3, 3, 2	2, 2, 2, 2	
9	3, 3, 3		

Rather than a spatial combination of colors, one can have a temporal combination. This means presenting the different colors at different times. Because the succession time of the colors is very short, the eye will not be able to pick out individual colors, and they will be mixed in the proportions in which they were presented. However, the drawback of this type of color presentation is that it gives considerably less well-defined colors, which are difficult to identify.

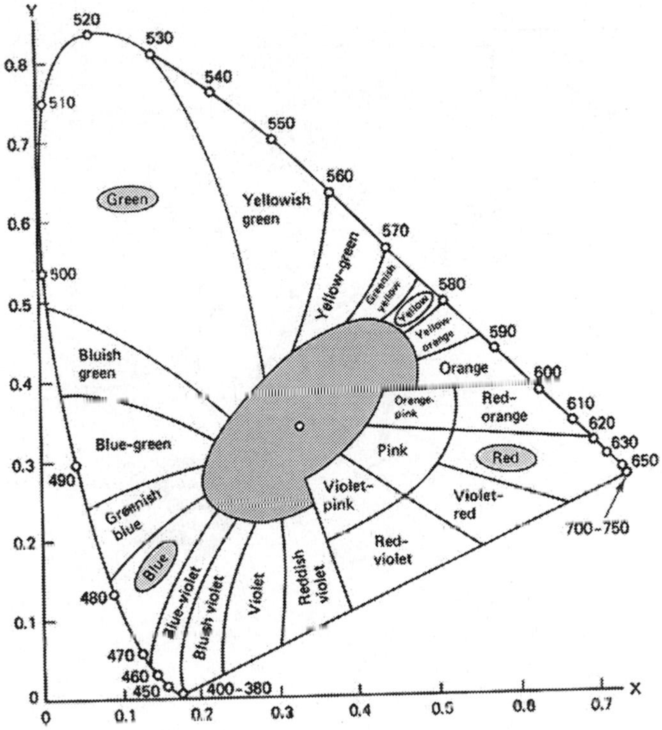

FIGURE 8.15 Chromatic area for common colors shown in the CIE 1931 standard system. The figures on the outer curve (430 to 780) are the wavelength of the light. The colors can be specified by giving their x–y coordinators. The colors which should be used primarily are circled.

In the CIE system, a color mixture will always lie on a line between the two colors of which it is a mixture. If the mixture is made up of three colors, the perceived color will lie within a triangle formed by the lines between pairs of colors from the three. In practice this means that the colors on a CRT will never fall outside the triangle shown in Figure 8.15. This ability to define colors on a CRT fairly simply using the CIE system is its major advantage over other descriptive classification systems. However, an obvious disadvantage of the CIE system is that colors cannot be classified in comprehensive descriptive terms such as color, saturation, brightness, etc. Other descriptive classification systems do have this advantage.

There are a large number of factors, which affect the perceived color. The following are the factors which must be taken into account:

1. Contrast
2. Size
3. Background color
4. Absolute relative discrimination
5. Number of colors
6. Surrounding lighting

When working only with black and white, it is sufficient to use luminance contrast (CR), which is normally defined as:

$$CR = \frac{L_1}{L_2}$$

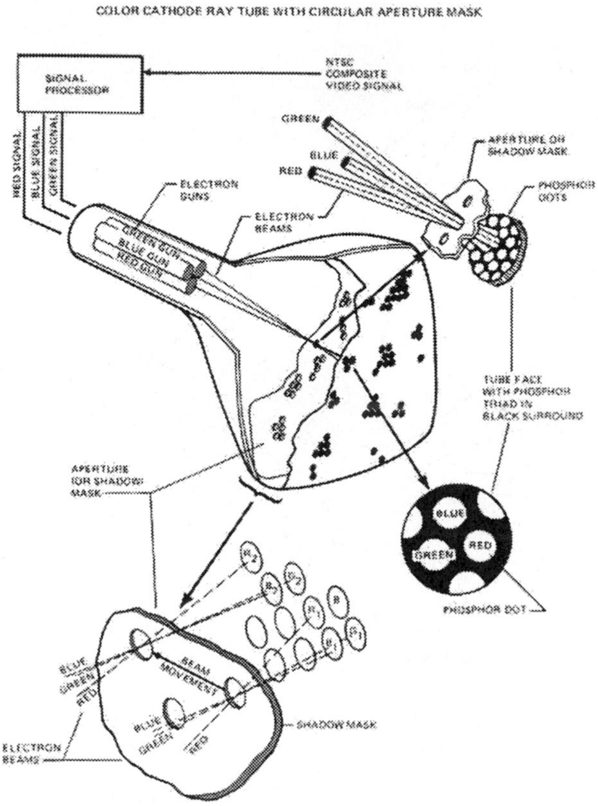

FIGURE 8.16 The colors on a CRT are created by the electronic guns shown in this figure. (From Farrell, R.J. and Booth, J.M., (1975), *Design Handbook for Imagery Interpretation Equipment,* (Seattle: Boeing Aerospace Company). With permission.)

where L_1, is the luminance of the object, and L_2, is the luminance of the surroundings. When color also has to be taken into account, the concept of color contrast is used. A special discrimination index has been worked out which takes into account both luminance and color contrast (see Silverstein, 1982).

Sensing of color is highly dependent on the size and brightness of detail within the picture. Very small details appear less saturated and sometimes change, depending on background color. Also, the ability to discriminate between colors is severely reduced for small pictures, especially along the blue–yellow axis (see Figure 8.15). The background color is also very important because the light from a small object is experienced as though it changes color toward its complementary color in the surroundings.

As the number of colors increases, color discrimination becomes more difficult, and greater demand is made for color coding. Research on the use of colors on VDU screens has shown that a maximum of four to six colors can be used. The number of colors, that can be used is of course highly dependent on whether the discrimination is absolute or relative. In absolute discrimination, the number of colors is further reduced. When color is used for decorative purposes, more colors can be used. The reasons for using colors to any great extent must be powerful ones — it is far easier to produce a poor color picture than a poor black-and-white one.

Comprehensive research has been carried out in recent years into the use of color screens, and the level of knowledge has increased considerably. Unfortunately, some important knowledge is still missing, which makes it difficult to make final recommendations. In particular, the effects of interacting factors are not well understood. For those who wish to learn more about the analytical method used in determining colors on a color screen, Silverstein's (1982) *Human Factors for Colour Display Systems* is to be recommended.

Why should a color display be chosen? In most cases, color displays are just a more costly alternative to the monochrome version. When there is an advantage in using a color display, it does not seem to have an economic basis. It is probable that a color display is *not* preferable to a monochrome display for most types of common office applications. Where there are special office applications needing more complex pictures, or in special applications such as control rooms in the process industries and traffic monitoring situations, color displays are certainly preferable.

Selection of a color display does not depend only on the fact that the operator prefers to have a color screen for aesthetic reasons. The primary reason for choosing a color display is for increased coding potential and the reduction in search time it brings about in complex pictures. It is also important to remember, as demonstrated later, that color contrast increases visibility and thereby decreases the need for luminance contrast. This, too, has many advantages, because color screens can be utilized in rooms with an ordinary level of lighting which fulfils the requirements for other types of visual tasks. Under special visual conditions, for example on planes and ships, the color screen is almost obligatory, although the requirement for well-shielded lights still remains.

Color VDUs drastically increase the requirement for specialist knowledge in the design of the display system and the choice of coding for pictures. It is difficult to make a good picture on a color display. A color picture can easily be made worse than a black-and-white picture if one does not have access to the right knowledge. A poorly designed color display decreases efficiency of performance as compared with a black-and-white screen.

Choice of Colors

A number of general ergonomic rules pertaining to absolute discrimination indicate that for recognizing and naming a color, a maximum of about seven colors is desirable. When it comes to seeing the difference between colors, many different colors and shades can be distinguished, probably well over twenty. Research on color screens has shown, however, that this is unrealistic. One cannot normally have more than three to seven colors on a CRT, depending on a number of different factors (Kinney, 1979; Teichner, 1979; Silverstein and Mayfield, 1981). For an operational color display, where absolute color discrimination is required, it is suitable to have three to four colors, and these should preferably be green, red, blue-green, and possibly a purple-red (plus white or black, depending on the background color used).

Where comparison between colors is possible, six to seven colors can be allowed, preferably red, yellow, green, blue-green (cyan), reddish-blue, and perhaps magenta (plus white or black depending on the background).

Color Screen Character and Symbol Design

It is fairly clear that the requirements will be far more stringent for the design of characters and symbols on a color screen than for those on a monochrome display with a dark background and light text. Research has shown that color sensitivity increases as the color field increases in angle, up to 10 minutes of arc. Small light fields have reduced color saturation, which means that, in practice, light colors tend to be seen as white.

On this basis, characters, symbols, or critical details in the picture which subtend less than 20 minutes of arc should not be used. On larger fields, where greater accuracy in color discrimination is required, this angle should be at least 1°. The lines, which make up the characters, should be thicker on a color screen than on a conventional one. If there is more than one minute of arc between lines, color separation will be distinguishable. The characters on a color screen should not be placed more than 15 minutes of arc from the line of sight. Characters that appear toward the periphery of the screen will not be color-coded to the same accuracy.

Research has also been performed on legibility and luminance contrast of color screens under different environmental lighting conditions (Silverstein, 1982). This has shown that there is no further improvement in legibility when the luminance contrast is increased beyond 5. It should also be remembered that the colors seen on the screen would change depending on the room lighting. The room lighting should therefore not be changed once the color screens are installed.

VDU Screen and Background Requirements

The technical requirements for a color screen are considerably greater than for an ordinary monochrome VDU screen. One must, for example, set considerably greater demands on sharpness of edges and the ability of the guns to send their rays to the correct place at the edges and corners of the screen, so that no convergence problems occur.

It has been shown that increased luminance on a VDU screen increases the color contrast up to a luminance on the screen of 3000 cd/m^2. Increasing the operator's light adaptation level also increases his sensitivity to color (Silverstein, 1982). A light background gives the effect of a higher degree of color saturation. In practice this means that a higher level of environmental light can be used when working with color screens. In many cases it may also be desirable to select a light background color on the VDU screen in order to achieve an even higher level of saturation which, in turn, reduces the risk of both diffused and reflected glare.

Standardization of Colors

There are various conventions that have been specified as standards. There is, for example, an 150 standard (1964) where various safety colors are recommended (see Figure 8.17). There is also an electrical standard (IEC, 1975) describing the recommended use of red, yellow, green, blue, and white colors for signal lamps and buttons. These recommendations are summarized in Table 8.6

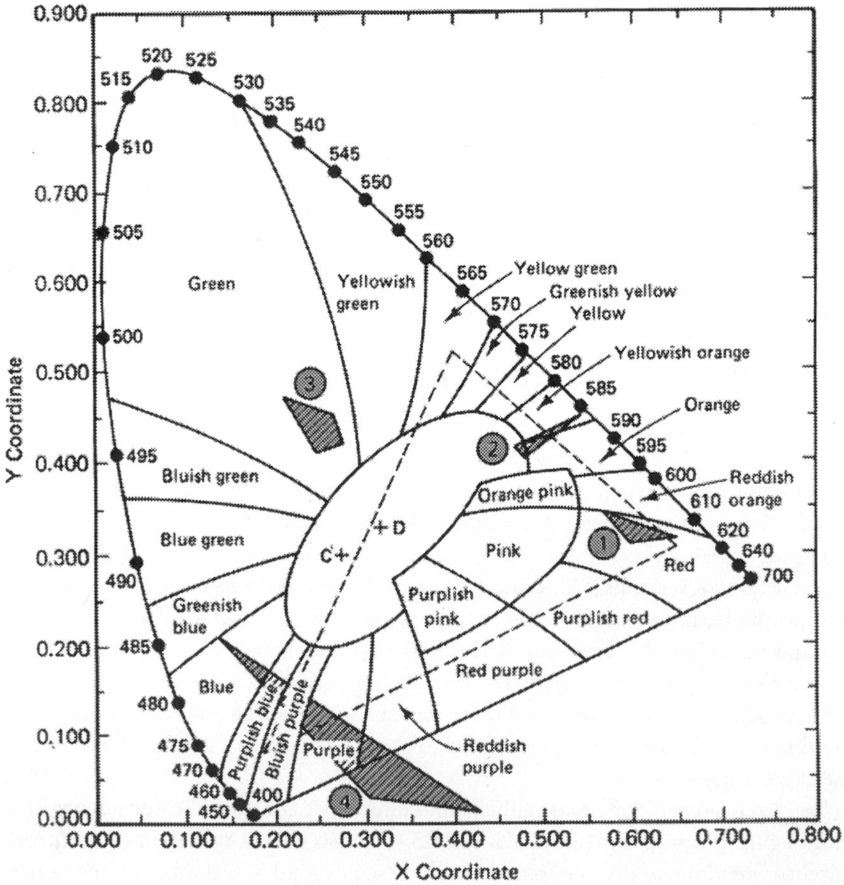

FIGURE 8.17 ISO recommendation for safety colors (R408, Dec. 1964) (1) safety red, (2) safety yellow, (3) safety green, (4) auxiliary blue.

TABLE 8.6 Examples of the Use of Different Colors for Signal Lamps and Control Buttons (SEN 280801)

	Signal lamps	Control buttons
Red	Danger, alarm	Stop, off, emergency
Yellow		Care, caution
Green	Safety	On, start
Blue	Any application	Any application
White	Any application	Any application

Red is used for danger and alarms; yellow illustrates caution, while green is used to signify safety. Control devices (buttons) use red to denote "Stop," "Off," and "Emergency"; yellow for "Action" (to avoid unwanted changes); green for "On" and "Start." Blue and white can be used for any other instructions as required.

There are, unfortunately, risks of misunderstandings with this electrical standard. A switch in an electrical circuit which is *closed* is marked by a red signal lamp outside the door to the switchroom to signify danger. A switch which is open is marked, according to Table 8.6, with red to indicate its "off" condition, i.e., the electrical circuit is "open." So a switch can be marked with red both when it is closed and when it is open! This example illustrates the difficulty of using colors to carry information. Colors must always be complemented with other information in order to avoid misunderstandings. Colors should only be used as a complement and to simplify readings.

In the IEC (1975) standard, it is suggested that in technical applications within the process industries the following practices should be adopted for the use of colors:

Red	Abnormal conditions in variables, and warnings
Green	Static plant information (e.g., pumps, lines, valves, etc.)
Blue-green (cyan)	Alphanumeric information and scales (e.g., legends)
Yellow, orange	Variables (e.g., histograms, curves, etc.)
Violet/red (magenta)	A reserve color to replace red if the red gun fails.

8.6 Speech Generation and Recognition

Language is probably the most distinguishable ability that man possesses. Man/machine systems have long included written language (e.g., in alphanumeric displays and keyboards), but spoken language has only been used for communication between people. Automatic speech generation and recognition by machine now offers an alternative to other forms of input and output to computers. An interactive speech system consists of speech recognition devices for control or information input and speech generation devices as a form of information display.

Automatic speech technology is of great interest for the future. Simpson et al. (1985) gives a comprehensive review of *system design for speech recognition and generation.* From the human factors/ergonomic perspective, a speech recognition system consists of a human speaker, recognition algorithms, and a device that responds appropriately to the recognized speech:

Human speech ~ Recognition algorithms ~ Response device

A speech generation system is the mirror image of a recognition system. It consists of a device to generate messages in the form of symbol strings, a speech generation algorithm to convert the symbol strings to an acoustic imitation of human speech, and a human listener. A speech generation system operates within the context of the user's working environment:

Message generator ~ Transformation of message codes to acoustic imitation of voice ~ Human listener

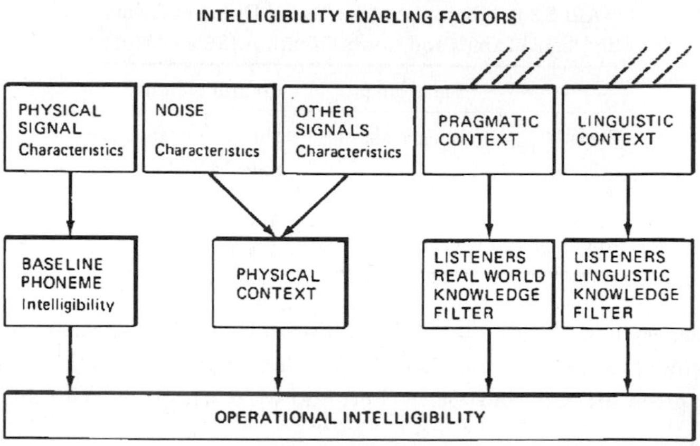

FIGURE 8.18 Factors that contribute to operational intelligibility (after Simpson et al., 1985). (Redrawn from *Human Factors,* 23, p. 131. © The Human Factors Society, Inc., and reproduced by permission.)

There are many factors that contribute to the operational intelligibility of speech. Simpson et al. (1985) proposed a model for operation intelligibility (Figure 8.18).

It is important to note that intelligibility and human speech are not necessarily correlated. A radio announcer may sound natural despite a background of static noise, but may have low intelligibility. Conversely, synthesized speech warning messages in an aircraft cockpit may sound mechanical, but pilots consider them to be more intelligible than messages received over the aircraft radio.

There are still no comprehensive design guidelines in this area, but some general points can be made. Although the ergonomics/human factors literature includes reports of research that supports certain principles of speech design, this knowledge has not yet been formulated into design guidelines. Human factors methodology is sufficiently well-developed to permit comparison of task-specific speech systems experimentally, but the tools required for producing generic design guidelines for speech systems are not yet available. In the short term, simulation of speech system capabilities in conjunction with the development of improved system performance should prove a productive methodology to achieve these aims.

Speech generation algorithms seem to be more advanced than speech recognition algorithms. Reasonably intelligible text-to-speech from standard English spelling is now available commercially. The recognition counterpart, speech-to-text (not to be confused with speaker identification systems), will probably not be commercially available in the near future.

In the short term, the current recognition algorithms appear adequate for use in favorable environments characterized by low-to-moderate noise levels, and for applications that only require small vocabularies and that do not place the operator under stress. Great caution must be exercised in the use of current technology in stressful situations.

Speech generation algorithms, on the other hand, have demonstrated acceptable performance even under conditions of severe noise and high workload. This technology is sufficiently advanced to be applied appropriately, with careful attention being paid to the integration of ergonomics. Simpson and Williams (1980) have studied the use of synthesized voice for cockpit warnings. They concluded that voice warnings do not need to be preceded by an alerting tone, but can be recommended for practical use.

8.7 Design of Controls

This section covers the design of the more traditional controls and the design of specific controls for communication with computers. Traditional control panels have the advantage that they give feedback to the operator of the maneuvers, that have been carried out. Examples of the advantages and disadvantages of each type of control are presented with design recommendations for these controls.

Keyboards are the traditional input devices used in data processing (DP) applications to communicate with computers. In the process control situation, it can often be advantageous to use other types of controls such as multi-way joysticks or light pens. The advantages and disadvantages of each type of control are described.

8.8 Functional Aspects of Controls

The control device is the means by which information on a decision made by man is transferred to the machine. The decision may, for example, be taken on the basis of previously read information devices or on the basis of information from other sources, or from some form of cognitive process.

Functionally, controls may be divided into the following categories:

1. Switching on/off, start or stop.
2. Increase and reduction (quantitative changes).
3. Spatial control (e.g., continuous control upward, downward, to the left or right).
4. Symbol/character production (e.g., alphanumeric keyboards).
5. Special tasks (e.g., producing sound or speech).
6. Multi-function (e.g., controls for communicating with computers).

Examples of control type 1 include the starting or stopping of motors, or switching lamps on or off. Type 2 may consist of an accelerator pedal to increase and reduce the flow of fuel to the engine. Traditionally, the best-known example of spatial control (3) is the steering of a car. Examples of character production (4) include typewriting and telegraphy. Different forms of control 5 are used for the production of sound. Of special interest here are the machines that are beginning to appear for the production and transmission of speech (as discussed in the previous section).

Of particular importance in control room design are the types of control 6 used in conjunction with computers. Controls operated by hand are of particular use where great accuracy is required in the control movement. Hands are considerably better at carrying out precision movements than feet. Where a very high degree of accuracy of movement is required, it is best for only the fingers to be used.

Other alternatives are also possible for the design of special controls. For example, if one has a large crank, or two cranks coupled in parallel which have to be controlled by both hands, very fine control movements can be made.

Because the power available from the leg is considerably greater than that from the hands, foot controls are suitable for maneuvering over long periods or continuously. Foot controls are also valuable where very large pressures are needed, because the body weight can be added to the force of the strong leg musculature. It may also be necessary to use foot control devices when the hands are occupied in other tasks. However, it should be noted that it may be necessary to use hand controls when there is insufficient space to accommodate foot controls.

When designing traditional types of controls, it is possible to design them in such a way that they naturally represent the changes one wishes to bring about in the process. For example, a lever which is pushed forward may determine the forward direction of movement of a digger bucket. Or the flow in a pipe can be stopped by turning a knob that lies on a line drawn on the panel. In this way the design of the control increases the understanding of the current state of the process.

For communication with computers, the keyboard is often chosen for carrying out all the different control functions. Technically, it is often easy to connect a keyboard to a computer system. Other control devices also exist for communicating with computers, such as light pens. However, a particular failing of this type of multipurpose control device is that the control movements in themselves have no natural analogy with the changes that they aim to bring about in the process.

8.9 Anatomical and Anthropometric Aspects of Control Design

Some of the principal anatomical and anthropometric aspects will be considered. For detailed specifications, some excellent handbooks are recommended: for example, Morgan et al., 1963 and Grandjean (1988). One important limitation of the recommendations available today is that they are based on the Caucasian races. For the Japanese population, for example, the measurements must be adapted for their proportionally shorter leg lengths.

The following rules can be applied in the design of all types of control:

1. The maximum strength, speed, precision, or body movement required to operate a control must not exceed the ability of any possible operator.
2. The number of controls must be kept to a minimum.
3. Control movements that are natural for the operator are the best and the least tiring.
4. Control movements must be as short as possible, while still maintaining the requirement for "feel."
5. The controls must have enough resistance to prevent their activation by mistake. For controls that are only used occasionally and for short periods, the resistance should be about half the maximum strength of the operator. Controls that are used for longer periods must have a much lower resistance.
6. The control must be designed to cope with misuse. In panic or emergency situations, very great forces are often applied and the control must be able to withstand these.
7. The control must give feedback so that the operator knows when it has been activated, even when it has been done by mistake.
8. The control must be designed so that the hand/foot does not slide off or lose its grip.

Table 8.7 gives a summary of the areas of use and the design recommendations for different controls. The controls are discussed in more detail later. Figure 8.19a–c gives the optimal areas for the different controls.

Press-Buttons and Keys

These are suitable for starting and stopping and for switching on or off. This type of control is also suitable for foot control, where it should be operated by the "ball" of the foot. The following recommendations apply to both hand- and foot-operated controls:

1. The resistance of the push-button should increase gradually, and then disappear suddenly to indicate that the button has been activated.
2. The top of the button should have a high coefficient of friction to stop the fingers/feet from sliding off (see Figure 8.20). Where press-buttons are to be activated by the fingers, the concave form is preferable.
3. In order to indicate that the button has been activated, a sound should be emitted if the workplace has low light levels.

Table 8.8 gives detailed recommendations for push buttons.

Toggle Switches

Toggle switches can be used to show two or three positions. Where there are three positions, one should be up, the middle one straight out, and the other one downward. Toggle switches take up very little room. The following recommendations also apply to toggle switches:

1. A sound should be heard to indicate activation of the switch.
2. If a number of switches are used, they should be placed in a horizontal row. Vertical positioning requires more space in order to avoid accidental operation.

Table 8.9 gives detailed design recommendations for toggle switches.

TABLE 8.7 Recommendations for Controls

	Stepwise Adjustments				Continuous adjustments				
	Rotary switch	Hand push-button	Foot press-button	Toggle switch	Small wheel	Wheel	Crank	Pedal	Lever
Large forces can be developed					No	Yes	No	Yes	Yes
Time constraint for adjustment	Medium	Fast	Fast	Very fast					
Recommended number of positions	3–24	2	2	2–3					
Space requirements for placing and using	Medium	Small	Large	Small	Small to medium	Large	Medium to large	Large	Medium to large
Activation by accident	Small	Medium	Large	Medium	Medium	Large	Medium	Medium	Large
Limits of control movement	270°	3.2 × 38 mm	12.7 × 100 mm	120°	None	±60°	None	Small[1]	±45°
Legibility	Good	Acceptable	Bad	Acceptable	Bad	Acceptable	Acceptable	Bad	Good
Visual identification of control position	Acceptable	Bad[2]	Bad	Acceptable	Acceptable[3]	Bad to acceptable	Bad	Bad	Acceptable
Checking control position on panel together	Good	Bad	Bad	Good	Good	Bad	Bad	Bad	Good
Usability as part of a combination of controls		Good	Bad	Good	Good	Good	Bad	Bad	Good

[1] The exception is "cycle" pedals, which have no limit

[2] The exception is when the control is back-lit and the light goes off when the control is activated

[3] Only usable when control cannot be turned more than one revolution Round wheels/knobs must be marked

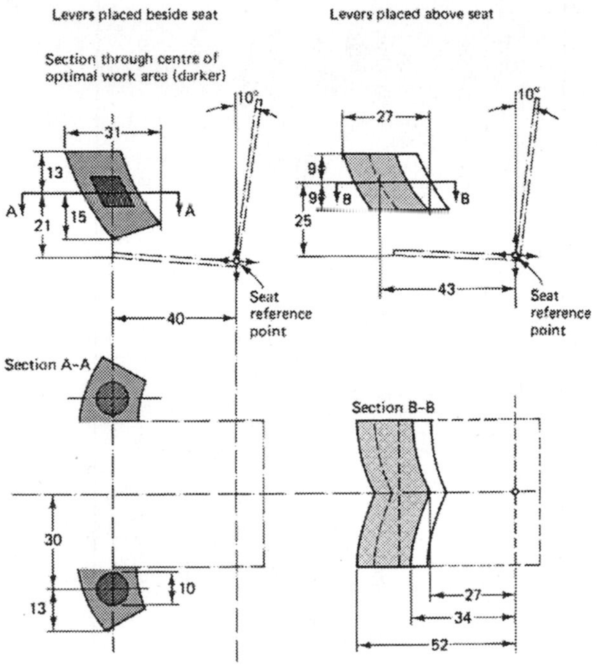

FIGURE 8.19 Preferred vertical surface areas and limits for different classes of manual controls. (Modified from McCormick, E.J. and Sanders, M.S. 1982, *Human Factors in Engineering and Design,* 5th edition (New York: McGraw-Hill.).

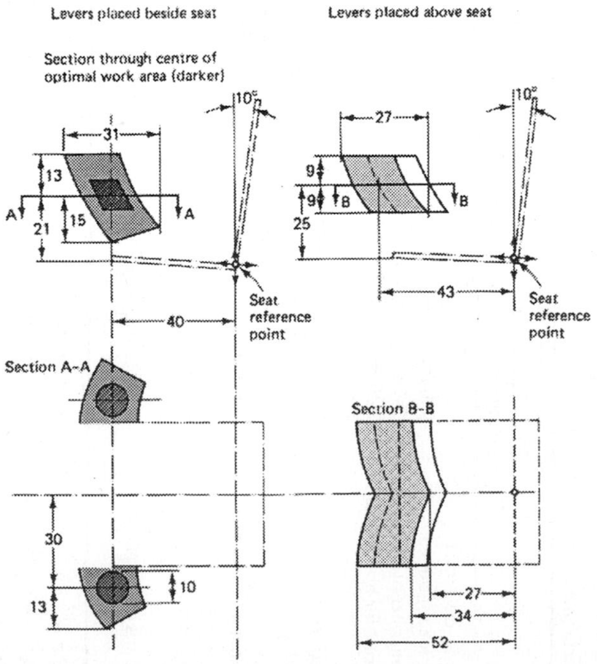

FIGURE 8.19(c) Optimal working areas for hands moving controls.

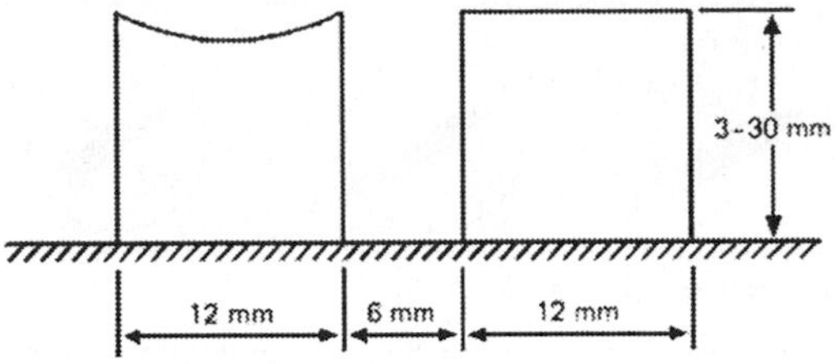

FIGURE 8.20 Press-buttons should be designed so that the fingers will not slide off them. The surface can be made concave or with some form of increased friction.

TABLE 8.8 Recommendations for Design of Press-Buttons

	Diameter (mm)		Travel (mm)		Resistance		Distance between push buttons (mm)	
	Min	Max	Min	Max	Min	Max	Min	Max
Finger								
One finger at random	13		3.2	38	280 g	1130 g	13	50
One finger in order							6.5	25
Different fingers in random order					140 g	560 g	13	13
Thumb, nail	19		3.2	38				
Foot								
Normal	13		13		1.8–4.5 kg			
Heavy shoes		25				9 kg		
Stretching ankle				64				
Leg movement				100				

TABLE 8.9 Recommendation for Design of Toggle Switches

Variables	Minimum	Maximum
Size (mm)		
Toggle switch		
for fingers	3.2	25
for hands	13	50
Travel (degrees)		
between positions	30°	
total travel		120°
Resistance (g)	280	1130
Number of positions	2	3
	Minimum	**Desirable**
Distance between control (mm)		
One finger — random	19	50
One finger — in order	13	25
Different fingers random or in order	16	19

Rotary Switches

These can be divided into two categories — cylindrical and winged. The primary difference between these is that the winged version has a pair of "wings" above the cylindrical part. The wings function both as a positional marker and as a finger grip. Rotary switches may have from 3 to 24 different positions. They require a relatively large amount of space, because the whole hand has to have room to turn around

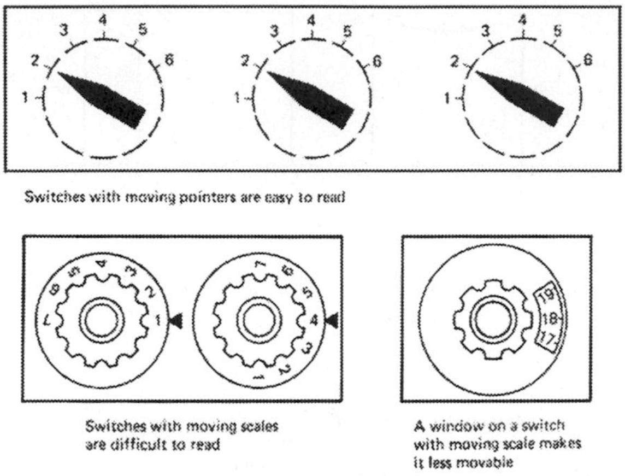

Switches with moving pointers are easy to read
Switches with moving scales are difficult to read
A window on a switch with moving scale makes it less movable

FIGURE 8.21 Design factors for rotary switches.

the switch. However, where multiple position switches are used, they take up less space than the number of push buttons or toggle switches required to fulfill the same function. Rotary switches can either have a fixed scale and moving pointer or a moving scale and fixed pointer. A variant on the moving scale is to have a window, which only shows a small part of the scale. Various models are shown in Figure 8.21.

The following recommendations apply to rotary switches:

1. In most applications, rotary switches should have a fixed scale and moving pointer.
2. There should be a detent in every position.
3. The turning resistance should steadily increase and then suddenly decrease as the next position is approached.
4. Cylindrical switches (knobs) should not be used if the resistance has to be high. In these cases, wing knobs are preferable.
5. Where only a few positions (2 to 5) are needed, they should be separated by 30 to 40 degrees.
6. Where fewer than 24 positions are used, the beginning and end of the scale should be separated by a greater space than between the different positions.
7. Where the workplace has low lighting levels, a sound should be made to denote that the switch has been activated. In these cases, there should also be a definite stop position at the beginning of the scale, so that the positions can be counted out.
8. The scale should always increase clockwise.
9. The hand should not shield the scale.
10. The surface of the switch should have a high coefficient of friction so that the hand does not slip.
11. The distance between panel and knob should be at least 3 mm.
12. The maximum amount of slope on the sides of the knob should be 5 degrees.

Levers

Levers are activated either by the whole hand or just by the fingers. In general, where fine control is needed, only the fingers should be used. The following recommendations apply to levers:

1. The maximum resistance (force) for push–pull movements with one hand, with the control placed centrally in front of the body, is between 12 and 22 kg, depending on how far from the body the control is positioned.
2. The maximum resistance for push–pull movements for two hands is double that for one hand.
3. The maximum resistance for one hand moving in the left–right direction is about 9 kg, and is considerably lower in the opposite direction.
4. The maximum resistance for two-handed movements in the left–right direction is about 13 kg.

5. The lever movement should never be greater than the arm's reach without moving the body.
6. Where precision is required, a supporting surface should be provided for the part of the body used; an elbow rest for large hand movements and a hand rest for finger movements.
7. When levers are used for step-wise control (e.g., gear levers), the distance between positions should be one third of the length of the lever.
8. Where the lever also acts as a visual indicator, the distance between positions can be reduced. The critical distance is then the operator's ability to see the markings.
9. The surface of the lever handle should have a high friction coefficient, so that the hand does not slip.

Cranks

These are suited to continuous control where there are high demands for speed. Cranks can be used for both fine and coarse control depending on the degree of gearing selected. The following recommendations apply to cranks:

1. Cranks are preferable to wheels where two or more revolutions are to be made.
2. For small cranks less than 8 cm in radius, the resistance should be at least 9 N and a maximum of 22 N when rapid movement is required.
3. Large cranks of 125–200 cm radius should have a resistance between 22 and 45 N.
4. Large cranks should be used when precision is required (accuracy between a half to one revolution), with the resistance between 10 and 35 N.
5. The handle should have a high surface friction to prevent the hand slipping.

Wheels

Wheels are used for two-handed operations. Identification of the position is very important if the wheel can be rotated through several revolutions. In addition, the following recommendations apply:

1. The turning angle should not exceed + 60 degrees from the zero position.
2. The diameter of the ring forming the outside of the wheel should be between 18 and 50 mm, and should increase as the size of the wheel increases.
3. The wheel should have a high surface friction so that the hand does not slip.

Table 8.10 shows the relative advantages of different forms of control device for computerized process systems for four common tasks.

8.10 Controls for Communication with Computers

The traditional controls in administrative computer systems are various different types of keyboard. There is often a numerical keyboard as well as the traditional typewriter keyboard. These types of keyboard have been tested over a long period of time and can be well specified. They are also thought to be well suited to most forms of administrative computer system.

For more specialized application, for example, computer systems for control and monitoring of process industries, the situation is often very different. The requirement for control devices is unique to each type of process industry, depending on the process to be controlled and the type of computer system installed. It is thus impossible to give any specific guidelines for the control devices on computerized control systems. However, some overall guidelines may be given. The advantages and disadvantages of different control devices for computerized systems will be examined. Many of the devices are new and have not yet been subjected to ergonomic evaluation. It is therefore difficult to give detailed guidelines for all controls. Finally, some more detailed design recommendations for the different types of keyboard will be presented.

TABLE 8.10 Relative Advantages of Different Forms of Control Device for Computerized Process Systems for Four Common Tasks

	Task			
	1	2	3	4
Control	Numeric data	Alphabetic data	Position cursor	Graphic information
Fixed-function keyboard	X	XX		
Variable-function keyboard	XX			
Lever			(X)	
Wheel			(X)	
Light pen			X	
Electronic data board				X
Touch screen			X	(X)
Mouse			X	(X)
Joystick (Track ball)			(X)	

(X), may sometimes be usable; X, usable; XX, very good; —, not recommended. Empty columns mean that the advantages and disadvantages of the application are not known.

Advantages and Disadvantages of Different Controls for Computers

The more traditional types of control can of course also be used for computerized systems. Controls, which have been produced specifically for communicating with computers, include:

1. Keyboard with predetermined functions for various keys
2. Keyboard with variable functions for the keys
3. Light pen
4. Touch screen
5. Electronic data board
6. Voice identification
7. Trackball and joystick (multi-position lever)
8. Mouse

Keyboards with Predetermined Functions for the Keys

Keyboards with predetermined functions for all keys normally have two main parts — an alphanumeric or a numeric part and a function key part. The traditional keyboard — numeric or alphanumeric — will be discussed later. The function key part has different keys for different predetermined tasks, such as starting, stopping, process a, b, c, etc. The keyboard works by the operator pressing the keys in a certain order, which he or she either remembers, or with the aid of some form of crib-sheet. The sequence in which the keys are to be pressed is thus often predetermined both by the system and by the design of the keyboard.

This type of keyboard is characterized by the need for a large number of keys, usually one per function. Where there are many functions and several subfunctions within every main function, problems arise with grouping the keys in the proper way and in positioning the keys in a mutually logical way, which is consistent in terms of movements.

It is unusual to be successful with this at the first attempt; the keyboard will need to be redesigned when it has been operational for long enough for the designer to build up enough experience with it to determine its optimal design. Making changes to the keyboard are often costly, but if it is not redesigned at a later stage, it means that large and frequent arm movements become tiresome and time consuming. The advantage of this type of keyboard is that it needs relatively little computer programming and, to a certain extent, a standard board can be used, at least for the alphanumeric part.

The alternative to having a large number of function keys is to have just a few, and to use particular codes instead, which can be entered numerically or alphanumerically. This type of keyboard is best when the operator is spending a large part of his working time at the keyboard. However, this is relatively uncommon in process industries.

Keyboards with Variable Functions for the Keys

Keyboards on which there is a variety of functions for each of the different keys are relatively uncommon but often exist as part of the more traditional keyboard (e.g., the top row of keys on the keyboard). Keyboards with a variety of functions per key are often particularly useful in process industries. A common form is to have a row of unmarked keys under the monitor screen, and to have squares representing the different keys directly above them on the screen. Depending on the picture being shown on the screen, text appears in different windows showing the functions that the keys have for each frame. There are more advanced systems for this type of keyboard in which there are several rows of unmarked keys and parallel pictures are projected down from the screen onto the keyboard using an arrangement of mirrors in order to show the current function of the keys. Because considerably fewer keys are used on this type of keyboard, fewer hand and arm movements are required by the operator, and the risk of errors occurring is also reduced.

Another application for this type of keyboard is to build lights into the keys. The relevant keys light up for each particular function. The lights in the keys are lit or extinguished when particular keys are pressed, depending on the sequence of operations required. In this way the operator is guided through the correct operation sequence.

Nonilluminated keys are then disconnected from the system. The risk of errors occurring with this type of system is very small, and work on this keyboard is also faster, particularly if the operator is not accustomed to the work. It is important, however, that if the lamps in the keys break, a warning signal must be produced.

There are also applications where keys can be pressed with different pressures. A light pressure on the key causes the function associated with that key to be written out on the screen, and the action is taken if the key is pressed harder. If it is fully depressed a signal is sent to the computer dictating changes to be made to the process. The operator can also receive new information on the screen that informs him which new keys can be used.

Depending on its design, the keyboard can be preprogrammed to lead the operator naturally through the work. This type of programming of the keyboard functions may be an advantage for especially important types of operations, in which errors could have serious consequences. A major disadvantage from the operators' perspective is that they may feel their work is being too highly controlled. Another disadvantage with this type of keyboard is that it requires a lot of programming, and this takes up a large part of the computer's capacity. An advantage is that the hardware does not need to be changed (rebuilding or extending the keyboard, etc.) to any great extent even if a major change is to be made in the function of the control. In other words, this form of control is very flexible.

Light Pens

The light pen consists of a photocell that senses the light radiated from the phosphor on a CRT screen. The light pen reacts every time a pixel on the screen is lit up by the electron ray within the tube. The signal passes from the light pen to the computer, which at the same time receives information on where and when the spot passes different places on the screen. In this way, the computer can identify where the pen is on the screen.

The light pen can be used for pointing to the parts of the screen one wishes to know more about. It can also be used to activate different functions. If, for example, one pointed to a valve and at the same time pressed a button on the side of the pen, this may cause the valve to close. The light pen is suitable for moving cursors on a screen. However, it is difficult to see any operational advantages of light pens over other controls.

If a light pen is used over a long period, it is necessary to have a specially designed armrest to prevent discomfort. The light pen has to come close up against the screen, which means that it is impossible to have any form of reflection shield or filter on the screen, and this can give rise to visual problems. Positioning of the light pen must be exact, which makes considerable demands on vision and also contributes to bad working posture in many instances.

Touch Screens

Touch screens involve moving the finger, a pen, a pointer or some other object within an active matrix placed over the screen This active matrix may be designed in several ways. It could be composed of a thin metal net, for example, on which an electrical circuit is made when it is touched. Electrical bridges and infrared beams can also be used to determine touch on the screen. Another type of touch screen is based on the use of a transparent material, which senses the pressure of the touch on the screen. Special measurement bridges are used to determine how the pressure field is distributed over the screen, and the position of the touch is deduced from this.

Functionally, the touch screen is very similar to the light pen and has similar advantages and disadvantages, although an additional disadvantage is that the screen becomes dirty. An advantage is that it is sometimes faster to point with the finger than with a light pen, however, the technical reliability of the touch screen is usually considered to be lower than that of the light pen.

There is limited research to show that the touch screen can be used for numeric and alphabetical data where there are a limited number of functions, as may be the case in the process industries. If it is necessary to send a large number of different types of words and information to the computer, the traditional keyboard is preferable.

Electronic Data Boards

Electronic data boards consist of a rectangular plate, which represents the surface of the screen. Some form of electric field is created over the plate. When a sensor is run over the board's surface, it "senses" its position on the board. One common form of board is placed directly onto the screen, and in this case, functions very like a light pen or touch screen. Another form of board is placed beside the screen, and one can work with a transparency of a picture. One of the advantages of the electronic data board is that one can very quickly make drawings or change them.

Voice Identification Instruments

As discussed earlier, instruments for voice recognition and identification have been connected to computers for a long time. Recognition of speech, however, is much more difficult. There are many apparent advantages of this type of device. It is, for example, very flexible and requires no special motor skill. The problems are that the equipment available today requires specially trained operators who have to use a very limited vocabulary and have to speak at a particular speed. In the future, however, this type of control may well be more widely applicable. Its present applications are primarily for different forms of emergency and alarm situations. There are many interesting development possibilities for this form of control device within the process industries. Development should progress in such a way that natural words and sentences will be able to be used directly. In an emergency situation, which requires immediate response, the operator should be able to shout "Stop" to control the process if he or she considers this the appropriate action for the situation.

The Trackball

The trackball is a mounted sphere that can be rotated in all directions and can be placed on a table or special fixture. The ball is usually used for moving the cursor on a screen. The cursor moves a certain distance (x/y directions) or with a speed proportional to the movement of the ball.

The Joystick

The joystick, a lever movable in all directions, has a function similar to the trackball.

The Mouse

The mouse is a small device with wheels or a ball mounted on the underside. If the mouse is moved to the left or right, this represents a corresponding movement on the screen. The mouse is especially suited for moving the cursor and for transferring graphic information.

There is no conclusive evidence to produce recommendations for the use of the trackball, mouse, or joystick. In practice, most people seem to prefer the mouse if one has access to a free table surface, otherwise the trackball is generally preferred.

Other Traditional Computer Controls

There are also many traditional types of controls, for example small wheels, levers, or joysticks (control levers that can be moved in two or three dimensions). These more traditional types of control devices are usually used for moving the cursor on the screen. However, in the future there will be a need for new types of controls that better suit the computer applications within the process industries. Traditional control panels, as simulated on a CRT, can sometimes be an improvement on a keyboard, giving direct feedback of different control settings.

The Keyboard

The keyboard is still the most common computer input device. The design of the keyboard has a significant effect on the operator's performance in terms of speed and accuracy. The most common keyboard layout is the QWERTY layout. Where two hands are used on the keyboard, 57% of the workload is on the left hand, even though 80% of the population is right-handed. This is advantageous for the type of job in which the right hand alternates between handwriting and typing. It is also important to have one standard keyboard layout. Although the QWERTY layout is not the most efficient (it is said to have been designed for slowness, so that early mechanical typewriters did not become overloaded), it is now the best compromise because it has become the standard keyboard.

Keyboards are usually designed so that the alphanumeric section is in the center, with the cursor, editing keys and numeric keypad to the right. Function keys may be placed anywhere on the keyboard, but in order to give an aesthetically pleasing design they are often placed on the left-hand side. Lateral hand movements also require less energy than longitudinal (front to back) ones.

There are no specific recommendations for keyboard layout, as their design is extremely sensitive to the task being carried out. However, a degree of flexibility must be incorporated in their design to cater to all variations in user requirements. One solution to this would be to develop a modular keyboard consisting of several units. Each unit would be made up of a different set of keys. The units could be arranged in the desired layout based on the results of the task analysis. However, care must be taken in using a flexible keyboard configuration due to the risk of a negative transfer of training. For example, if an operator carries out a number of different tasks and different keyboards are used for each of the different tasks, then high error rates must be expected.

Chord keyboards are a combination of a keyboard and a coding system. In a similar way to keys on a piano, one can press several keys at the same time. The advantage of this type of keyboard is that key-pressing speed compared with a standard typewriter is considerably greater than 50% faster. There are, however, no special design recommendations for this type of keyboard. In general, it may be said that this type of keyboard needs further study before any firm recommendations can be given regarding suitable areas of use and suitable design.

The numeric keyboard appears in two different designs. The accepted layout is a $3 \times 3 + 1$ key set, but there are two alternatives within this. Adding machines have the 7, 8, and 9 keys on the top row, while push-button telephones have the 1, 2, and 3 keys at the top (see Figure 8.22). In the future, all telephones will use the 123 keypad. Once this happens, it will be recommended that all numeric keyboards are of this design. Uniformity is important, and the user should not have to switch from one keyboard design to the other while working.

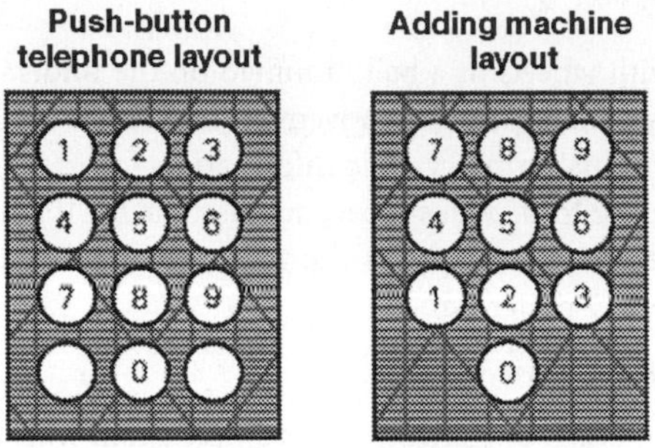

FIGURE 8.22 Layout of numerical keyboards.

The height of the keyboard is largely determined by its physical design, for example, electrical contacts and activating mechanism. Thicker keyboards (greater than 30 mm thick) should be lowered into the table surface to ensure a correct user posture. This, unfortunately, does not allow for flexible workplace design. Ideally, keyboards should be as thin as possible (less than 30 mm thick from the desk surface to the top of the second key row (ASDF …) and not need to be lowered into the surface. Recent product development, particularly by ERGOLAB in Sweden, has resulted in keyboards tending to become thinner, and this allows for a more flexible workplace design.

Keyboards can be stepped, sloped, or dished. There is no evidence on the relative advantages of any of these profiles. The most important factor is for the keyboard to be able to be angled between 0 to 15 degrees up at the back and, if the keyboard operator is standing, it is advantageous if it can be raised at the front from 0 to 30 degrees.

The size of the key tops is a compromise between producing enough space for the finger on the key, while at the same time keeping the total size of the keyboard as small as possible. Key tops should be square and 12 to 15 mm in size. This size is quite sufficient for touch typing, but in cases where keyboards are used for other tasks, for example on the shop floor, key sizes can be larger. The spacing of keys is standardized 19.5 mm between key top centers. This is within the ergonomic recommendations of between 18 to 20 mm. The force required for key displacement should be the same on all keys.

For skilled users, the actuating force should be 0.25 to 0.5 N, and the key displacement (travel) 0.8 to 10 mm (from rest to activation of system). For unskilled users, the force should be 1 to 2 N and the displacement 2 to 5 mm. The user requires feedback to indicate that the system has accepted the keystroke. This is an important keyboard characteristic, although the exact requirements vary according to the individual levels of user skill.

In normal typing and other key-pressing tasks, there is kinesthetic (muscle) and tactile (touch) feedback from the actual depression of the key, auditory feedback from the key press and/or activation of the print mechanism, and visual feedback from the keyboard or from the output display. For skilled operators, feedback from the keyboard (sound and pressure change) is of little importance. When learning, and for unskilled operators, this feedback is important. The operator should be able to remove the acoustic feedback.

The color of the keys is not generally regarded as important. A dark keyboard with light lettering is preferable when used in conjunction with light-on-dark image displays, and care should be taken not to cause any distracting reflections on the screen by light key colors. Matte finishes should be used where possible. The recommended reflectance factors for keyboards used in conjunction with negative-image (light on a dark background) VDUs are:

1. The lettering on the keys should be light and clearly defined. Its minimum height should be 25 mm in good lighting conditions. In the case of function keys, certain abbreviations may be required. These must follow a clear, logical pattern, easily identifiable by the operator.
2. Keyboards used with positive image (black on white) VDUs should be lighter in color with darker text. All keytop surfaces should have a matte finish.

Care should be taken when using colors to code various function keys. Attention should not be drawn to a red key or a group of red keys if their importance in the system is minimal. These principles concerning color may also be applied to any information lights found on the keyboard. There are international color standards (IEC 1975) which can apply to both keys and information lights. These standards should be adhered to whenever possible.

References

Berns, T. and Herring, V., 1985, Positive vs. negative image polarity of visual display screens, *ERGOLAB report*, 85:06 (Stockholm: ERGOLAB).

Cakir, A., Hart, D., and Stewart, T., 1979, *The VDT Manual*, (Damstadt: IFRA).

Farrell, R. J. and Booth, J. M., 1975, *Design Handbook for Imagery Interpretation Equipment*, (Seattle: Boeing Aerospace Company).

Grandjean, E., 1988, *Fitting the Task to the Man*, 4th edition (London: Taylor & Francis Ltd.).

IEC, 1975, International Standard, *Colours of Indicator Lights and Push Buttons*, Publ. no. 73 (Paris: International Electrotechnical Commission).

Ivergård, T., 1982, *Information Ergonomics*, (Lund, Sweden: Studentlitteratur).

Kinney, J. S., 1979, The use of color in wide-angle displays, *Proceedings of the Society for Information Display*, **20**, 33-40.

McCormick, E. J. and Sanders, M. S., 1982, *Human Factors in Engineering and Design*, 5th edition (New York: McGraw-Hill).

Morgan, C., Cook, J., Chapanis, A., and Lund, M., 1963, *Human Engineering Guide to Equipment Design*, (New York: McGraw-Hill Book Comp. Inc.).

Murrell, K. F. H., 1958, *Fitting the Job to the Worker. A Study of American and European Research into Working Conditions in Industrial Situations*, (Paris: Organisation for European Economic Co-operation–EPA).

Seminara, J. L., 1980, *Human Factors Considerations for Advanced Control Board Design*, Vol. 4 of *Human Factors Methods for Nuclear Control Room Design* (Palo Alto, CA: Electric Power Research Institute).

Silverstein, L. D. and Maryfield, R. M., 1981, *Color Selection and Verification Testing for Airborne Color CRT Displays*, Proceedings of the 5th Advanced Aircrew Display Symposium, Sept. (Naval Air Test Center).

Silverstein, L., 1982, *Human Factors for Color CRT Displays*, (San Diego, CA: The Society for Information Display).

Simpson, C. A. and Williams, D., 1980, Response time effects of alerting tone and semantic context for synthesized voice cockpit warnings, *Human Factors*, **22**, 319-330.

Simpson, C. A., McCouley, M. E., Poland, E. F., Ruth, J. C., and Williges, B. H., 1985, Systems design for speech recognition and generation, *Human Factors*, **27**, 115-141.

Stark, Lawrence, 1983, Personal Communication (reported in "Study Visit" US-1983 HF — Research in Automation," *Arbetslivsforden*, Stockholm).

Stewart, T. F. M., 1976, Displays & the software interface, *Applied Ergonomics*, **9**, 137-146.

Teichner, W. H., 1979, Color and information coding, *Proceedings of the Society for Information Display*, **20**, 3-9.

Wyszecki, G. and Stiles, W. S., 1967, *Colorscience — Concepts and Methods, Quantitative Data and Formulas*, (New York: John Wiley & Sons Inc.).

9

Engineering Anthropometry

Karl H. E. Kroemer
Virginia Tech

9.1 Overview

People come in a variety of sizes, and their bodies are not assembled in the same proportions. Thus, fitting equipment to suit the body requires careful consideration; design for the statistical "average" will not do. Instead, for each body segment to be fitted, the dimension(s) critical for design must be determined. A minimal or a maximal value, or a range may be critical. Often, a series of such decisions must be made to accommodate body segments or the whole body by clothing, workspace, and equipment.

The following text uses this procedure. It describes the steps involved, provides statistical tools, and supplies anthropometric data. For more detail see Kroemer, Kroemer, and Kroemer-Elbert (1997).

9.2 Terminology

Special terms often used in anthropometry are listed in Table 9.1. Together with the reference planes shown in Figure 9.1, they describe major aspects of anthropometric information used by designers and engineers.

9.3 Designing to Fit the Body

While all humans have heads and trunks, arms and legs, the body parts come in various sizes and are assembled in different proportions. The science of measuring human bodies is called anthropometry. The results of anthropometric surveys are described in statistical terms.

0-8493-2641-9/99/$0.00+$.50

TABLE 9.1 Terms Used in Engineering Anthropometry

Anthropometry —	measure of the human body. The term is derived from the Greek words "anthropos," human and "metrein," measure.
Height —	straight-line, point-to-point vertical measurement.
Breadth —	straight-line, point-to-point horizontal measurement running across the body or a segment.
Depth —	straight-line, point-to-point horizontal measurement running fore-aft the body.
Distance —	straight-line, point-to-point measurement between landmarks on the body.
Curvature —	point-to-point measurement following a contour; this measurement is neither closed nor usually circular.
Circumference —	closed measurement that follows a body contour; hence this measurement usually is not circular.
Reach —	point-to-point measurement following the long axis of the arm or leg.
	Terms Related to Body Reference Planes (see Figure 9.1)
Medial or *mid-sagittal* —	cutting body into left and right halves.
Frontal or *coronal* —	cutting body into fore-aft (anterior-posterior) sections.
Transverse —	cutting body into upper/lower (superior-inferior) sections.
Sagittal —	parallel to medial (occasionally used like *medial*).
	Anatomical Terms Related to Position
Anterior —	in front of, toward the front of the body.
Posterior —	behind, toward the back of the body.
Ventral —	toward the abdomen (occasionally used like *anterior*)
Dorsal —	toward the back or spine.
Medial —	near or toward the middle.
Lateral —	to the side, away from the middle.
Superior —	above, toward the top.
Inferior —	below, toward the bottom.
*Proximal** —	toward or near the center of the body.
*Distal** —	away from the center of the body.
Superficial —	on or near the surface.
Deep —	away from or below the surface.

* *Proximal* and *distal* usually refer to limbs, where the point of reference is the attachment to the body.

Most body data appear, statistically speaking, in a normal (Gaussian) distribution. Such distribution of data can be described by using the statistical descriptors *mean* (same as *average*), *standard deviation,* and *range,* if the sample size is large enough (see below for more detail). Misunderstanding and misuse have led to the false idea that one could "design for the average"; yet, the mean value is larger than half the data, and smaller than the other half. Consequently, the "average" does not describe the ranges of different statures, arm lengths, or hip breadths. Furthermore, one is unlikely ever to encounter a person who displays mean values in several, many, or all dimensions. The mythical "average person" is nothing but a statistical phantom.

> "A pioneer in the field of statistics, Sir Francis Galton (1822–1911) wrote years ago that 'it is difficult to understand why statisticians commonly limit their interests to averages. Their souls seem as dull to the charm of variety as that of a native of one of our flat English counties whose retrospect of Switzerland was that, if its mountains would be thrown into its lakes, two nuisances could be got rid at once.' Basic to virtually all design problems is the fact that mankind is far more like Switzerland than a flat English county, and that, whatever the charms of variety may be, we need statistics to quantify this variety." (Edmund Churchill on page IX-5 of NASA/Webb, 1978.)

Using Percentiles

Most body dimensions are normally distributed. A plot of their individual measures falls inside the well-known bell curve, shown in Figure 9.2. Only a few persons are very short, or very tall, but many cluster

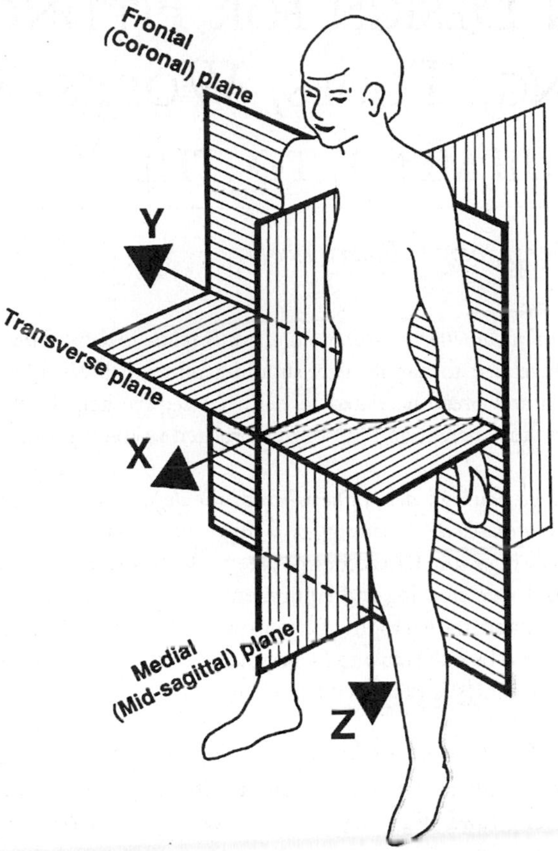

FIGURE 9.1 Reference planes used in conventional anthropometry.

around the center of the distribution (the mean or average). Figure 9.2 shows an approximate distribution of the stature of male Americans; only 2.5% are shorter than approximately 1,620 mm, and another 2.5% are taller than 1,880 mm. In other words: about 95% of all men are in the height range of 1,620 to 1,880 mm, because the 2.5th percentile value is at 1,620 mm and the 97.5th percentile is at 1,880 mm. The 50th percentile is at 1,750 mm. (In a normal — Gaussian — data distribution, mean (m), average, median, and mode coincide with the 50th percentile. The standard deviation (S) describes the peakedness or flatness of the data set. These statistical descriptors are discussed in some detail later in this chapter under "Estimation by Probability Statistics.")

There are two ways to determine given percentile values. One is simply to take a distribution of data, such as shown in Figure 9.2, and determine from the graph (measure, count, or estimate) critical percentile values. This works whether the distribution is normal, skewed, binomial, or in any other form. Fortunately, most anthropometric data are normally distributed, which allows the second, even easier (and usually more exact) approach: to calculate percentile values. This involves the standard deviation, S. If the distribution is flat (the data are widely scattered), the value of S is larger than when the data cluster close to the mean, m.

To calculate a percentile value, you simply multiply the standard deviation S by a factor k, selected from Table 9.2. Then you add the product to the mean m:

$$p = m + k * S \quad (1)$$

Steps in Design for Fitting Clothing, Tools, Workstations, and Equipment to the Body

(Kroemer, Kroemer, Kroemer-Elbert 1994)

Step 1: *Select those anthropometric measures that directly relate to defined design dimensions.* Examples are: hand length related to handle size; shoulder and hip breadth related to escape-hatch diameter; head length and breadth related to helmet size; eye height related to the heights of windows and displays; knee height and hip breadth related to the leg room in a console.

Step 2: *For each of these pairings, determine whether the design must fit only one given percentile (minimal or maximal) of the body dimension, or a range along that body dimension.* Examples are: the escape hatch must be big enough to accommodate the largest extreme value of shoulder breadth and hip breadth, considering clothing and equipment worn; the handle size of pliers is probably selected to fit a smallish hand; the leg room of a console must accommodate the tallest knee heights; the height of a seat should be adjustable to fit persons with short and with long lower legs. (How to use and calculate percentiles is explained below.)

Step 3: *Combine all selected design values in a careful drawing, mock-up, or computer model to ascertain that they are compatible.* For example, the required leg-room clearance height needed for sitting persons with long lower legs may be very close to the height of the working surface determined from elbow height.

Step 4: *Determine whether one design will fit all users.* If not, several sizes or adjustment must be provided to fit all users. Examples are one extra-large bed size fits all sleepers; gloves and shoes must come in different sizes; seat heights are adjustable.

If the desired percentile is above the 50th percentile, the factor k has a positive sign and the product $k * S$ is added to the mean m; if the p-value is below average, k is negative and the product $k * S$ is subtracted from the mean. Examples:

1st percentile is at $m–kS$	with $k = -2.33$ (see Table 9.2)
2nd percentile is at $m–kS$	with $k = -2.05$
2.5th percentile is at $m–kS$	with $k = -1.96$
5th percentile is at $m–kS$	with $k = -1.64$
10th percentile is at $m–kS$	with $k = -1.28$
50th percentile is at m	with $k = 0$
60th percentile is at $m+kS$	with $k = 1.28$
95th percentile is at $m+kS$	with $k = 1.64$

Percentiles serve the designer in several ways. First, they help to establish the portion of a user population that will be included in (or excluded from) a specific design solution. For example, a certain product may need to fit everybody who is taller than 5th percentile and smaller than the 60th percentile

TABLE 9.2 Percentile Values and Associated *k* Factors

BELOW MEAN				ABOVE MEAN			
percentile	factor k	percentile	factor k	percentile	factor k	percentile	factor k
0.001	−4.25	**25**	**−0.67**	**50**	**0**	76	0.71
0.01	−3.72	26	−0.64	51	0.03	77	0.74
0.1	−3.09	27	−0.61	52	0.05	78	0.77
0.5	−2.58	28	−0.58	53	0.08	79	0.81
1	−2.33	29	−0.55	54	0.10	**80**	**0.84**
2	−2.05	**30**	**−0.52**	**55**	**0.13**	81	0.88
2.5	−1.96	31	−0.50	56	0.15	82	0.92
3	−1.88	32	−0.47	57	0.18	83	0.95
4	−1.75	33	−0.44	58	0.20	84	0.99
5	**−1.64**	34	−0.41	59	0.23	**85**	**1.04**
6	−1.55	**35**	**−0.39**	**60**	**0.25**	86	1.08
7	−1.48	36	−0.36	61	0.28	87	1.13
8	−1.41	37	−0.33	62	0.31	88	1.18
9	−1.34	38	−0.31	63	0.33	89	1.23
10	**−1.28**	39	−0.28	64	0.36	**90**	**1.28**
11	−1.23	**40**	**−0.25**	**65**	**0.39**	91	1.34
12	−1.18	41	−0.23	66	0.41	92	1.41
13	−1.13	42	−0.20	67	0.44	93	1.48
14	−1.08	43	−0.18	68	0.47	94	1.55
15	**−1.04**	44	−0.15	69	0.50	**95**	**1.64**
16	−0.99	**45**	**−0.13**	**70**	**0.52**	96	1.75
17	−0.95	46	−0.10	71	0.55	97	1.88
18	−0.92	47	−0.08	72	0.58	98	2.05
19	−0.88	48	−0.05	73	0.61	99	2.33
20	**−0.84**	49	−0.03	74	0.64	99.5	2.58
21	−0.81	**50**	**0**	**75**	**0.67**	99.9	3.09
22	−0.77					99.99	3.72
23	−0.74					99.999	4.26
24	−0.71						

Any percentile value p can be calculated from the mean m and the standard deviation s (normal distribution assumed) by $p = m + ks$.

in hand size or arm reach. Thus, only the 5% having values smaller than the 5th percentile, and the 40% having values larger than the 60th percentile, will not be fitted, while 55% (60% – 5%) of all users will be accommodated.

Second, percentiles are easily used to select subjects for fit tests. For example, if the product needs to be tested, persons having 5th or 60th percentile values in the critical dimensions can be employed for use tests.

Third, any body dimension, design value, or score of a subject can be exactly located. For example, a certain foot length can be described as a given percentile value of that dimension, or a certain seat height can be described as fitting a certain percentile value of lower leg length (e.g., popliteal height), or a test score can be described as falling at a certain percentile value.

Fourth, the use of percentiles helps in the selection of persons to use a given product. For example, if a cockpit of an airplane is designed to fit the 5th to 95th percentiles, one can select cockpit crews whose body measures are at or between the 5th and 95th percentiles in the critical design dimensions.

To Determine a Single (Distinct) Percentile Point

a. Select the desired percentile value
b. Determine the associated k value from Table 9.2
c. Calculate the p value from $p = m$ plus k times S (Note that k, and hence the product, may be negative.)

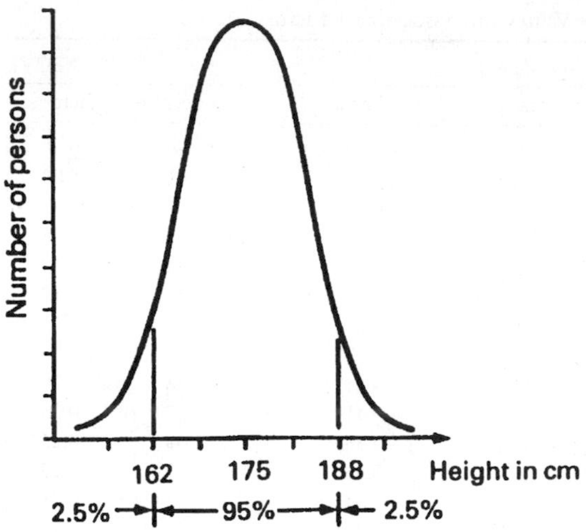

FIGURE 9.2 Distribution of body height (stature) in male Americans, as shown by Kroemer, Kroemer and Kroemer-Elbert (1997) and by Kroemer and Grandjean (1997). About 95% of all men are between 162 and 188 cm tall; about 2.5% are either shorter or taller.

To Determine a Range

1a. Select upper percentile p_{max}
1b. Find related k_{max} value in Table 9.2
1c. Calculate upper percentile value $p_{max} = m + k_{max} * S$.
2a. Select lower percentile p_{min}
(Note that the two percentile values need not be at the same distance from the 50th *p*, i.e., the range does not have to be "symmetrical to the mean.")
2b. Find related k_{min} value in Table 9.2
2c. Calculate lower percentile value $p_{min} = m + k_{min}*S$
3. Determine range $R = p_{max} - p_{min}$.

To Determine Tariffs

A distribution of body dimensions is often divided into certain sections, such as in establishing clothing tariffs. An example is the use of neck circumference to establish selected collar sizes for men's shirts. The first step is to establish the ranges (see above) which shall be covered by the tariff sections. The second step is to associate other body dimensions with the primary one, such as chest circumference, or sleeve length, with collar (neck) circumference. This can become a rather complex procedure, because the combination of body dimensions (and their derived equipment dimensions), depends on correlations among these dimensions, as discussed below. For more information, see McConville's Chapter VIII in NASA/Webb (1978) and the 1975 book by Roebuck, Kroemer, and Thompson.

9.4 Body Postures

To standardize measurements, the body is put into defined static postures:

Standing: the instruction is "stand erect; heels together; rears of heels, buttocks, and shoulders touching a vertical wall; head erect; look straight ahead; arms hang straight down (or upper arms hang, forearms are horizontal and extended forward); fingers extended."

Sitting: on a plane, horizontal, hard surface adjusted in height so that the thighs are horizontal; "sit with lower legs vertical, feet flat on the floor; trunk and head erect; look straight ahead; arms hang straight down (or upper arms hang, forearms horizontal and extended forward); fingers extended."

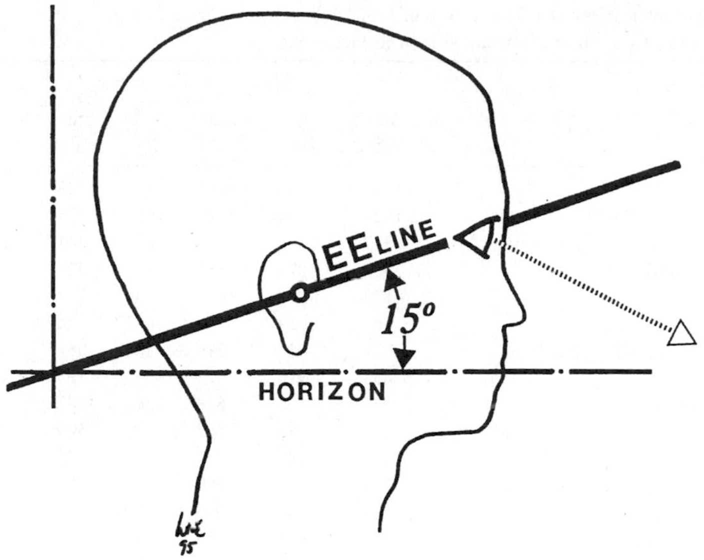

FIGURE 9.3 The ear–eye line serves as a reference to describe head posture and the line-of-sight angle.

Head (including the neck) is held erect (or "upright") when, *in the front view*, the pupils are aligned horizontally, and, *in the side view* the ear–eye line is angled about 15 degrees above the horizon (see Figure 9.3). The ear–eye (EE) line runs through the ear hole and the outside juncture of the eyelids.

People do not stand or sit in these postures naturally. Thus, the dimensions taken on the body in the standardized postures must be converted to reflect real postures. The postures assumed at work or leisure can vary greatly. Therefore, it is impossible to give "conversion factors" that apply to all conditions. The designer has to estimate the corrections that reflect the anticipated postures. Some general guidelines are presented in Table 9.3.

TABLE 9.3 Guidelines for the Conversion of Standard Measuring Postures to Real Work Conditions

Slumped standing or sitting:	deduct 5–10% from appropriate height measurements
Relaxed trunk:	add 5–10% to trunk circumferences and depths
Wearing shoes:	add approximately 25 mm to standing and sitting heights; more for "high heels"
Wearing light clothing:	add about 5% to appropriate dimensions
Wearing heavy clothing:	add 15% or more to appropriate dimensions (Note that mobility may be strongly reduced by heavy clothing.)
Extended reaches:	add 10% or more for strong motions of the trunk
Use of hand tools:	Center of handle is at about 40% hand length, measured from the wrist
Forward bent head (and neck) posture:	ear–eye line close to horizontal
Comfortable seat height:	add or subtract up to 10% to or from standard seat height

Adapted from Kroemer, Kroemer, and Kroemer-Elbert (1997).

9.5 Available Body Size Data

Recent information on body sizes of North American adults, females and males, is presented in Table 9.4. These data were derived from measurements of U.S. Army personnel taken in 1988. In spite of the military sampling bias due to selection, these data are the best available for the total North American adult population. The main reservation is with respect to body weight, which is more variable in the civilian population than in the military. Head, foot, and hand sizes should not differ appreciably between soldiers and civilians. Descriptive data of British, German, and Japanese adults are presented in Tables 9.5, 9.6, and 9.7. Each of these lists contains data, as and if available, measured in similar fashion, as illustrated in Figure 9.4.

TABLE 9.4 Anthropometric Measured Data in mm of U.S. Adults, 19 to 60 Years of Age. The reference numbers of the dimensions are shown in Figure 9.4.

	Men				Women			
Dimension	5th %ile	Mean	95th %ile	SD	5th %ile	Mean	95th %ile	SD
1. Stature [99]	1647	1756	1867	67	1528	1629	1737	64
2. Eye height, standing [D19]	1528	1634	1743	66	1415	1516	1621	63
3. Shoulder height (acromion), standing [2]	1342	1443	1546	62	1241	1334	1432	58
4. Elbow height, standing [D16]	995	1073	1153	48	926	998	1074	45
5. Hip height (trochanter) [107]	853	928	1009	48	789	862	938	45
6. Knuckle height, standing	na	na	na	na	na	na	na	na
7. Fingertip height, standing [D13]	591	653	716	40	551	610	670	36
8. Sitting height [93]	855	914	972	36	795	852	910	35
9. Sitting eye height [49]	735	792	848	34	685	739	794	33
10. Sitting shoulder height (acromion) [3]	549	598	646	30	509	556	604	29
11. Sitting elbow height [48]	184	231	274	27	176	221	264	27
12. Sitting thigh height (clearance) [104]	149	168	190	13	140	160	180	12
13. Sitting knee height [73]	514	559	606	28	474	515	560	26
14. Sitting popliteal height [86]	395	434	476	25	351	389	429	24
15. Shoulder-elbow length [91]	340	369	399	18	308	336	365	17
16. Elbow-fingertip length [54]	448	484	524	23	406	443	483	23
17. Overhead grip reach, sitting [D45]	1221	1310	1401	55	1127	1212	1296	51
18. Overhead grip reach, standing [D42]	1958	2107	2260	92	1808	1947	2094	87
19. Forward grip reach [D21]	693	751	813	37	632	686	744	34
20. Arm length, vertical [D3]	729	790	856	39	662	724	788	38
21. Downward grip reach [D43]	612	666	722	33	557	700	664	33
22. Chest depth [36]	210	243	280	22	209	239	279	21
23. Abdominal depth, sitting [1]	199	236	291	28	185	219	271	26
24. Buttock-knee depth, sitting [26]	569	616	667	30	542	589	640	30
25. Buttock-popliteal depth, sitting [27]	458	500	546	27	440	482	528	27
26. Shoulder breadth (biacromial) [10]	367	397	426	18	333	363	391	17
27. Shoulder breadth (bideltoid) [12]	450	492	535	26	397	433	472	23
28. Hip breadth, sitting [66]	329	367	412	25	343	385	432	27
29. Span [98]	1693	1823	1960	82	1542	1672	1809	81
30. Elbow span	na	na	na	na	na	na	na	na
31. Head length [62]	185	197	209	7	176	187	198	6
32. Head breadth [60]	143	152	161	5	137	144	153	5
33. Hand length [59]	179	194	211	10	165	181	197	10
34. Hand breadth [57]	84	90	98	4	73	79	86	4
35. Foot length [51]	249	270	292	13	224	244	265	12
36. Foot breadth [50]	92	101	110	5	82	90	98	5
37. Weight (kg), estimated by Kroemer	58	78	99	13	39	62	85	14

According to Gordon, Churchill, Clauser, et al. (1989), who used the numbers in brackets.

In both sets of male and female body dimensions the data are presented for the 5th and 95th percentiles. The mean (50th percentile) is also given, as is the standard deviation. This allows us to calculate other than 5th and 95th percentiles by using Equation 1 and the multiplication factors in Table 9.2.

Most populations have not been measured thoroughly and completely. Table 9.8 presents an overview of the most recent anthropometric data on national, ethnic, and geographic populations. It is unfortunate that, in most cases, only few people were measured; their small number n makes it unlikely that the statistical descriptors mean m and standard deviation S truly represent the underlying large population.

Table 9.9 contains general estimates for main regions of the earth. Given the paucity of existing data in 1990 and even today, it is not surprising that the estimates must be applied with great caution; note, for example, that the stature averages estimated for North Americans are several centimeters higher than actually measured, as reported in Table 9.4.

TABLE 9.5 Anthropometric Estimated Data in mm of British Adults, 19 to 35 Years of Age. The reference numbers of the dimensions are shown in Figure 9.4.

		Men				Women			
Dimension		5th %ile	Mean	95th %ile	SD	5th %ile	Mean	95th %ile	SD
1.	Stature	1625	1740	1855	70	1505	1610	1710	62
2.	Eye height, standing	1515	1630	1745	69	1405	1505	1610	61
3.	Shoulder height (acromion), standing	1315	1425	1535	66	1215	1310	1405	58
4.	Elbow height, standing	1005	1090	1180	52	930	1005	1085	46
5.	Hip height (trochanter)	840	920	1000	50	740	810	885	43
6.	Knuckle height, standing	690	755	825	41	660	720	780	36
7.	Fingertip height, standing	590	655	720	38	560	625	685	38
8.	Sitting height	850	910	965	36	795	850	910	35
9.	Sitting eye height	735	790	845	35	685	740	795	33
10.	Sitting shoulder height (acromion)	540	595	645	32	505	555	610	31
11.	Sitting elbow height	195	245	295	31	185	235	280	29
12.	Sitting thigh height (clearance)	135	160	185	15	125	155	180	17
13.	Sitting knee height	490	545	595	32	455	500	540	27
14.	Sitting popliteal height	395	440	490	29	355	400	445	27
15.	Shoulder-elbow length	330	365	395	20	300	330	360	17
16.	Elbow-fingertip length	440	475	510	21	400	430	460	19
17.	Overhead grip reach, sitting	1145	1245	1340	60	1060	1150	1235	53
18.	Overhead grip reach, standing	1925	2060	2190	80	1790	1905	2020	71
19.	Forward grip reach	720	780	835	34	650	705	755	31
20.	Arm length, vertical	720	780	840	36	655	705	760	32
21.	Downward grip reach	610	665	715	32	555	600	650	29
22.	Chest depth	215	250	285	22	210	250	295	27
23.	Abdominal depth, sitting	220	270	325	32	205	255	305	30
24.	Buttock-knee depth, sitting	540	595	645	31	520	570	620	30
25.	Buttock-popliteal depth, sitting	440	495	550	32	435	480	530	30
26.	Shoulder breadth (biacromial)	365	400	430	20	325	355	385	18
27.	Shoulder breadth (bideltoid)	420	465	510	28	355	395	435	24
28.	Hip breadth, sitting	310	360	405	29	310	370	435	38
29.	Span	1655	1790	1925	83	1490	1605	1725	71
30.	Elbow span	865	945	1020	47	780	850	920	43
31.	Head length	180	195	205	8	165	180	190	7
32.	Head breadth	145	155	165	6	135	145	150	6
33.	Hand length	175	190	205	10	160	175	190	9
34.	Hand breadth	80	85	95	5	70	75	85	4
35.	Foot length	240	265	285	14	215	235	255	12
36.	Foot breadth	85	95	110	6	80	90	100	6
37.	Weight (kg)	55	75	94	12	44	63	81	11

According to Pheasant (1986, 1996).

9.6 How to Get Missing Data

Often we design new products for users for whom we do not have exact information about their body size or strength. This is not a great problem if the product is similar to items already in use or if the users are fairly well known to us, such as our colleagues or at least people from our own country. In this case, we can probably take a few measurements on our acquaintances for a "rough guesstimate" of what the needed dimensions might be.

For exact and comprehensive information, however, more than such informal data gathering is necessary. Two avenues are open: one is to conduct a formal anthropometric survey; this is a major enterprise and best done by qualified anthropometrists (Kroemer, 1989; Roebuck, 1995). The other option is to deduce from existing data those that we need to know. There are several engineering approaches to estimate missing data.

TABLE 9.6 Anthropometric Measured Data in mm of East German Adults, 18 to 59 Years of Age. The reference numbers of the dimensions are shown in Figure 9.4.

		Men				Women			
Dimension		5th %ile	Mean	95th %ile	SD	5th %ile	Mean	95th %ile	SD
1.	Stature	1607	1715	1825	66	1514	1608	1707	59
2.	Eye height, standing	1498	1601	1705	64	1415	1504	1597	57
3.	Shoulder height (acromion), standing	1320	1414	1512	60	1232	1319	1403	53
4.	Elbow height, standing	na	na	na	na	na	na	na	na
5.	Hip height (trochanter)	na	na	na	na	na	na	na	na
6.	Knuckle height, standing	682	748	819	42	643	703	764	37
7.	Fingertip height, standing	588	652	717	39	557	616	672	35
8.	Sitting height	846	903	958	34	804	854	905	31
9.	Sitting eye height	719	775	831	34	684	733	782	30
10.	Sitting shoulder height (acromion)	552	601	650	31	517	562	609	29
11.	Sitting elbow height	198	244	293	29	190	234	282	28
12.	Sitting thigh height (clearance)	126	151	176	15	125	148	175	15
13.	Sitting knee height	490	531	575	27	458	497	538	24
14.	Sitting popliteal height	410	452	496	26	380	416	455	23
15.	Shoulder-elbow length	na	na	na	na	na	na	na	na
16.	Elbow-fingertip length	432	465	500	20	394	425	556	19
17.	Overhead grip reach, sitting	na	na	na	na	na	na	na	na
18.	Overhead grip reach, standing	1975	2121	2267	89	1843	1973	2103	79
19.	Forward grip reach	704	763	824	37	650	706	767	35
20.	Arm length, vertical	704	762	820	35	650	703	758	33
21.	Downward grip reach	na	na	na	na	na	na	na	na
22.	Chest depth	na	na	na	na	na	na	na	na
23.	Abdominal depth, sitting	na	na	na	na	na	na	na	na
24.	Buttock-knee depth, sitting	560	603	648	27	541	585	630	27
25.	Buttock-popliteal depth, sitting	444	486	527	25	437	479	521	26
26.	Shoulder breadth (biacromial)	365	399	430	20	336	365	393	17
27.	Shoulder breadth (bideltoid)	432	471	510	24	393	437	481	27
28.	Hip breadth, sitting	334	369	406	22	346	401	460	35
29.	Span	1640	1760	1885	75	1503	1616	1735	70
30.	Elbow span	833	895	911	39	757	817	881	38
31.	Head length	179	190	201	7	170	181	191	6
32.	Head breadth	148	158	168	6	141	151	160	6
33.	Hand length	174	189	205	9	161	174	189	9
34.	Hand breadth	81	88	96	5	71	78	85	44
35.	Foot length	243	264	285	13	222	241	260	12
36.	Foot breadth	91	102	113	6	83	93	104	6
37.	Weight (kg)	na	na	na	na	na	na	na	na

According to Fluegel, Greil and Sommer (1986).

Estimation by "Ratio Scaling"

"Ratio scaling" (as used by Pheasant 1986, 1996) is one technique for estimating data from known dimensions. It relies on the assumption that, though people vary greatly in size, they are likely to be similar in proportions. This premise holds true for body components that are related in size to each other, as discussed in detail by Roebuck, Kroemer, and Thomson (1975), by Pheasant (1982), and most recently by Roebuck (1995). For example, many body "lengths" are highly correlated with each other; also, groups of body "breadths" are related, as are "circumferences" as a group. However, not all body lengths (or breadths, or circumferences) are highly correlated with each other, and certainly many lengths are not related highly with breadths, nor with depths or circumferences. Thus, one has to be very careful in deriving one set of data from another.

TABLE 9.7 Anthropometric Measured Data in mm of Japanese Adults, 18-30 Years of Age. The reference numbers of the dimensions are shown in Figure 9.4.

		Men				Women			
Dimension		5th %ile	Mean	95th %ile	SD	5th %ile	Mean	95th %ile	SD
1.	Stature	1599	1688	1777	55	1510	1584	1671	50
2.	Eye height, standing	1489	1577	1664	53	1382	1460	1541	49
3.	Shoulder height (acromion), standing	1291	1370	1454	50	1208	1279	1367	48
4.	Elbow height, standing	970	1035	1098	39	909	967	1028	37
5.	Hip height (trochanter)	775	834	899	38	730	787	847	35
6.	Knuckle height, standing	na	na	na	na	na	na	na	na
7.	Fingertip height, standing	600	644	694	30	563	608	652	27
8.	Sitting height	859	910	958	30	810	855	902	28
9.	Sitting eye height	741	790	837	29	692	733	778	27
10.	Sitting shoulder height (acromion)	549	591	633	26	513	551	588	24
11.	Sitting elbow height	216	254	292	23	202	236	269	20
12.	Sitting thigh height (clearance)	138	156	176	12	130	143	162	10
13.	Sitting knee height	475	509	545	22	442	475	508	20
14.	Sitting popliteal height	371	402	434	19	345	372	402	17
15.	Shoulder-elbow length	307	337	366	18	289	315	339	15
16.	Elbow-fingertip length	418	448	479	18	390	416	445	17
17.	Overhead grip reach, sitting	na	na	na	na	na	na	na	na
18.	Overhead grip reach, standing	na	na	na	na	na	na	na	na
19.	Forward grip reach	na	na	na	na	na	na	na	na
20.	Arm length, vertical	na	na	na	na	na	na	na	na
21.	Downward grip reach	na	na	na	na	na	na	na	na
22.	Chest depth	190	217	246	18	190	215	250	19
23.	Abdominal depth, sitting	179	208	245	20	161	188	218	17
24.	Buttock-knee depth, sitting	530	567	604	23	511	550	586	22
25.	Buttock-popliteal depth, sitting	na	na	na	na	na	na	na	na
26.	Shoulder breadth (biacromial)	368	395	423	17	346	367	391	14
27.	Shoulder breadth (bideltoid)	na	na	na	na	na	na	na	na
28.	Hip breadth, sitting	318	349	380	19	331	358	386	17
29.	Span	1591	1690	1795	63	1483	1579	1693	62
30.	Elbow span	na	na	na	na	na	na	na	na
31.	Head length	178	190	203	7	168	177	187	6
32.	Head breadth	152	161	171	6	143	151	160	6
33.	Hand length	na	na	na	na	na	na	na	na
34.	Hand breadth	79	85	91	4	70	75	81	3
35.	Foot length	234	251	269	11	217	232	246	9
36.	Foot breadth	97	104	111	5	89	96	103	4
37.	Weight (kg)	54	66	80	8	45	54	65	6

According to Kagimoto (1990).

A good rule for ratio scaling is to use only pairings of data that are related to each other with a coefficient of correlation of at least 0.7. (This assures that the variability of the derived information is determined at least by 50% by the variability of the predictor, which derives from the squaring of the correlation coefficient, $0.7^2 = 0.49$.) However, never use ratio scaling if you must assume that the sample from which you want to scale has body proportions different from those of the other set; e.g., many Asian populations have proportionally shorter legs and longer trunks than Europeans or North-Americans.

For sets of highly correlated data, you can establish the estimate E of a ratio scaling factor for a desired dimension (d_y) in the population sample Y

- if you know the value of that dimension in sample X (d_x), and
- if you know the values of a reference dimension D in both samples X and Y (D_x and D_y).

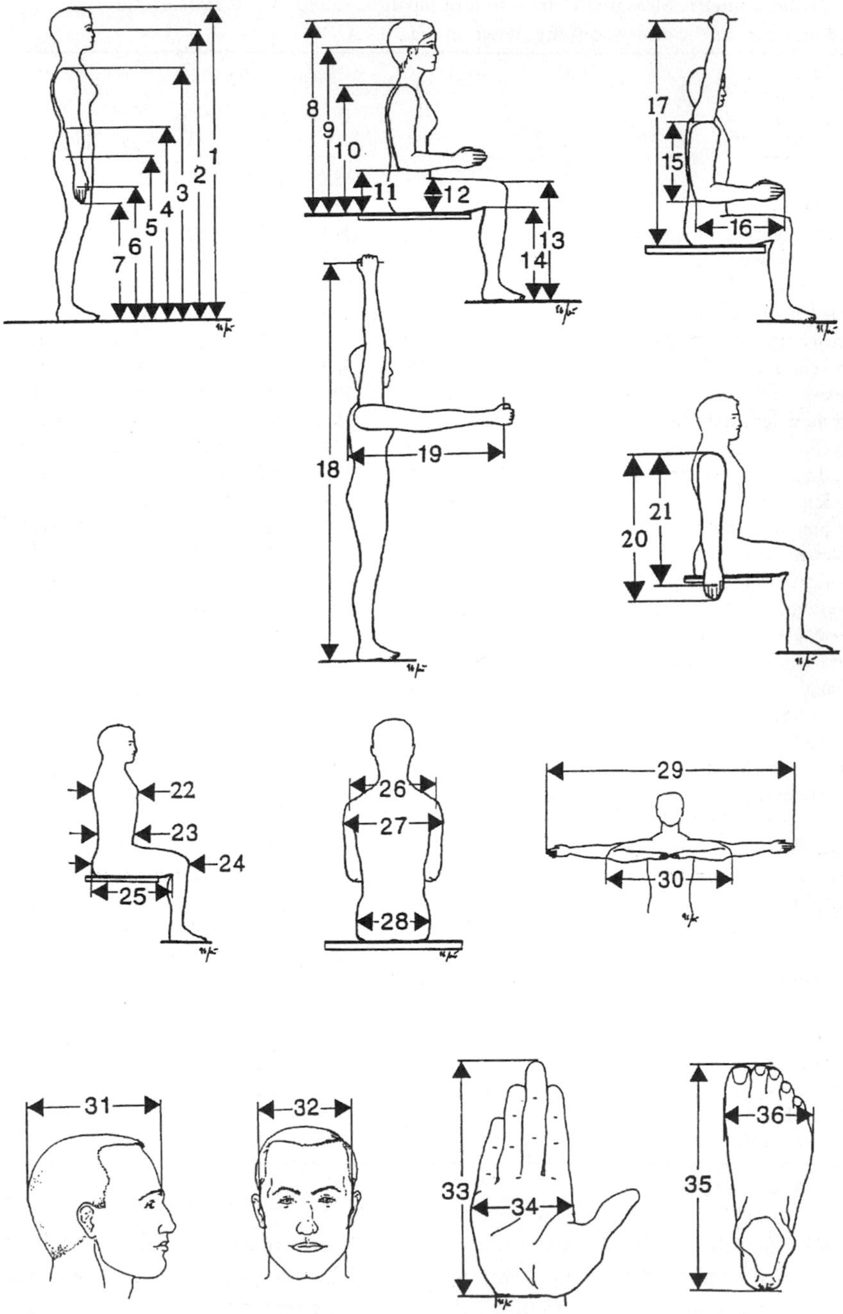

FIGURE 9.4 Illustrations of measured body dimensions.

The following commentary identifies measurements and design implications. The numbers in brackets are the same as used by Gordon et al. (1989), who also provide exact definitions of the anthropometric terms. Most of the listed dimensions are the same as listed by Pheasant (1986, 1996). The data listed in Tables 9.4 through 9.7 have been selected to include both traditional anthropometric sample descriptors (such as stature and weight) and body dimensions and reaches that are relevant to the sizing of workstations and equipment. However, in order to apply this information for engineering purposes, usually the data must be converted to reflect actual positions and motions instead of the "standardized frozen" postures (all body angles at either 0, 90, or 180 degrees) which the subject assumed for the measurements. Furthermore, allotments must be made for clothing, shoes, gloves, and other "real world" conditions. Table 9.3 provides some guidelines.

FIGURE 9.4 (continued)

1. **Stature:** The vertical distance from the floor to the top of the head, when standing [99]. A main reference for comparing population samples. Relates to the minimal height (clearance) of overhead obstructions. Add height for more clearance, hat, shoes, stride.
2. **Eye height, standing:** The vertical distance from the floor to the outer corner of the right eye, when standing [D19]. Origin of the visual field. Reference point for the location of vision obstructions and of visual targets such as displays; consider slump and motion of the standing person.
3. **Shoulder height (acromion), standing:** The vertical distance from the floor to the tip (acromion) of the shoulder, when standing [2]. Starting point for arm length measurements; near the center of rotation of the upper arm (shoulder joint), reference point for hand reaches; consider slump and motion of the standing person.
4. **Elbow height, standing:** The vertical distance from the floor to the lowest point of the right elbow, when standing, with the elbow flexed at 90 degrees [D16]. Reference point for height and distance of the work area of the hand and for the location of controls and fixtures; consider slump and motion of the standing person.
5. **Hip height (trochanter), standing:** The vertical distance from the floor to the trochanter landmark on the upper side of the right thigh, when standing [107]. Starting point for leg length measurement; near the center of the hip joint; reference point for leg reaches; consider slump and motion of the standing person.
6. **Knuckle height, standing:** The vertical distance from the floor to the knuckle (metacarpal bone) of the middle finger of the right hand, when standing. Reference point for lowest location of controls, handles, and handrails; consider slump and motion of the standing person.
7. **Fingertip height, standing:** The vertical distance from the floor to the tip of the index finger of the right hand, when standing [D13]. Reference point for lowest location of controls, handles, and handrails; consider slump and motion of the standing person.
8. **Sitting height:** The vertical distance from the sitting surface to the top of the head, when sitting [93]. The vertical distance from the floor to the underside of the thigh directly behind the right knee; when sitting, with the knee flexed at 90 degrees. Relates to the minimal height of overhead obstructions. Add height for more clearance, hat, trunk motion of the seated person.
9. **Sitting eye height:** The vertical distance from the sitting surface to the outer corner of the right eye, when sitting [49]. Origin of the visual field; reference point for the location of vision obstructions and of visual targets such as displays; consider slump and motion of the seated person.
10. **Sitting shoulder height (acromion):** The vertical distance from the sitting surface to the tip (acromion) of the shoulder, when sitting [3]. Starting point for arm length measurements; near the center of rotation of the upper arm (shoulder joint), reference point for hand reaches; consider slump and motion of the seated person.
11. **Sitting elbow height:** The vertical distance from the sitting surface to the lowest point of the right elbow, when sitting, with the elbow flexed at 90 degrees [48]. Reference point for height of an arm rest, of the work area of the hand, and of keyboard and controls; consider slump and motion of the seated person.
12. **Sitting thigh height (clearance):** The vertical distance from the sitting surface to the highest point on the top of the right thigh, when sitting, with the knee flexed at 90 degrees [104]. Minimal clearance needed between seat pan and the underside of a structure, such as a table.; add clearance for clothing and motions.
13. **Sitting knee height:** The vertical distance from the floor to the top of the right knee cap, when sitting, with the knees flexed at 90 degrees [73]. Minimal clearance needed below the underside of a structure, such as a table; add height for shoe.
14. **Sitting popliteal height:** The vertical distance from the floor to the underside of the thigh directly behind the right knee; when sitting, with the knees flexed at 90 degrees [86]. Reference for the height of a seat; add height for shoes, consider movement of the feet.
15. **Shoulder-elbow length:** The vertical distance from the underside of the right elbow to the right acromion, with the elbow flexed at 90 degrees and the upper arm hanging vertically [91]. A general reference for comparing population samples.
16. **Elbow-fingertip length:** The distance from the back of the right elbow to the tip of the middle finger, with the elbow flexed at 90 degrees [54]. Reference for fingertip reach when moving the forearm in the elbow.
17. **Overhead grip reach, sitting:** The vertical distance from the sitting surface to the center of a cylindrical rod firmly held in the palm of the right hand [D45]. Reference for height of overhead controls to be operated by the seated person. Consider ease of motion, reach, and finger/hand/arm strength.

FIGURE 9.4 (continued)

18. **Overhead grip reach, standing:** The vertical distance from the standing surface to the center of a cylindrical rod firmly held in the palm of the right hand [D42]. Reference for height of overhead controls to be operated by the standing person. Add shoe height. Consider ease of motion, reach, and finger/hand/arm strength.
19. **Forward grip reach:** The horizontal distance from the back of the right shoulder blade to the center of a cylindrical rod firmly held in the palm of the right hand [D21]. Reference for forward reach distance. Consider ease of motion, reach and finger/hand/arm strength.
20. **Arm length, vertical:** The vertical distance from the tip of the right middle finger to the right acromion, with the arm hanging vertically [D3]. A general reference for comparing population samples. Reference for the location of controls very low on the side of the operator. Consider ease of motion, reach and finger/hand/arm strength.
21. **Downward grip reach:** The vertical distance from the right acromion to the center of a cylindrical rod firmly held in the palm of the right hand, with the arm hanging vertically [D43]. Reference for the location of controls low on the side of the operator. Consider ease of motion, reach, and finger/hand/arm strength.
22. **Chest depth:** The horizontal distance from the back to the right nipple [36]. A general reference for comparing population samples. Reference for the clearance between seat backrest and the location of obstructions in front of the trunk.
23. **Abdominal depth, sitting:** The horizontal distance from the back to the most protruding point on the abdomen [1]. A general reference for comparing population samples. Reference for the clearance between seat backrest and the location of obstructions in front of the trunk.
24. **Buttock-knee depth, sitting:** The horizontal distance from the back of the buttocks to the most protruding point on the right knee, when sitting with the knees flexed at 90 degrees [26]. Reference for the clearance between seat backrest and the location of obstructions in front of the knees.
25. **Buttock-popliteal depth, sitting:** The horizontal distance from the back of the buttocks to back of the right knee just below the thigh, when sitting with the knees flexed at 90 degrees [27]. Reference for the depth of a seat.
26. **Shoulder breadth, biacromial:** The distance between the right and left acromion [10]. A general reference for comparing population samples. Indication of the distance between the centers of rotation (shoulder joints) of the upper arms.
27. **Shoulder breadth, bideltoid:** The maximal horizontal breadth across the shoulders between the lateral margins of the right and left deltoid muscles [12]. Reference for the clearance requirement at shoulder level. Add space for ease of motion, tool use.
28. **Hip breadth, sitting:** The maximal horizontal breadth across the hips or thighs, whatever is greater, when sitting [66]. Reference for seat width. Add space for clothing and ease of motion.
29. **Span:** The distance between the tips of the middle fingers of the horizontally outstretched arms and hands [98]. Reference for sideway reach.
30. **Elbow span:** The distance between the tips of the elbows of the horizontally outstretched upper arms with the elbows flexed so that the fingertips of the hands meet in front of the trunk. Reference for "elbow room."
31. **Head length:** The distance from the glabella (between the browridges) to the most rearward protrusion (the occiput) on the back, in the middle of the skull [62]. A general reference for comparing population samples. Reference for head gear size.
32. **Head breadth:** The maximal horizontal breadth of the head above the attachment of the ears [60]. A general reference for comparing population samples. Reference for head gear size.
33. **Hand length:** The length of the right hand between the crease of the wrist and the tip of the middle finger, with the hand flat [59]. A general reference for comparing population samples. Reference for hand tool and gear size. Consider changes due to manipulations, gloves, tool use.
34. **Hand breadth:** The breadth of the right hand across the knuckles of the four fingers [57]. A general reference for comparing population samples. Reference for hand tool and gear size, and for the opening through which a hand may (or may not) fit. Consider changes due to manipulations, gloves, tool use.
35. **Foot length:** The maximal length of the right foot, when standing [51]. A general reference for comparing population samples. Reference for shoe and pedal size.
36. **Foot breadth:** The maximal breadth of the right foot, at right angle to the long axis of the foot, when standing [50]. A general reference for comparing population samples. Reference for shoe size, spacing of pedals.
37. **Weight:** Nude body weight taken to the nearest tenth of a kilogram. A general reference for comparing population samples. Reference for body size, clothing, strength, health, etc. Add weight for clothing and equipment worn on the body.

TABLE 9.8 Recent Anthropometric Data on National and Ethnic Populations: Averages (and Standard Deviations), in mm but Weight in kg. Contact the author for details on the sources.

	Sample Size N	Stature	Sitting Height	Knee Height, sitting	Weight
Algerian females (Mebarki and Davies, 1990)	666	1576 (56)	795 (50)	487 (36)	61.3 (12.9)
Brazilian males (Ferreira, 1988; cited by Al-Haboubi, 1991)	3076	1699 (67)	—	—	—
Chinese females (Singapore) (Ong, Koh, Phoon, and Low, 1988)	46	1598 (58)	855 (31)	—	—
Chinese females (Taiwan) (Huang and You, 1994)	300	1582 (49)	—	—	51.2 (6.9)
Cantonese males (Evans, 1990)	41	1720 (63)	—	—	60.0 (6.2)
Egyptian females (Moustafa, Davies, Darwich, and Ibraheem, 1987)	4960	1606 (72)	838 (43)	499 (25)	62.6 (4.4)
Indian males (farmers) (Nag, Sebastian, and Mavlankar, 1980)	13	1576 (17)	—	—	44.6 (1.4)
Central Indian male farm workers (Gite and Yadav, 1989)	39	1620 (50)	739 (26)	509 (30)	49.3 (6.0)
South Indian males (workers) (Fernandez and Uppugonduri, 1992)	128	1607 (60)	791 (40)	542 (38)	56.6 (5.1)
Indonesian females	468	1516 (54)	719 (34)	—	—
Indonesian males (Sama'mur, 1985; cited by Intaranont, 1991)	949	1613 (56)	872 (37)	—	—
Iranian female students	74	1597 (58)	861 (36)	488 (23)	56.2 (10.1)
Iranian male students (Mououdi, 1997)	105	1725 (58)	912 (26)	531 (24)	65.7 (10.1)
Irish males (Gallwey and Fitzgibbon, 1991)	164	1731 (58)	911 (30)	508 (28)	73.9 (8.7)
Italian females	753	1610 (64)	850 (34)	495 (30)	58 (8.3)
Italian males	913	1733 (71)	896 (36)	541 (30)	75 (9.6)
(Coniglio, Fubini, Masali, Masiero, Pierlorenzi and Sagone, 1991)					
Jamaican females	123	1648	832	—	61.4
Jamaican males (Camey, Aghazadeh, and Nye, 1991)	30	1749	856	—	67.6
Korean female workers (Fernandez, Malzahn, Eyada, and Kim, 1989)	101	1580 (57)	833 (32)	460 (22)	53.9 (6.9)
Malay females (Ong, Koh, Phoon, and Low, 1988)	32	1559 (66)	831 (39)	—	—
Saudi-Arabian males (Dairi, 1986; cited by Al-Haboubi, 1991)	1440	1675 (61)	—	—	—
Singapore males (pilot trainees) (Singh, Pen, Lim, and Ong, 1995)	832	1685(53)	894(32)	—	—
Sri Lankan females	287	1523 (59)	774 (22)	—	—
Sri Lankan males (Abeysekera, 1985; cited by Intaranont, 1991)	435	1639 (63)	833 (27)	—	—
Sudanese Males					
Villagers	37*	1687 (63)	—	—	57.1 (7.6)
City dwellers	16*	1704 (72)	—	—	62.3 (13.1)
	48**	1668	—	—	51.3
Soldiers	21*	1735 (71)	—	—	71.1 (8.4)
	104**	1728	—	—	60.0
* (El-Karim, Sukkar, Collins, and Dore, 1981)					
**(Ballal et al., 1982; cited by Intaranont, 1991)					
Thai females	250*	1512 (48)	—	—	—
	711*	1540 (50)	817 (27)	—	—
Thai males	250*	1607 (20)	—	—	—
	1478**	1654 (59)	872 (32)	—	—
* (Intaranont, 1991)					
**(NICE; cited by Intaranont, 1991)					
Turkish females					
Villagers	47	1567 (52)	792 (38)	486 (27)	69.1 (13.8)
City dwellers	53	1563 (55)	786 (35)	471 (25)	65.9 (13.0)
(Goenen, Kalinkara, and Oezgen, 1991)					
Turkish males (soldiers) (Kayis and Oezok, 1991)	5108	1702 (60)	888 (34)	513 (28)	63.3 (7.3)
Vietnamese (American V.)					
Females	30	1559 (61)	—	—	48.6
Males	41	1646 (54)	—	—	58.9
(Imrhan, Nguyen and Nguyen (1993)					
U.S. Midwest workers, with shoes and light clothes					
Females	125	1637 (62)	—	—	64.7 (11.8)
Males	384	1778 (73)	—	—	84.2 (15.5)
(Marras and Kim, 1993)					
U.S. male miners (Kuenzi and Kennedy, 1993)	105	1803 (65)	—	—	89.4 (15.1)

Adapted from Kroemer, Kroemer, and Kroemer-Elbert (1997)

TABLE 9.9 Average Anthropometric Data (in mm) Estimated for 20 Regions of the Earth

	Stature		Sitting Height		Knee Height, Sitting	
	Females	Males	Females	Males	Females	Males
NORTH AMERICA	1650	1790	880	930	500	550
LATIN AMERICA						
Indian Population	1480	1620	800	850	445	495
European and Negroid population	1620	1750	860	930	480	540
EUROPE						
North	1690	1810	900	950	500	550
Central	1660	1770	880	940	500	550
East	1630	1750	870	910	510	550
Southeast	1620	1730	860	900	460	535
France	1630	1770	860	930	490	540
Iberia	1600	1710	850	890	480	520
AFRICA						
North	1610	1690	840	870	500	535
West	1530	1670	790	820	480	530
Southeast	1570	1680	820	860	495	540
NEAR EAST	1610	1710	850	890	490	520
INDIA						
North	1540	1670	820	870	490	530
South	1500	1620	800	820	470	510
ASIA						
North	1590	1690	850	900	475	515
Southeast	1530	1630	800	840	460	495
SOUTH CHINA	1520	1660	790	840	460	505
JAPAN	1590	1720	860	920	395	515
AUSTRALIA						
European extraction	1670	1770	880	930	525	570

Adapted from Juergens, Aune, and Pieper (1990).

In this case, you calculate the scaling factor E from

$$E = d_x / D_x \tag{2}$$

Since the basic assumption is that the two samples are similar in proportion, the same scaling factor E applies to both samples X and Y:

$$E = d_x / D_x = d_y / D_y \tag{3}$$

with

$$E = d_y / D_y \tag{3a}$$

known, you can calculate

$$d_y = E * D_y \tag{4}$$

in stepwise fashion, as shown in the following:

Step 1: In population sample X, establish a scaling factor E between the desired dimension and a known reference dimension. The reference parameter must be common for both population samples; stature is often used. For example, if shoulder height is to be estimated for sample Y, and is known in sample X, then calculate

$$E = \frac{\text{shoulder height in sample } X}{\text{stature in sample } X} \quad \text{see (2)}$$

Step 2: With E now known, the desired unknown dimension in population sample Y equals E times the reference parameter in sample Y. For example, shoulder height in sample $Y = E *$ stature in sample Y

$$\text{shoulder height in sample Y} = \text{E} * (\text{stature in sample Y}) \quad \text{see (4)}$$

The common parameter is often stature because its value is commonly and easily measured. Note, however, that stature is generally related well with other heights, but not necessarily with depths, breadths, circumferences, or weight (as discussed above). Thus, ratio scaling must be done with great caution and careful consideration of the circumstances, especially taking into account statistical correlations.

The technique of ratio scaling has been applied primarily to estimate the mean of a required dimension, and to estimate its standard deviation. For more detail on ratio scaling, read the books by Pheasant (1986, 1996) and Roebuck (1995).

Estimation by Regression Equations

Another way of estimating the relations among dimensions is through regression equations. Most regression equations are bivariate in nature, meaning two variables are involved, and it is presumed that the two variables are linearly related. (That linear relationship is seldom explicitly confirmed.) The general form is

$$y = a + b * x \quad (5)$$

where x is the known mean value and y the predicted mean. The constants a (the "intercept") and b (the "slope") must be determined (known) for the data set of interest. A recent example of this procedure is the estimation of body dimensions of American soldiers by Cheverud, Gordon, Walker, Jacquish, Kohn, Moore, and Yamashita (1990).

If you predict the mean value of y (for any value of x) using the regression equation shown above, you must remember that the actual values of y are scattered about the mean in a normal (Gaussian) probability distribution. The standard error SE of the estimate depends on the correlation r between x and y, and on the standard deviation of y (S_y) according to

$$SE_y = S_y\sqrt{1 - r^2} \quad (6)$$

Roebuck (1995) discussed this concept in some detail, including its extension to develop multivariate regression equations, principal component analyses, and boundary description analyses.

Estimation by Probability Statistics

In most cases, we are unable to measure every person of a user population with respect to body size or strength. If we were able to do so, we could describe the parameters of that total population by the mean (average) μ and standard deviation σ. (The terminology convention in statistics is to use Greek letters to indicate population parameters.) In reality, we can measure only a subgroup (sample), and from its parameters we infer or estimate what the actual population would have yielded. Using roman letters to describe the sample data, we say that

$$m = (\Sigma x)/n \quad (7)$$

where m is the mean (average), x is the individual measurement, and n is the number of measured individuals. The distribution of the data is described by the equation

$$S = \sqrt{\Sigma (x - m)^2 / n} \tag{8a}$$

with S called the standard deviation of the sample. If the sample size is small (conventionally, 30 or less) one makes an arbitrary correction by using $(n - 1)$ instead of n:

$$S = \sqrt{\Sigma (x - m)^2 / n - 1} \tag{8b}$$

The smaller n, the larger the standard error SE in sampling. The standard error SE of the mean is determined from

$$SE \text{ of the mean } = S/\sqrt{n} \tag{9}$$

The standard error SE of the standard deviation is determined from

$$SE \text{ of the standard deviation } = S/\sqrt{2n} = 0.71 \; SE \text{ of the mean} \tag{10}$$

As the number n increases, the mean m and the standard deviation S become more reliable estimates of the underlying general population (i.e., of μ and σ).

It is often useful to describe the variability of a sample by dividing the standard deviation S by the mean m (and multiplying the result by 100). This yields the coefficient of variation CV:

$$CV \text{ (in percent)} = 100 \; S/m \tag{11}$$

This expression is independent of the magnitude and of the unit of measurement. Groups of human measurements show characteristic variabilities. Typical coefficients of variation are listed in Table 9.10. This information can be used to judge the reliability of reported data.

Considering the *CV* is often very helpful when you try to determine the credibility of data published in the literature: unusually large or small *CV* values indicate that either the distribution of the measured population is indeed different from other populations, or that irregularities in measuring, or in data treatment, or in data reporting occurred. (See the more detailed discussion of possible data variations in the 1997 book by Kroemer, Kroemer, and Kroemer-Elbert, and in the 1995 book by Roebuck.) The *CV* may also help you make an estimate of the standard deviation of an unknown data set.

TABLE 9.10 Variability of Body Measurements

Variables Measured	CV in %
Body heights (stature, sitting height, elbow height, etc.)	3 to 5
Body breadths (hip, shoulder, etc.)	5 to 9
Body depths (abdominal, chest, etc.)	6 to 9
Reaches	4 to 10
Total body weights	10 to 20
Joint ranges	7 to 30
Muscular static strength	10 to 85

From Kroemer, Kroemer, and Kroemer-Elbert, 1990, *Engineering Physiology: Bases of Human Factors/Ergonomics*, 2nd ed. With permission.

9.7 Combining Anthropometric Data Sets

Occasionally, one must add or subtract anthropometric values; for example, total arm length is the sum of upper and lower arm lengths. If you want to add two measures, such as leg length and torso (with head) length, you generate a new combined distribution, stature. In doing so, you must take into account the covariation *COV* between the two measures of leg and torso: usually (but not always) a taller torso is associated with a taller head. This is mathematically described by the correlation coefficient r between the two data sets, x and y, and their standard deviations, S_x and S_y:

$$COV(x, y) = r_{x,y} * S_y * S_y \quad (12)$$

This allows us to calculate the *sum* of the two mean values of the x and y distributions from

$$m_z = m_x + m_y \quad (13)$$

and the estimated standard deviation of z from

$$S_z = \left[S_x^2 + S_y^2 + 2r * S_x * S_y\right]^{1/2} \quad (14)$$

The *difference* z between two mean values is

$$m_z = m_x - m_y \quad (15)$$

and its standard deviation

$$S_2 = \left[S_x^2 + S_y^2 - 2r * S_x * S_y\right]^{1/2} \quad (16)$$

Three Examples

EX 1: What is the 95p shoulder-to-fingertip length? You know the mean lower arm LA link length (with the hand) to be 442.9 mm with a standard deviation of 23.4 mm. You also know the mean upper arm UA link length of 335.8 mm and its standard deviation of 17.4 mm.

The multiplication factor of $k = 1.64$ (from Table 9.2) leads you to the 95th percentile. But you cannot calculate the sum of the two 95p lengths because this would disregard their covariance; instead, you calculate the sum of the mean values first:

$$m = m_{LA} + m_{UA} = 442.9 + 335.8 = 778.7 \text{ mm} \quad \text{see (13)}$$

The standard deviation is calculated next, using an assumed coefficient of correlation of 0.4:

$$S = \left[23.4^2 + 17.4^2 + 2 * 0.4 * 23.4 * 18.4\right]^{1/2} \text{ mm}$$
$$S = 34.6 \text{ mm} \quad \text{see (14)}$$

The 95p total arm length AL can now be calculated:

$$AL_{95} = 778.7 \text{ mm} + 1.64 * 34.6 \text{ mm} = 835.4 \text{ mm} \quad \text{see (1)}$$

EX 2: What is the average arm (acromion to wrist) length of an American pilot? You know that for a standing pilot the 90th percentile acromial (shoulder) height is 1532.0 mm and the wrist height is

905.6 mm; for the 10th percentile, the values are 1379.5 and 808.6 mm, respectively. The correlation between shoulder and wrist heights is estimated at 0.3. You first calculate the mean 90p and 10p acromion (A) and wrist (W) heights to be able to estimate the standard deviations:

$$m_A = (1532.0 + 1379.5)\,\text{mm}/2 = 1455.75\ \text{mm}$$

and, with $k = 1.28$ taken from Table 9.2,

$$S_A = (1532.0 - 1455.75)\,\text{mm}/1.28 = 59.6\ \text{mm} \qquad \text{see (1)}$$

$$\left[\text{or: } S_A = (1455.75 - 1379.5)\,\text{mm}/1.28 = 59.6\ \text{mm}\right]$$

Likewise,

$$m_w = (905.6 + 808.6)\,\text{mm}/2 = 857.1\ \text{mm}$$

$$S_w = (905.6 - 857.1)\,\text{mm}/1.28 = 37.9\ \text{mm} \qquad \text{see (1)}$$

$$\left[\text{or: } S_w = (857.1 - 808.6)\,\text{mm}/1.28 = 37.9\ \text{mm}\right]$$

The average arm length (acromion to wrist, AW) is

$$m_{AW} = m_A - m_W = 1455.75\ \text{mm} - 857.1\ \text{mm} = 598.65\ \text{mm} \qquad \text{see (15)}$$

The standard deviation of the arm length is

$$S_{AW} = \left(59.6^2 + 37.9^2 - 2*0.3*59.6*37.9\right)^{1/2}\text{mm} = 60.3\ \text{mm} \qquad \text{see (16)}$$

EX 3: What is the mass of the head of a 75p Japanese female? The mass of the total body has a mean of 54.0 kg with a standard deviation of 6.0 kg (see Table 9.7). The estimated mass of the head is 6.2% of body mass (from data compiled by Kroemer, Kroemer, and Kroemer-Elbert, 1997). Assume the correlation between total body and head masses to be 1. The mean head mass is

$$\text{mean}_{\text{head}} = 0.062 * 54.0\ \text{kg} = 3.35\ \text{kg}.$$

Given the assumed perfect correlation between head and total body masses, the standard deviation of the head mass may be calculated with

$$E = S_{\text{total body}}/m_{\text{total body}} = (6.0/54.0) = 0.11 \qquad \text{see (2)}$$

From

$$S_{\text{head}} = m_{\text{head}} * E = 3.35\ kg * 0.11 \qquad \text{see (4)}$$

you calculate

$$S_{\text{head}} = 0.37\ \text{kg}$$

The mass of a 75th percentile head is (with $k = 0.67$ taken from Table 9.2)

$$\text{mass}_{\text{head 75p}} = 3.35\ \text{kg} + 0.67 * 0.37\ \text{kg} = 3.6\ \text{kg} \qquad \text{see (1)}$$

9.8 The "Normative" Adult vs. "Real Persons"

The "Average Person" Phantom

Without formally stating so, even without consciously being aware of it, we commonly design for a group of "regular" people who are in the 25- to 45-year age bracket; who are of "normal" anthropometry, i.e., have body dimensions such as stature, hand reach, or weight close to the 50th percentile; who are "healthy" in their metabolic, circulatory, and respiratory subsystems; whose nervous control, sensory capabilities, and intelligence are all "near average," and who are able and willing to perform "normally." Thus, by default or for reasons, the normative stereotype of many human factor engineers is the "regular" adult woman or man. In fact, the proverbial "average person" appears only in newspapers, design guidelines, biomechanical models, and textbooks.

This mythical normative adult has become our user prototype to which we compare other subgroups, such as children, temporarily or permanently impaired persons, women during pregnancy, or aging people. Yet, most individuals and whole population subgroups deviate in size, strength, or other performance capabilities from the normative adult. Neither in a statistical sense nor in reality are there persons who are average in most or all respects, and products or processes "designed for the average" fit nobody well (Kroemer, Kroemer, and Kroemer-Elbert, 1994). To achieve ease, efficiency, and safety, it is mandatory to consider the ranges of, the variations in, and the combinations of physiologic and psychologic traits; the foregoing discussions showed ways to accommodate anthropometric variability.

Posture versus Motions

A similarly simplistic approach incorporates the idea of designing for body "posture." In part, this false concept may have been provoked by the standardized erect posture, sitting or standing, utilized in measuring body size, or static strength. Unfortunately, the "upright" posture has been employed as a design model, probably because it is easily visualized and made into a design template. This upright idol was promoted by orthopedists of the late 19th century who translated their postural concerns into the desire for an erect trunk posture, especially when sitting in school or office. Yet, over extended periods of time, the human is unable to maintain any given posture, upright or otherwise. Standing still, immobile sitting, even lying stiffly, quickly become uncomfortable and then, with time, physically impossible to maintain; if enforced by injury or sickness, circulatory and metabolic functions become impaired, bed sores appear. The human body is made to move.

Our bodies are designed for movement especially in the arms, with shoulder and elbow joints providing extensive angular freedom. The strong legs are able to propel the body on the ground, with major motions occurring in the knee and hip joints. Movements of the trunk occur mostly in flexion and extension at the lower back. However, these bending and unbending motions (in the medial plane) are rather limited, and often lead to overexertions, especially if combined with sideways twisting of the torso: low back pain has been reported throughout the history of mankind. Wrist problems have been associated with excessive motion requirements since the early 1700s. Head and neck have limited mobility in bending and twisting. Our thumbs and fingers have limited but finely controlled motion capability.

Ranges of motion (also called mobility or flexibility) depend much on age, health, fitness, training, and skill. Mobility ranges have been measured on dissimilar groups of people with various measuring instructions and techniques; hence, there is much diversity in reported results. However, at least one set of mobility measurements has been taken on groups of 100 females and of 100 males by the same researchers using the same techniques. These data are reported in Table 9.11. Note that the differences in mobility between males and females are negligible in most cases.

Designing to fit motion ranges, instead of fixed postures, is not difficult. The articulations in the human body have varying degrees of freedom for movements. These are shown in Figure 9.5 for major body joints, and the motion ranges are listed in Table 9.11. These maximal ranges were measured on students of physical education, hence, many people will have slightly less mobility than shown. "Convenient" mobility

TABLE 9.11 Comparison of Mobility Data (in degrees) for Females and Males

Joint	Movement	5th Percentile		50th Percentile		95th Percentile		Difference*	
		Female	Male	Female	Male	Female	Male	Female	Male
Neck	Ventral flexion	34.0	25.0	51.5	43.0	69.0	60.0	+8.5	
	Dorsal flexion	47.5	38.0	70.5	56.5	93.5	74.0	+14.0	
	Right rotation	67.0	56.0	81.0	74.0	95.0	85.0	+7.0	
	Left rotation	64.0	67.5	77.0	77.0	90.0	85.0	NS	
Shoulder	Flexion	169.5	161.0	184.5	178.0	199.5	193.5	+6.5	
	Extension	47.0	41.5	66.0	57.5	85.0	76.0	+8.5	
	Adduction	37.5	36.0	52.5	50.5	67.5	63.0	NS	
	Abduction	106.0	106.0	122.5	123.5	139.0	140.0	NS	
	Medial rotation	94.0	68.5	110.5	95.0	127.0	114.0	+15.5	
	Lateral rotation	19.5	16.0	37.0	31.5	54.5	46.0	+5.5	
Elbow-forearm	Flexion	135.5	122.51	148.0	138.0	160.5	150.0	+10.0	
	Supination	87.0	86.0	108.5	107.5	130.0	135.0	NS	
	Pronation	63.0	42.5	81.0	65.0	99.0	86.5	+16.0	
Wrist	Extension	56.5	47.0	72.0	62.0	87.5	76.0	+10.0	
	Flexion	53.5	50.5	71.5	67.5	89.5	85.0	+4.0	
	Adduction	16.5	14.0	26.5	22.0	36.5	30.0	+4.5	
	Abduction	19.0	22.0	28.0	30.5	37.0	40.0	–2.5	
Hip	Flexion	103.0	95.0	125.0	109.5	147.0	130.0	+15.5	
	Adduction	27.0	15.5	38.5	26.0	50.0	39.0	+12.5	
	Abduction	47.0	38.0	66.0	59.0	85.0	81.0	+7.0	
	Medial rotation (prone)	30.5	30.5	44.5	46.0	58.5	62.5	NS	
	Lateral rotation (prone)	29.0	21.5	45.5	33.0	62.0	46.0	+12.5	
	Medial rotation (sitting)	20.5	18.0	32.0	28.0	43.5	43.0	+4.0	
	Lateral rotation (sitting)	20.5	18.0	33.0	26.5	45.5	37.0	+6.5	
Knee	Flexion (standing)	99.5	87.0	113.5	103.5	127.5	122.0	+10.0	
	Flexion (prone)	116.0	99.5	130.0	117.0	144.0	130.0	+13.0	
	Medial rotation	18.5	14.5	31.5	23.0	44.5	35.0	+8.5	
	Lateral rotation	28.5	21.0	43.5	33.5	58.5	48.0	+10.0	
Ankle	Flexion	13.0	18.0	23.0	29.0	33.0	34.0	–6.0	
	Extension	30.5	21.0	41.0	35.5	51.5	51.5	+5.5	
	Adduction	13.0	15.0	23.5	25.0	34.0	38.0	NS	
	Abduction	11.5	11.0	24.0	19.0	36.5	30.0	+5.0	

* Listed are only differences at the 50th percentile, and if significant ($\alpha < 0.5$).

From Kroemer, Kroemer, and Kroemer-Elbert, 1990, *Engineering Physiology: Bases of Human Factors/Ergonomics,* 2nd ed. With permission.

is somewhere within the range of maximal values shown in Table 9.11, but not always in the middle of the ranges; not seldom, convenient motions are near the limits of mobility. Habits and skill as well as strength requirements may make different ranges preferred.

Design for motions starts by establishing the actual movement ranges. Convenient motions may cluster around the mean of mobility in a body joint, or may be close to the limits of flexibility. For example, a person walking about on a job, or standing, has the knees most of the time nearly extended, that is — in the sagittal view — the knee angles range close to the extreme value of about 180 degrees. The sagittal hip angle (between trunk and thigh) also varies in the neighborhood of 180 degrees. Both angles change to cluster about 90 degrees when sitting. (See Table 9.12.)

Preferred work areas of the hands and feet are in front of the body, within curved envelopes that reflect the mobility of the forearm in the elbow joint, or of the total arm in the shoulder joint; of the lower leg in the knee joint, and of the total leg in the hip joint. Thus, these reach envelopes are often described as

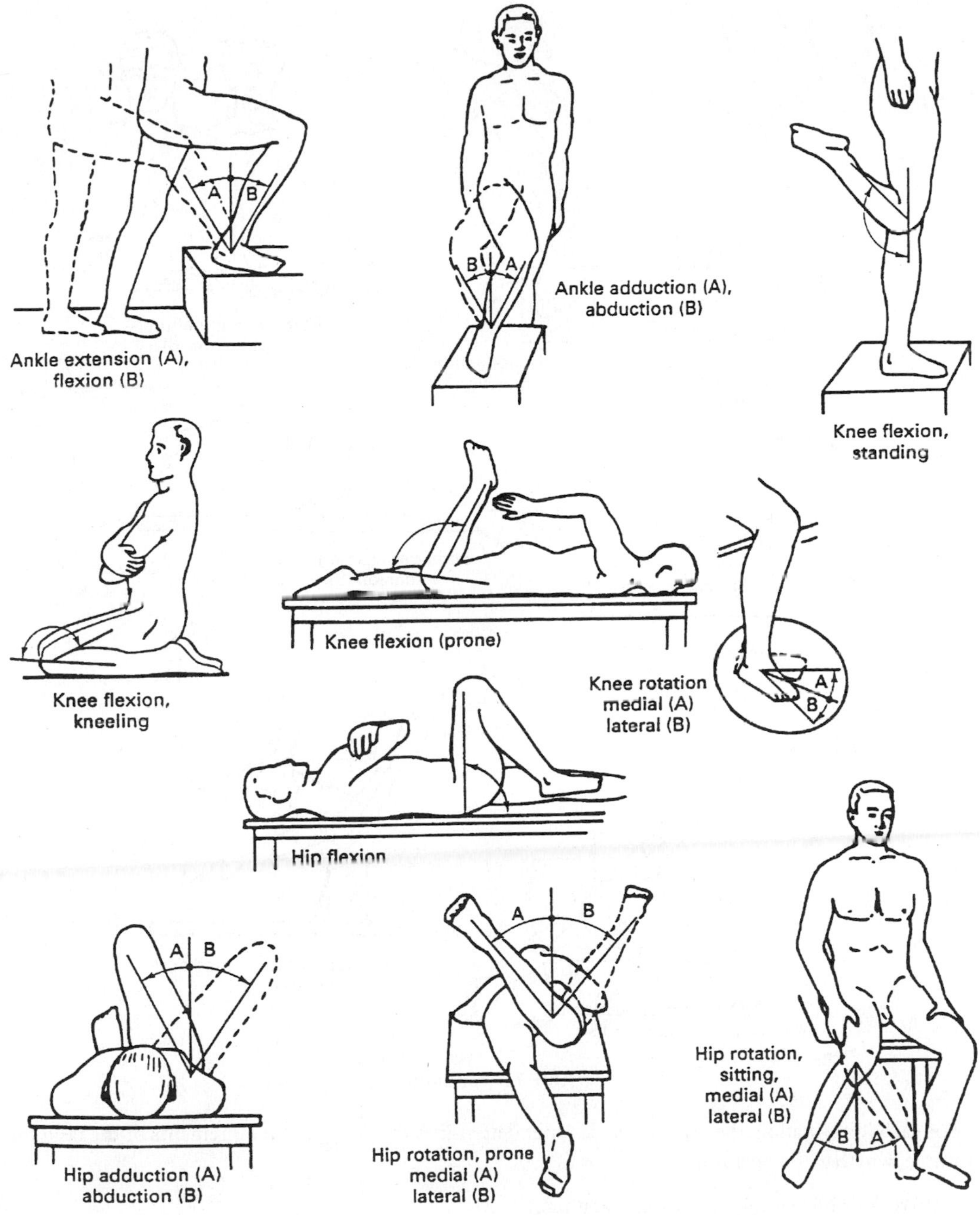

FIGURE 9.5 Maximal displacements in body joints. (From Kroemer, Kroemer, and Kroemer-Elbert, 1990, *Engineering Physiology: Bases of Human Factors/Ergonomics*, 2nd ed. With permission.)

(partial) spheres around the presumed locations of the body joints. However, the preferred ranges within the possible motion zones are different when the main requirements are strength, or speed, or accuracy, or vision — as discussed, in some detail, by Kroemer, Kroemer, and Kroemer-Elbert (1994). Thus, there is not one reach envelope, but different preferred envelopes.

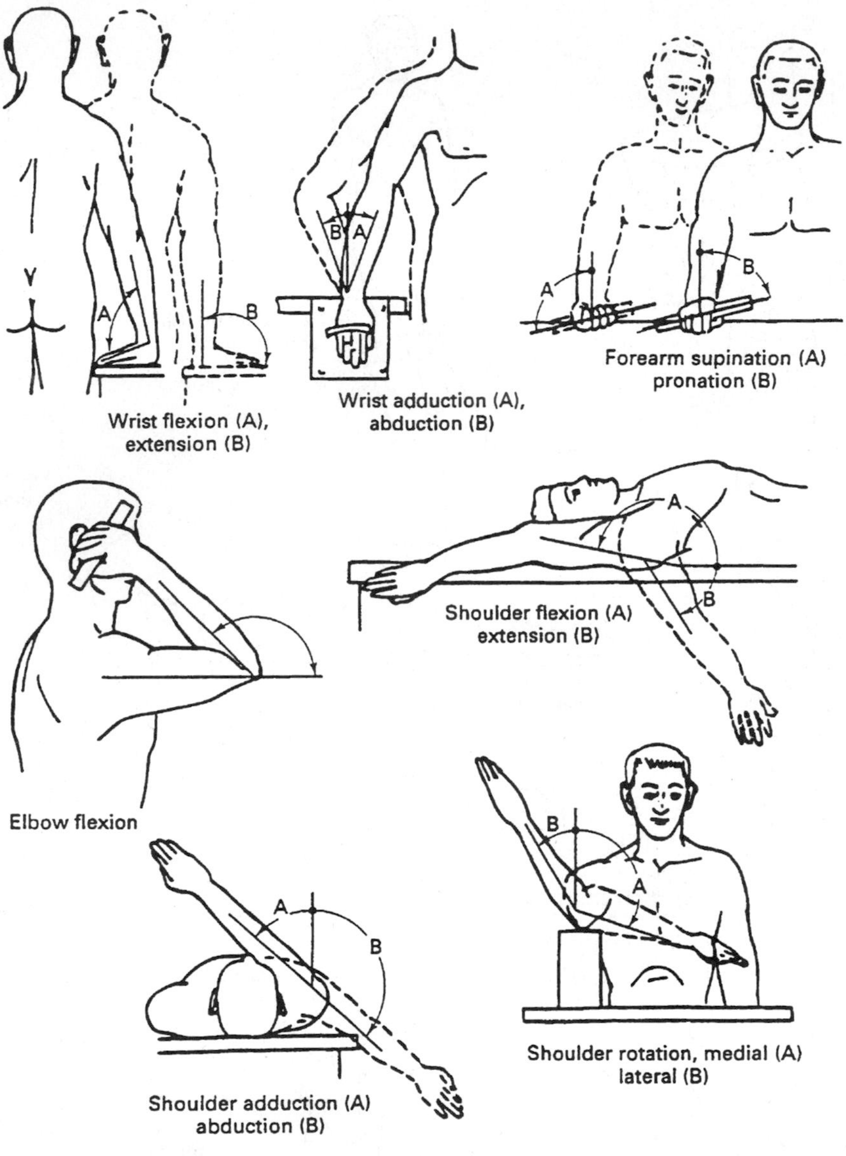

FIGURE 9.5 (continued)

For each job situation, the ergonomic designer determines the dominant requirements of the task; for example, whether the operator

- Works while sitting or walking (standing)
- Performs wide-ranging or specialized work
- Must exert large or small forces
- Executes fast and gross or slow and exact motions
- Needs high or low visual control.

Such circumstances affect the selection of the specific work envelope.

While sitting or moving about, the trunk is normally kept nearly erect, as are the neck and head. In most work situations, the upper arm hangs from the shoulder, while the elbow angle tends to be near 90 degrees; but the wrist is nearly straight. Table 9.12 lists typical body angles at work.

DESIGNING FOR MOTION IS DONE IN THESE STEPS:

Step 1: Select the major body joints involved.

Step 2: *Adjust body dimensions reported for standardized postures* (e.g., Tables 9.4 through 9.9) *to accommodate the real work conditions.* Use Table 9.3 for guidance.

Step 3: *Select appropriate motion ranges in the body joints.* The range can be depicted as the area between two positions, such as knee angles ranging between 60 and 105 degrees; or as a motion envelope, such as circumscribed by combined hand-and–arm movements, or by the clearance envelope under (through, within, beyond) which body parts must fit. Use Table 9.12 for guidance.

Basic work space design faults should be avoided. These include:

1. *Avoid twisted body positions,* especially of the trunk and neck. This results often from bad location of work objects, controls, and displays.
2. *Avoid forward bending of trunk, neck, and head.* This is frequently provoked by improperly positioned controls and visual targets including working surfaces that are too low.
3. *Avoid postures that must be maintained* for long periods of time, especially at the extreme limits of the range of motion. This is particularly important for the wrist and the back.
4. *Avoid holding the arms raised.* This results commonly from locating controls or objects too high, higher than the elbow when the upper arm hangs down. The upper limit for regular manipulation tasks is about chest height.

TABLE 9.12 Mobility Ranges at Work

Angles at	Walking About, Standing	Sitting
Knee	Near extreme stretch: 180 deg. or slightly less	Mostly mid-range: about 90 deg.
Hip (lateral new)	Near extreme stretch: about 180 deg.	Mostly mid-range: about 90 deg.
Shoulder	Mostly mid-range: Upper arm often hanging down	
Elbow	Mostly mid-range: about 90 deg.	
Wrist	Mostly mid-range: about straight	
Neck/Head	Mostly mid-range: about straight	
Back	Near extreme stretch: about erect	

From Kroemer, Kroemer, and Kroemer-Elbert, 1990, *Engineering Physiology: Bases of Human Factors/Ergonomics,* 2nd ed. With permission.

General criteria for work space layout relate to human strength, speed, effort, accuracy, importance, frequency, function and sequence of use, as listed in Table 9.13. Achieving the task while assuring safety for the human, avoiding overuse and unnecessary effort, and assuring ease and efficiency are primary design goals.

9.9 Summary

It is inexcusable to design tasks, tools or workstations for the phantom of "the average person" in a static position. No such person exists, and design for the average fits nobody well. Instead, ranges of body sizes,

TABLE 9.13 Guidelines for Workspace Design

Human Strength —	facilitate extension of strength (work, power) by object location and orientation.
Human Speed —	place items so that they can be reached and manipulated quickly.
Human Effort —	arrange work so that it can be performed with least effort.
Human Accuracy —	select and position objects so that they can be manipulated and seen with ease.
Importance —	the most important items should be in the most accessible locations.
Frequency of use —	the most frequently used items should be in the most accessible locations.
Function —	items with similar functions should be grouped together.
Sequence of use —	items which are commonly used in sequence should be laid out in that sequence.

of motions, and of strengths establish the design criteria. "Designing for function" is easy for the engineer who starts with proper anthropometric information and applies it ergonomically, that is with "ease and efficiency" as the guiding principles.

References

Bhattacharia, A. and McGlothlin, J.D. (Eds.) (1996). *Occupational Ergonomics.* New York, NY: Marcel Dekker.

Cheverud, J., Gordon, C.C., Walker, R.A., Jacquish, C., Kohn, L., Moore, A., and Yamashita, N. (1990). *1988 Anthropometric Survey of U.S. Army Personnel: Correlation Coefficients and Regression Equations.* (Natick TR 90/032-6). Natick, MA: U.S. Army Research, Development and Engineering Center.

Fluegel, F., Greil, H., and Sommer, K. (1986). *Anthropologischer Atlas.* Berlin, Germany: Tribuene.

Gordon, C.C., Churchill, T., Clauser, C.E., Bradtmiller, B., McConville, J.T., Tebbetts, I., and Walker, R.A. (1989). *1988 Anthropometric Survey of U.S. Army Personnel: Summary Statistics Interim Report.* (Natick-TR-89/027). Natick, MA: U.S. Army Natick Research, Development and Engineering Center.

Juergens, H.W., Aune, I.A., and Pieper, U. (1990). *International Data on Anthropometry.* (Occupational Safety and Health Series No. 65). Geneva, Switzerland: International Labour Office.

Kagimoto, Y. (Ed.) (1990). *Anthropometry of JASDF Personnel and Its Applications for Human Engineering.* Tokyo, Japan: Aeromedical Laboratory, Air Development and Test Wing JASDF.

Kroemer, K.H.E. (1983). Engineering anthropometry, in D.J. Oborne and M.M. Gruneberg, (Eds.) *The Physical Environment at Work* (pp. 39-68). London: Wiley.

Kroemer, K.H.E. (1989). Engineering anthropometry. *Ergonomics,* 32, 767-784

Kroemer, K.H.E. and Grandjean, E. (1997). *Fitting the Task to the Human* (5th ed.). London, U.K. and Bristol, PA: Taylor & Francis.

Kroemer, K.H.E., Kroemer, H.B., and Kroemer-Elbert, K.E. (1994). *Ergonomics: How to Design for Ease and Efficiency.* Englewood Cliffs, NJ: Prentice Hall.

Kroemer, K.H.E., Kroemer, H.J., and Kroemer-Elbert, K.E. (1997). *Engineering Physiology. Bases of Human Factors/Ergonomics* (3rd ed.). New York, NY: Van Nostrand Reinhold–Wiley.

NASA/WEBB (Ed.) (1978). *Anthropometric Sourcebook* (3 volumes). (NASA Reference Publication 1024). Houston, TX: NASA (NTIS, Springfield, VA 22161, Order No. 79 11 734).

Pheasant, S. (1986). *Bodyspace: Anthropometry, Ergonomics and Design.* London, U.K.: Taylor & Francis.

Pheasant, S. (1996). *Bodyspace: Anthropometry, Ergonomics and the Design of Work* (2nd ed.). London, UK: Taylor & Francis.

Pheasant, S.T. (1982). A technique for estimating anthropometric data from the parameters of the distribution of stature. *Ergonomics, 25,* 981-992

Roebuck, J.A. (1995). *Anthropometric Methods.* Santa Monica, CA: Human Factors and Ergonomics Society.

Roebuck, J.A., Kroemer, K.H.E., and Thomson, W.G. (1975). *Engineering Anthropometry Methods.* New York, NY: Wiley.

For Further Information

You are likely to encounter more involved considerations if data must be developed that describe composite populations which consist, for example, of subsamples such as *a*% females and *b*% males (Kroemer, 1983). Or you may have only mean values of body dimensions, but no information about the associated standard deviations that you must estimate. Also, a large set of design questions arises from link lengths and mass properties of body segments including locations of mass centers and definitions of joint centers, as well as mobility and motions characteristics. Related information and solutions for such challenges can be found in recent publications by Annis (in Bhattacharia and McGlothlin, 1996), Kroemer, Kroemer, and Kroemer-Elbert (1994, 1997), Pheasant (1986, 1996), or Roebuck (1995).

10

Occupational Biomechanics

William S. Marras
The Ohio State University

10.1 Biomechanic Analyses and Ergonomics

Definitions

Biomechanics may be defined as an interdisciplinary field in which information from both the biological sciences and engineering mechanics is used to assess the function of the body. *A major assumption of occupational biomechanics is that the body behaves according to the laws of Newtonian mechanics. By definition, "mechanics is the study of forces and their effects on masses"* (Kroemer, 1987). The function of interest in an occupational ergonomics context is most often a quantitative assessment of mechanical loading that occurs within the musculoskeletal system. The goal of an occupational biomechanics assessment is to quantitatively describe the musculoskeletal loading that occurs during work so that one can derive an appreciation for the degree of risk associated with an occupationally-related task. The characteristic that distinguishes occupational biomechanics analyses from other types of ergonomic analyses is that the comparison is quantitative in nature. The quantitative nature of occupational biomechanics permits ergonomists to address the question of how much exposure to the occupational risk factors is too much exposure.

The portion of biomechanics dealing with ergonomics issues is often labeled industrial or occupational biomechanics. Chaffin and Andersson (1991) have defined occupational biomechanics as "the study of the physical interaction of workers with their tools, machines, and materials so as to enhance the worker's performance while minimizing the risk of musculoskeletal disorders." This chapter will address occupational biomechanical issues exclusively in this ergonomics framework.

0-8493-2641-9/99/$0.00+$.50

Occupational Biomechanics Approach

In order to effectively address ergonomic issues in the workplace, one must develop an appreciation for the *trade-offs* associated with ergonomics. When one considers biomechanical rationale one finds that it is very difficult to accommodate all parts of the body in an ideal biomechanical environment. It is often the case that in attempting to accommodate one portion of the body the biomechanical situation at another body site is compromised. Therefore, the key to the proper employment of occupational biomechanical principles is to be able to consider the appropriate biomechanical trade-offs with various parts of the body associated with different workplace design options. For this reason this chapter will focus on the information required to develop proper biomechanical reasoning when considering a workplace. The chapter will first present and explain a series of key concepts that make up the heart of biomechanical reasoning. From there, these concepts will be applied to different parts of the body. Once this reasoning is developed, we will examine how the various biomechanical concepts must be considered collectively in terms of trade-off when designing a workplace from an ergonomic perspective under realistic conditions. This chapter will demonstrate that one cannot successfully practice ergonomics by simply memorizing a set of "ergonomic rules" (e.g., keep the wrist straight or don't bend from the waist when lifting). These types of rule-based design strategies ultimately result in suboptimizing the workplace ergonomic conditions.

10.2 Biomechanical Concepts

The Load–Tolerance Model

The fundamental concept in the application of occupational biomechanics to ergonomics is that one could design workplaces so that the load imposed on a structure does not exceed the tolerance of the structure. This basic concept is illustrated in Figure 10.1. This figure shows the traditional concept of biomechanical risk in occupational biomechanics. A loading pattern is developed on a body structure that is repeated as the work cycles are repeated during a job. The structure tolerance is also shown in this figure. If the tolerance far exceeds load, then the task is considered safe and the magnitude of the difference between the load and the tolerance is considered the safety margin. As implied in this figure, risk occurs when the imposed load exceeds the tolerance. As many of the industrial tasks become more repetitive, this model is beginning to change. As shown in Figure 10.2, occupational biomechanics logic is beginning to appreciate the fact that, with repetitive loading, the tolerance of the structure of interest may decrease over time to the point where it is more likely that the structure loading will exceed the structure tolerance and result in injury or illness. Thus, occupational biomechanical models and logic are beginning to build systems that consider observations in the workplace, such as cumulative trauma disorders.

Acute versus Cumulative Trauma

It is well recognized that in occupational settings two types of trauma can affect the human body and lead to musculoskeletal disorders. First, *acute* trauma can occur which refers to an application of force that is so large that it exceeds the tolerance of the body structure during an occupational task. Thus, acute trauma is typically associated with large exertions of force that occur infrequently. For example, an acute trauma can occur when a worker is asked to lift an extremely heavy object, as when moving a heavy part. This situation would relate to a peak load pattern that exceeded the load tolerance in Figure 10.1. *Cumulative* trauma, on the other hand, refers to the repeated application of force to a structure that tends to wear it down, thus lowering the structure tolerance to the point where the tolerance is exceeded through a reduction of the tolerance limit. This situation is illustrated in Figure 10.2. Cumulative trauma represents more of a "wear and tear" on the structure. This type of trauma is becoming far more common in the workplace because more repetitive jobs are becoming common in industry and are a concern for many ergonomics evaluations.

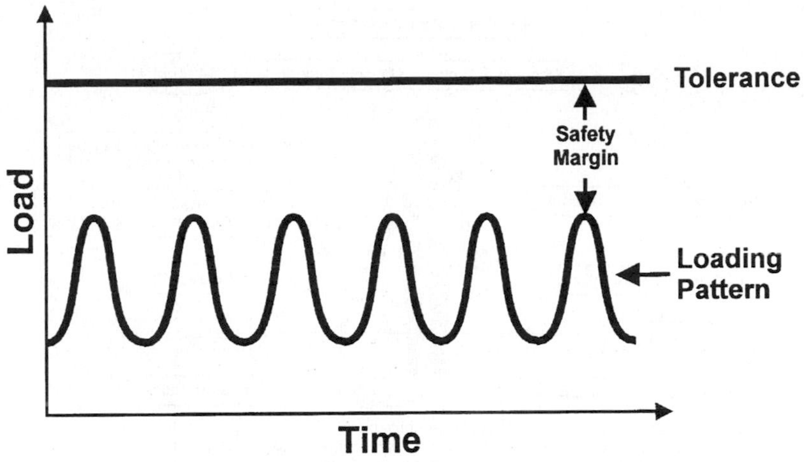

FIGURE 10.1 Traditional concept of biomechanical risk.

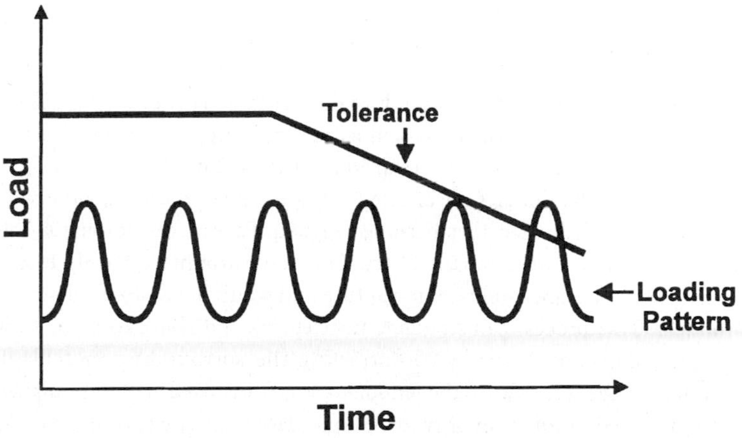

FIGURE 10.2 Realistic scenario of biomechanical risk.

Cumulative trauma can initiate a process that can result in a cycle of reaction of the body's structures that are very difficult to break. This process is illustrated in Figure 10.3. The cumulative trauma process begins by exposing the worker to manual exertions that are either frequent or prolonged. This repetitive or prolonged application of force can affect either the tendons or the muscles of the body. If the tendons are affected, the following sequence occurs. The tendons are subject to mechanical irritation when they are repeatedly exposed to high levels of tension, and groups of tendons may rub against each other. The physiologic reaction to this mechanical irritation can result in inflammation and swelling of the tendon. This swelling will stimulate the activities of the nociceptors surrounding the structure and signal the central control mechanism (brain), via pain perception, that a problem exists. In response to this pain, the body will attempt to control the problem via two mechanisms. First, the muscles surrounding the irritated area will coactivate in an attempt to minimize the motion of the tendon or stiffen the structure. Since motion will further stimulate the nociceptors and result in further pain, motion avoidance is often indicative of the start of a cumulative trauma disorder. Second, in an attempt to reduce the friction occurring within the tendon, the body will increase its production of synovial fluid within the tendon sheath. However, given the limited space available between the tendon and the tendon sheath, the increased production of synovial fluid often exacerbates the problem by further expanding the tendon

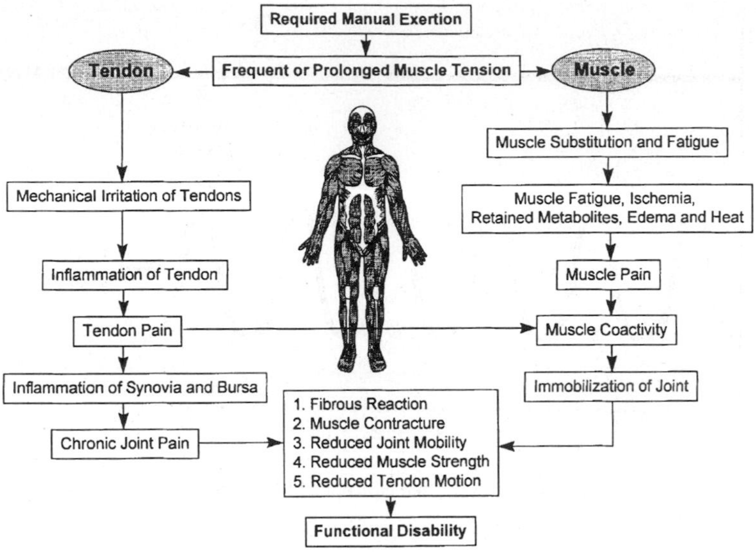

FIGURE 10.3 Sequence of events in cumultive trauma disorders. (Adapted from Chaffin, D.B. and Andersson, G.B. (1991) *Occupational Biomechanics*, John Wiley & Sons, Inc. New York, NY.)

sheath and, thus, further stimulating the surrounding nociceptors. This process results in chronic joint pain and a series of musculoskeletal reactions such as reduced strength, reduced tendon motion, and reduced mobility. Collectively, these reactions result in a functional disability.

A similar process occurs if the muscles are affected by cumulative trauma as opposed to the tendons. Muscles can be easily overloaded when they become fatigued. Fatigue can lower the tolerance to stress and can result in microtrauma to the muscle fibers. This microtrauma typically means the muscle is partially torn and the tear will cause capillaries to rupture and result in swelling, edema, or inflammation near the site of the tear. This process can stimulate nociceptors and result in pain. As with cumulative trauma to the tendons, the body reacts by cocontracting the surrounding musculature and thereby minimizing the motion of the joint. Since the tendons are not involved with cumulative trauma to the muscles, there is no increased production of synovial fluid. However, the end result is the same series of musculoskeletal reactions resulting from tendon irritation (i.e., reduced strength, reduced tendon motion, and reduced mobility). The ultimate result of this process is once again a functional disability.

Even though the stimulus associated with the cumulative trauma process is somewhat similar between tendons and muscles, there is a significant difference in the time required to heal from the damage to a tendon compared to a muscle. The mechanism of repair for both the tendons and muscles is dependent upon blood flow to the damaged structure. Blood provides nutrients for repair and dissipates waste materials. However, the blood supply to a tendon is just a fraction of that supplied to a muscle. Thus, given an equivalent strain to a muscle and a tendon, the muscle will heal rapidly (in about ten days if not reinjured), whereas the tendon could take months to accomplish the same level of repair. For this reason, ergonomists must be particularly vigilant in the assessment of workplaces that could pose a danger to the tendons of the body.

Moments and Levers

Biomechanical loads are *not* defined solely by the magnitude of weight supported by the body. The position of the weight relative to the axis of rotation of the body joint of interest defines the imposed load on the body and is referred to as a *moment*. Thus, a moment is defined as the product of force and distance. For example, a 50 Newton mass held at a horizontal distance of 75 cm (.75 meters) from the shoulder joint imposes a moment of 37.5 Nm (50 N × 0.75 m) on the shoulder joint, whereas the same weight held at a horizontal distance of 25 cm from the shoulder joint imposes a moment or load of only 12.5 Nm (50/n × 0.25m) on the shoulder. Thus, the load on a joint is a function of where the load is held relative to the joint and the mass of the weight held. Load is not simply a function of weight.

As implied by this example, moments are a function of the mechanical lever systems of the body. The musculoskeletal system can be represented by systems of levers, and these lever systems usually form the basis of most biomechanical assessments and models. Three types of lever systems are present in the human body. First-class levers are those that have a fulcrum in the middle of the system, an imposed load on one end of the system, and the restorative or internal load imposed on the opposite end of the system.

As will be discussed later (Figure 10.18), the trunk is an example of a first-class lever. In this example, the spine serves as the fulcrum. As the worker lifts, a moment is imposed anterior to the spine due to the object weight times the distance of the object from the spine. This moment is counterbalanced by the activity of the back musculature, but the mechanical advantage of the back muscles is much less than that of the object lifted. A second-class lever system can be found in the lower extremity. In a second-class lever system, the fulcrum is on one end of the lever, the restorative load is on the other end of the system, and the applied load is between these two. In the body, the lower leg is a good example of this lever system. In this example, the ball of the foot acts as the fulcrum, the load is applied through the tibia or bone of the lower leg. The restorative force is applied through gastrocnemius or calf muscle. In this manner, the muscle activates and causes the body to rotate about the fulcrum or ball of the foot and move the body forward. Finally, a third-class lever system is one where the fulcrum is on one end of the system, the applied load acts at the other end of the system, and the restorative force acts between the two. An example of this system in the human body is the elbow joint and is shown in Figure 10.4.

External and Internal Loading

Two types of forces can load the body during work. *External* loads refer to those forces that are imposed on the body as a result of gravity acting upon an external object being manipulated by the worker. For example, in Figure 10.4a the tool held in the worker's hand is subject to the forces of gravity and imposes a 44.5 N (10 pound) external load at a distance from the joint of 30.5 cm (12 inches) on the elbow joint. However, in order to maintain equilibrium, this external load must be counteracted by an *internal* load that is supplied by the muscles of the body. Figure 10.4a also shows that the internal load (muscle) acts at a distance relative to the elbow joint that is much closer to the fulcrum than the external load (tool). Thus, the internal load or force is at a biomechanical disadvantage and must be much larger (534 N or 120 lbs.) than the external load (44.5 N or 10 lbs.) in order to keep the musculoskeletal system in equilibrium. As shown in this example, it is not unusual for the magnitude of the internal load to be much greater (typically 10 times greater) than the external load. Thus, it is typically the internal loading that contributes most to cumulative trauma of the musculoskeletal system during work. The sum of the external load and the internal load define the total loading experienced at the joint. When evaluating a workstation, the ergonomist must not only consider the externally applied load but must be particularly sensitive to the magnitude of the internal forces that can load the musculoskeletal system.

Factors Affecting Internal Loading

The previous section has discussed the importance of understanding the relationship between the external loads imposed on the body and the internal loads generated by the force-generating mechanisms within the body. The key to proper ergonomic design involves designing workplaces so that the internal loads are minimized. Several properties of the work environment can be manipulated in order to facilitate this goal.

Posture and Length–Strength

The posture assumed when one works can affect the arrangement of the body's leverage system, and thus, can greatly affect the magnitude of the internal load required to support the external load. The

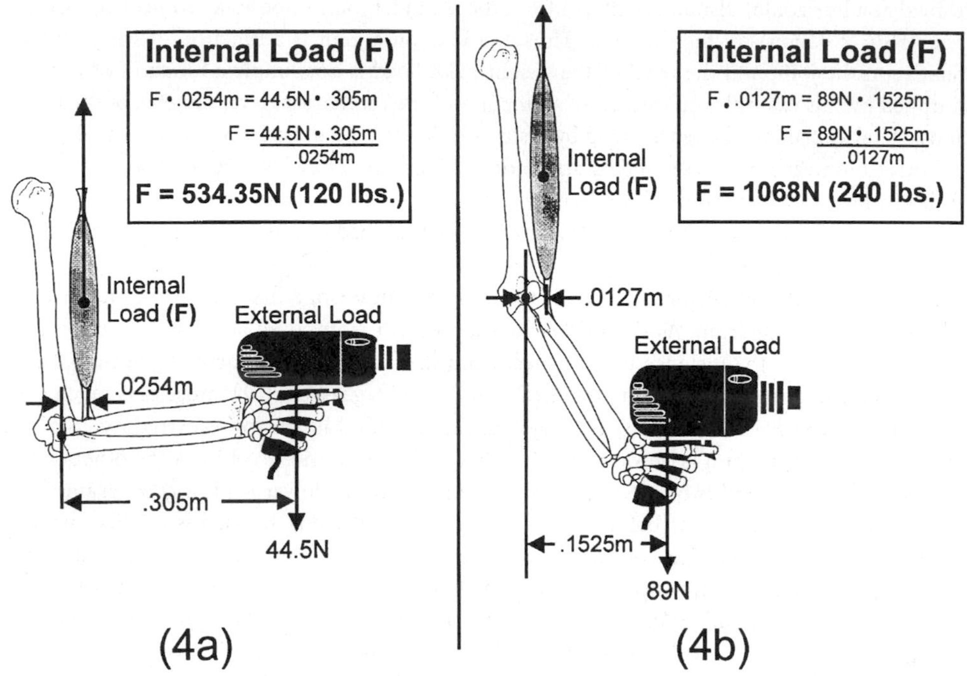

FIGURE 10.4 An example of an anatomical third class lever (a) demonstrating how the mechanical advantage changes as the elbow position changes (b).

arrangement of the lever system can influence the magnitude of the external moment (force × distance) imposed on the body as well as dictate the magnitude of the internal forces and the subsequent risk of cumulative trauma. Consider the biomechanical arrangement of the elbow joint that is shown in Figure 10.3. In Figure 10.4a, the mechanical advantage of the internal force generated by the biceps muscle and tendon is minimized by maintaining a posture keeping one's arm bent at a 90° angle. If one palpates the tendon and inserts the index finger between the joint center and the tendon one can gain an appreciation for the internal moment arm distance. One can also appreciate how this internal mechanical advantage can change with posture. With the index finger still inserted between the elbow joint and the tendon if the arm is slowly straightened one can appreciate how the distance between the tendon and the joint center of rotation is significantly reduced. If the moment imposed about the elbow joint is held constant as shown in Figure 10.4b, the mechanical advantage of the internal force generator is significantly reduced. In other words, since the internal moment or distance between the tendon and the joint center is reduced (compared to the situation where the elbow is positioned at a 90° angle), the muscle must produce more force in order to support the external load. This force is transmitted through the tendon and can increase the risk of cumulative trauma. Therefore, the positioning of the mechanical lever system can greatly affect the internal load transmission within the body. The same task can be performed in a variety of ways, but some of these positions are much more costly in terms of loading of the musculoskeletal system than others.

Another important relationship in defining the load on the musculoskeletal system is the length–strength relationship of the muscles. Figure 10.5 shows this relationship. The active portion of this figure refers to structures that actively generate force, such as muscles. The figure indicates that when muscles are close to their resting length they are in a position where they have the greatest capacity to generate force. However, when the muscle length deviates from this resting position, the capacity to generate force is greatly reduced. Hence, when a muscle stretches or becomes very short, the ability to generate force is greatly diminished. Note also, as indicated in Figure 10.5, that passive tissues in the muscle (and also ligaments) can generate tension when muscles are stretched. Thus, the orientation of

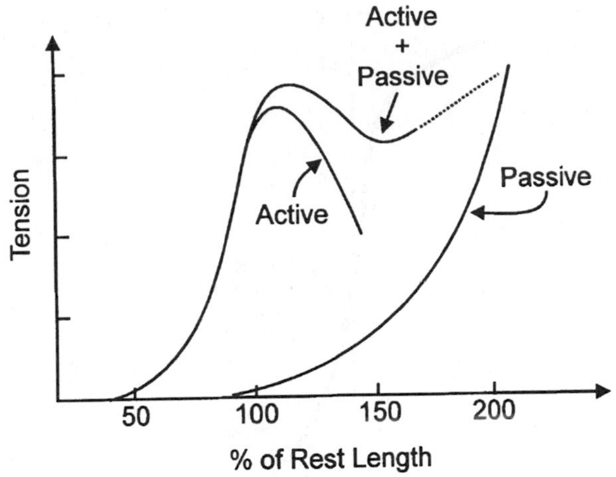

FIGURE 10.5 Length–tension relationship for a human muscle. (Adapted from Basmajian, J.V. and De Luca, C.J. (1985) *Muscles Alive: Their Functions Revealed by Electromyography* (5th edition), Williams and Wilkins, Baltimore, MD.)

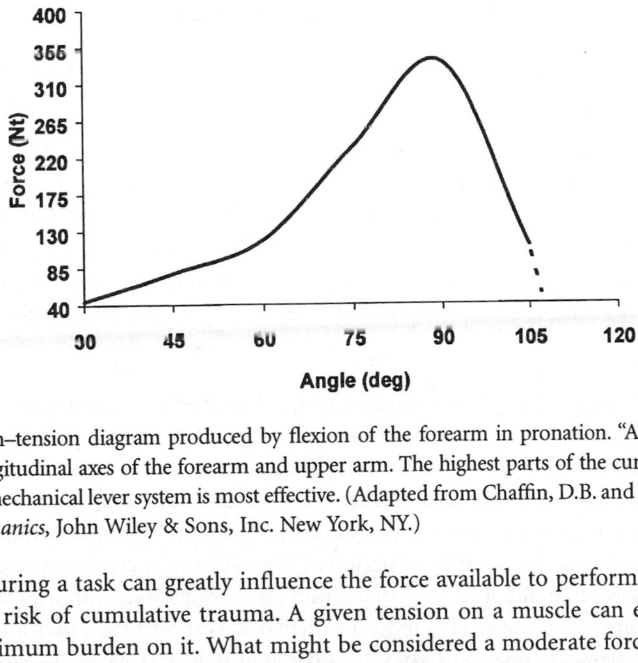

FIGURE 10.6 Length–tension diagram produced by flexion of the forearm in pronation. "Angle" refers to included angle between the longitudinal axes of the forearm and upper arm. The highest parts of the curve indicate the configurations where the biomechanical lever system is most effective. (Adapted from Chaffin, D.B. and Andersson, G.B. (1991) *Occupational Biomechanics*, John Wiley & Sons, Inc. New York, NY.)

the muscle fibers during a task can greatly influence the force available to perform a work task and can therefore influence risk of cumulative trauma. A given tension on a muscle can either tax the muscle greatly or be a minimum burden on it. What might be considered a moderate force for a muscle at the resting length can become the maximum force a muscle can produce when it is in a stretched or contracted position, thus increasing the risk of muscle strain. When this relationship is considered in conjunction with the mechanical load placed on the muscle and tendon via the arrangement of the lever system the position of the joint arrangement becomes a major factor in the design of the work environment. It is typically the case that the length–strength relationship interacts synergistically with the lever system. Figure 10.6 shows the effect of elbow position on the force-generation capability of the elbow. This figure indicates that position can have a dramatic effect on force generation. As already discussed, this position can also have a great effect on internal loading of the joint and the subsequent risk of cumulative trauma.

Force–Velocity

Motion can profoundly influence the ability of a muscle to generate force and load the biomechanical system. Motion can either be a benefit to the biomechanical system if momentum is properly used or it

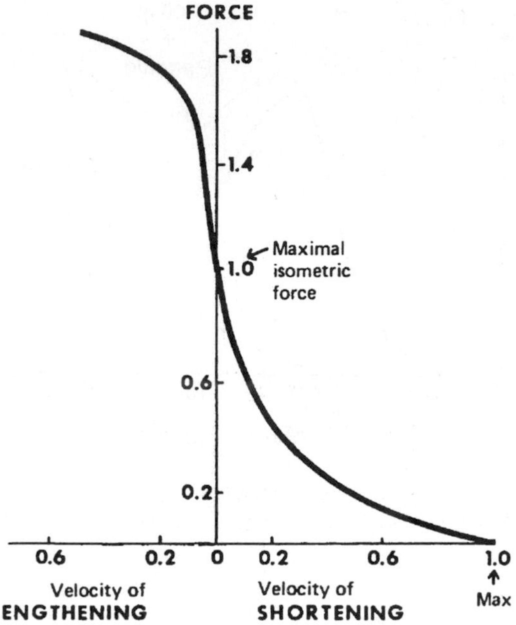

FIGURE 10.7 Influence of velocity upon muscle force. (Adapted from *The Textbook of Work Physiology*, McGraw-Hill, 1977)

can increase the load on the system if the worker is not taking advantage of momentum. The relationship between muscle velocity and force generation is shown in Figure 10.7. This figure shows that, in general, the faster the muscle is moving the greater the reduction in its force capability. As with most of the biomechanical principles mentioned in this chapter, this reduction in muscle capacity can result in the muscle strain occurring at a lower level of external loading and a subsequent increase in the risk of cumulative trauma. In addition, this effect is considered in many dynamic ergonomic biomechanical models.

Strength–Endurance

It is important to realize that strength is a transient factor. A worker may generate a great deal of strength during a one-time exertion. However, if the worker is required to exert large percentages of strength either repeatedly or for a prolonged period of time, the amount of force that the worker can generate is dramatically reduced. Figure 10.8 demonstrates this relationship during an isometric exertion. The broken line in this figure indicates that if a person is asked to generate maximum muscle force, maximum force output is only generated for a very brief period of time. As time increases, strength output decreases exponentially and levels off at about 20% of maximum after about seven minutes. Similar trends occur under repeated dynamic conditions. This indicates that if it is determined that a task requires a large portion of a worker's strength, one must consider how long that portion of the strength is required in order to ensure that the work does not strain the musculoskeletal system.

Rest Time

As has been mentioned several times in this chapter, the risk of cumulative trauma increases when the capacity to exert force is challenged by the external force requirements of the job. Another factor that can affect this strength capacity is rest time. Rest time has a profound effect on the ability to exert force. Figure 10.9 shows how energy for a muscular contraction is regenerated during work. Adenosine triphosphate (ATP) is required to produce a significant muscular contraction. ATP changes to adenosine diphosphate (ADP) once a muscular contraction has occurred. This ADP must be converted to ATP in order to enable another muscular contraction. This conversion can occur with the addition of oxygen to the system. If oxygen is not present, then the system goes into oxygen debt and there is insufficient

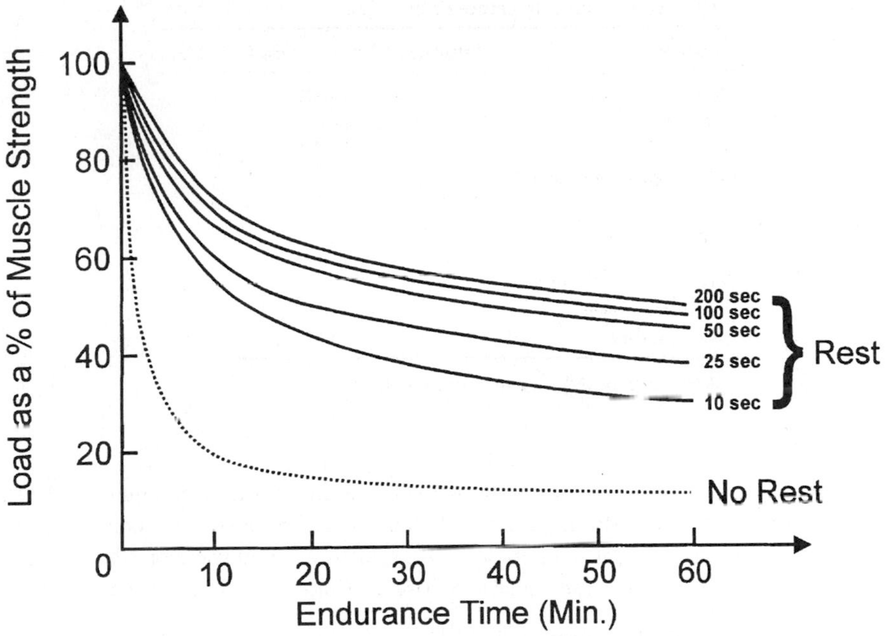

FIGURE 10.8 Forearm flexor muscle endurance times in consecutive static contractions of 2.5 sec duration with varied rest periods. (Adapted from Chaffin, D.B. and Andersson, G.B. (1991) *Occupational Biomechanics*, John Wiley & Sons, Inc. New York, NY.)

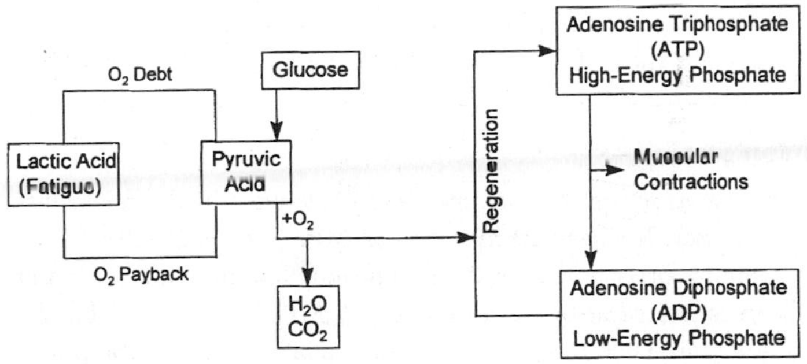

FIGURE 10.9 The body's energy system during work. (Adapted from Grandjean, E. (1982) *Fitting the Task to the Man: An Ergonomic Approach*, Taylor & Francis, Ltd. London.)

ATP for a muscular contraction. Thus, this flow chart indicates that oxygen is a key ingredient in maintaining a high level of muscular exertion. Oxygen is delivered to the target muscles via the blood flow. However, under static exertions the blood flow is reduced and there is a subsequent reduction in the blood flow to the muscle. This restriction of blood flow and subsequent oxygen deficit was responsible for the rapid decrease in force generation over time as shown in Figure 10.8. The solid lines shown in Figure 10.8 show how the force-generation capacity of the muscles increases when different amounts of rest are permitted in a fatiguing exertion. As more and more rest time is permitted, increases in force generation are achieved when more oxygen is delivered to the muscle and more ADP can be converted to ATP. This relationship also shows that any more than about 50 seconds of rest, under these conditions, does not result in a significant increase in force-generation capacity of the muscle. Practically, this indicates that in order to optimize the strength capacity of the worker and minimize the risk of muscle strain, a schedule of frequent and brief rest periods would be more beneficial than lengthy infrequent rest periods.

TABLE 10.1 Tissue Tolerance of the Musculoskeletal System

Structure	Estimated Ultimate Stress (σ_u) (MPa)
Muscle	32–60
Ligament	20
Tendon	60–100
Bone longitudinal loading	
Tension	133
Compression	193
Shear	68
Bone transverse loading	
Tension	51
Compression	133

Adapted from Ozkaya and Nordin, 1991

Load Tolerance

To this point this chapter has considered primarily factors that influence the loads applied to the structures of the body. As mentioned previously, occupational biomechanical analyses must consider not only the loads imposed upon a structure but also the ability of the structure to withstand or tolerate a load during work. This section will briefly review the knowledge base associated with body structure tolerances.

Muscle, Ligament, Tendon, and Bone Capacity

The exact tolerance characteristics of human tissues such as muscles, ligaments, tendons, and bones loaded under various working conditions is difficult to estimate. Tolerances of the structures in the body vary greatly under similar loading conditions. In addition, tolerance depends upon many other factors such as strain rate, age of the structure, frequency of loading, physiologic influences, heredity, conditioning, and many unknown factors. Furthermore, it is not possible to measure these tolerances under human *in vivo* conditions. Therefore, most of the published estimates of tissue tolerance have been derived from various animal and/or theoretical sources.

Muscle and Tendon Strain

Muscle appears to be the structure that has the lowest tolerance in the musculoskeletal system. The ultimate strength of a muscle has been estimated at 32 MPa (Hoy et al., 1990). It is generally believed that the muscle will rupture prior to the tendon in a healthy tendon (Nordin and Frankel, 1989), since tendon stress has been estimated at between 60 and 100 MPa (Nordin and Frankel, 1989; Hoy et al., 1990). Hence, as indicated in Table 10.1, there is a safety margin between the muscle failure point and the failure point of the tendon of about twofold (Nordin and Frankel, 1989) to threefold (Hoy et al., 1990).

Ligament and Bone Tolerance

Ligaments and bone tolerances within the musculoskeletal system have also been estimated. Ultimate ligament stress has been estimated at approximately 20 MPa. The ultimate stress of bone varies depending upon the direction of loading. Bone tolerance can range from as low as 51 MPa in transverse tension to over 190 MPa in longitudinal compression. Table 10.1 also indicates the ultimate stress of bone loaded in different loading conditions.

Disc/Endplate and Vertebrae Tolerance

The mechanism of cumulative trauma in the disc is thought to be related to repeated trauma to the vertebral endplate. The endplate is a very thin (about 1 mm thick) structure that facilitates nutrient flow to the disc fibers (anulus fibrosis). Repeated microfracture of this vertebral endplate is thought to impair the nutrient flow to the disc fibers and thereby lead to atrophy of the fiber and fiber degeneration. It is believed that if one can determine the level at which the endplate experiences a microfracture, one can then minimize the effects of cumulative trauma and disc degeneration within the spine.

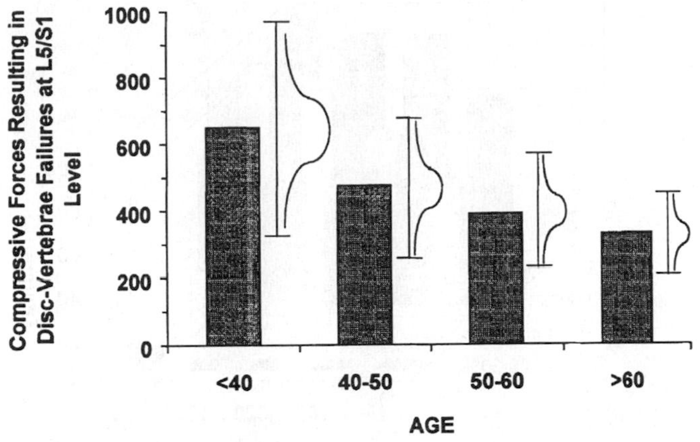

FIGURE 10.10 Mean and range of disc compression failures by age. (Adapted from National Institute for Occupational Safety and Health (NIOSH) (1981) Work practices guide for manual lifting. Department of Health and Human Services (DHHS), National Institute for Occupational Safety and Health (NIOSH), Publication No. 81-122.)

TABLE 10.2 Lumbar Spine Compressive Strength

		Strength in kN	
Population	n	Mean	s.d.
Females	132	3.97	1.50
Males	174	5.81	2.58
Total	507	4.96	2.20

Jager et al., 1991

Several studies of disc endplate tolerance have been performed. Figure 10.10 shows the levels of endplate compressive loading tolerance that have been used to establish safe lifting situations at the worksite (NIOSH, 1981). This figure shows the compressive force mean (column value) as well as the compression force distribution (thin line and normal distribution curve) that would result in vertebral endplate failure (microfracture). This figure indicates that for those under 40 years of age endplate microfracture damage begins to occur at about 350 kg (3432 N) of compressive load on the spine. If the compressive load is increased to 650 kg (6375 N), approximately 50% of those exposed to the load will experience vertebral endplate microfracture. When the compressive load on the spine reaches a value of 950 kg (9317 N), almost all of those exposed to the loading will experience a vertebral endplate microfracture. It should also be noted that the tolerance distribution shifts to lower levels with increasing age. In addition, it should be emphasized that this tolerance is based upon compression of the vertebral endplate alone. Shear and torsional forces in combination with compressive loading would be expected to further lower the tolerance of the endplate.

This distribution of risk has been widely used as the tolerance limits of the spine. However, it should be noted that others have identified different limits of vertebral endplate tolerance. Jager, Luttmann, and Laurig (1991) have reviewed 13 studies of spine compressive strength and suggested different compression value limits. Their summary of these spine tolerance limits is shown in Table 10.2. These researchers have also been able to describe the vertebral compressive strength based on an analysis of 262 values collected from 120 samples. They have related the compressive strength of the lumbar spine according to a regression equation:

$$\text{Compressive Strength (kN)} = (7.26 + 1.88\ G) - 0.494 + 0.468\ G) \times A + (0.042 + 0.106\ G) \times C - 0.145 \times L - 0.749 \times S$$

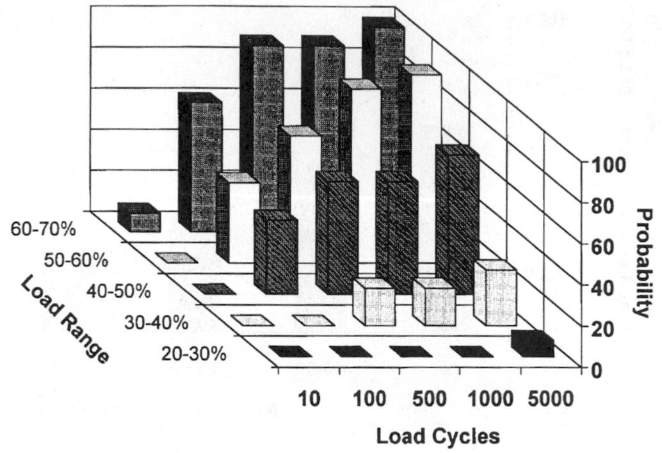

FIGURE 10.11 Probability of a motion segment to be fractured in dependence on the load range and the number of load cycles. (Adapted from Brinckmann, P., Biggemann, M., and Hilweg, D. (1988) Fatigue fracture of human lumbar vertebrae. *Clinical Biomechanics*, 3: Supplement 1, S1-S23.)

where

A = age in decade
G = gender coded as 0 for female or 1 for male
C = cross-sectional area of the vertebrae in cm^2
L = the lumbar level unit where 0 is the L5/S1 disc, 1 represents the L5 vertebrae, etc. through 10 which represents the T10/L1 disc
S = the structure of interest where 0 is a disc and 1 is a vertebrae

This analysis suggests that the decrease in strength within a lumbar level is about 0.15 kN of that of the adjacent vertebrae and that the strength of the vertebrae is about 0.8 kN lower than the strength of the discs (Jager et al., 1991). Using this equation, these researchers were able to account for 62% of the variability among the samples.

It has also been suggested that the tolerance limits of the spine varies as a function of frequency of loading (Brinckmann et al.; 1988). Figure 10.11 indicates that the spine tolerance varies as a function of spine load level and frequency of loading.

10.3 The Application of Biomechanics to the Workplace

Biomechanics of Commonly Injured Body Structures

Now that the basic concepts and principles of biomechanics relevant to ergonomics situations have been established we can apply these principles to various work situations. This section will show how one can apply these principles to various regions of the body that are typically affected by occupational tasks.

Shoulder

Shoulder pain is suspected of being one of the most under-recognized musculoskeletal disorders in the workplace. Second only to low back injury and neck pain in clinical frequency and reporting, shoulder region disorders are increasingly being recognized as a major workplace problem by those organizations that have reporting systems sensitive enough to detect such trends. The shoulder is one of the more complex structures of the body with numerous muscles and ligaments crossing the shoulder joint girdle complex. Because of its biomechanical complexity, surgical repair of the shoulder can be problematic. During many shoulder surgeries, it is often necessary to damage much of the surrounding tissue in an

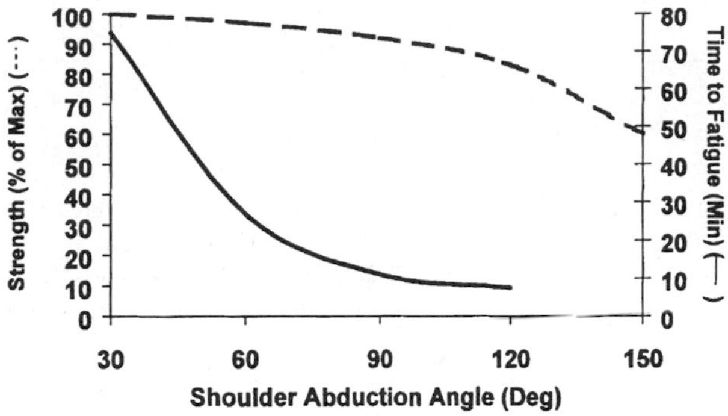

FIGURE 10.12 Shoulder abduction strength and fatigue time as a function of shoulder abducted from the torso. (Adapted from Chaffin, D.B. and Andersson, G.B. (1991) *Occupational Biomechanics,* John Wiley & Sons, Inc. New York, NY.)

attempt to reach the structure in need of repair. Often the target structure is small and difficult to reach. Thus, many times more damage is done to surrounding tissues than the benefit derived to the target tissue. Therefore, the best course of action is to ergonomically design workstations so that risk of initial injury is minimized.

Since the shoulder joint is so biomechanically complex, much of our biomechanical knowledge is derived from empirical evidence. The shoulder represents a statically indeterminate system in that we can typically measure six external moments and forces acting about the point of rotation, yet there are far more internal forces (over 30 muscles and ligaments that must counteract the external moments. Thus, quantitative estimates of shoulder joint loading are rare.

With respect to the shoulder, optimal workplace design is typically defined in terms of preferred posture during work. Shoulder *abduction,* defined as the elevation of the shoulder in the lateral direction, is of concern when work is performed overhead. Figure 10.12 indicates shoulder performance measures in terms of both available strength and perceived fatigue while the shoulder is held in varying degrees of abduction. This figure indicates that the shoulder can produce a considerable amount of strength throughout shoulder abduction angles of between 30 and 90 degrees. However, when we compare fatigue characteristics at these same abduction angles, it is apparent that fatigue increases rapidly as the shoulder is abducted above 30 degrees. Thus, even though strength is not a problem at shoulder abduction angles up to 90 degrees, fatigue becomes the limiting factor. Therefore, the only position of the shoulder that is acceptable from both a strength and fatigue standpoint is a shoulder abduction of, at most, 30 degrees.

Shoulder *flexion* has been examined almost exclusively as a function of fatigue. Chaffin (1973) has shown that even slight shoulder flexion can influence fatigue characteristics of the shoulder musculature. Figures 10.13 and 10.14 indicate the effects of vertical height of the work and horizontal distance, respectively, during shoulder flexion while seated, upon fatiguability of the shoulder musculature. During vertical flexion/extension (Figure 10.13), fatigue occurs more rapidly as the worker's arm becomes more elevated. This trend is most likely due to the fact that the muscles are farther from the neutral position as the shoulder becomes more elevated, thus affecting the length–strength relationship (Figure 10.5) of the shoulder muscles. Figure 10.14 shows that as the horizontal distance between the work and the body is increased, the time to reach significant fatigue is decreased. This trend is due to the fact that, as a load is held further from the body, more of the external moment (force × distance) must be supported by the shoulder. Thus, the shoulder muscles must produce a greater internal force when the load is held farther from the body, and they fatigue more quickly. Elbow supports have been shown to significantly increase the endurance time in these postures. In addition, an elbow support has the effect of changing the biomechanical situation by providing a fulcrum at the elbow. Thus, the axis of rotation becomes the elbow instead of the shoulder, and this makes the external moment much shorter. As shown in Figure 10.15, this not only increases the time one can maintain a posture, but also significantly increases the external load one can hold in the hand.

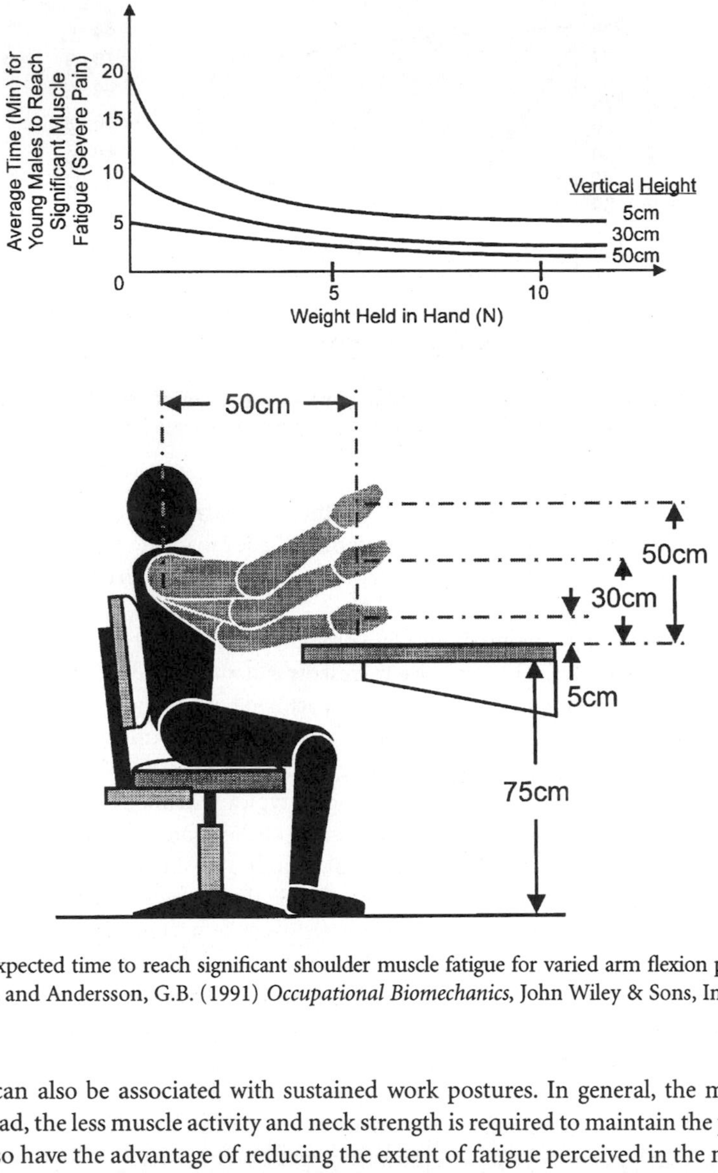

FIGURE 10.13 Expected time to reach significant shoulder muscle fatigue for varied arm flexion postures. (Adapted from Chaffin, D.B. and Andersson, G.B. (1991) *Occupational Biomechanics*, John Wiley & Sons, Inc. New York, NY.)

Neck

Neck disorders can also be associated with sustained work postures. In general, the more upright the posture of the head, the less muscle activity and neck strength is required to maintain the posture. Upright neck postures also have the advantage of reducing the extent of fatigue perceived in the neck region. This relationship is shown in Figure 10.16. This trend indicates that when the head is tilted forward 30 degrees or more from the vertical position, the time to experiencing significant neck fatigue increases rapidly. From a biomechanical standpoint, as the head is flexed forward, the center of mass of the head moves forward relative to the base of support of the head (spine). Therefore, as the head is moved forward, more of a moment is imposed about the spine, which necessitates increased activation of the neck musculature and greater risk probability of fatigue because a static posture is maintained by the neck muscles. When the head is not flexed forward and is relatively upright, the neck can be positioned in such a way that minimal muscle activity is required of the neck muscles and thus fatigue is minimized.

Neck–Shoulder Trade-offs and Work Height

As mentioned earlier, the key to proper ergonomic design of a workplace from a biomechanical standpoint is to consider the biomechanical trade-offs associated with a particular work situation. These trade-offs are necessary because often a situation that is advantageous for one part of the body is disadvantageous for another part of the body. Thus, many biomechanical considerations in the ergonomic design of the workplace require one to consider the various trade-offs and rationales for various design options.

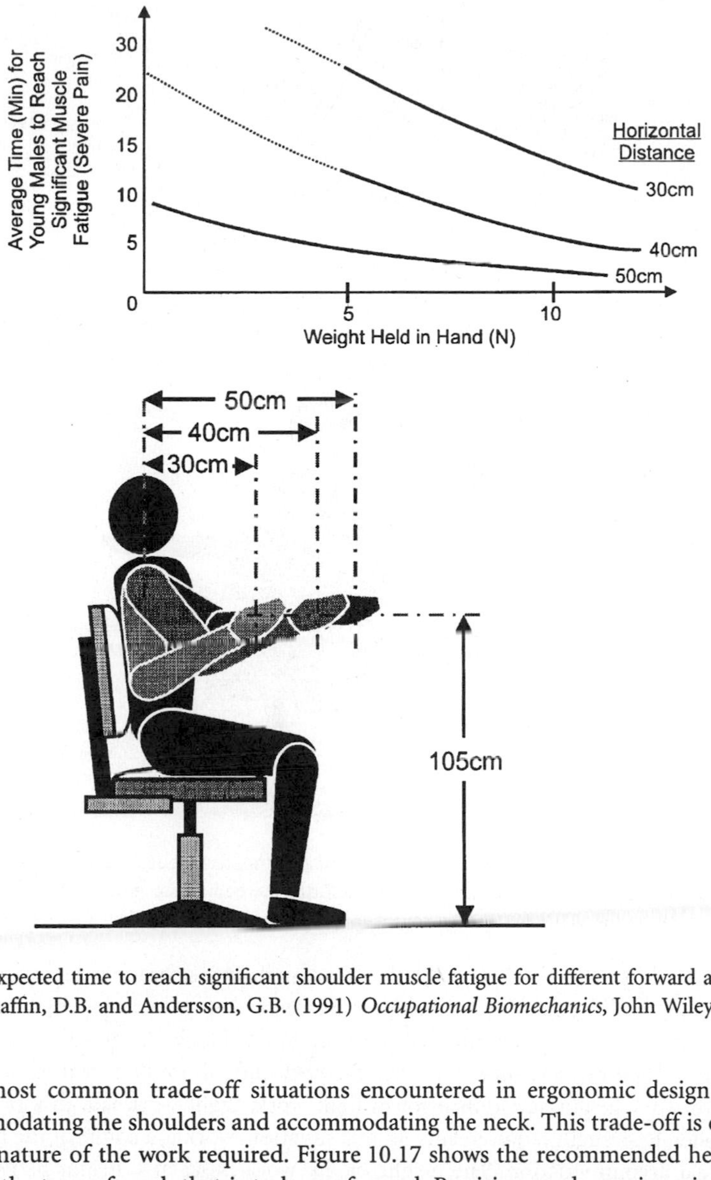

FIGURE 10.14 Expected time to reach significant shoulder muscle fatigue for different forward arm reach postures. (Adapted from Chaffin, D.B. and Andersson, G.B. (1991) *Occupational Biomechanics*, John Wiley & Sons, Inc. New York, NY.)

One of the most common trade-off situations encountered in ergonomic design is the trade-off between accommodating the shoulders and accommodating the neck. This trade-off is often resolved by considering the nature of the work required. Figure 10.17 shows the recommended height of the work as a function of the type of work that is to be performed. Precision work requires visual acuity, which is of prime importance in order for the worker to be able to accomplish the task. If the work is performed at too low a level, the head must be flexed in order to accommodate the visual requirements of the job. This situation could result in significant neck discomfort. Therefore, in this situation the proper work height is dictated by visual acuity requirements, and the work is typically raised to a relatively high level (95 to 110 cm above the floor). This position accommodates the neck but creates a problem for the shoulders because they must be abducted when the work level is high. Thus, a trade-off must be considered. Ideal shoulder posture is sacrificed in order to accommodate the neck because the visual requirements of the job are great. The logic associated with these trade-offs also dictates that the shoulder problems can be minimized by providing wrist or elbow supports at the workplace.

The other extreme of the working height situation involves heavy work. The greatest demand on the worker in heavy work is for a high degree of arm strength, whereas visual requirements in this type of work are typically minimal. Therefore, in this situation, ideal neck posture is typically sacrificed in favor of more favorable shoulder and arm postures. Hence, heavy work is performed at a height of 70 to 90 cm

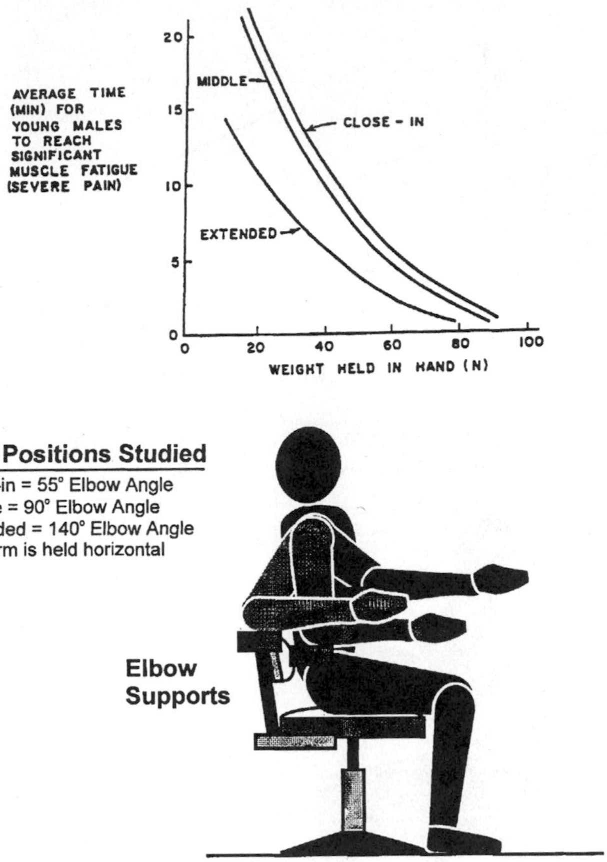

FIGURE 10.15 Expected time to reach significant shoulder and arm muscle fatigue for different arm postures and hand loads with the elbow supported. The greater the reach, the shorter the endurance time. (Adapted from Chaffin, D.B. and Andersson, G.B. (1991) *Occupational Biomechanics*, John Wiley & Sons, Inc. New York, NY.)

above the floor. With the work set at this height, the elbow angles are close to 90 degrees, which maximizes strength (Figure 10.6), and the shoulders are close to 30 degrees of abduction, which minimizes fatigue. In this situation, the neck is not in an optimal position, but the logic dictates that the visual demands of a heavy task would not be substantial and thus the neck should not be flexed for prolonged periods of time. The third work height situation involves light work. Light work is a mix of moderate visual demand with moderate strength requirements. In this situation, work is a compromise between shoulder position and visual accommodation. The height of the work is set at a height between those of the precision work height level and the heavy work height level. This dictates that the work is performed at a level of between 85 and 95 cm off the floor under light work conditions.

The Back

Low back disorders (LBDs) have been labeled as one of the most common and significant musculoskeletal problems in the U.S. that result in substantial amounts of morbidity, disability, and economic loss (Hollbrook et al., 1984; Praemer et al., 1992). Next to the common cold, low back disorders are the most common reason for workers to miss work. Back disorders were responsible for half a billion lost workdays in 1988, with 22 million cases reported that year (Guo, 1993). Among those under 45 years of age, LBD is the leading cause of activity limitations and can affect up to 47% of workers with physically demanding jobs (Andersson, 1991). The prevalence of LBD has also been observed to have increased by 2700% since 1980 (Pope, 1993). The costs associated with LBD are very significant. Estimates of lost wages alone amount to nearly four billion dollars annually (Frymoyer et al., 1983). Recent estimates of the total costs to society from low back pain range from $25 to $95 billion annually (Cats-Baril and Frymoyer, 1991).

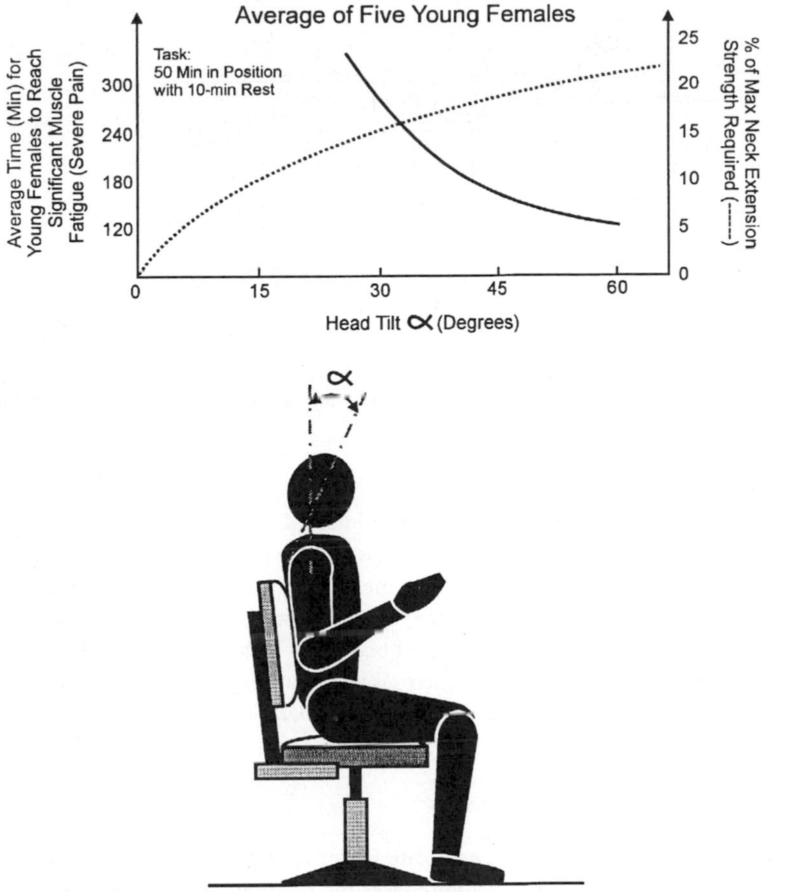

FIGURE 10.16 Neck extensor fatigue and muscle strength required vs. head tilt angle. (Adapted from Chaffin, D.B. and Andersson, G.B. (1991) *Occupational Biomechanics*, John Wiley & Sons, Inc. New York, NY.)

It is clear that the risk of LBD can be associated with industrial work (Andersson, 1981). Thirty percent of occupational injuries in the U.S. are caused by overexertion, lifting, throwing, holding, carrying, pushing, and or pulling objects that weigh 50 pounds or less (National Safety Council, 1989). Twenty percent of all workplace injuries and illnesses are back injuries, which account for up to 40% of compensation costs. Estimates of occupational low back disorder prevalence vary from one to 15% annually, depending upon occupation (Kelsey and White, 1980) and, over a career, can seriously affect 56% of workers (Rowe, 1981).

Manual materials handling (MMH) activities, specifically lifting, dominate occupationally related low back disorder risk. It has been estimated that lifting and MMH account for 50 to 75% of all back injuries (Bigos et al., 1986; Snook, 1989; Spengler et al., 1986). From a biomechanical standpoint, we assume that most serious and costly back pain is discogenic in nature and has a mechanical origin (Nachemson, 1975). Studies have found increased degeneration in the spines of cadaver specimens who had previously been exposed to physically heavy work (Videman et al., 1990). This suggests that occupationally related low back disorders are closely associated with spine loading.

Significance of Moments

The most important concept associated with occupationally related low back disorder risk is that of the external moments imposed about the spine (Marras et al., 1993). As with most body structures, the loading of the trunk is greatly influenced by the external moment imposed about the spine. However, because of the geometric arrangement of the trunk musculature relative to the trunk fulcrum during

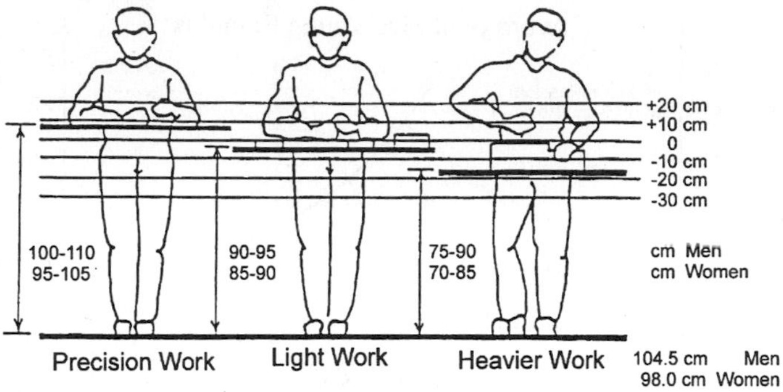

FIGURE 10.17 Recommended heights of bench for standing work. The reference line (+0) is the height of the elbows above the floor (From Grandjean, E. (1982) *Fitting the Task to the Man: An Ergonomic Approach,* Taylor & Francis, Ltd. London. With permission.)

lifting, very large loads can be generated by the muscles and imposed upon the spine. Figure 10.17 shows this biomechanical arrangement of lever system. As indicated here, the back musculature is at a severe biomechanical disadvantage in many manual materials handling situations. Supporting an external load of 222 N (about 50 pounds) at a distance of one meter from the spine imposes a load of 222 Nm of external moment about the spine. However, since the spine-supporting musculature is relatively close to the external load, the trunk musculature must exert extremely large forces (4440 N or 998 lbs.) to simply hold the external load in equilibrium. These internal loads can be far greater if dynamic motion of the body is considered (since force is a product of mass and acceleration). Thus, the most important concept to consider in workplace design from a back protection standpoint is to keep the moment arm at a minimum.

Lifting Style

The external moment concept has major implications for lifting styles, or the best "way" to lift. Since the externally applied moment significantly influences the internal loading, the lifting style is of far less concern compared to the magnitude of the applied moment. Some have suggested that proper lifting involves lifting by "using the legs" as opposed to "stoop" lifting (bending from the waist). However, spine loading has also been found to be a function of anthropometry as well as lifting style. Biomechanical analyses (Park and Chaffin, 1974) have demonstrated that no one lift style is correct for all body types. For this reason, the National Institute of Occupational Safety and Health (NIOSH, 1981) has concluded that lift style need not be a consideration when assessing the risk of occupationally related low back disorder. Some have suggested that the internal moment of the trunk has a greater mechanical advantage when lumbar lordosis is preserved during the lift (McGill, 1986; Anderson and Chaffin, 1986). Thus, from a biomechanical standpoint, the primary indicator of spine loading and, thus, the correct lifting style is whatever style permits the worker to bring the center of mass of the load as close to the spine as possible.

Seated versus Standing Workplaces

Seated workplaces have become more prominent of late, especially with the aging of the workforce and the introduction of service-oriented and data processing jobs. It has been well documented that loads on the lumbar spine are always greater when one is seated compared to standing. This is due to the tendency for the posterior (bony) elements of the spine to form an active load path when one is standing. When one is seated, these elements are disengaged and more of the load passes through the intervertebral disc. Thus, work performed in a seated position puts the worker at greater risk of loading and therefore damaging the disc. Given this situation, it is important to consider the design features of a chair since it may be possible to influence disc loading through chair design. Figure 10.19 shows the results of pressure measurements made in the intervertebral disc of workers as the back angle of the chair and magnitude

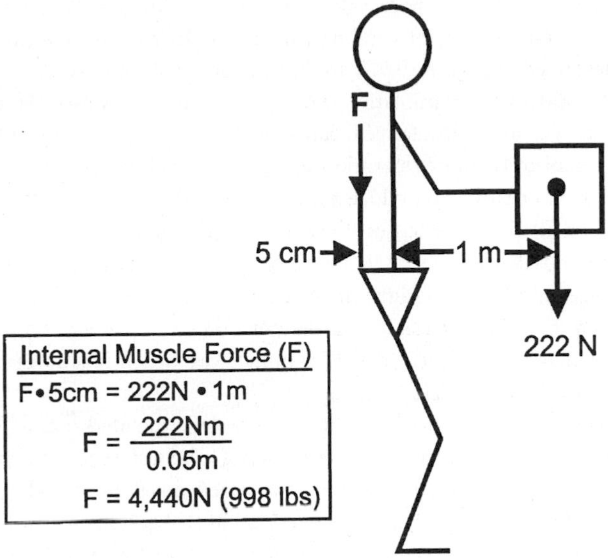

FIGURE 10.18 Internal muscle force required to counterbalance an external load during lifting.

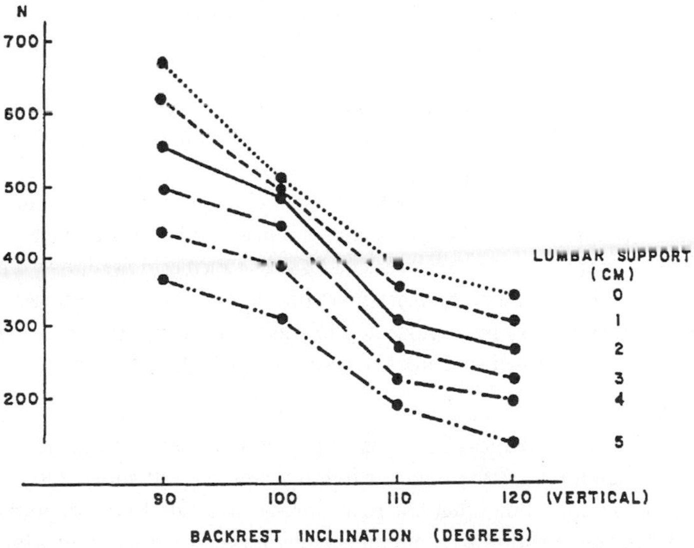

FIGURE 10.19 Disc pressures measured with different backrest inclinations and different size lumbar supports. (From Chaffin, D.B. and Andersson, G.B. (1991) *Occupational Biomechanics,* John Wiley & Sons, Inc. New York, NY. With permission.)

of the lumbar support were varied. This figure indicates that both the seat back angle and lumbar support features have a significant effect on disc pressure. Disc pressure is observed to decrease as the backrest angle is increased. However, increasing the backrest angle in the workplace is often not practical because it also moves the worker away from the work and thereby increases external moment. However, the figure also indicates that increasing lumbar support can also significantly reduce disc pressure. This reduction in pressure is most likely due to the fact that as lumbar curvature (lordosis) is reestablished (with lumbar support), the posterior elements play more of a role in providing an alternative load path, as is the case when standing in the upright position.

Less is known about risk to the low back associated with prolonged standing. It is known that the muscles experience low level static exertions and may be subject to the static overload through the muscle static fatigue process described in Figure 10.9. This fatigue can result in lowered muscle force-generation capacity and can, thus, initiate the cumulative trauma sequence of events (Figure 10.3). It has been demonstrated that this fatigue and cumulative trauma sequence can be minimized by two actions. First, foot rails provide a mechanism to allow relaxation of the large back muscles and, thus, increased blood flow to the muscle. This reduces the static load and fatigue in the muscle by the process described in Figure 10.9. When a leg is lifted and rested on the foot rest, the large back muscles are relaxed on one side of the body and the muscle can be supplied with oxygen. Alternating legs on the foot rest provides a mechanism to minimize back muscle fatigue throughout the day. Second, floor mats have been shown to decrease the fatigue in the back muscles provided that they have proper compression characteristics (Kim et al., 1993). Floor mats are believed to induce body sway which facilitates the pumping of blood through back muscles, thereby minimizing fatigue.

Our knowledge of when standing workplaces are preferable is dictated mainly by work performance criteria. In general, standing workplaces are preferred when: 1) the task requires a high degree of mobility (reaching and monitoring in positions that exceed the reach envelope or when performing tasks at different heights or different locations), 2) precise manual control actions are not required, 3) leg room is not available, (when leg room is not available the moment arm distance between the external load and the back is increased and thus greater internal back muscle force and spinal load result), and 4) heavy weights are handled or large forces are applied. When jobs must accommodate both sitting and standing, it is important to ensure that the positions and orientations of the body, especially the upper extremity, are in the same location under both standing and sitting conditions.

Wrists

The wrist has been of increased interest to ergonomists for the past two decades. The Bureau of Labor Statistics reports that repetitive trauma has increased from 18% of occupational illnesses in 1981 to 63% of occupational illnesses in 1993. Based upon these figures, repetitive trauma has been described as the *fastest growing* occupational problem. Even though these numbers and statements appear alarming, one must acknowledge that occupational illnesses represent 6% of all occupational injuries and illnesses. Furthermore, these figures for illness include illnesses unrelated to musculoskeletal disorders such as noise-induced hearing loss. Thus, the magnitude of the cumulative trauma problem must not be overstated. Nonetheless, there are specific industries (i.e., meat packing, poultry processing, etc.) in which cumulative trauma to the wrist is a major problem, and this problem has reached epidemic proportions within these industries.

Wrist Anatomy and Loading

In order to understand the biomechanics of the wrist and how cumulative trauma occurs in this structure, one must appreciate the anatomy of the upper extremity. Figure 10.20 shows a simplified anatomical drawing of the wrist. This figure shows that few power producing muscles reside in the hand itself. The thenar muscle which activates the thumb is one of the few power-producing muscles in the hand. The vast majority of the power-producing muscles of the hand are located in the forearm. Force is transmitted from these forearm muscles to the fingers through a network of tendons (tendons attach muscles to bone). These tendons originate at the muscles in the forearm, transverse the wrist (with many of them passing through the carpal canal), pass through the hand, and culminate at the fingers. These tendons are secured or "strapped down" at various points along this path with ligaments that keep the tendons in close proximity to the bones. This system results in a hand that is very small and compact yet capable of generating large amounts of force. The price the musculoskeletal system pays for this design is friction. The forearm muscles must transmit force over a very long distance in order to supply internal forces to the fingers. Thus, a great deal of tendon travel must occur, and this tendon travel can result in significant tendon friction under repetitive motion conditions, thereby initiating the events outlined in Figure 10.3. Thus, the key to controlling wrist cumulative trauma is rooted in an understanding of those workplace factors that adversely affect the internal force-generating (muscles) and transmitting (tendons) structures.

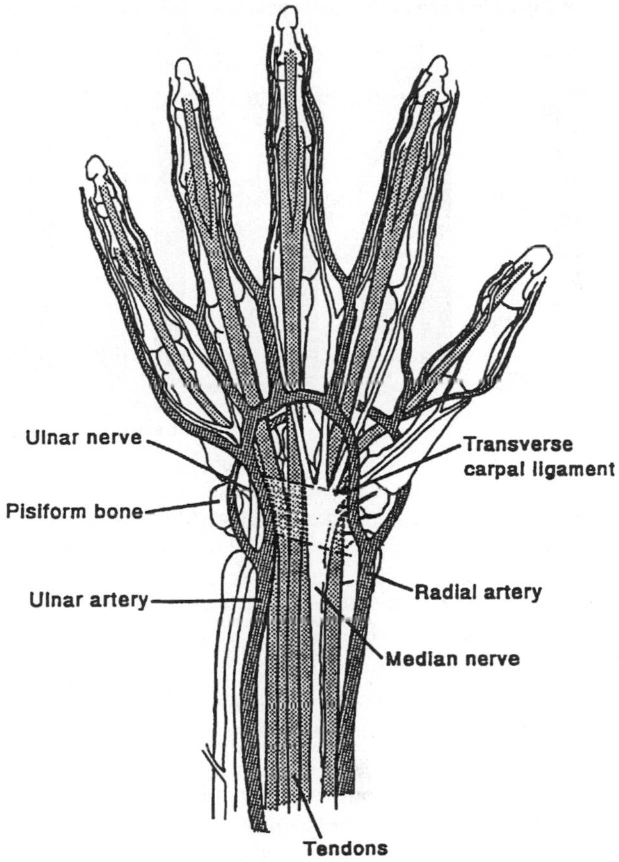

FIGURE 10.20 Important anatomical structures in the wrist.

Biomechanical Risk Factors for the Wrist

A number of risk factors for wrist cumulative trauma have been documented in the literature. First, deviated wrist postures are known to reduce the volume of the carpal tunnel and, thus, increase tendon friction. In addition, grip strength is dramatically reduced by deviations in the wrist posture. Figure 10.19 indicates that any deviation from the wrist's neutral position significantly decreases the grip strength of the hand. This reduction in strength is caused by a change in the length–strength relationship (Figure 10.5) of the forearm muscles once the wrist is bent. Hence, the muscles are working at a level that is greater than necessary. This reduced strength potential associated with deviated wrist positions can, therefore, more easily initiate the sequence of events associated with cumulative trauma (Figure 10.3). Thus, deviated wrist postures not only increase tendon travel and friction, but also for a given grip strength requirement, they increase the percentage of muscle activity and relative percentage of muscle force necessary to grip securely.

Second, increased frequency or repetition of the work cycle has been identified as a risk factor for cumulative trauma disorders (Silverstein et al., 1986, 1987). Studies have indicated that increased frequency of wrist motions increases the risk of developing a cumulative trauma disorder. Repeated motions with a cycle time of less than 30 seconds is considered a candidate for cumulative trauma disorder risk (Putz-Anderson, 1984).

Third, the force applied by the hands and fingers during a work cycle has been identified as a risk factor. In general, the greater the force required by the work the greater the risk of CTD. Greater hand forces result in greater tension within the tendons and result in greater tendon friction and tendon travel.

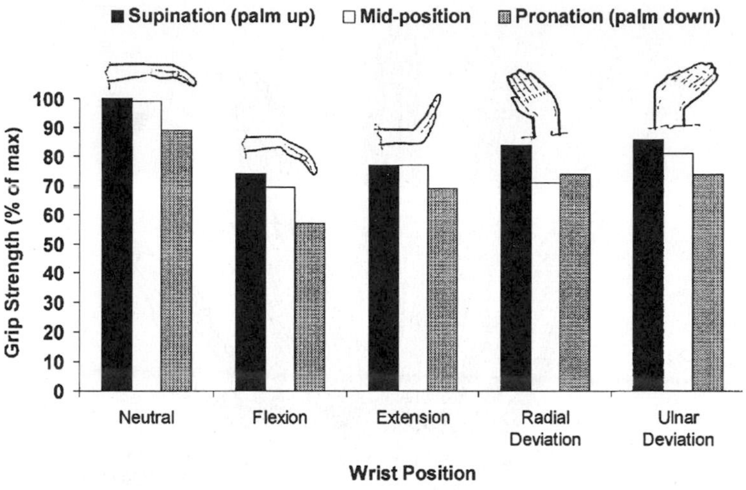

FIGURE 10.21 Grip strength as a function of wrist and forearm position. (Adapted from Sanders, M.S. and McCormick, E.J. (1993) *Human Factors in Engineering and Design*, McGraw-Hill Inc., New York, NY.)

Another factor related to force is wrist acceleration. Industrial surveillance studies have reported that repetitive jobs resulting in greater wrist acceleration are associated with greater cumulative trauma disorder incident rates (Marras and Schoenmarklin, 1993; Schoenmarklin et al., 1994). Since force is a product of mass and acceleration, jobs that increase the angular acceleration of the wrist joint result in greater tension and force transmitted through the tendons. Thus, wrist acceleration can be another mechanism of imposing force on the wrist structures.

Fourth, as shown in Figure 10.20, the anatomy of the hand is such that the median nerve becomes very superficial at the palm. Direct impact to the palm of the hand through pounding or striking an object with the palm, as is often done in assembly work, can directly stimulate the median nerve and initiate symptoms of cumulative trauma even though the work may not be repetitive.

Grip Design

The design of a tool's gripping surface can dramatically affect the activity of the internal force transmission system (tendon travel and tension). The grip opening and shape have a major influence on the available grip strength. Figure 10.21 shows how grip strength capacity changes as a function of the separation distance of the grip opening. This figure indicates that maximum grip strength occurs within a very narrow range of grip openings. If the grip opening deviates from this ideal range by as little as an inch (a couple of centimeters), then grip strength is dramatically reduced. This change in strength is also due to the length–strength relationship of the forearm muscles. Also indicated in Figure 10.22 are the effects of hand anthropometry. The worker's hand size as well as hand preference can influence grip strength and risk. Therefore, proper design of the handles is crucial in ergonomic workplace design.

Handle shape can also affect the strength of the wrist. Figure 10.23 shows how changes in the design of screwdriver handles can affect the maximum force that can be exerted. The biomechanical origin of these differences in strength capacity are most likely related to the length–strength relationship of the forearm muscles as well as contact area with the tool. The handle designs that result in less strength probably permit the wrist to twist or the grip to slip, resulting in a deviation from the ideal length–strength position in the forearm muscles.

Gloves

The use of gloves can significantly influence the generation of grip strength and may play a role in the development of cumulative trauma disorders. When gloves are worn, three effects must be considered. First, the grip strength generated is often reduced. There is typically a 10 to 20% reduction in grip strength when gloves are worn. When using gloves, the coefficient of friction between the hand and the tool can

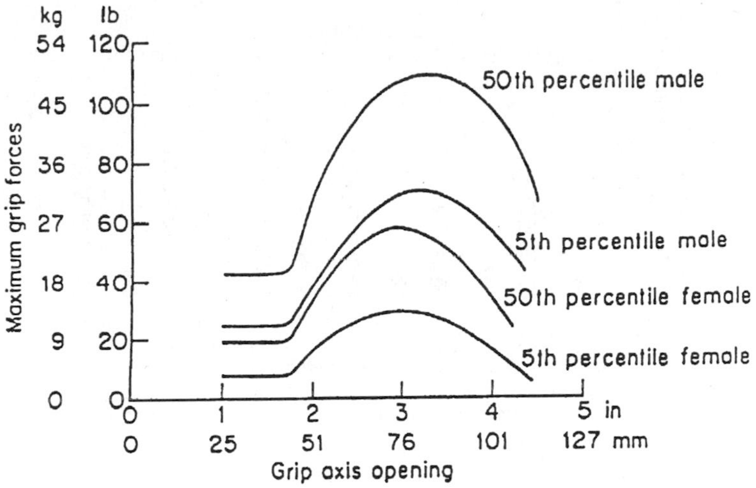

FIGURE 10.22 Grip strength as a function of grip opening and hand anthropometry. (Adapted from Sanders, M.S. and McCormick, E.J. (1993) *Human Factors in Engineering and Design*, McGraw-Hill Inc., New York, NY.)

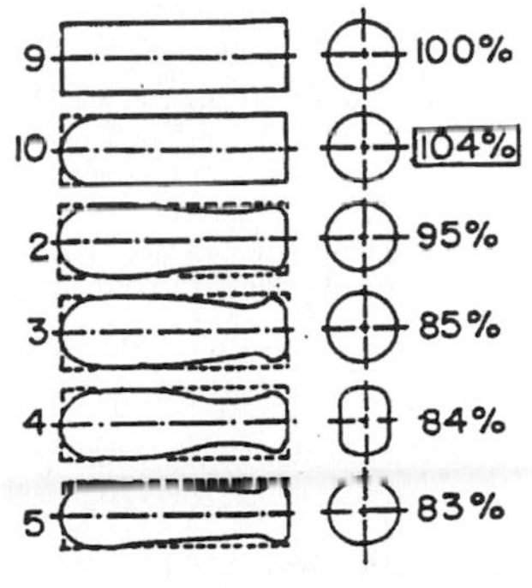

FIGURE 10.23 Maximum force which could be exerted on a screwdriver as a function of handle shape. (From Konz, S.A. (1983) *Work Design: Industrial Ergonomics*, Second Edition, Grid Publishing, Inc., Columbus, OH. With permission.)

be reduced, which in turn permits some slippage of the hand on the tool surface. This slippage can result in a deviation from the ideal muscle length and thus a reduction in available strength. The degree of slippage and the degree of strength lost depends upon how well the gloves fit the hand and the type of material used in the glove. Poorly fitting gloves result in greater strength loss.

Second, when wearing gloves, even though the externally applied force (grip strength) is often reduced, the internal forces are often very large compared to not using a glove. Studies have indicated that, for a given grip strength, the muscle activity is significantly greater when using gloves compared to a bare-handed condition (Sudhakar et al., 1988). Thus, the musculoskeletal system is less efficient when wearing a glove.

Third, the ability to perform a task is generally negatively affected when wearing gloves. Figure 10.24 shows the increase in time required to perform tasks when wearing gloves composed of different materials compared to performing the task bare-handed. The figure indicates that task performance time can increase up to 70% when wearing gloves.

These effects have indicated that there are biomechanical costs associated with glove usage. Less strength capacity is available to the worker, more internal force is generated, and worker productivity is affected. These negative effects of gloves do not mean that gloves should never be worn at work. When

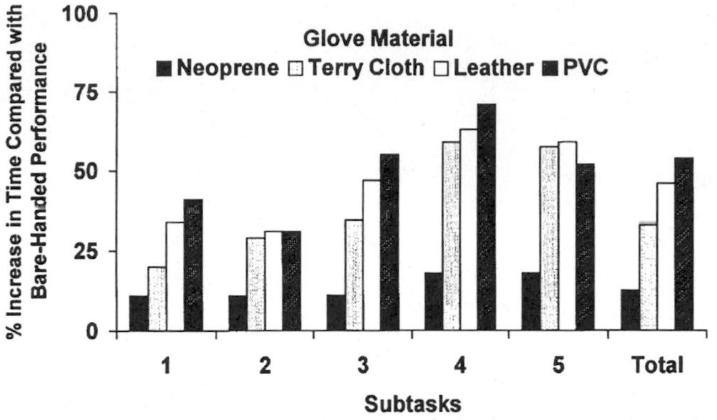

FIGURE 10.24 Performance (time to complete) on a maintenance-type task while wearing gloves constructed of five different material. (From Sanders, M.S. and McCormick, E.J. (1993) *Human Factors in Engineering and Design*, McGraw-Hill Inc., New York, NY. With permission.)

hand protection is needed, gloves should be considered as a potential solution. However, protection should only be provided to the parts of the hand that require it. For example, if the palm of the hand requires protection, fingerless gloves might provide an acceptable solution. If the fingers require protection but there is little risk to the palm of the hand, then grip tape wrapped around the fingers might be considered. In addition, different styles, materials, and sizes of gloves will fit workers differently. Thus, gloves produced by various manufacturers and of different sizes should be available to the worker to minimize the negative effects mentioned above.

Design Guidelines

This discussion has indicated that there are many factors that can affect the biomechanics of the wrist and the subsequent risk of cumulative trauma disorders. This suggests that proper ergonomic design of a work task cannot be accomplished by simply providing the worker with a "ergonomically designed" tool. Since ergonomics is associated with matching the workplace design to the worker's capabilities, it is not possible to design an "ergonomic tool" without considering the workplace design and task requirements simultaneously. What might be an "ergonomic" tool for one work situation may be improper while a worker is assuming another work posture. For example, using an *in-line* tool may keep the wrist straight when inserting a bolt into a horizontal surface. However, if the bolt is to be inserted into a vertical surface a *pistol grip* tool may be more appropriate. Using the in-line tool in this situation (inserting a bolt into a vertical surface) may cause the wrist to be significantly deviated. Hence, there are no ergonomic tools. There are just ergonomic *situations*. What may be an ergonomically correct tool in one situation may not be ergonomically correct in another. Thus, workplace design should be performed with care and trade-offs between different parts of the body must be considered by taking into consideration the various biomechanical trade-offs. Given these considerations, the following components of the workplace should be considered when designing a workplace so that cumulative trauma risk is minimized. First, maintain a neutral wrist posture. Second, minimize tissue compression. Third, avoid actions that repeatedly impose force on the internal structures. Fourth, minimize required wrist accelerations and motions. Fifth, consider the impact of glove use, hand size, and left-handed workers.

10.4 Analysis and Control Measures Used in the Workplace

Several analysis and control measures have been developed to evaluate and control biomechanical loading of body during work. Since low back disorders are often the objective of a biomechanical workplace analysis, most of these analysis methods have focused on spine risk. However, several of the measures also include analyses of risk to other body parts.

Lift Belts

Back support belts or lifting belts have been used with increasing frequency in the workplace. There exists a great deal of controversy as to whether use of these belts is a benefit or a liability during manual materials handling. A review of the literature related to lifting belts offers no clear answer as to the benefits of belt use. Reviews by McGill (1993) and NIOSH (1994) have concluded that there are so few well executed studies that one cannot unequivocally judge the benefits of lifting belts. Therefore, what is known about these devices can be summarized and used as a basis for an informed opinion.

Epidemiological studies have generally been limited in scope and often result in findings that were confounded by other factors such as training, the type of belt used, or the "Hawthorne Effect." Walsh and Schwartz (1990) reported a reduction in low back disorder (LBD) injury rate with the usage of back supports (hard shell corsets) and have recommended that they be used in controlling the risk of low back disorder. However, the data from this study suggest that back supports were only effective for those workers who had previously suffered a low back disorder. Mitchell et al. (1994) retrospectively evaluated injury data associated with belt use over a six year period at Tinker Air Force Base. Over this period, two different types of belts were used. Leather belts were used in the first two years of the study, and Velcro belts were used over the last four years. No relationship between belt usage and back injury could be established, but they did find that those who wore belts suffered more costly injuries once they occurred. Riddle et al. (1992) observed that when workers stopped wearing belts the risk of injury increased. However, this study suffers from small sample size, which makes it difficult to assess the strength of the association. More recently, Straus et al. (1996) in a large prospective study found that lift belts significantly reduced the risk of low back pain in a chain of home improvement stores. However, many unresolved study design questions are also associated with this study. Unfortunately, none of these epidemiologic studies may be considered conclusive because many of them suffer from low participation rates, inadequate observation periods, confounding with training, small sample size, low back-injury rates, improper counterbalancing of experimental treatments, uncontrolled work tasks, reporting bias, and/or previous back injury history (NIOSH, 1994).

Psychophysical studies have attempted to assess whether the magnitude of the weight a person was willing to lift changes when wearing a back belt. McCoy et al. (1988) found that subjects were willing to lift 19% more weight when belts were used but found no difference between belt types. Subjects reported that they preferred the elastic belt. However, this does not suggest that workers would be at lowered risk of back injury because it is not clear that spine tolerance to load would be increased with belt use.

Biomechanically based studies of lifting belts have documented their influence upon trunk motion, trunk muscle activity, and indirect indicators or predictions of trunk loading. The most consistent finding of these studies is that side bending and twisting trunk motion is significantly reduced with belt usage (McGill et al., 1994; Lavender et al., 1995; Lantz and Schultz, 1986). A recent study has indicated that for some subjects there may be a slight reduction in spine loading when lifting asymmetrically without moving the feet. However, this trend was not true for all subjects. Some subjects increased their spine loading under these conditions (Granata et al., 1996). Thus, if belts do have the potential to reduce spine loading, they do not appear to do so in all workers and may increase loading in some workers.

Perhaps the most important reason to be cautious of lifting belts is unrelated to biomechanical loading of the spine. There appear to be physiological reasons to be concerned within the use of lifting belts. One study has shown that lifting belts can significantly increase blood pressure. This could become problematic for workers who have a compromised cardiovascular system.

The brief review indicates that there is a large amount of conflicting evidence as to the benefits or liabilities associated with the use of back belts. A consistent finding among the studies is that if there is a benefit to back belts it is probably for those who have previously experienced a low back disorder. The literature also suggests that belts should only be used for a limited period of time. Until more definitive studies are available, it is prudent to use caution when recommending the use of back belts in a work environment. This includes a screening by an occupational physician who is familiar with the literature so that potential cardiovascular problems can be assessed.

TABLE 10.3 Fmax Table

	Average Vertical Location (cm) (in)	
Period	Standing V > 75 (3)	Stooped V ≤ 75 (3)
1 Hour	18	15
8 Hours	15	12

Reprinted from NIOSH, *Work Practices Guide for Manual Lifting*

1981 NIOSH Lifting Guide

The National Institute for Occupational Safety and Health (NIOSH) has developed two assessment tools or guides to help determine whether a manual materials handling task is safe or risky. The lifting guide was originally developed in 1981 (NIOSH, 1981) and applies to lifting situations in which the lifts are performed in the sagittal plane and to motions that are slow and smooth. Two benchmarks or limits are defined by this guide. The first limit is called the *action limit* (AL) and represents a magnitude of weight in a given lifting situation which would impose a spine load corresponding to the beginning of low back disorder risk along a risk continuum. The AL is associated with the point in Figure 10.10 at which people under 40 years of age just begin to experience a risk of vertebral endplate microfracture (350 kg of compressive load). The weight of an object to be lifted by a worker in a given task is compared to the AL. If the weight of the object is below that of the AL, the job is considered safe. If the weight lifted by the worker is larger than the AL, there is at least some level of risk associated with the task. The general form of the AL is defined according to equation (1).

$$AL = k\,(HF)(VF)(DF)(FF) \tag{1}$$

where

AL = The action limit in kg or pounds
k = Load constant (40 kg or 90 lbs.) which is the greatest weight a subject could lift if all lifting conditions are optimal.
HF = Horizontal factor defined as the horizontal distance from a point bisecting the ankles to the center of gravity of the load at the lift origin. Defined algebraically as 15/H (metric) or 6/H (U.S. units).
VF = Vertical factor or height of the load at lift origin. Defined algebraically as $(.004)\ |V - 75|$ (metric) or $1 - (.01)\,|V - 30|$ (U.S. units).
DF = Distance factor or the vertical travel distance of the load. Defined algebraically as .7 + 7.5/D (metric) or .7 + 3/D (U.S. units).
FF = Frequency factor or lifting rate defined algebraically as 1 – F/Fmax
F = average frequency of lift, Fmax is shown in Table 10.3

The logic associated with this equation assumes that if the lifting conditions are ideal, a worker could safely hold (and this implies lift) the load constant, *k*, (40 kg or 90 lbs.). If the lifting conditions are not ideal, the allowable weight is discounted according to the four factors HF, VF, DF, and FF. These four factors are shown in monogram form in Figures 10.25 through 10.28. According to the load discounting associated with these figures, the HF which is associated with the external moment has the most dramatic effect on acceptable lifting conditions. VF and DF are associated with the back muscle's length–strength relationship. FF attempts to account for the cumulative effects of repetitive lifting.

The second benchmark associated with this guide is the *maximum permissible limit* or MPL. The MPL represents the point at which significant risk, defined in part as a significant risk of vertebral endplate microfracture (Figure 10.10). The MPL is associated with a compressive load on the spine of 650 kg,

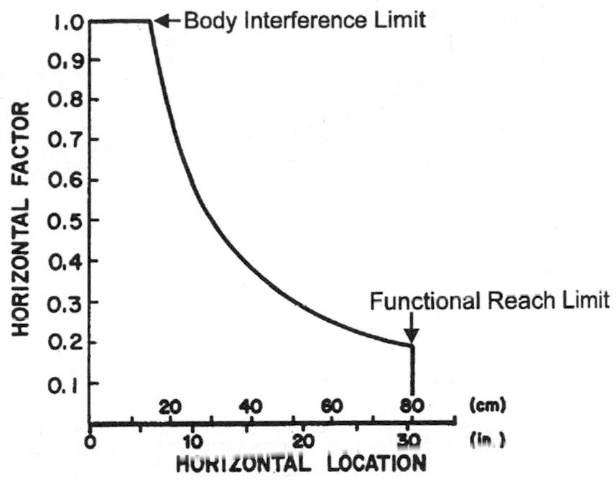

FIGURE 10.25 Horizontal Factor (HF) varies between the body interference limit and the limit of functional reach. (Adapted from National Institute for Occupational Safety and Health (NIOSH) (1981) Work practices guide for manual lifting. Department of Health and Human Services (DHHS), National Institute for Occupational Safety and Health (NIOSH), Publication No. 81-122.)

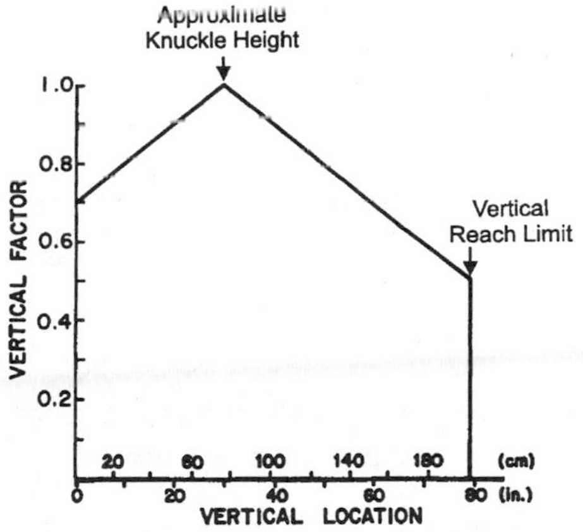

FIGURE 10.26 Vertical Factor (VF) varies both ways from knuckle height. (Adapted from National Institute for Occupational Safety and Health (NIOSH) (1981) Work practices guide for manual lifting. Department of Health and Human Services (DHHS), National Institute for Occupational Safety and Health (NIOSH), Publication No. 81-122.)

which corresponds to a point at which 50% of the people would be expected to suffer a vertebral endplate microfracture. Equation (2) indicates that the MPL is a function of the AL and is defined as follows:

$$\text{MPL} = 3\ (\text{AL}) \tag{2}$$

The weight that the worker expected to lift in a work situation is compared to the AL and MPL. If the magnitude of weight falls below the AL, the work is considered safe and no adjustments are necessary. If the magnitude of the weight falls above the MPL, then the work is considered risky and engineering changes involving the adjustment of HF, VF, and/or DF are required to reduce the AL and MPL. If the weight falls between the AL and MPL, then either engineering changes or administrative changes, defined as selecting workers who are less likely to be injured or rotating workers, are recommended. The AL and MPL were also indexed to non-biomechanical benchmarks. According to NIOSH (1981), these limits also correspond to strength, energy expenditure, and psychophysical acceptance points.

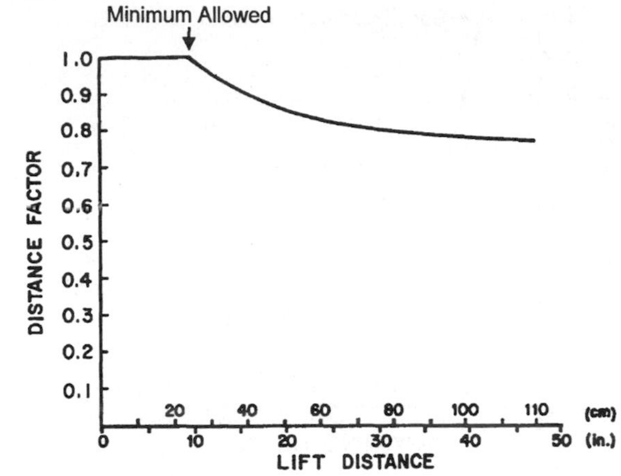

FIGURE 10.27 Distance factor (DF) varies between a minimum vertical distance moved of 25 cm (10 in.) to a maximum distance of 200 cm (80 in). (Adapted from National Institute for Occupational Safety and Health (NIOSH) (1981) Work practices guide for manual lifting. Department of Health and Human Services (DHHS), National Institute for Occupational Safety and Health (NIOSH), Publication No. 81-122.)

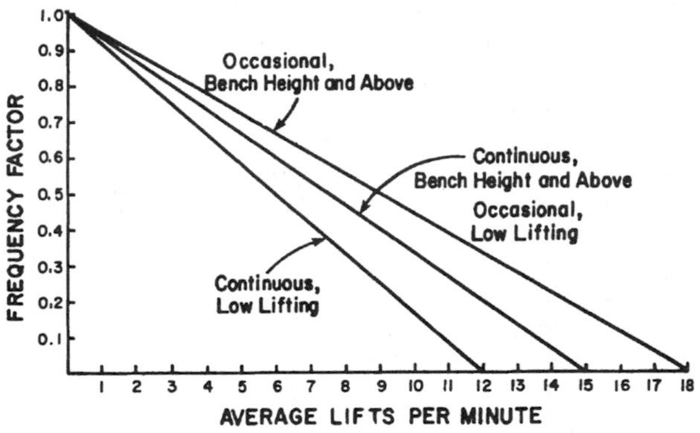

FIGURE 10.28 Frequency Factor (FF) varies with lifts/minute and the F_{max} curve. The F_{max} curve depends upon lifting posture and lifting time. (From National Institute for Occupational Safety and Health (NIOSH) (1981) Work practices guide for manual lifting. Department of Health and Human Services (DHHS), National Institute for Occupational Safety and Health (NIOSH), Publication No. 81-122.)

1991 Revised NIOSH Equation

The 1991 NIOSH revised lifting equation was introduce in order to address those lifting jobs that violate the sagittally symmetric lifting assumption (Waters et al., 1993). The concept of AL and MPL was replaced with a concept of a *lifting index* or LI. The LI is defined in equation (3).

$$LI = \frac{L}{RWL} \qquad (3)$$

where

L = Load weight or the weight of the object to be lifted.
RWL = Recommended Weight Limit for the particular lifting situation.
LI = Lifting Index used to estimate relative magnitude of physical stress for a particular job.

TABLE 10.4 Frequency Multiplier Table (FM)

Frequency Lifts/min (F)‡	Work Duration					
	≤1 hour		>1 but ≤2 hours		>2 but ≤8 hours	
	V < 30†	V ≥ 30	V < 30	V ≥ 30	V < 30	V ≥ 30
≥0.2	1.00	1.00	.95	.95	.85	.85
0.5	.97	.97	.92	.92	.81	.81
1	.94	.94	.88	.88	.75	.75
2	.91	.91	.84	.84	.65	.65
3	.88	.88	.79	.79	.55	.55
4	.84	.84	.72	.72	.45	.45
5	.80	.80	.60	.60	.35	.35
6	.75	.75	.50	.50	.27	.27
7	.70	.70	.42	.42	.22	.22
8	.60	.60	.35	.35	.18	.18
9	.52	.52	.30	.30	.00	.15
10	.45	.45	.26	.26	.00	.13
11	.41	.41	.00	.23	.00	.00
12	.37	.37	.00	.21	.00	.00
13	.00	.34	.00	.00	.00	.00
14	.00	.31	.00	.00	.00	.00
15	.00	.28	.00	.00	.00	.00
>15	.00	.00	.00	.00	.00	.00

†Values of V are in inches. ‡For lifting less frequently than once per 5 minutes, set F = .2 lifts/minute.

Reprinted from NIOSH, *Applications Manual for the Revised NIOSH Lifting Equation.*

If the LI is greater than 1.0, an increased risk for suffering a lifting-related low back disorder exists. The RWL is similar to the 1981 Lifting Guide AL equation (Equation 1) in that it contains factors that discount the allowable load according to the horizontal distance, vertical location of the load, vertical travel distance, and frequency of lift. However, the form of these discounting factors was changed. Moreover, two additional discounting factors have been included. These additional factors include a lift asymmetry factor to account for asymmetric lifting conditions and a coupling factor that accounts for whether or not the load lifted has handles. The RWL is represented algebraically in equations (4) (metric units) and (5) (U.S. units).

$$RWL(kg) = 23(25/H)(1-(0.003|V-75|))(.82+4.5/D))(FM)(1-(0.0032A))(CM) \quad (4)$$

$$RWL(lb) = 51(10/H)(1-(0.0075|V-30|))(.82+1.8/D))(FM)(1-(0.0032A))(CM) \quad (5)$$

where

H = Horizontal location forward of the midpoint between the ankles at the origin of the lift. If significant control is required at the destination, then H should be measured both at the origin and destination of the lift.
V = Vertical location at the origin of the lift.
D = Vertical travel distance between origin and destination of the lift.
FM = Frequency multiplier shown in Table 10.4.
A = Angle between the midpoint of the ankles and the midpoint between the hands at the origin of the lift.
CM = Coupling multiplier ranked as either good, fair, or poor and described in Table 10.5.

In this revised equation, the load constant has been significantly reduced compared to the 1981 equation. The adjustments for load moment, muscle length–strength relationships, and cumulative loading are still integral parts of this equation. However, these adjustments or discounting factors have

TABLE 10.5 Coupling Multiplier

	Coupling Multiplier	
Coupling Type	V < 30 inches (75 cm)	V ≥ 30 inches (75 cm)
Good	1.00	1.00
Fair	0.95	1.00
Poor	0.90	0.90

Reprinted from NIOSH, *Application Manual for Revised NIOSH Equation, 1994.*

been changed (compared to the 1981 guide) to reflect the most conservative value of the biomechanical, physiological, psychophysical, or strength data upon which they are based. Recent studies report that the 1991 revised equation yields a more conservative (protective) prediction of work-related low back disorder risk (Marras et al., 1997).

2D/3D Static Models

Biomechanically based spine models have been developed to help assess occupationally related manual materials handling tasks. These models assess the task based upon both spine loading criteria as well as through an evaluation of the strength required at the various major body joints in order to perform the task. One of the early static assessment models was developed by Chaffin at the University of Michigan (1969). Both two-dimensional (2D) as well as three-dimensional (3D) static models (Chaffin and Muzaffer, 1991) have been developed to help assess the risk of injury during manual materials handling activities. In both models, the moments imposed upon the various joints of the body due to the object lifted are evaluated assuming that a static posture is representative of the instantaneous loading of the body. These models then compare the imposed moments about each joint with the static strength capacity derived from a working population. The static strength capacity of the major joints assessed by this model have been documented in a database of over 3000 workers. In this manner the proportion of the population capable of performing a particular static exertion is predicted. In addition, the joint that limits the capacity to perform the task can be identified via this method. These models assume that a single equivalent muscle (internal force) supports the external moment about each joint. By considering the contribution of the externally applied load and the internally generated single muscle equivalent, spine compression acting on the lumbar discs is predicted. The predicted compression can then be compared to the tolerance limits of the vertebral endplate (Figure 10.10). The 2D version of this computer model assumes (as does the 1981 NIOSH Lifting Guide) that all lifts occur directly in front of the worker in the sagittal plane. Another important assumption of these models is that no significant motion occurs during the exertion because it is a static model. Figure 10.29 shows the output screen for this computer model where the lifting posture, lifting distances, strength predictions, and spine compression are shown. The 3D version of the computer model works in a similar manner, however, non-sagittal symmetric lifting assumptions are permitted.

Multiple Muscle System Models

One of the significant simplifying assumptions inherent in most static models is that the coactivation of the trunk musculature during a lift is negligible. The trunk is truly a multiple muscle system with many major muscle groups supporting and loading the spine. This can be seen in the cross-section of the trunk shown in Figure 10.30. Studies have shown that there is significant coactivation occurring in many of the major muscle groups in the trunk during realistic *dynamic* lifting (Marras and Mirka, 1993). This coactivation is important because all the trunk muscles have the ability to load the spine. Thus, assumptions regarding single equivalent muscles within the trunk can lead to erroneous conclusions about spine loading during a task. A recent study has indicated that ignoring the coactivation of trunk muscles during

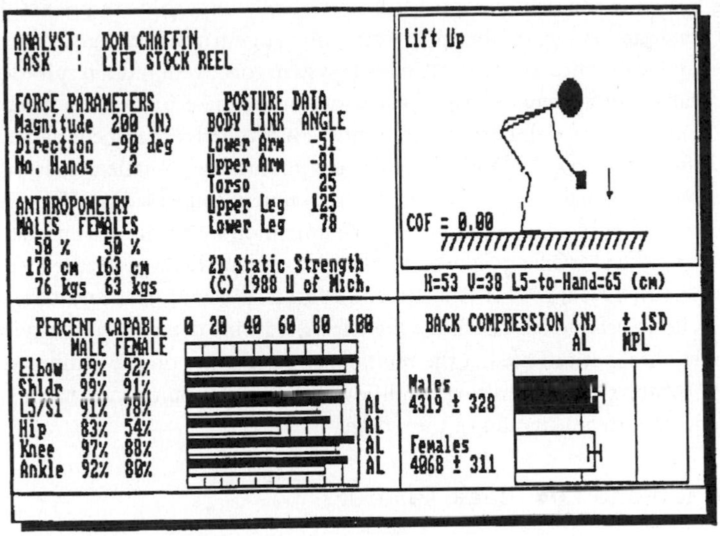

FIGURE 10.29 The 2D Static Strength Prediction Model. (From Chaffin, D.B. and Andersson, G.B. (1991) *Occupational Biomechanics*, John Wiley & Sons, Inc. New York, NY. With permission.)

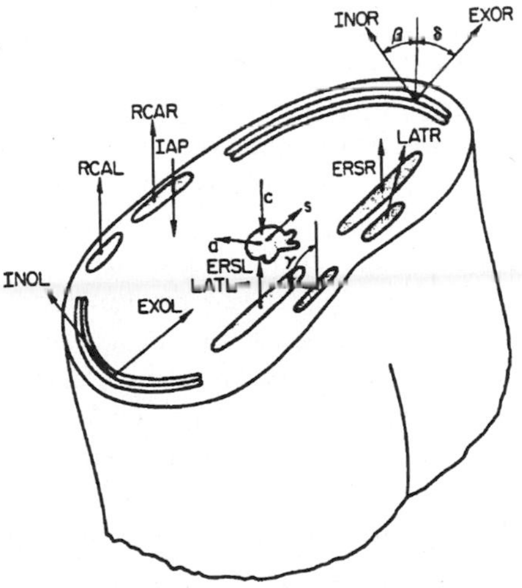

FIGURE 10.30 Cross-sectional view of the human trunk at the lumbrosacral junction. (Adapted from Schultz, A.B. and Andersson, G.B.J. (1981) Analysis of loads on the lumbar spine. *Spine*, 6:76-82.)

dynamic lifting can misrepresent spine loading by 45 to 70% (Granata and Marras, 1995). In an effort to more accurately estimate the loads on the lumbar spine, especially under complex, changing (dynamic) postures, multiple muscle system models of the trunk have been developed. Much of the research in the past decade has been centered around predicting how the multiple trunk muscles coactivation during dynamic lifting.

EMG-Assisted Multiple Muscle System Models

People recruit their muscles in various manners when moving dynamically. For example, when moving slowly the agonist muscle may dominate the muscles' activities during a lift. However, when moving cautiously, asymmetrically, or rapidly there may be a great deal of antagonistic coactivation present. During occupational lifting tasks, these latter dynamic conditions are typically the rule rather than the exception during lifting. For these reasons it has been virtually impossible to predict the instantaneous coactivation and resultant loading on the spine during dynamic trunk exertions. One of the few means to accurately account for the effect of the trunk muscle system coactivation upon spine loading is through the use of biologically assisted models. The most common of these models are electromyographic or EMG-assisted models. These models take into account the individual recruitment patterns of the muscles during a specific lift for a specific individual. By directly monitoring muscle activity, the EMG-assisted model can determine individual muscle force and the subsequent spine loading. These models have been developed and tested under bending and twisting dynamic motion conditions and have been validated (McGill and Norman, 1985; McGill and Norman, 1986; Marras and Reilly, 1988; Reilly and Marras, 1989; Marras and Sommerich, 1991a; 1991b; Granata and Marras, 1993; Marras and Granata, 1995, 1997). Figure 10.31 shows how such models can assess the effects of lifting dynamics upon spine loading. These models are the only ones that can predict the *multidimensional loads* on the lumbar spine under many three-dimensional complex dynamic lifting conditions. The limitation of such models is that it is that they require significant instrumentation of the worker.

Dynamic Motion and Low Back Disorder

As discussed throughout this chapter, it is clear that dynamic activity may increase the risk of low back disorder. However, in order to control this biomechanical situation at the worksite, one must know the type of motion that increases biomechanical load and how much is too much motion from a biomechanical standpoint. These issues were the focus of an industrial retrospective study performed over a six-year period in 68 different industrial environments. Trunk motion and workplace conditions were assessed in workers exposed to high risk of low back disorder jobs and compared to trunk motions and workplace conditions of low risk jobs (Marras et al., 1993, 1995). A trunk goniometer (lumbar motion monitor or LMM) that has been used to document the trunk motion patterns of workers in the workplace is shown in Figure 10.32. Trunk motion and workplace conditions associated with the high-risk and low-risk environments are listed in Table 6. Based on these findings, a five-factor multiple logistic regression model was developed that is capable of discriminating between tasks that indicate probability of high-risk group membership. These factors include: 1) frequency of lifting, 2) load moment (load weight multiplied by the distance of the load from the spine), 3) average twisting velocity (measured by the LMM), 4) maximum sagittal flexion angle through the job cycle (measured by the LMM), and 5) maximum lateral velocity (measured by the LMM). This LMM risk assessment model is the only model capable of assessing the *risk* of three-dimensional trunk motion on the job. This model has been shown to have a high degree of predictability (odds ratio = 10.7) compared to previous attempts to assess work-related low back disorder risk. The advantage of this assessment is that the evaluation provides information about risk that would take years to derive from historical accounts of incidence rates.

10.5 Summary

This chapter has shown that biomechanics provides one of the few means to *quantitatively* consider the implications of workplace design. Biomechanical design is important when a particular job is suspected of imposing large or repetitive forces on a particular structure of the body. It is particularly important to recognize that the internal structures of the body, such as muscles, are the primary loaders of the joint and tendon structures. In order to evaluate the risk of injury from a particular task, one must consider the contribution of both the external loads and internal loads upon the structure. Several quantitative models and assessment methods have been developed that systematically consider the internal loading imposed on the worker due to workplace layout and task requirements. Proper use of these models and methods involves recognizing the limitations and assumptions of each technique so that they are not applied inappropriately. When properly used, these assessments can help assess the risk of work-related injury and illness.

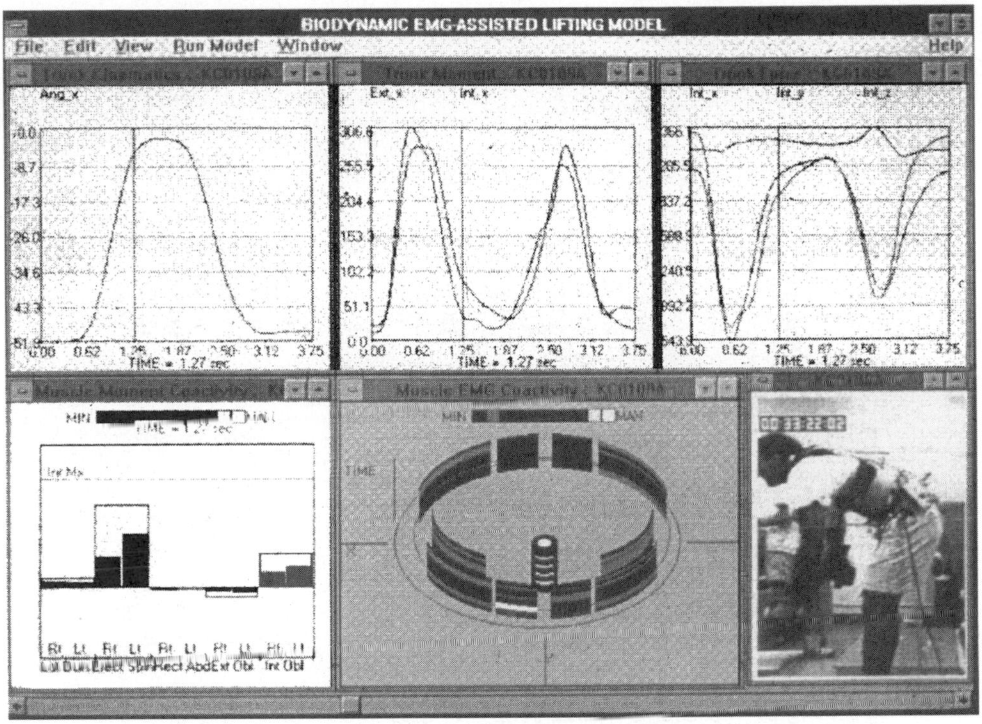

FIGURE 10.31 Windows EMG-assisted model.

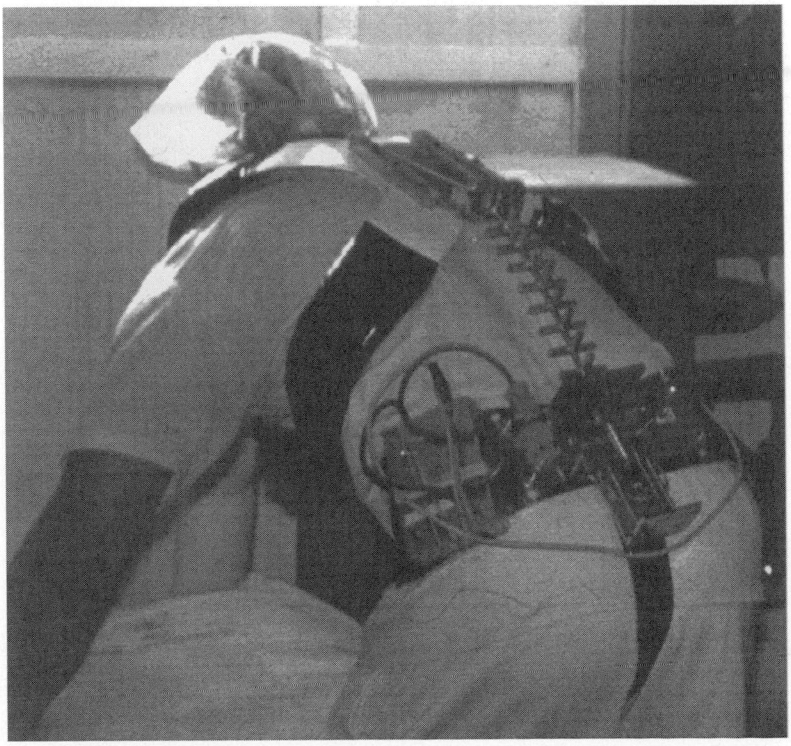

FIGURE 10.32 The lumbar motion monitor (LMM).

TABLE 10.6 Descriptive and t Statistics of the Workplace and Trunk Motion Factors in Each of the Risk Groups

	High Risk (N = 111)				Low Risk (N = 124)				
Factors	Mean	SD	Minimum	Maximum	Mean	SD	Minimum	Maximum	Statistics t
WORKPLACE FACTORS									
Lift rate (lifts/hr)	175.89	8.65	15.30	900.00	118.83	169.09	5.40	1500.00	2.1*
Vertical load location at origin (m)	1.00	0.21	0.38	1.80	1.05	0.27	0.18	2.18	1.4
Vertical load location at destination (m)	1.04	0.22	0.55	1.79	1.15	0.26	0.25	1.88	3.2†
Vertical distance traveled by load (m)	0.23	0.17	0.00	0.76	0.25	0.22	0.00	1.04	0.8
Average weight handled (N)	84.74	79.39	0.45	423.61	29.30	48.87	0.45	280.92	6.4†
Maximum weight handled (N)	104.36	88.81	0.45	423.61	37.15	60.83	0.45	325.51	6.7†
Average horizontal distance between load and L5-S1 (N)	0.66	0.12	0.30	0.99	0.61	0.14	0.33	1.12	2.5*
Maximum horizontal distance between load and L5-S1 (N)	0.76	0.17	0.38	1.24	0.67	0.19	0.33	1.17	3.7†
Average moment (Nm)	55.26	51.41	0.16	258.23	17.70	29.18	0.17	150.72	6.8†
Maximum moment (Nm)	73.65	60.65	0.19	275.90	23.64	38.62	0.17	198.21	7.4†
Job satisfaction	5.96	2.26	1.00	10.00	7.28	1.95	1.00	10.00	4.7†
TRUNK MOTION FACTORS									
Sagittal Plane									
Maximum extension position (°)	–8.30	9.10	–30.82	18.96	–10.19	10.58	–30.00	33.12	3.5†
Maximum flexion position (°)	17.85	16.63	–13.96	45.00	10.37	16.02	–25.23	45.00	1.5
Range of motion (°)	31.50	15.67	7.50	75.00	23.82	14.22	3.99	67.74	3.8†
Average velocity(°/sec)	11.74	8.14	3.27	48.88	6.55	4.28	1.40	35.73	6.0†
Maximum velocity (°/sec)	55.00	38.23	14.20	207.55	38.69	26.52	9.02	193.29	3.7†
Maximum acceleration (°/sec^2)	316.73	224.57	80.61	1341.92	226.04	173.88	59.1	1120.10	4.2†
Maximum deceleration (°/sec^2)	–92.45	63.55	–514.08	–18.45	–83.32	47.71	–227.12	–4.57	1.2

Lateral Plane									
Maximum left bend(°)	−1.47	6.02	−16.80	24.49	−2.54	5.45	−23.80	13.96	1.4
Maximum right bend (°)	15.60	7.61	3.65	43.11	13.24	6.32	0.34	34.14	2.6*
Range of motion (°)	24.44	9.77	7.10	47.54	21.59	10.34	5.42	62.41	2.2*
Average velocity (°/sec)	10.28	4.54	3.12	33.11	7.15	3.16	2.13	18.86	6.1†
Maximum velocity (°/sec)	46.36	19.12	13.51	119.94	35.45	12.88	11.97	76.25	4.9†
Maximum acceleration (°/sec²)	301.41	166.69	82.64	1030.29	229.29	90.9	66.72	495.88	4.1†
Maximum deceleration (°/sec²)	−103.65	60.31	−376.75	0.00	−106.20	58.27	−294.83	0.00	0.3
Twisting Plane									
Maximum left twist (°)	1.21	9.08	−27.56	29.54	−1.92	5.36	−30.00	11.44	3.2†
Maximum right twist (°)	13.95	8.69	−13.45	30.00	10.83	6.08	−11.20	30.00	2.2*
Range of motion (°)	20.71	10.61	3.28	53.30	17.08	8.13	1.74	38.59	2.9†
Average velocity (°/sec)	8.71	6.61	1.02	34.77	5.44	3.19	0.66	17.44	3.8†
Maximum velocity (°/sec)	46.36	25.61	8.06	136.72	38.04	17.51	5.93	91.97	4.7*
Maximum acceleration (°/sec²)	304.55	175.31	54.48	853.93	269.49	146.65	44.17	940.27	2.9†
Maximum deceleration (°/sec²)	−88.52	70.30	−428.94	−5.84	−100.32	72.40	−325.93	−2.74	1.6*

*Significant at $\alpha \le 0.05$ (two-sided). †Significant at $\alpha \le 0.01$ (two-sided).

Adapted from Marras et al., 1993.

References

Adams, M.A. and Dolan, P. (1995) Recent advances in lumbar spinal mechanics and their clinical significance, *Clinical Biomechanics*, 10(1), 3-19.

Anderson, C.K. and Chaffin, D.B. (1986) A biomechanical evaluation of five lifting techniques. *Applied Ergonomics*, 17(1): 2-8.

Andersson, G.B. (1981) Epidemiologic aspects of low back pain in industry. *Spine*, 6: 53-60.

Andersson, G.B. (1991), The epidemiology of spinal disorders, in *The Adult Spine*. Eds. Frymoyer, J.W., Ducker, T.B., Hadler, N.M., Kostuik, J.P., Weinstein, J.N., and Whitecloud, T.S. Raven Press, New York, pp. 107-146.

Basmajian, J.V. and De Luca, C.J. (1985) *Muscles Alive: Their Functions Revealed by Electromyography* (5th edition), Williams and Wilkins, Baltimore, MD.

Bean, J.C., Chaffin, D.B., and Schultz, A.B. (1988) Biomechanical model calculation of muscle forces: A double linear programming method. *J. Biomechanics*, 21 (1) 59-66.

Bigos, S.J., Spengler, D.M., Martin, N.A., Zeh, J., Fisher, L., Nachemson, A., and Wang, M.H. (1986) Back injuries in industry: A retrospective study. II. Injury factors. *Spine*, 11(3), 246-251.

Brinckmann, P., Biggemann, M., and Hilweg, D. (1988) Fatigue fracture of human lumbar vertebrae. *Clinical Biomechanics*, 3: Supplement 1, S1-S23.

Cats-Baril, W. and Frymoyer, J.W. (1991) The economics of spinal disorders, in *The Adult Spine*. Eds. Frymoyer, J.W., Ducker, T.B., Hadler, N.M., Kostuik, J.P., Weinstein, J.N., and Whitecloud, T.S. Raven Press, New York, pp. 85-105.

Chaffin, D.B.(1969) A computerized biomechanical model: development of and use in studying gross body actions. *Journal of Biomechanics*, 2, 429-441.

Chaffin, D.B. and Andersson, G.B. (1991) *Occupational Biomechanics*, John Wiley & Sons, Inc. New York, NY.

Chaffin D.B. and W.H. Baker (1970) A biomechanical model for analysis of symmetric sagittal plane lifting. *AIIE Transactions*, II(1): 16-27.

Chaffin, D.B. (1973) Localized muscle fatigue definition and measurement, *Journal of Occupational Medicine*, 15(4), 346-354.

Chaffin D.B. and Muzaffer, E. (1991) Three-dimensional biomechanical static strength prediction model sensitivity to postural and anthropometric inaccuracies. *IIE Transactions*, 23(3), 215-227.

Frymoyer, J.W., Pope, M.H., Clements, J.H., Wilder, D.G., MacPherson, B., and Ashikaga, T. (1983) Risk factors in low back pain: An epidemiologic survey. J. Bone Joint Surg., 65A, 213-216.

Granata K.P. and Marras, W.S. (1993) An EMG-assisted model of loads on the lumbar spine during asymmetric trunk extensions. *J. Biomechanics*, 26 (12), 1429-1438.

Grandjean, E. (1982) *Fitting the Task to the Man: An Ergonomic Approach*, Taylor & Francis, Ltd. London.

Guo, H.R., (1993) Back pain and U.S. workers (NIOSH report), presented at American Occupational Health Conference, April 29.

Hollbrook, T.L., Grazier, K., Kelsey, J.L., and Stauffer, R.N. (1984) The frequency of occurrence, impact and cost of selected musculoskeletal conditions in the United States, American Academy of Orthopaedic Surgeons, Chicago, IL, pp. 24-45.

Hoy, M.G., Zajac, F.E., and Gordon, M.E. (1990) A musculoskeletal model of the human lower extremity: The effect of muscle, tendon, and moment arm on the moment-angle relationship of the musculotendon actuators at the hip, knee, and ankle. *Journal of Biomechanics*, 23(2): 157-169.

Jager, M., Luttmann, A., and Laurig, W. (1991) Lumbar load during one-handed bricklaying, *International Journal of Industrial Ergonomics*, 8(3), 261-277.

Kelsey, J.L. and White, A.A. III (1980) Epidemiology and impact on low back pain. *Spine*, 5(2), 133-142.

Kelsey, K.L., Githens, P.B., White, A.A. III, Holford, T.R., Walter, S.D., O'Conner, T., Ostfeld, A.M., Weil, U., Southwick, W.O., and Calogero, J.A. (1984) An epidemiologic study of lifting and twisting on the job and risk for acute prolapsed lumbar intervertebral disc. *J Ortho Res.*, 2(1): 61-66.

Konz, S.A. (1983) *Work Design: Industrial Ergonomics*, Second Edition, Grid Publishing, Inc., Columbus, OH.

Kroemer, K.H.E. (1987) Biomechanics of the human body, in *Handbook of Human Factors*, (Ed. Salvendy, G.) John Wiley & Sons, New York, NY.

Lantz, S.A. and Schultz, A.B. (1986) Lumbar spine orthosis wearing. I. Restrictions of gross body motion. *Spine*, 11(8): 834-837.

Lavender, S.A., Thomas, J.S., Chang, D., and Andersson, G.B. (1995) The effects of lifting belts, foot movement and lifting asymmetry on trunk motions. *Human Factors*, (in press).

Marras, W.S. (1988) Predictions of forces acting upon the lumbar spine under isometric and isokinetic conditions: A model — experimental comparison. *Int. J. Ind. Ergonomics*, 3, 19-27.

Marras, W.S. and Granata, K.P. (1994) A biomechanical assessment and model of axial twisting in the thoraco-lumbar spine. *Spine*, 20(13): 1440-1451.

Marras, W.S., Lavender, S.A., Leurgans, S.E., Rajulu, S.L., Allread, W.G., Fathallah, F.A., and Ferguson, S.A. (1993) The role of dynamic three-dimensional trunk motion in occupationally-related low back disorders: The effects of workplace factors, trunk position and trunk motion characteristics on risk of injury. *Spine*, 18 (5), 617-628.

Marras, W.S., Lavender, S.A., Leurgans, S.E., Rajulu, S.L., Allread, W.G., Fathallah, F.A., and Ferguson, S.A. (1995) Biomechanical risk factors and trunk motion. *Ergonomics*, 38(2): 377-410.

Marras, W.S. and Reilly, C.H. (1988) Networks of internal trunk-loading activities under controlled trunk-motion conditions. *Spine*, 13(6), 661-667.

Marras, W.S. and Schoenmarklin, R.W. (1993) Wrist motion in industry, *Ergonomics*, 36(4), 341-351.

Marras, W.S. and Sommerich, C.M. (1991a) A three-dimensional motion model of loads on the lumbar spine: I. Model structure. *Human Factors*, 33 (2), 123-137.

Marras W.S. and Sommerich, C.M. (1991b) A three-dimensional motion model of loads on the lumbar spine: II. Model validation. *Human Factors*, 33 (2), 139-149.

McCoy, M.A., Congleton, W.L., Johnston, W.L., and Jiang, B.C. (1988) The role of lifting belts in manual lifting. *International Journal of Industrial Ergonomics*, 2: 259-256.

McGill, S.M. and Norman, R.W. (1985) Dynamically and statically determined low back moments during lifting. *J. Biomechanics*, 8 (12), 877-885.

McGill, S.M. and Norman, R. (1986) Partitioning the L4-L5 dynamic moment into disc, ligamentous, and muscular components during lifting. *Spine*, 11, 666-678.

McGill, S.M. (1993) Abdominal belts in industry: a position paper on their assets, liabilities and use. *American Industrial Hygiene Association Journal*, 54(12): 752-754.

McGill, S., Seguin, J., and Bennett, G. (1994) Passive stiffness of the torso in flexion, extension, lateral bending, and axial rotation: effects of belt wearing and breath holding. *Spine*, 19(6): 696-704.

Mitchell, L.V., Lawler, F.H., Bowen, D., Mote, W., Asundi, P., and Purswell, J. (1994) Effectiveness and cost-effectiveness of employer-issued back belts in areas of high risk for low back injury. *Journal of Medicine*, 36(1): 90-94.

Mirka G.A. and Marras, W.S. (1993) A stochastic model of trunk muscle coactivation during trunk bending. *Spine*, 18 (11), 1396-1409.

Mundt, D.J., Kelsey, J.L., Golden, A.L. et al. (1993) An epidemiologic study of non-occupational lifting as a risk factor for herniated lumbar intervertebral disc. *Spine*, 18(5): 595-602.

Nachemson, A. (1975) Towards a better understanding of low-back pain: a review of the mechanics of the lumbar disc. *Rheumatology and Rehabilitation*, 14, 129-143.

National Institute for Occupational Safety and Health (NIOSH) (1981) Work practices guide for manual lifting. Department of Health and Human Services (DHHS), National Institute for Occupational Safety and Health (NIOSH), Publication No. 81-122.

National Institute for Occupational Safety and Health (NIOSH) (1994) Workplace Use of Back Belts. Department of Health and Human Services (DHHS), National Institute for Occupational Safety and Health (NIOSH), Publication No. 94-122.

National Safety Council. (1989) *Accident Facts, 1989*, Chicago, IL.

National Safety Council. (1991) *Accident Facts*, Chicago, IL.

National Institute for Occupational Safety and Health (NIOSH), U.S. Department of Health and Human Services, Centers for Disease Control (1973) *The Industrial Environment-Its Evaluation and Control*, U.S. Government Printing Office, Washington, D.C.

Nordin, M. and Frankel, V.M. (1989) *Basic Biomechanics of the Musculoskeletal System,* 2nd edition, Lea and Febiger, Philadelphia, PA, pg 67.

Nussbaum, M.A., Chaffin, D.B., and Martin, B.J. (1995) A back-propagation neural network model of lumbar muscle recruitment during moderate static exertions. *Journal of Biomechanics,* 28(9): 1015-1024.

Ozkaya, N. and Nordin, M. (1991) *Fundamentals of Biomechanics, Equilibrium, Motion and Deformation.* Van Nostrand Reinhold, New York, NY.

Park, K.S. and Chaffin, D.B. (1974) A biomechanical evaluation of two methods of manual load lifting. *AIIE Transactions,* 6(2), 105-113.

Pope, M.H. (1993) Muybridge Lecture, International Society of Biomechanics XIVth Congress, Paris, France, July 5, 1993.

Praemer, A., Furner, S., and Rice, D.P. (1992) *Musculoskeletal Conditions in the United States,* American Academy of Orthopaedic Surgeons, Park Ridge, IL.

Reddell, C.R., Congleton, J.J., Huchingson, R.D., and Montgomery, J.F. (1992) An evaluation of a weightlifting belt and back injury prevention training class for airline baggage handlers. *Applied Ergonomics,* 23(5): 319-329.

Reilly, C. and Marras, W. (1989) Simulift: A simulation model of the human trunk motion. *Spine,* 14, (1), 5-11.

Rowe, M.L. (1981) Low back disability in industry: an updated position. *Journal of Occupational Medicine,* 13(10), 476-478.

Sanders, M.S. and McCormick, E.J. (1993) *Human Factors in Engineering and Design,* McGraw-Hill Inc., New York, NY.

Schoenmarklin, R.W., Marras, W.S., and Leurgans, S.E. (1994) Industrial wrist motions and risk of cumulative trauma disorders in industry. *Ergonomics,* 37(9), 1449-1459.

Schultz, A.B. and Andersson, G.B.J. (1981) Analysis of loads on the lumbar spine. *Spine,* 6: 76-82.

Schultz, A.B., Andersson, G.B.J., Haderspeck, K., Ortgren, R., Nordin, R., and Bjork, R. (1982a) Analysis and measurement of the lumbar trunk loads in tasks involving bends and twists. *J. Biomechanics,* 15, 669-675.

Silverstein, B.A., Fine, L.J., and Armstrong, T.J. (1986) Hand wrist cumulative trauma disorders in industry. *Journal of Industrial Medicine,* 43: 779-784.

Silverstein, B.A., Fine, L.J., and Armstrong, T.J. (1987) Occupational factors and carpal tunnel syndrome. *American Journal of Industrial Medicine,* 11: 343-358.

Snook, S.H. (1989) The control of low back disability: the role of management. *Manual Materials Handling: Understanding and Preventing Back Trauma.* American Industrial Hygiene Association, Akron, OH.

Spengler, D.M., Bigos, S.J., Martin, B.A., et al. (1986) Back injuries in industry: a retrospective study, I. Overview and costs analysis. *Spine,* 11: 241-245.

Sudhakar, L.R., Schoenmarklin, R.W., Lavender, S.A., and Marras, W.S. (1988) The effects of gloves on grip strength and muscle activity. *Proceedings of the Human Factors Society 32nd Annual Meeting,* October 24 to 28, Anaheim, CA.

UAW International Union (1982) *Strains and Sprains: A Worker's Guide to Job Design,* UAW, Detroit, MI.

Videman, T., Nurminen, M. and Troup, T.D.G. (1990) Lumbar spinal pathology in cadaveric material in relation to history of back pain, occupation, and physical loading. *Spine,* 15(8), 728-740.

Walsh, N.E. and Schwartz, R.K. (1990) The influence of prophylactic orthoses on abdominal strength and low back injury in the workplace. *American Journal of Physical Medicine and Rehabilitation,* 69(5): 245-250.

Waters, T.R., Putz-Anderson, V., Garg, A., and Fine, L.J. (1993) Revised NIOSH equation for the design and evaluation of manual lifting tasks. *Ergonomics,* 36(7): 749-776.

Webster, B.S. and Snook, S.H. (1989) The cost of compensable low back pain, Liberty Mutual internal report.

11
Human Strength Evaluation

Karl H. E. Kroemer
Virginia Tech

11.1 Overview

Skeletal muscles are able to move body segments with respect to each other against internal and external resistances. Muscle components can shorten dynamically, statically retain their length, or be lengthened. Various methods and techniques are available for assessing muscular strength. The engineering application of data on available body strength requires the determination of whether minimal or maximal exertions, static or dynamic, are critical. Data on body strength are presented for the design of tools, equipment, and work tasks.

11.2 Background and Terminology

Muscular efforts have been of special interest to physiological science; therefore, there is a long tradition of philosophical and experimental approaches and use of terminology. Of particular importance are Newton's three laws:

- The first explains that *unbalanced force acting on a mass changes its motion condition*
- The second states that *force f equals mass m multiplied by acceleration a:*

$$f = m * a.$$

- The third makes it clear that *force exertion requires the presence of an equally large counter force.*

0-8493-2641-9/99/$0.00+$.50

Physiology books published until the middle of the 20th century tended to divide muscle activities into either dynamic efforts lasting for minutes or hours, with work, energy, and endurance as typical topics; or short bursts of contractile exertion. Much research on muscle effort concentrated on the "isometric" condition in which muscle length (and hence body segment position) did not change. (The Greek term *iso* means unchanged or constant, and *metrein* refers to the measure, i.e., length of the muscle.) Consequently, most information on muscle strength was gathered for such static exertion. All other muscle activities were typically called "anisometric," often even falsely labeled "isotonic" or "kinetic," meant to cover all the many possible dynamic muscle uses. Table 11.1 lists and explains terms that correctly describe muscular events.

For the engineer, skeletal muscles are of primary interest because they pull on segments of the human body and generate energy for exertion to outside objects. Skeletal muscles connect two body links across their joint, as shown in Figure 11.1 in some cases muscles cross even two joints. Muscles are usually arranged in "functional pairs" so the contracting muscle is counteracted by its opponent. The muscle, or group of synergistic muscles, pulling in the intended direction, is called agonist (also called protagonist) and the opposite antagonist. Cocontraction, the simultaneous activation of paired opposing muscles, serves to control speed and strength exertion.

There are several hundred skeletal muscles in the human body, known by their Latin names. They are wrapped in connective tissue (fascia) which imbeds nerves and blood vessels. At the ends of the muscle, the tissues combine to form tendons which attach the muscle to bones.

Thousands of individual muscle fibers run, more or less parallel, the length of the muscle. Seen with a microscope, skeletal muscle fibers appear striped (striated): thin and thick, light and dark bands run across the fiber in regular patterns, which repeat along the length of the fiber. One such dark stripe appears to penetrate the fiber like a thick membrane or disc: this is the so-called z-disk (from the German *zwischen,* between). The distance between two adjacent z-lines defines the sarcomere. Its length at rest is approximately 250 Å (1Å = 10^{-10} m), meaning that there are about 40,000 sarcomeres in series within 1 mm of muscle fiber length.

Within each muscle fiber, thread-like myofibrils (from the Greek *mys,* muscle) are arranged by the hundreds or thousands in parallel. Each of these, in turn, consists of bundles of myofilaments. Spaces between them are filled with a network of tubular channels, sacs, and cisterns connected with a larger tubular system in the z-disks, which itself is part of the networks of blood vessels and nerves in the fascia. This is the sarcoplasmic reticulum, the "plumbing and control" system of the muscle. It provides the fluid transport between the cells inside and outside the muscle and also carries chemical and electrical messages.

Two of the myofibrils, myosin and actin, have the ability to slide along each other; this is the source of muscular contraction. Small projections, called cross-bridges, protrude from the myosin filaments toward neighboring actins. The actin filaments are twisted, double-stranded protein molecules, wrapped in a double helix around the myosin molecules. This is the "contracting microstructure" of the muscle.

The only *active* action a muscle can take is to contract; elongation is brought about by external forces that stretch the muscle. According to the "sliding filament theory," contraction is brought about by the heads of adjacent actin rods moving toward each other. This pulls the z-disks closer together: sarcomeres in series (and parallel) shorten, and with them the whole muscle. After a contraction, the muscle returns to its resting length, primarily through a recoiling of its shortened filaments, fibrils, fibers, and other connective tissues. Stretching the muscle beyond its resting length can be done by forces external to the muscle: either by gravity or other force acting from outside the body, or by the action of antagonistic muscle.

You may want to consult books by Asimov (1963), Åstrand and Rodahl (1986), Chaffin and Andersson (1991), Enoka (1988), Kroemer, Kroemer, and Kroemer-Elbert (1994, 1997), Schneck (1990, 1992), and Winter (1990), among others, for more information.

11.3 Relation between Muscle Length and Tension

Stimulation from the central nervous system (CNS) causes the muscle to contract to its smallest possible length which is about 60% of resting length with no external load. In this condition, the actin proteins

TABLE 11.1 Glossary of Muscle Terms

Activation of muscle — *See* contraction.
Co-contraction — Simultaneous contraction of two or more muscles.
Concentric (muscle effort) — Shortening of a muscle against a resistance.
Contraction — Literally, "pulling together" the Z lines delineating the length of a sarcomere, caused by the sliding action of actin and myosin filaments. Contraction develops muscle tension only if the shortening is resisted. *Note: during an isometric "contraction" no change in sarcomere length occurs, and in an eccentric "contraction" the sarcomere is actually lengthened. To avoid such contradiction in terms, it is often better to use the terms activation, effort, or exertion.*
Distal — Away from the center of the body.
Dynamics — A subdivision of mechanics that deals with forces and bodies in motion.
Eccentric (muscle effort) — Lengthening of a resisting muscle by external force.
Fiber — *See* muscle.
Fibril — *See* muscle fibers
Filament — *See* muscle fibers.
Force — As per Newton's Third Law, the product of mass and acceleration; the proper unit is the Newton, with 1 N = 1 kg m s^{-2}. On earth, one kg applies a (weight) force of 9.81 N (1 lb. exerts 4.44. N) to its support. Muscular force is defined as muscle tension multiplied by transmitting cross-sectional area.
Free dynamic — In this context, an experimental condition in which neither displacement and its time derivatives, nor force are manipulated as independent variables.
Iso — A prefix meaning constant or the same.
Isoacceleration — A condition in which the acceleration is kept constant.
Isoforce — A condition in which the muscular force (tension) is constant, i.e., isokinetic. This term is equivalent to isotonic.
Isoinertial — A condition in which muscle moves a constant mass.
Isojerk — A condition in which the time derivative of acceleration, jerk, is kept constant.
Isokinetic — A condition in which muscle tension (force) is kept constant. *See* isoforce and isotonic; compare with isokinematic.
Isokinematic — A condition in which the velocity of muscle shortening (or lengthening) is constant. (Depending on the given biomechanical conditions, this may or may not coincide with a constant angular speed of a body segment about its articulation.) Compare with isokinetic.
Isometric — A condition in which the length of the muscle remains constant.
Isotonic — A condition in which muscle tension (force) is kept constant — *see* isoforce. (In the past, this term was occasionally falsely applied to any condition other than isometric.).
Kinematics — A subdivision of dynamics that deals with the motions of bodies, but not the causing forces.
Kinetics — A subdivision of dynamics that deals with forces applied to masses.
Mechanical advantage — In this context, the lever arm (moment arm, leverage) at which a muscle pulls about a bony articulation.
Mechanics — The branch of physics that deals with forces applied to bodies and their ensuing motions.
Moment — The product of force and the length of the (perpendicular) lever arm at which it acts. Mechanically equivalent to torque.
Motor unit — All muscle filaments under the control of one efferent nerve axon.
Muscle — A bundle of fibers, able to contract or be lengthened. In this context, striated (skeletal) muscle that moves body segments about each other under voluntary control.
Muscle contraction — The result of contractions of motor units distributed through a muscle so that the muscle length is shortened. See contraction.
Muscle fibers — Elements of muscle, containing fibrils, which consist of filaments.
Muscle fibrils — Elements of muscle fibers, containing filaments.
Muscle filaments — Muscle fibril elements, especially actin and myosin (polymerized protein molecules), capable of sliding along each other, thus shortening the muscle and, if doing so against resistance, generating tension.
Muscle force — The product of tension within a muscle multiplied by the transmitting muscle cross-section.
Muscle strength — The ability of a muscle to generate and transmit tension in the direction of its fibers. *See also* body strength.
Muscle tension — The pull within a muscle expressed as force divided by transmitting cross-section.
Myo — A prefix referring to muscle (Greek *mys*, muscle).
Mys — A prefix referring to muscle (Greek *mys*, muscle).
Proximal — Toward the center of the body.
Repetition — Performing the same activity more than once. (One repetition indicates two exertions.)
Statics — A subdivision of mechanics that deals with bodies at rest.
Strength — *See* body strength and muscle strength.
Tension — force divided by the cross-sectional area through which it is transmitted.
Torque — The product of force and the length of the (perpendicular) lever arm at which it acts. Mechanically equivalent to moment.

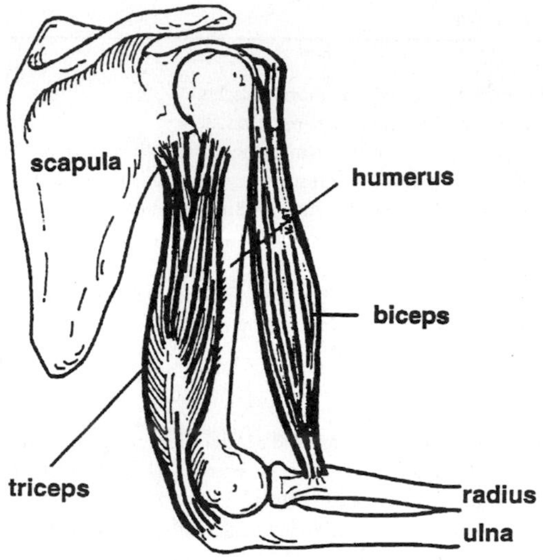

FIGURE 11.1 The biceps muscle reduces the elbow angle as agonist, counteracted by the triceps muscle as antagonist. Note the simplification of the actual conditions in modeling: in addition to the biceps, two other muscles (radialis and brachioradialis) also contribute to flexion about the elbow joint.

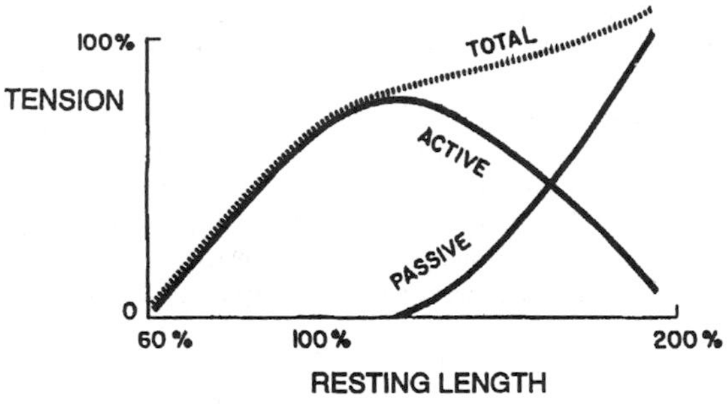

FIGURE 11.2 Active, passive, and total tension within a muscle at different lengths. (From Kroemer, K.H.E., Kroemer, H.J., and Kroemer-Elbert, K.E. (1997). *Engineering Physiology. Bases of Human Factors/Ergonomics* (3rd ed.). New York, NY: Van Nostrand Reinhold. With permission.)

are completely curled around the myosin rods. This is the shortest possible length of the sarcomeres, below which the muscle cannot develop any additional active contraction force.

Near resting length, the cross-bridges between the actin and myosin rods are in an optimal position to generate contact for contractile resistance. If the muscle is elongated further, the actin and myosin fibrils are slid along each other, which reduces the cross-bridge overlap between the protein rods. At about 160% of resting length, so little overlap remains that no active resistance can be developed internally. Thus, the curve of active contractile tension developed within a muscle in isometric twitch is as shown in Figure 11.2 minimal at approximately 60% resting length, rising to about 0.9 at resting length, at unit value at about 120 to 130% of resting length, and then falling back to minimum at about 160% resting length.

The muscle passively resists stretch, like a rubber band. This passive resistance becomes stronger the more the muscle is pulled from its resting length and is strongest near the point of muscle or tendon (attachment) breakage. This is also shown in Figure 11.2 above resting length, the tension in the muscle is the summation of active and passive strains. The summation effect explains why we stretch muscles

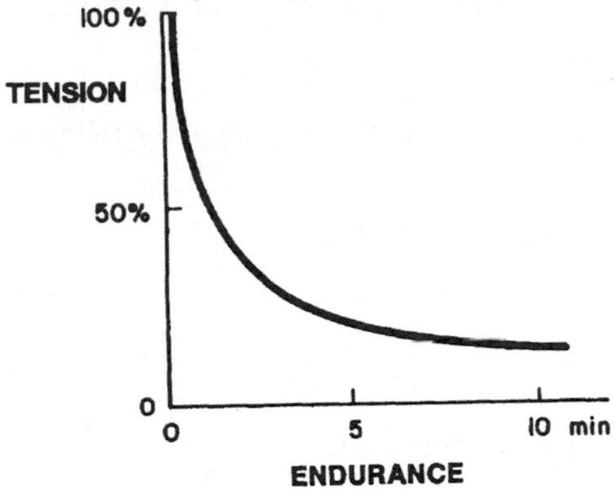

FIGURE 11.3 Muscle exertion and endurance. (From Kroemer, K.H.E., Kroemer, H.J., and Kroemer-Elbert, K.E. (1997). *Engineering Physiology. Bases of Human Factors/Ergonomics* (3rd ed.). New York, NY: Van Nostrand Reinhold. With permission.)

for a strong exertion, like in bringing the arm behind the shoulder before throwing a rock. This "preloading" tenses the muscle for a strong exertion.

In engineering terms it is said that muscles exhibit "viscoelastic" qualities. They are viscous in that their behavior depends both on the amount by which they are deformed, and on the rate of deformation. They are elastic in that, after deformation, they return to the original length and shape. These behaviors help to explain why the tension that can be developed isometrically ("statically," especially in a state of eccentric stretch) is the highest possible, while in active shortening (in a "dynamic" concentric movement) muscle tension is decidedly lower.

11.4 Muscle Endurance and Fatigue

Sufficient supply of arterial blood to the muscle and its unimpeded flow through the capillary bed into the venules and veins are crucial because they determine the ability of the contractile and metabolic processes of the muscle to continue. Blood brings needed energy-carriers and oxygen, and it removes metabolic byproducts, particularly lactic acid and potassium, as well as heat, carbon dioxide, and water liberated during metabolism.

The fine blood vessels permeating the muscle are easily compressed by pressure applied to them. A strongly contracting muscle generates pressure within itself, as can be felt by touching a tightened biceps or calf muscle. By this pressure, the muscle compresses its own blood vessels, thus reducing or even shutting off its own blood circulation. The interruption of blood flow through a muscle leads to muscle fatigue within seconds, forcing relaxation. Such fatigue, which occurs slowly when the muscle is not maximally contracting, is experienced painfully when one works overhead with raised arms, e.g., while fastening a screw in the ceiling of a room. Muscle fatigue in the shoulder muscles makes it impossible to keep one's arms raised after only a minute or so, even though nerve impulses from the CNS still arrive at the neuromuscular junctions, and the resulting action potentials continue to spread over the muscle fibers. Muscle fatigue is defined operationally as a "state of reduced physical ability which can be restored by rest."

Figure 11.3 shows the relation between static exertion and muscle endurance schematically: a maximal exertion can be maintained for just a few seconds; 50% of tension is available for about one minute; but less than 20% can be applied for long endurance periods.

11.5 Muscle Tension and Its Internal Transmission to the Point of Application

The term "strength" is often used in reference to any or all of the following:

- The tension *within* a muscle
- The *internal transmission* via body links across joints
- The *external application* of force or torque by a body segment to an outside object.

Confusion can be avoided by distinguishing these aspects and using proper terminology.

Muscle Tension — "Muscle Strength"

Within the muscle, all filament pulls combine to a resultant tension in the muscle. Its magnitude depends mostly on the involved number of parallel muscle fibers, i.e., on the cross-sectional thickness of the muscle. Maximal tensions reported on human skeletal muscle are within the range of 16 to 61 N/cm^2. Enoka (1988) uses 30 N/cm^2 as a typical value and calls it "specific (human muscle) tension." If the muscle cross-section area is known (such as from cadaver measurements or from MRI scans), one can calculate a resultant muscle force.

From the muscle, one tendon extends proximally to the origin and, in the opposite direction, another tendon extends outward to the insertion (see Figure 11.4). Like cables, tendons transmit the muscle tension, usually to the surface of a bone, but some end at strong connective tissues, such as in the fingers. The distal tendon may be quite long; for example, the tendons that reach from the muscles in the forearm to the digits of the hand can be 20 cm in length.

Internal Transmission

The tension inside a muscle–tendon unit tries to pull together the origin and the insertion of the muscle–tendon unit across a joint. This generates torque about the articulation with the long bones, the lever arms at which muscle pulls. The torque developed depends, hence, not only on the strength of the muscle, but also on the effective lever arm, or on the pull angle. Figure 11.4 depicts these conditions as

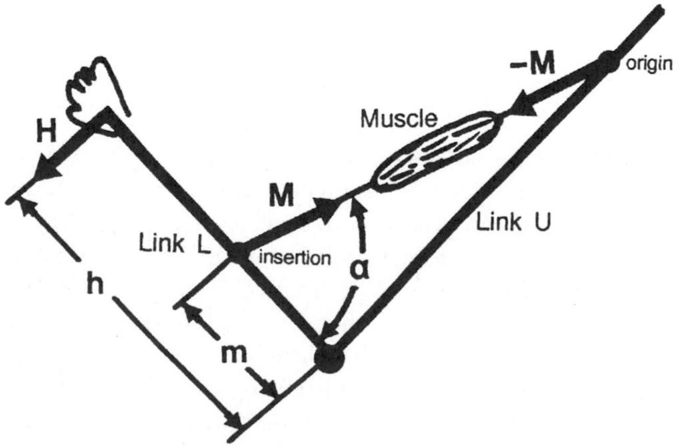

FIGURE 11.4 The muscle–tendon unit exerting pull forces *M* to links *L* and *U* at origin and insertion. Note the simplification in modeling the actual conditions at the elbow: in reality the biceps muscle has its origin proximal to the shoulder joint: two other muscles (radialis and brachioradialis) contribute to flexion about the elbow joint. Torque $T = m * M * \sin \alpha$ must be transmitted across the wrist joint to generate hand force H. (From Kroemer, K.H.E., Kroemer, H.J., and Kroemer-Elbert, K.E. (1997). *Engineering Physiology. Bases of Human Factors/Ergonomics* (3rd ed.). New York, NY: Van Nostrand Reinhold. With permission.)

a simplified scheme of the lower arm link *L* attached to the upper arm link *U*, articulated in the elbow joint. The muscle has its origin at the proximate end of *U*. Its insertion point is at the distance *m* from the joint on the arm link *L*. It generates a muscle force *M*, pulling at the angle α. The torque *T* generated by *M* depends on the lever arm *m* and the pull angle α according to

$$T = m * M * \sin \alpha \tag{1}$$

This torque *T* then counteracts an external force *H*, acting perpendicularly at its lever arm *h*, according to

$$T = m * M * \sin \alpha = \mathrm{h} * \mathrm{H} \tag{2}$$

External Application — "Body (Segment) Strength"

The final output of the biomechanical system is the torque or force (*H* in Figure 11.4) available at the hand, foot, or other body segment for exertion to a resisting object. This object is usually outside the body, but the resistance may be from an antagonistic muscle. The "body (segment) strength" available for application to an outside object is of primary importance to the engineer, designer, and manager.

The model depicted in Figure 11.4 shows that the amount of force (*H*) available at the body interface with an external object depends on

- Muscle force (*M*)
- Lever arms (*m* and *h*)
- Pull angle (α) which, in turn, depends on the angle between the two links. If all of these are known, the body segment force can be calculated from equation (2)

$$H = (m/h) * M * \sin \alpha \tag{3}$$

Definitions

To help distinguish among muscle tension, its internal transmission, and the exertion to an outside object, it is useful to define terms as follows:

MUSCLE STRENGTH is the maximal tension (or force) that muscle can develop voluntarily between its origin and insertion.

The best word to refer to this is "muscle tension" (in N/mm^2 or N/cm^2) but the term strength (force in N) is commonly used. If the variables *m, h,* α, and *H* in equation (2) are known, one can solve for the muscle force

$$M = (h/m) * H/\sin \alpha \tag{4}$$

INTERNAL TRANSMISSION is the manner in which muscle tension is transferred inside the body along links and across joint(s) as torque to the point of application to a resisting object.

If several link-joint systems in series constitute the internal path of torque (in Nm or Ncm) transmission, each transfers the arriving torque by the existent ratio of lever arms (*m* and *h* in the example above) until resistance is met, usually the point where the body interfaces with an outside object. This transfer of torques is more complicated under dynamic conditions than in the static case because of changes in muscle functions with motion and because of the effects of accelerations and decelerations on masses.

BODY SEGMENT STRENGTH is the force or torque that can be applied by a body segment to an object external to the body.

The segment is usually identified such as "hand," "elbow," "shoulder," "back," "foot." (Strength in N, torque in Ncm or Nm.)

The original muscle pull (after being internally transmitted to the appropriate body member and transformed in magnitude and direction during that transmission) finally results in a force or torque that can be applied to an outside object: often by hand to a hand tool, or to a handle on a box as in load lifting, or by shoulder or back in pushing or carrying; or by the feet in operating pedals, or in walking or running.

The quality and quantity of the force or torque transmitted to an outside object depends on many biomechanical and physical conditions including

- Body segment employed, e.g., hand or foot
- Type of body object attachment, e.g., by a touch or grasp
- Coupling type, e.g., by friction or interlocking
- Direction of force/torque vector
- Static posture or body motions in dynamic exertion

Within the field of ergonomics (*aka* human factors, *aka* human engineering) *muscle strength* is of particular interest to the engineering physiologist; *internal transmission* is of particular interest to the biomechanist and to the designer because of the implications for body segment posture and motion; and *body segment strength* is of particular interest to the designer of tools, equipment, and work tasks.

Figure 11.5 shows, in the form of a flow diagram, the feedforward of excitation signals from the CNS to muscle to generate tension. EEGs (electroencephalograms) and EMGs (electromyograms) can be recorded and measured, while muscle tension is calculated using biomechanical modeling. Torques are internally transmitted via bone leverages and across articulations to the body segment that applies energy to a resisting object (often a handle or pedal) for task execution. The internal torques are calculated biomechanically, while posture and motion as well as applied force and torque can be measured. Three feedback paths can be identified, although at present they do not provide convenient avenues for measurements. The first is a short reflex loop F_1 originating at interoceptors. The other two loops start at exteroceptors and lead to a comparator where they modify the input to the CNS. F_2 provides kinesthetic signals related to touch, body position, and motion. F_3 relates to sound and vision; its signal can be influenced by the experimenter, e.g., via verbal exhortation or by an instrument showing the applied force.

11.6 Assessment of Body Segment Strength

Generation of muscle strength is a complex procedure of myofilament activation through nervous feedforward and feedback control. It may involve substantial shortening or lengthening of muscle, i.e., a concentric or eccentric effort; or there may be no perceptible change in length, i.e., the effort is isometric. Mechanically, the main distinction between muscle actions is whether they are "dynamic" or "static."

Static Strength

In physiological terms, an isometric muscle contraction generates (usually after some initial sarcomere shortening) the static condition. When there is no change in muscle length during the isometric effort, then involved body segments do not move; in physics terms, all forces acting within the system are in equilibrium, as Newton's First Law requires. Therefore, the physiological "isometric" case is equivalent to the "static" condition in physics.

The static condition is theoretically simple and experimentally well controllable. It allows rather easy measurement of muscular effort. Therefore, most of the information currently available on "human strength" describes the outcomes of static (isometric) testing. Accordingly, most of the tables on body segment strength in this chapter and in other human engineering or physiologic literature contain static data.

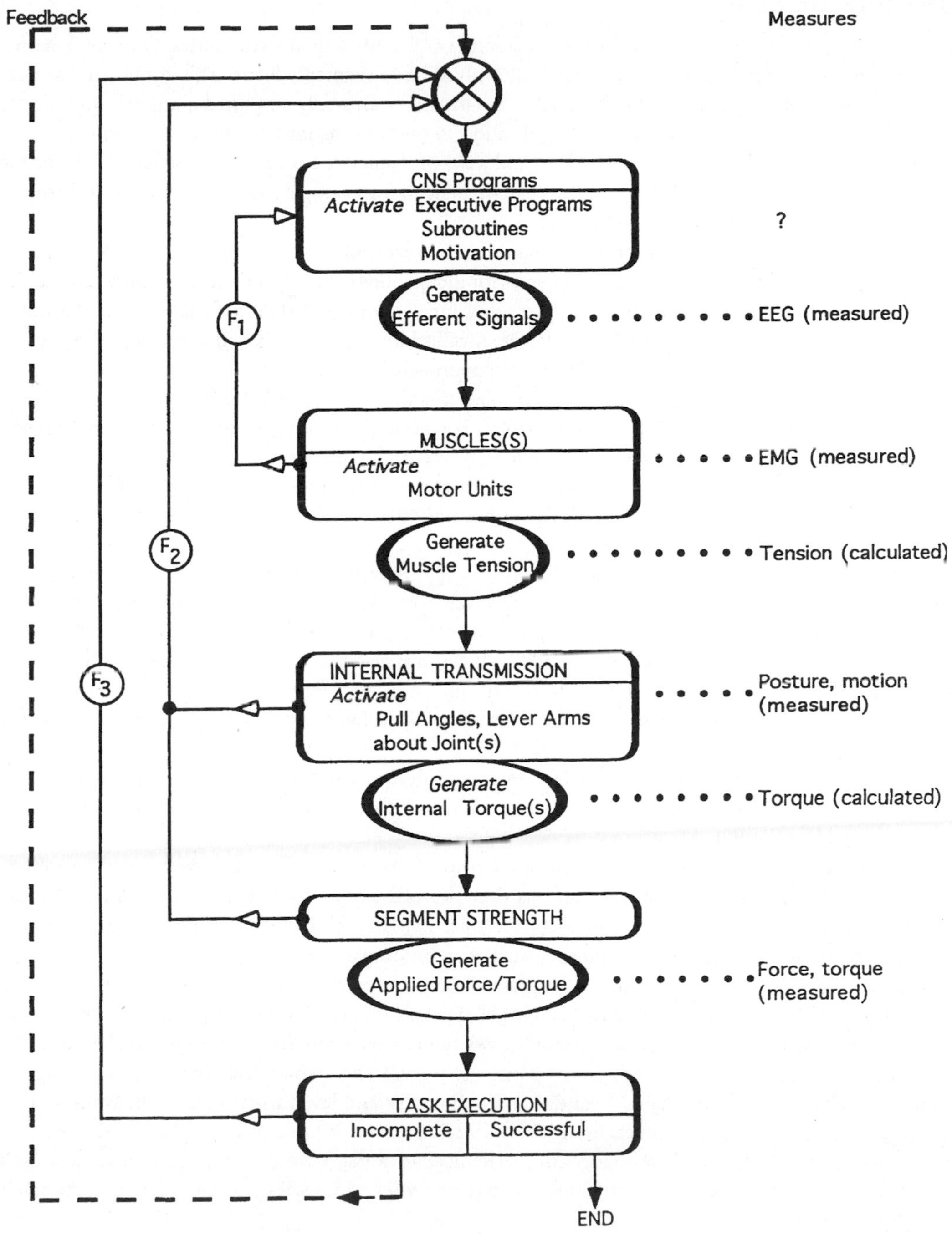

FIGURE 11.5 A conceptual model of the feedforward and feedback loops employed in generating and controlling muscle strength exertion. (Modified from Kroemer 1979, and Kroemer, Kroemer, and Kroemer-Elbert, 1994, 1997.)

Besides the simple convenience of dealing with statics, measurement of isometric strength appears to yield, for most cases, a reasonable estimate of the maximal possible exertion for most slow body link movements, especially if they are excentric. However, the data do not estimate fast exertions well, especially if they are concentric and of the ballistic-impulse type, such as throwing or hammering.

Dynamic Strength

Dynamic muscular efforts are more difficult to describe and control than static contractions. In dynamic activities, muscle length changes and, therefore, involved body segments move. This results in displacement. The amount of travel is relatively small at the muscle but usually amplified along the links of the internal transmission path to the point of application to the outside, for example at the hand or foot.

The time derivatives (velocity, acceleration, and jerk) of displacement are of importance both for the muscular effort (as discussed earlier) and the external effect: for example, change in velocity determines impact and force, as per Newton's Second Law.

Definition and experimental control of dynamic muscle exertions are much more complex tasks than in static testing. Various new classification schemes for independent and dependent experimental variables can be developed. Such a system has been presented for dynamic and static efforts (Kroemer, Marras, McGlothlin, McIntyre, and Nordin 1989; Marras, McGlothlin, McIntyre, Nordin, and Kroemer 1993). It includes the traditional isometric and isoinertial approaches.

Independent variables are those that are purposely manipulated during the experiment in order to assess resulting changes in the dependent variables. For example, if one sets the displacement (muscle length change) to zero — the *isometric* condition — one may either measure the magnitude of force generated, or the number of repetitions that can be performed until force is reduced because of muscular fatigue. This case is described in Table 11.2. Of course, since there is no displacement, its time derivatives, velocity, acceleration, and jerk, are also zero.

Alternatively, one may choose to control velocity as an independent variable, i.e., the rate at which muscle length changes. If velocity is set to a constant value, one speaks of an *isokinematic* muscle strength measurement. (Note that this *isovelocity* condition is often mislabeled "isokinetic.") Time derivatives of constant velocity, acceleration, and jerk, are zero. Mass properties are usually controlled in isokinematic tests. The variables displacement, force, and repetition can either be chosen as dependent variables or they may become controlled independent variables. Most likely, force and/or repetition are chosen as the dependent test variables.

Following the scheme laid out in Table 11.2, one also can devise tests in which acceleration or its time derivative, jerk, are kept constant. These test conditions are theoretically possible, but so far they have not been commonly applied.

For some tests, one sets the amount of muscle tension (force) to a constant value. In such an *isoforce* test, mass properties and displacement (and its time derivatives) are likely to become controlled independent variables, and repetition a dependent variable. This *isotonic* condition is, for practical reasons, often combined with an isometric condition, such as in holding a load motionless.

Note that the term isotonic has often been applied falsely. Some older textbooks described the examples of lifting or lowering of a constant mass (weight) as typical for isotonics. This is physically false for two reasons. The first is that according to Newton's Laws the change from acceleration to deceleration of a mass requires application of changing (not constant) forces. The second fault lies in overlooking the changes that occur in the mechanical conditions (pull angles and lever arms) under which the muscle functions during the activity. Hence, even if there were a constant force to be applied to the external object (which is not the case), the changes in mechanical advantages would result in changes in muscle tonus. It is certainly misleading to label all dynamic activities of muscles isotonic, as is unfortunately done occasionally.

In the *isoinertial* test, the external mass is set to a constant value. In this case, repetition of moving such constant mass (as in lifting) may either be a controlled independent or, more likely, a dependent variable. Also, displacement and its derivatives may become dependent outputs. Force (or torque) applied is likely to be a dependent value, according to Newton's Second Law. (Note that in the isoinertial test the external load is controlled, while in the previously described tests the conditions at the muscle were controlled.)

TABLE 11.2 Techniques to Measure Muscle Performance by Selecting Specific Independent and Dependent Variables

Names of Technique Variables	Isometric (Static) Indep. Dep.	Isovelocity (Dynamic) Indep. Dep.	Isoacceleration (Dynamic) Indep. Dep.	Isojerk (Dynamic) Indep. Dep.	Isoforce (Static or Dynamic) Indep. Dep.	Isoinertia (Static or Dynamic) Indep. Dep.	Free Dynamic Indep. Dep.
Displacement, linear/Angular	[Constant* (zero)]	C or X	C or X	C or X	C or X	C or X	X
Velocity, linear/angular	0	[constant]	C or X	C or X	C or X	C or X	X
Acceleration, linear/angular	0	0	[constant]	C or X	C or X	C or X	X
Jerk, linear/angular	0	0	0	[constant]	C or X	C or X	X
Force, Torque	C or X	C or X	C or X	C or X	[constant]	C or X	X
Mass, Moment of Inertia	C	C	C	C	C	[constant]	C or X
Repetition	C or X	C or X	C or X	C or X	C or X	C or X	C or X

Legend

Indep = independent
Dep = dependent
C = variable can be controlled
* = set to zero
0 = variable is not present (zero)
X = can be dependent variables

The [boxed] constant variable provides the descriptive name.

From Kroemer, K.H.E., Kroemer, H.J., and Kroemer-Elbert, K.E. (1997). *Engineering Physiology. Bases of Human Factors/Ergonomics* (3rd ed.). New York, NY: Van Nostrand Reinhold. With permission.

Table 11.2 also contains the most general case of motor performance measurement, labeled "free dynamic." Here the independent variables force and displacement (and its time derivatives) are left to the free choice of the subject. Only mass and repetition are usually controlled but may be used as dependent variables. Force, torque, or some other performance measure is likely to be chosen as a dependent output.

This discussion indicates that dynamic tests indeed require more effort to describe and control than static (isometric) measurements. This complexity explains why, in the past, dynamic measurements other than isokinematic and isoinertial testing have been rarely performed in the laboratory. On the other hand, if one is free to perform as one pleases, such as in the "free dynamic" test common in sports, very little experimental control can be executed. Nevertheless, Table 11.2 shows that it is possible to include both the traditional static and the important dynamic exertions in one systematic matrix of measurements.

11.7 Designing for Body Strength

The engineer or designer wanting to consider operator strength has to make a series of decisions. These include:

- Is the use mostly static or dynamic? If static, information about isometric capabilities, listed below, can be used. If dynamic, other considerations apply in addition, concerning for example physical endurance (circulatory, respiratory, metabolic) capabilities of the operator or prevailing environmental conditions. Physiologic and ergonomic texts (e.g., by Åstrand and Rodahl, 1986; Kroemer, Kroemer, and Kroemer-Elbert, 1994, 1997; Winter 1990) provide such information.
- Is the exertion by hand, by foot, or with other body segments? For each, specific design information is available. If a choice is possible, it must be based on physiologic and ergonomic considerations to achieve the safest, least strenuous, and most efficient performance. In comparison to hand movements over the same distance, foot motions consume more energy, are less accurate and slower, but they are stronger.
- Is a maximal or a minimal strength exertion the critical design factor?

  "Maximal" user strength usually relates to the structural strength of the object, so that a handle or a pedal may not be broken by the strongest operator. The design value is set, with a safety margin, above the highest perceivable strength application.

  "Minimal" user strength is that expected from the weakest operator which still yields the desired result, so that a door handle or brake pedal can be successfully operated or a heavy object be moved.

  A "range" of expected strength exertions is, obviously, that between the considered minimum and maximum. The infamous "average user" strength is usually of no design value (See Chapter 9).

- Most body segment strength data are available for static (isometric) exertions. They provide reasonable guidance also for slow motions, although they are probably too high for concentric motions and a bit too low for eccentric motions. Of the little information available for dynamic strength exertions, much is limited to isokinematic (constant velocity) cases. As a general rule, strength exerted in motion is less than measured in static positions located on the path of motion.
- Measured strength data are often treated, statistically, as if they were normally distributed and reported in terms of averages (means) and standard deviations. This allows the use of common statistical techniques to determine data points of special interest to the designer. In reality, body segment strength data are often in a skewed rather than in a bell-shaped distribution. This is not of great concern, however, because usually the data points of special interest are the extremes. The maximal forces or torques that the equipment must be able to bear without breaking are those above or near the strongest measured data points. The minimal exertions, which even "weak" persons are able to generate, can be identified as given percentile values at the low end of the distribution: often the 5th percentile is selected. This can be done either by calculation or by estimation (See Chapter 9 for details and instructions.)

Designing for Hand Strength

The human hand is able to perform a large variety of activities, ranging from those that require fine control to others that demand large forces. (But the feet and legs are capable of more forceful exertions than the hand, see below.)

One may divide hand tasks in this manner:

- Fine manipulation of objects, with little displacement and force. Examples are writing by hand, assembly of small parts, adjustment of controls.
- Fast movements to an object, requiring moderate accuracy to reach the target but fairly small force exertion there. An example is the movement to a switch and its operation.
- Frequent movements between targets, usually with some accuracy but little force; such as in an assembly task, where parts must be taken from bins and assembled.
- Forceful activities with little or moderate displacement (such as with many assembly or repair activities, for example when turning a hand tool against resistance).
- Forceful activities with large displacements (e.g., when hammering).

Accordingly, there are three major types of requirements: accuracy, displacement, and strength exertion. Design for the first two tasks is described in Chapters 8 through 11 of the book by Kroemer, Kroemer, and Kroemer-Elbert (1994).

Of the digits of the hand, the thumb is the strongest and the little finger the weakest. Gripping and grasping strengths of the whole hand are larger, but depend on the coupling between the hand and the handle (see Figure 11.6.) The forearm can develop fairly large twisting torques. Large force and torque vectors are exertable with the elbow at about a right angle, but the strongest pulling/pushing forces toward/away from the shoulder can be exerted with the extended arm, provided that the trunk can be braced against a solid structure. Torque about the elbow depends on the elbow angle as depicted in Figure 11.6 and, in more detail, in Figure 11.7.

Obviously, forces exerted with the arm and shoulder muscles are largely determined by body posture and body support. Likewise, finger forces depend on the finger joint angles, as listed in Tables 11.3 and 11.4. Table 11.5 provides detailed information about manual force capabilities measured in male students and machinists. Female students developed between 50 and 60% of the digit strength of their male peers, but achieved 80 to 90% in "pinches."

The Use of Tables of Exerted Torques and Forces

There are many sources for data on body strengths that operators can apply (see the tables and figures to follow). While these data indicate "orders of magnitude" of forces and torques, the exact numbers should be viewed with great caution because they were measured on various subject groups of rather small numbers under widely varying circumstances. It is advisable to take body strength measurements on a sample of the intended user population to verify that a new design is operable.

Note that thumb and finger forces, for example, depend decidedly on "skill and training" of the digits as well as the posture of the hand and wrist. Hand forces (and torques) also depend on wrist position, and on arm and shoulder posture. Exertions with arm, leg, and "body" (shoulder, backside) depend much on the posture of the body and on the support provided to the body (i.e., on the "reaction force" in the sense of Newton's Third Law) in terms of friction or bracing against solid structures. Figure 11.9 and Table 11.6 illustrate this: both were derived from the same set of empirical data but extrapolated to show the effects of

- Location of the point of force exertion
- Body posture
- Friction at the feet

on horizontal push (and pull) forces applied by male soldiers.

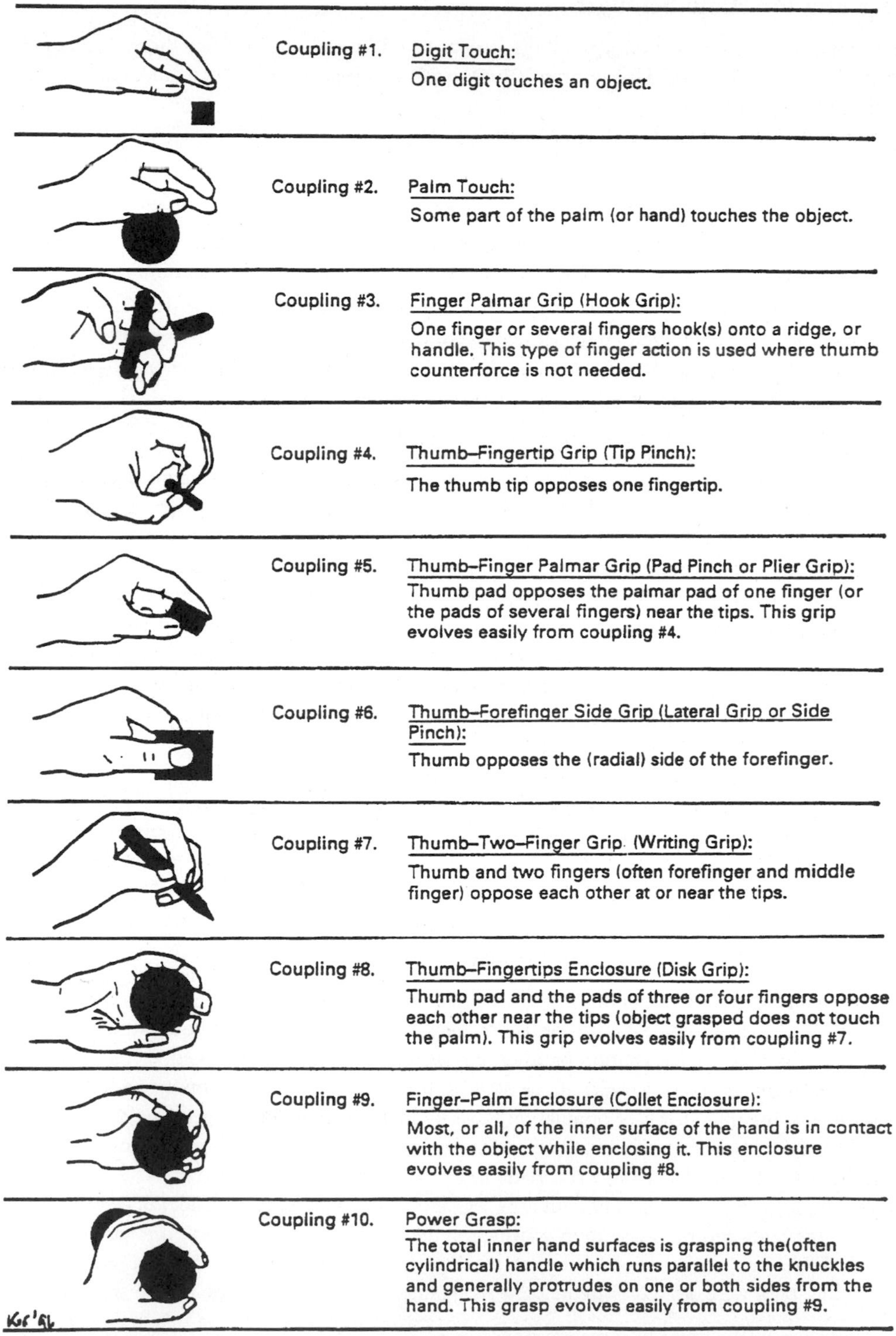
Coupling #1.
Digit Touch:
One digit touches an object.
Coupling #2.
Palm Touch:
Some part of the palm (or hand) touches the object.
Coupling #3.
Finger Palmar Grip (Hook Grip):
One finger or several fingers hook(s) onto a ridge, or handle. This type of finger action is used where thumb counterforce is not needed.
Coupling #4.
Thumb–Fingertip Grip (Tip Pinch):
The thumb tip opposes one fingertip.
Coupling #5.
Thumb–Finger Palmar Grip (Pad Pinch or Plier Grip):
Thumb pad opposes the palmar pad of one finger (or the pads of several fingers) near the tips. This grip evolves easily from coupling #4.
Coupling #6.
Thumb–Forefinger Side Grip (Lateral Grip or Side Pinch):
Thumb opposes the (radial) side of the forefinger.
Coupling #7.
Thumb–Two–Finger Grip (Writing Grip):
Thumb and two fingers (often forefinger and middle finger) oppose each other at or near the tips.
Coupling #8.
Thumb–Fingertips Enclosure (Disk Grip):
Thumb pad and the pads of three or four fingers oppose each other near the tips (object grasped does not touch the palm). This grip evolves easily from coupling #7.
Coupling #9.
Finger–Palm Enclosure (Collet Enclosure):
Most, or all, of the inner surface of the hand is in contact with the object while enclosing it. This enclosure evolves easily from coupling #8.
Coupling #10.
Power Grasp:
The total inner hand surfaces is grasping the(often cylindrical) handle which runs parallel to the knuckles and generally protrudes on one or both sides from the hand. This grasp evolves easily from coupling #9.

FIGURE 11.6 Couplings between hand and handle. (Adapted from Coupling the hand with the handle: an improved notation of touch, grip, and grasp, K.H.E. Kroemer, *Human Factors, 28,* 337-339.)

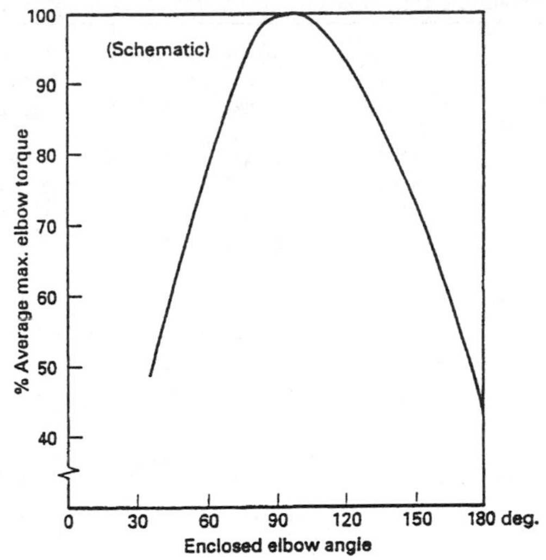

FIGURE 11.7 Relation of elbow angle and elbow torque (From Kroemer, K.H.E., Kroemer, H.B., and Kroemer-Elbert, K.E. (1994) *Ergonomics: How to Design for Ease and Efficiency.* Englewood Cliffs, NJ: Prentice Hall. With permission.)

TABLE 11.3 Mean Forces and Standard Deviations in N exerted by 9 Subjects in Fore, Aft, and Down Direction with the Fingertips, Depending on the Angle of the Proximal Interphalangeal Joint (PIP)

	PIP joint at 30 degrees			PIP joint at 60 degrees		
Direction:	Fore	Aft	Down	Fore	Aft	Down
DIGIT						
2 Index	5.4 (2.0)	5.3 (2.2)	27.4 (13.0)	5.2 (2.4)	6.8 (2.8)	24.4 (13.6)
2 n	4.8 (2.2)	6.1 (2.2)	21.7 (11.7)	5.6 (2.9)	5.3 (2.1)	25.1 (13.7)
3 Middle	4.8 (2.5)	5.4 (2.4)	24.0 (12.6)	4.2 (1.9)	6.5 (2.2)	21.3 (10.9)
4 Ring	4.3 (2.4)	5.2 (2.0)	19.1 (10.4)	3.7 (1.7)	5.2 (1.9)	19.5 (10.9)
5 Little	4.8 (1.9)	4.1 (1.6)	15.1 (8.0)	3.5 (1.6)	3.5 (2.2)	15.5 (8.5)

n: Nonpreferred hand

From Kroemer, K.H.E., Kroemer, H.B., and Kroemer-Elbert, K.E. (1994) *Ergonomics: How to Design for Ease and Efficiency.* Englewood Cliffs, NJ: Prentice Hall. With permission.

TABLE 11.4 Mean Poke Forces and Standard Deviations in N exerted by 30 subjects in the Direction of the Straight Digits.

Digit	10 Male Mechanics	10 Male Students	10 Female Students
1. Thumb	83.8 (25.19) A	46.7 (29.19) C	32.4 (15.36) D
2. Index Finger	60.4 (25.81) B	45.0 (29.99) C	25.4 (9.55) DE
3. Middle Finger	55.9 (31.85) B	41.3 (21.55) C	21.5 (6.46) E

Entries with different letters are significantly different from each other ($p \leq 0.05$).

From Kroemer, K.H.E., Kroemer, H.B., and Kroemer-Elbert, K.E. (1994) *Ergonomics: How to Design for Ease and Efficiency.* Englewood Cliffs, NJ: Prentice Hall. With permission.

TABLE 11.5 Mean Forces and Standard Deviations in *N* Exerted by 21 Male Students* and by 12 Male Machinists

Couplings (see Figure 11.6)	Digit 1 (thumb)	Digit 2 (index)	Digit 3 (middle)	Digit 4 (ring)	Digit 5 (little)	
Push with digit tip in direction of the extended digit ("Poke")	91 (39)* 138 (41)	52 (16)* 84 (35)	51 (20)* 86 (28)	35 (12)* 66 (22)	30 (10)* 52 (14)	See also Table 11.4
Digit Touch (Coupling #1) perpendicular to extended digit.	84 (33)* 131 (42)	43 (14)* 70 (17)	36 (13)* 76 (20)	30 (13)* 57 (17)	25 (10)* 55 (16)	
Same, but all fingers press on one bar	—	digits 2, 3, 4, 5 combined: 162 (33)				
Tip force (like in typing; angle between distal and proximal phalanges about 135 degrees)	— —	30 (12)* 65 (12)	29 (11)* 69 (22)	23 (9)* 50 (11)	19 (7)* 46 (14)	
Palm Touch (Coupling #2) perpendicular to palm (arm, hand, digits extended and horizontal)	—	—	—	—	—	233 (65)
Hook Force exerted with digit tip pad (Coupling #3, "Scratch")	61 (21)* 118 (24)	49 (17)* 89 (29)	48 (19)* 104 (26)	38 (13)* 77 (21)	34 (10)* 66 (17)	all digits combined: 108 (39)* 252 (63)
Thumb-Fingertip Grip (Coupling #4, "Tip Pinch")	—	1 on 2 50 (14)* 59 (15)	1 on 3 53 (14)* 63 (16)	1 on 4 38 (7)* 44 (12)	1 on 5 28 (7)* 30 (6)	
Thumb-Finger Palmar Grip (Coupling #5, "Pad Pinch")	1 on 2 and 3 85 (16)* 95 (19)	1 on 2 63 (12)* 34 (7)	1 on 3 61 (16)* 70 (15)	1 on 4 41 (12)* 54 (15)	1 on 5 31 (9)* 34 (7)	
Thumb-Forefinger Side Grip (Coupling #6, "Side Pinch")	—	1 on 2 98 (13)* 112 (16)	—	—	—	—
Power Grasp (Coupling #10, "Grip Strength")	—	—	—	—	—	318 (61)* 366 (53)

From Kroemer, K.H.E., Kroemer, H.B., and Kroemer-Elbert, K.E. (1994) *Ergonomics: How to Design for Ease and Efficiency.* Englewood Cliffs, NJ: Prentice Hall. With permission.

It is obvious that the amount of strength available for exertion to an object outside the body depends on the weakest part in the chain of strength-transmitting body parts. Hand pull force, for example, may be limited by finger strength, or shoulder strength, or low back strength; or it may be limited by the reaction force available to the body, as per Newton's Third Law. Figure 11.9 helps in determining where the "critical body segment "is in that sequence of torques about body joints. Starting at the point of external exertion, e.g., at the hand, one assesses the strength requirements joint-by-joint along the arm, shoulder, and back. Often, the lumbar back area is the "weak link" as evidenced by the large number of low-back pain cases reported in the literature.

To a sitting person, the reaction countering the forces actively exerted through upper body and arms is largely provided by the seat, although some support may be gathered from the floor via the legs. A walking or standing person receives all support from the ground up, of course, and not seldom hip or knee joint strength limits the ability to do hard efforts, such as lifting a load on the back. A slippery surface may make it impossible to push a heavy object sideways with the shoulder; one can experience this in the winter on icy ground when trying to push a car out of the ditch. These examples demonstrate how important is to provide proper body support at seat or ground.

Designing for Foot Strength

If a person stands at work, fairly little force and only infrequent operations of foot controls should be required because, during these exertions, the operator has to stand on the other leg. For a seated operator, however, operation of foot controls is much easier because the body is largely supported by the seat. Thus, the feet can move more freely and, given suitable conditions, can exert large forces and energies.

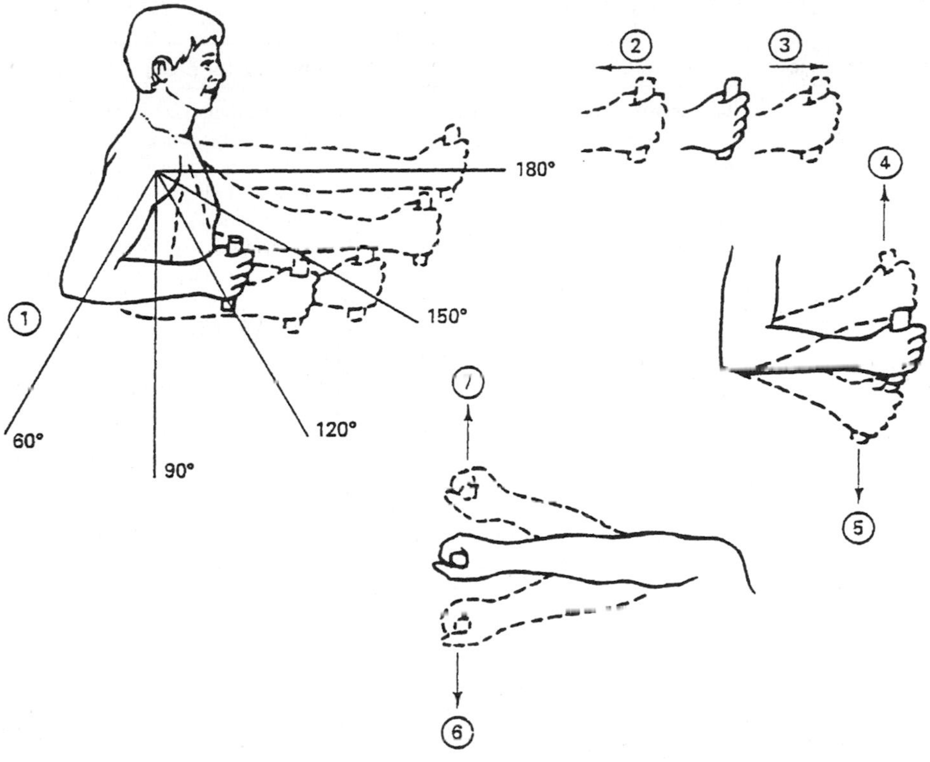

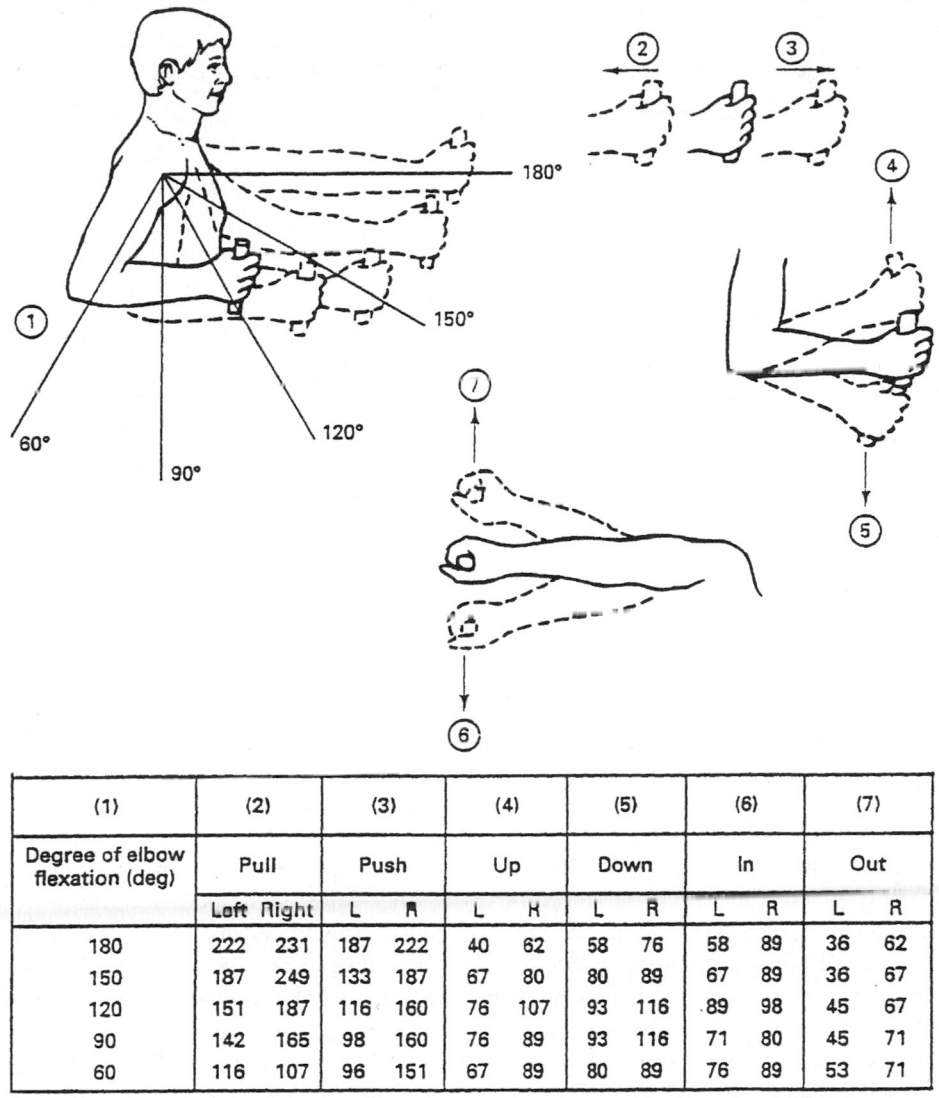

(1)	(2)		(3)		(4)		(5)		(6)		(7)	
Degree of elbow flexation (deg)	Pull		Push		Up		Down		In		Out	
	Left	Right	L	R	L	R	L	R	L	R	L	R
180	222	231	187	222	40	62	58	76	58	89	36	62
150	187	249	133	187	67	80	80	89	67	89	36	67
120	151	187	116	160	76	107	93	116	89	98	45	67
90	142	165	98	160	76	89	93	116	71	80	45	71
60	116	107	96	151	67	89	80	89	76	89	53	71

FIGURE 11.8 Fifth percentile arm forces in N exerted by sitting men. (Adapted from MIL HDBK 759.)

A typical example for such an exertion is pedaling a bicycle: all energy is transmitted from the leg muscles through the feet to the pedals. For normal use, these should be located directly underneath the body, so that the body weight above them provides the reactive force to the force transmitted to the pedal. Placing the pedals forward makes body weight less effective for generation of reaction force to the pedal effort, hence a suitable backrest should be provided against which the buttocks and low back press while the feet push forward on the pedal.

Small forces, such as for the operation of switches, can be generated in nearly all directions with the feet, with the downward or down-and-fore directions preferred. The largest forces can be generated with extended or nearly extended legs: in the downward direction limited by body inertia, in the more forward direction both by inertia and the provision of buttock and back support surfaces. These principles are typically applied in automobiles. For example, operation of a clutch or brake pedal can normally be performed easily with about a right angle at the knee. But if the power-assist system fails, very large

TABLE 11.6 Horizontal Push and Pull Forces in *N* that Male Soldiers Can Exert Intermittently or for Short Periods of Time

Horizontal force*; at least	Applied with**	Condition (μ: coefficient of friction at floor)
100 N push or pull	both hands or one shoulder or the back	with low traction, $0.2 < \mu < 0.3$
200 N push or pull	both hands or one shoulder or the back	with medium traction, $\mu \sim 0.6$
250 N push	one hand	if braced against a vertical wall 51 to 150 cm from and parallel to the push panel
300 N push or pull	both hands or one shoulder or the back	with high traction, $\mu > 0.9$
500 N push or pull	both hands or one shoulder or the back	if braced against a vertical wall 51 to 180 cm from and parallel to the panel or if anchoring the feet on a perfectly nonslip ground (like a footrest)
750 N push	the back	if braced against a vertical wall 600 to 110 cm from and parallel to the push panel or if anchoring the feet on a perfectly nonslip ground (like a footrest)

* May be nearly doubled for two and less than tripled for three operators pushing simultaneously. For the fourth and each additional operator, not more than 75% of their push capability should be added.

** See Figure 11.9 for examples.

Adapted from MIL STD 1472.

forces must be exerted with the feet: in this case, thrusting one's back against a strong backrest and extending the legs are necessary to generate the needed pedal force.

Figures 11.11 through 11.15 provide information about the forces that can be applied with legs and feet to a pedal. Of course, the forces depend on body support and hip and knee angles. The largest forward thrust force can be exerted with the nearly extended legs which leaves very little room to move the foot control further away from the hip.

Of course, the strength that can be exerted with the foot to an object, such as a pedal, depends on the joint strengths that can be transmitted along the "chain" ankle–knee–hip to the seat, which provides the reaction support needed according to Newton's Third Law. This is shown in Figure 11.16. The foot exertion loads all proximal segments according to the prevailing angles and leverarms. Following the diagram outward one sees that bad seat design, frail hip, knee or ankle, or low friction at the coupling of the shoe with the object may all make for a "weak kick."

Information on body strengths has been compiled, e.g., in NASA and U.S Military Standards; by Eastman-Kodak Company (1983, 1986); Kroemer, Kroemer, and Kroemer-Elbert (1994, 1997); Salvendy (1987); Weimer (1993); and Woodson, Tillman, and Tillman (1991). However, caution is necessary when applying these data because they were measured on different populations under varying conditions.

11.8 Summary

Muscle contraction is brought about by active shortening of muscle substructures. Elongation of the muscle is due to external forces. Maximal muscle tension depends on the individual's muscle size and exertion skill.

Prolonged strong contraction leads to muscular fatigue, which hinders the continuation of the effort and finally cuts it off. Hence, maximal voluntary contraction can be maintained for only a few seconds.

In isometric contraction, muscle length remains constant, which establishes a static condition for the body segments affected by the muscle. In an isotonic effort, the muscle tension remains constant, which usually coincides with a static (isometric) effort.

Dynamic activities result from changes in muscle length, which bring about motion of body segments. In an isokinematic effort, speed remains unchanged. In an isoinertial test, the mass properties remain constant.

Human body (segment) strength is measured routinely as the force (or torque) exerted to an instrument external to the body. This is information of great importance to the ergonomic designer/engineer.

	Force-plate[1] height	Distance[2]	Force, N Mean	Force, N SD
	50	80	664	177
	50	100	772	216
	50	120	780	165
	70	80	716	162
	70	100	731	233
	70	120	820	138
	90	80	625	147
	90	100	678	195
	90	120	863	141
	Percent of shoulder height		Both hands	
	60	70	761	172
	60	00	854	177
	60	90	792	141
	70	60	580	110
	70	70	698	124
	70	80	729	140
	80	60	621	130
	80	70	620	129
	80	80	636	133
	Percent of shoulder height			
	70	70	623	147
	70	80	088	154
	70	90	586	132
	80	70	545	127
	80	80	543	123
	80	90	533	81
	90	70	433	95
	90	80	448	93
	90	90	485	80
	Percent of shoulder height		Both hands	
Force plate 			Both hands	
	100 percent of shoulder height	50	581	143
		60	667	160
		70	981	271
		80	1285	398
		90	980	302
		100	646	254
			Preferred hand	
		50	262	67
		60	298	71
		70	360	98
		80	520	142
		90	494	169
		100	427	173
		Percent of thumb-tip reach*		
	100 percent of shoulder height	50	367	136
		60	346	125
		70	519	164
		80	707	190
		90	325	132
		Percent of span**		

(1) Height of the center of the force plate – 20 cm high by 25 cm long – upon which force is applied.
(2) Horizontal distance between the vertical surface of the force plate and the opposing vertical surface (wall or footrest, respectively) against which the subjects brace themselves.

*Thumb-tip reach – distance from backrest to tip of subject's thumb as arm and hand are extended forward.
**Span – the maximal distance between a person's fingertips when arms and hands are extended to each side.

FIGURE 11.9 Mean horizontal push forces and standard deviations in N exerted by standing men with their hands, the shoulder, and the back. (Adapted from NASA STD. 3000 A, 1989.)

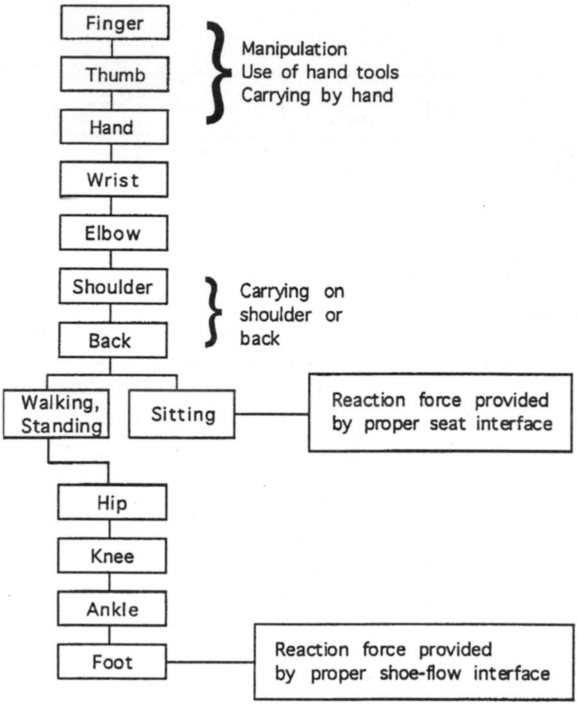

FIGURE 11.10 Determining the critical body segment strength for manipulating and carrying. (From Kroemer, K.H.E., Kroemer, H.J., and Kroemer-Elbert, K.E. (1997). *Engineering Physiology. Bases of Human Factors/Ergonomics* (3rd ed.). New York, NY: Van Nostrand Reinhold. With permission.)

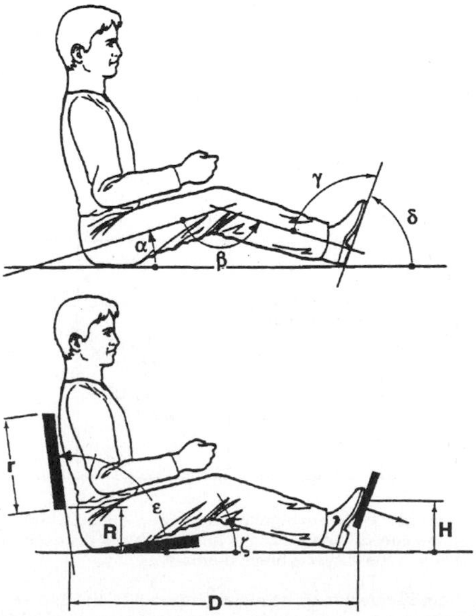

FIGURE 11.11 Conditions affecting the force that can be exerted on a pedal: body angles (upper illustration) and work space dimensions. (From Kroemer, K.H.E., Kroemer, H.B., and Kroemer-Elbert, K.E. (1994). *Ergonomics: How to Design for Ease and Efficiency.* Englewood Cliffs, NJ: Prentice Hall. With permission.)

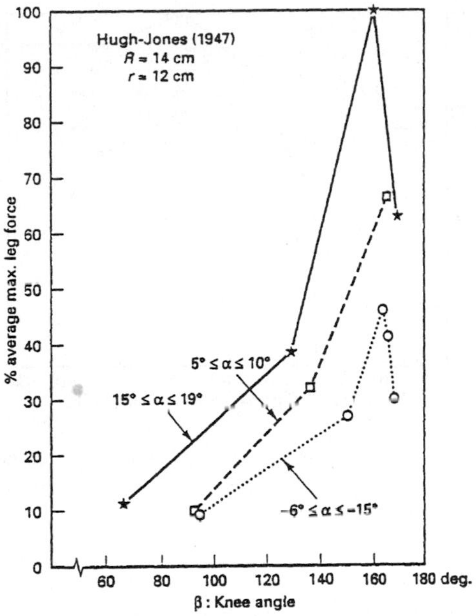

FIGURE 11.12 Effects of thigh angle α and knee angle β (see Figure 11.11) on pedal push force. (From Kroemer, K.H.E., Kroemer, H.B., and Kroemer-Elbert, K.E. (1994) *Ergonomics: How to Design for Ease and Efficiency.* Englewood Cliffs, NJ: Prentice Hall. With permission.)

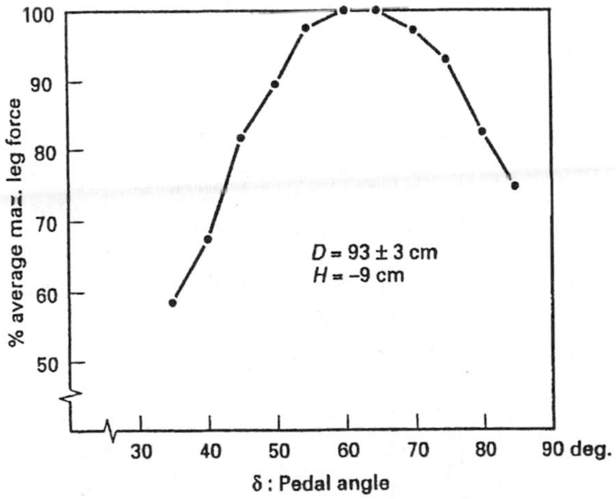

FIGURE 11.13 Effects of ankle (pedal) angle δ (see Figure 11.11) on foot force generated by ankle rotation. (From Kroemer, K.H.E., Kroemer, H.B., and Kroemer-Elbert, K.E. (1994) *Ergonomics: How to Design for Ease and Efficiency.* Englewood Cliffs, NJ: Prentice Hall. With permission.)

Design of equipment and work tasks for human body segment strength capabilities is done systematically by:

- Determining whether the exertion is static or dynamic
- Establishing with what body part the force or torque is exerted
- Following the chain of strength vectors through the involved body segments to find the "weak link" and to improve and rearrange the conditions, if possible
- Selecting the body strength percentile (minimum and/or maximum) that is critical for the operation.

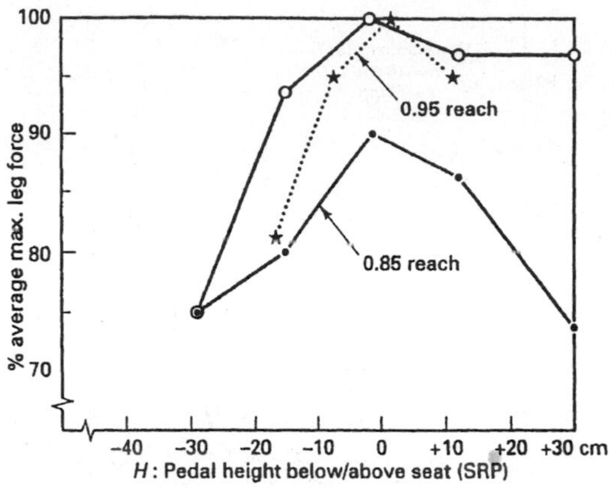

FIGURE 11.14 Effects of pedal height h and leg extension (see Figure 11.11) on pedal push force. (From Kroemer, K.H.E., Kroemer, H.B., and Kroemer-Elbert, K.E. (1994) *Ergonomics: How to Design for Ease and Efficiency.* Englewood Cliffs, NJ: Prentice Hall. With permission.)

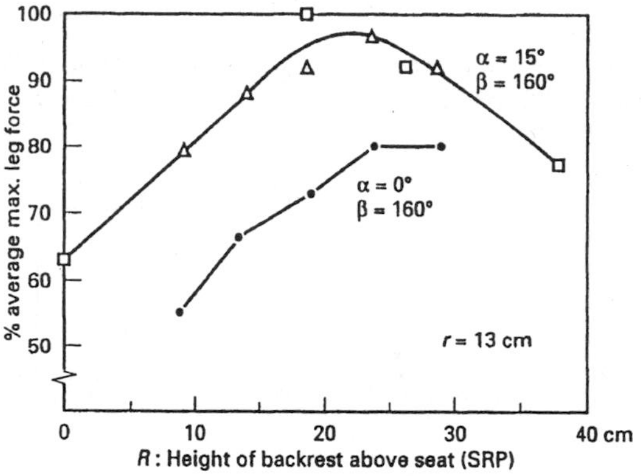

FIGURE 11.15 Effects of backrest height R (see Figure 11.11) on pedal push force. (From Kroemer, K.H.E., Kroemer, H.B., and Kroemer-Elbert, K.E. (1994) *Ergonomics: How to Design for Ease and Efficiency.* Englewood Cliffs, NJ: Prentice Hall. With permission.)

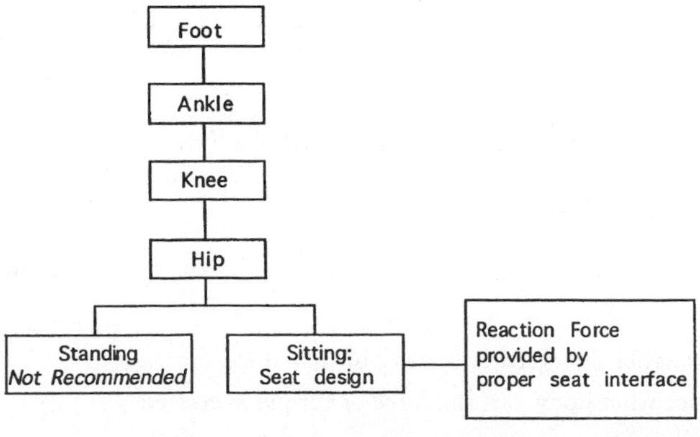

FIGURE 11.16 Determining the "critical body segment strength" for foot operation.

References

Asimov, I. (1963). *The Human Body. Its Structure and Operation.* New York, NY: The New American Library/Signet.

Åstrand, P. O. and Rodahl, K. (1977, 1986). *Textbook of Work Physiology.* (2nd, 3rd ed.). New York, NY: McGraw-Hill.

Chaffin, D.B. and Andersson, G.B.J. (1991). *Occupational Biomechanics* (2nd ed.) New York, NY: Wiley

Enoka, R.M. (1988). *Neuromechanical Basis of Kinesiology.* Champaign IL: Human Kinetics Books.

Eastman-Kodak Company (Ed.) (Vol. 1, 1983; Vol. 2,1986). *Ergonomic Design for People at Work.* New York, NY: Van Nostrand Reinhold.

Kroemer, K.H.E. (1979). A new model of muscle strength regulation, In *Proceedings, Annual Conference of the Human Factors Society,* (pp. 19-20). Santa Monica, CA: Human Factors Society.

Kroemer, K.H.E., Kroemer, H.B., and Kroemer-Elbert, K.E. (1994). *Ergonomics: How to Design for Ease and Efficiency.* Englewood Cliffs, NJ: Prentice Hall.

Kroemer, K.H.E., Kroemer, H.J., and Kroemer-Elbert, K.E. (1997). *Engineering Physiology. Bases of Human Factors/Ergonomics* (3rd ed.). New York, NY: Van Nostrand Reinhold.

Kroemer, K.H.E., Marras, W. S., McGlothlin, J. D., McIntyre, D. R., and Nordin, M. (1989). Assessing Human Dynamic Muscle Strength. (Technical Report, 8-30-89). Blacksburg, VA: Virginia Tech, Industrial Ergonomics Laboratory. Also published (1990) in *International Journal of Industrial Ergonomics,* 6, 199-210.

Marras, W.S., McGlothlin, J.D., McIntyre, D.R., Nordin, M., and Kroemer, K.H.E. (1993). *Dynamic Measures of Low Back Performance.* Fairfax, VA: American Industrial Hygiene Association.

Salvendy, G. (Ed.). (1987). *Handbook of Human Factors.* New York, NY: Wiley.

Schneck, D.J. (1990). *Engineering Principles of Physiologic Function.* New York, NY: New York University Press.

Schneck, D.J. (1992). *Mechanics of Muscle* (2nd ed.) New York, NY: New York University Press.

Weimer, J. (1993). *Handbook of Ergonomic and Human Factors Tables.* Englewood Cliffs, NJ: Prentice Hall.

Winter, D.A. (1990). *Biomechanics and Motor Control of Human Movement. (2nd ed.).* New York, NY: Wiley.

Woodson, W.E., Tillman, B., and Tillman, P. (1991). *Human Factors Design Handbook* (2nd ed.). New York, NY: McGraw-Hill.

12

Methods Based on Maximum Holding Time for Evaluation of Working Postures

Marjolein Douwes
TNO Prevention and Health
The Netherlands

Mathilde C. Miedema
TNO Prevention and Health
The Netherlands

J. Dul
TNO Prevention and Health
The Netherlands

12.1 Introduction

Background

Many work situations require postures which have to be maintained for a long period of time (e.g., machine operation, assembly work, VDU work). Forty-two percent of European workers adopt uncomfortable working postures for more than two hours a day (Paoli, 1992). This is illustrated in Figure 12.1. Depending on the position, of the load, the posture and the duration of holding the posture, there is a risk of acute discomfort and long-term health effects (musculoskeletal disorders) (Keyserling et al., 1988; Kilbom, 1988; Genaidy and Karwowski, 1993; Putz-Anderson and Galinsky, 1993). The percentage of workers on sick leave or unfit for work due to musculoskeletal disorders is enormous. This elicits the development of preventive programs.

Static work load can be diminished by improving the working posture (by optimization of the workplace and the equipment), by reducing the holding time of postures, and by supplying sufficient and properly distributed rest pauses. For these measures, standards and evaluation methods need to be

0-8493-2641-9/99/$0.00+$.50

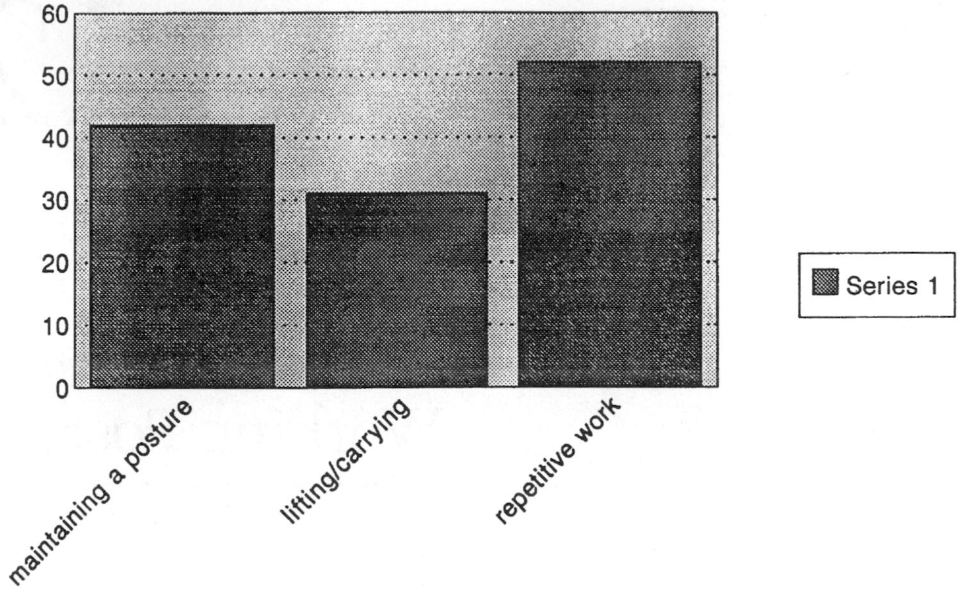

FIGURE 12.1 Percentage of the work force of different countries in the EC who work in uncomfortable postures for more than 2 hours per day (first column), who lift or carry loads for more than 2 hours (second column) or perform repetitive work for more than 2 hours per day (third column). (From Paoli, P. 1992. *First European Survey on the Work Environment: 1991–1992.* European Foundation for the Improvement of Living and Working Conditions Dublin. With permission.)

developed. Because of the lack of quantitative exposure–effect data, ergonomic recommendations to prevent risks for musculoskeletal disorders due to static load cannot yet be based on long-term health effects. However, acute discomfort can also be considered as an independent evaluation criteria for static postures (Miedema, 1992; Dul et al., 1994; Miedema et al., 1996).

Perceived Discomfort

An indicator of work load is the discomfort perceived by the worker. Static postures can cause strain on interior body structures, such as bones, joints, tendons, muscles, and ligaments. Subjective measurements like registration of body areas affected by discomfort and the intensity of such discomfort can be an indicator of the postural load. Van der Grinten and Smitt (1992) developed a method *localized musculoskeletal discomfort* (LMD) for recording the changing levels of discomfort over the working period. The method uses the 10 points category-ratio scale developed by Borg (1982) for recording the level of discomfort. To indicate the location of discomfort, the body diagram of Corlett and Bishop (1976) is used.

Since discomfort and musculoskeletal disorders are both related to exposure to biomechanical load on the musculoskeletal system (Milner, 1985; Nag, 1991; Putz-Anderson and Galinsky, 1993), reduction of discomfort will presumably contribute to reduction of the risk of musculoskeletal disorders as well (Dul et al., 1994).

Maximum Holding Time

A posture can be maintained for a limited period of time. The *maximum holding time* (MHT) is the maximum time that a posture, with or without external force exertion, can be maintained continuously until maximum discomfort, from a rested state. Because of the endless amount of different combinations of postures and force exertions, data on the MHT are only available for a limited amount of those combinations. To estimate the MHT for other combinations, the relationship between the MHT and the

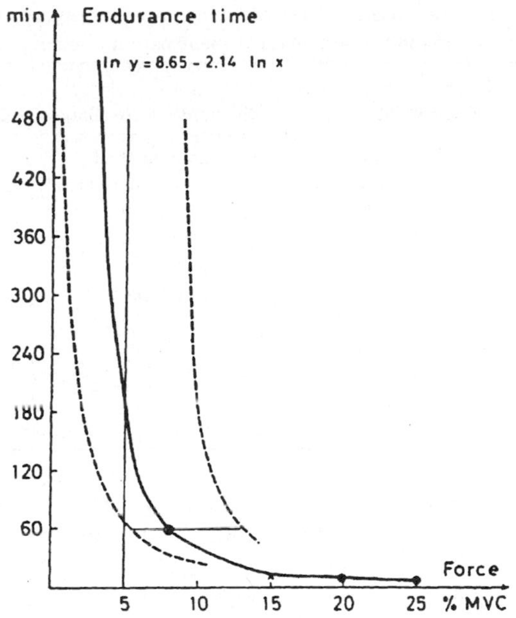

FIGURE 12.2 The relationship between muscle effort and MHT (Sjøgaard, 1986). ● from Rohmert (1960), X from Björkstén and Jonsson (1977), ○ from Hagberg (1981).

muscle effort can be used. The muscle effort of a posture is the force (or moment) that is needed to maintain the posture (with or without external force exertion) as a percentage of the maximal force that can be exerted in the same posture.

The relationship between muscle effort and MHT has been studied by many authors. Rohmert (1960) measured MHT for different values of muscle effort for different muscle groups. He found that despite large individual differences in maximal force exertion, equal relative loading (i.e., muscle effort) eliminated differences in MHT. Caldwell (1974) also found little variation in MHT among various levels of MVC when the muscle effort was equal for each individual. The studies of Rohmert (1960), Björkstén and Jonsson (1977), and Sjøgaard (1986) show that the MHT of contractions decreases exponentially as the muscle effort increases (see Figure 12.2). This relationship between muscle effort and MHT can be used to predict the MHT from the muscle effort of a certain posture.

Sjøgaard also found that for low-level static loads (less than 20% MVC) the variation in MHT is large, possibly because of differences between the various muscle groups (e.g., muscle structure and fiber composition).

The Relation Between MHT and Discomfort

The time that a static posture can still be maintained continuously after a period of loading (and resting) is called the Remaining Endurance Capacity (REC) and is expressed as a percentage of the MHT. Experiments have shown that at group level perceived discomfort, as measured with a 10-point rating scale (Borg, 1982), increases linearly in time, independent of the magnitude of MHT (Taksic, 1986; Manenica, 1986; Meijst et al., 1995; see Table 12.1). For example, a discomfort level of 5 after 10 minutes of holding a certain posture, means that the MHT of this posture is 20 minutes.

Maximum Acceptable Level of Discomfort

For standardization purposes agreement should be reached on the level of discomfort that can be considered acceptable. The choice of the maximum acceptable level of discomfort is not a scientific question but a matter of agreement between parties involved. Hagerup and Time (1992) have (arbitrarily) proposed a division of the Borg scale into three categories. They consider the mean score of a group of

TABLE 12.1 The linear relationship between the 10-point category-ratio scale for recording discomfort and the percentage of the maximum holding time (MHT).

Time (% MHT)	Remaining Endurance Capacity (% MHT)	Discomfort score (Borg CR-10 scale)	
0%	100%	0	nothing at all
10%	90%	1	very weak (just noticeable)
20%	80%	2	weak (light)
30%	70%	3	moderate
40%	60%	4	somewhat strong
50%	50%	5	strong (heavy)
60%	40%	6	
70%	30%	7	very strong
80%	20%	8	
90%	10%	9	
100%	0%	10	extremely strong (maximal)

individuals acceptable if this score is 1 to 3. Rose et al. (1992) found that when the subjects were allowed to decide on the duration of a static work task themselves, they stopped to pause at approximately 20% of the MHT. Until we have more data to make a better choice, we use a maximum acceptable mean discomfort level of 2 on the Borg scale (weak discomfort). Because of the linear relationship between discomfort and REC at group level, this implies that the holding time of a continuous static posture should be no more than 20% of the MHT of that posture. The duration of an intermittent exercise until the minimum acceptable mean discomfort score 2 can be considerably longer, depending on the work–rest schedule.

Ergonomic standards are meant to protect a certain percentage (e.g., 95%) of the population. If the maximum acceptable mean level of discomfort would be a score of 2, we estimate that at least 50% of the population will have less than "weak discomfort" (score 2), and 95% of the population will have less than "strong discomfort" (score 5). These estimations are based on the distribution of individual discomfort scores in our experimental data set (Dul et al., 1994).

Methods to Evaluate Static Load

As stated before, muscular fatigue and risk of musculoskeletal disorders can be reduced in a number of ways. The muscle effort during work can be reduced by improving the posture or reducing the external force exertion. The working posture and required external force can be controlled by variables such as the working height, reaching distance, and the force required to operate a machine. When postural and force changes are difficult to establish, sufficient and properly distributed rest pauses can be supplied. The duration and distribution of work and rest periods can be controlled by organizational (time based) factors. Both types of variables can be influenced by designers and manufacturers of machinery and by occupational health and safety personnel.

In this chapter, guidelines and methods are presented to help occupational health practitioners and designers bring muscle fatigue during work to an acceptable level. In the next paragraph an evaluation method for working postures based on hand positions is presented. A subsequent section describes an evaluation method for working postures alternated with rest periods ("The Work–Rest Model for Static Postures"). In the final section, guidelines for static load used in CEN and ISO standards are presented.

12.2 Evaluation Method for Hand Positions

Due to the increment of discomfort in time, holding time (%MHT) can be taken as a measure for making recommendations concerning the maximum duration of static postures. We studied endurance data of experiments found in the literature and ranked the postures (Dul et al., 1993). Based on this ranking we developed recommendations for the holding time of static standing postures. This section describes the

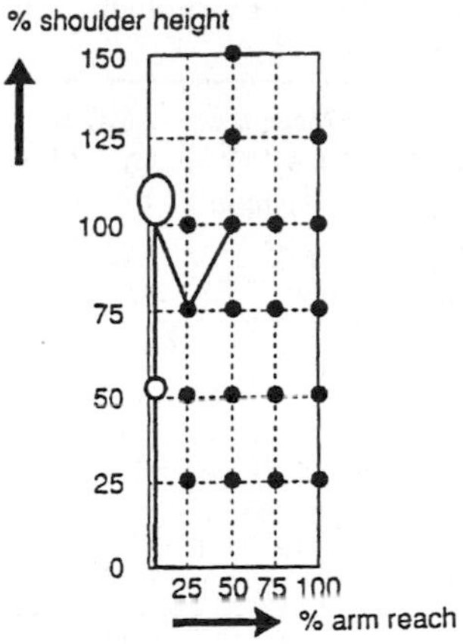

FIGURE 12.3 The 19 postures found in literature are indicated with ●. Posture is defined by the horizontal and vertical hand position. The vertical hand position is expressed as a percentage of the shoulder height and the horizontal hand position as a percentage of the arm reach in the upright standing posture.

development and results of a method to evaluate postures on the basis of hand position. In the last part of the paragraph an example of using this method in practice is included.

MHT Data from the Literature

Information was gathered about the MHT of 19 different standing postures which were maintained without rest pauses and without external load. These data were found in seven studies (Corlett and Manenica, 1980; Hagberg, 1981; Boussenna et al., 1982; Milner, 1985; Taksic, 1986; Manenica, 1986; Meijst et al., 1995). All postures were defined by two parameters, i.e., the horizontal distance (% shoulder height) and vertical distance (% arm reach) of the position of the hands with respect to the feet in upright standing posture. Shoulder height (SH) is defined as the distance from acromion to the floor in the upright position. The arm reach (AR) is defined as the maximum distance from the knuckles to the wall when standing upright with the back against the wall and the shoulder in 90° anteflexion. The 19 postures differ in the combination of 25, 50, 75, 100, 125, or 150% SH and 25, 50, 75, or 100% AR, and are shown in Figure 12.3. In all studies the participants were asked to maintain the posture as long as they could. In almost all studies the subjects had to perform a task while holding the posture. These tasks implied television games, spot-tracking, or tapping tasks. While maintaining the posture, location and amount of perceived discomfort was registered. The experiments ended when maximum discomfort was reached (score 10 on a 10-point rating scale; Borg, 1990). The MHTs that were recorded are summarized in Table 12.2.

Description of the Method

As can be seen in Table 12.2, there is much variation in MHT's of similar postures within and between studies. In spite of this variation, a ranking of the 19 postures was made, based on the mean MHT from all available data. These ranked postures are shown in Figure 12.4. The posture 75%SH/50%AR has the highest MHT (35.7 min.), and the posture 25%SH/100%AR has the lowest MHT (2.7 min.). The ranked postures can be arbitrarily classified into 3 groups; "comfortable," "moderate," and "uncomfortable," with relatively large, medium, and small MHT's, respectively. Uncomfortable postures are defined as postures with an MHT smaller than 5 minutes, which implies that maintaining an uncomfortable posture will lead to a relatively quickly increasing feeling of discomfort. All postures with an extremely low or high working height (25% and 150% SH) appear to be uncomfortable postures. According to the classification, the postures with a combination of a moderate working height (50%, 75%, 100%, and 125% SH) and

TABLE 12.2 Summary of the data from literature. For each study the postures investigated and their maximum holding times (MHT; in minutes) are given.

		Meijst n = 20 10♀/10♂ television game		Corlett n = ? tapping task		Manenica n = 15♀ tapping task		Milner n = 9♂ computer game		Boussenna n = 8♂ spot tracking		Douwes n = 12 6♂/6♀ no task		Hagberg n = 7♀ no task	
Posture SH/AR		mean	SD	mean	SD	mean	SD	mean	SD	mean	SD	mean	SD	mean	SD
25/25	males														
	females														
	all			4.5											
25/50	males														
	females														
	all			4.0											
25/75	males														
	females														
	all			2.9											
25/100	males									4.41	1.53				
	females					1.57	0.72								
	all			2.0											
50/25	males														
	females														
	all			13.5											
50/50	males	10.89	6.75												
	females	16.52	5.99												
	all	13.71	6.98	8.5											
50/75	males														
	females														
	all			5.0											
50/100	males	8.56	1.67					10.10		5.26	1.66				
	females	7.88	2.92			2.35	1.55	8.12	1.52						
	all	8.22	2.40	3.2				8.73	1.50						
75/25	males														
	females					4.27	2.04								
	all			30.0											
75/50	males	65.5	38.7												
	females	36.5	24.6												
	all	50.9	13.7	20.5											
75/75	males					4.25	1.75								
	females			7.5											
	all														
75/100	males	17.4	11.7							6.35	1.91				
	females	12.0	3.41												
	all	14.7	9.05	4.2											
100/25	males														
	females														
	all			9.0											
100/50	males	38.5	15.7												
	females	16.6	4.7												
	all	26.2	15.1	6.0											

TABLE 12.2 (continued) Summary of the data from literature. For each study the postures investigated and their maximum holding times (MHT; in minutes) are given.

		Meijst n = 20 10♀/10♂ television game		Corlett n = ? tapping task		Manenica n = 15♀ tapping task		Milner n = 9♂ computer game		Boussenna n = 8♂ spot tracking		Douwes n = 12 6♂/6♀ no task		Hagberg n = 7♀ no task	
Posture SH/AR		mean	SD	mean	SD	mean	SD	mean	SD	mean	SD	mean	SD	mean	SD
100/75	males														
	females														
	all			5.25											
100/100	males	11.42	2.76												
	females	8.45	2.18			3.57	1.38			7.85	2.09				
	all	9.93	2.90												
125/50	males	10.83	3.79									15.3	7.7	21.4	
	females	6.76	2.43			3.26	1.53					9.4	3.8		
	all	8.79	3.78									12.3	6.6		
125/100	males	8.94	2.38												
	females	6.51	3.40												
	all	7.72	3.18												
150/50	males														
	females					3.22	1.93								
	all														

Summary of the literature used in this study: posture is defined as the relative hand position with respect to the feet, i.e., the working height (as a percentage of shoulder height [%SH]) and working distance (as a percentage of arm reach [%AR]). For each study the postures investigated and resulting mean MHT (minutes) for males and females are given.

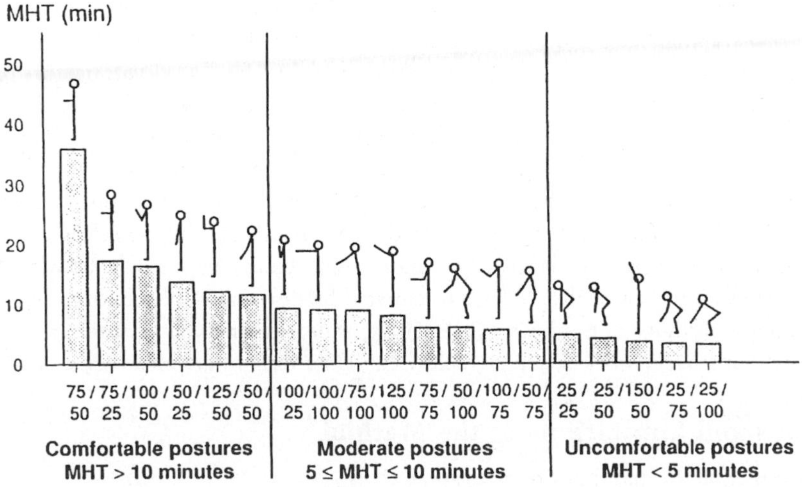

FIGURE 12.4 Ranking of the postures on the basis of the mean values of maximum holding time (MHT) of available data. The recommended holding time is 20% of MHT.

small working distance (25% and 50% AR) are comfortable postures (MHT longer than 10 minutes). Postures with a moderate working height (50%, 75%, 100%, and 125% SH) and a large working distance (75% and 100% AR) appear to be moderate postures (MHT ≥5 and ≤10 minutes).

Holding a comfortable, moderate, or uncomfortable posture for the maximum period of time causes extremely strong (maximal) discomfort in (a part of) the body. By limiting the actual holding time of a

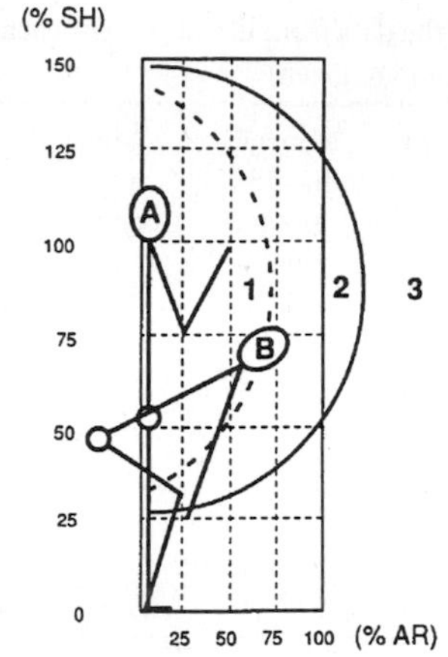

FIGURE 12.5 Recommendations concerning the maximum holding times of static postures. —— = demarcation line between uncomfortable and moderate postures. ---- = demarcation line between moderate and comfortable postures. 1 = hand positions of comfortable postures with MHT longer than 10 minutes and recommended maximum holding time of 2 minutes. 2 = hand positions of moderate postures with MHT between 5 and 10 minutes and recommended maximum holding time of 1 minute. 3 = Hand positions of uncomfortable postures with MHT less than 5 minutes, which are advised against. Person A adopts a comfortable posture, and person B adopts an uncomfortable posture.

posture, discomfort can be limited, even in uncomfortable postures. As mentioned earlier we propose that for a group of individuals the maximum acceptable holding time is 20% of the MHT, corresponding to a discomfort score of 2 of the Borg scale (weak discomfort). To calculate the maximum acceptable holding time of a posture, the MHT has to be divided by 5. To make the recommendations safe for all postures in the three classes, the maximum acceptable holding time valid for each class of hand position corresponds with the lowest maximum acceptable holding time of that class (most uncomfortable posture of that class). Thus, comfortable postures with an MHT of more than 10 minutes are recommended to be maintained 2 minutes maximally. Following the same procedure for a moderate posture the maximum acceptable holding time is 1 minute, and an uncomfortable posture is not acceptable. In Figure 12.5 the possible hand positions are divided into 3 areas corresponding to these recommendations (area 1: 2 minutes; area 2: 1 minute; area 3: 0 minutes). As stated before, we estimate that for a mean discomfort of 2, 95% of the population will have less than "strong discomfort" (score 5 on the 10-point scale). The body part(s) in which discomfort is felt, depends on the posture. All healthy subjects who adopt the same posture (independent of the study) perceive discomfort in approximately the same body part(s). Postures with hand positions at or below 50% SH are terminated by discomfort in the lower back and legs. In postures with hand positions at or above 100% SH, the shoulders and arms are the critical body parts. Also, it appears that a larger work distance results in a higher discomfort in shoulders and arms.

Applications and Limitations of the Method

The evaluation method for hand positions relates to:

- Standing postures
- Postures without external force
- Postures that are symmetric in the sagittal plane
- Postures that are maintained without rest pauses (static work)
- Healthy, young adults.

The classification can be a guidance in practical situations for occupational health officers, designers, labor inspectors, and ergonomists to match the working time to the working posture.

The recommendations are based on data with a large variation, caused by differences in intra- and interindividual characteristics and study design (including the task). One should be cautious when putting these recommendations into practice.

The method has been developed for pure static postures without body motions. In most working postures minor changes in posture and loading may occur. This may result in partial recovery due to changing the critical muscle group or variations of the muscle effort. Another point of attention is that in many work situations body parts are supported by a table, an armrest, or a machine. This support unloads the muscles and the joint. It can be assumed that the MHT of a "dynamic" or supported posture is longer than the MHT of a static posture. For these kinds of working situations, the recommendations are expected to be relatively safe.

In this study, posture was defined by the position of the hands with respect to the feet. This definition can influence the variation in MHT. The subjects were free to choose body angles of the knees, hips, back, shoulders, and elbows. Variation in these body angles within the same posture may have caused variation in MHT.

Using the data of Meijst et al. (1995), the relation between hand position and posture was studied (Miedema, 1992; Miedema et al., 1996). Also, we calculated the effect of the interindividual variation in posture on the muscle effort. We used the 2-Dimensional Static Strength Prediction Program (2DSSPP; Chaffin and Andersson, 1984) for calculation of the muscle effort. It appears that the body angles of the upper extremity show the largest variation; i.e., up to 30° variation in the shoulder. For postures with the shoulders as the critical body part, this implicates a range of muscle effort of 8 to 31% MVC. The angles of the lower extremity and trunk show a variation between 7° and 15°. The variation in body angles increases for lower working heights. In the postures with the hands on 50% shoulder height the interindividual variation in muscle effort (of the back) increases up to 50% MVC between subjects. Thus, the classification of standing working postures defined by hand position is mainly usable for the moderate and comfortable postures. For uncomfortable postures with a low working height (50%SH), one should be cautious about using the method.

Relationship with Other Methods

The evaluation method for hand positions was compared with biomechanical calculations. For all postures, the muscle effort of the critical muscle group was calculated by using the 2-Dimensional Static Strength Prediction Program (2DSSPP; Chaffin and Andersson, 1984). Both classifications are comparable. The muscle effort (posture and force) increases when the work distance increases and/or when the working height is very low or very high. In comfortable postures, the biomechanical load is relatively low compared with moderate and uncomfortable postures.

Figure 12.5 coincides also with anthropometric data. Recommendations concerning the optimal work area for the hands for upright postures result in lines similar to those in Figure 12.5 (Burandt, 1978). The evaluation method for hand positions was also compared with the classification of the Ovako Working Postures Analyzing System (OWAS; Karhu et al., 1977). The OWAS postures have been classified into four categories by experts, including physicians, work analysts, workers, and ergonomists. The postures of OWAS action category 1 (no improvement needed) are classified in our holding time classification as comfortable postures. Postures from OWAS action category 2 (improvements may be necessary in the future) are classified in our holding time classification as moderate postures. Postures from OWAS action categories 3 and 4 (improvement is needed as soon as possible, and immediately, respectively) are classified as uncomfortable postures. In this comparison, 10 postures (of the 19 postures) do not correspond with each other. It appeared that our classification of hand positions is more strict than the OWAS classification. This can be explained by the fact that OWAS is based on male workers, on more dynamic postures, and on more heavy work with external loads (steel industry).

TABLE 12.3 Calculated absolute hand positions (horizontal and vertical distance) in meters. It is assumed that shoulder height is 83% of the total body height.

	North Europe		Central Europe		East Europe	
	♂	♀	♂	♀	♂	♀
25% shoulder height	0.38	0.35	0.37	0.35	0.36	0.34
50% shoulder height	0.75	0.70	0.74	0.69	0.73	0.68
75% shoulder height	1.13	1.05	1.11	1.04	1.09	1.01
100 shoulder height	1.50	1.40	1.47	1.38	1.45	1.35
125% shoulder height	1.88	1.75	1.84	1.73	1.81	1.69
150% shoulder height	2.25	2.10	2.21	2.07	2.18	2.03
25% arm reach	0.22	0.20	0.21	0.20	0.21	0.20
50% arm reach	0.44	0.41	0.43	0.40	0.42	0.39
75% arm reach	0.65	0.61	0.64	0.60	0.63	0.59
100% arm reach	0.87	0.81	0.85	0.80	0.84	0.78

Example of an Application of the Hand Position Method

When a painter is painting a ceiling with a brush while standing on scaffolding, he lifts the right elbow to shoulder height. With flexion and extension movements in the elbow and wrist, he moves the brush. The shoulder is held in the same static position. This hand position varies between 125%SH/25%AR and 125%SH/75%AR. The posture 125%SH/25%AR is classified as a comfortable posture with a maximum acceptable holding time of 2 minutes. The posture 125%SH/75%AR is defined as a moderate posture with a maximum acceptable holding time of 1 minute. The final recommendation has to be based on the worst occurring posture. In this situation, the maximum acceptable holding time is 1 minute.

Absolute Hand Positions

For application of the recommendations, it may sometimes be easier to indicate the absolute hand position rather than the hand position as a percentage of shoulder height and arm reach. The absolute hand position was calculated from the relative hand position and anthropometric data. We assume that for the European population the shoulder height is 83% of the total body height (Molenbroek, personal communication). For the shoulder height and arm reach of the population, the international anthropometric data of Jürgens et al. (1989) and the Dutch data of Molenbroek and Dirken (1987) were used. The absolute hand positions are listed in Table 12.3.

12.3 The Work–Rest Model for Static Postures

Description of the Method

A mathematical work–rest model (WR model) for static postures has been developed to help designers and occupational health officers in selecting the most effective measures for reducing physical load by comparing different combinations of work–rest schemes and muscle load. The model can also be useful to develop standards and guidelines and to evaluate the acceptability of specific working conditions by comparing them with these standards or guidelines. Figure 12.6 shows a schematic representation of the WR-model.

The WR model predicts the course of muscular fatigue and recovery during work with static postures and rest. The prediction is based on four input variables. The first is the *muscle effort*, which is expressed as a percentage of the muscle strength (% MVC) of the critical muscle group. The critical muscle group determines the MHT of a given posture and force exertion from the Sjøgaard curve (see Figure 12.2). It is assumed that the muscle effort remains constant during all work periods. The muscle effort can be

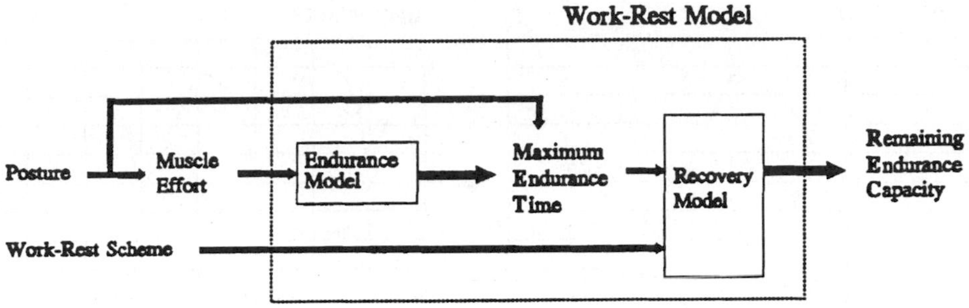

FIGURE 12.6 Schematic representation of the work rest model.

estimated with different techniques, such as EMG, biomechanical calculation, and possibly psychophysical methods.

The total work time and rest time are usually divided into a *number of work and rest periods* (n) with either constant or variable work and rest times. The *work time* (t_w) of a work–rest period is the duration of the muscle effort of the critical muscle group. The *rest time* (t_r) of a work–rest period is the length of time the critical muscle group can relax.

Based on the inputs, the WR model calculates the *maximum holding time* of the muscle effort. Also, the WR model calculates the course and minimum value of the *remaining endurance capacity (REC)*, which is the fraction of the MHT that a muscle effort can still be maintained continuously after a period of loading (and resting). The REC can be considered as the opposite of muscle fatigue and is considered to be the most important output variable of the model, because it indicates the (maximum) discomfort during that work–rest schedule (see Table 12.1 about the linear relationship between discomfort and time). For example, "no discomfort" (score 0 on the Borg scale) is felt at 100% REC, "strong discomfort" (score 5 on the Borg scale) is felt at 50% REC, and "extremely strong discomfort" is felt at 0% REC.

The model combines empirical studies on muscle fatigue and on muscle recovery during static contractions. The model equations were selected from the literature. The regression equation given by Sjøgaard (1986) was used for the relationship between muscle effort and MHT.

General Guidelines from the WR Model

From computer simulations and mathematical derivations, some general model predictions were formulated that can be considered general ergonomic guidelines for static postures. The first general guideline that can be given from the WR model predictions is that for a given total work and rest time and constant work and rest times, many short work–rest periods are better (i.e., generate less discomfort) than a few long work and rest periods.

For example, for a muscle effort of 20% MVC and total work time of 16 minutes and total rest time of 16 minutes, if the number of work and rest periods increases from 2 to 4, 8 and 16 minutes, the minimum REC increases from about 0 to about 40, 60, and 70%, respectively. This is illustrated by Figure 12.7. Furthermore, the model predicts that for a given number of variable work–rest cycles it is better to start with the longest work cycles.

Computer Program

Based on the model, a user-friendly program for a personal computer is being developed. With this program, the minimum REC (output) can be computed for a given combination of muscle effort, and number, duration, and distribution of work and rest periods (input). Instead of the minimum REC, the program can also calculate the required rest time for a given desired minimum REC. For each calculation a graph can be displayed, which shows the course of the REC during work and rest.

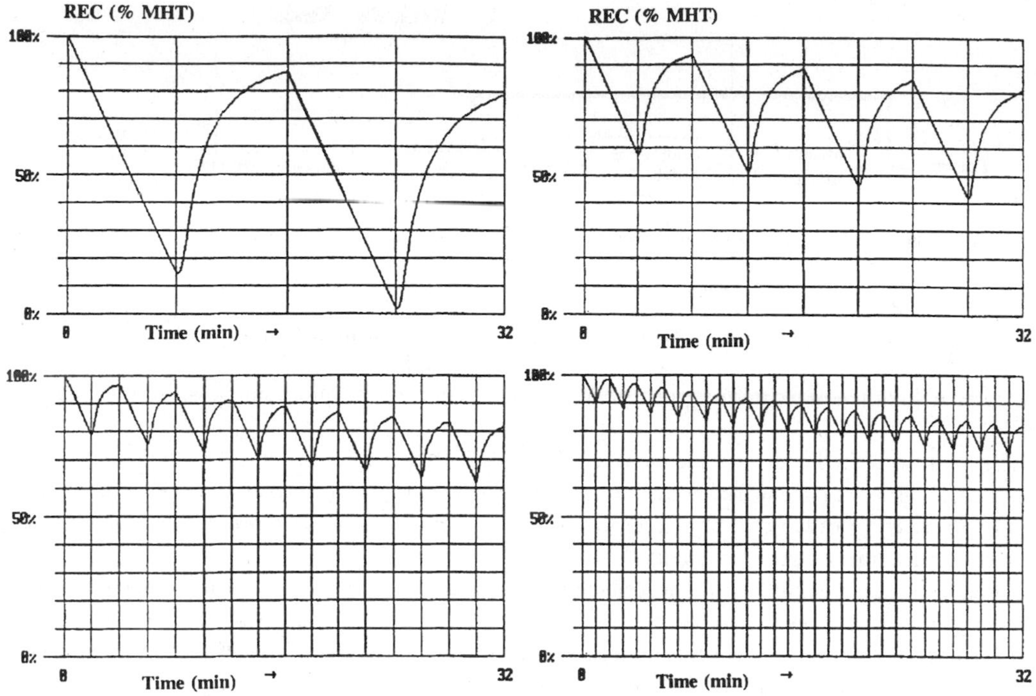

FIGURE 12.7 Illustration of the increasing REC with increasing number of work and rest periods (and constant total work and rest times).

The practical use of the program was tested by 13 occupational health officers. It appears that the program can be used by these practitioners to develop and support recommendations for changes of work and rest times during monotonous work with static postures. Before the program is released it will be tested more extensively.

Validity of the Method

The validity of the Work–Rest-model for static postures was tested in an experimental study with 10 subjects (Douwes and Dul, 1993). To validate the Work–Rest model predictions of REC, these predictions were compared with REC predictions from discomfort measurements during four different work–rest schedules, with the shoulder muscle effort varying from 20% to 36% MVC. During the work periods, the subjects were holding a weight in a seated posture with one arm elevated. Both the shoulder and elbow of that arm were 90° flexed forward.

Discomfort was measured using a method called *local musculoskeletal discomfort* (LMD), which has been developed and described by Van der Grinten and Smitt (1992). To record the level of discomfort, the LMD method uses the 10-point category ratio scale developed by Borg (1982). The location of discomfort is indicated on a modified body diagram of Corlett and Bishop (1976).

The results of the study (shown in Figure 12.8) indicate that the overall time pattern of REC during work–rest schedules is predicted well by the model. The maximum differences between model predictions and experimental REC values vary from 5% to 30% MHT. Both over- and underestimations occur. These results correspond well with the results of Serratos-Perez and Haslegrave (1992) in their study of the validity of the model of Milner. In this study ($n = 5$), eight different schedules of one work period and one rest period were performed. The muscle effort during work was 20%. Immediately after the rest periods, REC was measured. WR model predictions of REC appeared to overestimate the measured REC by 5% to 20% MHT.

Thus, it seems that WR model predictions are reasonably good on average but can differ up to 30% MHT in specific situations. These specific predictions could be improved by taking into account other

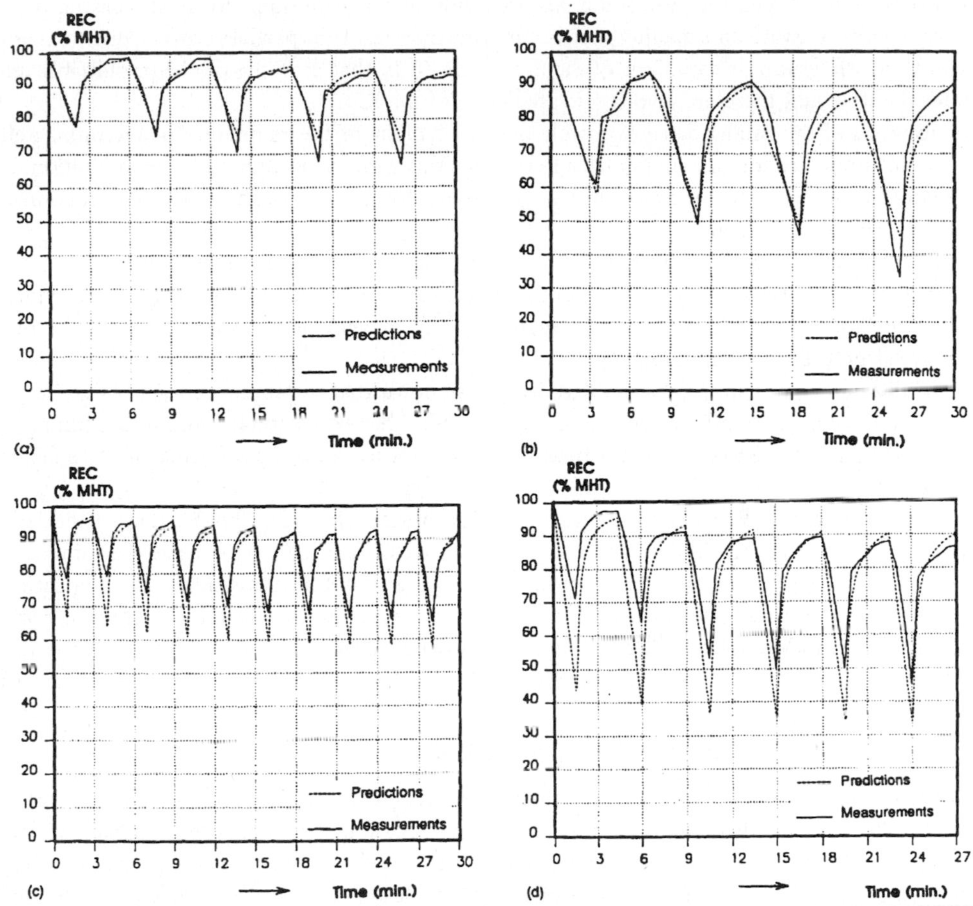

FIGURE 12.8 The course of experimental data (——) and model predictions (----) in the four validation experiments (n = 10): (a) muscle effort = 20% MVC; 2 min work, 4 min rest; (b) muscle effort = 21% MVC; 3.5 min work, 4 min rest; (c) muscle effort = 34% MVC; 1 min work, 2 min rest; (d) muscle effort = 36% MVC; 1.5 min work, 3 min rest (the total duration can be seen in the graphs).

factors that influence the REC. Therefore, the role of factors that may influence REC should be studied more thoroughly. In the meantime, it is advised to use the WR model only for comparing work–rest schedules and developing guidelines but not to evaluate specific situations in an absolute sense.

Possibilities of the WR Model

After further tests on the validity of the WR model, the model and its computer program may be a useful tool for evaluating static working postures. The model can be used to develop general ergonomic guidelines, such as those presented above. For specific working situations, the model can be used to select optimum combinations of muscle effort, work time(s), and rest time(s). Also, specific working situations may be assessed by comparing the minimum REC with a given standard, for example a limit value of 80% minimum REC such that "strong discomfort" is prevented.

Limitations of the WR Model

The WR model uses muscle fatigue as the only criterion to analyze static working postures. The load on the passive structures (i.e., ligaments, tendons) is not considered in the model. This load may be very important for the development of musculoskeletal disorders, in particular for extreme postures.

The model is developed for pure static postures without body motions. In most working postures minor changes in posture and loading may occur. This may result in partial recovery due to changing the critical muscle group or variations of the muscle effort. In that case, the model presumably underestimates the REC and therefore gives a safe prediction of the REC.

The accuracy of the model depends mainly on the accuracy of the estimation of the muscle effort. Presently, no simple techniques are available to estimate this variable accurately. If the estimation of the muscle effort is not accurate, the model may only be useful for a comparison of working situations, and not for an absolute assessment.

It is assumed that one muscle group is the critical one that determines the maximum holding time of the posture. It is also assumed that the relationship between muscle effort and MHT is independent of several factors, such as the critical muscle group, the task that is being performed, and the role of muscles and passive structures in maintaining a posture.

The validity of these assumptions is not known. It is known that certain muscle groups are relatively more fatigue resistant than others because of differences in fiber type composition. In the future, more detailed modeling may be necessary. It is therefore important to assess the influence of these factors in future studies:

In the model, the muscle effort during work remains constant. In reality, however, it is expected that during static load the MVC will decrease, and because of a constant load, the muscle effort increases.

The present model can only be used for groups of people and not for individuals. The empirical data on muscle fatigue and recovery which were used in the model are average data for a group of people.

To extend WR model applications to prediction of risks of musculoskeletal disorders, information is needed on the relationship between REC (or discomfort) and the incidence of musculoskeletal disorders.

Example of an Application of the WR Model

Imagine a painter who paints a ceiling with a paint sprayer and works above shoulder height for many hours a day. To evaluate his physical load and to compare possible measures to reduce his load, we use the WR model. Suppose that the sprayer weights 5 kg. We measure the posture and use a biomechanical model to calculate the muscle effort at the different joints. It appears that the shoulder has the highest muscle effort, namely 26% MVC. When we put in 26% muscle effort the WR model tells us that the MHT of this load is 5.35 minutes. After 4 minutes of painting in this posture, the REC is 25.27%.

Suppose that the painter needs 40 minutes to paint one ceiling. What would be the course of the REC if we divide the 40 minutes into 10 periods of 4 minutes' work and add 4 minutes of rest in between the working periods?

In Figure 12.9 we can see that in this case the REC is 0% before the end of the task. So we need to add more rest time. To find out how much rest time is needed for the painter to finish his job without the REC exceeding 10%, the output parameter of the WR model can be changed into "needed rest time." We fill in 10% for the minimal REC value, and the WR model calculates that the rest periods should be at least 6.61 minutes.

However, according to our guidelines of the discomfort not exceeding 2 on the Borg scale, we would prefer that the REC not exceed 80%. With 10 work periods of 4 minutes, the model tells us that there is no solution in adding rest time. To solve this problem we have to reduce the work times or reduce the muscle load, by improving the posture or reducing the weight of the paint sprayer.

12.4 Standards for Working Postures Based on MHT Data

The MHT can be used to compare static loads of different postures (with or without force exertion) and is also playing an important role in the development of standards and guidelines for acceptable working postures and duration of static load and rest periods by national standardization committees, as well as in Europe (CEN; Comité Européen de Normalisation) and worldwide (ISO; the International Standardization Organization) committees. ISO/CD 11226 (1995; *Ergonomics — Evaluation of Working Postures*)

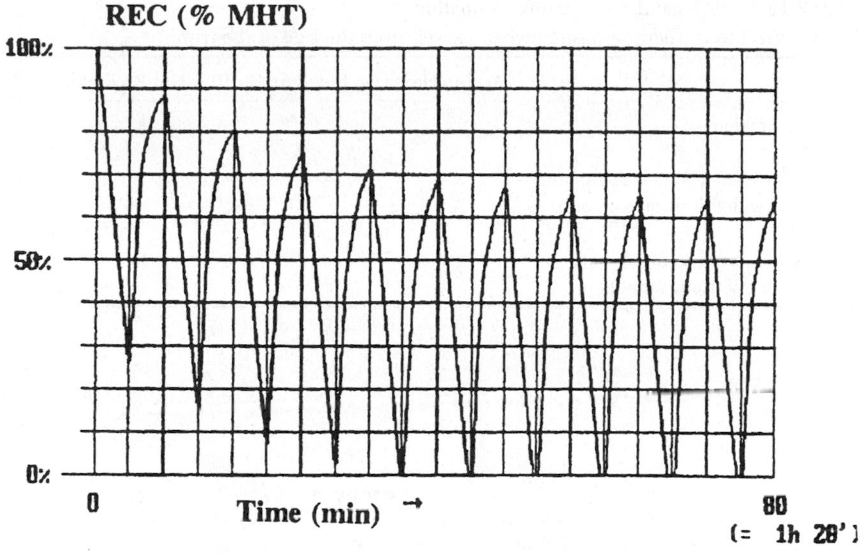

FIGURE 12.9 The course of the REC during 10 work and rest periods of 4 minutes, with a muscle effort of 26%.

contains an international standard to determine the acceptability of working postures. A comparable standard for working postures is being developed by the CEN under the machinery directive (prEN 1005-4).

Scope

The standard provides information for designers, employers, employees, and others involved in work, job, and product design. It specifies recommended limits for working postures with minimal external force exertion, while taking into account body angles and durations. The standard is meant to give reasonable protection to nearly the total healthy adult working population.

The Standard

The ISO standard for trunk inclination, head inclination, and upper arm elevation uses MHT data for judgment of acceptability. We will not present the complete ISO standard, but merely some examples of its use of MHT.

In the standard, trunk inclination is defined as the deviation angle from a neutral trunk position when viewed from the side. Head inclination is defined as the deviation angle from a neutral head position when viewed from the side. According to the standard, a trunk inclination of less than 20° is acceptable, a trunk inclination of between 20° and 60° needs to be evaluated with Table 12.4, and a trunk inclination of more than 60° is "not recommended." In Figure 12.10, the relationship between muscle effort and MHT is transferred to a relationship between trunk inclination and the maximum acceptable holding time, which is 20% of the MHT. When the trunk is fully supported, angles between 20° and 60° are acceptable without a time limit.

The inclination of the head is evaluated as acceptable when it is less than 25° and not recommended when it is more than 85°. A head inclination of between 25° and 85° needs to be evaluated with Figure 12.11. When the head is fully supported, these angles are acceptable without a time limit.

Depending on the purpose and situation different methods can be used for determining the working postures. These methods are observation, photography, video recordings, three-dimensional optoelectronic or ultrasound measuring systems, and body-mounted measuring devices such as inclinometers or goniometers are mentioned. Descriptions of these methods can be found in other parts of this book. For determining the MHT, the Sjøgaard curve or the WR model can be used.

TABLE 12.4 ISO-standard for trunk inclination
(with respect to the neutral posture when viewed from the side of the trunk).

	Acceptable	Go to Figure 12.10	Not Recommended
trunk inclination			
>60°	X	X	X
20°–60° without full trunk support	X		X
20°–60° with full trunk support	X		
0°–20°			
<0° without full trunk support			
<0° with full trunk support			

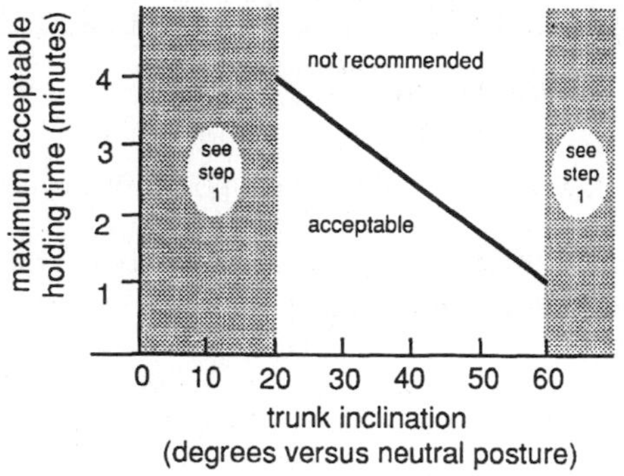

FIGURE 12.10 The relationship between trunk inclination and the maximum acceptable holding time.

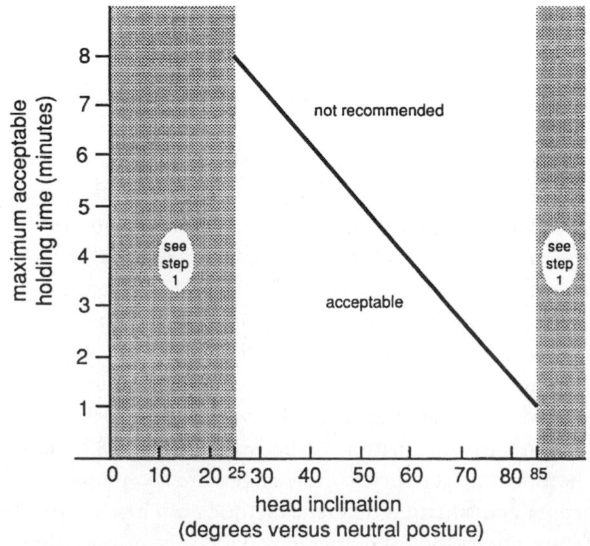

FIGURE 12.11 The relationship between head inclination and the maximum acceptable holding time.

Defining Terms

Critical muscle group: Muscle group with the largest muscle effort in a given posture.

Maximum Holding Time (MHT): Maximum duration of a muscle effort without rest until maximum discomfort, from a rested state.

Maximum Voluntary Contraction (MVC): Maximum isometric force exertion.

Muscle effort: Force (or moment) needed to maintain a posture (with or without external force exertion), as a percentage of the maximum force (MVC) or moment that can be exerted in the same posture.

Maximum acceptable holding time: Maximum acceptable time that a static posture can be held continuously without strong discomfort.

Rest time (t_r): Duration of one relaxation period, or period in which the muscle is not loaded.

Remaining Endurance Capacity (REC): Time that a muscle effort can still be maintained continuously after a period of loading (and resting), expressed as a percentage of the MHT of the muscle effort during loading periods.

Work time (t_w): Duration of one work period, or period in which the muscle is loaded.

References

Björkstén, M. and Jonsson, B. 1977. Endurance limit of force in long-term intermittent static contractions. *Scand. J. Work Environ. and Health* 3: 23-27.

Borg, G.A.V. 1982. A category scale with ratio properties for intermodal and interindividual comparisons, in *Psychophysical Judgment and the Process of Perception*, eds. H.G. Geissler and P. Petzold, VEB Deutscher Verlag der Wissenschaften, Berlin.

Borg, G.A.V. 1990. Psychophysical scaling with applications in physical work and perception of exertion. *Scand. J. Work, Environ. and Health* 16(suppl): 55-58.

Boussenna, M., Corlett, E.N., and Pheasant, S.T. 1982. The relation between discomfort and postural loading at the joints. *Ergonomics* 25: 315-322.

Burandt, U. 1978. *Ergonomie für Design und Entwicklung*, pp. 34-37, Schmidt, Köln.

Caldwell, L.S. 1974. The load-endurance relationship for a static manual response. *Human Factors* 6(1): 71-79.

Chaffin, D.B. and Andersson, G.B.J. 1984. *Occupational Biomechanics.* John Wiley & Sons, New York.

Corlett, E.N. and Bishop, R.P. 1976. A technique for assessing postural discomfort. *Ergonomics* 19: 175-182.

Corlett, E.N. and Manenica, I. 1980. The effects and measurement of working postures. *Applied Ergonomics* 11: 7-16.

Douwes, M. and Dul, J. 1993. Studies on the validity of a work–rest model for static postures. *Proceedings of the International Association World Congress 1993. Ergonomics of Materials Handling*, Warsaw, Poland.

Dul J., Douwes M., and Miedema M.C. 1993. A guideline for the prevention of discomfort of static postures, in *Advances in Industrial Ergonomics and Safety V.* eds. R. Nielsen and K. Jorgensen, pp. 3-5. Taylor & Francis.

Dul, J., Douwes, M., and Smitt, P. 1994. Ergonomic guidelines for the prevention of discomfort of static postures can be based on endurance data. *Ergonomics* 37: 807-815.

Genaidy, A.M. and Karwowski, W. 1993. The effects of neutral posture deviation on perceived joint discomfort ratings in sitting and standing postures. *Ergonomics* 36: 785-792.

Grinten, M. Van der and Smitt, P. 1992. Development of a practical method for measuring body part discomfort, in *Advances in Industrial Ergonomics and Safety*, ed. S. Kumar, Taylor & Francis, Denver.

Hagberg, M. 1981. Electromyographic signs of shoulder muscular fatigue in two elevated arm positions. *Am. J. of Phys. Med.*, 60: 111-121.

Hagerup, A.B. and Time, K. 1992. Felt load on shoulder in the handling of 3 milking units with one- and two-handgrips in various heights. *Proceedings of International Scientific Conference on Prevention of Work-Related Musculoskeletal Disorders (PREMUS)*, pp. 105-107, Sweden.

ISO Document N 62. 1995. Final version of ISO/CD 11226 — Ergonomics — Evaluation of working postures, International Organization for Standardization, Delft, The Netherlands (not published).

Jürgens, H.W., Aune, I.A., and Pieper, U. 1989. *Internationale anthropometrischer Datenatlas. Bundesanstalt für Arbeidsschutz*, pp. 36-38, Dortmund.

Karhu, O., Kansi, P., and Kuorinka, I. 1977. Correcting working postures in industry: a practical method for analysis. *Appl. Ergonomics* 8: 199-201.

Keyserling, W.M., Punnett, L., and Fine, L.J. 1988. Trunk posture and back pain: identification and control of occupational risk factors. *Applied Ind. Hyg.* 3: 87-92.

Kilbom, Å. 1988. Intervention programmes for work related neck and upper limb disorders — strategies and evaluations. *Proceedings of the 10th Congress of the IEA*, pp. 33-47.

Manenica, I. 1986. *The Ergonomics of Working Postures: A Technique for Postural Load Assessment*, pp. 270-277, Taylor & Francis, London.

Meijst, W., Dul, J., and Haslegrave, C. 1995. Maximum holding times of static standing postures. Thesis of extended essay. TNO Institute of Preventive Health Care, Leiden, The Netherlands.

Miedema, M.C. 1992. Static working postures. *Part 1: Classification of static working postures on the basis of maximum holding time. Part 2: Secondary analysis on endurance data.* Thesis of extended essay. TNO Institute of Preventive Health Care, Leiden, The Netherlands.

Miedema, M.C., Douwes, M., and Dul, J. 1996. Recommended holding times for prevention of discomfort of static standing postures. *Industrial Ergonomics* 19: p 9-18.

Milner, N. 1985. *Modelling Fatigue and Recovery in Static Postural Exercise*, University of Nottingham, Nottingham. Ph.D. Thesis.

Molenbroek, J.F.M. and Dirken, J.M. 1987. Nederlandse lichaamsmaten voor ontwerpen, DINED-tabel (3e herziene versie). *Nederlands Tijdschrift voor Ergonomie* 12: 23-24.

Nag, P.K. 1991. Endurance limits in different modes of load holding. *Appl. Ergonomics* 22: 185-188.

Paoli, P. 1992. *First European Survey on the Work Environment 1991-1992.* European Foundation for the Improvement of Living and Working Conditions, Dublin.

Putz-Anderson, V. and Galinsky, T.L. 1993. Psychophysically determined work durations for limiting shoulder girdle fatigue from elevated manual work. *Int. J. of Industrial Ergonomics 11* 19-28.

Rohmert, W. 1960, Ermittlung von Erhohlungspausen für statische Arbeit des Menschen. *International Zeitschrift für angewandte Physiologie einschliesslich Arbeitsphysiologie* 18: 123-164.

Rose, L., Ericson, M., Glimskär, B., Nordgren, B., and Örtengren, R. 1992. Ergo-Index: Development of a model to determine pause needs after fatigue and pain reactions during work, in *Computer Applications in Ergonomics Occupational Safety and Health*, eds. M. Mattila and W. Karwowski pp. 461-468. Elsevier Science Publishers B.V., Amsterdam.

Serratos-Perez, J.N. and Haslegrave, C.M. 1992. In *Contemporary Ergonomics* ed. E.J. Lovesey, pp. 66-71. Taylor & Francis, London.

Sjøgaard, G. 1986. Intramuscular changes during long-term contraction, in *The Ergonomics of Working Postures*, eds. N. Corlett et al. Chapter 14. Taylor & Francis, London.

Taksic, V. 1986. *The Ergonomics of Working Postures: Comparison of Some Indices of Postural Load Assessment.* Taylor & Francis, London, pp. 278-282.

13

Low-Level Static Exertions

Gisela Sjøgaard
National Institute of Occupational Health
Denmark

Bente Rona Jensen
National Institute of Occupational Health
Denmark

13.1 Low-Level Static Exertions in the Workplace

Low-level static exertions have been identified as a risk factor for the development of cumulative trauma disorders or repetitive strain injuries from epidemiological studies. The exposure in terms of static exertions in the workplace has been assessed for different jobs and/or tasks based on electromyographic recordings from specific muscle groups (Table 13.1). Jobs characterized by relatively high static levels in neck and shoulder showed health outcomes in terms of musculoskeletal disorders in these body regions (Table 13.2). In the 1970s, static contractions of 15% MVC (maximum voluntary contraction) were considered to be tolerated for an "unlimited" period of time for a muscle.[1] However, later studies showed that if a contraction is to be maintained for just one hour, it may have to be as low as 8% MVC.[2] A permissible level of static muscle load of 2 to 5% MVC was then suggested.[3] However, it was observed that musculoskeletal disorders were frequent even in jobs with static levels of this magnitude, and it was suggested to reduce the acceptable static level, e.g., by job rotation.[4] Static levels as low as 0.5 to 1% MVC may relate to troubles in the shoulder region,[5,6] and most recently, statements have been brought forward that static loads are not acceptable at all if sustained frequently or over a long period of time. Actually, "working hours as a risk factor in the development of musculoskeletal complaints" has been proposed.[7] Such continuous revision of recommendations can be foreseen if we do not understand why low-level static exertions cause disorders. The acceptable limits or interventions recommended in the workplace will only reduce cumulative trauma disorders if the true risk factors that elicit adverse health outcome, are eliminated or minimized. Therefore, it is important to identify which aspect of these so-called low-level static exertions may be the risk factors. In this context, plausibility also plays an important role in risk identification, that is, possible physiological mechanisms of tissue degradation which may be causally related to the identified risk aspect. The term *low-level static exertions* will be discussed, followed by a presentation of possible short- and long-term physiological responses. Based on this, preventive strategies are presented.

0-8493-2641-9/99/$0.00+$.50

TABLE 13.1 Electromyographic Data on Static (P = 0.1), Mean (P = 0.5), and Peak (P = 0.9) Muscle Load in the Shoulder Region During Different Work Tasks Expressed in Percentage of Maximal Electromyographic Activity or Percentage of Maximal Voluntary Force Development (%MAX).

Job	Muscles	P = 0.1 %MAX	P = 0.5 %MAX	P = 0.9 %MAX	References
Typewriting	m. trapezius	4	7	10	(8)
	m. deltoideus				
Office work	m. trapezius	1	4	—	(6)
CAD-work	m. trapezius	2	5	9	(9)
Industrial sewing	m. trapezius (r)	9	14	21	(10)
	m. trapezius (l)	9	16	25	
	m. infraspinatus	4	9	20	
Floor cleaning	m. trapezius	10	25	54	(11)
Assembly plant	m. trapezius	8	16	27	(12)
electronic work	m. deltoideus	7	13	28	
	m. infraspinatus	13	20	33	
Meat cutting	m. trapezius	6	10	17	(13)
Dental work	m. trapezius	9	13	18	(14)
Flight loading/unloading	m. trapezius	5	14	45	(15)
Letter sorting	m. trapezius	5	10	27	(16)
	m. deltoideus	5	14	19	
	m. infraspinatus	5	10	16	
Post office work	m. trapezius	6	14	33	(17)
(stamping)	m. infraspinatus	7	17	28	
Chocolate manufacturing work	m. trapezius	2	5	—	(6)

13.2 What Are "Static Exertions"?

Within the area of mechanics "static," in the strict sense means "no motion." In the workplace, truly static work postures are quite rare because most jobs include a number of movements to be performed often by the upper limbs. Even in supervision jobs, a number of objects have to be handled now and then.

According to the strict definition of static, one might suggest to use observation techniques to quantify how long a time a certain posture is maintained without any movement. But most likely this variable would fail to show a relationship to musculoskeletal disorders. For instance, lying in bed or sitting relaxed in a well-supporting chair is hardly considered a risk, although highly static. The reason is, of course, that no muscle exertions or contractions need to be performed in these conditions. Therefore, quantifying the true variable "static" when trying to identify risk factors is not sufficient. What we are looking for is the static muscle contraction that may induce an overload on the musculoskeletal system.

A profile of the muscular load during a period of work may be obtained by analyzing the amplitude probability distribution function (APDF) of the electromyographic signal (EMG). Such measurements for analysis of static muscle activity have been widely used in workplace studies, where the static level is defined as the probability level P = 0.1. For instance, a static level of 5% MVC means that the contraction level of the muscle is 5% MVC or above for 90% of the time, or in other words, only for 10% of the time is the muscle contraction below 5% MVC. This implies that muscular rest may occur for 10% of the recording time or less. The interpretation of a static contraction according to the APDF analysis has caused some confusion because the static level is actually defined in the time domain. Also, this variable does not give the information that the muscle is really performing a 5% MVC throughout the recording period; indeed, larger contraction forces may occur. Finally, this variable does not control for length changes of the muscle which means that the muscle contractions may well be dynamic. Nevertheless, redefining "static" in occupational settings has been a great "success." This is probably due to the time variable in essence being the real risk. But this was not intentional and no awareness has been paid to this fact by practitioners. Of note is that the risk factor probably is the *sustained* contraction.

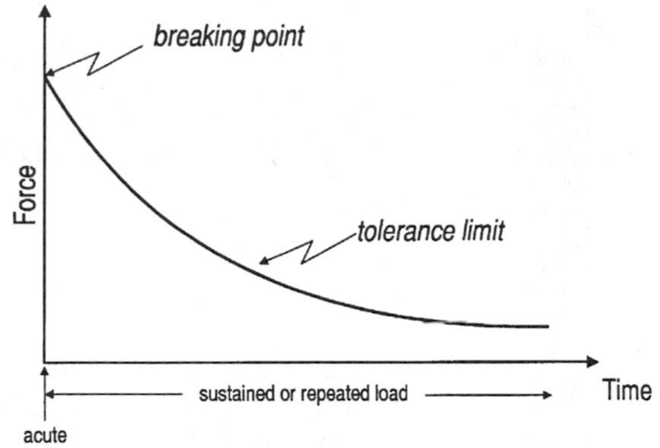

FIGURE 13.1 Tolerance limit for maximum force at breaking point depends on the type of muscle, e.g., cross-sectional area, state of training, and age. Further, the contraction mode, i.e., static or dynamic, is significant. For submaximal forces, the tolerance limit decreases with time, the rate of decrease depending on force magnitude and repetition frequency.

13.3 What Is a "Low Level"?

It may be surprising that low rather than high-level contraction forces seem to imply a risk. Of course, high forces can cause ruptures as seen in accidents where bone, ligament, tendon, or muscle are exerted beyond their breaking point, and in this sense, high forces imply a risk. However, low forces constitute a corresponding risk if repeated or sustained for a prolonged time. All structures, inert materials as well as biological tissues, are able to withstand a force characteristic to their structure. At high forces, disruption will occur when the breaking point is exceeded, and lesser forces repeated over time will eventually cause fatigue fracture (Figure 13.1). Repetitive force exertions are accumulated and cause eventual disruption, possibly not of the tissue as a whole but in terms of micro ruptures.

First of all, when evaluating the force level, the maximum strength of the muscle must be taken into account. This relates to the muscle's cross-sectional area, age, and state of training; and different muscle groups and subjects show highly different muscle strength. Therefore, exposure assessment in terms of force recordings in absolute numbers in Newtons (N) will not give sufficient information regarding the level of exertion. The maximum voluntary contraction (MVC) force must be recorded as well, and data must be presented in percentage of MVC as mentioned above regarding the EMG data.

Second, endurance time for muscles plays a significant role in this context. The relationship between force level and the time for which it can be sustained is depicted in the endurance time curve (Figure 13.2). At low force levels relative to the maximum strength, the muscle is capable of developing such force for long periods of time before being exhausted. Different muscles show highly different endurance capacity depending on muscle fiber type, anatomy, and state of training. But for every muscle there is a limit.

Third, in industry many low-level exertions include repeated static exertions or movements at quite high speed but with little displacement. When observing such tasks, often little attention is payed to the displacement, which is the cause for such exertions to be assessed as static. Also for intermittent static as well as dynamic contractions, endurance time curves exist.[2,18] It is for the dynamic contractions that the contraction level cannot be described only by the force in N or % MVC. The force–velocity relationship must also be taken into account, the relative load being higher when a specific force is developed with increasing speed (Figure 13.2). For instance, keyboard operators may press 200,000 keys a day or 500 per minute and a piano player strikes the tangents with finger movements at very high, maybe sometimes maximum, speed. For the unloaded limb, the EMG activity increases linearly with the velocity of the movement.[19] If, in addition to the movement velocity, there is an external force to overcome during the work tasks, this could imply maximum effort at high velocities even if the force level is low. Thus, low level cannot be assessed only in terms of % MVC, but the mode of contractions must also be taken

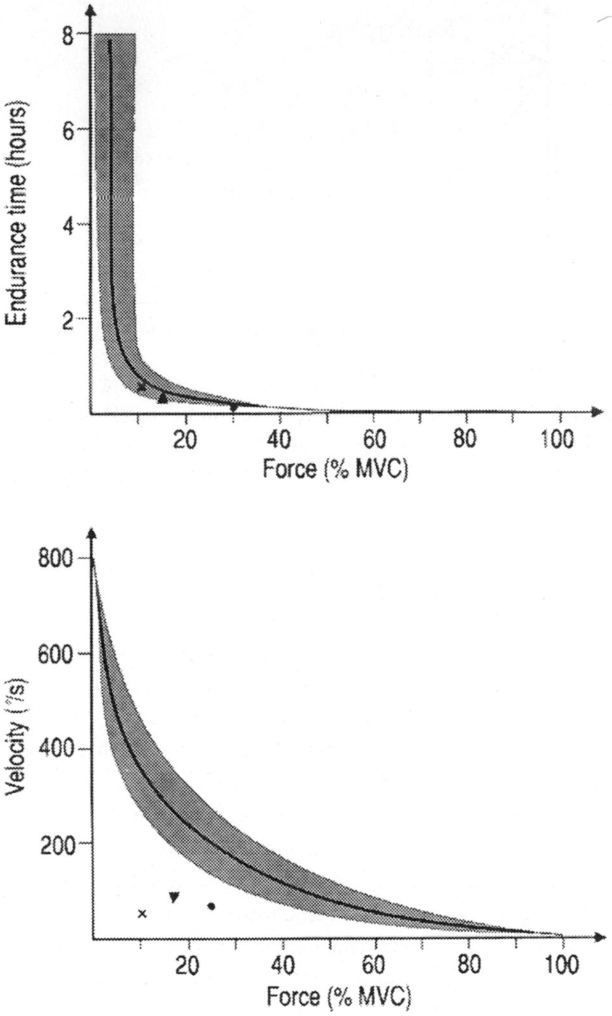

FIGURE 13.2 Upper part shows an endurance time curve for static contractions, which are defined as contractions at constant muscle length. Of note is the large range of endurance time at low forces, where the end point of exhaustion varies significantly, e.g., due to the level of motivation. Examples are given for handgrip (▲), shoulder abduction (x), and trunk extension (●). Lower part shows a force-velocity curve for dynamic contractions, only shown for concentric contractions, i.e., during muscle shortening. The maximum muscle force decreases with increasing velocity of shortening. Examples are given for shoulder movements during floor cleaning (●), shoulder movements during forking in agriculture (▼), and wrist movements during meat cutting (x). For the latter, it is seen that although the load is only 10% MVC, it is about 20% of the dynamic strength, since the maximal dynamic strength at this velocity is approximately 50% MVC.

into account. Ideally, the maximum dynamic voluntary contraction forces should be assessed and the work task evaluated in relation to the corresponding maximum force–velocity relationship.

In short, the term "low level" in the context of work-related static exertions refers to a working condition in which a muscle is activated at a level that can be maintained for a long period. This may be a true static contraction, sustaining a constant force and posture or varying in force within a limited range and without any movement. But even performing intermittent static or dynamic contractions (concentric/eccentric) at submaximal force velocities with small displacements and at intensities that can be maintained for a long time may be considered low-level static exertions in occupational

TABLE 13.2 One-Year Prevalence of Musculoskeletal Symptoms in Neck, Shoulder, Elbow/Forearm, and Hand/Wrist According to the Standardized Nordic Questionnaire[21]

Job	Sex	Number of workers	Neck (%)	Shoulder (%)	Elbow/forearm (%)	Hand/wrist (%)	References
Computer work	F	1285	49	48	13	25	(22)
	M	433	25	23	7	9	
Office work	F	643	48	48	12	22	(23)
	M	35	18	18	6	9	
CAD-work	F+M	149	70	—	41	52	(9)
Sewing machine operators	F	77	55	51	7	26	(24)
	F	303	57	53	7	28	
Cleaners	F	737	63	63	27	46	(25)
Assembly plant electronic work	F+M	25	64	56	—	—	(12)
Meat cutting	F	16	67	54	7	47	(23)
	M	114	39	62	15	47	
Meat cutting	(M)	2463	52	66	28	60	(13)
Dental work	F	43	49	40	19	40	(14)
	M	56	57	39	13	5	
Flight loading	M	808	30	31	14	22	(15)

From Kuorinka, I., Jonsson, B., Kilbom, Å., Vinterberg, H., Biering-Sørensen, F., Andersson, G., Jørgensen, K. Standardised Nordic questionnaires for the analysis of musculoskeletal symptoms. *Applied Ergonomics* 18(3):233-237, 1987.

settings.[20] Actually, when such exertions are measured by electromyography and analyzed by the above-mentioned APDF of the EMG, "static" levels of 5% MVC or more may be found. This means that a low-level static exertion is to be considered in the time domain and is characterized by workers being able to perform it for hours. The main feature is that the exertion is sufficiently low so that it can be sustained for a *prolonged* time, and the duration probably implies the risk.

13.4 Which Work Requirements Induce "Low-Level Static Exertions"?

Examples of jobs in which low-level static exertions are frequent are presented in Table 13.1 and 13.2. Additional job titles are numerous in the literature.[26] It is important in risk assessment to identify generic work requirements that induce these exertions. At random, requirements such as precision, speed, visual demand, and mental load can be mentioned. Also monotony or lack of variation is a characteristic that concerns working posture and movement as well as mental challenge. The same task is repeated over and over again most often by the hands. When operating with fast precise movements with the hands, there is a demand to stabilize the shoulder girdle. One reason is that the shoulders are the reference point for the upper limbs and if they move, the hands will be repositioned with respect to the motor control pattern for the upper limbs. Similarly, to control the position of the eyes, fixation of the neck is needed, and stable eye position is a prerequisite for most visual demands in industry. Interestingly, the fastest repositioning of the eyes can be performed when the neck and shoulder muscles are contracted up to 30% MVC.[27] When performing tasks at high speed, the stiffness of the musculoskeletal system must be increased. For this purpose, *co-contractions* are performed, which means that antagonistic muscles, i.e., muscles on each side of a joint, are contracting. This is especially common for the shoulder muscles. One reason is, as mentioned above, that the shoulder must be the stable fix point and "take-off" for arms and hands. Also the anatomy of the shoulder is such that is has the greatest mobility of all the joints in the body. It is a joint highly dependent on muscle stabilization, including *co-contractions*. These *co-contractions* have been shown to increase with increasing speed and precision demands.[19,28]

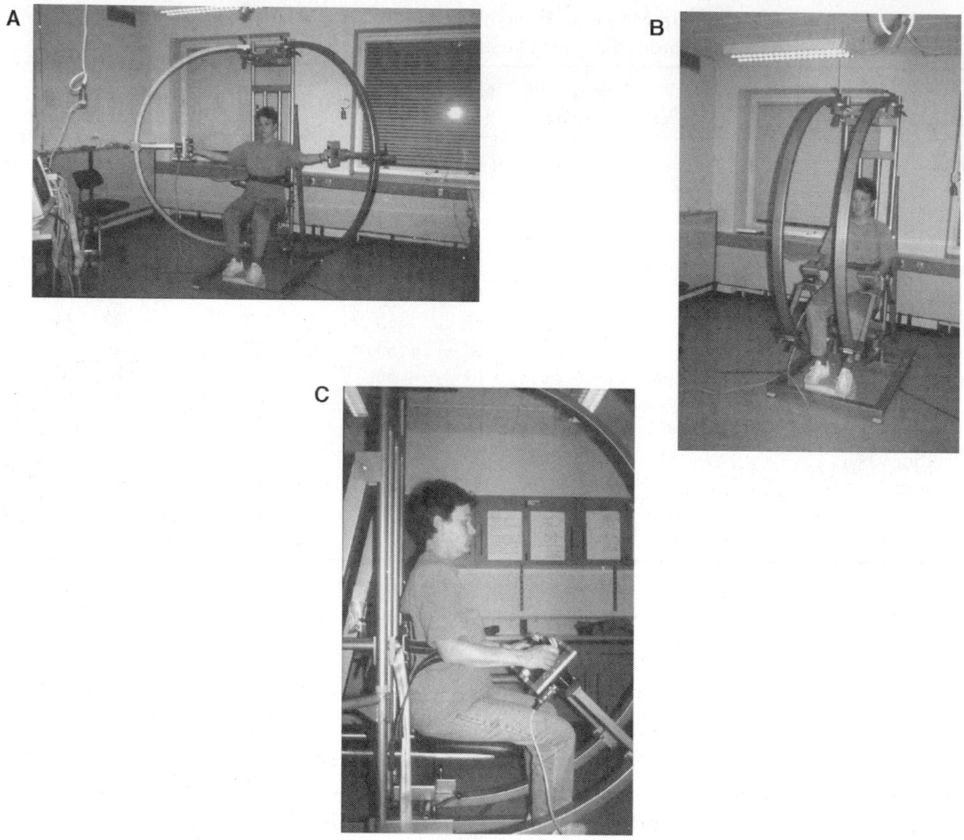

FIGURE 13.3 Experimental chair where the arm posture can be adjusted in any position of abduction (a) and flexion (b). The hands are grasping handles connected to 3-dimensional force transducers (c). Professor Bjørn Quistorff, University of Copenhagen, is acknowledged for the design.

13.5 Why Do "Low-Level Static Exertions" Imply a Risk?

As discussed above, it is not necessarily because exertions are static or at a low level that they imply a risk, but because such exertions are often sustained for prolonged periods of time. Additionally, often no sufficient recovery periods are allowed during such work tasks. A more informative term for the related risk factor would be *prolonged sustained or repeated muscle contractions.* According to the endurance curve, it is possible to sustain low-level exertions for a longer time than high-level exertions. It is likely, that this time factor is the risk. This hypothesis is supported by the physiological responses to such exertions, which constitute the plausibility. Standardized muscle contractions have been studied in combination with detailed physiological responses. In the following discussion, focus will be on mechanisms which may induce muscle damage.

An example of a standardized setup for studying muscle contractions is shown in Figure 13.3. The test chair can be regulated for the subject to adopt any working posture and the force transducers connected to the hand grips allow for three-dimensional recordings. During specific work tasks, biomechanical calculations may then assess the relative load on various muscles or muscle parts/groups based on maximum contractions performed in identical postures and directions.[29]

Intramuscular pressure and blood flow: With each muscle contraction, the tissue pressure (hydrostatic pressure) in the muscle increases in proportion to the force development. The absolute level in terms of mmHg varies widely between muscles and depends, among other things on the anatomy of the muscle itself as well as its surroundings. A bulky muscle attains higher pressures than a thin muscle, and a muscle

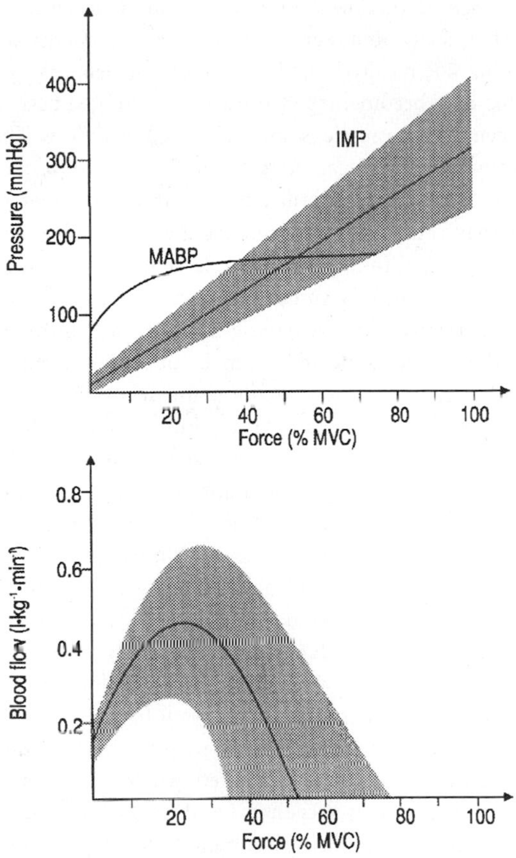

FIGURE 13.4 Upper part shows mean arterial blood pressure (MABP) and intramuscular pressure (IMP) with increasing contraction force. Lower part shows corresponding blood flow.

with bony surroundings or tight fascia shows relatively large increases because of the low compliance of these surroundings. At high contraction forces, the intramuscular pressure may attain values far above blood pressure (Figure 13.4) and obviously cause muscle blood flow to be occluded in areas where intramuscular pressure exceeds blood pressure, the highest pressures normally occurring deep in the muscles. However, even at low-level contractions, the complex microcirculatory regulation may become impeded. First of all, at low blood flow velocities it is not the mean blood pressure but the diastolic pressure that is decisive for maintenance of blood flow.[30] Further, with prolonged contractions, the muscle water content will increase[31] and correspondingly, the thickness of the muscle has been shown to increase.[32] Such a state of edematic tissue with increased volume will *per se* increase tissue pressure in a delimited closed muscle compartment with low compliance. At contraction levels in the order of 5 to 10% MVC, intramuscular pressures of 40 to 60 mmHg or more have been reported in muscles such as the *m. supraspinatus* in the shoulder.[33,34] Causal relationships between prolonged moderately increased tissue pressure and pathogenic changes have been studied extensively in relation to compartment syndromes.[35] Pressures above 30 mmHg maintained for eight hours have been shown to induce necrotic changes in the muscle even if no active contraction was performed and energy demand therefore was minimal.[36] One possible mechanism is that although initially blood flow is sufficient during low-level contractions, this may not be the case when the contraction is maintained for prolonged periods. Conditions with low flow and low perfusion pressure may provoke granulocyte plugging in the capillaries, which affects microcirculation, and may also facilitate formation of free radicals, which have a highly toxic effect.[37,38]

Metabolism: Adequate muscle blood flow is essential for muscle function because force development relies on the conversion of chemically bound energy to mechanical energy, a process also called energy turnover or metabolism. Some chemically bound energy or substrate is located in the muscle tissue (especially glycogen), but this may become depleted during prolonged activities. Therefore, the supply of substrates (including oxygen) to the muscle is crucial for such activities. The ultimate substrate in the conversion of chemical energy to mechanical energy is ATP, which is broken down in the myofibrils during the actin–myosin reaction. ATP is significant for the detachment of actin and myosin, and insufficiency of this process may cause rigor or contracture with massive pain. In normal muscle contractions, the actin–myosin reaction is initiated by the release of Ca^{2+} from the sarcoplasmic reticulum into the cytosol, and has been the focus in a number of studies on muscle fatigue. However, during the last decade, attention has been drawn also to the pathogenesis of Ca^{2+} -induced damage of muscle cells.[39] The reuptake of Ca^{2+} into the sarcoplasmic reticulum is an ATP-dependent process, which may be insufficient during prolonged activity because it accounts for up to 30% of the energy turnover during muscle activity. Further, energy crisis may result in an influx of Ca^{2+} from the extracellular space. Consequently, the cytosolic free Ca^{2+} is likely to be increased above normal for a prolonged time. This has serious implications for the phospholipids, including those in the muscle membrane. Ca^{2+} has a direct effect on phospholipase activity and, in addition, increases the susceptibility of the membrane lipids to free radicals, which have a highly toxic effect as mentioned above. Both these processes promote breakdown of the muscle membrane.[40] Finally, prolonged increased cytosolic Ca^{2+} concentration induces a Ca^{2+} load on the mitochondria and may eventually impair ATP formation, a sufficient concentration of which is a prerequisite for active force production. For more details see reference 41.

Motor control: Another important aspect during low force development is that although the muscle as a whole may not be metabolically exhausted, this may well be the case for single muscle fibers. The muscles are composed of different muscle fiber types and motor units with different recruitment thresholds. A stereotype recruitment order has been documented, which means that with increasing force, the low threshold motor units are always being recruited first.[42] Within a motor unit pool, various motor units may be alternating in activity pattern during a submaximal muscle contraction postponing fatigue to develop in each of the involved fibers.[30] However, in performing highly skilled movements and accurate manipulations, it is likely that the very same motor units are being recruited continuously. This holds true for pure static as well as slow force-varying and low-velocity dynamic contractions.[43,44] Additionally, contractions may be elicited due to reflexes, causing even more stereotype recruitment than during voluntary contractions. Mental load has been demonstrated to generate nonpostural muscle tension in shoulder muscles, and the same holds true for visual demands and neck muscles.[27,45-48] Also reflexes originating in the muscle itself from chemo- as well as mechanoreceptors may play a role, and recently the gamma-loop has been proposed to play a role in developing a potentially vicious circle.[49,50] The muscle fibers being continuously activated have been termed Cinderella fibers, because they are working from early to late.[51] A high energy turnover occurs in these fibers, and most likely they receive the least blood flow because tissue pressure increases in their vicinity due to the mechanical contraction impeding blood flow.[30] The pathogenic mechanisms described above regarding accumulation of Ca^{2+} and free radicals may be a concern, especially at the single muscle fiber level. Prolonged activity of specific motor units throughout an eight-hour working day may cause insufficient time for full recovery of these motor units due to a long-lasting element of fatigue,[52] which has been shown to occur in simulated occupational settings.[53] This may cause necrosis and, finally, cell destruction in these fibers. In line with this, fibers with marked degenerative characteristics have been found more frequently in muscle biopsies from patients with work-related chronic myalgia than in normal subjects in the trapezius muscle.[54] Interestingly, the degenerative fibers identified are slow twitch fibers, which connect with low threshold motor nerves.[55]

Perception of fatigue: When muscular work is performed over a prolonged period, fatigue develops. Fatigue may cause the work to be performed with less care or precision, and an accident can result. A fatigued worker is more likely to make a wrong movement, such as a slip and fall, leading to injury. However, even if an accident does not occur, prolonged fatigue without adequate time for recovery can lead to the development of musculoskeletal disorders. From the beginning of every muscle activity, the muscle is fatiguing, and muscle function is decreasing.[56] This condition is normally perceived as muscle fatigue. The perception of fatigue is a very useful mechanism for protecting the muscle against overload. Among other factors, the work-induced increase in potassium concentration in the interstitial space can help mediate the perception of fatigue in the central nervous system (CNS).[57] However, during very low level contractions, the accumulated increase in interstitial potassium may be subliminal to the threshold of the sensory afferents mediating the information to the CNS.[58] Also, in situations of machine-controlled work or heavy work pressure, the fatigue message is depressed, when it is not possible for the employee to take a rest. In other words, fatigue is ignored — consciously or subconsciously — which in the long term can have serious consequences.

The processes that take place in relation to fatigue are normally reversible for biological tissues, which is in contrast to inert tissues and a reason why we normally do not consider fatigue as dangerous. This means that muscles recover when resting after exertion. A rest period following muscular activity is therefore essential to enable the muscle to recover its full functional potential with regard to strength and endurance. Even an improvement or training effect of these variables may be obtained if optimal performance of activity and recovery periods is planned. There is no simple time equation for length of work and adequate length of a subsequent resting period. The process of recovery depends on the type of work that caused the muscle fatigue. For instance, so-called low-frequency fatigue and high-frequency fatigue are caused by fundamentally different biochemical changes in the muscle.[58] If fatigue is due to relatively high loads over a short time, the necessary recovery will be quicker than if fatigue is due to prolonged working at low load levels. Thus, if the same muscle group or group of fibers is activated continuously for a full working day of 7 or 8 hours, there is a risk that the muscle will not even be fully recovered by the next day. If such conditions persist for months or even years, they can ultimately inflict irreversible or chronic changes that may result in pain and impaired function.

13.6 How to Prevent Musculoskeletal Disorders from "Low-Level Static Exertions"

A prerequisite for effective prevention is knowledge of the cause of the disorder. The documentation so far of the time factor being essential gives the simple answer to this question: limit the time for each specific sustained muscle effort. Each single muscle cell and corresponding motor nerve and tendon demand recovery periods sufficiently long to attain full recovery. Time for recovery is not linearly related to time for activity. Rather, it increases exponentially. For example, if exhaustion is elicited by a high contraction force for 1 minute, then recovery is very fast, and after 2 to 3 minutes the same force can be performed again. But when a muscle is fatigued for 1 hour, it may take many hours for full recovery, and if the fatiguing process has lasted for an 8-hour working day, full recovery may not even have occurred the next morning when the next working day starts and the same tasks are to be managed. Interestingly, such sports activities as a marathon (lasting about 2 to 5 hours, depending on the state of training) are only performed a few times a year even by top athletes. Limits to prolonged activities are also seen in sport events such as the Tour de France or other endurance activities. Normally in the workplace, the worker is not totally exerted and often only part of the body is exerted. This means that somewhat shorter recovery will be acceptable, but still it is essential that the activity period is followed by a recovery period and that the duration of both is matched to the intensity and mode of contraction in the activity period (Figure 13.5).

In summary, it can be stated in concordance with an earlier discussion:[59] Human skeletal muscles are not adapted for continuous long-lasting activity. Indeed, no matter how low the exertion level is, rest periods are needed for the muscle to recover. Guidelines for low-level static exertions should therefore deal not just with the acceptable static level in percentage MVC. To recommend a reduction in exertion level from, for instance, 5 to 2% MVC will not help much physiologically; also it is not practical. Instead, we need guidelines for maximum acceptable *time limits* for prolonged sustained or repeated muscle

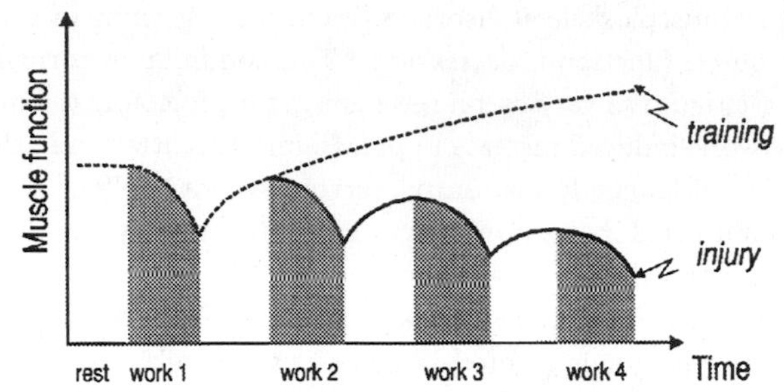

FIGURE 13.5 Muscle function in relation to work and recovery (duty circle).

contractions. This means that we must focus our attention on how long a muscle or group of muscles can tolerate maintaining or repeating the same low-level exertions. This is especially true when these same low-level exertions are imposed on a muscle day after day.

13.7 Recommendation for the Practitioner Regarding Job Profile and Workplace Design

- The workplace must allow for variation in working postures. This means, for example, that table and chair are easily adjustable. For instance, shifting frequently between sitting and standing is recommended, and instructions should be given to implement adjustments at frequent *time* intervals.
- The workplace must be designed based on principles of optimization and not minimization of mechanical workload. Therefore, it is recommended that work cycles include loads ranging from complete relaxation to moderately high contraction forces and velocities. Workers should be given the opportunity to optimize the phases of a duty cycle according to their capacity. It is important that the level or intensity of exertion changes over a wide range continuously over *time.*
- The job profile must allow for performing a variety of different tasks. The task variation should include variation regarding mental load as well as physical (mechanical) load on the musculoskeletal system. If only specialized work tasks can be performed at each workstation, a job must include tasks at different workstations. For the use of tools, it is recommended that a variety of tools with different designs are used interchangeably. In combination, these variations should cause loading of different body regions and muscle groups regularly over *time.*

References

1. Rohmert, W. Problems of determination of rest allowances. *Applied Ergonomics* 4(3):158-162, 1973.
2. Björkstén, M. and Jonsson, B. Endurance limit of force in long-term intermittent static contractions. *Scand J Work Environ Health* 3:23-27, 1977.
3. Jonsson, B. Kinesiology. With special reference to electromyographic kinesiology, pp. 417-428, in *Contemporary Clinical Neurophysiology,* edited by Cobb, W.A., Van Duijn, H., Amsterdam, Elsevier Scientific Publishing Company, 1978.
4. Jonsson, B. The static load component in muscle work. *Eur J Appl Physiol* 57:305-310, 1988.
5. Veiersted K.B., Westgaard, R.H., and Andersen, P. Pattern of muscle activity during stereotyped work and its relation to muscle pain. *Int Arch Occup Environ Health* 62:31-41, 1990.
6. Jensen, C.; Nilsen, K., Hansen, K., and Westgaard, R.H. Trapezius muscle load as a risk indicator for occupational shoulder-neck complaints. *Int Arch Occup Environ Health* 64:415-423, 1993.
7. Wærsted, M. and Westgaard, R.H. Working hours as a risk factor in the development of musculoskeletal complaints. *Ergonomics* 34(3):265-276, 1991.

8. Björkstén, M., Itani. T., Jonsson, B., and Yoshizawa, M. Evaluation of muscular load in shoulder and forearm muscles among medical secretaries during occupational typing and some nonoccupational activities, pp. 35-39, in *Biomechanics X-A*, edited by Jonsson, B., 1987, p. 35.
9. Jensen, C., Borg, V., Burr, H., Finsen, L., Olsen, H.B., Hansen, K., Juul-Kristensen, B., Vestergaard, K.B., Jensen, H.B., and Christensen, H. *Fysiske og psykosociale påvirkninger ved arbejde med computer-aided design (CAD). RAMBØLL. Rapport nr. 3.* København, Arbejdsmiljøinstituttet, 1996.
10. Jensen, B.R., Schibye, B., Søgaard, K., Simonsen, E.B., and Sjøgaard, G. Shoulder muscle load and muscle fatigue among industrial sewing-machine operators. *Eur J Appl Physiol* 67:467-475, 1993.
11. Søgaard, K., Fallentin, N., and Nielsen, J. Work load during floor cleaning. The effect of cleaning methods and work technique. *Eur J Appl Physiol* 73:73-81, 1996.
12. Christensen, H. Muscle activity and fatigue in the shoulder muscles of assembly-plant employees. *Scand J Work Environ Health* 12(6):582-587, 1986.
13. Christensen, H. (ed). *Udbeningsarbejde i svineslagterier.* København, Arbejdsmiljøinstituttet, 1996.
14. Finsen, L. *Biomechanical analyses of occupational work loads in the neck and shoulder. A study in dentistry* (Ph.D. thesis). Copenhagen, National Institute of Occupational Health, University of Copenhagen, 1995.
15. Jørgensen, K., Jensen, B, and Stokholm, J. Postural strain and discomfort during loading and unloading flight. An ergonomic intervention study, pp. 663-673. In *Trends in Ergonomics/Human Factors IV*, edited by Asfour, S.S., North-Holland, Elsevier Science Publishers B.V. 1987.
16. Jørgensen, K., Fallentin, N., and Sidenius, B. The strain on the shoulder and neck muscles during letter sorting. *Int J Ind Erg* 3:243-248, 1989.
17. Jørgensen, K. and Fallentin, N. *Lokal muskelbelastning og bevægeapparatssymptomer blandt ekspeditionspersonale på danske postkontorer.* København, August Krogh Institutet, University of Copenhagen, 1986.
18. Sjøgaard, G., Sejersted, O.M., Winkel, J., Smolander, J., Jørgensen, K., and Westgaard, R. Exposure assessment and mechanisms of pathogenesis in work-related musculoskeletal disorders: Significant aspects in the documentation of risk factors, in *Work and health. Scientific basis of progress in the working environment.* edited by Svane, O., Johansen, C., Luxembourg, European Commission, Directorate-General V, 1995.
19. Carpentier, A., Duchateau, J., and Hainaut, K. Velocity-dependent muscle strategy during plantarflexion in humans. *J Electromyogr Kinesiol* 6:225-233, 1996.
20. Jørgensen, K., Fallentin, N., Krogh-Lund, C., and Jensen, B.R. Electromyography and fatigue during prolonged, low-level static contractions. *Eur J Appl Physiol* 57:316-321, 1988.
21. Kuorinka, I., Jonsson, B., Kilbom, Å., Vinterberg, H., Biering-Sørensen, F., Andersson, G., and Jørgensen, K. Standardised Nordic questionnaires for the analysis of musculoskeletal symptoms. *Applied Ergonomics* 18(3):233-237, 1987.
22. Aronsson, G., Åborg C., Örelius, M., Datoriseringens vinnare och förlorare. En studie av arbetsförhållanden inom statliga myndigheter och verk. *Arbete och Hälsa* 27:1-87, 1988.
23. Ydreborg, B., Bryngelsson, I., and Gustafsson, C. *Referensdata till Örebroformulären FHV 001 D (200 D), 002 D (202 D), 003 D, 004 D och 007 D. Data från 95 yrkesgrupper insamlade åren 1984-1989.* Örebro, Stiftelsen för yrkes- och miljömedicinsk forskning och utveckling i Örebro, 1989.
24. Schibye, B., Skov, T., Ekner, D., Christiansen, J.U., and Sjøgaard, G. Musculoskeletal symptoms among sewing machine operators. *Scand J Work Environ Health* 21:426-433, 1995.
25. Nielsen, J. *Occupational health among cleaners. (In Danish with English summary.) (Ph.D. thesis).* Copenhagen, National Institute of Occupational Health, University of Copenhagen, 1995.
26. Armstrong, T.J., Buckle, P., Fine, L.J., Hagberg, M., Jonsson, B., Kilbom, Å., Kuorinka, I.A.A., Silverstein, B.A., Sjøgaard, G., and Viikari-Juntura, E.R.A. A conceptual model for work-related neck and upper-limb musculoskeletal disorders. *Scand J Work Environ Health* 19(2):73-84, 1993.
27. Kunita, K. and Fujiwara, K. Relationship between reaction time of eye movement and activity of the neck extensors. *Eur J Appl Physiol* 74:553-557, 1996.

28. Sjøgaard, G., Laursen, B., Németh, G., and Jensen, B. High speed precision tasks increase muscle activity. *Book of Abstracts XVth Congress of the International Society of Biomechanics, Jyväskylä, Finland* 858-859, 1995(abstract).
29. Laursen, B. *Shoulder muscle forces during work. EMG-based biomechanical models (Ph.D. thesis).* Copenhagen, National Institute of Occupational Health, Technical University of Denmark, 1996.
30. Sjøgaard, G., Kiens, B., Jørgensen, K., and Saltin, B. Intramuscular pressure, EMG and blood flow during low-level prolonged static contraction in man. *Acta Physiol Scand* 128:475-484, 1986.
31. Sjøgaard, G. Muscle energy metabolism and electrolyte shifts during low-level prolonged static contraction in man. *Acta Physiol Scand* 134:181-187, 1988.
32. Jensen, B.R., Jørgensen, K., and Sjøgaard, G. The effect of prolonged isometric contractions on muscle fluid balance. *Eur J Appl Physiol* 69:439-444, 1994.
33. Jensen, B.R., Jørgensen, K., Huijing, P.A., and Sjøgaard, G. Soft tissue architecture and intramuscular pressure in the shoulder region. *Eur J Morphol* 33(3):205-220, 1995.
34. Järvholm, U., Palmerud, G., Herberts, P., Högfors, C., and Kadefors, R. Intramuscular pressure and electromyography in the supraspinatus muscle at shoulder abduction. *Clin Orthop* 245:102-109, 1989.
35. Pedowitz, R.A., Hargens, A.R., Mubarak, S.J., and Gershuni, D.H. Modified criteria for the objective diagnosis of chronic compartment syndrome of the leg. *Am J Sports Med* 18(1):35-40, 1990.
36. Hargens, A.R., Schmidt, D.A., Evans, K.L., Gonsalves, M.R., Cologne, J.B., Garfin, S.R., Mubarak, S.J., Hagan, P.L., and Akeson, W.H. Quantitation of skeletal-muscle necrosis in a model compartment syndrome. *Bone Joint Surg (Am)* 63-A(4):631-636, 1981.
37. Schmid-Schönbein, G.W. Capillary plugging by granulocytes and the no-reflow phenomenon in the microcirculation. *Fed Proc* 46(7):2397-2401, 1987.
38. Jensen, B.R., Sjøgaard, G., Bornmyr, S., Arborelius, M., and Jørgensen, K. Intramuscular laser-Doppler flowmetry in the supraspinatus muscle during isometric contractions. *Eur J Appl Physiol* 71(4):373-378, 1995.
39. Jackson, M.J., Jones, D.A., and Edwards, R.H.T. Experimental skeletal muscle damage: The nature of the calcium-activated degenerative processes. *Eur J Clin Invest* 14:369-374, 1984.
40. Das, D.K. and Essman, W.B. *Oxygen Radicals: Systemic events and disease processes.* Karger, 1990.
41. Sjøgaard, G. and Jensen, B.R. Muscle pathology with overuse, in *Chronic upper limb musculoskeletal injuries in the workplace*, edited by Ranney, D., Philadelphia, U.S.A., W.B. Saunders Company, 1997.
42. Henneman, E. and Olson, C.B. Relations between structure and function in the design of skeletal muscles. *J Neurophysiol* 28:581-598, 1965.
43. Søgaard, K., Christensen, H., Jensen, B.R., Finsen, L., and Sjøgaard, G. Motor control and kinetics during low level concentric and eccentric contractions in man. *Electroenceph Clin Neurophysiol* 101:453-460, 1996.
44. Christensen, H., Søgaard, K., Jensen, B.R., Finsen, L., and Sjøgaard, G. Intramuscular and surface EMG power spectrum from dynamic and static contractions. *J Electromyogr Kinesiol* 5(1):27-36, 1995.
45. Westgaard, R.H. and Bjørklund, R. Generation of muscle tension additional to postural muscle load. *Ergonomics* 30(6):911-923, 1987.
46. Lie, I. and Watten, R.G. Oculomotor factors in the aetiology of occupational cervicobrachial diseases (OCD). *Eur J Appl Physiol* 56(2):151-156, 1987.
47. Wærsted, M. and Westgaard, R.H. Attention-related muscle activity in different body regions during VDU work with minimal physical activity. *Ergonomics* 39:661-676, 1996.
48. Wærsted, M., Eken, T., and Westgaard, R.H. Activity of single motor unit in attention-demanding tasks: firing pattern in the human trapezius muscle. *Eur J Appl Physiol* 72:323-329, 1996.
49. Johansson, H. and Sojka, P. Pathophysiological mechanisms involved in genesis and spread of muscular tension in occupational muscle pain and in chronic musculoskeletal pain syndromes: A hypothesis. *Med Hypotheses* 35:196-203, 1991.

50. Mense, S. Considerations concerning the neurobiological basis of muscle pain. *Can J Physiol Pharmacol* 69:610-616, 1991.
51. Hägg, G.M. Static work loads and occupational myalgia — a new explanation model, pp. 141-144. In *Electromyographical Kinesiology*, edited by Anderson, P.A., Hobart, D.J., Danoff, J.V., Elsevier Science Publishers B.V. 1991.
52. Edwards, R.H.T., Hill, D.K., Jones, D.A., and Merton, P.A. Fatigue of long duration in human skeletal muscle after exercise. *J Physiol (Lond)* 272:769-778, 1977.
53. Byström, S. Physiological response and acceptability of isometric intermittent handgrip contractions. *Arbete och Hälsa* 38:1-108, 1991.
54. Larsson, S., Bengtsson, A., Bodegård, L., Henriksson, K.G., and Larsson, J. Muscle changes in work-related chronic myalgia. *Acta Orthop Scand* 59(5):552-556, 1988.
55. Henriksson, K.G. Muscle pain in neuromuscular disorders and primary fibromyalgia. *Eur J Appl Physiol* 57(3):348-352, 1988.
56. Bigland-Ritchie, B., Cafarelli, E., and Vøllestad, N.K. Fatigue of submaximal static contractions. *Acta Physiol Scand* 128 (Suppl 556):137-148, 1986.
57. Sjøgaard, G. Exercise-induced muscle fatigue: The significance of potassium. *Acta Physiol Scand* 140 (Suppl. 593):1-64, 1990.
58. Sjøgaard, G. Potassium and fatigue: the pros and cons. *Acta Physiol Scand* 156:257-264, 1996.
59. Sjøgaard, G. Intramuscular changes during long-term contraction, pp. 136-143. In *The ergonomics of working postures. Models, methods and cases*, edited by Corlett, N., Wilson, J., Manenica, I., London and Philadelphia, Taylor & Francis, 1986.

14

Job Demands and Physical Fitness

Veikko Louhevaara
Finnish Institute of Occupational Health and University of Kuopio

14.1 Introduction

Job demands include actual physical, mental, and social loading factors, all of which are needed for working in a productive and qualified manner. When job demands are considered in relation to physical fitness, the physical aspect of the work load is most important. Without the help of external power, the physical work load can only be handled by a worker's muscular performance in terms of dynamic and static muscle contractions.

In both industrialized and developing countries, there are numerous jobs requiring physical work in spite of rapid technological developments. Rutenfranz et al. (1990) estimated that about 20% of the workforce in industrialized countries are regularly exposed to heavy muscular work even though the proportion of conventional dynamic jobs with simple manual tools has decreased. On the other hand, static and repetitive tasks have increased in many jobs (Smolander and Louhevaara, 1996).

Various control measures affecting muscular work performance via a worker's physical fitness can be categorized as secondary preventive measures. These are the most relevant in physically heavy occupations in which primary preventive ergonomic measures involving technical and organizational arrangements at work prove to be insufficient. Usually the control measures on physical fitness include, broadly speaking, individual health promotion, a healthy and satisfying lifestyle, and the maintenance of work ability or productive aging (WHO, 1993; Ilmarinen and Louhevaara, 1994; Louhevaara and Ilmarinen,

0-8493-2641-9/99/$0.00+$.50

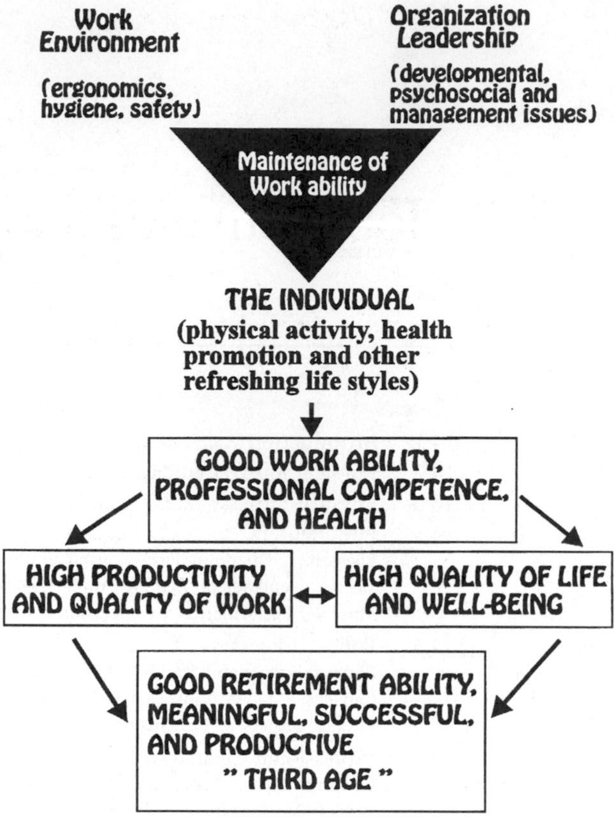

FIGURE 14.1 A concept for the maintenance of work ability and professional competence developed at the Finnish Institute of Occupational Health. The combination of various work-related and individual measures results in good work ability, professional competence, and health, as well as high productivity and quality of work. Simultaneously, the quality of life and the well-being of the workers improve. Thus, they also have better chances for a meaningful, satisfactory, and active "third age" after retirement.

1994). Of these measures, the most common ones are physical fitness training (exercise) and nutrition (Blair et al., 1996).

Individual physical fitness is one of the key elements of work ability which covers all capacities for coping with job demands without excessive over- or understrain. The concept of the maintenance of work ability involves measures in three areas, according to the triangle strategy developed at the Finnish Institute of Occupational Health. These areas are work and the environment (ergonomics, industrial hygiene, and safety on the job), organizational culture (psychosocial and management issues related to the job), and the individual worker (physical fitness training, health promotion, satisfying lifestyle) (Ilmarinen and Louhevaara, 1994; Ilmarinen et al., 1995) (Figure 14.1).

14.2 Physical Job Demands

According to the stress–strain concept introduced by Rutenfranz (1981), physical job demands i.e., muscular work load (stress, exposure, burden, exertion, effort) can be categorized as heavy dynamic work, manual materials handling (MMH), static postural work and repetitive work (Louhevaara, 1992) (Figure 14.2).

Heavy dynamic work with large muscle groups consists mainly of activities requiring the moving of a worker's own body mass, and his or her strain responses are mostly cardiorespiratory (overall) in nature. The load of heavy dynamic work increases in relation to moving speed, distance, the degree of ascent at the covered distance, and the amount of a worker's own body mass as well as the additional mass of

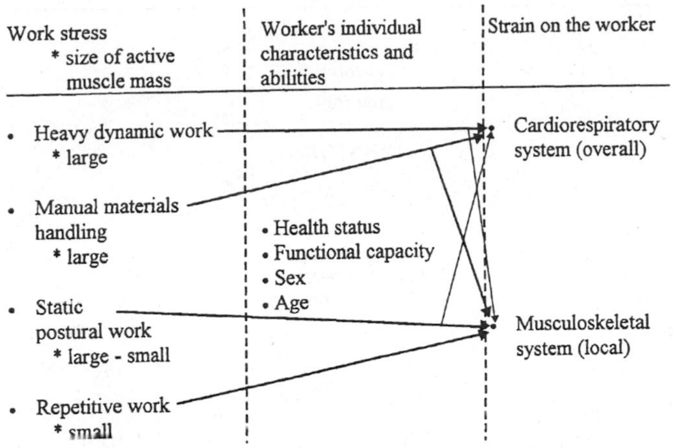

FIGURE 14.2 The effects of different types of physical (muscular) load on the cardiorespiratory and musculoskeletal system. The model has been adopted from the stress–strain concept, and individual characteristics and capacities are considered as intervening factors modifying strain responses due to physical work load.

personal protective equipment which must be worn in many heavy physical tasks. MMH involves mixed dynamic and static work with large muscle groups. The ordinary components of MMH are lifting, carrying, pushing, pulling, and holding external loads of various weights and sizes. Physical work load in MMH equally affects the cardiorespiratory and musculoskeletal system, whereas static postural and repetitive types of physical work loads predominantly produce musculoskeletal (local) strain responses (Louhevaara, 1992). Thus, the type of muscle contraction (dynamic versus static) and the amount of active muscle mass are very important factors as regards physical work load as well as the strain responses of a worker (Aminoff et al., 1996).

In addition to the above-mentioned aspects of muscular work, physical work load is greatly affected by the use of strength, the frequency of sudden peak load efforts, and work–rest regimens, as well as environmental factors such as basic thermal parameters (ambient temperature, relative humidity, and air velocity) (e.g., Kähkönen, 1993), and work rate or the intensity of work (e.g., Louhevaara et al., 1988).

Heavy dynamic work and MMH with large muscle groups are most often needed in jobs in forestry and agriculture, building, installation, transport, manual sorting, health care and home care, and cleaning, as well as in the work of fire fighters, police officers, and soldiers (Ilmarinen, 1984; Smolander et al., 1984; Ahonen et al., 1990; Louhevaara et al., 1990; Hopsu, 1993; Lusa, 1994; Soininen, 1995). Typical jobs with static postural and/or dynamic repetitive muscular work with small muscle masses are, for instance, electrical assembly and meat-processing (e.g., Jonsson et al., 1988; Viikari-Juntura et al., 1991).

The level of physical strain on a worker depends both on job demands and on his or her individual characteristics, capacities, skills, and motivation. Therefore, when considering optimal or acceptable physical job demands for different types of muscular work, one must base the criteria on cardiorespiratory, musculoskeletal (biomechanical), and subjective (psychophysical) strain responses. These may involve overall physiological changes, fatigue, symptoms and disorders in the whole body, or merely specific local changes, for instance, in a single small muscle group or joint.

The individual-based and multi-response-based evaluation of acceptable job demands arises from the principle that there are individual limits for overstrain and understrain as well as damage for each type of physical work load (Louhevaara, 1992). When physical job demands do not exceed the worker's individual capacities, his or her physiological organs adapt to the demands, and recovery is quick after the termination of work. If the job demands are too high, fatigue and various symptoms occur, work capacity and productivity diminish, and recovery is slow. Prolonged or repetitive overload or sudden peak loads may result in organic damage, i.e., injury, as well as work-related or occupational disease. On the other hand, muscular work at a specific intensity, frequency, and duration may also produce fitness training effects; similarly, muscular inactivity may cause detraining effects (Rohmert, 1983).

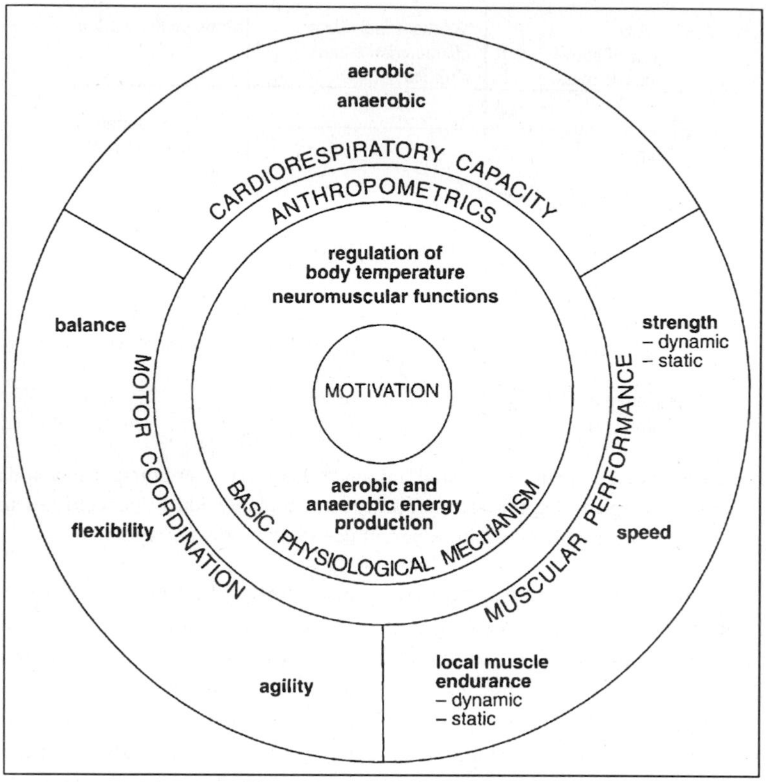

FIGURE 14.3 The main dimensions of physical fitness. The quality and quantity of output variables within cardiorespiratory capacity, muscular performance, and motor coordination depend on the basic physiological mechanisms and are affected by individual characteristics.

14.3 Physical Fitness

Physical fitness in terms of cardiorespiratory (aerobic) capacity, muscular performance (muscle strength and endurance), and motor coordination (body control) is based on basic physiological mechanisms: aerobic and anaerobic energy production, neuromuscular functions, and the regulation of body temperature. Anthropometric characteristics can be regarded as intervening factors in association with the output parameters of physical fitness. The utilization of different physical fitness capacities is done by means of voluntary muscle contractions which are impossible without an adequate level of motivation (Hollman and Hettinger, 1980, Åstrand and Rodahl, 1986) (Figure 14.3).

Physical work capacity is based on physical fitness. If the worker's profile of physical fitness and his or her anthropometrics are in harmony with the physical job demands, the situation is optimal or at least acceptable. The balanced situation is neither static nor permanent, but highly dynamic, and depends on changes both in job demands and individual capacities. Disturbances in the production process or the blackout of automatic data processing systems may change physical job demands completely within a few seconds. Similarly, sudden illness or injury may decrease a worker's level of physical fitness and work capacity dramatically. With the exception of these quick and often unexpected changes, the implementation of advanced technology and the unavoidable aging of workers will affect the sensitive balance between physical job demands and physical work capacity (WHO, 1993; de Zwart et al., 1995).

In this chapter, the physical job demands and fitness requirements are considered more thoroughly in the work of a fire fighter, a police officer, and a professional cleaner. The first two jobs can be classified as special occupations which occasionally encompass physically extreme and hazardous situations. These require near maximal or maximal physical exertion without any possibilities to apply even basic primary ergonomic controls. Professional cleaning belongs to basic service jobs and includes a lot of both heavy dynamic work and MMH activities. In many countries, and particularly in the Nordic countries, fire fighters taking part in operative tasks are almost without exception men. Also, the majority of the police officers on street duty are men. In professional cleaning, the situation is almost the reverse, as the workers are usually women. In the European Union, they are also often aging i.e., over 45 years old (WHO, 1993; Krueger et al., 1995).

14.4 Fire Fighters

Physically Demanding Fire-Fighting and Rescue Tasks

In the questionnaire study of Lusa et al. (1994), fire-fighting and rescue tasks were rated by 200 full-time fire fighters according to the job demands on physical fitness in terms of aerobic power (cardiorespiratory capacity), muscular performance (strength and endurance), and motor coordination (flexibility, agility, balance). Most of the respondents (79%) rated smoke-diving (entry into a smoke-filled space) with the use of the personal protective clothing system and a self-contained breathing apparatus (SCBA) as the task demanding most aerobic power. Almost half of the respondents (43%) felt that clearing passages with heavy manual tools required the highest demands on muscular performance. About three fourths (72%) of the respondents felt that the need for motor coordination was the greatest in fire-fighting and rescue operations on the roof. During the past five years, 54 to 71% of the respondents had carried out smoke-diving, clearing tasks, and roof work at least four times a year

Demands on Aerobic Power

The oxygen consumption (energy expenditure) is linearly related to the amount and intensity of dynamic muscle work. The oxygen consumption, i.e., the oxygen supply to active muscles and tissues, is high in heavy dynamic tasks and many MMH jobs. This also sets high demands on the fitness or efficiency of the cardiorespiratory system.

In various fire-fighting and rescue operations with the protective clothing system and SCBA, such as smoke-diving, fire-suppression, ladder climbing, rescuing a victim, dragging a hose, and raising a ladder, mean oxygen consumption levels of 2.1 to 2.8 l/min have been reported by Lemon and Hermiston (1977), Louhevaara et al. (1985), Sothmann et al. (1990), and Lusa et al. (1993). Maximal exertion is frequent in peak loads and has been reported to correspond to the average oxygen consumption of 3.8 l/min and heart rate of 180 beats/min in young and healthy fire fighters (Lusa et al. 1993). In the heat, the effective regulation of body temperature is necessary for carrying out potential fire-fighting and rescue tasks. In prolonged or repetitive exposure to heat, heavy dynamic muscle work, and the wearing of personal protective equipment, the needed release of body heat is impossible, leading to the termination of work or to health hazards (White and Hodus, 1987; Ilmarinen et al., 1994).

In actual field operations with the protective clothing system and SCBA, the mean and particularly peak demands on the cardiorespiratory capacity have consistently been noted to be very high without any possibilities to alleviate the work load. Therefore, secondary preventive control measures are needed to guarantee the health and safety of fire fighters. In practice, this means that every fire fighter, regardless of age, exposed to physically demanding fire-fighting and rescue operations must be at the adequate level of cardiorespiratory fitness, i.e., maximal oxygen consumption.

Lemon and Hermiston (1977), Louhevaara et al. (1985), Sothmann et al. (1990), and Louhevaara et al. (1994) have recommended that a fire fighter should have a maximal oxygen consumption of 2.7 to 3.0 l/min and/or 34 to 45 ml/min per kilogram of body weight. For instance, if a fire fighter weights 75 kg and his maximal oxygen consumption is 3.0 l/min, his maximal oxygen consumption related to body weight is 40 ml/min/kg.

In Finland, the *Guide on Smoke-diving* (1991) gives a recommendation for a four-stage scale to classify fire fighters' maximal oxygen consumption, based on the research done at the Finnish Institute of Occupational Health (Table 14.1). According to the *Guide,* the cardiorespiratory fitness of male and female

TABLE 14.1 Tests and Their Classification for Assessing Male and Female Fire Fighters' Cardiorespiratory (Aerobic) Capacity and Muscular Performance (Strength and Endurance) (Maximal oxygen consumption = VO2max)

	Classification			
Test	Poor	Moderate	Good	Excellent
VO2max[a]				
(l/min)	≤2.4	2.5–2.9	3.0–3.9	≥4.0
(ml/min/kg)	≤29	30–35	36	≥50
Bench press (45 kg)				
(reps/60 s)	≤9	10–17	18–29	≥30
Sit-up				
(reps/60 s)	≤20	21–28	29–40	≥41
Squatting (45 kg)				
(reps/60 s)	≤9	10–17	18–26	≥37
Pull-up				
(max reps)	≤2	3–4	5–9	≥10

[a] Bicycle-ergometer or treadmill test

fire fighters in the age range of 20 to 63 years is considered sufficient for smoke-diving tasks when they attain a "good" result in the tests for maximal oxygen consumption (3.0 l/min and/or 36 ml/min/kg) (Lusa, 1994).

Demands on Muscular Performance

In the study of Lusa et al. (1991), the biomechanical features were evaluated in a simulated rescue–clearing task, in which a 9-kg power saw was lifted from the floor to ceiling level (211 cm) by seven young and six older fire fighters. The maximal muscular capacity of the subjects did not differ in regard to their maximal isokinetic muscle strength or to the results of the repetitive muscle performance tests.

In the task, the mean dynamic compression force at the L5/S1 disc was 6228 N, being equal for the young and older subjects. The peak torque for the back and knee extension was also similar for the subjects, and was, on average, 242 Nm and 120 Nm, respectively. The peak values corresponded to over 90% of their maximal isokinetic muscle strengths. The results showed that the lifting and handling of a heavy power saw produced a high load on the musculoskeletal system.

The tests on muscular performance in Table 14.1 are based on the study of Lusa et al. (1991) and the results of several pilot studies carried out at the Emergency Service Institute of Finland, Rescue Department of the City of Helsinki, Fire and Rescue Department of the City of Oulu, and the Fire and Rescue Department of the City of Jyväskylä (Louhevaara et al. unpublished results). The tests are mainly designed to follow up the fire fighters' physical fitness with special reference to their physical work capacity and muscular job demands. These tests have also been used since 1993 for the selection of applicants for the training courses of the Emergency Service Institute of Finland. The applicants should attain the minimum results determined by the Institute. In the actual selection of applicants for basic training courses, the required minimal level has been between "good" and "excellent" in each test presented in Table 14.1.

Conclusions

The fire fighters' job demands on physical fitness and work capacity are occasionally extremely high during their entire occupational career. Their cardiorespiratory and muscular fitness should be tested regularly in order to guarantee their work capacity for physically extreme operations. The use of fitness tests is necessary in planning and carrying out preventive measures for maintaining the health and work ability of fire fighters. The most effective and necessary measure is regular physical fitness training, which should be done either during working hours or in their leisure time.

14.5 Police Officers

Physical Fitness and Health

In operative police work, it is almost impossible to influence job demands or to change the work environment. Due to the strict selection criteria, a young police officer is healthy and physically fit when starting his or her career. However, in middle-age, the lifestyle of many police officers is sedentary and often associated with various health problems. These factors accelerate the unavoidable age-related reduction of physical fitness and, particularly after the age of 40 years, the physical fitness as well as work capacity of police officers often decline rapidly. Obesity, elevated blood pressure, high serum lipid levels, and musculoskeletal disorders and diseases are the most common health problems that also reduce the physical fitness needed to carry out job demands without excessive strain and increased hazards on health or safety (Soininen 1995).

Maintenance of Work Capacity

Soininen (1995) investigated in his intervention study the feasibility and effects of 8-month-long intensive and medium-intensive worksite fitness programs in 111 male police officers aged 40 to 54 years who were serving in three middle-sized cities. The intensive program (n = 48) included guided fitness training two hours per week during working hours and three training sessions per week during leisure time. The subjects (n = 24) in the medium intensive program participated in three training sessions per week during their leisure time. Both interventions were separately carried out in two cities, whereas the subjects (n = 39) were the controls in the third city. The effects of the fitness programs were evaluated in regard to health, physical fitness, and subjective work ability, and were assessed with a questionnaire and laboratory and field measurements. The intensive fitness program was more feasible and efficient than the medium-intensive program. In the intensive group, the cardiovascular risk factors fell significantly, and both the maximal oxygen consumption and the index of the muscular fitness and flexibility improved 8%, on average. The corresponding improvements were smaller in the medium-intensive group. No positive changes were observed in the control group.

Physical Job Demands

In the field part of his intervention study, Soininen (1995) confirmed the previous observations of Smolander et al. (1985) on the physical job demands associated with police work. The mean physical work level was low, and the police officers spent, on average, about 60% of their working hours sitting. The most common workplaces were an office, a patrol car, and an alarm center. However, the questionnaire revealed that almost all police officers had a few physically extreme peak loads in their work every year. The peak loads usually occurred when the police officer had to catch or transport criminals, or intoxicated or mentally ill persons. The situations leading to physical peak load were usually very sudden and unexpected. They usually demanded a high level of muscular strength and endurance (carrying, lifting, pressing, wrestling, running, wrenching), often in unfavorable (dark, cold, and/or cramped space) environmental conditions.

Adequate Level of Physical Fitness

In order to manage the physical peak load situations successfully, police officers need to have a sufficient level of physical fitness. Smolander et al. (1984) suggested that the necessary level of the police officers operating in the street should be at least the same as the typical clients, who were 20- to 40-year-old men in Finland. Smolander et al. (1985) recommended that the test battery for evaluating the physical fitness of police officers should consist of items on anthropometrics, aerobic power (assessment of maximal oxygen consumption), muscular performance (assessment of muscle strength and endurance with repetitive tests), and flexibility. Later Soininen (1995) concluded that the sufficient level of each physical fitness dimension (aerobic power, muscle performance, and motor coordination) should be higher than the mean sex- and age-related reference values obtained from the population studies. The assessment of physical fitness should be done with reliable and feasible tests that could be completed in the occupational health services.

Based on the studies of Smolander et al. (1984, 1985) and Soininen (1995), the Finnish Ministry of Internal Affairs is preparing a guide for testing the work ability of police officers. Based on preliminary information, the guide will include tests on physical fitness and professional competence. Physical fitness will be determined by an indirect assessment of maximal oxygen consumption (submaximal bicycle-ergometer test or 2-km walking test), the evaluation of the strength and endurance of the trunk and arm muscles (sit-up test and the dynamic endurance and strength test for the upper arms with 5 or 10 kg barbells), as well as the determination of the flexibility of the hip and low back (sit and reach test) (Laukkanen, 1993; Pollock and Wilmore, 1994).

Conclusions

The selection criteria for police training seems to guarantee an acceptable level of physical fitness of police officers up to the age of 40 years. The physical demands of the activities of the police job are not sufficient to maintain the necessary level of physical fitness. Therefore, it must be obtained through regular and effective fitness training either during leisure time or working hours. Furthermore, after the age of 35 years the police officers' physical fitness should be followed and tested regularly by the occupational health services.

14.6 Professional Cleaners

Common Service Occupation

A professional cleaner is one of the basic service occupations, including e.g., hospital maids, home care workers, and kitchen and laundry workers. The professional cleaner is one of the most common occupations in the world. Krueger et al. (1995) estimated that in the European Union the number of full-time or part-time cleaners is about three million, and the great majority of them (95%) are women.

In many industrialized countries, the average age of the workforce is rapidly increasing due to the aging of the large post-war age cohort. For instance, the present Finnish professional cleaners are mostly over 45 years of age, i.e., aging workers. They have numerous problems in their work ability, resulting in a high incidence of sickness and a high frequency of early retirement (Hopsu, 1993; Hopsu et al., 1994). Moreover, during the past few years the serious economic depression has hampered the availability of adequate detergents, cleaning tools, and machinery and, on the other hand, the spaces to be cleaned have become substantially larger. Therefore, it is essential to maintain the work ability and to promote the professional competence and work motivation of the current professional cleaners with all possible means (Hopsu 1993).

Physical Fitness in Cleaning Work

Hopsu et al. (1994) reported the results of their intervention study on professional cleaners aimed at supporting their individual physical fitness and work capacity with physical fitness exercise and reducing their physical work load with ergonomic redesign measures. The intervention process was carried out using a participatory approach in the work units. Its main characteristic was to motivate workers and work units to participate in teamwork for developing their own physical fitness and work performance with individual health promotion and ergonomic measures.

The intervention and control group consisted of 30 and 32 female cleaners, respectively. The intervention group took part in individual physical fitness programs and also developed ergonomics of their work. The mean age of the subjects was 46 years. The length of the intervention period was one year. After the intervention, the subjects' cardiorespiratory and muscular performance had increased markedly and the work methods and tools had improved, resulting in a reduction of poor work postures. Due to the increased individual physical fitness level and work capacity, and the developments in ergonomics, the subjects' strain at work decreased according to the results of the heart rate. The improvements also positively affected subjective work ability. The intervention did not lower productivity.

Maintenance of Ability to Work and Physical Fitness

According to their intervention study, Hopsu et al. (1994) concluded that the maintenance of work ability, professional competence, and health of cleaners always needs to include a combination of measures which must affect work and the environment (ergonomics), work organization (teamwork and leadership), and the capacities of the individual workers. In this combination, the maintenance of physical fitness is an essential element. Furthermore, Hopsu et al. (1994) recommended that female professional cleaners should have at least average cardiorespiratory and muscular fitness without excessive weight in order to avoid overstrain. The fitness of cleaners should be evaluated regularly at intervals of five years by applying simple fitness tests in the occupational health services. The suggested test battery consisted of tests on anthropometrics (weight, height, and body mass index), aerobic power (2-km walking test), and the strength and endurance of the main muscle groups of the trunk (sit-up test, static endurance of the back muscles), legs (squatting test), arms (strength and endurance of upper arms with 5-kg barbells), and the flexibility of the trunk (bending forward and sideways in a standing position) (Laukkanen, 1993; Pollock and Wilmore, 1994).

Conclusions

Professional cleaning is a physically heavy job for women and requires physical fitness in terms of cardiorespiratory capacity, muscular performance, and motor coordination. The physical fitness level of professional cleaners should be higher than the average age-related values for women and should be tested in a pre-employment health examination, as well as followed regularly in periodic health examinations of the occupational health services. The participatory process based on the triangle strategy of the maintenance of work ability (Figure 14.1) proved to be efficient and feasible for improving the physical fitness and work capacity of cleaners; physical exercises and consideration of the ergonomic features of cleaning work by a teamwork-based developmental process were central elements.

14.7 Physical Exercise and Fitness

Several reports indicate that a relevant measure affecting individual health, functional capacity, and work ability is regular, moderate, and versatile physical exercise (fitness training) or activity. Together with moderate nutrition, physical exercise is effective in reducing body weight and lowering blood pressure and cholesterol levels, and it prevents ischemic heart disease (e.g., Fletcher, 1992; Lakka, 1994; Blair et al., 1996). Physical exercise is essential for sustaining cardiorespiratory, musculoskeletal, and nervous fitness and is, to a great extent, linked with good work ability regardless of the type of work (WHO, 1993). Physical exercise also has a positive effect on productivity, quality of work, absenteeism, and the turnover rate (Shephard, 1992). Therefore, physical exercise can be regarded as one of the basic elements in the maintenance of work ability (Figure 14.1).

Work ability is supported by physical fitness, and a variety of physical exercise modes is available for maintaining or improving different cardiorespiratory, musculoskeletal, and neuromuscular capacities involved in physical fitness (Figure 14.3). The quality and quantity of physical exercise, when aiming to improve work ability, depends on real job demands on physical fitness (see Figure 14.2) because training effects are highly specific in nature. It is obvious that jobs which encompass a lot of heavy dynamic work, i.e., high aerobic demands on the cardiorespiratory system, will not be benefited by a physical exercise program consisting of, for instance, near maximal weight lifting with a low number of repetitions. Instead, such jobs need physical exercise like brisk walking or jogging, which improve the capacity of the cardiorespiratory system. Moreover, both aerobic and strength exercise should be supplemented by training elements that aim for the promotion of sufficient motor coordination. Therefore, it is essential that physical exercise, targeting improved work ability, must be tailored to actual job demands. On the other hand, in jobs having low physical activity and high psychosocial demands, the main benefit of physical exercise may include its effects on mood, well-being, and mental work capacity.

Before starting physical exercise it is sometimes necessary to perform assessments of physical job demands (Louhevaara 1995). The assessments may help the prescription of appropriate training programs. However, in many cases just a careful observation of work and a brief worker interview may give the information needed for planning an adequate training program. If some standardized methods are necessary, the following basic work physiology and ergonomic methods can be used for this purpose:

- Job load and hazard analysis (Mattila 1985)
- Measurement of heart rate (Louhevaara et al., 1990)
- Basic Edholm Scale for estimation of energy expenditure (Ilmarinen et al., 1979)
- Rating of overall perceived exertion (Borg, 1970)
- Rating of local perceived exertion for the back, arms, and legs (Borg, 1970, Hultman et al., 1984).

Often, physical job demands are affected by environmental conditions and work rate. Thermal load increases muscular demands at work if the temperature, humidity, and air velocity substantially deviate from temperate ones, or if wearing heavy personal protective clothing and equipment is necessary. An unusually slow or rapid work rate may severely bias the reliable assessments of physical job demands.

The effects of work-site physical exercise interventions on physical fitness, work ability, health, and different work-related characteristics have been studied intensively at the Finnish Institute of Occupational Health during the past 10 years. The studies have focused on six occupational groups, including professional cleaners, nurses, home care workers, metal workers, fire fighters, and police officers.

The interventions lasted for 2 to 12 months, and musculoskeletal (dynamic and static muscle performance and motor coordination) and cardiorespiratory capacity (maximal oxygen consumption) were found to improve, on average, by 7 to 136% and 4 to 10%, respectively. Positive effects were also observed in subjective work ability and health, musculoskeletal symptoms, need for physiotherapy, absenteeism, strain at work, risk factors for ischemic heart disease (body fat, blood pressure, cholesterol, smoking), tolerance of shift work, and the mastering of work. The only drawback of the exercises was musculoskeletal injuries, which were usually minor, however (Louhevaara and Ilmarinen, 1995).

The results and experiences obtained from the interventions emphasize the following prerequisites for feasible physical exercise programs at the company level:

- Commitment and support of top management
- Commitment of the whole work unit
- Implementation entirely or partly during working hours
- Quick feedback on improvements in physical fitness
- Providing and strengthening of motivation
- Meaningful, versatile, and positive experiences from the exercise
- Skillful instruction and guidance (Louhevaara, 1995).

It is also very important that exercise programs are strictly confidential, voluntary, available for evaluation, and do not arouse feelings of guilt for anyone (WHO, 1988). Furthermore, all exercise programs should be based on close integration with other measures related to the triangle concept of the maintenance of work ability (Figure 14.1) (Louhevaara and Ilmarinen, 1995).

14.8 General Conclusions

In many jobs, heavy physical work requiring continuous or repetitive moving and MMH activities will remain indispensable in spite of rapid technological developments. In these types of jobs and particularly in a few specific occupations (e.g., fire fighters and police officers) in which primary preventive ergonomic controls on job demands are often difficult or impossible to apply, the workers need to have an adequate level of physical fitness in terms of cardiorespiratory capacity, muscular performance, and motor coordination. This is necessary in order to avoid over-strain and prevent health and safety hazards, as well as to attain good productivity and quality of work.

The required level of physical fitness depends on job demands and should be tested with reliable and feasible tests. In physically demanding and hazardous occupations like a fire fighter and a police officer, fitness tests should be done in the pre-employment situation and also regularly during the entire occupational career.

The only efficient way to improve physical fitness is regular fitness training. However, it is essential that physical training, targeting improved work ability, must be tailored to actual job demands. Both aerobic and strength exercise should be supplemented by training elements that aim for the promotion of sufficient motor coordination. On the other hand, in jobs having low physical activity and high psychosocial demands, the main benefit of physical exercise may include its effects on mood, well-being, and mental work capacity. Together with other individual health promotion measures physical training is an essential part of the concept for maintaining work ability, professional competence, and health. The best results are attained if individual measures (physical exercise) are carried out simultaneously with other developmental measures on work and the environment (ergonomics), as well as on organizational features of work (leadership).

References

Ahonen, E., Venäläinen, J.M., Könönen, U., and Klen, T. 1990. The physical strain of dairy farming. *Ergonomics* 33:1549-1555.

Aminoff, T., Smolander, J., Korhonen, O., and Louhevaara, V. 1996. Physical work capacity in dynamic exercise with differing muscle masses in healthy and older men. *Eur. J. Appl. Physiol.* 73:180-185.

Åstrand, P. O. and Rodahl, K. 1986. *Textbook of work physiology,* 3rd ed. McGraw-Hill, New York, NY.

Blair, S.N., Horton, E., Leon, A.S., et al. 1996. Physical activity, nutrition, and chronic disease. *Med. Sci. Sports Exerc.* 28:335-349.

Borg, G. 1970. Perceived exertion as an indicator of somatic stress. *Scand. J. Rehab. Med.* 2:92-98.

Fletcher, G.F. 1993. Statement of exercise: benefits and recommendations for physical activity programs for all Americans. *Circulation* 86:340-344.

Guide for Smoke-Diving. 1991. Finnish Ministry of Internal Affairs. Helsinki.

Hollmann, W. and Hettinger, Th. 1980. *Sportmedizin — Arbeits- und Trainingsgrundlagen.* 1. Lag. Schattauer Verlag, Stuttgart.

Hopsu, L. 1993. *Working conditions in jobs where the majority of workers are women: Cleaning work.* OECD panel group on women, work and health: National Report. Ministry of Social Affairs and Health. Helsinki.

Hopsu, L., Louhevaara, V., Korhonen, O., and Miettinen, M. 1994. Ergonomic and developmental intervention in cleaning work, in *Proceedings of the 12th Triennial Congress of the International Ergonomics Association.* Vol. 6: General Issues, pp. 159-160. University Press, Toronto, Canada.

Hultman, G., Nordin, M., and Örtengren, R. 1984. The influence of preventive educational programs on trunk flexion in janitors. *Appl. Ergonomics* 15:127-133.

Ilmarinen, J. 1984. Physical work load on cardiovascular system in different work tasks. *Scand. J. Work Environ. Health* 10:403-408.

Ilmarinen, J., Huuhtanen, P., Louhevaara, V., and Näsman, O. 1995. A new concept for maintaining work ability during aging, in *From Research to Prevention,* eds. Rantanen, J., Lehtinen, S., Hernberg, S., et al., pp. 123-127. Finnish Institute of Occupational Health, Helsinki, Finland.

Ilmarinen, J., Knauth, P., Klimmer, F., and Rutenfranz, J. 1979. The applicability of the Edholm scale for activity studies in industry. *Ergonomics* 22:369-376.

Ilmarinen, J. and Louhevaara, V. 1994. Preserving the capacity to work. *Ageing International* 21:34-36.

Ilmarinen, R., Griefahn, B., Mäkinen, H., and Kunemund, C. 1993. Physiological responses to wearing a fire fighter's turnout suit with and without microporous membrane in the heat, in *Proceedings of the Sixth International Conference on Environmental Ergonomics,* eds. Frim, J., Ducharne, M.B., and Tikuisis, P., pp. 78-79. Government of Canada, Montebello, Canada.

Jonsson, B., Persson, J., and Kilbom, Å. 1988. Disorders of the cerviobrachial region among female workers in the electronics industry. A two-year follow up. *Int. J. Ind. Erg.* 3:1-12.

Krueger, D. 1995. *Prevention of health and safety risks in professional cleaning and the work environment.* European Commission, Biomed 2 programme. Brussels, Belgium.

Kähkönen, E. 1993. *Comparison and error analysis of instrumentation and methods for assessment of neutral and hot environments on the basis of ISO standards.* Doctoral dissertation. University of Kuopio, Kuopio, Finland.

Lakka, T. 1994. *Leisure time physical activity, cardiorespiratory fitness, biological coronary risk factors and coronary heart disease: a population study in men in eastern Finland.* Doctoral dissertation. University of Kuopio, Kuopio, Finland.

Laukkanen, R. 1993. *Development and evaluation of a 2-km walking test for assessing maximal aerobic power of adults in field conditions.* Doctoral dissertation. University of Kuopio, Kuopio, Finland.

Lemon, P.W. and Hermiston, R.T. 1977. Physiological profile of professional fire fighters. *J. Occup. Med.* 19:337-340.

Louhevaara, V. 1992. Cardiorespiratory and muscle strain during manual sorting of postal parcels: A review. *J. Occup. Med.* (Singapore) 4:9-17.

Louhevaara, V. 1995. Assessment of physical work load at worksites: A Finnish-German concept. *Int. J. Occup. Safety Ergonomics* 1:144-152.

Louhevaara, V. 1995. *Feasibility of physical exercise in early rehabilitation of work ability.* 5th European Congress on Research in Rehabilitation, p. 105, Helsinki, Finland.

Louhevaara, V., Hakola, T., and Ollila H. 1990. Physical strain and work involved in manual sorting of postal parcels. *Ergonomics* 33:1115-1130.

Louhevaara, V. and Ilmarinen, J. 1995. Physical exercise as a measure for maintaining work ability during aging, in *The Paths to Productive Aging,* ed. Kumashiro, M., pp. 289-293. Taylor & Francis. Basingstoke.

Louhevaara, V., Teräslinna, P., Piirilä, P., et al. 1988. Physiological responses during and after intermittent sorting of postal parcels. *Ergonomics* 31:1165-1175.

Louhevaara, V., Tuomi, T., Smolander, J., et al. 1985. Cardiorespiratory strain in jobs that require respiratory protection, *Int. Arch. Occup. Health* 55:195-206.

Lusa, S. 1993. *Job demands and assessment of the physical work capacity of fire fighters.* Doctoral dissertation. University of Jyväskylä, Jyväskylä, Finland.

Lusa, S., Louhevaara, V., Smolander, J., et al. 1991. Biomechanical evaluation of heavy tool-handling in two age groups of firemen. *Ergonomics* 34:1429-1432.

Lusa, S., Louhevaara, V., and Kinnunen, K. 1994. Are the job demands on physical work capacity equal for young and aging firefighters? *J. Occup. Med.* 36:70-74.

Lusa, S., Louhevaara V., Smolander, J., et al. 1993. Physiological responses of firefighting students during simulated smoke-diving in the heat. *Am. Ind. Hyg. Assoc. J.* 54:228-231.

Pollock, M.L. and Wilmore, J.H. 1990. Exercise in health and disease. Evaluation and prescription for prevention and rehabilitation. 1st ed. W.B. Saunders Company, Philadelphia, NY.

Rohmert, W. 1983. Formen menslicher Arbeit, in *Praktische Arbeitsphysiologie,* eds. Rohmert, W. and Rutenfranz, J., 1. Lag. Georg Thieme Verlag, Stuttgart.

Rutenfranz, J. 1981. Arbeitsmedizinische Aspekts des Stressproblems, in *Stress, Theorien, Untersuchungen, Massnahmen,* ed. Nitsch, J.R. 1. Lag. Verlag Hans Huber, Bonn.

Rutenfranz, J., Ilmarinen, J., Klimmer, F., and Kylian, H. 1990. Work load and demanded physical performance capacity under different industrial conditions, in *Fitness for Aged, Disabled, and Industrial Worker,* ed. Kanenko, M. pp. 217-235. International Series on Sport Sciences, Vol. 20. Human Kinetics Books, Champaign.

Shephard, R.J. 1992. A critical analysis of work-site fitness programs and their postulated economic benefits. *Med. Sci. Sports and Exerc.* 24:354-370.

Smolander, J., Louhevaara, V., and Oja, P. 1984. Policemen's physical fitness in relation to the frequency of leisure time physical exercise. *Int. Arch. Occup. Environ. Health* 54:261-270.

Smolander, J., Louhevaara, V., Nygård, C-H., et al. 1985. Job demands and assessment of physical working capacity in policemen's occupational health service, in *Proceedings of the Ninth Congress of the International Ergonomics Association,* eds. Brown, I.D., Goldsmith, R., Coombes, K., and Sinclair, M.A., pp. 613-615, Taylor & Francis, Bournemouth, England.

Smolander, J. and Louhevaara V. 1998. Muscular work, in Stellman J.M., et al., eds. *Encyclopaedia of Occupational Health and Safety.* 4th ed. International Labour Office, pp. 1:29.28–29.31, Geneva, Switzerland.

Soininen, H. 1995. *The feasibility of worksite fitness programs and their effects on the health, physical capacity and work ability of aging police officers.* Doctoral dissertation. University of Kuopio, Kuopio, Finland.

Sothmann, M.S., Saupe, K.W., Jasenof, D., et al. 1990. Advancing age and the cardiorespiratory stress of fire suppression: Determining a minimum standard for aerobic fitness. *Human Performance* 3:217-236.

de Zwart, B.C.H., Frings-Dresen, M.H.W., and Dijk, F.J.H. 1995. Physical workload and the ageing worker: a review of the literature. *Int. Arch. Occup. Environ. Health* 68:1-12.

Viikari-Juntura, E., Kurppa, K., Kuosma, E., et al. 1991. Prevalence of epicondytis and elbow pain in the meat-processing industry. *Scand. J. Work Environ. Health* 17:38-45.

White, M.K. and Hodus, T.K. 1987. Reduced work tolerance associated with wearing protective clothing and respirators. *Am. Ind. Hyg. Assoc. J.* 48:304-310.

WHO (World Health Organization). 1988. *Health promotion for working populations.* Report of a WHO Expert Committee. WHO Technical Report Series 765, pp. 1-52, Geneva, Switzerland.

WHO (World Health Organization). 1993. *Ageing and work capacity.* Report of a WHO Expert Committee. WHO Technical Report Series 835, pp. 1-49, Geneva, Switzerland.

15

Psychosocial Work Factors

Pascale Carayon
Ecole des Mines de Nancy
France

Soo-Yee Lim
NIOSH

15.1 Introduction

This chapter examines the concept of psychosocial work factors and its relationship to occupational ergonomics. First, we provide a brief historical perspective of the development of theories and models of work organization and psychosocial work factors. Definitions and examples are then presented. Several explanations are given for the importance of psychosocial work factors in occupational ergonomics. Finally, measurement issues and methods for controlling and managing psychosocial work factors are discussed.

The role of "psychosocial work factors" in influencing individual and organizational health can be traced back to the early days of work mechanization and specialization, and the emergence of the concept of division of labor. Taylor (1911) expanded the principle of division of labor by designing efficient work systems accounting for proper job design, providing the right tools, motivating the individuals, and sharing of responsibilities between management and labor, and sharing of profits. This is known as the era of scientific management in which scientific methods are used to objectively measure work with the aim of improving its efficiency. These scientific methods involved breaking the tasks into small components or units, thus making work requirements and performance evaluations easy to define and monitor. Under these methods, work is simplified and standardized, therefore having a great impact on job and work processes. An analysis of psychosocial work factors in a job in this system would reveal that skill variety is minimal, workers have no control of the work processes, and the job is highly repetitive and monotonous. Such work system design can still be found in numerous workplaces.

As the workforce became more educated, individuals became more aware of their working conditions and environment, and began to seek avenues for improving their quality of working life. This is when the human relations movement emerged (Mayo, 1945), which raised the issue of the potential influence of the work environment on an individual's motivation, productivity, and well-being. Individual needs and wants were emphasized (Maslow, 1970). Thus, job design theorists incorporated worker behavior and work factors in their theories. The two theories of job enlargement and job enrichment formed the basis for many job design theories thereafter. These theories conceptualize the role of worker behavior

0-8493-2641-9/99/$0.00+$.50

and perception of the work environment in influencing personal and organizational outcomes. Job enlargement theory emphasized giving a larger variety of tasks or activities to the worker. While this was an improvement from the era of scientific management, the additional tasks or activities could be of a similar skill level and content: workers were performing multiple tasks of the same "kind." This has been called "horizontal loading" of the job, and is the opposite of job enrichment, which focused on the "vertical loading" of the job. Job enrichment aims at expanding the skills used by workers, while at the same time increasing their responsibility. Herzberg (1966), the father of the job enrichment theory, defined intrinsic and extrinsic factors (or motivation versus hygiene factors) that are important to worker motivation, thus leading to satisfaction or dissatisfaction, and psychological well-being. Intrinsic factors are related to the work (or job) conditions, such as having additional control over work schedules or resources, feedback, client relationships, skill use and development, better work content, direct communications, and personal accountability (Herzberg, 1974). Extrinsic factors are related to aspects of financial rewards and benefits and also to the physical environment. Herzberg indicated that extrinsic factors could lead to dissatisfaction with work, but not to satisfaction, while intrinsic factors could increase satisfaction with work. Herzberg's work demonstrated the complex relationships of job conditions, the individual's motivation, satisfaction, dissatisfaction, and psychological well-being. In a way similar to Herzberg's job enrichment theory, the Job Characteristics Theory (Hackman and Oldham, 1976) focused on the idea that specific characteristics of the job (i.e., skill variety, task identity, task significance, autonomy, and feedback) in combination with individual characteristics (growth need strength) would determine personal and work outcomes.

The Sociotechnical Systems Theory recognized two inter-related systems in an organization: the social system and the technical system. The main principle of the Sociotechnical Systems Theory is that the social and technical systems interact with each other, and that the joint optimization of both systems can lead to increased satisfaction and performance. The social system focused on the workers' perception of the work environment (i.e., job design factors) and the technical system emphasized the technology and the work processes used in the work (for example, automation, paced systems, and monitoring systems). In a study of coal mining (Trist and Bamforth, 1951), it was demonstrated that the technical system could impact the social system. In this study where semi-autonomous work groups were set up, workers were given opportunities to make decisions related to their work, and experienced better interactions with workers in their group, as well as task significance and completeness (see also Trist, 1981). Work by Trist and his colleagues showed that technological factors could influence both organizational and job factors. However, it was Davis (Davis 1980) who provided a conceptual framework and a set of principles that formulated the Sociotechnical theory. His framework called for a flattened management structure that would promote participation, interaction between and across groups of workers, enriched jobs, and most important, meeting individual needs. The Sociotechnical Systems Theory laid down the groundwork for the current understanding of how psychosocial work factors can be related to ergonomic factors by examining the interplay between the social and technical systems in organizations. Other recent theories and models of psychosocial work factors will be discussed later.

This rapid overview of the development of job design theories in the 20th century demonstrates the increasing role of psychological, social, and organizational factors in the design of work.

15.2 Definitions

Within the last decade, the role of psychosocial work factors on worker health has gained much popularity. However, the term of "psychosocial work factors" has been used loosely to define and represent many factors that are a part of, attached to or associated with the individuals. Some would consider what has been traditionally termed socioeconomic factors such as income, education level, and demographic or individual factors (e.g., age and marital status) as part of the psychosocial factors (Hogstedt, Vingard et al. 1995; Ong, Jeyaratnam et al. 1995). In order to understand psychosocial factors in the workplace, one needs to take into account the ability of an individual to make a psychological connection to his or her job, thus formulating the relationship between the person and the job. For instance, the International

Labour Office (ILO, 1986) defines psychosocial work factors as "interactions between and among work environment, job content, organizational conditions and workers' capacities, needs, culture, personal extra-job considerations that may, through perceptions and experience, influence health, work performance, and job satisfaction." Thus, the underlying premise in defining psychosocial work factors is the inclusion of the behavioral and psychological components of job factors. In the rest of the chapter, we will use the definitions proposed by Hagberg and his colleagues (Hagberg et al., 1995) because they are most highly relevant for occupational ergonomics.

Work organization is defined as the way work is structured, distributed, processed, and supervised (Hagberg et al., 1995). It is an "objective" characteristic of the work environment, and depends on many factors, including management style, type of product or service, characteristics of the workforce, level and type of technology, and market conditions. Psychosocial work factors are "perceived" characteristics of the work environment that have an emotional connotation for workers and managers, and that can result in stress and strain (Hagberg et al., 1995). Examples of psychosocial work factors include overload, lack of control, social support, and job future ambiguity. Other examples are described in the following section.

The concept of psychosocial work factors raises the issue of objectivity–subjectivity. Objectivity has multiple meanings and levels in the literature. According to Kasl (1978), objective data is not supplied by the self-same respondent who is also describing his distress, strain, or discomfort. At another level, Kasl (1987) feels that "psychosocial factor perception" can be less subjective when the main source of information is the employee but that this self-reported exposure is devoid of evaluation and reaction. Similarly, Frese and Zapf (1988) conceptualize and operationalize "objective stressors" (i.e., work organization) as not being influenced by an individual's cognitive and emotional processing. Based on this, it is more appropriate to conceptualize a continuum of objectivity and subjectivity. Work organization can be placed at one extreme of the continuum (that is the objective nature of work) whereas psychosocial work factors have some degree of subjectivity (see definitions above).

Psychosocial work factors result from the interplay between the work organization and the individual. Given our definitions, psychosocial work factors have a *subjective*, perceptual dimension, which is related to the *objective* dimension of work organization. Different work organizations will 'produce' different psychosocial work factors. The work organization determines to a large extent the type and degree of psychosocial work factors experienced by workers. For instance, electronic performance monitoring, or the on-line, continuous computer recording of employee performance-related activities, is a type of work organization that has been related to a range of negative psychosocial work factors, including lack of control, high work pressure, and low social support (Smith et al. 1992). In a study of office workers, information on psychosocial work factors was related to objective information on job title (Sainfort, 1990). Therefore, psychosocial work factors are very much anchored in the objective work situation, and are related to the work organization.

15.3 Examples of Psychosocial Work Factors

Psychosocial work factors are multiple and various, and are produced by different, interacting aspects of work. The Balance Theory of Job Design (Smith and Carayon-Sainfort, 1989) proposed a conceptualization of the work system with five elements interacting to produce a "stress load." The five elements of the work system are: (1) the individual, (2) tasks, (3) technology and tools, (4) environment, and (5) organizational factors. The interplay and interactions between these different factors can produce various stressors on the individual which then produce a "stress load" which has both physical and psychological components. The stress load, if sustained over time and depending on the individual resources, can produce adverse effects, such as health problems and lack of performance. The models and theories of job design reviewed at the beginning of the chapter tended to emphasize a small set of psychosocial work factors. For instance, the human relations movement (Mayo, 1945) focused on the social aspects of work, whereas the job characteristics theory (Hackman and Oldham, 1976) lists five job characteristics, i.e., skill variety, task identity, task significance, autonomy, and feedback. However, research and practice in

TABLE 15.1 Selected Psychosocial Work Factors and their Facets

1.	Job demands	Quantitative workload Variance in workload Work pressure Cognitive demands
2.	Job content	Repetitiveness Challenge Utilization and development of skills
3.	Job control	Task/instrumental control Decision/organizational control Control over physical environment Resource control Control over work pace: machine-pacing
4.	Social interactions	Social support from supervisor and colleagues Supervisor complaint, praise, monitoring Dealing with (difficult) clients/customers
5.	Role factors	Role ambiguity Role conflict
6.	Job future and career issues	Job future ambiguity Fear of job loss
7.	Technology issues	Computer-related problems Electronic performance monitoring
8.	Organizational and management issues	Participation Management style

the field of work organization has demonstrated that considering only a small number of work factors can be misleading and inefficient in solving job design problems. The balance theory proposes a systematic, global approach to the diagnosis and design or redesign of work systems that does not emphasize any one aspect of work. According to the balance theory, psychosocial work factors are multiple and of diverse nature.

Table 15.1 lists eight categories of psychosocial work factors and specific facets in each category. This list cannot be considered as exhaustive, but is representative of the most often studied psychosocial work factors.

The study of psychosocial work factors needs to be tuned in to the changes in society. Changes in the economic, social, technological, legal, and physical environment can produce new psychosocial work factors. For instance, in the context of office automation, four emerging issues are appearing (Carayon and Lim, 1994): (1) electronic monitoring of worker performance, (2) computer-supported work groups, (3) links between the physical and psychosocial aspects of work in automated offices, and (4) technological changes. The issue of technological changes applies nowadays to a large segment of the work population. Employees are asked to learn new technologies on a frequent, sometimes continuous, basis. Other trends in work organization include the development of teamwork and other work arrangements, such as telecommuting. These new trends may produce new psychosocial work factors, such as high dependency on technology, lack of socialization on the job and identity with the organization, and pressures from teamwork. Two APA publications review psychosocial stress issues related to changes in the workforce in terms of gender, diversity, and family issues (Keita and Hurrell, 1994), and some of the emergent psychosocial risk factors and selected occupations at risk of psychosocial stress (Sauter and Murphy, 1995).

15.4 Occupational Ergonomics and Psychosocial Work Factors

The emergence of macroergonomics has strongly contributed to the increasing interest in psychosocial work factors in the occupational ergonomics field (Hendrick, 1991; Hendrick, 1996). As shown above, the work factors can be categorized into the individual, task, tools and technologies, physical environment, and the organization (Smith and Carayon-Sainfort, 1989). They can also be described as either physical or psychosocial (Cox and Ferguson, 1994). Cox and Ferguson (1994) developed a model of the effects of physical and psychosocial factors on health. According to this model, the effects of work factors on health are mediated by two pathways: (1) a direct physicochemical pathway, and (2) an indirect psychophysiological pathway. These pathways are present at the same time, and interact in different ways to affect health. Physical work factors can have direct effects on health via the physicochemical pathway, and indirect effects on health via the psychophysiological pathway, but can also moderate the effect of psychosocial work factors on health via the psychophysiological pathway. This model demonstrates the close relationship between physical and psychosocial work factors in their influence on health and well-being.

The importance of psychosocial work factors in the field of occupational ergonomics emerges from several considerations.

1. Physical and psychosocial ergonomics are interested in the same job factors.
2. Physical and psychosocial work factors are related to each other.
3. Psychosocial work factors play an important role in physical ergonomics interventions.
4. Physical and psychosocial work factors are related to the same outcome, for instance, work-related musculoskeletal disorders.

First, some of the concepts examined in the physical ergonomics literature are similar to concepts examined in the psychosocial ergonomics literature. For instance, the degree of repetitiveness of a task is very important from both physical and psychosocial points of view. Physical ergonomists are more interested in the effect of the task repetitiveness on motions and force exerted on certain body parts, such as hands; whereas psychosocial ergonomists are concerned about the effect of task repetitiveness on monotony, boredom, and dissatisfaction with one's work (Cox 1985). In the physical ergonomics literature, an important job redesign strategy for dealing with repetitiveness is job rotation: workers are rotated between tasks which require effort from different body parts and muscles, therefore reducing the negative effects of repetitiveness of motions in a single task. From a psychosocial point of view, job rotation is one form of job enlargement (see above for a discussion of job enlargement). However, as discussed earlier, the psychosocial benefits of job rotation are limited because workers may be simply performing a range of similar, nonchallenging tasks. From a physical ergonomics point of view, job rotation is effective only if the physical variety of the tasks is increased; whereas from a psychosocial ergonomics point of view, job rotation is effective only to the extent of the content and meaningfulness of the tasks.

Second, physical and psychosocial work factors can be related to each other. For instance, the model proposed by Lim (1994) states that the psychosocial factor of work pressure can influence the physical factors of force and speed of motions. According to this model, workers may change their behaviors under the influence of work pressure, and, therefore, tend to exert more force or to speed up their work. Empirical evidence tends to confirm this relationship between work pressure (i.e., a psychosocial work factor) and physical work factors (Lim 1994). Another form of relationship between physical and psychosocial work factors is evident in the literature on control over one's physical environment. In this case, the psychosocial work factor of control is applied to one particular facet of the work, that is the physical environment. Control over one's physical environment can, therefore, have benefits from a physical point of view (i.e., being able to adapt one's physical environment to one's physical characteristics and task requirements), but also from a psychosocial point of view (i.e., having control is known to have many psychosocial benefits [Sauter et al., 1989]).

Third, psychosocial work factors are a crucial component of physical ergonomics interventions. In particular, the concept of participatory ergonomics uses the benefits of one psychosocial work factor, that is participation, in the process of implementing physical ergonomics changes (Noro and Imada, 1991). From a psychosocial point of view, using participation is important to improve the process and outcomes of ergonomic interventions. In addition, any type of organizational interventions, including ergonomic interventions, can be stressful because of the emergence of negative psychosocial work factors, such as uncertainty and increased workload (i.e., having more work during the intervention or the transitory period). Therefore, in any physical ergonomics intervention, attention should be paid to psychosocial work factors in order to improve the effectiveness of the intervention and to reduce or minimize its negative effects on workers.

Fourth, physical and psychosocial work factors can be related to the same outcome. One of these outcomes is work-related musculoskeletal disorders (WMSDs). There is increasing theoretical and empirical evidence that both physical and psychosocial work factors play a role in the experience and development of WMSDs (Hagberg et al. 1995; Moon and Sauter, 1996). Several mechanisms for the joint influence of physical and psychosocial work factors on WMSDs have been presented (Smith and Carayon, 1996). Therefore, in order to fully prevent or reduce WMSDs, both physical and psychosocial work factors need to be considered.

15.5 Measurement of Psychosocial Work Factors

From the occupational ergonomics point of view, the purpose of examining psychosocial work factors is to investigate their influence on and role in worker health and well-being. Thus, psychosocial work factors can be considered as predictors (i.e., independent variables), while worker health and well-being serve as the dependent variables or outcomes. The measurement or assessment of well-being can be classified into two levels of measures in terms of "context-free" (that is, life in general or general satisfaction) and "context-specific" (for example, job-related well-being) (Warr, 1994). It is the latter level of measure, "context-specific" that is relevant to the assessment of psychosocial work factors in the workplace. Table 15.1 shows a selected sample of the many different dimensions of jobs (for example, job demands, control, social support) that have been studied extensively. Furthermore, each dimension is made up of different facets that define and operationalize that particular dimension. For example, as shown in Table 15.1, the dimension of job demands consists of various facets, such as quantitative workload, variance in workload, work pressure, and cognitive demands; the dimension of job content includes repetitiveness, challenge on the job, and utilization of skills. It should be noted that Table 15.1 is not an exhaustive list of psychosocial work factors.

The most often used method for measuring psychosocial work factors in applied settings is the questionnaire survey. Difficulties with questionnaire data on psychosocial work factors are often due to the lack of clarity of the definitions of the measured factors or poorly designed questionnaire items that measure "overlapping" conceptual dimensions of the psychosocial work factor of interest. Measures of any one facet typically include several items that can be grouped in a "scale." Reliability of the scale is often being assessed by the Cronbach-alpha score method in which the intercorrelations among the scale items are examined for internal consistency. In general, it is recommended that existing, well-established scales be used in order to ensure the "quality" of the data (i.e., reliability and validity) and to be able to compare the newly collected data with other groups for which data has been collected with the same instrument (benchmarking).

The level of objectivity/subjectivity of the measures of psychosocial work factors will depend on the degree of influence of cognitive and emotional processing. For example, ratings of work factors by an observer cannot be considered as purely objective because of the potential influence of the observer's cognitive and emotional processing. However, ratings of work factors by an outside observer can be considered as more objective than an evaluative question answered by an employee about his/her work environment (e.g., "How stressful is your work environment?"). However, self-reported measures of psychosocial work factors can be more objective when devoid of evaluation and reaction (Kasl and Cooper, 1987). As discussed earlier, any kind of data can be placed somewhere on this objectivity/subjectivity continuum from "low in dependency on cognitive and emotional processing" (e.g., objective) to "high in dependency on cognitive and emotional processing" (e.g., subjective).

We discuss three different questionnaires which include numerous scales of psychosocial work factors. In addition, validity and reliability analyses have been performed on all three questionnaires. Two of these questionnaires have been developed and used to measure psychosocial work factors in various groups of workers or large samples of workers: (1) the NIOSH Job Stress questionnaire (Hurrell and

McLaney 1988), and (2) the Job Content Questionnaire (JCQ) (Karasek, 1979). The NIOSH Job Stress questionnaire is often used in the Health Hazard Evaluations performed by NIOSH. Translations of Karasek's JCQ exist in many different languages, including Dutch and French. The University of Wisconsin Office Worker Survey (OWS) is a questionnaire developed to measure psychosocial work factors in office/computer work (Carayon, 1991). This questionnaire covers a wide range of psychosocial work factors of importance in office and computer work. In addition to many of the psychosocial work factors measured by the NIOSH Job Stress Questionnaire or Karasek's JCQ, the OWS measures psychosocial work factors related to computer technology, such as computer-related problems (Carayon-Sainfort, 1992). The OWS questionnaire has been translated into Finnish, Swedish, and German. For all three questionnaires, data exist for various groups of workers in numerous organizations of multiple countries. This data can serve as a comparison to newly collected data and for benchmarking. Numerous other questionnaires for measuring psychosocial work factors exist, such as the Occupational Stress Questionnaire in Finland (Elo et al., 1994) and the Occupational Stress Indicator in England (Cooper et al., 1988). Other questionnaires are listed in Cook et al. (1981).

15.6 Managing and Controlling Psychosocial Work Factors

It is clear from the job design and occupational stress literature that jobs with negative psychosocial work factors, such as repetitiveness, no opportunity to develop skills, and low control, can have adverse effects on job performance and mental and physical health. Various approaches have been proposed to improve the design of jobs, such as job rotation and other forms of job enlargement, and job enrichment (see above). These strategies can be efficient to increase the variety in a job, to reduce the dependence on a particular technology or tool, and to increase worker control and responsibility. In particular, lack of job control is seen as a critical psychosocial work factor (Sauter et al., 1989). Providing a greater amount of control can be achieved by, for instance, allowing workers to determine their work schedules in accordance with organizational policies and production requirements, by allowing workers to give input into decisions that affect their jobs, by letting workers choose the best work procedures and task order, and by increasing worker participation in the production process. An experimental field study of a participation program showed the positive effects of participation on emotional distress and turnover (Jackson, 1983). According to the Sociotechnical Systems theory, autonomous work groups can be an effective strategy for increasing worker control and enriching jobs. Beyond increased control and improved job content, some forms of teamwork can have other positive psychosocial benefits, such as increased opportunity for socialization and learning.

Achieving the perfect job without any negative psychosocial work factors may not be feasible or realistic, given individual, organizational, or technological constraints and requirements. The balance theory (Smith and Carayon-Sainfort, 1989) proposes a job redesign strategy that aims to achieve an optimal job design. In this process, negative psychosocial work factors need to be eliminated or reduced as much as possible. However, when this is not possible, positive psychosocial work factors can be used to reduce the impact of negative psychosocial work factors. This balancing, or compensating, effect is based on the concept of the work system of the balance theory. The five elements of the work system (the individual, tasks, technology and tools, environment, and organizational factors) are interrelated: they can influence each other, and they can also influence the impact or effect of each other or their interactions. In this systems approach, negative psychosocial work factors can be balanced out or compensated by positive work factors.

Some trends in the field of organizational design and management may have positive characteristics from a psychosocial point of view. For instance, under certain conditions, the use of quality engineering and management methods can positively affect the psychosocial work environment, such as increased opportunity for participation, and learning and development of quality-related skills (Smith et al., 1989). However, other trends in the business world can have negative effects on the psychosocial work environment. For instance, downsizing and other organizational restructuring and reengineering may create highly stressful situations of uncertainty and loss of control (DOL 1995).

15.7 Conclusion

This chapter has demonstrated the importance of psychosocial work factors in the research and practice of occupational ergonomics. In order to clarify the issue at hand, we presented definitions of work organization and psychosocial work factors. It is important to understand the long research tradition on psychosocial work factors that has produced numerous models and theories, but also valid and reliable methods for measuring psychosocial work factors. At the end of the chapter, we presented examples of methods for managing and controlling psychosocial work factors.

Psychosocial work factors need to be taken into account in the research on and practice of occupational ergonomics. We have discussed the important role of psychosocial work factors with regard to physical ergonomics. In addition, given the constantly changing world of work and organizations, we need to pay even more attention to the multiple aspects of people at work, including psychosocial work factors.

References

Carayon, P. (1991). *The Office Worker Survey.* Madison, WI, Department of Industrial Engineering, University of Wisconsin-Madison.

Carayon, P. and Lim, S.-Y. (1994). Stress in automated offices. *The Encyclopedia of Library and Information Science.* A. Kent. New York, Marcel Dekker. Vol. 53, Supplement 16: 314-354.

Carayon-Sainfort, P. (1992). The use of computers in offices: impact on task characteristics and worker stress. *International Journal of Human Computer Interaction* 4(3): 245-261.

Cook, J. D., Hepworth, S. J. et al. (1981). *The Experience of Work.* London, Academic Press.

Cooper, C. L., Sloan, S. J. et al. (1988). *Occupational Stress Indicator.* Windsor, England, NFER-Nelson.

Cox, T. (1985). Repetitive work: Occupational stress and health. *Job Stress and Blue-Collar Work.* C. L. Cooper and M. J. Smith. New York, John Wiley & Sons: 85-112.

Cox, T. and Ferguson, E. (1994). Measurement of the subjective work environment. *Work and Stress* 8(2): 98-109.

Davis, L. E. (1980). Individuals and the organization. *California Management Review* 22(2): 5-14.

DOL (1995). *Guide to responsible restructuring.* Washington, D.C. 20210, U.S. Department of Labor, Office of the American Workplace.

Elo, A.-L., Leppanen, A., et al. (1994). The Occupational Stress Questionnaire. *Occupational Medicine.* C. Zenz, O. B. Dickerson and E. P. Horvarth. St. Louis, Mosby: 1234-1237.

Frese, M. and Zapf, D. (1988). Methodological issues in the study of work stress. *Causes, Coping and Consequences of Stress at Work.* C. L. Cooper and R. Payne. Chichester, John Wiley & Sons.

Hackman, J. R. and Oldham, G. R. (1976). Motivation through the design of work: test of a theory. *Organizational Behavior and Human Performance* 16: 250-279.

Hagberg, M., Silverstein, B., et al. (1995). *Work-Related Musculoskeletal Disorders (WMSDs): A Reference Book for Prevention.* London, Taylor & Francis.

Hendrick, H. W. (1991). Human Factors in organizational design and management. *Ergonomics* 34: 743-756.

Hendrick, H. W. (1996). Human factors in ODAM: an historical perspective. *Human Factors in Organizational Design and Management -V.* O. J. Brown and H. W. Hendrick. Amsterdam, The Netherlands, Elsevier Science Publishers: 429-434.

Herzberg, F. (1966). *Work and the Nature of Man.* New York, Thomas Y. Crowell Company.

Herzberg, F. (1974). The wise old turk. *Harvard Business Review*(September/October): 70-80.

Hogstedt, C., E. Vingard, et al. (1995). *The Norrtalje-MUSIC Study — An ongoing epidemiological study on risk and health factors for low back and neck-shoulder disorders.* PREMUS'95-Second International Scientific Conference and Prevention of Work-Related Musculoskeletal Disorders, Montreal, Canada.

Hurrell, J. J. J. and M. A. McLaney (1988). "Exposure to job stress — A new psychometric instrument." *Scandinavian Journal of Work Environment and Health* 14(suppl.1): 27-28.

ILO (1986). *Psychosocial Factors at Work: Recognition and Control.* Geneva, Switzerland, International Labour Office.

Jackson, W. E. (1983). Participation in decision-making as a strategy for reducing job-related strain. *Journal of Applied Psychology* 68: 3-19.

Karasek, R. A. (1979). Job demands, job decision latitude, and mental strain: implications for job redesign. *Administrative Science Quarterly* 24: 285-308.

Kasl, S. V. (1987). Methodologies in stress and health: past difficulties, present dilemmas, future directions. *Stress and Health: Issues in Research and Methodology.* S. V. Kasl and C. L. Cooper. Chichester, John Wiley & Sons: 307-318.

Kasl, S. V. and C. L. Cooper, Eds. (1987). *Stress and Health: Issues in Research and Methodology.* Chichester, John Wiley & Sons.

Keita, G. P. and Hurrell, J. J. (1994). *Job Stress in a Changing Workforce — Investigating Gender, Diversity, and Family Issues.* Washington, D.C., APA.

Lim, S. (1994). An integrated approach to cumulative trauma disorders in computerized offices: the role of psychosocial work factors, psychological stress and ergonomic risk factors. *IE.* Madison, WI, University of Wisconsin-Madison.

Maslow, A. H. (1970). *Motivation and Personality.* New York, Harper and Row.

Mayo, E. (1945). *The Social Problems of an Industrial Civilization.* Andover, MA, The Andover Press.

Moon, S. D. and Sauter, S. L. Eds. (1996). *Beyond Biomechanics — Psychosocial Aspects of Musculoskeletal Disorders in Office Work.* London, Taylor & Francis.

Noro, K. and Imada, A. (1991). *Participatory Ergonomics.* London, Taylor & Francis.

Ong, C. N., Jeyaratnam, J., et al. (1995). "Musculoskeletal disorders among operators of video display terminals." *Scandinavian Journal of Work Environment and Health* 21(1): 60-64.

Sainfort, P. C. (1990). *Perceptions of Work Environment and Psychological Strain Across Categories of Office Jobs.* The Human Factors Society 34th Annual Meeting.

Sauter, S. L., Hurrell, J. J. Jr., et al., Eds. (1989). *Job Control and Worker Health.* Chichester, John Wiley & Sons.

Sauter, S. L. and Murphy, L. R. (1995). *Organizational Risk Factors for Job Stress.* Washington, D.C., APA.

Smith, M. J. and Carayon, P. (1996). Work organization, stress, and cumulative trauma disorders. *Beyond Biomechanics — Psychosocial Aspects of Musculoskeletal Disorders in Office Work.* S. D. Moon and S. L. Sauter. London, Taylor & Francis: 23-41.

Smith, M. J., Carayon, P., et al. (1992). Employee stress and health complaints in jobs with and without electronic performance monitoring. *Applied Ergonomics* 23(1): 17-27.

Smith, M. J. and Carayon-Sainfort, P. (1989). A balance theory of job design for stress reduction. *International Journal of Industrial Ergonomics* 4: 67-79.

Smith, M. J., Sainfort, F., et al., Eds. (1989). *Efforts to Solve Quality Problems,* Secretary's Commission on Workforce Quality and Labor Market Efficiency, U.S. Department of Labor, Washington, D.C.

Taylor, F. (1911). *The Principles of Scientific Management.* New York, Norton and Company.

Trist, E. (1981). *The Evaluation of Sociotechnical Systems.* Toronto, Quality of Working Life Center.

Trist, E. L. and Bamforth, K. (1951). Some social and psychological consequences of the long-wall method of coal getting. *Human Relations* 4: 3-39.

16

Cognitive Factors

Philip J. Smith
The Ohio State University

The field of ergonomics can be divided into two broad categories: physical ergonomics and cognitive ergonomics (Kroemer et al., 1994; Van Der Verr, Bagnara, and Kempen, 1992). Given this division, the list of cognitive factors goes well beyond purely cognitive functions to encompass all mental activity. Thus, this list includes:

1. Psychomotor skills
2. Sensory and perceptual skills
3. Affective responses and motivation
4. Attention
5. Learning and memory
6. Language and communication
7. Problem solving and decision making
8. Group dynamics and teamwork

In terms of our understanding of these basic mental functions, cognitive ergonomics draws heavily upon the fields of psychology and linguistics (Anderson, 1993; Fleishman and Quaintance, 1989; Proctor and Dutta, 1995; Proctor and VanZandt, 1994; Rasmussen, 1988; Sheridan and Ferrell, 1974).

The cognitive issues within industrial ergonomics go beyond modeling these mental processes, however. The goal is to understand how the work environment interacts with the strengths and limitations of the worker to determine performance. For example, in considering the design of a set of controls and displays, it is critical to consider not only the characteristics of the worker, but also the impact of such factors as:

1. Intended functions of the controls and displays (supported tasks)
2. Relative and absolute positions of the controls and displays
3. The broader task environment
4. Consistency within the system
5. Consistency with other systems (population stereotypes or expectancies)

These task and environmental factors interact with characteristics of the operator, ranging from visual acuity and contrast sensitivity to the determinants of attention and the problem-solving strategies used to direct the access to and interpretation of information provided by displays.

0-8493-2641-9/99/$0.00+$.50

It is, therefore, the interaction of human abilities with task and environmental factors that characterizes ergonomics. Based on this emphasis on the performance of the worker in some context, important design concepts and principles have been developed. For example, in the design of displays and controls, basic psychological findings such as the Gestalt "Laws of Good Form" (Gibson, 1986; Rock, 1995) have been incorporated in recommendations for achieving effective designs, such as the use of proximity or color coding to indicate functional groupings of displays and controls.

16.1 Example 1

To illustrate the role of cognitive factors in ergonomic evaluations, a series of examples are presented below. As a first simple example, consider the cognitive ergonomic concerns associated with designing a visual warning sign.

When designing a warning sign, one critical factor is its salience. The underlying cognitive factor is the selectiveness of human attention (which limits the ability of people to simultaneously attend to many different sources of information in the environment). Such *selective attention* is important in two senses. First, it is critical that the worker's attention be drawn to the warning sign. Second, it is equally important that attention be focused on the important contents of the warning sign.

A second cognitive factor (in the broad sense of the term) concerns legibility. Considerations like character size, stroke width and font, viewing distance, illumination and contrast are important, as well as characteristics of the worker, such as *visual acuity.*

A third factor is intelligibility: Will the worker correctly interpret the contents of the warning? And a fourth major factor is motivation: Even if the individual's attention is drawn to important contents of the warning and it is correctly read and interpreted, will behavior change? (Since the goal of ergonomics is to provide practical guidance, consideration of this last factor has lead to guidelines that recommend that, to avoid *design-induced error,* warning signs should be treated as a last resort and that, wherever feasible, designers should engineer out a predicted hazard rather than relying on a warning sign to prevent hazardous behavior.) Furthermore, such guidelines emphasize a major theme in cognitive ergonomics: **Designers need to play "psychologist" as part of the design process, predicting possible behaviors by prospective users of a system.**

Example 1, then, describes a range of cognitive factors that need to be considered in the design of a relatively simple artifact. Examples 2 through 4 further demonstrate the approach of cognitive ergonomics to design in more complex settings.

16.2 Example 2

Example 1 illustrated a fundamental concept in cognitive ergonomics: **Design is a prediction task.**

Application of this concept is further demonstrated in the design of documentation and help for computer software. Here, basic research has provided a model for how (many) software users approach documentation and help (Doheny-Farina, 1988; Kearsley, 1988).

According to this model, such assistance tends to be accessed when the user encounters difficulty, rather than through reading the software documentation and help prior to starting to use a new software system. This means the user accesses the help with a particular task or goal in mind. Then the user scans salient landmarks in the text (such as the bold subheadings on a help screen) looking for something that looks relevant. When the label for some section is judged relevant, the user then processes that localized portion of the text, using perceptual cues (such as spacing) to decide what constitutes that "chunk" of text, and reads only enough to develop a mental model of how to accomplish the desired task or goal (Janosky, Smith, and Hildreth, 1986).

From a designer's perspective, this qualitative model provides structure in predicting user performance:

1. Predict the alternative goals that users might have when accessing a given section of the documentation or help
2. Structure the text so easily identified headings can be scanned, using labels that will be judged as relevant to the appropriate user goals
3. Display the message so that all of its components appear to be visually related to the heading
4. Structure the contents of the text and graphics so that the reader is likely to process all of the critical contents. (For instance, a fact or instruction that is critical to forming a correct model of performance should not be placed last in a chunk of text, as it is less likely to be read in that position.)

Example 2, then illustrates how qualitative task-specific models can be developed to expand upon basic cognitive functions like perception and the focus of attention, supporting completion of detailed cognitive task analyses. It further illustrates how such models are used to identify specific steps or procedures to aid designers.

16.3 Example 3

The key importance of taking a broad systems view is even clearer when an industrial task such as visual inspection is considered (Drury and Prabhu, 1994). As an example, companies manufacturing glassware (such as glass bottles) traditionally have had workers view the bottles as they moved along a conveyor, trying to watch for and reject defective items. Defects can range from stones and bubbles in the glass to a crack in the lip of a jar.

The range of relevant factors includes:

1. Sensory and perceptual demands and skills (task characteristics such as *illuminance, contrast, effective size or visual angle, rate of movement,* and *time available for viewing,* as well as worker characteristics such as *acuity* and *contrast sensitivity*)
2. Attentional demands (particularly the impact of *vigilance decrements* over time on the task)
3. Memory and information processing rates (recalling and checking for different defects)
4. Decision making (deciding — given a very limited viewing time — whether or not to call a bottle defective)
5. Motivation

Physical and environmental factors are relevant as well, as discomfort can affect attentiveness.

In addition, the influence of a poorly designed inspection workstation can have an important impact not only on the inspector's attitude and motivation, but also on co-workers performing other types of jobs, as a result of group norms propagating throughout the plant. Thus, the designer of such a job needs to consider not only the direct impact of factors such as lighting and line speed, but also the impact of poor job design on worker motivation. In short, another important ergonomic perspective is that: **Design is a communication task.**

The message communicated by a poorly designed job is that the company isn't too concerned with the quality of the worker's performance. Such a message can induce low motivation on that job, which may then spread to workers elsewhere in the plant.

16.4 Example 4

Similarly, rich issues arise in considering the cognitive factors involved in designing decision support tools and systems to support training. As an illustration, consider the design of the Antibody IDentification Assistant (AIDA). This is a critiquing system (Fischer, 1991; Miller, 1986; Silverman, 1992) developed to assist with problem solving in blood banks and with embedded training both in the lab and as part of a college course on transfusion medicine.

AIDA as a Decision-Support System

AIDA was designed to provide assistance to blood bankers in performing a complex *abductive reasoning* task (Josephson and Josephson, 1994), the identification of antibodies in a patient's blood. This is an important step in determining compatible blood for a transfusion. As an abduction task, antibody identification exhibits a number of complexities, including a combinatorial explosion in the number of possible combinations of solutions (combinations of different antibodies), the potential for masking when multiple antibodies are present, and noise in the data.

Cognitive ergonomics plays a role in a number of steps when attempting to improve performance on such a task. These steps are illustrated below.

Problem Identification. One application where an understanding of cognitive ergonomics is useful is in identifying the areas where performance needs to be improved. In particular, review of the literature on the causes of human error provides a sizable list of generic cognitive processes that characterize human performance of abduction or diagnosis tasks, and that produce errors in certain identifiable task environments (Bell, Raiffa, and Tversky, 1988; Chi, Glaser, and Farr, 1988; Fraser, Smith, and Smith, 1994; Hollnagel, 1993; Reason, 1990; Tversky and Kahneman, 1974). (For example, the use of a positive test strategy will produce a confirmation bias in certain situations [Klayman and Ha, 1987; Mynatt et al., 1978], leading to hypothesis fixation.) Such knowledge can be used to more efficiently and exhaustively identify important errors in a particular application by providing a top-down approach to problem identification.

In the case of antibody identification, empirical studies have demonstrated that a wide range of such predictable cognitive errors are made, including:

1. Slips
2. Perceptual distortions
3. Hypothesis fixation
4. Ignoring base rates
5. Biased assimilation

Some of these errors result from the use of simplifying (but fallible) domain-specific heuristics to reduce the cognitive load of the problem-solving task, but many are due to fundamental cognitive processes.

Cognitive Modeling. In system design, cognitive ergonomics not only helps provide a top-down approach to identifying problems or sources of errors, it also provides guidance in modeling the cognitive processes and strategies involved in expert performance (Backland and McDermott, 1994; Clancey, 1986; Hoffman, 1987; Kolodner, 1993; Kuipers, 1994; Miller, 1984; Newell, 1990; Shortliffe, 1990; Wenger, 1987; Wyatt, Spiegel and Halter, 1992). A great deal is known about expert/novice differences in problem solving on tasks like abductive reasoning or diagnosis (Chi, Glaser, and Farr, 1988). This knowledge makes it possible to more efficiently and effectively model performance in a particular domain. For example, in performing antibody identification, experts use heuristic methods in order to keep the cognitive load manageable. However, since these heuristics are fallible, these experts also employ a more global strategy to reduce error. This metastrategy involves always collecting converging evidence using two or more independent strategies and data sources (Guerlain et al., 1994; Obradovich et al., 1996; Smith, Galdes, et al., 1991; Smith, Miller, et al., 1992).

Design. Based on an understanding of needs of the user population and current problem-solving methods, cognitive ergonomics plays a role in answering a number of questions. First, there is the question of: **What general approach should be used to improve performance?**

This could range from training and education (Bailey, 1993; Fleming and Levie, 1993), to the design of an information system, to the design of an active decision support system, to full automation (eliminating the person from the job).

In the case of AIDA, the decision was to develop a system to provide both training and decision support. This decision was based on two ergonomic considerations. First, many of the errors observed in empirical studies of practitioners in the current environment were the result of slips (Norman, 1981).

While some of those might be reduced by better design of data displays, it is likely that a significant number would remain because of the nature of the task. This argues for use of the computer to avoid or catch such errors. Second, many other errors were the result of ignorance or misconceptions. This similarly supports the need for some type of intervention, such as a combination of training and active decision support. The third consideration, however, concerns the fallibility of the designer rather than the user. Evidence to date suggests that we do not know how to fully capture the expertise of people on such complex tasks, with the result that decision support systems (whether based on expert systems technology or optimization techniques) exhibit *brittleness* (Guerlain et al., 1994; Layton, Smith, and McCoy, 1994; Roth, Bennet, and Woods, 1987; Smith, McCoy, and Layton, 1997). This suggests the need to keep a well-trained person involved in the task.

This raises two additional questions if computers are to be used for decision support: **What technology should be used? What role should the computer play?**

Regarding the first question, human factors studies suggest that it is important that the user be able to develop an accurate mental model (Gentner and Stevens, 1983) of the functioning of the system, so that the user can work as an effective "partner" (Lehnert and Zirk, 1987). Such *cognitive compatibility* is important if the user is to compensate for the limitations of the designer (and the resultant brittleness of the technology). In the case of AIDA, based on such considerations, an expert systems approach was taken, modeling the computer's reasoning after the problem-solving strategies exhibited by human experts.

Regarding the second question, there are numerous variations, ranging from designing a system where the computer critiques the human user (which was the choice made for AIDA), to a system where the human critiques the computer's answer, to a system where the computer automatically completes subtasks at the request of the user. The best solution clearly depends upon the costs of different types of errors and the level of confidence that can be placed on the technology. Ergonomic studies clearly indicate, though, that systems based on either expert systems technology or optimization can induce errors in their users' reasoning when the user is asked to critique the computer's reasoning rather than vice versa. Such studies also show that the use of the computer as a critic can avoid such design-induced errors (as an example of such a *critiquing system* [Miller, 1986; Silverman, 1991], AIDA reduces errors by 33 to 62% on cases where it is fully competent and even reduces errors by 30% on cases where it is less than fully competent, because it prods the user to make more effective use of her own expertise [Obradovich, et al., 1996].) In short, consideration of the psychology of the user and of the designer, as well as the nature of the particular application, is necessary to make decisions about the appropriate roles for the user and for technology.

Ergonomics also has contributions to make at a more detailed level in developing technological support, specifically the design of the human–computer interface (Baecker, et al., 1995; Carroll, 1995; Gardiner and Christie, 1987; Norman, 1990; Treu, 1994; Tufte, 1990; Tufte, 1993). In the case of AIDA, this involved decisions about how to provide access to and display data more effectively, how to provide memory aids to reduce the memory load for the user, and about how to design an interface where the computer could unobtrusively monitor the user's performance and make meaningful, context sensitive inferences about when and where the user likely has made an error.

Evaluation. Finally, cognitive ergonomics has important contributions to make in system evaluation and usability testing. This includes the use of analytic methods, such as heuristic evaluation, as well as the design and analysis of empirical studies. For the latter approach, this includes guiding decisions about what tasks to use in testing, and the selection of what data to collect (such as concurrent verbal reports [Ericsson and Simon, 1993]).

AIDA as a Decision-Support System — Summary. In short, cognitive ergonomics provides a principled approach to the design of technological systems such as AIDA. It provides direction through the use of cognitive models of human problem solving and human error, as well as through the application of general design concepts and principles. It also prescribes the use of empirical methods in recognition of the fallibility of designers. In short, the role of cognitive ergonomics is to ensure that certain broad questions (such as the appropriate role for technology) are adequately addressed, as well as to provide guidance at a detailed level in the implementation of a design concept.

16.5 Summary

As illustrated above, one of the hallmarks of cognitive ergonomics is an emphasis on a broad systems approach. In evaluating cognitive factors, it is not enough to consider the strengths and limitations of the worker. The task and work environment must be considered as well as the broader organizational context. Furthermore, even when the focus of an analysis is on some cognitive task, physical ergonomics issues must be considered because the health and physical comfort of the worker can have a major impact on cognitive functions. Thus, cognitive ergonomics supports design by providing a link between our knowledge of basic psychological processes and their interaction with the tasks and tools within a work setting. The goal of this chapter has been to indicate the range of cognitive factors that need to be considered in designing different types of systems, and to illustrate how an understanding of these factors influences design. More details are found on specific topics in the following sections of this book:

Chapter 8	Design of Information Devices
Chapter 34	How Complex Human–Machine Systems Fail: Putting Human Error in Context
Chapter 35	Human and System Reliability Analysis
Chapter 36	Some Developments in Human Reliability Assessment
Chapter 38	Managing the Speed–Accuracy Trade-off
Chapter 40	Design of Industrial Warnings
Chapter 99	Human-Computer Interface Requirements
Chapter 100	Psychosocial Factors and Musculoskeletal Disorders in Computer Work
Chapter 111	Ergonomic Issues in Air Traffic Management

Defining Terms

Abductive reasoning: Reasoning or problem solving to determine the "best" explanation for a set of data.

Brittleness: A characteristic of computerized problem-solving systems (based on either expert systems technology or optimization techniques) where they provide inadequate or incorrect advice because they have incomplete or incorrect knowledge or an incomplete model of the domain.

Cognitive compatibility: In reference to computer systems, this term deals with the ability of the user to understand and effectively interact with the computer.

Contrast: A measure of the relative luminances of some visual target and its background.

Critiquing system: A problem-solving system (typically based on expert systems technologies) that plays the role of a critic, monitoring the performance of a person for potential errors and providing feedback when a potential problem is detected.

Design-induced errors: Errors made by a person using some product or system that could have been anticipated or predicted by the designer and could have been feasibly engineered out so that the resultant hazard no longer existed.

Illuminance: A measure of the amount of light falling on a surface.

Selective attention: A limitation in the ability of a person to attend to a large number of sources of information (such as multiple data displays) simultaneously.

Vigilance decrement: A decrease in performance over time on tasks such as inspection of a product for defects or monitoring of a sonar screen.

Visual acuity: A measure of the ability of a person to discriminate visual stimuli (such as characters in text strings).

Visual angle: A measure of the effective "size" of some visual target (such as a character in a text string). This effective size is a function of both the actual physical size of components of the target (such as the stroke width of the lines forming a character) and the distance of the viewer from that target.

References

Anderson, J.R. 1993. *Rules of the Mind.* Lawrence Erlbaum Associates: Hillsdale, NJ.

Bachant, J. and McDermott, J. 1994. R1 revisited: Four years in the trenches. *The AI Magazine,* 21-32.

Baecker, R.M., Grudin, J., Buxton, W.A., and Greenberg, S. 1995. *Readings in Human–Computer Interaction: Toward the Year 2000.* Morgan Kaufman: San Francisco, CA.

Bailey, G. D. (ed.). 1993. *Computer-Based Integrated Learning Systems.* Educational Technology: Englewood Cliffs, NJ.

Bell, D., Raiffa, H., and Tversky, A. 1988. Descriptive normative, and prescriptive interactions in decision making, in *Decision Making: Descriptive, Normative, and Prescriptive Interactions,* ed. D. Bell, H. Raiffa, and A. Tversky, pp. 9-30. Cambridge University Press: New York.

Carroll, J. (ed.) 1995. *Scenario-Based Design.* John Wiley & Sons, Inc.: New York.

Chi, M. T., Glaser, R., and Farr, M.J. (eds.). 1988. *The Nature of Expertise.* Lawrence Erlbaum: Hillsdale, NJ.

Clancey, W.J. 1986. From GUIDON to NEOMYCIN to HERACLES in twenty short lessons: ORN Final Report 1979-1985. *The AI Magazine,* 40-60.

Doheny-Farina, S., ed. 1988. *Effective Documentation: What We Have Learned from Research.* MIT Press: Cambridge, MA.

Drury, C. and Prabhu, P. 1994. Human factors in testing and inspection, in *Design of Work and Development of Personnel in Advanced Manufacturing,* ed. Salvendy, G. and Karwowski, pp. 331-354. W. Wiley: New York.

Ericsson, K.A. and Simon, H. 1993. *Protocol Analysis: Verbal Reports as Data.* MIT Press: Cambridge, MA.

Fischer, G., Lemke, A.C., Mastaglio, T., and Morch, A.I. (1991). The role of critiquing in cooperative problem solving. *ACM Transactions on Information Systems,* 9(3), 123- 151.

Fleishman, E. and Quaintance, M. (1984). *Taxonomies of Human Performance.* Academic Press: Orlando, FL.

Fleming, M. and Levie, H.H. (eds.). 1993. *Instructional Message Design: Principles from the Behavioral and Cognitive Sciences.* Educational Technology: Englewood Cliffs, NJ.

Gardiner, M. and Christie, B. (eds.). 1987. *Applying Cognitive Psychology to User Interface Design.* John Wiley & Sons: New York.

Gentner, D. and Stevens, A.C. (eds.). 1983. *Mental Models.* Lawrence Erlbaum: Hillsdale, NJ.

Gibson, J.J. 1986. *The Ecological Approach to Perception.* Lawrence Erlbaum: Hillsdale, NJ.

Guerlain, S., Smith, P.J., Gross, S.M., Miller, T.E., Smith, J.W., Svirbely, J.R., Rudmann, S., and Strohm, P. 1994. Critiquing versus partial automation: How the role of the computer affects human-computer cooperative problem solving, in *Human Performance in Automated Systems: Current Research and Trends,* M. Mouloua and R. Parasuraman, pp. 73-80. Lawrence Erlbaum Associates: Hillsdale, NJ.

Hoffman, R. 1987. The problem of extracting the knowledge of experts from experts from the perspective of experimental psychology. *The AI Magazine,* 8(2), 53-67.

Hollnagel, E. 1993. *Human Reliability Analysis Context and Control.* Academic Press: New York.

Janosky, B., Smith, P.J., and Hildreth, C. 1986. Online library catalog systems: An analysis of user errors. *Internatl. J. of Man-Machine Studies,* 25, 573-592.

Josephson, J. and Josephson, S. 1994. *Abductive Inference: Computation, Philosophy. Technology,* Cambridge University Press: New York.

Kearsley, G. 1988. *Online Help Systems: Design and Implementation.* Ablex: Norwood NJ.

Klayman, J. and Ha, Y.W. 1987. Confirmation, disconfirmation and information in hypothesis testing. *Psychological Review.* 94: 211-228.

Kolodner, J. 1993. *Case-Based Reasoning.* Morgan Kaufman: San Mateo, CA.

Kroemer, K., Kroemer, H., and Kroemer-Elbert, K. 1994. *Ergonomics: How to Design for Ease and Efficiency.* Prentice Hall: Englewood Cliffs, NJ.

Kuipers, B. 1994. *Qualitative Reasoning: Modeling and Simulation with Incomplete Knowledge.* MIT Press: Cambridge, MA.

Layton, C., Smith, P.J., and McCoy, E. 1994. Design of a cooperative problem-solving system for en-route flight planning: An empirical evaluation. *Human Factors,* 36, 94-119.

Lehner, P.E. and Zirk, D.A. 1987. Cognitive factors in user/expert-system interaction. *Human Factors,* 29, 97-109.

Miller, P. 1986. *Expert Critiquing Systems: Practice-Based Medical Consultation by Computer.* Springer-Verlag: New York.

Miller, R.A. 1984. INTERNIST-/CADUCEUS: Problems facing expert consultant programs. *Meth. Inform. Med.,* 23, 9-14.

Mynatt, C., Douherty, M., and Tweeney, R. 1978. Consequences of confirmation and disconfirmation in a simulated research environment. Q. J. *Expt. Psych.* 30:395-406.

Newell, A. 1990. *Unified Theories of Cognition.* Harvard University Press: Cambridge, MA.

Norman, D.A. 1981. Categorization of action slips. *Psychological Review.* 88, 1-15.

Norman, D.A. 1990. *The Design of Everyday Things.* Doubleday: New York.

Obradovich, J., Guerlain, S., Smith, P.J., Smith, J., Rudmann, S., Sachs, L., Svirbely, J., Kennedy, M., and Strohm, P. 1996. The Transfusion Medicine Tutor: The use of expert-systems technology to teach students and provide support to practitioners in antibody identification. *Proceedings of the 1996 International Conference on the Learning Sciences.* 249-255.

Proctor, R.W. and Dutta, A. 1995. *Skill Acquisition and Human Performance.* Sage: Thousand Oaks, CA.

Proctor, R.W. and VanZandt, T. 1994. *Human Factors in Simple and Complex Systems.* Allyn and Bacon: Boston.

Rasmussen, J. 1988. *Information Processing and Human–Machine Interaction: An Approach to Cognitive Engineering.* North-Holland: New York.

Reason, J. 1990. *Human Error.* Cambridge University Press: New York.

Rock, I. 1995. *Perception.* W.H. Freeman: New York.

Roth, E., Bennett, K., and Woods, D. 1987. Human interaction with an "intelligent" machine, *Internatl. J. of Man-Machine Studies,* 27, 479-525.

Sheridan, T.B. and Ferrell, W.R. 1974. *Man–Machine Studies: Information, Control, and Decision Models of Human Performance.* MIT Press: Cambridge, MA.

Shortliffe, E. 1990. Clinical decision-support systems, in *Medical Informatics: Computer Applications in Health Care,* ed. E. Shortliffe and L. Perreault, pp. 466-500. Addison-Wesley Publishing Company: New York.

Silverman, B.G. 1992. Building a better critic: Recent empirical results. I*EEE Expert,* 18-25.

Silverman, B.G. 1992. Survey of expert critiquing systems: Practical and theoretical frontiers. *Communications of the ACM,* 35(4), 106-128.

Smith, P.J., Galdes, D., Fraser, J., Miller, T., Smith, J.W., Svirbely, J.R. Blazina, J., Kennedy, M., Rudmann, S., and Thomas, D.L. 1991. Coping with the complexities of multiple-solution problems: a case study. *Internatl. J. of Man-Machine Studies,* 35, 429-453.

Smith, P.J., Giffin, Rockwell, T., and Thomas, M. 1986. Modeling fault diagnosis as the activation and use of a frame system. *Human Factors,* 28(6), 703-716.

Smith, P.J., McCoy, C.E., and Layton, C. 1997. Brittleness in the design of cooperative problem-solving systems: The effects on user performance. *IEEE Trans. on Syst., Man, Cybern.,* 27(3): 360-371.

Smith, P.J., Miller, T., Gross, S. Guerlain, S., Smith, J., Svirbely, J., Rudmann, S., and Strohm, P. 1992. The transfusion medicine tutor: A case study in the design of an intelligent tutoring system. *Proceedings of the 1992 Annual Meeting of the IEEE Society on Systems, Man and Cybernetics,* 515-520.

Treu, S. (1994). *User Interface Design: A Structured Approach.* New York: Plenum Press.

Tufte, E.R. 1993. *The Visual Display of Quantitative Information.* Graphics Press: Chesire, CT.

Tufte, E.R. 1990. *Envisioning Information.* Graphics Press: Chesire, CT.

Tversky, A. and Kahneman, D. 1974. Judgment under uncertainty: Heuristics and biases. *Science,* 185, 1124-1131.

Van der Veer, G.C., Bagnara, S., and Kempen, G.A.M. (eds.). 1992. *Cognitive Ergonomics: Contributions from Experimental Psychology.* North-Holland: Amsterdam.

Wenger, E. 1987. *Artificial Intelligence and Tutoring Systems: Computational and Cognitive Approaches to the Communication of Knowledge.* Kaufmann: Los Altos, CA.

Wyatt, J. and Spiegelhalter, D. 1992. Field trials of medical decision-aids: potential problems and solutions. *Proceedings of the 14th Annual Symposium on Computer Application in Medical Care,* 3-7.

For Further Information

Bogner, M.S., ed. 1994. *Human Error in Medicine.* Erlbaum: Hillsdale NJ.

Card, S., Moran, T., and Newell, A. 1983. *The Psychology of Human–Computer Interaction.* Erlbaum: Hillsdale, NJ.

Crane, J.G. 1992. *Field Projects in Anthropology: A Student Handbook* (3rd ed.). Waveland Press. Prospect Heights, IL.

Gaines, B. and Boose, J. 1989. *Knowledge Acquisition for Knowledge-Based Systems.* Academic Press: London.

Klein, G.A., Calderwood, R., and MacGregor, D. 1989. Critical decision method for eliciting knowledge. *IEEE Trans. Syst., Man, Cybern.,* 19(3).

Lajoie, S.P. and Lesgold, A. 1989. Apprenticeship training in the workplace: Computer- coached practice environment as a new form of apprenticeship. *Machine-Mediated Learning,* 3, 7-28.

Johnson-Laird, P. 1993. *Human and Machine Thinking.* Erlbaum: Hillsdale NJ.

Larkin, J. and Rainard, B. (1984). A research methodology for studying how people think. *Journal of Research in Science Teaching,* 21, 235-254.

Michie, D. 1986. *On Machine Intelligence,* 2nd ed. Ellis Horwood, Ltd.: Chichester.

Mitchell, C. and Saisi, D. 1987. Use of model-based qualitative icons and adaptive windows in workstations for supervisory control. *IEEE Trans. Syst., Man, Cybern.* 17, 573-593.

Moran, T. and Carroll, J., Eds. 1996. *Design Rationale: Concepts, Techniques, and Use.* Erlbaum: Hillsdale, NJ.

Moray N. 1986. Monitoring behavior and supervisory control, in *Handbook of Perception and Human Performance: Vol. 2, Cognitive Processes and Performance,* ed. K.R. Boff, L. Kaufman, and J.P. Thomas, pp. 40-51. John Wiley & Sons: New York.

Newell, A. and Simon, H.A. 1972. *Human Problem Solving.* Prentice-Hall: Englewood Cliffs, NJ.

Poulton, E. 1989. *Bias in Quantifying Judgments.* Lawrence Erlbaum Associates: Hillsdale NJ.

Rasmussen, J., Brehner, B., and Leplat, J. (eds.). 1991. *Distributed Decision Making: Cognitive Models for Cooperative Work.* John Wiley & Sons: New York.

Rasmussen, J., Pejtersen, A., and Goodstein, L. 1994. *Cognitive Systems Engineering.* John Wiley & Sons: New York.

Rouse, W. 1980. *Systems Engineering Models of Human–Machine Interaction.* North-Holland: New York.

Rubenstein, R. and Hersh, H. 1984. *The Human Factor: Designing Computer Systems for People.* Digital Press: Boston, MA.

Sanders, M. and McCormick, E. 1987. *Human Factors in Engineering and Design.* McGraw-Hill: New York.

Sanjek, R., ed. 1990. *Fieldnotes: The Making of Anthropology.* Cornell University Press: Ithaca, NY.

Shneiderman, B. 1987. *Designing the User Interface: Strategies for Effective Human–Computer Interaction.* Addison-Wesley: Reading MA.

Shafer, G. and Pearl, J. (Ed.). 1990. *Readings in Uncertain Reasoning.* Morgan Kaufman: San Mateo, CA.

Smith, J.B. 1994. *Collective Intelligence in Computer-Based Collaboration.* Lawrence Erlbaum: Hillsdale, NJ.

Tversky, A. 1982. *Judgment Under Uncertainty: Heuristics and Biases.* Cambridge University Press: Cambridge.

Wickens, C. 1992. *Engineering Psychology and Human Performance,* 2nd ed. Harper-Collins: New York.

Section II
Fundamentals of Work Analysis

17

Task Analysis

Anand Gramopadhye
Clemson University

Jatin Thaker
Clemson University

17.1 Background

Introduction

The need to improve the systems we work in has been the driving force behind human factors. Traditionally, studies in this area have been carried out by observing humans at work and by analyzing their work environments. The term "task analysis" refers to the formal approach of analyzing human performance in systems, and is comprised of the systematic recording and analysis of human work to identify human/system mismatches with an eye to ultimately designing superior systems. The goal, therefore, of any task analysis is to examine the existing human/machine systems in order to provide a basis for designing more efficient, effective systems that are based on known human capabilities.

This chapter contains a look at how task analysis has evolved to its current state, a discussion of the definitions of commonly used terms and descriptions of the task analysis procedure, an explanation of several different approaches to task analysis, and finally, a case study in which task analytic methodology is used to identify ergonomic interventions to minimize human error.

Historical Perspective

There are a number of different approaches to analyzing observable human behavior, each with its own name and purpose, but all sharing the broad goal of measuring human performance. For example, a time study is used primarily to determine the standard time needed to perform a job; an activity study is the more effective method for establishing an improved method of accomplishing a job; and link analysis is used to determine the best physical layout for a work area. Initially, the analysis of tasks focused on observable human behavior, but because automation has increased the number of cognitive and decision-making activities required of many operators, many task analyses now attempt to measure the cognitive activities which drive the observable behavior. Newer approaches, such as GOMS analysis, task

0-8493-2641-9/99/$0.00+$.50

knowledge structures, mental model development, and cognitive simulations, concentrate on describing and analyzing the operator's cognitive activities. Regardless of whether the objective of a particular analysis is to measure manual or cognitive tasks, it is undertaken to determine the performance requirements imposed by the task–hence the term task analysis.

Task analysis as we know it today did not develop in any predictable, or step-by-step fashion, but rather emerged from a variety of theories which developed somewhat independently from one another to address various contexts. This somewhat unstructured development has left us with a number of theoretical approaches to the study of human performance as well as with a number of terms, the definitions of which sometimes overlap. In a recent article summarizing the current status of task analysis, Stammers (1995) concluded that there is limited consensus on the terminology surrounding task analyses. It seems though, that the term "task analysis" is generally understood to refer to all activities involving the analyses of tasks and is, therefore, the term we will use throughout this chapter to mean just that (Drury et al., 1987; Kirwan and Ainsworth, 1992; Singleton, 1974; Stammers, 1995).

The growth of task analysis has closely followed that of human factors/ergonomics, because task analysis is central to the study of human factors and has become the expected means for beginning almost any human factors effort (Sheridan, 1997).

Task analysis as we know it today began early in this century with the work of Gilbreth (1911) and Taylor (1929). Taylor's work focused on describing a process and identifying ways to improve a job, but his interest was efficiency, not human wellness. Taylor's time-study approach, developed to determine time standards, did not address the ergonomic content or appropriateness of a particular task based on anthropometry or on human capabilities and limitations. Taylor's work was criticized in several quarters, but

> "Taylor's greatest and lasting contribution to the science of industry is the approach he adopted. He approached problems which had been thought either not to exist or to be easily solved by common sense, in the spirit of scientific enquiry."
>
> (Farmer, in Barnes, 1980)

Taylor used his scientific approach to describe, analyze, and improve the process of shoveling work at Bethlehem Steel Works. In the same vein, the Gilbreths expanded on this scientific approach, developing the motion study, a process that tried to determine the preferred method of doing work. As part of his work on motion study, Gilbreth identified 17 basic motions, called therbligs, which were common to all kinds of manual work, then used these basic motions to analyze the sequence of actions in a task. Today we use the combined term "motion and time study," as well as terms such as work measurement and work methods and design, to represent these initial techniques intended to analyze human work.

The U.S. Department of Labor also pursued job and task analysis in the 1930s. The intention of the Department of Labor was to formulate a consistent repertoire of personnel skills that could then be used in hiring, placement, and promotion.

The approaches developed by Taylor and Gilbreth were fine for measuring manual and repetitive tasks, but failed to address the cognitive components that made up those tasks. Realizing this limitation, Crossman (1956) proposed "mental therbligs," including planning and controlling activities, and developed the "sensori-motor" chart which drew links between planning/controlling and executing activities. Only much later were cognitive and information-processing tasks analyzed in a similar fashion (e.g., Card et al., 1983).

Work by Taylor, the Gilbreths, the Department of Labor, and Crossman, were important in lending credence to the potential for the systematic study of human work, but any look back at the development of task analysis would be remiss if it ignored the impact of the studies conducted by the U.S. military in the 1950s. Research by the military led to the development of a systematic task analytic process that defined performance requirements, training needs, and equipment design specifications. This process was especially helpful as personnel were becoming involved with increasingly sophisticated aircraft systems and complex weapon systems. In fact, the complexity of these systems led to changes in the very nature of the tasks humans would be expected to perform. The most commonly referenced work in this area is Miller's (1953) report "Method for Man–Machine Task Analysis," which used an approach that analyzed operators' tasks, then "linked" input to output. Later Miller developed a classification of tasks which used a codifying scheme based on the temporal patterns of a task. This classification system led to a taxonomic approach of classifying human performance. Since that time, a number of taxonomies of human performance have been developed (e.g., Fleishman and Quaintance, 1984). The advent of task analysis in the military provided a standardized approach to catalog skill requirements, and training and support needs. It also provided a basis for determining the requirements of new systems (e.g., Qualitative and Quantitative Personnel Requirements Information (QQPRI; Swain, 1962).

Outside of the United States though, specifically in the United Kingdom, the impetus for task analysis was driven by training needs. Early efforts in this direction were driven by the skills-based training movement (Crossman, 1956; Seymour, 1966). The focus here was to work with human operators to determine the knowledge an operator needed to perform subtasks, such as operating, inspecting, and disposing. Analysts tried to determine "how," "when," "why," etc., about a particular task, by observing a skilled operator over several cycles and using a questioning approach, or by performing the tasks themselves. The risk here, of course, was that analysts might assume that their own experience was, in all ways, a typical experience. Task information could, therefore, be misinterpreted using this technique. Another example where training has driven the growth of task analysis can be found in the works of Annett and Duncan (1967). They used a hierarchical task analytic (HTA) approach to establish training requirements. In an HTA approach, a task can be broken down to different levels, with each level covering more detail about the task. The number of levels to which a task was broken down was decided based on the degree of the analysis and various stopping rules. These rules are used to help the analyst determine the levels of hierarchy in a particular case (Stammers and Shepherd, 1995). The HTA approach has been used to define the training procedures and ergonomic requirements in complex process and nuclear industries (e.g., Duncan, 1974; Piso, 1981; Umbers and Reiersen, 1995). Examples of the use of HTA in other contexts can be found in the works of Shepherd (1989, 1993) and Carey, Stammers, and Astley (1989). Other approaches to task analysis within the human–computer interaction arena include Task Analysis for Knowledge Descriptions (TAKD) (Johnson et al., 1984), Task Action Grammars (TAG) (Payne and Green, 1986), and Task Knowledge Structure (TKS) (Johnson et al., 1988). Diaper (1989) provides a comprehensive coverage of various approaches to task analysis in the human–computer interaction arena.

Extensive use of task analysis outside of the military was first documented in Singleton's early work (1974). In the twenty-some years since these studies were conducted, we have seen an increase in the use of task analytic approaches to resolve ergonomic issues. In 1983, Drury et al. proposed a task analysis of aligning a lamp in the lamp holder of a copy machine using a column format. They found that it was very helpful to formally state the requirements for a good design between the task analysis and the ergonomic redesign. Stating the design requirements helped other members join the design phase, resulting in a participatory effort and greater ownership of the resulting design. In another effort, Armstrong et al. (1986) used task analysis based on traditional time and motion studies to identify risk factors in repetitive motion tasks. A similar approach had been used by Drury and Wick (1984) to analyze repetitive tasks in a shoe plant to identify potential ergonomic improvements. In their analyses, the tasks were observed and videotaped, then elementary job motions (based on therbligs) were determined for each body member. Their objective was to determine ergonomic interventions that would minimize musculoskeletal injuries. In this case, the investigators kept a detailed log of the body angles using the task description form. (Table 17.1 shows the task description form used for recording body angles data from Drury, 1987.) They administered both the body-part discomfort form developed by Corlett and Bishop (1976) and the general comfort rating scale developed by Shackel et al. (1979) several times during a shift to collect data on postural discomfort. They also recorded measures of productivity and quality. Their task analysis consisted of analyses of the raw descriptive data on body angles, forces, and discomfort. A major portion of the analysis involved categorizing the angles involved at each joint to determine just how close to the extreme value they came. For example, a measure that has proven to be effective in estimating exposure to Repeated Motion Injuries (RMI) problems of the wrist is simply the number of

TABLE 17.1 Task Description Form (Drury, 1987).

Job title:			Task 1	Task 2	Task 3	Task 4	Task 5	Task 6	Task 7	Task 8
FREQUENCY/BUNDLE										
BACK	Rotation									
	Lateral Bend									
	Flex/Ext									
NECK	Rotation									
	Lateral Bend									
	Flex/Ext									
SHOULDER	Rotation	R								
		L								
	Abd/Add	R								
		L								
	Flew/Ext	R								
		L								
ELBOW	Flexion	R								
		L								
FOREARM	Pron/Sup	R								
		L								
WRIST	Flex/Ext	R								
		L								
	Rad/Ulnar	R								
		L								
LEGS	Thigh to H	R								
		L								
	Shin to V	R								
		L								
	Foot to H	R								
		L								
	Rotation	R								
		L								
	Force	R								
		L								
POSTURE	Sit/Stand									
	Armrest									
	Foot Pedal									
	Backrest									
GRIP	Power	R								
		L								
	Tip pinch	R								
		L								
	Pulp pinch	R								
		L								
	Lat pinch	R								
		L								
FORCES	Push/Pull	R								
		L								
	Up/Down	R								
		L								
	In/Out	R								
		L								
	Fingers	R								
		L								
VIBRATION		R								
		L								
SHOCK		R								
		L								
LIGHTING	Luminance									
	Glare ?									

From Drury, C. G. (1987) A bio-mechanical evaluation of the repetitive motion injury potential of industrial jobs. *Seminars in Occupational Medicine,* Volume 2, No. 1, 41-49.

daily damaging wrist motions, i.e., any combination of a grip or external force with any nonzero wrist exposure. As part of the analysis, data from the body-part discomfort scale was summarized by determining the incidence of any nonzero body-part discomfort reading for each body part divided by the number of times the scale was administered. The data from the analysis was used to compare the demands placed on the operator with the operator's ability to meet those demands. Following this step, ergonomic interventions that would minimize ergonomic risks were identified. Finally, task analysis was applied as a recursive procedure to justify the choice of interventions and to measure the effect of those interventions.

Another method widely used for ergonomic analysis in Europe is the AET method (Landau and Rohmert, 1981). This method is based on the man-at-work system and the concept of stress and strain. The AET method initially analyzes an activity by determining the objects of work, the equipment, and the working environment. The objects of work are analyzed under material, energetic, and informational aspects, then analyzed by demands.

A more recent application of the use of a task analytic methodology for a specific industrial task is given by Gramopadhye et al. (1995). Gramopadhye et al. were faced with the task of improving inspection performance for a contact lens manufacturing company. The task description was designed to collect information on different aspects of the inspection task. Following this step, the different tasks within the inspection process were analyzed using Rasmussen's (1983) Skills-Rules-Knowledge (SRK) framework. The SRK framework was used to identify errors and to develop a taxonomy of errors. Following the development of the error taxonomy, a number of interventions, each expected to minimize inspection errors, were identified. Consequently, a computer-based inspection training program using simulated images of the contact lenses was developed to train inspectors to minimize inspection errors (Gramopadhye et al., 1998).

We have seen that a variety of approaches have been proposed to conduct task analysis. Unfortunately, there is no predetermined formula to ensure that investigators select the appropriate task analysis approach for any specific situation. The variety of approaches that we have, have emerged because of the variety of theoretical and methodological backgrounds and the range of contexts in which specific approaches have been developed (Stammers, 1995). Often the selection of any approach for task analysis depends upon the application domain, and more often (though not correctly), on the expertise the practitioner has in using a specific approach. In addition to the plethora of approaches available, another source of confusion about task analysis can be attributed to the various terms and related definitions that have appeared in the literature over the years. Despite the extensive use of task analysis in various domains, there is limited agreement among practitioners as to what constitutes a "task," or even what basic elements comprise a task analysis. Unless we resolve these theoretical issues, progress on task analysis, or at least our ability to communicate about it, will lag. The following section is intended to provide a better understanding of the theoretical issues underpinning task analysis.

17.2 Discussion of the Terms "Task" and "Task Analysis"

Task

A task can be defined as the unit element of analysis. In the context of human/machine systems, each task is considered to have an objective, and a specific input produces a system output. The definition of a "task" inherently defines the boundaries of a task. Based on the objectives and the level of analysis detail desired, the boundaries of a task can be expanded or collapsed. However, caution must be exercised by the practitioner when defining the boundaries of different tasks so that within any study, all tasks are comparable in terms of size (often defined by the scope, time, and level of effort required to perform them). Ensuring comparability between various tasks facilitates their comparison in terms of human performance requirements. The defining elements of a task are listed below.

1. Every task has a definite objective.
2. A task is a unit of activity conducted by one or more individuals.

3. An analyst must articulate each task's beginning and its end. The beginning is defined by an input to the system (e.g., information input, manual input, or a combination). The end of a task is defined by a system response (indication that the task objective has been completed).
4. Various tasks are associated with each other based on their objectives, functional relationship, and on space and time.

A task can be classified as continuous, discrete, or rule-based. Tracking tasks are examples of continuous tasks by which the operator has to control a system so that the system operates within predefined parameters. Most procedural tasks are discrete in nature. A typical example of such a discrete and procedural task is an assembly task. Rule-based tasks are a subset of the discrete task category. Rule-based tasks are found, for example, in diagnostic maintenance wherein the operator's next task is determined by the previous one.

Task Analysis

A task analysis is a study that aims to provide a systematic and comprehensive description of a task that is executed to achieve an overall system goal. Over the years, various definitions of task analysis have been proposed. McCormick (1976) considers task analysis to be the division of human work into component tasks and then the analysis of those components. Kirwan and Ainsworth (1992) define task analysis as the study of what an operator is required to do, in terms of actions and/or cognitive processes, to realize a system goal. A more general definition with a broader appeal in ergonomics is the following one, espoused by Drury (1983): "Task analysis is the comparison of the demands placed on the human operator (task demands) with the human's ability to meet those demands."

The task analytic approach has been used by system designers over a number of years (Meister, 1985; Johnson and Johnson, 1989). Figure 17.1 shows the role of task analysis in the overall ergonomic system design process (Pitkaar, Lenior, and Rijnsdorp, 1990). In the context of system design, task analysis has two main purposes, to design a new system, or to redesign an existing system (see Figure 17.2).

When designing a new system, the analyst must determine the system goals and the system functions that are used to achieve those goals. Following this step, the tasks needed to accomplish system functions are determined. Third, the analyst describes each task and determines the skills necessary to accomplish the task. Finally, task demands are compared with human capabilities and various function allocation alternatives (i.e., whether the task should be performed by the human alone, by the machine alone, or by a human who is assisted by a machine, etc.) are considered. The task analysis at this stage is a matter of articulating the task constraints and making them visible. These are the independent variables (inputs) that must be considered while performing the task, and the dependent variables which measure task performance (Sheridan, 1997).

In the redesign of an existing system, task analysis is used to identify problems and to suggest potential interventions that might remedy any problems. In these cases, task analysis is used to analyze complete systems or portions of existing systems in order to identify modifications to an existing system or to propose a completely new, better system. When viewed within the system design context, task analysis can be used for the following applications (adapted from Kirwan and Ainsworth, 1992):

1. System function allocation — decide on human–machine function allocation issues
2. Organizational issues — personnel selection, personnel qualification, skill requirements
3. Task design — identify the skills, procedures, and knowledge necessary to perform a task
4. Human–machine interface — workplace design, equipment/tool design
5. Human support requirements — design training and job aids
6. System reliability analysis — using data on human error to determine system reliability

The term "task analysis" has, unfortunately, been accepted to mean the simple description of a "task," in addition to the analysis of one. To resolve the resulting confusion, one needs to consider the various steps involved in conducting a task analysis. Referring to Figure 17.2, note that task analysis can be represented as consisting of the following steps:

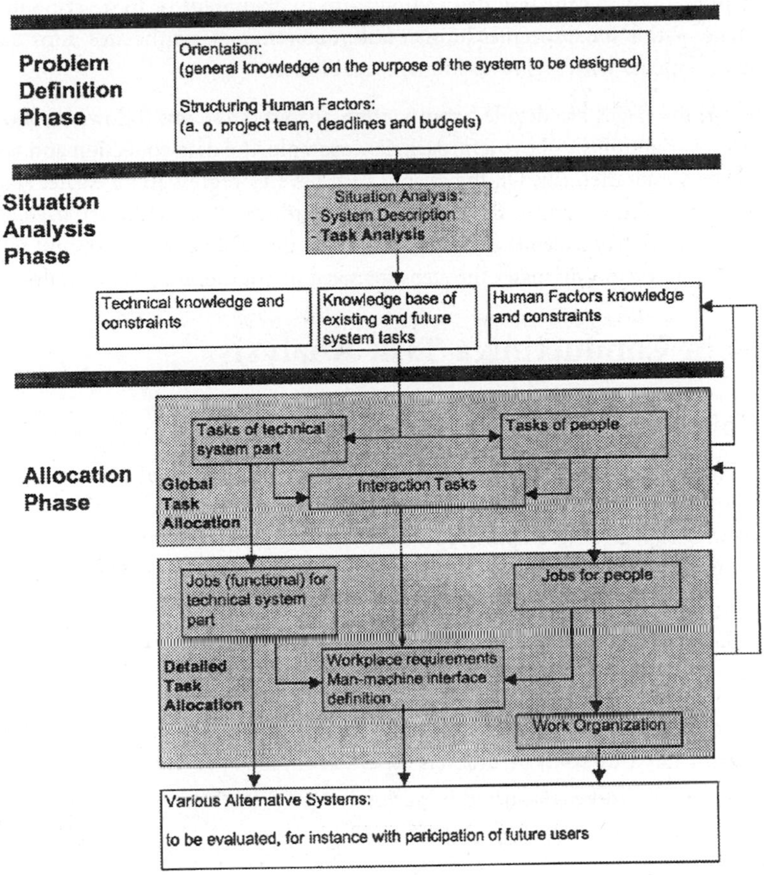

FIGURE 17.1 Role of task analysis within system design. (From Pitkaar, R. N., Lenior, T. M. J., and Rijnedorp, J. E. (1990) Implementation of ergonomics in design practice: outline of an approach and some discussion points. *Ergonomics*, 1990, 33, 5, 583-587. With permission.)

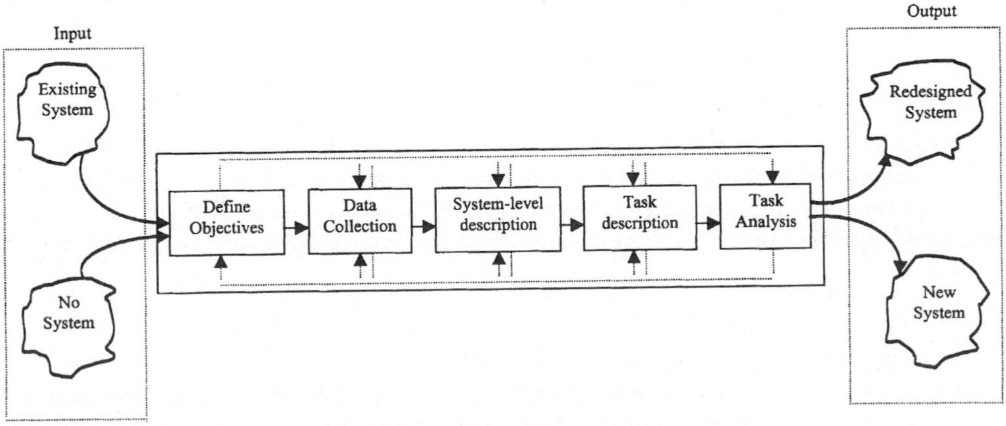

FIGURE 17.2 Steps in conducting task analysis.

1. Define objectives — establishing the scope of the task analysis
2. Data collection stage — collecting data on the task in question
3. System description stage — developing a system-level description

4. Task description — using the task description form to document or to describe the task
5. Analysis stage — this stage specifies human task requirements, synthesizes information, analyzes the task requirements with regard to human capabilities

Although, the various stages are detailed above, often analysts may not follow the above-mentioned steps exactly or use a variant of the above approach. For example, the data collection and task description stages might be done simultaneously. Or, the analyst may have to regress to an earlier stage or conduct several iterations at one or more stages. Finally, though, regardless of the approach used, it is the fourth stage that is most appropriately called the task analysis. Now that the various pertinent terms have been described, the following section discusses the steps involved in conducting a task analysis.

17.3 Steps in Conducting a Task Analysis

Define the Objectives and Scope of the Task Analysis

The first step in conducting a task analysis is to define the objectives/goals of the analysis to establish its scope. Doing this helps investigators identify the interactions between various personnel involved in all the phases of system design and development. At this stage, the analyst should also develop the detailed schedule required to successfully complete the task analysis and other system development activities. The following specific steps should be followed (in order):

- List the overall objectives of the task analysis
- Identify the personnel involved with the tasks to be studied
- Develop a detailed plan and schedule, identifying interactions with other personnel
- Obtain support for your study at all levels (operator level, supervisory level, management level)
- Identify the activities to be conducted by various members of the investigating team
- Develop a plan to accomplish these activities

Data Collection

Identifying the exact data to be collected is critical to conducting a task analysis. This phase involves defining the components and the subcomponents of the task. The type of analysis desired, the cost of the analysis, and the time available to conduct the analysis will often determine the level of detail required in the information to be collected.

System-Level Description

As a first step, an overall system level description should be developed to identify the goals of the system. Given the system-level goals, the inputs and outputs to the system should be determined. Depending upon the system under consideration, this process can be extensive. Various system description tools can be used to describe the system. The steps typically adopted to accomplish a system-level description (in order) are:

- Define the objective(s) of the system
- Develop a detailed verbal description of the system
- Use graphical approaches to clearly identify system components, the goals of each component, and the links among components; it is important to keep the stated objectives in mind while working on this step
- Identify inputs for each component of the system

Once the overall inputs, outputs, and system objectives are understood, it is critical to identify the various tasks required to achieve overall system objectives.

Task Description

After identifying the system components, the next step is to identify the different tasks within each system component. Detailed task information must be collected to ensure a comprehensive analysis. The first step in collecting task information is to develop a detailed verbal description of the task. Information is typically collected on the following items:

Task Name: A statement identifying the human performance requirement.

Task Objective: Outlines the goal of the task, so the analyst can develop links between this and other tasks and to identify relationships and commonalties with other tasks that have similar objectives and use similar methods, procedures, tools, or information.

Task Environment: Any specialized environment necessary to perform a particular task (e.g., specialized lighting to perform inspection, physical environment).

Time Required: The time required to perform the task (measured from when input is initiated to when output is obtained).

Tools/Equipment: The tools and type of equipment used to execute the task. The materials processed and products made.

Incidence/Frequency: Number of times per unit time (e.g., three times in an hour).

Control Actions/Input: Inputs or control actions necessary to execute a particular task. These could be manual, automated, or semi-automated.

Criticality of Task: Effect of task failure on system performance (often based on expert opinion).

Error potential: Significant human errors that are likely to occur and have an impact on task objective(s) and system performance. Often based on observations and expert opinion.

Information: Information required by each task and the source of the information input (e.g., electronic, oral).

Knowledge: Cognitive skills (perception, attention, decision making) required to perform the task (useful in selection and training of appropriate personnel).

Outputs/Feedback: Response that tells the human that a particular task was completed. This feedback is often obtained by a change in the system state.

Rules: Specific procedures, rules, and guidelines needed to perform the task.

Sequence: The order in which the various subtasks are organized (e.g., serial or branching).

Skills: Human manual skills necessary to perform the task (useful in selecting and training of appropriate personnel).

Support Systems: Support systems to execute a task (e.g., decision-aiding tools used to perform diagnostic maintenance tasks or other personnel/equipment that supplies information or a service).

A variety of data collection techniques may be employed to collect task information. The cost of data collection, the type of task information, and the task analysis approach may often prescribe the choice of a specific data collection technique. A variety of these techniques are described in greater detail below.

Activity Sampling: Activity sampling was initially developed for time and work-study measurements. It can be conducted either by direct observation of the human or through video recordings. The method is used to collect data on the percentage of time spent on different tasks by a human in a system. By observing human activities at different time intervals, an analyst collects information on how humans spend their time. An important requirement in conducting activity sampling is that all activities are observable and discrete (distinguishable from the next activity), and occur for sufficient time periods so as to make measurement possible. Activity sampling is especially suitable to situations in which an operator has to do several different things but in no fixed order.

Critical Incident Technique: This technique is used to collect data on critical events that have the potential to significantly impact system performance. The critical incident technique is most useful in a system in which problems are suspected, but the source, nature, and severity of these problems are not known or completely understood. Data on critical incidents can be collected from operators because they work closely with the system and have first-hand knowledge of the system.

Observation: The direct observation of humans performing a task is a time-honored technique for collecting task information. The observer can watch an actual performance, a simulated performance, or a video recording of the performance of a task. This technique is useful for collecting initial information on the performance of a task and for verifying information collected through other data collection techniques. Another technique, closely linked to observation, is shadowing, wherein the analyst follows the operator around as the operator performs his daily work. The analyst is concerned not only with the tasks he/she is expecting the operator to perform, but also with any other task(s) the operator performs. Shadowing is particularly useful when an operator is required to perform a myriad of tasks and to interact with a number of personnel to perform these tasks. In these cases, the operator typically cannot recollect task-specific information when asked in an interview or in response to a questionnaire. It seems that the best way for an analyst to collect accurate information in these cases is by following the operators as they conduct their daily tasks. In collecting information by direct observation, though, analysts must be sensitive to the possible effect of their observing on the operator. This intrusion, caused by the operator's awareness of the analyst, might alter the operator's attitude or even the actual performance on the task.

Interviews: Interviewing domain experts is a popular technique for collecting information. Both structured and unstructured interviews can be used to collect task information. In a structured interview, the questions themselves and the order in which the questions are asked is predetermined. In an unstructured interview, the interviewer may have only a rough idea of the exact questions to ask, and may allow answers to a few initial questions to determine the direction of the interview. These unstructured or less structured interviews can be particularly helpful during the initial stages of data collection, whereas a structured interview may be more appropriate for collecting specific information. Interviews can also help analysts collect information that was missed during the direct observation of a person performing a task. When interviewing, analysts should be concerned with (1) asking probing questions to obtain more details about the task (e.g., Is the way the task is typically performed the "right," or "textbook" fashion, or is it a "quicker," way developed by the operator?) (2) identifying the interrelationships among various subtasks, and (3) identifying other ways the task might be performed. Interviews of both individuals and groups of personnel can be helpful.

Surveys/Questionnaires: Asking domain experts to complete questionnaires is another method of collecting task information. The advantage to using questionnaires is that information can be collected inexpensively from large numbers of participants in various geographical locations. The disadvantage is the often low response rate and the reduced amount of information obtained per respondent. Surveys and questionnaires are most popularly used for collecting opinion and attitude data and may be best used as a follow-up to an interview.

Verbal Protocols: The increase in automation has changed the nature of many jobs from manual to cognitive. Information on tasks with large cognitive components can often be best captured by asking the subject to "think aloud" while carrying out his actions. This technique, wherein subjects are asked to verbalize their actions (i.e., to explain what they are doing it, how they are doing and why they are doing it) while carrying out their task is called "verbal protocol." Here again, as with the shadowing technique, there is a risk that asking the subjects to verbalize their actions will interfere with task performance and/or change the way the task gets executed. It can be helpful, however, to compare the findings obtained from a verbal protocol with information gathered through other data collection techniques in an effort to corroborate the data.

Task Documentation: Information on the task can often be collected through task documentation, i.e., typical procedural steps are often clearly described in system documents. Drawings, specifications, log reports, company-wide procedures, operation manuals, and training and instructional material are a sampling of the variety of documents that can be used to collect information on a task.

It is important that a task description be comprehensive, have integrity and validity, and be in a format that is clear, useful, and easy for the analyst to understand. Alternate formats can be used to transcribe task descriptive information (e.g., flow charts, column format). Table 17.2. shows a sample of a column format used for describing a real-time shop floor control system (Anne and Greenstein, 1993).

TABLE 17.2 Task Description: Column Format

				JOB DESIGN QUESTIONNAIRE															
Task no.	Function	Task	Allocation	Information Required	Information Presented	Human Input	Comp. Input	Coordination	Cognitive Demand	Possible Errors	Consequences	Task Duration	Frequency	Who	Knowledge employed	Skill level	Task complexity	Task criticality	Comments
	System Management																		
1.1	Activate RTS																		
1.1.1		Place ON/OFF switch in "ON" position	Human	Position of ON/OFF switch	Switch label	Updated switch position		None	Low	Failure to locate switch	Inability to run the system	<1 min.	Once a day	Real time clerk or Supervisor	Basic operation of a PC	Low	Simple	High	
1.1.2		Access RTS directory	Human	Current directory	Prompt of current directory	Change directory command		None	Low	1. Specifying wrong directory 2. Issuing wrong command	Inability to run the system	<1 min.	Once a day	Real time clerk or Supervisor	Basic knowledge of DOS	Medium	Simple	High	Knowledge of when and how to perform this task must be currently memorized and is not displayed. Could display activation procedure above terminal and/or replace individual commands with one command that invokes a batch file.
1.1.3		Activate memory record manager	Human	Description of activation sequence	None currently	Btrieve command		None	Moderate	Issuing wrong command	Inability to run the system	<1 min.	Once a day	Real time clerk or Supervisor	Knowledge of RTS activation requirement and DOS	Low	Simple	High	
1.1.4		Load RTS and generate main menu	Human	Description of activation sequence	None currently	Btrieve command		None	Moderate	Issuing wrong command	Generation of no display or another display	<1 min.	Once a day	Real time clerk or Supervisor	Knowledge of RTS activation requirement and DOS	Medium	Simple	High	

Anne, M. and Greenstein, J. S. (1993) The design of a computer-based tool to aid the manager of a real-time manufacturing control system (*Tech. Report*). Clemson, SC: Clemson University, Department of Industrial Engineering.

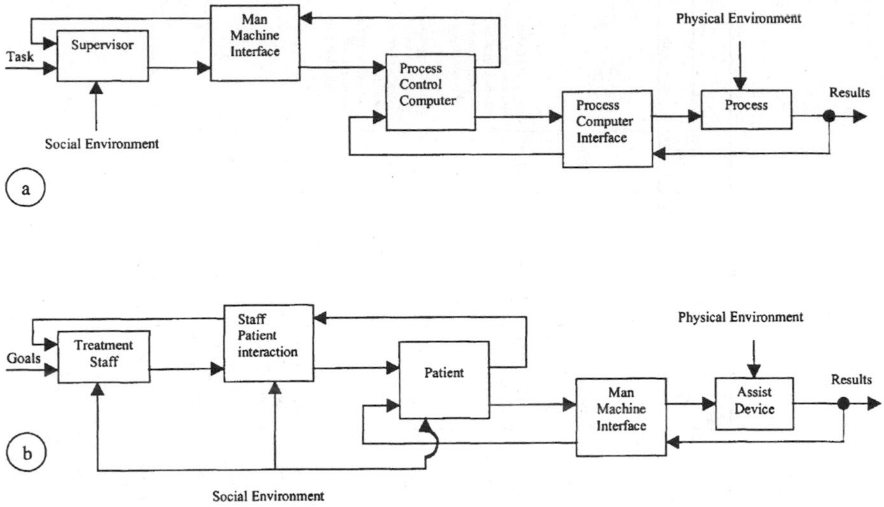

FIGURE 17.3 Flow chart showing the environment of a supervisor in process control and a patient in a rehabilitation center. (From Stassen, H. G., Steele, R. D., and Lymam (1989) A man-machine system approach in the field of rehabilitation: a challenge or necessity, in Baosheng Hu (ed.) *IFAC Analysis, Design and Evaluation of Man–Machine Systems*, PRC, 79-86. With permission.)

Task Analyses

Information collected during the task description stage is analyzed using one of the several approaches described below. The analysis stage applies various theories, models, and results of empirical ergonomic studies to the problem at hand. Although the list of approaches described below is not exhaustive, it does provide a representative list of some commonly used approaches. In addition, appropriate references have been included to lead the reader to opportunities for further study.

Flow Charting Approaches: Over the past several decades, various flow charting techniques have been used for task analysis. Flow charts or networks are graphical descriptions that can be used to describe and show links between the tasks and subtasks within a system. Figure 17.3 is a flow chart showing the environment of a supervisor in a process industry and a patient in a rehabilitation center (Stassen et al., 1989). A number of flow charting techniques are currently in use. The more popular of these are operation sequence diagrams, functional flow analysis, and decision diagrams. Figure 17.4 shows an operation sequence diagram (OSD) for baking soda crackers (Barnes, 1980). In this diagram, the various steps in the assembly process are outlined using operation sequence symbols. OSD can be depicted using alternate formats. Figure 17.5, an example of this, uses operation sequence symbols to depict the flow of information for the simple task of requesting a purchase order. The summary table shows the number of activities, times, and distance traveled, information which can be used to compare the method currently in use with the proposed method. Whatever the graphical representation, an operation sequence diagram assists by providing the analyst with a graphical representation of the tasks as they relate to both machine elements and other operators. This information is useful in helping analysts understand the links between various activities and then, consequently, to design more efficient systems. Similarly, functional analysis depicts the sequence of actions/functions that need to be performed by the system (Greer, 1981), and decision diagrams are often used to represent activities with decision components. Figure 17.6 shows a decision flow chart developed by Lock and Strutt (1985) to analyze error-likely situations in an inspection system.

Link Analysis: The objective of link analysis is to assist the analyst in identifying the relationships (links) between different system components (Chapanis, 1962). This approach can be used for arranging the physical layout of a screen display, instrument panel, workstations, and office, and help in understanding the communication links between individuals. Figure 17.7, adapted from Chapanis (1991), shows how

Baking Soda Crackers

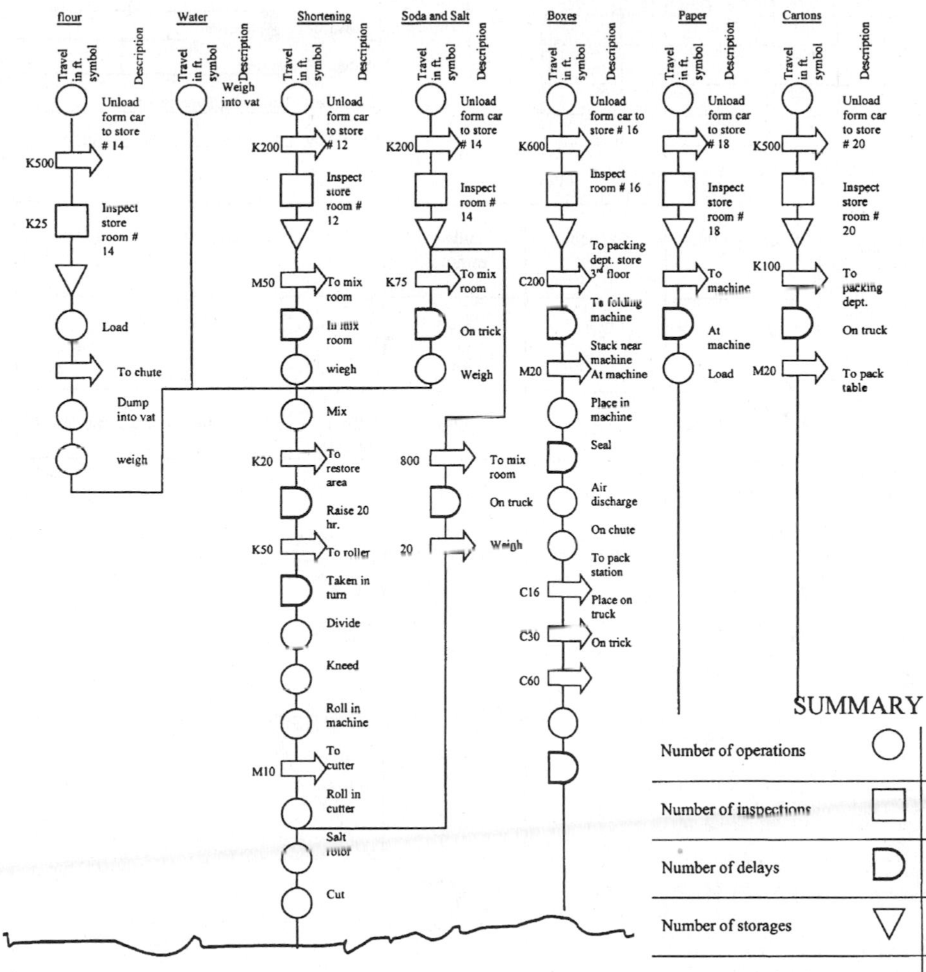

SUMMARY

Number of operations	○	37	
Number of inspections	□	6	
Number of delays	D	12	
Number of storages	▽	6	
Transportations		No.	Dist. Ft.
By carrying----------------M	⇨	8	140
By truck--------------------K	⇨	15	3820
By conveyor -------------- C	⇨	11	816
Total		34	4776

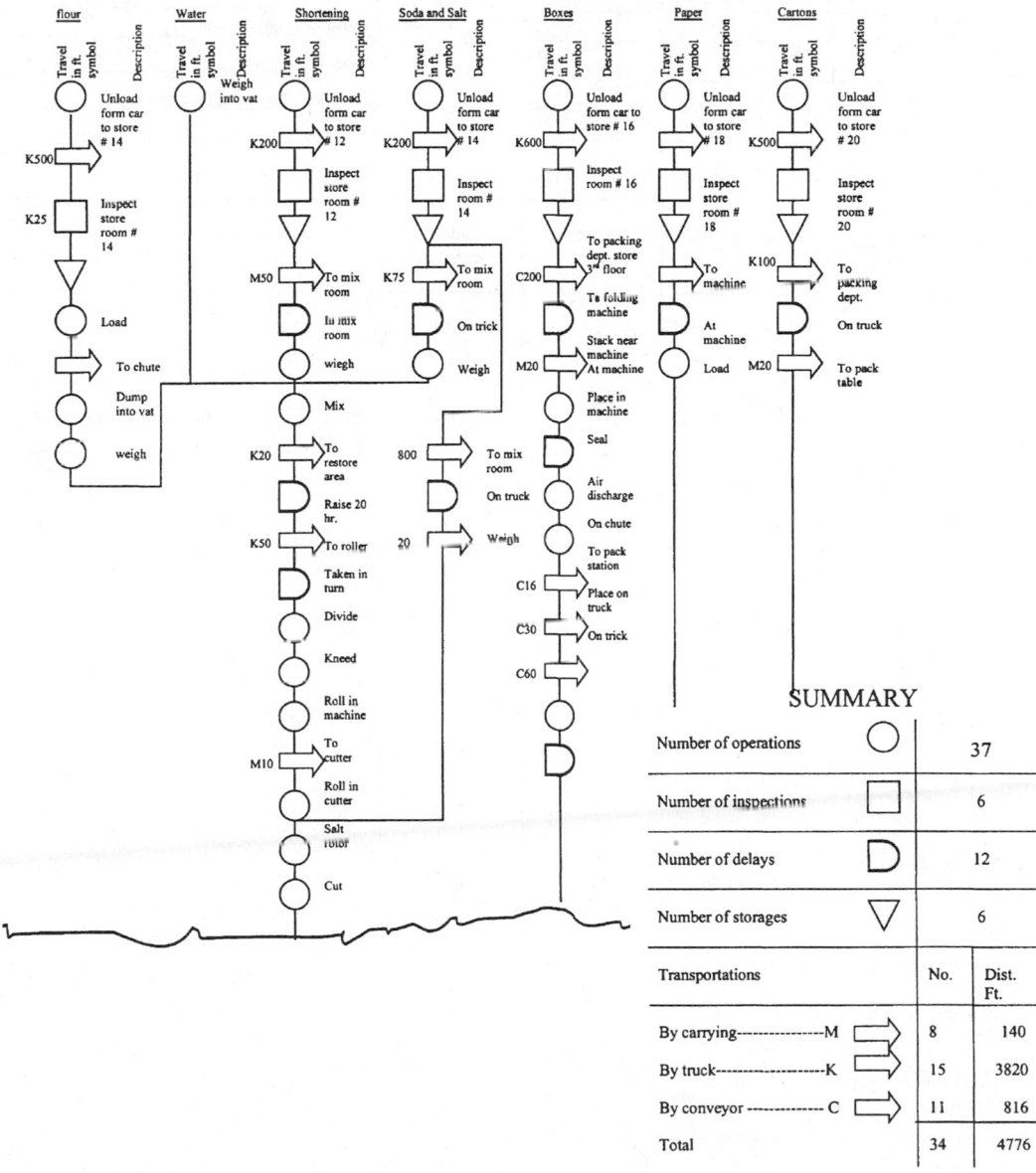

FIGURE 17.4 Assembly process chart: baking soda crackers. (From Barnes, R. (1980) *Motion and Time Study Design and Measurement of Work.* John Wiley & Sons Inc., New York. With permission.)

link analysis was used successfully to design an alternate physical layout for the *U.S.S. Louisville.* Initially a link table (see Table 17.3) was created by counting the number of times each system component interacted with the other. Then the associations between the different system components were identified. This information was also represented graphically using the schematic link diagram. In the link diagram, more frequently linked components are indicated by the greater number of lines shown connecting them. The information from Figure 17.7 and Table 17.3 was later used to devise an alternate schematic diagram which was sensitive to the links between the different system components (Figure 17.8).

FLOW PROCESS CHART

SUMMARY

	PRESENT		PROPOSED		DIFFERENCE	
	No	Time	No	Time	No	Time
○ Operations	3	7.00				
⇨ Transactions	5	1.25				
□ Inspections	1	0.50				
D Delay	1	5.00				
▽ Storages	1	1.25				
Distance Traveled		95 FT.		FT.		FT.

Job Requisition for Purchase

□ Man or ■ Material

Model Present

Department Purchasing Department

Charted By J. Thaker Date 07/21/95

Details of Present/ Proposed Method	Chart symbols	Distance in feet	Time in minutes	Comments
Supervisor picks up order from administrative department.	○ ➡ □ D ▽	20	0.25	
Supervisor makes revisions to the order.	● ⇨ □ D ▽		3.00	Often the order does not contain the latest updates.
Supervisor drops the order at the Secretary's desk.	○ ➡ □ D ▽	10	0.15	
Secretary types the revisions.	● ⇨ □ D ▽		2.00	
Secretary drops the order at the Supervisors desk.	○ ➡ □ D ▽	20	0.25	
Supervisor inspects the typed order.	○ ⇨ ■ D ▽		0.50	
Supervisor delivers the inspected order at the Manager's desk.	○ ➡ □ D ▽	25	0.30	
Waits for approval from the Manager.	○ ⇨ □ ◗ ▽		~5	
Supervisor delivers the order to the Secretary.	○ ➡ □ D ▽	20		
Secretary executes the order.	● ⇨ □ D ▽			
Secretary files a copy of the order.	○ ⇨ □ D ▼			

FIGURE 17.5 Operation sequence diagram.

Critical Incident Analysis: The critical incident analysis technique is used widely with operational systems to study human error and has been described in great detail by Flanagan (1954). The technique has also found extensive use in the personnel and skills-related areas (Meister, 1985). This technique documents accidents, misses, and near misses using first-hand information from operators. Once such information is documented, human factors knowledge is applied to analyze accidents. The goal of this analysis is to lead analysts to hypothesize about the source of the errors, and to help them identify interventions to minimize errors in the future.

Workload Analysis: Workload analysis focuses on determining whether a human has been successful in completing assigned tasks within an allocated time. The technique focuses on measuring or estimating

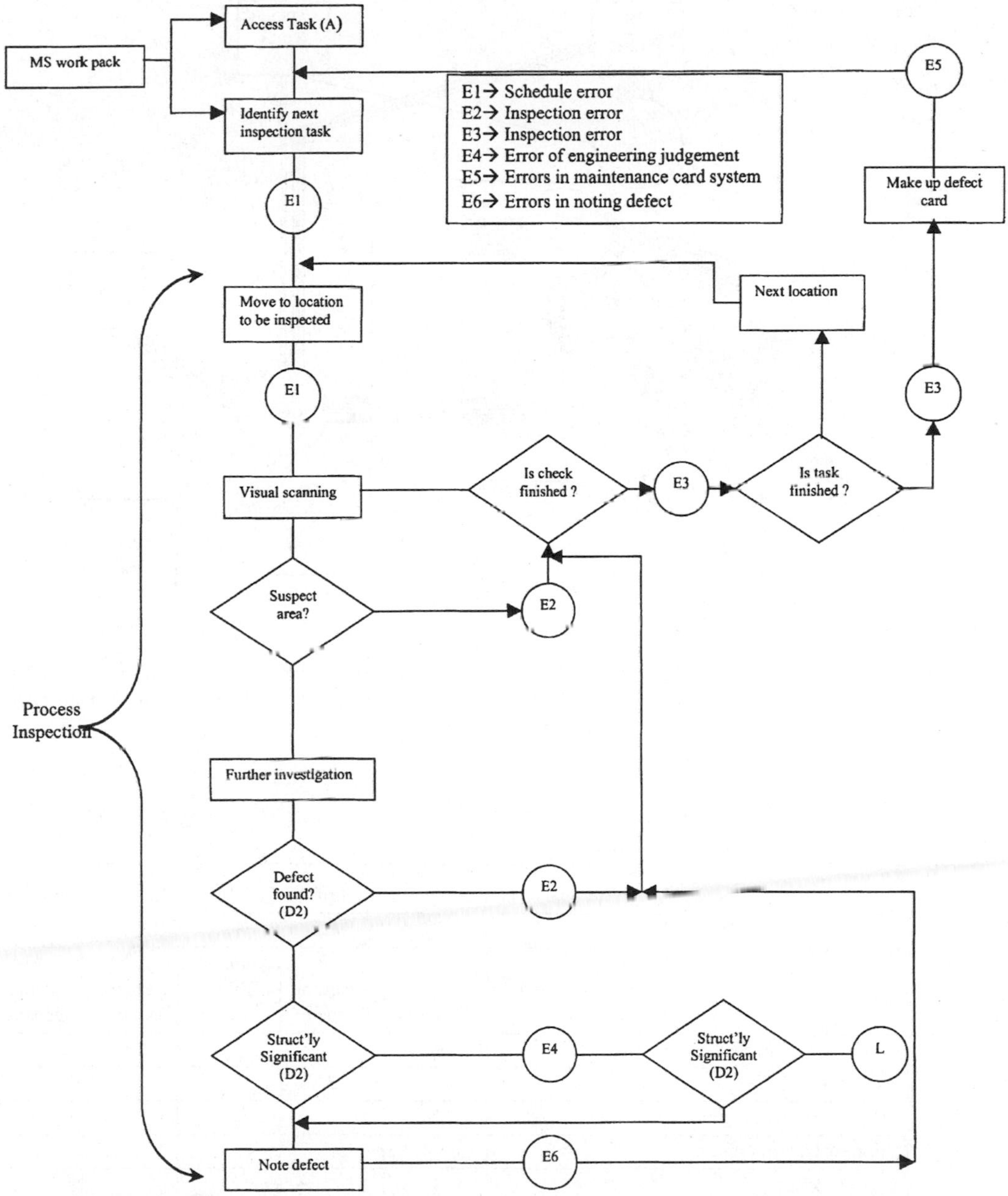

FIGURE 17.6 Inspection model flowchart. (From Lock, M. W. B. and Strutt, J. E. (1985) Reliability of in-service inspection of transport aircraft structures. *CAA Paper 85013*, London. With permission.)

the workload for different task segments as a function of time. Thus, knowing the workloads and human capabilities for different task segments, the analyst considers alternate function allocation strategies or provides additional resources for those tasks which have been identified as likely to overload the operator. In addition, the information on workloads and human capabilities can be used to specify or design hardware and software requirements. In a recent study, Wilson (1993) evaluated the workloads on aircraft pilots and the weapons systems officer during air-to-ground training missions using different workload measures. One such measure was the cardiac inter-beat intervals plotted as a function of different mission segment codes (see Figure 17.9).

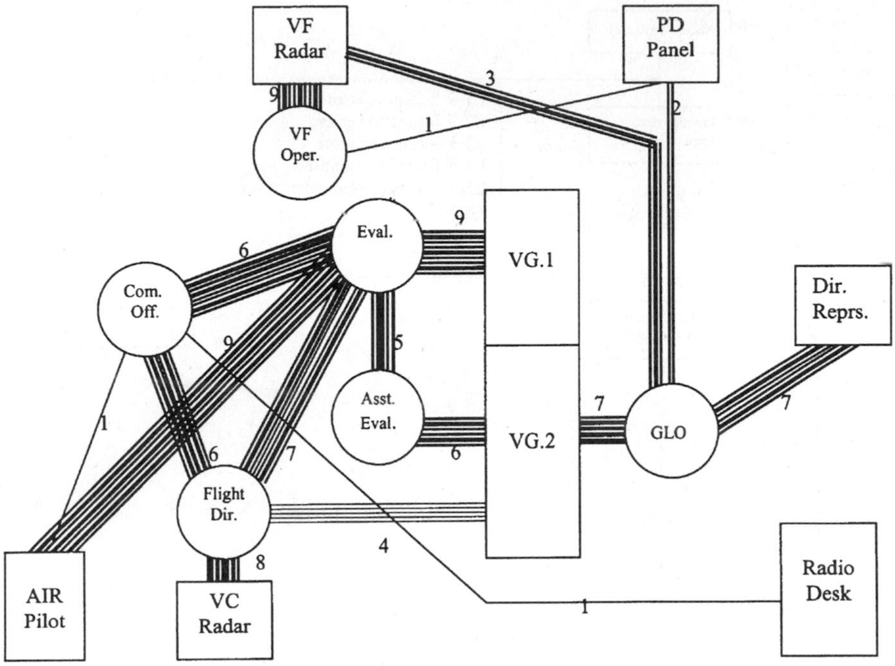

FIGURE 17.7 Schematic layout of the original arrangement of critical men and machines and the linkages between them. (From Chapanis, A. and Shafer, J. (1991) A workshop on human factors methods. *Annual Human Factors Society Meeting*, San Francisco, CA. With permission.)

TABLE 17.3 Linkages Between Men and Machine

An Example of Link Analysis: Layout of the CIC Aboard the U.S.S. Louisville-2

		MEN					
		Communications Officer	Evaluator	Assistant Evaluator	Gunnery Liaison Officer	Flight Director	VF Radar Operator
Men	Communications Officer		6			6	
	Evaluator	6		5		7	
	Assistant Evaluator		5				
	Gunnery Liaison Officer						
	Flight Director	6	7				
	VF Radar Operator						
Machines	VC Radar					8	
	Air Pilot	1	9				
	Radio Desk	1					
	PD Panel				2		1
	VG Radar No. 1		9				
	VG Radar No. 2			8	7	4	
	VF Radar				3		9
	Director Repeaters				7		

Chapanis, A. and Shafer, J. (1991) A workshop on human factors methods. *Annual Human Factors Society Meeting*, San Francisco, CA.

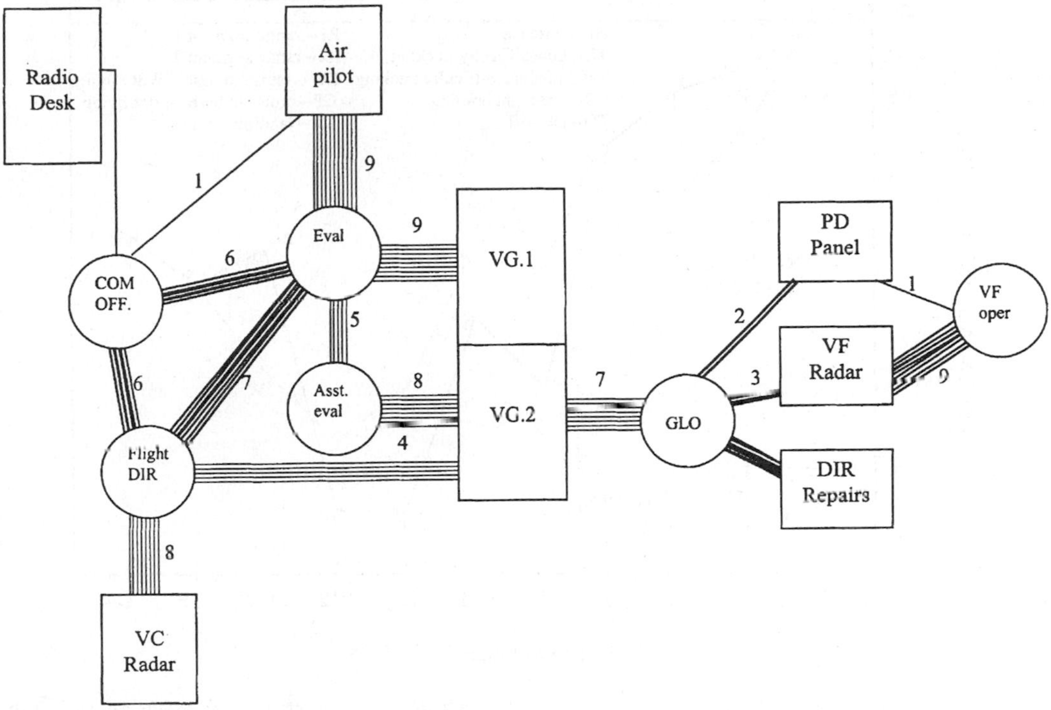

FIGURE 17.8 Schematic layout of the revised arrangement of critical men and machines and the linkages between them. (From Chapanis, A. and Shafer, J. (1991) A workshop on human factors methods. *Annual Human Factors Society Meeting*, San Francisco, CA. With permission.)

Hierarchical Task Analysis: The hierarchical task analysis (HTA) approach was originally an outgrowth of the work of Annett and Duncan (1967) and since then has been used extensively to study human error, human–machine interface design, human–computer interaction, training, job design, allocation of function, and assessment (Stammers and Shepherd, 1995). Simply stated, the HTA approach represents the task hierarchically.

To begin with, the HTA approach defines the overall system objectives, and later the tasks that are required to achieve system objectives. The tasks are redescribed in terms of a set of subtasks and subgoals and a plan that governs how the subtasks should be executed. Thus the final task analysis is hierarchical, with the level of detail increasing with each level. Each subtask can be examined to determine if it is defined to a sufficient level of detail; if not, the analyst can define it further. The level of detail is one of the critical issues in HTA, and various rules can be used to guide the analyst as to when the necessary level of detail has been attained. A commonly used rule is the P × C rule espoused by Annett and Duncan (1967). This rule states that further redefinition of the task is not necessary if the product of the probability (P) of inadequate performance and the cost (C) of inadequate performance is acceptable. The P × C rule can be applied to each task/subtask at a specific level. If the P × C value is unacceptable, then the task is broken down to greater levels of detail until an acceptable value of P × C is obtained for the lower level subtasks. Figure 17.10 shows an example of hierarchical task analysis applied to aircraft ramp maintenance activities (Mitchell, Bright, and Rickman, 1996). In this case, a human error classification scheme was developed and applied to a sample of operations in the HTA. This assisted the analysts in identifying and classifying errors and provided examples of potential human errors that might be expected to occur as operators perform generic maintenance tasks.

Fault Tree Analysis (FTA): The FTA is a quantitative top down approach that uses the and/or logic to estimate errors, representing tasks via a tree-like structure (Green, 1983; Stammers and Shepherd, 1995). The fault tree approach analyzes undesirable events often referred to as the "top event" by determining

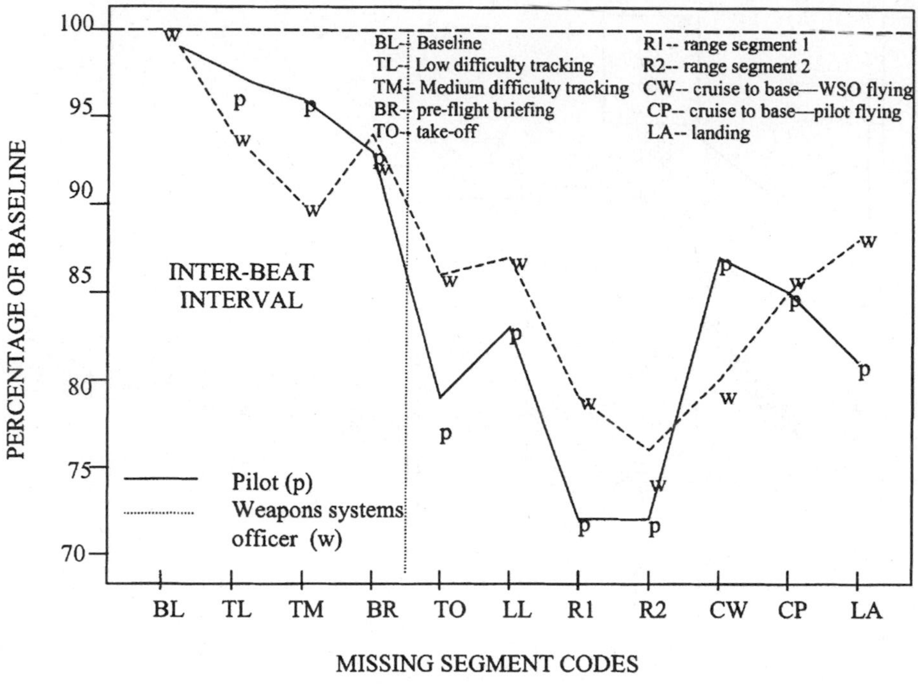

FIGURE 17.9 Mean Inter-beat Interval (IBI) percentage change from baseline for pilots and WSOs for each of the 11 segments. The vertical dotted line separates ground from flight segments. (From Wilson, G. F. (1993) Air-to-ground training missions: a psycho physiological workload analysis. *Ergonomics*, 36, 9, 1071-1087. With permission.)

what could cause it, either alone or in combination with other events. Herein "and" and "or" indicate the relationship between events. When two events are linked by "and," both events must occur in order for the parent event to occur, whereas when two events are linked by the term "or," the occurrence of one of the two events is sufficient to cause the parent event to occur. Probabilities can be assigned to individual events, enabling the analyst to estimate the probabilities of specific failures and those of the undesirable top event. The FTA has been used extensively to analyze the reliability and safety of complex systems (e.g., process industry, nuclear plant). Figure 17.11 shows the fault tree of an undesirable event in a nuclear power plant situation (Amendola et al., 1982). The FTA has also been used to a great degree to predict human error probabilities in complex systems — THERP approach (Swain and Guttman, 1980).

Failure Modes and Effects Analysis (FMEA): Unlike the FTA, which is a top-down approach, the FMEA is a bottom-up approach that has been successful in human reliability and error analysis studies (Hammer, 1985; Kirwan and Ainsworth, 1992). The FMEA approach is relatively straightforward. The analysis starts at the lowest level (i.e., task) and determines what effects a failure/error can have on system performance. The analyst typically starts with tasks and subtasks and identifies typical errors the human operator would be expected to make while executing the subtasks. Probability estimates or frequencies for each kind of error are estimated, following which the consequences of errors on system performance are deduced. If the consequences can be expected to be serious, further investigation is undertaken to identify interventions that can minimize errors or to design an error-tolerant system.

Behavior Taxonomy Approach: This approach focuses on developing behavioral taxonomies and classifying tasks based on different dimensions of performance. Over the years, various taxonomies have been developed to classify tasks (Miller, 1967; Fleishman and Quaintance, 1984). In addition, Gagne's (1977) taxonomy for learning tasks and the Position Analysis Questionnaire (McCormick, Jeanneret and Mecham, 1969) are commonly used in the areas of personnel selection and training.

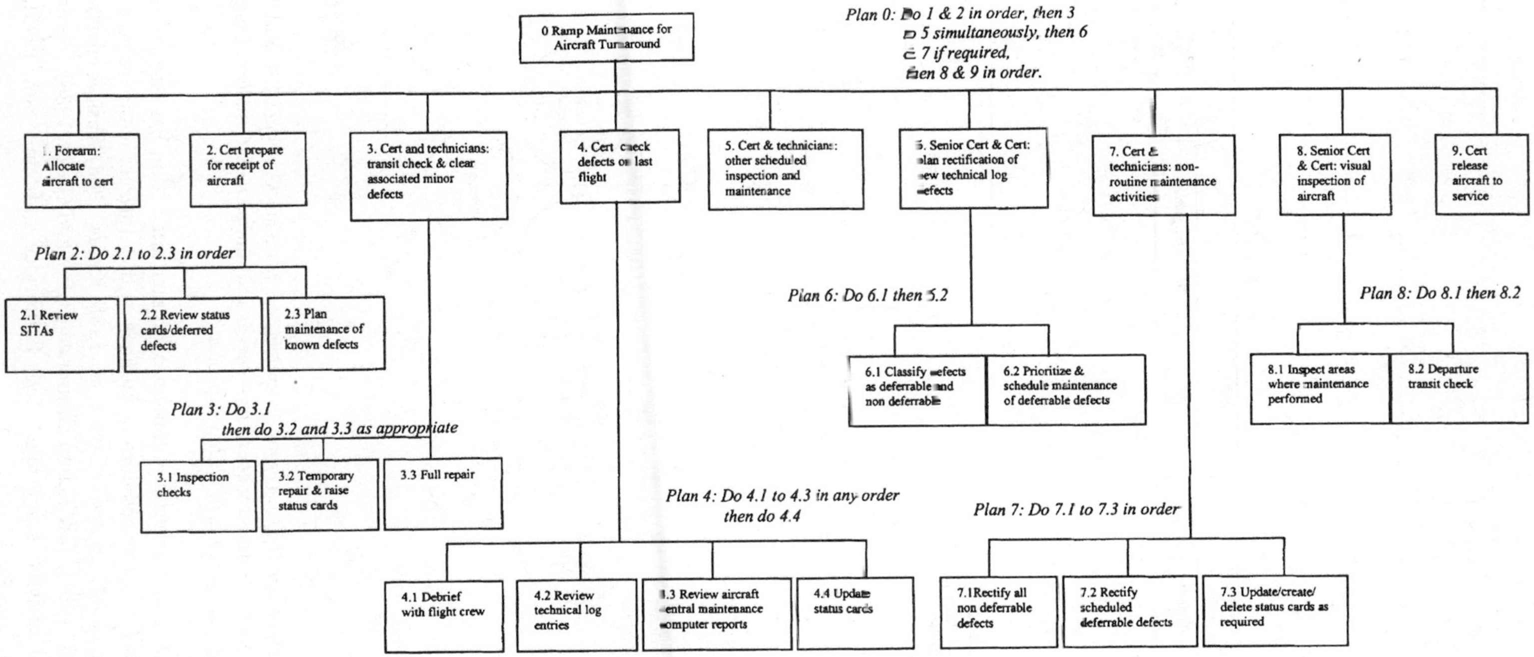

FIGURE 17.10 A hierarchical task analysis of aircraft ramp activities. (From Mitchell, K., Bright, C. K., and Rickman, J. K. (1996) A study into potential sources of human error in the maintenance of large civil transport aircraft. *CAA Paper 96004*, Lloyd's Register, Civil Aviation Authority, London. With permission.)

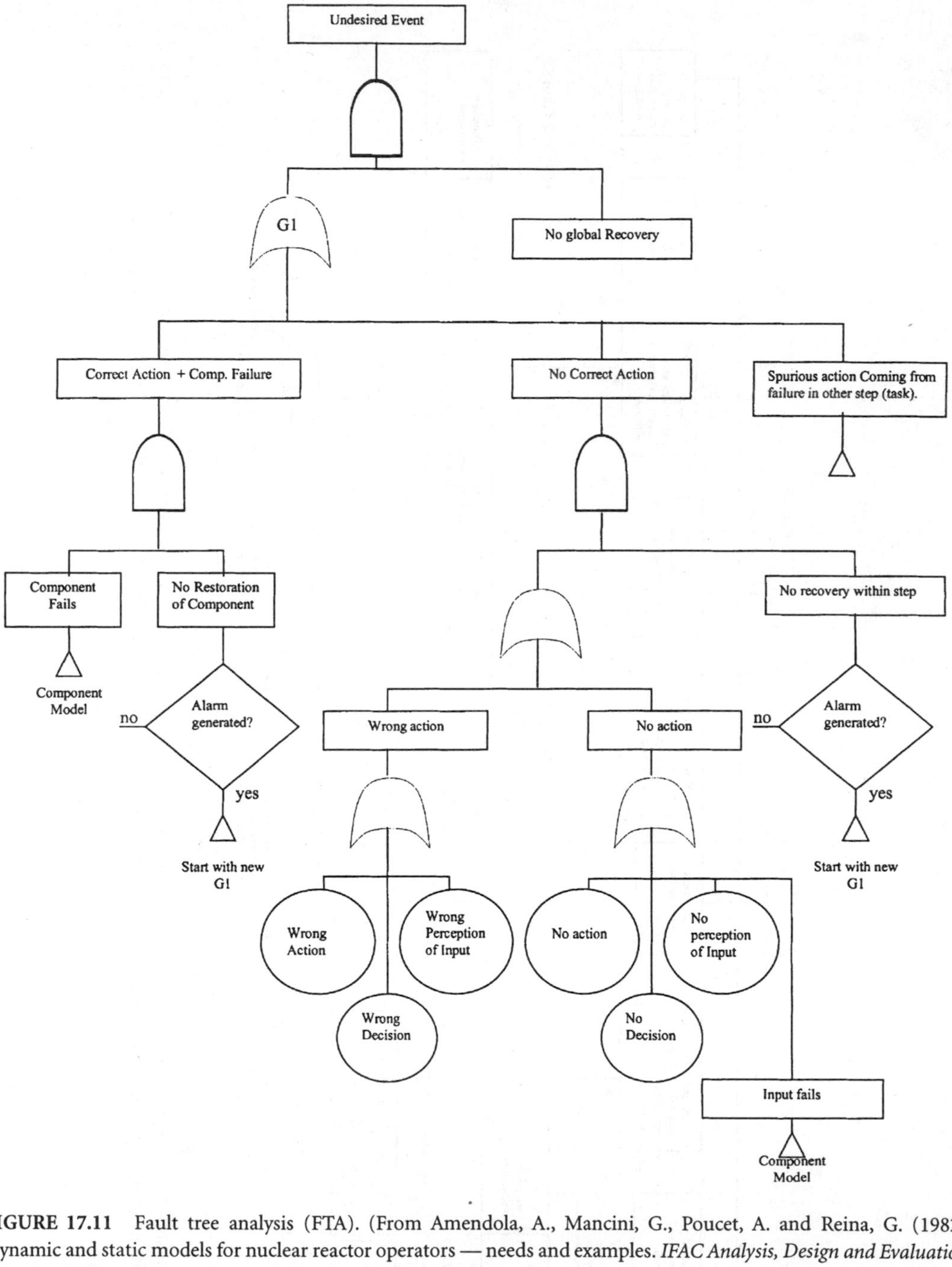

FIGURE 17.11 Fault tree analysis (FTA). (From Amendola, A., Mancini, G., Poucet, A. and Reina, G. (1982) Dynamic and static models for nuclear reactor operators — needs and examples. *IFAC Analysis, Design and Evaluation of Man-Machine Systems*, Germany, 103-110. With permission.)

Analysis of Cognitive Activities: Interest in human/computer interaction has resulted in several newer approaches to analyze the cognitive activities of humans as they interact with computers. Examples of such approaches include GOMS analysis (Card et al., 1983; Kieras, 1988), Task Analysis for Knowledge Structures (Johnson et al., 1984), Task Action Grammar (Payne and Green, 1986), mental model development, verbal protocol analysis (Bainbridge, 1991; Bainbridge and Sanderson, 1995), and cognitive simulation (Roth, Woods, and Pople, 1992).

17.4 Task Analysis: The Aircraft Inspection System

This section describes a study of the U.S. commercial aircraft inspection and maintenance system that was accomplished via a task analytic approach (Drury, Prabhu, and Gramopadhye, 1990; FAA, 1991; FAA, 1993a; Gramopadhye, Drury, and Prabhu, 1997). In this study, researchers analyzed the existing system, then identified potential ergonomic interventions. As is typical, following the identification of various interventions, the investigators undertook a detailed task analysis of specific subsystems with the intention that this analysis would lead to the implementation of specific interventions. A greater-than-usual level of detail is included here, since this document is intended to serve as a practical guide to others embarking on projects that require task analysis.

Introduction

The aircraft inspection/maintenance system is a complex one (Drury, Prabhu, and Gramopadhye, 1990; Drury, 1991; FAA, 1991) and is affected by a variety of geographically dispersed entities. These entities include large international carriers, regional and commuter airlines, repair and maintenance facilities, as well as the fixed-based operators associated with general aviation. An effective inspection is seen as a necessary prerequisite to public safety, so both inspection and maintenance procedures are regulated by the U.S. federal government via the Federal Aviation Administration (FAA). Investigators conducting this study found that while adherence to inspection procedures and protocols is relatively easy to monitor, tracking the efficacy of these procedures is not.

The maintenance process begins when a team that includes representatives from the FAA, aircraft manufacturers, and start-up operators schedule the maintenance for a particular aircraft. These schedules may be, and often are, later modified by individual carriers to suit their own scheduling requirements. These maintenance schedules are comprised of a variety of checks that must be conducted at various intervals. Such checks or inspections include flight line checks, overnight checks, and four different inspections of increasing thoroughness — the A, B, C, and most thorough and most time-consuming, D check. In each of these inspections, the inspector checks both the routine and non-routine maintenance of the aircraft. If a defect is discovered during one of these inspections, the necessary repairs are scheduled. Following these inspections, maintenance is scheduled to (1) repair known problems, (2) replace items because the prescribed amount of air time, number of cycles, or calendar time has elapsed, (3) repair previously documented defects (e.g., reports logged by pilot and crew, line inspection, items deferred from previous maintenance), and (4) perform the scheduled repairs (those scheduled by the team including the FAA representatives). In the context of an aging fleet, inspection takes an increasingly vital role. Scheduled repairs to an older fleet account for only 30% of all maintenance compared with the 60 to 80% in a newer fleet. This difference can be attributed to the increase in the number of age-related defects (FAA, 1991). In such an environment, the importance of inspection cannot be overemphasized. It is critical that these visual inspections be performed effectively, efficiently, and consistently over time. Moreover, 90% of all inspection in aircraft maintenance is visual in nature and is conducted by inspectors, so inspector reliability is fundamental to an effective inspection.

When the aircraft arrives at a maintenance site, the scheduled maintenance is translated into a set of job cards or work cards (i.e., instructions for inspection and maintenance). Initially, the aircraft is cleaned and access hatches opened so inspectors can view these various areas. This activity is followed by a thorough inspection — again, the inspection being primarily visual in nature. Since such a large part of the maintenance workload is dependent on the discovery of defects during inspection, it is imperative that this "incoming" inspection is completed as soon as possible after the aircraft arrives at the inspection maintenance site. At this point in the inspection process, inspectors are expected to discover those critical defects that will necessitate long follow-up maintenance times so this maintenance can be scheduled. Thus, there is a heavy inspection workload at the commencement of each inspection or check. It is only after the discovery of defects that the planning group can estimate the expected maintenance workload, order replacement parts, and schedule the maintenance. Frequently, maintenance facilities resort to overtime to accomplish these inspections. This overtime results in an increase in the total number of inspection hours, and often leads to prolonged work hours for the inspector. In addition, much of the inspection, including routine inspections on the flight line, are carried out during the night shift, because that is the time between the last flight of the day and first flight of the next.

During inspection, each defect is written up as a Non-Routine Card. This is translated into a set of work cards that identify the specific work needed to rectify the defect. Each of these defects is rectified by the maintenance crew. Rectifying each defect generates an additional inspection, typically called a "buy-back" inspection, conducted by the inspector to ensure that the work done by the maintenance crew meets the necessary standards.

The previous paragraph has pointed out that when an aircraft initially arrives at a maintenance site the inspection workload is quite large. As the service on the aircraft progresses, i.e., as maintenance crews begin work on the repairs, the inspection workload decreases. The inspection load increases again as maintenance tasks are completed. However, at this time the rhythm of the inspector's work is different; inspectors at this time are frequently interrupted as aircraft maintenance technicians request that inspectors conduct the required "buy-back" inspections of the completed work.

As in any system that is highly dependent on human performance, efforts made to reduce human errors by identifying human/system mismatches can have an impact on the overall effectiveness and the efficiency of the system. Given the backdrop of the inspection system, the objective of this particular study was to identify human/system mismatches and to design interventions that would improve human inspection performance in the aircraft maintenance system.

Objectives of the Study

The objectives of the study were fourfold:

1. To describe and analyze the existing aircraft inspection maintenance system
2. To identify human errors and develop a taxonomy of errors
3. To outline ergonomic interventions
4. To implement the specific ergonomic interventions expected to have the greatest impact on the system

Data Collection

The first stage of the assessment was to collect detailed information about the inspection process. A variety of data collection techniques were used to collect not only the process and procedural information and the idealized way of completing inspection tasks, but also information about the way tasks actually get accomplished. The main information sources for collection of task-descriptive data were the following:

1. *Observation and Shadowing:* Data was collected by observing aircraft inspection and maintenance operations at various sites, ranging from large international carriers to startup and regional operators of general aviation. This involved watching various inspectors and maintenance technicians accomplishing various tasks over several shifts.
2. *Interviewing:* Personal interviews conducted with inspectors, supervisors, aircraft maintenance technicians, lead mechanics and foremen, managers, planners, and other personnel associated with aircraft maintenance, were used to collect data on aircraft maintenance tasks. Interviews with system participants at all levels helped investigators to collect data on the structure and functioning of the system as well as to collect data on rare events such as system errors. This process was used to identify not only the prescribed way of working on the task but also the "quick and dirty" way that those tasks often really get completed.
3. *Documentation:* Information on aircraft inspection and maintenance procedures was obtained through company-wide procedures, Federal Aviation Authority mandated procedures (Federal Aviation Regulations — FARs), airworthiness directives, aircraft manufacturers manuals, and other documents.

Information from the above sources was used to obtain a basic understanding of the task(s) involved and to identify problem areas.

Task Description

The information on the inspection process was represented in various formats: (1) an inspection flow chart, (2) a task description based on generic inspection, and (3) a task description form. Following are more detailed descriptions of each of these formats.

Inspection Flow Chart: As a first step, a flow chart of inspection and maintenance activities was developed to illustrate the relationship between inspection and maintenance activities and the relationships among the personnel involved. A modified inspection/maintenance diagram is shown in Figure 17.12 (Kraus and Gramopadhye, in press)

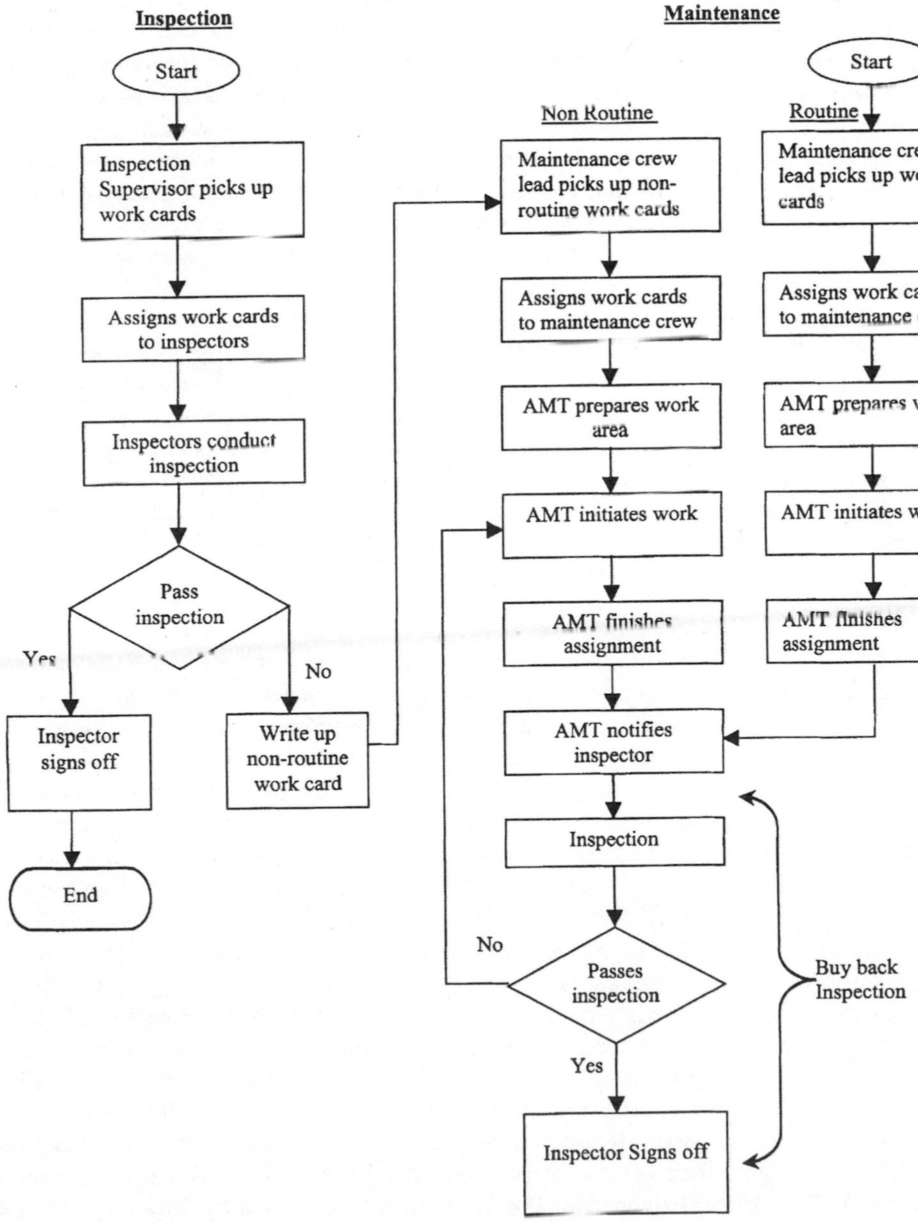

FIGURE 17.12 Inspection flow diagram. (From Kraus, D. and Gramopadhye, A. K. Transfer effects of computer-based team training. [To appear in the special issue on Human Factors in Aviation Maintenance in the International Journal of Industrial Ergonomics.]

Task Description Based on Generic Inspection Activities: Although general task description approaches are widely available (see Drury, 1983; Kirwan and Ainsworth, 1992), investigators in this case decided it would be advantageous to use an approach directly related to inspection. Literature describing human factors in inspections has produced the following generic list of inspection activities (e.g., Sinclair, 1984):

1. Present item to inspector
2. Search for flaws
3. Decide on rejection/acceptance of each flaw
4. Respond by taking necessary action.

It should be noted that not all steps may be required for all inspections. For example, some inspections may not necessitate search (e.g., color matching) and some may not necessitate decision making (e.g., absence of a rivet head on a lap splice). The description of various tasks within the aircraft inspection context is as shown in Table 17.4. The unit of task description was the work card (instructions outlining the specific inspection activity). A task was seen as continuing until a repair was completed and had passed airworthiness requirements. The work card was the unit of work assigned to one particular inspector on one physical assignment, but the actual time required to complete work shown on a work card is not consistent from work card to work card. Typically, inspectors were expected to complete the work outlined by the work card within one shift. However, work might have to be continued across shifts. The work listed on each work card could, potentially, require several inspections. Detailed task descriptions were obtained through site visits to various aircraft maintenance facilities.

Task Description Form: Data collected through interviews and site visits were transcribed using a standard working form (see Table 17.5), with a separate form for each of the five steps (initiate, access, search, decision making, and respond) in the generic task description. During the data collection phase, the human factors analysts remained with the inspectors, asking probing questions while these inspectors were actually at work. Following this step, knowledge of human factors models of inspection performance and the functioning of individual subsystems were used to identify specific subsystems and potential human/system mismatches under the Observations column in Table 17.5.

Task Analysis

Following this step, a detailed taxonomy of errors was developed from the failure modes of each task in aircraft inspection (Table 17.6.). This taxonomy, based on the failure modes and effects analysis (FMEA) approach, was developed due to the realization that a proactive approach to error control is necessary to identify potential errors. Thus, the taxonomy was aimed at the phenotypes of error (Hollnagel, 1989), that is, the observed errors. Using the generic task description of the inspection system, the goal or outcome of each task was postulated as shown in Table 17.6 (third column). These outcomes then formed the basis for identifying the failure modes of each task and included operational error data gained from the observations of inspectors and from discussions with various aircraft maintenance personnel, which were collected over a period of two years. Later, the frequencies of each kind of error were estimated, following which the consequence of errors on system performance was deduced. The error taxonomy provided the analysts a systematic framework to suggest appropriate intervention strategies. Following the classification of errors, the investigators, using their knowledge of human factors, identified interventions expected to minimize the errors. The role of the human factors effort was to change the human/machine system, thereby reducing the error incidence and making the system more reliable. Specifically, two types of interventions were considered: changing the system to fit the human, or changing the human inspector to fit the system. The specific interventions for five steps of the inspection process are listed in Table 17.7. Since then, various interventions for improving inspection performance have been implemented (FAA 1993b; Gramopadhye, Drury, and Sharit, 1993; Gramopadhye et al., 1997; Patel, Drury, and Prabhu 1993).

TABLE 17.4 Detailed Breakdown of Aircraft Visual Inspection by Task Step

Task Description	Visual Example	Task Step
1. Initiate	Get work card. Read and understand area to be covered.	1.1 Correct instructions written. 1.2 Correct equipment procured. 1.3 Inspector gets instructions. 1.4 Inspector reads instructions. 1.5 Inspector understands instructions. 1.6 Correct instructions available. 1.7 Inspector gets equipment. 1.8 Inspector checks/calibrates equipment
2. Access	Locate area on aircraft. Get into correct position.	2.1 Locate area to inspect. 2.2 Area to inspect. 2.3 Access area to inspect.
3. Search	Move eyes across area systematically. Stop if any indication.	3.1 Move to next lobe. 3.2 Enhance lobe (e.g., illuminate, magnify for vision, use dye penetrant, tap for auditory inspection). 3.3 Examine lobe. 3.4 Sense indication in lobe. 3.5 Match indication against list. 3.6 Remember matched indication 3.7 Remember lobe location. 3.8 Remember access area location. 3.9 Move to next access area.
4. Decision making	Examine indication against remembered standards (e.g., for dishing or corrosion).	4.1 Interpret indication. 4.2 Access comparison standard. 4.3 Access measuring equipment. 4.4 Decide on if it is a fault. 4.5 Decide on action. 4.6 Remember decision/action.
5. Respond	Mark defect. Write up repair sheet or if no defect, return to search.	5.1 Mark fault on aircraft. 5.2 Record fault. 5.3 Write repair action.
6. Repair	Repair area (drill out and repair rivet).	6.1 Repair fault
7. Buy back inspect	Visually inspect marked area.	7.1 Initiate. 7.2 Access. 7.3 Search. 7.4 Decision. 7.5 Respond.

From FAA (1991) *Human Factors in Aviation Maintenance — Phase One Progress Report.* DOT/FAA/AM-91/16, Office of Aviation Medicine, Washington, D.C.

17.5 Conclusion

This review has shown that task analysis has evolved over the years, with its growth linked closely to advances in technology and the growth of new application domains. A common theme that resonates throughout the chapter is that task analysis needs to be the basis for any human factors design effort. Ergonomists and human factors engineers need to bear this in mind as they embark upon any system design efforts. Furthermore, as practitioners continue to use different task analytic approaches, it is critical that they communicate the success and pitfalls in using a specific approach for a particular application. This will ultimately lead to the development of detailed principles or guidelines in using various task analytic approaches based on inputs available and outputs desired.

TABLE 17.5 Task Description Form

Task: Wing and Leading Edge Inspection **Location:** Right Wing

	Task Analysis									
	Subsystems									
Task Description	A	S	P	D	M	C	F	P	O	Observations
Search										
1.0 Inspect slat structure, wiring and installation.										
1.1 Check wear on male duct of telescopic shaft duct.										
1.1.1 Check wear by moving it and seeing if it is loose.			X		X					No prescribed force.
1.2 Inspect slatwell area for corrosion and cracks.								X		No standards for wear.
1.2.1 Hold flashlight such that light falls perpendicular to the surface.								X		Holding flashlight for a long period at odd positions – strenuous.
1.2.2 Visually look for cracks and corrosion.			X	X	X					Lack of information on type of cracks and figures.
1.2.3 The visual indication confirmed by tactile and moves scrutinized inspection.			X	X	X					
1.3 Look for play in slat actuator nut.										

Attention:	Number of time-shared tasks	Perception:	
Memory:	SSTS, Working, Long-term	Feedback:	Quality, Amount, Timing
Senses:	Visual, Tactile, Auditory	Decision:	Sensitivity, Criterion, Timing
Control:	Continuous, Discrete	Posture:	Reading, Forces, Balances, Extreme Angles

From FAA (1991) *Human Factors in Aviation Maintenance — Phase One Progress Report.* DOT/FAA/AM-91/16, Office of Aviation Medicine, Washington, D.C.

TABLE 17.6 Task and Error Taxonomy for Visual Inspection (Tasks: Initiate and Access)

Task	Errors	Outcome
	1. INITIATE	
1.1 Correct instructions.	1.1.1 Incorrect instructions. 1.1.2 Incomplete instructions. 1.1.3 No instructions available.	Inspector has correct and correctly working equipment, and understands instructions.
1.2 Correct equipment procured.	1.2.1 Incorrect equipment. 1.2.2 Equipment not procured.	
1.3 Inspector gets instructions.	1.3.1 Fails to get instructions.	
1.4 Inspector reads instructions.	1.4.1 Fails to read instructions. 1.4.2 Partially reads instructions.	
	2. ACCESS	
1.5 Inspector understands instructions.	1.5.1 Fails to understand instructions. 1.5.2 Misinterprets instructions. 1.5.3 Does not act on instructions.	
1.6 Correct equipment available.	1.6.1 Correct equipment not available. 1.6.2 Equipment not available. 1.6.3 Equipment not working.	
1.7 Inspector gets equipment.	1.7.1 Gets wrong equipment. 1.7.2 Gets incomplete equipment. 1.7.3 Gets non-working equipment.	
1.8 Inspector checks/calibrates equipment.	1.8.1 Fails to check/calibrate. 1.8.2 Checks/calibrates incorrectly.	
2.1 Locate area to inspect.	2.1.1 Locate wrong aircraft. 2.1.2 Locate wrong area on aircraft. 2.1.3 Mis-locate boundaries of area.	Inspector with correct equipment at correct inspection site, is ready to begin inspection.
2.2 Area ready to inspect.	2.2.1 Cleaning work not completed. 2.2.2 Cleaning work incorrect. 2.2.3 Mtc. access tasks not completed. 2.2.4 Mtc. access tasks incorrect. 2.2.5 Parallel work prevents access. 2.2.6 Parallel work impedes inspection.	
2.3 Access area to inspect.	2.3.1 Access equipment not available. 2.3.2 Incorrect access equipment. 2.3.3 Access equipment poorly designed. 2.3.4 Access not physically possible. 2.3.5 Access discouragingly difficult. 2.3.6 Access dangerous to inspection.	

TABLE 17.6 (continued) Task and Error Taxonomy for Visual Inspection (Tasks: Initiate and Access)

Task	Errors	Outcome
	3 SEARCH	
3.1 Move to next lobe.	3.1.1 Misses parts of access area. 3.1.2 Multiple searches of parts. 3.1.3 Too close/far between lobes. 3.1.4 Move to non-required area.	All indications located in all access areas.
3.2 Enhance lobe (e.g., illuminate, magnify for vision, use dye penetrant, tap for auditory inspection).	3.2.1 Enhance wrong area. 3.2.2 Enhance area inadequately. 3.2.3 Fail to use enhancing equipment.	
3.3 Examine lobe.	3.3.1 Fail to examine lobe. 3.3.2 Examine too short/long time. 3.3.3 Incorrect depth of examination. 3.3.4 Incomplete examination of lobe. 3.3.5 Fatigue from fixed posture.	
3.4 Sense indication in lobe.	3.4.1 Fail to attend lobe. 3.4.2 Fail to use cues present. 3.4.3 Fail to sense indication. 3.4.4 Sense wrong indication.	
3.5 Match indication against list.	3.5.1 Match against faults not listed. 3.5.2 Fail to match against full list. 3.5.3 Incorrect match.	
3.6 Remember matched indication.	3.6.1 Fail to record matched indication. 3.6.2 Forget matched indication.	
3.7 Remember lobe location.	3.7.1 Fail to record lobe location. 3.7.2 Forget lobe location.	
3.8 Remember access area location.	3.8.1 Fail to record access area location. 3.8.2 Forget access area location.	
3.9 Move to next access area.	3.9.1 Miss parts of area. 3.9.2 Multiple searches of parts. 3.9.3 Move to non-required area.	

Note: Search proceeds by successively examining each small area, called here a lobe, within a single area accessible without performing a new access, called here an access area. When all lobes have been examined in that access area, a new access is performed followed by a new search. The concept of a lobe comes from visual search where it is called a visual lobe.

TABLE 17.6 (continued) Task and Error Taxonomy for Visual Inspection (Tasks: Initiate and Access)

Task	Errors	Outcome
	4 DECISION	
4.1 Interpret indication.	4.1.1 Classify as wrong fault type.	All indications located are correctly classified, correctly labeled as fault or no fault, and actions correctly planned for each indication.
4.2 Access measuring equipment	4.2.1 Choose wrong measuring equipment. 4.2.2 Measuring equipment not available. 4.2.3 Measuring equipment not working. 4.2.4 Measuring equipment not calibrated. 4.2.5 Measuring equipment wrongly calibrated. 4.2.6 Does not use measuring equipment.	
4.3 Access comparison standard.	4.3.1 Choose wrong comparison standards. 4.3.2 Comparison standard not available. 4.3.3 Comparison standard not correct. 4.3.4 Comparison incomplete. 4.3.5 Does not use comparison standard.	
4.4 Decide on if fault.	4.4.1 Type I error, false alarm. 4.4.2 Type II error, missed fault.	
4.5 Decide on action.	4.5.1 Choose wrong action. 4.5.2 Second opinion if not needed. 4.5.3 No second opinion if needed. 4.5.4 Call for buy-back when not required. 4.5.5 Fail to call for required buy-back.	
4.6 Remember decision/action.	4.6.1 Forget decision/action. 4.6.2 Fail to record decision/action.	
	5 RESPOND	
5.1 Mark fault on aircraft	5.1.1 Fail to mark fault 5.1.2 Mark non-fault. 5.1.3 Mark fault in wrong place. 5.1.4 Mark fault with wrong tag. 5.1.5 Mark fault with wrong marker.	All faults and repair items are correctly recorded.
5.2 Record fault.	5.2.1 Fail to record fault. 5.2.2 Record non-fault. 5.2.3 Record fault in wrong place. 5.2.4 Record fault incorrectly.	
5.3 Write repair action.	5.3.1 Fail to write repair action. 5.3.2 Write repair action for non-fault. 5.3.3 Write repair action for wrong place. 5.3.4 Mis-write repair action. 5.3.5 Specify buy-back if not needed. 5.3.6 Fail to specify needed buy-back.	
	6. REPAIR	
6.1 Repair fault.	6.1.1 Fail to repair fault. 6.1.2 Repair non-fault. 6.1.3 Mis-repair fault. 6.1.4 Prevent access for buy-back.	All recorded faults correctly repaired and accessible for buy-back inspection.

From FAA (1991) *Human Factors in Aviation Maintenance — Phase One Progress Report.* DOT/FAA/AM-91/16, Office of Aviation Medicine, Washington, D.C.

TABLE 17.7 Potential Strategies for Improving Inspection (FAA, 1991).

	Strategy	
Task Step	Changing Inspector	Changing System
Initiate	Training in visual inspection procedures (procedural training).	Redesign of job cards Feedforward of expected flaws
Access	Training in area location (knowledge and recognition training).	Better support stands Better area location systems Location for visual inspection equipment
Search	Training in visual search (cueing, progressive-part).	Task lighting Optical aids Improved visual inspection templates
Decision	Decision training (cueing feedback, understanding of standards).	Standards at the work point Pattern recognition off job aids Improved feedback to inspection
Response	Training in writing skills.	Improved fault marking Hands-free fault recording

From FAA (1991) *Human Factors in Aviation Maintenance — Phase One Progress Report.* DOT/FAA/AM-91/16, Office of Aviation Medicine, Washington, D.C.

References

Amendola, A., Mancini, G., Poucet, A. and Reina, G. (1982) Dynamic and static models for nuclear reactor operators — needs and examples. *IFAC Analysis, Design and Evaluation of Man-Machine Systems*, Germany, 103-110.

Anne, M. and Greenstein, J. S. (1993) The design of a computer-based tool to aid the manager of a real-time manufacturing control system (*Tech. Report*). Clemson, SC: Clemson University, Department of Industrial Engineering.

Armstrong, T. J., Radwin, R. Hansen, D. J., and Kennedy, K. W. (1986) Repetitive trauma disorders: job evaluation and design. *Human Factors*, 28(3), 325-336.

Annett, J. and Duncan, K. D. (1967) Task analysis and training design. *Occupational Psychology*, 41, 211-212.

Bainbridge, L. and Sanderson, P. (1995) Verbal protocol analysis, in J. R. Wilson and E. N. Corlett (eds.) *Evaluation of Human Work: A Practical Ergonomic Methodology.* Taylor & Francis: London, 169-201.

Bainbridge, L. (1991) Mental models in cognitive skill: the example of industrial process operation, in P. Bibby et al. (eds.) *Models in the Mind.* Academic Press: London.

Barnes, R. (1980) *Motion and Time Study Design and Measurement of Work.* John Wiley & Sons Inc., New York.

Card, S. K., Moran, T. P., and Newell, A. L. (1983) *The Psychology of Human Computer Interaction.* Hillsdale, NJ: Erlbaum.

Carey, M. S., Stammers, R. B., and Astley, J. A. (1989) Human computer interaction design: the potential and pitfalls of hierarchical task analysis, in D. Diaper (ed.) *Task Analysis for Human–Computer Interaction.* Ellis Horwood: Chichester, 56-70.

Chapanis, A. (1962) *Research Techniques in Human Engineering.* John Hopkins University Press: Baltimore.

Chapanis, A. and Shafer, J. (1991) A workshop on human factors methods. *Annual Human Factors Society Meeting*, San Francisco, CA.

Corlett, E. N. and Bishop, R. P. (1976) A technique for assessing postural discomfort. *Ergonomics*, 19, 175-182.

Crossman, E. R. F. W. (1956) Perceptual activity in manual work, *Research*, 9, 42-49.

Diaper, D. (1989) *Task Analysis for Human–Computer Interaction.* Ellis Horwood Limited: Chichester, England.

Drury, C. G., Paramore, B., Van Cott, H. P., Grey, S., and Corlett, E. N. (1987) Task analysis, Chapter 3.4, In G. Salvendy (ed.), *Handbook of Human Factors.* John Wiley & Sons Inc.

Drury, C. G. (1991) The maintenance technician in inspection. Chapter 3, in FAA (1991) *Human Factors in Aviation Maintenance — Phase One Progress Report.* DOT/FAA/AM-91/16, Office of Aviation Medicine, Washington, D.C., 45-103.

Drury, C. G., Prabhu, P., and Gramopadhye, A. K. (1990) Task analysis of aircraft inspection activities: methods and findings. *Proceedings of the Human Factors Society 34th Annual Meeting,* 1181-1184.

Drury, C. G. (1987) A bio-mechanical evaluation of the repetitive motion injury potential of industrial jobs. *Seminars in Occupational Medicine,* Volume 2, No. 1, 41-49.

Drury, C. G. and Wick, J. (1984) Ergonomic applications in the shoe industry. *Proceedings of the International Conference on Occupational Ergonomics,* Toronto, Canada, 489-493.

Drury, C. G. (1983) Task analysis methods in industry. *Applied Ergonomics,* 14(1), 19-28.

Duncan, K. D. (1974) Analysis techniques in training design, in E. Edwards and F. P. Lees (eds.) *The Human Operator in Process Control.* Taylor & Francis: London.

FAA (1991) *Human Factors in Aviation Maintenance — Phase One Progress Report.* DOT/FAA/AM-91/16, Office of Aviation Medicine, Washington, D.C.

FAA (1993a) *Human Factors in Aviation Maintenance — Phase Two Progress Report.* DOT/FAA/AM-93/5, Office of Aviation Medicine, Washington, D.C.

FAA (1993b) *Human Factors in Aviation Maintenance — Phase Two Progress Report.* DOT/FAA/AM-93/15, Office of Aviation Medicine, Washington, D.C.

Flanagan, J. C. (1954) The critical incident technique. *Psychological Bulletin,* 51, 327-358.

Fleishman, E. A. and Quaintance, M. K. (1984) *Taxonomies of Human Performance.* Academic Press: New York.

Gilbreth, F. B. (1911) *Motion Study.* D. Van Nostrand Co. Princeton.

Gagne, R. M. (1977) *The Conditions of Learning.* 3rd edition. Holt, Rinehart and Winston: New York.

Gramopadhye, A. K., Drury, C. G., and Prabhu, P. V. (1997) Training strategies for visual inspection. *International Journal of Human Factors in Manufacturing,* 7(3), 171-196.

Gramopadhye, A. K., Kimbler, D., Kimbler, E., Bhagwat, S, and Rao, P. (1995) Application of advanced technology to training for visual inspection. *Proceedings of the Human Factors and Ergonomics Society 39th Annual Meeting,* 1299-1304.

Gramopadhye, A. K., Drury, C. G., and Sharit, J. (1993) Training for decision making in aircraft inspection. *Proceedings of the Human Factors and Ergonomics Society 37th Annual Meeting,* 1267-1271.

Gramopadhye, A. K., Bhagwat, S., Kimbler, D., and Greenstein, J. (1998) The use of advanced technology for visual inspection training. *Applied Ergonomics,* Vol. 29, No. 3.

Green, A. E. (1983) *Safety Systems Reliability.* John Wiley: Chichester.

Greer, C. W. (1981) Human engineering procedure guide. *Report AFAMRL-TR-81-35.* Wright-Patterson Air Force Base, Ohio, USA,.

Hammer, W. (1985) *Occupational Safety Management and Engineering.* Prentice Hall, New Jersey.

Hollnagel, E. (1989) The phenotypes of erroneous actions: implications for HCI design, in G. R. S. Weir and J. L. Alty (eds.) *Human–Computer Interaction and Complex Systems.* Academic Press: London.

Johnson, P., Diaper, D., and Long, J. B. (1984) Task, skills and knowledge: task analysis for knowledge based descriptions, in Shackel, B. (ed.) *Interact'84 — Proceedings of the First IFIP Conference on Human Computer Interaction.* Amsterdam: North Holland, 23-27.

Johnson, H. and Johnson, P. (1989) Integrating task analysis into system design: surveying designers needs. *Ergonomics,* Vol. 32, No. 11, 1451-1467.

Johnson, P., Johnson, H., Waddington, R., and Shouls, A. (1988) Task related knowledge structures: analysis, modeling and application, in Jones, D. M. and Winder, R. (eds.) *People and Computers: From Research to Implementation.* Cambridge University Press: Cambridge, 35-62.

Kieras, D. (1988) Towards a practical goms model methodology for user interface design, in M. Helander (ed.), *Handbook of Human–Computer Interaction.* Elsevier Science Publishers B. V., North Holland, 135-157.

Kirwan, B. and Ainsworth, L. K. (1992) *A Guide to Task Analysis.* Taylor & Francis: London.

Kraus, D. and Gramopadhye, A. K. Transfer effects of computer-based team training. (To appear in the special issue on Human Factors in Aviation Maintenance in the International Journal of Industrial Ergonomics).

Landau, K. and Rohmert, W. (1981) AET — A new job analysis method, *Spring Annual Conference Proceedings.* Bern — Suttgart: Hans Huber.

Lock, M. W. B. and Strutt, J. E. (1985) Reliability of in-service inspection of transport aircraft structures. *CAA Paper 85013,* London.

McCormick, E. J., Jeanneret, P. R., and Machan, R. C. (1969) A study of job characteristics and job dimensions as based on the position analysis questionnaire. *Report No. 6,* Occupational Research Center, Purdue University, West Lafayette, Indiana, USA.

McCormick, E. J. (1976) Job and task analysis, in M. D. Dunette (ed.) *Handbook of Organizational and Industrial Psychology.* Rand McNally: Chicago.

Meister, D. (1985) *Behavioral Analysis and Measurement Methods.* Wiley: New York.

Miller, R. B. (1953) *A Method for Man–Machine Task Analysis.* Wright-Patterson Air Force Base, OH: Wright Air Development Center (DTIC AD-15721).

Miller, R. B. (1967) Task taxonomy: science or technology? *Ergonomics,* 10, 167-176.

Mitchell, K., Bright, C. K., and Rickman, J. K. (1996) A study into potential sources of human error in the maintenance of large civil transport aircraft. *CAA Paper 96004,* Lloyd's Register, Civil Aviation Authority, London.

Patel, S. C., Drury, C. G., and Prabhu, P. (1993) Design and usability evaluation of work control documentation. *Proceedings of the Human Factors and Ergonomics Society 37th Annual Meeting,* 1156-1160.

Payne, S. J. and Green, T. R. G. (1986) Task-action grammars: a model of the mental representation of task languages. *Human–Computer Interaction,* 2, 93-133.

Piso, E. (1981) Task analysis for process control tasks: The method of Annett et al., applied. *Journal of Occupational Psychology,* 54, 247-254.

Pitkaar, R. N., Lenior, T. M. J., and Rijnsdorp, J. E. (1990) Implementation of ergonomics in design practice: outline of an approach and some discussion points. *Ergonomics,* 1990, 33, 5, 583-587.

Rasmussen, J. (1983) Skills, rules, knowledge: signals, signs and symbols and other distinctions in human performance models. *IEEE Transactions: Systems, Man and Cybernetics,* Vol. SMC-13, 257-267.

Roth, E. M., Woods, D. D., and Pople, Jr., H. E. (1992) Cognitive simulation as a tool for cognitive task analysis. *Ergonomics,* 35, 10, 1163-1198.

Seymour, W. D. (1966) *Industrial Skills.* Isaac Pitman: London.

Shackel, B., Chidsey, K. S., and Shipley, P. (1979) The assessment of chair comfort. *Ergonomics,* 12, 269-306.

Shepherd, A. (1989) Analysis and training in information technology tasks, in D. Diaper (ed.) *Task Analysis for Human–Computer Interaction.* Ellis Horwood: Chichester, 15-55.

Shepherd, A. (1993) An approach to information requirements specification for process control tasks. *Ergonomics,* 36, 1425-1437.

Sheridan, T. (1997) Task analysis, task allocation and supervisory control, in M. Helander, T. K. Landauer, and P. Prabhu (eds.) *Handbook of Human–Computer Interaction.* Elsevier Science B. V., 87-105.

Sinclair, M. (1984) Ergonomics of quality control. Workshop document, *International Conference on Occupational Ergonomics,* Toronto.

Singleton, W. T. (1974) *Man–Machine Systems.* London: Penguin.

Stassen, H. G., Steele, R. D., and Lyman, L. (1989) A man-machine system approach in the field of rehabilitation: a challenge or necessity, in Baosheng Hu (ed.) *IFAC Analysis, Design and Evaluation of Man–Machine Systems,* PRC, 79-86.

Stammers, R. B. and Shepherd, A. (1995) Task analysis, in J. R. Wilson and E. N. Corlett (eds.) *Evaluation of Human Work: A Practical Ergonomic Methodology.* Taylor & Francis: London, 144-169.

Stammers, R. B. (1995) Factors limiting the development of task analysis. *Ergonomics*, Vol. 38, No. 3, 588-594.

Swain, A. D. (1962) System and task analysis, a major tool for designing the personnel subsystem. (*Report SCR-457*), Sandia Corp.: Albuquerque, New Mexico.

Swain, A. D. (1964) THERP (*Report SC-R-64-1338*) Sandia Corp.: Albuquerque, New Mexico.

Swain, A. D. and Guttman, H. E. (1980) Handbook of Human Reliability Analysis and Emphasis on nuclear power plant application (*NUREG/CR-1278*). Washington, D.C.: U.S.A.

Taylor, F. W. (1929) *The Principles of Scientific Management.* Harper and Bros. New York.

Umbers, I. G. and Reiersen, C. S. (1995) Task analysis in support of the design and development of a nuclear power plant safety system. *Ergonomics*, 38, 3, 443-454.

Wilson, G. F. (1993) Air-to-ground training missions: a psycho physiological workload analysis. *Ergonomics*, 36, 9, 1071-1087.

18

A Computer-Based Tool for Practical Ergonomic Job Analysis

Steven L. Johnson
University of Arkansas

18.1 Introduction

One of the simplest definitions of ergonomics is matching the physical, physiological, and psychological requirements of the job with the capabilities of the human operator. This goal is obviously not new and has long been an important component of ensuring the operational effectiveness of both military and commercial operations from the perspectives of productivity and product quality. Traditional industrial engineering efforts in methods and work measurement have also addressed job analysis and documentation from the time of Taylor and the Gilbreths. The adequacy of the match between operator capability and job requirements also affects a company's indirect costs related to absenteeism, turn-over rates, and training costs. In addition to reducing the quality of working life, a mismatch contributes to lost time, accidents and injuries, restricted work assignments, as well as workers' compensation costs. A factor that has increased industry's attention to job requirements is the need to be in regulatory compliance.

Management, labor, and governmental agencies have all increased their attention to ensuring that job characteristics are compatible with the abilities of operators because of two federal initiatives. The Americans with Disabilities Act (ADA) addresses the analysis of job requirements in the contexts of both "essential job functions" and "reasonable accommodations." The Occupational Health and Safety Administration's recent attention to musculoskeletal disorders to the upper extremities has also increased the focus on the characteristics of tasks that increase the risk of injuries and illnesses (i.e., risk factors), particularly in manufacturing, assembly, and processing facilities.

The direct and indirect financial costs to both the company and the employee, as well as the regulatory compliance implications of a mismatch between the job requirements and operator capabilities, necessitate an accurate method of analyzing and documenting the job characteristics. The focus of this chapter is on the various methods of performing an ergonomic analysis and documenting the job characteristics that can affect the occupational safety and health of the operators.

0-8493-2641-9/99/$0.00+$.50

18.2 Documenting the Risk of Work-Related Musculoskeletal Disorders

There are three general approaches to appraising the characteristics of tasks that have been associated with increased risk of work-related injuries and illnesses. These are passive surveillance, active surveillance, and job-site analysis. *Passive surveillance* involves a review of the archival occupational safety and health records (i.e., OSHA logs, workers' compensation records, etc.). Additional data that can contribute to an assessment of potential discomfort that does not result in a reportable event, or even a visit to the medical department, are the requests for a job transfer and the absenteeism records for particular operations. Although this information can illuminate the extreme cases in a facility, the data are very noisy (i.e., incorrect names for operations, an injury that actually occurred as a result of a previous operation, etc.). In addition, for most organizations, it can be anticipated that the number of occurrences is small enough to render any statistical trend analysis erroneous. In fact, the potential for misinterpretation of such data can severely hinder an effective ergonomics effort.

Active surveillance generally involves soliciting information from the current employees pertaining to any discomfort or disorder they have experienced because of their work. The term "symptom survey" is often used in this context; although it is very likely that the term itself predisposes the employee to symptoms. Instead, labeling the instrument that requests information from the operator as to potential improvements as a "job improvement survey," prior to their indicating discomfort or pain, can have a very positive effect on the quality and usability of the information collected. It is the change in the responses over subsequent surveys that generally provides more useful information than that from a single application of an active surveillance instrument.

For both the active and passive surveillance methods, the obvious difficulty is differentiating between soreness that can occur naturally as someone begins or returns to a task and persistent discomfort that could indicate the onset of a disorder or injury. Another issue that relates to the effectiveness of active surveillance methods is the confidentiality of the data. That is, it is uncomfortable for many companies to collect data on discomfort or pain and not be able to identify the person in order to follow up on the reported problem. However, to obtain unrestrained information, and because the data could be interpreted as a form of medical record, confidentiality can be critical. One method of addressing this issue is to number the form and a corresponding tear-off signature area. The health care provider (i.e., usually the company nurse) is the only person who would have access to the cross reference of names with surveys. The health care provider can then use this information to follow up on medical problems that are reported.

A third general method of assessing the adequacy of the workplace design, work methods, tools, and equipment from an ergonomic perspective is the *job-site analysis*. There are two general approaches to job-site analysis: checklists and narrative reports. The narrative approach generally involves a qualified ergonomist with extensive training and experience in job-site analysis, evaluating each production operation in a facility. The product of the ergonomic analysis is a written report that discusses each operation, often accompanied with photographs and/or video recordings to illustrate the observations. It is often as important for operational personnel to understand the reasons some tasks do not experience problems as it is to understand the risks posed by problematic tasks. The major disadvantages of this narrative method of job-site analysis are the access to a qualified analyst, the time required to conduct the analysis and develop the report, and the resulting high cost. The primary benefit of this type of approach is that recommendations are provided that have the potential of reducing or eliminating the risk of fatigue, discomfort, and injury.

There have been some methodologies developed to assist in the analysis of injuries, specifically related to cumulative trauma disorders to the upper extremities. Drury and Wick (1984) developed a task analysis method to specifically document the posture, force, and frequency of tasks that could lead to injury. Jobs were videotaped and broken down into task steps (i.e., elements). The characteristics of the task elements were recorded on a task analysis form that addressed both the upper extremities and overall body posture.

The frequency of task characteristics that are considered to increase risk (i.e., the combination of radial deviation and pinch grip) were calculated. Each task element was considered "damaging" if it required a grip with deviation, flexion, or extension of the wrist. The compressive forces on the spinal discs were also calculated from the body angles and the amount of weight moved.

Drury (1987) developed a job analysis method that was broken down into two sections: task description and task analysis. The task description section involved analyzing jobs by using video recordings of the worker from five angles to develop an element breakdown documenting the body angles, posture, grip, forces, vibration, shock, and lighting. The frequency of each task was estimated by dividing the task working time over the day by the cycle time. The task analysis also included documenting the angles at each joint with respect to the maximum range of motion. The joint range-of-motion data were divided into zones: (1) no exposure (neutral to ± 10% of range), (2) low exposure (± 10% to ± 25% of range), (3) moderate exposure (± 25% to ± 50% of range), and (4) severe exposure (more than ± 50% of range). From the task description, each angle was replaced with a corresponding zone number. Drury used the term "Daily Damaging Wrist Motions (DDWM)" as a metric of exposure where a damaging wrist motion was defined as the frequency of any combination of a grip or external force with any nonzero wrist exposure.

Armstrong, Radwin, Hansen, and Kennedy (1986) developed a methodology for analysis, identification, and elimination of cumulative trauma disorders that involved fundamental industrial engineering work-methods procedures. The work content of each task element was evaluated with respect to risk factors: (1) repetitive or forceful sustained exertions, (2) shoulder posture (elbow above mid torso reaching down and behind), forearm (inward or outward rotation with a bent wrist), wrist (palmar flexion, full extension, ulnar or radial deviation), hand (pinching), (3) mechanical stress concentrations, (4) vibration, (5) cold, and (6) gloves. If any risk factor was identified, it was recommended that modifications be made.

A structured job analysis procedure was developed by Keyserling, Armstrong, and Punnett (1991) to assist safety professionals in recognizing and evaluating exposures to risk factors. The method consisted of: basic job documentation, identification and evaluation of exposure to risk factors, and methods of controlling exposures to work-related risk, eliminating or reducing the risk factors to acceptable levels.

These various methods of job-site analysis span a continuum from recommendations based on expert judgment with little or no objective or quantitative documentation to very detailed measurement and analysis methods that characterize the biomechanical considerations of the job in great detail. However, in each case, there can be a significant amount of time and money devoted to the analysis procedure.

An alternative approach, that is generally quicker to implement, involves the use of a *checklist* that helps in the identification of the task characteristics that are associated with increased fatigue, discomfort, or injuries. Armstrong and Lifshitz (1986) developed a checklist that included physical stress, force, posture, workstation hardware, repetitiveness, and tool design. The final score was calculated as the fraction or percentage of the responses scored by a "yes." Subscores for each category of risk factors were calculated to indicate where the attention should be focused to control the problem.

The Occupational Safety and Health Administration (OSHA) has distributed a draft checklist associated with a proposed ergonomic standard. The objective of the checklist is to assist employers in the evaluation of the jobs in their facilities. The checklist includes a set of risk factors related to three categories: (1) the upper extremities, (2) back and lower extremities, and (3) manual handling of materials. Within each category, points are assigned, depending upon the importance of the factor (i.e., pinch grip more than two pounds) and the daily exposure time (2 to 4 hours, 4 to 8 hours, or more than 8 hours). The points are combined within each category, and the total score is compared to a trigger value. If the score is greater than the trigger value, the job was designated as a "problem job" that deserves a more detailed analysis.

Keyserling, Stetson, Silverstein, and Brower (1993) developed a checklist that included the following categories: environment, posture, metabolic rate, manual lifting, and the use of the upper extremities. Within each category, questions were asked that characterized the job. Each question was designed to evaluate the presence and/or duration of the risk factors. Each response resulted in a stress rating as follows: (1) zero (exposures were insignificant), (2) check (moderate exposures were present), and (3) star (substantial exposures were present). The number of elements within which it occurred was also recorded. An overall score from ergonomic stresses was calculated by summing the total number of checks and stars.

Both the narrative and the checklist methods of job-site analysis have advantages and disadvantages. The checklist is relatively quick and easily performed; however, it is generally quite incomplete, particularly with respect to the temporal aspects of the job, across task elements, tasks and recovery periods. In addition, even the checklist procedure generally requires some training on the administration of the checklist and the interpretation of the results. Most important, it is descriptive in that it indicates when a problem exists rather than being prescriptive in terms of providing recommended modifications to reduce the risks.

The more narrative approach to job analysis and documentation is generally much more complete, and the analyst generally includes recommended modifications, in addition to a description of the risks. However, this method requires a significant amount of training and experience in ergonomics, which results in the time required and the cost being relatively high. Another serious disadvantage of utilizing either an external or internal ergonomic expert to conduct the analysis and document the job characteristic is that the analyst often does not have a very complete knowledge of the job itself. That is, the impact of extraneous variables (different incoming material, alternative process conditions, etc.) can make an apparently accurate ergonomic analysis totally invalid because the conditions are not representative. In general, a somewhat incomplete knowledge of ergonomics is less dangerous than an incomplete knowledge of the operational process.

An alternative method of conducting a job-site analysis is to use a *computer-based system* to document and evaluate the task requirements. This approach has the advantage of being interactive and easily understood by operational personnel in the context of an active surveillance process or independently. The computer-based approach also provides a structured method of evaluating the postures and motions across task elements and tasks to determine the total amount of exposure and the time provided for recovery. The system can be used to simulate operations that have not yet been implemented, and it can be used to evaluate alternative configurations for the workplace design, work methods, tools, and equipment. As changes occur in the production process, the documentation can easily be updated by addressing only the individual tasks or task elements that have changed. Possibly the most important feature of the computer-based system is that it provides recommendations that can be used to reduce the characteristics of the tasks that increase postural and biomechanical stress, along with the potential for work-related musculoskeletal disorders.

Many of the modifications that have the potential of reducing or eliminating cumulative trauma disorders are not difficult to recognize or implement. The personnel already in place often have both the skills and knowledge to properly design the systems, given the proper analysis tools and resources.It is not sufficient to simply provide operational personnel with only a method of evaluating jobs using a rating scale that indicates that a hazard may exist. The computer-based job analysis system provides the resource necessary for individuals with little or no training in ergonomics to both evaluate jobs and suggest modifications that can reduce or eliminate the incidence of cumulative trauma disorders. In particular, the system is designed for production supervisors who have a full understanding of the job, but no training in ergonomics.

18.3 Format Used in the Job Analysis and Documentation System

The computer-based ergonomic job analysis system has two primary functions. First, it is an evaluation tool to analyze and document the characteristics of the job that are associated with musculoskeletal disorders (i.e., risk factors). Second, it is used by operational personnel (i.e., first-line supervisors and engineers) to prescriptively evaluate the workplace design, work methods, tools and equipment by suggesting effective modifications that can lead to more efficient operations, in addition to controlling work-related occupational injuries and illnesses. The completeness of the job description is of primary importance to the validity of an analysis system. One of the reasons that occupational safety records (i.e., OSHA 200 logs) are of less value than might be expected is that job titles are not applied consistently by all people. For example, the terminology used by production supervisors is often different from that used by the human resources department. Operators often do not know the name of the job that they were performing when they visit the company medical facility. In addition to ensuring that the terminology used to label the job is consistent, it is also important to consider the individual being analyzed. Obviously, the posture that occurs at a particular workplace can be different for a very short individual as opposed to a very tall person. The height and gender of the person being analyzed is entered into the system so that subsequent graphical presentations are adjusted to accurately represent the anthropometric considerations. This allows the workplace geometry to be entered relative to body landmarks (i.e., knee or waist height) rather than requiring absolute measurements.

Entering Job Characteristics

To ensure that the documentation is accurate and complete, it is often beneficial to have a complete list of all jobs for specific workstations. These are not necessarily the same as the job categories defined by the human resources department. The supervisor can subsequently identify and describe the tasks and task elements that constitute the job. In addition, information about the total work time, frequency and duration of the breaks, and job rotation are required to determine a valid evaluation of the exposure. If the time allocated to the various task elements and tasks do not correspond relatively closely to the total work time, the analysis will be invalid and the conclusions can be very misleading.

Figure 18.1 illustrates a task and task element manager that is used to enter the frequency and duration of the various task elements. The task duration is subsequently derived by the system based on the element times and frequencies. If the cumulative time differs from the total work time by more than 5%, a message is provided that informs the analyst of the magnitude of the discrepancy. If the total time differs by more than 15%, the resulting errors are considered excessive, and the analyst is not allowed to proceed until corrections are made.

The next portion of the data entry process involves the work environment and organizational characteristics of the task. In particular, the room and product temperature, task pacing (i.e., machine paced, incentive paced, etc.), and personal protective equipment (including glove types) can affect the postural and biomechanical stress experienced during a workday.

For a production supervisor to be able to characterize the postures and motions that occur during a task element, it is true that a picture is worth a thousand words. The graphical representation shown in Figure 18.2 illustrates the method of entering information related to the position and movement of the upper arm and shoulder. The beginning of each element is, by default, the ending position of the previous element. The term *repetitive* in this context does not refer to whether the task element occurs repetitively, but whether the motion occurs repetitively within the task element (i.e., multiple rotations when driving a single screw). Often, the left and right arms are performing the same activities, and this can be indicated without reentering the information.

Figures 18.3 and 18.4 illustrate the screens used to enter the information for the position and rotation of the forearm and the position and action of the wrist and hand. For each of the alternative choices provided to the analyst (i.e., production supervisor), a help screen can be accessed to define the terms and give examples that illustrate the concepts (i.e., pinch grip).

Frequently, much of the fatigue and discomfort experienced in industrial operations is due to the posture required. Figure 18.5 illustrates the screen that is used to indicate bending and/or twisting of the torso. Other postural considerations relating to sitting and standing are entered on the screen shown in Figure 18.6.

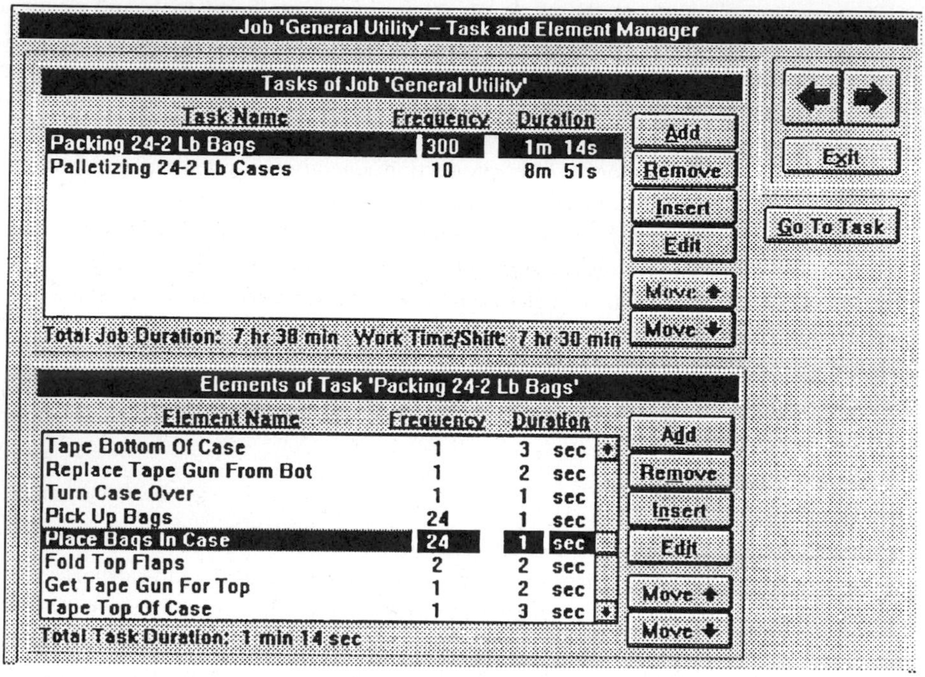

FIGURE 18.1 Screen used to enter task and element sequence, frequencies, and durations.

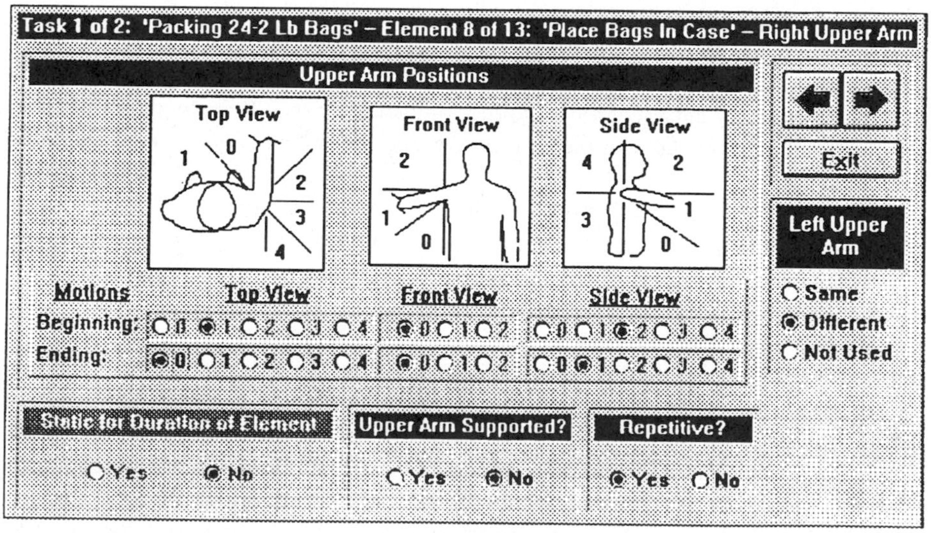

FIGURE 18.2 Screen used to enter shoulder and upper arm positions and movements.

Manual Material Handling Task Characteristics

Two approaches are included in the system to evaluate manual material handling tasks. First, the 1991 revision of the NIOSH Lifting Guidelines (Waters, Putz-Anderson, Garg and Fine, 1993) are used to develop the *recommended weight limit* (RWL) and the *lifting index* (LI). The second approach utilizes the tables developed by Liberty Mutual Insurance Company (Snook, 1978) that provide guidelines for lifting, lowering, pushing, pulling, and carrying. In general, the Liberty Mutual tables are more useful than the NIOSH guidelines as an engineering tool to evaluate and modify workplaces. The percent of the population capable of performing the task is more understandable and usable by operational personnel than is the

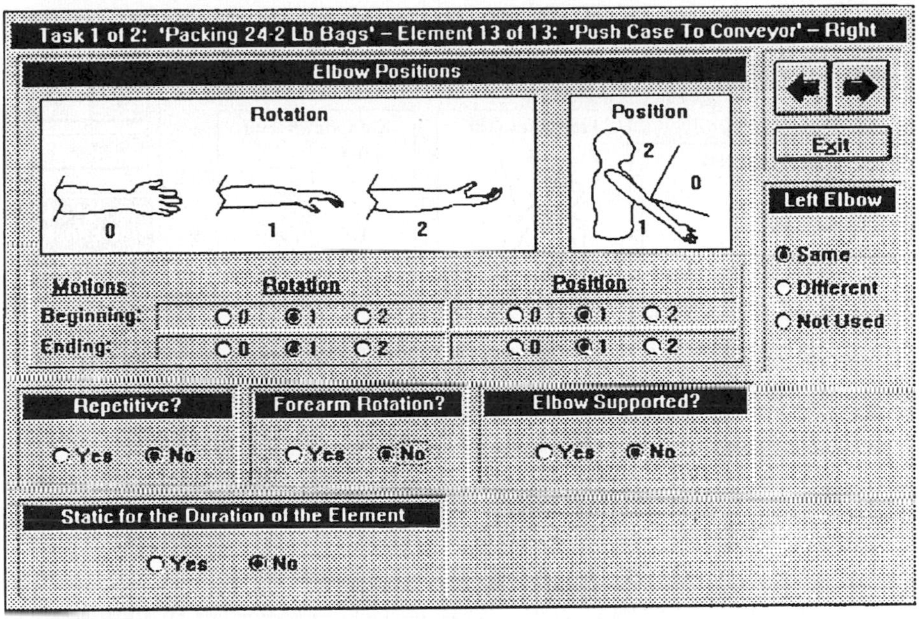

FIGURE 18.3 Screen used to enter elbow and forearm positions and movements.

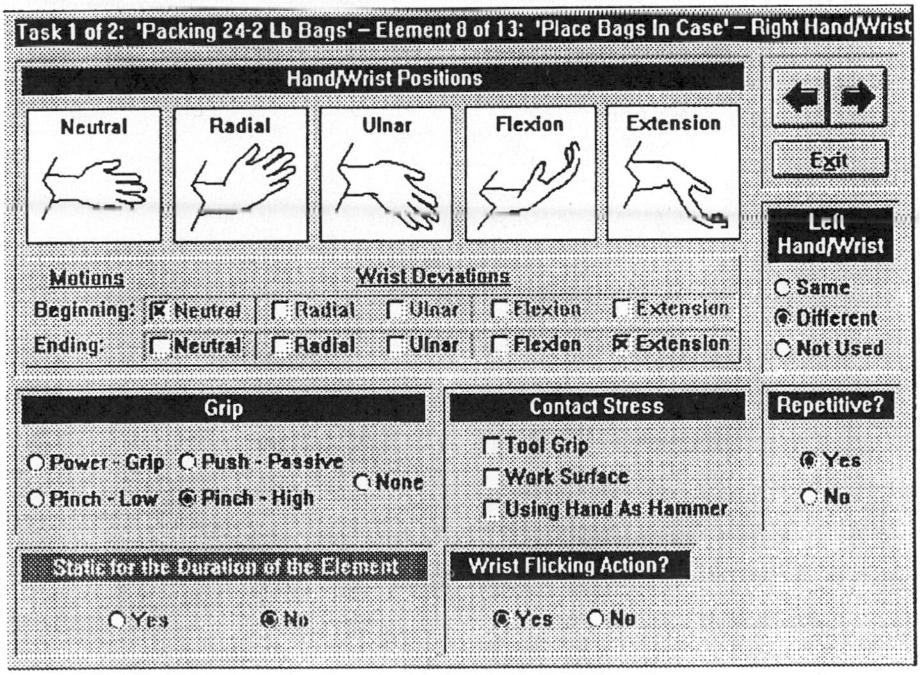

FIGURE 18.4 Screen used to enter wrist and hand positions and movements.

concept of a lifting index. Figures 18.7 and 18.8 illustrate the screens used to enter the data for the manual material handling aspects of each task element. Because it is generally easier for the production supervisor to document the geometry of the task using anatomical landmarks (i.e., knee or waist height) than to make measurements, a graphical method of entering the information is used. To eliminate the fact that knee heights are different for different people, the graphic is adjusted for the height of the person being analyzed, based on information entered previously.

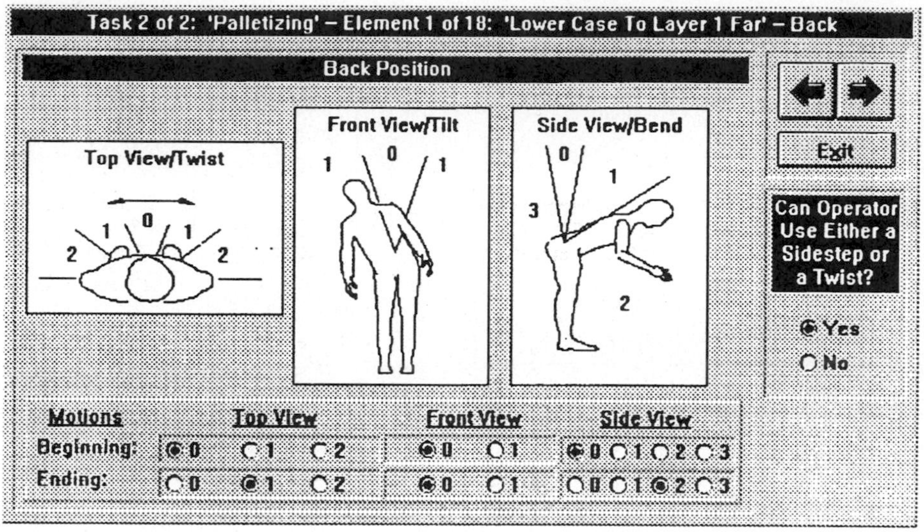

FIGURE 18.5 Screen used to enter back posture.

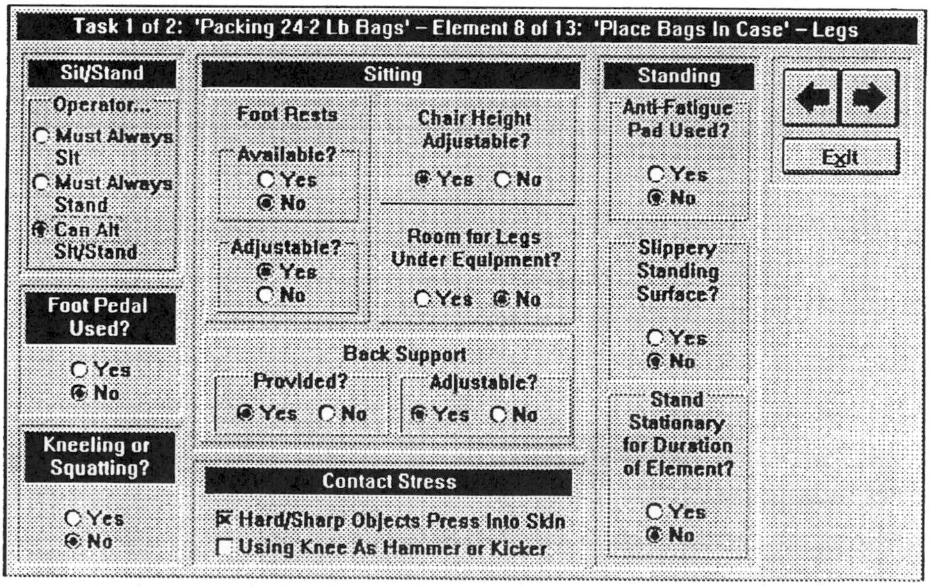

FIGURE 18.6 Screen used to enter lower extremity task characteristics.

Presentation of the Results and Recommendations of the Analysis

The common motions and postures are combined across task elements and tasks to represent the total exposure. The first set of screens present the motions per day and the exposure time for the various task characteristics (i.e., forearm rotated with arm extended). Figure 18.9 shows an example of a results screen for the arm. There are similar screens that report the results for the shoulder and the wrist/hand. Both the motions per day and the total exposure time are important in evaluating exposure to risk factors. The individual task elements that contribute to the particular activity are indicated by number.

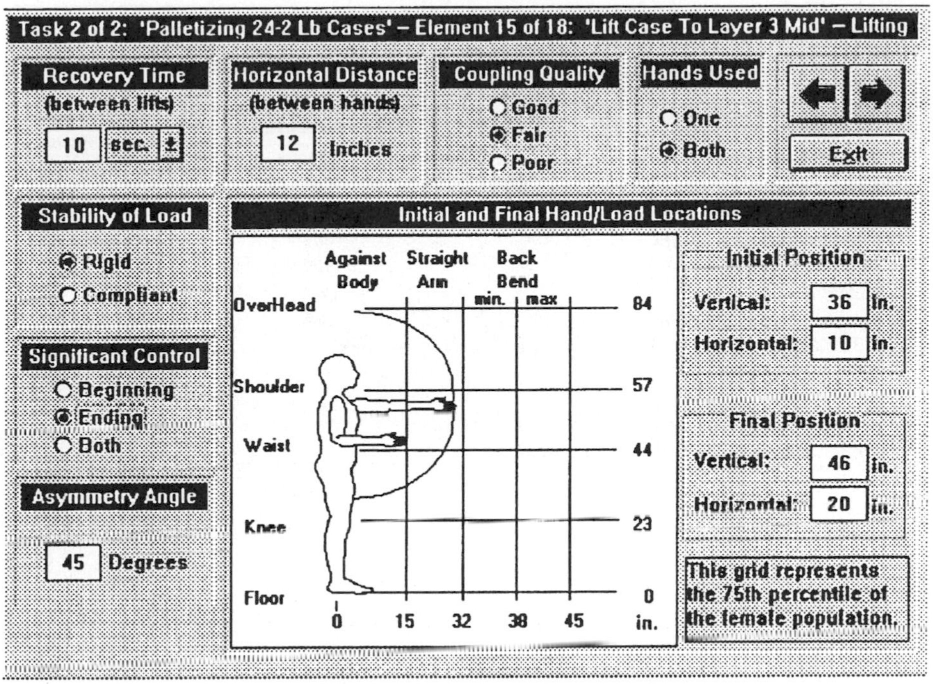

FIGURE 18.7 Screen used to enter lifting task characteristics.

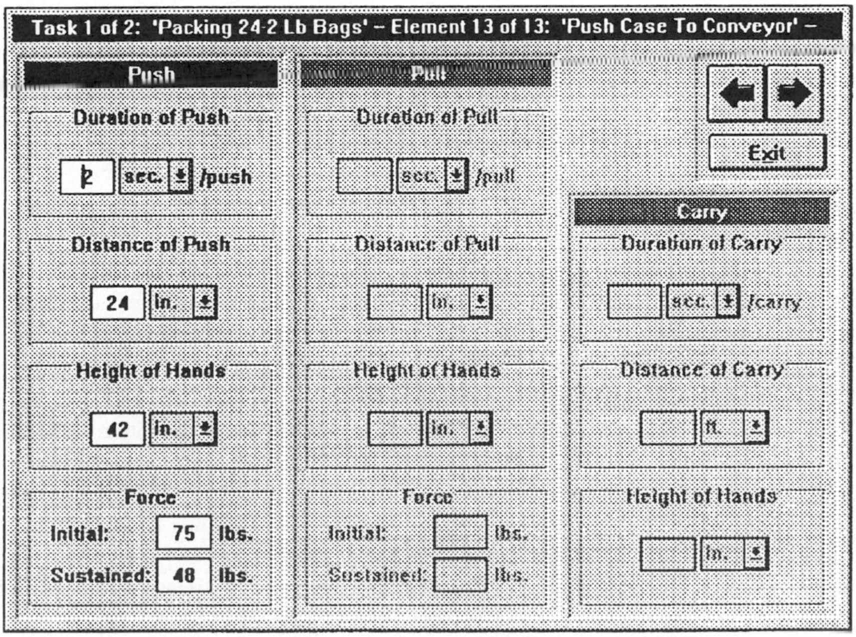

FIGURE 18.8 Screen used to enter push, pull and carry task characteristics.

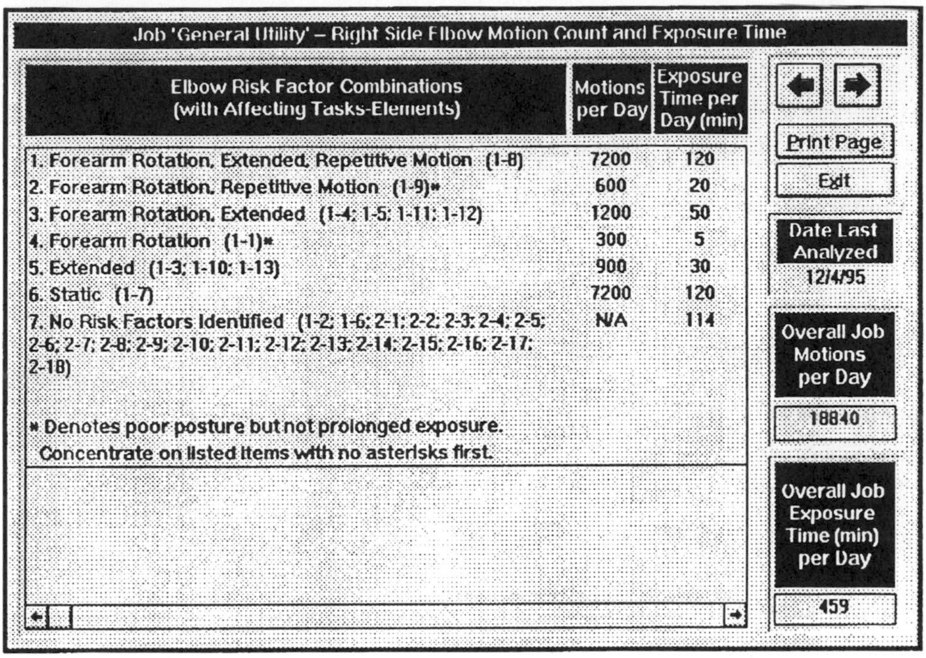
Job 'General Utility' – Right Side Elbow Motion Count and Exposure Time
Elbow Risk Factor Combinations (with Affecting Tasks-Elements)
Motions per Day
Exposure Time per Day (min)
1. Forearm Rotation, Extended, Repetitive Motion (1-8) 7200 120
2. Forearm Rotation, Repetitive Motion (1-9)* 600 20
3. Forearm Rotation, Extended (1-4; 1-5; 1-11; 1-12) 1200 50
4. Forearm Rotation (1-1)* 300 5
5. Extended (1-3; 1-10; 1-13) 900 30
6. Static (1-7) 7200 120
7. No Risk Factors Identified (1-2; 1-6; 2-1; 2-2; 2-3; 2-4; 2-5; 2-6; 2-7; 2-8; 2-9; 2-10; 2-11; 2-12; 2-13; 2-14; 2-15; 2-16; 2-17; 2-18) N/A 114
* Denotes poor posture but not prolonged exposure.
Concentrate on listed items with no asterisks first.
Print Page
Exit
Date Last Analyzed
12/4/95
Overall Job Motions per Day
18840
Overall Job Exposure Time (min) per Day
459

FIGURE 18.9 Results screen for the right-side elbow and forearm risk factors.

Although an indication of the task characteristics that surpass a threshold of risk (i.e., "problem" jobs) would obviously be helpful, the information necessary to provide such a conclusion is not currently available. It is the philosophy of the job analysis system that any and all awkward postures and motions should be addressed with the intention of reducing unnecessary fatigue and improving efficiency, as well as reducing work-related occupational injuries and illnesses. Therefore, the task of the analyst (production supervisor) is to continuously improve the workplace design, work methods, tools, and equipment to improve the effectiveness of the operations. The results screens can be used to identify the task elements contributing to risk and quantify the magnitude of the problem and the potential for improvement through modifications. As more information becomes available as to the dose–response relationship between the risk factors and musculoskeletal disorders, these can easily be incorporated into the system.

The screens that document the job characteristics that affect the operator's back and legs present the resulting exposure time in the particular postures (Figure 18.10). Whereas the previously discussed screens address the exposure of the body part, across task elements and tasks, the screen shown in Figure 18.11 documents the risk factors within a task element for the various parts of the body.

The results and recommendations for the manual material handling portions of the task are given together. Figures 18.12 and 18.13 illustrate a lifting and pushing task, respectively. The NIOSH lifting index and the recommended weight limit are included for the lifting tasks. In addition, the percent of males and females capable of performing the task element as determined by the Liberty Mutual Insurance data are included for lifting, lowering, pushing, pulling, and carrying. Figure 18.14 illustrates a screen that can be used by the analyst to evaluate the effects of task modifications. By changing the characteristics of the task and noting the changes in the results, alternative modifications can be evaluated, recommended, and supported.

The recommendations related to posture and biomechanical stress associated with non-material handling tasks also indicate both the number of motions and the total exposure time. Figure 18.15 illustrates an example of a recommendations screen for the hand and wrist.

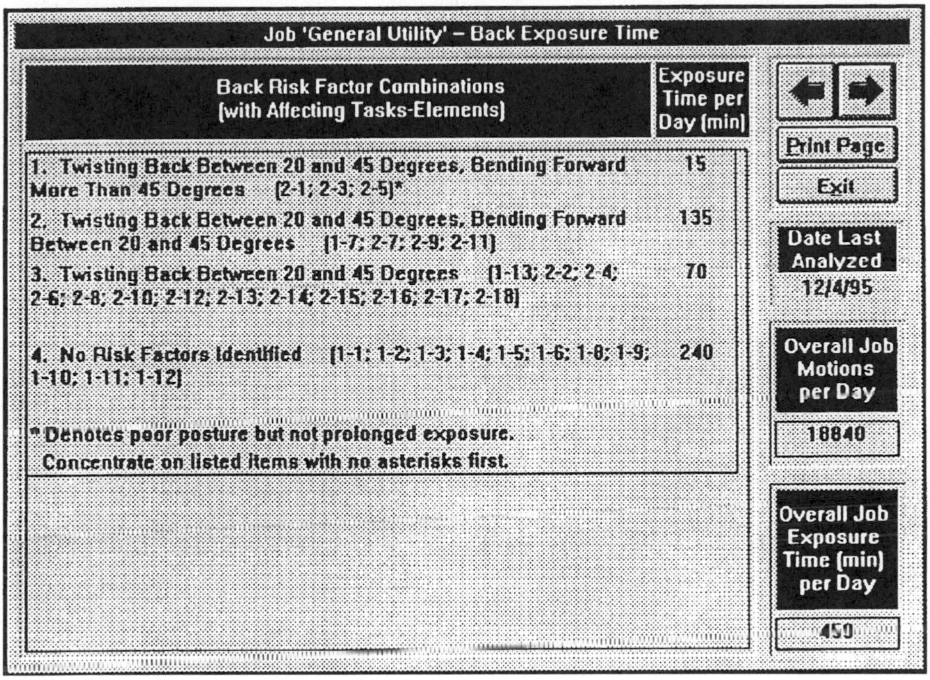

FIGURE 18.10 Results screen for the back posture factors.

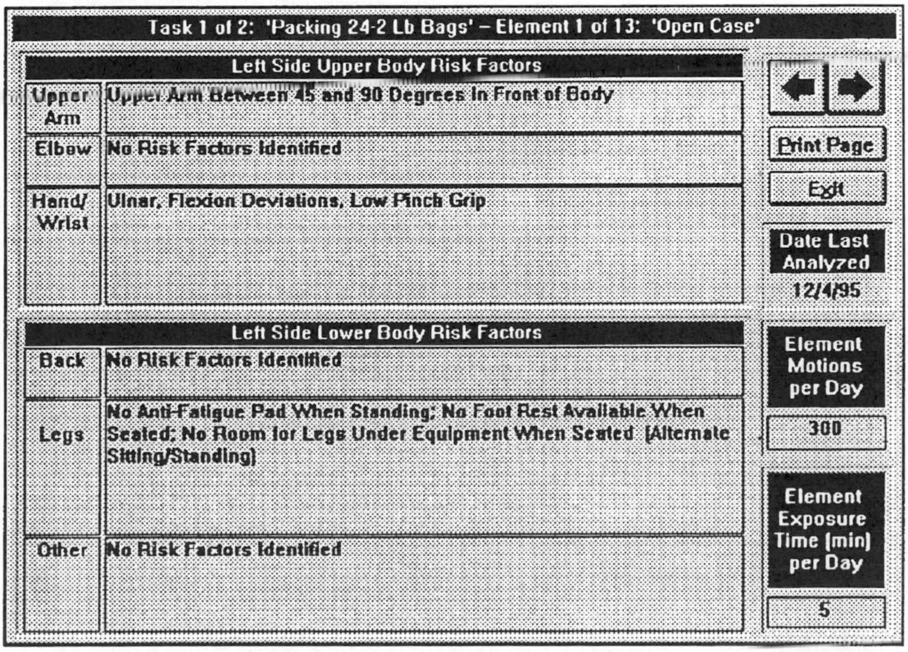

FIGURE 18.11 Results screen for an individual task element.

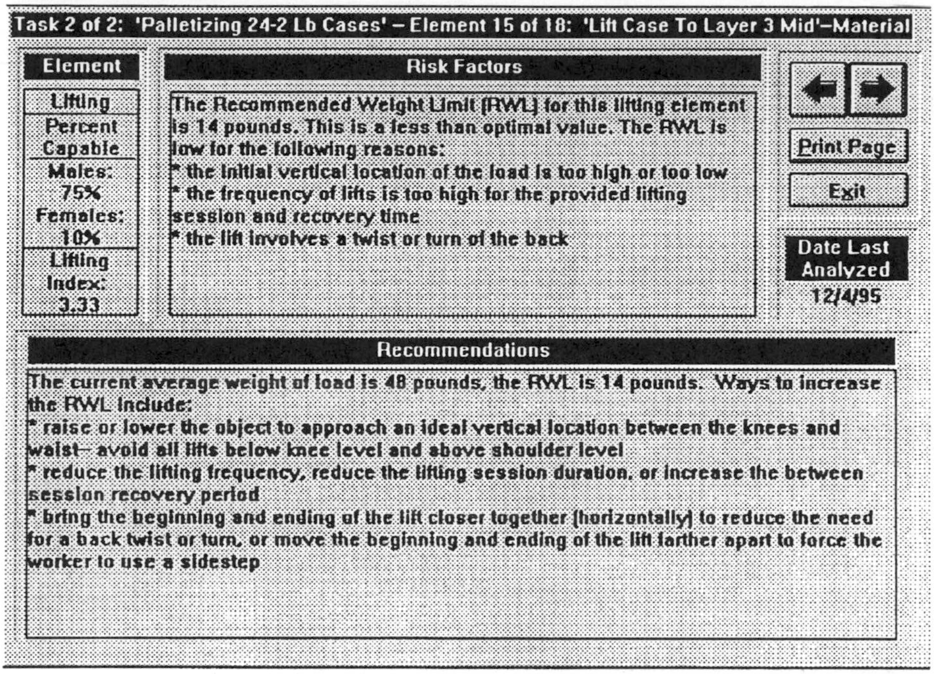

FIGURE 18.12 Results screen for a lifting task.

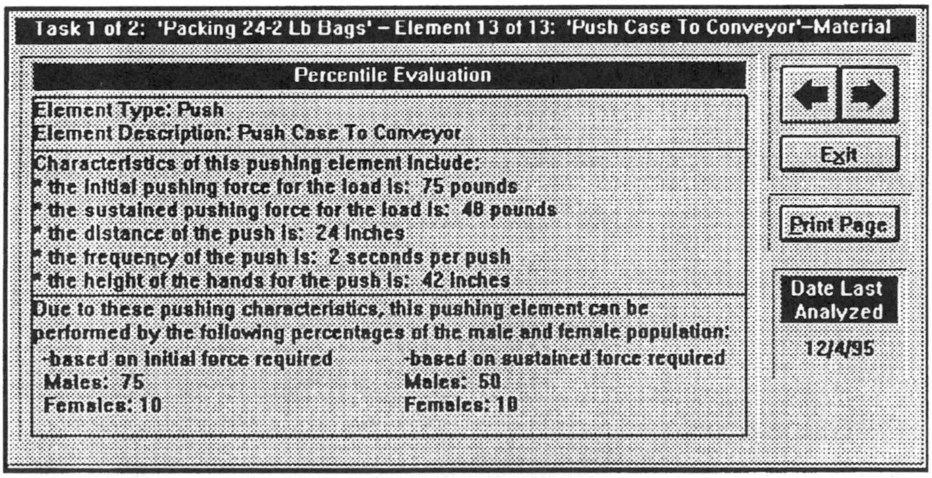

FIGURE 18.13 Results screen for a pushing task.

Many organizations prefer to have an assessment of how their jobs would be assessed using the most current OSHA checklist. Therefore, the computer-based system attempted to translate the job characteristics into the score format used for the checklist. The checklist is divided into the upper extremities, lower extremities and manual material handling. Figure 18.16 illustrates part of the checklist.

18.4 Discussion

As with any analysis process, the quality of the results and recommendations are highly dependent on the completeness and validity of the information entered by the analyst. That is, if the descriptions of

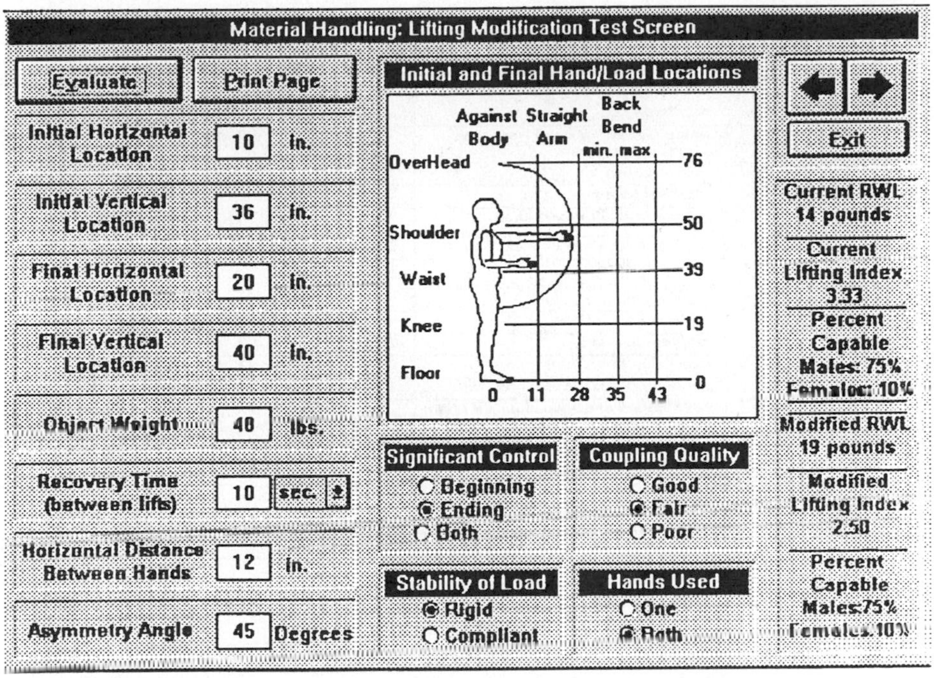

FIGURE 18.14 Lifting test screen to evaluate alternative task characteristics.

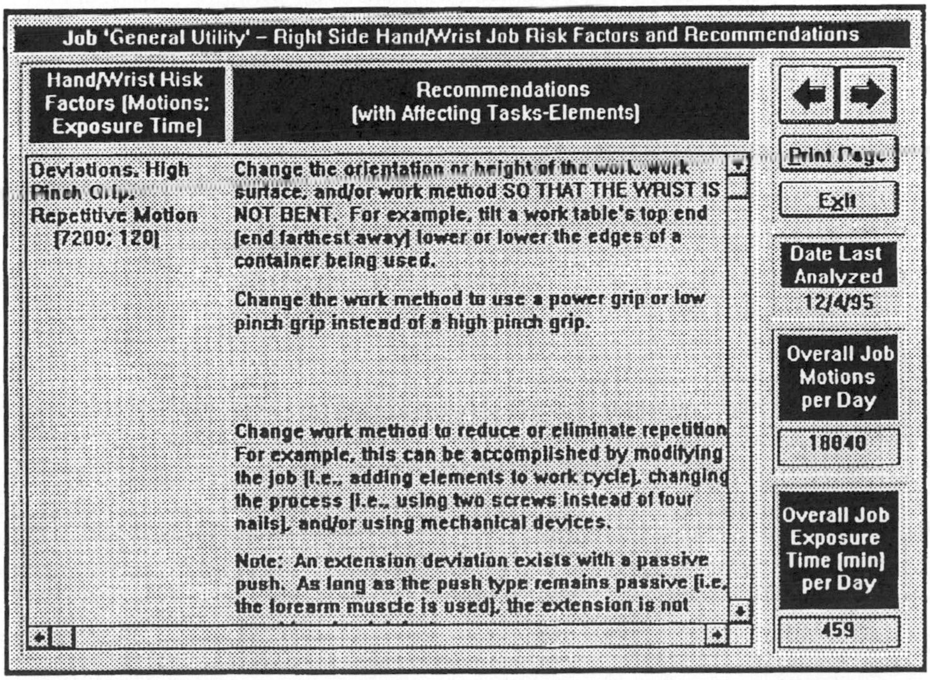

FIGURE 18.15 Recommendations screen to address risk factors related to the wrist and hand.

the task and task elements do not accurately represent the true job characteristics (physically or temporally), the results of the analysis will not be valid. Subsequently, the recommendations could be very inappropriate. It is also important to note that the results of the analysis provided by this system constitute

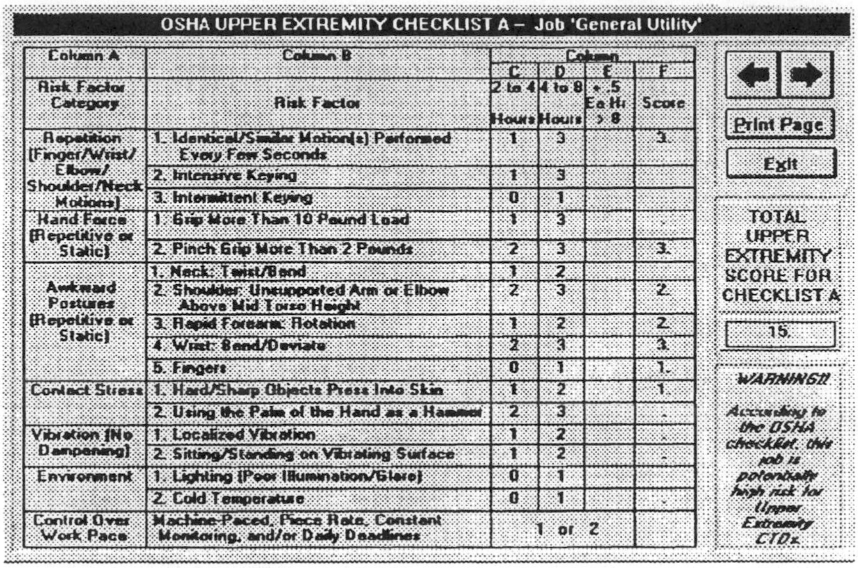
OSHA UPPER EXTREMITY CHECKLIST A – Job 'General Utility'
Column A
Column B
Column
C
D
E
F
Risk Factor Category
Risk Factor
2 to 4 Hours
4 to 8 Hours
+.5 Ea Hr > 8
Score
Repetition (Finger/Wrist/Elbow/Shoulder/Neck Motions)
1. Identical/Similar Motion(s) Performed Every Few Seconds
1
3
3.
2. Intensive Keying
1
3
3. Intermittent Keying
0
1
Hand Force (Repetitive or Static)
1. Grip More Than 10 Pound Load
1
3
2. Pinch Grip More Than 2 Pounds
2
3
3.
Awkward Postures (Repetitive or Static)
1. Neck: Twist/Bend
1
2
2. Shoulder: Unsupported Arm or Elbow Above Mid Torso Height
2
3
2.
3. Rapid Forearm Rotation
1
2
2.
4. Wrist: Bend/Deviate
2
3
3.
5. Fingers
0
1
1.
Contact Stress
1. Hard/Sharp Objects Press Into Skin
1
2
1.
2. Using the Palm of the Hand as a Hammer
2
3
Vibration (No Dampening)
1. Localized Vibration
1
2
2. Sitting/Standing on Vibrating Surface
1
2
Environment
1. Lighting (Poor Illumination/Glare)
0
1
2. Cold Temperature
0
1
Control Over Work Pace
Machine-Paced, Piece Rate, Constant Monitoring, and/or Daily Deadlines
1 or 2
Print Page
Exit
TOTAL UPPER EXTREMITY SCORE FOR CHECKLIST A
15.
WARNING!!
According to the OSHA checklist, this job is potentially high risk for Upper Extremity CTDs

FIGURE 18.16 Example of an OSHA checklist screen.

the beginning of the problem-solving process, not its completion. That is, there is no substitute for the knowledge and experience of operational personnel in the process of establishing effective and efficient methods of modifying the workplace design, work methods, tools, and equipment. In general, the operational personnel can usually develop applicable solutions to the problems, given that the problems are recognized in the first place. That is the goal of this analysis system. In addition, this system can be very useful in evaluating alternative designs before they are installed, thus reducing both the time and cost involved with retrofitting. By using the system to evaluate alternative configurations, the relative effects can both be documented and communicated in the decision-making process.

References

Armstrong, T. and Lifshitz, Y. (1986). Evaluation and design of jobs for control of cumulative trauma disorders. Presented at *Symposium on Ergonomic Interventions to Prevent Musculo-skeletal Injuries in Industry*, Denver, CO, October 9-10, 1986.

Armstrong, T., Radwin, R., and Hansen, D. (1986). Repetitive trauma disorders: job evaluation and design. *Human Factors*, 28(3), 325-336.

Drury, C. (1987). A biomechanical evaluation of the repetitive motion injury potential of industrial jobs. *Seminars in Occupational Medicine*, 2(1), 41-49.

Drury, C. and Wick, J. (1984). Ergonomic applications in the shoe industry. *Proceedings of the 1984 International Conference on Occupational Ergonomics*, 489-493.

Keyserling, W., Armstrong, T., and Punnett, L. (1991). Ergonomic job analysis: A structured approach for identifying risk factors associated with overexertion injuries and disorders. *Applied Occupational Environmental Hygiene*, 6(5), 353-363.

Keyserling, W., Stetson, D., Silverstein, B., and Brower, M. (1993). A checklist for evaluating ergonomic risk factors associated with upper extremity cumulative trauma disorders. *Ergonomics*, 36(7), 807-831.

Snook, S. (1978). The design of manual handling tasks. *Ergonomics*, 21(12), 963-985.

Waters, T.R., Putz-Anderson, V., Garg, A., and Fine, L. (1993). Revised NIOSH equation for the design and evaluation of manual lifting tasks. *Ergonomics*, 36(7), 749-776.

19

Job or Task Analysis for Risk Factors of Musculoskeletal Disorders

David J. Cochran
University of Nebraska-Lincoln
Terry L. Stentz
University of Nebraska-Lincoln
Brian L. Stonecipher
University of Nebraska-Lincoln
M. Susan Hallbeck
University of Nebraska-Lincoln

19.1 Introduction

There are numerous methods of analyzing jobs for ergonomic risk factors or ergonomic stressors (Keyserling et al., 1991; Ulin et al., 1992). Many companies, consultants, and others have devised job analysis methods. Most of these are proprietary or protected under copyright and as such are not available to everyone. There are some that are available in book form (Burke, 1992). These seem to be regimented and somewhat simplistic. Job analyses generally take one of three forms — checklist, interactive form-based, and narrative or open-ended methods. The checklist method is the simplest to conduct, but it has some very real drawbacks. Some will believe it can be used in place of training and experience, but in the hands of inadequately trained people, checklists can be and have been grossly misused. They do not typically allow for flexibility and may not ask the appropriate questions for a particular job unless they are extremely thorough and therefore long. The interactive form-based method takes considerable time and money to develop and to verify its efficacy. These sometimes have the same drawbacks that checklists have. Namely, they may not allow for flexibility and may not ask the appropriate questions for a particular job unless they are extremely thorough and therefore long. The narrative method is thorough, straightforward, relatively easy to learn, versatile, easy to use, and is generally accepted. On top of all this, it is an intuitively pleasing method.

0-8493-2641-9/99/$0.00+$.50

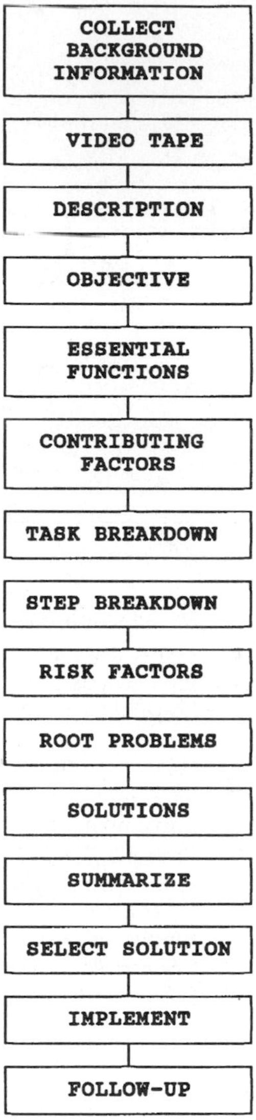

FIGURE 19.1

To conduct a narrative job analysis, background information is collected, the job in question is observed (live or preferably on videotape), a narrative description is written, the job objective is written, the essential functions are identified and written, and the contributing factors are identified and written. The job is then broken into tasks and then into steps. The steps are described; risk factors are identified for each step; root problems that cause these risk factors to be present are identified for each step; solutions are created and evaluated and a follow-up evaluation is conducted. This is nothing more than a general and logical procedure to thoroughly examine a job, and it is pictured in the flow chart in Figure 19.1. It is not tremendously difficult, but it does require some training, practice, and knowledge and awareness of the risk factors for musculoskeletal disorders. Experience and practice will improve the results.

19.2 Preparation

Prior to actually analyzing the job, collect background information and videotape the job. This should include all of the information necessary to be knowledgeable enough of the job to analyze it and propose feasible solutions. This is covered in some detail in Cochran et al., 1999. As a bare minimum, include the job title, work objective, work standard, major and minor tasks, the product, tools, equipment, materials, personal protective equipment, workstation layout, environmental conditions, any rotation scheme, preceding job, succeeding job, worker attributes, and required maintenance and repair. Also become knowledgeable in the physical, psychological, and medical problems experienced by workers on this job.

19.3 Job Analysis

Job Description

This section should inform the reader as to what the job consists of and how it is done. Describe the job in a narrative form. Cover each component in the sequence in which it occurs.

Job Objective

A short statement of what is to be accomplished by the job is sufficient here. This statement has the effect of setting a tone for the job analysis and can be referred to when questions arise about the necessity of particular parts of the job.

Job Essential Functions

Essential functions are basic and fundamental to the performance of a specific job. Essential functions are not marginal to the job or its performance. Enumeration of the essential functions is very useful for getting a job down to its essence. It can also be useful in complying with the Americans with Disabilities Act (see U.S. Equal Opportunity Commission, 1992).

Describe or list the functions of the job that are absolutely essential to its successful performance. The first consideration is whether employees in the position are actually required to perform the function. Reasons a function could be considered essential include that the position exists to perform the function; there are a limited number of other employees available to perform the function or among whom the function can be distributed; the function is highly specialized and the person in the position is hired for special expertise or ability to perform it. Examples of types of essential functions identified are activity tasks which have a process objective, physical movements and/or force exertions, body postures or positioning, cognitive operations and/or judgment-making, use of special knowledge, training, abilities or skills, or forms of communication and/or interaction with others or other programmed operational units.

Contributing Factors

At this point in the analysis, those things that are present in the job that might have an impact on the ergonomic problems of that job but are not identified in specific steps are identified. These are things such as incentive or piece work systems, overtime or unusual workdays, strictly controlled pace of work, intimidating management style, fear of job loss, and many others. Environmental factors that are present and may have an effect on the worker may also be included here.

Task Breakdown

Many jobs have more than one task. When these tasks are separate or dissimilar it is advisable to break the job into its separate tasks and treat each as a separate job.

Step Breakdown

Breaking a job into steps or subtasks is the first action to be taken. The whole job is observed (live or on tape). If it is an assembly line job, it will generally be very routine and easily described in steps. If it is a more complicated job and has been subdivided into tasks that are each analyzed separately, an overall summarization of the job, including all tasks, is required.

Another possibility exists when the job is complicated and has many different activities that do not occur in a set order or pattern. In this case, it is helpful to discern similar job activities and analyze each separately. Once again, an overall summarization of the job, including all activities, is required.

Step Description

Next, describe the job steps in ordinary language. This should describe what the person does — which hands are used, what tools are used, and what activities are conducted. The postures assumed and an estimate of the forces involved should be included. If the description of the step is longer than several sentences, the breakdown into steps may not be adequate or the description may be unnecessarily detailed.

Risk Factors

After the step description has been created, the risk factors present are noted and listed. It is useful for the body part involved to be identified and the risk factor described. It is not the purpose of this paper to present the risk factors, any quantification scheme, or their merits and demerits for predicting musculoskeletal disorders. That has been done in numerous other papers and will continue to be done as the science of ergonomics advances. Some recommended references are Armstrong (1983), Armstrong et al. (1986a), Rogers (1992), Marras and Schoenmarklin (1993), Marras and Lavender (1995), Putz-Anderson (1988), and Sommerich et al., (1993).

Root Problems

The root problems of the job that cause the risk factor or factors to be present are identified. There is an almost irresistible tendency to go from the risk factors to the solutions. This can be counterproductive. As an example, a risk factor may be extreme wrist flexion. The root problem may be improper working height, product orientation, lack of a jig or fixture, poorly designed tool, or a combination of these. The root problem is the cause of the risk factor being present, and it is important to specifically identify it. This step better points to the appropriate solutions than just identifying the risk factors that are present.

Solutions

Solutions are developed to reduce or eliminate as many of the problems as possible. Modifications of the workstation, tools, product, work methods, or the work organization are proposed. Creativity is emphasized here. The input and opinions of the workers doing the job are invaluable in creating and evaluating solutions. This also creates a "buy-in" situation, in which the workers are part of the solution and will try to make it work.

In order to facilitate the step breakdown, to enforce the idea of identifying the root problem, and to associate the solutions with the risk factors and root problems that they address, the authors have created a very simple form that forces the juxtaposition of these components of the job analysis. These are illustrated in Figures 19.2 through 19.5. The form asks for the rated importance of each risk factor with a scale from 0 to 4. Zero indicates that the rated importance to musculoskeletal disorders is none or negligible, and 4 indicates that the rated importance of the risk factor is high. The form also asks for the rated potential of each solution for a particular risk factor with a scale from 0 to 4. Zero indicates that the rated potential for reducing the root problem or risk factor, and therefore the musculoskeletal disorders, is none or negligible, and 4 indicates that the rated potential is high.

Job Analysis Summary

Risk Factors and Root Problems

This is a narrative section that summarizes the findings. It can be broken into two parts — step related and whole job related.

JOB: Loaf Mold Extractor Operator. STEP NUMBER: 1		
DESCRIPTION: Operator pushes the cart of 10 molds into position to unload the molds. Approximate weight of the loaded cart is 800 pounds.		
RISK FACTOR (*)	**ROOT PROBLEM(S)**	**SOLUTION (**)**
Ulnar deviation, extreme extension, pronation, of both wrists. (3)	No handle.	Provide well designed handle(s). (2)
High pushing forces.(3)	Small wheels on cart	Larger wheels on cart(2)
	Large weight of cart	Larger wheels on cart(2)
	Uneven floors.	Floor maintenance.(2)
Slippery floor.(2)	Wet floor.	Better drainage.(2)
		Housekeeping.(1)
	Wrong shoe sole.	Appropriate shoes.(1)

* Include the rated importance.
** Include the rated potential.
Use a 0 to 4 scale. Zero indicates none and 4 the highest importance or potential.

FIGURE 19.2

JOB: Loaf Mold Extractor Operator. STEP NUMBER: 2		
DESCRIPTION: The operator reaches into the storage rack, removes the mold lid, throws the lid about 4 feet to a bin. Lid weighs about 1 pound.		
RISK FACTOR (*)	**ROOT PROBLEM(S)**	**SOLUTION (**)**
Shoulder flexion, 45 degrees (0)	Height of mold in the cart.	Lower the mold height in the cart. (0)
Possible high wrist acceleration.	Location of the bin for lids.	Relocate the bin for a drop rather than a throw. (3)

* Include the rated importance.
** Include the rated potential.
Use a 0 to 4 scale. Zero indicates none and 4 the highest importance or potential.

FIGURE 19.3

The risk factors and the associated root problems found in the job steps are summarized in one or more paragraphs. Additionally, at this stage of the job analysis it may be useful to summarize overall levels of some factors for the entire job. In particular, the factors of repetition and static loading sometimes make more sense for the job and for the workday as a frequency or percentage of the time, respectively. Also, when the job involves different tasks, it is useful to determine the amount and percent of time spent on each. A summarization of the average cycle time, the range of cycle times, the active or working time, and the non-working time are often useful. From this, daily statistics can be developed and compared with production specifications and requirements.

JOB: Loaf Mold Extractor Operator.	STEP NUMBER: 3	
DESCRIPTION: The operator reaches into the storage rack and removes the double loaf mold, and carries mold to the tilt/extraction machine approximately 5 feet away. Approximate weight of the loaded mold is 70 to 75 pounds.		
RISK FACTOR (*)	**ROOT PROBLEM(S)**	**SOLUTION (**)**
High grasp force (3)	Mold weight	Eliminate lifting by slide transfer designed into the cart and T/E table (3)
	Very poor transfer design	
Ulnar deviation (3)	Mold design - bad handles	"
	Mold height/cart design	"
	Very poor transfer design	"
Shoulder flexion with high force, 45 degrees. (3)	Very poor transfer design	"
Lifting force 70 - 75 lb (4)	Mold weight	"
	Very poor transfer design	"

* Include the rated importance.
** Include the rated potential.
Use a 0 to 4 scale. Zero indicates none and 4 the highest importance or potential.

FIGURE 19.4

JOB: Loaf Mold Extractor Operator.	STEP NUMBER: 3 (continued)	
DESCRIPTION:		
RISK FACTOR (*)	**ROOT PROBLEM(S)**	**SOLUTION (**)**
Compression on calves from mold hitting while walking	Very poor transfer design	"
Slippery floor (3)	Same as Step 1 above.	Same as Step 1 above.

* Include the rated importance.
** Include the rated potential.
Use a 0 to 4 scale. Zero indicates none and 4 the highest importance or potential.

FIGURE 19.5

Finally, it is important to relate the risk factors found in the job and the injuries and illnesses found on the job. If these do not correspond, the analysis has been inadequate and further analysis is required.

Solutions

This is a narrative section that summarizes the possible solutions. It can also be broken into two parts — step related and whole job related. It states the possible solutions and which of the risk factors and root problems they will address. It can also give an appraisal of the sufficiency of change that these solutions will bring about, separately or as groups.

FOLLOW-UP EVALUATION			
JOB NAME: Loaf Mold Extractor Operator.			DATE:
CHANGE IMPLEMENTED	(#)RISK FACTORS REDUCED (**)	RISK FACTORS CREATED (*)	RISK FACTORS UNCHANGED (*)
Cart handles	(1)Ulnar deviation, extreme extension, and pronation (3)		
New floor surface	(1,2)High pushing forces (2)		
	(1)Slippery floor (1)		
Larger wheels	(1)High pushing forces (2)		
Shoe change	(1,3)Slippery floor (1)		
Drop chute for lids	(2)High wrist acceleration (3)		

\# Step numbers in which the risk factor occurred.
* Include the rated importance. Use a 0 to 4 scale. Zero indicates none and 4 the highest importance or potential.
** Include the rated change. Use a -1 to 4 scale. Minus one indicates worse and 4 complete elimination.

FIGURE 19.6

19.4 Implementation and Follow-Up

The solution or solutions selected are tested and then implemented if they are found to be feasible, if they address the problems, and if they have the potential to improve the workers' safety and health. Follow-up is absolutely necessary to fine-tune the solution and to ensure that the problems have been addressed and that significant new ones have not been created. In order to facilitate the evaluation of changes that have been implemented, the authors have created another simple form that gives the change implemented, the risk factors reduced, the risk factors created, and the risk factors that remain unchanged. This form is illustrated in Figures 19.6 and 19.7. The form asks for the rated importance of each risk factor created with a scale from 0 to 4. Zero indicates that the rated importance to musculoskeletal disorders is none or negligible, and 4 indicates that the rated importance of the risk factor is very high. The form also asks for a rating of each changed risk factor with a scale from –1 to 4. Minus one indicates that the change made the root problem or risk factor worse, and 4 indicates that the root problem and its associated risk factor is completely eliminated. Implied in all of this is that a new job analysis is likely. These ratings are based on the analyst's judgment and experience. They can be improved upon by getting input from the people doing the job.

Worker input is critical in the follow-up. No matter how good the job may look to someone else, the person doing the job truly knows that job and the problems encountered. One example is that the job is fine as long as periodic maintenance is performed, but it becomes a real problem job when the maintenance is not performed. Also, worker input allows for fine-tuning and new suggestions.

19.5 Example Job — Loaf Mold Extractor Operator (100–500)

Job Description

In this job, the operator pushes a cart weighing hundreds of pounds into position (approximately 10 feet). Next, he removes a one-pound mold top and throws it into a bin about 5 feet away. He then lifts a heavy (70 to 75 lbs.) mold full of cooked product (lunch meat) from the cart, carries it to a table about 5 feet away, places it into a tilt/extraction table (T/E table), activates the T/E table, removes the endplates and throws them into a bin, removes the plastic coating and places it in a receptacle under the table, lifts

FOLLOW-UP EVALUATION			
JOB NAME: Loaf Mold Extractor Operator.		DATE:	
CHANGE IMPLEMENTED	RISK FACTORS REDUCED (**)	RISK FACTORS CREATED (*)	RISK FACTORS UNCHANGED (*)
Slide transfer designed into the T/E table	(3)High grasp force (4)		
	(3)Ulnar deviation (4)		
	(3)Shoulder flexion, high force 45 degrees (4)		
	(3)Lifting force (4)		
	(3)Compression on the calves from mold (4)		
		(4)Shoulder flexion, 45 degrees (0)	
		(4)Push force (1)	

\# Step numbers in which the risk factor occurred.
* Include the rated importance. Use a 0 to 4 scale. Zero indicates none and 4 the highest importance or potential.
** Include the rated change. Use a -1 to 4 scale. Minus one indicates worse and 4 complete elimination.

FIGURE 19.7

and carries one of the two loaves 4 feet to the slice/package cart, returns to the tilt/extraction table, carries the other of the two loaves 4 feet to the slice/package cart, returns to the T/E table, removes the mold, carries the mold back to the original cart, and places the mold in the cart. Each mold contains two logs or columns of lunch meat approximately 6" by 6" by 36". When the mold cart is empty, he pushes it (150 lbs.) 15 feet to a staging area. When the slice/package cart is full, he pushes it 5 feet to a staging area.

Job Objective

The objective of this job is to get a mold cart with full molds, remove the cooked product from the molds, place the product on the slice/package cart, move the empty mold cart with empty molds, and move the full slice/packaging cart to a staging area.

Job Essential Functions

The essential functions involve:

Pushing a very heavy cart — 800 pounds
Removing and disposing of mold lids
Transferring full molds to a tilt/extraction table
Removing plastic and an endplate from the product
Transferring cooked product to a cart
Transferring empty molds to a cart
Pushing a low-to-moderately heavy cart — 100 pounds
Pushing a heavy cart — 500 pounds

Contributing Factors

This job rarely involves overtime or high pressure. It is a secure job, and the worker doing it is usually left alone to do his job. The temperature is cool, but no drafts seem to be present. The worker does wear a rubber-type glove. Looking at all of this, there appear to be no significant contributing factors.

Task Breakdown

There is only one task in this job. It involves getting a mold cart with full molds, removing the cooked product from the molds, placing the product on the slice/package cart, returning the mold cart with empty molds, and moving the slice/packaging cart to a staging area. It is repeated throughout the workday.

Step Breakdown

The first three steps of the job analysis for this job are contained in Figures 19.2 through 19.5.

Risk Factor and Root Problem Summary

The stressors found in this job are ulnar deviation, extreme wrist extension, pronation of both wrists, high pushing forces, high grip forces, ulnar deviation, wrist flexion and extension, shoulder flexion and abduction, high arm force, inefficient side pull, trunk flexion, and trunk torsion, carrying heavy loads, toss or flip actions probably involving high wrist accelerations, molds that hit upper legs and knees as the worker walks, and wet floors.

These risk factors relate closely with the problems suffered by the workers doing this job. They have had tendinitis of the wrist, pain in the neck and shoulder, low back pain, and one incidence of a slip and fall resulting in a sprained ankle and bruises.

Solutions

Several solutions have been developed and are considered feasible. They are handles for the mold cart, larger wheels for the cart, different shoe soles, resurfaced floor, redesigned mold cart, redesigned T/E table, have the plastic stripped off mechanically, and a redesigned transfer to the slice/packaging cart. Well-designed handles for pushing or pulling the mold cart, larger wheels for the cart, better shoe sole, and a better floor surface will significantly reduce the risk factors associated with moving the mold cart. A well-designed cart and T/E table will significantly reduce the risk factors associated with mold transfer. A well-designed ramp, chute, or conveyor from the T/E table to the slice/packaging cart will significantly reduce the risk factors associated with product transfer.

Implementation and Follow-Up

Numerous solutions were implemented. Well-designed handles were installed on the cart for pushing. Larger wheels were installed on the cart. The floor was repaired such that it is less slippery and does not cause problems in pushing the cart. Appropriate shoe sole material was recommended to the employee for his next pair of boots. The cart carrying the molds and the T/E table were redesigned to allow pushing the molds from one to the other, thereby eliminating the lifting of molds. A convenient drop bin was provided to dispose of the mold lids. A hole was provided in the counter top to dispose of the endplates. An adjustable chute was installed so that the extracted loaf could be pushed onto the slice/package cart. Follow-up evaluation forms provided in Figures 19.6 and 19.7 show changes implemented, the risk factors reduced, the risk factors created, and the risk factors that are unchanged. Overall evaluation determined that this job was changed from one with considerable problems to one with very few problems. Even though the potential for musculoskeletal disorders is greatly reduced, periodic monitoring is recommended.

References

A Technical Assistance Manual on the Employment Provisions (Title I) of the Americans With Disabilities Act. (January 1992.) U.S. Equal Opportunity Commission, U.S. Government Printing Office, Superintendent of Documents, Washington, D.C.

Armstrong, T. J. (1983) *An Ergonomics Guide to Carpal Syndrome;* American Industrial Hygiene Association, Akron, OH.

Armstrong, T. J. (1986a) Ergonomics and cumulative trauma disorders, *Hand Clinics,* 1, 3.

Armstrong, T. J. (1986b) Repetitive trauma disorders: job evaluation and design, *Human Factors,* 28(3), 325-336.

Burke, M. (1992) *Applied Ergonomics Handbook,* Lewis Publishers, Inc., Boca Raton.

Cochran, D., Stentz, T., Stonecipher, B., and Hallbeck, S. Guide for videotaping and gathering data on jobs for analysis for risks of musculoskeletal disorders, in *Handbook of Industrial Ergonomics,* 1999.

Keyserling, W. M., Armstrong, T. J., and Punnett, L. (May 1991) Ergonomic job analysis: a structured approach for identifying risk factors associated with overexertion injuries and disorders, *Applied Occupational Environmental Hygiene,* 353-363, 6(5).

Marras, W. S. and Lavender, S. L. (1995) Biomechanical risk factors for occupationally related low back disorders, *Ergonomics,* 377, 38(2).

Marras, W. S. and Schoenmarklin, R. W. (1993) Wrist motions in industry, *Ergonomics,* 341, 36(4).

Putz-Anderson, V. (1988) *Cumulative Trauma Disorders, A Manual for Musculoskeletal Diseases of the Upper Limbs,* Taylor & Francis.

Rogers, S. H., A functional job analysis technique, *Occupational Medicine: State of the Art Reviews,* 7, 4.

Sommerich, C. M., McGlothlin, J. D., and Marras, W. S. (1993) Occupational risk factors associated with soft tissue disorders of the shoulder: a review of recent investigations in the literature; *Ergonomics,* 36(6), 697-718.

Ulin, S. S. and Armstrong, T. J. (1992) A strategy for evaluating occupational risk factors of musculoskeletal disorders, *Journal of Occupational Rehabilitation,* 2, 1, 35-49.

20

The AET Method of Job Evaluation

Kurt Landau
University of Technology–Darmstadt

Regina Brauchler
University of Stuttgart–Hohenheim

Walter Rohmert
University of Technology–Darmstadt

20.1 Objectives of Job Analysis

Planning, design, and evaluation of work should be preceded by an analysis of the job, the work tasks, and the resulting demands that are placed on the worker. This type of systematic analysis following a standard pattern is performed in only very few cases, mainly manual jobs in industry. Instead, *ad hoc* procedures relating to the individual case are used. No further use is made of the data after the immediate problem has been solved. This would, in any case, be impossible because the analytical instrument is either totally or, at least, partially inapplicable outside the confines of the company that used it. This means that companies regularly "reinvent the wheel." Analytical data that could be further evaluated for general purposes is not passed on, and the opportunity to further develop the discipline of ergonomics is lost. No taxonomies of jobs and tasks can be compiled, and questions relating to occupational research are left unanswered.

This raises the question of whether it would be possible to develop job analysis procedures that are universally applicable. Such procedures should cover the whole spectrum from heavy physical work to mental work, and they should be equally suitable for use in large and small operations and in different branches of industry.

For the evaluation of job analysis procedures, the following criteria, most of which have been developed by Frieling and Graf Hoyos, can be used (Frieling, 1975; Graf Hoyos, 1974):

The procedure should:

- Be based on a theoretical model that allows a practical interpretation of the results obtained with the job analysis
- Offer complete coverage of all demands that are present within a specific person-at-work system
- Offer maximum cost-effectiveness with regard to application, data processing, and data evaluation; the application of the procedure should allow standardization
- Go beyond a merely verbal work description and allow quantitative statements, at least at ordinal numbers

0-8493-2641-9/99/$0.00+$.50

When applying the procedures, it should be possible to make statements as to:

- The reliability with which several raters analyze a person-at-work system at the same time (inter-rater-reliability)
- The reliability with which all items of the job analysis procedure can be rated (item reliability)

If job analysis is seen as an analysis of stress determinants, it can be assumed that it will be possible to make a quantitative evaluation of stress factors (generally rated on an ordinal scale) by duration, intensity, sequence, overlap, and time of occurrence within a job. If it is also claimed that the analysis procedure will provide information on the strains resulting from the stress patterns, this implies that the procedure is capable of:

- Producing repeatable qualitative and quantitative analyses of the strains arising (with the exception of emotional strains) (Luczak, 1975)
- Allocating psychological or physical strain to selected items qualitatively
- Rating specific items for the psychological or physical strains produced by them
- Helping to make quantitative evaluations of strains (ratings on a set scale) based on the results of physical or psychological examinations of the workers involved

This distribution describes the AET, a job analysis procedure which seeks to fulfill the requirements listed above.

The origins of AET (Arbeitswissenschaftliches Erhebungsverfahren zur Tätigkeitsanalyse; ergonomic job analysis procedure) date back to a study (Rohmert and Rutenfranz, 1975) ordered by the German government to investigate discrimination against women at work with respect to pay. A job analysis procedure was required that allowed a detailed investigation of workload and strain within a given person-at-work system. At that time, there was no job analysis procedure that could readily be used, although the PAQ (McCormick et al., 1969) seemed to provide a basis for the psychological items.

AET (Arbeitswissenschaftliches Erhebungsverfahren zur Tätigkeitsanalyse) job analysis (job evaluation) procedure was developed in 1978 (Landau et al., 1975; Landau, 1978; Rohmert and Landau, 1979); it has been continuously improved since then. An important characteristic is that when a firm applies the AET procedure to a specific job, the data are recorded in a central data bank at Darmstadt. The data bank now holds the analyses of over 7,000 jobs. Another important characteristic is that the AET procedure is applicable over a very wide range of jobs, blue collar as well as white collar, manual as well as engineering and executive, manufacturing as well as retail, office as well as factory.

For an analyst, the general procedure is to evaluate a job using 216 characteristics. The individual doing the job is not evaluated. The points for each characteristic are obtained and totaled. The pay for a worker is a function of the number of points, but the shape of the function varies with the firm.

20.2 AET — The Ergonomic Job Description Questionnaire

Theoretical Basis

The AET procedure is based, on the one hand, on the model of the person-at-work system and, on the other hand, on the concept of stress and strain. Concerning its criteria of classification of activities, in order to facilitate logical and deductive testing, AET is oriented toward the elements and flows of the person-at-work system.

The stress and strain concept of human work (see Rohmert et al., 1975; Luczak, 1975) assigns the work task to the "object area" (see Figure 20.1). Both the work task and the conditions of the work environment are included in the object area, from which job-specific and situation-specific demands result. These demands characterize the energetic-effective heaviness of work and the informative-mental difficulty of work. Partial stresses related to the work content are the result of duration and time distribution of work heaviness and difficulty.

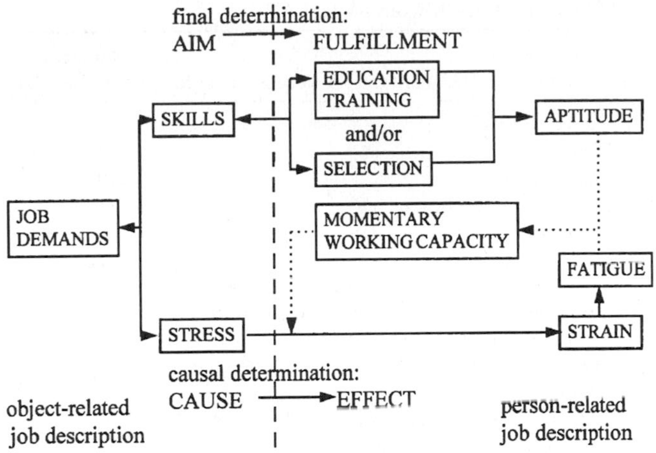

FIGURE 20.1 Stress-strain concept. (From Landau, K. and Rohmert, W. 1989. Introduction of the problems of job analysis, in Landau, K. and Rohmert, W. (eds.) *Recent Developments in Job Analysis, Proceedings of an International Symposium in Job Analysis*. University of Hohenheim, March 1989. London, New York, Philadelphia: Taylor & Francis, pp. 1-24. With permission.)

The partial stresses are expressed in factors and quantities describing the external effects of the person-at-work system on the working person (Luczak, 1975; Rohmert, 1983). Together with the situation-specific partial stresses that result from the physical and social working conditions, the work content-specific partial stresses determine the subjective activities. Activities are also influenced by the motivations and disposition of the working person; this can be partly derived from the required skill. These relationships form the "object area" of the enlarged stress–strain concept and the starting point of the job analysis with AET. The simultaneous consideration of the categories "tasks," "demands," "environmental conditions," and "required skills" as determinants of stress is the cornerstone of AET (Landau and Rohmert, 1989).

AET Structure with Regard to the Contents

To begin with, the person-at-work system is described. The description and scaling refers to the objects of work and to the equipment and the working environment. The objects of work are analyzed under material, energy, and information aspects. When a person is an object of work, the characteristics of the group of people he came from must be investigated. Finally, the qualities of material objects of work (like raw materials) are analyzed. The equipment, the work instruments (like tools, implements), and other operating materials (like hard- and software) are of interest. In this case, the ergonomic system of classification is completed by technical aspects. The analysis of the environment is related to the physicochemical conditions in the workroom, the organizational and social working conditions and to principles and methods of pay. There are 36 items for working and operating materials, 50 for the physical, organizational, and social working conditions, and 24 for the economic working conditions (see Table 20.1).

The object of investigation is the human activity in the person-at-work system; this is composed of the tasks which have to be carried out. The tasks ensue from the purpose of the person-at-work system. In this context part B of the analysis procedure represents a link between the tasks to be fulfilled and the resulting demands exerted on the working person. So, the analysis of the person-at-work system is followed by the task analysis, an evaluation by means of 31 items, subdivided according to the objects of work (see Table 20.2).

The task analysis is finally followed by the analysis of demands (see Table 20.3). Within this part the items were chosen in such a way as to take as many functions of the body as possible into account. This

TABLE 20.1 AET-Structure — Part A

A: ANALYSIS OF THE PERSON-AT-WORK SYSTEM	Number of characteristics	
1. Object of Work		33
Kind	(4)	
Characteristics	(28)	
Person as object of work	(1)	
2. Equipment		36
Means of production	(17)	
Changes in the state of the objects of work		
Change in the location of the objects of work		
Other means of production		
Other equipment	(19)	
for controlling the state		
for supporting human sense		
seat, worktable, workroom		
3. Working Environment		74
Physiochemical working environment	(12)	
Organizational and social conditions	(38)	
Temporal work organization		
Position on the work within the:		
operation process organization		
organization structure		
communication system		
Principals and methods of pay	(24)	
		143

TABLE 20.2 AET-Structure — Part B

B: ANALYSIS OF TASKS	Number of Characteristics
1. Related to Material Objects of Work	13
(Preparing, assembling, equipping, inserting, transporting, measuring, operating, checking, supervising ...)	
2. Related to Abstract Objects of Work	8
(Planning, organizing, coding information, transcribing information, combining information, analyzing information ...)	
3. Person-Related	8
(Representing, instructing, serving, attending ...)	
4. Number and Frequency of Task Repetitions	2
	31

is true for the function of power and energy generation in the different organs (stress during postural work and static work, heavy dynamic work and active light work), as well as for the functions of perception, decision, and action in different mechanisms of information processing. The classification depends on the question of which area is required concerning the reception of information, decision, and action. In the case of the analysis of demands, one distinguishes between the demands in connection with the reception of information (17 items), with information processing (8 items), and with information output or activity (17 items). Besides the usual test–statistical requirements of an analysis procedure, a fundamental problem arises as a result of the possibility of theoretical justification. That is namely asking for the selection of characteristics within the outlined classification of the procedure and for the scale level of the selected characteristics.

TABLE 20.3 AET-Structure — Part C

C: ANALYSIS OF DEMANDS		Number of Characteristics
1. Range of Demands: Reception of Information		17
Organs of Sense for the Reception of Information	(6)	
Dimensions of Identification	(7)	
Forms of Identification	(2)	
Accuracy Necessary for the Reception of Information	(1)	
Vigilance	(1)	
2. Range of Demans: Decision		8
Complexity of decision	(1)	
Temporal scope of decision	(1)	
Necessary knowledge	(6)	
3. Range of Demands: Activity		17
Organs of activity and accuracy of activity for stress while acting:		
Caused by postural work	(6)	
Caused by static work	(4)	
Caused by heavy dynamic muscular work	(3)	
Caused by active light work	(4)	
		42

Particularly when selecting the items, the following basic facts have to be recognized and considered:

1. It is not possible that a fully completed catalogue of items can be expected as part of the selected theoretical concept, especially in view of the economy of the procedure.
2. Only those items that are important for numerous people-at-work systems should be included.
3. The selection of items implies a certain judgment; hence, "user-specific traits" — such as experience, standards of values, opinions — influence the system of analysis.

Thus, it must be noted that as the selection of the theoretical model is at the basis of the procedure, the selection, scaling, and description of items within this model entail a series of subjective assumptions. However, these assumptions are acceptable as long as several analysts achieve reliable results by applying the procedure.

Parallel to the development of AET, an attempt was made to limit the subjective influences on the construction of items by means of the concept of an iterative procedure development with respect to the experience of both company specialists and ergonomists.

If one acknowledges the subjective, author-specific influence of the selection of a theoretical model concept substantiating AET on the selection of items and on the determination of scale levels, codes, and aids of classification, then this means also recognizing the need to include technical criticism to achieve a progressive development of AET. The goals of the iterative AET development consisted of increasing its accessibility to users while safeguarding or improving the test–statistical quality criteria.

The relatively vague notion of "ease of use of AET" can be operationalized by the following desired qualities:

- A clear analysis
- Clear and comprehensible formulas
- Conformity of the definitions of AET with the meaning of terms in ergonomic literature (as far as possible)
- A time-saving and economical analysis and evaluation

The development of AET was oriented toward certain groups of users. In addition to ergonomically trained groups of users in research and practice, AET addresses ergonomists. Another category at which AET is aimed, and which is more a group of persons interested in AET rather than a group of users, includes people such as industrial psychologists, representatives of labor and management, and so forth.

TABLE 20.4 AET Versions and Reliability

AET Version	Number of Examined Jobs	Item Reliabilities	Position Reliabilities
AET (A) - Draft	26	0.57	0.64
AET (A)	42	0.65	0.87
AET (B)	4	0.71	0.71
	17 Analysts (Seminar classification)		
AET (C)	62	0.71	0.89
AET (D)	2	0.79	0.74
	22 Analysts (Seminar classification)		

Incorporating the practical experience and the extent of knowledge of these target groups into the iterative development of AET was accomplished by means of three AET seminars and several meetings of ergonomists with representatives of labor and management. These seminars were designed to take account of the broad area of AET's validity and application. The aims of the seminars were:

- To discuss basic questions on matters of the form and contents of AET
- To determine how long it takes to train experienced analysts in using AET
- To ascertain the reliability of use of AET in the corresponding version
- To state difficulties of application and to initiate suggestions for improvement

The iterative elaboration of AET, based on AET version (A) and carried out both on this basis and on the grounds of the results of the reliability studies, altogether resulted in three further versions (B,C,D) with corresponding drafts (see Table 20.4). The number of items was able to be reduced considerably from 390 to 216.

AET Structure with Regard to Form

The analysis of the job is done in the form of an observation interview, which means that necessary analytical data are collected first by observation of the job and working environment and second by interviewing the incumbent and the incumbent's superior.

Each AET item consists of an (underlined) question outlining the state of affairs to be grasped and indicates the code for classifying this item. In certain circumstances, examples are given as classification aids. The explanations clarify the questioning of the AET characteristic in view of extent, delineation, and classification, but they cannot be taken as a complete and binding instruction for the rating.

The items of the analysis of demands — which might be difficult to answer for somebody who is not sufficiently trained in ergonomics — contain additional classification aids in the form of "activity scales." The activity scale, based on previously investigated data, contains a series of grades of mainly illustrative activities. Equivalent to an ascending rating scale, we may suppose, at least approximately, an intensified demand (see Figure 20.2).

However, the selected job examples only represent an optional part of reality; an equidistant reproduction of the job examples an ordinal scales of the item concerned is not possible.

The classification of characteristics can only be done by means of the corresponding code, which follows the suggestions of McCormick et al. (1969). The different codes are as follows:

- Significance Code (S)

  The importance or significance of this aspect for the task should be estimated in relation to other tasks or activities. Use a range from 0 to 5.

- Duration Code (D)

  The code "duration" is based on a shift lasting eight hours. This is assumed even if the incumbent is a part-time worker. This is done in order to be able to compare the work content of jobs with different shift hours. Use a range from under 1/3 of shift time to whole shift time.

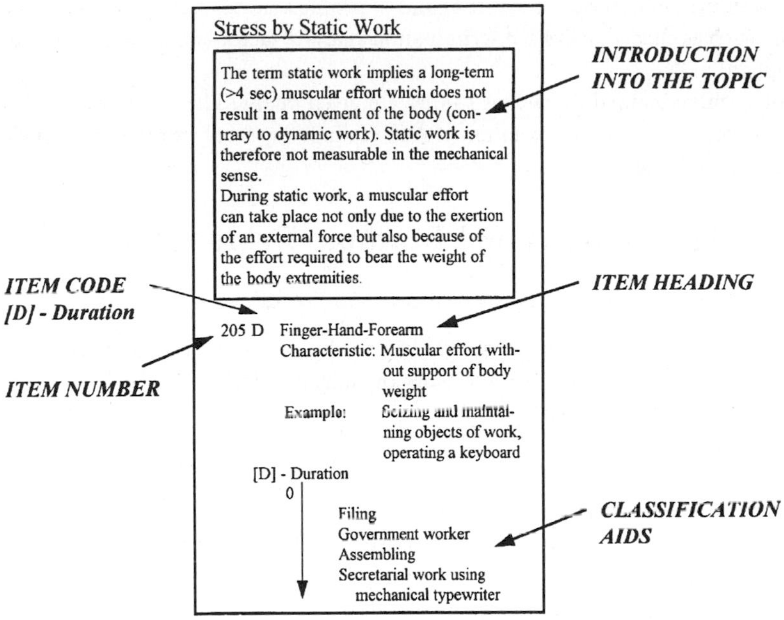

FIGURE 20.2 Example of an AET item.

- Frequency Code (F)

  This code characterizes the temporal distribution and position of stress sections. Use a range from 0 to 5.
- Alternative Code (A)

  The alternative code only asks about the presence of a characteristic: Does the work characteristic in question apply 1, or does it not apply 0?
- Exclusive Code (E)

  The exclusive code is always related to only one specific question, e.g., multiple properties of a working instrument. Use a range from 0 to 5.

An AET analysis in the field is directly followed by the coding of the AET items. This coding has to be done on standardized paper or in direct dialogue with the computer. The time required for an observation interview is usually 1 to 3 hours, depending on the analyst`s practical experience, the type of job and the repetition rate of work processes.

AET Method of Evaluation

As a result of the various opportunities that exist to use AET data, the data can be used to answer fundamental questions arising in nearly all potential fields of application:

1. The AET codings can be used to characterize a person-at-work system.
2. The AET codings can be used to describe people-at-work systems or groups of people-at-work systems.
3. The characteristics revealed by all or most of the AET codings can be used to classify a person-at-work system.
4. Conversely, people-at-work systems can be grouped together according to common AET characteristics.

For these interrelated and basic requests, the best solution is through using univariate evaluation methods that are supplemented by a series of multivariate methods. Concerning univariate methods, well-known procedures of descriptive statistics and of profile analysis can be used. Procedures of multivariate analysis, such as cluster analysis, discrimination analysis, factor analysis, multidimensional scaling, etc. can be applied.

Examples of results obtained in various studies are given below. These are intended to give the user guidance in the evaluation and interpretation of AET codings, especially for the types of evaluation listed under the fundamental questions 2 and 3 above.

20.3 AET Applications and Example Evaluations

The evaluation of AET codings has to provide answers to the following questions:

1. What are the differences in the jobs regarding work content, objects of work, work instruments, workplace, working environment, and work organization in:
   different branches of industry
   different enterprises
   different departments
   different wage groups?
2. What are the differences in the job characteristics of jobs requiring different education and job-related training?
3. To what extent do the job characteristics of native and foreign workers differ?
4. To what extent do the job characteristics of industrial employees, office workers, executive personnel, and government workers differ?
5. Which jobs are particularly similar or differ markedly in view of different strain-relevant stress components?

Profile Analysis

A representation of the scores obtained by the groups of items in the form of profiles is suitable to give a graphic survey of the extent or the duration of stress experienced during the execution of jobs or groups of activities. This type of evaluation of the data derived from job analysis is designated "profile analysis," regardless of the fact that this term may possibly be used in a different sense by other disciplines. Figure 20.3 shows an example of a job profile obtained from AET analyses. It shows the characterization of the individual person-at-work system in laying bricks on the basis of one actual coding of the 216 AET-items. The grouping of the characteristics necessary for the execution of a profile analysis used by AET is derived from the structural items of the job analysis procedure. The method used for computing the profiles is explained in Landau and Rohmert (1981) and Rohmert and Landau (1983). The vertical plane shows the characteristics of the workplace and the types of demand, while the horizontal scale shows the maximum AET classification in percent.

Figure 20.4 uses the AET procedure to analyze the incidence of the various types of demand in 2,838 jobs usually occupied by males and 866 usually occupied by females in German industry. The upper bar represents the female jobs, the lower bar the male jobs. The analysis shows that the most important tasks for males involve operating, controlling, supervising, planning, organizing, and analyzing. The main tasks performed by females are checking, and also a variety of general, people-oriented service tasks. The tasks are divided into the stereotyped patterns of "typically male" and "typically female." Males perform more (complex) operating, controlling, and assembly tasks in which they are required to plan and organize their own work, while females, in addition to their additional job as mother or housewife, are employed in industry mainly for simple checking activities and also for people-oriented services. The jobs occupied by males were exposed to far higher levels of physical or chemical stress from the working environment. This applies both to factors like illumination, climatic conditions, vibration, and noise as well as to other environmental influences like noxious materials. The work hazards, including the frequency or probability of a work accident or an occupational disease, are rated higher for the male jobs.

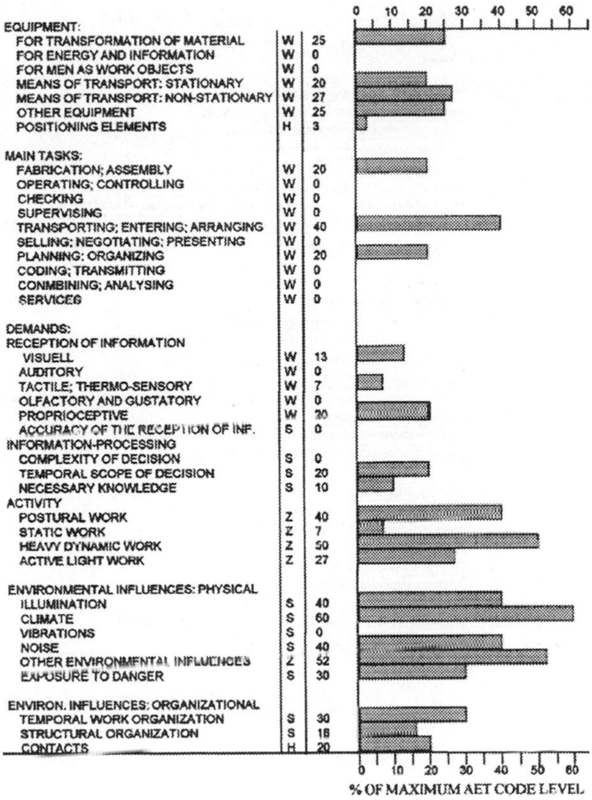

FIGURE 20.3 AET job profile: laying bricks.

The male jobs also show a higher level of demands in both the organization of working time (shift work), the sequence of operations, and overall planning. As night work by females was severely restricted by German law, it was and is largely a male preserve in industry, and the demands in this respect are therefore much higher for men.

Closer investigation shows that information reception and information processing place substantially higher degrees of certain types of demand on the male than on the female jobs. This led to a higher classification against the AET criteria for demands involving the reception of visual information, of auditory information, and proprioception. Similar levels for both male and female jobs were registered only for information reception via the senses of smell, taste, touch, and temperature sensitivity of the skin. There are only very slight differences between the sexes in the demands for accuracy of information reception. The male jobs involve tasks of greater complexity and, in some cases, greater critical stress. The level of knowledge required in male jobs is rated higher than in female jobs.

The proportion of physical work demands is very similar for both sexes. They are identical for static handling work and only insignificantly higher for males in the case of static holding work. In male jobs, longer periods of the shift are devoted to heavy dynamic work and, in female jobs, to active light work. This corresponds to the German role expectations in the division of work between men and women in industry, i.e., heavy dynamic work for men and active light work involving monotonous procedures for women.

Frequency Distributions

Tables 20.5 and 20.6 explain how company-, branch-, and sex-related position groups that were analyzed by means of the AET procedure can be evaluated in a simple way by analyzing the frequency distributions of levels. Table 20.5 shows percentages of the item levels "high" and "extreme" for selected characteristics of the AET procedure.

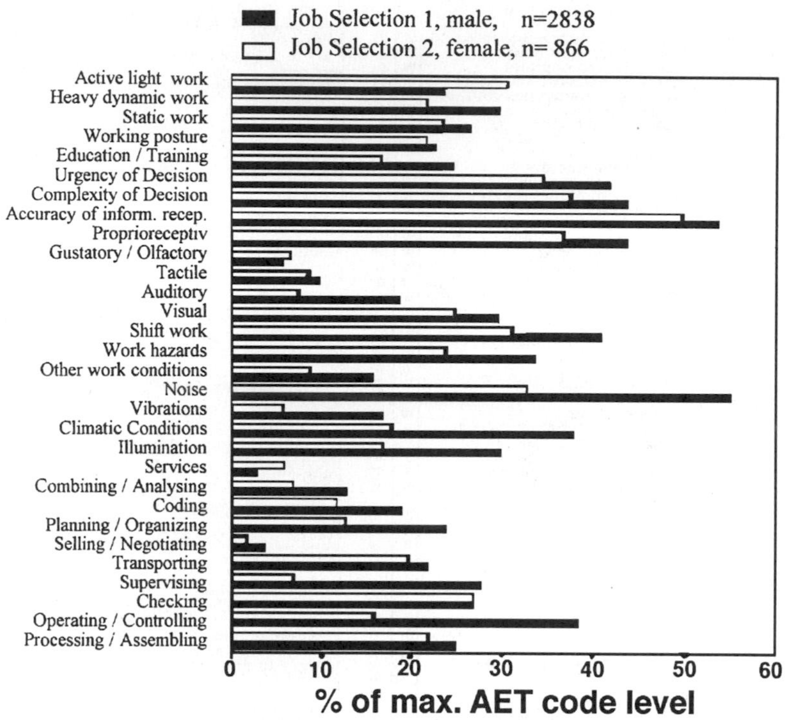

FIGURE 20.4 Analysis of job tasks and demands for 2,838 males and 866 females in German industry (216 AET items). (From Landau, K. and Rohmert, W. 1992. Evaluation of worker workload in flexible manufacturing industry. *The International Journal of Human Factors in Manufacturing,* 2 (4), pp. 369-388. With permission.)

TABLE 20.5 Frequency Distributions of High [4] and Extreme [5] AET Codings of Selected AET Items for In-Company Applications

	% Level for AET Characteristics "HIGH" (4) and "EXTREME" (5)					
	Fabrication		Preassembly		Assembly	
AET Item	Male	Female	Male	Female	Male	Female
Climate (71)	7	7	13	1	0	0
Vibrations (78)	87	25	2	0	18	17
Accuracy of Perception (17)	61	57	0	0	0	0
Decision Making Under Pressure of Time (19)	0	0	39	39	91	89

From Rohmert, W. and Landau, K. 1983. *A New Technique for Job Analysis.* London, Taylor & Francis. With permission.

For 7% of the workplaces of male workers, the climatic stress is rated "high" or "extreme." In the field of preassembly, 13% of the workplaces of male workers shows a high or extreme climatic stress. In the assembly department, an extraordinary climatic stress is to be found. Effects of mechanical vibrations in the fabrication department are rated "high" or "extreme" at 87% of the workplaces occupied by men and at 28% of the workplaces occupied by women. This method of evaluation can be continued in the same way for all characteristics of the procedure. If large-scale data collections related to branches are available, then it is possible to carry out an analysis of frequency distributions of levels of AET data due to conditions in the particular trade. This is shown in Table 20.6.

TABLE 20.6 Sex- and Branch-Related List of the Classifications High [4] and Extreme [5] for Selected AET Items as an Example of Frequency Distribution of AET Codings

	% Level for AET Characteristics "High" (4) and "Extreme" (5)			
	Occupied by Women	Occupied by Men	Metal Working Industry	Chemistry
Active Light Work (211)	32	6	24	22
Visual Identif. of Surface Structures (175)	34	25	34	39
Combining (161)	2	20	3	0
Noise (73)	95	40	35	43
Responsibility (109–112)	1	51	22	21

From Rohmert, W. and Landau, K. 1983. *A New Technique for Job Analysis.* London, Taylor & Francis. With permission.

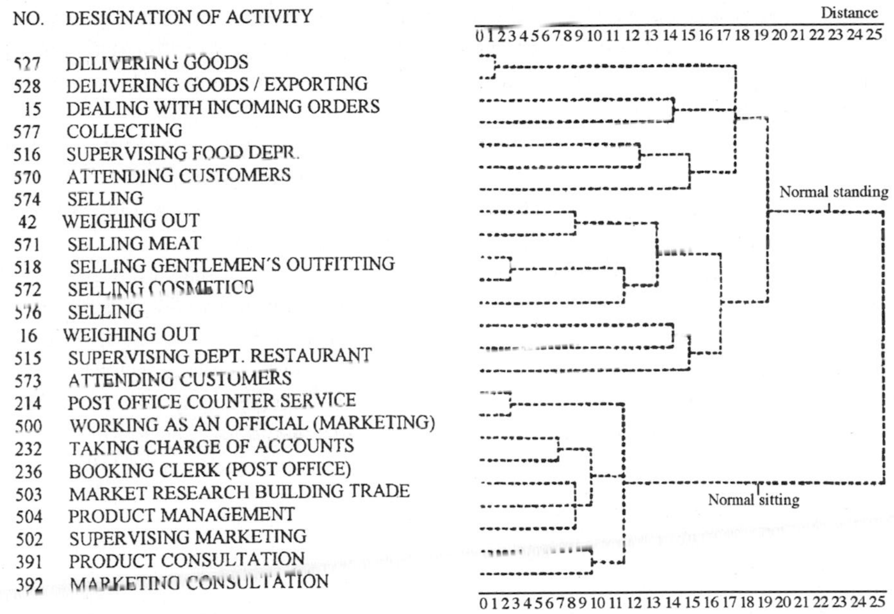

FIGURE 20.5 Example of an AET cluster analysis. (From Rohmert, W. and Landau, K. 1983. *A New Technique for Job Analysis.* London, Taylor & Francis. With permission.)

Cluster Analysis

The main aim of applying cluster-analytic methods to job analysis data is a reduction in the quantity of data. Apart from the generation of job groups, numerical relationships between the job groups must be established and interpreted. When carrying out the classification of activities, one must examine which conclusions can be drawn from the numerical relationship for the elaboration of a taxonomy of human activities. As far as ergonomics is concerned, the effects of specific work content on group composition and on numerical relationships between the groups are of particular interest. With the classification of large ergonomic data tools into smaller and comprehensive groups arises the possibility of investigating group scores further, i.e., the positions that are most representative for the group as well as for the respective workers, using (rather expensive) methods of occupational physiology and psychology.

Figure 20.5 shows one application of cluster analysis methods to AET data. A hierarchic cluster analysis has been carried out for 24 buying, selling, or trading jobs and is represented in the form of an activity dendrogram. In this analysis it is possible to distinguish that, with respect to working posture, two groups are clearly separated from the others.

1. Prevailing working posture is normal standing, with a small quota of the postures "bent standing," "crouching" and "kneeling." This group covers the working postures typical of selling activities in department stores and retail shops.
2. Prevailing body posture is sitting, with a small quota of standing. This posture is typical of commercial activities not to be classed with the retail trade. Significant activities in this group are the fields of marketing consultation and sales promotion.

The "dissection" of the results of a cluster analysis makes it possible to identify the job clusters at a given hierarchical level. In cases where jobs are grouped by sectors of industry, the mean duration of the observed postures per shift is shown graphically. As this involves the calculation of mean arithmetical values from data rated on an ordinal scale, it is advisable to interpret the results with caution.

Comparing different industries, Landau and Rohmert (1992) estimated, for example, the percentage of shift time involving static work (their Figure 6.1). The jobs in the iron and steel industry received the highest rating for static work, followed by the chemical industry, the automotive industry, and the services sector in that order. Static work mainly involves the use of the finger/hand/forearm region or the arm/shoulder/back region. Static work using the leg or foot region is of only minor importance in all the industries covered by this study.

The chemical industry has the highest percentages of heavy dynamic work and active light work. This is followed by the iron and steel industry in the case of heavy dynamic work, and by the automotive industry in the case of active light work. The percentages of heavy dynamic work and active light work are lowest in the services sector (Landau and Rohmert [1992] Figure 6.2).

Heavy dynamic work can involve either the arms and upper body muscles or the legs and pelvic muscles. Heavy dynamic stress on the legs and pelvic muscles was caused by walking, climbing, etc., in some cases with loads. Walking and climbing are still important work factors in the chemical industry, followed by the iron and steel industry, the automotive industry, and the service sector.

Active light work in the chemical industry involves mainly the finger/hand system. In the automotive industry, there are more gross motor activities using the hand/arm system. The foot/leg system is not used to any significant degree in any of the industries investigated (Landau and Rohmert [1992] Figure 6.3).

Shifting of Demands

The AET is capable of analyzing exceptionally high stress levels affecting specific body organs in people-at-work systems and also of quantifying tasks and demands at the workplace and, by extension, identifying shifts in demands such as:

- Cessation or addition of specific types of demand
- Changes in intensity of one or more types of demand
- Changes in duration of specific types of demand
- Changes in time spread of specific types of demand
- Changes in association between different types of demand.

Using the AET data, Landau and Rohmert (1992) compared eight jobs in mechanized assembly with 67 traditional assembly jobs in the automotive industry (Figure 20.7). The small size of the sample populations makes it necessary to interpret the present results with caution.

Visual information reception increases from 24% to 37% of the maximum AET score. There is an even higher increase, from 39% to 60% of the maximum score, in the requirement for accuracy of information reception. Proprioceptive information reception remains almost unchanged, while demands involving information reception by touch and via the thermosensors of the skin have been eliminated, because gripping actions are now carried out by the mechanized assembly. Demands involving auditory information reception, especially in cases where problems are starting to develop, increase because of the high noise level in the work environment.

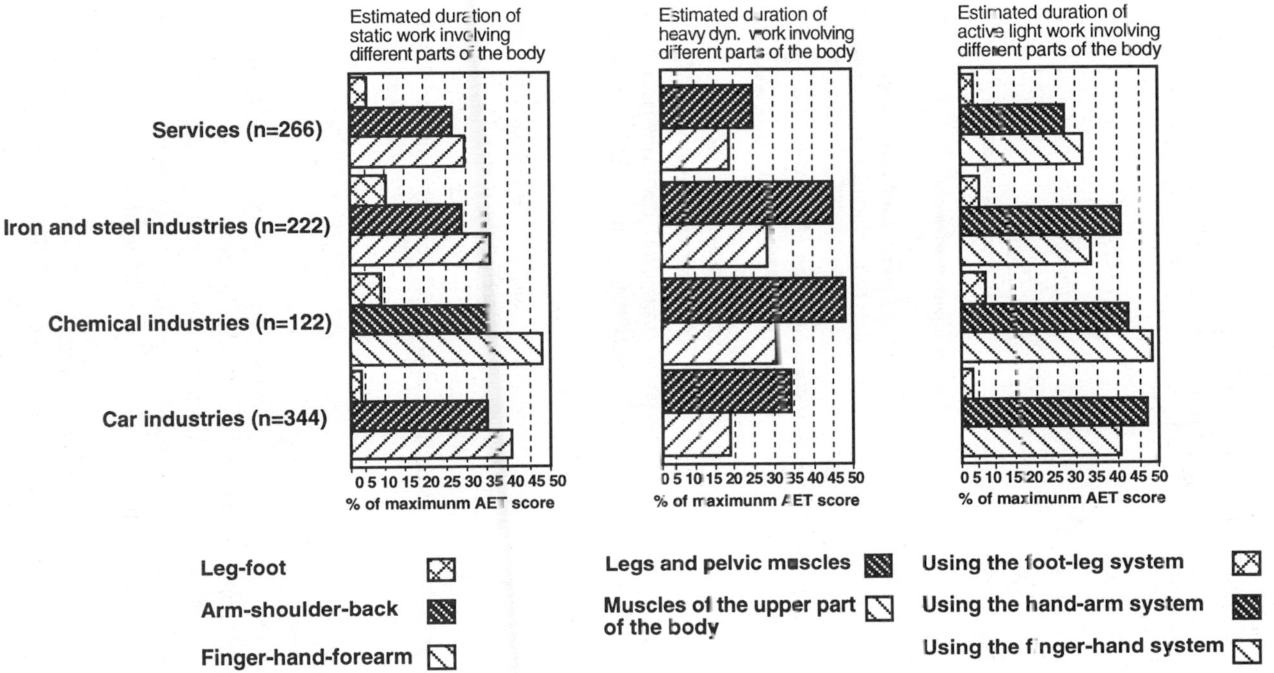

FIGURE 20.6 Estimated duration of physical work forms. An example of task analysis results obtained with AET (From Landau, K. and Rohmert, W. 1992. Evaluation of Worker Workload in Flexible Manufacturing Industry. *The International Journal of Human Factors in Manufacturing*, 2 (4), pp. 369-388. With permission.)

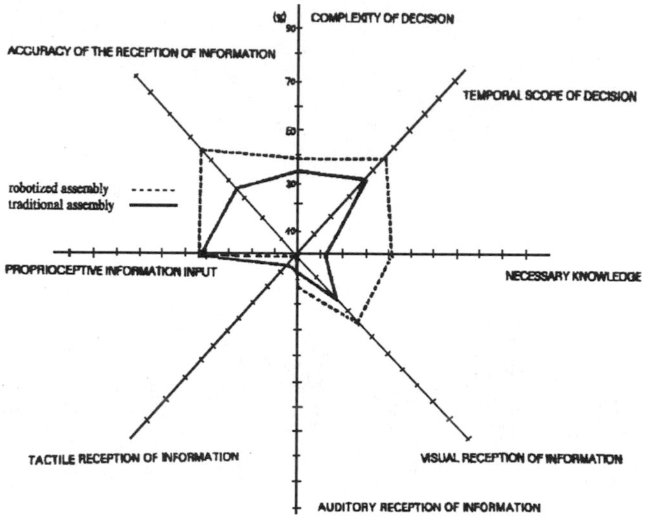

FIGURE 20.7 Shifts in demand involving information reception and information processing following robotization. An example of job analysis results obtained with AET. n1: solid line: 67 traditional assembly jobs; n2: dashed line: 8 jobs in mechanized assembly. (From Landau, K. and Rohmert, W. 1992. Evaluation of Worker Workload in Flexible Manufacturing Industry. *The International Journal of Human Factors in Manufacturing,* 2 (4), pp. 369-388. With permission.)

Demands relating to information processing produce an increase in qualification requirements. The amount of knowledge required increases from 12% to 41% of the maximum AET score. Decision complexity shows a small increase of 5%. However, the urgency of decisions rises from 43% to 55% because of the highly integrated assembly processes in which defects in the equipment have to be rectified without delay.

In conclusion, it can be stated that technological change *per se* does not necessarily bring an improvement in working conditions or a more humane work pattern. Industrial robots actually have both positive and negative effects for the workers. Their advantage is merely that they offer the option of a new production process better adapted to the needs of the workers. Whether this option and the option of improving the qualifications of the workers is exercised, depends very much on the attitudes of management and engineering departments.

20.4 Conclusion

Job analysis procedures are often criticized because of their unsuitability for a wide range of applications. The problem is that the development of universally applicable job analysis procedures is a costly business. The optimal solution is to design the procedure to be as universal as possible within the financial limitations imposed. Universality is defined in this case as suitability for the interpretation and generalization of data obtained from either different investigations, different sectors of industry or different companies.

Appropriation of the contributed AET procedure was done in several fields of applications in the last 20 years. The data obtained from 7,000 AET analyses checked and filed by us forms an available tool that allows detailed examinations, in particular the comparison of work tasks in different sectors of industry. The comparison of tasks and demands of work before and after the introduction of robotized, mechanized assembly is also suitable. Moreover, the use of statistical data analysis procedures for AET data provides significant aid in interpretation and data reduction. The cluster analysis in particular allows an economical application of job analysis in the fields of work design, demand analysis, personnel management, work and occupational research, detection of accident causes, and so forth.

To provide answers to ergonomic, technical, and epidemiological questions, it is necessary to collect all available data on potential deficiency and inadequate conditions at the workplace. Consequently, the main aim of further developments of job analysis procedures should be to maintain the number of items examined and to achieve maximum cost-effectiveness by applying the database obtained from successive investigations as a forecasting and indicative aid in the widest possible range of work situations and to achieve maximum cost-effectiveness by evaluating and interpreting the data.

References

Frieling, E. 1975. *Psychologische Arbeitsanalyse,* Stuttgart: Verlag W. Kohlhammer.

Frieling, E. and Hoyos, C. Graf. 1978. *Fragebogen zur Arbeitsanalyse–FAA.* Bern: Hans Huber Verlag.

Hoyos, Graf 1974. *Arbeitspsychologie.* Stuttgart: Verlag W. Kohlhammer GmbH.

Jeanneret, P.R. 1969. *A study of the job dimensions of "worker oriented" job variables and of their attribute profiles.* Lafayette: Unpublished doctoral dissertation, Purdue University.

Landau, K. 1978. *Das Arbeitswissenschaftliche Erhebungsverfahren zur Tätigkeitsanalyse — AET,* Dissertation, TH Darmstadt.

Landau, K. 1978. Das Arbeitswissenschaftliche Erhebungsverfahren zur Tätigkeitsanalyse — AET im Vergleich zu Verfahren der analytischen Arbeitsbewertung, *Fortschrittliche Betriebsführung,* 27, 1, 33-38.

Landau, K., Luczak, H., and Rohmert, W. 1975. Arbeitswissenschftliche Erhebungsbogen zur Tätigkeitsanalyse, in Rohmert, W. and Rutenfranz, J. (eds.) *Arbeitswissenschaftliche Beurteilung der Belastung und Beanspruchung an unterschiedlichen Arbeitsplätzen.* Der Bundesminister für Arbeit und Sozialordnung, Bonn.

Landau, K. and Rohmert, W. (eds.) 1981. *Fallbeispiele zur Arbeitsanalyse — Ergebnisse zum AET-Einsatz.* Bern: Huber.

Landau, K. and Rohmert, W. 1989. Introduction to the problems of job analysis, in Landau, K. and Rohmert, W., (eds.) *Recent Developments in Job Analysis,* Proceedings of an International Symposium on Job Analysis. University of Hohenheim, March, 14-15, 1989. London, New York, Philadelphia: Taylor & Francis, pp. 1-24.

Landau, K. and Rohmert, W. 1992. Evaluation of worker workload in flexible manufacturing industry. In: *The International Journal of Human Factors in Manufacturing,* 2 (4), 369-388.

Luczak, H. 1975. *Untersuchungen informatorischer Belastung und Beanspruchung des Menschen.* Fortschritts-Berichte der VDI-Zeitschriften, Reihe 10, 2, Düsseldorf, VDI-Verlag.

Luczak, H., Landau, K., and Rohmert, W. 1976. Faktoranalytische Untersuchungen zum Arbeitswissenschaftlichen Erhebungsbogen zur Tätigkeitsanalyse — AET. *Zeitschrift für Arbeitswissenschaft,* 30, 1976, 1, 22-30.

McCormick, E.J., Jeanneret, P.R., and Mecham, R.C. 1969. *The Development and Background of the Position Analysis Questionnaire (PQA).* Occupational Research Center, Purdue University, Report No. 5.

McCormick, E.J., Mecham, R.C., and Jeanneret, P.R. 1972. *Technical Manual for the Position Analysis Questionnaire* (PAQ). Lafayette: PAQ Services, Inc.

Rohmert, W., Rutenfranz, J., Luczak, H., Landau, K., and Wucherpfennig, D. 1975. Arbeitswissenschaftliche Beurteilung der Belastung und Beanspruchung an unterschiedlichen industriellen Arbeitsplätzen, in Rohmert, W. and Rutenfranz, J. (eds.) *Arbeitswissenschaftliche Beurteilung der Belastung und Beanspruchung an unterschiedlichen industriellen Arbeitsplätzen.* Der Bundesminister für Arbeit und Sozialordnung, Bonn, 15-250.

Rohmert, W. and Rutenfranz, J. 1975. *Arbeitswissenschaftliche Beurteilung der Belastung und Beanspruchung an unterschiedlichen industriellen Arbeitsplätzen.* Der Bundesminister für Arbeit und Sozialordnung, Bonn.

Rohmert, W. and Landau, K. 1979. *Das Arbeitswissenschaftliche Erhebungsverfahren zur Tätigkeitsanalyse (AET). Handbuch und Merkmalheft.* Bern, Stuttgart, Wien: Hans Huber Verlag.

Rohmert, W. 1983. Determination of stress and strain at real work places: Methods and results of field studies with air traffic control officers, in Moray, N. (Hrsg.): *Mental Workload, Its Theory and Measurement.* NATO Conference Series, Series III: Human Factors Col. 8 New York: Plenum Press 1979, 423-443.

Rohmert, W. and Landau, K. 1983. *A New Technique for Job Analysis.* London, Taylor & Francis.

For Further Information

For definitions and terminology used in job and task analysis, please refer to Landau, K., Rohmert, W. and Brauchler, R. 1997, *Task analysis: Part I — guidelines for the practitioner,* in Mital, A. and Kumar, S. 1997. One special Issue of the *International Journal of Industrial Ergonomics.* Elsevier Science Publishers North Holland (about 15 pages). Also refer to Landau, K. and Brauchler, R. 1997. *Task analysis: Part II — the scientific basis (knowledge base) for the guide,* in Mital, A. and Kumar, S. 1997. The same special issue of the *International Journal of Industrial Ergonomics.* Elsevier Science Publishers North Holland (about 30 pages).

Recent developments in job analysis in particular applications of job analysis procedures were discussed within the scope of an international symposium at the University of Hohenheim in March 1989. Please refer to Landau, K. and Rohmert, W. (eds.) 1989. *Recent Developments in Job Analysis — Proceedings of the International Symposium on Job Analysis.* (ISBN 0-85066-790-9) London, New York, Philadelphia: Taylor & Francis.

Actual investigation out of the AET data collection (7000 AET analysis) assessing physical load and musculoskeletal disorders was published in 1996. Please refer to Landau, K., Rohmert, W., Imhof-Gildein, B., Mücke, St., and Brauchler, R. 1996, *Risikoindikatoren für Wirbelsäulenerkrankungen (Schlußbericht) — Auswertung des Arbeitswissenschaftlichen Erhebungsverfahrens zur Tätigkeitsanalyse (AET-Datenbank) und Validierung eines neuen Arbeitsanalyseverfahrens.* (ISBN 3-89429-725-5) Bremerhaven, Wirtschaftsverlag NW. An abstract of this contribution is published by the German Bundesanstalt für Arbeitsschutz (eds.) 1996, *Problems and Progress in Assessing Physical Load and Musculoskeletal Disorders — Workshop vom 6. Oktober 1995 in der BAfAM (Bundesanstalt für Arbeitsmedizin).* (ISBN 3-89429-713-1) Bremerhaven, Wirtschaftsverlag NW.

In accordance with the basic European Guideline on Safety and Protection of Workers (89/391/EWG), the new European Union work safety laws stipulate an analysis of dangers at the workplace. These require the employer to carry out systematic workplace analysis to identify and eliminate the stresses caused by the work and the risks resulting from them. Computerized processing and evaluation of the data help to make the ABBA (Inspection of Workplace and Analysis of Stresses) system extremely economical. The methods used in the ABBA system are a further development of the Ergonomic Job Analysis Procedure (AET). The K-AET checklist comprises 102 items covering all the tasks and demands arising in a real person-at-work system. Supplements are used to collect data on special types of stress. Please refer to Landau, K., Maas, C., Schaub, Kh., Mücke, St., and Fischer, T. 1996. New software tools for developing job stress registers, in BAfAM (eds): *Tagungsband Nr. 10, Schriftenreihe der Bundesanstalt für Arbeitsmedizin 1996,* pp. 31-48

The use of ergonomic job analysis procedures and, in particular, a further development of the AET, to predict work-related diseases has already been successfully implemented in an expert system shell. The developed job analysis procedure used to predict work-related diseases was tested in a study of driving and control activities in the mining industry. Please refer to Landau, K. and Brauchler, R. 1994, Disease prediction using ergonomic knowledge bases, in Proceedings of the 12th Triennial Congress of the International Ergonomics Association, Volume 2, *Ergonomics in Occupational Health and Safety,* Toronto 1994, p. 159ff.

Brauchler, R. and Landau, K. 1992. Implementation of an epidemiological early-warning system by using rule induction algorithms, in Karwowski, W. and Mattila, M. *Computer Applications in Ergonomics, Occupational Safety and Health — Proceedings of the International Conference on Computer Aided Ergonomics and Safety,* Tampere, Finland, S. 249-254.

Landau, K. and Brauchler, R. 1992. Theoretical background and applications of epidemiological expert systems, in Karwowski, W. and Mattila, M. *Computer Applications in Ergonomics, Occupational Safety and Health — Proceedings of the International Conference on Computer Aided Ergonomics and Safety,* Tampere, Finland, S. 469-476.

21

Worker Strength Evaluation: Job Design and Worker Selection

Sean Gallagher
National Institute for Occupational Safety and Health

J. Steven Moore
University of Texas Health Center

21.1 Introduction

Many jobs in industry severely tax the worker's musculoskeletal system and may approach or exceed the worker's maximum voluntary strength capabilities. When this occurs, there is evidence that the worker is at higher risk of experiencing a musculoskeletal disorder (Chaffin, 1978; Keyserling et al., 1980). It is for this reason that efforts have been taken over the last couple of decades to provide a means of evaluating the muscular strength capabilities of workers, so that jobs can be designed to eliminate taxing exertions, and to ensure that workers performing physically demanding jobs have the strength to safely perform required tasks.

The effectiveness of worker strength evaluation in reducing work-related musculoskeletal disorders (WMSDs) depends in large part on the purpose of the evaluation. Assessment of physical strength has been used for two primary purposes in the field of ergonomics: job design and worker selection. Job design has been the focus of the psychophysical method of strength assessment and is the technique most likely to have a positive impact in reducing WMSDs. In this technique, the strength of a population of workers is used to design the job so that the majority of workers find the exertion to be acceptable. Studies have indicated that designing tasks by this approach may reduce back injuries by up to 33% (Snook et al., 1978). Strength testing has also been used for the purpose of worker selection, that is,

0-8493-2641-9/99/$0.00+$.50

making sure that workers have sufficient strength to perform physically demanding jobs. In a sense, this approach to controlling WMSDs is antithetical to one of the primary tenets of ergonomics: design the job to fit the worker. Instead, worker selection seeks to "fit" a strong worker into a physically demanding job. Predictably, this technique does not result in nearly the same magnitude of reduction in injuries compared to the job design approach. However, some studies have indicated a partial success using this technique. It should be noted that such an effect has only been evident when the procedure is employed in an environment known to place workers at very high risk of injury. Furthermore, it must be noted that most studies that have examined worker selection procedures have been short term (usually a follow-up period of one year or less). There can be no guarantee that this approach will be successful in protecting workers over the long term.

Muscular strength is a very complex function that can vary greatly depending upon the methods of assessment (Gallagher et al., 1998). As a result, there is often a great deal of confusion and misunderstanding with regard to the appropriate uses of strength testing in the context of ergonomics. It is not uncommon to see techniques misapplied by those unfamiliar with the caveats and limitations associated with various strength-testing procedures. The purposes of this chapter will be threefold: to provide the reader with a basic understanding of human strength, to characterize various methods of strength testing, and to describe ways that these techniques have been used in the attempt to control work-related musculoskeletal disorders (WMSDs).

21.2 Definition of Muscular Strength

Before describing the various strength-testing procedures available, one must first understand what is meant by the term *muscular strength.* For the purposes of this paper, muscular strength will be defined as *the capacity to produce a force or torque with a voluntary muscle contraction* (Gallagher et al., 1998). It is important to note that the strength or force output measured is that which the subject is willing to produce, and is probably somewhat lower than what the muscle is capable of producing in absolute terms (Chaffin and Andersson, 1991). It has been estimated that the maximal voluntary strength a subject is willing to put forth may be as much as 30% lower than the physiological tolerance of the muscle–tendon–bone system (Hettinger, 1961).

21.3 Measurement of Human Strength

We do not currently have the ability to directly measure the force or tension developed within the muscle of a living person (Kroemer et al., 1994). If this were possible, it might greatly simplify the analysis of worker strength. Lacking this ability, we must use indirect measurement techniques in which we measure (externally) the forces or torques generated at some interface between the person and a measurement device. This is important to realize because there are a multitude of ways that such an interface can be constructed, each of which can (and will) influence the resulting strength measure.

Consider the isometric elbow flexion measurement depicted in Figure 21.1 (Gallagher et al., 1998). Were we able to measure the muscle force directly, we would find that the muscle was developing a force of 1,000 Newtons (N). Being unable to do so, we must measure the forces external to the body using a force cuff. But where should we place the force cuff, close to the elbow joint or near the wrist? As will be demonstrated, the force reading will be dramatically affected depending on where the cuff is placed.

In this figure, the tension developed by the muscle acts through a lever arm of distance a. In so doing, it creates a torque about the elbow joint equal to $F_m \times a$. Assuming that the exertion is static (nothing moves), measured forces (on the gauge) will equal the elbow flexor torque divided by the distance that the gauge's associated force cuff is from the elbow joint. That is,

$$Q = (F_m \times a)/b \qquad (1)$$

or

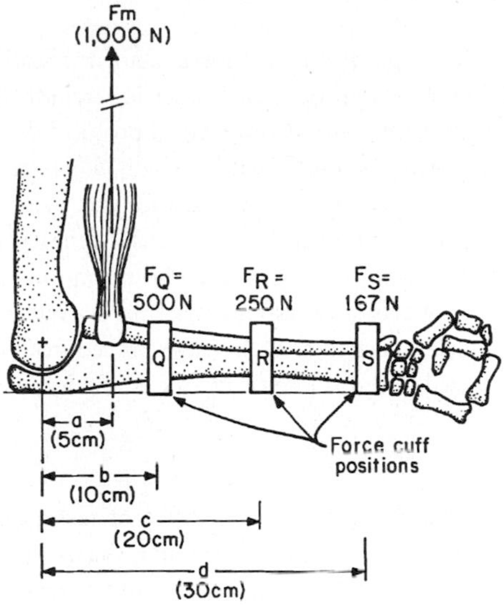

FIGURE 21.1 Given a constant muscle force (F_m), forces measured at various distances from the elbow will result in different force readings (F_Q, F_R, or F_S).

$$R = (F_m \times a)/c \tag{2}$$

or

$$S = (F_m \times a)/d \tag{3}$$

As we move the interface (a force cuff) from the elbow to the hand, the measured force will decrease. So, what can we say is the maximal force that can be generated in elbow flexion?

The answer is that it depends on how and where the forces are being measured.

This example highlights several important points. One central idea that should be understood is that "muscular strength is what is measured by an instrument" (Kroemer et al., 1994). One can see from the example given above that it would be entirely possible to have a case where two groups of subjects have (in actuality) identical muscle strength, but where differences in measurement techniques indicate wildly different strength capabilities. People using strength data must understand in detail how the measurements were done. Thus, a record of a person's strength describes what the instrumentation measured when the person voluntarily produced a muscle contraction in a specific set of circumstances with a specific interface and instrumentation (Gallagher et al., 1998).

21.4 Types of Muscular Strength

Muscular exertions can be divided into those which produce motion about a joint (*dynamic* exertions), and those which do not (*isometric* or *static* exertions). The vast majority of occupational tasks involve dynamic exertions. Unfortunately, the complexity of dynamic tasks (where one has to deal with factors such as velocity and acceleration) makes quantification of this type of strength more difficult (Chaffin and Andersson, 1991; Kroemer et al., 1994). For example, there may be great variability in speed of contraction with different people performing a given task. This, in turn, has a large bearing on the forces that can be produced by the muscles (Åstrand and Rodahl, 1977). Static exertions, on the other hand, are easier to quantify, but do not accurately represent muscle forces where the activity is very dynamic in nature (Kroemer et al., 1994). Neither of these types of strength testing is inherently better than the other — the important thing is to make sure that the test that is used is appropriate for the application being studied.

Isometric Strength

Tests of isometric strength involve application of a force against a stationary load-measuring device (Chaffin, 1996). Because of the relative simplicity of isometric strength tests, standardized procedures have been developed (Caldwell et al., 1974). The recommended protocol describes several control measures to standardize the execution and reporting of tests of static strength. For example, the recommended exertion duration is four to six seconds, with 30 seconds' to two minutes' rest provided between tests. Instructions are to be carefully stated to inform subjects of potential risk and use of the test results, and to prevent coercion or undue incentives to the subject during the exertions. The recommendations also detail methods of standardizing test postures, body supports, and restraint systems, as well as the control of environmental factors (temperature, humidity, noise, spectators, etc.). These procedures have been widely accepted, and have helped unify the techniques used to test isometric muscle strength by researchers around the world.

Dynamic Strength

In contrast to isometric strength testing, a number of different techniques exist to examine dynamic strength capabilities. One type of dynamic strength assessment is that involving measurement of *isoinertial* strength. Kroemer (1983) and Kroemer et al. (1990) define the isoinertial technique of strength assessment as one in which *mass properties of an object are held constant,* as in lifting a given weight over a predetermined distance. Several strength assessment procedures possess the attribute in this definition. Most commonly associated with the term is a specific test developed to provide a relatively quick assessment of a subject's maximal lifting capacity using a modified weight-lifting device. Another is a technique where the subject is asked to provide an estimate of an acceptable (submaximal) load, under set conditions (frequency and duration of lift, a specified lifting task, etc.). This technique is called the *psychophysical* methodology (Snook, 1978). Both will be discussed in greater detail later in the chapter.

Dynamic strength can also be evaluated using tests of *isokinetic* strength (Hislop and Perrine, 1969). This procedure evaluates dynamic strength *throughout a range of motion and at a constant velocity.* Such an exercise allows the muscle to contract at its maximum capability at all points throughout the range of motion. At the extremes of the range of motion of a joint, the muscle has the least mechanical advantage, and the resistance offered by the machine is correspondingly lower. Similarly, as the muscle reaches its optimal mechanical advantage, the resistance of the machine increases proportionally.

These are the most common tests of dynamic strength used in ergonomics; however, others are available. For example, there are devices that can measure force exerted during a constant acceleration exertion, those measuring strengths in an eccentric (muscle lengthening) mode, and several others (Kroemer et al., 1990). However, most of these have been used primarily for research purposes and not for worker strength evaluations. These devices are beyond the scope of the present chapter.

21.5 Factors Affecting Muscular Strength

Before discussing the use of physical strength assessment in job design and worker selection, it is important for the reader to understand some of the factors that can influence muscular strength. These may include personal factors, variables related to the task, or environmental factors (Ayoub and Mital, 1989). The following sections describe some of the major factors known to have a significant influence on strength test performance.

Personal Factors

Gender

There is a distinct difference between males and females in terms of muscular strength. On the average, the muscle strength of women is about two thirds that of men; however, the difference is variable according to which muscle group is examined. For example, for certain muscle groups women may have only 35% (on the average) the strength of the same muscle group in men. For other muscle groups, the difference

TABLE 21.1 Psychological Factors Affecting Maximal Muscular Strength and Their Likely Effects

Factor	Likely effect
Feedback of results	Positive
Instructions on how to exert strength	Positive
Arousal of ego involvement, aspiration	Positive
Pharmaceutical agents	Positive
Startling noise, subject's outcry	Positive
Hypnosis	Positive
Setting of goals, incentives	Positive or negative
Competition, contest	Positive or negative
Verbal encouragement	Positive or negative
Spectators	?
Deception by researcher	?
Fear of injury	Negative
Deception by subject	Negative

From Kroemer, K.H.E. and Marras, W.S., 1981. Evaluation of maximal and submaximal static muscle exertions. *Human Factors,* 25: 643-653. With permission.

between genders may be as little as 15% (Chaffin and Andersson, 1991). Women tend to perform relatively better when the task involves lower extremity muscle groups, and relatively poorer when the exertion requires a great deal of upper body strength. Some of the difference in strength between men and women can be accounted for by the difference in body size between the genders. However, it seems that even when one accounts for the difference in size, there remains a 20% difference in strength between males and females (Åstrand and Rodahl, 1977).

Age

Muscle strength generally reaches a peak in an individual's late 20s or early 30s, and begins a gradual decline thereafter. In general, the strength of the 40-year-old is approximately 5% less than that achieved at its peak. By the time the individual is 65, strength is 20% below its peak. However, it should be duly noted that a regimen of strength training can significantly influence the rate of decline (Åstrand and Rodahl, 1977).

Anthropometry

Anthropometry is the study of the physical dimensions and composition of the human body (Stramler, 1993). Certain anthropometric measures appear to be related to the amount of strength of which a subject is capable. The measures that most highly relate to strength are lean body mass (body mass corrected for fat), and limb cross-sectional data (obtained from measurements of circumference) (Chaffin and Andersson, 1991). Stature (a subject's height) does not appear to be highly related to strength.

Motivation

Many psychological factors, especially subject motivation, can have a marked influence on measured strength. These factors can have both positive and negative effects. Table 21.1 illustrates certain factors that may influence subject motivation, and the expected effect (Kroemer and Marras, 1981).

Task Influences

Posture

The posture adopted by the body can have a major impact on the expression of human strength. For example, the angle of a joint during an exertion can profoundly affect measured muscle strength. As illustrated in Figure 21.2, elbow flexion strength is highest when the joint is at 90 degrees. As the joint deviates from that angle, less force can be developed. Whole-body posture has also been shown to have a large effect on strength. For example, lifting strength is much lower when a subject is kneeling or seated as compared with standing (Gallagher et al., 1988; Yates and Karwowski, 1987).

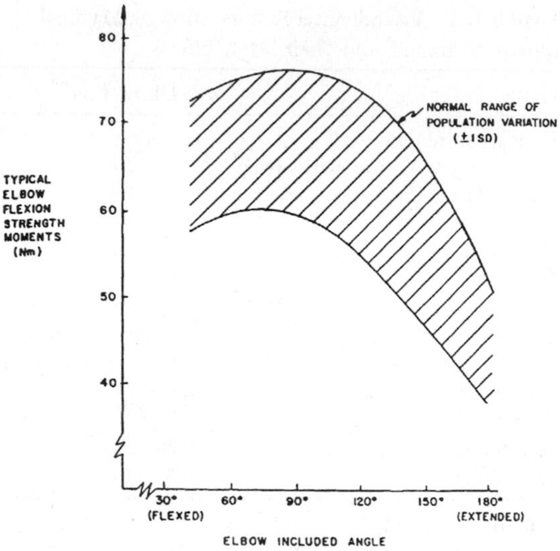

FIGURE 21.2 Change in elbow flexion strength as a result of changes in the angle of the elbow. (From *Occupational Biomechanics,* Chaffin, D.B. and G.B.J. Andersson, © 1991 by John Wiley & Sons, Inc. Reprinted by permission of John Wiley & Sons, Inc.)

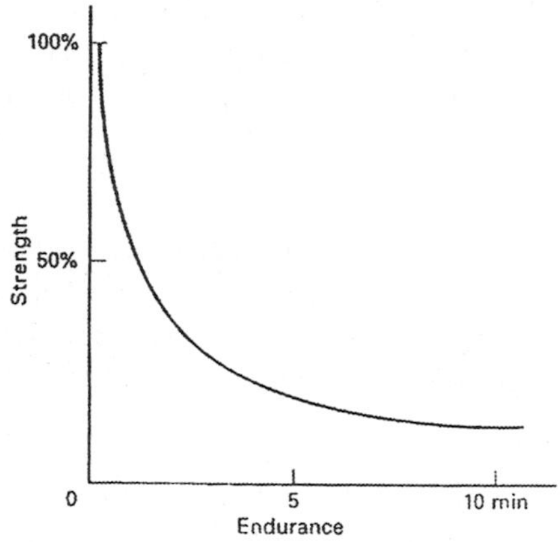

FIGURE 21.3 Endurance time as a function or required strength. (From Rohmert, W., 1966. *Maximal Forces of Men Within the Reach Envelope of the Arms and Legs* (in German), Research Report No. 1616, State of Northrhine — Westfalia, Westdeuscher Verlag Koeln-Opladen. With permission.)

Duration of Exertion

The amount of force that can be sustained during an exertion depends on the length of time of that exertion. Figure 21.3 illustrates this point. An exertion requiring 100% of a subject's maximal voluntary strength can only be sustained for a short period of time. However, as strength requirements are reduced, the exertion can be maintained for longer periods (Rohmert, 1966).

Velocity of Contraction

In work activities, muscular forces are usually applied through a range of motion and may be performed at a wide range of velocity of movement. As illustrated in Figure 21.4, the peak force (or torque) generated by a muscle decreases with increasing velocity of movement. In other words, higher forces can be produced at slow movement speeds. Muscles cannot generate as much force during high-velocity movements (Fox and Mathews, 1981).

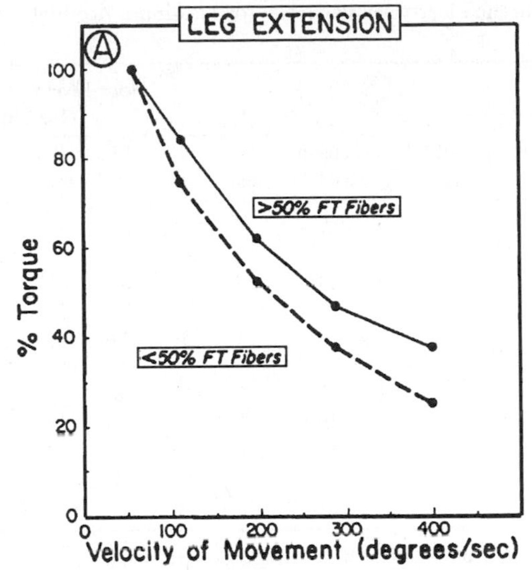

FIGURE 21.4 An example of the force–velocity relationship of muscle: forces generated by a muscle decrease with increasing velocity of movement. (From *The Physiological Basis of Physical Education and Athletics,* Fox, E.L. and D.K. Mathews, © 1981 By CBS College Publishing. With permission.)

Muscle Fatigue

Muscles that are highly stressed become fatigued, which correspondingly reduces the amount of strength that can be produced. Muscular fatigue appears to be dependent on the blood flow through the muscle. When a muscle is tightly contracted, blood flow is impeded, and the delivery of oxygen to the muscle is thereby reduced. Very low levels of exertion (less that 15% of maximal contraction) can be performed for long periods of time without excessive muscular fatigue (Åstrand and Rodahl, 1977).

Environmental Influences on Strength

Temperature and Humidity

Changes in temperature and humidity can affect strength capabilities, particularly at higher levels. Snook and Ciriello (1974) reported that an increase in Wet Bulb Globe Temperature (an index of heat stress) from 20 degrees C to 27 degrees C resulted in a 20% decrease in lifting capacity, a 16% reduction in pushing strength and an 11% reduction in carrying capacity. The effect of cold environments of strength capacity has not been well studied.

21.6 Purposes of Strength Measurement in Ergonomics

There are a number of reasons people may want to collect human strength data. Among the most common is collecting population strength data which can be used to build an anthropometric database; create design data for products, tasks, equipment, etc.; and for basic research into the strength phenomenon. This chapter will discuss two common uses of physical strength assessment in ergonomics: job design and worker selection and placement.

Strength Assessment for Job Design

Perhaps the most effective use of worker strength evaluations is in the area of job design (Snook, 1978). Job design has been a primary focus of the psychophysical method of determining acceptable weights and forces. In this technique, subjects are typically asked to adjust the weight or force associated with a

TABLE 21.2 Excerpt from the Liberty Mutual Tables for Maximum Acceptable Weight of Lift (kg) for Males and Females

				Floor Level to Knuckle Height One Lift Every							
Gender	Box Width (cm)	Distance of Lift (cm)	Percent Capable	5 sec	9 sec	14 sec	1 min	2 min	5 min	30 min	8 hr
			90	7	9	10	14	16	17	18	20
			75	10	13	15	20	23	25	25	30
Males	49	51	50	*14*	17	20	27	30	33	34	40
			25	*18*	*21*	25	34	38	42	43	50
			10	*21*	*25*	*29*	40	45	49	50	59
			90	*6*	7	8	9	10	10	11	15
			75	*7*	9	9	11	12	12	14	18
Females	49	51	50	*9*	10	11	13	15	15	16	22
			25	*10*	*12*	13	16	17	17	19	26
			10	*11*	*14*	15	18	19	20	22	30

Italicized values exceed 8-hour physiological criteria (energy expenditure).

Source: Snook, S.H. and Ciriello, V.M., 1991. The design of manual handling tasks: revised tables of maximum acceptable weights and forces. *Ergonomics* 34(9):1197-1213. With permission.

task in accordance with their own perception of what is an *acceptable workload* under specified test conditions (Snook, 1985). It can be seen from this description that this technique does not attempt to evaluate the maximum forces a subject is capable of producing. Instead, it evaluates a type of "submaximal," endurance-based estimate of acceptable weights or forces.

In the context of lifting tasks, the following procedure is usually used in psychophysical strength assessments. The subject is given control of one variable, typically the amount of weight contained in a lifting box. There will usually be two 20-minute periods of lifting for each specified task: one starting with a light box (to which the subject will add weight), the other starting with a heavy box (from which the subject will extract weight). The box will have a hidden compartment containing an unknown (to the subject) amount of weight, varied before each test, to prevent visual cues to the subject regarding how much weight is being lifted. The amount of weight selected during these two sessions is averaged and is taken as the maximum acceptable weight of lift for the specified conditions. In psychophysical assessments, the subject is instructed to work consistently according to the concept of "a fair day's pay for a fair day's work": working as hard as he or she can without straining or becoming unusually tired, weakened, overheated, or out of breath (Snook and Ciriello, 1974).

As psychophysical strength data are collected on large numbers of subjects, it becomes possible to design jobs so that they are well within the strength capabilities of the vast majority of workers. One criterion that is often used is to design the job so that 75% of workers rate the load as acceptable (Snook et al., 1978). Studies have indicated that if workers lift more than this amount, they may be three times more likely to experience a low back injury. On the other hand, designing jobs in accordance with this criterion has the potential to reduce the occurrence of low back injuries by up to 33% (Snook et al., 1978).

Several authors have published comprehensive tables on loads deemed acceptable by workers over a wide variety of industrial tasks. Of these, the most comprehensive are those developed by Snook and colleagues (Snook et al., 1978; Snook and Ciriello, 1991), which detail maximum acceptable loads for lifting, lowering, pushing, pulling, and carrying for both male and female workers. Table 21.2 presents an excerpt from one such tabulation, dealing with acceptable weights of lift for male and female workers for a floor level to knuckle height lift (Snook and Ciriello, 1991). If one wanted to design a lifting task requiring 4.1 lifts/min (one lift every 14 seconds) so that it was acceptable to 75% of females and 90% of males for the given box dimensions, one could go to these tables and determine that the acceptable weights of lift are 9 kg and 10 kg, respectively. Such a job should be designed so that the weight lifted is no more than 9 kg, or approximately 20 pounds.

As with all strength evaluation techniques, there are both advantages and disadvantages to the psychophysical approach. The major advantages of the psychophysical approach include the fact that it allows a realistic simulation of industrial tasks, and is capable of simulating intermittent as well as continuous types of lifting tasks. Furthermore, psychophysical results appear to be very reproducible, and are related to the occurrence of low back pain (Snook, 1985). However, disadvantages can also be identified. Perhaps the most important of these is that psychophysical results sometimes exceed what experts feel are safe according to other criteria, such as biomechanical stress or physiological cost (Snook, 1985). Furthermore, the technique is subjective. That is to say, it relies on self-reporting by the subject. Given the large number of overexertion injuries that occur every year, it is not clear that workers can always tell how much weight is safe for them to lift.

Despite the disadvantages that may be present with this technique, it is the only strength evaluation procedure that focuses on using the acquired data to develop a permanent engineering (job design) solution to the control of low back pain. It would be expected that this approach would afford a greater level of protection to workers than the worker selection techniques that follow.

Worker Selection and Placement

The purpose of worker selection and placement programs is to ensure that jobs which involve heavy physical demands are not performed by those lacking the necessary strength capabilities (Chaffin, 1996). It should be noted that this method is not the preferred strategy of the ergonomist, and there continues to be some controversy regarding the effectiveness of the approach, but there is some support for worker selection procedures in the literature (Chaffin et al., 1978; Keyserling et al., 1980). However, it is clear that specific conditions need to be met for this procedure to have a chance of success.

The process of strength evaluation for worker selection should be approached with caution on many fronts. In the first place, issues of unfair discrimination may be raised if appropriate testing procedures are not used (Chaffin and Andersson, 1991). If strength is to be used as a screening criterion, it is critical that the strength test employed is directly related to specific work requirements. Accurate representations of working postures are also important, as strength in one posture cannot accurately predict strength in another posture. Chaffin and Andersson (1991) suggested that the following criteria be used for methods of worker selection:

1. Is it safe to administer?
2. Does it give reliable, quantitative values?
3. Is it related to specific job requirements?
4. Is it practical?
5. Does it predict risk of future injury or illness?

Prior to testing, a history should be taken to ensure the worker does not have any cardiovascular or musculoskeletal problems that would increase the risk of taking the test.

Research has shown that worker selection cannot consist of general tests of strength (Battié et al., 1989; Troup et al., 1981; Mostardi et al., 1992). Such tests do not appear to be helpful in identifying those at risk of overexertion injury (strong workers seem to experience injury rates similar to those less strong). Instead, worker strength measures must be tied to a biomechanical analysis of workplace demands in order to predict those having increased risk of injury (Chaffin et al., 1978; Keyserling et al., 1980). There are two key principles that must be considered regarding the use of strength assessment for purposes of worker selection. These principles deal with the job relatedness of the test employed, and use of strength tests only under conditions where they have shown the ability to identify workers at high risk of injury.

The literature has shown that worker selection is only effective when a worker's strength capacity is equated with the demands of the job. All too often, emphasis is placed on collecting data on the former attribute, while the latter receives little or no attention (Chaffin, 1996). As will be illustrated, strength data in the absence of information regarding job demands is insufficient for purposes of worker selection. Consider the following scenario: an employer has an opening for a physically demanding job and wishes to hire an individual with strength sufficient for the task. This employer decides to base his employment decision on a strength test given to a group of applicants. Naturally, the employer selects the applicant with the highest strength score to perform the job. The employer may have hired the strongest job applicant; however, what this employer must understand is that he may not have decreased the risk of injury to his employee if, for example, the demands of the job still exceed this individual's capacity. This illustration should make it clear that only through knowing about the person's capabilities *and* the job demands might worker selection protect workers from WMSDs.

A second issue that must be considered when worker selection is to be implemented is that of the test's predictive value. The predictive value of a test is a measure of its ability to determine who is at risk of future WMSD. In the case of job-related strength testing, the predictive value appears to hold only when testing individuals for jobs where high risk is known (Chaffin, 1996). Strength testing does not appear to predict the risk of injury or disease to an individual when job demands are low or moderate. Furthermore, as noted previously, the effectiveness of worker selection techniques has not been demonstrated in long-term studies, only in relatively short-term investigations. It is unclear whether such tests will predict workers at risk of injury over the long term.

Finally, it should be noted that muscular strength is only one factor in a complicated and poorly understood mechanism of injury. A host of other tissues (such as the tendons, ligaments, and joint surfaces) may be deformed or injured by the stresses they experience, whether or not the muscles are able to develop sufficient strength for the job. Thus, it can be said that adequate muscular strength is necessary for safe performance of physical work but is not in itself sufficient for protection against injury.

21.7 Isometric Analysis

When a worker is called upon to perform a physically demanding lifting task, moments (or torques) are produced about various joints of the body by the external load (Chaffin and Andersson, 1991). Often these moments are augmented by the force of gravity acting on the mass of various body segments. For example, in a biceps curl exercise, the moment produced by the forearm flexors must counteract the moment of the weight held in the hands, as well as the moment caused by gravity acting on the center of mass of the forearm. In order to successfully perform the task, the muscles responsible for moving the joint must develop a greater moment than that imposed by the combined moment of the external load and body segment. It should be clear that for each joint of the body, there exists a limit to the strength that can be produced by the muscle to move ever-increasing external loads. This concept has formed the basis of isometric muscle strength prediction modeling (Chaffin and Andersson, 1991).

The following procedures are generally used in this biomechanical analysis technique. First, workers are observed (and usually photographed or videotaped) during the performance of physically demanding tasks. For each task, the posture of the torso and the extremities is documented at the time of peak exertion. The postures are then recreated using a computerized software package, which calculates the load moments produced at various joints of the body during the performance of the task (Chaffin and Andersson, 1991). The values obtained during this analysis are then compared to population norms for isometric strength obtained from a population of industrial workers. In this manner, the model can estimate the proportion of the population capable of performing the exertion, as well as the predicted compression forces acting on the lumbar discs resulting from the task.

Figure 21.5 shows an example of the workplace analysis necessary for this type of approach. Direct observations of the worker performing the task provide the necessary data. For example, the load magnitude and direction must be known (in this case a 200 N load acting downward), the size of the worker, the postural angles of the body (obtained from photographs or videotape), and whether the task requires one or two hands. Furthermore, the analysis requires accurate measurement of the load center relative to the ankles and the low back. A computer analysis program can be used to calculate the strength requirements for the task, and the percentage of workers who would be likely to have sufficient strength capabilities to perform it. Results of this particular analysis indicate that the muscles at the hip are most

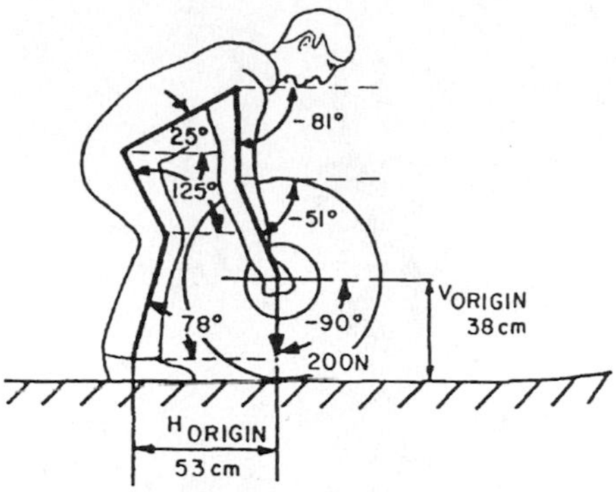

FIGURE 21.5 Postural data required for analysis of joint moment strengths using the isometric technique. (From *Occupational Biomechanics,* Chaffin, D.B. and G.B.J. Andersson, © 1991 by John Wiley & Sons, Inc. Reprinted by permission of John Wiley & Sons, Inc.)

stressed, with 83% of men having the necessary capabilities but only slightly more than half of women would have the necessary strength in this region. These results can then be used as the basis for determining those workers who have adequate strength for the job. However, such results can also serve as ammunition for recommending changes in job design (Chaffin and Andersson, 1991).

21.8 Isoinertial Testing

The Strength Aptitude Test

The Strength Aptitude Test (SAT) is a classification tool for matching the physical strength abilities of individuals with the physical strength requirements of jobs in the Air Force (McDaniel et al., 1983). The SAT is given to all Air Force recruits as part of their preinduction examinations. Results of the SAT are used to determine whether an individual has the minimum strength criterion which is a prerequisite for admission to various Air Force Specialties (AFSs). The physical demands of each AFS are objectively computed from an average physical demand weighted by the frequency of performance and the percent of the AFS members performing the task. Objects weighing less than 10 pounds are not considered physically demanding and are not considered in the job analysis. Prior to averaging the physical demands of the AFS, the actual weights of objects handled are converted into equivalent performance on the incremental weight lift test using statistical procedures developed over years of testing. These relationships consider the type of task (lifting, carrying, pushing, etc.), the size and weight of the object handled, and the type and height of the lift. Thus, the physical job demands are related to, but are not identical to, the ability to lift an object to a certain height. Job demands for various AFSs are reanalyzed periodically for purposes of updating the SAT (McDaniel, 1994).

In this technique, a preselected mass, constant in each test, is lifted by the subject (typically from knee height to knuckle height, elbow height, or to overhead reach height). The amount of weight to be lifted is relatively light at first, but the amount of mass is continually increased in succeeding tests until it reaches the maximal amount that the subject voluntarily indicates he/she can handle. Figure 21.6 shows an example of an isoinertial strength testing device. At the time of this writing, over 2 million Air Force personnel have been tested using this procedure.

A unique aspect of this technique is that it is the only strength measurement procedure discussed in this document where results are based on the success or failure to perform a prescribed criterion task (Kroemer, 1983). The criterion tasks studied have typically included lifting to shoulder height, elbow height, or knuckle height.

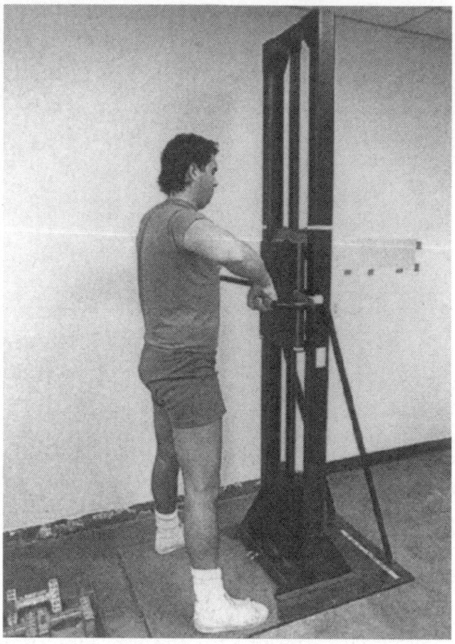

FIGURE 21.6 An isoinertial weight-lifting device. (From Kroemer, K.H.E., 1983. An isoinertial technique to assess individual lifting capacity. *Human Factors,* 25: 493–506. With permission.)

When developing the SAT, the Air Force examined more than 60 candidate tests in an extensive, four-year research program and found the incremental weight lift to 1.83 m to be the single best test of overall dynamic strength capability, which was both safe and reliable (McDaniel, 1994). This finding was confirmed by an independent study funded by the U.S. Army (Myers et al., 1984). This study compared the SAT to a battery of tests developed by the Army (including isometric and dynamic tests) and compared these with representative heavy demand tasks performed within the Army. Results showed the SAT to be superior to all others in predicting performance on the criterion tasks.

The Progressive Inertial Lifting Evaluation (PILE)

Another variety of isoinertial strength test is the Progressive Isoinertial Lifting Evaluation (PILE) (Mayer et al., 1988a, b). Instead of using a weight rack as shown in Figure 21.6, the Progressive Isoinertial Lifting Evaluation (PILE) is performed using a lifting box with handles and increasing weight in the box as it is lifted and lowered. Subjects perform two isoinertial lifting/lowering tests: one from floor to 30" (LUMBAR) and one from 30" to 54" (CERVICAL). Unlike the isoinertial procedures described above, there are three possible criteria for termination of the test: (1) voluntary termination due to fatigue, excessive discomfort, or inability to complete the specified lifting task; (2) achievement of a target heart rate (usually 85% of age predicted maximal heart rate); or (3) when the subject lifts a "safe limit" of 55 to 60% of his/her body weight. Thus, contrary to the tests described above, the PILE test may be terminated due to cardiovascular factors, rather than when an acceptable load limit is reached.

Since the PILE was developed as a means of evaluating the degree of restoration of functional capacity of individuals complaining of chronic low back pain (LBP), the initial weight lifted by subjects using this procedure is somewhat lower than the tests described above. The initial starting weight is 3.6 kg for women and 5.9 kg for men. Weight is incremented upward at a rate of 2.3 kg every 20 seconds for women, and 4.6 kg every 20 seconds for men. During each 20-second period, four lifting movements (box lift or box lower) are performed. The lifting sequence is repeated until one of the three endpoints is reached. The vast majority of subjects are stopped by the "psychophysical" endpoint, indicating the subject has a perception of fatigue or overexertion. The target heart rate endpoint is typically reached in older or large individuals. The "safe limit" endpoint is typically encountered only by very thin or small individuals.

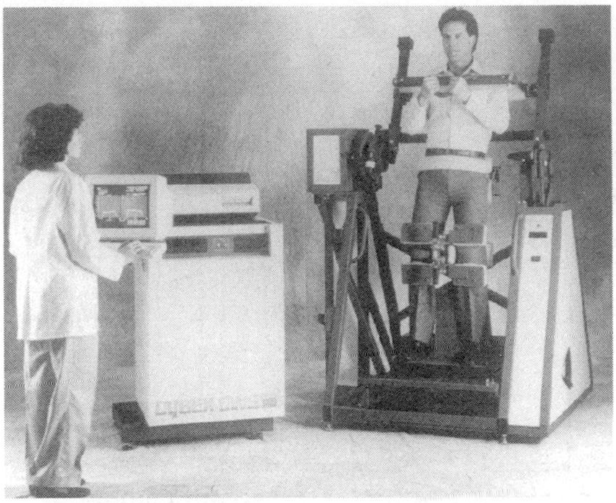

FIGURE 21.7 An isokinetic trunk flexion and extension device used to evaluate lumbar muscle strength. (Cybex Medical, Division of Henley Healthcare, Sugarland, TX.) (Photo courtesy of Henley Healthcare.)

Mayer et al. (1988b) developed a normative database for the PILE, consisting of 61 males and 31 females. Both total work (TW) and force in lbs. (F) were normalized according to age, gender, and a body weight variable. The body weight variable, the adjusted weight (AW), was taken as actual body weight in slim individuals, but was taken as the ideal weight in overweight individuals. This was done to prevent skewing the normalization in overweight individuals.

21.9 Isokinetic Tests

A technique of dynamic testing that has been growing in popularity is that dealing with the measurement of isokinetic strength (Hislop and Perrine, 1969). As defined previously, this technique evaluates muscular strength throughout a range of motion and at a constant velocity. It is important to realize that people do not normally move at a constant velocity (Kroemer et al., 1990). Instead, human movement is usually associated with significant acceleration and deceleration of body segments. Thus, there is a perceptible difference between isokinetic strength and free dynamic lifting. In the latter instance, subjects may use rapid acceleration to gain a weight lifting advantage, as in the Strength Aptitude Test described above. Acceleration is not permitted in isokinetic tests of strength.

The majority of isokinetic devices available on the market focus on quantifying strength about isolated joints or body segments, for example, trunk extension and flexion (see Figure 21.7). This may be useful for rehabilitation or clinical use, but isolated joint testing is generally not appropriate for evaluating an individual's ability to perform occupational lifting tasks. One should not make the mistake of assuming, for instance, that isolated trunk extension strength is representative of an individual's ability to perform a lift. In fact, lifting strength for a task may be almost entirely unrelated to trunk muscle strength (Himmelstein and Andersson, 1988). Strength of the arms or legs (and not the trunk) may be the limiting factor in an individual's lifting strength. For this reason, machines that measure isokinetic strengths of isolated joints or body segments should not be used as a method of evaluating worker capabilities related to job demands in most instances.

Many investigators have used dynamic isokinetic lifting devices specifically designed to measure whole-body lifting strength (Pytel and Kamon, 1981; Kishino et al., 1985) (see Figure 21.8). These devices typically have a handle connected by a rope to a winch which rotates at a specified isokinetic velocity when the handle is pulled. Studies using this type of device have demonstrated good correlations between isokinetic Dynamic Lift Strength (i.e., a lift from floor to chest height) and the maximum weights individuals were willing to lift for infrequent tasks using the psychophysical approach (Pytel and Kamon, 1981). Thus, under certain circumstances, this device appears to have some validity for assessment of job-related dynamic lifting strength capabilities of individuals. Some investigators have attempted to

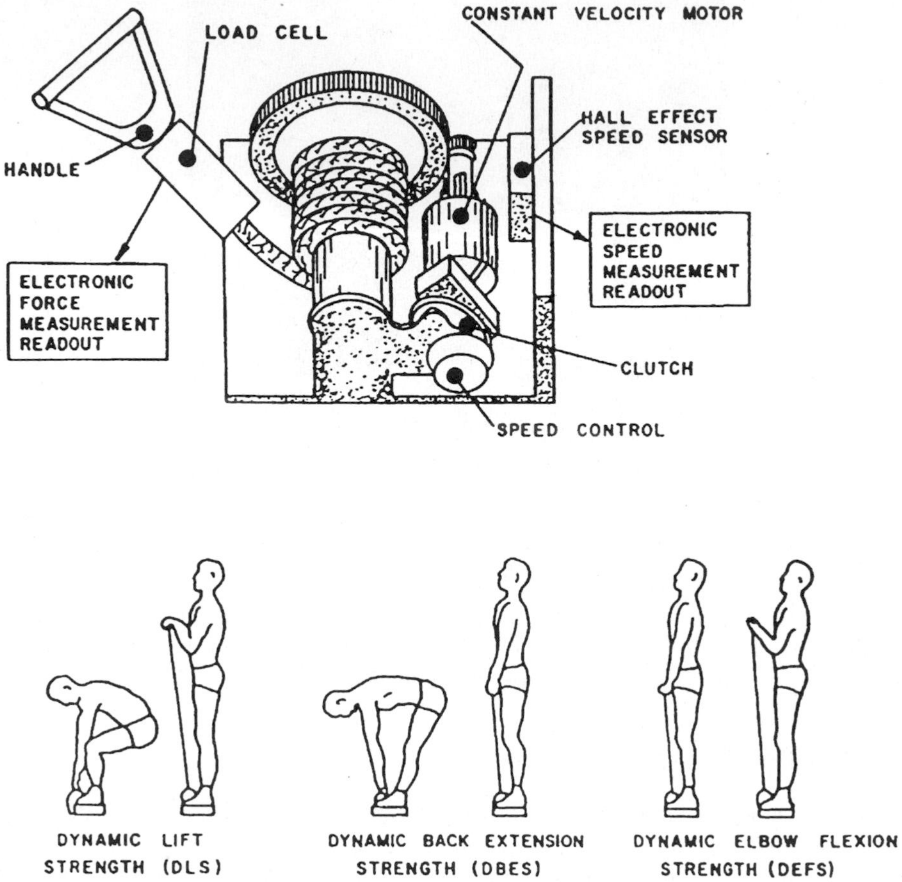

FIGURE 21.8 An isokinetic device used to evaluate whole-body lifting strengths. (From Pytel, J.L. and Kamon, E. Dynamic strength test as a predictor for maximal and acceptable lift, *Ergonomics,* 24: 663–672. With permission.)

modify this type of instrument by providing a means to mount it so that isokinetic strength can be measured in vertical, horizontal, and transverse planes (Mital and Vingararamoothy, 1984). However, while advances have been made in the use of isokinetic devices for worker strength evaluation, this procedure cannot be considered fully developed in the context of worker selection procedures.

21.10 Conclusions

In spite of advances in measurement techniques and an explosive increase in the volume of research, our understanding of human strength remains in its preliminary stages. It is clear that muscle strength is a highly complex and variable function dependent on a large number of factors. It is not surprising, therefore, that there are not only substantial differences in strength between individuals, or that strength measurements for a single individual can vary a great deal even during the course of a single day. Strength is not a fixed attribute — strength training regimens can increase an individual's capability by 30 to 40% or more. Disuse can lead to muscle atrophy (Åstrand and Rodahl, 1977).

The use of physical strength assessment in ergonomics has focused on both job design and worker selection techniques. Of these, the former has a much greater potential to significantly reduce WMSDs. Worker selection techniques must be considered a method of last resort — where engineering changes or administrative controls cannot be used to reduce worker exposure to WMSD risk factors. This technique has only shown a moderate effect in truly high risk environments, and only in short-term studies. It is not known whether worker selection procedures have a protective effect over the long term.

References

Åstrand, P.O. and Rodahl, K., 1977. *Textbook of Work Physiology*, McGraw-Hill Book Company, New York, 681 pp.

Ayoub, M.M. and Mital, A., 1989. *Manual Materials Handling*. London: Taylor & Francis, 324 pp.

Battié, M.C., Bigos, S.J., Fisher, L.D., Hansson, T.H., Jones, M.E., and Wortley, M.D., 1989. Isometric lifting strength as a predictor of low back pain. *Spine* 14:851-856.

Caldwell, L.S., Chaffin, D.B., Dukes-Dobos, F.N., Kroemer, K.H.E., Laubach, L.L., Snook, S.H., et al., 1974. A proposed standard procedure for static muscle strength testing. *American Industrial Hygiene Association Journal* 35:201-206.

Chaffin, D.B., 1996. Ergonomic basis for job-related strength testing, in *Disability Evaluation* (1st edition, Demeter, S.L., Andersson, G.B.J., Smith, G.M. eds.), Mosby, St. Louis, MO, Chapter 17, 159-167.

Chaffin, D.B. and Andersson, G.B.J., 1991. *Occupational Biomechanics* (2nd ed.), New York: John Wiley & Sons, 518 pp.

Chaffin, D.B., Herrin, G.D., and Keyserling, W.M., 1978. Pre-employment strength testing: an updated position. *Journal of Occupational Medicine* 20(6):403-408.

Fox, E.L. and Mathews, E.L., 1981. *The Physiological Basis of Physical Education and Athletics* (3rd ed.), Philadelphia: Saunders College Publishing, 677 pp.

Gallagher, S., Marras, W.S. and Bobick, T.G., 1988. Lifting in stooped and kneeling postures: Effects on lifting capacity, metabolic costs, and electromyography of eight trunk muscles. *Int. J. Ind. Erg.* 3:65-76.

Gallagher, S., Moore, J.S., and Stobbe, T.J., 1998. *Physical Strength Assessment in Ergonomics*, American Industrial Hygiene Association Ergonomics Guide, AIHA Press, Fairfax, VA, 60 pp.

Hettinger, T., 1961. *Physiology of Strength,* Springfield, IL: Charles C Thomas.

Himmelstein, J.S. and Andersson, G.B.J., 1988. Low back pain: risk evaluation and preplacement screening. *Occupational Medicine: State of the Art Reviews.* 3(2):255-269.

Hislop, H. and Perrine, J.J., 1967. The isokinetic concept of exercise. *Physical Therapy* 47:114-117.

Keyserling, W.M., Herrin, G.D., and Chaffin, D.B., 1980. Isometric strength testing as a means of controlling medical incidents on strenuous jobs. *Journal of Occupational Medicine 22(5)*:332-336.

Kishino, N.D., Mayer, T.G., Gatchel, R.J., Parish, M.M., Anderson, C., Gustin, L., and Mooney, V., 1985. Quantification of lumbar function: Part 4: Isometric and isokinetic lifting simulation in normal subjects and low-back dysfunction patients, *Spine 10* (10): 921-927.

Kroemer, K.H.E., 1983. An isoinertial technique to assess individual lifting capability. *Human Factors* 25(5): 493-506.

Kroemer, K.H.E. and Marras, W.S., 1981. Evaluation of maximal and submaximal static muscle exertions. *Human Factors,* 25: 643-653.

Kroemer, K.H.E., Marras, W.S., McGlothlin, J.D., McIntyre, D.R., and Nordin, M., 1990. On the measurement of human strength. *International Journal of Industrial Ergonomics,* 6: 199-210.

Kroemer, K.H.E., Kroemer, H.B., and Kroemer-Elbert, K.E., 1994. *Ergonomics: How to Design for Ease and Efficiency.* Englewood Cliffs, NJ: Prentice-Hall.

Mayer, T.G., Barnes, D., Kishino, N.D., Nichols, G., Gatchell, R.J., Mayer, H., and Mooney, V. (1988a). Progressive isoinertial lifting evaluation — I. A standardized protocol and normative database. *Spine* 13(8): 993-997.

Mayer, T.G., Barnes, D., Nichols, G., Kishino, N.D., Coval, K., Piel, B., Hoshino, D., and Gatchell, R.J. (1988b): Progressive isoinertial lifting evaluation — II. A comparison with isokinetic lifting in a chronic low-back pain industrial population. *Spine* 13(8): 998-1002.

McDaniel, J.W., 1994. Personal communication.

McDaniel, J.W., Shandis, R.J., and Madole, S.W., 1983. *Weight Lifting Capabilities of Air Force Basic Trainees.* AFAMRL-TR-83-0001. Wright-Patterson AFBDH, Air Force Aerospace Medical Research Laboratory.

Mostardi, R.A., Noe, D.A., Kovacik, M.W., and Porterfield, J.A., 1992. Isokinetic lifting strength and occupational injury: A prospective study. *Spine* 17(2): 189-193.

Myers, D.O., Gebhardt, D.L., Crump, C.E., and Fleishman, E.A., 1984. *Validation of the Military Entrance Physical Strength Capacity Test (MEPSCAT).* U.S. Army Research Institute Technical Report 610, NTIS No. AD-A142 169.

Pytel, J.L and Kamon, E., 1981. Dynamic strength test as a predictor for maximal and acceptable lift. *Ergonomics 24(9)*:663-672.

Rohmert, W., 1966. *Maximal Forces of Men Within the Reach Envelope of the Arms and Legs* (in German), Research Report No. 1616, State of Northrhine — Westfalia, Westdeuscher Verlag Koeln-Opladen.

Snook, S.H., 1978. The design of manual handling tasks. *Ergonomics 21*:963-985.

Snook, S.H., 1985. Psychophysical considerations in permissible loads. *Ergonomics 28(1)*:327- 330.

Snook, S.H. and Ciriello, V.M., 1974. Maximum weights and work loads acceptable to female workers. *J. Occup. Med.,* 16: 527-534.

Snook, S.H. and Ciriello, V.M., 1991. The design of manual handling tasks: revised tables of maximum acceptable weights and forces. *Ergonomics* 34(9):1197-1213.

Snook, S.H., Campanelli, R.A., and Hart, J.W., 1978. A study of three preventive approaches to low back injury. *J. Occup. Med* 20(7):478-481.

Stramler, Jr., J. H., 1993. *The Dictionary for Human Factors/Ergonomics.* Boca Raton, FL, CRC Press, 413 pp.

Troup, J.D.G., Martin, J.W., and Lloyd, D.C.E.F., 1981. Back pain in industry. A prospective Study. *Spine,* 6: 61-69.

Yates, J.W., Karwowski, W., 1987. Maximum acceptable lifting loads during seated and standing work. *Applied Ergonomics* 18: 239-243.

22

Dynamic Workplace Factors in Manual Lifting

William S. Marras
The Ohio State University

22.1 Introduction

One need only observe a worker perform an industrial materials handling task to appreciate the fact that dynamics or motion is an integral part of any lifting task. The significance of this dynamic component of the task has been recognized since the days of Isaac Newton, who demonstrated that force is a product of mass and acceleration ($F = m \times a$). Until recently, the field of ergonomics has not had many tools that were capable of assessing the effects of lifting dynamics. The significance of lifting dynamics has been suggested in much of the literature, however, the importance of considering motion has been underappreciated.

Several studies indicate that risk of a low back disorder may increase significantly when motion occurs during a lift. Punnet et al. (1991) have shown that the vast majority of low back disorders in automobile assembly plants cannot be explained simply by the weight of the object lifted or the instantaneous posture in many high-risk jobs that are dynamic. Bigos and associates (1986) also suggested that those workers who performed dynamics lifting tasks were at three times the risk of low back disorder compared to those who were exposed to static awkward postures. Marras and colleagues (1993) identified the levels of exposure to dynamic trunk motions that were associated with high and low risk of occupationally related low back disorders. Thus, this recent evidence suggests that dynamics can clearly influence the ability to safely lift objects.

There are few ergonomic tools currently available to evaluate the dynamic aspects of lifting tasks. For the most part, dynamic analyses and controls have been developed along the lines of: (1) biomechanical loading principles, (2) psychophysical and dynamic strength assessments, and (3) kinematic evaluations based upon epidemiologic trends. This chapter will review the development of dynamic analytic techniques in these three areas as well as demonstrate their application.

0-8493-2641-9/99/$0.00+$.50

22.2 Biomechanical Loading of the Spine During Dynamic Lifting

Techniques to assess and control occupational low back disorder exposure risk have evolved over the past three decades. Early attempts at controlling low back disorders (LBDs) at work were based on biomechanical logic. This logic assumes LBD risk can be assessed by comparing the load imposed upon the spine to the tolerance limits for the vertebral endplates. (See Chapter 10 for a description of the LBD injury process.) Spinal loading is a function of both *external* forces and *internal* forces acting on the body. External forces are the result of load mass characteristics and the body's system dynamics imposing loads on the spine. For example, the object mass being lifted (or the mass of the arms and trunk) is acted upon by the forces of gravity in order to load the spine. Internal forces refer to forces generated internal to the musculoskeletal system in response to these external forces. For example, the muscles and passive tissues (e.g., ligaments) respond to an external load by generating force within the musculoskeletal system that counteract the external load. Since the internal force-generating structures vary in their geometric orientation, it is important that their vector orientation is realistic in biomechanical modeling efforts. This is essential so that the nature and magnitude of spine loading (compression versus shear) during a task can be accurately assessed. Since the tolerance limits of the spine vary dramatically in compression versus shear versus torsion, accurate assessments of spine loading are crucial to an understanding of occupational risk. This review will focus on the ability of the various assessment tools to accurately evaluate internal forces and the subsequent spine loading which will define the subsequent risk of LBD.

The early biomechanical models of occupational low back loading simplified the analyses of the musculoskeletal system in two ways. First, these models assumed that the internal force-generating system could be simplified by representing the body as a cantilever system. This assumption permitted the internal force to be simplified by assuming the musculoskeletal system generates force using one equivalent muscle that best counteracts the external load. Under realistic conditions, the internal forces are generated by a complex activation of many muscles. This complex activation pattern creates vectors of loading on the spine that can be quite complex. Assumptions regarding a single equivalent internal force-generation system greatly simplify this situation.

The second major assumption in these early models was that the lifting activity could be represented by a static system. Early models assumed that the loads imposed upon the musculoskeletal system during lifting can be represented by observing the body at one instant in time while it was in a static (frozen) posture. These models were thought to be appropriate for situations where no appreciable acceleration was present during the lift. In other words, lifts were assumed to be slow and smooth.

One of the early attempts to quantitatively assess the loads on the spine as a function of an occupational lifting task was developed by Chaffin and Baker (1969). This model assumed that the body could be considered a cantilever system with a single equivalent muscle counterbalancing the externally applied moment. It also assumed the lift could be represented by a static model of the lifting situation. Chaffin and Baker developed a two-dimensional (coplanar) model that predicted the load imposed upon the spine as well as the moments at the major joints that would be required to balance the external load. When the moments were compared to the working population's static strength, a strength "benchmark" was available which would describe the percent of the population who would be expected to have adequate strength to perform the task. Later, this model was expanded to a three-dimensional assessment (Chaffin and Muzaffer, 1991). This model used the same logic as the two-dimensional coplanar model but permitted the body to be represented in static asymmetric postures. This model logic was also extended to include the influence of dynamic motion of the object lifted. Frievalds and associates (1984) were able to account for the effects of body segment dynamics and made predictions about the activity of single equivalent muscles in the musculoskeletal system in response to dynamic motion. In this model, the assumption that the lift could be represented by a static system was relaxed, and motion during lifting was allowed. However, the assumption that the body behaved as a cantilever system with one equivalent muscle supplying the internal force was maintained.

The internal force-generation system of the trunk has been recognized as a multiple muscle system, especially during motion (Winter and Woo, 1990). In order to accurately understand the nature of the loading on the spine during dynamic lifting, the coactivation pattern of the muscles and other internal force-generation mechanisms must be understood. Studies have demonstrated that the muscles that are active and the timing of the muscle activation change dramatically as dynamic movement conditions and asymmetry conditions of the body change (Marras and Reilly, 1988; Marras et al., 1986; Marras and Mirka, 1993). Hence, the significance of muscle coactivation upon spine loading predictions has been duly recognized in the literature.

Several attempts have been made to predict the activity of the multiple muscle system so that this information can be used in lifting models. Schultz and Andersson (1981) described a multiple muscle system consisting of the 10 major muscle groups that would be identifiable when passing a cutting plane through the lumbar level of the trunk. Using classical engineering techniques, they were able to describe the three forces and three moments acting upon the spine due to the external load being supported. In order to predict internal muscle activity, they tried two approaches. First, they simply made assumptions about which muscles would be active in a given task. This typically involved limiting the muscle activity to agonist muscles with the greatest mechanical advantage. Second, they employed linear optimization techniques to deal with the indeterminate nature of the problem (6 force and moment equations and 10 unknowns [muscles]). This attempt provided a solution to the problem without having to make simplifying assumptions, however, the solutions often did not match observations, especially under dynamic lifting conditions. Other optimization techniques have been attempted (e.g., Bean et al., 1988; Rathske, 1995), however, most suffer from the basic problem that linear optimization is incapable of providing more non-boundary solutions than the number of functional constraints. In most optimization models, the force and moment equations provide 5 functional constraints (plus an optimization function), and there are at least 10 unknowns (muscles and internal forces). Thus, the solution to the problem is destined to be indeterminate (Marras, 1988). In addition, optimization has only been attempted and has had limited success for static trunk loading conditions. Other techniques have also attempted to predict the activity of the multiple muscle system. Neural network techniques (Nussbaum et al., 1997) have been employed to predict the activity of the muscles under static conditions. This technique appears more valuable for classifying muscle responses as typical or nontypical. In addition, once the lifting situation changes, networks must be retrained for specific conditions. Stochastic models of muscle activities have also been produced (Mirka and Marras, 1993). These models indicate the probability that a certain combination of muscles would be performed given the dynamic characteristics of the lift. The limitation with this approach is that the lifting situation must be simulated many times for each given lifting task in order to accurately assess the loading characteristics (and probabilities) of the spine.

As demonstrated by these modeling efforts, it is a very difficult task to predict the activity of the trunk musculature during manual lifting tasks. Muscle activity must be determined in order to reasonably predict spine loading. Under static loading conditions, significant variability in muscle recruitment has been documented, and this variability becomes even greater when dynamic motion is present in the lift (Marras and Mirka, 1993). Specifically, large amounts of trunk muscle coactivity occur during dynamic activity that must be assessed in order to reasonably assess spinal loading characteristics.

This complexity in muscle recruitment has prompted several researchers to develop biologically assisted models of the trunk. In biologically assisted models, the person performing the task of interest is biologically monitored, and this recorded activity is used as model input in order to assist in the assessment of internal force development in the trunk. The most commonly used biological signal for these purposes is electromyography (EMG) of the trunk musculature. McGill and Norman (1985) were the first to develop an EMG-assisted model of the trunk. They developed an elegant anatomically detailed EMG-assisted model of bending motions containing 50 muscles and 12 ligaments. Other models (Thelan et al., 1994) have demonstrated that it is possible to assess simultaneously spinal loads in three dimensions under static loading conditions. The degree of complexity in these models also becomes very large and necessitates many simplifying assumptions about how many of the nonmonitored muscles behave.

An EMG-assisted model as been developed that is specifically designed to assess spine loading under dynamic manual lifting conditions. This model was specifically designed as a tool to evaluate manual lifting under actual lifting conditions. The goal in model development was to make the model only as

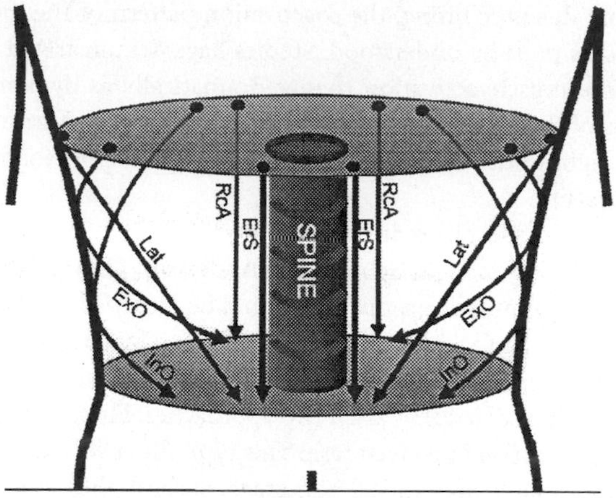

FIGURE 22.1 Geometric representation of the trunk used in the EMG-assisted model. The trunk is modeled as two plates that move dynamically throughout an exertion.

complex and detailed as needed to accurately reflect spinal loading. The model has been developed and validated so that it can accurately assess dynamic trunk loading under symmetric and asymmetric lifting conditions (Marras and Sommerich, 1991a, b; Granata and Marras, 1993; 1995), twisting force generation conditions (Marras and Granata, 1995), and lateral bending lifting conditions (Marras and Granata, 1997). The general structure of the model is shown in Figure 22.1. Geometrically, the model assumes that the trunk can be represented by two plates, one passed through the thorax, the other passed through the pelvis. Trunk muscle forces are represented by 10 vectors connecting the two plates. EMG electrodes monitor the activity of 10 major muscle groups in the trunk and each EMG signal is adjusted for muscle velocity, length, cross-sectional area, and force capacity before the muscle force is calculated. The contributions of each force vector is summed in each cardinal direction within the trunk to predict spine compression and shears as well as the moments imposed upon the spine. The moment calculation is used as a validation check for model fidelity. A person performing a materials handling task while standing on a force plate can provide an independent measure of spinal moments. If the measured moment imposed about the spine correlates well with the model predicted moment, then it is likely that the model is performing well. This model has been used extensively to evaluate dynamic materials handling tasks in industry. It has the advantage of being able to simultaneously evaluate many aspects of the lifts such as muscle activities, trunk kinematics, muscle coactivation, trunk muscle internal moment contributions, spinal loading, etc. Since the model resides in a Windows environment it makes it possible to evaluate multiple trunk parameters (e.g., muscle activities, kinematics, loading, etc.) while the lifting task is observed. Figure 22.2 shows an example of such an evaluation.

One should also consider the limitation associated with biomechanical assessments of dynamic lifting situations. Accurate biomechanical analyses can provide a measure of loading imposed upon the spine and when compared with spine load tolerance data (Brinckman, 1990; Evans and Lissner, 1965; Sonoda, 1962) can offer an objective measure of the risk associated with a one-time performance of a dynamic lift. However, in order to assess the risk of repeated lifts as would occur in the workplace, the loading models must be performed over multiple observations of the lifting situation throughout the workday. In addition, the loading must be compared to either spine tolerances observed during repetitive loading of the spine (Brinckmann, 1990) or a finite element model of the spine must be employed to predict when the spine structures would fail under repetitive loading conditions. Such tolerance data are still under development, and therefore, accurate risk predictions based upon biomechanical analyses may be premature.

Current investigations on spine tolerance limits have changed to explore more of the cumulative trauma aspects of spine loading. The current trend in tolerance research involves investigations relative to the

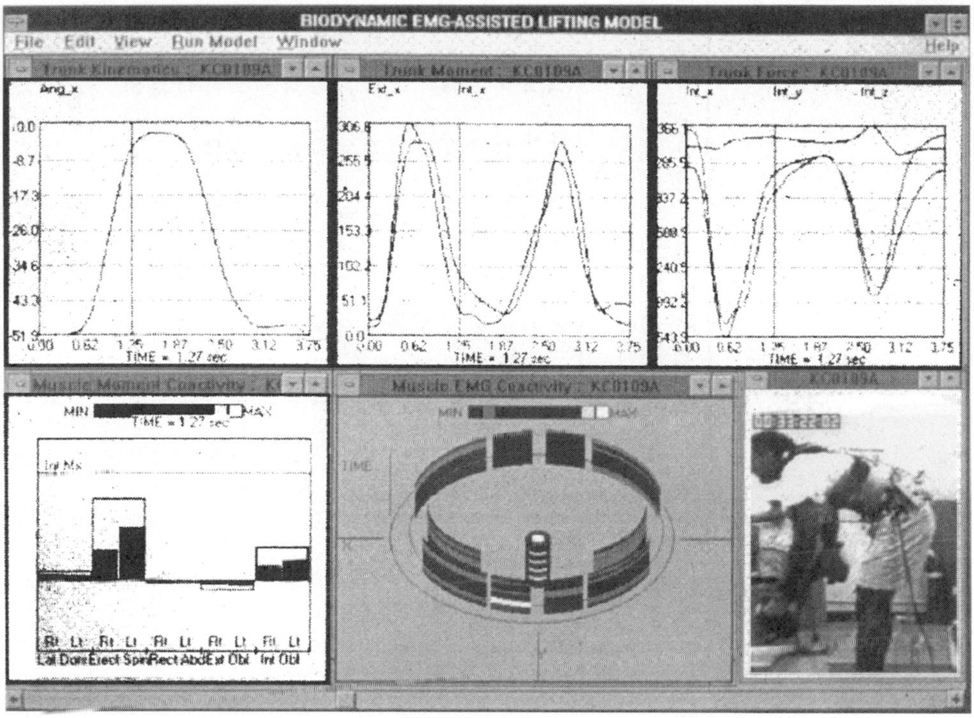

FIGURE 22.2 Example of a model "window" used to evaluate dynamic spinal loading.

biochemical changes that occur in spinal tissue as a function of age, spine load, and repetitive loading even at low level. This work is in stark contrast to the early investigations that explored strictly the mechanical limits of loading. However, most of this work is still in its early stages. In addition, such discussions are outside the scope of this chapter.

22.3 Psychophysical and Strength Assessments of Dynamic Lifts

Another approach used to assess dynamic lifting has been psychophysical analyses. In a psychophysical analysis, the weight of the object lifted is adjusted until the weight is judged as subjectively acceptable to the lifter for a given lifting situation. Using this logic, data sets have been collected and tabulated that describe the amount of weight that is acceptable to both males and females given lifting task variables. Task variables documented include the frequency of the lift, the duration of the lifting period, box size and dimensions, height of the lift, movement distance, number of people involved in the lift, the symmetric or asymmetric conditions of the lift, shape of the object lifted, load distribution (number of hands used for lifting), coupling conditions, load stability, and direction of applied force (pushing versus pulling). Typically, workplaces are designed according to a criteria that the workplace would be acceptable to 75% of the females in the population based upon these data.

An example of this type of data is shown in Table 22.1. These data were collected by Snook and Ciriello (1991) under very precisely controlled conditions. The effects of changing the task dynamics can be appreciated through an examination of this table. As shown here, for any given task conditions the weight judged acceptable decreases as the frequency of the lift increases. Greater frequency implies that workers are moving faster. Thus, these analyses indirectly account for dynamic workplace factors in manual lifting.

A related measure for the assessment of dynamic workplace factors is that of dynamic strength assessment. Dynamic strength has been evaluated for the whole body as well as for the back isolated from the rest of the body. In whole body dynamic strength testing, the dynamic characteristics of the object lifted are controlled or measured, and the force imposed on the object is the variable of interest. In

TABLE 22.1 An Example of Psychophysical Data Used to Evaluate the Acceptability of a Lifting Condition.

Width‡	Distance§	Percent¶	Floor level to knuckle height One lift every								Knuckle height to shoulder height One lift every								Shoulder height to arm reach One lift every							
			5	9	14	1	2	5	30	8	5	9	14	1	2	5	30	8	5	9	14	1	2	5	30	8
				s			min			h		s			min			h		s			min			h
		90	*6*	7	9	11	13	14	14	17	8	10	12	13	14	14	16	17	6	8	9	10	10	11	12	13
		75	*9*	11	13	16	19	20	21	24	*10*	14	16	18	18	19	21	23	8	10	12	14	14	14	16	17
	76	50	*12*	*15*	17	22	25	27	28	32	13	17	20	22	23	24	26	29	10	13	15	17	17	18	20	22
		25	*15*	*18*	*21*	28	31	34	35	41	16	*21*	*24*	27	27	28	32	35	11	16	18	21	21	22	24	27
		10	*18*	*22*	*25*	33	37	40	41	48	*19*	*24*	*28*	31	32	33	37	40	*14*	18	21	24	24	25	28	31
		90	6	8	9	12	13	15	15	17	8	11	13	15	15	16	18	19	6	8	9	12	12	12	14	15
		75	*9*	11	13	17	19	21	22	25	11	15	17	20	20	21	23	25	8	11	12	15	15	16	18	20
75	51	50	*13*	15	18	23	26	28	29	34	14	19	21	25	25	26	29	32	10	14	16	19	20	20	23	25
		25	*16*	*19*	22	29	33	35	36	42	*17*	*23*	26	30	31	32	36	39	13	17	19	23	24	25	27	30
		10	*19*	*22*	*26*	34	38	42	43	50	*20*	*26*	*30*	35	36	37	41	45	*15*	19	22	27	27	29	32	35
		90	8	9	11	13	15	16	17	20	10	13	15	18	18	19	21	23	7	10	11	14	14	14	16	18
		75	*11*	13	15	19	22	24	24	28	13	17	20	23	24	25	27	30	10	13	15	18	18	19	21	23
	25	50	*15*	18	21	26	29	32	33	38	17	22	25	30	30	31	35	38	12	16	19	23	23	24	27	29
		25	*18*	*22*	26	33	37	40	41	48	*20*	*27*	*30*	36	36	38	42	46	15	20	22	28	28	29	32	35
		10	22	*26*	*31*	38	44	47	49	57	*23*	*31*	*35*	42	42	44	49	53	*17*	*23*	26	32	32	34	38	41
		90	7	8	10	13	15	16	17	20	8	10	12	13	14	14	16	17	7	9	10	12	12	13	14	16
		75	*10*	12	14	19	22	24	24	28	10	14	16	18	18	19	21	23	9	11	13	16	16	17	19	21
	76	50	*14*	*16*	19	26	29	32	33	38	13	17	20	22	23	24	26	29	11	15	17	20	21	21	24	26
		25	*17*	*20*	*24*	33	37	40	41	48	16	21	24	27	27	28	32	35	13	18	20	25	25	26	29	31
		10	*20*	*24*	*28*	38	43	47	48	57	*19*	*24*	*28*	31	32	33	37	40	*15*	*21*	23	28	29	30	33	36

		90	7	9	10	14	16	17	18	20	8	11	13	15	15	16	18	19	7	9	11	14	14	14	16	18
		75	10	13	15	20	23	25	25	30	11	15	17	20	20	21	23	25	9	12	14	18	18	19	21	23
49	51	50	*14*	17	20	27	30	33	34	40	14	19	21	25	25	26	29	32	12	15	18	23	23	24	27	29
		25	*18*	*21*	25	34	38	42	43	50	17	23	26	30	31	32	36	39	14	19	21	28	28	29	32	35
		10	*21*	*25*	*29*	40	45	49	50	59	*20*	*26*	*30*	35	36	37	41	45	*16*	*22*	25	32	32	34	37	41
		90	8	10	12	16	18	19	20	23	10	13	15	18	18	19	21	23	9	11	12	16	16	17	19	21
		75	12	15	17	23	26	28	29	33	13	17	20	23	24	25	27	30	11	14	16	21	21	22	25	27
	25	50	*16*	20	23	30	34	37	38	45	17	22	25	30	30	31	35	38	14	18	21	27	27	28	32	35
		25	*21*	*25*	29	38	43	47	48	56	*20*	*27*	30	36	36	38	42	46	16	22	25	33	33	34	38	42
		10	*24*	*29*	*34*	45	51	56	57	67	*23*	*31*	*35*	42	42	44	49	53	*19*	*25*	29	38	38	40	44	48
		90	8	10	11	15	17	19	19	23	8	11	13	15	15	16	18	19	8	10	12	14	14	15	16	18
	76	75	*12*	14	17	22	25	28	28	33	11	15	17	20	20	21	23	25	10	14	16	18	19	19	22	24
		50	*16*	*19*	22	30	34	37	38	44	14	19	21	25	25	26	29	32	13	17	20	23	24	25	27	30
		25	*20*	*24*	*28*	37	42	47	47	55	17	*23*	26	30	31	32	36	39	*16*	*21*	24	28	29	30	33	36
		10	*24*	*29*	*33*	44	50	54	56	65	*20*	*26*	*30*	35	36	37	41	45	*18*	*24*	*28*	33	33	34	38	42
		90	9	10	12	16	18	20	20	24	9	12	14	17	17	18	20	22	8	11	13	16	16	17	18	20
		75	*12*	15	18	23	26	28	29	34	12	16	18	22	23	23	26	29	11	14	17	21	21	22	24	26
34	51	50	*17*	20	24	31	35	38	39	46	15	20	23	28	29	30	33	36	14	18	21	26	27	28	31	34
		25	*21*	*25*	*30*	39	44	48	49	57	18	24	27	34	35	36	40	44	*17*	22	25	32	32	33	37	41
		10	*25*	*30*	*35*	46	52	57	58	68	*21*	*28*	*32*	40	40	42	46	51	*19*	*26*	29	37	37	39	43	47
		90	10	12	14	18	20	22	23	27	11	14	16	20	20	21	23	26	10	13	15	19	19	19	22	24
		75	*15*	18	21	26	30	32	33	38	14	18	21	26	27	28	31	34	13	17	20	24	25	26	29	31
	25	50	*20*	24	28	35	40	43	44	52	18	23	27	33	34	35	39	43	16	22	25	31	31	33	36	40
		25	*26*	*30*	*35*	44	50	54	55	65	*21*	*28*	32	40	41	42	47	52	*20*	*26*	*30*	37	38	39	44	46
		10	*29*	*35*	*41*	52	59	64	66	76	*25*	*33*	*37*	47	47	49	55	60	*23*	*30*	*35*	43	44	45	51	55

‡Box width (the dimension away fromt he body) (cm).
§Vertical distance of lift (cm).
¶Percentage of industrial population.
Italicized values exceed 8 h physiological criteria (see text).

Adapted from Snook, S.H. and Ciriello, V.M. (1991) The design of manual handling tasks: revised tables of maximum acceptable weights and forces, *Ergonomics*, 34(9), 1197-1213.

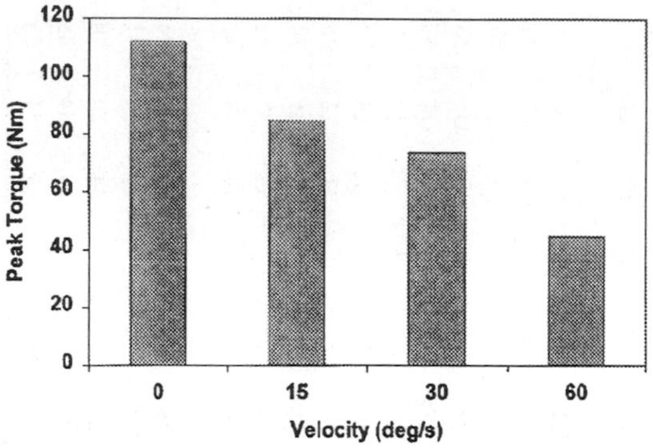

FIGURE 22.3 Trunk torque production as a function of increasing trunk velocity.

isolated back testing a dynamometer is aligned with a point of rotation on the back and the rotational motion characteristics are either controlled or documented while the torque-generation capacity of the trunk is monitored. The difference in the information derived between these two approaches is related to the specificity of the analysis desired. If one wants to assess dynamic lifting capacity without concern for the specific level of capacity of any given joint, then one might be interested in whole body strength. In this situation, strength is limited by whatever joint is weakest in the kinetic chain in the body. For example, if one is limited by shoulder strength in a lift, the strength of the shoulder will define the limit of the whole body lift. However, we will not have any idea about which joint limits the lift or how much capacity and relative level of protection is present in the other joints in the kinetic chain.

Whole body dynamic strength assessments have been assessed in several ways. Pytel and Kamon (1981) as well as Kamon et al. (1982) have used a dynamometer to document peak force capacity during dynamic whole body lifting. They limited the isokinetic speed of the object lifted to 0.73 m/s and observed the force that could be generated by a subject. In these studies, the peak force generated in the vertical direction was documented as both students and industrial workers lifted a handle from floor to chest height, from the floor to knuckle height, and when pulling a handle from knuckle height to chest height. These test positions were similar to some of the positions tested for psychophysical analyses. However, as expected, the force documented was much greater in these tests because the maximum force generation was of interest as compared to the acceptable weight of lift, as is documented in psychophysical studies. When compared to whole body isometric lifting strength, these dynamic forces were lower by about one third.

Isolated dynamic trunk strength testing has also been well documented in the literature. In this situation, the subject is placed in a dynamometer and the strength generated about the L5/S1 joint has been recorded through motions with different dynamic characteristics. Marras and associates (1987) have demonstrated the capacity to generate torque with the back changes as a function of the increasing velocity, increasing asymmetry, and as a function of forward trunk flexion angle. Figure 22.3 demonstrates how trunk strength changes with trunk angular velocity. This type of information can be compared to the strength required by a task (as assessed by dynamic models) to estimate the risk of trunk muscle overexertion given the motion characteristics of a dynamic lifting task. In this manner, the percent of the population that would exceed their trunk strength could be predicted and this, in turn, could provide a measure of back injury risk.

One needs to consider the advantages and disadvantages of psychophysical and dynamic strength information when evaluating and designing a manual lifting task. In particular, one must distinguish the objectives of evaluating whether one is capable of *performing* a task and whether one is at *risk of injury* when performing a materials handling task. The advantages of both psychophysical and dynamic strength data is that one has an objective measure of task performance and can therefore predict the percentage of the population that is capable of performing a task. When using psychophysical data to evaluate dynamic tasks, one must assume that the task will be performed at a speed and pace similar to that of the subjects used in the development of the database. However, in realistic work situations, it is a common observation that workers will work very rapidly to finish a materials handling task so that they can maximize their rest time. Thus, assumptions about the dynamic characteristics matching those of the database may not hold. When using psychophysical data, one also assumes that if lifting a load that is judged acceptable to a person they are less likely to suffer an injury. On one hand, there is evidence that the psychological perception of risk is a significant factor in the reporting of back injuries on the job (Bigos et al., 1986). On the other hand, there is also a belief that perceived acceptability may not be associated with risk over the long run. For example, many people choose to smoke and would call that risk acceptable. However, the literature is clear in that smoking dramatically increases the chances of physiological problems. Thus, one can make the argument that people are not good judges of risk perception. Therefore, psychophysical acceptance may not provide a level of protection for the working population. The argument can also be made that back injury is cumulative, thus, the load level that is acceptable at one point in time may not be acceptable as cumulative trauma occurs. Similar arguments can be made for the cautious application of dynamic strength data. Even though one has the capacity to perform a task, that does not mean they will not become injured, especially when the task is performed repeatedly. Thus, psychophysical and dynamic strength data must be used with caution and should be one of many tools used for the consideration of workplace design.

22.4 Assessments Based on Trunk Kinematics and Historical Observations of Risk

A unique approach to the assessment of occupationally related low back disorder risk associated with dynamic lifting at the worksite has recently been developed (Marras et al., 1993, 1995). This approach considers trunk motion characteristics in three-dimensional space along with traditionally documented characteristics of the workplace in a multiple logistic regression model of occupationally related LBD risk. This model requires documentation of trunk motions during a given occupational task. One means to facilitate these measurements is through the use of a trunk goniometer. A trunk goniometer, called the lumbar motion monitor or LMM, was developed to document the rotational position, velocity, and acceleration characteristics of the trunk during work tasks.

The LMM was used to document the trunk motion patterns and workplace characteristics of over 400 workers in jobs that have been associated historically with varying degrees of risk. Jobs with at least three years of historical LBD risk data (derived from medical records) were documented for trunk motion and workplace characteristics and used to form two risk groups (Marras et al. 1993). One group consisted of materials handling jobs that result in tasks associated with no recordable LBDs (0 incidences per 200,000 hours of exposure). The other group consisted of jobs that could be classified as having a high risk of LBD (average of 26.4 LBDs per 200,000 hours exposure and at least 12 incidences per 200,000 exposures). The analyses of the data indicated that high-risk jobs were more likely to involve rapid trunk motions that occurred between several planes of the body. Figure 22.4a and b show examples of low-risk and high-risk work cycles, respectively. These figures show the three-dimensional position of the thorax relative to the pelvis. The points in the figure are spaced one-sixtieth of a second apart. Thus, the farther apart the points, the faster the trunk is moving. Note how the low-risk motions shown in Figure 22.4a are confined to one or two planes of the body, and the speed of motion is low. On the other hand, Figure 22.4b indicates motions that are occurring in all three planes of the body and much more rapidly.

Multiple logistic regression models were developed that best discriminate between high-risk and low-risk groups of low back disorder based upon characteristics of the job and worker trunk motion patterns. Of the 114 variables associated and documented for each job, five factors were used to create a multiple logistic regression model that best discriminates between membership in the high- and low-risk groups. These five factors consist of: (1) frequency of lifting, (2) load moment (load weight multiplied by the distance of the load from the spine), (3) average twisting velocity (measured by the LMM), (4) maximum sagittal flexion angle through the job cycle (measured by the LMM), and (5) maximum lateral velocity

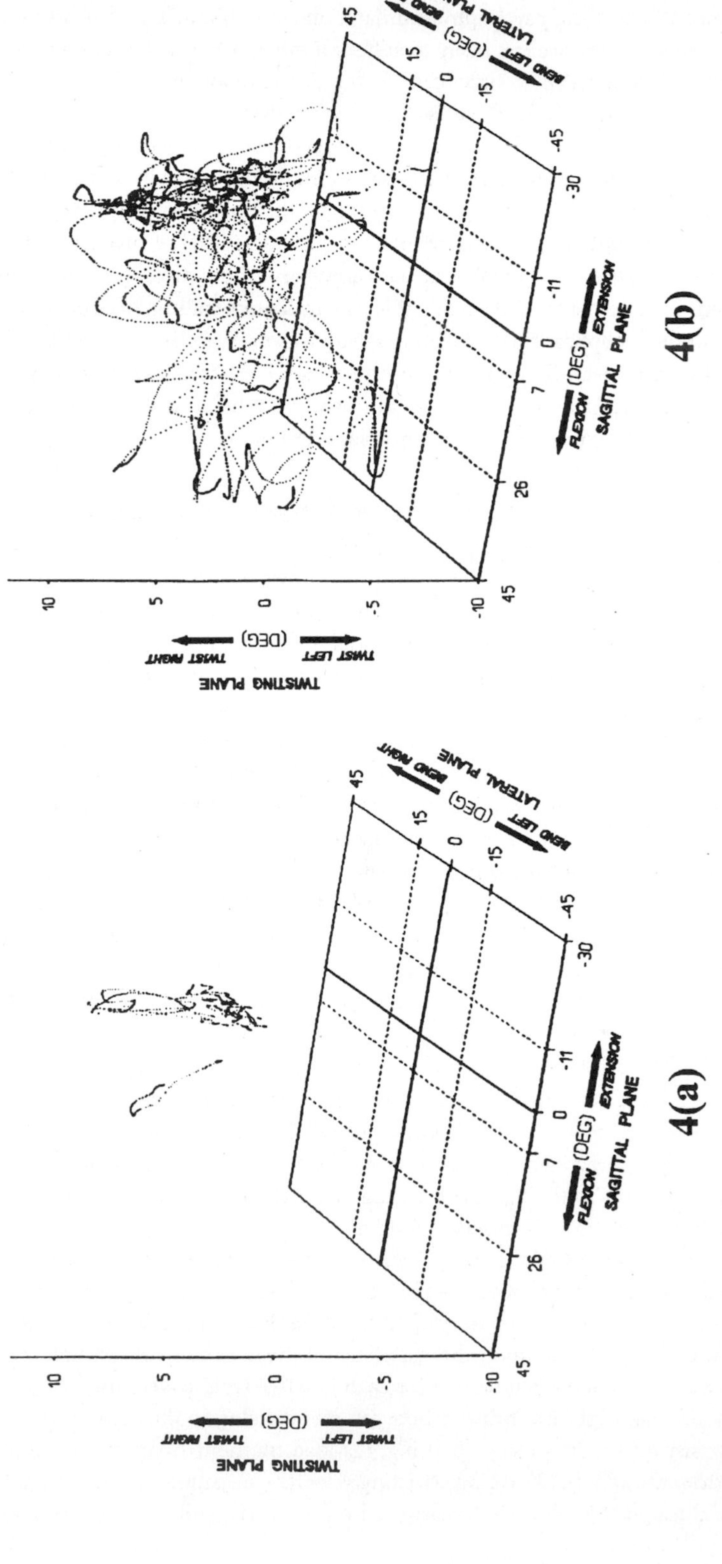

FIGURE 22.4 An example of trunk motion in a (a) low-risk and (b) high-risk job. Each point in the figure indicates the position of the thorax relative to the pelvis in three-dimensional space. The points are spaced one-sixtieth of a second apart. Thus, the more space between points the faster the motion.

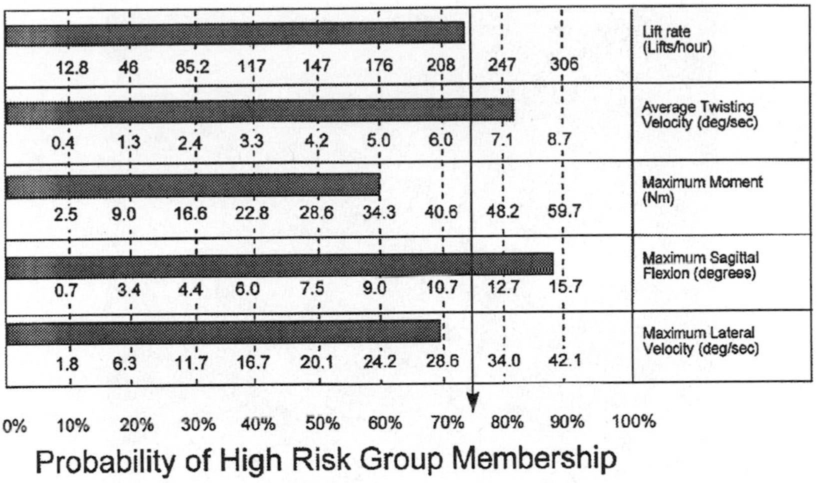

FIGURE 22.5 The LMM risk model. The five factors in the model are scaled according to the relative risk associated with each variable. A combination of the variables indicates overall probability of increased LBD risk that is shown at the bottom of the scale.

(measured by the LMM). These five variables have been scaled relative to their association with high-risk group membership. Figure 22.5 shows the risk model along with the appropriate scaling that can be used for workplace design and assessment purposes. The figure shows that by considering the combination of these five factors the probability of high-risk group membership can be predicted. Using this technique it is possible to address the issue of how much exposure to a risk factor is too much exposure, while considering the interactive effects of the other risk factors. This model has been shown to have a high degree of predictability (odds ratio = 10.7) compared to previous attempts to assess work-related low back disorder risk. Validity studies currently under way have indicated that the model is at least as sensitive as originally documented (Marras et al., 1993) and most likely more predictive than documented.

A similar assessment has been performed of the LMM database that indicates it is possible to assess the *severity* of the risk associated with dynamic lifting tasks (Marras et al. 1995). This assessment is very similar to the one just described with the exception that the scaling of the five risk variables (as shown in Figure 22.5) is different. Ongoing validity studies have also indicated that this is a very sensitive technique for dynamic manual lifting assessment.

This model can be used to assess current risk associated with the design of material handling tasks or it can be used to assess the expected risk associated with modifications or redesign of a job. In these cases it would be necessary to "mock up" the workplace and test a worker performing 5 or 6 repetitions of the job. The advantage of this assessment is that the evaluation provides information about risk that would take years to derive from historical accounts of incidence rates.

22.5 Applications of Dynamic Assessment Tools

This section will provide an example of how some of the dynamic assessment tools discussed in this chapter can be used for the evaluation of dynamic manual lifting assessments. The example shows how the three dynamic tools can be used to assess the job of an order selector in food distribution warehouses.

It is widely known that the order selection task in food distribution centers places the worker at risk of occupationally related low back disorders (LBDs). This job is associated with one of the greatest incidence rates of LBD in the United States. The National Association of Wholesale Grocers of America (NAWGA) and the International Foodservice Distribution Association (IFDA) disclosed that 30% of the injuries reported by food distribution warehouse workers were attributable to back sprains/strains

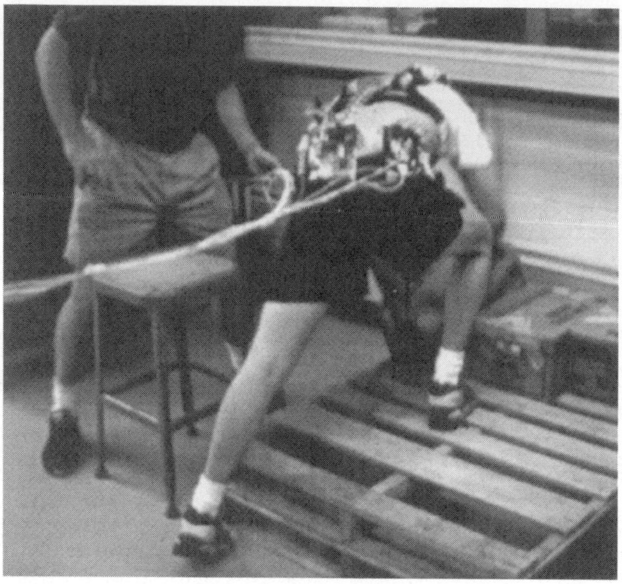

FIGURE 22.6 An order selector performing a stock picking task and wearing the experimental apparatus required to perform an LMM risk analysis and EMG-assisted model analysis.

(Waters, 1993). In addition, over a five-year period, it was found that back injuries could account for nearly 60% of lost workdays (NIOSH, 1992). Hence, grocery item selectors have an incidence of low back pain that is at least as severe as other manual materials handling jobs.

One approach to controlling this risk consists of manipulating the characteristics of the object or box to be handled in the food distribution center. A committee organized by the Food Marketing Institute (FMI) was interested in considering the various options available to them in order to help mediate the risk of work-related LBD in these food distribution centers. Among the options considered are: (1) reducing the weight of the boxes, (2) reducing the size of the boxes, or (3) incorporating handles into the boxes. However, it is currently unknown what effect these changes to the box characteristics would have on the loading of the spine and the subsequent risk of low back disorder. Such decisions would have a significant financial impact on the manufacturers of the items in the food distribution centers because it would require them to significantly change the packaging system for all products. Therefore, the FMI committee was interested in assessments that could *realistically* assess the significance of the contemplated changes to the box or case design.

The objective of this assessment was to determine the change in LBD risk and spine loading associated with selecting cases (in a warehouse environment) that varied as a function of: (1) weight (40, 50, and 60 lbs.), (2) size (2681 or 1584 cu. in.), and (3) the existence of handles or hand holds. In addition, in order to assess the problem in context, these variables were explored as a function of where the case was on a pallet. Ten experienced order selectors were evaluated as they selected cases from a slot (storage bin) on to a pallet jack. During the different experimental trials the case weight, case size, and case coupling (handles) conditions were varied. Workers were instructed to pick the entire complement of cases from the pallet so that they could be observed picking from all locations on a pallet. While the workers were lifting cases, they were being continuously monitored so that LBD risk could be assessed. Figure 22.6 shows the workplace environment of an order selector performing an order selection task.

Risk of LBD was evaluated through three assessment tools. First, a lumbar motion monitor (LMM) risk model was used that compared work conditions to those identified historically as being associated with a high risk of LBD injury. Second, the EMG-assisted biodynamic model described earlier was used to more specifically evaluate spine loading characteristics associated with the various potential changes in the case design. Third, the psychophysical acceptance was assessed as a function of each condition.

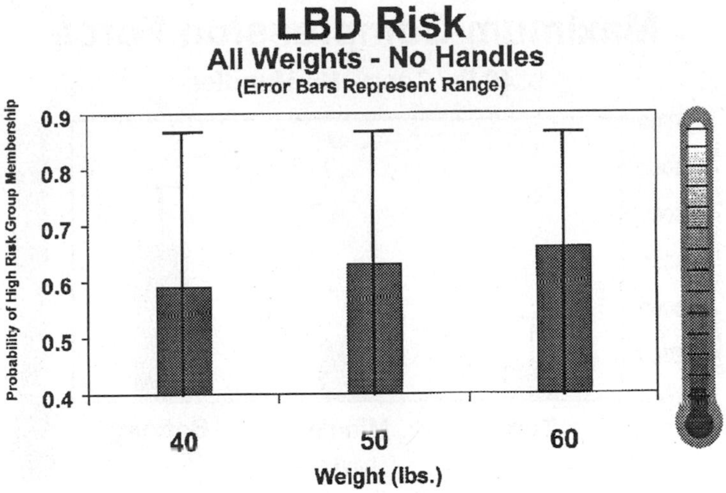

FIGURE 22.7 LMM risk as a function of case weight lifted.

Collectively, this information provided a rich understanding of the cost-benefits associated with the various case parameters under consideration.

As discussed above, the assessment tools assess risk by very different means. The LMM risk model *represents risk* in terms of the probability (based upon historical trends in industry) that the conditions of the task (job profile) resemble a high-risk situation. The EMG-assisted biodynamic model assesses *spine loading* that is assumed to be associated with risk of developing LBD. Finally, the psychophysical tool determines the percentage of the population that would be expected to find a load acceptable. These methods were used to assess the risk of LBD as a function of each case characteristic considered. In order to facilitate interpretation of these results a risk "thermometer" concept was used along with the graphical representation of the results. This risk thermometer can be used to assist in risk assessment regardless of the evaluation system used to determine risk. When the values on the graph are within the upper regions of the thermometer a highly risky (dangerous) condition is indicated, whereas when the values on the figure are aligned with the lower region of the thermometer, the conditions are safe. A continuous spectrum between these extremes permits one to make relative judgments between conditions.

Case Weights

The effect of case weight upon risk of LBD is indicated in Figure 22.7. This figure shows the results of the LMM risk model, however, similar trends were indicated when spine loading was assessed. Generally, these results indicate that as the weight of the case increases the risk increases in a rather linear fashion. The *average* risk associated with a 40-lb. case is close to the acceptable range, whereas, the 60-lb. case approaches the danger zone on the risk thermometer. Of particular significance is the fact that under each condition the range of the data (indicated by the error bars) is extremely large. Thus, for any given case weight, the risk associated with the condition can be either extremely safe or extremely risky. Further analyses indicated that the position of the case on the pallet more specifically defined risk than did case weight.

Position on the Pallet

These analyses have shown that the case position on the pallet is instrumental in defining LBD risk. Figure 22.8 indicates spine loading as a function of pallet position for the 40-lb. cases. Note that for this condition most of the spine loadings associated with the top and middle layers of the risk are within the safe region on the risk thermometer. However, the data indicate that the vast majority of the observations

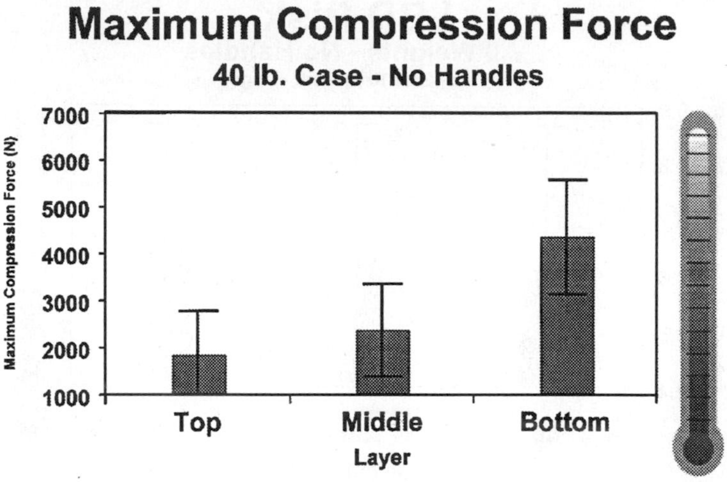

FIGURE 22.8 Spine loading (compression) associated with the location of a 40-lb. case on a pallet.

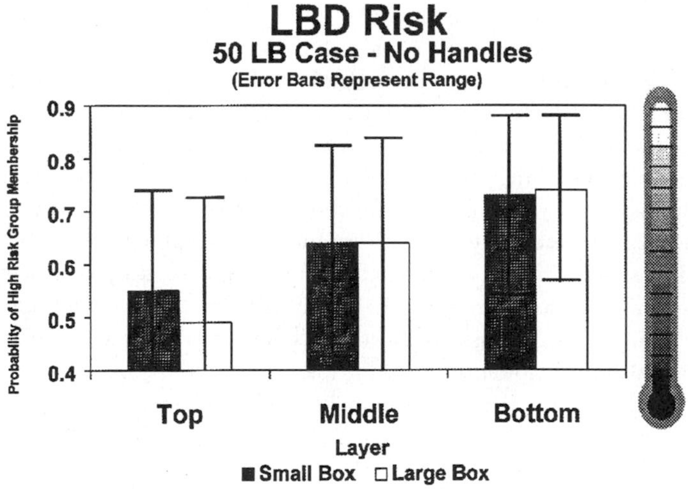

FIGURE 22.9 LMM risk associated with two different case sizes while lifting a 50-lb. case from different regions of a pallet.

associated with lifting from the bottom layer of the pallet imposed a significant risk of LBD to the workers. Similar trends were noted when the LMM risk model was used as a measure. The risk increased substantially for the bottom layer of the pallet when the 50- and 60-lb. cases were lifted. Thus, this analysis shows that the risk was significant only for lifts from the bottom layer of the pallet.

Case Size

The LMM risk analyses did indicate that case size was a significant factor in determining risk. However, the practical significance of this finding was questionable. Figure 22.9 shows the effects of case size for the 50-lb. case as a function of location on the pallet. This indicates that the only benefit of varying case size occurs at the top layer of the pallet and the evaluation of risk as a function of location has shown that little risk exists at the top layer of the pallet. The differences noted at the bottom (problematic) layer of the pallet were not significant enough to justify the control of case size. Therefore, there is no practical benefit of controlling case size among the sizes considered in this study.

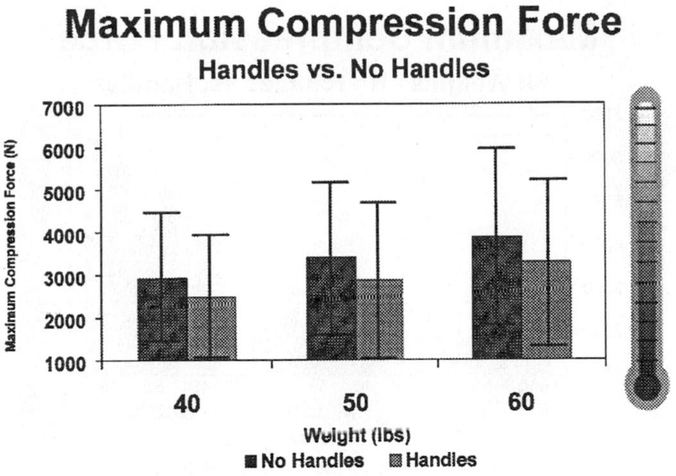

FIGURE 22.10 Spine compression as a function of handles and case weights.

Handles

The effects of case handles on spine loading compared to no handle conditions is illustrated in Figure 22.10. This figure indicates that *on the average* the effect upon spine loading of incorporating handles into a box is approximately equivalent to reducing the case weight by 10 pounds. However, as mentioned previously, serious risk only occurs at the bottom layer of the pallet. Therefore, if one considers the effects of handles upon spine loading as a function of the different levels of the pallet, the situation shown in Figure 22.11 becomes apparent. This figure shows that none of the conditions yield all of the data within the safe potion of the risk thermometer. However, at the bottom layer the 40-lb case with handles yields the lowest risk situation, with approximately 40% of the observations within the safe range of the risk thermometer and only 3% of the observations within the danger range. By contrast, the 40-lb box without handles and the 50-lb. box with handles increase the percent of the observation within the dangerous risk range to about 7 to 10% of the observations. Furthermore, a 50 lb. box without handles further increases risk to where lifting from the bottom layer would yield less than about 12% of the observations in the acceptable zone and over 20% of the observations within the danger zone. A summary of the percentage of data within the biomechanical benchmark zones of the risk thermometer is shown in Table 22.2. This table provides a means to quantitatively consider the trade-offs associated with the various workplace variables under consideration. For comparison purposes, similar information was provided in tabular form in Tables 22.3 and 22.4 for the LMM risk assessment and psychophysical acceptance, respectively. Table 22.3 indicates that subtle differences can be noted in risk between the different box conditions. Table 22.4 indicates that from a psychophysical perspective none of the conditions would be acceptable because none of the conditions would be judged acceptable to 75% of the females. Note the different sensitivities of information that can be derived given the different approaches. One should also note the dramatically different levels of effort required to perform the various analyses. EMG-assisted biomechanical models require significant time, effort, and resources, whereas LMM risk assessments and psychophysical assessments are quite quick and relatively easy to perform.

This analysis has demonstrated how one can pinpoint which case parameter variables or features are worthy of consideration for inclusion in the food distribution environment for the purpose of reducing the risk of work-related LBD. In general, the following conclusions were drawn.

- Risk of LBD increases linearly as case weight increases.
- The greatest risk and loading of the spine occur during lifts from the bottom layers of the pallet. The other layers of the pallet pose acceptable lifts regardless of the case weight.
- Case size has a significant effect on risk, but the difference has no practical meaning. There is no reason to control case size based upon the range of sizes explored in this study.

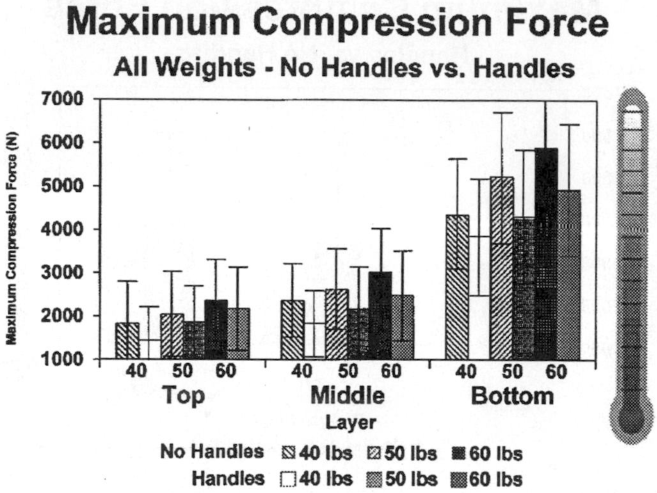

FIGURE 22.11 The effects of handles and the location of the case on a pallet on spine loading.

TABLE 22.2 Summary of Spine Compression Loading Compared to Spine Tolerance "Benchmarks" as a Function of Case Weight, Handles, and Location on a Pallet.

		Case Weight					
		40 lbs		50 lbs		60 lbs	
		Handles	No Handles	Handles	No Handles	Handles	No Handles
Top Layer	≤3400 N	99.1	94.6	92.5	88.2	91.7	84.3
	3400 to 6400 N	0.9	5.4	7.5	11.8	8.3	15.7
	>6400 N	0.0	0.0	0.0	0.0	0.0	0.0
Middle Layer	<3400 N	93.5	86.7	87.4	79.0	84.6	70.2
	3400 to 6400 N	6.5	13.3	12.6	21.0	15.4	29.1
	>6400 N	0.0	0.0	0.0	0.0	0.0	0.7
Bottom Layer	<3400 N	40.3	27.4	29.7	12.3	12.7	2.7
	3400 to 6400 N	56.3	65.5	59.7	65.3	71.7	64.4
	>6400 N	3.3	7.1	10.7	22.3	15.7	32.9

Safe (<3400 N)
Caution (3400–6400 N)
Danger (>6400)

- Handles have a significant effect upon spine loading. The effect is particularly significant when lifting from the lowest levels of the pallet. The 40-lb box, when combined with handles, represents the condition with the lowest level of risk even when lifting from the lowest level of the pallet. Handles have the effect upon spine loading of lowering the case weight by 10 lbs. Thus, a 50-lb case with handles can reduce spine loading to the level of a 40-lb case without handles.

These findings provided some practical solutions to FMI for the design of the distribution center, and these recommendations were incorporated into Food Industry Guidelines. This example demonstrates the range of information that is obtainable from these dynamic assessment tools. Psychophysical methods are easiest to apply. However, as this example shows, the specific conditions of a work situation may not match the psychophysical database well as was the case for the pallet middle level order selecting in this example. The LMM risk model is straightforward and easy to use and provides a realistic assessment of the risk of LBD risk probability. This risk model can relate to specific work conditions since it interprets the kinematics associated with a particular work condition. Finally, in order to derive more precise information about the task, such as actual spinal loading information, a much more sophisticated

TABLE 22.3 Summary of LMM Risk Assessment of an Order Selecting Task as a Function of Case Weight, Handles, and Location on a Pallet.

		Case Weight					
	Probability of High Risk	40 lbs		50 lbs		60 lbs	
	Group Membership	Handles	No Handles	Handles	No Handles	Handles	No Handles
Top Layer	<30%	0.0%	5.0%	0.0%	0.0%	0.0%	0.0%
	30% to 70%	100.0%	95.0%	90.0%	90.0%	90.0%	85.0%
	>70%	0.0%	0.0%	10.0%	10.0%	10.0%	15.0%
Middle Layer	<30%	5.0%	0.0%	0.0%	0.0%	0.0%	0.0%
	30% to 70%	85.0%	85.0%	85.0%	70.0%	60.0%	65.0%
	>70%	10.0%	15.0%	15.0%	30.0%	40.0%	35.0%
Bottom Layer	<30%	0.0%	0.0%	0.0%	0.0%	0.0%	0.0%
	30% to 70%	50.0%	50.0%	55.0%	45.0%	35.0%	45.0%
	>70%	50.0%	50.0%	45.0%	55.0%	65.0%	55.0%

Safe (<30%)
Caution (30% to 70%)
Danger (>70%)

TABLE 22.4 Summary of Psychophysical Assessment of Order Selection Task. The Values in the Table Indicate the Percentage of the Industrial Population that Would Find the Specific Conditions Acceptable.

	Case Weight											
	40 lbs				50 lbs				60 lbs			
	Handles		No Handles		Handles		No Handles		Handles		No Handles	
Gender	M	F	M	F	M	F	M	F	M	F	M	F
Top Layer	75%	0%	75%	0%	50%	0%	50%	0%	25%	0%	25%	0%
Middle Layer	n/a	n/a	n/a	n/a	n/a	n/a	n/a	n/a	n/a	n/a	n/a	n/a
Bottom Layer	50%	0%	25%	0%	25%	0%	10%	0%	10%	0%	0%	0%

Safe (<30%)
Caution (30% to 70%
Danger (>70%)

assessment is necessary (EMG-assisted model). However, this requires a great deal of effort and may not be practical for "on the job" workplace assessments. The correct level of assessment depends upon the consequences of information. In this example, the FMI wanted to make sure that the answer to their problem was realistic because it would require many manufacturers to make significant, expensive changes to their production process and would affect national guidelines. Therefore, they elected to seek the most complete and richest information possible. Had the magnitude of the problem not been so great and the consequences so costly, less complex analyses (the LMM risk model or psychophysical assessments) may have been adequate.

References

Bean, J.C., Chaffin, D.B., and Schultz, A.B. (1988) Biomechanical model calculation of muscle forces: A double linear programming method. *J. Biomechanics*, 21 (1) 59-66.

Bigos, S.J., Spengler, D.M., Martin, N.A., Zeh, J., Fisher, L., Nachemson, A., and Wang, M.H. (1986) Back injuries in industry: A retrospective study. II. Injury factors. *Spine*, 11(3), 246-251.

Brinckmann, P., Biggemann, M., and Hilweg, D. (1988) Fatigue fracture of human lumbar vertebrae. *Clinical Biomechanics*, Supplement No. 1, 1988.

Chaffin, D.B. (1969) A computerized biomechanical model: development of and use in studying gross body actions. *J. Biomechanics*, 2, 429-441.

Chaffin, D.B. and Muzaffer, E. (1991) Three-dimensional biomechanical static strength prediction model sensitivity to postural and anthropometric inaccuracies. *IIE Transactions*, 23(3), 215-227.

Evans, F.G. and Lissner, H.R. (1965) Studies on the energy absorbing capacity of human lumber intervertebral discs. *Proceedings of the Seventh Strapp Car Crash Conference*, Springfield, IL.

Frievalds, A., Chaffin, D.B., Garg, A., and Lee, K. (1984) A dynamic evaluation of lifting maximum acceptable loads. *J. Biomechanics*, 17, 251-262.

Granata, K.P. and Marras, W.S. (1993) An EMG-assisted model of loads on the lumbar spine during asymmetric trunk extensions, *J. Biomechanics*, 26(12), 1429-1438.

Granata, K.P. and Marras, W.S. (1995), An EMG-assisted model of trunk loading during free-dynamic lifting. *J. Biomechanics*, 28(11), 1309-1317.

Kaymon, E., Kiser, D., and Pytel, J. (1982) Dynamic and static lifting capacity and muscular strength of steelmill workers. *Am. Ind. Hygiene Assoc. J.*, 43, 853-857.

Marras, W.S. and Granata, K.P. (1995) A biomechanical assessment and model of axial twisting in the thoraco-lumbar spine. *Spine*, 20(13), 1440-1451.

Marras, W.S. and Granata, K.P. (1996) An assessment of spine loading as a function of lateral trunk velocity. *J. Biomechanics*, (30)7, 697-703.

Marras, W.S., Lavender, S.A, Leurgans, S., Rajulu, S., Allread, W.G., Fathallah F., and Ferguson, S.A., (1993) The role of dynamic three dimensional trunk motion in occupationally-related low back disorders: the effects of workplace factors, trunk position and trunk motion characteristics on injury. *Spine*, 18(5), 617-628.

Marras, W.S., Lavender, S.A., Leurgans, S., Fathallah F., Allread, W.G., Ferguson, S.A., and Rajulu, S. (1995) Biomechanical risk factors for occupationally related low back disorder risk. *Ergonomics*, 38(2), 377-410.

Marras, W.S. and Mirka, G.A.(1992) A comprehensive evaluation of asymmetric trunk motions. *Spine*, 17(3), 318-326.

Marras, W.S. and Reilly, C.H.(1988) Networks of internal trunk loading activities under controlled trunk motion conditions. *Spine*, 13(6), 661-667.

Marras, W.S. (1988) Predictions of forces acting upon the lumbar spine under isometric and isokinetic conditions: A model-experiment comparison. *International Journal of Industrial Ergonomics*, 3(1), 19-27.

Marras, W.S., Rangarajulu, S.L., and Wongsam, P.E. (1987) Trunk force development during static and dynamic lifts. *Human Factors*, 29(1), 19-29.

Marras, W.S. and Sommerich, C.M., (1991) A three dimensional motion model of loads on the lumbar spine, Part I: Model structure. *Human Factors*, 33(2), 123- 137.

Marras, W.S. and Sommerich, C.M. (1991) A three dimensional motion model of loads on the lumbar spine, Part II: Model validation. *Human Factors*, 33(2), 139-149.

Marras, W.S., Wongsam, P.E., and Rangarajulu, S.L.(1986) Trunk motion during lifting: The relative cost. *International Journal of Industrial Ergonomics*, 1(2), 103-113.

McGill, S.M. and Norman, R.W. (1985) Dynamically and statically determined low back moments during lifting. *J. Biomechanics*, 18, 877-885.

NIOSH 1992 Interim Report, HETA 91-405, March 1992.

Nussbaum, M.A., Martin, B.J., and Chaffin, D.B. (1997) A neural network model for simulation of torso muscle coordination. *J. Biomechanics*, 30(3), 251-258.

Punnett, L., Fine, L.J., Kyserling, W.M., Herrin, G.O., and Chaffin, D.B. (1991) Back disorders and nonneutral trunk postures of automotive assembly workers. *Scand. J. Work Environ. Health*, 17:337-346.

Pytel, J.L. and Kamon, E. (1981) Dynamic strength test as a predictor for maximal and acceptable lifting. *Ergonomics*, 24, 663-672.

Raschke, U., Martin, B.J., and Chaffin, D.B. (1996) Distributed moment histogram: A neurophysiology based method of agonist and antagonist trunk muscle activity prediction. *J. Biomechanics*, 29(12), 1587-1596.

Schultz, A.B. and Andersson, G.B.J. (1981) Analysis of loads on the lumbar spine. *Spine*, 6:76-82.

Snook, S.H. and Ciriello, V.M. (1991) The design of manual handling tasks: revised tables of maximum acceptable weights and forces, *Ergonomics*, 34(9), 1197-1213.

Sonoda, T. (1962) Studies of the strength for compression, tension, and torsion of the human vertebral column. *J. Kyoto Perfect. Med. Univ.*, 71, 659-702.

Thelan, D.G. and Schultz, A.B. (1994) Identification of dynamic myoelectric signal-to-force models during isometric lumbar muscle contractions. *J. Biomechanics*, 27(7), 907-919.

Waters, T.R., Putz-Anderson, V., Garg, A., and Fine, L.J. (1993) Revised NIOSH equation for the design and evaluation of manual lifting tasks. *Ergonomics*, 36(7), 749-776.

Winter, J.M. and Woo, S.L.-Y. (1990) *Multiple Muscle Systems*, Springer-Verlag, New York.

23

Push–Pull Force Limits

Sheik N. Imrhan
University of Texas at Arlington

23.1 Introduction

Pushing and pulling (push–pull) activities involves static or dynamic muscular force exertions for moving or stabilizing objects. Neither the direction of the force nor the motion of the object need be perfectly linear or in the horizontal plane so long as deviation is small. In cases where the deviation is great, the exertion can be considered a hybrid of two types, such as lift–push, lift–pull, etc. as used by Pheasant et al. (1982). Push–pull activities occur in many types of work environments — shipping and receiving, moving, warehousing, agriculture and farming, retailing, etc. — and are becoming more common as a result of efforts to minimize lifting, lowering, holding, and carrying, the most debilitating and costly manual materials handling (MMH) activities. Baril-Gingras and Lortie (1990), in studying over 900 tasks, estimated that nearly half of all materials handling activities were push–pull ones. However, push–pull activities also account for a significant amount of overexertion musculoskeletal disorders (NIOSH, 1981; Klein, Jensen, and Sanderson, 1984; Troup and Edwards, 1985), accounting for approximately 20% of all back injuries from MMH activities.

Most push–pull activities can be distinguished by the general body posture, the number of hands involved, and whether the exertion is static or dynamic. The choice depends on a number of factors: standing posture and two-handed exertions for strong forces and significant control, and sitting posture and one-handed exertions for moderate and weak forces; dynamic activity for moving loads (e.g., industrial carts) from one point to another and for activating electromechanical mechanisms (e.g., in lawn mower and outboard motor engines), and static activity for supporting objects and operating certain mechanisms, such as levers in transportation vehicles.

Pulling and pushing can be achieved with both the arms and legs, but this chapter is concerned only with the arm activities because they constitute almost all of the pulling and pushing in the workplace

0-8493-2641-9/99/$0.00+$.50

and are they ones associated with health concerns. In the workplace, both push and pull forces are applied for moving objects out of their positions to create space for other objects or to allow passage of other objects, though this may imply inefficiency in workplace design; moving objects from one position to another where they can be stored or further manipulated; and stabilizing an object, as in preventing a box from sliding along an incline. In addition, pull forces only are applied for activating mechanical devices, such as engines on small boats and lawn mowers, and opening packages, as in pinching and tearing plastic wrappers on consumer products; and push forces only are applied for shutting lids on consumer products, and activating electrical and electronic switches with the fingers.

The choice of using a push or pull may also depend on the designs of the workplace and object being handled. For example, it is difficult and often impossible to pull a large box that has no handles, especially if it is heavy. Instead, a worker may simply change position and push it, assuming the workplace can accommodate such change of position. Similarly, a cart with defective steering in the front wheels may be pulled on a handle instead of being pushed.

The choice of a hand–object contact depends on the size and shape of handle, the intended activity, or the amount of force desired. Power grips should be the best choice when great push–pull forces are desired. However, such forces may not be achieved without a handle of adequate size and shape to accommodate a power grip. As mentioned by Kumar (1995), handle orientation may also be important. For some objects, such as such as a large television box or a panel with no handles, a flat-palm contact may be the only recourse. When contact area is small — not more than about a few square centimeters — only pinch grips may be possible (Imrhan and Sundararajan, 1992), but in these cases the expected force application is usually relatively small. Hook grips are applicable where handles are not large enough for power grips and the expected forces are only of moderate or small magnitudes. Unlike pulling, great push forces can be achieved without a firm grip on a handle and even without a handle. In the latter case, a flat palm contact may be appropriate.

23.2 Factors Influencing Push–Pull Strengths

Studies in pushing or pulling have shown that wide differences in strength capacities are obtained according to the type of exertion (static or dynamic), number of hands performing the exertion (one or two), and general body posture (standing, sitting, kneeling, or lying down). Therefore, in discussing push–pull force limits and applying data from the literature to assess task, equipment, or workplace design, one must identify the exertions according to these or similar classifications. However, many other independent variables have been shown to affect people's capacity to exert pull or push forces. They are discussed below.

Handles

The type of handle influences the maximal attainable force. In general, for maximal pulling the person should have a comfortable power grip on the handle, with maximum contact along the width of the hands (i.e., no overlapping of the hands), and with adequate clearance at the sides of the objects. For cylindrical handles, a diameter of about 1.5 to 2 in. (3.8 to 5.1 cm) (Ayoub and LoPreisti, 1971) and length of 5 in. (13 cm), which are recommended for other manual materials tasks, are also suitable for pushing and pulling. A poor hand–handle or hand–object interface acts like a weak link in the chain of force transmission and can limit push–pull force capacity severely. Fothergill, Grieve, and Pheasant (1991), for example, found up to 65% decrease in strength when poor hand–handle interfaces were used for pulling and pushing. A poor hand–handle interface also fails to capture advantageous handle-handle locations (Fothergill et al., 1991). A handle is always necessary for effective pulling but not always for pushing. When, instead of a handle, a flat surface is available for pushing, great forces can still be achieved, as demonstrated by Kroemer (1969 and 1974). Push–pull tasks and work equipment should also be designed, when possible, to position the forearm in mid-supination/pronation where the muscles in the forearm are in least tension.

One-Handed versus Two-Handed Exertions

The difference between one- and two-handed maximal efforts is much smaller than is generally believed and is dependent on the height of force application and body posture. Data from Fothergill et al. (1991) and Chaffin, Andres, and Garg (1983) indicate that the ratio of one- to two-handed pushes and pulls varies from 0.64 to 1.04, depending on the conditions of force exertion. The ratio is high (well above 0.50) because (1) push–pull strengths come not only from the arms but also the trunk, which is active in one-handed exertions, and (2) one-handed exertions afford greater freedom in a person's posture and, hence, more effective use of the body's weight and center of gravity (Fothergill et al., 1991). The ratio is greater when arm strength is the limiting factor in the exertions, as when pushing or pulling at high positions, such as above the shoulder, and standing erect.

Body Posture

The strength of a muscular exertion depends, to a large extent, on body posture, which defines the magnitude of the biomechanical advantages of the various lever systems contributing to muscular torques about the body's joints. Standing, sitting, and kneeling are probably the three most common of these postures. Davis and Stubbs (1980) investigated two-handed horizontal push–pull forces but their data are not definitive — standing forces were stronger than kneeling ones only in some situations. For one-handed exertions, Mital and Fard (1990) showed that standing one-handed dynamic pull strength is about 37% stronger than sitting strength.

Strength in any of these three whole-body postures is dependent on the geometrical configuration of other body segments, such as foot position, hand–handle height (height of force application), and the angles in the arms and legs. These factors are discussed in later sections in this chapter. It must be noted also that the best posture for static exertions is not necessarily the best for dynamic exertions (Lee, Chaffin, Herrin, and Waikar, 1991), and dynamic postures are difficult to monitor because they change continuously as the body moves (Resnick and Chaffin, 1995).

Foot Position

In standing exertions, foot and hand placements determine the effective posture. The placement of the feet relative to the hand–handle contact influences the stability of the body and, hence, the leverage for pulling and pushing efforts. In general, static standing exertions are enhanced when the feet are separated, one foot in front the other (Ayoub and McDaniel, 1974). For pushing, one should lean forward with the rear (pivoting) foot positioned behind the body's center of mass; and for pulling, with the front (pivoting) foot in front of the center of mass. The leaning postures allow people to use their body weight more effectively in counteracting the push or pull force at the hands and enhance strength (Kroemer and Robinson, 1971; Ayoub and McDaniel, 1974; Warwick, Novack, Schultz, and Berkeson, 1980; Pheasant et al., 1982; Chaffin et al., 1983). Even when not leaning, stronger forces are realized when the feet are apart, one in front of the other, than when placed side by side (Chaffin et al., 1993; Daams, 1993), and pushing or pulling in free-style posture has been shown to yield considerably greater forces than in standardized postures with the feet together or one in front of the other. The amount of leaning and shoe–floor traction influence the distance of the foot from the hand–handle contact. In general, static MVC push force is greater than pull force when the feet are separated in the fore–aft plane and the body is leaning, but are of about the same strength when the feet are close together, side by side (Chaffin et al., 1983). In addition, one can achieve an optimal push–pull posture over a wider range of angles of force application when the feet are separated compared to when they are together (Fothergill et al., 1991).

Standing, Sitting, and Kneeling

Given the stability of the body when standing and the added strength from using the legs, one would expect that standing push–pull strengths are greater than kneeling ones. However, the data from Davis and Stubbs (1980) indicate that this depends on the arm position when pulling or pushing horizontally.

In some cases, and for some age groups, kneeling generated MVC forces of about the same magnitudes as when standing. There is no clear pattern from their study, however, to warrant a simple generalization. Moreover, Gallagher (1989) found that kneeling generated greater MVC forces than standing. Sitting and standing push–pull strengths are difficult to compare directly because of the great differences in the geometry in the arms and legs and in other factors. However, it is worth mentioning that Mital and Faard (1990) found isokinetic pulls while sitting unrestrained to be 73% as strong as while standing.

Height of Application of Forces

The optimal height for application of push or pull forces depends on general body posture (especially the angles in the arms and knees), and the degree of leaning forward (for pushing) or backward (for pulling). There seems to be wide agreement that best height for static pushing or pulling in the horizontal direction while standing is between chest and knee, with pulling height being lower than pushing height (Martin and Chaffin, 1972; Ayoub and McDaniel, 1974; Chaffin et al., 1983; Kumar, Narayan, and Baccus, 1995). Recommending an exact height would be unwise since the optimal height has been shown to depend on such conditions as type of exertion (static or dynamic), angle at the elbow, frictional characteristics at the shoe–floor contact, etc. However, various recommendations seem to indicate that two-handed pushing while standing is strongest at about elbow to hip height,and pulling at about hip to knee height. The direction of the applied force modifies this relationship. Pheasant and Grieve (1981) showed that the optimal height for pulls gets lower (to 25 cm from 63 cm), below the knee, and for pushes gets higher (to 175 cm from 100 cm), above the shoulder, when the forces are exerted at an angle upward (lift–pull and lift–push).

Horizontal Foot Distance and Reach Distance

The horizontal foot distance (HFD) is the perpendicular distance between the ankle (of the rear foot, for separated feet) and hand–handle contact, when standing. For pushing, the pivoting foot is the rear one, and for pulling it is the front one. The reach distance (RD) is the distance from the shoulder to the hand–handle contact, when sitting. Both HFD and RD define the configuration of the upper body, and HFD also defines the configuration of the lower body. They influence push–pull strengths significantly, but in different ways. In standing, strength increases with RD (Martin and Chaffin, 1972), but reach distance is itself influenced by a number of factors: amount of space available, foot– or shoe–floor contact, height of the handle, etc. (Kroemer, 1974; Chaffin et al., 1983). Ayoub and McDaniel (1974) have shown that the best HFD for static exertions is about 90 to 100% of shoulder height behind the hand for pushing and 10% in front of the hand for pulling. There is evidence that, when sitting, peak static push strength for a given starting position of the arm follows an increasing–decreasing trend with reach distance (Lower, et al., 1977) and both dynamic and static pull strength increases with inrcreasing reach distance (Mital and Faard; 1990). An analysis of data from VanCott (1972) also indicates that static pull strength increases with reach distance. Dynamic (isokinetic) pull strength in a single pull follows an increasing–decreasing trend over the range of pull (Imrhan and Ayoub 1985; Garg and Beller, 1990; Imrhan and Ramakrishnan, 1992), and there is some evidence that, for any given velocity of dynamic pull, there is an optimal arm configuration for maximal pull strength (Imrhan and Ayoub, 1990).

Direction of Push/Pull Force Exertion

Direction of pushing and pulling can be described with respect to (1) the transverse plane and (2) the sagittal or coronal plane. The strongest direction of force exertion should be that in which body's reaction force is maximized at the contact of the body with the floor, seat, etc. When standing and pulling from below the horizontal or pushing from above the horizontal (that is, at an angle to the horizontal), the reaction at the shoe–floor contact can be increased and, therefore, push–pull force increased, as found by Garg and Beller (1990) and Imrhan and Ramakrishnan (1992). When sitting, however, and especially if restrictions are placed on motion of the upper body (e.g., chest harness), this relationship may be modified and some other direction may be the strongest, as found by Imrhan and Ayoub (1988). In their study, horizontal pulls at shoulder level were stronger that angled pulls (toward the shoulder) above and below the horizontal. Pushing on a high handle upward at an angle to the horizontal or pulling on a low handle upward at an angle can produce a horizontal force component that is stronger than if these exertions were directed along the horizontal (Fothergill et al., 1991).

Pushing or pulling in the horizontal is strongest with the hand directly in front of the shoulder, for one-handed exertion, (Lower et al., 1977) or with both hands directly in front of the body, for two-handed exertions (Kumar, 1994). However, when pulling or pushing at an angle to the sagittal plane (sideways), reacton force and, hence, muscular force is less. One- or two-handed push–pull strength decreases rapidly as the arm moves across the body to the left or right (Mital and Fard, 1990; Kumar and Garand, 1992; Kumar et al., 1995). Mital and Faard (1990) showed that the decrease is more rapid to the left of the body than to the right for right-handed exertions. Pulling with the hand directly toward the body has been found to be about 10% stronger than across the body, (Mital and Fard 1990; Imrhan and Ramakrishnan, 1992); and pulling with the fingers (pinch–pulling) across the body horizontally is not significantly different in strength from pulling across the body obliquely (e.g., from left waist to right shoulder) (Imrhan and Sundararajan, 1992).

Body Support

Research has shown that a body support can enhance push–pull strengths by as much as 50% (Kroemer, 1974). The support helps to enhance the reaction force to the muscular exertions and is best positioned perpendicular to the exerted force. Supports are suitable only for static exertions, however. When standing, a panel, wall, or fixed footrest is effective, especially for pushing (Kroemer, 1974). When sitting, a stable chair backrest or footrest (Caldwell, 1964) is effective for pushing, and a harness is effective for pulling (Imrhan and Ayoub, 1985). Pheasant et al. (1982) also found that a low ceiling improved push–press strength because people were able to brace themselves on it, but it weakened lift strength because it constituted an obstacle.

Gender Differences

A female's absolute push–pull strengths are significantly weaker than a man's, and the female/male strength ratio varies considerably depending on posture and type of strength. The ratios fall mostly in the range 0.5 to 0.9 (Fothergill et al., 1991; Kumar et al., 1995), which is similar to the ratios found in other MMH activities. The higher ratios (less inequality) occur when the legs are more influential in the exertion, such as in pulling at low to medium height (Fothergill et al., 1991) and when strengths are weaker in general, such as exertions in planes away from the sagittal plane (Kumar et al., 1995).

Back Muscle Forces and L5/S1 Disc Compression Forces

MMH tasks, including pulling and pushing, that produce compression at the L5/S1 spinal disc in excess of 3400 N are considered hazardous (NIOSH, 1981). The results of certain studies (Lee et al., 1989; Chaffin et al., 1983; Resnick and Chaffin, 1995) show that maximal dynamic pushing and pulling create hazardous L5/S1 compression, especially when the exertions are performed at low heights (hip to knee), and that pulling creates greater compression than pushing. Disc compression also increases as push–pull forces or speed of walking increases (Lee et al., 1989).

Body-Support Traction

The shoe–floor or seat–buttocks coefficient of friction (COF) when standing or sitting is an important determinant of MVC or maximal acceptable force. (Fox, 1967; Kroemer and Robinson, 1971; Kroemer, 1974). Kroemer (1974) demonstrated that push force can increase by as much as 50% when the COF of shoe–floor contact increases from 0.3 (poor traction) to 0.6 (good traction). One can compensate for low traction by using foot and back supports.

Coefficient of friction can influence posture and self-selected speed of movement dramatically and is also influenced by the height of the push–pull exertions (Resnick and Chaffin, 1995; Lee et al., 1991). Differences in the effects of posture on strength are more pronounced when the coefficient of friction is low. Designers can avoid this situation by ensuring high traction on walking and sitting surfaces where pushing and pulling are common.

Distance of Movement of Body

The maximal acceptable dynamic push–pull loads (limits) that can be attained depends on the distance the load is moved. Clearly, the greater the distance the lighter the load. Maximal acceptable dynamic push–pull force limits over various distances have been established by Snook and Ciriello (1991).

Speed of Push/Pull

Resnick and Chaffin (1995) found that speeds of push selected voluntarily by subjects for short walking distances (1.5 m or 5 ft.) were much slower (0.2 to 1.1 m/s) than those proposed by the methods and time measurements (MTM) system for longer distances (1.8 m/s). These authors concluded that slower speeds than the MTM ones should be used for pushing heavy loads, especially over short distances.

Maximum walking speed for pulling and pushing loads depends on a number of factors including the shoe–floor COF, person's strength, load, type of handle, and height of handle. Stronger two-handed pushes while walking are executed at greater speeds than weaker ones, and heavier loads are pushed at slower speeds than lighter loads, regardless of people's strengths. Resnick and Chaffin (1995) found the speeds for stronger pushes on a light load (45 kg) to average 0.75 m/s for a voluntarily selected "hard push" and 0.4 m/s for "easy push." With a 450-kg cart load, subjects pushed at 0.44 m/s and 0.3 m/s, respectively. These authors also found that push speed increased as handle height decreased from shoulder to elbow to knee. Peak dynamic pull strength is achieved at a later position in the pull range as velocity of pull increases (Imrhan and Ayoub, 1985 and 1990).

Anthropometry

Body weight is the most influential anthropometric variable on push–pull strength, especially when it is used to enhance force directly, as when leaning forward in pushing and leaning backward in pulling (Kroemer, 1969; Ayoub and MacDaniel, 1974). Imrhan and Sundararajan (1992) also found body weight to be the most highly correlated anthropometric variable, even for strengths (finger pull) that are not affected by posture. As for almost all types of working strengths, no single anthropometric variable or combination of variables is suitable for predicting push or pull force accurately (Imrhan, 1983; Kumar et al., 1995). However, anthropometric variables of the upper body, especially the arm, can be used in combination with task variables, such as speed, arm position, etc., to significantly improve the predictive power of models of push or pull strength (Imrhan, 1983 and 1988).

Environmental Stress

One study (Snook and Cirello, 1974b) has found that workload for push decreased by 16% as environmental temperature changed from WBGT temperature of 17.2°C (63.0°F) to 27.0°C (80.6°F). Until more specific data becomes available, one may apply these findings to the design of jobs and administrative controls for protecting workers subjected to heat stress.

Endurance

The body posture and other conditions associated with enhanced maximal strength seem to be the same conditions that are associated with enhanced endurance. Caldwell (1964) showed that the sitting body posture that generated the greatest pull strength was the one in which endurance at submaximal strength (at 80% MVC) was also greatest. Presumably, in this posture, the body's levers are at the best overall mechanical advantage and require the least effort for eliciting and maintaining a specified level of strength. However, there is no other push–pull study that substantiates this relationship.

Frequency

Maximal push or pull forces are expected to decrease as frequency of task performance increases. Snook and Ciriello (1991) provide data from psychophysical push–pull experiments over several frequencies (from once per 6 seconds to once per 8 hours), showing that maximal acceptable forces decrease with increasing frequency. As for other MMH activities, this kind of decrease is nonlinear.

23.3 Push–Pull Magnitudes and Safe Force Limits

Safe push–pull force exertion limits may be interpreted as the maximum force magnitudes that people can exert without injuries (for single strong muscular exertions) or cumulative trauma disorders (for repeated exertions) of the upper extremities under specified conditions. Safe limits for static exertions should not be the same as for dynamic exertions.

Static Standing Forces

Many factors influence the magnitude of a static MVC force (single exertion) and, therefore, it would be unwise to recommend a single value for either push or pull force limits for task or workplace design. In addition, even for a given set of conditions, such as handle height, arm and leg posture, etc., MVC values differ considerably across published studies. Average static two-handed MVC push forces have ranged from about 400 to 620 N in males and from about 180 to 335 N in females when there is no bracing of the body, and pull forces from about 310 to 370 N in males and 180 to 270 N in females. Bracing, as mentioned before, can enhance force by as much as 50%.

Dynamic Standing Forces

Dynamic two-handed push–pull forces are not as strong as static ones. Dynamic push forces (mostly in moving industrial carts) have ranged from 170 to 430 N in males and 200 to 290 in females, and push forces from 225 to 500 N in males and 160 to 180 N in females. Initial forces (required to set a stationary object in motion) are generally lower than sustained forces (required to keep an object moving) in pushing and pulling tasks. Snook and Ciriello (1991) observed that maximal acceptable initial pulling force was 13% lower than pushing force, and maximal acceptable sustained force, 20% lower.

The most useful guidelines on dynamic push–pull force limits have been published by Snook and Ciriello (1991). These authors have proposed maximal acceptable force limits (MAFs) for males and females (four tables) in comfortable work conditions by combining two sets of data: the first from several early studies (Snook, Irvine, and Bass, 1970; Snook and Ciriello, 1974a, b; Snook, 1978; Ciriello and Snook, 1978), and the second from four subsequent studies found in Ciriello and Snook (1983), Ciriello, Snook, Blick, and Wilkinson (1990), and Ciriello, Snook, and Hughes (1991). Partial reproductions of the final four tables for pushing and pulling in males and females are given in Tables 22.1 through 22.4 in this chapter. Given the large number of ways in which a person can move an object from one location to another, and that many are not embodied in these tables, it is necessary that the tables be interpreted carefully to apply their values to task and workplace design.

The tables are the psychophysically determined maximal forces that people are willing to accept if they were to perform the push or pull activities as a normal eight-hour job. The forces are stated as a function of other work-related independent variables for both males and females. These are:

1. Distance of push/pull: 2.1, 7.6, 15.2, 30.5, 45.7, and 61.0 m (or 7, 25, 50, 75 and 100 ft.)
2. Frequency of push/pull: The same frequencies are not given for all distances, but each distance has force limits for one exertion per 8 hr., 30 min., 5 min., and 2 min. Greater frequencies are quoted for the smaller distances.

TABLE 23.1 Maximum Acceptable Forces of Pull for Females (kg)

Height	Percent	2.1 m pull							45.7 m pull				
		6	12	1	2	5	30	8	1	2	5	30	8
		s		min				h	min				h
		Initial forces											
	90	13	16	17	18	20	21	22	12	13	14	15	17
	75	16	19	20	21	24	25	26	14	16	17	18	20
135	50	19	22	24	25	28	29	31	17	18	20	21	24
	25	21	25	28	29	32	33	35	19	21	23	24	27
	10	24	28	31	32	36	37	39	22	24	25	27	31
	90	15	17	19	20	22	23	24	13	14	15	17	19
	75	17	20	22	23	26	27	28	16	17	18	20	22
57	50	20	24	26	27	30	32	33	18	20	22	23	26
	25	23	27	30	31	35	36	38	21	23	25	27	30
	10	26	31	34	35	39	40	43	24	26	28	30	34
		Sustained forces											
	90	6	9	10	10	11	12	15	6	6	7	7	9
	75	8	12	13	14	15	16	20	8	9	9	9	12
135	50	10	16	17	18	19	21	25	10	11	11	12	16
	25	13	19	21	21	23	25	31	12	13	14	14	19
	10	15	22	24	25	27	29	36	14	15	16	17	23
	90	5	8	9	9	10	11	13	5	6	6	6	8
	75	7	11	12	12	13	14	18	7	8	8	8	11
57	50	9	14	15	16	17	18	23	9	10	10	11	14
	25	11	17	18	19	21	22	27	11	12	12	13	19
	10	13	20	21	22	24	26	32	12	14	14	15	20

Height = vertical distance from floor to hand–object (handle) contact.

Percent = percentage of industrial workers capable of exerting the stated forces in work situations.

From Snook, S.H. and Ciriello, V.M. 1991. The design of manual handling tasks: revised tables of maximum acceptable weights and forces. *Ergonomics,* 34(9): 1197–1213. With permission.

3. Height (vertical distance from floor to hands) at which push/pull was exerted. Different heights are given for males and females — 144, 95, and 64 cm (or 57, 37.4, 25.2 in.) for males; and 135, 89, and 57 (53.1, 35.0, and 22.4 in.) for females.
4. The percentage of workers (10, 25, 50, 75, and 90%) who are capable of sustaining the particular force during a typical eight-hour job.

The data are also given for both initial force (force required to get object in motion) and sustained force (force required to keep object in motion). Each number in a table, therefore, corresponds to *the maximum initial or sustained push or pull force which a given percent percentage of the population can exert for a specified distance, at a specified frequency, and at a specified height without a significant chance of being injured or developing cumulative disorders.* Profound judgment must be used when applying the results of these tables in the workplace because the tables do not represent all possible push–pull conditions, and because the tabled conditions are represented by discrete states. It may be necessary to interpolate force limits for variable values not stated in the tables but which are within the stated ranges, for example, pulling or pushing over a distance of 9.1 m (30 ft.); but, this must be done with caution since the relationships are not linear.

Note that body posture is not a variable in these tables, even though it is one of the most influential variables in force exertion. This is because, as stated earlier, posture changes continuously during dynamic activities. The best description of posture relevant to these tables and the appropriate assumptions for work conditions are:

TABLE 23.2 Maximum Acceptable Forces of Push for Females (kg)

Height	Percent	2.1 m push							45.7 m push				
		6 s	12 s	1 min	2 min	5 min	30 min	8 h	1 min	2 min	5 min	30 min	8 h
		Initial forces											
	90	14	15	17	18	20	21	22	12	13	14	15	17
	75	17	18	21	22	24	25	27	15	16	17	19	21
135	50	20	22	25	26	29	30	32	18	19	21	22	25
	25	24	25	29	30	33	35	37	20	22	24	26	29
	10	26	28	33	34	38	39	41	23	25	27	29	33
	90	11	12	14	14	16	17	18	11	12	12	13	15
	75	14	15	17	17	19	20	21	13	14	15	16	18
57	50	16	17	20	21	23	24	25	15	17	18	19	22
	25	79	20	23	24	27	28	30	18	19	21	22	25
	10	21	23	26	27	30	31	33	20	22	23	25	28
		Sustained forces											
	90	6	8	10	10	11	12	14	5	5	5	6	8
	75	9	12	14	14	16	17	21	7	8	8	8	11
135	50	12	16	19	20	21	23	28	9	10	11	11	15
	25	16	20	24	25	27	29	36	11	13	13	14	19
	10	18	23	28	29	32	34	42	14	15	16	17	22
	90	5	6	8	8	9	9	12	5	5	5	6	7
	75	7	9	11	12	13	14	17	7	7	8	8	11
57	50	10	13	15	16	17	18	23	9	10	10	11	15
	25	12	16	19	20	22	23	29	11	13	13	14	19
	10	15	19	23	23	26	28	34	13	15	16	16	22

Height = vertical distance from floor to hand–object (handle) contact.

Percent = percentage of industrial workers capable of exerting the stated forces in work situations.

From Snook, S.H. and Ciriello, V.M. 1991. The design of manual handling tasks: revised tables of maximum acceptable weights and forces. *Ergonomics*, 34(9): 1197–1213. With permission.

1. The force limits apply to subjects walking in unrestricted postures and exerting two-handed symmetrical forces on handles (with comfortable grips) with the hands in approximately the same horizontal plane.
2. The walking surface is flat and level, and provides enough traction to prevent slipping of the feet. In the experiments in which the tabled data were developed, subjects wore shoes that provided high friction with their contact surface (a treadmill walkway).
3. The forces in the tables are the horizontal components of the applied muscular forces. In the experiments, subjects directed their exertions in the horizontal.
4. Environmental conditions are considered comfortable. The experiments were conducted in a climate controlled chamber at 21°C (80.6°F) and 45% humidity.
5. The people pushing and pulling are physically fit and accustomed to manual labor.
6. There is no obstruction to body movement due to clothing or obstacles along the path of movement.

If we wished to evaluate an existing MMH system, we can compare the load being pushed or pulled with the value in the table corresponding most closely to the actual task conditions. Tasks with high significant risks for injuries can then be identified. If we wished to design a task to prevent injuries we can use the tabled values as force limits for the tasks. Where there are differences in task conditions from the experimental conditions on which the tables are based, as listed above, we should use our judgment and make appropriate adjustments, if necessary, to the table values.

TABLE 23.3 Maximum Acceptable Forces of Pull for Males (kg)

Height	Percent	2.1 m pull							45.7 m pull				
		6	12	1	2	5	30	8	1	2	5	30	8
		s		min				h	min				h
		Initial forces											
	90	14	16	18	18	19	19	23	10	11	13	13	16
	75	17	19	22	22	23	24	28	12	14	16	16	20
144	50	20	23	26	26	28	28	33	15	16	19	19	24
	25	24	27	31	31	32	33	39	17	19	22	22	28
	10	26	30	34	34	36	37	44	20	22	25	25	31
	90	22	25	28	28	30	30	36	16	18	21	21	26
	75	27	30	34	34	37	37	44	19	22	25	25	31
64	50	32	36	41	41	44	44	53	23	26	30	30	37
	25	37	42	48	48	51	51	61	27	30	35	35	43
	10	42	48	54	54	57	58	69	30	34	39	39	49
		Sustained forces											
	90	8	10	12	13	15	15	18	6	7	8	9	10
	75	10	13	16	17	19	20	23	7	9	10	11	14
144	50	13	16	20	21	23	24	28	9	11	12	14	17
	25	15	20	24	25	28	29	34	11	13	15	17	20
	10	17	22	27	28	32	33	39	12	14	17	19	23
	90	11	14	17	18	20	21	25	8	9	11	12	15
	75	14	19	23	23	26	27	32	10	12	14	16	19
64	50	17	23	28	29	32	34	40	13	15	17	20	23
	25	20	27	33	35	39	40	48	15	18	21	24	28
	10	23	31	38	40	45	46	54	17	20	24	27	32

Height = vertical distance from floor to hand–object (handle) contact.

Percent = percentage of industrial workers capable of exerting the stated forces in work situations.

From Snook, S.H. and Ciriello, V.M. 1991. The design of manual handling tasks: revised tables of maximum acceptable weights and forces. *Ergonomics,* 34(9): 1197–1213. With permission.

One-Handed Force Magnitudes

All sitting push–pull forces described in this chapter are one-handed ones. Daams (1993) provides data for standing one-handed exertions. As for two-handed standing forces, one-handed forces vary considerably among studies with similar variables, and within individual studies depending on the test conditions or variables. An examination of published studies of one-handed MVC forces achievable by people range widely among studies and even among test conditions within individual studies. Thus, generalizations on recommended forces are not easy to promote. Two are mentioned below. It would be more appropriate to state the ranges of these forces. Average static standing push–pull forces have ranged from 70 to 134 N and sitting forces from 350 to 540 N. Dynamic pull forces have ranged from 170 to 380 N in females and from 335 to 673 N in males when sitting, for almost all studies. Average pull forces in males, while lying down prone, have ranged from 270 to 383 N and push forces, 285 to 330 N (Hunsicker and Greey, 1957). Davis and Stubbs (1980) and Mital, Nicholson, and Ayoub (1993) have published guidelines for one-handed push–pull forces, but they are not specific enough and must be interpreted with caution. They do not cover the wide range of conditions (posture, reach, static or dynamic contraction, handles, etc.) that can occur in a typical push-pull task. David and Stubbs (1980) recommended one-handed occasional standing static push forces for three age groups (under 40 yrs., 41 to 50 yrs., and 51 to 60 yrs.) for different reach distances in the range 5 to 70 cm. The values from their graphs range from 30 kg at 5 cm to 15 kg at 70 cm for under 40-yrs. males, with the values decreasing by 1 to 2 kg

TABLE 23.4 Maximum Acceptable Forces of Push for Males (kg)

Height	Percent	2.1 m pull: 6 s	12 s	1 min	2 min	5 min	30 min	8 h	45.7 m pull: 1 min	2 min	5 min	30 min	8 h
							Initial forces						
	90	20	22	25	25	26	26	31	13	14	16	16	20
	75	26	29	32	32	34	34	41	16	18	21	21	26
144	50	32	36	40	40	42	42	51	20	23	26	26	33
	25	38	43	47	47	50	51	61	24	27	32	32	39
	10	44	49	55	55	58	58	70	28	31	36	36	45
	90	19	22	24	24	25	26	31	12	14	16	16	20
	75	25	28	31	31	33	33	40	16	18	21	21	26
64	50	31	35	39	39	41	41	50	20	22	26	26	32
	25	38	42	46	46	49	50	59	24	27	31	31	39
	10	43	48	53	53	57	57	68	27	31	36	36	44
							Sustained forces						
	90	10	13	15	16	18	18	22	7	8	10	11	13
	75	13	17	21	22	24	25	30	10	11	13	15	18
144	50	17	225	27	28	31	32	38	12	14	17	19	23
	25	21	27	33	34	38	40	47	15	18	21	24	28
	10	25	31	38	40	45	46	54	18	21	24	28	33
	90	10	13	16	16	18	19	23	7	8	9	11	13
	75	14	18	21	22	25	26	31	9	11	12	14	17
64	50	18	23	28	29	32	33	39	12	14	16	18	22
	25	22	28	34	35	39	41	48	14	17	20	23	27
	10	26	32	39	41	46	48	56	17	20	23	26	31

Height = vertical distance from floor to hand–object (handle) contact.
Percent = percentage of industrial workers capable of exerting the stated forces in work situations.
From Snook, S.H. and Ciriello, V.M. 1991. The design of manual handling tasks: revised tables of maximum acceptable weights and forces. *Ergonomics*, 34(9): 1197–1213. With permission.

for 41 to 50-yrs. males and by 4 to 8 kg for 51 to 60-yrs. males. They also recommended a 30% decrease in these values for pushing–pulling at frequencies greater than once per minute. Mital, Nicholson, and Ayoub (1993) recommend the following guidelines for a typical workday for standing work: A push force of 107 N (24 lbs.) for an exertion of less than once per minute, and 73 N (16.5 lbs.) for greater frequencies; a pull force of 98 N (22 lbs.) for less than once per minute and 67 N (15 lbs.) for greater frequencies.

Pinch–Pull Force Magnitudes

Pinching and pulling with one hand while stabilizing the object with the other hand has been observed in male adults to yield forces of 100, 68, and 50 N when using the lateral, chuck, and pulp pinches, respectively (Imrhan and Sundararajan, 1992; Imrhan and Alhaery, 1994).

23.4 Conclusions

Push–pull strengths depend on numerous task-related factors. The strength variations have been examined from controlled laboratory experiments and have yielded profound insights into the characteristics of these strengths. However, much is still not known about push–pull strengths or acceptable loads (forces). The use of these strengths for establishing force limits for the design of tasks, equipment, and workplace still depend strongly on data gathered from simulation of typical occupational activities. The best compromise, at present, is to use the results of simulated experiments, as published by Snook and

Ciriello (1991), and modify the data according to the information available on the nature of push–pull strengths.

Defining Terms

Body support: An object which a person can brace on to enhance muscular force exertion.

Coefficient of (static) friction (μ): The ratio of the maximum force (F) acting along the area of contact between an object in contact with another body to the weight (N) of the object (or the force pressing the two bodies together); that is, $\mu = F/N$. Its value depends on the types of materials in contact with each other. Coefficient of friction for wet (slippery) surfaces may be below 0.2 and for dry surfaces with very good traction, above 0.8.

Horizontal foot distance: The perpendicular distance between the ankle of the pivoting foot (for separated feet) and hand–handle contact, when standing and pushing or pulling.

Manual materials handling: Moving or stabilizing loads with the hand(s) mostly by pulling, pushing, lifting, lowering, or carrying.

Maximum acceptable force: The maximal force (in pushing, pulling, or any other activity) a person is willing to exert voluntarily under work conditions during a workday with the opinion that the force will not cause undue discomfort or strain. It is usually determined experimentally by simulating work conditions in a laboratory.

Maximal voluntary contraction (MVC): A muscular contraction in which a person applies the strongest effort (for lifting, pushing, pulling, etc.) to the point where he or she does not suffer from significant muscular discomfort or pain. The resulting force or torque, measured with an appropriate instrument, is called the person's MVC strength for that particular task.

Push: A muscular effort applied to an object such that the object moves or tends to move away from the body. Neither the direction of movement nor the applied force need be linear or directly toward the body, as long as the deviation is not sharp.

Pull: A muscular effort applied to an object such that the object moves or tends to move closer to the body. The same conditions for direction as in "push" apply.

Push–pull strength (force): The maximal force (peak or 3-second mean) exerted on an object at the hand–object interface in a single MVC push or pull exertion, under specified conditions.

Reach distance: The perpendicular distance from the shoulder to the hand at the hand–object contact.

Safe force limits: Safe push–pull force exertion limits are the maximum force magnitudes that people can exert without injuries (for single strong muscular exertions) or cumulative trauma disorders (for repeated exertions) of the upper extremities under specified conditions.

Wet bulb globe temperature: A temperature value (in degrees Celsius or Fahrenheit) that represents the combined effect of air temperature, natural wet bulb temperature (reflecting humidity), and radiant temperature (solar load). It is used for evaluating environmental heat stress conditions.

References

Ayoub, M.M. and McDaniel, J.W. 1974. Effect of operator stance and pushing and pulling tasks. *Transactions of American Institute of Industrial Engineers,* 6, 185-195.

Ayoub, M.M. and LoPreisti, P. 1971. The determination of an optimum size handle by use of EMG. *Ergonomics,* 14: 509-518.

Caldwell, L.S. 1964. Body position and strength and endurance of manual pull. *Human Factors,* 6: 479-484.

Baril-Gingras, G. and Lortie, M. 1990. Les modes operatoires et leur determinants: Etudes des activites de manutention dans une grand entreprise de transport, in *Proceedings of the 23rd Annual Conference of HFAC,* Ottawa, pp. 137-142.

Chaffin, D.B., Andres, R.O., and Garg, A. 1983. Volitional postures during maximal push/pull exertions in the sagittal plane. *Human Factors,* 25: 541-550.

Ciriello, V.M. and Snook, S.H. 1978. The effects of size distance, height, and frequency on manual handling performance, in *Proceedings of the Human Facttors Society 22nd Annual Meeting,* Detroit, MI, pp. 318-322.

Ciriello, V.M. and Snook, S.H. 1983. A study of size, distance, height, and frequency effects on manual handling tasks. *Human Factors,* 25: 473-483.

Ciriello, V.M., Snook, S.H., Blick, A.C., and Wilkinson, P.L. 1990. The effect of task duration on psychophysically determined maximum acceptable weights and forces. *Ergonomics,* 33: 187-200.

Daams, B.J. 1993. Static force exertion in postures with different degrees of freedom. *Ergonomics,* 36: 397-406.

Davis, P.R. and Stubbs, D.A. 1980. *Force Limits in Manual Work.* Guilford: IPC Science and Technology Press.

Fothergill, D.M., Grieve, D.W., and Pheasant, S.T. 1991. Human strength capabilities during one handed maximum voluntary exertions in the fore and aft plane. *Ergonomics,* 35: 203-212.

Gallagher, S. 1989. Isometric pushing, pulling and lifting strengths in three postures, in *Proceeding of the HFS 33rd Annual Meeting,* Santa Monica, CA, pp. 637-640.

Garg, A. and Beller, D. 1990. One handed dynamic pulling strength with special reference to speed, handle height and angles of pulling. *Intl. J. Ind. Ergon.,* 6: 231-240.

Hunsicker, P.A. and Greey, G. 1957. Studies in human strength. *Research Quarterly,* 28:109

Imrhan, S.N. 1983. *Modeling Isokinetic Strength of the Upper Extremity.* Ph.D. Dissertation, Texas Tech University Lubbock, TX.

Imrhan, S.N. and Ayoub, M.M., 1985, An analysis of rotary and pull strength of the upper extremity, in *Trends in Ergonomics/Human Factors II,* ed. R.E. Eberts and C.G. Eberts, Elsevier Science Publishers, B.V., North Holland.

Imrhan, S.N. and Ayoub, M.M. 1988. Predictive models of upper extremity rotary and linear pull strength. *Human Factors,* 30(1): 83-94.

Imrhan, S.N. and Ayoub, M.M. 1990. The arm configuration at the point of peak dynamic pull strength. *Intl. J. Ind. Ergon.,* 6: 9-15.

Imrhan, S.N. and Alhaery, M. 1994. Finger pinch-pull strengths: large sample statistics, in *Advances in Industrial Ergonomics and Safety VI,* ed. F. Aghazadeh. Taylor & Francis, London, pp. 595-597.

Imrhan, S.N. and Ramakrishnan, U. 1992. The effects of arm elevation, direction of pull and speed of pull on isokinetic pull strength. *Intl. J. Ind. Ergon.,* 9: 265-273.

Imrhan, S.N. and Sundararajan, K. 1992. An investigation of finger pull strengths. *Ergonomics,* 35(3): 289-299.

Klein, B.P., Jensen, R.C., and Sanderson, L.M. 1984. Assessment of workers' compensation claims for back strains/sprains. *J. Occup. Med.,* 26(6): 443-448.

Kroemer, K.H.E. 1969. *Push Forces Exerted in Sixty-Five Common Working Positions. ARML*-TR, WPAFB, Ohio, 68-143.

Kroemer, K.H.E. 1974. Horizontal push and pull forces. *Applied Ergonomics,* 5(2): 94-102.

Kroemer, K.H.E. and Robinson, D.E. 1971. *Horizontal Static Forces Exerted by Men Standing in Common Working Postures on Surfaces of Various Tractions.* Aerospace AMRL-TR, WPAFB, Ohio, 70-114.

Kumar, S. and Garand, D. 1992. Static and dynamic strength at different reach distances in symmetrical and asymmetrical planes. *Ergonomics,* 35: 861-880.

Kumar, S. 1994. The back compressive forces during maximal push–pull activities in the sagittal plane. *J. Human Ergol.,* 23: 133-150.

Kumar, S., Narayan, Y., and Bacchus, C. 1995. Symmetric and assymetric two-handed pull-push strength of young adults. *Human Factors,* 37(4): 854-865.

Lee, K.S., Chaffin, D.B., Herrin, G.D. and Waikar, A.M. 1991. Effect on handle height on lower back loading in cart pushing and pulling. *Applied Ergonomics,* 22(2):117-123.

Lee, K.S., Chaffin, D.B., Waikar, A.M., and Chung, M.K. 1989. Lower back muscle forces in pushing and pulling. *Ergonomics,* 32: 1551-1563.

Lower, R.S., Schutz, R.K., and Sadosky, T.S. 1977. A prediction model of arm push strength in the transverse plane, in *Proceedings of the Human Factors Society 21st Annual Meeting,* Human Factors Society, Santa Monica, CA, pp. 132-136.

Martin, J.B. and Chaffin, D.B. 1972. Biomechanical computerized simulation of human strength in sagittal-plane activities. *AIIE Trans.* 4(1): 19-28.

Mital, A., Nicholson, A.S., and Ayoub, M.M. 1993. *A Guide to Manual Materials Handling.* Taylor & Francis, Washington, D.C.

Mital, A. and Faard, H.F. 1990. Effects of sitting and standing, reach distance, and arm orientation on isokinetic pull strengths in the horizontal plane. *Intl. J. Ind. Ergon.* 6(3): 241-248.

NIOSH 1981. *Work Practices Guide for Manual Lifting.* U.S., D.H.S.S., Pub. No. 81-122.

Pheasant, S.T. and Grieve, D.W. 1981. The principal features of maximal exertions in the sagittal plane. *Ergonomics,* 24(5): 327-338.

Pheasant, S.T., Grieve, D.W., Rubin, T. and Thompson, S.J. 1982. Vector representations of human strength in whole body exertion. *Applied Ergonomics,* 13(2), 139-144.

Resnick, M.L. and Chaffin, D.B. 1995. An ergonomic evaluation of handle height and load in maximal and submaximal cart pushing. *Applied Ergonomics,* 26(3): 173-178.

Snook, S.H. and Ciriello, V.M. 1974a. Maximum weights and workloads acceptable to female workers. *J. Occ. Med.,* 16: 527-534.

Snook, S.H. and Ciriello, V.M. 1974b. The effects on heat stress on manual handling tasks. *Am. Ind. Hyg. Assn. J.,* 31: 681-685.

Snook, S.H., Irvine, C.H., and Bass, S.F. 1970. Maximum weights and workloads acceptable to male industrial workers. *Am. Ind. Hyg. Assn. J.,* 31: 579-586.

Snook, S.H. 1978. The design of manual tasks. *Ergonomics,* 21: 963-985.

Snook, S.H. and Ciriello, V.M. 1991. The design of manual handling tasks: revised tables of maximum acceptable weights and forces. *Ergonomics,* 34(9): 1197-1213.

Troup, J.D.G. and Edwards, F.C. 1985. *Manual Handling and Lifting. An Information and Literature Review with Special Reference to the Back,* Her Majesty's Stationery Office, London.

Van Cott, H.P. and Kinkade, R.G. (eds.) 1972. *Human Engineering Guide to Equipment Design,* McGraw-Hill Book Company, New York.

Warwick, D., Novack, G., Schultz, A., and Berkeson, M. 1980. Maximal voluntary strengths of male adults in some lifting pushing and pulling activities. *Ergonomics,* 23, 49-54.

For Further Information

The design of manual handling tasks: revised tables of maximum acceptable weights and forces. by S.H. Snook and V.M.Ciriello, 1991, *Ergonomics,* 34(9) 1197-1213. This paper gives detailed tables of maximum acceptable force limits (for eight hours of work) for many manual materials handling tasks, including pushing and pulling.

Human Engineering Guide to Equipment Design edited by H.P Van Cott and R.G. Kinkade, R.G., 1972, Washington, D.C., U.S. Govt. Printing Office. This book gives several tables of data on push–pull static strengths in standing, sitting, and lying down postures.

Manual Materials Handling by M.M. Ayoub and A. Mital, 1989, Taylor & Francis, NY. This book gives tables of some recent data on push–pull activities.

24

Force Exertion in User–Product Interaction

Brechtje J. Daams
Delft University of Technology

24.1 Introduction

How to crack a nut? How to design a nutcracker with which to crack a nut? What force needs to be applied to a nutcracker designed to crack that nut? What force can be applied by the intended user of that nutcracker designed to crack that nut?

Questions like these arise in the process of product design. The forces users can and will exert in the use of a product are important criteria for the design of that product. It is usual to distinguish consumer products from professional products. This is a useful distinction in so far as it relates to the differences between the composition of a group of professional users and a group of normal consumers, and to the differences in the way both groups use the products. Consumer products are usually bought by their prospective users. These users are usually nonspecialists with diverse backgrounds, different levels of education, and of varying strength. Consumers include children, the elderly, the physically disabled, and the world's largest minority group, women.

Professional products are, as a rule, used frequently, possibly up to 8 hours a day, five days a week, all year round. A group of professional users is more homogeneous than a group of consumers. Users of professional products are aged between about 18 and 65 and are generally healthy. Even so, their strength may vary substantially. Both types of products and users are considered in this chapter.

In general the design limits for a product are dictated by the comfort, efficiency, and safety of its operation. Products should be designed so that even the weakest of the intended users are able to operate, use, or handle them. Maximal forces may be required of users for a short time in cases like emergencies, but in general a comfortable level of force is preferred.

Products should also be designed to withstand the largest forces strong users may exert. Accidental damage or breakage must be avoided, as this may lead to anything from minor annoyance to serious injury.

0-8493-2641-9/99/$0.00+$.50

Consumer products that cause discomfort will be difficult to sell. Professional products that cause health damages, either in the short term or in the long term, will be banned from the workplace. Both from an ethical and an economical point of view it is therefore important for designers to adapt products to the force capabilities of their prospective users.

The use product designers may make of the results from research on force exertion is the central issue of this chapter. First, a guideline is given on how to consider force exertion in the design process. A short survey of results pertaining to relevant aspects of force exertion is summarized in a list of rules of thumb. Finally, recommendations are given on how to determine the relevance of results of literature and research to a particular design.

24.2 Considering Force Exertion in Product Design

The Design Process

Design is an iterative process. This implies that there is no clear-cut path from a problem posed to a product designed to solve that problem. Nevertheless, specific activities can be discerned in the design process and some order in these activities is advisable (Roozenburg and Eekels, 1991). The starting point of a design is the definition of the problem to be solved, an assessment of its setting, and the collection and analysis of information. On the basis of this, a program of requirements for the future product is formulated. Ideas are generated, sketches are made, good ones are worked out, and of these the best one is selected through comparison with the requirements. Details are then dealt with and technical drawings and/or a model of the design is made. If possible, a test version of the product is evaluated through user trials. The results of these trials may be incorporated in an improved version of the design.

In the next section an attempt is made to outline how considerations of the force capabilities of users may be incorporated in the design process as it has been described above.

Considering Force Exertion in the Design Process

Problem definition. A product fulfills a need and solves a problem. Nuts need to be cracked because people wish to eat nuts. Grass should not be left to grow too long. These are both problems for which a number of solutions already exist. A good and accurate definition of the problem is essential to an appropriate solution. The problem definition should refer to the problem to which the product should be the solution. Defining the problem by stating a solution is not correct. It remains, however, common practice, in particular with principals. They may tell the designer, for example, that the problem is that she has to design a nutcracker, or that the problem is that no lawn mower suitable for elderly people exists. Such a problem definition is unnecessarily restrictive. It will result in yet another nutcracker and the umpteenth lawn mower. To define a problem as a solution (or a product) limits creative thought. Due consideration of users' force capabilities at the problem definition stage may lead to new, unexpected, and unconventional solutions that would be better in the given situation. These solutions do not necessarily involve designed products. Thus the sale of peeled and cracked nuts is an alternative to the sale of nutcrackers. For the elderly, perhaps a walking frame with lawn-mowing option would be a good solution to the problem of long grass. More generally, keeping a sheep or goat, or even changing the lawn into maintenance-free paving, may be a good solution. Although these are extreme examples, they illustrate the point.

Assessment of the setting. The function(s) of the product, the target user group, the circumstances of use, and the relevant behavior of the users should be established or estimated. For example, it makes a great difference whether the future users will be children or adults, male or female, and whether they will wear gloves or not. Will they be able to brace themselves to exert force? On a product that is hand-held and can be manipulated in the best position (like a jam jar), the maximal force that can be exerted is different from that on a fixed object where obstructions can hinder the adoption of the optimal posture. If the product is to be used frequently, this aspect should certainly be taken into account when establishing the maximal force. A product used by a person in a state of panic may demand a different force (and less precision!) of the user than one used for leisure purposes.

For example, after assessment it may be clear that the target group consists of the general public aged ten and upward (all minorities who are able to walk included); that the function of the product is to mow the lawn; that the situation of use is outdoors on a lawn, often in the sun and sometimes in wind and rain; that the users may be wearing a minimum of clothing and footwear, that the product will be used for a period of a few minutes to an hour once every week; that the exerted force will be dynamic; that there is no possibility for the users to brace themselves while exerting force; and that, if the force that has to be exerted is too great, people will buy a motorized lawn mower right away or next time.

Choosing design principles. The proposed function of the product must be translated into an action which can be performed optimally by the user. In this view, the product is the concrete intermediate between the two. At this stage, a designer should not be concerned with concrete products, but rather with selecting the best working principle for force exertion. Knowledge of the principles of physics and human force exertion and essential logical thinking are indispensable. The question a designer should ask is not "How much force can be exerted this way?" but "Is this the optimal way to exert force, or is there a better one?" A thorough evaluation of principles will make the selection of relevant information on force exertion in the subsequent stages of the design process more effective. It will help to avoid the risk of being bogged down by an excess of information, resulting in the dreaded "designer's block."

Assessment of information needed. Once the design principles have been established, it can be inferred which information on force exertion is needed. The users, their postures, the direction of the exerted force, and bracing possibilities are known within certain limits. The force capabilities are distributed over a wide range. The average or the median of this distribution is generally not a very useful measure for product design. Products designed for the average force cannot be operated by the weaker end of the population distribution, and may be damaged or broken by the strongest users. For design purposes, the weakest users are often more relevant than the strongest, and beginners more so than experienced users. The same applies to the postures and the directions of the exerted forces. It depends on the requirements of the product whether the lower maximally exertable forces, or the higher maximally exertable forces, or maybe both, should indicate the design limits. In addition to deciding whether the weakest or the strongest users are the most relevant for specific design purposes, the designer ought to decide and clearly define which percentile of the chosen population he or she is designing for: P_5, P_1, or P_{99}? Especially with extreme percentiles this can make an enormous difference for the values involved, and consequently for the design.

Here a comment should be made on the widespread habit of quoting the P_5 as a "normal" design maximum. To the annoyance of millions of people, many designers think it is accepted, standard, or even good practice, to exclude the upper or lower 5% (or both) of a population. This "P_5-P_{95} syndrome" results in products that cannot be used effectively, or comfortably, by 10% of the users.

It is important to note that products that are designed for the lower percentiles of the population (where forces are concerned, these are the weaker persons) can be easier to use or operate by the average user. Those products may sell even better to "normal" users than standard products, because they too appreciate clear, simple features and light operation. Unless, of course, these products get stigmatized as being specially made for the disabled and the elderly. Neither strong nor weak users want to be seen with a product that visibly classifies them as weak. If this negative image can be avoided, products designed for the weak can be a (commercial) success with everyone.

Taking again the example of the lawn mower, one can now look up which force can be exerted, for instance, comfortably for a few minutes up to an hour by 99% of the general public aged 10 years and older (all minorities that are able to walk included), pushing horizontally with two hands on a handle with a diameter of, for example, 3 cm, at elbow height while walking (dynamic force), without bracing themselves. External factors which should be taken into account include clothing, footwear, and the weather.

Gathering of information. The most relevant and exact information is obtained by custom research using a representative sample of the target user group in the required postures. This will best serve to reduce design uncertainty.

One should never take one's own strength as a measure "to get an indication." The large dispersion of data on force exertion (Sanchez and Grieve, 1992; Sanders and McCormick, 1993) goes unnoticed when measurements are limited to one or a few persons. We are none of us the average person, but still there are many designers who presume that if they can exert a certain force, anybody else must be able to do so too.

If custom research, even on a small scale, is impossible, literature is the alternative, though at best it will give only an indication of the range of forces involved. Caution is required with the application of data from literature. Generally the subjects and the experimental setting will not be similar, or even relevant, to those of the product. Consequently, such figures will be difficult to use. They should only be applied with the necessary comments and safety margins.

Definition of requirements. When it is known within which limits the product is to function, the program of requirements can now be extended with requirements concerning force exertion. As with all requirements, those on force exertion should preferably be operational. Clearly defined limits for compliance (preferably in hard numbers) must be included in the requirements. Only thus can a design later be tested for compliance with these requirements.

Design. In the design stage, the requirements are translated into a design for a product, in which they are assimilated as much as possible. Conflicting requirements sometimes lead to the inevitable compromise, a process inherent to design. Obviously, some requirements, especially those pertaining to safety, should not be compromised. It is the art of the designer to come up with a good product, despite any conflicts that may have arisen.

Follow-up. The first prototypes should be tested with subjects and evaluated, and the product should, if possible, be improved accordingly. This is important, especially if the information forming the basis for the program of requirements is obtained from literature.

Examples

The outline of the way considerations of force exertion are incorporated in the design process is best illuminated with some examples from practical experience. They illustrate the point that questions on force exertion in user–product interaction can never receive a standard, ready-to-use answer.

A first example of a product to be (re)designed is a large professional cheese slicer, as used in supermarkets (see Figure 24.1). Present cheese slicers are fitted with a handgrip at the end of a blade that rotates around a pivot. Enlargement of the blade (or the arm of the moment) reduces the force needed to slice, but increases the movement of arm and hand. The handgrip is positioned in the same direction as the blade, so that the wrist is in an uncomfortable position when exerting force. The slicer is usually positioned on a table or bench, so that the force is exerted on the handle from about shoulder height to about elbow height. The users are professional women and men between 18 and 65. Instead of trying to find out how much force can be exerted in such a situation, thought is given to a more comfortable way to slice. Suggested improvements include lower positioning of the equipment, so that the force is exerted with the hand at elbow height and lower, and body weight can be utilized. A further suggestion is to change the handgrip so that the wrist need not be flexed to extreme angles. If possible, the movement should be a translation instead of a rotation, so that force needs to be exerted in one direction only. The optimal length of the arm, weighing the length of the stroke against the force needed to operate the cheese slicer, may be determined experimentally.

A second product is a wheelchair for children in Sri Lanka who suffer from the consequences of polio. Their legs are paralyzed, so the vehicle has to be moved by using the arms. For some mechanical reason, force is to be exerted in one direction only: either pushing or pulling. Deciding on pulling horizontally, based solely on the fact that in this direction the greatest force can be exerted, would be wrong. When pulling in a horizontal direction, the child will tend to pull himself/herself out of the seat, which is uncomfortable and prevents maximal force exertion, even if the child were to be strapped to the back of the seat. When pushing, however, the child will be able to brace himself/herself against the back of the seat, which is more comfortable, and allows more force to be exerted. Therefore, pushing against the

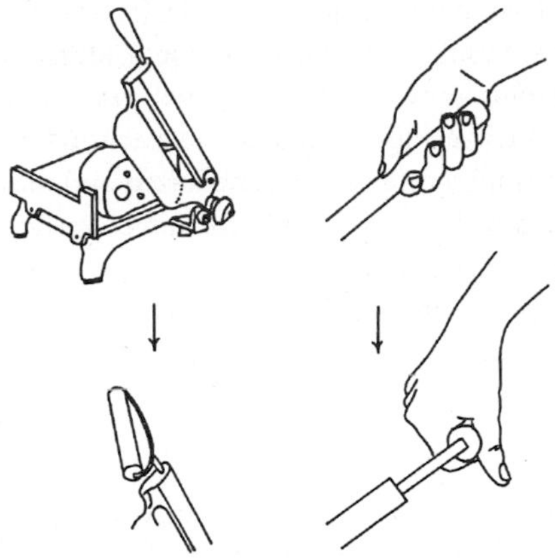

FIGURE 24.1 Suggestion for improvement of a professional cheese slicer. (From Daams, B.J. 1994. *Human Force Exertion in User–Product Interaction*. Delft University Press, Delft. With permission.)

back of the seat is preferred. It should be noted, however, that to exert force in both directions in a cyclic movement is better for the development of the muscles. Uneven development of muscle groups may lead to incorrect loading of the joints and related problems in the future. The use of two arms instead of one will, of course, allow more force to be exerted, and will also stimulate more symmetrical physical development in the child. To know the maximal forces, these should be measured for the actual children concerned. Although there are some data on maximal pushing and pulling abilities of European children in literature (e.g., Steenbekkers, 1993), this information cannot be used, because it cannot be assumed that children who are disabled and from a different ethnic group exert maximal forces equal to those exerted by able-bodied European children.

For the design of a portable or rolling easel, information on maximal exertable push, pull, and carry forces were asked for. The easel is intended to be used for outdoor painting and should be easily transportable by a person on foot. The target group consists of elderly people, so the required force forms an important aspect of the design. Carrying is no option, because it will certainly require more force and energy than simply rolling the easel along. Rather than pushing it, a wheeled object is preferably pulled along, because this makes it easier to negotiate ramps and curbs, as a straightforward analysis of the physics involved shows. To decrease rolling resistance, few and large wheels are advised. The force necessary to stabilize and maneuver the easel should be as small as possible, because energy and attention should not be diverted from the main activity, pulling the easel along. Therefore, two wheels are preferred over one.

The original question thus reduces to how much force elderly people can exert when pulling something along. If the lowest part of the distribution (say the P_1) of elderly women is included in the target group, the force that can be exerted is practically zero, for these people have barely sufficient strength to walk about unsupported, and will not have much force left to pull easels around. These considerations lead to the conclusion that exact information on the strength of the users is no longer relevant. The easel should be designed to be pulled with the minimal possible force to allow the largest possible proportion of the target user group to use it.

24.3 Summary of Literature

Literature is the alternative if custom research tailored to a specific product design, even on a small scale, is impossible. Finding the relevant information can be difficult though, because literature on force exertion is sparse and far between, many variables play a role, and research is generally aimed at working conditions, rehabilitation, and sports rather than the use of products.

Results quoted in literature have been obtained mainly for standardized and static postures, and concern maximal forces, which are usually measured for about four seconds. In practice, however, only the position of a handle or control is known and subjects will exert static or dynamic submaximal force for any time and in any posture they feel like, and to some extent with different muscle groups.

Nevertheless, some results from literature can be generalized and may serve as rules of thumb. These rules can be useful to find a good technical principle for force exertion in the early stages of a design process. Their validity should be checked at a later stage of the design process when product ideas are tested in more detail. The summary of these rules of thumb we give here has been arranged according to the variables person, posture, product, environment, and force.

Person

Gender. The physical strength of women is roughly half to two thirds that of men. The ratio of mean female to mean male maximal force varies for various adult subject groups between 35% (Sing and Karpovich, 1968) and 88% (Fothergill et al., 1992); for elderly people over 75 years of age between 19% (Page, 1981) and 68% (Mathiowetz et al., 1985); and for children between 74% and 103% (Mathiowetz et al., 1986 and Steenbekkers, 1993). Although males and females differ significantly in strength, the effect of gender is small after allowing for body size and composition.

Age. Maximal strength is attained between 20 and 30 years, it is relatively stable from 20 up to 59 years and starts to decline between approximately 35 to 60 years, but there is no close agreement on the exact ages in literature. The large dispersion in the quoted age limits reflects the disagreement on the exact ages found in literature. There is also little agreement on how rapidly strength declines with age. This may be related to the increase of inter-individual variance with age.

Anthropometric variables. The correlation between maximal force exertion and body height and weight is in general significant but not very high, except for children.

Laterality. The difference in strength between the preferred and the nonpreferred side of the body ranges between "not significant" for finger pull strength of right-handed subjects (Imrhan and Sundararajan, 1992) and the nonpreferred side exerting 87% of the hand grip strength of the preferred side (Mathiowetz et al., 1985).

Motivation. With auditory encouragement and visual feedback, maximal force exertion increases about 10% (Peacock et al., 1981).

Gloves. When gloves are worn by users, the maximally exertable force may decrease. Gloves do not enhance grip strength and torque, and in general reduce both up to 79% (Swain et al., 1970; Chen et al., 1989; Vincent and Tipton, 1988; McMullin and Hallbeck, 1991; Wang, 1991).

Product

Torque. Maximal torque, but not maximal force, increases with increasing diameter of knobs and lids. The optimal diameter is determined by the hand size of the subjects and roughness of the surface. Imrhan and Loo (1988) found an optimal diameter of 83 mm for a screw top container with a smooth surface. More torque can be exerted on a control, handle, or lid which is not round but offers an opportunity for a good grip (like a paddle, Bordett et al., 1988). The longer the lever, the larger the resulting moment.

Pull, lift, and carry. A good handle for lifting and carrying by adults should be at least 115 mm long, about 25 to 50 mm in diameter, with a hand clearance of 30 to 50 mm (Drury, 1980 and Hsia and Drury, 1986). For various force directions, various cross-sectional shapes of handle are recommended (Cochran and Riley, 1986). Sharp corners, edges, ridges, finger grooves, and curvatures should be avoided (Drury, 1980). However, if a handle is curved, it should be convex, following the natural curve of the gripping hand. Soft, smooth, nonslip surfaces are preferred. Cold and hard surfaces and vibration are to be avoided. The handle should be oriented such that it can be used without undue deviation of the wrist. Depending on the task, handle height can influence force exertion.

Grip force. Grip force appears to be maximal with a handle separation of about 55 to 60 mm. This separation is related to hand size (Fransson and Winkel, 1991; Radwin and Oh, 1991).

Environment

Support. More push and pull force is exerted maximally when the subjects can brace themselves, for example against a backrest, armrest, or footrest (Rohmert et al., 1987; Caldwell, 1962). For exerting large foot forces, a lumbar support is recommended. For subjects who are standing and reaching far forward, a support at the level of the pelvis can reduce the moment at the lower back with 30%, which results in a more comfortable posture (Frankel et al., 1984).

Temperature. The optimum ambient air temperature for sustained contractions of grip force appears to be 18°C. At other temperatures, endurance is shorter (Clarke et al., 1954).

Acceleration. Accelerations up to 5 *g* affect endurance but do not affect strength. Arm movements are effective up to 6 *g*, wrist and finger movements up to 12 *g* (Morgan et al., 1963; Woodson, 1981).

Posture

Free posture. Force can be exerted optimally in a freely adopted posture (Daams, 1993). Even small constraints in posture are found to have a considerable effect on the measured strength (Haslegrave, 1992).

Statics. Nature may assist in finding the most efficient way to exert force through the laws of statics. Moment is the product of lever and force. Adapting the size of the lever will change the resulting moment or change the required force.

Exerting force with use of body weight can make a task more pleasant.

The maximal force in a certain posture allowed by the laws of statics can be calculated from the height of the handle, the position of the center of mass, the pivot around which rotation would start, and sometimes the coefficient of friction with the environment. If this calculated force exceeds the force that physiologically can be exerted by the muscles of the subject, his or her maximal force is not limited by the posture. If, however, the calculated maximal force is less than the physiologically exertable force, it is clear that the posture will limit the force that the subject can maximally exert.

One/two handed. Two-handed strength commonly exceeds one-handed strength but is found to be less than twice the value. The ratio of one- versus two-handed force exertion in the sagittal plane (pushing, pulling, lifting, pressing, and all their combinations), ranges from 0.61 to 1.04, as measured by Fothergill et al. (1991) and Sanchez and Grieve (1992).

Joint angles. Extreme joint angles should be avoided during force exertion, especially if repeated often or maintained for sustained periods of time. Maximal force can be limited at extreme joint angles. Grip force and pulp, chuck, lateral, and three-jaw chuck pinch force are maximal with the wrist in neutral position (Imrhan, 1991; Hallbeck and McMullin, 1991; McMullin and Hallbeck, 1991; Fernandez et al., 1991).

Maximum leg force occurs when the knee is slightly bent.

Task distance. Hand grip and torque forces generally are greater if the task is close to the individual's body rather than at arm's length (Woodson, 1981). When pushing with one hand and lifting with one or both hands, the closer to the body, the more force can be exerted (The Materials Handling Research Unit, 1980). Static and dynamic lifting strengths are inversely related to reach distance (Kumar, 1990). For pulling, it seems to be the reverse: one-armed isokinetic pulling strength increases with reach distance (Mital and Faard, 1990). For comfortable force exertion, the arms should stay below shoulder height (Wiker et al., 1990).

Finger position. Maximal push and slide forces of the thumb are larger than those of the other fingers, and all types of force exerted with the pad of the fingertip are larger than those exerted with the tip of the finger. Maximal forces exerted with the flat of the hand exceed all finger forces (Bandera et al., 1985).

Asymmetric postures. Maximal lifting strength decreases with increasing asymmetry of posture (Kumar, 1990).

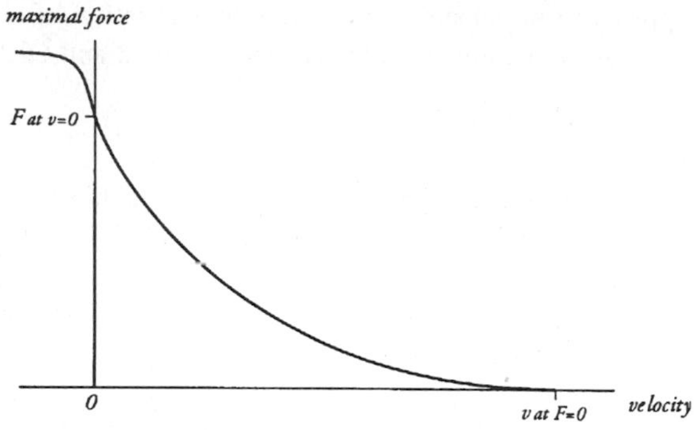

FIGURE 24.2 Force-velocity characteristics of skeletal muscle, showing a decrease of maximal tension (maximal force) as the muscle shortens and an increase as it lengthens (Adapted from Hof, A.L. 1987. Spiermechanica, in *Biomechanica.* Edited by R. Huiskes. Samson Stafleu, Alphen aan de Rijn. [in Dutch])

Endurance. The weight of limbs should be taken into account when considering the exertion of smaller forces over extended periods of time; hence sometimes the support of limbs is favored.

Force

Rest. A long and heavy task can be relieved by including periods of rest between start and finish. Frequent and short rest periods are much more effective than a few of long duration, the total time of rest being equal.

Precision. Maintained isometric muscle work reduces the precision of manual performance (Hammarskjöld et al., 1989).

Dynamic force. Maximal dynamic force decreases with increasing speed. Figure 24.2 shows the theoretical relation between the exerted force and velocity of contraction for a separate muscle (Hof, 1987). Maximal static force and maximal eccentric force in general exceed maximal concentric force. The smaller the range of motion, the smaller the maximal exerted force (Ayoub et al., 1981 and 1982; Hafez et al., 1982). With cycling, peak force is exerted at 90° past the top dead center in each revolution (Hoes et al., 1968; Sargeant et al., 1981). Optimal power output is not attained at maximum velocity nor at zero velocity (static force), but somewhere in between. With cycling, maximal power is generated at a velocity of 110 rpm (Sargeant et al., 1981).

Endurance time. Maximal endurance time increases consistently with lower force levels (see Figure 24.3). There is no agreement on the influence of age, sex, muscle group, and subject on maximal endurance time (Elbel, 1949; Rohmert, 1960 and 1965; Caldwell, 1963 and 1964; Byrd and Jeness, 1982; Sato et al., 1984; Bishu et al., 1990; Deeb et al., 1992; Deeb and Drury, 1992; Daams, 1994). When significant differences are found, they are small.

Although long endurance of low force levels is possible, it is not desirable nor advisable. The maximal endurance time at very low force levels should not be considered infinite. Force is easier to maintain if the relative force level is low, if the subject is not subject to boredom, and may change his posture from time to time. These last two conditions apply even more to the lower force levels. Consequently, dynamic and/or cyclic force exertion should be considered preferable. This corresponds with old physiological insights and with the findings of Sjøgaard (1986) and Ulmer et al. (1989).

(Dis)comfort. Discomfort increases with increasing force, work/rest ratio, and task duration (Wiker, 1991).

For dynamic one-handed pulling, it is suggested that exertion of maximal force at high speed is more comfortable than at low speed (Garg and Beller, 1990).

The ratio of "comfortable" to maximal endurance time of force exertion may range from 0.50 at 80% and 0.19 at 15% of maximal force (see Figure 24.3 and next paragraph).

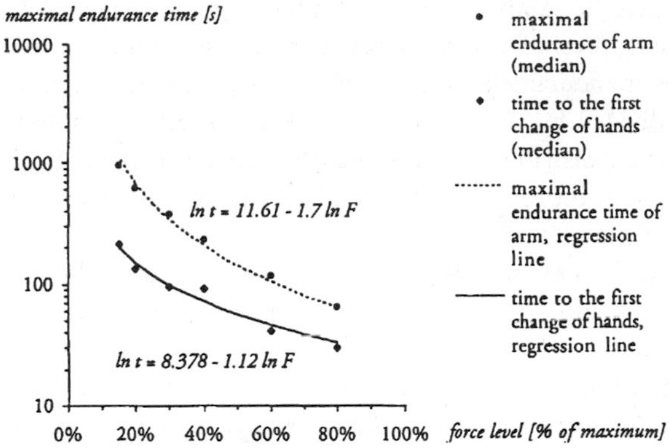

FIGURE 24.3 Medians and interpolations of maximal endurance time and time to the first change of hand, for arms only. (From Daams, B.J. 1994. *Human Force Exertion in User–Product Interaction.* Delft University Press, Delft. With permission.)

24.4 Recommendations for Research and the Use of Literature

Using Literature

General guidelines as outlined in the previous paragraph are useful tools in the early stages of the design process. They help to define the principal directions of design solutions. As the designs are worked out in more detail in subsequent stages of the design process, the need for more exact information arises. This information is best obtained from custom research tailored to the design at hand. Failing that, a designer will have to resort to literature. In this paragraph requirements are formulated that will guide the design of custom-tailored research and will help to determine the relevance, representativeness, and validity of data obtained from literature.

Physical strength and force exertion are important research topics in industrial ergonomics, military ergonomics, rehabilitation, and sports. Results of this research are reported in a vast and growing body of literature. Research in the different areas of ergonomics is diverse with respect to purpose, method, and subject group. The literature in the list of references with this chapter reflects this diversity. Relatively little of this research is directed toward force exertion in user–product interaction, but what there is tends to be too specific, aimed at one particular product or a select group of products. Designers would profit from research according to a more general product-oriented principle with a large variety of subjects. Such an approach ideally would lead to an atlas of human force exertion as outlined by Daams (1994).

For force exertion data from literature to be useful in product design, two essential requirements must be fulfilled at least. First, the experiment must be specified in all relevant detail. Journals and handbooks can make an important contribution here by requiring, and allowing for, extensive descriptions of the experiments. A second important requirement is that the experiment is similar to the real-life situation considered. In other words, the outcome of the experiment must be relevant to and valid for this intended use of the data. This is often neglected by both designers and researchers. For example, standardized postures inflicted by researchers on their subjects tend to be artificial and reduce the relevance of experiments for real-life situations (Daams, 1993).

Regarding the subjects, it is important to know their number, sex, age, and other characteristics that may be of importance, such as level of training or type of employment. For consumer products, it is imperative to include children, elderly, and disabled people in the sample. In any case, both females and males should be investigated.

A minimum number of subjects is needed to ensure a reproducible and sometimes representative result. To establish whether forces exerted in different situations are significantly different, a dozen or more subjects are needed. To establish a force exertion which is representative for a whole population, the main requirement is that the sample should match the population regarding the distribution of sex, age, build, profession, health, level of training, level of ability, etc. In total, several hundred subjects may be necessary. In both cases, the actual minimum sample size depends on the size of the variation coefficient (the standard deviation divided by the average force) which affects the power of the test, and on the accuracy with which the actual distribution of forces in the population needs to be known. The calculation of the sample size depends on the sort of test (paired or unpaired). For more information, we refer to standard textbooks on statistics.

It is important to know the posture, the direction of force exertion, the type of force, the exertion time, the laterality, the shape and size of the handle, the position of the handle, and other factors that possibly influence the result.

At present, most literature on force exertion concerns static, maximal force exertion in standardized postures. Characteristic of forces exerted on products, however, is that they are most often dynamic, they last for any time between a second and a few minutes, that usually they are not maximal, and that the users adopt a "natural" posture while exerting force.

Thus, there is a need for more information on dynamic, submaximal (comfortable) force, on endurance time and on force exerted in a "natural" (or "free") posture.

Conducting Research

Accounting for the factors "posture," "endurance time of sub-maximal force," and "(dis)comfort" in an appropriate manner will benefit research for product design and increase its relevance. In the following paragraphs, these aspects along with measuring methods are discussed.

Posture. In assessing human force exertion, the use of standardized postures can lead to inaccurate prediction of the forces and postures that occur in everyday life. A standardized posture is generally considered to yield more reliable data (implicit in Caldwell et al., 1974). Even if this were so, force data acquired from measurements in standardized postures may not have predictive validity.

For pushing, pulling, and exerting torque in various postures, research shows that the forces measured in free postures are highly reproducible, even with extreme handle heights and when the place to which the force is to be applied is not fixed (Daams, 1993). The exerted forces are equally reproducible in free and standardized postures. The difference in average force, though, is considerable and significant: much more force can be exerted in free posture. Furthermore, postures of subjects during force exertion in free posture show a remarkable intra-individual reproducibility. This measurement method thus not only yields information on force, but also on the standing room needed by the subjects, which is also relevant to design.

Research on force exertion in free posture, therefore, may be considered most suitable for product design research.

Endurance. Several investigations have been conducted into the endurance of submaximal force. There is, however, no standardized way to measure endurance. In general, endurance time is measured as a function of a percentage of maximal force. To enable the use of the results in product design and allow a comparison with literature or a reproduction of the experiment, several additional aspects of the measuring method should be known. In particular the limitation of the movements subjects are allowed to make, the limits between which the force has to be maintained, the definition of the point at which the measurement ends, and the way maximal force is determined should be registered. This last aspect not only influences maximal endurance time, but even affects the influence of other variables on the results (Daams, 1994).

In general, these aspects are not mentioned in present literature, which makes the results unsuitable for application to product design.

For a reproducible measurement of endurance time during force exertion, the following method was found to yield good results:

- First, maximal force is measured according to the method of Caldwell et al. (1974): 2 seconds buildup and 4 seconds maximal force. No feedback or encouragement is given.
- The subject is then asked to exert the required force (unaware of the level) and maintain it for as long as possible. The limb which exerts force is supported. During measurement of endurance time, again no encouragement is given and no conversation is allowed, as it appears that talking and the presence of other people both influence the endurance time positively.
- The range within which the force is allowed to vary during a measurement should be fairly wide to prevent unintentional, premature ending of the measurement, e.g., by muscle tremors, sneezing, flagging attention, and other minor actions by the subject. Such an unintentional end of a measurement proves to be very frustrating to subjects. X-t recordings of measurements show that, with a narrow limit at which warning beeps sound (10% above and below the required force level) and a wide limit at which the measurement ends (50% above and below the required force level), forces are maintained sufficiently close to the required level.

There is no need to limit the movement of subjects, because other than slight movements cause the exerted force to exceed one of the warning or even end limits. This motivates the subject to return to the previous posture, or it ends the measurement altogether.

Reproducible results obtained with this method are shown in Figure 24.3.

Comfort. It is hard to define what constitutes a comfortable force. In ergonomics, comfort is generally defined as the "absence of discomfort." In literature, various methods are used to measure (un)comfortable force exertion, resulting in different values (Arnold, 1991; Berns, 1981; Bordett et al., 1988; Kanis, 1993; Schoorlemmer and Kanis, 1992; Garg and Beller, 1990; Wiker, 1991; Schutz, 1972; Sato et al., 1984). Apparently, there is not one single discomfort level, and which discomfort level is measured depends on the measuring method. In all experiments, however, there is one common factor: the subjects are asked to think about and indicate their feelings of (dis)comfort. This will inevitably make them concentrate on registering and comparing any feelings of discomfort, thereby disturbing the measurement.

Product designers are interested in discomfort as the "absence of comfort." They wish to prevent a situation in which a user operating a product starts to note discomfort and gets annoyed. The following measurement is intended to catch the moment at which the endurance of force exertion is no longer comfortable, without involving subjects consciously in the judging process. The underlying assumption is that the moment at which the user starts to note discomfort is indicated by a spontaneous change of posture. Here this method is illustrated for horizontal push, but it is assumed to be applicable to other postures and forces, keeping in mind both the influence of posture on the results, especially at lower force levels, and the reproducibility of the results.

Subjects are asked to exert a certain percentage of maximal push force with one hand (see Figure 24.4), for a time which is three times the expected maximal endurance time at that force level, according to the formula by Rohmert (1960 and 1965). They are told they can change hands whenever they feel like it. The time to the first change of hands is taken as an indication of the moment at which discomfort is noted. Figure 24.3 shows the results of an actual experiment. The ratio to the maximal endurance time ranges from 0.50 at 80% to 0.19 at 15% of maximal force. This kind of "comfortable endurance time" has proven to be reproducible. It may serve product designers as a useful and valid measure for the onset of discomfort. Apart from probably being more relevant to product design than maximal endurance time it is also a good deal more comfortable to measure both for subject and experimenter.

24.5 Conclusion

Comfort, efficiency and safety of operation are important properties of a product. To incorporate these properties into products, designers have to be aware of the forces the intended users can and will exert on these products. This awareness should permeate the whole design process, from the conceptual stage

FIGURE 24.4 Posture during research on endurance time and (dis)comfort: pushing with one hand. (From Daams, B.J. 1994. *Human Force Exertion in User–Product Interaction.* Delft University Press, Delft. With permission.)

where rules of thumb may assist in finding the right direction for a design, to the detailing of the final design, where exact information on the expected forces determines the design limits. This information can be derived from custom-tailored research or from literature. The information from both these sources should not be used blindly, but it should be judged time and again on its relevance to and validity for a particular design. Such an intelligent and critical attitude is the key to successful integration of force exertion data in product design.

Defining Terms

Comfort: Comfort is generally defined as "absence of discomfort."

Comfortable endurance (time): The time during which a force can be maintained without discomfort by a subject. There is no standard procedure to measure comfortable endurance. A usable and reproducible method is proposed in this chapter. In general, comfortable endurance is measured as a function of a percentage of maximal force.

Concentric force exertion: Dynamic force exertion during which the muscles shorten.

Dynamic force exertion: Force exertion during which the length of the muscles changes. Consequently, the segments of the body that are involved rotate relative to each other. There is no standard procedure to measure dynamic force. Dynamic force exertion is expressed in Newtons (N).

Eccentric force exertion: Dynamic force exertion during which the muscles lengthen.

Endurance (time): See *Comfortable endurance (time)* and *maximal endurance (time).*

Laterality: The difference between the preferred and the nonpreferred side of the body.

Maximal force: The maximal force which subjects can exert in an experiment. Static maximal force is generally measured according to the method of Caldwell et al. (1974). Static force exertion is built up for a second, without jerk, and the maximum is maintained for a few (3 to 5) seconds. No feedback or encouragement is given.

Maximal endurance (time): The maximal time during which a force can be maintained by a subject. There is no standard procedure to measure maximal endurance. A usable and reproducible method is proposed in this chapter. In general, maximal endurance is measured as a function of a percentage of maximal force.

P_5-P_{95} syndrome: The habit of quoting the P_5 and/or the P_{95} as a "normal" design maximum, thus excluding 5 or even 10% of the population from using a product effectively or comfortably.

Power: A measure of work. It is expressed in watts, as the product of exerted force and velocity (Nm/s).

Static force exertion: Force exertion during which the length of the muscles does not change. Consequently, the speed of the movement during force exertion is zero. Force exertion is expressed in Newtons (N). When the force exertion is maximal, some authors refer to this measured variable as "Maximal Voluntary Contraction" or MVC.

References

Arnold, A.-K. 1991. An ergonomic approach to the design of consumer packaging, in *Interface '91, Proceedings of the 7th symposium on Human Factors and Industrial Design in Consumer Products* (pp. 138-143). Edited by D. Boyer and J. Pollack. The Human Factors Society, Santa Monica, CA.

Ayoub, M.M., Gidcumb, C.F., Beshir, M.Y., Hafez, H.A., Aghazadeh, F., and Bethea, N.J. 1981. *Development of an Atlas of Strengths and Establishment of an Appropriate Model Structure.* Institute for Ergonomics Research, Texas Tech University, Lubbock, Texas.

Ayoub, M.M., Gidcumb, C.F., Reeder, M.J., Beshir, M.Y., Hafez, H.A., Aghazadeh, F., and Bethea, N.J. 1982. *Development of a Female Atlas of Strengths.* Institute for Ergonomics Research, Texas Tech University, Lubbock, Texas.

Bandera, J.E., Kern, P., and Solf, J.J. 1985. *Ergonomische Kenngrößen für Kontaktgreifarten.* Bundesanstalt für Arbeitsschutz, Dortmund. [in German]

Berns, T. 1981. The handling of consumer packaging. *Applied Ergonomics,* 12: 153-161.

Bishu, R.R., Myung, R.H., and Deeb, J.M., 1990. Evaluation of handle positions using force/endurance relationship of an isometric holding task, in *Proceedings of the Human Factors Society 34th Annual Meeting* (pp. 684-687). The Human Factors Society, Santa Monica, CA.

Bordett, H.M., Koppa, R.J., and Congelton, J.J. 1988. Torque required from elderly females to operate faucet handles of various shapes. *Human Factors,* 20 (3): 339-346.

Byrd, R. and Jeness, M.E., 1982. Effect of maximal grip strength and initial grip on contraction time and on areas under force-time curves during isometric contractions. *Ergonomics,* 25 (5): 387-392.

Caldwell, L.S. 1962. Body stabilisation and the strength of arm extension. *Human Factors,* 4: 125-130.

Caldwell, L.S., 1963. Relative muscle loading and endurance. *Journal of Engineering Psychology,* 2: 155-161.

Caldwell, L.S., 1964. The load-endurance relationship for a static manual response. *Human Factors,* 6: 479-484.

Caldwell, L.S., Chaffin, D.B., Dukes-Dobos, F.N., Kroemer, K.H.E., Laubach, L.L., Snook, S.H., and Wasserman, D.E. 1974. A proposed standard procedure for static muscle strength testing. *American Industrial Hygiene Association Journal,* 35: 201-206.

Chen, Y., Cochran, D.J., Bishu, R.R., and Riley, M.W. 1989. Glove size and material effects on task performance, in *Proceedings of the Human Factors Society 33rd Annual Meeting* (pp. 708-712). The Human Factors Society, Santa Monica, CA.

Clarke, R.S.J., Hellon, R.F., and Lind, A.R. 1954. The duration of sustained contractions of the human forearm at different muscle temperatures. *J. Physiol.,* 143: 454-473.

Cochran, D.J. and Riley, M.W. 1986. The effects of handle shape and size on exerted forces. *Human Factors,* 28 (3): 253-265.

Daams, B.J. 1993. Static force exertion in postures with different degrees of freedom. *Ergonomics,* 36 (4): 397-406.

Daams, B.J. 1994. *Human Force Exertion in User–Product Interaction.* Delft University Press, Delft.

Deeb, J.M. and Drury, C.G., 1992. Perceived exertion in isometric uscular contractions related to age, muscle, force level and duration in *Proceedings of the Human Factors Society 36th Annual Meeting* (pp. 712-716). The Human Factors Society, Santa Monica, CA.

Deeb, J.M., Drury, C.G., and Pendergast, D.R., 1992. An exponential model of isometric muscular fatigue as a function of age and muscle groups. *Ergonomics,* 35 (7/8): 899-918.

Drury, C.G. 1980. Handles for manual materials handling. *Applied Ergonomics,* 11 (1): 35-42.

Elbel, E.R., 1949. Relationship between leg strength, leg endurance and other body measurements. *Journal of Applied Physiology,* 2 (4): 197-207.

Fernandez, J.E., Dahalan, J.B., Halpern, C.A. and Viswanath, V. 1991. The effect of wrist posture on pinch strength, in *Proceedings of the Human Factors Society 35th Annual Meeting* (pp. 748-752). The Human Factors Society, Santa Monica, CA.

Fothergill, D.M., Grieve, D.W., and Pheasant, S.T. 1991. Human strength capabilities during one-handed maximum voluntary exertions in the fore and aft plane. *Ergonomics,* 34 (5): 563-573.

Fothergill, D.M., Grieve, D.W., and Pheasant, S.T. 1992. The influence of some handle designs and handle height on the strength of the horizontal pulling action. *Ergonomics,* 35 (2): 203-212.

Frankel, V.H., Nordin, M., and Snijders, C.J. 1984. *Biomechanica van het skeletsysteem.* De Tijdstroom, Lochem. [in Dutch]

Fransson, C. and Winkel, J. 1991. Hand strength: the influence of grip span and grip type. *Ergonomics,* 34 (7): 881-892.

Garg, A. and Beller, D. 1990. One-handed dynamic pulling strength with special reference to speed, handle heights and angles of pulling. *International Journal of Industrial Ergonomics,* 6: 231-240.

Hafez, H.A., Gidcumb, C.F., Reeder, M.J., Beshir, M.Y. and Ayoub, M.M. 1982. Development of a human atlas of strengths, in *Proceedings of the Human Factors Society 26th Annual Meeting* (pp. 575-579). The Human Factors Society, Santa Monica, CA.

Hallbeck, M.S. and McMullin, D.L. 1991. The effect of gloves, wrist position, and age on peak three-jaw chuck pinch force: a pilot study, in *Proceedings of the Human Factors Society 35th Annual Meeting* (pp. 753-757). The Human Factors Society, Santa Monica, CA.

Hammarskjöld, E., Ekholm, J., and Harms-Ringdahl, K. 1989. Reproducibility of work movements with carpenters' hand tools. *Ergonomics,* 32 (8): 1005-1018.

Haslegrave, C.M. 1992. Predicting postures adopted for force exertion: thesis summary. *Clin. Biomech.,* 7 (4): 249-250.

Hoes, M.J.A.J.M., Binkhorst, R.A., Smeekes-Kuyl, A.E.M.C., and Vissers, A.C.A. 1968. Measurement of forces exerted on pedal and crank during work on a bicycle ergometer at different loads. *Int. Z. angew. Physiol. einschl. Arbeitsphysiol.,* 26: 33-42.

Hof, A.L. 1987. Spiermechanica, in *Biomechanica.* Edited by R. Huiskes. Samson Stafleu, Alphen aan de Rijn. [in Dutch]

Hsia, P.T. and Drury, C.G. 1986. A simple method of evaluating handle design. *Applied Ergonomics,* 17 (3): 209-213.

Imrhan, S.N., 1991. The influence of wrist position on different types of pinch strength. *Applied Ergonomics,* 22 (6): 379-384.

Imrhan, S.N. and Loo, C.H. 1988. Modelling wrist-twisting strength of the elderly. *Ergonomics,* 31 (12): 1807-1819.

Imrhan, S.N. and Sundararajan, K. 1992. An investigation of finger pull strengths. *Ergonomics,* 35 (3): 289-299.

Kanis, H. 1993. Operation of controls on consumer products by physically impaired users. *Human Factors,* 35 (2): 305-328.

Kumar, S. 1990. Symmetric and asymmetric stoop-lifting strength, in *Proceedings of the Human Factors Society 34th Annual Meeting* (pp. 762-766). The Human Factors Society, Santa Monica, CA.

Materials Handling Research Unit, the, 1980. *Grenzen van de voor handenarbeid benodigde lichaamskracht.* IPC Science & Technology Press Ltd., Luxembourg. [in Dutch]

Mathiowetz, V., Kashman, N., Volland, G., Weber, K., Dowe, M., and Rogers, S. 1985. Grip and pinch strength: normative data for adults. *Arch. Phys. Med. Rehabil.,* 66: 68-74.

Mathiowetz, V., Wiemer, D.M., and Federman, S.M. 1986. Grip and pinch strength: norms for 6- to 19-year-olds. *The American Journal of Occupational Therapy,* 40 (10): 705-711.

McMullin, D.L. and Hallbeck, M.S. 1991. Maximal power grasp force as a function of wrist position, age and glove type: a pilot study, in *Proceedings of the Human Factors Society 35th Annual Meeting* (pp. 733-737). The Human Factors Society, Santa Monica, CA.

Mital, A. and Faard, H.F. 1990. Effects of sitting and standing, reach distance, and arm orientation on isokinetic pull strengths in the horizontal plane. *International Journal of Industrial Ergonomics,* 6: 241-248.

Morgan, C.T., Cook, J.S., Chapanis, A., and Lund, M.W. (eds.) 1963. *Human Engineering Guide to Equipment Design.* McGraw-Hill, New York.

Page, M. 1981. An ergonomics evaluation of a reclosable pharmaceutical container with special reference to the elderly. *Ergonomics,* 24 (11): 847-862.

Peacock, B., Westers, T., Walsh, S., and Nicholson, K. 1981. Feedback and maximum voluntary contraction, *Ergonomics,* 24 (3): 223-228.

Radwin, R.G. and Oh, S. 1991. Handle and trigger size effects on power tool operation, in *Proceedings of the Human Factors Society 35th Annual Meeting* (pp. 843-847). The Human Factors Society, Santa Monica, CA.

Rohmert, W. 1960. Ermittlung von Erholungspausen für statische Arbeit des Menschen. *Int. Z. angew. Physiol. enschl. Arbeitsphysiol.,* 18: 123-164. [in German]

Rohmert, W. 1965. Physiologische Grundlagen der Erholungszeitbestimmung. *Arbeit und Leistung,* 19 (1): 1-28. [in German]

Rohmert, W., Mainzer, J., and Kahabka, G. 1987. Analyse biomechanischer und physiologischer Engpässer beim Ausüben von Stellungskräften. *Zeitschrift für Arbeitswissenschaft,* 41 (2): 114-130. [in German]

Roozenburg, N.F.M. and Eekels, J. 1991. *Produktontwerpen, struktuur en methoden.* Lemma, Utrecht. [in Dutch]

Sanchez, D. and Grieve, D.W. 1992. The measurement and prediction of isometric lifting strength in symmetrical and asymmetrical postures. *Ergonomics,* 35 (1): 49-64.

Sanders, M.S. and McCormick, E.J. 1993. *Human factors in engineering and design,* 7th edition. McGraw-Hill, New York.

Sargeant, A.J., Hoinville, E., and Young, A. 1981. Maximum leg force and power output during short-term dynamic exercise. *Journal of Applied Physiology: Respirat. Environ. Exercise Physiol.,* 51 (5): 1175-1182.

Sato, H., Ohashi, J., Iwanaga, K., Yoshitake, R., and Shimada, K. 1984. Endurance time and fatigue in static contractions. *Journal of Human Ergology,* 13: 147-154.

Schoorlemmer, W. and Kanis, H. 1992. Operation of controls on everyday products, in *Proceedings of the Human Factors Society 36th Annual Meeting* (pp. 509-513). The Human Factors Society, Santa Monica, CA.

Schutz, R.K. 1972. *Cyclic Work–Rest Exercise's Effect on Continuous Hold Endurance Capacity.* Unpublished Ph.D.-dissertation. University of Michigan, Ann Arbor.

Singh, M. and Karpovich, P. 1968. Strength of forearm flexors and extensors in men and women. *Journal of Applied Physiology,* 25 (2), 177-180.

Sjøgaard, G. 1986. Intramuscular changes during long-term contraction, in *The Ergonomics of Working Postures.* (pp. 136-143). Edited by E.J. Corlett, I. Manenica and J. Wilson. Taylor & Francis, London.

Steenbekkers, L.P.A. 1993. *Child Development, Design Implications and Accident Prevention.* Physical Ergonomics Series. Delft University Press, Delft.

Swain, A.D., Shelton, G.C., and Rigby, L.V. 1970. Maximum torque for small knobs operated with and without gloves. *Ergonomics,* 13 (2): 201-208.

Ulmer, H.-V., Knieriemen, W., Warlo, T., and Zech, B. 1989. Interindividual variability of isometric endurance with regard to the endurance performance limit for static work. *Biomed. Biochem. Acta,* 48 (5/6), 504-508.

Vincent, M.J. and Tipton, M.J. 1988. The effect of hand protection and cold immersion on grip strength, in *Contemporary Ergonomics 1988, Proceedings of the Ergonomics Society Conference* (pp. 323-327). Edited by E.J. Lovesey. Taylor & Francis, London.

Wang, M.-J.J. 1991. The effect of six different kinds of gloves on grip strength, in *Towards Human Work: Solutions to Problems in Occupational Health and Human Work.* (pp. 164-169). Edited by M. Kumashiro and E.D. Megaw. Taylor & Francis, London.

Wiker, S.E. 1991. Fatigue, discomfort and changes in the psychometric function with repetitive pinch grasps, in *Designing for Everyone. Proceedings of the 11th Congress of the International Ergonomics Association* (pp. 368-370). Edited by Y. Queinnec and F. Daniellou. Taylor & Francis, London.

Wiker, S.E., Chaffin, D.B., and Langolf, G.D. 1990. Shoulder postural fatigue and discomfort. *International Journal of Industrial Ergonomics,* 5 (2):133-146.

Woodson, W.E. 1981. *Human Factors Design Handbook.* McGraw-Hill, New York.

25

Rapid Upper Limb Assessment (RULA)

E. N. Corlett
University of Nottingham

25.1 Introduction

During the last two decades there has been an increasing awareness of the problems arising in the upper limb, wrist, elbow, or shoulder, as a consequence of the continuous use of the muscles and joints during work. First signs were evident in the users of computers. Due to a poor understanding of the tasks involved and inadequate task analyses, these were ascribed to users being work shy, bored, or in one famous phrase in Britain, "egg shell personalities."

Detailed studies in, for example, Australia and the United States revealed that although psychosocial factors could contribute to the problems being experienced it was the mechanical component that predominated. This was compounded from the frequency of use of the joint involved, the posture adopted by the limb, and the forces exerted during the tasks. Where these factors were appropriately analyzed and changes introduced, the incidence of injuries could be significantly reduced or even eliminated. Changes typically included the reorganization of the task.

The urgent requirement for methods to investigate what had become a major problem across the world gave rise to many reports and papers. Health and safety bodies demanded measures and in the European Union the international requirement was that employers were required to assess the risks of the problems arising, and to make the necessary changes. These requirements were embodied in a "daughter" directive of the main European Union Health and Safety Directive No. 89/391/EEC, and it is a requirement of these directives that they are incorporated into the laws of the nation states that make up the European Union. In Britain the "daughter" directive concerning work with display screen equipment was published by the government as *"Work-Related Upper Limb Disorders — A Guide to Prevention,"* (*Health and Safety Executive,* 1990).

For organizations unused to risk assessment, the demand to undertake such a process was formidable, although national health and safety bodies in the various European countries issued guidance notes to assist with the activity. However, many of these just discussed some possible sources of upper limb problems but gave no methodology for assessment, use of the results, or indications of the most desirable directions for change. The RULA procedure was developed as a consequence of this lack.

0-8493-2641-9/99/$0.00+$.50

25.2 Requirements

What was needed was a procedure by which an individual, with relatively little training, could assess a workplace, recognize from the assessment the major points that implied a risk to the user, and know from the results of the analysis what actions to take to reduce the risk. The system developed was not a system for diagnosing the presence or otherwise of upper limb disorder in the worker, it was to assess whether the workplace could present a hazard to that worker whereby he or she may be at risk of suffering upper limb problems. Thus the technique was devised as a proactive risk assessment tool, rather than a response to already reported injuries.

The procedure uses workers themselves, from whom the measurements are taken. Since it is workers, not machines, who suffer the problems, it is necessary to do more than pursue a relatively impersonal observation. The experience of these workers is relevant to probe the problem. Hence, apart from the measurements arising from the use of the RULA technique, some additional information is needed if the risks are to be more reliably assessed.

The aim was to produce a system that used observation of the workers doing their jobs, and a minimum of other questions and measurements, to identify the most probable points presenting the risks. These procedures had to give their results in such a form that they would demonstrate to the analyst where changes had to be made, and in what directions. The same procedures had to be usable after the changes had been made, so as to provide measures of improvement in risk more rapidly than retrospective counts of injuries.

25.3 The RULA Procedure

The model on which the RULA procedure was based was the OWAS method, developed by the Finnish Institute of Occupational Health and the OVAKO Steel Company to investigate problems of lifting and of back injuries. This process required that segments of the body be judged on a simple scale, producing a sequence of numbers which were matched against a grid. The value of the numbers and their position on the grid told the analyst the severity of the posture, while taking actions that reduced the magnitude of the numbers improved the situation. RULA extends this assessment process to the control of upper limb disorders.

As indicated earlier, the factors contributing to upper limb disorders are more than posture. They include frequency of use, forces, and the availability of changes in the work cycle. Hence, each of these must be part of the assessment if a reasonable judgment of the likely risk is to be obtained. Furthermore, posture is not just that of the wrist and arm; the whole body posture can tell us something about the adequacy, or otherwise, of the task. (See Figure 25.1.)

To perform a RULA assessment, observations are made of the limb and body postures for those parts of the work cycle the analyst considers to give the most frequent use of the joint, or the most extreme joint angles. For each of the chosen parts of the work cycle, the position of the upper and lower arm and the wrist are assessed, using the diagrams of Group A, and entered on the score sheet in the boxes marked A. The positions of the neck, trunk, and legs are then assessed, using the information in the diagram marked Group B, and entered on the score sheet under B. (See Figure 25.2) Note that for posture score A, the left and the right arms are assessed separately, and both a score A and a Grand Score derived for each. Of course, both arms may not require assessing if, in the judgment of the assessor, one of them is not at risk.

From each of these sets of observations a score can be obtained, using Table A or Table B in Figure 25.3 for the observation sets A and B, respectively. These tables combine the effects of the postures to give an initial estimate of risk, but as yet we have not taken into account the possible contributions of force and frequency.

To include these we now turn to the boxes dealing with the Muscle Use score and the Forces or Load scores (Figure 25.4). The former allows us to modify the initial estimate, which is based on posture, to account for aspects of muscle use known to influence upper limb problems, while the latter takes into account the effects of loadings imposed on the limb, which also affect whether or not some injury will be suffered.

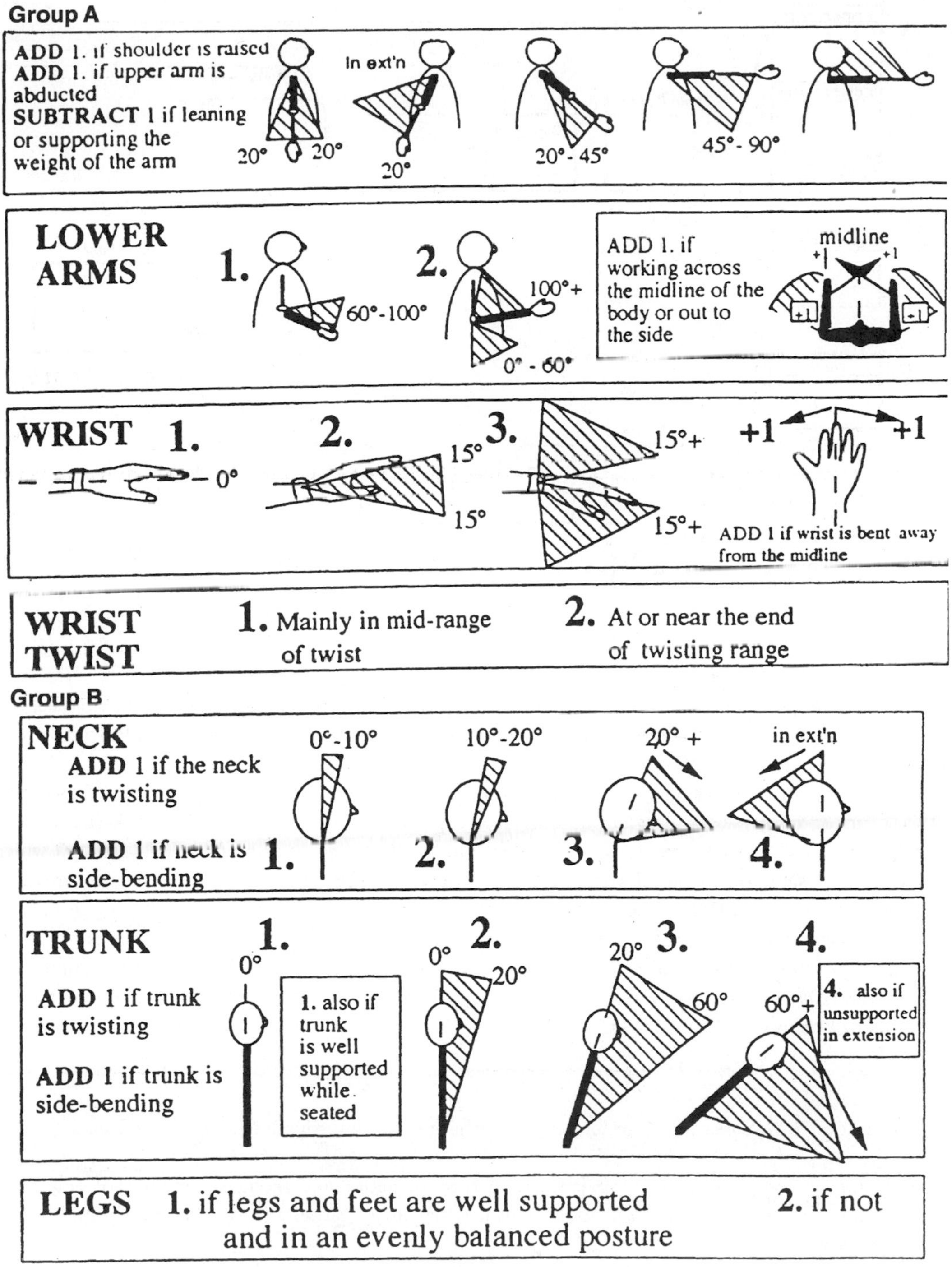
Group A
ADD 1. if shoulder is raised
ADD 1. if upper arm is abducted
SUBTRACT 1 if leaning or supporting the weight of the arm
20°
20°
In ext'n
20°
20°- 45°
45°- 90°
LOWER ARMS
1.
60°-100°
2.
100°+
0° - 60°
ADD 1. if working across the midline of the body or out to the side
midline
+1
+1
+1
+1
WRIST
1.
0°
2.
15°
15°
3.
15°+
15°+
+1
+1
ADD 1 if wrist is bent away from the midline
WRIST TWIST
1. Mainly in mid-range of twist
2. At or near the end of twisting range
Group B
NECK
ADD 1 if the neck is twisting
ADD 1 if neck is side-bending
0°-10°
1.
10°-20°
2.
20° +
3.
in ext'n
4.
TRUNK
ADD 1 if trunk is twisting
ADD 1 if trunk is side-bending
1.
0°
1. also if trunk is well supported while seated
2.
0°
20°
20°
3.
60°
4.
60°+
4. also if unsupported in extension
LEGS
1. if legs and feet are well supported and in an evenly balanced posture
2. if not

FIGURE 25.1

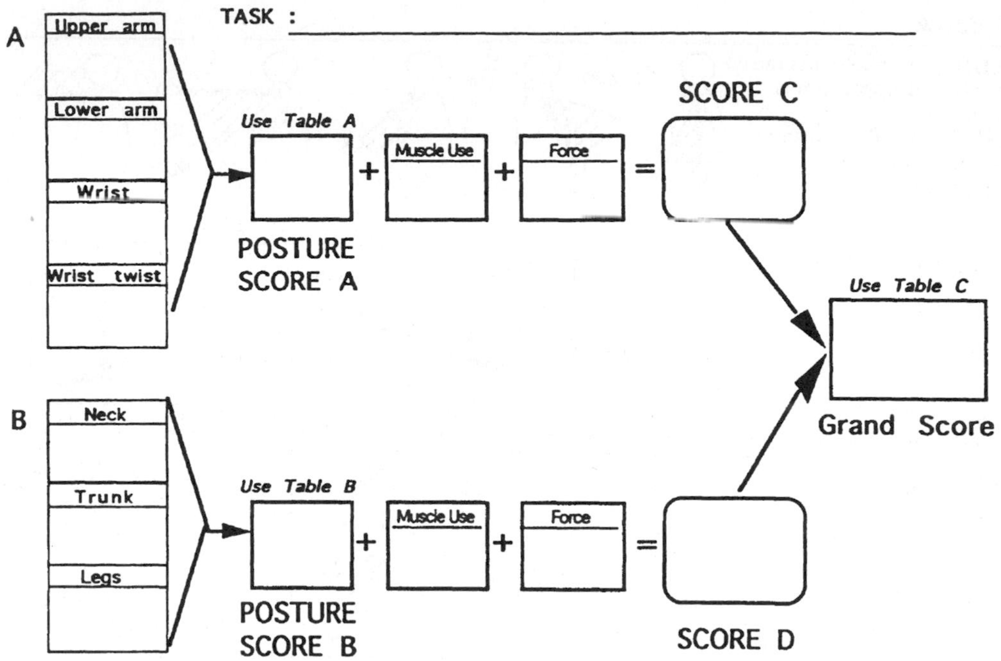

FIGURE 25.2

Having incorporated these two additional factors, we now sum the three values for each of sections A and B of the score sheet, arriving at score C and score D (Figure 25.5). It is with these two scores (C and D) that we now enter the Grand Score table, to find the final score (Figure 25.6).

It is this final score which gives us an estimate of the risk potential for the task. This Grand Score table was developed in a research program that involved experiments in which the various risk factors were manipulated, and an extensive study of the literature, resulting in a minimum set of numbers which would give a usable indication of severity without implying a greater precision than the system would achieve (McAtamney 1994).

As we move across the Grand Score table from the top left to the bottom right, the numbers get larger. This is emphasized by the shading, which darkens as we take this direction. The higher the number, the greater the risk of musculoskeletal symptoms arising. To put some numbers on this, we have proposed a number of action levels, as follows:

Action level 1: A score of one or two indicates that the posture is acceptable if it is not maintained or repeated for long periods.

Action level 2: A score of three or four indicates that further investigation is needed and changes may be required.

Action level 3: A score of five or six indicates that investigation and changes are required soon.

Action level 4: A score of seven or more indicates that investigation and change are required immediately.

Having found the Action level, it is then necessary to decide the most appropriate changes to reduce the risk. It will be evident, from an inspection of the score sheets, that as the situation worsens the numbers get higher. Hence, a first approach to improvement is to identify changes that will reduce the value of the scores, aiming to reduce the high numbers first. This will improve the total score, although a change in one of the values is unlikely to reduce the total scores by enough; several changes are usually necessary.

TABLE A Upper Limb Posture Score

UPPER ARM	LOWER ARM	WRIST POSTURE SCORE 1		2		3		4	
		TWIST 1	TWIST 2	TWIST 1	TWIST 2	TWIST 1	TWIST 2	TWIST 1	TWIST 2
1	1	1	2	2	2	2	3	3	3
	2	2	2	2	2	3	3	3	3
	3	2	3	3	3	3	3	4	4
2	1	2	3	3	3	3	4	4	4
	2	3	3	3	3	3	4	4	4
	3	3	4	4	4	4	4	5	5
3	1	3	3	4	4	4	4	5	5
	2	3	4	4	4	4	4	5	5
	3	4	4	4	4	4	5	5	5
4	1	4	4	4	4	4	5	5	5
	2	4	4	4	4	4	5	5	5
	3	4	4	4	5	5	5	6	6
5	1	5	5	5	5	5	6	6	7
	2	5	6	6	6	6	6	7	7
	3	6	6	6	7	7	7	7	8
6	1	7	7	7	7	7	8	8	9
	2	8	8	8	8	8	9	9	9
	3	9	9	9	9	9	9	9	9

TABLE B Neck, Trunk, Legs Posture Score

NECK POSTURE SCORE	TRUNK POSTURE SCORE 1		2		3		4		5		6	
	LEGS 1	LEGS 2	LEGS 1	LEGS 2	LEGS 1	LEGS 2	LEGS 1	LEGS 2	LEGS 1	LEGS 2	LEGS 1	LEGS 2
1	1	3	2	3	3	4	5	5	6	6	7	7
2	2	3	2	3	4	5	5	5	6	7	7	7
3	3	3	3	4	4	5	5	6	6	7	7	7
4	5	5	5	6	6	7	7	7	7	7	8	8
5	7	7	7	7	7	8	8	8	8	8	8	8
6	8	8	8	8	8	8	8	9	9	9	9	9

FIGURE 25.3

25.4 Some Complementary Measures

All possible actions of the hands are not represented in the RULA tables, so it is necessary to record certain other possible actions. Investigators can add to the ones listed here from their experiences of the particular requirements of the businesses they are engaged in. Other aspects of bodily constraints are also listed here, as many of them influence the ability of operators to change positions and to rest various parts of their bodies during the working day.

Hands

Are the hands or fingers doing pounding, gripping, exerting a force, or performing stretching or twisting actions? Is a pinch grip required? If a handle is used, is it too large, or too small, to be gripped easily? Is

MUSCLE USE SCORE

Give a score of 1 if the posture is; **mainly static, e.g. held for longer than 1 minute** **repeated more than 4 times/minute**

Static posture, held longer than 1 minute: score = 1
Moderate posture, not static, not highly repetitive: score = 0
Highly repetitive posture, repeated more than 6 times/min.: score = 1

FORCES OR LOAD SCORE

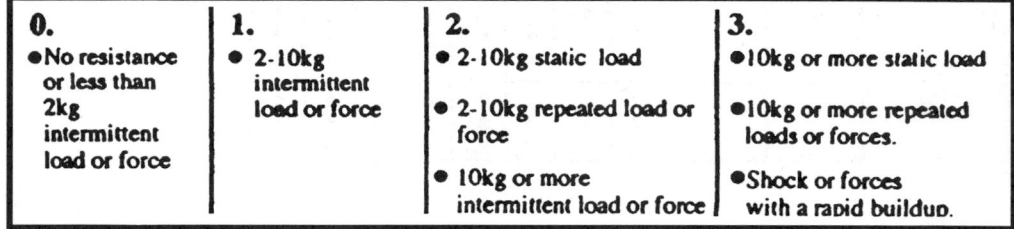

0.	1.	2.	3.
• No resistance or less than 2kg intermittent load or force	• 2-10kg intermittent load or force	• 2-10kg static load • 2-10kg repeated load or force • 10kg or more intermittent load or force	• 10kg or more static load • 10kg or more repeated loads or forces. • Shock or forces with a rapid buildup.

FIGURE 25.4

it too short or does it have pressure points that press into the palm? Is it difficult to grasp due to gloves, or a smooth surface?

Workpoint

Useful approximations for suitable work heights are:

Precision work: 100 to 200 mm above elbow height
Assembly work, with hand support: 50 to 70 mm above elbow height
Work requiring free hand movement: at, or just below, elbow height
Heavy work, requiring force: 100 to 300 mm below elbow height

Possible sources of obstructions, which may force inadequate postures, are awkwardly placed controls or visual requirements requiring reaching or bending, obstructions to the legs and feet, structures that prevent the operators from sitting, such as inadequate knee room.

Common problems with seating, which increase the difficulties of workers achieving good postures and changes of postures, are seats which are not adjustable, which have no padding, which dig into the users' legs, which prevent the users from swinging round to reach to the side, with a poor backrest which does not allow good lumbar support when in use.

Discomfort

Apart from the physical dimensions of the workpoint, as mentioned above, the responses of the users can give good guidance to assist with locating hazards. For full details of Body Part Discomfort recording, reference should be made to an appropriate textbook, such as that listed in the bibliography to this chapter. But even brief inquiries among workers concerning body discomforts can reveal the sources of

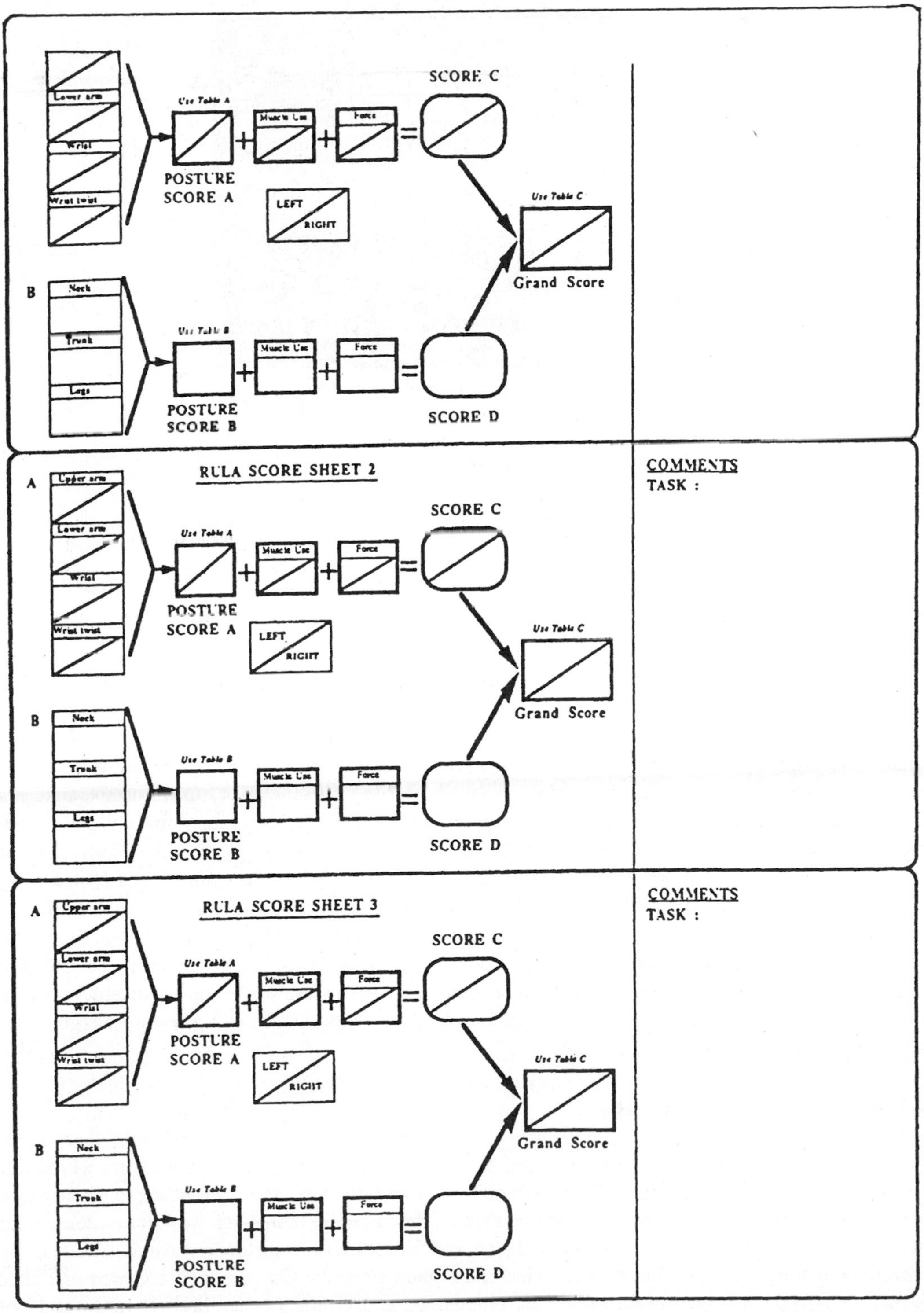
Lower arm
Wrist
Wrist twist
Use Table A
Muscle Use
Force
SCORE C
POSTURE SCORE A
LEFT
RIGHT
Use Table C
Grand Score
B
Neck
Trunk
Legs
Use Table B
Muscle Use
Force
POSTURE SCORE B
SCORE D
RULA SCORE SHEET 2
COMMENTS
TASK :
A
Upper arm
Lower arm
Wrist
Wrist twist
Use Table A
Muscle Use
Force
SCORE C
POSTURE SCORE A
LEFT
RIGHT
Use Table C
Grand Score
B
Neck
Trunk
Legs
Use Table B
Muscle Use
Force
POSTURE SCORE B
SCORE D
RULA SCORE SHEET 3
COMMENTS
TASK :
A
Upper arm
Lower arm
Wrist
Wrist twist
Use Table A
Muscle Use
Force
SCORE C
POSTURE SCORE A
LEFT
RIGHT
Use Table C
Grand Score
B
Neck
Trunk
Legs
Use Table B
Muscle Use
Force
POSTURE SCORE B
SCORE D

FIGURE 25.5

SCORE D (NECK, TRUNK, LEGS)

SCORE C (UPPER LIMB)	1	2	3	4	5	6	7+
1	1	2	3	3	4	5	5
2	2	2	3	4	4	5	5
3	3	3	3	4	4	5	6
4	3	3	3	4	5	6	6
5	4	4	4	5	6	7	7
6	4	4	5	6	6	7	7
7	5	5	6	6	7	7	7
8	5	5	6	7	7	7	7

TABLE C Grand Score Table

- **ACTION LEVEL 1** A score of one or two indicates that posture is acceptable if it is not maintained or repeated for long periods.
- **ACTION LEVEL 2** A score of three or four indicates further investigation is needed and changes may be required.
- **ACTION LEVEL 3** A score of five or six indicates investigation and changes are required soon.
- **ACTION LEVEL 4** A score of seven or more indicates investigation and changes are required immediately.

FIGURE 25.6

difficulties. Most musculoskeletal injuries give early warning by the growth of discomfort at the site, or in neighboring muscles. It is thus useful to ask workers, perhaps at intervals during the day, to point out those sites where discomfort has appeared during work, starting with the most uncomfortable first.

The results of such inquiries should complement the RULA analyses. Least discomfort will arise when the limbs are in positions represented by a value of one in each of the diagrams in the illustrations of Groups A and B. An increase in the value of the score will probably be accompanied by an increase in the discomfort levels reported, although this is not invariably true. If discomforts are found, they should be taken seriously as representing early warning of potential injuries, and action for change should be taken early. Sometimes management will not be too comfortable with inquiries of this sort, but it is very rare to find other than accurate responses to these inquiries about discomfort. What is more, they provide specific information relating to each individual's working condition, which will give valuable guidance for taking effective action.

25.5 Final Comments

The use of RULA is primarily as a survey tool. Although many, if not most, of the changes which will be required as a result of its use will be straightforward and obvious, there will be cases where more detailed investigations will be required. One user found that, in more than 8,000 workplaces, something like 80% of the changes could be done immediately and for little cost, while the rest could include some changes which were difficult and required time and money. Hence, it should not be assumed that after RULA all is well! An open mind should be maintained concerning further investigations, and the technique should also be used after changes to see if there is evidence of improvement in the scores, as well as in the experiences of work.

What is very useful about the technique is that relatively unskilled personnel can use the method after modest training; indeed, it has been taught to workers, who have then gone on to improve their own workplaces. One development has been the computerization of the procedure so that it can be available to all the computer users in a company, who can assess themselves using the display on their own screens. On completing their assessment, the scores are shown and suggestions for improvement are offered (Lueder,1996).

Methods engineers can find it a valuable tool that can be combined with their other techniques without any serious increase in the time needed to investigate working methods. It can be used, too, when new workplaces are being proposed, to see that they do not create unsuitable work situations. It will be evident that it can be used in mock-ups for new workplaces, giving assurance that the various features of the workplace will suit all users.

References

For information on the OWAS method, consult The Centre for Occupational Safety, Lonnrotinkatu 4B, SF-00120 Helsinki, FINLAND.

Health and Safety Executive. (1990) *Work-Related Upper Limb Disorders — A Guide to Prevention.* HSE Books, Sudbury, UK.

Lueder, R. K. (1996) A proposed RULA for computer users. *Proceedings of the Ergonomics Summer Workshop,* UC Berkeley Center for Occupational and Environmental Health, San Francisco.

McAtamney, L. (1994) *Interrelationship of Risk Factors Associated with Upper Limb Disorders in VDU Users.* PhD thesis, University of Nottingham, UK.

McAtamney, L. and Corlett, E. N. (1993) RULA: a survey method for the investigation of work-related upper limb disorders. *Applied Ergonomics,* 24,(2) 91-99.

McAtamney, L. and Corlett, E. N. (1992) *Reducing the Risks of Work-Related Upper Limb Disorders: A Guide and Methods.* The Institute for Occupational Ergonomics, The University of Nottingham, UK.

Wilson, J. R. and Corlett, E. N. (1995) *Evaluation of Human Work: A Practical Ergonomics Methodology.* (2nd ed.) Taylor & Francis, London.

26

OWAS Methods

Markku Mattila
Tampere University of Technology Finland

Mika Vilkki
Tampere University of Technology Finland

26.1 Introduction

Musculoskeletal disorders are one of the biggest occupational health problems in industrialized countries. According to national statistics, the proportion of musculoskeletal diseases of all occupational diseases in Finland was 31% (Kauppinen et al., 1994) and in the United States, 44% (Bureau of Labor Statistics, 1996). It was also noted that approximately 10% of occupational accidents resulted from sudden movements, lifting, repetitive motions, or overuse (Federation of Accident Insurance Institutions, 1995). Physical workload has been recognized as a factor affecting workers' health at several jobs. For example, about 33% of occupational diseases attributed to construction sites in Finland were linked to ergonomic factors associated with manual tasks (Federation of Accident Insurance Institutions, 1995).

Poor working postures constitute one of the main risk factors for musculoskeletal disorders (Burdorf et al., 1991), ranging from minor back problems to severe handicapping (Keyserling, 1986; Åaras et al., 1988). The effects of poor postures will continue unless proactive steps are taken to evaluate and reduce the problem. Therefore, it is essential to recognize early the patterns of work-related musculoskeletal symptoms and disorders and their risk factors in the workplace (Kuorinka and Forcier, 1995). More suitable working postures may have a positive effect on workers' musculoskeletal systems, and may allow for more effective control of work performance and reduction in the number of occupational accidents (Corlett et al., 1979; Wangenheim et al., 1986; Genaidy et al., 1990).

One practical method for analyzing and controlling poor working postures in industry is OWAS, the Ovako Working Posture Analysis System (Karhu et al., 1977).

0-8493-2641-9/99/$0.00+$.50

26.2 Theoretical Background For OWAS

Development of the OWAS Method

The Ovako Working Posture Analysis System (OWAS) was first developed in the Finnish steel industry in the 1970s. The need to identify and assess poor working posture arose from the fact that many jobs at the steel mill included physically stressful tasks. These problems had led to an increasing number of sick leaves and early retirements due to musculoskeletal disorders (Heinsalmi, 1985).

The project for working posture improvement was started at the plant. The jobs at the steel mill were studied and 680 photographs of different working postures were taken. This material represented nearly all existing typical work situations at the steel mill (Karhu et al., 1977). This material was analyzed and sorted by the researchers in order to create a classification system for the postures. The researchers were able to identify 84 typical posture combinations of back, arms, and legs. In practical experiments these proved to cover the most common postures in the steel industry. The typical working postures were combinations of four back postures, three arm postures, and seven leg postures.

After the typical postures were chosen, the usefulness of the system was tested. Twelve work design engineers were taught the system, and they then analyzed 28 tasks in the steel plant. The results were promising and led to the precise reliability testing of the method. During this assignment, 52 tasks were analyzed and over 36,000 postures were recorded. This analysis showed high inter-observer reliability but slightly lower inter-worker reliability (Karhu et al., 1977).

In order to evaluate the discomfort and health effects of the different postures, 32 experienced steel workers evaluated each posture in a four-point rating scale ranging from "normal posture with no discomfort and health effects" to "extremely bad posture, short exposure leads to discomfort, ill effects on health possible." The postures were then evaluated and their risks to the musculoskeletal system were assessed by a group of specialists consisting of international ergonomists. Based on these evaluations, the final classification of postures into different action categories for preventive measures was made.

Working Posture Classification in the OWAS Method

The 84 working postures classified in the OWAS system cover the most common and easily identifiable work postures for the back, arms, and legs. An estimate for load handled by the person observed is also made in connection with the posture.

Each classified posture of the OWAS is determined by a four-digit code in which the numbers indicate the postures of the back, the arms, and the legs, as well as the load/effort needed.

Back

In the OWAS system the first digit in the posture code indicates the posture of the back. There are four choices for the different back postures: (1) back straight, (2) back bent, (3) back twisted, and (4) back bent and twisted (Table 26.1).

Arms

The second digit in the observation code indicates the posture of the arms. There are three choices for the arm postures in the OWAS system: (1) both arms below shoulder level, (2) one arm at or above shoulder level, and (3) both arms at or above shoulder level (Table 26.2).

Legs

The third digit in the posture code indicates the posture of legs. There are seven choices for the postures of the legs in the OWAS system: (1) sitting, (2) standing on two straight legs, (3) standing on one straight leg, (4) standing or squatting on two bent legs, (5) standing or squatting on one bent leg, (6) kneeling, and (7) walking (Table 26.3).

TABLE 26.1 Definition of Four Codes for the Back Postures in the OWAS System

1 BACK STRAIGHT

"Back straight" means that worker's back is less than 20° (the angle of the lines which go between head–hips and legs) bent forward or sideways or less than 20° twisted (the angle between shoulders and hips).

2 BACK BENT

"Back bent" means that worker is in a posture in which the upper body is bent forward or backward 20° (the angle of the lines which go between head–hips and legs) or more.

3 BACK TWISTED (OR BENT SIDEWAYS)

"Back twisted" means that the back is twisted 20° or more (as defined above) or bent sideways 20° or more.

4 BACK BENT AND TWISTED

"Back bent and twisted" means a situation where back is bent (like in case 2) and simultaneously twisted (like in case 3).

TABLE 26.2 Definition of Three Codes for the Arm Postures in the OWAS Method

1 BOTH ARMS BELOW SHOULDER LEVEL

"Both arms below shoulder level" means a situation in which both arms are completely below shoulder level.

2 ONE ARM AT OR ABOVE SHOULDER LEVEL

"One arm at or above shoulder level" means that one arm or part of it is at or above shoulder level.

3 BOTH ARMS AT OR ABOVE SHOULDER LEVEL

In "Both arms at or above shoulder level" both arms are fully or partly at or above shoulder level.

TABLE 26.3 Definition of the Seven Postures for Legs in the OWAS Method

1 SITTING

"Sitting" means that the weight of the body is supported on the buttocks. In this posture the legs are also below the buttocks.

2 STANDING ON BOTH STRAIGHT LEGS

"Standing on both straight legs" means that the weight of the body is supported on two straight legs. The knee angle is more than 150°.

3 STANDING ON ONE STRAIGHT LEG

"Standing on one straight leg" is a situation in which one leg is straight and the weight of the body is completely supported by that leg. The knee angle is more than 150°.

4 STANDING OR SQUATTING ON BOTH FEET, KNEES BENT

In this posture the weight of the body is on both legs and both knees are bent on a 150° or smaller angle.

5 STANDING OR SQUATTING ON ONE FOOT, KNEE BENT

In this posture the weight of the body is on one leg, and it is also bent from the knee. The knee angle is 150° or smaller.

6 KNEELING ON ONE OR BOTH KNEES

In this posture the person is kneeling either on both knees or one knee.

7 WALKING OR MOVING

In this posture the person is walking or moving around at the workplace.

TABLE 26.4 Definition of Three Codes for the Load Handled in the OWAS Method

1	LOAD/USE OF FORCE ≤ 10 KG
	Weight handled or force needed is 10 kg or less.
2	LOAD/USE OF FORCE >10 KG ≤ 20 KG
	Weight handled or force needed is exceeds 10 kg but is less than 20 kg.
3	LOAD/USE OF FORCE > 20 KG
	Weight handled or force needed exceeds 20 kg.

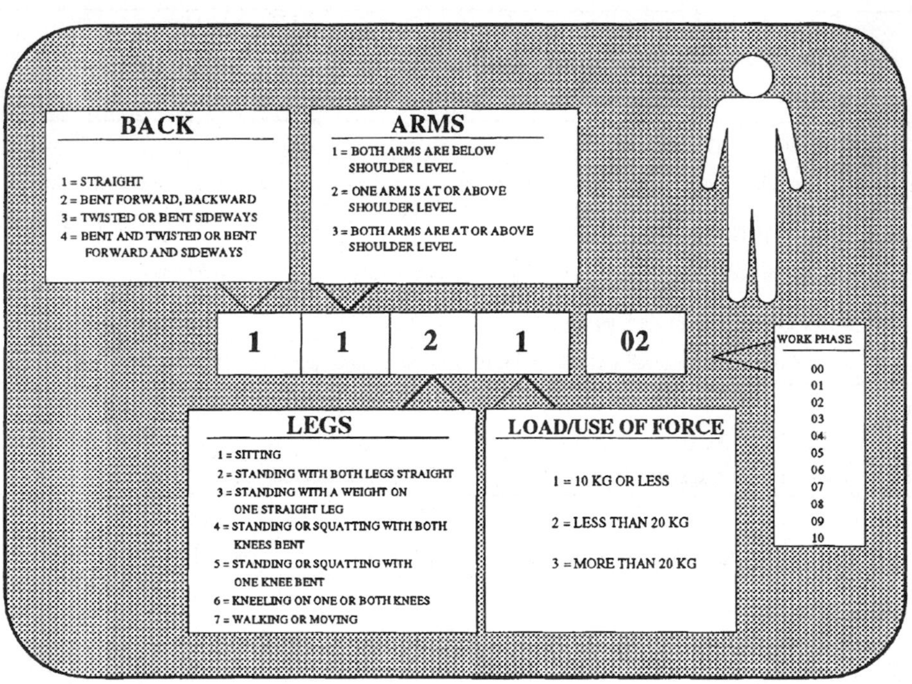

FIGURE 26.1 Example of a posture classification and codes for different body parts in the OWAS system.

Load/Use of Force

The fourth digit in the OWAS code indicates how big a load the person is handling or how much force must be used in the operation. Load/use of force has three alternatives: (1) less than 10 kilograms, (2) between 10 and 20 kilograms, and (3) more than 20 kilograms (Table 26.4).

The OWAS posture classification is presented in Figure 26.1. In addition to the posture code, it is possible to use the fifth digit to indicate the work phase or task the person was doing when the posture was observed. An example of the numerical code 1121 shows the usage of the OWAS system (Figure 26.1). Code 1121 indicates that the worker's back is straight (back: 1). He works with both arms below shoulder level (arms: 1), has his weight on two straight legs (legs: 2), and handles a load less than 10 kg (load/use of force: 1). The fifth digit indicates the work phase in which the posture occurs, for instance, carrying. The use of the work phase code and the analysis of different tasks among the observed material became possible through the computerized OWAS applications.

TABLE 26.5 The OWAS Action Categories for Prevention

Action Category	Explanation
1	Normal and natural postures with no harmful effect on the musculoskeletal system — **No actions required**
2	Postures with some harmful effect on the musculoskeletal system — **Corrective actions required in the near future**
3	Postures have a harmful effect on the musculoskeletal system — **Corrective actions should be done as soon as possible**
4	The load caused by these postures has a very harmful effect on the musculoskeletal system — **Corrective actions for improvement required immediately**

Making Observations for OWAS Data Gathering

The frequency of different postures and their proportional share of the working time are determined by observations. The basic idea in the observational technique is to collect material through observations made at set intervals over a given period of time. This provides an overall picture of the job studied. In the OWAS analysis, the observations are made using visual, split-second observations at the moment when the observer glances at the worker. The observations are collected in the original OWAS method using special forms made for this purpose. When the computer program for OWAS data gathering (OWASCO) is used, a portable laptop computer is utilized for the data input.

The observations should be done in the actual work situation, field conditions. In some cases, videotapes can be used. The advantage of using videotapes is that the observer has as much time as he/she wants to look at the observed posture. The videotapes can also easily and effectively be used in recalling the actual work situations when providing feedback from the posture study or when teaching new work methods (Mattila et al., 1992).

The observations are made using an equal interval system where the interval between observations normally is either 30 or 60 seconds. The reason for this is that in field conditions it is often too hard for the observer to use shorter intervals. Shorter intervals are suggested in cases when it is possible to make the analysis using the videotapes, or the nature of the task requires it, e.g., short work cycles. The observation period shouldn't exceed 40 minutes without 10 minutes for resting.

The frequencies of work postures and their relative proportion (%) of the working time are calculated from the observation results. The error limits associated with the mean relative proportions of work posture are calculated for 95% probability, using a random system formula. The error limits become smaller as the total number of observations increases. The error limits in mean values based on 100 observations are ±10%, and with 400 observations, ±5%. Mean values obtained through observation can be considered sufficiently reliable when the error limits are under 10% (Louhevaara and Suurnäkki, 1991).

The Analysis of Recognized Work Postures in the OWAS System

The posture combinations and the relative proportions of certain postures are classified into four action categories for improvement needs. The classification of the postures is based on the risk assessment of musculoskeletal disorders and the physical load on the subjects musculoskeletal system. The action category indicates the urgency and priority for corrective measures. The action categories for prevention range from 1, no actions required, to 4, corrective measures needed immediately (Table 26.5). This categorization based on risk assessment was originally constructed by physicians, work analysts, and workers and then revised and validated by an international group of experts (Karhu et al., 1977).

The classification for individual posture combinations indicates the level of risk of injury or harmful effects caused by that classified posture (combination of the postures of back, arms, and legs and the load handled) for the musculoskeletal system. If the risk for musculoskeletal disorder is high, then the action category indicates the need and urgency for corrective actions. The action categories for each individual postures are presented in Figure 26.2.

back	arms	legs 1			legs 2			legs 3			legs 4			legs 5			legs 6			legs 7		
	use of force	1	2	3	1	2	3	1	2	3	1	2	3	1	2	3	1	2	3	1	2	3
1	1	1	1	1	1	1	1	1	1	1	2	2	2	2	2	2	1	1	1	1	1	1
	2	1	1	1	1	1	1	1	1	1	2	2	2	2	2	2	1	1	1	1	1	1
	3	1	1	1	1	1	1	1	1	1	2	2	3	2	2	3	1	1	1	1	1	2
2	1	2	2	3	2	2	3	2	2	3	3	3	3	3	3	3	2	2	2	2	3	3
	2	2	2	3	2	2	3	2	3	3	3	4	4	3	4	4	3	3	4	2	3	4
	3	3	3	4	2	2	3	3	3	3	3	4	4	4	4	4	4	4	4	2	3	4
3	1	1	1	1	1	1	1	1	1	2	3	3	3	4	4	4	1	1	1	1	1	1
	2	2	2	3	1	1	1	1	1	2	4	4	4	4	4	4	3	3	3	1	1	1
	3	2	2	3	1	1	1	2	3	3	4	4	4	4	4	4	4	4	4	1	1	1
4	1	2	3	3	2	2	3	2	2	3	4	4	4	4	4	4	4	4	4	2	3	4
	2	3	3	4	2	3	4	3	3	4	4	4	4	4	4	4	4	4	4	2	3	4
	3	4	4	4	2	3	4	3	3	4	4	4	4	4	4	4	4	4	4	2	3	4

FIGURE 26.2 Action category for each individual OWAS classified posture combination.

The second classification is based on the time spent in different postures for each body part. This classification examines the relative proportion postures of the back, the arms, and the legs during the observation period (Figure 26.3). The same four action categories used in the classification of individual posture combinations are used here. The postures for each body part are counted together and when the relative proportion of certain posture during the observation period exceeds fixed limits, the action category changes from lower to higher. This indicates that the urgency of corrective actions is increasing. The OWAS system doesn't have a classification for the relative proportion of the use of force/load handled. In cases when heavy materials handling occurs, the situation must be evaluated separately in each case. For such evaluations, a biomechanical analysis is useful.

The Data Analysis and Reporting

When an adequate amount of postural data has been gathered, the results of the OWAS study are analyzed and reported. The use of a computerized OWAS application makes the analysis fast and more versatile than the traditional pen-and-paper method. The use of a computerized application is strongly recommended, but the traditional method is functional if the OWAS software is not available.

In the data analysis, all the collected postures are counted together and processed on the summary form. The basic analysis of the collected data includes two items: (1) the calculation of the percentages of posture combinations and (2) the proportional shares of postures of different body parts, falling into different action categories.

If a more detailed analysis is needed, the data may be analyzed according to different work phases. This helps in finding the most difficult tasks and operations in the work analyzed. It is also possible to analyze the most difficult postures (categories 3 and 4) separately and study these situations more carefully. This analysis makes it possible to quickly examine which tasks need corrective actions most urgently.

In the reporting of the OWAS study, the results are printed out and graphs are drawn from the distributions. If videotapes are available they can be very helpful in reporting the results of the postural study.

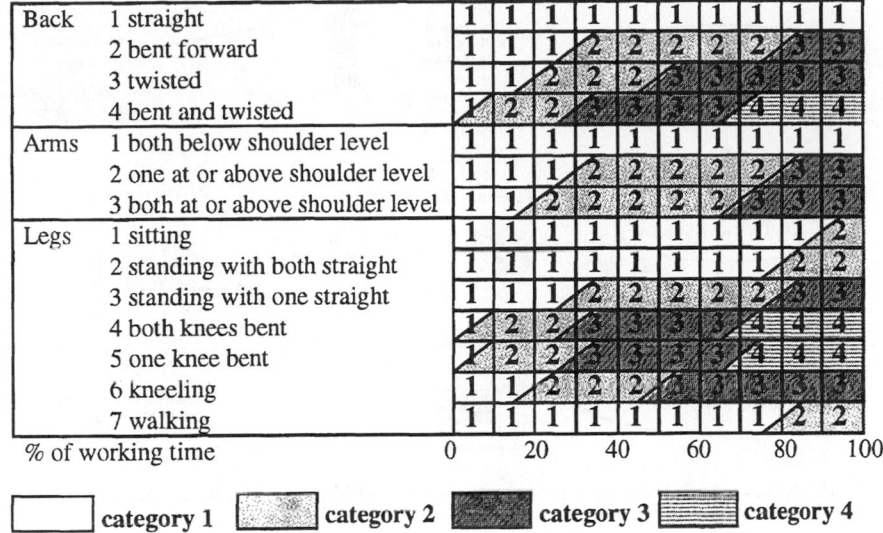

Back	1 straight	1	1	1	1	1	1	1	1	1	1
	2 bent forward	1	1	1	2	2	2	2	2	3	3
	3 twisted	1	1	2	2	2	3	3	3	3	3
	4 bent and twisted	1	2	2	3	3	3	3	4	4	4
Arms	1 both below shoulder level	1	1	1	1	1	1	1	1	1	1
	2 one at or above shoulder level	1	1	1	2	2	2	2	2	3	3
	3 both at or above shoulder level	1	1	2	2	2	2	2	3	3	3
Legs	1 sitting	1	1	1	1	1	1	1	1	1	2
	2 standing with both straight	1	1	1	1	1	1	1	1	2	2
	3 standing with one straight	1	1	1	2	2	2	2	2	3	3
	4 both knees bent	1	2	2	3	3	3	3	4	4	4
	5 one knee bent	1	2	2	3	3	3	3	4	4	4
	6 kneeling	1	1	2	2	2	3	3	3	3	3
	7 walking	1	1	1	1	1	1	1	1	2	2

% of working time: 0, 20, 40, 60, 80, 100

category 1 | category 2 | category 3 | category 4

FIGURE 26.3 Action categories for the relative proportions of the postures of the different body parts.

Procedure to Handle OWAS Results

It is highly recommended to establish a cooperative team to handle the results of OWAS analysis. This group could consist of: for example, the analyst, representatives of the workers, management, occupational health care specialist, and other persons who may be involved in job redesign and development. When the team has the results and identifies the problematic task associated with the poor work postures from the videotapes, it is easier and more effective to provide corrective measures and redesign the working methods (Kivi and Mattila, 1991; Mattila et al., 1992).

Reliability and Validity

The inter-observer reliability of the OWAS method has been tested in many jobs in different industries. In all reported cases the inter-observer reliability was high, averaging over 90% (Karhu et al., 1977; Louhevaara and Suurnäkki, 1991; Mattila et al., 1993). The posture of the back is most difficult to for observers to distinguish.

Leskinen and Tönnes (1993) studied the validity of the OWAS system by comparing the postures collected using the OWAS system with the accurate work postures recorded using the electronic Selspot II camera system. The OWAS system proved to give nearly the right picture about the postural load. Kuusela (1994) investigated the biomechanical load caused by typical OWAS postures, using a special biomechanical software. This study showed that the biomechanical basis of the OWAS system is correct with the exception of a few situations.

26.3 Computer-Aided Applications

The first application was introduced in the late 1980s by Tampere University of Technology in Finland. This system was based on a portable programmable HP-71B calculator used for data collection. The data was then transferred to a PC computer through a special interface linkage. The analysis of the collected data was then done by the analysis program in the PC computer (Kivi and Mattila, 1991). The next step in the software development was introduced in 1991 (Mattila et al., 1992). This system used a portable Canon X-07 computer for data collection. The main improvement compared to the previous system was that the data collection unit was already capable of performing a limited amount of data analysis. In order to achieve a comprehensive data analysis the data had to be transferred to the PC (Figure 26.4).

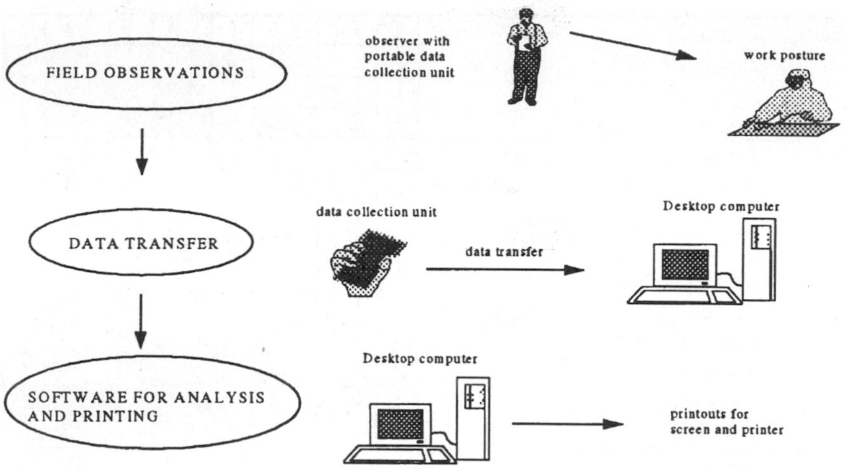

FIGURE 26.4 The OWAS data processing system used in the early computerized versions.

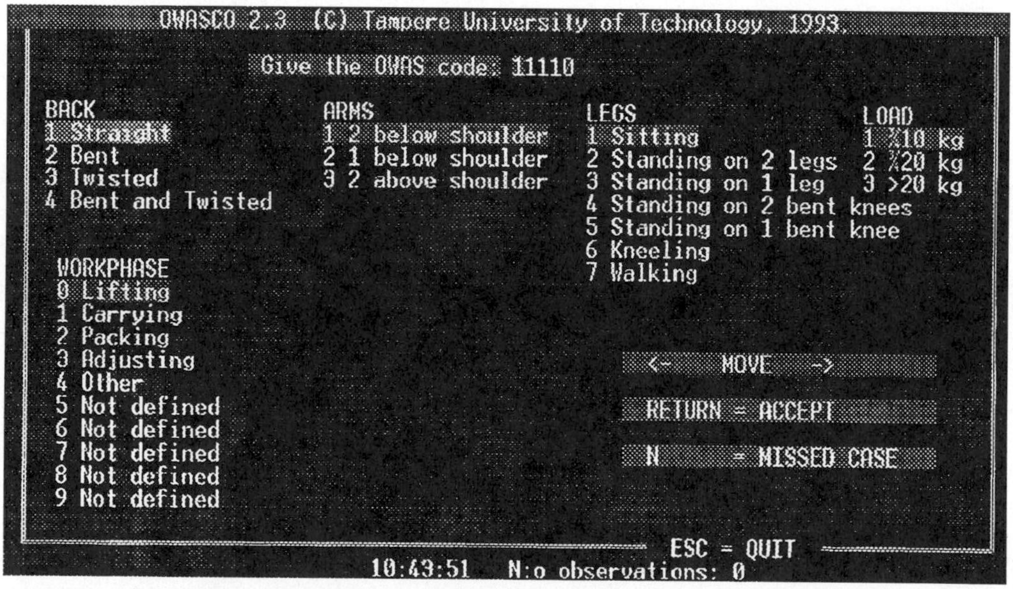

FIGURE 26.5 Data input interface of the OWASCO computer program.

The third-generation OWAS software was developed in 1992 (Vilkki et al., 1993). The light weight and reduced prices of PC notebook computers made it possible to use the same computer for data collection and analysis. This allowed faster analysis of data and immediate feedback at the worksite. The software program package is called OWASCO & OWASAN. In this system the OWASCO program is used for data collection and the OWASAN program is used for the analysis of data.

In the data analysis the background information for the work to be observed and the chosen analysis interval are given before starting the observation. In this stage it is also necessary to divide the observed job into work phases. If the observations are made properly using the work phase coding, it makes it easier in data analysis to identify the problematic tasks and to create measures for job redesign. The data input screen of the OWASCO program allows the choice of body postures from the menus to be always visible on the screen (Figure 26.5). A built-in clock with a "beep" sound indicates the moment when the observation should be done.

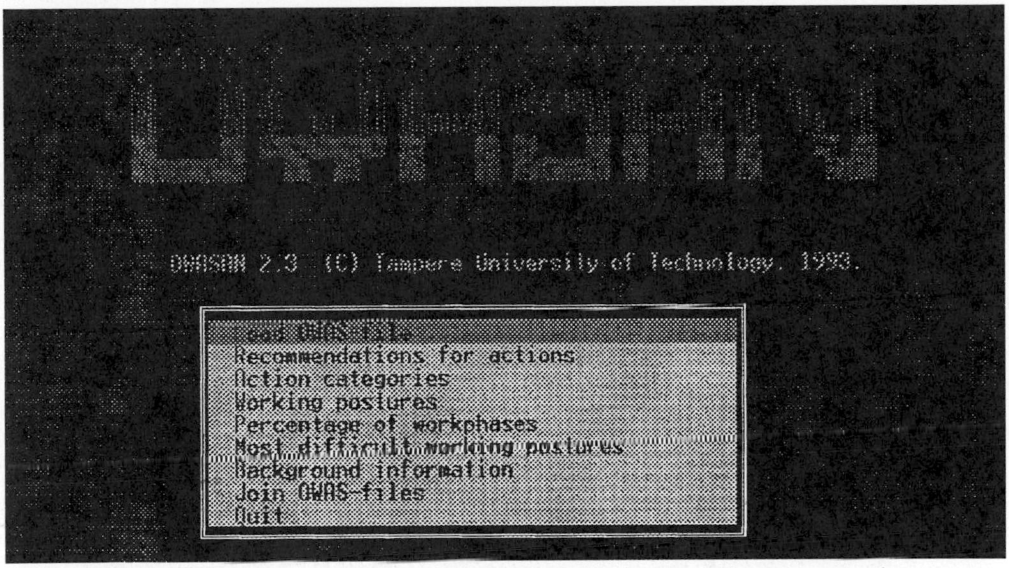

FIGURE 26.6 The main menu of the OWASAN computer program.

The OWASAN data analysis program calculates different types of analysis from the collected data. The functions of this program include: (1) recommendations for actions, (2) action categories, (3) listing of working postures, (4) percentage of work phases, (5) most difficult working postures, (6) printing the background information sheet, and (7) joining the OWAS files together (Figure 26.6). It is possible get the analysis from one particular work phase or the whole analyzed OWAS data.

The results can be printed on the screen, on a file for export to other software, e.g., word processing, or for a printer. An example of a printout of the "Recommendations for Actions" sheet is presented in Figure 26.7.

Other OWAS applications have also been introduced. Kant et al. (1990) introduced a computerized version of the OWAS system and tested it in garages. Pinzke (1992) introduced a software analysis method more precise than the OWAS method but using the same action categories.

26.4 Experiences from OWAS Usage in Different Fields

The OWAS method was originally developed in the steel industry. The jobs at the steel mill included a large variety of working postures combined with heavy materials handling. Due to the nature of the work for which the system was developed, it is most suitable in the analysis of physical work with clearly observable postures of different body segments.

The OWAS method has been applied over the years in various industries in many countries. Most of the reported cases are analyses of jobs including heavy physical work and dynamic working postures.

Steel Mill

Karhu et al. (1981) reported an OWAS study of two jobs: bricklaying and installation of a roll unit in a steel mill. In this study, poor working postures were indicated and corrective measures were suggested. The main result of this study was decreased number of poor working postures and considerably higher productivity.

Mining Industry

Heinsalmi (1985) reported a working posture study in the Finnish mining industry. In this study, the postures of a drilling unit were examined. The proportion of working postures in the OWAS categories

```
RECOMMENDATIONS FOR ACTIONS

COMPANY: Factory Inc.
DEPARTMENT: Assembly
THE JOB OBSERVED: Welding
OBSERVER: Ilkka
INTERVAL: 20

Whole material.   Total: 20

                0%  10% 20% 30% 40%  50%  60% 70%  80%  90% 100%
------------------------------ BACK ----------------------------------
Straight      11111111111111111111111111111111111■1111111111   75.00 %

Bent          111111111■11/2222222222222222222222/333333333   20.00 %

Twisted       111■1111/2222222222222/3333333333333333333333    5.00 %

Bent&Twisted  ■1/2222222222/333333333333333333/4444444444444   0.00 %
------------------------------ ARMS ----------------------------------
Free          11111■11111111111111111111111111111111111111111   10.00 %

One up        1111111111111/2222222222222222222222/33■333333   85.00 %

Both up       111■11111/2222222222222222222222/33333333333333   5.00 %
------------------------------ LEGS ----------------------------------
Sitting       111■1111111111111111111111111111111111111/2222    5.00 %

Stand.with 2  111111111111111111111111111111■11111/2222222222   60.00 %

Stand.with 1  1111111111111■2222222222222222222222/333333333   30.00 %

Both bent     11/■2222222222/3333333333333333/4444444444444    5.00 %

One bent      ■1/22222222222/3333333333333333/4444444444444    0.00 %

On knees      ■11111111/2222222222222/3333333333333333333333    0.00 %

Walking       ■1111111111111111111111111111111111111/222222222   0.00 %
------------------------------ LOAD ----------------------------------
 < 10 kg      Find out the reasons for differe■t loads . . .   65.00 %

 < 20 kg      Find out the reason■ for different loads . . .   35.00 %

 > 20 kg      ■ind out the reasons for different loads . . .    0.00 %
----------------------------------------------------------------------
```

FIGURE 26.7 An example of the "Recommendations for Actions" sheet printed with the OWASAN program.

2 through 4 was 5.4%. Changes in the machinery layout in the workplace decreased the share of poor postures to 1.0%.

Cleaning

Hopsu and Louhevaara (1991) used the OWAS method to measure postural load in cleaners work during an intervention study. The intervention included educational training and ergonomic job redesign. The results show a decrease from 39% to 25% in the amount of postures in categories 2 through 4. The constancy of the improvement was evaluated in a follow-up study a year after the intervention.

Garage

Kant et al. (1990) used the OWAS system in the analysis of the jobs of 84 mechanics in 42 garages. In this study, the use of the computerized OWAS system made it possible to easily identify the most difficult work phases. Some tasks, e.g., working at the side of the car, had 18% of posture combinations in categories 3 and 4. But while using the vehicle lift in the same job, the number of category 3 and 4 postures was only 4%.

Construction

Kivi and Mattila (1991) used the OWAS method to analyze twelve jobs at construction sites. This study showed that many jobs at construction sites include a high number of poor working postures. The most difficult jobs found in this study were cement worker and repair worker. The proportion of OWAS category 3 and 4 postures in these jobs was 28% and 18%, respectively. In this study, the computerized OWASCO & OWASAN program package was found useful in pinpointing the needs for improvement.

Mattila et al. (1993) analyzed the working postures of carpenters doing hammering tasks. The OWAS method showed a high incidence of poor postures in many tasks, but it was not possible to find differences in working postures caused by different types of hammers.

Rohmert et al. (1993) did an extensive study of tilers work. In this study, about 12,000 working postures were analyzed. The results showed that in tiling work more than 60% of the postures require corrective actions.

Paper Mill

Mattila et al. (1992) analyzed jobs in the paper mill industry. In the maintenance and adjustment tasks performed while the machine was stopped, the proportion of the OWAS category 3 and 4 postures was highest, in some cases up to 50%. The OWAS method proved to be suitable for analysis of this type of job also.

Railways

Peereboom (1993) presented a strategy for how to use the OWAS method. He used this approach to study the postures of train mechanics and railway maintenance workers.

Manufacturing

Vilkki et al. (1993) studied manual materials handling tasks in manufacturing. The percentages of postures in categories 3 and 4 were relatively low, but heavy lifting occurred frequently. Therefore, a biomechanical analysis was combined with the OWAS analysis. The computerized OWAS method proved to be an effective way to analyze poor postures during the redesign process of a production line.

Nursing

Nurses' working postures were studied by Engels et al. (1994). The nurses jobs were analyzed in two wards. This study showed that working postures of nurses were slightly harmful. Differences in postural load between wards were found.

Farming

Nevala-Puranen (1995) reported a study of farmers' postural load. In this study, poor postures were identified by the OWAS method and new techniques were taught. For example, in milking the proportion of category 3 and 4 postures decreased from about 50% to 20%. The study showed positive and permanent changes in farmers' working postures.

26.5 Conclusions and Future Prospects

The OWAS method has shown its functionality in many studies in different types of industries over the years. It is a useful way to get an overall picture of the postural load caused by different postures in different jobs and to direct the improvements in the working methods used at work. The method is feasible and the computer-aided applications have made the analysis more effective and more versatile.

The OWAS method effectively supports the modern safety management at the plant and the preventive occupational health care especially, while it includes a risk assessment procedure to show the most problematic postures and to pinpoint the priorities for corrective actions.

The OWAS analysis has proved to give good background information for participatory ergonomics. It is possible to use the results at the workplace immediately after the analysis in discussions to improve the working postures and to redesign the working methods. The utilization of the results is most effective in a cooperative team which consists of, e.g., the occupational health care professional, management and worker representatives, the person who did the analysis, and others who are involved with the analyzed work. The analysis of the OWAS results by a cooperative team gives a good basis for open discussion at the workplace. In this way, the OWAS method is a successful tool to support continuous improvement at the workplace. Such an approach allows the improvement of tasks, job redesign, and new working methods development. The videotapes and photographs taken during the study help the team to visualize the actual situations.

In the future there seems to be the possibility for automatic posture recording which will improve the accuracy and the speed needed for the analysis. In Manual materials handling tasks and jobs where loads handled play an essential role, it would be beneficial if biomechanical analysis could be integrated in the OWAS method. These are some challenges for future development for the OWAS method.

References

Åaras, A., Westgaard, R.H., and Strandew, E. 1988. Postural angles as an indicator of postural load and muscular injury in occupational work situations, *Ergonomics* 31, 915-933.

Burdorf, A., Govaert, G., and Elders, L. 1991. Postural load and back pain of workers in the manufacturing of prefabricated concrete elements et al., *Ergonomics* 34, 909-918

Bureau of Labor Statistics, 1996. *"Characteristics of injuries and illnesses resulting in absences for work, 1994."* ftp://stats.bls.gov/pub/news.release/osh2.txt/, U.S. Department of Labor.

Corlett, E.N., Madeley, S.J., and Manenica, J. 1979. Posture targeting: a technique for recording work postures, *Ergonomics* 24, 795-806.

Engels, J.A., Landeweerd, J.A., and Kant, Y. 1994, An OWAS-based analysis of nurses' working postures. *Ergonomics*, 37, 5, 909-919.

Federation of Accident Insurance Institutions, 1995. *Työtapaturma- ja ammattitautitilasto 1993 ("Occupational accidents and diseases 1993").* Tapaturmavakuutuslaitosten liitto, Helsinki, p. 96.

Genaidy, A., Karwowski, W., and Musavinezhad, S.H. 1990. Computer-aided ergonomics: a tool for control of musculoskeletal injury, in Karwowski, W., Genaidy, A., and Asfour, S.S. (eds.), *Computer-Aided Ergonomics*, Taylor & Francis, 8-28.

Heinsalmi, P. 1985. Method to measure working posture loads at working sites (OWAS), in Corlett, N., Wilson, J., and Manenica, I. (ed.), *The Ergonomics of Working Postures.* Taylor & Francis, London, pp. 100-104.

Hopsu, L. and Louhevaara, V. 1991. The influence of educational training and ergonomic job redesign intervention on the cleaners' work: a follow up study, in Quéinnec, Y. and Daniellou, F. (eds.), *Designing for Everyone.* Taylor & Francis, pp. 534-536.

Kant, I., Notermans, J.H.V., Borm, P.J.A. 1990, Observation of working postures in garages using the Ovako Working Posture Analyzing System (OWAS) and consequent workload reduction requirements. *Ergonomics* 33, 209-220.

Karhu, O., Kansi, P., and Kuorinka, I. 1977. Correcting working postures in industry. A practical method for analysis. *Applied Ergonomics* 8, 199-201.

Karhu, O., Härkönen, R., Sorvali, P., and Vepsäläinen, P. 1981. Observing working postures in industry: Examples of OWAS application. *Applied Ergonomics* 12, 13 — 17.

Kauppinen, T., Vaaranen, V., Vasama, M., Tokkanen, J., and Jolanki, R. 1994. *Occupational Diseases in Finland in 1993*, The Finnish Institute of Occupational Health, Helsinki, p. 102 (in Finnish).

Kivi, P. and Mattila, M. 1991. Analysis and improvement of work postures in the building industry: application of the computerized OWAS method. *Applied Ergonomics* 22 (1), 43-48.

Kuorinka, I. and Forcier L. (eds.). 1995. *Work Related Musculoskeletal Disorders (WMSDs)*, Taylor & Francis, London.

Kuusela, J. 1994. *Working Postures and Their Biomechanical Loading*, Tampere University of Technology, Finland, Unpublished technical report (in Finnish).

Leskinen, T. and Tönnes, M. 1993. Validity of observation methods used for the evaluation of working postures, *Työ ja Ihminen*, 7(1993):4, 299-314. (in Finnish, with English summary).

Long, A.F. 1992. A computerized system for OWAS field collection and analysis, in Mattila, M. and Karwowski, W. (eds.), *Computer Applications in Ergonomics, Occupational Safety and Health*. Elsevier, Amsterdam, pp. 353-358.

Louhevaara, V. and Suurnäkki, T. 1992. *OWAS: A Method for the Evaluation of Postural Load During Work*. Institute of Occupational Health and Centre for Occupational Safety, Helsinki, p. 23.

Mattila, M. and Kivi, P. 1991. Analysis of problematic working postures and manual lifting in building tasks, in Quéinnec, Y. and Daniellou, F. (eds.), *Designing for Everyone*. Taylor & Francis, London.

Mattila, M., Vilkki, M. and Tiilikainen, I. 1992. A Computerized OWAS analysis of work postures in the papermill industry, in Mattila, M. and Karwowski, W. (eds.), *Computer Applications in Ergonomics, Occupational Safety and Health*. Elsevier, Amsterdam, pp. 365-372.

Mattila, M., Karwowski, W., and Vilkki, M. 1993. Analysis of working postures in hammering tasks on building construction sites using the computerized OWAS method. *Applied Ergonomics* 24(6), 405-412.

Nevala-Puranen, N. 1995. Reduction of farmers' postural load during occupationally oriented medical rehabilitation. *Applied Ergonomics* 26, 6, 411-415.

Peereboom, K.J. 1993. A strategy for using the OVAKO working posture analyzing system (OWAS) to determine the physical load of actions, in Marras, W.S., Karwowski, W., Smith, J.L., and Pacholski, L. (eds.), *The Ergonomics of Manual Work*. Taylor & Francis, London, pp. 245-248.

Pintzke, S. 1992. A computerized method of observation used to demonstrate injurious work operations, in Mattila, M. and Karwowski, W. (eds.), *Computer Applications in Ergonomics, Occupational Safety and Health*. Elsevier, Amsterdam, pp. 359-364.

Rohmert, W., Wakula, J., and Schildge, B. 1993. Analysis of working postures of tilers, in Marras, W.S., Karwowski, W., Smith, J.L., and Pacholski, L. (eds.), *The Ergonomics of Manual Work*. Taylor & Francis, London, pp. 33-40.

Wangenheim, M., Samuelson, B., and Wos, H. 1986. ARBAN — a force ergonomic analysis method, in Corlett, E.N., Wilson J.R., and Manenica, I. (eds.) *The Ergonomics of Working Postures*, Taylor and Francis, 243-255.

Vilkki, M., Mattila, M., and Siuko, M. 1993. Improving work postures and manual materials handling tasks in manufacturing: A case study, in Marras, W.S., Karwowski, W., Smith, J.L. and Pacholski, L. (eds.), *The Ergonomics of Manual Work*. Taylor & Francis, London, pp. 273-276.

27

Ergonomics of Hand Tools

Andris Freivalds
The Pennsylvania State University

27.1 Introduction

Tools are as old as the human race itself. The hands and feet could be considered tools given to the human by nature. However, tools as we know them were developed as extensions of the hands and feet to amplify the range, strength, and effectiveness of these limbs. Thus, the early human, by picking up a stone, could make the fist heavier and harder, producing a more effective blow. Similarly, by using a stick, a longer and stronger arm was created.

The exact time when humans began to use and to make tools is not known. Leakey (1960), during his excavations in Africa, uncovered evidence that more than a million years ago the prehistoric human was already a tool-maker, using stones for chipping and bones for leather work. Similarly, Napier (1962) indicated that with changing tasks, such as converting from the power to precision grip, there was similar change in the anatomy of the hand as well as development of tools. An important milestone occurred when the stone tools were provided with handles some 35,000 years ago. The addition of the handle increased the range and speed of action and increased the kinetic energy for striking tasks. A still later change in tool development occurred with the change in tasks from food gathering to food production. New tools were required and developed accordingly. Surprisingly, many of these tools, with minor improvements and refinements, are still in use today. The reasons for such stagnation could be twofold: either the tool reached an optimal form very quickly with no room for improvement or there was no impetus for further improvement. The latter is the resigned view that since a tool has been used by so many people for so many years, no further improvement is possible. The former view is obviously not true since Lehmann (1953) noted the existence of over 12,000 different styles of shovels in Germany in the 1930s, all essentially used for the same task. Indeed, the last great

0-8493-2641-9/99/$0.00+$.50

change in tool development occurred with the start of the Industrial Revolution, with a change in task, from food production to manufacturing of goods.

The parallel development of tools with changing technology has given rise to another problem. The current technology explosion has proceeded too quickly to permit the gradual development of tools appropriate for the new industrial tasks. The instant demands for new and specialized tools to match the needs of technology has, in many cases, bypassed the testing needed to fit these tools to human users. This has resulted in a variety of hand-tool-generated work stressors and an increased incidence of cumulative trauma of the hand, wrist, and forearm, reducing productivity, disabling individuals, and increasing the medical costs for industry.

Cumulative trauma disorders (CTDs) are injuries to the musculoskeletal system that develop gradually as a result of repeated microtrauma. Because of the slow onset and relatively mild nature of the trauma, the condition is often ignored until the symptoms become chronic and more severe injury occurs. These problems are a collection of a variety of problems including repetitive motion disorders, carpal tunnel syndrome, tendinitis, ganglionitis, tenosynovitis, and bursitis, with these terms sometimes being used interchangeably. There are four major work-related factors that seem to lead to the development of CTD: (1) use of excessive force during normal motions, (2) awkward or extreme joint motions, (3) high amounts of repetition of the same movement, and (4) the lack of sufficient rest to allow the traumatized joint to recover. The most common symptoms associated with CTD include: pain, restriction of joint movement, and soft tissue swelling. In the early stages there may be few visible signs, however, if the nerves are affected, sensory responses and motor control may be impaired. If left untreated, CTD can result in permanent disability (Putz-Anderson, 1988).

The cost of CTD in U.S. industry, although not all due to improper tool design, is quite high. Data from the National Safety Council (1993) suggest that 15 to 20% of workers in key industries (meatpacking, poultry processing, auto assembly, and garment manufacturing) are at potential risk for CTD and that in 1991 some 223,600 cases or 61% of all occupational injuries were associated with repetitive actions. The worst industry was manufacturing, while the worst occupational title was butchering with 222 CTD claims per 100,000 workers (Putz-Anderson, 1988). With such high rates and average costs of $30,000 per case, NIOSH, in its *Year 2000 Objectives,* has targeted the reduction of CTD incidence from 82 to 60 cases per 100,000 overall workers and from 285 to 150 in certain manufacturing industries (NIOSH, 1989a).

The proper selection, evaluation, and use of hand tools is a major ergonomic concern. The following review will discuss the basic principles involved in tool design and the desirable attributes for specific tools.

27.2 Principles and Problems of Tool Design

General Principles

An efficient tool has to fulfill some basic requirements (Drillis, 1963):

1. It must effectively perform the function for which it is intended. Thus, an axe should convert a maximum amount of its kinetic energy into useful chopping work, cleanly separate wood fibers, and be easily withdrawn.
2. It must be properly proportioned to the body dimensions of the operator to maximize efficiency of human involvement.
3. It must be designed to match the strength and work capacity of the operator. Thus, allowances have to be made for the gender, age, training, and physical fitness of the operator.
4. It should not cause undue fatigue, i.e., it should not demand unusual postures or practices that will require more energy expenditure than necessary.
5. It must provide sensory feedback in the form of pressure, some shock, texture, temperature, etc., to the user.
6. The capital and maintenance costs should be reasonable.

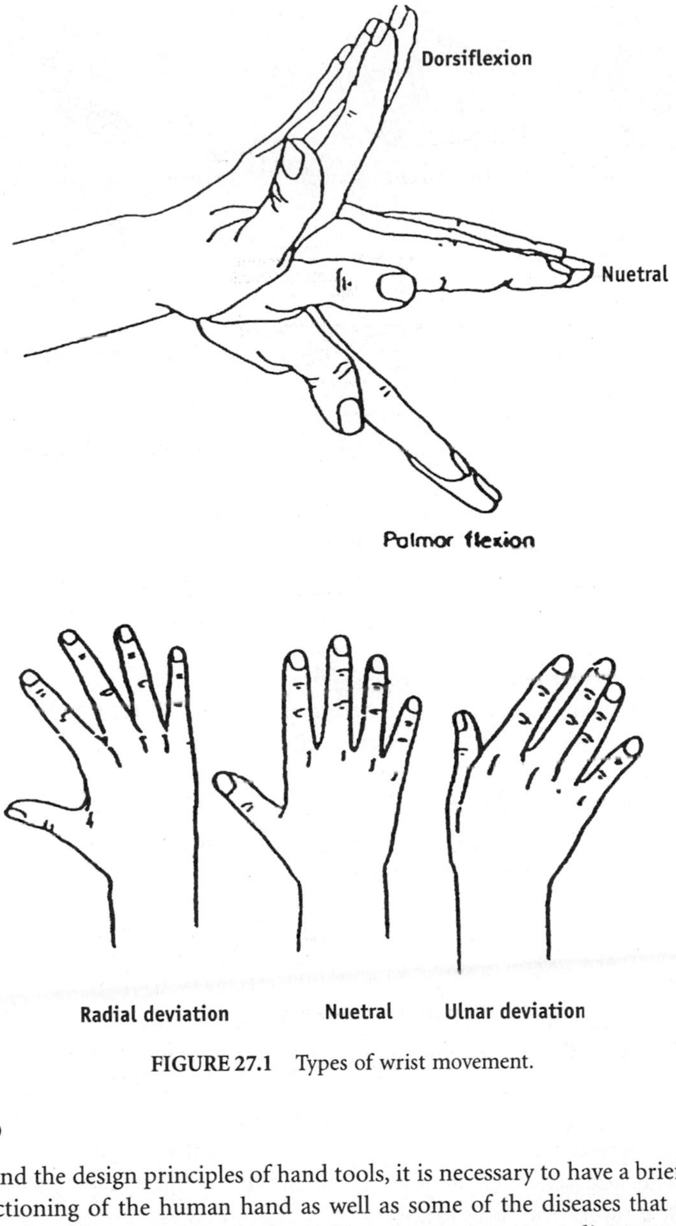

FIGURE 27.1 Types of wrist movement.

Types of Grip

To better understand the design principles of hand tools, it is necessary to have a brief description of the anatomy and functioning of the human hand as well as some of the diseases that can result from its misuse. The human hand is a complex structure of bones, arteries, nerves, ligaments, and tendons. The fingers are controlled by the extensor carpi and flexor carpi muscles in the forearm. The muscles are connected to the fingers by tendons which pass through a channel in the wrist, formed by the bones of the back of the hand on one side and the transverse carpal ligament on the other. Through this channel, called the carpal tunnel, pass also various arteries and nerves. The bones of the wrist connect to two long bones in the forearm, the ulna and the radius. The radius connects to the thumb side of the wrist and the ulna connects to the little finger side of the wrist. The orientation of the wrist joint allows movement in only two planes, each at 90° to the other (Figure 27.1). The first gives rise to *palmar flexion* and *dorsiflexion* (or *extension*). The second movement plane gives *ulnar* and *radial deviation.*

The manual dexterity produced by the hand can be defined in terms of a *power grip* and a *precision grip*. In a power grip, the tool, whose axis is more or less perpendicular to the forearm, is held in a clamp formed by the partly flexed fingers and the palm, with opposing pressure being applied by the thumb (Figure 27.2). There are three subcategories of the power grip differentiated by the line of action of force:

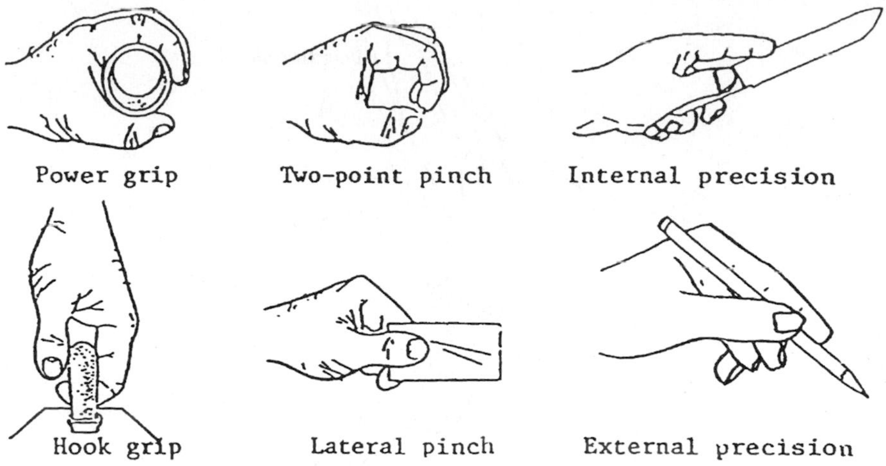

FIGURE 27.2 Types of grip.

(1) force parallel to the forearm, as in sawing; (2) force at an angle to the forearm, as in hammering; and (3) torque about the forearm, as when using a screwdriver. As the name implies, the power grip is used for power or for holding heavy objects.

In a precision grip, the tool is pinched between the flexor aspects of the finger and the opposing thumb. The relative position of the thumb and fingers determines how much force is to be applied and provides a sensory surface for receiving feedback necessary to give the precision needed. There are two types of precision grip: (1) internal, in which the shaft of the tool (e.g., knife) passes under the thumb and is thus internal to the hand; and (2) external, in which the shaft (e.g., pencil) passes over the thumb and is thus external to the hand. The precision grip is used for control. Other grips are just variations of the power or precision grip and include the hook grip, for holding a box or handle, a two-point pinch, and a lateral pinch (Figure 27.2).

Static Muscle Loading

When tools are used in situations in which the arms must be elevated or tools have to be held for extended periods, muscles of the shoulders, arms, and hands may be loaded statically, resulting in fatigue, reduced work capacity, and soreness. Abduction of the shoulder with corresponding elevation of the elbow will occur if work has to be done with a pistol-grip tool on a horizontal workplace. An in-line or straight tool reduces the need to raise the arm and also allows for a neutral wrist posture.

Prolonged work with arms extended can produce soreness in the forearm for assembly tasks done with force. By rearranging the workplace so as to keep the elbows at 90°, most of the problem can be eliminated (Figure 27.3). Of course, the orientation of the tool as related to the work surface can modify this posture. This will be discussed further in the next section. Similarly, continuous holding of an activation switch can result in fatigue of the fingers and reduced flexibility.

Awkward Wrist Position

As the wrist is moved from its neutral position, there is loss of grip strength. Starting from a neutral wrist position, full pronation decreases grip strength by 12%, full flexion/extension by 25%, and full radial/ulnar deviation by 15% (Terrell and Pursewell, 1970). The percent of maximum grip strength available can be quantified by ($r^2 = .854$, $p<.001$):

$$\%Grip = 98.6 - 8.8 \times PS - 25.2 \times FE - 16.3 \times RU \qquad (1)$$

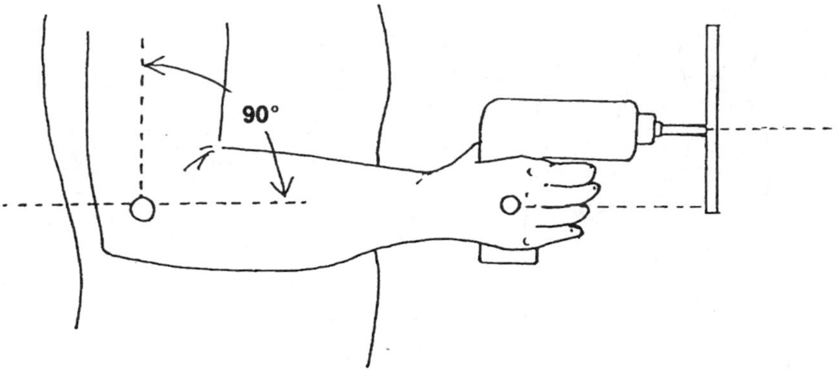

FIGURE 27.3 Optimum working posture with elbow bent at 90°.

where

PS = 1 if the wrist is fully pronated and 0 if in a neutral position or supinated
FE = 1 if the wrist is fully flexed or extended and 0 if in a neutral position
RU = 1 if the wrist is fully in radial or ulnar deviate and 0 if in a neutral position

Furthermore, awkward hand positions may result in soreness of the wrist, loss of grip, and, if sustained for extended periods of time, occurrence of *carpal tunnel syndrome*. To reduce this problem, the workplace or tools should be redesigned to allow for a straight wrist, i.e., lowering work surface and edges of containers, tilting jigs toward the user (Figure 27.4), using a pistol handle on power tools for vertical surfaces and in-line handles for horizontal surfaces (Figure 27.5), etc. Similarly, the tool handle should reflect the axis of grasp, such as a pistol grip on knives (Figure 27.6) (Armstrong et al., 1982).

Tissue Compression

Often, in the operation of hand tools, considerable force is applied by the hand. Such actions can concentrate considerable compressive force on the palm of the hand or the fingers, resulting in *ischemia*, obstruction of blood flow to the tissues, and eventual numbness and tingling of the fingers. Handles should be designed to have large contact surfaces to distribute the force over a larger area (Figure 27.7) or to direct it to less sensitive areas such as the tissue between the thumb and index finger. Similarly, finger grooves or recesses in tool handles should be avoided. Since hands vary considerably in size, the grooves will accommodate only a fraction of the population.

Gender

Female grip strength typically ranges from 50 to 67% of male strength (Konz, 1990; Chaffin and Andersson, 1984), i.e., the average male can be expected to exert approximately 500 N, while the average female can be expected to exert approximately 250 N. An interesting survey by Ducharme (1975) examined how tools and equipment that were physically inadequate for female workers hampered their performance. The worst offenders were crimpers, wire strippers and soldering irons. Females have a twofold disadvantage — an average lower strength and an average smaller grip span. Ducharme (1975) concluded that women could be integrated more quickly and safely into the work force if tools were designed to accommodate their smaller hand dimensions.

Handedness

Alternating hands permits reduction of local muscle fatigue. However, in many situations this is not possible because the tool use is one-handed. Furthermore, if the tool is designated for the user's preferred

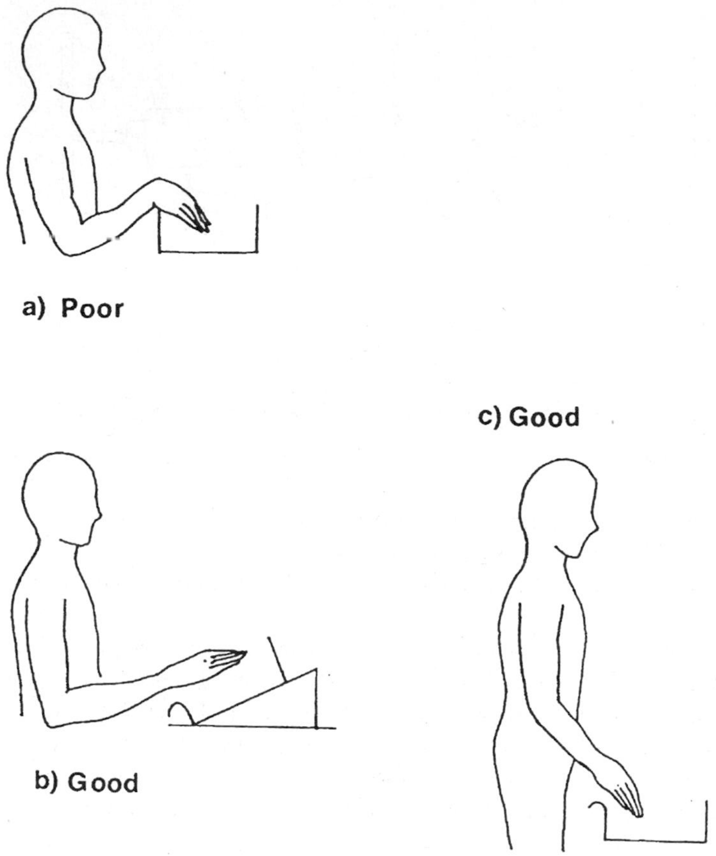

FIGURE 27.4 Proper orientation of jigs and containers.

hand — which for 90% of the population is the right hand — then 10% are left out (Konz, 1974). Laveson and Meyer (1976) gave several good examples of right-handed tools that cannot be used by a left-handed person, e.g., a power drill with side handle on the left side only, a circular saw, and a serrated knife beveled on one side only. Miller and Freivalds (1987) found right-handed males to show a 12% strength decrement in the left hand, while right-handed females showed a 7% strength decrement. Both left-handed males and females had nearly equal strengths in both hands. They concluded that left-handed subjects were forced to adapt to a right-handed world. Using time study, Konz and Warraich (1985) found decrements, ranging from 9% for an electric drill to 48% for manual scissors, for using the nonpreferred hand as opposed to the preferred hand.

Posture

In general, unless the posture is extreme, i.e., standing versus lying, torque exertion capability is not affected substantially by posture (Mital, 1986). The height at which torque was applied had no influence on peak torque exertion capability.

Repetitive Finger Action

If the index finger is used excessively for operating triggers, symptoms of *trigger finger* develop. Thus, trigger forces should be kept low, preferable below 10 N, to reduce the load on the index finger. Two- or three-finger operated controls are preferable (Figure 27.8); finger strip controls or a power grip bar are even better. For a two-handled tool, a spring-loaded return saves the fingers from having to return the

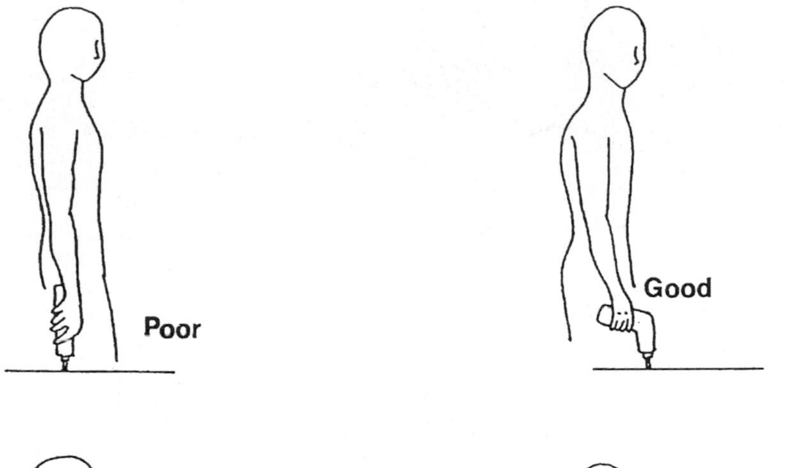

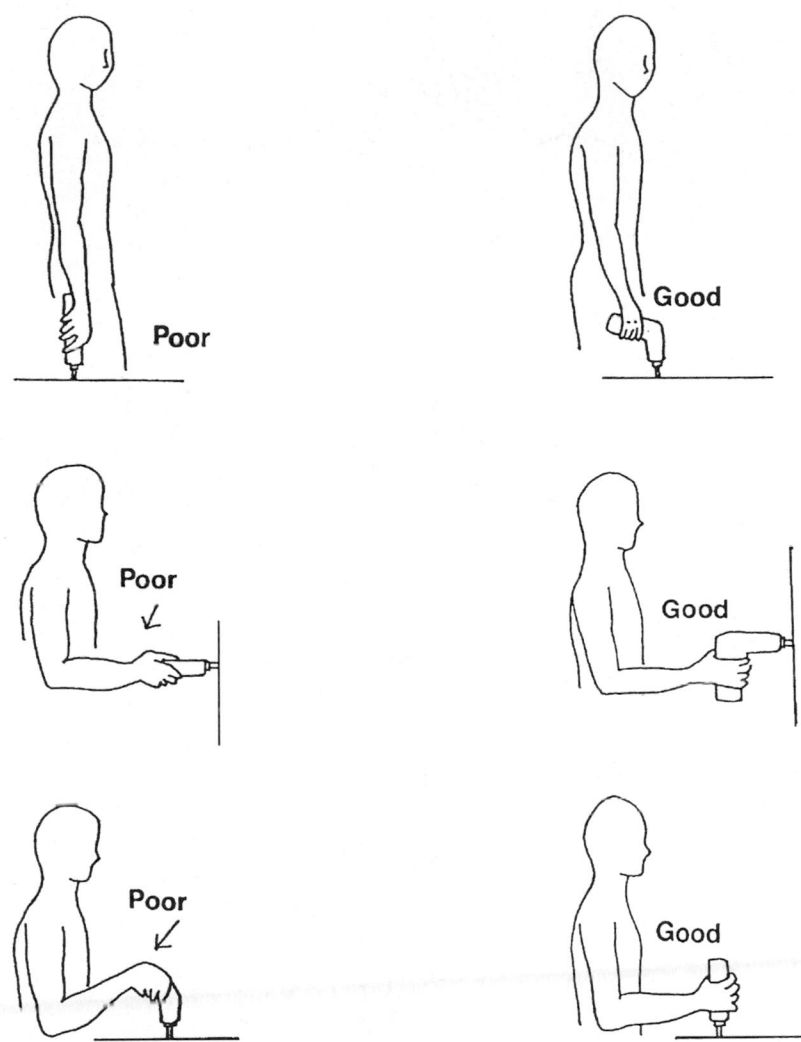

FIGURE 27.5 Proper orientation of power tools in the workplace.

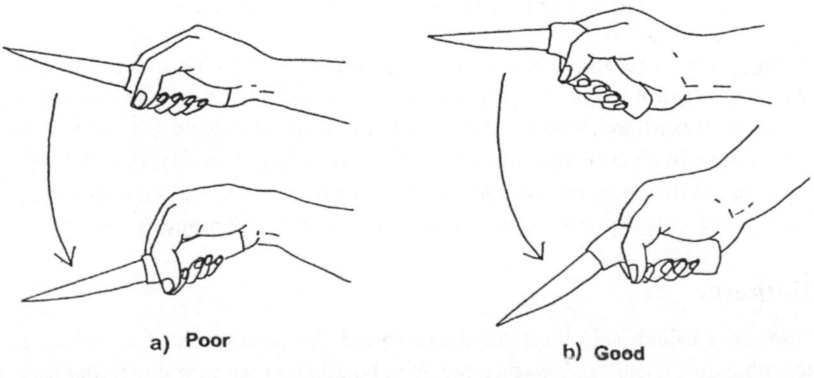

FIGURE 27.6 Pistol-grip knife.

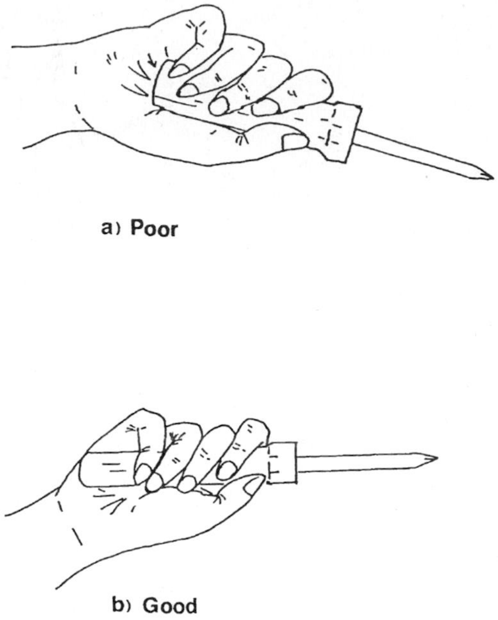

FIGURE 27.7 Avoiding tissue compression in tool design.

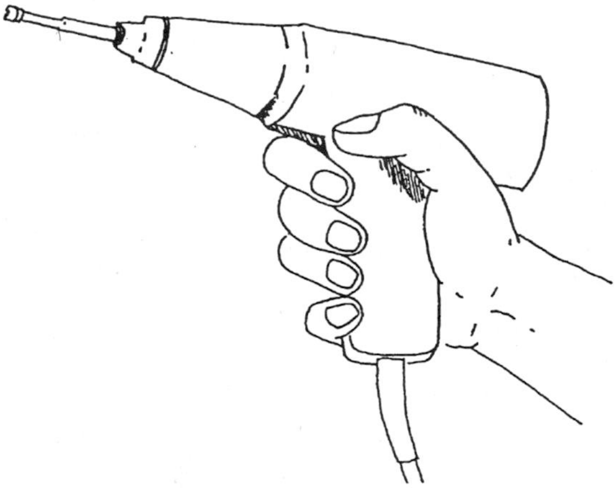

FIGURE 27.8 Three-finger trigger for power tools.

tool to its starting position (Eastman Kodak, 1983). In addition, the high number of repetitions must be reduced. Although critical levels of repetitions are not known, attempts have been made to identify the maximal number of exertions per hour or shift that can be tolerated. Most of these have been regarding wrist movements, but could reasonably apply to the fingers as well. Luopajarvi et al. (1979) found high rates of muscle–tendon disorders in assembly line packers with over 25,000 movements per day, while NIOSH (1989b) found similar problems in workers exceeding 10,000 motions per day.

Handle Diameter

Power grips around a cylindrical object should surround the circumference of the cylinder, with the fingers and thumb barely touching. However, the actual diameter may vary due to the task as well as the hand size of the operator. Thus, for a power grip on screwdrivers, Rubarth (1928) recommended a diameter of 40 mm. For minimum *EMG* activity, Ayoub and LoPresti (1971) found a 51-mm handle diameter to be best. However, based on the maximum number of work cycles completed before fatigue and on the ratio of grip force to EMG activity, they suggested a 38-mm diameter. For handles on boxes, Drury (1980) found diameters of 31 to 38 mm to be best in terms of the least reduction in grip strength. Using various handles of noncircular cross section, Cochran and Riley (1986) found the largest thrust forces in handles of 41-mm equivalent circular diameter (based on a total 130 mm circumference) for both males and females. For manipulation, however, smaller handles of 22 mm were found to be the best. Eastman Kodak (1983), based on company experience, recommended 30 to 40 mm with an optimum of 40 mm for power grips. Thus, one can summarize that handle diameters should be in the range of 30 to 50 mm, with the upper end best for maximum torque and the lower end for dexterity and speed.

Handle Length

The length of the handle has been studied to a lesser extent. For cut-out as well as normal handles, there should be enough space to admit all four fingers. Hand breadth across the metacarpals ranges from 71 mm for a 5th percentile female to 97 mm for a 95th percentile male (Garrett, 1971). Thus, 100 mm may be a reasonable minimum, but 120 mm may be more comfortable. Eastman Kodak (1983) recommended 120 mm. If the grip is enclosed or gloves are used, even larger openings are recommended. For an external precision grip, the tool shaft must be long enough to be supported at the base of the first finger or thumb. A minimum value of 100 mm is suggested (Konz,1990). For an internal precision grip, the tool should extend past the palm, but not so far as to hit the wrist (Figure 27.7) (Konz, 1990).

Handle Shape

Rubarth (1928) investigated handle shape and concluded that, for a power grip, one should design for maximum surface contact so as to minimize unit pressure of the hand. A tool with a circular cross section was found to give largest torque. Pheasant and O'Neill (1975) concluded that the precise shape of the handle was irrelevant and recommended simple knurled cylinders. Similarly, Cochran and Riley (1986) found that no one shape may be perfect, and that shape may be more dependent on the type of task and motions involved than initially thought. The circular cross section was found to be worst and a triangular one best for thrusting forces; triangular shape was slowest for a rolling type of manipulation, while a rectangular shape of height/width ratios from .67 to .8 appeared to be a good compromise for many tasks. A further advantage of a noncircular cross section is that the tool does not roll when placed on a table. It should also be noted that handles may not always have the shape of a true cylinder except for a hook grip. For screwdriver-type tools, the handle end is rounded to prevent undue pressure at the palm. For hammer type tools, the handle may have some flattening curving to indicate the end of the handle.

A final note on shape is that T-handles yield much better performance than straight screwdriver handles. Pheasant and O'Neill (1975) reported as much as a 50% increase in torque. Optimum handle diameter was found to be 25 mm and optimum angle was 60°, i.e., a slanted T (Saran, 1973). The slant allows the wrist to remain straight and, thus, generate larger forces.

Grip Surface, Texture, and Materials

For centuries wood was the material of choice for tool handles. Wood was readily available and easily worked. It has good resistance to shock and thermal and electrical conductivity, and has good frictional qualities even when wet. Since wooden handles can break and stain with grease and oil, there has been a shift to plastic and even metal. However, metal should be covered with rubber or leather to reduce shock and electrical conductivity and increase friction. Such compressible materials also dampen vibration and allow a better distribution of pressure, reducing the feeling of fatigue and hand tenderness (Fellows and Freivalds, 1991). The grip material, however, should not be too soft, otherwise sharp objects, such as metal chips, will get embedded in the grip and make it difficult to use. Grip surface area should be maximized to ensure a pressure distribution over as large an area as possible. Excessive localized pressure sometimes causes pain that forces workers to interrupt their work. Pressure/pain thresholds of around 500 kPa for females and 700 kPa for males have been found, with the thenar and os pisiforme areas being most sensitive (Fransson-Hall and Kilbom, 1993). During maximal power grips these values are greatly exceeded.

The frictional characteristics of the tool surface vary with the pressure exerted by the hand, the smoothness and porosity of the surface, and the type of contamination (Buchholz et al., 1988; Bobjer et al., 1993). Sweat increases the coefficient of friction, while oil and fat reduce it. Adhesive tape and suede provide good friction when moisture is present (Buchholz et al., 1988). The type of surface pattern as defined by the ratio of ridge area to groove area shows some interesting characteristics. When the hand is clean or sweaty, the maximum frictions were obtained with high ratios (i.e., maximizing the hand–surface contact area), while when the hand is contaminated, maximum frictions were obtained with low ratios (i.e., maximizing the capacity to channel away contaminants) (Bobjer et al., 1993).

Angulation of Handle

As discussed previously, deviations of the wrist from the neutral position under repetitive load can lead to a variety of cumulative trauma disorders as well as decreased performance. Therefore, angulation of tool handles, e.g., power tools, may be necessary so as to maintain a straight wrist. The handle should reflect the axis of grasp, i.e., about 78° from the horizontal, and should be oriented so the eventual tool axis is in line with the index finger (Fraser, 1980). This principle has been applied to various tools such as pliers and soldering irons, as mentioned previously.

An interesting extension of this concept has been promoted as Bennett's handle (Emanual et al., 1980). Bennett developed this concept based on the angle formed by the index finger and the life line under the thumb. This angle of 19°, used for his handles, is claimed to maintain a straight wrist, generate increased strength and control, and reduce stress, shock, and fatigue. Konz (1986) conducted a variety of tests to evaluate the effectiveness of Bennett's handle on a hammer in comparison with a standard hammer. Subjects preferred a 10° bend but performed no better than with a standard hammer. Knowlton and Gilbert (1983) used cinematography to evaluate a curved and conventional claw hammer. Bilateral grip strength was measured before and after a task, nail driving. The curved hammer produced a smaller strength decrement and caused less ulnar deviation than the conventional hammer. Thus, a bent handle may give some benefits.

Grip Span for Two-Handled Tools

Grip strength and the resulting stress on finger flexor tendons vary with the size of the object being grasped. A maximum grip strength is achieved at about 45 to 50 mm on a dynamometer with parallel sides, or about 75 to 80 mm on a dynamometer with handles angled inward (Chaffin and Andersson, 1984). At distances different from the optimum, percent grip strength decreases (Figure 27.9) as defined by ($r^2 = .99$, $p<.001$):

$$\%\text{Grip} = 100 - .11 \times S - 10.2 \times S^2 \tag{2}$$

where S = given grip span minus optimum grip span in cm.

Because of the large variation in individual strength capacities, and to accommodate 95% of the population, maximal grip requirements should be limited to less than 90 N.

Weight

For non-striking applications, the weight of the hand tool will determine how long it can be held or used and how precisely it can be manipulated. For tools held in one hand with the elbow at 90° for extended periods of time, Eastman Kodak (1983) recommended a load of no more than 2.3 kg. For precision operations, tool weights greater than .4 kg are not recommended unless a counter-balanced system is

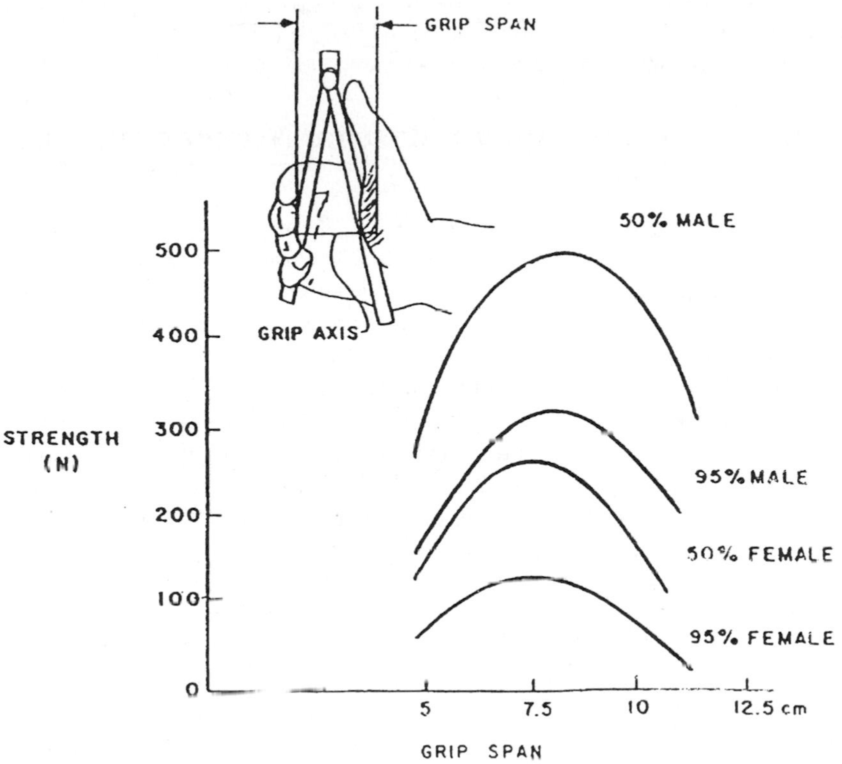

FIGURE 27.9 Grip strength as a function of grip span. (From Chaffin, D.B. and Andersson, G. 1984. *Occupational Biomechanics*. John Wiley & Sons, New York. With permission.)

used. Heavy tools, used to absorb impact or vibration, should be mounted on a truck to reduce effort for the operator. In addition, the tool should be well balanced, with its center of gravity as close as possible to the center of gravity of the hand (unless the purpose of the tool is to transfer force, as in a hammer). Thus, the hand or arm muscles do not need to oppose any torque development by an unbalanced tool.

Gloves

Gloves are often used with hand tools for safety and comfort. Safety gloves are seldom bulky, but gloves worn in subfreezing climates can be very heavy and interfere with grasping ability. Wearing woolen or leather gloves may add 5 mm to the hand thickness and 8 mm to the hand breadth at the thumb, while heavy mittens add 25 mm and 40 mm, respectively (Damon et al., 1966). More important, gloves reduce grip strength 10 to 20% (Cochran et al., 1986) and manual dexterity performance times 12 to 64% (Weidman, 1970). Neoprene gloves slowed performance times by 13% over barehanded performance, over terry cloth by 36%, over leather by 45%, and over PVC by 64%. Thus, there is a trade-off to be considered between increased injury and reduced performance without gloves and reduced performance with gloves. Perhaps the tool can be redesigned even more to reduce the need for gloves.

Vibration

Vibration is a separate and very complex problem with powered hand tools. Vibration can induce *white finger* syndrome, the primary symptom of which is a reduction in blood flow to the fingers and hand due to *vasoconstriction* of the blood vessels. As a result, there is a loss of sensory feedback and decreased performance. In addition, vibration may contribute to the development of carpal tunnel syndrome, especially in jobs with a combination of forceful and repetitive exertions (Silverstein et al., 1987). It is generally recommended that vibrations in the critical range of 2 to 200 Hz (Lundstrom and Johansson, 1986) be avoided. The exposure to vibration can be reduced through a reduction in the driving force and the use of vibration-damping materials on either the tool or the hand (gloves) (Andersson, 1990).

27.3 Attributes of Common Industrial Nonpowered Tools

Shovels

Shovels are used to lift, move, and toss loose dirt or sand. The blade is fastened to the shaft through a socket which, if stamped from a flat sheet, is generally rolled over to form a crimp known as a frog. The shaft may either taper to an end or have a handle. The handle traditionally has been of a T form, but more lately is of a D form. Long shafts are generally 1.2 to 1.7 m long, while the short D handle is about .7 m long. However, for an unconstrained posture and task, long-handled shovels are 18% more efficient (in terms of energy expenditure per sand shoveled) than short-handled shovels (Freivalds, 1986b). The external angle of the shaft with respect to the blade is called the lift and provides the tool with added leverage. Based on a trade-off between smaller low-back compressive forces and lower energy cost, Freivalds (1986b) found the optimum lift angle to be approximately 32°. Shovel weight should be as small as possible, especially if one considers that this is unproductive weight that has to be lifted along with the load moved. Weights should be below 1.5 kg (Freivalds and Kim, 1990). Blade size depends very much on the density of the material being handled; the less dense the material the larger the blade size. For a given material (foundry sand), Freivalds and Kim (1990) found that a blade-size/weight ratio of .068 m^2/kg is optimum, i.e., for a 1.5 kg shovel the blade size is .1 m^2 with a resulting load weighing 4.4 kg.

In terms of maximum amount of sand shoveled per given time period, the optimum shoveling rate is in the range of 18 to 21 scoops/min. The optimum shovel load ranges from 5 to 11 kg, depending on the decision criterion to be used, i.e., for high rates of shoveling (18 to 20 scoops/min), the lower end of the load range (5 to 7 kg) may be more appropriate (which follows the principle of reducing static loading), while for lower rates (6 to 8 scoops/min), the higher end of the load range (8 to 11 kg) may be acceptable (which follows the principles of increasing efficiency with larger loads).

This was determined by F. W. Taylor in the first scientific study on shoveling at the Bethlehem Steel Works in 1898, in which he was able to reduce the manpower requirements for material handling from 600 to 140 men (Taylor, 1913).

Shoveling throw height is a trade-off between increased efficiency for higher heights at a cost of an increased energy expenditure. Since shoveling performance stays reasonably constant up to a height of 1.3 m, an acceptable throw height may be as high as 1 to 1.3 m. Similarly, shoveling performance remains fairly constant up to a distance of 1.2 m (Freivalds, 1986a).

Hammers

Hammers are striking tools designed to transmit a force to an object by direct contact and thereby change its shape or drive it forward. The tool's efficiency in doing this may be defined as the ratio of the energy utilized in striking to the energy available in the stroke. This efficiency is maximized by placing the tool mass center as close as possible to the center of action, i.e., increasing the mass of the tool head relative to the handle. Another aim is to transform as much of the kinetic energy of the hammer into deforming an object's shape as possible. Thus, the mass of the hammer should be small relative to the mass of the forging and anvil. On the other hand, in driving a nail, the intent is to transform the kinetic energy of the hammer into the kinetic energy of the nail. Then the mass of the hammer should be great in relation to the mass of the nail. The overall mechanical efficiency for hammering a 15-cm nail into a wooden block can be as high as 57% (Drillis, 1963). However, there is a limit to the weight that can be placed in the head of the hammer. Increasing the head weight decreases angular velocity and ultimately the total kinetic energy, in addition to increasing physiological energy costs (Widule et al., 1978; Corrigan et al., 1981).

Saws

The action of heavy sawing requires a power grip with repetitive flexion and extension at the elbow, while the action of light sawing involves a precision grip with manipulation of the wrist. For the former, pistol grips are utilized, while for the latter a cylindrical screwdriver-type handle provides the best precision grip. Gläser (1933) found that for forestry work, a two-handed action provided more force and better performance but at a higher energy cost. Most efficient was the kneeling posture in which less torso support was needed and less energy was expended. Typically, Western saws cut as they are pushed through the wood, while Oriental saws cut as they are pulled through the wood. Although sawing times were not significantly different, energy expenditure was significantly lower (26%) for the Oriental saws (Bleed et al., 1982).

Pliers

Pliers and related tools — wire strippers, pincers, and nippers — are tools with a head in the form of jaws; jaws can have a variety of configurations, i.e., joints which may be simple or complex. Sometimes the handles are straight, but more typically they are curved outward to conform roughly to the shape of the grasp. The grasp, depending on use, can be of the precision or power type. In their simple form, pliers are a very common tool and, if used casually for short periods of time, will give reasonable performance with little fatigue. However, the relationship of the handles to the head forces the wrist into ulnar deviation, a posture which cannot be held repeatedly or for prolonged periods of time without fatigue or occurrence of cumulative trauma disorders. A further problem is that ulnar deviation reduces the range of wrist rotation by 50%, possibly reducing productivity. By bending the handles of the pliers instead of the wrist, Tichauer (1976) was able to reduce the stress on the operator's wrists and overall injury rates by a factor of six at Western Electric Co.

Other factors to be considered in the design of pliers were detailed by Lindstrom (1973): a working grip (subjectively chosen on an adjustable grip) width of 90 mm for men and 80 mm for women and a handle length of 110 mm for men and 100 m for women. For repeated or continuous operation, the required working strength should not exceed 33 to 50% of the individual's maximum strength. To minimize the applied pressure to the soft tissue of the palm (keep less than 200 to 400 kPa), the handles should be enlarged and flattened. Thus, indentation of the handles for the fingers is undesirable. Encasing the basic metal handles in a rubber or plastic sheath provides insulation and improves the tactile feel. Also, spring-loaded handles eliminate the need to manually open the handles.

Screwdrivers

The handles of screwdrivers (and similar tools: files, chisels, etc.) can either be used with a precision grip for stabilization or a power grip for torque. Crucial factors are also the size, shape, and texture of the handle. Applied torque increases with an increase in the diameter of the handle (Pheasant and O'Neill, 1975). Differences in the precise shape of handles appears to be less significant, as long as the hand does not slip around the handle (Cochran and Riley, 1982). Thus, knurled cylinders allow for significantly greater torque production than smooth cylinders. Further details can be found in the previous section on handle shape.

Knives

Although a very old tool, the knife has recently appeared in the literature as a possible cause for the increase in cumulative trauma disorders in food processing due to the long hours of static loading on the forearm flexors and the slippery handles requiring high grip forces (Armstrong et al., 1982; Karlqvist, 1984). For poultry processing, Armstrong et al. (1982) suggested a pistol-type grip (Figure 27.6) to allow the operator to hold the blade and the forearm horizontal so as to eliminate ulnar deviation and wrist flexion. A circular or elliptical handle with a circumference of 99 mm, as well as a strap, were recommended to allow the hand to relax between exertions without losing the grip on the knife. Similarly, for the fish canning industry, Karlqvist (1984) fitted some knives with a pistol-grip handle and others with larger diameter handles for better balance and movement. Note that one knife design will not be best for all jobs.

27.4 Attributes of Common Industrial Power Tools

Power Drills

In a power drill, or other power tools, the major function of the operator is to hold, stabilize, and monitor the tool against a workpiece, while the tool performs the main effort of the job. Although the operator may at times need to shift or orient the tool, the main function for the operator is to effectively grasp and hold the tool. A drill is comprised of a head, body, and handle, with all three, ideally, being in line. The line of action is from the line of the extended index finger so that in the ideal drill, the head is off-center with respect to the central axis of the body. Handle configuration is important, with the choices being pistol-grip, in-line, or right-angle. As a rule of thumb, in-line and right-angle are best for tightening downward on a horizontal surface, while pistol-grips are best for tightening on a vertical surface (Figure 27.5) with the aim being to obtain a standing posture with a straight back, upper arms hanging down, and a straight wrist (Figure 27.3). For the pistol grip, this results in the handle being at an angle of approximately 78° with the horizontal (Fraser, 1980).

Another important factor is the center of gravity. If it is too far forward in the body of the tool, a turning moment is created, which must be overcome by the muscle of the hand and forearm, creating muscular effort additional to that required for holding, positioning, and pushing the drill into the workpiece. The primary handle should be placed directly under the center of gravity, such that the body juts out behind the handle as well as in front. For heavy drills, a secondary supportive handle may be needed, either to the side or preferably below the tool, such that the supporting arm can be tucked in against the body rather than being abducted. Also tool balancers should be utilized for heavy tools (Fraser, 1980).

Nutrunners

Nutrunners, especially common in the automobile industry, are used to tighten nuts, screws, and other fasteners. They come in a variety of handle configurations, torque outputs, shut-off mechanisms, speeds, weights, and spindle diameters. Torque levels range from .1 to 5000 Nm and, for pneumatic tools, are generally lumped into approximately 22 power levels (M1.6 to M45), depending on motor size and gearing required to drive the tool. The torque is transferred from the motor to the spindle through a variety of mechanisms such that the power (often air) can be quickly shut off once the nut or other fastener is tight. The simplest and cheapest mechanism is a direct-drive, which is under the operator's control, but, because of the relatively long time it takes to release the trigger once the nut is tightened, transfers a very large *reaction torque* to the operator's arm. Mechanical friction clutches will allow the spindle to slip, reducing some of this reaction torque. A better mechanism for reducing the reaction torque is the air-flow shut-off which automatically senses when to cut off the air supply as the nut is tightened. A still faster mechanism is an automatic mechanical clutch shut-off. The most recent mechanisms include the hydraulic pulse system where the rotational energy from the motor is transferred over a pulse unit containing an oil cushion (filtering off the high-frequency pulses as well as noise) and a similar electrical pulse system, both of which, to a large extent, reduce the reaction torque (Freivalds and Eklund, 1993).

Variation of torque delivered to the nut depends on a variety of conditions including properties of the tool, the operator of the tool, properties of the joint, i.e., the combination of the fastener and material being fastened (ranging from soft, with the materials having elastic properties such as body panels, to hard, when two stiff surfaces, such as pulleys on a crankshaft, are brought together), stability of the air supply, etc. The torque experienced by the user (the reaction torque) depends on the above factors plus the torque shut-off system and may contribute to the development of cumulative trauma disorders. In general, using electrical tools at lower than normal rpm levels or underpowering pneumatic tools resulted in larger reaction torques and more stressful ratings. Pulse-type tools produced the lowest reaction torques and were rated as less stressful. It was hypothesized that the short pulses "chop up" or allow the inertia of the tool to resist the reaction torque. Another possibility is to provide reaction torque bars (Freivalds and Eklund, 1993).

27.5 Discussion

Currently, the most important issue regarding tool design is the reduction in the potential for the development of cumulative trauma disorders. Until about 10 years ago, tool usage was little changed from the days of the Industrial Revolution. The operator used tools manually in the manufacturing of goods. The operations required considerable forces, which was somewhat leveraged by the appropriate tool. Because of the manual nature of the tasks and the forces involved, the operations were fairly slow. With the advent of automation, the excessive force levels were eliminated, and many task elements relegated to the human could be eliminated. The operator performs a smaller part of the original task, which now can be speeded up because the machine handles most of the work. Unfortunately, the elements still left to the human operator become more limited in scope and thus more repetitious in nature. This incomplete and unergonomic automation has led to an upsurge in CTD cases, especially if the repetition is combined with excessive wrist deviations and forceful exertions. The threshold for injury for any single risk factor is not known, let alone for a combination of factors. There is also, probably, a trade-off between each of those factors that need to be quantified for exposure and threshold levels.

Tied in with frequency is the trade-off with productivity. Any reduction in frequency of tool usage may have a direct result in decreasing productivity. One alternative to maintaining constant productivity is to rotate operators for a critically repetitive task. But then again it is necessary to know threshold levels of frequency and how much rest, at which intervals, must be provided to recover from the trauma induced from repetitive tool usage. Also, it is important to know whether performing a greater variety of movements from those in the injurious task will allow the body to recover or will only delay recovery. These are all issues that haven't been fully addressed.

Another issue not fully resolved is the trade-off between manual and power tool usage. Most researchers, based on the limited human force capacity and greater fatiguability as compared to machines, have advocated the use of power tools. Unfortunately, power tools, whether powered electrically or pneumatically, produce some vibration. Vibration damping typically requires either an increase in the inertial mass (at the cost of increasing the weight of the tool and increasing the fatigue of the user) or vibration-absorbing systems which introduce a bit of "slop" in the hand–handle interface absorbing the vibrations (again at the cost of reducing control of the tool). Power tools, also, have a tendency to produce reaction torques, which can be reduced by using pulse-type tools (at the cost of increasing vibration) or by using reaction bars (again at the cost of limiting the control or maneuverability of the tool). These are issues that need to be clarified further.

Recent research indicates that power grip capabilities can be increased by a better understanding of the pressure distribution of the hand while using a tool or by improving the frictional characteristics of the tool handle surface. Perhaps the development or application of new polymers to tool handles can improve the efficiency of tool usage. Also new ways of measuring the hand–handle interface, such as the "data glove" of Yun et al. (1996), can provide more accurate information on this topic.

Most current work has addressed the power grip for tools. However, most power requirements are being fulfilled by machines, leaving the human operator to perform more precise tasks that currently cannot be easily replicated by machines. Unfortunately, there is very little information on precision or pinch grips and the precision aspects of tools. Questions on grip design and force exertion capabilities for precision grips, as well as occupational injury risk during work with high demands on precision, need to be studied further.

Epidemiological considerations are also important in substantiating proper ergonomic designs. Unfortunately, at present, there are few good studies that support good ergonomic tool design or clearly indicate the deficiencies in such designs. More injury data for both hand and powered tools are needed.

A final but very important consideration is the adaptation of tools for a more diverse population. With the aging of the worker population and the passage of the Americans with Disabilities Act, it is imperative that tools also be usable by individuals with a wide range of capabilities. This is both a challenge and an opportunity for ergonomists and tool designers to put their skills to effective use.

Defining Terms

Carpal tunnel syndrome: Compression of the median nerve in the carpal tunnel of the wrist resulting in pain, tingling, and/or paralysis in the fingers.

Cumulative trauma disorders: The collection of problems occurring in the upper extremities due to repetitive motion.

Dorsiflexion: *See Extension.*

EMG (electromyography): Electrical activity of the muscle.

Extension: Movement that decreases the angle between two adjacent bones.

Flexion: Movement that decreases the angle between two adjacent bones.

Ischemia: Occlusion of blood flow in an artery.

Palmar flexion: *See Flexion.*

Power grip: Hand grip such that the thumb opposes partly flexed fingers, barely overlapping. Provides maximum power.

Precision grip: Hand grip such that the thumb opposes only the first or second fingers, resulting in a much lower gripping force, but greater precision.

Radial deviation: Bending the wrist in the direction of the thumb.

Reaction torque: The transfer of torque to the operator's arms as a nut tightens and before the tool cuts power.

Trigger finger: Swollen tendon sheath resulting in the tendon being locked, such that attempts to move the finger cause a snapping or jerking movement.

Ulnar deviation: Bending the wrist in the direction of the little finger.

Vasoconstriction: Ischemia of the peripheral blood flow.

White finger syndrome: Occupation vibration syndrome characterized by finger blanching due to ischemia of the digital arteries.

References

Andersson, E.R. 1990. Design and testing of a vibration attenuating handle, *Int. J. Indus. Erg.* 6:119-125.

Armstrong, T.J., Foulke, J.A., Joseph, B.S., and Goldstein, S.A. 1982. Investigation of cumulative trauma disorders in a poultry processing plant. *Am. Indus. Hygiene Assoc. J.* 43:103-116.

Ayoub, M. and LoPresti, P. 1971. The determination of an optimum size cylindrical handle by use of electromyography. *Ergonomics* 14:509-518.

Bleed, A.S., Bleed, P., Cochran, D.J., and Riley, M.W. 1982. A performance comparison of Japanese and American hand saws, *Proc. Human Factors Soc.* Santa Monica, CA, 26:403-407.

Bobjer, O., Johansson, S.E., and Piguet, S. 1993. Friction between hand and handle. Effects of oil and lard on textured and non-textured surfaces; perception of discomfort. *Appl. Erg.* 24:190-202.

Buchholz, B., Frederick, L.J., and Armstrong, T.J. 1988. An investigation of human palmar skin friction and the effects of materials, pinch force and moisture. *Ergonomics* 31:317-325.

Chaffin, D.B. and Andersson, G. 1984. *Occupational Biomechanics.* John Wiley & Sons, New York.

Cochran, D.J., Albin, T.J., Riley, M.W., and Bishu, R.R. 1986. Analysis of grasp force degradation with commercially available gloves. *Proc. Human Factors Soc.* Santa Monica, CA. 30:852-855.

Cochran, D.J. and Riley, M.W. 1986. An evaluation of knife handle guarding. *Human Factors* 28:295-301.

Corrigan, D.L., Foley, V., and Widule, C.J. 1981. Axe use efficiency — a work theory explanation of an historical trend. *Ergonomics* 24:103-109.

Damon, A., Stoudt, H.W., and McFarland, R.A. 1966. *The Human Body in Equipment Design.* Harvard University Press, Cambridge, MA.

Drillis, R.J. 1963. Folk norms and biomechanics. *Human Factors.* 5:427-441.

Drury, C.G. 1980. Handles for manual materials handling. *Appl. Erg.,* 11:35-42.

Ducharme, R.E. 1975. Problem tools for women. *Indus. Eng.* Sep:46-50.

Eastman Kodak Co. 1983. *Ergonomic Design for People at Work.* Lifetime Learning Pub., Belmont, CA.

Emanual, J.T., Mills, S.J., and Bennett, J.F. 1980. In search of a better handle. *Proc. Symp.* Human Factors Indus. Design Consumer Products. Tufts University, Medford, MA, pp. 34-40.

Fellows, G.L. and Freivalds, A. 1991. Ergonomics evaluation of a foam rubber grip for tool handles. *Appl. Erg.* 22:225-230.

Fraser, T.M. 1980. *Ergonomic Principles in the Design of Hand Tools.* International Labour Office. Geneva, Switzerland.

Fransson-Hall, C. and Kilbom, Å. 1993. Sensitivity of the hand to surface pressure. *Appl. Erg.* 24:181-189.

Freivalds, A. 1986a. The ergonomics of shovelling and shovel design — a review of the literature. *Ergonomics* 29:3-18.

Freivalds, A. 1986b. The ergonomics of shovelling and shovel design — an experimental study, *Ergonomics* 29:19-30.

Freivalds, A. and Eklund, J. 1993. Reaction torques and operator stress while using powered nutrunners, *Appl. Erg.* 24:158-164.

Freivalds, A. and Kim, Y.J. 1990. Blade size and weight effects in shovel design, *Appl. Erg.* 21:39-42.

Garrett, J. 1971. The adult human hand: some anthropometric and biomechanical considerations. *Human Factors.* 13:117-131.

Gläser, H. 1933. *Beiträge zur Form der Waldsäge und zur Technik des Sägens,* Ph.D. Dissertation, Everswalde, Germany.

Karlqvist, L. 1984. Cutting operations at canning bench — a case study of handtool design, *Proc. 1984 Inter. Conf. Occup. Erg.* Human Factors Association of Canada, Rexdale, Ont. pp. 452-456.

Knowlton, R.G. and Gilbert, J.C. 1983. Ulnar deviation and short term strength reductions as affected by a curve handled ripping hammer and a conventional claw hammer, *Ergonomics* 26:173-179.

Konz, S. 1974. Design of handtools. *Proceedings Human Factors Soc.* Santa Monica, CA, 18:292-300.

Konz, S. 1986. Bent hammer handles. *Human Factors* 27:317-323.

Konz, S. 1995. *Work Design.* 4th ed. Publishing Horizons. Worthington, OH.

Konz, S. and Warraich, M. 1985. Performance differences between the preferred and non-preferred hand when using various tools. *Ergonomics International '85* (I.D. Brown, R. Goldsmith, K. Coombes and M.A. Sinclair, eds.), Taylor & Francis, London, pp. 451-453.

Laveson, J.K. and Meyer, R.P. 1976. Left out "lefties" in design. *Proc. Human Factors Soc.* 20:122-125.

Leakey, L.S.B. 1960. Finding the world's earliest man. *Nat. Geog.* 118:420-435.

Lehmann, G. 1953. *Praktische Arbeitsphysiologie.* Thieme Verlag, Stuttgart, Germany.

Lindstrom, F.E. 1973. *Modern Pliers.* Bahco Verktyg, Enköping, Sweden.

Lundstrom, R. and Johansson, R.S. 1986. Acute impairment of the sensitivity of skin mechanoreceptive units caused by vibration exposure of the hand. *Ergonomics* 29:687-698.

Luopajarvi, T., Kuorinka, I., Virolainen, M., and Holmberg, M. 1979. Prevalence of tenosynovitis and other injuries of the upper extremities in repetitive work. *Scand. J. Work Env. Health.* 5, Sup. 3:48-55.

Miller, G. and Freivalds, A. 1987. Gender and handedness in grip strength. *Proc. Human Factors Soc.* Santa Monica, CA, 31:906-909.

Mital, A. 1986. Effects of body posture and common hand tools on peak torque exertion capabilities, *Appl. Erg.* 17:87-96.

Napier, J. 1962. The evolution of the hand. *Sci. Am.* 207:56-62.

National Safety Council. 1993. *Accident Facts.* Chicago, IL.

NIOSH. 1989a. *Occupational Safety and Health, Year 2000 Objectives.* National Institute for Occupational Safety and Health, Center for Disease Control, Atlanta, GA.

NIOSH. 1989b. *Health Hazard Evaluation — Eagle Convex Glass, Co.,* HETA-89-137-2005, National Institute for Occupational Safety and Health, Cincinnati, OH.

Pheasant. S.T. and O'Neill, D. 1975. Performance in gripping and turning — a study in hand\handle effectiveness. *Appl. Erg.* 6:205-208.

Putz-Anderson, V. 1988. *Cumulative Trauma Disorders.* Taylor & Francis, London.

Rubarth, B. 1928. Untersuchung zur Festgestaltung von Handheften für Schraubenzieher und ähnliche Werkzeuge. *Industrielle Psychotechnik* 5:129-142.

Saran, C. 1973. Biomechanical evaluation of T-handles for a pronation supination task. *J. Occup. Med.* 15:712-716.

Silverstein, B.A., Fine, L.J., and Armstrong, T.J. 1987. Occupational factors and carpal tunnel syndrome. *Am. J. Indus. Med.* 11:343-358.

Taylor, F.W. 1913. *The Principles of Scientific Management.* Harper & Bros. New York.

Terrell, R. and Purswell, J. 1976. The influence of forearm and wrist orientation on static grip strength as a design criterion for hand tools. *Proc. Human Factors Soc.* Santa Monica, CA, 20:28-32.

Tichauer, E.R. 1976. Biomechanics sustains occupational safety and health. *Indust. Eng.* pp. 46-56 (Feb).

Weidman, B. 1970. *Effect of Safety Gloves on Simulated Work Tasks.* AD 738981, National Technical Information Service, Springfield, VA.

Widule, C.J., Foley, V., and Demo, F. 1978. Dynamics of the axe swing, *Ergonomics* 21:925-930.

Yun, M.H., Cannon, D., Freivalds, A., and Thomas, G. 1997. An instrumented glove for grasp specification in virtual reality based point and direct telerobotics, *IEEE Trans. Sys. Man Cyber.* 27: 835-847.

For Further Information

The article by R. J. Drillis, *Folk norms and biomechanics* is a good introduction for the historical evolution of tools.

Ergonomic Principles in the Design of Hand Tools by T.M. Fraser is a very good overall reference for tool design.

V. Putz Anderson's *Cumulative Trauma Disorders* is a good introduction into cumulative trauma disorders that may result from poor job practices and poor tool design.

Ergonomic Design for People at Work by the Human Factors Section at Eastman Kodak Co. has many very good practical applications of tool design as related to industrial practice.

(See References for complete citations.)

The Human Factors and Ergonomics Society (P.O. Box 1369, Santa Monica, CA 90406, USA) and The Ergonomics Society (Devonshire House, Devonshire Sq., Loughborough, Leic. LE11 3DW, UK) are good sources of information, various publications, newsletters, annual conferences, etc. that relate to ergonomics in general.

28

Computer-Aided Design and Human Models

J. Mark Porter
Loughborough University

Keith Case
Loughborough University

Martin T. Freer
Loughborough University

28.1 Introduction

With the CAD/CAM systems available today it is quite possible for some products to progress from concept design through production without requiring full-size physical models or mock-ups to perform any necessary evaluations. This helps to reduce the time scale of product design considerably. However, as most CAD/CAM systems provide little or no information concerning the needs of the end user, there is considerable danger that design decisions may only consider the engineering, styling, legislative, and financial constraints for the product, with the ergonomics issues comparatively unassessed until the design is completed.

It is essential that the ergonomics input take place throughout the design process, but nowhere is it more important than at the concept and early development stages. Basic ergonomics criteria, such as the adoption of healthy and efficient postures for the range of future users, need to be satisfied very early because there is usually only limited scope for modification later on without considerable financial and time penalties. Traditionally, the various ergonomics criteria were assessed during a product's development by conducting user trials with hand-built prototypes. These prototypes were not constructed just in order to perform ergonomics trials, they were often made first and foremost to visualize the product in 3D and to determine how to efficiently deal with the legal and manufacturing issues. As the visualization and engineering functions are now increasingly being performed with digital prototypes or mock-ups displayed on computer graphics terminals, it is clear that there is an urgent need to provide computer-based ergonomics assessment functions as well. The computer modeling of people (known as *man-modeling* since its inception in the 1960s but now increasingly termed *human modeling* in the 1990s) provides the ability to construct 2D or 3D models from anthropometric data which can be articulated between the body segments to simulate a wide variety of postures. These human models can then be used, in conjunction with the CAD model of the product being designed, to conduct computer-based user trials to assess criteria such as fit, reach, vision, and the resulting constraints upon posture. Such predictions enable the ergonomist to be more proactive in the design process and to be able to work

0-8493-2641-9/99/$0.00+$.50

closely with the other design team members to achieve ergonomic solutions to the design within the various financial, legal, engineering, and aesthetic constraints.

28.2 Human Models, Past and Present

Kinematic modeling enables the spatial evaluation of workplaces where either the human or parts of the physical environment are placed in different positions over time. This type of modeling is the focus of this chapter, while kinetic (or dynamic) modeling, usually associated with assessing the body's response to large external forces such as those experienced in car crash simulations, is not covered.

Kinematic human modeling and global systems, past and present, include ADAPS (Delft University, The Netherlands, see Post and Smeets, 1981), ANYBODY and ANTHROPOS (IST GmbH, Germany), APOLIN (Grobelny et al., 1992), BOEMAN (Boeing Co., USA), BUFORD (Rockwell International, USA), CAR (Naval Air Development Centre, USA), COMBIMAN and CREW CHIEF (Armstrong Aerospace Medical Research Laboratory, USA, see McDaniel, 1990), CYBERMAN (Chrysler Co., USA), Envision/ERGO (Deneb Robotics Inc., USA), ERGODATA (Laboratoire d'Anthropologie Appliquée. France), ERGOMAN (see Coblenz et al., 1991), ergoSHAPE (Institute of Occupational Health, Finland, see Launis and Lehtela, 1992, 2D only), ergoSPACE (Institute of Occupational Health, Finland, see Launis and Lehtela, 1990), FRANKY (G.I.T., Germany, see Elias and Lux, 1986), JACK (University of Pennsylvania, USA, see Badler et al., 1993), MDHMS (McDonnell Douglas, USA), MANNEQUIN (Biomechanics Corporation of America), MINTAC (Kuopio Regional Institute of Occupational Health and the University of Oulu, Finland, see Kuusisto and Mattila, 1990), RAMSIS (BMW and other car manufacturers, Germany), SAFEWORK (Genicom Consultants, Canada, see Fortin et al., 1990), SAMMIE (SAMMIE CAD Ltd. and Loughborough University, UK, see Porter et al., 1995), TADAPS (University of Twente, The Netherlands, see Westerink et al., 1990) and WERNER (Institute of Occupational Health, University of Dortmund, Germany, see Kloke, 1990).

Comparisons between some of these systems can be found in Dooley (1982), Rothwell (1985), Porter et al. (1993, 1995), and Das and Sengupta (1995), and it is not the intent of this chapter to present a detailed description of each human modeling system. It is important to appreciate that the quality of a product's ergonomics has more to do with the design team's judgment and ability to incorporate sound ergonomics principles in the design than to the use of any specific human modeling system (Das and Sengupta, 1995). Such systems do not automate the design process by creating ergonomics solutions to a set of specified inputs, rather they should be regarded as tools to be used by the design team. An increasing number of system developers and users have created information pages on the world wide web. Figures 28.1 through 28.5 show plots of the human models from JACK (as used by MIDAS, a U.S. Army–NASA product), MDHMS, and SAMMIE. These plots were taken from their respective web sites listed at the end of this chapter. Figures 28.6 and 28.7 show plots from SAFEWORK and Envision/ERGO.

The differences between the systems can be examined in terms of a number of features.

- The complexity of the human model (e.g., 2D or 3D, number of body segments, surface details for visualization or specified only by anthropometry)
- Whether the joint angles are constrained to possible angles or whether impossible postures could be inadvertently set
- The anthropometric databases available
- The extent of control over the size of individual body segments (fixed, linear scaling possible or direct control with data
- Whether there are extensive CAD facilities integrated fully with the human model or whether the design model has to be ported to and from another CAD system
- The ability of the workplace modeler to provide functional modeling (e.g., several components can be modeled and collectively called "driver's door", which can be rotated as one unit around the hinge point), hidden line views (lines behind solids removed), color surface shading, reflections, shadows and textures

FIGURE 28.1 MDHMS model of an aircraft maintenance operation. (Source: MDHMS web page.)

FIGURE 28.2 JACK models showing the use of textures and shadowing for imparting added realism. (Source: University of Pennsylvania web page.)

- How the system can assess the human model's reach, fit, or vision (the flexibility of these assessments varies considerably from simple reach to a point to automated volumetric reach, from visually inspected clearance to automated intersecting solid detection routines, and from the display of eye point location to the display of perspective views and mirror views from either eye)
- Whether the system provides strength data or calculates torque loads on selected joints.

The various systems cost from as little as a few hundred U.S. dollars up to $60,000 for a software license. Some systems can run on a PC, but many require a Sun or Silicon Graphics workstation or equivalent. Usability is a key requirement of such systems, and a fast response time often requires a high specification computer.

FIGURE 28.3 JACK performing a simple reach task in a cockpit. (Source: MIDAS (US Army-NASA) web page.)

FIGURE 28.4 JACK wearing protective clothing and equipment for aircrew. (Source: MIDAS (US Army-NASA) web page.)

Basic Functionality of the Systems

These systems are intended to be used as a predictive tool for the assessment of the capabilities of people when interacting with the designed physical environment. The basic functionality that is required is listed below in bold, in each case followed by a brief discussion of the relevant issues:

- **3D modeling of people** of the selected sex, age, nationality, and occupational groups. This is achieved using published anthropometric data, if indeed it exists for the population being examined.

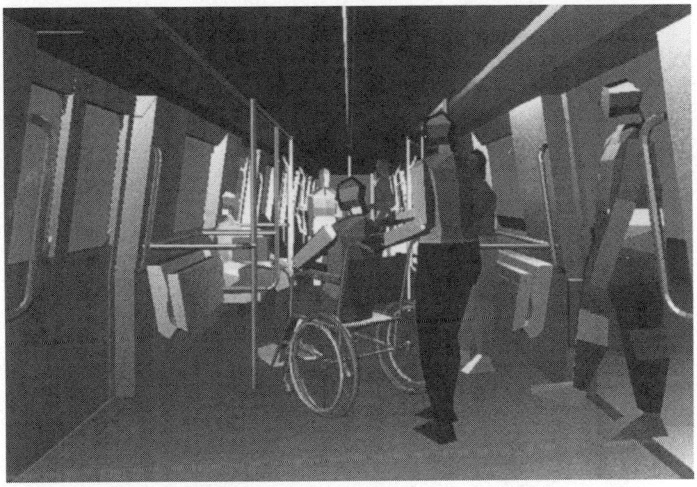

FIGURE 28.5 SAMMIE models used to evaluate accommodation in a railway carriage. (Source: Loughborough University web page.)

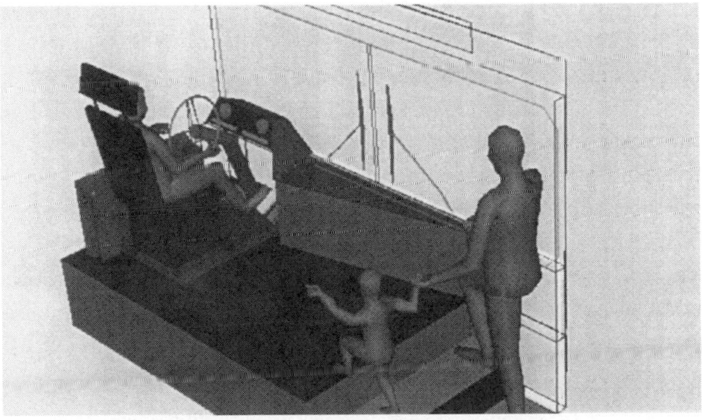

FIGURE 28.6 Design of a bus driver's cabin using SAFEWORK. (Courtesy of Safework Incorporated.)

FIGURE 28.7 Simulation of fuel tank assembly using Envision/ERGO. (Courtesy of Deneb Robotics Incorporated.)

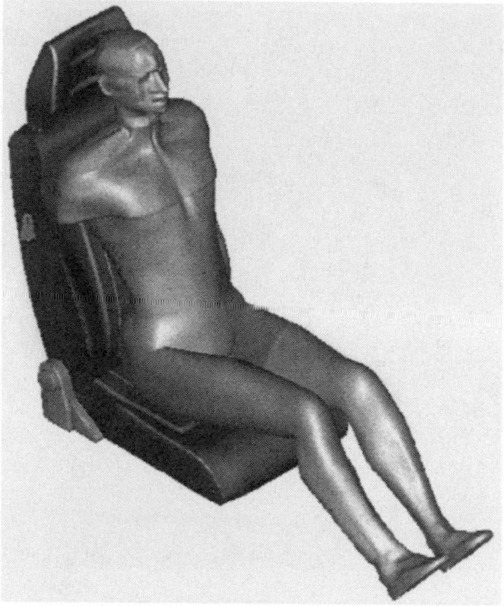

FIGURE 28.8 SAMMIE model of an individual person recorded by the LASS bodyscanner system.

The current databases have several shortcomings, basically because they were established with little consideration for the needs of 3D human modeling systems. For example, surveys record external body dimensions, whereas computer models need joint-to-joint dimensions in addition. The limited number of anthropometric dimensions recorded in surveys leave many gaps when having to fully define a 3D computer model. Should the human model remain as true as possible to the real data or should artistic license be granted to model more "realistic" models? The danger with the latter approach is that the designer (be it a stylist, industrial designer, engineer, or ergonomist) may come to believe the "added" data and, for example, feel confident that it is possible to design seat profiles based upon such highly detailed, but fictitious, models.

The relatively recent technique of body scanning, whereby thousands of data points can be recorded from the surface of the body, makes it possible to model individual people with considerable accuracy. Figure 28.8 shows a SAMMIE model constructed using scanned data from the LASS system (e.g., Jones et al., 1989).

- **knowledge base of comfort angles** for the major joints of the body

Human models come with various numbers of joints. Those with relatively few (e.g., fewer than 20) do not have detailed models of the hands or spine. With such details the number of joints can be well over 100. This large number of degrees of freedom in the human model's posture poses problems for the user who has to decide how to position the model realistically. The problem is made easier in some systems with the provision of automated reach tests, inverse kinematics, and grasping behaviors such that the model's hand can reach, grasp, and operate specified handles. This is done automatically, ensuring that the various joint angles do not exceed maximum or comfortable ranges as specified in the published literature. Such data on comfort angles are widely available for application areas such as computer workstations and cars. However, closer examination often reveals disagreement in the literature or the recognition that the recommended postural angles are based only on theoretical analysis. For example, when assessing a computer workstation design, should the human models be positioned sitting upright with a 90 degree trunk–thigh angle as generally recommended by many sources, or should the seat have a reclined backrest (Grandjean et al., 1984) or a forward tilting seat cushion (Mandal, 1984)?

The interrelationships between joints, such as the knee and hip, are typically not considered when using comfort angle recommendations. For example, the range of comfortable backrest angles is affected by any constraints to the knee angle, such as is experienced in a low sports car seat.

- ability to **model the proposed workstation in 3D, together with the simulation of ranges of adjustment** to be incorporated into the design.

FIGURE 28.9 SAMMIE evaluation of the Fiat Punto before full-size prototypes were available for road trials.

This "working" model of the product being developed is an essential part of a human modeling ergonomics design system because the human model needs to interact with the design in order to assess the physical characteristics of the interface. The requirements for an ergonomics model of a prototype design are, however, substantially different from the needs of other forms of CAD systems because the extremely detailed geometric information from an engineering CAD package is rarely required for ergonomics evaluations. Furthermore, human modeling systems should be used at the concept stages in design in order to help define the initial design specification, rather than just evaluating it at a later stage after the engineering criteria have been satisfied. Engineering CAD models rarely have the functionality of the various components under investigation embedded in their data structure (e.g., seat adjustment ranges, mirror rotation constraints), so these must be added to the ergonomics model. In many cases, it will be easier to create specific models for an ergonomics evaluation rather than simply transfer in detailed engineering models. Some human modeling systems have their own integrated CAD facilities which permit modifications to both the human models and the product being designed to be easily carried out. Other human modeling systems have only very crude CAD facilities, and these work in association with established commercial CAD systems, requiring the porting from system to system of either the human model or the product model. Software standards are continuing to be developed to improve the porting of geometric and functional information across different makes of computer hardware and software (Case et al., 1991).

- ability to **assess the kinematic interaction between the models of people and the workstation,** specifically in terms of the issues of user fit (e.g., headroom and legroom in a car), reach (e.g., to the steering wheel, gear selector, and pedals), and vision (e.g., of the road environment, both directly and in the mirrors, and the instrument binnacle).

The assessments focus on whether or not the people modeled can work efficiently at the workstation and can adopt a "comfortable" posture (i.e., within the ranges of joint angles considered acceptable). Figure 28.9 shows a SAMMIE model of the prototype Fiat Punto car in which a large male driver is simulating reversing the car, simultaneously assessing reach to the clutch pedal with one foot, reach to the steering wheel with one hand, and reach to the gear selector with the other, twisting in the seat and assessing vision around the head restraint, past the rear seat occupants, and through the rear window to the window environment. The same analysis can be conducted with a small female driver with the seat, steering wheel, and head restraint adjusted to suit her needs, within the ranges specified by the prototype design.

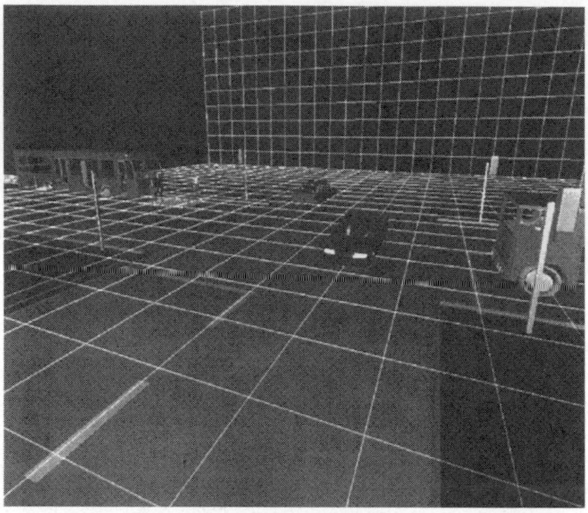

FIGURE 28.10 General view showing layout of SAMMIE models of vehicles, pedestrians, and road infrastructure.

FIGURE 28.11 Small male driver's view (5th percentile sitting eye height) from the coach shown on the left of Figure 28.10. Note that only two pedestrians are seen crossing in front of the cab.

Figures 28.10 through 28.12 show how simply and clearly the interactions between people, machines, and the physical environment can be presented. Figure 28.10 shows a general perspective view of a SAMMIE model of a road junction with various vehicles and pedestrians in position for the subsequent evaluation of driver vision. Figures 28.11 and 28.12 show a small and large male driver's view from one of the vehicles (which required the prior positioning of these drivers within the cab using the adjustment ranges as appropriate). This evaluation revealed that small drivers are more at risk of not being able to see children crossing directly in front of the cab.

- ability to make **iterative modifications to the design** to achieve optimum compromises.

Figure 28.13 shows one possible solution to the problem of poor driver vision directly in front of the cab. This solution involved the specification of the size, orientation, and radius of curvature of a supplementary mirror so that the range of drivers could see a wide-angle view across the front of the cab. Iterative modifications to the workstation model can be easily made to improve the human models' posture. Some systems provide information on static strength or calculate torque loads on certain joints,

FIGURE 28.12 Large male coach driver's view (99th percentile sitting eye height) of the same scene as shown in Figure 28.12. Note that the large driver can see the child crossing in front of the cab, but the small driver cannot.

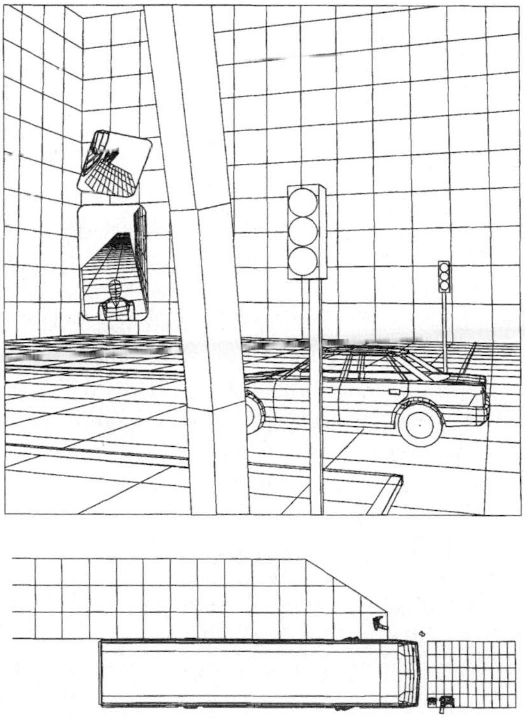

FIGURE 28.13 Coach driver's view in the exterior mirrors. The lower mirror shows the reflected field of view alongside the coach, including a legislative field of view at ground level and a pedestrian. The upper mirror reflects a wide-angle view across the front of the cab showing a child with an adult.

providing information to help identify more efficient designs in this respect. Human modeling systems have most to offer at the concept stage of design when they can be used to explore possible options for a design. Design is all about working within constraints, and sometimes challenging these constraints, to achieve the best compromises.

28.3 Benefits of Using Human Modeling CAD Systems

SAMMIE CAD has operated an ergonomics design consultancy service since 1978 and has now completed more than 150 commercial projects for more than 40 national and international clients. Some of these projects are discussed to illustrate the benefits derived from the use of a 3D human modeling CAD system.

The Formal Specification of the Future Users

It is crucial to the success of a design project to determine exactly who the intended users of a design will be. While seemingly an obvious starting point, it is often not at all clear in the client's mind. Any SAMMIE evaluation requires the investigation of how well human variability, in terms of size and shape, is accommodated by that design. This forces the client to make important decisions about acceptable accommodation range (e.g., 5th to 95th percentile or wider for a particular dimension) and the user population in terms of nationality, sex, and age groups at the earliest stage of design. For example, in an evaluation of a helicopter redesign we were able to demonstrate to the client that the existing aircraft chosen as a starting point, initially without particular regard to the users, was not capable of accommodating the population extremes (97.5th percentile Dutch male pilots and 25th percentile female pilots of other European nationalities) without structural changes so great as to warrant an almost completely new airframe. As a consequence, the project was aborted at an early stage, well before any full-size mock-ups were constructed or other major development costs incurred.

The Formal Specification of the Tasks

The next step is to help the client to establish a clear definition of all the tasks the users will be required to perform so they can be simulated in the evaluation. Often, this process identifies conflicts between various task functions. For example, SAMMIE was used in the design of the Brussels Tram 2000 (Figures 28.14 through 28.17). It was established that the driver had two equally important but conflicting tasks, namely driving the vehicle and selling tickets to passengers. A cab designed to allow ease of operation, optimum visibility and comfortable postures while driving was found to be severely compromised by the requirement to have the driver swivel around and sell tickets while remaining seated (given insufficient space for the driver to stand during ticketing operations). Since SAMMIE is a visual medium, it was possible to clearly demonstrate the problem to the rest of the design team and together look for solutions by quickly developing and investigating a variety of alternative seat swivel mechanisms and rotation points in the SAMMIE model. By group effort, a mechanism was developed that allowed the seat to move and swivel so that both tasks could be easily accomplished and that was feasible, cost effective, and did not require major changes to the cab or console structure.

The Formal Consideration of Other Factors

Because human models are used as predictive tools, it is important to have many other factors specified while conducting any evaluations. Traditionally, these other factors were often identified at the working prototype testing stage. This creates the need for an early dialogue with all stakeholders in the design process. These other factors are wide ranging, covering all aspects of the people involved, their job design, as well as the organizational and psychosocial issues. Another important consideration often overlooked by the client is the physical environment and its possible effects upon user task performance. A recent project examined control design for a new European fighter aircraft in which the control would only be used when the aircraft was "out of control". This posed several issues which the engineers had not considered because the pilot had always expected to be "in control" when considering the design of other controls. The motion conditions under which this particular control might be used are so severe that the "normal" usability criteria for acceptable reach and vision identified were totally inappropriate.

FIGURE 28.14 Stylist's concept sketches for the Brussels Tram 2000. (Courtesy of Design Triangle, UK).

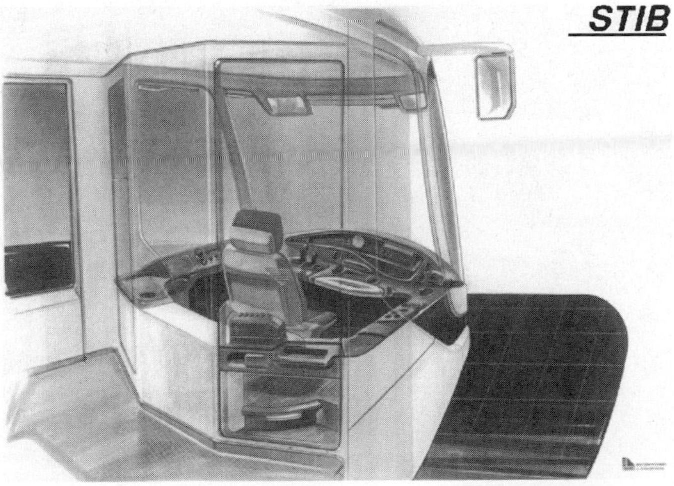

FIGURE 28.15 Stylist's rendering of a SAMMIE model showing a prototype design for the driver's cab in the Tram 2000. Note the ticketing desk behind the seat, which posed postural problems for the driver. The identified solution was to provide a seat which swivelled with offset centers of rotation so that it swung closer to the desk.

Proactive Ergonomics

Having specified who the future users will be, the tasks they will perform, and under what circumstances, it is then possible to construct appropriate human models and position them in "working" postures with only the simplest of workstation models. Areas of common reach and vision for various sized human models can be identified for the placement of the primary controls and displays before these items have

FIGURE 28.16 Advertising poster showing the new Tram 2000 in operation.

FIGURE 28.17 SAMMIE model used in the development of the driver's cab and the passenger areas.

been actually designed in detail. The minimum volume of space required by the human models to adopt their various task postures is easily observed at the earliest stages of design, and this provides the possibility of ensuring that sufficient clearances are provided as the design progresses. Such information and feedback can be provided throughout the development process, allowing the ergonomics issues to be considered proactively. This simultaneous consideration of people issues and engineering issues promotes the identification of optimum compromises, which are essential for a successful design. Such an approach was used in the design of the Lightweight Sports Car, a prototype car exhibited at the 1996 U.K. Motor Show. SAMMIE models of the defined future drivers and passengers (ranging from 99th percentile U.S.

FIGURE 28.18 SAMMIE model ported into Alias for developing the styling and engineering around the defined future occupants. (Photograph courtesy of Dr. C.S. Porter, Coventry University, UK.)

dimensions for both occupants down to 5th percentile Japanese male dimensions) were positioned in comfortable and appropriate postures before being ported into Alias, a 3D computer-aided styling system (Figure 28.18). This enabled the styling and engineering to develop around the human models (Porter and Porter, 1998), ensuring that the final design was both functional and aesthetic.

Reduction in Project Time-Scale

Another recent project involved the development of the driver's cab for the new Amsterdam tram, which provides a good example of the time savings achievable. The SAMMIE model was based on the bare minimum of engineering hard points (including the floor plan, the external body work, the crash resistant structure, and the rear wall) as soon as they were established. With a detailed ergonomics specification of the users and their vision and posture requirements, it was possible to quickly determine the required seat movement envelope and begin to develop a set of surfaces for controls and displays based upon the reach and vision capabilities of the user population (Figure 28.19). The engineers were provided with 3D coordinate and modeling data for an ergonomically designed workstation from which they could build their own CAD model within a matter of days. A mock-up was in fact built directly from the SAMMIE model of the driver's cab (Figure 28.20) and evaluated by a sample of Dutch tram drivers. The design was fully accepted without any changes being made to the SAMMIE design.

Iterative Design and Evaluation

SAMMIE CAD was involved in the development of the driver's cab for the new Lantau express train for Hong Kong's new airport. The designers (Design Triangle, U.K.) sketched a number of exterior forms for initial development (Figure 28.21). The SAMMIE model of the cab structure was built within a single day, and work was started to develop a suitable driver's workstation well in advance of any engineering drawing or engineering CAD work (Figures 28.22 and 28.23). The client subsequently had several changes of mind regarding the external form, which required major changes to the cab body and reduced the space available inside. In addition, the cab space was further reduced as the electrical equipment requirement for space grew, and the carriages were also shortened to enable them to pass through tunnels with a tighter radius of turn than usual. These changes were made to the model as they arose, allowing the assessment of their effect on the workstation ergonomics immediately. Indeed the client later decided that the passenger emergency evacuation route had to be through the front of the train, which effectively cut the cab into three parts. SAMMIE was used to explore how this requirement might be accommodated

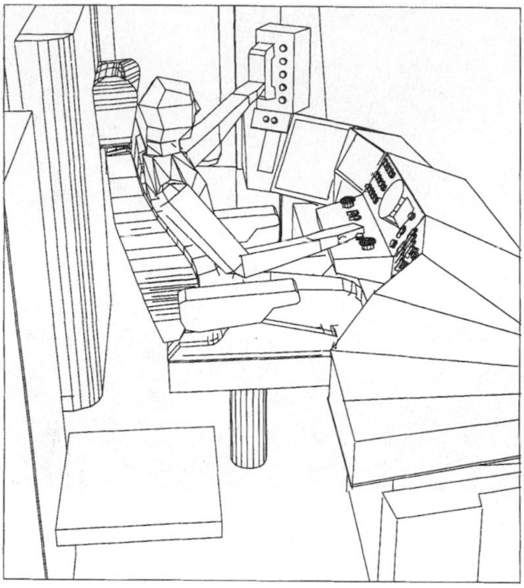

FIGURE 28.19 SAMMIE model of the driver's cab for the Amsterdam Tram was developed before any detailed engineering took place.

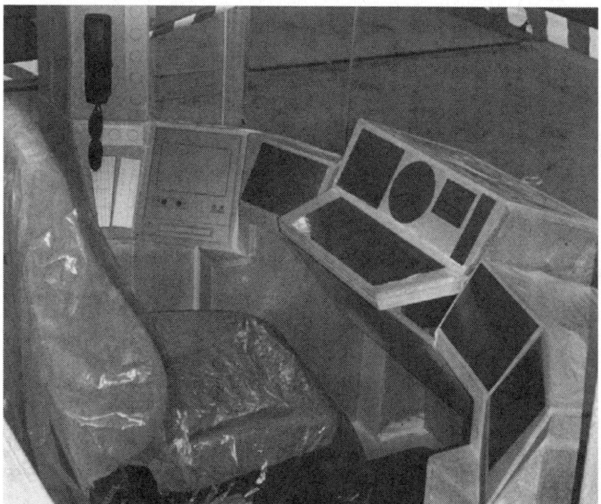

FIGURE 28.20 This mock-up of the Amsterdam Tram was made exactly to the SAMMIE specification and successful trials with Dutch drivers were conducted.

with the minimum number of compromises, mostly by reducing the amount and size of equipment required by the driver in order to fit a usable workstation into the smaller space. One novel solution that arose from this was the provision of a chair that can be slid away from the workstation and into a recess in the rear wall (Figure 28.24) to improve cross-cab access and allow sit/stand operation, while still providing a high-quality seat system. (There was insufficient space to swivel a seat, and commercially available flip-up seating would not satisfy the stated comfort criteria.) A full-size mock-up was built from the SAMMIE design (see Figure 28.25) which was used to confirm the easy evacuation of passengers in an emergency.

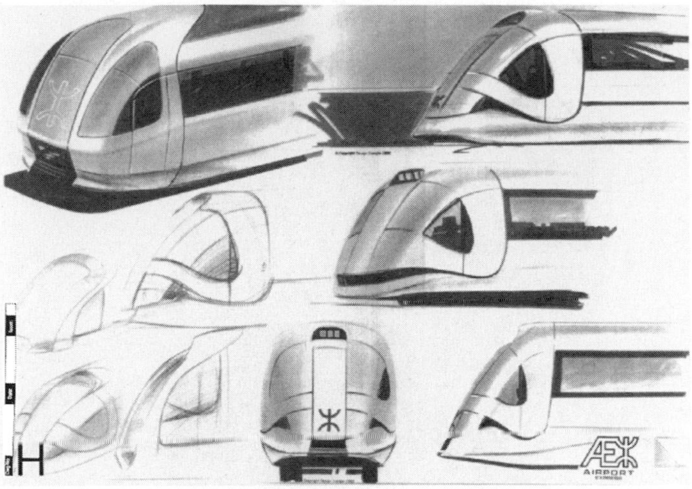

FIGURE 28.21 Stylist's concept sketches for the Lantau Line Express Train. (Courtesy of Design Triangle Ltd., UK.)

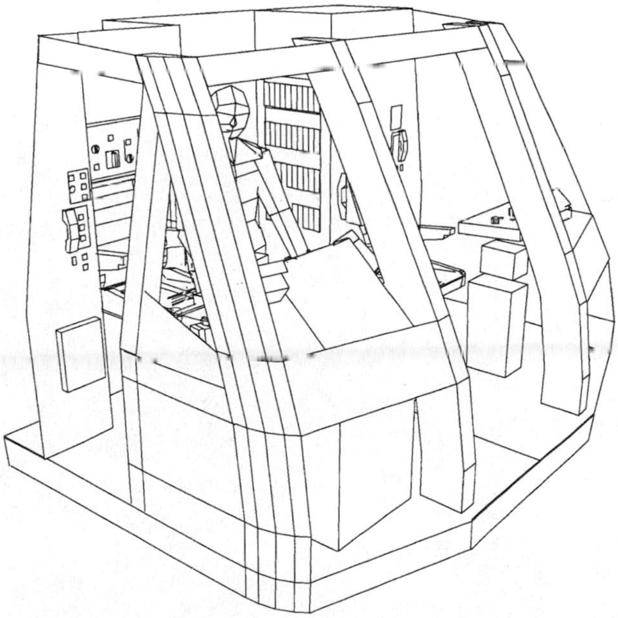

FIGURE 28.22 SAMMIE model of the Express Train cab showing the effects of the structural design and electrical requirements upon the accommodation space.

Cost-Effective Ergonomics

Any reduction in development times should be advantageous from a financial point of view. In addition, the cost of a sophisticated human modeling system is often less than the cost of making just one full-size mock-up using simple materials such as wood and glass fiber. Fitting trials or similar studies also incur extra costs for space rental, staff costs, subject payment, and so on. Because human modeling CAD systems enable the ergonomics input to be provided much earlier in the design process, this reduces the likelihood of expensive modifications being necessary at later stages.

FIGURE 28.23 Stylist's rendering of a SAMMIE model of the driver's workstation for the Express Train. (Courtesy of Design Triangle, UK.)

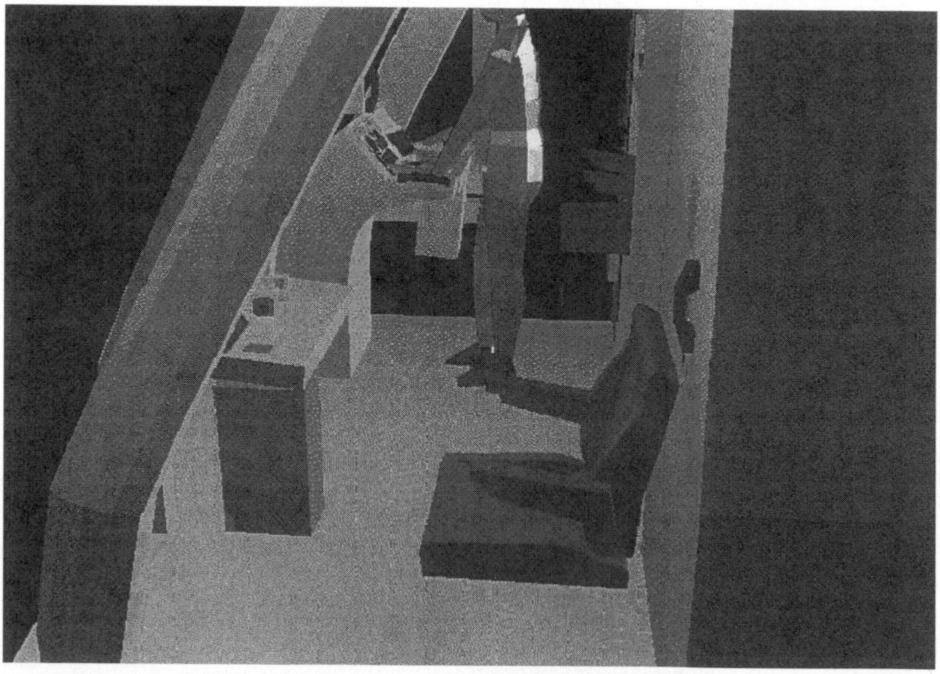

FIGURE 28.24 SAMMIE model of the Express Train cab showing the sit/stand workstation with the seat recessed to improve cross cab access.

Improved Communication

The ergonomist can work most effectively when in collaboration with other members rather than acting as a critic assessing the efforts of others. Collaboration is encouraged by human modeling CAD systems as the system can act as a focal point for the design team even before detailed engineering

FIGURE 28.25 Full size mock-up of the Express Train cab constructed for user trials in Hong Kong.

development work commences. The ergonomics problems with a proposed design can be presented visually and accurately, both in perspective with color surface rendering or as engineering drawings. This ensures efficient communication within the design team and leads naturally to solution-oriented action. The use of a human modeling CAD system is systematic and objective in its approach, which enables all stakeholders in a project (such as the designers, manufacturers, installers, operators, maintainers, recyclers) to examine any assumptions and constraints and to question the conclusions drawn. They can easily visualize any design problems identified and also have a direct involvement with the investigation of alternatives. This gives the ergonomist the opportunity to be proactive and to support the other design team members using communication methods (i.e., computer graphics) that are completely natural for them.

SAMMIE uses CAD techniques both as a tool for ergonomics analyses and as a medium for communication. The work is conducted "on screen" and it requires a high degree of interaction between the members of the design team. The working computer models are often very simple, and this helps to focus on the important ergonomics issues. However, at the completion of this work, we often spend nearly as much time again constructing a more detailed and aesthetically pleasing model, and this helps to impart a greater sense of validity. It is much more persuasive to the client to embody a good ergonomics specification into a CAD model than it is to present a written report listing ergonomics recommendations.

28.4 Issues of Validity

We have always advocated that human modeling systems should not replace user trials with full-size mock-ups, unless the design or the design modifications are so simple as to not warrant concern. In-depth user trials can reveal problems with so many more issues including long-term discomfort, effects of fatigue, negative transfer of training, error rate, performance, and the acceptance of the product. Many designers, engineers and ergonomists are expectantly waiting for the all-singing, all-dancing human modeling system to appear. The likelihood of such a system being developed in the near future seems remote. However, the advantage of using human modeling systems is that it is possible to build full-size mock-ups with the confidence that few, if any, modifications will be necessary to physically accommodate the users. The detailed evaluation of criteria such as those above can proceed without delay and without the extra costs of getting the basics right.

Human-modeling CAD systems have the potential to offer considerably more validity as a simulation of people than the traditional 2D manikins that are overlaid on drawings. This is because they can be made to simulate specific user groups and tasks in three dimensions. Also, 2D manikins are often used in a very simplistic way without a knowledge of their origin in terms of user type (e.g., military/civilian data, age range, nationality, size, date of survey) and application (e.g., erect or slumped sitting height, percentile values for individual body segments). For example, designers may have a 5th percentile adult female manikin and 50th and 95th percentile adult male manikins. The nature of these manikins gives support to the notion that people come either short and short limbed, tall and long limbed, or somewhere in between. It has been repeatedly demonstrated that this is not true and that the intercorrelation between body dimensions is rather poor (e.g., Haslegrave, 1980). Statistically, it is not possible for an individual to be 5th or 95th percentile in all vertical body dimensions and still be 5th or 95th percentile in stature. The manikin designer can resort to other techniques to ensure that the manikins are statistically correct, for example by calculating median values or using regression equations to describe component body dimensions for groups of men or women of a given stature and weight. Whichever method is chosen to define a variety of statistically correct manikins, there is still the problem of estimating the percentage of people accommodated by a particular design. A common mistake made by many designers is to use the 5th percentile female stature and 95th percentile male stature manikins to assess a workstation, assuming that if both of these manikins can be accommodated then so can 95% of the adult population. This is an incorrect assumption because it implies that those people designed out, because either their sitting height, hip breadth, or leg length, for example, are greater than 95th percentile male values, are all the same people. Similarly, all those with sitting eye height, arm length, or leg length smaller than 5th percentile female values are assumed to be the same individuals. Because these dimensions are not strongly correlated, these assumptions are incorrect. A study of air crew selection standards and design criteria analysis reported by Roebuck, Kroemer, and Thomson (1975, p. 268) illustrates the problem perfectly. It was shown that nearly half of the air crew were designed out when the 5th to 95th percentile range was used on a large number of body dimensions (in this case 15). Even limiting the number of dimensions to just 7 (sitting height, eye height-sitting, shoulder height-sitting, elbow rest height, knee height, forearm-hand length, and buttock-leg length) designed out over 30% of the available air crew.

This brief overview of the inherent problems with the use of manikins indicates the benefits that can arise from using 3D human modeling CAD systems with variable anthropometry. For example, the CAD operator has complete control of the dimensions or percentile values of each segment of the human model (assuming the system offers this facility) and can interactively change them in a matter of seconds. For example, a model of 99th percentile male stature with short arms and long legs for such a stature can be used to determine the rearmost position required of an adjustable steering wheel. Unfortunately data at this level of detail are not commonly presented in surveys, one of the exceptions being the anthropometric survey of Royal Air Force Aircrew by Simpson and Hartley (1981).

There is an important distinction between the evaluation of which percentile values are accommodated for a particular design dimension and the evaluation of what percentage of the population will be accommodated in all respects. Three-dimensional human modeling systems offer significant advantages in both respects and, in particular, the latter. Roebuck (1995) discusses two statistical methods by which some human modeling systems predict the percentage of the target population that will be accommodated by a particular workstation design. CAR and MDHMS use Monte Carlo methods to generate a large number of theoretical human models, each one of which represents a possible case that could occur in a population of people without violating any of the underlying anthropometric statistics for that population. Another statistical approach is Principal Component Analysis, a version of which is used in the SAFEWORK system.

SAMMIE is currently developing a dataset of body scans and anthropometric dimensions, including joint centers and joint mobility, from a carefully selected sample of people representative of the British and European population. Each person can then be modeled individually within SAMMIE and automatically positioned in a prototype workstation model according to a range of predefined criteria. Evaluations of fit, reach, vision, and the required postures will then be conducted, automatically resulting in the identification of those individuals who failed to successfully complete any of these tests.

Whenever possible we aim to combine the use of SAMMIE with the more traditional ergonomics methods. For example, our involvement with a major supermarket chain commenced with a survey of the musculoskeletal discomfort reported by staff in every work area (Porter et al., 1991). This allowed us to expose the high-risk workstations, namely the delicatessen and the cashier's workstation, and to subsequently model these using SAMMIE. This computer analysis simulated the users, both staff and customers, and their tasks, and the detailed postural analysis revealed several casual factors for the reported discomfort. Modifications were then made to the computer models of these workstations in order to improve the working postures. These designs were subsequently mocked-up and fine-tuned in terms of other more subjective attributes, such as the aesthetic issues.

In a similar way, SAMMIE was more recently used in the development of the Fiat Punto. The system was used to model the prototype Punto from engineering drawings and then to investigate driver accommodation for a variety of nationalities (Porter, 1995; Figure 28.9). It was considered to be essential that the Punto would accommodate drivers of all sizes. Two prototype Puntos were subsequently made available for the development of the seat with detailed assessments of the prototypes and competitor cars from 20 carefully selected members of the public driving over a 60-mile test route. This attention to detail by the manufacturer was rewarded when the Punto was voted European Car of the Year in 1995.

Human modeling systems are increasingly being used by designers and engineers who may not all have a thorough training in ergonomics issues relevant to the design of equipment and workplaces. This can result in the human models being regarded as just another CAD model, the size and shape of which can be specified and positioned within set constraints. It is all too easy to forget the important differences between designing and manufacturing the product, where variables can be specified, and evaluating the design using human models. These human models are also specified by the designer, but it is essential that they cover the wide range of sizes, shapes, functional mobility, and postural preferences that are exhibited by the population of future users of the product. Ergonomists learn the hard way that people are all different. Their experiences are based on trials, interviews, and accident reports. There is a concern that the use of human modeling systems by operators without this appreciation will lead to standardized procedures being developed taking little account of such differences. It does not necessarily follow that people will hold dangerous pieces of equipment by the handle as intended. Computer people may do as they are instructed, but real people, particularly when poorly trained, fatigued, under stress, working to a tight schedule, and so on, must not be expected to be so disciplined.

References

Badler, N.I., Phillips, C.B., and Webber, B.L., (1993). *Simulating Humans: Computer Graphics Animation and Control,* Oxford University Press, N.Y.

Case, K., Bonney, M.C., and Porter, J.M., 1991. Computer graphics standards for man modelling, *Computer Aided Design,* 23, 4, 257-268.

Coblenz, A., Mollard, R., and Renaud, C., 1991. Ergoman: 3-D representations of human operator and man-machine systems. *International Journal of Human Factors in Manufacturing*, 167-178.

Das, B. and Sengupta, A.K., 1995. Computer-aided human modeling programs for workstation design, *Ergonomics*, 38, 9, 1958-1972.

Dooley, M., 1982. Anthropometric modelling programme — a survey. *IEEE Computer Graphics and Applications*, 2, 17-25.

Elias, H.J. and Lux, C., 1986. Gestatung ergonomisch optimierter Arbeitsplatze und Produkte mit Franky und CAD (The design of ergonomically optimized workstations and products using Franky and CAD). *REFA Nachrichten*, 3, 5-12.

Fortin, C., Gilbert, R., Beuter, A., Laurent, F., Schiettekatte, J., Carrier, R., and Dechamplain, B., 1990. SAFEWORK: A micro-computer aided workstation design and analysis, new advances and future developments. Genicom Inc., Montreal Quebec.

Grandjean, E., Hunting, W., and Nishiyama, K., 1984. Preferred VDT workstation settings, body posture and physical impairments, *Applied Ergonomics*, 15, 99-104.

Grobelny, J., Cyewski, P., Karwowski, W., and Zurada, J., 1992. APOLIN: a 3-dimensional ergonomic design and analysis system, in *Computer Applications in Ergonomics, Occupational Safety and Health*, eds. M. Mattila and W. Karwowski, pp. 129-135. Elsevier Science Publishers BV, Amsterdam.

Haslegrave, C.M., 1980. Anthropometric profile of the British car driver, *Ergonomics*, 23, 436-67.

Jones, P.R.M., West, G.M., Harris, D.H., and Read, J.B., 1989. The Loughborough anthropometric shadow scanner, *Endeavour, New Series*, 13, 4, 162-168.

Kloke, W.B., 1990. WERNER: a personal computer implementation of an extensive anthropometric workplace design tool, in *Computer-Aided Ergonomics*, eds. W. Karwowski, A.M. Genaidy, and S.S. Asfour, pp 57-67. Taylor & Francis, London.

Kuusisto, A. and Mattila, M., 1990. Anthropometric and biomechanical man models in computer-aided ergonomic design structure and experiences of some programs, in *Computer-Aided Ergonomics*, eds. W. Karwowski, A.M. Genaidy, and S.S. Asfour, pp 104-114. Taylor & Francis, London.

Launis, M. and Lehtelä, J., 1990. Man models in the ergonomic design of workplaces with the micro-computer, in *Computer-Aided Ergonomics*, eds. W. Karwowski, A.M. Genaidy and S.S. Asfour, pp 68-79. Taylor & Francis, London.

Mandal, A.C., 1984. What is the correct height of furniture?, in *Ergonomics and Health in Modern Offices*, ed. E. Grandjean, pp. 471-476, Taylor & Francis, London.

McDaniel, J.W., 1990. Models for ergonomic analysis and design: COMBIMAN & CREW CHIEF, in *Computer-Aided Ergonomics*, eds. W. Karwowski, A.M. Genaidy, and S.S. Asfour, pp. 138-156, Taylor & Francis, London.

Porter, J.M., 1995. The ergonomics development of the Fiat Punto — European Car of the Year 1995, *Proceedings of the IEA World Conference 1995*, Rio de Janeiro, Brazil, eds. A. de Moraes and S. Marino, pp. 73-76. Associacao Brasileira de Ergonomia.

Porter, J.M., Almeida, G.M., Freer, M.T., and Case, K., 1991. The design of supermarket workstations to reduce the incidence of musculo-skeletal discomfort, in *Designing for Everyone and Everybody*, eds. Y. Queinnec and F. Daniellou pp. 1122-1124. Taylor & Francis, London.

Porter, J.M., Case, K., Freer, M.T., and Bonney, M.C., 1993. Computer-aided ergonomics design of automobiles, in *Automotive Ergonomics*, eds. B. Peacock and W. Karwowski, pp. 43-78, Taylor & Francis, London.

Porter, J.M., Freer, M., Case, K., and Bonney, M.C., 1995. Computer aided ergonomics and workspace design, in *Evaluation of Human Work: A Practical Ergonomics Methodology*, 2nd edition, eds. J.A. Wilson and E.N. Corlett, pp. 574-620. Taylor & Francis, London.

Porter, J.M. and Porter, C.S., 1998. Turning automotive design "inside-out". *International Journal of Vehicle Design*, Vol. 19, No. 4, 385-401.

Post, F.H. and Smeets, J.W., 1981. ADAPS: Computer aided anthropomerical design. *Tijdschrift voor Ergonomic*, 6, (4), 11-18 (in Dutch).

Roebuck, J.A., 1995. *Anthropometric Methods: Designing to Fit the Human Body*, Human Factors and Ergonomics Society, USA.

Roebuck, J.A., Kroemer, K.H.E., and Thomson, W.G., 1975. *Engineering Anthropometry Methods*, John Wiley & Sons, New York.

Rothwell, P.L., 1985. Use of man-modelling CAD systems by the ergonomist, in *People and Computers: Designing the Interface*, eds. P. Johnson and S. Cook, pp. 199-208. Cambridge University Press, U.K.

Simpson, R.E. and Hartley, E.V., 1981. Scatter diagrams based on the anthropometric survey of 2000 Royal Air Force Aircrew (1970/71), *Royal Aircraft Establishment Technical Report 81017*, Farnborough, Hampshire, England.

Westerink, J., Tragter, H., Van Der Star, A., and Rookmaaker, D.P., 1990. TADAPS: a three-dimensional CAD man model, in *Computer-Aided Ergonomics*, Taylor & Francis, London, pp 90-103.

For Further Information

Books

A wide selection of human modeling systems is individually presented in the following two books: *Computer-Aided Ergonomics*, eds. Karwowski, W., Genaidy, A. and Asfour, S.S., 1990. Taylor & Francis Ltd., London; and *Computer Applications in Ergonomics, Occupational Safety and Health*, eds. Mattila, M. and Karwowski, W., 1992. Elsevier Science Publishers B.V., The Netherlands.

A detailed description of the development of the Jack system is presented in: *Simulating Humans: Computer Graphics Animation and Control*, Badler, N.I., Phillips, C.B., and Webber, B.L., 1993. Oxford University Press, N.Y.

Database

An up-to-data database concerning the technical specifications of the major commercially available human modeling systems is available from the CSERIAC Program Office, AL/CFH/CSERIAC Bldg. #248, 2255 H Street, Wright-Patterson AFB, OH 45433-7022, USA. Tel: +513 255-4842, Fax: +513 255-4823. Acknowledgment is given to Aaron Gayman and Chris Sharbaugh for kindly supplying information for this chapter.

Web Sites

Several system developers and users have created web pages providing graphic images and current information. These include:

Deneb/ERGO, Auburn Hills, Michigan, USA: http://www.deneb.com/ergo.html

JACK at the University of Pennsylvania, USA: http://www.cis.upenn.edu/~hms/jack.html

McDonnell Douglas Human modeling system (MDHMS), Long Beach, California, USA: http://pat.mdc.com/LB/LB.html

MIDAS application of Jack at NASA Ames Research Center, Moffett Field, California, USA: http://ccf.arc.nasa.gov/af/aff/midas/MIDAS_home_page.html

SAFEWORK at Genicom Consultants Inc., Montreal (Quebec): http://www.safework.com

SAMMIE at Department of Design and Technology, Loughborough University, Leicestershire, UK: http://www.lboro.ac.uk/departments/cd/docs_dandt/staff/Porter/sammie.html

29

A Guide to Computer Software for Ergonomics

Jari Järvinen
Motorola

Hongzheng Lu
Lucent Technologies

29.1 Introduction

Usually ergonomics or human factors specialists attempt to optimize work with computers and software. The purpose of this chapter is to review the computer software that can help ergonomists, researchers, and practitioners at their work.

A vast amount of ergonomics knowledge has been generated by researchers, but it is not always effectively communicated to the practitioners who design, redesign, and evaluate products, tools, equipment, facilities, and environments. Tight schedules of the design projects, constantly changing design specifications, and the increasing complexity of work systems and equipment create situations in which the ergonomic considerations may be ignored by the designers. Moreover, the diversity of the available ergonomic literature may confuse a designer who does not have a comprehensive understanding of the field.

Recent developments in ergonomics software and other computer-aided tools make specialized knowledge accessible and present it in a suitable form to practitioners. The designers can now more easily obtain feedback needed to design products for human use and to optimize manufacturing and maintenance processes already at the design phase, when the cost of making changes is still relatively low. The incorporation of ergonomic knowledge through the use of ergonomics software at the early stages of the design process also reduces the number of prototypes needed to bring a product to market, and enables the manufacturer to create products that are cheaper to make and provide a better fit for the user.

New software applications in the field of ergonomics are being developed at an accelerating pace. This guide cannot be considered a complete catalog of ergonomics software. Selected, and most applicable, commercially available software from various application areas are described. Although ergonomics software can provide an access to applicable ergonomics knowledge and data, the human is still the main component in the design and analysis processes. The computer excels in computation, data processing, and storage, while creativity is the human's strength.

0-8493-2641-9/99/$0.00+$.50

29.2 Fundamentals

Since the 1960s several computer programs have been created to assist ergonomics practitioners. The first ergonomics computer programs were based on geometric human models. More recently, a variety of analysis modules and databases have been incorporated into these human modeling systems. Examples of other currently available software include ergonomics analysis programs (without human modeling capabilities), computerized database programs, and programs that provide a computer operator with guidance in ergonomic work techniques and reminders about rest breaks. A general description of different types of ergonomics software is presented in this section.

Human Modeling Systems

Human modeling is not a new design tool. For decades, wooden and plastic dimensionally accurate templates have been used in the design of automobiles and aircraft. The increase in the computing power of the workstations and personal computers has made it possible to use computerized human modeling tools at the design stage of workplaces and equipment.

The human modeling programs use anthropometrically accurate human models to simulate human size and motions as they do their work or use a product or tool. The product or environment model is usually created using a CAD modeling software. The design can then be evaluated using a human modeling system that is either a modular component of the CAD software or an independent, stand-alone human modeling system. Some of the stand-alone systems incorporate basic CAD modeling capabilities, and/or the CAD models can be imported to the human modeling system, or in some cases, the human models can be exported to the CAD system.

Different human models have been developed for different needs. The simple models are more or less static, easy to use, do not require much computer memory, and thus, can be used in almost any computer for routine design tasks. The more sophisticated human models are more versatile and can animate human motions in real time. However, these systems are complex and require special knowledge, skills, and training to be used effectively. Some of the sophisticated human modeling programs require powerful graphic workstations, or minicomputers, and are used mainly in the automotive and aerospace industries, as well as in academia and the military.

With the increasing computing power of personal computers, the sophistication of the PC-based human modeling systems can also be increased. Recently, relatively sophisticated human modeling systems have been developed that do not require investments in high-end hardware and software, and thus, are easily accessible for small and midsize companies. Some organizations, such as NASA, have adopted a two-tier approach that uses PC-based human modeling systems (such as custom version of MQPro) for routine design tasks, and high-end workstation programs (such as Jack) for modeling complex human–machine systems and for *virtual reality* applications.

Some of the ergonomics software use *virtual reality* (*VR*) to simulate human responses and fit to a proposed design. VR is a high-end user interface that involves real-time simulation and interactions through multiple sensorial channels (Burdea and Coiffet, 1994). The VR user is immersed in a computer-generated world via a boom or head-mounted display, and may "fly" or "walk" through the virtual world and interact with it (Mourant, 1994). In so-called desktop VR, the user wears stereo glasses and/or a head tracking system. This is sometimes called nonimmersion or "fishbowl" VR.

Other Ergonomics Software

Ergonomics analysis programs usually consist of computerized algorithms or equations used in ergonomics analyses, such as biomechanical analysis or energy expenditure prediction. Some established analysis methods, such as OWAS, have also been computerized. These programs do not usually include graphical human or environment modeling capabilities.

Other types of commercially available ergonomics software designed for ergonomics practitioners include computerized databases consisting of anthropometric information, design guidelines, and other information. The databases usually combine data from several sources, and the sophisticated user interfaces and database management procedures of the modern programs can efficiently provide the required design information for the practitioner.

Recently, due to the rapidly increased use of computers and the recognition of the problems related to excessive computer use, new types of ergonomics programs have emerged. Instead of assisting a designer or an ergonomics practitioner, these programs provide the computer user with guidance on ergonomic work techniques and stretching exercises, and reminds him/her about rest breaks.

29.3 Applications and Examples

Selected examples of ergonomics software from various application areas are described in this section. The examples were selected from the following areas: human modeling systems for personal computers (MQPro, ergoSHAPE) and workstations (SAMMIE, Deneb/ERGO, Jack, Safework, MDHMS), ergonomics analysis programs (ErgoEASER, ErgoMOST, NIOSH Lifting Equation, OWASwin, 3D SSPP, Energy Expenditure Prediction Program), database programs (PeopleSize), and training/stretching software (WorkSmart).

Human Modeling Systems for Personal Computers

MQPro

MQPro is a PC-based human modeling system and ergonomic design software. With a few mouse clicks, the program creates anthropometrically accurate, three-dimensional human models representing several ethnic groups, *percentiles,* and *somatotypes.* These models can be manipulated to any human-compatible position and viewed from any angle, distance, or perspective. The views can be printed, plotted, or exported to other graphics software for further enhancement of the image. The human models can walk, bend, reach, and grasp objects. Although equipped with a set of basic three-dimensional modeling and editing tools, MQPro is designed with import/export capabilities so it can be used with other graphics software, such as AutoCAD and 3D Studio.

MQPro has evolved from the human modeling package Mannequin. It has been developed by BCAM International and used for ergonomics consulting services in product and workplace evaluation and design projects. Custom versions of the software have been provided to select organizations, such as NASA and university research centers. The custom applications have ranged from specific interface needs to the integration of motion capture hardware (Flock of Birds).

Main features: anthropometric database of eleven populations, including 1988 U.S. Army (Natick) and NASA-STD-3000; adult male, adult female, and child human models with three somatotypes; 2.5th, 5th, 50th, 95th, 97.5th percentile human models or customize option for each body part; normal, customized, or free joint range of motion; human model representation in five levels of detail ranging from stick-figure to skeleton and high-resolution humanoid; physically challenged human models with wheelchairs, canes, and crutches; field of vision cones; reach envelopes for hands and feet; reach analysis; 2D and 3D drawing and editing tools; wire frame, shading, and hidden line removal capabilities; ability to "see" through selected human model's eyes; placement of camera, field of view, and light source for any three-dimensional perspective view; *Revised NIOSH 1991 lifting equation*; simulation of lifting, pushing, pulling by adding external forces and torques in any direction on any body part; calculation of reaction joint forces and torques due to external load and body posture; presentation of anthropometric information, joint angles, joint forces, and torques in tabular form; frame-by-frame animation; and animated walking on specified path. Some of the MQPro's capabilities are illustrated in Figures 29.1 and 29.2.

System requirements: 486 PC with co-processor or Pentium, Windows 3.1 or higher, Windows 95, or Windows NT, 4MB RAM, 4MB of available hard disk space, VGA display. Price range: less than $1,000.

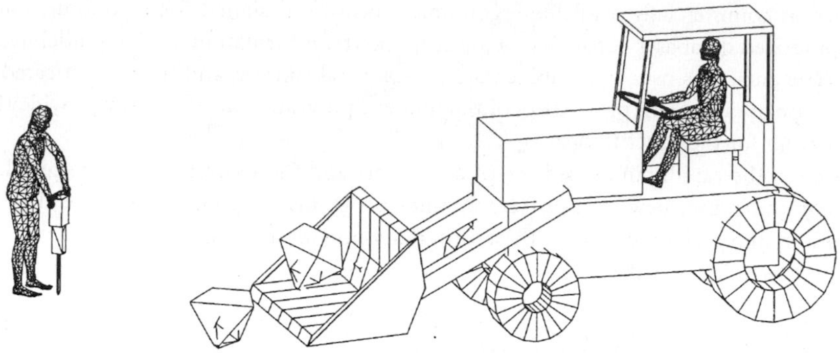

FIGURE 29.1 Two human models and three-dimensional models of a backhoe and jackhammer created using MQPro. (Courtesy of HumanCAD Systems)

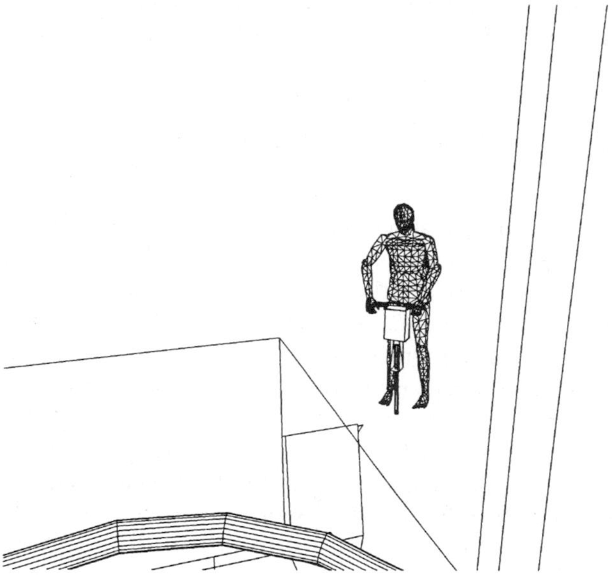

FIGURE 29.2 A field of view as seen by the MQPro human model sitting in the backhoe seat. (Courtesy of HumanCAD Systems)

ErgoSHAPE

ErgoSHAPE is a two-dimensional human modeling system developed by the Institute of Occupational Health in Finland to supplement a widely used PC-based AutoCAD design program. The system is used as a module of AutoCAD and consists of drawing files, menu files, and AutoLISP program files. Recently, a version of ergoSHAPE for the Designer computer-aided design program was developed.

The ergoSHAPE system consists of the following parts: (1) two-dimensional 50th percentile male and female human models based on the anthropometric dimensions of the northern European and North American populations, (2) biomechanical calculations in two dimensions; the stress is calculated as a percentage of the maximal static muscle strength, (3) recommendation charts providing design guidelines, such as dimensions for seated and standing workstations, and (4) ErgoTEXT text files, that include ergonomic design guidelines covering several areas of ergonomics.

ErgoSHAPE human models can also illustrate the curves indicating the viewing angles and distances and the reach zones in two dimensions. A more detailed description can be found in Launis and Lehtelä (1990).

System requirements: IBM-compatible PC, AutoCAD or Designer software. Price range: less than $1,000.

Human Modeling Systems for Workstation Computers

SAMMIE

SAMMIE (System for Aiding Man–Machine Interaction Evaluation), developed by the University of Nottingham, is one of the first commercially available human modeling systems. A detailed description of SAMMIE can be found in Case et al. (1990).

The human model of SAMMIE was originally based on data from Dreyfus's anthropometric measurements (Dreyfus, 1966). The limb and body segment lengths of the model can be varied by percentiles or by explicitly defining the dimensions for each body segment. Anthropometry data files can also be created and modified by the designer. In addition, the somatotypes, which define the flesh outline of the model, can be varied.

The user can build a three dimensional environment around the human model by solid modeling techniques. The user can specify primitive geometric shapes and assemble these to models of equipment or environments.

The human model can interact with the created environment model in several ways. Logical relationships between the body segments are included. The limbs can only be moved within normal human ranges of motion. The joint constraints can be defined by the user. The following ergonomic evaluations can be performed: reach assessment, fit assessment, visual field assessment, and posture evaluation against "normal" constraints.

SAMMIE has been used for the following applications, among others: power station control consoles, automobile design, visualization of car driver views in the English Channel Tunnel, financial trading facilities, helicopter concept design, and underground design.

System requirements: mainframe or workstation computer (Apollo or Sun), UNIX operating system.

Deneb/ERGO

Deneb/ERGO human-factors analysis software has been developed to be used in a workstation platform, such as Silicon Graphics, with a UNIX operating system. It is used in conjunction with Deneb's IGRIP modeling software.

Main features: 5th, 50th, and 95th percentile anthropometric male and female models that move realistically; enables creation of generic assembly motion routines and sequences; human models have strength capabilities, i.e., they can "lift" as much as average humans; NIOSH 1991 lifting equation; energy expenditure prediction of the simulated tasks (Garg's model); time measurements of the simulated tasks (MTM-UAS); evaluation of the injury potential of the simulated tasks; and immersive VR capability.

System requirements: workstation computer (Silicon Graphics), UNIX operating system, Deneb's IGRIP modeling software package.

Jack

Jack is a human modeling software package developed at the Center for Human Modeling and Simulation at the University of Pennsylvania.

Jack provides a three-dimensional interactive environment for controlling articulated figures. It features a detailed human model and includes realistic behavioral controls, anthropometric scaling, task animation and evaluation systems, strength-guided motion, view analysis including viewing the environment model through the human model's eyes, automatic reach and grasp, collision detection and avoidance, and other tools for a wide range of applications. Dynamic torques can be computed throughout animated motions and compared against strength data to assess the validity of motions and postures.

Jack also provides an interface to other hardware such as Cyberglove and Ascension Flock of Birds, which enable virtual reality applications.

System Requirements: Silicon Graphics IRIS 4D, INDIGO, or INDY workstations, Z-buffer, three-button mouse, SGI Operating System 4.0.1 or higher, 16 MB RAM. Price range: $10,000 to 20,000.

Safework

Safework is a modular human modeling system that runs in Silicon Graphics workstations. Safework is available in four versions with increasing functionality. The following features are included in the most advanced package, which contains all available functions and options: six basic anthropometric three-dimensional human models (5th, 50th, and 95th percentile male and female); user access to anthropometric variables and data (1988 U.S. Army Natick laboratories); seven different somatotypes; manipulation through inverse and direct kinematics; fully articulated hand and spine models; normal and restricted joint mobility; coupled range of motion; vision analysis, including binocular, ambinocular, and monocular vision, and various vision cones; animation; postural, comfort angle and reach analysis modules; collision detection; files can be imported from most CAD programs; basic geometric modeling capabilities.

System requirements: Silicon Graphics workstation Indy or Indigo2 R4400 (recommended), 32 MB RAM or higher (recommended: 64 MB or higher), Graphics Engine XZ (recommended: Graphics Engine Impact), IRIX 5.3. Price range: $30,000 to $60,000 depending on the version.

McDonnell Douglas Human Modeling System

In addition to three-dimensional biomechanical human models and virtual models of equipment and environments, the McDonnell Douglas Human Modeling System (MDHMS) offers detailed eye articulation, hand and finger articulation, and shoulder articulation. The model can simulate rotating eyeballs, which enables the illustration of what a human can see more realistically than most virtual reality applications can.

This software package has been used to study human-machine interactions and other human activities in the crew stations of a fighter/attack aircraft, crew-return vehicle, commercial aircraft, and extravehicular mobility unit. McDonnell Douglas' suppliers are also using MDHMS to evaluate the assembly and maintenance of their aircraft components before they are sent to McDonnell Douglas. This software requires a powerful workstation computer with a UNIX-based operation system.

Ergonomics Analysis Programs

ErgoEASER

ErgoEASER (Ergonomics Education, Awareness, System Evaluation & Recording) is a set of PC-based interactive tools for evaluating computer workstations and lifting tasks, and recommending appropriate controls.

ErgoEASER has been developed by Pacific Northwest National Laboratory (PNNL), the U.S. Department of Energy (DoE), and the U.S. Department of Defense (DoD), in consultation with the U.S. Department of Labor/Occupational Safety and Health Administration (DOL/OSHA). ErgoEASER is a work in progress, and is being shared with federal agencies and made available to the public.

ErgoEASER allows the user to evaluate workplaces and tasks to identify risk factors that should be addressed. Users may then modify variables affecting the risk factors and determine which variables will eliminate or reduce the hazards when adjusted.

Currently, ErgoEASER consists of three components: (1) getting started, (2) awareness and reporting, and (3) analysis modules for lifting and VDT workstations.

The "Getting Started" module includes background information on occupational ergonomics. The "Awareness and Reporting" module consists of examples and photographs of hazardous postures. A risk factor checklist is included to help the user assess his/her own company's work situations. Based on entered information, the software may recommend additional analyses. The analysis modules assist the user in more detailed evaluations to support the design of VDT workstations and lifting tasks. The user interactively enters variables that describe specific work situations, such as key VDT workstation and operator dimensions. The user can then view the simulated postures and observe the ergonomic effects resulting from workstation configurations. The software highlights the body parts with an increased stress level. The user can make adjustments by changing the values of the variables. Finally, the software generates a report that documents task specifications and recommended solutions. The Lifting Analysis module incorporates the 1991 NIOSH Lifting Equation.

ErgoMOST

ErgoMOST software was developed to analyze physical risk factors associated with repetitive motion. It is a module of the MOST (Maynard Operation Sequence Technique) Work Measurement Systems software package. ErgoMOST focuses on methods for improvements in the workplace to minimize ergonomic stress and maximize motion efficiency, and it can be used as a stand-alone application, or it can interact with other MOST software products. Relative Ergonomic Stress Indices are assigned based on used force, posture, repetition, grip, and vibration. Twelve body joints are evaluated by combining ergonomic analysis with defined methods using MOST. ErgoMOST also provides a range of reports and parameters for displaying results. Users can select a step- or job-level report, a report by body parts or ergonomic areas, and text or graphic formats.

System requirements: IBM PC-compatible 486 33 MHz or higher, 8MB RAM, Windows 3.1 or higher.

NIOSH Lifting Equation

The revised NIOSH Lifting Equation can be used to evaluate simple and well-defined lifting tasks. Specific lifting parameters are entered as inputs, and the equation provides a Recommended Weight Limit (RWL) for the lift. The revised NIOSH Lifting Equation is discussed in detail elsewhere in this book.

The NIOSH Lifting Equation is included as an analysis module in some of the human modeling systems described above, such as MQPro and Deneb/Ergo. In addition, numerous software sources that include the equation in computerized form are listed in the NIOSH web page (http://www.cdc.gov/niosh/homepage.html).

OWASwin

OWAS (Ovako Working posture Analyzing System) is a method for the evaluation of postural load during work based on a systematic observation and classification of working postures (Karhu et al., 1977). It is an observational sampling technique in which a four-digit code is used to describe postures of the body parts and the required force. Postures are observed at a set time interval. The coded posture combinations are classified into four action categories which allow the most stressful activities to be identified and indicate the priorities for corrective measures.

The action categories along with recommended actions are defined as follows: (1) normal posture: *no actions required,* (2) the posture is slightly harmful: actions to change the posture should be taken *in the near future,* (3) the posture is distinctly harmful: actions to change the posture should be taken *as soon as possible,* and (4) the posture is extremely harmful: actions to change the posture should be taken *immediately.* More information about the OWAS method is presented in Chapter 26.

The computerized version of OWAS consists of data collection and data analysis modules. The data collection module is used either in field conditions, using a portable computer and observing "live" work, or in the laboratory, where the videotaped work is observed. The postural data are collected using visual, split-second observations. The postures of the back, arms, and legs, as well as weight of load or used force are identified and recorded using a predefined coding system. During the posture coding the subtask or activity can also be recorded along with the posture codes. The subtask information can be used to focus interventions on the most demanding subtasks.

The data analysis module allows the postural data to be analyzed in two ways. One way is to examine the combined posture of the back, arms, legs, and exerted effort, and determine its effect on the musculoskeletal system. The other way is to examine the relative time spent in a particular posture for each body part and determine the time effect on the musculoskeletal system.

The analysis will result in the following output: number of observations, percentage of individual postures in each action category, relative time spent in each posture, relative time spent in each subtask, number of observations of each posture, and remedial action recommendations.

System requirements: Windows 3.1 or higher. Price range: less than $1,000.

3D Static Strength Prediction Program

The 3D Static Strength Prediction Program (3D SSPP) can be used to evaluate the physical demands of a prescribed job. Both DOS and Windows versions of the program are available. The program is based on years of research at the Center for Ergonomics at the University of Michigan concerning work-related human biomechanical and static strength capabilities. The program can be applied to a worker's motions in three-dimensional space. However, the effects of acceleration and momentum must be negligible. The program is most useful in the analysis of the slow movements used in heavy materials handling tasks that can be described as a sequence of static postures.

The first estimation of the working posture to be evaluated can be created by entering the position of the hands and the load on the hands. The posture can be modified by changing or entering joint angles for selected joints. The posture and force direction are presented using a stick figure in the analysis screen. A three-dimensional illustration of the posture is also available. The modeling of the work environment is not possible.

The analysis results are presented using the following output screens: (1) analysis summary; (2) anthropometric data, including link lengths and weights, joint angles, and balancing status of the analyzed posture; (3) moments at different joints; (4) moments about joint movement hinges including percentage of population with sufficient strength capability; (5) low back muscle and disc forces (L5/S1 disc); (6) spinal analysis summary including resultant moments and forces at L2/L3 and L4/L5 spinal segments; and (7) low back compression optimization summary including resultant forces of all muscles involved in the lumbar area.

Energy Expenditure Prediction Program

The Energy Expenditure Prediction Program (EEPP) software is used to predict the total energy expenditure and energy expenditure rate in performing a job. It is based on the assumption that a job can be divided into simple tasks, or activity elements, and that the average metabolic energy rate of the job can be predicted by knowing the energy expenditure of the simple tasks and the time duration of the job.

Other Ergonomics Software

PeopleSize

PeopleSize is a computerized database with a graphical mouse-driven interface consisting of most human dimensions that are relevant to design. The database is a comprehensive and validated anthropometry reference compiled from over 70 sources. Each body dimension in the database is based on the average of several available sources for each major racial group, weighted by sample size. The data sources are listed in the Help file of the software.

Main features: dimension-specific data for any percentile; adjustments for clothing (gloves, shoes, head gear, winter clothes, summer clothes, etc.) and sitting posture (slumped or erect); calculates percentiles for given measurements; program output can be stored to the "log" to be used later; displays detailed descriptions of measurement methods and terminology.

System requirements: Windows 3.1 or higher, 2 MB RAM, 1.5 MB free disk space, EGA or better graphics, mouse. Mac System 7 or higher, 2 MB RAM + 2 MB virtual memory, 2 MB free disk space.

WorkSmart Stretch Software

WorkSmart stretch software reminds the user about a stretch break every 50 minutes during the workday. WorkSmart offers four groups of stretches designed to benefit different areas of the body affected by daily computer-related muscle stress: Neck and shoulders, upper extremities, lower body, and back. The software displays instructions and graphic demonstrations for the stretch exercises. The user can also reschedule the break by postponing it by 10 to 50 minutes.

System requirements: Windows 3.1 or above, 3.5" Floppy drive, mouse, 386 IBM compatible or above, 16 color VGA with palletized VGA display driver, or 256 color VGA.

Defining Terms

Percentile: The frequency distributions for each measurement of population size are expressed in percentiles. A percentile indicates the percentage of population at or below certain measure.

Revised NIOSH Lifting Equation: An equation, developed by NIOSH, for calculating a recommended weight for specified two-handed lifting tasks.

Somatotype: The morphological type of a human body.

Virtual reality (VR): A high-end user interface that involves real-time simulation and interactions through multiple sensorial channels. A VR user is immersed into a computer-generated world via boom or head-mounted display, and may "fly" or "walk" through the virtual world and interact with it.

References

Burdea, G. and Coiffet, P. 1994. *Virtual Reality Technology*, John Wiley & Sons, New York.

Dreyfus, H. 1966. *The Measure of Man — Human Factors in Design*, Whitney Library of Design, New York.

Case, K., Porter, J.M., and Bonney, M.C. 1990. SAMMIE: a man and workplace modelling system. In *Computer-Aided Ergonomics*, eds. W. Karwowski, A.M. Genaidy and S.S. Asfour, p. 31-56. Taylor & Francis, New York.

Karhu, O., Kansi, P., and Kuorinka, I. 1977. Correcting working postures in industry: A practical method for analysis. *Applied Ergonomics*. 8:199-201.

Launis, M. and Lehtelä, J. 1990. Man models in the ergonomic design of workplaces with the microcomputer, in *Computer-Aided Ergonomics*, eds. W. Karwowski, A.M. Genaidy, and S.S. Asfour, p. 68-79. Taylor & Francis, New York.

Mourant, R.R. 1994. *Virtual Reality and Ergonomics*. Workshop presented at IEA 12th Triennial Congress.

T.R. Waters, V. Putz-Anderson, and A. Garg, 1994. *Application Manual for the Revised NIOSH Lifting Equation*, DHHS (NIOSH) Publication No. 94-110, NTIS, Springfield, VA.

For Further Information

Deneb/ERGO: Deneb Robotics, Inc., 3285 Lapeer Road West, P.O. Box 214687, Auburn Hills, MI 48321-4687.

Energy Expenditure Prediction Program: University of Michigan Software, 3003 South State Street, Suite 2071, Ann Arbor, Michigan 48109-1280, USA.

ErgoEASER®: Pacific Northwest National Laboratory, P.O. Box 999, Richland, WA 99352.

ErgoMOST: H.B. Maynard and Company, Inc., Eight Parkway Center, Pittsburgh, PA 15220, Internet: www.hbmaynard.com.

ErgoSHAPE: Institute of Occupational Health, Ergonomics Unit, Topeliuksenkatu 41 a A, 00250 Helsinki, Finland.

Jack®: Transom Technologies, Inc., 201 South Main St., Suite 1000, Ann Arbor, MI 48104. Internet: http://www.cis.upenn.edu/~hms/jack.html.

Karwowski, W., Genaidy, A.M., and Asfour, S.S. (eds.) 1990. *Computer-Aided Ergonomics*, Taylor & Francis, New York.

MQPro: HumanCAD Systems, 3100 Steeles Avenue West, Concord, Ontario L4K 3R1, Canada. Internet: http://www.mqpro.com.

PeopleSize: Friendly Systems Ltd., 443 Walton Lane, Loughborough, LE12 8JX, England.

Safework: Les Consultants Genicom, Inc., 3400 de Maisonneuve West, Suite 1430, Montreal (PQ), H3Z 3B8, Canada. Internet: http://www.safework.com/index.html.

SAMMIE: SAMMIE CAD Ltd., Quorn, Loughborough, Leicestershire, UK.

WorkSmart Stretch Software: Ergodyne Corporation, 1410 Energy Park Dr., STE 1, ST Paul, MN 55108-9950.

30

Guide for Videotaping and Gathering Data on Jobs for Analysis for Risks of Musculoskeletal Disorders

David J. Cochran
University of Nebraska-Lincoln

Terry L. Stentz
University of Nebraska-Lincoln

Brian L. Stonecipher
University of Nebraska-Lincoln

M. Susan Hallbeck
University of Nebraska-Lincoln

30.1 Introduction

Videotaping jobs for a detailed ergonomic analysis at a later time is common practice. This is an especially effective method for documenting and evaluating the risk factors for musculoskeletal disorders (MSD). The quality of the analysis cannot be any better than the quality of the videotape and related data gathered. Therefore, it is critical that the videotaping be done correctly. The authors have videotaped and/or analyzed hundreds of jobs in industry. In that process we have learned what works and what does not. The following is a compilation of that knowledge. This guide used unpublished information developed by NIOSH (V. Putz-Anderson and D. Habes) circa 1991 but is primarily based on the experience of the authors. The methodology applies to a very wide variety of industry and job types. The flow or sequence of activities is presented in Figure 30.1.

30.2 Preparation and Equipment

Prior to actually taping the job, it is advisable to collect background information. This will normally include job title, work objective, work standard, major and minor tasks, the product, tools, equipment, workstation layout, environmental conditions, any rotation scheme, worker attributes, work flow, and the jobs just preceding and just following this job. It is also important to know the injury and illness history of the job so that the videotape will feature the appropriate views of the workstation and worker. This information is all included in the job analysis sheets included in this article. Schedule the taping with appropriate managers

0-8493-2641-9/99/$0.00+$.50

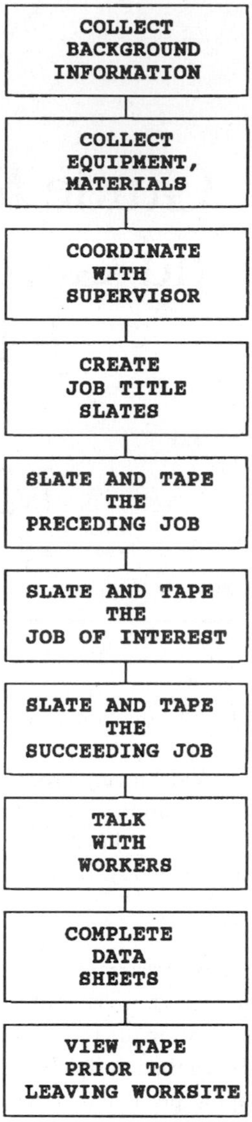

FIGURE 30.1. Flow chart of the activities involved in videotaping a job.

and supervisors and ask them to inform their workers that videotaping will take place. Make certain that the jobs to be taped will be performed during the scheduled taping session and will be typical of all of the work normally done. It is often constructive to have a meeting with the supervisor prior to taping the job to get as much of the information gathering done as possible and to iron out any potential difficulties. Prior contact with workers to be taped is advised to make them comfortable with the process and to assure them of the objectives. In some cases, the workers' permission is required before taping can begin.

Prior to entering the work area, it is advisable to create job title information slates or visual starters to introduce each job to be taped. This includes making a job title information slate for the job immediately before the job of interest and one for the job immediately after. With a dark marker in very large print (computer printing in large print also works very well) write the company, date, location, job name or title, shift, and any other pertinent information desired (see Figure 30.2). It is important that the job title information slate contain all of the pertinent information and be easily readable in the video.

XYZ COMPANY
DATE 6/6/66
TIME - 16:30
LOCATION - Z LINE
JOB NAME - HACKER
SHIFT - DAY

FIGURE 30.2. Job Title Information Slate.

In videotaping, two records are being created. The first is the videotape. The second is a log that is on paper or a voice recording. Maintain the log in the same sequence as the jobs on the videotape. This log is to contain information and notes about the job that will not be on the videotape. The log should have data sheets (discussed later) associated with each job attached to it. The person analyzing the tape must be able to follow the videotape and the log easily. Make certain that the job name is consistent on the videotape and the log.

Equipment that is necessary for taping includes:

- Video camera with at least two charged batteries good for 2 hours each
  - Date and time, including seconds, on the tape is highly recommended
- An adequate supply of videotape
- Objects of known length
  - Dollar bills (6 inches)
  - Yard stick with obviously alternating colors every 6 inches
- Tape measure
- Clipboard or tablet

Equipment that is useful but may not be absolutely required:

- Dictating recorder
- Force-measuring equipment
- Stop watch
- Light meter
- Sound pressure meter
- Thermometer
- Air movement measuring equipment
- Laptop computer
- Reflective tape or dots to put on worker joints
- Tripod or monopod

30.3 Videotaping Procedure

Be sure the light is adequate. If you are taping a worker wearing dark clothing against a light background the picture quality will generally be inadequate. To remedy this you may select another worker wearing lighter clothing, have the worker wear lighter clothing, or provide additional lighting.

Videotape the job title information slate for the job. If possible, verbally record all of the slate information on the audio portion of the tape while taping the slate. On-screen date, time, job name, or other useful information on the tape should be used if available. If this capability is limited, always put the time, including seconds, in the picture because the date is already on the job title information slate and these times can be useful when the tape is analyzed. Note on the log the job name and the first individual to be taped. If more than one individual is taped, list them in the order of taping with a very short description for future identification.

The objective is to get a good representation of all aspects of the job. Tape all tasks and subtasks of the job. Tape each task for 5 to 10 minutes and try to include at least 10 repetitions of all of its parts.

This is especially true for those angles that you consider most revealing of the task or the most important for future analysis. Sometimes when the task is long, there will be repeated cycles within it. In these cases it is important to get videotape of approximately 10 repetitions or cycles of everything included in each task and/or the entire job. Record all the idle time or other time between the cycles and/or the task activities. It is important that the camera be held as steady as possible. Where possible, use a tripod or at least a monopod. Another possibility is to lean or brace yourself or the camera against a solid, nonvibrating, object. Walking with the camera while recording is very detrimental to the quality of the videotape produced. Therefore, avoid walking with the camera while recording unless absolutely necessary. When walking and taping, slowness and steadiness are critical.

The sequence of activities in the actual taping is presented in Figure 30.3. Begin taping each task with a view of the entire workstation and then go to a whole-body view of the worker (Figure 30.4a). The seat or chair along with the standing surface should be included in this view. Two or more cycles are recommended before zooming in on the body area of interest. Resist the temptation to zoom in and out as the task progresses. A single view is almost always better. If the hands, arms, shoulders, neck, back, and/or hips are important, it is usually advisable to keep the whole upper body, including the buttocks, in view (Figure 30.4b). If the feet, legs, and/or hips are important, it is usually advisable to keep the whole lower body, including the buttocks, in view (Figure 30.4c). A more restricted viewing area is rarely useful. If it is useful, it should be supplemental to the larger body views. Videotape from angles which will allow determination of hand, wrist, arm, back, or other important postures. Tape from both sides and the front if possible. A view from above, as in Figure 30.4c, is helpful but often not possible.

Videotape the hand tools, jigs, fixtures, materials, parts, and personal protective equipment involved in the job. Videotape one or two cycles of the tasks immediately preceding and following the task in question. Where more than one worker is doing a job, record several of them. Try to tape workers who exhibit different experience or skill levels and those with a variety of body sizes and types.

In our experience, the same problems with the quality of videotapes occur repeatedly. The most frequently encountered bad characteristics are:

Too close
No complete workstation view
Not steady
Insufficient time or cycles
Move from one view to another too quickly or frequently
Frequent zoom in and out
No slate
No job before and after
Poor light or contrast
Audio narrative that is unintelligible
Back view
Walking or moving with the line
Lack of dimensional cues

30.4 Additional Data to Be Gathered

Even though the videotape lets the analyst see the job, it does not have all of the information needed to properly analyze a job and to propose viable solutions. Therefore, before, after, or during (if another person is available) the taping, it is useful to collect as much information about the job as possible. For this purpose six data sheets, contained in Figures 30.5 through 30.10, are included in this article. These data sheets attempt to structure all of the information the analyst might need. For any specific job, there may be blanks and whole data sheets that are not necessary or are not appropriate. If there is a certainty

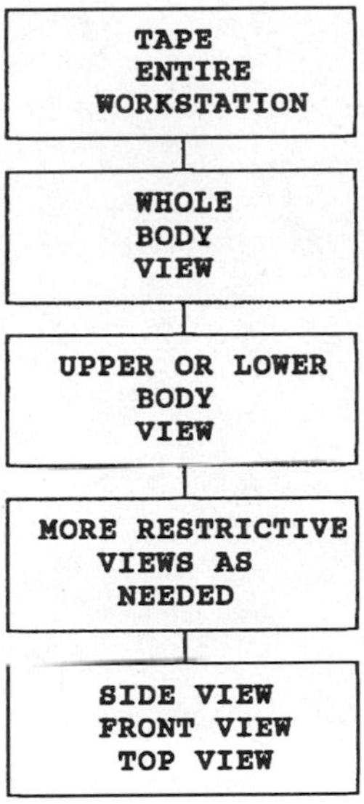

FIGURE 30.3. Flow chart of videotaping a job.

that the information is not going to be used or useful then don't collect it. Also there will be information that is pertinent to that job that is not addressed in the data sheets. These sheets are as complete as we can make them at this time. We recommend modifying or improving them for your specific needs.

The data sheets are, for the most part, self explanatory, but some of the items may need a little explanation. The Job/Task Information Data Sheet contains blanks for the basic information needed in evaluating the job. The "Work Typical" asks for a yes or no response as to whether the work being taped is what is normally done for that job.

For the Workstation Dimensions Data Sheet the forces can sometimes be measured directly. When this is not possible, they can be arrived at indirectly. This is done by having the workers exert a force on force-measuring equipment that they feel is equivalent to that exerted on the job or by having them estimate the forces and give this information verbally. When getting the force estimates indirectly it is best to get them from as many workers as possible and sometimes to do the job yourself and estimate the forces. It is also worthwhile to estimate what the percent of the maximum force is for those workers doing the job. This can be done by having the workers exert their maximum force on force-measuring equipment or by having them estimate the percent of their maximum force this job requires.

The Tool Data Sheet; Materials, Parts, Bins, Pallets, etc. Data Sheet; and Personal Protective Equipment Data Sheet are self explanatory. When possible, collect examples of tools, materials, parts, bins, and personal protective equipment to accompany the videotape. Having examples of the items used in a job accompany the videotape is preferable to completing the data sheet. However, listing them on the appropriate data sheets is still advisable. The sketches called for in these data sheets should have the dimensions on them.

The Maintenance Data Sheet is also self explanatory. The information gathered on this data sheet is very important. This information is rarely gathered and may have a tremendous impact on the job analysis and redesign. In our experience, lack of or poor maintenance has been the major cause of job or workstation problems, and dealing with that has been the only remedy necessary.

During or just after the videotaping process, it is useful to get input from the supervisor and employees. This should include their thoughts as to what parts of the job are causing difficulty, what the real problems are, and their ideas as to what the possible solution might be. Additionally, the Rated Perceived Exertion (RPE) (Borg, 1982) and a body part discomfort survey (Corlett 1976), can be used to determine how strenuous the job is and which parts of the workers bodies are affected.

FIGURE 30.4 Video frames of an individual at work. a) Seated view of the whole body that includes the workstation. b) Seated view of the upper body. c) Standing lower body view. d) Seated overhead and forward view.

30.5 Expected Results

The expected results of this process should be a videotape that is easy for a competent person to analyze. In addition, all of the details and information that cannot be put on a videotape will be available for the analyst. With this complete package the analysis will go smoothly with a minimum of delay and confusion. The analysis produced will not be deficient because of the documentation of the job. Additionally, the documentation created will be useful in follow-up and for demonstrating the original job. It is also useful to videotape and collect data about the job after it is changed. These materials can be used in training and demonstration.

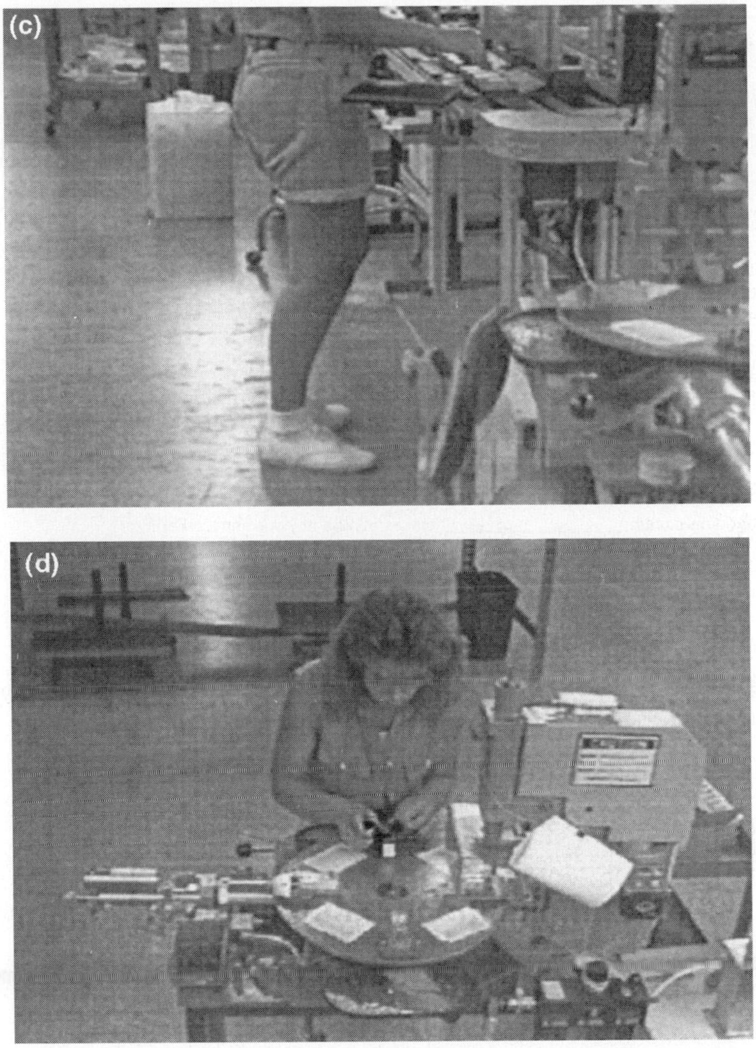

FIGURE 30.4 (continued)

30.6 Summary

Video taping a job for ergonomic analysis is not a hit or miss operation. Care and planning are required and a systematic approach is recommended. The steps recommended in this paper are stated in the flow chart of Figure 30.1. Proper taping and data gathering will make job analysis for ergonomic purposes much easier and more valid. The care and time spent properly videotaping and gathering data about the job will pay off in the long run.

DATA SHEET #1 - JOB/TASK INFORMATION

JOB NAME(S): ______________________________.

JOB LOCATION: Bldg.________. Floor:________. Department:________.

Line:__________. Work typical?_____. Exceptions ____________________

Brief Job Description:____

____

____.

Preceding job:______________ Succeeding job:________________

Number of employees on this job:____________. Shifts:______________.

Control of work pace: Worker:____ Line_____

Line speed :___________________________________.

Effective line speed:______________ (pieces/minute or hour per worker).

Incentive or piece work? ____________________________________.

Work Time: (hours and or minutes) Per Day:________________.

Per Week:________ Schedule:______________________________.

Jobs rotated with____________ at ________ intervals. If the rotation scheme is more complicated describe it more completely here or on a separate page.

Temperature: Area:__________ Product:_________ Drafts:_________

PPE (gloves etc.):______________________________________.

Tools:____.

Materials:____.

Training? ____

LIST THE TAPED WORKER'S NAMES OR IDENTIFICATION HERE, ON THE BACK, OR ANOTHER PAGE.

FIGURE 30.5 Data Sheet #1 — Job/Task Information.

DATA SHEET #2 - WORKSTATION DIMENSIONS

JOB NAME(S): ____________________.

WORK SURFACE AND/OR WORKING AREA

1. Height:______. 2. Depth:______. 3. Width:______.

4. Angle:______. 5. Product Height:______. 6. Hand Height:______.

SEATING

7. Chair make and model:____________________.

8. Adjustable?______. Height:______ Range:______.

Seat Pan:______. Range:______.

Back:______. Range:______.

Arm Rests:______. Range:______.

9. Foot Rest: Needed?______. Available?______. Adequate:______.

10. Thickness of work surface:______. 11. Leg room adequate?______.

REACHES - Repeat 12 through 18 for all major reaches.

12. Description:____________________.

13. Distance:______. 14. Height:______.

15. Angle (away from straight ahead):______.

16. Angle (+up and -down):______.

17. Load Carried (actual weight when possible):______.

18. Frequency:______.

THROWS:

19. Distance_____ Height_____ Weight_____ Angle (away from straight ahead)_____.

FORCES (It may be necessary to use another sheet or the back of this one):

20. Push:______ 21. Pull:______

22. Pinch:______ 23. Grasp:______

23. Lifting or carrying:______.

SKETCH THE WORKSTATION WITH DIMENSIONS ON ANOTHER PAGE:

FIGURE 30.6 Data Sheet #2 — Workstation Dimensions.

DATA SHEET #3 - TOOL

1. Name of tool:______________________________.
2. Job(s) used on:______________________________.
3. Tool weight?________lbs. 4. Balance?____________.
5. Handle(s): Span:_________inches. Length:______inches.
   Material:__________. Texture:__________.
   Pressure points?______________________________.
6. Counterbalanced?_______. Needed?_________. Appropriate tension?_______.
7. Place for the tool in the workplace (e.g., holster, fixture)?________________.
8. Is the tool powered?_________.
   Type? Torque:__________. Reciprocating/Vibrating:_____.
   Other (describe):______________________________.
   Power source? Air?_________. Exhaust away from the hand?_____.
   Electric?______. Hydraulic?______.
   Other (describe):______________________________.
9. Vibration? Present:_________. Measured:____________________.
10. Heat Source:______ Temperature:______________.
11. Cold Source:______ Temperature:______________.
12. Forces:______________________________

SKETCH TOOL WITH DIMENSIONS HERE:

FIGURE 30.7 Data Sheet #3 — Tool.

DATA SHEET #4 - MATERIALS, PARTS, BINS, PALLETS, ETC.

If the material or part described here is of a reasonable size and inexpensive ask for an example piece.

JOB NAME(S):______________________________.

1. Name of object or material other than tools handled?______________

____

2. Describe:

    Name:__________________________

    Material:_______________________

    Size:__________________________

    Sharp Edges:____________________

    Weight:_______________________lbs.

3. Hot?_____. Temperature:_________.

4. Cold?____. Temperature:_________.

5 Comments or additional description:

SKETCH THE PART WITH DIMENSIONS HERE.

FIGURE 30.8 Data Sheet #4 — Materials, Parts, Bins, Pallets, Etc.

DATA SHEET #5 - PERSONAL PROTECTIVE EQUIPMENT

1. Name of equipment:________________________.
2. Purpose:________________________.
3. Job(s) used on:________________________.
4. Description:________________________.
5. Available Sizes:________________________.
6. Fit:________________________.
7. Material:________________________.
8. Texture:________________________.
9. Interference?________________________.

Comments:________________________

____

______________________________.

SKETCH ITEM WITH DIMENSIONS HERE.

FIGURE 30.9 Data Sheet #5 — Personal Protective Equipment.

DATA SHEET #6 - MAINTENANCE

WORKSTATION

Nature: ______________________________

____

Frequency (now/required): ______________________________

____

Impact on the work/worker: ______________________________

____

____

TOOLS

Nature: ______________________________

____

Frequency (now/required): ______________________________

____

Impact on the work/worker: ______________________________

____

____

PPE

Nature: ______________________________

____

Frequency (now/required): ______________________________

____

Impact on the work/worker: ______________________________

____

____

FIGURE 30.10 Data Sheet #6 — Maintenance.

References

Borg, G. A. V. 1982. Psychological bases of perceived exertion. *Medical Science Sports Exercise* 14(5): 377

Corlett, E. N. and Bishop, R. P. 1976. A technique for assessing postural discomfort. *Ergonomics* 19(2): 175-182.

31

Physiological Instrumentation

Danuta Koradecka
Central Institute for Labour Protection Warsaw

J. Bugajska
Central Institute for Labour Protection Warsaw

31.1 Introduction

Physiological load of occupational physical work can be assessed on the basis of a rate of energy expenditure or oxygen consumption by the body. In many industrial tasks, it is also important to determine

0-8493-2641-9/99/$0.00+$.50

how much force must be generated by muscles to lift or support the weight of objects or to maintain the steady body position. However, these data are unsatisfactory when individual work tolerance has to be predicted. For this purpose, it is necessary to establish a relationship between the oxygen requirement for a given job to the individual maximal oxygen uptake or relation of force required to perform any task to the individual muscle strength (maximal voluntary contraction force). Moreover, many additional factors in the working environment, such as high or low ambient temperature, noise, etc., or psychological stress can influence work tolerance. Thus, the assessment of strain imposed by work and accompanying environmental factors should also include direct evaluation of the body responses to the work load or exposure to specific factors. Among them are indices of cardiovascular and respiratory system function, changes in body temperature, rate of sweating, muscle strength, muscle electrical activity (EMG), or indices of fatigue on the central nervous system such as disturbances in movement coordination, lowered level of arousal, etc.

31.2 Heart Rate (HR)

Description of the Parameter

Heart beat frequency (heart rate) is the most important and the simplest parameter of cardiovascular system function. Physical work, as well as many environmental stimuli, evoke an immediate acceleration of HR and, subsequently, an increase in the volume of blood ejected by the heart per unit of time (cardiac output). This response is usually followed by an increment in the amount of blood ejected during a single contraction of cardiac ventricles and changes in the distribution of blood flow among various organs due to the local constriction or dilatation of blood vessels. The changes in the cardiovascular system function allow the delivery of more oxygen and nutrients to active tissues, e.g., skeletal muscles, and the removal of metabolic waste products from the cells. The system also plays an important role in hormone transport from endocrine glands to target organs and in thermoregulation.

Specialized cells in the heart play a role of internal pacemakers due to their unique ability to produce electrical impulses initiating rhythmic cardiac muscle contractions. An acceleration of HR may be simply caused by increased blood flow to the heart, e.g., due to the dynamic contraction of skeletal muscles acting as pumps promoting venous return. However, the changes in HR may also be evoked by neural impulses from the autonomic nervous system and adrenal hormones, i.e., adrenaline and noradrenaline. There are two branches of the autonomic nervous system affecting the internal pacemaker of the heart: one of them — called sympathetic — increases HR, while the other — parasympathetic — exerts an opposite effect. Both branches of the autonomic system are under the control of the central nervous system and receive information from other brain centers and peripheral tissues. Circulating in blood, noradrenaline and adrenaline act in concert with the sympathetic nerves enhancing HR. These mechanisms adjust HR and, subsequently, cardiac output to the current needs. It is worth noting, however, that during psychological stress the increases in HR may exceed the real needs considerably.

At rest, under comfortable conditions, HR is usually 60 to 80 beats/min. It shows, however, spontaneous fluctuations of various frequency, known as physiological HR variability (HRV). Fast fluctuations of a frequency corresponding to the respiratory rhythm are caused mainly by the changes in the parasympathetic nervous system activity, while the slow fluctuations depend on both sympathetic and parasympathetic activities.

During dynamic exercise, HR is directly proportional to the intensity of the exercise. The increases in HR are greater during small muscle group exercise, e.g., arms, than during large muscle group involvement, e.g., legs, in spite of the same load. The maximal HR attained during exhausting efforts in young, healthy people usually exceeds resting values three times. The ability to reach the highest heart rate during maximal exercise decreases with age (Table 31.1). As a result, the same heart rate for a younger and older person implies different physiological load, imposed on the circulatory system in these people.

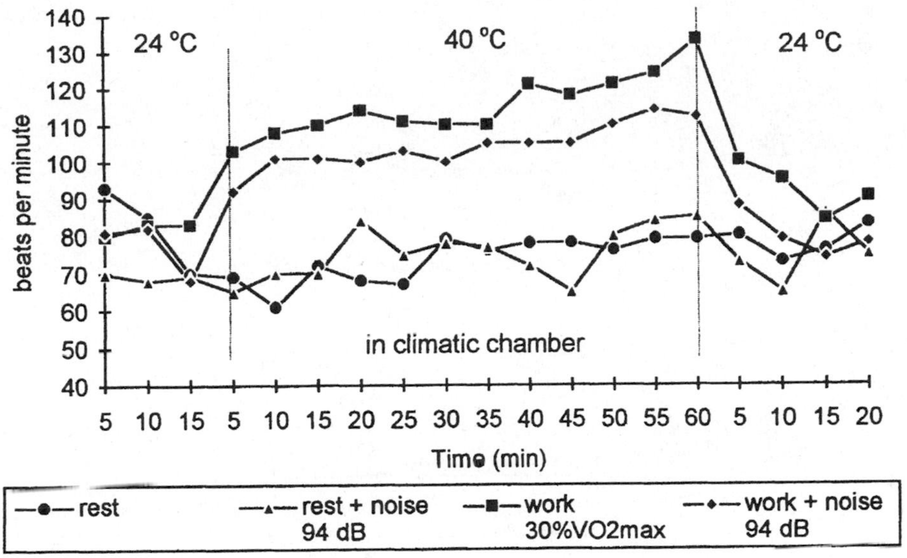

FIGURE 31.1 Comparison of heart rate at rest and during physical work (30% VO_{2max}) performed at 40°C (climatic chamber) accompanied by noise of 94 dB. (From Kurkus-Rozowska B. et al., Changes in blood pressure observed as a combined effect of heat and noise during physical load, *Arch. Complex Environmental Studies,* 8, 3-4, 11-16, 1996. With permission.)

Heart rate during exercise shows a high correlation with relative physical load expressed as the percentage of an individual maximal oxygen uptake (VO_{2max}). It has been shown that at 50% of maximal load (50% VO_{2max}), the heart rate of young, healthy men is 130 beats per minute, and at 30%, it is approximately 110 beats per minute. In women, a load of 50% VO_{2max} results in a heart rate increase up to 140 beats per minute. Stress and environmental stimuli, such as high or low temperature or noise, increase HR at rest and may exaggerate its response to exercise (Figure 31.1).

The increase in heart rate in response to acute stimuli is a result of an increase in sympathetic activity, a decrease in parasympathetic activity, or a combination of the two. The deceleration of heart rate occurring, e.g., during face cooling (diving reflex), results from an increase in parasympathetic activity, a decrease in sympathetic activity, or both (Papillo and Shapiro, 1990). Recently, spectral analysis of heart rate variability has come into use as a useful and sensitive index of autonomic function, particularly for the evaluation of mental stress (Pagani et al., 1991).

Description of the Parameter Measurement

Measurement of HR under various field conditions is quite easy. It can be made palpable, i.e., by putting a finger tip on an artery (e.g., carotid or radial artery) and counting the number of beats per minute. The method is based on the artery wall pulsation phenomenon. This is the most common method. However, it is not very precise and is not always possible to administer, e.g., during work requiring arm movements.

It is also possible to count heart rate by listening to the sound of the working heart with a stethoscope or using a microphone registering the acoustic signal, processing it, and displaying HR. This method can be used during physical effort.

Automatic manometers (of the so-called capacitive type) use the phenomenon of capacity changes connected with the filling of vessels according to the rhythm of the pulse rate. This information is electronically processed and HR is shown on a screen.

The electrocardiograph detects, amplifies, and prints out changes in the electrical potential between electrodes attached to the skin surface. These changes are caused by electrical impulses generated by

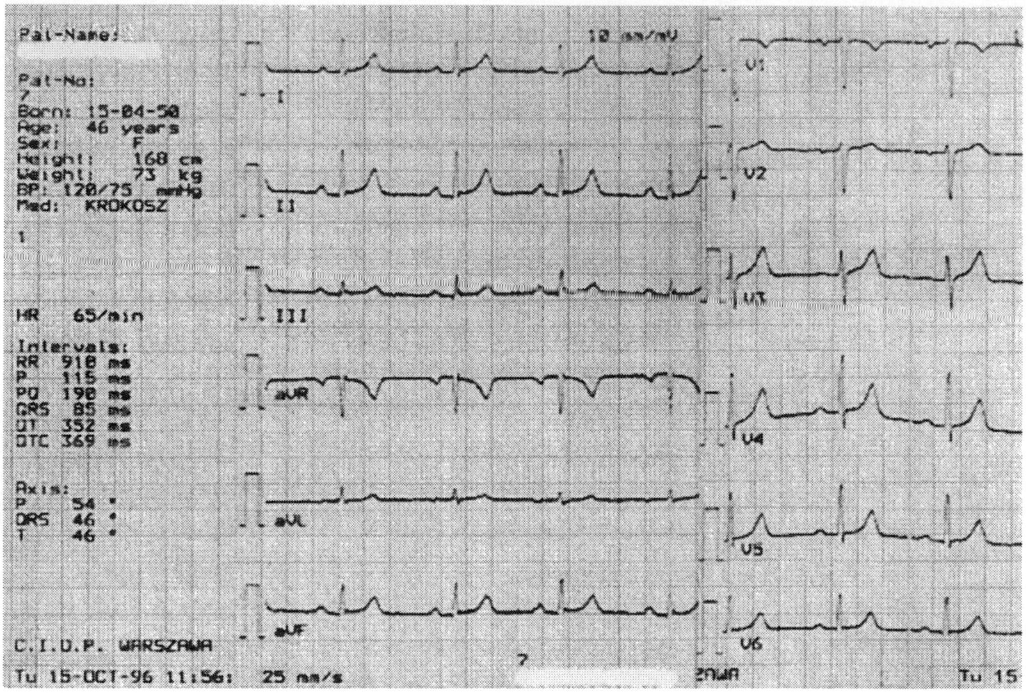

FIGURE 31.2 Sample ECG record with automatic measurement of heart rate calculated on the basis of R-R waves (in milliseconds).

cardiac pacemakers and spread through the heart. They show a regular pattern with the peak potential, called R waves, occurring as the impulse spreads through the heart ventricles. The time interval(s) between the two successive peak R waves (R-R interval) corresponds to the period of beat cycle and is the reciprocal of HR (60/R-R interval = beats/min). The electrocardiograph can be easily recorded during various activities using electrodes attached to the chest.

In the case of older ECG equipment, the measurement of the R-R intervals has to be performed manually (putting a special ruler to the ECG curve). In newer ECG equipment, this function is taken over by a processor and heart rate is displayed on a screen (Figure 31.2). In the ECG equipment which works in the Holter system, ECG signals are monitored and "remembered" for 24 hours. Statistical data can be processed by a computer, pathological changes can be assessed, and the course of this parameter over time can be shown (Figure 31.3).

Another possibility is to register heart rate in a given time with equipment in the form of a watch, which — via radio waves — registers signals from an electrode placed on the chest. This information can be sent through an appropriate interface to a computer. A computer program processes the data statistically and graphically (Figure 31.4).

Interpretation of the Parameter

Mean heart rate at rest is 60 to 80 beats per minute. If daytime mean heart rate falls below 50 beats per minute, it is a case of so-called bradycardia, which is considered a pathological symptom in people of average physical capacity. The phenomenon of functional bradycardia takes place in the case of people with high physical capacity, especially in endurance-trained athletes. It is connected with a great volume of blood ejected from the heart during each contraction, which — even with lower heart rate — allows the delivery of an adequate amount of blood to the tissues (the so-called athlete's heart). On the other hand, heart rate at rest exceeding 100 beats per minute should be considered a symptom of tachycardia. It is often connected with myocardial ischemia or neurohormonal hyperexcitability related to stress or some illnesses (e.g., hyperthyreosis).

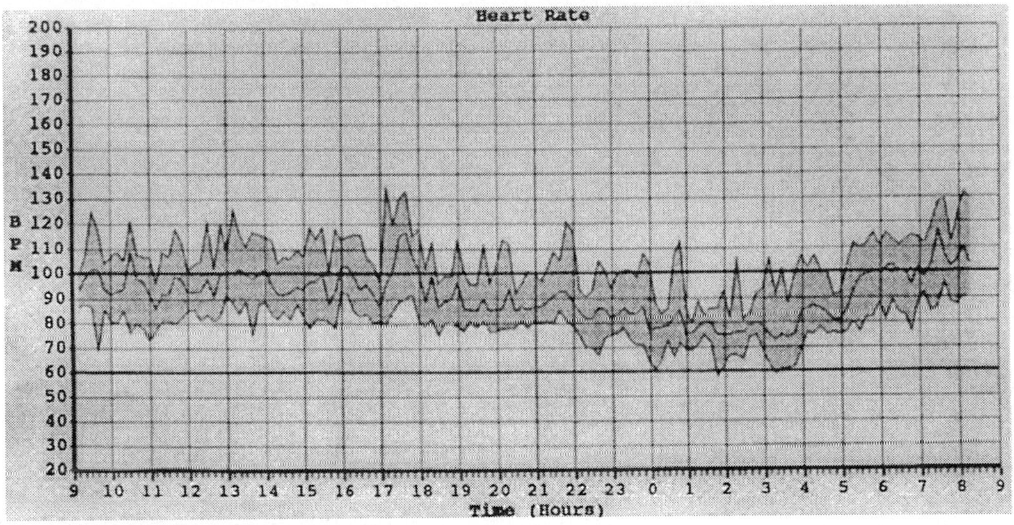

FIGURE 31.3 Sample record of the measurement of heart rate over 24 hours with the use of an electrocardiograph.

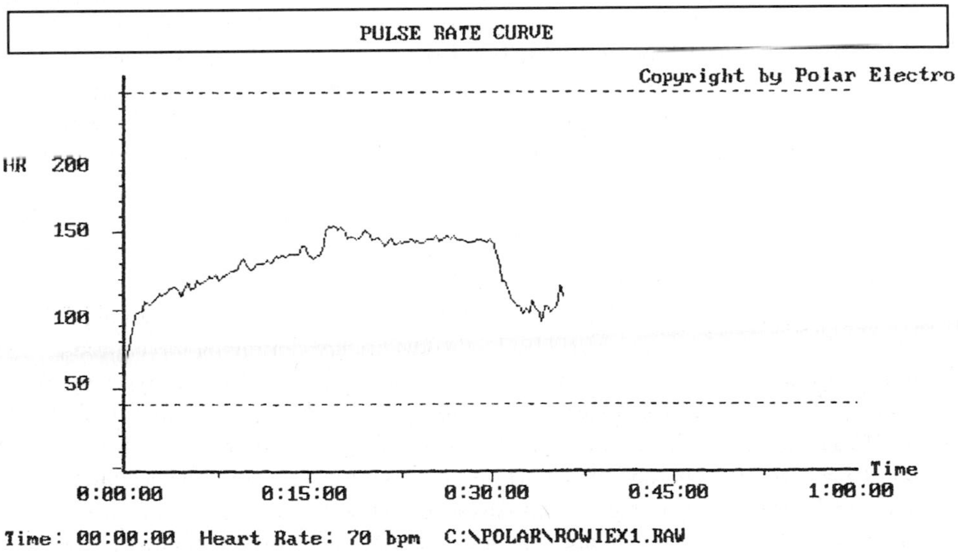

FIGURE 31.4 Sample record of a heart rate measurement performed during exercise with a watch-like device collecting the signal via radio waves from an electrode placed on the chest.

Maximal heart rate during exercise for healthy people can be estimated by subtracting the age of the examined person from the number 220.

31.3 Minute Pulmonary Ventilation (VE) and Respiratory Rate (RR)

Description of the Parameter

The respiratory and cardiovascular systems act in concert to provide a system of delivering oxygen to tissues and remove carbon dioxide from them. Respiratory rate (number of breaths during 1 min), tidal

TABLE 31.1 Correlation Between Maximal Heart Rate and Age

age (years)	Maximal Heart Rate per minute (HR_{max}/min)	
	mean	range
10	210	190–215
15	203	185–218
20–29	193	173–213
30–39	185	165–205
40–49	176	156–196
50–59	168	148–188
60–69	162	141–181
70–79	153	133–173
80–89	145	125–165

From Kozłowski St., Nazar K., *Introduction to Clinical Physiology*, PZWL 1995 (in Polish).

volume (TV), that is the volume of air inspired during a single breath (depth of breathing), and the product of these two variables — minute pulmonary ventilation — are the main parameters of respiratory system function. The diffusion of gases occurs in the pulmonary alveoli. It depends on the differences in partial pressures of oxygen and carbon dioxide between the alveoli and blood. The maintenance of balance in partial pressures of blood gases and blood pH requires coordination of the respiratory and cardiovascular systems. This is accomplished by involuntary regulation of rate and depth of breathing by the respiratory centers in the brain. These centers establish the breathing pattern by sending out impulses to the respiratory muscles. The respiratory centers receive information on the chemical environment in the body, from the areas of the brain sensitive to carbon dioxide and pH, and from the special receptors located in the aorta and carotid artery that are sensitive to partial pressure of oxygen and carbon dioxide in blood and its pH. Accumulation of carbon dioxide and the subsequent increase in body fluid acidity (drop of pH) appear to be the strongest stimulators of respiration. However, the respiratory system function may also be influenced by other neural factors, among them sensory impulses from working muscles, emotional distress, etc. There is also a possibility of voluntary control of breathing through the cerebral cortex. This can be, however, overridden by the involuntary controlling mechanisms.

Minute ventilation (V_E) increases in direct proportion to the effort intensity and oxygen uptake (VO_2) (Figure 31.5): up to 70% of maximal oxygen consumption (VO_{2max}) in people with low physical capacity, and up to 85% in people with high physical capacity. Exceeding this level results in hyperventilation, which is an increase in the proportion of ventilation to oxygen consumption. The beginning of hyperventilation coincides with an increase in carbon dioxide percentage in the expired air, an abrupt rise of the level of lactic acid in blood, and a decrease of blood pH. The above-mentioned phenomenon is illustrated in Figure the 31.6.

Description of the Parameter Measurement

Respiratory rate can be determined approximately by watching chest movements. However, this method is not applicable if there is little chest mobility during breathing. In order to precisely determine respiratory rate, spirometers — which measure changes of mechanical parameters on the basis of the air movement during inspiration or expiration — can be used.

Recently, electronic spirometers, in which air flow sensors measure breathing, have been introduced. Devices that measure man's metabolism by analyzing the expiratory gases can also measure respiratory rate.

In order to measure minute ventilation of the lungs, it is necessary to put on a face mask connected to a gas meter and to measure the volume of gas expired over a minute.

The measurement of ventilation in newer types of devices can be made by monitoring the respiration rate (RR) and tidal volume (TV) with an air flow sensor.

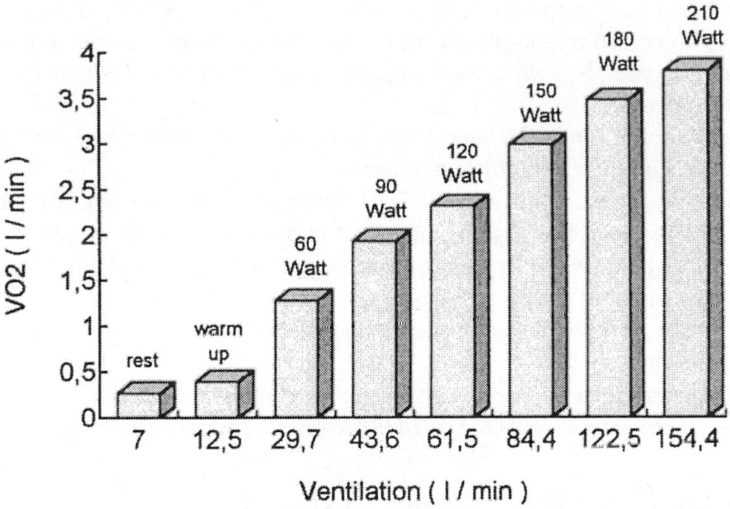

FIGURE 31.5 Sample of relationship between oxygen uptake (VO_2) and minute ventilation of the lungs (V_E) in a male subject during different physical load (in Watts). (From Konarska, M., Kurkus-Rozowska, B., Krokosz, A., and Furmanik, M., Application of pulmonary ventilation measurements to assess energy expenditure during normal and massive muscular work, *Proceedings of the 12th Triennial Congress of the IEA,* 3, Human Factors Association of Canada, 1994. With permission.)

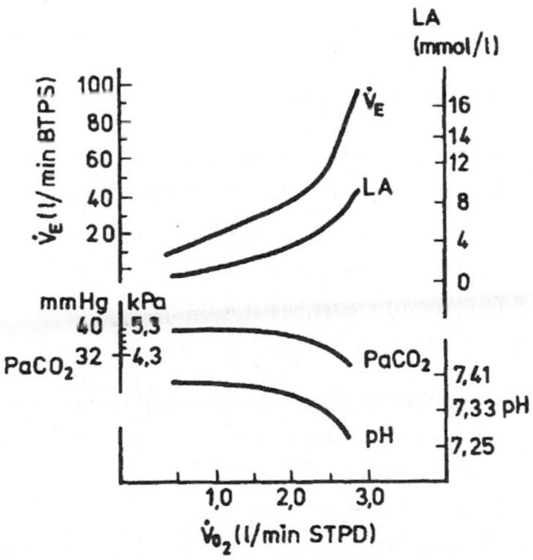

FIGURE 31.6 Changes in pulmonary ventilation, blood lactic acid, blood pH, and blood $PaCO_2$ in relation to oxygen uptake, during exercise of increasing intensity. (Adapted from Kozłowski St., Nazar K., *Introduction to Clinical Physiology,* PZWL 1995 [in Polish]. With permission.)

Interpretation of the Parameter

Mean respiratory rate at rest in healthy people is 10 to 20 breaths per minute. In small children, respiratory rate is higher. In adults, respiratory rate decreases with an increase of tidal volume.

During exercise, respiratory rate increases to 30 to 60 breaths per minute. The number of breaths is directly proportional to the intensity of the performed exercise, and it stabilizes at a fixed level after a few minutes of effort.

Respiratory rate exceeding 60 per minute has low efficiency from the point of view of pulmonary ventilation, while causing rapid fatigue of the respiratory muscles.

There is often an increase of respiratory rate per minute at rest in people with changes in the respiratory system related to a lower tidal volume, which can be caused by constriction of the respiratory tract, (e.g., asthma). This can also take place while breathing hot air and during illnesses that are accompanied by higher internal temperature.

The mean minute ventilation of the lungs at rest is 6 to 12 l/min. During exercise, minute ventilation can increase 25-fold in relation to ventilation at rest.

Maximal minute ventilation of the lungs (VE_{max}) during exercise is different for people of different physical capacity. In people of low physical capacity it is 70 to 90 l/min; in people of average physical capacity it is 110 to 130 l/min, in people of high physical capacity it is 15 to 160 l/min. In athletes maximal minute ventilation of the lungs can reach 200 l/min. In women, the maximal minute ventilation is lower than in men of the same age because of smaller respiratory volume. Maximal minute ventilation of the lungs decreases with age.

Minute ventilation of the lungs below the suggested standards can be the first signal of respiratory obduration related to the pathological changes in the respiratory system (e.g., among cigarette smokers).

31.4 Maximal Oxygen Uptake (VO_{2max})

Description of the Parameter

Maximal oxygen uptake is the ability to take oxygen during maximal physical exercise. It is the best indicator of the efficiency of the oxygen transport system in the body and the ability of tissues to utilize oxygen in metabolic processes. Oxygen consumption depends mainly on maximal cardiac output, the mass of skeletal muscles, and the capacity of aerobic biochemical processes yielding energy for muscular work. Maximal pulmonary ventilation, blood volume, the number of red blood cells, and hemoglobin content also play an important role in VO_{2max} limitation.

Maximal oxygen uptake (VO_{2max}) is a basic parameter for evaluating physical capacity because:

- It determines the amount of oxygen per minute which man is able to consume in order to satisfy the demand for oxygen during physical load. Exercise during which oxygen requirement exceeds VO_{2max} can be maintained only for a few minutes,.
- It makes it possible to calculate the relative load expressed as % VO_{2max} (i.e., it makes it possible to calculate the load, which permits work for a long time without developing muscle fatigue and the accumulation of lactic acid in the blood).

Maximal oxygen uptake of a given person, expressed per kg of body mass per minute, allows evaluation of individual physical work capacity and classification of it as low, average, high, or very high (Table 31.2). The ability of oxygen uptake decreases with age (Figure 31.7).

The amount of oxygen uptake during submaximal effort correlates with relative load and, as a result, also with physiological parameters, e.g., heart rate (Figure 31.8). Physiological responses to exercise such as an increase of minute pulmonary ventilation, increase of heart rate, blood pressure and changes in peripheral blood flow are similar in various persons if the relative work load expressed as percentage of VO_{2max} is the same.

Description of the Parameter Measurement

The amount of maximal oxygen consumption is defined during maximal or supramaximal exercise as a result of a difference in oxygen concentrations in inspired air (oxygen concentration in the surrounding air) and in expired air. However, most often the value of VO_{2max} is predicted on the basis of the correlation of oxygen consumption and heart rate during submaximal exercise, according to Åstrand's nomogram (Åstrand and Rodahl, 1986).

Interpretation of the Parameter

The evaluation of maximal oxygen consumption according to sex, age, and physical capacity expressed as milliliters of O_2 per minute per kilogram is shown in Table 31.2.

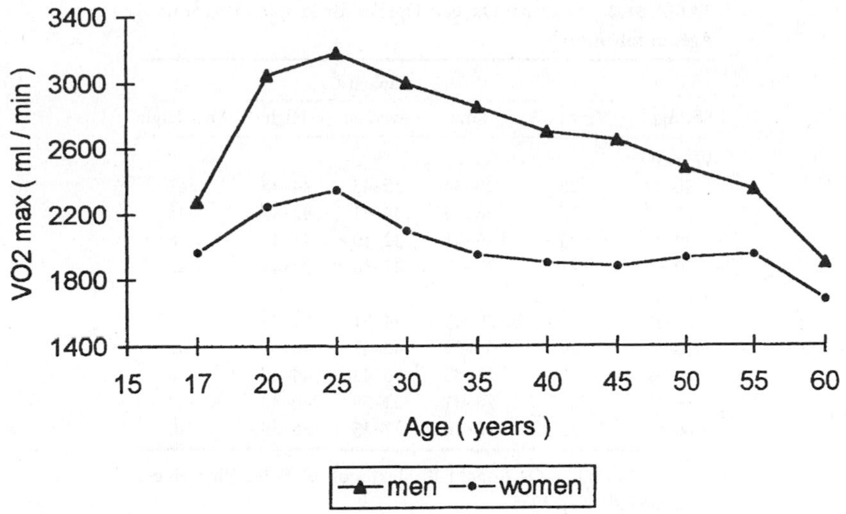

FIGURE 31.7 Maximal oxygen uptake (VO_{2max}) calculated on the basis of heart rate during exercise (submaximal) in male (a) and female (b) subjects. (From Kozłowski St., Nazar K., *Introduction to Clinical Physiology,* PZWL 1995 [in Polish]. With permission.)

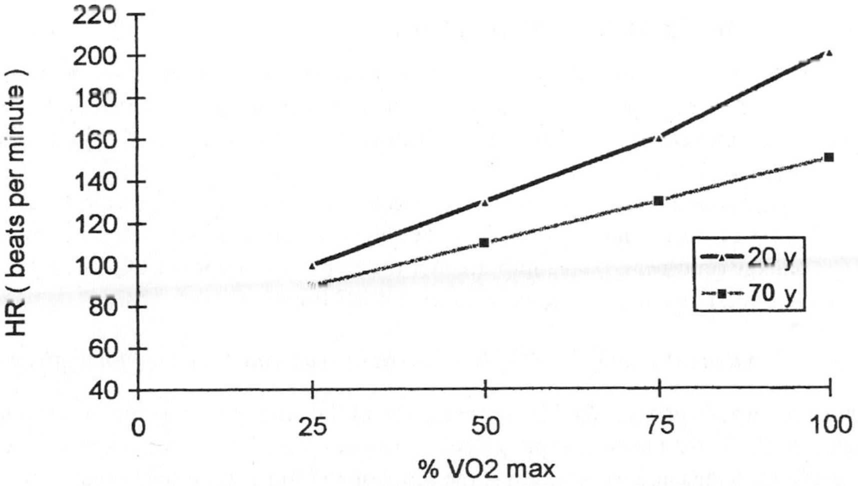

FIGURE 31.8 Relationship between heart rate (HR) and relative physical load (given as percent of maximum oxygen uptake by the body, $\%VO_{2max}$) in subjects 20 and 70 years old. (From Kozłowski St., Nazar K., *Introduction to Clinical Physiology,* PZWL 1995 [in Polish]. With permission.)

31.5 Energy Expenditure

Description of the Parameter

Energy expenditure is the total energy liberated by the body in order to maintain life functions (basic or rest metabolism), body temperature at rest, and support muscle work during exercise. The amount of the energy used for basic metabolism depends on body mass, sex, age, and health. The amount of energy required to perform physical work depends on the mass of the involved muscles, body position during work, and work intensity.

TABLE 31.2 Maximal Oxygen Uptake for People of Various Ages in ml/min/kg

	Capacity				
Sex/Age	Very Low	Low	Average	High	Very High
Women					
20–29	28	29–34	35–43	44–48	>49
30–39	27	28–33	34–41	42–47	>48
40–49	25	26–31	32–40	41–45	>46
50–65	21	22–28	29–36	37–41	>42
Men					
20–29	38	39–43	44–51	52–56	>57
30–39	34	35–39	40–47	48–51	>52
40–49	30	31–35	36–43	44–47	>48
50–59	25	26–31	32–39	40–43	>44
60–69	22	22–26	27–35	36–39	>40

From Åstrand P-O., Rodahl K.: Textbook of Work Physiology, McGraw-Hill, N.Y. 1986.

Energy expenditure can be used to evaluate the physical load of work and to determine balance, i.e., an equilibrium between intake in the form of food and energy expenditure, e.g., when treating obesity.

Energy expenditure is expressed in W/m^2 per body surface of the subject (previously in kcal/min and kJ/min).

Description of the Parameter Measurement

Energy expenditure can be measured directly in calorimetric chambers or indirectly on the basis of oxygen consumption during exercise. The amount of energy expended can also be estimated on the basis of the measurement of minute ventilation of the lungs, which is in direct proportion to oxygen consumption (up to 75% VO_{2max} of a given person).

At present, oxygen consumption is measured with electronic oximeters. Energy expenditure calculated on the basis of oxygen consumption is the result of multiplying oxygen consumption (in liters per minute) by the oxygen energy equivalent, taking into account the respiratory quotient RQ (the ratio of carbon dioxide to oxygen in expired air). The result is expressed in kJ/min.

$$\textit{Energy expenditure}\,(\text{kJ/min}) = \text{VO}_2(\text{l/min}) \times \textit{energy equivalent of}\ \text{O}_2\ (\textit{for a given}\ \text{RQ})$$

By using equipment for measuring minute ventilation of the lungs, it is possible to determine energy expenditure on the basis of the simple relation between ventilation and oxygen consumption (Figure 31.9). Datta-Ramanathan's equation is the simplest and most commonly used:

$$\textit{energy expenditure}\,(\text{kcal/min}) = \textit{VE}\,(\text{STPD}) \times 0.21$$

where VE = minute ventilation of the lungs in l/min, STPD = standard conditions of dry gas volume for 0°C and for atmospheric pressure of 760 mm Hg.

Interpretation of the Parameter

Energy expenditure is an absolute measure of work load. However, the physiological load and, subsequently, work tolerance depend on the individual working capacity. The relation of oxygen uptake or the known oxygen requirement for a given task to VO_{2max} of the working person is often used for the assessment of physical strain in occupational work. The suggested physical work load is 30% VO_{2max} during a shift. Load which exceeds 50% VO_{2max} per shift can adversely affect the worker's health in the long run. Energy expenditure was measured and evaluated in industry, in a large number of subjects

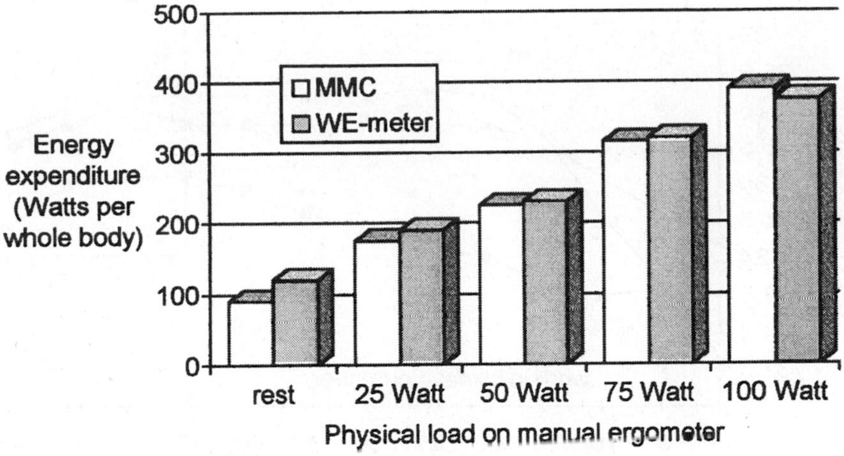

FIGURE 31.9 Comparison of the energy expenditure evaluation using MMC (Metabolic Measurement Card, on the basis of oxygen uptake) and WE-meter (electronic ventilation meter, on the basis of minute ventilation) during exercise on a manual ergometer. (From Konarska, M., Kurkus-Rozowska, B., Krokosz, A., and Furmanik, M., Application of pulmonary ventilation measurements to assess energy expenditure during normal and massive muscular work, *Proceedings of the 12th Triennial Congress of the IEA*, 3, Human Factors Association of Canada, 1994. With permission.)

TABLE 31.3 Work Load Classification on the Basis of Net Energy Expenditure in kJ/min

Physical Work	Energy Expenditure (kJ/min) Men	Women
Light	8.4–20.5	6.3–14.2
Moderately hard	20.9–30.0	14.7–22.6
Hard	31.4–41.4	23.0–30.0
Very hard	41.9–51.9	31.4–39.3
Extremely hard	>52.0	>39.8

Christensen E. H.: Physiological evaluation of work in the Nykroppa iron works, in *Ergonomics Society Symposium on Fatique.* Floyd W.F. ed. Lewis, London 1953.

performing various tasks. The data can be found in the available literature (Passmore and Durnin, 1967). Because of that, energy expenditure evaluation and physical work load classification mostly apply to a "standard person." Net energy or oxygen requirement can be calculated for a shift (after subtracting the value of basic metabolism) and compared with the data in the work load classification table (Table 31.3).

31.6 Skin Temperature and Core Temperature

Description of the Parameter

Skin temperature and core temperature allow us to evaluate the efficiency of the thermoregulatory system in relation to changes in ambient temperature, exercise, etc. (Figure 31.10). This efficiency is important because of the necessity to maintain constant core temperature within a narrow range of temperatures for particular organs (e.g., approximately 36°C for muscles and approximately 37.2°C for the brain). The average body core temperature is 37°C. Core temperature varies according to the circadian cycle (approximately 37°C during the day and approximately 36°C at night). In small children, the temperature is slightly higher because of the smaller number of sweat glands. The average core temperature changes as a result of metabolism changes in some diseases (e.g., thyropathy).

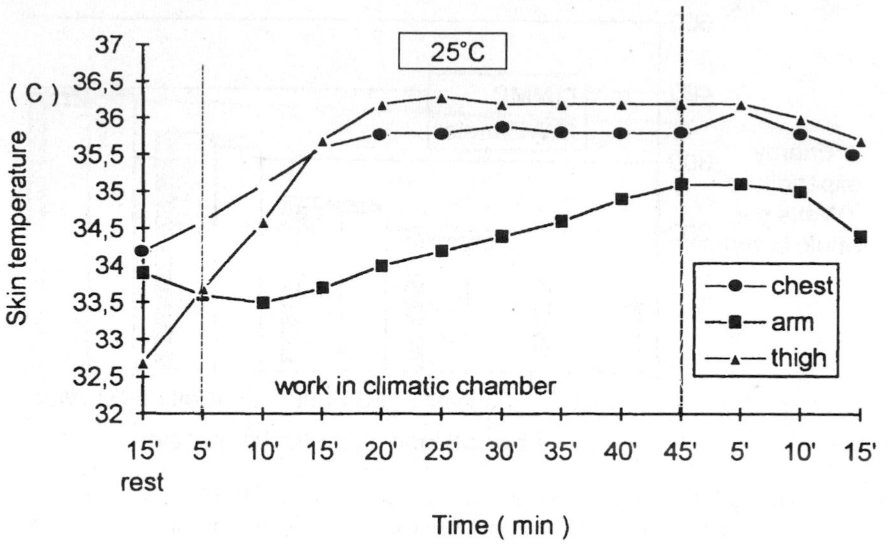

FIGURE 31.10 Sample of the changes of skin surface temperature during work (30% VO_{2max}) performed in oil protective clothing, at 25°C. (From Marszałek A., Sawicka A., Necessity to limit the duration of work performed in oil protective clothing, *Proceedings of 8th Congress of Polish Association of Occupational Medicine,* 1996 (in Polish).

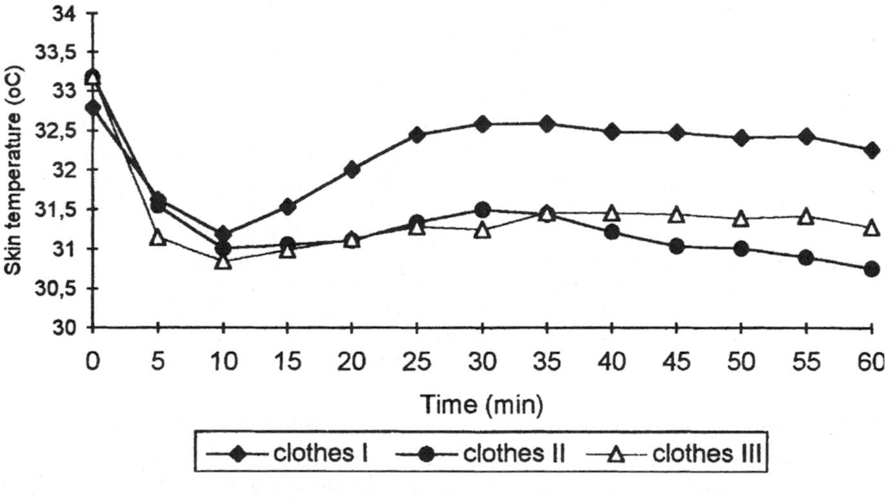

FIGURE 31.11 Time course of mean weighted skin temperature during exercise of 30% VO_{2max}, performed in three kinds of protective clothing. Clothes I = cold weather clothing for working in cold store; clothes II = cold weather and rain clothing for airport mechanics; clothes III = cold weather and rain clothing for construction workers. (From Marszałek A., Sołtyński K., Sawicka A.: Physiological evaluation of cold protective clothing, *Int. J. Occup. Safety & Ergonomics,* 1995, 1, 3, 235. With permission.)

Skin temperature is measured on the body surface. Local differences in skin temperature result from the skin blood flow difference. Because of that, skin temperature is measured in several places (from 3 to 10 points). The result in every spot is multiplied by the ratio characteristic for blood supply of a given body part. In this way, mean weighted skin temperature is calculated in accordance with the 8-point formula (ISO 9886, 1992). (Figure 31.11).

$$T_{sk} = 0.07T_{s1} + 0.175T_{s3} + 0.175T_{s4} + 0.07T_{s5} + 0.07T_{s6} + 0.05T_{s7} + 0.19T_{s10} + 0.2T_{s13}$$

where
T_{s1} = temperature of the forehead
T_{s3} = temperature of the right scapula
T_{s4} = temperature of upper left front part of the chest
T_{s5} = temperature of the upper right arm
T_{s6} = temperature of lower left arm
T_{s7} = temperature of the dorsum of the left palm
T_{s10} = temperature of the front part of the right thigh
T_{s13} = temperature of the back part of the left calf

Core temperature is influenced by ambient temperature and physical exercise. During physical exercise a considerable amount of thermal energy is produced. Out of 5 kcal produced during metabolism, only about 1 kcal is converted into work, and the rest into heat that has to be dispersed. Otherwise, core temperature would rise, and that would make further work impossible. It could also pose a life hazard (heat stroke).

The thermoregulatory mechanism utilizes the following physical phenomena:

- Redistribution of heat produced inside the body to skin through conduction and convection
- Redistribution of body heat to the environment or inversely through radiation, convection, and conduction
- Evaporation of sweat from skin and evaporation of water from pulmonary alveoli and mucous membranes

Description of the Parameter Measurement

Core temperature of sick adults is usually measured with a mercury thermometer in the armpit or in the mouth under the tongue. In the case of small children, a thermometer is usually put in the rectum.

Physiologists measure core temperature in the rectum, 5 to 8 cm deep, or in the esophagus or near the tympanic membrane using appropriate thermistor sensors or thermocouples. Skin temperature is measured with special thermistor sensors or thermocouples placed in specific places of the body.

Interpretation of the Parameter

Average core temperature measured with a mercury thermometer in the mouth is 37°C, in the armpit, 36.6°C.

Man's core temperatures below 27°C and over 42°C most often result in death. The rise of core temperature by more than 3°C, resulting from the impairment of the sweating mechanism or heat exchange problems, can be hazardous because of the disturbance of the functions of the circulatory system.

Average skin temperature measured in thermoneutral conditions is 34 to 36°C.

31.7 Sweating

Description of the Parameter

Sweating and sweat evaporation is the most effective way to remove heat surplus from the body. The efficiency of thermoregulation depends on the number of sweat glands in skin, the pace of sweat secretion, environmental conditions influencing sweat evaporation, and on the efficiency of the circulatory system responsible for transporting heat from working muscles to the skin. Sweat secretion increases in the 3rd to 5th second of exercise. First, in a reflex way, sweat glands placed on the trunk are activated. After 5 to 10 minutes of work, the number of active glands increases due to the rise of core temperature. After exceeding 60 to 70% maximal oxygen uptake (VO_{2max}), sweat secretion decreases as a result of lower flow of blood to the skin, which is related to the increase of blood flow in the working muscles.

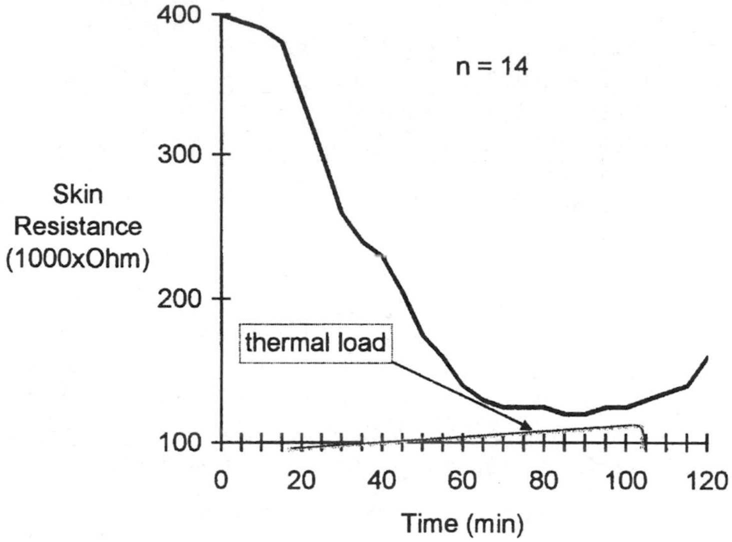

FIGURE 31.12 Changes of chest skin resistance R_s in subjects with exogenous thermal load (Adapted from Grucza R.: *Thermoregulation Model in Case of Endogenic and Exogenic Thermal Load of the Organism,* CMDiK PAN Warszawa, 1979, [in Polish].)

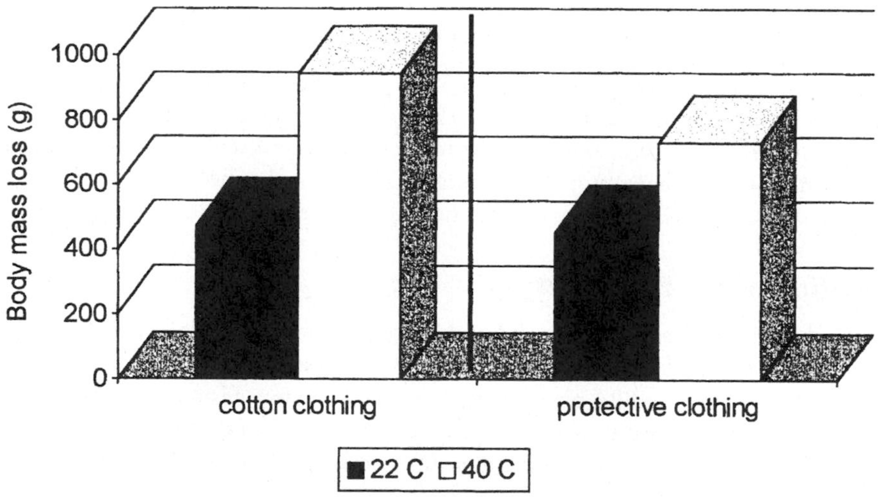

FIGURE 31.13 Sample of body mass loss (g) in subjects as an indicator of water loss when sweating during work at 30% VO_{2max}, in two kinds of protective clothing was compared with cotton clothing (control). (From Bugajska J. et al. Development of methods for evaluation of protective clothing for compliance with occupational physiology and hygiene, *International Journal of Occupational Safety and Ergonomics,* submitted for publication.

Description of the Measurement

Sweat can be measured with a device registering changes of skin resistance when sweating is low (Figure 31.12) or with hygrometers in the case of heavy sweating or high ambient temperature. Sweat rate can also be calculated indirectly on the basis of body mass loss measurement (Figure 31.13).

Interpretation of the Measurement

The ability to sweat depends on acclimatization to high temperature, age, race, etc. During maximal exercise the acclimatized person can secrete up to 1 liter of sweat per hour and 3 to 4 liters of sweat per working shift.

31.8 Tremometry

Description of the Parameter

Tremometry is a method used for measuring motor efficiency as far as hand tremor (trembling) and visual–motor coordination are concerned.

Method of Measurement

Testing is performed with a tremometer consisting of a desk in which holes of various shapes and sizes are cut out and a metal pen. The subject is to contour all cut-out shapes with a pen in such a way that side and bottom areas of the holes are not touched. The examination is made in the same position (standing or sitting) without resting the hand on the device's desk.

The result of the examination: time of performing the task, number of errors, time of errors.

Interpretation of the Measurement

Interpretation consists in comparing the results before and after a particular load factor is applied.

31.9 Critical Flicker Frequency Threshold

Description of the Parameter

Critical flicker frequency (CFF) threshold is considered to be an indicator of the level of stimulation of cortical centers, tiredness caused by work, and physiological changes resulting from the influence of alcohol or some pharmacological agents.

Method of Measurement

CFF is measured with the Flicker Test. The test is conducted with a device consisting of a measuring instrument and a black tube allowing both-eyed observation of the source of flickering light without any inflow of light from the outside. In the tube there will be a central flickering red light of 1 cm^2, shining with the intensity of 7.8 milliamperes. This light is surrounded by eight diodes, also red, shining with continuous light.

The range of flicker frequency changes is 0 to 99.9 Hz. The change of flicker frequency takes place automatically. The accuracy of the result is ± 0.1 Hz.

The subject is asked to react (by pressing a button) when he/she notices that the flickering light becomes continuous (in the case of ascending threshold) or when the person notices that the continuous light has started to flicker (in the case of descending threshold). Ascending threshold of light flickering is measured during the increase of light flicker frequency from 30 Hz upward; descending threshold, on the other hand, is measured when flicker frequency decreases from 99.9 Hz downward.

The number of measurements depends on the examination; most frequently a six-fold measurement is used.

The result of the examination: light flicker frequency when the subject notices the beginning of flickering (descending threshold) or cessation of flickering (ascending threshold) measured in Hz/s.

Interpretation of the Measurement

Interpretation consists in comparing the results before and after a particular load factor is applied.

31.10 Muscle Fatigue

Description of the Parameter

The term fatigue is used to describe sensations of general tiredness and accompanying decrement in the muscle ability to generate force or power. There are many causes of fatigue, among them a failure of the muscle fiber contractile mechanism, exhaustion of muscle energy substrates, such as high-energy phosphates and glycogen, the accumulation of metabolic waste products (lactic acid, hydrogen ion, ammonia), and failure of the near-muscular impulse transmission are considered. The contribution of various mechanisms to the limitation of muscle working ability differ in various types of work and none of the above-mentioned causes alone can explain all aspects of fatigue. It is generally accepted that the central nervous system is also the site of fatigue. In fact, perceived discomfort, decrease of alertness, and coordination disturbances always accompany strenuous muscular work and even precede the physiological limitation in muscle performance.

Muscle fatigue can be estimated by methods which register the electrical activity of muscle (electromyographic signal — EMG) and the exerted force. A comparison of the measured parameters before, during, and after exercise can be used as an indicator of fatigue. EMG is used for the evaluation of fatigue of particular muscles, whereas the measurement of force is used for the evaluation of whole muscle groups, e.g., muscles of the upper or lower limb, or dorsal muscles.

Description of the Measurement

Electromyography

Muscle fatigue may be indicated on the EMG signal record as a shift of the signal power spectrum toward lower frequencies (Bugajska et al., 1993; Lindstrom et al., 1970). In such a case, the parameter determining the median power frequency (MPF) is used to evaluate muscle fatigue.

Moreover, muscle fatigue can be evaluated on the basis of the changes in the following EMG parameters: ZC (the number of signal crossing of zero, the so-called Zero Crossing) (Hagg, 1981) and the signal amplitude value (Lindstrom et al., 1970).

The values of the coefficients of slope of the regression line between parameters indicating muscle fatigue and the time of the measurement constitute an indicator of muscle fatigue (Roman-Liu, 1996; Sundelin et al., 1992). The cause of the changes of the value of the EMG signal parameters is not clearly determined. Production of lactic acid resulting in pH change in the studied muscles is indicated. The rise in potassium concentration in the intercellular fluid (Bigland-Ritchie et al., 1979) connected with fatigue is given as another cause.

Decrease of Maximal Force or Time of Its Maintenance

As a result of fatigue, the maximal exerted force or the time of maintenance of this force decreases. Dynamometric force measurements at certain time intervals can be used to obtain (as in the case of EMG signal) the coefficient of inclination of regression lines between the measured value and the time indicating muscle fatigue.

During dynamic activities such as a jump, a power indicator can also be used. The subject performs, with no breaks, several maximal vertical jumps. The reaction force of the base (measured on a dynamometric platform) is registered and on this basis the power of each bounce is calculated.

Power is defined by an equation of simple regression:

$$P = a + b \times t$$

where P = power of each cycle, t = time in which a given power is reached, and a, b = equation coefficients. Coefficient b, the angle of the slope of the line, is an indicator of fatigue of the lower limb muscles.

31.11 Muscle Strength

Definition of the Parameter

Muscle strength depends on the muscle size (the sum of the cross-sectional areas of every muscle fiber) and neural mechanisms that recruit fibers during muscle contraction. The more fibers are recruited simultaneously, the greater the force that can be generated. Moreover, the frequency of each fiber stimulation results in the force increment. The pattern of muscle fiber stimulation can be assessed by EMG. There is a close correlation between the force of contraction and integrated muscle electrical activity.

Muscle strength (Ważny, 1977; Zorski, 1985) indicates man's ability to generate force during muscle contractions. It can be also defined as the ability to overcome and oppose external resistance.

Muscle strength is characterized by structural and geometric as well as energy and information factors. All these factors also have an influence on muscle strength effect during performance. According to the equation of muscle contribution understood as the value of the moment of muscle strength (Fidelus, 1989):

$$M_z = \rho \sum_{i=1}^{n} p_i r_i(\alpha) \left\{ \frac{F_i}{F_{io}} \left[\frac{l_i}{l_{io}}(\alpha) \right] \right\} \frac{U_i}{U_{i\,\max}}, \ [\text{Nm}],$$

muscle strength effect is influenced by:

M_z = moment of external forces in relation to a human body
ρ = muscle tension identical for a given muscle group, $[\text{Nm}^{-2}]$
p_i = physiological cross-section of the i-th muscle, $[\text{m}^2]$
r_i = arm of the force of i-th muscle, [m]
α = articular angle [°]
$\left\{ \frac{F_i}{F_{io}} \left[\frac{l_i}{l_{io}}(\alpha) \right] \right\}$ = dependence of the active force on the length of the i-th muscle, which is a function of the articular angle
F_o = value of the muscle active force when its length at rest is l_o
$U_i/U_{i\,\max}$ = ratio of the current EMG value [V] of the i-th muscle to maximal value of EMG for a given muscle.

In order to solve the equation, all values, except for tension, must be known. The calculated value of muscular tension also contains a certain factor of proportionality between the mean values of the figures used in the equation and the actual figures of the subject.

In mechanics, force is understood as a vector of force acting on a given particle. According to Newton's second principle of dynamics, force:

$$m \times a = F$$

where m = mass (of the particle), a = acceleration of the body, F = force (Zorski, 1985).

In order to estimate muscular strength, it is best to measure the moments of strength developed by muscle groups in relation to a particular center of rotation; this results from the analysis of correlation between muscle strength and geometric parameters (Buśko et al., 1991).

Method of Measurement

Measurement can be performed in two ways: indirectly and directly. In the case of an indirect measurement, the force and its arm are measured, and then the following multiplication takes place:

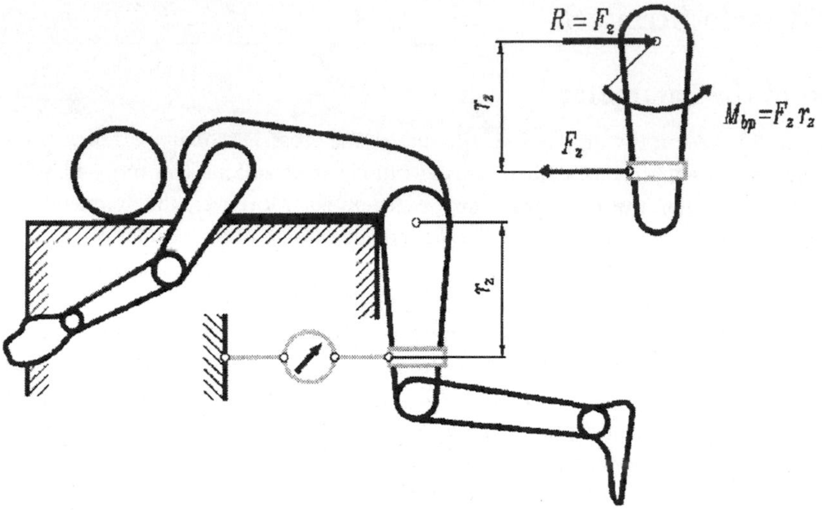

FIGURE 31.14 Measurement of the torque of extensors of the hip joint (M_{bp} — torque, F_z — external force, r_z — arm of external force, R — reaction force). (From Tokarski T. Dependence between muscle force and dynamic parameters developed on a dynamometric platform and bicycle, Doctoral thesis, Warsaw 1994. With permission.)

$$M = r \times F$$

where M = torque, r = arm of force (measured with a ruler), F = force (read from the dynamometer).

In the case of a direct measurement, the torque is measured with a torque meter.

During the measurements, the conditions in which the length of muscles allows the development of maximal torque are established. In most cases, muscles develop maximal torque in the range of 60 to 120° (Komi, 1979; Pieter et al., 1989). Not all authors, however, agree. This can result from using various methods of measuring the torque (Jensen et al., 1971; Kowalk et al., 1993). Most frequently, the measurement of moments of muscle forces is performed at 90° (Fidelus et al., 1984; Fugl-Meyer et al., 1980), in such a way that the center of gravity of the free rotating part (e.g., shank or forearm) is situated under or above the axis of rotation in a joint, so the torque resulting from weight of this part of the body is zero. It is easier to set the angle at 90° than, e.g., 60° or 120°; besides, when using the indirect method of calculating the torque, only one operation of multiplication of force and arm (sin 90° = 1) read from the dynamometer is made. (see Figure 31.14.)

Static conditions during the measurement are ensured by stabilizing the neighboring body parts.

Measuring Equipment

Indirect measurement of torque (Figure 31.14): stabilizing frame, stabilizing belts, dynamometer, ruler.

Direct measurement of torque (Figure 31.15): stabilizing armchair, torque meter, amplifier, digital voltmeter.

31.12 Body Position

Description of the Parameter

Body position is the position of individual body parts in relation to one another. Body position is defined by the angles of individual joints.

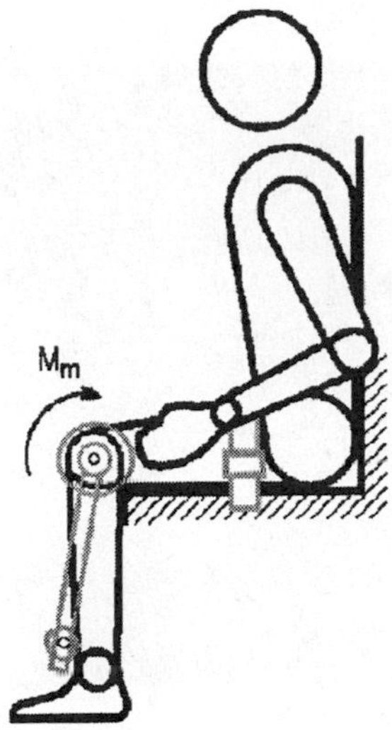

FIGURE 31.15 Measurement of torque of knee joint extensors (M_m). (From Tokarski T. Dependence between muscle force and dynamic parameters developed on a dynamometric platform and bicycle, Doctoral thesis, Warsaw 1994. With permission.)

There are several methods to evaluate the position at work. One is analysis of the posture at work with the use of audiovisual techniques. Such systems can analyze the movement of the human body recorded on a videotape. The system can consist of the following parts: video camera, videocassette recorder, computer, card for image processing, and a computer program for movement analysis (Macellari et al., 1983). Movement is recorded with a camera, which films the movement of individual body parts to which markers have been attached. Then, the image from the camera is converted into a digital form that can be read by a computer. As a result, it is possible to determine the position of all the markers, that is, individual body parts in space and time (Normand et al., 1983) (Figure 31.16). On the basis of the obtained values, computer analysis allows calculation of, among others, articular angles and the speed and angle acceleration in individual joints.

Another method is recording with position sensors placed on the body.

Description of the Measurement

Equipment with pendulum potentiometer sensors can be used for such measurements (Fig, 31.17). Sensors placed on the head, back, and arm make it possible to record and then analyze the deflection of the head, back, or arm from the position indicated as initial (Aaras and Stranden, 1988). The movement is analyzed in two planes, frontal and sagittal.

OWAS (Ovako Working Posture Analysis System)

OWAS (Karhu et al., 1986) can be used to make a direct and objective evaluation of load caused by the work posture. With the help of OWAS, a quantitative analysis of standard work postures can be performed, taking into consideration the external forces by classifying the position of the back, arms, and legs, and the external load. Combinations of positions of individual body parts (back, arms, legs), taking into consideration the external load, are grouped into four categories. This method can be used in examinations conducted directly at the workstation. By conducting an observation with a device for coding the work posture or with a video camera, it is possible to evaluate the time of holding and the frequency of changing a given work posture and to estimate the muscular load related to it.

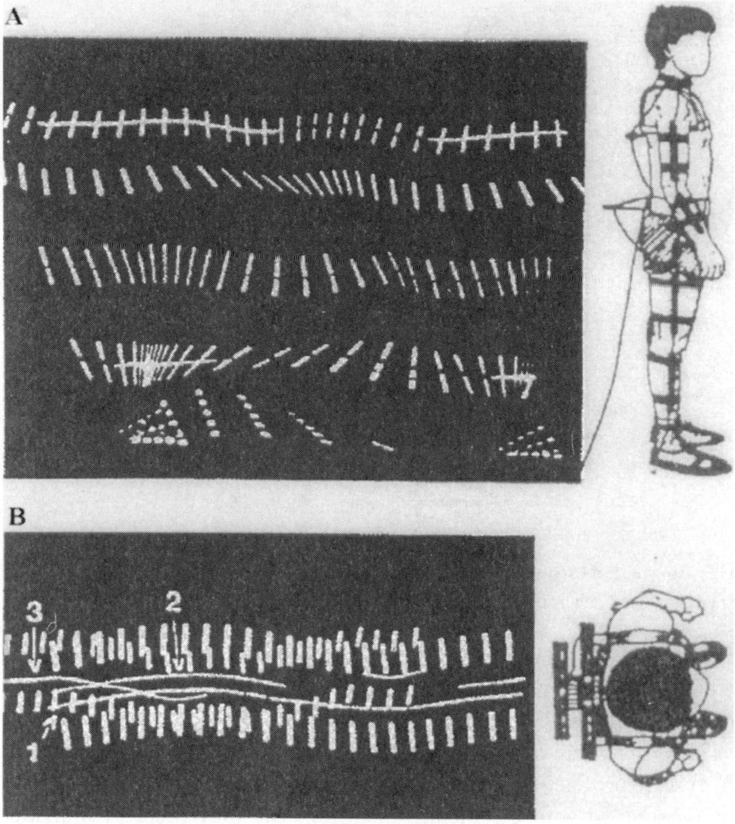

FIGURE 31.16 Position of individual body parts when moving, seen from the side (A) and from above (B) recorded with a light-emitted diodes and video cameras. (From Normand M. C., Richards C. L., Filion M., Dumas F., Tardif D.: A simplified method for tridimensional analysis of gait movements, in *Biomechanics IX-B,* Winter A. et al. (eds.), Illinois, 1983, 255-259.)

Maximum Time of Holding a Work Posture

For a full analysis of the working conditions, after defining and evaluating the work posture, holding time should be estimated for each of those postures. Maximum Holding Time (MHT) has been defined for several chosen positions (Karhu et al., 1986). MHT is the time a specific posture can be held (nonstop or with intervals) without the risk of fatigue or musculoskeletal system problems.

When there is a great variability of positions, MHT can be much higher than for a position held continuously. By comparing the holding time of a given position with MHT, it is possible to evaluate working conditions. Green, yellow, and red zones in a three-zone rating system of evaluation were marked in the following way to indicate safe holding time for each position:

Green zone — holding time at 0 to 30% MHT
Yellow zone — holding time at 30 to 50% MHT
Red zone — holding time at over 50% MHT

UniTOR

UniTOR is a computer-aided method of recording and analyzing the work process. This is a computer program made for recording, statistically processing data, and documenting the work process in accordance with the TOR (Time — Object — Recording) method (Gedliczka et al., 1993). It is especially used in recording the process of current work.

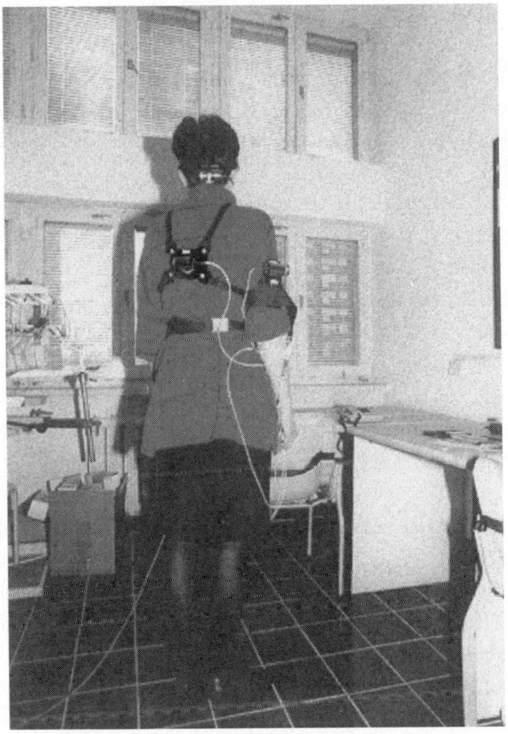

FIGURE 31.17 Pendulum potentiometer sensors placed on the head, back, and arm.

References

Aaras A., Stranden E.: Measurement of postural angels during work. *Ergonomics*, 31, 5, 1988, 935-944.

Åstrand P-O., Rodahl K.: *Textbook of Work Physiology*, McGraw-Hill, N.Y. 1986.

Bigland-Ritchie B., Jones D., Woods J.: Excitation frequency and muscle fatigue, electrical responses during human voluntary and stimulated contractions, *Experimental Neurology*, 64, 1979, 414-427.

Bugajska J., Roman D., Koradecka D.: Analysis of fatique during repetitive manual work (hand-grip), in *The Ergonomics of Manual Work*, Marras W., Karwowski W., Smith J. L., Pacholski L., (eds.) Taylor & Francis, London, 1993.

Bugajska J., Szmauz-Dybko M., Sawicka A., Tokarski T.: Development of methodologies for evaluation of protective clothing for compliance with occupational physiology and hygiene. *International Journal of Occupational Safety and Ergonomics*, submitted for publication.

Buśko K., Musiał W., Wychowański M.: *Instructions to Exercises in Biomechanics*, Academy of Physical Education edition, Warszawa 1991.

Christensen E. H.: Physiological evaluation of work in the Nykroppa iron works, in *Ergonomics Society Symposium on Fatique*. Floyd W.F. (ed.) Lewis, London 1953.

Fidelus K., Urbanik Cz.: The influence of various types of muscle effort on the effect of strength and speed training, *Biol. Sport*, 1984, 1, 186-198.

Fidelus K.: *Biomechanic Outline of Physical Exercises*, part 1, Academy of Physical Education edition, Warszawa 1989.

Fugl-Meyer A. R., Gustafsson L., Burstedt Y.: Isokinetic and static plantar flexion characteristics, *Eur. J. Appl. Physiol.*, 1980, 44, 221-234.

Gedliczka A., Goralczyk A., Otręba R., Wolska A.: New method of timekeeping, the power of time-object-recording, in *The Ergonomics of Manual Work*, Marras W. et al. (eds.), Taylor & Francis, London 1993.

Greenleaf J. E.: Hyperthermia and exercise. *Internat. Rev. Physiol.: Environmental Physiology III T.*, 20, Robertson D. (ed.) University Park Press, Baltimore, 1979.

Grucza R. et al.: Dynamics of sweating in men and women during passive heating, *Europ. J. Appl. Physiol.*, 1987, 51, 309-314.

Grucza R.: *Thermoregulation Model in Case of Endogenic and Exogenetic Thermal Load of the Organism*, CMDiK PAN Warszawa, 1979, (in Polish).

Hagg G.: Electromyographic fatigue analysis based on the Number of Zero Crossings, *European Journal of Physiology*, 1981, 391, 78-80.

Hagg G., Suurkula J.: Zero crossing rate of electromyograms during occupational work and endurance tests as predictors for work related myalgia in the shoulder/neck region, *European Journal of Applied Physiology*, 62, 1991, 436-444.

Ikeda M., Sato K., Oshima M., Physiological workload of train drivers on a suburban commuter railway, *RTTI Raport*, 1988, Mae., 2, 3, 13-16.

Ishibashi Y., On the degree of fatique of workers in a sewing factory for three different ages, *Jap. J. of Science of Clothing*, 1982, Oct., 26, 1, 19-26.

Iwasaki T., Kurimoto S., Noro K., The change in colour critical flicker fusion (CFF) values and accommodation times during experimental repetitive tasks with CRT display screens, *Ergonomics*, 1989, 32 (3), 293-305.

Jansen A. A., de Gier J. J., Slanger J. L., Alcohol effects on signal detection performance, *Neuropsychobiology*, 1985, 14/2/, 83-87.

Jansen A. A., de Gier J. J., Slanger J. L., Diazepam-induced changes in signal detection performance: a comparison with the effects on the critical flicker-fusion frequency and the digit symbol substitution test, *Neuropsychobiology*, 1986, 16/4/, 193-197.

Jensen R. H., Smith G. L., Johnston R. C.: A technique for obtaining measurements of force generated by the hip muscle, *Arch. Phys. Med. Rehabil.*, 1971, 52, 201-215.

Karhu U., Kansi P., Kuorinka I.: Correcting working postures in industry: a practical method for analysis, *Applied Ergonomics*, 1986, 8, 199-201.

Komi P V.: Neuromuscular performance factors influencing force and speed production, *Scand. J. Sports Sci.*, 1979, 12, 417-466.

Komi P.: Electromyographic, mechanical and metabolic changes during static dynamic fatigue, in Knuttgen G. H., Vogel J. A. and Poortmans J., *Biochemistry of Exercise*, Vol. 13, pp. 197-215. International Series on Sport Science. Human Kinetics Publishers Inc., Champaign, IL, 1983.

Konarska M. et al. Application of pulmonary ventilation measurements to assess energy expenditure during manual and massive muscular work, *Proceedings of the 12th Triennial Congress of the International Ergonomics Association*, Vol. 3, Human Factors Association of Canada, 1994.

Kowalk D. J., Besser M. P., Vaughan Ch. L., Bowsher K. F.: Abduction — adduction moments of the knee during stair ascent and descent, *Abstracts of International Society of Biomechanics XIVth Congress*, 1993, Vol. 1, pp. 716-717.

Kozłowski St., Nazar K., *Introduction to Clinical Physiology*, PZWL 1995 (in Polish).

Kurkus-Rozowska B. et al. Changes in blood pressure observed as a combined effect of heat and noise during physical load, *Archives of Complex Environmental Studies*, Vol. 8, No. 3-4, pp 11-16, 1996.

Lindstrom L., Magnusson R., Petersen J.: Muscular fatigue and action potential conduction velocity changes studied with frequency analysis of EMG signals, *Electromyography*, 1970, 4, 341-356.

Luczak A., Sobolewski A., 1995, The results of long-term observations of critical flicker frequency threshold (CFF), *Ergonomia*, 18, 2, 179-187.

Macellari V., Rossi M., Bugarini M.: Human motion monitoring using the CoSTEL system with reflective markers, in *Biomechanics IX-B*, Winter A. et al. (eds.), Illinois 1983, 260-264.

Marek T., Pieczonka- Błaszczyk W., Method of critical flicker frequency measurement and its accuracy, *Zeszyty Naukowe U. J.*, 1979, 29.

Marszałek A., Sołtyński K., Sawicka A.: Physiological evaluation of cold protective clothing, *Int. J. Occup. Safety & Ergonomics*, 1995, 1, 3, 235.

Matsumoto K., Sasagawa N., Kawamori M., Studies of fatigue of hospital nurses due to shiftwork, *Jap. J. of Industrial Health*, 1978, 20, 81-93.

Misawa T., Shigeta S., An experimental study of work load on VDT performance, part 1, Effects of polarity of screen and colour of display, *Jap. J. of Industrial Health,* 1986, Nov., 28, 6, 420-427.

Misawa T., Shigeta S., An experimental study of work load on VDT performance, part 2, Effects of difference in input devices, *Jap. J. of Industrial Health,* Nov., 1986, 28, 6, 462-469.

Ogiński A., Koźlakowska-Swigoń L., Pokorski J., Iskra-Golec, I., CFF as the strain indicator of control desks working in continuous motion, *Przegląd Lekarski,* 1981, 38, 9, 695-700.

Nadel E. R.: *Problems with Temperature Regulation During Exercise.* Academic Press, London 1977.

Normand M. C., Richards C. L., Filion M., Dumas F., Tardif D.: A simplified method for tridimensional analysis of gait movements, in *Biomechanics IX-B,* Winter A. et al. (eds.), Illinois, 1983, 255-259.

Nygaard E.: Woman and exercise–with special reference to muscle morphology and metabolism, in *Biochemistry of Exercise IV B,* Poortmans J. R., Niset G. (eds.) Univ. Park Press, Baltimore 1981.

Pagani M., Mazzuero G., Ferrari A., Liberati D., et al.: Sympathovagal Interaction During Mental Stress, *Circulation,* 1991, 83, (suppl. II).

Papillo J. F., Shapiro D.: The cardiovascular system, in *Principles of Psychophysiology-Physical, Social and Inferential Elements,* Cacioppo J. T., Tassinary L. G., (eds.), CUP, Cambridge, 1990.

Passmore R., Durnin J. U. G. A., *Energy, Work and Leisure,* Heinemann Educational Books, London, 1967.

Pieter W., Hijmans J., Taffe D.: Isokinetic leg strength of taekwondo practitioners, *Asian J. Phys. Education,* 1989, 12, 3, 55-64.

Rewerski W., Kozłowski St., Wróblewski T., Korolkiewicz K.: *Thermoregulation,* PZWL, Warszawa 1973.

Roman-Liu, D., Wittek, A., Kędzior, K., Musculoskeletal Load assessment of the upper limb positions subjectively chosen as the most convenient, *International Journal of Occupational Safety and Ergonomics,* 1996, 2, 4.

Seki K., Hugon M., Fatigue subjective et degradations de performance en environnement hyperbarea saturation, *Ergonomics,* 1977, 20, 2, 103-119.

Shephard R. J.: *Human Physiological Work Capacity.* CUP, Cambridge 1978.

Smolander J., Kolari P., Korhonen O., Ilmarinen R.: Skin blood flow during incremental exercise in a thermoneutral and hot dry environmental, *Europ. J. Appl. Physiol.,* 1987, 56, 273-280.

Sołtyński K. et al.: A method for determining thermal load in a high intensity radiation field of short duration, *Proceedings of the 12 Triennial Congress of the International Ergonomics Association,* Toronto, 1994.

Sołtyński K., Konarska M.: Body heat balance a man with deficient sweat rate subjected to physical work in hot environment, *International Journal of Occupational Safety & Ergonomics,* 1996, submitted for publication.

Sundelin G., Hagberg M.: Electromyographic signs of shoulder muscle fatigue in repetitive arm work paced by the methods-time measurement system. *Scand. J. Work, Environ. Health,* 1992, 18, 262-268.

Ważny Z.: *Training of Muscle Strength,* Sport i Turystyka, Warszawa 1977.

Ważny Z.: *Glossary of Athletic Training,* Academy of Physical Education edition, Warszawa 1994.

Zorski H. red.: *Technical Mechanics,* Ed. PWN, Warszawa 1985.

prEN 1005-4, Safety of Machinery — Human physical performance — Part 4: Working postures during machinery operation.

32

Video-Based Measurements of Human Movement

Peter M. Quesada
University of Louisville

32.1 Introduction

Quantitative analysis of human movement, via video motion measurement, can be a powerful means for addressing important ergonomics issues (e.g., potential modifications of occupational tasks, directed at preventing acute and/or cumulative trauma injury) (Berguer, 1997; Boston, 1995; Garg, 1991; Gracovetsky, 1990; Kumar, 1990, 1994; Lee, 1994; Pascarelli, 1993; Rudy, 1995). Statistics and correlations pertaining to injury incidence can suggest potential hazards of certain occupational tasks; however, statistical inferences cannot explain mechanisms by which injuries occur due to task performance. Identifying such mechanisms generally requires quantitative understanding of the kinematics, and often the kinetics, of a task. Actual knowledge of injury mechanisms is typically vital for effecting preventative modifications.

Video-based motion measurement is an effective and advantageous method of obtaining quantitative human movement data for a variety of ergonomics and biomechanics applications (Asato, 1993; Boninger, 1997; Cooper, 1996; Gracovetsky, 1989, 1995; Peterson, 1996; Robinson, 1993; Roosmon, 1993). Video-based motion measurements, which can provide absolute positional data for individual body segments, are commonly preferable to electrogoniometric measurements, which quantify only relative orientations of adjacent segments. Video motion measurement can also be less cumbersome for human subjects, and can allow individuals to perform tasks more naturally.

0-8493-2641-9/99/$0.00+$.50

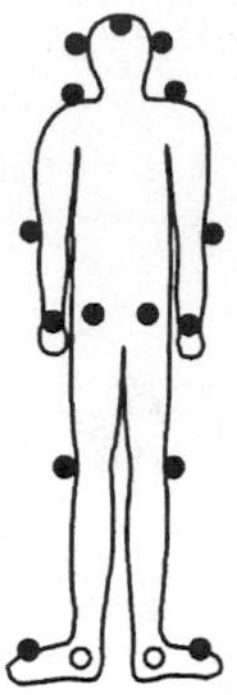

FIGURE 32.1 Depiction of subject with markers placed at anatomic locations. Solid circles represent markers that are visible in the view shown, while empty circles represent markers that are hidden in the view shown.

32.2 Overview of Video Motion Measurement Equipment

Cameras

Video cameras are the most obvious, if not most essential, items required for video motion measurement. A wide range of cameras are, or can be made, suitable for collecting motion data. Suitable cameras include both analog and digital types, as well as cameras with short-range, medium-range, and long-range lenses. The types of cameras available can influence the range of movement tasks that can be measured; or, alternatively, the need to measure a specific movement task can influence the selection of cameras to be used.

Image Capture Devices

For each frame of video data, two-dimensional camera coordinates for markers, placed on the body surface are obtained from each camera used. Two-dimensional data from a single camera may be sufficient to draw meaningful conclusions for a limited number of applications. More commonly, however, two-dimensional data from multiple cameras are integrated to determine three-dimensional positional data, for each marker, with respect to a global coordinate system (GCS), fixed to a given point in the testing object space. As with the types of cameras, the number of cameras available also affects the range of movement tasks that can be measured.

Body Surface Markers

Preparation of subjects involves placement of markers on a subject's body surface (Figure 32.1). Such markers generally do not substantially restrict subject movement, thereby permitting measurement of a wide range of movements. Markers are either positioned on palpable bony landmarks (e.g., lateral maleoli, anterior superior iliac spines, acromion processes, etc.) or attached to rigid fixtures that are placed on body segments (often in the vicinity of segment mass centers). Specific sites of marker or fixture placement are generally related to the biomechanical model that will be used to compute kinematic variables. Conversely, biomechanical model development can be affected by ease or difficulty of identifying specific marker placement locations.

Markers can be characterized as being either passive or active. Passive markers reflect light from sources attached, or in close proximity, to each camera. Both infrared and incandescent light are commonly used to illuminate passive markers. Lights can be positioned slightly above and behind a camera, or around the periphery of a camera lens (Figure 32.2). In these or other light/camera arrangements, the light source

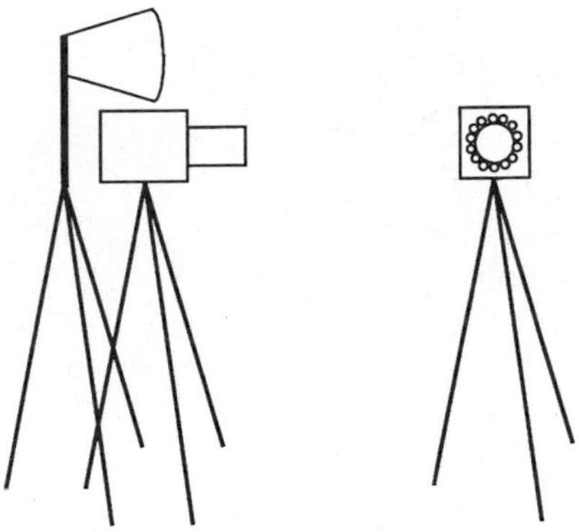

FIGURE 32.2 Depiction of light/camera arrangements for illuminating passive markers. On the left is a single light, positioned just above and behind a camera. The right side depicts an array of lights around the periphery of a camera lens.

must be out of the view of the camera in whose vicinity the source is positioned. Suitable arrangements do not necessarily require, however, that each light source be out of view for all cameras. Incandescent lighting is generally held constant, while infrared lighting is often strobed. Active markers generate their own light to be viewed by cameras. Active markers can be more cumbersome than passive markers if lead wires are needed to deliver power. With some video motion systems, active markers are strobed as a means of providing automated marker identification. Colored lights have also been used to assist or automate marker identification. In addition to colored lights, light emitting diodes and infrared lights have been used as active markers.

Load Measurement Devices

Computation of kinetic variables associated with human movement commonly involves use of various load measuring devices and application of inverse dynamic formulations. Force platforms are among the most common devices used for collecting raw kinetic data. These devices measure the forces imposed on them, generally through the feet, in three component directions, as well as the moments of force about three axes oriented along the measured force directions. Platform force and moment measurements can be used to determine the location of the resultant force's point of application (Figure 32.3). The location of this point is essential if resultant force data is to be integrated with video-generated kinematic data to estimate internal loading variables such as joint moments and contact forces. A variety of devices are available for collecting kinetic data associated with body segments other than the feet. Uniaxial load cells can measure simple tensile and compressive loads, such as the net load on the hands due to pulling on a rope. Pressure sensitive mats can be used to record the load distribution at a seating interface. Additionally, and somewhat more indirectly, accelerometers placed on an object of known mass can record object acceleration data that are suitable for determining loads placed on an object by an individual.

32.3 Overview of Video Motion Measurement Tasks

Collection of three-dimensional video motion data begins with a camera calibration procedure. The purpose of this process is to determine the position and orientation, with respect to the GCS (Figure 32.4), of each camera to be used. Placement of body surface markers generally follows camera calibration or proceeds simultaneously, if sufficient personnel are available. With cameras calibrated and markers placed, video data collection proceeds as the subject performs the task(s) to be studied. Initial processing then reduces raw video data (i.e., two-dimensional coordinates for each marker, in each camera's image plane)

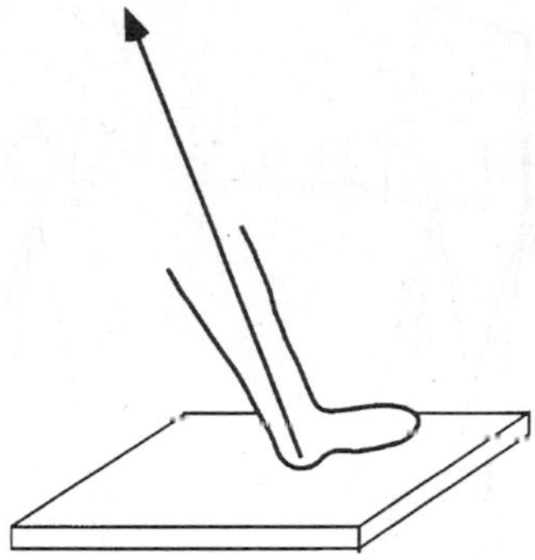

FIGURE 32.3 Depiction of a resultant force measurement with a force platform. The length of the vector indicates the magnitude of the force, the orientation indicates the force's direction, and the origin of the vector indicates the point of load application.

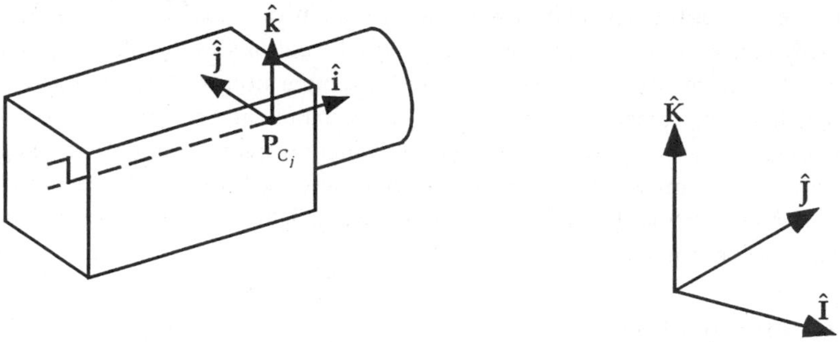

FIGURE 32.4 Depiction of a camera with its position, P_{C_j}, and LCS axes directions indicated with respect to a GCS.

to generate three-dimensional coordinates, with respect to a GCS, for each marker at every sampling interval. Based on these three-dimensional marker coordinates, as well as estimated coordinates of body locations without markers, biomechanical variables are then computed. Computation of motion variables is generally the final step in the video measurement process. Variables can be formulated to quantify both temporal and spatial characteristics of a movement task.

32.4 Calibration of Object Space

Successful collection and reduction of video motion data requires a quantitative knowledge of the location and orientation of each camera to be used. Consequently, execution of a calibration procedure invariably accompanies sessions of video motion data collection, to determine camera location and orientation information (Figure 32.4). Camera location can be described as the position, in the GCS, of a characteristic point (e.g., the focal point). Similarly, camera orientation can be quantified by the orientation/direction of a local coordinate system (LCS), which typically has one axis perpendicular to the image plane. Traditionally, camera calibration has involved imaging of several markers, whose three-dimensional coordinates are known with respect to the GCS, and that are arranged to fill the object volume (Figure 32.5). Early variations of this technique require that each marker be identified in each camera view, while more current applications have markers identified automatically. In either case, camera locations and orientations are determined once marker identification has been completed.

The mechanical process of calibrating video motion systems has changed little (until very recently) as technology for video motion measurement has progressed from using cine film and other similar photographic means, to using video cameras for acquiring raw movement data. Advancements in computational algorithms, however, have provided substantial improvements in both accuracy and data processing

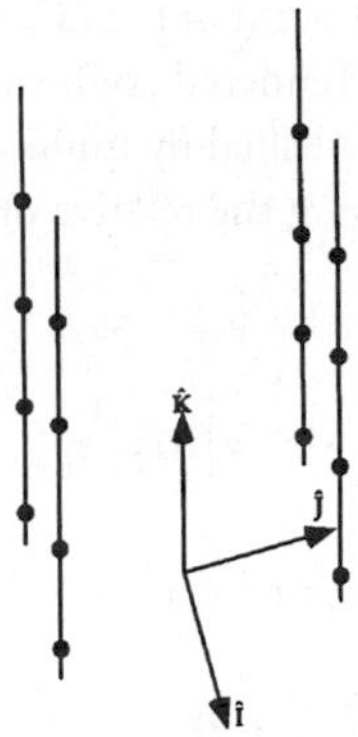

FIGURE 32.5 Depiction of a calibration, reference marker arrangement.

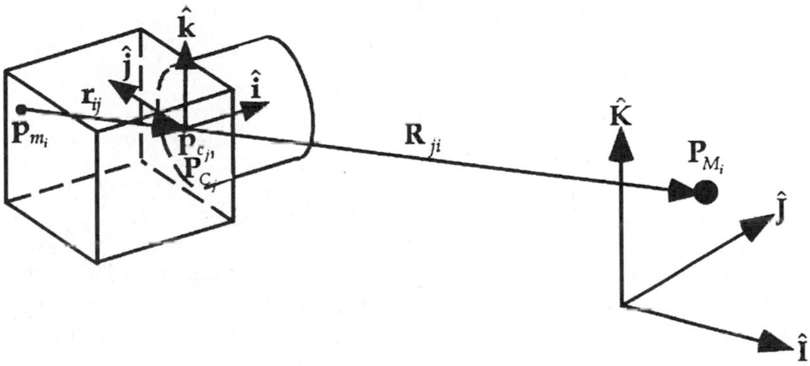

FIGURE 32.6 Depiction of LCS-based vector, $\mathbf{r}_{ij}$, from marker image to camera focal point, and GCS-based vector, $\mathbf{R}_{ji}$, from focal point to marker location.

time associated with camera calibration (Abdel-Aziz, 1971; Marzan, 1975; Woltring, 1976; Whittle, 1982). A substantial volume of software, that incorporates the more basic algorithms, is available in the public domain. Many of the more sophisticated algorithms, however, have been commercialized as the demand for software products to perform camera calibration as well as marker coordinate generation has grown. Consequently, details of current calibration and coordinate generation algorithms will not be discussed here.

Complete mastery of camera calibration details is unnecessary for most exercises in video motion measurement. An overview of the principles involved with the process, however, should assist in arranging appropriate camera configurations and calibration object volumes. To understand, fundamentally, the camera calibration process, one can visualize a pair of vectors, associated with marker, *i*, that is of interest (Figure 32.6). The first vector, $\mathbf{r}_{ij}$, originates from the projected image of the marker, $\mathbf{p}_{m_i}$, on the backplane of camera, *j* (i.e., image plane, *j*), and terminates at the camera focal point. The second vector, $\mathbf{R}_{ji}$, originates from the focal point and terminates at the marker, $\mathbf{P}_{M_i}$. Vector $\mathbf{r}_{ij}$ and point $\mathbf{p}_{m_i}$ are described with respect to an LCS, which is attached to the camera and defined by three mutually perpendicular unit vectors, two unit vectors lying in the image plane and the third being perpendicular to the image plane. Vector $\mathbf{R}_{ji}$ and point $\mathbf{P}_{M_i}$ are described for a GCS, fixed to the laboratory space. These vectors are thus expressed as:

$$\mathbf{r}_{ij} = \mathbf{p}_{c_j} - \mathbf{p}_{m_i} \quad \text{(in LCS)} \tag{1.a}$$

$$\mathbf{R}_{ji} = \mathbf{P}_{M_i} - \mathbf{P}_{C_j} \quad \text{(in GCS)} \tag{1.b}$$

where $\mathbf{p}_{c_j}$ and $\mathbf{P}_{C_j}$ designate the camera focal point in the LCS and GCS, respectively. For ideal flat camera lenses and backplanes, these vectors can be rendered colinear by rotating the LCS to be parallel with the GCS. This process is mathematically accomplished by multiplying $\mathbf{r}_{ij}$ by an appropriate rotation matrix, $[\mathbf{M}_{R_j}]$(that is a function of variables describing the relative orientation of the of the LCS with respect to the GCS):

$$\mathbf{r}'_{ij} = \left[\mathbf{M}_{R_j}\right]\mathbf{r}_{ij} \tag{2}$$

Subsequently, the GCS vector is directly proportional to the rotated LCS vector:

$$\mathbf{r}'_{ij} = k_{ij}\mathbf{R}_{ji} \tag{3.a}$$

$$\left[\mathbf{M}_{R_j}\right]\mathbf{r}_{ij} = k_{ij}\mathbf{R}_{ji} \tag{3.b}$$

$$\mathbf{r}_{ij} = k_{ij}\left[\mathbf{M}_{R_j}\right]^{-1}\mathbf{R}_{ji} \tag{3.c}$$

$$\left[\mathbf{p}_{c_j} - \mathbf{p}_{m_i}\right] = k_{ij}\left[\mathbf{M}_{R_j}\right]^{-1}\left[\mathbf{P}_{M_i} - \mathbf{P}_{C_j}\right] \tag{3.d}$$

where k_{ij} is a constant pertaining to the combination of marker, *i*, with camera, *j*. This vector relationship yields three component equations for a given marker, *i*, and camera, *j*. Marker position, $\mathbf{P}_{M_i}$ (in the GCS) is known, while location of the projected marker image, $\mathbf{p}_{m_i}$ (in the LCS), is measured for most traditional camera calibration protocols. Unknown variables include: global camera position components (constituting $\mathbf{P}_{C_j}$), the camera orientation parameters (contained within $[\mathbf{M}_{R_j}]$) and the proportionality constant, k_{ij}. One or more of the LCS components of $\mathbf{p}_{c_j}$ may also be unknown in some calibration algorithms. Consequently, the equations generated by a single marker are insufficient to determine the unknown variables associated with a camera's location and orientation. Fortunately, each additional reference marker will generate three similar vector component equations, while introducing an additional unknown, k_{ij}. Continued addition of reference markers will, ultimately, yield a sufficient number of equations to estimate the location and orientation of camera, *j*. In most applications the number of reference markers exceeds the minimum required, and, thus provides some redundancy to the process. The large number of equations, typically involved, generally dictate the use of numerical algorithms to effect a solution. The process is repeatable for each camera to be used in the desired video motion measurement task.

Estimations of GCS coordinates for unknown marker positions tend to be better when markers are near the positions of reference markers used during calibration (Figure 32.7). Consequently, reference markers are typically arranged to fill, as much as possible, the space to be occupied by markers during data collection.

The process of determining unknown GCS coordinates of markers, imaged during video motion measurement, utilizes the same vector relationship (Equation 3) described for camera calibration. Application of this relationship at each desired sampling interval will yield, as unknown variables, the marker GCS coordinates as well as a value of k_{ij} for a given camera, *j*, with a view of marker, *i*. The number of unknowns generated, thus exceeds the number of equations available for a single application of this relationship. Generation of sufficient equations to solve for all unknowns associated with a marker at a specific sampling interval, requires at least two applications of the vector relationship, with one occurring for each camera with a view of the marker. Consequently, estimation of a marker's GCS coordinates requires that the marker be visible to a minimum of two cameras.

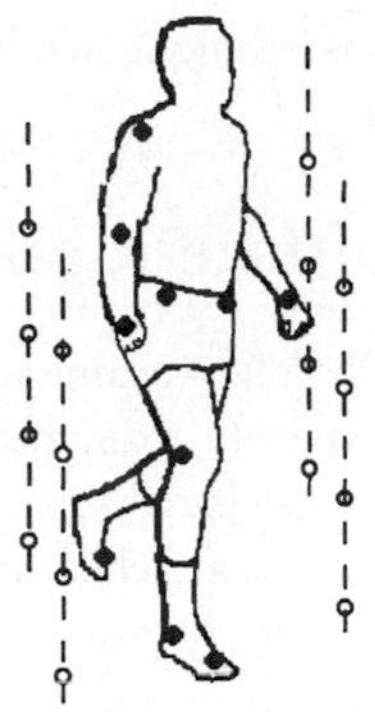

FIGURE 32.7 Depiction of subject with markers moving through a calibrated object space. Empty circles represent locations of reference markers that were present during calibration. Some body markers (e.g., left wrist) are close to reference marker locations, while others (e.g., right shoulder) are rather far from reference locations.

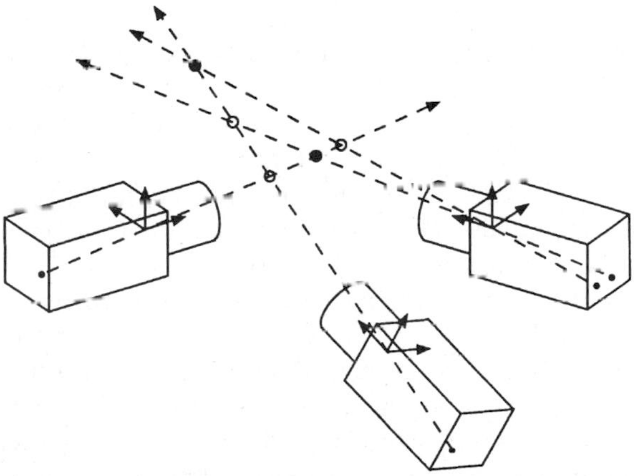

FIGURE 32.8 Depiction of markers (larger solid circles) imaged by multiple cameras. Vectors originating from images of markers (smaller solid circles), on camera image planes, cross at marker locations; however they can also cross at other "ghost" marker locations, indicated by empty circles.

The process of determining the GCS coordinates, for a given marker at a specified sampling interval, can be described graphically as identifying the intersection of two or more vectors in three-dimensional space (Figure 32.8). Each vector originates from a camera's backplane image of the marker, passes through the camera focal point, and continues indefinitely through the marker and into three-dimensional space.

This graphical representation visually demonstrates the process of determining GCS marker coordinates; however, it can also suggest the potential for identifying false, or "ghost" markers when vectors passing through different markers unintentionally intersect with each other (Figure 32.8). Generation of ghost markers tends to complicate further processing of marker coordinate data to compute kinematic variables associated with the movement tasks being measured. Ghost markers can occur when multiple cameras are positioned and directed in a common plane that multiple markers tend to occupy during measured movements. This undesirable effect can be largely avoided by arranging cameras to be positioned and directed in different planes.

Recently, a different calibration technique has been developed in which a rigid rod of known length and with two markers attached at the ends is moved throughout the desired object volume while camera data are collected. Considering some of the basic concepts developed previously, one can perhaps appreciate the potential strengths of this technique, such as the potential to generate a large number of calibration points and the capacity for those points to fill a large portion of the object volume. The computational algorithms for this calibration technique involve somewhat more sophisticated programming logic and will not be discussed further.

32.5 Placement of Body Surface Marker

Placement of body surface markers generally follows camera calibration or proceeds simultaneously if sufficient personnel are available. Two approaches to marker placement are commonly used. In one approach, markers are placed individually at specific anatomical locations (Kadaba, 1990). In the other approach, a cluster of markers (three or more) is attached to a rigid or semirigid fixture that is placed on a body segment (Antonsson, 1989). Selection of locations for marker placement is dictated by the ease with which such locations can be reproduced and/or the biomechanical modeling for which marker coordinate data will serve as input. When applying individual markers, reproducibility of marker placements is particularly essential if comparisons are to be made between data from different subjects or subject groups, or between data from a single subject acquired under different conditions or on separate occasions. Typically, individual marker placements are most repeatable when they are located directly over bony prominences, such as the lateral maleolus.

32.6 Collection and Initial Processing of Video Motion Data

With cameras calibrated and markers placed, video data collection proceeds as the subject performs the task(s) to be studied. Once motion data are collected, the process for reducing raw video data to generate marker coordinate information will depend on the type of system used to acquire the data. The data reduction process will typically include marker identification (either manual or automated identification, or a combination of the two) and reconstruction of image plane marker coordinates (performed via software). The order in which these two elements are performed is often the factor that determines the amount of time needed to complete initial data reduction. Earlier motion systems required that markers be identified from the perspective of each camera used, with three dimensional reconstruction following marker identification. For many of the more recent systems, however, three dimensional reconstruction is performed first, followed by marker identification from any arbitrary perspective. The computational basis for determining three-dimensional marker coordinates from two-dimensional image plane data is essentially the same whether marker identification precedes or follows reconstruction.

32.7 Computation of Biomechanical Variables

With three-dimensional marker coordinates determined for each frame of video data, biomechanical variables can be computed. For some applications computations are based upon surface movement (i.e., coordinate trajectories of the markers placed on the body surface), while for other purposes, calculations are based upon skeletal movement. This latter approach requires that three-dimensional coordinates be determined for skeletal locations (typically joint centers) at which markers cannot be placed. Surface based calculations may be sufficient when motion is desired in just a single place (e.g., sagittal plane). Quantification of movement in multiple planes, however, generally requires application of skeletal-based computational techniques (Kadaba, 1990; Vaughan, 1992).

Estimation of coordinate data for skeletal locations, often referred to as "virtual markers," is commonly based on mathematical models involving surface marker coordinates, at times in conjunction with previously estimated coordinates for other virtual markers. Such models may take the form of statistical regression equations that can be evaluated directly, or multiple constraint equations requiring sequential or simultaneous solution (Kadaba, 1990; Vaughan, 1992). Some of these approaches permit estimation of virtual marker coordinates for a given frame using other marker data for the same frame only. Other techniques exist, however, for which virtual marker coordinates are determined for a given frame using marker data from both preceding and succeeding frames.

A commonly employed technique for computing hip joint center coordinates is a typical example of using regression-based relationships to approximate virtual marker coordinates. This technique involves placement of a marker at each of the anterior superior iliac spines (ASISs) and placement of a third marker at the sacrum or at the end of a wand, such that the marker lies in the plane formed by sacrum and the ASIS markers. Three unit vectors are then determined as follows:

$$\hat{\mathbf{j}}_{pelvis} = \frac{\mathbf{P}_{M_{left\ ASIS}} - \mathbf{P}_{M_{right\ ASIS}}}{\left|\mathbf{P}_{M_{left\ ASIS}} - \mathbf{P}_{M_{right\ ASIS}}\right|} \tag{4.a}$$

$$\hat{\mathbf{k}}_{pelvis} = \frac{\left(\mathbf{P}_{M_{left\ ASIS}} - \mathbf{P}_{M_{right\ ASIS}}\right) \times \left(\mathbf{P}_{M_{sacral}} - \mathbf{P}_{M_{right\ ASIS}}\right)}{\left|\left(\mathbf{P}_{M_{left\ ASIS}} - \mathbf{P}_{M_{right\ ASIS}}\right) \times \left(\mathbf{P}_{M_{right\ ASIS}}\right) \times \left(\mathbf{P}_{M_{sacral}} - \mathbf{P}_{M_{right\ ASIS}}\right)\right|} \tag{4.b}$$

$$\hat{\mathbf{i}}_{pelvis} = \hat{\mathbf{j}}_{pelvis} \times \hat{\mathbf{k}}_{pelvis} \tag{4.c}$$

where $\mathbf{P}_{M_{right\ ASIS}}$, $\mathbf{P}_{M_{left\ ASIS}}$, and $\mathbf{P}_{M_{sacral}}$ are vectors containing three-dimensional coordinates in the GCS of the right ASIS, left ASIS, and sacral markers, respectively. The vectors, $\hat{\mathbf{i}}_{pelvis}$, $\hat{\mathbf{j}}_{pelvis}$, and $\hat{\mathbf{k}}_{pelvis}$, are unit vectors (i.e., magnitude of each equals unity) that form a cartesian coordinate system describing the orientation of the pelvis with respect to the GCS. The positions of the hip joint centers, $\mathbf{P}_{M_{right\ hip\ center}}$ and $\mathbf{P}_{M_{left\ hip\ center}}$, can be described as

$$\mathbf{P}_{M_{right\ hip\ center}} = \mathbf{P}_{M_{right\ ASIS}} + \left|\mathbf{P}_{M_{right\ ASIS}} - \mathbf{P}_{M_{left\ ASIS}}\right|\left(-a_1\hat{\mathbf{i}}_{pelvis} + a_2\hat{\mathbf{j}}_{pelvis} - a_3\hat{\mathbf{k}}_{pelvis}\right) \tag{5.a}$$

$$\mathbf{P}_{M_{left\ hip\ center}} = \mathbf{P}_{M_{left\ ASIS}} + \left|\mathbf{P}_{M_{right\ ASIS}} - \mathbf{P}_{M_{left\ ASIS}}\right|\left(-a_1\hat{\mathbf{i}}_{pelvis} - a_2\hat{\mathbf{j}}_{pelvis} - a_3\hat{\mathbf{k}}_{pelvis}\right) \tag{5.b}$$

where a_1, a_2, and a_3 are constants equal to the fractions of the inter-ASIS distance that an ASIS marker position must be translated in the respective unit vector directions, in order to arrive at a hip joint center.

As indicated, determination of virtual marker locations via application of multiple constraints is also possible. Knee joint centers are often approximated in such a manner, following identification of hip joint centers as described previously. This approach typically involves coordinate data for two markers in addition to hip joint center coordinates. One marker (with coordinate vector, $\mathbf{p}_{M_{right/left\ knee\ marker}}$) is placed on the lateral knee joint such that it lies on the assumed joint axis extended. The other marker (with coordinate vector, $\mathbf{p}_{M_{right/left\ femoral\ marker}}$) is placed on the femur or at the end of a femoral wand such that it lies approximately in a plane defined by the assumed rotational axis and the hip joint center (recall that a vector and a point off the vector uniquely define a plane). Considering the right lower limb for this discussion, a unit vector, $\hat{\mathbf{u}}_I$, perpendicular to the plane is first computed as

$$\hat{\mathbf{u}}_I = \frac{\left(\mathbf{p}_{M_{right\ hip\ center}} - \mathbf{p}_{M_{right\ knee\ marker}}\right) \times \left(\mathbf{p}_{M_{right\ femoral\ marker}} - \mathbf{p}_{M_{right\ knee\ marker}}\right)}{\left|\left(\mathbf{p}_{M_{right\ hip\ center}} - \mathbf{p}_{M_{right\ knee\ marker}}\right) \times \left(\mathbf{p}_{M_{right\ femoral\ marker}} - \mathbf{p}_{M_{right\ knee\ marker}}\right)\right|} \tag{6}$$

Next, two intermediate unit vectors, $\mathbf{u}'_{II}$ and $\mathbf{u}'_{III}$ are determined as

$$\hat{\mathbf{u}}'_{II} = \frac{\left(\mathbf{p}_{M_{right\ hip\ center}} - \mathbf{p}_{M_{right\ knee\ marker}}\right)}{\left|\left(\mathbf{p}_{M_{right\ hip\ center}} - \mathbf{p}_{M_{right\ knee\ marker}}\right)\right|} \tag{7.a}$$

$$\hat{\mathbf{u}}'_{III} = \hat{\mathbf{u}}'_{II} \times \hat{\mathbf{u}}_{I} \tag{7.b}$$

Based upon an assumption that the knee joint axis is perpendicular to a vector from knee center to hip center, $\hat{\mathbf{u}}'_{III}$ can be rotated about $\hat{\mathbf{u}}_I$ by some angle, ϕ, to obtain vector, $\hat{\mathbf{u}}_{III}$, in the direction of the joint axis. The angle ϕ, vector $\hat{\mathbf{u}}_{III}$, and knee center location, $\mathbf{p}_{M_{right\ knee\ center}}$, are computed as

$$\phi = \sin^{-1}\left[\frac{\frac{1}{2}(knee\ width + marker\ diameter)}{\left|\mathbf{p}_{M_{right\ hip\ center}} - \mathbf{p}_{M_{right\ knee\ marker}}\right|}\right] \tag{8.a}$$

$$\hat{\mathbf{u}}_{III} = \hat{\mathbf{u}}'_{II}\sin(\phi) + \hat{\mathbf{u}}'_{III}\cos(\phi) \tag{8.b}$$

$$\mathbf{p}_{M_{right\ knee\ center}} = \mathbf{p}_{M_{right\ knee\ marker}} + \frac{1}{2}(knee\ width + marker\ diameter)\hat{\mathbf{u}}_{III} \tag{8.c}$$

where knee width is measured from lateral to medial femoral condyle.

Computation of motion variables is generally the final step in the video measurement process. Variables can be formulated to quantify both temporal and spatial characteristics of a movement task. Temporal variables describe times or durations associated with the task being measured (e.g., time to initiate movement following a stimulus) and are typically single values. Angular variables that quantify absolute or relative orientation of body segments (e.g., elbow flexion angle) represent typical spatial variables. Rates of changes of relative segment angles are potential variables that incorporate both temporal and spatial attributes.

By identifying the video frames associated with initiation and completion of a task, the task duration, t_{task}, is readily computed as

$$t_{task} = (frame_f - frame_i)\Delta t \tag{9}$$

where $frame_f$ and $frame_i$ are the frame numbers for the initiation and finish of the task, respectively, and Δt is the video sampling rate. Similar expressions can quantify the time required to perform a specific portion of a task. Alternatively, it is often useful to quantify the duration of a task component as a percentage of the overall task duration:

$$\%_{task\ component} = \frac{frame_b - frame_a}{frame_f - frame_i} \times 100\% \tag{10}$$

where $frame_a$ and $frame_b$ are the frame numbers for the start and end of the task component. Such normalized time values are often useful for comparing task component durations when overall task durations differ.

Kinematic variables computed from marker coordinate data can quantify both translational and rotational movement of one or more body segments with respect to either a GCS or one another. Marker coordinate data are often filtered prior to calculating kinematic variables. Absolute translation of a body segment (e.g., forearm) between two sampling intervals, i and f (e.g., the beginning and end of a gear shift task) is computed as

$$\Delta\mathbf{P}_A = \mathbf{P}_{A,f} - \mathbf{P}_{A,i} \tag{11}$$

where $\mathbf{P}_{A,i}$ and $\mathbf{P}_{A,f}$ are the position vectors at frames i and f, respectively, for location A (e.g., the mass center) on the body segment of interest. Translational velocity, $\mathbf{v}_{A,i}$, and acceleration, $\mathbf{a}_{A,i}$, of point A at frame, i, can be discretely approximated as

$$\mathbf{v}_{A,i} = \frac{1}{2(\Delta t)}\left(\mathbf{P}_{A,i+1} - \mathbf{P}_{A,i-1}\right) \tag{12.a}$$

$$\mathbf{a}_{A,i} = \frac{1}{(\Delta t)^2}\left(\mathbf{P}_{A,i+1} - 2\mathbf{P}_{A,i} + \mathbf{P}_{A,i-1}\right) \tag{12.b}$$

Estimations for velocity and acceleration can also be obtained as the first and second time derivatives, respectively, of time functions for, i.e., $\mathbf{P}_A$:

$$\mathbf{v}_A(t) = \frac{d}{dt}\mathbf{P}_A \tag{13.a}$$

$$\mathbf{a}_A(t) = \frac{d^2}{dt^2}\mathbf{P}_A(t) \tag{13.b}$$

Often the functions fitted to the experimental data are piecewise polynomials.

Kinematic variables that quantify the orientation or rotation of body segments are often of greater interest than translational variables. Marker coordinate data can be used to compute either two- or three-dimensional body segment orientations or rotations, with respect to a GCS or with respect to one another. Most of these computations inherently assume that body segments behave as rigid bodies (i.e., that the distance between any two points on a segment remains constant).

Measurements of two marker (or virtual marker) positions ($\mathbf{P}_A$ and $\mathbf{P}_B$) per body segment (a marker need not be unique to a particular body segment) permit orientations and rotations of each segment to be computed based upon a single unit vector per segment (i.e., two-dimensional computation). The absolute orientation of a body segment (e.g., upper arm) can be quantified by a unit vector, $\hat{\mathbf{u}}_s$, directed from one marker to the other:

$$\hat{\mathbf{u}}_s = \frac{\mathbf{P}_B - \mathbf{P}_A}{\left|\mathbf{P}_B - \mathbf{P}_A\right|} \tag{14}$$

The relative orientation of two segments, i and j (e.g., upper arm and lower arm), can be described by an angle, ϕ, between the projection of the unit vectors, $\hat{\mathbf{u}}_{s,i}$ and $\hat{\mathbf{u}}_{s,j}$:

$$\phi = \cos^{-1}\left(\hat{\mathbf{u}}_{s,i} \bullet \hat{\mathbf{u}}_{s,j}\right) \tag{15}$$

These expressions are generally intended to be evaluated with three-dimensional marker coordinate data; however, they can be applied to two-dimensional coordinates as well. Two-dimensional applications assume the axis of rotation to be perpendicular to the coordinate plane. This requirement, unfortunately, limits the utility of two-dimensional data to movements occurring in a single plane. Occupational movements adhering to this restriction are rather limited.

Three-dimensional computation of absolute and relative segment orientations and rotations involves associating a three-dimensional local coordinate system (LCS) with each body segment of interest. Each LCS is described by three mutually orthogonal unit vectors, $\hat{\mathbf{u}}_1$, $\hat{\mathbf{u}}_2$, and $\hat{\mathbf{u}}_3$ (Figure 32.9). Calculation of these unit vectors requires measurement of at least three marker (or virtual marker) positions ($\mathbf{P}_A$, $\mathbf{P}_B$,

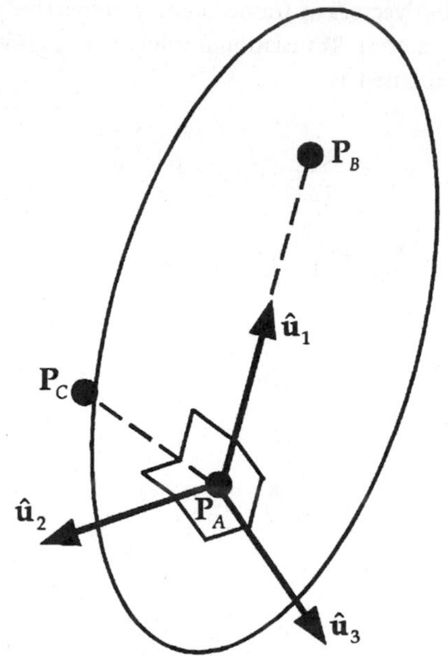

FIGURE 32.9 Depiction of segment LCS determined from segment marker positions.

and $\mathbf{P}_C$) per segment for each sampling interval. A given marker position can potentially be used in the computations of more than one segment LCS. For example, a virtual marker at the knee joint center could be used in determining LCSs for both thigh and lower leg segments. As with previously described two-dimensional computations, rigid body assumptions are also inherent to these three-dimensional kinematic calculations. The LCS unit vectors for a segment are computed as

$$\hat{\mathbf{u}}_1 = \frac{\mathbf{P}_B - \mathbf{P}_A}{\left|\mathbf{P}_B - \mathbf{P}_A\right|} \tag{16.a}$$

$$\hat{\mathbf{u}}_2 = \frac{(\mathbf{P}_B - \mathbf{P}_A)\times(\mathbf{P}_C - \mathbf{P}_A)}{\left|(\mathbf{P}_B - \mathbf{P}_A)\times(\mathbf{P}_C - \mathbf{P}_A)\right|} \tag{16.b}$$

$$\hat{\mathbf{u}}_3 = \hat{\mathbf{u}}_1 \times \hat{\mathbf{u}}_2 \tag{16.c}$$

The unit vectors, $\hat{\mathbf{u}}_1$, $\hat{\mathbf{u}}_2$, and $\hat{\mathbf{u}}_3$, and, can be designated, for a segment, as $\hat{\mathbf{i}}_s$, $\hat{\mathbf{j}}_s$, and $\hat{\mathbf{k}}_s$, though not necessarily in that order. The order of designation, while somewhat arbitrary, generally follows the convention of forming a right-hand coordinate system (i.e., $\hat{\mathbf{i}}_s \times \hat{\mathbf{j}}_s = \hat{\mathbf{k}}_s$; $\hat{\mathbf{j}}_s \times \hat{\mathbf{i}}_s = -\hat{\mathbf{k}}_s$).

With an LCS determined for each body segment of interest, orientation of any given segment can be described with respect to any other coordinate system. Consequently, one can quantify a segment's absolute orientation (i.e., with respect to a GCS) or relative orientation (i.e., with respect to another LCS). One technique, for quantifying orientation of an LCS, involves rotating a reference coordinate system, $\hat{\mathbf{i}}_r$, $\hat{\mathbf{j}}_r$, and $\hat{\mathbf{k}}_r$, (either the GCS or another LCS) through angles ϕ, θ, and ψ about its three coordinate system axes, in succession so as to render it parallel to the LCS of interest ($\hat{\mathbf{i}}_s$, $\hat{\mathbf{j}}_s$, $\hat{\mathbf{k}}_s$). Many rotation sequences are possible, however, as an example one can consider rotation through angle, ϕ, about $\hat{\mathbf{i}}_r$ (which transforms $\hat{\mathbf{j}}_r$ and $\hat{\mathbf{k}}_r$ to $\hat{\mathbf{j}}_r'$ and $\hat{\mathbf{k}}_r'$); then rotation through angle, θ, about (which transforms $\hat{\mathbf{i}}_r$ and $\hat{\mathbf{k}}_r'$ to $\hat{\mathbf{i}}_r'$ and $\hat{\mathbf{k}}_r''$); and last, rotation through angle, ψ, about $\hat{\mathbf{k}}_r''$ (which transforms $\hat{\mathbf{i}}_r'$ and $\hat{\mathbf{j}}_r'$ to $\hat{\mathbf{i}}_r''$ and $\hat{\mathbf{j}}_r''$). Parallelism between the LCS of interest and the rotated reference coordinate system dictates that

$$\begin{bmatrix} \hat{\mathbf{i}}''_r \\ \hat{\mathbf{j}}''_r \\ \hat{\mathbf{k}}''_r \end{bmatrix} = \begin{bmatrix} \hat{\mathbf{i}}_s \\ \hat{\mathbf{j}}_s \\ \hat{\mathbf{k}}_s \end{bmatrix} \tag{17}$$

Transformation of the reference coordinate system via three successive rotations can be computed by multiplying a rotation matrix, $[{}_r\mathbf{M}_s]$, with the original LCS unit vectors:

$$\left[{}_r\mathbf{M}_s\right] \begin{bmatrix} \hat{\mathbf{i}}_r \\ \hat{\mathbf{j}}_r \\ \hat{\mathbf{k}}_r \end{bmatrix} = \begin{bmatrix} \hat{\mathbf{i}}_s \\ \hat{\mathbf{j}}_s \\ \hat{\mathbf{k}}_s \end{bmatrix} \tag{18}$$

For the rotation sequence described above, the rotation matrix takes the form

$$\left[{}_r\mathbf{M}_s\right] = \begin{bmatrix} \cos\psi\cos\theta & \sin\psi\cos\phi + \cos\psi\sin\theta\sin\phi & \sin\psi\sin\phi - \cos\psi\sin\theta\cos\phi \\ -\sin\psi\cos\theta & \cos\psi\cos\phi - \sin\psi\sin\theta\sin\phi & \cos\psi\sin\phi - \sin\psi\sin\theta\cos\phi \\ \sin\theta & -\cos\theta\sin\phi & \cos\theta\cos\phi \end{bmatrix} \tag{19}$$

From the indicated matrix multiplication, three of nine potential expressions can be chosen, from which the three angular rotations, ϕ, θ, and ψ, can be determined. One possible set of expressions is

$$\sin\theta = \hat{\mathbf{k}}_s \bullet \hat{\mathbf{i}}_r \tag{20.a}$$

$$\cos\theta\cos\phi = \hat{\mathbf{k}}_s \bullet \hat{\mathbf{k}}_r \tag{20.b}$$

$$\cos\psi\cos\theta = \hat{\mathbf{i}}_s \bullet \hat{\mathbf{i}}_r \tag{20.c}$$

From these relationships, the rotations can be determined as

$$\phi = \cos^{-1}\left\{ \frac{\hat{\mathbf{k}}_s \bullet \hat{\mathbf{k}}_r}{\cos\left[\sin^{-1}\left(\hat{\mathbf{k}}_s \bullet \hat{\mathbf{i}}_r\right)\right]} \right\} \tag{21.a}$$

$$\theta = \left[\sin^{-1}\left(\hat{\mathbf{k}}_s \bullet \hat{\mathbf{i}}_r\right)\right] \tag{21.b}$$

$$\psi = \cos^{-1}\left\{ \frac{\hat{\mathbf{i}}_s \bullet \hat{\mathbf{i}}_r}{\cos\left[\sin^{-1}\left(\hat{\mathbf{k}}_s \bullet \hat{\mathbf{i}}_r\right)\right]} \right\} \tag{21.c}$$

Angles computed for rotations about each of the LCS axes are often referred to as "Cardan" angles. Alternatively, rotations could be performed such that the first and third rotations are about the same axis, and the second rotation is about one of the remaining axes. Such a rotation scheme results in angles that are commonly termed "Euler" angles.

For some applications, particularly those involving rotational kinetics, calculation of segment angular velocities and acceleration may be desirable. Segment angular velocity and acceleration calculations are somewhat more complicated than translational velocity and acceleration computations. Angular velocity, $\boldsymbol{\Omega}$, and acceleration, $\dot{\boldsymbol{\Omega}}$, of a segment are determined as vectors with components Ω_x, Ω_y, Ω_z and $\dot{\Omega}_x$, $\dot{\Omega}_y$, $\dot{\Omega}_z$, in the $\hat{\mathbf{i}}_s$, $\hat{\mathbf{j}}_s$, and $\hat{\mathbf{k}}_s$ directions, respectively. For the rotation sequence described above the angular velocity and acceleration components can be expressed as:

$$\Omega_x = \dot{\phi}\cos\theta\cos\psi + \dot{\theta}\sin\psi \tag{22.a}$$

$$\Omega_y = -\dot{\phi}\cos\theta\sin\psi + \dot{\theta}\cos\psi \tag{22.b}$$

$$\Omega_z = \dot{\phi}\sin\theta + \dot{\psi} \tag{22.c}$$

$$\dot{\Omega}_x = \ddot{\phi}\cos\theta\cos\psi - \dot{\phi}\dot{\theta}\sin\theta\cos\psi - \dot{\phi}\dot{\psi}\cos\theta\sin\psi + \ddot{\theta}\sin\psi + \dot{\theta}\dot{\psi}\cos\psi \tag{22.d}$$

$$\dot{\Omega}_y = -\ddot{\phi}\cos\theta\sin\psi + \dot{\phi}\dot{\theta}\sin\theta\sin\psi - \dot{\phi}\dot{\psi}\cos\theta\cos\psi + \ddot{\theta}\cos\psi - \dot{\theta}\dot{\psi}\sin\psi \tag{22.e}$$

$$\Omega_z = \ddot{\phi}\sin\theta + \dot{\phi}\dot{\theta}\cos\theta + \ddot{\psi} \tag{22.f}$$

The first and second time derivatives of ϕ, θ, and ψ can be approximated using the same techniques (i.e., discrete differentiation, or differentiation of time function approximations) used to estimate translational velocity and acceleration.

Computation of rotational kinematics can be readily applied to a wide range of video movement measurements; however, the occupational ergonomist must provide meaningful interpretation for these angles. For example, the designations "flexion/extension," "abduction/adduction," and "internal/external rotation" could be assigned to the angles describing relative orientation of a femoral LCS with respect to a pelvic LCS. Assignment of relevant designations, however, can be considerably more complicated for some body segments.

Construction of plots of rotation angles versus time are often useful with tasks for which cycles can be defined (e.g., a reach across a workspace). In such plots, time is often expressed as a percentage of the time to complete one cycle. Averages, maximums, minimums, and ranges of rotation angles, as well as angular values at specific instances, are all potential variables of interest to the ergonomist. Selection of the most appropriate variables for particular video movement measurements, however, is a matter of the ergonomist's discretion.

The techniques, described in this discussion, for video movement data collection and processing can be generally applied to a variety of occupational tasks. Ergonomists using these techniques for specific applications will need to adapt them for the particular nuances of the occupational tasks of interest. In doing so, they should consider that any aspect of video movement measurement (calibration, marker placement, image reconstruction, biomechanical modeling, and kinematic computation) can represent a weak link in the process if not appropriately considered.

References

Abdel-Aziz YI, and Karara HM: Direct Linear Transformation from comparator coordinates into object space coordinates in close range photogrammetry. Proceedings of the Symposium on Close Range Photogrammetry, Falls Church, VA, 1-18, 1971.

Antonsson EK, and Mann, RW: Automatic 6-d.o.f. kinematic trajectory acquisition and analysis. *Journal of Dynamic Systems, Measurement and Control,* 111:31-39, 1989.

Asato KT, Cooper RA, Robertson RN, and Ster JF: SMARTWheels: development and testing of a system for measuring manual wheelchair propulsion dynamics. *IEEE Transactions on Biomedical Engineering,* 40(12):1320-4, 1993.

Berguer R, Rab GT, Abu-Ghaida H, Alarcon A, and Chung J: A comparison of surgeons' posture during laparoscopic and open surgical procedures. *Surgical Endoscopy,* 11(2):139-42, 1997.

Boninger ML, Cooper RA, Robertson RN, and Rudy TE: Wrist biomechanics during two speeds of wheelchair propulsion: an analysis using a local coordinate system. *Archives of Physical Medicine & Rehabilitation,* 78(4):364-72, 1997.

Boston JR, Rudy TE, Lieber SJ, and Stacey BR: Measuring treatment effects on repetitive lifting for patients with chronic low back pain: speed, style, and coordination. *Journal of Spinal Disorders,* 8(5):342-51, 1995.

Cooper RA, Robertson RN, VanSickle DP, Boninger ML, and Shimada SD: Projection of the point of force application onto a palmar plane of the hand during wheelchair propulsion. *IEEE Transactions on Rehabilitation Engineering,* 4(3):133-42, 1996.

Garg A, Owen B, Beller D, and Banaag J: A biomechanical and ergonomic evaluation of patient transferring tasks: wheelchair to shower chair and shower chair to wheelchair. *Ergonomics,* 34(4):407-19, 1991.

Gracovetsky S, Kary M, Pitchen I, Levy S, and Ben Said R: The importance of pelvic tilt in reducing compressive stress in the spine during flexion-extension exercises. *Spine,* 14(4):412-6, 1989.

Gracovetsky S, Kary M, Levy S, Ben Said R, Pitchen I, and Helie J: Analysis of spinal and muscular activity during flexion/extension and free lifts. *Spine,* 15(12):1333-9, 1990.

Gracovetsky S, Newman N, Pawlowsky M, Lanzo V, Davey B, and Robinson L: A database for estimating normal spinal motion derived from noninvasive measurements. *Spine,* 20(9):1036-46, 1995.

Kadaba MP, Ramakrishnan HK, Wootten ME: Measurement of lower extremity kinematics during level walking. *Journal of Orthopaedic Research,* 8(3):383-92, 1990.

Kumar S, and Cheng CK: Spinal stresses in simulated raking with various rake handles. *Ergonomics,* 33(1):1-11, 1990.

Kumar S: Lumbosacral compression in maximal lifting efforts in sagittal plane with varying mechanical disadvantage in isometric and isokinetic modes. *Ergonomics,* 37(12):1975-83, 1994.

Lee YH, Cheng CK, and Tsuang YH: Biomechanical analysis in ladder climbing: the effect of slant angle and climbing speed. Proceedings of the National Science Council, Republic of China — Part B, Life Sciences, 18(4):170-8, 1994.

Marzan GT: Rational design for close-photogrammetry. Doctoral dissertation, University of Illinois at Urbana-Champaign, 1975.

Pascarelli EF, and Kella JJ: Soft-tissue injuries related to use of the computer keyboard. A clinical study of 53 severely injured persons. *Journal of Occupational Medicine,* 35(5):522-32, 1993.

Peterson B, and Palmerud G: Measurement of upper extremity orientation by video stereometry system. *Medical & Biological Engineering & Computing,* 34(2):149-54, 1996.

Robinson ME, O'Connor PD, Shirley FR, and MacMillan M: Intrasubject reliability of spinal range of motion and velocity determined by video motion analysis. *Physical Therapy,* 73(9):626-31, 1993.

Roozmon P, Gracovetsky SA, Gouw GJ, and Newman N: Examining motion in the cervical spine. I: Imaging systems and measurement techniques. *Journal of Biomedical Engineering,* 15(1):5-12, 1993.

Roozmon P, Gracovetsky SA, Gouw GJ, and Newman N: Examining motion in the cervical spine. II: Characterization of coupled joint motion using an opto-electronic device to track skin markers. *Journal of Biomedical Engineering,* 15(1):13-22, 1993.

Rudy TE, Boston JR, Lieber SJ, Kubinski JA, and Delitto A: Body motion patterns during a novel repetitive wheel-rotation task. A comparative study of healthy subjects and patients with low back pain. *Spine,* 20(23):2547-54, 1995.

Vaughan CL, Davis BL, and O'Connor JC: *Dynamics of Human Gait.* Human Kinetics Publishers, Champaign, Illinois, 1992.

Whittle MW: Calibration and performance of a three-dimensional television system for kinematic analysis. *Journal of Biomechanics,* 15:185-196, 1982.

Wickstrom G, Laine M, Pentti J, Hyytiainen K, and Salminen JJ: A video-based method for evaluation of low-back load in long-cycle jobs. *Ergonomics,* 39(6):826-41, 1996.

Woltring HJ: Calibration and measurement in 3-dimensional monitoring of human motion by optoelectronic means II. *Biotelemetry,* 3:65-97, 1976.

33

Force Dynamometers and Accelerometers

Robert G. Radwin
University of Wisconsin–Madison

Thomas Y. Yen
University of Wisconsin–Madison

33.1 Introduction

Forces considered in industrial ergonomics are usually classified as external or internal. External forces act against the human body, and they may be produced by an external object or in reaction to the human body exerting forces against an external object. It is possible to directly measure external forces using mechanical or electromechanical force measurement instruments. Internal forces are tension, compression, torsion, or shearing within muscles, tendons, bones, or other anatomical structures. Voluntary motions and exertions are produced through the generation of internal forces through active muscle contraction and passive action of connective tissues. Internal forces produce torque or rotation about the joints. External forces often result from internal force actions. Internal forces are usually not measured directly but by using indirect electrophysiological correlates such as electromyography (EMG).

The means of measuring external force will vary greatly depending on the circumstances of the task and practical considerations such as the accuracy required and the equipment and expertise available. Strength or maximal voluntary exertions represent the maximum force an individual is capable of producing. Forces associated with industrial tasks are usually less than maximal and are sometimes estimated from indirect measurements of the task requirements rather than measuring the exertions of individuals performing the task. These include measuring the weight of objects carried or lifted, or measuring the force necessary to do work, such as pushing or pulling a control. Direct force measurements should consider variability among individuals.

Human vibration is quantified from acceleration of objects that transmit vibrational energy to the body by contact either through the seat or feet (whole-body vibration) or by grasping vibrating objects (hand–arm vibration). Whole-body vibration is associated with vibration from riding in a vehicle or from standing on a moving platform. Hand–arm vibration may be introduced by using power hand tools or operating controls such as steering wheels on off-road vehicles.

Force and acceleration are expressed as a vector having magnitude and direction. The unit for force in the MKS system is the Newton ($1\ N = 1\ kg\ m/s^2$). The corresponding unit for acceleration is meters per second per second (m/s^2). Sometimes acceleration is expressed as multiples of the acceleration of gravity in units of g ($1\ g = 9.8\ m/s^2$). Force and acceleration measurements should specify both magnitude and direction. Usually force and acceleration vectors are decomposed into orthogonal components in a reference coordinate system.

0-8493-2641-9/99/$0.00+$.50

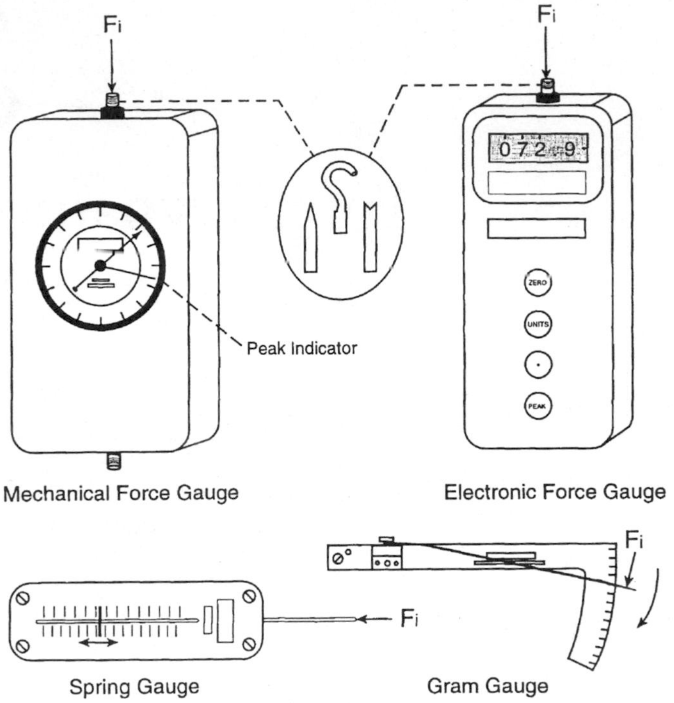

FIGURE 33.1 Hand-held force transducers with a variety of attachments.

Instruments applicable to force and acceleration measurements range from simple mechanical instruments to electromechanical devices. This chapter presents the theory and practice of force and acceleration measurement devices in industrial ergonomics. Mechanical and electromechanical force measurement devices are described, and force measurement applications using specialized dynamometers are discussed. The chapter also describes accelerometer theory and applications in human vibration measurement.

33.2 Mechanical Measurement Devices

Simple mechanical devices such as spring scales are used in many instances for estimating forces and loads, and are adequate for numerous industrial ergonomics applications. An assortment of these instruments is illustrated in Figure 33.1. Spring scales are available in various load levels, ranging from just a few grams to thousands of kilograms. These devices typically have a precision of 1% full scale. Most spring scales actually display units of mass (kg) rather than force because they are usually calibrated against a known mass. Force, which is measured in units of Newtons, can be determined by multiplying kilograms by the acceleration of gravity (9.81 m/s^2). Ergonomics practitioners find that instruments that are threaded for attachments such as various hooks and points are convenient for force measurement applications in the field (see Figure 33.1).

Mechanical force measurement instruments operate like a "fish scale" on the principle that applied force displaces a spring that is mechanically coupled by a spring-lever or a spring-cable system to a pointer and scale display. These simple devices often act as second-order mechanical systems consisting of a single mass, spring, and damping element as illustrated in Figure 33.2. The relationship between the input force and the output displacement is established as a simple second-order differential equation of the following form:

$$F_i - kx_o - c\dot{x}_o = m\ddot{x}_o \tag{1}$$

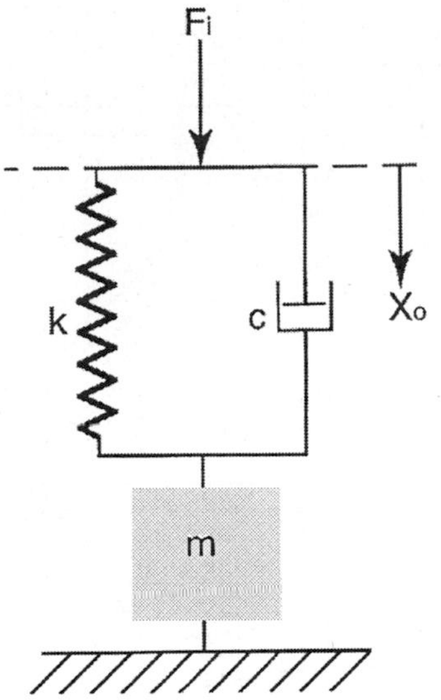

FIGURE 33.2 Second-order mechanical system consisting of a single mass, spring, and damping element.

where F_i is the applied force, k_s is the spring constant, c is the damping factor, m is the mass of elastic member, and x_o is the resulting displacement.

It is usually sufficient to assume that the mass of the elastic member does not affect the spring displacement, so it can considered zero. A damping element is used for damping the mass-spring system and preventing it from being excited into resonance, but in the static or quasi-static case where the excitation frequency is low it can be ignored. Consequently, the spring scale displacement can be simplified to:

$$x_o = F_i/k \tag{2}$$

indicating that the displacement is proportional to the applied force. A stiffer spring results in a less sensitive instrument, yielding a greater force range.

Many common mechanical force measurement instruments used in ergonomics practice have a continuously moving pointer and a peak force indicator. Peak forces are recorded by the peak indicator, while sustaining forces are read directly from the indicator dial. The peak force indicator in mechanical force transducers is sometimes used for measuring isometric strength. This may be accomplished by anchoring one end of the transducer while exerting force against the free end of the transducer with an appropriate handle. Grip dynamometers such as the Jamar and Smedely dynamometers are spring force measurement instruments specifically designed for pinch and power grip strength.

Mechanical force transducers are most suitable for static force measurements such as determining the weight of a stationary object or for measuring quasistatic, or very slowly changing forces such as the force needed to overcome friction and push or pull a rolling cart along the floor. Although these simple mechanical devices are easy to use and they do not require an external power source, they are somewhat limited in their application and if not used cautiously they can yield erroneous measurements. For example, when measuring the force needed to push a hand truck, forces when starting and stopping the truck are usually greater due to friction and inertia of the truck. But since the measured force is read directly off the display dial, it is difficult to continuously record fluctuating forces, particularly when the force rapidly changes. Care must also be taken to ensure the alignment of the scale or dynamometer with the axis of the exertion. Because springs can become permanently deformed when stretched beyond their elastic limits, their calibration should be periodically tested using known loads.

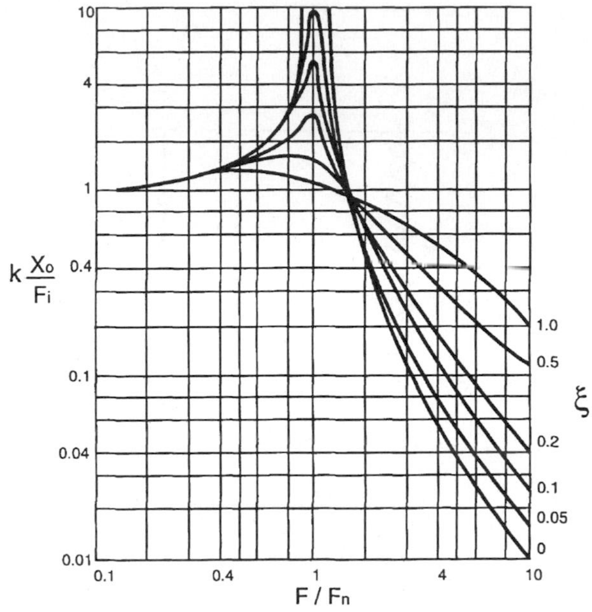

FIGURE 33.3 Second-order system response characteristics over a range of frequencies.

As is the case with any measurement, it is critical that the measurement instrument doesn't affect the system it is measuring. When dynamic or changing force is applied to spring devices, the mechanical spring instrument does in fact become part of the mechanical system. Second-order systems respond to changing inputs as depicted in Figure 33.3. When the system input excitation frequency is near or at resonance, the ratio of output displacement to input force greatly increases and the system becomes unstable. Since spring scales behave as second-order systems with resonant frequencies typically less than 1 Hz, they can be put into resonance and become unstable even by a relative low frequency input or by an impulsive force. Furthermore, if the changing force input has a frequency that is greater than the natural frequency of the instrument, the force measurements may read much less than its actual value due to attenuation by the spring-mass system (See Figure 33.3).

33.3 Electromechanical Measurement Devices

Electronic force transducers overcome many of the limitations of mechanical force transducers. Strain gauge load cells are capable of measuring static force, and they are much better suited for measuring forces that change with time than mechanical spring scales. Since electronic force transducers are much stiffer than mechanical force instruments, their response is much less affected by the system being measured. They also have better accuracy, and they can provide continuous force-time data. There are many self-contained electronic force measurement devices commercially available for use in industrial ergonomics studies (see Figure 33.1). These devices often contain circuitry for recording continuous force, peak force, and average force. It may also be possible to record the force output using an external instrument, and the device may have provisions to connect it directly to a digital computer. A summary of the advantages and disadvantages for each type of force transducers is shown in Table 33.1.

Many electromechanical sensors operate on the principle that mechanical inputs can alter the electrical resistance of a resistive element. The resistance (Ω) of a cylindrical electrical conductor is $R = \rho L/A$, where L is its length (m), A is its cross-sectional area (m^2), and ρ is the resistivity of the particular material (Ωm). The greater resistivity a material has, the more it acts as an insulator. Mechanical actions affect resistive sensors by either changing L or A (for metals) or by changing ρ (for semiconductors), consequently changing the sensor's resistance and the voltage across it. Metal strain gauges are made from

TABLE 33.1 Force Transducer Comparison

	Mechanical Force Transducers	Electronic Force Transducers
Advantages	Relatively inexpensive Simple to use Requires no power	Highly linear Continuous recording is possible Large force range with single transducer Small size Durable Fast response time
Disadvantages	Coarse resolution Continuous force recording is difficult Large size Limited dynamic capability	Calibration needed often Expensive transducer and instrumentation

lengths of very fine wire (<25 μm diameter). When the wire is stretched, its resistance changes, mainly due to changes in its cross-sectional area A and length L. The resistivity of metals increases with increasing temperatures because of the increased number of collisions that electrons make, thus increasing their electrical resistance. Consequently, resistive sensor accuracy may be affected by temperature.

Commercial strain gauge transducers of various types are available with a capacity for measuring loads of a few grams up to hundreds of thousands of kilograms and in several accuracy grades. The lowest accuracy grade typically has an overall (combined nonlinearity, hysteresis, nonrepeatability, etc.) error of about 1% of full scale, while the best accuracy grade has about 0.15% of full scale overall error.

Strain gauge transducers are often constructed using a calibrated metal plate or beam that undergoes a very small change (strain) in one of its dimensions. This mechanical deflection, usually a fraction of 1%, causes a small change in electrical resistance in the gauge wire. Low modulus materials such as aluminum are used to increase strain per unit force. Strain gauges are usually used in pairs. A simple load cell consists of a cantilever beam with strain gauges bonded on two opposing sides. The load cell can be made more sensitive and accurate when two gauges are placed on the top of the beam and an additional two gauges are attached to the bottom. The folded cantilever beam permits four gauges to be placed on the top surface, which eases manufacture and increases performance (Doebelin, 1990).

Silicon strain gauges are made of diffused resistors integrated into a silicon substrate. Both silicon and metal strain gauges are used in a similar manner and provide a very linear response within the elastic limits of the material they are fastened to. Silicon strain gauges exhibit even higher temperature effects than metal because deformation affects ρ. These temperature effects can often be controlled using special compensation circuits.

The gauge factor $G = (\Delta R/R_0)/(\Delta L/L_0)$, specifies a strain gauge's sensitivity to mechanical deformation. In this proportion, $\Delta R/R_0$ is the fractional change in resistance due to strain, and $\Delta L/L_0$ is the fractional change in strain gauge length. The gauge factor for metal strain gauges is typically between 2 and 5, while the gauge factor for silicon can be as high as 170. This mechanical deflection, usually a fraction of 1%, causes a small but measurable change in resistance.

When strain gauges are arranged in a Wheatstone bridge circuit, their sensitivity can be increased. The resulting imbalance in voltages in the bridge is proportional to the applied force. Bridge circuits also provide temperature compensation which prevents output shifts due to temperature effects on the transducers. The bridge circuit configuration increases sensitivity by $2(1 + \nu)$ over a single strain gauge configuration, where ν is Poisson's ratio. Bridge circuits also provide temperature compensation which prevents output shifts due to temperature effects on the transducers. A low modulus material such as aluminum is often used to increase strain per unit force. When four strain gauges are used in a Wheatstone bridge they yield an electrical output that is insensitive to bending stresses due to the force being applied off center or at an angle, and they can be temperature compensated (Doebelin, 1990). The circuit in Figure 33.4 uses two strain gauges in adjacent legs of the bridge. The output voltage of the bridge V_o is given by the equation:

$$V_o = V_e\left[\frac{R_2}{R_2 + R_4} - \frac{R_1}{R_1 + R_3}\right] \tag{3}$$

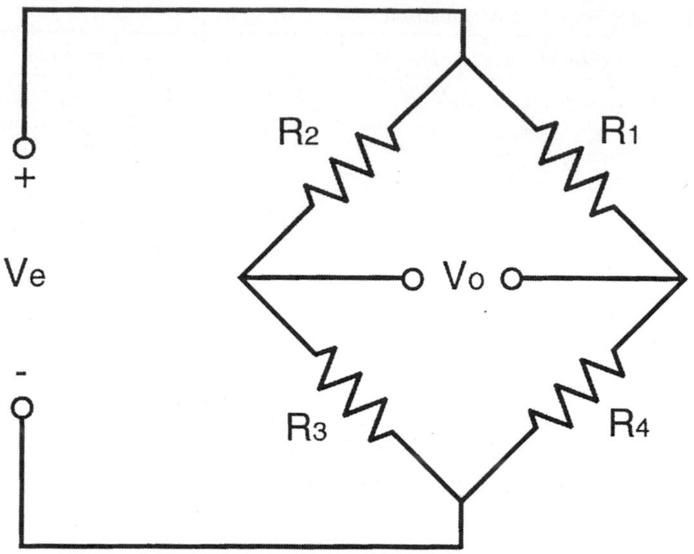

FIGURE 33.4 Strain gauges in Wheatstone bridge circuit.

When the beam bends, gauge 1 elongates and increases ΔR, while gauge 2 is compressed and decreases ΔR. The two fixed resistors in the opposite two legs of the bridge have identical resistance to the undeformed strain gauges, R_0, but they do not deform. This arrangement is known as a half-bridge circuit. If ΔR is much less than R_0 then

$$V_o \approx -\frac{V_e}{2R_0}\Delta R, \qquad (4)$$

and the output is proportional to the change in resistance. A full bridge configuration in which all four legs contain strain gauges yields even higher sensitivity with an output given by $V_o \approx \Delta R/R_0$. Load cell amplifiers are commercially available that already contain the necessary Wheatstone bridge circuitry and excitation power sources.

Piezoelectric and piezoresistive load cells require minute deformations of their atomic structure within a block of crystalline material. Piezoelectric materials, such as quartz and barium titanate, produce a change in charge distribution when subjected to a mechanical stress. Quartz is a naturally found piezoelectric material, and deformation of its crystalline structure changes the electrical characteristics such that the electrical charge across its surfaces is altered. The charges collect on metal electrodes deposited onto the surface of piezoelectric material. Special amplifiers are used called charge amplifiers that output a voltage proportional to a charge at its input. Since piezoelectric sensors operate on changes in charge distribution, the most notable drawback to piezoelectric sensors is their inability to respond to static loads.

Piezo material can be quite small in size, which allows for easy mounting in many applications. Rapidly changing forces can be measured using piezoelectric force sensors. Because of their small size, many commercial piezo load cells have the necessary electronic circuitry already built into the package. A single piezo load cell is useful over a very wide range of forces because the 1% nonlinearity applies to any calibration range. Piezoelectric load cells can measure forces as great as 16,000 N. Piezo transducers are excellent for dynamic force measurements because of their very fast time constant. Typical piezo transducers have very high stiffness, and natural frequency ranging from 10 KHz to 300 KHz (Doebelin, 1990). Because piezoelectric load cells are very stiff, they are suitable for measuring isometric forces. Since they only respond to changing and impulsive forces, piezoelectric load cells are unsuitable for measuring steady state forces.

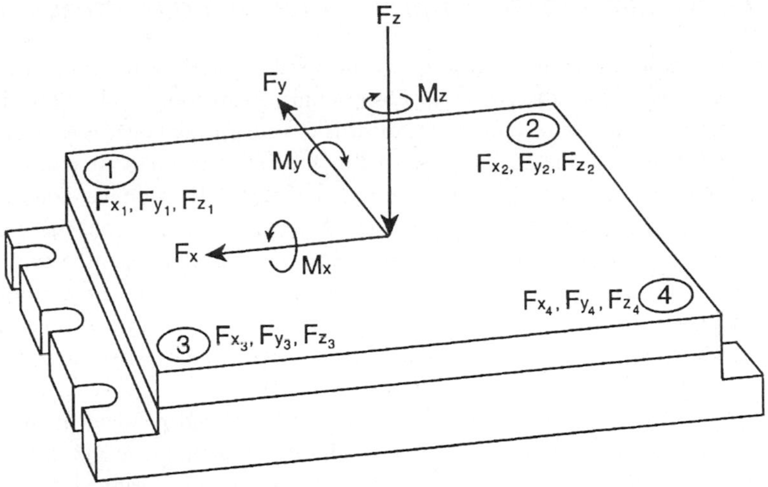

FIGURE 33.5 A force platform for measuring the three orthogonal force components (F_x, F_y, F_z) and three moments (M_x, M_y, M_z) from triaxial force transducers 1 through 4.

Unlike piezo*electric* sensors that produce changes in charge when stressed, piezo*resistive* sensors change resistance when they are stressed. Consequently, piezoresistive sensors are not limited to changing forces and are suitable for static loads. Piezoresistive sensors are usually configured in a Wheatstone bridge, and like strain gauges they require an external excitation voltage and a sensitive instrumentation amplifier.

Force platforms are used for measuring forces in two or more directions, such as ground reaction force acting on the feet during standing or walking. Ground reaction force contains a vertical component plus two shear components acting along the surface (Winter, 1990). This is accomplished using three or more force sensors that are arranged at right angles to each other. Force platforms are available with strain gauge or piezo force transducers. A common force plate configuration contains a flat plate supported by four triaxial force transducers as shown in Figure 33.5. This configuration measures three orthogonal force components in the x, y, and z axes for each of the four transducers (1, 2, 3, 4). Orthogonal force components (F_x, F_y, F_z) and moments (M_x, M_y, M_z) are resolved by using the equation:

$$\begin{bmatrix} F_x \\ F_y \\ F_z \\ M_x \\ M_y \\ M_z \end{bmatrix} = \begin{bmatrix} 1 & 1 & 1 & 1 & 0 & 0 & 0 & 0 & 0 & 0 & 0 & 0 \\ 0 & 0 & 0 & 0 & 1 & 1 & 1 & 1 & 0 & 0 & 0 & 0 \\ 0 & 0 & 0 & 0 & 0 & 0 & 0 & 0 & 1 & 1 & 1 & 1 \\ 0 & 0 & 0 & 0 & -1 & -1 & 1 & 1 & 0 & 0 & 0 & 0 \\ 0 & 0 & 0 & 0 & 0 & 0 & 0 & 0 & 1 & -1 & 1 & -1 \\ -1 & -1 & 1 & 1 & 1 & -1 & 1 & -1 & 0 & 0 & 0 & 0 \end{bmatrix} \cdot \begin{bmatrix} F_{x_1} \\ F_{x_2} \\ F_{x_3} \\ F_{x_4} \\ F_{y_1} \\ F_{y_2} \\ F_{y_3} \\ F_{y_4} \\ F_{z_1} \\ F_{z_2} \\ F_{z_3} \\ F_{z_4} \end{bmatrix} \qquad (5)$$

33.4 Force Measurement Applications in Ergonomics

One way of directly measuring hand and grip force is by installing strain gauge force sensors directly in handles and objects grasped in industrial tasks. For example, Armstrong, et al. (1994) installed strain gauge load cells directly underneath computer keyboards for measuring finger exertions during typing tasks. Grip measurements are sometimes complicated by the fact that forces are unevenly applied and distributed throughout the palmar and finger surfaces and often involve multiple digits. A conventional strain gauge force instrument that measures force using the strain produced from the bending moment of a cantilever beam is extremely limited because the point of application must be controlled in order to know the particular bending moment arm. Furthermore, these instruments cannot linearly sum forces applied at arbitrary locations along the beam. Because of these constraints, the simple cantilever beam strain gauge system will not suffice for practical hand force measurements when using an instrumented handle.

A strain gauge dynamometer was developed that has sensitivity independent of the point of force application and linearly sums forces applied at multiple locations along the length of the active area. It is based on the principle of shearing stress acting in the cross-section of a beam when a transverse force is applied (Pronk and Niesing, 1981; Radwin et al., 1991). Instead of basing the dynamometer on sensing bending stresses produced when an applied force creates a bending moment in a cantilever beam, which is commonly used in many strain gauge instruments and is highly dependent on point of application, this instrument employs the principle of measuring beam shear stresses acting in the cross section of the beam when a transverse force is applied. This is accomplished by measuring shearing stress acting in the cross section of the beam. Strain gauges are mounted on a thin web machined into the central longitudinal plane and aligned at 45° with respect to the long axis (Pronk and Niesing, 1981; Radwin, Masters and Lupton, 1991). By selecting a measurement point at the neutral axis of the beam, the effect of bending stresses are completely removed from the strain gauges and all strain at the measurement point is strictly due to shear stress. Shear strain is totally independent of the point of application. A schematic diagram of the dynamometer design and the critical dimensions are given in Figure 33.6. This instrument is highly linear and has a typical force sensitivity of 2 mV/N.

The strain gauge dynamometer was used in a number of ergonomics investigations involving both maximal and submaximal exertions. A dynamometer for measuring grip strength when grasping a cylindrical handle (Oh and Radwin, 1993) was constructed for an average force of 250 N and a maximum load of 1000 N force. Aluminum caps were attached to both beams for producing a cylindrical surface. The handle length was 145 mm in order to accommodate hands of various sizes. The beams were mounted on a track so that they were capable of being separated arbitrary distances in order to provide a variable grip span.

A similar dynamometer was designed and constructed for measuring submaximal finger forces during five-finger prehension tasks when weights of various sizes were suspended from the dynamometer (Radwin et al., 1992). Another version of this instrument was attached to an electromagnetic shaker stage for directly measuring forces exerted when the handle was set into vibration resembling a vibrating power hand tool (Radwin et al., 1987). A smaller version with greater sensitivity was constructed for measuring pinch strength (Jeng et al., 1994). The pinch force instrument was 85 mm in length and was designed for an average force of 10 N and a maximum force of 200 N. Two opposing active dynamometer beams were used for independently measuring forces exerted by the thumb in opposition to one of the four fingers. Since the dynamometer was insensitive to the point of application, it was useful for clinical evaluations where it may be difficult to control the location where the fingers apply forces against the bars, particularly when using different fingers.

Another version of this instrument was used for investigating the forces involved in operating a pistol grip power hand tool (Oh and Radwin, 1993). The apparatus is shown in Figure 33.7. Two strain gauge instrumented beams were constructed for the handles of the power hand tool and an in-line pneumatic hand tool was mounted perpendicular to the handle to resemble a pistol grip hand tool. The tool was completely operational and capable of measuring palmar feed force and finger exertions during power

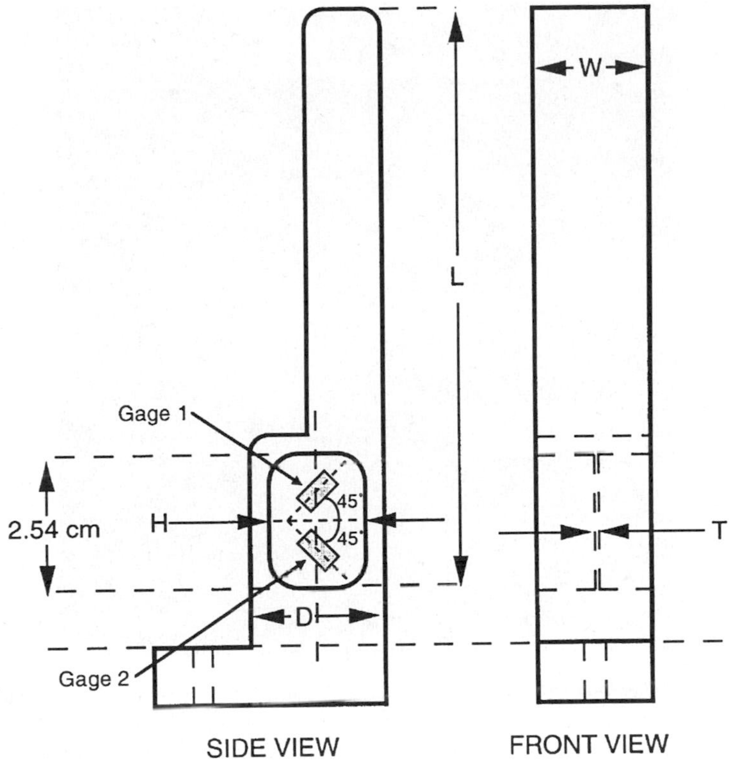

FIGURE 33.6 Schematic diagram containing dimensional variables for general dynamometer design. Dimensions that are varied include length (*L*), width (*W*), depth (*D*), pocket length (*H*), and web thickness (*T*). (From Radwin, R. G., Masters, G. P., and Lupton, F. W. 1991. A linear force-summing hand dynamometer independent of point of application, *Applied Ergonomics*, 22(5): 339-345. Copyright (1991) with kind permission from Elsevier Science Ltd., The Boulevard Langford Lane, Kidlington, U.K.)

hand tool operation. Contoured plastic caps were constructed and attached to the handle for simulating the shape of an actual hand tool. A trigger for activating the power hand tool was integrated into the finger side of the handle. The trigger contained a leaf spring switch used for activating a relay and solenoid valve for controlling the tool air supply.

Although strain gauge load cells can measure force with great accuracy, load cells with sufficient range are sometimes too large and bulky for attaching to handles or to the body, and in many instances it is difficult to use load cells for directly measuring exerted hand force in industrial tasks. A durable and thin conductive polymer force transducer was found useful for measuring external forces on regions of the body where conventional force sensors were far too large (Jensen et al., 1991). Because of its very small size and high durability, the sensor can be easily attached to the skin, which makes fingertip and palmar forces easy to measure. The conductive polymer sensing elements are composed of two conducting interdigitated patterns deposited on a thermoplastic sheet facing against another sheet containing a conductive polyetherimide film (see Figure 33.8). A spacer between the two plastic layers prevents contact, causing the sensor to have infinite impedance. As applied force increases, the two layers compress together, increasing the contact area. This subsequently decreases the electrical resistance of the sensor. An increased force leads to a decrease in resistance.

A dome for distributing force over the active sensing area is necessary for these elements to operate as force sensors (see Figure 33.8). Without the dome, the measurements are erroneous as shown in Figure 33.9. These sensors are very limited; their useful range is up to 30 N with an accuracy of 1 N; however, there are few alternatives available for directly measuring finger and hand forces.

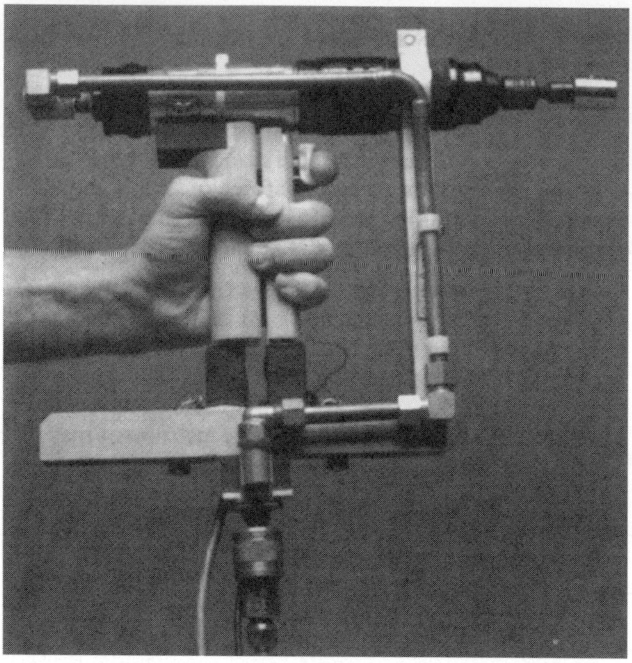

FIGURE 33.7 Dynamometer used for measuring finger and palm forces exerted when operating a completely functional simulation of a pistol-grip pneumatic power hand tool. (From Radwin, R. G., Masters, G. P., and Lupton, F. W. 1991. A linear force-summing hand dynamometer independent of point of application, *Applied Ergonomics,* 22(5): 339-345. Copyright (1991) with kind permission from Elsevier Science Ltd., The Boulevard Langford Lane, Kidlington, U.K.)

The small size and flexibility of the transducer materials, greatly increases the mountability of the transducers on a large variety of objects, and tool handles and grips. Due to the physical properties or the transducer mounting configuration, most electronic force transducers are only sensitive to forces acting normal or perpendicular to the surface plane of the transducer. This allows the identification of directional components of the forces acting on or by an object by using a different force transducer for each force direction of interest.

33.5 Acceleration Measurements

Acceleration is measured directly using devices called accelerometers. Accelerometers consist of a small mass and a piezoelectric element that measures the resulting force when mass accelerates. Recall that piezoelectric force sensors operate on the principle that the electrical charge measured across a piezoelectric material is proportional to its deformation when force is applied. Piezoelectric accelerometers operate the same way except a small mass is mounted on top of the piezoelectric material. The mass weighs usually no more than several grams. When the device is accelerated, the mass exerts a force against the piezoelectric material which produces a signal proportional to its acceleration. A typical accelerometer design is shown in Figure 33.10.

Piezoelectric accelerometer sensitivity is usually expressed in terms of coulomb charge per unit of acceleration (typically pC/g). The outputs are amplified using a charge amplifier that converts charge into voltage. Accelerometers may also be made from piezoresistive material, and their sensitivity is expressed in terms of millivolts per unit of acceleration (mV/g). Like piezoresistive force sensors, piezoresistive accelerometers require an excitation voltage, and they require an instrumentation amplifier similar to a load cell.

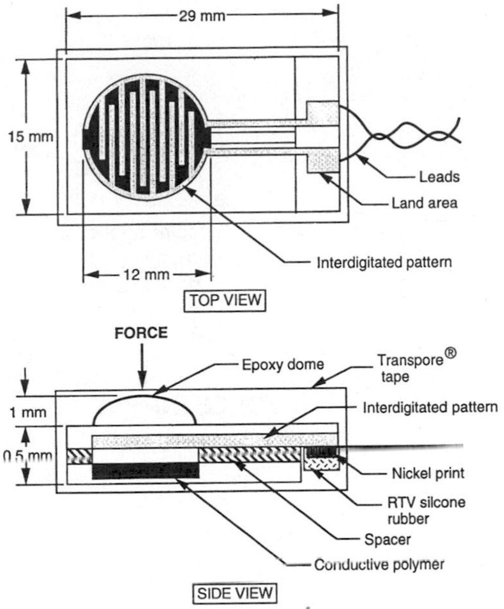

FIGURE 33.8 Schematic diagram of the conductive polymer finger force sensor showing top and side views. After a dome was placed over the conductive polymer sensing area the sensor was encased in Transpore® tape. (Reprinted from Jensen, T. R., Radwin, R. G., and Webster, J. G. 1991. A conductive polymer sensor for measuring external finger forces, *Journal of Biomechanics*, 24 (9): 851-858. Copyright (1991) with kind permission from Elsevier Science Ltd., The Boulevard Langford Lane, Kidlington, U.K.)

Accelerometer sensitivities are proportional to their mass. Consequently, the smaller the accelerometer, the less sensitive it is. Typical accelerometer frequency response characteristics are shown in Figure 33.11. It is important that an accelerometer not be excited by frequencies near or at its resonant frequency. This would produce erroneously large measurements. Accelerometer resonant frequencies vary inversely proportional to the square root of their mass. Therefore, smaller accelerometers have higher resonant frequencies and are usable over a greater frequency range. Accelerometers are also influenced by temperature changes, humidity, and other harsh environmental conditions.

It is important that the total mass of an accelerometer is sufficiently small not to interfere with the measurement by loading the vibrating body. Commercial accelerometers are small enough and light enough to attach directly to the limbs for measuring body motions. Accelerometers weighing more than 15 g are typically unsuitable for vibration measurements made by mounting them on a human body.

Usually accelerometers are sensitive to motion in a single direction. Triaxial accelerometers are available for simultaneously measuring acceleration in three orthogonal directions. Angular acceleration may be measured by mounting an accelerometer tangential to the rotating object. The angular acceleration α is the tangential acceleration a_p divided by the radius of rotation r: $\alpha = a_p/r$.

When accelerometers are mounted on the body, they are usually located near bony eminencies and surfaces. More commonly, accelerometers are mounted on objects that transmit vibration to the body. This may be a seat or a platform. In that case, accelerometer mass is not as critical for measuring whole-body vibration (WBV), although size might be a consideration when mounting accelerometers under a seated operator. Accelerometer resonant frequencies should be greater than 300 Hz and be capable of sustaining instantaneous acceleration levels up to 100 m/s^2 without damage. A triaxial seat disk accelerometer may be inserted between a vehicle seat and a passenger's buttocks. Vibration at the feet of a vehicle passenger can be measured by mounting an accelerometer directly to the floor. SAE J1013 specifies a

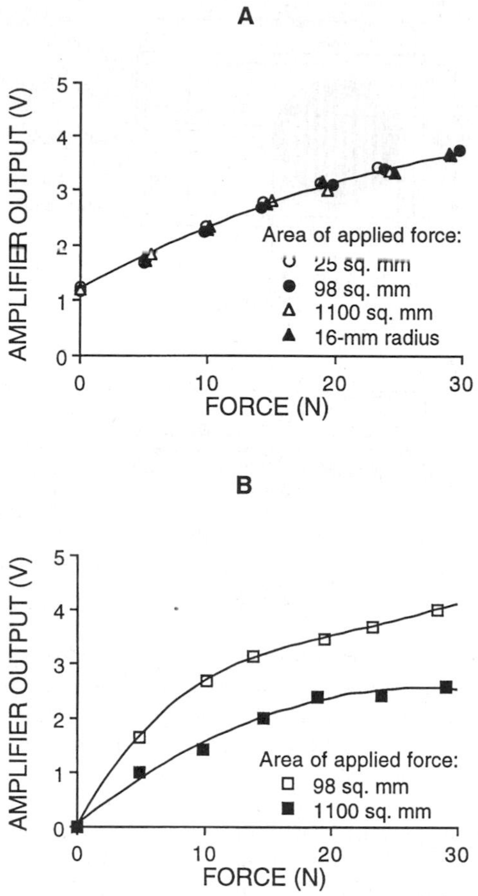

FIGURE 33.9 (A) Sensor response with an epoxy dome for input forces applied using different size surface areas, including a curved surface, showing insensitivity to area of application. (B) Sensor response without an epoxy dome for forces applied using different size surface areas, showing high sensitivity to the contact area of force application when the dome is not included. (Reprinted from Jensen, T. R., Radwin, R. G., and Webster, J. G. 1991. A conductive polymer sensor for measuring external finger forces, *Journal of Biomechanics,* 24 (9): 851-858. Copyright (1991) with kind permission from Elsevier Science Ltd., The Boulevard Langford Lane, Kidlington, U.K.)

transducer mounting for measuring seated vibration using either a 200-mm-diameter by 6-cm-thick disc placed between the operator and the seat cushion, or using a semirigid disc made of 80 to 90 durometer molded rubber or plastic. The semirigid disc is the most practical, and it is recommended particularly for soft or highly contoured cushions. The disc should be placed on the seat so that the triaxial accelerometers are located midway between the ischial tuberosities and are aligned parallel to the ISO basicentric seated operator coordinate axes. The disc should be taped or similarly attached to the cushion to maintain its location.

The frequencies of interest for whole-body vibration are between 0 Hz and 80 Hz. Piezoresistive accelerometers are the most common for WBV because of their low frequency response. Calibration is simplest for piezoresistive accelerometers with a DC response because they can be calibrated by just inverting the accelerometer in the direction of gravity. A 180° tilt represents a 19.61 m/s^2 (2g) peak-to-peak change in acceleration. WBV acceleration measurements are referenced to the ISO biodynamic coordinate system (ISO 2631) as shown in Figure 33.12.

Hand–arm vibration (HAV) usually is transmitted through the handles of manually operated equipment, such as power hand tools. Small piezoelectric accelerometers are the most common type of

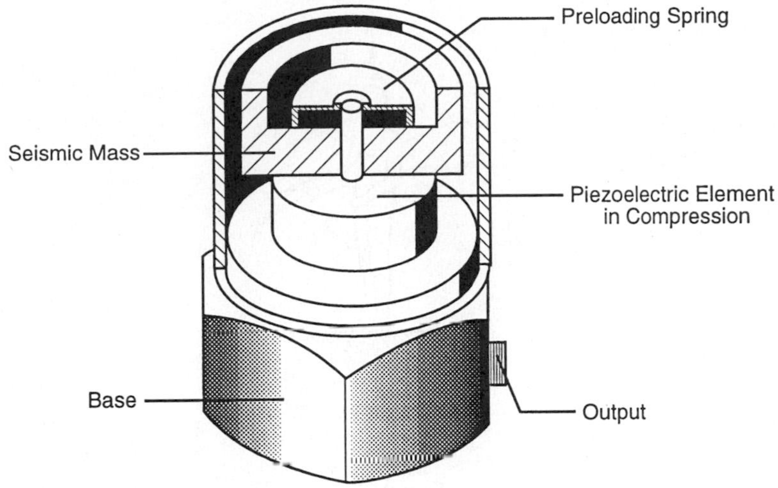

FIGURE 33.10 Piezoelectric accelerometer.

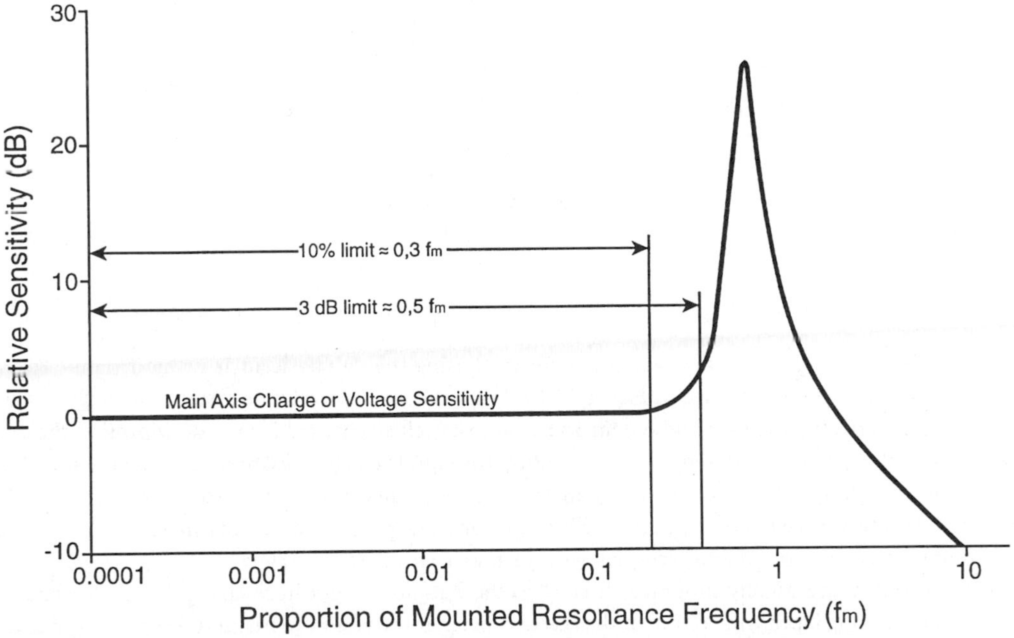

FIGURE 33.11 Representative accelerometer frequency characteristics. The usable frequency range is typically limited to the frequency where the output becomes amplified less than 10% or the frequency where the output becomes amplified no more than 3 dB.

transducer used for HAV measurements. Size and durability are critical considerations since the accelerometers are mounted on tool handles as operators use the tools for work. Accelerometer mass should be as small as possible. Accelerometers weighing 10 to 15 grams or less with a cross-axis sensitivity less than 10% are most suitable for HAV measurements. Piezoelectric accelerometers should be calibrated using a portable vibration exciter (such as the Brüel & Kjær 4294) that can verify accelerometer condition at a study site, because accelerometers are easily damaged during measurements in the field.

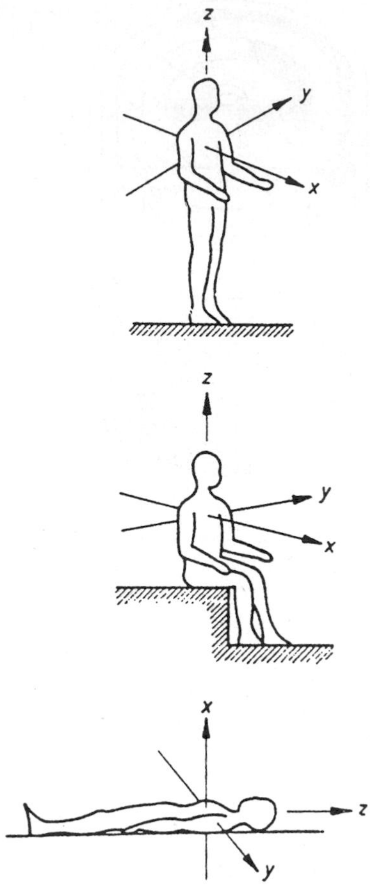

FIGURE 33.12 ISO Biodynamic coordinate system as defined in ISO 2631.

Vibration measurements of the hand are performed using the ISO basicentric or biodynamic coordinate system in ISO 5349 as shown in Figure 33.13. The basicentric coordinate system is usually easier to use than the biodynamic coordinate system and is almost exclusively used for handle vibration measurements. The three linear axes designated as x (motion through the palm), y (motion across the knuckles), and z (motion along the axis of the arm) are measured simultaneously. The accelerometer should be mounted directly on the handle nearest to where the hand grasps the handle. All three linear axes should be simultaneously measured and recorded, with each axis analyzed separately.

Accelerometers are usually mounted directly to the handle using a hose clamp or similar strap, or welding a mounting block with a stud for the accelerometer. If the vibration magnitude varies significantly over different parts of the handle, then the maximum value at the point of contact should be measured. Accelerometers are commonly screwed to a small aluminum block with a radius on the bottom and strapped to the handle using a hose clamp or a Panduit cable tie. If a resilient element is interposed between the hand and the vibrating surface, a thin and suitably shaped piece of metal may be placed between the hand and the handle.

Triaxial accelerometers are the most convenient because remounting an accelerometer in different orientations is often time-consuming and more obtrusive in the field. Single axis accelerometers may be used for measuring vibration as long as the event is relatively repeatable; however, error might be introduced if each successive measurement involves considerably different tool orientations and motions, as is possible in field measurements.

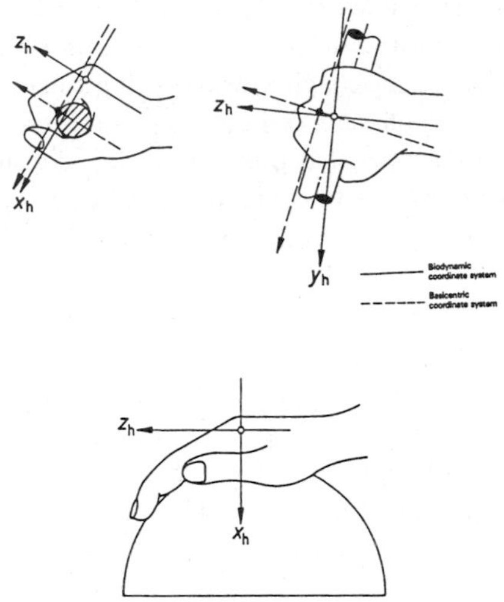

FIGURE 33.13 Biodynamic and basicentric coordinate system as defined in ISO 5349.

Root-mean-squared (RMS) linear acceleration (in meters/second/second or "g," where 1g = 9.81 m/sec/sec) is the measurement parameter of choice for vibration magnitude. The hand–arm vibration frequency range for most standards cover the one third octave bands having center frequencies from 6.3 to 1250 Hz, octave bands having center frequencies from 8 to 1000 Hz, and frequency weighted measurements covering the frequency range 5.6 Hz to 1400 Hz. A frequency range as high as 5 KHz for unweighted acceleration measurements has been advocated by some researchers.

Vibration signals with very high peak acceleration, such as percussive tools, require a more durable transducer with a resonant frequency above 25 KHz and cross axis sensitivity at least 20 dB below the sensitivity of the axis to be measured. One source of artifact error comes from DC shifts that result when an accelerometer is excited at its resonant frequency. It sometimes may be necessary to mechanically attenuate vibration in order to avoid accelerometer resonances, which should be made as near to the transducer as possible. The preferred method of measurement involves inserting a mechanical low-pass filter with a cut-off frequency no less than 2000 Hz between the transducer and the vibrating surface.

Human occupational vibration exposure is usually assessed by measuring vibration acceleration and determining acceleration magnitude and frequency characteristics, in addition to vibration exposure time. There are no specified methods for obtaining accurate exposure time (duration) data. Exposure time could be based on a time study, work sampling, historic data, or determined directly from the vibration record. Exposure time is critical in most standards. Statistical methods may be useful for determining duration from sampled data if repetitive events are within a specified error for a certain accuracy (e.g., 95% probability that sampling error is less than 5%).

Vibration exposure time is difficult to predict based on production or work standards and may vary significantly depending on the specific operation, individual work methods, and operator experience. A study was undertaken to evaluate individual worker's exposure to hand–arm vibration from power hand tool use in an electric appliance assembly plant where workers used small in-line pneumatic screwdrivers to perform highly repetitive tasks on an assembly line (Radwin and Armstrong, 1985). The average daily exposure was predicted from observed vibration samples recorded for each operator on the assembly line.

Accelerometers were attached to the power hand tools used by each operator, and vibration measurements were recorded for an observation period ranging from 4 to 18 min. The analysis consisted of measuring the average air tool operation time for each operator. Vibration and frequency measurements were made at a later time in the laboratory. There were two phases associated with the screwdriver operation. The first, or "run-down," phase corresponded to running the tool during screw driving. The second "clutch-slippage" phase corresponded to slippage of the clutch mechanism when the screw was tightened to the set torque. Visual inspection of the acceleration waveform clearly revealed that the vibration produced during the clutch-slippage phase was much greater in amplitude and fundamentally lower in frequency than the run-down phase. A computer program was used to recognize the run-down and chutch-slippage phases from data sampled using an analog–digital converter, on the basis of an empirically derived digital moving average filter. Total exposure for each operator was computed by measuring samples of their screwdriver operation times and multiplying the cumulative time by the number of operations per shift.

Current HAV standards frequency weightings were arrived at by consensus and originate from early studies by Miwa (1967, 1968) of subjective perception of vibration and discomfort and tolerance levels. The ISO 5349 and ANSI S3.34 specify similar frequency weighting networks. These weighted curves were revised in the 1980s and issued in 1986 as ISO 5349. Frequency weighting is accomplished by applying specified one third octave band weightings or by passing the data through the specified filter network. Discrete spectra may also be weighted according to the filter network characteristics. Because early HAV standards were based solely on subjective vibration perception and the paucity of epidemiological data available, NIOSH (1989) and others have advocated measuring unweighted HAV.

Use of linear integrating vibration exposure measurement equipment is best suited for measuring vibration exposure when the vibration signal is of short duration and its operation varies substantially with time. When task cycle times are relatively long, lasting more than one minute or longer, linear integrating vibration spectrum analyzers or software having integration periods greater than one minute are not always available. Vibration exposure is difficult to assess directly using many fast Fourier transform (FFT) spectrum analyzers because of long tool operating times. Providing that the vibration signal is stationary (i.e., the signal is time-invariant) during tool operation, linear integration over the entire tool operating period is not always necessary. A study showed that if tool operation time can be accurately determined independently of vibration acceleration, vibration exposure levels can then be computed using transient vibration measurements (Radwin, et al., 1990). The vibration measurements can be taken off-line where conditions can be controlled and not interfere with ongoing operations.

References

American National Standards Institute (1986). *S3.34: Guide for the Measurement and Evaluation of Human Exposure to Vibration Transmitted to the Hand*, New York, NY.

Armstrong, T. J., Foulke, J. A., Martin, B. J., Gerson, J., and Rempel, D. M. 1994. Investigation of applied forces in alphanumeric keyboard work, *American Industrial Hygiene Association Journal*, 55: 30–35.

Doebelin E. O. 1990. *Measurement Systems: Application and Design*, 4th ed. New York: McGraw-Hill.

Eastman Kodak Company 1986. *Ergonomic Design for People at Work: Volume 2*, New York: Van Nostrand Reinhold.

International Organization for Standardization 1985. *ISO 2631: Evaluation of Human Exposure to Whole-body Vibration–Part 1: General Requirements*, Geneva.

International Organization for Standardization 1986. *ISO 5349: Guidelines for the Measurement and Assessment of Human Exposure to Hand-Transmitted Vibration*, Geneva.

Jeng, O. J., Radwin, R. G., and Rodriquez, A. A. 1994. Functional psychomotor deficits associated with carpal tunnel syndrome, *Ergonomics*, 37(6): 1055-1069.

Jensen, T. R., Radwin, R. G., and Webster, J. G. 1991. A conductive polymer sensor for measuring external finger forces, *Journal of Biomechanics*, 24 (9): 851-858.

Miwa, T. 1967. Evaluation methods for vibration effect: Part 3: Measurements of threshold and equal sensation contours on hand for vertical and horizontal sinusoidal vibrations, *Industrial Health*, 5: 213-220.

Miwa, T. 1968. Evaluation methods for vibration effect: Part 6. Measurements of unpleasant and tolerance limit levels for sinusoidal vibrations, *Industrial Health*, 6, 18-27.

National Institute for Occupational Safety and Health 1989. *Criteria for a Recommended Standard: Occupational Exposure to Hand–arm Vibration*, DHHS/NIOSH Publication No. 89-106.

Oh, S. and Radwin, R. G. 1993. Pistol grip power tool handle and trigger size effects on grip exertions and operator preference, *Human Factors*, 35(3), 551-569.

Pronk, C. N. A. and Niesing, R. 1981. Measuring hand-grip force using a new application of strain gauges, *Medical & Biological Engineering and Computing*, 19, 127-128.

Radwin, R. G. and Armstrong, T. J. 1985. Assessment of hand vibration exposure on an assembly line, *American Industrial Hygiene Association Journal*, 46(4), 211-219.

Radwin, R. G., Armstrong, T. J., and Chaffin, D. B. 1987. Power hand tool vibration effects on grip exertions, *Ergonomics*, 30 (5), 833-855.

Radwin, R. G., Armstrong, T. J., and VanBergeijk, E. 1990. Vibration exposure for selected power hand tools used in automobile assembly, *American Industrial Hygiene Association Journal*, 51(9): 510-518.

Radwin, R. G., Masters, G. P., and Lupton, F. W. 1991. A linear force-summing hand dynamometer independent of point of application, *Applied Ergonomics*, 22(5): 339-345.

Radwin, R. G., Oh, S., Jensen, T. R., and Webster, J. G. 1992. External finger forces in submaximal static prehension, *Ergonomics*, 35(3): 275-288.

Winter, D. A. (1990). *Biomechanics and Motor Control of Human Movement*, New York: John Wiley & Sons.

Section III
Cognitive Environment Issues

34

How Complex Human-Machine Systems Fail: Putting "Human Error" in Context

Klaus Christoffersen
The Ohio State University

David D. Woods
The Ohio State University

"Rather than being the main instigators of an accident, operators tend to be the inheritors of system defects… Their part is that of adding the final garnish to a lethal brew whose ingredients have already been long in the cooking"

(Reason, 1990; p. 173).

34.1 Introduction

Research on human error is as old as the field of human factors (Fitts and Jones, 1947). Despite this long record of investigation into the factors that lead to erroneous actions and assessments, the belief remains commonplace that the verdict "human error" is a meaningful statement of the cause of failures. The belief that human errors are random acts constituting a basic category of human performance blocks our ability to understand and therefore control the factors that lie behind failures of complex systems. Human factors has always probed for the systematic factors behind the label "human error" in the nature of the problems people face, in the design of the artifacts that people use, and in the organization that

0-8493-2641-9/99/$0.00+$.50

provides resources and sets goals. In other words, the label human error is not a conclusion, but rather a starting point for investigation (Woods et al., 1994).

A fundamental premise of human factors research on errors, therefore, is that human performance is shaped by systematic factors. The scientific study of failure is not possible unless one understands this basic tenet of social and behavioral science. We do not yet understand all of the factors and how they interact. Individual differences are always a prominent fact of human behavior. But there are regularities which shape human cognition, collaboration and performance, and we do understand how these factors can make certain *kinds* of erroneous actions and assessments predictable. Again, finding points where these design-induced and organization-induced errors will occur and developing countermeasures has been among the major topics of human factors from its origins. Our ability to predict the timing and number of erroneous actions is very weak, but our ability to predict the kind of errors that will occur, when people do err, is often good or very good.

A number of serious, widely publicized accidents, such as the Three Mile Island nuclear power accident in 1979, the Bhopal chemical plant accident in 1984, the Challenger space shuttle accident in 1986, the Strasbourg automated airliner crash in 1992, and numerous less celebrated but no less catastrophic events led to a new wave of research into the factors that lead complex systems to fail (Perrow, 1984; Senders and Moray, 1991; Reason, 1990; Hollnagel, 1993). This research has shown how popular beliefs that such accidents are due simply to isolated acts of human error mask the deeper story. In pursuing this story, the research has shown that the processes that lead complex systems to fail are much more complex than a single "error" by a single human. The opportunity to learn from accidents and incidents, and the ability to make systems more reliable and robust, depends on pursuing the underlying factors beyond the label "human error" (Woods et al., 1994). This chapter presents a portion of the deeper story of how complex systems fail.

34.2 Hindsight Bias

Why is human error so persistently viewed as a legitimate explanation for system failures? In part, the problem is that attributions of "human error" occur after the fact, in the aftermath of a failure, through a process of social judgment. After an accident or incident, it is easy for us with the benefit of hindsight to say, "How could they have missed x?" or "How could they have not realized that x would obviously lead to y?" This is because our knowledge of the bad outcome makes it seem that participants failed to account for information or conditions that "should have been obvious" or behaved in ways that were inconsistent with the (now known to be) significant information. (See Woods et al., 1994, Chapter 6 for an extensive discussion of how the hindsight bias degrades our ability to learn from accidents.) Fundamentally, this omniscience is not available to any of us before we know the results of our actions. To react, after the fact, as if this knowledge were available to operators, trivializes the situation confronting the practitioner, and masks the processes affecting practitioner behavior before-the-fact.

Hindsight bias is the tendency for people to "consistently exaggerate what could have been anticipated in foresight" (Fischhoff, 1975). Studies have consistently shown that people have a tendency to judge the quality of a process by its outcome. The information about outcome biases their evaluation of the process that was followed. Decisions and actions followed by a negative outcome will be judged more harshly than if the *same* decisions had resulted in a neutral or positive outcome. Indeed, this effect is present even when those making the judgments have been warned about the phenomenon and been advised to guard against it (Fischhoff, 1975, 1982).

Given knowledge of outcome, reviewers will tend to *simplify* the problem-solving situation that was actually faced by the practitioner. The dilemmas, the uncertainties, the trade-offs, the attentional demands, and double binds faced by practitioners may be missed or under-emphasized when an incident is viewed in hindsight. Because the hindsight bias masks the real dilemmas, uncertainties, and demands practitioners confront, we have a distorted view of the factors contributing to the incident or accident. In this vacuum, we only see human performance after an accident or near miss as irrational, willing disregard (for what is now obvious to us and to them), or even diabolical. This fuels the traditional responses to punish the individuals associated most closely with the outcome in time and space.

34.3 Design and Organizational Factors Shape Human Performance

If we accept human error as a symptom rather than as a cause of problems, then what is human error a symptom of? Three of the underlying factors which most significantly influence human performance and create opportunities for errors by front-line operators are the design of tasks, the design of human–computer interfaces, and organizational characteristics of systems. A comprehensive discussion of the ways in which design and organizational factors can negatively impact human cognition, collaboration, and performance is outside the scope of this chapter, but some examples will serve to illustrate the point.

Task Design: Omissions of Isolated Acts

One example of a kind of slip of action is the omission of an isolated act (often the last action in a prescribed sequence or post-completion slip; Byrne and Bovair, 1997). In slips of action, the user intends to act in the appropriate way, but the process of translating that correct intention into the specific sequence of detailed action needed to carry out that intention is disturbed (cf. Norman, 1981). One example of this class of erroneous actions can occur when one of the actions that makes up an action sequence is unconnected physically or functionally to previous or successive actions. As a result of how the task has been designed, there are no cues in the structure of the task to act as an external memory or remind the actor to carry out the step. This characteristic increases the user's working memory load. When this increased memory burden is combined with the occurrence of other factors that challenge user's working memory (e.g., disruptions, multiple tasks, high work load, fatigue), the isolated step is easily omitted (Byrne and Bovair, 1997). The design of the task influences the memory load on the actor, i.e., it shapes the cognitive activities of the people in the system. If other factors are present that also challenge working memory, a specific kind of erroneous action can result.

This is only one kind of erroneous action that is affected by high memory loads on users. Many tasks and devices are designed in ways that place high demands on user memory; so many in typical human–computer interfaces that Norman (1988) refers to a conspiracy against human memory. Symptomatic of the loads on user memory are adaptations, such as the creation of external memory aids, which users often devise in response to this threat to their performance. For example, people may develop paper reminders or "crib notes" and attach them to a computer workstation to aid in various tasks.

Human–Computer Interface Design: Mode Error

A classic example of how design factors can create the potential for poor human performance is mode error (see Woods et al., 1994, Chapter 5 for a summary). Mode error is one kind of breakdown in the interaction between humans and machines, especially computerized devices. Norman (1988, p. 179) summarizes the source of mode error quite simply by suggesting that one way to create or increase the possibilities for erroneous actions is to "… change the rules. Let something be done one way in one mode and another way in another mode." When this is the case, a human user can commit an erroneous action by executing an intention in the way appropriate to one mode of the device when the device is actually in another mode. Put simply, multiple modes in devices create the potential for mode errors. The consequences of mode errors depending on the context in which they occur. Mode errors in human–computer interaction have been critical contributors to accidents in aviation (Billings, 1996).

Mode error is inherently a human–machine system breakdown. It requires a user who loses track of the system's active mode configuration and a machine that interprets user input differently depending on the current mode of operation. Mode error has been identified and studied as a systematic form or user error created by design factors since at least 1981 (e.g., Norman, 1981; Lewis and Norman, 1986; Monk, 1986; Sellen et al., 1992; Sarter and Woods, 1995; Obradovich and Woods, 1996). Interestingly, characteristics of the computer medium and common pressures on the design process have made it easy for designers to proliferate modes and to create more complex interactions across modes. The result is an epidemic of new opportunities for mode errors to occur and new kinds of mode-related problems in today's computerized devices. These studies provide methods to identify when computerized devices will produce mode errors and suggest several design techniques as countermeasures.

Organizational Factors: Goal Conflicts

Organizational factors also shape human cognition and collaboration in ways that lead predictably to certain forms of erroneous actions (e.g., Reason, 1997). Organizations provide resources but also create or sharpen the dilemmas practitioners face. Organizational pressures to meet some goals (generally throughput or economic goals) without taking into account potential conflicts (typically with goals that provide a safety margin against the potential for failure) can place operators in "double-" or "N-tuple-" binds. (See Woods et al., 1994, Chapter 4 for an in depth treatment of goal conflicts and organizational pressures.) In these situations, any attempt to achieve one goal involves sacrificing achievement of another goal. Multiple conflicting goals result in situations where all of the operator's degrees of freedom for action are consumed by the various demands and there is no course of action left which does not violate some goal. In these situations, operators are essentially forced to choose among the lesser of N evils, thereby committing some "error" by default.

In one tragic aviation disaster, the Dryden Ontario crash (Moshansky, 1992), several different organizational pressures to meet economic goals and organizational decisions to reduce resources created a situation where a pilot faced this kind of double bind. Deciding not to take off in deteriorating weather conditions would strand a full plane of passengers, disrupt schedules, and lose money for the carrier. Such a decision would be regarded as an economic failure with potential sanctions for the pilot. On the other hand, the means to accommodate the weather threat (de-icing equipment) were not available due to organizational choices not to invest or to cut back on equipment at peripheral airports such as Dryden. In the end, the pilot attempted to take off after an overlengthy delay despite the risk of icing, and the aircraft crashed.

The folk models that lead us to think that human error is the cause of accidents mislead us. These folk models create an environment where accidents are followed by a search for a culprit and solutions that consist of punishment and exile for the apparent culprit and increased regimentation or remediation for other practitioners as if the cause resided in defects inside people. However, these countermeasures are ineffective or counterproductive because they completely miss the systematic deeper factors that produced the multiple conditions necessary for failure. Other practitioners, regardless of motivation levels or skill levels remain vulnerable to the same systematic factors. If the erroneous action was an omission of an isolated act, the memory burdens imposed by task mis-design are still present as a latent factor ready to contribute to the same type of error. If a mode error was part of the failure chain, the computer interface design still creates the potential for others to commit the same kind of error. If a double bind was behind the actions that contributed to the failure, that goal conflict remains to perplex other practitioners.

If we examine deeper organizational and design factors and understand how they shape the cognitive and collaborative activities of people who work in the field of practice, then we can predict the kinds of human performance problems that will arise and we can learn where and how to invest to improve the system.

34.4 How Do Complex Systems Fail?

The Nature of Complex Systems

The research on disasters and the role of human performance over the last 20 years have revealed the basic form of failure in complex systems. However, before we can enter into a discussion of the characteristic signature of complex system failures, we must clarify what we mean by a complex system. Specifically, we will outline those features of modern systems which shape and contribute to the nature of the failures we see. These characteristics are becoming increasingly prevalent today as the forces of economy and safety exert ever greater pressure on system designers, managers, and operators.

Complexity

The defining characteristic of complex systems refers to the degree of interconnection and interdependence one finds among the system components (Perrow, 1984). While it is difficult to objectively measure complexity, systems that rate highly in this dimension tend to exhibit a high number of common mode connections, highly interconnected subsystems, numerous feedback loops, and interacting control parameters (Perrow, 1984; p. 88). In part, these characteristics are design responses to pressures to make systems efficient. These qualities can be exploited to make systems highly responsive to changes in demands and to certain classes of internal disturbances.

Highly integrated manufacturing systems offer a prime example of these characteristics. Individual machines and machine cells are informationally linked such that their activities co-determine one another. Thus, the arrival of a new production order or a machine breakdown somewhere in the system can produce a wave of effects in the form of rescheduling production, rerouting part flows, and reconfiguring machines and machine cells, thus allowing the system to dynamically redesign itself in response to changing conditions.

The negative aspect of these qualities is a tendency to result in system dynamics that are highly nonlinear and very difficult for operators to understand or predict accurately, especially in the presence of disturbances (Kugler and Lintern, 1995). Because of the interrelationships among parts of the system, disruptions or anomalies can produce effects that are "distant" (in a physical or functional way) from their source. Single faults can have consequences for multiple system elements. Multiple faults can simultaneously influence individual elements. All of these factors add significantly to the difficulty confronting operators attempting to assess and respond to disturbances in the system.

A typical "error" in complex systems is for operators to miss side-effects of their actions. Because control parameters and systems can interact, operator actions can often have unintended effects on other parts of the system. When these effects are not anticipated or accounted for, operators can be drawn into erroneous situation assessments. For example, an operator may judge that a new and independent fault has occurred when in fact the new indications are a result of the operator's previous actions. In the presence of one or multiple faults, system functioning may be altered such that it becomes impossible for operators to anticipate the full effects of their response actions.

Coupling

Often associated with complexity, coupling represents another dimension of modern complex systems which has direct implications for the way in which these systems fail. Perrow (1984) describes coupling as the amount of "slack" or buffering between system elements. In a tightly coupled system, effects will propagate very quickly between neighboring parts of the system. Again, this is a highly desirable property for systems that need to be sensitive and responsive to changes in demands. Such systems are also more efficient in that they shed superfluous capacity and adaptability in exchange for a more streamlined process. Just-in-time (JIT) inventory management practices represent an example where increasing the coupling in systems can result in demonstrably more efficient performance.

The price of the increased responsiveness and efficiency observed in tightly coupled systems is that the effects of disturbances propagate very quickly, with limited opportunities for intervention, and can rapidly impact overall system functions. The margins between successful system performance and system breakdown become significantly narrowed. For example, the United Parcel Service strike of August 1997 caused rapid, widespread shutdowns in the manufacturing sector due to shortages in parts supplies (a direct result of JIT inventory practices). Under exceptional circumstances (see Latent Failure model below), these characteristics can create a window of opportunity for individual events (e.g., a disturbance or erroneous operator action) to lead quickly and decisively to a large-scale failure of the system.

Uncertainty

There is an irreducible level of uncertainty inherent in complex human–machine systems. For designers, managers, and operators, the complexity of these systems is such that it is difficult if not impossible to be entirely certain of what the effects of certain decisions will be. In the context of operator decision making, uncertainty arises due to the indirect nature of most information. Operators must take into account the possibility of failed sensors, noisy information channels, inaccuracies in human-reported data, stale data, the presence of faults, etc. The complexity of the system also means that there is often a many to one mapping between root causes and the observable symptoms in data. One of the ways these factors manifest themselves is in the form of a trade-off operators must negotiate between their confidence in their assessment of a situation and the value of using resources (time, cognitive effort) to gather more information. Because system elements are highly interdependent and tightly coupled, the effects of a delay in taking action can quickly spread throughout the system. Therefore, operators will often be under considerable pressure to act to preserve the integrity of the system, even if decisions about how to act must be made on the basis of highly uncertain data.

Variability

Ashby's (1956) Law of Requisite Variety states that the possibility for functionally significant variation, or "variety," that the environment presents to a controlled system must be matched by the variety of the control system if effective regulation over the "essential variables" (e.g., productivity, safety) is to be maintained. There are two fundamental ways to ensure that the "essential variables" of a system are unaffected by unwanted variability in the environment. The first is to insulate the system by restricting and controlling the input channels which can transmit variety from the external environment to the system. Examples include the building in which a production system is housed, which serves to block variety due to the weather, to control access by the general public, etc. The second approach to protecting the essential variables is to design variety into the control system to allow it to respond to both externally and internally generated variability. This is most commonly done by attempting to anticipate significant classes of variability and designing preplanned measures specifically to recognize and respond to those types of conditions. Automated safety systems and the development of detailed standard operating procedures (SOPs) are two examples of this method.

However, this second method is necessarily limited because it is impossible, even in principle, for designers to exhaustively foresee every anomaly, every twist of circumstances, and every combination of events that may confront the system. That is, it is impossible to eliminate the potential for *unanticipated variability* in the system. In the extreme case, genuinely novel events can occur, i.e., events for which no planned responses have been developed in advance, for example, the explosion of an oxygen tank on Apollo 13. In these cases, operators must improvise a response with whatever resources are available. At the other extreme, there will be cases that match very well with previously developed plans and that can thus be handled routinely (so called "textbook cases"). However, even in these instances, the presence of SOPs, while reasonable and important, cannot reduce activity to the status of simple rote procedure following, no matter how detailed they are. (See Suchman, 1987 for an extensive discussion of the practical and theoretical limits of plans to specify completely all needed actions in advance.)

Moreover, the most frequent manifestation of unanticipated variability consists of small complicating factors (e.g., multiple faults that mask each other or suggest conflicting responses; see Roth et al., 1992 for a list of examples) that present subtle variations on "textbook cases." These situations challenge operational personnel to consider how previously developed plans and automatic responses are relevant to the unique circumstances they confront and to consider how they should be modified if critical goals are to be achieved (Woods et al., 1990). For example, despite massive efforts by the nuclear power industry to develop comprehensive and fully detailed procedures to guide operators through any conceivable scenario, Roth et al. (1992) were able to show that highly plausible scenarios involving complicating factors could be generated that presented serious challenges to the procedures.

Thus, helping the operational (human–machine) system to confront and absorb the variability of the domain is critical to producing highly robust systems. The notion that a system is reliable in the sense that it performs as designed a high proportion of the time is incomplete — first, because "a high proportion" can never be 100% and second, because circumstances *will* arise that circumvent or challenge the designed features of the system.

The Signature of Complex System Failures

Our usual conception is that a system fails when some single catastrophic event occurs that overwhelms the coping ability of the people on the scene. But because engineers and others are aware of the potential for disaster they develop multiple redundant mechanisms, safety systems, and elaborate policies and procedures to keep them from failing in ways that produce bad outcomes. The results of combined operational and engineering measures make these systems relatively safe from single point failures; that is, they are protected against the failure of a single component or procedure directly leading to a bad outcome.

The scale, complexity, and coupling of these systems create a different pattern for serious failures where incidents develop or *evolve* through a *conjunction* of several small failures, both machine and human (e.g., Turner, 1978; Pew et al., 1981; Perrow, 1984; Wagenaar and Groeneweg, 1987; Reason, 1990). This pattern has been seen in multiple disasters or incidents in a variety of different industries, and despite the fact that each critical incident is unique in many respects.

1. Incidents *evolve* toward failure.

   These incidents evolve through a series of interactions between the people responsible for system integrity and the behavior of the technical systems themselves (the engineered or physiological processes under control). One acts, the other responds, which generates a response from the first and so forth. The incident evolution can be stopped or redirected away from undesirable outcomes at various points.
2. Multiple contributors, each necessary but only jointly sufficient.

   System failures are characterized by a concatenation of several small failures and contributing events rather than a single large failure (e.g., Pew et al., 1981; Reason, 1990). The multiple contributors are all necessary but individually insufficient for the system failure to have occurred. If any of the contributing factors were missing, the failure would have been avoided. Similarly, a contributing disturbance or fault can occur without producing negative outcomes if other potential factors are not present.
3. Human–machine interaction.

   Often the multiple contributing factors include aspects of human–machine interaction.
4. Latent factors.

   Some of the factors that combine to produce a disaster are latent in the sense that they were present before the incident began (Turner, 1978). Reason (1990) uses the term *latent failures* or factors to refer to conditions resident in a system that can produce a negative effect but whose consequences are not revealed or activated until some other enabling condition is met. These conditions are latent or hidden because their consequences are not manifest until the enabling conditions occur. A typical example is a condition that makes safety systems unable to function properly if called on, such as the maintenance problem that resulted in the emergency feedwater system being unavailable during the Three Mile Island incident (The Kemeny Commission, 1979). Latent failures require a trigger, i.e., an initiating or enabling event, that activates its effects or consequences. For example, in the Space Shuttle Challenger disaster, the decision to launch in cold weather was the initiating event that activated the consequences of the latent failure in booster seal design (Rogers et al., 1986).

   Latent failures are typically associated with managers, designers, maintainers, or regulators; people who are generally not directly involved in routine operations and handling incidents and accidents. The latent failure model thus highlights the fact that the causes of system failure are much broader than simply erroneous actions on the part of front-line operators. Organizational activities such as goal-setting, planning, maintaining, and communicating shape and interact with task and environmental conditions, individual unsafe acts, and failed system defenses to produce failures.

An Example: The "Going Sour" Scenario

In the "going sour" class of accidents, an event occurs or a set of circumstances come together that appear to be minor and unproblematic, at least when viewed in isolation or from hindsight. This event triggers an evolving situation that is, in principle, possible to recover from. But through a series of commissions and omissions, misassessments and miscommunications, the human team or the human–machine team manages the situation into a serious and risky incident or even accident. In effect, the situation is managed into hazard (originally, Cook et al., 1991; cf., Woods and Sarter, 1997) Several recent accidents in aviation involving highly automated aircraft show this signature (Billings, 1996).

After the fact, going sour incidents look mysterious and dreadful to outsiders who have complete knowledge of the actual state of affairs (Woods et al., 1994). Since the system is managed into hazard, in hindsight, it is easy to see opportunities to break the progression toward disaster. The benefits of hindsight allow reviewers to comment (Woods et al., 1994, Chapter 6),

- "How could they have missed X; it was *the* critical piece of information?"
- "How could they have misunderstood Y; it is so logical to us?"
- "Why didn't they understand that X would lead to Y, given the inputs, past instructions and internal logic of the system?"

In fact, one test for whether an incident is a going sour scenario is to ask whether reviewers, with the advantage of hindsight, make comments such as, "All of the necessary data was available, why was no one able to put it all together to see what it meant?"

Luckily, going sour accidents are relatively rare even in complex systems. The going sour progression is usually blocked because of two factors:

- The expertise embodied in operational systems and personnel allows practitioners to avoid or stop the incident progression
- The problems that can erode human expertise and trigger this kind of scenario are significant only when a collection of factors or exceptional circumstances come together.

The going sour accident is one kind of latent failure scenario and illustrates how latent failures are a side effect of complexity. The latent failure signature is an evolving process in which there are several points or opportunities to detect that the system is being managed into hazard and to act to recover the situation.

34.5 Adaptation in Human–Machine Systems

Unanticipated variability is a primary example of the factors that can force operational systems to adapt, structurally and behaviorally, away from standard or canonical practices in order to cope with the potential for change. In general, adaptation is a potent concept for describing and interpreting the nature of human–machine systems, particularly with respect to its implications for how systems evolve over time and how we ought to interpret the meaning of "human error" in these systems. Rasmussen (e.g., 1990) has argued convincingly for the need to more fully understand the forces that serve to drive and limit adaptation in complex work environments, both bottom-up through the nature of the technology (e.g., human–machine interfaces) used to support work, and top-down through management practices and organizational structures. The ubiquity and power of adaptation makes it important to understand how it expresses itself in human machine systems, what factors shape the forms of adaptation we see, and particularly how adaptive processes can lead to weaknesses. By examining adaptation more closely we can hope to learn more about how to support it and encourage it in ways that are effective, coherent, and consistent with global goals.

Adaptation can be interpreted as a sort of equilibrium-seeking process, guided by subjective criteria (e.g., speed, quality, robustness against unanticipated variability, etc.), in which people attempt to dynamically match their behavior to the state of their environment. That is, based on their experience and knowledge of their environment, and on available feedback, people will actively try to locate an acceptable balance among their criteria by modifying their behavior, modifying their environment, or both. In any sufficiently complex and dynamic environment, this process will take place continually as demands and constraints shift, appear, and disappear. There will always be forces compelling people to explore new or modified ways of performing tasks. This is especially true in industries such as manufacturing, where the drive to be "flexible" and "agile" (Karwowski et al. 1997) means that the system must constantly reinvent itself to accommodate novel production runs, integrate new process technologies, and dynamically manage breakdowns in the system. Total quality management and continuous improvement strategies attempt to harness people's adaptive tendencies by promoting and capturing positive adaptations.

We take it as a basic premise then that adaptation *does* occur. In fact, it *must* occur to allow complex human–machine systems to function at all. Regardless of how well thought out any system or piece of technology is beforehand, there will always be a gap between the description of system functioning as designed on paper and the complete, situated reality of what must be done to make the system work (cf. the "irremediable incompleteness" of plans; Suchman, 1987). The job of designers and planners is to create a system that *can* work; the function of operators is to resolve the remaining degrees of freedom in ways such that the system *does* work (Rasmussen et al., 1994).

One of the most important ways in which adaptation is influenced is by the nature of the technology with which operators must interact in the performance of their tasks. Embedded within the broader environment of dynamically changing demands and constraints, an ongoing dialogue of co-adaptation between operator and technology continually takes place. This is especially apparent in the context of the introduction of new information technology into the workplace. As Rasmussen (1995) notes, "When the system is put to work, the human elements change their characteristics; they adapt to the functional characteristics of the working system, and they modify system characteristics to serve their particular needs and preferences" (p. 4). Note the dual forms of adaptation implied: not only do operators adapt to the system characteristics, they actively alter the system characteristics to suit their own criteria (Woods et al., 1994, Chapter 5). Cook and Woods (1996) have referred to these co-adaptive processes as *task tailoring* and *system tailoring*. These are fundamental processes in any human–machine work system, and are significant in shaping opportunities for "error" and system failure.

Task and System Tailoring

Operators are not passive in the process of accommodating to changes in technology. Rather, they are actively adaptive. Multiple studies have shown that practitioners adapt information technology provided for them to the immediate tasks at hand in a *locally* pragmatic way, usually in ways not anticipated by the designers of the information technology (Flores et al., 1988; Hutchins, 1990; Cook and Woods, 1996; Obradovich and Woods, 1996). In fact, human adaptive processes can often obscure the effects of poorly designed information technology because humans compensate for the weaknesses of the technology. The point is that the artifacts of information technology that designers introduce to the workplace are shaped by their users until they become useful "tools."

System Tailoring

System-tailoring types of adaptations tend to focus on shaping the technology itself to fit the pre-existing strategies of operators and the demands of the field of activity (e.g., adaptation focuses on the setup of the device, device configuration, how the device is situated in the larger context). For example, in one study (Cook and Woods, 1996), operators set up the new device in a particular way to minimize their need to interact with the new technology during high-criticality, high-tempo periods. This occurred despite the fact that the operators' configurations neutralized many of the putative advantages of the new system (the flexibility to perform greater numbers and kinds of data manipulation). Note that system tailoring frequently results in only a small portion of the "in principle" device functionality actually being used operationally. That is, operators will throw away or alter functionality in order to achieve simplicity and ease of use.

Task Tailoring

In task tailoring, operators adapt their strategies, especially cognitive processing strategies, for carrying out tasks to accommodate constraints imposed by new technology. Thus, task-tailoring types of adaptations tend to focus on how operators adjust their activities and strategies given the constraints imposed by the characteristics of the device. For example, information systems that force operators to access related data serially instead of in parallel result in a proliferation of windows and new window management tasks (e.g., searching for related data, decluttering displays as windows accumulate, etc.). Operators may tailor the device itself, for example, by trying to configure windows so that related data are available in parallel, but they may still need to tailor their activities. For example, the may need to learn when to schedule the new decluttering task (e.g., by devising external reminders) to avoid being caught in a high criticality situation where their first need is to reconfigure the display so that they can "see" what is going on in the monitored process.

Brittle Adaptations

Task and system tailoring represent examples of operators' adaptive coping strategies for dealing with clumsy aspects of new technology, usually in response to criteria such as work load, cognitive effort, ease of use, robustness to common errors, etc. The danger associated with these strategies is that adaptation based on operators' locally defined criteria can lead to "brittle" features in the larger system. In the language of adaptation, local work practices become "overspecialized" with respect to the prevailing conditions, and thus become highly sensitive and prone to failure when these conditions change. For example, data monitoring strategies developed in the context of routine operations may cause critical data to be missed in the context of a fault detection scenario. Such adaptations can thereby become a form of latent failure (Reason, 1990) within the system which can then be triggered by critical changes in the local conditions which originally shaped the adaptation.

The problem is that operator adaptation is a fundamentally local phenomenon, in terms of both time and space. In the absence of influences to the contrary, the criteria which serve to shape adaptation will tend to be applied with respect to *recent* experience and the *current* state of the *immediate* environment. The tendency towards overspecialization occurs when operators (knowingly or not) trade the ability to successfully adapt to novel conditions in exchange for better adaptation to the current, prevailing conditions. That is, there is a tendency to trade long-term adaptability for short-term efficiency and simplicity. Adaptation at higher organizational levels is subject to the same processes.

Thus, while adaptation can be a powerful force for positive change, there is no guarantee that the individually local adaptive activities of operators will result in emergently adaptive changes at a global level. Tailoring can be clever or it can be brittle. Adaptations can conflict, compete, and lead to weaknesses in the system. Therefore, to be successful in a global sense, adaptation at local levels must be guided by the provision of appropriate criteria, informational and material resources, and feedback (Rasmussen et al., 1994). In a sense, this can be seen as a primary function of management. Because operators have privileged access to the dynamic details and demands of their situated work context, management should provide the freedom and resources for operators to adapt to local conditions. But at the same time management must provide ways to constructively constrain and coordinate that adaptation such that global goals such as productivity and safety are protected.

Adaptation and Error

There is a fundamental sense in which the adaptive processes that lead to highly skilled, highly robust operator performance are precisely the same as those that lead to errors. Adaptation is basically a process of exploring the space of possible behaviors in search of stable and efficient modes of performance. Occasionally, exploration may happen unintentionally (e.g., in the case of action slips; Norman, 1981). Other times, the exploration will be a deliberate modification to existing behaviors due to pressures or changes in the environment. Such explorations are not random, but represent "educated guesses," normally based on minor modifications to existing strategies or heuristically guided generation of novel strategies and methods.

"Errors" in this context represent information about the limits of successful adaptation. Adaptation thus relies on the feedback or learning about the system which can be derived from failures, near misses, and incidents. A simple example is the ubiquitous "speed–accuracy trade-off" observed in manual tasks. All other things being equal, it is generally true that as the speed at which a manual task is attempted increases, there comes a point at which the accuracy or quality of the task performance begins to degrade. This point may change as skill increases, but within local bounds this relationship places a practical limit on the speed with which the task can be performed. If the process of adaptation in a manual task is guided by the simultaneous criteria of speed and accuracy, then each individual must attempt to locate an acceptable balance between these two factors. Generally though, the *only* way to locate this point is to continue to increase speed until accuracy begins to suffer; i.e., "errors" begin to occur.

Adaptation with respect to more complex cognitive activities and work practices are subject to the same basic processes. Operators' limited access to the state of the environment, compounded by the inherent variability and uncertainty of the world means that every modification to established behaviors has the potential to result in a negative outcome, particularly if the system is operating at or near acceptable performance limits (see below). If the result is positive, we tend to call it "successful adaptation" and reward the operator. If the result is negative, we tend to call it "human error" and begin remedial action. The point is that in order to reap the rewards of the power of human adaptive processes, *people must be allowed to be wrong.* "Zero tolerance" attitudes toward error, or policies that enforce strict adherence to SOPs in an attempt to eliminate errors, create a double bind by forcing operators to persist in established practices, despite whatever forces of change in the environment may be pushing them to adapt.

Boundary-Seeking Behavior

Rasmussen et al. (1994) have described how complex human–machine systems will naturally tend to migrate toward the boundaries of safe/acceptable performance (i.e., boundaries beyond which significant negative consequences become imminent). In response to pressures of efficiency and work load, managers and operators will naturally adapt in ways that push the system nearer to the edges of its performance envelope. (See Rasmussen et al., 1994, p. 149 for a graphical depiction of this phenomenon.) In a resource-constrained, competitive environment, organizations are rewarded for operating as closely as possible to the edges of system performance limits in pursuit of maximum efficiency, productivity, etc. However, consistent operation near the boundaries of performance means that the system is continually vulnerable to the effects of critical changes in the environment or to erroneous operator actions. In critical situations, events that might otherwise be relatively benign can in fact nudge the system into an unsafe or unacceptable region of performance.

When changes in technology and organizational structures are introduced, the shape of the performance envelope is changed. Some weak points are eliminated; new weak points are created; the nature of weak points can change. The result is not simply a system that is equally safe but more productive. For example, technological changes designed to increase efficiency or safety often do so at the price of increased complexity. This can drastically change the nature and location of the system's performance limits. Moreover, competitive stresses mean that the benefits of changes tend to be taken in productivity gains rather than as increased safety margins. Thus, the system remains vulnerable to failures, but not necessarily the same failures. The system may fail in new, unexpected ways; it may fail more suddenly; the consequences of failure may be more severe. The point is that system performance boundaries are always changing, and that the forces of adaptation will tend to seek out those boundaries. The key to a system that is both efficient and robust in the long run lies not in artificially constraining operators to established work practices, but in allowing them to sense and become familiar with system performance boundaries so that they can recognize when performance limits are being approached and know how to recover when they have been crossed.

34.6 Error Tolerance, Error Recovery, and High Robustness Systems

The implication of the preceding discussion is that systems must be able to cope effectively with the nature of operating at the boundaries of system performance (Rasmussen, 1997). Ultimately, system reliability is measured by outcomes. That is, evaluations of system performance depend not so much on whether the system operated optimally in every detail, but whether the "essential variables" (Ashby, 1956), such as those defined by production or safety goals were satisfactorily kept in their desired ranges. One must accept that systems will experience disturbances, some due to operator actions, which push the system into unacceptable states where negative consequences become imminent. Efforts to improve system reliability cannot focus exclusively on preventing these disturbances because of our limited ability to predict for all eventualities. To achieve a truly robust system, the key is to incorporate mechanisms that allow the system to detect, recover from, and absorb the effects of errors.

Rasmussen (1986) has described the concept of "unkind" work environments, where the effects of errors lead quickly and decisively to failures. Recovery intervals (the period during which the effects of errors can be reversed) are short, errors are difficult to detect, the system degrades quickly, and the consequences are severe. The variabilities in behavior and performance which normally drive adaptive processes become "an unsuccessful experiment with unacceptable consequences" (Rasmussen, 1986). In these sorts of environments, the natural tolerance to errors is very low. What is needed are measures that make the environment "kinder" by supporting not only avoidance, but the detection of and recovery from errors.

> *"The ultimate error frequency largely depends upon the features of the work interface which support immediate error recovery, which in turn depends on the observability and reversibility of the emerging unacceptable effects. The feature of reversibility largely depends upon the dynamics and linearity of the system properties, whereas observability depends on the properties of the task interface which will be dramatically influenced by the modern information technology"*
>
> (Rasmussen, 1985; p. 1188).

Human–Machine Cooperation

Studies of highly reliable, high-performance organizations (e.g., Rochlin et al., 1987; Seifert and Hutchins, 1992; Hutchins, 1990, 1995) have shown that the processes that support error detection and recovery are fundamentally *cooperative* and *distributed.* These studies reveal that reliability *emerges,* not because the individual agents never commit errors, but because the cooperative structures in the system (e.g., shared workspaces, cross-checking strategies) provide mechanisms for catching errors and correcting them before negative consequences begin to accrue. For example, miscommunications between air traffic control and aviation crews are relatively common, but the air transport system has evolved robust cooperative processes such as crew cross-checks and readbacks which help ensure that these errors are revealed and corrected quickly. The reliability of the larger system is not a function of the reliability of the agents but rather of the *ways in which the agents interact.*

The prevalence of advanced automation in human–machine systems makes it important to ask how well automated agents support the cooperative processes which produce highly robust performance in distributed human teams. We must recognize that introducing advanced automation into a system is not simply a matter of substituting machine activity for human activity. Automation changes the cooperative structure of the system and the patterns of communication and coordination that occur among the human and machine agents in the system. Patterns of errors will be changed; some types of error will be eliminated, some will be created. What is needed are automated systems which consider the larger cooperative structure in which they are implemented (Sarter et al., 1997). To make automation support robust systems we must pay close attention to the question of how well such systems support the interactive activities that form the basis for error detection and recovery.

The answer, unfortunately, is often not very well. The properties of advanced automation have tended to hinder cooperation with human partners rather than help it. Woods (1996) has observed that automated systems tend to be:

Strong. Much advanced automation has the ability to act with considerable autonomy. That is, it is capable of performing extended sequences of tasks without any direct intervention from human operators. Closely related to autonomy is the issue of authority. This refers to the automation's ability to *independently* assess when a situation calls for intervention based on its own internal criteria, and to take control of the situation if it finds it to be warranted.

Silent. Many automated systems are "silent" in the sense that they provide little feedback about their activities. This makes it very easy for operators to lose track of the state of the automation. Particularly when combined with the properties of "strong" automation as outlined above, this creates the potential for "automation surprises" (Sarter et al., 1997). These are situations where operators have lost track of the activities of their automated partners and are "surprised" by the actions of the automated system. Operators can be left asking questions of the automation such as "What is it doing?" "Why did it do that?" "What is it going to do next?" (Weiner, 1989). All of these are symptoms of a general breakdown in coordination among the human and the automated system.

Clumsy. One of the putative benefits of the use of automation is an expected decrease in work load for human operators. However, rather than smoothing out the peaks and troughs in operator work load or reducing it in an absolute sense, many times automation has been found to simply amplify patterns of work load in a syndrome termed "clumsy automation" by Wiener (1989). That is, work load reductions tend to occur during periods in which work load was already low, while additional work load tends to appear in periods where work load is already high and the consequences of breakdowns are greatest. It is at these times that the automated system demands the most of the operator in terms of providing input and coordinating activities, leaving the user in the paradoxical position of needing the automated system's help but not having the time or attentional resources to help the automation do so.

Difficult to Direct. Uncooperative automated systems make it difficult for users to interrupt and/or redirect the automation's activities if the operator recognizes a need to do so. If the automation is not designed to support cooperative problem solving, a typical result is that the user's only option for intervening is to essentially "turn off" the automated system and take over the problem in its entirety and full complexity.

Strong, silent, clumsy, difficult to direct automated systems act as uncooperative partners rather than resources adapted to support people as situations vary in tempo and criticality. Such uncooperative interactions in both human–human and human–machine interaction have been recognized as a latent factor in disasters (Billings, 1996). Uncooperative partners lead to miscommunications and misassessments which push situations toward greater hazard, which retard the detection of the deteriorating situation, and impair initiation of recovery strategies.

34.7 The Complexity of Human Error

We have attempted to convey some of the flavor of the results of the past two decades of research into the ways that complex human–machine systems fail and the role of human error. Fundamentally, research has shown that human performance, including errors, is shaped by systematic factors such as task design, human computer interaction, and organizational influences. Reacting to failures as if they were strictly a function of inherent human fallibility prevents us from examining the deeper factors that shape performance and lead to errors and system failures. Although we do not yet understand precisely how all of these factors interact, our ability to predict the sorts of errors that will occur given certain system characteristics is very good.

Pressures of economy and safety have driven up the complexity of modern human–machine systems and have led to a new signature of failure characterized by the presence of multiple contributors, including factors latent in the system. The pressures of the environment, unanticipated variability, and the potential for failure cause people to adapt their behavior and available technology in ways that can be both positive and negative. While adaptation drives much of the behavior we see, it is based on feedback about the limits of successful performance. Moreover, competitive stresses will encourage systems to migrate toward these limits.

A constructive response to failures, near misses, and incidents involves a search for the vulnerabilities, constraints, pressures, and dilemmas behind the label "human error." In the end, improving the system lies in helping people in various roles and at different levels of an organization adapt successfully to these demands of their field of practice. This is best done by making systems error tolerant and by supporting detection and recovery from errors before negative consequences result.

The Paradox of Simultaneous Success and Vulnerability

Ultimately, the difficulties and controversies in dealing with human error arise from seemingly paradoxical effects of our efforts to improve system performance. The highly successful efforts of engineers in protecting systems against catastrophic single-point failures have lead to systems that are "safer" in an actuarial sense, but which at the same time are vulnerable in new ways — the latent failure model and going sour incidents. The efforts to improve systems seem to produce unanticipated side effects on human performance as the systems increase in complexity. The characteristics of complex system failures are perplexing — although individual failures become less frequent, the consequences of failures are more severe. Failures may be fewer, but they more visible and more dreadful to stakeholders.

Effort after Success

It is common for us to think of complex technical systems as inherently safe. When problems occur, we are lead to believe that the reason is that the system was somehow interfered with or prevented from operating *as designed*. This belief leads to a search for the (presumably human) guilty parties, and simultaneously blinds us to the tremendous effort invested by human operators in the (normally) successful performance of the system. The truth is that such systems are inherently hazardous. People and organizations are aware of these hazards and actively work to devise defenses against them. This effort after success is needed continuously. When these efforts break down and we see a failure, we gain information, not about the innate fallibilities of people, but about the nature of the threats to complex systems and the limits of the defenses we have put in place.

We can thus begin to see how attributing complex system failures to mere "human error" represents a basic misunderstanding of the nature of complex systems and the relationship between errors and system failure. The story behind human error is every bit as complex as the environments in which it occurs. Human operators behave in ways that are rational given the demands imposed upon them and the combination of cognitive, technological, and organizational resources at their disposal. The typical mismatches between these demands and resources cause operators to develop coping strategies; they structure their environment; they adjust and adapt their behavior to make systems work. Errors are not symptoms of random variability in human performance, but rather represent what is simply the most visible evidence of the mismatches between demands and resources which human ingenuity and diligence hide from us most of the time.

References

Ashby, W. R. (1956). *An Introduction to Cybernetics.* New York, NY: Wiley.

Billings, C. E. (1996). *Aviation Automation: The Search for a Human-Centered Approach.* Hillsdale, NJ: Erlbaum.

Byrne, M. D. and Bovair, S. A working memory model of a common procedural error. *Cognitive Science,* 21(1), 31-61, 1997.

Cook, R. I. and Woods, D. D. (1996). Adapting to new technology in the operating room. *Human Factors,* 38(4), 593-613.

Cook, R. I., Woods, D. D., and McDonald, J. S. (1991). *Human Performance in Anesthesia: A Corpus of Cases.* (CSEL Technical Report 91-TR-03). Columbus, OH: The Ohio State University, Department of Industrial and Systems Engineering, Cognitive Systems Engineering Laboratory.

Fischhoff, B. (1975). Hindsight-foresight: The effect of outcome knowledge on judgment under uncertainty. *Journal of Experimental Psychology: Human Perception and Performance,* 1(3), 288-299.

Fischhoff, B. (1982). For those condemned to study the past: Heuristics and biases in hindsight, in D. Kahneman, P. Slovic, and A. Tversky (Eds.), *Judgment under Uncertainty: Heuristics and Biases.* Cambridge, England: Cambridge University Press.

Fitts, P. M, and Jones, R. E. (1947). *Analysis of Factors Contributing to 460 "Pilot-Error" Experiences in Operating Aircraft Controls* (Memorandum Report TSEAA-694-12). Wright Field, OH: U.S. Air Force Air Materiel Command, Aero Medical Laboratory.

Flores, F., Graves, M., Hartfield, B., and Winograd, T. (1988). Computer systems and the design of organizational interaction. *ACM Transactions on Office Information Systems,* 6, 153-172.

Hollnagel, E. (1993). *Human Reliability Analysis: Context and Control.* London: Academic Press.

Hutchins, E. (1990). The technology of team navigation, in J. Galegher, R. Kraut, and C. Egido (Eds.), *Intellectual Teamwork. Social and Technical Bases of Cooperative Work.* Hillsdale, NJ: Erlbaum.

Hutchins, E. (1995). *Cognition in the Wild.* Cambridge, MA: MIT Press.

Karwowski, W., Warnecke, H. J., Hueser, M., and Salvendy, G. (1997). Human factors in manufacturing, in G. Salvendy (Ed.), *Handbook of Human Factors and Ergonomics (2ed.),* (pp. 1865-1925). New York, NY: Wiley.

Kemeny, J. G. et al. (1979). *Report of the President's Commission on the Accident at Three Mile Island.* New York: Pergamon Press.

Kugler, P. N. and Lintern, G. (1995). Risk Management and the Evolution of Instability in Large-Scale, Industrial Systems, in P. Hancock, J. Flach, J. Caird, and K. Vicente (Eds.), *Local Applications of the Ecological Approach to Human–Machine Systems* (*Vol. 2*), (pp. 416-450). Hillsdale, NJ: Erlbaum.

Lewis, C. and Norman, D. A. (1986). Designing for error, in D. A. Norman and S. W. Draper (Eds.), *User-Centered System Design: New Perspectives on Human–Computer Interaction* (pp. 411-432). Hillsdale, NJ: Erlbaum.

Monk, A. (1986). Mode errors: A user centered analysis and some preventative measures using keying-contingent sound. *International Journal of Man-Machine Studies,* 24, 313-327.

Moshansky, V. P. (1992). *Final Report of the Commission of Inquiry into the Air Ontario Crash at Dryden Ontario.* Ottawa, Canada: Ministry of Supply and Services.

Norman, D. A. (1981). Categorization of action slips. *Psychological Review,* 88, 1-15.

Norman, D. A. (1988). *The Psychology of Everyday Things.* New York: Basic Books.

Obradovich, J. H. and Woods, D. D. (1996). Users as designers: how people cope with poor HCI design in computer-based medical devices. *Human Factors,* 38(4), 574-592.

Perrow, C. (1984). *Normal Accidents: Living with High Risk Technologies.* New York, NY: Basic Books.

Pew, R. W., Miller, D. C., and Feehrer, C. E. (1981). *Evaluation of Proposed Control Room Improvements Through Analysis of Critical Operator Decisions* (NP-1982). Palo Alto, CA: Electric Power Research Institute.

Rasmussen, J. (1990). The role of error in organizing behavior. *Ergonomics,* 33(10/11), 1185-1199.

Rasmussen, J. (1985). Trends in human reliability analysis. *Ergonomics,* 28(8), 1185-1196.

Rasmussen, J. (1986). *Information Processing and Human–Machine Interaction: An Approach to Cognitive Engineering.* New York: North-Holland.

Rasmussen, J. (1995). The concept of human error and the design of reliable human–machine systems. (Invited address) *1st Berliner Workshop on Man-Machine Systems,* Berlin, October 1995.

Rasmussen, J. (1997). Risk management in a dynamic society: A modeling problem. *Safety Science,* 27(2/3), 183-213.

Rasmussen, J., Pejtersen, A. M., and Goodstein, L. P. (1994). *Cognitive Systems Engineering.* New York, NY: Wiley.

Reason, J. (1997). *Managing the Risks of Organizational Accidents.* Aldershot, U.K.: Ashgate.

Reason, J. (1990). *Human Error.* Cambridge, England: Cambridge University Press.

Rochlin, G., Laporte, T. R., and Roberts, K. H. (1987). The self-designing high reliability organization: Aircraft carrier flight operations at sea. *Naval War College Review,* (Autumn), 76-90.

Rogers, W. P. et al. (1986). *Report of the Presidential Commission on the Space Shuttle Challenger Accident.* Washington, D.C.: Government Printing Office.

Roth, E. M., Mumaw, R. J., and Pople, H. E. (1992). Enhancing the training of cognitive skills for improved human reliability: lessons learned from the cognitive environment simulation project, in *Proceedings of the IEEE Fifth Conference on Human Factors in Power Plants.* Monterey, CA: IEEE.

Seifert, C. M. and Hutchins, E. (1992). Error as opportunity: Learning in a cooperative task. *Human-Computer-Interaction,* 7, 409-435.

Sellen, A. J., Kurtenbach, G. P. and Buxton, W. A. S. (1992). The prevention of mode errors through sensory feedback. *Human-Computer-Interaction,* 7, 141-164.

Sarter, N. B. and Woods, D. D. (1995). "How in the world did we get into that mode?" Mode error and awareness in supervisory control. *Human Factors,* 37, 5-19.

Sarter, N. B., Woods, D. D., and Billings, C. E. (1997). Automation surprises, in G. Salvendy (Ed.), *Handbook of Human Factors and Ergonomics (2ed.),* (pp. 1926-1943). New York, NY: Wiley.

Senders, J. and Moray, N. (1991). *Human Error: Cause, Prediction, and Reduction.* Hillsdale, NJ: Erlbaum.

Suchman, L. A. (1987). *Plans and Situated Actions: The Problem of Human–Machine Communication.* New York, NY: Cambridge University Press.

Turner, B. A. (1978). *Man-Made Disasters.* London: Wykeham.

Wagenaar, W. and Groeneweg, J. (1987). Accidents at sea: Multiple causes and impossible consequences. *International Journal of Man-Machine Studies,* 27, 587-598.

Weiner, E. L. (1989). *Human Factors of Advanced Technology ("Glass Cockpit") Transport Aircraft.* (NASA Contractor Report 117528). Moffett Field, CA: NASA-Ames Research Center.

Woods, D. D. (1996). Decomposing automation: apparent simplicity, real complexity, In R. Parasuraman and M. Mouloula, (Eds.), *Automation Technology and Human Performance: Theory and Applications,* (pp. 3-17). Hillsdale, NJ: Erlbaum.

Woods, D. D. and Sarter, N. B. (1997) *Learning from Automation Surprises and Going Sour Accidents.* (CSEL Technical Report 97-TR-05). Columbus, OH: The Ohio State University, Institute for Ergonomics, Cognitive Systems Engineering Laboratory.

Woods, D. D., Roth, E. M., and Bennett, K. B. (1990). Explorations in joint human–machine cognitive systems, in S. Robertson, W. Zachary, and J. Black, (Eds.), *Cognition, Computing and Cooperation.* Norwood, NJ: Ablex.

Woods, D. D., Johannesen, L. J., Cook, R. I., and Sarter, N. B. (1994). *Behind Human Error: Cognitive Systems, Computers, and Hindsight.* CSERIAC State-of-the-Art-Report. Crew Systems Ergonomics Information Analysis Center: Wright-Patterson AFB, OH.

35

Human and System Reliability Analysis

Joseph Sharit
The University of Miami

35.1 The Evolution of Perspectives to Human and System Reliability Analysis

In most scientific applications, the reliability of a system is formally expressed as the probability that the system performs its intended objective. When systems consist only of machine or material components, mathematical methods exist for computing or estimating component or system reliability, either in terms of the probability of functioning normally each time it is used or in terms of the probability that it will not fail during some prescribed time of use (Kapur and Lamberson, 1977; Dhillon and Singh, 1981).

However, when humans are integral to system operation, as is the case in many of today's systems, the problem of reliability assessment is altered. First, there is the need to assess human reliability. Second, the assessment of system reliability must now take into account the effects, both immediate and delayed, of human behaviors on other components of the system, including other humans. Although a reasonable understanding of human reliability is pivotal to evaluating system reliability, as human–system interactions become more complex the balance in emphasis will likely shift to methods that focus on how these interactions can adversely affect system reliability, even in the absence of identifiable human reliability problems or human errors.

With engineering reliability methods having matured much earlier than methods of human reliability analysis (HRA), it is not surprising that approaches to HRA initially emphasized computing probabilities of human error. This perspective enabled computations of system reliability based on probability assessments of all individual system components — mechanical and human components alike. However,

0-8493-2641-9/99/$0.00+$.50

improved understanding of cognitive and sociotechnical processes and development of approaches to the analysis of work settings has led to a shift in emphasis, away from computing human error probabilities (HEPs) and toward understanding how and why human errors occur. The underlying premise of this qualitative perspective to HRA is that with this understanding we can better predict the potential for particular types of human errors as well as better understand why accidents or adverse system consequences occur, thereby providing insights into system reliability and the design of countermeasures that are not necessarily apparent from application of quantitative techniques. However, this perspective also places much greater demands on analysts to identify and analyze the complex texturings underlying work contexts that are capable of triggering human errors or propagating human actions into adverse system outcomes.

Paralleling this qualitative or systems perspective was a change in safety philosophy adopted by some industries. The traditional safety perspective emphasized a blame culture whereby accidents were considered the fault of the worker who either violated procedures or was not exercising proper caution. This perspective was gradually being replaced in some corporations by safety philosophies that recognized the role of organizational structure and management in shaping work environments that promote human errors and violations (Reason, 1990, 1995). Although quantitative approaches to HRA also recognized such factors, they were used either to modify or directly assess the probability of human error. In contrast, the qualitative perspective to HRA was primarily interested in understanding the link between these factors and human behaviors that result in errors, violations, and inadequate performance.

35.2 The Current Dilemma Facing Industries

As a result of these developments, there are currently many options available to industries who are interested in human and system reliability, including the consideration of hybrid approaches that combine both quantitative and qualitative perspectives. First, industries must determine the perspective that is most consistent with their goals and interests. If the industry is primarily concerned with meeting regulatory standards that require "quantitative risk assessments" — i.e., that require the derivation of probabilities associated with potentially hazardous system events — then a quantitative perspective to human and system reliability will be required. However, if an industry is concerned about a growing culture of procedural violations that have potentially important implications for worker safety and system productivity, then a qualitative perspective emphasizing sociotechnical considerations will likely be more appropriate.

Even if an appropriate or dominant perspective can be identified, decisions still need to be made concerning the degree of depth to which human (and system) reliability analysis should be pursued. For example, with quantitative approaches to HRA, methods range from relatively quick assessments to very involved analyses. These decisions will be based, in part, on the availability of resources and the expertise of analysts. Decisions also need to be made concerning the degree to which other perspectives are to be incorporated. The current pressures industries face, which include the need to reduce accidents, litigation, and worker compensation costs while increasing both the quantity and quality of the product, often demand a well-balanced, multifaceted perspective to the problem of human and system reliability analysis.

Industries also need to consider issues of training, data collection, and data organization required for maintaining and improving HRA programs that are put in place. For example, if an industry's objective is to improve the design of written work procedures in order to minimize human errors that may occur during execution of these procedures, some form of training program will likely be necessary that enables designers to anticipate the potential for such errors. Such anticipation will necessarily require some understanding of the interrelationships between work contexts and tendencies humans have in processing information that form the basis for various types of errors. Finally, the ability of an industry to determine whether they have adopted an appropriate perspective (or "model") and have the capability to continuously improve human and system reliability through the implementation of design strategies requires having mechanisms in place that dictate what information should be collected, how it should be organized, and how it should be evaluated in order to determine what adjustments may be needed to the human and system reliability programs (Center for Chemical Process Safety, 1994).

35.3 Hazard Analysis Techniques and Quantitative Risk Assessments

It is perhaps misleading to suggest that qualitative and quantitative approaches to HRA are necessarily mutually exclusive. When one considers that HRA generally encompasses human error identification, prediction, and reduction, some degree of qualitative emphasis is necessarily implied. The extent to which a quantitative perspective is adopted will likely depend on the extent to which the HRA will serve as input to a quantitative risk assessment. Depending on the industry, this type of system reliability assessment is typically referred to as a probabilistic safety analysis (PSA) or probabilistic risk assessment (PRA). Most industries that carry out such assessments, such as the chemical processing or nuclear power industries, are concerned about hazards arising from interactions between hardware and software failures, environmental events, and human errors and violations that can result in injuries, fatalities, and damage to the plant and environment.

A variety of well-known hazard analysis techniques have been developed that can be used to identify hazards (Center for Chemical Process Safety, 1992). These methods are usually differentiated based on the stage in the system design life cycle on which the analysis is being performed. At the conceptual and preliminary design stages of system development, Preliminary Hazard Analysis, Failure Mode and Effects Analysis, and HAZOP techniques are often used. These approaches, by and large, enable design solutions to be implemented, including those incorporating human factors and ergonomics design principles, for the various undesirable consequences of hazards that were identified. PRAs, however, are usually more concerned with undesirable events at the later stages of the system's life cycle that cannot be resolved by these methods.

The two primary hazard analysis techniques that have become associated with PSAs and PRAs are Fault Tree (FT) Analysis and Event Tree (ET) Analysis. Both have as starting points undesirable events to which hazards contribute. These events may have been identified through informal methods such as expert opinion, or from previous hazard analysis techniques such as HAZOP which were unable to further resolve the hazard. The FT, however, represents a deductive, top-down decomposition of this event, whereas the ET represents an inductive analysis that determines how this event can propagate, and consequently the ways in which the event can be circumvented. In both cases, human interventions in combination with other system components and environmental factors can be assessed.

Figure 35.1 illustrates an ET for an offshore emergency shutdown scenario. This particular ET addresses the sequence of human actions in response to the initiating event and is therefore often referred to as an Operator Action Event Tree. Each branch of the tree represents either success (the upper branch) or failure (represented in this diagram as a HEP) in achieving the required actions specified along the top. The probability of each failure state on the right is the product of the error and/or success probabilities of each branch leading to that failure state, where the overall probability of failure is the sum of the individual failure states. Dashed lines indicate paths through which recovery from previous errors can occur.

A portion of an FT developed by Ozog (1985) associated with loading a flammable liquid storage tank from a tank truck is illustrated in Figure 35.2 (Center for Chemical Process Safety, 1989). As indicated in this figure, FTs are essentially Boolean logic models that depict the relationships between events in a system — either hardware, human, or environmental — that lead to a final outcome or top event which, in principle, could be either desirable or undesirable. Qualitative applications of FTs emphasize the different combinations of events that could lead to the top event. For many applications, this information is revealing in its own right. As a quantitative method, "basic events" or events for which no further analysis of the cause is carried out, are assigned probabilities which can either be static or in the form of rate measures that take into account dynamic considerations. Methods are then employed to propagate these values into either a probability or rate measure associated with the top event (Dhillon and Singh, 1981). These methods also enable the individual contributions of events to the top event to be computed, making this technique very suitable for performing cost-benefit analyses that can lead to design interventions.

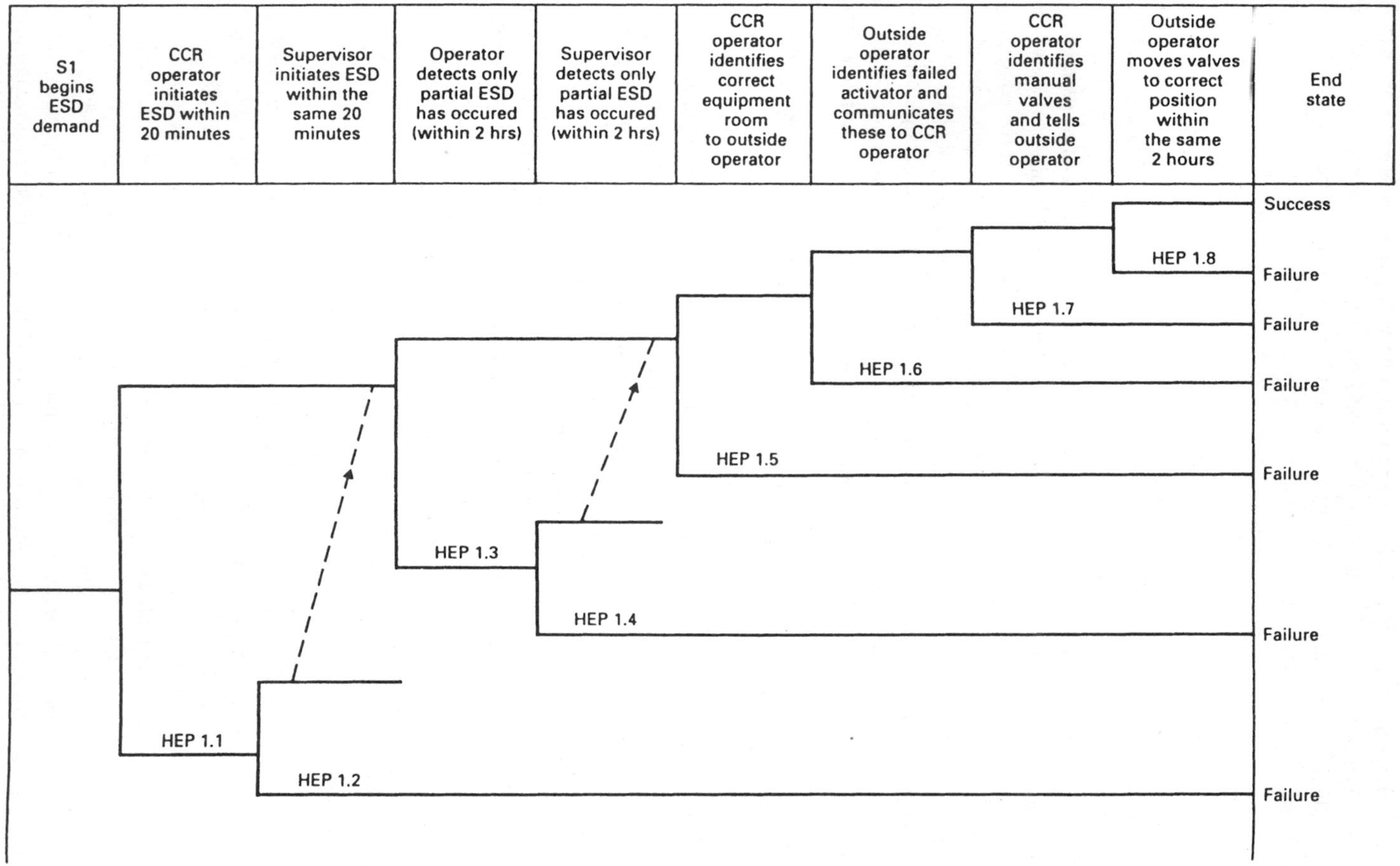

FIGURE 35.1 An Operator Action Event Tree. ESD refers to emergency shutdown procedure, and CCR refers to chemical control room. (From Kirwan, B. (1994). *A Guide to Practical Human Reliability Assessment.* London: Taylor & Francis. With permission.)

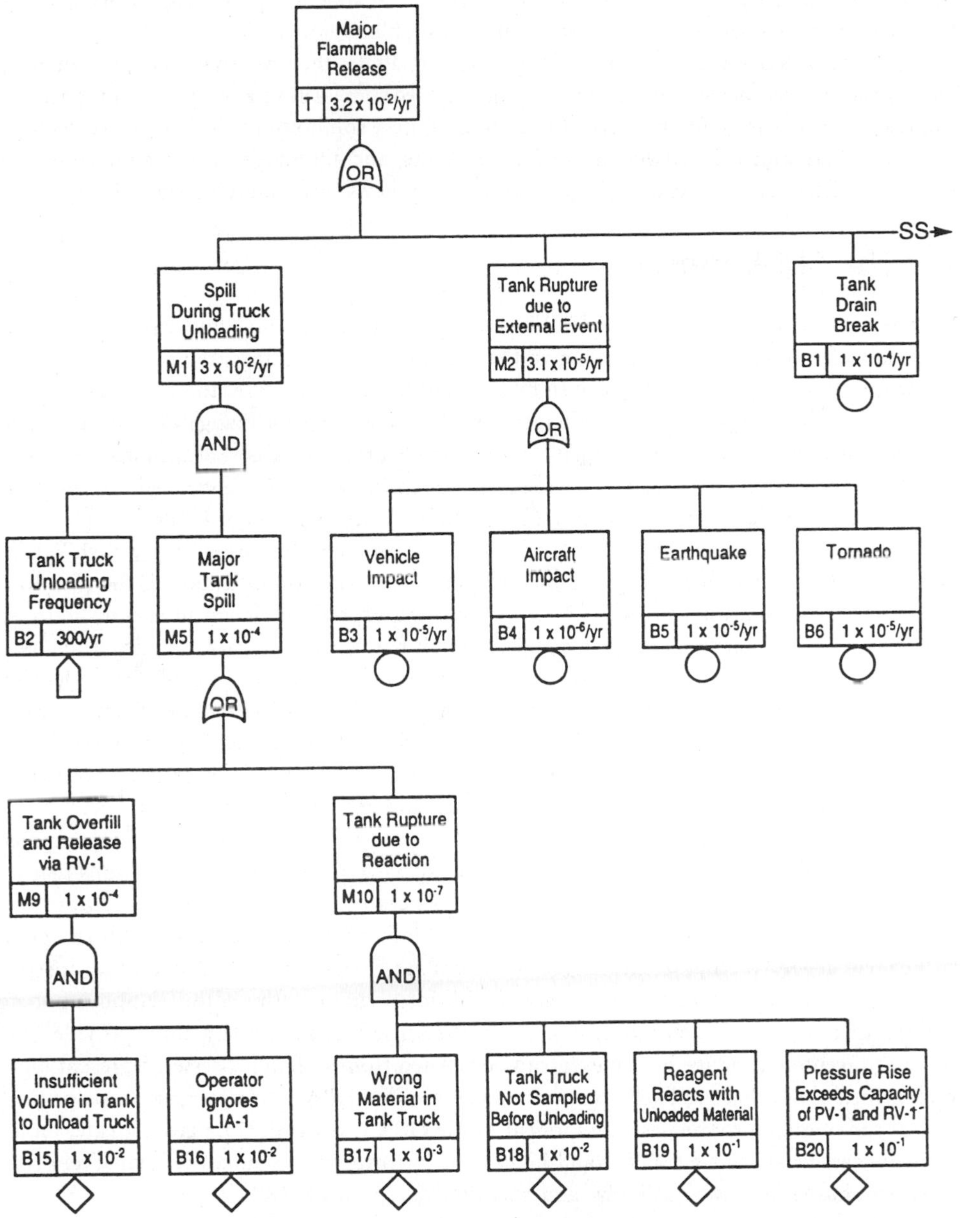

FIGURE 35.2 A fault tree for analysis of a major release of a flammable liquid. (From Ozog, H. (1985) Hazard identification, analysis, and control. *Chemical Engineering*, February 18, 161-170. With permission.)

FTs often differ in their depictions due to subtle distinctions in symbology (Roland and Moriarty, 1983). In Figure 35.2, the circles under the rectangles denote basic events; the house symbol denotes a normal event that either occurs or does not; and the diamond symbol denotes an undeveloped event that the analyst chooses not to analyze further but one that, in theory, is at a sufficiently high level in the decomposition process that it can be further analyzed. Events for which these symbols are not present represent intermediate events that are further analyzed through successive levels of causality as described by logical AND and OR relationships between events.

In those cases where the top event in an FT represents the initiating event potentially leading to accidental conditions in an ET, Cause Consequence Charts which combine FTs and ETs can be used

(CCPS, 1992). Although potentially cumbersome, they are capable of representing both the deductive and inductive analyses associated with a particular undesirable event.

When FTs and ETs are used as a basis for performing PSAs, they necessarily require quantitative solutions. However, regardless of whether FTs or ETs are used there will be a need to quantify the human error contributions to the event of interest. In addition to these solutions, PSAs also involve determining the risks associated with the consequences of these events, and ultimately whether the results of the analysis are consistent with the industry's (or regulatory agency's) acceptable risk criteria.

35.4 The HRA Process

The HRA process as viewed by Kirwan (1994) is presented in Figure 35.3. In this depiction, the qualitative–quantitative emphasis issue is addressed in the problem definition stage, representing the first step in the HRA process. If the HRA is not driven by the need to perform a PSA, the emphasis of the HRA will likely be qualitative. A quantitative component may be useful for prioritizing different human errors, especially in meeting the objectives of safety improvement programs that have limited resources for implementing risk reduction strategies. However, such quantitative components need not invoke many of the assumptions that underlie approaches that emphasize quantification of HEPs.

Clearly, the most important aspect of HRA following problem definition, and which qualitative approaches are extending to increasingly greater levels of depth and sophistication, is the identification and modeling of human errors (Figure 35.3). In general, this aspect remains insufficiently addressed in HRAs driven by PSAs, where the balance in emphasis on human errors is toward quantification. As noted earlier, there is a growing concern in many high-risk systems (such as health care) for understanding adverse system outcomes that may not arise from human error *per se,* but rather from the complex couplings between organizational, individual, collaborative work group, and environmental factors, including human tendencies for errors and violations (Sharit et al., 1996). In these cases, HRAs and not PSAs will determine which situations should be analyzed and how they should be assessed.

For HRAs that are driven by PSAs, a number of issues need to be addressed, including specification of PSA criteria (such as the target criteria for the PSA) and the types of human interactions that will be dealt with (e.g., whether human-initiated accident sequences will be considered in addition to those initiated by hardware failures or environmental events). Many of these issues relate to the scope of the HRA, which will depend on the system's degree of vulnerability to human error. Systems that are complex in terms of the underlying system design processes, highly interactive in terms of the relationships between subsystems and components, and tightly coupled in the sense that such interactions provide minimal flexibility in preventing negative system events to be isolated from system processes, represent high-risk systems (Perrow, 1984) that are likely to require more detailed HRAs. This increase in scope, in turn, implies that the human error analysis component of the HRA (Figure 35.3) be given greater emphasis. It is in these cases where justification would likely exist for investing resources in high-level or hybrid HRAs that emphasize both the qualitative and quantitative aspects of HRA.

Task Analysis, Cognitive Task Analysis, and System Analysis

Most HRAs, whether quantitative or qualitative in emphasis, require some form of organizing structure for describing human involvement with other system components in order to identify human behaviors that could potentially lead to undesirable consequences. This would also provide a basis for suggesting error reduction design interventions and for conducting follow-up assessments of these interventions. Task analysis (TA) represents the most formal approach to this problem. Many TA methods exist that differ primarily on the basis of the types of information acquired and in the ways in which this information is represented. These distinctions, in turn, depend on the types of problems the analyst is interested in. A particular type of TA technique referred to as hierarchical task analysis (HTA) can readily organize large amounts of task-based information (Stammers and Shepherd, 1995), making it particularly suitable for HRAs involving high-risk, but relatively routine tasks. An illustration of the use of HTA in HRA will

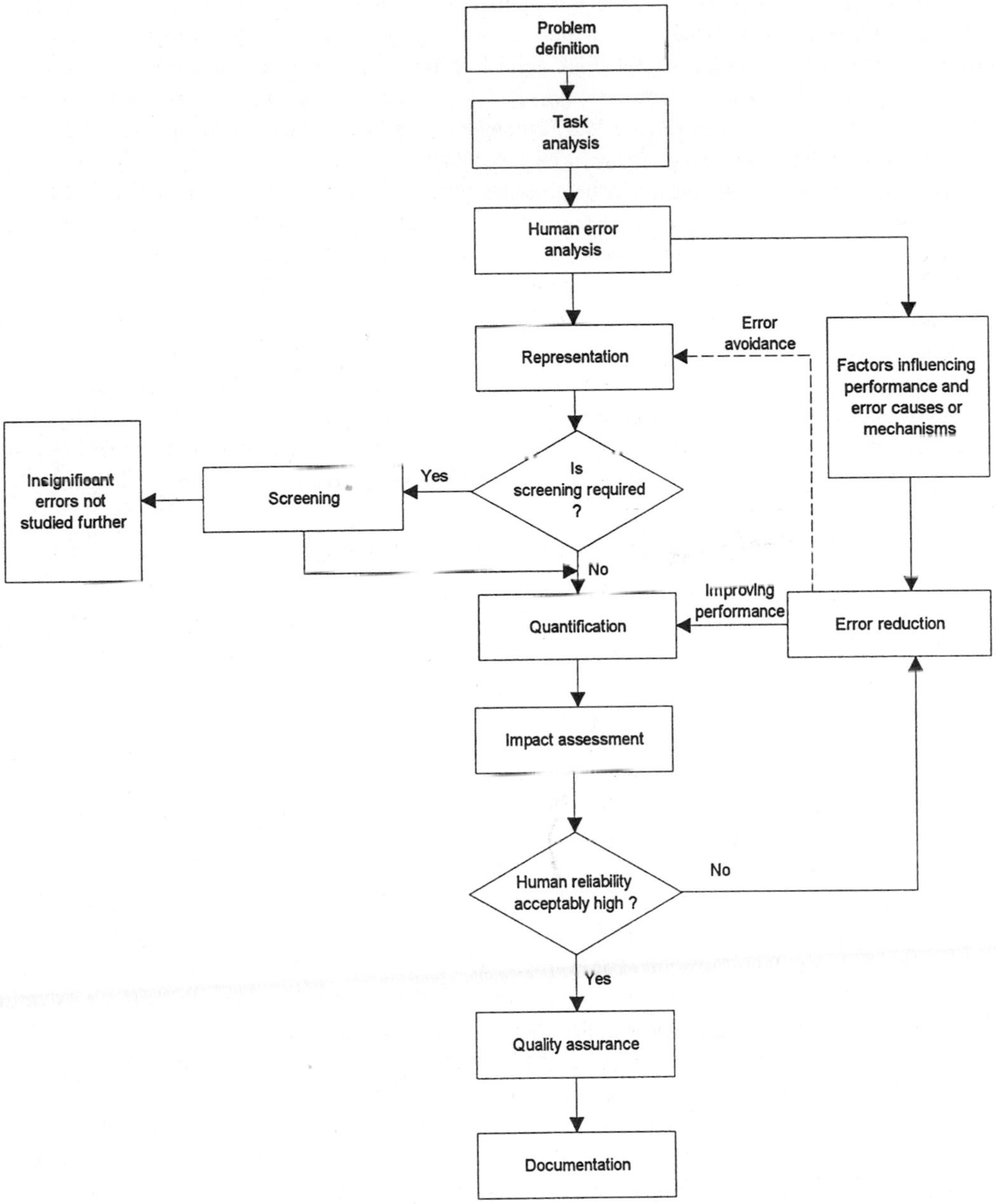

FIGURE 35.3 The HRA process. (From Kirwan, B. (1994). *A Guide to Practical Human Reliability Assessment.* London: Taylor & Francis. With permission.)

be presented in the next section. Reviews of task analysis techniques can be found in Kirwan and Ainsworth (1992), CCPS (1994), and Luczak (1997).

While TA remains essential in almost any study of human reliability, qualitative approaches to HRA benefit significantly from cognitive task analysis and systems analysis. Cognitive task analysis (Rasmussen, 1986; Rasmussen et al., 1994) focuses on cognitive processes that support human planning, problem solving, and decision making and control activities that are often expressed through interactions with computer-based interfaces to complex systems. Examples of such systems include expert system based fault-diagnostic systems, nuclear power systems, intelligent vehicle highway systems, and intelligent information management systems that act as pilot assistants in advanced aviation. Cognitive task analysis often relies on "modeling frameworks" for deriving insights concerning human performance. For example, the distinctions between skill-based, rule-based, and knowledge-based levels of performance (discussed in the next section), and the use of a "stepladder" model of human performance developed by Rasmussen (1986) for depicting the relationships between various stages of information processing that occur in response to process disturbances (Figure 35.4) can, when used in conjunction, enable predictions to be made concerning the types of errors that can be expected.

Cognitive task analysis would generally be preceded by a systems analysis (Rasmussen et al., 1994). In HRA, systems analysis entails describing the overall system, in terms of its various characteristics, in ways that can provide insights into work contexts that are most relevant to the problem of human and system reliability analysis. A brief illustration of systems analysis is provided in the following section. When systems analysis is combined with cognitive task analysis, the tendencies for performing at either the skill-, rule-, or knowledge-based levels can be determined. In addition, analysis at the systems level can reveal tendencies for errors and violations arising due to complex "higher-level" system factors. Cognitive task analysis and the more traditional task analysis would then provide additional layers of texture in the analysis of work contexts that could enhance predictions concerning these errors, including their types and the potential for their recovery, and also provide more detailed design recommendations for their reduction.

35.5 The Qualitative Perspective to Human and System Reliability Analysis

The qualitative perspective to HRA emphasizes system factors — specifically, the contexts that arise from the interplay of system factors. The identification and analysis of system contexts can potentially provide insight not only into how situational factors can trigger particular types of human errors and how these errors can propagate into adverse outcomes, but also how human interventions, even if not considered erroneous, can lead to negative consequences. In addition, the qualitative perspective is also potentially capable of predicting worker violations, depending on the extent to which factors such as organizational structure and work culture are considered.

In adopting this perspective, it is important that the analyst choose an appropriate system description that is consistent with the general problem definition. System descriptions can take many forms (cf. Sharit, 1997, pp. 303-305), and ultimately the analyst may choose to explore the degree to which different descriptions provide different insights. For example, in a large-scale trauma center, system descriptions in the form of: (1) defining relevant subsystems and the links between the different subsystems within which a trauma patient may directly or indirectly receive care; (2) the types of collaborations that exist between health care providers both within and across these subsystems; (3) the temporal constraints governing these collaborations; and (4) the communication channels between these providers, provide a basis for understanding the possibility for corruption of relevant patient-based information, and thus the occurrence of adverse system consequences. This analysis, in turn, would dictate the methods that would be adopted for analyzing human–system interactions in more detail in order to better evaluate human and system reliability. In contrast, in the case of a nuclear power plant control room, a different approach to systems description may be advised. For example, when the concern is for a nuclear power plant operator's capabilities in handling abnormal events through interaction with a control room computer interface, a more appropriate system description may be one based on an abstraction hierarchy (Rasmussen, 1986; Vicente and Rasmussen, 1992).

Unlike the quantitative perspective, which is driven by quantitative risk assessments and design recommendations that are primarily influenced by risk potential, the qualitative perspective is compatible with the "proactive" problem of predicting human and system reliability as well as with the "retrospective" problem of analysis of accidents. In the latter case, a clear starting point will almost invariably result in the identification of accident causes. However, when the objective is prediction, there are many "contextual paths" that could conceivably lead to reduced human and system reliability. These distinct cases are illustrated in Figure 35.5.

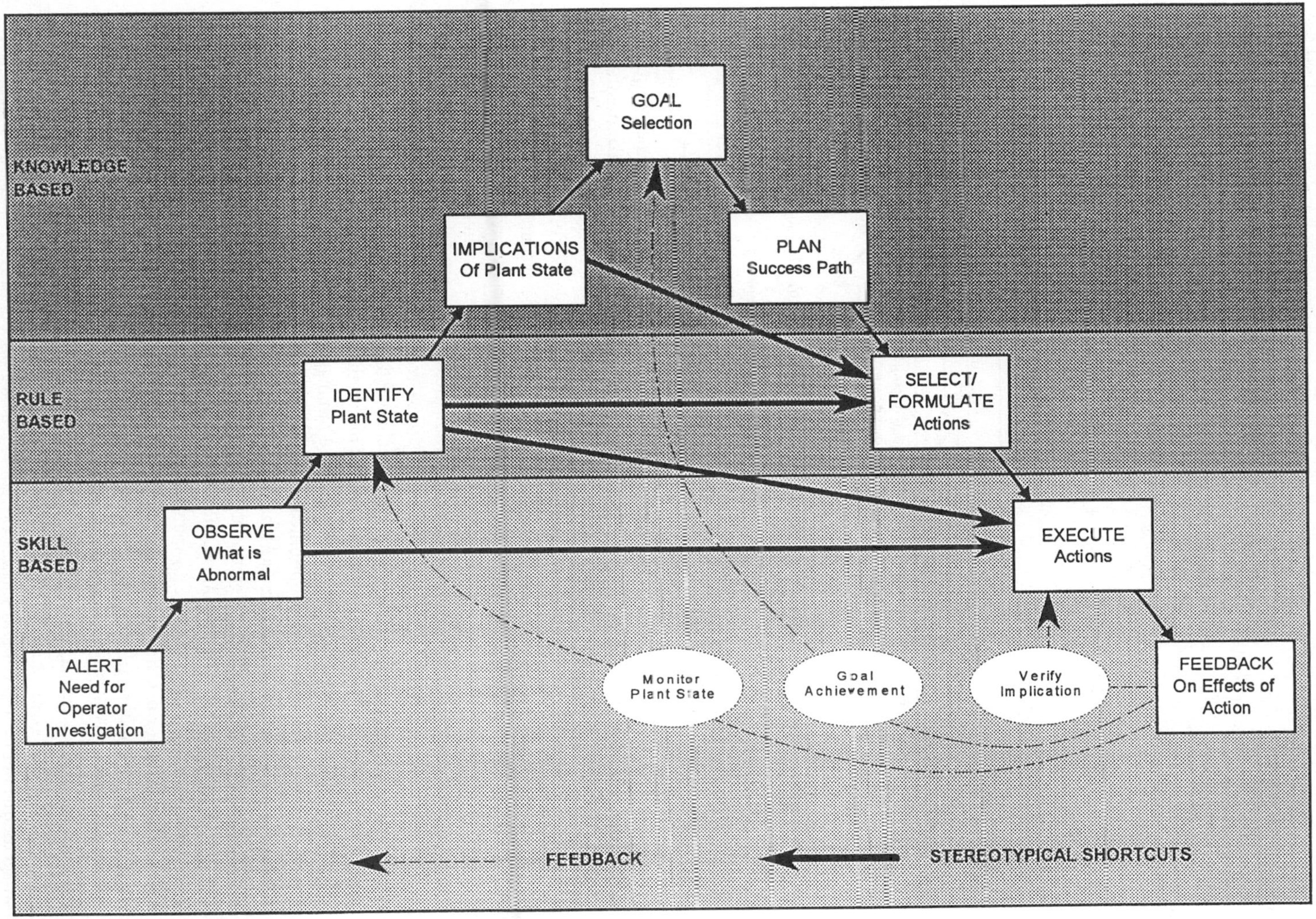

FIGURE 35.4 A "stepladder" model for depicting the relationships between various stages of information processing that occur in response to process disturbances. (From Center for Chemical Process Safety (CCPS). (1994). *Guidelines for Preventing Human Error in Process Safety.* New York: American Institute of Chemical Engineers. With permission. Adapted from Rasmussen, J. (1986). *Information Processing and Human-Machine Interaction.* Amsterdam.)

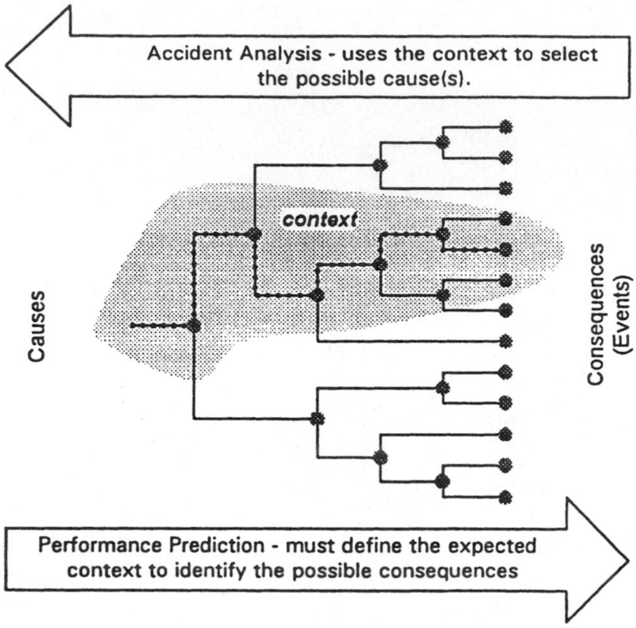

FIGURE 35.5 Accident analysis and performance or human error prediction. (From Hollnagel, E. (1993). *Human Reliability Analysis: Context and Control.* London: Academic Press. With permission.)

Prediction of Human Error

Prediction of human error and violations, and more generally, of adverse system outcomes, requires understanding the interplay between three factors: (1) the general error-inducing environment; (2) particular types of events characteristic to the system that could promote or trigger human behaviors leading to adverse system consequences; and (3) the tendencies humans have for committing errors and violations. These factors are summarized in Figure 35.6 for a trauma care system application (Sharit et al., 1996). As discussed earlier, the particular system analysis framework adopted will influence the characterization of the error-inducing environment. This analysis can be further aided by tools or methods that focus on identifying and elaborating on relevant sociotechnical factors, and by analysis of Performance Shaping Factors (PSFs). Discussion of each of these concepts follows.

An example of a "systems method" that considers sociotechnical factors is the Human Factors Analysis Methodology or HFAM (Pennycook and Embrey, 1993). This methodology is comprised of 20 groups of factors that are subdivided into three broad categories: (1) management-level factors (such as degree of worker participation, effectiveness of communications, and effectiveness of procedures development system); (2) operational-level generic factors (such as work group factors, training, process management, and job aids and procedures); and (3) operational-level job-specific factors (such as computer-based systems, control panel design, and maintenance). HFAM first invokes a screening process to identify the major areas vulnerable to human error (e.g., maintenance operations involving steam generators in nuclear power plant operations); the generic and appropriate job-specific factors are then applied to these areas. The components of each factor that applies can then be evaluated at two levels of detail as illustrated in Figure 35.7. The problems that are identified ultimately reflect failures at the management control level. Corresponding management-level factors would then be evaluated to identify the nature of the management-based error; these types of errors are often referred to as "latent" errors (Reason, 1990). Management-level factors fall into various categories, including: (1) those that can be specifically linked to operational-level factors (e.g., training, procedures); (2) those that are indicators of the quality of the safety culture and therefore can affect the potential for both errors and violations; and (3) those

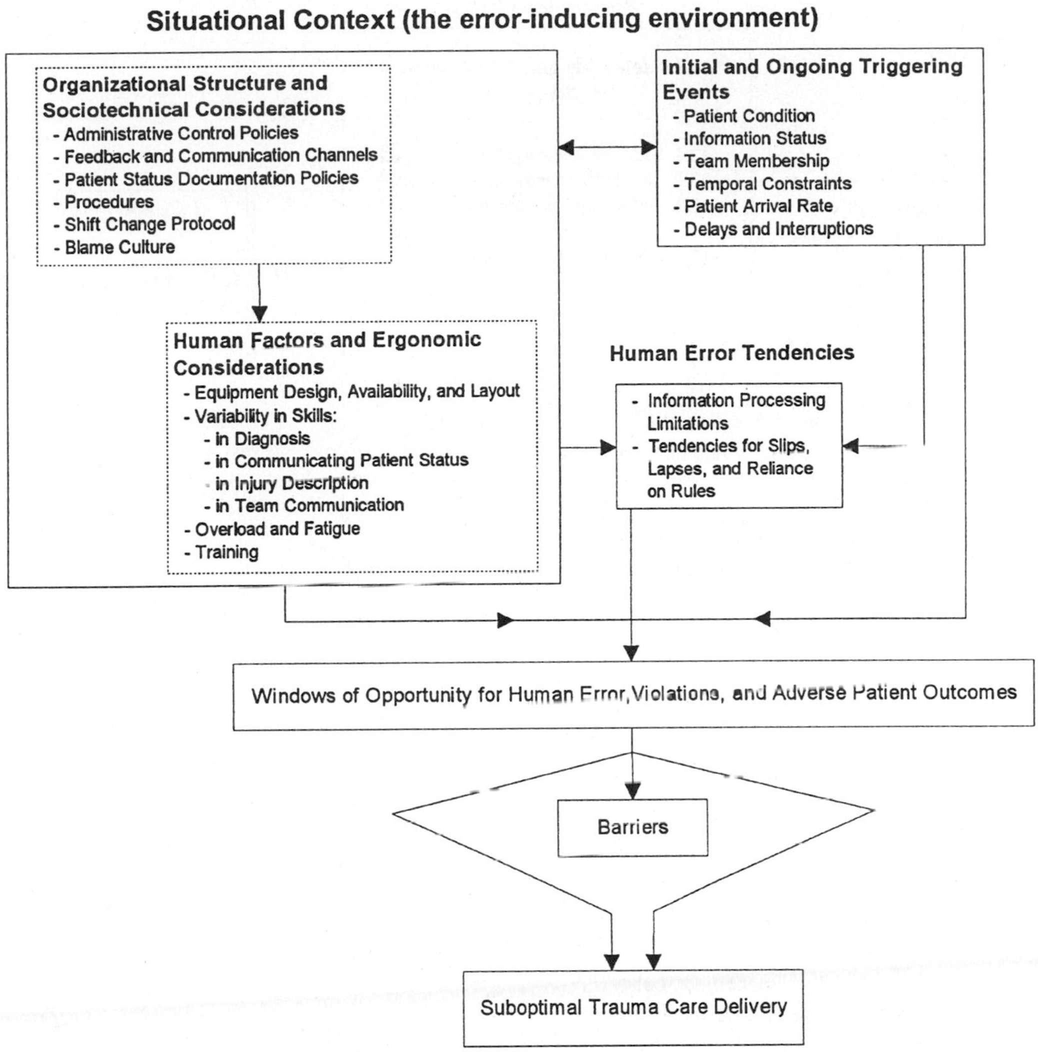

FIGURE 35.6 The interplay between situational context and human error tendencies in generating windows of opportunity for human error and adverse system outcomes. (From Sharit, J., Czaja, S. J., Augenstein, J., and Dilsen, K. (1996). A systems analysis of a trauma center: a methodology for predicting human error, in A. F. Ozog and G. Salvendy (eds.) *Advances in Applied Ergonomics*, 996-1101. Indiana: U.S.A. Publishing Corporation. With permission.)

that reflect communication of information throughout the organization, including the capability for learning lessons from operational experience based on various forms of feedback channels.

In principle, Performance Shaping Factors (PSFs), also referred to as Performance Influencing Factors (PIFs), represent any factors that can influence the potential for human error or violations. As will be discussed, in quantitative HRA they are used to modify or directly compute estimates of HEPs. In qualitative human and system reliability assessments, PSFs can serve a number of purposes. For example, their identification and assessment are important components of human factors and ergonomics audits that are used for establishing which design features might be susceptible to human error. This application of PSFs can also be used by workers as part of a participative error-reduction program. When used for *predicting* human error and adverse system outcomes, PSFs essentially shape the system context — their interplay with initiating and ongoing system events and with human error tendencies define contexts conducive to error and adverse outcomes (Figure 35.6).

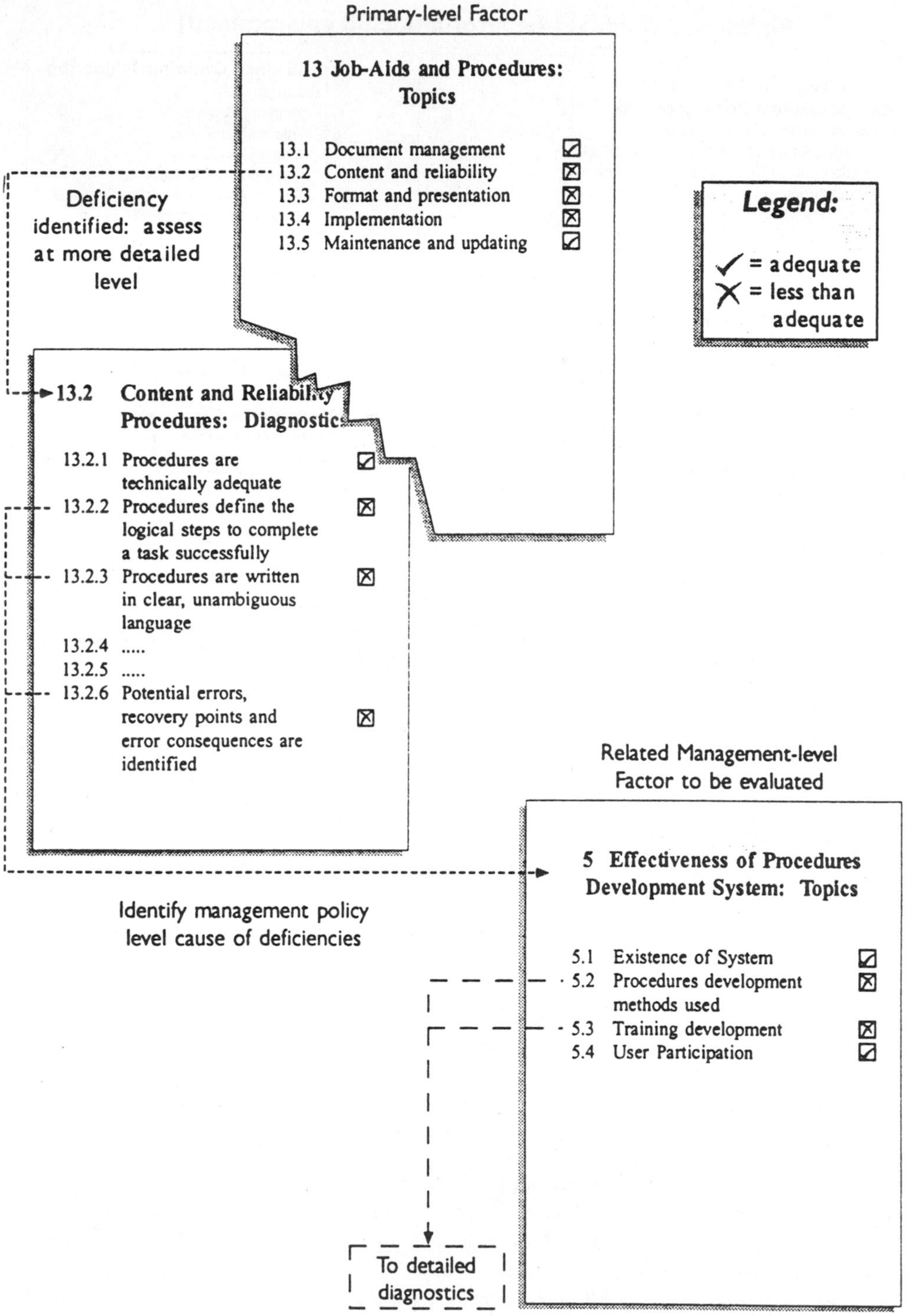

FIGURE 35.7 Illustration of the structure applied in one form of the HFAM tool. (From Pennycook, W. A. and Embrey, D. E. (1993). An operating approach to error analysis, in *Proceedings of the First Biennial Canadian Conference on Process Safety and Loss Management.* Edmonton, Alberta, Canada. Waterloo, Ontario, Canada: Institute for Risk Research, University of Waterloo. With permission.)

Table 35.1 presents a classification structure for PSFs compiled by Swain and Guttmann (1983) for application to the nuclear power industry. It is not uncommon for analysts to devise their own classification schemes based on the characteristics of the particular operating environment or organization. In general, at least three broad categories of PSFs need to be considered: those related to demands (e.g., task and physical work environment factors), resources (e.g., job aids and training), and management policies. The assumption generally adopted by both the quantitative and qualitative perspectives is that when all PSFs relevant to a particular situation are optimal, error likelihood will be minimized. However, errors will still occur due to a phenomenon known as "stochastic variability" in human performance that can derive, for instance, from movement variability or from unique intentions and biases. Note that many of the sets of factors in HFAM are essentially PSFs which are expressed at increasingly finer levels of definition.

While PSFs are essential for constructing a framework for error prediction, as alluded to earlier many critical links between human behavioral tendencies and the situational context will still likely be "under-specified." The additional level of articulation generally required is typically provided by task and cognitive task analysis. As with systems analysis, however, it is important that the analyst choose an appropriate method of task analysis (TA) in order to expose the subtleties that best satisfy the contextual systems model. A number of taxonomies exist for classifying human performance (Fleishman and Quaintance, 1984), and many variations and hybrid schemes can be derived from these fundamental structures. Returning to the example of complex trauma care systems, a TA method that, among other considerations, emphasizes: (1) how teams of workers coordinate activities over time and different locations; (2) the constraints imposed on workers relating to how information is documented; (3) the variability of worker skills for various areas of health care expertise; (4) the types of communication protocol; and (5) the opportunities for interruptions and delays, are examples of the types of descriptions required for elaboration of the error-inducing environment.

Areas where cognitive task analysis could help further establish the potential for error should now be more readily identifiable. For example, it can be used to analyze how the constraints imposed by a computer-based information system on information documentation can affect information processing activities by other workers who must assimilate this documented patient information with other data during information acquisition activities. At this point, an understanding of human error tendencies based on models of human information processing and human error is needed.

Human Error: Information Processing, Classification Schemes, and Models

The Traditional Human Factors Approach

Predicting human error requires some understanding of the relationships between the various attentional processes or components comprising the human's information processing system (Figure 35.8). Even with only a very fundamental appreciation of these processes it may, for example, be possible to predict that: (1) a worker does not have enough time to input information accurately given the design of the interface; (2) the design of displays is likely to evoke control responses that are contraindicated; (3) equipment is positioned in a manner that makes it likely that a poor position will be adopted when performing some activity or that other operations will be interfered with; and (4) decision making will take place without the benefits of complete or unambiguous information.

This perspective to prediction essentially translates the contextually rich system-based and task-based information into task demands. Mismatches between these demands and the human's capabilities for meeting these demands, which are largely reflected in information processing considerations (Figure 35.8), help map out areas with increased vulnerability to human error. Depending on the objectives of the human and system reliability assessment, this type of analysis may be sufficient. It is, however, suboptimal in that it is generally incapable of predicting the *types* of errors that might occur, and in this sense it is likely to underutilize contextually rich descriptions of the error inducing environment.

Systematic Approaches to Error Prediction

Systematic approaches to error prediction provide several advantages, including the ability to more easily rationalize and document the consequences of errors and error reduction strategies. This capability, in

TABLE 35.1 Examples of Performance Shaping Factors

External PSFs		Stressor PSFs	Internal PSFs
Situational Characteristics **Those PSFs General to One or More Jobs in a Work Situation**	**Task and Equipment Characteristics:** **Those PSFs Specific to Tasks in a Job**	**Psychological Stressors:** **PSFs which Directly Affect Mental Stress**	**Organismic Factors:** **Characteristics of People Resulting from Internal & External Influences**
Architectural features Quality of environment: Temperature, humidity, air quality, and radiation Lighting Noise and vibration Degree of general cleanliness Work hours/work breaks Shift rotation Availability/adequacy of special equipment, tools, and supplies Manning parameters Organizational structure (e.g., authority, responsibility, communication channels) Actions by supervisors, co-workers, union representatives, and regulatory personnel Rewards, recognition, benefits	Perceptual requirements Motor requirements (speed, strength, precision) Control–Display relationships Anticipatory requirements Interpretation Decision-making Complexity (information load) Narrowness of task Frequency and repetitiveness Task criticality Long- and short-term memory Calculational requirements Feedback (knowledge of results) Dynamic vs. step-by-step activities Team structure and communication Man–machine interface factors: design of prime equipment, test equipment, manufacturing equipment, job aids, tools, fixtures	Suddenness of onset Duration of stress Task speed Task load High jeopardy risk Threats (of failure, loss of job) Monotonous, degrading, or meaningless work Long, uneventful vigilance periods Conflicts of motives about job performance Reinforcement absent or negative Sensory deprivation Distractions (noise, glare, movement, flicker, color) Inconsistent cueing	Previous training/experience State of current practice or skill Personality and intelligence variables Motivation and attitudes Emotional state Stress (mental or bodily tension) Knowledge of required performance standards Sex differences Physical condition Attitudes based on influence of family and other outside persons or agencies Group identifications
Job and Task Instructions: **Single Most Important Tool for Most Tasks**		**Physiological Stressors:** **PSFs which Directly Affect Physical Stress**	
Procedures required (written or not written) Written or oral communications Cautions and warnings Work methods Plant policies (shop practices)		Duration of stress Fatigue Pain or discomfort Hunger or thirst Temperature extremes Radiation G-Force extremes Atmospheric pressure extremes Oxygen insufficiency Vibration Movement constriction Lack of physical exercise Disruption of circadian rhythm	

Source: Swain, A. D. and Guttmann, H. E. (1983). *Handbook of Human Reliability Analysis with Emphasis on Nuclear Power Plant Applications.* NUREG/CR-1278, U.S. Nuclear Regulatory Commission, Washington, D.C.

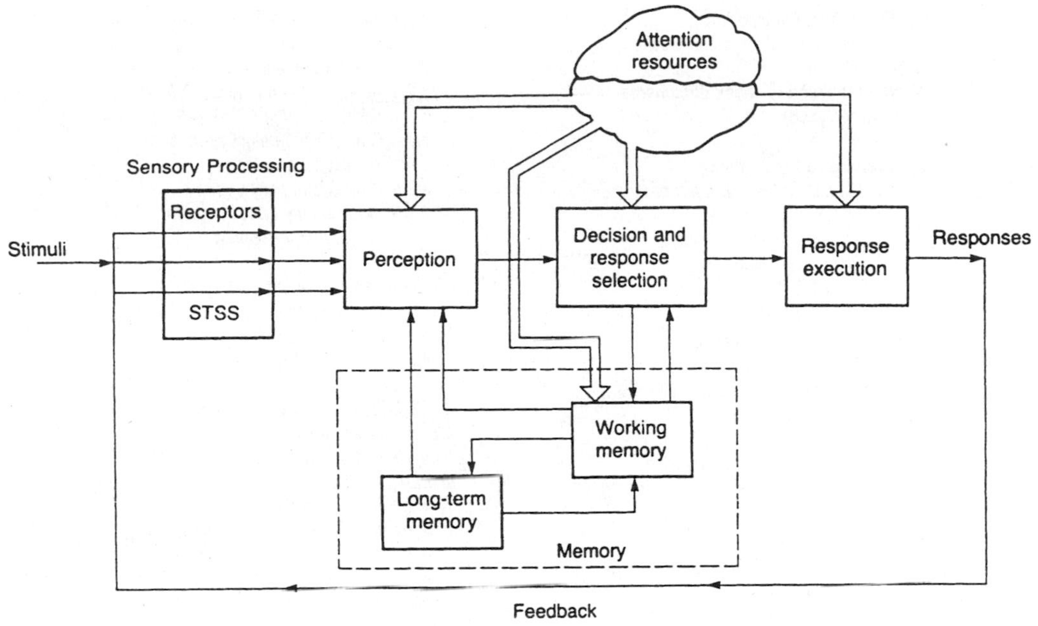

FIGURE 35.8 A model of human information processing (STSS refers to the short-term sensory store). (From Wickens, C. D. (1992). *Engineering Psychology and Human Performance*, Second Edition. New York: HarperCollins. With permission.)

Action		**Retrieval**	
A1	Action too long / short	R1	Information not obtained
A2	Action mistimed	R2	Wrong information
A3	Action in wrong direction	R3	Information retrieval incomplete
A4	Action too little / too much		
A5	Misalign	**Transmission**	
A6	Right action on right object	T1	Information not transmitted
A7	Wrong action on right object	T2	Wrong information transmitted
A8	Action omitted	T3	Information transmission incomplete
A9	Action incomplete		
A10	Wrong action on wrong object	**Selection**	
		S1	Selection omitted
Checking		S2	Wrong selection made
C1	Checking omitted		
C2	Check incomplete	**Plan**	
C3	Right check on wrong object	P1	Plan preconditions ignored
C4	Wrong check on right object	P2	Incorrect plan executed
C5	Check mistimed		
C6	Wrong check on wrong object		

FIGURE 35.9 An example of an error taxonomy used as part of a systematic methodology for predicting human error for the chlorine tanker filling problem. (From Center for Chemical Process Safety (CCPS). (1994). *Guidelines for Preventing Human Error in Process Safety.* New York: American Institute of Chemical Engineers. With permission.)

turn, increases the chances of management's acceptance of the HRA program. An example of one such approach is SPEAR — System for Predictive Error Analysis and Reduction (CCPS, 1993). SPEAR consists of five steps: task analysis (TA), PSF analysis, predictive human error analysis, consequence analysis, and error reduction analysis. For each step of the TA, and taking the influence of PSFs into consideration, the occurrence of one or more errors derived from some type of "error taxonomy," such as the one presented in Figure 35.9, is evaluated.

An example of this approach taken from CCPS (1993) involves the filling of a storage tank with chlorine from a tank truck. The primary purpose of this analysis was to identify potential human errors that could

0. Fill tanker with chlorine
Plan: Do tasks 1 to 5 in order.

1. Park tanker and check documents (not analyzed)

2. Prepare tanker for filling
Plan: Do 2.1 or 2.2 in any order then do 2.3 to 2.5 in order.
2.1 Verify tanker is empty
Plan: Do in order.
2.1.1 Open test valve
2.1.2 Test for Cl_2
2.1.3 Close test valve
2.2 Check weight of tanker
2.3 Enter tanker target weight
2.4 Prepare fill line
Plan: Do in order.
2.4.1 Vent and purge line
2.4.2 Ensure main Cl_2 valve closed
2.5 Connect main Cl_2 fill line

3. Initiate and monitor tanker filling operation
Plan: Do in order.
3.1 Initiate filling operation
Plan: Do in order.
3.1.1 Open supply line valves
3.1.2 Ensure tanker is filling with chlorine
3.2 Monitor tanker filling operation
Plan: Do 3.2.1, do 3.2.2 every 20 minutes on initial weight alarm, do 3.2.3 and 3.2.4. On final weight alarm, do 3.2.5 and 3.2.6.
3.2.1 Remain within earshot while tanker is filling
3.2.2 Check road tanker
3.2.3 Attend tanker during last 2-3 ton filling
3.2.4 Cancel initial weight alarm and remain at controls
3.2.5 Cancel final weight alarm
3.2.6 Close supply valve A when target weight reached

4. Terminate filling and release tanker
4.1 Stop filling operation
Plan: Do in order.
4.1.1 Close supply valve B
4.1.2 Clear lines
4.1.3 Close tanker valve
4.2 Disconnect tanker
Plan: Repeat 4.2.1 five times then do 4.2.2 to 4.2.4 in order.
4.2.1 Vent and purge lines
4.2.2 Remove instrument air from valves
4.2.3 Secure blocking device on valves
4.2.4 Break tanker connections
4.3 Store hoses
4.4 Secure tanker
Plan: Do in order.
4.4.1 Check valves for leakage
4.4.2 Secure log-in nuts
4.4.3 Close and secure dome
4.5 Secure panel (not analyzed)

5. Document and report (not analyzed)

FIGURE 35.10 Part of a hierarchical task analysis for the chlorine tanker filling problem. (From Center for Chemical Process Safety (CCPS). (1994). *Guidelines for Preventing Human Error in Process Safety.* New York: American Institute of Chemical Engineers. With permission.)

contribute to a major flammable release resulting from either a spill during unloading of the truck or from a tank rupture. Prior to this HRA, an FTA revealed that the frequency of such a release is largely due to human errors. Figure 35.10 illustrates a portion of a HTA — the TA method used in this case — that addresses the operations that could lead to a spill during tanker loading. Note that with an HTA, task operations can be described to whatever level of detail is required. Figure 35.11 illustrates some of the results of this analysis, where errors are coded according to their classification (Figure 35.9). The appeal of such a systematic method should be obvious. Moreover, this procedure does not preclude quantification of human errors considered important for satisfying PSA requirements.

Cognitive Engineering Approaches to Error Prediction

Perspectives that emphasize mental or cognitive processes that underlie human error can potentially reveal the causes of errors. This knowledge, in turn, provides a stronger basis for design countermeasures, and for making use of data on "near misses" — incidents that are precursors of more serious events—that industries would be well advised to keep. These "cognitive engineering system" perspectives, however, require descriptions of error-inducing environments that are generally more detailed and less systematic than approaches that emphasize the external form of the error — i.e., *what* happened (in terms of its observable manifestation) — rather than *how* (from the standpoint of information processing mechanisms) and *why* (from the interplay between behavioral tendencies and the situational context) it happened.

A modeling framework that is consistent with the goal of understanding underlying causes of human error is based on distinguishing fundamentally different categories of human information processing. These categories are referred to as skill-based, rule-based, and knowledge-based levels of performance

STEP	ERROR TYPE	ERROR DESCRIPTION	RECOVERY	CONSEQUENCES AND COMMENTS	ERROR REDUCTION RECOMMENDATIONS		
					PROCEDURES	TRAINING	EQUIPMENT
2.3 Enter Tanker target weight	Wrong information obtained (R2)	Wrong weight entered	On check	Alarm does not sound before tanker overfills	Independent validation of target weight	Ensure operator double checks entered date. Recording of values in checklist.	Automatic setting of weight alarms from unladen weight. Computerize logging system and build in checks on tanker reg. No. and unladen weight linked to warning system. Display differences
3.2.2 Check Tanker while filling	Check omitted (C1)	Tanker not monitored while filling	On initial weight alarm	Alarm will alert the operator if correctly set. Equipment fault, e.g., leaks not detected early and remedial action delayed.	Provide secondary task involving other personnel. Supervisor periodically checks operation.	Stress importance of regular checks for safety	Provide automatic log-in procedure
3.2.3 Attend tanker during last 2-3 ton filling	Operation omitted (O8)	Operator fails to attend	On step 3.2.5	If alarm not detected within 10 minutes tanker will overfill	Ensure work schedule allows operator to do this without pressure	Illustrate consequences of not attending	Repeat alarm in secondary area. Automatic interlock to terminate loading if alarm not acknowledged. Visual indication of alarm.
3.2.5 Cancel final weight alarm	Operation omitted (O8)	Final weight alarm taken as initial weight alarm	No recovery	Tanker overfills	Note differences between the sound of the two alarms in checklist.	Alert operators during training about differences in sounds of alarms	Use completely different tones for initial and final weight alarms
4.1.3 Close tanker valve	Operation omitted (O8)	Tanker valve not closed	4.2.1	Failure to close tanker valve would result in pressure not being detected during the pressure check in 4.2.1	Independent check on action. Use checklist	Ensure operator is aware of consequences of failure	Valve position indicator would reduce probability of error
4.2.1 Vent and purge lines	Operation omitted (O8) Operation incomplete (O9)	Lines not fully purged	4.2.4	Failure of operator to detect pressure in lines could lead to leak when tanker connections broken	Procedure to indicate how to check if fully purged	Ensure training covers symptoms of pressure in line	Line pressure indicator at controls. Interlock device on line pressure.
4.4.2 Secure locking nuts	Operation omitted (O8)	Locking nuts left unsecured	None	Failure to secure locking nuts could result in leakage during transportation	Use checklist	Stress safety implication of training	Locking nuts to give tactile feedback when secure

FIGURE 35.11 Results from the application of a systematic approach to error prediction and analysis for the chlorine tanker filling problem. (From Center for Chemical Process Safety (CCPS). (1994). *Guidelines for Preventing Human Error in Process Safety.* New York: American Institute of Chemical Engineers. With permission.)

(Rasmussen, 1986). Work activities at the skill-based level are highly practiced routines that require little conscious attention. The rule-based level involves the use of rules that the worker invokes from memory, or obtains from other sources such as reference manuals or co-workers. For example, a maintenance worker might conclude that a certain fault is present in the equipment based on symptoms revealed by diagnostic tests. This conclusion may then trigger another rule that addresses actions that need to be taken in response to that fault. Knowledge-based performance typically occurs when workers are attempting to solve problems in relatively unfamiliar situations, and demands the greatest use of information processing resources.

The distinctions implicit to this "SRK framework" can be used in accident analysis to trace the observed or "external error form" to its underlying causes. For error prediction, while models of human error have been proposed that utilize this framework (e.g., Reason, 1990), the usefulness of these models is not likely to be readily apparent to analysts. Much more useful are the distinctions Reason (1990) makes between different types of errors, specifically, between skill-based slips and lapses, rule-based mistakes, and knowledge-based mistakes. In his view, the potential for error begins with "cognitive underspecification," implying that at some point in the processing of task-related information the specification of information is incomplete. This underspecification, which to some extent results from information processing and memory limitations, can promote one of two forms of biases: "frequency bias" or "similarity bias." These biases, respectively, reflect tendencies to process information and act based on the frequency with which a behavior has been performed, or the degree to which information currently being perceived or processed appears similar to patterns of information the person is readily tuned to. The manifestation of these biases in terms of different types of errors will depend largely on the particular level of performance, the situational context, and characteristics unique to the person.

For example, assume a worker is performing a series of operations on a machine that are fairly well-practiced, requiring little attention. A modification is made to the machine that requires the operations ADB to be performed instead of ABCD. Following the modification, the worker immediately proceeded to perform operation B after performing operation A, even though the worker was aware of the new sequence and intended to perform it. This error is referred to as a "double-capture slip." More generally, this error is classified as a "skill-based slip" due to "inattention" (Table 35.2): had the worker invested more attention at the critical point where, due to the sheer frequency with which the previous routine was performed, one would expect the worker to slip into the old routine, then this error would likely have been avoided. An example of a skill-based lapse resulting from this same absence of attentional control is when a worker intends to initiate a sequence of operations but is interrupted by an alarm. After addressing the source of the alarm the worker goes on to other activities, perhaps because initiating the intended sequence of events has not in the past been generally associated with corrective actions in response to alarms. This type of error is referred to as "an omission following an interruption" (Table 35.2). Excessive attentional control can also lead to skill-based slips, as when a worker disrupts activities being performed in order to analyze the situation. Disruption of the "preprogrammed" or automatic sequence of activities typical of skill-based work can result in the worker picking up the task at a point further along than it is (an "omission due to overattention"), or repeating steps already taken ("repetition due to overattention").

Unlike errors at the skill-based level, errors at the rule-based level represent actions that are intentional; they just turn out to be wrong, and thus are viewed as "mistakes." As with the skill-based level, cognitive underspecification serves to set in motion frequency and similarity biases that shape the form errors assume at the rule-based level. Consider a job-shop situation where machine setup operations are based on factors such as part type and processing operations. The current work part has features that do not completely match the conditions associated with the IF or antecedent portion of any of the rules available to the worker. However, despite these inconsistencies (i.e., cognitive underspecification) that result in only partial matching of the rule, the worker may choose to use a rule that has been successful many times before, but one that is incorrect for the current situation. This form of error or "failure mode" is referred to as "rule strength" (Table 35.2), and represents the misapplication of a good rule in that under the appropriate conditions this rule would have been successful.

TABLE 35.2 Human Error Failure Modes at Each of Three Levels of Human Performance Associated with the SRK Framework

Skill-based performance	
Inattention	*Overattention*
Double-capture slips	Omissions
Omissions following interruptions	Repetitions
Reduced intentionality	Reversals
Perceptual confusions	
Interference errors	
Rule-based performance	
Misapplication of good rules	*Application of bad rules*
First exceptions	Encoding deficiencies
Countersigns and nonsigns	Action deficiencies
Informational overload	Wrong rules
Rule strength	Inelegant rules
General rules	Inadvisable rules
Redundancy	
Rigidity	
Knowledge-based performance	
Selectivity	*Problems with causality*
Workspace limitations	*Problems with complexity*
Out of sight out of mind	Problems with delayed feedback
Confirmation bias	Insufficient consideration of processes in time
Overconfidence	Difficulties with exponential developments
Biased reviewing	Thinking in causal series not causal nets
Illusory correlation	Thematic vagabonding
Halo effects	Encysting

Source: Reason, J. (1990). *Human Error.* Cambridge: Cambridge University Press.

As indicated in Table 35.2, there are many forms of knowledge-based mistakes, which is not surprising given the enormous variance in human behavioral tendencies manifest during performance at the knowledge-based level. The challenge for the analyst is to recognize the potential for these situations and to recommend design solutions such as decision and memory aids. The use of principles of ecological interface design (Vicente and Rasmussen, 1992) can provide a means for handling abnormal or unanticipated situations characteristic of this level of performance. Through these methods and principles, system representations on human–computer interfaces can potentially be devised that can transform knowledge-based problem solving to the rule-based level, where the worker has a more realistic opportunity to succeed.

Rasmussen (1982) has provided, in flow-chart form, a guide for answering questions concerning *what, how,* and *why* an error occurred. Figure 35.12 illustrates the flow chart corresponding to identification of the internal error mechanism — i.e., *how* an error occurred. Although this procedure is more easily applied to the analysis of accidents, it can also be used to predict errors if work contexts have been analyzed to sufficient detail. In that case, the links between various factors comprising the error-inducing environment and the possible internal error mechanisms will represent the underlying causes of the error. Note that the end points in Figure 35.12 related to information processing can, in principle, be further resolved based on Reason's (1990) analysis of failure modes. However, to extend the analysis of causality to this level of detail requires methods for: (1) identifying cognitive underspecification in work situations; and (2) translating this underspecification into particular failure modes.

For a cognitive-based analysis of human error that still maintains the appeal of systematic methods, one can substitute the predictive human error analysis stage of SPEAR with a classification scheme that

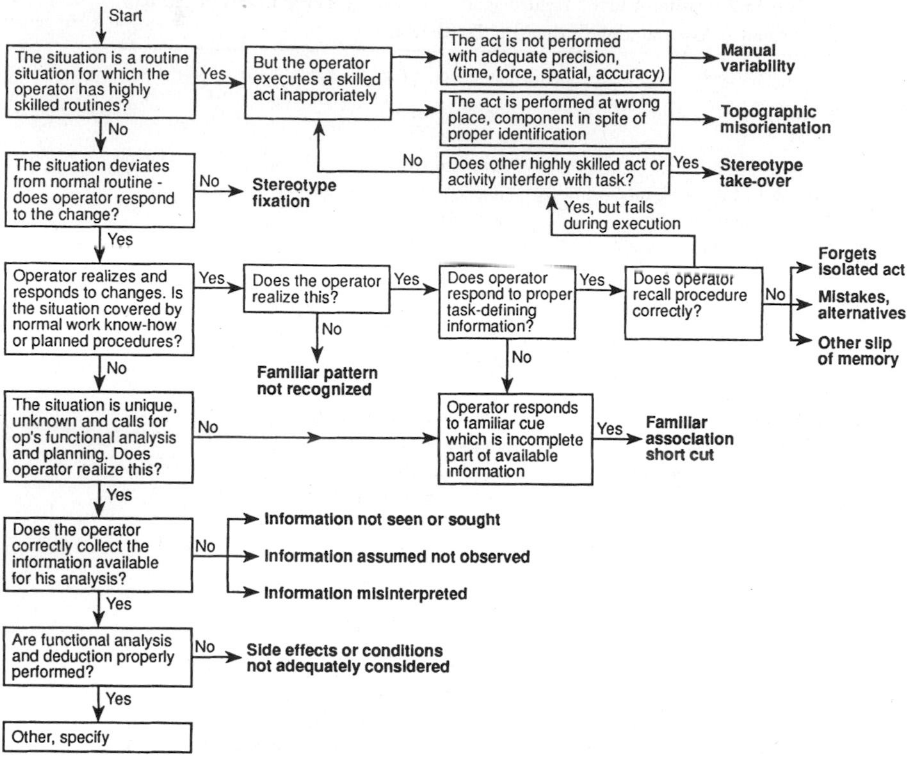

FIGURE 35.12 Guide for identifying the mechanisms concerning "how" a human error occurred. (From Rasmussen, J. (1982). Human errors: a taxonomy for describing human malfunction in industrial installations, *Journal of Occupational Accidents*, 4, 311-333. With permission.)

adds considerable depth to the one illustrated in Figure 35.9. An example of such a scheme is the "human error identification in systems tool" (HEIST). Based on an analysis of PSFs, this tool first identifies the external error mode. A listing of external error modes classified according to eight stages of human information processing is presented in Table 35.3. The first column in a "HEIST table" would then consist of a code whose first letter (or first two letters) refers to one of these eight stages, and whose next letter refers to one of six general or "global" PSFs: time (T); interface (I); training/experience/familiarity (E); procedures (P); task organization (O); and task complexity (C).

These external error modes are then linked to underlying psychological error mechanisms. Many of these mechanisms are consistent with modes of failure or types of errors listed in Table 35.2. Table 35.4 presents an extract from a HEIST table corresponding to two of the eight stages of human information processing referred to above — specifically, "Activation/detection" and "Observation/data collection." The linking to psychological error mechanisms is depicted in column four of the HEIST table (Table 35.4). Table 35.5 presents a listing of these mechanisms for the two stages of information processing included in Table 35.4. Note that the error-reduction guidelines contained in the HEIST table can be used for making specific error reduction recommendations in methods such as SPEAR (Figure 35.11). A complete HEIST table and associated listing of psychological error mechanisms can be found in Kirwan (1994).

35.6 Simulators and Computer Simulation

A simulator is a facility consisting of a mock-up of a system and its human interface, and containing a mathematical representation of the system. Although typically used for training purposes, the potential exists for investigating various types of errors on the part of workers and worker crews, depending on

TABLE 35.3 A Listing of External Error Modes Classified According to Stages of Human Information Processing

1. Activation/detection
   - 1.1 Fails to detect signal/cue
   - 1.2 Incomplete/partial detection
   - 1.3 Ignore signal
   - 1.4 Signal absent
   - 1.5 Fails to detect deterioration of situation

2. Observation/data collection
   - 2.1 Insufficient information gathered
   - 2.2 Confusing information gathered
   - 2.3 Monitoring/observation omitted

3. Identification of system state
   - 3.1 Plant-state-identification failure
   - 3.2 Incomplete-state identification
   - 3.3 Incorrect-state identification

4. Interpretation
   - 4.1 Incorrect interpretation
   - 4.2 Incomplete interpretation
   - 4.3 Problem solving (other)

5. Evaluation
   - 5.1 Judgment error
   - 5.2 Problem-solving error (evaluation)
   - 5.3 Fails to define criteria
   - 5.4 Fails to carry out evaluation

6. Goal selection and task definition
   - 6.1 Fails to define goal/task
   - 6.2 Defines incomplete goal/task
   - 6.3 Defines incorrect or inappropriate goal/task

7. Procedure selection
   - 7.1 Selects wrong procedure
   - 7.2 Procedure inadequately formulated/short-cut invoked
   - 7.3 Procedure contains rule violation
   - 7.4 Fails to select or identify procedure

8. Procedure execution
   - 8.1 Too early/late
   - 8.2 Too much/little
   - 8.3 Wrong sequence
   - 8.4 Repeated action
   - 8.5 Substitution/intrusion error
   - 8.6 Orientation/misalignment error
   - 8.7 Right action on wrong object
   - 8.8 Wrong action on right object
   - 8.9 Check omitted
   - 8.10 Check fails/wrong check
   - 8.11 Check mistimed
   - 8.12 Communication error
   - 8.13 Act performed wrongly
   - 8.14 Part of act performed
   - 8.15 Forgets isolated act at end of task
   - 8.16 Accidental timing with other event/circumstance
   - 8.17 Latent error prevents execution
   - 8.18 Action omitted
   - 8.19 Information not obtained/transmitted
   - 8.20 Wrong information obtained/transmitted
   - 8.21 Other

Source: Kirwan, B. (1994). *A Guide to Practical Human Reliability Assessment.* London: Taylor & Francis.

the apparatus available for tracking human performance. When properly run, simulator trials can also be used as a basis for estimating HEPs for quantitative PRA-driven HRAs.

Computer simulation attempts to capture the interaction between the human and environment through software. When applied to work contexts, one can define worker tasks, environmental events, and performance variables, and, through the simulation's software logic, determine various outcomes of interest. An example of such a simulation is MicroMAPPS (Gertman and Blackman, 1994), a PC-based version of the Maintenance Personnel Performance Simulation (Siegel et al., 1984). Sponsored by the Nuclear Regulatory Council, this simulation enables the user to define various subtasks for particular maintenance crew tasks and rate a number of variables that impact performance (e.g., time since task was last performed, time available, crew's initial ability). Given its underlying probabilistic basis, over many simulation runs MAPPS can provide the average HEP for any given subtask. When using this simulation model, analysts have the opportunity to evaluate changes in worker and work crew reliability as a function of changes in a variety of relevant parameters. In addition, HEP estimates can be used as inputs into PRAs.

The Cognitive Environment Simulation (CES) is a computer simulation model that is more consistent with the objective of qualitative approaches to HRA that emphasize causality of error (Woods et al., 1987). CES uses as inputs the data produced by a nuclear power plant (NPP) simulator. It responds to those inputs through an array of human information processing mechanisms and a knowledge base derived from models of what humans would do in facing the circumstances unfolding over time in the CES model. This simulation model primarily focuses on modeling the processes by which humans assess

TABLE 35.4 Extract from a HEIST Table

Code	Error-identifier prompt	External error mode	System cause/ psychological error-mechanism	Error-reduction guidelines
		1. Activation/Detection		
AT1	Does the signal occur at the appropriate time? Could it be delayed?	Action omitted; performed too early or too late	Signal timing deficiency; failure of prospective memory	Alter system configuration to present signal appropriately; generate hard copy to aid prospective memory; repeat signal until action has occurred
AI1	Could the signal source fail?	Action omitted or performed too late	Signal failure	Use diverse/redundant signal sources; use a higher-reliability signal system; give training and ensure procedures incorporate investigation checks on 'no signal'
AI2	Can the signal be perceived as unreliable?	Action omitted	Signal ignored	Use diverse signal sources; ensure higher signal reliability; retrain if signal is more reliable than it is perceived to be
AI3	Is the signal a strong one, and is it in a prominent location? Could the signal be confused with another?	Action omitted; or performed too late; or wrong act performed	Signal-detection failure	Prioritize signals; place signals in primary (and unobscured) location; use diverse signals; use multiple-signal coding; give training in signal priorities; make procedures cross-reference the relevant signals; increase signal intensity
AI4	Does the signal rely on oral communication?	Action omitted or performed too late	Communication failure; lapse of memory	Provide physical back-up/substitute signal; build required communications requirements into procedures
AE1	Is the signal very rare?	Action omitted or performed too late	Signal ignored (false alarm); stereotype fixation	Give training for low-frequency events; ensure diversity of signals; prioritize signals into a hierarchy of several levels
AE2	Does the operator understand the significance of the signal?	Action omitted or performed too late	Inadequate mental model	Training and procedures should be amended to ensure significance is understood
AP1	Are procedures clear about action following the signal or the previous step, or when to start the task?	Action omitted or performed either too early or too late	Incorrect mental model	Procedures must be rendered accurate, or at least made more precise; give training if judgment is required on when to act
AO1	Does activation rely on prospective memory (i.e., remembering to do something at a future time, with no specific cue or signal at that later time)?	Action omitted or performed either too late or too early	Prospective memory failure	Proceduralize task, noting calling conditions, timings of actions, etc...; utilize an interlock system preventing task occurring at undesirable times; provide a later cue; emphasize this aspect during training
AO2	Will the operator have other duties to perform concurrently? Are there likely to be distractions? Could the operator become incapacitated?	Action omitted or performed too late	Lapse of memory; memory failure; signal-detection failure	Training should prioritize signal importances; improve task organization for crew; use memory aids; use a recurring signal; consider automation; utilize flexible crewing
AO3	Will the operator have a very high or low work load?	Action omitted or performed either too late or too early	Lapse of memory; other memory failure; signal-detection failure	Improve task and crew organization; use a recurring signal; consider automation; utilize flexible crewing; enhance signal salience

TABLE 35.4 (continued) Extract from a HEIST Table

Code	Error-identifier prompt	External error mode	System cause/ psychological error-mechanism	Error-reduction guidelines
AO4	Will it be clear who must respond?	Action omitted or performed too late	Crew-coordination failure	Emphasize task responsibility in training and task allocation among crew; utilize team training
AC1	Is the signal highly complex?	Action omitted, or wrong act performed, or act performed either too late or too early	Cognitive overload; inadequate mental model	Simplify signal; automate system response; give adequate training in the nature of the signal; provide on-line, automated, diagnostic support; develop procedures which allow rapid analysis of the signal (e.g., use of flow charts)
AC2	Is the signal in conflict with the current diagnostic "mindset"?	Action omitted or wrong act performed	Confirmation bias; signal ignored	Procedures should emphasize disconfirming as well as confirmatory signals; utilize a shift technical advisor in the shift-structure; carry out problem-solving training and team training; utilize diverse signals; implement automation
AC3	Could the signal be seen as part of a different signal set? Or is, in fact, the signal part of a series of signals which the operator needs to respond to?	Action performed too early or wrong act performed	Familiar-association shortcut/stereo-type take-over	Training and procedures could involve display of signals embedded within mimics or other representations showing their true contexts or range of possible contexts; use fault-symptom matrix aids, etc.
		2. Observation/Data Collection		
OT1	Could the information or check occur at the wrong time?	Failure to act; or action performed too late or too early; or wrong act performed	Inadequate mental model/ inexperience/ crew coordination failure	Procedure and training should specify the priority and timing of checks; present key information centrally; utilize trend displays and predictor displays if possible; implement team training
OI1	Could important information be missing due to instrument failure?	Action omitted or performed either too late or too early; or wrong act performed	Signal failure	Use diverse signal sources; maintain back-up power supplies for signals; have periodic manual checks; procedures should specify action to be taken in event of signal failure; engineer automatic protection/action; use a higher-reliability system
OI2	Could information sources be erroneous?	Action omitted or performed either too late or too early; or wrong act performed	Erroneous signal	Use diverse signal sources; procedures should specify cross-checking; design system-self-integrity monitoring; use higher-reliability signals
OI3	Could the operator select a wrong but similar information source?	Action omitted or performed either too late or too early; or wrong act performed	Mistakes alternatives; spatial misorientation; topographic misorientation	Ensure unique coding of displays, cross-referenced in procedures; enhance discriminability via coding; improve training

TABLE 35.4 (continued) Extract from a HEIST Table

Code	Error-identifier prompt	External error mode	System cause/ psychological error-mechanism	Error-reduction guidelines
OI4	Is an information source accessed only via oral communication?	Action omitted or performed either too late or too early; or wrong act performed	Communication failure	Use diverse signals from hardwired or softwired displays; ensure back-up human corroboration; design communication protocols
OI5	Are any information sources ambiguous?	Action omitted or performed either too late or too early; or wrong act performed	Misinterpretation; mistakes alternatives	Use task-based displays; design symptom-based diagnostic aids; utilize diverse information sources; ensure clarity of information displayed; utilize alarm conditioning
OI6	Is an information source difficult or time-consuming to access?	Action omitted or performed too late; or wrong act performed	Information assumed	Centralize key data; enhance data access; provide training on importance of verification of signals; enhance procedures
OI7	Is there an abundance of information in the scenario, some of which is irrelevant, or a large part of which is redundant?	Action omitted or performed too late	Information overload	Prioritize information displays (especially alarms); utilize overview mimics (VDU or hard wired); put training and procedural emphasis on data-collection priorities and data management
OE1	Could the operator focus on key indication(s) related to a potential event while ignoring other information sources?	Action omitted or performed too late; or wrong act performed	Confirmation bias; tunnel vision	Training in diagnostic skills; enhance procedural structuring of diagnosis, emphasizing checks on disconfirming evidence; implement a staff-technical-advisor role; present overview mimics of key parameters showing whether system integrity is improving or worsening or adequate
OE2	Could the operator interrogate too many information sources for too long, so that progress toward stating identification or action is not achieved?	Action omitted or performed too late	Thematic vagabonding; risk-recognition failure; inadequate mental model	Training in fault diagnosis; team training; put procedural emphasis on required data-collection time frames; implement high-level indicators (alarms) of system-integrity deterioration
OE3	Could the operator fail to realize the need to check a particular source? Is there an adequate cue prompting the operator?	Action omitted or performed either too late or too early; or wrong act performed	Need for information not prompted; prospective memory failure	Procedural guidance on checks required; training; use of memory aids; use of attention-gaining devices (flash; alarms; central displays and messages)
OE4	Could the operator terminate the data collection/observation early?	Action omitted or performed either too early or too late; or wrong act performed	Overconfidence; inadequate mental model; incorrect mental model; familiar-association short-cut	Training in diagnostic procedures and verification; procedural specification of required checks, etc.; implement a shift-technical-advisor role

TABLE 35.4 (continued) Extract from a HEIST Table

Code	Error-identifier prompt	External error mode	System cause/ psychological error-mechanism	Error-reduction guidelines
OE5	Could the operator fail to recognize that special circumstances apply?	Action omitted or performed either too late or too early; or wrong act performed	Fail to consider special circumstances; slip of memory; inadequate mental model	Ensure training for, as well as procedural noting of, special circumstances; STA; give local warnings in the interface displays/controls
OP1	Could the operator fail to follow the procedures entirely?	Action omitted or wrong act performed	Rule violation; risk-recognition failure; production–safety conflict; safety-culture deficiency	Training in use of procedures; operator involvement in development and verification of procedures
OP2	Could the operator forget one or more items in the procedures?	Action omitted or performed either too early or too late; or wrong act performed	Forget isolated act; slip of memory; place-losing error	Ensure an ergonomic procedure design; utilize tick-off sheets, place keeping aids, etc.; team training to emphasize checking by other team member(s)
OO1 (AO2)	Will the operator have other duties to perform concurrently? Are there likely to be distractions? Could the operator become incapacitated?	Action omitted or performed too late	Lapse of memory; memory failure; signal-detection failure	Training should prioritize signal importances; develop better task organization for crew; use memory aids; use a recurring signal; consider automation; use flexible crewing
OO2 (AO3)	Will the operator have a very high or low work load?	Action omitted or performed either too late or too early	Lapse of memory; other memory failure; signal-detection failure	Better task and crew organization; utilize a recurring signal; consider automation; use flexible crewing; enhance signal salience
OO3 (AO4)	Will it be clear who must respond?	Action omitted or performed too late	Crew-coordination failure	Improve training and task allocation among crew; team training
OO4	Could information collected fail to be transmitted effectively across shift-hand-over boundaries?	Failure to act; or wrong action performed; or action performed either too late or too early; or an error of quality (too little or too much)	Crew-coordination failure	Develop robust shift-hand-over procedures; training; team training across shift boundaries; develop robust and auditable data-recording systems (logs)
OC1	Does the scenario involve multiple events, thus causing a high level of complexity or a high work load?	Failure to act; or wrong action performed; or action performed too early or too late	Cognitive overload	Emergency-response training; design crash-shutdown facilities; use flexible crewing strategies; implement shift-technical-advisor role; develop emergency operating procedures able to deal with multiple transients; engineer automatic information recording (trends, logs, printouts); generate decision/diagnostic support facilities

Source: Kirwan, B. (1994). *A Guide to Practical Human Reliability Assessment.* London: Taylor & Francis.

TABLE 35.5 Listing of Psychological Error Mechanisms for the Two Stages of Information Processing in Table 35.4

Activation/detection

1. Vigilance failure: lapse of attention.
   Ergonomic design of interface to allow provision of effective attention-gaining measures; supervision and checking; task-organization optimization, so that the operators are not inactive for long periods, and are not isolated.
2. Cognitive/stimulus overload: too many signals present for the operator to cope with.
   Prioritization of signals (e.g., high-, medium-, and low-level alarms); overview displays; decision-support systems; simplification of signals; flow chart procedures; simulator training; automation.
3. Stereotype fixation: operator fails to realize that situation has deviated from norm.
   Training and procedural emphasis on range of possible symptoms/causes; fault-symptom matrix as a job-aid; decision support system; shift technical advisor/supervision.
4. Signal unreliable: operator treats signal as false due to its unreliability.
   Improved signal reliability; diversity of signals; increased level of tolerance on the part of the system, or delay in effects of error, which allows error detection and correction (decreases "coupling"); training in consequences associated with incorrect false-alarm diagnosis.
5. Signal absent: signal absent due to a maintenance/calibration failure or a hardware/software error.
   Provide signal; redundancy/diversity in signaling-design approach; procedures/training to allow operator to recognize when signal is absent.
6. Signal-discrimination failure: operator fails to realize the signal is different.
   Improved ergonomics in the interface design; enhanced training and procedural support in the area of signal differentiation; supervision and checking.

Observation/data collection

7. Attention failure: lapse of attention.
   Multiple signal coding; enhanced alarm salience; improved task organization with respect to back-up crew and rest pauses.
8. Inaccurate recall: operator remembers data incorrectly (usually quantitative data).
   Non-reliance on memorized data, which would necessitate better interface design — as data are received, they can either be acted on while still present on a display (controls and displays are co-located) or at least be logged onto a "scratch pad"; sufficient displays for presenting all information necessary for a decision/action simultaneously; printer usage; training in non-reliance on memorized data.
9. Confirmation bias: operator only selects data that confirms given hypothesis, and ignores other disconfirming data sources.
   Problem-solving training; team training (including training in the need to question decisions, and in the ability of the team leader(s) to take constructive criticism); shift technical advisor (diverse, highly qualified operator who can "stand back" and consider alternative diagnoses); functional procedures; high-level information displays; simulator training; high-level alarms for system-integrity degradation; automatic protection.
10. Thematic vagabonding: operator flits from datum to datum, never actually collating it meaningfully.
    Problem-solving training; team training; simulator training; functional-procedure specification for decision-timing requirements; high-level alarms for system-integrity degradation.
11. Encystment: operator focuses exclusively on only one data source.
    Problem-solving training; team training (including training in the need to question decisions, and in the ability of the team leader(s) to take constructive criticism); shift technical advisor; functional procedures; high-level information displays; simulator training; high-level alarms for system-integrity degradation.
12. Stereotype fixation revisited: need for information is not prompted (by either memory or procedures).
    Emergency procedure enhancements, and emphasis of key symptoms and indicators to be checked; team training; problem-solving training; alarm re-prioritization; simulator training.
13. Crew-functioning problem: allocation of responsibility or priorities is unclear, with the result that data collection/observation fails.
    Improved crew coordination, and allocation of responsibilities; team training; emergency training; accident-management procedures; remote-incident-monitoring/back-up center; high-level displays; "crash-shutdown" facilities.
14. Cognitive/stimulus overload: operator too busy, or being bombarded by signals, with the result that effective data collection/observation fails. See 2 above.

Source: Kirwan, B. (1994). *A Guide to Practical Human Reliability Assessment.* London: Taylor & Francis.

situations and form intentions to act in emergency NPP operations; inappropriate intentions therefore define human error. CES allows the analyst to investigate situational characteristics conducive to intention failures and the form and consequences of these failures. It does so by allowing the analyst to make various adjustments in the processing mechanisms and knowledge base underlying CES's problem-solving model, as well as in the input stream of plant data. A relatively sophisticated understanding of human information processing and situational modeling is obviously required of the analyst in utilizing this model.

Data from both simulators and simulations must be interpreted with caution (Sharit, 1993). With simulators the particular application may determine the usefulness of data on human error. For example, in NPP simulators real-world conditions of stress can rarely be duplicated, whereas in flight simulators pilots can not only experience "simulator sickness," but, unlike NPP operators, are also likely to have experienced many of the emergency conditions being simulated, and therefore may respond with different strategies. In any simulator application, however, it is unlikely that many of the more subtle factors that contribute to establishing the context within which the human performs can be captured. With simulation models, the requirements for verifiability and validity, fundamental to any simulation modeling application, must be satisfied. In human and system reliability applications, validating simulation outcomes based on assumptions concerning human performance variables and their multiple interactions is far from trivial.

35.7 Accident and Incident Analysis

While the primary goal in human and system reliability should be on preventing adverse outcomes, the occurrence of accidents demands that mechanisms be in place for determining the causes of such events, and these mechanisms will necessarily be qualitative in emphasis. As indicated in Figure 35.5, tracing paths from accidents to causes is much more constrained, and therefore feasible; not surprisingly, these endeavors tend to be successful, though often expensive. Depending on the method used, however, these backward paths can be traced to levels of causality that are relatively shallow (e.g., the worker forgot to check protective equipment properly prior to entering a hazardous area), to those that are much more revealing. A modeling framework consisting of an error causation chain (Figure 35.13) and corresponding flow charts corresponding to different stages in this chain (e.g., Figure 35.12) is, in principle, more effective at analyzing adverse consequences such as an injury, near miss, or poor product quality that directly result from the *external* error or mode of malfunction (Figure 35.13), than when used for the purpose of error prediction. A hypothetical case study illustrating the use of this modeling framework in a process control scenario is presented in CCPS (1994, p. 100).

An intuitively appealing and cost-effective method for investigating accidents is "change analysis." Change analysis techniques are based on the well-documented general relationship between change and increased risk. These techniques were developed at the Rand Corporation and improved by Kepner and Tregoe (1981). The primary objective of these techniques is to establish accident-free reference bases and then systematically identify changes or differences relative to the accident/incident situation. A worksheet is typically used to explore potential changes contributory to adverse outcomes. The various factors that the analyst feels should be included in the analysis are listed under the first column. These are designed to ask the questions: who (e.g., operator, fellow worker, supervisor), where, what (e.g., protective equipment), when, how, and why in terms of task factors, working conditions, initiating events, and management control factors. The next columns, respectively, address each of these factors in terms of the present (accident/incident) situation, prior situation, a comparison of these two situations in order to identify changes or differences, and a listing of all the differences. Finally, differences are analyzed for their effect on the accident in terms of both their independent and interactive contributions, and the resulting analysis is integrated into the overall accident system evaluation process.

If integrated with a contextual modeling framework, change analysis can contribute to identification of underlying causes of human error and/or adverse system consequences. For example, a change in a work procedure may result in workers violating the new procedure due to a perceived increase in risk or work load associated with the task. An organizational structure characterized by poor feedback channels between workers and management, and by policies that emphasize productivity, reinforce continuation of these violations which, under other changes instituted in the system, can result in increased exposure to hazards. Note that change analysis techniques can also be used *proactively* to predict adverse consequences

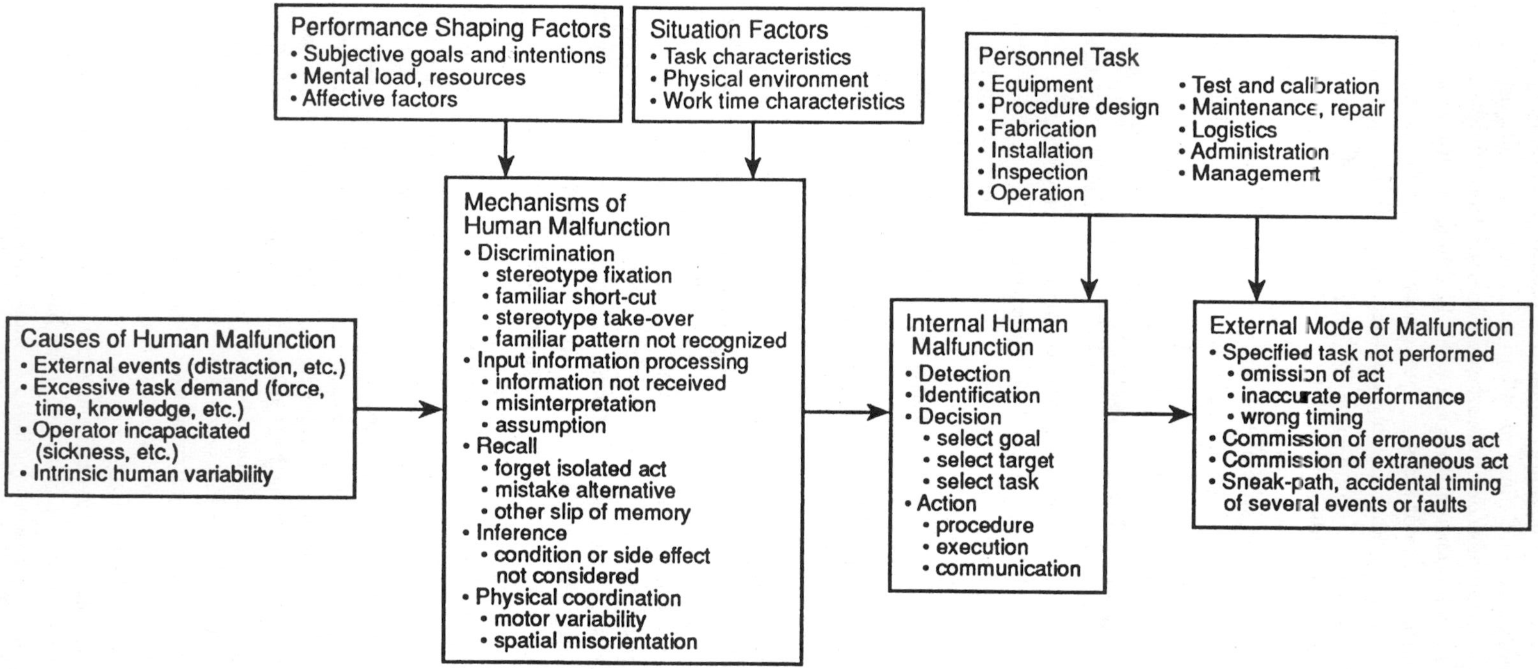

FIGURE 35.13 A model of an error causation chain tracing human error in a backward sequence from its consequences, through "what" happened (the external error mode), "how" it happened (the internal error mode), and "why" it happened (an interplay between PSFs, error-inducing factors, and the internal error mechanisms or error tendencies). (From Rasmussen, J. (1982). Human errors: a taxonomy for describing human malfunction in industrial installations, *Journal of Occupational Accidents*, 4, 311-333. With permission.)

by investigating potential problems associated with proposed changes in normal or stable functioning systems. In this "change control" application, mechanisms for detecting and monitoring change are essential.

The Management Oversight and Risk Tree (MORT) relies on a logic diagram for investigating the various factors contributing to an accident (Johnson, 1980). Factors considered by MORT include lines of responsibility, barriers to unwanted energy, and management factors. The emphasis on barriers essentially constitutes a "barrier analysis" which, like change analysis, is an accident analysis technique. In barrier analysis, the various energy sources to be considered (e.g., the energy associated with heights as derived from scaffolds or the energy associated with gamma radiation) and barriers to these energy sources (e.g., lanyards or lead-shielded body suits) are analyzed to determine the availability and adequacy of barrier safeguards.

By reasoning backward through a sequence of contributory factors, posing "yes" and "no" to questions along the way, and through the availability of accompanying text that aids the analyst in judging whether a factor is adequate or less than adequate (LTA), MORT assists the analyst in detecting omissions, oversights, or defects. MORT is especially powerful in identifying organizational root causes, which makes it useful for PRAs as well as accident analysis (Gertman and Blackman, 1994).

A generic MORT diagram is illustrated in Figure 35.14. The three flow charts in Figures 35.15 through 35.17 correspond to elaboration of the "OR" gate in Figure 35.14 labeled "S" arising from "Specific Control Factors." Specifically, they expand on the breakdown of the "Accident" event resulting from specific control factors being LTA, into "Barriers" (Figure 35.15), "Persons or Objects in Energy Channel" (Figure 35.16), and "Incident" (Figure 35.17). The codes in these diagrams correspond to further elaborations depicted in a series of MORT flow charts that can be found in Gertman and Blackman (1994).

Other relatively well-known accident analysis techniques include Events and Causal Factors Charting (Ontario Hydro, 1978), the Sequentially Timed Events Plotting Procedure or STEP (Hendrick and Benner, 1987), and Root Cause Coding (Armstrong et al., 1988).

35.8 The Quantitative Perspective to Human and System Reliability Analysis

The primary objective of all quantitative approaches to HRA is to derive estimates of the probability of error associated with various facets of performance. While some behavioral scientists may accept the validity of such estimates for relatively simplistic and repetitive behaviors, they would not be inclined to attach any significance to these probability estimates for behaviors that are even moderately complex. Such behaviors are considered too multidimensional to be summarized by single numerical estimates. However, with this information the opportunity exists for PRAs to be performed that enable system reliability to be evaluated, and ultimately, for the most cost-effective design interventions to be determined from the standpoint of reducing system risks.

The most classical approach to quantitative human reliability assessment is generally associated with engineering reliability techniques (Kapur and Lamberson, 1977). When applying these techniques, a fundamental distinction exists between static and dynamic reliability models. In the static case, a prescribed time period of operation is implied, whereas in the dynamic case the reliability of the component is expressed as a function of time. Some of the early quantitative approaches to modeling human reliability closely paralleled the classical paradigm (Askren and Regulinski, 1969, 1971). In the dynamic case, this involved deriving an instantaneous error rate, $h_e(t)$ (analogous to the hazard function in classical reliability modeling), from which the human reliability function, $R_e(t)$, corresponding to the probability that the human will not have committed an error by time t, could be derived.

This distinction between the static and dynamic cases also exists in many of the current quantitative perspectives to HRA. However, in contrast to approaches to modeling human error that have adopted classical engineering reliability methods, these perspectives address real-world tasks and a variety of task-related behavioral issues. The models corresponding to the dynamic case are often referred to as "time

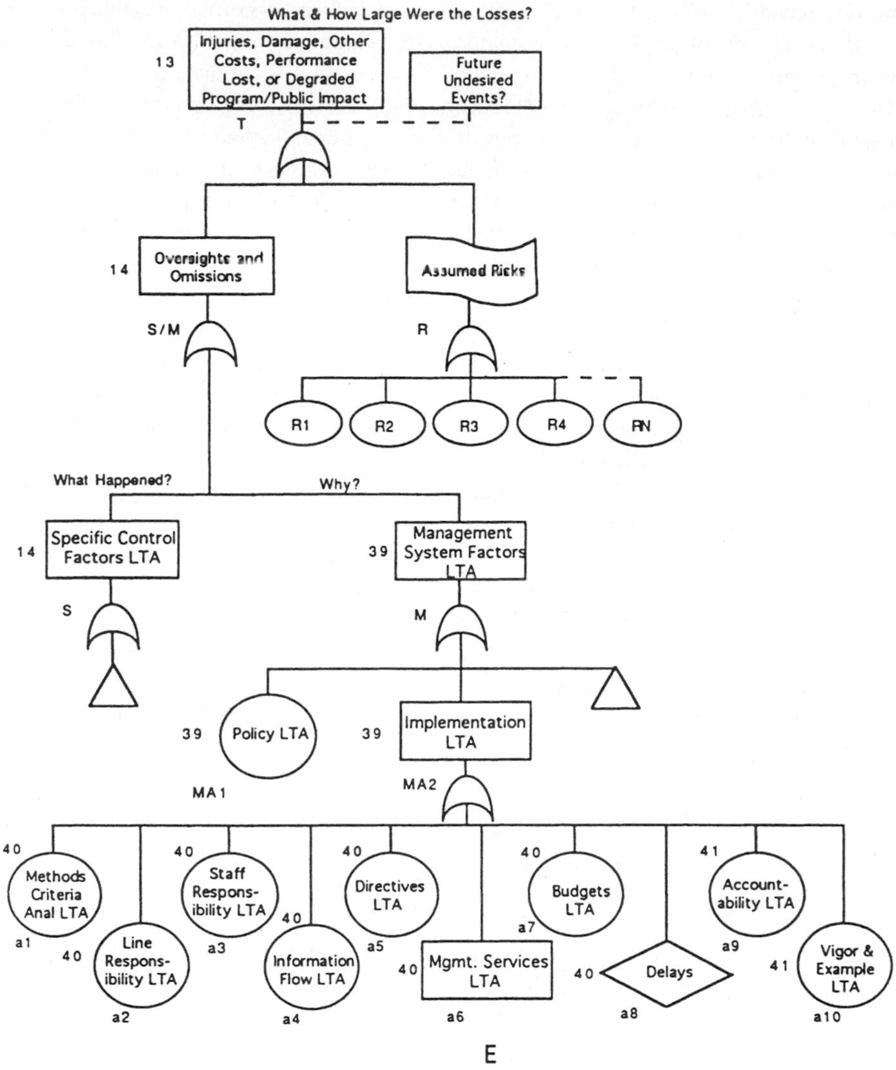

FIGURE 35.14 A Generic MORT diagram. (From Gertman, D. I. and Blackman, H. S. (1994). *Human Reliability & Safety Analysis Data Handbook*. New York: John Wiley & Sons. With permission.)

reliability correlations" (TRC). These models constitute mathematical relationships between the failure of an operator or crew to appropriately deal with some compelling situation and the time t (in minutes) after the initiation of this situation, and are usually depicted in the form of probability vs. time plots. An example of a TRC is the human cognitive reliability model (Hannaman and Spurgeon, 1988). This model is characterized by a three-parameter Weibull distribution (a type of probability distribution often used in reliability modeling) whose parameter values depend on whether the crew response to a problem can be best described by either the skill-, rule-, or the knowledge-based level of performance. Figure 35.18 illustrates the predictions of this model for a normalized time variable (that controls for contributions to crew response times that are unrelated to human activities). More general TRC models capable of reflecting a wide variety of influences are discussed by Dougherty and Fragola (1988).

In the remainder of this section, the focus will be on two techniques used to determine HEPs. Although each of these techniques relies heavily on the concept of PSFs (as do most other currently used quantitative approaches to HRA), they illustrate the different types of approaches that can be taken to predicting human error.

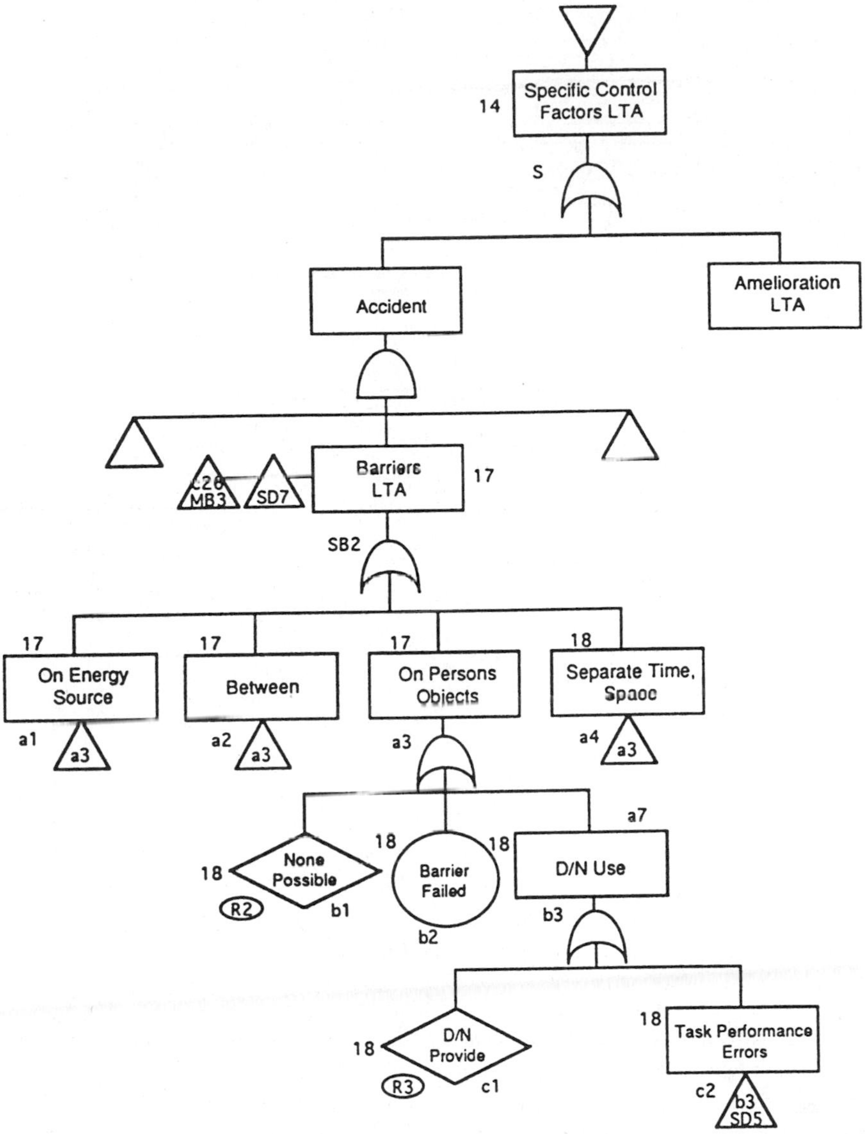

FIGURE 35.15 The Specific Control Factors MORT tree deriving from Figure 35.14. (From Gertman, D. I. and Blackman, H. S. (1994). *Human Reliability & Safety Analysis Data Handbook.* New York: John Wiley & Sons. With permission.)

THERP

The Technique for Human Error Rate Prediction, generally referred to as THERP, is perhaps the most well-known of all HRA techniques. The motivation for its development was twofold: (1) to derive estimates of human error that could be used in system fault trees in combination with other failure event data in order to evaluate system reliability; and (2) to aid in making decisions involving design trade-offs by evaluating the contribution of human error to subsystem or system reliability. The basic steps comprising THERP are outlined in Figure 35.19 and detailed in a work by Swain and Guttmann (1983) sponsored by the U.S. Nuclear Regulatory Commission. The overall objective of the method is to (1) *decompose* human tasks to a level of description whereby human error probabilities (HEPs) can be

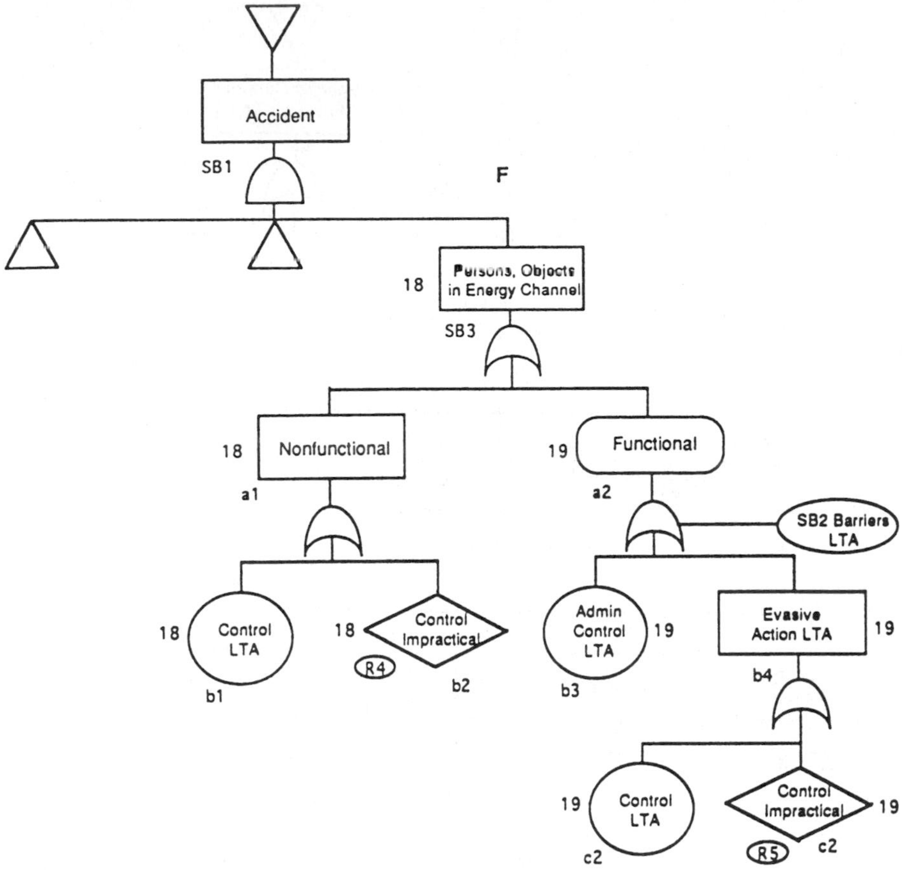

FIGURE 35.16 The Accidents with Persons, Objects in Energy Channel MORT tree derived from Figure 35.15. (From Gertman, D. I. and Blackman, H. S. (1994). *Human Reliability & Safety Analysis Data Handbook.* New York: John Wiley & Sons. With permission.)

estimated for the constituent sequential subtasks, and then to (2) *aggregate* these estimates to derive probabilities of task failure.

The first four steps of this method establish which work activities will be emphasized, the concerns for human error associated with these operations, time and skill requirements, and factors related to error detection and the potential for error recovery. The results of this effort are represented by a type of event tree referred to as a "probability tree." Each relevant subtask in a probability tree is characterized by two limbs representing either successful or unsuccessful performance, as illustrated in Figure 35.20. In this hypothetical task, a worker checks the calibration of a series of set points consisting of three comparators that operate by OR logic to detect abnormal pressure of a plant stream. This task requires that test equipment be set up correctly; an uncorrected error would result in miscalibration of all three comparators and consequently a failure (denoted by F_1 and F_2). Note that path "a" has been arbitrarily designated a success path (S_1) under the assumption that if the test equipment was set up correctly, adopting a conservative estimate of 10^{-2} for the probability of miscalibration would result in a probability of 10^{-6} of miscalibrating all three comparators, a value considered negligible in most PRAs for which this type of analysis would serve as input.

The next five steps (Figure 35.20) constitute the quantitative assessment stage. First, HEPs are assigned to each of the limbs of the tree corresponding to incorrect performance. These probabilities are referred to as *nominal* HEPs and, in theory, are presumed to represent medians of lognormal distributions.

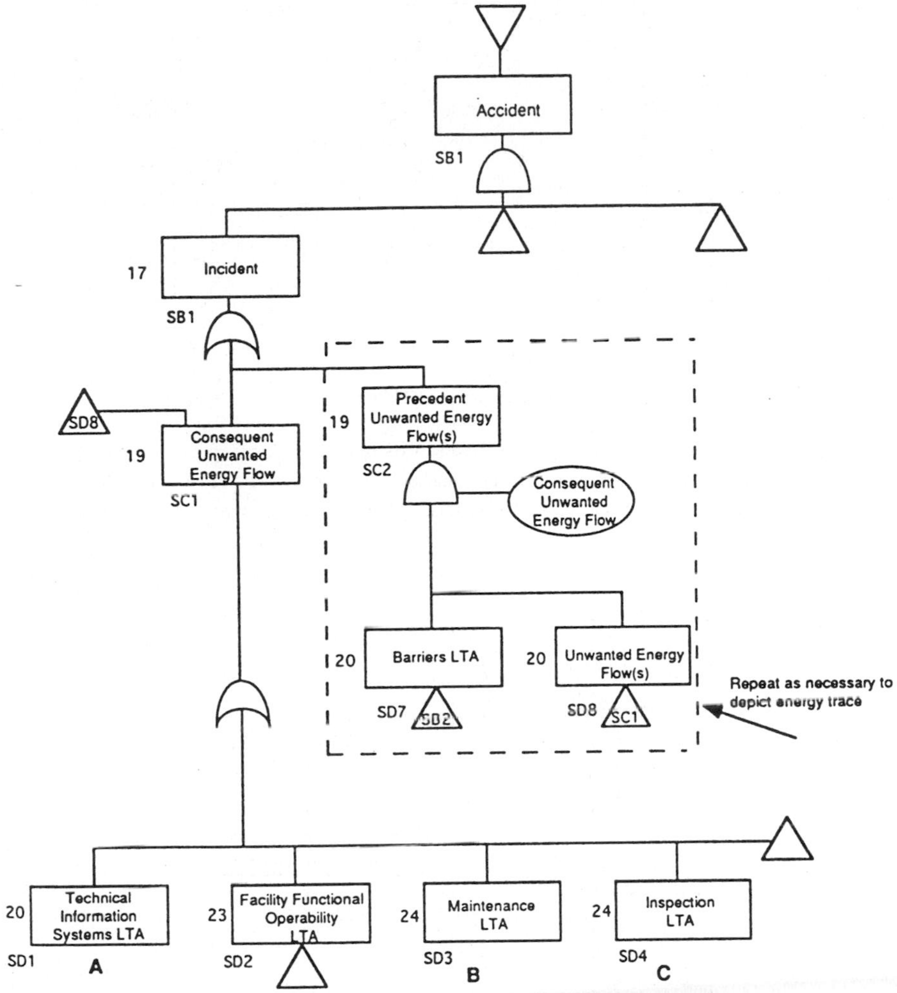

FIGURE 35.17 The Accident, Incident MORT tree derived from Figure 35.15. (From Gertman, D. I. and Blackman, H. S. (1994). *Human Reliability & Safety Analysis Data Handbook*. New York: John Wiley & Sons. With permission.)

Associated with each nominal HEP are upper and lower uncertainty bounds (UCBs) that reflect the variance associated with any given error distribution. The square root of the ratio of the upper to the lower UCB defines the *error factor*, the choice of which reflects the variability associated with the distribution for a particular error. Large error factors reflect variance arising from the assignment of nominal HEPs in addition to the variance associated with individual differences in worker performance. Swain and Guttmann (1983) provide a variety of nominal HEPs and their associated error factors for a variety of NPP tasks. Tables 35.6 and 35.7 illustrate these values for two different tasks; the values in Table 35.7 refer to *joint* HEPs in that the performance of a crew rather than an individual worker is being evaluated. In general, however, the absence of existing data from the operations in question will require that nominal HEPs be derived from other sources such as: (1) expert judgment using a variety of techniques such as absolute probability judgment and paired comparisons (Kirwan, 1994); (2) simulators (Gertman and Blackman, 1994); and (3) data from other jobs similar in psychological content to the operations of interest.

In order to account for more specific individual, environmental, and task-related influences on performance, nominal HEPs are subjected to a series of refinements (Figure 35.19). First, nominal HEPs are modified based on the influence of PSFs, resulting in *basic* HEPs (BHEPs). In some cases, guidelines are

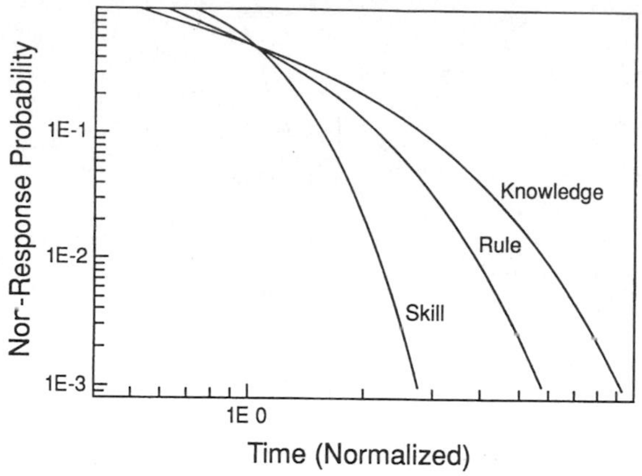

FIGURE 35.18 An example of a time reliability correlation model corresponding to skill-based, rule-based, and knowledge-based processing for work crews. (From Hannaman, G. W. and Worledge, D. H. (1988). Some Developments in Human Reliability Analysis Approaches and Tools. *Reliability Engineering and System Safety,* 22, 235-256. With permission.)

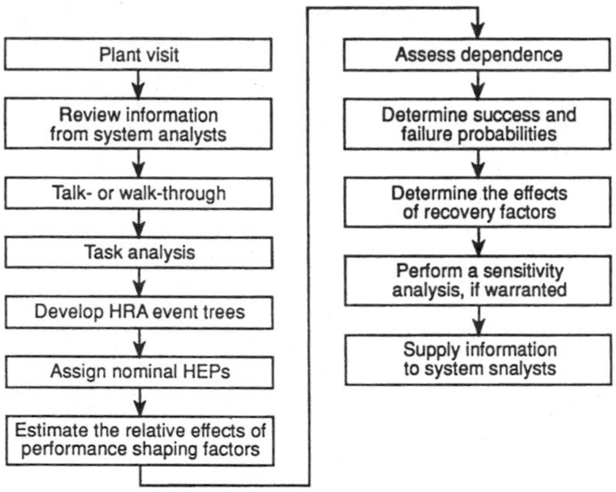

FIGURE 35.19 The steps comprising THERP. (From Swain, A. D. and Guttmann, H. E. (1983). *Handbook of Human Reliability Analysis with Emphasis on Nuclear Power Plant Applications.* NUREG/CR-1278, U.S. Nuclear Regulatory Commission, Washington, D.C. With permission.)

provided in the form of tables indicating the direction and extent of influence on nominal HEPs of particular PSFs. For example, Table 35.8 illustrates the influence of stress on nominal HEPs as a function of type of task and worker experience. Next, a nonlinear *dependency model* is incorporated that considers *positive* dependencies that exist between adjacent limbs of the tree, resulting in *conditional* HEPs (CHEPs). In a positive dependency model, failure on a subtask increases the probability of failure on the following subtask, and successful performance of a subtask decreases the probability of failure in performing the subsequent task element. Instances of negative dependence can be accounted for, but require the discretion of the analyst. In the case of positive dependence, THERP provides equations for modifying BHEPs to CHEPs based on the extent of dependence assumed by the analyst.

For example, in Figure 35.20, for the case of a small setup error (α), *complete* dependence was assumed between the setup task element and the calibration of the first set point, as well as between the subtasks of calibrating the second and third set points. The .9 probability that the operator will be alerted by the misalignment of the second setup and therefore recheck the test setup represents an instance of negative

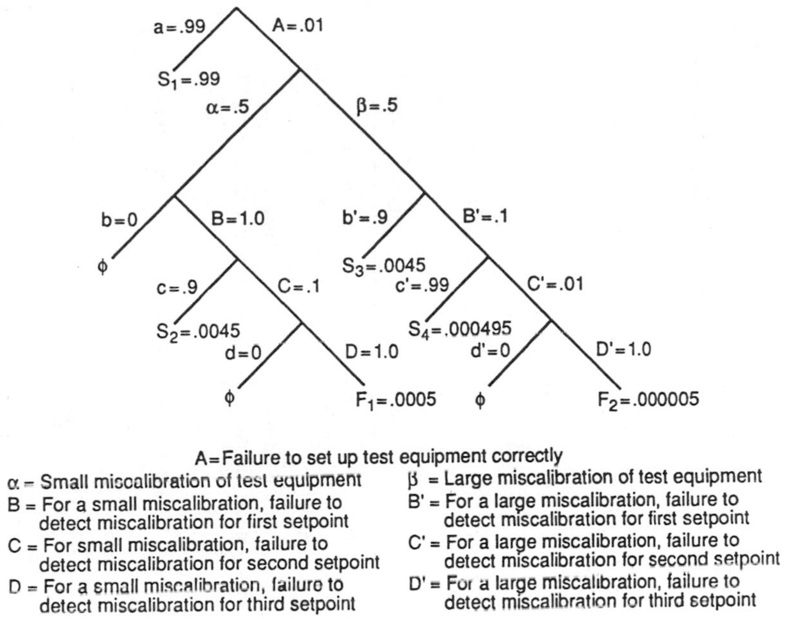

FIGURE 35.20 An example of a probability tree used in THERP for a hypothetical calibration task. (From Swain, A. D. and Guttmann, H. E. (1983). *Handbook of Human Reliability Analysis with Emphasis on Nuclear Power Plant Applications*. NUREG/CR-1278, U.S. Nuclear Regulatory Commission, Washington, D.C. With permission.)

TABLE 35.6 Estimated HEPs for Errors of Commission in Reading and Recording Quantitative Information from Unannunciated Displays

Item	Display of Task	HEP[a]	EF
1	Analog meter	0.003	3
2	Digital readout (less than four digits)	0.001	3
3	Chart recorder	0.006	3
4	Printing recorder with large number of parameters	0.05	5
5	Graphs	0.01	3
6	Values from indicator lamps that are used as quantitative displays	0.001	3
7	Recognition that an instrument being read is jammed, if there are no indicators to alert the user	0.1	5
8	Less than three	—[b]	—
9	More than three	0.01	3
10	Simple arithmetic calculations with or without calculators	0.01	3
11	Detect out-of-range arithmetic calculations	0.05	5

[a] Multiply HEPs by 10 for reading quantitative values under a high level of stress if the design violates a strong populational stereotype (e.g., a horizontal analog meter in which values increase from right to left). In this case, "letters" refer to those that convey no meaning. Groups of letters such as motor-operated value (MOV) do convey meaning, and the recording HEP is considered to be negligible.

[b] Negligible 0.001 (per symbol).

Source: Gertman, D.I. and Blackman, H.S. (1994). *Human Reliability & Safety Analysis Data Handbook.* New York: John Wiley & Sons.

dependence; for most PRA applications, however, the assumption of *zero* dependence as opposed to negative dependence will lead to more conservative estimates of HEPs. In addition to zero and complete dependence, the dependency model also accounts for low, medium, and high levels of dependence between adjacent task elements.

TABLE 35.7 Nominal Model of Estimated HEPs and EFs for Diagnosis within Time (T) by Control Room Personnel of Abnormal Events Annunciated Closely in Time[a]

Item	*T* (minutes[b] after T_0)[c]	Median joint HEP[d] for diagnosis of single or first event	EF	Item	*T* (minutes[b] after T_0)[c]	Median joint HEP[d] for diagnosis of single or second event	EF	Item	*T* (minutes[b] after T_0)[c]	Median joint HEP[d] for diagnosis of single or third event	EF
1	1	1.0	—	7	1	1.0	—	14	1	1.0	—
2	10	0.1	—	8	10	1.0	—	15	10	1.0	—
3	20	0.01	10	9	20	0.1	10	16	20	1.0	—
4	30	0.001	10	10	30	0.01	10	17	30	0.1	10
5	60	0.0001	30	11	40	0.001	10	18	50	0.01	10
6	1500	0.00001	30	12	70	0.0001	30	19	50	0.001	10
				13	1510	0.00001	30	20	80	0.0001	30
								21	1520	0.00001	30

[a] Closely in time refers to cases in which the annunciation of the second event occurs while the control room personnel are still actively engaged in diagnosing and/or planning the responses to cope with the first event. This is situation specific, but for the initial analysis, use within 10 minutes as a working definition of closely in time. Note that this model pertains to the control room crew rather than to one individual. Note that this nominal model for diagnosis includes the activities listed in Table 12-1 of NUREG/CR-1278, as perceive, discriminate, interpret, diagnosis, and the first level of decision making. The modeling includes those aspects of behavior included in the Annunciator Response Model in NUREG/CR-1278, Table 20-23; therefore, when the nominal model for diagnosis is used, the annunciator model should not be used for the initial diagnosis. The annunciator model may be used for estimating recovery factors for an incorrect diagnosis.

[b] For points between the times shown, the medians and EFs may be chosen from NUREG/CR-1278, Figure 12-4.

[c] T_0 is a compelling signal of an abnormal situation and is usually taken as a pattern of annunciators. A probability of 1.0 is assumed for observing that there is some abnormal situation.

[d] NUREG/CR-1278, Table 12-5, presents some guidelines to use in adjusting or retaining the nominal HEPs presented above.

Source: Gertman, D.I. and Blackman, H.S. (1994). *Human Reliability & Safety Analysis Data Handbook.* New York: John Wiley & Sons.

TABLE 35.8 Modifications of Estimated HEPs for Step-by-Step and Dynamic Processing as a Function of Stress

		Modifiers for nominal HEPs[a]	
Item	Stress level	Skilled[b]	Novice[b]
1	Very low (very low task load)	×2	×2
	Optimum (optimum task load)		
2	Step-by-step[c]	×1	×1
3	Dynamic[c] moderately high (heavy task load)	×1	×2
4	Step-by-step[c]	×2	×4
5	Dynamic[c] extremely high (threat stress)	×5	×10
6	Step-by-step	×5	×10
7	Dynamic[d]	0.25 (EF = 5)	0.50 (EF = 5)
	Diagnosis[d]	These are the actual HEPs to use with dynamic tasks or diagnosis. They are NOT modifiers.	

[a] The nominal HEPs are those in the data tables in Part III and in Chapter 20 of NUREG/CR-1278. Error factors are listed in NUREG/CR-1278, Table 5-20.

[b] A skilled person is one with 6 months or more experience in the tasks being assessed. A novice is one with 6 months or less experience. Both levels have the required licensing or certificates.

[c] Step-by-step tasks are routine, procedurally guided tasks, such as carrying out written calibration procedures. Dynamic tasks require a higher degree of man–machine interaction, such as decision making, keeping track of several functions, controlling several functions, or any combination of these. These requirements are the basis of the distinction between step-by-step tasks and dynamic tasks, which are often involved in responding to an abnormal event.

[d] Diagnosis may be carried out under varying degrees of stress, ranging from optimum to extremely high (threat stress). For threat stress, the HEP of 0.25 is used to estimate performance of an individual. Ordinarily, more than one person will be involved. NUREG/CR-1278, Tables 5-6 and 5-8, lists joint HEPs based on the number of control room personnel presumed to be involved in the diagnosis of an abnormal event for various times after annunciation of the event and their presumed dependence levels, as presented in the staffing model in NUREG/CR-1278, Table 5-9.

Source: Gertman, D.I. and Blackman, H.S. (1994). *Human Reliability & Safety Analysis Data Handbook.* New York: John Wiley & Sons.

At this point, success and failure probabilities for the entire task are computed. Various approaches to these computations can be taken. The simplest approach is to multiply the individual CHEPs associated with any path on the tree leading to failure and then to sum these individual failure probabilities to arrive at the probability of failure for the total task, and then assign UCBs to this probability. More complex approaches to these computations take into account the variability associated with the combinations of events comprising the probability tree (Swain and Guttmann, 1983).

Following these computations, the analyst can choose to consider ways in which errors can be recovered. Common "recovery factors" include: (1) alerting the operator to the occurrence of an error through an annunciator, in which case the HEP associated with correctly responding to the annunciator would also have to be considered in the HRA; (2) the presence of co-workers who can potentially catch a fellow worker's errors, especially during work crew performance; and (3) various types of walk through inspections that are scheduled. As with event trees, these "recovery paths" can easily be represented on the original HRA probability tree. In the case of annunciators or inspectors, the relevant failure limb is extended into two (one failure and one success) additional limbs, with the probability of the operator being alerted to the annunciator or the inspector spotting the error, respectively, feeding back into the success path of the original tree. In the case of recovery through fellow workers, BHEPs are modified to CHEPs by considering the degree of dependency between the operator and one or more fellow workers who are in a position to notice the error. The computations for total task failure can now be repeated to determine the effects of recovery factors.

In addition to considering error recovery factors, the analyst can choose to perform sensitivity analysis. An example of one approach to sensitivity analysis is to identify the most probable errors on the tree and determine the degree to which design modifications corresponding to those task elements, which would reduce the magnitudes of those errors accordingly, affect the total failure probabilities previously computed. Finally, the results of the HRA are incorporated into system risk assessments such as PRAs.

An obvious deficiency of THERP, even among proponents of this approach, is its limitations in addressing decision-based errors, or what some have referred to as "errors of commission." One approach proposed for identifying these types of errors in situations, such as NPP applications, where the system's operations are flow oriented, is SNEAK analysis (Hahn et al., 1991), which is based on methods used to identify faults in electrical circuits. A computer-interactive "Sneak Analysis Tool" is available to the analyst for identifying these errors or "sneak conditions." These errors can then be used as starting points (i.e., initiating events) in event trees to determine whether these errors could be recovered (Blackman, 1991). Although these tools potentially provide valuable qualitative information to HRA analysts, if probabilities for these errors of commission were available the modification of this error based on the use of event trees for determining recovery paths would also provide useful inputs into PRAs. A method proposed for estimating HEPs for these types of errors is referred to as INTENT (Gertman et al., 1992). This method is based on subjective assessments from experts and borrows heavily from a method for deriving HEPs discussed below.

SLIM-MAUD

SLIM refers to the Success Likelihood Index Methodology (Embrey et al., 1984), a procedure for deriving HEPs; MAUD (Multi-Attribute Utility Decomposition) refers to a computer-interactive implementation of SLIM. In contrast to THERP, SLIM allows the analyst to focus on any human action or task, including those that can lead to highly infrequent events. Consequently, this method can provide inputs into PRAs at various system levels; that is, the HEPs can reflect relatively low-level actions that cannot be further decomposed as well as more broadly defined actions that encompass many of these lower-level actions. This increased flexibility, however, comes at the expense of a greatly reduced emphasis on task analysis and an increased reliance on subjective assessments.

SLIM assumes that the probability a human will carry out a particular task successfully depends on the combined effects of a number of PSFs that can be identified and appropriately evaluated through expert judgment. Task domain experts are assumed to be capable of assessing the relative importance (or weights) of each PSF with respect to the likelihood of human error in the task being evaluated and, independently of this assessment, rating how good or bad each PSF actually is for each of these tasks. The likelihood of success for each human action under consideration is determined by summing the product of the weights and ratings for each PSF, resulting in numbers (SLIs) that represent a scale of likelihood of success. These SLIs are useful in their own right, especially when the actions under consideration represent alternative modes of response in an emergency scenario, and the analyst is interested in determining which types of responses are least or most likely to succeed. However, for the purpose of conducting PRAs, SLIM converts the SLIs to HEPs.

The basic procedures for implementing SLIM are summarized as follows. First, an appropriate group of task-domain experts are identified. These experts help identify the potential error modes associated with the human actions of interest, and the set of PSFs most relevant to performance of these actions. The identification of all possible error modes is essential, and is generally arrived at through in-depth analysis and discussions that could include task analyses and reviews of documentation concerning emergency operating procedures. Relative importance weights for the PSFs are then derived by asking each "judge" to assign a weight of 100 to the most important PSF, and then assign weights to the remaining PSFs as a ratio of the one assigned the value of 100. Each individual weight is then divided by the sum of the weights for all the PSFs, resulting in normalized weights. The judges then rate each PSF on each task, with the lowest scale value indicating that the PSF is as poor as it is likely to be under real operating conditions, and the highest value indicating that the PSF is as good as it is likely to be in terms of promoting successful task performance. The ranges of values associated with the rating scale will dictate the range of possible SLI values that are subsequently computed.

SLIs are computed for each task or action by summing the product of the normalized weights with the ratings for each PSF. An estimate of the HEP, which equals one minus the probability of success ($P(S)$), can then be derived using the relationship $log\ P(S) = a \times SLI + b$. To derive the constants a and b, the probabilities of success must be available for at least two tasks taken from the cluster of tasks for which the relevant set of PSFs were identified. However, even if information on such "reference" tasks is not available, methods exist for deriving HEPs for the tasks of interest. Methods also exist for deriving upper and lower uncertainty bounds for these HEPs that PRAs typically require.

MAUD represents a user-friendly computer-interactive environment for implementing SLIM. This feature ensures that many of the assumptions that are critical to the theoretical underpinnings of this methodology are met. For example, MAUD ensures that the ratings for the various PSFs by a given judge (or analyst) are independent of one another, and that the relative importance weights elicited for the PSFs are consistent with the judge's preferences. In addition, MAUD provides procedures for assisting the expert in identifying the relevant PSFs. Further details concerning SLIM-MAUD can be found in Embrey et al. (1984) and Kirwan (1994).

35.9 Summary

In summary, the task facing a human and system reliability analyst has become increasingly more complex. The analyst must first understand the problem from the standpoint of determining whether a qualitative or quantitative perspective should be emphasized. Given the various conceptual and methodological tools at the analyst's disposal, a variety of options exist in adopting either perspective. In applying qualitative approaches, the analyst may have to consider the extent to which effort is to be invested in generating appropriate system descriptions, and in identifying or developing appropriate sociotechnical systems analysis, task analysis, and cognitive task analysis techniques. At the next level, the analyst must decide on the extent to which scenarios or contexts, and the propagation of both human errors and "reasonable actions" given these contexts, are to be modeled and analyzed. Contextual analysis requires that the sociotechnical systems analysis and task and cognitive task analyses be linked; this process, in turn, is governed by the relevant system descriptions. Furthermore, this contextual modeling and analysis process implies that a suitable characterization or model of human error (and of human violations) be available and integrated into the systems model.

In adopting a quantitative emphasis, the analyst must have a clear understanding of the motivation for deriving quantitative estimates of human error. Often, this requirement will be equivalent to understanding the motivation and objectives associated with performing quantitative risk assessments such as PRAs and PSAs. The analyst must then choose from a number of methods available for deriving estimates of human error probabilities (HEPs). In addition to understanding the assumptions and inner workings of each of these methods, the analyst must also be able to address issues related to criteria such as cost, practicality, usefulness, face validity, and training requirements. Moreover, the analyst must decide to what extent qualitative techniques need to be integrated into the HRA, giving rise to a potentially large number of hybrid HRA methods.

Finally, a more complete treatment of human and system reliability analysis, which is beyond the scope of this chapter, entails consideration of issues related to data collection and information systems organization in order to establish a reliability program dedicated to continuous improvement. However, understanding what data to collect, how to organize data so that it answers the questions of interest, and how to update these efforts based on ongoing human and system reliability analysis, requires an understanding of many of the fundamental qualitative and quantitative considerations discussed to this point. Finally, risk management or error reduction strategies need to be integrated into any human and system reliability analysis program. This requires understanding the effects of various design interventions, both in terms of how well they serve as barriers to human error and violations and to the adverse system outcomes they can induce, and in terms of how the design interventions can contribute to new contexts that can give rise to adverse outcomes.

This summary, which essentially defines the boundaries within which the ideal human and system reliability analyst operates, is presented in Figure 35.21. In the coming years we can expect many of these concepts and methods to become further refined and new techniques to be introduced, especially as analysts continue to gain insights into this problem.

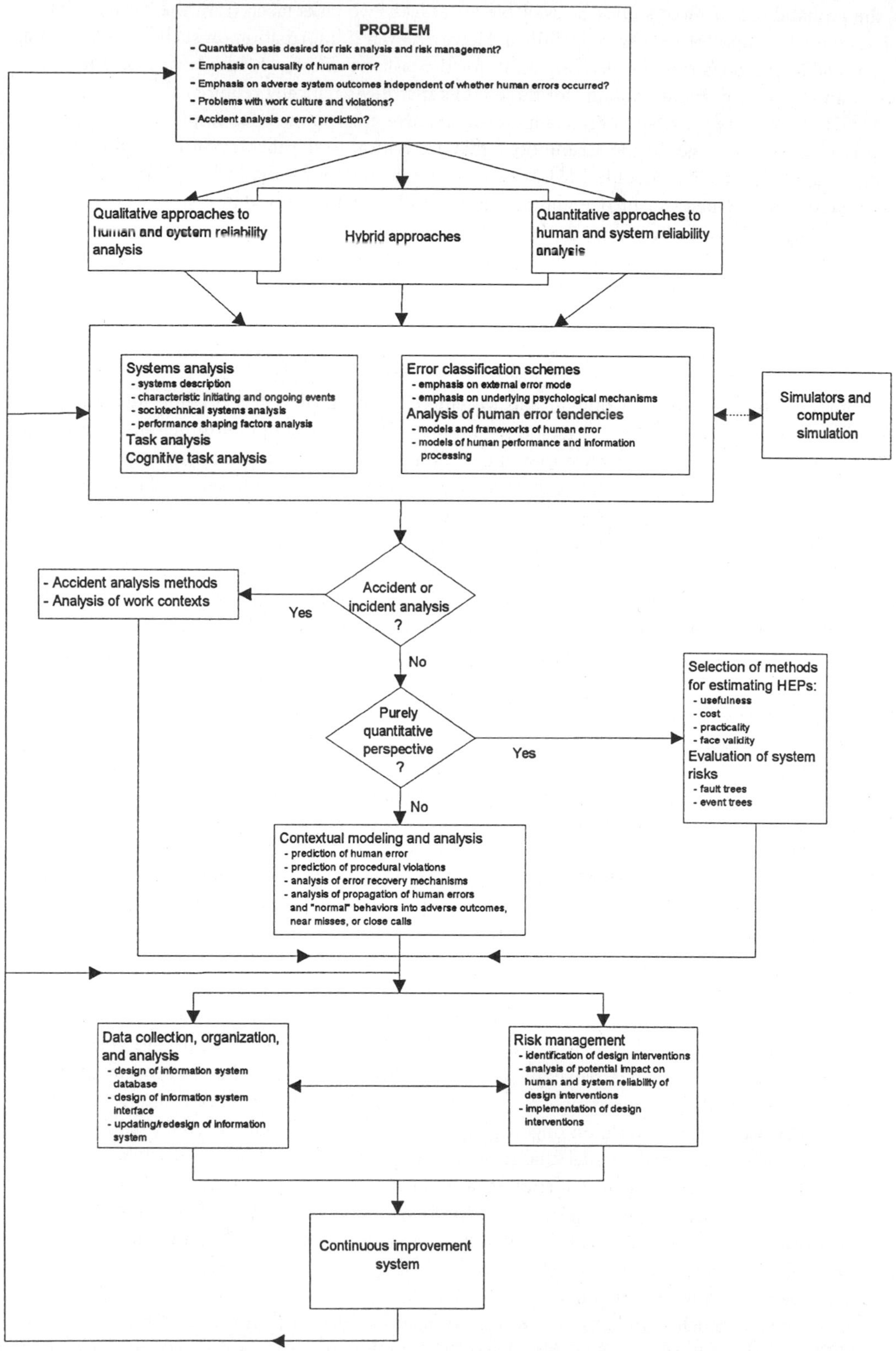

FIGURE 35.21 Defining the boundaries within which the human and system reliability analyst operates.

References

Armstrong, M. E., Cecil, W. L., and Taylor, K. (1988). *Root Cause Analysis Handbook.* Report No. DPSTOM-81, E.I. Dupont De Nemours & Company, Savannah River Laboratory, Aiken, SC.

Askren, W. B. and Regulinski, T. L. (1969). *Mathematical Modeling of Human Performance Errors for Reliability Analysis of Systems.* Aerospace Medical Research Laboratory, AMRL-TR-68-93.

Askren, B. W., and Regulinski, T. L. (1971). *Quantifying Human Performance Reliability.* Technical Report AFHRL-TR-71-22, Air Force Systems Command, Brooks Air Force Base, Texas.

Blackman, H. S. (1991). Modeling the influence of errors of commission on success probability, *Proceedings of the Human Factors Society 35th Annual Meeting,* pp. 1085-1089.

Center for Chemical Process Safety (CCPS). (1989). *Guidelines for Chemical Process Quantitative Risk Analysis.* New York: American Institute of Chemical Engineers.

Center for Chemical Process Safety (CCPS). (1992). *Guidelines for Hazard Evaluation Procedures, Second Edition, with Worked Examples.* New York: American Institute of Chemical Engineers.

Center for Chemical Process Safety (CCPS). (1994). *Guidelines for Preventing Human Error in Process Safety.* New York: American Institute of Chemical Engineers.

Dhillon, B. S. and Singh, C. (1988). *Engineering Reliability.* New York: John Wiley & Sons.

Dougherty, E. M. and Fragola, J. R. (1988). Foundations for a time reliability correlation system to quantify human reliability. *4th IEEE Conference on Human Factors and Power Plants,* pp. 268-278.

Embrey, D. E., Humphreys, P., Rosa, E. A., Kirwan, B., and Rea, K. (1984). *SLIM-MAUD: An Approach to Assessing Human Error Probabilities Using Structured Expert Judgment.* NUREG/CR-3518, U.S. Nuclear Regulatory Commission, Washington, D.C.

Fleishmann, E. and Quaintance, M. (1984). *Taxonomies of Human Performance: The Description of Human Tasks,* Orlando, FL, Academic Press.

Gertman, D. I., Blackman, H. S., Haney, L. N., Seidler, K. S., and Hahn, H. A. (1992). INTENT: A method for estimating human error probabilities for decision-based errors. *Reliability Engineering and System Safety,* 35, 127-137.

Gertman, D. I. and Blackman, H. S. (1994). *Human Reliability & Safety Analysis Data Handbook.* New York: John Wiley & Sons.

Hahn, A. H. and deVries II, J. A. (1991). Identification of Human Errors of Commission Using Sneak Analysis, *Proceedings of the Human Factors Society 35th Annual Meeting,* pp. 1080-1084.

Hannaman, G. W. and Worledge, D. H. (1988). Some Developments in Human Reliability Analysis Approaches and Tools. *Reliability Engineering and System Safety,* 22, 235-256.

Hendrick, K. and Benner, L. Jr. (1987). *Investigating Accidents with STEP.* New York: Marcel Dekker.

Hollnagel, E. (1993). *Human Reliability Analysis: Context and Control.* London: Academic Press.

Johnson, W. G. (1980). *MORT Safety Assurance Systems.* New York: Marcel Dekker.

Kepner, C. H. and Tregoe, B. B. (1981). *The New Rational Manager.* Princeton, NJ: Kepner-Tregoe Inc.

Kapur, K. C. and Lamberson, L. R. (1977). *Reliability in Engineering and Design.* New York: John Wiley & Sons.

Kirwan, B. (1994). *A Guide to Practical Human Reliability Assessment.* London: Taylor & Francis.

Kirwan, B. and Ainsworth, L. K. (1992). *Guide to Task Analysis.* London: Taylor & Francis.

Luczak, H. (1997). Task analysis, in G. Salvendy (ed.) *Handbook of Human Factors and Ergonomics, Second Edition.* New York: John Wiley & Sons.

Ontario Hydro (1977). *Events and Causal Factors Charting.* U.S. Department of Energy 76-45/14, SSDC-14, Ontario Hydro Toronto, Canada.

Ozog, H. (1985) Hazard identification, analysis, and control. *Chemical Engineering,* February 18, 161-170.

Pennycook, W. A. and Embrey, D. E. (1993). An operating approach to error analysis, in *Proceedings of the First Biennial Canadian Conference on Process Safety and Loss Management.* Edmonton, Alberta, Canada. Waterloo, Ontario, Canada: Institute for Risk Research, University of Waterloo.

Perrow, C. (1984). *Normal Accidents: Living with High-Risk Technologies.* New York: Basic Books.

Rasmussen, J. (1982). Human errors: a taxonomy for describing human malfunction in industrial installations, *Journal of Occupational Accidents*, 4, 311-333.

Rasmussen, J. (1986). *Information Processing and Human-Machine Interaction: An Approach to Cognitive Engineering*, New York: North-Holland.

Reason, J. (1990). *Human Error.* Cambridge: Cambridge University Press.

Reason, J. (1995) A systems approach to organizational error. *Ergonomics*, 38, 1708-1721

Roland, H. E. and Moriarty, B. (1983). System Safety Engineering and Management. New York: John Wiley & Sons.

Sharit, J. (1993). Human reliability modeling, in K.B. Misra (ed.) New Trends in System Reliability Evaluation. Amsterdam: Elsevier, 369-410.

Sharit, J. (1997). Allocation of functions, in G. Salvendy (ed.) *Handbook of Human Factors and Ergonomics,* Second Edition. New York: John Wiley & Sons.

Sharit, J., Czaja, S.J., Augenstein, J., and Dilsen, K. (1996). A systems analysis of a trauma center: a methodology for predicting human error, in A.F. Ozok and G. Salvendy (eds.) *Advances in Applied Ergonomics*, 996-1101. Indiana: U.S.A. Publishing Corporation.

Siegel, A. I., Bartter, W. D., Wolf, J. J., Knee, H. E., and Haas, P. M. (1984). *Maintenance Personnel Performance Simulation (MAPPS) Model: Summary Description.* NUREG/CR-3626, U.S Nuclear Regulatory Commission, Washington, D.C.

Stammers, R. B. and Shephard, A. (1995). Task analysis, in J. R. Wilson and E. N. Corlett (eds.) *Evaluation of Human Work, Second Edition.* London: Taylor & Francis, 144-168.

Swain, A. D. and Guttmann, H. E. (1983). *Handbook of Human Reliability Analysis with Emphasis on Nuclear Power Plant Applications.* NUREG/CR-1278, U.S. Nuclear Regulatory Commission, Washington, D.C.

Vicente, K. J. and Rasmussen, J. (1992). Ecological interface design: theoretical foundations, *IEEE Transactions on Systems, Man, and Cybernetics*, 22, 589-606.

Wickens, C.D. (1992). *Engineering Psychology and Human Performance*, Second Edition. New York: HarperCollins.

Woods, D. D., Roth, E., and Pople, H. (1987). *Cognitive Environment: System for Human Performance Assessment.* NUREG-CR-4862, U.S. Nuclear Regulatory Commission, Washington, D.C.

36

Some Developments in Human Reliability Assessment

Barry Kirwan
University of Birmingham

36.1 Introduction

This chapter deals with the subject of human reliability assessment (HRA). HRA may be considered a subdiscipline of ergonomics or human factors (these terms are used interchangeably in this chapter), but it emanates also from the fields of reliability engineering and risk assessment, and is therefore a hybrid discipline. HRA is fundamentally the analysis of human failures. Unlike accident analysis, however, HRA is prospective or predictive — it is concerned with determining what can go wrong, before it happens. This is no trivial task. HRA also not only tries to determine what can go wrong (i.e., human errors), but also how likely it is to go wrong, i.e., it predicts the probabilities of different errors and failures occurring. Furthermore, since HRA has become more linked to psychology and ergonomics over the last decade and a half, it has focused on how human failures occur, and what factors cause them or increase their likelihood of occurrence. Therefore, based on such analysis, it then becomes possible to determine how to prevent such errors from occurring at all, or at least to decrease their likelihood. HRA, broadly speaking, can therefore be seen to have three interlinked functions:

0-8493-2641-9/99/$0.00+$.50

1. Determination of what can go wrong (human error identification)
2. Quantification of the probabilities of errors (human reliability quantification)
3. Reduction of error likelihood (error reduction analysis)

HRA is most commonly used in a risk assessment format, essentially determining how frequently accidental outcomes (e.g., fatalities) will occur in a given period of operation of a system (usually such predicted frequencies are very small, e.g., once in one hundred thousand years of operation). When utilized within risk assessment, HRA is effectively assessing the human contribution to risk. This contribution is integrated within the overall risk assessment framework, so that the human contribution to risk can be seen in conjunction with other contributions to risk: hardware and software failures, and environmental events. Therefore, when total risk is estimated for a system such as a chemical plant or an offshore platform, the relative contribution of human error (and human recovery capabilities) to risk can be judged by the owners, designers, and/or regulators of such a system. Sometimes human error will be seen as a major contributor to risk, and other times its role may be negligible, or at least tolerable. If, however, risk assessment and HRA do show that human error is of significant concern, there will be the need for more human factors effort to improve the designed operator support systems (interfaces, training, procedures, etc.). HRA can therefore lead to the determination of the adequacy, from a safety perspective, of the human factors considerations designed into a system.

A typical question that HRA might be used to address, therefore, is the following:

Is the human error contribution to risk for an offshore drilling system (or nuclear power plant, or chemical plant, or transportation system, etc.) acceptable?

Such a question might be posed by a regulatory body, and the oil and gas company would then be obliged to carry out an HRA/risk assessment to answer it. Sometimes, however, HRAs are carried out to compare two different designs, e.g., an automated process vs. a semiautomated process, and then the risks of the two systems are compared, in order to determine which design to utilize.

Whether HRA is used for risk assessment purposes or for design evaluations and comparisons, it is clearly relevant to human factors on two major counts. First, the prime content or subject matter of HRA is human performance (and particularly human error in actions or decisions) in the working environment, leading to system failures, often as a function of poor original attention to ergonomics factors in the design of the system. Second, the impact of many HRAs leads to the determination that more human factors input into the design process is required. Therefore, HRA is obviously related to human factors by its content and can act as a mechanism for enhancing the incorporation of more human factors into system designs.

A number of questions arise concerning the approach of HRA, and among them are the following, which will be at least partly addressed within the confines of this chapter:

- How does HRA work? (the HRA process)
- What is the scope of real HRAs, i.e., how big are they, how long do they take, what do they achieve, etc.?
- What contemporary developments are occurring in HRA?
- Is there evidence to support the validity of HRA approaches, given that they ultimately have a rather ambitious aim of predicting human performance in non-trivial industrial tasks?

This chapter therefore attempts to define the HRA process, and briefly outline some recent practical HRAs and research initiatives,* which show in more detail how HRA may be applied and how it is developing as an approach. A recent validation study is also summarized, which supports the quantitative part of the HRA process, thereby at least in part supporting the empirical validity of the approach of HRA.

*Much of the work reported in this chapter is of U.K. origin. This is due to the accessibility of this work to the author and, as the referee for this chapter noted, due to a current decrease in work in HRA in the U.S. at the present time. Although there is therefore a slight bias to U.K. work, these studies are still relatively representative of work ongoing in other countries.

36.2 Objectives

The objectives of the chapter can be stated formally as follows:

1. To give an overview of the HRA process
2. To outline the scope and some of the impacts of two recent large-scale (U.K.) HRAs, and a U.K. HRA assessment program
3. To consider recent developments in two major areas of HRA
   A new human error analysis system
   A human error probability (HEP) data bank
4. To summarize a recent validation exercise

36.3 Scope

In the limits of this chapter, a good many short-cuts are necessary. For the more interested reader, there are a number of texts on HRA which describe the overall process, how it can be used, and its impacts (Swain and Guttmann, 1983; Dhillon, 1986; Park, 1987; Dougherty and Fragola, 1987; Kirwan et al., 1988; Swain, 1989; Embrey et al., 1994; Gertman and Blackman, 1994, and Kirwan, 1994). There are also two other texts of interest which take a more pessimistic view of HRA (Reason, 1990; Hollnagel, 1993), but which are also important for their discussion of the nature of human error, as is a recent review of the nature of human error (Woods et al., 1994). The reader interested in finding out more about risk assessment may consult a number of texts (e.g., Henley and Kumamoto, 1981; Green, 1983; Cox and Tait, 1991).

Since this chapter is written by a U.K.-based author, its contents are biased to developments in HRA approaches in the U.K. However, it is believed that these developments are also relevant to other international spheres of HRA activity.

36.4 Human Reliability Assessment — The HRA Process

HRA usually progresses through a number of stages, embodied in the HRA process as shown in Figure 36.1. These stages are described below.

Problem Definition

This refers to scoping the HRA, i.e., deciding what tasks or human involvements will be addressed by the HRA (very few HRAs can address all human involvements, due to resource constraints). In some cases, many tasks will be analyzed in detail, whereas in other HRAs only a representative sample of tasks will be assessed, and many of these will be assessed only using basic rather than detailed methods. There are no fixed criteria for determining the depth and breadth of an HRA, but the following encapsulate some of the major considerations:

- The nature of the plant being assessed and the cost of failure
- The criticality of the role of the operator
- The novelty of the plant design
- The system life cycle stage

The more hazardous the plant, the more critical the role of the operator, and the more novel the plant design, then the deeper or more exhaustive the assessment must be (as with novel systems, there will inevitably be more unpredicted events or situations that the operator will have to deal with). Therefore, a hazardous, human-critical, novel plant design will warrant full error analysis of all types of errors (and a corresponding and interrelated human factors assessment). Alternatively, a well known and tested plant system that has already been operational for 30 years, has a good (i.e., low) accident record, and has been

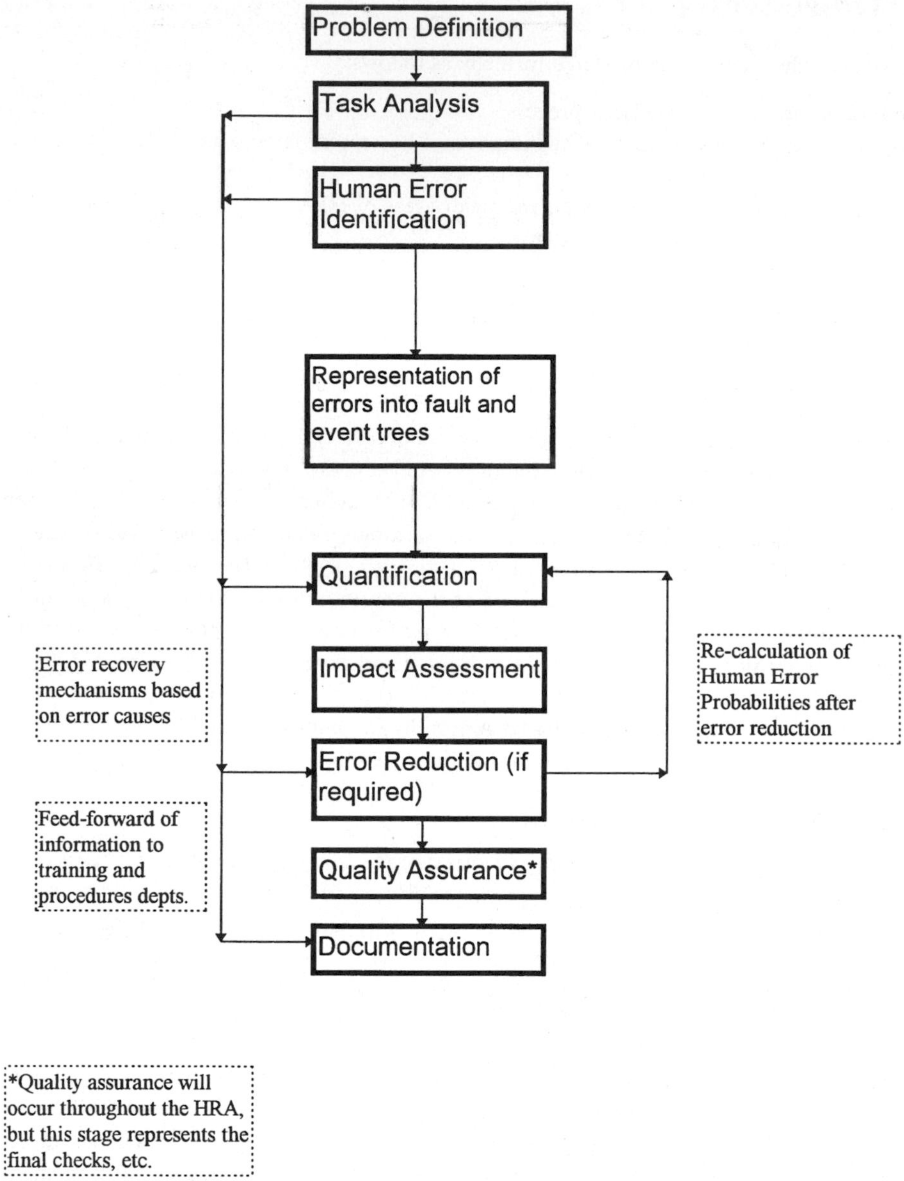

FIGURE 36.1 The HRA process. (Adapted from Kirwan, B. (1990) — A resources flexible approach to human reliability assessment for PRA, Safety and Reliability Symposium, Altrincham, September. London: Elsevier Applied Sciences, pp. 114-135.)

analyzed several times previously will warrant much less assessment effort. However, even an old plant may warrant special attention, e.g., if rule violations are starting to appear in incident records, or if new control systems are being installed, etc. Last, the earlier the life cycle stage, the more the plant can be favorably influenced as a result of error reduction, but the less powerful are the error identification and prediction techniques, and the more difficult task analysis (see below) becomes due to lack of detailed information. Furthermore, some analyses will be difficult at an early design stage, e.g., rule violation assessment (see below) in particular.

The assessment team must also decide upon the scope in terms of which operational stages are to be addressed. Typically, the major focus will be on emergency events, such as post-trip recovery actions in a nuclear power plant following a reactor trip, or the equivalent recovery actions following an event in a chemical plant. However, a number of phases of operation may be considered, namely the following:

- **Normal operation** — Actions and errors may occur in normal operation which lead to hazardous events or situations
- **Abnormal operation** — Conditions which are off-normal must be detected by operators and prevented from developing into incidents, or if the conditions are intentional, special provisions will have been made to maintain safety integrity, and the operators must act accordingly
- **Emergency operation** — An event or incident has happened, and the operators must maintain or restore safety barriers, or mitigate accident consequences (this is often the focus of HRA)
- **Beyond design basis accident** (BDBA) — A severe accident has occurred and operators must limit the accidental consequences (this will involve usage of emergency provisions and procedures, and will probably involve offsite organizations and resources)
- **Maintenance, test, and calibration** — Errors may occur during these phases which may then lead to initiating events or to the unavailability or erroneous operation of safety-related systems or instrumentation during a later event or incident (a so-called *latent failure*)
- **Outage/shutdown** — Certain operations will occur during outages or shutdown (or partial shut-down) conditions, when the usual interlocks and protective systems may be in a partly disabled state. Operations during these phases may still be hazardous, and safety integrity may rely more on administrative controls (e.g., both *ad hoc* and formal procedures), which are ultimately a human barrier form, rather than the usual physical protective systems
- **Start-up/shut-down** — Errors may occur during start-up/shut-down which will cause complications later on, or which could lead directly to events during start up or shut-down
- **Other** (specific events related to a specific plant)

Once the scope of the analysis has been determined, the next question becomes what can go wrong in the identified spheres of activities, and this is the area of human error analysis. However, before identifying what can go wrong, it is (logically) necessary to determine how tasks and operations *should* proceed, i.e., to determine what is correct or normative performance. This is the human factors subject area of task analysis, discussed next.

Task Analysis

Task analysis is a fundamental approach describing and analyzing how the operator interacts with a system and with other personnel in that system (Kirwan and Ainsworth, 1992). In particular, task analysis describes what an operator is required to do, in terms of actions and/or cognitive processes, to achieve a system goal. Task analysis methods can also detail the displays which cue the operator to perform or cease an operation, and the controls with which such operations are achieved. The first primary aim of task analysis is to create a detailed picture of human involvement, with all the necessary information for analysis of the adequacy of that involvement. Thus, although all task analysis techniques aim to describe the task, they differ in the information that is encoded, depending on the aim of the evaluation.

Prior to representing tasks in some format, data on the operator's task must be collected in some way. There are a number of data collection approaches, the most commonly used ones in HRA being the following (see Kirwan and Ainsworth, 1992; and also Sinclair, 1995):

Observation
Interviews
Critical incident technique (CIT)
Operational experience review (OER) (of incident and near-misses, etc.)
Walk-through and talk-through (WT/TT)

The most frequently used formal approach in task analysis for HRA purposes is *hierarchical task analysis* (HTA) (see Shepherd, 1989; Kirwan and Ainsworth, 1992). This is a useful way of eliciting the *goals* driving behavior, the *tasks* used to achieve the goal or goals, the tasks' subordinate *operations* (actions and behaviors), and the operational sequence options, called *plans,* in a human operator task. HTA is a powerful and flexible method and can represent the task either graphically or in a tabular fashion, and both representations are concise and amenable to usage for the next stage of the HRA process (error identification). There are a number of other more detailed task analysis techniques, which can be useful in the support of error identification, as follows:

***Tabular Task Analysis* (*TTA*)** — To consider interface-related errors, i.e., associated with problems in knowing when to act or whether an act has been performed correctly, as a function of the feedback available to the operators via the interface.

***Tabular Scenario Analysis* (*TSA*)** — To define the sequence of tasks in an abnormal/emergency event-driven situation (see Kirwan, 1994).

***Timeline Analysis* (*TLA*)** — To determine overall task feasibility with respect to time constraints (usually used for emergency response tasks — *horizontal timeline analysis*), and the overall coordination of personnel required (including communications and personnel locations) and individual personnel involvement in tasks (thus, problems of work load, in a crude sense, can be identified and addressed — *vertical timeline analysis*).

***Link Analysis* (*LA*)** — To be used if it is required to consider detailed movements in a confined location, e.g., to determine if task interference will occur, or if signals will be missed due to poor location of instruments etc., or to consider accidental activation of critical controls, or to consider information flow-paths.

***Decision-Action-Diagrams* (*DADs*)** — These are used to model decision making sequences and are a useful representation method for all but simple/straightforward decisions.

The usage of these various approaches for HRA purposes is shown in Figure 36.2.

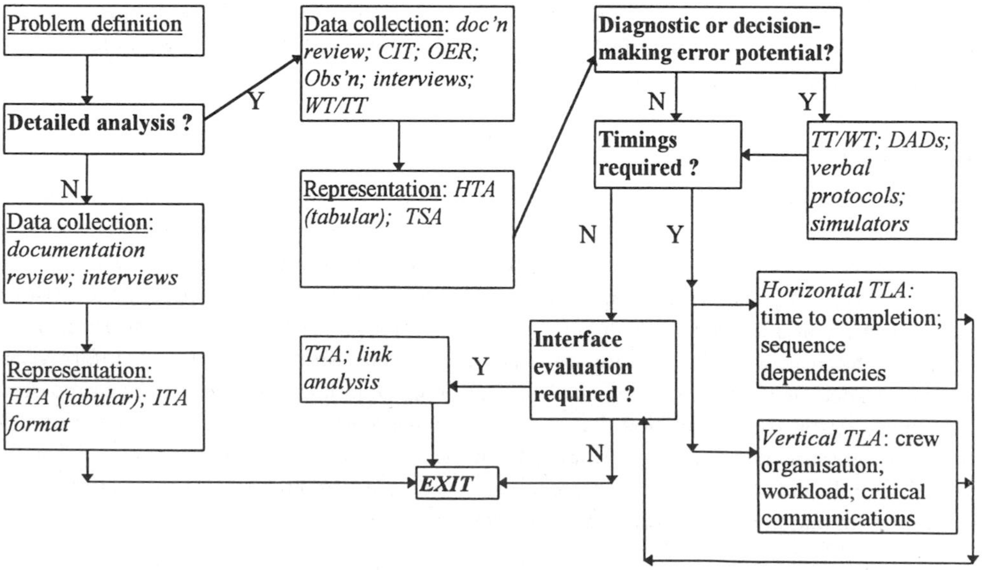

FIGURE 36.2 HRA task analysis technique selection. (Adapted from Kirwan, B. (1994) A guide to practical human reliability assessment. London: Taylor & Francis.)

Human Error Analysis

Once task analysis has been completed, human error identification may then be applied to consider what can go wrong. At the least, this error identification process will consider the following types of error (adapted from Swain and Guttmann, 1983):

- Omission error — Failing to carry out a required act
- Action error — Failing to carry out a required act adequately:
  - Act performed without required precision, or with too much/little force
  - Act performed at the wrong time
  - Acts performed in the wrong sequence
- Extraneous act — Unrequired act performed instead of or in addition to the required act (also called error of commission, or EOC)

While this is a basic and useful set of error modes, a more refined, recent, and cognitive process control-related set is shown in Table 36.1 (Meister, 1995). The human error identification phase can identify many errors, and there are many techniques available for identifying errors (see Kirwan, 1992a; 1992b; 1995). Not all of these errors will be important for the study, as can be determined by reviewing their consequences on the system's performance. This is done during the error identification stage, usually by completing a human error analysis table, as shown in Table 36.2. This simplified table shows the task step (or operation, or goal, or plan) from the task analysis, and then the error(s) associated with that step. Consequences are determined (often in conjunction with reliability engineers assessing hardware, software, and environmental failures), and possible recovery steps (e.g., checks by supervisor, etc.) are considered. The assessor may also at this stage identify how the error could be prevented, or how its effects could be mitigated.

The errors which most clearly contribute to a degraded system state, alone or in conjunction with other hardware/software failures and/or environmental events, must next be *represented* in the risk analysis framework.

Representation

Having defined what the operators should do (via task analysis) and what can go wrong (at a detailed error level or simply at the overall task execution level), the next step is to represent this information in a form in which quantitative evaluation of human error impact on the system, in conjunction with other failure events, can take place. It is usual that the human error impact be seen in the context of all other potential contributions to system risk, such as hardware and software failures, and environmental events. This enables total risk to be calculated from all sources, and enables interactions between different sources of failure to be assessed. It also means that when risk is calculated, human error will be seen in its proper perspective, as one component factor affecting risk. Sometimes human error will be found to dominate risk, and sometimes it will have less importance than other failure types (e.g., hardware, software, or equipment failures).

Risk assessments typically use logic trees, called fault and event trees, to determine risk, and human errors and recoveries are usually embedded within such logical frameworks (see Henley and Kumamoto, 1981; Green, 1983; Cox and Tait, 1991; Kirwan, 1994). Fault trees look at how various failures and combinations of failures can lead to a "top event" of interest (e.g., loss of core cooling in a nuclear power reactor). Event trees are more sequential in nature, and look at how an event, once occurred, develops and proceeds toward accidental circumstances (e.g., loss of cooling can eventually lead to core melt if certain hardware and human functions fail, which in turn could lead to leakage of radioactivity into the biosphere, and to the need for public evacuation, etc.). Both fault trees and event trees can be very large, and so are usually developed and mathematically evaluated using computerized methods and tools.

TABLE 36.1 Possible Cognitive Errors in Process Control

Information Gathering

1. Failure to detect a disturbance
2. Misreading system status information
3. Misinterpretation of system status information: a general term for instances not covered by one of the more specific interpretation errors below
4. Failure to associate two or more items of information, when their combined effect should be noted
5. Wrongly associate two or more items of information
6. Interpret the situation as having changed, when in fact it hasn't
7. Failure to realize that the situation has changed, as has been indicated by new information
8. Over/under-estimation of situation severity
9. Use excessive time acquiring information
10. Spend too little time acquiring information about system status
11. Great difficulty in interpreting symptoms of system status
12. Concentration of attention on one deviation to the exclusion of a second concurrent deviation

Stabilization Requirements

1. Failure to attempt stabilization
2. Incomplete stabilization requirements
3. Incorrect stabilization requirements
4. Over/under-estimate basic requirements for stabilization

Hypothesis Generation

1. Inability to develop a hypothesis
2. Generate too few possible hypotheses
3. Difficulty in deciding between competing hypotheses
4. Selection of an incorrect hypothesis
5. Considers correct hypothesis of fault cause, but rejects it without even testing
6. Perseveres with working hypothesis of fault cause, despite contrary evidence

Hypothesis Testing

1. Failure to test hypothesis
2. Select incorrect test of hypothesis
3. Perform selected test incorrectly, due to skill-based/procedural errors
4. Misinterpretation of test result

Performing Corrective Action

1. Selection of corrective action inconsistent with information gathered
2. Skill-based/procedural errors during execution of the corrective action
3. Failure to complete execution of corrective action
4. Reacts to system status information in excessively rapid manner when the need to do so is not apparent
5. Operator continues with corrective action after it is clear it has not rectified the situation
6. Proceed to initiate selected action, even though it is no longer applicable because the situation has changed
7. Failure to monitor the effects of corrective actions
8. Inability to decide whether or not the corrective action has been effective, and has subsequent problems in deciding whether the action should continue
9. Misinterpretation of effects of corrective action

From Meister, D. (1995) Cognitive behaviour of nuclear reactor operators. *International Journal of Industrial Ergonomics,* 16, 109-122. With permission.

TABLE 36.2 Simple HEA Tabular Format

Task Step (from the HTA)	Error	Consequence	Recovery	Error Reduction
4.1.3 Close valve	Omits task step	Line will rupture leading to release	Pressure build up alarm in CCR	Emphasis in procedures; spring-returned valve

A further representation issue is that of dependence between two or more errors or tasks, where for example, failure on one task will increase the likelihood of failure on a subsequent task (e.g., misdiagnosis may affect the successful outcome of a number of subsequent tasks). It is important that such dependencies are included in the representation, and that their effects are given appropriate mathematical weighting, as otherwise, risk may be seriously under-estimated. Currently only one HRA technique really deals with dependence (THERP — see later), and most other techniques borrow from this method when required.

Human Error Quantification

Once the human error potential has been represented, the next step is to quantify the likelihood of the errors to determine the overall effect of human error on system safety or reliability. Human reliability quantification techniques all quantify the human error probability (HEP), which is the metric of human reliability assessment. The HEP is defined as follows:

$$\text{HEP} = \frac{\text{number of errors occurred}}{\text{number of opportunities for error to occur}}$$

Thus, if when buying a cup of coffee from a vending machine, on average one time in a hundred tea is accidentally purchased, the HEP is taken as 0.01 (it is somewhat educational to try and identify HEPs in everyday life with a value of less than once in a thousand opportunities, or even as low as once in ten thousand). In an ideal world, there would be many studies and experiments in which HEPs were recorded, (e.g., operator fails to fully close a valve once every five thousand times (s)he is required to close a valve). In reality there are few such recorded data. The ideal source of human error "data" would be from industrial studies of performance and accidents, but at least three reasons can be deduced for the lack of such data:

- Difficulties in estimating the number of opportunities for error in realistically complex tasks (the so-called denominator problem)
- Confidentiality and unwillingness to publish data on poor performance
- Lack of awareness of why it would be useful to collect in the first place (and hence lack of financial incentive for such data collection)

There are other potential reasons (see Kirwan et al., 1990) but the net result is a scarcity of HEP data. HRA therefore uses quantification techniques, which either rely on expert judgment, or a mixture of data and psychologically based models which evaluate the effects of major influences on human performance.

Below are listed the major techniques in existence in the field of human reliability quantification, arguably the most developed field within human reliability assessment today. These are categorized below into four classes, depending on their data sources, and mode of operation.

1. Unstructured Expert Opinion Techniques
   - Absolute Probability Judgment or Direct Numerical Estimation (APJ or DNE: Seaver and Stillwell, 1983)
   - Paired Comparisons (PC: Hunns, 1982; Comer et al., 1983)
2. Data-Driven Techniques
   - Human Error Assessment and Reduction Technique (HEART: Williams, 1986; 1988; 1992)
   - Technique for Human Error Rate Prediction (THERP: Swain and Guttmann, 1983)
   - Human Reliability Management System (HRMS: Kirwan and James, 1989; Kirwan, 1994)
   - Justification of Human Error Data Information (JHEDI: Kirwan 1990; 1994)
3. Structured Expert Opinion Techniques
   - Success Likelihood Index Method using Multi Attribute Utility Decomposition (SLIM-MAUD: Embrey et al., 1984)
   - Socio-technical Approach to Assessing Human Reliability (STAHR: Phillips et al., 1983)

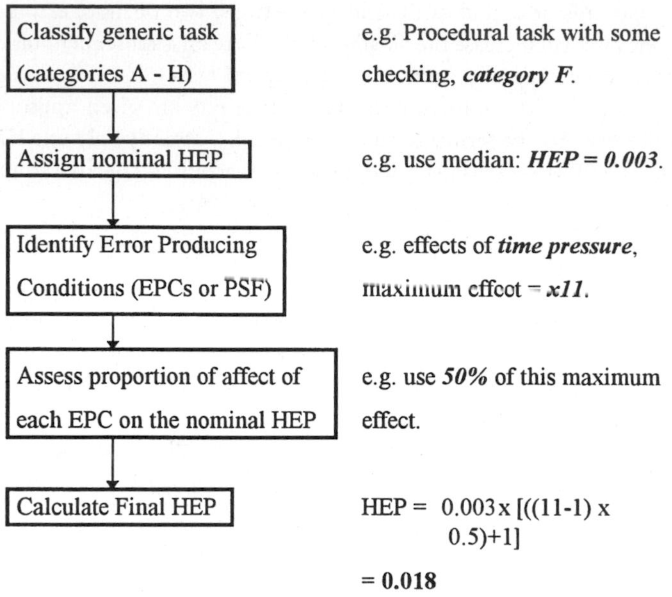

FIGURE 36.3 HEART quantification process.

4. Accident Sequence Data Driven Techniques
   - Human Cognitive Reliability (HCR: e.g., Hannaman et al., 1984)
   - Accident Sequence Evaluation Program (ASEP: Swain, 1987)

All of these techniques (and others) generate human error probabilities. Swain (1989) and Kirwan et al. (1988) discuss the relative advantages and disadvantages of these techniques, and Kirwan (1988a) gives the results of a comparative evaluation of five techniques. Kirwan et al. (1988) and Kirwan (1994) also give selection guidance to help practitioners decide which one(s) to use for a particular assessment problem.

To give a simplified indication of how data-driven techniques work, the THERP technique uses a database of "nominal" human error probabilities, e.g., failure to respond to a single annunciator alarm. Performance shaping factors (PSF) such as the *quality of the interface design* (e.g., whether alarms are prioritized, adequately color-coded, near to the operator and in the normal viewing range, etc.), or *time pressure,* are then considered with respect to this error. If such factors are indeed evident in the scenario under investigation, then the nominal human error probability may be modified by the assessor (e.g., in this case increased to reflect poor quality of interface) by using an error factor (EF) of, say, 10. Thus, if an initial nominal HEP is 0.001, an EF of 10 can be used to increase the actual estimated HEP to a value of 0.01. This is then the HEP that could be inserted into the fault or event tree.

Expert judgment techniques, (1) and (3) above, on the other hand, use personnel with relevant operational experience (e.g., more than 10 years) to estimate HEPs. The rationale is that these personnel will have had significant opportunities for error and will have also committed certain errors (and seen others commit errors), and hence have information in their memories which can be used to generate HEPs. Such expert opinion methods may either ask experts directly for such estimates, or may use more subtle and indirect methods, to avoid the various biases associated with human recall which can occur (see Tversky and Kahneman, 1974).

Another technique, currently popular in the U.K., is the HEART approach. The HEART quantification process is shown in Figure 36.3. The first stage is to consider what category the task in question belongs to, from a range of nine categories. Each category has an associated nominal or baseline error probability range, for example as follows:

TABLE 36.3 HEART Calculational Formula and Example Calculation (task = carry out valve closure sequence)

Type of Task F	Generic Error Probability — 0.003		
Error Producing Conditions (EPCs)	Maximum effect	Assessed Proportion of Effect	Calculation
Inexperience	×3	0.4	((3-1).0.4) +1 = 1.8
Opposite technique	×6	1.0	((6-1).1.0) +1 = 6.0
Low morale	×1.2	0.6	((1.2-1).0.6)+1 = 1.12

HEP = 0.003 × 1.8 × 6.0 × 1.12 = 0.036

Category A: Totally unfamiliar task, performed at speed with no real idea of likely consequences. HEP range = 0.35–0.97, median HEP = 0.55.

Category F: Restore or shift a system to original or new state following procedures, with some checking. HEP range = 0.0008–0.007, median HEP = 0.003.

Once the assessor has decided which task category to use, the next stage is to consider whether there are any negative factors that could lead to an increased HEP. More generally known as performance shaping factors (PSF), in HEART these factors are called error producing conditions (EPCs). There are over 30 EPCs, though most assessors use a range of approximately 20 EPCs. Each EPC has a derived associated maximum effect it can have on the nominal HEP, for example, as follows:

EPC	Maximum Effect on Nominal HEP
Unfamiliarity	x17
Time pressure	x11
Operator inexperience	x3

The next step with HEART, the calculation of the final HEP, is fairly straightforward (see example in Table 36.3). It should be noted that the formula uses a mathematical "fix" to avoid low-maximum-effect EPCs inadvertently resulting in decreasing the HEP, rather than increasing it. The effectiveness of this fix can be seen in Table 36.3 with respect to the EPC of motivation: if the assessed proportion of effect was applied directly to the maximum effect for this EPC, then the resulting multiplier would be less than unity (0.72 in this example), and this would then reduce the final HEP instead of increasing it.

In summary, therefore, HEART begins the quantification process by determining which of a number of generic error probabilities is appropriate to the scenario under investigation. The assessor then considers PSF effects on the task, though these are called error producing conditions (EPCs) in HEART terminology. Having selected EPCs from a range of 32 EPCs, which are weighted in terms of the maximum effect they can each have on performance, the assessor then judges how much of that maximum effect should be applied in the scenario. The overall HEP is then calculated as a function of the original generic probability, the number of EPCs, and each EPC's weighting factored by the assessor's assessed rating of the EPC's level in the scenario.

One important advantage of techniques like HEART (and SLIM, and HRMS/JHEDI) is that they can give insights into avenues for error reduction. This facility is available because of their utilization of performance shaping factors in the HEP estimation process. Thus, for example, if the above HEP was considered unacceptably high, it can be seen that the EPC of *opposite technique* is contributing significantly to the final derived HEP. This EPC may have been used by the assessor because one of the valves closes in the opposite direction to the others (also called a stereotype violation). If the plant designers/owners were to change that valve to make it consistent with others in the plant, the EPC of opposite technique could be removed from the HEART calculation. This would lead to a revised HEP of 0.006 (an error reduction factor of 6). This facility of some of the HRA techniques is useful for determining risk reduction measures and priorities.

Impact Assessment, Error Reduction Assessment, Quality Assurance, and Documentation

Following such quantifications, system risk will be calculated. It will then be determinable whether the installation's overall risk is acceptable. If not, it may be necessary or desirable to try to reduce the level of risk. If human error is contributing significantly to risk levels, then it will be desirable to attempt to find ways of reducing the human error impact. This may or may not utilize the quantification technique (as shown above) in devising error reduction mechanisms (ERMs) which will reduce the system vulnerability to human error. If ERMs are derived (whether by the quantification means or other qualitative means, or both), then the quantification technique will be used to recalculate the HEPs, for the system as it would perform with ERMs in place. Following this stage (which can run through several iterations), the results will be documented and quality assurance systems should ensure that ERMs are effectively implemented, and that assumptions made during the analysis remain valid throughout the lifetime of the system (see Kirwan, 1994). This completes the HRA process. The next section gives a brief overview of some applications of HRA.

36.5 Human Reliability Assessment in Practice

The following examples are aimed at giving insight into large HRAs in practice. There are three examples as follows (this section is adapted from Kirwan, 1996a):

1. Two recent plant design HRAs, discussing briefly their scope and impact
2. An HRA program for a number of existing and aging nuclear power plants, showing the impact of HRA on the human factors in the system design

HRA for THORP (Thermal Oxide Reprocessing Plant, BNFL, Sellafield)

THORP is a very large nuclear fuel reprocessing plant, sometimes described as a number of conventionally sized plant modules joined together under one roof, and has a staff complement approaching 800 personnel. It has process complexity similar to that of a nuclear power plant. The HRA approach started in earnest in the detailed design stage for THORP, and was predicated upon a large human factors assessment exercise (Kirwan, 1988b; 1989), amounting to approximately 15 person-years of effort. This assessment addressed the safety adequacy (from a human factors perspective) of the central control room, local control rooms, control and instrumentation panels local in the plant, staffing and organization issues, training, and emergency preparedness. The human factors reviews were used to determine the effect of certain performance-shaping factors on performance (e.g., the adequacy of the interface in the central control room during emergency conditions). Such information significantly influenced the two computerized quantification systems developed for the THORP HRA, namely the Human Reliability Management System (HRMS: Kirwan and James, 1989; Kirwan, 1990) and the Justification of Human Error Data Information system (JHEDI: Kirwan, 1990).

JHEDI was designed to rapidly but conservatively assess all identified human involvements in the THORP risk assessment, and assessed over 800 errors. JHEDI requires a simplified task analysis, error identification via keyword prompting, quantification according to PSF, and noting of training or procedural implications to be fed forward to the respective THORP operations departments.

HRMS is a more intensive system and is used for those errors that are found to be risk significant in the PSA (e.g., for the errors that have a potentially major risk impact on the THORP system: see Kirwan, 1990). HRMS requires detailed task analysis, error identification and quantification, and computer-supported error reduction is also carried out based on the PSF assessment. Approximately 20 tasks were the subject of the more detailed HRMS assessment approach.

All assessments have been rigorously documented, and information arising out of the assessments is fed forward to the operational departments that are now running THORP and assessing its safety performance. JHEDI is still being applied to other plant designs at BNFL Risley, U.K., the design center for BNFL.

Sizewell B Pressurized Water Reactor HRA*

The Sizewell B risk assessment and HRA (Whitworth, 1987; Whitfield, 1991; 1995) for the U.K.'s first and only PWR were also to an extent predicated upon extensive human factors assessment of the design of the interface and other systems (e.g., see Umbers and Reiersen, 1995; Ainsworth and Pendlebury, 1995), although the linkage between the human factors assessments and the HRA inputs was less formalized than for THORP. The HRA approach involved a very large amount of initial task analysis and error analysis. The Human Error Assessment and Reduction Technique (HEART: Williams, 1986) was used as the main quantification tool, supplemented by the Technique for Human Error Rate Prediction (THERP: Swain and Guttmann, 1983), and error reduction was carried out as required.

One of the early impacts of the human reliability assessment work for Sizewell B was the recognition of the need for more automation to support the operator in a particular accidental event scenario (Fewins et al., 1992).

In the early HRA phase, a great deal of human error analysis and task analysis was carried out, whereas later on, as the design became more detailed and construction started, the focus shifted to particular significant scenarios and the quantification of HEPs for these scenarios. The balance of effort in the HRA was therefore largely qualitative in the early design stages, and more quantitative later on. This is a sensible approach, since early in the design phase is when most design impact can be easily achieved, and the numbers (HEPs and risk calculations) may be seen as secondary when compared to the goal of achieving a good (safe) and operable (ergonomic) design. Later on, having a strong safety case so that operation will be allowed by the regulators, becomes the primary driving force behind assessment. Therefore, the HRA effort becomes more focused on quantitative predictions and fault and event trees, as the resultant risk estimates will determine when the plant may become operable. This shift of emphasis within the life cycle of a large HRA for a novel and complex plant is probably typical of such projects.

The Sizewell B detailed design and assessment program spanned over a decade, following an exhaustive public inquiry, during which one of the key recommendations was for a high degree of human factors support for the station design and development. (This inquiry spawned the Advisory Committee for the Safety of Nuclear Installations, ACSNI, which still advises the nuclear power industry in the U.K.). As a consequence, the project received a good deal of guided human factors and HRA effort, as it was seen as a high-profile project.

Continued Operation Risk Assessments/HRAs

These risk assessments and associated HRAs are part of a required program of work to determine whether the aging gas-cooled reactor plants are safe to continue operating beyond their original sanctioned lifetimes (e.g., whether the Magnox stations can operate beyond 30 years). So far, two Magnox plants have been shut down (due probably more to economic reasons than safety concerns), and the results of the continued operation HRAs for the other stations are still being reviewed by the relevant U.K. regulatory body, HM Nuclear Installations Inspectorate.

These HRAs have used a significant amount of task analysis and have generally each followed a similar basic format: detailed task analysis for a small number of key scenarios, and less detailed analysis of the remainder of the scenarios. No detailed error analysis is utilized, and task failure likelihood is calculated by using the HEART quantification method. Error reduction measures are identified either based on the HEART calculations or on the task analysis. An example of the methodology from one of the continued operation HRAs is given in Kirwan, Scannali, and Robinson (1995). In this particular study, which was carried out over a period of two years, four tasks were assessed in exhaustive depth using a range of task analysis approaches (hierarchical, tabular, and timeline analyses), and approximately 40 other tasks were

*It should be noted that the insights into the Sizewell "B" PWR nuclear power plant project are based largely on the author's observations as an outsider to that program of HRA activities, and discussions with colleagues working on the project at the time. These views may therefore not be representative of the owners' or project team's views.

analyzed in less (though still substantial) depth. The HEART approach was used for quantification, with THERP used to independently corroborate the quantifications for three of the tasks (the results using the two techniques agreed within a factor of three of each other, and this was considered a positive result). Fault trees were the predominant representation method used, and THERP's dependence model was adapted for the HRA. The HRA led to a number of impacts on the existing plant, which have since been implemented. Some examples are as follows:

- Procedural recommendations (i.e., changes to existing procedural documentation)
- Training recommendations (concerning simulator training)
- Emergency lighting recommendations
- Recommended changes to the VDU interface
- Recommendations for certain local alarms being made more central

As a result of the risk assessment and HRA, and the implementation of certain human factors and other risk assessment-identified design recommendations, the plant is still currently operating and producing power.

36.6 Developments in HRA

This section discusses two recent developments in HRA in the U.K. The first is the development of a HEP database to support HRA activities, and the second is to support error identification activities. Both of these areas were seen as primary and urgent areas for development of practical tools in the U.K. The first was necessary to lend credibility to the whole HRA quantification process (i.e., so that at least some real and robust HEPs were in evidence, among all the expert judgment that enshrouded the area of HRA), and to provide at least enough HEPs to be useful for validation and training efforts. The second was to provide much needed support to assessors in the difficult area of determining what can go wrong (error identification). This latter project on error identification was also pursued since industry and academia alike realized that, whereas prediction of HEPs was reasonably robust and a range of tools existed for quantifying HEPs, the logical precursor to quantification, namely deciding what needed to be quantified, was an immature area in terms of technique development and was probably a large source of inconsistency in practical HRA. Although there have been other developments in HRA, therefore, these two are seen as significant in enhancing this applied methodology.

Human Error Database (CORE-DATA) Development

Although much human reliability assessment (HRA) is carried out internationally, in nuclear power and other industries (e.g., chemical and offshore), there has been a paucity of HEP data useful for predicting how often errors will occur. This lack of data has been despite a number of attempts to develop such a database over the past 30 years (e.g., see Topmiller et al., 1984). However, in the late 1980s, certain developments in the theory of human error suggested that a database could be constructed. In the U.S. this led to the NUCLARR database (e.g., see Gertman and Blackman, 1994).

Following an Advisory Committee for the Safety of Nuclear Installations (ACSNI: 1991) report in the U.K. recommending the construction of a HEP database, a three-year project was set up to develop the Computerised Operator Reliability and Error Database (CORE-DATA: see Taylor-Adams and Kirwan, 1995; Taylor-Adams, 1995). The work was funded by the U.K. Nuclear Power and Reprocessing Generic Nuclear Safety Research (GNSR) Programme.

The main objectives of the project were as follows: to aggregate existing HEP data and to structure the database to be theoretically satisfactory yet usable for assessors; to collect new data for the database; and to produce a computerized prototype of the database for demonstration purposes. There was also a more long term objective, namely the investigation of the feasibility of developing extrapolation rules to render the technique usable as a human reliability assessment tool in its own right, i.e., so that data could be manipulated to be applicable to new situations, scenarios, and even new industrial contexts.

The development of a sound theoretical structure for the database relies on the development of a number of taxonomies or classification systems, which enable categorization of the data in psychologically meaningful, robust, and mutually exclusive terms. This was not a trivial task, and five taxonomies were developed, analyzed, and evaluated in the context of the current dominant models in human factors and human reliability, namely the Information Processing Model (Wickens, 1992), and the Skill, Rule and Knowledge Model (Rasmussen et al., 1981). Once these taxonomies had been developed, existing data could be incorporated into a working database.

CORE-DATA currently contains about 250 HEPs, and a further 900 or so, of varying pedigree, have been put into hard copy CORE-DATA format. There have also been two recent and successful data collection exercises (one offshore evacuation study [Basra and Kirwan, 1996], and one manufacturing study), which have produced new data for the CORE-DATA system. An example of the CORE-DATA interface is shown in Figure 36.4.

Human Error Analysis

A recent research program has produced a prototype human error analysis system for dealing with the following types of error:

- Skill- and rule-based error forms (slips and lapses, and rule-following errors)
- Cognitive errors (diagnosis and decision-making errors)
- Errors of commission
- Rule violations
- Teamwork and communication errors

This subsection gives some insights into the human error analysis tool-kit that has been developed, called the Human Error and Recovery Analysis (HERA) system (Kirwan, 1998), with respect to the most detailed and computerized element, the skill and rule-based error analysis section.

The skill and rule-based error analysis section has seven sets of error identification prompts or questions, as follows:

1. Mission — High level mission-oriented questions, e.g., Could the task fail to be achieved in time?
2. Operations — Still fairly high-level questions, e.g., Could a previous latent maintenance error lead to errors or difficulties in the current task?
3. Goals — Concerned with high-level planning of the team, e.g., Could the team fail to realize the need to shift to a higher goal?
4. Plans — Concerned with the timing and sequencing of the task, e.g., Will the pre-conditions for the plan be met?
5. Error — External error mode prompts, e.g., omission; right action on wrong object; etc.
6. Performance shaping factors (PSF) — macro human factors considerations, e.g., Is the alarm reliable? How often do operators expect to see this event?
7. Psychological error mechanisms — The internal mechanisms of failure, e.g., manual variability; substitution error; information transmission error, etc.

It is up to the assessor as to which of the above seven modules are utilized, and several or even all can be applied to the hierarchical task analysis that is entered in a tabular format into the computerized system. Once errors are identified, they are automatically encoded into a human error analysis table. The assessor can then add the consequences of the errors and any likely error recovery possibilities. The tabulated results can then be printed out, ready for the representation phase of the HRA process.

An example of the usage of the computerized HERA system is shown in Figures 36.5 and 36.6, and Table 36.4. In this example, goals analysis is being applied to a (UK) Nuclear Power Plant Loss of Offsite Power (Loss of Grid) scenario. Figure 36.5 shows the goals analysis questions in full, with their abbreviated short-form highlighted in bold. Figure 36.6 shows a screen dump from the HERA system, showing a goals analysis in progress. Table 36.4 shows part of the resultant human error analysis table based on goals analysis of this scenario.

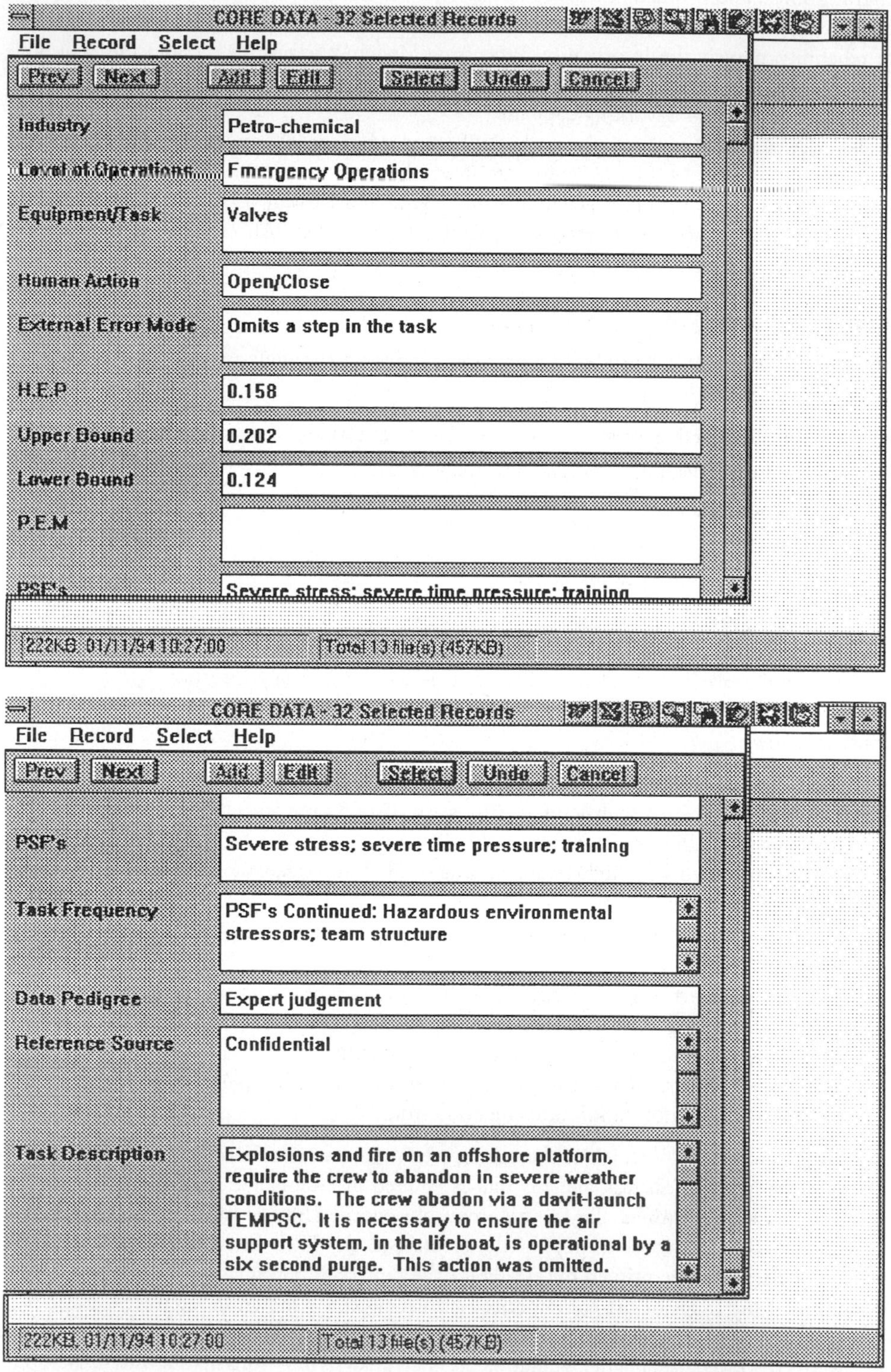

FIGURE 36.4 Screen of first part of CORE-DATA data screen (split here into two halves).

GOALS ANALYSIS
1. Could the operators have no goal, e.g. due to a flood of conflicting information; the sudden onset of an unanticipated situation; a rapidly evolving and worsening situation; or due to disagreement or other decision-making failure to develop a goal ? **[no goal]**
2. Could operators find themselves without any clear strategy, i.e. they would be 'outside' the procedures ? {as in a Beyond Design Basis Accident or BDBA} **[outside procedures]**
3. Could the key decision-maker(s) panic or become unavailable, and hence obstruct goal-setting? {chain of command is broken - also known as 'decapitation' if the task leader is lost} **[no goal]**
4. Could the operators have the wrong goal, e.g. due to a misperception of the circumstances, or due to over-familiarity with that goal (e.g. due to training bias)? **[wrong goal]**
5. Could the operators have the wrong goal due to production pressure, or due to practical difficulties in implementing the stated goal, leading to apparent violations ? **[wrong goal]**
6. Could there be a goal conflict, e.g. between production and safety goals (e.g. shutting down plant when the plant safety margin is narrow), or between competing safety goals (e.g. rescuing operational personnel from fire versus protecting the public from contamination) **[goal conflict - note: this may result in goal delayed, wrong goal, no goal, or goal inadequate]**
7. Could the operators fail to realise the need to shift to a higher goal, as the scenario worsens, e.g. due to detection failure or disbelief of indications, or due to operators getting 'locked in' to a lower level goal or task (e.g. diagnosis), ignoring the deterioration of events, and persevering with a goal when it should be aborted in favour of another? **[wrong goal]** ?

FIGURE 36.5 HERA Goals Analysis.

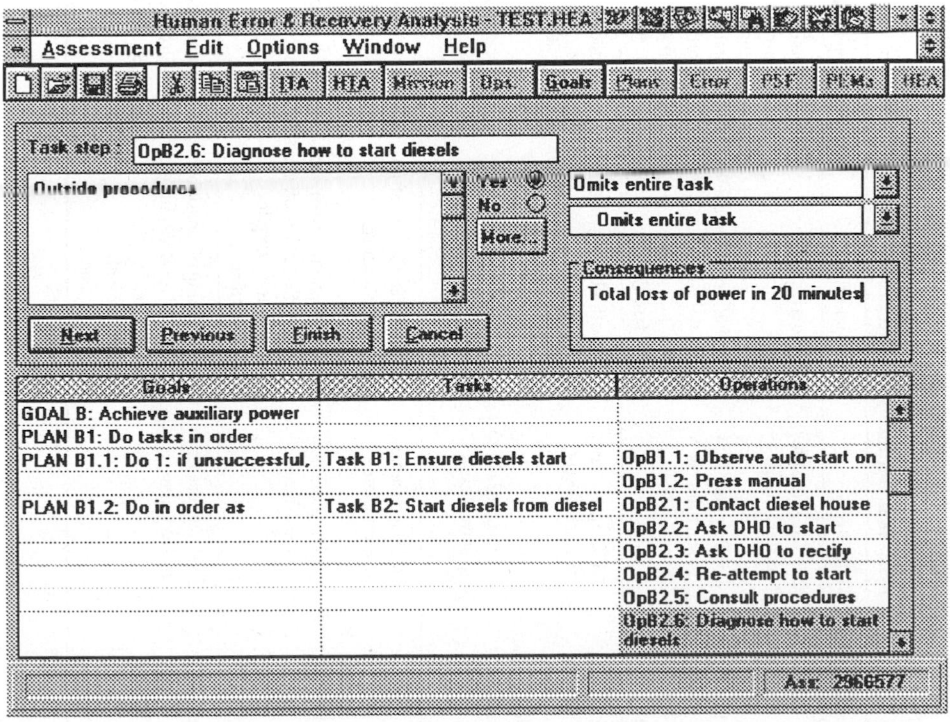

FIGURE 36.6 HERA Goals Analysis Screen.

TABLE 36.4 Goals Analysis: Skill and Rule Identified Errors for Loss of Offsite Power (Grid) Scenario

Identifier from HERA	Task Step	Error Identified	Consequence	Recovery	Comments
1. No goal	2.6:. Diagnose how to start diesels	Fail to derive plan	Total loss of power and forced cooling in 20 minutes.	Restoration of grid within 20 minutes (unlikely).	Overall failure to diagnose.
2. Outside procedures	2.6: Diagnose how to start diesels	No *ad hoc* procedures derived	Total loss of power and forced cooling in 20 minutes.	Restoration of grid within 20 minutes (unlikely).	
3. Decapitation		None identified			
4. Wrong goal	2.6: Diagnose how to start diesels	Switch off all pony motors	Undesirable to switch off all pony motors. Cooling will suffer.	When diesels started, will restart pony motors.	
5. Production pressure		N/A			
6. Goal conflict	2.6: Diagnose how to start diesels	Unwilling to turn off pony motors	Total loss of power and forced cooling in 20 minutes.	Restoration of grid within 20 minutes (unlikely).	Also failure to start diesels, a form of reluctance.
7. Goal shift failure	2.6: Diagnose how to start diesels	Fail to realize they must go beyond procedures	Total loss of power and forced cooling in 20 minutes.	Restoration of grid within 20 minutes (unlikely).	
8. Premature shift	2.6: Diagnose how to start diesels	Switch off cooling when not required (i.e., before off-loading non-essential supplies)	Loss of cooling when not necessary	Diesels start and pony motors restarted.	A partial diagnostic failure — it is safer to do this than do nothing, but will compromise short-term cooling.
9. Too many goals		None identified			
10. Goal delayed	2.6: Diagnose how to start diesels	Delay opening of circuit breakers until too late (>20 minutes)	Total loss of power and forced cooling in 20 minutes.	Restoration of grid within 20 minutes (unlikely).	
11. Incomplete goal		None identified			
12. Violating goal		None identified			

36.7 Validation of Human Reliability Quantification Techniques

A recent large-scale validation has taken place in the U.K. to test the three quantification techniques THERP, HEART, and JHEDI. Thirty U.K. practitioners took part in the exercise, and each assessor independently used only one of the three techniques (very few assessors were practitioners in more than one technique). Each assessor quantified HEPs for 30 tasks, and for each task there was the following information:

- General description of the scenario
- Inclusion of relevant PSF information in the description
- Provision of simple linear task analysis
- Provision of diagrams where necessary and relevant
- Statement of exact human error requiring quantification

Each assessor had two days to carry out the assessments, and experimental controls were exercised, so that the assessors were working effectively under invigilated examination conditions. For each of the 30 tasks the HEP was known to the experimenter, but not to any of the assessors. Tasks were chosen to be relevant to nuclear power and reprocessing industries, since all of the assessors were currently working in these two areas. A large proportion of the data were from real recorded incidents,* with the data range spanning five orders of magnitude (i.e., from 1.0 to 1E-5). The results are summarized below. (For a fuller analysis and presentation/discussion of results, see Kirwan et al., 1996.)

Predictive Validity

The analysis of all the data (i.e., all 895 estimated HEPs — there were five missing values) showed a significant correlation between estimates and their corresponding true values (Kendall's coefficient of concordance: $Z = 11.807$, $p < 0.01$). This supports the validity of the HRA quantification approach as a whole, especially as no assessors or outliers have been excluded from these results. The analysis of all data for individual techniques shows a significant correlation in each case (using Kendall's Coefficient of Concordance): THERP $Z = 6.86$; HEART $Z = 6.29$; JHEDI $Z = 8.14$; all significant at $p < 0.01$.

Individual correlations for all subjects are shown in Table 36.5. There are 23 significant correlations (some significant at the $p < 0.01$ level) out of a possible 30 correlations. This is a very positive result, again supporting the validity of the HRA quantification approach.

Precision

Table 36.6 shows that there is an overall average of 72% precision (estimates within a factor of 10) for all assessors, irrespective of whether they were significantly correlated or not. This figure includes all data estimates, even the apparent outliers that have been identified in the study. This is therefore a reasonably good result, supporting HRA quantification as a whole. Furthermore, no single assessor dropped below 60% precision in the study. The precision within a factor of 3 is approximately 38% for all techniques. This is a fairly high percentage given the required precision level of a factor of 3. The degree of optimism and pessimism is not too large, as also shown in the histogram in Figure 36.7, with only a small percentage of estimates at the extreme optimistic and pessimistic ends of the histogram (i.e., greater than a factor of a hundred from the actual estimate). Certainly there is room for improvement, but the optimism and pessimism are not in themselves dominating the results, and estimates were more likely to be pessimistic (17.5% of the total number of estimates) than optimistic (9.7% of the total number of estimates).

The highest and lowest precision values for the techniques within a factor of 3 and within a factor of 10 are shown in Table 36.7.

---

*Real data came largely from actual recorded incidents in the nuclear chemical and other heavy industry sectors for which the number of opportunities for error could be calculated robustly. See Kirwan et al., 1990 for similar data collection and generation activities.

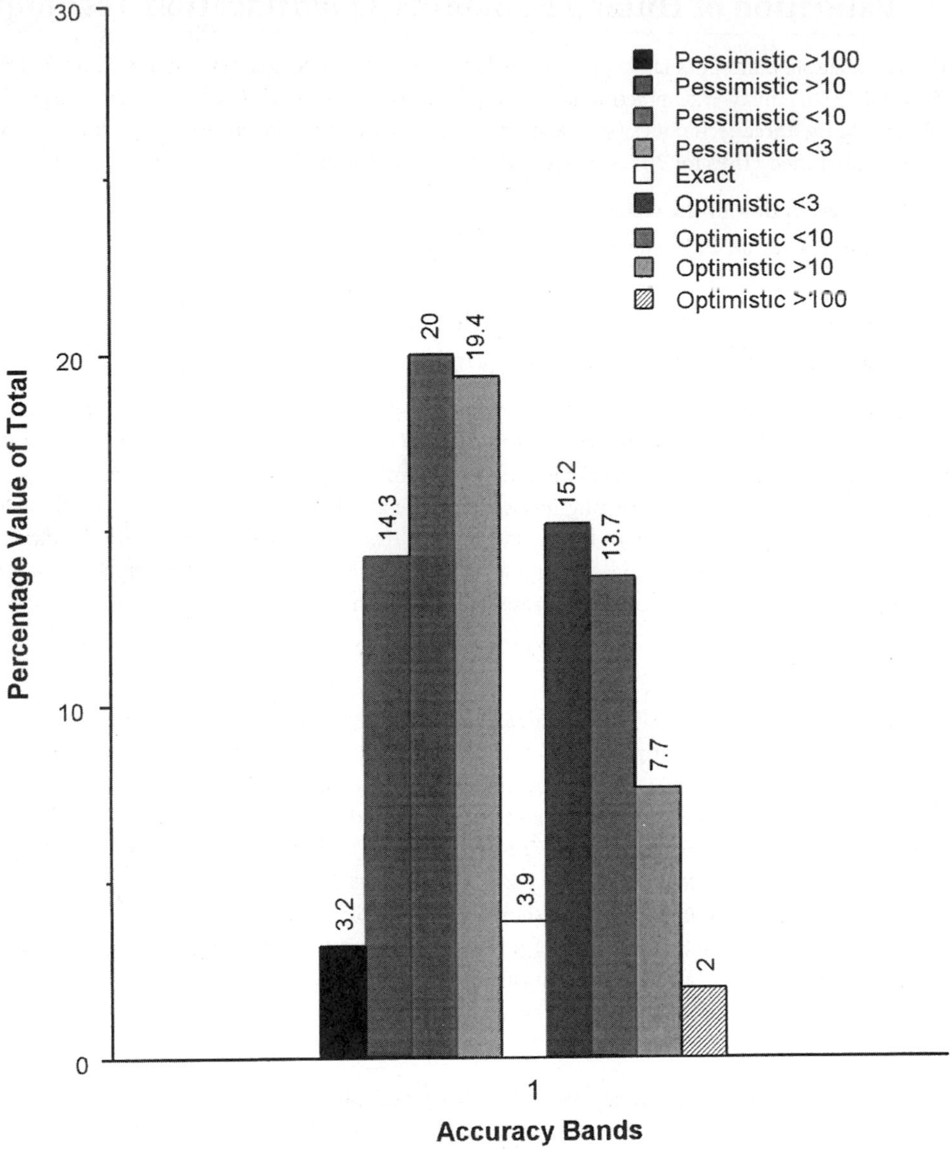

FIGURE 36.7 Overall HRA optimism and pessimism.

The overall results were therefore positive, with a significant overall correlation of all estimates with the known true values, with 23 individual significant correlations, and a general precision range of 60 to 87%, the average precision being 72%. These results lend support for the HRA quantification part of the HRA process.

36.8 Concluding Comments

This chapter has attempted to outline the subject of Human Reliability Assessment, by describing the HRA process, and citing some recent large-scale applications in the U.K. The two recent developments and the validation exercise are also encouraging for HRA practitioners, since they represent improvements in HRA technology, and also demonstrate that such projects themselves are considered worth funding by industry. In the U.K. HRA is also seen as a means of evaluating the human factors adequacy of a

TABLE 36.5 Correlations for Each Subject for the Three Techniques

	THERP	HEART	JHEDI
1	.615**	.577**	.633**
2	.581**	.558**	.551**
3	.540**	.473**	.533**
4	.521**	.440**	.452**
5	.437**	.389*	.436**
6	.311*	.370*	.423*
7	.298	.351*	.418*
8	.297	.347*	.401*
9	.254	.217	.386*
10	.078	.124	.275

* SIGNIFICANT P < 0.05.
** SIGNIFICANT P < 0.01.

TABLE 36.6 Precision for the Three Techniques for all HRA Assessors

	Factor 3	Factor 10	Optimistic	Pessimistic
THERP	38.33%	72%	13.67%	13.33%
HEART	32.67%	70.33%	11%	18.33%
JHEDI	43.67%	75%	3.33%	21.33%

TABLE 36.7 Precision for the Three Techniques

	Highest Precision		Lowest Precision	
	Factor of 3	Factor of 10	Factor of 3	Factor of 10
THERP	56.67%	80%	10%	63.33%
HEART	46.67%	76.67%	16.67%	60%
JHEDI	66.67%	86.67%	26.67%	70%

system, albeit from a safety or risk perspective, and many HRA assessors, particularly on large HRA projects, achieve significant human factors improvements in systems via the HRA process. Human factors and HRA are therefore not only clearly related, but can operate synergistically, with benefits to both camps. It is hoped that in the future such collaboration between HRA and Human Factors will continue and grow. Human error and human performance, after all, are simply two sides of the same coin.

References

ACSNI (1991) *Human Reliability Assessment — a Critical Overview.* Advisory Committee on the Safety of Nuclear Installations, Health and Safety Commission, London: HMSO.

Ainsworth, L. and Pendlebury, G. (1995) Task-based contributions to the design and assessment of the man-machine interfaces for a pressurised water reactor. *Ergonomics,* 38, 3, pp. 462-474.

Basra, G. and Kirwan, B. (1998) Collection of offshore human error probability data. *Reliability Engineering & System Safety,* 61, 77-93.

Comer, M. K., Seaver, D. A., Stillwell, W. G. and Gaddy, C. D. (1984) *Generating Human Reliability Estimates Using Expert Judgment.* Vols. 1 and 2. NUREG/CR-3688 (SAND 84-7115). Sandia National Laboratory, Albuquerque, New Mexico, 87185 for Office of Nuclear Regulatory Research, U.S. Nuclear Regulatory Commission, Washington, D.C. 20555.

Cox, S.J. and Tait, N.R.S. (1991) *Reliability, Safety and Risk Management.* Oxford: Butterworth-Heinemann.

Dhillon, B.S. (1986) *Human Reliability with Human Factors.* Oxford: Pergamon.

Dougherty, E.M. and Fragola, J.R. (1988) *Human Reliability Analysis: a Systems Engineering Approach with Nuclear Power Plant Applications,* New York: John Wiley & Sons.

Embrey, D.E., Humphreys, P.C., Rosa, E.A., Kirwan, B. and Rea, K. (1984). *SLIM-MAUD: An Approach to Assessing Human Error Probabilities Using Structured Expert Judgment.* NUREG/CR-3518, Volumes 1 and 2, U.S. Nuclear Regulatory Commission, Washington, D.C. 20555, 180 pages.

Embrey, D.E., Kontogiannis, T. and Green, M. (1994) *Preventing Human Error in Process Safety.* Centre for Chemical Process Safety (CCPS), American Institute of Chemical Engineers. New York: CCPS.

Fewins, A., Mitchell, K., and Williams, J.C. (1992) Balancing automation and human actin through task analysis, in Kirwan, B., and Ainsworth, L.K. (Eds.) *A Guide to Task Analysis.* London: Taylor & Francis, pp. 241-251.

Gertman, D.I. and Blackman, H. (1994) *Human Reliability and Safety Analysis Data Handbook.* Chichester: John Wiley & Sons.

Green, A.E. (1983) *Safety Systems Reliability.* Chichester: John Wiley & Sons.

Hannaman, G.W., Spurgin, A.J., and Lukic, Y.D. (1984) *Human Cognitive Reliability Model for PRA Analysis.* Report NUS-4531, Electric Power Research Institute, Palo Alto, California.

Henley, E.J. and Kumamoto, H. (1981) *Reliability Engineering and Risk Assessment.* New Jersey: Prentice-Hall.

Hollnagel, E. (1993) *Human Reliability Analysis: Context and Control.* London: Academic Press.

Hunns, D.M. (1982) The method of paired comparisons, in Green, A.E. (Ed.) *High Risk Safety Technology.* Chichester: John Wiley & Sons.

Kirwan, B. (1988) Integrating human factors and reliability into the plant design and assessment process, in *Contemporary Ergonomics,* Megaw, E.D. (Ed.). London: Taylor & Francis, pp. 154-162.

Kirwan, B. (1988a) A comparative evaluation of five human reliability assessment techniques, in *Human Factors and Decision Making.* Sayers, B.A. (Ed.). London: Elsevier, pp. 87-109.

Kirwan, B., Embrey D.E., and Rea, K. (1988b) *The Human Reliability Assessors Guide,* Report RTS 88/95Q, NCSR, UKAEA, Culcheth, Cheshire, 271 pages.

Kirwan, B. (1989) A human factors and reliability programme for the design of a large nuclear chemical plant. *Human Factors Annual Conference,* Denver, Colorado, October, pp. 1009-1013.

Kirwan, B. and James, N.J. (1989) Development of a human reliability assessment system for the management of human error in complex systems. *Reliability '89,* Brighton, June 14–16, pp. 5A/2/1-5A/2/11.

Kirwan, B. (1990) A resources flexible approach to human reliability assessment for PRA, *Safety and Reliability Symposium,* Altrincham, September. London: Elsevier Applied Sciences, pp. 114-135.

Kirwan, B., Martin, B.R., Rycraft, H., and Smith, A. (1990) Human error data collection and data generation, in *International Journal of Quality and Reliability Management,* 7.4, pp. 34-66.

Kirwan, B. (1992a) Human error identification in human reliability assessment. Part 1: overview of approaches. *Applied Ergonomics,* 23, 5, 299-318.

Kirwan, B. (1992b) Human error identification in HRA. Part 2: Detailed comparison of techniques. *Applied Ergonomics,* 23, 6, 371-381.

Kirwan, B. (1992c) A task analysis programme for THORP, in Kirwan B and Ainsworth L K (eds.), *A Guide to Task Analysis,* pp. 363-388. London: Taylor & Francis.

Kirwan, B. and Ainsworth, L.K. (Eds.) (1992) *A Guide to Task Analysis.* London: Taylor & Francis.

Kirwan, B. (1994) *A Guide to Practical Human Reliability Assessment.* London: Taylor & Francis.

Kirwan, B. (1995) Current trends in human error analysis technique development, in *Contemporary Ergonomics,* Robertson, S.A. (Ed.), London: Taylor & Francis, pp. 111-117.

Kirwan, B., Scannali, S., and Robinson, L. (1995) Practical HRA in PSA — a case study. *European Safety and Reliability Conference,* ESREL '95, Bournemouth, June 26–28. Institute of Quality Assurance.

Kirwan, B. (1996) Human Reliability Assessment in the U.K. Nuclear Power and Reprocessing Industries, in Stanton, N (Ed.) *Human Factors in Nuclear Safety.* London: Taylor & Francis.

Kirwan, B. (1998a) Human error identification techniques for risk assessment of high risk systems — Part 1: Review and evaluation of techniques. *Applied Ergonomics,* 29, 3, 157-177.

Kirwan, B. (1998b) Human error identification techniques for risk assessment of high risk systems — Part 2: Towards a framework approach. *Applied Ergonomics,* 29, 5, 299-318.

Meister, D. (1995) Cognitive behaviour of nuclear reactor operators. *International Journal of Industrial Ergonomics,* 16, 109-122.

Park, K.S. (1987) *Human Reliability: Analysis, Prediction, and Prevention of Human Errors.* Oxford: Elsevier.

Phillips, L.D., Humphreys, P., and Embrey, D.E. (1983) *A Socio-Technical Approach to Assessing Human Reliability.* London School of Economics, Decision Analysis Unit, Technical Report 83-4.

Rasmussen, J., Pedersen, O.M., Carnino, A., Griffon, M., Mancini, C., and Gagnolet, P. (1981) *Classification System for Reporting Events Involving Human Malfunctions,* RISO-M-2240, DK-4000, Riso National Laboratories, Roskilde, Denmark.

Reason, J.T. (1990) *Human Error.* Cambridge: Cambridge University Press.

Seaver, D.A. and Stillwell, W.G. (1983) *Procedures for Using Expert Judgment to Estimate Human Error Probabilities in Nuclear Power Plant Operations.* NUREG/CR-2743, Washington, D.C. 20555.

Shepherd, A. (1989) Analysis and training of information technology tasks, in Diaper, D. (Ed.) *Task Analysis for Human–Computer Interaction.* Chichester: Ellis Horwood, pp. 15-54.

Sinclair, M. (1995) Subjective assessment, in Wilson, J.R., and Corlett, N.E. (Eds.) *Evaluation of Human Work.* London: Taylor & Francis, pp. 69-100.

Swain, A.D. and Guttmann, H.E. (1983) *Human Reliability Analysis with Emphasis on Nuclear Power Plant Applications.* NUREG/CR-1278, USNRC, Washington, D.C. 20555.

Swain, A.D. (1987) *Accident Sequence Evaluation Program Human Reliability Analysis Procedure.* NUREG/CR 4722. Washington, D.C.-20555: USNRC.

Swain, A.D. (1989) *Comparative Evaluation of Methods for Human Reliability Analysis.* Gessellschaft fur reaktorsicherheit, GRS-71. Schwertnergasse 1, 5000 Koln.

Taylor-Adams, S.E. (1995) The use of the Computerised Operator Reliability and Error Database (CORE-DATA) in the Nuclear Power and Electrical Industries, in the *IBC Conference on Human Factors in the Electrical Supply Industries,* Copthorne Tara Hotel, London, 17/18 October. 16 pages.

Taylor-Adams, S.T., and Kirwan, B. (1995) Human reliability data requirements. *International Journal of Quality and Reliability Management,* 12, 1, 24-46.

Topmiller, D.A., Eckel, J.S., and Kozinsky, E.J. (1984) *Human reliability databank for nuclear power plant operations.* USNRC Report NUREG/CR-2744, Washington, D.C.-20555

Tversky, A. and Kahneman, D. (1974) Judgment under uncertainty: heuristics and biases. *Science,* 185, 1124-1131.

Umbers, I. and Reiersen, C.S. (1995) Task analysis in support of the design and development of a nuclear power plant safety system. *Ergonomics,* 38, 3, pp. 443-454.

Whitfield, D. (1991) An overview of human factors principles for the development and support of nuclear power station personnel and their tasks, in the Conference *Quality management in the nuclear industry: the Human Factor.* London: Institute of Mechanical Engineers.

Whitfield, D. (1995) Ergonomics in the design and operation of Sizewell B nuclear power station. *Ergonomics,* 38, 3, pp. 455-461.

Whitworth, D. (1987) Application of operator error analysis in the design of Sizewell "B", in *Reliability '87,* NEC, Birmingham, April. 14–16. London: Institute of Quality Assurance, pp. 5A/1/1-5A/1/14

Wickens, C. (1992) *Engineering Psychology and Human Performance.* New York: Harper-Collins.

Williams, J.C., 1986, HEART — A proposed method for assessing and reducing human error, *Proceedings of the 9th "Advances in Reliability Technology"* Symposium, University of Bradford.

Williams, J.C. (1988) A data-based method for assessing and reducing human error to improve operational performance, in *IEEE Conference on Human Factors in Power Plants,* pp. 436-450. Monterey, California, June 5–9.

Williams, J.C., 1992, Toward an improved evaluation analysis tool for users of HEART, *Proceedings of the International Conference on Hazard Identification, Risk Analysis, Human Factors & Human Reliability in Process Safety,* Orlando.

Woods, D.D. et al. (1994) *Behind Human Error: Cognitive Systems, Computers and Hindsight.* CSERIAC state of the art report. Ohio: Wright Patterson Air Force Base. October.

ACRONYMS

ACSNI	(UK) Advisory Committee for the Safety of Nuclear Installations
APJ	Absolute Probability Judgment
BDBA	Beyond Design Basis Accident
BNFL	British Nuclear Fuels plc
CIT	Critical Incident Technique
CORE-DATA	Computerised Operator Reliability and Error Database
DADs	Decision Action Diagrams
DHO	Diesel House Operator
EEM	External Error Mode
ERM	Error Reduction Mechanism
GNSR	Generic Nuclear Safety Research
HEART	Human Error Assessment and Reduction Technique
HEP	Human Error Probability
HERA	Human Error and Recovery Assessment
HF	Human Factors
HRA	Human Reliability Assessment
HRMS	Human Reliability Management System
HRQ	Human Reliability Quantification
HTA	Hierarchical Task Analysis
JHEDI	Justification of Human Error Data Information
LA	Link Analysis
MAGNOX	Magnesium Oxide (cladded fuel reactor)
NUCLARR	Nuclear Computerised Library for Assessing Reactor Reliability
OER	Operational Experience Review
PC	Paired Comparisons
PEM	Psychological Error Mechanism
PSFs	Performance Shaping Factors
SLIM-MAUD	Success Likelihood Index Method using Multi-Attribute UtilityDecomposition
THERP	Technique for Human Error Rate Prediction
TLA	Timeline Analysis
TSA	Tabular Scenario Analysis
TTA	Tabular Task Analysis
WT/TT	Walk-through/talk-through
UK	United Kingdom

37

DIALOG: A Computer-Based System for Development of Experimental Data on Human Reliability

Heiner Bubb
Technische Universität München

Iwona Jastrzebska-Fraczek
Technische Universität München

37.1 System Ergonomics and Human Error

Most of the conventionally applied human reliability assessment (HRA) methods are based on an event-orientated categorizing system, as it is given by, for example, every accident statistic. A most important system of this category is THERP (Technique for Human Error Rate Prediction) prepared by Swain and Guttmann (1983). Systems of this kind provide reliable prognosis for so-called skill-based actions, which are often enough repeated. In the case of cognitive tasks they cannot support sufficient data. The HCR (Human Cognitive Reliability, Hannaman and Spurgin, 1984) methods classify human error as belonging only to the ratio of time available to time required. This assumption is often challenged. As system ergonomics deals with the information flow in a man–machine system (Figure 37.1) as well as with the information flow between different systems, this would serve as a base to assess human reliability independent of the level of information processing.

By the information flow of a singular man–machine system, the quality of human work can be calculated as the relation of the result of work to the task. Performance in this context is: Quality per time. Limits of quality are explicitly or implicitly defined in every system. If any human action results in a jumping over such predefined limits, we call that *human error* (HE, Rigby, 1970). The probability of nonoccurrence of such HEs defines *human reliability* (HR). This consideration is important not only for a simple singular man–machine system but also for any complicated combinations of different singular man–machine systems, which describes in its totality any arbitrary system, as for example a power plant, an assembly plant, or a traffic system. Apart from technical defects, human errors influence the reliability

0-8493-2641-9/99/$0.00+$.50

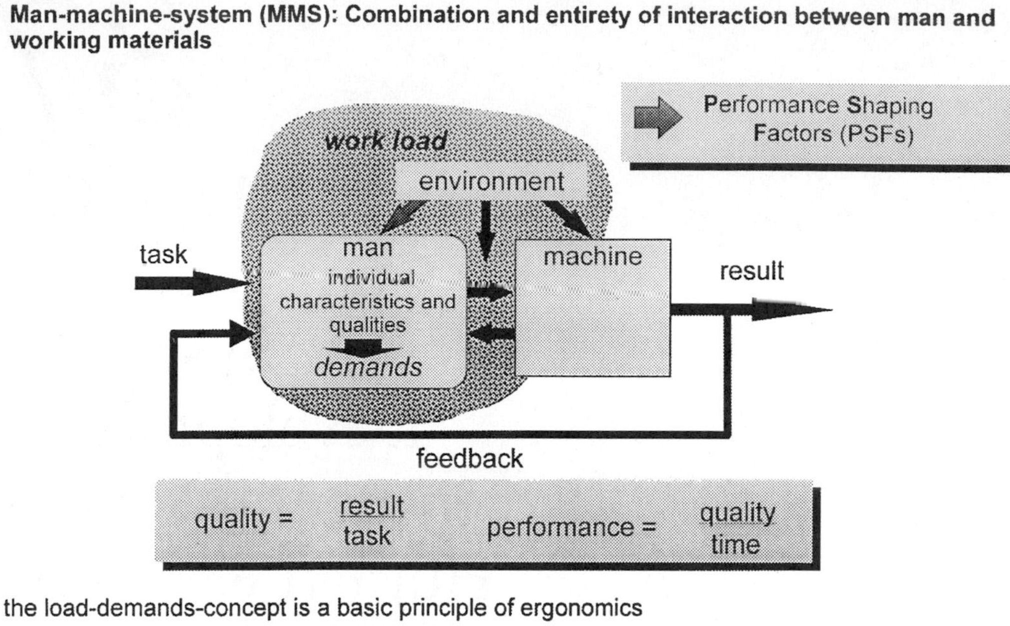

FIGURE 37.1 Structural block diagram of human work.

and dependability of such systems. Primary technical, often unimportant, failures occur and initiate inadequate human reactions. Because such effects cannot be treated only by measurement in the technical area, more profound treatments of human operation in the technological context and its possibilities of deviation are necessary. System ergonomics can be used as a base.

37.2 The Idea of System Ergonomics

System ergonomics starts with a general description of properties of every task and assigns experimental experience, and from that, ergonomic recommendations to the partial aspects of the task can be made. The fundamental idea is that by the knowledge of the information transfer by the subsystems man and machine, the tasks to be performed by the operator may be designed. For this designing of the task that results from the system mission and from the specifically chosen lay-out of the system and the system components (e.g., the machine), the following fundamental rules are to be considered, formulated as questions (see also Figure 37.2):

1. Function: "What has the operator in view and how far is he assisted by the technical system?"
2. Feedback: "Is the operator allowed to recognize if he has effected something and what was the success of it?"
3. Compatibility: "How much effort has the operator to make in order to convert the code system between the different technical information channels?"

The *function* may be separated into the intrinsic task contents and the influencable task design. Under this aspect the *task contents* are essentially defined by the *temporal and spatial order* of the activities which are to be carried out to perform the total task. It may be described by the terms operation (= *temporal* organization), dimensionality (= *spatial order* of the task), and manner of control (= the *time and location window* within which the task must be finished (see Bubb, 1988). The *task design* refers to the system structure which may be chosen to a large extent by the system planner. In this area the degree of difficulty may be influenced by the specifically designed layout. It can be distinguished between the manner of presenting task and result to the operator — the so-called *display* (*compensatory* or *pursuit*

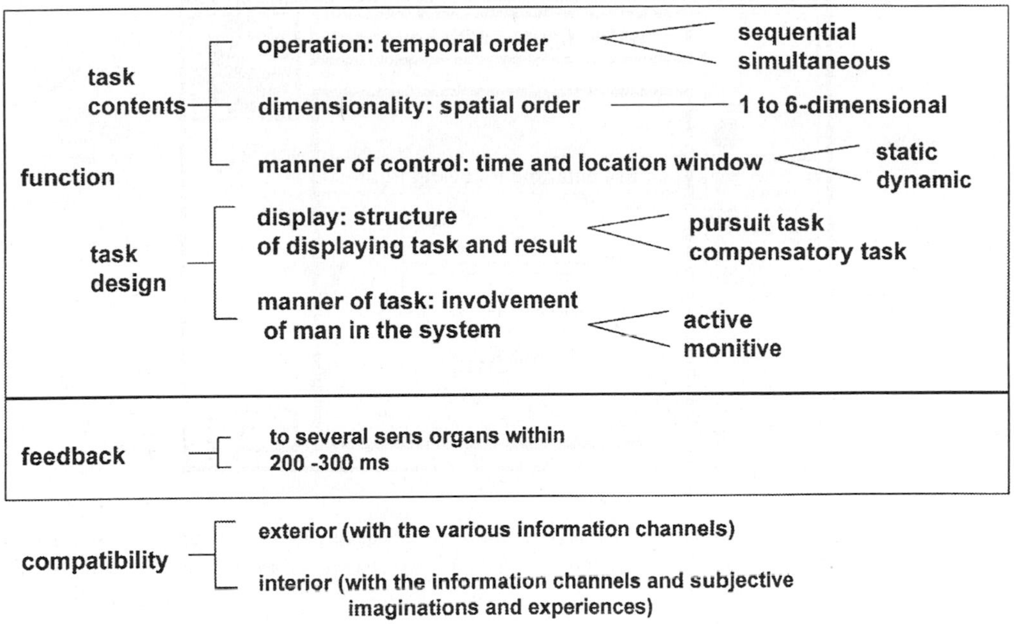

FIGURE 37.2 System ergonomic elements.

task), and the manner of involving the operator in the total system — the so-called *manner of task* (*active* or *monitive task*).

Feedback calls for a certain kind of redundancy which utilizes a *number of sensory organs*. A further aspect is the *time* that elapses between the input of information on the control element and the reaction of the system on the output side. Considering the relevant recommendations a well-designed feedback allows the operator to answer the questions:

- What have I done?
- In what a state is the system?

Compatibility describes the facility that enables the operator to convert the code system between the different information channels that he has to handle. It characterizes the obvious between different areas of information, as indication by instruments, control elements, and internal models of the operator. Neglecting the rules of compatibility decreases human reliability immensely (Spanner, 1993).

37.3 The Software Tool DIALOG

With DIALOG, a great part of system ergonomic aspects can be investigated. For instance, tasks of differing difficulties are realized by entering figures which appear as a combination of figures. Additionally, continuous tasks in the form of simple tracking experiments are possible, simultaneous tasks can be simulated, and time delay, concerning feedback, can also be presented. With another feature of the program, the subject can be put under time pressure.

The Task "Entering Figures"

In the "entering figures" mode the task for the subject is to rewrite a combination of several numbers between 1 and 20. This is an example for *sequential operation* of different degree of difficulty. Simultaneously further tasks can be displayed in form of traffic lights (with the color sequence green, yellow,

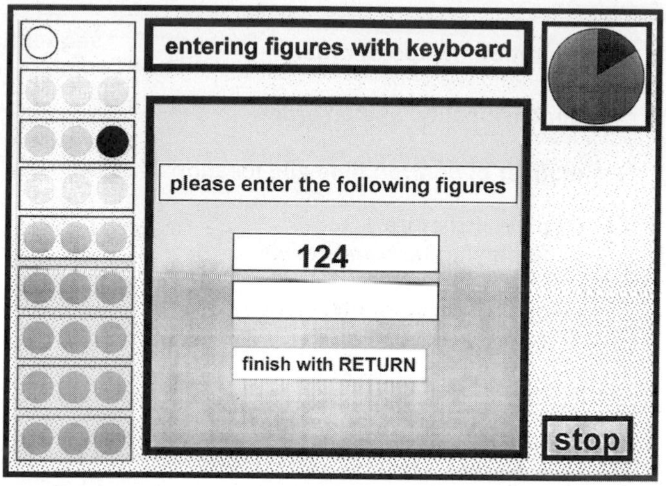

FIGURE 37.3 Example for displayed tasks by DIALOG. In this case: Sequential task of 3 numbers, 2 simultaneous tasks (rewriting of numbers and switching off of the traffic lights), restricted time window (indicated by the green/red disk right above).

red), which are to be switched off by a mouse click before the appearance of the color red. Up to 9 traffic lights can be presented. That is an example for *simultaneous tasks* of different degrees of difficulty. By an additional indicated disk different time windows can be investigated. This disk is originally green. During the task a red segment appears increasing with time. That is a representation of a *static task* (manner of control). Figure 37.3 shows an example of the video screen as it is presented to the subject during the experimental run.

The designer of the experiment can adjust the subject's task by a special input mask (see Figure 37.4). As it can be seen from this figure, the software tool DIALOG allows us to choose the number of test runs (here 30) and to adjust the time limit in steps of milliseconds. By a mouse click the time limit may be activated (indicated by a cross). The box "feedback" allows us to choose a feedback delay and to activate or deactivate optical feedback (= indication of the numbers) and/or acoustic feedback (= a clicking sound). In the box "determination of figures" the minimum and maximum number of figures may be determined. If the same number is not taken for both, by arbitrary numbers with different figures appear in the task window in the chosen range. The box "simultaneous operation" allows us to determine the number of control actuators (= traffic lights) and independent of it the number of actual activated operation during a test run. In the field "speed" the time between the appearance of two traffic lights is determined. Of course, this time must be chosen in such a manner, that the amount of "number of operations" can run within the general chosen "time limit."

The program modus "entering of figures" offers an example of a one-dimensional, static task presented as a pursuit task. The manner of task is active. In this combination the following elements of system ergonomics can be treated by experiment: Sequential and simultaneous operations of different degrees of difficulty and different conditions of feedback.

The Task "Following a Line"

In a second modus, DIALOG allows us to investigate continuous tasks. This is a kind of tracking task. In this case the subject has to redraw a line with the mouse, which is presented under a defined angle. This task can be demanded with and without time pressure and with and without optical feedback. By changing the allowed tolerance field, the difficulty of task can be varied. The described task is presented to the subject by a screen picture as shown in Figure 37.5.

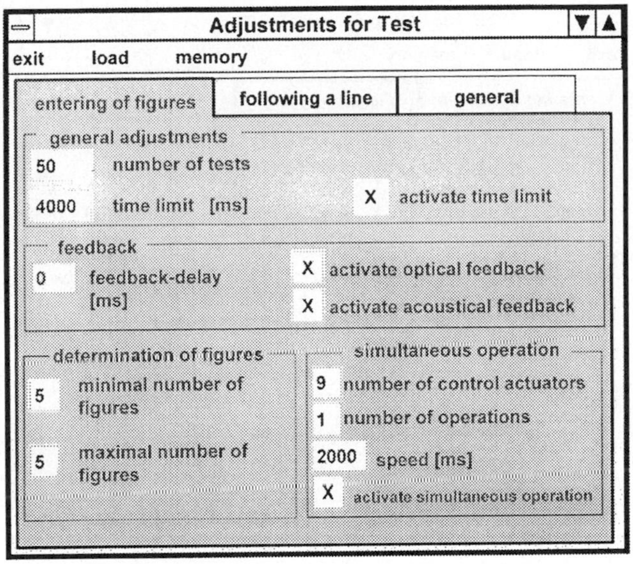

FIGURE 37.4 Possibilities of adjustment for the test "entering of figures."

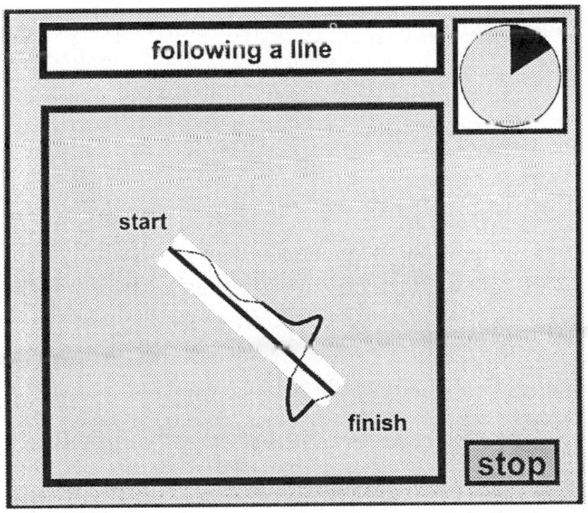

FIGURE 37.5 Example for displayed tasks by DIALOG. In this case: Sequential (Following the line), optical feedback, restricted time window (indicated by the green/red disk right above).

In order to adjust the different experimental conditions the user surface "following a line" of Figure 37.6 is applied. Also in this case by "general adjustments" the number of tests (here: 30) is chosen and the time limit, which determines the time pressure, is adjusted and in a given case selected ("activate time limit"). The box "description of range" allows us to determine the width of tolerance (radius). This can be done even asymmetrically (tolerance: positive, negative, or symmetric). By the box "feedback" the optical replay of the drawn line can be switched on and off. The box "general adjustments" allows us to define the starting position of the displayed line, the angle under which it is inclined, and its length.

The program modus "following a line" represents a two-dimensional static task with a location window given by the width of tolerance. Under the view point of operation it is a singular task. The display again creates a pursuit task.

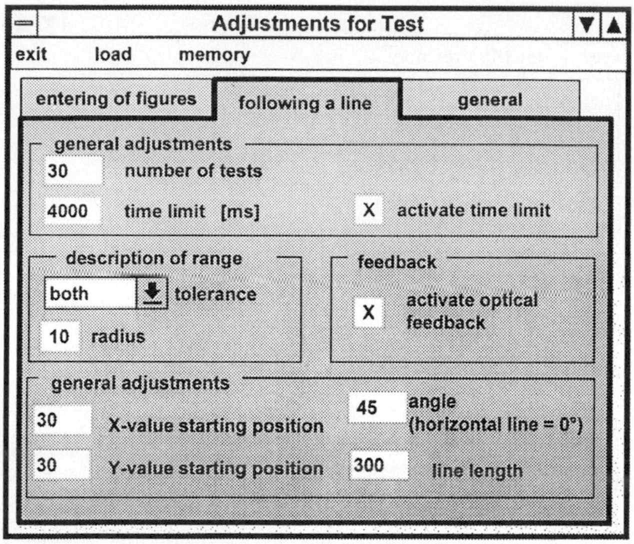

FIGURE 37.6 Possibilities of adjustment for the test "following a line."

37.4 Results

Using the described equipment, experiments were carried out. Thirty-two subjects had to accomplish different tasks presented by DIALOG during a session of about 1.5 hours. All reactions of the subjects were recorded by DIALOG. In the following only a part of the possible analysis is reported as an example. All results are statistically safe on a significance level of 0.95.

Entering Figures

In a first investigation the areas "operation" (sequential task of the difficulty "5 numbers" and "7 numbers" in combination with and without simultaneous switching off of traffic lights) and manner of control (with and without a temporal window of 4s or 8s, within which the tasks had to be accomplished) have been the objective. In this case the decision "task correctly accomplished" or "not" can be made, and the usual definition of human error probability (HEP) can be used:

$$\text{HEP} = \text{number of observed errors/number of opportunities for error}$$

We found out that time pressure results in more increased error probability than additional simultaneous tasks. In the case of recording 5 numbers, the average error probability was about 0.03. This value increased with the factor 14 to 0.43 under the condition of time pressure (Figure 37.7).

The observation shows that a more difficult (sequential) task under the condition of time pressure has an under-proportional ascent of error probability in relation to the behavior of easier tasks (Figure 37.7).

The simultaneous task of switching off traffic lights had no additional effect to the error probability, with a time window of 8s. Without time pressure, the increase of the difficulty of a sequential task of 5 to 7 numbers had an effect of the factor 2.5. Under the condition of time pressure of 4s, the same increase had only an effect of the factor 1.3. A more difficult task seems to stimulate the attention (see Figure 37.8).

This suggestion is substituted by a more detailed investigation of errors, so we could separate the following kinds of error (see also Swain and Guttmann, 1983):

- Error of quality, z_1 (writing a wrong number, e.g., 123 instead of 124)
- Error of part omission, z_2 (omitting one number, e.g., 12 instead of 124)
- Error of omission, or negligence, z_3 (the task is totally neglected)
- Error of addition, z_4 (writing an additional number not asked for, e.g., 1234 instead of 124)

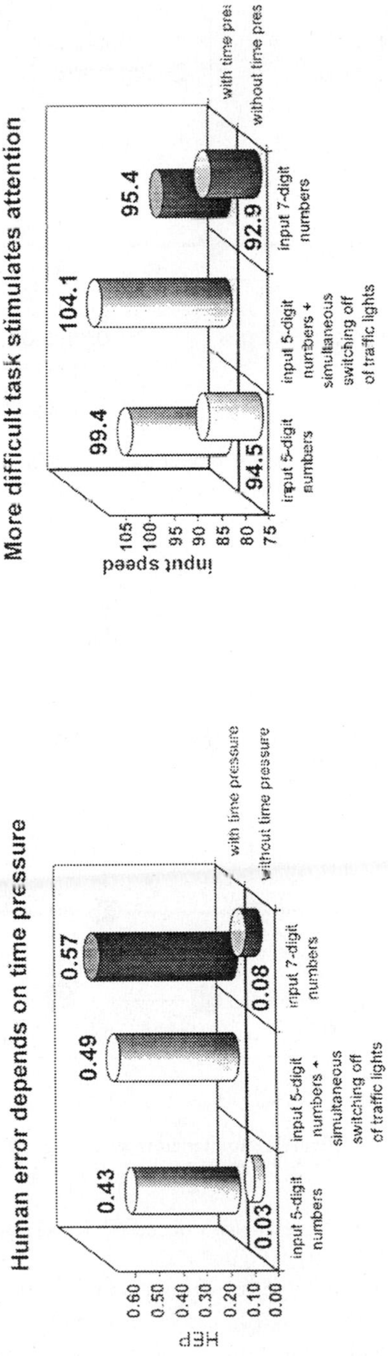

FIGURE 37.7 Human error depends on time pressure.

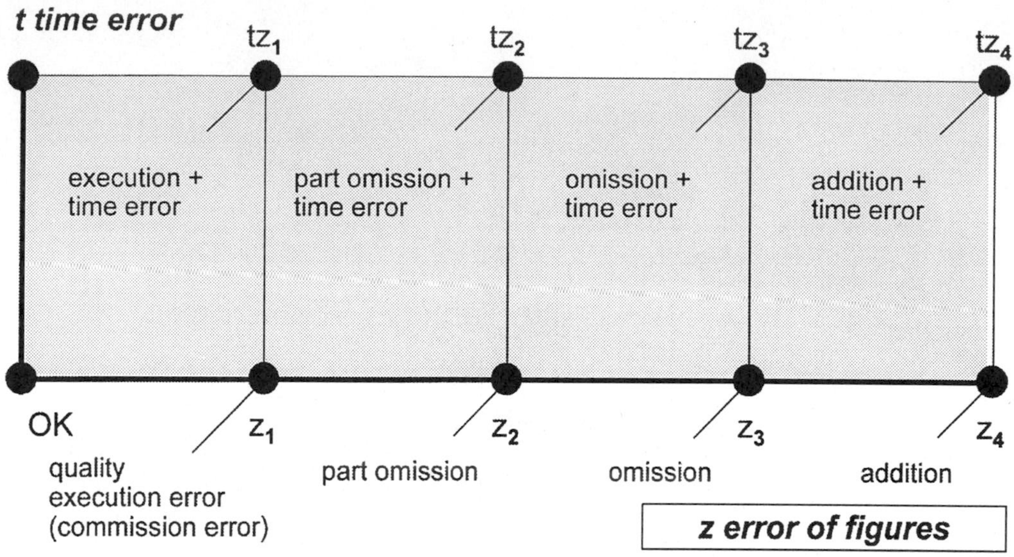

FIGURE 37.8 Errors made at "entering of numbers" with time pressure.

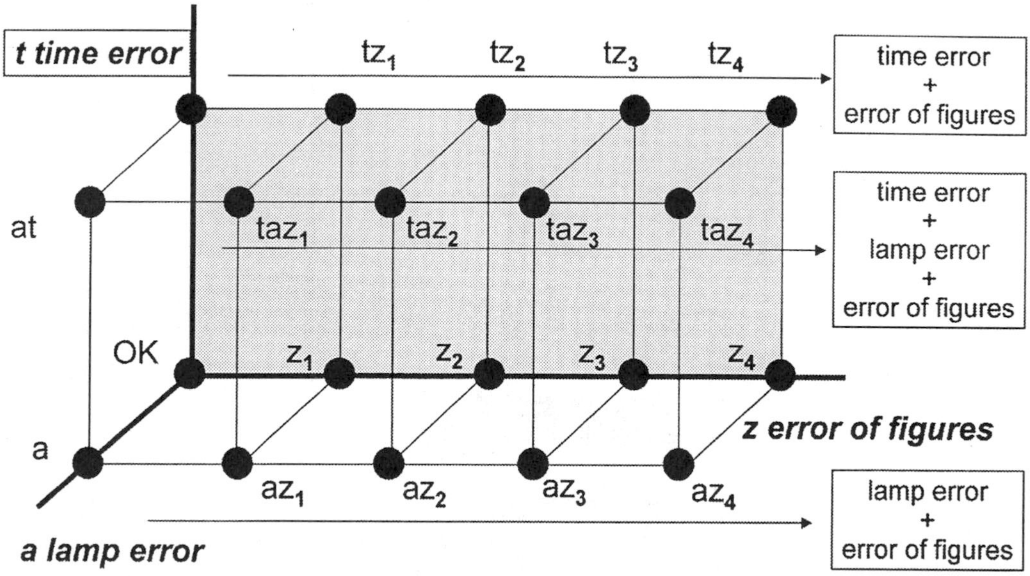

FIGURE 37.9 Errors made at "entering of figures, simultaneous switching off of the traffic lights" under time pressure.

In the case of time pressure, all the types of error can occur in combination with a time error, i.e., the task is not finished within the time window, t (Figure 37.9).

Whereas the error of negligence does not appear without time pressure, this type crops up with time pressure and gets a value of 0.076 (see Figure 37.10). Under the condition of time pressure the time error appears with a probability independent of the difficulty of the task of 0.07 to 0.09. Of course, all types of error are increasing with time pressure. There is one exception: the error of quality disappears under this condition. The probability of all kinds of error becomes smaller when these appear in combination with time error. There is again one exception: the error of omission increases in connection with time error with the factor 16 for 5 numbers and 12 for 7 numbers.

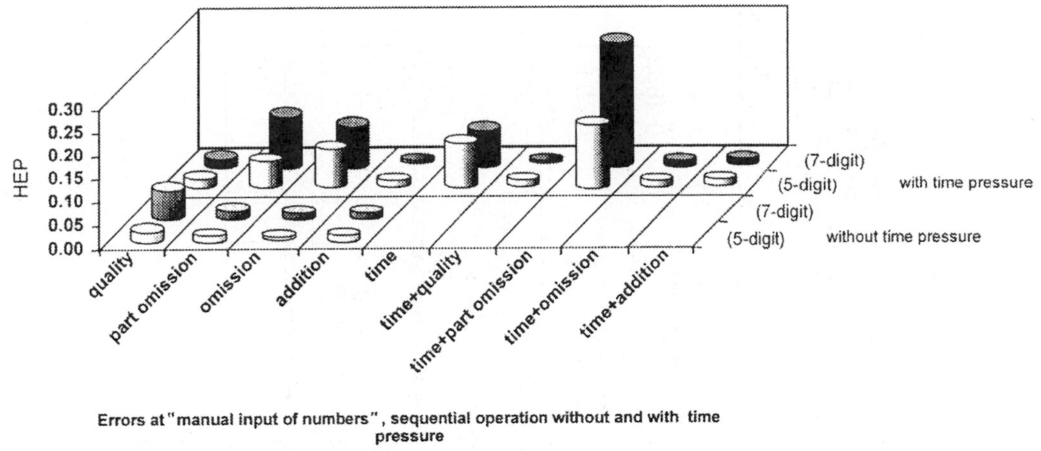

FIGURE 37.10 In the case of time pressure, all the types of errors can occur in combination with the time error.

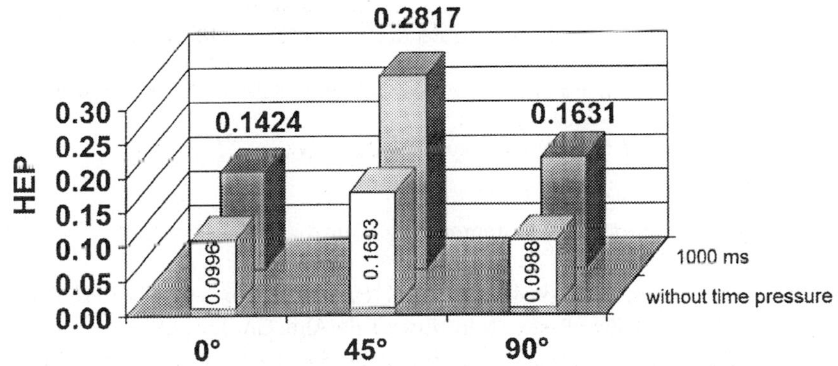

FIGURE 37.11 HEPs during "following a line" under the condition of time pressure and without time pressure and the orientation of the line of 00, 45° and 900.

Following a Line

As during continuous tasks the number of tasks and thereby the number of errors cannot be defined, the definition was carried out considering the time:

$$\text{HEP} = \text{total of time quota status “error”/total time}$$

In the case of these results, the tolerance width was adjusted to 10 pixels. Figure 37.11 shows the results of the HEP values without time pressure and time pressure of 1s. Different orientations of the line were used. We observed that under the condition of 0° and 90° the same error probability occurs, whereas in connection with an inclination of 45° the error probability is nearly doubled. Under this condition the time pressure plays a more important role. However, as Figure 37.12 shows, the jump of error probability from without time pressure to any time pressure is much higher than the increase from time pressure of 3s over 2.25s to 1.5 s. Figure 37.12 shows also that the inclination of the line is altogether unimportant.

The results reported here only show examples of the possibilities of experiments that can be done by DIALOG. In future, by using this program, a more systematic investigation in the area of system ergonomics will be possible in regard to error probability. There is a certain hope, that experiences observed by DIALOG will complement and add to observations of errors during real situations. In future, there might be a system in which a prediction of human error probability will be possible in cognitive tasks as well.

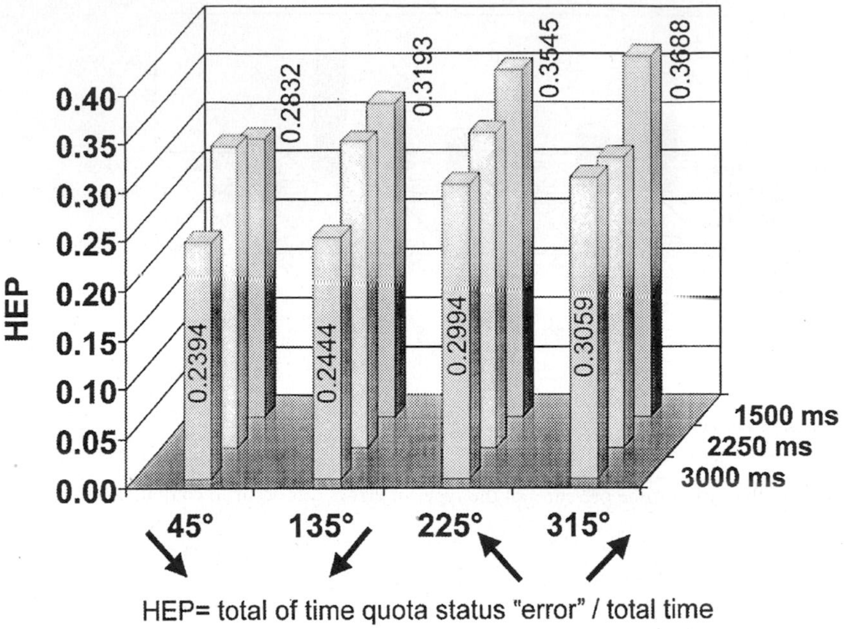

FIGURE 37.12 HEPs during "following a line" under the condition of different time pressure and different inclination of the line.

References

Bubb, H. (1988): System ergonomics as an approach to improve human reliability. *Nuclear Engineering and Design* 110, S. 233-245.

Hannaman, G.W. and Spurgin, A. J. (1984): *Systematic Human Action Reliability Procedure (SHARP).* ERPI NP-3583, Electric Power Research Institute, Palo Alto, CA, June 1984.

Rigby, L. V. (1970): The nature of human error, in: *Amer. Soc. for Quality Control, Annual Technical Conference,* 24th, 457-466.

Spanner, B. (1993): Einfluß der Kompatibilität von Stellteilen auf die menschliche Zuverlässigkeit. Fortschritts-Berichte VDI, Reihe 17 "Biotechnik," VDI-Verlag Düsseldorf.

Sträter, O. (1997): *Beurteilung der menschlichen Zuverlässigkeit auf der Basis von Betriebserfahrungen.* Dissertation an der Technischen Universität München.

Swain, A.D. and Guttmann, H.E. (1983): *Handbook of Human Reliability Analysis with Emphasis on Nuclear Power Plant Applications.* NUREG/CR-1278, Scandia Laboratories, Albuquerque, NM, 97185.

38

Managing the Speed–Accuracy Trade-Off

Colin G. Drury
State University of New York at Buffalo

38.1 Introduction: Speed, Accuracy and Performance

This chapter examines two aspects of task performance, speed and accuracy, and shows how they interact with each other. The best-known interaction is a speed/accuracy trade-off (SATO) — as speed increases accuracy decreases. But this is not the only interaction between these two performance measures, so managing their interaction needs more detailed knowledge. Speed and accuracy are first defined, and their modes of interaction shown as three levels. Finally, a practical way of managing this interaction is presented with a worked example.

Under conditions of increasingly global competition, industry must remain focused on performance in order to survive and prosper. We use many measures of "performance" based on our different perspectives and on the level of aggregation of our data. Thus at a low level, a supervisor might define performance as labor hours to produce the week's output. At a somewhat higher level, a production manager may define it as percentage of orders shipped on time. At the board level, an appropriate performance measure may be how the stock price has increased in the past quarter.

While all are useful performance measures, at the plant level an augmented set may be more appropriate, for example:

Quality:	e.g., fraction of acceptable product shipped
Timeliness:	e.g., fraction of product shipped on time
Efficiency:	e.g., resources used to achieve the quality and timeliness levels.
Plant Health:	e.g., how fit the plant remains to achieve its performance in subsequent time periods

In fact, these are all examples of the three basic measures of performance — effectiveness, efficiency, and well-being — which can be applied to an individual employee, a production unit, or the whole plant:

1. *Effectiveness:* How well does the outcome match expectations? Was the refrigerator well designed? Was the error rate in hospital prescriptions low? Effectiveness is measured by accuracy or quality, or by their complement: errors.

0-8493-2641-9/99/$0.00+$.50

2. *Efficiency:* How much resource is consumed to achieve the effectiveness? Did the design of the refrigerator take too many designer-months? What staffing level was needed to achieve the low prescription error rate? Efficiency is measured traditionally by throughput or speed, but in fact includes other resource use, such as capital and lead-time.
3. *Well-being:* How did the achievement of *that* level of effectiveness with *that* efficiency affect the worker and the organization? What was the operator's stress level? How many injuries were sustained? Did job satisfaction measures increase? Well-being can be measured by the cost to the operator and the organization of achieving the desired performance.

Note that the first two, effectiveness and efficiency, are those typically measured and used by business people and together define "performance." Effectiveness, particularly customer satisfaction, has been promoted as the key measure. Indeed, customer satisfaction is seen by many as the price of entry into the competitive global marketplace. Efficiency is closely scrutinized by both managers and market analysts to determine whether the costs are outstripping the sales base. Corporate well-being is a more recent consideration, but is becoming more readily measured by satisfaction surveys. Most companies' annual reports talk about "Our people are our greatest asset" and "We put safety before all other considerations." Some companies, such as chemical and continuous process operations, have espoused the philosophy that if a process is understood well enough to make it safe, then it will also be effective and efficient. Thus the modern view is that effectiveness, efficiency, and well-being are all required to define "performance."

These three aspects can, of course, co-vary. Under some conditions, high efficiency can mean low effectiveness. This is the classic speed/accuracy trade-off where lack of time reduces quality. Under others, a plant-wide change can give both high effectiveness and high efficiency. Finally, some means of achieving high effectiveness and efficiency (for example, piece-rate payment systems) can adversely affect the well-being of plant and workers. Figure 38.1 shows some of the possible interactions between efficiency, effectiveness, and well-being applied to the individual human performing a task. It is these covariations, specifically between effectiveness and efficiency, that constitute the speed–accuracy trade-off (SATO). This chapter explains how different SATO functions can arise, and how to manage processes that may have SATO characteristics.

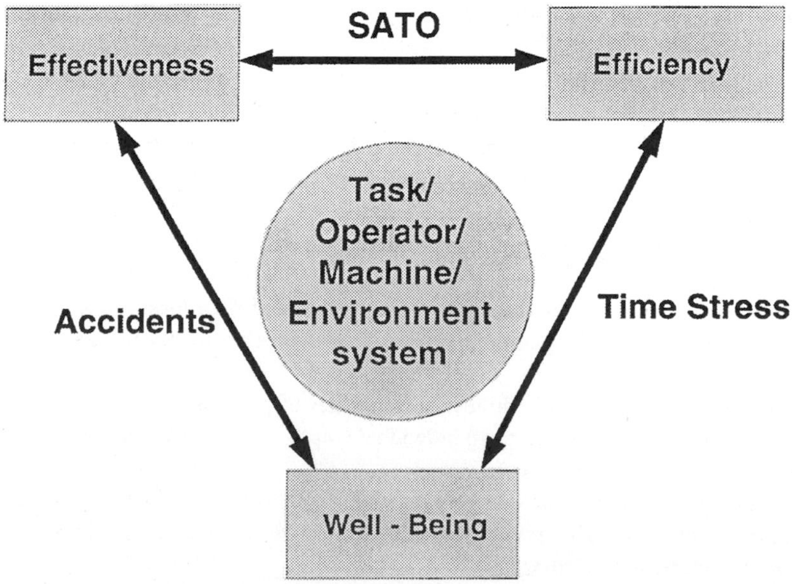

FIGURE 38.1 Interactions between effectiveness, efficiency, and well-being for the individual operator.

To understand SATO, we need to make the connections of speed with efficiency and accuracy with effectiveness. Speed is the rate at which a task is performed. At times it is measured as a rate, for example throughput of a manufacturing cell in parts per day or speed of a conveyor or forklift truck in meters per second (ms^{-1}). Often, however, speed is only implied and its reciprocal, time per unit, is measured. Thus a sewing task has standard allowed hours for each garment, or an aircraft engine maintenance shop has a throughput time measured in person days per engine. In research studies, speed or time is a basic parameter measured or controlled. A very early study of human motions (Woodworth, 1899) had people make rapid movements at different rates from 20 to over 200 per minute, while accuracy was measured to give one of the earliest examples of SATO. Since that date, studies of reaction times, decision times, movement times, search times, and so on have provided the basis for many of our quantitative models of human performance. Time and rate measurements have an equally long history in industry, stemming from scientific management, and evolving into time study and methods study (e.g., Konz, 1990; Mundel and Danner, 1994).

Accuracy is measured as "freedom from error in discrete tasks" (Drury, 1994), with analogous measures for continuous (e.g., tracking) tasks. Thus to define accuracy, we must also define error. Rasmussen (1986) considered errors as unsuccessful experiments in an unkind environment (p. 150). In a major compilation of current positions on error, Senders and Moray (1991) suggest that an error is a human action that fails to meet an implicit or explicit standard. More formally they characterize errors as (p. 25):

Actions not intended by the actor
Actions not desired by a set of rules or an external observer
Actions that led the task or system outside acceptable limits

Responding to costly and visible errors, such as the Bhopal disaster, the study of human errors is undergoing something of a renaissance in ergonomics (e.g., Reason, 1990; Norman, 1981), with numerous classification schemes being proposed. However, for the purpose of this chapter a simple classification into slips and mistakes (Norman, 1981) is appropriate. While both can be affected by speed, it is the skill-based slips that most obviously show this effect. Rule-based and knowledge-based mistakes represent incorrect intentions, which could be the result of insufficient time for information processing, but correct intentions incorrectly executed (slips) are the most typical speed-affected errors.

At the beginning of this introduction, we emphasized that modern manufacturing requires both speed *and* accuracy to survive. Now that speed and accuracy have been defined, we can proceed to see how they interact with each other, i.e., SATO. We are not forgetting well-being, merely holding it constant. We will only consider SATO which leaves well-being unchanged, meaning that we do not allow the plant to deteriorate, or its workforce to develop chronic injuries in our pursuit of speed and accuracy.

38.2 The Speed–Accuracy Trade-Off

Not all tasks exhibit a speed–accuracy trade-off. In some tasks, the accuracy attainable is an inherent function of the quality of the data used to perform the task, rather than a function of how long the operator can consider that data. For example, in a difficult quality-control discrimination task, such as judging the noisiness of electric motor bearings, more time to decide does not produce increased accuracy. Such tasks are called *data-limited* (Norman and Bobrow, 1979). "Data limited" means that the inherent quality of the data limits performance, no matter how much time is devoted to the task. Tasks which *do* exhibit a SATO are ones where evidence in favor of a correct response is accumulated over time, for example, searching a circuit board for defects. These tasks are called *resource-limited* in that the limitation on accuracy is how much resource (in this case, time) is expended on the task.

Examining the shape of the speed–accuracy trade-off function of many tasks suggests that most are resource-limited at very short times and data-limited at very long times. Figure 38.2, for example, shows a task that is resource limited at short durations (500 ms or so), but data limited for durations longer than about 500 ms. In the remainder of this chapter, we shall consider tasks that do exhibit SATO, i.e., those that are resource-limited over a practical range of times.

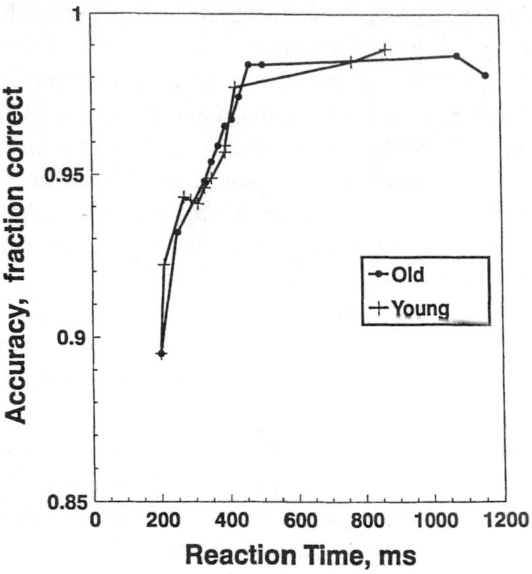

FIGURE 38.2 Level 1 speed–accuracy trade-off in reaction times for two age groups. (Adapted from Smith, G.A. and Brewer, N. 1995. Slowness and age: speed–accuracy mechanisms. *British Journal of Psychology,* 76:199-203.)

Woodworth's early movement study showed that as speed increased, accuracy worsened. This classic inverse relationship between speed and accuracy is what most consider "SATO." Perhaps the first comprehensive SATO studies were Garrett's (1922) inquiries into SATO for human perception and coordination. Since then, major advances were made by Pachella (1974) and Pew (1969). The latter coined the term "Speed–Accuracy Operating Characteristic," or SAOC, to describe the plot of speed *versus* accuracy for a particular task or experiment. It is one type of isoperformance curve (Jones and Kennedy, 1996) showing a limit on possible combinations of performance. We will examine at a "task" level where SATO arises and the different forms it can take.

A common-sense view of human performance suggests that people will choose levels of speed and accuracy that do them the most good. Such ideas have indeed been influential in the study of SATO, under the general title of economic maximization models. They postulate that human operators consider the rewards for speed and the penalties for error, finally choosing an operating point that maximizes their expected payoff given a fixed level of ability. Examples of this thinking include:

1. A model of choice reaction time (Fitts, 1966) that postulates that the operator gradually builds up evidence for one or the other alternative until the amount of evidence passes some threshold that the operator considers enough to start a response. Because the evidence also contains noise, the operator's path toward the final alternative is a "random walk." Time to respond can be reduced by reducing the threshold amount of evidence needed for a response, but the result will also mean more errors.
2. Signal Detection Theory Models of the decision process as a choice between alternatives based on inherently noisy data. In, for example, a quality control task, the inspector has to decide on whether or not to reject an item of product based upon whether it is judged to meet a standard. Choice of the criterion of what to accept depends on the relative costs and probabilities of the various outcomes of the decision (McNichol, 1972).
3. Optimal stopping models of visual search (e.g., Karwan et al., 1995) where a visual inspector is modeled as having to choose when to stop searching one item of a product for possible defects and move on to the next item. Stopping earlier on each item will save time but lead to the inspector missing some defects. Stopping time is optimum when the costs of increased search time begin to outweigh the value of finding a defect.

TABLE 38.1 Possible Interventions by SATO Level

SATO Level	Interventions	Intervention Label
1. Micro-SATO	Keep strategy at constant point on the SAOC	1. Reduce SATO variability
2. Macro-SATO	Choose the best operating point on the existing SAOC	2. Optimize SATO point
3. Performance intervention	(a) Reduce the fixed time associated with task	3. (a) Reduce overhead
	(b) Improve the slope of the SAOC, i.e., improve efficiency of process	3. (b) Improve process capability

4. Driving along a roadway of fixed width (Montazer et al., 1987). During a fixed visual sampling interval, a driver covers a distance proportional to road speed. There is value to covering the distance as quickly as possible, but large costs associated with deviating into a fixed boundary, e.g., drifting off the road. The optimization model correctly predicts the speed and accuracy as a function of roadway width.

While such models have been useful and influential within laboratory tasks, their use in shaping long-term human behavior at work may be limited. People tend to be poor at estimating the costs and probabilities needed in everyday life, and often act as "satisficers" who select a solution that is good enough rather than as "optimizers" who seek the best solution. In particular, using rewards and punishments to control the speed–accuracy trade-off may well distort the very system it seeks to optimize. Even Deming (1986) has removal of speed incentives as one of his 14-points for improving quality.

In parallel, we are being urged by modern management techniques to stress the achievement of quality as a way to improve speed of performance. This is taking place at a higher level of aggregation than the task or its detailed components. We are here considering the introduction of dedicated cells, just-in-time manufacturing, statistical process control procedures, or even plant-wide total quality management programs.

There is a clear disparity between the known negative SATO at a task level and the known positive SATO at a more highly aggregated level. This disparity is mainly the result of the multiple ways in which SATO can take place. We can classify where SATO occurs at three levels as a micro-SATO (Level 1), where speed and accuracy vary from trial to trial in a repetitive task, a macro-SATO (Level 2), where the task may be performed under different speed and accuracy reward structures or a performance-intervention-SATO (Level 3), where deliberate changes are made to improve both the speed and accuracy of task performance. We shall later examine the possible interventions at each level, as shown in Table 38.1. First, we consider each level in turn.

Level 1: Micro-SATO: In a repeated task, the repetition-to-repetition performance changes can show a speed–accuracy trade-off. In laboratory reaction-time tasks, where the criteria are for high speed with high accuracy, both speed and accuracy vary from trial to trial. Trials with the shortest reaction times tend to exhibit the highest error rates. This was demonstrated by Pew (1969) and has since been a consistent finding in reaction time research. For example, Smith and Brewer (1995) found very similar speed/accuracy relationships for two groups of subjects 18 to 30 years old and 58 to 75 years old (Figure 38.2), despite the older subjects generally responding more slowly. Note that in these carefully controlled laboratory tasks, the time from stimulus to response (reaction time) may be very short compared with typical industrial task completion times.

Studies of human performance within a sequence of repetitions of a task (e.g., Drury and Corlett, 1975; Rabbitt, 1981) have shown that people control their level of effort on a trial-to-trial basis. For control over speed and accuracy, people try to maintain speed performance within a narrow band which gives a sufficiently low error rate. When speeds increase beyond this band, errors occur. An error (at least a detected error) is followed by a slowing of performance for a time until the desired speed band is achieved once again. This effect can be seen in the Smith and Brewer study of SATO and aging. Figure 38.3 shows mean reaction times in the trials before and after an error occurred.

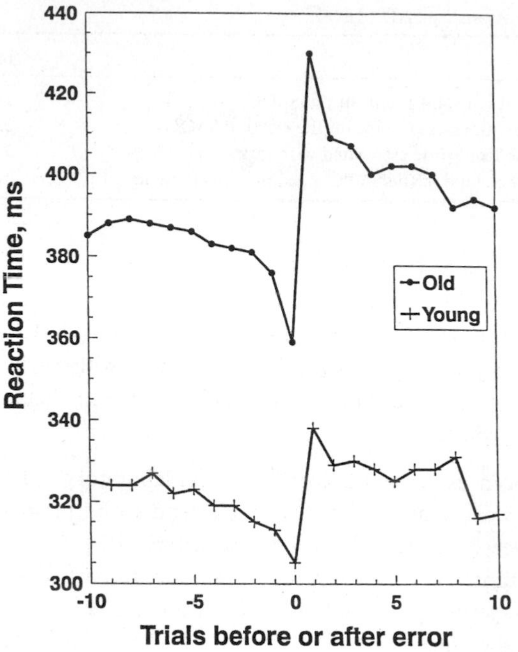

FIGURE 38.3 Speed-up prior to error and slow down following error. (Adapted from Smith, G.A. and Brewer, N. 1995. Slowness and age: speed–accuracy mechanisms. *British Journal of Psychology,* 76:199-203.)

Level 2: Macro-SATO: When the same task is repeated under different reward conditions for speed and accuracy, the mean speed and accuracy over each reward condition show a SATO. In a classic experiment, Hick (1952) rewarded subjects in his choice reaction time experiment either for speed or for accuracy. He found that when encouraged to react faster, more errors occurred. In fact the increase in mean errors balanced the decrease in mean reaction times exactly as predicted from information theory, i.e., the subjects' rate of transmission of information remained constant across all reward conditions.

This macro-SATO level is the most typical lay interpretation of SATO: if you are encouraged to increase speed, you may be expected to decrease accuracy. Perhaps because of this popularity, there have been many studies that have demonstrated a macro-SATO. Wickelgren (1977) reviews earlier studies and examines how the SAOC can be measured. Wickens (1994) reviews SATO in an information theory context, while Meyer, Smith, Abrams and Wright (1990) provide a comprehensive historical view of SATO applied to accurate movements. Measurement of macro-SATO in visual search is described in Drury and Forsman (1996).

Much of the SATO literature comes from the study of accurate movements, e.g., Gan and Hoffmann (1988). For low accuracy movements, known as ballistic movements, the movement time is a function only of the distance traveled. For high accuracy movements, time depends both on the distance traveled and the accuracy demanded. This is known as Fitts' Law, with the formula given in Table 38.2.

In an industrial context, the same effects are observed. Omissions and errors increase with pacing speed in industrial assembly (Dudley, 1958) and in textile machine servicing (Conrad, 1954). Accurate movement of a computer's mouse obeys Fitts' Law (Card et al., 1983). Forklift trucks are driven more slowly as aisle width decreases (Drury and Dawson, 1972). Inspection of minted coins for defects shows decreased accuracy at higher pacing speeds (Fox, 1977).

The macro-level of SATO has involved a number of models of human performance. These predict the levels of both speed and accuracy that are possible, provide some explanation of *why* the SATO effects occur, and give a rationale for measuring and plotting the SAOC. It is not the intention of this chapter to present these models in any detail, nor to continue the arguments for and against alternative models.

TABLE 38.2 Speed–Accuracy Functions for Selected Tasks

Name	Task	SAOC Form	Time Range, s
Target detection (Bloch's Law)	Detect dim targets of intensity I on a dark field with exposure time T	$IT = a$	<0.1
Visual search (random strategy)	Search for and locate a target in an extended, often cluttered, background	$p(det) = 1 - \exp(-aT)$	<100
Signal detection (SD theory)	Detect signal in the presence of noise. p(det) = prob. of detection; p (FA) = prob. of false alarm. z(p) is normal deviate for prob. p	$z(p(det)) + z(p(FA)) = a\sqrt{T}$	<2
Choice reaction time (Hick's Law)	Respond correctly to one of a number N of different events	$\text{Log}\left(\frac{p(correct)}{p(error)}\right) = aT$	<2
Accurate movements (Fitts' Law)	Move the hand or an object a distance A to a target of width W	$T + a + b\,\text{Log}\left(\frac{2A}{W}\right)$	<2
Driving/path following	Control body or vehicle along a path with lateral boundaries giving clearance = W	$\text{Speed} = \frac{k}{T} = a + bW$	—

T = time allowed or taken; a, b are constants

Rather, some models of specific tasks are given in Table 38.2 to indicate how much assistance is available in the literature in understanding SATO.

A single example is instructive, however. Figure 38.4 shows a SAOC for the task of automobile driving, from DeFazio et al. (1992). Here, the issue was the effect of an accuracy restriction on chosen speed. Subjects drove an automobile around a circular course of radius 8.13 m as fast as possible without exceeding the course boundaries. A model of this process (Drury, 1971; Montazer et al. 1987) showed that if the driver adjusts the speed to maintain a constant (low) probability of leaving the road, then a linear relationship would be expected between speed and car/boundary clearance. The model tells us what to control and measure in this experiment (control clearance, measure speed), and what shape of SAOC to expect. In this experiment, the aim was to determine whether car width and boundary width have the same effect on clearance, and hence speed. They did not, and the authors were able to calculate effective car width which differed from the actual car width.

Level 3: Performance Intervention SATO: When a change (other than just a change of reward structure) has been made to a task, then both speed and accuracy can be affected. The particular relationship between speed and accuracy seen in Levels 1 and 2 will probably not hold when changes are made to the task. Thus an improved task may well give:

1. Improved accuracy at the same speed
2. Improved speed at the same accuracy
3. Simultaneous improvements in speed and accuracy

The point is that changes to the task move us to an entirely new SAOC, rather than merely moving *along* a single SAOC as was the case for Levels 1 and 2. For example, Figure 38.4 shows SAOCs for two tasks in circuit board inspection: finding a wrong IC chip and finding a bent component lead (Drury and Chi, 1995). Both SAOCs are examples of systematic visual search, but at any chosen search time the probability of detection is always higher for the Wrong IC condition.

Change of conditions can be due to changing the operator (e.g., selection, motivation training), the task (as in the above example), the equipment (e.g., hybrid automation in inspection: Hou et al., 1993) or the environment (e.g., lighting, Drury, 1987). Such changes have costs attached, but also payoffs for increased performance. For example, Drury (1991a) extended the data from Hasslequist (1981) on a before-and-after study of ergonomics changes at an assembly plant. Quality improved — 50% fewer errors — at the same time as productivity was up 6%.

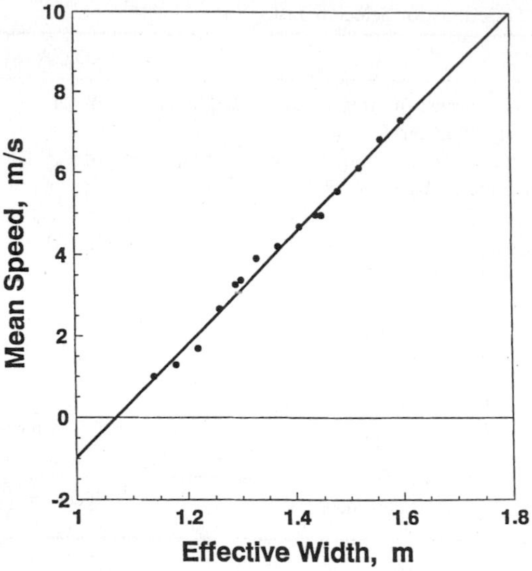

FIGURE 38.4 Speed–accuracy operating characteristic for car driving along narrow road. Chosen speed varies with the effective width of clearance. (Adapted from DeFazio, K., Wittman, D., and Drury, C.G. 1992. Effective vehicle width in self-paced tracking, *Applied Ergonomics,* 23(6): 382-386.)

Much of the data supporting the concept that improving quality will also improve productivity comes from popular publications rather than scientific journals. From Crosby's classic book, *Quality Is Free* (1979) comes the idea that the true cost of quality is no greater than that of producing defects, but productivity is not addressed. In other similar books, executives from successful quality companies, such as Motorola, do make the connection, saying "Quality is quantity; quality is low cost" (Dobyas and Crawford-Mason, 1991, p. 253). In the same source an executive at Siemens shows 30,000 fuel injectors being produced in a month (1985), in a week (1987), and now in a day (1990). Plans were for them to be produced in one shift in 1992. "That has been and is being done through quality improvement" is the quote.

An excellent special issue of *Business Week* on quality in 1991 provided many more examples. Toyo Ltd., a maker of bathroom fixtures, started a quality program in 1984, and credits it with halving inventories and lead time, and also with a 50% increase in productivity at the same time as a 25% decrease in customer complaints (p. 23). Similarly, at NEC's Tohoku printer plant a quality function deployment program reduced defect rates from 26% to 1.2%, while doubling productivity (p. 61). As a final example, a quality vice president at IBM is quoted as saying "The longer you work on something, the more time you have to interject defects." (p. 68).

Some examples have more traditional references. The compilation of interventions in *Work in America* (Anon., 1973) showed many that reported both throughput and quality increases. A more recent example (Drury, 1991a) introduced a just-in-time manufacturing cell for machining end-caps for automotive alternators. In just-in-time (JIT), the aim is to improve performance by reducing inventory levels between processes. When one process fails, it brings the whole production system to a halt, which focuses management attention on finding permanent cures for process failures. This eventually improves quality by reducing many of the failure modes. With a higher-quality product and system, the efficiency can also increase, thus giving simultaneous speed and accuracy benefits. In the manufacturing cell (Drury, 1991a), cycle time decreased from 7 days to 2.5 hours, while quality improved so much that no defects were measured during the test period of several months after the change.

Incidentally, changing to a different person performing the task can change both speed and accuracy, but rarely in any predictable manner. Because people differ in their overall capabilities, they can display better or worse speed and accuracy. Thus a plot of speed against accuracy where each point is the average performance of an individual will, in general, produce just a cloud of points with no correlation. We cannot expect that faster individuals will be any more, or any less, accurate than slower individuals. This concept has been used to classify people on the two dimensions of speed and accuracy using a standard test, the Matching Familiar Figures Test (MFFT). The four groups are:

Slow, inaccurate
Slow, accurate ("Reflective")
Fast, inaccurate ("Impulsive")
Fast, accurate

We have used this categorization in two inspection tasks (Schwabish and Drury, 1984; Chi and Drury, 1996) without much success. In fact, performance on an inspection task was best predicted in both cases by using only the accuracy dimension of MFFT.

38.3 Practical Intervention

Corresponding to each level of SATO, there are appropriate interventions to manage those jobs where SATO has an influence. Drury (1994) developed four possible interventions, provided a methodology for managing SATO, and gave an example of its use. In the current chapter, his interventions, methodology, and example will be adapted to the three levels of SATO defined in the previous section.

The micro-SATO of Level 1 showed how trial-to-trial variability in SATO can arise. For example, in Figure 38.2, if the mean reaction time had been chosen for each reaction, the error would probably not have occurred. A suitable intervention for Level 1 is thus to ensure that a consistent strategy is used, i.e., the operator does not drift along the SAOC toward either very slow performance or frequent errors.

Level 2 introduced macro-SATO, showing how the operating point chosen on the SAOC moves as the rewards for speed and accuracy are manipulated. Here, the intervention strategy should be to ensure that operators choose the "best point" on the SAOC to operate. "Best" has to be defined in systems terms depending upon the shape of the underlying SAOC and the relative value to the company of efficiency and effectiveness.

At the performance intervention level (Level 3) the aim should be to move to a better SAOC, rather than merely choosing and sticking to the best point on the existing SAOC. Figure 38.5 shows two SAOCs that illustrate the ways in which a SAOC can be "better." First, the two fitted lines cut the time axis at different points. This point is where effectiveness (here, probability of detection) is zero, and represents the time to perform the task when no searching is taking place. A suitable name for this would be the overhead time. Reduction of overhead time is thus one aspect of a Level 3 intervention. The other difference between the SAOCs of Figure 38.5 is in their slopes. The "better" SAOC has a steeper slope, i.e., probability of detection increases more rapidly with every second spent searching. In general, this represents improved performance efficiency, or improved process capability.

To summarize the interventions, Table 38.1 shows the levels, interventions and appropriate labels. Note that most of the simple views on SATO are at Levels 1 and 2 which emphasize trading off speed for accuracy (e.g., haste makes waste). However, most of the quality-oriented management philosophies emphasize the simultaneous improvement of quality *and* productivity, largely by using Level 3 strategies.

The objective of an ergonomic analysis of a task is not just to understand the task, but to generate beneficial interventions. How can we improve speed and accuracy aspects of productivity? If a task is expected to exhibit SATO effects, the four generic intervention strategies can be employed to generate specific interventions. All that is required is a task description, and a task analysis which determines whether SATO is likely. Such a task analysis should be the first step in almost any ergonomic project, so that its production does not represent additional ergonomics work load. Thus a four-step process is recommended (Drury, 1994):

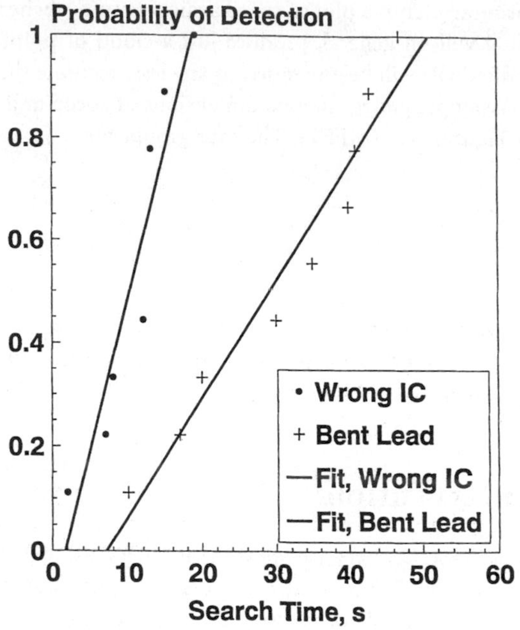

FIGURE 38.5 Speed–accuracy operating characteristics for two circuit board search tasks. (Adapted from Drury, C.G. and Chi, C.-F. 1995. A test of economic models of stopping policy in visual search. *IIE Transactions* (1995), 27: 382-393.)

1. *Perform Task Description and Task Analysis*: Use standard methods (e.g., Kirwan and Ainsworth, 1992; Drury et al., 1987) to produce a list of tasks, with the ergonomic models or database underlying each.
2. *Select SATO Tasks*: Determine whether each task is potentially resource-limited. Prototypical tasks are those given in Table 38.2.
3. *Form Task/Intervention Matrix*: List all SATO-prone tasks and, for each, list the four generic intervention types.
4. *Use Matrix to Generate Specific Interventions*: For each cell in the Task/Intervention matrix, use standard techniques (e.g., group processes) to allow the design team to determine specific interventions.

This will now be applied to a task about which much has been written recently (Drury et al., 1990), the inspection of aircraft structures.

The world's fleet of airliners has built up over several decades. With excellent safety records and a shortage of new aircraft in the 1980s, older aircraft are often replaced at much longer time intervals. The mean age of several aircraft types in the U.S. is now over 20 years (Shepherd, 1990). Continuing airworthiness has demanded increased inspection load to discover faults (e.g., cracks, corrosion) before they affect flight safety. While the system has several beneficial redundancies, people (inspectors) are still required to detect small targets in complex structures, visually or with electronic aids.

As the first part of a study of improvements to aircraft inspection, task analyses were undertaken of many inspection activities in several airlines (Drury et al., 1990). From those came a generic list of inspection functions which can serve here as the required task description. Although this same list covers both visual and non-destructive inspection (electronically aided), only the former will be considered here because it represents the majority of inspection activities in the airline practice.

It should be noted that for this task, errors of failure to detect defects would have enormous cost. This is recognized by inspectors and management whose philosophy represents the extreme accuracy end of the SAOC. However, there are time pressures involved, and (more commonly) adverse working conditions

TABLE 38.3 Generic Tasks in Aircraft Inspection, with Simple Task Analysis, Task Type, and SATO Potential

Task	Description	Type	SATO Potential
1. Initiate	Get workcard, read and understand area to be covered. Get equipment.	Accurate, movement Reading	Small
2. Access	Locate area on aircraft. Move to position to inspect.	Self-paced movement	Large
3. Search	More field of view (FOV) across area. More fixation across FOV. Stop if any indication.	Search	Large
4. Decision	Examine indication against standards. Reach conclusion.	Discrimination Response selection	Medium
5. Respond	Mark defect on aircraft. Record data onto repair record.	Accurate movement	Medium

of cramped spaces and awkward postures which can create the same effects as time pressures. The four step process is as follows:

Perform Task Description and Task Analysis. Table 38.3 shows the generic functions, with a brief task description for each and a note about the type of task.

Select SATO Task. The major tasks where SATO is to be expected in Table 38.3 are in the continuous movement to the area inspected (access) and the serial search for defects (search), although all five tasks have some SATO potential. Search is the most time consuming of the tasks and potentially the most error prone (Drury, 1991b), so that it is the obvious place to concentrate efforts at intervention. However, other tasks cannot be neglected. For example, although the SATO for the response selection task (in decision) is over periods of less than 1.0 s from the literature, interventions to improve the task should lead to higher decision reliability. Equally, although the recording of defects (respond) is relatively brief, it will necessarily interrupt search activities. Thus any improvement here will decrease the probability of forgetting the current search point.

Form Task/Intervention Matrix. Table 38.4 shows this matrix, with all tasks listed. To save space, a completed matrix (step 4) is presented.

Use Matrix to Generate Specific Interventions. Examples of relatively obvious interventions are given in Table 38.4. They were generated by considering the form of the SAOC for each of the tasks, and postulating practical methods of affecting each aspect of the SAOC. In a design (or redesign) setting, a team would generate ideas, whether practical or not, within this matrix for later careful evaluation.

Some of the interventions given in Table 38.4 are already being implemented. More readable workcards (Patel et al., 1992), better lighting (Reynolds et al., 1992), and training interventions for search strategy (Drury and Gramopadhye, 1992) have all been evaluated.

38.4 Summary and Suggestions for Implementation

This chapter has defined how speed and accuracy can be measured and the various ways in which speed and accuracy can be traded off. Level 1, or micro-SATO, is where successive repetitions of a task can either be performed slowly to gain accuracy or rapidly, in which case accuracy suffers. An appropriate strategy to manage Level 1 is to train operators to keep a consistent strategy from trial to trial. At the level of macro-SATO (Level 2) operators respond to overall pressures by changing their mean speed, and hence accuracy. Here, management should ensure that operators have a clear idea of the costs of both slow and erroneous performance. Management should also not insist on either very high speeds, which have a disproportionate effect on errors (Figure 38.1), or on unrealistic error rates, which lead operators to become overly slow (Figure 38.1 again). Level 3 SATO is performance intervention where the task, operator, machinery, or environment is changed to move to a new and better SAOC. This is the level at which most quality-oriented management interventions take place.

TABLE 38.4 Tasks/Interventions Matrix for Aircraft Structural Inspection

Task	Intervention Level			
	Level 1	Level 2	Level 3	
	1. Reduce SATO Variability	2. Optimize SATO Point	3.a. Reduce Overhead	3.b. Improve Capability
1. Initiate	Train to complete understanding before accessing area		Accessible workcard Accessible equipment	More readable workcard Checklist for equipment
2. Access	Provide performance feedback on using safe access practices	Make walkways as wide as possible Train not to hurry	Use fixed walkways If not, use easily movable equipment Ensure equipment available	Design walkways for easy walking, e.g., clear boundaries, good surface Ensure correct footwear
3. Search	Train to use correct overlap of field of view (FOVs)	Determine the correct degree of overlap of fixations/FOVs Minimize awkward posture constraints to discourage early terminations	Allow easy FOV moves (e.g., lightweight flashlight) Well-defined boundaries to search area	Train to increase size of visual lobe Improve lighting for indication detection Workcard should list expected fault types
4. Decision		Minimize space and posture constraints as above	Clear listing of possible defects and their standards Accessible comparison standards	Correct lighting to discriminate faults Training in judgment against standards
5. Respond	Training to complete information	Standards for information completeness	Accessible recording device	Voice recording at inspection site Easy transcribing of voice recording to repair records (e.g., barcodes)

The common factor in all of these interventions is that they cost money, at least more money than the occasional talk to the workforce. Most have a single one-time cost, although others, such as training, have a continuing cost element. Replacing ongoing costs by one-time costs is, however, the basic strategy of industry. Much innovation, and practically all automation, has a single investment which produces long-term savings. Standard engineering texts have chapters on return-on-investment calculations, while management texts elevate these simple formulae to an impressive level of real-world complexity. It is not that industry has no precedents for such interventions of substance, but the will has been somewhat lacking. Presumably, the "new logic" of insistence on high quality, better design, and continuous improvement will allow more scope in the future for implementation.

But to follow the methodology outlined here places the onus on both the ergonomist and the manager to interpret tasks in terms of models so that predictions of benefits can be made. Practicing managers will not have to dismiss models as "academic," and the ergonomics research community will have to consider that their models may be used to evaluate investment decisions. The benefits of such changes in approach are that we utilize detailed, quantitative knowledge of human performance to make more rational use of the time and material resources of society.

References

Anon 1973. *Work in America, Report of a Special Task Force to the Secretary of Health, Education, and Welfare*, MIT Press, Cambridge, MA.

Brewer, N. and Smith, G.A. 1989. Developmental changes in processing speed: influence of speed–accuracy regulation. *Journal of Experimental Psychology: General*, 118(3): 298-310.

Business Week 1991. Special Issue, October 1991, The Quality Imperative.

Card, S.K., Moran, T.P. and Newall, A. 1983. *The Psychology of Human-Computer Interaction*, Lawrence Erlbaum Associates, Hillsdale, NJ.

Chi, C.-F. and Drury, C.G. 1998. Do people choose an optimal response criterion in an inspection task? *IIE Transactions*, 30, 257-266.

Conrad, R. 1954. Speed stress, in *Human Factors in Equipment Design*, eds. W.F. Floyd and A.T. Welford, Lewis Publishers, Chelsea, Michigan.

Craig, A. and Condon, R. 1985. Speed–accuracy trade-off and time of day. *Acta Psychologica* 58, p. 115-122, Elsevier Science Publishers B.V., North-Holland.

Crosby, 1979. *Quality is Free. The Art of Making Quality Certain*. McGraw-Hill, New York.

Dar-el, E.M. and Vollichman, R. 1996. Speed v/s accuracy under continuous and intermittent learning, in *Advances in Applied Ergonomics, Proceedings of the 1st International Conference on Applied Ergonomics (ICAE'96)*, ed. A.F. Ozok, Istanbul, Turkey, 676-682.

DeFazio, K., Wittman, D., and Drury, C.G. 1992. Effective vehicle width in self-paced tracking, *Applied Ergonomics*, 23(6): 382-386.

Deming, W.E. 1986. *Out of the Crisis*, Massachusetts Institute of Technology, Cambridge, Mass.

Dobyns, L. and Crawford-Mason, C. 1991. *Quality or Else. The Revolution in World Business*, Houghton Mifflin Company, Boston.

Drury, C.G. 1971. Movements with lateral constraint. *Ergonomics*, 14: 293-305.

Drury, C.G. 1987. Inspection performance and quality assurance, Chapter 65, *Job Analysis Handbook*, John Wiley & Sons, New York.

Drury, C.G. 1991a. Ergonomics practice in manufacturing. *Ergonomics*, 34: 825-839.

Drury, C.G. 1991b. Errors in aviation maintenance: taxonomy and control, In Proceedings of the 35th Annual Meeting of the Human Factors Society, San Francisco, CA, 42-46.

Drury, C.G. 1994. The speed–accuracy trade-off in industry. *Ergonomics*, 37: 747-763.

Drury, C.G. and Chi, C.-F. 1995. A test of economic models of stopping policy in visual search. *IIE Transactions* (1995), 27: 382-393.

Drury, C.G., and Corlett, E.N. 1975. Control of performance in multi-element repetitive tasks, *Ergonomics*, 18: 279-298.

Drury, C.G., and Dawson, P. 1974. Human factors limitations in fork-lift truck performance. *Ergonomics,* 17: 447-456.

Drury, C.G. and Forsman, D.R. 1996. Measurement of the speed accuracy operating characteristic for visual search. *Ergonomics,* 39(1): 41-45.

Drury, C.G. and Gramopadhye, A. 1992. Training for visual inspection: controlled studies and field implications, in *Meeting Proceedings of the Seventh Federal Aviation Administration Meeting on Human Factors Issues in Aircraft Maintenance and Inspection,* Atlanta, GA, 135-146.

Drury, C.G., Paramore, B., Van Cott, H.P., Grey, S.M. and Corlett, E.N. 1987. Task Analysis, Chapter 3.4, in *Handbook of Human Factors,* ed. G. Salvendy, p. 370-401, John Wiley & Sons, New York.

Drury, C.G., Prabhu, P. and Gramopadhye, A. 1990. Task analysis of aircraft inspection activities: methods and findings, in *Proceedings of the Human Factors Society 34th Annual Conference,* Santa Monica, California, 1181-1185.

Dudley, N.A. 1958. Output pattern in repetitive tasks, *Institute of Production Engineers' Journal,* 37: 303-313.

Fitts, P.M. 1966. Cognitive aspects of information processing: III. Set for speed vs. accuracy. *Journal of Experimental Psychology,* 71(6): 849-857.

Fox, J.G. 1977. Quality control of coins, in *Case Studies in Ergonomics Practice,* eds. H.G. Maule, and J.S. Weiner, Vol. 1, Taylor & Francis, London.

Gan, K.C. and Hoffmann, E.R. 1988. Geometrical conditions for ballistic and visually-controlled movements. *Ergonomics,* 31: 829-840.

Garrett, H.E. 1922. A study of the relation of accuracy to speed, in *Archives of Psychology,* ed. R.S. Woodworth, No. 56, G.E. Stechert & Co., London.

Hasslequist, R.J. 1981. Increasing manufacturing productivity using human factors principles. *Proceedings of the Human Factors Society,* 25th Annual Conference, Santa Monica, CA, 204-206.

Hick, W.E. 1952. On the rate of gain of information, *Quart. J. Exp. Psychol.,* 4: 11-26.

Hou, T.-S., Lin, L., and Drury, C.G. 1993. An empirical study of hybrid inspection systems and allocation of inspection function. *International Journal of Human Factors in Manufacturing,* 3: 351-367.

Jones, M.B. and Kennedy, R.S. 1996. Isoperformance curves in applied psychology. *Human Factors,* 38(1): 167-182.

Karwan, M., Morawski, T.B., and Drury, C.G. (1995). Optimum speed of visual inspection using a systematic search strategy. *IIE Transactions (1995),* 27: 291-299.

Kirwan, B. and Ainsworth, C.K. 1992. *A Guide to Task Analysis,* Taylor & Francis, London.

Konz, S. 1990. *Work Design: Industrial Ergonomics,* p. 219-236, Publishing Horizons, Inc., Worthington, Ohio.

Meyer, D.E., Irwin, D.E., Osman, A.M., and Kounios, J. 1988. The dynamics of cognition and action: mental processes inferred from speed–accuracy decomposition. *Psychological Review,* 95(2): 183-237.

Meyer, D.E., Smith, J.E.K., Abrams, R.A. and Wright, C.E. 1990. Speed–accuracy tradeoffs in aimed movements: toward a theory of rapid voluntary action, in *Attention and Performance XIII, Motor Representation and Control,* ed. M. Jeannerod, Chapter 6, 173-226.

McNichol, D. 1972. *A Primer of Signal Detection Theory,* Allen and Unwin, Sydney.

Montazer, M.A., Drury, C.G., and Karwan, M. 1987. Self-paced path control as an optimization task, *IEEE Transactions,* SMC-17.3: 455-464.

Mundel, M.E. and Danner, D.L. 1994. *Motion and Time Study, Improving Productivity,* Seventh Edition, Prentice Hall, London.

Myerson, J., Hall, S., Wagstaff, D., Pon, L.W., and Smith, G.A. 1990. The information-loss model: a mathematical theory of age-related cognitive slowing. *Psychological Review,* 97(4): 475-487.

Norman, D.A. 1981. Categorization of action slips. *Psychological Review,* 88(1): 1-15.

Norman, D.A. and Bobrow, D.J. 1979. On data limited and resource limited processes. *Cognitive Psychology,* 5: 44-64.

Pacella, R.G. 1974. The interpretation of reaction time in information processing research, in *Human Information Processing: Tutorials in Performance and Cognition,* ed. B.H. Kantowitz, p. 41-82, Lawrence Erlbaum Associates, Hillsdale, NJ.

Patel, S., Prabhu, P., and Drury, C.G. 1992. Design of work control cards, in *Meeting Proceedings of the Seventh Federal Aviation Administration Meeting on Human Factors Issues in Aircraft Maintenance and Inspection,* Atlanta, GA, 163-172.

Pew, R.W. 1969. The speed–accuracy operating characteristic, in *Acta Psychologica,* 3: 16-26.

Rabbit, P.M.A. 1981. Sequential reactions, in *Human Skills,* ed. D.H. Holding, John Wiley & Sons, Chichester.

Rasmussen, J. 1986. *Information Processing and Human-Machine Interaction, An Approach to Cognitive Engineering,* North-Holland, New York.

Reason, J. 1990. *Human Error,* Cambridge University Press, Cambridge, U.K.

Reynolds, J.L., Gramopadhye, A., and Drury, C.G. 1992. Design of the aircraft inspection/maintenance visual environment, in *Meeting Proceedings of the Seventh Federal Aviation Administration Meeting on Human Factors Issues in Aircraft Maintenance and Inspection,* Atlanta, GA, 151-162.

Salvendy, G. and Smith, M.J. (eds.) 1981. *Machine Pacing and Occupational Stress,* Taylor & Francis, London.

Schwabish, S.D. and Drury, C.G. 1984. The influence of the reflective-impulsive cognitive style on visual inspection, *Human Factors,* 26.6: 641-647.

Senders, J.W. and Moray, N.P. 1991. *Human Error: Cause, Prediction and Reduction,* Lawrence Erlbaum Associates, Hillsdale, New Jersey.

Shepherd, W.T. 1990. Human factors in the maintenance and inspection of aircraft. *Proceedings of the Human Factors Society 34th Annual Meeting,* Santa Monica, CA, 1167-1170.

Smith, G.A. and Brewer, N. 1995. Slowness and age: speed–accuracy mechanisms. *British Journal of Psychology,* 76:199-203.

Wickelgren, W.A. 1977. Speed–accuracy tradeoff and information processing dynamics. *Acta Psychologica* 41: 67-85.

Wickens, C.D. 1994. *Engineering Psychology and Human Performance,* Harper Collins, New York.

Woodworth, R.S. 1899. The accuracy of voluntary movement. *Psychol. Rev. Monogr. Suppl.,* 3, No. 3: 1-114.

For Further Information

For an excellent review of SATO in accurate movements, see Meyer, Hall, Wagstaff, Pon, and Smith (1990).

For a discussion of the methodology of measuring the SAOC, see Wickelgren (1977).

A new way of measuring the SAOC in the laboratory has been developed by Meyer, Irwin, Osman, and Kounios (1988).

The issue of paced performance has not been covered in this chapter. A useful book on the subject is Salvendy and Smith (1981).

For an example of current studies of SATO and learning, see Dar-el and Vollichman (1996).

Time-of-day effects on SATO are covered in Craig and Condon (1985), while development and aging processes affecting SATO can be found in Brewer and Smith (1989) and Meyerson, Hall, Wagstaff, Pon, and Smith (1990).

39

Receiver Characteristics in Safety Communications

Stephen L. Young
Liberty Mutual Research Center for Safety & Health

Kenneth R. Laughery
Rice University

Michael S. Wogalter
North Carolina State University

David R. Lovvoll
Rice University

39.1 Introduction

An interesting aspect of the warning process occurs when people pick up a prescription drug at a pharmacy. In many cases, these medications are accompanied by a patient package insert (PPI). PPIs contain detailed information about the nature of a drug, potential side effects, prescriptions and proscriptions for use, and a wealth of other details about the chemical makeup of the drug. PPIs are similar in scope and detail to the information contained in drug reference books (e.g., *Physician's Desk Reference*). One look at these documents will tell you that they are not designed for the layperson, but rather they are designed to provide the kinds of information and the level of detail that would primarily benefit an individual with substantial medical training. Because of the level of sophistication required to acquire relevant information from PPIs, pharmacies will often provide a briefer and simpler summary of the relevant information for use by the lay customer. The purpose of the summary is to provide the end user with the most important information necessary to use the drug properly.

While a central tenet of warning theory is that it is important to provide people with information so they can make informed choices regarding their behavior, it is not necessarily true that more information is better. Table 39.1 shows the information provided in a PPI and in a pharmacy summary for the same drug. It is clear that the information is targeted for two different audiences. Physicians are provided detailed information, because they need it to make proper prescribing decisions. However, patients (for the most part) will not find much of the detail helpful, and they could find it difficult and confusing. It may actually make the extraction of relevant information *more* difficult.

This example demonstrates, in a very basic way, that one must consider who is the audience when designing, producing, and delivering safety-related information. Other examples might include the presentation of information in material safety data sheets (MSDSs) and in OSHA regulations.

0-8493-2641-9/99/$0.00+$.50

TABLE 39.1 Example of Pharmaceutical Information Provided to Physicians and Patients

Information Category	Physicians	Patients
Dosage	Adult: 3-4 g/day in evenly divided doses Child (>2): 40–60 mg/kg/day in 3–6 evenly divided doses	Take two tablets twice daily. Take this medicine with meals or a snack. Try to space your doses evenly over each 24 hour period. If you miss a dose of this medicine, take it as soon as possible. If it is almost time for your next dose, skip the missed dose and go back to your regular dosing schedule. Do not take 2 doses at once.
Adverse Reactions	Cardiovascular — Vasculitis, pericarditis with or without tamponade CNS — Headache, transverse myelitis, convulsions, meningitis, transient lesions of the posterior spinal column, cauda equina syndrome, Guillain-Barre syndrome, peripheral neuropathy, mental depression, vertigo, hearing loss, insomnia, ataxia, hallucinations, tinnitus, drowsiness Genitourinary — Oligospermia, infertility Gastrointestinal — Anorexia, nausea, vomiting, gastric distress, hepatitis, pancreatitis, diarrhea, stomatitis, abdominal pains, neutropenic enterocolitis Hematological — Heinz body anemia and hemolytic anemia, aplastic anemia, agranulocytosis, leukopenia, megaloblastic (macrocytic) anemia, purpura, thrombocytopenia, hypoprothrombinemia, methemoglobinemia, congenital neutropenia	Side effects, that may go away during treatment, include headache, nausea, loss of appetite, or indigestion. If they continue or are bothersome, check with your doctor. CHECK WITH YOUR DOCTOR AS SOON AS POSSIBLE if you experience sore throat, fever, rash, tightness of chest or difficulty breathing. If you notice other effects not listed above, contact your doctor, nurse or pharmacist.
Cautions	Drug-induced hypersensitivity reactions, blood dyscrasias, neuromuscular and CNS reactions, hepatotoxicity, nephrotoxicity, or fibrosing alveolitis may result in death. Watch for clinical signs suggesting a serious blood dyscrasia; obtain a CBC frequently. To prevent crystalluria and lithiasis, perform a urinalysis, including a careful microscopic examination, frequently and instruct patients to maintain an adequate fluid intake. Caution patients that their skin or urine may turn orange-yellow during therapy. Contraindications include hypersensitivity to sulfasalizine, sulfapyridine, or other sulfonamides or to 5-aminosalicylic acid or other salicylates.	TELL YOUR DOCTOR OR PHARMACIST if you are allergic to sulfa drugs. Keep all medical appointments while receiving this medicine. Drinking extra fluids while you are taking this medicine is recommended. Check with your doctor or nurse for instructions. This medicine may cause increased sensitivity to the sun. Avoid exposure to the sun or sunlamps until you know how you react to this medicine. Use a sunscreen or protective clothing if you must be outside for a prolonged period. This medicine may cause a harmless, yellow-orange discoloration of the urine or skin.

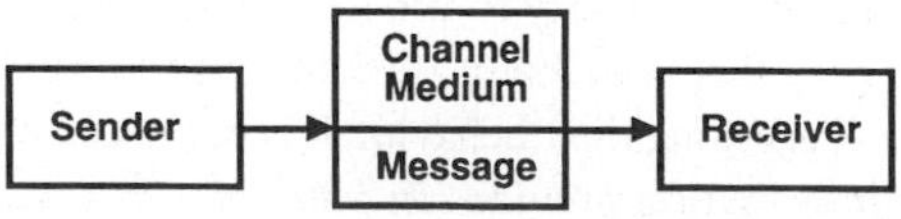

FIGURE 39.1 Basic communication model.

The task of communicating warning information to an individual, whether through product warnings, safety signs, auditory warnings, etc., can be described in terms of communication theory (Lehto and Miller, 1986). Using this theoretical context, McGuire (1980) defined warnings as communications designed to influence behavior with respect to a product. Figure 39.1 shows a simple, generic communications model, which includes four primary components: the sender, the message, the medium, and the receiver.

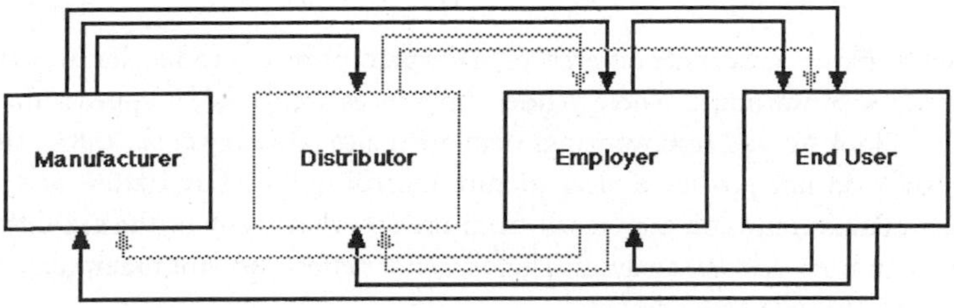

FIGURE 39.2 Complex warning communication system.

The sender, or source, represents the originator of the communication. With respect to warnings, the message is the relevant information that is to be transmitted via some medium. The message could (and preferably would) contain information about the nature of the hazard, consequences of exposure to the hazard, and/or instructions on how to avoid the hazard. The medium refers to the channel or route by which information moves from the source to the receiver. Media for warnings can include MSDSs, on-product labels, package inserts, signs, oral instructions, and so forth. The receiver refers to any and all persons who are at risk and to whom the warning should be directed. Characteristics of each of these components may, and often do, play a critical role in the effectiveness of a warning.

In the most simple application of this model to the warning process, a manufacturer of a product (the source) attempts to relay some warning message using one or more media to an end user of the product (the receiver). However, the process of conveying warning information is not always so straightforward. For example, Figure 39.2 represents the elements of a more complex warning communication system for a product being used in an industrial setting. Here the product might be marketed through a distributor (or a series of distributors) to an employer (the business and its management) who in turn communicates in various ways with the end user (employee). Communication from the manufacturer to the end user may be direct, such as labels on the product, or indirect, through various intermediaries (e.g., distributors). The media through which the information is communicated may also be quite varied. Feedback between various components may be involved, such as an employer notifying the manufacturer about a safety problem associated with the use of the product.

Even more complex warning systems could have several receivers, including distributors, employers, and end users. These receivers might differ markedly in several important respects. For example, an employer's industrial toxicologist who may be a receiver in the communication process will probably have a great deal more technical knowledge than a laborer working in the plant who is the end user of the product. This knowledge difference may have implications for the warning system associated with the product. A parallel example, is that given at the outset of the chapter, where there are at least two kinds of users with respect to medications manufactured by a drug company (e.g., a prescribing physician, pharmacists, and a patient who are all targeted receivers of safety information).

Whether the circumstances are simple or complex, the success of a warning communication system depends on accounting for the properties of the various system components. Previous research has examined issues related to the source, the medium, and the message. In this chapter, we focus on the receiver. We review the literature on the most commonly addressed receiver characteristics and present warning-design implications that stem from these characteristics. This chapter is organized into four sections: demographic variables, competence, familiarity, and risk perception. Finally, we offer observations that warning designers should consider with respect to the receiver.

39.2 Demographic Variables

Demographics are statistical characteristics of individuals that can be used for the purpose of grouping. It is easy to collect such data and many warning-related studies have done so in the past. Two common demographic variables are gender and age.

Gender

Research indicates that there are gender differences in the perception of product hazards and in willingness to read and comply with warnings. Where gender differences are found, it appears that females report being more likely to look for and read warnings than are males (Godfrey et al., 1983). However, in many instances, the results do not provide a clear picture regarding the exact nature and cause of gender differences. Much of this confusion may result from the fact that many of the sex differences reported in the literature come from *post hoc* analyses. That is, gender effects are not manipulated *a priori* but are simply analyzed after the data are collected.

Post Hoc Analyses of Gender Effects

For example, Donner (1990) manipulated warning modality (written versus oral versus both) and formality (informal versus formal) for two different products (fabric protector versus a bench grinder). Compliance with the warning information, as well as noticing and reading the warning, was reported. The results showed a strong product effect, with no significant effect of warning modality or formality. After these analyses (which were the primary focus of the study), an evaluation of gender effects was conducted. Donner found gender differences in previous experience with the product (e.g., females had greater experience with the fabric protector), but this variable did not affect compliance with the product warning. This type of analysis is common among studies that report gender effects. After the primary analyses are performed, a *post hoc* analysis is conducted to determine how gender interacted with or influenced the findings.

Studies employing *post hoc* analyses of gender have produced mixed results. Dorris and Tabrizi (1978) had male and female subjects rate the hazardousness of different products and found very small effects of gender. Godfrey et al. (1983) showed that females were more likely to look at warnings than were males. Silver and Braun (1993) found no gender differences in preference for font size and style in warning labels. Green and Pew (1978) and Laux, Mayer, and Thompson (1989) found that females produced lower accuracy in safety symbol comprehension than males, but only under certain conditions (e.g., gender effects interacted with previous training) and with certain symbols. Leonard, Hill, and Karnes (1989) reported that females were more likely to wear seatbelts than were males. Other research shows that females would be more likely to report complying with warnings than would males (Desaulniers, 1991; Viscusi et al., 1986).

Note that in these studies, gender was not a formal variable under study, nor was gender crossed with the other experimental variables. In such *post hoc* analyses, there is often little reasoning provided as to why gender might influence the results. Gender analyses are presented because the authors thought to do them and because they could be done. It is possible that many studies do not report *post hoc* analyses of gender because of a lack of observed differences. Thus, it is not really surprising that gender has produced mixed results.

A Priori Analyses of Gender Effects

Several studies have examined sex differences as a primary focus of the study. These studies have, for the most part, demonstrated differences between males and females with regard to several safety-related variables. For example, Goldhaber and deTurck (1988a; 1988b) reported that, while females were less likely to recall a "No Diving" warning sign posted around a swimming pool, they were significantly less likely to actually dive than were males. LaRue and Cohen (1987) had males and females rate different consumer products according to several questions. Reported familiarity and perceptions of product hazardousness were statistically similar between males and females in this study. However, females reported being significantly more likely to read warnings for the products than males. Young, Martin, and Wogalter (1989) had males and females rate consumer products according to several questions. The products in this study were classified from the ratings as being either masculine or feminine. Product ratings were then examined as a function of subject gender. The results demonstrated a main effect of product's masculinity/femininity and an interaction between this variable and the subject's gender. According to the interaction, males and females judged the hazardousness of feminine products similarly, but females rated the masculine products as significantly more hazardous than did the males.

Although gender was the variable of interest in these studies, it remains unclear as to whether gender is truly responsible for the observed results. Specifically, gender may simply act as a proxy or alias for another variable or group of variables (e.g., familiarity). Young et al. (1989) demonstrated that greater hazard ratings were given by females to products that they used infrequently. Other studies have shown that familiarity and product hazard ratings are negatively correlated (Otsubo, 1988; Young, 1996; Wogalter et al., 1987; Young et al., 1990). It may be that females and males would rate product hazards similarly if females only used the products more frequently and/or had greater knowledge about them generally or their hazards specifically. However, other studies suggest that this hypothesis may not be true. For example, Godfrey and Laughery (1984) showed that females underestimated the risks associated with tampon use due, in part, to their familiarity with the product. Karnes and Leonard (1986) examined male and female knowledge of hazards associated with a contraceptive device (an IUD) based on a safety-related pamphlet. Males were included in the subject sample specifically because it was assumed (whether true or not) that they would serve as a low-familiarity control to the females. Females in this study perceived the risks to be significantly higher than did the males.

Conclusions: Gender

These studies suggest that gender effects may or may not be observed in research that does not specifically address gender issues. Evidence tends to suggest that females are more willing to act safely with products (e.g., read or comply with warnings) than are males. However, it is unclear whether this trend would be observed in all situations and with all products. It is also unclear whether gender is the true source of observed differences in product perceptions or whether those differences are related to more basic considerations (e.g., knowledge of the hazards, familiarity, frequency of use, etc.). Further systematic research, which accounts for confounding variables (e.g., familiarity, etc.), is needed before design considerations with regard to warnings can be provided.

Age

When considering age, it is generally believed that (a) older people tend to be more risk averse and (b) younger people (especially younger males) tend to be predisposed to taking greater risks. Smith and Watzke (1990) demonstrated that older people (30 to 59 years, 60 to 75 years, and over 75 years old) more carefully consider the risks and are more cautious than younger adults (under 30 years of age). The authors suggest that cautiousness may be a characteristic of older adults and that this characteristic may start to exhibit itself in the middle years of life (between 30 and 59 years of age). If it is true that people are more or less risk averse depending on their age, then this should manifest itself in safety-related behaviors (e.g., looking for, reading, and complying with warnings). For example, older people (> 25 years old) are more likely to wear seatbelts than younger individuals (< 25 years old) (Leonard et al., 1989). Desaulniers (1991) found that older people, 40 and above, reported being more likely to take precautions in response to warnings.

There are many hypotheses as to why age may influence risk perceptions and/or safety-related behaviors, but the most reasonable ones surmise that younger people do not actually take risks in a formal sense. Specifically, they do not consider an action with a conscious view toward the costs and benefits of acting one way over another (see Lehto, 1991; Wagenaar, 1992). As such, their behavior may appear to be nonrational and more dangerous than the types of judgment-based behavior exhibited by older individuals. While a great deal of evidence for risk-taking in younger people can be found in the literature on traffic accidents (see Edwards and Ellis, 1976; Leonard et al., 1989), there is somewhat less evidence of this phenomenon in day-to-day behaviors. Thus, it is not surprising that the literature is not entirely consistent in demonstrating age effects.

For example, Mazis, Morris, and Gordon (1978) showed that a sample of premenopausal females preferred longer and more detailed information regarding the risks associated with oral contraceptives, but that this trend was more pronounced for younger subjects (e.g., college-age students) than for older subjects. Purswell, Schlegel, and Kejriwal (1986) demonstrated that older people (>30 years old) exhibited safer behaviors than younger subjects (<30 years old) with a router, but that the opposite pattern was observed with an electric knife. Leonard, Ponsi, Silver, and Wogalter (1989) found minor differences between older (M = 37 years) and younger (M = 18 years) subjects on willingness to read warnings for pest-control products. Wright, Creighton, and Threlfall (1982) observed no effect of age (ages in the sample ranged from under 30 years old to over 50) with regard to willingness to read instructions.

As with the evidence regarding gender effects, findings associated with age in warnings research have not been entirely consistent. Much of the conflicting data may result from the fact that a majority of analyses are *post hoc* evaluations. There is an additional problem associated with research on age: defining "younger" and "older" categories in research. These terms can vary widely in meaning from one study to the next. The "older" group of subjects in one study could be the "younger" group in another. Because of these issues, the effect of age on risk perceptions and safety-related behaviors is inconclusive. It appears as though younger adults may be a more difficult group to warn because of their lack of (formal) consideration of risks and benefits in the decision-making process. Relative to younger people (and possibly younger males in particular) older adults may be more likely to comply with safety-related information. However, this hypothesis needs additional testing, with systematic research that controls for confounding variables (e.g., knowledge and familiarity with products, etc.).

39.3 Competence

Competence can be defined as possessing the capacity to meet the demands of a particular task. There are many dimensions of receiver competence that may be relevant to the design of warnings. We discuss three here: sensory, physical, and cognitive capabilities.

Sensory Capabilities

It is obvious that the blind person cannot see a written warning, nor would the deaf person receive an auditory warning. Although these extreme examples are obvious, we also know that sensory capabilities lie along a continuum. Consider that many older adults, who use more medications as a group, cannot read medication labels because of age-related visual decrements. Yet many over-the-counter pharmaceutical labels have print that is too small for older adults to read. One way to solve this problem is to design product labels to accommodate larger type (see Wogalter and Young, 1994). Another way is to provide relevant safety information in the form of pictorial symbols. However, several studies have demonstrated that older adults have greater difficulty in interpreting safety-related symbols (Collins and Lerner, 1982; Easterby and Hakiel, 1981; Ringseis and Caird, 1995).

Physical Capabilities

This topic deals with the extent to which the user will be physically capable of carrying out a task. For example, older adults may not have the dexterity to grab hold of a three-point manual seat belt or they may not be able to generate the torque necessary to open small medicine containers. Wogalter and Young (1994) demonstrated that different label designs, while increasing the size of warning information (and thereby making it easier to read), could provide the user with a greater surface area on which to apply force. If special equipment is required to comply with the warning, it must be available or obtainable. For example, some hair dyes contain warnings that direct the use of gloves during application. Rather than imposing on the user to find and/or purchase gloves separately, plastic gloves are generally included in the package with the dye. If special skills are required, they must be present in the receiver population. To some extent, as with the sensory limitations of receiver populations, the behavioral limitations that may be involved could be considered rather obvious, although we are constantly amazed at the number of warnings that violate such considerations — especially in the behavioral domain where basic product instructions (e.g., for assembly/installation) are often difficult to carry out.

Cognitive Capabilities

Examples of cognitive competence include requisite technical capacity, language, and reading ability.

Technical Capacity

One of the primary issues in warning design with respect to competence concerns the level of technical information to be communicated. Comprehension of such information is generally a function of the receiver's existing technical knowledge of the domain. Here we are referring to conceptual knowledge that includes both factual information and process understanding (the receiver's mental model). Some examples include: (a) medications where knowledge of physiology may be relevant, (b) chemical reactions that require an understanding of what not to mix with what, and (c) mechanical properties where knowledge is needed to understand the hazards of handling certain kinds of equipment. In formulating warnings, it is important to take into account the relevant technical knowledge of the receiver. Further, the problem may be more complicated in the sense that warnings regarding a particular product hazard may be directed to multiple groups (or receivers) differing in knowledge.

The point to be emphasized here is that the level or levels of knowledge and understanding must be considered. Of course, it is also a valid concern that variability in knowledge about facts and processes exists within the target audience for a particular product warning. There may be various approaches to address these concerns. One approach is to construct a single warning system that will be understood at a range that reaches as many people in the target audience as possible. Another approach is to develop a multiple-component warning system where different components are directed at subgroups varying in technical knowledge. The second approach, as shown in the example provided at the beginning of the chapter, is the one selected for presentation of drug-related information by pharmacies. Physicians receive detailed information about prescription drugs and patients receive summaries of that information.

Language

A second cognitive issue with respect to competence is language. The target audience may know a language different from the majority. A warning printed in only one language is much less likely to be accessible to all potential users. Attempts to deal with this problem include the use of pictorial symbols and printing the message in multiple languages. The latter technique is commonly employed in instruction booklets that accompany various electronic products such as watches and calculators. Signs printed in multiple languages must be either increased in size to accommodate the extra material or, if the size is held constant, they must be more cluttered or dense. Neither of these alternatives is desirable. Also, selection of languages to appear on these signs may not be so straightforward. How many languages does one need in order to cover all potential users? The number could be prohibitively high.

Symbols, on the other hand, provide the promise of non-verbal communication — a method of conveying safety-related information regardless of the language spoken in the population. Research has demonstrated that symbols are effective in attracting user attention (Young, 1991; Laughery and Young, 1991; Wogalter et al., 1996; Young and Wogalter, 1990) and in conveying safety information (Collins, 1983; Collins and Lerner, 1982; Laux et al., 1989). However, the promise of completely non-verbal communication has not been and may not be fully realized. Symbols necessarily involve an abstraction of some message. This method of information display is easier for certain safety-related concepts (e.g., slippery floor) than with others (e.g., biohazard, cancer). Designing symbols to convey information that can be interpreted accurately under many different circumstances can be difficult.

Reading Ability

Many warnings require high levels of reading ability on the part of the receiver. The usual recommendation for general target audiences is a reading level near the elementary school range. An exception to this rule is found in Leonard et al. (1989), who found that college students and other highly educated individuals reported being more likely to read complex warnings than simple ones. The complex warnings in this study were used primarily for more hazardous products. Perceived hazardousness is a factor that (as we will see in the next section) has a strong relationship to willingness/likelihood of reading warnings. Obviously, if comprehension of a warning is to be achieved, the material should be written at the level that accommodates the readership. One way to evaluate the readability of safety-related information is to conduct some type of readability analysis of the materials. A discussion of reading level measures and their application in the design of instructions and warnings can be found in Duffy (1985). It should be noted that readability formulas are only indications of comprehensibility, and they may be less suitable with short messages like those that commonly appear in warning signs. Readability scores and comprehension measures are not always highly correlated. Thus, readability formulas should be used with caution and probably as only a first step in determining the material's understandability.

The problem of warning readability may require more than simply keeping reading levels to a minimum. There are a very large number of functionally illiterate adults in the population who cannot read written (verbal) warnings at any level. We offer no simple solutions to this problem, but certainly pictorial symbols, oral warnings, special training programs, etc. may be important ingredients of warning systems for such populations.

Conclusions: Competence

There are many factors that influence the capacity of individuals to meet the task of acting safely around products. Sensory, physical, and cognitive capabilities can influence whether users are capable of accessing, understanding, and using safety-related information. These characteristics of potential users must be considered when designing warnings.

39.4 Familiarity and Experience

One of the issues that has received substantial attention in research concerns the familiarity and/or experience that people have with products and how such factors influence the effectiveness of warnings. Familiarity has been defined in many ways, but we define it here as a state of being intimate or closely acquainted with a product and its hazards. Familiarity is not a dichotomous state. People may be unfamiliar with a product. They may be familiar with a product generally (i.e., they have heard *about* the product) or they may be familiar with it specifically (i.e., they have used the product). Users can become familiar with a product indirectly (i.e., through the acquisition of knowledge about it) or directly (i.e., through direct use). Familiarity is a belief and it is most commonly measured through ratings where people express their familiarity with a product on a Likert-type scale (e.g., 0 = "not at all familiar" to 7 = "extremely familiar"). Experience, on the other hand, can be operationally defined in terms of time and/or frequency of use. A distinction between familiarity and experience has been noted by Wogalter et al. (1986, 1987). We do not consider the familiarity and experience to be synonymous, but we will discuss them together here for the purpose of dealing with users' knowledge-based product perceptions.

Familiarity

Numerous studies have explored the effect of familiarity on safety-related product perceptions and behavior. In general, higher levels of product familiarity or experience are associated with decreases in the probability that warnings will influence user behavior. The reasons for this conclusion are varied, but they tend to revolve around the notion that as people use a product and become more familiar with and knowledgeable about it, they perceive it to be less dangerous. Desaulniers (1989) showed that people perceived more familiar products as less hazardous. Karnes and Leonard (1986) showed that subjects with greater experience riding ATVs considered them to be less dangerous than did subjects with less experience.

The utility of warning information may be reduced as people come to see the products as less hazardous. Thus, users may not seek out or read relevant information. Godfrey and Laughery (1984) showed that females reported being less likely to read warnings for products with which they are familiar (e.g., tampons). Johnson (1992) showed that willingness to look for and read warning information for scaffolds was negatively related to the number of times workers had previously used scaffolding. Morris, Mazis, and Gordon (1977) showed that about 78% of females read a PPI for oral contraceptives the first time they used the drug, but that less than 11% read the insert when it accompanied subsequent prescriptions. However, other research suggests that unwillingness to look for and read warnings is not exclusively related to familiarity. Leonard et al. (1989) demonstrated that willingness to read warnings for a pest-control product was unrelated to familiarity with that product. In addition, Godfrey et al. (1983) demonstrated that familiarity was not related to subjects' reported willingness to look for warning labels on products perceived as hazardous. However, they also reported being more willing to look for warnings on unfamiliar products when the hazard level was perceived as being low. Thus, familiarity may be one factor, along with other perceptions, that influence the extent to which users may seek information.

Behavioral effects of familiarity have also been demonstrated. Goldhaber and deTurck (1988a, b) showed that previous experience with diving into pools was related to lower likelihood of noticing a "No Diving" sign, a higher likelihood of diving into shallow water, and a lower perception of the risks associated with such activities. Otsubo (1988) showed that people with less experience were more likely to read the warnings for two types of saws.

The above review suggests that the more people become familiar with a product, the less likely they will be to engage in safe behaviors (and vice versa). While this relationship may be linear (or at least monotonic), there is some evidence to suggest that the relationship is nonlinear. Bettman and Park (1980) found that subjects with a moderate level of knowledge or experience relied most on external information when making a product-related decision. People with low and high levels of previous knowledge relied to a greater extent on this external information. The authors suggested that users with high levels of experience did not need the information and that users with low levels did not have the capacity to use it properly. Johnson and Russo (1980) demonstrated that both the linear and nonlinear ("inverted-U") functions were observed in different decision-making tasks.

Conclusions: Familiarity

Research generally suggests that lower levels of familiarity are associated with higher levels of perceived hazard and greater reported willingness to act with caution (and vice versa). The most common explanation for this finding is that greater familiarity is associated with greater knowledge of and appreciation for the product's hazards. However, familiarity with a product is not synonymous with knowledge of the hazards associated with it. People may report being familiar with a product and yet have little or no knowledge of its hazards. Familiarity lies along a continuum — people can have indirect, general familiarity with a product or they can have more direct, specific familiarity (i.e., from lower to higher forms of familiarity). Subjects who provide ratings of familiarity in research studies may not make the distinction between the two types. Thus, they might report a high degree of familiarity with a product that they have very little personal knowledge of or experience with (e.g., they may have heard a lot about a product, but have no direct experience with it).

High levels of perceived familiarity may lull people into thinking that they have greater knowledge of and control over product hazards than they actually have and/or that the products are less hazardous. This perception, coupled with the fact that familiarity may reduce information-seeking behaviors on the part of users, can produce a dangerous situation and a special challenge to safety professionals. That challenge is to make warning information salient (i.e., to attract the attention of familiar users) and to make the information seem relevant. A considerable body of research has addressed various forms of salience. However, there is much less research dealing with relevance issues. Ways to make warnings more relevant can include prioritizing warning information based on users' needs and presentation of information to specific users at intermittent schedules.

39.5 Risk Perception

Risk perception in the present context refers to the way people understand and consider the hazards associated with products and the ways in which these perceptions influence people's behavior when using them. A consistent finding in warning research is that people's perception of the risk associated with a product or situation is an important determinant of warning effectiveness — the greater the risk, the more likely people will look for, read, and comply with warnings (Donner and Brelsford, 1988; Friedmann, 1988; Godfrey et al., 1983; LaRue and Cohen, 1987; Leonard et al., 1986; Otsubo, 1988; Wogalter et al., 1991). While most of the research on risk perception has evaluated the nature of the products themselves, some research has examined subject characteristics.

One study (Young, 1996) had subjects rate a list of consumer products according to several different rating questions:

- How hazardous is this product?
- How frequently do you use this product?
- How likely are you to be injured while using this product?

As in other studies of risk perception, the results demonstrated that subjects' risk perceptions varied as a function of the product being evaluated, with some products being considered more hazardous (as a whole) than others. However, the risk ratings also varied as a function of the subject. Specifically, there were differences in the way different subjects perceived the hazard for individual products (e.g., chain saw).

Based on the ratings in this study, Young (1996) partitioned the subjects into three distinct groups which varied in terms of how they perceived the products in general and in terms of the information they accessed when evaluating product risks. The first group of subjects was labeled *Fearful,* because they subjects considered the products as a whole to be quite hazardous while having only average knowledge of the risks and average familiarity with the products. The second group of subjects, labeled *Fearless,* considered the products as a whole to be nonhazardous despite having little knowledge of the products' risks and little familiarity with them. The third group of subjects was labeled the *Informed* group. These subjects considered the products as a whole to be nonhazardous, but they also knew a great deal about the risks associated with the products and they were very familiar with them as well. This group is similar to the internal locus-of-control subjects reported in Laux and Brelsford (1989) in that these subjects believe they are capable of controlling the hazards. One interesting demographic relationship was the finding that subjects in the *Fearless* group were significantly younger than were subjects in the *Fearful* group.

Young (1996) also demonstrated that these subject groups considered different kinds of information when forming risk perceptions. The *Fearful* group considered products to be risky if they had severe potential injury consequences, if an injury was likely, if the number of different risks associated with the product was high, and/or if the subject had been injured or had known someone who had been injured with the product in the past. The *Fearless* group considered products to be dangerous only if they had the potential to injure or kill many people at a time and if the product hazards were encountered involuntarily. When considering product hazards, the *Informed* group not only looked at the potential for catastrophe, but they also weighed information about the benefits provided by the product and the degree of control they exercised over the hazards. Thus, subjects not only perceived the products differently, but they accessed different information when forming perceptions of the risk associated with consumer products.

The results of this work demonstrated that at least some variance of risk perceptions could be attributed to how people perceive consumer products as a whole. The results suggest that information in warnings could be designed to suit the informational needs of the targeted audience. For instance, with the *Fearful* group, one could provide information about the nature of the hazard and the potential severity of injury associated with it. One way to accomplish this is to provide explicit information regarding injury consequences. Research has demonstrated that the explicitness with which the consequence information is expressed is an important determinant of perceived hazard and of recall of warning information (Laughery and Stanush, 1989; Sherer and Rogers, 1984). As expected, the more explicit the consequence information, the greater the perceived hazard and the more information recalled. This would hold true for the *Fearful* group of subjects. However, different information may be needed for the other groups. For the *Fearless* and *Informed* groups, information about the potential catastrophic nature of the hazard, about the extent to which exposure to the hazard is voluntary, and/or about the degree of personal control over the hazard may be necessary for these subjects to develop a proper appreciation of the risks.

39.6 Conclusions and Recommendations

In this chapter we have focused on characteristics of receivers that are important in the design of warnings. There are several principles or guidelines that appear warranted on the basis of the analyses presented.

Principle #1 — Know thy receiver. This statement may seem trivial and obvious; yet, as noted earlier, warnings are often designed with little or no regard for characteristics of the people to whom they are directed. Examples include warnings that require reading levels greater that the receiver's capability and that contain unfamiliar, technical terminology. Gathering knowledge and data about relevant characteristics of target audiences may require time, effort and money, but without such information, the warning designer and ultimately the receiver will be at a serious disadvantage. Analyzing existing data, such as demographic information, or collecting new data by conducting surveys may be necessary.

Principle #2 — When variability exists in the target audience, design for the low end of that audience. Whether the variability exists in competence, technical knowledge, familiarity, perception of hazardousness, or other receiver characteristics, it is important that warnings not be designed for the average. While it would be desirable to choose a criterion for warning designs that would include up to 99% of the population, there are several instances in which this may not be possible. For example, warning about such hazards as radon gas will necessarily involve information that may not be understood by all people. It is inappropriate to suggest that warning information should not be provided simply because the information may not be understood by 100% of the population. The point is to consider the variability in the target audience and to design the safety information so that it can be used by as many people in the target audience as is practical.

Principle #3 — When the target audience consists of subgroups that differ in relevant characteristics, consider employing a warning system that includes different components designed for the different subgroups. As in the prescription drug example provided at the beginning of the chapter, different types of information and different levels of detail are provided to different groups of receivers. This information is tailored to the needs and capabilities of these receivers.

A corollary to this principle is: do not try to accomplish too much with a single warning. Consider the current OSHA guidelines regarding the variety of subgroups in the target audience for material safety data sheets (MSDSs). These subgroups include toxicologists, safety engineers, managers, physicians, and end users (such as the laborer using the product). It is unlikely that one warning or pamphlet will be sufficient to meet the informational needs and capabilities of all these users. If the warning system does not include communications designed for the capabilities (both the strengths and weaknesses) of each group, it is probably destined to fail.

Principle #4 — Warnings should be tested using samples of potential receivers. Warning design guidelines (e.g., ANSI, 1991; FMC, 1985; Westinghouse Electric Corporation, 1985) can be used to develop candidate warnings for testing, thereby limiting the number of items that need to be tested. However, it is not always possible to use these guidelines to design a perfect warning system. The guidelines presented here can enable one to develop a preliminary warning. Testing of the warning system could assist the designer in refining and developing an effective system by providing information on ways to modify and improve the warnings. These tests might consist of "trying it out" on a target audience sample to assess comprehension and/or behavioral intentions. Our experience indicates that even such *minimal efforts* are seldom part of the warning design process, but would benefit the produced warning had they been taken.

Last, warnings should be viewed within the context of a communication *system* that includes the message, the medium, and the receiver. This chapter has sought to demonstrate that it is important to consider the capabilities and limitations of the receiver when designing the warning message. The most well-researched receiver characteristics were discussed in this chapter. Other receiver characteristics have been reported in the literature (e.g., locus of control, risk taking as a personality trait), but thus far they have received much less attention. The essential point of the chapter is that receiver characteristics should be considered in the warning design process in order to maximize effectiveness for their intended target audience.

References

ANSI (1991). *American National Standard for Product Safety Signs and Labels, ANSI Z535.4-1991.* Washington, D.C.: National Electrical Manufacturers Association.

Bettman, J.R. and Park, C.W. (1980). Effects of prior knowledge and experience and phase of the choice process on consumer decision processes: A protocol analysis. *Journal of Consumer Research, 7,* 234-248.

Collins, B.L. (1983). Evaluation of mine-safety symbols, in *Proceedings of the Human Factors Society 27th Annual Meeting* (pp. 947-949). Santa Monica, CA: The Human Factors Society.

Collins, B.L. and Lerner, N.D. (1982). Assessment of fire-safety symbols. *Human Factors,* 24, 75-84.

Desaulniers, D.R. (1989). Consumer product hazards: What will we think of next? In *Interface '89: The Sixth Symposium on Human Factors and Industrial Design in Consumer Products* (pp. 115-120). Santa Monica, CA: The Human Factors Society.

Desaulniers, D.R. (1991). *An Examination of Consequence Probability as a Determinant of Precautionary Intent.* Doctoral dissertation, Rice University, Houston, TX.

deTurck, M.A. and Goldhaber, G.M. (1991). A developmental analysis of warning signs: The case of familiarity and gender. *Journal of Products Liability, 13,* 65-78.

Donner, K.A. (1990). *The Effects of Warning Modality, Warning Formality, and Product on Safety Behavior.* Masters thesis, Rice University, Houston, TX.

Donner, K.A. and Brelsford, J.W. (1988). Cueing hazard information for consumer products, in *Proceedings of the Human Factors Society 32nd Annual Meeting* (pp. 532-535). Santa Monica, CA: The Human Factors Society.

Dorris, A.L. and Tabrizi, M.F. (1978). An empirical investigation of consumer perception of product safety. *Journal of Products Liability,* 2, 155-163.

Duffy, T.M. (1985). Chapter 6: Readability Formulas: What's the Use? In T.M. Duffy and R. Waller (Eds.), *Designing Usable Texts.* (pp. 113-140). Orlando: Academic Press, Inc.

Easterby, R.S. and Hakiel, S.R. (1981). The comprehension of pictorially presented messages. *Applied Ergonomics,* 12, 143-152.

Edwards, M.L. and Ellis, N.C. (1976). An evaluation of the Texas driver improvement training program. *Human Factors,* 18, 327-334.

FMC (1985). *Product Safety Sign and Label System.* Santa Clara, CA: FMC Corporation.

Friedmann, K. (1988). The effect of adding symbols to written warning labels on user behavior and recall. *Human Factors,* 30, 507-515.

Godfrey, S.S., Allender, L., Laughery, K.R. and Smith, V.L. (1983). Warning messages: Will the consumer bother to look? In *Proceedings of the Human Factors Society 27th Annual Meeting* (pp. 950-954). Santa Monica, CA: Human Factors Society.

Godfrey, S.S. and Laughery, K.R. (1984). The biasing effect of familiarity on consumer's awareness of hazard, in *Proceedings of the Human Factors Society 28th Annual Meeting* (pp. 483-486). Santa Monica, CA: Human Factors Society.

Goldhaber, G.M. and DeTurck, M.A. (1988a). Effects of consumers' familiarity with a product on attention to and compliance with warnings. *Journal of Products Liability,* 11, 29-37.

Goldhaber, G.M. and deTurck, M.A. (1988b). Effectiveness of warning signs: Gender and familiarity effects. *Journal of Products Liability,* 11, 271-284.

Green, P. and Pew, R.W. (1978). Evaluating pictographic symbols: An automotive application. *Human Factors,* 20, 103-114.

Johnson, D. (1992). A warning label for scaffold users, in *Proceedings of the Human Factors Society 36th Annual Meeting* (pp. 611-615). Santa Monica, CA: The Human Factors Society.

Johnson, E.J. and Russo, J.E. (1980). Product Familiarity and Learning New Information, in K.E. Monroe (Ed.) *Advances in Consumer Research.*

Karnes, E.W. and Leonard, S.D. (1986). Consumer product warnings: reception and understanding of warning information by final users, in W. Karwowski (Ed.) *Trends in Ergonomics/Human Factors III, Part B: Proceedings of the Annual International Industrial Ergonomics and Safety Conference* (pp. 995-1003).

LaRue, C. and Cohen, H. (1987). Factors influencing consumers' perceptions of warning: an examination of the differences between male and female consumers, in *Proceedings of the Human Factors Society 31st Annual Meeting* (pp. 610-614). Santa Monica, CA: The Human Factors Society.

Laughery, K.R. and Stanush, J.A. (1989). Effects of warning explicitness on product perceptions, in *Proceedings of the Human Factors Society 33rd Annual Meeting* (pp. 431-435). Santa Monica, CA: The Human Factors Society.

Laughery, K.R. and Young, S.L. (1991). An eye scan analysis of accessing product warning information, in *Proceedings of the Human Factors Society 35th Annual Meeting* (pp. 585-589). Santa Monica, CA: The Human Factors Society.

Laux, L.F. and Brelsford, J.W. (1989). Locus of control, risk perception and precautionary behavior, in *Interface 89: The Sixth Symposium on Human Factors and Industrial Design in Consumer Products* (pp. 121-121). Santa Monica, CA: Human Factors Society.

Laux, L.F., Mayer, D.L. and Thompson, D.B. (1989). Usefulness of symbols and pictorials to communicate hazard information, in *Interface 89: The Sixth Symposium on Human Factors and Industrial Design in Consumer Products* (pp. 79-83). Santa Monica, CA: Human Factors Society.

Lehto, M.R. (1991). A proposed conceptual model of human behavior and its implications for design of warnings. *Perceptual and Motor Skills,* 73, 595-611.

Lehto, M.R. and Miller, J.M. (1986). *Warnings, Volume 1.* Ann Arbor, MI: Fuller Technical Publications (p. 18).

Leonard, S.D., Hill, G.W. and Karnes, E.W. (1989). Risk perception and use of warnings, in *Proceedings of the Human Factors Society 33rd Annual Meeting* (pp. 550-554). Santa Monica, CA: The Human Factors Society.

Leonard, S.D., Matthews, D., and Karnes, E.W. (1986). How does the population interpret warning signals? In *Proceedings of the Human Factors Society 30th Annual Meeting* (pp. 116-120). Santa Monica, CA: Human Factors Society.

Leonard, D.C., Ponsi, K.A., Silver, N.C. and Wogalter, M.S. (1989). Pest-control products: Reading warnings and purchasing intentions, in *Proceedings of the Human Factors Society 33rd Annual Meeting* (pp. 436-440). Santa Monica, CA: The Human Factors Society.

Mazis, M., Morris, L.A., and Gordon, E. (1978). Patient attitudes about two forms of printed oral contraceptive information. *Medical Care,* 16, 1045-1054.

McGuire, W.J. (1980). The communication-persuasion model and health-risk labeling. (pp. 99-122), in L.A. Morris, M.B Mazis and I. Barofsky (Eds.) *Product Labeling and Health Risks: Banbury Report 6,* Cold Spring Harbor Laboratory.

Morris, L.A., Mazis, M.B. and Gordon, E. (1977). A survey of the effects of oral contraceptive patient information. *The Journal of the American Medical Association,* 238, 2504-2508.

Otsubo, S.M. (1988). A behavioral study of warning labels for consumer products: Perceived danger and use of pictographs, in *Proceedings of the Human Factors Society 32nd Annual Meeting* (pp. 536-540). Santa Monica, CA: The Human Factors Society.

Purswell, J.L., Schlegel, R.E. and Kejriwal, S.K. (1986). A prediction model for consumer behavior regarding product safety, in *Proceedings of the Human Factors Society 30th Annual Meeting* (pp. 1202-1205). Santa Monica, CA: The Human Factors Society.

Ringseis, E.L. and Caird, J.K. (1995). The comprehensibility and legibility of twenty pharmaceutical warning pictograms, in *Proceedings of the Human Factors & Ergonomics Society 39th Annual Meeting* (pp. 974-978). Santa Monica, CA: The Human Factors & Ergonomics Society.

Sherer, M. and Rogers, R.W. (1984). The role of vivid information in fear appeals and attitude change. *Journal of Research in Personality*, 18, 321-334.

Silver, N.C. and Braun, C.C. (1993). Perceived readability of warning labels with varied font sizes and styles. *Safety Science*, 16, 615-626.

Smith, D.B.D. and Watzke, J.R. (1990). Perception of safety hazards across the adult life-span, in *Proceedings of the Human Factors Society 34th Annual Meeting* (pp. 141-145). Santa Monica, CA: The Human Factors Society.

Viscusi, W.K., Magat, W.A. and Huber, J. (1986). Informational regulation of consumer health risks: an empirical evaluation of hazard warnings. *Rand Journal of Economics*, 17, 351-365.

Wagenaar, W.A. (1992). Risk taking and accident causation, in Yates, J.F. (Ed.) *Risk-Taking Behavior.* (pp. 257-281). New York: John Wiley & Sons.

Westinghouse Electric Corporation (1985). *Product Safety Label Handbook.* 2nd Edition. Trafford, PA: Westinghouse Printing Division.

Wogalter, M.S. and Silver, N.S. (1990). Arousal strength of signal words. *Forensic Reports*, 3, 407-420.

Wogalter, M.S. and Silver, N.C. (1995). Warning signal words: Connoted strength and understandability by children, elders, and non-native English speakers. *Ergonomics*, 38, 2188-2206.

Wogalter, M.S. and Young, S.L. (1994). The effect of alternative product-label design on warning compliance. *Applied Ergonomics*, 25, 53-57.

Wogalter, M.S., Desaulniers, D.R., and Brelsford, J.W. (1986). Perceptions of consumer product hazards: Implications for the need to warn, in *Proceedings of the Human Factors Society 30th Annual Meeting* (pp. 1197-1201). Santa Monica, CA: Human Factors Society.

Wogalter, M.S., Desaulniers, D.R. and Brelsford, J.W. (1987). Consumer products: How are the hazards perceived? In *Proceedings of the Human Factors Society 31st Annual Meeting* (pp. 615-619). Santa Monica, CA: The Human Factors Society.

Wogalter, M.S., Sojourner, R.J. and Brelsford, J.W. (1997). Comprehension and retention of safety pictorials. *Ergonomics*, 40, 531-542.

Wogalter, M.S., Brelsford, J.W., Desaulniers, D.R., and Laughery, K.R. (1991). Consumer product warnings: The role of hazard perception. *Journal of Safety Research*, 22, 71-82.

Wright, P., Creighton, P. and Threlfall, S.M. (1982). Some factors determining when instructions will be read. *Ergonomics*, 25, 225-237.

Young, S.L. (1991). Increasing the noticeability of warnings: effects of pictorial, color, signal icon and border, in *Proceedings of the Human Factors Society 35th Annual Meeting* (pp. 580-584). Santa Monica: Human Factors Society.

Young, S.L. (1996). Subject differences in the perception of risk, in *Proceedings of the Human Factors and Ergonomics Society 40th Annual Meeting* (pp. 503-507). Santa Monica, CA: The Human Factors and Ergonomics Society.

Young, S.L. and Wogalter, M.S. (1990). Comprehension and memory of instruction manual warnings; Conspicuous print and pictorial icons. *Human Factors, 32*, 637-649.

Young, S.L., Brelsford, J.W., and Wogalter, M.S. (1990). Judgments of hazard, risk and danger: Do they differ? In *Proceedings of the Human Factors Society 34th Annual Meeting* (pp. 503-507). Santa Monica, CA: Human Factors Society.

Young, S.L., Martin, E.G. and Wogalter, M.S. (1989). Gender differences in consumer product hazard perceptions, in *Interface '89: The Sixth Symposium on Human Factors and Industrial Design in Consumer Products* (pp. 73-78). Santa Monica, CA: The Human Factors Society.

40

Design of Industrial Warnings

David R. Clark
GMI Engineering & Management Institute

Susan A. H. Benysh
Purdue University

Mark R. Lehto
Purdue University

40.1 Introduction

The design of an effective warning sign or label is a complex and difficult task. For the working professional in ergonomics, it can be perplexing to decipher the vast amount of research that has been, and is being, performed in the field, as well as the regulations and standards that affect the design and use of warning signs or labels. The purpose of this chapter is to provide industrial professionals with a succinct, but comprehensible, overview of the design of warning signs and labels for application in their own work environment.

This chapter is divided into four main sections: (1) the hierarchy of hazard control, where the role and place of warnings in a comprehensive program of hazard control is discussed, (2) generic guidelines for warning signs and labels, which covers the underlying human factors principles that guide the development of warnings, (3) effectiveness of warning signs and labels, which provides guidance on assessing whether or not the labels or signs are affecting behavior as desired, and (4) specific requirements and guidelines for warning signs and labels, where sources and provisions of existing design standards are outlined.

40.2 The Hierarchy of Hazard Control

Due to the increase of product liability litigation in the United States, especially those lawsuits citing "the failure to warn" as grounds for seeking damages, preventive criteria may be developed by the litigators of such cases that are not based on research. One example of this is the preventive criteria for companies to warn against ALL hazards with explicit warning signs and labels. The dangers of overwarning have been noted by several authors, including Lehto and Miller (1986). Overwarning might reduce the overall effectiveness of warnings, and thus increase product accidents. Too many warnings may cause habituation. Information overload is another possible problem. Therefore, warnings should be considered a supplement to, but not a substitute for, engineering of the product itself. The placement of warning signs and

0-8493-2641-9/99/$0.00+$.50

labels to products should be a final step in the attempt to control hazards associated with a product (Lehto and Salvendy, 1995). The engineering approaches to increase safety in the industrial environment related to product use that will be discussed in this section are product design (Norman, 1988; Rouse, 1992), job design (Campion and Medsker, 1992), personnel selection (Conoley and Kramer, 1989), training (Phillips, 1983), and supervision (Peterson, 1975; 1984).

Figure 40.1 (Lehto and Salvendy, 1995) illustrates the requirements for safe product design and use within workplaces in terms of the approaches discussed above. The most crucial stage for safe product design and use is product and process design. Essentially, this stage is fundamental for ensuring safe use of the product. During this stage, designers should be considering not only the intended use of the product, but also the many different ways in which the product might be misused. For example, an employee might use a chair as a ladder. Consideration of potential incorrect product-usage scenarios is vital to the design of a safer product. It is up to the employer to ensure that the employee uses the product for what it is designed to do, thus minimizing accidents associated with it.

Personnel selection is the next critical stage. This stage is important due to the individualistic nature of workers and the variety of skills and abilities they may possess. The selection process must ensure that employees have the prerequisite abilities to effectively and safely use a product. Training may be essential to ensure that the workers are qualified to perform tasks involving the product. Job design is also important since it determines exactly how, when, and for how long the operator will be using or exposed to the product. Through job design, the correct product will be chosen for the job, ensuring that misapplication does not occur. Forms of job performance aids, such as written instructions and operator manuals, perform a vital function at this stage.

Employee supervision is an essential method of ensuring that safe behavior is followed during employee training and job design. As well, supervision is a necessary deterrent when the employee can rationalize unsafe behavior or protocol, i.e., the behavior is an inconvenience or is unnecessary.

Warning signs and labels are the final resort, when all else fails, to produce reasonably safe behavior. The purpose of the warning is to alert operators to the presence of hazard, in hopes of avoiding unsafe or risky behavior. Warnings can enter the employee's field of attention in a variety of modes, verbal or nonverbal, through the auditory channel, the tactile channel, making use of the olfactory sense, or through the visual senses. Also, they can be presented through many different types of communications, including product packaging, the instruction manual, and the product itself.

From this discussion of accident prevention comes the question: At what stage is the most effective intervention performed? Lehto and Salvendy (1995) suggest starting by determining the error scenarios that involve the product. From there, determine intervention strategies for each stage mentioned in Figure 40.1. Brainstorming will suggest many options, and the preferred intervention strategy can then be determined and initiated. Generally, error scenarios are performed case by case, but intervention strategies for generic error categories can also be determined. This approach can be seen in Table 40.1.

Employee errors can be classified into three broad categories: errors that occur when performing routine behavior, when performing nonroutine behavior, and when the employee commits an error through an intentional violation (Reason, 1990). According to Rasmussen (1986), experienced workers performing routine tasks make errors when they fail to notice changes from ordinary conditions. When employees perform nonroutine actions, errors usually occur because of the user not knowing the proper procedure. Intentional violations occur when the user knowingly performs an unsafe act.

From the earlier discussion of Figure 40.1, product design is the preferred intervention strategy for any error scenario, especially with respect to errors occurring from routine behavior. One approach using product design to reduce potential routine behavior errors is to have the product provide warnings signals when the product is in a hazardous state (Lehto, 1991; 1992). An example of such a warning signal is when hazardous chemicals have a distinctive odor. This type of warning ensures that the hazard is immediately presented to the worker, so an effective action may be taken to possibly reduce the severity of the accident. Another approach is to design the products so that they do not provide affordances for, or trigger, inappropriate responses (Gibson, 1979). An example of this type of intervention would be to design an oven door out of glass to discourage users from standing on it. A final approach is to design

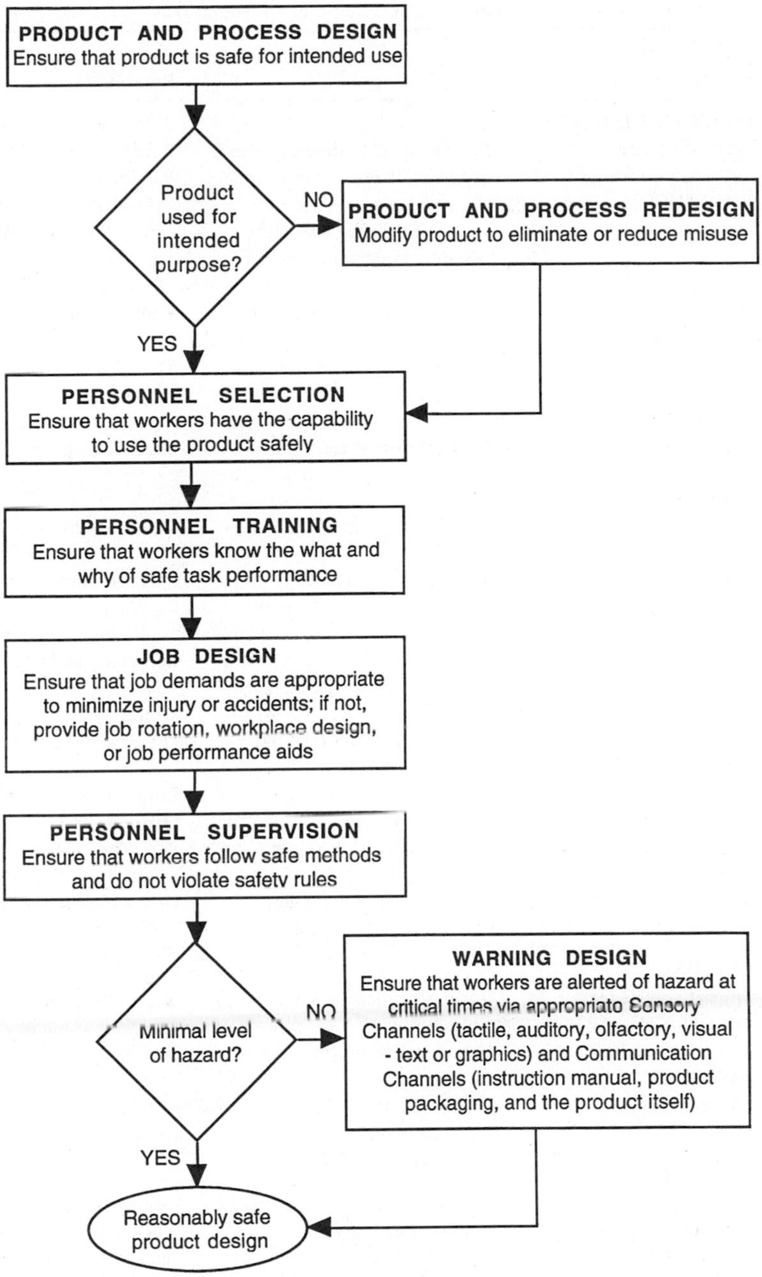

FIGURE 40.1 Intervention strategies and requirements for safe product design and use. (From Lehto, M.R. and Salvendy, G. 1995. Warnings: A supplement not a substitute for other approaches to safety. *Ergonomics.* 38 (11):2155-2163. With permission.)

the product to tolerate a minor range of errors, allowing for variability of work behavior. An example would be to have a knob click to certain values, instead of a free spinning dial which requires the worker to dial to an exact value.

Personnel selection is the next preferred strategy for error prevention. Again, this strategy seems to only be applied to errors associated with routine behavior. The only approach that these authors can foresee for the reduction of accidents is to ensure that the employees have adequate abilities to use the

TABLE 40.1 Intervention Strategies Based on Specific and Generic Error Scenarios

Error Scenario	Preferred Intervention Strategies	Objectives	Examples
ROUTINE BEHAVIOR	**PRODUCT DESIGN**		
Fail to perceive hazard condition or forget hazard condition	hazard signals	Provide signal at time of hazard.	Bumps marking border of roadway.
	interruptive features	Trigger shift away from routine.	Barricade between street curb and exit of building.
	WARNINGS		
	interactive	Combine warning text with product interruptive feature.	Taping the top drawer of a file cabinet shut with a warning label. Placing a sign on barricade.
	selective, nonvisual	Provide signal at time of hazard.	Auditory warning triggered when person enters hazard area.
	TRAINING		
	on product signals	Develop necessary skill to interpret product hazard signals.	The odor of ethyl mercaptan indicates a gas leak.
Forget intended action	**JOB DESIGN**		
	checklists	Confirmation of each action taken.	Flight start up and shut down checklist.
	PRODUCT DESIGN		
	hazard signal	Response relevant at time of signal.	Vibrations from speed bumps.
	WARNINGS reminders	Stimulate error recovery.	Seat belt buzzer/chimes.
Activate incorrect script	**PRODUCT DESIGN**		
	affordances and constraints	Eliminate design features that afford inappropriate responses.	Eliminate stair-like ladder features that encourage users to face away instead of toward rungs.
		Deactivate product features.	Use of interlocks and lockouts.
Psycho-Motor variability	**PRODUCT DESIGN**		
	hazard signals	Increase signal strength.	Ethyl mercaptan (odor) added to natural gas.
	error tolerance	Tolerates normal response range.	Adequate separation distance between controls.
	TRAINING		
	skill development	Reduce response variability.	Driver training on simulator.
	EMPLOYEE SELECTION	Ensure adequate abilities.	Testing driver visual acuity.
NON-ROUTINE BEHAVIOR	**TRAINING**		
	impart knowledge	Teach safe procedures, safety rules, etc.	HAZCOM program in U.S.
	JOB DESIGN		
	written procedures	Change task to routine following of instructions.	Step-by-step maintenance or repair procedures.
	WARNINGS		
	procedures	Change task to routine following of instructions.	Instruction manuals or step-by-step procedures on product labels.
INTENTIONAL VIOLATIONS	**SUPERVISION**		
	enforcement	Make sure safety rules followed.	Visible enforcement and supervision.
	information gathering	Identify extent of violations.	Work sampling, employee monitoring, etc.
	JOB DESIGN		
	compliance and noncompliance costs and benefits	Reduce the cost of compliance. Eliminate benefits of non-compliance.	Provide personal protective equipment (PPE) near point of hazard. Make PPE comfortable. Make sure PPE doesn't interfere with task.
	TRAINING		
	impart knowledge	Teach about hazards, reasons for safety rules, etc.	HAZCOM program in U.S.

Adapted from Lehto, M.R. and Salvendy, G. 1995. Warnings: A supplement not a substitute for other approaches to safety. *Ergonomics.* 38 (11):2155-2163. With permission.

product. This would include selecting workers for both current abilities and their potential for learning new ones. The next stage for accident intervention is personnel training. Training can be employed for all three types of behavior — routine, nonroutine, and intentional violations — to reduce errors. Training can be used to ensure that the worker has the necessary skills to interpret hazard signals and understand the reasons for safety rules. Safe behavior can also be taught to the worker. Finally, training to reduce response variability by employees can lead to a minimization of accidents by the worker.

Designing the job to minimize errors is the next level of priority from Figure 40.1. As with employee training, this level can be applied to all three categories of behavior. For routine behavior, a set of checklists can be created to ensure the employee is following a set routine. A checklist will also provide for validation of a correctly completed task. For nonroutine behavior, written procedures will alter the task from a nonroutine task to a routine of following instructions. Finally, job design can be employed to reduce intentional violations by reducing the compliance costs and increasing the benefits associated with safe behavior. An example of this would be to provide personal protective equipment near the point of a hazard.

Supervision is useful when attempting to reduce errors that occur due to intentional violations. This reduction of errors is a result of enforcing safety rules. Supervision is also useful in reminding employees of the negative repercussions of improper behavior. The final step in accident intervention is to use warning signs or labels. Warnings are useful by interrupting the task, making certain that the employee is reminded to not perform incorrect behavior before it occurs (as found by Frantz and Rhoades, 1993). As well, warnings can stimulate recovery from errors. One example of this type of warning is to have a seatbelt buzzer reminding users to secure their seatbelts before driving.

Warnings, though, should be used with care, especially when considering the level of experience of the employees to which they are directed. Dorris and Purswell (1977) and others (Frantz and Rhoades, 1993; Otsubo, 1988) found that experienced users, who were familiar with a task and the product used to complete that task, often failed to read warning labels. One potential reason may be that the user is highly focused on task-related goals, only looking for information that will satisfy the end-goal of task completion. Interruptions may also take the user out of an automatic form of behavior and encourage him or her to enter a less routine information-seeking mode of behavior.

40.3 Generic Guidelines

From the research regarding the effectiveness of warning signs and labels, many generic guidelines can be inferred. These guidelines are: "(#1) conform to standards and population stereotypes when possible; (#2) make warning signs and labels conspicuous and each component legible; (#3) simplify the syntax of text and combinations of symbols; (#4) make symbols and text as concrete as possible; (#5) make sure that the cost of compliance is within a reasonable level; (#6) be selective; (#7) integrate the warning into the task and hazard related context; and (#8) match the warning to the level of performance at which critical errors occur for the relevant population" (Lehto, 1992, p. 108). These general guidelines are expanded upon and classified into three broad categories: format, content, and mode of presentation. Though much research has been performed on the format of warnings, content and mode of presentation are still, if not more, important to the overall effectiveness of warning signs and labels.

Format

The format of a warning sign or label refers to the physical aspects of the warning, which include text, iconic, and/or graphic information, both in individual and in collective form. Other format aspects of warning signs and labels are the size of the label, methods of color coding, and the arrangement of the warning components. The two generic guidelines that are of the greatest relevance to format specification are (#1) conform to standards and population stereotypes when possible; and (#2) make warning signs and labels conspicuous and each component legible. Usually, the normal variations in typography are acceptable, but there exist at least four circumstances (according to Sanders and McCormick, 1993) in which it may be preferred to have set guidelines for labels, one being when working with warning messages. There are twelve specific guidelines that have surfaced from the literature. These will be presented, with examples when appropriate, below.

Provide a signal word or words in capital letters at the top of a warning sign or label to indicate the sign or label is a warning. Examples of signal words are NOTICE, WARNING, CAUTION, and DANGER (ANSI Z535.2, 1996; ANSI Z535.4, 1996; and ANSI Z129.1, 1994). A study by Bresnahan and Bryk (1975) addressed the perception of the terms DANGER and CAUTION by industrial workers. They found that greater levels of hazard were associated with the term DANGER. Lirtzmann (1984) also found these results, but when he tested WARNING and CAUTION, there was found no significant difference between perceived danger levels. Hadden (1986) provides support for a new system of signal words, EXTREME-DANGER, SERIOUS-DANGER, and MODERATE-DANGER.

Use lowercase for text within a warning sign or label. Poulton (1967) found that lowercase is easier to read and faster to comprehend than when the text is ALL UPPERCASE. This seems to be the general consensus of the literature in this area (Sanders and McCormick, 1993).

For text within smaller labels, a visual angle of at least 10 minutes of arc for the intended viewing distance should be provided (Heglin, 1973; Woodson, 1963). For text within signs, a visual angle of a minimum of 25 minutes of arc should be present for the intended viewing distance (Heglin, 1973; Woodson, 1963). Other factors, such as luminance and the population, should be factored into determining the proper visual angle (Sanders and McCormick, 1993). The National Bureau of Standards (Howett, 1983) developed a formula for determining the size of alphanumeric characters to be read at various distances by persons with various Snellen acuity scores. The following formulas should be used:

$$W_S = 1.45 \times 10^{-5} \times S \times d \tag{1}$$

$$H_L = W_S/R \tag{2}$$

where W_s, d, and H_L are in the same units (inches or mm) and:

W_s = stroke width
S = denominator of Snellen acuity score (i.e., if Snellen acuity = 20/40, then S = 40)
d = reading distance
H_L = letter height
R = stroke width-to-height ratio of the font, expressed as a decimal proportion (i.e., for a ratio of 1:10, R = 1/10 = 0.10)

For short messages or signal words, text characters should be sans serif (ANSI Z35.1, 1972; SAE Recommended Practice J115a Safety Signs, 1979; Westinghouse, 1981; FMC Guidelines, 1980).

Test characters should be dark against a light background. This recommendation is based on research that shows that the contents of a label or sign need to be larger when they are presented as white on a dark or black background (Heglin, 1973). Also, irradiation may occur, which is the phenomenon in which white features on a black background tend to "spread" into adjacent dark areas (though the reverse is not true).

The stroke width-to-height ratio of text characters should be between 1:6 and 1:10 (Heglin, 1973). The stroke ratio can also be expressed as a proportion (i.e., 1:10 = 0.10).

The width-to-height ratio of alphanumeric characters should be between 1:1 and 1:3.5 (Heglin, 1973; Sanders and McCormick, 1993). Alphanumeric characters are a mix of both text and numeric characters. A ratio that is often used is 3:5. This ratio has its roots in the fact that most letters of the English alphabet have five elements in height and three elements in width. This can easily be seen in the numeric 8 (two vertical strokes plus the space in-between: three horizontal strokes plus two spaces between these).

Provide for a brightness contrast of at least 50% between the warning text and its background. The perceived brightness of a warning sign or label depends on the energy spectrum of the emitted or reflected light and the type of color that is used on the warning. One should not depend on color contrasts, since printing inks and fading may cause the contrast to lessen. The brightness of a particular stimulus can be described using photometric units. Data regarding the energy spectrum of various forms of lighting on different types of signs and labels is available in the *IES Lighting Handbook* (1981).

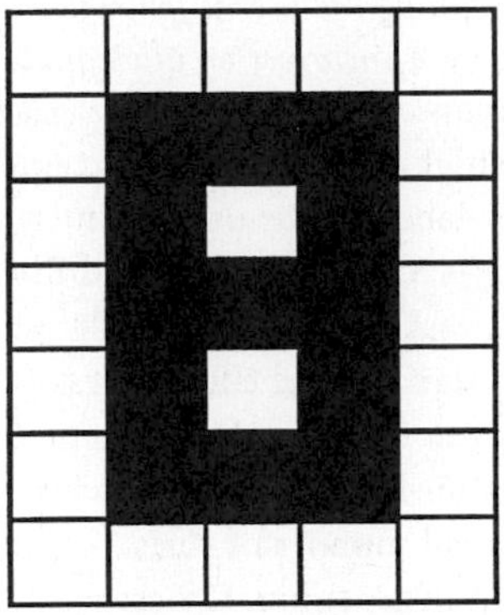

FIGURE 40.2 Example of a stroke height-to-width ratio.

Consider color coding schemes consistent with ANSI Z35.1 (1996). If color is used, consider the energy spectrum of foreseeable lighting sources when specifying the color mix. Avoid using the extremes of the color spectrum, such as reds and blues (Matthews, 1987), since some subjects with various types of color blindness may not be able to read these signs or labels.

Avoid crowding of components on the sign or label, assuring that each component is legible. Increased density of the sign or label (even when the necessary information is present) increases both reaction times and errors (Tullis, 1988). Avoid crowding by eliminating unnecessary information and use concise wording.

The presence of adverse viewing environments, such as the presence of dirt, grease, and other contaminants, can degrade the legibility of warning components. Compensation should take place to counteract these variables by testing the legibility under realistic conditions and replacing damaged signs and labels.

Consider the use of symbols/pictographs when users are performing routine behavior. Consider using text when employees need to make decisions or are performing non-routine behavior. FMC (1980) and Westinghouse (1981) encourage the use of symbols/pictographs. Conversely, the ANSI Z535 (1996) standards recommend using symbols as a supplement to words.

Content

The content of the warning sign or label refers to the warning message itself, its level of abstraction, and its syntactic structure. The generic guidelines applicable here are (#3) simplify the syntax of text and combinations of symbols, (#4) make sure that the cost of compliance is within a reasonable level, (#5) be selective, and (#8) match the warning to the level of performance at which critical errors occur for the relevant population. Some specific guidelines that can be derived from these generic guidelines and from the research follow.

Messages should focus on critical errors that cause a significant safety problem. When choosing warnings, one should be selective to avoid habituation and warning overload, and thus reduce the potential for ignoring the warning. Therefore, avoid long lists of hazards and messages that describe trivial hazards or hazards that are obvious to the intended audience. Such information is best provided in media other than warning signs or labels (Lehto and Miller, 1986; Scammon, 1977).

Focus on developing messages for the following two types of error situations: (1) forgetting to perform an action ordinarily performed (i.e., the sign or label reminds), and (2) not knowing the consequences of performing or failing to perform some action.

When a user's performance is skill-based, meaning they are performing an automatized set of procedures, and they commit an error based on their failure to perceive a condition or motor variability, provide a warning signal and consider training. An example of this type of error is when a person is walking in a dry area and does not change his or her gait before stepping on a wet spot. Rhoades et al. (1990) provide many case studies of such errors.

When performance is rule-based, meaning the user is following a set of rules and the behavior is not yet automatized, and the error is caused by a incorrect or inadequate rule, determine whether the rule was originally developed on the basis of knowledge or judgment-based behavior. If it seems to be judgment-based, like when people speed on the highway, focus on enforcement. If it seems to be knowledge-based, determine whether a warning sign or label can be used to interrupt the task (i.e., to place its message into short term memory at the time it is relevant). Frantz and Rhoades (1991) showed that interrupting a task increased the compliance in a task from 13% to 73%. Other studies have also concluded that interrupting a task increases warning sign or label effectiveness (Dingus et al., 1991).

When performance is knowledge-based, meaning the user is problem-solving, and the error is caused by inadequate knowledge, the amount of knowledge necessary to prevent the error should be determined. If the knowledge can be described with a small number of rules, consider a warning sign or label containing these rules in the form of step-by-step instructions. Otherwise, focus on training, instruction manuals, or other forms of education.

When performance is judgment-based, meaning the user is experiencing an affective reaction of some sort, and the error is caused by inappropriate priorities, evaluate the user's behavior pattern. If the undesired behavior pattern appears to have significant value to the user (i.e., pleasure, comfort, convenience, etc.) or is likely to be entrenched, focus on enforcement through supervision.

Regardless of the level of a user's performance, consider messages that minimize the cost and increase the benefits of compliance. Therefore, the behavior most desired by the user will be the safer and more effective behavior.

If a large number of potential warnings are present after applying these guidelines, increasing the probability of overloading the user, other means of providing the information (such as instruction manuals or training courses) should be considered to reduce warning overload.

Given that a message satisfies the preceding rules for content, further recommendations pertain to the *level of abstraction* and are based on the generic guidelines (#8) match the warning to the level of performance at which critical errors occur for the relevant population and (#4) make symbols and text as concrete as possible. Specific recommendations inferred from these guidelines are:

When subjects are inexperienced, consider pictographs (having a more detailed design) instead of symbols. The abstractness of the symbols have been found to cause a decrease in warning comprehension (Jacobs et al., 1975; Lerner and Collins, 1980).

When performance is at a skill-based or rule-based level, consider brief messages that describe conditions or actions. Also, consider symbols or pictographs instead of text, since these are found to make the warning more memorable (Young and Wogalter, 1988).

When performance is at a knowledge-based level, that is the user needs to understand "why" an action or behavior needs to be performed, consider more detailed messages that describe both conditions and actions.

When performance is at a judgment-based level, in that the incoming information is processed as values that evoke an effective reaction, consider messages that describe the hazard and the benefits of compliance. Also, consider citing highly credible sources. Craig and McCann (1981) found that compliance to warnings was higher when the message came from a government source versus a public utility company.

When the hazard is complex or occurs in different manifestations, consider abstract text, which better covers hazard contingencies, rather than a long list of concrete examples or symbols. It was found by Lehto and DeSalvo (1992) and Morris and Kanouse (1980) that these longer messages created problems for some subjects. Also, increasing the number of items on a label has been shown to decrease overall recall performance (Scammon, 1977; Reder and Anderson, 1982). It was found in numerous studies (Cahill, 1976; Johnson, 1980; Cairney and Sless, 1982) that symbols and pictographs were poorly comprehended when describing a complex hazard (as opposed to reading and understanding text). However, it was also found that concrete text may be better comprehended than abstract text (Lehto and DeSalvo, 1992; Laughery et al., 1991; Leonard and Matthews, 1986). An example of concrete text is "Using this product without the use of a mask will give you lung cancer." An example of abstract text is "Using this product incorrectly will increase your chances of illness."

When knowledge or understanding of a product or task is low, consider concrete text instead of abstract symbols or pictographs. Concrete text is easier to comprehend, and the interpretation of symbols has been found to vary across different cultures (Easterby and Zwaga, 1976; Sinaiko, 1975; Cairney and Sless, 1982).

Use text and symbols that people in the intended user population can comprehend. Consider language, reading level, and cultural effects.

The *syntactic structure* of warning signs and labels is addressed by generic guideline (#3) simplify the syntax of text and combinations of symbols. Specific recommendations derived from this guideline and research in the area are:

Use short simple sentences; complex conditional sentences, particularly those containing negations, should be avoided. Longer messages are not necessarily comprehended better (Morris and Kanouse, 1980; Lehto and DeSalvo, 1992). The latter study showed that action statements ALONE consistently received the highest comprehensibility ratings.

Symbolic signs or labels should focus on describing conditions (i.e., flammable). With few exceptions (i.e., a slash/bar to indicate negation) they should not combine multiple meanings or be used to describe complicated sequences of actions. This should lead to an increase in the comprehension of the warning (Young and Wogalter, 1988).

Mode of Presentation

The mode of presentation of a warning sign or label refers to the location and task-specific timing of contact with a warning sign or label. The following generic guidelines apply here: (#5) make sure that the cost of compliance is within a reasonable level, (#7) integrate the warning into the task and hazard related context, and (#8) match the warning to the level of performance at which critical errors occur for the relevant population. Specifically, the following guidelines should be followed:

The warning sign or label should be presented at a location and time in which the danger is still avoidable. It was shown by Wogalter et al. (1987) that the location of warnings in instructions can be a major determinant of effectiveness. Another example is the use of a seatbelt buzzer that alerts the driver of an automobile to secure his or her seatbelt when starting the automobile.

The location and timing of presentation should minimize the cost or difficulty of compliance (i.e., a sign to wear goggles should be close to an available source of goggles). Wogalter et al. (1989) found that subjects were more likely to use goggles and gloves in a chemistry lab task when they were provided with the equipment. Several other studies have shown that the likelihood of following a warning is influenced by the perceived cost of compliance with respect to cost to the worker, effort, time, comfort, productivity, and more (Godfrey et al., 1985).

Make an attempt to present the warning sign or label at a time when the person has available attentional capacity. Several studies have shown that people have trouble remembering warnings shortly after completing a task (Wright, 1979; Strawbridge, 1986; Ursic, 1984).

When performance is skill-based (Rasmussen, 1986), determine if the task can be interrupted to bring attention to the label. If this can be done, consider a warning sign or label which describes the condition and prescribes an action. If this cannot be done, consider providing a warning signal, training, or modifications of the product.

Avoid embedding the sign or label in a cluttered background. Several studies have shown the adverse effects of visual clutter on the perception of signs (Holahan, 1977; Boersema and Zwaga, 1985). Holahan was able to show that traffic accidents increased at a stop sign where the presence of commercial signs increased.

40.4 Evaluating Effectiveness

To measure the effectiveness of a warning, one must first understand the steps that must occur before a sign or label can prevent an accident. These include: (1) attending to the warning, (2) comprehending the warning, (3) deciding to perform the proper action to avoid an accident, and (4) taking the appropriate behavior. Finally, (5) the action decided upon must be sufficient to avoid the accident. Since all the steps must be taken in order to avoid an accident, the probability of achieving correct behavior, or avoiding an accident, can never be greater than the probability of successfully completing a single step.

A simple accident illustrates this argument. Consider the hypothetical situation in which (1) 50% of the population will read the warning, (2) 80% of those individuals understand the warning once they have read it, (3) 95% of those individuals decide to perform the proper action to avoid an accident, (4) 100% of those individuals take the appropriate behavior, and (5) that action is sufficient to avoid the accident 75% of the time. When the probability of successfully completing the entire sequence of events is defined as the effectiveness of the warning, then effectiveness is the multiplication of the conditional probabilities of completing each step. This example results in an effectiveness of an 0.285, or the warning will be 28.5% effective for the population of users.

Recent research has confirmed the basis of this model, that the end result of compliance cannot be greater than any probability of completing a single step. Otsubo (1988) observed that warning labels created to encourage workers to wear gloves when operating a circular saw were complied with by 38% of the workers, though 74% of the subjects noticed the warning and 52% read the warning. The same results were found when Otsubo performed the same experiment with a jigsaw. It was found that 54% of the subjects noticed the warning, 25% read it, and only 13% complied with the warning to wear gloves.

There are many dependent variables related to the effectiveness of warnings. A majority of these dependent variables can be grouped into three areas, perceptual factors, comprehension levels, and those factors associated with behavior patterns. The evaluation of the perception of warnings should give insight as to the conspicuity, or attention-attracting ability of the sign. There are numerous variables related to the perception of warning signs or labels. Some of these measures include reaction time, accuracy of task completion, attention to different elements of the warning (accomplished through the study of eye movement), the use of tachistoscopic procedures, and measures of legibility distance. Table 40.2 gives the approach taken for each procedure and their respective advantages and disadvantages.

When evaluating the comprehensibility of a warning, four variables are often used: symbol recognition or matching, message recall, psychometric (rating scales), and readability indexes. Symbol recognition is the most commonly applied technique used to measure warning sign and label comprehension. It is usually applied via open-ended questions, in which the user is asked to describe the symbol. When symbol matching is used, the user is asked to match the symbol to possible meanings. Recognition gives a more accurate understanding of comprehension, though matching allows for more quantifiable and, thus, analytical results. Message recall is an approach in which subjects are asked to remember the message after a task has been completed. Obviously, the user's memory may introduce error, but this measure provides the best indication as to how the contextual material is related to perception of the warning. Psychometric scales are often used when requesting users to rate their perceptions of different aspects of the warning. From this rating, comprehension can be inferred using statistical approaches such as factor analysis or cluster analysis. A lot of pre-experiment work should be performed to ensure that the scales give the information that is desired by the experimenter. Finally, readability indexes are applied to the warning signs and labels to describe the difficulty of written material in terms of word length, sentence length, or other variables. Many indexes have been developed, but due to the nature of most warnings, being in terse fragments vs. prose, readability indexes have little value.

Evaluating behavior patterns also offers some insight as to the effectiveness of warning signs and labels. When evaluating a subject's behavior, this allows the warning to be evaluated under realistic conditions as to how it affects overall behavior. Since the ultimate objective of a warning sign or label is to change behavior, this technique will offer insight into the realization of this goal. The most common approach is to set up an experiment in which two groups of subjects are tested on the same task. One group is the control without the warning, and the other group receives the warning. The prevalence of safe behavior can then be compared between the groups. Field observations are also useful in determining warning effectiveness. This will allow even more realistic conditions, and should reduce potential error introduced due to the unrealistic situation of a controlled experiment.

TABLE 40.2 The Evaluation of Perceptual Factors for Warning Signs and Labels

Procedure	Approach	Advantages	Disadvantages
1. Reaction time	Measures the reaction time to different warnings. Assumes that a symbol quickly reacted to is more salient than one reacted to more slowly	Useful when time to react is needed for the task	Reaction time not always related to sign or label quality
2. Accuracy of product use	Conversely, errors can be measured to determine possible perceptual problems associated with the sign or label. A confusion matrix can be created to establish the correct and incorrect responses	Shows areas that need to be investigated further to decrease errors An absolute measure is obtained and can be used for comparison to other warning designs	May not give insight into the area that has caused accuracy to decrease
3. Attention to items within the warning	Usually, measuring attention is achieved through the measurement of the placement and time of eye movements. This gives insight into the elements that capture a user's attention	This allows the specific aspects of the warning that is causing trouble or decreasing accuracy to be identified	The amount of data produced by this method makes it difficult to analyze Equipment used to gather this data can be quite intrusive and not allow for an accurate setting for the study
4. Tachistoscopic procedures	A sign or label is presented to the subject with a tachistoscope for precisely timed intervals. After viewing the sign or label, the subjects are asked questions regarding it's content	Allows for the amount of time that is needed to perceive a warning to be correctly measured Gives an accurate measure of the attention needed to perceive a warning	This measure not based on a realistic setting and therefore, may confound errors The results of this study rely on the user's memory, which may introduce error
5. Measures of legibility distance	Determines the distance in which a warning can be perceived and read	Gives an absolute values as to the distance required to read a warning	Questionable measure of symbol effectiveness when quick perception is not important Studies have found this measure to not be related to measures of comprehension or even other measures of legibility

There are four methods commonly used to measure the dependent variables needed to evaluate the effectiveness of warning signs and labels. These methods include interviews, questionnaires, behavioral observations, and ratings of warning effectiveness. Each method has advantages and disadvantages. Therefore, the best method will depend upon the objectives of the study.

Interviewing is a good method for understanding the reasoning behind user actions. The interview is normally performed some time after the task is performed. Therefore, the basic errors that can occur in knowledge elicitation will be compounded with memory errors. A structured interview begins by the interviewer having a transcript of questions to ask the subject. These questions are general, such as "What did you do to reach this goal?" and "What did you do next?" The interviewer will then ask these questions over and over again until all relevant information is extracted from the subject. Obviously, this process can take a very long time. The main concern with applying this technique is that the subject can only produce what can be verbalized, so automatic actions are not given as much detail as the processes that have non-proceduralized actions (Wilson and Corlett, 1990). Also, subjects tend to justify their decision when in an interview situation, creating false confidence (Wilson and Corlett, 1990).

The advantages of questionnaires and rating scales are often the same as for an interview, but the amount of knowledge obtained from the questionnaire and rating scales will be limited to the amount of forethought by the researcher. Since a questionnaire can contain either open- or closed-ended questions, one must decide on this as well. Closed-ended questions (such as "On a scale of 1 to 10, with 10 being the highest, how safe do you view this product?") are easy to analyze, but may not have the construct validity that is needed. Open-ended questionnaires can be difficult to analyze and can be open to interpretation error by the researcher. Also, while some subjects will explain their reasoning, others will give very short and, quite often, uninformative answers.

Behavioral observations can be used to describe the use of the product. In some cases, interpretation on the part of the researcher may need to be employed, possibly introducing error. Also, it may not be apparent from the user's actions what mental actions were taken to perform a consequential action. Therefore, sometimes a verbal protocol is performed along with the observation. A verbal protocol is a knowledge elicitation technique in which subjects are requested to "think aloud" as they perform a task (Shadbolt and Burton, 1990; Belkin et al., 1987; Kuipers and Kassirer, 1983). During this time, the analyst records the process (usually on video). After the task is completed, the video is reviewed by the analyst and the subject, with the subject giving details about how the task was performed or why certain actions were taken. Also, a review of the video by experts could offer insights regarding certain focal areas that the subject did not, or could not, elaborate upon. The first step to performing a verbal protocol involves some preliminary work, so that the analyst has basic knowledge of the task for best time utilization (Wilson and Corlett, 1990). Also, a few training sessions in "thinking aloud" should be held so as to avoid embarrassment to the subject. The main problem that may occur due to verbal protocol technique is that the verbalization may interfere with the subject's performance (Wilson and Corlett, 1990). But, with enough training sessions, that problem should be minimized.

The foregoing provides a basis for the use of warnings within the hierarchy of hazard control, and outlines general guidelines for the construction and evaluation of warning signs and labels based on research findings and principles of human factors. As a practical matter, the designer of warnings must also consider an array of government and consensus standards that can either (1) provide explicit guidelines on how to accomplish that which has been generically discussed above, or (2) require compromise between incompatible generic guidelines and explicit requirements (and not infrequently, between different explicit requirements). The remainder of this chapter identifies sources of some of the most important or commonly used standards and illustrates some of the provisions of those standards.

40.5 Government and Consensus Standards

Consider the typical industrial signs shown in Figure 40.3. Are these warning signs or are they not? Are they instructions instead? Are they combinations of both? A warning may be defined as something that (1) gives or serves to give (a) notice beforehand, especially of danger or evil, or (b) admonishing advice, or (2) calls or serves to call one's attention; an instruction is a direction calling for compliance (Webster 1988). Consequently, warnings do not need to include instructions; or conversely, instructions do not constitute a warning. On the other hand, certain guidelines for warning signs and labels recommend that provisions be made for information on how to avoid the hazard, that is, instructions. Depending on the nature of the hazard and the target audiences of the warning, information of differing types may be warranted. Therefore, in the following review of specific warning standards, it is not possible to totally separate the giving of instructions in the construction of warnings. Perhaps it would be better to think of the combination of elements as hazard communication. Indeed, the current American National Standards Institute warning sign and label standards (ANSI Z535.1 *et seq.*, 1996) refer to themselves as a Hazard Communication System.

FIGURE 40.3 Warnings or instructions?

Sources of Standards

It is natural to divide warnings standards into two categories: mandatory requirements and voluntary guidelines. Each is discussed below.

Mandatory Requirements

Sources of mandatory design guidelines include government regulatory agencies at all levels. The primary source of interest to industry is the U.S. Department of Labor's Occupational Safety and Health Administration, although other important sources include the U.S. Department of Transportation, the Environmental Protection Agency, and even the Consumer Product Safety Commission. Many states have analogs of these agencies whose regulations supplement or supersede the federal guidelines, although they must be at least as rigorous as the federal versions.

In many cases, the information available in these regulations to the designer or specifier of an industrial warning is generally not of a systems nature. This means that certain components of warnings designed to address certain specific hazards are given, but guidelines on how to address new or unanticipated hazards are not.

Occupational Safety and Health Administration (*OSHA*). The primary OSHA regulations on warnings are found in 29 CFR 1910.144 (color coding) and .145 (accident prevention signs and tags), and in 29 CFR 1910.1200 (hazard communication standard, or "right to know"); these will be covered in more detail later in this chapter. Other provisions address specific hazards such as egress (29 CFR 1910.36 *et seq.*), radiation (.96 *et seq.*), and flammables or explosives (.103 *et seq.*), while others focus on processes/industries, such as sawmills (.265) or telecommunications (.268). Similar provisions are found in OSHA's construction (29 CFR 1926) and maritime standards (29 CFR 1915 *et seq.*).

Department of Transportation (*DOT*). The DOT has extensive labeling, marking, and placarding regulations that cover materials determined to be hazardous when transported. Inasmuch as these regulations provide cradle-to-grave coverage, they are applicable to all parts of industry, whether in the preparation, transportation, or use of these materials.

Environmental Protection Agency (*EPA*). The hazard labeling authority of the EPA arises out of the Federal Insecticide, Fungicide and Rodenticide Act of 1978 (FIFRA), the Toxic Substances Control Act of 1976 (TSCA), and the Resource Conservation and Recovery Act of 1976 (RCRA). FIFRA requires that pesticide labels include information on product identification, ingredients, warnings, and use.

Consumer Product Safety Commission (*CPSC*). Although not directly applicable to the industrial workplace, many products that find their way into the workplace would be considered consumer products under other circumstances. Indeed, the general availability of Material Safety Data Sheets (MSDS) through the OSHA Right-to-Know Standard has been extended to consumer products. Therefore, the role of the Consumer Product Safety Commission should be appreciated, although no specific CPSC regulations will be treated herein.

The Courts. Also bear in mind that the courts have influenced the role and design of warnings, although the decisions typically give very little in the way of engineering design criteria.

Voluntary Guidelines

Several voluntary systems-based warning sign and label guidelines are provided by private sector standards-making organizations. These systems promote consistency in design and allow designers to leverage the considerable experience of other organizations in the development of warning signs and labels.

Consensus Organizations. These organizations develop voluntary requirements that represent a consensus, or agreement, among its membership and the affected industries or public. This process is somewhat similar to that which the federal government follows under the rules of the Administrative Procedures Act. Such organizations include: (1) international groups, such as the United Nations, the European Community, the International Standardization Organization (ISO), and the International Electrotechnical Commission (IEC), (2) national organizations, such as the American National Standards Institute (ANSI), the British Standards Institute, the Canadian Standards Association, the German Institute for Normalization (DIN), and the Japanese Industrial Standards Committee, and (3) independent agencies which often submit their standards to ANSI for approval as consensus standards, such the National Fire Protection Association (NFPA). Table 3 identifies examples of various U.S. consensus

TABLE 40.3 Examples of Warnings or Warnings-Related U.S. Consensus Standards

ANSI A13.1	Piping
ANSI C95.2	Radio Frequency Radiation
ANSI D6.1, 6.1b, 10.1	Traffic Control
ANSI N12.1	Fissile Material
ANSI N2.1	Radiation
ANSI Z129.1	Industrial Chemicals
ANSI Z138.2	Color
ANSI Z241.1	Sand Foundry Industry
ANSI Z244.1	Lock Out/Tag Out
ANSI Z35.1, 535.2	Accident Prevention Signs
ANSI Z35.2, 535.5	Accident Prevention Tags
ANSI Z35.4	Informational Signs
ANSI Z35.5	Biological Hazard
ANSI Z53.1, 535.1	Color Code
ANSI Z535.2	Environmental and Facility
ANSI Z535.3	Symbols
ANSI Z535.4	Consumer Product
ANSI/ASAE S338.1	Towed Equipment
ANSI/ASTM D1535	Color
ANSI/ASTM D56, 93	Flash Point
ANSI/ISA S5.5	Process Display Symbols
ANSI/MH11.3	Powered Industrial Trucks
ANSI/NEMA ICS1, 6	Industrial Control & Systems
ANSI/NFPA 30	Flammable and Combustible Liquids
ANSI/NFPA 70	Electrical
ANSI/NFPA 101	Safety to Life from Fire
ANSI/NFPA 178	Fire Fighting Operations
ANSI/NFPA 1901	Automotive Fire Apparatus
ANSI/SAE J1116	Off-Road Work Machines
ANSI/SAE J115	Safety Signs
ANSI/SAE J1164	ROPS and FOPS
ANSI/SAE J137	Agricultural Equipment on Highways
ANSI/SAE J1500	Operator Controls
ANSI/SAE J208, 389, 841, 1170	Agricultural Equipment
ANSI/SAE J284	Agricultural, Construction, and Industrial Equipment
ANSI/SAE J298	Industrial Equipment
ANSI/SAE J575	Motor Vehicle Lighting
ANSI/SAE J594	Reflex Reflectors
ANSI/SAE J674	Motor Vehicle Glazing
ANSI/SAE J725, 943	Slow-Moving Vehicles
ANSI/SAE J99	Industrial Equipment on Highways
ASAE S441	Safety Signs
ASTM C1023, ES6	Ceramic Art Material
ASTM D1014	Paints on Steel
ASTM D1729, 2244, E308	Color
ASTM D1788	ABS Plastic
ASTM D2794	Organic Coatings
ASTM D3278	Flash Point
ASTM D4086	Metamerism
ASTM D4257	Coatings & Lining Industry
ASTM D4267	Parenteral Drug Containers
ASTM E239	Paint, Varnish, Lacquer, and Related Products
ASTM E42, 188, 822, G23, 26	Nonmetallic Material
ASTM E991	Fluorescent Color
ASTM ES9, F926	Kerosine Containers
ASTM F406	Play Yards
ASTM F839	Gasoline Containers
EIA RS257	Mercury
NEMA 260	Switchgear & Transformers

TABLE 40.3 Examples of Warnings or Warnings-Related U.S. Consensus Standards (continued)

NEMA EW6	Arc-Welding and Cutting
NEMA IB1	Lead-Acid Batteries
NFPA 291	Fire Hydrants
NFPA 704	Fire Hazards
SAE J1048	Motor Vehicles
SAE J107	Motorcycles
SAE J1150	Agricultural Equipment
SAE J179	Truck Wheel Rims
TAPPI UM586	Label & Tape Aging Testing

standards. Two sources of particular interest to U.S. industry are the ANSI Z535 Committee on Safety Signs and Colors and the ANSI Z129 Committee on Labeling of Hazardous Industrial Chemicals.

The ANSI Z535 Committee on Safety Signs and Colors committee was formed in 1979 by combining the previous Z35 Committee on Safety Signs with the Z53 Committee on Safety Colors. The Z35.1 (Specification for Accident Prevention Signs) and Z53.1 (Safety Color Code for Marking Physical Hazards) standards (along with Z129.1) existing at that time were the primary standards for industrial warning systems. The new Z535 Committee was "to develop standards for the design, application, and use of signs, colors, and symbols intended to identify and warn against specific hazards and or other accident prevention purposes" (ANSI Z535.1). In 1991, a series of five standards were published by the Z535 co-secretariat National Electrical Manufacturers Association after being approved by ANSI. These addressed color coding (Z535.1, which replaced Z53.1), environmental and facility safety signs (Z535.2, which replaced Z35.1 and Z35.4), safety symbols (Z535.3), product safety signs and labels (Z535.4), and accident prevention tags for temporary hazards (Z535.5, which replaced Z35.2). A 1996 version of each of these standards is now in the approval process. Note, however, that Z35.1-1968 and Z53.1-1967 were used in 1972 as the basis for the current OSHA regulations on accident prevention signs, and most sign manufacturers still provide Z35.1/Z53.1-compliant signs. New design efforts should normally conform to the Z535 series of standards.

Notable among the exceptions in the scope of the ANSI Z535 standards is that of chemical products and mixtures. These are addressed by the ANSI Z129 Committee on Labeling of Hazardous Industrial Chemicals and the provisions of a standard that was first constituted in 1946 as an industry guide under the Manufacturing Chemists Association and later approved as ANSI Z129.1 in 1976, now sponsored by the Chemical Manufacturers Association. The current standard was approved in 1994.

Professional Organizations. This includes such organizations as the Society of Automotive Engineers (SAE), the American Society of Mechanical Engineers (ASME), the National Safety Council (NSC), and the American Society of Safety Engineers (ASSE).

Industrial Trade Associations. Typically based on a product or service focus, these organizations include such organizations as the National Electrical Manufacturers Association (NEMA), the Chemical Manufacturers Association (CMA), and the Material Handling Institute, (MHI). They are often the initial source of standards approved by consensus organizations.

Companies. Individual companies often develop internal systems for warning design, some quite extensive. And these efforts are sometimes made available outside the company. Two notable examples of this type are the *Product Safety Sign and Label System* developed by the FMC Corporation (1980) and the *Product Safety Label Handbook* developed by the Westinghouse Electric Corporation (1981). Elements of the FMC system were the basis of the current ANSI Z535 labeling standards.

Commercial. There are a number of "sign" companies that provide warning signs and labels that are marketed as conforming to various and specific standards set by many of the organizations mentioned above.

Other Sources. There are a number of reference books, handbooks, and textbooks that cover warning issues to varying degrees. Most books on safety, human factors, and ergonomics include requisite sections on warnings as a hazard control measure, and often discuss general issues about design and effectiveness.

A few books focus on warnings, albeit from a theoretical or research perspective, including Miller, Lehto, and Frantz (1994), Lehto and Miller (1986), and Edworthy and Adams (1996), and offer little in terms of specific design guidelines. Then there are examples of books designed to address the needs of specific users, such as the chemical industry (O'Connor and Lirtzman, 1984).

Provisions Within Selected Standards

According to ANSI Z535.2 (1996), a safety sign is "a visual alerting device in the form of a sign, label, decal, placard, or other marking which advises the observer of the nature and degree of the potential hazard(s) which can cause injury or death. ..." The key here is that a sign (or label) is intended to be a permanently mounted device. The difference is normally in terms of where the device is placed. If it is on the hazard-producing or -containing object, especially as manufactured or distributed, it is usually referred to as a label. On the other hand, if it is mounted not on the object but nearby or in the general environment, or refers to hazards not specifically "attached" to an object, it is usually referred to as a sign. Temporary devices (or tags) used in either place are also used in industry for similar purposes and are addressed by their own standard (i.e., Z535.5). There are a fairly large number of existing warning sign and label standard "systems," not all of which are compatible. These differ in the choice of words, symbols, colors, and layout (see Table 40.4). The following is based on a selection of the most important of these standard systems. The reader should note that, due to space limitations, not all of the provisions of these standards can be included here. Further, standards are time-sensitive entities. While every effort has been made to use the latest information, it is a certainty that parts of these standards will change over time. Consequently, one should always consider obtaining the latest version of the standard(s) in their full text.

ANSI Z535.2-1996 Environmental and Facility Safety Signs

ANSI Z535.2 establishes requirements for a uniform system of visual identification of potential hazards in the environment, such as industrial facilities. It specifically excludes applicability to product labels (see ANSI Z535.4) and chemicals (see ANSI Z129.1). As ANSI Z535.2 signs include the use of safety colors and symbols, it incorporates by reference the Safety Color Code (ANSI Z535.1) and Criteria for Safety Symbols (ANSI Z535.3). As noted in Table 40.5, there are seven types of safety signs in ANSI Z535.2: (1) danger, (2) warning, (3) caution, (4) notice, (5) general safety, (6) fire, and (7) directional. The selection of the appropriate type of sign for a given hazard and the design of its panel are also illustrated in Table 40.5, and further information on the colors used is shown in Table 40.6. A typical ANSI Z535.2 sign consists of multiple panels for signal words, messages, and symbols (see Table 40.7). Finally, the use of symbols is encouraged, provided that they are understood. ANSI Z535.3 describes methodology for the development of such symbols. In general, symbols must be correctly identifiable 85% of the time, with no more than 5% critical confusions (i.e., results in an action that is opposite of the desired response). ANSI Z535.3 provides 22 symbols that meet this criteria (see Table 40.8) along with 6 appropriate surround shapes (see Table 40.9).

OSHA 29 CFR 1910.145 Specification for Accident Prevention Signs and Tags, and 29 CFR 1910.144 Safety Color Code for Marking Physical Hazards

For all practical purposes, one can follow either the old ANSI Z35.1/Z53.1 standards or the new ANSI Z535 standards to meet the requirements of OSHA for hazard warning signs. Consequently, there is no need to replace existing compliant signs. Also, specific reference is made within the regulations to conforming with the ANSI Z53.1 color code. However, conformance with the new ANSI Z535.1 color code will result in the same design. Therefore, when designing or specifying new signs, follow the new ANSI standards.

ANSI Z535.5–1996 Accident Prevention Tags

These tags are intended to be used to identify temporary hazards. Therefore, they are to be used only until such time as the hazard is eliminated. They should never be used as a permanent substitute for

PRODUCT NAME OR IDENTIFICATION
(Identity of Hazardous Component(s), Where Appropriate)

SIGNAL WORD
IMMEDIATE HAZARD(S) STATEMENT
DELAYED HAZARD(S) LABEL STATEMENT

Precautionary Measures

Instructions in Case of Contact or Exposure
(First Aid Statements and Antidotes Where Appropriate)

Fire Instructions

Spill or Leak Instructions

Container Handling and Storage Instructions

References

Additional Useful Statements

Name and Address of Company

Telephone Number

FIGURE 40.4 ANSI Z129.1-1994 general example of label format.

ANSI Z535.2 signs. The format follows that of the two-panel designs of ANSI Z535.2 (see Table 40.7) with the lower panel reserved for word messages and/or symbols.

ANSI Z129.1-1994, Hazardous Industrial Chemicals – Precautionary Labeling

This standard is applicable to chemical products and mixtures and is complementary in scope to ANSI Z535.2 in the environmental or facility area, and to ANSI Z535.4 in the product area. It is directed toward manufacturers, distributors, and employers who wish to alert persons to hazards inherent with chemicals. The required contents of an ANSI Z129.1 label is shown in Table 40.10, with an example arrangement shown in Figure 40.4.

OSHA 29 CFR 1910.1200 Hazard Communication Standard

This section of the OSHA General Industry Standards, also known as the Right-to-Know Standard, requires employers to provide information to employees about the hazardous chemicals they are or may be exposed to during normal conditions, or during any foreseeable emergency. This is to be done through hazard communication including on-container labels and Material Safety Data Sheets (MSDS). The information necessary for such communication must be provided by the chemical manufacturers or importers through their distributors.

Labels — Manufacturers' Responsibility. Manufacturers' container labels (or tags, other markings) must include: (a) the identity of the chemical, (b) appropriate hazard warnings, and (c) the name and address of the manufacturer, importer, etc.

Labels — Employers' Responsibility. Employers must ensure that all labels on incoming containers of hazardous chemicals are maintained. Further, it is the employer's responsibility that all containers in the workplace are labeled, tagged, or marked with: (a) the identity of the chemical, and (b) appropriate hazard warnings. As an alternative, and provided that the same information is provided and is readily accessible to all employees, the employer may use signs, placards, or other written materials. Where material is being transferred from labeled containers to portable containers, and for immediate use by the same employee who performed the transfer, labeling is not required.

Labels — Format. The only specifications for label format and placement is that the labels must be: (a) legible, (b) in English, and (c) prominently displayed or readily available. Other standards with more

TABLE 40.4 Summary of Recommendations Within Selected Warning Systems

System	Signal Words	Color Coding	Typography	Symbols	Arrangement	Hazard ID
ANSI Z35.1 Specifications for Accident Prevention Signs	Danger Caution	Red Yellow	Sans serif typeface. All upper case or upper and lower case.	Symbols only as supplement to words.	Defines signal word, message, symbol panels (optional, attached to side of label).	Not specified
(replaced by ANSI Z535.2, but 1968 version is still basis for OSHA regulations)						
ANSI Z129.1 Precautionary Labeling of Hazardous Chemicals	Danger Warning Caution Notice* Attention* * optional words for "delayed" hazards	Not specified	Not specified	Exclamation point may follow signal word for emphasis. Skull-and-crossbones as supplement to words.	Label arrangement not specified; examples given.	Provides guidance about how to select signal words and precautionary text.
ANSI Z535.2 Environmental and Facility Safety Signs	Danger Warning Caution Notice (general safety) (arrows)	Red Orange Yellow Blue Green as above when combined; black & white otherwise per ANSI Z535.1	Sans serif, upper case only for signal word, acceptable typefaces, letter heights	Symbols and pictographs per ANSI Z535.3; safety alert symbol	Defines signal word, word message, symbol panels in 1-3 panel designs. 4 shapes for special use.	Provides guidance
(use ANSI Z129.1 for chemical hazards)						
ANSI Z535.4 Product Safety Signs and Labels	Danger Warning Caution	Red Orange Yellow per ANSI Z535.1	Sans serif, upper case only for signal word, acceptable typefaces, letter heights	Symbols and pictographs per ANSI Z535.3; safety alert symbol	Defines signal word, message, pictorial panels in order of general to specific. Can use ANSI Z535.2 for uniformity.	Provides guidance
(use ANSI Z129.1 for chemical hazards)						
ANSI Z535.5 Accident Prevention Tags (for Temporary Hazards)	Danger Warning Caution	Red Orange Yellow per ANSI Z535.1	Sans serif, upper case only for signal word, acceptable typefaces, letter heights	Symbols and pictographs (to replace or supplement word messages) per ANSI Z535.3; safety alert symbol	Defines signal word and message panels.	Provides guidance

NEMA Guidelines: NEMA 260	Danger Warning	Red Red	Not specified	Electric shock symbol	Defines signal word, hazard, consequences, instructions, symbol. Does not specify order.	Not specified
SAE J115 Safety Signs	Danger Warning Caution	Red Yellow Yellow	Sans serif typeface, upper case	Layout to accommodate symbols; specific symbols/pictographs not prescribed	Defines 3 areas: signal word panel, pictorial panel, message panel. Arrange in order of general to specific.	Provides guidance
ISO R557, 3864	None. 3 kinds of labels: Stop/prohibition Mandatory action Warning	 Red Blue Yellow	Message panel is added below if necessary	Symbols and pictographs	Pictograph or symbol is placed inside appropriate shape with message panel below if necessary	Not specified
OSHA 1910.145 Specification for Accident Prevention Signs and Tags OSHA 1926.200 Accident Prevention Signs and Tags	Danger Warning (tags) Caution Biological Hazard, BIOHAZARD, or symbol (safety instruction) (slow-moving Vehicle)	Red Yellow Yellow Fluorescent orange/orange-red Green Fluorescent yellow-orange & dark red per OSHA 1910.144 and ANSI Z53.1	Readable at 5 feet or as required by task	Biological hazard symbol. Major message can be supplied by pictograph (tags only). Slow-Moving Vehicle (SAE J943)	Signal word and major message (tags only)	Provides guidance
OSHA 1910.1200 Hazard Communication Standard			In English		Only as Material Safety Data Sheet	Provides guidance
	Per applicable requirements of EPA, FDA, BATF, and CPSC; not otherwise specified.					
Westinghouse Handbook; FMC Guidelines	Danger Warning Caution Notice	Red Orange Yellow Blue	Helvetica bold and regular weights, upper/lower case	Symbols and pictographs	Recommends 5 components: Signal word, Symbol/pictograph hazard, Result of ignoring warning, Avoiding hazard	Provides guidance about how to select signal words

TABLE 40.5 ANSI Z535.2-1996 Signal Word Panels

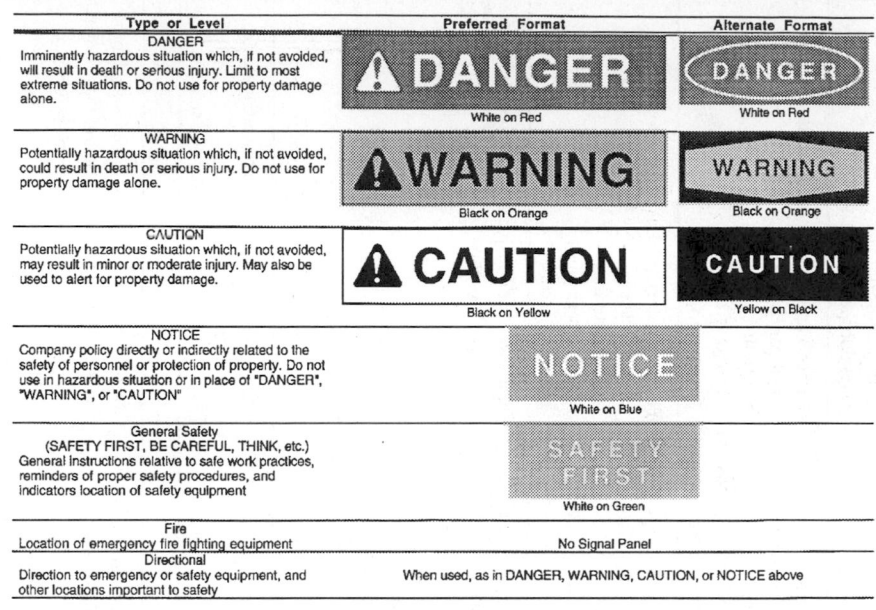

Type or Level	Preferred Format	Alternate Format
DANGER Imminently hazardous situation which, if not avoided, will result in death or serious injury. Limit to most extreme situations. Do not use for property damage alone.	DANGER White on Red	DANGER White on Red
WARNING Potentially hazardous situation which, if not avoided, could result in death or serious injury. Do not use for property damage alone.	WARNING Black on Orange	WARNING Black on Orange
CAUTION Potentially hazardous situation which, if not avoided, may result in minor or moderate injury. May also be used to alert for property damage.	CAUTION Black on Yellow	CAUTION Yellow on Black
NOTICE Company policy directly or indirectly related to the safety of personnel or protection of property. Do not use in hazardous situation or in place of "DANGER", "WARNING", or "CAUTION"	NOTICE White on Blue	
General Safety (SAFETY FIRST, BE CAREFUL, THINK, etc.) General instructions relative to safe work practices, reminders of proper safety procedures, and indicators location of safety equipment	SAFETY FIRST White on Green	
Fire Location of emergency fire fighting equipment	No Signal Panel	
Directional Direction to emergency or safety equipment, and other locations important to safety	When used, as in DANGER, WARNING, CAUTION, or NOTICE above	

TABLE 40.6 ANSI Z535.1-1996 Safety Colors, Meanings, and Examples of Use

Color	Meaning	Commonly Used Examples
Red	Identification of DANGER or STOP.	Flammable liquid containers Emergency machine-stop bars/buttons/switches Fire protection equipment
Orange	Identification of WARNING. Identification of hazardous parts of machines.	Machine parts that may cut, crush, or otherwise injure, and emphasizing such when guards are open Inside of moveable guards Exposed edges of power transmission devices
Yellow	Identification of CAUTION. May be striped or checkered with black for enhancing contrast with background.	Hazards of striking against, stumbling, falling, tripping, or being caught in-between Flammable material storage cabinets Corrosive or unstable material containers
Green	Identification of SAFETY, emergency egress, and location of first aid and safety equipment	Gas masks First aid kits or dispensary Stretchers Safety deluge showers Safety bulletin boards and signs Emergency egress routes
Blue	Identification of SAFETY INFORMATION used on informational signs and bulletin boards	Mandatory action signs for wearing of personal protective gear such as hard hats
Other	Unassigned	

specific design guidelines that can be applied include: (a) OSHA 29 CFR 145 *et seq*., (b) ANSI Z535.2, and (c) ANSI Z129.1, all of which are covered elsewhere herein.

Material Safety Data Sheets — Manufacturers' Responsibility. Chemical manufacturers and importers must obtain or develop an MSDS for every hazardous chemical produced or imported by them.

TABLE 40.7 ANSI Z535.2-1996 Multi-Panel Sign Layouts by Type Classification

Type or Level	Layouts Permitted
All except Fire Safety and Directional	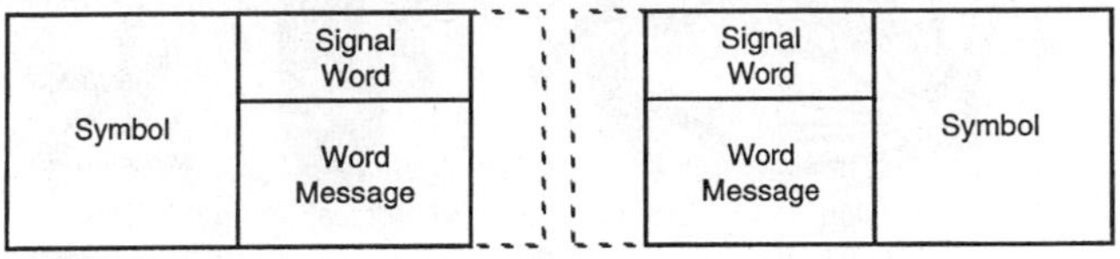
All except Fire Safety	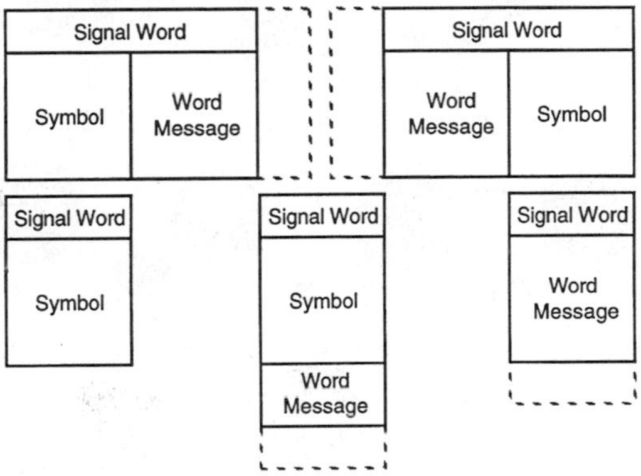
Only General Safety and Fire Safety	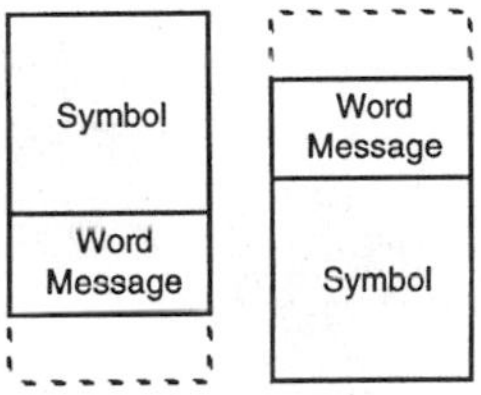

Material Safety Data Sheets — Employers' Responsibility. Employers must obtain, maintain, and make readily accessible MSDSs for every hazardous chemical they use.

Material Safety Data Sheets — Format. While there is no required format for OSHA MSDSs, they must be in English and contain the information outlined in Table 40.11.

The physical layout of an MSDS is not specified by OSHA. However, two guidelines exist. The first is a nonmandatory form by OSHA itself, shown in Figure 40.5. The second is a standard proposed by the Chemical Manufacturers Association and ANSI, Z400.1. An example from this draft standard is shown in Figure 40.6.

ANSI Z535.4 Product Safety Signs and Labels

Part of the ANSI Z535 series, this standard addresses product signs and labels, with their inherent characteristics of generally being smaller and observed at a closer distance than environmental or facility signs or labels. The other primary differences between Z535.2 and Z535.4 signs and labels are that there are fewer types for products, with fewer signal words and panel layouts (see Table 40.12 and Figure 40.7).

DOT 49 CFR 170 Hazardous Materials Transportation Regulations

The probably familiar "diamond" signs on the outside of trucks, especially tankers, is due to U.S. Department of Transportation regulations. The primary purpose is to alert emergency response personnel to potential hazards present within and in the vicinity of the vehicle, especially when there has been an

TABLE 40.8 ANSI Z535.3-1996 Symbol Examples

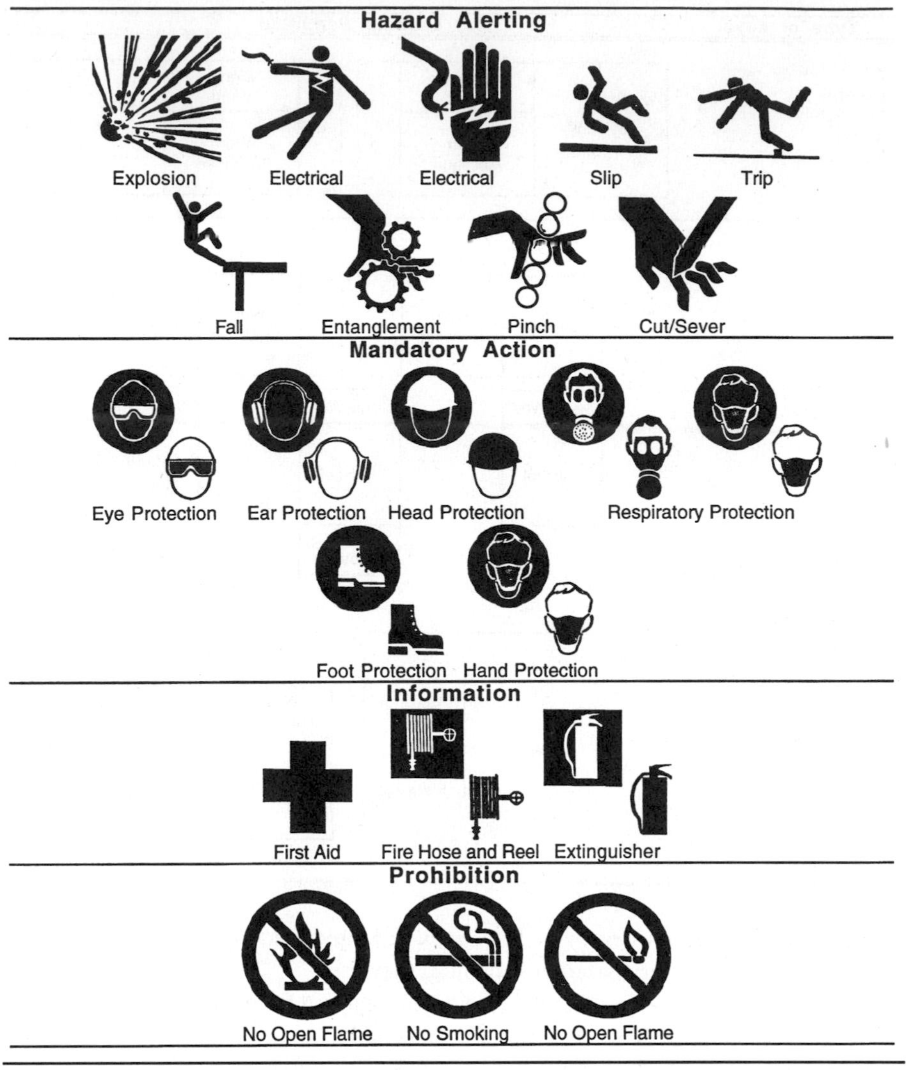

accident. These signs are comprised of nine classes of hazards, with corresponding colors, symbols, and text, with one version using standardized four-digit code numbers to identify the hazard (see Figure 40.8).

NFPA 704 Standard System for the Identification of Fire Hazards

The National Fire Protection Association developed the NFPA 704 system to aid fire emergency response personnel in the identification of hazards and the appropriate procedural response. It addresses fire, health, reactivity levels, and specific hazard concerns (see Figure 40.9). These signs or labels may be commonly found both in the environment and on specific product containers.

Hazardous Material Identification System

The HMIS® was developed by the National Paint and Coatings Association. It is similar to another system, the ASTM Safety Alert System (ASTM D 4257, 1984). In both systems (and like NFPA 704), coded indices indicating health, flammability, and reactivity are given. In addition, they both indicate the types of personal protective equipment that should be used. Figure 40.10 illustrates the sign or label format and the meanings of the four indices.

TABLE 40.9 ANSI Z535.3-1996 Safety Symbol Surround Shapes and Use

Symbol Type and Use	Shape
Hazard Alerting Usually not recommended. Reduces symbol size on on Z535.2, .4, and .5 formats. Preferred colors: Black on white. Red on white for some symbols. White may be replaced by orange or yellow if used with Z535.2	
Mandatory Action Generally used on Z535.2 Notice signs, although optional. Colors: White symbol on solid blue/black circular surround, in turn on white symbol panel background. Reverse when no surround.	
Information Generally used on Z535.2 General Safety or Fire Safety signs. Colors: Green or black image on white background. For fire-related symbols, red image on white background, or reverse.	
Prohibition Use is mandatory. Colors: Black symbol on white background with red or black circle and slash. White border on slash if both symbol and slash are black.	

40.6 Summary

This chapter has endeavored to give the reader some background on the fundamentals of warnings research and practice. The conclusions reached give guidance on how to construct warnings that will hopefully have a better chance at achieving the goals of eliciting appropriate behavioral response for accident avoidance. This was followed by a review of the major warning sign and label standards and their provisions that are applicable or likely to be found in common industrial environments.

The reader is cautioned that both knowledge from warnings research and mandatory requirements and voluntary guidelines are constantly evolving, resulting in changing criteria for the design and assessment of warning signs and labels. Since most standards are revised every five years or less, one should always try to obtain the latest versions of the references given for the standards in this chapter.

TABLE 40.10 ANSI Z129.1-1994 Precautionary Label Content

Category	Criteria or Guidelines
Identification of chemical product or its hazardous components	Adequate to allow proper action in case of exposure. Chemical name(s) of substances contributing substantially to hazards (if mixture proprietary, informative not required on label, but procedure to obtain such information in an emergency situtation must be provided). Nondescriptive code or trade name only not permitted
Signal word	Based on greatest immediate hazard. DANGER, WARNING, or CAUTION for immediate hazards. NOTICE or ATTENTION for delayed hazards. May be followed by exclamation mark (!) for emphasis.
Immediate hazards	Reasonably foreseeable immediate physical and health hazards associated with reasonably foreseeable handling, use, or misuse. Order by seriousness of hazard. Standard provides guidelines for selection of statement of hazard., 6.2, 7.1, Table 1
Delayed hazards	Reasonably foreseeable delayed physical and health hazards associated with reasonably foreseeable handling, use, or misuse. Order by seriousness of hazard. No need to repeat hazards included as immediate hazards., 6.2, 6.3, 7.1, Table 2
Precautionary measures	Brief supplement to hazard statement(s). Measures to avoid injury., 7.2, Table 1
Contact or exposure instructions (first aid or antidotes)	Section captioned "FIRST AID": Use when immediate treatment warranted and simple remedial measures may be taken; limit to procedures that do not require special training. Section captioned "ANTIDOTE": Use when specific antidoes known and administration does not required special training. Section captioned "NOTES TO PHYSICIANS": Recommended medical practices or antidotes to be administered by a physician., 7.2, Table 1
Fire instructions	Intended for persons who handle containers during shipment and storage. Instructions for confining and extinguishing fire. Simple and brief as possible. Advise suitable control material. When appropriate and personnel and property not at risk, may advise to allow fire to burn out rather than risk contamination., 6.7, 7.1, 7.2, Table 3
Spill or leak instructions	Methods to use, in the absence of fire, to contain spills or leaks to minimize exposures and prevent personal injury and environmental contamination., 6.8, 7.1, 7.2, Table 4
Container handling and storage instructions	Special or unusual handling and storage procedures. Where flash point of flammable or combustible liquids determines storage requirements, include this information as pertinent to storage code (ANSI/NFPA 30)., 6.9, 7.1, 7.2
References	Reference to MSDS, if available.
Additional useful statements	5.2, 6.9, 7.1, 7.2
Name and address of company	Manufacturer, importer, or distributor. Actual corporate or business name. Include street address, city, state, and Zip code. Country and mail code if foreign.
Telephone number	For additional information on product. Indicate hours or type of information restrictions.

TABLE 40.11 Information Required on Material Safety Data Sheet (MSDS) by OSHA (29 CFR 1910.1200(g)(2))

Identity of chemical as used on the container label
Physical and chemical charteristics
Physcial hazards, such as fire, explosion, or reactivity
Health hazards, signs and symptoms, and medical conditions that may be aggravated by exposure
Primary routes of entry
OSHA permissible exposure limit (PEL), ACGIH Threshold Limit Value (TLV), and common or recommended exposure limits
Potential as carcinogen
Precautions for safe handling and use
Control measures
Emergency and first aid procedures
Date of preparation or last revision to MSDS
Name, address, and telephone number of party responsible for MSDS who can provide additional information

Material Safety Data Sheet May be used to comply with OSHA's Hazard Communication Standard, 29 CFR 1910.1200. Standard must be consulted for specific requirements.	**U.S. Department of Labor** Occupational Safety and Health Administration (Non-Mandatory Form) Form Approved OMB No. 1218-0072
IDENTITY *(As Used on Label and List)*	*Note:* *Blank spaces are not permitted. If any item is not applicable, or no information is available, the space must be marked to indicate that.*

Section I

Manufacturer's Name	Emergency Telephone Number
Address *(Number, Street, City, State, and ZIP Code)*	Telephone Number for Information
	Date Prepared
	Signature of Preparer *(optional)*

Section II – Hazardous Ingredients/Identity Information

Hazardous Components (Specific Chemical Identity: Common Name(s))	OSHA PEL	ACGIH TLV	Other Limits	% *(optional)*

Section III – Physical/Chemical Characteristics

Boiling Point		Specific Gravity (H_2O = 1)	
Vapor Pressure (mm Hg)		Melting Point	
Vapor Density (AIR = 1)		Evaporation Rate (Butyl Acetate = 1)	
Solubility in Water			
Appearance and Odor			

Section IV – Fire and Explosion Hazard Data

Flash Point (Method Used)	Flammable Limits	LEL	UEL
Extinguishing Media			
Special Fire Fighting Procedures			
Unusual Fire and Explosion Hazards			

FIGURE 40.5a & 5b OSHA nonmandatory Material Safety Data Sheet (MSDS) format.

Finally, bear in mind that, in a court of law, compliance with standards is generally not a sufficient sole defense because standards are often viewed as minimum requirements (often due to a "one size fits all" strategy in their development). Therefore, it is necessary to fully understand the potential hazards of any system being analyzed (i.e., environment, facility, product, etc.), apply the hierarchy of hazard control to achieve the most effective protection for those who use or are exposed to the hazards, and apply warnings as a sole remedy sparingly and in a rigorous manner.

Section V – Reactivity Data

Stability	Unstable		Conditions to Avoid
	Stable		

Incompatibility *(Materials to Avoid)*

Hazardous Decomposition or Byproducts

Hazardous Polymerization	May Occur		Conditions to Avoid
	Will Not Occur		

Section IV – Health Hazard Data

Route(s) of Entry	Inhalation?	Skin?	Ingestion?

Health Hazards *(Acute and Chronic)*

Carcinogenicity	NTP?	IARC Monographs?	OSHA Regulated?

Signs and Symptoms of Exposure

Medical Conditions
Generally Aggravated by Exposure

Emergency and First Aid Procedures

Section VII – Precautions for Safe Handling and Use

Steps to Be Taken in Case Material is Released or Spilled

Waste Disposal Method

Precautions to Be Taken in Handling and Storing

Other Precautions

Section VIII – Control Measures

Respiratory Protection *(Specify Type)*

Ventilation	Local Exhaust	Special
	Mechanical *(General)*	Other

Protective Gloves	Eye Protection

Other Protective Clothing or Equipment

Work/Hygienic Practices

FIGURE 40.5a & 5b (continued)

1. **CHEMICAL PRODUCT AND COMPANY INFORMATION**
Company name and address — **PRODUCT NAME:**
Emergency Phone — **PRODUCT CODE:**
Effective Date: — Print Date:

2. **COMPOSITION/INFORMATION ON INGREDIENTS**

Chemical Ingredients (% by wt.)		
Component A	CAS#	xx–xx%
Component B	CAS#	xx–xx%
Impurity C	CAS#	xx ppm max

3. **HAZARDS IDENTIFICATION**

EMERGENCY OVERVIEW

[short description of characteristics and hazards]

POTENTIAL HEALTH EFFECTS

EYE:
SKIN CONTACT:
SKIN ABSORPTION:
INGESTION:
INHALATION

CHRONIC EFFECTS/CARCINOGENICITY:

4. **FIRST AID MEASURES**

EYE:
SKIN:
INGESTION:
INHALATION

5. **FIRE FIGHTING MEASURES**

FLAMMABLE PROPERTIES
FLASH POINT:
METHOD USED:
FLAMMABLE LIMITS
LFL:
UFL:
EXTINGUISHING MEDIA:
FIRE & EXPLOSION HAZARDS:
FIRE-FIGHTING EQUIPMENT:

6. **ACCIDENTAL RELEASE MEASURES**

7. **HANDLING AND STORAGE**

8. **EXPOSURE CONTROLS/PERSONAL PROTECTION**

RESPIRATORY PROTECTION:
SKIN PROTECTION:
EYE PROTECTION:
EXPOSURE GUIDELINE(S):
ENGINEERING CONTROLS:

9. **PHYSICAL AND CHEMICAL PROPERTIES**

APPEARANCE:
ODOR:
BOILING POINT:
VAPOR PRESSURE:
VAPOR DENSITY:
SOLUBILITY IN WATER:
SPECIFIC GRAVITY:
FREEZING POINT:
pH:
VOLATILE:

10. **STABILITY AND REACTIVITY**

STABILITY: (CONDITIONS TO AVOID)
INCOMPATIBILITY: (SPECIFIC MATERIALS TO AVOID)
HAZARDOUS DECOMPOSITION PRODUCTS:
HAZARDOUS POLYMERIZATION:

11. **TOXICOLOGICAL INFORMATION**

12. **ECOLOGICAL INFORMATION**

13. **DISPOSAL CONSIDERATIONS**

14. **TRANSPORT INFORMATION**

TRANSPORTATION AND HAZARDOUS MATERIALS DESCRIPTION:

15. **REGULATORY INFORMATION**

OSHA HAZARD COMMUNICATION RULE, 29 CFR 1910.1200:
CERCLA/SUPERFUND, 40 CFR 117, 302:
SARA HAZARD CATEGORY:
SARA 313 INFORMATION:
TOXIC SUBSTANCES CONTROL ACT (TSCA):
CALIFORNIA PROPOSITION 65:

16. **OTHER INFORMATION**

MSDS STATUS:

FIGURE 40.6 Example of Material Safety Data Sheet (MSDS) format per proposed ANSI/CMA Z400.1.

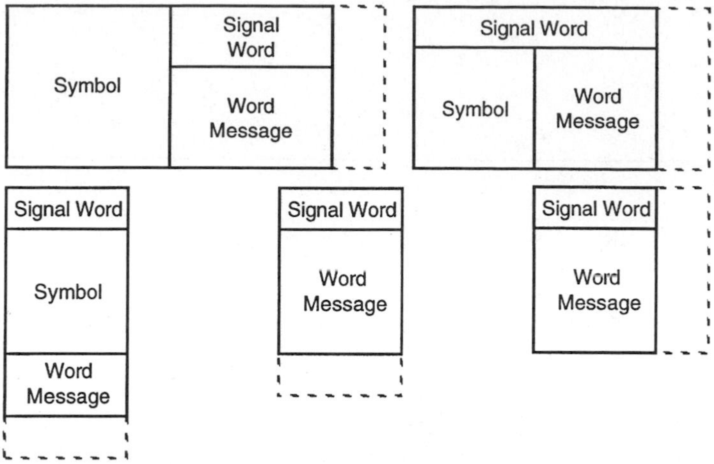

FIGURE 40.7 ANSI Z535.4-1996 multi-panel sign layouts.

TABLE 40.12 ANSI Z535.4-1996 Signal Word Panels

Type or Level	Format
DANGER Imminently hazardous situation which, if not avoided, will result in death or serious injury. Limit to most extreme situations. Do not use for property damage alone.	DANGER White on Red
WARNING Potentially hazardous situation which, if not avoided, could result in death or serious injury. Do not use for property damage alone.	WARNING Black on Orange
CAUTION Potentially hazardous situation which, if not avoided, may result in minor or moderate injury. May also be used to alert against unsafe practices.	CAUTION Black on Yellow
CAUTION without Safety Alert Symbol Propery-damage-only accidents.	CAUTION Black on Yellow

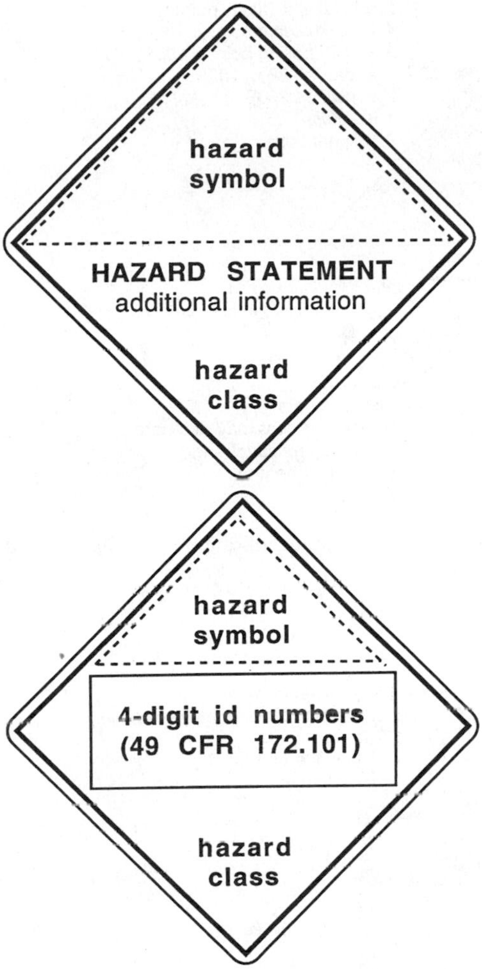

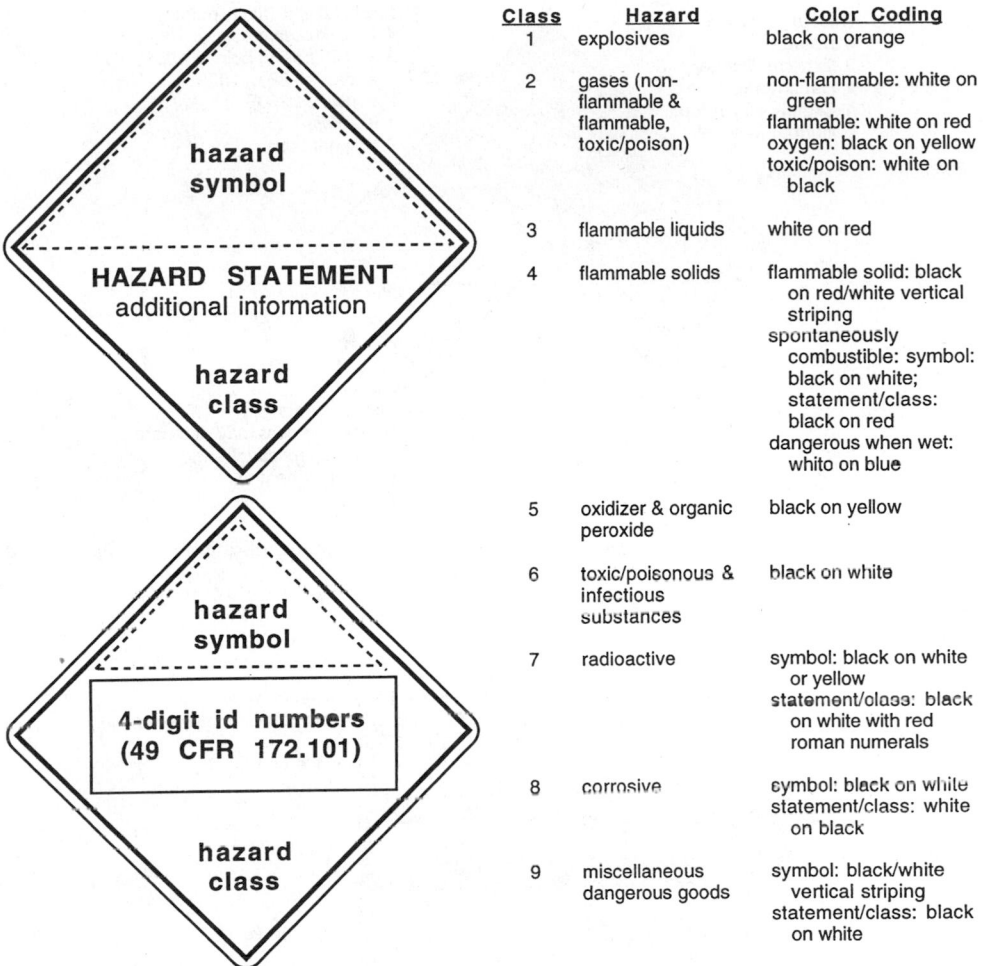

Class	Hazard	Color Coding
1	explosives	black on orange
2	gases (non-flammable & flammable, toxic/poison)	non-flammable: white on green flammable: white on red oxygen: black on yellow toxic/poison: white on black
3	flammable liquids	white on red
4	flammable solids	flammable solid: black on red/white vertical striping spontaneously combustible: symbol: black on white; statement/class: black on red dangerous when wet: whito on blue
5	oxidizer & organic peroxide	black on yellow
6	toxic/poisonous & infectious substances	black on white
7	radioactive	symbol: black on white or yellow statement/class: black on white with red roman numerals
8	corrosive	symbol: black on white statement/class: white on black
9	miscellaneous dangerous goods	symbol: black/white vertical striping statement/class: black on white

FIGURE 40.8 U.S. Department of Transportation (49 CFR 172) Hazard Class label formats.

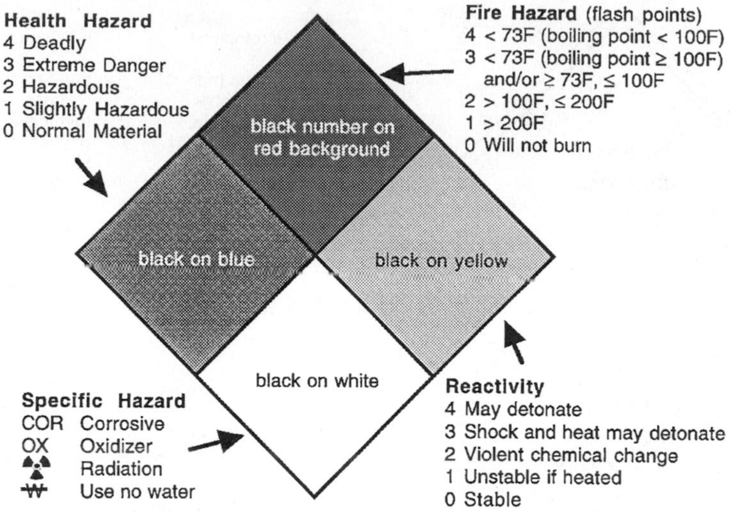

FIGURE 40.9 Label format from NFPA 704-1990, Standard System for the Identification of the Fire Hazards of Materials.

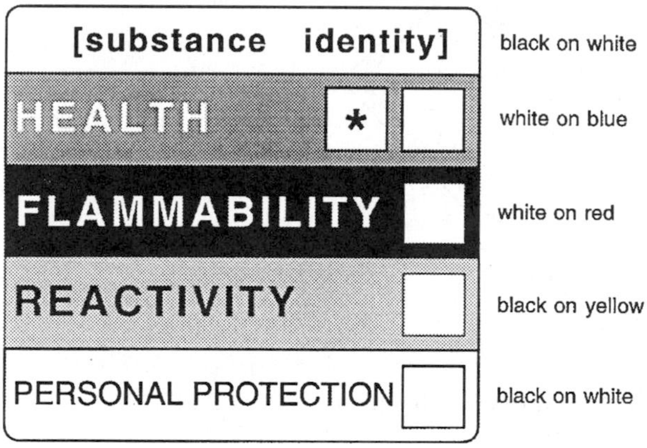

Index	Health Hazard*	Flammability or Reactivity
0	No significant	Minimal
1	Slight	Slight
2	Moderate	Moderate
3	High	Serious
4	Extreme	Severe

* Asterisk indicates a chronic/target organ hazard

Personal Protection	Index*										
	A	B	C	D	E	F	G	H	I	J	K
Safety glasses	x	x	x		x	x	x		x		
Gloves		x	x	x	x	x	x	x	x	x	x
Splash goggles								x		x	
Face shield				x							
Airline hood or mask											x
Dust respirator					x	x					
Vapor respirator											
Combination dust & vapor respirator							x	x	x	x	
Synthetic apron			x	x		x		x		x	
Boots											x
Full protective suit											x

* Index **X** = Ask your supervisor for specialized handling directivers

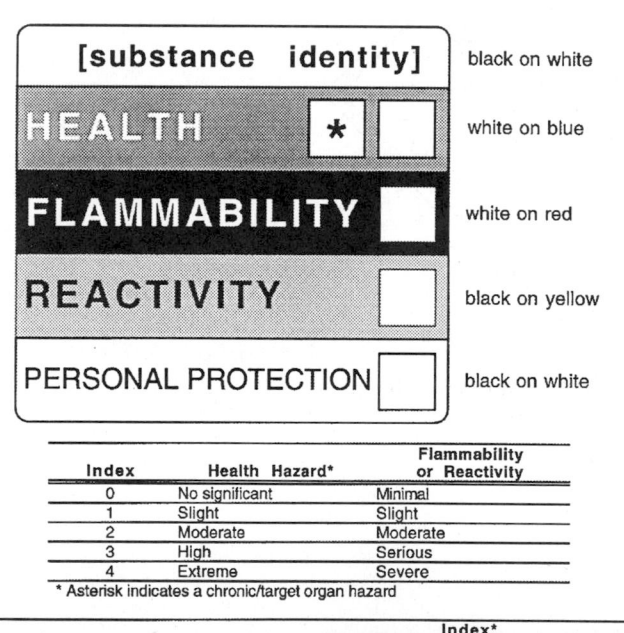

FIGURE 40.10 Hazardous Materials Identification System (HMIS®) (From National Paint and Coatings Association. *Hazardous Materials Identification System.* With permission.)

References

American National Standards Institute/Chemical Manufacturers Association. 1994. *American National Standard for Hazardous Industrial Chemicals – Precautionary Labeling*, ANSI Z129.1.

American National Standards Institute/Chemical Manufacturers Association. 1995, *American National Standard for Material Safety Data Sheets*, ANSI Z400.1 (draft version).

American National Standards Institute/National Electrical Manufacturers Association. 1996. *American National Standard for Accident Prevention Tags (for Temporary Hazards)*, ANSI Z535.5 (draft revision).

American National Standards Institute/National Electrical Manufacturers Association. 1996. *American National Standard for Criteria for Safety Symbols*, ANSI Z535.3 (draft revision).

American National Standards Institute/National Electrical Manufacturers Association. 1996. *American National Standard for Environmental and Facility Safety Signs*, ANSI Z535.2 (draft revision).

American National Standards Institute/National Electrical Manufacturers Association. 1996. *American National Standard for Product Safety Signs and Labels*, ANSI Z535.4 (draft revision).

American National Standards Institute/National Electrical Manufacturers Association. 1996. *American National Standard Safety Color Code*, ANSI Z535.1 (draft revision).

American National Standards Institute/National Bureau of Standards. 1979. *American National Standard Safety Color Code for Marking Physical Hazards*, ANSI Z53.1.

American National Standards Institute/National Electrical Manufacturers Association/National Safety Council. 1972. *American National Standard Specifications for Accident Prevention Signs*, ANSI Z35.1.

Belkin, N.J., Brooks, H.M., and Daniels, P.J. 1987. Knowledge elicitation using discourse analysis. *International Journal of Man-Machine Studies*. 27:127-144.

Boersema, T. and Zwaga, H. 1985. The influence of advertisements on the conspicuity of routing information. *Applied Ergonomics*. 16(4):267-273.

Bresnahan, T.F. and Bryk, J. 1975. The hazard association values of accident-prevention signs. *Professional Safety*. 20(1):17-25.

Cahill, M.C. 1976. Design features of graphic symbols varying in interpretability. *Perceptual and Motor Skills*. 42(2):647-653.

Cairney, P.T. and Sless, D. 1982. Communication effectiveness of symbolic safety signs with different user groups. *Applied Ergonomics*. 13(2):91-97.

Campion, M.A. and Medsker, G.J. 1992. Job design, in *Handbook of Industrial Engineering*, 2nd edition, ed. G. Salvendy, p. 845-881. John Wiley & Sons, New York, NY.

Conoley, J.C. and Kramer, J.J., (Eds.) 1989. *The Mental Measurements Yearbook*. Buros Institute of Mental Measurements of the University of Nebraska, Lincoln.

Craig, C.S. and McCann, J.M. 1978. Assessing communication effects on energy conservation. *Journal of Consumer Research*. 5:82-88.

Dorris, A.L. and Purswell, J.P. 1977. Human factors in the design of effective product warnings. *Proceedings of the Human Factors Society*. 22:343-346.

Easterby, R.S. and Hakiel, S.R. 1981. Field testing of consumer safety signs: The comprehension of pictorially presented messages. *Applied Ergonomics*. 12(3):143-152.

Easterby, R.S. and Zwaga, H.J.G. 1976. Evaluation of public information symbols ISO tests: 1975 Series, A.P.Report No. 60. University of Aston in Birmingham, UK.

Edworthy, J. and Adams, A. 1996. *Warning Design: A Research Perspective*. Taylor & Francis, London, UK.

FMC Corporation. 1980. *Product Safety Sign and Label System*, 2nd ed. Santa Clara, CA.

Frantz, J.P. and Rhoades, T.P. 1993. A task analytic approach to the temporal and spacial placement of product warnings. *Human Factors*. 35:719-730.

Gibson, J.J. 1979. *The Ecological Approach to Visual Perception*. Houghton-Mifflin, Boston, MA.

Godfrey, S.S., Rothstein, P.R., and Laughery, K.R. 1985. Warnings: Do they make a difference. *Proceedings of the Human Factors Society*, 29th Annual Meeting. 669-673.

Hadden, S.G. 1986. *Read the Label*. Westview Press, Boulder, CO.

Heglin, H. 1973, July. NAVSHIPS display illumination design guide, Volume 2: Human Factors (NELC-TD223), Naval Electronics Laboratory Center, San Diego, CA.

Holahan, C.J. 1977. Relationship Between Roadside Signs and Traffic Accidents: A Field Investigation, Research Report 54. Council for Advanced Transportation Studies, Austin, TX.

Illuminating Engineering Society. 1981. *IES Lighting Handbook: Application Volume.* New York, NY.

Jacobs, R.J., Johnston, A.W., and Cole, B L. 1975. The visibility of alphabetic and symbolic traffic signs. *Australian Road Research.* 5(7):68-86.

Johnson, D.A. 1980. The design of effective safety information displays. *Proceedings of the Symposium: Human Factors and Industrial Design in Consumer Products.* 314-328.

Kuipers, B. and Kassirer, J.P. 1983. How to discover a knowledge representation for causal reasoning by studying an expert physician. *IJCAI-83: Proceedings of the 8th International Conference on Artificial Intelligence.*

Laughery, K.R., Rowe-Hallbert, A.L., Young, S.L., Vaubel, K.P., and Laux, L.F. 1991. Effects of explicitness in conveying severity information in product warnings. *Proceedings of the Human Factors Society,* 35th Annual Meeting. 481-485.

Lehto, M.R., and Miller, J.M. 1986. *Warnings: Volume I: Fundamentals, Design, and Evaluation Methodologies.* Fuller Technical Publications, Ann Arbor, MI.

Lehto, M.R. and Papastavrou, J.D. 1993. Models of the warning process: Important implications towards effectiveness. *Safety Science.* 16:569-595.

Lehto, M.R. and Salvendy, G. 1995. Warnings: A supplement not a substitute for other approaches to safety. *Ergonomics.* 38 (11):2155-2163.

Lehto, M.R. 1991. A proposed conceptual model of human behavior and its implications for the design of product warnings. *Perceptual and Motor Skills.* 73:595-611.

Lehto, M.R. 1992. Designing warning signs and warning labels: Scientific basis for initial guidelines. *International Journal of Industrial Ergonomics.* 10:115-138.

Leonard, S.D. and Matthews, D. 1986. How does the population interpret warning signals? *Proceedings of the Human Factors Society,* 30th Annual Meeting. 116-120.

Lerner, N.D. and Collins, B.L. 1980. The assessment of safety symbol understandability by different testing methods, PB81-185647. National Bureau of Standards, Washington, D.C.

Lirtzman, S.I. 1984. Labels, perception, and psychometrics, in *Handbook of Chemical Industry Labeling,* Ed. C.J. O'Connor and S.I. Lirtzman. Noyes Publications, Park Ridge, NJ.

Matthews, M. 1987. The influence of color on CRT reading performance and subjective comfort under operational conditions. *Applied Ergonomics.* 18:259-271.

Miller, J.M., Lehto, M.R., and Frantz, J.P. 1994. *Warnings & Safety Instructions.* Fuller Technical, Ann Arbor, MI.

Morris, L.A. and Kanouse, D.E. 1981. Consumer reaction to the tone of written drug information. *American Journal of Hospital Pharmacy.* 38(5): 667-671.

National Electronic Manufacturers Association. 1982. *Safety Labels for Padmounted Switchgear and Transformers Sited in Public Areas,* NEMA 260.

National Fire Protection Association. 1990. *Standard System for the Identification of the Fire Hazards of Materials,* NFPA 704.

National Paint and Coatings Association. *Hazardous Materials Identification System.*

Norman, D.A. 1988. *The Psychology of Everyday Things.* Basic Books, NY.

O'Connor, C.J. and Lirtzman, S.I. 1984. *Handbook of Chemical Industry Labeling.* Noyes Publications, Park Ridge, NJ.

Otsubo, S.M. 1988. A behavioral study of warning labels for consumer products: Perceived danger and use of pictographs. *Proceedings of the Human Factors Society,* 30th Annual meeting. 1202-1205.

Peterson, D. 1975. *Safety Management — A Human Approach.* Aloray, Deer Park, NY.

Peterson, D. 1984. *Analyzing Safety Performance.* Aloray, Deer Park, NY.

Phillips, J.J. 1983. *Handbook of Training Evaluation and Measurement Methods.* Gulf Publishing Company, Houston.

Poulton, E. 1967. Searching for newspaper headlines printed in capitals or lower-case letters. *Journal of Applied Psychology.* 51:417-425.

Rasmussen, J. 1986. *Information Processing and Human-Machine Interaction.* North-Holland.

Reason, J. 1990. *Human Error.* Cambridge University Press, Cambridge, UK.

Reder, L.M. and Anderson, J.R. 1982. Effects of spacing and embellishment on memory for the main points of a text. *Memory and Cognition.* 10(2):97-102.

Rhoades, T.P., Frantz, J.P., and Miller, J.M. 1991. Emerging strategies for the assessment of safety related product communications. *Proceedings of the Human Factors Society,* 35th Annual Meeting, San Francisco, CA. 998-1002.

Rouse, W.B. 1992. Human-centered product planning and design, in *Handbook of Industrial Engineering,* 2nd edition, Ed. G. Salvendy. John Wiley & Sons, New York, NY. 1220-1240.

Sanders, M.S., and McCormick, E.J. 1993. *Human Factors in Engineering and Design,* 7th edition. McGraw-Hill, New York, NY.

Scammon, D.L. 1977. Information load And consumers. *The Journal of Consumer Research.* 4:148-155.

Shadbolt, N. and Burton, M. 1990. Knowledge elicitation, in *Evaluation of Human Work,* Ed. J.R. Wilson and E.N. Corlett, Taylor & Francis, London, UK.

Sinaiko, H.W. 1975. Verbal factors in human engineering: some cultural and psychological data, in *Verbal Factors in Human Engineering.* The Smithsonian Institution, Washington, D.C. 159-177.

Strawbridge, J.A. 1986. The influence of position, highlighting, and imbedding on warning effectiveness. *Proceedings of the Human Factors Society,* 30th Annual Meeting. Human Factors Society. Santa Monica, CA. 716-720.

Tullis, T. 1988. Screen design, in *Handbook of Human-Computer Interaction,* Ed. M. Helander. Elsevier Science, Amsterdam. 377-411.

U.S. Consumer Product Safety Commission. Hazardous Substances and Articles, 16 CFR 1500.

U.S. Department of Labor, Occupational Health and Safety Administration. Accident Prevention Signs and Tags, 29 CFR 1926.200, in *Construction Standards,* 29 CFR 1926.

U.S. Department of Labor, Occupational Health and Safety Administration. *Hazard Communication Standard,* 29 CFR 1910.1200, in *General Industry Standards,* 29 CFR 1910.

U.S. Department of Labor, Occupational Health and Safety Administration. *Safety Color Code for Marking Physical Hazards,* 29 CFR 1910.144, in *General Industry Standards,* 29 CFR 1910.

U.S. Department of Labor, Occupational Health and Safety Administration. *Specification for Accident Prevention Signs and Tags,* 29 CFR 1910.145, in *General Industry Standards,* 29 CFR 1910.

U.S. Department of Transportation. *Hazardous Materials Transportation Regulations,* 49 CFR 170 *et seq.*

Ursic, M. 1984. The impact of safety warnings on perception and memory, *Human Factors.* 16(6):677-682.

Webster's Ninth New Collegiate Dictionary. 1988. Merriam-Webster, Springfield, MA.

Westinghouse Electric Corporation. 1981. *Product Safety Label Handbook.* Westinghouse Printing Division. Trafford, PA.

Wilson, J.R. and Corlett, E.N. 1990. *Evaluation of Human Work.* Taylor & Francis, London, UK.

Wogalter, M.S., Allison, S.T., and McKenna, N.A. 1989. Effects of cost and social influence on warning compliance. *Human Factors.* 31:133-140.

Wogalter, M.S., Godfrey, S.S., Fontenelle, G.A., Desaulniers, D.R., Rothstein, P.R., and Laughery, K.R. 1987. Effectiveness of warnings. *Human Factors.* 29(5):599-622.

Woodson, W. 1963. Human engineering design standards for spacecraft controls and displays (General Dynamics Aeronautics Report GDS-63-0894-1). National Aeronautics and Space Administration, Orlando, FL.

Wright, P. 1979. Concrete actions plans in TV messages to increase reading of drug warnings. *Journal of Consumer Research.* 6:256-259.

Young, S.L. and Wogalter, M.S. 1988. Memory of instruction manual warnings: Effects of pictorial icons and conspicuous print. *Proceedings of the Human Factors Society,* 32nd Annual Meeting. Human Factors Society, Santa Monica, CA. 905-909.

41

Ergonomics Methods in the Design of Consumer Products

Neville A. Stanton
University of Southampton

Mark S. Young
University of Southampton

41.1 Introduction

There is an immense variety in the range of methods that can be applied to the design of consumer products. The methods differ in what they address (i.e., the human element, the device element, or the interaction) and what they produce (e.g., task descriptions, predicted errors, and performance times). Ergonomics has a practical role to play in device design, on the basis that use of the methods should improve design by: reducing device interaction time, reducing user errors, improving user satisfaction, and improving device usability. Although there are many different design processes, most can be reduced to six main phases:

Concept: in which the idea for the device is considered in a largely informal manner, many implementations are considered, and many degrees of freedom remain.

Flowsheeting: in which the ideas for the device become formalized and the alternatives considered become very limited.

Design: in which the design solution becomes crystallized and blueprints are devised.

Prototyping: in which soft- and hard-built prototype devices are developed for evaluation.

Commissioning: in which the final design solution is implemented and the product enters the marketplace.

Operation and maintenance: when the device is supported in the marketplace.

0-8493-2641-9/99/$0.00+$.50

We believe that ergonomics methods may have the greatest impact at the prototyping stage, particularly *analytic prototyping* (i.e., when the device exists as a paper-based or computer-based model). Although in the past, it may have been costly to alter design at structural prototyping, and perhaps even impossible, with the advent of computer-aided design such retooling is made much simpler. It may even be possible to compare alternative designs at this stage with such technology. These ideas have yet to be proved in practice; however, given the nature of most ergonomics methods, it would seem most sensible to apply the methods at the analytic prototyping stage. We have selected 12 methods for consideration, based upon our analysis that these are a representative spread of methods that are currently being used to evaluate human–machine performance and assess the demands and effects upon people (Diaper, 1989; Kirwan and Ainsworth, 1992; Kirwan, 1994; Corlett and Clarke, 1995; Wilson and Corlett, 1995; Jordan et al., 1996). They were also chosen because of the appropriateness to the assessment of consumer products and user activity. Methods selected were as follows:

- Heuristics
- Checklists/Guidelines
- Observation
- Interviews
- Questionnaires
- Link analysis
- Layout analysis
- Hierarchical task analysis
- Systematic human error reduction and prediction approach
- Task analysis for error identification
- Repertory grids
- Keystroke level model

In terms of the analytic prototyping of human interfaces, we feel that there are three main forms: functional analysis (i.e., consideration of the range of functions the device supports), scenario analysis (i.e., consideration of the device with regard to a particular sequence of activities), and structural analysis (i.e., nondestructive testing of the interface from a user-centered perspective). We have classified the methods in this chapter into each of these types, as follows:

Functional Analysis	Scenario Analysis	Structural Analysis
Interviews	Link analysis	KLM
Questionnaires	Layout analysis	PHEA
Checklists	HTA	TAFEI
Repertory grids	Heuristics	Observation

We hope this chapter will help to highlight the different contributions each of the methods makes to analytic prototyping.

41.2 Radio-Cassette Players

Given the number of methods to be covered, emphasis will be given to providing an example of each approach with some accompanying text and reference to source material on the approach. The review is based upon the application of the methods to the evaluation of two radio-cassette machines taken from a car, as shown in Figures 41.1 and 41.2.

Although these analyses were undertaken as part of a research project concerned with operation of in-car devices, we believe that the methods would be equally appropriate to all kinds of consumer products, e.g., lawn mowers, hairdryers, kettles, videocassette recorders, cookers, washing machines, freezers, power tools, and vacuum cleaners. A fuller evaluation of devices may be found in Stanton (1998).

41.3 Examples of Ergonomics Methods

Heuristics (Nielsen, 1992)

Heuristics require the analyst to use judgment, intuition, and experience to guide product evaluation. This method is wholly subjective and the output is likely to be extremely variable. In favor of the heuristic

Ford 7000 RDS-EON

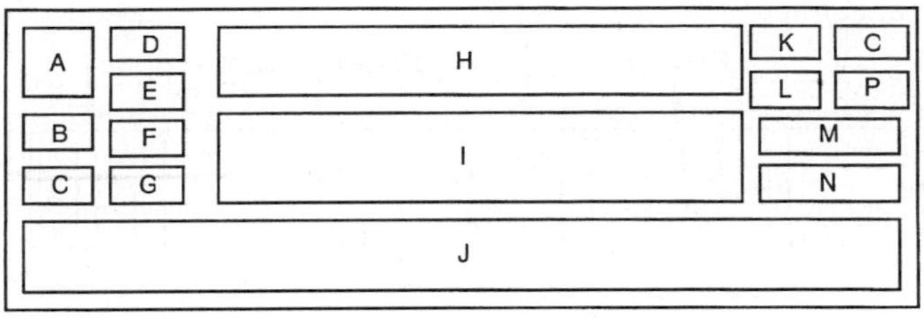

A - On/Off/Vol.	I - Display
B - Bass/Treb.	J - Presets
C - Fade/Bal.	K - Tape
D - Eject	L - PTY
E - Dolby	M - Menu
F - News	N - Seek
G - TA	O - CD
H - Cassette Door	P - AM/FM

This is a schematic diagram of the Ford radio cassette referred to in the examples. It is quite an advanced device, with RDS facilities, automatic volume control, automatic music search and programme type tuning amongst its functions. It is generally very mode-driven, with most of the more advanced functions being hidden in a menu. It is about twice the height of a standard radio, and apart from the On/Off/Volume control, all of the controls are buttons. This means that simple tasks such as adjusting the bass or treble involve more than one element (i.e., press Bass button, then turn Volume control).

FIGURE 41.1 The Ford car radio-cassette player (7000 RDS-EON).

approach is the ease and speed with which it may be applied. Several techniques incorporate the heuristic approach (e.g., checklists, guidelines, SHERPA) but serve to structure heuristic judgment (see Figure 41.3).

An heuristic analysis was applied to the Sharp car radio. Needless to say, the analysis was very quick to execute and did not require any special knowledge on the analyst's part. Indeed, in many ways it resembled a simple walkthrough. Problems encountered in the analysis were few; the only real dissatisfaction resided in the fact that most of the output pertained to very similar faults with the device. Regarding the output, it may be observed that it is largely similar to the checklist approach in form and content, particularly as here we also see many items concerned with anthropometrics (e.g., button sizes). In addition, one of the remedial suggestions was also proposed in the SHERPA analysis (although, this particular section of SHERPA is also subjective). Thus there is some overlap between these approaches, which can only be encouraging for heuristics.

Checklists/Guidelines (Ravden and Johnson, 1989; Woodson et al., 1992)

Checklists and guidelines would seem to be a useful *aide memoir*, to make sure that the full range of ergonomics issues have been considered. However, the approach may suffer from a problem of situational sensitivity, i.e., the discrimination of an appropriate item from a nonappropriate item largely depends upon the expertise of the analyst. Nevertheless, checklists offer a quick and relatively easy method for device evaluation (see Figure 41.4).

Woodson, Tillman, and Tillman (1992) Transport Checklist turned out to be more of a set of guidelines, so could not be used as an example in this analysis. A cursory inspection of the vehicular-related guidelines revealed that such recommendations are largely concerned with anthropometric improvements and safety (e.g., ingress and egress; protruding units etc.). Thus, these can only deal with issues such as consequentiality of accidents, comfort, and satisfaction, rather than error prediction and usability. (Although

Sharp RG-F832E

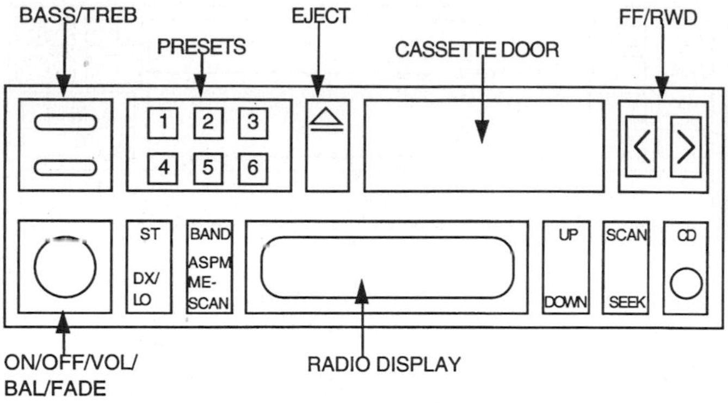

This is a schematic diagram of the Sharp radio cassette referred to in the examples. It is a rather standard radio and has no RDS facilities. Some of the controls may need further elaboration.

On/Off:	This is a knob-twist control - turn clockwise for On, then further to increase volume. Push and turn to adjust fade; a collar adjusts balance.
ST:	Push the top of this button to toggle stereo/mono radio reception.
DX/LO:	Toggles local or distance reception when scanning for radio stations.
BAND:	Switches between wavebands.
ASPM ME-SCAN:	Scans the preset stations.
UP/DOWN:	Manual radio tuning.
SCAN:	Scans the current waveband for radio signals; continues until interrupted by user.
SEEK:	Looks for next radio signal on current waveband and locks onto it.
CD:	CD/auxilliary input socket.
BASS/TREB:	Sliding controls for bass and treble.

FIGURE 41.2 The Sharp car radio-cassette player (RG-F832E).

comfort and satisfaction are elements of usability, there are also more cognitive components.) The Human Engineering Design Checklist proved more useful. Only one section was deemed relevant to assessing a car radio; this was the section on console and panel design. Even so, it was clear that this was constructed for control room assessments, and many of the items were simply irrelevant. However, it would not be an arduous task to extract the relevant items and thus construct a checklist for assessing in-car devices. Furthermore, some items that were not relevant to a car radio (e.g., those concerned with CRTs) may be applicable to other devices (e.g., navigation aids). Again, though, the checklist was, for the most part, concerned with anthropometric issues.

Observation
(Drury, 1990; Kirwan and Ainsworth, 1992; Baber and Stanton, 1996a)

Observation is perhaps the most obvious way of collecting information about a person's interaction with a device; watching and recording the interaction will undoubtedly inform the analyst of what occurred on the occasion observed (see Figure 41.5). Observation is also a deceptively simple method. One simply watches, participates in, and/or records the interaction. However, the quality of the observation will largely depend upon the method of recording and analyzing the data. There are concerns about the intrusiveness of observation, the amount of effort required in analyzing the data, the objectivity of the analysis, and the comprehensiveness of the observational method. Despite these concerns, it is difficult

Example of Heuristic Output
(Sharp RG-F832E)

- On/Off/Volume control is a tad small and awkward, combined with difficult Balance control
- Pushbutton operation would be more satisfactory for On/Off, as Volume stays at preferred level
- Fader Control is particularly small and awkward
- Both of the above points are related to the fact that a single button location has multiple functions - this is too complex
- Treble and Bass Controls also difficult and stiff; although these functions are rarely adjusted once set
- Station Preset Buttons are satisfactory; quite large and clear
- Band Selector Button and FM Mono-Stereo Button should not have 2 functions on each button - could result in confusion if wrong function occurs. These buttons are the only buttons on the radio which are not self-explanatory - the user must consult the manual to discover their function
- Tuning Seek and Tuning Scan Buttons are easier to understand and use, although there are still two functions on the same button. These are probably used more than the aforementioned buttons
- Cassette FF, RWD and Eject Buttons are self-explanatory; the same accepted style that is on all car radio designs. FF and RWD Buttons could be a little larger
- Auto-reverse function is not so obvious, although it is an accepted standard (pressing FF and RWD Buttons simultaneously)
- Illumination - is daytime/nighttime illumination satisfactory? A dimmer control would probably aid matters

FIGURE 41.3 Heuristic analysis of the radio-cassette for Sharp RG-F832E.

to manage without some form of observational data, as most ergonomics methods rely upon it, e.g., hierarchical task analysis and link analysis.

The observational studies show data from 30 participants performing a range of tasks on two occasions. These data on actual performance show errors and response times. These data may seem to be highly credible in the eyes of designers, but observational studies are very resource-intensive and provide little output regarding cognitive mechanisms. They are applicable only late in the design process. The wide use of the observational technique suggests that it is both reliable and valid, as well as useful.

Interviews (Cook, 1988; Sinclair, 1990; Kirwan and Ainsworth, 1992)

Like observation, the interview has a high degree of ecological validity associated with it. If you want to find out what people think of a device you simply ask them (see Figure 41.6). Interviewing has many forms, ranging from highly unstructured (free-form discussion) through focused (a situational interview), to highly structured (an oral questionnaire). For the purposes of device evaluation, a focused approach would seem most appropriate. The interview is good at addressing issues beyond direct interaction with devices, such as the adequacy of manuals and other forms of support. The strengths of the interview are the flexibility and thoroughness it offers. For the purposes of this review we undertook the interview within the Ravden and Johnson (1989) framework of 11 areas of usability. This served as an interview agenda to focus the interviewer and respondent on usability issues associated with the operation of the radio-cassette.

The most striking aspect about the interview was its speed of administration — the whole process lasted around 30 minutes. As its structure was based on the sections of an HCI checklist, we can be quite confident that it thoroughly covered all aspects of device interaction. Admittedly, some aspects of the checklist were simply inapplicable to a car radio, but this just affirmed one advantage of the interview — its flexibility in adapting to changing scenarios. The output of the interview was on the whole unsurprising;

Example of Checklist Output
(Sharp RG-F832E)

The following are selected items from the Human Engineering Design Checklist (Woodson, 1981) which are relevant and/or marginal (unsatisfactory) for the car radio under analysis.

4. Console and panel design
4.1 Displays
4.1.1 Principles
e. Crucial visual checks identified by attention-getting devices (e.g. visual or aural signals).
g. Probability of confusion among instruments is minimal.
4.1.2 Labeling
a. Trade names and other irrelevant information deleted.
b. Easy to read under expected conditions of illumination.
4.1.4 Scales, dials, counters
a. Numbers and letters are large enough for accurate reading at normal distance.
b. Reflected light does not create illusion warning is "ON" or obscure reading.
4.1.6 Indicator and legend lights
k. Displays are arranged in relation to one another to reflect the sequence of use or the functional relations of the components they represent, in that order of preference.
l. Distinct, functional areas set apart for purposes of ready identification are outlined by black lines...
w. Button surfaces are concave to fit the finger, or provide a high degree of frictional resistance to prevent slipping.
x. Buttons provide "snap feel" or an audible click to indicate that the control has been activated.
y. A channel or cover guard is provided when prevention of accidental activation is imperative.
4.2 Control/Display Relationships
4.2.1 Arrangements
a. All controls having sequential relations, or having to do with a particular function or operation, or which are operated together, are grouped together, along with the associated displays.
d. If a control knob is adjacent to the instrument it controls, it is located so that the control or the hand normally used for setting does not obscure the indicator.
4.2.2 Precautions
a. The control is located or oriented so that the operator is not likely to hit it or move it accidentally in the normal sequence of control movements.
d. Interlocks are provided so that extra movement of the prior operation of a related or locking control is required.
e. Resistance is built into the control so that definite or sustained effort is required to actuate it.

FIGURE 41.4 Checklist applied to the radio-cassette for Sharp RG-F832E.

much of it resembling the output of the heuristic analysis. Again, though, there are a number of advantages to the structured approach. Thorough coverage has been mentioned; in addition, it was possible to relate the responses to psychological issues of design (e.g., cognitive compatibility) as a domain expert was present. It could be argued that the subjective responses, combined with professional wisdom, make this a very strong technique. In particular, one aspect emerged from this analysis which had not been covered by any of the other techniques — the usability of the instruction manual.

Questionnaires (Brooke, 1996)

There are few examples of standardized questionnaires appropriate for the evaluation of in-car devices. However, the Software Usability Scale (SUS) may, with some minor adaptation, be appropriate. SUS was developed as part of the usability engineering program in integrated office systems developed at the

Task	Errors Observed	F1	F2	T1 (s) mean (sd)	T2 (s) mean (sd)
1				4.64 (4.38)	4.05 (3.02)
2	Didn't turn knob enough to adjust	3	1	10.5 (7.67)	6.14 (1.88)
	Pressed Seek	1			
3	Adjusted Treble	1		20.4 (13.3)	10.1 (3.75)
	Adjusted Volume	2			
	Pressed On/Off	1			
4				15.7 (15.5)	8.55 (4.50)
5	Adjusted Fade	2	5	18.1 (9.61)	11.9 (5.53)
	Adjusted Bass	1			
	Didn't attempt - forgot how		1		
6	Used Seek	1		7.14 (9.88)	3.86 (1.73)
7	Pressed Preset and Seek together	1	1	31.5 (24.7)	23.6 (10.7)
	Held Seek button down	2	1		
	Interrupted Seek by pressing Preset		1		
	Pressed preset		1		
	Used Manual tuning		1		
	Failed to store - didn't know how	10			
	Didn't hold preset long enough	10	4		
	Pressed Seek to store	1			
8	Didn't know function	4		44.2 (23.8)	29.5 (13.6)
	Used Seek	15	3		
	Held Seek button down	2			
	Pressed Preset	1			
	Failed to store - didn't know how	7			
	Didn't hold preset long enough	10	4		
	Hit 2 Presets and storage failed		1		
9				3.64 (1.87)	3.18 (1.33)
10	Failed to stop FF/RWD	1	2	36.6 (19.2)	30.9 (16.1)
	Pressed Seek instead of autoreverse	3			
	Pressed wrong direction	3	2		
	Turned tape over manually	6			
	Failed to Seek	1	1		
11	Pressed twice		1	4.55 (2.77)	3.91 (1.69)
12				2.86 (1.21)	2.05 (0.576)

Task list:

1. Switch On
2. Adjust Volume
3. Adjust Bass
4. Adjust Treble
5. Adjust Balance
6. Choose a new Preset station
7. Choose a new station using Seek and store it
8. Choose a new station using Manual search and store it
9. Insert cassette
10. Find next track on other side of cassette
11. Eject cassette
12. Switch Off

FIGURE 41.5 Observation of user activity with the radio-cassette for Ford 7000 RDS-EON.

Digital Equipment Company. SUS comprises 10 items that relate to the usability of the device. Originally conceived as a measure of software usability, it has some evidence of proven success. The distinct advantage of this approach is the ease with which the measure may be applied. It takes less than a minute to complete the questionnaire, and no training is required.

The SUS score is determined by taking 1 from all the scores on items with odd numbers and subtracting the scores from 5 for all the items with even numbers. The resultant value is multiplied by 2.5 to give an overall SUS rating of between 0 (extremely poor usability) and 100 (excellent usability). In the example shown in Figure 41.7, this resulted in the value 29 multiplied by 2.5. giving a SUS rating of 72.5. On its own, this just offers a subjective value, but the rating could be of most use when a dozen or more participants are rating two or more products. Statistical analysis of the results could be used to indicate real differences between the products.

Given the brevity of the approach, it is likely to serve as a useful adjunct to other methods. Brooke (1996) reports that SUS is a reliable measure and correlates well with other subjective measures.

Example of Interview output (Sharp RG-F832E)

SECTION 1: VISUAL CLARITY *Information displayed on the screen should be clear, well-organised and easy to read.* • There is a certain amount of visual clutter on the LCD • Writing (labelling) is small but readable • Ambiguous abbreviations (e.g., DX/LO; ASPM ME-SCAN)
SECTION 2: CONSISTENCY *The way the system looks and works should be consistent at all times* • Tuning buttons (especially Scan and Seek functions) present inconsistent labelling • Moded functions create problems in knowing how to initiate the function
SECTION 3: COMPATIBILITY *The way the system looks and works should be compatible with user expectations* • 4 functions on 'On/Off' switch makes it somewhat incompatible • Auto-reverse function could cause cognitive compatibility problems
SECTION 4: INFORMATIVE FEEDBACK *Users should be given clear, informative feedback on where they are in the system* • Tactile feedback is poor, particularly for the 'On/Off' switch • Operational feedback poor when programming a preset station
SECTION 5: EXPLICITNESS *The way the system works and is structured should be clear to the user* • Novice users may not understand station programming without instruction • Resuming normal cassette playback after FF or RWD is not clear
SECTION 6: APPROPRIATE FUNCTIONALITY *The system should meet the needs and requirements of users when carrying out tasks* • Rotating dial is not appropriate for front/rear fader control • Prompts for task steps may be useful when programming stations
SECTION 7: FLEXIBILITY AND CONTROL *The interface should be sufficiently flexible in structure, information presentation and in terms of what the user can do, to suit the needs and requirements of all users* • Users with larger fingers may find controls fiddly • Radio is inaudible whilst winding cassette - this is inflexible
SECTION 8: ERROR PREVENTION AND CORRECTION *The system should be designed to minimise the possibility of user error, users should be able to check their inputs and to correct errors* • There is no 'undo' function for stored stations • Separate functions would be better initiated from separate buttons
SECTION 9: USER GUIDANCE AND SUPPORT *Informative, easy-to-use and relevant guidance and support should be provided* • Manual is not well structured, relevant sections are difficult to find • Instructions in the manual are matched to the task
SECTION 10: SYSTEM USABILITY PROBLEMS • Minor problems in understanding function of 2 or 3 buttons • Treble and bass controls are tiny
SECTION 11: GENERAL SYSTEM USABILITY • Best aspect: This radio is *not* mode-dependent • Worst aspect: Ambiguity in button labelling • Common mistakes: Adjusting balance instead of volume • Recommended changes: Substitute pushbutton operation for 'On/Off' control

FIGURE 41.6 Interviewing users about the radio-cassette for Sharp RG-F832E.

Link Analysis (Stammers et al., 1990; Kirwan and Ainsworth, 1992; Drury, 1995)

Link analysis represents the sequence in which device elements are used in a given task or scenario. The sequence provides the links between elements of the device interface. This may be used to determine if the current relationship between device elements is optimal in terms of the task sequence. Time data recorded on duration of attentional gaze may also be recorded in order to determine if display elements are laid out in the most efficient manner. The link data may be used to evaluate a range of alternatives before the most appropriate arrangement is accepted. The following diagrams represent the relevant link diagrams and tables for the Ford 7000 RDS EON. The analyses are abbreviated and based on a standard subset of tasks (see below). Redesign is offered on the basis of the analyses. As can be seen, very little is changed on the Ford radio, suggesting that the original satisfied the principles of link analysis well (see Figure 41.8).

Example of SUS output and scoring (Ford 7000 RDS-EON)

	Strongly disagree 1	2	3	4	Strongly agree 5
1. I think that I would like to use this system frequently					✓
2. I found the system unnecessarily complex		✓			
3. I thought the system was easy to use					✓
4. I think that I would need the support of a technical person to be able to use this system	✓				
5. I found the various functions in this system were well integrated				✓	
6. I thought there was too much inconsistency in this system	✓				
7. I would imagine that most people would learn to use this system very quickly				✓	
8. I found the system very cumbersome to use		✓			
9. I felt very confident using the system					✓
10. I needed to learn a lot of things before I could get going with this system			✓		

Scoring SUS

Odd-numbered items score = scale position - 1	Even-numbered items score = 5 - scale position
1. 5 - 1 = 4	2. 5 - 2 = 3
3. 5 - 1 = 4	4. 5 - 1 = 4
5. 4 - 1 = 3	6. 5 - 1 = 4
7. 4 - 1 = 3	8. 5 - 2 = 3
9. 5 - 1 = 4	10. 5 - 3 = 2
Total for odd-numbered items = 18	Total for even-numbered items = 16

Grand Total = 34 (multiply this by 2.5 to obtain usability score)

SUS overall usability score = 34 * 2.5 = **85**

FIGURE 41.7 Rating the radio-cassette using SUS for Ford 7000 RDS-EON.

Link Analysis - Ford 7000 RDS EON

Initial design:

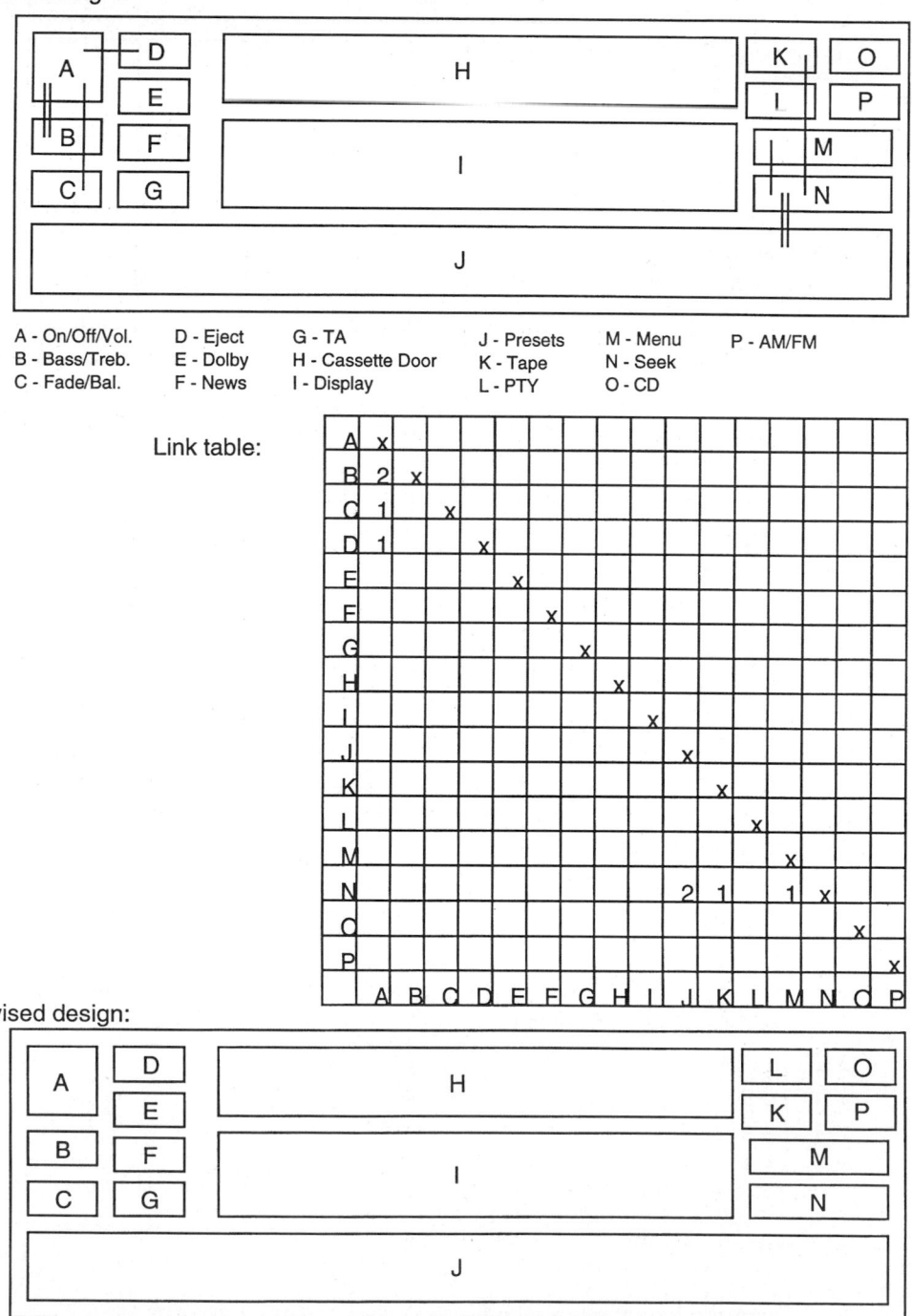

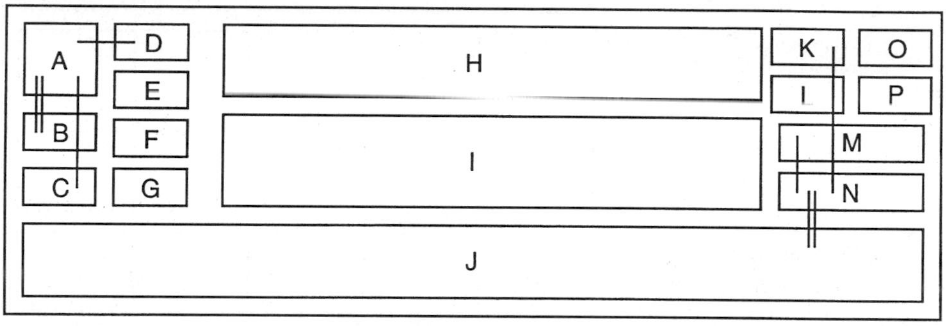

A - On/Off/Vol.
B - Bass/Treb.
C - Fade/Bal.
D - Eject
E - Dolby
F - News
G - TA
H - Cassette Door
I - Display
J - Presets
K - Tape
L - PTY
M - Menu
N - Seek
O - CD
P - AM/FM

Link table:

	A	B	C	D	E	F	G	H	I	J	K	L	M	N	O	P
A	x															
B	2	x														
C	1		x													
D	1			x												
E					x											
F						x										
G							x									
H								x								
I									x							
J										x						
K											x					
L												x				
M													x			
N										2	1		1	x		
O															x	
P																x

Revised design:

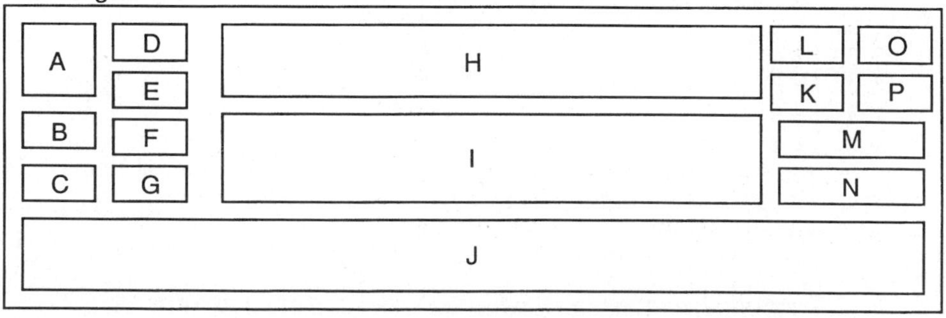

FIGURE 41.8 A link analysis for Ford 7000 RDS-EON.

A task list follows:

1. Switch On
2. Adjust Volume
3. Adjust Bass
4. Adjust Treble
5. Adjust Balance
6. Choose New Preset
7. Use Seek, then store station
8. Use Manual Search, then store station
9. Insert cassette
10. Autoreverse, then Fast Forward
11. Eject cassette and Switch Off

An initial dilemma was encountered in link analysis in whether to study hand or eye movements. For simplicity, the analysis was restricted to hand movements. A basic walk-through was used for the data collection. One particular problem here was concerned with the fact that operating a radio is far from being a set procedure, leading to a possibly infinite set of links. This was circumvented by analyzing a single run of a typical task.

Layout Analysis (Easterby, 1984)

Layout analysis builds upon link analysis to consider functional groupings of device elements (see Figure 41.9). Within functional groupings, elements are sorted according to optimum trade-off of three criteria: frequency of use, sequence of use, and importance of element.

Layout analysis was undoubtedly easier to execute. Functional groupings on a radio are obvious, and their importance, sequence, and frequency of use were also easily determined. The analysis also maintains something of a hierarchical structure, as it progresses from general categories to specific functions. Overall, this was a very straightforward and seemingly effective technique. Both techniques (link and layout analysis) lead to suggested improvements for interface layout.

Hierarchical Task Analysis (Annett et al., 1971; Stammers and Shepherd, 1995)

Hierarchical task analysis (HTA) has been a technique central to the discipline of ergonomics in the U.K. for over 2 decades. Application of the technique breaks tasks down into goals, plans, and operations in a hierarchical structure (see Figure 41.10). While the technique offers little more than a task description, it serves as the input into other predictive methods, for example SHERPA and KLM. The concepts of HTA are relatively straightforward, but the approach requires some practice and reiteration before HTA can be applied with confidence.

The structure of the task presented itself immediately; however, some detailed aspects later in the analysis (such as the logic involved in decisions and plans) were slightly more problematic. This, though, merely illustrates one of the characteristics (and some may say benefits) of HTA — that it is an iterative technique. This was certainly borne out in the analysis conducted.

Systematic Human Error Reduction and Prediction Approach (Embrey, 1993; Stanton, 1995; Baber and Stanton, 1996b)

Systematic human error reduction and prediction approach (SHERPA) is a semistructured human error identification technique. It is based upon hierarchical task analysis (HTA) and an error taxonomy. Briefly, each task step from the bottom level in HTA is taken in turn and potential error modes associated with that activity are identified (see Figure 41.11). From this the consequences of those errors are determined. SHERPA appears to offer reasonable predictions of performance but may have some limitations in its comprehensiveness and generalizability.

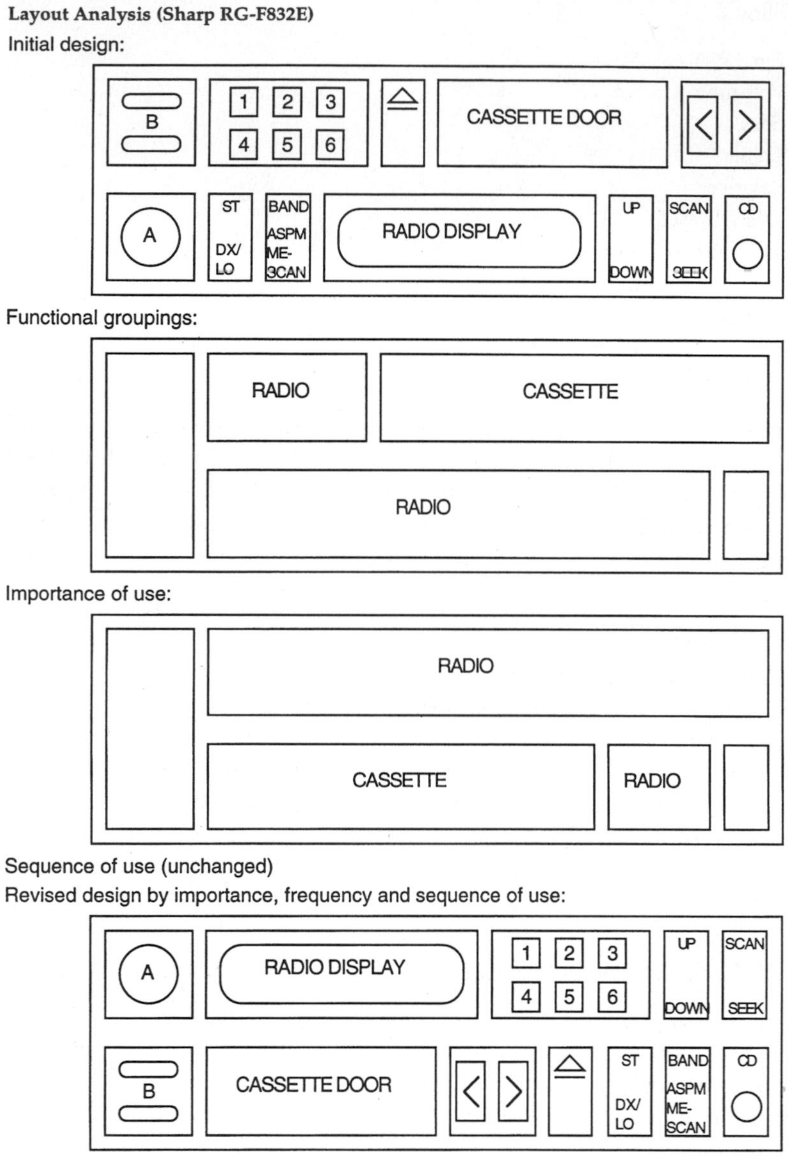

FIGURE 41.9 A layout analysis for Sharp RG-F832E.

While the human error taxonomy incorporated in SHERPA is certainly a handy prompt in identifying errors, it does have limitations in generalization across tasks. As SHERPA was originally designed for process control room tasks, some sections of the taxonomy are quite inappropriate for product design and evaluation. Thus, as stated above, there is room for improvement in this area. The error predictions and remedies offered by SHERPA are certainly of applied value if proved to be valid. Indeed, there are some recent studies (e.g., Baber and Stanton, 1996b; Stanton and Stevenage, 1997) that suggest such predictions are, to a large extent, accurate and reliable. However, the credibility and salience of some of the predicted errors may be questioned. Even after a thorough SHERPA, error reduction strategies must be evaluated along dimensions such as practicability and cost-effectiveness to determine whether they are worth applying. Moreover, cognitive aspects of errors, which may aid in the above determination, are largely avoided by SHERPA (although differing error mechanisms were intimated when multiple

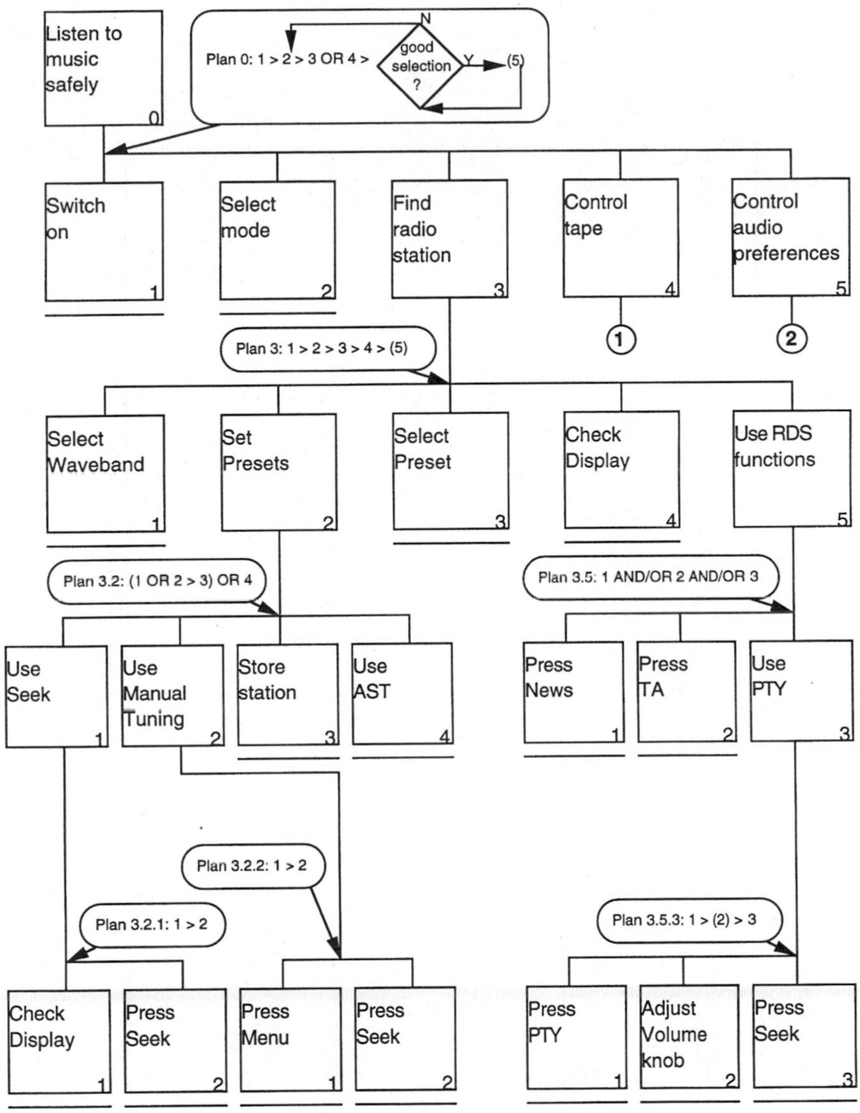

FIGURE 41.10 Hierarchical task analysis for Ford 7000 RDS-EON.

error modes lead to similar consequences). Finally, an observation worthy of note is that executing the SHERPA served to highlight deficiencies in the HTA — undoubtedly a bonus for task analysis.

Task Analysis For Error Identification (Baber and Stanton, 1994; Stanton and Baber, 1996b)

Task analysis for error identification (TAFEI) is an approach for modeling the interaction between device and user (see Figure 41.12). TAFEI is based upon hierarchical task analysis (HTA) and state space diagrams (SSDs), both established techniques with a pedigree of over 25 years. By mapping HTA onto SSDs and employing a transition matrix, it is possible to start to consider what may go wrong in the interaction. Essentially, TAFEI is a human error identification method (like SHERPA).

Once one has understood HTA and SSDs, executing a TAFEI analysis is not difficult, although it is still a little time consuming. It is certainly an advantage to possess either or both HTA and SSD before

Abridged example of HTA tabular format (Ford 7000 RDS-EON)

Super-ordinate	Subtask/plan	Notes
0	LISTEN TO IN-CAR ENTERTAINMENT P0: 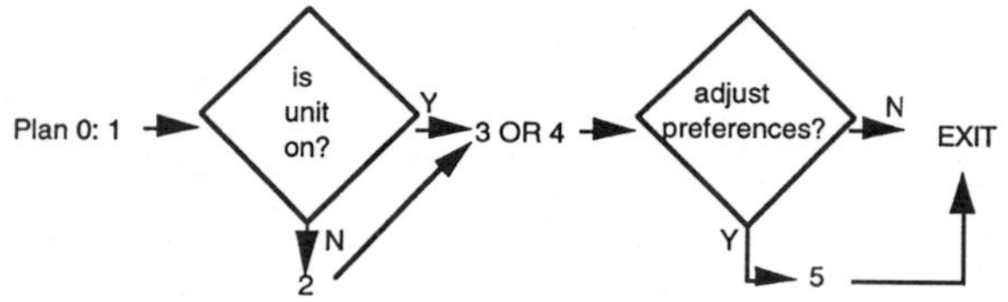 1. Check unit status // 2. Press on/off button // 3. Listen to radio 4. Listen to cassette 5. Adjust audio preferences	User checks whether radio is already on. If not, user switches on by pressing on/off knob. User may then decide to listen to radio or cassette, and may wish to adjust their audio preferences.
5	ADJUST AUDIO PREFERENCES P5: Plan 5: 1 AND/OR 2 AND/OR 3 AND/OR 4 AND/OR 5 1. Adjust volume 2. Adjust bass 3. Adjust treble 4. Adjust fade 5. Adjust balance	User may decide to adjust any or all of these preferences.
etc.	...	HTA table continues in this manner.

FIGURE 41.10 (continued)

Abridged example of PHEA output (Sharp RG-F832E)

Task Step	Error Mode	Description	Consequence	Recovery	P	C	Remedies
1	A4	Volume level adjusted inappropriately	Volume is at undesirable level	Immediate /4.2	M		Separate vol/on/off Preset startup volume
	A7	Balance adjusted instead of on/off/vol	Unit is not switched on; balance settings altered	Immediate /4.3	H		Separate balance/on/off Lockout mechanism
2.1	C1	Omit check of station	Listening to wrong station	2.2	L		Untuned on startup Reminder to check
	C2	Misidentify station	Listening to wrong station	2.2	M		RDS
2.2.1	S2	Select wrong preset	Listen to wrong station	2.2.2 on	H		Label buttons clearly Aide-mémoire
2.2.2	A1	Preset button held too long	Undesired station storage	2.3	L		Confirm before storage
	A6	Press wrong button	Desired station not found	Immediate	H		Label buttons clearly Aide-mémoire
2.3.1	C1	Omit check of wavelength	Unnecessary retuning	2.3.2	L		Display conspicuity (e.g., RDS)
	C3	Check wrong display	Wavelength not identified	2.3.2	L		Display conspicuity (e.g., RDS)
etc...							

FIGURE 41.11 SHERPA for Sharp RG-F832E.

beginning, as this saves a great deal of time. The most difficult part of this trial analysis was in constructing the SSDs for a car radio, as these have to be quite accurate for the remainder of the analysis to be effective. Completing the transition matrix is then a straightforward affair.

Repertory Grids (Kelly, 1955; Baber, 1996)

Repertory grids may be used to determine people's perception of a device. In essence, the procedure requires the analyst to determine the elements (the forms of the product) and the constructs (the aspects of the product that are important to its operation) (see Figure 41.13). Each version of the product is then rated against each construct. This approach seems to offer a way of gaining insight into consumer perception of the device, but does not necessarily offer predictive information.

The repertory grid is not a difficult technique to execute, once the concept has been grasped. Constructing a thorough grid should take no more than an hour. However, analysis is a different story. Initially in this example, the revised analysis technique of Baber (1996) was attempted. While it certainly seemed an easier method than the usual factor analysis, there also appeared to be weaknesses in its approach. These weaknesses were particularly borne out when it came to the factor extraction stage, for no constructs were significantly related to be grouped as a factor. Thus while the process of constructing the repertory grid was useful in itself, no useful quantification of the grid could be gleaned. Thus, the analysis turned to more conventional methods. It is quite possible to execute a number of analyses on a repertory grid, such as factor analysis, multidimensional scaling, and even analyses of variance (although none of these methods were actually carried out in the present example). In summary, the repertory grid did provide a useful insight into perception of this product, such that the output could be useful in design. However, the constructs elicited also seemed to mirror somewhat the output of some of the other techniques reviewed so far, such as checklists and heuristics.

Keystroke Level Model (Card et al., 1983)

The keystroke level model (KLM) is a technique that is used to predict task performance time for error-free operation of a device. The technique works by breaking tasks down into component activities, e.g., mental operations, motor operations, and device operations, then determining response times for each of these operations and summing them. The resultant value is the estimated performance time for the whole operation. While there are some obvious limitations to this approach (such as the analysis of

Abridged example of TAFEI

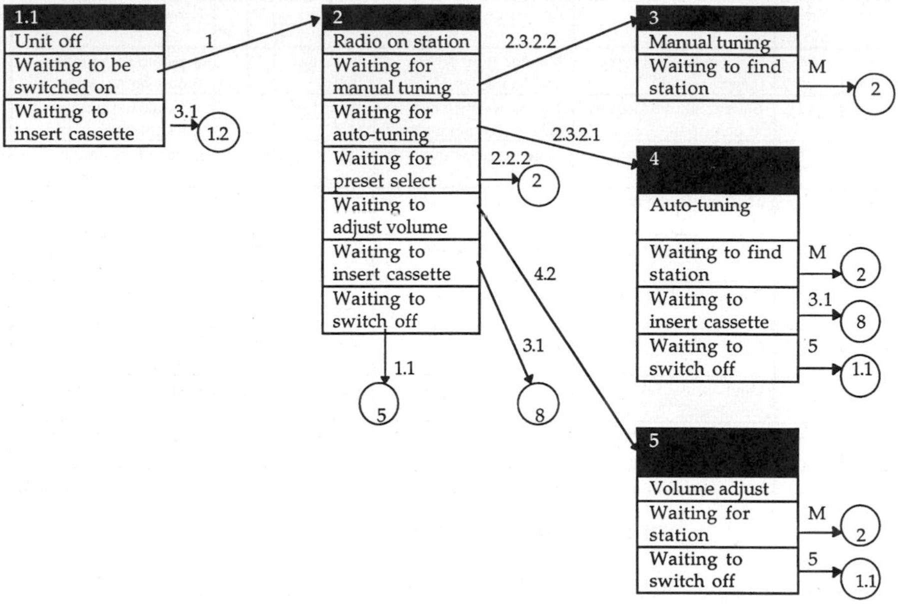

Transition matrix

	1.1	1.2	1.3	1.4	2	3	4	5	6	7	8	9	10	11	12	13	14
1.1	-	I	-	-	L	-	-	-	-	-	-	-	-	-	-	-	-
1.2	L	-	I	I	-	-	-	-	-	-	L	-	-	-	-	-	-
1.3	-	I	-	-	-	-	-	-	-	-	-	L	-	-	-	-	-
1.4	-	I	-	-	-	-	-	-	-	-	-	-	L	-	-	-	-
2	L	-	-	-	-	L	L	L	L	L	L	-	-	-	-	-	-
3	-	-	-	-	L	-	-	-	-	-	-	-	-	-	-	-	-
4	L	-	-	-	L	-	-	-	-	-	L	-	-	-	-	-	-
5	I	-	-	-	L	-	-	-	-	-	-	-	-	-	-	-	-
6	-	-	-	-	L	-	-	-	-	-	-	-	-	-	-	-	-
7	-	-	-	-	L	-	-	-	-	-	-	-	-	-	-	-	-
8	-	I	-	-	L	-	-	-	-	-	-	L	L	L	L	L	L
9	-	-	I	-	-	-	-	-	-	-	L	-	-	-	-	-	-
10	-	-	-	I	-	-	-	-	-	-	L	-	-	-	-	-	-
11	-	-	-	-	-	-	-	-	-	-	L	-	-	-	-	-	-
12	-	I	-	-	-	-	-	-	-	-	L	-	-	-	-	-	-
13	-	-	-	-	-	-	-	-	-	-	L	-	-	-	-	-	-
14	-	-	-	-	-	-	-	-	-	-	L	-	-	-	-	-	-

FIGURE 41.12 TAFEI for Sharp RG-F832E.

cognitive operations) and some ambiguity in determining the number of mental operations to be included in the equation, the approach does appear to have some support. There are four motor operators in KLM — keystroking, pointing, homing and drawing; one mental operator; and one operator for system response. Each of these operators has an associated nominal time, derived by experiment (although drawing and response times are variable). It is thus a simple matter of determining the components of the task in question and summing the times of the associated operators to arrive at an overall task time prediction (see Figure 41.14).

Although KLM is indeed a simplistic method, it was designed for human–computer interaction (HCI), and this is evident when attempting to apply it. With a car radio, an immediate stumbling block was encountered because there is no operator accounting for "turning a knob." Thus, while some operators have no place outside HCI (e.g., pointing, drawing), there are others that are not foreseen within HCI.

Example of Repertory Grid Output
(Ford 7000 RDS-EON and Sharp RG-F832E)

	Elements						
Constructs	Rover	Ford	Vaux-hall	New Rover	Worst	Best	Opposites
Mode dependent	1	5	4	1	5	1	Separate functions
Pushbutton operation	2	5	4	2	1	5	Knob-turn operation
Bad labelling	2	5	4	2	5	1	Clear labelling
Easy controls	1	5	5	2	1	5	Fiddly controls
Poor functional grouping	4	5	2	2	5	1	Good functional grouping
Good illumination	2	4	5	2	1	5	Poor illumination

5 = left side very much applicable (right side not applicable at all)
4 = left side somewhat applicable (right side not really applicable)
3 = in between
2 = left side not really applicable (right side somewhat applicable)
1 = left side not applicable at all (right side very much applicable)
0 = characteristic irrelevant

FIGURE 41.13 Repertory grids for Sharp RG-F832E and Ford 7000 RDS-EON.

Worked example for comparing two alternative car radio designs using the Keystroke Level Model (KLM)

Task	Time - Design 1 (s)	Time - Design 2 (s)	Difference +/-
Switch On	MHKR=2.65+1=3.65	MHKR=2.65+1=3.65	0
Adjust Volume	MHKR=2.65+0.1=2.75	MHKR=2.65+0=2.65	+0.1
Adjust Bass	MHKHKR=3.95+0.2=4.15	MHKR=2.65+0=2.65	+1.5
Adjust Treble	MHKKHKR=4.15+0.3=4.45	MHKR=2.65+0=2.65	+1.8
Adjust Balance	MHKKHKR=4.15+0.3=4.45	MHKKR=2.85+0.1=2.95	+1.5
Choose new Preset	MHKR=2.65+0.2=2.85	MHKR=2.65+0.2=2.85	0
Use Seek	MHKR=2.65+1=3.65	MHKR=2.65+1=3.65	0
Use Manual search	MHKHKR=3.95+1=4.95	MHKR=2.65+1=3.65	1.3
Store station	MHKR=2.65+1=3.65	MHKR=2.65+3=5.65	-2
Insert Cassette	MHKR=2.65+1=3.65	MHKR=2.65+1=3.65	0
Autoreverse and FF	MHKRHKRKR=4.15+5=9.15	MHKRKRK=3.05+5=8.05	1.1
Eject Cassette	MHKR=2.65+0.5=3.15	MHKR=2.65+0.3=2.95	0.2
Switch Off	MHKR=2.65+0.5=3.15	MHKR=2.65+0.7=3.35	-0.2
Total time	**53.65**	**48.35**	**5.3**

This table represents the calculations for execution times of standard tasks across the two different radio designs. Design 1 is the Ford 7000 RDS EON, design 2 is a Sharp RG-F832E. As is evident, on this set of tasks, the Ford design takes around 5 seconds longer to complete. Analysing the operators suggests that this extra time is very much taken up by the moded nature of the device.

N.B.: System response times have been largely estimated in the above table. For purposes of equality, standard response times (e.g., for radio tuning) were applied to both designs.

FIGURE 41.14 KLM for Sharp RG-F832E and Ford 7000 RDS-EON.

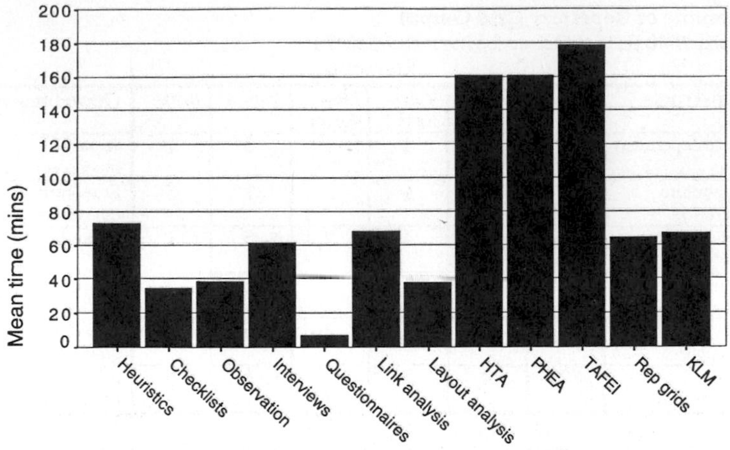

FIGURE 41.15 Time to apply each method to the evaluation of a radio-cassette (Ford 7000 RDS-EON).

A consequence of this is that we are immediately limited as to the tasks we can analyze in the automobile. Further restrictions are imposed regarding the fixed times associated with most operators, leading to some errant predictions — a time of 2 seconds simply for selecting a preset station seems generous. A final limitation of KLM is that it does not predict anything over and above task execution time, nor does it claim to.

41.4 Conclusions

The detailed review of ergonomics methods led to a greater insight into the demands and outputs of the methods under scrutiny. A study by Stanton and Young (1997) indicated that link analysis, layout analysis, repertory grids, and KLM appear to offer good utility when compared with other, more commonly used methods. These data also seem to reinforce the reason for the popularity of questionnaires, interviews, observations, checklists, and heuristics, as they take relatively little time to apply when compared with HTA, TAFEI, and SHERPA. Perhaps it is surprising that link and layout analysis are not more popular given that they are also relatively quick to use. Similarly, repertory grids and the keystroke level model seem to be no more time consuming to use than the focused interview. However, these techniques are rather more specialized in their output, like link and layout analysis (see Figure 41.15).

These analyses seem to suggest that some methods will be more acceptable than others because of the time required to apply them to an evaluation. However, we hope that ergonomists and designers would explore the utility of other methods rather than always relying on three or four of their favorite approaches.

Acknowledgments

The research reported in this chapter was supported by the EPSRC under the LINK Transport Infrastructure and Operations Programme in the U.K. The authors are grateful to Taylor & Francis for allowing them to reproduce some material from *Human Factors in Consumer Products.*

References

Annett, J., Duncan, K. D., Stammers, R., and Grey, M. J. (1971) *Task Analysis.* Department of Employment Training information paper 6. HMSO, London.

Baber, C. (1996) Repertory grid theory and its application to product evaluation, in P. W. Jordan; B. Thomas; B. A. Weerdmeester and I. L. McClelland (eds.) *Usability Evaluation in Industry* Taylor & Francis, London, pp. 157-165.

Baber, C. and Stanton, N. A. (1994) Task analysis for error identification: a methodology for designing error-tolerant consumer products. *Ergonomics,* 37, 11, 1923-1941.

Baber, C. and Stanton, N. A. (1996a) Observation as a usability method, in P. W. Jordan; B. Thomas; B. A. Weerdmeester and I. L. McClelland (eds.) *Usability Evaluation in Industry* Taylor & Francis, London, pp. 85-94.

Baber, C. and Stanton, N. A. (1996b) Human error identification techniques applied to public technology: predictions compared with observed use, *Applied Ergonomics,* 27 (2) pp. 119-131.

Brooke, J. (1996) SUS: a "quick and dirty" usability scale, in P. W. Jordan; B. Thomas; B. A. Weerdmeester and I. L. McClelland (eds.) *Usability Evaluation in Industry* Taylor & Francis, London, pp. 189-194.

Card, S. K., Moran, T. P., and Newell, A. (1983) *The Psychology of Human-Computer Interaction* Erlbaum, Hillsdale, NJ.

Corlett, E. N. & Clarke, T. S. (1995) *The Ergonomics of Workspaces and Machines.* 2nd Edition Taylor & Francis, London.

Diaper, D. (1989) *Task Analysis in Human Computer Interaction.* Ellis Horwood, Chichester.

Drury, C. G. (1995) Methods for direct observation of performance, in J. Wilson and N. Corlett (eds.) *Evaluation of Human Work.* 2nd Edition Taylor & Francis, London, pp. 45-68.

Easterby, R. (1984) Tasks, processes and display design, in R. Easterby and H. Zwaga (eds.) *Information Design* Taylor & Francis, London.

Embrey, D. (1983) Quantitative and qualitative prediction of human error in safety assessments, in *The Institution of Chemical Engineers symposium Series* 130 pp. 329-350.

Jordan, P. W., Thomas, B., Weerdmeester, B. A., and McClelland, I. L. (1996) *Usability Evaluation in Industry* Taylor & Francis, London.

Kelly, G. A. (1955) *The Psychology of Personal Constructs* Norton, New York.

Kirwan, B. and Ainsworth, L. (1992) *A Guide to Task Analysis* Taylor & Francis, London.

Kirwan, B. (1994) *A Guide to Practical Human Reliability Assessment* Taylor & Francis, London.

Nielsen, J. (1992) Finding usability problems through heuristic evaluation, in *Proceedings of the ACM Conference on Human Factors in Computing Systems* ACM Press, Monterey, CA, pp. 373-380.

Ravden, S. J. and Johnson, G. I. (1989) *Evaluating Usability of Human-Computer Interfaces: A Practical Method.* Ellis Horwood, Chichester.

Sinclair, M. (1995) Subjective assessment, in J. Wilson and N. Corlett (eds.) *Evaluation of Human Work.* 2nd Edition Taylor & Francis, London, pp. 69-100.

Stammers, R. B. Carey, M., and Astley, J. A. (1990) Task analysis, in J. Wilson and N. Corlett (eds.) *Evaluation of Human Work* Taylor & Francis, London, pp. 134-160.

Stammers, R. B. and Shepherd, A. (1995) Task analysis, in J. Wilson and N. Corlett (eds.) *Evaluation of Human Work.* 2nd Edition Taylor & Francis, London, pp. 144-168.

Stanton, N. A. (1995) Analysing worker activity: a new approach to risk assessment, *Health and Safety Bulletin,* 240 pp. 9-11.

Stanton, N. A. (1998) *Human Factors in Consumer Products.* Taylor & Francis, London.

Stanton, N. A. and Baber, C. (1996a) Factors affecting the selection of methods and techniques prior to conducting a usability evaluation, in P. W. Jordan; B. Thomas; B. A. Weerdmeester and I. L. McClelland (eds.) *Usability Evaluation in Industry.* Taylor & Francis, London pp. 39-48.

Stanton, N. A. and Baber, C. (1996b) A systems approach to human error identification, *Safety Science,* 22 (1-3) 215-228.

Stanton, N. A. and Stevenage, S. (1998) Learning to predict human error: issues of reliability, validity and acceptability. *Ergonomics,* (in press)

Stanton, N. A. and Young, M. (1995) *Development of a Methodology for Improving Safety in the Operation of In-Car Devices* EPSRC/DOT LINK Report 1. University of Southampton, Southampton.

Stanton, N. A. and Young, M. (1997) Is utility in the mind of the beholder? A study of ergonomics methods, *Applied Ergonomics,* 29 (1) 41-54.

Wilson, J. (1995) A framework and context for ergonomics methodology, in J. Wilson and N. Corlett (eds.) *Evaluation of Human Work.* 2nd Edition Taylor & Francis, London pp. 1-39

Wilson, J. and Corlett, N. (1995) *Evaluation of Human Work.* 2nd Edition. Taylor & Francis, London.

Woodson, W. E., Tillman, B., and Tillman, P. (1992) *Human Factors Design Handbook.* 2nd edition McGraw-Hill, New York.

Part III
Musculoskeletal Disorders — Engineering Factors

Section I
Disorders of the Extremities

42

Epidemiology of Upper Extremity Disorders

Bradley Evanoff
Washington University School of Medicine

David Rempel
University of California San Francisco

This chapter summarizes findings from epidemiologic studies that address workplace and individual factors associated with upper extremity musculoskeletal disorders. These disorders are not new: epidemics and clinical case series of work-related upper extremity problems were reported throughout the 1800s and early 1900s (Conn, 1931; Thompson, 1951). Although there are almost no prospective studies in this area, within the last 20 years a number of well-designed, cross-sectional studies have focused on disorders of the hand, wrist, and elbow as related to work. These studies point to the multifactorial nature of work-related upper extremity disorders. The severity of these disorders is influenced not only by biomechanical factors, but also by other work organizational factors, the worker's perception of the work environment, and medical management.

From an epidemiologic point of view, this topic is problematic because there are many specific disorders that can occur in the hand, arm, and shoulder, ranging from arthritis to nerve entrapments. To complicate the matter further, there are few accepted criteria for case definitions for these many disorders. In their early stages, these disorders usually present with nonspecific symptoms without physical examination or laboratory findings. In fact, the only laboratory tests consistently of value in diagnosing these disorders are nerve conduction studies for nerve entrapment disorders and radiographs for osteoarthritis. Finally, symptoms at the hand or wrist may be due to nerve compression or vascular pathology in the neck or shoulder.

42.1 Frequency, Rates, and Costs

Rates of hand and wrist symptoms and associated disability among working adults were assessed by a 1988 national interview survey of 44,000 randomly selected U.S. adults (National Health Interview Survey) (Park, 1993). Of those who had worked anytime in the past 12 months, 22% reported some finger, hand, or wrist discomfort that fit the category "pain, burning, stiffness, numbness, or tingling" for one or more days in the past 12 months. Only one-quarter were due to an acute injury such as a cut, sprain, or broken bone. Nine percent reported having prolonged hand discomfort that was not due to an acute injury; that is, discomfort of 20 or more days or 7 or more consecutive days during the last 12 months. Of those with prolonged hand discomfort, 6% changed work activities and 5% changed jobs due to the hand discomfort.

0-8493-2641-9/99/$0.00+$.50

TABLE 42.1 Examples of Disorders of the Hand, Wrist, and Elbow Observed in Workplace Studies

Non-specific hand and wrist pain	Hand Arm Vibration Syndrome
Tendinitis	Osteoarthritis
Tenosynovitis	Hypothenar hammer syndrome
Finger tendinitis	Gamekeeper's thumb
Wrist tendinitis	Digital neuritis
Stenosing tenosynovitis	Nerve entrapments
Lateral epicondylitis	Carpal Tunnel Syndrome
Medial epicondylitis	Ulnar neuropathy at the wrist
Ganglion cysts	Ulnar neuropathy at the elbow

Elbow pain and epicondylitis are common in working populations. Symptoms of elbow pain are reported by 7 to 21% of workers in industrial populations (Chiang, 1993; Ohlsson, 1989; Buckle, 1987). Epicondylitis is seen in 0.7 to 2.0% of workers in jobs with low levels of physical demands to the arms and hands, and in 2 to 33% of worker groups with high levels of demands.

In the U.S., hand and wrist disorders account for 55% of all work-related repeated motion disorders reported by U.S. private employers (Bureau of Labor Statistics, 1993). This category excludes low back pain. A similar percentage is also reported in industrial (McCormick, 1990) and other national studies (Kivi, 1984). A similar rise in work-related hand/forearm problems has been observed in other countries such as Finland (Kivi, 1984), Australia (Bammer, 1987), and Japan (Ohara, 1976).

Costs for work-related musculoskeletal disorders are difficult to estimate reliably. Webster and Snook (1994) analyzed 1989 insurance claims data from 45 states, restricting their analysis to upper extremity claims classified as cumulative trauma disorders. They estimated that the total compensable cost for upper extremity cumulative trauma disorders in the U.S. was $563 million in 1989. The National Institute for Occupational Safety and Health has estimated that the annual workers' compensation costs for neck and upper extremity disorders is $2.1 billion, plus $90 million in indirect costs (NIOSH, 1996).

42.2 Disorder Types and their Natural History

Table 42.1 lists the most common workplace hand, wrist, and elbow problems. Nonspecific hand/wrist pain is the most common problem, followed by tendinitis, ganglion cysts, and carpal tunnel syndrome (Silverstein, 1987; McCormack, 1990; Hales, 1994). In many workplace studies, rates of nonspecific symptoms, tendinitis, and CTS appear to track each other, that is, a number of specific disorders typically occur together. For example, in a pork processing plant, the rank order of hand and wrist problems, as a percentage of all morbidity, was: nonspecific hand/wrist pain (39%), CTS (26%), trigger finger (23%), trigger thumb (17%), and DeQuervain's tenosynovitis (17%) (Moore, 1994). Similar ratios of disorders have been observed in manufacturing (Armstrong, 1982; Silverstein, 1986; McCormack, 1990), food processors (Kurppa, 1991; Luopajarvi, 1979), and among computer operators (Hales, 1994; Bernard, 1993).

Tendinitis is the most common specific, work-related hand disorder (McCormack, 1990; Luopajarvi, 1979). For the purposes of this chapter tendinitis will include hand, wrist, and distal forearm tendinitis or tenosynovitis, and trigger finger. Tendinitis occurs at discrete locations; the most common site is the first extensor compartment (De Quervain's Disease), followed by the five other pulley sites on the extensor side of the hand and three on the flexor side. The diagnosis is based on history, symptom location, and palpation and provocative maneuvers on physical exam. There has been no association of tendinitis with age or gender, but work-related tendinitis is higher among workers with less than 3 years of employment (McCormack, 1990).

Lateral epicondylitis is the most common specific elbow disorder; medial epicondylitis is less common. The diagnosis is based on pain and tenderness over the lateral or medial elbow and pain on movement of the wrist or fingers against resistance. Other disorders of the elbow which may be related to occupational activities include olecranon bursitis, triceps tendinitis, and osteoarthritis.

Studies of carpal tunnel syndrome have generated considerable controversy. While there is agreement that this disorder results from compression of the median nerve at the wrist, there are no universally accepted diagnostic criteria for carpal tunnel syndrome. Some consider an abnormal nerve conduction study a gold standard (Katz, 1991; Nathan, 1992; Heller, 1986). However, relying exclusively on nerve conduction studies can lead to reporting very high prevalence rates — 28% (Nathan, 1992) and 19% (Barnhart, 1991) in low-risk working populations. A case definition incorporating typical symptoms and signs has been proposed by NIOSH for surveillance purposes (CDC, 1989); however, the usual signs have relatively poor sensitivities and specificities (Katz, 1991; Heller, 1986; Franzblau, 1993). Therefore, this definition may have limited value in distinguishing CTS from other hand disorders. Hand diagrams completed by patients are reproducible and sensitive, but may lack specificity (Katz, 1990; Franzblau, 1994). Only in the later stages are weakness and thenar atrophy a noticeable feature. In approximately 25% of cases, CTS is accompanied by other disorders of the hand or wrist (Phalen, 1966).

Few studies have evaluated the work-relatedness of osteoarthritis of the hand and wrist (Hadler, 1978; Williams, 1987). Hadler et al. (1978) assessed the hands of 67 workers at a textile plant in Virginia. Significant differences in finger and wrist joint range of motion, joint swelling, and X-ray patterns of degenerative joint disease were observed between three different hand intensive jobs; the observed differences matched the pattern of hand usage.

Hand arm vibration syndrome or Vibration White Finger disease occurs in occupations involving many years of exposure to vibrating hand tools (NIOSH, 1989). This is a disorder of the small vessels and nerves in the fingers and hands presenting as localized blanching at the fingertips with numbness on exposure to cold or vibration. The symptoms are largely self-limited if vibration exposure is eliminated at an early stage (Ekenval, 1987; Futatsuka, 1986).

Hypothenar hammer syndrome or occlusion of the superficial palmar branch of the ulnar artery has been associated in clinical series and case-control studies with habitually using the hand for hammering (Little, 1972; Nilsson, 1989) and with exposure to vibrating hand tools (Kaji, 1993). The mean years of exposure before presentation were 20 to 30 years.

Small case-control studies or clinical series have described factors associated with less common disorders such as Gamekeeper's thumb (Campbell, 1955; Newland, 1992), digital neuritis, and ulnar neuropathy at the wrist (Silverstein, 1986).

42.3 Individual Factors

Some data on individual risk factors, such as age and gender, are available for carpal tunnel syndrome but not for other disorders of the hand and wrist. The risk of CTS increases with age (Stevens, 1988), but in a cross-sectional study of an industrial cohort, age explained only 3% of the variability in median nerve latency (Nathan, 1992). Although CTS is more common among women in the general population, in workplace studies, when employees perform similar hand activities, the ratio of female to male rates is close to 1.2:1 (Franklin, 1991; Nathan, 1992; Silverstein, 1986). Certain female specific factors, such as pregnancy (Eckman-Ordeberg, 1987) are clearly associated with pregnancy; however, the role of other female factors such oophorectomy, hysterectomy (Cannon, 1981; Bjorkquist, 1977; de Krom, 1990), or use of oral contraceptives (Sabour, 1970), is less certain. Other individual factors have strong associations with carpal tunnel syndrome based on multiple studies: diabetes mellitus (Phalen, 1966; Yamaguchi, 1965; Stevens, 1987), rheumatoid arthritis (Phalen, 1966; Yamaguchi, 1965; Stevens, 1987), and obesity (Nathan, 1992; DeKrom, 1990; Falck, 1983; Vessey, 1990; Werner, 1994). For some putative risk factors, the associations are based on single studies on studies presenting conflicting results: thyroid disorders (Phalen, 1966; Hales, 1994), vitamin B6 deficiency (Amadio, 1985; Ellis, 1982; McCann, 1978), wrist size and shape (Johnson, 1983; Armstrong, 1979; Bleeker, 1985), and general de-conditioning (Nathan, 1988, 1992).

TABLE 42.2 Work-Related Factors Associated with Disorders of the Hands and Wrists

Repetition	Mechanical contact
Force	Duration
Posture extremes	Work organization
Vibration	

TABLE 42.3 Controlled Epidemiologic Workplace Studies Evaluating the Association Between Work and Wrist, Hand or Distal Forearm Tendinitis*

Authors	Exposed Population	Control Population	Rate in Exposed	Rate in Control
Luopajarvi et al., 1979[5]	152 bread packaging	133 shop attendants	53%[1]	14%
Silverstein et al., 1986[2,3]	industrial	industrial		
	143 low force/high repetition	136 low force/low repetition	3%	1.5%
	153 high force/low repetition	136 low force/low repetition	4%[1]	1.5%
	142 high force/high repetition	136 low force/low repetition	20%[1]	1.5%
McCormack et al., 1990	manufacturing	manufacturing		
	369 packers/folders	352 knitting workers	3.3%[1]	0.9%
	562 sewers	352 knitting workers	4.4%[1]	0.9%
	296 boarding workers	352 knitting workers	6.4%[1]	0.9%
Kurppa et al., 1991[4,5]	102 meat cutters	141 office workers	12.5%	0.9%
	107 sausage makers	197 office workers	16.3%[1]	0.7%
	118 packers	197 office workers	25.3%[1]	0.7%

* Case criteria are based on history and physical examination.
[1] significant difference from control
[2] adjusted for age, sex, and plant
[3] analysis includes other disorders, although tendinitis was most common
[4] cohort study with 31-month follow-up
[5] all exposed and control subjects are female
From Rempel, D. and Punnet, L., Epidemiology of wrist and hand disorders, in *Musculoskeletal Disorders in the Workplace: Principles and Practice,* eds. M. Nordin et al., Mosby-Year Book, Inc., St. Louis, Missouri, 1997. With permission.

42.4 Work-Related Factors

Table 42.2 summarizes the characteristics of work that have been associated with elevated rates of upper extremity symptoms and specific disorders, including carpal tunnel syndrome and tendinitis. These associations have been observed in multiple studies and in different population groups, while dose–response trends have been seen in several studies. Most studies have been cross-sectional in design, limiting our ability to draw conclusions about causation. The preponderance of evidence, however, suggests strongly that there is a causal relationship between work exposures and upper extremity disorders. Carpal tunnel syndrome and hand–wrist tendinitis have been the best studied; several recent reviews have evaluated the work-relatedness of these disorders and concluded that there is a causal relationship (Stock, 1991; Hagberg, 1992; Kuorinka and Forcier, 1995). Tables 42.3, 42.4, and 42.5 summarize selected studies of wrist and hand tendinitis, carpal tunnel syndrome, and epicondylitis.

Studies using crude measures of exposure have reported associations between repetition and hand/wrist pain and disorders. In a study relying exclusively on nerve conduction measurements, median nerve slowing occurred at a higher rate among assembly line workers than among administrative controls (Nathan, 1992; Hagberg, 1992). Although no systematic assessment of exposure was carried out, the assembly line work was considered more repetitive than the control group. Rate of persistent wrist and hand pain was higher in garment workers performing repetitive hand tasks than in the control group, hospital employees (Punnett, 1985). Persistent wrist pain, or that lasting most of the day for at least one

TABLE 42.4 Selected Controlled Epidemiologic Workplace Studies Evaluating the Association Between Work and Carpal Tunnel Syndrome*

Authors	Exposed Population	Control Population	Criteria	Rate in Exposed	Rate in Control
Silverstein et al., 1987[2]	industrial high force, high repetition	industrial low force, low repetition	history & physical exam	5.1%[1]	0.6%
Nathan, 1988[3,4]	22 keyboard operators	147 administrative/clerical	electrodiagnostic	27%	28%
	164 industrial assembly line	147 administrative/clerical	electrodiagnostic	47%[1]	28%
	115 general plant	147 administrative/clerical	electrodiagnostic	38%	28%
	23 grinders	147 administrative/clerical	electrodiagnostic	61%[1]	28%
Chiang, 1990[2]	frozen food factory	frozen food factory	history and signs		
	37 high repetition	49 low repetition & cold		46%	4%
	121 high repetition & cold	49 low repetition & cold		47%[1]	4%
Barnhart, 1991[3]	106 ski manufacturing repetitive jobs	67 ski manufacturing nonrepetitive jobs	electrodiagnostic and signs	15.4%[1]	3.1%
Osorio, 1994[2,3]	supermarket workers	supermarket workers	history & signs	63%[1]	0.0%
	high exposure	low exposure	electrodiagnostic	33%[1]	0.0%

* Diagnosis based on history and physical exam or nerve conduction study.
[1] significantly different from control group
[2] control for age, gender, years on job
[3] control for age and gender
[4] low participation rate and limited exposure assessment
From Rempel, D. and Punnet, L., Epidemiology of wrist and hand disorders, in *Musculoskeletal Disorders in the Workplace: Principles and Practice,* eds. M. Nordin et al., Mosby-Year Book, Inc., St. Louis, Missouri, 1997. With permission.

TABLE 42.5 Selected Epidemiologic Workplace Studies Evaluating the Association Between Work and Epicondylitis*

Authors	Exposed Population	Control Population Criteria	Rate in Exposed	Rate in Controls
Kurppa, 1991[1]	107 female sausage makers	197 female office workers and supervisors	11.1	1.1
	118 female meatpackers	197 female office workers and supervisors	7.0	1.1
	102 male meat cutters	141 male office workers, maintenance men an d supervisors	6.4	0.9
Chiang, 1993[2]	28 fish processors with high repetition and high force movements of the arms	61 fish processors without high repetition or high force	21.4%	9.8%
	118 fish processors with high repetition or high force movements of the arms	61 fish processors without high repetition or high force	15.3%	9.8%
Roto and Kivi, 1984[2]	90 male meat cutters	77 male construction foremen	8.9%	1.4%
McCormack, 1990[2]	369 manufacturing workers	352 knitting workers	2.2%	1.4%
	562 manufacturing workers	352 knitting workers	2.1%	1.4%
	468 manufacturing workers	352 knitting workers	1.9%	1.4%
	296 manufacturing workers	352 knitting workers	1.0%	1.4%
Viikari-Juntura, 1991[2]	332 meat plant workers	288 office workers, maintenance workers and supervisors	0.6%	0.5%
Luopajärvi, 1979[2]	152 female packers	133 female shop assistants	2.6%	2.3%

* Diagnosis based on history and physical exam.
[1] prospective cohort study: rates are incidence of epicondylitis per 100 workers/yr.
[2] cross-sectional study: rates are prevalence of epicondylitis observed in active workers

month in the last year, occurred in 17% of garment workers and 4% of hospital controls, while persistent hand pain occurred in 27% of garment workers and 10% of controls. Others have observed a similar link between high hand/wrist repetition and carpal tunnel syndrome (Chiang, 1990; Barnhart, 1991) and tendinitis (Kurppa, 1991). The link to repetition may be that these are jobs that require high velocity or accelerations of the wrist (Marras and Schoenmarklin, 1993).

Rates of wrist tendinitis among scissors makers was compared to shop attendants in department stores in Finland. Examinations and histories were systematized and performed by one person. The rates between the groups were not significantly different; however, among the scissors makers the rate of tendinitis increased with increasing number of scissors handled (Kuorinka, 1979). Luopajärvi et al. (1979) compared packers in a bread factory to the same control group. The packers' work involved repetitive gripping, up to 25,000 cycles per day, with maximum extension of thumb and fingers to handle wide bread packages. Approximately half of the packers had wrist/hand tenosynovitis compared to 14% among the controls. The most common disorder of the hand or wrist was thumb tenosynovitis followed by finger/wrist extensor tenosynovitis. CTS was diagnosed in four packers and no controls.

The force applied to a tool or materials during repeated or sustained gripping are also predictors of risk for tendinitis and carpal tunnel syndrome. For example, in a study of the textile industry the risk of hand and wrist tendinitis was 3.9 times higher among packaging and folding workers than among knitters (McCormack, 1990). The packing and folding workers were considered to be performing physically demanding work compared to the knitting workers. Armstrong et al. (1979) observed that women with carpal tunnel syndrome applied more pinch force during production sewing than did their job- and sex-matched controls. It is possible that those with carpal tunnel syndrome altered their working style as the carpal tunnel syndrome progressed; however, it is unlikely that they would increase the pinch force because this would also trigger symptoms. In a study by Moore et al. (1994) at a pork processing plant, the jobs that involved high grip force or long grip durations, such as Wizard knife operator, snipper, feeder, scaler, bagger, packer, hanger, and stuffer, affected almost every employee. Others have observed a similar relationship with work involving sustained or high-force grip in grinders (Nathan, 1992), meatpackers and butchers (Kurppa, 1991; Falck, 1983), and other industrial workers (Thompson, 1951; Welch, 1972).

The most comprehensive study of the combined factors of repetition and force was a cross-sectional study of 574 industrial workers by Silverstein et al. (1986, 1987; Armstrong, 1987). Disorders were assessed by physical exam and history and were primarily tendinitis followed by carpal tunnel syndrome, Guyon tunnel syndrome, and digital neuritis. Subjects were classified into four exposure groups based on force and repetition. The "high-force" work was that requiring a grip force on average of more than 4 kg-force, while "low-force" work required less than 1 kg of grip force. The "high-repetition" work involved a repetitive task in which either the cycle time was less than 30 seconds (greater than 900 times in a work day) or more than 50% of cycle time was spent performing the same kind of fundamental movements. The high-risk groups were compared to the low-risk group after adjusting for plant, age, gender, and years on the job. The odds ratio of all hand/wrist disorders for just high force was 4.9, and it increased to 30 for jobs which required both high-force and high-repetition. The identical analysis of just carpal tunnel syndrome revealed an odds ratio of 1.8 for force and 14 for the combined high-force and high-repetition group. A meta-analysis of Silverstein's data and Luopajarvi study concluded that for high-force and high-repetition work the common odds ratio for carpal tunnel syndrome was 15.5 (95% C.I. 1.7–141) and for hand/wrist tendinitis it was 9.1 (95% C.I. 5–16) (Stock, 1991). Estimates of the percentage of CTS cases among workers who perform repetitive or forceful hand activity that can be attributed to work range from 50 to 90% (Hagberg, 1992; Cummins, 1992; Tenaka, 1994).

With regard to epicondylitis, the individual roles played by force and repetition are less clear. One cohort study and six cross-sectional studies have evaluated the incidence or prevalence of epicondylitis in relation to specific jobs, which were characterized by high force, high repetition, or both. Kurppa (1991) found a relative risk of 6.4 for epicondylitis in jobs with high repetition, some of which also involved high force. One cross-sectional study found a significantly elevated risk of epicondylitis only among recently employed workers in high-repetition or high-repetition/high-force jobs (Chiang, 1993).

Another cross-sectional study found an odds ratio of 6.9 epicondylitis in a high-repetition, high-force job (Roto and Kivi, 1984). This odds ratio was not statistically significant. Four other cross-sectional studies found little or no increase in risk for epicondylitis in workers involved in jobs characterized by high force and/or high repetition. (McCormack, 1990; Viikari-Juntura, 1991; Luopajärvi, 1979; Dimberg, 1987)

Work involving increased wrist deviation from a neutral posture in either the extension, flexion or ulnar, radial direction has been associated with carpal tunnel syndrome and other hand and wrist problems (Thompson, 1951; Hoffman, 1981; Tichauer, 1966). De Krom et al. (1990) conducted a case-control study of 156 subjects with carpal tunnel syndrome compared to 473 controls randomly sampled from the hospital and population registers in a region of the Netherlands. After adjusting for age and sex, a dose–response relationship was observed for increasing hours of work with the wrist in extension or flexion. No risk was observed for increasing hours performing a pinch grasp or typing. Some studies of computer operators have linked awkward wrist postures to severity of hand symptoms (Faucett, 1994), risk of tendinitis or carpal tunnel syndrome (Seligman, 1986), arm and hand discomfort (Sauter, 1991; Duncan, 1974; Hunting, 1981).

Prolonged exposure to vibrating hand tools, such as chain saws, has been linked in prospective studies to Hand Arm Vibration Syndrome (Ekenval, 1987; Futatsuka, 1986). The risks are primarily vibration acceleration amplitude, frequency, hand coupling to tool, hours per day of exposure, and years of exposure. However, based on existing studies, there is no clear vibration acceleration/frequency/duration threshold that would protect most workers. Therefore, medical surveillance is recommended to identify cases early while the disease can still be reversed (NIOSH, 1989). Use of vibrating hand tools may also increase the risk of CTS (Seppalainen, 1970; Cannon, 1981; Rothfleisch, 1978) indirectly by increasing applied grip force through a reflex pathway (Radwin, 1987).

Prolonged or high-load localized mechanical stress over tendons or nerves from tools or resting the hand on hard objects have been associated with tendinitis (Tichauer, 1966) and nerve entrapments (Phalen, 1966; Hoffman, 1985) in case studies.

The average total hours per day that a task is repeated or sustained has been a factor in predicting hand problems (Margolis, 1987; Macdonald, 1988). Among computer operators increasing self-reported hours of computer use has been a predictor of symptom intensity or disorder rate in all (Faucett, 1994; Burt, Bernard, 1993; Oxenburgh, 1985; Maeda, 1982; Hunting, 1981). DeKrom et al. (1990) did not observe a relationship between CTS and hours of computer use.

Work organizational (work structure, decision control, work load, deadline work, supervision) and psychosocial factors (job satisfaction, social support, relationship with supervisor) appear to have some influence on hand and wrist symptoms among computer users. Among newspaper reporters and editors, work organizational factors modified the expected relationship between workstation design and hand and wrist symptoms. Symptom intensity increased as keyboard height increased among those with low decision latitude but not among those with high decision latitude (Faucett, 1994). In another study of newspaper employees, the risk of hand and wrist symptoms was increased among those with increasing hours on deadline work and less support from the immediate supervisor (Bernard, 1993). Among directory assistance operators at a telephone company, high information processing demands were associated with an elevated rate of hand and wrist disorders (Hales, 1994). On the other hand, in the industrial setting, Silverstein et al. (1986) observed no effect on job satisfaction.

42.5 Summary

The lack of prospective studies and an uncertainty about the precise pathophysiologic mechanisms involved limits our ability to definitively identify causative factors. Nonetheless, current studies point to a multifactorial relationship between work exposures and disorders of the hand, wrist, and elbow. Symptom severity and disorder rate appear to be influenced by work organizational factors, such as decision latitude and cognitive demands. Some disorders, such as tendinitis and carpal tunnel syndrome, are clearly associated with work involving repetitive and forceful use of the hands. It seems likely that there is a causal relationship between some work exposures and these disorders. For other disorders, such as epicondylitis and osteoarthritis, the relationship to work exposures is less clear, although current data are suggestive. Carpal tunnel syndrome has been linked to individual factors in population-based studies and in clinical case series. However, in workplace studies where workplace exposures are adequately quantified, individual factors play a limited role relative to workplace factors (Cannon, 1981; Silverstein, 1987; Armstrong, 1979; Franklin, 1991; Faucett, 1994; Hales, 1994).

References

Adams ML, Franklin GM, Barnhart S. Outcome of carpal tunnel surgery in Washington State workers' compensation. *Am J Ind Med* 1994; 25:527-536.

Al-Qattan MM, Bowen V, Manktelow RT. Factors associated with poor outcome following primary carpal tunnel release in non-diabetic patients. *J Hand Surg* (Br Volume) 1994; 19B:622-625.

Amadio PC. Pyridoxine as an adjunct in the treatment of carpal tunnel syndrome. *J Hand Surg* 1985; 10A:237-241.

Armstrong TJ, Langolf GD. Ergonomics and occupational safety and health, in *Environmental and Occupational Medicine,* ed WN Rom, Little, Brown Co, Boston, 1982, pp. 765-784.

Armstrong TJ, Chaffin DB. Carpal tunnel syndrome and selected personal attributes. *J Occup Med* 1979; 21:481-486.

Armstrong TJ, Foulke JA, Joseph BS, Goldstein SA. Investigation of cumulative trauma disorders in a poultry processing plant. *Am Ind Hygiene Assoc J* 1982; (43)2:103-116.

Armstrong TJ, Buckle P, Fine LJ et al. A conceptual model for work-related neck and upper-limb musculoskeletal disorders. *Scand J Work Environ Health* 1993; 19:73-84.

Bammer G. VDUs and musculoskeletal problems at the Australian National University, in *Work with Display Units 86,* Eds Knave B and Wideback PG, Elsevier Science Publishers B.V. North-Holland, 1987.

Barnhart S, Demers PA, Miller M, Longstreth WE, Rosenstock L. Carpal tunnel syndrome among ski manufacturing workers. *Scand J Work Environ Health* 1991; 17:46-52.

Bernard B, Sauter S, Peterson M, Fine L, Hales T. Health Hazard Evaluation Report: Los Angeles Times. U.S. Department of Health and Human Services, Public Health Service, Centers for Disease Control, National Institute for Occupational Safety and Health, NIOSH Report No. 90-013-2277. 1993.

Birkbeck MQ, Beer TC. Occupation in relation to the carpal tunnel syndrome. *Rheumatol Rehabil* 1975; 14:218-221.

Bjorkqvist SE, Lang AH, Punnonen R, Rauramo L. Carpal tunnel syndrome in ovariectomized women. *Acta Obstet Gynecol Scand* 1977; 56:127-130.

Bleeker MQ, Bohlman M, Moreland R, Tipton A. Carpal tunnel syndrome: role of carpal canal size. *Neurology* 1985; 35:1599-1604.

Burt S, Hornung R, Fine L, Silverstein B, Armstrong T. Health hazard evaluation report: Newsday. U.S. Department of Health and Human Services, Public Health Service, Centers for Disease Control, National Institute for Occupational Safety and Health, NIOSH Report No. 89-250-2046. 1990.

Campbell, CS. Gamekeeper's Thumb. *Journal of Bone and Joint Surgery* 1955; 37(B) 1:148-149.

Cannon LJ, Bernacki EJ, Walter SD. Personal and occupational factors associated with carpal tunnel syndrome. *J Occup Med* 1981; 23:255-258.

Centers for Disease Control: Occupational disease surveillance — carpal tunnel syndrome. *MMWR* 1989; 38:485-489.

Cheadle A, Franklin G, Wolfhagen C, Savarino J, Liu PY, Salley C, Weaver M. Factors influencing the duration of work-related disability: a population-based study of Washington State Workers' Compensation. *Am J Pub Health* 1994; 84:190-196.

Chiang HC, Chen SS, Yu HS, Ko YC. The occurrence of carpal tunnel syndrome in frozen food factory employees. *Kaohsiung J Med Sci* 1990; 6:73-80.

Chiang HC, Ko YC, Chen SS, Yu HS, Wu TN, Chang PY. Prevalence of shoulder and upper-limb disorders among workers in the fish-processing industry. *Scand J Work Environ Health* 1993; 19:126-131.

Conn HR. Tenosynovitis. *Ohio State Med J* 1931; 27:713-716.

de Krom M, Kester A, Knipschild P, Spaans F. Risk factors for carpal tunnel syndrome. *Am J Epi* 1990, 132:1102-1110.

Duncan J, Ferguson D. Keyboard operating posture and symptoms in operating. *Ergonomics* 1974; 17:651-662.

Ekenvall L, Carlsson A. Vibration white finger: a follow up study. *Br J Ind Med* 1987; 44:476-478.

Ekman-Ordeberg G, Salgeback S, Ordeberg G. Carpal tunnel syndrome in pregnancy: A prospective study. *Acta Obstet Gynec Scand* 1987; 66:233-235.

Ellis J, Folkers K, Watanabe T et al. Clinical results of a cross-over treatment with pyridoxine and placebo of the carpal tunnel syndrome. *J Clin Nutr* 1979; 2046-2070.

Falck B and Aarnio P. Left-sided carpal tunnel syndrome in butchers. *Scand J Work Environ Health* 1983; 9:291-297.

Faucett J and Rempel D. VDT-related musculoskeletal symptoms: Interactions between work posture and psychosocial work factors. *Am J Ind Med* 1994; 26:597-612.

Fine LJ, Silverstein BA, Armstrong TJ et al. Detection of cumulative trauma disorders of upper extremities in the workplace. *J Occup Med* 1986; 28:674-678.

Franklin GM, Haug J, Heyer N, Checkoway H, Peck N. Occupational carpal tunnel syndrome in Washington State, 1984-1988. *Am J Pub Health* 1991, 81:741-746.

Franzblau A, Werner R, Valle J, Johnston E. Workplace surveillance for carpal tunnel syndrome: a comparison of methods. *J Occup Rehab* 1993; 3:1-14.

Franzblau A, Werner RA, Albers JW, Grant CL, Olinski D, Johnston E. Workplace surveillance for carpal tunnel syndrome using hand diagrams. *J Occup Rehab* 1994; 4:185-198.

Futatsuka M, Ueno T. A follow-up study of vibration-induced white finger due to chain-saw operation. *Scand J Work Environ Health* 1986; 12:304-306.

Hadler N, Gillings D, Imbus H et al. Hand structure and function in an industrial setting. *Arthritis and Rheum* 1978, 21:210-220.

Hagberg M, Morgenstern H, Kelsh M. Impact of occupations and job tasks on the prevalence of carpal tunnel syndrome. *Scand J Work Environ Health* 1992; 18: 337-345.

Hales TR, Sauter SL, Peterson MR, Fine LJ, Putz-Anderson V, Schleifer LR, Ochs TT, Bernard BP. Musculoskeletal disorders among visual display terminal users in a telecommunications company. *Ergonomics* 1994; 10:1603-1621.

Heller L, Ring H, Costeff H, Solzi. Evaluation of Tinel's and Phalen's signs in diagnosis of carpal tunnel syndrome. *Eur Neurol* 1986; 25:40-42.

Hoffman J, Hoffman PL. Staple gun carpal tunnel syndrome. *J Occup Med* 1985; 27:848-849.

Hünting W, Läubli T, Grandjean E. Postural and visual loads at VDT workplaces. *Ergonomics* 1981; 24:917-931.

Johnson EW, Gatens T, Poindexter D, Bowers D. Wrist dimensions: correlation with median sensory latencies. *Arch Phys Med Rehab* 1983; 64:556-557.

Katz JN, Stirrat CR, Larson MG, Fossel AN, Eaton HM, Liang MH. A self-administered hand symptom diagram for the diagnosis and epidemiologic study of carpal tunnel syndrome. *J Rheumatol* 1990; 3:1-14.

Katz JN, Larson MG, Fossel AH, Liang MH. Validation of a surveillance case definition of carpal tunnel syndrome. *Am J Public Health* 1991; 81:189-193.

Kaji H, Honma H, Usui M, Yasuno Y, Saito K. Hypothenar Hammer Syndrome in workers occupationally exposed to vibrating tools. *J Hand Surg* (Br Volume) 1993; 18B:761-766.

Kivi P. Rheumatic disorders of the upper limbs associated with repetitive occupational tasks in Finald in 1975-1979. *Scand J Rheum* 1984; 13:101-107.

Kuorinka I, Koskinen P. Occupational rheumatic diseases and upper limb strain in manual jobs in a light mechanical industry. *Scand J Work Environ Health* 1979, 5:39-47.

Kourinka I, Forcier L. (eds.) *Work Related Musculoskeletal Disorders: A Reference Book for Prevention.* London: Taylor & Francis, 1995.

Kurppa K, Viikari-Juntura E, Kuosma E, Huuskonen M, Kivi P. Incidence of tenosynovitis or peritendinitis and epicondylitis in a meat processing factory. *Scand J Work Environ Health* 1991; 17:32-37.

Little JM, Ferguson DA. The incidence of Hypothenar Hammer Syndrome. *Arch Surg* 1972; 105:684-685.

Luopajarvi T, Kuorinka I, Virolainen M, Holmberg M. Prevalence of tenosynovitis and other injuries of the upper extremities in repetitive work. *Scand J Work Environ Health* 1979. 5:48-55.

Marras WS, Schoenmarklin RW. Wrist motions in industry. *Ergonomics* 1993; 36: 341-351.

Maeda K, Hunting W, Grandjean E. Factor analysis of localized fatigue complaints of accounting-machine operators. *J Human Ergol* 1982; 11:37-43.

Magnusson M, Ortengren R. Investigation of optimal table height and surface angle in meatcutting. *Applied Ergonomics* 1987; 18.2:146-152.

Masear VR, Hayes JM, Hyde AG. An industrial cause of carpal tunnel syndrome. *J Hand Surg* 1986; 11A:222-227.

McCormack RR Jr., Inman RD, Wells A, Berntsen C, Imbus HR. Prevalence of tendinitis and related disorders of the upper extremity in a manufacturing workforce. *J Rheumatol* 1990; 17:958-964.

Moore JS, Garg A. Upper extremity disorders in a pork processing plant: relationships between job risk factors and morbidity. *Am Ind Hyg Assoc J* 1994, 55:703-715.

Muffly-Elsey D, Flinn-Wagner S. Proposed screening tool for the detection of cumulative trauma disorders of the upper extremity. *J Hand Surg* 1987, 12A: 2(2), 931-935.

Nathan PA, Keniston RC, Myers LD, Meadows KD. Obesity as a risk factor for slowing of sensory conduction of the median nerve in industry. *J Occup Med* 1992; 34:379-383.

Nathan PA, Meadows KD, Doyle LS. Occupation as a risk factor for impaired sensory conduction of the median nerve at the carpal tunnel. *J Hand Surg* 1988; 13B: 167-170.

NIOSH Criteria for a recommended standard. Occupational exposure to hand-arm vibration. DDHS Publication No. 89-106. 1989. National Institute for Occupational Safety and Health. Cincinnati, Ohio.

NIOSH National Occupational Research Agenda. DDHS Publication No. 96-1115. 1996. National Institute for Occupational Safety and Health. Cincinnati, Ohio.

Newland, CC. Gamekeeper's Thumb. Orthopedic Clinics of North America 1992; 23(1):41-48.

Nilsson T, Burström L, Hagberg M. Risk assessment of vibration exposure and white fingers among platers. *Int Arch Occup Environ Health* 1989; 61: 473-481.

Ohara H, Aoyama H, Itani T. Health hazards among cash register operators and the effects of improved working conditions. *J Human Ergology* 1976; 5:31-40.

Osorio AM, Ames RG, Jones JR, Rempel D, Castorina J, Estrin W, Thompson D. Carpal tunnel syndrome among grocery store workers. *Am J Ind Med* 1994, 25:229-245.

Oxenburgh M, Rowe S, Douglas D. Repetitive strain injury in keyboard operators. *J Occup Health and Safety — Australia and New Zealand* 1985; 1:106-112.

Park CH, Wagener DK, Winn DM, Pierce JP. Health conditions among the currently employed: United States, 1988. National Center for Health Statistics. *Vital Health Stat* 1993. 10(186).

Phalen GS. The carpal-tunnel syndrome. *J Bone Joint Surg* 1966; 48A:211-228.

Punnett L, Robins JM, Wegmen DH, Keyserling WM. Soft tissue disorders in the upper limbs of female garment workers. *Scand J Work Environ Health* 1985; 11:417-425.

Radwin RG, Armstrong TJ, Chaffin DB. Power hand tool vibration effects on grip exertions. *Ergonomics* 1987; 30:833-855.

Rempel D, Punnet L, Epidemiology of wrist and hand disorders, in *Musculoskeletal Disorders in the Workplace: Principles and Practice,* eds M Nordin et al., Mosby-Year Book, Inc., St. Louis, Missouri, 1997.

Roto P, Kivi P. Prevalence of epicondylitis and tenosynovitis among meatcutters. *Scand J Work Environ Health* 1984; 10:203-205.

Sauter SL, Schleifer LM, Knutson SJ. Work posture, workstation design, and musculoskeletal discomfort in a VDT data entry task. *Human Factors* 1991; 33:151-167.

Seligman P, Boiano J, Anderson C. Health Hazard Evaluation of the Minneapolis Police Department. NIOSH HETA 84-417-1745, 1986. U.S. Department of Commerce, NTIS, Springfield, Virginia.

Seppalainen AM. Nerve conduction in the vibration syndrome. *Scand J Work Environ Health* 1970; 7:82-84.

Silverstein BA, Armstrong T, Longmate A, Woody D. Can in-plant exercise control musculoskeletal symptoms? *J Occup Med* 1988; 30: 922-927.

Silverstein BA, Fine LJ, Armstrong TJ. Hand wrist cumulative trauma disorders in industry. *Br J Ind Med* 1986. 43:779-784.

Silverstein BA, Fine LJ, Armstrong TJ. Occupational factors and carpal tunnel syndrome. *Am J Ind Med* 1987. 11:343-358.

Silverstein BA, Fine LJ, Stetson D. Hand-wrist disorders among investment casting plant workers. *J Hand Surg* 1987; 12A (5 part 2): 838-844.

Stevens JC, Sun S, Beard CM, O'Fallon WM, Kurland LT. Carpal tunnel syndrome in Rochester, Minnesota. 1961 to 1980. *Neurology* 1988; 38:134-138.

Stock SR. Workplace ergonomic factors and the development of musculoskeletal disorders of the neck and upper limbs: A meta-analysis. *Am J Ind Med* 1991; 19:87-107.

Tanaka S, Wild DK, Seligman PJ, Behrens V, Cameron L, Putz-Anderson V. The U.S. prevalence of self-reported carpal tunnel syndrome. *Am J Public Health* 1994; 84:1846-1848.

Thompson A, Plewes L, Shaw E. Peritendinitis crepitans and simple tenosynovitis: a clinical study of 544 cases in industry. *Br J Ind Med* 1951; 8:150-160.

Tichauer E. Some aspects of stress on forearm and hand in industry. *J Occup Med* 1966; 8:63-71.

Viikari-Juntura E. Neck and upperlimb disorders among slaughterhouse workers. *Scand J Work Environ Health,* 1983; 9:283-290.

Vessy MP, Villard-MacIntosh L, Yeates D. Epidemiology of carpal tunnel syndrome in women of child-bearing age. Finding in a large cohort study. *Am J Epi* 1990; 19:655-659.

Webster BS, Snook SH. The cost of compensable upper extremity cumulative trauma disorders. *J Occup Med* 1994; 36:713-7.

Welch R. The causes of tenosynovitis in industry. *Indust Med* 1972; 41:16-19.

Werner RA, Albers JW, Franzblau A, Armstrong TJ. The relationship between body mass index and the diagnosis of carpal tunnel syndrome. *Muscle & Nerve* 1994; 17:632-636.

Williams, Cope et al. Metacarpo-phalangeal arthropathy associated with manual labor (Missouri metacarpal syndrome). *Arthritis and Rheum* 1987; 30:1362-1371.

Yamaguchi D, Liscomb P, Soule E. Carpal tunnel syndrome. *Minn Med J* 1965; January 22-23.

43

Integrated Analysis of Upper Extremity Disorders

Richard Wells
University of Waterloo

43.1 Introduction

Work and activity-related musculoskeletal disorders (WMSD) have a complex multifactorial etiology including not only the physical aspects of the activities that people perform but also the psychosocial aspects. These disorders may involve muscular, tendinous, ligamentous, nervous tissues and include both acute (overexertion) as well as chronic (overuse) onset. A number of sources of information ranging from biomechanics, epidemiology, and clinical case series have identified a number of major risk factors associated with the development of upper limb musculoskeletal disorders. (For reviews, see Stock, 1992 or Hagberg et al., 1995.) These include forcefulness, adverse posture, repetition or continuous activity, angular velocity and acceleration, or joints and duration of exposure. Plausible biological mechanisms by which these risk factors may result in disorders of the musculoskeletal system have been proposed. Despite this, our best evidence points to a complex interaction of physical, psychosocial, and individual factors in the development of musculoskeletal disorders at work.

An integrated approach to the causation of WMSD helps us understand the many simultaneous and interacting physical stressors which act on the upper limb during activity. These approaches help form a bridge between the performance of work and the cellular and other descriptions of the degenerative/inflammatory processes involved in work and activity-related musculoskeletal disorders. An integrated approach

0-8493-2641-9/99/$0.00+$.50

also guides us in the construction and evaluation of workplace assessment tools. In the sections which follow, concepts important to the assessment of the risk factors of force and posture are reviewed prior to analyzing the features of a number of assessment tools.

43.2 Site and Types of Upper Limb Work-Related Musculoskeletal Disorders

Terminology describing work-related musculoskeletal disorders (WMSD) has become extremely convoluted; for example, in the U.S., where the term of preference is cumulative trauma disorders (CTD), disorders in visual display terminal (VDT) operators are called repetitive strain injuries (RSI). In this chapter WMSD refers to all disorders of the musculoskeletal system (both upper extremity and low back and limbs) both to specific tissue as well as nonspecific symptoms and syndromes where associations with work have been found (cf. Hagberg et al., 1995).

An examination of Figure 43.1 reveals that a large number of types of tissues have been identified as being affected by work: tendon, muscle, nerve, and joint. The disorders identified are found in a wide variety of locations in the hand, forearm, arm, shoulder, and neck. How can we possibly devise methods which will allow us to predict injury risk in such a wide variety of sites and tissues? Fortunately, quantification of the external loads applied to the upper limb and its posture have been successful in describing the differences between jobs and tasks with high versus low risks of developing WMSD. Technically this is known as a low specificity of effect; a specific work factor can cause a number of different musculoskeletal disorders in a number of anatomic sites (Hagberg et al., 1995). This is likely so because increasing the external demand, in terms of increased force or frequency of exertion, increases the demands on most of the tissues (internal) of the arm and shoulder. While this makes the development of workplace assessment tools simpler, it makes it more difficult to examine causation and the mechanisms of disorders.

43.3 Risk Factors for Upper Limb Work-Related Musculoskeletal Disorders

Sources of information ranging from biomechanics, epidemiology and clinical case series have identified a number of major extrinsic (external) risk factors associated with the development of upper limb musculoskeletal disorders. These include forcefulness, adverse posture, repetition or continuous activity, joint angular velocity and acceleration, and duration of exposure. In addition, there are a number of potentiating factors which are commonly mentioned including, cold, vibration, and use of gloves (Hagberg et al., 1995). The following sections explore some concepts useful in the quantification of time and posture.

Time as an Integral Part of Risk Factor Description

The time or frequency characteristics of tasks have typically been described by the term "repetitiveness." Unfortunately, this word is so often used and overused as to make such terms as "repetitive job" and "highly repetitive" almost meaningless. No clear definition of the term is usually offered, which compounds the lack of clarity.

In general the word is used in three main ways. First, it is used as a qualitative term to describe both the high frequency of actions as well as the sameness or monotony of the job. Second, it has been used to describe fast manual work with little apparent rest between movements. Third, repetitive work can be quantified by the number of parts, efforts, keystrokes or wrist movements/per unit time. Perhaps the most widely used operational definition of repetitive is that of Silverstein and colleagues: work with a cycle time of less than 30 seconds or having a repeated sub-cycle lasting more than 50% of the main cycle was categorized as being highly repetitive (Silverstein et al., 1986). Marras and colleagues have

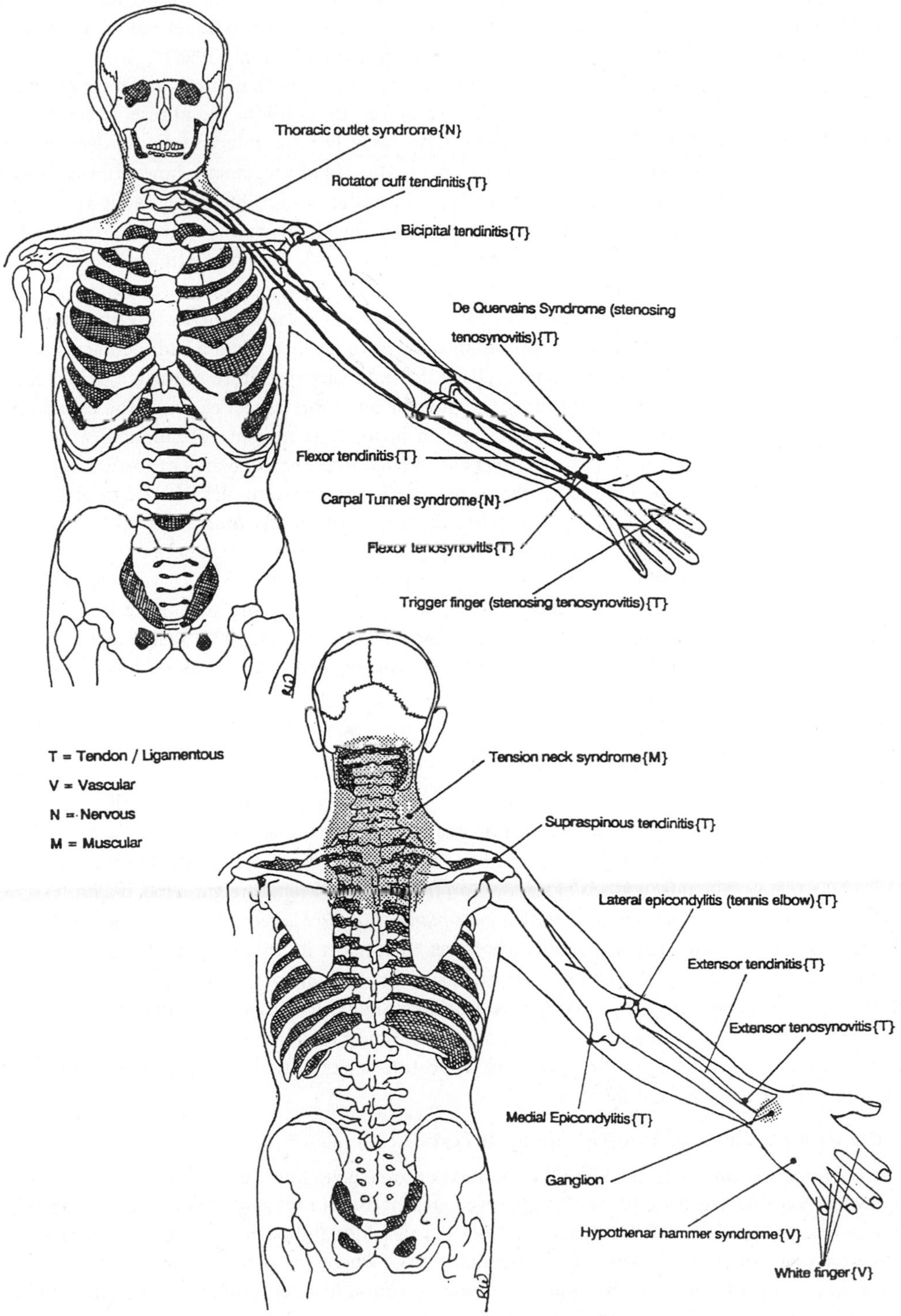

FIGURE 43.1 Schematic of the upper limb showing examples of the sites and tissues potentially involved in work- and activity-related musculoskeletal disorders: {T} = tendon-related disorders; {N} = nerve-related disorders; {M} = muscle-related disorders; {V} = vascular disorders.

developed approaches to quantifying the time-varying nature of body motions using angular velocity and acceleration of both the wrist and trunk which have also been shown to be related to risk of injury, (Marras and Schoenmarklin, 1993; Marras et al., 1993; Schoenmarklin et al., 1994)

It can be noted that many of these definitions rate the frequency of both motions and force generation; in general, each risk factor has an associated time variation. In addition, the phrase, "repetitive job," ignores the various functions of the different parts of the body. For example, a keyboard data entry task is thought of as repetitive; true, the fingers have a high frequency of movement; however, the forearms, shoulders, and back have almost constant and unchanging (static) posture and muscle activity. Westgaard and Winkel have argued that each risk factor should be described by its intensity, time variation, and its duration (Westgaard and Winkel, 1994; Winkel and Mathiassen, 1994). Ideally, the time dimension should allow the effects of different work organizations to be predicted: the effects of micropauses, of different work/rest ratios, of different break schedules, of rotation and work enlargement. At this time our knowledge does not permit us to deal with this important dimension at more than a rudimentary level. This argues for research into better ways of characterizing the time-varying nature of the major risk factors.

In the assessment of injury risk, tools must account for one more aspect of time. The estimation of risk factors which relate either to the highest demand or to some measure of cumulative or average loading. Despite the common notion that WMSD's are related to the accumulated exposure over months or years, i.e., cumulative trauma disorders, there are surprisingly few examples where cumulative exposure-response relationships have been demonstrated; most associations found are between exposure intensity and WMSD. For example, Stenlund et al., (1992) found relationships between the cumulative load lifted and arthrosis of the acromioclavicular joint in the shoulder, and Kumar (1990) found that workers with back pain had higher cumulative loads on the spine. In a similar manner, long periods in non-neutral postures have been associated with back pain (Punnett et al., 1991) Relationships have also been found for maximum loads; for example, one of the stronger predictors found by Marras et al. (1993) for low back pain was from maximum hand load.

Posture as a Risk Factor

Postures of the limbs and trunk have a long history in characterizing tasks because, unlike many other risk factors, they are often observable and quantifiable without instrumentation. Posture is an important element of task analysis because it can be related to a number of injury mechanisms. In general, posture can give information about four kinds of stressors on the musculoskeletal system. First, if a limb segment is inclined with respect to the line of gravity a *joint moment of force* is required about the proximal end with the necessity for muscular or ligamentous forces to support it. Second, a joint angle close to the end range of motion ("extreme" posture) will load ligaments and may compress blood vessels and nerves. Third, joint angles away from the joint's optimal working range will change the geometry of the muscles crossing the joint, possibly impairing the optimal functioning of joints or tendons around the joint. Fourth and last, the change (or lack of change) in posture may be used to characterize the frequency (repetitiveness) or the static nature of the task.

Posture as a Predictor of Joint Moment of Force

As the previous section illustrated, the joint moment of force gives important insights into tissue loads. As body segments deviate from the vertical, the ever-present force of gravity acts on the mass of the body segment: the "hidden load" of the arm mass about the shoulder and particularly the trunk mass about the low back are important, especially in sedentary tasks where the posture may be maintained for substantial periods of time. This is frequently termed postural load. If weights are held in the hand, a moment is usually created by the load in non-neutral postures.

"Extreme" Posture as a Predictor of Soft Tissue Loads

Usually the extremes of a joint's motion are constrained by ligaments: use of extreme posture during work may not be desirable. For example, in the low back during "stoop" lifts, flexion of the lumbar spine creates tension in the posterior ligaments of the spine, and in many people a "flexion relaxation" phenomenon is seen whereby the extensor muscles of the spine become inactive and the ligaments support the moment (McGill and Norman, 1993). The drawback and potential risk in this for low back injury is that if there are unexpected loads or slipping, the only structures which can support the extra loads are the ligaments. If the posture is held for long periods of time, for example in steel reinforcement workers or gardeners, creep of the spinal ligaments and a change in the stability of the spine may result.

Joint Posture and Optimal Musculoskeletal Geometry

For each joint there is a range of posture which minimizes possible adverse features of work and which allows effective force application with minimum fatigue and injury potential. Even before an "extreme" posture is reached, there are changes in the function of the musculoskeletal system which usually make the postures less than optimal and which may elevate tissue loads.

For example, at the wrist, extension of greater than about 30 degrees increases intracarpal pressure, even in normal people, above 30 mmHg (Rempel et al., 1995). This pressure, if maintained for substantial periods of time, likely decreases microcirculation of the structures in the tunnel, including the median nerve. This may be one of the mechanisms by which work activities cause carpal tunnel syndrome. Another example at the wrist involves grasping a small object with the wrist in flexion. This can require large effort, and forcing the wrist into maximal flexion will usually cause the object to be dropped. This is the basis of a number of actions in self-defense. This example shows that nonoptimal postures require higher efforts to perform a given task. Large deviation from approximately neutral postures can also affect blood supply: looking upward, as during the picking of fruit, can compromise cerebral blood flow especially if coupled with neck twist (Sakakibara et al., 1987).

Each joint has an optimal position for different work activities; it is often near the midpoint of the range of motion, but this rule has sufficient exceptions to make it unreliable. For example, the knee functions very well close to the extreme straight position during most locomotor tasks.

Change of Posture

Work involves changes in posture, and the changes can be used to quantify the frequency of movements. Frequency of activity is described further in a later section. If postures do not change for long periods of time, such as shoulder and trunk posture during computer (VDT) work, the task may be called static.

Force as a Risk Factor

Despite the existence of a large number of external risk factors (Figure 43.2), it can be argued that the final common pathway by which work causes or contributes to the development of WMSD is force. External loads and postures give rise to "internal exposures" in the tissues of the upper limb. Thus a fingertip force may give rise to tensile force in the finger flexor tendons; simultaneous wrist flexion stretches the wrist and finger extensors, increasing their passive tensile force (Keir et al., 1996); the wrist flexion also increases the hydrostatic pressure in the carpal canal. These forces have a different effect on each tissue which will be discussed in more detail shortly.

Externally we may wish to measure the force required or the force exerted by the hand (which may be considerably more, depending on the friction and size of the object). We may wish to measure the absolute force or the force relative to an individual's capacity.

It can be seen that posture is most frequently a modifier of or a predictor of the loads experienced by the tissue. The foregoing shows why posture is such a valuable measure of workplace risks: only when external loads are applied is it of limited use in workplace evaluation. The internal exposure to force, however, remains central; other risk factors affect this directly or indirectly.

43.4 Integrated Approach to Evaluate Potential for Upper Limb WMSD

Figure 43.2 illustrates an integrated approach to evaluate the potential for upper limb WMSD: the "external" factors of force, posture, time variation (repetitiveness), and duration of exposure act on the

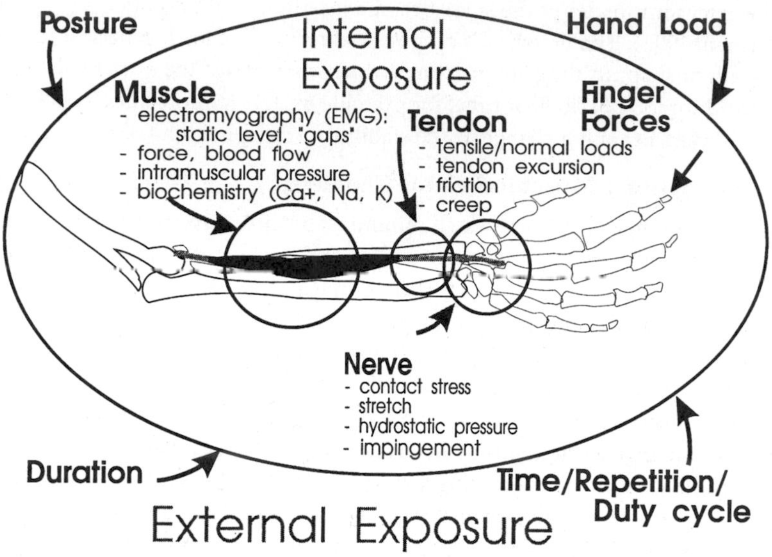

FIGURE 43.2 Illustration of the distinction between "external exposures" to the body and "internal exposures" to individual tissues. Under each tissue some of the measures for evaluating the potential for developing work-related musculoskeletal disorders are listed.

musculoskeletal system. These create "internal exposures" to the tissues of the body. It is these internal exposures which stress the individual tissues and which must be resisted: it is at this level which injury mechanisms can be tested using histological, physiological, or electrophysiological techniques. For example, through the use of external forces (An et al., 1987) and limb accelerations (Marras and Schoenmarklin, 1993) moments of force at the elbow during work can be predicted. Because the musculoskeletal system is mechanically indeterminate, i.e., there are more force-producing structures than equations of equilibrium to describe the system, analysis to predict the load in the individual tissues demands either that assumptions be made concerning how muscles are recruited be made or some criterion measure is minimized through optimization approaches (An et al., 1987; Wells et al., 1995). In addition, through the incorporation of the biological materials properties the response of the tissue to load can be grossly predicted. The models available to study activity-related musculoskeletal disorders are in preliminary stages of development. This is complicated by the lack of a good animal model of these disorders and the delicate balance of physical strain and restorative responses.

In general these "internal exposures" are the subject of laboratory-based research rather than workplace assessment, but they are important because they help us conceptualize the best external variables to measure and the best way in which to evaluate them. For example, in the depiction of manual material handling tasks to elucidate the link between work and low back pain, one could describe the load lifted by a person, and separately, the distance away from the body of this mass. It has been found more useful to compute the joint moment; the product of the force and the distance (Marras et al., 1993). This is done based on a biomechanical model which demonstrates that tissue loads (internal exposures) are better reflected by moment than either load or posture separately. Similar arguments can be made at the shoulder.

Figure 43.3 depicts a conceptual model of the factors influencing the development of work-related carpal tunnel syndrome. The pathways indicated a way in which the multiple demands of work may combine to reduce or elevate potential for trauma to the tissue. Take, for example, work in the cold with a vibrating tool; gloves will also likely be worn. The diagram illustrates how the presence of gloves may increase grip force; the presence of cold and vibration may decrease tactile sensation and further increase grip force.

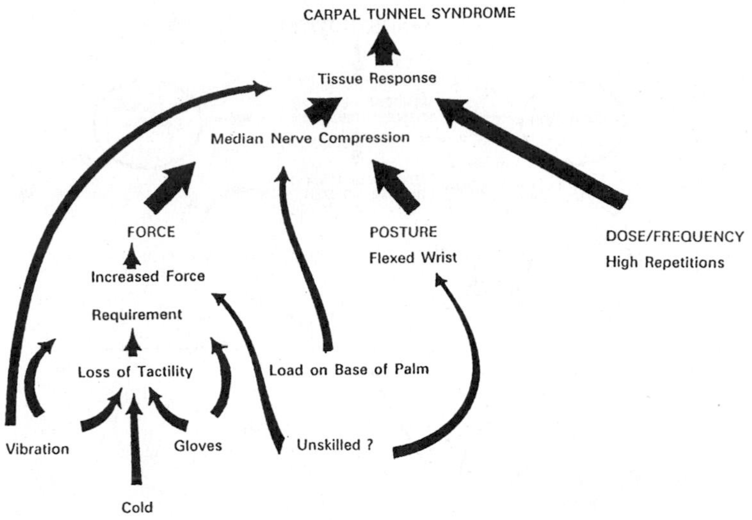

FIGURE 43.3 Conceptual model of the relationship between external risk factors and the development of carpal tunnel syndrome due to both compression of the median nerve by the flexor tendons and changes in carpal tunnel geometry due to posture alone. Across the center are listed the major (external) risk factors of force, posture and time. Below are listed risk factors which may elevate the effect of these primary factors, and above are noted possible pathways by which the external exposures (risk factors) create internal exposures that could plausibly lead to the development or aggravation of CTS. Many additional factors can alter the forces in the flexor tendons such as the grip type, number of digits used, and the angular acceleration of the wrist. Lack of skill may lead to poorer postures, higher grip forces, or increased coactivation of muscles, more movements to perform a given task, or jerkier (higher acceleration) motions.

Building on this concept, Moore et al., (1991) described an integrated approach using biomechanical modeling to predict internal exposure variables likely related to injury. This is achieved by the synthesis of posture, force, movement, and muscle loading data. The model produced a profile of measures for use in industrial settings, that reflect the loading on the different tissues affected by WMSD (nerves, tendons, muscles) and the different loading mechanisms (e.g., highly static postures, repeated extreme postures, dynamic movements). The measures involve continuous monitoring of hand/wrist postures, forces, and muscle activations (electromyographic signals) over the duration of the task both in the arms and shoulders. Using the previously described measures as input to a biomechanical model of the forearm and hand, a profile of 12 risk factors was created which characterized the demand of the task on the distal upper limb. The 12 variables were: peak tendon force, cumulative tendon force, cumulative tendon excursion, peak tendon excursion velocity, average pressure on the flexor retinaculum (and thus the median nerve), peak pressure on the flexor retinaculum, cumulative pressure on the flexor retinaculum, cumulative frictional work on the flexor tendons, peak frictional power, and three measures of the flexor myographic signal, the 10th or static, 50th, and 90th percentiles of the amplitude probability distribution function (APDF).

The models above integrated information from anatomical, biomechanical, and epidemiological studies to produce a profile of measures to characterize tasks and which reflected injury mechanisms for different tissue types. These will now be briefly reviewed.

Approaches to Investigating Tendon Disorders

Etiologically, reduced lubrication between tendons and tendon sheaths due to excess relative movement has been suggested in tenosynovitis (Rowe, 1987) while high peak loads and cumulative strain have been suggested for tendinitis (Goldstein et al., 1987). Norman and Wells (1990) have proposed a model for assessment of tenosynovitis which was operationalized by Moore et al. (1991). This is seen in Figure 43.4.

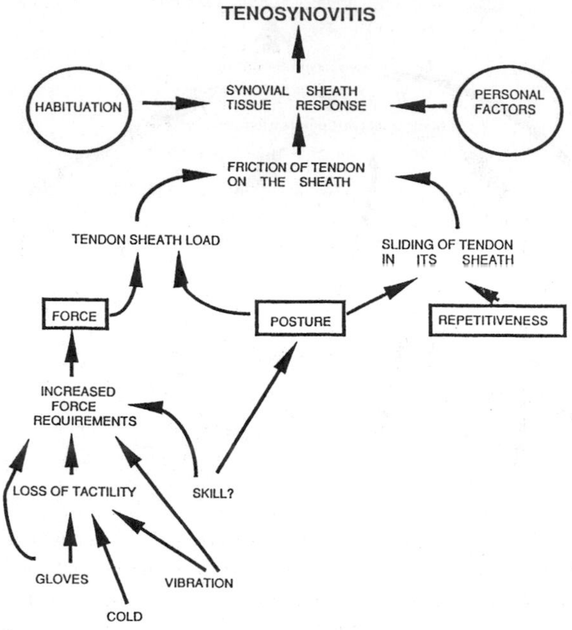

FIGURE 43.4 Conceptual model of the relationship between external risk factors and the development of tenosynovitis. Across the center are listed the major risk factors of force and posture. Below are listed risk factors which may elevate the effect of these primary factors, and above are noted possible pathways by which the external exposures (risk factors) create internal exposures that could plausibly lead to the development or aggravation of tenosynovitis. Many additional factors can alter the forces in the flexor tendons, such as the grip type, number of digits used, and the angular acceleration of the wrist. (Adapted from Norman, R.K. and Wells, R.P. 1990. Biomechanical aspects of occupational injury, *Proceedings of the 23rd Annual Conference of the Human Factors Association of Canada.* With permission.)

In this approach the frictional work done by the tendon sliding through its sheath is calculated. One type of frictional work is present due to a "belt-pulley" interaction when the wrist deviates from a straight position (Armstrong and Chaffin, 1979). In addition to this type of frictional work, it is suggested that a non-negligible resistance to movement is present to move the tendons through the carpal tunnel even in the straight position. Estimates from the work of Goldstein et al. (1987) and Smutz et al. (1995) put this resistance at around 5N in the neutral position. Excursion of the tendons at the wrist (caused by finger and wrist movement) in both deviated, and straight postures will therefore create an energy input, possibly beyond the recovery capability of the tissue.

Approaches to Investigating Nerve Disorders

Insult to the median nerve, whether due to increased hydrostatic pressures in the carpal canal (Rempel et al., 1995) or due to mechanical insult upon the nerve by overlying tendon (Keir and Wells, 1995) has often been suggested as a likely mechanism of work-induced carpal tunnel syndrome. More controversially, it may be caused by hypertrophy or edema of the synovial sheaths and thus be secondary to tenosynovitis.

Mechanical stress to the median nerve can be predicted by modified belt-pulley models of the wrist and is an output of a biomechanical model (Moore et al., 1991). Hydrostatic pressures are measurable using *in vivo* and *in vitro* techniques (Keir and Wells, 1995).

Methodological Approaches to Predicting Muscle Disorders in the Upper Limb

Recent clinical findings have suggested that forearm muscle pain may be an overlooked problem in studying work-related chronic musculoskeletal injuries (Ranney et al., 1995). While work-related muscle pain is well accepted in the shoulder area (e.g., Veiersted et al., 1993), pain in the forearm is usually attributed to tendinitis or epicondylitis. The major approach to investigating muscle-related occupational disorders is electromyography. Jonsson (1982) described a technique in which the frequency of any particular level of EMG occurring is calculated. From this, an amplitude probability curve is developed. The static level describes the ability of the muscle to rest at least 10% of the time and appears important in the development of chronic work-related muscle problems. If the value is greater than zero, the muscle is not given a chance to completely rest at least 10% of the time during a task. While this is a useful technique for quantifying muscle usage throughout the duration of a task, it gives no indication of the duration of each rest pause, i.e., whether the rests came as numerous pauses or one big pause. Veiersted et al. (1990, 1993) addressed this by using a "gaps" analysis. This analysis looks at the number of times the muscle is turned off (an EMG "gap" is defined as a muscle activation of less than 0.5% MVC lasting for more than 0.2 seconds), and it appeared that people with pain had fewer gaps. More recently it has been shown that workers likely to require neck/shoulder sick leave can be predicted from measures of "gap" frequency. (Veiersted et al., 1993).

43.5 Workplace Assessment Tools

Scope of Workplace Assessment Tools Reviewed

The previous sections have described some of the main risk factors which increase the risk of developing upper limb musculoskeletal disorders and how these risk factors may produce internal exposures to tissues of the body potentially leading to premature fatigue or WMSD. It falls to this section to describe and review some of the tools in the literature which have been developed to assess workplace risk of WMSD. Tools included met two conditions: (1) the tool or method not only recorded risk factors but rated them and (2) rated the injury risk to the upper limbs or the intervention priority.

The workplace assessment tools reviewed fell into two main categories: those using mainly observational methods to identify, rate, and combine a number of risk factors to produce an estimate of risk for upper limb musculoskeletal disorders (e.g., Strain Index) or intervention priority (e.g., RULA) and which are often suggested as screening tools, and those which measured (sometimes called technical methods) the time course of a more complex risk factor (e.g., the trapezius electromyogram) and produced an estimate of risk. The tools are described in Table 43.1 and evaluated on a number of criteria which are important either from a conceptual, measurement, or usability viewpoint. The following sections describe these criteria.

Stated Purpose

The tools can only be reviewed on the basis of the stated purpose; this purpose differed between the tools. For example, the RULA output is in terms of an intervention priority, whereas trapezius EMG has been associated with neck/shoulder musculoskeletal sick leave. Some tools were designed for screening purposes whereas others were intended to produce more definitive analyses. The precise area of the body for which the tool predicts risk also differs between tools and methods; for example, RULA has upper limb and trunk scores and defines an intervention priority based upon the combined whole body score, whereas the Strain Index is specific to the distal upper limb.

Input Variables

As noted in the previous sections, there is a wide range of potential stressors for the upper limb; how many there are and how they are measured is important information concerning the potential generalizability of the method. The handling of the time-varying nature of the risk factors is, to this author, of prime importance both for its usefulness and its ability to answer the frequently asked questions concerning line balancing, job rotation, and break schedules. The inputs are described by the forces and postures adopted, the characterization of their time course, as well as the treatment of other risk factors.

TABLE 43.1 Comparison of Workplace Assessment Tools in the Literature which Have Been Developed to Assess Workplace Risk of WMSD and Priority for Intervention

Workplace Assessment Tool	Reference(s)	Stated Purpose	Input Variables			
			Force	Posture	Time	Other
Cumulative Trauma Disorder Checklist *Upper Extremity CTD*	Lifschitz and Armstrong (1986)	"…to indicate easily perceived problems in the workplace from the design point of view with respect to upper extremity cumulative trauma disorders; … to rapidly give readable feedback to the user."	"Weight of the tool below 10 lbs"	"…bending the wrist"	"is the cycle time above 30 seconds"	contact stress, vibration, temperature, seating, adjustability, tool characteristics.
Electromyography *Neck/shoulder*	Aaras and Westgaard (1987), Veiersted et al., (1990,1993) and Veiersted (1996)	Use of electromyographic evaluation as a predictor of neck/shoulder sick leave and myalgia.	Muscle activation in %Maximum Voluntary Contraction (MVC)	—	Continuous record	—
Wrist Goniometry *Upper Limb CTD*	Marras and Schoenmarklin (1993), Schoenmarklin, Marass and Leurgans., (1994)	"… to determine *quantitatively* the association between specific wrist motion parameters and the incidence of CTD as a group."	—	Wrist flexion/extension, ulnar/radial deviation and pronation/ supination	Continuous record	—
Psychophysical *Hand/wrist*	Snook et al., (1995)	"…to investigate the feasibility of using psychophysical methods to determine maximum acceptable forces for various types and frequencies of repetitive wrist motion."	Wrist moment	Testing performed over ROM, ± 45 deg.	Frequency (#/min) and duration (hrs)	—

Workplace Assessment Tool	Evaluation of Risk Factors		Equipment Required	Measurement Characteristics	Validity	Study base/ Generalizability
	Rated	Combined				
Cumulative Trauma Disorder Checklist	Factor present or absent	Sum of number of risk factors present (NO's).	—	—	Number of NO scores increased with the incidence of reported CTD	Auto plant
Electromyography	Raw surface EMG processed to form APDF or "gaps."	—	Portable EMG system	Reliability; "static" APDF (0.59) and gaps (0.85) in a field setting	Static level of trapezius EMG predictive of neck/shoulder sick leave in an inception cohort of packers	Packaging workers and workers from telephone assembly
Goniometry	Wrist posture differentiated to form angular velocity and acceleration	—	3-D goniometer system and software	—	In a comparison of high and low risk jobs flexion accelerations were significantly different between groups.	40 workers in 8 automotive sector manufacturers
Psychophysical	Acceptability of a combination of wrist moment and frequency judged by psychophysical adjustment experimentation	—	Applied force and its moment arm required	—	Such a method has been validated for manual materials handling, Snook (1978).	Laboratory study, 29 female participants drawn from a working population

TABLE 43.1 (continued) Comparison of Workplace Assessment Tools in the Literature which Have Been Developed to Assess Workplace Risk of WMSD and Priority for Intervention

Workplace Assessment Tool	Limit/Guideline Level Proposed	Information for Intervention	Limitations	Other
Cumulative Trauma Disorder Checklist	No	Address NO items in checklist	—	
Electromyography	Yes, some suggestion for static load of 2–5% and less than 1%.	No clear link to help work station redesign except reducing static loading. May be valuable in evaluation of prototypes. Could be used as EMG biofeedback to improve work methods	May not be sensitive to some postural effects, impingement and postures likely to promote nerve entrapment. Not tested for sensitivity to short duration, high load situations	Could be generalized to other areas of the upper limb, (e.g., Moore et al., 1991)
Goniometry	Yes; acceleration characteristics of high and low risk jobs described	Reduce acceleration	Not tested for activities involving constant gripping	One of the few approaches to allow evaluation of rapid dynamic manual tasks
Psychophysical	Yes, acceptable for x % of the population	Can determine a combination of wrist moment and frequency to be acceptable to any percentage of the population.	Unknown relationship between acceptability and risk of musculoskeletal disorders. A "library" of tasks is needed to allow analysis of many jobs.	This approach uses psychophysical experimentation to determine acceptable (guideline) values for wrist moment. Other tasks are planned to be added to the "library" of tasks.

Workplace Assessment Tool	Reference(s)	Stated Purpose	Input Variables			
			Force	Posture	Time	Other
OWAS (Ovako Working Posture Analysis System) *Whole Body*	Karhu et al., (1977), Mattilla et al., (1992), Leskinen and Tönnes (1994)	"…OWAS has been planned to meet the criteria. (a) simple enough to be used by ergonomically untrained personnel, (b) it must provide unambiguous answers even if it results in over-simplification, (c) it must also offer possibilities for correcting the over-simplified ergonomic approach…so that a systematic guide to corrective action can be constructed."	Effort of <10kg, <20kg and >20kg.	Observed shoulder posture, limb above or below shoulder. (Other categories for trunk and lower limbs)	Posture sampling. OWASCO/OWASAN allow calculation of sampling statistics	—
RULA (Rapid Upper Limb Assessment) *Upper Limbs + Trunk and Lower Limbs*	McAttamney and Corlett (1993)	"…investigate the exposure of individual workers to risk factors associated with work-related musculoskeletal disorders."	Categories for 0–2kg, 2–10kg and more than 10kg.	Observed posture at wrist, lower arm and shoulder from predefined postures.	Action repeated more than 4/min or mainly static and held for more than 1 minute. Analysis performed at an instant "…held for the greatest amount of time or where highest loads occur."	—
Strain Index *Distal Upper Extremity*	Moore and Garg (1995)	"…semi-quantitative job analysis that … appears to identify accurately jobs associated with distal upper extremity disorders…."	Observed intensity of exertion on a five point scale from light to near maximal	Hand/wrist posture on a scale from very good to very bad	Three temporal descriptors: Duration of cycle(%), efforts/min and duration/day all on a five point scale.	—
Upper Extremity Checklist *Upper Limb*	Keyserling et al., (1993)	"…designed to function as a rapid screening tool which could be used by persons with relatively little training and experience in ergonomics to identify jobs with potentially harmful exposures to upper extremity risk factors."	Presence of forceful exertions; manual materials handling >4.5kg, poor grip friction, pressing with fingertip, hold object >2.7kg.	Postures of the arms and shoulder; pinch, wrist deviation, twisting 2 adverse shoulder postures.	Duration of risk factor as none, up 1/3 and greater than a 1/3 of a cycle. Presence of a cycle time of 30sec or less.	Contact stresses, gloves, jerk.

TABLE 43.1 (continued) Comparison of Workplace Assessment Tools in the Literature which Have Been Developed to Assess Workplace Risk of WMSD and Priority for Intervention

Workplace Assessment Tool	Evaluation of Risk Factors		Equipment Required	Measurement Characteristics	Validity	Study base/ Generalizability
	Rated	Combined				
OWAS (Ovako Working Posture Analysis System)	Postures rated by 32 experienced steel workers from "…no discomfort and no effect on health …" to "…short exposure leads to discomfort, ill effect on posture possible." Ranking also by small group of international ergonomists	—	Paper form or video/computer system for OWASCO/ OWASAN	Inter-observer (2 observers); 74–99% agreement. Test-retest (morning/afternoon) (2 observers) 70–100% agreement	—	Steel Works.
RULA (Rapid Upper Limb Assessment)	Individual risk factor weights by expert judgment	Additive model. " Look-up" tables to obtain intervention priority by expert judgment.	Clipboard and pen	"….high consistency of scoring among …" 120 persons	Statistically significant association between upper and lower body scores (A and B) and discomfort in that area in 16 VDT operators	Ratings developed from production line, VDT data entry and sewing tasks. Validity data determined on VDT operators
Strain Index	Intensity of effort raised to power of 1.6 (psycho-physical power law) Other rating using professional judgment	Multiplicative model	None	Sensitivity analysis	Increase in mean incidence rate with Strain Index in 25 jobs	One pork processing plant
Upper Limb Checklist	By time duration and expert judgment as minor(√) or significant(*)	Potential(_) and significant risk (*) factors summed separately.	None	Inter-observer; moderate, intra-observer; moderate	Agreement between trained Health and Safety Personnel and experts:	Auto-industry

Workplace Assessment Tool	Limit or Guideline Level Proposed?	Information for Intervention?	Limitations	Other
OWAS (Ovako Working Posture Analysis System)	Posture ranking used to create 4 action categories; Class1 = Normal postures which do not need any special attention Class 4 = postures need immediate attention.	"…postures can be evaluated in terms of their desirability and need of attention…"	Only shoulder posture rated and at only two postures. No information on distal upper limbs	
RULA (Rapid Upper Limb Assessment)	Intervention priority indicated (for whole body)	Address highest upper limb risk factor scores	Little effect of time course of risk factors on ratings	Scores for the upper limbs as well as the trunk and lower limbs. Both are combined to form a grand score which is used to indicate an intervention priority.
Strain Index	Preliminary suggestion of 5 as limit value	Address highest upper limb risk scores. Equation also allows impact of changing risk factor levels to be assessed.		Contact stresses, cold and vibration not accounted for
Upper Limb Checklist	"…any job receiving one or more 'stars' was considered to have a high priority for additional investigation and analysis."	"…any job receiving one or more 'stars' was considered to have a high priority for additional investigation and analysis."		The period between the number of √ and * is a separator with no meaning. A job with ten * is not necessarily ten times as risky as a job with one *; it simply has more separate risk factors. These risk factors may or may not be synergistic or act at the same body joint.

The tools differed considerably in their treatment of the time course of the input variables; some used only a single instant while others used mathematical or electronic processing to extract information about the time variation of the risk factor.

Rating of Individual Risk Factors

As previously noted, only tools which evaluated the size of the risk factor(s) have been reviewed here; how this is done and the quality of this assessment is key to the usefulness of the tool. It is difficult to imagine any work without some risk factors present and one can quickly fall into the mindset that work is inherently dangerous and the observation of a bent wrist during work implies hazard. A distinction is made in industrial hygiene between toxicity and hazard. Benzene is highly toxic, yet if used infrequently where the concentration is small (a person's exposure is low), the hazard is small. Similarly, even for wrist flexion close to an individual's range of motion (ROM), the risk is also low if the motion is infrequent. In fact the adoption of "extreme postures" for short periods of time is probably beneficial; they are called stretch breaks.

A number of approaches are seen ranging from statistical treatments based on epidemiological studies to expert and consensus judgments. For single risk factors epidemiological approaches are possible; however, some element of expert judgment becomes necessary to "fill-in" the holes in the epidemiological literature.

Combination of Risk Factors

Very few epidemiological studies allow the interactions of a number of risk factors to be examined. For example, Silverstein and colleagues did study a simple 2 × 2 interaction of force and repetitiveness (Silverstein et al., 1986), while the psychophysical approach allows combinations of multiple dimensions to be rated. These studies are not common and so the majority of studies combine rating of the individual risk factors with additive or multiplicative models to arrive at a risk estimate.

The combination of risk factors to produce an estimate of risk is perhaps the most difficult issue in workplace evaluation and tool construction. Should the individual risk factor ratings be added or multiplied or even considered completely separately? For example, in evaluating the risk of low back pain on a job one could measure the risk factor of posture and load separately. Does one add the posture and load score or multiply them? Biomechanical models indicate that multiplying the load and its moment arm about the low back (in effect, posture) gives the low back moment of force (or torque). This has been found to give the best single prediction of low back pain risk (Marras et al., 1993). Clearly the integrated approach advocated here uses biological and mechanical arguments to help in this decision.

In industrial hygiene exposure levels are considered separately except when the agent of interest has the same target organ or pathway. For example, in a given job there is exposure to work overhead and hand/arm vibration from a hand tool. Does this mean the job has two risk factors which need improvement or is there more risk than if either of these exposures occurred separately? The first approach is supported by different target "organs," the shoulder and forearm, while the low specificity of effect (Hagberg et al., 1995) could argue for the second interpretation.

Based on the discussion of the importance of time as a descriptor of risk factors, it would appear that time must be considered along with the primary risk factor, force. In some cases both force and posture are considered and in these cases, both force and posture may be considered with their time variation. In some cases posture may be used as a surrogate of force and in such cases its time variation must be considered. In all these cases, this may be done additively or, more commonly, multiplicatively (cf. NIOSH, 1981; Moore and Garg, 1995).

Definitive answers to the above conceptual issues are not available, and so the magnitude of the total score calculated and its relation to risk must be cautiously interpreted. In the case of tools whose purpose is to calculate intervention priority, these questions are not as critical because the scores are used as a summary of the size of the individual risk factors and their number combined.

Equipment Required

The equipment required is important in the choice of a tool; some methods utilize observations of work, while others use various "technical" methods, such as electromyography or goniometry, in the measurement process. This obviously affects the time and cost of the assessment. While it is sometimes heard that ergonomics assessments must be simple and cheap, the value of more costly yet precise technical measures with epidemiologically determined relationships to risk must not be undervalued. The time, training required, and cost of using the tools are rarely reported.

Measurement Characteristics

The tools developer can arrive at an estimate of risk in a large number of ways; what is important, however, is the quality of the tool's predictions. This can be assessed in terms of the measurement characteristics and validity. The measurement characteristics refer to such qualities as intra-observer reliability (or test–retest reliability) as well as inter-observer agreement or reliability. A tool with poor reliability will be of limited usefulness. It may still, however, be able to distinguish jobs with many risk factors and high risks from those with few risk factors and low risks, but it may not reliably distinguish between jobs with less extreme contrasts. Good measurement characteristics become of even greater concern if the tool is used for guidelines or legislative purposes.

Validity

The term validity can be used in a number of senses. Content validity refers to the completeness of the assessment. Questions such as "are all important risk factors rated" are asked here. Most tools reviewed, however, used some variant of criterion-related validity. The output of the tool was compared to some health-related output on jobs or individuals. The stated purpose is important here in evaluating the appropriate comparisons.

Study Base/Generalizability

It is not possible to test a tool under all possible conditions; the range of workplaces used to develop and test the tool are useful in judging the applicability of the tool to a given target workplace. For example, it is likely problematic to use a tool developed in an office environment to apply to a construction site.

Proposed Limit of Guideline Level?

Although not universal, many of the tools produced some recommendations or guidelines in terms of the score or output of the tool. A number of tools have screening as their stated purpose. In this case a two- (or more) step process is assumed, and those jobs exceeding some criterion score are further analyzed. In this case a high sensitivity is desirable; a moderate number of "false positive" findings are accepted so as not to miss potentially risky jobs in the first step.

For those tools whose purpose is to define risk, it is unreasonable to imagine that a single threshold divides risky jobs from non-risky ones. Where the risk of developing various WMSDs has been produced against a continuous exposure measure, it has been found that the risk increases steadily from the nonexposed state, i.e., there is no obvious step or threshold evident (e.g., Punnett et al., 1991). The threshold chosen is then dependent on the increased risk which is to be accepted (a societal judgment). While a guideline value can be useful in the interpretation of an instrument's score, the measurement characteristics (the validity and the generalizability) of the tool must be of high quality before reliance can be placed on these recommendations.

Information for Intervention

The assessment of risk is but one step in process of workplace improvement; a good tool provides direction on which risk factors need addressing and also provides material and suggestions for solutions. Ideally, it might also allow "what if" scenarios to be explored and predict what the level of risk will be for the new combination of risk factors.

Limitations

No tool is perfect; the limitations of a given tool need to be understood, however, so that undue reliance on the output is not made where the tool's predictions are likely not to be of high quality. Each tool could have a large number of limitations; the focus of this section is on major areas where the predictive power of the tool is suspect or untested.

43.6 Summary

This chapter has reviewed some concepts important in the measurement and evaluation of the major risk factors for the development of WMSDs in the upper extremity. An "integrative" approach is followed whereby anatomical, physiological, and biomechanical information is used to conceptualize upper limb function and as a possible means to learn how work might cause WMSDs. These concepts are then used to inform a review of some of the major upper limb workplace evaluation tools; each has different purposes, strengths, and weaknesses. As these tools are further developed, the framework used in this chapter should provide a springboard from which the reader can assess existing and new tools in the light of their needs and available resources.

References

Åaras, A. 1987. Postural load and the development of musculo-skeletal illness: *Scand J. Rehab Med;* Suppl. 18, 1-35.

Åaras, A. and Westgaard, R.H. 1987. Further studies of postural load and musculo-skeletal injuries of workers at an electro-mechanical assembly plant. *Applied Ergonomics;* 18(3): 211-219.

An, K-N, Chao, E.Y., Cooney, W.P., and Lischeid, R.L. 1985. Forces in the normal and abnormal hand. *J. Orthop. Res.*; 3:202-211.

Armstrong, T.J., Castelli, W.A., Evans, F.G., and Diaz-Perez, R. 1984. Some histological changes in the carpal tunnel contents and their biomechanical implications. *J. Occup. Med*; 26(3), 197-201.

Armstrong, T.J. and Chaffin, D.B. 1979. Some biomechanical aspects of the carpal tunnel. *J. of Biomechanics*; 12:567-570.

Goldstein, S.A. 1981. *Biomechanical Aspects of Cumulative Trauma to Tendons and Tendon Sheaths.* PhD Thesis, University of Michigan.

Goldstein, S.A., Armstrong, T.J., Chaffin, D.B., and Matthews, L.S. 1987. Analysis of cumulative strain in tendons and tendon sheaths. *J. of Biomechanics;* 20(1):1-6.

Hagberg, M., Silverstein, B., Wells, R., Smith, R., Carayon, Hendrick, H.P., Perusse, M., Kourinka, I., and Forcier, L. (eds.). 1995. *Work-Related Musculoskeletal Disorders (WMSD): A Handbook for Prevention*, Taylor & Francis, London.

Jonsson B. 1982. Measurement and evaluation of local muscular strain in the shoulder during constrained work. *J. Human Ergol.*; 11, 73-88.

Karhu, O., Kansi, P., and Kuorinka, I. 1977. Correcting working postures in industry: a practical method for analysis. *Applied Ergonomics*; 8(4): 199-201.

Keir, P., Wells, R., and Ranney, D. 1996. Passive stiffness of the forearm musculature and functional implications; A pilot study, in press *Clin Biomech.*, 11(7): 401-409.

Keir, P.J. and Wells R. 1995. The effect of tendon loading and wrist posture on carpal tunnel pressure in cadavers. *Proceedings of the 19th Annual Meeting of the American Society of Biomechanics*, Stanford University, August 1, 1995, pp.129-130.

Keyserling, W.M., Stetson, D.S., Silverstein, B.A., and Brouwer, M.L. 1993. A checklist for evaluating ergonomic risk factors associated with upper extremity cumulative trauma disorders. *Ergonomics*; 36, (7); 807-831.

Kumar, S., 1990. Cumulative load as a risk factor for low-back pain. *Spine* 15, 1311-1316.

Lifshitz, Y. and Armstrong, J. 1986. A design checklist for control and prediction of cumulative trauma disorder in intensive manual jobs. *Proceedings of the Human Factors Society*, 30th Annual Meeting; 837-841.

Larsson S.E., Bengtsson, A Bodegård, L., Henriksson, K.G., and Larsson, J. 1988. Muscle changes in work related chronic myalgia, *Acta Orthop Scand*; 59(5): 552-6.

Leskinen, T. and Tönnes, M. 1994. Utilization of a video-computer system for analyzing postural load-evaluation of observation. *Proceedings of the 12th Triennial Congress of the International Ergonomics Association*, Toronto, Canada, August 15, 1994; Vol. 2 pp. 383-385.

Marras, W.S. and Schoenmarklin, R.W. 1993. Wrist motions in industry. *Ergonomics*; 36(4):341-351.

Marras, W.S., Lavender, S.A., Leurgans, S.E., Rajulu, S.L., Allread, W.G., Fathallah, F.A., and Ferguson, S.A. 1993. The role of dynamic three-dimensional trunk motion in occupational-related low back disorders: the effects of workplace factors trunk position and trunk motion characteristics on risk of injury. *Spine*; 18(5), pp. 617-28.

Mattilla, M., Vilkki, M., and Tiilikainen, I. 1992. A computerized OWAS analysis of work postures in the papermill industry, in: Mattila, M. and Karwowski, W. (eds.), *Computer Applications in Ergonomics, Occupational Safety and Health* Elsevier, Amsterdam), 1-11.

Mathiassen, R. and Winkel, J. 1991. Quantifying variation in physical load using exposure-vs-time data. *Ergonomics*; 34(12):1455-68.

McGill, S.M. and Norman, R.W. 1992. Low back biomechanics in industry: the prevention of injury through safer lifting, in Grabiner, M. (ed.), *Current Issues in Biomechanics*, Human Kinetics Publishers, Champaign, Ill, 69-120.

McAtamney, L. and Corlett, E.N. 1993. RULA: a survey method for the investigation of work-related upper limb disorders. *Applied Ergonomics*; 24(2): 91-99.

Moore, J.S. and Garg, A. 1995. The strain index: A proposed method to analyze jobs for risk of distal upper extremity disorders. *Am. Ind. Hyg. Assoc. J.*; 56: 443-458.

Moore, A., Wells, R., and Ranney, D. 1991. Quantifying exposure in occupational manual tasks with cumulative trauma disorder potential. *Ergonomics*; 34(12):1433-1453.

NIOSH 1981. *Work Practices Guide for Manual Lifting*, Cincinnati, OH: U.S. Department of Health and Human Services. Technical Report No. 81-122).

Norman, R.K. and Wells, R.P. 1990. Biomechanical aspects of occupational injury, *Proceedings of the 23rd Annual Conference of the Human Factors Association of Canada.*

Punnett, L., Robins, J.M., Keyserling, W.M., Herrin, G., and Chaffin, D.B. 1991. Back disorders and non-neutral trunk postures of automobile assembly workers. *Scand. J. Work Environ. Health*, 17(5):337-346.

Ranney, D., Wells, R., and Moore, A. 1995. Upper limb musculoskeletal disorders in highly repetitive industries: precise anatomical physical findings. *Ergonomics*, 38(7): 1408-1423.

Rempel, D. 1995. Musculoskeletal loading and carpal tunnel pressure, in Gordon, S., Blair, S., and Fine L. (eds.), *Repetitive Motion Disorders of the Upper Extremity*, American Academy of Orthopedic Surgeons, Rosemont Il, 123-133.

Rowe M. 1987. The diagnosis of tendon and tendon sheath injuries: *Sem. Occup. Med.*; 2(1):1-6.

Sakakibara, H., Miyao, M., Kondo, T., Yamada, S., Nakagawa, T., and Kobayashi, F. 1987. Relationship between overhead work and complaints of pear and apple orchard workers. *Ergonomics;* 30(5): 805-815.

Schoenmarklin, R.W, Marass, W.S., and Leurgans, S. 1994. Industrial wrist motions and incidence of hand/wrist cumulative trauma disorders, *Ergonomics*; 37(9): 1449-1460.

Silverstein, B.A, Fine, L.J., and Armstrong, T.J. 1986. Hand wrist cumulative trauma disorders in industry. *Br. J. Ind. Med.;* 43:779-784.

Skie, M., Zeiss, J., Ebraheim, N.A., and Jackson, W.T. 1990. Carpal tunnel changes and median nerve compression during wrist flexion and extension seen by magnetic resonance imaging. *J. Hand Surg*; 15-A(6), 934-939.

Smutz, W.P, Bishop, A., Niblock, H., and Drexler, M. 1995. Measurement of creep strain in flexor tendons during low-force, high-frequency activities such as computer keyboard use. *Clinical Biomechanics*; 10(2): 67-72.

Snook, S., 1978. The design of manual materials handling tasks. *Ergonomics*; 21:963-985.

Snook, S., Vaillancourt, D.R., Ciriello, V.M., and Webster, B.S. 1995. Psychophysical studies of repetitive wrist flexion and extension. *Ergonomics;* 38(7):1488-1507.

Stenlund, B., Goldie, I., Hagberg, M., Hogstedt, C., and Marions, O., 1992, Radiographic osteoarthrosis in the acromioclavicular joint resulting from manual work or exposure to vibration. *Brit. J. Ind. Med.;* 49:588-593.

Stock, S.R., 1991, Workplace ergonomic factors and the development of musculoskeletal disorders of the neck and upper limbs: A meta-analysis, *Am. J. Ind. Med.*, 19:87-107.

Veiersted, K., Westgaard, R., and Andersen, P. 1990. Pattern of muscle activity during stereotyped work and its relation to muscle pain. *Int. Arch. Occup. Environ. Health*; 62:31-41.

Veiersted, K.B., Westgaard, R.H., and Andersen, P. 1993. Electromyographic evaluation of muscular work pattern as a predictor of trapezius myalgia. *Scand. J. Work Environ. Health*; 19:284-290.

Veiersted, K.B. 1996. Reliability of myoelectric trapezius muscle activity in repetitive light work. *Ergonomics*; 39(5): 797-807.

Wells, R., Ranney, D., and Keir, P. 1994. Passive force length properties of cadaveric human forearm musculature, in *Advances in the Biomechanics of the Hand and Wrist.* Schuind, F., An, K-N., Cooney, W.P., and Garcia-Elias, M. (eds.), pp. 31-40.

Wells, R., Keir, P.J., and Moore, A.E, 1995. Applications of biomechanical hand and wrist models to work-related musculoskeletal disorders of the upper extremity, in Gordon, S.L., Blair, S.J., and Fine, L.J. (eds.) *Repetitive Motion Disorders of the Upper Extremity*, American Academy of Orthopaedic Surgeons, Rosemount, IL.

Winkel, J. and Westgaard, R., 1992. Occupational and individual risk factors for shoulder-neck complaints, part II: the scientific basis (literature review) for the guide. *Int J Ind Erg; 10*, 85-104.

Winkel, J and Mathiasen, S-E. 1994. Assessment of physical work load in epidemiologic studies, concepts, issues and operational considerations. *Ergonomics*; 37, 979-988.

44

Biomechanical Aspects of CTDs

Richard W. Marklin
Marquette University

44.1 Introduction

The purpose of this chapter is to explain the biomechanical *etiology* of *cumulative trauma disorders* (*CTDs*) that affect the hand, wrist, elbow, and shoulder. The assumption that these CTDs are caused, in part, by work-related activity is based on biomechanical mechanisms that are consistent with epidemiological findings. CTDs affect the soft tissues in the body, namely tendons, ligaments, muscles, and nerves, and in general not bone tissue. Although some authors include bone tissue within the umbrella of CTDs (Kuorinka and Forcier, 1995), this chapter will focus only on those CTDs affecting soft tissue. A brief description of the anatomy of the upper extremity will be provided to familiarize the reader with anatomical terms. Then the three major classes of CTDs, namely those involving muscle, the muscle–tendon unit, and nerve compression, will be discussed.

44.2 Anatomy of the Upper Extremity

Skeletal System

The bones of the upper extremity, which are illustrated in Figure 44.1, are of two types, long and short bones. The long bones connecting the shoulder to the elbow (humerus), the elbow to the wrist (radius

0-8493-2641-9/99/$0.00+$.50

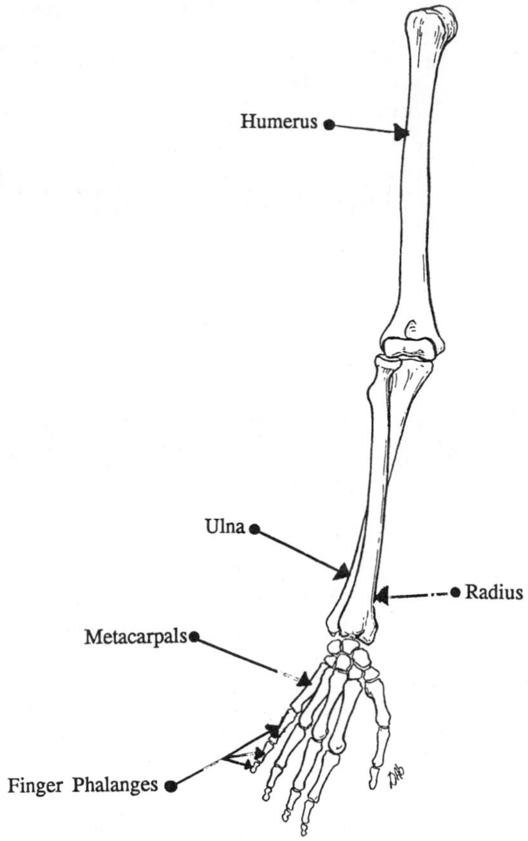

FIGURE 44.1 The long and short bones of the right upper extremity. (From Basmajian, J.V. 1982. *Primary Anatomy,* 8th edition, p. 57, Williams and Wilkins, Inc. With permission.)

and ulna), and the wrist to the fingers (metacarpals and phalanges) are adapted for weight bearing and for sweeping, speedy movements that allow the hand to move in space and grasp and touch objects (Rasch, 1989). The movement of the radius (thumb side) around the ulna (little finger side) in the forearm permits the hand to be turned up (supination) or down (pronation), as illustrated in Figure 44.2. The *proximal* and *distal* parts of the long bones display flared ends that act as attachment points for other bones and for connective tissue, such as *tendons* and *ligaments.*

The cluster of small cubical bones comprising the wrist are the eight carpal bones, which are categorized as short bones (Rasch, 1989). The carpal bones move with respect to each other to *flex* (palm side) and *extend* (back side of hand) the wrist joint, while also allowing the wrist to move side to side, from a neutral position to *radial deviation* (thumb side) and to *ulnar deviation* (little finger side), as shown in Figure 44.3.

Muscular System

The *muscles* of the body are the generators of internal force that convert energy chemically stored in the body into mechanical work (Rasch, 1989). Skeletal muscle, also called *striated muscle,* is composed of longitudinal fibers that follow the direction in which a muscle exerts a force, as seen in Figure 44.4. A muscle is like a rope in that it can only pull or exert a force in tension, and it cannot push or exert a weight-bearing force (compression force). As shown in Figure 44.5, a muscle exerts a tensile force by contracting its thread-like fibers, which shortens the length of the muscle and in fusiform muscles creates a bulge at its center.

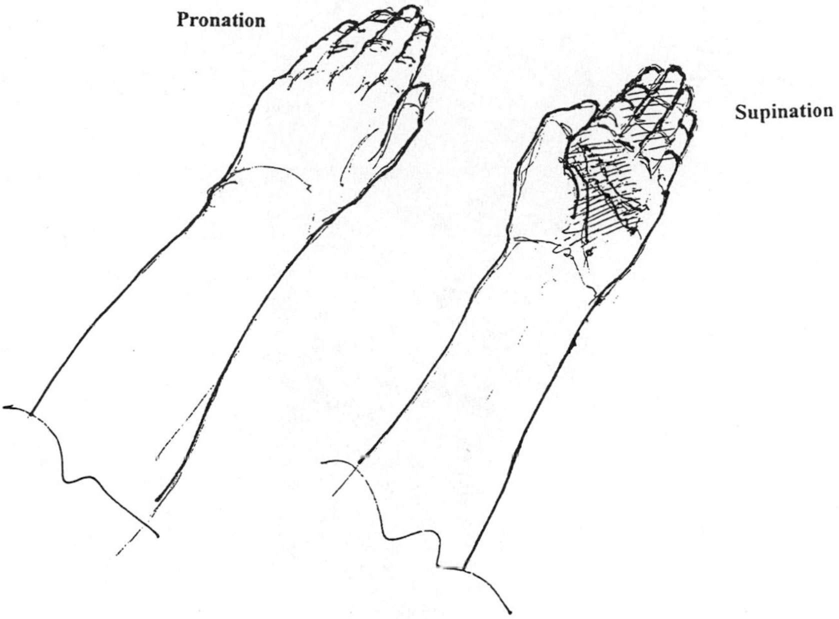

FIGURE 44.2 The forearm in a pronated and supinated posture. (From Marklin, R.W. original artwork. With permission.)

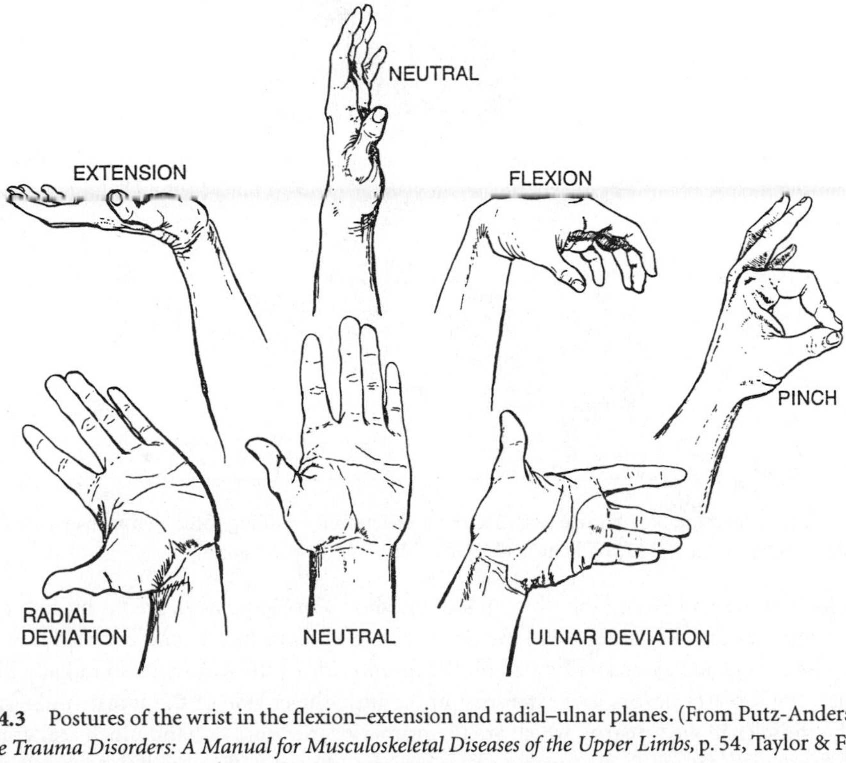

FIGURE 44.3 Postures of the wrist in the flexion–extension and radial–ulnar planes. (From Putz-Anderson, V. 1988. *Cumulative Trauma Disorders: A Manual for Musculoskeletal Diseases of the Upper Limbs*, p. 54, Taylor & Francis. With permission.)

FIGURE 44.4 A photograph of the left shoulder muscles from a cadaver, as seen from the side. (From McMinn, R.M.H. and Hutchings, R.T. 1977. *Color Atlas of Human Anatomy,* p. 113, Year Book Medical Publishers, Inc., Chicago, IL. With permission.)

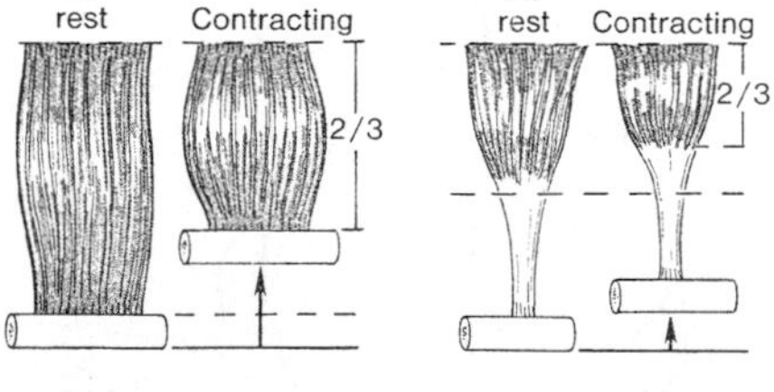

FIGURE 44.5 The shortening of a muscle as it contracts, generating a pulling force. (From Basmajian, J.V. 1982. *Primary Anatomy,* 8th edition, p. 113, Williams and Wilkins, Inc. With permission.)

The muscles that flex and extend the elbow, which are shown in Figure 44.6, are the biceps and triceps. The group of muscles that flex and extend the wrist are the forearm flexors and extensors, as shown in Figure 44.7. The flexors and extensors located on the thumb side of the forearm also radially deviate the wrist; likewise, the forearm flexors and extensors on the little finger side of the forearm ulnarly deviate the wrist. The muscles in the forearm, which are the primary generators of hand pinch and grasp forces, are called *extrinsic muscles,* while the much smaller muscles located within the hand are called *intrinsic muscles.* One of the main functions of the intrinsic muscles in the hand is to cooperate with the extrinsic muscles to generate hand movements that require dexterity and fine motor control.

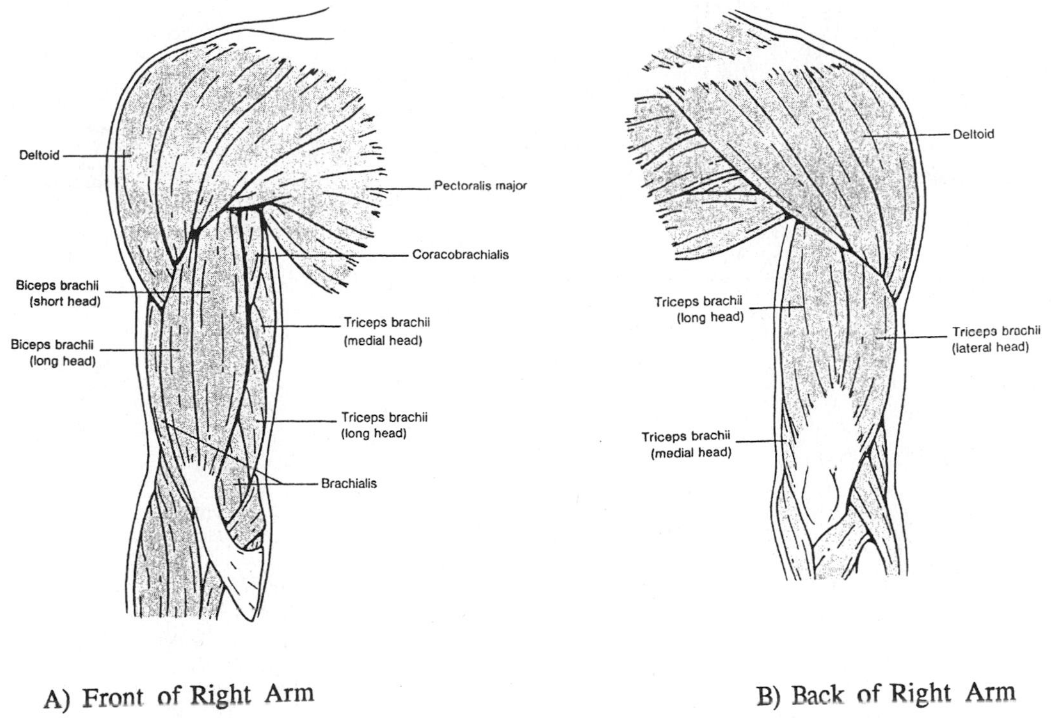

FIGURE 44.6 The muscles that flex (a) and extend (b) the elbow. Figures (a) and (b) are the front and back views of the right arm, respectively. (From Van de Graaff, K.M. and Rhees, R.W. 1987. *Human Anatomy and Physiology*, p. 107, Schaum's Outline Series, McGraw-Hill Book Co. With permission.)

Connective Tissue and Carpal Tunnel

As shown in Figure 44.7, the extrinsic muscles of the forearm are attached to the fingers with strong cord-like collagen structures called tendons. The tendons attached to the flexor and extensor forearm muscles are constrained within the wrist area by thick bands called the *flexor retinaculum* and *extensor retinaculum*, as illustrated in Figure 44.8. The flexor and extensor retinacula are ligaments that attach carpal bones on one side of the wrist to bones on the other side. The flexor retinaculum and carpal bones form a canal called the *carpal tunnel*, through which nine tendons from the forearm flexor muscles and the *median nerve* pass, as shown in Figure 44.8. As the flexor tendons course through the carpal tunnel on their way to the fingers, they travel through a network of *synovial sheaths*, as shown in Figures 44.9 and 44.10. These sheaths reduce the friction between the tendons and their adjacent structures as they wrap around tendons in articulating joints of the wrist and fingers. The structure of a synovial sheath is an elongated and double-walled *bursa* that contains *synovial fluid*, as illustrated in Figures 44.11 and 44.12. The inner wall of the sheath is attached to the tendon, and the outer wall is attached to a fibrous sheath moored to a bone or ligament. The inside surfaces of the sheath's inner and outer walls are lined with synovial fluid, which acts as a lubricant as the tendon traverses inside the tunnel formed by the fibrous sheath.

Nervous System

The primary purposes of the *peripheral nervous system* (*PNS*), which serves voluntary skeletal muscles of the extremities, head, neck, and torso, are first, to receive sensory information from outlying parts of the body and relay this information to the *central nervous system* (*CNS*), which consists of the brain and spinal cord. The second major purpose of the peripheral nervous system is to send motor signals that activate muscles in the outlying area(s) in response to the sensory input. Of the several nerves traveling

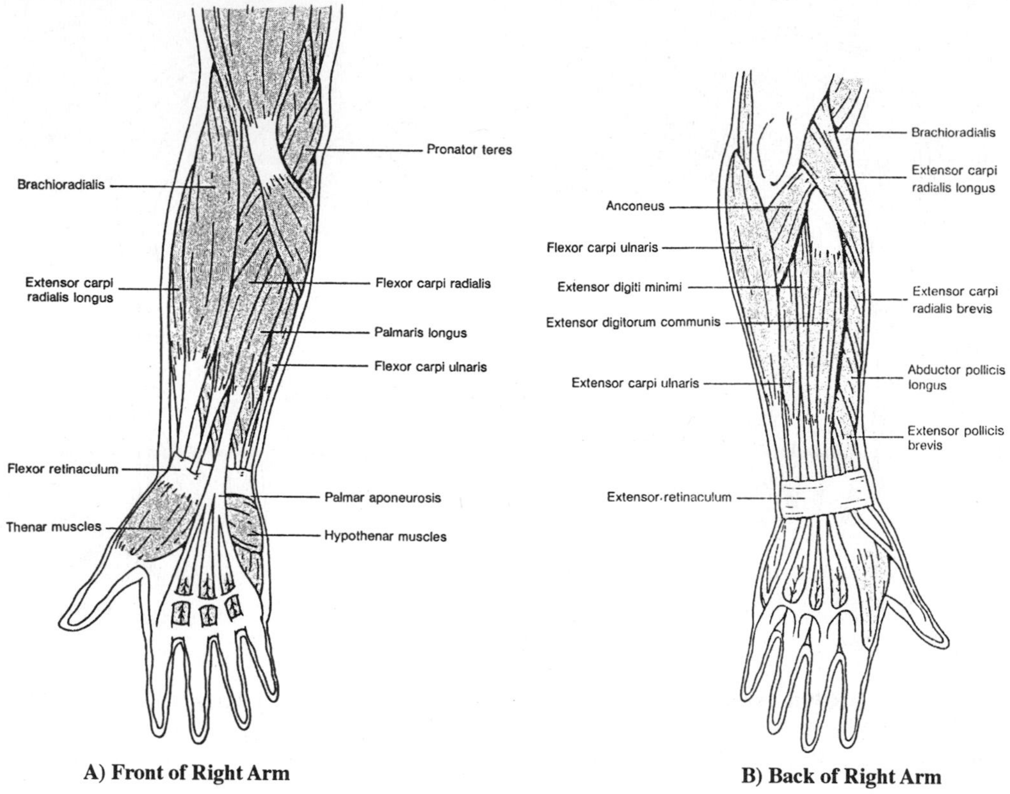

FIGURE 44.7 The muscles that flex (a) and extend (b) the wrist. Figures (a) and (b) are the front and rear views of the right arm, respectively. (From Van de Graaff, K.M. and Rhees, R.W. 1987. *Human Anatomy and Physiology*, p. 109, Schaum's Outline Series, McGraw-Hill Book Co. With permission.)

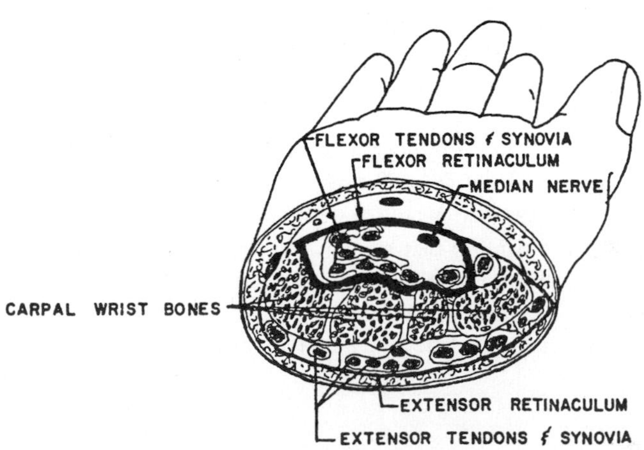

FIGURE 44.8 Cross-sectional anatomy of the wrist. The area highlighted is the carpal tunnel. (From Chaffin, D.B. and Andersson, G.B.J. 1991. *Occupational Biomechanics*, 2nd edition, p. 240, John Wiley & Sons Publishers. With permission.)

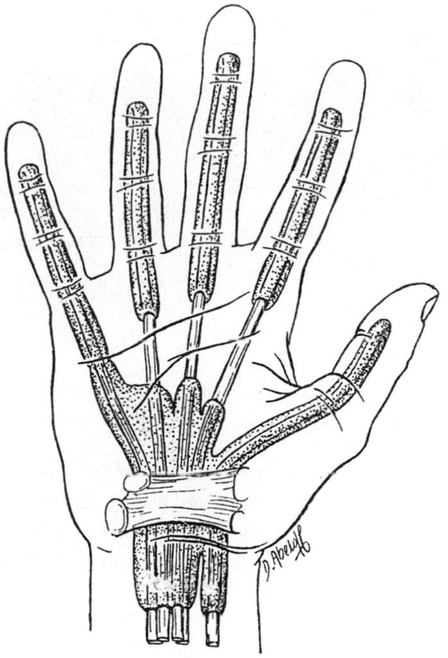

FIGURE 44.9 The system of synovial sheaths that lubricate the flexor tendons as they bend around the wrist and finger joints (palmar view of the right hand). (From Basmajian, J.V. 1982. *Primary Anatomy,* 8th edition, p. 158, Williams and Wilkins, Inc. With permission.)

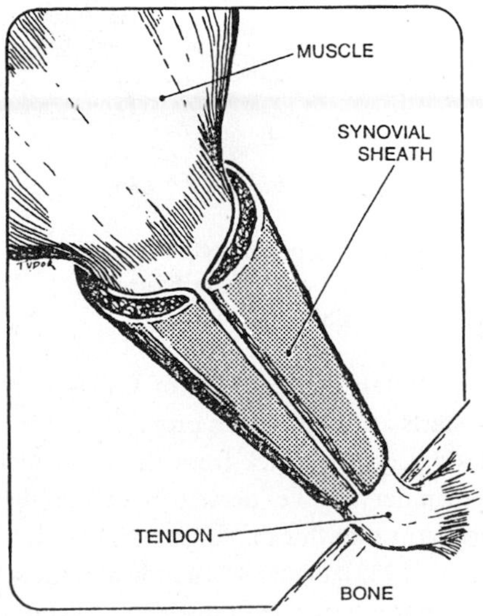

FIGURE 44.10 A magnified end view of a muscle, tendon, sheath, and bony attachment point. (From Putz-Anderson, V. 1988. *Cumulative Trauma Disorders: A Manual for Musculoskeletal Diseases of the Upper Limbs,* p. 12, Taylor & Francis. With permission.)

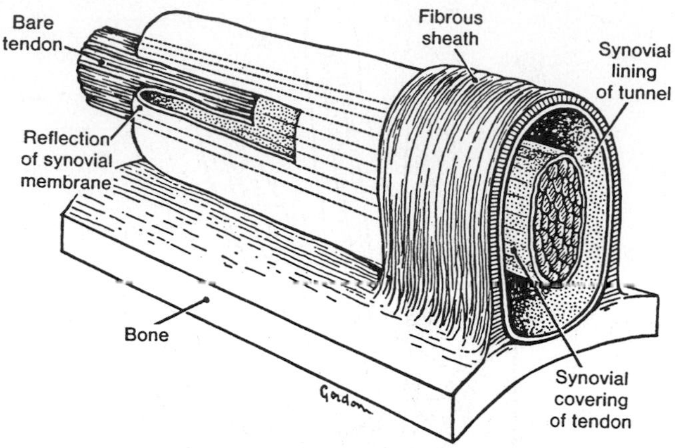

FIGURE 44.11 Structure of a synovial sheath. An area of the sheath has been cutaway to expose its double-walled structure. Synovial fluid lines the inside of the sheath's inner and outer walls and reduces friction as the tendon moves within its tunnel. **Note:** normally the tendon fits snugly in its tunnel, but is shown having a loose fit in this figure for illustration purposes. (From Basmajian, J.V. 1982. *Primary Anatomy,* 8th edition, p. 119, Williams and Wilkins, Inc. With permission.)

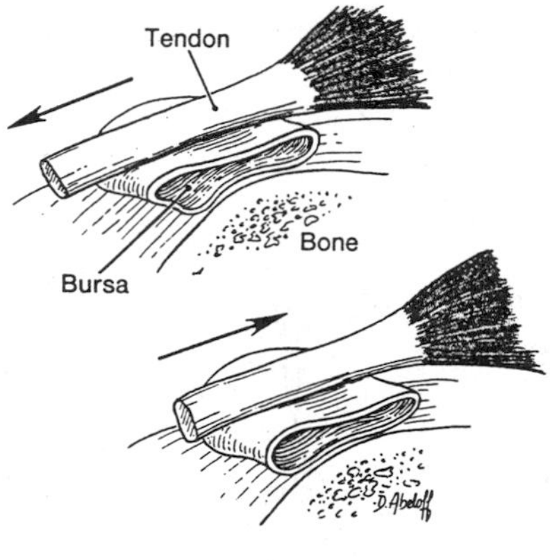

FIGURE 44.12 A bursa, which is a collapsed bag of connective tissue filled with synovial fluid, forms whenever a tendon rubs against a hard structure, such as a bone. (From Basmajian, J.V. 1982. *Primary Anatomy,* 8th edition, p. 118, Williams and Wilkins, Inc. With permission.)

through the arm, the nerve most often associated with CTDs is the median nerve. As shown in Figure 44.13a, the median nerve starts at the shoulder, provides motor inputs to muscles in the forearm and thumb region, and provides sensory feedback from the palm region and from the thumb to the center of the ring finger. The median nerve is the "nerve of precision" because it supplies motor function to the extrinsic muscles in the forearm that flex the fingers and the intrinsic muscles in the thumb that exert a precision grip (Feldman et al., 1983). Figure 44.14 indicates the sensory regions of the hand served by the median nerve and the radial and ulnar nerves, which travel down the radial and ulnar sides of the forearm, respectively, as illustrated in Figures 44.13b and 44.13c. The radial nerve is the "nerve of stability" because it *innervates* the forearm extensor muscles that oppose and stabilize the precision and power muscles on the forearm's flexor side. The ulnar nerve is the "nerve of power" because it innervates the muscles that provide wrist flexor power, but little precision (Feldman et al., 1983).

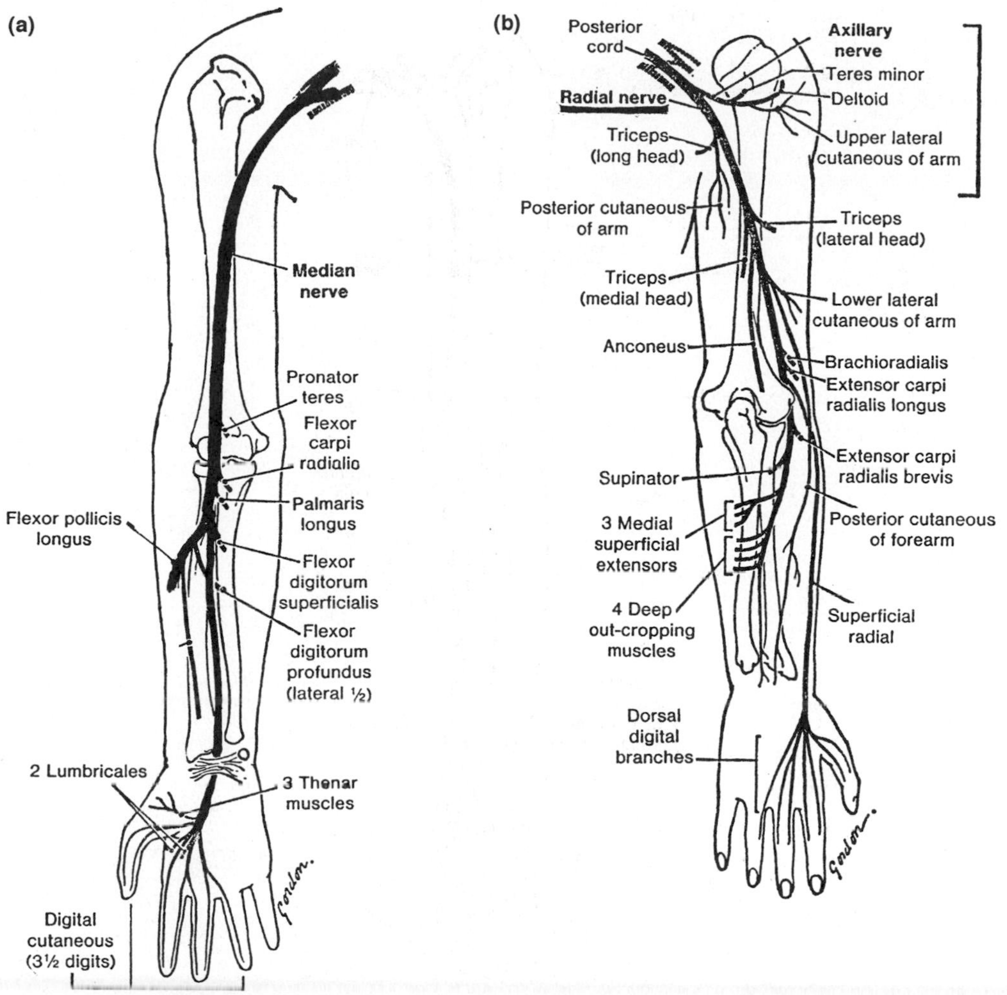

FIGURE 44.13 a) Front view of the paths of the *median nerve* as it travels down the right upper extremity. b) Rear view of the paths of the *radial nerve*. c) Front view of the paths of the *ulnar nerve*. (From Basmajian, J.V. 1982. *Primary Anatomy*, 8th edition, p. 340, Williams and Wilkins, Inc. With permission.)

44.3 Work-Related Muscle Disorders

After frequent or prolonged contractions, a muscle can feel painful for a relatively short period of time and recover to full function, or it could develop a more serious chronic condition. If the pain disappears after a relatively short period of time, the cause was probably temporary fatigue of the muscular tissues. However, if the pain persists, the worker could have developed a muscle CTD.

The medical term describing muscle pain is *myalgia*, which includes a few specific muscle pain syndromes. Myalgia can occur after vigorous or unaccustomed exercise, and also from work-related activity. A worker can develop a *myopathy* called *myofascial syndrome*, which is characterized by "the presence of one or more discrete areas (or trigger points) that are tender and hypersensitive and from which pain may radiate when pressure is applied" (Kuorinka and Forcier, 1995, p. 81). Myofascial syndrome could be associated with work-related activity, and a common work-related myofascial syndrome is *tension neck syndrome* (also called *shoulder–neck myofascial syndrome*). Tension neck syndrome

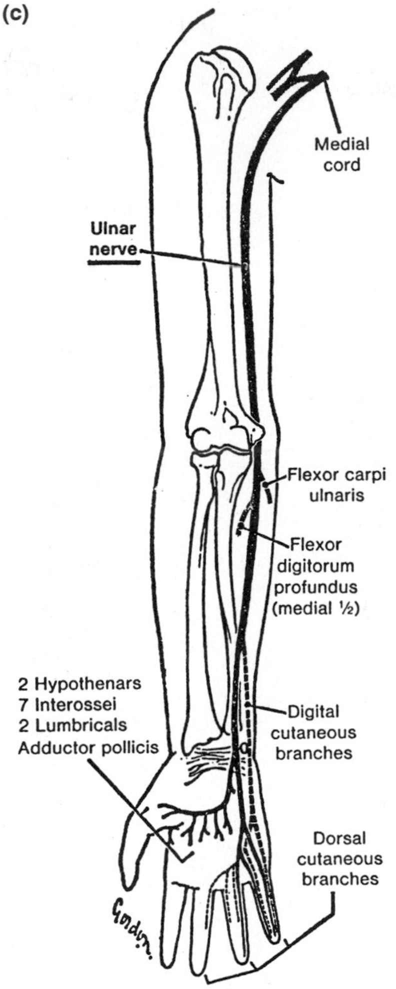

FIGURE 44.13 (continued)

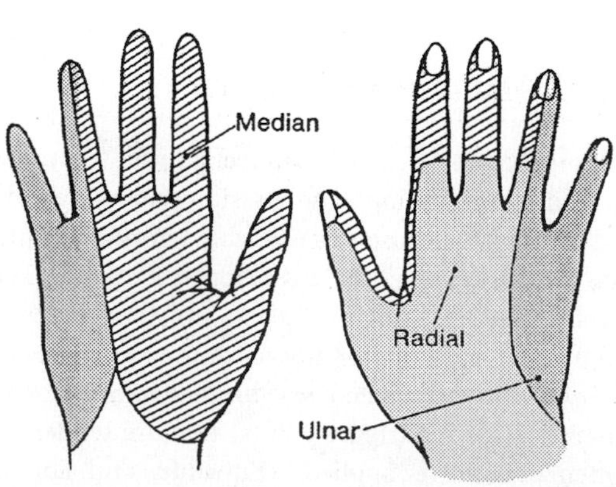

FIGURE 44.14 Sensory regions of the right hand served by the median, radial, and ulnar nerves. (From Basmajian, J.V. 1982. *Primary Anatomy,* 8th edition, p. 341, Williams and Wilkins, Inc. With permission.)

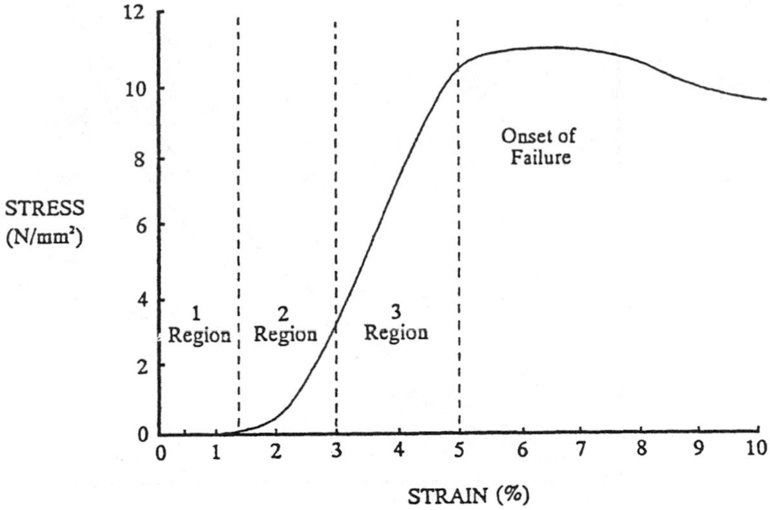

FIGURE 44.15 A typical stress-strain curve of a tendon. (From Abrahams, M. 1967. Mechanical analysis of tendon *in vitro*: A preliminary report, *Med Biol Eng*, Vol. 5, p. 435. With permission.)

is a myofascial syndrome localized in the shoulder and neck region with tenderness descending into the trapezius muscle. Occupational groups cited in the literature that have been associated with high rates of tension neck syndrome are those requiring repetitive arm movements and constrained postures (Kuorinka and Forcier, 1995).

The *pathogenesis* of myofascial syndromes is unknown; however, several hypotheses have been offered in the literature, which include a lower capillary-to-fiber ratio for the slow twitch fibers (Type I), severe depletion of ATP in the muscle, and dysfunctional energy metabolism (Kuorinka and Forcier, 1995). For a more thorough discussion of muscle CTDs, the reader is referred to Kuorinka's and Forcier's (1995) book, which has a comprehensive description and discussion of biomechanical mechanisms of muscle CTDs.

44.4 Biomechanical Aspects of Muscle–Tendon Disorders

As a muscle shortens during contraction and lengthens during stretching, its tendon acts like a rope and transmits the muscle force to the bony attachment site. As a tendon moves with a muscle, the length of the tendon does not necessarily stay constant. A tendon has elastic properties and is analogous to a rubber band. The muscle force applied to the tendon is a tensile force, which is commonly converted to the units of stress (force divided by cross-sectional area of tendon). As the tensile force increases, the tendon elongates, which is measured by strain (the percentage of change in length). Figure 44.15 shows a typical stress–strain curve of a tendon with its three characteristic regions (Abrahams, 1967). In Region 1, the crumpled collagen fibers of a relaxed tendon merely straighten under negligible loads. Then, as the tensile force increases, the tendon passes through the "toe" region (region 2), and then has a linear relationship between stress and strain in region 3. Although the tendon can elongate up to 5% strain before onset of failure, normal tendon strain is below 3% (Abrahams, 1967; Elliott, 1965; Rigby et al., 1959).

The loading and unloading of a tendon can change its elastic properties depending on whether the tensile force is increasing or decreasing. As shown in Figure 44.16, the amount of stress required to elongate a tendon to a specific strain level is greater when a tendon is loaded (increase in tensile load) then when it is unloaded (decrease in tensile load). This change in stress–strain curve is called *hysteresis*, which results from a loss of energy, probably as heat, during the unloading phase (Moore, 1992).

Several CTDs reported in the literature occur at sites where a tendon wraps around a deviated joint. As a muscle contracts and moves its tendon accordingly, the tendon can rub against its adjacent surface,

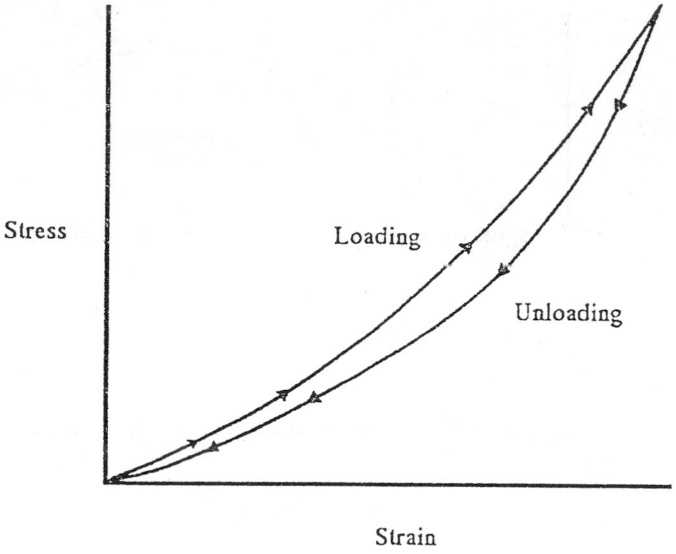

FIGURE 44.16 The stress–stress curve of a tendon depends on whether the tensile force is increasing or decreasing. The difference in the stress–strain curves is called hysteresis. The units of stress and strain are N/mm^2 and % elongation, respectively. (From Moore, J.S. 1992. Function, structure, and responses of components of the muscle–tendon unit, in *State of the Art Reviews in Occupational Medicine*, Vol. 7, No. 4, p. 721. With permission.)

usually a bone or ligament, as a rope rubs against a nonrotating pulley. Likewise, when the muscle lengthens, the tendon moves in the opposite direction against its adjacent structures. In the wrist area, the repetitive rubbing of the tendons against the carpal bones and flexor retinaculum can cause CTDs known as *tendinitis* and *tenosynovitis*, which are inflammation of the tendon and its sheath, respectively.

Based on Landsmeer's (1962) model, Armstrong and Chaffin (1979) developed a static model of a tendon wrapping around a joint. Figure 44.17 depicts Landsmeer's model of a tendon, which is analogous to a rope bent around a nonrotating pulley, and Figure 44.18 illustrates the Armstrong and Chaffin (1979) model as a reasonable representation of Landsmeer's model. When the wrist is flexed, the flexor tendons bend around the flexor retinaculum that is assumed to have a constant radius. When the wrist is extended, the flexor tendons are supported on the *dorsal* side by the carpal bones that are assumed to have a constant radius. Armstrong and Chaffin (1978) found that the radius in a flexed posture is larger than in an extended posture.

The arc length of the tendon wrapping around the pulley is defined in equation (1).

$$X = R \times \theta \tag{1}$$

where
X = tendon arc length around pulley (mm)
R = radius of curvature of supporting tissues (mm)
θ = angle of deviation of wrist from neutral (in radians)

The reaction forces acting normal to the tendon are shown in Figure 44.18 and defined in equation (2).

$$F_n = \left(F_t \times e^{[\mu \times \theta]}\right) / R \tag{2}$$

where
F_n = normal supporting force per unit of arc length (N/mm)
F_t = average tendon force in tension (N)

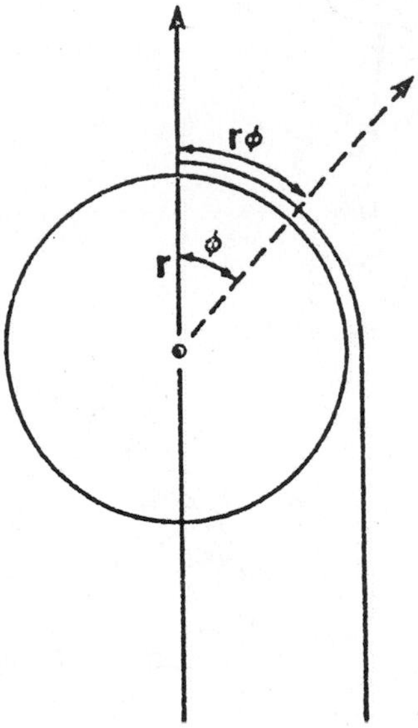

FIGURE 44.17 Landsmeer's (1962) model of a tendon wrapping around a joint. (From Chao, Y.S., An, K.N., Cooney, W.P, and Linscheid, R.L. 1982. *Biomechanics of the Hand: A Basic Research Study,* p. 15, Biomechanics Laboratory, Department of Orthopaedic Surgery, Mayo Clinic/Mayo Foundation, Rochester, MN. With permission.)

μ = coefficient of friction between tendon and supporting synovia
θ = wrist deviation angle (radians)
R = radius of curvature of supporting tissues (mm)

Since μ is considered small (approximately 0.0032 [Fung, 1981]), it can be approximated by zero. This changes equation (2) to equation (3).

$$F_n = F_t / R \tag{3}$$

Equation (3) reveals F_n is a function of the tendon force and radius of curvature. As the radius of curvature decreases, the normal supporting force per unit of arc length increases. The normal supporting force for women would be greater than for men because women have smaller wrists. Also, as the tendon force increases, the normal supporting force increases.

The total supporting force F_r in Figure 44.18 is the force of the ligaments, bones, and median nerve in the carpal tunnel acting on the flexor tendons. F_r is defined in equation (4).

$$F_r = 2 \times F_t \times \sin(\theta/2) \tag{4}$$

where
F_r = resultant force exerted by adjacent wrist structures on the flexor tendons (N)
F_t = tendon force (N)
θ = wrist deviation angle (in radians)

Equation (4) indicates that F_r is a function of the tendon force and wrist deviation angle, but is independent of radius of curvature. Figure 44.19 illustrates this relationship in that as the tendon force and wrist angle increase, the resultant force F_r increases linearly.

The significance of F_n and F_r is based on the theory that increased normal forces place greater stress on the tendon and its surrounding structures. The increase in normal force could cause the tendon and

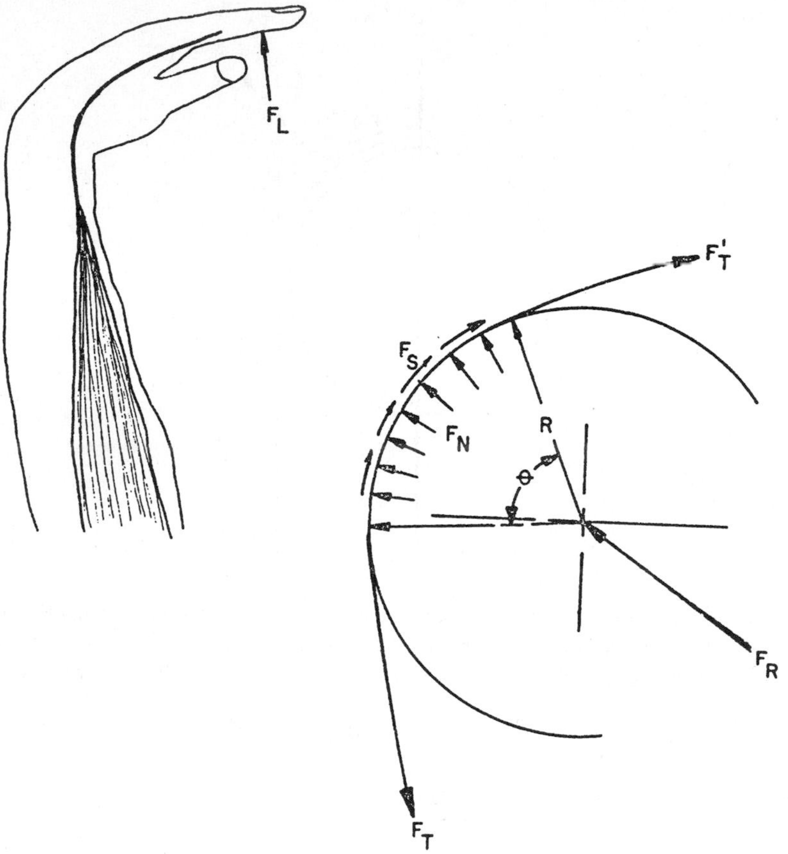

FIGURE 44.18 Armstrong's and Chaffin's (1979) biomechanical model of a flexor tendon wrapping around the flexor retinaculum. F_t is the tendon force, and F_r is the resultant reaction force exerted against the tendon. (From Chaffin, D.B. and Andersson, G.B.J. 1991. *Occupational Biomechanics,* 2nd edition, p. 243, John Wiley & Sons Publishers. With permission.)

its sheath or the fibrous sheath moored to bone or ligament to *hypertrophy* or *inflame.* If these structures were to hypertrophy or inflame, then the coefficient of friction (μ in equation (2)) would increase, thereby placing even greater F_n on the tendons.

Dynamic movements that accelerate and decelerate the tendons around a nonrotating pulley could exacerbate the trauma imposed on the tendons. Schoenmarklin and Marras (1990) developed a dynamic model of a flexor tendon bent around the carpal bones or flexor retinaculum, taking into account the acceleration and deceleration of a tendon's movements. This model analyzes the effects of peak angular acceleration on the resultant reaction force that the wrist bones and ligaments exert on tendons and their sheaths in the flexion/extension plane. Like the Landsmeer (1962) and Armstrong and Chaffin (1979) models, Schoenmarklin and Marras (1990) model the tendon as a rope bent around a fixed pulley.

The quantitative effects of the wrist's peak angular acceleration on resultant reaction forces were based on the free body diagram (FBD) and mass × acceleration diagram (MAD) approach in engineering dynamics (Meriam and Kraige, 1986). Figure 44.20 illustrates the FBD and MAD approach applied to a wrist and hand in midposition (neither pronated or supinated). There is no externally applied load in the hand. The hand is rotated in the horizontal plane around a vertical **z** axis, so the effects of gravity do not play a role in this example. All the flexor tendons are grouped together as one tendon force vector in order to maintain static determinacy. The hand is assumed to accelerate from a stationary posture, so the angular velocity is theoretically zero, resulting in zero centripetal force.

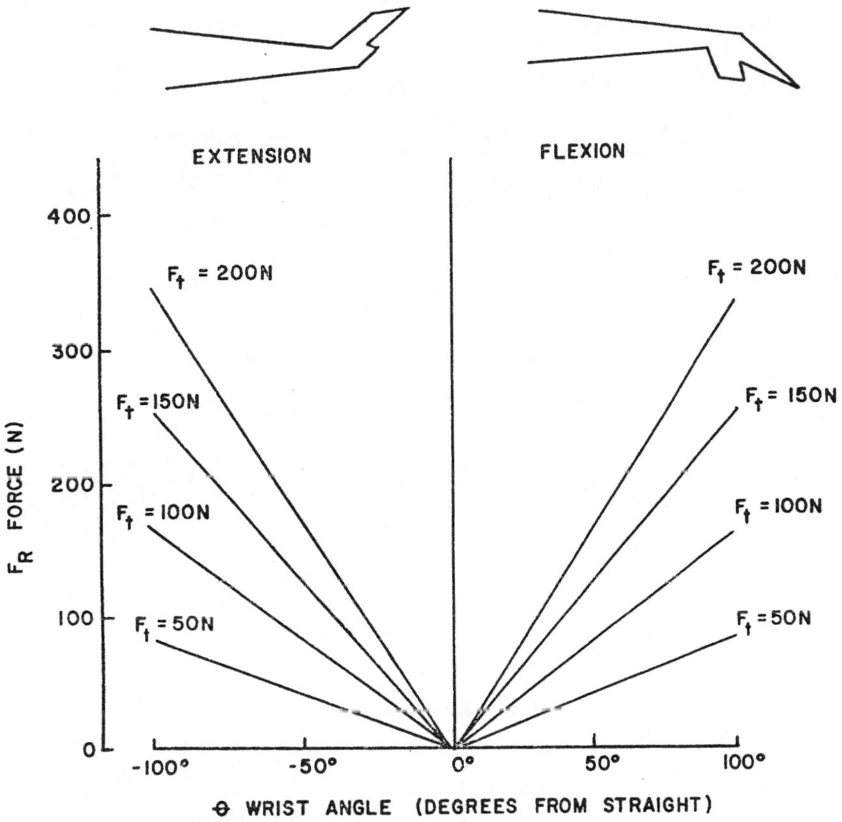

FIGURE 44.19 The resultant reaction force (F_r), as modeled by Armstrong and Chaffin (1979), that is exerted against the flexor tendons as a function of wrist angle and tendon force. (From Chaffin, D.B. and Andersson, G.B.J. 1991. *Occupational Biomechanics*, 2nd edition, p. 247, John Wiley & Sons Publishers. With permission.)

The maximum tendon force ($F_{t\text{-max}}$) was computed as a function of five peak angular accelerations ($\ddot{\theta}$ = 3000, 6000, 9000, 12000, and 15000°/sec²). Based on empirical data from normal subjects, 15000°/sec² was found to be about 50% of peak wrist acceleration in the flexion/extension plane (Schoenmarklin and Marras, 1993). The $F_{t\text{-max}}$ is depicted in Figure 44.21 and is derived from equation (5) (LeVeau, 1977).

$$F_{t\text{-max}} = F_{t\text{-min}} \times e^{[\mu \times \theta]} \tag{5}$$

where

$F_{t\text{-max}}$ = maximum force in flexor tendons, which is the force that the extrinsic flexor muscles in the forearm exert on their tendons (N)
$F_{t\text{-min}}$ = minimum force in flexor tendons, which is the force that the flexor tendons transmit to the hand and fingers (N)
μ = coefficient of friction between tendons and their sheaths
θ = wrist deviation angle (radians)

Since the coefficient of friction for human synovial joints bones is estimated to be very low (0.0032 according to Fung, 1981), then the calculation of the $F_{t\text{-max}}$ force is very close to $F_{t\text{-min}}$. The F_r depicted in Figure 44.21 and expressed in equation (4) was calculated as the resultant force necessary to resist $F_{t\text{-max}}$ and $F_{t\text{-min}}$.

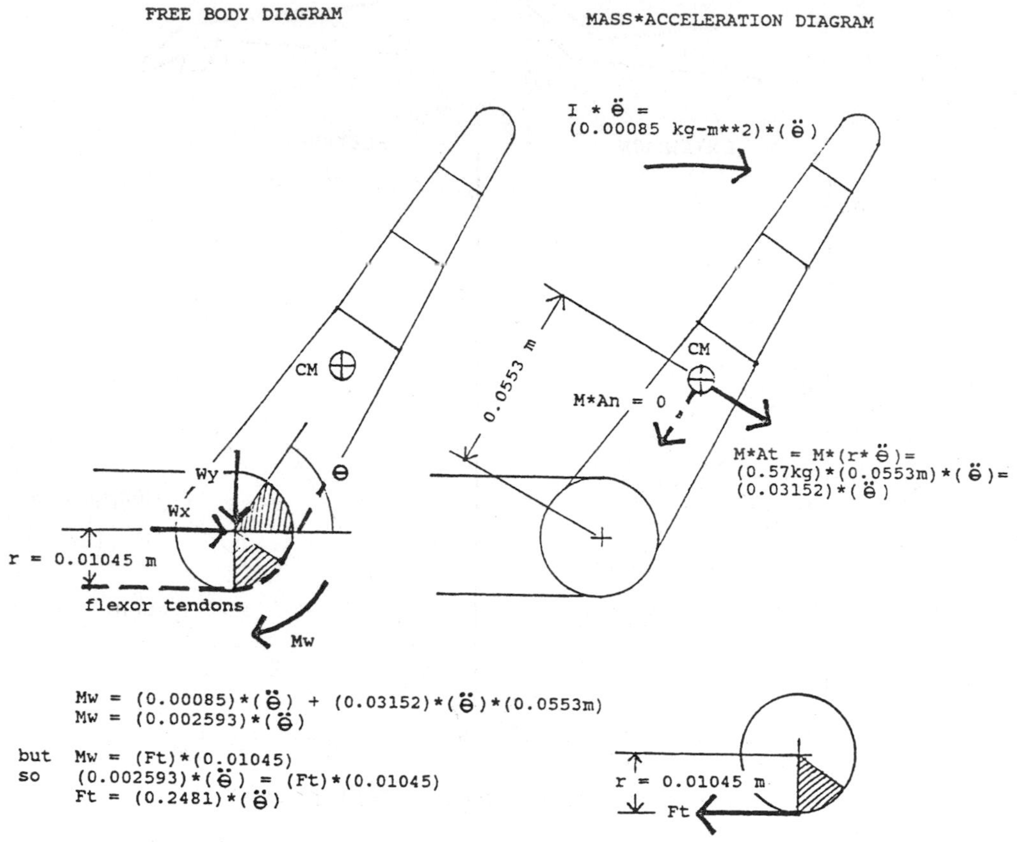

FIGURE 44.20 The free body diagram (FBD) and mass acceleration diagram (MAD) approach used by Schoenmarklin and Marras (1990) to calculate the peak reaction force (F_r in Figures 44.21 and 44.22) on the wrist when the wrist is accelerated (in the flexion direction) at an extension angle of θ. (From Schoenmarklin, R.W. and Marras, W.S. 1990. In *Proceedings of the 34th Meeting of the Human Factors Society*, p. 807. With permission.)

As shown in Figure 44.22, F_r increases approximately linearly as wrist angle or angular acceleration increases, resulting in a curved plane that signifies an interactive effect between wrist angle and angular acceleration. The greatest F_r occurs when the wrist is accelerated at a deviated wrist posture. The large peak reaction forces exerted on the flexor tendons and their sheaths are due solely to wrist motion without any externally applied load in the hand. If loads were applied in the hand (e.g., power grip or pinch grip) while the hand was accelerated in deviated postures, then F_r would increase even more, resulting in even more stress on flexor tendon tissue. The large peak F_r in Figure 44.22 could possibly cause the tendon and its sheath or the fibrous sheath moored to bone or ligament to hypertrophy or inflame, which could result in tendinitis or tenosynovitis. The occurrence of either tendinitis or tenosynovitis would most likely increase μ in equation (5), thereby increasing $F_{t\text{-}max}$ and F_r even more (refer to Figure 44.21 and equation (4).

The large resultant reaction forces on the tendons from wrist deviation and accelerations could possibly explain the findings of Armstrong et al. (1984), who investigated the histological changes in the flexor tendons as they pass through the carpal tunnel. These investigators found hypertrophy and increased density in the synovial tissue in the carpal tunnel area. These authors suggested that biomechanical factors, such as repeated exertions with a flexed or extended wrist posture, could have partially caused degenerative changes in tendon tissue. In addition to reaction forces from supporting structures, the hypertrophy of the tendon tissue could have been caused by differences in strain within a tendon. In an investigation of the viscoelastic properties of tendons and their sheaths, Goldstein et al. (1987) found

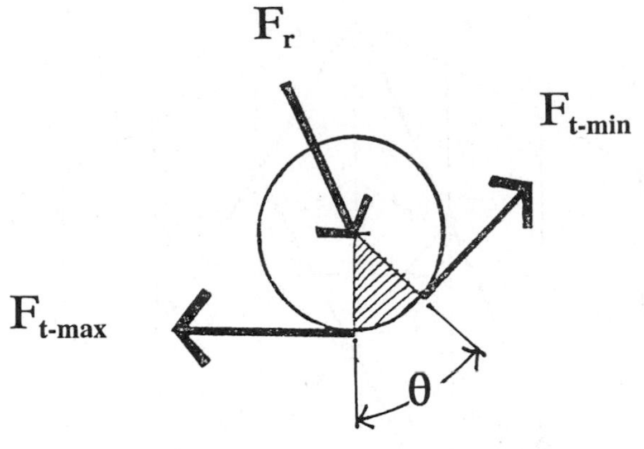

$$F_{t\text{-max}} = F_{t\text{-min}} \bullet e^{[\mu \bullet \theta]}$$

FIGURE 44.21 Relationship between the maximum ($F_{t\text{-max}}$) and minimum ($F_{t\text{-min}}$) forces of a tendon and the resultant reaction force (F_r). $F_{t\text{-max}}$ is the force emanating from the forearm flexor muscles, and $F_{t\text{-min}}$ is the force transmitted to the hand. The flexor tendon is wrapped around the wrist's carpal bones. $F_{t\text{-max}}$ and $F_{t\text{-min}}$ are the maximum and minimum tendon forces in Schoenmarklin's and Marras's (1990) dynamic model of a flexor tendon passing through the wrist joint. The equation for $F_{t\text{-max}}$ and $F_{t\text{-min}}$ is from Leveau (1977). (From Schoenmarklin, R.W. and Marras, W.S. 1990. In *Proceedings of the 34th Meeting of the Human Factors Society*, p. 807. With permission.)

that flexion/extension wrist angle increased the shear traction forces between tendons, their sheaths, and bones and ligaments that form the anatomical pulley. As depicted in Figure 44.23, when the wrist is extended approximately 10°, the strain in the flexor digitorum profundus (FDP) tendons, which pass through the carpal tunnel and move the fingers, is approximately 10% to 15% lower on the side distal (hand side) to the flexor retinaculum than the proximal side (forearm side). This difference in strain within a tendon creates shear traction forces, which are magnified when the wrist angle is deviated to 65° flexion or extension.

44.5 Work-Related CTDs Involving the Muscle–Tendon Unit

The pathogenesis of the most frequently studied CTDs involving the muscle–tendon unit will be discussed below.

Tendinitis

Although tendinitis is defined as inflammation of the tendon, Moore (1992) contends there is scant scientific evidence that the collagenous fibers that comprise the tendon actually inflame. According to Moore (1992), tendinitis is often used as a term that implies soreness localized to a muscle–tendon unit that increases with tensile load from either muscle contraction or passive stretch. Moore (1992) further states that these clinical findings of soreness "may represent no more than a normal pattern to varying degrees of use, rather than inflammation." Because clinicians do not have sensitive diagnostic tools to differentiate between tendinitis and tenosynovitis (inflammation of the tendon's sheath), soreness in joints where the tendons do not have sheaths, such as in the elbow and shoulder, is usually diagnosed as tendinitis, whereas soreness in joints with sheathed tendons is commonly diagnosed as tenosynovitis.

One theory of the pathogenesis of tendinitis is the physical disruption of a small number of collagen fibers within a tendon and the ensuing repair process. According to Moore (1992), the body responds to this disruption in a manner similar to that of a partial tendon laceration, which is the partial cutting or severing of a tendon. The healing process of the tendon occurs in three stages: inflammatory stage,

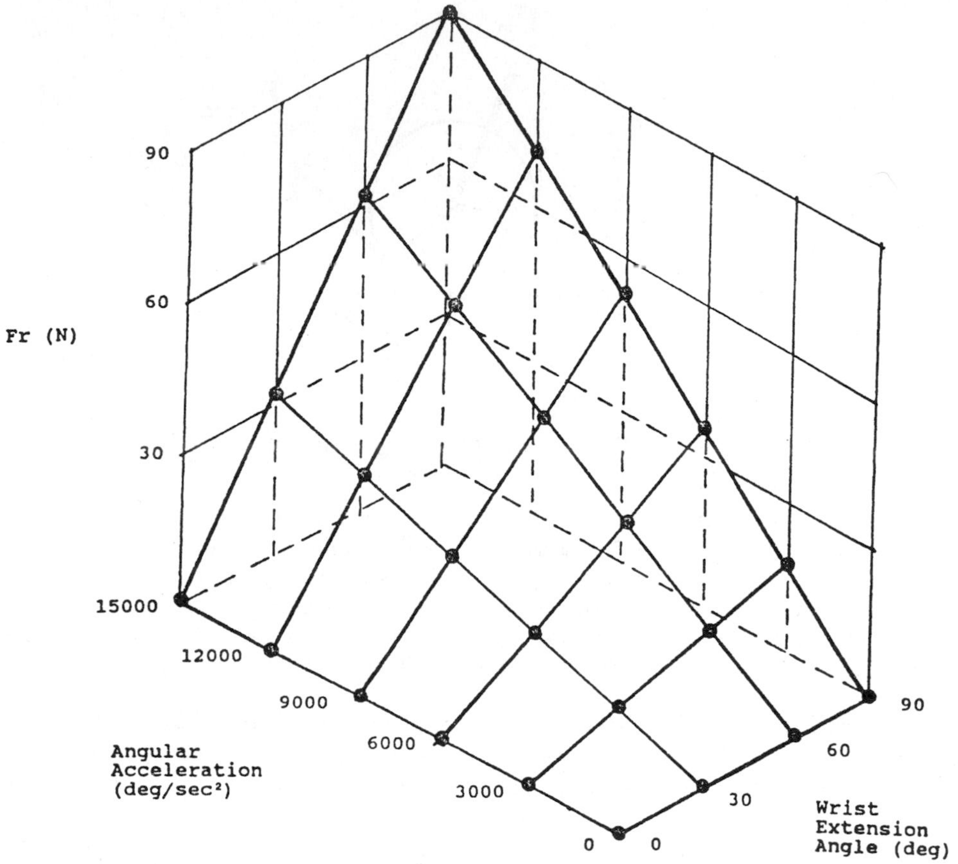

FIGURE 44.22 The resultant reaction force (F_r) exerted by the carpal bones or flexor retinaculum against a flexor tendon and its sheath as a function of wrist angle and acceleration. (From Schoenmarklin, R.W. and Marras, W.S. 1990. In *Proceedings of the 34th Meeting of the Human Factors Society,* p. 809. With permission.)

reparative or collagen-production stage, and a remodeling stage (Gelberman et al., 1988). As the body rebuilds its tendon tissue, the collagen content increases and the tendon could hypertrophy, or increase in size. In addition, the body may not repair all of the disrupted fibers, resulting in permanent fraying of the tendon. Due to hypertrophy and fraying, the tendon may be biomechanically different, possibly deficient, after the body's attempt to completely restore the disrupted tissues.

Moore (1992) relates the effects of partial tendon laceration to CTDs in that when a joint is deviated from a neutral position, the tendons could react to resulting reaction forces from the joint structures (F_r in equation (4) and Figures 44.21 and 44.22) in a manner similar to a partial tendon laceration. As shown by Armstrong and Chaffin's (1979) static model, as the wrist deviates, the resulting reaction forces from the flexor retinaculum or carpal bones could cause physical disruption to the tendons, much like the effects of partial tendon laceration. Theoretically, physical disruption of the tendon tissue would be exacerbated by even greater reaction forces if the wrist were accelerated or decelerated, particularly at extreme wrist deviation angles (Schoenmarklin and Marras, 1990). If the wrist and fingers were deviated excessively in repetitive motions, hypertrophy of the tendons from the healing process or permanent fraying of the tendons could cause soreness at the wrist. Depending on the specific tendon, this soreness may be diagnosed as tendinitis or tenosynovitis. Soreness in the wrist flexor muscles' tendons (flexor carpi radialis and ulnaris) would probably be diagnosed as tendinitis because these tendons do not have sheaths, whereas soreness in tendons passing through the carpal tunnel (flexor digitorum superficialis and profundus) would commonly be diagnosed as tenosynovitis these tendons are sheathed.

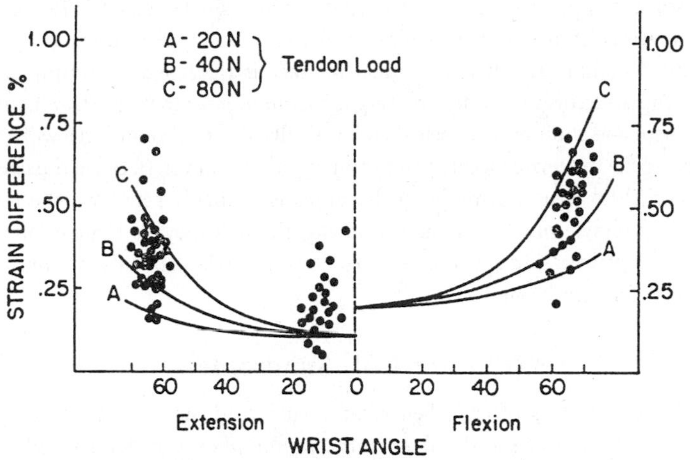

FIGURE 44.23 The difference in strain within the flexor digitorum profundus tendon between measurements taken proximal and distal to the flexor retinaculum. Even at an extended wrist angle of 10°, there is a 10% to 15% difference in strain between the FDP tendon proximal and distal to the flexor retinaculum. The difference in strain is magnified as wrist deviation angle increases. (From Goldstein, S.A., Armstrong, T.J., Chaffin, D.B., and Matthews, L.S. 1987. Analysis of cumulative strain in tendons and tendon sheaths. *J Biomed,* 20, p. 4. With permission.)

Lateral Epicondylitis (Tennis Elbow)

Lateral epicondylitis, which is also called tennis elbow in lay parlance, is tendinitis of the forearm extensor and supinator muscles at the *lateral epicondyle* of the elbow. The lateral epicondyle is the small bony attachment point on the outside of the elbow where the group of forearm extensor and supinator muscles originate. The extrinsic extensor and supinator muscles fuse into an *aponeurosis,* or a broad, flat tendon, which is attached to the lateral epicondyle in the elbow. Soreness and pain occur at the point where the aponeurosis of the extensor and supinator muscles pull on the lateral epicondyle. The small size of the lateral epicondyle and the relatively large mass of extensor and supinator muscles create high stresses on the lateral epicondyle and its attached aponeurosis. Patients who have lateral epicondylitis report their pain is particularly acute when they extend their wrist or supinate the forearm against resistance, which occurs when one is hitting a tennis ball with a backhand stroke.

Lateral epicondylitis is a CTD in that it is directly related to the motions that tense the wrist's extensor and supinator muscles (Nirschl, 1983). In a study of 113 patients, Goldie (1964) found that repeated wrist extensions or alternating pronating and supinating movements of the forearm were causal factors in 83 of the cases.

Review of the medical literature reveals several hypotheses regarding the pathogenesis of lateral epicondylitis, although all of them do agree that the basic mechanism is deterioration of the aponeurotic tendinous tissue at the lateral epicondyle. Cyriax (1936), who treated 20 patients with lateral epicondylitis, concluded it is caused by a tear between the tendinous origin of the extensor muscles and the *periosteum* of the lateral epicondyle. Goldie (1964) suggested that lateral epicondylitis is due to a buildup of lesions in a space under the tendon and distal to the epicondyle. Microscopically, Nirschl (1985) found that the affected tendon in lateral epicondylitis had a characteristic appearance of hypertrophy that was grayish, edematous, and friable. Nirschl (1985) interpreted this medical description as a "thick unhappy gray tendon, weeping with *edema.*" A normal tendon has collagen fibers that run parallel, but the tendons afflicted with lateral epicondylitis look coarse and granular.

Often, a patient with lateral epicondylitis will wear a brace around the forearm near the elbow (Froimson, 1971; Moore, 1992), as is often seen on tennis players. Although it has not been validated experimentally, one of the theories of why the forearm brace is beneficial is based on biomechanics. Figure 44.24 shows a free body diagram of the elbow as viewed from the head position. The tendons of the forearm extensor and supinator muscles are modeled as a single vector. When tightened, the brace may keep the aponeurosis from vibrating against underlying bony tissue during repeated extending or supinating exertions. In addition, when the forearm brace is tightened it compresses the aponeurosis against the underlying structures, thereby creating a *frictional force* that resists, albeit partially, the pull of the forearm extensor and supinator muscles. Theoretically, this frictional force would relieve the lateral epicondyle of carrying the full tensile load of the aponeurosis. However, because the coefficient of friction among the musculoskeletal tissues underlying the brace is probably very low, the reduction in tensile load on the lateral epicondyle may be small and negligible or it may be large enough to retard lateral epicondylitis. Experimental research is needed to determine whether this biomechanical theory can explain the efficacy of forearm braces.

Supraspinatus Tendinitis (Rotator Cuff Syndrome)

Supraspinatus tendinitis, which is often called rotator cuff syndrome, is tendinitis of the muscle that elevates the shoulder. Elevation of the shoulder in the *frontal plane* is called shoulder *abduction*. Pain is felt on the *acromion process*, or bony top of the shoulder, when one abducts the shoulder, particularly when the arm is holding a load or exerting a pushing force.

The pathogenesis of supraspinatus tendinitis is impingement of the bursa and supraspinatus tendon as the shoulder is abducted. The superficial and deep muscles of the shoulder are shown in Figure 44.25. The deltoid muscle, which covers the outside of the shoulder, and the supraspinatus muscle, which is a smaller muscle under the deltoid and trapezius muscles and acromion, are the major abductors of the shoulder. As shown in Figures 44.26a and 44.26b, abduction of the shoulder compresses the acromion downward, thereby pinching the underlying bursa and supraspinatus tendon (Chaffin and Andersson, 1991). As illustrated in Figure 44.12, the bursa is a tubular synovial sheath whose purpose is to lubricate the contact between the deltoid muscle and acromion and the supraspinatus tendon. However, if the shoulder is abducted repeatedly, and particularly under heavy loads, the resulting impingement could damage the bursa and supraspinatus tendon fibrils and produce fraying of the tendon. The relative *avascular* nature of the supraspinatus tendon diminishes its capability to repair itself, thereby leading to degeneration, as shown in Figure 44.26c (Moore, 1992). In addition, intramuscular pressure from muscle fibers attached to the tendon can also diminish the reparative process of the tendon.

Tenosynovitis

Although tenosynovitis is defined as inflammation of the tendon sheath (Stedman, 1982), any tendon sheath disorder is called tenosynovitis, regardless of the presence or absence of inflammation (Moore, 1992). As shown in Figures 44.9 and 44.10, the tendon sheath is a tubular structure that wraps around a tendon and contains synovial fluid to "provide lubrication, protection, and repair assistance for the surrounded tendon" (Moore, 1992). Tenosynovitis is diagnosed only where tendons are sheathed, whereas tendinitis could occur in a tendon regardless of whether it is sheathed. Usually, soreness in a sheathed tendon area is diagnosed as tenosynovitis, whereas soreness in a tendon without sheathing is diagnosed as tendinitis.

DeQuervain's Tenosynovitis

DeQuervain's disease is the *stenosing tenosynovitis* of the tendons that abduct and extend the thumb (abductor pollicis longus [APL] and extensor pollicis brevis [EPB]) (Williams and Ward, 1983). This disease was named after a Swiss surgeon who observed the condition in 1895. The practical importance of the muscles that flex, extend, and abduct the thumb cannot be overestimated. According to Bunnel (1956), "a hand without a thumb is no more than a hook." The APL and EPB are two of the thumb's muscles that are necessary for dexterity and fine manipulations.

As shown in Figure 44.27, the tendons of the APL and EPB pass underneath the extensor retinaculum of the wrist, and then they share the same synovial sheath on their way to the dorsal and lateral side of

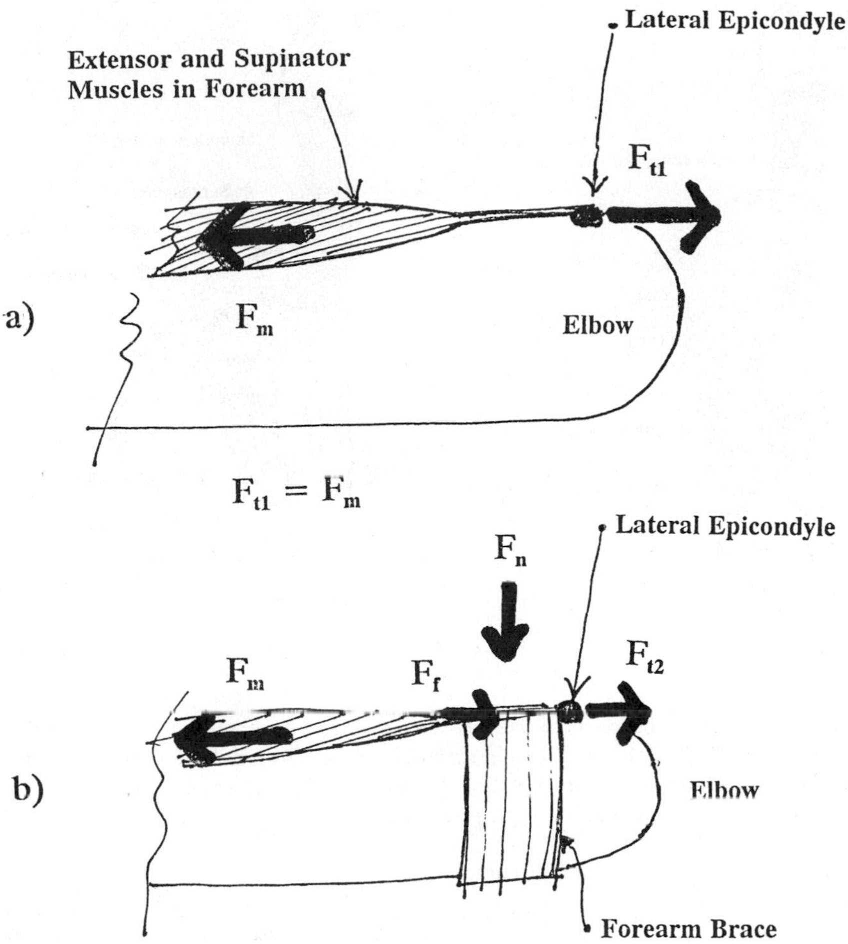

$$F_m = F_f + F_{t2}$$

$$F_f = F_n \times \mu$$

$$\therefore\ F_{t2} < F_m$$

FIGURE 44.24 Free body diagram (FBD) analysis of the forearm extensor and supinator muscles and their tendinous attachment (aponeurosis) to the lateral epicondyle. The view of the elbow is from the head looking down, with the forearm in midposition (neither supinated or pronated). a) FBD of the elbow without a forearm brace. The force the aponeurosis has to exert, F_{t1}, is equal to the tensile pull of the extensor and supinator muscles, F_m. b) The tightening of the forearm brace around the forearm creates a frictional force, F_f, that opposes F_m, thereby lessening the force on the aponeurosis, F_{t2}. (From Marklin, R.W. original artwork. With permission.)

the thumb (Lamphier, 1965). The APL's and EPB's common sheath, which is about 5 cm long, passes over a bony depression called the radial styloid.

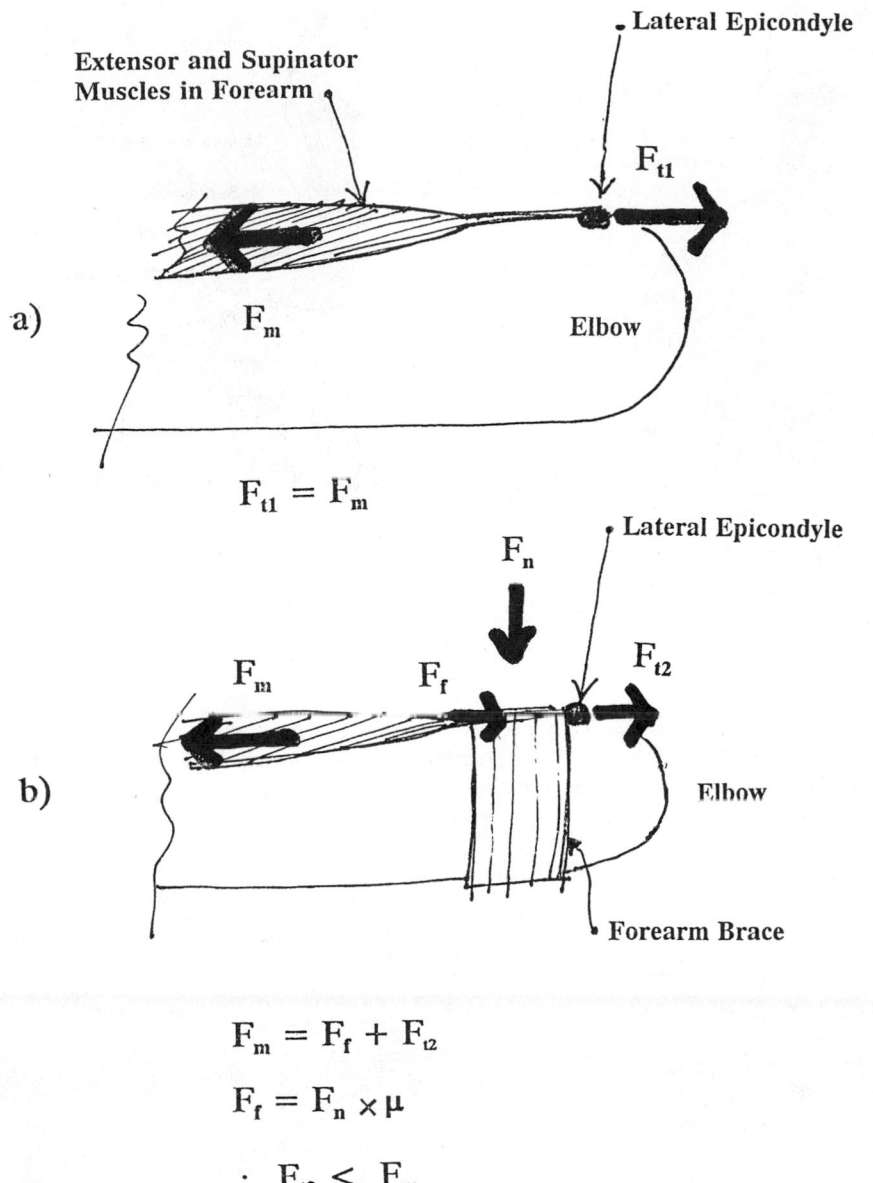

The APL and EPB tendons and their common sheath are subject to cumulative trauma because of their position in the bony groove in the radial styloid. DeQuervain's disease is caused by the friction of the two tendons rubbing against each and against the long bony groove (Lamphier et al., 1965). DeQuervain's disease is a stenosing tenosynovitis in that the common synovial sheath thickens (refer to

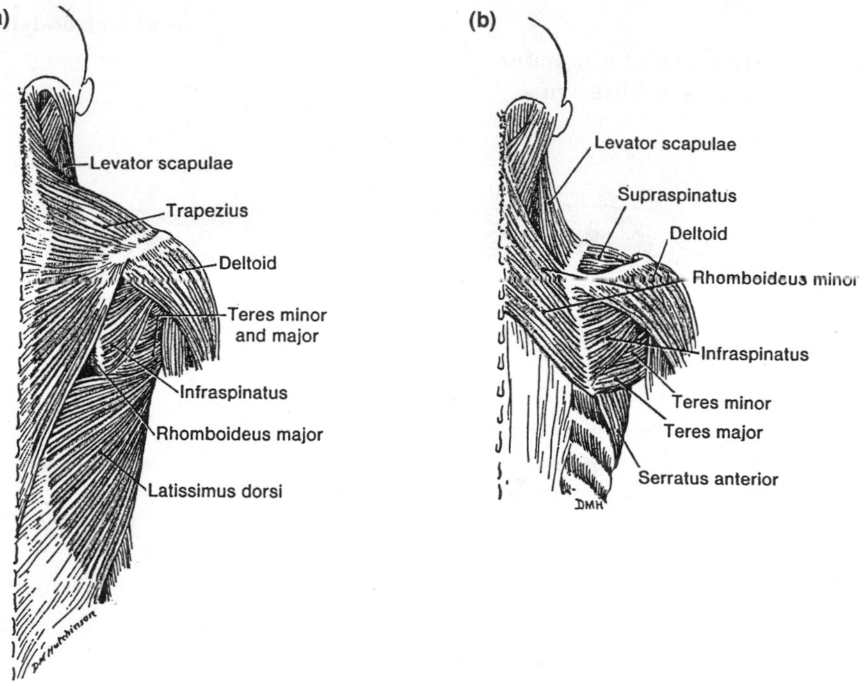

FIGURE 44.25 a) Superficial muscles of the shoulder complex. The deltoid muscle is a major shoulder abductor. b) Deep muscles of the shoulder complex. The supraspinatus muscle is responsible for initiating abduction of the shoulder, after which the deltoid provides most of the abduction force. (From Basmajian, J.V. 1982. *Primary Anatomy,* 8th edition, p. 141, Williams and Wilkins, Inc. With permission.)

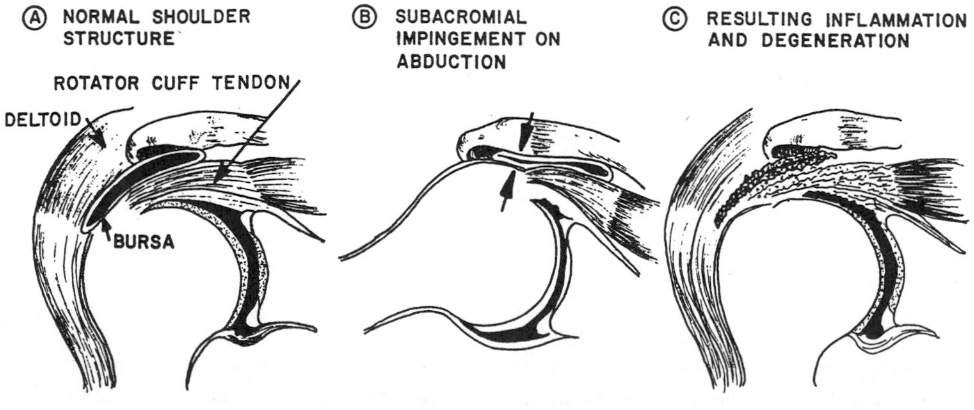

FIGURE 44.26 a) Normal shoulder structure with the arm hanging at the side. Bursa separates the deltoid muscle and acromion from the supraspinatus tendon (rotator cuff tendon). b) When the shoulder is abducted, bursa and supraspinatus tendon are pinched between the acromion and humerus bone (arm bone). c) With repeated abductions, both the bursa and tendon could swell and degenerate and the tendon could fray. (From Chaffin, D.B. and Andersson, G.B.J. 1991. *Occupational Biomechanics,* 2nd edition, p. 381, John Wiley & Sons Publishers. With permission.)

Figure 44.27), thereby increasing the friction between the APL and EPB tendons within their common sheath. In his review of the medical literature, Moore (1992) described the pathogenesis of DeQuervain's disease. In mild cases, the synovial layer within the synovial sheath thickens up to twice the normal

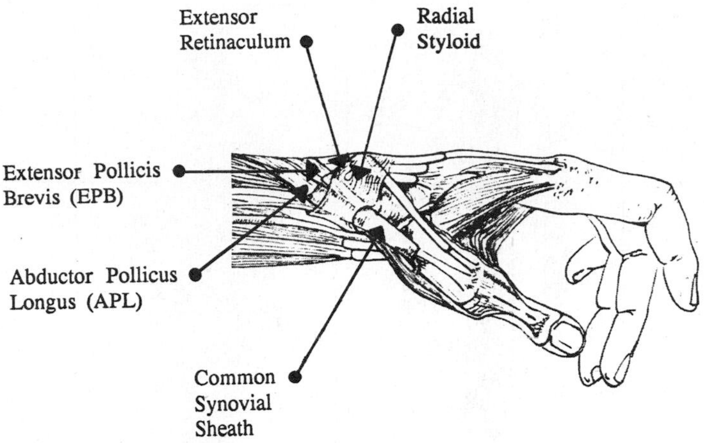

FIGURE 44.27 Abductor pollicis longus (APL) and extensor pollicis brevis (EPB) tendons as they proceed under the extensor retinaculum and through their common synovial sheath. DeQuervain's disease is stenosing tenosynovitis of the APL and EPB in their common sheath. (From Lamphier, T.A., Crooker, C., and Crooker, J.L. 1965. DeQuervain's disease. *Industrial Medicine and Surgery*, p. 848. With permission.)

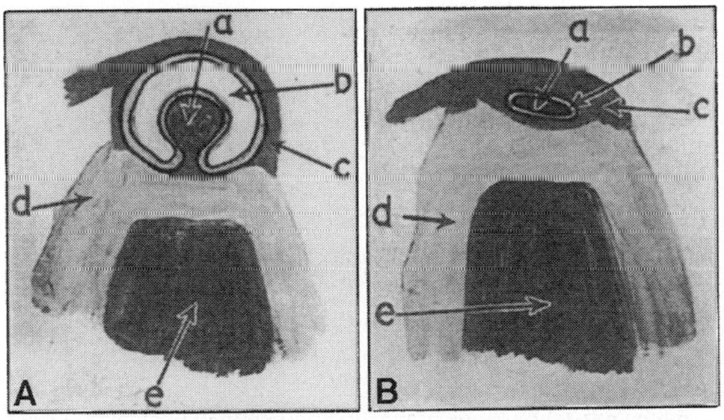

FIGURE 44.28 A) Cross-section of a normal fibroosseus canal as it passes over the radial styloid. The tendon (a) and its synovial sheaths (b) are moored to the radius bone (d) with a fibrous ligamentous sheath (c). B) Cross-section of a fibroosseus canal with stenosing tenosynovitis. The tendons (a) are flattened, the synovial sheath (b) is thinned, and the fibrous ligamentous sheath (c) is thickened. (From Finkelstein, H. 1930. Stenosing tendovaginitis at the radial styloid process. *J Bone Jt Surg*, Vol. 12, p. 515. With permission.)

thickness. However, at the point of constriction where the tendons rub against the radial styloid, the synovial sheath of the APL and EPB tendons thin and the tendons flatten (Finkelstein, 1930), as illustrated in Figure 44.28. The thinning of the tendon sheath and flattening of the tendons is caused by hypertrophy of the fibrous ligamentous sheath, namely the extensor retinaculum, that holds the APB and EPB tendons and their common sheath to the radial styloid bone. In severe cases, the fibrous ligamentous sheath thickens three to four times (Lamphier et al., 1965), and the tendon could swell, forming a bulbous shape adjacent to the site of constriction, as illustrated in Figure 44.29. This bulbous swelling could cause popping of the APL and EPB tendons as the wrist is ulnarly deviated with the thumb flexed inside the palm.

Trigger Finger

Trigger finger is stenosing tenosynovitis of the tendons that flex the fingers and is manifested by painful locking of the finger during finger flexion. As illustrated in Figure 44.30, the finger flexor tendons and

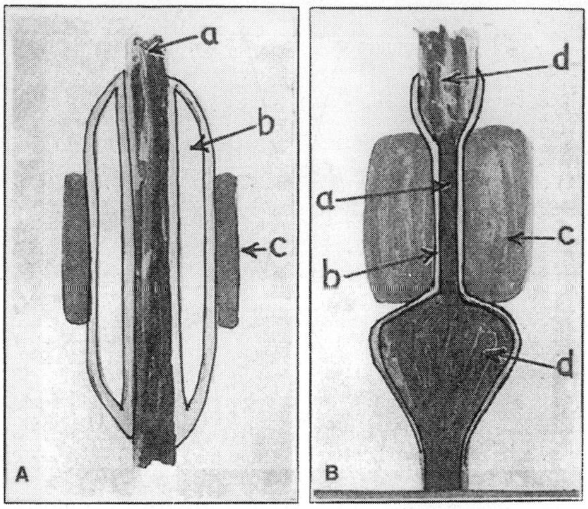

FIGURE 44.29 A) Longitudinal section of a normal fibroosseus canal. The tendon (a) is covered by it sheath (b). The fibrous ligamentous sheath (c) forms a canal through which the tendon can travel unimpeded. B) Longitudinal section of a fibroosseus canal with stenosing tenosynovitis. The tendon (a) and its synovial sheath (b) are thinned by a thickened fibrous ligamentous sheath (c), forming a nodule in the tendon (d). (From Finkelstein, H. 1930. Stenosing tendovaginitis at the radial styloid process. *J Bone Jt Surg*, Vol. 12, pp. 514 and 515. With permission.)

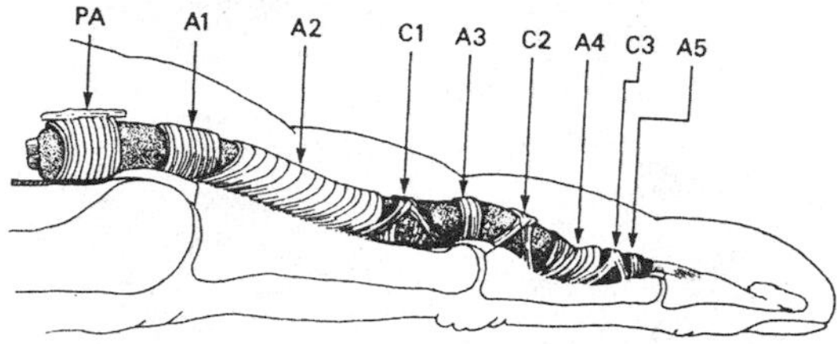

FIGURE 44.30 A finger flexor tendon as it traverses from the knuckle (left) to the tip (right). Fibrous ligamentous sheaths comprise the five annular pulleys (A1–A5) and three cruciate pulleys (C1–C3). These pulleys hold the tendons close to the finger joints to avoid bowstringing during finger flexion. (From Doyle, J.R. 1989. Anatomy of the finger flexor tendon sheath and pulley system: A current review. *J Hand Surg*, 14A, p. 350. With permission.)

their synovial sheaths are moored to the bones of the finger with fibrous ligamentous sheaths to avoid bowstringing during finger flexion. These fibrous ligamentous sheaths are called pulleys, of which there are two types: annular (A1–A5) and cruciate (C1–C3). Repetitive and forceful flexing of the fingers could cause trigger finger, whose pathogenesis is related to that of DeQuervain's disease. As depicted in Figure 44.29A, a normal fibroosseus canal allows the tendon to glide with no obstruction. However, in the case of trigger finger, bulbous swellings, such as those shown in Figure 44.29B, will restrict the finger flexor tendons from traversing through their fibrous ligamentous sheaths. If large enough, the bulbous swellings could immobilize the tendon, thereby locking the finger in a fixed flexed position. In order to "unlock" the finger, the bulbous swelling has to snap to move beyond the constriction in the fibroosseus canal. External aid from the other hand may be required to extend the finger back to a straight, neutral position (Rowe, 1985).

Although trigger finger tends to occur at the creases of the finger (Caillet, 1984), it occurs most frequently near the knuckle (metacarpophalangeal joint) (Quinnel, 1980). The middle and ring fingers are the predominant sites of trigger finger in the dominant hand, accounting for over 40% of the cases in Quinnel's (1980) study.

44.6 Biomechanical Aspects of Nerve Compression Disorders

Nerve compression disorders are a group of disorders in which a peripheral nerve is compressed or pinched, causing trauma to the immediate area served by the nerve and sometimes distal to the site of impingement. The effects from the trauma could be temporary or long term. An example of a widely known nerve disorder is sciatica, which is compression of the sciatic nerve in the lower back but whose pain is felt throughout the areas of the lower leg served by the sciatic nerve.

Nerve compression disorders of the upper extremity can affect all three major nerves of the upper extremity, namely the radial, ulnar, and median nerves (refer to Figure 44.13a). The most widely publicized nerve compression disorder of the upper extremity is *carpal tunnel syndrome*, which is compression of the median nerve at the site where it passes through the carpal tunnel in the wrist. Examples of lesser known nerve compression disorders of the upper extremity are cubital tunnel syndrome and posterior interosseous nerve syndrome, which are caused by compression of the ulnar and radial nerves, respectively.

The anatomy of a *neuron* is illustrated in Figure 44.31. Although neurons vary in size and shape, they generally consist of a cell body (soma), *dendrites*, and an *axon*. The axon is the shaft of the nerve through which electrical impulses travel. When the impulse reaches the axon terminal, it then crosses over to the dendrites of an adjacent neuron. As shown in Figure 44.31, there are small gaps, called the *nodes of Ranvier*, between segments of the axon. These segments are covered with a *myelin sheath* and *neurilemma* (or Schwann cells), which insulate the nerve fibers from adjacent cellular compartments and allow the electrical impulse to travel from node to node (Van de Graaf and Rhees, 1987). The speed of an impulse traveling through a neuron is called *conduction velocity*. A cross-sectional view of the axon of a neuron reveals three layers of connective tissue that hold the nerve fibers together and protect them. As illustrated in Figure 44.32, the epineurium is the most external layer, holding together several *fasciculi*. Several fasciculi surrounded by epineurium is called a single *nerve* (Spence, 1986). The perineurium, which consists of fibrous collagen, surrounds each fasciculus or bundle of nerve fibers. The endoneurium is the connective tissue within each fasciculus and forms a tube-like membrane around each *nerve fiber* (Szabo and Gelberman, 1987).

Although the pathogenesis of nerve entrapment syndromes is controversial, two prominent theories have been hypothesized (Moore, 1992). These two theories are first, mechanical compression of the nerve, and second, inadequate blood supply serving the nerve. The first theory, mechanical compression, is described in detail in Feldman et al. (1983) and Szabo and Gelberman (1987) and is summarized below. Any physical disturbance to the nerve can cause motor or sensory dysfunction. An impingement upon the *efferent* portion of the nerve — carrying impulses to the peripheral nervous system from the brain and spinal cord — can result in loss of muscular strength. Likewise a disturbance to the *afferent portion* — carrying impulses to the central nervous system from receptors located throughout the body — can decrease sensory feedback. Myelinated nerves are more susceptible to the effects of pressure than unmyelinated nerves. Since motor nerves consist predominantly of thick myelinated fibers, the motor nerves are theoretically more susceptible to the effects of compression than sensory cutaneous nerves (Feldman et al., 1983). However, based on clinical observations of nerve entrapment syndromes, the sensory function appears to show decrements before motor function (Szabo and Gelberman, 1987). The pathogenesis of why sensory decrements are manifested before motor decrements is still unclear.

Compression can cause neural dysfunction in the following manner. First, compression causes bulbous swellings on the fibers (Aguayo, 1975), which can block conduction of electrical impulses. If compression continues, the myelin between nodes on the nerve becomes thinner, and the fibers start to segmentally demyelinate, which can further decrease the conduction velocity of nerve impulses (Feldman et al., 1983).

The second prominent theory posed to explain the pathogenesis of nerve entrapment syndromes is insufficient blood supply to the nerves (Moore, 1992). *Ischemia*, or inadequate circulation due to mechanical obstruction in the nerve's path, can cause symptoms typical of nerve entrapment syndromes, such as *paresthesia* or acute pain and possibly a reduction in conduction velocity. The mechanical obstruction could take the form of pinching or entrapment of the nerve as the nerve travels around tendons, ligaments,

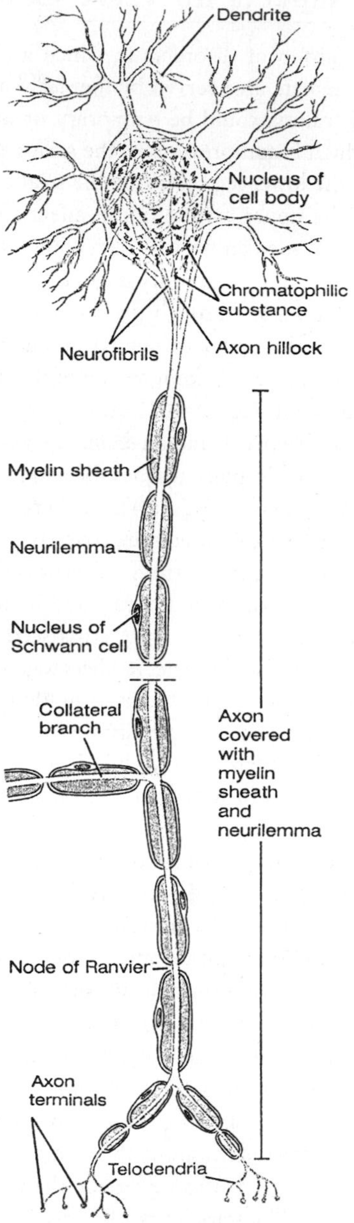

FIGURE 44.31 Structure of a typical motor neuron. (From Spence, A.P. 1986. *Basic Human Anatomy,* The Benjamin Cummings Publishing, Inc., p. 348. With permission.)

or bones in a joint. Vascular deficiencies and changes in blood pressure could account for the variation in symptoms noted in patients with approximately equal levels of nerve conduction delay (Shivde et al., 1981; Moore, 1992). In one study, vascular *sclerosis* was observed in 98% of the carpal tunnel cases (Fuchs et al., 1991).

The nerve fiber distal to the site of compression or physical trauma, whether by compression or inadequate blood supply, can also be detrimentally affected. *Wallerian degeneration* is the deterioration of the myelin sheath distal to the site of trauma and can lead to *atrophy* and destruction of a neuron. Wallerian degeneration is caused by impaired flow of electrical impulses down the nerve's axon and ischemia (Feldman et al., 1983). Ogata and Naito (1986) investigated the effects of compression and

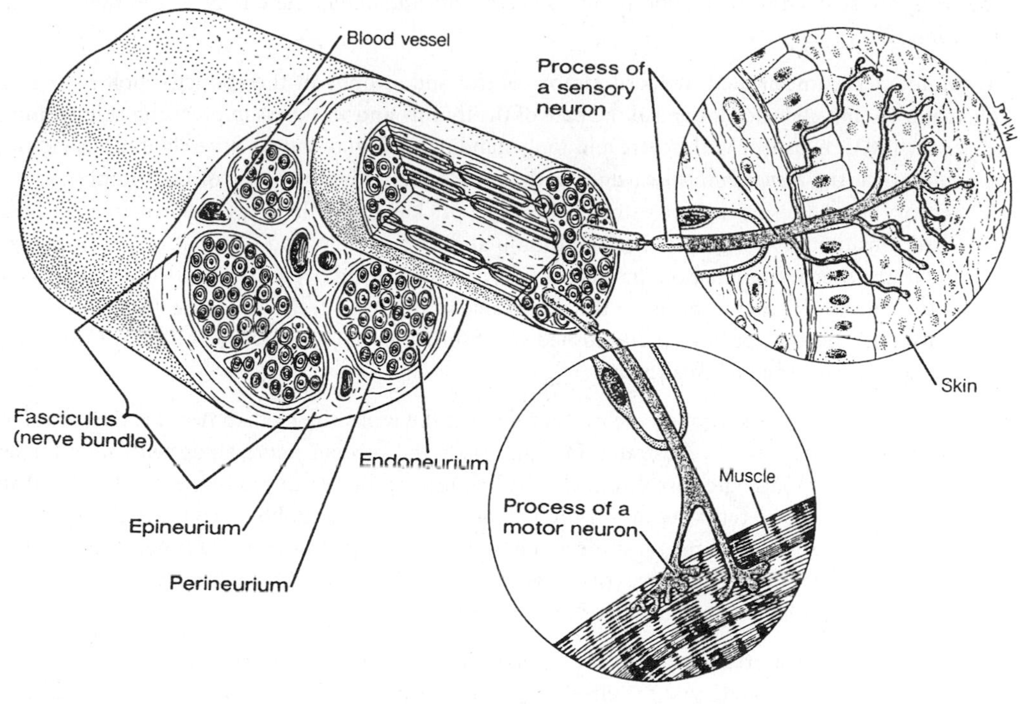

FIGURE 44.32 Cross section of a single peripheral nerve, which is composed of several fasciculi wrapped in epineurium connective tissue. The perineurium wraps each fasciculus, and the endoneurium encases each single nerve fiber. (From Spence, A.P. 1986. *Basic Human Anatomy,* The Benjamin Cummings Publishing, Inc., p. 353. With permission.)

stretching of nerves, and they found that compression and stretching do restrict blood flow to the nerves, even to the point of arresting blood flow. In addition to Wallerian degeneration, compression of a nerve can impede nerve conduction and lower conduction velocity distal to the site of trauma, resulting in decreased motor and sensory function such as muscle atrophy and paresthesia, respectively (Feldman et al., 1983).

In the upper extremity, peripheral nerves can be compressed at any point along its path, ranging from the wrist (carpal tunnel syndrome) to the shoulder (thoracic outlet syndrome). These two syndromes are described and discussed below.

44.7 Work-Related CTDs Involving the Nerve

Carpal Tunnel Syndrome

Carpal tunnel syndrome (CTS) is a nerve entrapment disorder arising from compression of the median nerve at the site where it travels through the carpal tunnel in the wrist. Although CTS was first recognized by Sir James Paget more than a century ago (1860), the name CTS was not uniformly applied to compression of the median nerve until the late 1950s (Phalen, 1981). Prior to and through the early 1950s, CTS was referred to with long, clumsy names, such as "spontaneous compression of the median nerve in the carpal tunnel" and "compression secondary to trauma, tumor, or systemic disease." The much shorter, more facile name, "carpal tunnel syndrome," started to gain acceptance in the late 1950s to include all conditions that might cause compression of the median nerve, regardless of the source of compression. In 1957, Phalen and Kendrick stated, "The term carpal tunnel syndrome is now used to describe all cases of compression neuropathy of the median nerve at the wrist" (Moore, 1992).

The symptoms of CTS involve the motor, sensory, and *autonomic* functions of the median nerve (Armstrong, 1983):

1. Motor nerve impairment: reduced motor control and atrophy of the abductor pollicis brevis, a major thumb abductor located at the base of the thumb, and weakness in precision grip (Feldman et al., 1983). A patient with severe and long-standing CTS may show a severely diminished musculature at the *thenar eminence*, which is the protruding area at the base of the thumb.
2. Sensory nerve impairment: paresthesia (burning, prickling, and tingling) and *hypesthesia* (diminished sensitivity to stimulation) in the thumb and fore, middle, and one half of the ring fingers and in the thenar eminence on the palmar side; paresthesia and hypesthesia on the distal *phalanges* of the thumb and same fingers on the dorsal side of the hand (refer to Figure 44.14).
3. Autonomic nerve impairment: diminished sweat function, resulting in dry and shiny skin in areas noted in sensory nerve impairment.

Of all the tissues that pass through the carpal tunnel at the wrist, the median nerve is the softest and most vulnerable to pressure. As illustrated in Figure 44.8, the median nerve along with nine forearm flexor tendons pass through the carpal tunnel, which is formed by the carpal bones on the dorsal side of the hand and the flexor retinaculum on the palmar side. In 1963, Robbins conducted a systematic study, considered by some to be the most thorough, of the anatomy of the carpal tunnel (Moore, 1992). He analyzed cross sections of cadaveric wrists dissected at distances 2 to 4 cm proximal and distal to the wrist crease. Among Robbins's (1963) major findings were:

1. The cross-sectional area of the carpal tunnel decreases from its proximal entrance to a point 2 cm distal from its origin. He said the effect is "to have a canal with a slightly narrowed waist."
2. All the structures are crowded in the canal, and in 6 of the 7 specimens, the median nerve was flattened and directly beneath the flexor retinaculum.
3. When the wrist was in a neutral position in the cadaveric specimens, Robbins (1963) inserted and withdrew a piece of rubber easily underneath the flexor retinaculum. However, when the wrist was flexed or extended, the resistance increased concomitantly with the angle of deviation. Considerable force was required to withdraw the rubber proximally in either of the extreme flexed or extended positions. Robbins (1963) concluded the volume of the carpal tunnel decreases as the wrist is deviated in either flexion or extension.

Patients with CTS show a significant elevation of pressure in the carpal tunnel. At a neutral posture, Gelberman et al. (1981) found that the mean pressure in the carpal tunnel of patients with CTS was 32 mmHg, compared to 2.5 mmHg for control subjects. These researchers also showed that 90° of wrist flexion and extension increased the pressure precipitously to approximately 100 mmHg and 32 mmHg for the CTS patients and healthy subjects, respectively.

Although the pathophysiology of CTS is unknown, researchers have postulated thickening of the flexor tendon sheaths in CTS patients as contributing to the increase in pressure in the carpal tunnel. In a study of 212 wrists surgically treated for CTS, Phalen (1966) observed thickening or fibrosis of the flexor synovium in 203 of the cases. Biopsy specimens of the flexor synovium from 181 of the 212 wrists revealed chronic fibrosis or thickening in 91 specimens, chronic inflammation compatible with symptoms of rheumatoid arthritis in 64, and no pathologic change in 26. Yamaguchi et al. (1965) observed microscopically an aging effect on the flexor tendon sheaths, manifested by fibrous thickening of the sheaths. They compared the sheath anatomy of CTS patients vs. healthy controls, and they found that almost 90% of the patients exhibited a greater increase in thickening and fibrosis of the sheaths than the healthy control subjects. Kerr et al. (1992) observed hypertrophy of the synovium of the flexor tendons, with little or no evidence of inflammation, in patients with CTS. Schuind et al. (1990) observed fibrous hypertrophy and necrotic lesions in flexor synovium "typical of a connective tissue undergoing degeneration under repeated mechanical stresses." The experimental findings of Armstrong et al. (1984) support

Schuind's (1990) association between degeneration of tendon tissue and mechanical stress. Synovial hypertrophy and the mean densities of subsynovium and adjacent connective tissue were significantly greater at the wrist crease as compared to locations proximal and distal to the crease. Armstrong et al. (1984) concluded that repeated exertions with a flexed or extended wrist are an important factor in the etiology of the degeneration and hypertrophy of tissue surrounding the tendon.

Occupational Sources of Median Nerve Compression in CTS

The three main occupational risk factors of CTS (wrist repetition, deviated wrist angle, and tendon force), along with the basic anatomical structure of the wrist, can increase the pressure in the carpal tunnel, compress the median nerve, and result in the following:

1. *Increase in carpal tunnel pressure at deviated wrist angles due to reduction in tunnel volume.* According to Robbins (1963), extreme flexion and extension of the wrist reduced the volume of the carpal tunnel, thereby augmenting the pressure on the median nerve. This increase in tunnel pressure would, theoretically, affect the median nerve first because it is the softest and most vulnerable tissue in the carpal tunnel.
2. *General increase in tunnel pressure from wrist deviation.* Deviation of the wrist in the flexion/extension plane has been shown repeatedly in the anatomical and physiological literature to increase the pressure in the carpal canal (Phalen, 1966; Smith et al., 1977). In a flexed or extended wrist posture, the median nerve is squeezed between the flexor retinaculum and the overlying flexor tendons, thereby exposing a worker to CTS. Recently, Rempel et al. (1994) measured the carpal tunnel pressure of subjects typing on a computer keyboard elevated at various slopes to extend the wrist at five different angles. As shown in Figure 44.33, these researchers found that the pressure in the carpal tunnel was lowest at a neutral position compared to postures up to 50° extension and 20° flexion. The approximately 100 mmHg maximum pressures Rempel et al. (1994) measured were in the same range as in Gelberman's et al. (1981) study. In addition to the flexion/extension plane, carpal tunnel pressure has been shown to increase as the wrist radially and ulnarly deviates from a neutral posture. Sommerich (1994) measured the carpal tunnel pressures of four subjects typing on a standard QWERTY computer keyboard and an alternative keyboard split and angled to reduce ulnar deviation. She found that all subjects showed a decrease in carpal tunnel pressure with a concomitant decrease in ulnar deviation. The maximum carpal tunnel pressures of 80 mmHg measured in Sommerich's (1994) study were similar to those measured in the studies of Gelberman et al. (1981) and Rempel et al. (1994).
3. *Thickening and fibrosis of synovium and hypertrophy of synovial sheaths.* The well-documented reporting of thickening of the flexor sheaths and synovium in the carpal tunnel (Phalen, 1966; Yamaguchi et al., 1965; Armstrong et al., 1984; Schuind et al., 1990; and Kerr et al., 1992) could possibly be explained by the biomechanical models of Armstrong and Chaffin (1979) and Schoenmarklin and Marras (1990). The resultant reaction force on the flexor tendons and the median nerve passing through the carpal tunnel increases concomitantly, not only with deviation angle (F_r in Figures 44.18 and 44.19 from Armstrong and Chaffin [1979]), but also with acceleration of the wrist (F_r in Figures 44.21 and 44.22 from Schoenmarklin and Marras [1990]). Wrist deviation and acceleration are the static and dynamic components of repetitive movements of the wrist, which have been associated with CTS.

In order to accelerate the wrist, the extrinsic muscles in the forearm have to exert force which is transmitted to the tendons. As modeled by Schoenmarklin and Marras (1990), some of the force transmitted through the tendon is lost to friction against the ligaments and bones that form the carpal tunnel (refer to Figure 44.21). This frictional force could irritate the tendons and their sheaths and possibly cause the synovitis and hypertrophy found experimentally in the carpal tunnel. Armstrong et al. (1984)

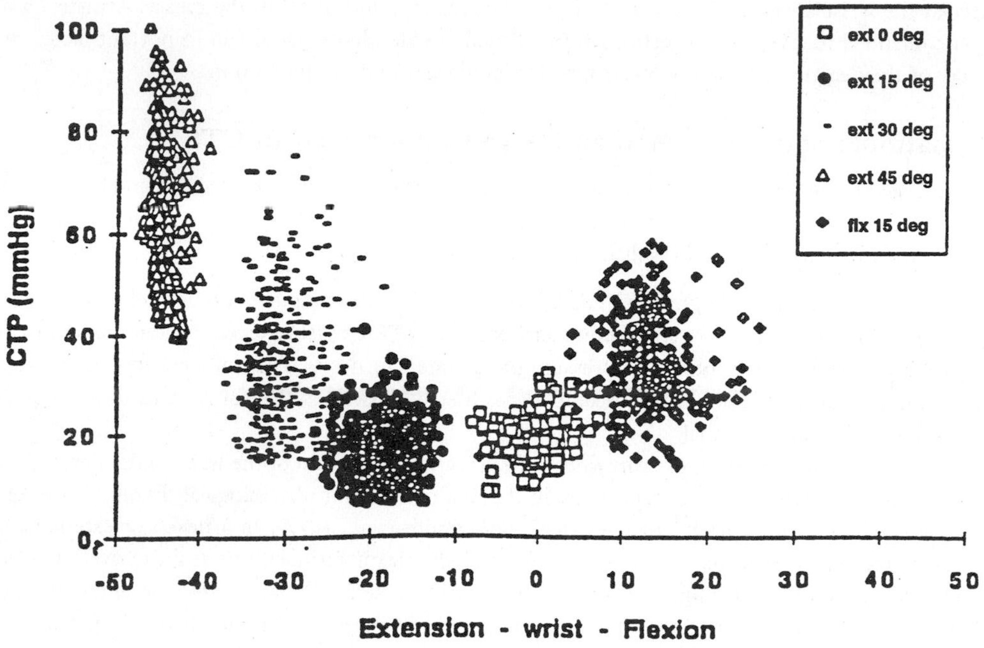

FIGURE 44.33 Carpal tunnel pressure in one subject's wrist while typing on a computer keyboard set at different wrist extension angles. (From Rempel, D. and Horie, S. 1994. Effect of wrist posture during typing on carpal tunnel pressure. In *Proceedings of Working With Display Units: Fourth International Scientific Conference,* Milan, Italy, p. C27. With permission.)

found sizeable increases in synovium and synovial density in the carpal tunnel area, which they attributed to repeated flexion/extension exertions. From a modeling point of view, Tanaka and McGlothlin (1989) hypothesized that the friction between tendons and adjacent structures is a major cause of CTDs and CTS, and Moore and Wells (1989) and Moore et al. (1991) showed that the frictional work generated in the carpal tunnel supported Silverstein's et al. (1986, 1987) dose–response relationship between repetition and CTD risk.

In theory, the deleterious effects of frictional work generated between the tendons and their sheaths or supporting structures is exacerbated by coactivation of the forearm extensor muscles during movements of the wrist and hand. In order for the hand to maintain the same flexor torque or power/pinch force, the flexor muscles have to exert more force to overcome the extensor force. Greater forces in the flexor muscles will generate, in theory, increased frictional work between the flexor tendons and their adjacent structures, thereby exposing workers to an increased risk of CTS.

Coactivation of *antagonist* muscles during static and dynamic contractions of the *agonist* muscles have been found experimentally and modeled (Schoenmarklin and Marras, 1992; Marras and Sommerich, 1991a,b; Marras, 1992). With regard to hand grip exertions, Grant et al. (1992) measured the electromyographic (EMG) signal of the forearm flexor and extensor musculature while subjects gripped various diameters of handle, and these researchers found contractions of the extensor muscles (which act as antagonists in this case) up to 30% maximum voluntary contraction (MVC). Coactivation of the extensors stabilizes the wrist during flexion movements (and vice versa, coactivation of the flexors stabilizes the wrist during extension movements), and coactivation of the extensors also helps guide and stabilize the hand while it exerts a power or pinch force.

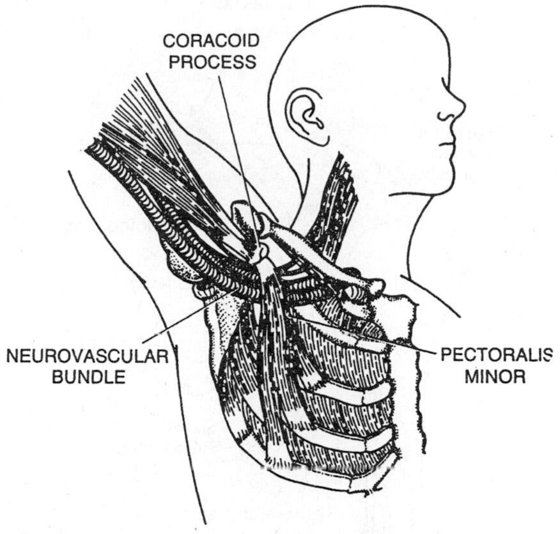

FIGURE 44.34 Abduction of the shoulder causes the neurovascular bundle, containing the brachial plexus nerve and subclavian artery, to be stretched under the pectoral muscles. Hypertrophy of the pectoral muscles would impinge the neurovascular bundle, thereby causing deficiencies in the neural and circulatory systems of the upper extremity. (From Putz-Anderson, V. 1988. *Cumulative Trauma Disorders: A Manual for Musculoskeletal Diseases of the Upper Limbs*, p. 20, Taylor & Francis. With permission.)

Thoracic Outlet Syndrome

Thoracic outlet syndrome (TOS) is a *neurovascular* disorder that affects the bundle carrying nerves, arteries, and veins from the neck through the shoulder and into the arm, as illustrated in Figure 44.34. This neurovascular bundle, which contains the brachial plexus nerve and the subclavian artery and vein, could be compressed by adjacent muscles in the shoulder region, such as the scalene, subclavius, or pectoralis minor, or the bones forming the thoracic outlet in the shoulder. The motions and tasks that are associated with TOS are repetitive shoulder abduction and *adduction*, carrying heavy loads on the shoulder, and working overhead (Feldman et al., 1983).

The pathogenesis of TOS is hypertrophy of the subclavius or pectoral muscles, which can pinch the neurovascular bundle and produce symptoms in the neural and circulatory system throughout the arm. Compression of the brachial plexus, which branches out into the median and ulnar nerves, can cause numbness or paresthesia in the lower parts of arm and hand served by the median and ulnar nerves. An impinged subclavian artery, which is the major artery supplying the upper extremity (refer to Figure 44.35), will result in ischemia, a reduction of blood flow to shoulder and arm (Basmajian, 1982). This diminished blood flow will reduce the amount of oxygen and nutrients available for dynamic arm movements and exacerbate the fatiguing effect from static (anaerobic) contractions by reducing the amount of blood needed to carry away lactate and metabolites.

44.8 Summary

In this chapter, major CTDs that affect the soft tissues of the body and have been associated with work-related activity are categorized into three groups: CTDs affecting muscle, the muscle–tendon unit, and nerve. The anatomy of the upper extremity's musculoskeletal system was described, followed by a discussion of biomechanical mechanism(s) that theoretically explain, in part, the epidemiological associations between respective categories of CTD and work-related activity. In addition, each CTDs is described in detail along with its pathogenesis.

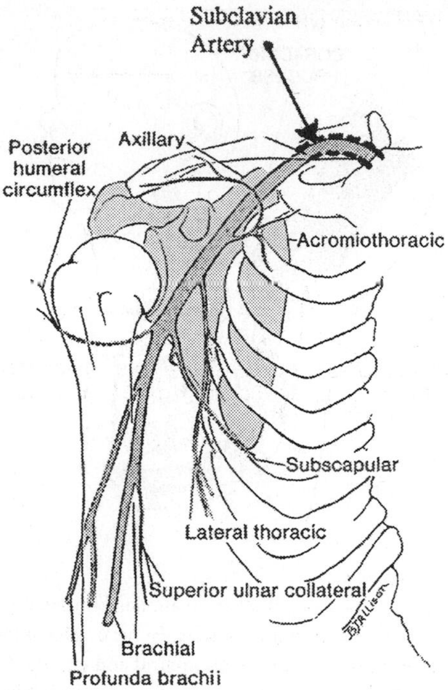

FIGURE 44.35 The subclavian artery, which branches off into several arteries, is the major artery that serves the upper extremity. Compression of the subclavian artery can cause ischemia and paresthesia in the arm and hand. (From Basmajian, J.V. 1982. *Primary Anatomy*, 8th edition, p. 274, Williams and Wilkins, Inc. With permission.)

Defining Terms

Abduct (abduction): Literally means to "to lead away from." In anatomical parlance, abduction is moving a joint away from the center of the body. Shoulder abduction is elevating the arm in the frontal plane.

Acromion process: The bony top of the shoulder. The acromion is actually the highest part of the scapula bone, which spans the back of the shoulder.

Adduct (adduction): Literally means "to lead towards." Adduction of a joint is moving the joint toward the center of the body. Shoulder adduction is moving the shoulder towards the side of the torso.

Afferent nerves: Nerves that travel from the peripheral nervous system to the central nervous system, namely the brain and spinal cord.

Agonist muscle: Muscle that initiates and carries out motion (Chaffin and Andersson, 1991).

Antagonist muscle: Muscle that opposes the action of the agonist muscle (Chaffin and Andersson, 1991).

Aponeurosis: A broad, flat tendon. An aponeurosis usually occurs at the point where the tendons from several muscles fuse and connect to a bone.

Atrophy: A wasting of tissues or organs. A severe case of carpal tunnel syndrome will show atrophy of the muscles in the thenar eminence.

Autonomic: Relating to the autonomic nervous system, which is part of the efferent division of the peripheral nervous system. The autonomic nervous system is the involuntary system functioning below the conscious level, that controls the heart, organs, and glands of the body (Spence, 1982). Patients with carpal tunnel syndrome may display dry and shiny palms, resulting from insufficient activity of the sweat glands in the hand, which are controlled by the autonomic nervous system.

Avascular: Nonvascular; without blood vessels (Stedman, 1982).

Axon: The shaft of a neuron that transmits electrical impulses (Spencer, 1986).

Bursa: A structure resembling a collapsed bag with cellophane-thin walls whose inner surfaces are extremely smooth, moist, and slippery. A bursa usually forms where a tendon rubs back and forth on a hard structure. For example, rotator cuff syndrome is the deterioration of the bursa that lubricates the shoulder bone from a tendon whose muscle raises the shoulder (Basmajian, 1982).

Carpal tunnel: The tunnel formed by eight small bones of the wrist and the transverse carpal ligament (flexor retinaculum). The tendons that connect the forearm flexor muscles to the fingers pass through the carpal tunnel.

Carpal tunnel syndrome: Compression of the median nerve as it passes through the carpal tunnel.

Central nervous system (CNS): The nervous system consisting of the brain and spinal cord. The CNS is the integrative and control center of the body. The CNS receives sensory input from the peripheral nervous system (PNS) and develops response strategies to the input (Spence, 1986).

Conduction velocity: The velocity at which an electrical impulse travels through a nerve. Conduction velocity, which for a motor nerve is normally 50 to 60 m/s (Johnson, 1989). When the impulse reaches muscle fiber, its conduction velocity slows down to about 3 to 6 m/s (Basmajian and DeLuca, 1985).

Cumulative trauma disorders (CTDs): Any of a class of pathologies affecting soft tissues (muscles, tendons, and nerves) created from excessively frequent use of a particular joint or tissue, especially in combination with awkward positioning, inadequate or no rest periods, or excessive loads. Also called repetitive strain injury (RSI), repetitive stress injury, repetitive motion injury, and overuse syndrome (Stramler, 1993).

Dendrites: Thin extensions emanating from the cell body of a neuron that receive electrical impulses from adjoining neurons (Spence, 1986).

DeQuervain's tenosynovitis: A narrowing of the passage (stenosing tenosynovitis) of the tendons and their sheaths that extend the thumb in the palmar plane.

Disease: Illness, sickness, or cessation of bodily functions, systems, or organs. A disease is characterized usually by at least two of the following criteria: a recognized causal agent(s), an identifiable group of signs and symptoms, or consistent anatomical alterations (Stedman, 1982). Compare to *Syndrome* and *Disorder.*

Distal: Located away from the center of the body or point of origin. For example, the wrist is distal to the elbow, and the elbow is distal to the shoulder (Stedman, 1982).

Disorder: A disturbance of the functions, structure, or both resulting from a failure in development or from external factors such as physical contact, injury, or disease (Stedman, 1982). Compare to *Disease* and *Syndrome.*

Dorsal: Pertaining to the back of an anatomical structure, as in the dorsal (back) side of the hand.

Edema: An accumulation of an excessive amount of watery fluid in cells, tissues, or cavities (Stedman, 1982).

Efferent nerves: Nerves that travel from the brain and spinal cord in the central nervous system to the outlying areas in the peripheral nervous system.

Endoneurium: Thin connective-tissue sheath that wraps each individual nerve fiber (Spence, 1986).

Epicondyle: A projection from a long bone above an articulating joint.

Epineurium: Connective-tissue sheath that surrounds several fasciculi in a nerve. Several fasciculi surrounded by epineurium constitute a single nerve (Spence, 1986).

Etiology: The science and study of the causes of disease and their mode of operation (Stedman, 1982). Compare to *Pathogenesis.*

Extend (extension): Movement of a body part that increases the angle of its adjacent joint. For example, moving your hand away from your shoulder requires elbow extension.

Extensor retinaculum: A strong fibrous ligament that stretches across the back of the hand at the wrist. The extensor retinaculum holds the forearm extensor tendons close to the carpal bones of the wrist.

Extrinsic muscles: Flexor and extensor muscles in the forearm that generate much of the movement and force production of the hand and fingers. The extrinsic muscles can also be recruited for hand movements and forces requiring fine motor control.

Fasciculus (fasciculi): A bundle of nerve fibers surrounded by connective tissue (perineurium) (Spence, 1986).

Fibroosseus canal: A canal formed by a fibrous ligamentous sheath and an underlying bone. DeQuervain's disease and trigger finger are thickening of the fibrous ligamentous sheath, resulting in a thinning of the tendon and its synovial sheath that pass through the canal.

Flex (flexion): Movement of a body part that decreases the angle of its adjacent joint. For example, moving your hand toward your shoulder requires elbow flexion.

Flexor retinaculum: The transverse carpal ligament that stretches across the palmar side of the wrist. The flexor retinaculum, which forms the top of the carpal tunnel, holds the flexor tendons and their sheaths inside the wrist.

Frictional force: A force that impedes impending motion.

Frontal plane: The plane of the body that travels through the chest and arms. The frontal plane shows the front of the body. Also called coronal plane.

Humerus: The arm bone connecting the shoulder to the elbow.

Hypertrophy: General increase in the bulk of a part or organ that is not due to tumor formation (Stedman, 1982). Compare to *Inflame.*

Hypesthesia (hypoesthesia): An abnormal sensation characterized by diminished sensitivity to stimulation (Stedman, 1982).

Hysteresis: The difference in the stress–strain response of a material to an increasing load or decreasing load. With tendons, the stress required to maintain a certain amount of strain is less when the force is decreasing than when increasing.

Innervate: To supply nerve function to a specific muscle group.

Insert (insertion): The distal attachment point of a muscle to bone, via a tendon.

Inflame: Pathologic process consisting of a histologic reaction to affected blood vessels and adjacent tissues in response to an injury or abnormal stimulation caused by a physical, chemical, or biologic agent (Stedman, 1982). Compare to *Hypertrophy.*

Intrinsic muscles: The small muscles located within the hand that move the thumb and fingers. The intrinsic muscles, which are much smaller than the extrinsic muscles in the forearm, are used primarily for motions and forces requiring dexterity and fine motor control.

Ischemia: Inadequate circulation of the blood due to mechanical obstruction, mainly arterial narrowing (Stedman, 1982).

Lateral epicondyle: A bony protrusion located on the lateral side of the elbow. Viewed from the side, the lateral epicondyle is located close to the pivot point of the elbow as it flexes and extends.

Lateral epicondylitis (tennis elbow): Tendinitis of the forearm extensor and supinator muscles at the lateral epicondyle of the elbow. Lateral epicondylitis is colloquially dubbed "tennis elbow" because patients with lateral epicondylitis report pain when the wrist is extended and supinated — a movement similar to that of a back stroke in tennis.

Ligament: Connective tissue resembling a tendon except a ligament attaches bone to bone (Basmajian, 1982).

Median nerve: The great "flexor" nerve of the upper extremity. It supplies motor function to the forearm flexor muscles and the thumb's (thenar) muscles and sensory function to the palm and the digits from the thumb to the center of the ring finger (Basmajian, 1982).

Morbidity: A diseased state (Stedman, 1982).

Muscle: A contractile organ of the body that moves various body parts and internal organs. A muscle can only pull (tensile force) and not push (compression force). The origin of a muscle is the end that is more fixed and the insertion of a muscle is the end that is more movable. For example, the flexor muscles of the forearm originate at the elbow and insert at the wrist and fingers (Stedman, 1982).

Myalgia: Muscular pain.

Myelin sheath: A sheath that surrounds the axon of a neuron. The purpose of the myelin sheath is to insulate the neuron from adjacent cellular fluids so an electrical impulse can travel down the axon jumping from one node of Ranvier to the next.

Myofascial syndrome: Term referring to regional muscle pain syndromes (Kuorinka and Forcier, 1995).

Myopathy: Term for measurable pathological changes in a muscle with or without symptoms (Kuorinka and Forcier, 1995).

Nerve: A collection of nerve fibers in the peripheral nervous system (Spence, 1986).

Nerve fiber: Any long process of a neuron. The term usually refers to axons, but also includes the peripheral processes of sensory neurons (Spence, 1986).

Neurilemma: The thin membrane between the myelin and connective tissue (endoneurium) in a neuron. Neurilemma is also called the sheath of Schwann (Spence, 1986).

Neuron: A nerve cell (Spence, 1986).

Neurovascular: Pertaining to the nervous and circulatory systems of the body.

Nociceptor: A peripheral nerve organ or mechanism that senses pain or injurious stimuli and transmits them (Stedman, 1982).

Nodes of Ranvier: Gaps in the myelin sheath and neurilemma of a neuron's axon. Electrical impulses known as potentials travel down the axon from one node to the next, which is called "saltatory conduction." (The etymology of "saltatory" is from the Latin word "saltare," meaning "to dance.")

Origin (originate): The proximal attachment point of a muscle to bone, via a tendon.

Paresthesia: An abnormal sensation, such as of burning, prickling, tickling, or tingling (Stedman, 1982).

Pathogenesis: The mode of origin or development of any disease or morbid process (Stedman, 1982). Compare to *Etiology.*

Periosteum: Thick fibrous membrane covering the surface of a bone (Stedman, 1982).

Peripheral nervous system: All nervous structures located outside the central nervous system (CNS), which consists of the brain and spinal cord. The PNS consists of nerves that connect the outlying parts of the body and their receptors with the brain and spinal cord (Spence, 1986).

Phalanx (phalanges): One of the long bones of the fingers (Stedman, 1982). A finger has three phalanges: distal (tip), middle, and proximal (connected to the knuckle).

Proximal: Located toward the center of the body or point of origin. For example, elbow is proximal to the wrist, and the shoulder is proximal to the elbow (Stedman, 1982).

Radial deviation: Rotation of the wrist joint toward the radius bone or thumb side. Sometimes radial deviation is referred to as abduction of the wrist joint.

Sclerosis: The process of becoming hard or firm. In the case of arteriosclerosis, the walls of the arteries harden and become less elastic. The blood vessels are unable to expand and recoil in response to pressure changes, thereby elevating one's maximum blood pressure (Spence, 1986).

Shoulder-neck myofascial syndrome: See *Tension neck syndrome.*

Stenosing tenosynovitis: A narrowing of the canal through which a tendon and its sheath pass.

Striated muscle: Muscle that appears striped with dark and light bands under a microscope. Also called skeletal muscle, which are the muscles that move the limbs, head, neck, and torso.

Supraspinatus tendinitis (rotator cuff syndrome): Tendinitis of the supraspinatus muscle, which abducts the shoulder from the side of the trunk. Supraspinatus tendinitis is often called rotator cuff syndrome.

Syndrome: The collection of signs and symptoms associated with any disease process and constituting the picture of the disease (Stedman, 1982). Compare to *Disease* and *Disorder.*

Synovial fluid: A clear fluid whose purpose is to lubricate a tendon within a sheath or a joint (Stedman, 1982).

Synovial sheath: An elongated and double-walled tubular structure (bursa) that surrounds a tendon and allows the tendon to travel with little friction. Synovial sheaths are located where tendons move around joints, such as in the wrist (Basmajian, 1982).

Tendon: Connective tissue resembling a tough cord or band and always part of a muscle, usually forming an attachment of muscle to bone (Basmajian, 1982).

Tendinitis (also spelled **tendonitis**): Inflammation of a tendon (Stedman, 1982).

Tenosynovitis: Inflammation of a tendon's sheath (Stedman, 1982). Clinically, any tendon sheath disorder is called tenosynovitis, regardless of the presence or absence of inflammation (Moore, 1992).

Tension neck syndrome (TNS): Myalgia in the shoulder–neck region of the body. TNS is synonymous with shoulder–neck myofascial syndrome, and TNS is defined by symptoms of pain in the shoulder–neck region with simultaneous findings of tenderness over the shoulder–neck muscles (Kuorinka and Forcier, 1995).

Thenar eminence: The area at the base of the thumb that is raised above the general level of the palm. The thenar eminence contains the intrinsic musculature that controls the thumb.

Thoracic outlet syndrome: A neurovascular disorder that affects the brachial plexus nerve and the subclavian artery and vein as they traverse through the thoracic outlet in the shoulder. The early symptoms of thoracic outlet syndrome are found in the areas in the forearm and hand served by the median and ulnar nerves.

Trigger finger: Painful locking of a finger caused by narrowing of the canal through which a finger flexor tendon and its sheath pass.

Ulnar deviation: Rotation of the wrist joint toward the ulna bone or little finger side. Sometimes ulnar deviation is referred to as adduction of the wrist joint.

Wallerian degeneration: Degeneration of the myelin sheath of a neuron's axon caused by compression of the nerve and ischemia. Wallerian degeneration leads to the atrophy and destruction of the neuron (Stedman, 1982).

References

Abrahams, M. Mechanical analysis of tendon *in vitro*: A preliminary report. *Med Biol Eng*, 5:433-443, 1967.

Aguayo, A.J. Neuropathy due to compression and entrapment, in Peripheral Neuropathy, Dyck, P.J. and Lambert, E.H., editors, 688-713, W.B. Saunders Co., 1975.

Armstrong, T.J. An ergonomics guide to carpal tunnel syndrome. *Ergonomics Guides*, American Industrial Hygiene Association, 1983.

Armstrong, T.J., Castelli, W.A., Evans, F.G., and Diaz-Perez, R.D. Some histological changes in carpal tunnel contents and their biomechanical implications. *Journal of Occupational Medicine*, 26, No. 3, 197-201, 1984.

Armstrong, T.J. and Chaffin, D.B. Some biomechanical aspects of the carpal tunnel. *Journal of Biomechanics*, 12, 567-570, 1979.

Basmajian, J.V. Primary Anatomy, Eighth edition. Williams and Wilkins, 1982.

Basmajian, J.V. and DeLuca, C.J. *Muscles Alive: Their Functions Revealed by Electromyography*, Fifth edition. Williams and Wilkins, 1985.

Brand, P.W. *Clinical Mechanics of the Hand.* The C.V. Mosby Co., 1985.

Bunnel, S. *Surgery of the Hand*, J.B. Lippincott Co., 1956.

Caillet, R. (1984). *Hand Pain and Impairment.* Philadelphia, F.A. Davis Co., 1984.

Chaffin, D.B. and Andersson, G.B.J. *Occupational Biomechanics*, Second Edition. John Wiley & Sons, 1991.

Cyriax, J.H. The pathology and treatment of tennis elbow, *Journal of Bone and Joint Surgery*, 18, 921-940, 1936.

Elliot, D.H. Structure and function of mammalian tendon. *Biol Rev*, 40:392-421, 1965.

Feldman, R.G., Goldman, R., and Keyserling, W.M. Peripheral nerve entrapment syndromes and ergonomic factors. *American Journal of Industrial Medicine*, 4, 661-681, 1983.

Finkelstein, H. Stenosing tendovaginitis at the radial styloid process. *Journal of Bone and Joint Surgery*, 12, 509-540, 1930.

Froimson, A.I. Treatment of tennis elbow with forearm support band. *Journal of Bone and Joint Surgery*, 53A:183-184, 1971.

Fuchs, P.C., Nathan, P.A., and Myers, L.D. Synovial histology in carpal tunnel syndrome. *Journal of Hand Surgery*, 16A, 753-758, 1991.

Fung, Y.C. *Biomechanics: Mechanical Properties of Living Tissues.* Springer-Verlag Publishers, 1981.

Gelberman, R.H., Hergenroeder, P.T., Hargens, A.R., Lundbor, G.N., and Akeson, W.H. The carpal tunnel syndrome: A study of carpal canal pressures. *Journal of Bone and Joint Surgery*, 63A(3), 380-383, 1981.

Gelberman, R., Goldberg, V., and An, K.N. Tendon, in *Injury and Repair of the Musculoskeletal Soft Tissues*, Woo, S.L.Y. and Buckwalter, J.A., editors, pp. 1-40, American Academy of Orthopaedic Surgeons, 1988.

Goldie, I. Epicondylitis lateralis humeri (Epicondylitis of tennis elbow): A pathogenetic study. *Acta Chir Scand Suppl*, 339, 1-119, 1964.

Goldstein, S.A., Armstrong, T.J., Chaffin, D.B., and Matthews, L.S. Analysis of cumulative strain in tendons and tendon sheaths. *Journal of Biomechanics*, 20, No. 1, 1-6, 1987.

Grant, K.A., Habes, D.J., and Steward, L.L. An analysis of handle designs for reducing manual effort: The influence of grip diameter. *International Journal of Industrial Ergonomics*, 10, 199-206, 1992.

Johnson, E.W. *Practical Electromyography*, 2nd edition, Williams and Wilkins, 1989.

Kerr, C., Sybert, D.R., and Albarracin, N.S. An analysis of the flexor synovium in idiopathic carpal tunnel syndrome: Report of 625 cases. *Journal of Hand Surgery*, 17A(6), 1028-1030, 1992.

Kuorinka, I. and Forcier, L., editors. *Work-Related Musculoskeletal Disorders (WMSDs): A Reference Book for Prevention*. Taylor & Francis, 1995.

Lamphier, T.A., Crooker, C., and Crooker, J.L. *Industrial Medicine and Surgery*, 847-856, 1965.

Landsmeer, J.M.F. Power grip and precision handling. *Annals Rheumatoid Diseases*, 21, 164-170, 1962.

LeVeau, B. *Biomechanics of Human Motion*. Baltimore: Williams and Lissner, 1977.

Marras, W.S. Towards an understanding of dynamic variables in ergonomics. *Occupational Medicine: State of the Art Reviews*, Vol. 7, No. 4, 655-677, 1992.

Marras, W.S. and Sommerich, C.M. A three-dimensional motion model of loads on the lumbar spine: I. Model structure. *Human Factors*, 33(2), 123-137, 1991a.

Marras, W.S. and Sommerich, C.M. A three-dimensional motion model of loads on the lumbar spine: II. Model validation. *Human Factors*, 33(2), 139-149, 1991b.

Merriam, J.L. and Kraige, L.G. *Engineering Mechanics: Dynamics*, Second edition, John Wiley & Sons, 1984.

Moore, J.S. and Garg, A. State of the art reviews: Ergonomics: low-back pain, carpal tunnel syndrome, and upper extremity disorders in the workplace. *Occupational Medicine*, Vol. 7, No. 4, 1992.

Moore, A.E. A system to predict internal load factors related to the development of cumulative trauma disorders of the carpal tunnel and extrinsic flexor musculature during grasping. Master's thesis, University of Waterloo, Waterloo, Canada, 1988.

Moore, A., Wells, R., and Ranney, D. Quantifying exposure in occupational manual tasks with cumulative trauma disorder potential. *Ergonomics*, 34, 1433-1453, 1991.

Nirschl, R. Muscle and tendon trauma: tennis elbow. Chapter 28 in *The Elbow and Its Disorders*, W.B. Saunders Co., 1985.

Ogata, K. and Naito, M. Blood flow of peripheral nerve effects of dissection, stretching, and compression. *The Journal of Hand Surgery*, 11B(1), 10-14, 1986.

Paget, J. *Lectures in Surgical Pathology*, Philadelphia: Lindsay & Blakiston, p. 42, 1860.

Phalen, G.S. The carpal tunnel syndrome: seventeen years' experience in diagnosis and treatment of six hundred fifty-four hands. *Journal of Bone and Joint Surgery*, 48-A(2), 211-228, 1966.

Phalen, G.S. The birth of a syndrome, or carpal tunnel revisited, Guest Editorial. *Journal of Hand Surgery*, 109-110, 1981.

Phalen, G.S. and Kendrick, J.I. Compression neuropathy of the median nerve in the carpal tunnel, *JAMA*, 164:524-530, 1957.

Putz-Anderson, V., Editor. *Cumulative Trauma Disorders: A Manual for Musculoskeletal Diseases of the Upper Limbs*. Taylor & Francis, 1988.

Quinnel, R.C. Conservative management of trigger finger. *The Practitioner*, 224, 187-190, 1980.

Rasch, P.J. *Kinesiology and Applied Anatomy*, 7th edition, Lea and Febiger, 1989.

Rempel, D. and Horie, S. Effect of wrist posture during typing on carpal tunnel pressure, in *Proceedings of Working With Display Units: 4th International Scientific Conference*, Milan, Italy, p. C27, 1994.

Rigby, B.J., Hirai, N., and Spikes, J.D. The mechanical properties of rat tail tendon. *J Gen Physiol*, 43:265-283, 1959.

Robbins, H. Anatomical study of the median nerve in the carpal tunnel and the etiologies of carpal tunnel syndrome. *Journal of Bone and Joint Surgery*, 45A, 953-966, 1963.

Rowe, M.L. *Orthopaedic Problems at Work,* Perinton Press, 1985.
Schoenmarklin, R.W. and Marras, W.S. A dynamic biomechanical model of the wrist joint, in *Proceedings of the 34th meeting of the Human Factors Society,* Orlando, FL., 805-809, 1990.
Schoenmarklin, R.W. and Marras, W.S. Dynamic capabilities of the wrist joint in industrial workers, *International Journal of Industrial Ergonomics,* 11, 207-224, 1993.
Schoenmarklin, R.W. and Marras, W.S. An EMG-assisted biomechanical model of the wrist joint. *Advances in Industrial Ergonomics and Safety IV,* edited by Kumar, S., 777-781, Taylor & Francis Publishers, 1992.
Schivde, A.J., Dreizen I., and Fisher, M.A. The carpal tunnel syndrome: A clinical-electrodiagnostic analysis. *Electromyogr Clin Neurophysiol,* 21, 143-153, 1981.
Schuind, P.F., Garcia-Elias, M., Cooney, W.P., and An, K.N. Flexor tendon forces: *In vivo* measurements. *Journal of Hand Surgery,* 17A(2), 291-298, 1992.
Silverstein, B.A., Fine, L.J., and Armstrong, T.J. Hand wrist cumulative trauma disorders in industry. *British Journal of Industrial Medicine,* 43, 779-784, 1986.
Silverstein, B.A., Fine, L.J., and Armstrong, T.J. Occupational factors and carpal tunnel syndrome. *American Journal of Industrial Medicine,* 11, 343-358, 1987.
Smith, E.M., Sonstegard, D.A., and Anderson, W.H. Contribution of flexor tendons to the carpal tunnel syndrome. *Archives of Physical Medicine and Medical Rehabilitation,* 58, 379-385, 1977.
Sommerich, C.M. Carpal tunnel pressure during typing: Effects of wrist posture and typing speed, in *Proceedings of the Human Factors and Ergonomics Society 38th Annual Meeting,* 611-615, 1994.
Spence, A.P. *Basic Human Anatomy,* Second Edition, The Benjamin/Cummings Publishing Co., 1986.
Stedman, T.L. *Stedman's Medical Dictionary,* 24th edition, Williams and Wilkins, 1982.
Stramler, J.H. *The Dictionary for Human Factors and Ergonomics,* CRC Press, 1993.
Szabo, R.M. and Gelberman, R.H. The pathophysiology of nerve entrapment syndromes. *The Journal of Hand Surgery,* 12A(5), 880-884, 1987.
Tanaka, S. and McGlothlin, J.D. A conceptual model to assess musculoskeletal stress of manual work for establishment of quantitative guidelines to prevent hand and wrist cumulative trauma disorders (CTDs), in *Advances in Industrial Ergonomics and Safety I,* Mital, A., editor, 419-426, Taylor & Francis, 1989.
Van de Graaff, K.M. and Rhees, R.W. *Human Anatomy and Physiology,* Schaum's Outline Series in Biology, McGraw-Hill Book Co., 1987.
Williams, H.J. and Ward, J.R. Musculoskeletal occupational syndromes. Chapter 29 in *Environmental and Occupational Medicine,* Little, Brown, & Co., 1983.
Yamaguchi, D.M., Lipscomb, P.R., Soule, E.H. Carpal tunnel syndrome. *Minn Med,* 22-23, Jan. 1965.

For Further Information

Although replete with medical terminology, an excellent reference for the biomechanical pathogenesis of cumulative trauma disorders of the upper extremity are chapters on the muscle–tendon unit and carpal tunnel syndrome in J.S. Moore's and A. Garg's *State of the Art Reviews: Ergonomics: Low- Back Pain, Carpal Tunnel Syndrome, and Upper Extremity Disorders in the Workplace.*

A good discussion of biomechanical models of the wrist is presented in Section 6.5.2 of *Occupational Biomechanics, 2nd Edition* by Don B. Chaffin and Gunnar B.J. Andersson. This book also has a chapter on handtool design (Chapter 11).

Basmajian's *Primary Anatomy, 8th edition* provides easy-to-read and well-illustrated depictions of the body's soft tissues (muscles, nerves, and tendons) and their structures and functions.

The *Clinical Mechanics of the Hand* by Paul W. Brand is particularly helpful in describing the function of specific muscles and tendons in the hand and forearm and how these interact to configure the hand in common pinch and grip postures.

The *Work-Related Musculoskeletal Disorders (WMSDs): A Reference Book for Prevention* (edited by Kuorinka and Forcier) is an excellent book that describes specific WMSDs and provides evidence for the association between work-related activity and each respective WMSD.

45

Occupational Risk Factors for Shoulder Disorders

Eira Viikari-Juntura
Finnish Institute of Occupational Health

45.1 Introduction

This chapter deals with occupational risk factors of shoulder disorders. The shoulder is structurally and functionally intimately linked with the neck, and often neck–shoulder disorders have been dealt with as one group of disorders. In this chapter, the main emphasis is on the disorders of shoulder structures, although some overlap to the neck area cannot be avoided.

The shoulder is a complex system of bones, muscles, tendons, and ligaments that attach the upper extremity to the torso. The glenohumeral joint is the joint with the largest range of motion in the human body, allowing large mobility for the upper extremity and enabling the body to reach far in all directions. The primary function of the shoulder is to direct and support the hand in its activities.

Because of its supporting function, high forces are imposed on the shoulder, especially if the hand is holding a heavy object. Due to the long moment arm of the extended upper arm, fairly light objects, weighing about 1 kg, impose high mechanical stress on the shoulder. To be able to withstand such forces great stability is demanded of the structures of the shoulder. On the other hand, the long moment arm of the upper arm and the large mobility and relatively poor protection of the shoulder joint render various tissues liable to injuries associated with falls and other sudden movements. Such injuries may heal only partially and decrease the strength and stability of the structures permanently.

Physical load factors occurring at work associated with various shoulder disorders include manually strenuous activities, postural factors of the arm and torso, static work, repetitive work, lack of rest pauses, vibration from handheld tools, environmental factors, and work organizational factors. Only a small amount of data exist on which to base any reference values for acceptable load intensities, frequencies, and durations of such factors. Because of the great liability of the shoulder to acute injuries, nonoccupational

0-8493-2641-9/99/$0.00+$.50

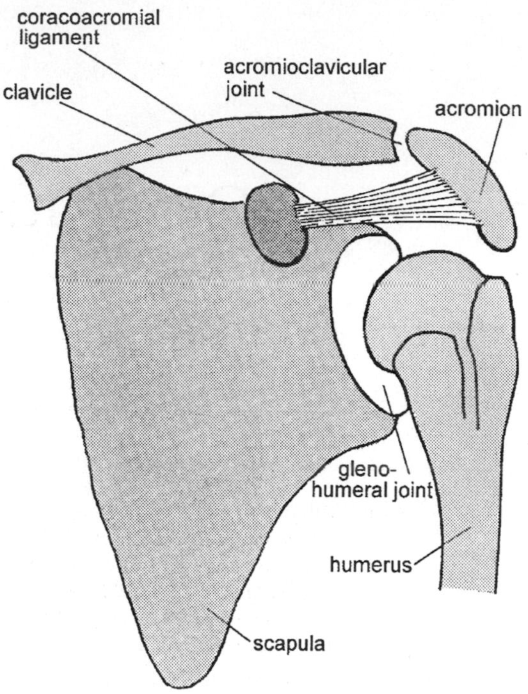

FIGURE 45.1 Bony structures in the shoulder, front view.

activities, especially hand-intensive sports, should be considered when investigating the etiology of shoulder pain in individual workers.

Due to the complexity of the shoulder, it has long been difficult to estimate the stresses to the different parts in the shoulder complex when exposed to various physical load factors. Recent advances in biomechanical modeling have markedly increased our knowledge of the stresses on the different structures of the shoulder and also of the capacity of shoulder structures.

45.2 Structure and Function of the Shoulder

Bony Structures

The most important bones in the shoulder are the shoulder blade (scapula), the humerus, and the clavicle (Figure 45.1). The humerus is attached to the scapula by the glenohumeral joint at the lateral aspect. It is in this joint that the primary motion occurs during movements of the arm. The clavicle is attached to the acromion in the upper lateral aspect of the shoulder, and the other end attaches to the sternum. The scapula covers part of the dorsal aspect of the rib cage.

Muscles and Tendons

The prime movers of the shoulder are the deltoid and four so-called rotator cuff muscles. The deltoid has its origin at the lateral part of the clavicle, the acromion process of the scapula, and the back of the scapula and inserts at the lateral aspect of the humerus. The rotator cuff muscles have their origin in the scapula and insert at the head of the humerus. Before the insertion, their tendons merge around the head of the humerus, thereby forming the rotator cuff (Figure 45.2). The trapezius is a flat muscle on the surface that has its origin at the occiput, neck, and thoracic vertebrae and inserts at the clavicle, acromion process of the scapula and spine of the scapula.

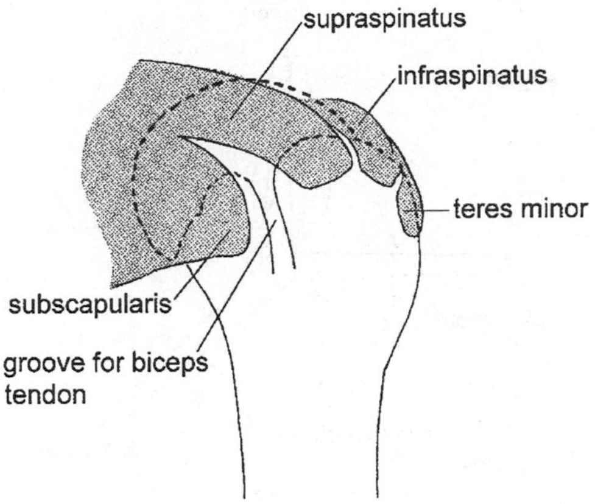

FIGURE 45.2 Insertion of rotator cuff tendons at the head of humerus, front view.

Nerves and Vessels Around the Shoulder

In addition to the nerves and vessels that supply the shoulder muscles and bones, all vessels and nerves supplying the arm and hand pass by the shoulder, forming a neurovascular bundle at the so-called thoracic outlet area. Due to postural factors, structural anomalies, and tight bands of the muscles at the thoracic outlet, the course of these latter nerves and vessels may be interfered with.

Movements of the Shoulder

The upper arm has a wide range of motion in all three planes: sagittal, transversal, and horizontal. The movement in the sagittal plane is the flexion–extension movement, the movement in the transversal plane is the abduction–adduction movement; and the movement in the horizontal plane is the horizontal abduction and adduction (Figure 45.3). Moreover, the arm has a wide range of rotation around its longitudinal axis. Most work activities demand varying degrees of flexion and abduction of the shoulder, often with the forearm in flexion at the elbow.

Loading of Shoulder Muscles in Different Activities

Shoulder muscle loading may be estimated by recording the electrical activity of the muscles by electromyography and measuring the intramuscular pressure. The trapezius, deltoid, supraspinatus, and infraspinatus muscles have been measured most commonly. The abduction movement in the plane of the scapula creates more loading than the flexion movement, especially for the supraspinatus muscle. The intramuscular pressure shows a linear increase with increasing abduction or flexion angle. In the supraspinatus, a flexion or abduction angle of 30° is enough to raise the intramuscular pressure above the level where muscle blood flow is impeded. Adding hand load increases the intramuscular pressure markedly. The increase is somewhat higher in shoulder abduction than in flexion. Flexing the elbow by 90° reduces the intramuscular pressure by about 30% in shoulder abduction.

Mechanisms of Injury in the Shoulder

Shoulder disorders may have their pathological process in the muscles, tendons, nerves, or joints. Commonly painful muscles are the trapezius, especially the descending part, supra- and infraspinatus, and the levator scapulae. *Tendinitis of the rotator cuff tendons* (e.g., *supraspinatous or infraspinatous tendinitis*) is the most common tendon disorder at the shoulder. In the *thoracic outlet syndrome,* the nerves and/or vessels of the neurovascular bundle are compressed at the thoracic outlet. The joints of the shoulder differ as to how prone they are to *degenerative changes.* In the glenohumeral joint, degenerative changes are uncommon, whereas in the acromioclavicular joint, such changes occur much more frequently.

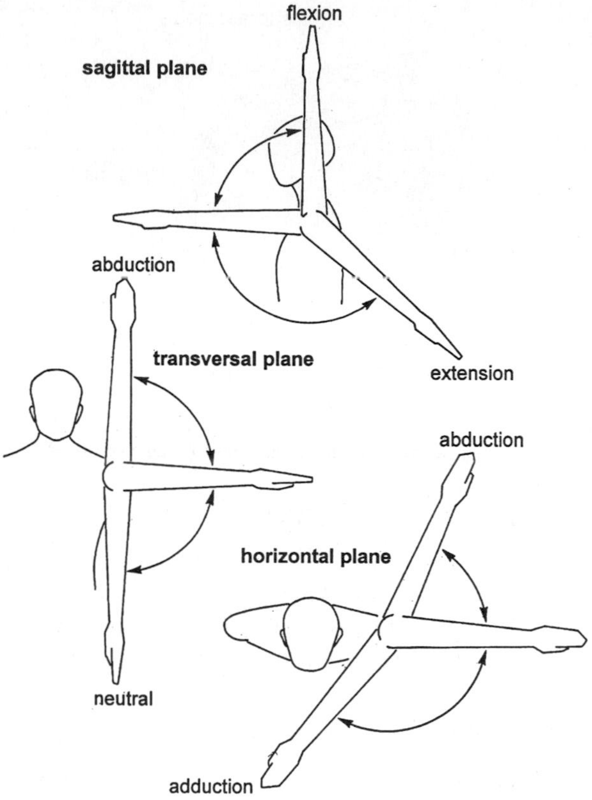

FIGURE 45.3 Movements of the shoulder in the three planes.

As described above, flexion and abduction of the arm increase the intramuscular pressure, which in turn interferes with blood circulation and may produce local ischemia and cause muscle pain. Other possible mechanisms for muscle pain are disturbances in energy metabolism and mechanical failure, especially after heavy physical exercise.

Rotator cuff tendinitis is usually associated with degenerative changes in the tendons of the rotator cuff. Such changes may be caused by impaired circulation in the tendon due to high tension in the muscle. When the hand is exposed to local low-frequency vibration, the muscles of the arm contract involuntarily as a consequence of the tonic vibration reflex. The tendons of the rotator cuff may also be mechanically injured due to compression under the acromion and the coracoacromial arch. The tendons may also be injured in falls and other accidents. A degenerated tendon is more likely to tear in an accident than is a healthy tendon.

45.3 Occupational Risk Factors

Knowledge of occupational risk factors for shoulder disorders is based on epidemiological studies in the field and experimental studies in the laboratory. Epidemiological studies have investigated the associations between physical load factors and clinically defined shoulder tendinitis, radiographically assessed degenerative joint disease, or reported shoulder pain. In many epidemiological studies, certain occupations have been selected to represent a certain type of work, e.g., welders for overhead work. In some other studies, exposure assessment has been based on self-assessment. Only rarely has the validity of such self-assessment been investigated. This means that the information on physical load factors in the epidemiological studies is usually crude and may have considerable inaccuracy and even bias. Only exceptionally have direct measurements been carried out.

In experimental studies, the intensity, frequency, and duration of the exposure can be determined and measured with high precision, but the outcome is different from that in the epidemiological studies. In addition to subjectively assessed discomfort, various physiological responses have been measured, such as myoelectrical activity of muscles, intramuscular pressure, and blood metabolites. Only some data exist on the associations between different types of physiological responses and the development of shoulder disorders. Based on epidemiological and experimental evidence, 10 work-related risk factors may be recognized:

- Heavy physical work
- Manual handling
- Elevated postures of the arm
- Nonneutral trunk postures
- Static postures
- Repetitive work
- Lack of pauses
- Vibration
- Draft
- Work organizational factors

Each of these risk factors will be discussed below. A compilation of selected epidemiological studies (Table 45.1), studies in which various physiological responses have been measured during real or simulated work (Table 45.2), and studies that have measured various physiological responses during basic movements and loading situations of the shoulder in the laboratory (Table 45.3) serve as a source of original data for the text. In all studies, various physical load factors (exposure) have been the independent variables under study. The outcome or health effect has usually been more long term by nature in the epidemiologic studies and a short-term response in the experimental studies.

Heavy Physical Work

Several studies have shown an association between heavy physical work and shoulder problems. The association has been found for shoulder tendinitis (Stenlund et al., 1993), radiographically defined acromioclavicular arthrosis (Stenlund et al., 1992), and reported shoulder pain (Viikari-Juntura et al., 1993). An increased risk of acromioclavicular arthrosis was observed for 10 to 28 years of manual work, and an even higher risk for more than 28 years of manual work, i.e., an exposure–response relationship was observed for cumulative exposure to manual work. Examples of occupational groups are rockblasters, bricklayers, and various jobs in forestry.

Heavy physical work may involve manual handling of heavy loads, nonneutral trunk postures, and elevated postures of the arm. Such work may also be associated with repeated minor traumas and a risk of major trauma.

Manual Handling

Manual handling of loads has been associated with shoulder disorders in some studies. Wells et al. (1983) found a higher prevalence (13%) of recurrent shoulder pain for letter carriers than for meter readers (7%) and postal clerks (5%). The maximum bag weight of the letter carriers was 11.4 kg. Letter carriers whose maximum bag weight had been increased by 4.4 kg had a prevalence of 23% of recurrent shoulder pain. Stenlund et al. (1992) determined life-long lifting as lifted tonnes, and found an increased risk of radiographically assessed osteoarthrosis of the acromioclavicular joint for 710 to 26,000 tonnes, and an even higher risk for more than 26,000 tonnes. This means that an exposure–response relationship was also found for lifted load and acromioclavicular joint arthrosis. Unfortunately, it is not known whether intensive lifting for a short period is associated with a different risk than less intensive lifting for a longer

TABLE 45.1 Selection of Epidemiological Studies on Shoulder Disorders

Study Population	Type of Study	Outcome	Exposure	Validity of Exposure Assessment Method	Adjustment for Confounders	Findings	Reference
17 cases (all men) with shoulder pain for at least 3 months seeking medical advice, 34 age, sex, and workplace matched controls	Case-control	Prolonged shoulder pain (clinical signs of shoulder tendinitis in 12 cases)	Physical heaviness of work, shoulder load (work height of arms), based on interview of the cases and referents	Not reported	Not necessary (matched design)	No difference between cases and controls in physical heaviness of work 11/17 cases and 5/34 controls performed their work at or above acromion height	Bjelle et al., 1979
152 female assembly-line packers 133 female shop assistants	Cross-sectional	Humeral tendinitis (supraspinatous or bicipital tendinitis; clinically verified)	Job title (packer, shop assistant) used in analysis Assembly-line: repetitive motions up to 2,500/day; static muscle work; extreme work postures of the hands; varying amount of lifting	Not necessary	Not performed	Prevalence of humeral tendinitis 9.2% among the packers and 3.8% among the shop assistants.	Luopajärvi et al., 1979
131 shipyard welders with more than 5 years of welding experience 56 male office clerks above 40 years of age	Cross-sectional + case-control among welders	Supraspinatus tendinitis (clinically verified)	Job title (welding work) Rating of shoulder load by expert	Not reported	Not performed in cross-sectional study. Not necessary in case-control study (matched design)	Estimated prevalence of supraspinatus tendinitis 18.3% among welders. Uncertainty of true occurrence of disease because questionnaire and clinical study 1 year apart and only symptomatic were examined. In case-control study, tendency toward higher load among those with supraspinatus tendinitis	Herberts et al., 1981

92 letter carriers with 11.4 kg maximum weight, 104 letter carriers with 15.8 kg maximum weight (weight increase occurred on an average of 5 months prior to study), 76 gas meter readers, 127 postal clerks	Cross-sectional	Occurrence of and disability due to shoulder pain by telephone interview	Maximum weight carried at work either 11.4 or 15.8 kg	Not necessary	Age, body mass index, years in current job, previous heavy work	Prevalence of "significant" shoulder problems 23, 13, 7, and 5% among letter carriers with weight increase, letter carriers without weight increase, meter readers, and postal clerks, respectively. Shoulder problems associated with weight carried.	Wells et al., 1983
96 female employees in the electronics industry	Cross-sectional	Neck, shoulder, and arm disorders (clinically verified) and symptoms (during preceding 12 months)	Observations from video taken from two projections (individually for each worker): work cycle time, number and duration of rest periods (>2 s) for arm, shoulder, and head, duration of different upper arm and head postures, frequency of changes between postural categories	Not reported	Age, anthropometry, maximum muscle strength and endurance, work history, leisure physical activity, hobbies, perceived psychological stress at work, work satisfaction, number of breaks and rest pauses at work, individual productivity	Long duration of employment in the electronics industry and % of work cycle time with upper arm abducted 0 to 30° associated with symptoms in the shoulder during preceding 12 months (slight or more severe). High stature and a high number of upper arm flexions per hour inversely related to shoulder symptoms.	Kilbom et al., 1986
54 bricklayers, 55 rock blasters, 98 construction foremen (all men)	Cross-sectional	Radiographical arthrosis of the acromioclavicular joints	Job title, life-long lifted load, life-long exposure to vibration, years of manual work based on questionnaire and interview	Not reported	Age, dexterity, smoking	Prevalence of osteoarthrosis in the right vs. left shoulder: rockblasters 61.8%; 56.4% bricklayers 59.3%; 40.7% foremen 36.7%; 23.4% Increased risk for years of manual work, cumulative lifted load and vibration (exposure-response relationship for all)	Stenlund et al., 1992

TABLE 45.1 (continued) Selection of Epidemiological Studies on Shoulder Disorders

Study Population	Type of Study	Outcome	Exposure	Validity of Exposure Assessment Method	Adjustment for Confounders	Findings	Reference
As above	As above	Shoulder tendinitis (rotator cuff tendinitis; clinically verified)	As above	Not reported	Age, dexterity, smoking, sports activities	Prevalence of rotator cuff tendinitis in the right vs. left shoulder: rockblasters 23.6%; 14.5% bricklayers 1.8%; 1.8% foremen 9.2%; 3.1% Increased risk for cumulative exposure to vibration	Stenlund et al., 1993
157 cases and 2,565 controls (incidence study) in a forest industry company, 352 cases and 122 controls (persistence study)	1-year follow-up	Incidence of severe shoulder pain, persistence of severe shoulder pain	Trunk and shoulder postures and movements, physical heaviness of work, work characteristics (self-administered questionnaire)	Fair for most items	Sex, age, body mass index, smoking, sleep disorder	Frequent twisting of the torso associated with incidence of severe shoulder pain Physically heavy work and mental overload at work associated with persistently severe shoulder pain	Viikari-Juntura et al., 1993
52 female orchard farmers	Two cross-sections	Prevalence of neck and shoulder stiffness and pain; pain in joint motion and muscle tenderness in the neck and shoulder	Pear bagging (2,455/day) in the first cross-section, apple bagging (1,887/day) in the second cross-section. Upper arm flexion >90° for 75% of time in pear bagging and 41% of time in apple bagging (based on goniometer measurements of one worker)	Not reported	Not necessary, because the same subjects were investigated at two time points	Neck and shoulder stiffness, neck pain, shoulder muscle tenderness, and pain in movements of the neck more prevalent during pear bagging than apple bagging. It is not, however, clear whether the higher prevalence of pain could be attributed to higher work pace or higher work height, or both when bagging pears.	Sakakibara et al., 1995

TABLE 45.2 Selected Studies on Physiological Responses During Real or Simulated Work Tasks Involving Activation of Shoulder Muscles

Study Group	Task	Physiological Response	Muscles	Findings*	Reference
41 female office workers (nonsymptomatic and symptomatic)	Various keyboard activities in the field: conventional typewriter, telex machine, visual display terminal, telephone exchanger, insurance calculating machine	Myoelectrical activity in forearm and shoulder muscles by surface electrodes	Trapezius, deltoid, rhomboideus, forearm extensors	The activities of the right trapezius ranged from 10 to 30% of maximal activity. Static activity in the trapezius measured also at times when the forearm was not active.	Onishi et al., 1982
25 assembly workers: 16 women and 9 men (nonsymptomatic and symptomatic)	Light assembly and soldering tasks in the field	Myoelectrical activity of shoulder muscles by surface electrodes	Anterior deltoid, upper trapezius, infraspinatus	High static activity levels of about 7–14% of maximal activity in all muscles; median activity levels high (about 16–20%) in infraspinatus and trapezius; acceptable maximum contraction levels (about 30%) in all muscles.	Christensen, 1986
6 healthy female secretaries	Wordprocessing: 5h, 3h, and 3h work period with 15 s pauses every 6th minute in the field	Myoelectrical activity by surface electrodes, discomfort in the eyes, neck, shoulders, elbows, hands, back	Upper trapezius	Median of static load 3.2% and 3.0%, of median load 7.3% and 5.5%, and of peak load 11.7% and 9.2% of maximal activity on right and left side, respectively. No difference in electrical activity between the three work periods. Discomfort ratings slightly lower after work with rest pauses than without rest pauses.	Hagberg and Sundelin, 1986
8 healthy female cashiers	Simulation of cash register operations in the laboratory using conventional keyboard, horizontal scanner, vertical scanner, and pen reader in the sitting posture. Horizontal scanner used also in the standing posture.	Myoelectrical activity of neck and shoulder muscles by surface electrodes	Upper trapezius, cervical erector spinae, levator scapulae, thoracal erector spinae/rhomboidei, infraspinatus	Static load between 2–5% of maximal activity; median load below 10%; peak load below 30% in all muscles.	Lannersten and Harms-Ringdahl, 1990

TABLE 45.2 (continued) Selected Studies on Physiological Responses During Real or Simulated Work Tasks Involving Activation of Shoulder Muscles

Study Group	Task	Physiological Response	Muscles	Findings*	Reference
6 healthy female physiotherapy students	Work simulation in the laboratory of grasping a cylinder (weight 15 g) and releasing it through a hole in the table, paced as 2,466 cycles per hour (MTM-110).	Myoelectrical activity by surface electrodes	Lateral and cervical parts of the upper trapezius, infraspinatus	Static and peak loads were 4.4 and 31%, 17 and 37%, and 12 and 55% of the maximal activity for the lateral and cervical part of the upper trapezius, and the infraspinatus, respectively. Signs of fatigue seen in all subjects. The localization and time for these signs varied between different subjects.	Sundelin and Hagberg, 1992
12 healthy female secretaries	Wordprocessing task in the field with 3 different air velocities (0.96, 1.4, 1.96 m/s)	Myoelectrical activity by surface electrodes	Cervical and lateral part of upper trapezius, levator scapulae	Increased myoelectrical activity (increase in both mean power frequency and root-mean-square amplitude) in the lateral part of upper trapezius, possibly due to reflex recruitment of motor units. In the cervical part of upper trapezius, increase of root-mean-square amplitude and a decrease in mean power frequency, possibly due to cooling of the muscle.	Sundelin and Hagberg, 1992
12 healthy female physiotherapy students	2-hour work simulation in the laboratory of grasping a cylinder and releasing it through a hole: one hour continuously (MTM-110), and another hour with 1-minute pause every 6th minute, consisting of lifting of a 2-kg box on a shelf slowly 5-6 times (MTM-132 for the repetitive task)	Myoelectrical activity by surface electrodes, rating of discomfort	Lateral and cervical parts of the upper trapezius, infraspinatus	Electromyographic signs of fatigue less pronounced with pause activities than without. Fatigue patterns lower during the second hour, indicating adaptation to work. Discomfort ratings did not differ between work with and without pauses and were higher during the second hour.	Sundelin, 1993

* Activity figures from different studies, given as % of maximal activity, can be compared only roughly, due to different calibration procedures in different studies.

TABLE 45.3 Selected Studies on Physiological Responses Associated with Controlled Postures and Loading of the Shoulder Muscles in the Laboratory

Study Group	Task	Physiological Response	Muscles	Findings*	Reference
6 healthy female students	Repetitive shoulder flexions 0–90° with hand loads between 0 and 3.1 kg. Repetition rate 15/min, duration of each trial 60 min	Myoelectrical activity of shoulder muscles, tenderness of shoulder muscles, clinical tests for shoulder tendinitis, heart rate, rating of perceived exertion	Upper trapezius, anterior deltoid, biceps brachii	Peak load ranged between 13 and 60% of maximal activity. Ratings of perceived exertion increased more than heart rate, indicating the importance of local factors. After two days, tenderness in the descending part of the trapezius and supraspinatus in all subjects, signs of shoulder tendinitis in 2 subjects.	Hagberg, 1981
9 healthy men; 5 with welding experience	Laboratory experiment on arm posture and hand tool weight (upper arm flexion and abduction, elbow flexion, and upper arm rotation in 21 combinations. Hand load 0, 1, and 2 kg)	Myoelectrical activity of shoulder muscles by intramuscular electrodes	Anterior, middle, and posterior deltoid, supraspinatus, infraspinatus, upper trapezius	Degree of upper arm elevation most important determinant of shoulder muscle load. Muscles of the rotator cuff more hand-load dependent than the deltoid muscle.	Sigholm et al., 1984
10 healthy females (skilled in assembly work)	Laboratory experiment on different sitting postures and tilting the work object from the horizontal plane. Experimental work cycle of simulated soldering at an electronics plant.	Myoelectrical activity of neck and shoulder muscles	Cervical erector spinae/trapezius, upper trapezius, middle trapezius/supra-spinatus, thoracal erector spinae/rhomboidei, levator scapulae, sternocleidomastoid	Sitting posture with the spine slightly tilted backward and the cervical spine vertical was associated with the lowest activity in posterior neck and shoulder muscles. The posture with the whole spine straight and vertical resulted in a higher myoelectrical activity, and the posture with the whole spine flexed was associated with the highest myoelectrical activity. A backward inclination of the spine generally requires that the work object should be tilted from the horizontal plane.	Schüldt et al., 1986
10 healthy females (skilled in assembly work)	Laboratory experiment on the effects of arm support and suspension on myoelectrical activity of posterior neck and shoulder muscles in different sitting postures and tilting the work object from the horizontal plane. Simulated work cycle of soldering at an electronics plant.	Myoelectrical activity of neck and shoulder muscles	Cervical erector spinae/trapezius, upper trapezius, middle trapezius/supra-spinatus, thoracal erector spinae/rhomboidei, levator scapulae, sternocleidomastoid	Both arm support and suspension reduced the myoelectrical activity of the muscles in the whole-spine-straight-and-vertical posture. Activity values were in most cases below 10% of maximal activity. Arm suspension might be more efficient than arm support in the posture with the trunk slightly inclined backward. Arm support might be more efficient than arm suspension in the whole-spine-flexed posture which even with support was associated with relatively high activities.	Schüldt et al., 1987

TABLE 45.3 (continued) Selected Studies on Physiological Responses Associated with Controlled Postures and Loading of the Shoulder Muscles in the Laboratory

Study Group	Task	Physiological Response	Muscles	Findings*	Reference
12 healthy students	Laboratory experiment on arm posture and hand tool weight (upper arm flexion and abduction, elbow flexion in 15 combinations; hand load 0, 1, and 2 kg)	Intramuscular pressure	Supraspinatus	High intramuscular pressure in abductions ≥30° with and without elbow flexion and with and without shoulder load. Hand load increased intramuscular pressure in most postures.	Järvholm et al., 1988
6 healthy females	Continuous and intermittent holding of the arm horizontally at 60° to the sagittal plane (cycle time 10, 60, and 360 s and duty cycle 0.33, 0.50, 0.67, and 0.83)	Myoelectrical activity by surface electrodes, arterial blood pressure, heart rate, perceived fatigue during exercise; venous blood potassium, lactate, and ammonia pre- and postexercise; maximal voluntary contraction, pressure pain threshold, proprioceptive performance, and 1-minute arm holding at 25% MVC before and up to 4 h postexercise	Upper trapezius	Duty cycle influenced all variables, cycle time only blood pressure and fatigue perception. Cardiovascular and neuromuscular recovery incomplete for hours. Ranking of protocols differed according to the criterion variable	Mathiassen, 1993
72 healthy subjects with assembly work experience, 35 men and 37 women	Simulation of a repetitive assembly job by a work simulator, consisting of lifting and lowering of a tool handle and striking a metal pointer to a plate at the end of each excursion (in the sitting posture). Repetition rate (20, 24, 35 lifts/min), required force (10, 20, 30% of maximal activity), tool weight (2.1, 2.5, 3.0 kg) and reach height (109, 120, 131 cm) were varied.	Duration of work until given rating of perceived muscle discomfort was reached	Muscles in the neck, shoulder, or upper arm	Increase in force, repetition rate, tool weight, and height of upper target all reduced work trial duration, repetition rate and force having the largest effects. Interaction between repetition rate and force, so that an increase in each variable led to an attenuation of the other variable's effect.	Putz-Anderson and Galinsky, 1993

* Activity figures from different studies, given as % of maximal activity, can be compared only roughly, due to different calibration procedures in different studies.

period. It should also be noted that the data on lifting were based on questionnaires and interviews with no assessment of validity.

Manual handling activities impose high loads on probably all shoulder structures of which the rotator cuff muscles and the deltoid have been the most investigated (Sigholm et al., 1984; Järvholm et al., 1988). Hand load has a strong effect on shoulder muscle activity, especially in the rotator cuff muscles, and on intramuscular pressure in the supraspinatus. Manual handling activities may also run a risk of trauma.

Elevated Postures of the Arm

There is some epidemiological evidence to support an association between elevated postures of the arm and shoulder pain (Bjelle et al., 1979; Sakakibara et al., 1995) as well as supraspinatus tendinitis (Herberts et al., 1981). The occupations involved in the studies have been shipyard welders and orchard farmers. Experimental studies have shown that the activity of shoulder muscles increases with increasing elevation (flexion and abduction) of the arm (Sigholm et al., 1984). A flexion angle of ≥30° without hand load raises the intramuscular pressure at a level where blood circulation is disturbed (Sigholm et al., 1984; Järvholm et al., 1988). After longer periods of intensive shoulder muscle exercise, cardiovascular and neuromuscular recovery may be incomplete for hours (Mathiassen, 1993). Elevated arm postures may also be associated with mechanical irritation of the rotator cuff tendons under the acromion and coracoacromial arch.

The loads imposed on shoulder structures in various activities with elevated arms may be decreased by suspending the arms. The results from the simulated work of welders showed that arm suspension reduced shoulder muscle load, but the intramuscular pressure of the supraspinatus remained at a level where muscle blood flow would still be compromised (Järvholm et al., 1991).

Nonneutral Trunk Postures

In sedentary work, the workplace layout largely determines the posture of the torso, neck, and limbs. Schüldt et al. (1987) investigated the myoelectrical activity of several neck and shoulder muscles in different postures during simulated soldering work in the laboratory. Sitting with the spine slightly tilted backward and the cervical spine vertical was associated with the lowest activity. The posture with the whole spine straight and vertical resulted in a higher myoelectrical activity, and the posture with the whole spine flexed was associated with the highest activity. In the latter experiment, the work object was horizontal in the whole-spine-flexed posture, tilted 35° in the whole-spine-straight-and-vertical posture, and tilted 75° in the posture with the spine tilted backward and neck vertical. As stated by these authors, a backward inclination of the spine generally requires that the work object should be tilted from the horizontal plane.

Static Work

According to epidemiological studies, shipyard welders (Herberts et al., 1981), orchard harvesters (Sakakibara et al., 1995), packers (Luopajärvi et al., 1979), garment workers (Punnett et al., 1985), workers in light assembly tasks (Kvarnström, 1983), and office workers with intensive use of the mouse (Hagberg and Karlqvist, 1994) have shown a high risk for shoulder disorders. Common to the tasks in these occupations is static exertion of shoulder muscles with or without elevation of the arm.

Measurements of myoelectrical activity in the field and simulated work in the laboratory have shown static activity levels ranging from 4 to 17% of maximal activity in different shoulder muscles (Christensen, 1986; Sundelin and Hagberg, 1992), the upper range being far above the 2 to 5% of maximal activity recommended by Jonsson (Jonsson, 1982). A simulation of cash register operation in the laboratory (Lannersten and Harms-Ringdahl, 1990) and word processing operations, both with spontaneous and forced pauses (Hagberg and Sundelin, 1986), was associated with a lower static load of 2 to 5% of maximal activity.

Measurements of various keyboard activities in the field showed activity in the trapezius also at times when no activity was performed by the forearm (Onishi et al., 1982). The importance of interruptions of activity was shown in a follow-up of workers at a chocolate manufacturing plant. The workers who had a smaller number of short unconscious interruptions of electromyographic activity (so-called emg-gaps) had a higher risk of contracting trapezius myalgia than those with a higher number of emg-gaps during a follow-up time of six months (Veiersted et al., 1993).

The load of the shoulder muscles may be reduced by arm support or arm suspension in activities involving static exertion of shoulder and arm muscles. A laboratory experiment of simulated soldering work in different postures of the trunk and neck suggested that arm suspension might be more efficient than arm support in the posture with the trunk slightly inclined backward. Arm support might be more effective than arm suspension in the whole-spine-flexed posture. This latter posture, however, showed generally high myoelectrical activities and should not be adopted for longer periods of work (Schüldt et al., 1987).

Repetitive Work

A typical pattern of repetitive work is that the fingers perform quick movements while the shoulder muscles perform mostly static exertions to fulfil their primary task in supporting the arms. Word processing, packing, and light assembly tasks are examples of such repetitive work, and high risk for shoulder disorders has been shown in these tasks. In a simulation of assembly work in the laboratory, the introduction of a one-minute pause with dynamic lifting activity every 6th minute resulted in less pronounced myoelectrical signs of fatigue in the shoulder muscles, suggesting that dynamic activity might be effective in counteracting the effects of static loading of shoulder muscles (Sundelin, 1993). No difference was seen in discomfort ratings, however.

Certain work tasks may demand repetitive movements of the upper arm, but the health effects of repetitive arm flexions or abductions are not well known. In an experimental study in which six healthy students performed repetitive shoulder flexions with a frequency of 15/min for 60 minutes with loads varying from 0 to 3.1 kg, all subjects had tenderness in the descending part of the trapezius and supraspinatus, and two of them also had other signs of supraspinatus tendinitis after two days (Hagberg, 1981). A study among electronics assembly workers showed that the duration of mild upper arm flexion was associated with shoulder disorders, but the number of upper arm flexions was inversely related to shoulder disorders (Kilbom et al., 1986). This suggests that a more dynamic working style decreases the risk of shoulder disorders.

An experimental study used the psychophysical approach to investigate the effects of various repetition rates, forces, tool weights, and reach heights on work durations until a given degree of subjectively rated fatigue was achieved in repeated arm flexions. The repetition rate was the prime determinant for work duration, followed by force, height of upper target, and tool weight. Repetition rate and force showed an interaction, so that increases in each variable led to a slight attenuation of the other variable's effect (Putz-Anderson and Galinsky, 1993). This study is among the few sources of data on which reference values for the frequencies of shoulder elevations with given loads and elevation angles may be established.

Work-Rest Schedule: Lack of Pauses

It is conceivable that the frequency, duration, and quality of pauses are crucial determinants for the development of fatigue in the muscles during forceful, repetitive, or static exertions. Work–rest schedules have been investigated in some studies in the field (Hagberg and Sundelin, 1986) and in the laboratory. In the aforementioned simulation of light industrial work in the laboratory, an MTM-110 pacing without pauses was compared with an MTM-132 pacing with one-minute pause every 6th minute, consisting of the dynamic lifting activity, the production rate of the task itself being equal in both schemes. As mentioned, electromyographic signs of fatigue were less pronounced with pause activities than without, despite the higher repetition rate and extra work of lifting during the pauses (Sundelin, 1993). In another experimental study, a range of physiological responses was measured when holding the arm continuously and intermittently in the horizontal plane at 60° to the sagittal plane. Duty cycle had a more pronounced effect on the various physiological responses than cycle time. A ranking of the protocols with different combinations of cycle time and duty cycle differed according to which physiological response was used (Mathiassen, 1993).

In conclusion, while the need for pauses is evident in tasks demanding forceful, repetitive, or static shoulder activities, only few data exist upon which to base any recommendation. Moreover, the recommendations will differ depending on what kind of physiological criterion is used. Sometimes objective and subjective criteria seem to contradict each other.

Vibration

Vibration from hand-held tools has been shown to be associated with both radiographically assessed arthrosis of the acromioclavicular joint (Stenlund et al., 1992) and shoulder tendinitis (Stenlund et al., 1993). For both conditions, an exposure–response relationship between cumulative exposure to vibration and the disease have been observed. The assessment of cumulative exposure took into consideration the hours that each vibrating tool had been used and the energy emission from the tool.

Draft

High air velocities in the work environment are perceived as draft and traditionally considered to increase neck and shoulder discomfort. Only some epidemiological evidence exists for the association of draft with neck and shoulder pain (Tola et al., 1988). The behavior of shoulder muscles was studied in an experiment with different air velocities in the office environment. The myoelectrical activity changes suggested increased recruitment of motor units in some muscles, and a possible cooling effect in others associated with increasing air velocity (Sundelin and Hagberg, 1992).

Work Organizational Factors

Demands, control, and social support are work organizational factors that have been most often investigated in association with shoulder disorders. An association between high job demands such as time pressure, high concentration, high work load and shoulder disorders has been shown in many studies. Also low control and little autonomy has been associated with shoulder disorders, but the results concerning social support are conflicting (Bongers et al., 1993). In a follow-up study, mental overload (difficult phases at work, the need to hurry to get work done) was associated with the persistence of severe shoulder pain but not with the incidence of pain, indicating that work organizational factors might have a greater role in the prognosis than in the genesis of shoulder disorders (Viikari-Juntura et al., 1993).

Summary of Occupational Risk Factors

The text above deals with individual occupational risk factors for shoulder disorders. In real work situations, many risk factors are present simultaneously and may have combined effects on the risk of shoulder disorders. In dynamic work, the overall risk of shoulder disorders is probably a function of arm posture, weight of load, and frequency of repetitions of arm movements. There is evidence that a high cumulative exposure of heavy work increases the risk of shoulder disorders, but whether a certain duration of heavy work per day could be tolerated for longer times is not known. Traumas to the shoulder may increase the risk of shoulder disorders in heavy work. There is convincing evidence of harmful effects of low-frequency vibration from handheld tools to the shoulder. Therefore, low-frequency vibration from tools should be eliminated or kept to the minimum.

In static work tasks with lower force demands, the elevation angle of the arm, the overall posture of the body, and the rest pauses largely determine the loading pattern of shoulder muscles. In this kind of work, optimal body and arm postures should be enabled by proper workplace layout and possibilities to support or suspend the arm according to the preference of the worker.

Work organizational factors may have an effect on the risk of shoulder disorders by influencing the intensity, frequency, or duration of physical load factors. They may also affect the reporting of the disorders or the recovery from them.

References

Bjelle, A., Hagberg, M., and Michaelsson, G. 1979. Clinical and ergonomic factors in prolonged shoulder pain among industrial workers. *Scand. J. Work Environ. Health* 5(3):205-210.

Bongers, P.M., De Winter, C., Kompier, M.A.J., and Hildebrandt, V.H. 1993. Psychosocial factors at work and musculoskeletal disease. *Scand. J. Work Environ. Health* 19(5):297-312.

Christensen, H. 1986. Muscle activity and fatigue in the shoulder muscles of assembly-plant employees. *Scand. J. Work Environ. Health* 12(6):582-587.

Hagberg, M. 1981. Work load and fatigue in repetitive arm elevations. *Ergonomics* 24(7):543-555.

Hagberg, M. and Karlqvist, L. 1994. Symptoms and disorders related to keyboard and computer mouse use. International Conference on Occupational Disorders of the Upper Extremities, December 1-2, 1994, Miyako Hotel, San Francisco, California.

Hagberg, M. and Sundelin, G. 1986. Discomfort and load on the upper trapezius muscle when operating a wordprocessor. *Ergonomics* 29(12):1637-1645.

Herberts, P., Kadefors, R., Andersson, G., and Petersén, I. 1981. Shoulder pain in industry: An epidemiological study on welders. *Acta Orthop. Scand.* 52(3):299-306.

Jonsson, B. 1982. Measurement and evaluation of local muscular strain in the shoulder during constrained work. *J. Human Ergol.* 11(1):73-88.

Järvholm U., Palmerud, G., Kadefors, R., and Herberts, P. 1991. The effect of arm support on the supraspinatus muscle during simulated assembly work and welding. *Ergonomics* 34(1):57-66.

Järvholm, U., Palmerud, G., Styf, J., Herberts, P., and Kadefors, R. 1988. Intramuscular pressure in the supraspinatus muscle. *J. Orthop. Res.* 6(2):230-238.

Kilbom, Å., Persson, J., and Jonsson, B.G. 1986. Disorders of the cervicobrachial region among female workers in the electronics industry. *Int. J. Ind. Ergonomics* 1(1):37-47.

Kvarnström, S. 1983. Occurrence of musculoskeletal disorders in a manufacturing industry, with special attention to occupational shoulder disorders. *Scand. J. Rehab. Med.* (Suppl. 8):6-101.

Lannersten, L. and Harms-Ringdahl, K. 1990. Neck and shoulder muscle activity during work with different cash register systems. *Ergonomics* 33(1):49-65.

Luopajärvi, T., Kuorinka, I., Virolainen, M., and Holmberg, M. 1979. Prevalence of tenosynovitis and other injuries of the upper extremities in repetitive work. *Scand. J. Work Environ. Health* 5(Suppl. 3):48-55.

Mathiassen, S.-E. 1993. The influence of exercise/rest schedule on the physiological and psychophysical response to isometric shoulder-neck exercise. *Eur. J. Appl. Physiol.* 67(6):528-539.

Onishi, N., Sakai, K., and Kogi, K. 1982. Arm and shoulder muscle load in various keyboard operating jobs of women. *J. Human Ergol.* 11(1):89-97.

Punnett, L., Robins, J.M., Wegman, D.H., and Keyserling, W.M. 1985. Soft tissue disorders in the upper limbs of female garment workers. *Scand. J. Work Environ. Health* 11(6):417-425.

Putz-Anderson, V. and Galinsky, T.L. Psychophysically determined work durations for limiting shoulder girdle fatigue from elevated manual work. *Int. J. Ind. Ergon.* 11(1):19-28.

Sakakibara, H., Miyao, M., Kondo, T., and Yamada, S. 1995. Overhead work and shoulder-neck pain in orchard farmers harvesting pears and apples. *Ergonomics* 38(4):700-706.

Schüldt, K., Ekholm, J., Harms-Ringdahl, K., Németh, G., and Arborelius, U.P. 1986. Effects of changes in sitting work posture on static neck and shoulder muscle activity. *Ergonomics* 29(12):1525-1537.

Schüldt, K., Ekholm, J., Harms-Ringdahl, K., Németh, G., and Arborelius, U.P. 1987. Effects of arm support or suspension on neck and shoulder muscle activity during sedentary work. *Scand. J. Rehab. Med.* 19(2):77-84.

Sigholm, G., Herberts, P., Almström, C., and Kadefors, R. 1984. Electromyographic analysis of shoulder muscle load. *J. Orthop. Res.* 1(4):379-386.

Stenlund, B., Goldie, I., Hagberg, M., and Hogstedt, C. 1993. Shoulder tendinitis and its relation to heavy manual work and exposure to vibration. *Scand. J. Work Environ. Health* 19(1):43-49.

Stenlund, B., Goldie, I., Hagberg, M., Hogstedt, C., and Marions, O. 1992. Radiographic osteoarthrosis in the acromioclavicular joint resulting from manual work or exposure to vibration. *Br. J. Ind. Med.* 49(8):588-593.

Sundelin, G. 1993. Patterns of electromyographic shoulder muscle fatigue during MTM-paced repetitive arm work with and without pauses. *Int. Arch. Occup. Environ. Health* 64(7):485-493.

Sundelin, G. 1993. Patterns of electromyographic shoulder muscle fatigue during MTM-paced repetitive arm work with and without pauses. *Int. Arch. Occup. Environ. Health* 64(7):485-493.

Sundelin, G. and Hagberg, M. 1992. Electromyographic signs of shoulder muscle fatigue in repetitive arm work paced by the Methods-Time-Measurement system. *Scand. J. Work Environ. Health* 18(4):262-268.

Sundelin, G., and Hagberg, M. Effects of exposure to excessive drafts on myoelectric activity in shoulder muscles. *J. Electromyogr. Kinesiol.* 2:36-41.

Tola, S., Riihimäki, H., Videman, T., Viikari-Juntura, E., and Hänninen, K. 1988. Neck and shoulder symptoms among men in machine operating, dynamic physical work and sedentary work. *Scand. J. Work Environ. Health* 14(5):299-305.

Veiersted, K.B., Westgaard, R., and Andersen, P. 1993. Electromyographic evaluation of muscular work pattern as a predictor of trapezius myalgia. *Scand. J. Work Environ. Health* 19:284-290.

Viikari-Juntura, E., Riihimäki, H., Takala, E.-P., Rauas, S., Leppänen, A., Malmivaara, A., Grönqvist, R., Härmä, M., Martikainen, R., Saarenmaa, K., and Kuosma, E. 1993. Niska-hartiaseudun ja yläraajan oireita ennustavat tekijät metsäteollisuudessa (Factors predicting pain in the neck, shoulders, and upper limbs in forestry work). *Työ ja ihminen* 7(4):233-253 (in Finnish with English summary).

Wells, J.A., Zipp, J.F., Schuette, P.T., and McEleney, J. 1983. Musculoskeletal disorders among letter carriers A comparison of weight carrying, walking & sedentary occupations. *J. Occup. Med.* 25(11):814-820.

For Further Information

A recent scientific review on occupational risk factors of soft tissue disorders of the shoulder is presented by Sommerich, McGlothlin, and Marras in *Ergonomics* (1993, 36:697-717).

An extensive overview of the work-relatedness of shoulder disorders as well as other disorders of the neck and upper limbs with a practical approach to prevention is presented in Hagberg, Silverstein, Wells et al.'s *Work-Related Musculoskeletal Disorders (WMSDs): A Handbook for Prevention,* Taylor & Francis, 1995.

Results on extensive investigations using electromyographic recordings in simulated sedentary work in different postures have been presented in a supplement (No. 19, 1988) of *Scandinavian Journal of Rehabilitation Medicine* by Kristina Schüldt "On Neck Muscle Activity and Load Reduction in Sitting Postures."

Basic biomechanics of the shoulder are included in *Occupational Biomechanics* by Chaffin and Andersson (John Wiley & Sons, Inc., 1991).

46

Hand Tools: Design and Evaluation

Robert G. Radwin
University of Wisconsin–Madison

46.1 Introduction

This chapter describes specific hand tool design features that help minimize physical stress and maximize task performance in jobs involving the continuous or repetitive use of hand tools. An important objective of ergonomics in the design, selection, installation, and use of hand tools is the reduction of muscle fatigue onset and the prevention of musculoskeletal disorders of the upper limb. It is not just the tool design, but how a tool is used for a specific task and workstation that imparts physical stress upon the tool operator. Consequently, there is no "ergonomic hand tool" *per se*. What makes sense in one situation can produce unnecessary stress in another.

It is generally agreed that physical stress, fatigue, and musculoskeletal disorders can be reduced and prevented by selecting the proper tool for the task. Tools used so that physical stress factors are minimized, such as reducing stress concentrations in the fingers and hands, producing low force demands on the operator, or minimizing shock, recoil, and vibration are usually the best tools for the job. Control of these factors depends on the tool and the specific tool application. Selection of tools should, therefore, be viewed within the context of the specific job being performed.

Tool selection should be based on (1) process engineering requirements, (2) human operator limitations, and (3) workstation and task factors. Some factors considered for each of these requirements are summarized in Table 46.1. A detailed description of each of these factors is contained in Radwin and Haney (1996). Manufacturing engineers often specify the process requirements with little regard for the operator and the workstation. Hand tool selection should therefore consider how the particular task and workstation relate to the capabilities and limitations of the human operator for a particular tool design. The process is not always simple and often involves an iterative approach, considering individual tool design features and their role in augmenting and mitigating physical stress. This chapter will describe some hand tool design features and the research leading to an understanding of how tool design can help reduce physical stress in hand tool operation.

0-8493-2641-9/99/$0.00+$.50

TABLE 46.1 Requirements for Ergonomic Hand Tool Selection

Process Engineering Requirements	Requirements specified in terms of the production process, such as how fast a drill bit should turn, or how much torque should be applied to a screw being tightened. Manufacturing process requirements are often based on the product design and parameters needed for accomplishing the task quickly and reliably at the desired level of quality.
Identify Human Operator Limitations	Consider how process and workstation requirements affect the tool operator's ability to perform the task. Human capabilities are limited by strength, fatigue, anthropometry, and manual dexterity.
Workstation and Task Factors	Consider the particular task and workstation where the tool is being used. Requirements may include work location, work orientation, tool shape, tool weight, gloves, frequency of operation, tool accessories, work methods and standards.

46.2 Power Tool Triggers and Grip Force

Extended-length triggers (see Figure 46.1) that distribute force among two or more fingers are often suggested for minimizing stress concentrations at the volar aspects of the fingers (Lindquist et al., 1986; Putz-Anderson, 1988). The rationale is that the force for squeezing the trigger and grasping the handle will be distributed among several fingers to reduce the stress in the index finger. Following is a description of a study that investigated how this particular design feature affects the force in the hands.

In order to directly measure finger and hand force exerted during actual tool operation, an apparatus was constructed for simulating a functioning pistol grip pneumatic nutrunner (Oh and Radwin, 1993). Strain gages were installed in two aluminum bars that were used as the handle for measuring force exerted against the fingers and palm. The instrumented bars were constructed so they were insensitive to the point of force application and linearly summed force applied along the length of the handle (Pronk and Niesing, 1981; Radwin et al., 1991). This was accomplished by measuring shearing stress acting in the cross section of the beam. Strain gages were mounted on a thin web that was machined into the central longitudinal plane and aligned at 45° with respect to the long axis. The effect of bending stresses were completely removed from the strain gages by selecting a measurement point at the neutral axis of the beam, so that all the strain at the measurement point is strictly due to shear stress. Shear strain is totally independent of the point of application.

The strain gage instrumented handle was mounted on a rigid frame and attached perpendicular to a modified in-line pneumatic nutrunner motor in a configuration resembling a pistol-grip power tool (see Figure 46.2). The two dynamometers were mounted in parallel on a track so the handle span could be continuously adjusted. The apparatus was completely functional. The air motor contained an automatic air shut-off torque control mechanism and was operated at a 6.8 Nm target torque setting.

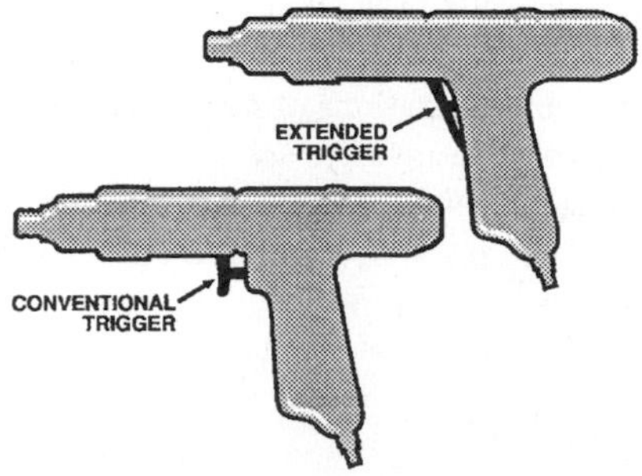

FIGURE 46.1 A conventional trigger and an extended-length trigger on pistol grip power hand tools. (Reprinted with permission from *Human Factors*, 35, 3, 1993. Copyright 1993 by the Human Factors and Ergonomics Society.)

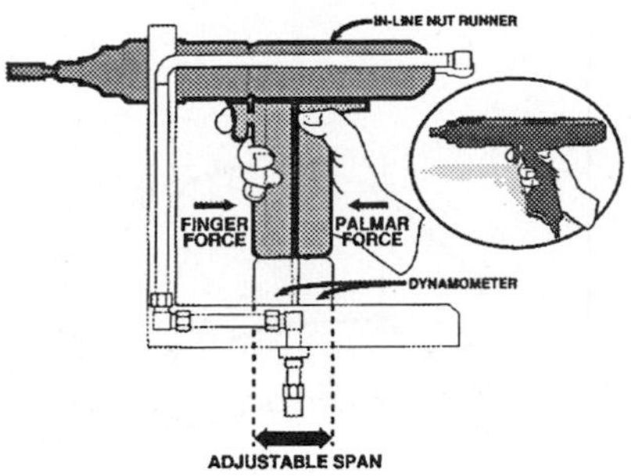

FIGURE 46.2 Dynamometer used for measuring finger and palm forces exerted when operating a completely functional simulation of a pistol-grip pneumatic power hand tool. (Reprinted with permission from *Human Factors*, 35, 3, 1993. Copyright 1993 by the Human Factors and Ergonomics Society.)

Plastic caps were formed and attached to each end of the dynamometer so the contours resembled a power tool handle. The handle circumference was 12 cm for a 4 cm span, measured between two points tangent to the handle contact surfaces. The handle circumference increased an additional 2 cm as the handle span was increased by 1 cm. A trigger was mounted on the finger side cap (see Figure 46.2), and a contact switch was installed inside the trigger. A leaf spring was used for controlling trigger tension. When the trigger was squeezed, the switch tripped a relay and a solenoid valve for supplying air to the pneumatic power tool motor.

Two different trigger types were tested. One was a conventional power tool trigger, activated using only the index finger. The second was longer than the conventional trigger and was activated using both the index and middle fingers (see Figure 46.1). The conventional trigger was 21 mm long and the extended trigger was 48 mm long. The conventional trigger required 8 N, and the extended trigger required 11 N for activation.

Use of the extended trigger was found beneficial for reducing grip force and exertion levels during tool operation. Average peak finger and palm forces were, respectively, 9% and 8% less for the extended trigger than for the conventional trigger. Eleven of eighteen subjects (61%) indicated that they preferred using the handle with the extended trigger after just an hour of use in the laboratory. The average finger and palmar holding force was 65% and 48%, respectively, less for the extended trigger, than for the conventional trigger. Since subjects spent 65% to 76% of the operating time holding the tool, using an extended trigger may have an important effect on reducing exposure to forceful exertions in the hand during power hand tool operation.

46.3 Handle Size and Grip Force

Research on handle design has typically focused on finding the optimal handle dimensions. Grip strength is affected through the biomechanics of grip from the relative position of the joints of the hand and by the position and length of the muscles involved. Consequently, grip strength is affected by the handle size. Recommendations for handle size are usually based on the span that maximizes grip strength, or the span that minimizes fatigue.

Hertzberg (1955), in an early Air Force study, reported that a handle span of 6.4 cm maximized power grip strength. Greenberg and Chaffin (1975) recommended that a tool handle span should be in the range between 6.4 cm and 8.9 cm in order to achieve high grip forces. Ayoub and Lo Presti (1971) found that a 3.8 cm diameter was optimum for a cylindrical handle. This was based on maximizing the ratio between strength and EMG activity, and on the number of work cycles before onset of fatigue. Another study by Petrofsky et al. (1980) showed that the greatest grip strength occurred at a handle span between 5 cm and 6 cm.

Grip strength is affected by hand size. Fitzhugh (1973) showed that the handle span resulting in maximum grip strength for a 95 percentile male hand length is larger than the handle span for a 50

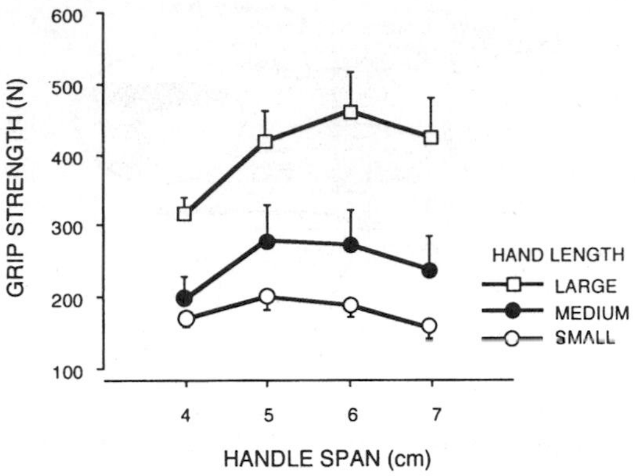

FIGURE 46.3 Average grip strength plotted against handle span for three hand size categories. Error bars represent standard error of the mean. (Reprinted with permission from *Human Factors,* 35, 3, 1993. Copyright 1993 by the Human Factors and Ergonomics Society.)

percentile female. Consequently, a person with a small hand might benefit from using a smaller handle, and a person with a large hand might benefit from using a larger one.

Grip strength data often used for handle design are based on population measurements made using instruments like the Jamar or Smedley dynamometers (Schmidt and Toews, 1970; Young et al., 1989) rather than using handle dimensions representative of an actual tool. In most cases, only one dimension (handle span) has been controlled, while the other handle dimensions were not necessarily similar to a tool handle.

A power hand tool manufacturer considered offering a power hand tool that provided a handle that was adjustable in size. An investigation of grip strength using handle dimensions similar to power hand tool handles was conducted in order to explore the differences against published grip strength data (Oh and Radwin, 1993). Hand length up to 17 cm was classified as small, between 17 cm and 19 cm as medium, and greater than 19 cm as large. Average grip strength is plotted against handle span and hand size in Figure 46.3. Grip strength increased as hand length increased. Large hand subjects produced their maximum grip strength (mean = 463 N, SD = 128 N) for a handle span of 6 cm, while medium hand (mean = 280 N, SD = 122 N) and small hand (mean = 203 N, SD = 51 N) subjects produced their maximum strength for a handle span of 5 cm.

The span resulting in maximum grip strength agreed with the findings of previous strength studies. Hertzberg (1955) found that subjects exerted more force at a 6.4 cm span than among 3.8 cm, 6.4 cm, 10.2 cm, and 12.7 cm handle spans. Petrofsky et al. (1980) reported that on the average, subjects produced maximum grip force for a handle span between 5 cm and 6 cm.

Although the span resulting in maximum grip strength and the grip strength function agreed with previous findings, the maximum grip strength for both student and industrial worker subjects was markedly less than what has been previously reported in the literature. Schmidt and Toews (1970) collected grip strength data from 1,128 male and 80 female Kaiser Steel Corporation employee applicants, using a Jamar dynamometer. They reported for a handle span of 3.8 cm, an average of 499 N for the dominant male hand and 308 N for the dominant female hand. Swanson et al. (1970) measured the grip strength of 50 females and 50 males using a Jamar dynamometer. Among these subjects, 36 were light manual workers, 16 were sedentary workers, and 48 were manual workers. They reported for a handle span of 6.4 cm, 467 N for the male dominant hand and 241 N for the female dominant hand. These all exceeded the strength levels observed (see Figure 46.3).

A major difference between grip strength measured for tool handles by Oh and Radwin (1993) and previously reported strength data is in the handle dimensions. The Jamar and Smedely dynamometers have smaller circumferences and narrower widths than the tool handle used in this study. The tool handle curvature was also straight while the Jamar dynamometer has a curved surface at the grip center. The handle used in this study closely represented an actual tool handle in circumference and width. These

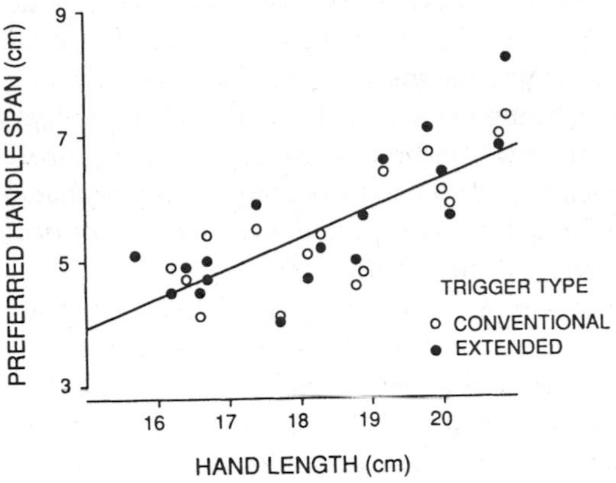

FIGURE 46.4 Preferred handle span plotted against hand length. (Reprinted with permission from *Human Factors,* 35, 3, 1993. Copyright 1993 by the Human Factors and Ergonomics Society.)

size and curvature differences can affect the position of the fingers and grip posture. These dimensional differences must be considered when designing handles based on strength using published grip strength data.

The investigation also found a difference in grip strength between student subjects and industrial workers. Grip strength, averaged over handle span, was 279 N (SD = 133 N) for the students and 327 N (SD = 90 N) for the workers. No significant grip strength differences, however, between the student and worker groups was observed within each hand size.

The underlying assumption in designing handles based on maximum strength is that the actual force exerted is independent of handle size. Exertion level is the ratio of the actual grip force used, to the maximum voluntary force generating capacity. If the grip force used during tool operation is the same for all handle sizes, then the handle span associated with the greatest grip strength should result in the lowest exertion level. If grip force, however is affected by handle size, then the handle span associated with the greatest grip strength may not be the handle span resulting in the minimum exertion level.

A series of experiments were performed using the pistol grip power hand tool with strain gage instrumented handles and an adjustable handle span as described above. Handle span affected peak finger and palmar force. Peak finger force increased 24% for a student subject group, and 30% for an industrial worker group, as handle span increased from 4 cm to 7 cm. Similarly, peak palmar force increased 21% for the student group and 22% for the worker group, as handle span increased from 4 cm to 7 cm. Handle span also influenced finger and palmar holding forces. Finger holding force increased 20%, and palmar holding force increased 16%, as handle span increased from 4 cm to 7 cm for the student subjects.

The study found that hand size was proportional to the handle span operators preferred when offered the opportunity to adjust the handle size to any size they desired. Operators with larger hand sizes reported they preferred using a tool with a larger handle. Preferred handle span is plotted against hand length in Figure 46.4 for both trigger types. There was no difference between the preferred handle span for the conventional trigger and the preferred handle span for the extended trigger. No anthropometric measurements, however, were related to the span resulting in the minimum peak exertion level. Exertion level when holding the tool was less for the large size hands than for the small size hands. Holding exertion level for the large hands was maximum for the 4 cm handle span, while holding exertion level for the small hands was maximum for the 7 cm handle span. In addition, the tendency for large hand subjects to prefer larger handle spans suggests that selectable size handles may be more desirable than having only a single size handle for power hand tools.

46.4 Static Hand Force

Safe power hand tool operation requires that an operator possess the ability to adequately support the tool in a particular position, apply the necessary forces, while reacting against the forces generated by the tool. Force demands that exceed an operator's strength capabilities can cause loss of control, resulting in an accident or an injury. Design and selection of power hand tools that minimize static grip and hand force will help reduce muscle fatigue and prevent upper limb disorders.

The force necessary for supporting a power hand tool depends on the tool weight, its center of gravity, the length of the tool, and air hose attachments. Power hand tools should be well balanced with all attachments installed. As a general rule, a hand tool center of gravity should be aligned with the center of the grasping hand so the hand does not have to overcome moments that cause the tool to rotate the operator's wrist and arm (Greenberg and Chaffin, 1977).

Psychophysical experiments have provided some insight into the load that power tool operators prefer. When experienced hand tool operators were asked to rate the mass of the power tools they operated, tools weighing 0.9 to 1.75 kg mass were rated "just right" (Armstrong et al. 1989). Other psychophysical experiments showed that perceived exertion for a tool mass of 1 kg was significantly less than for tools with a mass of 2 kg and 3 kg (Ulin and Armstrong, 1992).

There is a tradeoff between selecting a light tool and the benefit of the added weight for performing operations that require high feed force. The power available for a grinding task increases with increasing mass of the grinder. Reducing the weight of the grinder can increase the feed force the operator must provide and may increase the amount of time necessary for accomplishing the task, consequently subjecting the operator to more stressful work and greater vibration exposure. Heavy grinding tasks should be performed on horizontal surfaces so the weight of the tool does not have to be supported by the operator. Heavy power tools should be suspended using counterbalancing accessories.

In addition to supporting the tool load, power hand tool operators often have to exert push or feed force, or act against reaction forces. Feed force is necessary for starting a threaded fastener, advancing a bit or keeping a bit or socket engaged during the securing cycle. Feed force is affected by the work material and design of the tool, bit, or fastener. Large feed forces are sometimes needed when operating power tools such as drills and screwdrivers. Repetitive or sustained exertions associated with these operations should be minimized. Drill feed force is affected by the drill power and speed, bit type, material, and diameter of the hole drilled. Power screwdriver feed force may be affected by the fastener head and screw tip used. Feed force for a slotted or Phillips head screw generally requires more feed force than for a torx head screw. Self-tapping screws require more force than screws tightened through pre-tapped holes. Material hardness is also a factor for self-tapping screws and drilling. Feed force requirements also increase as torque level increases for cross recess screws.

Power hand tools such as screwdrivers or nutrunners, used for tightening threaded fasteners, are commonly configured as (1) in-line, (2) pistol grip, and (3) right angle. A mechanical model of a nutrunner was developed for static equilibrium (no movement) conditions (Radwin et al., 1995). Hand force, reaction force from the workpiece, tool orientation, weight, and output torque were included in this model. This chapter will describe the model developed for pistol grip nutrunners.

The model uses a Cartesian coordinate system relative to the orientation of the handle grasped in the hand using a power grip. This coordinate system has the x-axis perpendicular to the axial direction of the handle; the y-axis is parallel to the long axis of the handle; and the z-axis is parallel to the tool spindle. The origin is the end of the tool bit or socket. Hand forces are described in relation to these coordinate axes. To simplify the model, an initial assumption is that orthogonal forces can be applied along the handle without producing coupling moments. This assumption allows force to be considered as having a single point of application. The resultant hand force $\mathbf{F}_H$ at the grip center is the vector sum of the three orthogonal force components

$$\mathbf{F}_H = F_{H_x}\mathbf{i} + F_{H_y}\mathbf{j} + F_{H_z}\mathbf{k} \tag{1}$$

where the hand force magnitude is:

$$|\mathbf{F}_H| = \sqrt{{F_{H_x}}^2 + {F_{H_y}}^2 + {F_{H_z}}^2} \tag{2}$$

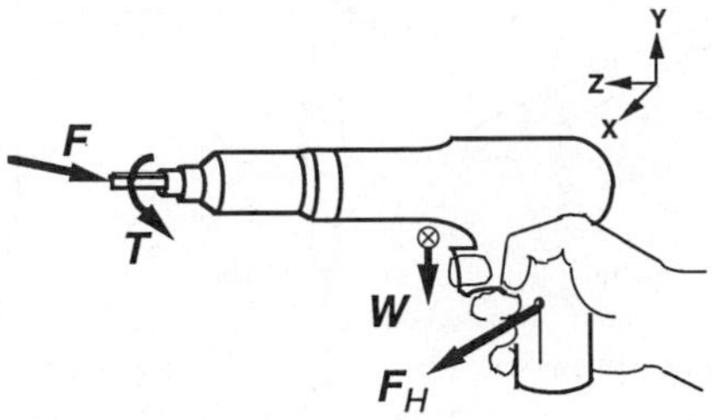

FIGURE 46.5 Free body diagram and orthogonal force components considered in the pistol grip force model.

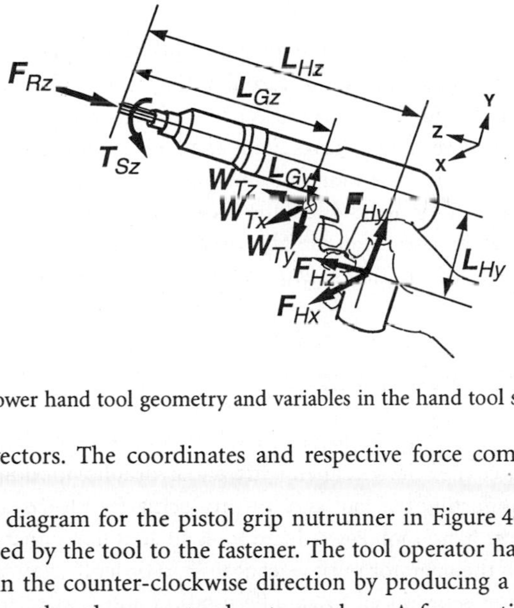

FIGURE 46.6 Power hand tool geometry and variables in the hand tool static force model.

and **i**, **j**, **k** are the unit vectors. The coordinates and respective force components are illustrated in Figure 46.5.

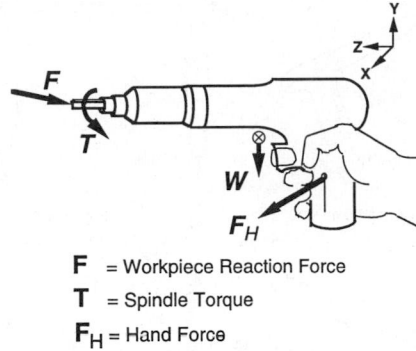

Consider the free-body diagram for the pistol grip nutrunner in Figure 46.6. The torque, T_S acts in reaction to torque T, applied by the tool to the fastener. The tool operator has to oppose this equal and opposite reaction torque in the counter-clockwise direction by producing a reaction force F_{Hx}. That is not the only force, however, that the operator has to produce. A force acting in the z direction, F_{Hz}, provides feed force and produces an equal and opposite reaction force, F_{Rz}. In addition, the operator has to react against the tool weight in order to support and position the tool by providing a vertical force component, F_{Hy}. The tool weight, W_T and push force, F_{Hz} tend to produce a clockwise moment about the tool spindle in the yz-plane which is countered by this vertical support force.

When a body is in static equilibrium, the sum of the external forces and the sum of the moments are equal to zero. Using that relationship, the following system of equations was developed for the pistol-grip nutrunner to describe these static forces:

$$\begin{bmatrix} F_{Hx} \\ F_{Hy} \\ F_{Hz} \end{bmatrix} = \begin{bmatrix} -L_{Gz}/L_{Hz} & 0 & 0 & -1/L_{Hz} & 0 \\ 0 & -L_{Gz}/L_{Hz} & \left(-L_{Gy}-L_{Hy}\right)/L_{Hz} & 0 & -L_{Gy}/L_{Hz} \\ 0 & 0 & -1 & 0 & -1 \end{bmatrix} \cdot \begin{bmatrix} W_{Tx} \\ W_{Ty} \\ W_{Tz} \\ T_{Sz} \\ F_{Rz} \end{bmatrix} \tag{3}$$

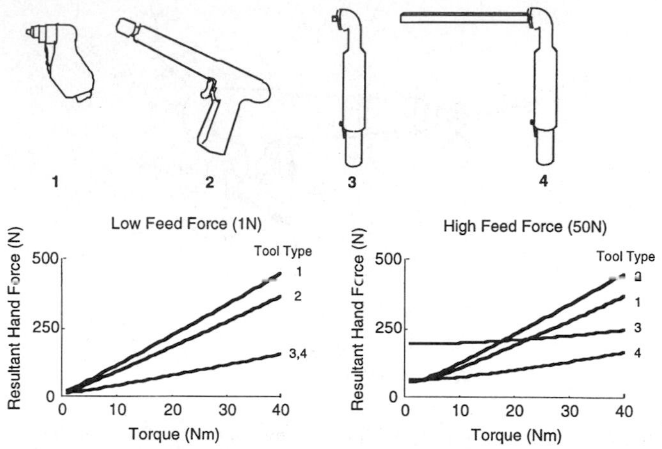

FIGURE 46.7 Comparison of predicted hand forces for four different tool configurations performing the same task for one-hand tool operation.

Assuming one-hand operation, resultant hand force magnitude was predicted using the model for the four different tools and plotted as a function of torque in Figure 46.7. Hand force was determined for both low feed force (1 N) and high feed force (50 N) conditions, when operating these tools against a vertical surface. When feed force was small, the resultant hand force was mostly affected by torque reaction force, which increased as torque increased for all four tools. Since the greatest force component in this case was torque reaction force, Tools 3 and 4 had the least resultant hand force since they both had the longest handles. Tool 3, however, had a considerably greater resultant hand force when feed force was high. This effect was not observed for Tool 4, which also had a similar handle, but contained a spindle extension shaft.

46.5 Dynamic Reaction Force

Whereas manual hand tools rely on the human operator for generating forces, power hand tools operate from an external energy source (i.e., electric, pneumatic, and hydraulic) for doing work. The tool operator provides static force for supporting the tool and for producing feed force, and must react against the forces generated by the power hand tool. Power hand tools such as nutrunners produce rapidly building torque reaction forces which the operator must react against in order to maintain full control of the tool.

Nutrunner reaction torque is produced by spindle rotation and is affected primarily by the spindle torque output and tool size. Nutrunner spindle torque can range from less than 0.8 Newton-meters (Nm) to more than 700 Nm. This torque is transmitted to the operator as a reaction force through the moment arm created by the tool and tool handle. A tool operator opposes reaction torque while supporting the tool and preventing it from losing control.

The three major operating modes for nutrunners include (1) mechanical clutch, (2) stall, or (3) automatic shut-off. When a stall tool is used, maximum reaction torque time is directly under operator control by releasing the throttle, which can last as long as several seconds. Stall tools tend to expose an operator to reaction torque the longest. Although clutch tools limit reaction torque exposure, ratcheting clutch tools can expose workers to significant levels of vibration if used frequently (Radwin and Armstrong, 1985). The speed of the shut-off mechanism controls exposure to peak reaction force for automatic shut-off tools. Consequently, automatic shut-off tools have the shortest torque reaction time because these tools cease operating immediately after the desired peak torque is achieved.

As torque is applied to a threaded fastener, it rotates at a relatively low spindle torque until the clamped pieces come into intimate contact. This torque can approach zero with free running nuts or can be rather significant as in the case where locking nuts, thread interference bolts, or thread-forming type fasteners are used. After the fastener brings the clamped members of the joint into initial intimate contact, it continues to draw the parts together until they form a solid joint. When the joint becomes solid, continued turning of the nut results in a proportionally increasing torque. This is the elastic portion of the cycle and is the time when reaction torque forces are produced. Torque build-up, and consequently torque reaction force, continues rising at a fixed rate until peak torque is achieved, which is the clamping force of the joint. Forearm muscle reflex responses when operating automatic air shut-off right angle nutrunners during the torque-reaction phase was more than four times greater than the muscle activity used for holding the tool and two times greater than the run-down phase (Radwin et al., 1989). Flexor EMG activity during the torque–reaction phase increased for tools having increasing peak spindle torque.

Threaded fastener joints are classified as "soft" or "hard" depending on the relationship between torque build-up and spindle angle. The International Standardization Organization (ISO) specifies that a hard joint has an angular displacement less than 30 degrees when torque increases from 50% to 100% of target torque, and a soft joint has an angular displacement greater than 360 degrees (ISO-6544).

Nutrunner torque reaction force is a function of several factors including target torque, spindle speed, joint hardness, and torque build-up time. Some of these factors are interdependent. Faster spindle speed results in shorter torque build-up time, and softer joints are related to longer build-up times. The duration of exertion is directly related to torque build-up time rather than just the speed of the tool or joint hardness.

Studies have shown that torque build-up time as well as the magnitude of torque reaction force has a significant influence on human operators during power nutrunner use. Kihlberg et al. (1995) studied right angle nutrunners having different shut-off mechanisms (fast, slow, and delayed) and found a strong correlation between perceived discomfort, handle displacement, and reaction forces. Radwin et al. (1989) investigated the effects of target torque and torque build-up time using right-angle pneumatic nutrunners and found that average flexor rms electromyography (EMG) activity scaled for grip force increased from 372 N for a low target torque (30 Nm) to 449 N for a high target torque (100 Nm), and that average grip force was 390 N for a long build-up time (2 s), and increased to 440 N for a medium build-up time (0.5 s). They also reported that EMG latency between tool torque onset and peak flexor rms EMG for the long torque build-up time (2 s) was 294 ms and decreased to 161 ms for the short build-up time (0.5 s). The findings suggested that torque reaction force can affect extrinsic hand muscles in the forearm, and hence grip exertions, by way of a reflex response. Johnson and Childress (1988) showed that low torque was associated with less muscular activity and reduced subjective evaluations of exertion.

Representative torque reaction force, handle kinematics, and EMG muscle activity are illustrated in Figure 46.8. Since torque builds up in a clockwise direction, the reaction torque has a tendency to rotate the tool counterclockwise with respect to the operator. When the operator has sufficient strength to react against the reaction torque, the tool remains stationary or rotates clockwise and the operator exerts concentric muscle contractions against the tool (positive work). However, when the tool overpowers the operator, it tends to move in a counterclockwise direction and the operator exerts eccentric muscle contractions against the tool (negative work). Therefore, measures of handle movement that occur (handle velocity and displacement) and the direction of rotation can indicate relative tool controllability. Handle movement direction was defined as positive when the handle moved in the direction of tool reaction torque (see Figure 46.1). If handle velocity increases after shutoff, it means that the tool and hand are unstable. The work done on the tool–hand system and the power involved in doing work during torque build-up were also assessed. If the operator has the capacity to successfully react against the torque build-up (positive work), then the tool is considered stable. This occurs when handle displacement and velocity were less than zero. If the handle become unstable and the net handle displacement occurs in the direction of torque reaction away from the operator, then work and power are negative.

A computer-controlled right angle nutrunner was used to study power hand tool reaction forces (Oh and Radwin, 1997). A torque transducer and an angle encoder were integrated into the tool spindle head which outputted analog torque and digital angular rotation signals. A threaded fastener joint simulator that could be oriented horizontally or vertically was mounted on a height adjustable platform. The longitudinal axis of the joint head was oriented perpendicular to the ground for the horizontal workstation setting, and oriented parallel to the ground for the vertical workstation setting.

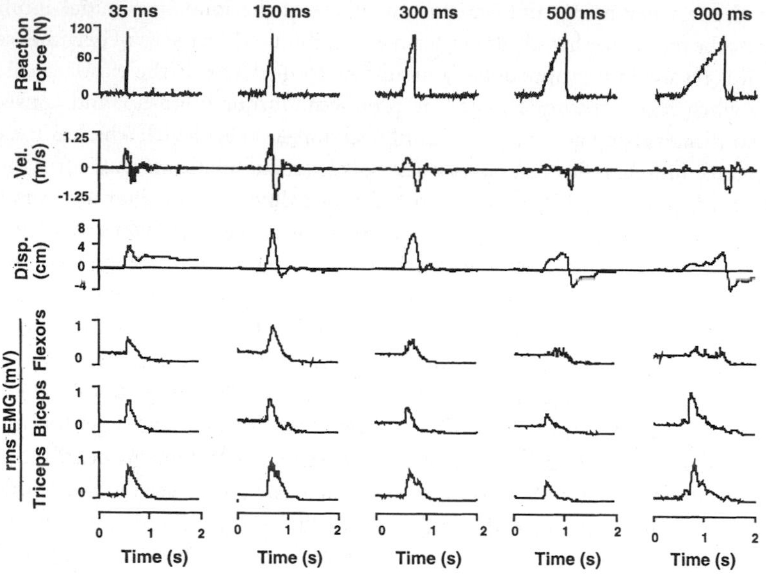

FIGURE 46.8 Representative torque reaction force, handle kinematics and EMG muscle activity for different torque build-up times.

The study showed that workstation orientation and tool dynamics (torque reaction force and torque build-up time) influenced operator muscular exertion and handle stability. In general, handle instability increased when the tool was operated on a vertical workstation (rather than a horizontal workstation), when torque reaction force was high (88.3 and 114.6 N), and for a 150 ms torque build-up time, regardless of torque reaction force.

As torque reaction force increased from 52.1 N to 114.6 N, peak hand velocity 89%, and peak hand displacement increased 113%. Peak hand velocity was greatest for a 150 ms build-up time and the least for a 900 ms build-up time. The effect of target torque was consistent with previous studies that showed that target torque was related to muscular exertion, subjective perceived exertion, and handle instability (Johnson and Childress, 1988; Lindqvist, 1993; Oh and Radwin, 1994; Radwin et al., 1989). As torque reaction force increased from 52.1 N to 114.6 N, the magnitude of negative work increased by 35%, and the magnitude of average power against the operator increased by 30%. Under these conditions, perceived exertion also increased from 2.7 to 4.3 (as rated on Borg's 10-point scale), and task acceptance rate decreased from 73% to 28%. When the tool was operated on a horizontal workstation, average finger flexor EMG was significantly influenced by torque reaction force. As torque reaction force increased from 52.1 N to 114.6 N, the average flexor EMG increased by 14%.

The effect of torque build-up time on power hand tool operators has been studied in terms of perceived exertion, muscular activity, and handle stability (Armstrong et al., 1994; Freivalds and Eklund, 1991; Lindqvist et al., 1986; Oh and Radwin, 1994; Radwin et al., 1989). Torque build-up time is a concern because it is directly related to assembly time and exertion duration. Increased duration may lead to earlier fatigue onset. Although longer build-up time results in longer duration exertions and increases the operation cycle time, it may provide an opportunity for better tool control since it gives the operator a longer time to react.

Peak hand velocity was 46.7% less for horizontal workstations (mean = 0.46 m/s, SD = 0.26 m/s) than for vertical workstations (mean = 0.67 m/s, SD = 0.34 m/s). A similar trend was observed for peak hand displacement. Peak hand displacement for horizontal workstations (mean = 4.0 cm, SD = 2.2 cm) was 90.2% less than peak hand displacement for vertical workstations (mean = 7.6 cm, SD = 4.6 cm). Previous findings agree that a horizontal workstation is preferable for right angle tool use. Ulin et al. (1992) showed that average subjective ratings of perceived exertion were significantly less when the tool was operated on horizontal workstations rather than vertical workstations. Also, 88% more negative work and 58% more power against the operator were recorded while the tool was operated on a vertical workstation. However, subjective ratings of perceived exertion and task acceptance rates did not differ between horizontal and vertical workstations. This might come from the fact that the torque levels in the current study were much greater than the torque level used for the Ulin et al. study (1992).

Although perceived exertion was less and task acceptance rate was greater for a 35 ms build-up time than for longer build-up times, the operator might not have sufficient time to voluntarily react against torque build-up with the 35 ms build-up time. On the average, the onset of the EMG burst occurred 40 ms after the onset of torque build-up for the 35 ms build-up time. This indicated that the muscles were not activated until a significant amount of torque had built up for the 35 ms build-up time. Lack of muscular contraction during torque build-up might explain why the peak handle velocity was higher for short build-up times. Without muscular contractions, the inertia of the tool and hand had to absorb all of the reaction force. Short exertion duration and lack of muscular contractions due to EMG latencies might contribute to lower subjective ratings of perceived exertion for the 35 ms build-up.

The larger torque variance that occurred for the 35 ms build-up time indicated that even though subjective perceived exertion was less, this condition might result in more target torque error. Also, the probability of increased handle instability after shutoff was significantly greater for the 35 ms build-up time. This suggests that even after shutoff, operators did not have sufficient capacity to control the tool reaction torque. Therefore, the 35 ms build-up time increased handle stability in terms of peak handle displacement and negative work, and reduced subjective perceived exertion, however, the lack of muscular contraction during torque build-up reduced tightening quality.

Methods for limiting reaction force include (1) use of torque reaction bars, (2) installing torque absorbing suspension balancers, (3) providing tool mounted nut holding devices, and (4) using tool support reaction arms. A torque reaction bar sometimes can be used to transfer loads back to the work piece. Tools that can be equipped with a stationary reaction bar adapted to a specific operation so reaction force can be absorbed by a convenient solid object can completely eliminate reaction torque from the operator's hand. These bars can be installed on in-line and pistol-grip tools. Right angle tools can react against a solid object instead of relying on the hand and arm. Reaction devices (1) remove reaction forces from the operator, (2) permit pistol-grip and in-line reaction bar tools to be operated using two hands, (3) free the operator from restricting postures, (4) provide weight improvements over right angle nutrunners, and (5) improve tool fastening performance. The disadvantages are that reaction bars must be custom made for each operation, and the combination of several attachments for one tool can be difficult Torque reaction bars may also add weight to the tool and can make the tool more cumbersome to handle.

46.6 Vibration

Vibration can be a by-product of power hand tool operation, or it can even be the desired action as is the case with abrasive tools like sanders or grinders. Vibration levels depend on tool size, weight, method of propulsion, and the tool drive mechanism. It is affected by work material properties, disk abrasives, and abrasive surface area. Continuous vibration is inherent in reciprocating and rotary power tools. Impulsive vibration is produced by tools operating by shock and impact action, such as impact wrenches or chippers. The tool power source, such as air power, electricity, or hydraulics can also affect vibration. Vibration is also generated at the tool-material interface by cutting, grinding, drilling, or other actions.

Pneumatic hammer recoil was observed producing a stretch reflex and muscular contractions in the elbow and wrist flexors (Carlsöö and Mayr, 1974). Studies of the short-term neuromuscular effects of hand tool vibration have demonstrated that hand tool vibration can introduce disturbances in neuromuscular force control resulting in excessive grip exertions when holding a vibrating handle (Radwin et al., 1987). The results of these studies demonstrated that grip exertions increased with tool vibration. Average grip force increased for low frequencies (40 Hz) vibration but did not change for higher frequencies (160 Hz) vibration. Since forceful exertions are a commonly cited factor for chronic upper extremity muscle, tendon, and nerve disorders, vibrating hand tool operation may increase the risk of CTDs through increased grip force.

Vibration has also been shown to produce temporary sensory impairments (Streeter, 1970; Radwin et al., 1989). Recovery is exponential and can require more than 20 minutes (Kume et al., 1984). Workers often sand or grind surfaces and periodically inspect their work using tactile inspection to determine if the surface was sanded to the desired level of smoothness. Diminished tactility may result in a surface feeling smoother than it actually is, resulting in a rougher surface than is actually desired.

Vibration has not been shown to be significantly reduced by using resilient mounts on handles. Vibration isolation techniques have been generally unsuccessful for limiting vibration transmission from power tools to the hands and arms. Isolation has been particularly difficult for vibration frequencies less than 100 Hz. This is because attenuation only occurs when the vibration spectrum falls above the resonant frequency of the isolation system or material. When the vibration frequency is less than the resonant frequency of the isolating material, the handle acts as a rigid body and no vibration is attenuated. Grinding tools typically run at speeds near 6000 rpm (100 Hz) making it difficult to have a resilient vibration isolating handle. Furthermore, if the vibration frequency is approximately equivalent to the isolator resonant frequency, the system will actually intensify vibration levels. Weaker suspension systems have lower resonant frequencies, but are often impractical because such a system is usually too flexible for the heavily loaded handles of tools like grinders. Handles loaded with high forces must be very rigid.

References

Armstrong, T.J., Bir, C., Finsen, L., Foulke, J., Martin, B., Sjøgaard, G., and Tseng, K. 1994. Muscle responses to torques of hand held power tools, *Journal of Biomechanics,* 26(6): 711-718.

Ayoub, M.M. and Lo Presti, P. 1971. The determination of an optimum size cylindrical handle by use of electromyography. *Ergonomics,* 14(4): 509-518.

Fitzhugh, F.E. 1973. *Dynamic aspects of grip strength.* (Tech Report). Department of Industrial & Operations Engineering, Ann Arbor: The University of Michigan.

Freivalds, A. and Eklund, J. 1991. Subjective ratings of stress levels while using powered nutrunners, in W. Karwowski and J.W. Yates (Eds.), *Advances in Industrial Ergonomics and Safety III,* New York, Taylor & Francis, 379-386.

Greenberg, L. and Chaffin, D.B. 1975. *Workers and Their Tools: A Guide to the Ergonomic Design of Hand Tools and Small Presses.* Midland, MI: Pendell.

Hertzberg, H.T.E. 1955. Some contributions of applied physical anthropology to human engineering. *Annals of NY Academy of Science, 63*(4): 616-629.

International Organization for Standardization 1981. *Hand-held Pneumatic Assembly Tools for Installing Threaded Fasteners — Reaction Torque Reaction Force and Torque Reaction Force Impulse Measurements.* ISO-6544.

Johnson, S.L. and Childress, L.J. 1988. Powered screwdriver design and use: tool, task, and operator effects, *International Journal of Industrial Ergonomics,* 2: 183-191.

Kihlberg, S., Kjellberg, A., and Lindbeck, L. 1995. Discomfort from pneumatic tool torque reaction force reaction: acceptability limits, *International Journal of Industrial Ergonomics,* 15: 417-426.

Lindqvist, B. 1993. Torque reaction force reaction in angled nutrunners, *Applied Ergonomics,* 24(3): 174-180.

Lindquist, B., Ahlberg, E., and Skogsberg, L. 1986. *Ergonomic Tools in Our Time.* Atlas Copco Tools, Stockholm.

Oh, S. and Radwin, R.G. 1993. Pistol grip power tool handle and trigger size effects on grip exertions and operator preference, *Human Factors,* 35(3): 551-569.

Oh, S. and Radwin, R.G., 1994, Dynamics of power hand tools on operator hand and arm stability, in *Proceedings of the Human Factors and Ergonomics Society 38th Annual Meeting,* 602-606, Santa Monica, CA: Human Factors and Ergonomics Society.

Oh, S. and Radwin, R.G. 1998. The influence of target torque and torque build-up time on physical stress in right angle nutrunner operation, *Ergonomics,* 41(2): 188-206.

Petrofsky, J.S., Williams, C., Kamen, G., and Lind, A.R. 1980. The effect of handgrip span on isometric exercise performance, *Ergonomics, 23*(12): 1129-1135.

Pronk, C.N.A. and Niesing, R. 1981. Measuring hand grip force using an application of strain gages, *Medical, Biological Engineering and Computing,* 19:127-128.

Putz-Anderson, V. 1988. *Cumulative Trauma Disorders.* New York: Taylor & Francis.

Radwin, R.G., Masters, G., and Lupton, F.W. 1991. A linear force summing hand dynamometer independent of point of application, *Applied Ergonomics,* 22(5): 339-345, 1991.

Radwin, R.G. and Haney, J.T. 1996. *An Ergonomics Guide to Hand Tools,* Fairfax, VA: American Industrial Hygiene Association.

Radwin, R.G., VanBergeijk, E., and Armstrong, T.J. 1989. Muscle response to pneumatic hand tool torque reaction force reaction forces, *Ergonomics,* 32(6): 655-673.

Radwin, R.G., Oh, S., and Fronczak. 1995. A mechanical model of hand force in power hand tool operation, *Proceedings of the Human Factors and Ergonomics Society 39th Annual Meeting,* Santa Monica: Human Factors and Ergonomics Society: 348-352.

Schmidt, R.T. and Toews, J.V. 1970. Grip strength as measured by the Jamar dynamometer, *Archives of Physical Medicine & Rehabilitation, 51*(6): 321-327.

Swanson, A.B., Matev, I.B., and Groot, G. 1970. The strength of the hand, *Bulletin of Prosthetics Research,* Fall: 145-153.

Ulin, S.S., Snook, S.H., Armstrong, T.J., and Herrin, G.D. 1992. Preferred tool shapes for various horizontal and vertical work locations, *Applied Occupational and Environmental Hygiene,* 7(5): 327-337.

Young, V.L., Pin, P., Kraemer, B.A., Gould, R.B., Nemergut, L., and Pellowski, M. 1989. Fluctuation in grip and pinch strength among normal subjects, *The Journal of Hand Surgery,* 14A(1): 125-129.

47 Gloves

Ram Bishu
University of Nebraska–Lincoln

A. Muralidhar
University of Nebraska–Lincoln

47.1 Importance of the Hand

The hand is probably the most complex of all anatomical structures in the human body. Along with the brain, it is the most important organ for accomplishing the tasks of exploration, prehension, perception, and manipulation, unique to humans. The importance of the hand to human culture is emphasized by its depiction in art and sculpture, its reference frequency in vocabulary and phraseology, and its importance in communication and expression (Chao et al., 1989). The human hand is distinguished from that of the primates by the presence of a strong opposable thumb, which enables humans to accomplish tasks requiring precision and fine control. The hand provides humans with both mechanical and sensory capabilities.

47.2 Prehensile Capabilities of the Hand

Napier (1956) divides hand movements into two main groups — prehensile movements, in which an object is seized and held partly or wholly within the compass of the hand, and nonprehensile movements, where no grasping and seizing is involved but by which objects can be manipulated by pushing or lifting motions of the hand as a whole or of the digits individually.

Landsmeer (1962) further classifies human grasping capabilities as *power grip,* where a dynamic initial phase can be distinguished from a static terminal phase, and *precision handling,* where there is no static terminal phase. The dynamic phase as defined by Landsmeer includes the opening of the hand, positioning of the fingers, and the grasping of the object. Westling and Johansson (1984) state that the factors that influence force control during precision grip are friction, weight, and a safety margin factor related to the individual subject. They also found that in multiple trials, the frictional conditions during a previous trial could affect the grip force. They also showed that the grip employed when holding small objects stationary in space was critically balanced such that neither accidental slipping between the skin

0-8493-2641-9/99/$0.00+$.50

TABLE 47.1 Comparison of Bare Hand–Gloved Hand Capabilities

Indices	Bare Hand	Gloved Hand
Thermal Tolerance	Poor	Good
Tactile Perception	Excellent	Poor
Grip Strength	Good	Reduced
Range of Motion	Excellent	Poor
Manipulative Ability	Excellent	Reduced
Prehension	Excellent	Poor
Torque Capability	Poor	Improved
Vibration Tolerance	Poor	Good
Dexterity	Excellent	Reduced
Chemical Resistance	Poor	Excellent
Electrical Energy	Poor	Excellent
Radiation (all kinds)	Poor	Excellent
Biohazard Risk	Poor	Excellent
Abrasive Trauma	Poor	Improved

and the object occurred, nor did the grip force reach exceedingly high values. This sense of critical balance as to the amount of force applied while gripping is important, as too firm a grip could result in the destruction of a fragile object, causing possible injury to the hand, or lead to muscle fatigue and interfere with further manipulative activity imposed upon the hand. Sensory perception in the hand is due to the presence of mechanoreceptors distributed all over the palmar area, especially at the tips of the fingers.

Thus, feedback from the hand is a critical component of the gripping task enabling the amount of force to be controlled. Anything that blocks the transmission of impulses from the hand interferes with the feedback cycle and affects grip force control. Gloves do affect the feedback cycle.

47.3 Need for Protection of the Hand

The hand, which provides humans with both mechanical and sensory capabilities, needs to be protected from the environment. Protection is needed from mechanical trauma (abrasions, cuts, pinches, punctures, crush injuries), thermal extremes (heat and cold), radiation (nuclear, ultraviolet, X-ray, and thermal), chemical hazards, blood-borne pathogens, electrical energy, and vibration.

There exist several forms of hand protection, which can be used as stand-alone protection, or in combination with other personal protective equipment. The commonly available hand protection are gloves, mittens, finger cots, and gauntlets made of several materials such as leather, cotton, rubber, nylon, latex, metal, and in combinations of the same, to provide maximum protection against the specific condition being guarded against. The use of gloves, although a necessity in many workplaces, has some associated disadvantages. Gloves have been found to affect hand performance adversely, and the performance parameters affected are dexterity, task time, grip strength, and range of motion. Table 47.1 provides a summary of bare hand and gloved hand capabilities.

Facilitation of these activities, with simultaneous protection from the hazards of the work environment, are often conflicting objectives of glove design. The conflicts associated with providing primary hand protection through the use of a glove while permitting adequate hand functioning has been widely recognized.

It will be relevant to give a brief description of a variety of gloves that are available today. It will also be relevant to discuss performance effects of gloves, before detailing the challenges of glove design.

47.4 Types of Gloves

There are a wide variety of gloves available today. Starting from a garden glove at 50 cents a pair from the local grocery store to the custom-fit shuttle gloves donned by astronauts for extra vehicular activities

(EVA), which cost a few hundred thousand dollars a pair, the variety among gloves can be so overwhelming as to defy easy categorization. Gloves can be categorized along a number of dimensions, such as materials, design, and location of use. According to the National Safety Council (1975, 1976), hand protection can be job-rated or general purpose. Job-rated hand protection is designed to protect against the hazards of specific operations, while general purpose gloves protect against many hazards. Materials used in gloves are cotton, nylon, duck, jersey, canvas, terry, flannel, lisle, leather, rubber, synthetic rubber, wire mesh, aluminized fabric, asbestos, plastic and synthetic coatings, impregnated fabrics, polyvinyl chloride, nitrile, neoprene, and many man-made fibers with identifiable brand names (Dionne, 1979; Riley and Cochran, 1988). Glove styles include liners, reversibles, open back, gloves or mittens with reinforced nubby palms and fingers, and double-thumb gloves. Certain tasks may need double or more gloves. For example, shuttle gloves are an assemblage of three layers of gloves, while latex-sensitive people in the medical community wear an inner liner with an outer shell. The length of glove may be wrist-, elbow-, or shoulder-length with exact dimensions depending on the manufacturer. In summary, the gloves range from easily available general purpose ones to highly task-specific and job-rated ones.

47.5 Glove Effect on Strength

Grip strength: Published evidence exists for glove effect on grip strength, grasp strength, pinch strength, grasp at submaximal levels of exertion, torque capabilities, and on endurance time. Reduction in grip and grasp force when gloves are donned has been reported by a number of investigators (Hertzberg, 1955; Lyman and Groth, 1958; Cochran et al. 1986; Wang et al. 1987; and Sudhakar et al. 1988). Hertzberg (1955) using a Smedley hand dynamometer determined that grip strength was reduced by about 20% among gloved airplane pilots. Reduction in strength may be as much as 30% or more, according to Lyman and Groth (1958). Cochran, Albin, Bishu, and Riley (1986) performed an experiment which examined the differences in grasp force degradation among five different types of commercially available gloves as compared to a bare-handed condition. The results indicated that the "no glove" condition was significantly higher in grasp force than any of the glove conditions. Wang, Bishu, and Rodgers, (1987) performed an experiment on strength decrements with three different types of gloves. The results of the study showed that there was a reduction in grip strength when comparing gloved performance to bare-handed performance. Bishu et al. (1995a, b) studied the effects of EVA gloves at different pressures on human hand capabilities. A factorial experiment was performed in which three types of EVA gloves were tested at five pressure differentials. The independent variables tested in this experiment were gender, glove type, pressure differential, and glove make. Six subjects participated in an experiment where a number of performance measures, namely grip strength, pinch strength, time to tie a rope, and the time to assemble a nut and bolt, were recorded. Tactile sensitivity was also measured through a two-point discrimination test. The salient results were that with EVA gloves strength is reduced by nearly 50%, and that performance decrements increase with increasing pressure differential. McMullin and Hallbeck (1991) studied the effect of wrist position, age, and glove type on the maximal power grasp force, and their findings indicate that a single-layer glove is better than several layers, as the bunching of glove material at the joints could cause strength decrement. More recently Muralidhar, Bishu, and Hallbeck (1996) evaluated two prototype gloves (contour and laminated) with a single layer and a double layered glove. Bare-hand performance was measured to assess the exact glove effect. Considerable reduction in grip strength with gloves was found. Figure 47.1 shows the effect of gloves on grip strength. Similar results were also reported by Bronkema and Bishu (1996).

In summary, most of the research evidence on gloves indicates that gloves reduce grip and grasp capabilities.

Torque strength: A number of studies have reported an increase in strength capabilities with gloves. Riley, Cochran, and Schanbacher (1985) examined forward handle pull, backward handle pull, maximum wrist flexion torque, and maximum wrist extension torque while using no-glove, one-glove, and double-glove conditions. The results of this study showed that the one-glove condition was superior to both the no-glove and two-glove conditions. Similar results have been reported by Adams and Peterson (1988),

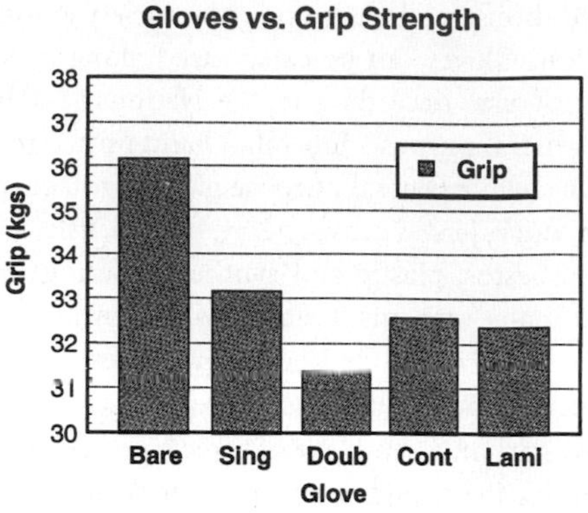

FIGURE 47.1 Gloves vs. grip strength.

who investigated the effects of two types of gloves on torque capabilities. In this study, a two-layer work glove and a three-layer chemical defense glove were found to enhance tightening performance, while only the work glove aided the loosening performance. Mital, Kuo, and Faard (1994) have reported an increase in peak torquing exertion capabilities when gloves are donned, with the extent of increase being dependent on the type of gloves donned. In contrast, Cochran, Batra, Bishu, and Riley (1988) found that gloves reduce torquing force. They had subjects perform a flexion torque task using four sizes of cylindrical handles (7 cm, 9 cm, 11 cm, and 13 cm) while wearing three types of gloves (cotton smooth leather, and suede leather). Using cotton gloves yielded the lowest torquing force, while bare-handed had the highest, and the two leather gloves were in between and were not significantly different from each other. These results are supported by Chen, Cochran, Bishu, and Riley (1989), who found the forces generated using cotton gloves of all sizes were significantly lower than the leather or deerskin gloves of different sizes in a similar torquing task. The effect of gloves on torque capabilities is far less clear. However, it is reasonable to assume that gloves would aid torquing tasks.

Pinch strength: As compared to grip or grasp capabilities, studies on glove effect on pinch strength are few and far between. Kamal, Moore, and Hallbeck (1992) report that gloves do not affect lateral pinch capabilities. Hallbeck and McMullin (1991, 1993) found similar results for three jaw chuck pinch. Overall, gloves do not affect pinch strength.

Endurance time: Almost all activities with a gloved hand involve certain levels of hand exertions for periods of time. Therefore, two issues are relevant here: the extent of exertion and the time of exertion. Most of the published studies on gloves have addressed the issue of extent of exertion. Bishu, Klute, and Kim (1995b) addressed the question of how long a person can sustain a level of exertion in the gloved-hand condition. This deals with muscular fatigue and related issues. They reported that the endurance time at any exertion level depended, not on the glove, but just on the level of exertion expressed as a percentage of maximum exertion possible at that condition. There is, however, a glove effect for the maximum exertion. Figure 47.2 shows the plot of the exertion level effect on the endurance time, across all glove and pressure configurations. The endurance time is least at 100% exertion level, while it is greatest at 25% exertion level.

47.6 Glove Effect on Dexterity

Bradley (1969) showed that control operation time was affected while wearing gloves. Banks and Goehring (1979), while studying the effects of degraded visual and tactile information in diver performance, found that the use of gloves increased task time by 50 to 60%. McGinnis, Bensel, and Lockhar (1973) investigated the effect of six different hand conditions on dexterity and torque capability. They used bare hand, leather glove, leather glove with inserts, impermeable glove, impermeable glove with inserts, and an impermeable glove with built-in insulation. They found that under dry conditions, the impermeable glove had the

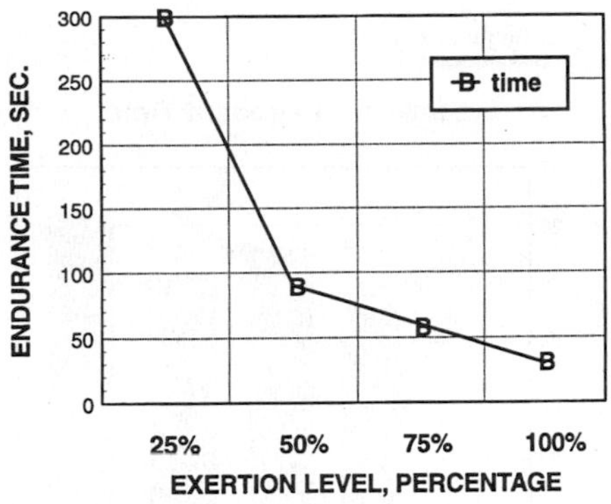

FIGURE 47.2 Endurance time.

best torque capability, and that the bare-handed dexterity performance was superior to that of gloved-hand performance. Plummer et al. (1985) studied the effects of nine glove combinations (six double and three single) on performance of the Bennett Hand Tool Dexterity Test apparatus. Results of the study indicated that subjects, with gloves donned, took longer to complete the task, with the double glove causing longer completion times. Cochran and Riley (1986) found that gloves generally reduce dexterity and force capability. Bensel (1993) conducted an experiment in which the effects of three thicknesses (0.18 mm, 0.36 mm, and 0.64 mm) of chemical protective gloves on five dexterity tests (the Minnesota rate of manipulation-turning; the O'Connor finger dexterity test; a cord and cylinder manipulation; the Bennet hand-tool dexterity test; and a rifle disassembly/assembly task) were investigated. Mean performance times were shortest for the bare-handed condition and longest for the thickest (0.64 mm) glove. Nelson and Mital (1995) found no appreciable differences in dexterity and tactility among latex gloves of five different thicknesses: 0.2083 mm; 0.5131 mm; 0.6452 mm; 0.7569 mm; and 0.8280 mm. The authors found the thickest latex glove (0.8280 mm) to be puncture resistant, with no loss in dexterity and tactility as compared to the thinner gloves. Bellinger and Slocum (1993) investigated the effect of protective gloves on hand movement and found that gloves decreased the range of motion in adduction/abduction and supination/pronation, while extension/flexion was not affected. Their findings suggest that there is an overall reduction in the kinematic abilities of the hand while wearing gloves. More recently Muralidhar, Bishu, and Hallbeck (1996) evaluated two prototype gloves (contour and laminated) with a single layer, and a double-layered glove. Bare-hand performance was measured to assess the exact glove effect. A battery of tests consisted of the Pennsylvania Bi-Manual Worksample Assembly Test (PBWAT), Minnesota Rate of Manipulation Test-Turning (MRMTT), a rope-tying task to evaluate dexterity for flexible object manipulation, and a manipulability test. Figure 47.3 shows the glove effect on MRMTT. Figure 47.4 shows the plot of the glove effect on PBWAT. Figure 47.5 shows the glove effect on the rope tying time, while Figure 47.6 shows the glove effect on the manipulation time. It is seen that gloves reduce dexterity. The reduction in gloved performance is seen consistently in all the measures.

Overall, gloves reduce finger dexterity, and manipulability.

47.7 Glove Effect on Tactility

Although intuitively most obvious, the effect of gloves on tactile sensitivity has not been well documented. The evidence on this matter is somewhat confusing mainly due to inadequacies of measures and inadequacies of instruments. The monofilament test (Weinstein, 1993) is by far the most popular to assess tactile sensitivity. Used in clinical testing, filaments with predetermined force are pressed against the fingers of the subjects by the experimenter until the sensation of touch is felt. The force is recorded as the tactile sensitivity. The two-point discrimination test used by O'Hara et al. (1988) and by Bishu and Klute (1995a) failed to give a clear indication of loss of tactile sensitivity. Bronkema et al. (1994) have used grasp force degradation at submaximal levels of exertion with gloves as a measure of the loss of

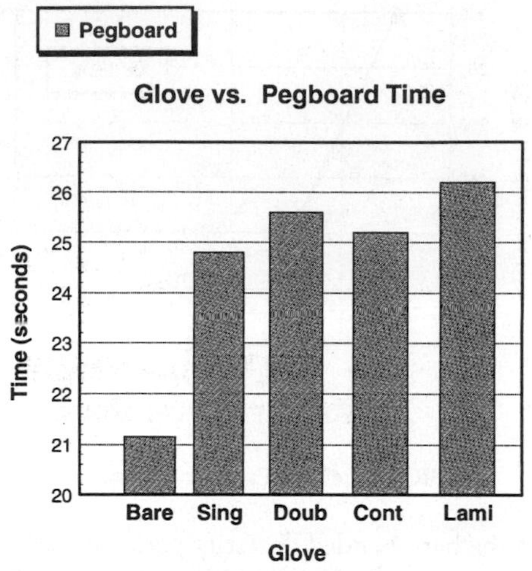

FIGURE 47.3 Gloves vs. pegboard time.

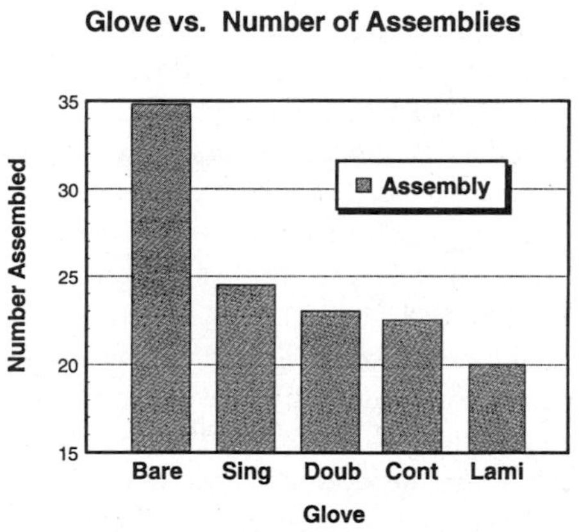

FIGURE 47.4 Gloves vs. number of assemblies.

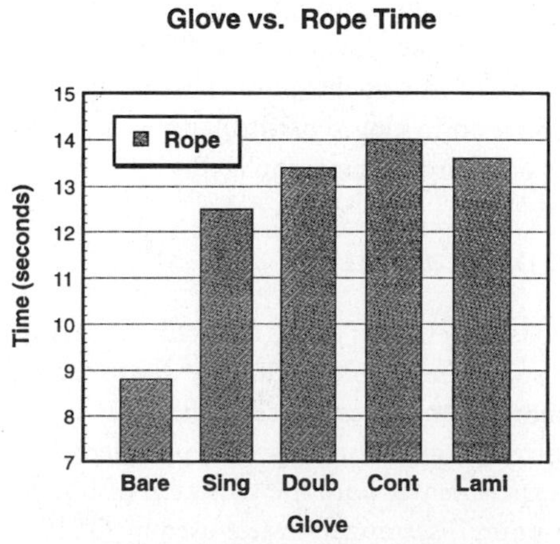

FIGURE 47.5 Gloves vs. rope knotting time.

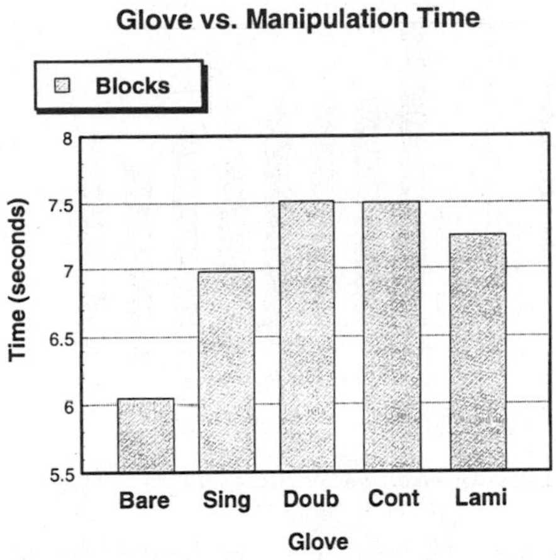

FIGURE 47.6 Gloves vs. blocks manipulation time.

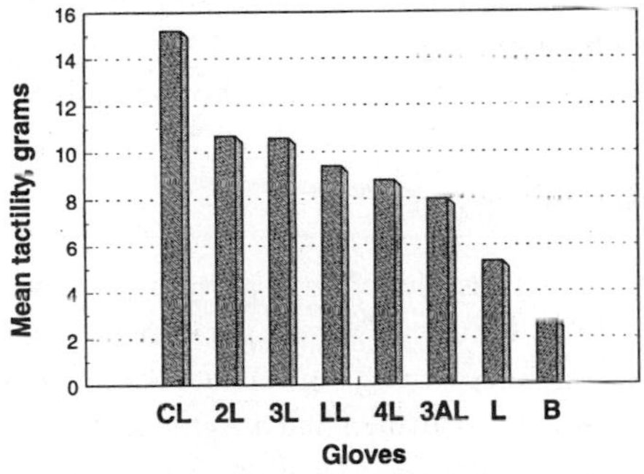

FIGURE 47.7 Glove effect on tactility.

tactility. Their results indicate that gloves do reduce tactile sensitivity. Desai and Konz (1983) studied the effect of gloves on tactile inspection performance and found that gloves had no significant effect on the inspection performance. In fact, they recommend that gloves be worn during tactile inspection tasks to protect the inspectors' hands from abrasion and to help in the detection of small surface irregularities. Nelson and Mital (1995) found no appreciable differences in dexterity and tactility among latex gloves of five different thicknesses. In summary, the effect of gloves on tactility has not been clearly understood and should be the focus of glove research in the future.

47.8 Liners

Today there is a growing trend toward the use of inner gloves or glove liners. For example, health care professionals often tend to use glove liners to prevent outer glove/skin interaction. Similarly, meat processors usually wear glove liners, while astronauts use multiple layers of liners. Almost all of the research efforts on gloves has focused on the outer glove, while liners have drawn little research attention. Using a standardized glove testing protocol, Bishu, Chin, and Goodwin (1997) investigated the effect of a number of glove liners.

The study compared three types of liners: liners made from PTFE, cotton, and latex. A battery of evaluation tests, comprising some standardized tests and certain functional tests, was designed. The tests assessed the following capabilities: tactile sensitivity, dexterity, manipulability, strength, and effect of continuous use. The actual tests performed were the pinch test, finger strength test, monofilament test,

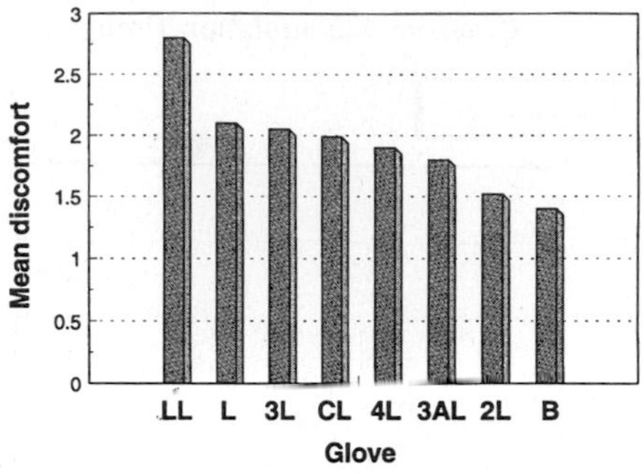

FIGURE 47.8 Glove effect on overall fatigue.

pain threshold test, rope-tying test, peg board test, and fatigue test. In the fatigue test the effect of continuous use of gloves was measured with Borg's RPE scale. Continuous use was simulated by making the subjects perform keyboard tasks and peg board tasks alternately for an hour. Discomfort measures were recorded every ten minutes. Important findings were (a) liners made a distinct contribution to the performance decrements with gloves, and (b) liners had a significant effect on the overall comfort (discomfort) during extended periods of glove usage. Figure 47.7 shows the graph of effect of liners on tactility as measured with monofilament test. Figure 47.8 shows the liner effect on overall fatigue.

47.9 Glove Attributes

Bradley (1969) also investigated dexterity as a function of glove attributes such as snugness of fit, tenacity, and suppleness in a wide variety of 18 industrial gloves. The conclusions reached are that various glove attributes influence dexterity performance to varying extents. Bishu, Batra, Cochran, and Riley (1987) found that glove attributes and the task performed had a significant effect on the force exertion. Wang, Bishu, and Rodgers (1987) concluded that altered feedback from a gloved hand caused strength degradation. Batra, Bronkema, Wang, and Bishu (1994) found grip strength reduction to be significantly correlated with glove thickness and subjective rating discomfort, and suggest that glove thickness should be minimized, while increasing the tenacity. In spite of these studies, no comprehensive model linking performance degradation with glove characteristics exists.

47.10 Challenges of Glove Design

In summary, gloves do reduce performance, but provide a vital protective function. Facilitation of performance, with simultaneous protection from the hazards of the work environment, are often conflicting objectives of glove design. The conflicts associated with providing primary hand protection through the use of a glove while permitting adequate hand functioning has been widely recognized. Looking at the glove attributes that cause performance differences, attributes or level of attributes that facilitate performance deteriorates safety function. This conflict poses certain challenges for the glove designer. Before attempting to design any kind of protection for the hand, it is necessary to first identify what it needs to be protected against. Human capability is limited to a narrow bandwidth of acceptable environmental conditions in which performance is not affected. There are a number of environmental hazards, often in combination, that are likely to pose a threat to the hands of workers interfacing with their workplace.

Glove material is often fixed by the environment. Environments that expose the worker to radiation, electrical, biological, fire, chemical, and extreme thermal hazards warrant that specific materials be incorporated into the hand protection, irrespective of the design. The hazard-specificity of such materials

TABLE 47.2 Glove Attributes as a Function of Design Parameters

Prehension	Function (Design, Material)
Torque Capability	Function (Design, Material)
Dexterity	Function (Design, Material)
Tactile Feedback	Function (Material, User)
Grip Strength	Function (User, Material, Design)
Fit	Function (Design, Material)
Pressure-Pain Threshold	Function (User, Material, Design)
Range of Motion	Function (User, Design, Material)
Abrasion Resistance	Function (Material)
Puncture Resistance	Function(Material)
Cut & Tear Resistance	Function (Material)
Thermal Protection	Function (Material)
Chemical Resistance	Function (Material)
Biohazard Protection	Function (Material)
Radiation Protection	Function (Material)
Electrical Protection	Function (Material)
Vibration Protection	Function (Material, Design)

also poses a problem for the glove designer, because the minimum thickness of the material required to provide adequate protection is usually a fixed value, limiting the designers' choice of variable parameters.

For example, a glove designed for use in an operating room or by a dental hygienist has to be capable of being sterilized either by steam or chemical disinfectants, impermeable to any potentially dangerous fluids, and of sufficient thickness and strength to maintain its integrity for a reasonable period of time.

However, in spite of the inflexibility in the materials' parameters, the glove is expected to enable the user to function without significant loss of desirable hand functions like grip strength, dexterity, range of motion, and tactile feedback. When these requirements are combined with multiple hazard condition protection requirements, glove design becomes a complex task. Table 47.2 shows the glove attributes as a function of design parameters. Muralidhar, Bishu, and Hallbeck (1996) suggest an ergonomic approach for glove design. Basing on published literature on force distribution in the hand during any task, they recommend that gloves should have variable thickness, with more thickness in regions where more force is exerted and less thickness in regions where force exertion is minimal. They argue that such an approach would yield gloves with minimal performance degradation and maximal protection.

47.11 Glove Evaluation Protocol

The question of how to evaluate a glove has always interested the designer, manufacturer, and the user of gloves. Standard evaluation protocols do not generally exist. Even in cases where they do exist, as in cases of rubber gloves used by utility people or fire fighters' gloves, the protocols are inadequate. It is recommended that a typical glove evaluation protocol include the following:

1. Strength tests including grip and pinch tests.
2. A battery of standardized tests to assess dexterity, tactility, and manipulability. Typical standard tests for these include Pennsylvania Bi-Manual Worksample Assembly Test (PBWAT), Minnesota Rate of Manipulation Test-Turning (MRMTT), Purdue peg board test, O'Connor dexterity test, and Monofilament test.
3. A battery of functional tests. This is what most of the existing glove evaluation protocols lack. Functional tests are task specific and should be appropriately designed to simulate actual tasks to be performed with the concerned gloves.

Evaluation protocols similar to one listed above have been used by O'Hara, Briganti, Cleland, and Winfield (1988) and Bishu and Klute (1995a) in the evaluation of EVA gloves. Similar protocols have been used for evaluating liners by Bishu, Chin, and Goodwin (1997).

47.12 Glove Standards

Existing glove standards are of three types. The standards generally describe protective requirements as in some of the U.S. Occupational Safety and Health Administration standards; or describe the protection the gloves must provide for safety as in gloves in the chemical industry or for the utility personnel; or specifically describe glove testing requirements.

47.13 Conclusion

In summary, the following statements can be made with regards to gloves.

1. Gloves protect the hand from the environment, but affect the performance.
2. Gloves range widely in size, type, and cost. They range from general purpose gloves to highly specialized task-specific gloves.
3. Gloves reduce grip or grasp strength capabilities, while they do not affect torque or pinch capabilities.
4. Glove reduce hand dexterity, tactility, and manipulability.
5. Providing protection without compromising performance is a continuous challenge for glove designers.

References

Adams, S.K. and Peterson, P.J. (1988). Maximum voluntary hand grip torque for circular electrical connectors. *Human Factors*, 30 (6), pp.733-745.

Banks, W.W. and Goehring, G.S. (1979). The effects of degraded visual and tactile information on diver work performance. *Human Factors*, 21 (4), pp. 409-415.

Batra, S., Bronkema, L.A., Wang, M., and Bishu, R.R. (1994). Glove attributes: can they predict performance? *International Journal of Industrial Ergonomics*, 14, pp. 201- 209.

Bellingar, T.A. and Slocum, A.C. (1993). Effect of protective gloves on hand movement: An exploratory study. *Applied Ergonomics*, 24 (4), pp. 244-250.

Bensel, C.K. (1993). The effects of various thicknesses of chemical protective gloves on manual dexterity. *Ergonomics*, 36 (6), pp. 687-696.

Bishu, R.R., Batra, S., Cochran, D.J., and Riley, M.W. (1987). Glove effect on strength: an investigation of glove attributes. *Proceedings of the 31st Annual Meeting of the Human Factors Society*, pp. 901-905.

Bishu, R.R., Chin, A., and Goodwin, B. (1997). Inner gloves: how good are they? To be published in *Advances in Occupational Ergonomics and Safety II*, (Editors: Das and Karwowski) Ohio.

Bishu, R.R. and Klute, G. (1995a). The effects of extra vehicular activity gloves on human performance. *International Journal of Industrial Ergonomics*, 16, pp. 165-174.

Bishu, R.R., Klute, G., and Kim, B. (1995b). Force endurance relationship: does it matter if gloves are donned? *Applied Ergonomics*, 26(3), pp. 179-185.

Bradley, J.V. (1969). Effect of gloves on control operation time. *Human Factors*, 11(1), pp. 13-20.

Bronkema, L. and Bishu R.R. (1996). The effects of glove frictional characteristics and load on grasp force and grasp control. *Proceedings of the 40th Annual Meeting of the Human Factors and Ergonomic Society*, Philadelphia, PA., pp. 702-706.

Bronkema, L., Bishu, R.R., Garcia, D., Klute, G., and Rajulu, S. (1994). Tactility as a function of grasp force: the effect of glove, pressure and load. *Advances in Ergonomics and Safety VI* (Editor: Aghazadeh), Taylor & Francis Ltd., London, pp. 627-632.

Chao, E.Y.S., An, K.N., Cooney, W.P., and Linschied, R.L. (1989). *Biomechanics of the Hand — A Basic Research Study*, World Scientific, Singapore.

Chen, Y., Cochran, D.J., Bishu, R.R., and Riley, M.W. (1989). Glove size and material effect on task performance. *Proceedings of the 33rd Annual Meeting of the Human Factors Society,* Denver, Colorado, pp. 708-712.

Cochran, D.J. and Riley, M. (1986). The effects of handle shape and size on exerted forces. *Human Factors,* 28 (3), pp. 253-265.

Cochran, D.J., Albin, T.J., Bishu, R.R., and Riley, M.W. (1986). An analysis of grasp force degradation with commercially available gloves. *Proceedings of the 30th Annual Meeting of the Human Factors Society,* pp. 852-855.

Cochran, D.J., Batra, S., Bishu, R.R., and Riley, M.W. (1988). The effects of gloves and handle size on maximum torque. *Proceedings of the 10th Congress of the International Ergonomics Association,* pp. 254-256.

Desai, S. and Konz, S. (1983). Tactile inspection performance with and without gloves. *Proceedings of the Human Factors Society,* pp. 782-785.

Dionne, E.D. (1979). How to select proper hand protection. *National Safety News,* 119, pp. 44-53.

Hallbeck, M.S. and McMullin, D.L. (1993). Maximal power grasp and three jaw chuck pinch as a function of wrist position, age and glove type. *International Journal of Industrial Ergonomics,* 11, (3) pp. 195-206.

Hall beck, M.S. and McMullin D.L. (1991). The effect of gloves, wrist position, and age on peak three-jaw chuck pinch force: a pilot study. *Proceedings of the Human Factors Society, 35th Annual Meeting,* pp. 753-757.

Hertzberg, T. (1955). Some contributions of applied physical anthropometry to human engineering. *Annals of New York Academy of Sciences,* 63, pp. 621-623.

Kamal, A.H., Moore, B.J., and Hallbeck, M.S. (1992). The effects of wrist position/glove type on maximal peak lateral pinch force. *Advances in Industrial Ergonomics and Safety IV* (Editor S. Kumar) London: Taylor & Francis, pp. 701-708.

Landsmeer, J.M.F. (1962). Power Grip and Precision Handling. *Annals of Rheumatic Disease,* 21, pp. 164-169.

Lyman, J. and Groth, H. (1958). Prehension force as measure of psychomotor skill for bare and gloved hands. *Journal of Applied Psychology,* 42:1, pp. 18-21.

McGinnis, J.S., Bensel, C.K., and Lockhar, J.M. (1973). *Dexterity Afforded by CB Protective Gloves.* U.S. Army Natick Laboratories, Natick, Massachusetts, Report No. 73-35-PR.

McMullin, D.L. and Hallbeck, M.S. (1991). Maximal power grasp force as a function of wrist position, age, and glove type: A pilot study. *Proceedings of Human Factors Society 35th Annual Meeting,* pp. 733-737.

Mital, A., Kuo, T., and Faard, H.F. (1994). A quantitative evaluation of gloves used with non-powered hand tools in routine maintenance tasks. *Ergonomics,* 37, (2), pp. 333-343.

Muralidhar, A. and Bishu, R.R., (1994) Glove evaluation: a lesson from impaired hand testing. *Advances in Industrial Ergonomics and Safety VI,* Editor F. Aghazadeh. Taylor & Francis. pp. 619-625.

Muralidhar, A., Bishu, R.R., and Hallbeck, M.S. (1996). The development of an ergonomic glove. Submitted to *Applied Ergonomics* for review and publication.

National Safety Council, (1975). Programming personal protection: hands and fingers, *National Safety News,* 114, pp. 56-58.

National Safety Council, (1976). Keeping hands safe. *National Safety News,* 119, pp. 44- 53.

Napier, J.R. (1956). The prehensile movements of the human hand. *The Journal of Bone and Joint Surgery,* 38B, pp. 902-913.

Nelson, J.B. and Mital, A., (1995). An ergonomic evaluation of dexterity and tactility with increase in examination/surgical glove thickness. *Ergonomics,* 38, (4) pp. 723-733.

O'Hara, J.M., Briganti, M., Cleland, J., and Winfield, D. (1988). Extravehicular activities limitations study. Volume II: establishment of physiological and performance criteria for eva gloves — final report (*Report number AS-EVALS-FR-8701, NASA Contract no NAS-9-17702*).

Plummer, R., Stobbe, T., Ronk, R., Myers, W., Kim, H., and Jaraiedi, M. (1985). Manual dexterity evaluation of gloves used in handling hazardous materials. *Proceedings of the 29th Annual Meeting of the Human Factors Society*, pp. 819-823.

Riley, M.W. and Cochran, D.J. (1988). Ergonomic aspects of gloves: design and use. *International Reviews of Ergonomics*, 2, Editor David J. Oborne, Taylor & Francis, pp. 233-250.

Riley, M.W., Cochran, D.J., and Schanbacher, C.A. (1985). Force capability differences due to gloves. *Ergonomics*, 28 (2) pp. 441-447.

Sudhakar, L.R., Schoenmarklin, R.W., Lavender, S.A., and Marras, W.S., (1988). The effects of gloves on muscle activity. *Proceedings of the 32nd Annual Meeting of the Human Factors Society*, pp. 647-650.

Wang, M.J., Bishu, R.R., and Rodgers, S.H. (1987). Grip strength changes when wearing three types of gloves. *Proceedings of the Fifth Symposium on Human Factors and Industrial Design in Consumer Products*, Interface 87, Rochester, NY.

Weinstein, S. (1993). Fifty years of somatosensory research: from the Semmes-Weinstein monofilaments to the enhanced sensory test. *Journal of Hand Therapy*, January- March, pp. 11-22.

Westling, G. and Johansson, R.S. (1984). Factors influencing the force control during precision grip. *Experimental Brain Research*, 53, pp. 277-284.

48

Industrial Mats

Jung-Yong Kim
Hanyang University

48.1 Introduction

Prolonged standing on one's feet is very common in the workplace. Workers who are exposed to prolonged standing often experience fatigue, discomfort, and swelling of the legs and feet (Winkel, 1981; Rys and Konz, 1989). Ryan (1989) showed that supermarket cashiers, who stood 90% of their working hours, experienced discomfort mostly in the lower back area. Redfern and Chaffin (1988) reported a significant level of fatigue and discomfort in various areas of the body after prolonged standing.

Many studies have shown that the lack of venous return in the lower extremities increased discomfort during prolonged standing (Brantingham et al., 1970; Winkel and Jorgensen, 1986; Konz et al., 1990). Local muscle fatigue in the lower back area has also been observed after two hours of standing (Kim et al., 1994). Likewise, the cause of discomfort and fatigue can be different depending upon the related body parts.

In industry, floor mats have been widely distributed as a quick remedy to help reduce the discomfort and fatigue of workers. However, there is no documented guideline to choose the proper matting for an individual's working condition. This made it difficult for ergonomists or safety managers to objectively evaluate mats for their own workplaces. Therefore, in this chapter, various studies are introduced and compared to help readers understand different approaches and testing methods. Furthermore, a few tips or guidelines in the selection of a proper mat are summarized based upon the results of these studies.

48.2 Psychophysical Approach

A subjective rating technique for postural discomfort (Corlett and Bishop, 1976) has been employed to examine the level of discomfort in various body parts after prolonged standing. In this technique, workers can score the level of discomfort by using a body diagram, even though it only provides subjective opinions. Redfern and Chaffin (1988) used this technique to examine the overall body fatigue and leg fatigue in nine different floor conditions at the end of the workday. They asked about the discomfort level of the feet, ankle, shank, knee, thigh, hips, lower back, and upper back. They found that all the

0-8493-2641-9/99/$0.00+$.50

body parts except for the legs and hips indicated significant differences in a discomfort rating with a change in the floor conditions. In their study, the feet showed the highest discomfort rating followed by the ankle and shank. Regarding the floor conditions, the relatively soft mats consistently showed less discomfort than the concrete floor or the hard mat. However, the uneven soft surface showed a relatively higher rating of tiredness despite its softness. No quantitative data were reported to specify the proper compressibility of mats in this study. It was concluded that the different hardness and depth, as well as the viscoelastic property of the mat are the main factors in determining the effectiveness of the mat.

Konz et al. (1990) investigated three different mats and a concrete surface. Twenty college students stood on mats for 90 minutes, and discomfort levels were examined from the neck, shoulder, upper back, mid back, lower back, buttocks, upper leg, lower leg, ankle, hind foot, mid foot, and fore foot. As a result, all of the lower parts of the body from the buttocks down were significantly affected by the floor surfaces. All three mats in this study showed a significant reduction of discomfort compared to the concrete surface. Importantly, the compressibility of the mats was quantitatively reported based upon the technique developed by Konz and Subramanian (1989). In this study, the comfort level was inversely related to the mat compressibility. In other words, the harder the mat was, the more effective it was in reducing discomfort, which was the reverse outcome compared to the previous study by Redfern and Chaffin (1988).

Hinnen and Konz (1994) tested five different mats by using a compressibility measure (Konz and Subramanian, 1989). They tested sixteen female subjects standing an entire shift for two days. Each hour they measured the discomfort levels of nine body regions. After they compared five mats in terms of discomfort level and compressibility, they found that there was an optimal range of thickness and compressibility of the mats to maximize the reduction of discomfort. They concluded that the most comfort can be provided when the mat is at least 1/2 inch thick and 3 to 4% compressible.

From the results of these studies, it was found that the discomfort level increased with time and appeared to be the greatest at the feet, and became progressively less and less from the feet up. Also, the surface type, thickness, and compressibility of the mat have been recognized as important factors in determining the anti-fatiguing effect during prolonged standing.

48.3 Physiological Approach

During a period of prolonged standing or sitting, the hampered venous and arterial circulation of the lower leg (Basmajian, 1979; Brantingham, 1970; Winkel and Jorgensen, 1986) can cause foot swelling or skin temperature change, which are signs of discomfort and tiredness of the leg.

Changes in foot dimension and skin temperature were measured by Rys and Konz (1989) and Konz et al. (1990). They examined the changes in foot length, width, thickness, and ankle thickness after standing for 90 minutes and found no significant differences with a variety of floor surfaces. The calf circumference was also measured by Kuorinka et al. (1978) after one hour of standing. The result showed 3.5 to 1 mm increase in calf circumference in an hour, but the increase somewhat leveled off toward the end of the session for half of the participants. Eventually, no difference in calf circumference was found on the three surfaces.

Skin temperature was measured from the calf and instep by Konz et al. (1990). They observed that the calf skin temperature had increased by 0.3° C for the concrete surface, while small or negative increases were recorded for the mat. Conversely, Rys and Konz (1989) reported that the calf skin temperature measured on the concrete was significantly lower, by 1.5 and 1.9° C, compared to the rubber mat after an hour of standing. No skin temperature change was found in the instep as the floor surface varied.

48.4 Postural Approach

The frequency of posture shifting was observed by a video recorder, and the center of gravity of body sway was measured by force platform by (Zhang et al., 1991). They found that the frequency of posture shift is a sensitive measure to show the effect of prolonged hours of standing, but it is a poor indicator of showing the difference in floor types. Kuorinka et al. (1978) also measured the frequency of a postural sway at the beginning and the end of one hour of standing. An increase of frequency was found at the end of the session, but no difference was found between the three different surfaces.

48.5 Biomechanical Approach

An electromyographic (EMG) study (Basmajian 1979) showed that the standing posture can be maintained by muscle activities in the solius, iliopsoas, sacrospinalis, and neck extensor. Kuorinka et al. (1978) examined the EMG signals of the solius muscle and showed a slight trend for the integrated EMG (IEMG) to rise on a concrete surface, although the initial IEMG was the lowest. Zhang et al. (1991) measured EMG from the tibialis anterior and gastrocnemius muscles during and after two hours of standing and found no differences in the fatigue level as the thickness of the floor changed.

Marras (1992) pointed out that the processed EMG used in previous studies simply indicated the level of muscle contraction that could be a very weak indicator of muscle fatigue during static standing. He suggested that the EMG power spectrum could be a more effective tool for detecting the localized muscle fatigue after prolonged standing than the processed EMG. LeVeau and Andersson (1992) used a spectral analysis to derive EMG power frequency distribution that is the second-order information of an IEMG signal. They showed that the shift of mean or median value of a frequency distribution could be used as a sign of local muscle fatigue.

Kim et al. (1994) studied muscle fatigue after a period of two hours of standing by using the EMG power spectrum. The erector spinae muscle as well as tibialis anterior and gastrocnemius muscle were examined. Three floor conditions including concrete, thin mat, and thick mat were tested. The compressibility of the mat was also quantified based on the method used by Konz and Subramanian (1989). The Kin/Com dynamometer was used to measure the 75% of maximum voluntary contraction (MVC) before and after prolonged standing. The IEMG immediately recorded after two hours of standing and the median frequency shift of the EMG power spectrum was computed. In the study, a more local muscle fatigue in the erector spinae muscle rather than the lower leg muscles was found. The significant increase of the processed IEMG signal was also observed in the leg muscles, however, that was not necessarily the sign of local muscle fatigue. It was stated that the decreased discomfort on the softer mat could be due to the active venous return following lower leg contraction on a soft surface.

48.6 Characteristics of Tested Mats

Various types of mats have been tested in different studies. Currently, individual mats cannot be quantitatively compared to each other because test protocols are not yet standardized for the studies. In spite of this, test results in different studies are summarized in Table 48.1. The comfort score in this table should not be used for direct comparison between mats.

48.7 A Standardized Protocol

A standardized protocol needs to be developed to quantitatively evaluate mats. The protocol is expected to specify the method and apparatus to measure the thickness and compressibility of material. The material can be further specified in terms of resiliency and elasticity if an apparatus such as an Instron machine is available. Moreover, surface type can be specified in terms of the shape and friction coefficient. In general, the standing period needs to be longer than an hour to see signs of discomfort. Since the standing posture may change the result greatly, the posture needs to be standardized and strictly instructed during the test.

Standardized Compressibility Measure

Konz and Subramanian (1989) computed an average foot pressure based upon the average male body weight of 170 lb. (77.3 kg), which is 0.35 kg/cm^2. If one uses a 7 × 7 cm mat specimen, 17.15 kg of force

TABLE 48.1 Characteristics of Various Mats

Mat Type	Description and Material	Thickness, mm (in)	Compression[1] (mm)	Compression[2] (%)	Source	Comfort score
Foam plastic	Dense surface (low resiliency, high elasticity)	10 (0.4")	n/a	n/a	Kuorinka et al. (1978)	high
San-EZE	Resilient rubber	22 (0.87")	0.9	4.1	Konz et al. (1990) Rys & Konz (1989)	high
Optimat	Honeycomb	12 (0.5")	0.7	5.8	Konz et al. (1990) Rys & Konz (1990)	high
Traction mat	Rubber	9.5 (0.37")	0.7	7.4	Rys & Konz (1989)	high
San-EZE II	n/a	18 (0.7")	1.7	9.4	Konz et al. (1990)	intermediate
Footsaver	Thick solid surface	12 (0.5")	n/a	n/a	Rys & Konz (1990)	low
Footsaver	n/a	7 (0.3")	1.3	18.6	Konz et al. (1990)	low
Thin (Blue)	Thin hard rubber with padding	8 (0.31")	0.55	6.9	Kim et al. (1994)	high
Thick (Black)	Thick hard rubber w/uneven surface	22 (0.87")	0.49	2.2	Kim et al. (1994)	low
Thin	Rubber	1.6 (0.06")	n/a	n/a	Redfern & Chaffin (1988)	low
Medium	Rubber	6.4 (0.25")	n/a	n/a	Redfern & Chaffin (1988)	intermediate
Thick	Rubber	9.5 (0.37")	n/a	n/a	Redfern & Chaffin (1988)	high
Hard w/trilaminate padding	Trilaminate padding	n/a	n/a	n/a	Redfern & Chaffin (1988)	high
Hard w/o trilaminate padding	No padding	n/a	n/a	n/a	Redfern & Chaffin (1988)	low
Viscoelastic mat	Viscoelastic material	n/a	n/a	n/a	Redfern & Chaffin (1988)	low
Uneven	Soft, uneven surface	n/a	n/a	n/a	Redfern & Chaffin (1988)	low
Workstation mat	PVC vinyl w/1/2" dia. hole	22 (0.87")	n/a	2.2	Hinnen & Konz (1994)	high
Super sponge cushion matting	Rubber tile w/sponge base	12.7 (0.5")	n/a	3.3	Hinnen & Konz (1994)	high
Comfort-EZE	Rubber w/5/16" dia. knobs	9.5 (0.37")	n/a	5.2	Hinnen & Konz (1994)	low
Cushion-rib runner tread	Vinyl sponge/corrugated top	9.5 (0.37")	n/a	8.9	Hinnen & Konz (1994)	low
Interlocking rubber matting	Rubber w/smooth top	12.7 (0.5")	n/a	3.5	Hinnen & Konz (1994)	intermediate

n/a: not available.

[1] The absolute amount of compressed part under the given pressure (Equation 1)

[2] The amount of compressed part relative to the original depth of mat

should be applied to generate the same average foot pressure on the specimen. In actual testing, 18 kg instead of 17.15 kg was used to record the compressibility of the mat. The duration of compression should be short enough (about a second) to simulate the actual foot-stepping situation. The equation computing the average adult foot pressure (Konz and Subramanian, 1989) is as follows:

$$\text{pressure} = 0.15 + 0.0026 \times \text{body weight (kg)} \tag{1}$$

A proper press machine should be used to control the pressure level precisely. The following are examples of devices used in previous studies:

1. Instron machine (Redfern and Chaffin, 1988), Instron model 1122 universal testing machine (Konz et al., 1990).
2. MTS Bionix 858 servo hydraulic materials testing system (Kim et al., 1994)

48.8 Foot Wear Conditions

The discomfort level can be greatly affected by footwear as well as the characteristics of the mat itself. That is, the standardized test also needs to be specified in terms of footwear conditions, and the results should be interpreted accordingly. The following are examples of the various footwear used in previous studies.

1. Working shoes (Redfern and Chaffin, 1988)
2. Thin-soled sneakers (Kuorinka et al., 1978)
3. Dress shoes with hard insole and sneaker with soft insole (Zhang et al., 1991)
4. Cotton socks and slippers (Konz et al., 1990)
5. Cotton socks without shoes (Kim et al., 1994)

Moreover, the final selection of mats should be made after considering the interactive effect between shoes and mats. The right combination of the two materials not only increases the comfort level but can also prevent slips and falls in the workplace. To quantitatively assess the interaction between shoes and mats, Redfern and Bidanda (1994) measured the friction level in terms of proper parameters including dynamic coefficient of friction. Leclercq et al. (1995) also suggested the use of reference conditions including lubricant, floor surface and footwear model to accurately measure the slipping resistance.

48.9 Suggestions for the Ergonomist

Selecting the best mat is not a simple task. Commercialized mats sometimes use materials with different compressibility for each foot. This type of mat may be able to stimulate the venous circulation of one leg better than the other. However, it has not been adequately studied to determine how effective those specially designed mats are. Presently, the simple and safe way of selecting a mat is to choose one that is "not too hard, not too soft." For example, workers who stand and walk around may need a good hard and even surface to minimize unnecessary ankle action to balance their posture. At the same time, the softness of the mat helps the blood pumping mechanism while standing quietly to reduce the discomfort level. Therefore, the combination of solid surface and a soft padding may meet the needs of workers who both walk and stand. In summary, the surface shape, softness, thickness, kind of padding, and material should be carefully considered in selecting the right mat to meet the specific requirements of an individual's workplace.

Further research should be conducted under the standardized protocol to acquire information on the anti-fatiguing and anti-slipping effect of various mats. Then, a quantitative ergonomic guideline can be developed based upon the standardized data. Eventually, such guidelines will help ergonomists and safety managers select the best mat for the individual worker.

References

Basmajian, J.V. and DeLuca, C. 1979. Their functions revealed by electromyography, in *Muscle Alive*. Williams & Wilkins, Baltimore, MD.

Brantingham, C.R., Beekman, B.E., Moss, C.N., and Gorden, R.B. 1970. Enhanced venous blood pump activity as a result of standing on a varied terrain floor surface. *Journal of Occupational Medicine*, 12: 164-169.

Corlett, E.N. and Bishop, R.P. 1976. A technique for assessing postural discomfort. *Ergonomics*, 19(2): 175-182.

Hinnen, P. and Konz, S. 1994. Fatigue mats. *Advances in Industrial Ergonomics and Safety VI*, Taylor & Francis. 323-327.

Kim, J.Y., Stuart-Buttle, C., and Marras, W.S. 1994. The effects of mats on back and leg fatigue. *Applied Ergonomics*, 25(1): 29-34.

Konz, S., Bandla, V., Rys, M., and Sambasivan, J. 1990. Standing on concrete vs. floor mats. *Advances in Industrial Ergonomics and Safety II*, Taylor & Francis. 991-998.

Konz, S. and Subramanian, V. 1989. Footprints. *Advances in Industrial Ergonomics and Safety I*, Taylor & Francis. 203-205.

Kuorinka, I., Hakkanen, S., Nieminen, K., and Saari, J. 1978. Comparison of floor surfaces for standing work. *Biomechanics VI-B*, 207-211.

Leclercq, S., Tisserand, M., and Saulnier, H. 1995. Assessment of slipping resistance of footwear and floor surfaces. influence of manufacture and utilization of the products. *Ergonomics*, 38(2): 209-219.

LeVeau B. and Andersson, G. 1992. Output forms: data analysis and applications. Interpretation of electromyographic signals. *Selected Topics in Surface Electromyography for Use in the Occupational Setting: Expert Perspectives*. U.S. Department of Health and Human Services. 70-102.

Marras, W.S. 1992. Applications of electromyography in ergonomics. *Selected Topics in Surface Electromyography for Use in the Occupational Setting: Expert Perspectives*. U.S. Department of Health and Human Services. 122-143.

Redfern, M.S. and Bidanda, B. 1994. Slip resistance of the shoe-floor interface under biomechanically-relevant conditions. *Ergonomics*, 37(3): 511-524.

Redfern, M.S. and Chaffin, D.B. 1988. The effect of floor types on standing tolerance in industry. *Trends in Ergonomics/Human Factors V*, (Ed.) F. Aghazadeh, Elsevier Science Pub.

Ryan, G.A. 1989. The prevalence of musculoskeletal symptoms in supermarket workers. *Ergonomics*, 32: 359-371.

Rys, M.J. and Konz, S. 1989. Standing with one foot forward. *Advances in Industrial Ergonomics and Safety I*, (Ed.) A. Mital, Taylor & Francis.

Rys, M.J. and Konz, S. 1989. An evaluation of floor surfaces. *Proceedings of the Human Factors Society 33rd Annual Meeting*. 517-520.

Rys, M. and Konz, S. 1990. Floor surfaces. *Proceedings of the Human Factors Society 34th Annual Meeting*. 575-579.

Winkel, J. 1981. Swelling of the lower leg in sedentary work — pilot study. *Journal of Human Ergology*, 10: 139-149.

Winkel, J. and Jorgensen, K. 1986. Evaluation of foot swelling and lower-limb temperature in relation to leg activity during long-term seated office work. *Ergonomics*, 29(2): 313-328.

Zhang, L. Drury, C.G., and Woollet, S.M. 1991. Constrained standing: evaluation of the foot/floor interface. *Ergonomics*, 34: 175-192.

49

Ergonomic Principles Applied to the Prevention of Injuries to the Lower Extremity

Steven A. Lavender
Rush-Presbyterian-St. Luke's Medical Center

Gunnar B.J. Andersson
Rush-Presbyterian-St. Luke's Medical Center

Most of the ergonomics literature dealing with the prevention and control of musculoskeletal disorders in the workplace has focused on the upper extremity and the back. Comparatively little attention has been given to lower extremity musculoskeletal disorders which occur in the workplace. One could argue that since the lower extremity problems are not well documented in ergonomic journals, the problems may not be of much practical significance. The first objective of this chapter is to review the current literature regarding occupational musculoskeletal disorders affecting the lower extremities and to demonstrate the significance of the problem. The second objective is to describe what types of intervention strategies are available to minimize the likelihood of future or recurrent injuries to the feet, ankles, knees, and hips.

49.1 Lower Extremity Injuries: Is There an Occupational Problem?

The sports medicine literature is full of lower extremity overuse injuries in athletes. All too often we have seen athletes relegated to the sidelines following some sort of soft tissue injury that is likely to be the effect of not just a single incidence, but rather a cumulative loading pattern during practice and competition. Luckily, in most occupational environments the intensity of the exercise is greatly diminished, however, the cumulative exposure problem still persists. Recent studies have begun to report the relationship between occupational factors and knee, hip, and foot trauma.

Lindberg and Axmacher (1988) reported the prevalence of coxarthosis in the hip to be greater in male farmers than in an age-matched group of urban dwellers. Vingard et al. (1991) classified blue-collar occupations as to whether static or dynamic forces could be expected to act on the lower extremity. The

0-8493-2641-9/99/$0.00+$.50

authors found that those employed in occupations that experienced greater loads on the lower extremity, namely farmers, construction workers, firefighters, grain mill workers, butchers, and meat preparation workers, had an increased risk of osteoarthrosis of the hip. Similarly, Vingard et al. (1992) found that disability pensions for hip osteoarthrosis were significantly more likely to be received by males employed as farmers, forest workers, and construction workers.

Lindberg and Montgomery (1987) reported that knee gonarthrosis (osteoarthritis), as defined by a "narrowing of the joint space with a loss of distance between the tibia and the femur in one compartment, of one-half or more of the distance in the other compartment of the same knee joint or the same compartment of the other knee, or less than 3 mm," was more common in those who had performed jobs that required heavy physical labor for a long time. Kohatsu and Schurman (1990) found that, relative to controls, the individuals with severe OA were two to three times more likely to have worked in occupations requiring moderate to heavy physical work. Anderson and Felson (1988) reported a relationship between the frequency of knee bending required in a respondent's occupation and osteoarthritis in the older working population (55 to 64 years). Moreover, these same authors have shown that the strength demands of the job were predictive of knee OA in the women from this older age group (Anderson and Felson, 1988). The authors suggest that the increased OA in those with long exposure to occupational tasks is indicative of the role of repetitive occupational exposure. Further supporting the link between material handling jobs and knee problems is the finding by McGlothlin (1996), who recently reported that beverage delivery personnel were experiencing discomfort in the knees, in addition to the anticipated discomfort in the back and shoulders. It should be recognized, though, that personal risk factors for osteoarthrosis of the knee include obesity and significant knee injury (Kohatsu and Schurman, 1990). These same authors found no relationship between leisure time activities and knee OA.

Torner et al. (1990) reported that chronic prepatellar bursitis was the predominant knee disorder in 120 fishermen who underwent an orthopedic physical examination. Forty-eight percent of the men examined showed this disorder. Interestingly, the finding was as common among younger men as in older men. The authors believe that this disorder is a secondary effect of the boat's motion. The knees are used to stabilize the body by pressing against gunwales or machinery as tasks are performed with the upper extremities. Furthermore, just standing in mild sea conditions (maximum roll angles of 8 degrees) has been shown to considerably elevate the moments at the knees as the motion in the lower extremities and the trunk are the primary means for counteracting a ship's motions (Torner et al., 1991).

The etiology of "beat knee" was described by Sharrard (1963). He reported on the examination of 579 coal miners. Forty percent of those examined were symptomatic or had previously experienced symptoms. Most of the injuries could be characterized as acute simple bursitis or chronic simple bursitis. The majority of the affected miners were colliers whose job requires constant kneeling at the mine face. There was a strong relationship between the coal seam height (directly related to roof height in a mine) and the incidence of beat knee. The incidence rates were much higher in mines with a roof height under four feet as compared with those with greater roof heights. Obviously, this factor greatly affects the work posture of the miners. With higher roof heights miners can alternate between stooped and kneeling postures, but when seams are one meter or less, the stooped posture is no longer an alternative. Gallagher and Unger (1990), for example, present recommendations for weight limits of handled materials in underground mines. Below 1.02 m these are based on miners in kneeling postures. Sharrard (1963) also speculated on the individual factors attributable to the disorder and found a higher incidence among younger men. However, this may be due to the "healthy worker effect" (Andersson, 1991) in which older miners with severe "beat knee" have left the mining occupation.

Tanaka et al. (1982) reported that the occupational morbidity ratios for workers' compensation claims of knee-joint inflammation among carpet installers was twice that found in tile setters and floor layers, and was over 13 times greater than that of carpenters, sheet metal workers, and tinsmiths. Others have shown the knees of those involved with carpet and flooring installation were more likely to have fluid collections in the superficial infrapatellar bursa, have a subcutaneous thickening in the anterior wall of the superficial infrapatellar bursa, and have an increased thickness in the subcutaneous prepatellar region (Myllymaki et al., 1993).

Thun et al. (1987) determined the incidence of repetitive knee trauma in the flooring installation professions. While all flooring installers spend a large amount of time kneeling, the authors divided the 154 survey respondents into two groups, "tilesetters" and "floor layers," based on their use of a "knee-kicker." This device is used to stretch the carpet during the installation process. These respondents were compared with a group of millwrights and brick layers whose jobs did not require extended kneeling and/or the use of a knee kicker. Of the 112 floor layers (those who used the knee kicker) the prevalence rate of bursitis was approximately twice that found in the 42 tilesetters, and over three times that found in the 243 millwrights and brick layers. However, the prevalence in both groups of flooring workers of having required needle aspiration of the knee was almost five times that of millwrights and bricklayers. These results suggest that long durations of occupational kneeling is related to fluid accumulation, yet the bursitis is due to the repetitive trauma endured by the floor layers using the knee kicker. Village et al. (1991) found that the peak impulse forces generated in the knees of carpet-layers when using the "knee-kicker" were on the order of 3000 N. The opposite knee which was supporting the body during this action had an average peak force of 893 N. Bhattacharya et al. (1985) reported knee impact forces of 2469 N (about three times body weight) for a light kick and 3019 N (or about four times body weight) for a hard kick. These light and hard kicks resulted in impact decelerations of 12.3 *g* and 20 *g*, respectively. The authors observed that the knee kicking action during flooring installation occurred at a rate of 141 kicks per hour. Putting the knee injuries in perspective, pain was reported by 22% of questionnaire respondents in the tufting job at a carpet manufacturer. However, knees were only listed in 2.4% of the accident records. Thus, the knee is frequently the site of discomfort, although, there may be few lost days associated with knee pain (Tellier and Montreuil, 1991).

Cumulative trauma injuries can take the form of stress or fatigue fractures. Linenger and Shwayhat (1992) reported training-related injuries to the foot occurred in military personnel undergoing basic training at a rate of three new injuries per 1,000 recruit days. These authors found that stress fractures to the foot, ankle sprains, and achilles tendinitis accounted for the bulk of the injuries. Anderson (1990) found the stress fractures to be most common in the distal second and third metatarsal bones but could occur in any of the bones in the foot. Giladi et al.'s (1985) findings indicated that 71% of the stress fractures in their sample of military recruits occurred in the tibia and 25% in the femoral shaft. Moreover, they found the fractures to occur later in the training process than reported by others. Jordaan and Schwellnus (1994) reported that overuse injuries, when normalized according to training hours per week, decreased from week 1 to week 4, showed a resurgence in week 5, and a large peak in the final week of training. The injury rates corresponded to the weeks in which there was increased marching and less field training.

With regard to the overuse injuries found in military recruits, some investigators have looked into aspects of lower limb morphology that may indicate which individuals will be more susceptible to injury while performing the tasks associated with military training. Giladi et al. (1991) reported the influence of individual factors on the incidence of fatigue fractures, specifically, they found that individuals with narrow tibiae, and/or a greater external rotation of the hip were more likely to experience fatigue fractures. Cowan and colleagues (1996) reported the relative risk of "overuse" injuries was significantly higher in military recruits with the most valgus knees. In addition, these authors showed that the "Q" angle, which defines the degree of deviation in the patellar tendon from the line of pull on the patella by the quadriceps muscles, was shown to be predictive of stress fractures.

In summary, several occupational risk factors have been identified which place an employee at increased risk for disorders in the lower extremity. The literature has shown that heavy physical labor and frequent knee bending are factors, especially in the older component of the work force, thereby suggesting an interaction between the age degenerative processes and cumulative work experience. In other occupations the risk of lower extremity disorders is increased through poor footing conditions. And clearly, the role of direct cumulative trauma in those employees who must maintain kneeling postures and use their knees to strike objects (knee kicker) cannot be overlooked when considering preventive measures.

49.2 Preventing Injury: Types of Ergonomic Controls

Several types of control mechanisms to prevent or accommodate lower extremity disorders are available. This section will focus on the techniques whereby the foot–floor interface can be optimized. This includes measures to prevent slips and falls, stress fractures, as well as improving circulation and comfort in the lower extremities for those who remain in relatively static work postures throughout the day.

Floor Mats

Floor mats are often used for local slip protection. While inexpensive, they create a possible trip hazard, interfere with operations or cleanliness, and wear excessively (Andres et al., 1992). Several investigators have looked into the use of floor mats to reduce the fatigue effects observed in jobs that require prolonged standing. The subjects tested by Kuorinka et al. (1978) indicated through subjective ratings that they preferred to work on softer surfaces as opposed to harder surfaces. A foam plastic surface was rated the best and concrete the worst. These authors reported a moderate correlation between the subjective comfort ratings of the five surfaces tested and the order of surface hardness. However, integrated electromyographic (EMG) signals, median frequency of the EMG, measures of postural sway, and measures of calf circumference did not show any significant difference due to the floor covering. Hinnen and Konz (1994) asked employees in a distribution center to stand for two 8-hour shifts on each of five mats tested in the study. Approximately every hour the employees rated their comfort in several body regions including the upper leg, lower leg, ankle, and back. A scale of 0 (no discomfort at all) to 10 (extreme discomfort) was used. While these workers experienced relatively little discomfort, the mats with compressibility between 3 and 4% did best in the upper leg discomfort rankings as well as subject preference rankings. Marginally significant changes in the discomfort ratings were reported for the ankle. The ratings of lower leg and back discomfort showed no significant differences.

Rys and Konz (1989, 1990) reported on several anthropometric and physiological measures including changes in foot size and skin temperature at the instep and the calf. In general, the mats included in this study were significantly different from concrete in that there was greater skin temperature at both measured locations and greater comfort ratings. These authors report that the comfort was inversely related to mat compressibility.

Cook et al. (1993) used surface EMG to study the recruitment of the anterior tibialis and paraspinal muscles when standing on linoleum-covered concrete vs. an expanded vinyl 9.5 mm-thick surgical mat. After subjects stood for two sessions, of two-hour duration, on the mat and on the linoleum, it was concluded that there were no significant changes in the mean of the rectified EMG signals in either muscle due to the mats. As in the studies above, subjective data support the use of the mat.

Kim et al. (1994) tested two types of floor mats and a control condition in which subjects stood on concrete. While these authors observed muscular fatigue, as determined by a shift in the EMG median frequencies in the gastrocnemius and anterior tibialis muscle, the EMG median frequencies in these muscles were not affected by the use of floor mats. The median frequency shift in the erector spinae was reduced when subjects stood on the thinner and more compressible mat. The authors hypothesized that greater compressibility would have made for a less stable base of support, thereby, requiring more frequent postural changes in the trunk to overcome the destabilization associated with postural sway. Thus, the dynamic use of erector spinae muscles to correct for postural sway would facilitate the oxygen delivery and the removal of contractile by-products through increased blood flow. A further test of this hypothesis would evaluate whether this motion occurred only in the trunk, or if it occurred in the lower extremities which did not show the spectral shift due to the floor condition.

Shoe Insoles

It is widely recognized that shoe design plays a critical role in the development of overuse syndromes in runners (Lehman, 1984; McKenzie et al., 1985; Pinshaw et al., 1984). Moreover, the role of the shoe in controlling lower extremity kinematics has been reviewed by Frederick (1986) and discussed by McKenzie et al. (1985). Similarly, the use of wedged insoles has been shown to alter the static posture of the lower extremity (Yasuda and Sasaki, 1987). Sasaki and Yasuda (1987) have shown the use of wedged insoles to be a good conservative treatment for medial osteoarthritis of the knee in the early stages. These authors reported that patients with early radiographic stages of osteoarthritis and who were provided a wedged insole had reduced pain and improved walking ability relative to controls without the insole.

Clearly, the lower extremity disorders reported by runners represent extreme overuse, however, the treatment and prevention mechanisms may be applicable to occupational settings where employees must stand, walk, run, or even jump during their normal work activities. Padded insoles have been investigated for the shock-abating effects on the skeletal system. Loy and Voloshin (1991) used lightweight accelerometers for measuring the shock waves as subjects walked, ascended and descended stairs, and jumped off platforms of a fixed height. The peak magnitude of the shock waves during jumping activities were approximately eight times that seen during normal walking. The results indicated that the insoles reduced the amplitude of the shock wave by between 9 and 41% depending upon the activity performed. The insoles were most effective at reducing heel strike impacts and had the largest effect with the jumping activities.

Milgrom et al. (1985) tested the effects of shock attenuation on the incidence of overuse injuries in infantry recruits. Earlier studies conducted by fixing accelerometers to the tibial tubercle showed that soldiers wearing modified basketball shoes had mean accelerations that were 19% less than soldiers wearing lightweight infantry boots. These authors also found that over the 14 weeks of basic training the modified basketball shoes reduced the metatarsal stress fractures. However, the tibial and femoral stress fractures were not affected by the shoes worn. Gardner et al. (1988) compared viscoelastic polymer insole and a standard mesh insole that were issued by platoon to over 3,000 marine recruits. While the polymer insole had good shock absorbing properties, the incidence of lower extremity stress injuries over the 12-week basic training program were unaffected by the insole used.

Several studies have been conducted to evaluate variations in insole materials. Leber and Evanski (1986) describe the characteristics of the following seven insole materials: Plastazote, Latex foam, Dynafoam, Ortho felt, Spenco, Molo, and PPT. These authors measured the plantar pressures in 26 patients with forefoot pain. All insole materials reduced the plantar pressure by between 28 and 53% relative to a control condition. However, PPT, Plastazote, and Spenco were the superior products. Viscolas and Poron were found to have the best shock absorbency of the five insole materials tested by Pratt et al. (1986). Maximum plantar pressures were found to be significantly reduced in the forefoot region with PPT, Spenco, and Viscolas, although the three materials were not significantly different (McPoil and Cornwall, 1992). In the rear foot region, however, McPoil and Cornwall (1992) report that only the PPT and the Spenco reduced the maximum plantar pressure relative to the barefoot condition. The plantar pressure in the rearfoot region was not significantly reduced with the Viscolas. Interestingly, based on the shock absorbency data from Pratt's (1988) 30-day durability test, the resilience of Viscolas, PPT, and Plastazote could be described as excellent, good, and poor, respectively. Sanfilippo et al. (1992) also reported the change in foot-to-ground contact area as a function of insole material. Plastazote, Spenco, and PPT led to a significantly greater contact area than the other materials tested.

In summary, insoles appear to be effective at modifying the lower extremity kinematics and reducing the peak plantar pressures, although their effectiveness is dependent upon the material used. Additional research is needed to clarify the effectiveness of insoles in controlling lower extremity stress injuries. Based on the previous discussion it should be clear that the effectiveness of this control strategy will be dependent upon shock absorbing capacity, the pressure dispersion, and the durability properties of the insole materials selected.

The Foot–Floor Interface

Controlling slip and fall injuries requires a multifaceted approach. The foot–floor interface is analogous to the four-legged stool shown in Figure 49.1. To optimize the postural stability all four legs need to be

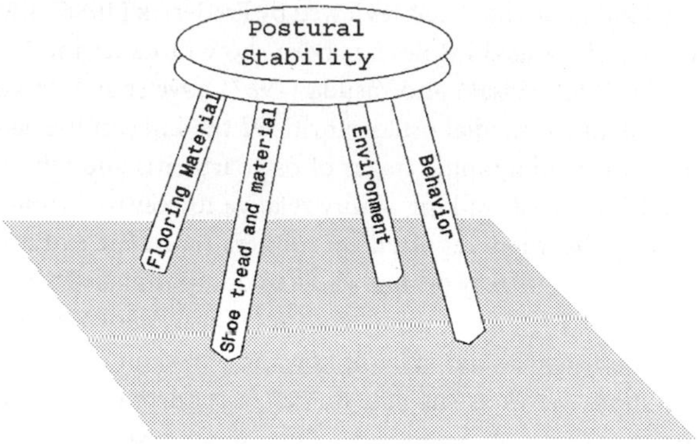

FIGURE 49.1 The four legged stool showing the interdependence of the factors affecting postural stability at the foot–floor interface.

in place and of equal length. The obvious legs are the flooring material and the shoe tread material and design. The environmental conditions represent the third leg as these affect the coefficient of friction between the shoe and the floor. And the fourth leg of the stool pertains to the behavior of the individual wearing the shoe. This behavioral component includes an individual's locomotion pattern, perception of environmental conditions, and allocation of the attentional resources necessary for adaptive behavior. If any of the four primary components are missing, at least to some degree, a leg of the stool is cut off by some random amount. For example, if the environmental conditions result in an oil film on the floor surface, the stool may still stand on the three remaining legs provided the shoe design and the floor material are adequate, and that the individual perceives the environmental conditions and adapts his or her behavior accordingly. Thus, the stool remains standing, although precariously, in spite of a shortened leg. If the individual did not attend to the environmental conditions the psychomotor leg would have been shortened, thereby making it unlikely that the stool will remain standing. In summary, the prevention of lower extremity injuries due to slips and falls requires attention be paid to each component or leg of the stool responsible for maintaining the body's stability.

In considering slip and trip prevention it is the dynamic friction of the interface, as opposed to static friction which is considered more critical in determining slip potential as most slips occur when the heal initially contacts the ground (Redfern et al., 1992; Strandberg, 1983). Gronqvist et al. (1992) has quantified slip resistance by determining the coefficient of friction between the interacting surfaces and possible contaminants. These authors reported that the important counter measures against floor slipperiness are the microscopic porosity and roughness of floor coverings. Flooring materials which have rough, unglazed, raised patterns, or are made from porous ceramic tiles are best for reducing slipping hazards in areas which must maintain very high standards for hygiene. In environments where the hygiene standards can be relaxed, very rough epoxy or acrylic resin floor materials should be used. Floor surface issues become even more critical on ramps and other inclined work surfaces. Redfern and McVay (1993) reported that when walking down ramps, the *required* coefficient of friction increased in a nearly linear fashion as ramp angle increased. This was due to the high shear forces encountered during heal strike as an individual walks down a ramp.

Gronqvist and Hirvonen (1994) studied slip resistance properties of footwear on iced surfaces. They found that the shoe material significantly affected the slip resistance. Shoe heels and soles constructed from thermoplastic rubber with a large cleated area were best for dry ice conditions (–10° C). Very few of the shoes tested functioned well on wet ice (0° C) where there is a boundary layer of water on top of the ice. In fact, the shoes which worked best for dry ice conditions were among the worst when tested in the wet ice condition. Shoes with the sharpest cleats yielded the greatest friction readings under this condition. Hardness of the heel and sole material was not a significant factor on wet ice.

Little consensus is found in sole hardness data, and what differences exist are not of practical significance (Leclercq, 1994). In general, microcellular PU (polyurethane) heels and soles are recommended, although, footwear showed less of an effect on the kinetic coefficient of friction than did variations in floor surfaces (Gronqvist et al., 1992). Relative to rubber soles and heels, Manning, Jones, and Bruce (1985) reported that boots with microcellular polyurethane (PU) soles and heels better resisted polishing caused by smooth, wet, or oily floors; had a longer life; and had a greater average coefficient of friction. The COF of the PU improved over time, indicating that the initial smooth surface of the boots should be roughened prior to use. Tisserand (1985) recommends avoiding the use of "micro-treads" or small bumps. However, tread design and material are not independent factors when it comes to the coefficient of dynamic friction. Leclercq et al. (1994) stress that the ridges in the tread need to be as sharp as possible to wipe away contaminants on the surface. In addition, these authors point out that the tread design needs to provide channels for the surface contaminants to flow through. Slipping risk is greatest at heel-strike, and therefore, the heel of the shoe requires considerable attention. There is controversy regarding whether the use of a bevelled heel to increase contact surface area is superior to the use of a small contact surface that would result in high contact pressure (Leclercq et al., 1994).

Tisserand (1985) has promoted the concept of a mental model whereby an individual has constant input as to the friction available from the surface being walked upon. The model is updated as new information is received indicating a change in the surface conditions. Slips occur where there is a discrepancy between the mental model of the surface and the actual conditions. Thus, as Tisserand (1985) points out: "The risk of slipping lies more in the gradient of the friction coefficient of the surface than in its absolute value: such as when in a car, a small patch of ground with a low coefficient (an isolated patch of ice, for example, on a large surface with a high coefficient of friction) is more dangerous than a surface with a medium but constant coefficient of friction" (pp. 1039).

Swensen assessed the subjective judgments of surface slipperiness by having groups of iron workers and students walk across steel beams. The subjects were asked to rank the slipperiness of four types of steel coatings and four levels of surface contaminant: none, water, clay, plastic covering oil. Static coefficient of friction (COF) values ranged between .98 and .20. For the experienced and inexperienced subjects the correlation between the subjective surface ratings and the actual COF measured following the test were .75 and .90, respectively. The subjects exposed to the very slippery conditions (COF = .20) created by the oil and plastic compensated by shortening their stride length, thereby lessening the foot velocity and shear forces, and maintained the body's center of gravity within the smaller region of stability. Thus, people can detect the COF and adapt their gait accordingly. When this adaptation fails to take place, an individual is much more likely to slip and possibly fall.

Ideally, with the perfect shoe–floor interface, one with no environmental contaminants, the individual's behavior would be not be a factor. But in few cases would this exist, and then, even a small fluctuation in the environment would be enough to disrupt the balance of the now three-legged stool. Therefore, administrative control measures need to be considered as a means for maintaining the behavior necessary for stable work postures. Employees should be trained to recognize where slippery conditions are likely within a facility, and that they be alert to changing environmental conditions. Further, employees should be encouraged to report maintenance problems with machines that affect the flooring conditions.

Stair Design

Pauls (1985) reports that only 6% of stair accidents entail slips, more often accidents are due to "overstepping." This author suggests that the overstepping occurs because the individual descending the stairs does not accurately perceive the stair width (also known as tread length). Thus, the foot is placed too far forward on the step. This scenario accounts for 19% of all stair accidents. This work highlights the interplay between engineering design factors (stair size) and behavioral factors. Research has suggested that stairs should have risers no higher than 178 mm (7 inches) and treads no shorter than 279 mm (11").

A secondary issue in stair design is the complex effect of visual distractions. Pauls (1985) reported that when visual distractions were present people actually focused harder on descending the stairs. When no distractions were present people exhibited less caution. Similarly, patients commonly reported that their falls leading to femoral neck fractures were initiated with missing a step down, for example unexpectedly stepping off a curb (Citron et al., 1985). This suggests that when the brain recognizes the potential distractions attention resources are allocated to the task at hand, whereas without some overt distraction the brain may allocate adequate attentional resources, or it may not. Archea (1985) has stressed that an older population that may not have the perceptual and motor capabilities found in younger individuals are more vulnerable to accidents on stairs. These findings suggest that the effects of distractions around stairs changes through the aging process.

Help for Those in Kneeling Postures

Sharrard (1963) reported that there was no relationship between the type of knee pads used and the incidence of beat knee in miners. This author recorded peak pressures on the order 35.7 kg per square cm as simulated mining tasks were performed. These compression forces were shown to vary widely throughout the 2.5 second cycle time for a shoveling task. Unfortunately, the author had no instrumentation capable of determining the shear forces and the torsional moments placed on the knee during the simulated tasks. At the time of Sharrard's paper a "bursa pad" had been designed that allowed perspiration to escape, pushed coal particles away from the skin, and provided satisfactory cushioning. Although no control group was used, the author reported that of the 24 previously affected men selected to test the pad under working conditions only two reported a recurrence of beat knee after a twelve-month period.

Ringen et al. (1995) reported on a new tool to reduce the knee and back trauma in those who tie rebar rods together in preparation for pouring concrete. No longer will concrete workers need to kneel or stoop for extended periods to interconnect the iron rods as this tool allows the operator to work in a standing posture.

Powered carpet stretching tools are available to remove the repeated trauma experienced by carpet layers. However, their widespread implementation depends upon educating flooring workers on the trade-offs between the additional time necessary to operate the tool and the knee disorders associated with the conventional technique.

49.3 Summary

Ergonomic texts historically have focused relatively little attention on the prevention of lower extremity disorders or the accommodation of individuals returning to work who have experienced a lower extremity disorder. In part this may be due to an under appreciation of the frequency and severity of occupational lower extremity disorders. Unlike many back or upper extremity disorders which have their origins in the repeated stresses placed on muscular, tendinous, and ligamentous tissues, many of the occupational lower extremity disorders occur through direct compression of the body tissues by a surface in the environment. As a result, the occupational lower extremity disorders often involve cartilaginous tissue and bone. Therefore, accommodation and prevention of these disorders occurs primarily through the optimizing the body's contact with surfaces in the environment. This chapter has illustrated some of the key ways in which this can be accomplished.

References

Anderson, E.G. (1990). Fatigue fractures of the foot. *Injury,* 21, 275-279.

Anderson, J.J. and Felson, D.T. (1988). Factors associated with osteoarthritis of the knee in the first national health and nutrition examination survey (HANES I). *American J. of Epidemiology,* 128, 179-189.

Andersson, G.B.J. (1991). The epidemiology of spinal disorders, in J.W. Frymoyer (Ed.) *The Adult Spine: Principles and Practice.* New York: Raven, 107-146.

Andres, R.O., O'Conner, D., and Eng, T. (1992). A practical synthesis of biomechanical results to prevent slips and falls in the workplace, in S. Kumar (Ed.) *Advances in Industrial Ergonomics and Safety IV,* London: Taylor & Francis, 1001-1006.

Archea, J.C. (1985). Environmental factors associated with stair accidents by the elderly. *Clinics in Geriatric Medicine*, 1, 555-569.

Bhattacharya, A., Mueller, M., and Putz-Anderson, V. (1985). Traumatogenic factors affecting the knees of carpet installers. *Applied Ergonomics*, 16, 243-250.

Citron, N. (1985). Femoral neck fractures: are some preventable? *Ergonomics*, 28, 993-997.

Cook, J., Branch, T.P., Baranowski, T.J., and Hutton, W.C. (1993). The effect of surgical floor mats in prolonged standing: an EMG study of the lumbar paraspinal and anterior tibialis muscles. *J. Biomedical Engineering*, 15, 247-250.

Cowan, D.N., Jones, B.H., Frykman, P.N., Polly, D.W., Harman, E.A., Rosenstein, R.M., and Rosenstein, M.T. (1996). Lower limb morphology and risk of overuse injury among male infantry trainees. *Medicine and Science in Sports and Exercise*, 28, 945-952.

Frederick, E.C. (1986). Kinematically mediated effects of sport shoe design: a review. *Journal of Sports Sciences*, 4, 169-184.

Gallagher, S. and Unger, R.L. (1990). Lifting in four restricted lifting conditions. *Applied Ergonomics*, 21, 237-245.

Gardner, L.I., Dziados, J.E., Jones, B.H., and Brundage, J.F. (1988). Prevention of lower extremity stress fractures: a controlled trial of a shock absorbent insole. *American Journal of Public Health*, 78, 1663-1567.

Giladi, M., Ahronson, Z., Stein, M., Danon, Y.L., and Milgrom, C. (1985). Unusual distribution and onset of stress fractures in soldiers. *Clinical Orthopaedics & Related Research*, 192, 142-146.

Giladi, M., Milgrom, C., Simkin, A., and Danon, Y. (1991). Stress fractures. Identifiable risk factors. *American Journal of Sports Medicine*, 19, 647-652.

Gronqvist, R., Hirvonen M., and Skytta, E. (1992). Countermeasures against floor slipperiness in the food industry, in S. Kumar (Ed.) *Advances in Industrial Ergonomics and Safety IV*, London: Taylor & Francis, 989-996.

Gronqvist, R. and Hirvonen, M. (1994). Pedestrian safety on icy surfaces: Anti-slip properties of footwear, in F. Aghazadeh (Ed.) *Advances in Industrial Ergonomics and Safety VI*, London: Taylor & Francis, 315-322.

Hinnen, P. and Konz, S. (1994). Fatigue mats, in F. Aghazadeh (ed.) *Advances in Industrial Ergonomics and Safety VI*, London: Taylor & Francis, 323-327.

Jordaan, G. and Schwellnus, M.P. (1994). The incidence of overuse injuries in military recruits during basic military training. *Military Medicine*, 159, 421-426.

Kim, J.Y., Stuart-Buttle C., and Marras, W.S. (1994). The effects of mats on back and leg fatigue. *Applied Ergonomics*, 25, 29-34.

Kuorinka, I., Hakkanen, S., Nieminen, K., and Saari, J. (1978). Comparison of floor surfaces for standing work, in E. Asmussen and K. Jorgensen (Eds.) *Biomechanics VI-B, Proceedings of the Sixth International Congress of Biomechanics*, Baltimore, University Park Press, 207-211.

Kohatsu, N.D. and Schurman, D.J. (1990). Risk factors for the development of osteoarthrosis of the knee. *Clinical Orthopaedics and Related Research*, 261, 242-246.

Kuorinka, I, Hakkanen, S, Nieminen, K., and Saari, J. (1978). Comparison of floor surfaces for standing work, in E. Asmussen and K. Jorgensen (Eds.) *Biomechanics VI-B*, Baltimore: University Park Press, 207-210.

Leber, C. and Evanski, P.M. (1986). A comparison of shoe insole materials in plantar pressure relief. *Prosthetics and Orthotics International*, 10, 135-138.

Leclercq, S., Tisserand, M., and Saulnier, H. (1994). Slip resistant footwear: A means for the prevention of slipping, in F. Aghazadeh (Ed.) *Advances in Industrial Ergonomics and Safety VI*, London: Taylor & Francis, 329-337.

Lehman, W.L. (1984). Overuse syndromes in runners. *American Family Physician*, 29, 157-161.

Lindberg, H. and Axmacher, B. (1988) Coxarthrosis in farmers. *Acta Orthop. Scand.*, 59, 607.

Lindberg H. and Montgomery, F. (1987). Heavy labor and the occurrence of gonarthrosis. *Clinical Orthopaedics and Related Research*, 214, 235-236.

Linenger, J.M. and Shwayhat, A.F. (1992). Epidemiology of podiatric injuries in US Marine recruits undergoing basic training. *Journal of the American Podiatric Medical Association*, 82, 269-271.

Loy, D.J. and Voloshin, A.S. (1991). Biomechanics of stair walking and jumping. *Journal of Sports Sciences*, 9, 136-149.

Manning, D., Jones, C., and Bruce, M. (1985). Boots for oily surfaces. *Ergonomics*, 28, 1011-1019.

McKenzie, D.C., Clement, D.B., and Taunton, J.E. (1985). Running shoes, orthotics, and injuries. *Sports Medicine*, 2, 334-47.

McGlothlin, J.D. (1996). *Ergonomic Interventions for the Soft Drink Beverage Delivery Industry.* U.S. Department of Health and Human Services (NIOSH) Publication No. 96-109.

McPoil, T.G. and Cornwall, M.W. (1992). Effect of insole material on force and plantar pressures during walking. *Journal of the American Podiatric Medical Association*, 82, 412-416.

Milgrom, C., Finestone, A., Shlamkovitch, N., Wosk, J., Laor, A., Voloshin, A., and Eldad, A. (1992). Prevention of overuse injuries of the foot by improved shoe shock attenuation. A randomized prospective study. *Clinical Orthopaedics & Related Research*, 281, 189-192.

Myllymaki, T., Tikkakoski, T., Typpo, T., Kivimaki, J., and Suramo, I. (1993). Carpet layer's knee. An ultrasonographic study. *Acta Radiol.*, 34, 496-499.

Pinshaw, R., Atlas, V., and Noakes, T.D. (1984). The nature and response to therapy of 196 consecutive injuries seen at a runners' clinic. *South African Medical Journal*, 65, 291-298.

Pratt, D.J., Rees, P.H., and Rodgers, C. (1986). Assessment of some shock absorbing insoles. *Prosthetics & Orthotics International*, 19, 43-45.

Pratt, D.J. (1988). Medium term comparison of shock attenuating insoles using a spectral analysis technique. *Journal of Biomedical Engineering*, 10, 426-429.

Redfern, M.S. and Bidanda, B. (1992). The effects of shoe angle, velocity, and vertical force on shoe/floor slip resistance, in S. Kumar (Ed.) *Advances in Industrial Ergonomics and Safety IV*, London: Taylor & Francis, 997-1000.

Redfern, M.S. and McVay, E.J. (1993). Slip potentials on ramps, in *Proceedings of the Human Factors and Ergonomics Society 37th Annual Meeting*, 2, 701-703.

Ringen, K., Englund, A., and Seegal, J. (1995). Construction workers, in B.S. Levey, and D.H. Wegman (Eds.) *Occupational Health: Recognizing and Preventing Work-Related Disease*, Boston: Little, Brown, and Company, 685-701.

Rys, M. and Konz, S. (1990). Floor mats, in the *Proceedings of the Human Factors Society 34th Annual Meeting*, 1, 575-579.

Rys, M. and Konz, S. (1989). An evaluation of floor surfaces, in the *Proceedings of the Human Factors Society 33rd Annual Meeting*, 1, 517-520.

Sanfilippo, P.B., Stess, R.M. and Moss, K.M. (1992). Dynamic plantar pressure analysis. Comparing common insole materials. *Journal of the American Podiatric Medical Association*, 82, 502-513.

Sasaki. T. and Yasuda, K. (1987). Clinical evaluation of the treatment of osteoarthritic knees using a newly designed wedged insole. *Clinical Orthopaedics & Related Research*, 221, 181-187.

Sharrard, W.J.W. (1963). Aetiology and pathology of beat knee. *British Journal of Industrial Medicine*, 20, 24-31.

Strandberg, L. (1983). On accident analysis and slip-resistance measurement. *Ergonomics*, 26, 1983.

Swensen, E.E., Purswell, J.L., Schlegel, R.E., and Stanevich, R.L. (1992). Coefficient of friction and subjective assessment of slippery work surfaces. *Human Factors*, 34, 67-77.

Tanaka, S., Smith, A.B., Halperin, W., Jensen, R. (1982). Carpet-layers knee. *The New England J. of Medicine*, 307, 1276-1277.

Tellier, C. and Montreuil, S. (1991) Pain felt by workers and musculoskeletal injuries: assessment relating to tufting shops in the carpet industry, in Y. Queinnec and F. Daniellou (Eds.) *Designing for Everyone*, 1, 287-289.

Thun, M., Tanaka, S., Smith, A.B., Haperin, W.E., Lee, S.T., Luggen, M.E., and Hess, E.V. (1987). Morbidity from repetitive knee trauma in carpet and floor layers. *British Journal of Medicine*, 44, 611-620.

Tisserand, M. (1985). Progress in the prevention of falls caused by slipping. *Ergonomics*, 28, 1027-1042.

Torner, M., Zetterberg, C., Hansson, T., and Lindell, V. (1990). Musculo-skeletal symptoms and signs and isometric strength among fishermen. *Ergonomics*, 33, 1155-1170.

Torner, M., Almstrom, C., Karlsson, R., and Kadefors, R. (1991). Biomechanical calculations of musculo-skeletal load caused by ship motions, in combination with work, on board a fishing vessel, in Y. Queinnec and F. Daniellou (Eds.) *Designing for Everyone*, 1, 293-295.

Village, J., Morrison, J.B., and Leyland, A. (1991). Carpetlayers and typesetters ergonomic analysis of work procedures and equipment, in Y. Queinnec and F. Daniellou (Eds.) *Designing for Everyone*, 1, 320-322.

Vingard, E., Alfredsson, L, Goldie, I., and Hogstedt, C. (1991). Occupation and osteoarthrosis of the hip and knee: A register-based cohort study. *International Journal of Epidemiology*, 20, 1025-1031.

Vingard, E., Alfredsson, L., Fellenius, E., and Christer, H. (1992). Disability pensions due to musculo-skeletal disorders among men in heavy occupations. *Scand. Journal Social Med.*, 20, 31-36.

Yasuda, K. and Sasaki, T. (1987). The mechanics of treatment of the osteoarthritic knee with a wedged insole. *Clinical Orthopaedics & Related Research*, 215, 162-172.

50

Ergonomics of the Foot

Stephan Konz
Kansas State University

This chapter is divided into five sections. Section 1 (Foot/Leg) gives the anatomy, physiology, and dimensions of the foot. Section 2 (Activities) describes the activities of standing, walking, running, and stepping. Section 3 (Accidents) discusses falls, their causes and solutions. Section 4 (Fatigue/Comfort) discusses walking and standing. Section 5 (Foot Controls) briefly describes pedals and switches.

50.1 Foot/Leg

Anatomy

Figure 50.1 shows the bones of the foot and ankle. The toes (foot fingers) are divided into *metatarsals* and three *phalanges* (except for the big toe, which only has two phalanges). In supporting the body, the *calcaneus* (heel) supports 50% of the weight, the lst and 2nd metatarsal 25%, and the 3rd, 4th, and 5th metatarsal 25%. In between are two arches:(1) the medial arch (calcaneus, the talus, the navicular, the cuneiform bones, and the 1st, 2nd, and 3rd metatarsals) and (2) the lateral arch (calcaneus, talus, cuboid, and the 4th and 5th metatarsals).

Under the heel (calcaneus) is a very important shock absorber, the heel pad (about 1.8 cm thick). The bottom of the calcaneus is not spherical but has two small "mountains"; the pad reduces the pressure on these mountains, and thus on the ankle, knee, and back.

The foot is connected to the ankle with a *mortise and tenon* joint. The vertical leg of the mortise is short on the outside (*lateral* side); in addition, the ligaments holding the bottom of the fibula (lateral malleolus) to the talus and calcaneus are relatively weak. In contrast, the vertical leg of the inside (*medial*) mortise is longer, and the ligaments holding the bottom of the tibia (medial malleolus) to the talus are relatively strong.

Inward rotation (*inversion*) of the foot tends to pull the ligaments from the bone; with proper treatment, healing is usually complete in about three weeks. There is a danger that the injured person may not seek medical advice even with a complete tear of the ligaments (connecting either the malleolus and the talus or the tibia and fibula). Then there would be need for surgical repair and rigid fixation in a cast for 2 to 3 months.

0-8493-2641-9/99/$0.00+$.50

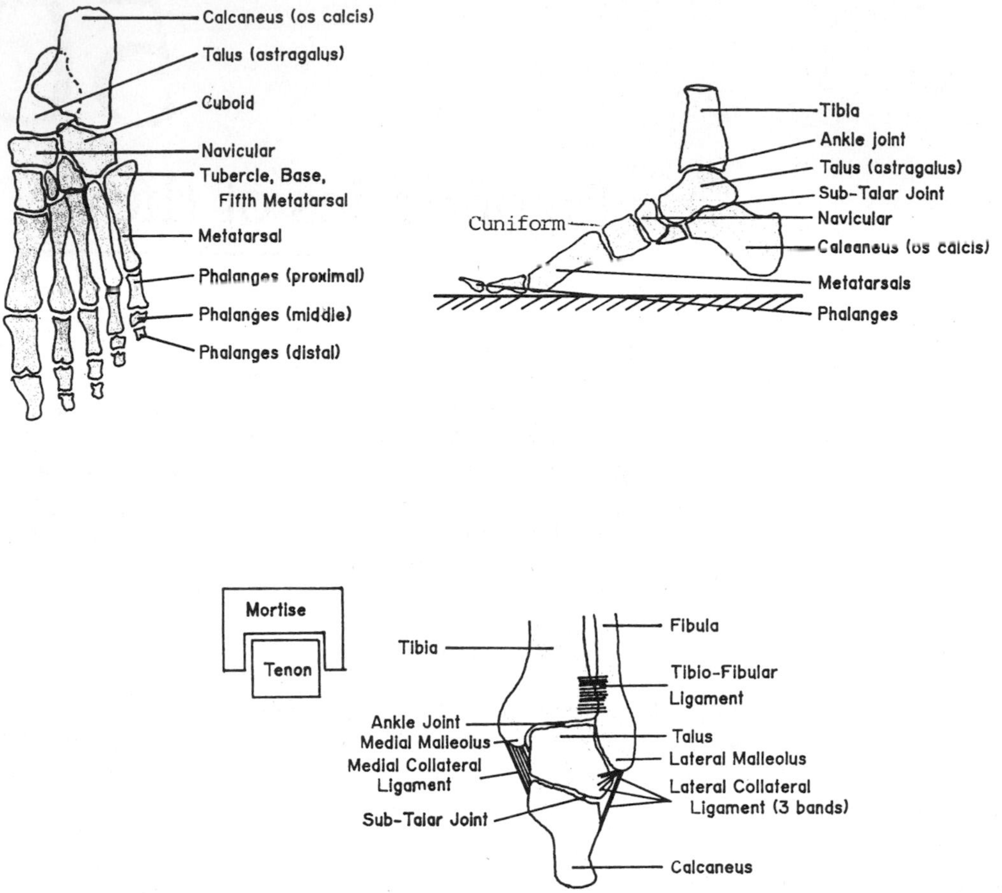

FIGURE 50.1 The foot and ankle. The right foot is viewed from below (top left) and the outside (top right); the left ankle (bottom) is viewed from the front.

External rotation (*eversion*) of the foot tends to break one of the malleoli bones (vertical part of the mortise). These serious injuries tend to be recognized and the injured person goes to a physician.

Approximately 80% of all foot fractures involve the toes; almost all of them could be prevented by safety shoes since they lie within the area protected by the metal toe cap (Rowe, 1985).

Three venous systems drain the lower limbs: (1) a deep central system drains the muscles, (2) a superficial system drains the foot and the skin of the leg, and (3) a perforating system connects the deep and superficial systems.

Physiology

The veins are the body's blood storage location. If the legs don't move, the blood from the heart tends to go down to the legs and stay there (*venous pooling*). This causes more work for the heart, as, for a constant supply of blood, when there is a lower ml of blood per beat, then there must be more beats. Venous pooling causes swelling of the legs (*edema*) and varicose veins. The foot swelling during stationary seated desk work can be overcome by modest leg activity (such as rolling the chair about the workstation) (Winkel and Jorgensen, 1986).

Venous pressure in the ankle of sedentary people is approximately equal to hydrostatic pressure from the right auricle. Pollack and Wood (1949) gave a mean ankle venous pressure of 56 mmHg for sitting and 87 for standing. Nodeland et al. (1983) gave 48 for sitting and 80 for standing. Pollack and Wood reported walking drops ankle venous pressure to about 23 mmHg (Nodeland et al. reported 21) in about

TABLE 50.1 Selected Dimensions (cm) of Nude U.S. Adult Civilians

	Mean		Std. Deviation	
	Females	Males	Females	Males
Stature	162.9 (100%)	175.6 (100%)	6.4	6.7
Crotch height	74.1 (45%)	83.7 (48%)	4.4	4.6
Knee height	51.5 (32%)	55.9 (32%)	2.6	2.8
Foot length	24.4 (15%)	27.0 (15%)	1.2	1.3
Foot breadth	9.0 (6%)	10.1 (6%)	.5	.5

The percentages show the dimension as a percent of stature height. Shoes add 25 mm height for males and 15 mm for females. Shoes add .9 kg to body weight.

Source: Konz, S. 1995. Work Design: Industrial Ergonomics, 4th ed., p. 111. Publishing Horizons, Scottsdale, AZ. With permission.

10 steps. The fall occurs as the calf muscles contract in taking the next step before venous filling has been completed; thus additional blood is pumped out of the leg, causing a further drop in pressure when the calf muscles relax. The drop stabilizes in about 10 steps when the incoming flow to the vein from the capillaries equals the flow out of the leg. Thus, walking can partially compensate for posture; for example, Nodeland et al. reported standing bench work (i.e., with occasional steps around the area) had ankle pressure approximately equal to sitting at a desk (48 mmHg).

Because of vasoconstriction, foot skin temperature (without shoes) usually is the lowest body skin temperature. Normal skin foot temperature = 33.3°C for males but 31.2° for females (Oleson and Fanger, 1973).

Dimensions

Table 50.1 gives some dimensions for U.S. adults. A large portion of the variation in human stature is in leg length; the torso is relatively constant in height. Figure 50.2 shows the mean difference, when standing, between the inside of the two feet is about 107 mm. The distance between foot centerlines is about 107 + 90 = 197 mm (200 mm in round numbers). The distance between outside edges is 107 + 90 + 89 = 286 (300 mm in round numbers). Yet mean height for males is 1756 mm! Thus there is a base of only 200 to 300 mm for a structure of 1756 mm.

The mean pressure (Rys and Konz, 1994) on the feet can be estimated from:

$$MP = .15 + .002\,6\,WT \quad \left(r^2 = .49\right) \tag{1}$$

where
MP = mean pressure, kg/cm^2
WT = body weight, kg

Thus a 70 kg person would have an MP = .33 kg/cm^2. But the peak pressure could be much higher (say 10 kg/cm^2). Diabetics (who may have neuropathic feet) can have pressures of 20 to 30 kg/cm^2, leading to recurrent ulceration and eventual amputation (Boulton et al., 1984).

There is no significant difference between the left and right foot. However, for specific individuals, there often is considerable difference between the left and right foot — especially in width (see Figure 50.3.)

The technical name for differences in leg length in the same person is *leg length discrepancy* (*LLD*). Contreras et al. (1993), summarizing studies with N = 2377, reported that 40% of people had LLD ≤ 5 mm, 30% had LLD ≤ 9 mm, 20% had LLD ≤ 11 mm, and 10% had LLD ≤ 14 mm.

Weight of leg segments (Clauser et al., 1969), as a percent of body weight, are: 1.47 for foot, 4.35 for calf, and 10.27 for thigh; a total leg is 16.10 and both legs are 32.2. For example, the weight of both legs for a 70 kg person would average 70 (.322) = 22.5 kg.

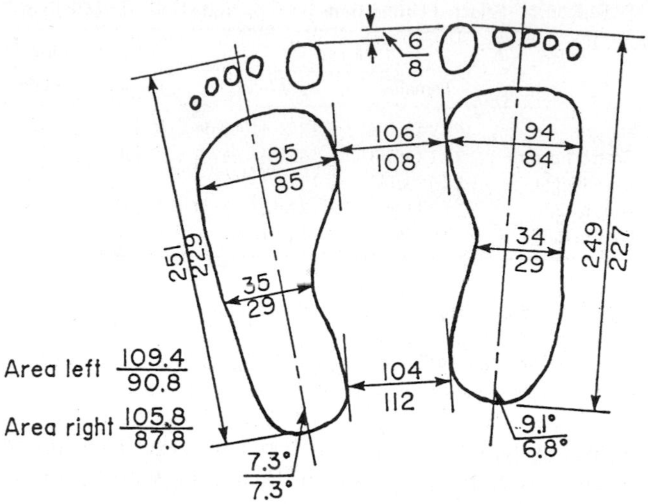

FIGURE 50.2 Footprint dimensions in mm (males above line; females below line); areas in mm²; angles in degrees (Rys and Konz, 1994). Toe area is 10% of contact area. They stood with the right foot slightly (6-8 mm) ahead of the left foot. The left foot (for males) averaged 7.3° to the left of the medial plane; the right foot averaged 9.1° to the right.

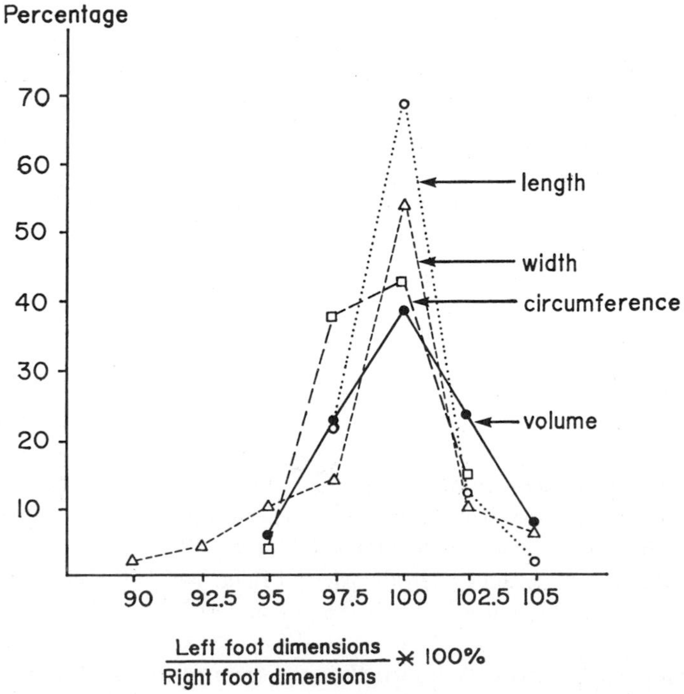

FIGURE 50.3 Distributions of left/right percentage of 84 Americans for foot length, width, circumference and volume (Rys and Konz, 1994). Although the mean of the left does not differ significantly from the right (i.e., the ratio does not differ from 100), for an individual the left can differ from the right.

When people stand at a work surface, there needs to be an indentation for their toes so they can stand close to the worksurface. Rys and Konz (1994) recommend a space at least 150 mm deep, 150 mm high, and 500 mm wide.

50.2 Activities of the Foot

Standing

During standing, the legs will generally move occasionally. Satzler et al. (1993) recorded foot movements for 120 min of standing; people moved a foot approximately every 90 s.

Walking

When walking, the activity of one leg has a shorter swing phase (when the foot is being passed forward) and a longer support (stepping, contact) phase (when the foot is on the ground). The support phase starts at heel strike and ends at toe-off; it has an early, passive section and a later active (propulsion) section (Davis, 1983).

At heel strike, the forward-moving heel hits the ground (causing deceleration). Continued forward motion of the body results in the forefoot contacting the ground; propulsion (acceleration) begins. The heel rises and the foot is pushed backward under the body. This tendency is resisted by friction under the sole; the body is propelled forward. The foot is everted, increasing forefoot contact area on the inner side, until only the skin around the big toe is in ground contact. Finally, contact ceases and the cycle repeats. At heel strike, horizontal velocity decreases from about 450 cm/s to 20 cm/s; heel angle to the floor changes from about 20° prior to heel contact to 0° at 100 ms after contact (Redfern and Rhoades, 1996). During a slip, instead of stopping, the heel continues to move and the leading foot moves out in front of the body.

Since the swing phase is shorter than the support phase, heel strike of the opposite limb occurs during the propulsion section of the support phase.

The length of stride (L) divided by stature height (h) varies linearly with velocity; L/h = .67 at v = .8 m/s and L/h = .9 at 1.7 m/s (Alexander, 1984).

Running

Walking changes to running, for normal size adults, at about 2.5 m/s (6 miles/h) (Alexander, 1984) since it uses less energy (for the same speed). Running differs from walking in that both feet are off the ground for part of the stride. In addition, the heel strike should be renamed the foot strike, since the initial contact probably will be forward of the heel. Peak force is about 3 × body weight at about .1 s after contact. For walking, heel touchdown to toe push-off is .48 s, while, for running, the average contact duration is .29 s (Scanton and McMaster, 1976). After foot strike (usually on the outside edge of the foot), the foot rolls inward and flattens out (*pronation*). Then the foot rolls through the ball and rotates outward (*supination*).

Stepping

Descending stairs demands a gait quite different from ascent (Templer, 1992). For descent, the leading foot swings forward over the nosing edge and stops its forward motion when it is directly over the tread below; the toe is pointed downward. Meanwhile the heel of the rear foot begins to rise, starting a controlled fall downward toward the tread. The heel of the forward foot then is lowered and the weight transferred to the forward foot. The rear foot then begins to swing forward. We tend to hold our center of gravity as far back as possible by leaning backward. Problems are overstepping the nosing with the forward foot, catching the toe of the forward foot, and snagging the heel of the rear foot on the nosing as it swings past. Falls tend to be down the stairs.

For ascent, the leading foot has a toe-off, swing, and first contact with the upper step. The foot is roughly horizontal. The ball of the foot is well forward on the tread; the heel may or may not be on the tread. The rear foot then rises on tiptoe, pushing down and back. The rear leg then begins the swing phase. The primary problem is catching the toe, foot, or heel of either foot on the stair nosing. Another problem is slipping by the rear foot when it pushes backward. Falls tend to be upward.

50.3 Accidents

Falls

The annual death rate from falls in the United States is about 11,700; about 6,500 of these are in the home (especially affecting elderly women). About 15% of the population will have hospital treatment in their lifetime because of injuries from a stair accident (Pauls, 1985). Occupational exposures result in about 1,500 deaths and 300,000 injuries (Leamon, 1992). In industrial fatalities, falls account for about 12%, which is greater than the total for electrical current, fires, burns, and poisons of all types (Leamon, 1992). Of workers injured in falls from heights, about 20% die (Eisma, 1990).

Not all underfoot accidents result in falls, and not all falls result in a lost-time injury. Some falls result in no lost time, some result in sprains and strains, some in broken bones, and some in death. In addition, falls often are not recorded by accident recording procedures (Leamon, 1992). Thus the accident reports tend to drastically underestimate the number of falls. It needs to be emphasized that the risk of a fall varies very much with occupation; all people do not have a dry, level indoor floor. Construction workers, cleaning personnel, transportation workers, and restaurant serving personnel have higher risks (Chaffin et al., 1992).

Andersson and Lagerlof (1983) analyzed 121,000 occupational accidents resulting in injuries; 20,600 had a fall. Falls on the same level were 2/3 and falls between levels were 1/3. For the same level, the main pre-events were slipping (55%) and tripping (19%). (Manning et al. (1988) reported, for falls on the same level, 62% for slips and 17% for trips.) For falls to a lower level (e.g., from stairs, ladders, roofs, vehicles), the main pre-events were loss of support of underlying surface (28%), slipping (28%), and stepping-on-air (8%). The lower-level problem is focused in job trades such as roofers, painters, and maintenance workers. Since the fall has a greater distance, the body velocity and resulting deceleration become greater.

Falls which occur when the person is carrying something are especially dangerous. The object carried decreases stability as a function of the torque above the ankle (weight × object height above ankle). Other problems are that the arms cannot be used for balance (to prevent a fall), to grab a railing, or to break the fall impact.

Causes of Falls

Falls can occur from slips (unexpected horizontal foot movement), trips (restriction of foot movement), and stepping-on-air (unexpected vertical foot movement). Loss of balance and falls can occur without a slip, trip, or stepping-on-air. Examples would be from alcohol or drugs, fainting, etc. Or, people who fall may have blood pressure problems or foot problems (Gabel et al., 1985).

The elderly may be more likely to fall because of deterioration in postural control mechanisms and decrements in visual acuity, strength, endurance, reaction time, and motor control. Furthermore, on falling, the elderly are more likely to sustain a bone fracture (due to osteoporosis) (Maki and Fernie, 1990).

Slips

Slips primarily occur during foot pushoff and heel strike. During pushoff, the person falls forward (less common and less dangerous). In addition, during pushoff, most of the weight has already been transferred to the other foot. If a slip occurs during heel strike, the person falls backward. The critical time is .05 to .1 s after heel strike. Leamon (1992) defines a microslip as a slip less than 2 cm, a slip as 8 to 10 cm, and a slide as uncontrolled movement of the heel. Microslips occur very often and normally are not perceived

TABLE 50.2 Minimizing Slips and Slip Effects

1. Eliminate the lubricant.
   A. Avoid spilling of lubricant.
   B. Quickly clean up lubricant. Note that the lubricant (water, mud) could be on the shoe. "Lubricants" also can be solid objects such as coins, paper clips, hairpins, screws, and metal chips.
   C. Don't add lubricants during cleaning (e.g., don't use an oil mop to clean a waxed floor).
2. Choose good flooring.
   A. Carpet is best (high friction, low effect of lubricants)
   B. For hard surfaces, stainless steel and ceramic are worst as they are the smoothest. Grooved or porous floors reduce lubricant problems but are hard to clean.
   C. Use mats and duckboards (elevated slatted flooring) for local areas where wet floors are common. For example, building entrances often are wet due to tracking in of water and snow. Machines using oil or coolant can be a problem. Mats should have beveled edges to reduce tripping and holes to encourage drainage.
3. Choose good shoes. The heel is critical.
   A. Bevel the rear of heel (reduce the contact angle during heel strike to 0 from 10 to 15°).
   B. Have a tread to penetrate (squeeze out) the lubricant.
   C. Have soft material to increase contact area, and thus grip. Slip resistance of shoes increases after about 5 km of walking so tests on new shoes underestimate the slip resistance (Leclercq, Tisserand, and Saulnier, 1995).
4. Walk carefully (short slow steps).
   A. Keep body center of gravity within stride.
   B. Reduce heel angle at heel strike.
5. If there is a slip, eliminate the fall (e.g., handrails).
6. If there is a fall, decrease consequence of the fall.
   A. Less distance to fall.
   B. Lower impact force/pressure (use soft surfaces such as carpet; minimize sharp objects).

by the person. A slip is perceived and the person typically jerks the upper body, moves the arms, etc.— but does not fall. A slide involves loss of control and usually a fall.

During a slip, there normally is a "lubricant" (water, oil, grease, dust, ice, snow) either on the surface or on the shoe heel (Leclercq et al., 1994).

Slips can also occur, with stationary feet, during pushing and pulling. Although the feet slip, there does not tend to be a fall and injury.

Slips also can occur when the "ground" slopes (front to back or side to side). Examples are ramps and ladder rungs. When moving a cart up or down a ramp, stay above the cart to prevent injury from the cart if it "gets loose." Outdoor walkways often are sloped, have poor illumination, and have water and ice as lubricants.

In the special outdoor circumstances of snow and ice, slipping can be very common; in Finland, slipping outdoors is 10 times more common in winter (Gronqvist and Hirvonen, 1995). The most danger occurs when the ice is "wet" (i.e., close to the freezing point) as the water is a lubricant. For this situation, the best shoes have a soft heel/sole (Shore A hardness <60) made of thermoplastic rubber and exhibiting a large apparent contact area (good tread). Cleats (spikes, studs) are effective if they can penetrate the ice (i.e., if ice is close to the freezing point); if they cannot penetrate (i.e., ice is too hard (say at –10° C), then shoes with spikes are very slippery. Strewing sand on ice is effective on wet ice (i.e., close to freezing) but has relatively little effect on dry ice. Adding salt to melt ice may work if the temperature is close to freezing; if the water then evaporates, this is good, but if the result is just a lubricant added to the ice, the result is bad.

Table 50.2 summarizes how to reduce slips.

Trips

Trips occur during swing. As the foot swings forward, it hits an obstacle and the person falls forward. Thus, in contrast to slips where the problem is excessive horizontal leg movement, with trips the problem is lack of leg movement. Outdoor trips often occur from uneven surfaces (walkways, parking lots) which the person expects to be even. Indoor trips tend to be from objects on the floor or stairs. Usually there is a visual problem.

Stepping-On-Air

Stepping-on-air occurs when the foot has unexpected vertical movement. This can occur on steps when the distance between stairs is unequal; it can occur when there is a hole in the ground; it can occur when there is no ground (i.e., "cliff," edge-of-scaffold, unexpected step, step on spiral stairs, unexpected curb or ramp). Very commonly, stepping-on-air occurs with "single steps" (small changes in elevation) such as curbs or one-step changes in floor level. Steps descending from large trucks and off-road vehicles often present problems. On steps, the fall usually occurs when descending; the fall can be for a considerable distance.

In some cases, the surface is there initially but breaks or moves (step breaks, floor mat slides, a chair used as a stepstool moves). Ladder "feet" need to be non-slip as slipping of the ladder base is the most common pre-event for portable ladder accidents (Alexsson and Carter, 1995). Often stepping-on-air has a visual cause.

Solutions for Falls

The goals are to: (1) prevent the fall and (2) reduce the consequences of the fall.

Prevent the Fall

Scaffolds and work platforms should have a waist-high (107 cm) guardrail as well as a 10 cm high toeboard (reduces slips over the edge as well as reducing falling objects). The top of the guardrail should discourage sitting.

Since the primary problem on stairs is overstepping (Pauls, 1985), for safe stairs: (1) have easily visible steps, (2) provide treads that are long enough, and (3) provide handrails that are both within reach and graspable.

Visual solutions consider both the quantity of light and the quality of light. Note that not everyone's vision is perfect. For example, consider people not wearing their glasses, vision of elderly, the "out of focus" of steps when people wearing bifocals descend stairs, etc.

The quantity of light typically is increased with fixed sources (ceiling lights, street lights, etc.). But lamps fail; the resulting lack of light is especially critical for stairs. One solution is to have two lamps illuminate critical areas. Portable sources such as flashlights are a temporary solution. Too much light causes glare; solutions include nonreflecting surfaces and glare shields for both natural and artificial illumination.

The quality of light is also important. Ideally, the light should give moderate shadows because shadows aid depth perception. Depth perception is improved by using multiple sources and considering the orientation (direction) of the light. Camouflage consists in obliterating contrasts; we are concerned with anticamouflage. Contrast is especially important on steps.

Because walking is automatic, attention must be drawn to steps, especially if the step is "camouflaged" so there is an "ambush." Do not distract attention from a stairs by providing a "view" as a person begins the descent. Call attention to steps by changing the color of the floor (e.g., red carpet on stairs vs. green carpet on approaches), having a handrail (especially for "one-step" stairs), changing wall color on stair walls (e.g., paint changes color and descends at the angle of the stairs), having the handrail color contrast with the wall and stair, and avoiding carpet patterns which confuse depth perception (e.g., narrow strips with strong contrast).

For stairs, the key for friction is the nosing, not the tread. Thus, have a high coefficient of friction for the nosing. Outdoor stairs often are lubricated by rain and snow. Such stairs should have a wash (slope of less than 1:60) to permit water to drain. Perhaps a roof is feasible — even if it doesn't give 100% protection. Prevent water (from the ground or a building) from draining onto the stairs.

Handrails on stairs help prevent falls. The handrail should permit a power grip (11 to 13 cm circumference), have a clearance from the wall of about 3.8 cm, and be 89 to 97 cm above the stair (Konz, 1994).

Box 1
Hard Surface Floors

Concrete floors can either be: (1) the wearing surface itself or (2) a base for other materials (e.g., terrazzo, plastic tiles or sheeting, or carpet).

If concrete is the wearing surface, it must be durable, have satisfactory slip resistance, and be easily cleaned.

Concrete typically is poured in slabs with joints. There are two types of joints: (1) contraction joints (5 to 10 mm wide) and (2) expansion (isolation) joints (about 20 mm wide, filled with compressible material). The expansion joints are not very necessary in a temperature-controlled building and are potential trip hazards as the slab shifts. Good design is to minimize contraction joints in a building. In addition, a horizontal reinforcing dowel between slabs will reduce slab tilting and thus trip hazards.

If tile or plastic is placed on concrete, the problem tends to be slips rather than trips.

If liquids are used or stored in an area, there will be spills. Install drains and slope the floor appropriately.

A handrail must be within reaching distance; at a minimum there should be a handrail on the right side descending. For detailed information on stair design, see Templer (1992) and NFPA (1991).

For mounting/dismounting vehicles, use the *three-contact rule* (at each phase of mounting/dismounting, at least three limbs should maintain contact with steps or handles at the same time).

Ramps for people should have a maximum angle of about 5°. If handtrucks are pushed up the ramp, there should be a landing at least every 3 m in elevation. Have a nonskid surface in the center of each lane. Ramps should have handrails and, if used by vehicles, heavy curbs. If vehicles use a ramp exposed to rain or snow, have a 60 cm strip of abrasive metal plate in the track of each wheel; attach it to the concrete with countersunk holes and flat-head expansion screws. Maximum ramp angles for vehicles are 3° for a power-operated hand truck, 7° for a powered platform truck, 10° for a low-lift pallet-skid truck, 10° for an electric fork truck, and 15° for a gasoline fork truck (Konz, 1994).

Don't use the hands to carry objects while on ladders and stairs.

If a person knows the surface is slippery, walking behavior can be changed. A short stride length reduces foot velocity, gives smaller foot shear force at the heel/ground interface, and keeps the body center of gravity between the feet. Leaning forward helps keep the center of gravity between the feet and, if you fall, it will be forward instead of backward. Sun et al. (1996) report that, when walking down a ramp, people over age 35 decrease stride length and steps/min.

Reduce the Consequences of the Fall

The solution depends upon the task and environment. For example, for workers on a scaffold or roof, use a full-body harness attached to an anchorage point. Another choice is safety nets.

A fall down some stairs is comparable to falling into a hole with jagged rocks at the bottom. For falls on stairs or on the same level, carpet can reduce peak body deceleration on the hip by 20% over hard floors (Makie and Fernie, 1990). (Of course, in addition, carpets have high coefficients of friction and thus have very few slips.) Stair landings reduce the distance of a fall. To minimize impact injuries, stairs, handrails, and balustrading should be free from hostile elements such as projecting elements, sharp edges, and corners.

Box 1 discusses hard surface floors.

50.4 Fatigue/Comfort

The discussion is divided into walking and standing.

Box 2
Shoes

Athletic shoes are divided into running shoes (designed for forward movement) and court shoes (designed for quick side to side movement).

For running, the main problem is the shock of foot strike, yielding arthritis of the knee and hip, Achilles tendonitis, low back pain, and shin splints. Overpronation gives knee pain.

For walking, heel strike is less forceful (1.5 × body weight) so cushioning is less critical (but still desirable); flexibility in the sole allows the normal heel-to-toe roll. For walking, deceleration properties of the cushioning material is important; for standing, time is not as critical and material stiffness and maximum compression is relevant to comfort (Goonetilleke and Himmelsbach, 1992).

At a given walking speed, the energy expenditure values increase by .7 to 1.0% per additional 100 g shoe weight; this increase in energy consumption is approximately 5 times higher than the same weight on the upper body (Smolander et al., 1989). Legg and Mahanty (1986) found it took 6.4 times as much energy to carry a kilogram on the feet as on the back.

Feet often swell so, for fit, buy shoes when your feet are swollen (late in the day). In addition, your left and right foot might vary slightly. Thus buy shoes with at least four pairs of eyelets as they increase the adjustment possibilities.

People with low arches (footprint has a broad connection of two areas) will be more comfortable with shoes with a straighter "inner line" (difficult to distinguish left from right shoe).

Boots support the ankle and calf as well as increasing insulation; they are especially useful for side support, such as when walking outdoors. Boots also protect against chemicals and animal products (such as fats and oils); some boot materials have a better life than others.

For impact protection, use a steel-toe shoe; in some industries (such as mining), metatarsal guards are also used.

Walking

The primary problem is the shock of heel strike being transmitted up the foot, leg and back. For shoe solutions, see Box 2.

The energy cost of walking depends upon the terrain, with a hard surface giving the minimum cost (Pandolf et al., 1976):

$$\text{WLKMET} = \text{C}\left(2.7 + 3.2\,(\text{v} - .7)^{1.65}\right) \qquad (2)$$

where

WLKMET = Walking metabolism, W/kg of body weight
C = Terrain coefficient
= 1.0 for treadmill, blacktop road
= 1.1 for dirt road
= 1.2 for light brush
= 1.3 for hard-packed snow; C = 1.3 + .082 (foot depression, cm)
= 1.5 for heavy brush
= 1.8 for swamp
= 2.1 for sand
v = velocity, m/s (for v > .7 m/s (2.5 km/h))

Standing

Problems can occur with floor temperature and static electricity. However, the primary problem is lack of circulation in the leg and static loading of the muscles.

When wearing normal shoes, 23° C is optimal comfort for floors for standing and walking people; use 25° C for sedentary people (ASHRAE, 1993). Heavy carpet will save about 1% of the total energy used to heat the building (Hager, 1977).

Static electricity is a problem in industries such as electronics. Some solutions are: (1) raise humidity in the air above 40%, (2) use conductive carpets (e.g., with carbon fibers), (3) use an antistatic floor mat, (4) connect the operator to ground with a static-bleed wrist strap, (5) use shoes with static-dissipating soles.

Teitelman et al. (1990) reported preterm births occurred more often (7.7%) when women had jobs with prolonged standing; the rate for sedentary jobs was 4.2% and for active jobs was 2.8%.

Avoid static standing by sitting, by walking, and by shifting posture while standing. If static standing is required, consider a cushioned floor or a footrest (Whistance et al., 1995).

Sitting

Perhaps the person can sit instead of stand. Sitting does tend to restrict movement of the shoulders and thus reach distance. A compromise is a sit-stand seat, such as that used by post office workers sorting mail into boxes. Nijboer and Dul (1987) reported a sit-stand seat was beneficial for upholstery workers even though they could use the seat for only part of the work cycle. Another technique for supporting part of the body weight is to have something to lean on — typically a counter. Another possibility is to sit part of the time. For example, service personnel often are required by management to stand when serving customers; they should be provided chairs for times when there are no customers. Seats should be available for factory personnel during breaks. One firm used swing-down benches on the wall of an aisle; during work they were up against the wall but during worker breaks they were pivoted into sitting position.

Walking

As mentioned earlier, blood circulation in the foot can be brought to normal in as few as 10 steps. Thus design the job so there is occasional walking (such as to get supplies, dispose of materials).

Shifting Posture

It also is possible to shift the posture while standing — remember the bar rail. Bar rails are designed to improve the comfort of those standing at the bar. Satzler et al. (1993) studied four conditions: (1) standing with one foot on a 100 mm high, flat platform, (2) standing with one foot on a 100 mm 15° angled platform, (3) standing with one foot on a 100 mm high, 50 mm diameter bar, and (4) standing with both feet flat on a concrete floor. The three standing aids were preferred over no aid; the two platforms were better than the bar. Note that bar patrons not only support their feet on bar rails but lean on the bar.

Cushioned Floor

The entire floor can be cushioned (carpeted) or the floor can be cushioned locally with a mat. Mats can also be used to raise the feet off the floor (raise above liquid) and act as a frictional surface (avoid slips).

Brantigham et al. (1970) studied a "varied terrain" floor mat with nonuniform resilience density; each placement of the foot caused a slight change in horizontal angle of the foot during weight-bearing. The concept is that many foot problems are due to the "over flat" nature of the built environment. The varied terrain mat enhanced circulation in the lower extremities (reduced venous pressure) and increased skin temperature of the calf .3 to 1.0° C (improved circulation to the surface); about 2/3 of the subjects reported they were less tired when using the special mat.

Additional studies have been done on floor mats (Kuorinka et al., 1978; Rys and Konz, 1988; Redfern and Chaffin, 1988; Rys and Konz, 1989; Rys and Konz, 1990; Konz et al., 1990; Zhang et al., 1990; Stuart-Buttle et al., 1993; Redfern, 1995). Summarizing:

- Floor mats improve comfort (over hard-surfaced floors). Comfort may be increased in the back as well as the legs.
- Mats should compress but not too much. Optimum is about 6% under the feet of a 70 kg adult.

- Mats should have beveled edges to reduce tripping and falling.
- Mats should have a nonslip surface; drain holes may be useful to aid drainage of fluids. In addition, the mat should not slip on the floor.
- Mats may have to be cleaned periodically (e.g., in food service environments). In these cases, large mats are difficult to handle.
- If a raised work platform is used to stand on, the surface should be resilient rather than rigid (i.e., wood or plastic, not steel). The platform also should have a high ratio of surface to holes (i.e., you are not standing on "knives").

50.5 Foot Controls

Although most controls are operated by the hands, some controls are operated by the foot. The foot does not have the dexterity of the hand, but it is connected to the leg instead of the arm so it can exert more force. A leg has approximately 3 times the strength of an arm. A foot control also reduces use of the hand/arm.

Foot controls can be divided into pedals and switches.

Pedals

Pedals can be used for power and control. Power generation can be continuous (bicycle) or discrete (nonpowered automobile brake pedal). For information on continuous power, see Whitt and Wilson (1982) and Brooks et al. (1986).

Discrete power generally is applied by one leg; there does not seem to be any advantage to using the left or right leg. Force using both feet is about 10% higher than using just one foot (Van Buseck, 1965).

A control example is an auto accelerator pedal.

Switches

A foot switch can actuate a machine (such as a punch press). Generally the foot remains on the switch so the time and effort of moving the foot/leg is not important.

On–off controls (such as faucets, clamping fixtures) can be actuated by lateral motion of the knee as well as vertical motion of the foot. The knee should not have to move more than 75 to 100 mm; force requirements should be light. Hospitals use knee switches to actuate faucets to improve germ control on the hands.

Avoid foot pedals/switches which must be operated while standing, because they tend to distort posture and cause back problems.

Defining Terms

Calcaneus: The heel bone; see Figure 50.1.
Edema: Swelling of legs due to fluid retention.
Eversion: External rotation of the foot.
Inversion: Inward rotation of the foot
Metatarsals: Bones in the foot; see Figure 50.1.
Mortise and tenon joint: A type of joint; see Figure 50.1.
Lateral: The outside (side farthest from the centerline).
Leg length discrepancy: Differences in leg length (in the same person).
Medial: The inside (side closest to the centerline).
Phalanges: Bones in the foot; see Figure 50.1.
Pronation: Rolling inward (toward the centerline) of the foot.
Supination: Rolling outward (away from centerline) of the foot.
Three-contact-rule: Rule used on ladders and steps. At least three limbs should be in contact with steps or handles at all times.
Venous pooling: Pooling of blood in the veins of the legs.

References

Alexander, R. 1984. Stride length and speed for adults, children and fossil hominids. *American J. of Physical Anthropology,* 63: 23-27.

Alexsson, P. and Carter, N. 1995. Measures to prevent portable ladder accidents in the construction industry. *Ergonomics,* 38 (2): 250-259.

Andersson, R. and Lagerloff, E. 1983. Accident data in the new Swedish information system on occupational injuries. *Ergonomics,* 26 (1): 33-42.

ASHRAE, 1993. *Handbook of Fundamentals,* ed., Parsons, R. Am. Society of Heating, Refrigeration and Air Conditioning Engineers, Atlanta, GA.

Boulton, A., Franks, C., Betts, R., Duckworth, T., and Ward, J. 1984. Reduction of abnormal foot pressures in diabetic neuropathy using a new polymer insole material. *Diabetes Care,* 7(1): 42-46. Jan-Feb.

Brantigham, R., Beekman, B., Moss, C., and Gordon, R. 1970. Enhanced venous pump activity as a result of standing on a varied terrain floor surface. *J. of Occupational Medicine,* 12(5): 164-169.

Brooks, A., Abbott, A., and Wilson, D. 1986. Human-powered watercraft. *Scientific American,* 256 (12):120-130, December.

Chaffin, D., Woldstad, J., and Trujillo, A. 1992. Floor/shoe slip resistance measurement. *American Industrial Hygiene Association J.,* 53 (5): 283-289.

Clauser, C., McConnville, J., and Young, J. 1969. *Weight, Volume and Center of Mass of the Human Body, AMRL-TR-70,* Aerospace Medical Research Laboratory, Dayton, OH.

Contreras, R., Rys, M., and Konz, S. 1993. Leg length discrepancy, in *The Ergonomics of Manual Work,* eds. W. Marras, W. Karwowski, and L. Pacholski, 199-202, Taylor & Francis, London.

Davis, P. 1983. Human factors contributing to slips, trips and falls. *Ergonomics,* 26 (1): 51-59.

Eisma, T. 1990. Rules changes, worker training help simplify fall prevention. *Occupational Health and Safety,* 52-55, March.

Gabell, A., Simons, M., and Nayak, U. 1985. Falls in the healthy elderly: predisposing causes. *Ergonomics,* 28 (7): 965-975.

Goonetilleke, R. and Himmelsbach, J. 1992. Shoe cushioning and related material properties. *Proceedings of the Human Factors Society,* 519-522.

Grondqvist, R. and Hirvonen, M. 1995. Slipperiness of footwear and mechanisms of walking friction on icy surfaces. *Int. J. of Industrial Ergonomics,* 16:191-200.

Hager, N. 1977. Energy conservation and floor covering materials. *ASHRAE Journal,* 34-39, September.

Konz, S., Bandla, V., Rys, M., and Sambasivan, J. 1990. Standing on concrete vs. floor mats, In *Advances in Industrial Ergonomics and Safety II,* 526-529, ed., B. Das, Taylor & Francis, London.

Konz, S. 1994. Change-in-level. *Facility Design: Manufacturing Engineering,* 118-122, Publishing Horizons, Scottsdale, AZ.

Konz, S. 1995. *Work Design: Industrial Ergonomics,* Publishing Horizons, Scottsdale, AZ.

Kourinka, I., Haakanen, S., Nieminen, K., and Saari, J. 1978. Comparison of floor surfaces for standing work. *Biomechanics VI-B; International Series on Biomechanics, Proceedings of 6th Int. Congress on Biomechanics,* 207-211, University Park Press, Baltimore, MD.

Leamon, T. 1992. The reduction of slip and fall injuries: Part 1--Guidelines for the practitioner and Part II — The scientific basis (knowledge base) for the guide. *Int. J. of Industrial Ergonomics,* 10: 23-34.

Leclercq, S., Tisserand, M., and Saulnier, H. 1994. Assessment of the slip-resistance of floors in the laboratory and in the field: Two complementary methods for two applications. *Int. J. of Industrial Ergonomics,* 13: 297-305.

Leclercq, S., Tisserand, M., and Saulnier, H. 1995. Assessment of slipping resistance of footwear and floor surfaces. *Ergonomics,* 38 (2): 209-219.

Legg, S. and Mahanty, A. 1986. Energy cost of backpacking in heavy boots. *Ergonomics,* 29(3): 433-438.

Manning, D., Ayers, I., Jones, C., Bruce, M., and Cohen, K. 1988. The incidence of underfoot accidents during 1985 in a working population of 10,000 Merseyside people. *J. of Occupied Accidents,* 10: 121-130.

Maki, B. and Fernie, G. 1990. Impact attenuation of floor coverings in simulated falling accidents. *Applied Ergonomics,* 21(2): 107-114.

NFPA, *NFPA Life Safety Code 1991,* NFPA, 1 Batterymarch Park, Quincy, MA 02269.

Nijboer, I. and Dul, J. 1987. Introduction of standing aids in the furniture industry, in *Musculoskeletal Disorders at Work,* ed. P. Buckle, 227-233, Taylor & Francis, London.

Nodeland, H., Ingemansen, R., Reed, R., and Aukland, K. 1983. A telemetric technique for studies of venous pressure in the human leg during different positions and activities. *Clinical Physiology,* 3: 573-576.

Oleson, B. and Fanger, P. 1973. The skin temperature distribution for resting man in comfort. *Archives des Sciences Physiologigues,* 27(4): A385-93.

Pandolf, K., Haisman, M., and Goldman, R. 1976. Metabolic energy expenditure and terrain coefficients for walking on snow. *Ergonomics,* 19: 683-690.

Pauls, J. 1985. Review of stair safety research with an emphasis on Canadian studies. *Ergonomics,* 28(7): 999-1010.

Pollack, A. and Wood, E. 1949. Venous pressure in the saphenous vein at the ankle in man during exercise and changes in posture. *J. Applied Physiology,* 1: 649-662, March.

Redfern, M. and Chaffin, D. 1988. The effects of floor types on standing tolerance in industry, in *Trends in Ergonomics/Human Factors V,* ed., F. Aghazadeh, 401-405, Elsevier, Amsterdam.

Redfern, M. 1995. Influence of flooring on standing fatigue. *Human Factors,* 37 (3): 570-581.

Redfern, M. and Rhoades, T. 1996. Fall prevention in industry using slip resistance testing, in *Occupational Ergonomics,* Eds., Bhattacharya, A. and McGloughlin, J., 463-476, Marcel Dekker, New York.

Rowe, M. 1985. *Orthopaedic Problems at Work,* Perinton Press, Fairport, NY.

Rys, M. and Konz, S. 1988. Standing work: carpet vs. concrete, *Proceedings of the Human Factors Society,* 522-526.

Rys, M. and Konz, S. 1989. An evaluation of floor surfaces, *Proceedings of the Human Factors Society,* 517-520.

Rys, M. and Konz, S. 1990. Floor mats. *Proceedings of the Human Factors Society,* 575-579.

Rys, M. and Konz, S. 1994. Standing. *Ergonomics,* 37(4): 677-687.

Satzler, C., Satzler, L., and Konz, S. 1993. Standing aids. *Proceedings of the Ayoub Symposium,* 29-31, Texas Tech. University, Lubbock, TX.

Scanton, P. and McMaster, J. 1976. Momentary distribution of forces under the foot. *J. Biomechanics,* 9:45-48.

Smolander, J., Louhevarra, V., and Hakola, T. 1989. Cardiorespiratory strain during walking in snow with boots of different weights. *Ergonomics,* 32(1): 3-13.

Stuart-Buttle, C., Marras, W., and Kim, J. 1993. The influence of anti-fatigue mats on back and leg fatigue. *Proceedings of the Human Factors and Ergonomics Society,* 769-773.

Sun, J., Walters, W., Svensson, N. and Lloyd, D. 1996. The influence of surface slope on human gait characteristics: a study of urban pedestrians walking on an inclined surface. *Ergonomics,* 39(4): 677-692.

Templer, J. 1992. *The Staircase: Studies of Hazards, Falls, and Safer Design,* MIT Press, Cambridge, MA.

Teitelman, A., Welch, L., Hellenbrand, K., and Bracken, M. 1990. Effect of maternal work activity on preterm birth and low birth weight. *Am. J. Epidemiology,* 131: 104-113.

Van Buseck, C. 1965. Excerpts from maximal brake pedal forces produced by male and female drivers. Research Report EM-18, Warren, MI: GM, Jan.

Whistance, R., Adams, L., van Geems, B., and Bridger, R., 1995. Postural adaptations to workbench modifications in standing workers. *Ergonomics,* 38(12): 2485-2503.

Whitt, F. and Wilson, D. 1982. *Bicycling Science,* MIT Press, Cambridge MA.

Winkel, J. and Jorgensen, K. 1986. Evaluation of foot swelling and lower-limb temperature in relation to leg activity during long-term seated office work. *Ergonomics,* 29(2): 313-328.

Zhang, L., Drury, C., Woollet, S. 1991. Constrained standing: evaluating the foot/floor interface. *Ergonomics,* 34: 175-192.

For Further Information

Konz, S. 1995. *Work Design: Industrial Ergonomics,* 4th edition, Holcomb Hathaway, Scottsdale AZ. This popular textbook concisely summarizes many aspects of job design and gives detailed design guidelines.

Konz, S. 1994. *Facility Design: Manufacturing Engineering,* 2nd edition, Holcomb Hathaway, Scottsdale, AZ. Gives many details and design recommendations for design and arrangement of industrial facilities.

Ergonomics. This journal, published in England, publishes articles on ergonomics from authors around the world.

Section II
Low Back Disorders

51

Epidemiology of Back Pain in Industry

Gunnar B. J. Andersson
Rush-Presbyterian-St. Luke's Medical Center

51.1 Introduction

Epidemiologic research in low back pain (LBP) has been, and still is, hampered by methodologic problems in definition, classification, and diagnosis. Objective evidence of existing low back pain is often lacking, and people's recall of previous episodes is poor. The intermittent nature of low back pain complicates prevalence studies, and studies of disability due to LBP are influenced by legal and socioeconomic factors. Methodologic problems also exist in the quantification of physical exposures that might be of etiologic importance.

In general, data about back pain may be obtained from official health registers or by retrospective, prospective, or cross-sectional surveys of general populations or of specific industrial populations. Such data are useful in defining the magnitude of the problem. Care must be taken when interpreting these data, however. As mentioned above there is no consensus on classification and diagnosis, making it difficult to rely on insurance and hospital data (Wood and Badley, 1987). Sickness absence and disability data are heavily influenced by work conditions and the legal and socioeconomic situation, and there is a poor correlation between tissue injury and disability.

Data from workers' compensation claims are affected by several inherent biases (Abenhaim and Suissa, 1987): (a) all workers are not covered by worker compensation programs; (b) the claims data are mainly administrative and therefore, while accurate on absence and cost, lack validity on symptoms and diagnosis; (c) all workers with back pain do not file a claim, and many do not stay away from work.

51.2 The Magnitude of the Problem

The magnitude of any health problem is measured by prevalence and incidence. In a prevalence study, the presence of LBP and other important variables is determined at one point in time (point prevalence)

0-8493-2641-9/99/$0.00+$.50

or during one period of time (period prevalence) for each member of the population studied or for a representative sample. Incidence may be defined as the number of people who develop LBP over a specified time period, such as their lifetimes (lifetime incidence, which is synonymous with lifetime prevalence) or in a single year (annual incidence). In short, prevalence means all cases of LBP, whereas incidence means all *new* cases of LBP.

51.3 National Studies

Information obtained from different countries is considered separately because the differing socioeconomic factors of these populations may influence the results. This is particularly true for disability data, which are significantly determined by local legal, social, and economic factors.

United States

Between 10 and 17% of adults in the U.S. have a back pain episode in a given year (Cunningham and Kelsey, 1984; Deyo and Tsui-Wu, 1987; Praemer et al. 1992). In about one-third of these the pain is severe and chronic. In a large National Health Survey performed from 1985 through 1988, about 4.1 million persons per year reported a "disc disorder," and another 4.6 million a "back strain" (Praemer, 1992). Other epidemiologic data show that back pain is the most frequent cause of activity limitation in people below age 45, the second most common reason for patient visits (over 14% of new visits are for back pain), the fifth ranking reason for hospitalization, and the third most common reason for surgical procedures (Praemer et al., 1992; Taylor et al., 1994; Hart et al., 1995; Andersson, 1997). About 2% of the U.S. workforce (500,000 workers), report compensation back injuries each year (National Safety Council, 1991). The frequency of surgical procedures for back related conditions has risen dramatically in the U.S. over the past two decades. In fact, it has more than doubled from 1979 to 1990. In that later year, 279,000 back operations were performed on adults; 232,500 without fusion and 46,500 fusions (Taylor et al., 1994).

United Kingdom

British surveys place low back pain at the top of the list of medical conditions as well. In 1992–93 there were 81 million certified back-related sick days, and about 7 million visits to general practitioners for back pain (National Back Pain Association, 1994). During the same period there were 33,000 work-related back injuries. Frank (1993) reported that back pain was the single largest cause of sick leave in 1988–89, responsible for 12.5% of all sickness absence days.

Sweden

Swedish national insurance data show a consistent sickness absence in percent of all annual sickness absence from the early sixties to the late eighties. In the 1961 to 1971 period, the average absence was 12.5% or 1% of all workdays (Helander, 1973). In 1983 the percent was 10.9, and in 1987, 13.5% (Nachemson, 1991). Unfortunately, the number of sickness absence days rose dramatically during that period so that the percent of insured sick listed for back pain rose from 1% of the working population in 1970 to 8% in 1987, the number of days per absence rose similarly from 20 to 34 days per year, and the cost in terms of lost production increased 16 fold (Table 15.1). Retirement and disability pensions caused by back pain rose by 6000% from 1952 to 1987.

During 1983 and 1984, a prospective Swedish study analyzed all patients who were sicklisted for LBP in a district of Gothenburg containing 49,000 subjects from 20 to 65 years of age (Choler et al. 1985). A total of 7,526 sickness absence episodes for LBP were reported over an 18-month period. Fifty-seven percent of patients recovered in 1 week, 90% in 6 weeks, and 95% after 12 weeks. At the end of a year 1.2% remained work disabled. Those with sciatica were out of work for longer periods of time than were

TABLE 51.1 Estimated Sicklisting for LBP and Associated Cost Due to Loss of Production in Sweden

Year	% of Insured[1]	Number of Days	Cost of Loss of Production[2] ($ Million[3])
1970	1	20	52
1975	3	22	179
1980	4	25	285
1987	8	34	806

[1]4.7 million
[2]Based on 1987 wages and social costs.
[3]Assuming $1 = 6 Sw. Cr.
Adapted from Nachemson AL (1991): Back Pain. Causes, diagnosis and treatment. The Swedish Council of Technology Assessment in Health Care. Stockholm.

patients who had back pain only. Recurrent pain and disability occurred in 12% over the 18-month period of observation.

Canada

Lee et al. (1985) analyzed data on musculoskeletal complaints based on the 1978 to 1979 Canada Health Survey. A prevalence of 4.4% with "serious back and spine problems" was calculated. The total number of disability days exceeded 21 million, and the average sickness absence period was 21.4 days. There was no difference in prevalence between men and women.

51.4 Cross-Sectional Studies

A cross-sectional epidemiologic study is one in which a population is studied at a single point in time, or over a defined period, in an attempt to evaluate all members of that population. In the past decades several cross-sectional studies have been performed. Table 51.2 presents the prevalence and lifetime incidence of LBP, as determined by some of these studies. The prevalence rates vary from a low of 12.0% to a high of 35.0%. Some authors report a higher prevalence in females, but others found no difference. The lifetime incidence rates are higher and range from 48.8% to 69.9%.

United States

In 1973, Nagi, Riley, and Newby determined the prevalence rates of persistent back pain of persons between 18 and 64 years residing in Columbus, Ohio (Nagi et al., 1973). A random sample of 1,135 subjects was studied, of whom 203 (18%) reported "often being bothered with pain in the back." Of those with back problems, 62% had had a spine radiograph; 26% had worn a back support; and 4% had had back operations.

Frymoyer et al. (1980, 1983) performed a retrospective and cross-sectional analysis of 1,221 males 18 to 55 years of age who had enrolled in a family practice facility from 1975 to 1978. Almost 70% had had LBP. When the data from that study were extrapolated to the 50 million working American males in the age group 18 to 55, it was calculated that 38.5 million workdays are lost annually. Patients with severe LBP had significantly more leg complaints, sought more medical care and treatment for LBP, and had lost more time from work for this reason when compared to subjects with no or moderate LBP. Sciatica-like symptoms were present in 28.9% of the males with moderate LBP and 54.5% of the males with severe LBP. Objective reports of numbness were present in 14.0% of the males with moderate LBP and 37.4% of those with severe LBP, while weakness was reported by 17.9% of those with moderate LBP and 44.0% with severe LBP.

TABLE 51.2 Prevalence and Lifetime Incidence of LBP in Difference Cross-Sectional Studies

Lifetime Incidence (%)	Prevalence (%) Point	Period	Study Group N	Age	Sex	Comment	Reference
62.6	12.0	—	449	30–60	M		Biering-Sorensen (1983)
61.4	15.2	—	479	30–60	F		
48.8	—	—	692	15–72	F		Hirsch, et al. (1969)
60.0	—	—	1193	25–59	M	Industrial Population	Hult (1954)
69.9	—	—	1221	28–55	M		Frymoyer, et al. (1980)
—	18.0	—	1135	18–64	M/F		Nagi, et al. (1973)
61.0	—	31	716	40–47	M	1-month period	Svensson and Andersson (1983)
67.0	—	35	1640	38–64	F	1-month period	Svensson, et al. (1988)
51.4	22.2	—	3091	20+	M		Valkenburg and Haanen (1982)
57.8	30.2	—	3493	20+	F		
—	12.9	—	3316	—	M/F	8 work groups	Magora (1972)
—	—	25	—	40–59	M	1-year period	Gyntelberg (1974)
—	29.0	—	575	55–	M/F		Bergenudd and Nilsson (1988)
75.0	—	21	7217	30–	M/F	1-month period	Heliovaara (1989)
58.0	—	36	2667	—	M/F	Adults, 1-year period	Walsh (1992)
59.0	33.0	—	4000	—	M/F	Adults	Skovrow, et al (1993)
—	18.0	—	4256	—	F	Family care employees	Moens, et al. (1993)

Andersson GBJ (1997): The epidemiology of spinal disorders, in *The Adult Spine:* Principles and Practice, 2nd edition, J.W. Frymoyer, Ed. Lippincott-Raven, Philadelphia, pp. 93-141.

Studies by Kelsey (1975a,b) and Kelsey and Hardy (1975) sampled 20- to 64-year-olds residing in the New Haven (Connecticut) area who had lumbar X-rays taken over a two-year period for suspected herniated nucleus pulposus. The researchers divided the sample into those with surgically confirmed herniated discs and those who had probable or possible herniated discs based on clinical signs and symptoms. She was able to define a variety of risk factors related to the diagnosis of herniated lumbar disc; including sedentary occupations, driving of motor vehicles, chronic cough and bronchitis, lack of physical exercise, participants in certain sports (baseball, golf, and bowling), suburban residence, and pregnancy.

Kelsey et al. (1984a,b) and Kelsey and Golden (1988) later performed another case-control study in Connecticut from 1979 to 1981 with minor methodologic modifications. The study population was 20- to 64-year-old women and men who had had X-rays and myelograms at various health centers in New Haven and Hartford. As in the previous study, they were divided into those with surgically confirmed disc herniations and those with probable or possible disc herniations. A control group of nonback patients admitted for in-hospital services was matched for sex and age. A number of possible risk factors were studied and odds ratios determined. Frequent lifting and twisting were both significant risk factors, as were driving and smoking.

The prevalence of low back pain in elderly people (over age 65) was determined in a survey of 3,097 persons living in rural parts of Iowa (Lavsky-Shulan et al., 1985). Twenty-four percent of the women and 18% of the men had low back pain in the year preceding the survey, and 40% had back pain at the time of the interview. Five percent of the population had been operated on.

Scandinavia

A number of studies have been performed in the city of Göteborg, Sweden (about 450,000 inhabitants). Four are reviewed here. Hirsch, Jonsson, and Lewin (1969) interviewed 692 women (15 to 72 years of age), selected at random to represent the adult Swedish female population. The lifetime incidence of LBP was 48.8% and increased with age up to 55 years, after which no further increase was noted. Horal (1969) and Westrin (1970, 1973) studied a random sample of subjects who in 1964 had been sicklisted for LBP by physicians in Göteborg, Sweden. They were compared to a control group matched with respect to sex, age, and sickness benefit but not previously sicklisted for LBP. Of the total group, Horal studied 212 pairs of probands and controls, and

TABLE 51.3 Data on Prevalence and Use of Medical Services from Two Retrospective Cross-Sectional Surveys in Göteborg, Sweden

	Men (Age 40–47) (N = 940)	Women (Age 38–64) (N = 1,760)
Lifetime prevalence	61	66
One-Month prevalence	31	35
Chronic pain	3.5–4	
Physician visit	40	38
Radiograph	23	30
Hospitalized	3.5	3.4
Operation	0.8	1.0

Adapted from Svensson HO, Andersson GBJ (1983): Low back pain in forty to forty-seven year old men: Work history and work environment factors. *Spine* 8:272-276. and Svensson HO, Andersson GBJ (1989): The relationship of low-back pain, work history, work environment, and stress: A retrospective cross-sectional study of 38- to 64-year-old women. *Spine* 14:517-522.

shortly thereafter Westrin studied 214 (78% of the base material). Ninety-five percent (95%) of the probands had had LBP in the preceding 3 to 4 years, and 52% had ongoing pain at the time of the interview. In the control group, the corresponding figures were 49% and 27%, respectively. This means that sickness absence statistics severely underestimate the true frequency of low back pain.

Svensson and Andersson (Andersson et al., 1983; Svensson, 1982; Svensson and Andersson, 1982, 1983; Svensson et al., 1983) studied a randomized sample of 940 40- to 47-year-old men in Göteborg, Sweden. Seven hundred and sixteen men were interviewed, and information about the remaining 234 was obtained from the Swedish National Health Insurance Office. Thirty-three percent of all sickness absence episodes experienced during their working life were spine related, constituting 47% of all sickness absence days; 3.6% were totally disabled and 4% had been off work more than 3 months because of LBP in the 3 years preceding the study. Forty percent had consulted a physician, 3.5% had been admitted to a hospital, and 0.8% had been operated on because of their LBP (Table 51.3).

The same study design was later used to survey 1,640 38- to 64-year-old women (Svensson et al., 1988; Svensson and Andersson, 1989). Of these, 19% had been off work because of LBP in the preceding three-year period, 3.5% for 3 months or longer. About 2.6% of 38- to 49-year old women had significant work disability, whereas the corresponding percentage among 50- to 64-year-olds was 5.9.

Biering-Sorensen (1982) sampled 82% of all 30- to 60-year-old inhabitants in Glostrup, Denmark. There were 449 men and 479 women. An extensive questionnaire regarding low back problems was administered along with objective measurements of spine function. Twelve months after the examination 99% of the study population completed a follow-up questionnaire on LBP occurring in the intervening period. The lifetime prevalence/incidence of LBP appears in Figure 51.1 along with the one-year period and point prevalence data. In general, increasing age was associated with increasing episodes of LBP. Work absence at some time was reported by 22.5% of those who had LBP, 10% had needed some job adjustment, and 63% had changed their jobs because of back pain. Of those who had experienced LBP, 60% had consulted a physician, 25% a specialist, and 15% a chiropractor (Biering-Sorenson, 1983). About 30% had had radiographs taken of the lumbar spine, 4.5% had been admitted to a hospital, and 1% had been operated on because of LBP.

The prevalence rate of sciatica and its impact on Finnish society was estimated by Heliovaara 1988 (see also Heliovaara et al., 1987), based on a sample of 8,000 persons representative of the Finnish population aged 30 or over. Sciatica was present in 5.3% of men and 3.7% of women. In both genders the prevalence rates were highest in the 45- to 64-year-old group. The prevalence of definite herniated discs was 1.9% for men and 1.3% for women. Low back syndrome other than sciatica was present in 12.5% of men and 17% of women. Disability due to lumbar disc syndrome was estimated at 3.5% in men, 4.5% in women.

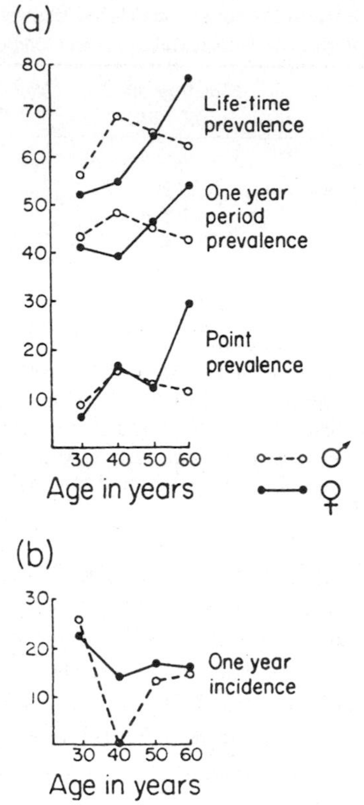

FIGURE 51.1 Top: Prevalence rates of "low back trouble" by age and sex. (Redrawn from Biering-Sorensen F (1982): Low back trouble in a general population of 30-, 40-, 50-, and 60-year-old men and women. Study design, representativeness, and basic results. *Dan Med Bull* 29:289.) Bottom: Incidence of "low back trouble" over one year in a general population. (From Biering-Sorensen F (1982): Low back trouble in a general population of 30-, 40-, 50-, and 60-year-old men and women. Study design, representativeness, and basic results. *Dan Med Bull* 29:289. With permission.)

Israel

Magora and Taustein (1969) studies a random sample of 3,316 individuals taken at random. Present and past LBP was determined. Four hundred twenty-nine (12.9%) were found to suffer from LBP at the time of the survey (point prevalence), and 92% (394) of those had pain on and off from 6 months to 11 years or more before the investigation. The majority of the subjects with LBP did not take sick leave (57.8%) and, of those who did, 29.4% had absence periods from 1 to 10 days.

The Netherlands

A study of 3,091 men and 3,493 women 20 years of age and older was performed by Valkenburg and Haanen (1982) in 1975 through 1978 in the Dutch city of Zoetermeer. A questionnaire, physical examination, and radiographs were obtained. The prevalence of LBP in men and women increased slightly with age up to 65 years, and thereafter decreased (Table 51.4). Disc prolapse, defined by clinical signs and symptoms, was found in 1.9% of men and 2.2% of women. Considerable disability was attributed to LBP: 85% had recurrences; 30% had LBP for more than 3 months, and 30% had become bedridden at some point by their symptoms. Nearly half of the men and one-third of the women reported that they had been unfit to work because of LBP at some time, and 8% of the men as well as 4% of the women had changed their jobs because of these complaints. Twenty-eight percent of the men and 42% of the women had consulted a physician for LBP.

TABLE 51.4 Low Back Complaints and Work Incapacity in the Zoetermeer Study

	Men (%)	Rel %	Women (%)	Rel %*
Point-prevalence	22.2		30.2	
Lifetime Incidence	51.4		57.8	
>3 Months	14.3	28	19.6	34
Unfit for Work	24.3	47	19.5	34
Work Change	4.2	8	2.4	4
Recurrences	—	85	—	85

*Rel % refers to proportion among those with a lifetime history of low back pain.

TABLE 51.5 Rates of Selected Back Operations in the United States per 100,000 of General Population

Procedure	1979	1981	1983	1985	1987	% Increase
Laminectomy	31	36	41	41	38	23
Discectomy	59	57	81	96	103	75
Lumbar Fusion	5	9	10	18	15	200

Deyo RA, Cherwin D, Conrad D, Volinn E (1991): Cost, controversy, crisis: Low Back Pain and the Health of the Public. *Ann Rev Publ Health* 12:141-56.

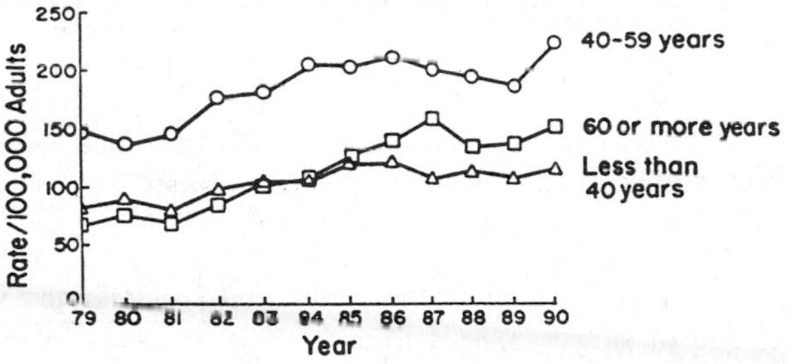

FIGURE 51.2 Low back surgery rates per 100,000 adults, by age, 1979-1990. (Adapted from Taylor VM, Deyo RA, Cherkin DC, Kreuter W (1994): Low back pain hospitalization. Recent United States Trends and Regional Variations. *Spine* 19:1207-1213.)

Belgium

About 4,000 individuals were studied to explore the influence of sociocultural and employment variables on back pain (Skovron et al., 1994). The lifetime incidence of back pain was 59%, and 33% had ongoing back pain (point prevalence). Age (OR = odds ratio, 2.0) and female gender (OR = 1.42) were associated with an increased risk of a LBP history. Sociocultural factors influenced the risk of first-time back pain episodes, but not of disability or severity.

Hospitalizations and Operations

Volinn et al. (1994) examined the National Hospital Discharge Survey for time trends (1979–1987), and for geographic variations (1987). The U.S. rate of lower back surgery increased 49% over the time period reviewed, while the rate of nonsurgical low back pain hospitalization decreased by 33%. Table 51.5 illustrates the dramatic increase in back operations in the U.S. and breaks it down into actual procedures (Deyo et al., 1991). A comparatively larger increase occurred in fusions than in laminectomies or discectomies. Volinn

TABLE 51.6 Hospitalizations in the U.S. for Back Conditions in 1988, Based on First Listed Diagnosis by ICD.9.CM.Code

Diagnosis	# Hospitalizations	ICD.9.CM.Code
Intervertebral Disc Disorders	417,000	722
Other and Unspecified Back Disorders	178,000	724
Fracture of the Vertebral Column	76,000	805
Spondylosis and Allied Disorders	75,000	721
Sprains and Strains of Other and Unspecified Parts of Back	55,000	847
Sprains and Strains of Sacroiliac Region	42,000	846

From National Hospital Discharge Survey, 1988.

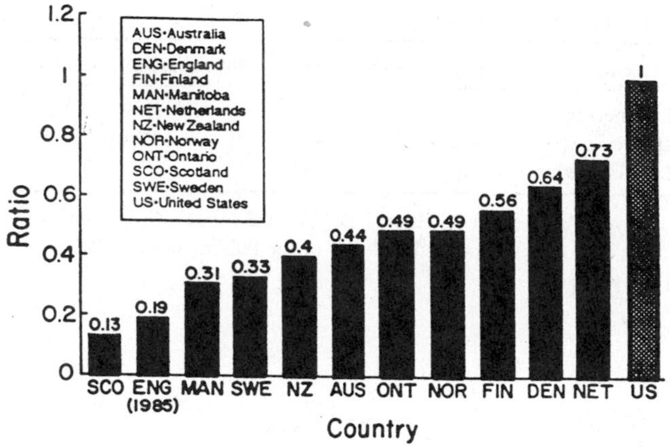

FIGURE 51.3 Ratios of back surgery rates in 11 countries or Canadian provinces compared to the back surgery rate in the U.S. (1988–1989). Adapted from Cherkin DC, Deyo RA, Loeser JD, Bush T, Waddell G (1994): An international comparison of back surgery rates. *Spine* 19:1201-1206.)

et al. (1994) also report large regional variation in hospitalization, surgical rates, and length of stay. They found that in 1987 the rate of surgery ranged from 77/100,000 adults in the Northeast region to 146/100,000 in the South region. Variations in surgical rates exist between areas of each region as well. Thus, among counties in the State of Washington rates of surgery for low back pain varied from 11.5/100,000 to 172/100,000, an almost 1.5-fold difference (Volinn, 1992). This indicates that cultural differences and practice patterns have a major influence on hospitalizations and procedures. Taylor et al. (1994) further analyzed the National Hospital Survey Data from 1979 through 1990. The increase in the number of surgical procedures continued from 1987 to 1990. Over the 11-year time period surveyed, adult low back operations increased by 55% from 147,500 (1979) to 279,000 (1990). This corresponds to an increase from 102 to 158 per 100,000 adults. The rise was particularly great for fusions which increased by 100% from 13 to 26/100,000. In 1990, there were 46,500 lumbar fusions and 232,500 low back pain operations without fusion. The upward trends occurred among all age groups, but were particularly great for patients aged 60 and older (Figure 51.2). Nonsurgical hospitalizations, on the other hand, decreased from 402/100,000 in 1979 to 150/100,000 in 1990. Table 51.6 shows the hospitalizations for back conditions in 1988.

There are marked international variations in rates of back surgery (Deyo et al., 1991, Cherkin et al., 1994). Nachemson (1991) has reported that the number of operative procedures for herniated nucleus pulposus (HNP) in Sweden has remained stable at 200/million inhabitants/year from the mid-1950s through the 1980s. The corresponding numbers in Finland (1967–1977) were 350 (Heliovaara, 1988) and in Great Britain 100 (Benn and Wood, 1975; Wood and Badley, 1987). The number of people operated on for HNP per million/year in the U.S. has been assessed in the past as ranging between 450 to 900

TABLE 51.7 Prevalence of Back and Spine Impairments in the United States, 1988

Impairment by		# in millions	per 1,000 population
Total		15.431	64.1
Gender	Males	6.701	57.4
	Females	8.730	70.3
Race	White	13.957	68.7
	Black	1.137	38.7
	Other	0.336	—
Age	0–17	0.714	11.2
	18–44	8.295	80.5
	45–64	4.105	90.1
	65–74	1.333	75.9
	75–84	0.780	87.2
	84–	0.203	93.6

Adapted from Praemer A, Furnes S, Rice DP (1992): Musculoskeletal conditions in the United States. *Amer Acad Ort Surg*, Park Ridge, IL pp. 1-199. Based on the National Health Interview Survey 1985–88.

(Kane, 1980; Frymoyer, 1988; Kelsey and White, 1980). In 1990, it was 1310 (Taylor et al. 1994). A recent study compares the back surgery rates of thirteen countries and provinces revealing that in the United States, it is at least 40% higher than in any other country, and more than five times higher than in Scotland and England (Figure 51.3). Differences in the underlying prevalence of back pain was not felt to explain the differences in surgical rates. Cultural differences, differences in practice patterns, and availability are more likely explanations.

Chronic Back Pain

Limited data is available about the prevalence of chronic back pain. The NHANES II data suggest that 33.7% of those reporting back pain lasting for at least two weeks had pain for six months or longer. This percentage is similar to the one reported by Valkenburg and Haanen for a Dutch population, where 34% of those with back pain had a duration of 3 months or longer (14.3% of all males, 19.6% of all females). Praemer et al. (1992) used the 1988 National Health Interview Survey (1985–1988) to determine the frequency of impairment in the U.S. Impairment in the survey was defined as a chronic or permanent defect representing a decrease or loss of ability to perform various functions. Musculoskeletal impairment was the most prevalent impairment in people up to age 65, and back and spine impairments the most frequently reported subcategory of musculoskeletal (51.7%). The annual rates vary significantly by gender and age group (Table 51.7). Back and spine impairments were more common in females (70.3/1,000) than males (57.4/1,000), and more common among whites (68.7/1,000) than blacks (38.7/1,000). In 1988, back and spine impairments resulted in over 185 million restricted activity days (21.0/impairment), including 83 million bed days (5.4/impairment) (Table 51.8). About 56% of restricted activity days occurred among women.

Rossignol et al. (1988) (see also Abenhaim et al., 1988) followed a cohort of 2,341 cases, randomly sampled to represent occupational back compensated individuals in the state of Quebec (Canada) in 1981. From the initial injury date, 6.7% of compensated workers were still absent from work after 6 months, accounting for 68% of work days lost and 76% of the total compensation cost for low back pain. When the cumulative absence was calculated over the three years, 9.7% of workers were absent for 6 months or longer. This increase illustrates the recurrent nature of back pain. A logistic regression model was used to calculate risk factors associated with absences of 6 months or more. Age and site of symptoms were the two most important variables. An increase in age of 23 years doubled the odds of accumulating at least 6 months of absence and lumbar symptoms were 2.86 times more likely than thoracic symptoms

TABLE 51.8 Restricted Activity Days and Bed Days/Impairment for Back or Spine Disorders in 1988 by Major Population Subgroup

		Rescheduled Activity Days/Impairment	Bed Days/ Impairment
Total		12.0	5.4
Gender	Male	14.5	5.9
	Female	10.1	5.0
Race	White	10.1	4.2
	Black	31.8	15.4
Age	>65	11.5	5.4
	65–	15.0	5.3

Adapted from Praemer A, Furnes S, Rice DP (1992): Musculoskeletal conditions in the United States. *Amer Acad Ort Surg,* Park Ridge, IL pp. 1-199.

to become chronic. The odds ratio for gender and occupation was not statistically significant. When occupation was eliminated, sex achieved statistical significance, however (OR 1.92;95% CI = 1.19-3.11).

van Doorn (1995) studied a self-employed subset of Dutch dentists, veterinarians, physicians, and physical therapists. Twenty-three percent of claims lasted longer than 6 months or were considered chronic (van Doorn 1995). The risk for chronicity increased with advancing age. Using a predictive cox-regression model, van Doorn found that specific diagnosis, older age, previous back pain, and psychosocial problems were all factors negatively influencing recovery.

51.5 Occupational Risk Factors

The relationship between occupational factors and low back pain is difficult to determine because exposure is usually difficult and sometimes impossible to quantify. Using job titles alone is not appropriate. Healthy workers may well stay in the same occupation and job, while workers with low back pain may leave a job and move to a less taxing one. The result is a shift in prevalence of back pain from heavy to light jobs. Another problem is the definition of what is heavy and what is light. Traditionally, heavy physical jobs have been defined as jobs with high energy demand, and light as jobs with low energy demand. Many low energy jobs are static in nature, however, which as such can be a risk factor for LBP. Burdorf (1992) reviewed 81 original epidemiologic articles to determine how well they assessed exposure to postural loading. In 58% there was no exposure data, while in the remaining subjects, exposure was based on questionnaire in 33%, observation in 9%, and direct measurement in 5% only. Throughout, it was felt that the quality of exposure data was poor.

The recall of exposure is even worse than the recall of previous pain and disability, and is often influenced by the insurance system. Subjects will tend to relate LBP to an "injury" or particular exposure if this results in compensation. Kelsey (1975a) found, for example, that 70.9% of men receiving compensation associated the onset of their problem with a specific event, compared to 35.5% of men not receiving compensation. Swedish and Canadian injury statistics would suggest the same.

Further complicating the situation is the fact that exposure to several occupational risk factors often occurs in the same job. For example, a truck driver may have to load and unload his truck (lifting), sit for many hours in an unchanged posture (static loading), and be exposed to whole-body vibration.

The seven most frequently discussed occupational risk factors are listed in Table 51.9. Among them, the first six are all physical risk factors which have been experimentally associated with the development of injuries in spinal tissues. Because they often occur together, the relative importance of each is difficult to determine. The seventh, "psychological and psychosocial work factors," is probably more related to disability claims than to occurrence of a specific pathology.

TABLE 51.9 The Seven Most Frequently Discussed Risk Factors for Lower Back Pain

Heavy Physical Work
Static Work Postures
Frequent Bending and Twisting
Lifting, Pushing and Pulling
Repetitive Work
Vibrations
Psychological and Psychosocial

Heavy Physical Work

Several investigations indicate an increase in sickness absence due to low back pain, and also an increase in low back symptoms in individuals performing physically heavy work (Andersson, 1981, 1997; Snook, 1982). A few will be discussed here. Magora (1970) found prevalence to vary greatly between occupations. Uyttendaele et al. (1981) found that university and hospital employees with occupations demanding high physical strains were absent from work significantly more often due to LBP than those with light physical work. Unskilled laborers had the highest prevalence rate for disc prolapse and lumbago in the Dutch study by Valkenburg and Haanen (1982). Svensson and Andersson (1983) found heavy physical work to be strongly associated with the occurrence of low back pain; and the highest prevalence of low back pain in their first cross-sectional study, which included men only, was in men with physically heavy professions (Table 51.9). In a subsequent study in women, however, Svensson and Andersson (1989) found only psychological variables, such as dissatisfaction with the work environment and the work itself, and fatigue and worry at the end of the work day to be directly associated with low back pain. Forward bending, lifting, standing, and monotonous work correlated to LBP, but in the univariate analysis.

Klein et al. (1984) found the highest rates of back sprain/strains among workers in physically heavy industries and with physically heavy occupations, as did Behrens et al. (1994). Lloyd et al. (1986) report a higher lifetime prevalence and three-month prevalence in miners than in office workers. Finnish employees in the metal industry were followed over a 10-year period (Leino et al., 1987). Low back pain was more common among blue collar workers than among white collar workers. Among blue collar workers, however, only weak associations were found between morbidity and indices of physical work load.

Mitchell (1985) surveyed low back pain occurring in RAF (Royal Air Force) ground trades personnel. The male prevalence was 9.1% overall, and increased with the severity of the job grade. In all job grades, the prevalence was highest in the 40 to 49 year age group. Both sickness absence and frequency of employment restriction increased with the severity of the job grade. Herrin et al. (1986) analyzed musculoskeletal injury rates among 6,900 workers in 55 industrial jobs with almost 3,000 different manual tasks. The manual exertion requirements were determined using various job stress indices, and 2-year retrospective as well as 1-year prospective medical reports were gathered. Musculoskeletal problems were twice as common if predicted peak lumbosacral disc compression forces exceeded 6,800 N (1,500 lb.). Back problems were about 2.5 times higher in workers with high physical performance requirements. Using the U.S. "Quality of Employment Survey" for 1972–1973, Leigh and Sheetz (1989) found physically heavy work to be associated with back pain, particularly farming (OR = 5.17). A cross-sectional survey at a U.S. oil company employing 10,350 full-time regular employees found the relative risk for a low-back injury to be 1.57 in physically demanding jobs (Tsai et al., 1992).

Similar to LBP, sciatica also has an increased prevalence in physically heavy occupations. Wickstrom et al. (1978) found a higher prevalence of sciatica in concrete reinforcement workers than in computer technicians, and higher rates of HNP have been reported in physically demanding occupations by Hrubec and Nashold (1975), Kelsey et al. (1984), and Videman et al. (1984). Riihimaki (1985) found higher prevalances of sciatica among concrete reinforcement workers compared to house painters. Hrubec and Nashold (1975) found a negative association between HNP and clerical work, while craftsmen and

TABLE 51.10 Annual Prevalence of Back Sprains/Strains (BS) and Unspecified Back Injuries (BI) in Selected Studies

Author & year	Prevalence (%)	Comment
Hult (1954)	2.9	BI. Worker samples
Blow/Jackson (1971)	3.6	BI. Dockworkers
Stubbs/Nicholson (1979)	0.9	BI. Construction workers
Klein et al. (1984)	0.7	BS. General population
Anderson (1986)	4.5	BI. Dockyard workers
Abenhaim/Suissa (1987)	1.4	BI. General population

TABLE 51.11 Prevalence of LBP in Studies Comparing Physically Heavy and Light Work

Reference	Physically heavy (%)	Physically light (%)	N	Comment
Hult (1954)				
LBP	64.2	52.7		
Severe LBP	10.6	6.8		
Work absence	43.5	22.5		
Lawrence (1955)	41.0	29.0	362	
Rowe (1963, 1969)	47.0	35.0		Medical visits
Ikata (1965)	22.4	5.2		Sciatica
Magora (1972)	21.6	10.4		
Lloyd et al. (1986)	69.0	58.0		
	35.0	26.0		3 months

Source: Rowe ML (1963): Preliminary statistical study of low back pain. *J Occup Med* 5(7): 336-341; Rowe ML (1969): Low back pain in industry. A position paper. *J Occup Med* 11(4): 161-169.

foremen had a significantly higher than average risk. Heliovaara (1987), in a large Finnish study, found that men had a significantly higher probability of being hospitalized for HNP if they were blue collar workers or motor vehicle drivers. A lesser difference between occupations was present in women.

All these studies would seem to support the idea that heavy physical work increases the risk of low back pain. Confounding factors may exist, however, and the level at which physical work load becomes a risk factor is not determined. Other studies are less clear. Lockshin et al. (1969), Sairanen et al. (1981), Porter (1987), and Bigos and Battié (1992) did not find any differences in prevalence between heavy and light work. And occupational factors did not predict the incidence of LBP over a 1-year follow-up period in a Danish cross-sectional sample (Biering-Sorensen and Thomsen, 1986). It is not surprising in epidemiologic studies of this type that some studies will be negative. The association between work load and LBP is relevant in time order, strong based on the majority of studies, dose-related (Heliovaara et al., 1991; Burdorf et al., 1991; Kumar, 1990; Punnett et al., 1989), consistent, and biologically coherent. Table 51.10 summarizes the annual rates per 100 workers reported by different investigators among different worker samples, using injury reports, registry statistics, questionnaires, and interviews. Although differences in research methods make data difficult to compare directly, the average prevalence rates range from a low of 0.75 to a high of 4.5, i.e., large injury rates are reported everywhere. Prevalence rates are higher for workers in physically demanding jobs (Table 51.11). This reflects not only the risk, but also the difficulty of performing a heavy job with a painful lower back. It is obvious from the data, however, that back pain is very frequent in physically light jobs as well. A change in the work environment, therefore, cannot be expected to solve the back injury problem completely.

Static Work Postures

Working in predominantly one posture, such as prolonged sitting, seems to carry an increased risk for back pain. But there is considerable disagreement. While many studies indicate an increased risk of low back pain in subjects with predominantly sitting work postures (Hult 1954; Kroemer and Robinette, 1969; Lawrence, 1955; Magora, 1972; Partridge, 1969), others do not (Bergquist-Ullman, 1977; Braun, 1969; Frymoyer et al., 1983; Damkot et al., 1984; Heliovaara, 1987; Riihimaki, 1989; Svensson and Andersson, 1983, 1989; Westrin, 1973). Kelsey (1975, 1978) and Kelsey and Hardy (1979) found that men who spend more than half their workday in a car have threefold increased risk of disc herniation. This could be due to the combined effect of sitting and vehicle vibration. Magora (1972) found that those who either sat or stood during most of the workday had an increased risk of low back pain. Frequent changes in posture were also found to increase the risk of back pain, however.

Frequent Bending and Twisting

The association between low back symptoms and frequent bending and twisting is difficult to evaluate as a separate activity because lifting is usually also involved. A large number of studies report an association between these movements in general and low back pain (Bergquist-Ullman and Larsson, 1977; Brown, 1975; Damkot et al., 1984; Daniel et al., 1980; Frymoyer et al., 1983; Frymoyer et al., 1980; Lloyd et al., 1986; Maeda et al., 1980; Troup, 1984; Troup et al., 1970; Wickstrom et al., 1978).

Magora (1973) established a connection between both excessive bending and occasional bending on the one hand and low back pain on the other, and a similar finding was made by Chaffin and Park (1973). Keyserling et al. (1988) found low back pain to be related to asymmetric postures in an automobile assembly plant, and Riihimaki et al. (1989) report a relationship between sciatica and twisting and bending work postures not only in physically heavy jobs, but in office workers as well. Further analysis of the data from the automobile assembly plant (Punnett et al., 1991) revealed the following odds ratios (OR) and confidence intervals (CI); mild trunk flexion (OR 4.9, CI 1.4–17.4), severe trunk flexion (OR 5.7, CI 1.6–20.4), and trunk twist or lateral bend (OR 5.9, CI 1.6–21.4). The risk increased with exposure to multiple postures and with increasing duration of exposure. Thus, a combination of mild flexion and twisting produced an OR of 7.4 (CI 1.8–29.4).

Lifting, Pushing, and Pulling

It has been clearly established that back pain can be triggered by lifting, but the frequency at which lifting is the main cause of back pain varies between studies (Bergquist-Ullman and Larsson, 1977; Bigos et al., 1986; Hult, 1954; Ikata, 1965; Kelsey, 1975; Klein et al., 1984; Lloyd et al., 1986; Magora, 1970, 1972). Sudden unexpected maximal efforts were found by Magora (1972) to be particularly harmful, and Glover (1960), Tichauer (1965), and Troup et al. (1970) express the same opinion about lifting in combination with lateral bending and twisting.

Chaffin and Park (1973) found that workers involved in heavy manual lifting had about eight times the number of lower back injuries as those with a more sedentary work situation. Svensson and Andersson (1983) found a direct association between occurrence of low back pain and frequent lifting, as did Frymoyer et al. (1980, 1983) and Hult (1954). Snook (1982) found that a worker was three times more susceptible to compensable low back injury if exposed to excessive manual handling tasks. The National Institute of Occupational Safety and Health (NIOSH) estimated in 1981 that one third of the U.S. workforce lifted in excess of what was considered acceptable, and that lifting was a major cause of low back pain. They also concluded that the severity of injury rate while lifting was proportional to the weight of the object, the bulk of the object, the location of the object at the start of the lift, and the frequency of lifting.

Kelsey (1975), on the other hand, in her first study found no indication that workers with herniated discs did more lifting on the job than workers without such symptoms. Further, there was no indication in that study that jobs requiring pushing, pulling, or carrying either increased or decreased the risk of herniated discs. In the second study, however, (Kelsey et al., 1984b) frequent lifting was identified as a risk factor for HNP, the risk increasing the heavier the weight lifted and the more frequent the lifts. Lifting while twisting increased the risk even further. The odds ratio for HNP in subjects performing frequent lifting of heavy weights while twisting was 3.4.

Troup et al. (1981) followed 802 workers over a two-year period. Half of all episodes of LBP were associated with a "back injury" and of those, 1 of 3 occurred from manual material handling. A cross-sectional survey in a small market town in the south of England correlated the life-time occupational history of 545 adults with the prevalence of low back pain (Walsh et al., 1989). The strongest associations were for lifting and moving weights over 25Kg (RR 2.0, CI 1.1–3.7). When considering those individuals with severe unremitting back pain, the risk ratio for lifting increased to 5.3 among the men, and 2.9 among the women.

Repetitive Work

Repetitive work increases, in general, the sickness absence rate. Low back pain seems to be no exception in this respect. This may explain, in part, why assembly line industries have a higher incidence of low back pain among their manual workers than among their office employees (Bergquist-Ullman and Larsson, 1977).

Duration of employment may also be associated with back pain, although the healthy worker effect makes this difficult to evaluate. Magora (1970) found an association with back pain among workers doing heavy work, and Astrand (1987) reports a direct relationship between back pain and time of employment among 391 male employees in a Swedish pulp and paper industry.

Vibrations

There are several studies suggesting an increasing risk of low back pain in drivers of tractors, (Christ, 1973, 1974; Christ and Dupuis, 1966; Christ and Dupuis, 1968; Damlund et al., 1986; Dupuis and Christ, 1966; Dupuis et al., 1972; Dupuis and Zerlett, 1986; Hulshof and van Zanten, 1987; Rosegger and Rosegger, 1960; Seidel et al., 1986; Seidel and Heide, 1986) of trucks, (Gruber, 1976; Kelsey and Hardy, 1975; Kristen et al., 1981; Wilder et al., 1982; Behrens et al., 1994) of buses, (Gruber and Ziperman, 1974; Kelsey and Hardy, 1975) and of airplanes (Fitzgerald and Crotty, 1972; Schulte-Wintrop and Knoche, 1978). These studies also suggest that low back pain occurs at an earlier age in subjects exposed to vibration.

Kelsey and Hardy (1975) found that truck driving increased the risk of disc herniation by a factor of four, while tractor driving and car commuting (20 miles or more per day) increased the risk by a factor of two. In a later study, the risk of HNP was related to the type of vehicle, indicating significant differences between different brands of cars (Kelsey et al., 1984a).

Hulshof and van Zanten (1987) have reviewed the epidemiologic data supporting a relationship between whole-body vibration and low back pain. They concluded that vibration was a probable risk factor in helicopter pilots, tractor drivers, construction machine operators, and transportation workers. They were critical of the data, however, concluding that none of the many studies reviewed was adequate in terms of the quality of exposure data, effect data, study design, and methodology. Most studies did not control for confounding variables, and only a few had control populations.

Dupuis and Zerlett (1986), in a 10-year prospective study (1961 to 1971), describe an increased incidence of backache reports from 47% to 58% among tractor drivers, while Hilfert et al. (1981) reported that 70% of 352 construction machine operators had periodic LBP compared to 54% in an unexposed control group. Gruber and Zipermann (1974) compared 1,448 male interstate bus drivers to three control groups. Experienced drivers had a higher prevalence of spinal disorders than controls. A significant correlation was found between prevalence rates and exposure level. In a later study Gruber (1976) found significantly higher back pain prevalence rates among 3,205 interstate truck drivers compared to 1,137 air traffic controllers. Behrens et al. (1994) found the highest prevalence estimates for back injuries among

U.S. occupational groups to occur among truck drivers. In a Danish study, 2,045 full-time male bus drivers in the three largest cities in Denmark were compared to 195 motormen (Netterstrom and Juel, 1989). The prevalence of low back pain was 57% vs. 40%. Burdorf and Zondervan (1990) found an odds ratio of 3.6 for low back pain among crane operators compared to controls.

Buckle et al. (1980), Frymoyer et al. (1980), Backman (1983), Damkot et al. (1984), Walsh et al. (1989), and Biering-Sorensen and Thomsen (1986) all report an association between automobile use and low back pain. The risk of being hospitalized because of HNP was high among motor vehicle drivers in the Finnish Study by Heliovaara (1987), and Pentinnen (1987) and Riihimaki et al. (1989) report an increased risk of sciatica with motor vehicle driving in other Finnish studies.

Studies of vibration-exposed populations have also indicated that radiographic changes occur in the spines of these subjects (Dupuis and Zerlett, 1986). These studies are retrospective and usually limited to selected subject groups. It is therefore difficult to make cause-effect conclusions. Further, the radiographic findings are diverse and cannot all be explained by mechanical theory. Nonetheless the prevalence rates of radiographic changes are very high.

Psychological and Psychosocial Work Factors

Several psychological work factors, including monotony at work, work dissatisfaction, and poor relationship to co-workers have been found to increase the risk of complaining of low back pain and report workers compensation claims. (Astrand, 1987; Battie, 1989; Cunningham and Kelsey, 1984; Deyo and Tsui-Wu, 1987; Bergquist-Ullman and Larsson, 1977; Bigos et al., 1996; Bigos and Spengler et al., 1986; Damkot et al., 1984; Svensson and Andersson, 1983 1989, 1988). Monotony had a direct relationship to low back pain in the study by Svensson and Andersson (1983), while Bergquist-Ullman and Larsson (1977) found that workers with monotonous jobs, requiring less concentration, had a longer sickness absence following low back pain than the others. Diminished work satisfaction has also been found to be related to an increased risk of low back pain by Westrin (1970), Magora (1973), and Svensson et al. (1983). Bergenudd and Nilsson (1988) found that middle-aged workers had an increased prevalence of back pain if they had physically heavy jobs and that the association increased further when the workers were dissatisfied with their work. Individuals with back pain had been less successful in a childhood intelligence test, and on average had a shorter education. Kelsey and Golden (1988) point out that since most of the studies are retrospective it is difficult to determine whether psychological factors are antecedents or consequences of pain. Bigos et al. (1986) and Battie et al. (1992) prospective studies concluded, however, that psychological work factors were more important than physical work factors as risk indicators of low back pain.

Acknowledgment

This manuscript is in part based on Chapter 7, The epidemiology of spinal disorders, by G. B. J. Andersson, in *The Adult Spine* (Ed. J. W. Frymoyer), Lippincott-Raven Publishers, Philadelphia, pp. 93-141, 1997.

References

Abenhaim L, Suissa S, Rossignol M (1988): Risk of recurrence of occupational back pain over three year follow-up. *Brit J Ind Med* 45:829-833.

Abenhaim LL, Suissa S (1987): Importance and economic burden of occupational back pain: A study of 2500 cases representative of Quebec. *J Occup Med* 29:670-674.

Andersson GBJ (1997): The epidemiology of spinal disorders, in *The Adult Spine:* Principles and Practice, 2nd edition, J.W. Frymoyer, Ed. Lippincott-Raven, Philadelphia, pp. 93-141.

Andersson GBJ, Svensson HO, Oden A (1983): The intensity of work recovery in low back pain. *Spine* 8:880-884.

Andersson GBJ (1981): Epidemiologic aspects on low back pain in industry. *Spine* 6:53-60.

Astrand NE (1987): Medical, psychological, and social factors associated with back abnormalities and self reported back pain. *Brit J Ind Med* 44:327-336.

Backman AL (1983): Health survey of professional drivers. *Scand J Work Environ Health* 9:30-35.

Battié MC (1989): The reliability of physical factors as predictors of the occurrence of back pain reports. A prospective study within industry. Thesis, University of Goteborg, Goteborg, Sweden.

Battié MC, Bigos SJ, Fisher LD, Spengler DM, Hansson TH, Nachemson AL, Wortley D (1990): Anthropometric and clinical measurements as predictors of industrial back pain complaints: A prospective study. *J Spinal Disorders* 3:195-204.

Behrens V, Seligman P, Cameron L, Mathias CGT, Fine L (1994): The prevalence of back pain, hand discomfort and dermatitis in the U.S. working population. *Am J Public Health* 84:1780-1785.

Benn RT, Wood PH (1975): Pain in the back: an attempt to estimate the size of the problem. *Rheumatol Rehabil* 14:121-128.

Bergenudd H, Nilsson B (1988): Back pain in middle age; occupational work load and psychologic factors: an epidemiologic survey. *Spine* 13:58-60.

Bergquist-Ullman M, Larsson U (1977): Acute low back pain in industry. A controlled prospective study with special reference to therapy and confounding factors. *Acta Orthop Scand* (Suppl) (170):1-117.

Biering-Sorensen F, Thomsen C (1986): Medical, social and occupational history as risk indicators for low-back trouble in a general population. *Spine* 11:720-725.

Biering-Sorensen F (1982): Low back trouble in a general population of 30-, 40-, 50-, and 60-year-old men and women. Study design, representativeness, and basic results. *Dan Med Bull* 29:289.

Biering-Sorensen F (1983): A prospective study of low back pain in a general population, III: Medical service-work consequence. *Scand J Rehab Med* 15:89.

Bigos SJ, Battie MC (1992): Risk factors for industrial back problems, in *Seminars in Spine Surgery*, Vol. 4 (Ed. S.W. Wiesel), W.B. Saunders, Philadelphia, pp. 2-11.

Bigos SJ, Battie MC, Fisher LD, Fordyce WE, Hansson TH, Nachemson AL, Spengler DM (1996): A longitudinal, prospective study of acute industrial back problems: The influence of work perceptions and psychosocial factors. *Spine.*

Bigos SJ, Spengler DM, Martin NA, Zeh J, Fisher L, Nachemson A (1986): Back injuries in industry: A retrospective study. III. Employee-related factors. *Spine* 11:252-256.

Braun W (1969): Ursachen des lumbalen Bandscheiberverfalls. Die Wirbelsaule in Forschung und Praxis 43.

Brown JR (1975): Factors contributing to the development of low back pain in industrial workers. *Amer Industr Hyg Assoc J* 36:26-31.

Buckle PW, Kember PA, Wood AD, Wood SN (1980): Factors influencing occupational back pain in Bedfordshire. *Spine* 5:254-258.

Burdorf A, Govaert G, Elders L (1991): Postural load and back pain of workers in the manufacturing of prefabricated concrete elements. *Ergonomics* 34:909-18.

Burdorf A, Zondervan H (1990): An epidemiological study of low-back pain in crane operators. *Ergonomics* 33:981-987.

Burdorf A (1992): Exposure assessment of risk factors for disorders of the back in occupational epidemiology. *Scand J Work Environ Health* 18:1-9.

Chaffin DB, Park KS (1973): A longitudinal study of low-back pain as associated with occupational weight lifting factors. *Amer Ind Hyg Assoc J* 34:513-525.

Cherkin DC, Deyo RA, Loeser JD, Bush T, Waddell G (1994): An international comparison of back surgery rates. *Spine* 19:1201-1206.

Choler U, Larsson R, Nachemson A, Peterson LE (1985): Back pain. Spri report 188 (in Swedish):1-100.

Christ W (1973): Beanspruchung und Leistungsfahigkeit des Menschen bei underbrochener und Langzeit-Exposition mit stochastischen Schwingungen. Dissertation, Technical University, Darmstadt. VDI Ber 11:1-85.

Christ W (1974): Belastung durch mechanische Schwingungen und mogliche Gesunheitsschadigungen im Bereich der Wirbelsaule. *Fortschr Med* 92:705-708.

Christ W, Dupuis H (1968): Untersuchung der Moglichkeit von gesundheitlichen Schadigungen im Bereich der Wirbelsaule. *Med Welt* 36:1919-1920; 37:1967-1972.

Christ W, Dupuis H (1966): Uber die Beanspruchung der Wirbelsaule unter dem Einfluss sinusformiger und stochastischer Schwingungen. *Int Z Angew Physiol Einschl Arbeitsphysiol* 22:258-278.

Cunningham LS, Kelsey JL (1984): Epidemiology of musculoskeletal impairments and associated disability. *Am J Public Health* 74:574-579.

Cust G, Pearson JC, Mair A (1972): The prevalence of low back pain in nurses. *Int Nurs Rev* 19:169-179.

Damkot DK, Pope MH, Lord J, Frymoyer JW (1984): The relationship between work history, work environment and low-back pain in men. *Spine* 9:395-399.

Damlund M, Goth S, Hasle P, Munk K (1982): Low-back pain and early retirement among Danish semi-skilled construction workers. *Scand J Work Environ Health* (Suppl) 8:100-104.

Damlund M, Goth S, Hasle P, Munk K (1986): Low back strain in Danish semi-skilled construction work. *Applied Ergonomics* 17:31-39.

Daniel JW, Fairbank JC, Vale PT, O'Brien JP (1980): Low back pain in the steel industry: A clinical, economic and occupational analysis at a North Wales integrated steelworks of the British Steel Corporation. *J Soc Occup Med* 30:49-56.

Deyo RA, Cherwin D, Conrad D, Volinn E (1991): Cost, controversy, crisis: Low Back Pain and the Health of the Public. *Ann Rev Publ Health* 12:141-56.

Deyo RA, Tsui-Wu Y-J (1987): Descriptive epidemiology of low-back pain and its related medical care in the United States. *Spine* 12:264-268.

Dupuis H, Christ W (1966): Untersuchung der Moglichkeit von Gesundheits-schadigungen im Bereich der Wirbelsaule bei Schlepperfahrern. Research report, Max-Planck-Institute fur Landarbeit und Landtechnik, Bad Kreuznach.

Dupuis H, Zerlett G (1986): *The Effects of Whole-Body Vibration.* New York: Springer-Verlag.

Dupuis H, Hartung E, Louda L (1972): Vergleich regelloser Schwingungen eines berenzten Frequensbereiches mit sinusformigen Schwingungen hinsichtlich der Einwirkung auf den Meschnen. *Ergonomics* 15:237-265.

Fitzgerald JG, Crotty J (1972): The incidence of backache among aircrew and groundcrew in the RAF.

Frank A (1993): Low Back Pain. *Brit Med Journal* 306:901-908.

Frymoyer JW (1988): Back pain and sciatica. *New Engl J Med* 318:291-300.

Frymoyer JW, Pope MH, Costanza MC, Rosen JC, Goggin JE, Wilder DG (1980): Epidemiologic studies of low-back pain. *Spine* 5:419-423.

Frymoyer JW, Pope MH, Clements JH et al. (1983): Risk factors in low back pain. *J Bone Joint Surg* 65:213.

Glover JR (1960): Back pain and hyperaesthesia. *Lancet* 1:1165-1169.

Gruber GJ (1976): Relationships between whole-body vibration and morbidity patterns among interstate truck drivers. U.S. Department of Health, Education and Welfare, DHEW (NIOSH) Publication No. 77-167.

Gruber GJ, Ziperman HH (1974): Relationship between whole-body vibration and morbidity patterns among motor coach operators. DHEW (NIOSH) Publication No. 75-104.

Hart LG, Deyo RA, cherkin DC (1995): Physician office visits for low back pain. *Spine* 20:11-19.

Helander E (1973): Back pain and work disability, (in Swedish). *Socialmed Tidskr* 50:398.

Heliovaara M (1987): Occupation and risk of herniated lumbar intervertebral disc or sciatica leading to hospitalization. *J Chronic Dis* 40:259-264.

Heliovaara M (1988): Epidemiology of sciatica and herniated lumbar intervertebral disc. Helsinki: The Research Institute for Social Security, pp. 1-147.

Heliovaara M, Makela M, Knekt P, Impivaara O, Aromaa A (1991): Determinants of sciatica and low-back pain. *Spine* 16:608-14.

Heliovaara M, Knekt P, Aroma A (1987): Incidence and risk factors of herniated lumbar disc or sciatica leading to hospitalization. *J Chron Dis* 3:251-285.

Herrin GD, Jaraiedi M, Anderson CK (1986): Prediction of overexertion injuries using biomechanical and psychophysical models. *Am Industr Hygiene Assoc J* 47:322-330.

Hilfert R, Kohne G, Toussaint R, Zerlett G (1981): Probleme der Ganzkorperschwingungs-belastung von Erdbaumaschinenfuhrern. Zentralblatt Arbeitsmedizin, Arbeitsschutz, Prophylaxe Ergonomie 31:152-155.

Hirsch C, Jonsson B, Lewin T (1969): Low-back symptoms in a Swedish female population. *Clin Orthop* 63:171.

Horal J (1969): The clinical appearance of low back pain disorders in the city of Gothenburg, Sweden. Comparisons of incapacitated probands and matched controls. *Acta Orthop Scand* (suppl) 118:1.

Hrubec Z, Nashold BS Jr. (1975): Epidemiology of lumbar disc lesions in the military in World War II. *Am J Epidem* 102(5):367-376.

Hulshof C, van Zanten BV (1987): Whole body vibration and low back pain. A review of epidemiological studies. *Int Arch Occup Environ Health* 59:205-220.

Hult L (1954): Cervical, dorsal, and lumbar spinal syndromes. *Acta Orthop Scand* (Suppl) 17:1-102.

Ikata T (1965): Statistical and dynamic studies of lesions due to overloading on the spine. *Shikoku Acta Med* 40:262.

Kane WJ (1980): Worldwide incidence rates of laminectomy for lumbar disc herniations. Presented at the annual meeting of ISSLS, New Orleans, LA.

Kelsey JL, Golden AL (1988): Occupational and workplace factors associated with low-back pain. *Occup Med* 3:7-16.

Kelsey JL, Hardy RJ (1975): Driving of motor vehicles as a risk factor for acute herniated lumbar intervertebral disc. *Am J Epidemiol* 102(1):63-73.

Kelsey JL (1975a): An epidemiological study of the relationship between occupations and acute herniated lumbar intervertebral discs. *Int J Epidemiol* 4(3):197-205.

Kelsey JL (1975b): An epidemiological study of acute herniated lumbar intervertebral discs. *Rheumatol Rehabil* 14(3):144-159.

Kelsey JL, White AA III (1980): Epidemiology and impact on low back pain. *Spine* 5(2):133-142.

Keyserling WM, Punnett L, Fine LJ (1988): Trunk posture and back pain: Identification and control of occupational risk factors. *Appl Ind Hyg* 3:87-92.

Klaukka T, Sievers K, Takala J (1982): Epidemiology of rheumatic diseases in Finland in 1967-76. *Scand J Rheumatol* (suppl) 47:5-15.

Klein BP, Jensen RC, Sanderson LM (1984): Assessment of workers' compensation claims for back strains/sprains. *J Occup Med* 26:443-448.

Kristen H, Lukeschitsch G, Ramach W (1981): Untersuchung der Lendenwirbelsaule bei Kleinlasttransportarbeitern. *Arb Med Soz Med Prav Med* 61:226-229.

Kroemer KH, Robinette JC (1969): Ergonomics in the design of office furniture. *Industr Med Surg* 38:115-125.

Kumar S (1990): Cumulative load as a risk factor for back pain. *Spine* 15:1311-16.

Lavsky-Shulan M, Wallace RB, Kohout FJ et al. (1985): Prevalence and Functional Correlates of Low Back Pain in the Elderly: The Iowa 65+ Rural Health Study. *J Am Geriatrics Soc* 33:23-28.

Lawrence JS (1955): Rheumatism in coal miners, Part III. Occupational factors. *Br J Indust Med* 12:249-261.

Lee P, Helewa A, Smythe HA et al. (1985): Epidemiology of musculoskeletal disorders (complaints) and related disability in Canada. *J Rheumatol* 12:1169-1173.

Leigh JP, Sheetz RM (1989): Prevalence of back pain among full-time United States workers. *Brit J Industr Med* 4:651-657.

Leino P, Aro S, Hasan J (1987): Trunk muscle function and low back disorders: A ten-year follow-up study. *J Chronic Dis* 40:289-296.

Lloyd MH, Gauld S, Soutar CA (1986): Epidemiologic study of back pain in miners and office workers. *Spine* 11:136-140.

Lockshin MD, Higgins IT, Higgins MW, Dodge HJ, Canale N (1969): Rheumatism in mining communities in Marion County, West Virginia. *Am J Epidemiol* 90:17-29.

Maeda K, Harada N, Takamatsu M (1980): Factor analysis of complaints of occupational cervicobrachial disorder in assembly lines of a cigarette factory. *Kurume Med J* 27:253-261.

Magora A (1970): Investigation of the relation between low back pain and occupation. 2. Work history. *Industr Med Surg* 39(12):504-510.

Magora A (1973): Investigation of the relation between low back pain and occupation. 5. Psychological aspects. *Scand J Rehab Med* 5:191-196.

Magora A (1972): Investigation of the relation between low back pain and occupation. 3. Physical requirements: Sitting, standing and weight lifting. *Industr Med Surg* 41:5-9.

Magora A, Taustein I (1969): An investigation of the problem of sick-leave in the patient suffering from low back pain. *Ind Med Surg* 38:398.

Mitchell JN (1985): Low back pain and the prospects for employment. *J Soc Occup Med* 35:91-94.

Nachemson AL (1991): Back Pain. Causes, diagnosis and treatment. The Swedish Council of Technology Assessment in Health Care. Stockholm.

Nagi SZ, Riley LE, Newby LG (1973): A social epidemiology of back pain in a general population. *J Chron Dis* 26:769.

National Safety Council (1991). *Accident Facts*, Chicago.

Netterstrom B, Juel K (1989): Low Back Trouble among Urban Bus Drivers in Denmark. *Scand J Soc Med* 17:203-206.

NIOSH (1981): Work practices guide for manual lifting. DHHS (NIOSH) Publication No 81-122.

Partridge REH, Anderson JAD (1969): Back pain in industrial workers. Proceedings of the International Rheumatology Congress, Prague, Czechoslovakia, Abstract 284.

Pentinnen J (1987): Back pain and sciatica in Finnish farmers. Helsinki: Publications of the Social Insurance Institution, ML:71.

Porter RW (1987): Does hard work prevent disc protrusion? *Clin Biomech* 2:196-198.

Praemer A, Furnes S, Rice DP (1992): Musculoskeletal conditions in the United States. *Amer Acad Ort Surg*, Park Ridge, IL pp. 1-199.

Punnett L, Fine LJ, Keyserling WM, Herrin GO, Chaffin DB (1991): Back Disorders and nonneutral trunk postures of automobile assembly workers. *Scand J Work Environ Health* 17:337-346.

Riihimaki H, Wickstrom G, Hanninen K, Luopajarvi T (1989): Predictors of sciatic pain among concrete reinforcement workers and house painters. A five year follow-up. *Scand J Work Environ Health* 15:415-423.

Riihimaki H (1985): Back pain and heavy physical work: A comparative study of concrete reinforcement workers and maintenance house painters. *Br J Ind Med* 42:226-232.

Rosegger R, Rosegger S (1960): Arbeitsmedizinische Erkenntnisse beim Schlepperfahren. *Arch Landtechn* 2:3-65.

Rossignol M, Suissa S, Abenhaim L (1988): Working disability due to occupational back pain: Three-year follow-up of 2300 compensated workers in Quebec. *J Occup Med* 30:502-505.

Sairanen E, Brushaber L, Kaskinen M (1981): Felling work, low back pain and osteoarthritis. *Scand J Work Environ Health* 7:18-30.

Schulte-Wintrop HC, Knoche H (1978): Backache in VH-ID helicopter crews. AGARD-CP-255.

Seidel H, Heide R (1986): Long-term effects of whole-body vibration: A critical survey of the literature. *Int Arch Occup Environ Health* 58:1-26.

Seidel H, Bluethner R, Hinz B (1986): Effects of sinusoidal whole-body vibration on the lumbar spine: The stress-strain relationship. *Int Arch Occup Environ Health* 57:207-223.

Skovron ML, Szpalski M, Nordin M, Melot C, Cukier D (1994): Sociocultural factors and back pain: a population-based study in Belgian adults. *Spine* 19:129-137.

Snook SH (1982): Low back pain in industry, in Symposium on Idiopathic Low Back Pain, AA White, SL Gordon, Eds. St. Louis: Mosby, pp. 23-28.

Svensson HO, Andersson GBJ (1983): Low back pain in forty to forty-seven year old men: Work history and work environment factors. *Spine* 8:272-276.

Svensson HO, Andersson GBJ, Johansson S, Wilhelmsson C, Vedin A (1988): A retrospective study of low back pain in 38- to 64-year-old women. Frequency and occurrence and impact on medical services. *Spine* 13:548-552.

Svensson HO, Andersson GBJ (1989): The relationship of low-back pain, work history, work environment, and stress: A retrospective cross-sectional study of 38- to 64-year-old women. *Spine* 14:517-522.

Svensson HO (1982): Low-back pain in 40-47 year old men: Some socioeconomic factors and previous sickness absence. *Scand J Rehabil Med* 14:54-59.

Svensson HO, Andersson GBJ (1982): Low back pain in 40-47 year old men. I: Frequency of occurrence and impact on medical services. *Scand J Rehabil Med* 14:47.

Svensson HO, Vedin A, Wilhelmsson C et al. (1983): Low back pain in relation to other diseases and cardiovascular risk factors. *Spine* 8:277.

Taylor VM, Deyo RA, Cherkin DC, Kreuter W (1994): Low back pain hospitalization. Recent United States Trends and Regional Variations. *Spine* 19:1207-1213.

Tichauer ER (1965): The biomechanics of the arm-back aggregate under industrial working conditions. ASME Rep No 65-WA/HUE-1.

Troup JD (1984): Causes, prediction and prevention of back pain at work. *Scand J Work Environ Health* 10:419-428.

Troup JD, Martin JW, Lloyd DC (1981): Back pain in industry: A prospective study. *Spine* 6:61-69.

Troup JDG, Roantree WB, Archibald RM (1970): Survey of cases of lumbar spinal disability. A methodological study. Med Officers' Broadsheet, National Coal Board.

Tsai SP, Gilstrap EL, Cowles SR, Waddell Jr., LC, Ross CE (1992): Personal and job characteristics of musculoskeletal injuries in an industrial population. *Journal Occupational Medicine* 34:606-612.

Uyttendaele D, Vandendriessche G, Vercauteren M, DeGroote W (1981): Sicklisting due to low back pain at the Ghent State University and University Hospital. *Acta Orthop Belgica* 47:523-546.

Valkenburg HA, Haanen HCM (1982): The epidemiology of low back pain, in *Symposium on Idiopathic Low Back Pain,* AA White, SL Gordon, Eds. St. Louis: Mosby, pp. 9-22.

vanDoorn TWC (1995): Low back disability among self-employed dentists, veterinarians, physicians and physical therapists in The Netherlands. *Acta Orthop Scand* 66(suppl 263) 1-64.

Videman T, Numminen T, Tola S, Kuorinka I, Vanharanta H, Troup JDG (1984): Low back pain in nurses and some loading factors of work. *Spine* 9:400-404.

Volinn E, Turczyn KM, Loeser JD (1994): Patterns in low back pain hospitalizations: Implications for the treatment of low back pain in an era of health care reform. *The Clinical Journal of Pain* 10:64-70.

Walsh K, Varnes N, Osmond C, Styles R, Coggon D (1989): Occupational causes of low-back pain. *Scand J Environ Health* 15:54-59.

Westrin C-G (1973): Low back sick-listing. A nosological and medical insurance investigation. *Scand J Soc Med* (suppl) 7:1-116.

Westrin C-G (1970): Low back sick-listing. A nosological and medical insurance investigation. *Acta Soc Med Scand* 2:127-134.

Wickstrom G, Hanninen K, Lehtinen M, Riihimaki H (1978): Previous back syndromes and present back symptoms in concrete reinforcement workers. *Scand J Work Environ Health* (Suppl 4) 1:20-29.

Wilder DG, Woodworth BB, Frymoyer JW, Pope MH (1982): Vibration and the human spine. *Spine* 7:243-254.

Wood PHN, Badley EM (1987): Epidemiology of back pain, in *The Lumbar Spine and Back Pain.* M Jayson, Ed. London: Churchill Livingstone, pp. 1-15.

52

Static Biomechanical Modeling in Manual Lifting

Don B. Chaffin
The University of Michigan

52.1 Introduction

Though most manual tasks in industry involve significant body motions, it continues to be very helpful to evaluate specific exertions within a manual task by performing a static biomechanical analysis. Such analyses are normally performed by combining the postural information (body angles) obtained from a stopped frame video image (or photograph) of a worker, and measured forces exerted at the hands. The latter is often obtained with a simple handheld force gauge.

What follows is a description of a computerized static biomechanical model which has been developed and used over the last 25 years to predict:

1. The percentage of men and women who would be capable of exerting specified hand forces in various work postures, and
2. The forces acting on various spinal motion segments.

Since these two different output predictions have specific criterion values referenced in the NIOSH Lifting Guideline, they are often used by professional ergonomists to determine the relative risk of injury associated with the performance of a manual exertion of interest (Chaffin, 1988). It also should be noted, that the prediction of the percent of the population capable of performing a specific exertion required on a job is often crucial to the determination of a job-specific strength test score for pre-employment and return to work purposes (Chaffin, 1996a). Finally, because the biomechanical population strengths and low back stresses are predicted by a computerized model which runs on common personal computer platforms, this has meant that job and product designers and engineers have been able to simulate various expected high exertion tasks during the early part of the design process, and thus avoid costly prototype evaluations and retrofits when the products and/or processes become operational (Chaffin, 1996b). It is this latter application of the biomechanical static strength prediction programs that provide perhaps the greatest potential benefit over other common job evaluation methods. Many other methods require a

0-8493-2641-9/99/$0.00+$.50

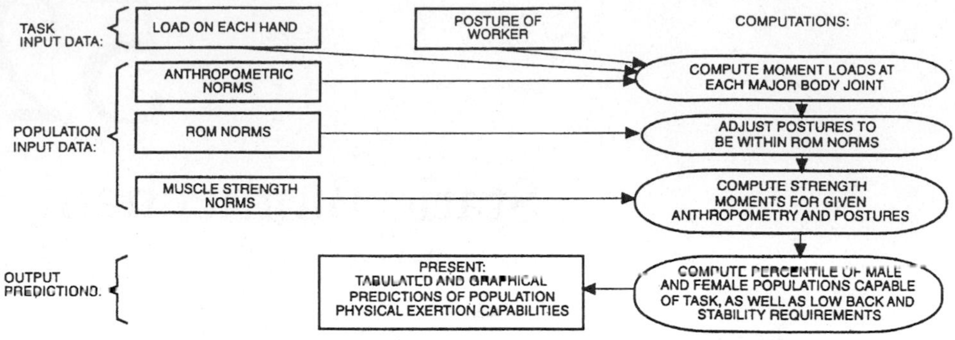

FIGURE 52.1 Biomechanical logic used to predict whole-body static exertion capabilities for given postures, hand force directions, and anthropometric groups.

person to be observed and measured, sometimes with expensive instrumentation. This precludes the use of these empirical methods for use in proscriptive job design, wherein the job exists only on paper or in a computer-rendered drawing of the workspace. By interfacing a computerized biomechanical model of a person (as described in the following) into a computerized rendering of the workspace, the designer can quickly perform a large number of simulated exertions to determine the human consequences of altering a proposed job design, much like that being done to accommodate various sized individuals using computerized anthropometric manikins.

What follows is a brief description of the development of a static biomechanical strength modeling technology, including illustrations of how it has been used to evaluate various manual lifting situations.

52.2 Development of Static Strength Prediction Programs

The general logic used to predict population static strengths in various jobs is depicted in Figure 52.1. In this model specific muscle group strength data and spinal vertebrae failure data are used as the limiting values for the reactive moments at various body joints created when a person of a designated stature and body weight attempts an exertion (i.e., lifts, pushes, or pulls in a specific direction with one or both hands while maintaining a known posture).

This logic has been well described for a sagittal, coplanar static strength analysis in Chaffin and Andersson (1991). When wishing to perform an analysis in three dimensions, the body is represented as a set of links with known mass, as depicted in Figure 52.2. The load moments M_j are computed by the cross products of the unit distance vectors to each joint and body segment weights and hand forces.

The static moment equilibrium equations for the elbow and shoulder in this linkage can be defined as:

$$\overline{F}_{R\,HAND} = F_{RX}i + F_{RY}j + F_{RZ}k \tag{1}$$

where

$\overline{F}_{R\,HAND}$ is the Right Hand Force with X, Y, Z unit vector (i, j, k) components for each,

F_{RX}: X component of R Hand Force,

F_{RY}: Y component of R Hand Force, and

F_{RZ}: Z component of R Hand Force.

$$\overline{V}_1 = \left(V_{1X}i + V_{1Y}j + V_{1Z}k\right) \bullet (\text{Lelink}) \tag{2}$$

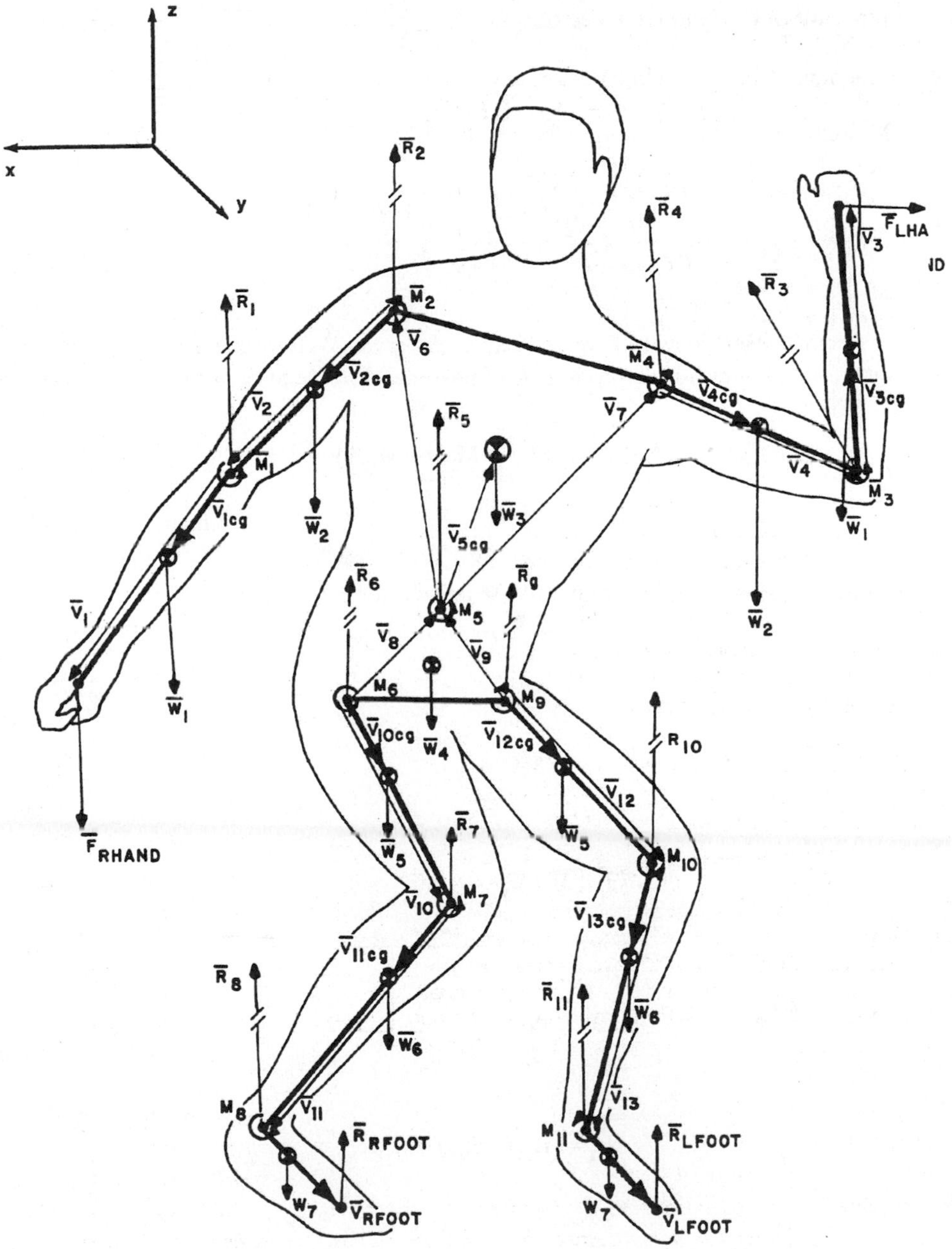

FIGURE 52.2 Three-D distance, force, and moment vectors used in 12 link biomechanical model for strength prediction. From Chaffin, D.B. and Erig, M. 1991. Three-dimensional biomechanical static strength prediction model sensitivity to postural and anthropometric inaccuracies. *IIE Trans.* 23(3):216-227. With permission.)

where

$\overline{V_1}$ is the Forearm Length with Link unit Vectors,

V_{1x}: X component of Forearm Unit Vector,

V_{1y}: Y component of Forearm Unit Vector,

V_{1z}: Z component of Forearm Unit Vector, and

Lelink: Magnitude of Forearm Length from Anthropometric Data.

$$\overline{V_{1cg}} = \left(V_{1X}i + V_{1Y}j + V_{1Z}k\right) \bullet (\text{cglink}) \tag{3}$$

where

V_{1cg} is the cg distance from elbow to Forearm Center of Gravity Vector expressed in unit vector form, multiplied by the cglink, which is the magnitude of proximal distance to cg of forearm from anthropometric data.

$$\overline{M_1} = \left(M_{1X}i + M_{1Y}j + M_{1Z}k\right) \tag{4}$$

where

$\overline{M_1}$ is the Elbow Resultant Moment with X, Y, Z unit vector components,

M_{1x}: Elbow Moment about X axis,

M_{1y}: Elbow Moment about Y axis,

M_{1z}: Elbow Moment about Z axis,

and

$$\overline{M_1} = \overline{V_1} * \overline{F_{R\,HAND}} + \overline{V_{1cg}} * \overline{W_1} \tag{5}$$

where

$\overline{W_1} = 0i + 0j - W_{1Z}k$, (which is Forearm Weight Vector),

and

$$\overline{R_1} = \left(R_{1X}i + R_{1Y}j + R_{1Z}k\right) \tag{6}$$

where

$\overline{R_1}$ is the Elbow Joint Reaction Force Vector with X,Y,Z unit vector components,

and

$$\overline{M_2} = \overline{M_1} + \overline{V_{2cg}} * \overline{W_2} + \overline{V_2} * \left(-\overline{R_1}\right) \tag{7}$$

where

$\overline{M}_2$ is the Right Shoulder Resultant Moment with:

$\overline{V_{2cg}}$: Upperarm Center of Gravity Vector,

$\overline{V}_2$: Upperarm Link Vector,

$\overline{W}_2$: Upperarm Weight Vector.

and

$$\overline{M}_2 = M_{2X}i + M_{2Y}j + M_{2Z}k \tag{8}$$

where

$\overline{M}_{2x}$: Shoulder Moment about X axis,

M_{2y}: Shoulder Moment about Y axis,

M_{2z}: Shoulder Moment about Z axis.

A recursive computational procedure is used to continue the analysis to compute external load moments and forces at the elbow and shoulder of the arm or arms doing the exertion, the lumbosacral joint, hip joints and knee and ankle joints.

The size and mass of the person (linkage) size is most often specified as a select strata of the population (i.e., a percentile of specific anthropometric dimensions is selected from population surveys). Thus, a small, medium, or large man or woman can be specified, or specific link anthropometry can be used if available. Link length-to-stature ratios from Drillis and Contini (1966) and link mass-to-bodyweight ratios from Dempster (1955) and Clauser et al. (1969) are used to simplify this procedure, if specific anthropometry is not available on a subject. Most often an average male or female anthropometry is chosen for assessing the strength requirements of a given task in industry.

The strength moment values used as population limit values in the program were measured by Stobbe (1982) for 25 men and 22 women employed in manual jobs in three different industries. These values have been combined with the earlier values from Chaffin and Baker (1970) and Schanne (1972) to form the statistical data for the population joint moment limits.

Once the size of the person has been specified or selected from a known anthropometric data source, the posture is entered with reference to either photographs or videos (or by manipulating a computer-generated hominoid) and then the hand forces of interest are keyed in. The program then computes the load moments at each joint of the linkage, and compares each to the corresponding strength moment capability obtained from the previously measured populations. This provides a prediction of the percent of the population that is capable of producing the necessary strength moments at each joint. The logic for computing the lumbar motion segment compression force is shown in Figure 52.3.

The logic assumes that once the lumbar moment is computed (as described in the preceding), torso muscles contract to stabilize the column. In the 2D Sagittal Plane logic a single torso muscle equivalent contraction force is included. When the necessary reactive torso muscle force is added to body segment weights and hand forces (with a minor adjustment for abdominal pressure effects) a prediction of the compression force on the L5/S1 disc results, as shown in Figure 52.4.

When an asymmetric exertion (e.g., one-handed force, or twisted or laterally bent torso) is being analyzed, many different torso muscle actions and passive supporting tissue reactions need to be considered. The first step in such a procedure requires that the position, orientation, cross-sectional size, and length of the various connective tissues be modeled at the lumbar spinal level. A geometric torso

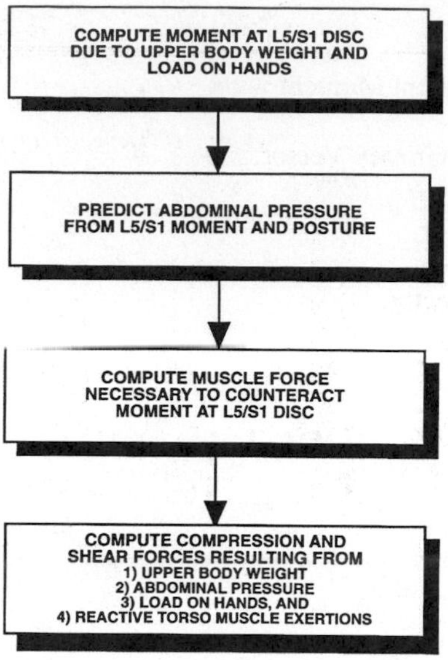

FIGURE 52.3 Logic for computing L5/S1 compression forces in 2D and 3D Static Strength Prediction Programs.

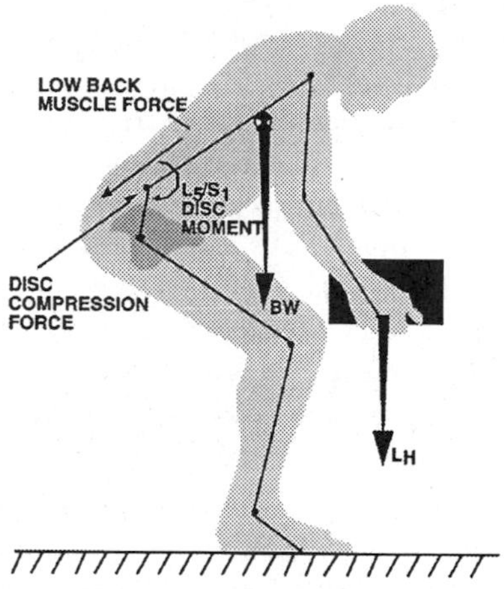

FIGURE 52.4 Simple low-back model of lifting for static coplanar lifting analyses. The load on the hands L_H and torso and arm weights BW act to create moments at the L5/S1 disc of the spine. The moments are resisted primarily by the back muscles. The high muscle forces required in such a task cause high disc compression forces. (From Chaffin, D.B. and Andersson, G.B.J. 1991. *Occupational Biomechanics*, 2nd ed. John Wiley & Sons, Inc., NY. With permission.)

model proposed by Nussbaum and Chaffin (1996) for this purpose is shown in Figure 52.5. This model includes estimates of specific tissue geometry acquired from various CT scans, (Tracy et al., 1989; Moga et al., 1993; Chaffin et al., 1990), along with passive tissue reaction forces estimated by McCully and Faulkner, (1983), Nachemson et al. (1979), Miller et al. (1986), and others.

The most important predictors of spinal column stress, however, are the muscle reaction forces required to stabilize the spine to external load moments. In the 3D torso models various approaches have been used to predict the required reactive muscle forces. Perhaps the most commonly cited torso biomechanical model for 3D Static Analysis is that developed by Schultz and Andersson (1981). It is depicted in Figure 52.6. A revised version of this model has been developed by Bean, Chaffin, and Schultz (1988).

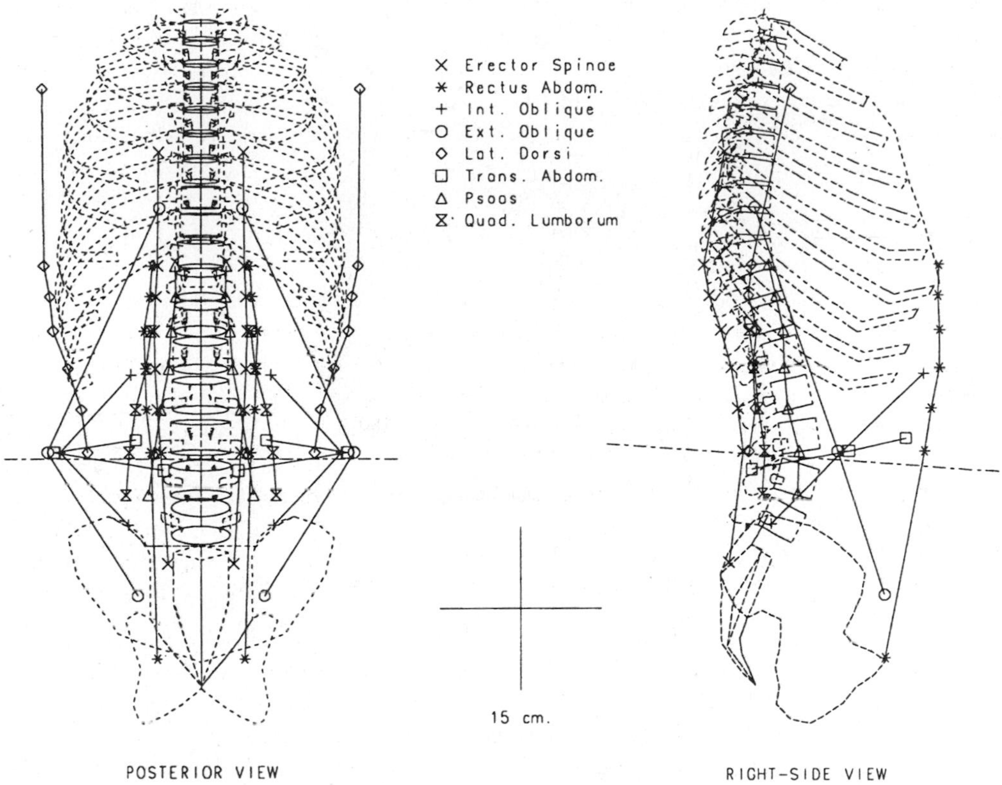

FIGURE 52.5 Muscle geometry illustrated for a 50th percentile male. Muscles are treated as pointwise connections from origin to insertion (see text). An imaginary cutting plane which bisects the L_3/L_4 motion segment is also shown. (From Nussbaum, M.A. and Chaffin, D.B. 1996. Development and evaluation of a scalable and deformable geometric model of the human torso. *Clin. Biomech.* 11(1):25-34. With permission.)

This latter model provides a more efficient computational method for solving the linear programs used to simultaneously minimize the torso muscle contraction intensities and motion segment compression forces. The present model predicts the minimum muscle force contractile intensities required to meet the moment equilibrium requirements about the three orthogonal axis-of-rotation of the motion segment. Given a set of optimal forces so computed, the model further seeks to minimize the disc compression forces. Because such an approach attempts to minimize *both* muscle intensity requirements and disc compression forces simultaneously, it is referred to as a "double linear optimization" approach.

More recently Hughes and Chaffin (1995) proposed that a nonlinear objective function be used as the basis for selecting the various muscle reaction forces during a given exertion. They referred to this as the sum of the cubed muscle intensity objective. Nussbaum, Chaffin, and Martin (1996) also have proposed a neural network model to predict torso muscle actions. And most recently, Raschke and Chaffin (1996) have proposed that the external moment is normally distributed about the torso, and activates several muscles simultaneously depending on the direction and magnitude of the external moment.

52.3 Computerization of Strength Prediction and Back Force Prediction Models

It should be clear from the preceding descriptions that the biomechanical models used for population strength and spinal motion segment force prediction are computationally intense, especially in the 3-dimensional form. For this reason a number of faculty, staff, and students associated with the Center

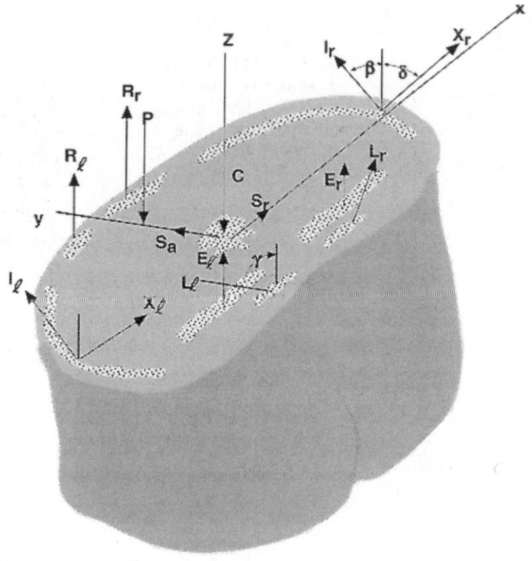

FIGURE 52.6 Schematic of 10 muscle model developed by Schultz and Andersson (1981).

for Ergonomics at the University of Michigan have worked to provide user friendly, computer programs of the models. These are referred to as the 2-dimensional and 3-dimensional Static Strength Prediction Programs™ (i.e., 2DSSPP™ and 3DSSPP™). Over 2,500 licenses for use of these programs have been granted by the University of Michigan's, IPO Software Group over the last 13 years.

The main screen of the 2DSSPP™ program is depicted in Figure 52.7. The input values, (i.e., body link angles, hand forces, and anthropometry) are shown in the upper left quadrant. A stick figure depicting the body posture, the hand location, and the hand force direction used as inputs is depicted in the upper right quadrant. The predicted percent of the male and female population having sufficient strength to perform the designated exertion (in this case lifting a 44-pound stock reel) is shown in tabular and graphical form in the lower left quadrant. The back compression force predictions for a man and women performing the 44-pound lift is shown in the lower right quadrant. From inspection of the percent capable predictions (left bottom) it is obvious that hip strengths are the most limiting muscle group strength, (only 66% of women and only 87% of men have sufficient hip strength to lift the 44-pound reel). These values are below that recommended by NIOSH, which believes that jobs should accommodate 99% of men's and 75% of women's strength (or 90% of a mixed gender population). It also is shown in the right bottom quadrant that the L5/S1 compression forces of 924 and 845 pounds are above that recommended by NIOSH of 760 pounds.

The 3DSSPP program requires more input data than the 2D version depicted in Figure 52.7. Three-dimensional exertions often involve two hand forces which can act in any direction. Also, a model of the human body in 3D has 12 body links (some with three postural angles). Two types of input presentations are used to assist in assuring the correct input data are used for 3D analyses. One presentation has three orthogonal views of a stick figure, and the other presents a shaded, enfleshed hominoid, which can be viewed from any direction. Figure 52.8 depicts these two presentations. The use of the 3D hominoid was found by Beck and Chaffin (1992) to allow postures viewed on a video or photograph to be accurately represented in a computer. An inverse kinematic model with preferred postural prediction capability is included to allow the user to easily manipulate the figure into the posture to be analyzed. The output screens are similar to that shown in Figure 52.7 for the 2DSSPP. These depict both percent capable strength predictions (for 21 muscle functions) and lumbar compression forces; the latter can include 3D predictions of a large number of individual muscle and spinal forces.

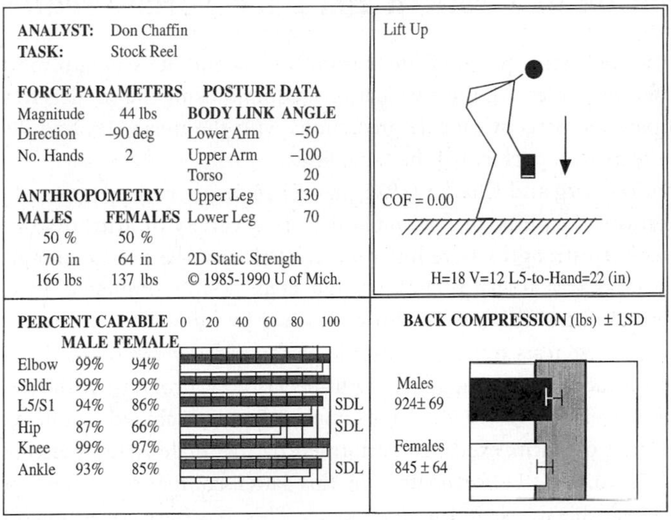

FIGURE 52.7 Main screen from the University of Michigan 2D Static Strength Prediction Program for personal computers. (From the University of Michigan, IPO Software, Ann Arbor, MI 48109.)

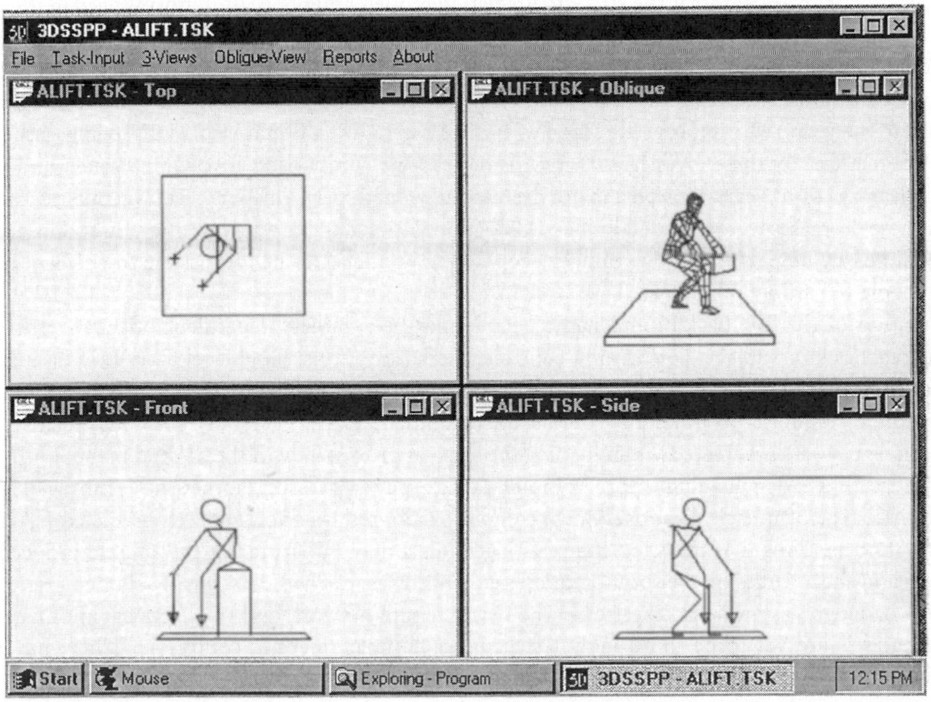

FIGURE 52.8 Input screen used in 3DSSPP™ to assure postural data are correct. (With permission, University of Michigan Regents).

52.4 Validation of Strength and Back Force Prediction Models

The validation of the static strength predictions from the 2D and 3D strength models has been accomplished in three different studies. All three validations required using the models to simulate whole-body exertions and compare the percent capable predictions with the mean, 10, and 90 percentile strengths of a group of volunteers who performed the same tasks.

In the first validation Garg and Chaffin (1975) had 71 male Air Force personnel perform 38 different maximum arm exertions (i.e., lifts, pushes, pulls, etc.) in a variety of arm/torso postures while seated. They found the predicted strengths were highly correlated with the group strengths when performing the 38 upper body tasks (r = 0.93 to 0.97). Chaffin et al. (1987) simulated 15 different whole-body exertions in the sagittal plane which were also performed by both men and women from a variety of industries. In some of these tests over 1,000 people performed the exertions, though on average about 200 people performed each one. Comparison with the 2DSSPP program with the group strength data revealed a very high correlation (r = 0.92). This same study also included simulations with the 3DSSPP program of 72 different one-arm exertions performed by five male Army personnel. The correlations ranged from r = 0.71 to 0.83. Unfortunately, in this latter comparison, exact postural and bracing conditions were not available to use in the simulations. This may have contributed to the lower correlations.

The last validation involved simulations of 56 one- and two-handed, whole-body exertions in 14 different symmetric, bent, and twisted-torso postures (Chaffin and Erig, 1991). The simulation results were compared with the group strengths of 29 young males. Photographs from several views were available to assist in replicating the postures used by these subjects. The results indicated that if care is taken to assure that the postures used in the model simulation is the same as that chosen by people performing the exertions, the prediction error standard deviation will be less than 6%.

In conclusion, it appears that the strength prediction models and population norms used in the present models are accurate in predicting the percent of the population capable of performing a large variety of different types of maximal static exertions. One caution should be noted, however. At present the strength norms used as limits in the models are based on male and female populations who are relatively young (i.e., 18 to 49 years). To improve the models further, strength values are currently being gathered on older populations by these investigators. In this regard, one comparison involving 98 men and women with a mean age of 73 years, showed a major decrease in strength performance in certain muscle functions. When these decreases were included in the 3DSSPP population database, it was found that some exertions that could easily be performed by younger people were predicted to be impossible to perform by most older people (Chaffin et al., 1994).

Validation of the low back biomechanical model has been largely dependent on EMG estimates of muscle reactions in subjects performing controlled torso exertions. Hughes et al. (1994) discusses this procedure, and the results of comparisons with four different optimization procedures used to predict torso muscle responses to different external torso moment loads. Generally speaking, relatively high correlation (r > 0.8) is achieved when loading the torso approximately in the sagittal plane. With greater asymmetric or sudden loading, more complex muscle patterns result, sometimes with a 10 to 30% antagonistic type of muscle response. These complex responses are often not well predicted (r < 0.6) by existing models. Thus it is expected that existing models may underpredict the muscle-induced compression and shear forces on the spinal motion segments by as much as 30% during sudden (i.e., jerking) motions or lateral, asymmetric exertions. The newer neural network and/or geometric moment distribution models are yet to be thoroughly tested under complex loading conditions. They may be less sensitive to this cocontraction phenomenon than existing optimization models.

52.5 Final Comments

The development of biomechanically based, static strength and low back force prediction models provide powerful tools for performing ergonomic assessments of high physical effort jobs. With the advent of

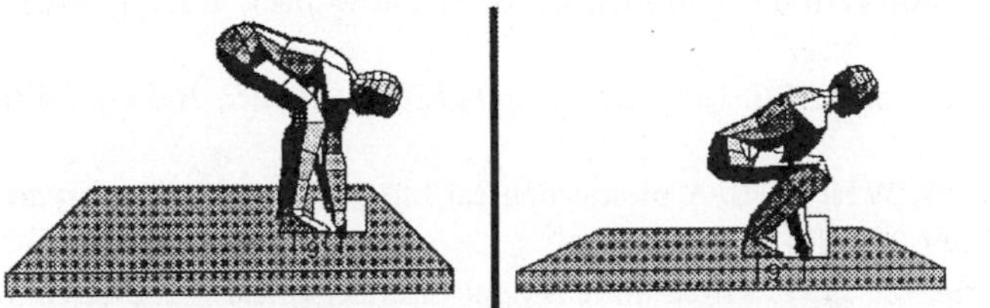

1A STOOP LIFT

25 LB. OBJECT:	MALE	FEMALE
Back Comp. (lbs)	621	538
Disc Shear (lbs)	96	83
% Pop. Strength	94 %	88 %
Limiting Muscles	Hip Ext.	Hip Ext.
50 LB. OBJECT		
Back Comp. (lbs)	790	706
Disc Shear (lbs)	121	108
% Pop. Strength	90 %	75 %
Limiting Muscles	Hip Ext.	Hip Ext.

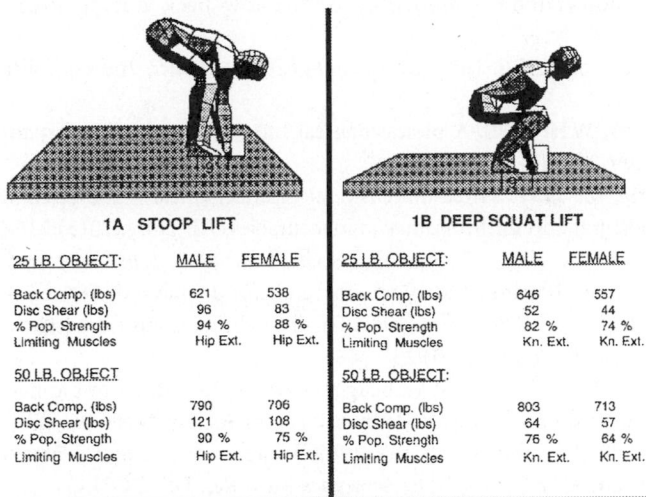

1B DEEP SQUAT LIFT

25 LB. OBJECT:	MALE	FEMALE
Back Comp. (lbs)	646	557
Disc Shear (lbs)	52	44
% Pop. Strength	82 %	74 %
Limiting Muscles	Kn. Ext.	Kn. Ext.
50 LB. OBJECT:		
Back Comp. (lbs)	803	713
Disc Shear (lbs)	64	57
% Pop. Strength	76 %	64 %
Limiting Muscles	Kn. Ext.	Kn. Ext.

FIGURE 52.9 Comparison of two different postures used to lift 25- and 50-pound objects from the floor close to feet using the Michigan 3DSSPP™.

faster personal computers these models are gaining popularity not only as a means to evaluate existing manual tasks, but also to aid in the design of new workplaces and equipment, as well as in the specification of personnel selection and training programs.

The inclusion of inverse kinematics, behaviorally based posture prediction methods, improved human form graphics, and direct human image video input data, have made the use of these models relatively easy for the ergonomics practitioner. By simply clicking and pointing to parts of the body, one can adjust postures and have a comprehensive biomechanical assessment of a specific exertion (such as depicted in Figure 52.9).

With the widely recognized need to "design workplaces right the first time," computerized biomechanical strength prediction models will become even more useful. It is hoped that this presentation assists those interested in knowing more about both the scientific basis for this rapidly expanding technology, and the potential benefits and limitations inherent in it.

Acknowledgment

I wish to acknowledge NIH Grant AR-39599 for partial support for some of the work described in this presentation.

References

Bean, J.C., Chaffin, D.B., and Schultz A.B. 1988. Biomechanical model calculation muscle contraction forces: a double linear programming method. *J. Biomech.* 21(1):59-66.

Beck, D.J. and Chaffin, D.B. 1992. An evaluation of inverse kinematics models for posture prediction. Proceedings of *CAES 1992*, Tampere, Finland. Center for Ergonomics, The University of Michigan, Ann Arbor, MI.

Chaffin, D.B. 1996a. Ergonomic basis for job-related strength testing, in *Disability Evaluations*, Eds. S.L. Demeter, G.B.J. Andersson, and G.M. Smith, p.159-167. American Medical Association, Mosby, St. Louis, MO.

Chaffin, D.B. 1997. Biomechanical aspects of workplace design, in *Handbook of Human Factors*, Ed. G. Salvendy. Wiley & Sons, Inc., New York, NY.

Chaffin, D.B. 1988. A biomechanical strength model for use in industry. *Appl. Ind. Hyg.* 3(3):79-86.

Chaffin, D.B. 1988. Biomechanical modelling of the low back during load lifting. *Ergonomics* 31(5):685-697.

Chaffin, D.B. and Andersson, G.B.J. 1991. *Occupational Biomechanics*, 2nd ed. John Wiley & Sons, Inc., NY.

Chaffin, D.B. and Baker, W.H. 1970. A biomechanical model for analysis of symmetric sagittal plane lifting. *AIIE Trans.* 2(1):16-27.

Chaffin, D.B. and Erig, M. 1991. Three-dimensional biomechanical static strength prediction model sensitivity to postural and anthropometric inaccuracies. *IIE Trans.* 23(3):216-227.

Chaffin, D.B., Redfern, M.S., Erig, M., and Goldstein, S.A. 1990. Lumbar muscle size and location measurements from CT scans of 96 older women. *Clin. Biomech.* 5(1):9-16.

Chaffin, D.B., Freivalds, A., and Evans, S.M. 1987. On the validity of an isometric biomechanical model of worker strengths. *IIE Trans.* 19(3):280-288.

Chaffin, D.B., Woolley, C.B., Buhr, T., and Verbrugge, L. 1994. Age effects in biomechanical modeling of static lifting strengths. Proceedings of Human Factors Society, Nashville, TN.

Clauser, C.E., McConville, J.T., and Young, J.W. 1969. Weight, volume and center of mass of segments of the human body. AMRL-TR-69-70, *Aerospace Med. Res. Labs.*, OH.

Dempster, W.T. 1955. Space requirements of the seated operator. WADC-TR-55-159, *Aerospace Med. Res. Lab.* Wright-Patterson AFB, OH.

Drillis, R. and Contini, R. 1966. *Body Segment Parameters.* BP174-945, Tech. Rep. No. 1166.03, S. of Engnrg. and Sci., NYU, NY.

Garg, A.D. and Chaffin, D.B. 1975. A biomechanic computerized simulation of human strength. *AIIE Trans.* 14:272-280.

Hughes, R.E. and Chaffin, D.B. 1995. The effect of strict muscle stress limits on abdominal muscle force predictions for combined torsion and extension loadings. *J. Biomech.* 28(5):527-533.

Hughes, R.E., Chaffin, D.B., Lavender, S.A., and Andersson, G.B.J. 1994. Evaluation of muscle force prediction models of the lumbar trunk using surface electromyography. *J. Orthop. Res.* 12:689-698.

Miller, J.A.A., Schultz, A.B., Warwick, D.N., and Spencer, D.L. 1986. Mechanical properties of lumbar spine motion segments under large loads. *J. Biomech.* 19:79-84.

Moga, P.J., Erig, M., Chaffin, D.B., and Nussbaum, M.A. 1993. Torso muscle moment arms at intervertebral levels T10 through L5 from CT scans on eleven male and eight female subjects. *Spine.* 18(15):2305-2309.

McCully, K.K. and Faulkner, J.A. 1983. Length-tension relationship of mammalian diaphragm muscles. *J. Appl. Physiol.* 54:1681-1686.

Nachemson, A.L., Schultz, A.B. and Berkson, M.H. 1979. Mechanical properties of human lumbar spine motion segments: influences of age, sex, disc level, and degeneration. *Spine.* 4:1-8.

Nussbaum, M.A. and Chaffin, D.B. 1996. Development and evaluation of a scalable and deformable geometric model of the human torso. *Clin. Biomech.* 11(1):25-34.

Nussbaum, M.A., Chaffin, D.B. and Martin, B.J. 1996. A back-propagation neural network model of lumbar muscle recruitment during moderate static exertions. *J. Biomech.* 28(9):1015-1024.

Raschke, U. and Chaffin, D.B. 1996. Trunk and hip muscle recruitment in response to external anterior lumbosacral shear and moment loads. *Clin. Biomech.* 11(3):145-152.

Schanne, F.T. 1972. Three dimensional hand force capability model for a seated person. An unpublished Ph.D. dissertation, University of Michigan, Ann Arbor, MI.

Schultz, A.B. and Andersson, B.J.G. 1981. Analysis of loads on the lumbar spine. *Spine.* 6(1):76-82.

Stobbe, T.J. 1982. The development of a practical strength testing program in industry. An unpublished Ph.D. dissertation, University of Michigan, Ann Arbor, MI.

Tracy, M.F., Gibson, M.J. Szypryt, E.P. et al. 1989. The geometry of the muscles of the lumbar spine determined by magnetic resonance imaging. *Spine.* 14:186-193.

53

Dynamic Low Back Models: Theory and Relevance in Assisting the Ergonomist to Reduce the Risk of Low Back Injury

Stuart M. McGill
University of Waterloo

0-8493-2641-9/99/$0.00+$.50

53.1 Introduction

This chapter will address some issues associated with "complex" models and highlight how ergonomists can take advantage of these special tools to reduce the risk of low back injury. Throughout the course of daily activity, the low back system is subjected to loading from external forces, and from forces produced by the internal tissues needed to create movement and maintain static postures. The fact that injury to the low back can only be caused by excessive loading of any given tissue cannot be easily dismissed. This overloading may occur during strenuous exertion or during prolonged postures; sitting may be such an example. On the other hand, too little tissue loading also leads to higher injury risk from atrophied tissues, detrained motor control systems and physiological impairment. Prevention of injury requires knowledge of the tissue loads during activity, which also enable testing of hypotheses designed to reduce the risk of injury. Because direct measurement of tissue loads *in vivo* is not feasible, the only tenable option for predicting tissue loads is to utilize modeling approaches.

Biomechanical models have been used to estimate loads in the low back tissues and identify high-risk jobs for approximately three decades. Several approaches to model development have been employed with each approach having a specific objective, and corresponding assets and liabilities. For example, some models were intended to reveal spine function and low back injury mechanisms which is background knowledge required to devise injury avoidance strategies in industry. Other model approaches were intended as simple tools for health and safety personnel to provide an approximate index of injury risk on the plant floor. The main issue boils down to the purpose, and necessary complexity, of the model. Complex models are required to reveal how tissues function inside a worker and to identify specific (and often subtle) injury mechanisms, while simpler models are needed to broaden utility of use and reduce the more overt risks on the plant floor. But the better "simple" models need the "complex" models to assess the many simplifying assumptions which affect accuracy and validity of output, depending on the type of application. Further, in most workplaces, the most blatant or overt ergonomic injury risks have been addressed, and only the more subtle and sometimes sublime risks remain. Ergonomists need "simple" models but also must be conversant with the more complex models which will assist in rectifying the more subtle injury risks, and assist in developing more effective intervention strategies.

The intent of this chapter is to first examine some issues associated with the development, interpretation, and application of "complex" models, then briefly describe a "complex" model, and finally apply the "complex" model to real occupational injury issues. Specifically, data and model output will be integrated into the formulation of a set of guidelines intended to reduce the risk of low back injury in a wide variety of occupational situations, ranging from heavy lifting to sedentary activities. Since some of these guidelines are based on recent research findings, they remain tentative until their efficacy under clinical trial is proven or disproven. While some may think it prudent to wait for the definitive study, ergonomists have not always the luxury of time to deal with injury issues, thus the intention in listing the tentative guidelines here is to provoke discussion and debate, and motivate the initiation of experiments to test their viability.

53.2 Issues Relevant for Complex Model Interpretation

Having made clear the need for "complex" models in the previous section, the ergonomist must understand the limitations, and conversely the most appropriate applications, associated with complex models. The issues of anatomical complexity and how tissue loading is determined will be discussed here. Simple and static models are described elsewhere in this handbook, together with a discussion of the static vs. dynamic modeling issue and when two-dimensional analysis or three-dimensional analysis is required (e.g., Chapter 54.).

The Asset of Anatomical Complexity

There is no doubt that the most overt violations of biomechanical injury risk reduction principles can be addressed by quite "simple" models. However, if the purpose of a biomechanical model is to provide insight into the functional role of various tissues and how they become injured, it is necessary that the model represent the structure of the anatomy as closely as possible. Unfortunately, some researchers have worked beyond the anatomical limitations of the "simple" models and have made erroneous and unjustified conclusions about the best choice of injury risk reduction strategy. Examples of questionable anatomical/mechanical simplification are as follows:

1. Muscle areas used in some models were obtained from cadavers. Dimensions obtained in this way are difficult to justify for use in models of healthy, young, and working individuals when atrophy and distortion from fluids would greatly under-predict the force potential of the musculature. These underestimations of muscle area, and force-producing potential, have been pointed out with CT scan data of younger, ambulatory adults presented by Nemeth and Ohlsen (1986), Reid and Costigan (1985), and McGill et al. (1988, 1993, 1997). Models for ergonomic application must give proper credit to the musculature to most accurately estimate the risk of injury for a given work load.
2. Muscles in the trunk do not pull in straight lines as represented in many lumbar spine models. Several important muscles within the trunk act around pulley systems of bone, other muscle bulk, and pressurized viscera, which alters length, force, and vector direction properties together with the resultant joint loading (cf. McGill and Norman, 1987; McGill, 1996).
3. Models which assume the extensor musculature can be represented by a single equivalent force vector acting parallel to the compressive axis produce controversial output. Excellent work by Langenberg (1970) and Macintosh and Bogduk (1987) provides clear descriptions of the connections for the prime extensors of longissimus thoracic pars thoracic and pars lumborum, iliocostalis lumborum pars thoracic and pars lumborum, and multifidus. Very few of these fibers run parallel to the axis of spinal compression, demonstrating that they exert both compression and shear forces on the spine (Figure 53.1). In addition, the laminated architecture of the muscle fascicles provides for a much larger muscle cross-sectional area to contribute to extensor moment production than would be observed in a single transverse section of the torso. For this reason, an estimate of extensor moment potential from a transverse scan at a single level of the spine would result in large error as only a small portion of the musculature would be measured. The relatively large bulk of thoracic fibers (shown in Figure 53.2) is often neglected as an important contributor to extension as the fibers produce extensor forces over the full lumbar spine through a moment arm often exceeding 10cm.
4. The extensor and torsional potential of latissimus dorsi is often neglected, yet it has the largest moment arm length of all of the posterior trunk muscles. Its association with the lumbodorsal fascia as a spine extensor is a contentious issue (cf. Tesh et al., 1987; McGill and Norman, 1988).
5. The passive force contributions of the supraspinous and interspinous ligaments are often modeled with a single equivalent element. However, interspinous fibers run obliquely to the supraspinous ligament, thus creating nonparallel forces (Figure 53.1a, vectors a,b,c). In fact, the interspinous acts to generate a shear force on the joint. However, the fiber direction of this ligament complex as depicted in most anatomical texts (e.g., *Gray's Anatomy* (1980)), is in error as pointed out by Heylings (1978). Instead, the interspinous ligament has been shown to contribute significant anterior not posterior, shear forces which increase, rather than reduce, facet load during large degrees of flexion (see Figure 53.1) (Shirazi-Adl and Drouin, 1987; McGill, 1988).
6. The diaphragm area and shape is fundamental to the calculation of the potential assistance provided by intra-abdominal pressure. It is suspected that the size of diaphragms that have been used in the past (up to 465 cm^2) is a gross overestimate; the normal surface area on which pressure is exerted is probably closer to 243 cm^2 (McGill and Norman, 1987), 276 cm^2 (Schultz et al., 1982), or 299cm^2 (Troup et al., 1983).
7. The psoas complex has long been considered and described in the textbooks as a flexor of the lumbar spine. Recent work has shown that while it is a flexor of the hip, its moment arm to flex the spine is extremely limited (Thorstensson et al., 1989; Santaguida and McGill, 1995), and its neural activation is not correlated with moments acting on the lumbar spine but rather with hip flexion moments (McGill et al., 1996).

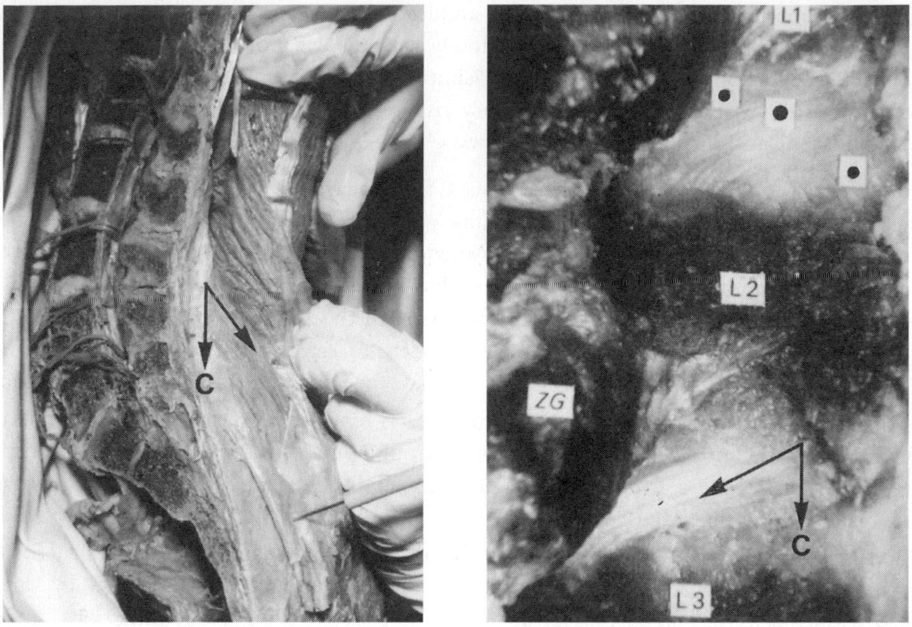

FIGURE 53.1 (Left panel) Pars lumborum fibers of iliocostalis lumborum and longissimus thoracis create a posterior shear force on the superior vertebra compared to the compressive axis (C), (right panel) while the interspinous ligament imposes an anterior shear when strained in flexion. The general oblique line of action of the muscle and ligament which causes shear loading, is shown compared to the compressive axis (C). (From Heylings, D.J.A. (1978) Supraspinous and interspinous ligaments of the human lumbar spine. *J. Anat.* 123:127-131. With permission.)

The Problem of Indeterminacy

Pain and disability have been documented to result from damage to ligaments, discs, vertebral end plates, vertebral bodies themselves, facet joints, various muscles, and to several other tissues. While investigation of these injuries requires a modeling approach that incorporates sufficient anatomical detail, a method is needed to solve for the inherent indeterminacy from so many unknown muscle, ligament, and bone forces. Because people differ in the way they perform work, causing only some of them to become injured, the modeling method should also be sensitive to the many different ways that individual people utilize their muscles and various ligaments to perform tasks. Historically, two basic approaches have been used to partition the supporting duties among the many components of the trunk musculature and ligamentous system — optimization approaches and biological approaches that utilize biological signals obtained from each subject (for example, measurements of muscle EMG and spine kinematics). The optimization approach attempts to satisfy the reaction moment requirements needed to support a posture by recruiting the various muscles based on an optimization criterion such as the minimization of joint compression and shear load (e.g., Gracovetsky et al., 1981)), or first minimization of muscle contraction intensity and then spine compression load (e.g., Bean et al., 1988). Generally optimization assumes that the motor control system operates to fulfill objectives that can be mathematically defined and in so doing ignores individual variability and predicts the same tissue load distribution for all subjects performing a certain task. Furthermore, most currently reported optimization approaches are very poor at predicting patterns of muscle cocontraction so characteristic of three-dimensional spine motion (cf. Hughes et al., 1994, 1995) and for predicting muscle activity in other areas of the body (e.g., Collins, 1995). It may turn out in the future that more robust and comprehensive optimization "cost functions," yet to be defined, will prove more effective in predicting patterns of muscle activation and their resulting forces.

On the other hand, a modeling approach that uses biological signals obtained directly from each subject is inherently sensitive to the individual ways that people load their low back tissues and, hence

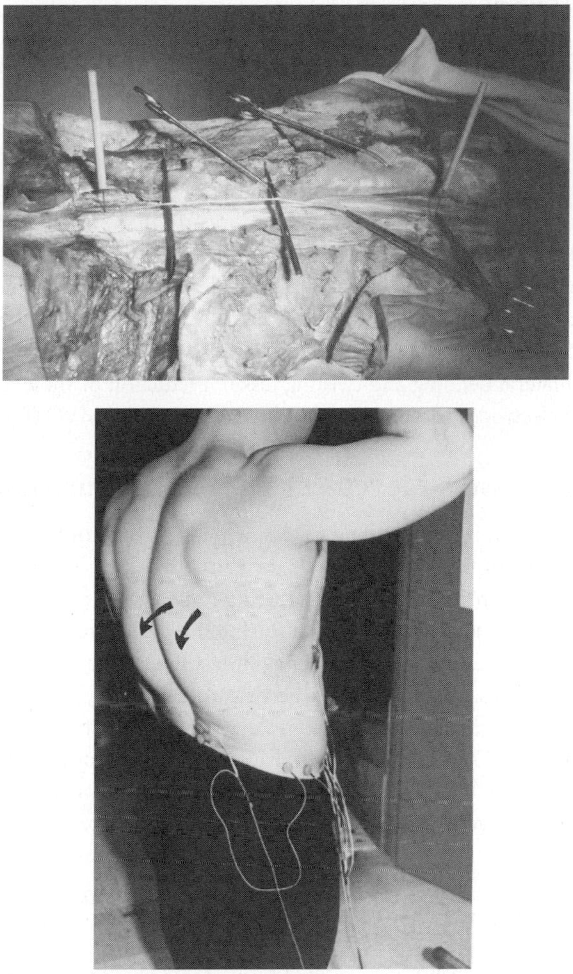

FIGURE 53.2 A bundle of fibers of longissimus thoracis has been dissected and the tendon isolated to show the insertion on the sixth and seventh thoracic (T6 and T7) ribs and sacral origin (a). Hence, these muscles create an extensor moment over the full length of the lumbar spine and minimize the compressive penalty due to their mechanical advantage. Longissimus thoracis bulk in a developed weight lifter (b). (From McGill, 1990. With permission.)

are better suited to investigations of injury. For example, McGill and Norman (1986) and Marras and Sommerich (1991) have proposed dynamic models of the lumbar spine that attempted to determine the significant forces in many muscles in the low back based in part on their neural activation measured through calibrated surface EMG and in the passive structures based on estimates of strain from directly measured spine kinematics. The major asset of the biologically based approach is that muscle cocontraction is fully accommodated together with the approach being sensitive to the differences in the way individuals perform a movement. However, estimations of muscle force, based in part on myoelectric signals, are problematic because the force potential per muscle cross-sectional area must be assumed, together with other variables that are known to modulate muscle force production. Furthermore, one must rely on anatomical accuracy to satisfy the moment requirements about all three joint axes and about several joints simultaneously so that errors in achieving moment equilibrium remain. Nonetheless, the ability to predict measured moments from biologically driven models adds some degree of validity to the modeling approach. In addition, animal studies involving the direct measurement of tendon forces compared with predictions of tendon forces, from EMG-based models, appear encouraging (see the work of Komi, 1990; Gregor et al., 1987; Norman et al., 1988). Recent work by Cholewicki et al. (1995)

compared the assets and liabilities of an EMG and spine kinematics approach with an optimization approach to obtain tissue force distribution profiles and concluded that the biological–EMG-based approach was more suitable for investigation of injury mechanisms.

Estimating Deep (and Inaccessible) Muscle Forces

A major drawback of the EMG-based approach using surface electrodes is the myoelectric inaccessibility of the deeper torso muscles (e.g., psoas, quadratus lumborum, three layers of the abdominal wall). In an attempt to address this criticism, recent work by McGill, Juker, and Kropf (1996) utilized in-dwelling intramuscular electrodes with simultaneous surface electrode sites to evaluate the possibility, and validity, of using surface activity profiles as surrogates to activate deeper muscles over a wide variety of tasks and exercises common in industry and in rehabilitation programs (e.g., situps, curlups, leg raises, pushups, some spine extensor tasks, and lateral bending and twisting tasks). Prediction of these deeper muscles is possible from well-chosen surface electrodes within an error criterion of 15% of MVC (RMS difference), or less.

Dynamic Models to Assess Movement Over Long Durations

Analysis of prolonged tasks requires assessment of tissue loads throughout the performance rather than trying to select a single event in time for analysis. Low back tissues are not always under static loading during static work, nor used in repeatable patterns during prolonged work (Potvin and Norman, 1992). For example, subtle shifts in tissue load distribution during the performance of lighter loading tasks, but of longer duration, can lead to situations where loads in a single tissue may rise to unreasonable levels (for example, during prolonged sitting or flexion where ligamentous creep transfers more load and strain to the posterior anulus). Once again, anatomical complexity is required to evaluate the interplay between muscle and passive tissues, but also a biologically driven model is required that is sensitive to subtle changes in spinal posture (and therefore sharing of ligament and disc loads) and sensitive to shifts in muscle activity (either between muscles or between muscles and passive tissues).

A Brief Description of a Dynamic, 3-Dimensional, Anatomically Complex, Biologically Driven, Modeling Approach

While two groups (the Marras Group, e.g., Granata and Marras, 1993, and the McGill Group) have devoted much effort to the development of biologically driven models, the McGill model will be described here, given its familiarity to the author.

Individual tissue loads have been predicted from a laboratory technique and model developed over the past 14 years. The model is composed of two distinct parts. First, a three-dimensional linked segment representation of the body was constructed using the dynamic load in the hands as input, and working through the arm and trunk linkage, reaction forces and moments about a joint in the low back (usually L4/L5) are computed (previously described in McGill and Norman, 1985) (see Figure 53.3). Joint displacements are recorded on two or more video cameras at 30 Hz to reconstruct the joints and body segments in three dimensions. This first model produces the three reaction forces and corresponding moments about the three orthopedic axes of the low back (flexion–extension, lateral bend, axial twist). The second anatomically detailed model enables the partitioning of the reaction moment obtained from the link segment model into the substantial restorative moment components (supporting tissues) using an anatomically detailed three-dimensional representation of the skeleton, muscles, ligaments, nonlinear elastic intervertebral discs, etc. This part of the model was first described in McGill and Norman (1986), with full three-dimensional methods described in McGill (1992) and the most recent update provided by Cholewicki and McGill (1996) where 90 low back and torso muscles are represented in total. Very briefly, first the passive tissue forces are predicted by assuming stress–strain or load deformation relationships for the individual passive tissues and calibrated for the differences in flexibility of each subject by normalizing the stress–strain curves to the passive range of motion of the subject, which is detected by electromagnetic instrumentation which monitors the relative lumbar angles in three dimensions. Then

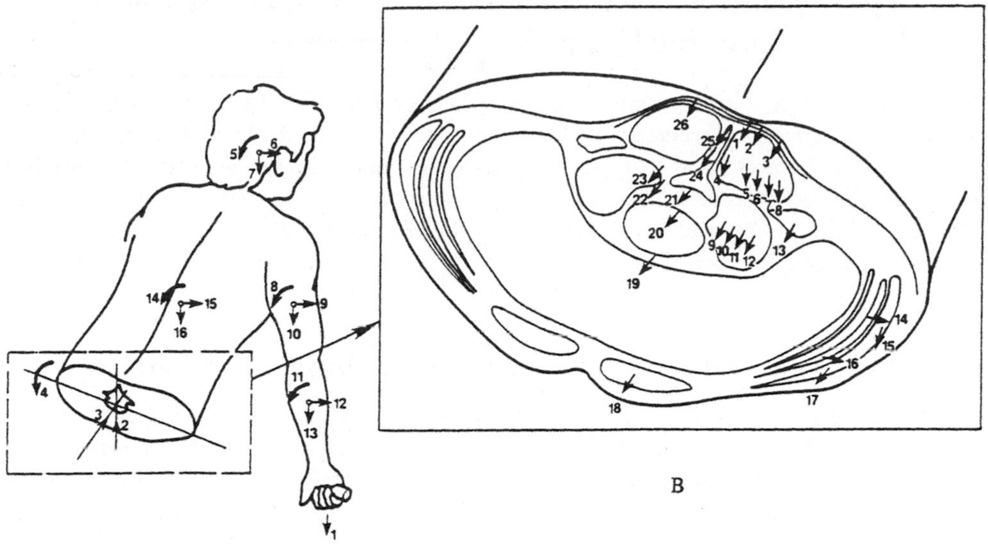

FIGURE 53.3 The tissue load prediction approach required two models: (A) the first is a dynamic, 3D linked segment model to obtain the three reaction moments about the low back; (B) the second model partitions the moments into tissue forces (muscle forces 1–18), ligaments 19–26, and moment contributions from deformed disc, gut, and skin in bending etc.).

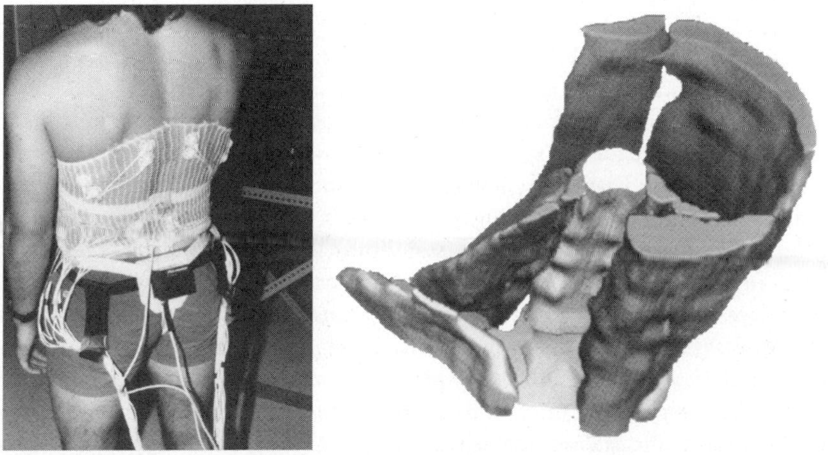

FIGURE 53.4 Subject monitored with EMG electrodes and electromagnetic instrumentation to directly measure 3D lumbar kinematics and muscle activity (left panel). The modeled spine (partially reconstructed for illustration purposes) moves in accordance with the subject's spine, whose muscles are activated by the subject.

the remaining moment is partitioned among the many laminae of muscle based on their myoelectric profile, their physiological cross-sectional area, and modulated with known relationships for instantaneous muscle length and of either shortening or lengthening velocity. Most recent improvements of the force velocity relationship have been described in Sutarno and McGill (1995). In this way, the modeled spine moves according to the movements of the subjects spine, and the virtual muscles are activated according to the activation measured directly from the subject (see Figure 53.4). This method of using biological signals to solve the indeterminacy of multiple load bearing tissues facilitates the assessment of the many ways that we choose to support loads, an objective that is necessary for evaluation of injury mechanisms and the formulation of injury avoidance initiatives.

53.3 Application of Complex Dynamic Models to Reduce the Risk of Low Back Injury

In the previous section a case was made, and hopefully justified, for the need for representing the musculature and ligaments as accurately as possible in spine models. Such models have enabled reassessment of many mechanical issues that pertain to spine function. The following section constitutes a discussion of some recent research findings as they relate to the issue of formulating guidelines for manual exertion for task assessment and implementation by the ergonomist. While not all issues were borne from the use of complex dynamic models, they were listed here to act as source material for the ergonomist.

Should One Avoid End Range of Spine Motion During Exertion?

It is recognized that very few lifting tasks in industry can be accomplished by "bending the knees and not the back." Furthermore, most workers rarely adhere to this technique when repetitive lifts are required — a fact which is quite probably due to the increased physiologic cost of squatting compared with stooping (Garg and Herrin, 1979). However, a case can be formulated for the preservation of neutral lumbar spine curvature while lifting, (specifically avoiding end range limits of spine motion about any of the three axes). This is a different concept than "trunk angle," as the posture of the lumbar spine can be maintained independent of thigh and trunk angles. Specifically, there has been much confusion in the literature between trunk angle or inclination, and the amount flexion in the lumbar spine. Bending over is accomplished by either hip flexion or spine flexion or both. It is the issue of specific lumbar spine flexion that is of importance here. Normal lordosis can be considered to be the curvature of the lumbar spine associated with the upright standing posture.

Using the tissue load distribution perspective, the following example demonstrates the shifts in tissue loading, predicted from our modeling approach, which has quite dramatic affects on shear loading of the intervertebral column and lends insight into the stoop–squat issue. First, the dominant direction of the pars lumborum fibers of longissimus thoracis and iliocostalis lumborum are noted to act obliquely to the compressive axis of the lumbar spine, producing a posterior shear force on the superior vertebra. In contrast, the interspinous ligament complex acts with the opposite obliquity to impose an anterior shear force on the superior vertebra (see Figure 53.1). Let's observe one example where spine posture determines the interplay between passive tissues and muscles which ultimately modulates the risk of several types of injury. For example, if a subject holds a load in the hands with the spine fully flexed sufficient to achieve myoelectric silence in the extensor muscles (reducing their tension), and with all joints held still so that the low back moment remains the same, then the recruited ligaments will add to the anterior shear to levels well over 1000 N, which is of great concern from an injury risk viewpoint (see Figure 53.5). However, when a more neutral lordotic posture is adopted, the extensor musculature is responsible for creating the extensor moment and at the same time will support the anterior shearing action of gravity on the upper body and handheld load. Disabling the ligaments by avoiding full flexion greatly reduced shear loading. Full flexion postures have injury implications on strained posterior tissues and also on structures affected by large shear loads (e.g., facet joints, neural arch, or conditions of spondylolisthesis).

Using knowledge of tissue loads, one could take the position that the important issue is not whether it is better to stoop lift or to squat lift but rather the emphasis should be to place the load close to the body to reduce the reaction moment (and the subsequent extensor forces and resultant compressive joint loading) and to avoid a fully flexed spine (i.e., maintain some lordosis) to minimize shear loading. In fact, sometimes it may be better to squat to achieve this, or in cases where the object is too large to fit between the knees, it may be better to stoop, flexing at the hips but always avoiding full lumbar flexion to minimize posterior ligamentous involvement. (For a more comprehensive discussion, see McGill and Norman, 1987; Potvin et al., 1991; McGill and Kippers, 1994.)

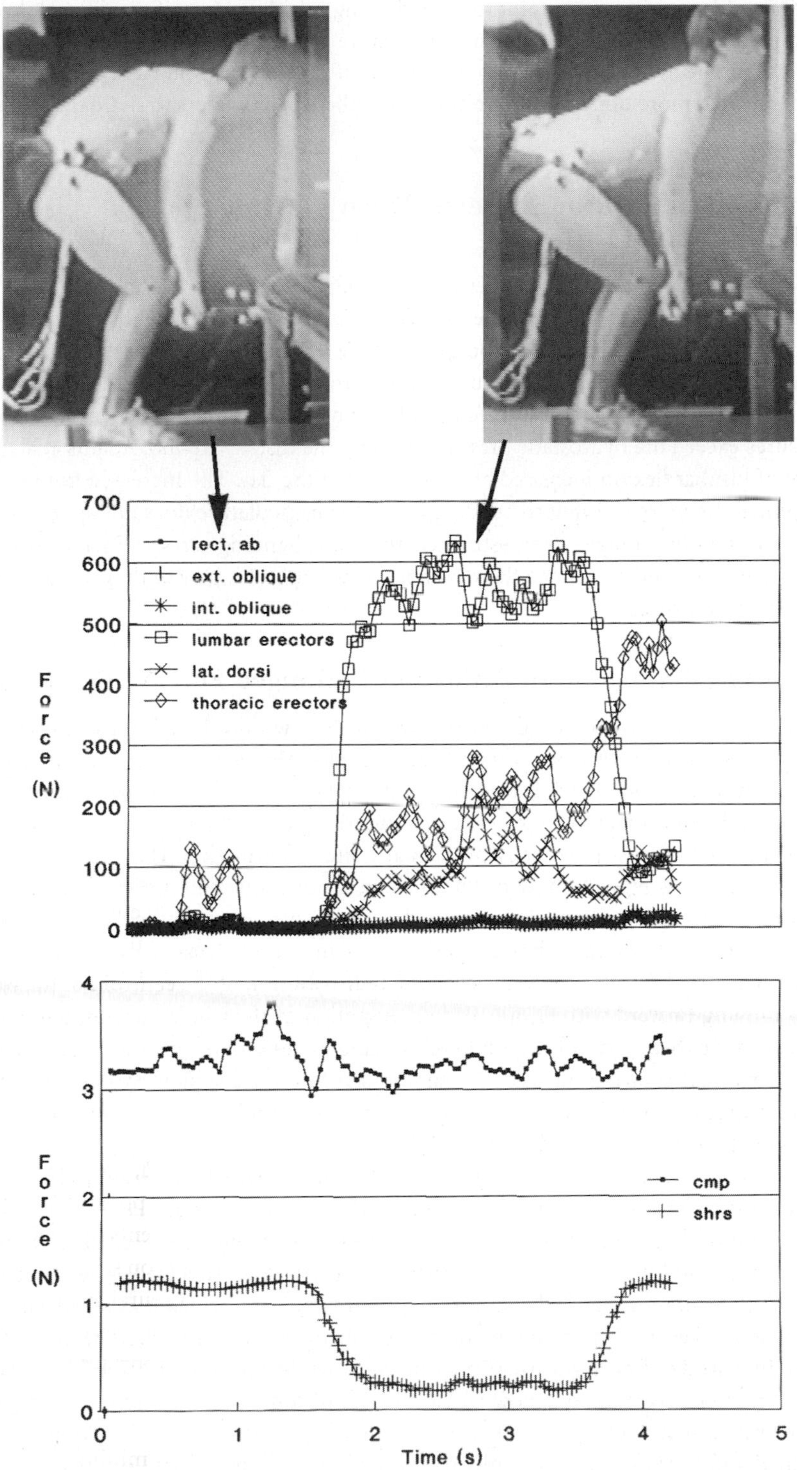

FIGURE 53.5 The fully flexed spine is associated with myoelectric silence in the back extensors and loaded posterior passive tissues, and high shearing forces on the lumbar spine (these are computer digitized images). A more neutral posture recruits the shear supporting pars lumborum extensors, disables the interspinous ligaments minimizing their shear producing forces, and reduces the net shear (Shrs) on the spine. Compression (CMP) remains essentially unchanged.

The scientific evidence appears to point to the following positive effects if one lifts while avoiding full lumbar flexion: conscious control of lumbar musculature is retained, reducing shear load on the facet joints and providing neurological protection not present if one depends on passive tissue to support the load; the disc is loaded more uniformly permitting all of the annular fibers to share the stress rather than disabling some.

Should One Lift or Perform Extreme Torso Bending Shortly After Rising from Bed?

The diurnal variation in spine length together with the ability to flex forward has been well documented. Losses in sitting height over a day have been measured up to 19mm by Reilly et al. (1984), who also noted that approximately 54% of this loss occurred in the first 30 minutes after rising. Over the course of a day, hydrostatic pressures cause a net fluid outflow from the disc, resulting in the narrowing of the space between the vertebrae, which in turn reduces tension in the ligaments. When lying down at night, osmotic pressures exceed the hydrostatic pressure, causing the disc to expand. Adams et al. (1987) noted that the range of lumbar flexion increased by 5° throughout the day. The increased fluid content caused the lumbar spine to be more resistant to bending, while the musculature does not appear to compensate by restricting the bending range. They estimated that disc bending stresses increased by 300% and ligament stresses by 80% and concluded there is an increased risk of injury to these tissues when bending forward early in the morning.

Should One Lift Immediately Following Prolonged Flexion Postures?

For a number of years, it has been proposed that the nucleus within the annulus "migrates" anteriorly during spinal extension and posteriorly during flexion (McKenzie, 1979). McKenzie's program of passive extension of the lumbar spine (which is presently popular in physical therapy) was based on the supposition that an anterior movement of the nucleus would decrease pressure on the posterior portions of the annulus, which is the most potentially problematic site of herniation. Due to viscous properties of the nuclear material, such repositioning of the nucleus is not immediate upon a postural change, but rather takes time. Krag et al. (1987) demonstrated anterior movement, albeit quite minute, from an elaborate experiment that placed radio-opaque markers in the nucleus of cadaveric lumbar motion segments. Whether this observation was due only to a redistribution of the centroid of the wedge-shaped nuclear cavity moving forward with flexion, or to a migration of the whole nucleus, remains to be seen. Nonetheless, hydraulic theory would suggest lower bulging forces on the posterior annulus if the nuclear centroid moved anteriorly during extension. If compressive forces are applied to a disc where the nuclear material is still posterior (as in lifting immediately after a prolonged period of flexion), then a concentration of stress will occur on the posterior annulus.

While this specific area of research needs more development, there does appear to be a time constant associated with this redistribution of nuclear material. If this is so, it would be unwise to lift an object immediately following prolonged flexion — such as sitting or stooping, as would a stooped gardener who may stand erect and lift a heavy object. Furthermore, it was suggested by Adams and Hutton (1988) that prolonged full flexion may cause the posterior ligaments to creep, which may allow damaging flexion postures to go unchecked if lordosis is not controlled during subsequent lifts. The data of McGill and Brown (1992), in a study of posterior passive tissue creep while slouching forward in a seated posture, showed that even after two minutes following 20 minutes of full flexion, subjects only regained half of their intervertebral joint stiffness, while even after 30 minutes of rest some residual joint laxity remained. This is of particular importance for those individuals whose work or movement patterns are characterized by cyclic bouts of full end range of motion postures followed by exertion.

These data suggest that the spine has a memory, since the mechanics of the joints are modulated by previous loading history. Before lifting, following a stooped posture, or after prolonged sitting, a case could be made for standing or even consciously extending the spine for a short period. Allowing the nuclear material to "equilibrate" or move anteriorly to a position associated with normal lordosis may decrease forces on the posterior nucleus. However, further research is required to assess this potentially important possibility.

Should Intra-abdominal Pressure Be Increased While Lifting?

It has been claimed for many years that intra-abdominal pressure (IAP) plays an important role in support of the lumbar spine, especially during strenuous lifting. This issue has been considered in lifting mechanics for years and, for some, has formed a cornerstone for prescription of abdominal belts to industrial workers and also has motivated various abdominal strengthening programs. Many have advocated the use of intra-abdominal pressure as a mechanism to reduce lumbar spine compression (Cyron et al., 1979; Troup et al., 1983; Thomson, 1988).

However, some have indicated that they believe the role of IAP in reducing spinal loads has been over emphasized (e.g., Bearn, 1961; Grew, 1980; Ekholm et al., 1982). In fact, some experimental evidence suggests that somehow, in the process of building up IAP, the net compressive load on the spine is increased! During squat lifts, it appears that the net effect of the involvement of the abdominal musculature and IAP is to increase compression rather than alleviate joint load (McGill and Norman, 1987). The size of the cross-sectional area of the diaphragm and the moment arm used to estimate force and moment produced by IAP had a major effect on conclusions reached about the role of IAP (see McGill and Norman, 1987). The diaphragm surface area was 243 cm^2 and the centroid of this area was 3.8 cm. anterior to the center of the T12 disc (compare with a 511 cm^2 pelvic floor and 465 cm^2 diaphragm together with moment arm distances [up to 11.4 cm] used in other studies). While these results were obtained with complex models, others have noted increased low back EMG activity with higher IAP during voluntary valsalva maneuvers (Krag and co-workers, 1986). Nachemson and Morris (1964) and Nachemson et al. (1986) showed an increase in intradiscal pressure during a valsalva maneuver indicating a net increase in spine compression with an increase in IAP, presumably a result of abdominal wall musculature activity.

The generation of appreciable IAP during load handling tasks is well documented; the role of IAP is not. Farfan (1973) has suggested that IAP creates a pressurized visceral cavity to maintain the hoop-like geometry of the abdominals. Recent work which measured the distance of the abdominals to the spine (their moment arms) was unable to confirm substantial changes in abdominal geometry when activated in a standing posture (McGill et al., 1996). However, the compression penalty of abdominal activity cannot be discounted. It appears that the spine prefers to sustain increased compression loads if intrinsic stability is increased. An unstabilized spine buckles under extremely low compressive load (e.g., approximately 20N, Lucas and Bresler, 1961). The geometry of the musculature suggests that individual components exert lateral and anterior–posterior forces on the spine, which perhaps can be thought of as guy wires on a mast to prevent bending and compressive buckling (Cholewicki et al., 1996). As well, activated abdominals create a rigid cylinder of the trunk resulting in a stiffer structure. Thus it appears that increased IAP, commonly observed during many activities including lifting, as well as in those people experiencing back pain, does not have a direct role to reduce spinal compression but rather is an agent used to stiffen the trunk and prevent tissue strain or failure from buckling.

Should Abdominal Belts Be Prescribed to Manual Materials Handlers?

Readers are directed to Chapter 74 on this topic in this handbook by McGill.

Should Workers Adopt a Lifting Strategy to Recruit the Lumbodorsal Fascia?

Studies have attributed various mechanical roles to the lumbodorsal fascia (LDF). In fact there have been some attempts to recommend lifting postures based on LDF hypotheses. Suggestions were originally made (Gracovetsky et al., 1981) that lateral forces generated by internal oblique and transverse abdominis are transmitted to the LDF via their attachments to the lateral border. This lateral tension was hypothesized to increase longitudinal tension, from Poisson's effect, pulling in the direction of the posterior midline of the lumbar spine, causing the posterior spinous processes to move together resulting in lumbar extension. This proposed sequence of events formed an attractive proposition because the LDF has the largest moment arm of all extensor tissues. As a result, any extensor forces within the LDF would impose the smallest compressive penalty to vertebral components of the spine.

However, this hypothesis was examined by three studies which used quite different methods, all published about the same time, and which, collectively, questioned its viability: Tesh et al. (1987), who performed mechanical tests on cadaveric material; Macintosh et al. (1987), who recognized anatomical inconsistencies with the abdominal activation; McGill and Norman (1988), who tested the viability of LDF involvement with latissimus dorsi as well as with the abdominals using a dynamic modeling approach. Regardless of the choice of LDF activation strategy, the LDF contribution to the restorative extension moment was negligible compared with the much larger low back reaction moment required to support the load in the hands. Although the LDF does not appear itself to be a significant active extensor of the spine, it is a strong tissue with a well-developed lattice of collagen fibers. Its function may be that of an extensor muscle retinaculum (Bogduk and Macintosh, 1984). The tendons of longissimus thoracis and iliocostalis lumborum pass under the LDF to their sacral and ilium attachments. Perhaps the LDF provides a form of "strapping" for the low back musculature. Hukins et al. (1990), on theoretical grounds only at this time, have proposed that the LDF acts to increase the force per unit cross-sectional area that muscle can produce by up to 30%. They suggest that it does this by constraining bulging of the muscles when they shorten. This contention remains to be proven. Tesh et al. (1987) have suggested that the LDF may be more important for supporting lateral bending. No doubt, this notion will be pursued in the future. Given the confused state of knowledge about the role, if any, of the LDF, the promotion of lifting strategies based on intentional LDF involvement cannot be justified at this time.

Should the Trunk Musculature Be Cocontracted to Stabilize the Spine?

The ability of the joints of the lumbar spine to bend in any direction is accomplished with large amounts of muscle coactivation. Such coactivation patterns are counter productive to generating the torque necessary to support the applied load in a way that minimizes the load penalty imposed on the spine from muscle contraction. Several ideas have been postulated to explain muscular coactivation: the abdominals are involved in the generation of intra-abdominal pressure (Davis, 1959), or in providing support forces to the lumbar spine via the lumbodorsal fascia (e.g., Gracovetsky et al., 1981); however, these ideas have not been without opposition (see previous sections).

It appears that another explanation for muscular co-activation is tenable. As noted above, ligamentous spine (one in which muscles are removed) will fail under compressive loading in a buckling mode, at very low forces of only about 20N (Lucas and Bresler, 1961). The spine can be likened to a flexible rod where under compressive loading it will buckle. However, if the rod has guy wires connected to it, like the rigging on a ship's mast, more compression is ultimately experienced by the rod, but it is able to bear much more compressive load as it is stiffened and more resistant to buckling. The cocontracting musculature of the lumbar spine can perform the role of stabilizing guy wires to each lumbar vertebrae bracing against buckling. Work by Crisco and Panjabi (1990) and Cholewicki and McGill (1996), has begun to quantify the influence of muscle architecture and the necessary coactivation on stability of the lumbar spine. The architecture of the lumbar erector spinae is especially suited for this role (see Macintosh and Bogduk [1987] and McGill and Norman [1987]). In order to invoke this antibuckling and stabilizing mechanism when lifting, one could justify lightly cocontracting the musculature to both minimize the potential of buckling and remove the possibility of any tissue having to bear a surprise load.

How Do People Hurt Their Backs Picking Up a Pencil?

While injury from large exertions is understandable, explanation of how people injure their backs performing rather benign-appearing tasks is more difficult — but the following is worth considering by

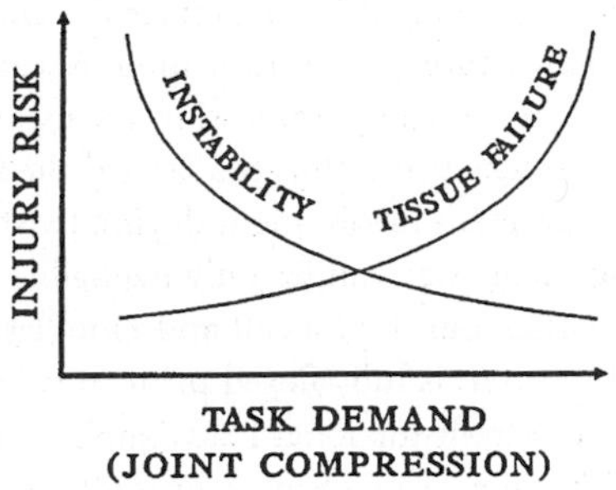

FIGURE 53.6 While injury from high loading tasks is easier to rationalize, injury from low loading tasks appears to reduce spine stability and increase the possibility of injury from errors in motor control, and the resulting joint displacement and tissue overload. (From Cholewicki, J. and McGill, S.M. (1996) Mechanical stability of the *in vivo* lumbar spine: Implications for injury and chronic low back pain, *Clin. Biomech.* 11(1):1-15. With permission.)

the ergonomist. Continuing the considerations about stabilization from the previous section — a number of years ago, we were investigating the mechanics of power lifters' spines while they lifted extremely heavy loads, using video fluoroscopy for a sagittal view of the lumbar spine (Cholewicki and McGill, 1992). The range of motion of the power lifters' spines was calibrated and normalized to full flexion by first asking them to flex at the waist and support the upper body against gravity with no load in the hands. During their lifts, although they outwardly appeared to have a very flexed spine, in fact, the lumbar joints were all two to three degrees per joint from full flexion, explaining how they could lift such huge loads (up to 210 kg) without sustaining the injuries suspected to be linked with full lumbar flexion. However, during the execution of a lift, one lifter reported discomfort and pain. Upon examination of the video fluoroscopy records, one of the lumbar joints (specifically, the L4/L5 joint) reached the full flexion calibrated angle, while all other joints maintained their static position (2 to 3 degrees from full flexion). This is the first observation that we know of reported in the scientific literature documenting proportionately more rotation occurring at a single lumbar joint than at the other joints, and it would appear that this unique occurrence was due to an inappropriate sequencing of muscle forces (or a temporary loss of the normal motor control pattern). This motivated the work of our colleague, Dr. Jacek Cholewicki, to investigate and continuously quantify stability of the lumbar spine throughout a reasonably wide variety of loading tasks (Cholewicki and McGill, 1996). Generally speaking, it appears that the occurrence of a motor control error which results in a temporary reduction in activation to one of the intersegmental muscles, perhaps for example a laminae of multifidus, could allow rotation at just a single joint to the point where passive tissue or other muscle tissue could become irritated or injured. Dr. Cholewicki noted that the risk of such an event was greatest when there are high forces in the large muscles with simultaneous low forces in the small intersegmental muscles (a possibility with our power lifter) or when all muscle forces are low, such as during a low-level exertion. Thus, a mechanism is proposed, based on motor control error resulting in temporary inappropriate neural activation, that explains how injury might occur during extremely low load situations, for example, picking a pencil up from the floor following a long day at work performing a very demanding job (see Figure 53.6).

Are Twisting Lifts Particularly Dangerous?

Twisting of the trunk has been identified as a factor in the incidence of occupational low back pain (Frymoyer et al., 1983; Troup et al., 1981), but the mechanisms of risk require some explanation. Some hypotheses have been based on an inertia argument in that twisting at speed will impose dangerous axial torques upon braking the axial rotation of the trunk at the end range of motion. Farfan and colleagues (1970) proposed that twisting of the disc is the only way to damage the collagenous fibers in the annulus leading to failure. They reported that distortions of the neural arch permitted such injurious rotations. More detailed analyses of the annulus under twist were conducted by Shirazi-Adl et al. (1986b) who supported Farfan's contention that twisting indeed can damage the annulus but also noted that twisting is not the sole mechanism of annulus failure. In contrast, some research has suggested that twisting *in vivo* is not dangerous to the disc as the facet in compression forms a mechanical stop to rotation well before the elastic limit of the disc is reached, and thus the facet is the first structure to sustain torsional failure (Adams and Hutton, 1981). Ligament involvement during twisting was studied by Ueno and Liu (1987), who concluded that the ligaments were under only negligible strain during a full physiological twist. However, an analysis of the L4/L5 joint by McGill and Hoodless (1990) suggested that posterior ligaments may become involved if the joint is fully flexed prior to twisting.

Certainly the mechanisms of injury from torsional loads applied under twisting conditions remain inconclusive. However, it is clear that the increase in compressive load on the spine is dramatic if a comparatively small amount of axial twist torque is required in addition to the dominant extensor torque. Using data from a combination of our previous studies, to support 50 N.m in extension imposes about 800 N of spinal compression, but 50 N.m in axial twist would impose over 2500 N, while 50 N.m of lateral bend imposes 1400 N of compression. These differences result from the difference in coactivation of the trunk musculature, combined with small moment arms in many cases, to generate the moments of force required. It appears that the joint pays dearly in order to support even small axial torques when extending during the lifting of a load.

Is "Lifting Smoothly" and Not Jerking the Load Always Best?

We have all heard that a load should be lifted smoothly and not "jerked." This recommendation was most likely rationalized on the basis that accelerating a load upwards increases its effective mass by virtue of an additional inertial force acting downward together with the gravitational vector. However, this may not always be the case as it is possible to lift a load by transferring momentum from an already moving segment. The concept of momentum transfer during lifting has been referred to by Troup and Chapman (1969) and by Grieve (1975), who coined the term "kinetic lift." Later, McGill and Norman (1985) documented that smaller low back moments were possible in certain cases compared to moments during static lifts in the same posture using skillful transfer of momentum. For example, if a load is awkwardly placed, perhaps placed on a worktable at a distance of 75 cm from the worker, a slow-smooth lift would necessitate the generation of a large lumbar extensor torque for a lengthy duration of time — a situation that is most strenuous on the back. However, this load could be lifted with a very low lumbar extensor moment or quite possibly no moment at all. For example, if the worker leaned forward and placed his hands on the load, with bent elbows, the elbow extensors and shoulder musculature could thrust upward initiating upward motion of the trunk to create both linear and angular momentum in the upper body (note that the load has not yet moved). As the arms straighten, coupling takes place between the load and the large trunk mass (as the hands then start to apply upward force on the load) transferring some, or all, of the body momentum to the load causing it to be lifted with a jerk. This highly skilled "inertial" technique is observed quite frequently throughout industry and in some athletic events such as competitive weight lifting, but it must be stressed that such lifts are conducted by highly practiced and skilled individuals. In most cases, acceleration of loads to decrease low back stress in the manner described is not suitable for the "lay" individual when conducting the lifting chores of daily living.

The momentum — transfer technique is a skilled movement that requires practice, is only feasible for awkwardly placed lighter loads, and could not be justified for heavy lifts. However, there may be another mechanical variable to be integrated into the analysis of a dynamic technique. The tissue property of viscoelasticity enables tissues to sustain higher loads when loaded at rate (Burstein and Frankel, 1968). Troup (1977) suggested that the margin of safety for spine injury may be increased during a higher strain rate but cautioned that incorporation of this principle into lifting technique depends on the rate of increase in spinal stress, the magnitude of peak stress, and its duration. The lifting instruction to always lift a load "smoothly" may not invariably result in the least risk of injury. Indeed, it is possible to skillfully transfer momentum to an awkwardly placed object to position the load as close to the body quickly and minimize the extensor torque required to support the load. There can be no argument that reduction of the extensor moment required to support the hand load is paramount in reducing the risk of injury and that this is best accomplished by keeping the load as close to the body as possible.

Is Sedentary-Seated Work Harmful?

Epidemiological evidence presented by Videman et al. (1990) documented the increased risk of disc herniation for those who perform sedentary jobs characterized by sitting. Known mechanical changes associated with the seated posture include the increase in intradiscal pressure when compared to standing postures (Andersson et al., 1975), increases in posterior annulus strain (Pope et al., 1977), creep in posterior passive tissues (McGill and Brown, 1992) which decreases anterior–posterior stiffness and increases shearing movement (Schultz et al., 1979), and posterior migration of the mechanical fulcrum (Wilder et al., 1988) which reduces the mechanical advantage of the extensor musculature (resulting in increased compressive loading). This has motivated occupational biomechanists to consider the duration of sitting as a risk factor when designing seated work in the interest of reducing the risk of injury. A recently proposed guideline has suggested a sitting limit of fifty minutes without a break, although this proposal will be tested and evaluated in the future.

53.4 Tentative Risk Reduction Guidelines for Occupational Injury

The following recommendations have been summarized from the biomechanical rationale developed in the previous section. Some are consistent with what has been advocated for years, while others contradict long-standing notions that were based on flawed, or unavailable, biomechanical understanding. They are more versatile and widely applicable than the commonly used instruction of "bend the knees, not the back" to reduce lifting stresses. However, they have not been subjected to rigorous scientific challenge and clinical trial. Due to the lack of conclusive evidence to support these recommendations, and acknowledging that the definitive studies will probably not occur in the near future, they are listed in the following form to provoke thought, stimulate discussion and generate research interest. Perhaps it is overly optimistic to expect that these recommendations will withstand time and remain intact. However, these recommendations may have the potential to reduce tissue loads during the performance of a wide range of industrial exertion tasks and they are able to accommodate all tasks including those outside the sagittal plane. Furthermore, the exact instructions issued to a specific worker should not be taken verbatim from the following list, but rather the biomechanical principle should be explained in a language and terminology which is familiar to the worker. In addition, often successful job incumbents have developed personal strategies for working that assist them in avoiding fatigue and injury. Their insights are the result of thousands of hours of performing the task, and they can be very perceptive — attempts should be made to accommodate them.

Recommendations for Safer Work — A Tentative List

1. Avoid a fully flexed or bent spine and rotate trunk using hips.
   - Strain in the annular fibers is equalized (Hickey and Hukins, 1980; Shirazi-Adl et al., 1986 a)
   - Posterior ligaments are not strained and cannot be injured (McGill, 1988).
   - Facets are in contact and can bear some load (e.g., Nachemson, 1960).
   - The anterior shearing effect from ligament involvement is minimized, and the posterior supporting shear of the musculature is maximized (McGill, 1989).
   - Compressive testing of lumbar motion units has shown increases in tolerance with partial flexion but decreased ability to withstand compressive load at full flexion (Adams and Hutton, 1988).
2. Choose a posture to minimize the reaction moment on low back so long as #1 is not compromised.
   - Neutral lordosis is still maintained but sometimes the load can be brought closer to the spine with bent knees (squat lift) or relatively straight knees (stoop lift). The key is to reduce the moment which has been shown to be a dominant risk factor (Marras et al., 1995).

3. Allow time for the disc nucleus to "equilibrate" and ligaments to regain stiffness after prolonged flexion.
   - After prolonged sitting or stooping, spend time standing to allow the nuclear material within the disc to equilibrate and equalize the stress on the annulus, and allow the ligaments to regain their rest length and provide protective stiffness to the lumbar spine (McGill and Brown, 1992).
4. Avoid lifting shortly after rising from bed.
   - Forward bending stresses on the disc and ligaments are higher in the early morning compared with later in the day (Adams et al., 1987).
5. Prestress system even during "light" tasks.
   - Lightly cocontract the stabilizing musculature to remove the slack from the system and stiffen the spine, even during "light" tasks such as picking up a pencil (Cholewicki and McGill, 1996).
   - Cocontraction and the corresponding increase in stability increases the margin of safety of material failure of the column under axial load (Crisco and Panjabi, 1990)
   - In this way, no tissue will have to bear a "surprise" load
6. Avoid twisting.
   - Twisting reduces the intrinsic strength of the annulus by disabling some of its supporting fibers while increasing the stress in the remaining fibers under load (Shirazi-Adl et al., 1986b)
   - Since there is no muscle designed to produce only axial torque, the collective ability of the muscles to resist axial torque is limited and may not be able to protect the spine in certain postures (McGill, 1991)
   - The additional compressive burden on the spine is substantial for even a low amount of axial torque production (McGill and Hoodless, 1990).
7. Exploit the acceleration profile of the load.
   - This is only for highly skilled individuals performing repetitive tasks.
   - Dangerous for heavy loads and should not be attempted.
   - It is possible that a transfer of momentum from the upper trunk to the load can start moving an awkwardly placed load without undue low back load (Grieve, 1975; Troup, 1977). Possibly, the viscoelastic property of biological material will safely absorb a momentary high load required to bring the load close to the trunk, which reduces the reaction moment.
8. Avoid prolonged sitting.
   - Prolonged sitting is associated with disc herniation (Videman et al., 1990).
   - When required to sit for long periods, adjust posture often, stand up, at least every 50 minutes, and extend spine and/or walk for a few minutes.
   - Organize work to break up bouts of prolonged sitting into shorter periods that are better tolerated by the spine.
9. Consider the best rest break strategies
   - Workers engaged in sedentary work would be best served by frequent, dynamic breaks to reduce tissue stress accumulation (Adams and Dolan, 1995) and migrate tissue load bearing (McGill, 1997).
   - Workers engaged in dynamic work may be better served with longer and more "restful" breaks.

53.5 Future Directions

The recent advances in low back research and ergonomic model sophistication has increased general understanding of lumbar mechanics, injury mechanisms, and planning injury avoidance strategies. However, the reader of this chapter becomes aware, quite quickly, that many issues pertaining to reducing the risk of industrial low back injury, and to the recommendation of injury risk reduction guidelines remain unsolved. The future of ergonomics and spinal biomechanics must address a range of issues that will demand the utmost effort in creating models that capture biological fidelity. Continued effort must be directed toward obtaining more sophisticated anatomy for model components, for it is the fine details that unlock the secrets of force generation, transmission, and sharing strategies among tissues. Tissue properties such as strength, viscoelasticity, and fatiguability must be better understood. The knowledge base of biomaterials is still in the developmental stage. Static behavior has been quite well documented for some tissues although not all. However, with the recent development of quite involved dynamic models, dynamic tissue behavior is desperately required. The property of viscoelasticity is paramount in the determination of dynamic tissue load due to its time and loading rate dependency. Examination of movement, particularly rapid movements of the trunk and limbs, is hindered by inadequate dynamic tissue information.

Muscle cocontraction and interplay with the passive tissues is far more prevalent under complex motion conditions which enforces the need to measure the activity of individual tissues. Both EMG techniques and optimization strategies to partition tissue loads require basic research to improve predictions of the cocontracting muscle forces. Perhaps major developments in artificial intelligence will provide the required interface with motor control in the future. However, at present only the mind is capable of such sophisticated processing, and research efforts would appear to be best directed toward the improvement of EMG techniques and the appropriate processing to obtain muscle force predictions. Work must continue to determine which muscles to monitor, where to place electrodes, and improve processing techniques to increase EMG reliability. Some international effort is ongoing, with the continual reporting of some quite impressive predictions of individual muscle force measures in animal preparations, from processed EMG.

Analysis of the single task, with no provision for repeated movements, has dominated the literature. However, most tasks in industry and those that are part of mundane daily events are repeated and prolonged. The effect of fatigue on the body system, intervertebral joints, muscles, and ligaments demand investigation. For example, at present the frequency of task repetition can only be recommended on the psychophysical criterion of what the individual "thinks" is appropriate. Data on repeated tissue loads is extremely scarce, although is appearing in the literature with greater frequency. Tissue fatigue must also be considered in the context of static holds that may occur in activities such as stooping for long periods of time while gardening. While ergonomic design of the workplace is of utmost importance in facilitating work postures that minimize joint loads, much basic research remains to test and refine low back hypotheses and related issues such as reducing the risk of low back injury for workers. Finally, there is a commonly held notion among ergonomists that minimal tissue loading is best. This is untrue. Biological tissues require repeated loading to be healthy. The trick is to develop a wise rest break strategy to facilitate optimal tissue adaption. The most healthy combination of work and rest will be achieved through first understanding the biomechanical, physiological, and psychological parameters of injury and human performance, followed with thoughtful application of this wisdom.

Acknowledgment

The author wishes to acknowledge the many colleagues who have made intellectual contributions to the series of works documented in this chapter and who have made the journey such a joy: Jacek Cholewicki, Ph.D., John Sequin, M.D., Lina Santaguida, M.Sc., Chrisanto Sutarno, M.Sc., Daniel Juker, M.D., Michael Sharratt, Ph.D. and, in particular, Robert Norman, Ph.D.

References

Adams, M.A. and Hutton, W.C. (1981) The relevance of torsion to the mechanical derangement of the lumbar spine, *Spine* 6:241-248.

Adams, M.A., Dolan, P., Hutton, W.C. (1987) Diurnal variations in the stresses on the lumbar spine, *Spine* 12 (2): 130:137.

Adams, M.A. and Hutton, W.C. (1988) Mechanics of the intervertebral disc, in *The Biology of the Intervertebral Disc* (Ed. P. Ghosh). CRC Press, Boca Raton.

Adams, M.A. and Dolan, P. (1995) Recent advances in lumbar spinal mechanics and their clinical significance, *Clin. Biomech.* 10(1): 3-19.

Andersson, G.B.J., Ortengren, R., Nachemson, A., Elfstrom, G., Broman, M. (1975) The seated posture: an electromyographic and discometric study, *Orthop. Clin. N. Am.* 6:105-120.

Bean, J.C., Chaffin, D.B., Schultz, A.B. (1988) Biomechanical model calculation of muscle contraction forces: a double linear programming method, *J. Biomech.* 21:59-66.

Bearn, J.G. (1961) The significance of the activity of the abdominal muscles in weight lifting, *Acta Anat.* 45:83.

Bogduk, N. and Macintosh, J.E. (1984) The applied anatomy of the thoracolumbar fascia. *Spine*, 9:164-170.

Burstein, A.H. and Frankel, W.H. (1968) The viscoelastic properties of some biological material, *Ann N.Y. Acad.* SCI., 146:158-165.

Cholewicki, J. and McGill, S.M. (1992) Lumbar posterior ligament involvement during extremely heavy lifts estimated from fluoroscopic measurements, *J. Biomech.* 25(1):17-28.

Cholewicki, J., McGill, S.M., Norman, R.W. (1995) Comparison of muscle forces and joint load from an optimization and EMG assisted lumbar spine model: towards development of a hybrid approach, *J. Biomech.* 28(3):321-331.

Cholewicki, J. and McGill, S.M. (1996) Mechanical stability of the *in vivo* lumbar spine: Implications for injury and chronic low back pain, *Clin. Biomech.* 11(1):1-15.

Collins, J.J. (1995) The redundant nature of locomotor optimization laws, *J. Biomech.* 28 (3): 251-167.

Crisco, J. J. and Panjabi, M.M. (1990) Postural biomechanical stability and gross muscular architecture in the spine, in *Multiple Muscle Systems* (Eds. J. Winters, S. Woo), Springer-Verlag, New York. pp. 438-450.

Cyron, B.M., Hutton, W.C., Stott, J.R. (1979) The mechanical properties of the lumbar spine, *Mech. Eng.* 8(2):63-68.

Davis, P.R. (1959) The causation of herniae by weight-lifting, *Lancet*, 2:155-157.

Ekholm, J., Arborelius, U.P., Nemeth, G. (1982) The load on the lumbosacral joint and trunk muscle activity during lifting, *Ergonomics* 25(2):145-161.

Farfan, H.F. (1973) Mechanical Disorders of the Low Back, Lea and Febiger, Philadelphia.

Farfan, H.F., Cossette, J.W., Robertson, G.H., Wells, R.V., Kraus, H. (1970) The effects of torsion on the lumbar intervertebral joints: the role of torsion in the production of disc degeneration, *J. Bone. Jt. Surg.* 52A (3):469-497

Frymoyer, J.W., Pope, M.H., Clements, J.H, Wilder, D.G., MacPherson, B., Ashikaga, T. (1983) Risk factors in low back pain, *J. Bone, Jt Surg.*, 65A:213-218.

Gracovetsky, S., Farfan, H.F., Lamy, C. (1981) Mechanism of the lumbar spine, *Spine* 6(1):249- 262.

Garg, A. and Herrin, G. (1979) Stoop or squat: a biomechanical and metabolic evaluation, *AIIE. Trans.* 11:293-302.

Granata, K.P. and Marras, W.S. (1993) An EMG-assisted model of loads on the lumbar spine during asymmetric, dynamic trunk extensions, *J. Biomech.* 26(12): 1429-1438.

Gray's Anatomy, Descriptive and Applied (1980) 36th ed. (Eds. Warwick, R., and Williams, P.L.) Longmans, London.

Gregor, R.J., Komi, P.V., Jarvinen, M. (1987) Achilles tendon forces during cycling, *Int. J. Sports Med.* 8:9-14.

Grew, N.D. (1980) Intrabdominal pressure response to loads applied to the torso in normal subjects, *Spine* 5(2):149-154.

Grieve, D.W. (1975) Dynamic characteristics of man during crouch and stoop lifting, in *Biomechanics iv*, (Eds. Nelson, R.C. and Morehouse, C.A.) pp. 19-29, University Park Press, Baltimore.

Heylings, D.J.A. (1978) Supraspinous and interspinous ligaments of the human lumbar spine. *J. Anat.* 123:127-131.

Hickey, D.S. and Hukins, D.W.L. (1980). Relation between the structure of the annulus fibrosus and the function and failure of the intervertebral disc, *Spine* 5(20):106-116.

Hughes, R.E., Chaffin, D.B., Lavender, S.A., Anderson, G.B.J. (1994) Evaluation of muscle force prediction models of the lumbar trunk using surface electromyography, *J. Orthop. Res.* 12:689-698.

Hughes, R.E., Bean, J.C., Chaffin, D.B. (1995) Evaluating the effect of coordination in optimization models, *J. Biomech.* 28(7):875-878.

Hukins, D.W.L., Aspden, R.M., Hickey, D.S. (1990). Thoracolumbar fascia can increase the efficiency of the erector spinae muscles, *Clin. Biomech.* 5(1): 30-34.

Komi, P.V. (1990) Relevance of in vivo force measurements to human biomechanics, *J. Biomech.* Suppl. 2:23-34.

Krag, M.H., Byrne, K.B., Gilbertson, L.G., Haugh, L.D. (1986) Failure of intraabdominal pressurization to reduce erector spinae loads during lifting tasks, in *Proceedings of the North American Congress on Biomechanics,* Montreal, 25-27 August pp. 87-88.

Krag, M.H., Seroussi, R.E., Wilder, D.G., Pope, M.H. (1987) Internal displacement distribution from *in vitro* loading of human thoracic and lumbar spinal motion segments: Experimental results and theoretical predictions, *Spine.* 12(10):1001-1007.

Langenberg, W (1970) Morphologic, Physiologischer Querschnitt and Kraft des M. erector spinae in Lumbalbereich des Menshen, *Z. Anat. Entwickl.* 132:158-190.

Lucas, D. and Bresler, B. (1961) *Stability of the Ligamentous Spine,* Tech. Report No. 40, Biomechanics Laboratory, University of California, San Francisco.

Macintosh, J.E., Bogduk, N., Gracovetsky, S. (1987) The biomechanics of the thoracolumbar fascia. *Clin. Biomech.* 2:78-93.

Macintosh, J.E. and Bogduk, N. (1987) The morphology of the lumbar erector Spinae, *Spine,* 12(7):658-668.

Marras, W.S. and Sommerich, C.M. (1991) A three-dimensional motion model of loads on the lumbar spine I and II, *Human Factors,* 33(2): 123-149.

Marras, W.S., Lavender, S.A., Leurgans, S.E., Fathallah, F.A., Ferguson, S.A., Allread, W.G., Rajulu, S.L. (1995) Biomechanical risk factors for occupationally-related low back disorders, *Ergonomics,* 36(2): 377-410.

McGill, S.M. and Norman, R.W. (1985) Dynamically and statically determined low back moments during lifting, *J. Biomech.* 18(12):877-885.

McGill, S.M. and Norman, R.W. (1986) Partitioning of the L4/L5 dynamic moment into disc, ligamentous and muscular components during lifting, *Spine* 11(7):666-678.

McGill, S.M. and Norman, R.W. (1987) Reassessment of the role of intraabdominal pressure in spinal compression, *Ergonomics* 30(11):1565-1588.

McGill, S.M. and Norman R.W. (1987) Effects of an anatomically detailed erector spinae model on L4/L5 disc compression and shear, *J. Biomech.* 20(6):591-600.

McGill, S.M. and Norman, R.W. (1988) The potential of lumbodorsal fascia forces to generate back extension moments during squat lifts, *J. Biomed Engng.*, 10:312-318.

McGill, S.M., Patt, N., Norman, R.W. (1988) Measurement of the trunk musculature of active males using CT Scan radiography: duplications for force and moment generating capacity about the L4/L5 joint, *J. Biomech.* 21(4):329-341.

McGill, S.M. (1988) Estimation of force and extensor moment contributions of the disc and ligaments at L4/L5, *Spine* 13:1395-1402.

McGill, S.M. (1989) Loads of the lumbar spine and associated tissues, in *Biomechanics of the Spine — Clinical and Surgical Perspective,* (Eds. Goel, V.K. and Weinstein, J.N.) CRC Press, Boca Raton.

McGill, S.M. and Hoodless, K. (1990) Measured and modelled static and dynamic axial trunk torsion during twisting in males and females, *J. Biomed. Engng.* 12:403-409.

McGill, S.M. (1991) Electromyographic activity of the abdominal and low back musculature during the generation of isometric and dynamic axial trunk torque: implications for lumbar mechanics, *J. Orthop. Res.* 9:91-103.

McGill, S.M. (1991) The kinetic potential of the lumbar trunk musculature about three orthogonal orthopaedic axis in extreme postures, *Spine* 16(7):809-815.

McGill, S.M. (1992) The influence of lordosis on axial trunk torque and trunk muscle myoelectric activity, *Spine* 17(10):1187-1193.

McGill, S.M. (1992) A myoelectrically based dynamic 3-D model to predict loads on lumbar spine tissues during lateral bending, *J. Biomech.* 25(4):395-414.

McGill, S.M. and Brown, S. (1992) Creep response of the lumbar spine to prolonged lumber flexion, *Clin. Biomech.* 7:43-46.

McGill, S.M., Santaguida, L., Stevens, J. (1993) Measurement of the trunk musculature from T6 to L5 using MRI scans of 15 young males corrected for muscle fiber orientation, *Clin. Biomech.* 8:171-178.

McGill, S.M. and Kippers, V. (1994) Transfer of loads between lumbar tissues during the flexion relaxation phenomenon, *Spine* 19(19):2190-2196.

McGill, S.M. (1996) A revised anatomical model of the abdominal musculature for torso flexion efforts, *J. Biomech.* 29(7):973-977.

McGill, S.M., Juker, D., Axler, C. (1996) Correcting trunk muscle geometry obtained from MRI and CT scans of supine postures for use in standing postures, *J. Biomech.* 29(5): 643-646.

McGill, S.M., Juker, D., Kropf, P. (1996) Appropriately placed surface EMG electrodes reflect deep muscle activities (psoas quadratus lumborum, abdominal wall) in the lumbar spine, *J. Biomech.* 29(11):1503-1507.

McGill, S.M. (1997) Biomechanics of low back injury: implications on current practice and the clinic. *J. Biomech.*, 30(5): 465-475.

McKenzie, R.A. (1979) Prophylaxis in recurrent low back pain, *Nz Med. J.* 89:22.

Nachemson, A. (1960) Lumbar intradiscal pressure. *Acta Orthop. Scand.* suppl. 43.

Nachemson, A.L. and Morris, J.M. (1964) *In vivo* measurements of intradiscal pressure, *J. Bone Jt. Surg.* 46A:1077-1092.

Nachemson, A., Andersson, G.B.J., Schultz, A.B. (1986) Valsalva manoeuvre biomechanics: effects on lumbar trunk loads of elevated intra-abdominal pressure, *Spine* 11(5):476-479

Nemeth, G. and Ohlsen, H. (1986) Moment arm lengths of trunk muscles to the lumbosacral joint obtained *in vivo* with computer Tomography, *Spine.* 11(2):158

Norman, R.W., Gregor, R., Dowling, J. (1988) The prediction of cat tendon force from EMG in dynamic muscular contractions, in proceedings of the fifth biennial conference of the Canadian Society for Biomechanics 1988, O'Hara, August 16-19, pp. 120-121.

Pope, M.H., Hanley, E.N., Matteri, R.E., Wilder, D.G., Frymoyer, J.W. (1977) Measurement of intervertebral disc space height, *Spine* 2:282-286.

Potvin, J., Norman, R.W., McGill, S. (1991) Reduction in anterior shear forces on the L4/L5 disc by the lumbar musculature, *Clin. Biomech.* 6:88-96.

Potvin, J. and Norman, R.W. (1992) Can fatigue compromise lifting safety? Proceedings of the Second North American Congress on Biomechanics, Chicago, August 24-28, pp. 513-514.

Reid, J.G. and Costigan, P.A. (1985) Geometry of adult rectus abdominis and erector spinae Muscles, *J. Orthop. Sports Phys. Ther.* 6:278-280.

Reilly, T., Tynell, A., Troup, J.D.G. (1984) Circadian variation in human stature, *Chronobiology Int.* 1:121-126

Santaguida, L. and McGill, S.M. (1995) The psoas major muscle: A three-dimensional mechanical modeling study with respect to the spine based on MRI measurement, *J. Biomech.* 28(3):339-345.

Schultz, A.B., Warwick, D.N., Berkson, M.H., Nachemson, A., (1979) Mechanical properties of the human lumbar spine motion segments — Part 1, Responses to flexion, extension, lateral bending and Torsion, *J. Biomech. Eng.* 101:46-52.

Schultz, A.B., Andersson, G.B.J., Ortengren, R., Haderspeck, K., Nachemson, A. (1982) Loads on the lumbar spine, *J. Bone. Jt. Surg.* 64A(5):713-720.

Shirazi-Adl, A., Ahmed, A.M., Shrivastava, S.C. (1986a) A finite element study of a lumbar motion segment subjected to pure sagittal plane moments, *J. Biomech.* 19(4):331-350

Shirazi-Adl, A., Ahmed, A.M., Shrivastava, S.C. (1986b) Mechanical response of a lumbar motion segment in axial torque alone and combined with compression, *Spine* 11(9): 914- 927

Shirazi-Adl, A. and Drouin, G. (1987) Load bearing role of the facets in the lumbar segment under sagittal plane Loadings *J. Biomech.* 20(6):601-603.

Sutarno, C. and McGill, S.M. (1995) Iso-velocity investigation of the lengthening behavior of the erector spinae muscles, *Eur. J. Appl. Physiol. Occup. Med.* 70(2):146-153.

Tesh, K.M., Dunn, J., Evans, J.H., (1987) The abdominal muscles and vertebral stability, *Spine* 12(5):501-508.

Thomson, K.D. (1988) On the bending moment capability of the pressurized abdominal cavity during human lifting Activity, *Ergonomics* 31(5):817-828

Thorstensson, A., Andersson, E., Cresswell, A., (1989) Lumbar spine and psoas muscle geometry revised with magnetic resonance imaging, in *Proceedings of the International Society of Biomechanics*, June, Los Angeles, abstract #251.

Troup, J.D.G. and Chapman, A.E. (1969) The strength of the flexor and extensor muscles of the trunk, *J. Biomech.* 2:49-62.

Troup, J.D.G. (1977) Dynamic factors in the analysis of stoop and crouch lifting methods: a methodological approach to the development of safe materials handling standards, *Orthop. Clin. N. Am.* 8(1):201-209.

Troup, J.D.G., Martin, J.W., Lloyd, D.C. (1981) Back pain in industry: a prospective survey, *Spine* 6:61-69

Troup, J.D.G., Leskinen, T.P.J., Stalhammear, H.R., Kuorinka, I.A. (1983) A comparison of intra-abdominal pressure increases, hip torque, and lumbar vertebral compression in different lifting techniques, *Human Factors* 25(5):517-525.

Ueno, K. and Liu, Y.K. (1987) A three-dimensional nonlinear finite element model of lumbar intervertebral joint in torsion, *J. Biomech. Eng.* 109:200-209.

Videman, T., Nurminen, J., Troup, J.D.G. (1990) Lumbar spinal pathology in cadaveric material in relation to history of back pain, occupation and physical loading, *Spine* 15(8):728-740.

Wilder, D.G., Pope, M.H., Frymoyer, J.W. (1988) The biomechanics of lumbar disc herniation and the effect of overload and instability, *J. Spine. Disorders* 1(1):16-32.

54

Selection of 2-D and 3-D Biomechanical Spine Models: Issues for Consideration by the Ergonomist

Robert W. Norman
University of Waterloo

Stuart M. McGill
University of Waterloo

54.1 Introduction

The usefulness and interpretation of complex biomechanical models have been explained in another chapter (McGill, 1999). Biomechanists continue to debate many contentious questions, such as whether there is a need for high levels of accuracy in the anatomical representation of spinal structures in a spine model, and whether the mathematical method used to solve large numbers of equations in these models reflects known physiological patterns of muscle activation in human movements. These authors believe that the answer to both of these questions is yes. However, the focus of this chapter is on simpler models intended for relatively easy use in workplace settings for the assessment of risk of back injury; but

0-8493-2641-9/99/$0.00+$.50

simplification raises another set of questions that ergonomists should ask before purchasing a biomechanical software package.

Some of the most pertinent questions are: How do biomechanists approach the assessment of level of risk of back injury in the workplace? Have biomechanical spine model estimates of risk factors (typically low back moments of force, spinal compression and shear forces are outputs from these models), been shown to be related to increased risk of injury? Are all spine models the same? Have spine models been validated, i.e., how close to reality are the model estimates of moments of force, spinal compression, and shear? How much anatomical detail in the model is necessary for adequate content validity? Is a model that will allow dynamic analysis of tasks always needed or are static models acceptable? Are single plane (2D) assessments still useful or must 3-dimensional analyses always be performed? If a 3-dimensional model is needed, how does the model estimate forces on joints and how are the forces and moments interpreted? Are there any spine models that produce outputs that reflect the effects of prolonged loading?

This chapter will address these issues and will then describe a family of 2D and 3D models into which the authors have attempted to incorporate as many important biological features as they could within the constraints of producing a method to assess low back injury risk that is practical for workplace use by ergonomists. The chapter is not a critical, comparative review of commercially available spine models suitable for ergonomic use, although several interesting models have been described in the past decade by spinal biomechanists (e.g., Chaffin et al., 1989; Marras and Sommerich, 1991).

54.2 The Biomechanical Approach to Assessment and Reduction of Risk of Back Injury

Biomechanists generally operate under the concept that a tissue will become irritated or will fail when the forces to which it is exposed at a particular point in time are greater than the tolerance of that tissue to force at that point in time. Both the size of the force acting on a tissue and the tolerance of tissue to force change with time. The forces on tissues change with changing demands of a task or with the way in which the worker does the same task from time to time; the tolerance of the tissue to force (or other mechanical input) varies with factors such as state of repair from previous injury, conditioning or deconditioning, age, gender, disease, and inherited individual differences in factors such as the cross-sectional area of tissues.

Typically, the biomechanics approach to injury risk assessment is to represent the body of the worker as a system of individual body segments linked at the joints (linked segment model). An external force acting on the linkage, for example the force of a weight held in the hand, requires the production of forces in muscles and in other tissues at all joints in the linkage to prevent the linkage from collapsing or to cause it to move in a desired direction. At the level of the lumbar spine these "reaction forces" and "reaction moments of force" are produced by some combination of muscles, ligaments, discs, bones in contact with each other, and by other body tissues, depending upon what the posture is and where in the joint range of motion each body segment is at each instant in time throughout the movement. The reaction forces in the supporting tissues can become very large for even relatively light loads held in the hands because the weights of body parts are very large (head, arms, and trunk above the pelvis are about one half body weight), and the mechanical advantage of body tissues in producing moments of force is very poor. If the forces in tissues gets too large, the tissues are damaged.

In all biomechanical spine models, reaction moments are produced by simplified representations of the real anatomical structures that actually produce the reaction moments that have to support the weight of the force on the hands and of the body parts and any inertial forces caused by accelerations of these masses, for example by combining forces of some or many small muscles to make one force vector. Some models are much more simplified anatomically and physiologically than others leading to the question as to whether they are over-simplified and produce unrealistic estimates of exposure variables; the reverse question can also be asked. Are some models more complex than is necessary for ergonomic use?

The forces calculated in these supporting structures are the measures of exposure used to estimate injury risk. For spinal models, typically the reaction (or support) moment about the lumbar spine required to support the load moment, and the consequent compressive and shear forces acting on the lumbar spine at the L4/L5 or L5/S1 level are calculated. Quantifying the risk of injury usually involves comparing these estimates of supporting tissue forces with estimates of the failure tolerance of these tissues (e.g., compressive or shear strength of the spinal tissues) or reaction moments of forces with back extension strength under the knowledge that operating at or near a person's strength limits is getting close to tissue fatigue limits, if not failure limits. Simply, as a force applied to a tissue approaches its capacity to bear force, the risk of injury increases to the point where the limit of force tolerance is reached and injury (tissue damage) occurs. More recently, epidemiological studies have appeared in which injury or pain reporting odds ratios have been calculated for high vs. low exposure to biomechanical variables such as trunk angle (Punnett et al., 1991; Marras et al., 1993), back extension moment of force (Marras et al., 1993), and peak and prolonged spine compression and shear soon to be published from work of our own group and colleagues from the Institute for Work & Health (Toronto). The establishment of target values for these variables to reduce risk epidemiologically is promising.

The advantages of the biomechanical approach are several: from a good model, any task should be analyzable — lifting, lowering, pushing, pulling, while interacting with any material, tool or machine; the analysis should be individualizable for gender, height, weight, strength etc.; and, in principle, any region of the body may be considered. The critical issue for the ergonomist is the ability to analyze the demands of a job using reasonably simple methods, to assess the risk of injury, and to redesign jobs in a way that can be shown to actually reduce forces on vulnerable tissues to levels below injury threshold.

54.3 Have Biomechanical Spine Model Outputs Been Shown to Be Related to Injury Risk?

Both biomechanical and epidemiological evidence show clearly that high moments of force about the lumbar spine which, in turn, are highly related to large compressive or shear forces on lumbar motion units, increase risk of injury. Biomechanical models that output lumbar moments of force, compression and shear, therefore, are useful as tools for measuring exposure to risk of injury.

Many studies on cadavers and on animal tissue in the laboratory have shown that excessive compressive or shear forces on the spine will result in failure of various tissues including the cartilaginous end plate of the vertebral body, the laminae of the annulus fibrosis of the intervertebral discs, muscles, ligaments, facet joints, and other structures. Original data and excellent compilations of many years of experimental results on the compressive tolerance of spinal motion units (two vertebral bodies and a disc) can be found in papers by Brinckmann et al. (1989) and Jager et al. (1991). The compilation by Jager et al. shows that older spines and female spines are weaker in compressive loading than younger spines and, for the same age, male spines on average but with large individual differences in tolerance to load. Figure 54.1 is an adaptation of data from regression equations presented by Jager et al. (1991).

Cripton et al. (1994) have shown failure of spinal motion units exposed to anterior shear forces of about 2000 N. Troup and Chapman (1969) reported maximum trunk extension moments of force (back strength) of 485 N.m for 98 men aged 18 to 39 years and 302 N.m for 132 women aged 18 to 23 years. Although people can produce maximum effort muscular contractions without injury, particularly well-conditioned people, well-motivated maximum muscular contractions, particularly in eccentric (active muscle lengthening) efforts, are approaching tissue tolerance levels. It is also well known that maximum muscular efforts can be maintained for only a few seconds before fatigue results in continual drops in force with efforts to sustain or repeat the contraction. Muscular fatigue may be a protective mechanism for active tissue, but it causes changes in the ways people perform tasks, in some cases loading tissues that would normally not be used in that task and that should not be used because of a lower injury tolerance (Potvin and Norman, 1992).

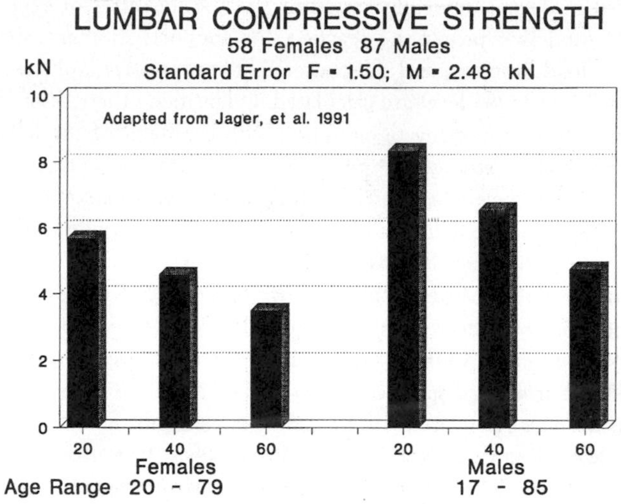

FIGURE 54.1 Lumbar spine compressive strength data compiled by Jager, Luttmann, and Laurig (1991). Means by age and sex calculated from their regression equation.

As noted earlier, not only is there biomechanical justification for assessing risk using biomechanical spine models, there is an increasing body of epidemiological studies that support this notion. Marras et al. (1993) showed, in an epidemiological study of 403 lifting jobs in 40 different industries, that those workers exposed to high average or peak lumbar moments of force were 4 to 5 times more likely to be in a high injury risk group than those exposed to low moments. Punnett et al. (1991), in a study of 219 auto assemblers, showed that those who spent more than 10% of the cycle time with the trunk flexed more than 20 degrees were 6 times more likely to report a back injury than those who worked in a more upright posture. If they spent more than 10% of the time flexed more than 45 degrees, the risk was 9 times higher than those who could work in an upright posture with less than 20 degrees of flexion. A flexed trunk increases the moment of force because of the weight of the head, arms, and trunk alone, even without loads in the hands.

In a recent study (unpublished) from our own group we showed, in data from 235 auto workers, that the correlation of trunk moment of force with L4/L5 compression was 0.98 and with reaction shear was 0.88. The 105 auto workers who were in the back pain group had peak moments of force of 182 N.m compared with 141 N.m for 130 control workers who reported no back pain, peak spinal compression of 3402 N vs. 2740 N and peak spinal shear of 462 N vs. 353 N, respectively. All of these differences were highly significant statistically ($p<0.0001$). These values were obtained from video analysis of all workers in the study while they worked combined with a quasidynamic, two-dimensional, biomechanical spine model (2DWATBAK to be discussed later). Interestingly, the mean spinal compression forces of the back pain group were very close to the Action Limit for spinal compression (3434 N) proposed as a limit by NIOSH (1981), below which most people are expected to have nominal risk of back injury. As noted earlier, high compressive forces, in the absence of impacts, are produced primarily because of very large forces in torso muscles, particularly back extensor muscles, which have relatively poor mechanical advantage but are required to support or produce the sizable moments of force required to do some tasks. The independent odds ratios on three biomechanical variables, over the spread in observed responses for the 25% of random control subjects (no reports of back pain) who had the highest exposures compared to the 25% with the lowest exposures, showed that there was a 232% greater risk of being a low back pain case (those who reported back pain to plant medical personnel) if the shear forces on the spine were high, 189% greater risk if the "usual" force on the hands was high, and 170% greater risk if the average degrees of torso flexion were high. If all three of these variables were high, the risk of being a case was

750% higher than if the exposures were low. The univariable odds ratio for peak spinal compression was 1.86 and for accumulated compression over a shift, 1.33.

While it is infrequent to see spinal motion unit failure associated with reported occupational low back pain, high spinal compression almost invariably indicates high forces on spinal muscles and, if the torso is flexed, laterally bent or twisted to very near the extremes of the range of motion, on ligaments and other spinal structures.

54.4 Are All Spine Models the Same?

All spine models are not the same. Several different models which implicitly or explicitly contain different assumptions regarding function and dramatically different levels of anatomical realism, have been presented in the literature. An earlier chapter by McGill (1999) (Chapter 53) addressed generic functional differences among complex, EMG-assisted spine models (e.g., McGill and Norman, 1986; Marras and Sommerich, 1991), optimization models (e.g., Gracovetsky et al., 1981; Schultz et al., 1982; Bean et al., 1988), and single muscle equivalent models (e.g., Frievalds et al., 1984; Norman et al., 1994). Cholewicki et al. (1995) argue that outputs from current optimization spine models are not physiologically realistic in that they do not account for the simultaneous activation of "antagonistic" muscles (cocontraction), such as those of the low back and abdomen, that is observed when people handle loads. Cocontraction of trunk muscles is extreme when a trunk twist moment of force is present, for example when one bends forward and lifts a garage door or some other load with one hand. In this case, spinal compression would be considerably underestimated by optimization models or other types of models that do not account for cocontraction of antagonistic torso muscles.

Not only are there differences in the physiological behavior of different spine models because of differences in how the models are activated (e.g., EMG vs. optimization), there are structural differences in the anatomical representation of the models. It is well known that the low back muscles, when activated, produce force in a slightly backward as well as downward direction (with respect to the spine) during a lift. The backward (posterior) component of their force reduces the shear forces acting on the facet joints and other structures but not the compressive force (Potvin et al., 1991). Excessive shear is a risk factor. It is also well known that if the normal curvature (curve seen when standing upright) of the low back is lost during a lift, the low back muscles reduce their contribution to the support moment and with extreme spine flexion these muscles may shut off completely. This causes the moment to be supported entirely by passive structures such as ligaments, and the shear force reducing contribution of the muscles is lost. Ligament injury is more serious than muscle strain, and shear forces can irritate facet joints without causing tissue damage if the joints are sensitive or result in severe damage to facet joints and other parts of the vertebral body if the forces are excessive.

It is possible to build these structural and functional features of spinal mechanics into even relatively simple models. Biomechanical spine models intended for the assessment of back injury risk in industry should include these known phenomena to increase the number of known injury mechanisms to which they are sensitive. Not all of them do this; therefore all spine models are not the same.

54.5 Have Spine Models Been Validated?

It is important to ask the question as to how close to reality the spine model estimates of exposure measure are, the measures that will be compared with tolerance levels (the lumbar moments of force, spine compression, and shear). The fact is that none of the spine models referred to above has been directly validated, although one has come close. The authors of some of the models claim that their model has been validated, but their claims are, at best, based on correlations of estimates of muscle forces from the model output with EMG recordings. This is only an indirect and not entirely convincing validation.

To be assured that outputs from models are reasonably accurate, direct validation is necessary. Therefore, if a model is to predict a muscle force or spine compression force in Newtons (N), direct validation requires that measurement of muscle forces or spine compression (N) must be made against which to compare the predictions. Technically it is not possible to directly measure spine muscle forces in living, active humans. Therefore, direct validation of spinal models has not been reported, except for the work of Schultz et al. (1982).

Schultz et al. claimed validation of a 22-muscle optimization model on the grounds that the disc compression force (N) predicted by the model was close to the pressure (mmHg) that was directly measured from a pressure transducer inserted into the lumbar discs of humans while they performed quite light, static tasks. This is a reasonable approach to validating the disc compression prediction part of the model although even in this study, Newtons of force were plotted against mmHg of disc pressure.

These authors also recorded EMG amplitude activity observed during static efforts and correlated these static amplitudes with the muscle forces predicted by the model. The correlations were far from perfect but were encouraging. However, EMG is not a direct measure of muscle force and consequently the validation of the distribution of muscle forces in this, and in all other models, particularly during asymmetric tasks, is still in question.

As noted above, correlating predictions of muscle forces in Newtons with electrical activity from muscle in microvolts is, at best, indirect. If some in the scientific community consider this type of validation of spine, or other models to be acceptable, then the EMG-assisted models are inherently valid because they use the electrical activity on a trial-to-trial and instant-to-instant basis. They do not predict it.

Direct validation of predicted forces against directly measured forces acting on most spinal tissues is virtually impossible. Therefore, content validity in the form of realistic representation of anatomical structures together with "biological" recruitment of passive tissues and simultaneous activation (cocontraction) of both agonist and antagonist muscles is important. The ergonomist or other users of commercially available spine models should take claims of validation of these models skeptically, demand detailed descriptions of the anatomical structures built into the models, and clear statements of the implicit and explicit assumptions and limitations of the model before purchasing it.

54.6 How Much Anatomical Detail in the Model is Necessary for Adequate Content Validity?

As much biological realism as possible should be incorporated into simple/industrial models so that content validity is at least present, even if direct validation of predicted forces in low back tissues is not possible. For example, the architecture of the extensor muscles and ligaments of the low back, and their dramatic effect on low back shear loads were explained briefly above and in detail in Chapter 53 (McGill, 1999). These features should be incorporated into industrial low back models. Another issue concerns the representation of contraction of the many muscles acting about the torso. For example, cocontraction of antagonistic torso muscles is commonly observed even during sagittal plane load handling and is particularly dominant in 3-D tasks. Cocontraction of antagonistic muscles results in higher muscle forces to satisfy the required moment, and in turn, higher spine compressive forces, than would be the case if there were no cocontractions (McGill and Norman, 1986; Marras and Granata, 1995). This effect can be easily incorporated into two-dimensional models by simply reducing the effective moment arm lengths of single equivalent muscles from anatomical dimensions observed from CT and MRI (McGill et al., 1988, 1993) to more physiologically effective lengths (McGill, 1991, McGill and Norman, 1992). Table 54.1 shows this comparison. It is highly desirable to incorporate this phenomenon into any 2D or 3D model of the lumbar spine because the cocontraction can have a dramatic effect on the sizes of forces acting on other structures within the system.

However, the problem of incorporating the effects of muscular cocontractions into three-dimensional spinal models is not trivial and is particularly problematic if a relatively simple model must be developed. In the opinion of the authors, optimization does not solve the problem. Indeed, one of the biggest

TABLE 54.1 Approximate moment arms (cm) of equivalent force generators based on measurements of muscle geometry and also estimated from an anatomically detailed, myoelectrically driven model that attempted to accommodate measured muscular cocontraction. Cocontraction has the effect of decreasing the mechanical advantage of an equivalent force generator and imposes higher compressive loads on the spine.

AXIS	Based on Muscle Geometry	Estimates from Model
Extension	6.0–7.5	5.5–6.5
Lateral bend	5.0–6.0	3.0–4.0
Axial twist	7.0–10.0	1.0–3.0
Flexion	8.5–10	4.0–4.5

weaknesses in current spine optimization models is that they are poor at predicting patterns of cocontractions of antagonistic muscles (Hughes et al., 1995; Cholewicki et al., 1995) particularly in complex 3-D tasks. Anatomically complex models are available (and described in Chapter 53) that have biological and mechanical content validity and can produce estimates of forces on tissues during dynamic three-dimensional tasks (e.g., McGill and Norman, 1986; Marras and Sommerich, 1991; McGill, 1992). Unfortunately, these models require the use of multichannel electromyography and direct measurement of spine kinematics and, thus, are too complex for ready application in industry although they are extremely useful in laboratory experiments in formulating and testing hypotheses designed to reduce low back injury.

In order to facilitate industrial application, anatomically simplified three-dimensional models have been reported in which the torso musculature is represented by a system of single equivalent force-moment generators (e.g., Schultz et al., 1982; Chaffin et al., 1989; Norman et al., 1994). During the generation of pure moment about any one axis, the single moment generators can be designed to quite closely predict the output from the detailed biomechanical models. The problem lies in tasks in which two or more moments must be satisfied. Should the three moment generators be activated separately, in spite of the fact that the human musculature does not work this way, or should they form a relationship that recognizes muscle coupling and allows the muscles to work together to satisfy the moments? One solution is to estimate spinal compression using a regression equation that relates compression, estimated from a complex EMG-assisted model, with the 3-D low back moments that resulted in the compression. while subjects performed dynamic, asymmetrical, three-dimensional tasks. An equation was proposed by McGill et al. (1996) that retained as much anatomical and physiological content validity as they could and, in particular, preserved the effects of antagonistic muscle cocontraction since it was based on subjects performing dynamic, asymmetrical, three-dimensional tasks. This is described in the 3DWATBAK section of this chapter.

54.7 Is a Model Needed that Will Allow Dynamic Analysis of Tasks or Are Static Models Acceptable?

A model is needed that will at least allow accounting of the inertial forces on the hands (the live loads) during lifts and lowers, pushes and pulls, not just the static load weights of objects held in the hands. Some static models can provide this versatility and a "quasidynamic" analysis of human lifting tasks is often a reasonable approach to estimating the size of the moments of force, joint reaction forces, and lumbar compression and shear forces of tasks which, in reality, are dynamic. The extent to which an entirely static model is acceptable depends upon the size of the inertial forces in a dynamic lift compared with the static load and body segment weights which create forces and moments. For example, heavy loads may be analyzed statically in many cases, simply because the lifters are incapable of appreciably accelerating the load. An exception from the athletic world is the competitive "snatch" or "clean and jerk" of a heavy load, which contains periods of very high accelerations and decelerations and corresponding high and low inertial forces. While a static analysis would be inappropriate in these cases, most heavy loads in industry are not lifted with a technique characterized by high accelerations. Of course, to analyze pushes or pulls, knowledge of the size and direction of forces acting on the hands is essential.

Freivalds et al. (1984) estimated that accelerations of loads lifted by their subjects could increase the static load by as much as 40% of its weight. The loads lifted were maximum loads the subjects were willing to lift repeatedly according to their own feelings of exertion and fatigue (not single maximum lifts). They varied from about 10 kg to 66 kg depending upon the subject and conditions of the awkward lifts. The authors of this chapter reported that dynamic analysis of lifts of 20 kg masses resulted in lumbar moments 19% higher on average than those statically determined by zeroing all accelerations of the body segments and load in the hands. Individual trials were up to 50% higher, but one skilled subject showed slightly lower dynamic than static moments on two trials. He exploited trunk accelerations which were transferred to the load to assist in the lift. A subsequent reanalysis of the data, in which only the load weight acceleration but not the body segment accelerations were accounted for, showed that this "quasi-dynamic" moment overestimated the actual dynamic moment by about 25% (McGill and Norman, 1985).

These studies indicated that caution must be exercised in interpreting statically determined moments from simple industrial models and all other static models in the context of lifts which are highly dynamic. On the other hand, if one knows the actual dynamic load weight in the posture analyzed, from an instrumented handle on the load for example, using this instantaneous value (or "live" load) as a hand force, instead of the static load weight, will usually give a liberal estimate of the actual dynamic joint moment. This is an error on the "safe" side as far as estimating demands of the lift on the low back, or other joints is concerned, avoids the complexity of a fully dynamic analysis for the ergonomist, and allows some accounting for dynamic tasks in methods suitable for workplace use. To reiterate, a model which cannot account for varying sizes and directions of forces acting on the hands cannot be used to analyze pushes or pulls.

54.8 Are Single Plane (2D) Models Useful or Must 3-Dimensional Analyses Always be Performed?

Many industrial tasks are dominated by joint loading in one plane even though there may be components of forces in other planes — the posture may be asymmetric but the loading is essentially planar and most often the dominant moment is in the flexion/extension (sagittal plane) of trunk movement. For many tasks, therefore, a 2-D model is satisfactory and many load handling postures that appear to be three dimensional do not result in large twisting moments around the long axis of the trunk. On the other hand, some efforts that do not appear as trunk twists around its long axis in fact have a substantial moment about this axis or the lateral bend axis; handling loads with one hand but with no actual torso twist or lateral bend displacement are examples. Sensitivity analyses have demonstrated that joint moments and low back compression estimated from a 2-D model are less than 10% different from those estimated from a 3-D model for two-handed lifts in which the load is placed toward the side of the body up to 30 degrees away from the midline (Bone et al., 1990). This posture looks far more twisted than calculation of torsional moments reveals (Figure 54.2). This is a far smaller difference than the trigonometry would explain and is the result of movement toward the side of the body to reach the load that was accomplished by shifting the position of the legs, pelvis, and torso together, even though the feet were constrained to remain facing forward. These small errors can be further reduced by orienting the camera axis orthogonal to the dominant loading plane and a 2-D analysis would not be out of the question. An example of the use of a 2-D and 3-D, quasidynamic spine model developed in our own laboratories and suitable for use in the workplace follows. Several models that have been reported in the literature by other authors were cited earlier.

54.9 A Simple 2D, Quasidynamic Model for Risk Assessment and Job Design (2DWATBAK)

The objective of this model, called 2DWATBAK, is to facilitate rather routine injury risk assessment of industrial tasks, but in a way that incorporates as much biological detail and content validity as possible, consistent with a method that is usable in the workplace. This model uses joint position data, obtained

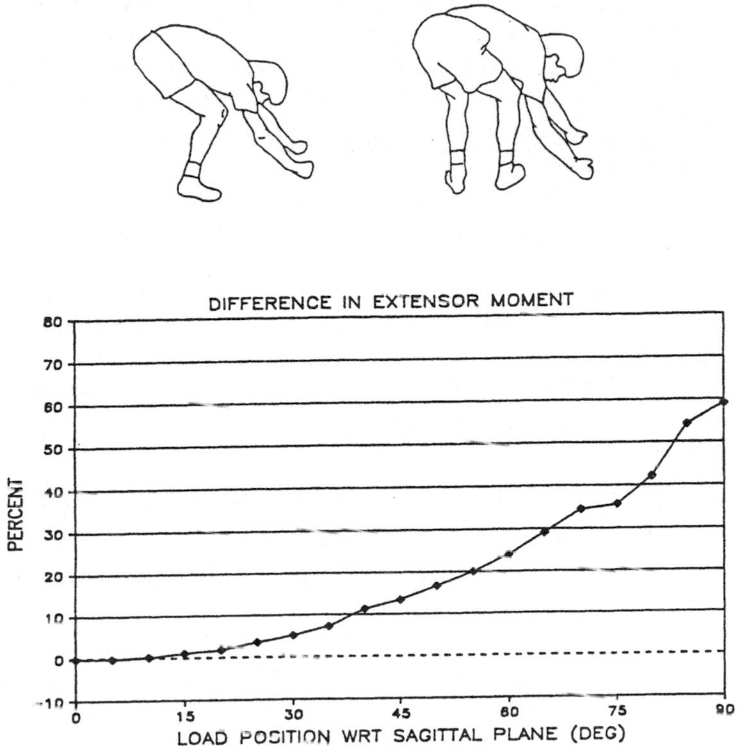

FIGURE 54.2 (Top panel) Illustration of two of the lifting postures performed with the load positioned in the sagittal plane and 90° from the sagittal plane; (middle and bottom panels) Differences in L4/L5 extensor moment predicted by a 2D model and the 3D model, expressed as a percentage of the 3D model prediction.

from a single frame of film, a slide or video, or from a manikin that the operator can move on the computer screen. The gender, body weight, and height of the worker and the sizes and directions of the static or dynamic forces acting on the hands must be input for the model to estimate moments of force about the elbows, shoulders, hips, knees, ankles, and the L4-L5 level of the low back. The compression and two shear forces, the reaction shear and the joint shear, acting at the L4/L5 level of the lumbar spine are then calculated.

To assess risk, the moments at all of the joints can be compared with strength data obtained from the literature or from the worker if measured, and the compression (e.g., Figure 54.1) and shear forces can be compared with tolerance data obtained on cadavers.

Several features are incorporated into 2DWATBAK in an attempt to increase biological content validity and to facilitate use:

1. Spinal posture (lordotic curvature) is monitored and entered by the operator to improve the accuracy of low back shear force predictions. Shear force injury risk has been underestimated in the past. For example, the size of the shear force supported by spinal structures depends upon whether the worker's lumbar spine is in a neutral posture (the curve observed during upright standing) or is flexed. A neutral posture requires erector spinae muscle activation to supply the supporting moment when the torso is inclined forward by rotating about the hips. When the posterior muscles are activated, they reduce the shear forces acting on the spinal motion unit joints (joint shear) below the level of anterior reaction shear. When the spine is severely flexed, the erector spinae muscles can be deactivated (flexion relaxation) and the moment has to be supported by ligaments and other passive tissues. When stretched, the interspinous ligament (Figure 54.3, force vectors 2a–d), described by Heylings (1978), adds undesirably to the anterior

shearing force produced by the weight of the head, arms, and trunk and forces in the hands; it does not offset this shearing force.

While most lumbar spines will tolerate 2000N of anterior shear (Cripton et al., 1994), one could argue for a workplace guideline of 1000N, half the average value, to provide a margin of safety for those in the bottom half of the shear strength distribution. Large individual differences typify tissue tolerance data.

2. Guidelines for interpreting low back compression values are improved with adjustment factors for gender differences and the effects of age (20, 40, 60 years) from data compiled by Jager et al. (1991).
3. Moments of force at each joint can be compared to joint strength data obtained from the literature (or from the worker) to identify joints at risk other than the low back (Figure 54.4).
4. The posture of a manikin can be altered on the computer screen by moving the body segment with the "arrow keys." 2DWATBAK then automatically recalculates the various exposure variable values to allow assessment of the redesigned task (Figure 54.5).

54.10 If a 3-Dimensional Model is Needed, How Does the Model Estimate Forces on Joints and How Are They Interpreted: An Example of a 3D Industrial Risk Assessment Model (3DWATBAK)

While many occupational tasks are performed essentially in the sagittal plane and can be reasonably analyzed using two-dimensional biomechanical models, many other tasks require three-dimensional analysis for approximation of the actual demands. However, obtaining input data for a 3-D model is a problem for industrial use in the absence of very expensive image reconstruction equipment or the need to use an often oversimplified look-up table of many possible 3-D postures. 3DWATBAK will take digitized x,y,z coordinates if the user has a 3-D image reconstruction system. An alternative to this expensive equipment for obtaining the necessary input coordinates of the body joints has been provided by means of a manipulable manikin on the computer screen.

First, a worker is captured in a frame of a video image or photograph. Then, 3DWATBAK provides three views (top, side, front) of a manikin which appears on the computer screen (Figure 54.6). The posture of the manikin is altered to approximate that of the worker by using a mouse. The body segments of any one of these views are moved one at a time. Movement of a segment in one view automatically moves the manikin in the other two views. When the posture of concern has been produced on the screen to the satisfaction of the analyst, the 3-D coordinates of the end points (body joints) of the manikin segments are automatically generated, scaled, and stored.

As for 2DWATBAK, the model can be individualized for a particular worker by providing data on height, weight, gender, and user-defined or standard anthropometrics. Control of anthropometrics (e.g., mass of thigh or any other segment) allows alteration of a body segment mass or location of mass center if necessary, for example to add the mass of a tool belt or to accommodate analysis of an amputee. Size and direction of forces acting on each hand are defined as a push, pull, lift, lower, or as a load mass.

Anatomical detail has inherently been incorporated through the development of a regression equation to enhance biological fidelity in estimations of low back compressive load, and, in particular, attempt to account for the cocontracting musculature. This regression equation was based on the data of three subjects, performing a variety of tasks which produced various combinations of 3-D moments about the low back. Compression was calculated using the anatomically detailed laboratory model which accommodated patterns of muscle/ligament interplay and muscle cocontraction measured in each subject-trial (see Chapter 53). The specific regression equation is as follows (R-square 0.936); the "shape" of the spinal compression/moment surface, from which the equation was derived, is seen in Figure 54.7.

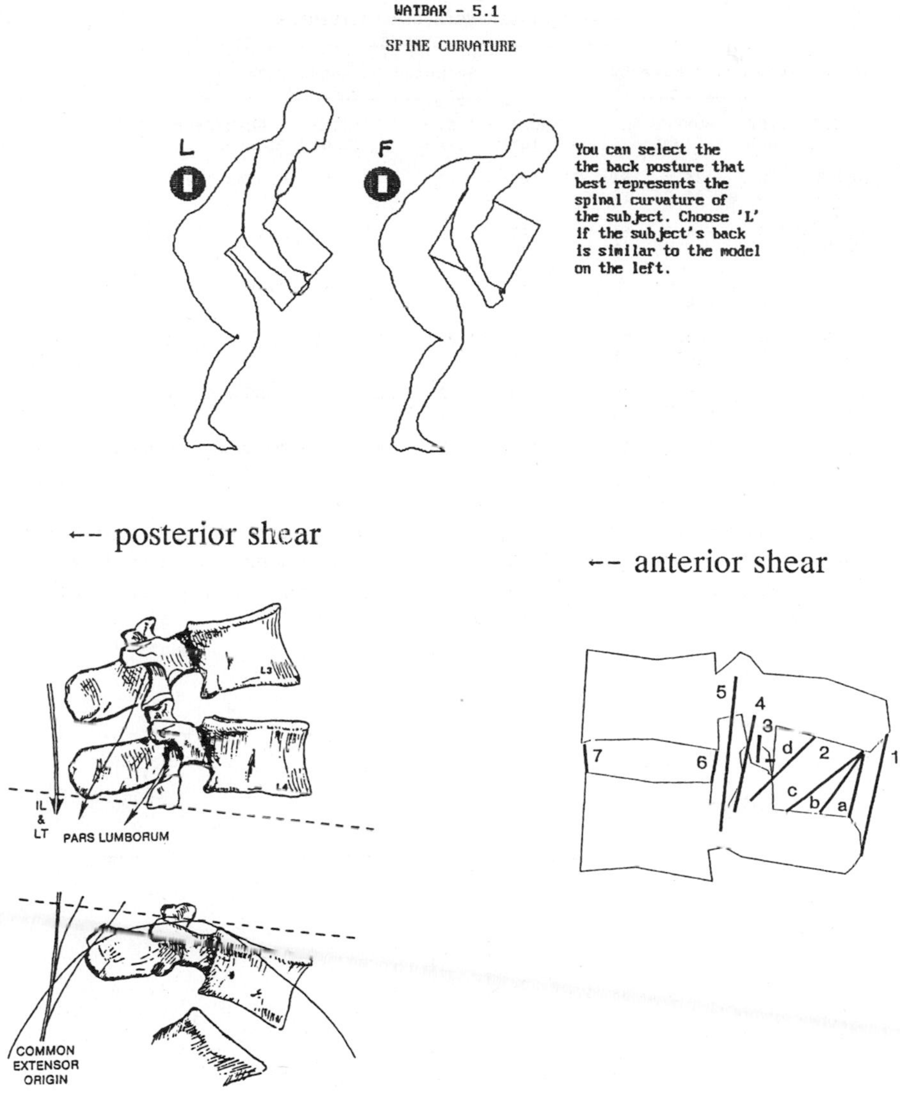

LUMBAR SPINE (L4/L5) PARAMETERS

MOMENT ARM		SHEAR Reaction (N)	SHEAR Joint (N)	COMPRESSION (N)	NIOSH AL (N)	NIOSH MPL (N)
ERECTOR SPINAE	5cm:	573.	1699.	5229.	3433	6376
ERECTOR SPINAE	6cm:	573.	1511.	4396.		

ABDOMINAL PRESSURE IS: 9.8 kPa.
EFFECT OF ABDOMINAL PRESSURE ON COMPRESSION, SUBTRACT: 457.0 N.
SUBJECT'S HEIGHT (m): 1.75 WEIGHT (kg): 80 GENDER: M
LOAD MASS (kg): 20 L4/L5 MOMENT (N.m): 250 .
FORCE ON RIGHT HAND (N): 98.1 ANGLE OF FORCE (DEG): -90
FORCE ON LEFT HAND (N): 98.1 ANGLE OF FORCE (DEG): -90

FIGURE 54.3 WATBAK allows the user to indicate the spine curvature (top panel), accommodates muscle and passive tissue load sharing (middle panel), and calculates the joint shear force — not only the reaction shear force (bottom panel). The joint shear force is the index of shear injury risk.

BODY JOINT STRENGTH DEMANDS (FEMALE)

Calculated Joint Moments				Selected Strength Data				
JOINT	SIDE (R/L)	MOMENT (Nm)		MOMENT (Nm)	S.D (+/-)		ANGLE (DEG)	REFERENCE
ELBOW	R	26.0	Fl	42	12	Fl	90	Stobbe (1982)
ELBOW	L	26.0	Fl					
SHOULDER	R	44.0	Fl	43	10	Fl	0	Koski and McGill (1992)
SHOULDER	L	44.0	Fl					
ANKLE	R	116.0	Ex	124	27	Ex	DNA	Fugl-Meyer et al (1980)
ANKLE	L	116.0	Ex					
KNEE	R	20.4	Fl	69	22	Fl	135	Stobbe (1982)
KNEE	L	20.4	Fl					
HIP	R	126.5	Ex	128	52	Ex	100	Stobbe (1982)
HIP	L	126.5	Ex					
TORSO		205.8	Ex	244	53	Ex	180	Troup and Chapman (1969)

* DNA: Data Not Available

FIGURE 54.4 Calculated joint moments from require by the task are compared with gender specific joint strength data from the literature. The average moment +/- 1 standard deviation at the joint angle reported in the studies cited are found under the "selected strength data" column. Data from a 50 kg woman holding a 20 kg load mass in front of her in a semi-squat with her torso inclined forward about 45°.

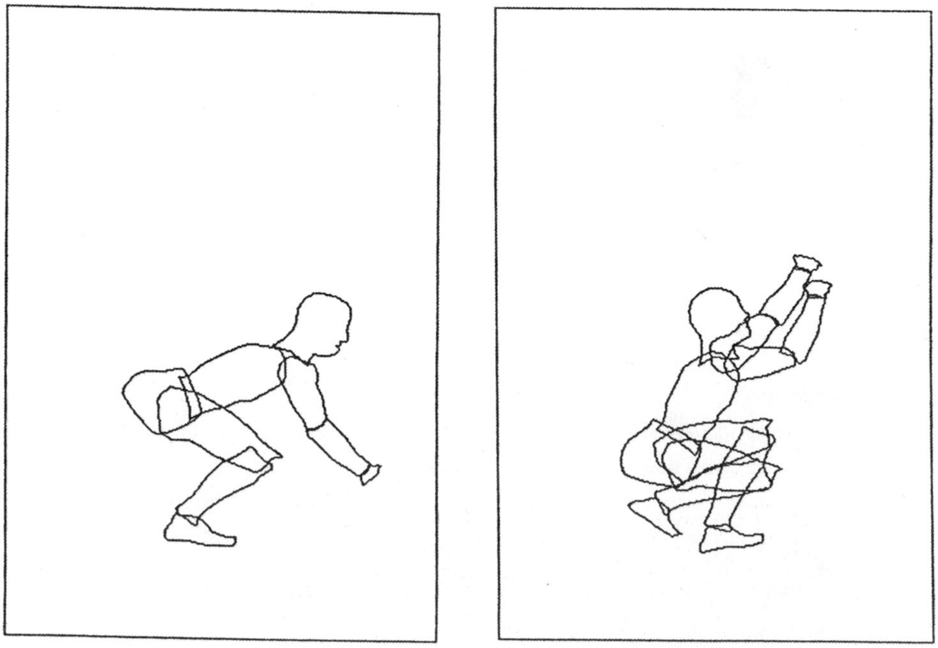

FIGURE 54.5 A manikin figure is moved into position in the sagittal plane, using arrow keys, which automatically creates posture data for joint loads. This method of data entry also facilitates rapid analysis of effects of redesign of tasks.

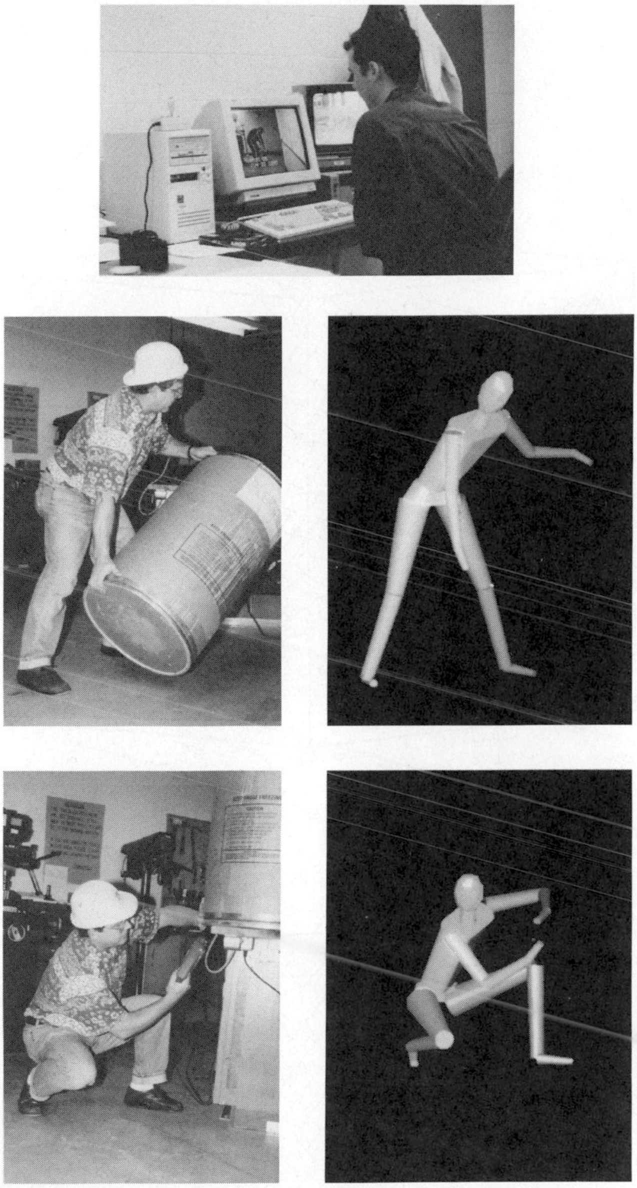

FIGURE 54.6 Input to 3DWATBAK can be obtained by manually moving a 3D manikin with a mouse into position to represent the posture of a worker seen on videotape (top panel). The manikin is automatically digitized and processed through the 3D computations to produce 3D output.

$$\begin{aligned} C = 1067.6 + 1.219F + 0.083F^2 - 0.0001F^3 + 3.299B + 0.119B^2 \\ - 0.0001B^3 + 0.862T + 0.393T^2 - 0.0001T^3 \end{aligned} \tag{1}$$

where

C = compression (N)

F = flexion–extension moment where negative values correspond to flexion (N.m)

B = lateral bending moment where bending to the right is positive (N.m)

T = axial twisting moment where CCW twist is positive (N.m)

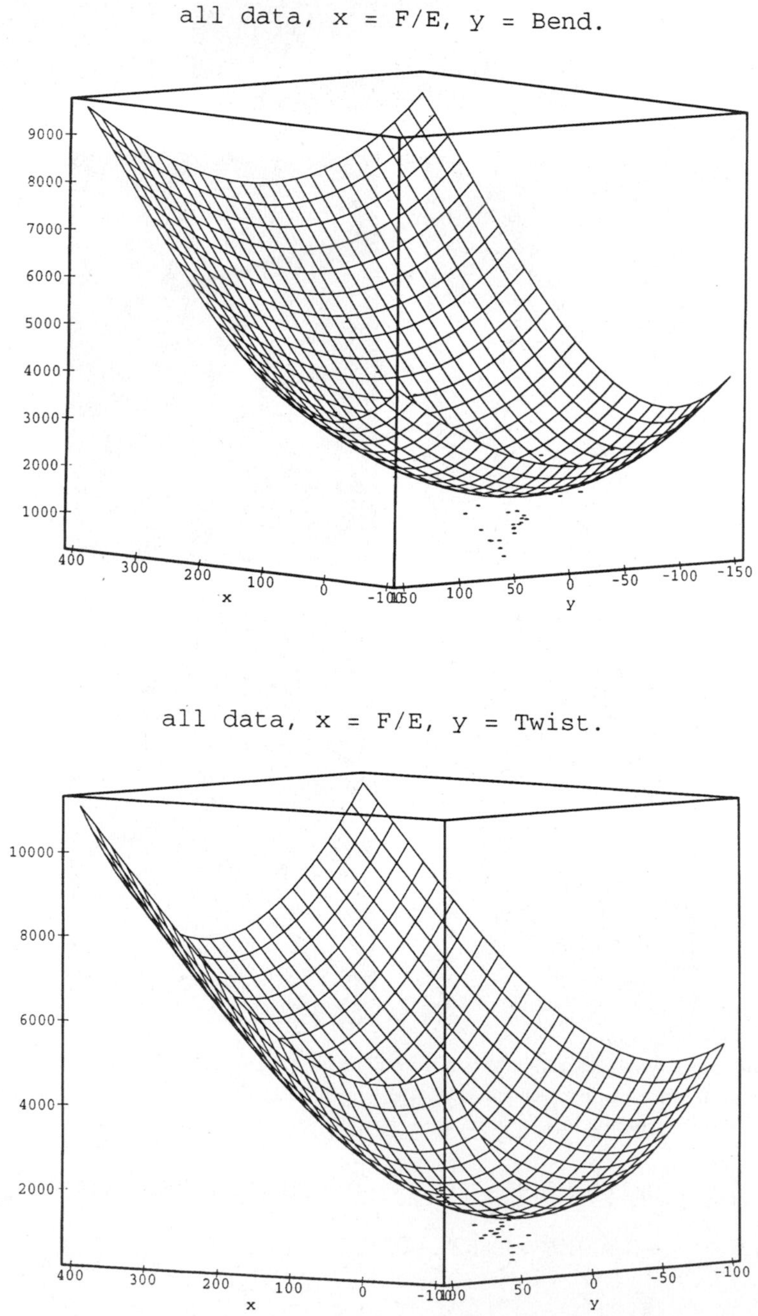

FIGURE 54.7 Compression as a function of three moments (flexion–extension, lateral bend, axial twist) is predicted using a 4D regression equation. Since only three dimensions can be plotted at one time, two plots are shown, compression (top panel) as a function of flexion–extension moment (x) and lateral bending moment (y), compression (bottom panel) as a function of flexion–extension moment (x) and axial twist moment (y).

TABLE 54.2 Low back loads and some examples of 3D joint moments produced during the barrel holding task shown in Figure 56.6. Typical moments about the x–y–z axes are automatically transformed into orthopedic orientations for each joint for functional interpretation.

Segment	Proximal Joint	Orthopedic Moment (N.m)	Direction
Right forearm	Right elbow	2. 2	Flexion
		0.6	Pronation
		15. 2	Adduction
Left forearm	Left elbow	3.9	Flexion
		0.6	Pronation
		4.6	Abduction
Right upperarm	Right shoulder	1.6	Flexion
		3.9	Internal rotation
		27. 9	Abduction
Left upperarm	Left shoulder	21. 2	Flexion
		4.6	External rotation
		2.8	Adduction
Thorax and abdomen	L4/L5	169	Extension
		26.3	Left axial twist
		6.9	Left lateral bend

Total compression: 3482 N

Anterior-posterior joint shear: 288 N

Anterior-posterior reaction shear: 534 N

Lateral reaction shear: 0 N

An example of the utility of 3DWATBAK is provided in Figure 54.6, where the real lifting task is observed and the ergonomist positions the manikin with the computer mouse to represent the posture, then 3DWATBAK uses linked segments to calculate the reaction moments (x, y, z), and then transforms the moments into specific joint axes (in this case flexion–extension, lateral bend, axial twist) and supports the moments with muscle and ligament forces (using the regression equation) to compute low back compression (Table 54.2).

It is interesting to compare the compressive cost of producing moments about the 3 lumbar axes. For example, a pure extensor moment of approximately 160 N.m results in a compression load close to 3000 N. On the other hand, an extensor moment of 100 N.m coupled with 50 N.m of left lateral bend and 50 Nm of CW axial twist produces the same level of compression. The spine pays dearly for supporting combinations of moments, particularly for those outside the sagittal plane.

In summary, the way in which 3DWATBAK represents the anatomy and biological muscle recruitment, represents a compromise between the relative simplicity that is needed in models for assessing the risk of injury in industry, and a rigorous and methodologically complex modeling approach that inherently contains quite high levels of biological fidelity in predicting low back loads but is impractical for industrial use. However, the equations within 3DWATBAK are based on model output that gives full credit to muscle cocontractions and also recognizes biological muscle coupling which occurs in three-dimensional spine loading. This method seems a reasonable way to examine risk of low back injury for three-dimensional occupational tasks.

54.11 Are There Any Spine Models that Reflect the Effects of Prolonged Loading?

Some experimental methods are under development but not yet routinely available to ergonomists. It is true that current biomechanical spine models provide an assessment of exposure to injury risk at a single

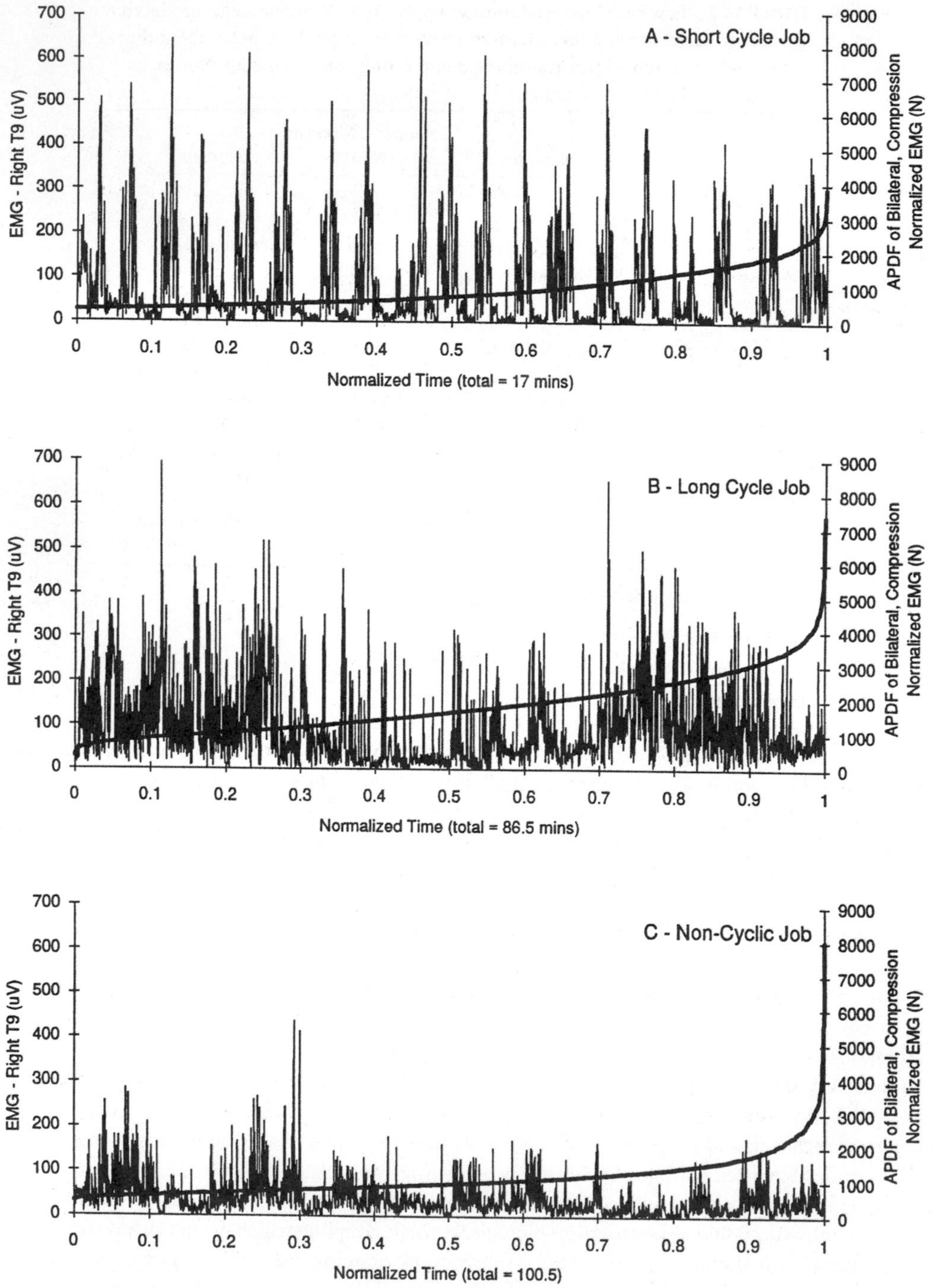

FIGURE 54.8 (A) Linear envelope and accompanying APDF of short cycle job (1 min.); (B) Long cycle job (5 min.); (C) Noncyclic job illustrating conversion of extensor muscle EMG to Newtons of estimated compression during prolonged monitoring of jobs (Compression Normalized EMG). Interpretation in text.

instant in time. Our work with the complex, EMG-assisted model has shown a high correlation of back muscle EMG with spinal compression in load-handling tasks that do not have a dominant axial twist component. Modern data loggers permit the collection of EMG for long periods of time. The question then arose as to whether it was possible to estimate spinal compression continuously as a person worked by using an electronically processed version of the EMG signal calibrated against the spine compression observed when a worker held a load statically in a standard posture and the back muscle EMG recorded during that calibration effort. In other words, could we approximate the compression estimate that a biomechanical model would yield if it were applied 100 times per second for periods of work of several hours by normalizing the EMG to Newtons of compression from the calibration effort? Without going into detail, the answer was yes, as long as the job did not involve a large part of the total time in which the dominant torso moment was spinal twisting. The method did not work well for pure torso twisting.

Most jobs that we have observed in our field research do not have large periods of time in which torso twist predominates. Even in asymmetrical tasks, if the worker is bent over to some degree, the dominant support moment is trunk extension. This method seems to work well in this case. The most useful presentation of data obtained from this so-called "compression normalized EMG" is in the form of a (compression) amplitude probability distribution function. Figure 54.8 shows compression normalized APDF curves recorded from three different auto workers, one worker (A) who did a short cycle assembly job (1 minute), one worker (B) who did a longer job cycle (5 minutes), and a third worker (C) whose job was highly varied over a 2-hour period. These workers were monitored for 17, 86, and 100 minutes, respectively. If one uses the 90% level as the estimate of peak compressive loading (90% of the cycle time the compression was below this level and above this level for only 10% of the time), a range from about 1800 to 3000 N for the three workers is observed. The median level of compression, representing 50% of the work cycle time, ranged from about 800 to 1800N during the period of observation, and the compression never dropped below 400 N (level at 10% on the normalized time axis). Standing upright with no muscular activity the compression on the L4/L5 motion unit is about one half body weight.

This method, based upon spine model output, has considerable potential for monitoring spinal loading over prolonged periods of time, but the limitations of the method need further exploration. The percentages of the work cycle time at various levels of compression can be compared with data on spine tolerance to acute loading. It would be helpful if epidemiological data could be obtained on the risk associated with lower level accumulated loads as well. This would extend the utility of this type of monitoring of exposure to back injury risk.

References

Bean, J.C., Chaffin, D.B., and Schulz, A.B. 1988. Biomechanical model calculation of muscle contraction forces: a double linear programming method. *J. Biomech.* 21:59-66.

Bone, B.C., Norman, R.W., McGill, S.M., and Ball, K.A. 1990. Comparison of 2D and 3D modelpredictions in analysing asymmetric lifting postures, in *Advances in Industrial Ergonomics and Safety II*, Taylor & Francis, pp. 543-550.

Brinckmann, P., Biggemann, M., and Hilweg, D. 1989. Prediction of the compressive strength of human lumbar vertebrae. *Clin. Biomech.* 4 (Suppl. 2).

Chaffin, D. Hughes, R., and Nussbaum, M. 1989. Towards a 3D biomechanical torso model for asymmetric loading. *Proceedings of the International Society for Biomechanics*, Los Angeles, abstract 132.

Cholewicki, J., McGill, S.M., and Norman, R.W. 1995. Comparison of muscle forces and joint load from an optimization and EMG assisted lumbar spine model: towards development of a hybrid approach. *J. Biomech.* 28(3):321-331.

Cripton, P., Berleman, U., Visarino, H., Begeman, P.C., Nolte, L.D., and Prasad, P. 1994. Response of the lumbar spine due to shear loading in: Injury prevention through biomechanics — *Symposium Proceedings*, May 4-5, Wayne State University.

Freivalds, A., Chaffin, D., Garg, A., and Lee, K.S. 1984. A dynamic biomechanical evaluation of lifting maximum acceptable loads. *J. Biomech.* 17(4):251-262.

Gracovetsky, S., Farfan, H.F., and Lamy, C. 1981. Mechanism of the lumbar spine. *Spine.* 6(1):249-262.

Heylings, D. 1978. Supraspinous and interspinous ligaments of the human lumbar spine. *J. Anat.* 123:127-131.

Hughes, R.E., Bean, J.C., and Chaffin, D.B. 1995. Evaluating the effect of cocontraction in optimization models. *J. Biomech.* 28(7):875-878.

Jager, M., Luttmann, A., and Laurig, W. 1991. Lumbar load during one-handed bricklaying. *Int. J. Ind. Ergonomics.* 8:261-277.

Marras, W.S. and Granata, K. 1995. A biomechanical assessment and model of axial twisting in the thoraco-lumbar spine. *Spine.* 20(13):1440-1451.

Marras, W. and Sommerich, C. 1991. A three-dimensional motion model of loads on the lumbar spine I: Model Structure. *Human Factors.* 33:123-137.

Marras, W.S., Lavender, S.A., Leurgans, S.E., Rajulu, S.L., Allread, W.G., Fathallah, F.A., and Ferguson, S.A. 1993. The role of dynamic three-dimensional trunk motion in occupationally-related low back disorders. *Spine.* 18(5):617-628.

McGill, S.M., and Norman R.W. 1985. Dynamically and statically determined low back moments during lifting. *J. Biomech.* 18(12):877-885.

McGill, S.M. 1999. Dynamic low back models: Theory and relevance in assisting the ergonomist to reduce the risk of low back injury, in *The Occupational Ergonomics Handbook* (eds. W. Karwowski, W. Marras) CRC Press, Boca Raton.

McGill, S.M. and Norman, R.W. 1986. Partitioning the L4-L5 dynamic moment into disc, ligamentous, and muscular components during lifting. *Spine.* 11(7):666-678.

McGill, S.M. and Norman, R.W. 1992. Loading of the low back during 3D moment generation. *Proceedings of the Human Factors Association of Canada*, Hamilton, October 25-18, pp. 73-79.

McGill, S.M. Norman, R.W., and Cholewicki, J. (1996). A Simple Polynomial that predicts low back compression during complex 3D tasks. *Ergonomics*, 39(9): 1107-1118.

McGill, S.M., Patt, N., and Norman R.W. 1988. Measurement of the trunk musculature of active males using CT Scan radiography: Implications for force and moment generating capacity about the L4/L5 joint. *J. Biomech.* 21(4):329-341.

McGill, S.M., Santaguida, L., and Stevens, J. 1993. Measurement of the trunk musculature from T6 to L5 using MRI scans of 15 young males connected for muscle fibre orientation. *Clin. Biomech.* 8:171-178.

McGill, S.M. 1991. Lumbar loads from moments about three orthopaedic axes: Developing the architecture of a 3-D occupational low back model, *Proceedings of the XIIIth International Congress on Biomechanics*, Perth, Australia, December 9-13, pp. 545-547.

McGill, S.M. 1992. A myoelectrically based dynamic 3-D model to predict loads on lumbar spine tissue during lateral bending. *J. Biomech.* 25:395-414.

NIOSH–1981. National Institute for Occupational Safety and Health. *A Work Practices Guide for Manual Lifting*, Cincinnati; Taft Industries.

Norman, R., McGill, S., Lu. W., and Frazer, M. 1994. Improvements in biological realism in an industrial low back model: 3DWATBAK. *Proceedings of the 12th Triennial Congress of the International Ergonomics Association*, Toronto, Volume 2, pp. 299-301.

Potvin, J.R., Norman, R.W., and McGill, S.M. 1991. Reduction in anterior shear forces on the L4/L5 disc by the lumbar musculature. *Clin. Biomech.* 6:88-96.

Potvin, J.R. and Norman, R.W. 1992. Can fatigue compromise lifting safety? *Proceedings of the Second North American Congress on Biomechanics*, Chicago, August 24-28, pp. 513-514.

Punnett, L., Fine, L.J., Keyserling, W.M., Herrin, G.D. and Chaffin, D.B. 1991. Back disorders and nonneutral trunk postures of automobile assembly workers. *Scand. J. Work Environ. Health.* 17:337-346.

Schultz, A.B., Andersson, G.B.J., Ortengren, R., Haderspeck, K., and Nachemson, A. 1982. Loads on the lumbar spine. *J. Bone Jt. Surg.* 64A (5): 713-720.

Troup, J.D.G. and Chapman, A.E. 1969. The strength of the flexor and extensor muscles of the trunk. *J. Biomech.* 2:49-62.

55

Quantitative Assessment of Trunk Performance

Mohamad Parnianpour
The Ohio State University

Aboulfazl Shirazi-Adl
Ecole Polytechnique, Montreal

55.1 Introduction

As early as 1700, Bernardino Ramazzini, one of the founders of occupational medicine, had associated certain physical activities with musculoskeletal disorders (MSD). He postulated that certain violent and irregular motions and unnatural postures of the body impair the internal structure (Snook et al., 1988; NIOSH, 1997). Presently, much effort is directed toward a better understanding of work-related musculoskeletal disorders involving the back, cervical spine, and upper extremities (Parnianpour et al., 1990). The World Health Organization (WHO) has defined occupational diseases as those work-related diseases where the relationship to specific causative factors at work has been fully established (WHO, 1985). Obtaining the occupational history is crucial to proper diagnosis and appropriate treatment of work-related disorders. The occupational physician must consider the conditions of both the workplace and the worker in the evaluation of the injured workers. Biomechanic and ergonomic evaluations have developed a series of techniques for quantification of the task demands and evaluation of the stresses in the workplace. Functional capacity evaluation has also been advanced to quantify the maximum performance capability of the workers. The motto of ergonomics is to avoid the mismatch between the task demand and the functional capacity of individuals. A multidisciplinary group of clinicians and engineers constitutes the rehabilitation team that will work together to implement the prevention measures. Through proper workplace design, workplace stressors and risk factors could be minimized. It is expected that one third of the compensable low back pain in industry could be prevented by proper ergonomic workplace or task design. In addition to reducing the probability of both the initial and recurring episodes, proper ergonomic design allows earlier return of injured workers by keeping the task demands at a lower

0-8493-2641-9/99/$0.00+$.50

level. Unfortunately, ergonomists are often asked to redesign the task or the workplace after a high incidence of injuries has already been experienced. The next preventive measure that has been suggested is preplacement of workers based on the medical history, strength, and physical examinations (Parnianpour et al., 1987; Snook et al., 1988; Parnianpour and Engin, 1994).

Title I of the Americans with Disabilities Act (ADA, 1990) prohibits discrimination in regard to any aspect of the employment process. Thus, the development of preplacement tests has been impeded by the possibility of discrimination against individuals based on gender, age, or medical conditions. The ADA requires physical tests to simulate the "essential functions" of the task. In addition, one must be aware of "reasonable accommodations," such as lifting aids, that may make an otherwise infeasible task possible for a disabled applicant to perform. Health care providers who perform physical examinations and provide recommendations for job applicants must consider the rights of disabled applicants. It is extremely crucial to quantify the specific physical requirements of the job to be performed, and to examine an applicant's capabilities to perform those specific tasks, taking into account any reasonable accommodations that may be provided. Hence, task analysis and functional capacity assessment are truly intertwined.

This chapter is intended to illustrate the application of some principles and practices of human performance engineering, especially quantification of human performance. The problem of low back pain has been selected to illustrate a series of concepts that are essential to evaluation of both the worker and the workplace, while realizing the importance of the disorders of the neck and upper extremities. By inference and generalization, most of these concepts can be extended to these situations. Manual material handling tasks have been the focus of our attention due to their prevalence in industry. In addition, the trunk muscles and spine were selected for the most detailed investigation due to the observation that a large proportion of ADA cases involve low back disability (Khalaf et al. 1997a).

55.2 Principles

Assessment of function across various dimensions of performance (i.e., strength, speed, endurance, and coordination) has provided the basis for a rational approach to clinical assessment, rehabilitation strategies, and determination of return to work potential for injured employees (Kondraske, 1990). To understand the complex problem of trunk performance evaluation of (low back pain) LBP patients, the terminology of muscle exertion must first be defined. However, it should be noted that a number of excellent reviews of trunk muscle function have been performed (Andersson, 1991; Beimborn and Morrissey, 1988; Newton and Waddell, 1993; Pope, 1992). We do not intend to reproduce this extensive literature here, as our motive is to provide a critical analysis that will lead the reader toward an understanding of the future of functional assessment techniques. A more extensive clinical application is presented elsewhere (Parnianpour and Tan, 1993; Parnianpour, 1995; Szpalski and Parnianpour, 1996).

Impairment, Disability, and Handicap

The tremendous human suffering and economic costs of disability present a formidable medical, social, and political challenge in the midst of growing health care costs and scarcity of resources. The WHO (1980) distinguished among impairment, disability, and handicap. Impairment is any loss or abnormality of psychological, physiological, or anatomical structure or function — impairment reflects disturbances at the organ level. Disability is any restriction or lack (resulting from impairment) of ability to perform an activity in the manner or within the range considered normal for a human being — disability reflects disturbances at the level of person. Handicap is a disadvantage for a given individual, resulting from an impairment or a disability, that limits or prevents the fulfillment of a role that is normal (depending on age, sex, and social and cultural factors) for that individual. As disability is the objectification of an impairment, handicap represents the socialization of an impairment or disability. Despite the immense improvement presented by the International Classification of Impairments, Disabilities, and Handicaps (ICIDH), it is limited from an industrial medicine or rehabilitation perspective. The hierarchical organization lacks the specificity required for evaluating the functional state of an individual with respect to task demands.

Kondraske (1990) has suggested an alternative approach, using the principles of resource economics. The resource economics paradigm is reflective of the principal goal of ergonomics: fitting the demands of the task to the functional capability of the worker. The Elemental Resource Model (ERM) is based on the application of general performance theory that is to present a unified theory for measurement, analysis, and modeling of human performance across all different aspects of performance, across all human subsystems, and at any hierarchical level. This approach uses the same bases to describe both the fundamental dimensions of performance capacity and task demand (available and utilized resources) of each functional unit involved in performance of the high-level tasks. The elegance of the ERM is due to its hierarchical organization, allowing causal models to be generated based on assessment of the task demands and performance capabilities across the same dimensions of performance (Kondraske 1990).

Muscle Action and Performance Quantification

The details of the complex processes of muscle contraction in terms of the bioelectrical, biochemical, and biophysical interactions are under intense research. Muscle tension is a function of muscle length and its rate of change, and can be scaled by the level of neural excitation. These relationships are called the length–tension and velocity–tension relationships. From a physiological point of view, the measured force or torque applied at the interface is a function of: (a) the individual's motivation (magnitude of the neural drive for excitation and activation processes); (b) environmental conditions (muscle length, rate of change of muscle length, nature of the external load, metabolic conditions, pH level, temperature, and so forth); (c) prior history of activation (fatigue); (d) instruction and descriptions of the tasks given to the subject; (e) the control strategies and motor programs employed to satisfy the demands of the task; and (f) the biophysical state of the muscles and fitness (fiber composition, physiological cross-sectional area of the muscle, cardiovascular capability). It cannot be overemphasized that these processes are complex and interrelated (Kroemer et al., 1994). Other factors that may affect the performance of the patients are: misunderstanding of the degree of effort needed in maximal testing, test anxiety, depression, nociception, fear of pain and re-injury, as well as unconscious and conscious symptom magnification.

In the following sections we review some methods to quantify performance and lifting capability of isolated trunk muscles during a multilink coordinated manual material handling task. Relevant factors that influence the static and dynamic strength and endurance measures of trunk muscles will be addressed, and the clinical applications of these assessment techniques will be illustrated.

The central nervous system (CNS) appropriately excites the muscle, and the generated tension is transferred to the skeletal system by the tendon to cause motion, stabilize the joint, and/or resist the effect of external forces on the body. Hence, the functional evaluation of muscles cannot be performed without the characterization of the interfaced mechanical environment. The four fundamental types of muscle exertion or action are isometric, isokinetic, isotonic, and isoinertial. In isometric exertion, the muscle length is kept constant and there is no movement. Although mechanical work is not achieved, physiological work, i.e., static work, is performed and energy is consumed. When the internal force exerted by the muscle is greater than the external force offered by the resistance, then concentric, i.e., shortening, muscle action occurs; whereas if the muscle is already activated and the external force exceeds the internal force of the muscle, then eccentric, i.e., lengthening, muscle action occurs. When the muscle moves, either concentrically or eccentrically, dynamic work is performed. If the rate of shortening or lengthening of the muscle is constant, the exertion is called isokinetic. When the muscle acts on a constant inertial mass, the exertion is called isoinertial. Isotonic action occurs when the muscle tension is constant throughout the range of motion.

These definitions are very clear when dealing with isolated muscles during physiological investigations. However, terminologies employed in the literature of strength evaluation are imprecise. The terms are intended to refer to the state of muscles, but they actually refer to the state of the mechanical interface, i.e., the dynamometer. Isotonic exertion, as defined, is not as realizable physiologically because muscular tensions change as its lever arm changes despite the constancy of external loads. Special designs may vary the resistance level in order to account for changes in mechanical efficiency of the muscles. In addition, the rate of muscle length change may not remain constant even when the joint angular velocity is regulated by the dynamometer during isokinetic exertions. During isoinertial action, the net external resistance is not only a function of the mass (inertia) but is also a function of the acceleration. The acceleration, however, is a function of the input energy to the mass. Hence, to fully characterize the net external resistance we need to have both the acceleration and the inertial parameters (mass and moment of inertia) of the load and body parts. Future research should better quantify the inertial effects of the dynamometers particularly during nonisometric and nonisokinetic exertions.

For any joint or joint complex, muscle performance can be quantified in terms of the basic dimensions of performance: strength, speed, endurance, steadiness, and coordination. Muscle strength is the capacity to produce torque or work by voluntary activation of the muscles, whereas muscle endurance is the ability to maintain a predetermined level of motor output — e.g., torque, velocity, range of motion, work, or energy — over a period of time. Fatigue is considered to be a process under which the capability of muscles diminishes. However, neuromuscular adjustments take place to meet the task demands (i.e., increase in neural excitation) until there is final performance breakdown — endurance time. Coordination, in this context, is the temporal and spatial organization of movement and the recruitment patterns of the muscle synergies.

Despite the proliferation of various technologies for measurement, basic questions such as: "What needs to be measured and how can it best be measured?" are still being investigated. However, there is a consensus on the need to measure objectively the performance capability along the following dimensions: range of motion, strength, endurance, coordination, speed, acceleration, etc. Strength is one of the most fundamental dimensions of human performance and has been the focus of many investigations. Despite the general consensus about the abstract definition of strength, there is no direct method for measurement of muscle tension *in vivo.* Strength has often been measured at the interface of a joint (or joints) with the mechanical environment. A dynamometer, which is an external apparatus onto which the body exerts force, is used to measure strength indirectly.

Different modes of strength testing have evolved based on different levels of technological sophistication. The practical implication of contextual dependencies on the provided mechanical environment of the strength measures must be considered during the selection of the appropriate mode of measurement (Khalaf et al., 1997a). In this regard, equipment that can measure strength in different modes is more efficient in terms of both initial capital investment, required floor space in the clinics or laboratories, and the amount of time it takes to get the person in and out of the dynamometer.

55.3 Low Back Pain and Trunk Performance

The problem of LBP is selected to present important models that could be utilized by the entire multidisciplinary rehabilitation team for the measurement, modeling, and analysis of human performance (Kondraske, 1990). The inability to relate low back pain (LBP) to anatomical findings and the difficulties in quantifying pain have directed much effort toward quantification of spinal performance. The problem is made even more complex by the increasing demand of the health care system to quantify the level of impairment of patients reporting back pain without objective findings. Indeed some studies have indicated that a precise cause of nociception cannot be recognized in more than 80% of patients with low back pain complaints. However, work-related disorders of upper extremity, unlike low back disorders, can better be related to specific anatomical sites such as a tendon or compressed nerve. Examples of the growing number of cumulative trauma disorders of the upper extremities and the neck are: carpal tunnel syndrome (CTS), DeQuervain's disease, trigger finger, lateral epicondylitis (tennis elbow), rotator cuff tendinitis, thoracic outlet syndrome, and tension neck syndrome (Kroemer et al., 1994).

There are three basic impairment evaluation systems, each having its merits and shortcomings: (1) anatomic, based on physical examination findings;)2) diagnostic, based on pathology; and (3) functional, based on performance or work capacity (Luck and Florence, 1988). The earlier systems were anatomical, based on amputation and ankylosis. Although this approach may be more applicable to the hand, it is very inappropriate for the spine. The diagnostic-based systems suffer from lack of correspondence between the degree of impairment for a given diagnosis and the resultant disability and even more from the lack of a clear diagnosis. A large percentage of symptom-free individuals have anatomical findings detectable by the imaging technologies, while some LBP patients have no structural anomalies.

The function-based systems are more desirable from an occupational medicine perspective for the following reasons. They allow the rehabilitation team to rationally evaluate the prospect for return to light duty work and the type of "reasonable accommodations" needed (such as assistive devices) that could reduce the task demand below the functional capability of the individual. By focusing on remaining ability and transferable skills rather than the disability or structural impairment of the injured worker, the set of feasible jobs can be identified. These points are extremely important, given the natural history of work disability after a single low back pain episode causing loss of work time: 40% to 50% of workers return to work by 2 weeks, 60% to 80% return by 4 weeks, and 85% to 90% return by 12 weeks. The small portion of disabled workers who become chronic are responsible for the majority of the economic cost of LBP. It is therefore the primary goal of the rehabilitation team to prevent the LBP, which is self correcting in most cases no matter what kind of therapy is used, from becoming a chronic disabling predicament. Injured workers should neither be returned to work too early nor too late, as both could complicate the prognosis. The results of functional capacity evaluation and task demand quantification should guide the timing for returning to work. It is clear that psycho-socioeconomic factors become more important than physical factors as the disability progresses into "chronicity syndrome" and play a major role in defining the evolution of a low back disability claim. Future research should further establish the reliability and reproducibility of performance assessment tools to expedite their widespread use (Luck and Florence, 1988; Newton and Waddell, 1993; Parnianpour et al., 1989a,b).

Maximal and Submaximal Protocols

Biomechanical strength models of the trunk are usually based on static maximal strength measurement. In real-life work situations, individuals rarely exert lengthy or maximum static effort. In most clinical situations, submaximal protocols are recommended, especially in patients with pain or with cardiovascular problems. Also, submaximal testing is less susceptible to fatigue and injury. The activity of daily living also has a great deal of submaximal efforts at the self-selected pace (Kim et al., 1996). Hence, it has been argued that testing at the preferred rate may be complementary to the maximal effort protocols. The preferred motion can be solicited by instructing the subject to perform repetitive movement at a pace and through the range of motion at which he/she feels the most comfortable. It has been shown that low back pain patients and normals have different resisted preferred flexion/extension motion characteristics. Having the subject perform against resistance is based on the hypothesis that, at higher resistance levels, the separation between the performance levels of patients and normal subjects becomes more evident. It has been shown, for example, that functional impairment of trunk extensors in LBP patients with respect to the normal population is larger at higher velocities during isokinetic trunk extension. However, the proponents of unconstrained testing have argued that separation of these groups can be performed based on the position, velocity, and acceleration profiles of the trunk during self-selected flexion/extension tasks. They have noted that pain and fear of re-injury may become the limiting factors. The sudden surge in acquiring performance measures of LBP patients during the initial rehabilitation process also underscores the validity of this concept.

Static and Dynamic Strength Measurements of Isolated Trunk Muscles

Weakness of the trunk extensor and abdominal muscles in patients with low back pain was demonstrated using the cable tensiometer to measure isometric strength. The disadvantage of the cable tensiometer (which records applied force) is that it neglects to measure the lever arm distance from the center of trunk motion. It is also recommended that the cable tensiometer be used to determine peak isometric torques rather than the stable average torque exerted over a 3-second period. Dynamometers used for testing dynamic muscle performances contain either hydraulic or servo motor systems to provide constant velocity (e.g., isokinetic devices) or constant resistance (e.g., isoinertial devices) The isokinetic devices can be further categorized into passive and active types. The robotics-based dynamometers can actively apply force on the body and hence allow eccentric muscle performance assessments, while only concentric exertions can be measured by the passive devices. Eccentric muscle action can simulate the lowering phase of a manual material handling task. Based on sport medicine literature, eccentric action has been implicated for its significant role in the muscle injury mechanism. Using isokinetic dynamometers, the isometric and isokinetic strength of trunk extensor and abdominal muscles were shown to be weaker in low back pain patients compared to healthy individuals. Dedicated trunk testing systems have become the cornerstone of objective functional evaluation and have been incorporated in the rehabilitation programs in many centers.

Two issues of importance for future research are the role of pelvic restraints and the significance of using newly developed triaxial dynamometers as opposed to more traditional uniaxial dynamometers. Studies on healthy volunteers have shown that trunk motions occur in more than one plane — lateral bending accompanies the primary motion of axial rotation. Numerous attempts have been made to measure the segmental range of motion three dimensionally in the lumbar spine with the purpose of quantifying abnormal coupling and diagnosing instabilities.

The effect of posture on the maximum strength capability can be described based on the length–tension relationship of muscle action and the changes in the moment arm of trunk muscles (Parnianpour et al., 1991; Parnianpour et al., 1993; Khalaf et al., 1997a). Marras and Mirka (1989) studied the effect of trunk postural asymmetry, flexion angle, and trunk velocity (eccentric, isometric, and concentric) on maximal trunk torque production. It was shown that trunk torque decreased by about 8.5% of the maximum for every 15° of asymmetric trunk angle. At higher trunk flexion angles, extensor strength increased. Complex, significant interaction effects of velocity, asymmetry, and sagittal posture were detected. The range of velocity studies were more limited (±30°/second) than those used customarily in spinal evaluation.

Tan et al. (1993) tested 31 healthy males for the effects of standing trunk–flexion positions (0, 15, 35 degrees) on triaxial torques and electromyogram (EMG) of ten trunk muscles during isometric trunk extension at 30%, 50%, 70%, and 100% of maximum voluntary exertions (MVE). Trunk muscle strength was significantly increased at a more flexed position. But the accessory torques in the transverse and coronal planes were not affected by trunk postures. The recorded lateral bending and rotation accessory torques were less than 5% and 16% of the primary extension torque, respectively. The rectus abdominus were inactive during all the tests. The EMG of erector spinae varied linearly with higher values of MVE, while the latissimus dorsi had a nonlinear behavior. The obliques were coactivated only during 100% MVE. The neuromuscular efficiency ratio (NMER) was constructed as the ratio of the extension torque over the processed (RMS) EMG of the extensor muscles. It was hoped that NMER could be used in a clinical setting where generation of the maximum exertion is not indicated. However, the NMER proved to have a limited clinical utility because it was significantly affected by both exertion level and posture. The NMER of the extensor muscles increased at more flexed position. Studies that have combined the EMG activities and dynamometric evaluations have the potential of discovering the neuromuscular adaptation during different phases of injury and rehabilitation processes.

Khalaf et al. (1997a) measured concentric trunk isokinetic extension/flexion strength for 20 male and female healthy volunteers in the range of upright to 40 degrees of trunk flexion at six levels of angular velocity (10, 20, 40, 60, 80, and 100 degrees/sec). Results indicated that trunk strength is significantly influenced by trunk angular position, trunk angular velocity, gender, direction, as well as the interaction between trunk angular position and velocity. Three-dimensional surfaces of trunk strength in response to trunk angular position and velocity were constructed for each subject per direction. Such data presentation is more accurate and gives better insight about an individual's strength profile as compared to the traditional use of a single strength value (Figure 55.1). In addition, when comparing the task demands of a particular MMH task and maximum strength capacity of an individual subject, the contextual

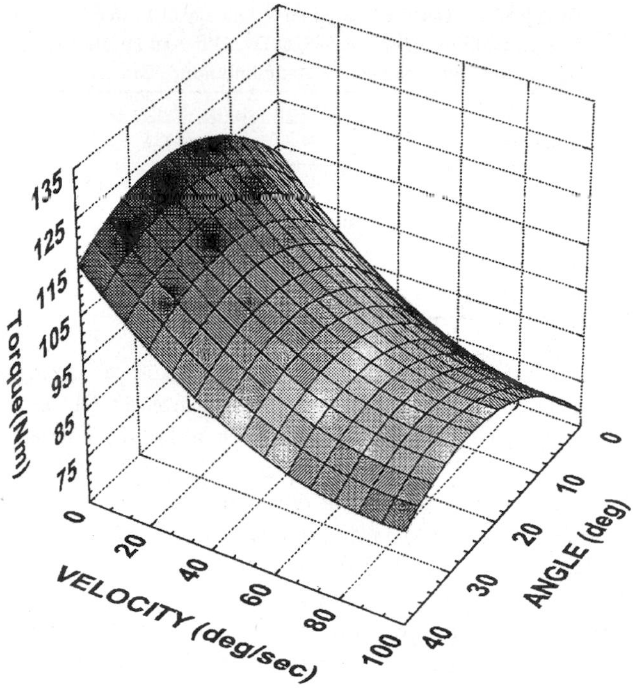

FIGURE 55.1 Trunk extension strength of a healthy female volunteer.

dependencies of fit or lack of it in terms of workplace design can better be illustrated. The workplace factors affect the required trunk speed, the muscular torque, and the range of motion of the worker. Concept of dynamic utilization ratio will be presented in the following section.

Static and Dynamic Trunk Muscle Endurance

The high percentage of type I fibers in the back muscles, in addition to the better vascularization of these muscle groups, contributes to their superior endurance (Parnianpour et al., 1987a,b; Zhu et al., 1989). Physiological studies indicate that at higher muscle utilization ratios (relative muscle loads), fatigue is detected earlier. Isometric endurance tests have been used to compute the median frequency (MF) of the myoelectrical activities of trunk muscles in both normal and LBP populations. The expected decline of the median frequency with fatigue is parameterized by the intercept (initial MF) and the slope of the fall. It has been shown that trunk range of motion (ROM) and isometric strength suffered from lower specificity and sensitivity than spectral parameters. Trunk muscle endurance does differ between healthy subjects and those reporting LBP. During isometric endurance testing, trunk flexors develop fatigue faster than extensors in symptom-free subjects. The flexor fatiguability appeared significantly higher in patients with low back pain as compared to controls. Chronicity also influences trunk muscle endurance. Chronic LBP patients showed reduced abdominal as well as back muscle endurance as compared to the healthy controls, and lower back muscle endurance as compared to the intermittent LBP group. Individuals with a history of debilitating low back pain demonstrated less isometric trunk extensor endurance than either normal individuals or patients with history of lesser low back pain.

We have developed the following testing protocols for the quantification of trunk muscle endurance during diverse sets of fatiguing isometric submaximal tasks: (a) constant torque isometric exertions (Sparto et al., 1997b); (b) sustained varying torque isometric exertions (Sparto et al., 1997c; 1998). These isometric submaximal extension exertions are analyzed by novel modeling approaches to quantify the state of muscular fatigue using spectral analysis of trunk muscle activity (EMG). In addition, new EMG-driven models predicted the forces developed in the muscles during these tests so that the severity of

TABLE 55.1 Decline in Median Frequency of Power Spectrum of Erector Spinae (ES) at Five Different Locations During the Performance of a Static Endurance Test

n = 10 Muscle Location	Median Frequency Decline (%/s) Mean (S.D.)
Medial ES @ L1	–0.36 (0.15)
Lateral ES @ L1	–0.22 (0.11)
Medial ES @ L3	–0.35 (0.13)
Lateral ES @ L3	–0.30 (0.17)
Medial ES @ L5	–0.39 (0.15)

spinal loading could be quantified. The safety of the proposed protocols can be assessed by comparing the developed disc compression and shear forces with tissue tolerances in the literature. A detailed finite element model of the spine has been developed that quantifies the stress and strain in the highly innervated tissues such as the annulus fibers, ligaments, and vertebral end plates (Wang et al., 1997a,b,c; 1998).

While quantifying the trunk muscle recruitment during the isometric tasks, it was observed that as the subjects fatigued, muscles other than the primary trunk extensors became more active. The implication of this finding is that if a worker has a deficit in either the primary or *secondary* muscles, risk of injury may increase. We have found that more caudal and medial muscles fatigue the most (Table 55.1). Hence, if experimental resources are limited, based on our results, we have identified the medial erector spinae muscle at the level of L5 to be the location of choice for quantifying trunk fatigue (Sparto et al., 1997b).

Soft tissues subjected to repetitive loading, due to their viscoelastic properties, demonstrate creep and load relaxation (Wang et al. 1997b,c; 1998). The loss of precision, speed, and control of the neuromuscular system induced by fatigue reduces the ability of muscles to protect the weakened passive structure, which may explain many industrial, clinical, and recreational injury mechanisms. These results further indicate the necessity of relating clinical protocols to the job and show how short-duration maximal isometric testing alone cannot provide the complex functional interaction of strength, endurance, control, and coordination (Parnianpour et al., 1988; Sparto et al., 1997c).

Parnianpour et al. (1988) studied the effect of isoinertial fatiguing of flexion and extension trunk movements on the movement pattern (angular position and velocity profile) and the motor output (torque) of the trunk. They showed that, with fatigue, there is a reduction of the functional capacity in the main sagittal plane. There is also a loss of motor control, enabling a greater range of motion in the transverse and coronal planes while performing the primary sagittal task. Association of sagittal with coronal and transverse movements is considered more likely to induce back injuries; thus the effect of fatigue and reduction of motor control and coordination may be an important risk factor leading to injury-prone working postures. The endurance limit is a more useful predictor of incidence and recurrence of low back disorders than the absolute strength values. Although physiological criteria used in the National Institute for Occupational Safety and Health Lifting Guide (NIOSH, 1981) considered cardiovascular demands of dynamic repetitive lifting tasks, the limits of muscular endurance were not explicitly addressed. Future research should fill this gap, since the maximum strength measures should not guide the design decisions. Maximum level of performance can only be maintained for short periods of time, and muscular fatigue should be avoided to prevent the development of MSD. This caveat should be applied to all dimensions of performance capability (Kondraske, 1990; Parnianpour, 1995).

Other dynamic endurance protocols use repetitive isoinertial and isokinetic lifting tasks (Sparto et al., 1997c; 1998b). These two variations provide low stress to the musculoskeletal system and high stress to the cardiovascular system, and vice versa. Decreased knee and hip motion and increased spine flexion were evident in these studies, during repetitive sagittal lifting and lowering tasks, suggesting that the effects of fatigue may increase the loading of the passive tissues of the spine (Sparto et al., 1997c,d; 1998b). The changes in the phase angles of hip and spine motion between the rested and fatigued states during a repetitive lifting task are shown in Figure 55.2. The implications of this altered coordination should be investigated further in future research. Additional analyses have demonstrated a decrease in the postural

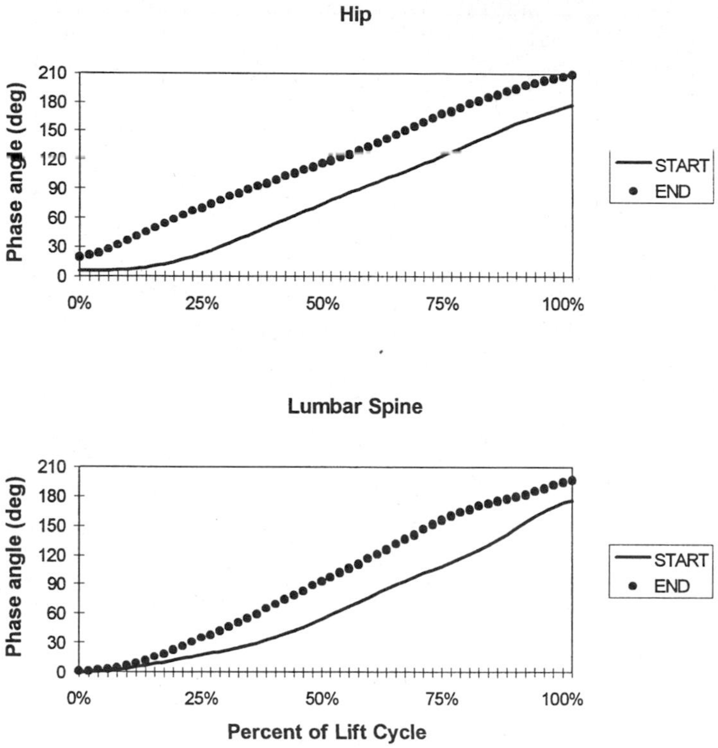

FIGURE 55.2 Change in the phase angles from the start to the end of the repetitive lifting test for one subject. The phase angles are shown for the hip and lumbar spine.

stability of subjects performing these tasks (Sparto et al., 1997d). The validation of the LIDOlift simulator was originally presented in Sparto et al. (1995).

Sparto and Parnianpour (1997) studied endurance of trunk extensors during repetitive submaximal trunk isokinetic extension. Trunk muscle electromyography (EMG) and torque output were measured in 16 healthy men [mean (s.d.) 23 (3) yrs, 178 (5) cm, 79 (9) kg] as they performed repetitive trunk extension exertions at 15 degrees/sec. A factorial design consisting of two load levels (35% and 70% of the maximum dynamic trunk extension torque) and two repetition rates (5 and 10 reps per minute) was implemented in order to cause different rates of fatigue. Each of the four load/rep combinations was tested during a single session, with at least one week in between. After the subjects stretched, bipolar surface electrodes were applied to the skin. Three bilateral trunk extensor muscles were sampled — latissimus dorsi (LAT), erector spinae (ERS), and posterior internal obliques (PINO) — as well as three trunk flexors — rectus abdomini (RAB), external obliques (EOB), and anterior internal obliques (AINO). The pelvis and lower extremities of the subjects were restrained within a trunk dynamometer. Maximal voluntary isometric exertions in each of the six cardinal directions were performed at 17.5 degrees of forward flexion to measure the maximum electromyographic activity of each muscle. The average of three dynamic maximal voluntary contractions (DMVC) in trunk extension was obtained, in order to measure the initial maximum torque-generating capability. During the dynamic exertions, subjects extended their trunks from a forward flexion posture of 35 degrees to upright standing, at a constant velocity of 15 degrees per second. The torque was measured using a load cell and knowing the distance from the point of force application to the center of rotation of the trunk. Two minutes of rest was provided between all maximum exertions.

During the repetitive endurance test, trunk extension was performed at 15 deg/sec while subjects controlled their torque output at 35% or 70% of the initial DMVC, by using visual feedback displayed on a computer monitor. The exertion rate, 5 or 10 reps per minute, was regulated by an audible tone.

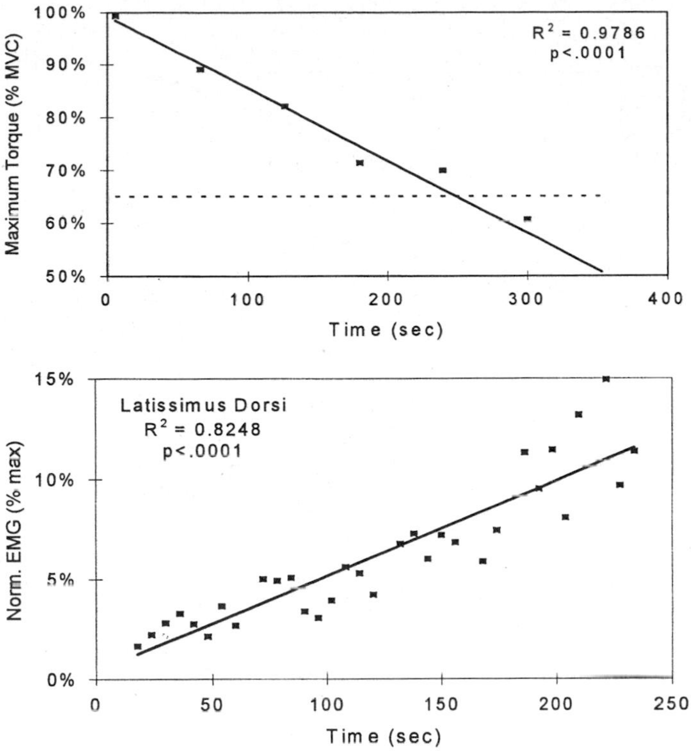

FIGURE 55.3 (Top) Decline in maximum dynamic torque generating capability during an endurance test performed at 35% MVC, 10 reps/minute. A 35% decline in torque is indicated by the dashed line. (Bottom) The normalized EMG of the left latissimus dorsi during the same test.

Once per minute, a DMVC was performed, to quantify the reduction in the maximal torque-generating capability (fatigue). The exertions continued until a 35% reduction in the MVC occurred, the subject could no longer continue, or 30 minutes elapsed.

The raw EMG from each exertion was sampled at 1536 Hz, rectified, and low pass filtered at 3 Hz. After normalizing with the maximum EMG obtained from the isometric MVCs, the root-mean-square (rms) value of the EMG between the trunk angles of 25 and 10 degrees of flexion was computed. From the exertions whose rms torque was within 5% of the designated load level, the change in EMG with time, or SLOPE, was quantified using linear regression. In order to obtain an estimate of the slope for each muscle group, the slopes were averaged for the right and left muscle pairs. The effect of the load magnitude and repetition rate on the slope was tested using repeated measures multiple analysis of variance (MANOVA).

The effect of the repetitive endurance test on the decline in maximum torque-generating capability and change in muscle activation are shown in Figure 55.3. For this test, the maximum torque declined by 35% in about 4 minutes, during which the EMG of the left latissimus dorsi muscle increased by 10%. The endurance time (point at which the predicted maximum torque declined by 35%) for each of the load/reps combinations is shown in Table 55.2. The effect of load, repetition rate, and their interaction on the endurance time were all highly significant ($p<.001$), indicating the study design was effective at eliciting different rates of fatigue. It can be seen that doubling the repetition rate from 5 to 10 reps/min had a greater effect in causing fatigue than doubling the load magnitude.

The results of the MANOVA indicated that as a group, the slopes of all muscles were significantly affected by the effects of load, repetition rate, and their interaction ($p<.005$). Furthermore, Table 55.3 demonstrates that significant differences in slope were found in each of the trunk extensor muscles (LAT,

TABLE 55.2 Mean (s.d.) Endurance Time, in Minutes, for the 4 Load/Reps Endurance Tests

	35% DMVC	70% DMVC
5 reps/min	24.1 (8.1)	9.8 (3.8)
10 reps/min	5.7 (3.9)	2.4 (0.9)

TABLE 55.3 P-values from the Univariate ANOVA, Testing the Effect of Load, Repetition Rate, and Interaction on the Slope

SLOPE	Load (L)	Reps (R)	L * R
LAT	.0001	.0001	.0011
ERS	.0001	.0001	.0001
PINO	.0001	.0001	.0007
RAB	.0209	.0094	.2441
EXO	.0636	.0836	.2947
AINO	.4040	.1338	.3949

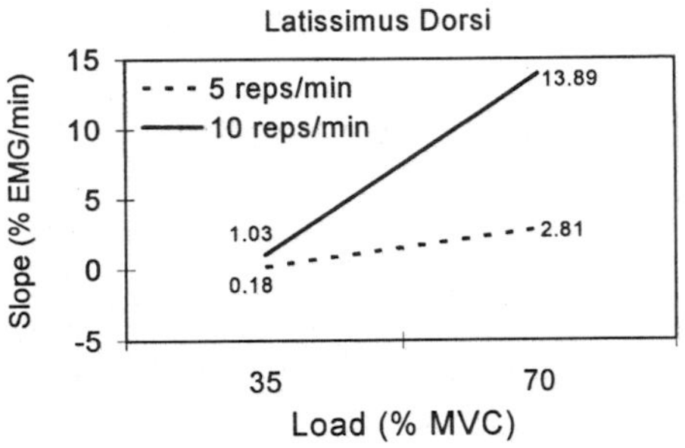

FIGURE 55.4 Increase in slope of the regression between the normalized EMG of the latissimus dorsi and time, due to the interaction between the load and repetition rate.

ERS, PINO). Figure 55.4 shows the mean slope as a result of the interaction between load and repetition rate for the latissimus dorsi. The other extensor muscles display a common pattern.

Because of the ambiguity in relating the amplitude of the processed EMG to muscle tension, one cannot speculate if the increase in activation resulted in greater muscle force, or represented a decreased efficiency in EMG/force production due to fatigue. Furthermore, it is not known if the function of the increased activation of the secondary trunk extensors (LAT and PINO), is to provide greater stability, or augment the torque generation. The increased and altered trunk extensor muscle activation due to fatigue requires further investigation to determine its effect on the spinal loading. EMG-driven models accounting for muscle fatigue have been developed to address these issues (Sparto, 1998).

In addition, models are being developed that are attempting to predict the endurance time based on parameters obtained from the noninvasive surface EMG of the trunk extensor muscles. Both short-time Fourier transform and wavelet transform techniques are being used to process the EMG (Sparto et al., 1997a). The use of the wavelet transform is more promising than the short-time Fourier transform because it does not depend on stationarity of the signal, the bases of analysis are not restricted to sinusoids, and it is not limited by the time-frequency resolution tradeoff. Furthermore, the analysis techniques can be incorporated into digital signal processing hardware that can be used to monitor the extent of back muscle fatigue in real time. The real time displays will allow the biofeedback application as well in the retraining of muscle endurance so essential in the conservative treatment of low back pain patients with spinal instability.

Lifting Strength Testing

The National Institute for Occupational Safety and Health (NIOSH, 1981) recommended static, i.e., isometric, strength measurement as its standard for lifting tasks. This was based on the evidence that associated low back pain with inadequate isometric strength. The incidence of an individual's sustaining an on-the-job back injury increases threefold when the task-lifting requirements approach or exceed the individual's strength capacity. However, lifting strength is not a true measure of trunk function but is a global measure taking into account arm, shoulder, and leg strength as well as the individual's lifting technique and overall fitness. It has been shown that strength tests were more valid and predictive of risk of low back disorders if they simulated the demands of the job. The clinicians must be aided with easy-to-use and validated instruments or questionnaires to gather information about the task demands in order to decide what testing protocol best simulates the applicant's spinal loading conditions.

Static strength measurements have been reported to underestimate significantly the loads on the spine during dynamic lifts. Comparing static and dynamic biomechanical models of the trunk, the predicted spinal loads under static conditions were 33% to 60% less than those under dynamic conditions, depending on the lifting technique. The recruitment patterns of trunk muscles (and thus the internal loading of the spine) are significantly different under isometric and dynamic conditions. General manual material handling tasks require a coordinated multilink activity which can be simulated using classical psychophysical techniques or the robotics-based lift task simulators. Various lifting tests, including static, dynamic, maximal, and submaximal, are currently available. The experimental results of correlational studies have confirmed the theoretical prediction that strength will be dependent upon the measurement technique (Balague et al., 1993; Parnianpour et al., 1987b; Parnianpour et al., 1993). Since muscle action requires external resistance, the effect of muscle action will depend on the nature of the resistance. These results refute the implicit assumption that a generic strength test exists that can be used for preplacing workers (pre-employment) and predicting the risk of injury or future occurrence of low back pain. The psychometric properties of isokinetic and isoresistive modes of strength testing were recently addressed. The quantification of the surface response of strength as function of joint angle and velocity was only possible for isokinetic testing, while isoresistive tests yielded a very sparse data set (Parnianpour, 1995).

The widely conflicting results found in the literature regarding the relationship of an individual's strength to the risk of developing LBP may be due to inappropriate modes of strength measurements, i.e., lack of job specificity (Parnianpour, 1995; Szpalski and Parnianpour, 1996; Parnianpour et al., 1977b). Isometric strength testing of the trunk is still widely used, especially in large-scale industrial or epidemiological studies, because it has been standardized and studied prospectively in industry. Compared to trunk dynamic strength testing protocols, the trunk isometric strength testing protocols are simpler and less expensive.

One outstanding issue during dynamic testing is the unresolved problem of how the wealth of information can be presented in a succinct and informative fashion. One approach has been to compare the statistical features of the data with the existing normal databases. This is particularly crucial because we do not have the option of comparing the results to the "contralateral asymptomatic joint," as we have with lower or upper extremity joints. Given the large differences between individuals, we recommend comparison be made to job-specific databases (Parnianpour et al., 1991; Parnianpour et al., 1994). For example, it is more appropriate for the trunk strength of an injured construction worker to compared to age- and gender-controlled healthy construction workers than to data from healthy college graduate students or office workers. However, given the scarcity of such data, we argue for comparison of performance capacity with job demand based on task analysis. The performance capacity evaluation is once again linked to task demand quantification (Parnianpour, 1995).

In evaluating any manual material handling (MMH) tasks such as lifting, the quantification of the various generated parameters (kinetic, kinematic, and electromyographic) is essential to assessing functional capacity and development of a biomechanical profile of task demands. We have developed a methodology for representing and quantifying performance data variability of the kinematic and kinetic motion profiles as a function of the different lift characteristics (load, mode, and speed) during MMH

TABLE 55.4 P-values of MANOVA to Test the Effect of Load (L), Mode (M), Speed (S), and Their Interactions on the Extracted Features (Coefficients) of the Joint Torque Profiles

N = 20	L	M	S	L*M	L*S	M*S
Ankle	.032	.001	.033	NS	NS	NS
Knee	.018	.001	NS	NS	NS	NS
L5/S1	.003	.003	.011	.001	NS	NS
Shoulder	.001	.001	.046	.005	NS	NS
Elbow	.045	.001	.039	NS	NS	NS

NS: Not Significant

tasks (Khalaf et al., 1997b,c). Using a database of motion profiles from a manual lifting experiment, the Karhunen–Loeve Expansion (KLE) was shown to be quite effective for representing the various motion profiles, where the number of basis vectors (eigenvectors) and coefficients needed to accurately represent the data were substantially smaller than the original data set resulting in lower order space or dimension. The factorial lifting experiment required subjects to lift 2 masses (6.8 and 13.6 kg), at three different speeds (2, 4, 6 seconds per lifting/lowering cycle), using three modes of lift (preferred, straight leg, bent leg). Table 55.4 demonstrates how the lift characteristics affect the extracted features of the joint torque profiles. We can see that the main effects of load (L), mode (M), and speed (S) significantly affect the torque profiles (Khalaf et al., 1997). Further inspection of the data revealed that the 13.6 kg load resulted in greater peak lumbosacral (L5/S1) torques; the greater speeds of lift resulted in greater peak lumbosacral torque; and that the preferred mode of lift was a compromise between the straight leg and bent leg modes, in terms of the peak lumbosacral torque. The following analysis also revealed that the patterns of joint movement was invariant with respect to the level of load and speed tested in this experiment (Khalaf et al., 1997c).

The effects of lift characteristics were also investigated using analysis of variance techniques which recognize the vectorial constitution of the waveforms as opposed to the traditional descriptive statistics representing the data over the lifting cycle (Khalaf et al., 1997). The application of these techniques will enhance the ability to document the effect of intervening measures such as education or physical training/exercise on the kinematic and kinetic patterns of performance (Parnianpour et al., 1987). In addition, the ADA-related applications are significant since this will aid in the assessment of the feasibility of task performance as a function of existing impairments, and assessment of limitations on functional capacity, allowing appropriate task assignment based on a worker's capabilities. Furthermore, the differential influence of lift characteristics on the variability of performance during different phases of lifting and lowering may provide added resolution in the analysis of MMH tasks. The most effective workplace modifications can be recommended based on the preceding results.

Inverse and Direct Dynamics

A major task of biomechanics has been to estimate the internal loading of musculoskeletal structure and establish the physiological loading during various daily activities. Kinematics studies deal with joint movement with no emphasis on the forces involved. However, kinetic studies address the effect of forces that generate such movements. Using sophisticated experimental and theoretical stress/strain analyses, hazardous/failure levels of loads have been determined. The estimated forces and stresses are used to estimate the level of deformation in the tissues. This technique allows us to assess the risk of over-exertion injury associated with any physical activity. Given repetitive motions and exertion levels much lower than the ultimate strength of the tissues, an alternative injury mechanism, the cumulative trauma model, has been used to describe much of the musculoskeletal disorders of the upper extremities (Wang et al., 1997a,b,c).

The experimental data on the joint trajectories are differentiated to obtain the angular velocity and acceleration. Appropriate inertial properties of the limb segments are used to compute the net external

moments about each joint. This mapping from joint kinematics to net moments is called inverse dynamics. Direct dynamics refers to studies that simulate the motion based on known actuator torques at each joint. The key issue in these investigations is understanding the control strategies underlying the trajectory planning and performance of purposeful motion. A highly multidisciplinary field has emerged to address these unsolved questions [see Berme and Cappozzo (1990) for a comprehensive treatment of these issues].

It should be pointed out that determination of the external moments about different joints during manual material handling tasks is based on the well-established laws of physics. However, the determination of human performance and assessment of functional capacity are based on other disciplines, e.g., psychophysics, that are not as exact or well developed. We can describe the job demand, in terms of the required moments about each joint, easily by analyzing the workers performing the tasks. However, we are unable to predict the ability to perform an arbitrary task based on our incomplete knowledge of functional capacities at the joint levels. A task is easily decomposed to its demands at the joint level, however, we cannot compose (construct) the set of feasible tasks based on our functional capacity knowledge. The mapping from high-level task demands to the joint-level functional capacity for a given performance trial is unique. However, the mapping from joint-level functional capacity to the high-level task demand is one to many (not unique). The challenge to the human performance research community is to establish this missing link. Much of the integration of ergonomics and functional analysis depends upon the removal of this obstacle. The question of whether a subject can perform a task based on the knowledge of his/her functional capacity at the joint level remains an area of open research. When ergonomists or occupational physicians evaluate the fitness of task demands and worker capability, the following clinical questions will be presented: (1) Which space should be explored for determining normalcy, fit, or equivalence? (2) Should we consider the performance of the multilink system in the joint space or end-effector (cartesian work space)? These issues have profound effects on both the development of new technologies and evaluation of trunk or lifting performance (Sparto et al., 1995).

The enormous degree of freedom existing in the neuromusculoskeletal system provides the control centers for both the kinematic and actuator redundancies. The redundancies provide optimization possibility. Since we can lift an object from point A to point B with infinite postural possibilities, it can be suggested that certain physical parameters may be optimized for the learned movements. The possible candidates for objective function to be optimized are: movement time, energy, smoothness, muscular activities, etc. This approach, though still in its early stage, may be very important for spine functional assessment. We could compare the given performance to the optimal performance that is predicted by the model. This approach provides specific goals and gives biofeedback with respect to the individual's performance.

Biomechanical Models

To complement the trunk strength and endurance testing protocols, a number of biomechanical simulation models have been developed which will be briefly discussed in the following section: (a) simulation of trunk strength based on anatomical cross-sectional area of muscles, line of action, and their lever arms (Parnianpour et al., 1997b); (b) simulation of trunk motion for healthy and impaired low back patients (Parnianpour et al., 1997a); (c) simulation of lifting tasks considering the strength and range of motion of each joint based on the load characteristics: initial and final load position, load magnitude, duration of lift (Khalaf et al., 1996).

Figure 55.5 represents the three-dimensional trunk strength boundaries based on the anatomical data of Hughes et al. (1995). Any task that requires moments that are interior to the surface are considered feasible, while those on the outside of the boundary are infeasible. These models can utilize the readily available CT scan or MRI data that the patients may already have. Any conceivable MMH task can be reduced to the three components of moment that it requires from the trunk muscles to keep it in equilibrium. Hence, the model is not limited to any specific task (pushing, lifting etc.), which is highly desirable. The preceding strength model can quickly identify the tasks that cannot be performed without any ergonomic modifications. It can easily be interfaced with the databases that may have been developed based on biomechanical analyses of the task in question (Parnianpour, 1995; Khalaf et al., 1997a).

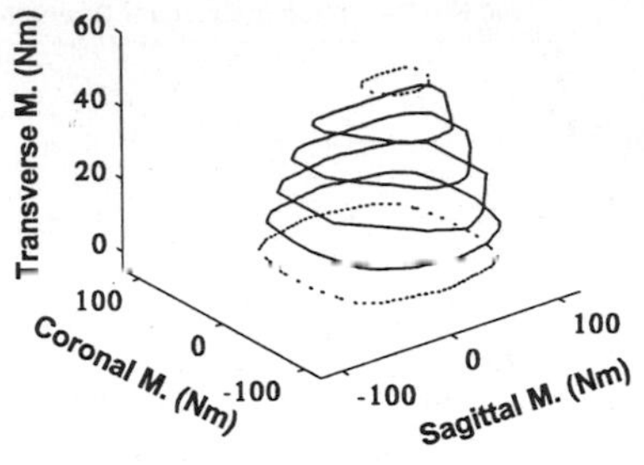

FIGURE 55.5 Three-dimensional boundaries for trunk strength, assuming muscle geometry from Hughes et al. (1995).

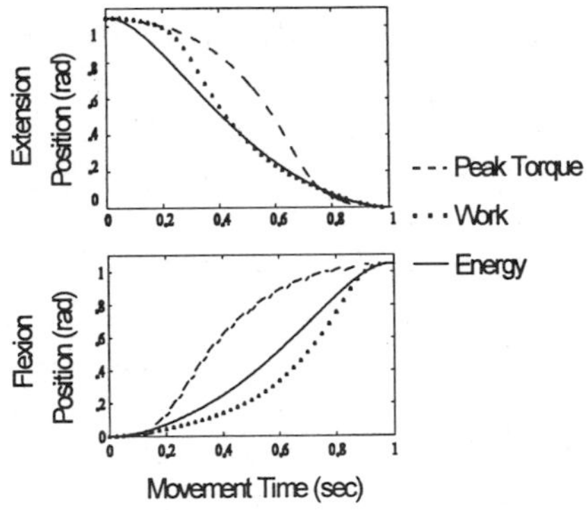

FIGURE 55.6 Effect of minimizing different cost functions (energy, work, peak torque) on the optimized trunk trajectory during simulation of trunk extension and flexion movements. The movement time has been specified at 1 second.

Figure 55.6 shows the simulated trajectory of flexion and extension of the trunk using a number of candidate cost functions in the range of upright to 60 degrees of flexion. The simulation models require specification of initial and final angular position, velocity and acceleration of the trunk and the subject's trunk flexion/extension strength. The movement time could be specified or kept as another free parameter. This model can indicate whether a task is feasible within the desired movement time or could provide a physical basis for real-time feedback to individual during the training. To simulate the impairment, we have limited the trunk extension strength to 200 Nm, and report the simulated trajectories for flexion task in Figure 55.7. It is clear that the imposed limitation on strength reduces the variability of predicted trunk trajectory (Parnianpour et al., 1997a). Although we may not be able to ever verify the cost function being used or whether CNS uses the optimization algorithms, the ability to simulate trajectories of movement is still invaluable. The numerical simulations may allow us to estimate the bounds on the required functional capacity or changes in the workplace to make the task feasible.

Biomechanical simulation models provide a time- and cost-effective tool for answering "what if" types of questions. In the light of the ADA, this is of great value in predicting the consequences of task modifications and/or workstation alterations without subjecting an injured worker or a disabled individual to unnecessary testing. Consequently, a computer-based simulation program of multilink coordinated lifting that predicts the optimum motion pattern(s) required to perform a wide range of lifting tasks subjected to constraints based on experimental strength profiles has been developed (Khalaf et al., 1996). The model uses nonlinear optimization techniques to investigate the feasibility of task performance as a function of existing impairments, and limitations on functional capacity such as the range of motion, strength, and speed of lifting.

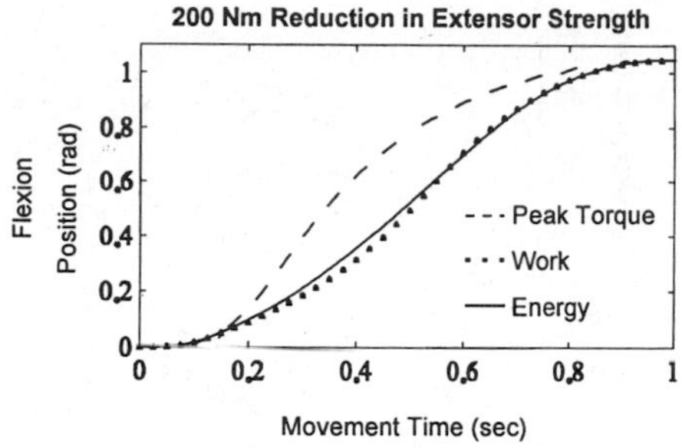

FIGURE 55.7 Effect of limiting trunk extension strength to 200 Nm on the optimized trunk flexion trajectory. The movement time has been specified at 1 second to allow comparison with unconstrained strength simulation presented in Figure 55.6.

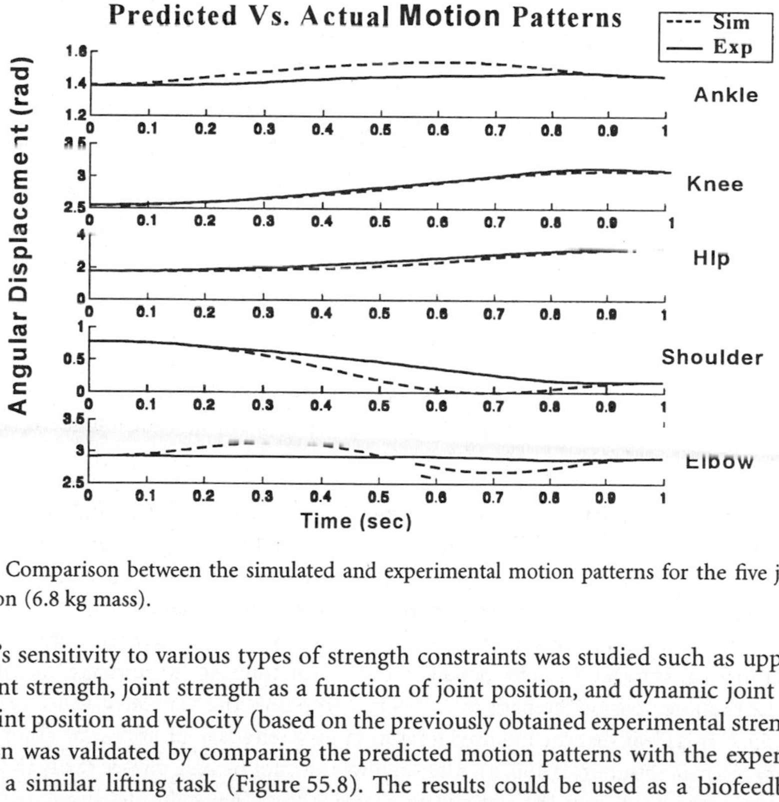

FIGURE 55.8 Comparison between the simulated and experimental motion patterns for the five joints during a lifting simulation (6.8 kg mass).

The model's sensitivity to various types of strength constraints was studied such as upper and lower bounds of joint strength, joint strength as a function of joint position, and dynamic joint strength as a function of joint position and velocity (based on the previously obtained experimental strength profiles). The simulation was validated by comparing the predicted motion patterns with the experimental data generated for a similar lifting task (Figure 55.8). The results could be used as a biofeedback tool for training injured workers during rehabilitation, return to work assessment, as well as workplace modifications or "reasonable accommodations" as dictated by the ADA. The proper choice of a cost function remains an open area of research. Present work includes further experimental validation for a diverse set of loading conditions. In addition, we are in the process of implementing a nonlinear feedback control scheme to enhance the model by providing stability.

Comparison of Task Demands and Performance Capacity

The regression analysis was used to model the dynamic torque, velocity, and power output as a function of resistance level during isoresistive flexion and extension using the B-200 Isostation (Parnianpour et al., 1990). Results indicated that the measured torque was not a good discriminator of the 10th, 50th, and 90th percentile populations. However, velocity and power were shown to effectively discriminate the

three populations. Based on these data, it was suggested that during clinical testing, sagittal plane resistance should not be set at higher than about 80 Nm in order to minimize the internal loading of spine while taxing trunk functional capacity. This presentation of data may be useful to the physician or ergonomist in evaluating the functional capacity requirements of workplace manual material handling tasks. For example, a manual material handling task that requires about 80 Nm (61 ft-lbs) of trunk extensor strength could be performed by 90% of the population in the normal database if the required average trunk velocity does not exceed 40°/sec, while only 50% could perform the task if the velocity requirement exceeds 70°/sec; but more importantly, only the top 10th percentile population could perform the task if the velocity requirement approaches 105°/sec (Parnianpour et al., 1990). A few versions of lumbar motion monitors that can record the triaxial motion in the workplace have been used to provide the trunk movement requirements. The following example also illustrates the importance of having the same bases for evaluation of both task and the functional capability of the worker.

Lifting Utilization Ratio

The utilization ratio provides a joint-specific unified scalar quantity representing the task demand normalized by an individual's maximum capacity. It would indicate whether a subject is capable of performing the task and how much of his/her maximum capacity is taxed by a given physical activity such as lifting. To illustrate this concept, Khalaf et al. (1997a) combined the trunk strength surface responses and the results of the task demand profiles of a lifting task performed by a typical female subject (Khalaf et al., 1997b). Very different results were predicted depending on whether the denominator of the utilization ratio was based on a single isometric strength value measured in upright position, posture dependent, or a function of both posture and trunk velocity. One of the reasons for the poor results of many epidemiological investigations for prediction of future low back injury based on isometric strength is shown here. To increase the predictive power, it has been suggested that the measured strength must simulate the essential functions of the task as closely as possible. The results of this analysis indicated that accurate estimation of the utilization ratio requires both the dynamic measurements of the external moment and strength in the range of joint positions being experienced during the performance of the lifting task. Moreover, the dynamic characteristics of the task should be considered. It is clear that the higher magnitudes of the dynamics components increase the task demand and reduce the maximum functional capacity, hence increasing the utilization ratio. The dynamic utilization ratio is computed considering the strength dependence on both the angular position and velocity. At the faster rate of lift, the numerator of the utilization ratio (task demand) is increased due to the increased acceleration. On the other hand, the denominator (maximum strength capacity) is decreased since the tension is inversely proportional to a muscle's speed of shortening (Hill's tension–velocity relationship). Hence, the utilization ratio approaches unity even for lifts that are only a fraction of the maximum lifting capacity. Moreover, the higher muscular activities (including higher coactivation) experienced at the faster lifting tasks may augment the risk of injury due to higher spinal loading (Ross et al., 1993).

55.4 Clinical Applications

Clinical studies have utilized quantitative human performance, i.e., strength and endurance measures, to predict the first incidence or recurrence of LBP, disability outcome, and also as a prognosis measure during the rehabilitation process. Training programs to enhance the endurance and strength of workers have been implemented in some industries. A prospective randomized study among employees in a geriatric hospital showed that exercising during work hours to improve back muscle strength, endurance, and coordination proved cost effective in preventing back symptoms and absence from work (Gundewall et al., 1993). Every hour spent by the physiotherapist on the exercise group reduced the work absence by 1.3 days. In this study both training and testing equipment were very modest. More studies on effectiveness of these programs are needed (Nordin et al., 1987; Kahanovitz et al., 1987). It can be hypothesized that these programs complement the stress management programs to enhance both worker satisfaction and coping strategies with regard to physical and nonphysical stressors at the workplace.

Functional-based impairment evaluation schemes have traditionally used spinal mobility (Buchalter et al., 1998; Elnaggar et al., 1991). Given the poor reliability of range of motion (ROM), its large variability among individuals, and the static psychometric nature of ROM, the use of continuous dynamic profiles of motion with the higher order derivatives has been suggested (Marras et al., 1993). Dynamic performances of 281 consecutive patients from the Impairment Evaluation Center at Mayo Clinic were used (Sharafeddin et al., 1996). As part of the comprehensive physical and psychological evaluation, 281 consecutive LBP patients underwent isometric and dynamic trunk testing using the B200 Isostation. Feature Extraction and Cluster Analysis techniques were used to find the main profiles in dynamic patient performances. The middle three cycles of movements were interpolated and averaged into 128 data points; thus the data were normalized with respect to cycle time. This allowed for comparison between individuals. The LBP patients in this study were shown to be heterogeneous with respect to their movement profile (Sharafeddin et al., 1996). Uniform treatment of these patients is questionable, and rehabilitation programs should consider their specific impairments. Future research should incorporate the clinical profiles with these movement profiles to further delineate the heterogeneity of low back patients (Parnianpour et al., 1994; Szpalski and Parnianpour, 1996).

Marras et al. (1993) used similar feature extraction techniques to characterize the movement profiles of 510 subjects belonging to normal (N = 339) and ten LBP patient groups (N = 171). Subjects were asked to perform flexion/extension trunk movement at five levels of asymmetry, while the three-dimensional movement of the spine was monitored by the Lumbar Motion Monitor (an exoskeleton goniometer developed at the Biodynamics Lab of The Ohio State University). Trunk motions were performed against no resistance, and no pelvic stabilization was required. The quadratic discriminant analysis was able to correctly classify over 80% of the subjects. The same technology was used to develop logistic regression models to identify the high-risk jobs in industrial workplaces. Hence, principles of human performance can successfully be applied to the worker and the task to avoid the mismatch between performance capability and task demand. A more detailed clinical implication of quantitative assessment of trunk performance is presented in Szpalski and Parnianpour (1996).

55.5 Conclusions

The outcome of trunk performance is affected by the many neural, mechanical, and environmental factors which must be considered during quantitative assessment. The objective evaluation of the critical dimensions of functional capacity and its comparison with the task demands is crucial to the decision-making processes in the different stages of the ergonomic prevention and rehabilitation process. Knowing the tissue tolerance limits from biomechanical studies, task demands from ergonomic analysis, and function capacities from performance evaluation, the rehabilitation team will optimize the changes to the workplace or the task that will maximize the functional reserves (unutilized resources) to reduce the occurrence of fatigue or overexertion. This will enhance worker satisfaction and productivity, while reducing the risk of MSD.

Based on ergonomic and motor control literature, the testing protocols that best simulate the loading conditions of the task will yield more valid results and better predictive ability. Ergonomic principles indicate that the ratio of the functional capacity to the task demand (utilization ratio) is critical to the development of muscular fatigue which may lead to more injurious muscle recruitment patterns and movement profiles due to loss of motor control and coordination. However, large prospective studies are still needed to verify this. The most promising application for these quantitative measures is to be used as a benchmark for the safe return to work of injured workers, given the enormous variations within the normal population.

The ability to identify subgroups of patients or high-risk individuals based on their functional performance will remain an open area of research to interested biomedical engineers within the multidisciplinary group of experts addressing neuromusculoskeletal occupational disorders. With the advent of technologies to monitor trunk performance in the workplace, we can obtain estimates of the injurious levels of task demands (kinematics and kinetic parameters) which can be used to guide our preplacement and rehabilitation strategies. The more functional the clinical tests become, the more clinicians need a

complex interpretation scheme. An increasingly complex interpretation scheme opens the possibility of using mathematical modeling with intelligent computer interfaces.

The integration of complex biomechanical modeling of viscoelastic elements of spine, EMG-driven models quantifying the muscular activity, and realistic anatomical models with the well-controlled experimental designs motivated by the task demand analyses in the workplace is crucial to further understanding of low back disorders and their work-related risk factors. We need to accelerate our research efforts toward understanding the stabilization role of muscles in spine and the consequences of various neuro-control strategies by applying detailed biomechanical models to quantify the tissue stresses and strains (Parnianpour, 1991; Parnianpour et al., 1997b; Shirazi-Adl and Parnianpour, 1993, 1996a,b, 1998; Kiefer et al., 1997) during physical exertions.

Acknowledgment

The authors acknowledge the support from OSURF and NIDRR H133E30009. The authors would like to thank, for their invaluable comments and contributions, George V. Kondraske, Margareta Nordin, Victor H. Frankel, Elen Ross, Jackson Tan, Robert R. Crowell, William Marras, Sheldon R. Simon, Ali Sheikhzadeh, Jung Yong Kim, Patrick Sparto, Kinda Khalaf, Jaw L. Wang, Alexander Kiefer, Marek Szpalski, and Sue Ferguson.

References

Andersson GBJ. 1991. Evaluation of muscle function, in *The Adult Spine: Principles and Practice*, Ed. J.W. Frymoyer, p. 241. Raven Press Ltd., New York, NY.

Buchalter D, Parnianpour M, Viola K, Nordin M, Kahanovitz N. 1988. Three-dimensional spinal motion measurements. Part 1: A technique for examining posture and functional spinal motion. *Journal of Spinal Disorders*, 1(4), 279-83.

Balague F, Damidot P, Nordin M, Parnianpour M, Waldburger M. 1993. Cross-sectional study of isokinetic muscle trunk strength among school children. *Spine*, 18, 1199-1205.

Beimborn DS and Morrissey, MC. 1988. A review of literature related to trunk muscle performance. *Spine*, 13(6): 655-660.

Berme N and Cappozzo, A. 1990. *Biomechanics of Human Movement: Applications in Rehabilitation, Sports and Ergonomics.* Bertec Corporation, Worthington, OH.

Elnaggar IM, Nordin M, Sheikhzadeh A, Parnianpour M, Kahanovitz N. 1991. Effects of spinal flexion and extension exercises on low-back pain and spinal mobility in chronic mechanical low-back pain patients. *Spine*, 16(8), 967-72.

Gundewall B, Liljeqvist M, Hansson T. 1993. Primary prevention of back symptoms and absence from work. *Spine*, 18(5): 587-594.

Hughes RE, Bean JC, Chaffin DB. 1995. Evaluating the effect of coactivation in optimization models. *J. Biomechanics, 28(7)*: 875-878.

Kahanovitz N, Nordin M, Verderame R, Yabut S, Parnianpour M, Viola K, Mulvihill M. 1987. Normal trunk muscle strength and endurance in women and the effect of exercises and electrical stimulation. Part 2: Comparative analysis of electrical stimulation and exercises to increase trunk muscle strength and endurance. *Spine*, 12(2), 112-118.

Khalaf KA, Parnianpour M, Wade L, Sparto PJ, Simon SR. 1996. Biomechanical simulation of manual multi-link coordinated lifting. *Proceedings of the 15th Southern Biomedical Engineering Conference*, Dayton, Ohio, 197-198.

Khalaf KA, Parnianpour M, Wade L. 1997. Feature extraction and modeling of the variability of performance in terms of biomechanical motion patterns during MMH tasks, in *Biomedical Sciences Instrumentation*, Eds. Benghuzzi, HA, Bajpai PK, Instrument Society of America, Vol. 33, 35-40.

Khalaf KA, Parnianpour M, Sparto PJ, Simon SR. 1997a. Modeling of functional trunk muscle performance: Interfacing ergonomics and spine rehabilitation in response to the ADA, *Journal of Rehabilitation Research and Development*, 34(4), 459-469.

Khalaf KA, Parnianpour M, Sparto PJ, Barin K, and Simon, SR. 1997b. Feature extraction and quantification of the variability of dynamic performance profiles at different sagittal lift characteristics. *IEEE Transactions on Rehabilitation Engineering*, in review.

Khalaf KA, Parnianpour M, Sparto PJ, Barin K. 1997c. The determination of the effect of lift characteristics on dynamic performance profiles during manual material handling (MMH) tasks. *Ergonomics*, in press.

Kiefer A, Shirazi-Adl A, Parnianpour M. 1997. On the stability of human spine in neutral postures. *European Spine Journal*, 6(1), 45-53.

Kim JY, Parnianpour M, Marras WS. 1996. Quantitative assessment of the control capability of the trunk muscles during oscillatory bending motion under a new experimental protocol. *Clinical Biomechanics*, 11(7), 385-391

Kondraske GV. 1990. Quantitative measurement and assessment of performance, in Smith RV and Leslie JH (Eds.) *Rehabilitation Engineering*, CRC Press, Boca Raton, FL.

Kroemer KE, Kroemer H, Kroemer-Elbert K. 1994. *Ergonomics: How to Design for Ease & Efficiency*. Prentice Hall, Inc., Englewood Cliffs, New Jersey.

Luck JV and Florence DW. 1988. A brief history and comparative analysis of disability systems and impairment evaluation guides. *Office Practice*, 19: 839-844.

Marras WS and Mirka GA. 1989. Trunk strength during asymmetric trunk motion. *Human Factors*. 31(6): 667-677.

Marras WS, Parnianpour M, Ferguson SA et al. 1993. Quantification and classification of low back disorders based on trunk motion. *European Journal of Medical Rehabilitation*, 3(6): 218-235.

National Institute for Occupational Safety and Health (NIOSH). 1981. Work practices guide for manual lifting (DHHS Publication No. 81122). Washington, D.C.: U.S. Government Printing Office.

National Institute for Occupational Safety and Health (NIOSH). 1997. Musculoskeletal Disorders and Workplace Factors (DHHS Publication No. 97-141). Cincinnati, Ohio: NIOSH Publication Dissemination.

Newton M and Waddell G. 1993. Trunk strength testing with iso-machines, Part 1: Review of a decade of scientific evidence. *Spine*, 18(7): 801-811.

Nordin M, Kahanovitz N, Verderame R, Parnianpour M, Yabut S, Viola K, Greenidge N, Mulvihill M. 1987. Normal trunk muscle strength and endurance in women and the effect of exercises and electrical stimulation. Part 1: Normal endurance and trunk muscle strength in 101 women. *Spine*, 12(2), 105-111.

Parnianpour M. 1991. Modeling of trunk muscles recruitment during isometric exertions. *IEEE Engineering in Medicine & Biology*, 10(2), 51-54.

Parnianpour M. 1995. Applications of quantitative assessment of human performance in occupational medicine, in Bronzino J. (Ed.). *Handbook of Biomedical Engineering*. CRC Press, Boca Raton, 1230-1239.

Parnianpour M, Bejjani FJ, Pavlidis L. 1987. Worker training: the fallacy of a single, correct lifting technique. *Ergonomics*. 30 (2), 331-334

Parnianpour M, Schecter S, Moritz U, Nordin M. 1987a. Back muscle endurance in response to external load. *Proceedings of the American Society of Biomechanics*, University of California Davis, pp. 41-42, 1987.

Parnianpour M, Nordin M, Moritz U et al. 1987b. Correlation between different tests of trunk strength. Buckle P. (Ed.): *Musculoskeletal Disorders at Work*. Taylor & Francis, London, pp. 234-238.

Parnianpour M, Nordin M, Kahanovitz N, Frankel VH. 1988. The triaxial coupling of torque generation of trunk muscles during isometric exertions and the effect of fatiguing isoinertial movements on the motor output and movement patterns. *Spine*, 13: 982-992.

Parnianpour M, Li F, Nordin M, Frankel VH. 1989. Reproducibility of trunk isoinertial performances in the sagittal, coronal, and transverse planes. *Bulletin of the Hospital for Joint Diseases Orthopaedic Institute*, 49(2), 148-154.

Parnianpour M, Li F, Nordin M, Kahanovitz N. 1989. A database of isoinertial trunk strength tests against three resistance levels in sagittal, frontal, and transverse planes in normal male subjects. *Spine*, 14(4), 409-411.

Parnianpour M, Nordin M, Skovron ML, Frankel VH. 1990. Environmentally induced disorders of the musculoskeletal system. *Medical Clinics of North America*, 74(2), 347-59.

Parnianpour M, Nordin M, Sheikhzadeh A. 1990. The relationship of torque, velocity and power with constant resistive load during sagittal trunk movement. *Spine*, 15: 639-643.

Parnianpour M, Campello M, Sheikhzadeh A. 1991. The effect of posture on triaxial trunk strength in different directions: Its biomechanical consideration with respect to incidence of low-back problem in construction industry. *International Journal of Industrial Ergonomics*, 8(3), 279-288.

Parnianpour M, Hasselquest L, Aaron A, Fagan L. 1993. The intercorrelation among isometric, isokinetic and isoinertial muscle performance during multi-joint coordinated exertions and isolated joint trunk exertion. *European Journal of Physical Medicine and Rehabilitation*, 3, 114-122.

Parnianpour M, Tan JC. 1993. Objective quantification of trunk performance, in D'Orazio B. (Ed.) *Back Pain Rehabilitation*. Andover Medical Publishers, Boston, 205-237.

Parnianpour M, Davoodi M, Forman M, Rose D. 1994. The normative database for the quantitative trunk performance of female dancers: Isometric and isoinertial strength and endurance. *Medical Problems of Performing Artists*, 9, 50-57.

Parnianpour M, Hanson T, Goldman S, Madson T, Sparto P. 1994a. The variability of trunk muscle performance in three distinct groups of females: impairment evaluation center, preplacement, and normal volunteers. *Proceedings of North American Spine Society Annual Meeting*, Minneapolis, MN, p. 165.

Parnianpour M, Engin AE. 1994. A more quantitative approach to classification of impairments, disabilities, and handicaps. *Journal of Rheum. Med. Rehab.*, 5(1), 52-64.

Parnianpour M, Wang JL, Shirazi-Adl A, Khayatian B, and Lafferriere G. 1997a. A computational method for simulation of trunk motion: Toward a theoretical based quantitative assessment of trunk performance, *IEEE Transactions on Rehabilitation Engineering*, in review.

Parnianpour M, Wang JL, Shirazi-Adl A, Wilke HJ. 1997b. The effect of trunk models in predicting muscle strength and spinal loading. *Journal of Musculoskeletal Research*, 1(1), 55-69.

Pope MH. 1992. A critical evaluation of functional muscle testing, in *Clinical Efficacy and Outcome in the Diagnosis and Treatment of Low Back Pain*, Ed. J.N. Weinstein, p. 101, Raven Press Ltd., New York, NY.

Ross EC, Parnianpour M, Martin D. 1993. The effects of resistance level on muscle coordination patterns and movement profile during trunk extension. *Spine*, 18, 1829-1839.

Sharafeddin H, Parnianpour M, Hemami H, Hanson T, Goldman S, Madson T. 1996. Computer aided diagnosis of low back disorders using the motion profile. *15th Southern Biomedical Engineering Conference*, Dayton, Ohio, pp. 431-432.

Shirazi-Adl A, Parnianpour M. 1993. Nonlinear response analysis of the human ligamentous lumbar spine in compression: On mechanisms affecting the postural stability. *Spine*, 18, 147-158.

Shirazi-Adl A, Parnianpour M. 1996a. Stabilizing role of moments and pelvic rotation on the human spine in compression. *ASME Journal of Biomechanical Engineering*, 118(1), 26-31.

Shirazi-Adl A, Parnianpour M. 1996b. Role of posture in mechanics of the lumbar spine. *Journal of Spinal Disorders*, 9, 277-286.

Shirazi-Adl A, Parnianpour M (1998) Finite element model studies in lumbar spine biomechanics, in Leondes C (Ed.) *Biomechanics Systems Techniques and Application*. Gordon and Breach Publishing Group, Newark, in press.

Snook SH, Fine LJ, Silverstein BA. 1988. Musculoskeletal disorders, in *Occupational Health: Recognizing and Preventing Work-related Disease*, B.S. Levy and D.H. Wegman (Eds.), p. 345-370. Little, Brown and Co., Boston/Toronto.

Sparto PJ. 1998. Trunk muscle electromyographic responses, wavelet detection of fatigue, and spinal loading during fatiguing repetitive trunk extension. Doctoral Dissertation, The Ohio State University, Columbus, Ohio.

Sparto PJ, Parnianpour M, Khalaf KA, Simon SR. 1995. The reliability and validity of a lift simulator and its functional equivalence with free weight lifting tasks, *IEEE Transactions on Rehabilitation Engineering*, 3(2), 155-165.

Sparto PJ, Parnianpour M. 1997. Changes in muscle recruitment patterns during fatiguing dynamic trunk extension exertions. *American Society of Biomechanics*, Clemson, South Carolina, September 24-27, 23-24.

Sparto PJ, Jagadeesh JM, Parnianpour M. 1997a. Real time wavelet analysis of electromyography for back muscle fatigue detection during dynamic work, in Benghuzzi HA, Bajpai PK (Eds.) Biomedical Sciences Instrumentation, Instrument Society of America, Vol. 33, 82-87.

Sparto PJ, Parnianpour M, Reinsel TE, Simon SR. 1997b. Spectral and temporal responses of trunk extensor emg to an isometric endurance test, *Spine*, 22(4), 418-425.

Sparto PJ, Parnianpour M, Marras WS, Granata KP, Reinsel TE, Simon SR. 1997c. Neuromuscular trunk performance and spinal loading during a fatiguing isometric trunk extension with varying torque requirements, *Journal of Spinal Disorders*, 10(2), 145-156.

Sparto PJ, Parnianpour M, Reinsel TE, Simon SR. 1997c. The effect of fatigue on multi-joint kinematics and load sharing during a repetitive lifting test, *Spine*, 22(22), 2647-2654.

Sparto PJ, Parnianpour M, Reinsel TE, Simon SR. 1997d. The effect of fatigue on multi-joint kinematics, coordination, and postural stability during a repetitive lifting test, *Journal of Orthopaedic and Sports Physical Therapy*, 25(1), 3-12.

Sparto PJ, Parnianpour M, Marras WS, Granata KP, Reinsel TE, Simon SR. 1998a. The effect of EMG-force relationships and method of gain estimation on the predictions of an EMG-driven model of spinal loading, *Spine*, 23(4), 423-429.

Sparto PJ, Parnianpour M, Reinsel TE, Simon SR. 1998b. The effect of lifting belt use on multi-joint motion and load bearing during repetitive and asymmetric lifting, *Journal of Spinal Disorders*, 11(1), 57-64.

Szpalski M, Parnianpour M. 1996. Trunk performance, strength and endurance: measurement techniques and application. In Weisel S., & Weinstein J. (Eds.): *The Lumbar Spine*, second edition. WB Saunders, Philadelphia, 1074-1105.

Tan JC, Parnianpour M, Nordin M, et al. 1993. Isometric maximal and submaximal trunk exertion at different flexed positions in standing: Triaxial torque output and EMG. *Spine*, 18(16), 2480-1011.

Wang JL, Parnianpour M, Shirazi-Adl A, Engin AE. 1997a. The review and evaluation of viscoelastic models for collagen fiber during constant strain rate loading. *Biomedical Engineering, Application, Basis, Communication*, 9 (1), 5-19.

Wang JL, Parnianpour M, Shirazi-Adl A, Engin AE. 1997b. The simulation of viscoelastic behaviors under experimental controlled loading and the failure criterion of collagen fiber. *Journal of Theoretical and Applied Fracture Mechanics*, 27(1), 1-12.

Wang JL, Parnianpour M, Shirazi-Adl A, Engin AE. 1997c. Development and validation of a viscoelastic finite element model of an L2-L3 motion segment. *Journal of Theoretical and Applied Fracture Mechanics*, 28(1), 81-93.

Wang JL, Parnianpour M, Shirazi-Adl A, Engin AE. 1998. The dynamic response of L2/L3 motion segment to cyclic axial compressive loading. *Clinical Biomechanics*, 13, 516-525.

World Health Organization. 1980. *International Classification of Impairments, Disabilities, and Handicaps.* Geneva.

World Health Organization. 1985. *Identification and Control of Work-related Diseases.* Technical report no. 174. Geneva.

Zhu XZ, Parnianpour M, Nordin M, Kahanovitz N. 1989. Histochemistry and morphology of erector spinae muscle in lumbar disc herniation. *Spine*, 14(4), 391-397.

56

Perspective on Industrial Low Back Pain

Malcolm H. Pope
University of Iowa

Donald R. McIntyre
Interlogics, Hillsborough, NC

56.1 Introduction

The first recorded case of occupational low back pain (LBP) was during the building of the pyramids. Although there is an extensive literature supporting a positive relationship between LBP and workplace factors (see reviews by Andersson et al., [1991]; Garg and Moore, 1992; Kelsey and Hochberg, 1988; Riihimaki, 1995), recent publications have claimed that psychosocial factors are much more important (Nachemson, 1991; Bigos et al., 1991; Hadler, 1994). This paper will address the important ergonomic principles.

The industrial risk factors that have been identified are:

1. Manual materials handling (both frequent and heavy lifting)
2. Awkward postures
3. Fixed postures (including sitting)
4. Whole body vibration
5. Slipping and tripping
6. Psychosocial factors

There is a lengthy epidemiological literature dealing with the relationship of these risk factors with LBP.

The epidemiological studies have several generic limitations:

1. Exposure data is generally limited (job title alone reduces risk toward null effects)
2. Confounding factors are often present
3. A definitive diagnosis is often absent

0-8493-2641-9/99/$0.00+$.50

Manual Materials Handling (MMH)

Numerous studies have found a positive relationship between MMH and LBP. Heliovaara (1987) found that the odds ration (OR) for herniated *nucleus pulposus* (HNP) for heavy MMH was 2.5 to 4.6. Riihimaki (1995) summarized reviews of the relationship of LBP and MMH. Of 26 studies since 1950 only two found no relationship. However, all of the studies were cross sectional rather than prospective and used job title as a surrogate of exposure.

Biomechanical studies, as summarized by Pope et al. (1991) show that a large load or one held at a distance from the body leads to extremely high forces in the disc due to contraction of the erector spinae. Likewise, flexion of the trunk increases compression forces. A simple means of addressing the issue of the allowable lift is to use the NIOSH lifting formula (Waters et al. 1993).

1. Recommended Weight Limit (RWL)

$$RWL = LC \times HM \times VM \times DM \times AM \times CM \times FM \tag{1}$$

where
LC = Load constant
HM = Horizontal multiplier
VM = Vertical multiplier
AM = Asymmetric multiplier
CM = Coupling multiplier
FM = Frequency multiplier

The six multipliers are the penalties for deviating from the ideal lifting situation. The ideal lifting situation is defined as lifting from a 30 in. (75 cm) height, to a distance of up to 10 in. (25 cm), with hands close to the body (up to 10 in. [25 cm] from the ankles), with no twisting, with good grasp, a lifting frequency of once every 5 minutes or less, and with the lifting duration not exceeding 1 hour.

Thus, the load constant (LC) is the maximum weight allowed to be lifted. Since the six multipliers are penalties, none of them can be more than 1. In other words, under ideal conditions, when all six multipliers are 1, the RWL = LC.

2. Lifting Index (LI)

The LI gives the ratio of the weight to be lifted to the recommended weight limit.

$$LI = \frac{\text{Load Weight (lbs) or (kg)}}{\text{Recommended Weight Limit (lbs) or (kg)}} \tag{2}$$

A convenient means of assessing the NIOSH lifting risk is via a tool such as Video Works (see example). The data can be digitized directly from the video into the NIOSH equations and the effects of ergonomic changes assessed.

Control of the risk can include reducing the:

1. Postural demands
2. Burden
3. Size of the load
4. Repetitions
5. Twisting

Other control strategies can include the use of:

1. Lifting equipment

2. Parts delivered at waist height
3. Two-man lift

A logical control technique would appear to be matching the worker to the task. Chaffin (1974) found that a mismatch increased the injury risk. Liles et al. (1984) found that the incidence of injury was related to the job severity index (JSI). The JSI is the weight of a lift divided by the acceptable weight of a lift as defined by the worker. A JSI of 0.6 increasing to 2.0 increases the risk of injury by 4.5 times. Repetitive lifting of greater than 5 kg increases the risk twofold.

Testing of workers to assess lifting ability is limited in the United States under the provisions of the ADA unless the lifting represents an essential function of the job.

Postural Demands

The literature contains numerous papers relating severe postural demands to LBP and HNP. These postural demands are often linked together with MMH. An exception is the static awkward postures normally associated with sedentary work. Punnett (1991), while studying auto workers, found that LBP was associated with awkward postures. Mild flexion has an OR = 4.9, high flexion 5.7 and twist or lateral bend 5.9. Pope et al. (1986) showed there is a higher level of antagonistic muscle activity in twisted postures, leading to high muscle and disc forces. This was supported by Kelsey et al. (1984) who found an OR of 3.0 between lifting while twisting and HNP. Adams and Hutton (1985), in laboratory studies, showed that loading a cadaver disc in a flexed, twisted orientation leads to an HNP.

Controls mean a redesign of the workplace to prevent high kinematic demands such as reaching, bending, and twisting. Conveyor belts, adjustable tables, and delivery of goods at waist height are some of the solutions that have been employed.

Whole Body Vibration (WBV)

There is considerable literature relating exposure to WBV through driving vehicles to LBP and HND. Hulshof and Van Zanton (1987) found OR to range from 1.6 to 7.0. Kelsey and Hardy (1975) found that spending over half the day in a car gives an OR of 3.0 for HND.

Studies of the occupational environment reveal that many vehicles subject the worker to levels of vibration greater than that recommended by the International Standards Organization (ISO, 1978). The human spinal system has a characteristic response to vibration in a seated posture. The first resonance occurs within a band of 4.5 to 5.5 Hz. Similar, but less tightly defined resonances, are also in the 9.4 to 13.1 Hz range (Pope et al., 1986, 1987, 1989). The resonance is markedly affected by the pelvis–buttocks system. The response of the human is due to a combination of a vertical subsystem and a rotational subsystem. The latter is by rocking of the pelvis.

The back muscle response with respect to the vibration stimulus was studied by Seroussi et al. (1989). The muscles are not able to protect the spine from adverse loads. At many frequencies, the muscle response is so far out of phase that muscle forces are added to those of the stimulus. The fatigue that was found in muscles, after vehicular vibration, is indicative of the loads in the muscles. Exposure of the seated subject to whole body vibrations of 5 Hz, in a position that ensured back muscle activity, increased the rate of development of fatigue in the erector spinae muscles, as compared to the same conditions without vibrations. After WBV, there is an increased latency in muscle recruitment (Wilder et al. 1996). This may be due to the fatigue inherent in WBV exposure. This suggests that if a worker unloads a vehicle after exposure to WBV this could present a problem for the back.

Many ergonomic interventions are possible. WBV should be attenuated at the cab or seat to comply with ISO 2631. The layout of the vehicle cab should be optimized to prevent awkward postures. This is particularly true for tractor and forklift drivers. Excessive driving shifts should be avoided and frequent stretch breaks encouraged. Lifting directly after driving should be avoided. The psychological environment should not be ignored and good health (i.e., aerobic strength, trunk strength, and endurance) encouraged.

Slipping and Tripping

It is reported in some studies that as much as 50% of LBP injuries occur in conjunction with slipping and tripping (Pope et al., 1991). In the absence of blunt trauma, the slip may cause a sudden load to be applied to the ligamentous spine or may cause the muscles to contract with high force (Wilder et al., 1996).

Psychosocial factors

Many studies support a positive relationship between LBP report or disability and psychosocial stressors. These stressors include:

1. Monotonous work
2. Poor social support
3. High work pressure
4. Lack of control
5. Job dissatisfaction
6. Mutual dislike of boss and/or co-workers

In the Boeing study (Bigos et al., 1991) found the OR for LBP disability was 1.7. Magnusson et al. (1996) found that mechanical stressors lead to LBP, but psychosocial factors are responsible for low back pain disability.

The challenge would seem to be to separate the issues of LBP and disability since the solutions may be different. Clearly, a systematic method needs to be employed to deal with this difficult problem.

56.2 An Ergonomics and Injury Evaluation System

With this in mind, we developed a systems approach to the ergonomics and injury evaluation. The overall system is shown in Figure 56.1. It will be noted that the usually disparate functions of the industrial and medical facilities now overlap, leading to vastly improved communications and efficiencies. The system can be thought of as having both baselining, or pre-injury phases and post-injury components. With the emphasis on evaluating and treating both the worker and workplace, the baselining or pre-injury phase includes task analysis, task screening, and training, baseline testing of the worker, ergonomics intervention/reengineering, while the post-injury phase includes testing of both the worker and workplace. The system is operated by Interlogics, Hillsborough, NC.

Baselining or Pre-Injury Phase

Software has been developed to download information from the standard OSHA 2000 or other customized databases in order to determine the incidence, nature, and severity of musculoskeletal-related injuries. These data are transferred to a central processing facility to identify jobs, activities, or sub populations with an increased injury risk (high-priority jobs). A typical data collection screen for a given injured worker is shown in Figure 56.2.

Task Analysis

The system for ergonomics assessment and intervention includes task analysis. Once the high-priority jobs have been identified, trained technicians travel to the workplace where they use novel portable measurement devices for data collection. Data are initially collected at the workplace using a pentop computer (Figure 56.3). The collection system is based on a time and motion analysis and identifies the essential functions of the job and selects tasks and workplace hazards leading to a higher risk of injury.

This process also provides detailed job descriptions of the jobs analyzed. Specific job data include:

- Posture data, including sustained postures, improper sitting positions, and documentation of the orientation of the ankles, hips, lower back, shoulders, wrists, neck, and elbows
- Lifting hazards data

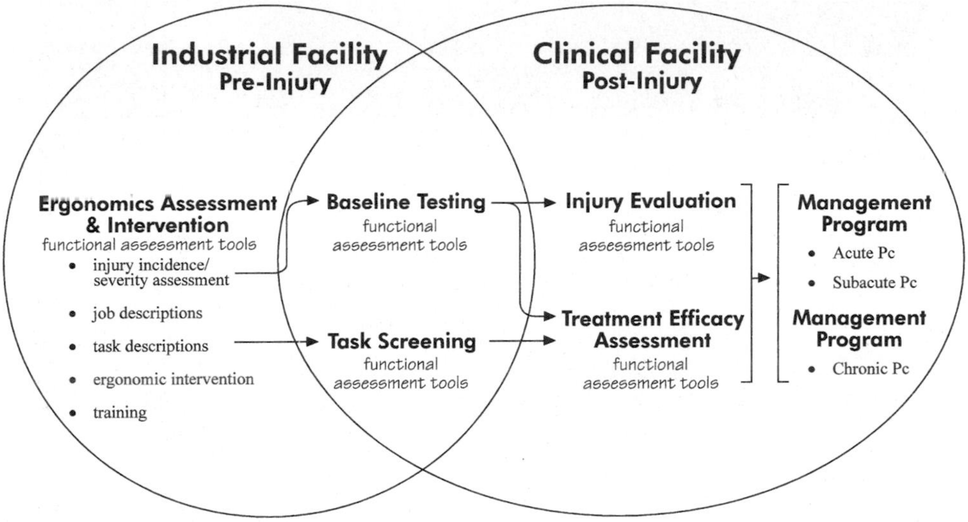

FIGURE 56.1 Overview of the system used to evaluate, reduce, and manage job-related injuries.

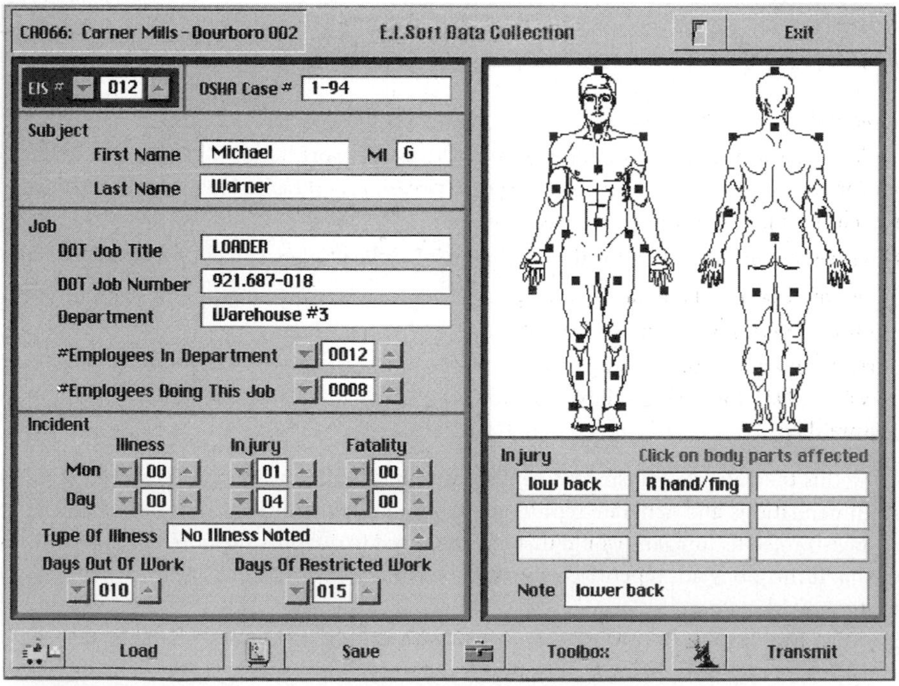

FIGURE 56.2 Injury incidence and severity data collection screen.

- Pushing/pulling hazards data
- Repetition hazards data, including constant work cycles, lack of stretching, and heavy duty cycles
- Fatigue hazards data, including abnormal speech patterns and worker behavior changes, such as looking for resting places and rubbing body parts used in performing a task
- Environmental hazards data, including flagging for hot or cold extremes, wetness, slippery conditions, sharp edges, noise, and obstructions

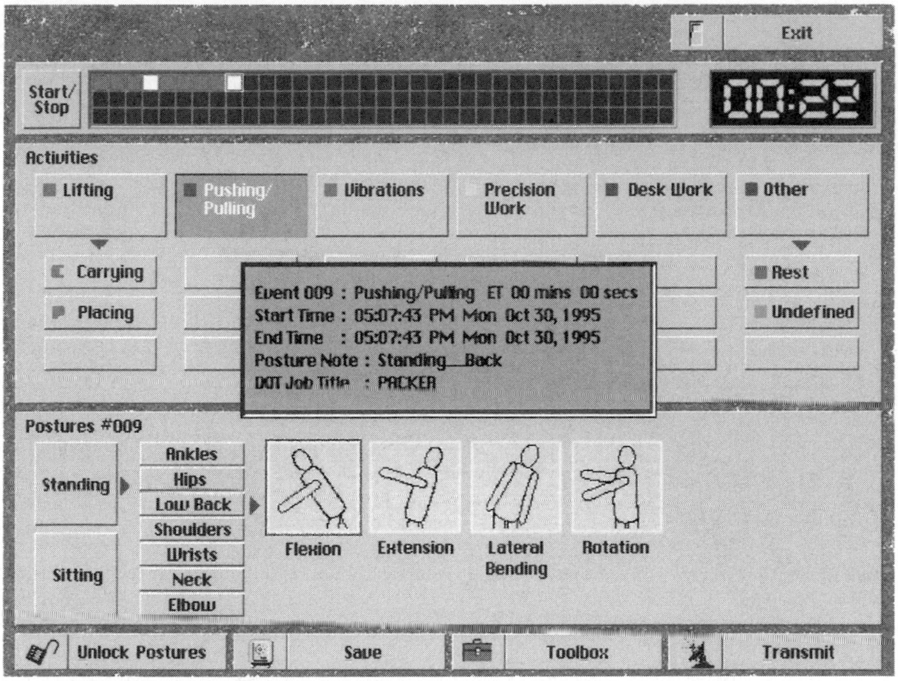

FIGURE 56.3 Example screen used during the process of collecting job description and job hazard information.

- Tool/equipment hazards data
- Desk job hazard data, including ergonomic risks for workstations, excessive reaching for tools and/or equipment, VDT height and distance violations, excessive low or high positions of chairs, and cluttering of desktop work spaces
- Time-based summary data for the *Dictionary of Occupational Titles* (1987) jobs performed
- Time-based summary data for lifting, carrying, placing, pushing/pulling, vibrations, precision work, desk work, locomotion (walking), and rest
- Correlation data of perceived pain vs. actual joint stresses observed, including Genaidy and Karwowski (1993) Joint Stress Tables, Nordic Questionnaire (Dickenson et al. 1992), and extreme posture flags

Following this the data are transmitted to the centralized facility where professional ergonomists will make recommendations and generate reports focusing on potentially injurious task characteristics.

At this stage trained technicians would then be instructed to quantify the potentially hazardous aspects of a given job. If the job is not repetitive or stereotyped, this may mean that measurements should occur throughout the working day.

For example, InterLogics' B-Tracker™ is used to measure the functional characteristics of low-back movement. At the central processing facility, professional ergonomists analyze the data to precisely define awkward and prolonged trunk postures and repetitive trunk movements. (An illustration of this application of the B-Tracker™ in the workplace is shown in Figure 56.4).

The report generated by the professional ergonomist with the help of the technicians would then make recommendations for workplace modification to reduce ergonomic risk or to hasten return to work for an injured worker.

Task Screening and Training

The same data transmitted to the central processing facility to define potentially injurious task characteristics can also be used to generate a task-specific protocol to test job candidates. Candidates can be

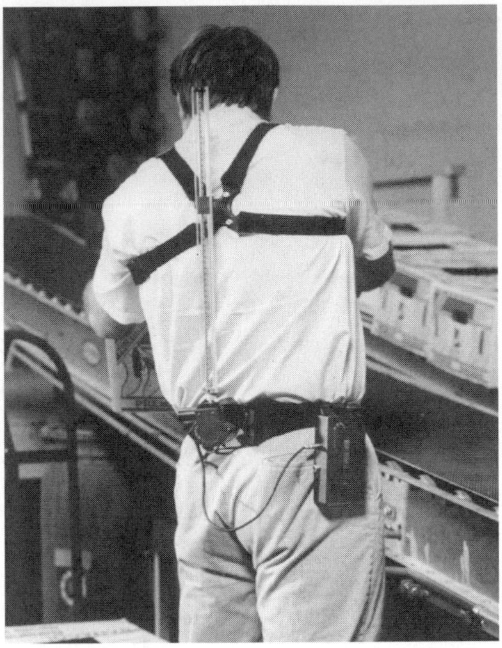

FIGURE 56.4 Example of a proprietary measurement tool used to collect task specific functional information at the worksite.

screened "post-offer and pre-placement," using the task-specific protocol to determine whether they can perform the extreme characteristics required of the task. This screening process, which is based on the actual demands of a particular job, complies with the Americans with Disabilities Act of 1990.

A typical screening test, allowable under the ADA, would be to have the prospective worker lift a load repetitively from the floor to a shelf across the room.

Task analysis information can also be used to create task-specific training programs. The actions of the "expert" worker can often be used as a model for a videotape. Risks of injury can therefore be reduced by ergonomic intervention and training program implementation, for example, by correcting hazardous lifting techniques, encouraging other productive noninjurious workplace behaviors, and teaching workers how to recognize and respond to potentially hazardous situations. To determine the effectiveness of ergonomics intervention and training, workers are subsequently analyzed performing their tasks.

Baseline Testing

Baseline testing of the worker, if economically feasible, can be a very useful tool for following a worker if he or she subsequently becomes injured. However, typically once injury incidence and severity data have been collected and analyzed, regardless of whether task-specific screening has occurred, testing can be conducted to generate baseline data for the identified "at risk" population. For instance, low-back baseline assessments are accomplished by using the B-Tracker, a portable stabilization system, and a standardized protocol consisting of software directed movements in all cardinal planes, circumductions, and stimulated sagittal plane lifts. (The testing set-up is shown in Figure 56.5) These baseline data serve as a record of the functional characteristics of the healthy worker's low back. Each worker's data are subsequently compared to healthy and low-back problem population profiles, and a certainty of group membership in either of those populations is assigned. Current work employs a neural network to refine this population.

Baseline data are gathered in the event a worker later becomes injured. The comparison of baseline and post-injury clinical data enables a determination of the precise nature and extent of changes in the worker's functional status, which in turn leads to the assignment of an appropriate treatment path.

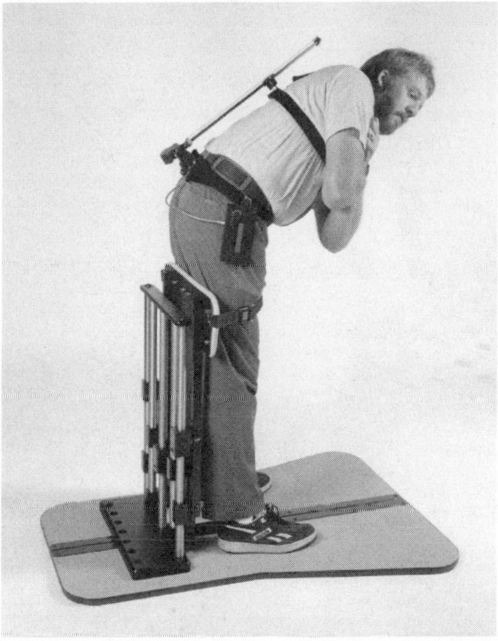

FIGURE 56.5 Testing set-up used for baselining healthy workers and for injury evaluations (Photo by Dan Crawford ©1996.)

Baseline data also inform the clinician when a worker has recovered to his or her pre-injury functional status.

Post Injury Analysis

The Worker

After an injury, the worker's functional status is assessed using the same protocols and devices used at baseline. In the case of a low-back injury, for instance, additional information is also collected, including a pain discomfort rating and a measure of nonorganic physical signs. This and the protocol data are transmitted to the central processing facility for analysis and report generation. The report identifies the functional status of the injured worker with respect to healthy and low-back problem populations, and it assigns a degree of certainty for the injured worker's membership in the low-back problem population. Work is under way to develop an expert system using neural networks that can learn to improve and refine its characterization. These data, especially the changes with time, will be helpful to the clinician in optimizing return to work and early activation while avoiding chronicity.

Comparing a worker's healthy baseline functional status with that derived from the clinical evaluation shows the extent of loss of functional ability. Subsequent testing of the injured worker can be used to track the recovery process.

Before release to return-to-work status, the injured worker can be screened for functional ability to perform specific task characteristics of his or her job. This is accomplished by using the task-specific screening protocol previously created from data collected at the work site. If the injured worker is found to be functionally unable to perform the specific task characteristics, the analyzed data are used to assign different job tasks suitable to the injured worker based on an analysis of the essential functions of alternative jobs.

In an effort to return injured workers to the workplace as soon as possible, separate programs are implemented for managing acute or chronic injuries. The acute low-back program is based on the clinical practice guidelines for acute low-back problems in adults as described by the Agency for Health Care

Policy and Research (1994). The chronic low-back pain management program was created and validated by the Iowa Spinal Research Center. An objective of the software-driven program for managing acutely injured patients is to begin the appropriate treatment regimens as early as possible in an effort to prevent any injuries from becoming chronic. The treatment costs are contained by selecting appropriate treatment protocols, maintaining contact with the injured worker, accurately tracking recovery progress, and objectively documenting the worker's return to pre-injury status.

56.3 Case Study

In many cases it would be helpful to evaluate the workplace (see task analysis) to assess whether ergonomic changes can be made to hasten return to work and prevent reinjury.

Injury Incidence/Severity Analysis

Bandy Boxing Group's OSHA Logs for the 6-month period Jan-Jun 1996 were input into the OCM software. The resulting injury incidence and severity report shows the Stacker job has the highest incidence/severity rate. (Table 56.1) The incidence rate was found to be 19%. This represents the number of injuries per 100 full-time workers. The industry incidence rate for the same time period is 6%. The high incidence rate at Bandy Boxing Group compared to the industry incidence rate indicates further investigation is warranted.

Severity is the number of average lost workdays per total lost workday cases. On average injured workers missed 5 days of work due to injury. Industry severity for the same time period is 5.4. In general it is desirable to reduce the number of days of missed work, i.e., assign alternate duties to keep workers at work.

Job Analysis

An ergonomist does an on-site analysis of the tasks associated with the Stacker job, using a pentop computer running TCM software. The resulting Job Analysis report pinpoints for further analysis the specific task that puts the Stacker at risk: the lifting task. The ergonomist's analysis of the Stacker job and related jobs at Bandy Boxing provides the data composing several job description reports. (Table 56.2A, 56.2B, 56.2C)

Task Analysis

A videotaped performance of the Stacker's lifting task generates a VideoWorks report, documenting the task's high lifting index. Simultaneous use of the B-Tracker provides data for the B-Tracker Job Analysis report, which shows the Stacker repetitively using extreme flexion and extreme lateral bending postures. (Table 56.3, A-E)

Recommendations

All reports are analyzed by the ergonomic experts at InterLogics' Central Processing Facility, yielding an Ergonomic Recommendations report with the following recommendations:

- Engineering Re-design
- Ergonomic Intervention
- Education and Training
- Screening
- Baseline Testing

(See Table 56.4.)

Injury Incidence/Severity Report

Bandy Boxing Group
July 14, 1996

Company Data

Company #:	39-0004-02
Address:	2912 Sousa Blvd. Muskegon, MI 49444
Contact:	Beth Fingers
Phone:	(616) 733-3800

Session Data

Period:	Jan 95 - Jun 95
# Job Titles:	15
Data Entry:	Tracy Marker
Phone:	(616) 733-3800

Incident Summary

Job Title	DOT Job Code	Injury	Lost WD	Rstrctd WD	Illness	Lost WD	Rstrctd WD	Death
Stacker	929.687-030	5	12	11	0	0	0	0
Gluer	795.687-014	4	10	10	0	0	0	0
Air-Drier-Machine Op	534.682-010	3	9	8	0	0	0	0
Rotary-Cutter Feeder	640.686-010	3	8	3	0	0	0	0
Tractor-Trailer-Truck Dr	904.383-010	2	3	1	0	0	0	0
Maintenance Mechan	638.281-014	2	1	1	0	0	0	0
Air Hammer Stripper	794.687-050	2	1	0	0	0	0	0
Flexographic-Press Op	651.682-010	2	1	0	0	0	0	0
Laminator	554.685-030	1	2	1	0	0	0	0
Packer	929.684-010	1	1	0	0	0	0	0
General Supervisor	183.167-018	1	0	1	0	0	0	0
Supplies Packer	919.687-022	1	0	1	0	0	0	0
Corrugator Operator	641.562-010	1	0	0	0	0	0	0
Printer-Slotter Helper	659.686-014	1	0	0	0	0	0	0
Quality-Control Tester	559.367-010	1	0	0	0	0	0	0
Totals		**30**	**48**	**37**	**0**	**0**	**0**	**0**

TABLE 56.1

Baseline Testing

All stackers at Bandy Boxing are baseline tested to provide benchmark data about the current functional status of their low backs. The resulting Baseline Testing reports indicate that all but one of the Stackers achieved scores that place them in the healthy low-back population (Table 56.7A–56.7D).

Clinical Evaluation

An injury to the Stacker Bobby Carrera immediately initiates a B-Tracker clinical evaluation, precisely the same protocol that was used during baseline testing. The Clinical Evaluation report score places Bobby Carrera in the problem low-back population, and also provides information about the relative extent of the injury. The attending physician prescribes a rehabilitation program and returns Bobby to work with alternative work duties, assigned on the basis of the Job Description reports previously compiled. Treatment and a subsequent Clinical Evaluation report show Bobby Carrera's improved functional status.

Job Analysis Report

Bandy Boxing Group
July 17, 1996

Company Data

Company #:	39-0004-02
Address:	2912 Sousa Blvd. Muskegon, MI 49444
Contact:	Beth Fingers
Phone:	(616) 733-3800
Ergonomist:	Tracy Marker

Job/Task Data

Job Title:	Stacker
Job Code:	929.687-030
Dept.:	Outgoing

Demographic Data

Worker Name:	Bobby Carrera
Worker ID:	123-45-6789
Height:	6′1″
Weight:	170#
Hand Dominance:	Right
Time on Job:	2 yrs

Job/Task Summary

Event	%Time	Time	#Events	Avg Duration
All	100	04:33	98	00:03
Lifting	75	03:24	68	00:03
Pushing/ Pulling	2	00:05	2	00:03
Precision Work	19	00:52	23	00:02
Passive Work	1	00:04	2	00:02
Locomotion (Walking)	2	00:05	2	00:03
Resting	1	00:03	1	00:03

TABLE 56.2A

At this point the physician, in conjunction with InterLogics, has Bobby return to work at full capacity as a Stacker. A later Baseline Testing report shows that Bobby Carrera is at or near pre-injury low-back functional status. (Table 56.6)

56.4 Conclusions

The cost and disability associated with LBP remain very high. The vast majority of studies show causality between mechanical loading factors and LBP. It is evident that psychosocial factors have an important role in the disability from LBP.

The ergonomics and injury evaluation system works well in practice and represents a novel, comprehensive system for health care cost containment. This program links the industrial and clinical arenas to

Job Analysis Report

Bandy Boxing Group
July 17, 1996 Page 2

Pre-Analysis Pain Locations†

Low Back

Relative Joint Stress Ratings*

	Neck	Shoulder	Elbow	Wrist	Back	Hip	Ankle
Lifting	68	204	136	0	340	0	0
Pushing/Pulling	0	0	0	0	0	0	0
Precision Work	69	46	115	23	69	0	0
Passive Work	0	0	0	0	0	0	0
Locomotion (Walking)	0	0	0	0	0	0	0
Totals	**137**	**250**	**251**	**23**	**409**	**0**	**0**

Observed Task Risks

Awkward positions
Sustained postures
Extreme temperatures
Inadequate rest period

Job Hazard Summary

Worker's pre-analysis pain location is matched with high stress rating at same joint (low back), suggesting further investigation warranted (VideoWorks, B-Tracker). Observed task risks suggest ergonomic intervention, administrative controls, or engineering re-design warranted.

Report Reviewed by: ____________________ **Date:** __________

Tracy Marker, Ergonomist

† from Nordic Musculoskeletal Questionnaire developed by Dickenson et al., 1992
* based on research by Genaidy & Karwowski, 1993

TABLE 56.2B

promote an integrated injury prevention and injury management program which effectively reduces injuries and more quickly returns injured workers to their jobs, all with the benefit of reduced expense. Injury prevention proceeds by initially identifying the causes of industry's job-related injuries and the at-risk worker populations; this then leads to baselining and screening procedures, to ergonomic intervention, or to training programs. On the clinical side, injury management proceeds by use of the information gathered at the workplace and by further evaluation with the same portable assessment tools. By use of such functional capability information, the injured worker can be placed in the chronic or acute treatment program, and his or her rehabilitation efficiently managed to expedite a speedy return-to- workplacement in a functionally appropriate job.

Job Description

Bandy Boxing Group
July 14, 1996

Job/Company Data

Company #:	39-0004-02
Address:	2912 Sousa Blvd Muskegon, MI 49444
Job Title:	Stacker
DOT Code:	929.687-030
Industry:	Packaging

Essential Job Functions	Essential Requirements
1. Re-stack paper stock onto pallet by flipping it over and evening it out in preparation for entry into printing press	1. a) Sufficient hand grip strength to grasp stack of paper stock b) Ability to lift, control, and flip large paper stock stacks up to 60 lbs each c) Work in noisy environment
2. Move re-stacked pallet with hand truck onto platform of printing press	2. a) Ability to push and maneuver hand truck in close quarters b) Properly position pallet in printing press

Secondary Job Functions	Secondary Requirements
1. Maintain appropriate flow of material into press	1. a) Verbally communicate with press operator b) Assist other stacker in flipping and restacking large sheets of paper stock
2. Assist printer operator as needed.	2. a) Fold cleaning rags b) Watch printing operation and signal to operator when problem arises

Strength Rating

0 10 20 50 100
Sed. Light Medium Heavy Very Heavy

DOT Physical Demand Strength Rating (PDSR) Scale for Occasional Weight

Job Rotation Options	DOT Job Code	PDSR
Scrapper	794.687-050	Heavy
Hand Packager	920.587-018	Medium

Job rotation reduces the continuous performance of a single task, preventing the hazards associated with work that is repetitive, static, or awkward in nature.

Report Reviewed by: ______________________ **Date:** ________

Tracy Marker, Ergonomist

TABLE 56.2C

The cost of injuries in the workplace will continue to be expensive, but the costs can be effectively contained. Integrating injury prevention and injury management programs reduces health care costs by reducing initial injuries, speeding recovery, and leveraging professional expertise via a central processing facility.

Defining Terms

Accelerometer: Instrument that measures acceleration.
Biomechanics: The application of principles of mechanics to the human body.

VideoWorks Report: Individual

Bandy Boxing Group
July 17, 1996

Company/Task Data

Company #:	39-0004-02
Address:	2912 Sousa Blvd. Muskegon, MI 49444
Job Title:	Stacker
Job Code:	929.687-030
PDSR:	Heavy Work

Performance Statement

On July 17, 1996, 3 lifting tasks associated with the job of Stacker at Bandy Boxing Group were performed by Bobby. An analysis of the performance was conducted with VideoWorks, yielding results based on the Revised NIOSH Lifting Equation (1991). **The greatest Lifting Index (LI) associated with the tasks performed was 4.46, indicating a high risk of injury**, i.e., this specific task was highly stressful and likely to put nearly all workers at increased risk of lifting-related injury. See the Recommended Adjustments section for specific ways in which to reduce the lifting index of the individual tasks performed.

Performance Description

	Name	Duration	Frequency (lifts/min)
Lift #1	Lift 1	Moder	4.30
Lift #2	Lift 2	Moder	4.30
Lift #3	Lift 3	Moder	4.30

Performance Evaluation

Name	L (lbs)	RWL(lbs)	FILI	LI(origin)	LI(dest)	Risk Level
Lift 1	50.00	11.20	3.05	2.19	4.46	high
Lift 2	50.00	12.26	2.79	1.95	4.08	high
Lift 3	50.00	13.71	2.49	2.01	3.65	high

Report Reviewed by: ____________________ **Date:** __________

Tracy Marker, Ergonomist

TABLE 56.3A

Electromyograph: Recording of electrical output of the contraction of a muscle.
Erector spinae: Back muscles that extend the spine.
Ergonomics: Science that seeks to adapt working conditions to suit the worker.
Extensor: Any muscle that performs extension.
Flexor: Any muscle that flexes a joint.
Intervertebral: Situated between two adjacent vertebrae of the spine.
Isometric: A type of muscle contraction in which the length of the muscle remains constant.
Kinematics: Science of motion, including movements of body.
Nucleus pulposus: The central, more viscous portion of the intervertebral disc.

VideoWorks Report: Individual

Bandy Boxing Group
July 17, 1996 Page 2

Recommended Adjustments

	Actual	NIOSH	Ideal 100%	Compromises 50%	25%
H	24.07	**10.0"**	-14.07	-7.03	-3.52
A	0.00	**0°**	0.00	0.00	0.00
V	41.51	**30.0"**	-11.51	-5.76	-2.88
D	15.04	**10.0"**	-5.04	-2.52	-1.26
C	poor	**good**	good	fair	poor
F	4.30	**0.2**	-4.10	-2.05	-1.03

H	22.97	**10.0**	-12.97	-6.49	-3.24
A	0.00	**0**	0.00	0.00	0.00
V	42.33	**30.0**	-12.33	-6.17	-3.08
D	10.66	**10.0**	-0.66	-0.33	-0.17
C	poor	**good**	good	fair	poor
F	4.30	**0.2**	-4.10	-2.05	-1.03

H	20.50	**10.0**	-10.50	-5.25	-2.63
A	0.00	**0**	0.00	0.00	0.00
V	43.69	**30.0**	-13.69	-6.84	-3.42
D	10.12	**10.0**	-0.12	-0.06	-0.03
C	poor	**good**	good	fair	poor
F	4.30	**0.2**	-4.10	-2.05	-1.03

Adjustment Results

	Posture Analyzed	Ideal 100%	Compromises 50%	25%
Lifting Indexes #1	4.46 (end)	0.98	2.19	2.54
Lifting Indexes #2	4.08 (end)	0.98	2.06	2.72
Lifting Indexes #3	3.65 (end)	0.98	1.91	2.47

Variable Key

H (distance of load from subject)
A (angular rotation while lifting)
V (distance of handhold from floor)
D (distance load moves vertically)
C (coupling; good/fair/poor load handhold)
F (frequency of lifts/minute)
L (load weight)

TABLE 56.3B

Static Dynamic Task: Characterized by both static and nonrepetitive low-back movements which show no obvious repeated pattern.

Repetitive/Excursion Task: Characterized by cyclical movement pattern where no posture is held continuously and the changes in trunk orientation are predictable.

Excursion: A continuous movement in a single direction with defined position end points (its length in degrees defines a range of motion for one or more axial movements).

Oscillation Angle: The angle which describes the symmetry of the excursion with respect to the subjects's neutral posture.

Repetitive/Excursion Parameters: Number of excursions, movement range (range of motion), velocity, oscillation angle, successive excursions (repetitions).

VideoWorks Report: Comparative

Bandy Boxing Group
July 17, 1996

Company/Task Data

Job Title:	Stacker
Job Code:	929.687-030
PDSR:	Heavy Work

Comparison Statement

This report is based on a single task analysis session from July 17, 1996, when lifting tasks for the position of Stacker at Bandy Boxing Group were performed by Bobby. Task #1 of the 3 tasks analyzed at that time has had parameters modified in order to make it more nearly comply with NIOSH guidelines. The results are presented below.

Selected Task with Modifications

	Actual	NIOSH	What If?
L	50.00	**50.00#**	40.00
H	10.00	**10.0"**	10.00
V	56.56	**30.0"**	56.56
D	15.04	**10.0"**	20.05
A	10.00	**0°**	10.00
F	4.30	**0.2**	3.30
C	poor	**good**	poor

	Actual	NIOSH	What If?
L	50.00	**50.00**	40.00
H	24.07	**10.0**	19.07
V	41.51	**30.0**	36.51
D	15.04	**10.0**	20.05
A	10.00	**0**	0.00
F	4.30	**0.2**	3.30
C	poor	**good**	poor

Modification Results

	Actual	NIOSH	What If?
RWL	11.20	**51.0**	16.02
LI	4.46 (end)	**0.98**	2.50 (end)
FILI	3.05	**0.98**	1.92

Variable Key

H	(distance of load from subject)	**D**	(distance load moves vertically)
A	(angular rotation while lifting)	**C**	(coupling; good/fair/poor load handhold)
V	(distance of handhold from floor)	**F**	(frequency of lifts/minute)
		L	(load weight)

TABLE 56.3C

Graphical Regions: Class 1: denotes normal velocities, oscillation angles, and ranges of motion for which no ergonomic/safety action is required (within 31% of mean values for each direction for each axis). Class 2: denotes velocities, oscillation angles, and ranges of motion for which postures need consideration in the near future (between 31% and 62% of mean values for each direction for each axis). Class 3: denotes velocities, oscillation angles, and ranges of motion for which postures need consideration immediately and job-redesign may be required (beyond 62% of mean values for each direction for each axis).

B-Tracker Task Analysis Report

Bobby Carrera, 123-45-6789
July 17, 1996

Company Data

Company:	Bandy Boxing Group
Company #:	39-0004-02
Address:	2912 Sousa Blvd. Muskegon, MI 49444

Job/Task Data

Job Title:	Stacker
Job Code:	929.687-030
Dept.:	Outgoing
PDSR:	Heavy

Session Summary

Duration:	0:04:09
Task Name:	Stacking
Task Type:	Repetitive/Excursion

Analysis Conclusions

This task involves repetitive trunk movements with biased flexion and lateral bending. Excursion analysis of the task reveals that the worker moved rapidly through large ranges with off-center orientations. These results suggest training, ergonomic intervention, engineering re-design or job-specific screening may be warranted.

Ergonomic/Safety Concerns

Average percentage of oscillation angles, velocities, and ranges of motion within each class for all axes.

Oscillation Angles:	Class 1:	44
	Class 2:	55
	Class 3:	1
Velocities	Class 1:	48
	Class 2:	51
	Class 3:	1
Ranges of Motion	Class 1:	46
	Class 2:	53
	Class 3:	1

Report Reviewed by: ____________________ **Date:** __________

Tracy Marker, Ergonomist

TABLE 56.3D

B-Tracker Task Analysis Report

Bobby Carrera, 123-45-6789
July 17, 1996 Page 2

Demographic Data

Worker Name:	Bobby Carrera
Worker ID:	123-45-6789
Height:	6'1"
Weight:	170#
Hand Dominance:	Right
Time on Job:	2 yrs

Oscillation Angles (by axis)

Flex/Ext	0	0	0	14	84	2
Lateral	1	67	30	1	1	0
Rotation	0	6	46	42	6	0

Percentage of oscillation angles in each class. Each axis is shown comprising two directions of movement, e.g. right and left lateral bending.

Velocities (by axis)

Flex/Ext	56	44	0
Lateral	25	75	0
Rotation	62	37	1

Percentage of velocities in each class.

Ranges of Motion (by axis)

Flex/Ext	35	64	1
Lateral	28	72	0
Rotation	75	25	0

Percentage of ranges of motion in each class.

TABLE 56.3E

Ergonomic Recommendations

Bandy Boxing Group
July 19, 1996

Re-Engineering

Eliminate the need to manually re-stack paper stock for printing press by re-engineering the previous operation to include mechanical stacking.

Ergonomic Intervention

Use a weight-dependent, adjustable pallet which maintains a constant height as stock is removed or added. Modify the orientation of the pallets with respect to each other to decrease awkward postures, e.g., butt pallets side by side to enable the worker to flip stock without lifting it completely off the pallet.

Education and Training

The trainer, a safety professional or a supervisor with knowledge of ergonomics, introduces the worker to the stacking job both verbally and by demonstration, and works side-by-side with worker for one to two days. The worker is monitored while performing job for compliance with job description, productivity, and safety/ergonomic guidelines (use B-Tracker or BackWorks Targeting/Avoidance software where appropriate). The worker is tested using a written scrambled job description. The worker is also tested practically by walking the trainer through the stacking job. The trainer calculates a learning curve and revisits training as needed.

Screening

BackWorks generates a protocol that can be used to test job candidates "post-offer and pre-placement." The protocol is based on the actual physical movements of the stacking task and complies with the Americans with Disabilities Act of 1990.

Baseline Testing

The B-Tracker is used to gather baseline or benchmark data on the current functional status of the low back. Baseline data directs managed care towards a known pre-injury functional status.

Report Reviewed by:____________________ **Date:**__________

Tracy Marker, Ergonomist

TABLE 56.4

Backability Baseline Report

Bobby Carrera, 123-45-6789
August 1, 1996

Level III

Demographic Data

Company: Bandy Boxing Group
Job Title: Stacker
Job Code: 929.687-030
Division: Packaging
Dept: Outgoing
Physical Demand Strength Rating: Heavy
Non-Work Activity Level: Heavy
Gender: Male
Age: 30
Birth Date: February 19, 1966
Height: 6 ft. 1 in.
Weight: 170 lbs.
Previous History of LBP: No
Date of last LBP Incident: NA

B-Tracker Analysis: 27

0 ———————————————— 100
▲ 27

The lower the B-Tracker Analysis Performance Score, the greater the certainty that this individual is a member of the healthy back population.

Conclusions

On August 1, 1996, Bobby Carrera participated in a functional low-back analysis. Results indicate that

- a low-back exercise program is recommended
- correct lifting techniques should be used; heavy lifting and prolonged and/or awkward trunk postures should be avoided
- performance of job tasks and work environment may be evaluated for potential improvement
- baseline testing should be repeated in 12 months

Test Site:
Administered by:
Title:

Account Charged: INT001
System Ref#: Oct3096 121428

TABLE 56.5

Low-Back Clinical Report

Bobby Carrera, 123-45-6789
August 29, 1996

Demographic Data

DOT Job Title: Stacker
DOT Job Code: 929.687-030

Conclusions

On August 29, 1996, Bobby Carrera participated in a functional low-back analysis. The test results confirm membership in a low-back problem population. Bobby's Waddell score was consistent with an organic etiology.

B-Tracker Analysis: 87

0 ———————————————— 100
▲ 87

The higher the B-Tracker Analysis Performance Score, the greater the certainty that the patient is a member of the low-back problem population. Scores between 0 and 5 indicate that the individual is a member of the healthy low-back population.

VAS Test Results: 65

0 ———————————————— 100
▲ 65

Visual Analog Scale (VAS) prior to the B-Tracker test: "Mark on the scale from 0 to 100 your level of pain discomfort with 0 being None and 100 being Unbearable."

Waddell's Signs: 0

Tenderness Simulation Distraction Regional Overreaction

0 - 2 of 5 Waddell's signs indicate a negative result;
3 - 5 of 5 Waddell's signs indicate a positive result.

Test Site Stewart PT--Burlington, NC #3954-1AE4
Signature ____________________
Title ____________________

TABLE 56.6A

Low-Back Clinical Report

Bobby Carrera, 123-45-6789
September 18, 1996

Demographic Data

DOT Job Title: Stacker
DOT Job Code: 929.687-030

Conclusions

On September 18, 1996, Bobby Carrera participated in a functional low-back analysis. The test results confirm membership in a low-back problem population. Bobby's Waddell score was consistent with an organic etiology.

B-Tracker Analysis: 35

0 — 100 (marker at 35)

The higher the B-Tracker Analysis Performance Score, the greater the certainty that the patient is a member of the low-back problem population. Scores between 0 and 5 indicate that the individual is a member of the healthy low-back population.

VAS Test Results: 11

0 — 100 (marker at 11)

Visual Analog Scale (VAS) prior to the B-Tracker test: "Mark on the scale from 0 to 100 your level of pain discomfort with 0 being None and 100 being Unbearable."

Waddell's Signs: 0

Tenderness Simulation Distraction Regional Overreaction

0 - 2 of 5 Waddell's signs indicate a negative result;
3 - 5 of 5 Waddell's signs indicate a positive result.

Test Site Stewart PT--Burlington, NC #3954-1AE4
Signature ______________________
Title ______________________

TABLE 56.6B

Backability Baseline Report

Bobby Carrera, 123-45-6789
September 20, 1996

Level III

Demographic Data

Company: Bandy Boxing Group
Job Title: Stacker
Job Code: 929.687-030
Division: Packaging
Dept: Outgoing
Physical Demand Strength Rating: Heavy
Non-Work Activity Level: Heavy
Gender: Male
Age: 30
Birth Date: February 19, 1966
Height: 6 ft. 1 in.
Weight: 170 lbs.
Previous History of LBP: No
Date of last LBP Incident: NA

B-Tracker Analysis: 31

0 ———————————— 100
▲ 31

The lower the B-Tracker Analysis Performance Score, the greater the certainty that this individual is a member of the healthy back population.

Conclusions

On September 20, 1996, Bobby Carrera participated in a functional low-back analysis. Results indicate that

- a low-back exercise program is recommended
- correct lifting techniques should be used; heavy lifting and prolonged and/or awkward trunk postures should be avoided
- performance of job tasks and work environment may be evaluated for potential improvement
- baseline testing should be repeated in 12 months

Test Site:
Administered by:
Title:

Account Charged: INT001
System Ref#: Oct3096_121428

TABLE 56.7A

BackAbility Baseline Summary

Bandy Boxing Group
August 1, 1996

Baseline Testing Summary

On August 1, 1996, Bandy Boxing Group had 18 individuals with 3 different job titles participate in the BackAbility Baseline Program.

Baseline Summary Data

	Job Code	#	Level IV	Level III	Level II	Level I
Stacker	929.687-030	11	4	6	0	1
Gluer	795.687-014	4	1	3	0	0
Air-Drier-Machine Op	534.682-010	3	0	3	0	0
		18	**5**	**12**	**0**	**1**

Baseline Level Descriptions

Level I

- a special medical intervention is necessary to determine if a low-back problem exists

Level II

- a low-back exercise program is strongly recommended
- correct lifting techniques should be used; heavy lifting and prolonged and/or awkward trunk postures should be avoided
- performance of job tasks and work environment may be evaluated for potential improvement
- baseline testing should be repeated in 6 months

Level III

- a low-back exercise program is recommended
- correct lifting techniques should be used; heavy lifting and prolonged and/or awkward trunk postures should be avoided
- performance of job tasks and work environment may be evaluated for potential improvement
- baseline testing should be repeated in 12 months

Level II

- current low-back exercise should be maintained
- correct lifting techniques should be continued; avoidance of heavy lifting and prolonged and/or awkward trunk postures should persist

Account Charged: INT001
System Ref#: Oct3096_121428

TABLE 56.7B

BackAbility Baseline Summary

Bandy Boxing Group
August 1, 1996 page 2

Individual Baseline Scores

Level I

Stacker (1)

Philbert, Yoakum	98				

Level II

NA

Level III

Stacker (6)

Carrera, Bobby	69	Barley, Filbert	57	Oliver, Clem	51
Philbert, Yoakum	49	Melbourne, Dan	49	Carrera, Bobby	22

Gluer (3)

Philbert, Yoakum	67	Laxaline, Martha	15	Bertram, Randy	7

Air-Drier-Machine (3)

Ballinger, Raymond	57	Mayonnaire, Lakisha	54	Blessingame, Chad	6

Level IV

Stacker (4)

Phillips, Tanya	5	Crosby, Willie	2	Boyles, Darlene	2
Fargo, Janelle	0				

Gluer (1)

Reynolds, Billy	0				

TABLE 56.7C

BackAbility Baseline Protocol

An Explanation of the Protocol and the Report

The B-Tracker BackAbility Baseline Protocol is used for testing subjects who have no low-back problem. Specific directions are provided about placement of the B-Tracker and use of a stabilization platform. The formalized protocol requires that the clinician demonstrate each baseline movement immediately preceding its performance. The following three movements compose the protocol:

Flexibility
: two repetitions of movements for each axis (flexion/extension, lateral bending, rotation). Subjects are instructed to move as far as possible.

Free Dynamic Movement
: seven repetitions of movements for each axis (flexion/extension, lateral bending, rotation). Subjects are instructed to move as far as feels comfortable at their preferred pace.

Circumduction
: four circumductions to the right, then four to the left. Subjects are instructed to move as far as is comfortable at their preferred pace.

The Baseline Score

The purpose of the baseline score is to confirm—by degrees of certainty—the subject's low-back health and to establish for future reference a quantifiable benchmark. There are four levels to which individuals are assigned on the basis of their B-Tracker baseline score:

Level I	95-100
Level II	70-94
Level III	6-69
Level IV	1-5

Use of the Baseline Report

In the event of injury, the subject's baseline data will serve as the goal to which recovery should be directed in order to achieve pre-injury functional status and return-to-work readiness. Categorization by level also enables recommendations which can assist individuals improve or maintain their low-back status. Of course, the recommendations should be tailored towards the functional demands placed on an individual's low back due to specific job requirements.

Level I
- a special medical intervention is necessary to determine if a low-back problem exists

Level II
- a low-back exercise program is strongly recommended
- correct lifting techniques should be used; heavy lifting and prolonged and/or awkward trunk postures should be avoided
- performance of job tasks and work environment may be evaluated for potential improvement
- baseline testing should be repeated in 6 months

Level III
- a low-back exercise program is recommended
- correct lifting techniques should be used; heavy lifting and prolonged and/or awkward trunk postures should be avoided
- performance of job tasks and work environment may be evaluated for potential improvement
- baseline testing should be repeated in 12 months

Level IV
- current low-back exercise should be maintained
- correct lifting techniques should be continued; avoidance of heavy lifting and prolonged and/or awkward trunk postures should persist

TABLE 56.7D

Low-Back Clinical Protocol

An Explanation of the Protocol and the Report

The B-Tracker Low-Back Clinical protocol is used when the patient asserts he/she has a low-back problem. The protocol comprises the following three parts:

- Waddell Signs Test
- VAS (Visual Analog Scale) Test
- B-Tracker movement exercises.

The Waddell Signs Test is administered in accordance with procedures documented by Waddell, et al. in *Spine* (5:2, 1980), "Nonorganic Physical Signs in Low-Back Pain."

The VAS requires patients to rate their pain just prior to beginning the B-Tracker movement exercises. As patients view a scale whose only markings are '0' and 'None' at one end and '100' and 'Unbearable' at the other end, they are asked to respond to the following question: "Mark on the scale from 0 to 100 your level of pain discomfort, with 0 being None and 100 being Unbearable."

The B-Tracker movement exercises are performed in strict order with specific directions. Further assuring consistent testing are specific directions about placement of the B-Tracker and use of a stabilization platform. The protocol also requires that the clinician demonstrate each exercise immediately preceding its performance. The following three exercises compose the protocol's movement exercises:

Flexibility

two repetitions of movements for each axis (flexion/extension, lateral bending, rotation). Patients are instructed to move as far as possible.

Free Dynamic Movement

seven repetitions of movements for each axis (flexion/extension, lateral bending, rotation. Patients are instructed to move as far as feels comfortable at their preferred pace.

Circumduction

four circumductions to the right, then four to the left. Patients are instructed to move as far as is comfortable at their preferred pace.

The Clinical Report Score

The purpose of the resulting B-Tracker score is to confirm—by degrees of certainty—the patient's assertion that he/she is experiencing a low-back problem without a specific diagnosis. To this end, there is a range of scores (6 - 100) which confirms a patient's low-back problem. A score falling in the range 0 - 5 suggests the patient's claim of a low-back problem should be further tested.

The Clinical Report score provides the means to make decisions regarding treatment paths. Consider the following two scenarios: confirmation can be used to initiate a treatment path involving exercise and other non-surgical procedures; or, confirmation can be used to initiate surgical intervention. In the first case, the treatment is conservative and in the second, aggressive. When considering an aggressive treatment path, greater certainty is desired.

An example may serve to illustrate: Consider a patient who obtains a score of 20. The score is sufficient to initiate a treatment path involving rehabilitation/exercise. However, the score may not be sufficient as the basis for consideration of surgical intervention due to the consequences associated with treatment if, in fact, the individual was not a member of the low-back problem population.

Use of the Low-Back Clinical Report

The B-Tracker Low-Back Clinical Report, when compared to a patient's earlier Low-Back Baseline Report, objectively documents the precise extent of low-back functional loss. The B-Tracker Low-Back Clinical Report further provides means to ensure a patient is directed to an appropriate patient management program. Subsequent assessments with the Low-Back Clinical protocol objectively indicate the patient's level of functional improvement, when a patient has returned to his/her pre-injury functional status, and the patient's return-to-work readiness.

TABLE 56.8

References

Adams, M.A. and Hutton, W.C. 1985. Gradual disc prolapse. *Spine.* 10:524-531.

Andersson, G.B.J., Chaffin, D.B., Pope, M.H. 1991. Occupational biomechanics of the lumbar spine, in *Occupational Low Back Pain,* Mosby-Year Book, Inc., St. Louis, MO.

Bigos, S.J., Battié, M.C., Spengler, D.M. et al. 1991. A prospective study of work perceptions and psychosocial factors affecting the report of back injury. *Spine.* 16:1-6.

Dickenson, C.E. et al. 1992. Questionnaire development: an examination of the Nordic musculoskeletal questionnaire. *Applied Ergonomics.* 23:197-201.

Dictionary of Occupational Titles. 1977. U.S. Department of Labor Employment and Training Administration, Washington, D.C.

Chaffin, D.B. 1974. Human strength capability and low-back pain. *J. Occup Med.* 16(4).

Garg A., Moore, S.J. 1992. Epidemiology of low-back pain in industry, in *Occupational Medicine, State of the Art Reviews: Ergonomics: Low-back Pain, Carpal Tunnel Syndrome, and Upper Extremity Disorders in the Workplace,* Eds. J. Moore and A. Garg, p. 599-608. Hanley & Belfus, Inc., Philadelphia, PA.

Genaidy, A.M., Karwowski W. 1993. The effects of neutral posture deviations on perceived joint discomfort ratings in sitting and standing postures. *Ergonomics.* 36:785-792.

Hadler, N.M. 1994. Backache and work incapacity in Japan. *J. Occup. Med.* 36(10):1110-1114.

Heliövaara, M., Knekt, P., Aromaa, A. 1987. Incidence and risk factors of herniated lumbar intervertebral disc or sciatica leading to hospitalization. *J. Chronic Dis.* 40:251-258.

Hulsof, C., Veldhuizen van Zantan, B. 1987. Whole-body vibration and low-back pain: A review of epidemiologic studies. *International Archives of Occupational and Environmental Health.* 59:205-220.

ISO: Evaluation of human response to whole body vibration. *Intl. Org. for Standardization* Ref. No: 2631 (E), 1978.

Kelsey, J.L., Hochberg, M.C. 1988. Epidemiology of chronic musculoskeletal disorders. *Ann. Rev. Public Health.* 9:379-401.

Kelsey, J.L., Githens, P.B., White, A.A. III et al. 1984. An epidemiologic study of lifting and twisting on the job and risk factors for acute prolapsed lumbar intervertebral disc. *J. Orthop. Research* 2:61.

Kelsey, J.L., Hardy, R.J. 1975. Driving of motor vehicles as a risk factor for acute herniated lumbar intervertebral disc. *Am. J. Epidemiology.* 102(1):63-73.

Liles, D.H., Deivanayagam, S., Ayoub, M.M., Mahajan, P. 1984. A job severity index for the evaluation and control of lifting injury. *Human Factors.* 26:683-694.

Magnusson, M.L., Pope, M.H., Wilder, D.G., Areskoug, B. 1996. Are occupational drivers at an increased risk for developing musculo-skeletal disorders? *Spine.* 21(6):710-717.

Nachemson, A.L. 1991. Spinal disorders: overall impact in society and the need for orthopaedic resources. *Acta Orthop. Scand. Suppl.* 241:17-22.

Pope, M.H., Frymoyer, J.W., Andersson, G.B.J., Chaffin, D. 1991. *Occupational Low Back Pain: Assessment, Treatment and Prevention,* 2nd ed. Mosby Press, St. Louis, MO.

Pope, M.H., Andersson, G.B.J., Broman, H., Svensson, M., Zetterberg, C. 1986. Electromyographic studies of the lumbar trunk musculature during the development of axial torques. *J. Ortho. Res.* 4(3):288-297.

Pope, M.H., Svensson, M., Broman, H., Andersson, G.B.J. 1986. Mounting of the transducer in measurements of sequential motion of the spine. *J. Biomech.* 19(8):675-677.

Pope, M.H., Wilder, D.G., Jorneus, L., Broman, H., Svensson, M., Andersson, G.B.J. 1987. The response of the seated human to sinusoidal vibration and impact. *J. Biomech. Eng.* 109:279-284.

Pope, M.H., Broman, H., Hansson, T. 1989. The dynamic response of a subject seated on various cushions. *Ergonomics.* 32(10):1155-1166.

Punnett, L., Fine, L.T., Keyserling, W.M. et al. 1991. Back disorders and non-neutral postures of automobile assembly workers. *Scandinavian J. of Work, Environ. and Health.* 17:337-346.

Riihimäki, H. 1995. Back and limb disorders, in *Epidemiology of Work Related Diseases,* Ed. J. C. McDonald, Chapter 10. BMJ Publishing Group, London, England.

Seroussi, R., Wilder, D., Pope, M.H. 1989. Trunk muscle electromyography and whole body vibration. *J. Biomech.* 22(3):219-229.

Waters, T.R., Putz-Anderson, V., Garg, A. et al. 1993. Revised NIOSH lifting equation for the design and evaluation of manual lifting tasks. *Ergonomics.* 36:749-776.

Wilder, D.G., Aleksiev, A., Magnusson, M., Pope, M.H., Spratt, K., Goel, V.K. 1996. Muscular response to sudden load: a tool to evaluate fatigue and rehabilitation. *Spine,* in press.

For Further Information

Contact Interlogics, 328 Elizabeth Brady Rd., Hillsborough, NC 27278.

57

Revised NIOSH Lifting Equation

Thomas R. Waters
National Institute for Occupational Safety and Health

Vern Putz-Anderson
National Institute for Occupational Safety and Health

57.1 Introduction

This chapter provides information about a revised equation for assessing the physical demands of certain two-handed manual lifting tasks that was developed by the National Institute for Occupational Safety and Health (NIOSH) and described earlier in an article by Waters et al.[1] We discuss what factors need to be measured, how they should be measured, what procedures should be used, and how the results can be used to ergonomically design new jobs or make decisions about redesigning existing jobs that may be hazardous. We define all pertinent terms and present the mathematical formulas and procedures needed to properly apply the NIOSH lifting equation. Several example problems are also provided to demonstrate how the equations should be used. An expanded version of this chapter is contained in a NIOSH document.[2]

Historically, NIOSH has recognized the problem of work-related back injuries and published the *Work Practices Guide for Manual Lifting* (WPG) in 1981.[3] The WPG contained a summary of the lifting-related literature up to 1981; analytical procedures and a lifting equation for calculating a recommended weight for specified two-handed, symmetrical lifting tasks; and an approach for controlling the hazards of low back injury from manual lifting. The approach to hazard control was coupled to the *action limit* (AL), a term that denoted the recommended weight derived from the lifting equation.

0-8493-2641-9/99/$0.00+$.50

In 1985, NIOSH convened an *ad hoc* committee of experts who reviewed the current literature on lifting, including the NIOSH WPG.* The literature review was summarized in a document containing updated information on the physiological, biomechanical, psychophysical, and epidemiological aspects of manual lifting.[4] Based on the results of the literature review, the *ad hoc* committee recommended criteria for defining the lifting capacity of healthy workers. The committee used the criteria to formulate the revised lifting equation.** Subsequently, NIOSH staff developed the documentation for the equation and played a prominent role in recommending methods for interpreting the results of the lifting equation. The revised lifting equation reflects new findings and provides methods for evaluating asymmetrical lifting tasks, and lifts of objects with less than optimal couplings between the object and the worker's hands. The revised lifting equation also provides guidelines for a more diverse range of lifting tasks than the earlier equation.[3]

The rationale and criterion for the development of the revised NIOSH lifting equation are provided in a journal article by Waters et al.[1] We suggest that those users who wish to achieve a better understanding of the data and decisions that were made in formulating the revised equation consult that article. It provides an explanation of the selection of the biomechanical, physiological, and psychophysical criterion as well as a description of the derivation of the individual components of the revised lifting equation. For those individuals, however, who are primarily concerned with the use and application of the revised lifting equation, this chapter provides a more complete description of the method and limitations.

Although the revised lifting equation has not been fully validated, the recommended weight limits derived from the revised equation are consistent with, or lower than, those generally reported in the literature. Moreover, the proper application of the revised equation is more likely to protect healthy workers for a wider variety of lifting tasks than methods that rely only on a single task factor or single criterion.

Finally, it should be stressed that the NIOSH lifting equation is only one tool in a comprehensive effort to prevent work-related low back pain and disability. Some examples of other approaches are described elsewhere.[5] Moreover, lifting is only one of the causes of work-related low back pain and disability. Other causes that have been hypothesized or established as risk factors include whole body vibration, static postures, prolonged sitting, and direct trauma to the back. Psychosocial factors, appropriate medical treatment, and job demands may also be particularly important in influencing the transition of acute low back pain to chronic disabling pain.

57.2 Definition of Terms

This section provides the basic technical information needed to properly use the revised lifting equation to evaluate a variety of two-handed manual lifting tasks. Definitions and data requirements for the revised lifting equation are also provided.

Recommended Weight Limit (RWL)

The *recommended weight limit* (RWL) is the principal product of the revised NIOSH lifting equation. The RWL is defined for a specific set of task conditions as the weight of the load that nearly all healthy workers could perform over a substantial period of time (e.g., up to 8 h) without an increased risk of developing lifting-related low back pain (LBP). By "healthy workers" we mean workers who are free of adverse health conditions that would increase their risk of musculoskeletal injury.

*The *ad hoc* 1991 NIOSH Lifting Committee members included: M.M. Ayoub, Donald B. Chaffin, Colin G. Drury, Arun Garg, and Suzanne Rodgers. NIOSH representatives included Vern Putz-Anderson and Thomas R. Waters.

**For this document, the revised 1991 NIOSH lifting equation will be identified simply as "the revised lifting equation."[1,2] The abbreviation WPG will continue to be used as the reference to the earlier NIOSH lifting equation, which was documented in a publication entitled *Work Practices Guide for Manual Lifting*.[3]

The concept behind the revised NIOSH lifting equation is to start with a recommended weight that is considered safe for an "ideal" lift (i.e., load constant equal to 51 lb) and then reduce the weight as the task becomes more stressful (i.e., as the task-related factors become less favorable). The precise formulation of the revised lifting equation for calculating the RWL is based on a multiplicative model that provides a weighting (multiplier) for each of six task variables:

1. Horizontal distance of the load from the worker (*H*)
2. Vertical height of the lift (*V*)
3. Vertical displacement during the lift (*D*)
4. Angle of asymmetry (*A*)
5. Frequency (*F*) and duration of lifting
6. Quality of the hand-to-object coupling (*C*)

The weightings are expressed as coefficients that serve to decrease the load constant, which represents the maximum recommended load weight to be lifted under ideal conditions. For example, as the horizontal distance between the load and the worker increases from 10 in, the recommended weight limit for that task would be reduced from the ideal starting weight.

The *recommended weight limit* (RWL) is defined as

$$RWL = LC \times HM \times VM \times DM \times AM \times FM \times CM$$

where

				METRIC	*U.S. CUSTOMARY*
LC	=	Load Constant	=	23 kg	51 lb
HM	=	Horizontal Multiplier	=	(25/H)	(10/H)
VM	=	Vertical Multiplier	=	1–(.003 |V-75|)	1–(.0075 |V-30|)
DM	=	Distance Multiplier	=	.82 + (4.5/D)	.82 + (1.8/D)
AM	=	Asymmetric Multiplier	=	1–(.0032A)	1–(.0032A)
FM	=	Frequency Multiplier	=	From Table 57.5	From Table 57.5
CM	=	Coupling Multiplier	=	From Table 57.7	From Table 57.7

The term *task variables* refers to the measurable task-related measurements that are used as input data for the formula (i.e., *H*, *V*, *D*, *A*, *F*, and *C*); whereas, the term *multipliers* refers to the reduction coefficients in the equation (i.e., HM, VM, DM, AM, FM, and CM).

Measurement Requirements

The following list briefly describes the measurements required to use the revised NIOSH lifting equation. Details for each of the variables are presented later in this chapter.

H = horizontal location of hands from midpoint between the inner ankle bones. Measure at the origin and the destination of the lift (cm or in).

V = vertical location of the hands from the floor. Measure at the origin and destination of the lift (cm or in).

D = vertical travel distance between the origin and the destination of the lift (cm or in).

A = angle of asymmetry or angular displacement of the load from the worker's sagittal plane. Measure at the origin and destination of the lift (degrees).

F = average frequency rate of lifting measured in lifts/min. Duration is defined to be ≤ 1 h; ≤ 2 h; or ≤ 8 h assuming appropriate recovery allowances (see Table 57.5).

C = quality of hand-to-object coupling (quality of interface between the worker and the load being lifted). The quality of the coupling is categorized as good, fair, or poor, depending upon the type and location of the coupling, the physical characteristics of load, and the vertical height of the lift.

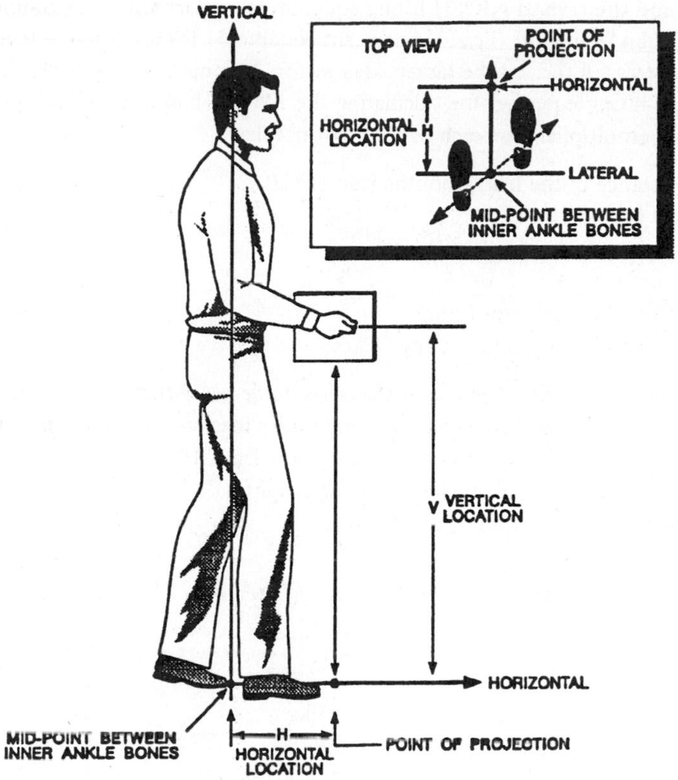

FIGURE 57.1 Graphic representation of hand location.

Lifting Index (LI)

The *lifting index* (LI) is a term that provides a relative estimate of the level of physical stress associated with a particular manual lifting task. The estimate of the level of physical stress is defined by the relationship between the weight of the load lifted and the recommended weight limit. The LI is defined by the equation

$$LI = \frac{Load\ Weight}{Recommended\ Weight\ Limit} = \frac{L}{RWL}$$

where Load Weight (L) = weight of the object lifted (lbs or kg).

Miscellaneous Terms

Lifting task The act of manually grasping an object of definable size and mass with two hands, and vertically moving the object without mechanical assistance.

Load weight (L) Weight of the object to be lifted, in pounds or kilograms, including the container.

Horizontal location (*H*) Distance of the hands away from the midpoint between the ankles, in inches or centimeters (measure at the origin and destination of lift). See Figure 57.1.

Vertical location (*V*) Distance of the hands above the floor, in inches or centimeters (measure at the origin and destination of lift). See Figure 57.1.

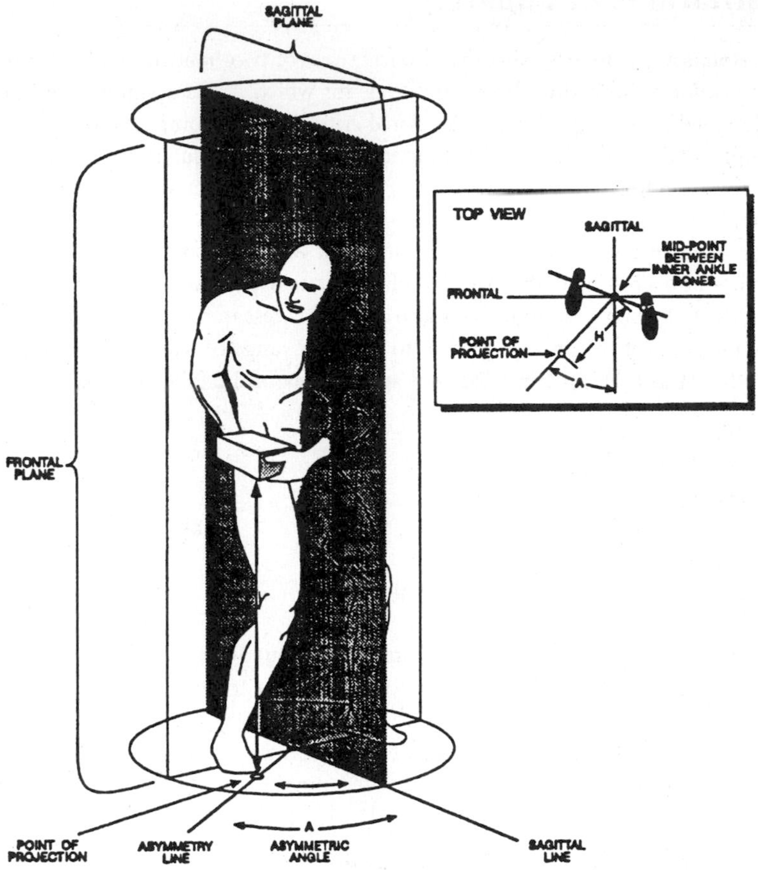

FIGURE 57.2 Graphic representation of angle of asymmetry (A).

Vertical travel distance (*D*) Absolute value of the difference between the vertical heights at the destination and origin of the lift, in inches or centimeters.

Angle of asymmetry (*A*) Angular measure of how far the *object* is displaced from the front (midsagittal plane) of the worker's body at the beginning or end of the lift, in degrees (measure at the origin and destination of lift). See Figure 57.2. The asymmetry angle is defined by the location of the load relative to the worker's midsagittal plane, as defined by the neutral body posture, rather than the position of the feet or the extent of body twist.

Neutral body position Position of the body when the hands are directly in front of the body and there is minimal twisting at the legs, torso, or shoulders.

Frequency of lifting (*F*) Average number of lifts per minute over a 15-minute period.

Duration of lifting Three-tiered classification of lifting duration specified by the distribution of work-time and recovery-time (work pattern). Duration is classified as either short (1 h), moderate (1 to 2 h), or long (2 to 8 h), depending on the work pattern.

Coupling classification Classification of the quality of the hand-to-object coupling (e.g., handle, cut-out, or grip). Coupling quality is classified as good, fair, or poor.

Significant control A condition requiring "precision placement" of the load at the destination of the lift. This is usually the case when (1) the worker has to regrasp the load near the destination of the lift, or (2) the worker has to momentarily hold the object at the destination, or (3) the worker has to carefully position or guide the load at the destination.

57.3 Limitations of Equation

The lifting equation is a tool for assessing the physical stress of two-handed manual lifting tasks. As with any tool, its application is limited to those conditions for which it was designed. Specifically, the lifting equation was designed to meet specific lifting-related criteria that encompass biomechanical, physiological, and psychophysical assumptions and data used to develop the equation. To the extent that a given lifting task accurately reflects these underlying conditions and criteria, this lifting equation may be appropriately applied.

The following list identifies a set of work conditions in which the application of the lifting equation could either under- or overestimate the extent of physical stress associated with a particular work-related activity. Each of the following task limitations also highlights research topics in need of further research to extend the application of the lifting equation to a greater range of real-world lifting tasks.

The revised NIOSH Lifting Equation does not apply if any of the following occur:

Lifting/lowering with one hand
Lifting/lowering for over 8 hours
Lifting/lowering while seated or kneeling
Lifting/lowering in a restricted work space
Lifting/lowering unstable objects
Lifting/lowering while carrying, pushing, or pulling
Lifting/lowering with wheelbarrows or shovels
Lifting/lowering with "high speed" motion (faster than about 30 in./second)
Lifting/lowering with unreasonable foot/floor coupling (<0.4 coefficient of friction between the sole and the floor)
Lifting/lowering in an unfavorable environment (temperature significantly outside 66 to 79°F [19 to 26°C] range; relative humidity outside 35 to 50% range)

57.4 Obtaining and Using the Data

Horizontal Component

Definition and Measurement

Horizontal location (*H*) is measured from the midpoint of the line joining the inner ankle bones to a point projected on the floor directly below the midpoint of the hand grasps (i.e., load center), as defined by the large middle knuckle of the hand (Figure 57.1). Typically, the worker's feet are not aligned with the midsagittal plane, as shown in Figure 57.1, but may be rotated inward or outward. If this is the case, then the midsagittal plane is defined by the worker's neutral body posture as defined above. If significant control is required at the destination (i.e., precision placement), then *H* should be measured at both the origin and destination of the lift.

Horizontal distance (*H*) should be measured. In those situations where the *H* value cannot be measured, then *H* may be approximated from the following equations:

Metric (All distances in cm)	U.S. Customary (All distances in inches)
$H = 20 + W/2$	$H = 8 + W/2$
for $V \geq 25$ cm	for $V \geq 10$ in.
$H = 25 + W/2$	$H = 10 + W/2$
for $V < 25$ cm	for $V < 10$ in.

where *W* is the width of the container in the sagittal plane and *V* is the vertical location of the hands from the floor.

TABLE 57.1 Horizontal Multiplier

H		H	
in	Hm	cm	Hm
≤10	1.00	≤25	1.00
11	.91	28	.89
12	.83	30	.83
13	.77	32	.78
14	.71	34	.74
15	.67	36	.69
16	.63	38	.66
17	.59	40	.63
18	.56	42	.60
19	.53	44	.57
20	.50	46	.54
21	.48	48	.52
22	.46	50	.50
23	.44	52	.48
24	.42	54	.46
25	.40	56	.45
>25	.00	58	.43
		60	.42
		63	.40
		>63	.00

Horizontal Restrictions

If the horizontal distance is less than 10 in. (25 cm), then *H* is set to 10 in. (25 cm). Although objects can be carried or held closer than 10 in. from the ankles, most objects that are closer than this cannot be lifted without encountering interference from the abdomen or hyperextending the shoulders. Although 25 in. (63 cm) was chosen as the maximum value for *H*, it is probably too great a distance for shorter workers, particularly when lifting asymmetrically. Furthermore, objects at a distance of more than 25 in. from the ankles normally cannot be lifted vertically without some loss of balance.

Horizontal Multiplier

The horizontal multiplier (HM) is $10/H$ for *H* measured in inches and $25/H$ for *H* measured in centimeters. If *H* is less than or equal to 10 in. (25 cm), the multiplier is 1.0. HM decreases with an increase in *H* value. The multiplier for *H* is reduced to 0.4 when *H* is 25 in. (63 cm). If *H* is greater than 25 in., then HM = 0. The HM value can be computed directly or determined from Table 57.1.

Vertical Component

Definition and Measurement

Vertical location (*V*) is defined as the vertical height of the hands above the floor. *V* is measured vertically from the floor to the midpoint between the hand grasps, as defined by the large middle knuckle. The coordinate system is illustrated in Figure 57.1.

Vertical Restrictions

The vertical location (*V*) is limited by the floor surface and the upper limit of vertical reach for lifting (i.e., 70 in. or 175 cm). The vertical location should be measured at the origin and the destination of the lift.

Vertical Multiplier

To determine the vertical multiplier (VM), the absolute value or deviation of *V* from an optimum height of 30 in. (75 cm) is calculated. A height of 30 in. above floor level is considered "knuckle height" for a

TABLE 57.2 Vertical Multiplier

V		V	
in	VM	cm	VM
0	.78	0	.78
5	.81	10	.81
10	.85	20	.84
15	.89	30	.87
20	.93	40	.90
25	.96	50	.93
30	1.00	60	.96
35	.96	70	.99
40	.93	80	.99
45	.89	90	.96
50	.85	100	.93
55	.81	110	.90
60	.78	120	.87
65	.74	130	.84
70	.70	140	.81
>70	.00	150	.78
		160	.75
		170	.75
		175	.70
		>175	.00

worker of average height (66 in. or 165 cm). The vertical multiplier (VM) is (1-(.0075 | V-30 |)) for *V* measured in inches, and VM is (1-(.003 | V-75 |)), for *V* measured in centimeters.

When *V* is at 30 in. (75 cm), the vertical multiplier (VM) is 1.0. The value of VM decreases linearly with an increase or decrease in height from this position. At floor level, VM = 0.78, and at 70 in. (175 cm) height, VM = 0.7. If *V* is greater than 70 in., then VM = 0. The VM value can be computed directly or determined from Table 57.2.

Distance Component

Definition and Measurement

The distance variable (*D*) is defined as the vertical travel distance of the hands between the origin and destination of the lift. For lifting, *D* can be computed by subtracting the vertical location (*V*) at the origin of the lift from the corresponding *V* at the destination of the lift (i.e., *D* is equal to *V* at the destination minus *V* at the origin). For a lowering task, *D* is equal to *V* at the origin minus *V* at the destination.

Distance Restrictions

The distance variable (*D*) is assumed to be at least 10 in. (25 cm), and no greater than 70 in. (175 cm). If the vertical travel distance is less than 10 in. (25 cm), then *D* should be set to the minimum distance of 10 in. (25 cm).

Distance Multiplier

The distance multiplier (DM) is (.82 + (1.8/D)) for *D* measured in inches and (.82 + (4.5/D)) for *D* measured in centimeters. For *D* less than 10 in. (25 cm) *D* is assumed to be 10 in. (25 cm), and DM is 1.0. The distance multiplier, therefore, decreases gradually with an increase in travel distance. The DM = 1.0 when *D* is set at 10 in., (25 cm); DM = 0.85 when *D* is 70 in. (175 cm). Thus, DM ranges from 1.0 to 0.85 as the *D* varies from 0 in. (0 cm) to 70 in. (175 cm). The DM value can be computed directly or determined from Table 57.3.

TABLE 57.3 Distance Multiplier

D		D	
≤10	DM	cm	DM
≤10	1.00	≤25	1.00
15	.94	40	.93
20	.91	55	.90
25	.89	70	.88
30	.88	85	.87
35	.87	100	.87
40	.87	115	.86
45	.86	130	.86
50	.86	145	.85
55	.85	160	.85
60	.85	175	.85
70	.85	>175	.00
>70	.00		

Asymmetry Component

Definition and Measurement

Asymmetry refers to a lift that begins or ends outside the midsagittal plane (See Figure 57.2). In general, asymmetric lifting should be avoided. If asymmetric lifting cannot be avoided, however, the recommended weight limits are significantly less than those used for symmetrical lifting.*

An asymmetric lift may be required under the following task or workplace conditions:

1. The origin and destination of the lift are oriented at an angle to each other.
2. The lifting motion is across the body, such as occurs in swinging bags or boxes from one location to another.
3. The lifting is done to maintain body balance in obstructed workplaces, on rough terrain, or on littered floors.
4. Productivity standards require reduced time per lift.

The asymmetric angle (A), which is depicted graphically in Figure 57.2, is operationally defined as the angle between the asymmetry line and the midsagittal line. The *asymmetry line* is defined as the line that joins the midpoint between the inner ankle bones and the point projected on the floor directly below the midpoint of the hand grasps, as defined by the large middle knuckle. The *sagittal line* is defined as the line passing through the midpoint between the inner ankle bones and lying in the midsagittal plane, as defined by the neutral body position (i.e., hands directly in front of the body, with no twisting at the legs, torso, or shoulders). *Note*: The asymmetry angle is not defined by foot position or the angle of torso twist, but by the location of the load relative to the worker's midsagittal plane.

In many cases of asymmetric lifting, the worker will pivot or use a step turn to complete the lift. Because this may vary significantly between workers and between lifts, we have assumed that no pivoting or stepping occurs. Although this assumption may overestimate the reduction in acceptable load weight, it will provide the greatest protection for the worker.

*It may not always be clear if asymmetry is an intrinsic element of the task or just a personal characteristic of the worker's lifting style. Regardless of the reason for the asymmetry, any observed asymmetric lifting should be considered an intrinsic element of the job design and should be considered in the assessment and subsequent redesign. Moreover, the design of the task should not rely on worker compliance, but rather the design should discourage or eliminate the need for asymmetric lifting.

TABLE 57.4 Asymmetric Multiplier

A deg	AM
0	1.00
15	.95
30	.90
45	.86
60	.81
75	.76
90	.71
105	.66
120	.62
135	.57
>135	.00

The asymmetry angle (*A*) must always be measured at the origin of the lift. If significant control is required at the destination, however, then angle *A* should be measured at both the origin and the destination of the lift.

Asymmetry Restrictions

The angle *A* is limited to the range from 0° to 135°. If $A > 135°$, then AM is set equal to zero, which results in an RWL of zero, or no load.

Asymmetric Multiplier

The asymmetric multiplier (AM) is 1–(.0032A). AM has a maximum value of 1.0 when the load is lifted directly in front of the body and decreases linearly as the angle of asymmetry (*A*) increases. The range is from a value of 0.57 at 135° of asymmetry to a value of 1.0 at 0° of asymmetry (i.e., symmetric lift). If *A* is greater than 135°, then AM = 0, and the RWL = 0.0. The AM value can be computed directly or determined from Table 57.4.

Frequency Component

Definition and Measurement

The frequency multiplier is defined by (1) the number of lifts per minute (frequency), (2) the amount of time engaged in the lifting activity (duration), and (3) the vertical height of the lift from the floor. Lifting frequency (*F*) refers to the average number of lifts made per minute, as measured over a 15-minute period. Because of the potential variation in work patterns, analysts may have difficulty obtaining an accurate or representative 15-minute work sample for computing *F.* If significant variation exists in the frequency of lifting over the course of the day, analysts should employ standard work sampling techniques to obtain a representative work sample for determining the number of lifts per minute. For those jobs where the frequency varies from session to session, each session should be analyzed separately, but the overall work pattern must still be considered. For more information, most standard industrial engineering or ergonomics texts provide guidance for establishing a representative job sampling strategy (e.g., Eastman Kodak Company).[6]

Lifting Duration

Lifting duration is classified into three categories based on the pattern of continuous work time and recovery time (i.e., light work) periods. A continuous work time (WT) period is defined as a period of uninterrupted work. Recovery time (RT) is defined as the duration of light work activity following a period of continuous lifting. Examples of light work include activities such as sitting at a desk or table, monitoring operations, light assembly work, etc. The three categories are short duration, moderate duration, and long duration.

TABLE 57.5 Frequency Multiplier (FM)

Frequency[‡] Lifts/min (F)	Work Duration ≤1 Hour		>1 but ≤ 2 Hours		>2 but ≤ 8 Hours	
	$V^{\dagger} < 30$	$V \geq 30$	$V < 30$	$V \geq 30$	$V < 30$	$V \geq 30$
≤0.2	1.00	1.00	.95	.95	.85	.85
0.5	.97	.97	.92	.92	.81	.81
1	.94	.94	.88	.88	.75	.75
2	.91	.91	.84	.84	.65	.65
3	.88	.88	.79	.79	.55	.55
4	.84	.84	.72	.72	.45	.45
5	.80	.80	.60	.60	.35	.35
6	.75	.75	.50	.50	.27	.27
7	.70	.70	.42	.42	.22	.22
8	.60	.60	.35	.35	.18	.18
9	.52	.52	.30	.30	.00	.15
10	.45	.45	.26	.26	.00	.13
11	.41	.41	.00	.23	.00	.00
12	.37	.37	.00	.21	.00	.00
13	.00	.34	.00	.00	.00	.00
14	.00	.31	.00	.00	.00	.00
15	.00	.28	.00	.00	.00	.00
>15	.00	.00	.00	.00	.00	.00

† Values of V are in inches.
‡ For lifting less frequently than once per 5 minutes, set F = .2 lifts/minute.

Short Duration. Short duration lifting tasks are those that have a work duration of 1 h or less, followed by a recovery time equal to 1.2 times the work time (i.e., at least a 1.2 recovery-time to work-time ratio [RT/WT]). For example, to be classified as short-duration, a 45-minute lifting job must be followed by at least a 54-minute recovery period prior to initiating a subsequent lifting session. If the required recovery time is not met for a job of one hour or less, and a subsequent lifting session is required, then the total lifting time must be combined to correctly determine the duration category. Moreover, if the recovery period does not meet the time requirement, it is disregarded for purposes of determining the appropriate duration category.

As another example, assume a worker lifts continuously for 30 min, then performs a light work task for 10 min, and then lifts for an additional 45-minute period. In this case, the recovery time between lifting sessions (10 min) is less than 1.2 times the initial 30-minute work time (36 min). Thus, the two work times (30 min and 45 min) must be added together to determine the duration. Since the total work time (75 min) exceeds 1 hour, the job is classified as moderate duration. On the other hand, if the recovery period between lifting sessions was increased to 36 min, then the short duration category would apply, which would result in a larger FM value.

A special procedure has been developed for determining the appropriate lifting frequency (*F*) for certain repetitive lifting tasks in which workers do not lift continuously during the 15-minute sampling period. This occurs when the work pattern is such that the worker lifts repetitively for a short time and then performs light work for a short time before starting another cycle. For work patterns such as this, *F* may be determined as follows, as long as the actual lifting frequency does not exceed 15 lifts/min:

1. Compute the total number of lifts performed for the 15-minute period (i.e., lift rate times work time).
2. Divide the total number of lifts by 15.
3. Use the resulting value as the frequency (*F*) to determine the frequency multiplier (FM) from Table 57.5.

For example, if the work pattern for a job consists of a series of cyclic sessions requiring 8 min of lifting followed by 7 min of light work, and the lifting rate during the work sessions is 10 lifts/min, then

the frequency rate (F) that is used to determine the frequency multiplier for this job is equal to $(10 \times 8)/15$ or 5.33 lifts/min. If the worker lifted continuously for more than 15 min, however, then the actual lifting frequency (10 lifts/min) would be used.

When using this special procedure, the duration category is based on the magnitude of the recovery periods *between* work sessions, not *within* work sessions. In other words, if the work pattern is intermittent and the special procedure applies, then the intermittent recovery periods that occur during the 15-minute sampling period are *not* considered as recovery periods for purposes of determining the duration category. For example, if the work pattern for a manual lifting job was composed of repetitive cycles consisting of 1 min of continuous lifting at a rate of 10 lifts/min, followed by 2 minutes of recovery, the correct procedure would be to adjust the frequency according to the special procedure [i.e., $F = (10 \text{ lifts/min} \times 5 \text{ min})/15 \text{ min} = 50/15 = 3.4$ lifts/min.] The 2-minute recovery periods would not count toward the RT/WT ratio, however, and additional recovery periods would have to be provided as described above.

Moderate Duration. Moderate duration lifting tasks are those that have a duration of more than 1 h but not more than 2 h, followed by a recovery period of at least 0.3 times the work time [i.e., at least a 0.3 recovery time to work time ratio (RT/WT)].

For example, a worker who continuously lifts for 2 h, would need at least a 36-minute recovery period before initiating a subsequent lifting session. If the recovery time requirement is not met, and a subsequent lifting session is required, then the total work time must be added together. If the total work time exceeds 2 h, then the job must be classified as a long duration lifting task.

Long Duration. Long duration lifting tasks are those that have a duration of between 2 and 8 h, with standard industrial rest allowances (e.g., morning, lunch, and afternoon rest breaks). *Note*: No weight limits are provided for more than 8 h of work.

The difference in the required RT/WT ratio for the short (<1 hour) duration category, which is 1.2, and the moderate (1 to 2 h) duration category, which is 0.3, is due to the difference in the magnitudes of the frequency multiplier values associated with each of the duration categories. Since the moderate category results in larger reductions in the RWL than the short category, there is less need for a recovery period between sessions than for the short duration category. In other words, the short duration category would result in higher weight limits than the moderate duration category, so larger recovery periods would be needed.

Frequency Restrictions

Lifting frequency (F) for repetitive lifting may range from 0.2 lifts/min to a maximum frequency that is dependent on the vertical location of the object (V) and the duration of lifting (Table 57.5). Lifting above the maximum frequency results in an RWL of 0.0 (except for the special case of discontinuous lifting discussed above, where the maximum frequency is 15 lifts/min)

Frequency Multiplier

The FM value depends upon the average number of lifts/min (F), the vertical location (V) of the hands at the origin, and the duration of continuous lifting. For lifting tasks with a frequency less than 0.2 lifts/min, set the frequency equal to 0.2 lifts/minute. Otherwise, the FM is determined from Table 57.5.

Coupling Component

Definition and Measurement

The nature of the hand-to-object coupling or gripping method can affect not only the maximum force a worker can or must exert on the object, but also the vertical location of the hands during the lift. A "good" coupling will reduce the maximum grasp forces required and increase the acceptable weight for lifting, while a "poor" coupling will generally require higher maximum grasp forces and decrease the acceptable weight for lifting.

The effectiveness of the coupling is not static, but may vary with the distance of the object from the ground, so that a good coupling could become a poor coupling during a single lift. The entire range of

TABLE 57.6 Hand-to-Container Coupling Classification

Good	Fair	Poor
1. For containers of optimal design, such as some boxes, crates, etc., a "Good" hand-to-object coupling would be defined as handles or hand-hold cut-outs of optimal design [see notes 1 to 3 below].	1. For containers of optimal design, a "Fair" hand-to-object coupling would be defined as handles or hand-hold cut-outs of less than optimal design [see notes 1 to 4 below].	1. Containers of less than optimal design or loose parts or irregular objects that are bulky, hard to handle, or have sharp edges [see note 5 below].
2. For loose parts or irregular objects, which are not usually containerized, such as castings, stock, and supply materials, a "Good" hand-to-object coupling would be defined as a comfortable grip in which the hand can be easily wrapped around the object [see note 6 below].	2. For containers of optimal design with no handles or hand-hold cut-outs or for loose parts or irregular objects, a "Fair" hand-to-object coupling is defined as a grip in which the hand can be flexed about 90 degrees [see note 4 below].	2. Lifting non-rigid bags (i.e., bags that sag in the middle).

1. An optimal handle design has .75 to 1.5 inches (1.9 to 3.8 cm) diameter, ≥4.5 inches (11.5 cm) length, 2 inches (5 cm) clearance, cylindrical shape, and a smooth, nonslip surface.
2. An optimal hand-hold cut-out has the following approximate characteristics: ≥1.5 inch (3.8 cm) height, 4.5 inch (11.5 cm) length, semi-oval shape, ≥2 inch (5 cm) clearance, smooth nonslip surface, and ≥ 0.25 inches (0.60 cm) container thickness (e.g., double thickness cardboard).
3. An optimal container design has ≤16 inches (40 cm) frontal length, ≤12 inches (30 cm) height, and a smooth nonslip surface.
4. A worker should be capable of clamping the fingers at nearly 90° under the container, such as required when lifting a cardboard box from the floor.
5. A container is considered less than optimal if it has a frontal length >16 inches (40 cm), height >12 inches (30 cm), rough or slippery surfaces, sharp edges, asymmetric center of mass, unstable contents, or requires the use of gloves.
6. A worker should be able to comfortably wrap the hand around the object without causing excessive wrist deviations or awkward postures, and the grip should not require excessive force.

the lift should be considered when classifying hand-to-object couplings, with classification based on overall effectiveness. The analyst must classify the coupling as good, fair, or poor. The three categories are defined in Table 57.6. If there is any doubt about classifying a particular coupling design, the more stressful classification should be selected.

The decision tree shown in Figure 57.3 may be helpful in classifying the hand-to-object coupling.

Limitations

There are no limitations for classifying the coupling, but both hands must be observed when assessing the coupling. If one hand is predominately used to lift the load, and the fingers are flexed at 90° under the load, then the coupling should be rated as fair, regardless of the position of the other hand.

Coupling Multiplier

Based on the coupling classification and vertical location of the lift, the Coupling Multiplier (CM) is determined from Table 57.7.

57.5 Procedures

Prior to data collection, the analyst must decide (1) if the job should be analyzed as a single-task or multitask manual lifting job, and (2) if significant control is required at the destination of the lift. This is necessary because the procedures differ according to the type of analysis required.

A manual lifting job may be analyzed as a single-task job if the task variables do not differ from task to task, or if only one task is of interest (e.g., single most stressful task). This may be the case if one of the tasks clearly has a dominant effect on strength demands, localized muscle fatigue, or whole-body

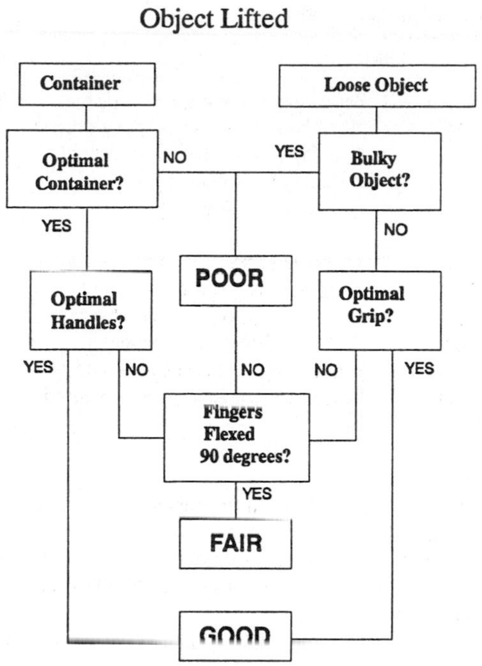

FIGURE 57.3 Decision tree for coupling quality.

TABLE 57.7 Coupling Multiplier

	Coupling Multiplier	
Coupling Type	V < 30 inches (75 cm)	V ≥ 30 inches (75 cm)
Good	1.00	1.00
Fair	0.95	1.00
Poor	0.90	0.90

fatigue. On the other hand, if the task variables differ significantly between tasks, it may be more appropriate to analyze a job as a multitask manual lifting job. A multitask analysis is more difficult to perform than a single-task analysis because additional data and computations are required. The multitask approach, however, will provide more detailed information about specific strength and physiological demands.

For many lifting jobs, it may be acceptable to use either the single- or multitask approach. The single-task analysis should be used when possible, but when a job consists of more than one task and detailed information is needed to specify engineering modifications, then the multitask approach provides a reasonable method of assessing the overall physical demands. The multitask procedure is more complicated than the single-task procedure, and requires a greater understanding of assessment terminology and mathematical concepts. Therefore, the decision to use the single- or multitask approach should be based on: (1) the need for detailed information about all facets of the multitask lifting job, (2) the need for accuracy and completeness of data regarding assessment of the physiological demands of the task, and (3) the analyst's level of understanding of the assessment procedures.

The decision about control at the destination is important because the physical demands on the worker may be greater at the destination of the lift than at the origin, especially when significant control is required. When significant control is required at the destination, for example, the physical stress is increased because the load will have to be accelerated upward to slow down its descent. This acceleration

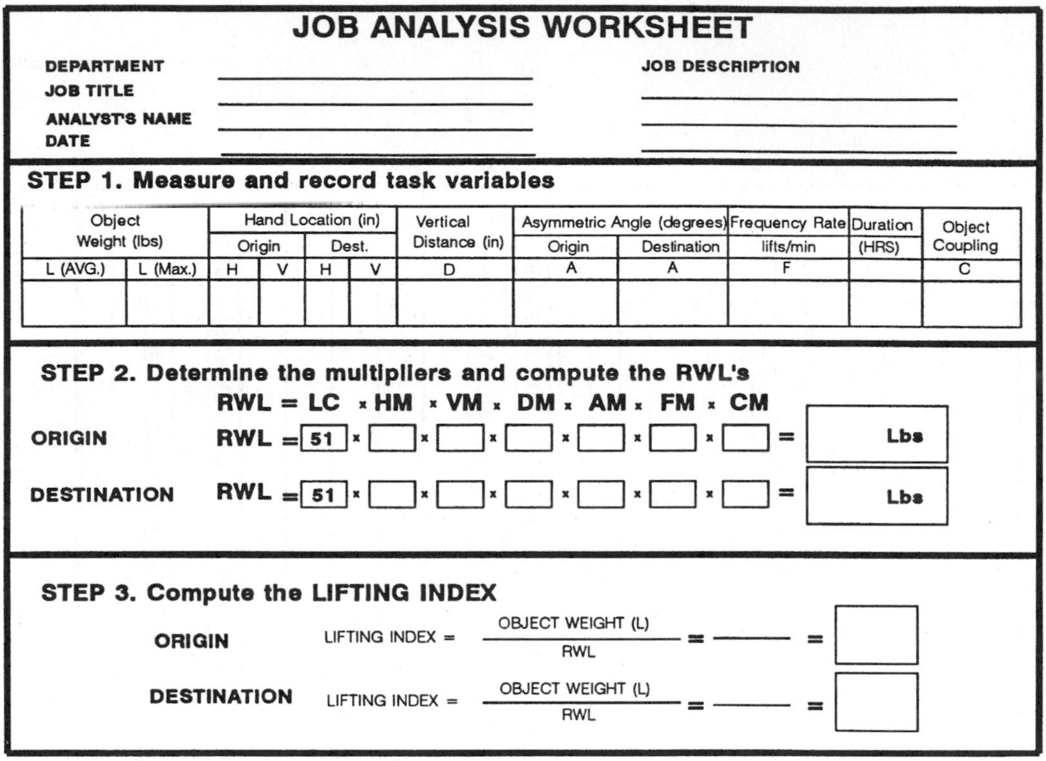

JOB ANALYSIS WORKSHEET

DEPARTMENT ____________ JOB DESCRIPTION

JOB TITLE ____________

ANALYST'S NAME ____________

DATE ____________

STEP 1. Measure and record task variables

Object Weight (lbs)		Hand Location (in)				Vertical Distance (in)	Asymmetric Angle (degrees)		Frequency Rate lifts/min	Duration (HRS)	Object Coupling
		Origin		Dest.			Origin	Destination			
L (AVG.)	L (Max.)	H	V	H	V	D	A	A	F		C

STEP 2. Determine the multipliers and compute the RWL's

RWL = LC × HM × VM × DM × AM × FM × CM

ORIGIN RWL = 51 × ☐ × ☐ × ☐ × ☐ × ☐ × ☐ = ☐ Lbs

DESTINATION RWL = 51 × ☐ × ☐ × ☐ × ☐ × ☐ × ☐ = ☐ Lbs

STEP 3. Compute the LIFTING INDEX

ORIGIN LIFTING INDEX = OBJECT WEIGHT (L) / RWL = ——— = ☐

DESTINATION LIFTING INDEX = OBJECT WEIGHT (L) / RWL = ——— = ☐

FIGURE 57.4 Single task job analysis worksheet.

may be as great as the acceleration at the origin of the lift and may create high loads on the spine. Therefore, if significant control is required, then the RWL and LI should be determined at both locations and the lower of the two values will specify the overall level of physical demand.

To perform a lifting analysis using the revised lifting equation, two steps are undertaken: (1) data are collected at the worksite as described in step 1 below; and, (2) the recommended weight limit and lifting index values are computed using the single-task or multitask analysis procedure described in step 2 below.

Step 1: Collect Data

The relevant task variables must be carefully measured and clearly recorded in a concise format. As mentioned previously, these variables include the horizontal location of the hands (*H*), vertical location of the hands (*V*), vertical displacement (*D*), asymmetric angle (*A*), lifting frequency (*F*), and coupling quality (*C*). A job analysis worksheet, as shown in Figure 57.4 for single-task jobs or Figure 57.5 for multitask jobs, provides a simple form for recording the task variables and the data needed to calculate the RWL and the LI values. A thorough job analysis is required to identify and catalog each independent lifting task in the worker's complete job. For multitask jobs, data must be collected for each task.

Step 2: Single-Task Procedure

For a single-task analysis, step 2 consists of computing the recommended weight limit (RWL) and the lifting index (LI). This is accomplished as follows.

Calculate the RWL at the origin for each lift. For lifting tasks that require significant control at the destination, calculate the RWL at *both* the origin and the destination of the lift. The latter procedure is required if (1) the worker has to regrasp the load near the destination of the lift, (2) the worker has to momentarily hold the object at the destination, or (3) the worker has to position or guide the load at

MULTI-TASK JOB ANALYSIS WORKSHEET

DEPARTMENT ______________________ JOB DESCRIPTION

JOB TITLE ______________________ ______________________

ANALYST'S NAME ______________________ ______________________

DATE ______________________ ______________________

STEP 1. Measure and Record Task Variable Data

Task No.	Object Weight (lbs)		Hand Location (in)				Vertical Distance (in)	Asymmetry Angle (degs)		Frequency Rate lifts/min	Duration Hrs	Coupling
			Origin		Dest.			Origin	Dest.			
	L (Avg.)	L (Max.)	H	V	H	V	D	A	A	F		C

STEP 2. Compute multipliers and FIRWL, STRWL, FILI, and STLI for Each Task

Task No.	LC	x HM	x VM	x DM	x AM	x CM	FIRWL	x FM	STRWL	FILI = L/FIRWL	STLI = L/STRWL	New Task No.	F
	51												
	51												
	51												
	51												
	51												

STEP 3. Compute the Composite Lifting Index for the Job (After renumbering tasks)

CLI =	$STLI_1$ +	$\Delta FILI_2$ +	$\Delta FILI_3$ +	$\Delta FILI_4$ +	$\Delta FILI_5$
		$FILI_2(1/FM_{1,2} - 1/FM_1)$	$FILI_3(1/FM_{1,2,3} - 1/FM_{1,2})$	$FILI_4(1/FM_{1,2,3,4} - 1/FM_{1,2,3})$	$FILI_5(1/FM_{1,2,3,4,5} - 1/FM_{1,2,3,4})$
CLI =					

FIGURE 57.5 Multitask job analysis worksheet.

the destination. The purpose of calculating the RWL at both the origin and destination of the lift is to identify the most stressful location of the lift. Therefore, the lower of the RWL values at the origin or destination should be used to compute the LI for the task, as this value would represent the limiting set of conditions.

The assessment is completed on the single-task worksheet by determining the LI for the task of interest. This is accomplished by comparing the actual weight of the load (*L*) lifted with the RWL value obtained from the lifting equation.

Step 2: Multitask Procedure

For a multitask analysis, step 2 comprises three steps:

1. Compute the frequency-independent recommended weight limit (FIRWL) and single-task recommended weight limit (STRWL) for each task.
2. Compute the frequency-independent lifting index (FILI) and single-task lifting index (STLI) for each task.
3. Compute the composite lifting index (CLI) for the overall job.

Compute the Frequency-Independent Recommended Weight Limits (FIRWLs). Compute the FIRWL value for each task by using the respective task variables and setting the frequency multiplier (FM) to a value of 1.0. The FIRWL for each task reflects the compressive force and muscle strength demands for a single repetition of that task. If significant control is required at the destination for any individual task, the FIRWL must be computed at both the origin and the destination of the lift, as described above for a single-task analysis.

Compute the Single-Task Recommended Weight Limit (STRWL). Compute the STRWL for each task by multiplying its FIRWL by its appropriate FM value. The STRWL for a task reflects the overall demands of that task, assuming it was the only task being performed. *Note*: This value does not reflect the overall demands of the task when the other tasks are considered. Nevertheless, it is helpful in determining the extent of excessive physical stress for an individual task.

Compute the Frequency-Independent Lifting Index (FILI). The FILI is computed for each task by dividing the *maximum* load weight (*L*) for that task by the respective FIRWL. The maximum weight is used to compute the FILI because the maximum weight determines the maximum biomechanical loads to which the body will be exposed, regardless of the frequency of occurrence. Thus, the FILI can identify individual tasks with potential strength problems for infrequent lifts. If any of the FILI values exceed a value of 1.0, then job design changes may be needed to decrease the strength demands.

Compute the Single-Task Lifting Index (STLI). The STLI is computed for each task by dividing the *average* load weight (*L*) for that task by the respective STRWL. The average weight is used to compute the STLI because the average weight provides a better representation of the metabolic demands, which are distributed across the tasks, rather than being dependent on individual tasks. The STLI can be used to identify individual tasks with excessive physical demands (i.e., tasks that would result in fatigue). The STLI values do not indicate the relative stress of the individual tasks in the context of the whole job, but they can be used to prioritize the individual tasks according to the magnitude of their physical stress. Thus, if any of the STLI values exceed a value of 1.0, then ergonomic changes may be needed to decrease the overall physical demands of the task. *Note*: It may be possible to have a job in which all of the individual tasks have an STLI less than 1.0 and yet is physically demanding due to the combined demands of the tasks. In cases where the FILI exceeds the STLI for any task, the maximum weights may represent a significant problem and careful evaluation is necessary.

Compute the Composite Lifting Index (CLI). The assessment is completed on the multitask worksheet by determining the composite lifting index (CLI) for the overall job. The CLI is computed as follows:

1. The tasks are renumbered in order of decreasing physical stress, from the task with the greatest STLI down to the task with the smallest STLI. The tasks are renumbered in this way so that the more difficult tasks are considered first.

TABLE 57.8 Computations from Multitask Example

Task #	Load Weight (L)	Task Frequency (F)	FIRWL	FM	STRWL	FILI	STLI	New Task #
1	30	1	20	.94	18.8	1.5	1.6	1
2	20	2	20	.91	18.2	1.0	1.1	2
3	10	4	15	.84	12.6	.67	.8	3

2. The CLI for the job is then computed according to the following formula:

$$CLI = STLI_1 + \sum \Delta LI$$

where:

$$\sum \Delta LI = \left(FILI_2\ X \left(\frac{1}{FM_{1,2}} - \frac{1}{FM_1} \right) \right)$$

$$+ \left(FILI_3\ X \left(\frac{1}{FM_{1,2,3}} - \frac{1}{FM_{1,2}} \right) \right)$$

$$+ \left(FILI_4\ X \left(\frac{1}{FM_{1,2,3,4}} - \frac{1}{FM_{1,2,3}} \right) \right)$$

$$\vdots$$

$$+ \left(FILI_n\ X \left(\frac{1}{FM_{1,2,3,4,\ldots,n}} - \frac{1}{FM_{1,2,3,\ldots,(n-1)}} \right) \right)$$

Note: (1) The numbers in the subscripts refer to the new task numbers; and, (2) the FM values are determined from Table 57.5, based on the sum of the frequencies for the tasks listed in the subscripts.

An Example. The following example is provided to demonstrate this step of the multi- task procedure. Assume that an analysis of a typical three-task job provided the results shown in Table 57.8.

To compute the Composite Lifting Index (CLI) for this job, the tasks are renumbered in order of decreasing physical stress, beginning with the task with the greatest STLI. In this case, as shown in Table 57.8, the task numbers do not change. Next, the CLI is computed according to the formula shown above. The task with the greatest CLI is Task 1 (STLI = 1.6). The sum of the frequencies for Tasks 1 and 2 is 1+2, or 3, and the sum of the frequencies for Tasks 1, 2 and 3 is 1+2+4, or 7. Then, from Table 57.5, FM_1 is 0.94, $FM_{1,2}$ is 0.88, and $FM_{1,2,3}$ is 0.70. Finally, the CLI = 1.6 + 1.0(1/.88 – 1/.94)+.67(1/.70 – 1/.88) = 1.6 + .07 + .20 = 1.9. *Note*: The FM values were based on the sum of the frequencies for the subscripts, the vertical height, and the duration of lifting.

57.6 Applying the Equations

Using the RWL and LI to Guide Ergonomic Design

The recommended weight limit (RWL) and lifting index (LI) can be used to guide ergonomic design in several ways.

1. The individual multipliers can be used to identify specific job-related problems. The relative magnitude of each multiplier indicates the relative contribution of each task factor (e.g., horizontal, vertical, frequency, etc.)
2. The RWL can be used to guide the redesign of existing manual lifting jobs or to design new manual lifting jobs. For example, if the task variables are fixed, then the maximum weight of the load could be selected so the RWL is not exceeded; if the weight is fixed, then the task variables could be optimized so that the weight does not exceed the RWL.
3. The LI can be used to estimate the relative magnitude of physical stress for a task or job. The greater the LI, the smaller the fraction of workers capable of safely sustaining the level of activity. Thus, two or more job designs could be compared.
4. The LI can be used to prioritize ergonomic redesign. For example, a series of suspected hazardous jobs could be rank ordered according to the LI and a control strategy could be developed according to the rank ordering (i.e., jobs with lifting indices above 1.0 or higher would benefit the most from redesign).

Rationale and Limitations for LI

The NIOSH recommended weight limit (RWL) and lifting index (LI) equations are based on the concept that the risk of lifting-related low back pain increases as the demands of the lifting task increase. In other words, as the magnitude of the LI increases, (1) the level of the risk for a given worker would be increased and (2) a greater percentage of the workforce is likely to be at risk for developing lifting-related low back pain. The shape of the risk function, however, is not known. Without additional data showing the relationship between low back pain and the LI, it is impossible to predict the magnitude of the risk for a given individual or the exact percent of the work population who would be at an elevated risk for low back pain.

To gain a better understanding of the rationale for the development of the RWL and LI, consult Waters et al.,[1] which provides a discussion of the criteria underlying the lifting equation and of the individual multipliers and identifies both the assumptions and uncertainties in the scientific studies that associate manual lifting and low back injuries.

Job-Related Intervention Strategy

The lifting index may be used to identify potentially hazardous lifting jobs or to compare the relative severity of two jobs for the purpose of evaluating and redesigning them. From the NIOSH perspective, it is likely that lifting tasks with an LI > 1.0 pose an increased risk for lifting-related low back pain for some fraction of the workforce.[1] Hence, to the extent possible, lifting jobs should be designed to achieve an LI of 1.0 or less.

Some experts believe, however, that worker selection criteria may be used to identify workers who can perform potentially stressful lifting tasks (i.e., lifting tasks that would exceed an LI of 1.0) without significantly increasing their risk of work-related injury above the baseline level.[6,7] Those who endorse the use of selection criteria believe that the criteria must be based on research studies, empirical observations, or theoretical considerations that include job-related strength testing and/or aerobic capacity testing. Even these experts agree, however, that many workers will be at a significant risk of a work-related injury when performing highly stressful lifting tasks (i.e., lifting tasks that would exceed an LI of

TABLE 57.9 General Design/Redesign Suggestions

If HM is less than 1.0	Bring the load closer to the worker by removing any horizontal barriers or reducing the size of the object. Lifts near the floor should be avoided; if unavoidable, the object should fit easily between the legs.
If VM is less than 1.0	Raise/lower the origin/destination of the lift. Avoid lifting near the floor or above the shoulders.
If DM is less than 1.0	Reduce the vertical distance between the origin and the destination of the lift.
If AM is less than 1.0	Move the origin and destination of the lift closer together to reduce the angle of twist, or move the origin and destination farther apart to force the worker to turn the feet and step, rather than twist the body.
If FM is less than 1.0	Reduce the lifting frequency rate, reduce the lifting duration, or provide longer recovery periods (i.e., light work period).
If CM is less than 1.0	Improve the hand-to-object coupling by providing optimal containers with handles or hand-hold cut-outs, or improve the hand-holds for irregular objects.
If the RWL at the destination is less than at the origin	Eliminate the need for significant control of the object at the destination by redesigning the job or modifying the container/object characteristics. (See requirements for significant control in text.)

3.0). Also, "informal" or "natural" selection of workers may occur in many jobs that require repetitive lifting tasks. According to some experts, this may result in a unique workforce that may be able to work above a lifting index of 1.0, at least in theory, without substantially increasing their risk of low back injuries above the baseline rate of injury.

Example Problems

Two example problems are provided to demonstrate the proper application of the lifting equation and procedures. The procedures provide a method for determining the level of physical stress associated with a specific set of lifting conditions and assist in identifying the contribution of each job-related factor. The examples also provide guidance in developing an ergonomic redesign strategy. Specifically, for each example, a job description, job analysis, hazard assessment, redesign suggestion, illustration, and completed worksheet are provided.

A series of general design/redesign suggestions for each job-related risk factor are provided in Table 57.9. These suggestions can be used to develop a practical ergonomic design/redesign strategy.

Loading Bags Into A Hopper, Example 1

Job Description. The worker positions himself midway between the handtruck and the mixing hopper, as illustrated in Figure 57.6. Without moving his feet, he twists to the right and picks up a bag off the handtruck. In one continuous motion he then twists to his left to place the bag on the rim of the hopper. A sharp-edged blade within the hopper cuts open the bag to allow the contents to fall into the hopper. This task is done infrequently (i.e., 1 to 12 times per shift) with large recovery periods between lifts (i.e., > 1.2 Recovery Time/Work Time ratio). In observing the worker perform the job, it was determined that the nonlifting activities could be disregarded because they require minimal force and energy expenditure. Significant control is not required at the destination, but the worker twists at the origin and destination of the lift. Although several bags are stacked on the hand truck, the highest risk of overexertion injury is associated with the bag on the bottom of the stack; therefore, only the lifting of the bottom bag will be examined. Note, however, that the frequency multiplier is based on the overall frequency of lifting for all of the bags.

Job Analysis. The task variable data are measured and recorded on the job analysis worksheet (Figure 57.7). The vertical location of the hands is 15 inches at the origin and 36 inches at the destination. The horizontal location of the hands is 18 inches at the origin and 10 inches at the destination. The asymmetric angle is 45° at the origin and 45° at the destination of the lift, and the frequency is less than .2 lifts/min for less than 1 hour (see Table 57.5).

Using Table 57.6, the coupling is classified as fair because the worker can flex the fingers about 90° and the bags are semi-rigid (i.e., they do not sag in the middle). Significant control of the object is not

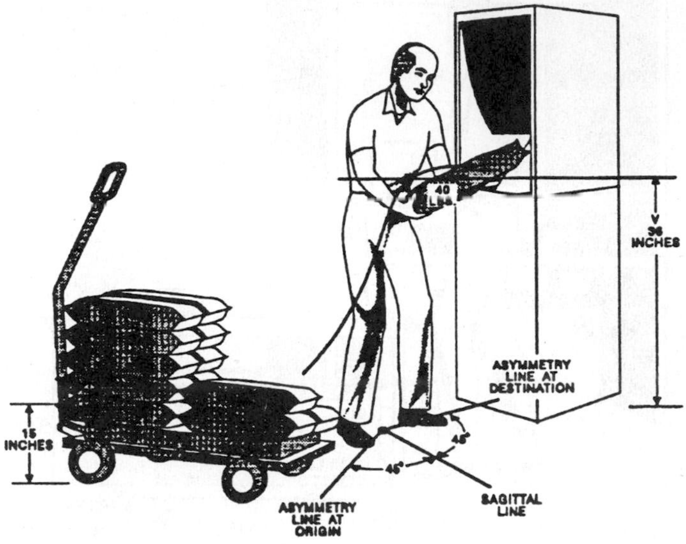

FIGURE 57.6 Loading bags into hopper, example 1.

JOB ANALYSIS WORKSHEET

DEPARTMENT Manufacturing
JOB TITLE Batch Processor
ANALYST'S NAME
DATE

JOB DESCRIPTION Dumping bags into mixing hopper
Example 1

STEP 1. Measure and record task variables

Object Weight (lbs)		Hand Location (in)				Vertical Distance (in)	Asymmetric Angle (degrees)		Frequency Rate lifts/min	Duration (HRS)	Object Coupling
		Origin		Dest.			Origin	Destination			
L (AVG.)	L (Max.)	H	V	H	V	D	A	A	F		C
40	40	18	15	10	36	21	45	45	<.2	<1	Fair

STEP 2. Determine the multipliers and compute the RWL's

		LC	HM	VM	DM	AM	FM	CM	
ORIGIN	RWL =	51	.56	.89	.91	.86	1.0	.95	= 18.9 Lbs
DESTINATION	RWL =	51							= Lbs

STEP 3. Compute the LIFTING INDEX

ORIGIN LIFTING INDEX = OBJECT WEIGHT (L) / RWL = 40 / 18.9 = 2.1

DESTINATION LIFTING INDEX = OBJECT WEIGHT (L) / RWL = ——— =

FIGURE 57.7 Job analysis worksheet, example 1.

required at the destination of the lift so the RWL is computed only at the origin. The multipliers are computed from the lifting equation or determined from the multiplier tables (Tables 57.1 to 57.5, and Table 57.7). As shown in Figure 57.7, the RWL for this activity is 18.9 lbs.

Hazard Assessment. The weight to be lifted (40 lbs) is greater than the RWL (18.9 lbs). Therefore, the LI is 40/18.9 or 2.1. This job would be physically stressful for many industrial workers.

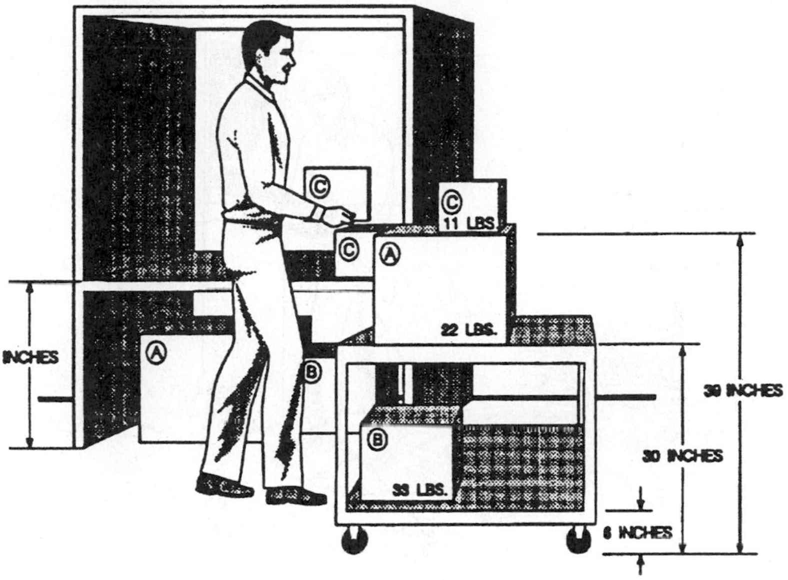

FIGURE 57.8 Warehouse order filling, example 2.

Redesign Suggestions. The worksheet shows that the smallest multipliers (i.e., the greatest penalties) are .56 for the HM, .86 for the AM, and .89 for the VM. Using Table 57.8, the following job modifications are suggested:

1. Bring the load closer to the worker to increase the HM.
2. Reduce the angle of asymmetry to increase AM. This could be accomplished either by moving the origin and destination points closer together or farther apart.
3. Raise the height at the origin to increase the VM.

If the worker could get closer to the bag before lifting, the H value could be decreased to 10 inches, which would increase the HM to 1.0, the RWL would be increased to 33.7 lbs, and the LI would be decreased to 1.2 (i.e., 40/33.7).

Comments. This example demonstrates that certain lifting jobs may be evaluated as a single-task or multitask job. In this case, only the most stressful component of the job was evaluated. For repetitive lifting jobs, the multitask approach may be more appropriate.

Warehouse Order Filling, Example 2

Job Description. A worker lifts cartons of various sizes from supply shelves onto a cart as illustrated in Figure 57.8. There are three box sizes (i.e., A, B, and C) of various weights. These lifting tasks are typical in warehousing, shipping, and receiving activities in which loads of varying weights and sizes are lifted at different frequencies. Assume that the following observations were made: (1) control of the load is not required at the destination of any lift; (2) the worker does not twist when picking up and putting down the cartons; (3) the worker can get close to each carton; and, (4) walking and carrying are minimized by keeping the cart close to the shelves.

Job Analysis. Since the job consists of more than one distinct task and the task variables often change, the multitask lifting analysis procedure should be used. This job can be divided into three tasks represented by cartons A, B, and C. The following measurements were made and recorded on the job analysis worksheet (Figure 57.9):

1. The horizontal locations (H) for each task at the origin and destination are as follows: Box A, 16 inches; Box B, 12 inches; and, Box C, 8 inches.
2. The vertical locations (V) at the origin are taken to be the position of the hands under the cartons as follows: Box A, 0 inches; Box B, 0 inches; and, Box C, 30 inches.

MULTI-TASK JOB ANALYSIS WORKSHEET

DEPARTMENT	Warehouse	**JOB DESCRIPTION**	
JOB TITLE	Shipping Clerk		Selecting an order for shipment
ANALYST'S NAME			Warehouse order filling
DATE			Example 2

STEP 1. Measure and Record Task Variable Data

Task No.	Object Weight (lbs) L (Avg.)	Object Weight (lbs) L (Max.)	Hand Location (in) Origin H	Origin V	Dest. H	Dest. V	Vertical Distance (in) D	Asymmetry Angle (degs) Origin A	Dest. A	Frequency Rate lifts/min F	Duration Hrs	Coupling C
1 (A)	22	33	16	0	16	30	30	0	0	1	8	Fair
2 (B)	33	44	12	0	12	6	6	0	0	2	8	Fair
3 (C)	11	22	8	30	8	39	9	0	0	5	8	Fair

STEP 2. Compute multipliers and FIRWL, STRWL, FILI, and STLI for Each Task

Task No.	LC	x HM	x VM	x DM	x AM	x CM	FIRWL	x FM	STRWL	FILI = L/FIRWL	STLI = L/STRWL	New Task No.	F
1	51	.63	.78	.88	1.0	.95	21.0	.75	15.8	1.6	1.4	2	1
2	51	.83	.78	1.0	1.0	.95	31.4	.65	20.4	1.4	1.6	1	2
3	51	1.0	1.0	1.0	1.0	1.0	51.0	.35	17.8	.4	.6	3	5
	51												
	51												

STEP 3. Compute the Composite Lifting Index for the Job (After renumbering tasks)

CLI =	$STLI_1$ +	$\Delta FILI_2$ +	$\Delta FILI_3$ +	$\Delta FILI_4$ +	$\Delta FILI_5$	
		$FILI_2(1/FM_{1+2} - 1/FM_1)$	$FILI_3(1/FM_{1+2+3} - 1/FM_{1+2})$	$FILI_4(1/FM_{1+2+3+4} - 1/FM_{1+2+3})$	$FILI_5(1/FM_{1+2+3+4+5} - 1/FM_{1+2+3+4})$	
		1.6(1/.55-1/.65)	.4(1/.18-1/.55)			
CLI =	1.6	.45	1.5			3.6

FIGURE 57.9 Job analysis worksheet, example 2.

3. The vertical locations (V) at the destination are the vertical position on the cart as follows: Box A, 30 inches; Box B, 6 inches; and, Box C, 39 inches.
4. The average weights lifted for each task are as follows: Box A, 22 lbs; Box B, 33 lbs; and, Box C, 11 lbs.
5. The maximum weights lifted for each task are as follows: Box A, 33 lbs; Box B, 44 lbs; and, Box C, 22 lbs.
6. No asymmetric lifting is involved (i.e., A = 0).
7. The lifting frequency rates for each task are as follows: Box A, 1 lift/min; Box B, 2 lifts/min; and Box C, 5 lifts/min.
8. The lifting duration for the job is 8 hours, however, the maximum weights are lifted infrequently (i.e., less than or equal to once every 5 minutes for 8 hours)
9. Using Table 57.6, the couplings are classified as fair.

The multitask lifting analysis consists of the following three steps:

1. Compute the frequency-independent-RWL (FIRWL) and frequency-independent- lifting index (FILI) values for each task using a default FM of 1.0.
2. Compute the single-task-RWL (STRWL) and single-task-lifting index (STLI) for each task.
3. Renumber the tasks in order of decreasing physical stress, as determined from the STLI value, starting with the task with the largest STLI.

Step 1

Compute the FIRWL and FILI values for each task using a default FM of 1.0. The other multipliers are computed from the lifting equation or determined from the multiplier tables (Table 57.1 to 57.5, and Table 57.7). Recall, that the FILI is computed for each task by dividing the *maximum* weight of that task by its FIRWL.

		FIRWL	*FILI*
Task 1	=	21.0 lbs	1.6
Task 2	=	31.4 lbs	1.4
Task 3	=	51.0 lbs	.4

These results indicate that *some of the tasks requires excessive strength.* Remember, however, that these results do not take the frequency of lifting into consideration.

Step 2
Compute the STRWL and STLI values for each task, where the STRWL for a task is equivalent to the product of the FIRWL and the FM for that task. Recall, that the STILI is computed for each task by dividing the *average* weight of that task by its STRWL. The appropriate FM values are determined from Table 57.5.

		STRWL	*STLI*
Task 1	=	15.8 lbs	1.4
Task 2	=	20.4 lbs	1.6
Task 3	=	17.8 lbs	.6

These results indicate that Tasks 1 and 2 *would be stressful* for some workers, if performed individually. Note, however, that these values do not consider the combined effects of all of the tasks.

Step 3
Renumber the tasks, starting with the task with the largest STLI value, and ending with the task with the smallest STLI value. If more than one task has the same STLI value, assign the lower task number to the task with the highest frequency.

Hazard Assessment. Compute the composite-lifting index (CLI) using the renumbered tasks. As shown in Figure 57.9, the CLI for this job is 3.6, which indicates that this job would be physically stressful for nearly all workers. Analysis of the results suggests that the combined effects of the three tasks are significantly more stressful than any individual task.

Redesign Suggestions. Developing a redesign strategy for a job depends on tangible and intangible factors that may be difficult to evaluate, including costs/benefits, feasibility, and practicality. No preferred procedure has been developed and tested. Therefore, the following suggestions represent only one approach to ergonomic job modification.

In this example, the magnitude of the FILI, STLI, and CLI values indicate that both strength and endurance would be a problem for many workers. Therefore, the redesign should attempt to decrease the physical demands by modifying the job layout and decrease the physiological demands by reducing the frequency rate or duration of continuous lifting. If the maximum weights were eliminated from the job, then the CLI would be significantly reduced, the job would be less stressful, and more workers could perform the job than before.

Those lifts with strength problems should be evaluated for specific engineering changes, such as (1) decreasing carton size or removing barriers to reduce the horizontal distance; (2) raising or lowering the origin of the lift; (3) reducing the vertical distance of the lift; improving carton couplings, and 4) decreasing the weight to be lifted. The redesign priority for this example is based on identifying interventions that provide the largest increase in the FIRWL for each task (Step 2 on worksheet). For example, the maximum weight lifted for carton A is unacceptable; however, if the carton at the origin were on the upper shelf, then the FIRWL for Task 1 would increase from 21.0 lbs to 27.0 lbs. The maximum weight lifted still exceeds the FIRWL, but lifts of average weight are now below the FIRWL. Additionally, providing handles, decreasing box size, or reducing the load to be lifted will decrease the stress of manual lifting.

Comments. This example demonstrates the complexity of analyzing multitask lifting jobs. Errors resulting from averaging, and errors introduced by ignoring other factors (e.g., walking, carrying, holding,

pushing and pulling activities, and environmental stressors), can only be resolved with detailed biomechanical, metabolic, cardiovascular, and psychophysical evaluations.

Several important application principles are illustrated in this example:

1. The horizontal distance (H) for Task 3 was less than the 10.0 inches minimum. Therefore, H was set equal to 10 inches (i.e., multipliers must be less than or equal to 1.0).
2. The vertical travel distance (D) in Task 2 was less than the 10 inches minimum. Therefore, D was set equal to 10 inches.

References

1. Waters T.R., Putz-Anderson V., Garg A., and Fine L.J., 1993, Revised NIOSH equation for the design and evaluation of manual lifting tasks. *Ergonomics.* Vol. 36(7), 749- 776.
2. Waters T.R., Putz-Anderson, and Garg A., *Applications Manual for the Revised NIOSH Lifting Equation.* 1994. National Institute for Occupational Safety and Health, Technical Report. DHHS(NIOSH) Pub. No. 94-110. Available from the National Technical Information Service (NTIS). NTIS document number PB94-176930 (1-800-553-NTIS).
3. NIOSH, 1981, *Work Practices Guide for Manual Lifting,* NIOSH Technical Report No. 81-122, (U.S. Department of Health and Human Services, National Institute for Occupational Safety and Health, Cincinnati, OH).
4. NIOSH, 1991, *Scientific Support Documentation for the Revised 1991 NIOSH Lifting Equation: Technical Contract Reports, May 8, 1991,* (U.S. Department of Health and Human Services, National Institute for Occupational Safety and Health, Cincinnati, OH). Available from the National Technical Information Service (NTIS No. PB-91-226-274).
5. ASPH/NIOSH, 1986, *Proposed National Strategies for the Prevention of Leading Work- Related Diseases and Injuries: Part 1.* Published by the Association of Schools of Public Health under a cooperative agreement with the National Institute for Occupational Safety and Health.
6. Eastman Kodak. 1986, *Ergonomic Design for People at Work, Vol. 2,* Van Nostrand Reinhold. New York. 1986
7. Ayoub, M.M. and Mital, A. 1989, *Manual Materials Handling,* Taylor & Francis, London.
8. Chaffin, D.B. and Andersson, G.B.J. 1984, *Occupational Biomechanics,* John Wiley & Sons, New York.

58

A Population-Based Load Threshold Limit (LTL) for Manual Lifting Tasks Performed by Males and Females

Waldemar Karwowski
University of Louisville

Paul Gaddie
University of Louisville

Renliu Jang
University of Louisville

Wook Gee Lee
University of Louisville

58.1 Introduction

In order to reduce the risk of low back pain and injuries due to manual lifting tasks, we need guidelines for load limits with respect to different percentiles of capable industrial population that are simple to understand by the practitioners, and easy to apply in practical settings. Unfortunately, the revised NIOSH (1991) method, which aims to protect 99% of males and 75% of females of the U.S. population, is quite cumbersome to use, and by design very restrictive in setting the acceptable load limits. This chapter introduces population-based load threshold limits (LTL) for evaluation of manual lifting tasks performed by males and females. The proposed limits are based on the simulation of the revised (1991) NIOSH lifting equation, and application of the threshold lifting index (scaling factors) developed in reference to the comprehensive database for manual lifting tasks reported by Snook and Ciriello (1991).

As described in Chapter 57 of this handbook, the National Institute for Occupational Safety and Health proposed the recommended weight limit (RWL) as the guide for evaluating manual tasks in industry (Waters et al., 1993). The revised lifting equation (RLE) and its predecessor (NIOSH, 1981) aim to

0-8493-2641-9/99/$0.00+$.50

simultaneously satisfy the biomechanical, physiological, and psychophysical criteria for setting limits in manual lifting tasks (Waters et al., 1993; Dempsey, 1998). The 1991 equation expands the previous guidelines, and it is claimed that it can be applied to a larger percentage of lifting tasks (Waters et al., 1993).

Because the RWL is intended to protect 90% of the mixed industrial working population (99% males and 75% females), it is designed for the least physically capable percent of the industrial workforce. Given that more capable populations may be available at the workplace to perform physically demanding lifting tasks (for example strong young males and females), the application of the RWL-based limits that accommodate the majority of mixed male/female population may result in some work inefficiency. As discussed by Waters et al. (1993), a critical point concerning the 1991 RLE concept is that the NIOSH perspective was very conservative in nature. Furthermore, as demonstrated by Karwowski (1992), an application of the 1991 RLE model leads to a fairly narrow range of RWL values.

According to Waters and Putz-Anderson (1998), the lifting index (LI) may be used to identify potentially hazardous lifting jobs. NIOSH believes that jobs with LI smaller than 1.0 would pose an increased risk for lifting-related injury to less than 25% of female and 1% of male industrial workers. However, it was also recognized, that worker selection criteria may be used to identify workers who can perform lifting tasks that would exceed an LI = 1.0 without significantly increasing their risk of work-related injury above the baseline level. It was also noted that the said selection criteria must be based on research studies, job-related strength testing, and/or aerobic capacity (Waters and Putz-Anderson, 1998). Although the risk function representing the relationship between the lifting index and probability of injury is mostly unknown today (for some preliminary data see Karwowski, Gaddie, and Caldewell, 1994), some experts hypothesize that many workers would be at significant risk of a work-related injury when performing lifting jobs with the LI values above 3.0, which are classified as highly stressful tasks (Waters et al., 1993; Waters and Putz-Anderson (1998). Such assumption is yet to be validated in the field.

58.2 Threshold RWL Values

Given the above discussion, until comprehensive epidemiological data that relate the LI values to the risk of lifting-related injury is developed, the paradigm of protecting most of the population represented by the RLE concept seems a plausible and desirable strategy. There is also, however, a pressing need, at least for some of the industrial applications, to identify lifting limits and the population of capable workers (both males and females), who are able to handle physically demanding jobs with a lifting index above 1.0 (but below 3.0). Such data could be developed with respect to the simulated threshold RWL (TRWL) values corresponding to different levels of lifting index.

The threshold RWL values were defined by Karwowski and Gaddie (1995) as those limiting values of RWL (corresponding to LI = 1.0) that would be expected to represent at least 99.5% of all of the potential industrial jobs. In other words, these values mean that only in 0.5% of all possible cases (lifting tasks), the expected recommended weight limit would exceed the specific threshold RWL value. From the practical standpoint that means that for the majority of industrial jobs (99.5%), the expected RWL would be at or below the specific TRWL value. This implies that when LI is set to 1.0 for task design or evaluation purposes, it can be expected that only half of one percent of the plausible industrial lifting tasks would have the recommended weight greater than the specific TRWL value. Assuming that the 0.5% is the additional safety factor, the TRWL indicates the maximum values of RWL that one would expect to get when using the RLE model for evaluating any of the real lifting jobs encountered in industry.

58.3 The 1991 Revised NIOSH Lifting Equation

Detailed description of the relevant terms utilized by the RLE (1991) is provided in Chapter 57. For the purpose of this discussion, it is noted that the recommended weight limit (RWL) is the product of the load constant (LC) and six multipliers (M) which account for seven risk factors, i.e.,: RWL (Kg) = LC *

HM * VM * DM * AM * FM * CM. The multipliers are defined in terms of the related risk factors, including the horizontal location (H), vertical location (V), vertical travel distance (D), frequency of lift (F), asymmetry angle (A), and coupling (C). The multipliers for frequency and coupling are defined using relevant tables. In addition to lifting frequency, the work duration and vertical distance factors are used to compute the frequency multiplier.

58.4 Computer Simulation of the NIOSH (1991) Lifting Equation

One way to investigate the practical implications of the RLE (1991) for industry is to determine the likely results of the equation when applying a realistic and practical range of values for the risk factors (Karwowski, 1992). Karwowski and Gaddie (1995) introduced a computer simulation modeling approach to derive the threshold RWL values for 90% of the mixed male/female population (i.e., 99% of male and 75% of female industrial workers). Shortly, a computer simulation was written in SLAM II, a Simulation Language for Alternative Modeling (Pritsker, 1986). The NIOSH 1991 Lifting Equation was modeled as the product of factor multipliers represented as attributes of an entity flowing through the network. The developed model is a sequence of operations which draws random values from each of the relevant factor distributions and tables, and then calculates the corresponding RWL values.

58.5 Summary of Simulated Characteristics of Lifting Factors and Multipliers

As much as possible, the probability distributions for the risk factors utilized by the RLE (1991) were designed to be representative of the real industrial workplaces (Ciriello et al., 1990; Brokaw, 1992; Karwowski and Brokaw, 1992; Marras et al., 1995). Except for the vertical travel distance factor, coupling and asymmetry multipliers, all factors were defined using lognormal or normal distributions. For all the factors defined as having lognormal distributions, the following procedure was used to adjust for the required range of real values whenever necessary. After the given distribution was generated, it was then adjusted for the range, i.e., truncated for values below the lower limit and above the upper limit. Specifically, the values outside the range were removed, and the new values that would fall within the desired range were generated. This process normalized the derived distributions to the area of 1.0.

Horizontal Factor

The horizontal factor (H) represents the distance of the hands from the midpoint of the ankles, measured in centimeters at the origin and the destination of lift. This horizontal distance was assumed to be the maximum of the two positions, and that no intermediate position was greater than this maximum. The range of horizontal distance as specified by NIOSH (Waters et al., 1993) is from 25 cm to 63 cm, and is used to define the horizontal multiplier (HM). A lognormal distribution for H (mean value of 44 cm, standard deviation of 6 cm) was used to generate random values within the required range. These values were chosen because $H = 44$ cm is the midpoint of the range from 25 cm to 63 cm, and the standard deviation of 6 cm allows almost all the values generated to fall within this range. It should be noted that the lognormal distribution gives greater weight to the values which are located near the mean.

Vertical Height Factor

The vertical height factor (V), indicates the vertical distance of the hands from the floor, with the range from the "floor position" (of 0 cm at a minimum) to the upper limit of vertical reach of 175 cm (Waters et al., 1993). This factor was defined using a lognormal distribution (mean = 100 cm, std = 25 cm). It should be noted that the vertical height factor is also used as an independent variable in the definition of vertical multiplier (VM), the coupling multiplier (CM), and the frequency multiplier (FM).

Distance Multiplier

The vertical travel distance factor (D), i.e., the difference between the origin and destination of the lift, is used by NIOSH (Waters et al., 1993) to define the distance multiplier (DM). For the purpose of this simulation, the vertical travel distance (D) was assumed uniform over the interval of (0, [175-V]). Since the vertical distance traveled is the difference between the origin and the destination heights, this distance can be no greater than the one defined by the origin of lift and the maximum vertical reach limit. A variable called the maximum vertical distance (MAVED) was created based on the vertical factor (V). The travel distance factor (D) was then drawn from a uniform distribution ranging from 0 to MAVED. The MAVED variable was then modified in order to reset all values of D which were less than 25 cm to 25 cm. This procedure guaranteed that the DM was set at 1.0 for such D values.

Asymmetry Factor

The asymmetry factor is defined for the (0 – 135 degrees) range of the upper body twisting angle. This factor was modeled using a lognormal distribution (45, 15), i.e., the mean of 45° and a standard deviation of 15°. This mean and standard deviation provide a natural range from 0° to 90° of twist for the underlying distribution. The mean of 45 degrees was chosen based on the conservative assumption that in most of the industrial lifting tasks the upper body twisting does not exceed 90 degrees. The standard deviation was chosen to ensure that almost all the lifting posture asymmetry values will fall within the NIOSH-specified range limit.

Work Duration Factor

The work duration factor is used to define the frequency multiplier (FM). For the purpose of this simulation, the work duration was conservatively modeled as a normal variable with the mean of 2.0 hours and standard deviation of 2.0 hours. This distribution was then adjusted for the range, i.e., truncated for values of work duration of zero or less, and those which were greater than 8 hours. For the purpose of simulation, the values outside the range of work duration were removed, and the new values that would fall within the desired range were generated. This process normalized the derived distribution to the area of 1.0. These random values were then used in the selection of the frequency multiplier from the NIOSH table for FM (Waters et al., 1993).

Frequency Factor

The frequency variable F (lifts/minute), for the range of lifting frequency from 0.2 lifts/min to 15 lifts/min, is used in the definition of the frequency multiplier (FM). The frequency multipliers (FMs) were selected from the NIOSH table based on randomization of three factors: (1) the frequency of lift, (2) work duration, and (3) vertical height (V). Due to the discrete nature of the values for frequency of lift, the probability distribution function for this factor was generated using normal distribution. The parameters for this distribution were defined as follows: a mean of 4.7 lifts/min, and a standard deviation 1.75 lifts/min. These values were derived based on the information provided by results of industrial surveys of manual lifting tasks. For example, Ciriello et al. (1990) reported that 94% of industrial lifting tasks are performed at frequencies of 4.3 lifts/min or slower. Brokaw (1992) analyzed 31 industrial lifting tasks and reported that in about 80% of these tasks the observed frequency was lower than 5 lifts/min. For the purpose of this simulation, however, it was decided that better representation of the higher lifting frequencies is warranted by the fact that such frequencies of lifting tasks are common today in the service industry, and by the attention given to such tasks by NIOSH (1993), as reflected by their inclusion in the frequency multiplier table. The randomly generated values of the frequency of lift which were less than or equal to zero were eliminated, and the truncated normal distribution was renormalized to equate the area under the curve to 1.0. The resulting distribution was used to generate the probability density function values for entries from the frequency of lift table.

Coupling Multiplier

Coupling describes the quality of coupling between the hand and the load. This quality has three linguistic values, i.e., *good, fair* and *poor.* The discrete values for the coupling multiplier (CM) were chosen based on the results of an industrial survey of 31 manual lifting tasks performed in three different companies reported by Brokaw (1992). This survey showed that slightly less than 10% of the observed lifting tasks had good couplings, about 64% had fair couplings, while about 26% had poor couplings. Due to the limited sample size of the above study, it was decided to allow for more liberal coupling assumption for the simulation purposes. Therefore, the probability of good hand-to-container coupling was set at 0.3, for fair coupling at 0.5, and the probability for poor coupling was set at 0.2.

58.6 Results of Simulation: Threshold Values of RWL

The results showed that over all conditions studied (represented by 100,000 randomly generated lifting task scenarios), the threshold RWL values (the RWL values which account for 99.5% of all simulated lifting tasks) were equal to or lower than: (1) 13.0 kg (or 28.6 lbs) for up to one hour of lifting, (2) 12.5 kg (or 26.4 lbs) for less than 2 hours of exposure, and (3) 10.5 kg (or 23.1 lbs) for lifting over an 8-hour shift. The above results indicate that under most of the examined lifting conditions (99.5% of the simulated cases), one can reasonably expect that an implementation of the RLE (1991) model at the level of Lifting Index of 1.0, which is designed to protect 90% of the mixed industrial working population, would necessitate redesign of those manual lifting tasks for which the load lifted exceeds the TRWL values reported above. From a practical point of view, these results define the threshold RWL values that can be used by practitioners for the purpose of immediate risk assessment of manual lifting tasks performed in industry.

58.7 Development of the Population-Based Design Load Threshold Limits

The next step in development of the population-based design load threshold limits (LTL) for manual lifting tasks was to define the threshold lifting index (TLI) values for different percentiles of capable male and female populations. This was done by using the scaling factors derived from the psychophysical data reported by Snook and Ciriello (1991), which account for different percentiles of the capable U.S. population. The scaling factors were defined in reference to 90% of male and 75% of female populations, at which their values were equated to 1.0 (e.g., LI = 1.0). These scaling factors (see Table 58.1) were derived based on the ratios between the 10th, 50th, and 90th percentile values of the maximum acceptable weights of lift (MAWL) under frequency of 1 lift per minute (box width

TABLE 58.1 Values of Threshold Lifting Index (TLI) for Male and Female Populations

Percent capable	Males V < 75	Males V > 75	Males SR	Females V < 75	Females V > 75	Females SR
90%	1.00	1.00	1.00	0.80	0.86	0.86
80%	1.33	1.24	1.22	0.95	0.96	0.96
70%	1.56	1.40	1.37	1.05	1.03	1.03
60%	1.76	1.54	1.50	1.13	1.10	1.09
50%	1.95	1.67	1.62	1.22	1.16	1.15
40%	2.14	1.81	1.75	1.30	1.22	1.21
30%	2.34	1.95	1.88	1.38	1.28	1.27
20%	2.57	2.11	2.03	1.48	1.35	1.34
10%	2.90	2.35	2.25	1.62	1.44	1.44

V = vertical height of lift (cm)
SR = shoulder-to-reach height

of 34 cm), assuming normal distribution of these values. Three lifting conditions with vertical heights of lift, i.e.,: (1) V < 75 cm, (2) V>75 cm, and the shoulder-to-reach, were adopted for the purpose of calculating the scaling factors for males and females based on the appropriate percentile distribution values (Snook and Ciriello, 1991).

The derived scaling factors were defined as the threshold lifting index (TLI) values (see Table 58.1). It should be noted that all of the TLI values are less than 3. For comparison purposes, the results of the field study of 24 lifting jobs (Karwowski et al., 1994) revealed the value of lifting index of 4.1 (at lift destination), which corresponds to the lumbar compressive strength of 4100 N (with one standard deviation) for 40-year-old males, as proposed by Jager and Luttman (1992).

The population-based load threshold limits (LTL) corresponding to different percentiles of capable workers were defined as follows:

$$\mathrm{LTL} = \mathrm{TRWL} \times \mathrm{TLI}\,[\mathrm{kg}]$$

The derived distributions of load threshold limits are illustrated in Figures 58.1 and 58.2, for males and females, respectively. Furthermore, Tables 58.2 and 58.3 show the LTL values as the function of work exposure and vertical lifting height. These values range from the most capable population of workers (10% value) to the least capable (90% value), and are expressed as a function of: (1) task exposure (up to one hour, one-to-two hours, and eight hours of work), and (2) vertical location of the load, i.e., the vertical height of the hands above the floor (V < 75 cm and V > 75 cm). In addition, the LTL values for lifting from shoulder-to-reach (SR) are also given. It should be remembered that the LTL values represent the threshold RWLs which apply to 99.5% of all possible simulated cases based on the NIOSH (1991) RLE model, and, therefore, represent the largest variation of all possible lifting tasks that can be encountered in industry today.

TABLE 58.2 Load Threshold Limits (LTL) for Male Population (kg) at Different Work Exposures

Percent	V < 75			V > 75			Shoulder-to-reach		
capable	1 hr	2 hrs	8 hrs	1 h	2 hrs	8 hrs	1 hr	2 hrs	8 hrs
90%	13.00	12.50	10.50	13.00	12.50	10.50	13.00	12.50	10.50
80%	17.29	16.63	13.97	16.12	15.50	13.02	15.86	15.25	12.81
70%	20.28	19.50	16.38	18.20	17.50	14.70	17.81	17.13	14.39
60%	22.88	22.00	18.48	20.02	19.25	16.17	19.50	18.75	15.75
50%	25.35	24.38	20.48	21.71	20.88	17.54	21.06	20.25	17.01
40%	27.82	26.75	22.47	23.53	22.63	19.01	22.75	21.88	18.38
30%	30.42	29.25	24.57	25.35	24.38	20.48	24.44	23.50	19.74
20%	33.41	32.13	26.99	27.43	26.38	22.16	26.39	25.38	21.32
10%	37.70	36.25	30.45	30.55	29.38	24.68	29.25	28.13	23.63

V = vertical height of lift (cm)

TABLE 58.3 Load Threshold Limits (LTL) for Female Population (kg) at Different Work Exposures

Percent	V < 75			V > 75			Shoulder-to-reach		
capable	1 hr	2 hrs	8 hrs	1 h	2 hrs	8 hrs	1 hr	2 hrs	8 hrs
90%	10.40	10.00	8.40	11.18	10.75	9.03	11.18	10.75	9.03
80%	12.35	11.88	9.98	12.48	12.00	10.08	12.48	12.00	10.08
70%	13.65	13.13	11.03	13.39	12.88	10.82	13.39	12.88	10.82
60%	14.69	14.13	11.87	14.30	13.75	11.55	14.17	13.63	11.45
50%	15.86	15.25	12.81	15.08	14.50	12.18	14.95	14.38	12.08
40%	16.90	16.25	13.65	15.86	15.25	12.81	15.73	15.13	12.71
30%	17.94	17.25	14.49	16.64	16.00	13.44	16.51	15.88	13.34
20%	19.24	18.50	15.54	17.55	16.88	14.18	17.42	16.75	14.07
10%	21.06	20.25	17.01	18.72	18.00	15.12	18.72	18.00	15.12

V = vertical height of lift (cm)

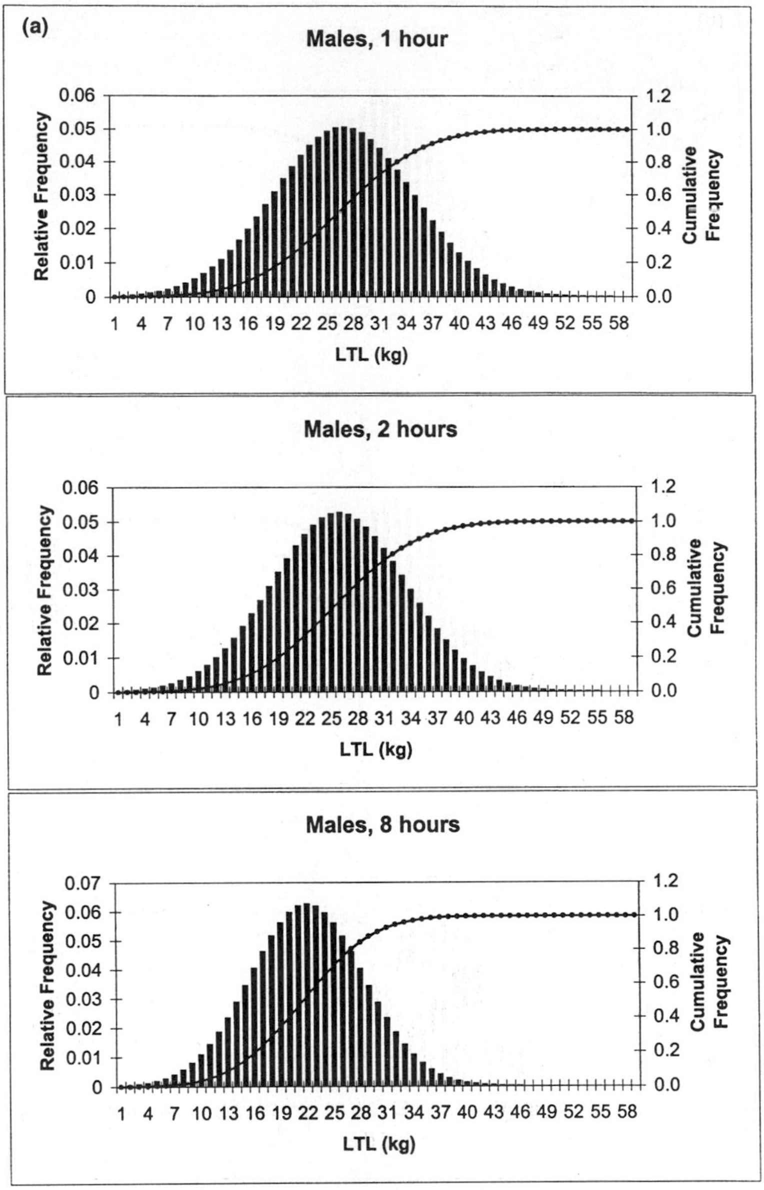

FIGURE 58.1 (a) Distributions of load threshold limits (LTL) for males (V < 75 cm). (b) Distributions of load threshold limits (LTL) for males (V > 75 cm).

58.8 The LTL Model Application

The population-based load threshold limits could be very useful for immediate risk assessment of manual lifting tasks performed in industry at large. Such values of lifted loads, if exceeded, provide an indication of the need for a more thorough examination of the manual lifting tasks performed in a given industrial setting, and, if appropriate, evaluation of the physical capacity of the exposed workers. It should be noted that the developed model does not imply that the LTL values are safe for any given individual performing a specific manual lifting task. Rather, these data should be used a guide for design and evaluation of manual lifting tasks performed in different types of industry by a given percentile of capable workers,

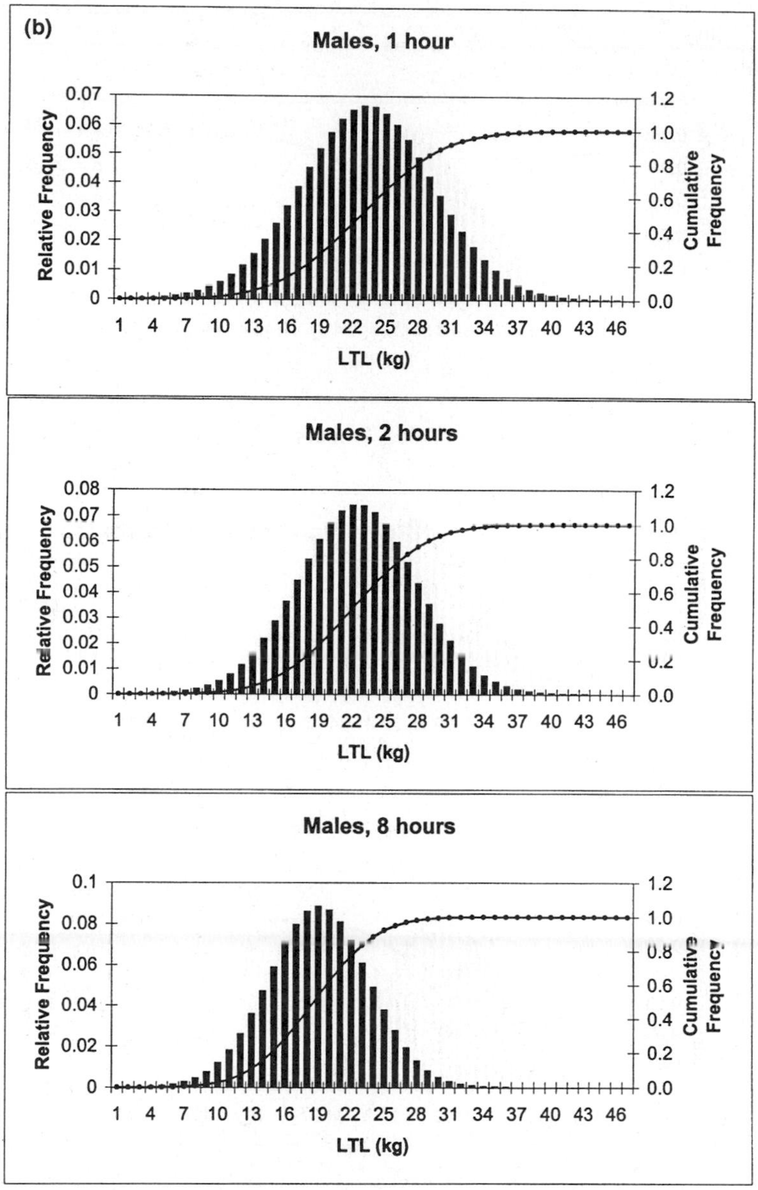

FIGURE 58.1 (continued)

until more objective epidemiological evidence regarding the causal relationship between the loads lifted on the job and related risk of low-back injury is available (Dempsey, 1998; Karwowski et al., 1998).

It should be remembered that the LTL data apply to 99.5% of the expected cases (i.e., potential lifting tasks that can be expected to occur in real industrial environments), and would be somewhat greater for 0.5% of the remaining cases. Since the TRWL concept, which integrates all aspects of the NIOSH revised lifting equation (1991), was used as the basis for development of the population-based design limits, the proposed LTL approach is applicable to a large variety of possible lifting task conditions as it integrates all lifting factors considered by the RLE model. Finally, the developed threshold load limit approach extends application of the original RWL concept to different percentiles of capable male and female populations, making it much more relevant and applicable to the real life situations.

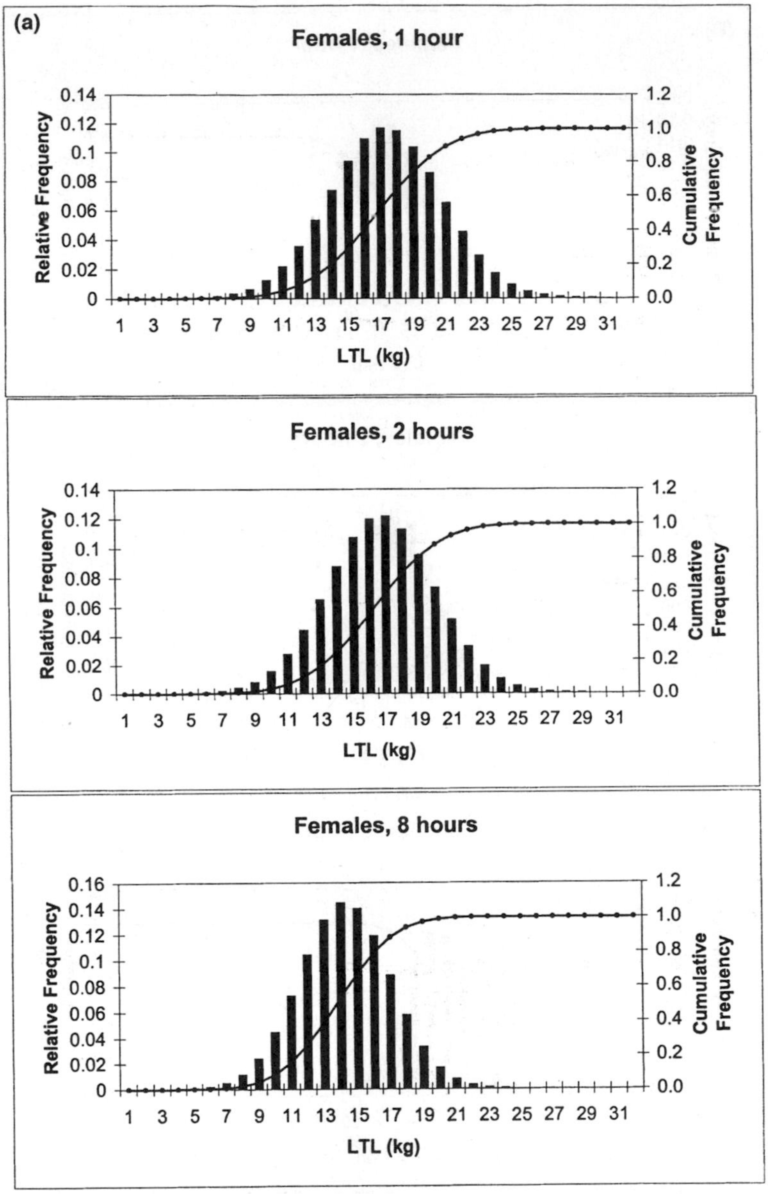

FIGURE 58.2 (a) Distributions of load threshold limits (LTL) for females (V < 75 cm). (b) Distributions of load threshold limits (LTL) for females (V > 75 cm).

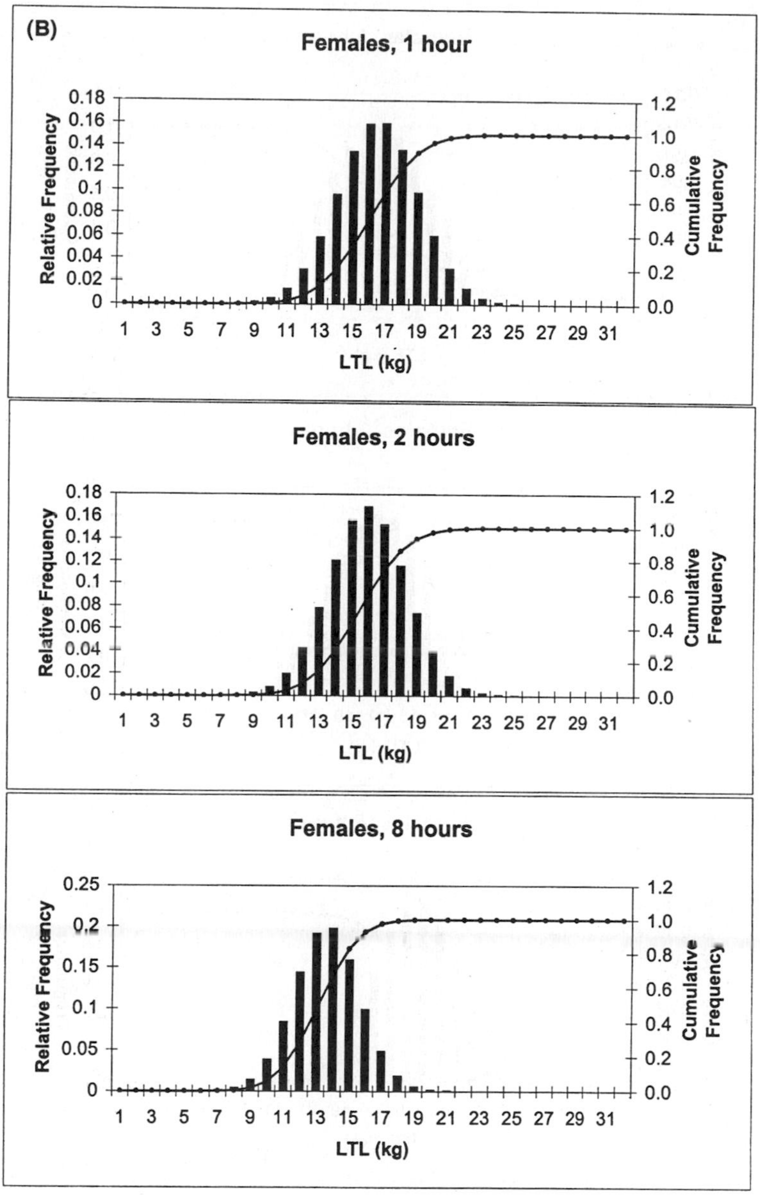

FIGURE 58.2 (continued)

References

Ayoub, M. M., Dempsey, P. G., and Karwowski, W. 1997, Manual materials handling, in *Handbook of Human Factors & Ergonomics*, G. Salvendy, Ed., John Wiley & Sons, New York, pp. 1085-1123.

Brokaw, N. 1992, Implications of the Revised NIOSH Lifting Guide of 1991: A Field Study. Unpublished M.S. Thesis, Department of Industrial Engineering, University of Louisville, Louisville, Kentucky.

Dempsey, P. 1998, A critical review of biomechanical, epidemiological, physiological and psychophysical criteria for designing manual materials handling tasks. *Ergonomics*, 41 (1) 73-88.

Jager, M. and Luttman, A. 1992, The load on the lumbar spine during asymmetrical bi-manual materials handling. *Ergonomics*, 35 (7/8), 783-805.

Karwowski, W. and Brokaw, N. 1992, Implications of the Proposed Revisions in a Draft of the Revised NIOSH Lifting Guide (1991) for Job Redesign: A Field Study, *Proceedings of the 36th Annual Meeting of the Human Factors Society*, Santa Monica, California, pp. 659-663.

Karwowski, W. 1992, Comments on the assumption of multiplicity of risk factors in the draft revisions to NIOSH Lifting Guide, in S. Kumar (Ed.), *Advances in Industrial Ergonomics and Safety IV*, Taylor & Francis, London, pp. 905-910.

Karwowski, W., Caldwell, M. and Gaddie, P. 1994, Relationship between the NIOSH lifting index, compressive and shear forces on the lumbosacral joint, and low back injury incidence rate based on industrial field study, in *Proceedings of the 36th Annual Meeting of the Human Factors Society*, Santa Monica, California, pp. 654-659.

Karwowski, W., Lee, W. G, Jamaldin, B., Gaddie, P., and Jang, R. 1998, Beyond psychophysics: a need for cognitive modeling approach to setting limits in manual lifting tasks, *Ergonomics*, in press.

Marras, W. S., Lavender, S. A., Leurgans, S. E., Fathallah, F. A., Ferguson, S. A., Allread, W. G., and Rajulu, S. L. 1995, Biomechanical risk factors for occupationally related low back disorders. *Ergonomics*, 38, 377-410.

Snook, S. H. and Ciriello, V. M. (1991). The design of manual handling tasks: Revised tables of maximum acceptable weights and forces. *Ergonomics*, 34 (9), 1197-1213.

Waters, T. R., Putz-Anderson, V., Garg, A., and Fine, L.J. 1993, Revised NIOSH equation for the design and evaluation of manual lifting tasks, *Ergonomics*, 36, 749-776.

Waters, T. R. and Putz-Anderson, V. 1998, Revised NIOSH Lifting Equation, in W. Karwowski and W. S. Marras (Eds.), *Handbook of Occupational Ergonomics*, CRC Press, Boca Raton.

59

Occupational Low Back Disorder Risk Assessment Using the Lumbar Motion Monitor

William S. Marras
The Ohio State University

W. Gary Allread
The Ohio State University

Richard G. Ried
The Ohio State University

59.1 Background

Occupational Back Injuries

Low back disorders (LBD) are among the most common occupationally related injuries. According to Andersson (1997), LBDs affect an estimated 80% of the population during their working career. The National Center for Health Statistics (1977) has documented that LBDs are the prime reason for activity limitation for those under 45 years of age. Cats-Baril (1996) has shown that LBDs cost society up to 100 billion dollars annually. Despite the prevalence and cost of these injuries, there are relatively few accurate methods available to predict the risk of occupationally related LBDs.

Tools for Analyzing Low Back Injury Risk

There is a host of tools available to evaluate LBD risk in industrial jobs. These tools vary widely in their assumptions and complexity. When choosing an analysis technique it is important to match the capabilities of tool to the characteristics of the job being evaluated.

0-8493-2641-9/99/$0.00+$.50

Two- and three-dimensional static models are easy to apply but assume that motion is not a significant factor in LBD risk. The 1981 and 1991 NIOSH lifting equations are also easy to apply but assume that all movements are slow and smooth. Within their limitations, these methods can provide insight on LBD risk. However, recent research has suggested that motion may play a role in LBD risk.

Numerous epidemiologic studies have specifically indicated that the risk of LBD increases when dynamic lifting occurs. Data from a retrospective study of 4,645 injuries by Bigos et al. (1986) suggest that there were greater reports of LBD with dynamic tasks relative to awkward static tasks. Magora (1973) concluded that lateral bending and twisting were only significant risk factors when they occurred simultaneously with sudden (quick) movements. Punnett et al. (1991) studied non-neutral postures in automobile assembly plants and reported that postural stress to the back was more dynamic than static. All of these studies have indicated that the risk of LBD increases with dynamic activity, especially when the body moves asymmetrically. Therefore, if the job has a dynamic component it is important to choose an analysis method that accounts for the additional risk that motion imparts.

Video-based motion analysis systems offer one way to study the dynamic component of a job, but they have several drawbacks when used in an industrial setting. Video assessments must take place in a calibrated space of usually no more than 2 to 3 cubic meters. Cameras must be carefully placed to obtain data for all three planes of motion, and time-consuming analysis is necessary to obtain usable data. In industrial environments, tasks often involve movement outside the calibrated space, work areas limit camera placement, and a great deal of labor must be expended to analyze a few minutes of usable data. These limitations often make video-based systems impractical for routinely evaluating a large number of industrial jobs.

59.2 Development of the Lumbar Motion Monitor

Physical Description

The Lumbar Motion Monitor (LMM) (Figure 59.1) was developed in the Biodynamics Laboratory at The Ohio State University in response to the need for a practical method of assessing the dynamic component of occupationally related LBD risk in industrial settings. The patented LMM is a triaxial electrogoniometer that acts as a lightweight exoskeleton of the lumbar spine. It is positioned on the back of a subject directly in line with the spine and attached by harnesses at the pelvis and thorax. Four potentiometers at the base of the LMM measure the instantaneous position of the spine (as a unit) in three-dimensional space relative to the pelvis. Position data from the potentiometers are recorded at 60 Hz, transmitted to an analog-to-digital (A/D) converter, and then recorded on a microcomputer. The data are then processed to calculate the position, velocity, and acceleration of the spine in each of the three planes of motion as a function of time.

Calibration and Validation

During data collection and analysis, each LMM is used with its matching calibration file. Before its first use, each LMM is calibrated individually, using a specially designed reference frame (Figure 59.2). During the calibration process the LMM is positioned on the calibration frame at 225 different positions in three-dimensional space and the voltage outputs recorded. The calibration process eliminates any individual variability in each LMM.

The LMM was validated to ensure its accuracy and sensitivity with a video-based motion analysis system (Marras et al., 1992). During the validation process the predicted velocities and accelerations of the LMM were compared relatively (to position accuracy) against the predicted values of the motion analysis system. The results of the validation process (Table 59.1) show high correlation coefficient values and significance levels ($r > .95$, $p < 0.0001$) for all three planes of motion. An independent group has also determined that the reproducibility of the LMM is suitably high for range of motion and velocity for the device to be used for evaluation in a clinical and research setting (Gill and Callaghan, 1996).

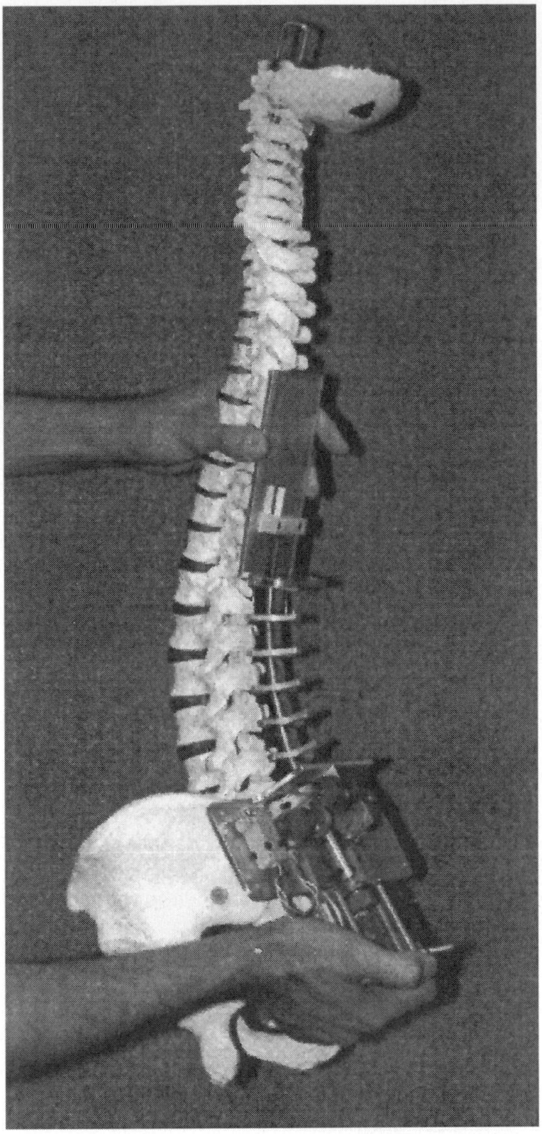

FIGURE 59.1 The Lumbar Motion Monitor (LMM) compared to an anatomical model of the spine.

Development of the LBD Risk Model

An *in vivo* study was undertaken to determine quantitatively whether dynamic trunk motions in combination with workplace and environmental factors may better describe the risk of LBD in repetitive manual materials handling (MMH) tasks.

Approach

The study involved an industrial surveillance of the trunk motions and workplace factors involved in high- and low-risk repetitive MMH jobs. The approach used in this phase of the project was to (1) identify industries involved with repetitive MMH work; (2) examine the company medical records as well as the health and safety records to identify those repetitive MMH jobs that were associated, historically, with either a high or low risk of occupationally related LBD; (3) quantitatively monitor the trunk motions and workplace factors associated with each of these jobs; and (4) evaluate the data to determine which combination of trunk motion and workplace factors was most closely associated with LBD risk.

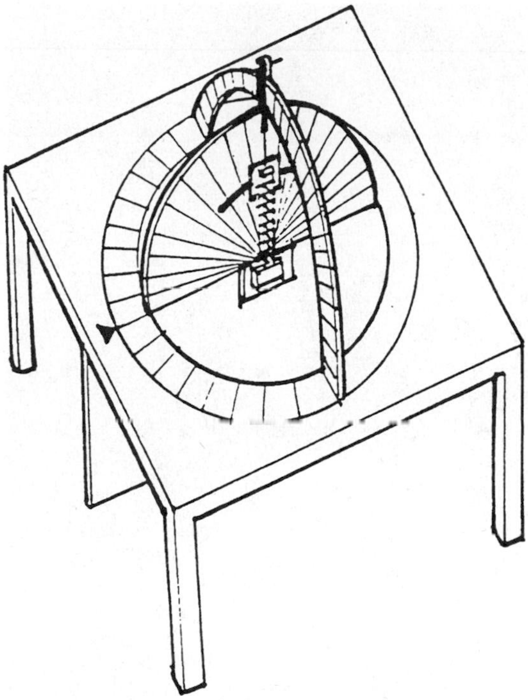

FIGURE 59.2 The three-dimensional reference frame.

TABLE 59.1 Correlations of the Velocities and Accelerations of the Motion Analysis System in the Three Planes of Motion with the Velocities and Accelerations of the LMM

	Correlation*	
Plane	Velocity	Acceleration
Lateral	0.95	0.95
Sagittal	0.99	0.96
Twisting	0.99	0.99

* $p < 0.0001$

Study Design

This study was a cross-sectional study of 403 industrial jobs from 48 manufacturing companies throughout the midwestern United States. Only repetitive jobs without job rotation were examined in this study. This was necessary to prevent the confounding effects created by alternate jobs. Jobs examined in this study were divided into two groups, high and low risk of LBD, based on examination of the injury and medical records. Whenever possible, company medical reports were used to categorize risk. In some cases only injury logs (OSHA 200 logs) were available. All medical reports, injury records and logs were scrutinized to ensure they were as accurate as possible. The outcome measure (LBD risk) derived from these medical and injury records consisted of the normalized rate of reported occupationally related LBD. Incidence of reported LBD was considered regardless of whether there was any restricted or lost time associated with the incident.

Thus, the dependent variable in this study consisted of two levels of job-related LBD risk categories. Low-risk group jobs were defined as those jobs with at least three years of records showing no injuries and

no turnover. Turnover is defined as the average number of employees who left a job per year. High-risk group jobs were those jobs associated with at least 12 injuries per 200,000 hours of exposure (average of 26/200,000 hours). The high-risk group category incidence rate corresponds to the 75th percentile value of risk for the 403 jobs examined. Of the 403 jobs examined, 124 of the jobs were categorized as low-risk and 111 were categorized as high-risk. The remainder of the jobs (168) were categorized as medium-risk and were not used in this particular analysis.

The independent variables in this study consisted of workplace, individual, and trunk motion characteristics that were indicative of each job. The workplace and individual characteristics consisted of variables typically considered in current workplace guidelines for materials handling (NIOSH, 1981; Putz-Anderson and Waters, 1991). Specifically, these variables were: (1) the maximum horizontal distance of the load from the spine; (2) the weight of the object lifted; (3) the height of the load at the origin of the lift; (4) the height of the load at the destination of the lift; (5) the frequency of lifting (lift rate); (6) the asymmetric angle of the lift (as defined by NIOSH, 1981); (7) employee anthropometry (12 measures); (8) employee injury history; (9) employee satisfaction; and (10) trunk motion. Trunk motion characteristics were those variables obtained using the LMM. These variables consisted of the trunk angular position, velocity, and acceleration characteristics (i.e., means, ranges, maximums, minimums, etc.) in each of the cardinal planes. Selected trunk motion factors along with selected workplace factors were used to develop a quantitative model of occupational risk factors.

Data Collection

Initially, data about employee health, employment history, and anthropometry were collected. Next, the employee was fitted with an LMM. A baseline reading from the LMM was then taken, while the individual stood erect and rigid. The employee then was asked to return to work and wore the LMM for at least ten job cycles. Thus, the length of time the employee wore the monitor depended upon the cycle time of the job. Monitoring of back motion was initiated as the employee began the MMH task and concluded when the employee completed the task. Extraneous activities not involving MMH were not monitored. Signals from the LMM were sampled at 60 Hz via an analog-to-digital converter and stored on a portable microcomputer. The data were further processed in the laboratory to determine position, velocity, and acceleration of the trunk as a function of time in the sagittal, lateral, and axial twisting planes of the body.

Analysis

The data were initially examined to determine whether the trunk motions were repeatable. This analysis indicated that more than half of the variation was attributable to the job. Hence, trunk motions were dictated largely by the design of the task, and repetitive trials resulted in motions that were fairly similar.

The various personal, environmental, and workplace factors from the database were analyzed using logistic regression techniques to determine if any single factor could distinguish jobs associated with high-risk group membership from those that were not. The most powerful single variable was maximum moment, which yielded an odds ratio of 5.17. Overall, however, the odds ratios were low, indicating that few of the individual variables discriminate well between high- and low-risk jobs. Of the trunk motion factors, the velocity variables generally produced greater odds ratios than maximum or minimum position, range of motion, or acceleration. Table 59.2 shows the descriptive statistics of the workplace and trunk motion factors for the high- and low-risk groups.

Next, multiple logistic regression was used to predict the probability of high-risk group membership as a function of the values of several workplace and trunk motion factors. A five-variable model incorporating the trunk motion and workplace factors was developed and further refined after examining a series of stepwise logistic regression models (containing different variables, e.g., velocity, acceleration) fitted to several intermediate data sets. A combination of five variables (moment, frequency of lift, sagittal flexion, twisting velocity, and lateral velocity) was found to have the greatest odds of predicting high-risk group membership. This combination of workplace and trunk motion factors forms the basis of the LBD Risk Model. The model was selected for the statistical importance of the predictors and for biomechanical plausibility. The model variables remained consistent when tested with the various intermediate

TABLE 59.2 Descriptive Statistics of the Workplace and Trunk Motion Factors in Each of the Risk Groups

Factors	High Risk (N = 111)				Low Risk (N = 124)			
	Mean	SD	Min	Max	Mean	SD	Min	Max
WORKPLACE FACTORS								
Lift Rate (lifts/hr)	175.89	8.65	15.30	900.00	118.83	169.09	5.40	1500.00
Vertical load location at origin (m)	1.00	0.21	0.38	1.80	1.05	0.27	0.18	2.18
Vertical load location at destination (m)	1.04	0.22	0.55	1.79	1.15	0.26	0.25	1.88
Vertical distance traveled by load (m)	0.23	0.17	0.00	0.76	0.25	0.22	0.00	1.04
Average weight handled (N)	84.74	79.39	0.45	423.61	29.30	48.87	0.45	280.92
Maximum weight handled (N)	104.36	88.81	0.45	423.61	37.15	60.83	0.45	325.51
Average horizontal distance between load and L_5/S_1 (N)	0.66	0.12	0.30	0.99	0.61	0.14	0.33	1.12
Maximum horizontal distance between load and L_5/S_1 (N)	0.76	0.17	0.38	1.24	0.67	0.19	0.33	1.17
Average moment (Nm)	55.26	51.41	0.16	258.23	17.70	29.18	0.17	150.72
Maximum moment (Nm)	73.65	60.65	0.19	275.90	23.64	38.62	0.17	198.21
Job satisfaction	5.96	2.26	1.00	10.00	7.28	1.95	1.00	10.00
TRUNK MOTION FACTORS								
Sagittal Plane								
Maximum extension position (°)	−8.30	9.10	−30.82	18.96	−10.19	10.58	−30.00	33.12
Maximum flexion position (°)	17.85	16.63	−13.96	45.00	10.37	16.02	−25.23	45.00
Range of motion (°)	31.50	15.67	7.50	75.00	23.82	14.22	399.00	67.74
Average velocity (°/sec)	11.74	8.14	3.27	48.88	6.55	4.28	1.40	35.73
Maximum velocity (°/sec)	55.00	38.23	14.20	207.55	38.69	26.52	9.02	193.29
Maximum acceleration (°/sec²)	316.73	224.57	80.61	1341.92	226.04	173.88	59.10	1120.10
Maximum deceleration (°/sec²)	−92.45	63.55	−514.08	−18.45	−83.32	47.71	−227.12	−4.57
Lateral Plane								
Maximum left bend (°)	−1.47	6.02	−16.80	24.49	−2.54	5.46	−23.80	13.96
Maximum right bend (°)	15.60	7.61	3.65	43.11	13.24	6.32	0.34	34.14
Range of motion (°)	24.44	9.77	7.10	47.54	21.59	10.34	5.42	62.41
Average velocity (°/sec)	10.28	4.54	3.12	33.11	7.15	3.16	2.13	18.86
Maximum velocity (t/sec)	46.36	19.12	13.51	119.94	35.45	12.88	11.97	76.25
Maximum acceleration (°/sec²)	301.41	166.69	82.64	1030.29	229.29	90.90	66.72	495.88
Maximum deceleration (°/sec²)	−103.65	60.31	−376.75	0.00	−106.20	58.27	−294.83	0.00
Twisting Plane								
Maximum left twist (°)	1.21	9.08	−27.56	29.54	−1.92	5.36	−30.00	11.44
Maximum right twist (°)	13.95	8.69	−13.45	30.00	10.83	6.08	−11.20	30.00
Range of motion (°)	20.71	10.61	3.28	53.30	17.08	8.13	1.74	38.59
Average velocity (°/sec)	8.71	6.61	1.02	34.77	5.44	3.19	0.66	17.44
Maximum velocity (°/sec)	46.36	25.61	8.06	136.72	38.04	17.51	5.93	91.97
Maximum acceleration (°/sec²)	304.55	175.31	54.48	853.93	269.49	146.65	44.17	940.27
Maximum deceleration (°/sec²)	−88.52	70.30	−428.94	−5.84	−100.32	72.40	−325.93	−2.74

data sets. The empirical stability of the model was checked by predicting the classification of 100 jobs based on the preliminary model. This model resulted in an odds ratio of 10.6.

By averaging individual probability values for moment, frequency of lift, sagittal flexion, twisting velocity, and lateral velocity, the LBD Risk Model is able to predict the probability of high-risk group membership (LBD risk) for any repetitive job (Figure 59.3). It is important to understand that the predictive power of the model is a result of the interaction of these five variables. Individually, each of these factors is unable to reliably distinguish between a high-risk and a low-risk situation, but when they are considered in combination the predictive power increases tenfold.

The LBD Risk Model is currently being validated in a longitudinal study. Jobs are prospectively tracked (over time) to determine if changes in injury rates after the implementation of ergonomic changes correspond with predicted value changes in the LBD Risk Model. Preliminary results indicate the changes

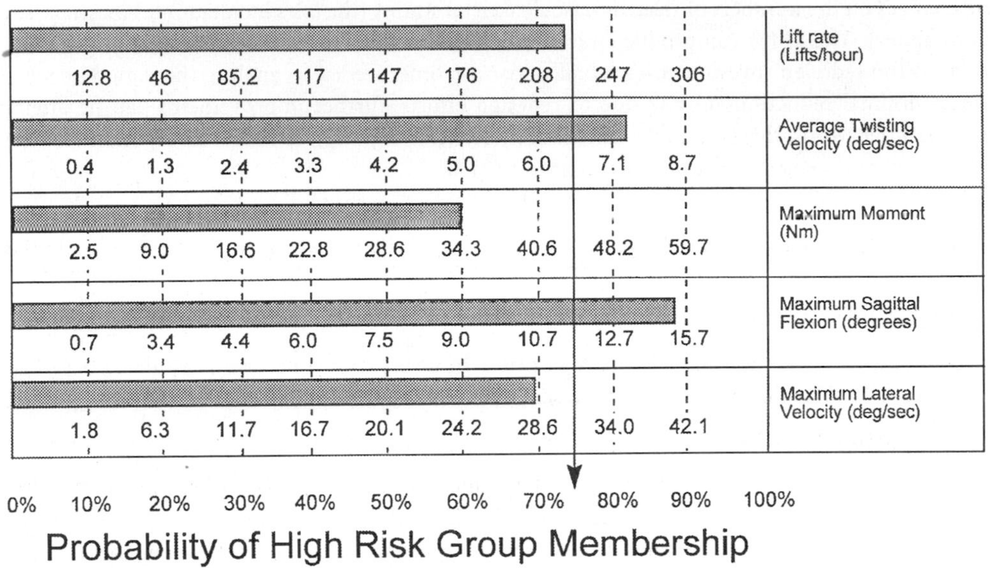

FIGURE 59.3 The LBD Risk Model showing the five risk variables scaled relative to risk. The arrow points to the overall probability of high risk of LBD group membership for a particular job.

in LMM probability of risk corresponds well with an observed change in actual injury rates after changes have been made to the work environment.

59.3 Benefits of the LMM and the LBD Risk Model

Use of the LMM and its computation of risk provides several advantages to the ergonomist studying injury risk for material handling activities. First, the LMM allows one to determine the instantaneous three-dimensional position of the trunk while individuals perform their actual job tasks, not work simulated instead in a laboratory. This ability eliminates the question of whether a material handling study conducted in an artificial setting can be generalized to the workplace. Also, the LMM data are gathered objectively. Resulting calculations of injury risk are determined irrespective of an investigator's (perhaps unintentionally biased) view of the work.

A second advantage of using the LMM system is that material handling jobs can be assessed relative to a large database covering jobs from diverse manufacturing environments and with different levels of risk. This enables the ergonomist to determine if a job has a high probability of injuring employees who perform that job. It also allows one to rank several jobs, based on their risk values, and to study solutions for those having the greatest chance of producing injury.

Third, the probability model enables the ergonomist to quantitatively assess and compare each task within a job. Specific factors that contribute to a task's probability are identified, as are the tasks that most contribute to the job's overall risk. This information pinpoints the specific tasks and the factors therein that must be addressed during job redesign to reduce the job's injury risk potential. This method of identifying and changing only those components that contribute to risk eliminates the need to change the entire job. This quantitative assessment permits the ergonomist to make decisions about what level of risk is acceptable for a job. It is not possible to totally reduce a job's risk to zero. However, the LBD Risk Model allows one to determine if a job's risk is above a criteria of acceptance set by the company or the ergonomist.

A fourth benefit of this LBD Risk Model is the assistance it can provide to the ergonomic intervention process. Modified jobs can be re-monitored using the LMM, and the effects of those changes can be quantified and compared with those values determined prior to the intervention. Traditionally, the effects

of job changes on the numbers of related musculoskeletal strains (the job's incident rate) may take several years to appear. The LMM can produce more timely feedback to the ergonomist regarding anticipated returns on the redesign investment — actually, just the time needed to analyze the data. For jobs that produce minimal reductions in risk due to redesign efforts, further improvements can be attempted sooner. In other words, this risk model can assist in determining whether or not an ergonomic intervention has produced *enough* of an improvement in the job.

59.4 Applications: How to Use the LMM and LBD Risk Model

Recommended Equipment

It is important to make sure that all equipment used in LMM data collection is available before going to a job site. The following list of items represents the essential equipment needed.

Lumbar Motion Monitors

Two sizes of LMMs are available commercially, large and small. These LMMs differ only in their length; the large LMM is suitable for taller individuals or individuals of moderate height who must bend forward extremely far in their work. The small LMM is more appropriate for shorter individuals or those of moderate height who do little forward bending.

LMM Harnesses

Elastic waist and shoulder harnesses are supplied with the commercial LMMs and consist of three sizes: large, medium, and small. The large harnesses are for taller or heavier people; the medium harnesses are for those with average body builds, and the small harnesses are for shorter individuals or those with slighter builds. All harnesses have adjustable Velcro straps to allow for individual differences. The shoulder harness is fitted over one's torso before the waist harness and LMM are attached. The waist harness attaches to the LMM before it is placed on the individual.

Laptop/Notebook Computer

The LMM software requires a PC-based operating system. (The software will be discussed in a later section.) The computer should have, as a minimum, a 486 processor. However, the faster the processing speed of the computer used, the more efficiently data will be collected and stored. Most industrial environments will have 110-volt outlets for computers, though battery-operated laptops may make data collection more flexible. Be aware of the length of battery life before recharge, however.

LMM/Computer Connections

A cable that connects the LMMs to the computer is supplied with the LMMs. This cable transmits information regarding trunk positions from the LMM to the computer for storage. It is important that this cable be in good working order, since data collection in an industrial environment can often put heavy wear on equipment. Also available is a digital telemetry system. This "wireless" system consists of a transmitter and battery pack, worn on the waist harness, and a receiver unit, placed alongside the computer. Both systems (the cable and the telemetry unit) provide the same information. The telemetry system, however, allows data to be collected without the cable connection between the monitored individual and the computer, providing greater freedom of movement.

Other Equipment

The LMM provides data regarding the trunk motions required of a job. However, other information is important and needs to be collected at the job site to fully assess the job's risk. Listed below is additional equipment that should be part of the LMM gear.

Scale and Push/Pull Gauge — It is critical to record the weights of items handled by individuals during data collection. A heavy-duty scale is needed, one that is capable of reading object weights from one to 100 pounds. Occasionally, during material handing, objects are not lifted but are pushed

or pulled instead. A gauge capable of recording these forces also is important to have. Both devices are available commercially.

Tape Measure— A heavy-duty tape measure is needed to record the reach distance and the vertical and horizontal locations of objects that are handled at the job site. The size of the objects themselves also may be of importance and can be measured using a tape.

Extension Cord/Outlet Strip— Power outlets in industry are not always conveniently located near where data collection is to take place. An extension cord permits greater flexibility in locating the computer closer to the job site and in full view of the individual being monitored. An outlet strip also allows the use of other electronic devices such as video recorders and battery chargers.

Wheeled Cart— Collapsible carts are useful to move equipment into the job site, and from one site to another, if multiple sites are to be monitored.

Selecting the Job(s) to Monitor

The LMM can be used to monitor the positions, velocities, and accelerations of the trunk in the three cardinal planes for any job in which trunk movements are required. This type of information can be important when one needs to describe the motion characteristics required of a job or activity. The LBD Risk Model is composed of five workplace factors that, when combined, assess a job's probability of having a high number of LBDs. This model is also useful for establishing relative comparisons between several tasks that comprise a job.

The industrial jobs used to develop the risk model had MMH frequencies ranging from approximately six to 1500 lifting tasks required per hour. Thus, the tasks to be assessed using this risk model should fall within this scope of lifting rates. Most MMH jobs fit this profile. For example, in automobile assembly, job cycles are repeated every one to two minutes, and the parts themselves often are standard in size and weight. Palletizing jobs may require very different types of objects to be handled, but the task of continual lifting from one storage area to another remains the same. These types of jobs are very consistent with those used to develop the risk model database, and the job's LBD risk can be easily assessed with the LMM's software.

Some jobs are less repetitive or have more job variation than assembly or palletizing tasks. However, the LBD risk model still can be used to make relative comparisons between the tasks that comprise the job. For jobs that require a larger number of tasks, the risk model is helpful in comparing the factors that make up the model. This will allow for the ergonomist to assess trade-offs between such factors as lifting frequency, object weights handled, and the trunk motions required for the different tasks. It should be noted that there is inherent variability in the way an individual may perform the same task repeatedly. Because of this difference, a minimum of five repetitions should be collected of any given task.

Another issue to be considered is job rotation. Employers often use a variety of rotation schemes for job processes. If the job to be monitored requires no rotation (employees perform the same job every day/week/month), then the risk assessment can be directly related to the tasks observed. Jobs in which individuals rotate regularly between a few work areas also may be used in assessing LBD risk, if this rotation schedule is fixed. When the job rotation requires employees to do completely different jobs on an hourly or weekly basis, it becomes difficult to relate a task's risk values to the overall job risk, since many tasks could contribute to the risk assessment. This issue is key to determining a job's suitability for LBD risk assessment. That is, does the job's work structure enable one to define the job in terms of a few repeatable, consistently performed tasks?

There may be some jobs that fit within the LBD Risk Model profile but still should not be monitored. Seated jobs may require repetitive activities, but they usually are not ones that require significant material handling. In any event, the LMM may rub against a chair's back or the waist harness will shift from its position on the hips during seated work, and erroneous LMM output will result. Also, jobs that require close contact of the LMM with a finished product could produce scratches on the product, and the employer may not want to risk product damage. Finally, exposure to water or other liquids may damage the LMM or its components.

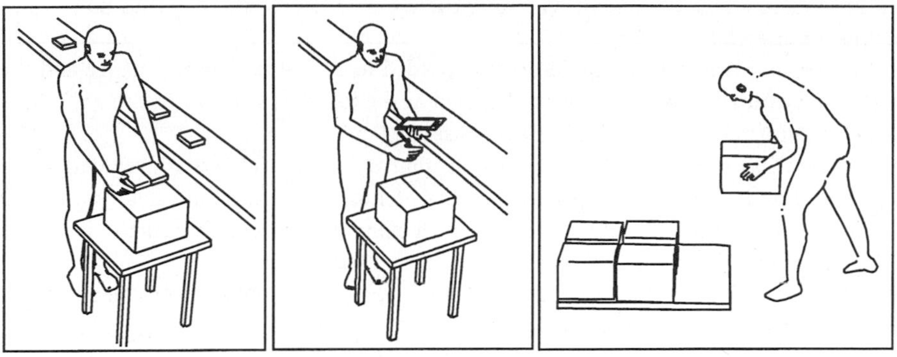

a b c

FIGURE 59.4 The three tasks of the frozen food packager job: (a) place two packages in box; (b) record date/time of packaging; (c) load full box onto pallet.

TABLE 59.3 Task Analysis of One Cycle of the Frozen Food Packager Job

Task	Length of Task	Notes
Place two food packages in box	42 seconds	Task time is seven seconds per two boxes packaged; twelve packages fit into each box.
Record date/time of packaging	10 seconds	No material handling required
Load full box onto pallet	8 seconds	Full pallet contains seven layers of boxes
Total time of job tasks	*60 seconds*	

Defining the Major Components of the Job through a Task Analysis

It is very important to properly define all relevant tasks that make up the job under investigation. These tasks should encompass the range of materials handling work that is required of the job — especially those that may present a risk of LBD. The tasks also should be defined so they are meaningful to those who will be interested in the LMM results. One way to define these job tasks is through a task analysis. A task analysis defines the discrete events of a job.

Take, for example, a job performed in a food processing plant (shown in Figure 59.4). An employee places twelve frozen food packages, two at a time, into a box (a), records the date and time of the packaging (b), then loads the packed box onto a pallet (c). A simple task analysis of this job is shown in Table 59.3. Two of the three tasks identified would qualify as material handling tasks for this job — loading the individual packages and placing the full box onto the pallet. Recording the date and time of the packaging would not be considered relevant material handling since low forces and exertions are required. Any trunk movement related to this task would not be considered in further analyses. However, trunk kinematics and the probability of risk could be determined for the two relevant MMH tasks identified.

It may be convenient to subdivide and redefine a task that is similar in all work dimensions except for one. For instance, in the above example, a fully loaded pallet may contain many boxes stacked several layers high. Tasks could be defined separately as "Place box on Layer 1," "Place box on Layer 2," etc. This may assist data interpretation. That is, differences in trunk kinematics could be interpreted not only as a function of the defined job tasks, but further in terms of workplace factors such as a box's location on a pallet.

Collecting and Recording Workplace Data for Risk Assessment

The trunk motion information that is used in the LBD Risk Model is automatically stored in the data collection software. Trunk motions include the position, the velocity of movement, and the related

acceleration of the trunk in all three directions of motion during a task— the lateral (side-bending) plane, the sagittal (forward bending) plane, and the rotational (twisting) plane. Two other components, lift rate and maximum moment, must be determined and input manually.

Lift Rate

Lift rate is defined as the *total* number of MMH actions that a job requires per hour. This number is related to the lifts for all tasks combined; it does not change from one task to another in a job. As the task analysis in Table 59.2 found, on average, one box is fully prepared and loaded onto the pallet per minute. Assuming this rate represents an average job cycle across the work shift, the packaging task alone would require 360 lifts per hour, since the product is placed in each of the 60 boxes six times (two packages per lift). For the palletizing task, 60 lifts are required per hour, for each fully packed box. Both tasks combine for a total of 420 lifts required per hour for this job. The non-MMH date/time recording task would not influence the lifting frequency of the job.

Because the lift rate value is directly input into the LBD Risk Model, it is very important to get an accurate estimate of the lifting frequency. It may be necessary to confirm the task analysis results by questioning employees familiar with the job or the job supervisor.

Maximum Moment

Maximum moment is defined as the external moment generated about the spine. A moment is composed of two factors, the weight of the object being handled and the horizontal distance from the spine at which it is handled.

Weight— Each object handled on a job must be weighed and recorded in the software or noted manually and input during data analysis. If objects of varying weights are handled, then each must be weighed individually and recorded. This often occurs in mail and freight delivery operations. In the aforementioned food processing example, the weights are constant for each task. The combined weight of the two food packages lifted together needs to be recorded, as should the weight of a fully packed box that is palletized.

Horizontal Distance—A tape measure is needed to determine how far from the spine objects are being handled for each task. With the tape measure held *horizontally,* one must measure the distance from the spine at the lumbosacral joint (near the top of the hips) to the center of the hands when the task is being performed. Obviously, as individuals handle objects, this distance changes as the object is moved. It is important to determine at what point during the task the distance is the greatest (i.e., generating the greatest moment about the spine) and to record this length. An ergonomist correctly measuring the horizontal distance is shown in Figure 59.5a. Here, the tape measure is kept level to determine the length from the individual's L_5/S_1 joint to the center of the hands. The *incorrect* approach is being used in Figure 59.5b, since the distance being measured is not horizontal.

For some MMH jobs, perhaps those on an assembly line, work requirements can be very repetitive and the employees' actions and movements rather consistent. In these cases, the horizontal distances may not vary much as identical objects are handled on each cycle. For other jobs, such as when pallets are loaded or unloaded, each cycle can produce very different trunk motions, since objects are being handled from different areas. This will likely change the horizontal distance at which an employee handles the load. Because the maximum moment value, a combination of an object's weight and its horizontal distance from the spine, is directly input into the LBD Risk Model, it is important that these distances be accurately measured and that changes in the distances for each task cycle being monitored by the LMM are recorded.

The measurement of these horizontal distances should not interfere with the work being done by the employee. The ergonomist who measures these distances should stand close enough to the individual to get accurate readings, but far enough away to not disturb the work. It is important that the individual be able to move naturally at the job site. A monitored employee being crowded by an investigator will move differently, change his/her trunk motions, and give erroneous information via the LMM.

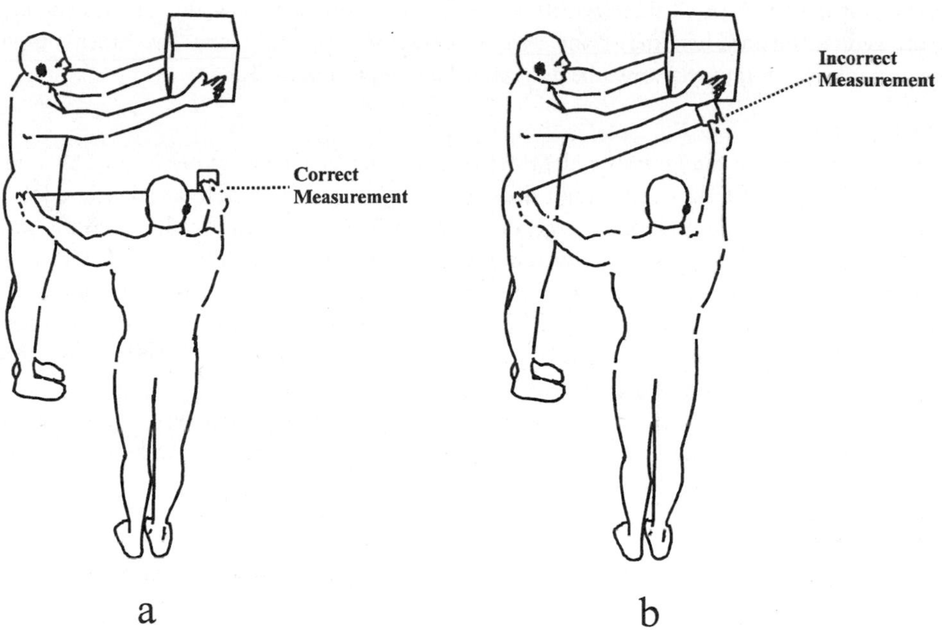

FIGURE 59.5 The correct (a) and incorrect (b) methods of measuring the distance a load is held from the L_5/S_1 joint.

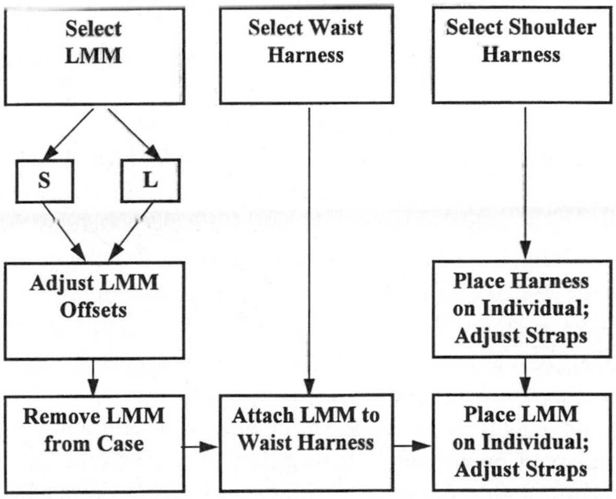

FIGURE 59.6 Flow chart of the LMM set-up procedure.

Setting Up the LMM for Data Collection

After the job to be monitored has been selected, the tasks identified, and basic workplace information has been gathered, the data collection process can start. To begin, the proper LMM and harnesses need to be chosen and prepared for use. This process is depicted in a flow chart in Figure 59.6 and is described below.

Selecting the Correct LMM

The standing height of the individual to be monitored is usually the best indicator of which LMM to use. As shown in Figure 59.7a, a small LMM placed on a taller individual will restrict his/her range of motion. A large LMM put on a shorter person will cause the LMM to buckle and give erroneous readings

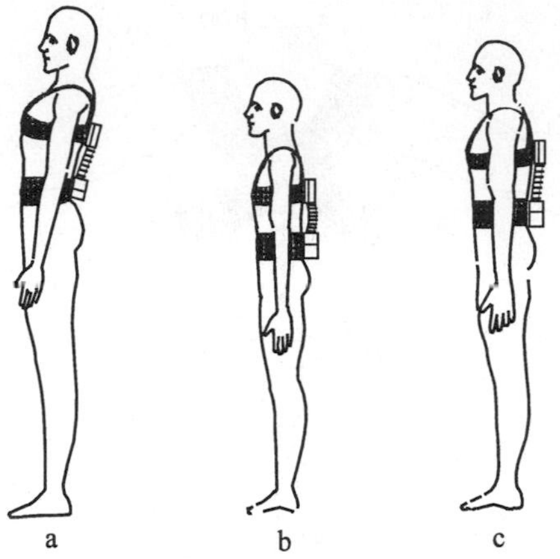

FIGURE 59.7 LMM sizing- (a) the small LMM incorrectly placed on a tall individual; (b) a large LMM incorrectly put on a small individual; (c) properly sized LMM for the individual.

(Figure 59.7b). Both situations could damage the LMM. A properly sized LMM (Figure 59.7c) will enable an individual to stand upright without the LMM pulling him/her backward or buckling due to the waist and shoulder harnesses being too close to one another. The shoulder harness also can be adjusted to create a proper fit.

Adjusting the LMM

LMMs are individually calibrated during their manufacture, and there is a specific position designated as its "neutral" position. During normal usage, this neutral position may change slightly and must be readjusted (through the software) before every subsequent use. To adjust the LMM, choose which LMM is to be used, but keep it in the LMM case. Attach the cable from the computer to the LMM (or use the telemetry device), and adjust the voltage offsets using the supplied LMM software. Every LMM should be adjusted in this manner before it is placed on an individual. Without disconnecting the LMM, use the software to monitor the LMM's three motion traces. Take the LMM out of the case and move the top of the LMM while keeping the bottom in place. The three traces on the computer screen should show realistic position values.

Putting the LMM on the Employee

After the LMM has been checked to be working properly, it is ready to be placed on an employee. When handling the LMM, note that it should never be bent into extreme angles; damage to the LMM could result. The following steps, diagrammed in Figure 59.6, will ensure proper placement.

Select the Shoulder Harness — The shoulder harness should fit snugly when worn but still enable the individual to breath normally. It's important that most of the Velcro strap that crosses the chest be overlapped during wear, so that it and the LMM remain in place during data collection. Place the harness on the individual to be monitored.

Attach LMM on Waist Harness — Three sets of harnesses are available. Select the size that will fit on the individual with the greatest amount of Velcro that will overlap when worn. Place the base of the LMM evenly on the waist harness and secure tightly.

Place Waist Harness/LMM on Individual — First slide the top of the LMM into place on the shoulder harness. Then, fit the waist harness over the hips and momentarily secure the harness with the Velcro strap. The base of the LMM (at the lowest t-section) should be aligned with the lumbosacral joint (L_5/S_1) of the trunk. This position can be found by first locating the tops of the hips (the iliac crest) with the fingers, then placing the thumbs about an inch lower than the fingers. See Figure 59.8 for proper LMM alignment on an individual. When the LMM base has been positioned, check that the LMM's t-sections are aligned vertically. If not, one or both of the following adjustments can be made:

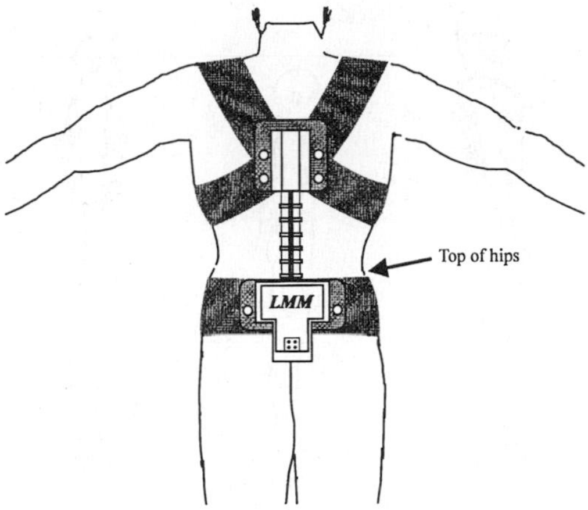

FIGURE 59.8 Proper placement of the LMM on the torso. The base of the LMM is located slightly below the top of the hips, and the LMM t-sections are aligned vertically.

1. The waist harness can be moved to the left or right.
2. The shoulder harness can be loosened and shifted. It can be adjusted by shortening or lengthening the Velcro straps that cross over the shoulders.

Once the LMM is properly positioned, securely tighten the Velcro strap on the waist harness, and insert each leg strap through the buckle. It is important to always use these leg straps. They prevent the LMM from moving up on the hips, which otherwise would move the base of the LMM from its L_5/S_1 location and result in erroneous data.

Collecting the Data

After the LMM is properly placed on an individual and is correctly hooked to the computer, the investigator is ready to gather job data. It is suggested that the individual be given some time (a few job cycles, for example) to become accustomed to the LMM before collecting data.

If not already done, the company, job, task, and employee data need to be entered into the LMM software. Follow the software instructions for this procedure. Examples of the job profile set-up screens for the Chattanooga™ DOS and Windows software are shown in Appendix A-59.1. After the task data are input and the data collection screen is open and showing the three traces (for the lateral, sagittal, and twisting planes of motion), data collection can begin.

The main goal of data collection is to gather information on trunk motions that is representative of all the work required of the job. For example, if the job requires handling objects of widely varying weights or from different locations (on a pallet, for example), then the data should be gathered of each of these tasks. The more data that are collected, the more it is likely to represent the requirements of the job. The amount of data collected should be reflective of the job and how it is structured. For example, if 80% of the job requires doing Task A and only 20% doing Task B, then a majority of the data should be of the employee performing Task A. In other words, data collection should be proportional to the job being studied. Example LMM data collection screens for DOS and Windows software packages are shown in Appendix A-59.2. Notice that only the motions associated with the designated tasks are marked using the computer keyboard's function keys.

There is no set number of job cycles that indicates when "enough" data have been collected. It is dependent on the nature of the tasks. Jobs that are highly consistent in their activities usually require fewer numbers of collected cycles than those with more widely varying requirements. The authors suggest, though, that a minimum of seven to ten cycles of any task be collected.

TABLE 59.4 Example Data Collection Form

Company: Job: Employee: Date:						
Trial #	Task	Object Weight	Horizontal Distance	Start Height	Finish Height	Comments

It's extremely important to keep a record of the weights handled and the measured horizontal lifting distances for *each task cycle collected*, so that these data are matched with the corresponding trunk motion data for that cycle. An example form is shown in Table 59.4. The ergonomist may wish to develop a customized data collection form beforehand to keep track of this information.

Analyzing and Interpreting the LMM Data

By following the guidelines in the previous sections of this chapter, data analysis and interpretation will be made easier. Tasks of a job that are carefully chosen will assist in the job's interpretation of its risk probability and lead to results having more practical significance. The software used to collect and analyze LMM data provides trunk kinematic information and risk probability charts for each task defined for a job and for each employee who was monitored. It is beyond the scope of this chapter to provide LMM software documentation. Instead, explanations of how these software outputs can be interpreted will be discussed.

Trunk Kinematic Information

Information on the specific trunk motions produced by employees during each task can be obtained from the software. This includes the trunk positions and the velocities and accelerations produced by the trunk for each plane of motion for those cycles of a task chosen by the ergonomist. This information can be useful for general descriptions of the material handling or for comparisons with other tasks or jobs. Use of these data can be valuable for investigators who have formed hypotheses about, for example, what tasks require more trunk motions than others.

Output from the frozen foods example job is shown in Table 59.5. As the header shows, these data include trials of data collected for the food packaging task of this job. Information includes minimum and maximum trunk positions, average and maximum velocities, and maximum accelerations and decelerations for each plane of motion. From these data, average motions over the six trials could be computed, as could the range of movement required and the amount of variation for each kinematic parameter.

Probability of High Risk Group Membership (LBD Risk)

The software will produce information that compares the job tasks of interest with a database of jobs previously determined to be at a "high" or "low" risk in terms of low back injury. The interpretation of data is best described through continued use of the food packager example.

TABLE 59.5 Output Data Showing Low-Back Kinematics for Six Trials of One Job Task

Company: Frosty Foods
Job: Frozen Food Packager
Employee: Allen Shaffer
Task: Place food packages in box
Date: 10/26/96
Run: 1

		Side-Bending (Lateral)						Forward-Bending (Sagittal)						Twisting (Rotational)					
Trial	Time	Min. Pos.	Max. Pos.	Avg. Vel.	Max. Vel.	Max. Acc.	Max. Dec.	Min. Pos.	Max. Pos.	Avg. Vel.	Max. Vel.	Max. Acc.	Max. Dec.	Min. Pos.	Max. Pos.	Avg. Vel.	Max. Vel.	Max. Acc.	Max. Dec.
1	09:11:33	−1.4	7.3	12.4	32.7	66.7	−228.5	−8.5	4.5	4.5	45.3	215.4	−97.4	−2.4	12.3	6.3	35.4	276.4	−101.2
2	09:11:42	−0.7	12.9	8.7	19.9	77.3	−197.3	−10.3	9.6	6.8	34.6	227.5	−101.2	−1.2	10.2	2.3	29.5	256.2	−98.6
3	09:11:49	−3.2	4.2	13.8	39.5	86.3	−207.8	−5.4	8.8	3.9	41.2	196.8	−125.5	−5.7	8.6	6.1	40.4	295.3	−112.5
4	09:11:59	−1.1	6.8	21.0	38.4	84.1	−216.7	−11.3	7.3	9.8	29.7	246.8	−89.4	−3.3	4.3	5.4	41.2	299.5	−95.2
5	09:12:06	0.2	11.0	17.7	29.0	56.8	−156.4	−9.5	9.0	7.2	39.4	209.5	−112.2	−6.7	9.4	5.8	36.5	255.4	−110.4
6	09:12:15	−3.8	3.2	22.5	35.7	80.0	−164.9	−4.3	6.7	5.5	42.1	198.5	−102.3	−0.2	10.0	4.8	32.1	267.4	−105.6

In the food packaging job, two tasks were defined. The first task involved placing individual frozen food packages into a box (twelve in all), and the second task involved loading the filled box of packages onto a pallet. Risk probability charts are shown for each task in Figure 59.9. In Figure 59.9a, The package loading task was found to have a risk probability of 40%. (Probability values are calculated by averaging the individual logits from each of the five risk factors — in Figure 59.9a, (99%+30%+3%+23%+45%)/5 = 40%.) The risk charts for this first task that were produced using the DOS and Windows LMM software are shown in Appendix A-59.3. The probability value for the palletizing task (Figure 59.9b) was calculated to be 64%, clearly the task more similar to those considered "high risk." These two probability values are to be used only for comparison purposes. It is the chart in Figure 59.9c that reflects the true risk value for the entire job. This chart summarizes the largest values for each risk factor across both tasks making up this job. The value shown on this chart, and thus the risk for this material handling job, is 66%.

A closer examination of the top two charts in Figure 59.9 shows which factors most contributed to the job summary values in Figure 59.9c. Each of the five factors is discussed separately.

Lift Rate — The lifting frequency for the entire job was 420 lifts/hour. Because this variable is composed of the total number of lifts from both tasks, this value is shown to be the same on all charts in Figure 59.9. As indicated by the length of the lift rate bar on the charts, this rate is very rapid and is comparable to some of the highest frequency material handling jobs found in industry.

Average Twisting Velocity — The amount of twisting velocity required for package loading was fairly low, but it was moderately high for the palletizing task, as indicated by the length of the bars on these charts in Figure 59.9. The greater value of the two is used in the job summary chart in Figure 59.9c, which was taken from the palletizing task.

Maximum Moment — As shown on the charts in Figure 59.9, the moment values were very low for both tasks comprising this job. The low weight of the individual packages (each at one pound) and the fully packed box being palletized (12 pounds) generated low maximum moment values. The greater moment value from the palletizing task was used in the job summary chart.

Maximum Sagittal Flexion — The package loading task was performed while employees were in relatively upright postures. This is reflected in Figure 59.9a by a short sagittal flexion bar on the chart. However, during box palletizing, those boxes placed on the lower layers of the stack required much forward trunk flexion. Figure 59.9b depicts these higher angles by the long bar for this factor. Subsequently, this higher value of the two tasks resulted in it being used in the job summary chart in Figure 59. 9c.

Maximum Lateral Velocity — The LMM determined that lateral velocities generated during the package loading task actually were higher than found during box palletizing. While the values for the previous factors all were larger during handling of the full box, and thus were used in the job summary chart, it is the lateral velocity value from the package loading task that must be used in the job summary, since it is the greater of the two tasks analyzed.

It is the job summary value of 66% (taken from the chart in Figure 59.9c) that represents the LBD risk for this example food processing job. The value indicates that, on the continuum of low-risk jobs (0%) to high-risk jobs (100%), this particular job has a 66% likelihood of being considered "high risk." As stated earlier in this chapter, a high-risk job was defined as one having twelve or more (with an average of 26.4) low back strains per 200,000 hours (or 100 workers/year) of employee exposure. Results here could be interpreted as indicating that this particular job has a moderately high chance of producing a large number of low back strain injuries among individuals who do this job.

The primary goal of the LMM analysis could be simply to determine whether or not a job presents a risk of low back strain to its employees. This can be determined by calculating the LBD risk value from a job summary. As Figure 59.10 shows, there is some overlap in risk values among jobs defined as low and high risk. However, our analyses (Marras et al., 1993) found that those jobs with fairly low probability values (below 30%, for example) were much more likely to be low-risk jobs. Similarly, very few jobs having a risk value over 50% were defined as "low risk," and no low-risk jobs had risk values over 70%. That is, jobs with probability values above 60% are virtually assured to have some low back injury risk associated with them.

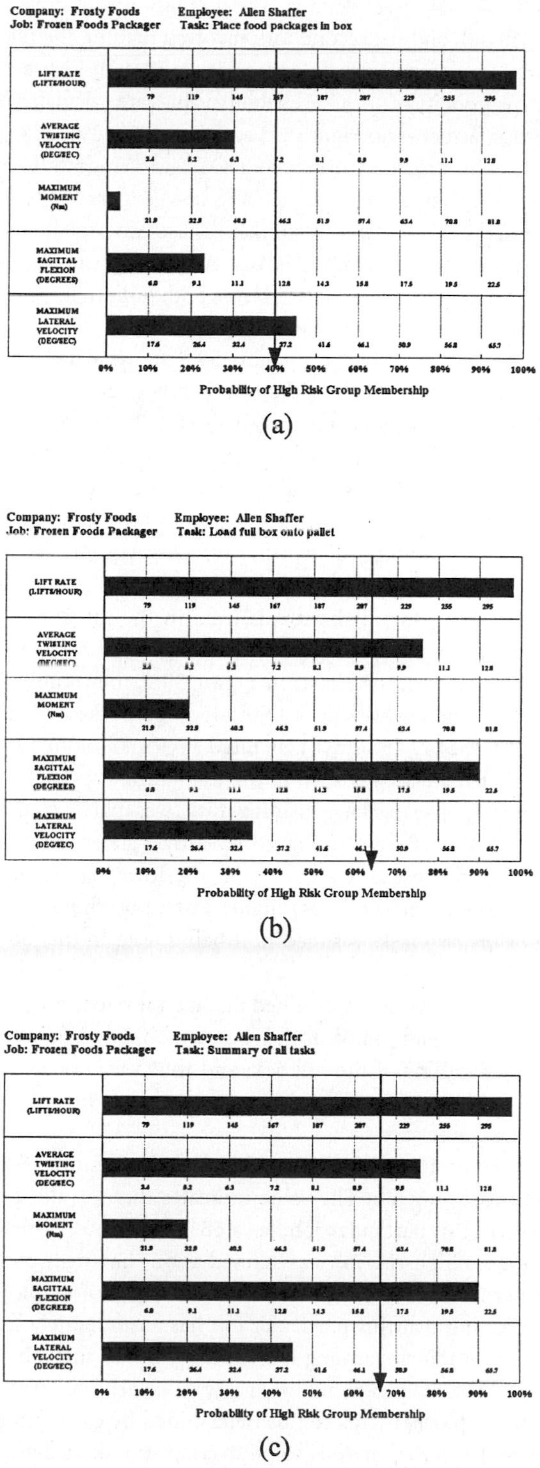

FIGURE 59.9 Risk value charts for the frozen food packager job: (a) "Place Two Food Packages in Box" task; (b) "Load Full Box onto Pallet" task; and (c) summary of all job tasks.

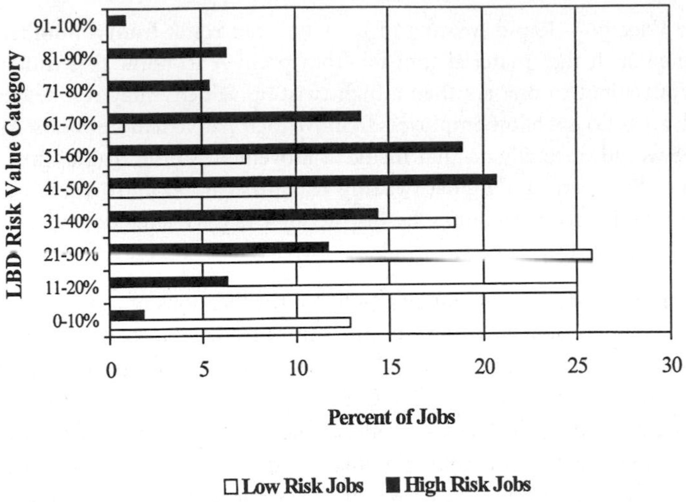

FIGURE 59.10 Distributions of the LBD risk values for jobs defined as "low risk" and those defined as "high risk" in Marras et al. (1993).

A second goal of the analysis may be to compare one job task with another, to determine which one(s) require more back motions and external moments about the trunk. This exercise can assist in learning which task(s) should be the focus of redesign efforts. Introducing ergonomic modifications to tasks already found to have low risk values probably will have little real impact on improving the job overall. However, making changes to tasks whose individual factors contribute most to the job's summary risk probability will reduce the probability for the entire job.

A third goal of the analysis may be to determine, for specific tasks, which individual components are most responsible for its composite probability value. This type of analysis provides direct information regarding how the job's requirements may affect those factors used in this probability model. Examples of ergonomic improvements that can be made on a job to affect each component are listed below. These examples should not be considered an exhaustive list of possible changes that can be implemented to improve working conditions.

Lift Rate — The total number of material handling actions required of the job affects this risk factor. Thus, reducing this rate will reduce the overall risk value. This type of change is not usually favored by management, since it will likely reduce productivity. However, there are ways to redesign jobs to reduce lifting frequency:

1. Rearrange Job Tasks — For jobs in which a number of tasks comprise the job, one method is to rearrange tasks with those of other jobs. This may more evenly distribute the lifting frequencies of several jobs, some of which may be considerably lower than the job monitored.
2. Rotate Jobs — Rotating employees between a job having a high lifting frequency with one having a much lower frequency also will reduce the rate for the job of interest. The effects of the job(s) into which employees are rotated must also be considered, however. It should be cautioned that this and the previous approach are most beneficial if the jobs that are rearranged or included in the rotation allow employees to use different muscle groups to perform the job. Rotating individuals into jobs that require the same muscles be used likely will have either no benefit or could produce greater musculoskeletal stress.
3. Add Employees — Dividing the job so that added personnel are available to perform the job will distribute the work across more people and lower the job's required lift rate. Of course, the cost of the additional employee(s) must be compared with the benefits of reduced low back strains and their related costs.
4. Automate — It may be possible that some job tasks can be automated through new equipment or robotics. This method of assisting the material handler will undoubtedly reduce the lifting frequency of the job and the overall job requirements.

Average Twisting Velocity— Rapid twisting of the trunk can result from a number of situations. If a work area is designed such that material transfer from point A to point B is difficult or if these two locations are not convenient to one another, a high twisting velocity may result. High velocities often result because work areas do not allow employees to move their feet to handle goods. As a result, a turning action an employee would normally do that included movement of the trunk, hips, legs, and ankles is more concentrated in the trunk, and higher twisting velocities can occur. It would be difficult to reduce the speed at which individuals twist simply by instructing them to "slow down." However, engineering controls that can reduce twisting velocity include:

1. Place Work in Front of Material Handler — Move the locations of point A relative to point B, so that they are more convenient to one another. In other words, create a workplace in which the material transfer requires moving in as few planes of motion as possible, thus allowing the employee to remain in a more neutral posture
2. Spread Out Congested Work Areas — Material handling areas that allow employees to walk or take at least one step between handling points A and B often reduces the twisting velocity of the job. This occurs because the added movement allows one to get into a position in which the entire body can assist in the transfer, rather than just the trunk.
3. Raise Working Heights — Depending on the location of goods to be handled, lower levels of work that require a great deal of sagittal flexion also may require additional trunk twisting. This is often the case when the work requires asymmetric lifting. By raising the work heights, not only will forward flexion be reduced, but twisting velocity can decrease because the handling distances are more appropriately located.

Maximum Moment— Because an external moment is the product of an object's weight and the distance from the body at which it is handled, reducing either of these two factors will reduce the moment. Various examples are described below.

1. Reduce Weight Requirements — Some work situations allow employees to handle as many units or as much weight as they feel is acceptable. However, to work faster, some employees may handle more goods at one time than is physically safe. A limit on the numbers/weights of objects handled in any given time can reduce Maximum Moment values. In some environments, raw materials handled are received from a supplier in bulk quantities. The weights of these materials may produce excessive moment values. By working with these suppliers, an arrangement may be possible to package materials in smaller, lower-weight containers. The weight changes just described likely will *increase* the lifting frequency of the job, however, so this trade-off in the risk model should be examined.
2. Install Material Handing Aids — For goods that are of uniform shape or size, several types of lifting aids are commercially available to provide handling assistance. These can be adapted to a wide range of work environments. Handling aids, such as lifting hoists, when incorporated successfully, greatly reduce the loading forces on the spine and result in much lower moment values. The device should be considered carefully, since a handling aid that is difficult to use or greatly slows the job process will likely be abandoned by the employees. In addition, some material handling aids can reduce the distance at which objects are handled. For example, lift/tilt tables are available commercially and able to be adapted to specific needs. These devices can raise or angle objects or bins of goods so that they can be more easily accessed. This can result in reduced horizontal reach distances.
3. Evaluate the Transfer Locations — The distance from the body at which individuals handle goods often is greatest during the initial or final contact with the product. This is often true during palletizing operations, in which cases need to be placed properly on a skid to ensure the load's stability. During carrying, for example, people tend to bring the product closer to their bodies. An evaluation of these locations may detect workplace arrangements that cause individuals to reach farther than is necessary to handle objects.

Maximum Sagittal Flexion — The more that individuals work in upright positions during material handling, the lower their trunk flexion will be. This reduces subsequent risk probabilities for jobs. Reducing sagittal flexion can be accomplished by eliminating tasks that, for example, require loads below knee level to be handled. Several interventions can prevent these situations.

1. Raise the Heights of Loads Placed near the Floor — Objects can be raised from the floor during material handling work a number of different ways. For palletizing tasks, stacking a skid underneath the pallet will raise the height of the bottom-most objects. Lift tables and self-raising devices placed underneath objects also will raise objects higher off the floor and reduce sagittal flexion. These changes should be evaluated for safety considerations and to ensure that objects at the top of the pallet can still be accessed. Other tables are available commercially that tilt or swivel objects, such as those on pallets, that bring the loads closer to the material handlers.
2. Adjust Working Heights Relative to an Individual's Standing Height — Work areas that are adjustable to accommodate those performing the work also can reduce trunk flexion. Work tables can be constructed or purchased that move vertically to a position most comfortable to the user. On assembly lines, conveyors can be adjusted that raise or lower objects depending on the work being performed at a specific location. Alternately, work areas alongside the lines, under the employees themselves, can be raised or lowered to produce the correct working height.
3. Train Employees on Proper Lifting Techniques — For some work situations, vertical adjustability may not be technically or economically feasible. In these cases, employees can be educated on proper lifting techniques aimed to reduce back strain and reduce the amount of sagittal flexion required.

Maximum Lateral Velocity — High lateral velocity values on the risk charts indicate that the material handling work requires rapid sideways bending. This motion may be difficult to visualize, but it usually indicates that work is not being performed in front of one's body, but asymmetrically instead. Workplace modifications that more conveniently locate or raise the work relative to the material handler (as already noted above) can assist in reducing this factor on the probability charts. A case study conducted by Stuart-Buttle (1995) found that reductions in lateral velocities and sagittal flexion can be achieved through workplace interventions in MMH tasks using lift tables. This paper also illustrated the importance of testing the impact of workplace modifications. The initial installation of this table actually produced higher LMM risk values and more employee dissatisfaction. The LMM analysis identified the problems with the new system and provided feedback about how the workplace needed to be further changed to produce actual risk reduction.

From the example job modifications just discussed, it is important to understand that these five workplace factors are interrelated. None of these factors responds independently from the others. For instance, adding a lift table to palletizing work may reduce sagittal flexion, because the load is being raised. However, is also can lower the maximum moment and average twisting velocity values, because the load may be held closer to the body during handling and be more easily accessed. If the work is self-paced, lift rate actually could *increase* since the work may be less physically demanding and those affected may be capable and willing to handle more material. This example illustrates the trade-offs that must be considered when evaluating the probability of risk for a job and implementing ergonomic interventions.

Interpreting Results from Several Individuals

Casual observation of any material handling activity will show that different people usually perform the same job slightly differently. Employees inherently vary in how they do manual work, and this will likely produce different trunk motions. To account for these differences, the ergonomist may wish to use the LMM to monitor several employees who perform the same job and then analyze and interpret the results in terms of average probabilities across these individuals. The authors know of no literature that specifies how many employees must be monitored to account for all variability in a job. This process is task-dependent, and the ergonomist should study the amount of variability of the job tasks and work layout before deciding how many employees to monitor. Assembly-line or machine-paced operations may limit the freedom employees have in performing job tasks in comparison with self-paced material handling work. The ergonomist may want to adjust the number of individuals monitored accordingly.

TABLE 59.6 Computational Example for Assessing LBD Risk from Three Employees Performing the Frozen Packager Job

Task 1: Place Two Food Packages in Box				
Risk Factor	Emp. 1	Emp. 2	Emp. 3	Average
Lift Rate	99.0%	99.0%	99.0%	99.0%
Average Twisting Velocity	30.0%	21.0%	42.0%	31.0%
Maximum Moment	3.0%	2.0%	4.0%	3.0%
Maximum Sagittal Flexion	23.0%	26.0%	17.0%	22.0%
Maximum Lateral Velocity	45.0%	32.0%	43.0%	40.0%
Average	**40.0%**	**36.0%**	**41.0%**	**39.0%**
Task 2: Load Full Box onto Pallet				
Risk Factor	Emp. 1	Emp. 2	Emp. 3	Average
Lift Rate	99.0%	99.0%	99.0%	99.0%
Average Twisting Velocity	76.0%	75.0%	80.0%	77.0%
Maximum Moment	20.0%	22.0%	24.0%	22.0%
Maximum Sagittal Flexion	90.0%	94.0%	89.0%	91.0%
Maximum Lateral Velocity	35.0%	30.0%	43.0%	36.0%
Average	**64.0%**	**64.0%**	**67.0%**	**65.0%**
Job Summary				
Risk Factor	Emp. 1	Emp. 2	Emp. 3	Average
Lift Rate	99.0%	99.0%	99.0%	99.0%
Average Twisting Velocity	76.0%	75.0%	80.0%	77.0%
Maximum Moment	20.0%	22.0%	24.0%	22.0%
Maximum Sagittal Flexion	90.0%	94.0%	89.0%	91.0%
Maximum Lateral Velocity	45.0%	32.0%	43.0%	40.0%
Average	**66.0%**	**64.4%**	**67.0%**	**65.8%**

If the LMM data collection and analysis software does not assess the average risk probability across employees doing the same tasks, then these values can be calculated manually, by following the steps listed below. A computational example of the results from three employees monitored who performed the frozen food packager job is shown in Table 59.6.

1. For Task 1, the average probabilities range from 36.0% to 41.0% for these three individuals. For Task 2, the range is 64.0% to 67.0%. To find the probability average for each of these tasks, average the values of *each* factor used in the model across all employees. That is, add the lift rate, average twisting velocity, maximum moment, maximum sagittal flexion, and maximum lateral velocity values separately and divide that number by the total number of employees monitored. Do this calculation for each job task. In Table 59.6, the right-most column shows the average values for each risk factor for the two material handling tasks of this job.
2. Compute the average probability for each task. To do this, sum the individual factor *probabilities* and divide by five. This value will be the average probability across employees. In the right-most column, Table 59.6 shows that the average probability for Task 1 is 39.0%, and it is 65% for Task 2. Also note that, due to individual differences in how the job is done, risk values vary from employee to employee.
3. Determine the *job* risk probability by reviewing all averaged task probabilities (again, those in the right-most column of Table 6 for Task 1 and Task 2). Find the highest probability value for each of the five factors and record it. This must be done for each employee and overall, to find the average risk for the job. As Table 59.6 shows, the largest average values for all five factors but maximum lateral velocity come from Task 2. These are shown in the bottom section of this table. Averaging these five values generates a job risk of 65.8%. This represents the risk for the entire job, averaged across all employees monitored.

References

Andersson, GBJ, *The Adult Spine. Principles and Practice*, 2nd ed. Raven, New York, 1997.

Andersson GBJ, Epidemiologic aspects on low back pain in industry. *Spine*, 1981, 6(1), pp. 53-60.

Bigos SJ, Spengler DM, Martin NA, Zeh J, Fisher L, Nachemson A, and Wang MH, Back injuries in industry: A retrospective study. II. Injury factors. *Spine*, 11(3):246-251, 1986.

Cats-Baril WL and Frymoyer JW, The economics of spinal disorders, in Frymoyer JW, Ducker TB, Hadlet NM, Kostuik JP, Weinstein JW, and TS Whitecloud III (Eds.): *The Adult Spine: Principles and Practice*. Raven Press, New York, 85-105, 1991.

Gill, KP, and Callaghan MJ, Intratester and intertester reproducibility of the lumbar motion monitor as a measure of range, velocity and acceleration of the thoracolumbar spine. *Clinical Biomechanics*, 11(7):418-421, 1996.

Magora A, Investigation of the relation between low back pain and occupation. 4. Physical requirements: Bending, rotation, reaching and sudden maximal effort. *Scand J Rehab Med*, 5(4):186-190, 1973.

Marras, WS, Fathallah FA, Miller RJ, Davis SW, and Mirka GA, Accuracy of a three dimensional lumbar motion monitor for recording dynamic trunk motion characteristics. *International Journal of Industrial Ergonomics*, 9(1):75-87, 1992.

Marras WS, Lavender SA, Leurgans SE, Rajulu SL, Allread WG, Fathallah FA, and Ferguson SA, The role of dynamic three-dimensional motion in occupationally-related low back disorders. *Spine*, 18(5):617-628, 1993.

National Center for Health Statistics, *Prevalence of Selected Impairments*, United States Government Printing Office, Series 10, No. 134, 1977.

National Institute for Occupational Safety and Health (NIOSH): Work practices guide for manual lifting. Department of Health and Human Services (DHHS), National Institute for Occupational Safety and Health (NIOSH), Publication No. 81-122, 1981.

Putz-Anderson V and Waters TR. (1991) Revisions in NIOSH guide to manual lifting. Paper presented at national conference entitled "A national strategy for occupational musculoskeletal injury prevention — Implementation issues and research needs." Ann Arbor, MI, University of Michigan, 1991.

Punnett L, Fine LJ, Keyserling WM, Herrin GD, and Chaffin DB. Back disorders and nonneutral trunk postures of automobile assembly workers. *Scand J Work Environ Health*, 17:337-346, 1991.

Stuart-Buttle C, A case study of factors influencing the effectiveness of scissor lifts for box palletizing. *American Industrial Hygiene Association Journal*, 56:1127-1132, 1995.

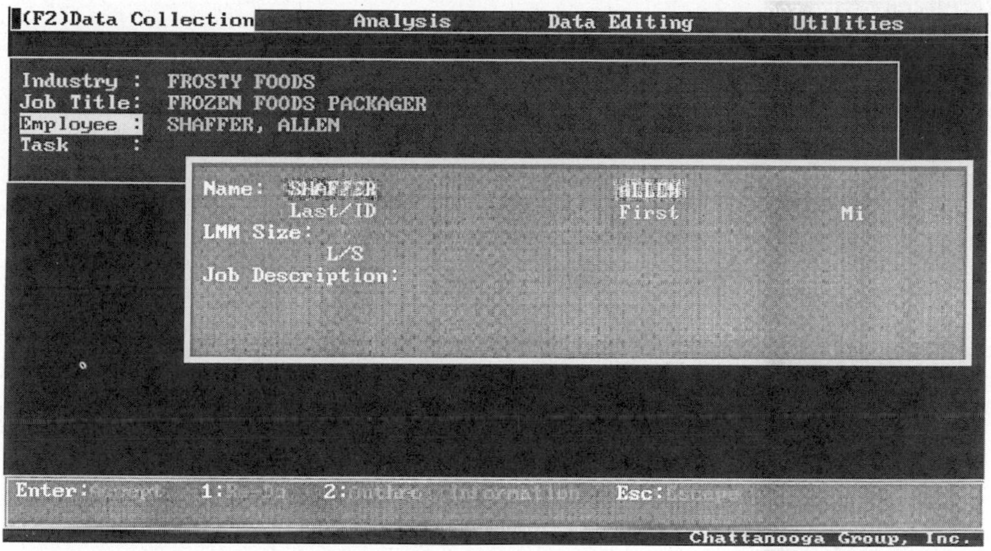

(a)

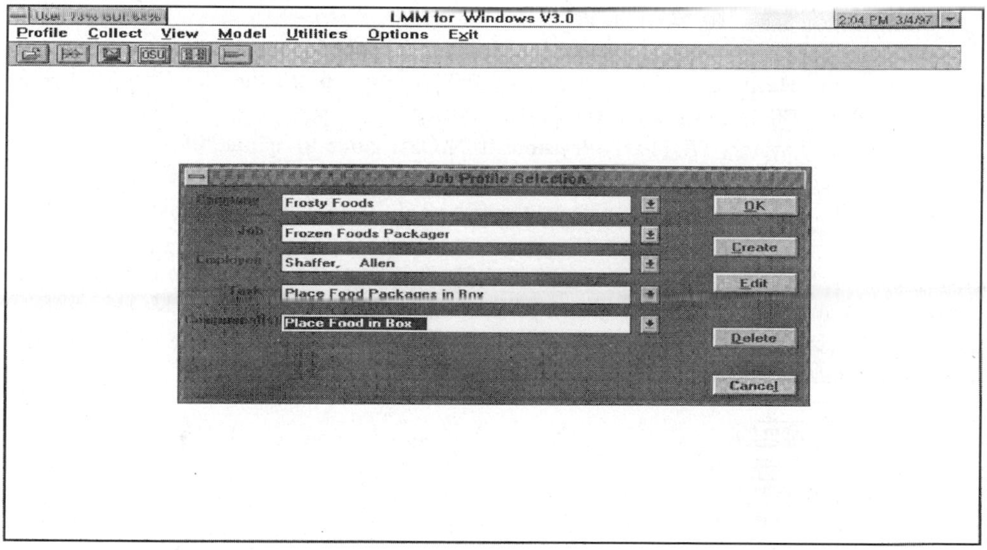

(b)

APPENDIX A-59.1 Example job profile selection screens for Chattanooga software; (a) DOS version; (b) Windows version.

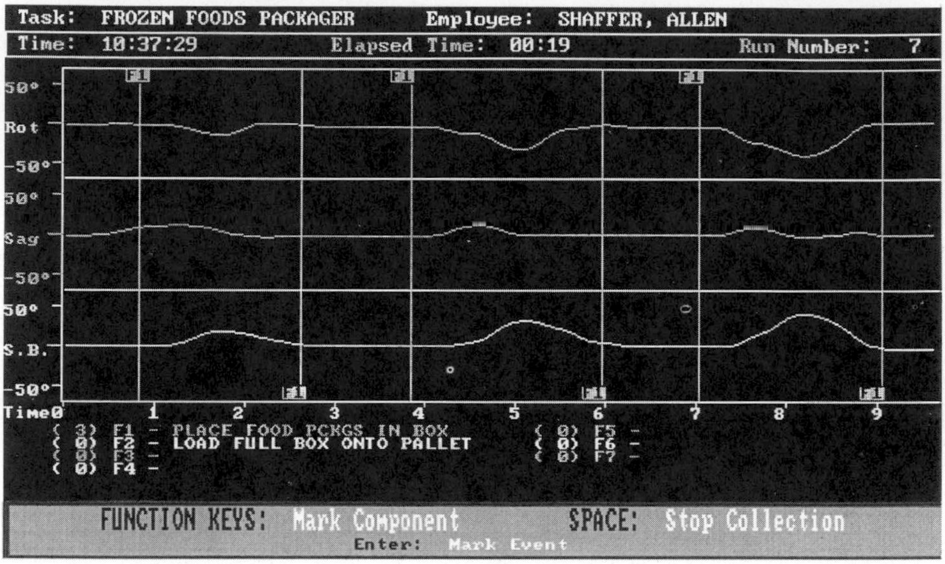

(a)

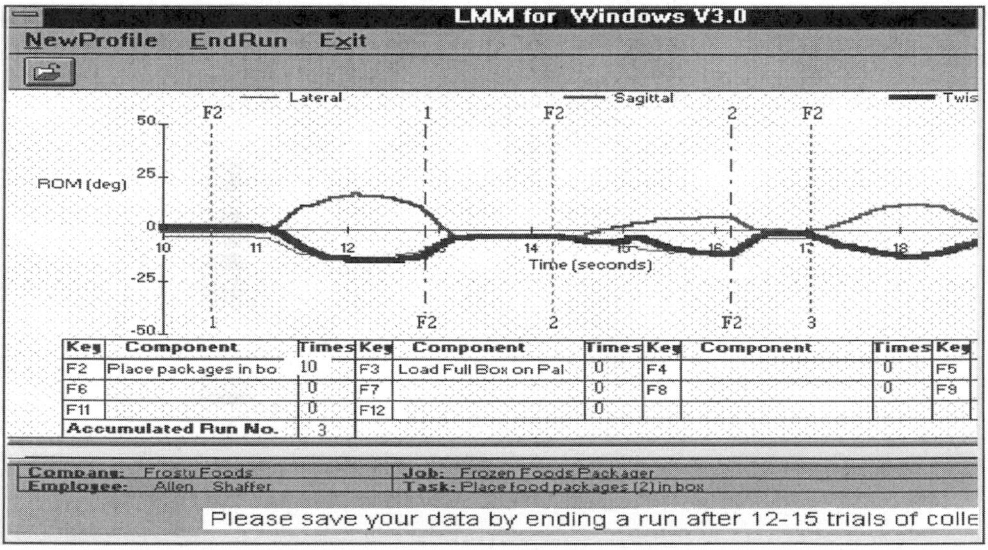

(b)

APPENDIX A-59.2 Example data collection screens for Chattanooga software; (a) DOS version; (b) Windows version.

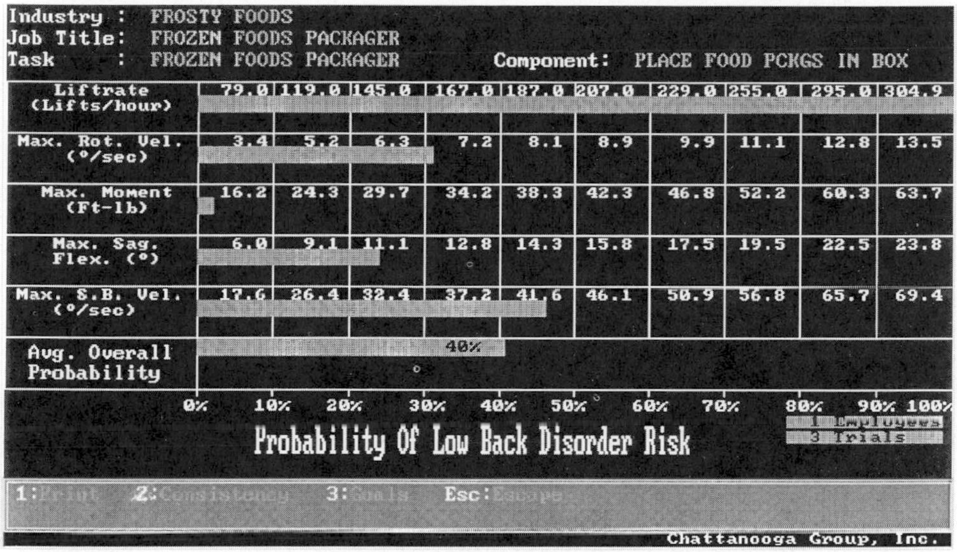

(a)

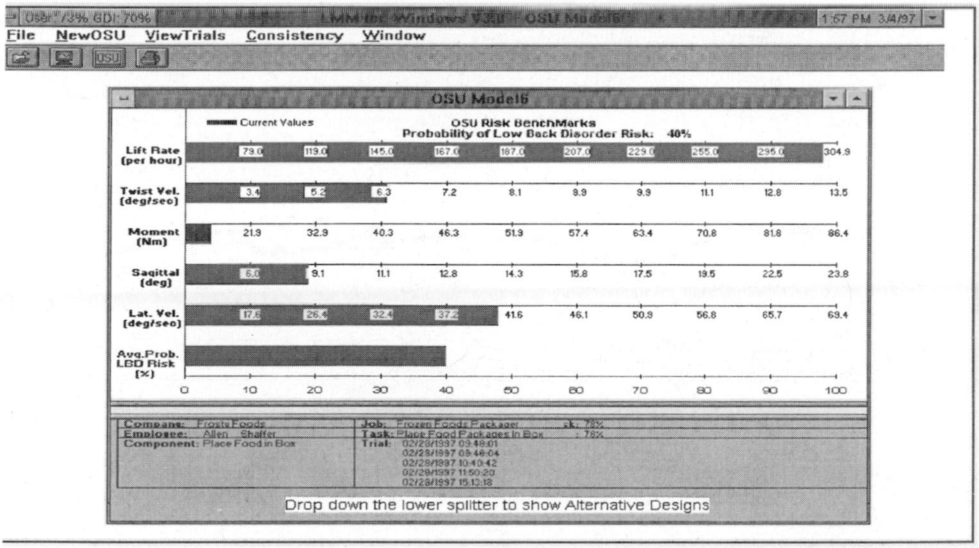

(b)

APPENDIX A-59.3 Example OSU Model screens for Chattanooga software; (a) DOS version; (b) Windows version.

60

Prevention of Musculoskeletal Disorders: Psychophysical Basis

Patrick G. Dempsey
Liberty Mutual Research Center for Safety & Health

60.1 Introduction

Occupational ergonomics is a discipline which seeks to maximize the performance of systems found in diverse settings such as production work, service industries, and health care. In this context, the system consists of the personnel, the equipment, the organizational and ambient environments, and the tasks required to produce a good or service. System performance refers to the ratio between system outputs (goods and services) and inputs to the system (personnel, capital, etc.). Not only is productivity at individual workstations important, but so is preventing injuries and illnesses associated with the tasks performed. A detriment to system performance in many settings in which manual work is performed is musculoskeletal disorders.

The direct costs (medical and indemnity payments) associated with work-related musculoskeletal disorders (WRMSDs) represent a significant source of financial losses. These losses have a direct economic impact on employers and insurers, which in turn are passed on to the customer. For example, customers who buy goods or services from an organization must bear the additional costs of the WRMSDs associated with producing those goods or providing those services.

An often overlooked aspect of WRMSDs is the negative impact of indirect costs (lost production time, training new workers, litigation, etc.) on system performance. For WRMSDs that develop over a period of time, there may be a gradual reduction in individual productivity. Once a disorder (whether it be the

0-8493-2641-9/99/$0.00+$.50

result of cumulative trauma or the direct result of a single overexertion) progresses to the point that the individual cannot continue to perform a job, there are additional indirect costs associated with lost production, replacing the individual, as well as potential performance decrements until the replacement worker becomes proficient at the task. The extent of indirect costs is partially a function of the injured individual's function in the system.

One approach to the prevention of WRMSDs is the psychophysical approach. This approach seeks to provide limits and guidelines for manual work that represent "maximum acceptable" work loads that minimize the injury potential of the work. To some extent, it is unknown if the psychophysical approach achieves this goal, as will be discussed in more detail later. In general, the extent to which many ergonomic criteria represent "optimal" limits with respect to optimizing the performance of systems comprised of manual work is unknown. Work loads and physical stresses have to be at or below levels which protect workers, but at the same time permit output levels which are economically sound. The psychophysical approach is an approach that seeks a balance between productivity and health and safety concerns.

Chapter Goal and Outline

The goal of this chapter is to provide the reader with information concerning basic theory behind the psychophysical approach, the availability of data for designing manual tasks, the methods of applying the data, and the limitations of the data. The primary focus will be on the application of psychophysical techniques rather than on empirical methodologies or a literature review of the theoretical underpinnings of the psychophysical approach. Where necessary, theoretical and empirical results will be used to justify specific application techniques or to explain caveats of psychophysical data. Manual materials handling (MMH) and upper-extremity intensive (UEI) tasks have been the focus of psychophysical research, and each will be considered separately due to the disparity in application methodologies.

Readers interested in the empirical and theoretical aspects of the psychophysical approach to MMH task design are referred to Snook[1,2] and Ayoub and Mital[3] for further reading. For specific information on the comparisons of the psychophysical approach to the biomechanical and physiological approaches to MMH task design, the reader should consult Ayoub,[4] Mital et al.,[5] and Nicholson.[6] Less thorough information on the empirical and theoretical aspects of the psychophysical approach to UEI task design is available. However, there are useful discussions in Kim et al.[7] and Fernandez et al.[8]

Introduction to Psychophysics

Psychophysics is a branch of psychology dealing with the relationships between stimuli and sensations. These relationships can be best described by the psychophysical power law. The psychological magnitude (sensation) ψ grows as a power function of the physical magnitude ϕ (stimulus) in the following manner:[9]

$$\psi = k\phi^n$$

The value of the constant k depends on the units of measure, while the exponent n has a value that varies for different sensations. The value of n may be lower than 1 for stimuli such as smell and brightness, or as high as 3.5 for electric shock.[9] Ljungberg et al.[10] found a value of $n = 1.86$ for a simulated brewery lifting task, whereas Gamberale et al.[11] found a value of 2.43 for a similar task. Gamberale et al. attributed the larger value in the latter study to more demanding lifting cycles.

In MMH experiments, the subject adjusts the magnitude of the stimulus (weight, force, or frequency) to correspond to a sensation which is "dictated" by the instructions given by the experimenter, i.e., "without straining yourself, or becoming unusually tired, weakened, overheated, or out of breath."[12] Adaptations of these instructions have been used to study UIE tasks. For UEI tasks, the instructions are directed at having subjects select work loads that do not result in "unusual discomfort in the hands, wrists, or forearms."[13] Subjects are also monitored during experimentation for signs of soreness, stiffness, and numbness.

The Current State of Psychophysical Data

Psychophysical data are currently available for designing MMH tasks as well as UEI tasks. For MMH tasks, there are fairly extensive data available for maximum acceptable weights and forces for lifting, lowering, pushing, pulling, holding, and carrying tasks. Research of the psychophysical approach to MMH task design has spanned approximately three decades, and current work continues to expand the range of task conditions for which psychophysical data are available.

For UEI tasks, data are available for maximum acceptable frequencies and forces for a variety of tasks. It should be noted that there are considerably more data available for designing MMH tasks than UEI tasks. The increased attention to cumulative trauma disorders (CTDs) in the past decade or so led researchers to adapt psychophysical techniques used in MMH research to the study of UEI tasks. In part, this was done because of the lack of quantitative guidelines for forces, durations, and postures associated with UEI work. Although there are not extensive data, there are data that can be used to design manual work involving the upper extremities.

60.2 The Psychophysical Approach to Designing Manual Materials Handling Tasks

One of the first approaches to the control of MMH injuries through specifying task limits was the psychophysical approach. Applications in the military which relied on subjective estimates of load handling limits[14,15] were followed by the psychophysical approach being utilized to set industrial materials handling limits.[16] The methodologies and database developed by Snook and his colleagues[1,16-18,19] at the Liberty Mutual Research Center have been used by researchers and practitioners for the past several decades. This has resulted in the availability of a wide range of data available for designing MMH tasks (e.g., References 5, 19–22).

Setting Weight and Force Limits

The application of the psychophysical approach to MMH task design is performed by using databases in the literature which provide limits specific to task conditions such as frequency, pushing or pulling distance, and load dimensions. Figure 60.1 presents a general model of the procedures associated with applying psychophysical data.

To use a database, the user records the relevant task parameters, then finds the value in the database applicable to the specific task. Since it is impossible to collect data for all combinations of tasks, one can either use interpolation to find the appropriate value, or one can use the closest value. In the latter case, the user should use the lower of the values in between which the task parameters fall. Next, the analyst determines if a task is acceptable. At a minimum, an MMH task should be designed to accommodate 75% of the population:[1] however, one should strive to accommodate at least 90% of the population whenever possible. If females perform the task, then the design should be based upon accommodating the female workers.

When a task does not accommodate at least 75% of the population, or if the task does accommodate 75% of the population and minor changes can be made to increase the acceptability of the task at little or no cost, then the task and/or workplace should be redesigned to accommodate at least 90% of the population. Options to increase the acceptability of a task will be described in more detail later. If a task cannot be redesigned to accommodate at least 75% of the population, then preplacement testing should be considered.

Psychophysical Data

Most psychophysical databases present maximum acceptable weights or forces. There is limited information available on maximum acceptable frequencies.[23-25] The data described in detail here are all related to force and weight data.

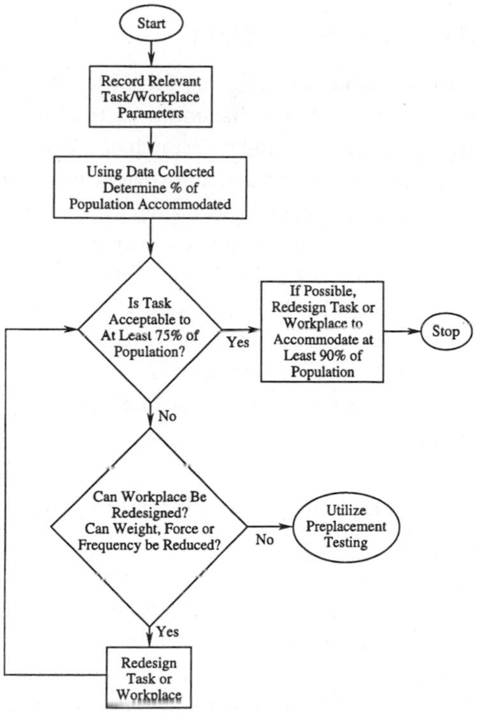

FIGURE 60.1 General model of the psychophysical approach to analyzing manual tasks.

The largest and most comprehensive single database for designing MMH tasks is that of Snook and Ciriello.[19] The data were collected from industrial subjects. The database includes maximum acceptable weights for lifting and lowering tasks, maximum acceptable initial and sustained forces for pushing and pulling tasks, and maximum acceptable weights for carrying tasks. The database provides values for males and females, as well as values that accommodate 10, 25, 50, 75, and 90% of the respective populations. An expanded computerized version of the database which covers a wider range of task conditions is available which performs analyses for both single- and multiple-component jobs (CompuTask™ *).

For lifting and lowering tasks, Snook and Ciriello's[19] database accommodates frequencies between one lift/lower every eight hours and 12 lifts/lowers per minute. Box dimensions in the sagittal plane between 34 and 75 cm are accommodated, as are lifts and lowers with vertical distances between 25 and 76 cm. Finally, data for floor to knuckle height, knuckle to shoulder height, and shoulder height to arm reach ranges of lift/lower are available.

Snook and Ciriello[19] presented maximum acceptable forces for pushing and pulling tasks, including forces required to start an object in motion (initial force) and forces required to keep an object in motion (sustained forces). Vertical handle heights between 57 and 135 cm are accommodated by the data. Frequencies for pushing/pulling are the same as those for lifting. The push/pull distances range between 2.1 m and 61.0 m.

Finally, Snook and Ciriello's[19] database also contains maximum acceptable weight of carry values for the same frequencies as the other MMH tasks mentioned above. Carry distances of 2.1 m, 4.3 m, and 8.5 m are available with vertical heights of 79 and 111 cm for males and 72 and 105 cm for females. Snook and Ciriello's[19] data are presented in Tables 60.1 through 60.9.

Modified Psychophysical Databases

Mital et al.[5] modified Snook and Ciriello's[19] database to conform with biomechanical, physiological, and epidemiological criteria. Mital et al. also present data from other sources for additional types of MMH

*Registered trademark of Liberty Mutual Insurance Group, Boston, Massachusetts

tasks. Snook and Ciriello's[19] database was modified to satisfy the following criteria: spinal compression limits of 2689 and 3920 N for females and males, respectively, in consideration of the results of Liles et al.;[26] an intra-abdominal pressure limit of 90 mmHg; and an energy expenditure limit of 21 to 23% of treadmill aerobic capacity or 28 to 29% of bicycle aerobic capacity. Mital et al.[5] also present maximum acceptable frequencies of lift for one-handed lifting, acceptable holding time data, and maximum acceptable forces and weights of lift for MMH tasks performed in nonstandard postures such as lying down and kneeling. In order to extend the range of applicability of the modified data, Mital et al.[5] presented seven multipliers to adjust the data for the following factors; work duration, limited headroom, asymmetrical lifting, load asymmetry, couplings, load placement clearance, and heat stress.

Mital[27] found that psychophysical data collected in short periods (i.e., 20 to 25 min.) assuming a longer work period (8 to 12 hours) should be reduced. Subsequently, Mital[21] presented psychophysical data for males and females performing lifting tasks for eight-hour work shifts based on the adjustments determined in the earlier study. The data were collected from 37 males and 37 females experienced in manual lifting. Mital[21] also presented a modified database representing the combined data from his study, Snook's[1] data, and data collected by Ayoub et al.[20] Although the modified database only accommodates lifting tasks, the combined sample size is considerable. Similarly, Mital[22] presented a psychophysical database for lifting tasks for males and females working 12-hour shifts. The database represents values valid for 12 hours based on adjustments of eight-hour data.

Data for Nonstandard MMH Tasks

One advantage of the psychophysical approach is that it allows for the realistic simulation of many types of materials handling tasks. Several such examples will be provided to illustrate how psychophysics has been used to develop guidelines for specific applications.

Smith et al.[28] presented a psychophysical database for evaluating MMH tasks performed in unusual postures. Maximum acceptable weight data for 99 different tasks were presented, including data for one- and two-handed lifting and lowering tasks performed in postures such as lifting on one knee, lifting on two knees, lifting while lying down, etc. These data are particularly appropriate for occasional maintenance tasks which impose postural constraints on the operator.

Like maintenance, mining is comprised of many activities for which standard psychophysical data are not applicable. Mining tasks are often performed under postural constraints, such as limited headroom. Gallagher[29] collected psychophysical data to address tasks performed under restricted headroom conditions and provided guidelines for tasks performed in low-seam coal mines requiring lifting while kneeling. Mining also requires handling of a variety of materials. Gallagher and Hamrick[30] provided psychophysical guidelines for the handling of rock dust bags, ventilation stopping blocks, and crib blocks.

Assessment of Multiple Component MMH Tasks

More often than not, workers who perform MMH tasks perform a number of tasks, often in sequence. For example, a common combination task in industry is where a worker lifts material, carries it for some distance, then lowers the material. In such situations, all of the tasks should be evaluated.

Snook and Ciriello[19] recommend using the weight or force limit for the task with lowest percentage of the population accommodated as the design criterion for multiple component tasks. This recommendation was based upon the findings of Ciriello et al.[17,18] Thus, for a combination where the worker lifts materials, carries materials, and then lowers material, the limiting component task would be the lift. Snook and Ciriello[19] do caution that this method of analysis may result in violation of recommended energy expenditure criteria for some multiple component tasks.

Straker et al.[31] disagree with the multiple component methodology described in the preceding paragraph, stating that "combination tasks should probably be assessed as whole entities and not separated into components for analysis." However, Straker et al.[31] recommend no alternative methodology to assess multiple component tasks as "whole entities." Ciriello et al.[17,18] did evaluate combination tasks as whole entities using psychophysics, and as stated above, there is typically one component that is the limiting component. In general, the design and evaluation of multiple component MMH jobs is one of the more underdeveloped areas of MMH research and practice.

TABLE 60.1 Maximum acceptable weight of lift for males (kg).

			Floor level to knuckle height One lift every								Knuckle height to shoulder height One lift every								Shoulder height to arm reach One lift every							
Width‡	Distance§	Percent¶	5	9	14	1	2	5	30	8	5	9	14	1	2	5	30	8	5	9	14	1	2	5	30	8
			s			min				h	s			min				h	s			min				h
		90	*6*	7	9	11	13	14	14	17	8	10	12	13	14	14	16	17	6	8	9	10	10	11	12	13
		75	*9*	11	13	16	19	20	21	24	10	14	16	18	18	19	21	23	8	10	12	14	14	14	16	17
	76	50	*12*	*15*	17	22	25	27	28	32	13	17	20	22	23	24	26	29	10	13	15	17	17	18	20	22
		25	*15*	*18*	*21*	28	31	34	35	41	16	*21*	*24*	27	27	28	32	35	11	16	18	21	21	22	24	27
		10	*18*	*22*	*25*	33	37	40	41	48	*19*	*24*	*28*	31	32	33	37	40	*14*	18	21	24	24	25	28	31
		90	6	8	9	12	13	15	15	17	8	11	13	15	15	16	18	19	6	8	9	12	12	12	14	15
		75	*9*	11	13	17	19	21	22	25	11	15	17	20	20	21	23	25	8	11	12	15	15	16	18	20
75	51	50	*13*	15	18	23	26	28	29	34	14	19	21	25	25	26	29	32	10	14	16	19	20	20	23	25
		25	*16*	*19*	22	29	33	35	35	42	*17*	*23*	26	30	31	32	36	39	13	17	19	23	24	25	27	30
		10	*19*	*22*	*26*	34	38	42	43	50	*20*	*26*	*30*	35	36	37	41	45	*15*	19	22	27	27	29	32	35
		90	8	9	11	13	15	16	17	20	10	13	15	18	18	19	21	23	7	10	11	14	14	14	16	18
		75	*11*	13	15	19	22	24	24	28	13	17	20	23	24	25	27	30	10	13	15	18	18	19	21	23
	25	50	*15*	18	21	26	29	32	33	38	17	22	25	30	30	31	35	38	12	16	19	23	23	24	27	29
		25	*18*	22	26	33	37	40	41	48	*20*	*27*	*30*	36	36	38	42	46	15	20	22	28	28	29	32	35
		10	*22*	*26*	*31*	38	44	47	49	57	*23*	*31*	*35*	42	42	44	49	53	*17*	*23*	26	32	32	34	38	41
		90	7	8	10	13	15	16	17	20	8	10	12	13	14	14	16	17	7	9	10	12	12	13	14	16
		75	*10*	12	14	19	22	24	24	28	10	14	16	18	18	19	21	23	9	11	13	16	16	17	19	21
	76	50	*14*	*16*	19	26	29	32	33	38	13	17	20	22	23	24	26	29	11	15	17	20	21	21	24	26
		25	*17*	*20*	*24*	33	37	40	41	48	16	21	24	27	27	28	32	35	13	18	20	25	25	26	29	31
		10	*20*	*24*	*28*	38	43	47	48	57	*19*	*24*	*28*	31	32	33	37	40	*15*	*21*	23	28	29	30	33	36

		90	7	9	10	14	16	17	18	20	8	11	13	15	15	16	18	19	7	9	11	14	14	14	16	18
		75	10	13	15	20	23	25	25	30	11	15	17	20	20	21	23	25	9	12	14	18	18	19	21	23
49	51	50	*14*	17	20	27	30	33	34	40	14	19	21	25	25	26	29	32	12	15	18	23	23	24	27	29
		25	*18*	*21*	25	34	38	42	43	50	17	23	26	30	31	32	36	39	14	19	21	28	28	29	32	35
		10	*21*	*25*	*29*	40	45	49	50	59	*20*	*26*	*30*	35	36	37	41	45	*16*	*22*	25	32	32	34	37	41
		90	8	10	12	16	18	19	20	23	10	13	15	18	18	19	21	23	9	11	12	16	16	17	19	21
		75	12	15	17	23	26	28	29	33	13	17	20	23	24	25	27	30	11	14	16	21	21	22	25	27
	25	50	*16*	20	23	30	34	37	38	45	17	22	25	30	30	31	35	38	14	18	21	27	27	28	32	35
		25	*21*	*25*	29	38	43	47	48	56	*20*	*27*	30	36	36	38	42	46	16	22	25	33	33	34	38	42
		10	*24*	*29*	*34*	45	51	56	57	67	*23*	*31*	*35*	42	42	44	49	53	*19*	*25*	*29*	38	38	40	44	48
		90	8	10	11	15	17	19	19	23	8	11	13	15	15	16	18	19	8	10	12	14	14	15	16	18
		75	*12*	14	17	22	25	28	28	33	11	15	17	20	20	21	23	25	10	14	16	18	19	19	22	24
	76	50	*16*	*19*	22	30	34	37	38	44	14	19	21	25	25	26	29	32	13	17	20	23	24	25	27	30
		25	*20*	*24*	*28*	37	42	47	47	55	17	*23*	26	30	31	32	36	39	*16*	*21*	24	28	29	30	33	36
		10	*24*	*29*	*33*	44	50	54	56	65	*20*	*26*	*30*	35	36	37	41	45	*18*	*24*	*28*	33	33	34	38	42
		90	9	10	12	16	18	20	20	24	9	12	14	17	17	18	20	22	8	11	13	16	16	17	18	20
		75	*12*	15	18	23	26	28	29	34	12	16	18	22	23	23	26	29	11	14	17	21	21	22	24	26
34	51	50	*17*	20	24	31	35	38	39	46	15	20	23	28	29	30	33	36	14	18	21	26	27	28	31	34
		25	*21*	*25*	*30*	39	44	48	49	57	18	24	27	34	35	36	40	44	*17*	22	25	32	32	33	37	41
		10	*25*	*30*	*35*	46	52	57	58	68	*21*	*28*	*32*	40	40	42	46	51	*19*	*26*	*29*	37	37	39	43	47
		90	10	12	14	18	20	22	23	27	11	14	16	20	20	21	23	26	10	13	15	19	19	19	22	24
		75	*15*	18	21	26	30	32	33	38	14	18	21	26	27	28	31	34	13	17	20	24	25	26	29	31
	25	50	*20*	24	28	35	40	43	44	52	18	23	27	33	34	35	39	43	16	22	25	31	31	33	36	40
		25	*26*	*30*	*35*	44	50	54	55	65	*21*	*28*	32	40	41	42	47	52	*20*	*26*	*30*	37	38	39	44	46
		10	*29*	*35*	*41*	52	59	64	66	76	*25*	*33*	*37*	47	47	49	55	60	*23*	*30*	*35*	43	44	45	51	55

‡Box width (the dimension away from the body)(cm).

§Vertical distance of lift (cm).

¶Percentage of industrial population.

Italicized values exceed 8 h physiological criteria.

From Snook, S. H. and Ciriello, V. M., The design of manual handling tasks: revised tables of maximum acceptable weights and forces, *Ergonomics*, 34, 1197, 1991. With permission.

TABLE 60.2 Maximum acceptable weight of lift for females (kg).

			Floor level to knuckle height One lift every								Knuckle height to shoulder height One lift every								Shoulder height to arm reach One lift every							
Width‡	Distance§	Percent¶	5 s	9 s	14 s	1 min	2 min	5 min	30 min	8 h	5 s	9 s	14 s	1 min	2 min	5 min	30 min	8 h	5 s	9 s	14 s	1 min	2 min	5 min	30 min	8 h
		90	*5*	6	7	7	8	8	9	12	5	6	7	9	9	9	10	12	4	5	5	6	7	7	7	8
		75	*7*	8	9	9	10	10	11	14	6	7	8	10	11	11	12	14	5	6	6	7	8	8	8	10
	76	50	*8*	10	10	11	12	12	13	17	7	8	9	11	12	12	13	16	6	7	7	8	9	9	10	11
		25	*9*	*11*	12	13	14	14	15	21	*8*	9	10	13	14	14	15	18	7	7	8	9	10	10	11	13
		10	*11*	*13*	*14*	14	15	16	17	23	*9*	10	11	14	15	15	17	20	7	8	9	10	11	11	12	14
		90	*6*	7	8	8	9	9	10	14	6	7	8	9	10	10	11	13	5	6	7	7	7	7	8	9
		75	*7*	*9*	9	10	11	11	13	17	7	8	9	11	12	12	13	15	6	7	8	8	9	9	9	11
75	51	50	*9*	*10*	11	12	13	14	15	21	*9*	9	11	13	14	14	15	17	7	8	9	9	10	10	11	13
		25	*10*	*12*	*13*	15	16	16	18	24	*10*	11	12	14	16	16	17	20	8	9	10	10	11	11	12	14
		10	*11*	*14*	*15*	17	18	18	20	27	*11*	12	14	16	17	17	19	22	9	10	11	12	13	13	14	16
		90	*6*	8	8	9	9	9	11	14	6	7	8	10	11	11	12	14	5	6	7	8	8	8	9	10
		75	*8*	10	11	11	12	12	13	18	7	8	9	12	13	13	14	17	6	7	8	9	9	9	10	12
	25	50	*10*	*12*	13	13	14	14	16	21	9	10	11	14	15	15	16	19	7	8	9	10	11	11	12	14
		25	*11*	*14*	15	15	16	17	19	25	10	11	12	16	17	17	19	22	8	9	10	12	12	12	14	16
		10	*13*	*16*	*17*	17	19	19	21	29	*11*	12	14	18	19	19	21	24	9	10	11	13	14	14	15	17
		90	5	6	7	8	8	8	9	13	5	6	7	9	9	9	10	12	4	5	5	7	7	7	8	9
		75	7	8	9	10	10	10	12	16	6	7	8	10	11	11	12	14	5	6	6	8	8	8	9	11
	76	50	*8*	10	10	12	12	13	14	19	7	8	9	11	12	12	13	16	6	7	7	9	10	10	11	12
		25	*9*	11	12	14	15	15	17	22	*8*	9	10	13	14	14	15	18	7	7	8	10	11	11	12	14
		10	*11*	*13*	14	15	17	17	19	25	*9*	10	11	14	15	15	17	20	7	8	9	11	12	12	13	15

		90	6	7	8	9	10	10	11	15	6	7	8	9	10	10	11	13	5	6	7	7	8	8	9	10
		75	7	9	9	11	12	12	14	18	7	8	9	11	12	12	13	15	6	7	8	9	9	9	10	12
49	51	50	*9*	10	11	13	15	15	16	22	*9*	9	11	13	14	14	15	17	7	8	9	10	11	11	12	14
		25	*10*	*12*	13	16	17	17	19	26	*10*	11	12	14	16	16	17	20	8	9	10	11	12	12	13	15
		10	*11*	*14*	15	18	19	20	22	30	*11*	12	14	16	17	17	19	22	*9*	10	11	13	14	14	15	17
		90	6	8	8	9	10	10	11	15	6	7	8	10	11	11	12	14	5	6	7	8	9	9	10	11
		75	*8*	10	11	12	12	13	14	19	7	8	9	12	13	13	14	17	6	7	8	9	10	10	11	13
	25	50	*10*	12	13	14	15	15	17	23	9	10	11	14	15	15	16	19	7	8	9	11	12	12	13	15
		25	*11*	14	15	16	18	18	20	27	10	11	12	16	17	17	19	22	8	9	10	12	13	13	15	17
		10	*13*	*16*	17	19	20	21	23	31	*11*	12	14	18	19	19	21	24	9	10	11	14	15	15	16	19
		90	7	8	9	9	10	10	11	15	6	7	8	9	10	10	11	13	5	6	7	8	9	9	10	11
		75	*8*	10	11	12	13	13	14	19	7	8	9	11	12	12	13	15	6	7	8	9	10	10	11	13
	76	50	*10*	*12*	13	14	15	16	17	23	*9*	9	11	13	14	14	15	17	7	8	9	11	12	12	13	15
		25	*12*	*14*	15	17	18	18	20	27	*10*	11	12	14	16	16	17	20	8	9	10	12	13	13	15	17
		10	*13*	*16*	*18*	19	20	21	23	31	*11*	12	13	16	17	17	19	22	*9*	10	11	14	15	15	16	19
		90	7	9	9	11	12	12	13	18	8	8	9	10	11	11	12	14	7	7	8	9	10	10	11	12
		75	*9*	*11*	12	14	15	15	16	22	9	10	11	12	13	13	14	17	8	8	9	11	11	11	12	14
34	51	50	*11*	*13*	*14*	16	18	18	20	27	*10*	11	13	14	15	15	17	19	*9*	10	11	12	13	13	14	17
		25	*13*	*15*	*17*	19	21	21	24	32	*12*	13	14	16	17	17	19	22	*10*	11	12	14	15	15	16	19
		10	*14*	*18*	*19*	22	24	24	27	36	*13*	14	16	18	19	19	21	24	*11*	*12*	14	15	16	16	18	21
		90	*8*	10	11	11	12	12	14	19	8	8	9	12	12	12	14	16	7	7	8	10	11	11	12	14
		75	*10*	12	13	14	15	15	17	23	9	10	11	13	14	14	16	18	8	8	9	12	12	12	14	16
	25	50	*12*	*15*	16	17	18	19	21	28	10	11	13	16	17	17	18	21	9	10	11	13	14	14	16	18
		25	*14*	*17*	*19*	20	22	22	24	33	12	13	14	18	19	19	21	24	10	11	12	15	16	16	18	21
		10	*16*	*20*	*21*	23	25	25	28	38	*13*	14	16	19	21	21	23	27	*11*	12	14	17	18	18	20	23

‡Box width (the dimension away from the body)(cm).
§Vertical distance of lift (cm).
¶Percentage of industrial population.
Italicized values exceed 8 h physiological criteria.

From Snook, S. H. and Ciriello, V. M., The design of manual handling tasks: revised tables of maximum acceptable weights and forces, *Ergonomics*, 34, 1197, 1991. With permission.

TABLE 60.3 Maximum acceptable weight of lower for males (kg).

			Knuckle height to floor level One lower every								Shoulder height to knuckle height One lower every								Arm reach to shoulder height One lower every							
Width‡	Distance§	Percent¶	5	9	14	1	2	5	30	8	5	9	14	1	2	5	30	8	5	9	14	1	2	5	30	8
				s			min			h		s			min			h		s			min			h
		90	7	9	10	12	14	15	16	20	10	11	14	14	15	15	16	19	6	7	9	9	10	10	11	13
		75	10	13	14	18	20	22	22	29	13	16	18	18	21	21	21	26	9	10	12	12	14	14	14	18
	76	50	*14*	17	19	23	27	29	30	38	18	20	24	24	27	27	28	34	11	13	15	16	18	18	19	23
		25	*17*	*21*	24	29	33	36	37	47	21	25	29	29	34	34	34	42	14	16	19	20	23	23	23	28
		10	*20*	*25*	*28*	34	39	42	44	56	*25*	29	34	34	39	39	39	49	16	19	22	23	26	26	27	33
		90	8	10	11	13	15	16	17	21	11	12	14	15	17	17	18	22	7	8	9	10	12	12	12	15
		75	11	14	15	18	21	23	23	30	14	17	20	21	24	24	24	30	9	11	13	14	16	16	16	20
75	51	50	*14*	18	20	24	28	30	31	40	19	21	25	27	31	31	31	38	12	14	16	18	21	21	21	26
		25	*18*	22	25	30	34	37	39	49	23	26	31	33	38	38	38	47	15	17	20	22	25	25	26	32
		10	*21*	*26*	29	36	41	44	46	58	*27*	31	36	38	44	44	44	55	17	20	24	26	30	30	30	37
		90	9	11	12	15	17	18	19	24	12	14	17	18	21	21	21	26	8	9	11	12	14	14	14	17
		75	13	16	17	21	24	25	26	34	17	20	23	24	28	28	28	35	11	13	15	16	19	19	19	24
	25	50	*17*	21	23	27	31	34	35	45	22	25	30	32	36	36	37	45	14	16	19	21	24	24	25	31
		25	*21*	26	29	34	39	42	44	56	27	31	37	39	44	44	45	56	17	20	24	26	30	30	30	38
		10	*24*	*31*	34	40	46	49	51	66	*31*	36	43	45	52	52	52	65	20	23	28	30	35	35	35	44
		90	8	10	11	15	17	18	19	24	10	11	14	14	15	15	16	19	7	8	10	11	12	12	12	15
		75	12	15	16	21	24	26	26	34	13	16	18	18	21	21	21	26	10	11	14	15	17	17	17	21
	76	50	*15*	19	21	27	31	34	35	45	18	20	24	24	27	27	28	34	13	15	17	19	22	22	22	27
		25	*19*	*24*	26	34	39	42	44	56	21	25	29	29	34	34	34	42	16	18	21	23	27	27	27	33
		10	*25*	*28*	*31*	40	46	49	51	65	25	29	34	34	39	39	39	49	18	21	25	27	31	31	31	39

		90	9	11	12	15	17	19	19	25	11	12	14	15	17	17	18	22	8	9	10	12	14	14	14	17
		75	12	15	17	22	25	26	28	35	14	17	20	21	24	24	24	30	10	12	14	16	19	19	19	24
49	51	50	16	20	22	29	33	35	37	47	19	21	25	27	31	31	31	38	14	16	18	21	24	24	25	31
		25	*20*	25	27	36	41	44	46	58	23	26	31	33	38	38	38	47	17	19	23	26	30	30	30	37
		10	*23*	*29*	32	42	48	51	54	68	27	31	36	38	44	44	44	55	19	22	26	30	35	35	35	44
		90	10	13	14	17	20	21	22	28	12	14	17	18	21	21	21	26	9	10	12	14	16	16	16	20
		75	14	18	19	24	28	30	31	40	17	20	23	24	28	28	28	35	12	14	17	19	22	22	22	28
	25	50	*19*	24	26	32	37	40	41	54	22	25	30	32	36	36	37	45	16	18	22	25	29	29	29	36
		25	*23*	*29*	32	40	46	49	51	65	27	31	37	39	44	44	45	56	20	23	27	31	35	35	36	44
		10	*27*	*34*	38	47	54	58	60	77	31	36	43	45	52	52	52	65	23	26	31	36	41	41	42	52
		90	10	12	13	17	19	21	21	27	11	12	14	15	17	17	18	22	9	10	12	12	14	14	14	18
		75	*14*	17	19	24	27	29	30	39	14	17	20	21	24	24	24	30	12	13	16	17	19	19	19	24
	76	50	*18*	*23*	25	32	36	39	40	51	19	21	25	27	31	31	31	38	15	17	21	22	25	25	25	31
		25	*23*	*29*	31	39	45	48	50	64	23	26	31	33	38	38	38	47	19	21	25	27	31	31	31	38
		10	*27*	*34*	*37*	46	53	57	59	75	27	31	36	38	44	44	44	55	22	25	30	31	36	36	36	45
		90	10	13	14	17	20	22	22	29	11	13	15	17	20	20	20	24	9	10	12	14	16	16	16	20
		75	14	18	20	25	28	30	32	40	15	18	21	23	27	27	27	33	12	14	17	19	22	22	22	27
34	51	50	*19*	24	26	33	37	40	42	53	20	23	27	30	35	35	35	43	16	19	22	24	28	28	28	35
		25	*24*	*30*	33	41	47	50	52	67	24	28	33	37	42	42	43	53	20	23	27	30	34	34	35	43
		10	*28*	*35*	*38*	48	55	59	62	78	28	33	39	43	49	49	50	62	23	27	31	35	40	40	40	50
		90	12	15	16	20	23	24	25	32	13	15	18	20	23	23	23	29	11	12	15	16	19	19	19	23
		75	17	21	23	28	32	34	36	46	18	21	25	27	31	31	32	39	15	17	20	22	26	26	26	32
	25	50	*23*	28	31	37	42	46	47	60	23	27	32	35	41	41	41	51	19	22	26	29	33	33	33	41
		25	*28*	*35*	38	46	53	57	59	75	29	33	39	42	50	50	50	63	23	27	32	35	41	41	41	51
		10	*33*	*41*	*45*	54	62	67	70	89	33	39	46	51	58	58	59	73	27	31	37	41	47	47	48	59

‡Box width (the dimension away from the body)(cm).

§Vertical distance of lower (cm).

¶Percentage of industrial population.

Italicize values exceed 8 h physiological criteria.

From Snook, S. H. and Ciriello, V. M., The design of manual handling tasks: revised tables of maximum acceptable weights and forces, *Ergonomics*, 34, 1197, 1991. With permission.

TABLE 60.4 Maximum acceptable weight of lower for females (kg).

Width‡	Distance§	Percent¶	Knuckle height to floor level, One lower every								Shoulder height to knuckle height, One lower every								Arm reach to shoulder height, One lower every							
			5 s	9 s	14 s	1 min	2 min	5 min	30 min	8 h	5 s	9 s	14 s	1 min	2 min	5 min	30 min	8 h	5 s	9 s	14 s	1 min	2 min	5 min	30 min	8 h
		90	5	6	7	7	8	8	9	12	6	6	7	8	9	10	10	13	5	5	5	6	7	7	7	9
		75	6	8	8	9	10	10	11	14	7	8	8	11	11	12	12	15	5	6	6	7	8	9	9	11
	76	50	7	9	10	11	12	12	13	17	8	9	10	12	13	14	14	18	7	8	8	8	10	10	10	13
		25	*9*	11	12	12	14	14	15	20	9	11	11	13	15	17	17	21	*8*	*9*	9	10	11	12	12	15
		10	*10*	13	13	14	15	16	17	23	*11*	12	13	15	17	19	19	24	*9*	*10*	10	11	12	14	14	17
		90	6	7	7	8	9	10	10	14	7	8	8	9	10	11	11	14	5	6	6	6	7	8	8	10
		75	7	8	9	10	11	12	13	17	8	9	9	11	12	13	13	17	7	7	8	8	9	10	10	12
75	51	50	*8*	10	11	12	14	14	15	20	10	11	11	13	15	16	16	20	8	9	9	9	11	12	12	15
		25	*10*	12	13	14	16	17	18	24	11	13	13	15	17	19	19	23	*9*	*10*	11	11	12	13	13	17
		10	*11*	*13*	14	16	18	19	20	27	*13*	15	15	17	19	21	21	26	*10*	*12*	12	12	14	15	15	19
		90	6	8	8	9	10	10	11	14	7	8	8	10	11	12	12	15	5	6	6	7	8	9	9	11
		75	8	10	10	11	12	12	13	17	8	9	9	12	13	15	15	19	7	7	8	9	10	11	11	13
	25	50	*9*	11	12	13	14	15	16	21	10	11	11	14	16	18	18	22	8	9	9	10	12	13	13	16
		25	*11*	13	14	15	17	17	19	25	11	13	13	16	19	20	20	26	9	10	11	12	13	15	15	19
		10	*12*	15	16	17	19	20	21	28	13	15	15	19	21	23	23	29	10	12	12	13	15	17	17	21
		90	5	6	7	8	8	9	10	13	6	6	7	8	9	10	10	13	5	5	5	6	7	8	8	10
		75	6	8	8	9	10	11	12	16	7	8	8	10	11	12	12	15	5	6	6	8	9	9	9	12
	76	50	8	9	10	11	13	13	14	19	8	9	10	12	13	14	14	18	7	8	8	9	10	11	11	14
		25	9	11	12	13	15	16	17	22	9	11	11	13	15	17	17	21	*8*	*9*	9	11	12	13	13	16
		10	*10*	13	13	15	17	18	19	25	*11*	12	13	15	17	19	19	24	*9*	*10*	10	12	13	15	15	19

		90	6	7	7	9	10	10	11	15	7	8	8	9	10	11	11	14	5	6	6	7	8	9	9	11
		75	7	8	9	11	12	13	14	18	8	9	10	11	12	13	13	17	7	7	8	8	10	10	10	13
49	51	50	*8*	10	11	13	15	15	16	22	10	11	11	13	15	16	16	20	8	9	9	10	11	13	13	16
		25	*10*	12	13	15	17	18	19	26	11	13	13	15	17	19	19	23	*9*	*10*	11	12	13	15	15	18
		10	*11*	13	14	17	19	20	22	29	*13*	15	15	17	19	21	21	26	*10*	*12*	12	13	15	16	16	21
		90	6	8	8	9	10	11	12	15	7	8	8	10	11	12	12	15	5	6	6	8	9	9	9	12
		75	8	10	10	11	13	13	14	19	8	9	10	12	13	15	15	19	7	7	8	9	10	12	12	14
	25	50	9	11	12	14	15	16	17	23	10	11	11	14	16	18	18	22	8	9	9	11	13	14	14	17
		25	*11*	13	14	16	18	19	20	27	11	13	13	16	19	20	20	26	9	10	11	13	15	16	16	20
		10	*12*	15	16	18	20	21	23	30	13	15	15	19	21	23	23	29	10	12	12	15	16	18	18	23
		90	6	8	9	9	10	11	12	15	7	8	8	9	10	11	11	14	6	6	7	8	9	9	9	12
		75	8	10	11	11	13	13	14	19	8	9	9	11	12	13	13	17	7	8	8	9	10	11	11	14
	76	50	*10*	12	13	14	15	16	17	23	10	11	11	13	15	16	16	20	8	9	10	11	13	14	14	17
		25	*11*	14	15	16	18	19	20	27	11	13	13	15	17	19	19	23	*9*	*11*	11	13	15	16	16	20
		10	*13*	*16*	17	18	20	21	23	30	*12*	14	15	17	19	21	21	26	*11*	*12*	13	14	16	18	18	23
		90	7	9	9	11	12	13	14	18	8	9	9	10	11	12	12	15	7	8	8	8	10	11	11	13
		75	*9*	11	11	13	15	16	17	22	9	11	11	12	14	15	15	19	8	9	10	10	12	13	13	16
34	51	50	*10*	13	14	16	18	19	20	27	11	13	13	14	16	18	18	22	*10*	11	11	12	14	15	15	19
		25	*12*	*15*	16	19	21	22	24	31	13	15	15	17	19	21	21	26	*11*	*13*	13	14	16	18	18	22
		10	*14*	*17*	*18*	21	24	25	27	35	*16*	*17*	17	19	21	23	23	29	*13*	*15*	15	16	18	20	20	25
		90	8	10	10	11	13	13	14	19	8	9	9	11	12	13	13	17	7	8	8	9	11	12	12	15
		75	*10*	12	13	14	15	16	17	23	9	11	11	13	15	16	16	21	8	9	10	11	13	14	14	18
	25	50	*12*	14	15	17	19	20	21	28	11	13	13	16	18	20	20	25	10	11	11	14	15	17	17	21
		25	*14*	*17*	18	20	22	23	24	33	13	15	15	18	21	23	23	29	11	13	13	16	18	19	19	24
		10	*15*	*19*	20	22	25	26	28	37	15	17	17	21	23	26	26	32	*13*	*15*	15	18	20	22	22	28

‡Box width (the dimension away from the body)(cm).
§Vertical distance of lower (cm).
¶Percentage of industrial population.
Italicized values exceed 8 h physiological criteria.

From Snook, S. H. and Ciriello, V. M., The design of manual handling tasks: revised tables of maximum acceptable weights and forces, *Ergonomics*, 34, 1197, 1991. With permission.

TABLE 60.5 Maximum acceptable forces of push for males (kg).

Height‡	Percent§	2.1 m push, One push every: 6 s	12 s	1 min	2 min	5 min	30 min	8 h	7.6 m push, One push every: 15 s	22 s	1 min	2 min	5 min	30 min	8 h
Initial forces¶															
144	90	20	22	25	25	26	26	31	14	16	21	21	22	22	26
	75	26	29	32	32	34	34	41	18	20	27	27	28	28	34
	50	32	36	40	40	42	42	51	23	25	33	33	35	35	42
	25	38	43	47	47	50	51	61	27	3	40	40	42	42	51
	10	44	49	55	55	58	58	70	31	35	46	46	48	49	58
95	90	21	24	26	26	28	28	34	16	18	23	23	25	25	30
	75	28	31	34	34	36	36	44	21	23	30	30	32	32	39
	50	34	38	43	43	45	45	54	26	29	38	38	40	40	48
	25	41	46	51	51	54	55	65	31	35	45	45	48	48	58
	10	47	53	59	59	62	63	75	35	40	52	52	55	56	66
64	90	19	22	24	24	25	26	31	13	14	20	20	21	21	26
	75	25	28	31	31	33	33	40	16	19	26	26	27	28	33
	50	31	35	39	39	41	41	50	20	23	32	32	34	35	41
	25	38	42	46	46	49	50	59	25	28	39	39	41	41	50
	10	43	48	53	53	57	57	68	28	32	45	45	47	48	57
Sustained forces*															
144	90	10	13	15	16	18	18	22	*8*	9	13	13	15	16	18
	75	*13*	17	21	22	24	25	30	*10*	*13*	17	18	20	21	25
	50	*17*	22	27	28	31	32	38	*13*	*16*	22	23	26	27	32
	25	*21*	*27*	33	34	38	40	47	*16*	*20*	28	29	32	33	39
	10	*25*	*31*	38	40	45	46	54	*19*	*23*	*32*	33	38	39	46
95	90	10	13	16	17	19	19	23	*8*	10	13	13	15	15	18
	75	*14*	18	22	22	25	26	31	*11*	*13*	17	18	20	21	25
	50	*18*	*23*	28	29	33	34	40	*14*	*17*	22	23	26	27	32
	25	*22*	*28*	34	35	40	41	49	*17*	*21*	27	29	32	33	39
	10	*26*	*33*	40	41	46	48	57	*20*	*24*	*32*	33	37	38	45
64	90	10	13	16	16	18	19	23	*8*	10	12	13	14	15	18
	75	*14*	18	21	22	25	26	31	*11*	*13*	17	17	19	20	24
	50	*18*	*23*	28	29	32	33	39	*14*	*17*	21	22	25	25	31
	25	*22*	*28*	34	35	39	41	48	*17*	*21*	26	27	31	32	37
	10	*26*	*32*	39	41	46	48	56	*20*	*25*	*30*	32	36	[illegible]	44

Height‡	Percent§	15.2 m push, One push every: 25 s	35 s	1 min	2 min	5 min	30 min	8 h	30.5 m push, One push every: 1 min	2 min	5 min	30 min	8 h
Initial forces¶													
144	90	16	18	19	19	20	21	25	15	16	19	19	24
	75	21	23	25	25	26	27	32	19	21	25	25	31
	50	26	29	31	31	33	33	40	24	27	31	31	38
	25	31	35	37	37	40	40	48	28	32	37	37	46
	10	36	40	43	43	45	46	55	32	37	42	42	53
95	90	18	21	22	22	23	24	28	17	19	22	22	27
	75	24	27	28	28	30	30	36	21	24	28	28	35
	50	29	33	35	35	37	38	45	27	30	35	35	44
	25	35	40	42	42	45	45	54	32	36	42	42	52
	10	40	46	49	49	52	52	62	37	41	48	48	60
64	90	15	17	19	19	20	20	24	14	16	19	19	23
	75	19	21	24	24	26	26	31	18	21	24	24	30
	50	23	27	30	30	32	33	39	23	26	30	30	37
	25	28	32	36	36	39	39	47	28	31	36	36	45
	10	32	37	42	42	44	45	54	32	36	41	41	52
Sustained forces*													
144	90	*8*	9	11	12	13	14	16	*8*	10	12	13	16
	75	*11*	*13*	15	16	18	18	22	*11*	13	16	18	21
	50	*14*	*17*	*20*	20	23	24	28	*15*	*17*	20	23	28
	25	*17*	*20*	*24*	25	28	29	34	*18*	*21*	25	29	34
	10	*20*	*24*	*28*	29	33	34	40	*21*	*25*	29	33	39
95	90	*8*	*10*	11	12	13	13	16	*8*	10	12	13	16
	75	*11*	*13*	15	16	18	18	21	*11*	13	16	18	21
	50	*14*	*17*	*19*	20	23	23	28	*15*	*17*	20	23	27
	25	*18*	*21*	*24*	25	28	29	34	*18*	*21*	25	28	33
	10	*20*	*25*	*28*	29	32	33	40	*21*	*25*	29	33	39
64	90	*8*	*10*	11	11	12	13	15	*8*	9	11	13	15
	75	*11*	*13*	14	15	17	17	21	*11*	13	15	17	20
	50	*14*	*17*	*19*	19	22	22	27	*14*	*16*	19	22	26
	25	*18*	*21*	*23*	24	27	28	33	*17*	*20*	24	27	32
	10	*21*	*25*	*27*	28	31	32	38	*20*	*24*	28	32	37

Height‡	Percent§	45.7 m push, One push every: 1 min	2 min	5 min	30 min	8 h	61.0 m push, One push every: 2 min	5 min	30 min	8 h
Initial forces¶										
144	90	13	14	16	16	20	12	14	14	18
	75	16	18	21	21	26	16	18	18	23
	50	20	23	26	26	33	20	22	22	28
	25	24	27	32	32	39	23	27	27	34
	10	28	31	36	36	45	27	31	31	39
95	90	14	16	19	19	23	14	16	16	20
	75	18	21	24	24	30	18	21	20	26
	50	23	26	30	30	37	22	26	26	32
	25	27	31	36	36	45	27	31	31	38
	10	32	36	41	41	52	31	35	35	44
64	90	12	14	16	16	20	12	14	14	17
	75	16	18	21	21	26	15	18	18	22
	50	20	22	26	26	32	19	22	22	28
	25	24	27	31	31	39	23	26	26	33
	10	27	31	36	36	44	26	30	30	38
Sustained forces*										
144	90	7	8	10	11	13	7	8	9	11
	75	*10*	*11*	13	15	18	*9*	11	13	15
	50	*12*	*14*	17	19	23	*12*	14	16	19
	25	*15*	*18*	21	24	28	*15*	17	20	24
	10	*18*	*21*	24	28	33	*19*	*20*	23	28
95	90	*7*	8	9	11	13	7	8	9	11
	75	*9*	*11*	13	15	18	*9*	11	12	15
	50	*12*	*14*	17	19	23	*12*	14	16	19
	25	*15*	*18*	21	24	28	*15*	17	20	23
	10	*17*	*20*	24	27	32	*17*	*20*	23	27
64	90	*7*	8	9	11	13	7	8	9	10
	75	*9*	*11*	12	14	17	*9*	10	12	14
	50	*12*	*14*	16	18	22	*12*	14	15	18
	25	*14*	*17*	20	23	27	*14*	17	19	22
	10	*17*	*20*	23	26	31	*16*	19	22	26

‡Vertical distance from floor to hands (cm).
§Percentage of industrial population.
¶The force required to get an object in motion.
*The force required to keep an object in motion.
Italicized values exceed 8 h physiological criteria.

From Snook, S. H. and Ciriello, V. M., The design of manual handling tasks: revised tables of maximum acceptable weights and forces, *Ergonomics*, 34, 1197, 1991. With permission.

TABLE 60.6 Maximum acceptable forces of push for females (kg).

		2.1 m push							7.6 m push							15.2 m push							30.5 m push					45.7 m push					61.0 m push			
		One push every							One push every							One push every							One push every					One push every					One push every			
Height‡	Percent§	6 s	12 s	1 min	2 min	5 min	30 min	8 h	15 s	22 s	1 min	2 min	5 min	30 min	8 h	25 s	35 s	1 min	2 min	5 min	30 min	8 h	1 min	2 min	5 min	30 min	8 h	1 min	2 min	5 min	30 min	8 h	2 min	5 min	30 min	8 h
																		Initial forces¶																		
	90	14	15	17	18	20	21	22	15	16	16	16	18	19	20	12	14	14	14	15	16	17	12	13	14	15	17	12	13	14	15	17	12	12	14	15
	75	17	18	21	22	24	25	27	18	19	19	20	22	23	24	15	17	17	17	19	20	21	15	16	17	19	21	15	16	17	19	21	14	15	17	19
135	50	20	22	25	26	29	30	32	21	23	23	24	26	27	29	18	20	20	20	22	23	25	18	19	21	22	25	18	19	21	22	25	17	18	20	22
	25	24	25	29	30	33	35	37	25	26	27	28	31	32	34	20	23	23	24	26	27	29	20	22	24	26	29	20	22	24	26	29	20	21	23	26
	10	26	28	33	34	38	39	41	28	30	30	31	34	36	38	23	26	26	26	29	31	32	23	25	27	29	33	23	25	27	29	33	22	24	26	29
	90	14	15	17	18	20	21	22	14	15	16	17	19	19	21	11	13	14	14	16	16	17	12	14	15	16	18	12	14	15	16	18	12	13	14	16
	75	17	18	21	22	24	25	27	17	18	20	20	22	23	25	14	16	17	17	19	20	21	15	16	18	19	21	15	16	18	19	21	15	16	17	19
89	50	20	22	25	26	29	30	32	20	21	23	24	27	28	30	16	19	20	21	23	24	25	18	20	21	23	26	18	20	21	23	26	18	19	20	23
	25	24	25	29	30	33	35	37	23	25	27	28	31	33	34	19	22	23	24	27	28	29	21	23	24	26	30	21	23	24	26	30	20	22	24	27
	10	26	28	33	34	38	39	41	29	28	31	32	35	37	39	22	24	26	27	30	31	33	24	26	28	30	33	24	26	28	30	33	23	25	26	30
	90	11	12	14	14	16	17	18	11	12	14	14	16	16	17	9	11	12	12	13	14	15	11	12	12	13	15	11	12	12	13	15	10	11	12	13
	75	14	15	17	17	19	20	21	14	15	17	17	19	20	21	11	13	14	15	16	17	18	13	14	15	16	18	13	14	15	16	18	12	103	14	16
57	50	16	17	20	21	23	24	25	16	18	20	21	23	24	25	14	15	17	18	19	20	21	15	17	18	19	22	15	17	18	19	22	15	16	17	19
	25	19	20	23	24	27	28	30	19	21	23	24	27	28	29	16	18	20	20	23	24	25	18	19	21	22	25	18	19	21	22	25	17	19	20	23
	10	21	23	26	27	30	31	33	22	23	26	27	30	31	33	18	20	22	23	25	26	28	20	22	23	25	28	20	22	23	25	28	19	21	23	25
																		Sustained forces*																		
	90	*6*	8	10	10	11	12	14	*6*	*7*	7	7	8	9	11	*5*	*6*	*6*	6	7	7	9	*5*	*6*	6	6	8	*5*	5	5	6	8	*4*	4	4	6
	75	*9*	*12*	14	14	16	17	21	*9*	*10*	11	11	12	13	16	*7*	*8*	*9*	9	10	11	13	*7*	*8*	9	9	12	*7*	*8*	8	8	11	*6*	6	6	9
135	50	*12*	*16*	19	20	21	23	28	*12*	*14*	*14*	15	16	17	21	*10*	*11*	*12*	*12*	14	14	18	*10*	*11*	12	12	16	*9*	*10*	*11*	11	15	*8*	*8*	9	12
	25	*16*	*20*	24	25	27	29	36	*15*	*17*	*18*	18	20	22	27	*12*	*14*	*15*	*16*	17	18	22	*13*	*14*	*15*	15	21	*11*	*13*	*13*	14	19	*10*	*10*	11	15
	10	*18*	*23*	28	29	32	34	42	*18*	*20*	*21*	22	24	26	32	*14*	*17*	*18*	*18*	20	22	27	*15*	*17*	*17*	18	25	*14*	*15*	*16*	17	22	*12*	*12*	13	17
	90	*6*	*7*	9	9	10	11	13	*6*	*7*	8	8	9	9	11	*5*	*6*	6	7	7	8	10	*5*	*6*	6	7	9	*5*	*6*	6	6	8	*4*	4	5	6
	75	*8*	*11*	13	13	15	16	19	*9*	*10*	11	11	13	13	17	*7*	*8*	9	10	11	11	14	*8*	*9*	9	10	13	*7*	*8*	8	9	12	*6*	6	7	9
89	50	*11*	*15*	18	18	20	21	26	*12*	*13*	*15*	15	17	18	22	*9*	*11*	*13*	*13*	14	15	19	*10*	*12*	12	13	17	*10*	*11*	*11*	12	16	*8*	*9*	9	12
	25	*14*	*18*	22	23	25	27	33	*15*	*17*	*19*	19	21	23	28	*12*	*14*	*16*	*16*	18	19	24	*13*	*15*	*15*	16	22	*12*	*14*	*14*	15	20	*11*	*11*	12	15
	10	*17*	*22*	26	27	30	32	39	*17*	*20*	*22*	*23*	25	27	33	*14*	*17*	*19*	*19*	21	23	28	*16*	*18*	*18*	19	26	*14*	*16*	*17*	18	24	*13*	*13*	14	18
	90	*5*	6	8	8	9	9	12	*6*	7	7	7	8	9	11	*5*	*6*	*6*	6	7	7	9	*5*	*6*	6	6	8	*5*	*5*	5	6	7	*4*	4	4	6
	75	*7*	*9*	11	12	13	14	17	*8*	*10*	10	11	12	12	15	*7*	*8*	*9*	*9*	10	10	13	*7*	*8*	8	9	12	*7*	*7*	8	8	11	*6*	6	6	8
57	50	*10*	*13*	15	16	17	18	23	*11*	*13*	*14*	14	16	17	21	*9*	*11*	*12*	*12*	13	14	17	*10*	*11*	11	12	16	*9*	*10*	*10*	11	15	*8*	*8*	8	11
	25	*12*	*16*	19	20	22	23	29	*14*	*17*	*18*	18	20	21	26	12	*14*	*15*	*15*	17	18	22	*12*	*14*	14	15	20	*11*	*13*	*13*	14	18	*10*	*10*	11	14
	10	*15*	*19*	23	23	26	28	34	*17*	*20*	*21*	21	23	25	31	*14*	*16*	*17*	*18*	20	21	26	*15*	*16*	*17*	18	24	*13*	*15*	*16*	16	22	*12*	*12*	13	17

‡Vertical distance from floor to hands (cm).
§Percentage of industrial population.
¶The force required to get an object in motion.
*The force required to keep an object in motion.
Italicized values exceed 8 h physiological criteria.

From Snook, S. H. and Ciriello, V. M., The design of manual handling tasks: revised tables of maximum acceptable weights and forces, *Ergonomics*, 34, 1197, 1991. With permission.

TABLE 60.7 Maximum acceptable forces of pull for males (kg).

Height‡	Percent§	2.1 m pull One pull every							7.6 m pull One pull every							15.2 m pull One pull every							30.5 m pull One pull every					45.7 m pull One pull every					61.0 m pull One pull every			
		6 s	12 s	1 min	2 min	5 min	30 min	8 h	15 s	22 s	1 min	2 min	5 min	30 min	8 h	25 s	35 s	1 min	2 min	5 min	30 min	8 h	1 min	2 min	5 min	30 min	8 h	1 min	2 min	5 min	30 min	8 h	2 min	5 min	30 min	8 h
																		Initial forces¶																		
144	90	14	16	18	18	19	19	23	11	13	16	16	17	18	21	13	15	15	15	16	17	20	12	13	15	15	19	10	11	13	13	16	10	11	11	14
	75	17	19	22	22	23	24	28	14	15	20	20	2[illegible]	21	26	16	18	19	19	20	20	24	14	16	19	19	23	12	14	16	16	20	12	14	14	17
	50	20	23	26	26	28	28	33	16	18	24	24	25	26	31	19	21	22	22	24	24	29	17	19	22	22	27	15	16	19	19	24	14	16	16	20
	25	24	27	31	31	32	33	39	19	21	28	28	29	30	36	22	25	26	26	28	28	33	20	22	26	26	32	17	19	22	22	28	16	19	19	24
	10	26	30	34	34	36	37	44	21	24	31	31	33	33	40	24	28	29	29	31	31	38	22	25	29	29	37	20	22	25	25	31	18	21	21	27
95	90	19	22	25	25	27	27	32	15	18	23	23	24	24	29	18	20	21	21	23	23	28	16	18	21	21	26	14	16	18	18	23	13	16	16	19
	75	23	27	31	31	32	33	39	19	21	28	28	29	30	36	22	25	26	26	28	28	33	20	22	26	26	32	17	19	22	22	28	16	19	19	24
	50	28	32	36	36	39	39	47	23	26	33	33	35	35	42	26	29	31	31	33	33	40	24	27	31	31	38	20	23	27	27	33	20	23	23	28
	25	33	37	42	42	45	45	54	26	30	39	39	41	41	49	30	34	36	36	38	39	46	27	31	36	36	45	24	27	31	31	38	23	26	26	33
	10	37	42	48	48	51	51	61	30	33	43	43	46	47	56	33	38	41	41	43	44	52	31	35	40	40	50	27	30	35	35	43	26	30	30	37
64	90	22	25	28	28	30	30	36	18	20	26	26	27	28	33	20	23	24	24	26	26	31	18	21	24	24	30	16	18	21	21	26	15	18	18	22
	75	27	30	34	34	37	37	44	21	24	31	31	33	34	40	24	28	29	29	31	32	38	22	25	29	29	36	19	22	25	25	31	19	21	21	27
	50	32	36	41	41	44	44	53	25	29	37	37	40	40	48	29	33	35	35	37	38	45	27	30	35	35	43	23	26	30	30	37	22	26	26	32
	25	37	42	48	48	51	51	61	30	34	44	44	46	47	56	34	39	41	41	43	44	52	31	35	41	41	50	27	30	35	35	43	26	30	30	37
	10	42	48	54	54	57	58	69	33	38	49	49	52	53	63	38	43	46	46	49	49	59	35	39	46	46	57	30	34	39	39	49	29	34	34	42
																		Sustained forces*																		
144	90	8	10	12	13	15	15	18	6	8	10	11	12	12	15	7	8	9	9	10	11	13	7	8	9	11	13	6	7	8	9	10	6	6	7	9
	75	10	13	16	17	19	20	23	*8*	10	13	14	16	16	19	*9*	*10*	12	12	14	14	17	*9*	10	12	14	16	7	9	10	11	14	7	8	10	11
	50	*13*	16	20	21	23	24	28	*10*	*13*	16	17	19	20	23	*11*	*13*	14	15	17	17	20	*11*	13	15	17	20	*9*	*11*	12	14	17	*9*	10	12	14
	25	*15*	20	24	25	28	29	34	*12*	*15*	20	20	23	24	28	*13*	*15*	*17*	18	20	21	24	*13*	15	18	20	24	*11*	*13*	15	17	20	*11*	12	14	17
	10	*17*	22	27	28	32	33	39	*14*	*17*	22	23	26	27	32	*14*	*17*	*19*	20	23	24	28	*15*	*17*	20	23	27	*12*	*14*	17	19	23	*12*	14	16	19
95	90	10	13	16	17	19	20	24	*8*	10	13	14	16	16	19	*9*	*10*	12	12	14	14	17	*9*	10	12	14	17	*7*	9	10	12	14	7	9	10	12
	75	*13*	17	21	22	25	26	30	*11*	*13*	17	18	20	21	25	*11*	*14*	15	15	18	18	22	*12*	13	16	18	21	*10*	*11*	13	15	18	*9*	11	13	15
	50	*16*	21	26	27	31	32	37	*13*	*17*	21	22	25	26	31	*14*	*17*	*19*	19	22	23	27	*14*	17	19	22	26	*12*	*14*	16	19	22	*12*	14	16	18
	25	*19*	*26*	31	33	37	38	45	*16*	*20*	26	27	30	31	37	*17*	*20*	*22*	23	26	27	32	*17*	*20*	23	27	32	*14*	*17*	19	22	26	*14*	16	19	22
	10	*22*	*29*	36	37	42	43	51	*18*	*23*	29	31	34	36	42	*19*	*23*	*26*	27	30	31	37	*19*	*23*	27	31	36	*16*	*19*	22	25	30	*16*	19	21	25
64	90	11	14	17	18	20	21	25	*9*	11	14	15	17	17	20	*9*	*11*	12	13	15	15	18	*9*	11	13	15	18	*8*	9	11	12	15	8	9	10	12
	75	*14*	19	23	23	26	27	32	*11*	*14*	19	19	22	22	26	*12*	*14*	16	17	19	19	23	*12*	14	17	19	23	*10*	*12*	14	16	19	*10*	12	13	16
	50	*17*	*23*	28	29	32	34	40	*14*	*18*	23	24	27	28	33	*15*	*18*	20	21	23	24	28	*15*	*18*	21	24	27	*13*	*15*	17	20	23	*12*	14	16	20
	25	*20*	*27*	33	35	39	40	48	*17*	*21*	27	28	32	33	39	*18*	*21*	*24*	25	28	29	34	*18*	*21*	25	28	33	*15*	*18*	21	24	28	*15*	17	20	23
	10	*23*	*31*	38	40	45	46	54	*19*	*24*	*31*	32	37	38	45	*20*	*24*	*27*	28	32	33	39	*21*	*24*	28	32	38	*17*	*20*	24	27	32	*17*	*20*	23	27

‡Vertical distance from floor to hands (cm).
§Percentage of industrial population.
¶The force required to get an object in motion.
*The force required to keep an object in motion.
Italicized values exceed 8 h physiological criteria.

From Snook, S. H. and Ciriello, V. M., The design of manual handling tasks: revised tables of maximum acceptable weights and forces, *Ergonomics*, 34, 1197, 1991. With permission.

TABLE 60.8 Maximum acceptable forces of pull for females (kg).

		2.1 m pull							7.6 m pull						
		One pull every							One pull every						
Height‡	Percent§	6 s	12 s	1 min	2 min	5 min	30 min	8 h	15 s	22 s	1 min	2 min	5 min	30 min	8 h
Initial forces¶															
	90	13	16	17	18	20	21	22	13	14	16	16	18	19	20
	75	16	19	20	21	24	25	26	16	17	19	19	21	22	24
135	50	19	22	24	25	28	29	31	19	20	22	23	25	26	28
	25	21	25	28	29	32	33	35	21	23	25	26	29	30	32
	10	24	28	31	32	36	37	39	24	26	28	29	32	34	36
	90	14	16	18	19	21	22	23	14	15	16	17	19	20	21
	75	16	19	21	22	25	26	27	17	18	19	20	22	23	25
89	50	19	23	25	26	29	30	32	19	21	23	24	26	27	29
	25	22	26	29	30	33	35	37	22	24	26	27	30	31	33
	10	25	29	32	33	37	39	41	25	27	29	30	33	35	37
	90	15	17	19	20	22	23	24	15	16	17	18	20	21	22
	75	17	20	22	23	26	27	28	17	19	20	21	23	24	26
57	50	20	24	26	27	30	32	33	20	22	24	25	28	29	30
	25	23	27	30	31	35	36	38	23	25	27	29	32	33	35
	10	26	31	34	35	39	40	43	26	28	31	32	35	37	39
Sustained forces*															
	90	*6*	*9*	10	10	11	12	15	*7*	*8*	9	9	10	11	13
	75	*8*	*12*	13	14	15	16	20	*9*	*11*	12	12	13	14	18
135	50	*10*	*16*	17	18	19	21	25	*12*	*13*	*15*	16	18	18	22
	25	*13*	*19*	21	21	23	25	31	*14*	*16*	*18*	19	21	22	27
	10	*15*	*22*	24	25	27	29	36	*16*	*19*	*21*	22	24	26	32
	90	*6*	*9*	10	10	11	12	14	*7*	*8*	9	9	10	10	13
	75	*8*	*12*	13	13	15	16	19	*9*	*10*	11	12	13	14	17
89	50	*10*	*15*	16	17	19	20	25	*11*	*13*	*15*	15	16	18	22
	25	*12*	*18*	20	21	23	24	30	*14*	*16*	*18*	18	20	22	27
	10	*14*	*21*	23	24	26	28	35	*16*	*18*	*21*	21	23	25	31
	90	*5*	8	9	9	1	11	13	*6*	*7*	8	8	9	10	12
	75	*7*	*11*	12	12	13	14	18	*8*	*9*	11	11	12	13	16
57	50	*9*	*14*	15	16	17	18	23	*10*	*12*	*13*	14	15	16	20
	25	*11*	*17*	18	19	21	22	27	*13*	*15*	*16*	17	19	20	24
	10	*13*	*20*	21	22	24	26	32	*15*	*17*	*19*	20	22	23	28

		15.2 m pull							30.5 m pull				
		One pull every							One pull every				
Height‡	Percent§	25 s	35 s	1 min	2 min	5 min	30 min	8 h	1 min	2 min	5 min	30 min	8 h
Initial forces¶													
	90	10	12	13	14	15	16	17	12	13	14	15	17
	75	12	14	16	16	18	19	20	14	16	17	18	20
135	50	14	16	19	19	21	22	24	17	18	20	21	24
	25	16	19	21	22	25	26	27	19	21	23	24	27
	10	18	21	24	25	27	29	30	22	24	25	27	31
	90	10	12	14	14	16	17	18	13	14	15	16	18
	75	12	15	17	17	19	20	21	15	16	18	19	21
89	50	14	17	19	20	22	23	25	18	19	21	22	25
	25	16	20	22	23	26	27	28	20	22	24	25	29
	10	18	22	25	26	29	30	32	23	25	26	28	32
	90	11	13	15	15	17	18	19	13	14	15	17	19
	75	13	15	17	18	20	21	22	16	17	18	20	22
57	50	15	18	20	21	23	24	26	18	20	22	23	26
	25	17	21	23	24	27	28	30	21	23	25	27	30
	10	19	23	26	27	30	31	33	24	26	28	30	34
Sustained forces*													
	90	*6*	*7*	*7*	8	8	9	11	*6*	*7*	7	8	10
	75	*7*	*9*	*10*	10	11	12	15	*8*	*9*	10	10	14
135	50	*9*	*11*	*13*	*13*	14	15	19	*11*	*12*	12	13	17
	25	*11*	*14*	*15*	*16*	17	19	23	*13*	*15*	*15*	16	21
	10	*13*	*16*	*18*	*18*	20	22	27	*15*	*17*	*17*	18	25
	90	*5*	*6*	*7*	7	8	9	11	*6*	*7*	7	7	10
	75	*7*	*8*	*10*	10	11	12	14	*8*	*9*	9	10	13
89	50	*9*	*11*	*12*	*13*	14	15	18	*10*	*12*	12	13	17
	25	*11*	*13*	*15*	*15*	17	18	22	*12*	*14*	*15*	15	21
	10	*13*	*15*	*17*	*18*	20	21	26	*15*	*16*	*17*	18	24
	90	*5*	*6*	*7*	7	7	8	10	*6*	*6*	6	7	9
	75	*7*	*8*	*9*	9	10	11	13	*7*	*8*	9	9	12
57	50	*8*	*10*	*11*	*12*	13	14	17	*9*	*11*	11	12	16
	25	*10*	12	*14*	*14*	16	17	21	*11*	*13*	13	14	19
	10	*12*	14	*16*	*16*	18	19	24	*13*	*15*	*16*	16	22

		45.7 m pull					61.0 m pull			
		One pull every					One pull every			
Height‡	Percent§	1 min	2 min	5 min	30 min	8 h	2 min	5 min	30 min	8 h
Initial forces¶										
	90	12	13	14	15	17	12	13	14	15
	75	14	16	17	18	20	14	15	16	18
135	50	17	18	20	21	24	16	18	19	21
	25	19	21	23	24	27	19	20	22	25
	10	22	24	25	27	31	21	23	24	27
	90	13	14	15	16	18	12	13	14	16
	75	15	16	18	19	21	15	16	17	19
89	50	18	19	21	22	25	17	18	20	22
	25	20	22	24	25	29	20	21	23	26
	10	23	25	26	28	32	22	24	25	29
	90	13	14	15	17	19	13	14	15	17
	75	16	17	18	20	22	15	16	18	20
57	50	18	20	22	23	26	18	19	21	23
	25	21	23	25	27	30	21	22	24	27
	10	24	26	28	30	34	23	25	27	30
Sustained forces*										
	90	*6*	*6*	7	7	9	*5*	5	5	7
	75	*8*	*9*	9	9	12	*7*	7	7	10
135	50	*10*	*11*	*11*	12	16	*8*	*9*	9	12
	25	*12*	*13*	*14*	14	19	*10*	*11*	11	15
	10	*14*	*15*	*16*	17	23	*12*	*12*	13	17
	90	*5*	*6*	6	7	9	*5*	5	5	7
	75	*7*	*8*	9	9	12	*6*	7	7	9
89	50	*9*	*11*	*11*	12	15	*8*	*8*	9	12
	25	*11*	*13*	*13*	14	19	*10*	*10*	11	15
	10	*13*	*15*	*16*	15	22	*12*	*12*	13	17
	90	*5*	*6*	6	5	8	*4*	5	5	6
	75	*7*	*8*	8	8	11	*6*	6	6	9
57	50	*9*	*10*	*10*	11	14	*8*	*8*	8	11
	25	*11*	*12*	*12*	13	17	*9*	*10*	10	13
	10	*12*	*14*	*14*	15	20	*11*	*11*	12	16

‡Vertical distance from floor to hands (cm).
§Percentage of industrial population.
¶The force required to get an object in motion.
*The force required to keep an object in motion.
Italicized values exceed 8 h physiological criteria.

From Snook, S. H. and Ciriello, V. M., The design of manual handling tasks: revised tables of maximum acceptable weights and forces, *Ergonomics*, 34, 1197, 1991. With permission.

TABLE 60.9 Maximum acceptable weight of carry (kg).

Height‡	Percent§	2.1 m carry, One carry every							4.3 m carry, One carry every							8.5 m carry, One carry every						
		6 s	12 s	1 min	2 min	5 min	30 min	8 h	10 s	16 s	1 min	2 min	5 min	30 min	8 h	18 s	24 s	1 min	2 min	5 min	30 min	8 h
Males																						
111	90	10	14	17	17	19	21	25	9	11	15	15	17	19	22	10	11	13	13	15	17	20
	75	14	19	23	23	26	29	34	13	16	21	21	23	26	30	13	15	18	18	20	23	27
	50	*19*	25	30	30	33	38	44	17	20	27	27	30	34	39	17	19	23	24	26	29	35
	25	*23*	30	37	37	41	46	54	*20*	25	33	33	37	41	48	*21*	*24*	29	29	32	36	43
	10	*27*	*35*	43	43	48	54	63	*24*	29	38	39	43	48	57	*24*	*28*	34	34	38	42	50
79	90	13	17	21	21	23	26	31	11	14	18	19	21	23	27	13	18	17	18	20	22	26
	75	18	23	28	29	32	36	42	16	19	25	25	28	32	37	17	20	24	24	27	30	35
	50	*23*	30	37	37	41	46	54	*20*	25	32	33	36	41	48	*22*	*26*	31	31	35	39	46
	25	*28*	*37*	45	46	51	57	67	*25*	30	40	40	45	50	59	*27*	*32*	38	38	42	48	56
	10	*33*	*43*	53	53	59	66	78	*29*	*35*	47	47	52	59	69	*32*	*38*	44	45	50	56	65
Females																						
105	90	11	12	13	13	13	13	18	9	10	13	13	13	13	18	10	11	12	12	12	12	16
	75	*13*	14	15	15	16	16	21	*11*	12	15	15	16	16	21	*12*	*13*	14	14	14	14	19
	50	*15*	16	18	18	18	18	25	*12*	13	18	18	18	18	24	*14*	*15*	16	16	16	16	22
	25	*17*	*18*	20	20	21	21	28	*14*	15	20	20	21	21	28	*15*	*17*	18	18	19	19	25
	10	*19*	*20*	22	22	23	23	31	*16*	17	22	22	23	23	31	*17*	*19*	20	20	21	21	28
72	90	*13*	14	16	16	16	16	22	10	11	14	14	14	14	20	*12*	12	14	14	14	14	19
	75	*15*	*17*	18	18	19	19	25	11	13	16	16	17	17	23	*14*	*15*	16	16	17	17	23
	50	*17*	*19*	21	21	22	22	29	*13*	15	19	19	20	20	26	*16*	*17*	19	19	20	20	26
	25	*20*	*22*	24	24	25	25	33	*15*	22	22	22	22	22	30	*18*	*19*	21	22	22	22	30
	10	*22*	*24*	27	27	28	28	37	*17*	*19*	24	24	25	25	33	*20*	*21*	24	24	25	25	33

‡Vertical distance from floor to hands (cm).
§Percentage of industrial population.
Italicized values exceed 8 h physiological criteria (see text).

From Snook, S. H. and Ciriello, V. M., The design of manual handling tasks: revised tables of maximum acceptable weights and forces, *Ergonomics*, 34, 1197, 1991. With permission.

The only other method for multitask assessment that incorporates MMH tasks in addition to lifting that the author was able to find was the method presented by Mital.[32] The data used with this methodology are from Mital et al.,[5] which are modified psychophysical data as described earlier. The method is similar to that developed by Jiang and Mital,[33] except that capacity is predicted using more contemporary data.

In general, this method requires that each MMH task is analyzed, and data regarding work duration, etc., are also needed. The analyst then determines the percentage of the population that the design should accommodate, which should be 75 or 90%. The next step involves calculating the recommended work rate (kg-m/min) for the percentage of the population being analyzed using the Mital et al.[5] data. The actual work rate is divided by the recommended work rate, which yields the risk potential. Any risk potential values greater than 1 signal the need for task redesign. This method focuses on the individual components that are unacceptable, as with the Snook and Ciriello[19] method.

Example

An example of Snook and Ciriello's[19] multicomponent task assessment will be used to illustrate how psychophysical data are used to analyze MMH tasks. The analysis was performed with the CompuTask computer program. The set of tasks is fairly simple and includes a worker bending over and lifting a box, carrying the box 5 feet, and lowering the box to the floor. The set of tasks is performed three times per minute for eight hours. The relevant data that need to be collected as well as the analysis are shown in Figure 60.2. Aside from psychophysical results, physiological analyses and NIOSH lifting equation computations (STRWL = single task recommended weight limit, FIRWL = frequency independent recommended weight limit, STLI = single task lifting index, FILI = frequency independent lifting index) are provided by the software.

The task with the lowest percentage of the population accommodated is the lifting task. This task accommodates 75% of the male population and <10% of the female population. Thus, the set of tasks is marginally acceptable for males and unacceptable for females. Also, the overall physiological evaluation shows that the task is not acceptable for eight hours. As was discussed earlier, the method of analysis being used may result in violation of energy expenditure criterion. Redesign efforts would be focused on eliminating the tasks through materials handling devices or a conveyor, or eliminating the need to lift and lower the boxes by increasing the vertical origin and destination of the lifting and lowering tasks, respectively.

Task and Workplace Design

An often overlooked use of psychophysical data is workplace design. In situations where unacceptable tasks cannot be eliminated, or loads, frequencies, or forces cannot be reduced to "acceptable" levels, psychophysical results can be used to suggest workplace design changes to increase the percentage of the population that a task or job will accommodate. For example, a common problem is that the dimensions and/or weight of the load cannot be reduced. In situations such as this, the task and/or workplace can be redesigned to decrease the physical demands of the task.

The following list provides several examples of task and workplace redesign principles to reduce physical demands when altering the material being handled is infeasible or mechanical aids cannot alleviate the need to handle materials manually. Application of these principles will increase the percentage of the population a task will accommodate.

1. For lifting or lowering tasks, bring the load closer to the body.
2. When lifting low loads, bring the vertical origin of the load as close to knuckle height as possible. Alternately, when lowering loads, the destination should be as close to knuckle height as possible. This principle will reduce bending. In general, try to avoid lifting or lowering to or from high and low locations.
3. Decrease the vertical distance that loads must be lifted/lowered and the distance which loads must be pushed, pulled, or carried.
4. Decrease the frequency of the task or increase the number or workers performing the task.

Shipping Department

LIBERTY MUTUAL.

Evaluation Results

COMPONENT	FREQUENCY ONE EVERY	FORCE (lbs)	HAND HEIGHT AT START (in)	HAND HEIGHT AT END (in)	DISTANCE MOVED	HAND DISTANCE FROM BODY (in)	BODY MOTION TWIST	BODY MOTION REACH	BODY MOTION BEND	POPULATION PERCENTAGES MALE	POPULATION PERCENTAGES FEMALE	SUGGESTED MAXIMUM DURATION (hrs) MALE / FEMALE
Lift	20 sec.	34	5	33	28 in.	10	Yes	Moderate	Considerable	75	<10	8 / EL
Carry	20 sec.	34	33		5 ft.		No		None	>90	77	8 / 8
Lower	20 sec.	34	33	5	28 in.	10	No	Moderate	Considerable	83	<10	8 / 8
EVALUATION FOR ENTIRE TASK							Yes	Moderate	Considerable	75	<10	EL / EL

EL —> Exceeds energy expenditure limit for the task duration specified

"POPULATION PERCENTAGES" are the percentages of the male and female population that can be expected to perform the task without excessive stress or excessive fatigue.

"SUGGESTED MAXIMUM DURATION" is the recommended continuous time the job can be performed during an 8-hour workday before exceeding the Energy Expenditure (kcal/Min) guidelines for males and females.

	STRWL	FIRWL	STLI	FILI
Component # 1 - Lift	11.6	21.1	2.9	1.6
Component # 3 - Lower	15.2	27.6	2.2	1.2

FIGURE 60.2 Example of psychophysical analysis of a multiple-component MMH job using CompuTask™.

5. For pushing and pulling tasks, provide equipment that provides the least resistance so that initial forces required to overcome inertia are as low as possible. Maintenance of mechanical assists is very important with regard to this principle.
6. For all MMH tasks, provide good hand-to-object coupling when possible, i.e., tote boxes with handles, carts with handle bars, etc.
7. Decrease the duration over which the task is performed.
8. Change pulling tasks to pushing tasks.

60.3 The Psychophysical Approach to Designing Upper Extremity Tasks

The primary risk factors for WRMSDs of the upper extremity are fairly well known.[34] Task-related risk factors include posture, force, and repetition. Vibration and cold are task-related risk factors for some disorders such as carpal tunnel syndrome. Duration of the task and rest periods are also important since these factors affect the acceptability of task. Altering work–rest relationships can alter the acceptability of a particular combination of posture, force, and repetition.

There are few quantitative guidelines for limits of posture, force, and repetition. Although general guidelines suggest maintaining a neutral wrist posture and reducing the force requirements and frequency of a task, these guidelines do not indicate acceptable levels of the variables. Once ergonomic analyses and task redesign are done, a decision as to the acceptability of a task is difficult.

The application of the psychophysical approach to the design of UEI tasks was a response to the need for establishing quantitative guidelines with which to assess tasks. Currently, quantitative dose–response relationships developed with epidemiological techniques that provide relationships between individual risk factors and their interactions and the risk of upper-extremity WRMSDs do not exist. In the absence of such relationships, psychophysical data will continue to be one option of setting task limits. The remainder of this section will provide an overview of the current state of psychophysical data as well as discussion of how these data are applied in the workplace.

One advantage of the psychophysical approach is that data can be developed which incorporate force, posture, and repetition into the development of data for different durations. This is important in that this approach allows for trade-offs between variables, i.e., for some tasks, it is not always possible to modify all factors.

Setting Acceptable Force and Frequency Limits

Fernandez and his colleagues have collected maximum acceptable frequency data for several types of tasks include drilling,[35-38] riveting,[39,40] and tasks requiring pinch and power grasps.[41,42] In these studies, factors such as wrist posture and duration were incorporated into the experimental protocol to provide frequency limits for a variety of UEI tasks. In a similar study, Abu-Ali et al.[43] had subjects control the length of rest periods for task combinations of varying wrist postures, exertion periods, and power grip forces.

In order to use these data, one would record relevant task parameters, find the data relevant for a particular task in the database, and determine if the task is acceptable to the majority of the population, just as with MMH data. If the task is not acceptable, then the frequency would need to be reduced, the duration would need to be reduced, or factors such as wrist deviation would need to be modified to increase the acceptability of a task.

While Fernandez and his colleagues chose frequency as the variable that subjects manipulate, Snook et al.[13] chose force as the manipulated variable. Also, Snook et al.[13] used a 7-hour adjustment period which was much longer than the shorter (20 to 25 min.) period used in the studies cited above. Snook et al.[13] studied tasks requiring wrist flexion with a power grasp, wrist flexion with a pinch grip, and extension with a power grasp. Frequencies between 2 and 20 repetitive motions per minute were studied. Female subjects adjusted wrist torque during the experiment by manipulating the resistance offered by

TABLE 60.10 Maximum acceptable forces for female wrist flexion (power grip) (N).

Percent of population	Repetition rate 2/min	5/min	10/min	15/min	20/min
90	14.9	14.9	13.5	12.0	10.2
75	23.2	23.2	20.9	18.6	15.8
50	32.3	32.3	29.0	26.0	22.1
25	41.5	41.5	37.2	33.5	28.4
10	49.8	49.8	44.6	40.1	34.0

From Snook, S. H., Vaillancourt, D. R., Ciriello, V. M., and Webster, B. S., Psychophysical studies of repetitive wrist flexion and extension, *Ergonomics*, 38, 1488, 1995. With permission.

TABLE 60.11 Maximum acceptable forces for female wrist flexion (pinch grip) (N).

Percent of population	Repetition rate 2/min	5/min	10/min	15/min	20/min
90	9.2	8.5	7.4	7.4	6.0
75	14.2	13.2	11.5	11.5	9.3
75	19.8	18.4	16.0	16.0	12.9
75	25.4	23.6	20.6	20.6	16.6
10	30.5	28.2	24.6	24.6	19.8

From Snook, S. H., Vaillancourt, D. R., Ciriello, V. M., and Webster, B. S., Psychophysical studies of repetitive wrist flexion and extension, *Ergonomics*, 38, 1488, 1995. With permission.

TABLE 60.12 Maximum acceptable forces for female wrist extension (power grip) (N).

Percent of population	Repetition rate 2/min	5/min	10/min	15/min	20/min
90	8.8	8.8	7.8	6.9	5.4
75	13.6	13.6	12.1	10.9	8.5
75	18.9	18.9	16.8	15.1	11.9
75	24.2	24.2	21.5	19.3	15.2
10	29.0	29.0	25.8	23.2	18.3

From Snook, S. H., Vaillancourt, D. R., Ciriello, V. M., and Webster, B. S., Psychophysical studies of repetitive wrist flexion and extension, *Ergonomics*, 38, 1488, 1995. With permission.

a magnetic particle brake. Aside from reporting the torques, forces were also reported which were computed by dividing the torques by the moment arms. The forces are reported in Tables 60.10 through 60.12. Currently, studies are being conducted that address other motions such as ulnar deviation, etc.

Snook et al.'s[13] data would be applied in a manner similar to that described earlier in this section. However, these data are more generic than some of the data collected by Fernandez and his colleagues. For example, the data collected by Snook et al.[13] do not apply only to specific tasks such as drilling, as do some of the data from other studies mentioned.[35-38]

Krawczyk et al.[14] presented preferred weights for manual transfer tasks for transfer distances of 0.5 and 1.0 m. and frequencies between 10 and 30 transfers per minute for an eight-hour work duration. Thus, depending on the situation, one could adjust frequency or transfer distance for a particular weight of object being transferred.

Tool and Workplace Design

Up to this point, the psychophysical results discussed have all involved experimentation where subjects control a variable. Another approach to the study of manual work using psychophysics is to elicit perceived exertion or discomfort ratings from subjects performing specific tasks. The results can then be used to select the task conditions with the lowest perceived exertion or discomfort.

Ulin et al.[45-47] utilized perceived exertions to study tool masses, tool shapes, and horizontal and vertical work locations. A very general summary of the results indicates that 114 cm was the preferred vertical height for driving screws with a variety of tools. At that height, a pistol shaped tool was the most preferred. The ratings of perceived exertion with respect to the horizontal distance of the workpiece from the body indicated that the workpiece should not be greater than 38 cm from the front of the worker. The optimal workpiece location was a vertical height of 114 cm and a horizontal distance of 13 cm. Likewise, increasing the tool mass increased the ratings, as would be expected.

In general, the following recommendations are examples of means to increase the psychophysical acceptability of a UEI task:

1. Maintain a neutral wrist posture.
2. Decrease the force requirements of a task.
3. Decrease the duration over which a task is performed. Worker rotation can be very helpful.
4. Decrease the frequency at which the task is performed. This will increase recovery time between exertions and help to prevent fatigue and possible CTDs.
5. Reduce the horizontal distance between the workpiece and the worker.
6. For work with hand tools performed while the operator is standing, position the workpiece at a vertical height of approximately 114 cm. Ideally, the location of the workpiece should be adjustable to accommodate different operators.

60.4 Advantages and Disadvantages of the Psychophysical Approach

Like any approach to setting limits for manual work, there are advantages and disadvantages of the psychophysical approach. This approach is only one of several approaches available for designing MMH and UEI tasks, and the advantages and disadvantages of each approach should be examined to determine the best fit for a particular situation. The advantages and disadvantages of the psychophysical approach to the design of MMH and UEI tasks are given below. When necessary, a distinction is made if the advantage or disadvantage is specific to MMH or UEI tasks.

The advantages of the psychophysical approach include:

- Psychophysics allows for the realistic simulation of industrial work.[2]
- Currently, there is a considerable amount of psychophysical data for MMH tasks available that were collected from industrial workers. Many physiological models were developed from limited samples of university students. Likewise, the representativeness of cadaver data used to set certain biomechanical criteria such as lumbosacral compression limits is questionable.
- Psychophysical results are consistent with the industrial engineering concept of a "fair day's work for a fair day's pay."[2]
- Psychophysics can be used to study intermittent MMH tasks which are common in industry.[2] Such tasks are not amenable to physiological analyses.
- Psychophysical results are very reproducible.[2]
- For MMH tasks, psychophysical judgments take into account the whole job, and integrate biomechanical and physiological factors.[48,49]
- Psychophysical results for MMH tasks appear to be related to low-back pain.[1,2,26]

- For MMH tasks, psychophysical data apply to a wider array of tasks than either the biomechanical or physiological approach.
- For MMH tasks that must necessarily be performed under postural restrictions (i.e., maintenance work and mining), psychophysics is one technique that can be used to develop handling limits specific to the tasks being examined.
- The psychophysical approach is less costly and time consuming to apply in industry than many of the biomechanical and physiological techniques.
- Currently, psychophysical data represent one of the only quantitative guides for the design of force limits for UEI work. In the absence of objective biomechanical or physiological criteria, psychophysics may be used to elicit acceptable task parameters for UEI work.[7]

The disadvantages and limitations of the psychophysical approach include:

- Psychophysics is a subjective method.[2]
- The assumption that the subjective work loads selected by subjects are below the threshold for injury has not been validated.[50] There is not extensive epidemiological support for psychophysical data for MMH tasks and no epidemiological support for using psychophysical data for the design of UEI data. However, the same is true of most of the other criteria currently in use for designing manual work.
- Psychophysical results for high-frequency MMH tasks exceed energy expenditure criteria.[2]
- Some psychophysical values for MMH tasks may violate the biomechanical spinal compression criterion of 3400 N.[51] However, this assumes that the spinal compression criterion of 3400 N is correct, for which there is not much support.
- Psychophysics does not appear to be sensitive to bending and twisting while performing MMH tasks, both of which have been related to compensable low-back pain cases.[2]
- The range of data for designing UEI tasks is somewhat limited at this time.

60.5 Conclusions

Psychophysical data are one option available to the ergonomist for designing manual tasks and assessing whether or not a task or set of tasks needs to be redesigned. These data have been applied in the workplace for many years with considerable success. As with any assessment tool, there are advantages and limitations associated with psychophysics as discussed in the previous section. When used properly, psychophysical data provide the analyst with a tool applicable to a diverse set of tasks involving manual work.

References

1. Snook, S. H., The design of manual handling tasks, *Ergonomics*, 21, 963, 1978.
2. Snook, S. H., Psychophysical considerations in permissible loads, *Ergonomics*, 28, 327, 1985.
3. Ayoub, M. M. and Mital, A., *Manual Materials Handling*, Taylor & Francis, London, 1989.
4. Ayoub, M. M., Problems and solutions in manual materials handling: the state of the art, *Ergonomics*, 35, 713, 1992.
5. Mital, A., Nicholson, A. S., and Ayoub, M. M., *A Guide to Manual Materials Handling*, Taylor and Francis, London, 1993.
6. Nicholson, L. M., A comparative study of methods for establishing load handling capabilities, *Ergonomics*, 32, 1125, 1989.
7. Kim, C. H., Marley, R. J., Fernandez, J. E., and Klein, M. G., Acceptable work limits for the upper extremities with the psychophysical approach, in *Proceedings of the 3rd Pan Pacific Conference on Occupational Ergonomics*, 1994, 312.

8. Fernandez, J. E., Fredericks, T. K., and Marley, R. J., The psychophysical approach in upper extremities work, in *Contemporary Ergonomics 1995*, Robertson, S. A., Ed., Taylor and Francis, London, 1995, 456.
9. Stevens, S. S., The psychophysics of sensory function, *American Scientist*, 48, 226, 1960.
10. Ljungberg, A. S., Gamberale, F., and Kilbom, Å., Horizontal lifting — physiological and psychological responses, *Ergonomics*, 25, 741, 1982.
11. Gamberale, F., Ljungberg, A. S., Annwall, G., and Kilbom, Å., An experimental evaluation of psychophysical criteria for repetitive lifting work, *Applied Ergonomics*, 18, 311, 1987.
12. Snook, S. H. and Ciriello, V. M., Maximum weights and work loads acceptable to female workers, *Journal of Occupational Medicine*, 16, 527, 1974.
13. Snook, S. H., Vaillancourt, D. R., Ciriello, V. M., and Webster, B. S., Psychophysical studies of repetitive wrist flexion and extension, *Ergonomics*, 38, 1488, 1995.
14. Emanuel, I., Chaffee, J. W., and Wing, J., A study of human weight lifting capabilities for loading ammunition into the F-86H aircraft, WADC Technical Report 56-367, Wright Air Development Center, Wright-Patterson Air Force Base, Ohio, 1956.
15. Switzer, S. A., Weight-Lifting Capabilities of a Selected Sample of Human Subjects, Technical Document Report No. MRL-TDR-62-57, Aerospace Medical Research Laboratories, Wright-Patterson Air Force Base, Ohio, 1962.
16. Snook, S. H. and Irvine, C. H., The evaluation of physical tasks in industry, *American Industrial Hygiene Association Journal*, 27, 228, 1966.
17. Ciriello, V. M., Snook, S. H., and Hughes, G. J. Further studies of psychophysically determined maximum acceptable weights and forces, *Human Factors*, 35, 175, 1993.
18. Ciriello, V. M., Snook, S. H., Blick, A. C., and Wilkinson, P. L., The effects of task duration on psychophysically-determined maximum acceptable weights and forces, *Ergonomics*, 33, 187, 1990.
19. Snook, S. H. and Ciriello, V. M., The design of manual handling tasks: revised tables of maximum acceptable weights and forces, *Ergonomics*, 34, 1197, 1991.
20. Ayoub, M. M., Bethea, N., Deivanayagam, S., Asfour, S., Bakken, G., Liles, D., Selan, J., and Sherif, M., Determination and modeling of lifting capacity, Final Report, HEW (NIOSH) Grant No. 5R010H00545-02, 1978.
21. Mital, A., Comprehensive maximum acceptable weight of lift database for regular 8-hour work shifts, *Ergonomics*, 27, 1127, 1984a.
22. Mital, A., Maximum weights of lift acceptable to male and female industrial workers for extended work shifts, *Ergonomics*, 27, 1115, 1984b.
23. Fox, R. R., A psychophysical study of high-frequency lifting, Unpublished doctoral dissertation, Texas Tech University, Lubbock, Texas, 1993.
24. Nicholson, L. M. and Legg, S. J., A psychophysical study of the effects of load and frequency upon selection of work load in repetitive lifting, *Ergonomics*, 29, 903, 1986.
25. Snook, S. H. and Irvine, C. H., Maximum frequency of lift acceptable to male industrial workers, *American Industrial Hygiene Association Journal*, 29, 531, 1968.
26. Liles, D. H., Deivanayagam, S., Ayoub, M. M., and Mahajan, P., A job severity index for the evaluation and control of lifting injury, *Human Factors*, 26, 683, 1984.
27. Mital, A., The psychophysical approach in manual lifting — a verification study, *Human Factors*, 25, 485, 1983.
28. Smith, J. L., Ayoub, M. M., and McDaniel, J. W., Manual materials handling capabilities in non-standard postures, *Ergonomics*, 35, 807, 1992.
29. Gallagher, S., Acceptable weights and physiological costs of performing combined manual handling tasks in restricted postures, *Ergonomics*, 34, 939, 1991.
30. Gallagher, S. and Hamrick, C. A., Acceptable work loads for three common mining materials, *Ergonomics*, 35, 1013, 1992.

31. Straker, L. M., Stevenson, M. G., and Twomey, L. T., A comparison of risk assessment of single and combination manual handling tasks: 1. Maximum acceptable weight measures, *Ergonomics*, 39, 128, 1996.
32. Mital, A., Using "A Guide to Manual Materials Handling" for designing/evaluating multiple activity manual materials handling tasks, in *Proceedings of the IEA World Conference on Ergonomic Design, Interfaces, Products, and Information*, 1995, 550.
33. Jiang, B. C., and Mital, A., A Procedure for designing/evaluating manual materials handling tasks, *Int J Prod Res*, 24, 913, 1986.
34. Putz-Anderson, V. (Ed.), *Cumulative Trauma Disorders: A Manual for Musculoskeletal Disorders of the Upper Limb*, Taylor and Francis, London, 1988.
35. Davis, P. J. and Fernandez, J. E., Maximum acceptable frequencies for females performing a drilling task in different wrist postures, *J Human Ergol*, 23, 81, 1994.
36. Fernandez, J. E., Dahalan, J. B., and Klein, M. G., Using the psychophysical approach in hand-wrist work, in *Proceedings of the M.M. Ayoub Occupational Ergonomics Symposium*, Institute for Ergonomics Research, Lubbock, Texas, 1993, 63.
37. Kim, C. H. and Fernandez, J. E., Psychophysical frequency for a drilling task, *International Journal of Industrial Ergonomics*, 12, 209, 1993.
38. Vaidyanathan, V. and Fernandez, J. E., MAF for males performing drilling tasks, in *Proceedings of the Human Factors Society 36th Annual Meeting*, Human Factors Society, Santa Monica, 1992, 692.
39. Fredericks, T. K., The effect of vibration on maximum acceptable frequency for a riveting task, Unpublished doctoral dissertation, The Wichita State University, Wichita, Kansas, 1995.
40. Fredericks, T. K. and Fernandez, J. E., The effect of vibration on maximum acceptable frequency for a riveting task, in *Proceedings of the Konz/Purswell Occupational Ergonomics Symposium*, Institute for Ergonomics Research, Lubbock, Texas, 1995, 27.
41. Dahalan, J. B. and Fernandez, J. E., Psychophysical frequency for a gripping task, *International Journal of Industrial Ergonomics*, 12, 219, 1993.
42. Klein, M. G. and Fernandez, J. E., The effects of posture, duration, and force on pinching frequency, *International Journal of Industrial Ergonomics*, 20, 267, 1997.
43. Abu-Ali, M., Purswell, J. L., and Schlegel, R. E., Psychophysically determined work-cycle parameters for repetitive hand gripping, *International Journal of Industrial Ergonomics*, 17, 35, 1996.
44. Krawczyk, S., Armstrong, T. J., and Snook, S. H., Preferred weights for hand transfer tasks for an eight hour day, in *Proceedings of International Scientific Conference on Prevention of Work-related Musculoskeletal Disorders*, Hagberg, M., and Kilbom, Å., Eds., 1992, 157.
45. Ulin, S. S., Armstrong, T. J., Snook, S. H., and Franzblau, A., Effect of tool shape and work location on perceived exertion for work on horizontal surfaces, *Am Ind Hyg Assoc* J, 54, 383, 1993a.
46. Ulin, S. S., Armstrong, T. J., Snook, S. H., and Keyserling, W. M., Examination of the effect of tool mass and work postures on perceived exertion for a screw driving task, *International Journal of Industrial Ergonomics*, 12, 105, 1993b.
47. Ulin, S. S., Snook, S. H., Armstrong, T. J., and Herrin, G. D., Preferred tool shapes for various horizontal and vertical work locations, *Appl Occup Environ Hyg*, 7, 327, 1992.
48. Haslegrave, C. M. and Corlett, E. N., Evaluating work conditions for risk of injury — techniques for field surveys, in *Evaluation of Human Work*, 2nd. ed., Wilson, J. R., and Corlett, E. N., Eds., Taylor and Francis, London, 1995, 892.
49. Karwowski, W. and Ayoub, M. M., Fuzzy modelling of stresses in manual lifting tasks, *Ergonomics*, 27, 641, 1984.
50. Gamberale, F. and Kilbom, Å., An experimental evaluation of psychophysically determined maximum acceptable work load for repetitive lifting work, in *Ergonomics International 88*, A. S. Adams, R. R. Hall, B. J. McPhee, and Oxenburgh, M. S., Eds., Ergonomics Society of Australia, Sydney, 1988, 233.
51. Chaffin, D. B. and Page, G. B., Postural effects on biomechanical and psychophysical weight-lifting limits, *Ergonomics*, 37, 663, 1994.

61

The Relative Importance of Biomechanical and Psychosocial Factors in Low Back Injuries

A. Kim Burton
University of Huddersfield

Michele C. Battié
University of Alberta

Chris J. Main
University of Manchester

61.1 Introduction

That low back trouble (LBT) is an increasing problem in industrialized society has become obvious. The realization that this increase is occurring despite the best efforts of ergonomists, clinicians, and legislators has not gone unnoticed (Nachemson, 1996; Waddell, 1992; Bigos and Battié, 1990; Winkel and Westgaard, 1996). Arguably the field of ergonomics has been thrown a little off balance by developments in the clinical arena over recent years. On the one hand, ergonomists and biomechanists strive to reduce physical stress at the workplace with the intent of lowering the risk of musculoskeletal problems. On the other hand, the clinicians and psychologists are suggesting that rehabilitation of the back-injured worker should involve not only activity but physical challenges to the musculoskeletal system (Kohles et al., 1990; Bendix et al., 1995). This apparent dichotomy cannot readily be dismissed; to understand how it has arisen requires an exploration of reports from the various fields of endeavor involved.

This chapter will attempt to bring together, from a variety of perspectives, evidence on the development, recurrence, and persistence of work-related low back trouble. Some of the issues raised are covered in detail elsewhere in this book; the intention here is to provide a rationale which links the considerations of ergonomics/biomechanics with those psychosocial aspects discussed in the two following chapters. Rather than present a systematic review, or attempt a meta-analysis, the starting point is the commonly held tenet that physically demanding work is detrimental to the back, i.e., it will cause injury leading to low back pain and consequent disability. From this it would follow logically that management of work-related low back disorders can best be achieved by reduction of occupational spinal stressors.

0-8493-2641-9/99/$0.00+$.50

The basic "injury/damage" model can be expressed as one where exposure to mechanical overload, whether a single event or cumulative stress, results in some form of damage to spinal tissues, and that further exposure leads to further damage and/or lack of recovery, which in turn leads to disabling consequences. On the face of it, this model would appear to be intuitively reasonable and valid. What follows is a selective exploration of the literature focusing on evidence which challenges the basic premise that ergonomic intervention is the key to reducing the burden of occupational back pain, and highlights the emergent role of psychosocial influences.

61.2 Background

There is no shortage of epidemiological reports that link heavy, strenuous work (physical stress) with back pain (Suadicani et al., 1994; Riihimaki, 1989), but this link is not universally reported (Burton et al., 1989; Spengler et al., 1986; Riihimaki et al., 1994); seemingly much depends on definitions for back pain and workload. Similarly, reports of an association between heavy work and absenteeism (Riihimaki et al., 1989; Andersson et al., 1983), are not entirely consistent (Lindstrom et al., 1994). Physical spinal stress from whole body vibration (WBV) also has been associated with an increased risk of occupational back injury (Wikstrom et al., 1994), and has been linked specifically with symptoms (Burton and Sandover, 1987), but a dose–response relationship has not been confirmed. There is certainly some experimental biomechanical evidence suggesting that strenuous work environments are likely to be detrimental; *in vitro* experiments, simulating physiological occupational loads, can result in fatigue damage of numerous spinal soft tissues (Adams and Dolan, 1995) and to the end plates of the vertebral bodies (Brinckmann et al., 1988).

The injury model suggests that if occupational loads are reduced there should be a concomitant reduction in occupational back trouble. It is highly likely that occupational physical stressors on workers' spines have progressively reduced over the last 20 years or so, due to the combined effects of increasing mechanization and ergonomics-driven legislative procedures. However, there is no evidence that back pain has decreased, and in fact the disability due to LBT continues to grow exponentially (Clinical Standards Advisory Group, 1994).

It might, at this point, be helpful to put the work-relatedness of LBT into perspective. Occupational low back trouble must be viewed in the light of a high prevalence of LBT in the general population, at least in the industrialized nations. The background prevalence is often quoted as 80% or more, but a more realistic figure for what might be termed "notable" back trouble (i.e., that which results in care-seeking or a period of disability, as opposed to the transient twinges suffered by most) is probably an average of around 60% (Papageorgiou et al., 1995). The lifetime prevalence is similar among males and females, but does seem to vary with age. Few adults can recall trouble before the age of 18 years (Burton, 1987), but then the prevalence rises through working years (from ~50% to ~65%), but does not increase thereafter (Papageorgiou et al., 1995). The 1-month prevalence rate amounts to 39% on average among adults (Papageorgiou et al., 1995), while the point prevalence rate is of the order of 14 to 30% (Clinical Standards Advisory Group. 1994). Such figures are derived from cross-sectional studies and are at the mercy of reporting errors; for instance it has been found that some will deny ever having back trouble despite the fact that previously they have been sick listed for the condition (Svennson and Andersson, 1982). Longitudinal studies of school children and adolescents have reported a surprisingly high lifetime prevalence of over 50% by 16 years of age (Burton et al., 1996b; Balagué et al., 1995). Undoubtedly back pain is a common life experience virtually irrespective of age, but adolescent back pain differs markedly from the adult experience in that, while a recurrent phenomenon, it is rarely disabling (Burton et al., 1996b).

Recent epidemiological studies have revealed that LBT can be as prevalent among sedentary workers as among manual workers (Hemmingway et al., 1994) but heavy jobs do seem to be associated with an increased work loss due to LBT (Riihimaki et al., 1989), though that increase may be due to longer spells rather than more spells (Andersson et al., 1983). However, the association between heavy work and absence rates is not a consistent finding (Lindstrom et al., 1994). A study of construction workers did

find a dose–response relationship between "severe" low back pain and stooping and kneeling, but back pain in general was associated more with a range of psychosocial parameters (Holmstrom et al., 1992).

The identification of risk factors for a condition with such a high background prevalence has long been recognized as problematic (Troup and Edwards, 1985). The etiology is clearly multifactorial (Troup and Videman, 1989) so it is exceedingly difficult to be certain that a particular job is involved in causation in individual cases. The problem may have arisen even if the individual had not been employed, there may have been a non-work etiology, or it may have resulted from a previous job. The high prevalence in adolescents, coupled with the fact that only 3 to 10% of episodes in adults are brought to medical attention (Wells, 1985), is a fair indication that much back pain is relatively benign and can be considered a normal life experience.

It should be remembered that LBT usually occurs simply as a complaint of pain (in back and/or leg, with or without sensory symptoms); very rarely is it possible to determine a physical reason for the symptoms; i.e., there is no objective evidence of tissue damage. A consequence of the symptoms is an inability (or reluctance) to perform activities of daily living, including work activities. In essence, LBT is a symptomatic complaint (usually) without evidence of specific tissue or structural damage, which may be associated with some degree of dysfunction; it is well recognized that there is poor correlation between the degree of symptoms and the extent of purported physical stress prior to onset.

Work-related LBT has to be viewed against this high background level of reporting, in the general population, of a symptom which has an undetermined pathology, a propensity for recurrence, and a variable tendency to progress to disability. It is the link between these features (damage, reporting/recurrence, and disability) which is the focus of the following discussion.

61.3 Spinal Damage

The structure presumed at greatest risk for damage from heavy work (including WBV) is the intervertebral disc, and this has been the major subject of investigation. While it has been indicated that heavy work is associated with lumbar disc degeneration (Riihimaki et al., 1990), much may depend on the measurement methods, imaging techniques, and the population studied. For instance, a review on this subject (Troup and Edwards, 1985) noted that reports of severe disc degeneration occurring in miners are not entirely consistent; numerous studies support the link, but at least one found that it was osteophytes rather than discal changes that were related to duration of mining work (Caplan et al., 1966). This association has been noticed in another mining population (author's unpublished data). Interestingly, Olympic weight lifters do not seem to show an excess of radiological changes (Hult, 1954). Assuming a link between work and disc degeneration, it might reasonably be assumed that disc degeneration will have an adverse effect on spine function (e.g., flexibility), and thus exposure to physical stressors (*via* occupation or leisure) will result in diminished flexibility. However, it would seem that lumbar flexibility, while being somewhat related to disc height, is not substantially influenced by either degeneration or exposure to physical stressors (Burton et al., 1996a). Recent investigation involving multivariable statistical techniques has confirmed that disc degeneration — as measured from magnetic resonance images (MRI) — is influenced only modestly by work history; the greatest proportion of the explained variances in degeneration score can be accounted for by genetic influences, though age did have some expected influence (Battié et al., 1995).

Disc herniations (as opposed to generalized degenerative changes) are not apparently associated with heavy work (Kelsey, 1975), but sitting work, particularly involving driving, may carry an increased risk (Kelsey et al., 1984). However, it should be borne in mind that the presence of herniations is poorly correlated with symptoms (Boos et al., 1995). Disc herniations were found in 76% of an asymptomatic control group (matched for age, sex, and work-related factors) compared with 96% in a symptomatic group; the presence of symptoms was related to neural compromise and psychosocial aspects of work, but not to the exposure to physical stressors (Boos et al., 1995).

Certainly we would expect to find, based on *in vitro* studies, signs of damage to the discs and vertebral bodies from exposure to mechanical overload (Brinckmann et al., 1988), but this has yet to be confirmed.

Perhaps the level of physical stressors commonly encountered in workers is simply below the threshold needed to create the predicted damage, or it may be that the measurement techniques currently available are insufficiently precise to identify it. Alternatively, it may be that the normal biological repair processes obscure the evidence (Brinckmann, 1985). A new method for quantifying overload damage from radiographs has become available (Brinckmann et al., 1994), by which it is possible to compare radiographs from cohorts exposed to heavy work or WBV with those exposed to light or sedentary work. Overload damage has only been detected in cohorts exposed to extremely heavy work, or to undamped vehicular vibration; cohorts performing strenuous work within current ergonomic guidelines did not show overload damage (Brinckmann et al, 1998). Also, certain strenuous sports activities have been shown to induce disc disruption (e.g., fast bowlers in cricket [Elliott et al., 1993]).

Severe trauma leading to vertebral body fractures, such as can happen in parachute landing (Murray-Leslie et al., 1977) or ejection from military aircraft (Laurell and Nachemson, 1963), may present few if any symptoms. Even when symptoms do ensue, recovery is relatively swift with no lasting disability or reduced effectiveness (Laurell and Nachemson, 1963).

Structures other than vertebrae and discs (i.e., muscles and ligaments) may, of course, be damaged (Adams and Dolan, 1995), but for the most part we do not have objective means for detecting such damage. Experience in areas other than the low back indicate that muscle and ligament injuries (in the absence of complete rupture) have a healing time of the order of 4 to 8 weeks. Perhaps 85% of those taking time off work with back trouble will have returned to work during that time (Clinical Standards Advisory Group. 1994), but that leaves a sizeable number for whom an obvious physiological explanation for the persisting disability is lacking.

Current knowledge is insufficient, in the vast majority of cases of LBT, for accurate identification of the structure or structures that are "injured," or the extent to which they might be injured. Some measure of strain to the soft tissues supporting the spine (possibly involving neuromuscular control mechanisms) are the most likely candidates. New modeling techniques are suggesting that spine stability is closely related to muscle activity. Deficient intrinsic spine muscles or a lack of motor control seem to reduce the mechanical stability of the spine which could increase the risk of straining muscles or ligaments (Cholewicki and McGill, 1996). The question remains whether back strains are most often the result of some specific element of work or the result of other circumstances in life. The evidence shows that the back can certainly be injured in various ways, but the injury model can be seen to be lacking explanatory power where prolonged disability is concerned.

Irrespective of whether damage to spinal structures can be identified or quantified, there is no doubt that workers do get painful backs and some will believe it is their work which is to blame. In a general survey of workers, 61% stated their back disorder was caused by their work, and 39% felt it was made worse by their work; workers report back disorders twice as frequently as any other disorder (Clinical Standards Advisory Group. 1994). A study, of sicklisted blue collar workers found that 60% of patients believed that work demands had caused their back trouble, but neither an assessment of workload (e.g., lifting, bending) nor calculated compression loads predicted the rate of return to work or sick leave during follow-up (Lindstrom et al., 1994). Low back pain attributed to work has been reported to be more common in female nurses than female teachers, but non-occupationally attributed back pain had the same prevalence in both groups (Cust et al., 1972). The finding that worker ratings of job-required physical abilities are correlated with back injury rates is interesting (Skovron et al., 1991) but, because both parameters are subject to possible reporting bias, the relationship may, in part, represent subjective attribution. Confounding of physical workload with psychosocial workload has been found among various branches of agricultural work with differing prevalence rates, rendering clear identification of risk factors difficult (Hildebrandt, 1995).

61.4 Injury, Recurrence, and Work Loss

The biomechanical approach to identification of occupational risk factors for LBT has shifted from static models of spinal loading to dynamic models which use accelerations and velocity to estimate forces acting

on the spine (Halpern, 1992), while the more sophisticated approaches are advocating methods which incorporate muscle synergy and cocontraction patterns (Cholewicki et al., 1995). This sort of modeling fits better with the epidemiology. Some data suggest that the risk of LBT is associated with dynamic lifting (Bigos et al., 1986), and this parameter has been studied in detail by Marras and colleagues (Marras et al., 1993). This latter study was the first to link epidemiological findings with quantitative biomechanical findings in a large working population. A multiple logistic regression model revealed that a combination of five trunk motion and workplace factors distinguished between high and low risk; the factors were lifting frequency, load moment, trunk lateral velocity, trunk twisting velocity, and trunk sagittal angle. The outcome variable (risk of LBT) was derived from medical and injury records, which admittedly suffer from confounding effects of reporting bias and inadequate information about previous trouble. While a causative link was not proved, the association between biomechanical factors and risk was clear. When looking at worker-rated job demands (including ratings of dynamic components) and back injury rates, it is apparent that high-risk jobs can have quite diverse demands, i.e., not all jobs with high injury rates require the same physical abilities (Halpern, 1992). A study of 1800 nurses in Belgium and The Netherlands, using a quasi-objective rating system for task demands, showed a significantly lower prevalence of back trouble (and other musculoskeletal complaints) in the Dutch nurses despite the fact that their average workload was substantially greater than their Belgian counterparts. Overall, symptoms and work loss in the previous 12 months were not related to workload, nor was attribution that work was causative. It transpired that the Dutch nurses differed strikingly from the Belgian nurses on a range of psychosocial variables; they reported less depressive symptoms and were significantly more positive about pain, work, and activity (Burton et al. 1997).

It is known that a previous history of back pain is highly predictive of future episodes (Troup et al., 1987; Burton et al., 1989; Bigos et al., 1991), so it might be more illuminating to investigate the relationship between occupational physical stressors and first-time back injury rather than with prevalence rates (i.e., study of workers without any back trouble history). Because of methodological difficulties, reports of such studies are rare. Videman and colleagues (Videman et al., 1984) studied nursing aides and trained nurses. They found that young nursing aides (with high work loads but low likelihood of previous LBT history) had a higher prevalence than the trained nurses who did less manual handling, suggesting that skill and training may be important. But it was also noted that the aides tended to have children at an earlier age, thus domestic spinal loading was possibly a confounding influence. While the annual incidence rate of LBT in nurses has been found to be higher than in teachers, the annual prevalence and point prevalence rates were the same (Leighton and Reilly, 1995), though first-onset at a younger age in nursing aides compared with teachers has been reported (Cust et al., 1972). It has been shown that experienced industrial workers do show a reduced risk for LBT compared with inexperienced workers, but this is not necessarily because they experience reduced spinal loads, rather, their smoother motions may be related to muscular coordination aiding spinal stability (Granata et al., 1996).

A large general population study (Croft et al., 1995) has found that new episodes of LBT are more likely for those who are psychologically distressed; a relationship that held true even for first onsets. In the large prospective study at the Boeing plant (Bigos et al., 1991) it was found that reported first injuries were not related specifically to job demands, rather to psychosocial factors (i.e., low job satisfaction and an elevated hysteria scale on the Minnesota Multiphasic Personality Inventory); the workers at Boeing, though, worked in an environment where job tasks were not particularly stressful for the back. Some clarification of the issue could emerge from study of first-onsets in workers exposed to the same substantial physical stressors over long periods; police officers in Northern Ireland have proved to be a useful group for study (Burton et al., 1996c). They compulsorily wear body armor (weighing >8kg) for up to 12 hours per day; they do so irrespective of rank, and they return to the same work on recovery from back trouble. This police force was compared with an English police force which did not wear armor. It was found that the physical stress of wearing body armor reduced the survival time to first-onset, and there was a slight increase in the hazard over time. It was also found that spending longer than two hours per day in vehicles comprised a separate risk for first-onset of LBT, but this hazard did not increase; it is interesting to note that the effect of exposure to armor and vehicles was not additive. Surprisingly, the

proportion of officers with persistent (chronic) back complaints did not depend on the length of exposure (to the same stressors) since first-onset, rather chronicity was associated with psychosocial factors (distress and blaming work) (Burton et al., 1996c).

Physical stress on the spine may actually be beneficial. It has been reported that a physical fitness program improved cardiovascular performance, strength, and flexibility, and was associated with reduced workers' compensation costs for back injuries (Cady et al., 1979), though this finding has not been confirmed by others (Battié et al., 1989). Porter has suggested that heavy manual work (in miners) may benefit the spine by developing ligamentous and annular strength, thus restraining encroachment of a disc protrusion into the spinal vertebral canal (Porter, 1987). A later cadaveric study gave some support to this notion; the compressive strength of the spines from young men who had been physically active tended to be greater than those who had been less active (Porter et al., 1989).

There may, of course, be factors other than task-induced stress that could be important in respect of causation. In a large prospective study of employees in various industries, Troup and colleagues (Troup et al., 1981) found that the current attack of back pain arose with no evidence of injury in over 50%. For those with an "injury," two thirds were truly accidental, i.e., they occurred in association with some identifiable event. Falls were actually as common a cause as handling injuries, and there was no indication that those currently injured by handling were more likely to have had a previous handling injury. Falls were also associated with longer periods of sick leave and a greater propensity for recurrence. Overall, recurrence was found to be common; a figure of 44% in the first year dropped to 31% in the second year, suggesting recurrence rates may be more a feature of the natural history of the disorder than a reflection of continued exposure to the work environment (Troup et al., 1981). Support for this notion has come from other studies, involving multivariable analyses, which have found little relationship between recurrence and work demands (Troup et al., 1987; Burton et al., 1989; Bigos et al., 1991). As mentioned above, the best predictor of future trouble seems to be a previous history, with perception of work demands being more important than objective measurement (Troup et al., 1987), and dissatisfaction with work being a significant factor; the term re-injury may be a misnomer (Bigos et al., 1991). More recently, an enhanced instrument, the Psychosocial Aspects of Work questionnaire, has been described which separately evaluates job satisfaction, social support, and mental stress (Symonds et al., 1996). It has been possible to show that workers with current LBT have a lower score for job satisfaction and social support but, surprisingly, absenteeism (Symonds et al., 1996) and work heaviness (Burton et al., 1997) were not related to these parameters.

While absenteeism may not be related to attitudes about work itself, other attitudes and beliefs do seem to be relevant. Psychosocial factors such as negative beliefs about the inevitable consequences of LBT (measured by the Back Beliefs Questionnaire [Symonds et al., 1996]), inadequate pain control strategies (Symonds et al., 1996), fear–avoidance beliefs (Waddell et al., 1993) and belief that work was causative (Burton et al., 1996c) have all been found to relate to absenteeism. The relationship between attribution of cause, job satisfaction, and pain perception is complex and, in part, related to job level; issues related to attribution are important factors for compliance with intervention strategies (Linton and Warg, 1993). Accordingly, it has been found that a simple educational intervention program (comprising workplace broadcasting of a pamphlet stressing the benign nature of LBT, the importance of activity and desirability of early work return) is capable of creating a positive shift in beliefs with a concomitant reduction in extended absence (Symonds et al., 1995).

The question of when and how to return workers with LBT to their job has attracted considerable attention. It has been popularly held that too early a return to the same task would risk recurrence of symptoms (or do further damage). Indeed back injured workers may perceive their back problem as lifelong trouble, and believe that their back injury has made them more vulnerable to reinjury and disability (Tarasuk and Eakin, 1994). There is accumulating evidence that this belief is not only false, but detrimental. The use of work restrictions does not necessarily correlate with reduced symptoms following return to work. A three-year follow-up of reported occupational musculoskeletal injuries (including LBT) found that those whose workloads had been reduced did not report fewer problems (Kemmlert et al., 1993). A one-year follow-up report concerning the use of restrictions for nonspecific back pain found

that restrictions were associated with a reduced probability of returning to the original work; they did not reduce work absence, and they did not significantly reduce recurrences (Skovron et al., 1997). In fact, a successful rehabilitation program for patients with subchronic back pain has advocated early return to unrestricted duties as part of a combined graded activity/behavioral therapy approach (Lindstrom et al., 1992). Prospective study of even the most severe chronic disabling spinal disorder workers' compensation patients who completed a functional restoration program and returned to work, found that they were at relatively low risk for recurrence (Garcy et al., 1996). When looking at predictors for chronicity, using multivariable analyses in a mixed population of workers, occupation was not found to be significant, rather chronicity was related to age and to the duration of the first spell (Burton et al., 1989). The study of working police officers revealed that change of duties after developing LBT was rare, and persistence of symptoms was found to be unrelated to the length of exposure (to the same physical stressors at work) following first-onset (Burton et al., 1996c).

Even when there is known damage to lumbar structures, activity restriction seems not to be necessary. A prospective study of 50 patients operated for lumbar disc herniation, who were urged to return to full activity as soon as possible, showed the following features: mean return to work time was 1.7 weeks with 25% returning within 1 to 2 days; 97% returned to their previous job and had returned to full work by 8 weeks; no patient changed jobs because of symptoms; only 6% had a recurrence (follow-up averaged 3.8 years) (Carragee et al., 1996). Clinical studies in workers' compensation back pain patients have found that delayed functional recovery was associated with psychosocial factors more than with perceived task demand (Hadler et al., 1995), and that longer spells off work were associated with a poor outcome (Lancourt and Kettelhut, 1992). In compensation cases, settlement of the claim, interestingly, does not result in a reduction in morbidity (Greenough and Fraser, 1989). Generally it would seem that workers over 50 years old return to work faster than younger ones; neither gender nor length of employment influenced recovery rates, but a lower hourly wage was associated with a longer return to work period (Reid et al., 1996).

The reluctance to confront normal physical challenges seen in back-disabled workers has been termed activity intolerance, which is variously linked to individual response to pain, the belief that a specific injury must be the cause of the pain, and behavioral roles such as suffering. This has led to the proposal that treatment and benefits should not, in general, be (pain) complaint-contingent but time-contingent such that there is a clear incentive to return to work within an allotted time (IASP, 1995).

The question obviously arises as to the origin of the various relevant psychosocial traits — do they develop before or after onset of symptoms? This issue will be addressed in the following chapters, but there is clinical evidence that psychological profiles predictive of chronicity are present early in the course of the back pain experience (within the first three weeks) (Burton et al., 1995) and will even precede it in a few people (Mannion et al., 1996). The profiles identified clinically (coping strategies and individual beliefs, as opposed to job satisfaction) are in the same domain as those found to be related to work loss (Symonds et al., 1996).

61.5 Effectiveness of Ergonomic Intervention

Most ergonomic interventions will focus on strategies to reduce spinal loading and have been discussed elsewhere in this book. What reports there are to support the belief that ergonomic intervention will reduce the impact of occupational low back pain have been considered largely anecdotal (Smedley and Coggon, 1994), and there are significant risks of ergonomics being confounded with rationalization leading to an "ergonomic pitfall" (Winkel and Westgaard, 1996). The only intervention which has been formally evaluated is worker training in manual handling techniques; while lifting techniques can be improved, the effect on injury rates has not been clearly demonstrated (Smedley and Coggon, 1994). Rigorous controlled trials of ergonomic intervention programs, with morbidity as the outcome, remain to be conducted. Meanwhile, there is evidence that personnel issues may be more important than physical ergonomics. A large industrial study has found that ergonomic solutions failed to reduce lost workdays, and that reliance on case monitoring and wellness orientation actually increased work loss. However,

diligent safety programs and an emphasis on systematic return-to-work programs did reduce lost workdays (Hunt and Habeck, 1993). Similarly, a comparison of a personnel program (designed to increase communication between claimants and their employer, their doctor and the Workers' Compensation Board) with a back program (designed to reduce back injuries through intensive feed-back training) found that it was the personnel program which was by far the most effective (Wood, 1987).

61.6 Concluding Remarks

There is now sufficient evidence to seriously challenge the power of the injury/damage model to explain the current phenomenon of occupational back trouble. Although epidemiological studies can link the incidence (first recalled onset) of back pain with certain physical stressors (notably spinal loading and vehicle use [Burton et al., 1996c]), the fact remains that these injuries should, by virtue of known physiological processes, resolve within the natural healing time of 4 to 8 weeks. Doubtless many such injuries do resolve naturally, but the problem (for science as well as society) is that a considerable number recur or progress to chronic disability. There is strong evidence that recurrence and disability are not generally related to physical stressors, rather they may be mediated by psychosocial phenomena. Biomechanics may help to explain initial injury mechanisms, but currently offers little in the way of explanation for persisting trouble.

It can be accepted that much work is physically demanding, and may (frequently) lead to some discomfort and pain. These transient symptoms may be a normal consequence of life, but if the worker believes that the job is to blame there is the potential for psychosocial factors to intervene. A proportion of back injured workers having inappropriate beliefs about the nature of their problem, and its relationship to work, will develop fear–avoidance behaviors because of inadequate pain coping strategies. They then begin to function in a disadvantageous way and drift into chronic disability. Once a worker has developed back trouble, it would seem that therapeutic programs combining physical challenges to the back with operant conditioning are more successful than traditional approaches involving rest. Just as bed rest is known to be detrimental to recovery from an episode of back trouble (Malmivaara et al., 1995), it may be that the same principle applies to prolonged work absence. Certainly, undue rest can be expected to result in the deficient motor control mechanisms which heighten biomechanical susceptibility to injury or reinjury (Cholewicki and McGill, 1996).

61.7 Summary

On balance, there is evidence to support the notion that biomechanics-based ergonomic improvements to the workplace may have some potential to limit first-time back injury; therefore they should be deployed where practicable. The possible role of ergonomics for reducing recurrence rates seems at best equivocal, but there is no convincing evidence that continuance of work is detrimental in respect of disability. Because it is likely that much back pain is only work-related inasmuch as people of working age get painful backs, it is becoming clear that reducing spinal loads or awkward postures is likely to have only a small impact on the overall pattern of back pain. Non-biomechanical approaches (organizational and psychosocial) seemingly are more effective in maintaining workability. Primary prevention may be an unrealistic goal, but reducing the costly burden of chronic disability may well be possible; strategies which involve psychosocial advice to overcome activity intolerance (IASP. 1995), reduce fear avoidance, and promote activity (along with therapeutic motor-control training) may well transpire to be effective.

References

Adams, M.A. and Dolan, P. 1995. Recent advances in lumbar spinal mechanics and their clinical significance. *Clin Biomech* 10:3-19.

Andersson, G.B.J., Svensson, H., and Oden, A. 1983. The intensity of work recovery in low back pain. *Spine* 8:880-884.

Balagué, F., Skovron, M.L., Nordin, M., Dutoit, G., and Waldburger, M. 1995. Low back pain in schoolchildren: a study of familial and psychological factors. *Spine* 20:1265-1270.

Battié, M.C., Bigos, S.J., and Fisher, L.D. 1989. A prospective study of the role of cardiovascular risk factors and fitness in industrial back pain complaints. *Spine* 14:141-147.

Battié, M.C., Videman, T., Gibbons, L., Fisher, L., Manninen, H., and Gill, K. 1995. Determinants of lumbar disc degeneration: a study relating lifetime exposures and MRI findings in identical twins. *Spine* 20:2601-2612.

Bendix, A.F., Bendix, T., Ostenfeld, S., Bush, E., and Andersen, A. 1995. Active treatment programs for patients with chronic low back pain: a prospective, randomized, observer-blinded study. *European Spine Journal* 4:148-152.

Bigos, S.J., Spengler, D.M., Martin, N.A., Zeh, J., Fisher, L., Nachemson, A., and Wang, M.H. 1986. Back injuries in industry: a retrospective study. II. Injury factors. *Spine* 11:246-251.

Bigos, S.J., Battié, M.C., Spengler, D.M., Fisher, L.D., Fordyce, W.E., Hansson, T., Nachemson, A.L., and Wortley, M.D. 1991. A prospective study of work perceptions and psychosocial factors affecting the report of back injury. *Spine* 16:1-6.

Bigos, S.J. and Battié, M.C. 1990. Industrial low back pain. Risk factors, in Weinstein, J.N. and Wiesel, S.W. (Eds.) *The Lumbar Spine*, pp. 846-859. Philadelphia: Saunders.

Boos, N., Reider, V., Schade, K., Spratt, N., Semmer, M., and Aebi, M. 1995. The diagnostic accuracy of magnetic resonance imaging, work perception, and psychosocial factors in identifying symptomatic disc herniations. *Spine* 20:2613-2625.

Brinckmann, P. 1985. Pathology of the vertebral column. *Ergonomics* 28:77-80.

Brinckmann, P., Biggemann, M., and Hilweg, D. 1988. Fatigue fracture of human lumbar vertebrae. *Clin Biomech* 3 (Suppl. 1):s1-s23.

Brinckmann, P., Frobin, W., Biggemann, M., Hilweg, D., Seidel, S., Burton, K., Tillotson, M., Sandover, J., Atha, J., and Quinell, R. 1994. Quantification of overload injuries to the thoracolumbar spine in persons exposed to heavy physical exertions or vibration at the workplace: Part I — The shape of vertebrae and intervertebral discs — study of a young, healthy population and a middleaged control group. *Clin Biomech* 9, (Suppl. 1):s5-s83.

Brinckmann, P., Frobin, W., Biggemann, M., Tillotson, K.M., Burton, A.K. 1998. Quantification of overload injuries to thoracolumbar vertebrae and discs in persons exposed to heavy physical exertions or vibration at the workplace. Part II. Occurrence and magnitude of overload injuries in exposed cohorts *Clin Biomech* 13, (Suppl. 2): s1-s42.

Burton, A.K. 1987. *Patterns of lumbar sagittal mobility and their predictive value in the natural history of back and sciatic pain*, Doctoral Thesis: Huddersfield Polytechnic/CNAA.

Burton, A.K., Tillotson, K.M., and Troup, J.D.G. 1989. Prediction of low-back trouble frequency in a working population. *Spine* 14:939-946.

Burton, A.K., Tillotson, K.M., Main, C.J., and Hollis, S. 1995. Psychosocial predictors of outcome in acute and subchronic low back trouble. *Spine* 20:722-728.

Burton, A.K., Battié, M.C., Gibbons, L., Videman, T., and Tillotson, K.M. 1996a. Lumbar disc degeneration and sagittal flexibility. *J Spinal Disord* 9:418-424.

Burton, A.K., Clarke, R.D., McClune, T.D., and Tillotson, K.M. 1996b. The natural history of low-back pain in adolescents. *Spine* 21:2323-2328.

Burton, A.K., Tillotson, K.M., Symonds, T.L., Burke, C., and Mathewson, T. 1996c. Occupational risk factors for the first-onset of low back trouble: a study of serving police officers. *Spine* 21: 2612-2620

Burton, A.K., Symonds, T.L., Zinzen, E., Tillotson, K.M., Caboor, D., Van Roy, P., and Clarys, J.P. 1997. Is ergonomics intervention alone sufficient to limit musculoskeletal problems in nurses? *Occup Med* 47: 25-32

Burton, A.K. and Sandover, J. 1987. Back pain in Grand Prix drivers: a "found" experiment. *Appl Erg* 18:3-8.

Cady, L.D., Bischoff, D.P., O'Connell, E.R., Thomas, P.C., and Allan, J.H. 1979. Strength and fitness and subsequent back injuries in firefighters. *J Occup Med* 21:269-272.

Caplan, P.S., Freedman, L.M.J., and Connelly, T.P. 1966. Degenerative joint disease of the lumbar spine in coal miners — a clinical and X-ray study. *Arthr Rheumatism* 9:693-701.

Carragee, E.J., Helms, E., and O'Sullivan, G.S. 1996. Are postoperative activity restrictions necessary after posterior lumbar discectomy? A prospective study of outcomes in 50 consecutive cases. *Spine* 21:1893-1897.

Cholewicki, J. and McGill, S.M. 1996. Mechanical stability of the *in vivo* lumbar spine: implications for injury and chronic low back pain. *Clin Biomech* 11:1-15.

Cholewicki, J., McGill, S.M., and Norman, R.W. 1995. Comparison of muscle forces and joint load from an optimization and EMG assisted lumbar spine model: towards development of a hybrid approach. *J Biomechanics* 28:321-332.

Clinical Standards Advisory Group. 1994. *Epidemiology review: the epidemiology and cost of back pain*, London: HMSO.

Croft, P.R., Papageorgiou, A.C., Ferry, S., Thomas, E., Jayson, M.I.V., and Silman, A.J. 1995. Psychologic distress and low back pain: evidence from a prospective study in the general population. *Spine* 20:2731-2737.

Cust, G., Pearson, J.C.G., and Mair, A. 1972. The prevalence of low back pain in nurses. *International Nursing Review* 19:169-179.

Elliott, B.C., Davis, J.W., Khangure, M.S., Hardcastle, P., and Foster, D. 1993. Disc degeneration and the young fast bowler in cricket. *Clin Biomech* 8:227-234.

Garcy, P., Mayer, T., and Gatchel, R.J. 1996. Recurrent or new injury outcomes after return to work in chronic disabling spinal disorders: tertiary preventative efficacy of functional restoration treatment. *Spine* 21:952-959.

Granata, K.P., Marras, W.S., and Kirking, B. (1996) Influence of experience on lifting kinematics and spinal loading, in Anonymous *20th Annual Meeting*, Georgia Tech, Atlanta, USA: American Society of Biomechanics.

Greenough, C.G. and Fraser, R.D. 1989. The effects of compensation on recovery from low-back injury. *Spine* 14:947-955.

Hadler, N.M., Carey, T.S., and Garrett, J. 1995. The influence of indemnification by works' compensation insurance on recovery from acute backache. *Spine* 20:2710-2715.

Halpern, M. (1992) Prevention of low back pain: basic ergonomics in the workplace and clinic, in Nordin, M. and Vischer, T.L. (Eds.) *Common Low Back Pain: Prevention of Chronicity*, pp. 705-730. London: Baillier Tindall.

Hemmingway, H., Shipley, M.J., Stansfield, S., and Marmot, M. 1994. *Society for Back Pain Research*, Leeds: November 11.

Hildebrandt, V.H. 1995. Musculoskeletal symptoms and workload in 12 branches of Dutch agriculture. *Ergonomics* 38:2576-2587.

Holmstrom, E.B., Lindell, J., and Mortiza, U. 1992. Low back and neck/shoulder pain in construction workers: Occupational workload and psychosocial risk factors. Part 2: Relationship to neck and shoulder pain. *Spine* 17:672-677.

Hult, L. 1954. Cervical, dorsal and lumbar spinal syndromes. *Acta Orthop Scand* Suppl. 17:1-102.

Hunt, A. and Habeck, R. 1993. *The Michigan Disability Prevention Study*, Kalamazoo, Mich. W.E Upjohn Institute for Employment Research.

IASP. 1995. *Back Pain in the Workplace: Management of Disability in Nonspecific Conditions*, Seattle: IASP Press.

Kelsey, J.L. 1975. An epidemiological study of the relationship between occupations and acute herniated lumbar intervertebral discs. *Int J Epidemiol* 3:197-205.

Kelsey, J.L., Githens, P.B., and O'Connor, T. 1984. Acute prolapsed lumbar intervertebral disc. An epidemiologic study with special reference to driving automobiles and cigarette smoking. *Spine* 9:608-613.

Kemmlert, K., Orelium-Dallner, M., Kilbom, Å. and Gamberale, F. 1993. A three-year follow-up of 195 reported occupational over-exertion injuries. *Scand J Rehabil Med* 25:16-24.

Kohles, S., Barnes, D., Gatchel, R.J., and Mayer, T.G. 1990. Improved physical performance outcomes after functional restoration treatment in patients with chronic low-back pain: Early vs. recent training results. *Spine* 15:1321-1324.

Lancourt, J. and Kettelhut, M. 1992. Predicting return to work for lower back pain patients receiving workers compensation. *Spine* 17:629-640.

Laurell, L. and Nachemson, A. 1963. Some factors influencing spinal injuries in seat ejected pilots. *Aerospace Med* 7:726-729.

Leighton, D.J. and Reilly, T. 1995. Epidemiological aspects of back pain: the incidence and prevalence of back pain in nurses compared to the general population. *Occup Med* 45:263-267.

Lindstrom, I., Ohlund, C., Eek, C., Wallin, L., Peterson, L., and Nachemson, A. 1992. Mobility strength and fitness after a graded activity program for patients with subacute low back pain: A randomized prospective clinical study with a behavioral therapy approach. *Spine* 17:641-652.

Lindstrom, I., Ohlund, C., and Nachemson, A. 1994. Validity of patient reporting and predictive value of industrial physical work demands. *Spine* 19:888-893.

Linton, S.J. and Warg, L. 1993. Attributions (beliefs) and job satisfaction associated with back pain in an industrial setting. *Percept Motor Skills* 76:51-62.

Malmivaara, A., Hakkinen, U., Heinrichs, M., Koskenniemi, L., Kuosma, E., Lappi, S., Paloheimo, R., Servo, C., Vaaranen, V., and Hernberg, S. 1995. The treatment of acute low back pain — bed rest, exercises, or ordinary activity? *New Eng J of Med* 332:351-355.

Mannion, A.F., Dolan, P., and Adams, M.A. 1996. Psychological questionnaires: do "abnormal" scores precede or follow first-time low back pain? *Spine* 21: 2603-2611.

Marras, W.S., Lavender, S.A., Leurgens, S.E., Rajulu, S.L., Allread, W.G., Farthallah, F.A., and Ferguson, S.A. 1993. The role of dynamic three-dimensional trunk motion in occupationally-related low back disorders: The effects of workplace factors trunk position and trunk motion characteristics on risk of injury. *Spine* 18:617-628.

Murray-Leslie, C.F., Lintott, D.J., and Wright, V. 1977. The spine in sport and veteran military parachutists. *Ann Rheum Dis* 36:332-342.

Nachemson, A.L. (1996) Low back pain in the industrial world, in Aspden, R.M. (Ed.) *Lumbar Spine Disorders*, pp. 1-5. Chesterfield: Arthritis & Rheumatism Council for Research.

Papageorgiou, A.C., Croft, P.R., Ferry, S., Jayson, M.I.V., and Silman, A.J. 1995. Estimating the prevalence of low back pain in the general population: Evidence from the South Manchester Back Pain Survey. *Spine* 20:1889-1894.

Porter, R.W. 1987. Does hard work prevent disc protrusion? *Clin Biomech* 2:196-198.

Porter, R.W., Adams, M.A., and Hutton, W.C. 1989. Physical activity and the strength of the lumbar spine. *Spine* 14:201-203.

Reid, S., Haugh, L.D., Hazard, R.G., and Tripathi, M. 1996. *Recovery Rates of Workers After Occupational Low Back Injury*, Burlington, VT. June 25-29: Int Soc Study Lumbar Spine.

Riihimaki, H. 1989. Radiographically detectable lumbar degenerative changes as risk indicators of back pain. A cross-sectional epidemiological study of concrete reinforcement workers and house painters. *Scand J Work Environ Health* 15:280-285.

Riihimaki, H., Tola, S., Videman, T., and Hanninen, K. 1989. Low-back pain and occupation. A cross-sectional questionnaire study of men in machine operating, dynamic physical work, and sedentary work. *Spine* 14:204-209.

Riihimaki, H., Mattison, T., Zitting, A., Wickstrom, G., Hanninen, K., and Waris, P. 1990. Radiologically detectable changes of the lumbar spine among concrete reinforcement workers and house painters. *Spine* 15:114-123.

Riihimaki, H., Viikari-Juntura, E., Moneta, G., Kuha, J., Videman, T., and Tola, S. 1994. Incidence of sciatic pain among men in machine operating dynamic physical work and sedentary work: A three-year follow up. *Spine* 19:138-142.

Skovron, M.L., Nordin, M., Halpern, N., and Cohen, H. 1991. *Do Worker Ratings of Job-Required Physical Abilities Correlate with Back Injury Rates?* Heidelberg, May 12-16: ISSLS.

Skovron, M.L., Hiebert, R., Nordin, M., Brisson, P.M., and Crane, M. 1997. Work restrictions and outcome of non-specific low back pain. *Spine,* in press.

Smedley, J. and Coggon, D. 1994. Will the manual handling regulations reduce the incidence of back disorders? *Occup Med* 44:63-65.

Spengler, D.M., Bigos, S.J., Martin, N.A., Zeh, J., Fisher, L., and Nachemson, A. 1986. Back injuries in industry: a retrospective study. I. Overview and cost analysis. *Spine* 11:241-245.

Suadicani, P., Hansen, K., Fenger, A.M., and Gyntelberg, F. 1994. Low back pain in steelplant workers. *Occup Med* 44:217-221.

Svensson, H. and Andersson, B.J. 1982. Low back pain in forty to forty-seven year old men. I. Frequency of occurrence and impact on medical services. *Scand J Rehabil Med* 14:47-53.

Symonds, T.L., Burton, A.K., Tillotson, K.M., and Main, C.J. 1995. Absence resulting from low back trouble can be reduced by psychosocial intervention at the workplace. *Spine* 20:2738-2745.

Symonds, T.L., Burton, A.K., Tillotson, K.M., and Main, C.J. 1996. Do attitudes and beliefs influence work loss due to low back trouble? *Occup Med* 46:25-32.

Tarasuk, V. and Eakin, J.M. 1994. Back problems are for life: perceived vulnerability and its implications fro chronic disability. *J Occup Rehabil* 4:55-64.

Troup, J.D.G., Martin, J.W., and Lloyd, D.C.E.F. 1981. Back pain in industry — a prospective survey. *Spine* 6:61-69.

Troup, J.D.G., Foreman, T.K., Baxter, C.E., and Brown, D. 1987. The perception of back pain and the role of psychophysical tests of lifting capacity. *Spine* 12:645-657.

Troup, J.D.G. and Edwards, F.C. 1985. *Manual Handling and Lifting: an Information and Literature Review with Special Reference to the Back,* London: Health & Safety Executive, HMSO.

Troup, J.D.G. and Videman, T. 1989. Inactivity and the aetiopathogenesis of musculoskeletal disorders. *Clin Biomech* 4:174-178.

Videman, T., Numinen, T., Tola, S., Kuorinka, I., Vanharranta, H., and Troup, J.D.G. 1984. Low back pain in nurses and some loading factors of work. *Spine* 9:400-404.

Waddell, G. (1992) Biopsychcosocial analysis of low back pain. In: Nordin, M. and Vischer, T.L. (Eds.) *Common Low Back Pain: Prevention of Chronicity,* pp. 523-557. London: Baillere Tindall.

Waddell, G., Newton, M., Henderson, I., Somerville, D., and Main, C.J. 1993. A fear avoidance belief questionnaire (FABQ) and the role of fear-avoidance beliefs in chronic low back pain and disability. *Pain* 52:157-168.

Wells, N. 1985. *Back Pain. Publication No. 78,* London: Office of Health Economics.

Wikstrom, B., Kjellberg, A., and Landstrom, U. 1994. Health effects of long-term occupational exposure to whole-body vibration: a review. *Int J Industrial Ergonomics* 14:273-292.

Winkel, J. and Westgaard, R.H. 1996. Editorial: a model for solving work related musculoskeletal problems in a profitable way. *Appl Erg* 27:71-78.

Wood, D.J. 1987. Design and evaluation of a back injury prevention program within a geriatric hospital. *Spine* 12:77-82.

62

Fall-Related Occupational Injuries

Stephen N. Robinovitch
San Francisco General Hospital

62.1 Introduction

In the United States, falls account for approximately one third of all injury-related hospitalizations and one tenth of all injury-related deaths, making them the leading cause of accidental injury and second leading cause of traumatic death (Rice, 1989; Public Health Service, 1993). The annual cost associated with such injuries has been estimated at a staggering $37.3 billion, over one third of which involves direct medical expenses (Rice, 1989). Therefore, in terms of both frequency and cost, falls closely rival motor vehicle accidents (MVA) as a leading cause of injury. However, aside perhaps for the problem of fall-related hip fractures in the elderly (which represent approximately one quarter of all fall-related injury costs [Praemar, 1992]), they have received considerably less attention from injury prevention experts.

To design effective strategies for preventing fall-related injuries, we must consider two sets of risk factors: those which lead to an increased incidence of falls, and those which increase risk for injury in the event of a fall. Research tools for identifying such risk factors generally fall under the categories of epidemiology and biomechanics. The former (discussed in Section 62.3 of this chapter) provides important information on the circumstances surrounding fall injuries, thereby identifying host and environmental targets for intervention. The latter (discussed in Section 62.4) complements epidemiological data by identifying, through the analysis of experimental or mathematical models of the injury-causing environment, how injury risk is influenced by factors such as motor control, Newtonian dynamics, and mechanical properties of biological tissues. Prevention strategies (discussed in Section 62.5) with the highest likelihood for success are often those derived from, and therefore compatible with, a combined epidemiological and biomechanical perspective.

62.2 General Considerations

Three general groups of individuals are at high risk for fall-related injuries: the elderly over age 65, children between ages 6 to 10, and individuals whose occupation requires them to work at heights (Figure 62.1). For each of these, injury results from the transfer of gravitational potential energy to tissue strain energy (and the corresponding generation of mechanical stresses and strains) that exceeds the tissue's energy-absorbing

0-8493-2641-9/99/$0.00+$.50

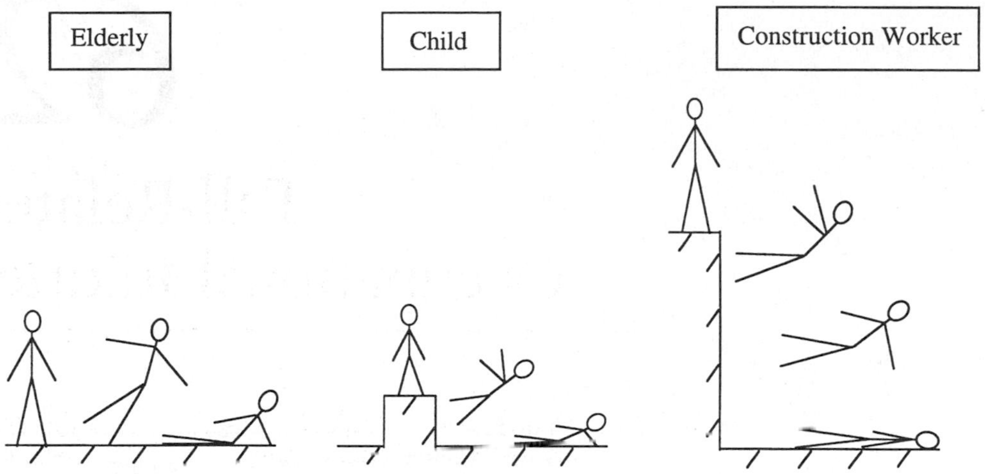

	Elderly	Child	Construction Worker
typical scenario:	fall from standing height	fall from play structure	fall from roof or scaffold
cause of fall:	intrinsic factors (poor balance, vision, etc.)	intrinsic factors (lack of caution, developing motor skills) and moderate-risk environmental factors (e.g. play structures with poor footing)	high-risk environmental factors and poor safety practices
energy content of fall:	low	moderate	high
resistance to trauma:	low (poor bone strength, impaired protective responses)	high	high
typical injury:	hip fracture, wrist fracture	abrasions, lacerations, wrist fracture	death, spinal cord injury, head injury, multiple fracture

FIGURE 62.1 Overview of factors influencing injury risk during typical falls in the elderly individual, child, and construction worker.

capacity. However, among the three groups, important differences exist in risk factors for fall-related injury. In particular, injury risk in the elderly is generally dominated by *host* factors such as balance, strength, reaction time, vision, and bone strength (Tinetti, 1994), while injury risk in the workplace is generally dominated by *environmental* factors such as poor lighting, slippery surfaces, workplace clutter, or lack of adequate fall prevention equipment (McVittie, 1995). Consequently, ergonomic or human factors-based strategies are likely to be highly effective in preventing fall-related injuries in the workplace, while this has, as yet, been untrue for the elderly (Fleming, 1993; Parker, 1996; Tinetti, 1993; Waller, 1978).

TABLE 62.1 Annual Fall-Related Occupational Injuries and Deaths in the United States, by Occupation

	Fatal Injuries[a]		Nonfatal Injuries[b]	
Industry	Occurrences	Percent of All Injuries	Occurrences (Thousands)	Percent of All Injuries
Agriculture/Forestry	63	7.9	6.9	16.7
Mining	10	6.4	3.9	18.9
Construction	335	32.0	43.0	19.7
Manufacturing	62	8.8	62.7	10.8
Transportation/Public Utilities	26	3.0	41.8	17.3
Wholesale Trade	14	5.5	24.2	14.6
Retail Trade	26	3.9	80.5	20.4
Services	71	9.6	99.5	19.3

[a] Source: Bureau of Labor Statistics, U.S. Department of Labor, Census of Fatal Occupational Injuries, 1995.
[b] Source: Bureau of Labor Statistics U.S. Department of Labor, Survey of Occupational Illnesses and Injuries, 1994. Data reflect a sum of injuries due to falls to a lower level and falls to the same level.

62.3 Epidemiology

Fatal Injuries

Incidence. In 1995, falls accounted for 10% of fatal work-related injuries in the United States (653 of 6210 total deaths), ranking them alongside MVA, violent assault, and being struck by an object as a leading cause of accidental work-related death (Bureau of Labor Statistics, 1996a). As shown in Table 62.1, approximately one half of these deaths occurred in the construction industry, where falls accounted for 32% of work-related fatalities, more than any other cause (Bureau of Labor Statistics, 1996a; National Safety Council, 1994; McVittie, 1995; Ore, 1996). Construction-related occupations with the highest risk for fall-related fatality include roofers, structural metal workers, carpenters, and painters. In each of these, falls account for about one half or more of total fatalities (Bureau of Labor Statistics, 1996a; Sorock, 1993; Suruda, 1992).

Circumstances. Not surprisingly, over 95% of fall fatalities in the workplace involve falls from elevation (Bureau of Labor Statistics, 1996c), and accordingly, working at elevation is the single greatest risk factor for a fatal fall in the workplace. A review of 63 fatal falls in the Quebec construction industry between 1989 and 1992 (McVittie, 1995) found that the most common working surfaces from which falls occurred were skeletal structures such as building frameworks and roof trusses (25%), unfinished floors (24%), scaffolds (24%), roofs (18%), and ladders (6%). OSHA reported slightly different results in the analysis of 1148 fall-related deaths in the U.S. construction industry between 1985 and 1989 (Occupational Safety and Health Administration, 1992), citing roofs as the most common working surface (26%), followed by scaffolds (21%), skeletal structures (14%), unfinished floors (10%), and ladders (6%).

A second important class of risk factors for fall fatalities is lack of proper protective equipment. Indeed, several investigators have observed that the vast majority of fall-related fatalities involve the absence or improper use of fall restraint equipment, and noncompliance with recommended equipment and job site standards. For example, in McVittie's study of fall-related injuries in the Ontario construction industry (McVittie, 1995), no obvious fall risk existed in 84% of deaths, which apparently could have been prevented with proper use of safety equipment. In such cases, safety harnesses were often available but not being used. In another 13% of cases, falls occurred through poorly guarded floor openings. Only one case resulted from failure of a fall restraint system, and this involved the worker fastening his safety harness to a hoisting line rather than a lifeline. Similarly, in Suruda's study of fatal falls to construction painters (Suruda, 1992), over one half involved falls from scaffolds, and the vast majority of these (88%) involved the painter falling off the scaffold, as opposed to the scaffold collapsing. In 74% of cases, OSHA issued safety violations, commonly citing the lack of scaffold guardrails, nets, lifelines, and safety belts. Indeed, the single most common cause of death was failure to connect a safety belt to a lifeline.

TABLE 62.2 Incidence Rates (Injuries per 10,000 Worker-Years) for Fall-Related Nonfatal Injury by Nature of Injury and Body Part Affected, 1994

Injury	Fall to Lower Level	Fall on Same Level
	Type	
Fractures	2.7	4.7
Dislocations	0.4	0.6
Sprains and Strains	4.6	11.7
Lacerations	0.3	0.8
Bruises	2.2	6.6
Intracranial Injury	0.1	0.3
Back Pain	0.3	1.0
Soreness, Non-Back	0.5	1.6
	Location	
Head	0.4	1.4
Neck	0.1	0.3
Trunk	4.3	11.3
Pelvis	0.4	1.1
Upper Extremity	1.6	4.6
Lower Extremity	4.4	9.4

Source: Bureau of Labor Statistics, U.S. Department of Labor, Survey of Occupational Illnesses and Injuries, 1994.

Nonfatal Injuries

Incidence. According to the Bureau of Labor Statistics, falls were the cause of 18% of all work-related injuries in 1994, trailing only overexertion and being struck by an object as a leading cause of injury (Bureau of Labor Statistics, 1996b). Between 1992 and 1994, the overall frequency for disabling fall-related injuries was fairly constant at 49 per 10,000 person-years. Sectors with the greatest number of fall-related injuries were the service and trade industries (Table 62.1). However, incidence rates were highest for the construction, transportation, agriculture, and mining industries, with respective risks of 96, 73, 64, and 62 injuries per 10,000 worker-years.

In contrast to fall fatalities, over two thirds of nonfatal fall injuries were due to falls onto the same level (Bureau of Labor Statistics, 1996b). However, we might generally expect that, the higher the fall height, the greater the injury severity, and indeed, lost workdays were considerably higher for injuries caused by falls to a lower level (10 days vs. 6 days for falls on the same level). As shown in Table 62.2, the most frequent types of injury were strains and sprains, followed by fractures and contusions, and the body part most affected was the trunk (including back and shoulders), followed by the lower extremity, upper extremity, and head. While a useful reflection of overall injury trends, Table 62.2 fails to show that falls represent the leading cause of work-related spinal cord injury (Rosenberg, 1993) and brain injury (Heyer, 1994), surely among the most devastating of injuries. Not surprisingly, workers in construction and agriculture are at greatest risk for these injuries.

Circumstances. When compared to fall fatalities, risk factors for nonfatal falls in the workplace are considerably more diverse, as reflected by the small variation among industries in injury incidence (Table 62.1). McVittie (McVittie, 1995) found that risk factors for falls in the construction industry generally fell under two categories: in-transit activities and poor housekeeping. Risk-factors associated with the former included uneven terrain and the use of steps and ladders, while those associated with the latter included the presence of waste materials, debris, clutter, poor lighting, and slippery surfaces.

62.4 Biomechanics

In considering the biomechanics of falls and fall-related injuries, it is useful to consider falls as having three stages (Hayes, 1993): (1) an *initiation* stage, involving a slip, trip, or loss of balance, and potential attempts to regain upright posture; (2) a *descent* stage, where movements may be attempted in preparation for landing; and (3) a *contact* stage, where impact occurs between the body parts and the ground, resulting in the generation of reaction forces and dissipation of the body's kinetic energy.

Fall Initiation

Numerous studies have been conducted on the control of balance, involving application of destabilizing perturbations to the feet or trunk (see reviews by Alexander, 1994; Prieto, 1993; Winter, 1990). Most have focused on balance correction through sway, or standing balance maneuvers (Alexander, 1992; Era, 1995; Kuo, 1993; Maki, 1987; Romick-Allen, 1988; Stelmach, 1989), which involves activation of lower extremity muscles in either a proximal-to-distal sequence (hip strategy), or distal-to-proximal sequence (ankle strategy) (Manchester, 1989; Woollacott, 1993). However, several investigators have observed that sway-based recovery is feasible only for small perturbations to posture (i.e., small displacements of the body's center-of-gravity from the base-of-support formed by the feet), such as those occurring during normal walking and standing. Consequently, in the event of a slip or trip, stepping emerges as the primary means for balance recovery (Maki, 1996; Nashner, 1980; Thelen, 1997; Wolfson, 1986).

To assess the factors which influence balance recovery by stepping after a slip, we recently measured body movements as subjects (aged 22 to 35 years) stood barefoot on a large gymnasium mat and attempted to prevent falls after the mattress was made to unexpectedly translate (Hsiao, 1998). Throughout the testing session, we randomly varied both the direction of the perturbation (by having the subject stand forward, backward, or sideways to the perturbation direction), and the strength of the perturbation (randomized between 4 acceleration levels). To evoke "natural" responses, no practice trials were allowed, and subjects were only instructed that (1) in the event of platform movement, they should "try to prevent themselves from falling," and (2) prior to platform movement, they should maintain their gaze directed forward and at eye level. In each trial, a three-dimensional motion measurement system acquired the positions of 20 skin surface markers located throughout the extremities and trunk.

As expected, stepping was the predominant balance recovery technique; only a single trial involved stabilization of posture through sway (Figure 62.2). However, the effectiveness of stepping in preventing a fall was highly directional-dependent, with subjects being more than twice as likely to fall when the perturbation was directed posteriorly (i.e., the mattress translated in the anterior direction), as opposed to anteriorly or laterally. Furthermore, falls due to posterior perturbations commonly involved trunk and/or pelvis impact, while this was rare in falls due to anterior or lateral perturbations (in these trials, impact was almost always restricted to the hands and/or knees). Finally, females were twice as likely to fall as males, and in all but one fall trial, a failed attempt at balance recovery by stepping was observed (Figure 62.3).

Implicit in the observation that stepping represents the primary means for balance recovery after a slip or trip is the suggestion that fall risk is inherently greater on scaffolds and ladders, since the small surface area of these environments nearly eliminates one's ability to recover balance by stepping. This highlights the importance of educating workers to correctly perceive this difference in fall risk, and the importance of utilizing fall restraint devices when working in such environments.

Fall Descent Kinematics

Even in a fall from standing height, sufficient energy exists to fracture the wrist (Chiu, 1996; Myers, 1993) or hip (Courtney, 1994; Robinovitch, 1991). However, only a small percentage of such falls actually result in serious injury (Melton, 1988). This suggests that highly effective movement strategies exist for preventing

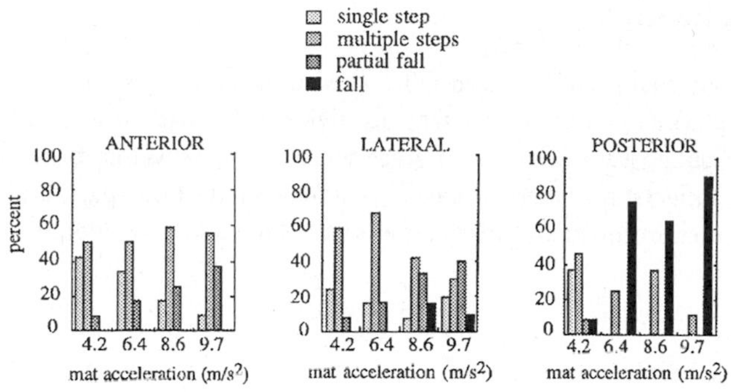

FIGURE 62.2 Balance recovery and falling responses during simulated slipping experiments. Seventy-seven percent of trials involved balance recovery through one or more steps. During anterior and lateral perturbations, loss-of-balance was more likely to lead to a "partial fall" (defined as contact to one or both knees and/or wrists) than a "fall" (defined as contact to the pelvis and trunk), while the reverse was true for posterior perturbations. (From Hsiao ET, Kearns M, Robinovitch SN. 1998. Analysis of movement strategies during unexpected falls. *J Biomechanics,* 31:1-9. With permission.)

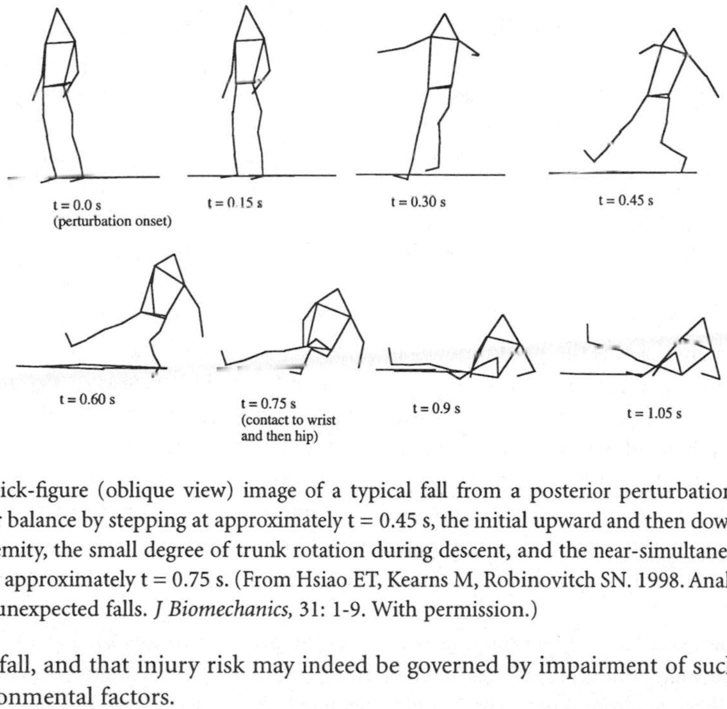

FIGURE 62.3 Stick-figure (oblique view) image of a typical fall from a posterior perturbation. Note the failed attempt to recover balance by stepping at approximately t = 0.45 s, the initial upward and then downward movement of the upper extremity, the small degree of trunk rotation during descent, and the near-simultaneous impact to the wrist and pelvis at approximately t = 0.75 s. (From Hsiao ET, Kearns M, Robinovitch SN. 1998. Analysis of movement strategies during unexpected falls. *J Biomechanics,* 31: 1-9. With permission.)

injury during a fall, and that injury risk may indeed be governed by impairment of such responses due to host or environmental factors.

To assess the nature of protective responses during falls, we analyzed body movements during the 28 falls occurring during the slipping experiments described in the previous section (Hsiao, 1997). Our results confirmed our suspicion that, rather than being random and unpredictable, body segment kinematics during falls are organized into specific movement sequences which facilitate safe landing. These could be classified into two major types. The first was a complex yet highly repeatable sequence of upper extremity movements which allowed subjects to impact their wrist at nearly the same instant as the pelvis (average time interval between contacts = 38 ms; Figure 62.4). This involved an immediate upward movement of the upper extremity (perhaps reflecting a startle response), followed by a rapid downward movement, and finally a second upward acceleration just prior to impact (Figure 62.3). This last deceleration substantially

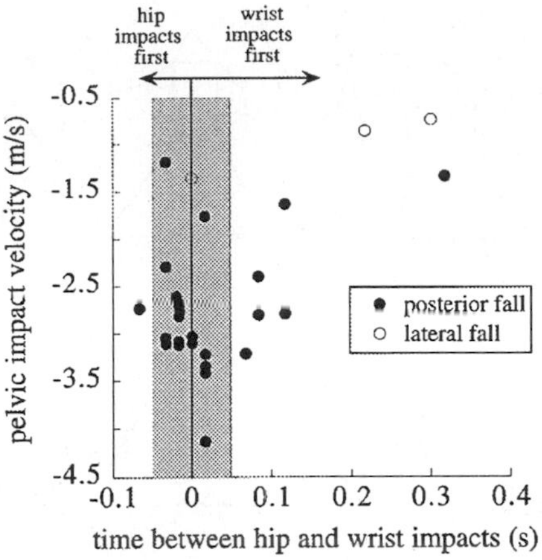

FIGURE 62.4 Vertical velocity of the pelvis at the moment of impact as a function of the time interval between wrist and pelvis contacts. Positive x-axis values reflect the wrist impacting before the pelvis, and negative values on the y-axis reflect downward movement. In approximately 70% of trials, the time difference between hip and wrist impacts was less than 50 ms (gray band), suggesting a sharing of impact energy between the two body regions. (From Hsiao ET, Kearns M, Robinovitch SN. 1998. Analysis of movement strategies during unexpected falls. *J Biomechanics,* 31: 1-9. With permission.)

decreased the vertical velocity of the wrist at contact velocity, which averaged 2.6 m/s, or 66% of its peak downward velocity during descent. Impact to both wrists was observed in 26 of the 28 falls, with an average interval of 70 ms between contacts. Average vertical pelvis velocity at impact was 2.5 m/s, or 83% of its peak downward velocity during descent. Head impact was observed in only five falls, three of which involved the same subject.

These results suggest (as does common experience) that the upper extremity plays a major role in allowing safe landing during a fall. For example, it is reasonable to assume that without the occurrence of upper extremity impact, all cases of failed balance recovery would have resulted in impact to the head and trunk. Therefore, braking the fall with the outstretched hand allowed for complete avoidance of upper body impact in anterior trials, and reduced the incidence of upper body impact by over fourfold in lateral trials. Second, over two thirds of falls involved wrist contact within 50 ms of pelvis contact (Figure 62.4), and previous studies have shown that approximately 50 ms is required to reach peak force during a fall on the hip (Robinovitch, 1991) or wrist (Chiu, 1996). This suggests that impact configurations are chosen to allow *sharing* of impact energy between these body parts, and a subsequent reduction in localized impact forces and injury risk.

The second class of protective response observed in these trials was marked trunk rotation in falls due to lateral (but not anterior or posterior) perturbations (Figure 62.5). This commenced late in the descent stage of the fall, and resulted in impact with the body anteriorly facing the contact surface. This, in turn, allowed subjects to contact the impact surface with both right and left upper extremities, and avoid impact to the lateral aspect of the hip (thereby decreasing one's risk for hip fracture [Greenspan, 1994; Nevitt, 1993]). In posterior perturbations, subjects apparently sensed the infeasibility of a 180-degree trunk rotation to anteriorly face the impact surface, and instead focused on safe landing on the buttocks and wrists.

Impact Mechanics

During a fall, the risk for injury to a specific anatomical structure depends on the ratio of the force applied to the structure divided by the force required to cause failure. To estimate failure forces of anatomical structures, cadaveric materials may be loaded to failure in a mechanical testing system which simulates the fall impact environment. Such studies have measured failure loads for the distal radius,

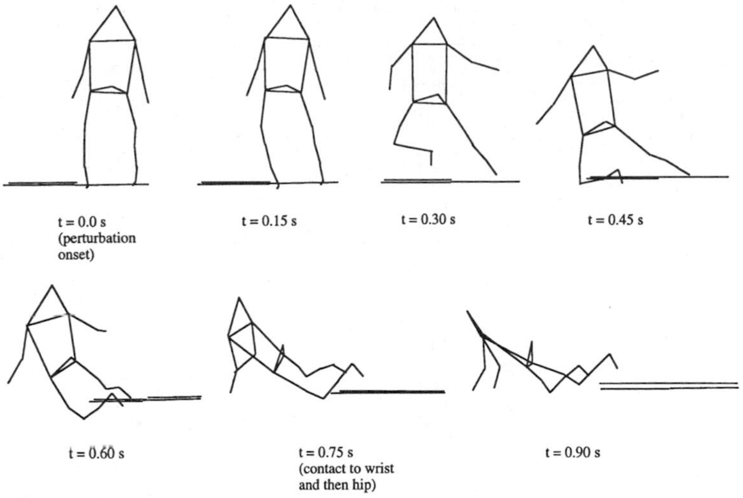

FIGURE 62.5 Stick figure image of a typical side fall. Note that trunk rotation about an inferior–superior axis allowed for impact to the anterior–lateral aspect of the pelvis rather than the hip, at nearly the same instant that contact occurred to the wrist. (From Hsiao ET, Kearns M, Robinovitch SN. 1998. Analysis of movement strategies during unexpected falls. *J Biomechanics,* 31: 1-9. With permission.)

which average 1800 ± 700 N (Myers, 1993; Spadaro, 1994), and the hip, which average 4200 ± 1600 (S.D.) N (Courtney, 1994; Weber, 1992).

In order to estimate fall impact forces, consider the mechanics of a simple mass–spring representation of the body, as shown in Figure 6. In configuration (a), the system is stationary at height h (measured in units of m), possessing zero kinetic energy and a gravitational potential energy of mgh, where m is the mass in kg, and g is the gravitational constant, 9.81 m/s^2. Configuration (b) represents the beginning of the impact phase. The system has descended a height h, and the spring just contacts the ground but has not yet undergone compression. Therefore, the system has zero potential energy and a kinetic energy of $mv^2/2$, where v is the downward velocity in m/s. By equating total energy between configurations (a) and (b), it is readily observed that $v = \sqrt{2\,gh}$ (recall that free-fall velocity is dependent only on descent height, and not system mass). Finally, in configuration (c), the spring has compressed maximally (by an amount x, measured in m), and is about to rebound upward. Therefore, the system has zero downward velocity and kinetic energy, and (neglecting the difference in gravitational potential energy between configurations (b) and (c)) a potential energy of $kx^2/2 = F^2/(2k)$. Equating this to the total energy in configurations (a) or (b), the peak impact force on the system is:

$$F = v\sqrt{mk} = \sqrt{2kmgh}. \tag{1}$$

Equation (1) illustrates that the main requirement in determining impact force is to derive reasonable estimates for the system's effective mass m and stiffness k. In previous studies, we estimated such values based on the measured response of human subjects during safe, simulated falls on the hip (Robinovitch, 1991) and outstretched hand (Chiu, 1996). Average values for m and k during falls onto the hip were 35 kg and 71 kN/m, respectively, while those for falls on the outstretched hand were 29 kg and 8.7 kN/m. Figure 62.7 shows resulting estimates of peak contact force when floor stiffness is infinitely stiff, and primary contact occurs to either the upper extremity (filled circles) or hip (filled squares). Under such conditions, hip contact force is over threefold greater than upper extremity contact force, due to the substantially higher effective stiffness of the body during a fall on the hip. Also shown in Figure 62.7 are peak contact forces for falls onto a rather soft surface, possessing a stiffness of 10 kN/m. These were

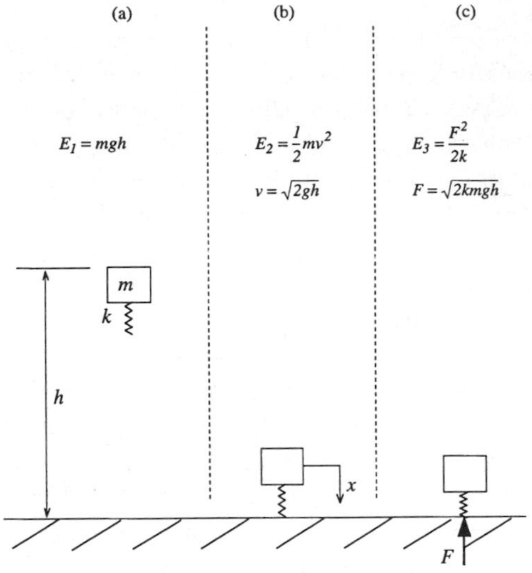

FIGURE 62.6 Idealized spring–mass model of impact during a fall. Simple energy considerations imply that contact force scales with the square-root of descent height, effective mass, and effective stiffness.

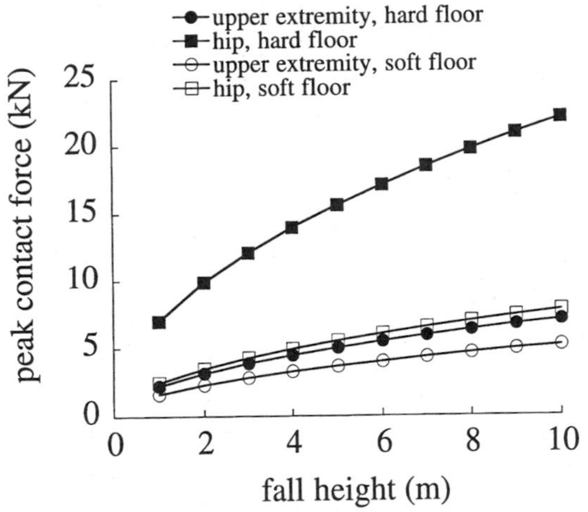

FIGURE 62.7 Effect of fall height and ground stiffness on estimated contact forces during falls. Contact force is based on $F = \sqrt{2k_{eq}mgh}$ (see Figure 62.6), where $k_{eq} = kk_f/(k + k_f)$ is the effective series stiffness (in kN/m) of the body spring k and ground spring k_f, m is the effective mass of the body (in m), $g = 9.81$ m/s^2, and h is descent height (in m). Estimated values for m and k are 35 kg and 71 kN/m for impact to the hip (Robinovitch, 1991) and 29 kg and 8.7 kN/m for impact to the outstretched hand (Chiu, 1996). Note that, during impact to an infinitely stiff surface ($k_f = \infty$; filled symbols), the much stiffer hip region experiences contact forces threefold higher than those generated during a fall on the upper extremity. However, during impact to a soft surface (k_f = 10 kN/m; open symbols), the series stiffness governing hip impact is greatly reduced, and peak hip impact forces exceed upper extremity forces by only 1.5-fold.

calculated by replacing k in Equation (1) with the equivalent series stiffness $kk_f/(k + k_f)$. Note that, under this scenario, hip contact force is reduced 65%, while upper extremity force is reduced by only 27%. Consequently, a soft impact surface substantially reduces one's risk for hip fracture during a fall (or injury to other "stiff" body parts, such as the head or back), but offers considerably less protection against upper extremity injury.

62.5 Prevention

The single most effective means for lowering the incidence of serious fall-related injuries in the workplace is to ensure appropriate use of personal fall protection equipment. This follows from the following facts discussed in Section 62.3: (1) that the vast majority of serious fall injuries and fatalities involve descents from high elevation; (2) that in the majority of cases, fall protection equipment was not being used at the time of the injury; and (3) that few fall-related deaths result from restraint system failure.

A major challenge for ergonomics researchers is to therefore engineer systems which ensure usage of fall protection equipment in high-risk environments. This requires the conversion of fall protection systems from an active form of protection, which relies on workers' choice in utilizing it, to an automatic (or passive) form of protection, in which user choice, and thus risk for noncompliance, is eliminated (Committee on Trauma Research, 1985). To illustrate the point here, consider the role of seatbelts in preventing injury during motor vehicle accidents. While it is well-established that seatbelts reduce injury risk, the existence of a seatbelt in itself provides no actual protection, unless the occupant chooses to use it (active protection). In contrast, an automobile designed so the transmission operates only when seatbelts are engaged (passive protection) is likely to be inherently safer, since it removes user choice as a variable. An analogous system for work at high elevation is, for example, one which prevents workers from ascending to elevated workspaces without the use of an effective fall protection system.

Simultaneous with efforts to develop such systems, improvements should be undertaken in job training and work site inspection for fall hazards. Equipment should be regularly tested for structural integrity, and inspected for evidence of wear or damage. Risk factors for standing height falls should also be minimized, with special attention to proper lighting, and the elimination of clutter and slippery surfaces. Workers and supervisors should also be alert to intrinsic risk factors for falls, such as alcohol or medication use, and impairments in vision, balance, strength, and joint range-of motion.

Education should focus on the principles of load-transfer and load-sharing among restraint system components, and the need to ensure that the load-bearing capacity of each component (e.g., anchoring sites, belts, lifeline attachment fixtures, ropes) meet appropriate factors of safety. Workers should also be educated regarding the importance of short lifelines, since the loads imposed on a tethering system increase with descent height. In using scaffolding, attention should be paid to manufacturer recommendations on the use of counterweights, safety brakes, tie-off of support devices, and floor load capacity. Different anchor points should be used to support scaffolding and harness or body belt lifelines. Workers should also be trained on proper use of ladders, including reaching practices, use of three-point contact, avoidance of unstable surfaces, and limitations on ladder inclination and ascent height.

Education on fall mechanics may also promote safe work practices, and eliminate misconceptions regarding risk for falls and fall-related injuries. For example, workers should be instructed on the relationship between injury tolerance limits and fall descent height, and be made aware that postural stability is inherently lower in the constrained environment of a scaffold or ladder, since reduced surface area impairs one's ability to recover balance by stepping.

References

Alexander NB, Shepard N, Gu MJ, Schultz A. 1992. Postural control in young and elderly adults when stance is perturbed: kinematics. *J Gerontol Med Sci* 47: M79-87.

Alexander NB. 1994. Postural control in older adults. *J Am Geriatr Soc* 42: 93-108.

Bureau of Labor Statistics. 1996a. *Census of Fatal Occupational Injuries,* 1995. Washington, D.C., U.S. Department of Labor.

Bureau of Labor Statistics. 1996b. *Characteristics of Injuries and Illnesses Resulting in Absences from Work, 1994.* Washington, D.C., U.S. Department of Labor.

Bureau of Labor Statistics. 1996c. *Issues in Labor Statistics: Issue 96-1: Construction Falls.* Washington, D.C., U.S. Department of Labor.

Chiu J, Robinovitch SN. 1996. Transient impact response of the body during a fall on the outstretched hand. BED-Volume 33, *1996 Advances in Bioengineering,* Proceedings from the ASME International Mechanical Engineering Congress and Exposition. New York: American Society of Mechanical Engineers. p. 269-270.

Committee on Trauma Research, Commission on Life Sciences, National Research Council and Institute of Medicine. 1985. *Injury in America: A Continuing Public Health Problem.* Washington, D.C.: National Academic Press.

Courtney AC, Wachtel EF, Myers BR, Hayes WC. 1994. Effects of loading rate on strength of the proximal femur. *Calcif Tissue Int* 55: 53-58.

Era P, Heikkinen E. 1995. Postural sway during standing and unexpected disturbance of balance in random samples of men of different ages. *J Gerontol* 40:287-295.

Fleming E, Pendergast DR. 1993. Physical condition, activity pattern, and environment as factors in falls by adult care facility residents. *Arch Phys Med Rehab* 74: 627-630.

Greenspan SL, Myers ER, Maitland LA, Resnick NM, Hayes WC 1994. Fall severity and bone mineral density as risk factors for hip fracture in ambulatory elderly. *JAMA* 271:128-133.

Hayes WC, Myers ER, Morris JN, Gerhart TN, Yett HS, Lipsitz LA. 1993. Impact near the hip dominates fracture risk in elderly nursing home residents who fall. *Calcif Tissue Int* 52: 192-198.

Heyer NJ, Franklin GM. 1994. Work-related traumatic brain injury in Washington State, 1988-90. *Am J Public Health* 84: 1106-1109.

Hsiao ET, Robinovitch SN. 1998. Common protective movements govern unexpected falls from standing height. *J Biomechanics,* 31: 1-9.

Kuo AD, Zajac FE. 1993. A biomechanical analysis of muscle strength as a limiting factor in standing posture. *J Biomechanics* 26S:137-150.

Maki BE, Holliday PJ, Fernie GR 1987. A posture control model and balance test for the prediction of relative postural stability. *IEEE Transactions on Biomedical Engineering* 34:797-810.

Maki BE, McIlroy WE, Perry SD. 1996. Influence of lateral destabilization on compensatory stepping responses. *J Biomechanics* 29:343-353.

Manchester D, Woollacott M, Zederbauer-Hylton N, Marin O. 1989. Visual, vestibular and somatosensory contributions to balance control in the older adult. *J Gerontol Med Sci* 44:M118-M127.

McVittie DJ. 1995. Fatalities and serious injuries. *Occup Med: State of the Art Rev* 10:285-293.

Melton LJ, Chao EYS, Lane J. 1988. Biomechanical aspects of fractures, in Riggs BL, Melton LJ. (Eds.), *Osteoporosis: Etiology, Diagnosis, and Management,* pp. 111-131, New York, NY, Raven Press.

Myers ER, Hecker AT, Rooks DS, Hipp JA, Hayes WC. 1993. Geometric variables from DXA of the Radius predict forearm fracture load *in vitro. Calcif Tissue Int* 52:199-204.

Nashner LM. 1980. Balance adjustments of humans perturbed while walking. *J Neurophysiol* 44:650-664.

National Safety Council. 1994. *Accident Facts,* 1994 Edition. Itasca, IL: National Safety Council.

Nevitt MC, Cummings SR. 1993. Type of fall and risk of hip and wrist fractures: the study of osteoporotic fractures. *J Am Geriatr Soc* 41:1226-1234.

Occupational Safety and Health Administration. *Construction Accidents: Workers Compensation Database 1985-88.* U.S. Department of Labor, Washington, D.C.

Ore T, Stout NA. 1996. Traumatic occupational fatalities in the U.S. and Australian construction industries. *Am J Indust Med* 30:202-206.

Parker MJ, Twemlow TR, Pryor GA. 1996. Environmental hazards and hip fractures. *Age and Ageing* 25:322-325.

Praemar A, Furner S, Rice DP. 1992. Costs of musculoskeletal conditions, in *Musculoskeletal Conditions in the United States,* pp. 143-170, Park Ridge, IL, American Academy of Orthopaedic Surgeons.

Prieto TE, Myklebust JB, Myklebust BM. 1993. Characterization and modeling of postural steadiness in the elderly: A review. *IEEE Trans Rehab Eng* 1:26-34.

Rice DP, MacKenzie EJ, and associates. 1989. *Cost of Injury in the United States: A Report to Congress.* Institute for Health and Aging, University of California San Francisco, and Injury Prevention Center, Johns Hopkins University.

Robinovitch SN, Hayes WC, McMahon TA. 1991. Prediction of femoral impact forces in falls on the hip. *J Biomech Eng* 113:366-374.

Romick-Allen R, Schultz AB. 1988. Biomechanics of reactions to impending falls. *J Biomechanics* 21:591-600.

Rosenberg NL, Gerhart K, Whiteneck G. 1993. Occupational spinal cord injury: demographic and etiologic differences from non-occupational injuries. *Neurology* 43 (July): 1385-1388.

Sorock GS, Smith EO, Goldoft M. 1993. Fatal occupational injuries in the New Jersey construction industry, 1983 to 1989. *J Occupational Med* 35:916-921.

Spadaro JA, Werner FW, Brenner RA, Fortino MD, Fay LA, Edwards WT. 1994. Cortical and trabecular bone contribute strength to the osteopenic distal radius. *J Orthop Res* 12:211-218.

Stelmach GE, Teasdale N, Di Fabio RP, Phillips J. 1989. Age-related declines in postural control mechanisms. *Intl J Aging Human Development* 29:205-223.

Suruda AJ. 1992. Work-related deaths in construction painting. *Scand J Work Environ Health* 18:30-33.

Thelen DG, Wojcik LA, Schultz AB, Ashton-Miller JA, Alexander NB. 1997. Age differences in using a rapid step to regain balance during a forward fall. *J Gerontol Med Sci* 52A:M8-M13.

Tinetti ME, Baker DI, Garrett PA, Gottschalk M, Koch ML, Horwitz RI. 1993. Yale FICSIT: risk factor abatement strategy for fall prevention. *J Am Geriatr Soc* 41:315-320.

Tinetti ME. 1994. Prevention of falls and fall injuries in elderly persons: A research agenda. *Prev Med* 23:756-762.

U.S. Department of Health and Human Services, Public Health Service. 1993. Healthy People 2000: National Health Promotion and Disease Prevention Objectives, Washington, D.C.

Waller JA. 1978. Falls among the elderly-human and environmental factors. *Accid Anal Prev.* 10:21-33.

Weber TG, Yang KH, Woo R, Fitzgerald RHJr. 1992. Proximal femur strength: Correlations of the rate of loading and bone mineral density. *ASME*, BED 22:111-114.

Winter DA, Patla AE, Frank JS. 1990. Assessment of balance control in humans. *Medical Prog Technol* 16:31-51.

Wolfson LI, Whipple R, Amerman P, Kleinberg A. 1986. Stressing the postural response: A quantitative method for testing balance. *J Am Geriatr Soc* 34:845-850.

Woollacott MH. 1993. Age-related changes in posture and movement. *J Gerontol* 48S:56-60.

63

Low Back Pain (LBP) Glossary: A Reference for Engineers and Ergonomists

Simon M. Hsiang
Liberty Mutual Research Center for Safety and Health

Raymond M. McGorry
Liberty Mutual Research Center for Safety and Health

63.1 Foreword

Low back pain (LBP) is a pervasive problem in many societies and low back pain disability (LBPD) is the most expensive and pervasive health and safety problem arising in industry (Spengler et al., 1986a; Webster and Snook, 1990). However, there is no clear definition, well-accepted pathology, diagnostic scheme or treatment protocol for the problem. With occurrences both in and out of the workplace and in both the presence and absence of biomechanical stressors arising from work, the problem is regarded as belonging to several academic domains. Consequently, ergonomists, sociologists, psychologists, and engineers are attempting to determine the relevant approaches from their individual disciplines. The large volume of LBP research papers produced by widely divergent disciplines is indicative of the scope, and the urgent need to control the problem.

In attempting to deal with the immense and sometimes controversial literature, many ergonomists and engineers interested or involved in the studies of LBP are often exposed to documents replete with medical vocabularies and unfamiliar theories.

For the benefit of these professionals, in their review of the literature, this glossary of LBP terminology and research has been compiled. The reader should be aware of the complexity and the subjective nature of the large volume of LBP literature. This document is not meant to substitute for a medical dictionary, or for review of the current literature, but rather as a desk reference for those requiring assistance with terms or hypotheses encountered during an article review.

63.2 Introduction

The terms compiled in this glossary can be classified into six highly interrelated categories. They are: *anatomy* of the vertebral column, the *mechanism* of disease process, *diagnosis, diagnostic tools, treatment* plans based on diagnosis, and terms specific to *low back pain* (see Figures 63.1 and 63.2).

0-8493-2641-9/99/$0.00+$.50

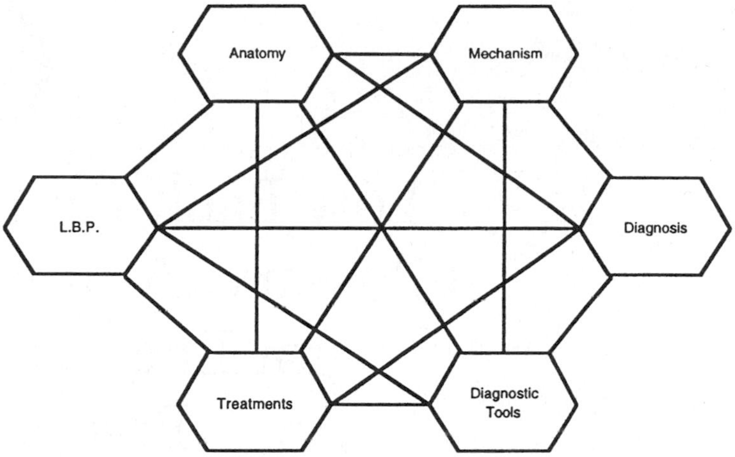

FIGURE 63.1 Schematic of the six categories of terms of low back pain.

Anatomy deals with terms that are descriptive of and specific to the structure and physical properties of the vertebrae, the intervertebral disc, the spinal cord and its nerve roots, and the muscles and ligaments that attach to it.

Mechanism describes the normal and abnormal function (physiology) of the anatomy, providing insight into the effect of disease process on the mechanics of the spinal column.

Diagnostic tools have been developed to provide insight into the anatomy and function to aid the medical practitioner in developing an appropriate diagnosis and subsequent treatment plan.

Diagnosis is a medical judgment based on the qualitative and quantitative findings produced by the diagnostic tools. Development of the appropriate diagnosis provides insight into the disease process and as a result provides goals and direction to the treatment plan.

Treatment covers a spectrum of techniques and procedures, performed by a wide array of medical professionals, based on any number of philosophies.

Low back pain is a category that encompasses terms related to the cause, nature, and description of the pain.

Two cases are provided to demonstrate the interrelationships that exist between the terms included in this glossary. In these cases the terms are linked to illustrate two different pathways from activity to the mechanism and finally the diagnosis and medical management of LBP. Consider the case where lifting and activities of daily living (ADL) create compressive and torsional forces acting on the spine. These forces are borne by both the anterior (disc) and posterior (apophysis) portion of the vertebral column. Case 1, as shown in Figure 63.3, uses the apophyseal (facet) joint as an example to trace the low back pain pathway for the posterior aspect of the spine. Case 2, as shown in Figure 63.4, uses the disc to trace the same pathway for the anterior spine. Note that both examples are for illustrative purposes only. They are *not* all-inclusive and are *not* intended as suggested limitations to the use of diagnostic tool or treatment protocols.

63.3 Terms and Definitions

Acupuncture — based on Chinese medicine techniques. Involves insertion of thin needles of various length through the skin at sites believed (1) to relieve symptoms, (2) to induce surgical anesthesia, and (3) to promote healing. Sometimes used to relieve low back pain.

Analgesic — (1) insensitive to pain, (2) relieving pain, and (3) an agent whose intended action is to alleviate pain (see pharmacological agents for low back pain).

Anaphylaxis (Anaphylactic) — a manifestation of sudden allergic reaction (see antigen).

Ankylosis — immobility or abnormal fixation of a joint. May be congenital, hereditary, or resulting from disease, injury, aging process (Sampson and Davis, 1988), or induced artificially by surgical fixation or fusion. Ankylosis reduces spinal segment range of motion, and often is painful. It can sometimes be asymptomatic.

Ankylosing spondylitis — inflammation of the spine that leads to stiffening or fusion of the vertebral joints.

Annulus — a ring or ring-like structure.

Annulus fibrosus — The fibrous outer ring-like portion of the intervertebral disc (Bogduk, 1983, 1991; Brinckmann, 1986; Hickey and Hukins, 1983).

Anatomy

- Can be classified into the inner portion and the outer portion. The inner portion of the annulus is mainly type II collagen (fibrocartilaginous). These collagen fibers are attached to the vertebral endplate, forming an internal capsule surrounding the nucleus pulposus. The outer portion is largely type I collagen (fibroelastic) and spans the rims of the vertebral bodies.
- The annulus fibrosis is comprised of 16 to 20 layers called lamellae. Collagen fibers in a single lamella of the annulus fibrosus have the same orientation. They are typically oriented at an angle of 65° to the long axis of the vertebral column. The orientation of the fibers alternates from 65° to –65° in successive lamellae.

Mechanism

- Both apophyseal joints and annulus fibrosus can resist shear force. The apophyseal joints have facets oriented to block forward translation by bony contact. The annulus fibrosus acts as a substantial ligament capable of resisting shear due to its high density of collagen.
- The orientation of collagen fibers may be biomechanically essential for the annulus to withstand: (1) the tensile force due to trunk flexion; (2) the radial load due to axial loading; (3) the torsional stress due to trunk rotation; (4) the shear stress due to anterior or posterior translation of the vertebrae; (5) or the combination of these forces.

Failure

- The annulus fibrosis can fail under conditions of low magnitude cyclic loading when exposed to a combined pattern of flexion and rotation (Gordon et al., 1991).
- The annulus may creep in time under compression with buckling and bulging of the annulus fibrosus and narrowing of the disc space. Creeping may become symptomatic as a result of foraminal or canal stenosis. In some cases of annular disruption, CT scan and discography provocation have demonstrated good correlation (Vanharanta et al., 1988b) (see creep and stenosis).
- The annulus fibrosus may creep due to the shear force when the mechanism of apophyseal joints is disabled by spondylolysis. Creeping may cause the upper vertebrae in the segment to slide progressively forward, as the condition progresses to spondylolisthesis (see spondylolysis and spondylolisthesis).

Anterior longitudinal ligament — the broad, strong band of longitudinal fibers that extends along the anterior surface of the vertebral bodies. It extends from the cervical spine to the sacrum (see ligament, and spinal ligament).

Antigen — any substance that the body regards as foreign and potentially dangerous and against which it initiates a specific immune response.

Apophyseal joint of the spine — commonly used as a synonym for the zygapophyseal or facet joint between adjacent vertebrae. The function of the apophyseal joint is to allow limited movement between vertebrae and to protect the disc from excessive shear forces, flexion, and axial rotation (Adams and Hutton, 1983b; Oxland et al., 1992). Excessive loading of this joint stretches the joint capsule and can itself be a cause of low back pain (Yang and King, 1984a). These joints contain rudimentary menisci which have been suspected of causing pain in entrapment disorders and may respond to treatment with manipulation (Bogduk and Engle, 1984).

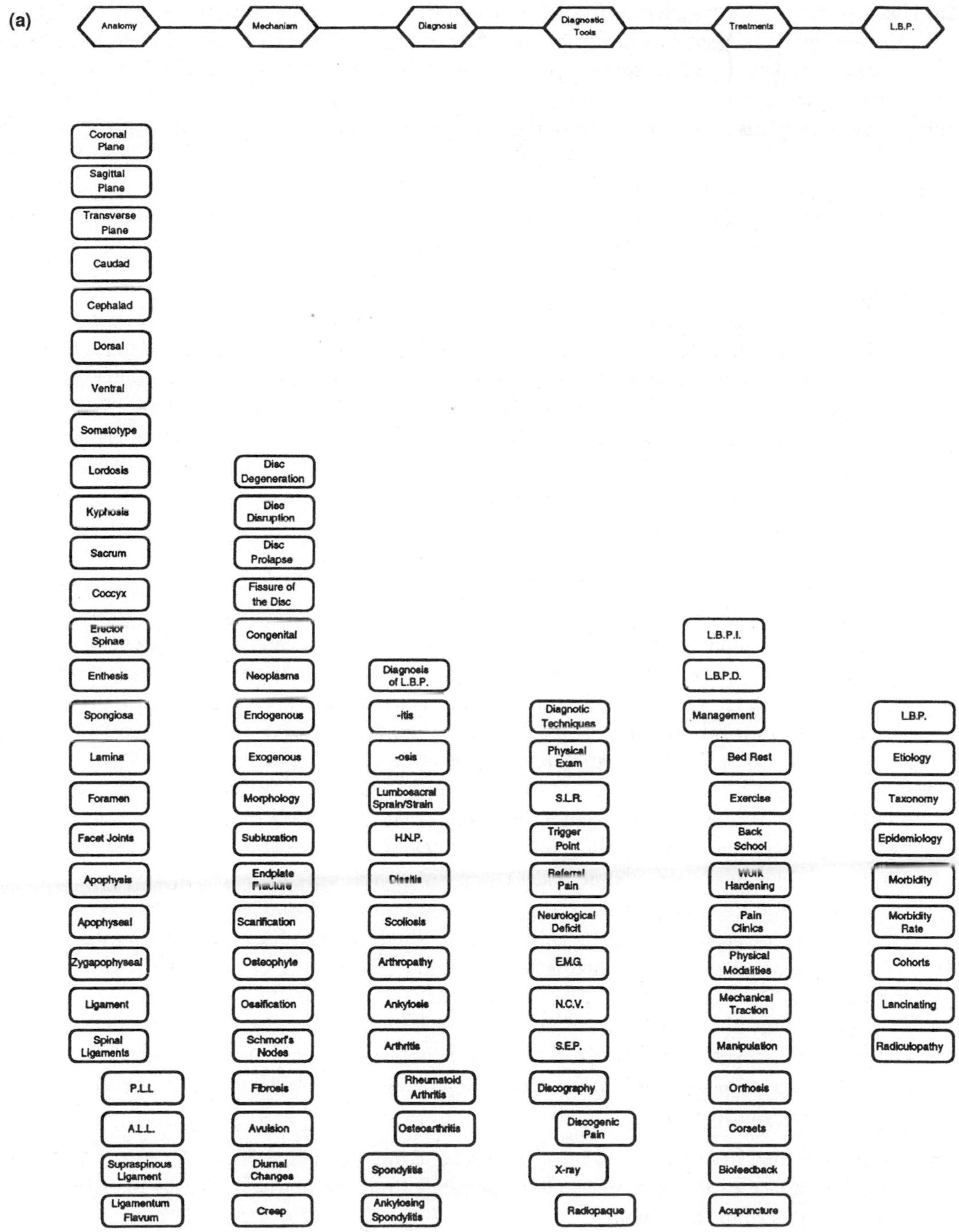

FIGURE 63.2 Compilation of glossary terms in each of the six categories.

Apophysis — projection or swelling from a bone.

Arachnoid — resembling a spider's web in consistency.

Arachnoid membrane — the middle of the three membranes (meninges) covering the spinal cord and brain (see meninges).

Arachnoiditis — inflammation of the arachnoid membrane; can progress to fibrosis or scarring that binds the nerve roots of the cauda equina (see cauda equina, sciatica).

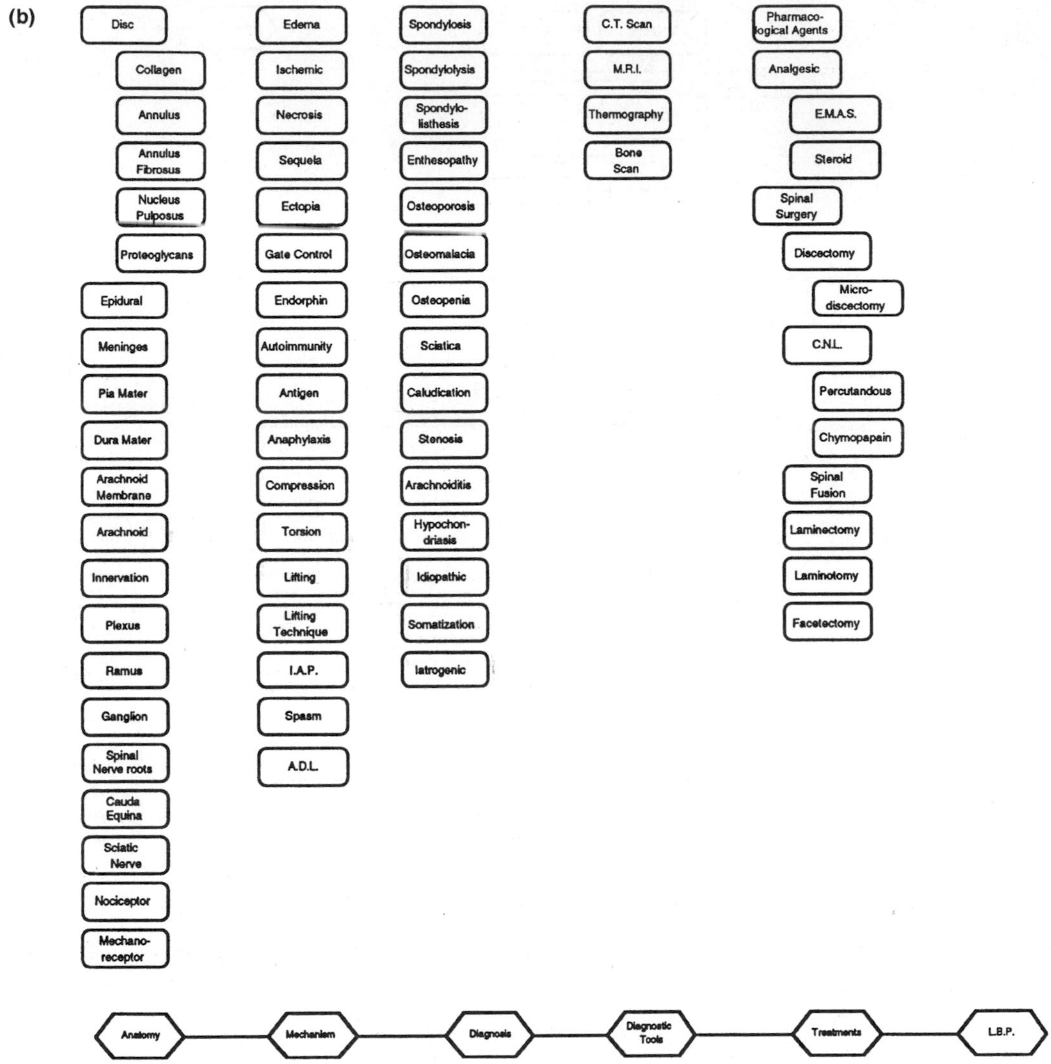

FIGURE 63.2 (continued)

Arthritis — inflammation of a joint (see osteoarthritis, rheumatoid arthritis).

Arthropathy — disease affecting a joint.

Autoimmunity — a condition characterized by an immune response directed against the body's own tissues.

Avulsion — the tearing or forcible separation of part of a structure.

Back school — a patient education program usually administered by health care professionals with the goal of achieving the greatest functional level for the patient. This usually includes education about the anatomy of the spine, mechanisms of back pain, proper body mechanics for lifting and in activities of daily living (ADLs), and safe exercise techniques (Saal and Saal, 1989).

Bed rest for low back pain — restriction of movement and activity during the acute phase of an LBP episode. Current medical management suggests bed rest for only a brief period (two days or less) (Deyo et al., 1986; Nachemson, 1985).

Biofeedback — the process of providing audio, visual, or tactile feedback of some physiological measure (e.g., skin temperature, heart rate, EMG, blood pressure) to the patient. The goal is to teach the patient self modulation or voluntary control of the measures, often with a resultant reduction or cessation of the associated symptoms (Asfour et al., 1989).

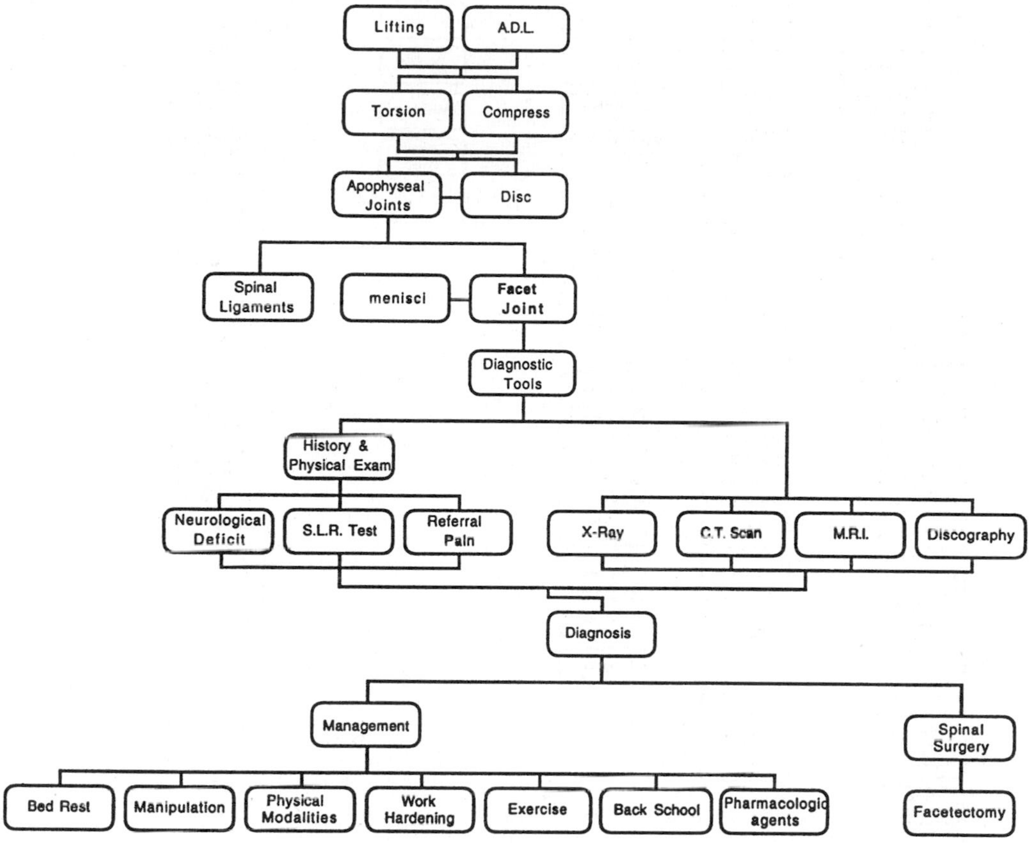

FIGURE 63.3 Schematic of Case 1: Facet joint.

Bone scan — an imaging technique during which a radioactively tagged metabolite is injected into the body and becomes incorporated into the skeletal system of the individual. Regions of increased metabolic activity and increased blood supply (e.g., inflammation, healing fractures, cancerous growths) can be visualized using nuclear medicine imaging techniques.

Cancellous — lattice like; applied to the bony tissue laid down by osteoblasts during development of bone and in the consolidation stage of fracture repair.

Caudal — toward the tail, opposite to cephalad.

Cauda equina — Latin for "mare's tail." The terminus of the spinal cord in the lower lumbar spine forming an array of nerve roots and occupying the vertebral canal below the cord.

Cephalad — toward the head, opposite to caudal.

Chemonucleolysis (CNL) — dissolution of the nucleus pulposus of an intervertebral disc by injection of a proteolytic agent (see chymopapain). Can be effective in the treatment of herniation of a disc (Farfan, 1985). Some of the complications or side effects of this procedure can be severe allergic reactions, migration of the enzyme to surrounding tissues, tissue destruction, postoperative LBP and spasm, and a slowing of nerve conduction (Wehling et al., 1989; Bromley et al., 1984; Fager, 1984; Agre et al., 1984).

Chymopapain — an enzyme that dissolves soft tissue, especially the intervertebral disc nucleus. The mechanism of the injection of chymopapain is the hydrolysis of proteoglycans resulting in a decrease in intradiscal pressure, and subsequent dissolution of disc components. A temporary change in the mechanical properties of the disc may result from the injection (Spencer et al., 1985; Nordby, 1983) (see chemonucleolysis).

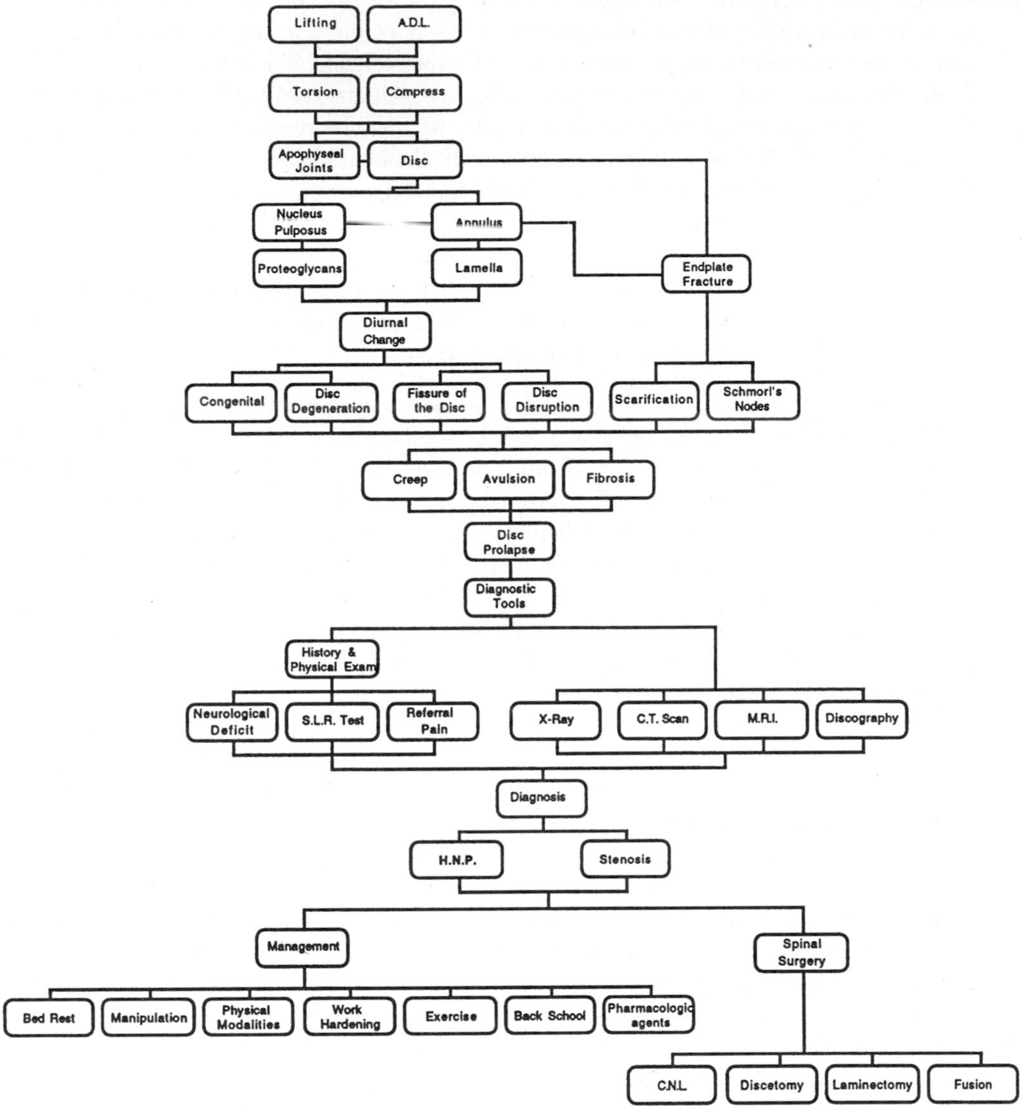

FIGURE 63.4 Schematic of Case 2: HNP.

Claudication (**neurogenic**) — pain due to neurologic or vascular compromise.

- Neurogenic claudication — pain of a neurologic origin occurring in the back, buttocks, and legs. It increases with walking or other weight-bearing activities (Dong and Porter, 1989) (see straight leg raise test).
- Vascular claudication — leg pain secondary to a circulatory insufficiency. No pain in a limb when at rest. The pain starts after walking has begun. Intensification of the pain will not occur without activity. Pain stops after a period of rest.

Coccyx — commonly known as the tail bone; the last four bones of the spine, generally fused.
Cohort — a group of individuals who share a common exposure or experience.
Collagen — a group of fibrous proteins that are major constituents of connective tissues (e.g., skin, tendon, bone, cartilage).

Compression loading of the disc — the load applied along the long axis of the vertebral column (Marras et al., 1984). It is the force vector normal to the body of the disc. Several mutually dependent factors can contribute to compressive loading of the intervertebral disc (Barnes et al., 1989; Dillard et al., 1991; Dolan et al., 1988; Marras and Reilly, 1988; Schultz, 1986): (1) weight of the body; (2) forces due to external loading; (3) forces secondary to muscular contraction (Bogduk et al., 1992; Cartas et al., 1992; Goel et al., 1993; Ladin et al., 1989; Poulsen, 1981; Schultz and Andersson, 1981; Schultz et al., 1981); (4) restorative forces of the spinal ligaments; (5) forces secondary to intra-abdominal pressure (Davis, 1981; Gracovetsky et al., 1981; Õrtengren et al., 1981; Bogduk and Macintosh, 1984). It may be that the L_3/L_4 lumbar disc is exposed to greater compressive loading than the L_5/S_1 disc. This is due to the greater mechanical advantage provided by the more pronounced L_5 spinous process which serves as the attachment for ligaments and extensor muscles (Gracovetsky et al., 1977; Gracovetsky and Farfan, 1986).

Capacity

- The forces required to produce failure purely by compression alone are quite large, on the order of 10 kN. Such forces are not incurred under normal circumstances (Hutton and Adams, 1982).
- There are studies suggesting that the maximal compressive force on the L_5/S_1 disc could be used to predict incidence rates of LBP (Chaffin and Park, 1973; Herrin et al., 1986). However, LBP due to annulus rupture from acute strain on the tissue appears improbable, and the ultimate compressive strength is not a reliable measure since it varies over a wide range.
- There is some evidence that an individual's level of physical activity may positively influence the compressive strength of the lumbar disc and vertebral body (Porter et al., 1989).

Mechanism (Lin et al., 1978)

- The annulus can passively transmit weight (compressive loads) between consecutive vertebral bodies because the sheer mass of its fibers renders them effective space fillers.
- Weight bearing between two vertebrae thus requires the cooperative action of both the nucleus pulposus and the annulus fibrosus.

Failure

- Sudden large magnitude compressive forces tend to cause anterior disc prolapse, while smaller magnitude cyclic compressive loading tends to result in bulging of the lamella of the posterior annulus (Adams and Hutton, 1985; Lavender et al., 1993, McNally et al., 1993).
- When a spinal joint is under compressive loading or flexed beyond its normal range, failure may result due to microfracture of a vertebral endplate (Adams and Hutton, 1982) and/or excessive strain on the underlying vertebra (Brinckmann et al., 1983).
- Interdependency has been documented between the surface area of the endplate and the density of the vertebral body (Brinckmann et al., 1989). Likewise, an interdependence has been reported between the vertebral bone density and the mechanical properties of the disc (Keller et al., 1989).
- Repeated microfractures and scarring of the bone underlying the vertebral endplates may induce a spot deterioration of the disc (Horst and Brinckmann, 1981). Uneven pressure gradients due to the spot deterioration propagate unrecoverable disc bulge. Deterioration is due to fluid loss and resultant change in viscoelastic properties. The mechanical failure of the endplate is highly dependent on the rigidity of the endplate and the underlying trabecular bone (Holmes et al., 1993). The consequences are decreased disc height and destabilization of the motion segments.
- The deterioration may continue with herniation of nuclear material into the vertebral canal, where it irritates or compresses nerve roots.
- Under compressive loading the facet joints are the foremost structure to yield at the limit of torsion. When compression of the disc is combined with torsion, the probability of failure increases.

Computer assisted tomography (CT or CAT) scan — an imaging technique involving a series of sequential X-ray exposures (slices) taken at discrete space intervals. The resultant tissue density information can be manipulated and stored by the computer. These data can be digitally enhanced to provide clinically useful information (Dvorak et al., 1987). The integration of a series of these exposures can be used to create three-dimensional images. CT images can provide evidence of soft tissue encroachment, such as disc herniation (Ninomiya et al., 1992); however, MRI has become the "gold standard" for identifying disc herniation except in some cases where CT with myelography is needed. MRI provides greater resolution of soft tissues and has become more widely used in the imaging and diagnosis of spinal pathology where soft tissue involvement is suspected (see MRI).

Congenital — referring to a condition present at the time of birth.

Coronal plane — relating to the plane that bisects the body into anterior and posterior sections.

Corsets — an orthopedic device that encircles and supports a part, as worn in certain spinal injuries or deformities (Lantz and Schultz, 1986a, 1986b) (see orthosis).

Creep — increasing strain (elongation) due to constant loading under the elastic limit of the material over a long period of time

Creep of the intervertebral disc — Under experimental conditions, the disc with its nucleus removed initially exhibits the same **compressive stiffness** as an intact disc (Adams and Hutton, 1983a). The problem with an isolated annulus fibrosus is that, in time, it will **creep; its lamellae** buckle inward and outward, and the disc space is decreased (Adams et al., 1987; Panjabi et al., 1984; Shirazi-Adl et al., 1984). Prolonged sitting or standing postures, and exposure to whole body vibration may cause creep detectable as a temporary decrease in stature (Brinckmann and Grootenboer, 1991).

Diagnosis of Low Back Pain — Diagnostic Terminology for LBP based on Fardon et al., (1993):

Nonspecific pain syndromes
- Deconditioning syndrome
- Facet syndrome
- Idiopathic lumbar pain
- Low back pain
- Lumbago
- Lumbar pain syndrome
- Lumbar syndrome
- Mechanical derangement
- Mechanical low back pain
- Sciatica

Disc specific disorders
- Degenerative disc disease
- Degenerative disc syndrome
- Disc disease
- Disc disruption
- Disc herniation
- Disc protrusion
- Herniated nucleus pulposus
- Internal disc derangement
- Internal disc disorder

Degenerative Arthritic
- Arthritis
- Degenerative arthritis
- Degenerative joint disease
- Degenerative spine disease
- Facet arthritis
- Facet arthrosis
- Facetitis
- Facet spondylosis
- Hypertrophicarthritis
- Spondylosis

Muscle / tendon disorder
- Acute and chronic pain
- Fibromyositis
- Sprain
- Sprain/strain
- Strain
- Tendinitis

Stability
- Degenerative instability
- Lumbar (spinal) instability
- Segmental instability

Stenosis
- Central stenosis
- Stenosis

Miscellaneous
- Failed low back surgery
- Fracture nerve root compression
- Osteoporosis
- Post laminectomy syndrome
- Radiculopathy
- Scoliosis
- Spondylolisthesis
- Spondylolysis

Diagnostic techniques/tools for low back pain (Wiesel et al., 1984; Nachemson, 1985; Hadler, 1984; Haig, 1992) —

- History, physical exam, and self report;
- Imaging techniques — X-rays (Panjabi et al., 1992a; Scavone et al., 1981), CT scan, MRI (Bernhardt et al., 1992), Thermograph (Mills et al., 1986);
- Enhanced imaging — discography, myelography, bone scan;
- Electrophysiologic testing — EMG (electromyography), NCV (nerve conduction velocity), and SEP (somatosensory evoked potentials) (Tonzola and Ackil, 1981; Haldeman, 1984; Eisen and Hoirch, 1983).

Disc — The intervertebral disc is intimately connected to two adjacent vertebral bodies. Composed of soft tissue, the disc serves to transmit loads down the vertebral column, as well as provide an energy absorptive capacity. Due to its flexibility and its interposition between the vertebrae, it allows for greater range of motion of the spine.

Gross anatomy (Inoue, 1981; Markolf and Morris, 1974)

- The disc consists of three parts: (1) a semifluid nucleus pulposus, consisting of water bound by a matrix of proteoglycans and collagen; (2) an annulus fibrosus, consisting of some 16 to 20 lamellae of collagen fibers surrounding the nucleus pulposus; and (3) vertebral endplates, covering the top and bottom surfaces of the disc and connecting it to the vertebral bodies.

Neuroanatomy (Bogduk et al., 1981; Goren et al., 1990; Panjabi et al., 1988b; Yoshizawa et al., 1980)

- Nerve fibers (including nociceptors) exist within the outer third of the annulus. The middle third of the annulus fibrosus may or may not be innervated. The inner third is not innervated (Bogduk, 1983, 1991).
- The sources of nerve fibers in the annulus fibrosus are the sinuvertebral nerves posteriorly, from the ventral rami posterolaterally, and from the gray rami communicants anteriorly and laterally (Bogduk et al., 1992).

Disc degeneration — a term describing the physical deterioration resulting from an injury to the disc (Thompson et al., 1990) or from the gradual wear and tear associated with the aging process (Hickey et al., 1986) and gender (Hsu et al., 1990; Kim et al., 1991; Miller et al., 1988; Pedrini-Mille et al., 1990; Tanaka et al., 1993; Vanharanta et al., 1989).

Mechanical property alterations

- The structural deterioration results in a change in viscoelastic properties (from resilient, with a high water concentration, to nonresilient, with a low water concentration) of the disc (Horst and Brinckmann, 1981; Panagiotacopulos et al., 1987; Hukins et al., 1990; McNally et al., 1992).
- The nucleus pulposus loses fluid and becomes less elastic (Adams and Hutton, 1983a).
- The normally firm and compact annulus fibrosis becomes fragmented and can surrender its structural integrity (Hickey and Hukins, 1982).
- The bone of the vertebral bodies adjacent to a degenerated disc exhibits a decrease in mechanical properties under compressive loading (Hansson et al., 1987). Relatively greater degeneration in the lumbar discs of cadavers occurs at lower spinal levels (Panjabi, 1988).

Diagnosis

- Measurement of disc height from radiographic studies has been found to be a poor method of detecting early, potentially symptomatic, degenerative changes (Vanharanta et al., 1988a).
- A relationship has been described between disc degeneration and osteoarthritis of the facet joint. It is believed that arthritis follows disc degeneration because of changes in the loading of the facet (Butler et al., 1990; Noren et al., 1991).

Disc disruption — degradation of the nucleus pulposus and inner annulus fibrosus without change in contour or size of the disc. There is a resultant degradation of the mechanical properties of the disc (Bogduk, 1983; Bogduk, 1991).

Mechanism of disc disruption

- Nuclear degradation may progress to involve erosion of the annulus fibrosus and fissures in the radial directions.
- The annulus fibrosus may be left to bear weight alone, because the bracing effect of the nucleus on the annulus fibrosus is reduced.
- The nerve endings in the annulus fibrosus can be sensitized by substances produced during the inflammatory response at thresholds lower than would be anticipated if the mechanical process operated alone (McCarron et al., 1987).

Physical evidence

- Very little physical evidence of disc disruption can be found directly. Since the outer perimeter (boundary) of the disc remains intact and normal in contour, no element of disc bulge or herniation is observed. Discography is required for definitive diagnosis (Sachs et al., 1987).

Symptoms (without frank herniation of the nucleus)

- Pain may be due to the chemical nociception and would be aggravated by any movement that mechanically stressed the affected disc.
- Inflammatory cells penetrate the annulus fibrosus of disrupted discs, whereupon chemical mediators may trigger nociceptive nerve endings.
- Nerve endings in the annulus fibrosus may be exposed to enzymes and breakdown products involved in the degradative process of the disc.

Diagnosis

- No abnormal neurologic signs will be present because the disc disruption does not involve nerve root irritation or compression.
- CT scans and myelography are usually normal because the outer perimeter of the disc is intact.

Discectomy — surgical removal or excision of all or part of the disc (Spengler, 1982a; Eismont and Currier, 1989). There is a change in disc contour and size following discectomy (Brinckmann and Horst, 1985). Disc height, radial bulging, and intradiscal pressure all decrease following discectomy (Brinckmann and Grootenboer, 1991). The inflammatory focus and the source of mechanical pain is often eliminated with excision of the problematic disc (see micro-discectomy). Total disc excision may destabilize the motion segment, and surgical fusion may be required (Goel et al., 1985, 1986; Frymoyer, 1981).

Disc prolapse or disc herniation — extrusion of the nucleus pulposus of an intervertebral disc. The extruded material may be mixed elements of the annulus fibrosus and nucleus pulposus (see HNP).

Discitis — inflammatory disease of the disc.

Discogenic pain — pain caused by derangement of an intervertebral disc.

Discography — an imaging technique for direct visualization of disc structure and pathology. This procedure involves X-ray photography of an intervertebral disc after direct injection of an absorbable radiopaque contrast medium (Sachs et al., 1987).

Pain provocation

- Discography pain provocation is usually conducted on the suspected disc and at least one adjacent disc (Weinstein et al., 1988a; Calhoun et al., 1988; Friedman and Goldner, 1955). During the procedure the patient is asked to distinguish between the painful and pain-free discs (Vanharanta et al., 1987).
- Patients undergoing discography sometimes suffer reproduction of their back pain.

Application

- Some clinicians and researchers question if discography provides information that cannot be obtained by other methods (Mooney et al., 1988; Nachemson, 1989). Others question the reliability and utility of the technique (Nachemson, 1989; Holt, 1968).
- Test specificity is reported to be markedly improved by combining analysis of the discographic image with information acquired during provocation (Walsh et al., 1990).
- A combined analysis using computed tomography (CT scan) and discography is reported to provide better documentation of the site and size of a prolapsed disc (Ninomiya et al., 1992; Anntti-Poika et al., 1990).
- Discography, like any invasive procedure, is not without its risks. One serious complication is discitis secondary to infection (Fraser et al., 1987).

Diurnal changes (in the disc) — It is hypothesized that the daily cyclic change in stature is due to expungement and reabsorbtion of fluid by the disc (Adams et al., 1986). This is thought to be related to the change of compressive loading of the disc associated with standing and sleep positioning (Tyrrell et al., 1985; Krag et al., 1990). Research indicates that the greatest rate of shrinkage or swelling occurs during the first hour of standing or lying, respectively. However, disc height measurements can vary greatly (Andersson et al., 1981; Adams and Hutton, 1983a; Adams et al., 1987; Adams et al., 1990; Adams and Dolan, 1991).

Dorsal — (1) related to the back; (2) posterior; a position more toward the back surface.

Dura mater — the tough outermost of the three membranes (meninges) covering the spine and brain (see epidural).

Ectopic — misplaced, occurring in an unnatural location, especially if congenital.

Edema (oedema) — excessive accumulation of fluid in the intercellular tissue spaces.

EMAS — endorphin mediated analgesia system (see endorphin).

Electromyography (EMG) — the measurement of the electrochemical potential produced during: (1) rest; (2) voluntary contraction; and (3) electrical stimulation of the skeletal muscle. Although EMG is used to estimate muscular exertion force, the limits of its accuracy and precision have not yet been clearly defined. EMG can also be used as an indirect measure of the function of the motor nerves which control the tested muscle (Schultz et al., 1982b; Khatri et al., 1984).

Endogenous — originating or produced within the organism.

Endorphin — a neurotransmitter produced in the central nervous system that acts as an analgesic (see EMAS).

Endplate fracture or disruption — refers to fissuring or loss of continuity of the cartilage interface between the disc and adjacent vertebral bodies. An endplate fracture may heal and cause no further problems. It is possible for an endplate fracture to set into motion a series of sequelae that manifest as pain and a variety of end stages. These symptoms may be demonstrated by pain provocation during discography (see discography) (Hsu et al., 1988).

Enthesis — (1) the junction of tendon and bone; (2) the use of a man-made substance as a substitute for lost or removed tissue.

Enthesopathy — any rheumatic disease resulting in inflammation of enthesis.

Epidemiology of low back pain — Epidemiology is the science concerned with the study of the factors influencing the frequency and distribution of disease or injury in human populations. Epidemiologic studies of low back pain have been limited due to the difficulty in defining exposures and outcomes. Three domains of risk factors for the epidemiology of low back pain have been established (Allan and Waddell, 1989, Bigos et al., 1986a, 1986b; Damkot et al., 1984; Frymoyer and Gordon, 1989; Nachemson et al., 1991; Sandover, 1988; Skovron, 1992).

Task factors (Pearcy et al., 1985; Troup et al., 1981)
- Vibration (Althoff et al., 1991; Kelsey, 1975; Kelsey et al., 1984b; Weinstein et al., 1988b; Wilder et al., 1982).
- Lifting (Kelsey et al., 1984b; Mirka and Marras, 1993; Mundt et al., 1993; Svensson and Andersson, 1983).
- Monotonous work (Magnusson et al., 1990; Svensson and Andersson, 1983).
- Work load (Sairanen et al., 1981; Stubbs, 1981; Videman et al., 1984).

Environmental factors
- Job satisfaction (Bigos et al., 1991a).
- Prolonged sitting/Sedentary jobs (Kelsey, 1975; Svensson and Andersson, 1983; Videman et al., 1990; Williams et al., 1991).

Individual factors (Bigos et al., 1991c)
- Cigarette smoking (Battié et al., 1989b; Kelsey et al., 1984a).
- Physical characteristics.
- Cardiovascular conditioning (Battié et al., 1989b).
- Age and gender (Videman et al., 1990; Videman et al., 1984).
- Pregnancy and menstruation (Ostgaard et al., 1993; Svensson et al., 1990; Videman et al., 1984).
- Strength (Battié et al., 1989b; Troup, 1991b).
- Psychological characteristics (Bigos et al., 1991b; Ostgaard et al., 1991).
- Psychosocial status (Troup, 1991a).

Epidural (extradural) — on or over the dura mater; often used in reference to the space outside the dura, a common site for administration of anesthetics (see dura mater).

Erector spinae — large rope-like muscle mass made up of many small, segmental muscles that run parallel to the spine. They act in concert to extend the trunk, creating an extension moment and compressive forces within the vertebral column (Schultz et al., 1982a; Macintosh and Bogduk, 1987, 1991; McGill, 1991). The extensor moment created by the erector spinae muscle force is a function of trunk position since the muscle moment arm varies with flexion angle (Cartas et al., 1992; Goel et al., 1993; Macinstosh et al., 1993; Tan et al., 1993; Triano and Schultz, 1987).

Etiology of low back pain — Etiology is the study of the factors and causes of disease. The potential causes of LBP can be categorized into mechanical, chemical, biological, infectious, genetic, degenerative, and psychosocial causes (Spengler, 1982b, 1983; Spengler and Freeman, 1979).

Exercise treatment for low back pain — usually prescribed to improve flexibility of the spine and strengthening of the trunk musculature. This may involve flexion or extension exercises, or both (Saal and Saal, 1989; Elnaggar et al., 1991; Nachemson, 1985).

Exogenous — originating or produced externally.

Facet joints of the spine — the paired synovial (moveable) joints between adjacent vertebrae, located on the apophyseal protuberances on the posterolateral aspect of the vertebral body. Its integrity and proper function depend on support from surrounding structures such as the joint capsule, ligaments, the intervertebral disc, and the muscles that cross the joint (Carrera, 1980a, 1980b; Davis and Carragee, 1993; Panjabi et al., 1992b, 1993; Van Schaik et al., 1985; Shirazi-Adl et al., 1987; Yang and King, 1984b).

Facetectomy — surgical excision of the articular facet of the spine (Wiltse, 1977). This procedure can be performed for correction of severe scoliosis. A major concern following this type of surgery is the stability of the spinal segment (Abumi et al., 1989b).

Fibrosis — formation of fibrous tissue, as in repair of or replacement of parenchymotous (organic, non-framework, non-supporting) material; pl. fibrosus.

Fissure of the disc — a channel, cleft, or groove through the annulus which forms secondary to injury. Degradation of the nuclear matrix destroys its cohesiveness. This may allow the nucleus to extrude through the fissure. Formation of a fissure is a prerequisite phenomenon for HNP to occur (see HNP).

Foramen — an opening or passage (Panjabi et al., 1983).

Ganglion — an aggregation of nerve cells.

Gate control theory — a theory on the modulation of pain perception by the central nervous system first proposed by Melzak and Wall (Melzack and Wall, 1965; Wall and Melzack, 1994). The main hypothesis of this theory is that sensory neurons trigger central control systems which inhibit or facilitate the input of nociceptors and their transmission pathways.

Herniated nucleus pulposus (HNP) — commonly known as a "slipped disc" or "herniated disc," or more technically as a prolapsed disc. Extrusion of material from the nucleus pulposus mixed with elements of the annulus fibrosus (Haig et al., 1993).

Failure mechanism

- Antecedent injury may damage the nucleus and produce a passageway through the annulus fibrosus. Heavy load lifting or long-term exposure to excessive vibration may be contributing factors to development of HNP (Wilder et al., 1982; McCarron et al., 1987).
- HNP may start when a radial fissure completely erodes the annulus. Spinal nerve roots may be compressed mechanically by the prolapsed material (Spencer et al., 1984; Adams and Hutton, 1985). The prolapsed material may elicit an inflammatory response that secondarily caused edema of the nerve roots. The blood vessels which supply the nerve may be compressed either mechanically or as a result of the edema, thereby rendering the nerve root ischemic.
- Recent research supports the idea that the annulus has a limited capacity for healing. Surgically created defects in the lumbar discs of dogs seem to show that large lesions can heal by filling with a solid plug of fibrous tissue and prevent leakage of nuclear fluid. Small wounds (comparable to a radial fissure), tend to heal more slowly, and continue to leak (Hampton et al., 1989). Other researchers have reported on the spontaneous resolution of extruded disc material over time without surgery (Saal and Saal, 1989).
- Disc herniations are categorized based on the degree of displacement of the nuclear material. The four levels ranked in order of increasing severity are (Spengler, 1982a):
    1. Intraspongy herniation (herniation through the endplate into the underlying cancellous, or spongy, bone of the vertebral body),
    2. Protrusion (displacement beyond a line drawn between the margin of two adjacent vertebral bodies),
    3. Extrusion (displacement into the foramina or canal,
    4. Sequestration (separation of the extruded material).

Symptoms

- HNP is present in a small percentage of incidence of low back pain, and conversely not all instances of HNP produce low back pain and neurologic symptoms. It is hypothesized that chemical nociception starts when a radial fissure reaches the middle and outer thirds of the annulus fibrosus and encounters nerve endings (Mooney, 1987). It may be that a combination of pressure and chemical irritation is necessary for HNP to produce pain.
- HNP patients often present with lancinating pain in the lower limb, severely limited straight-leg raising, and objective neurologic signs of numbness or weakness in the distribution of the affected nerve root.

- Stenosis of the disc space occurs when there is prolapse into the spinal canal or near the nerve roots. This can cause pressure and can become symptomatic if it compromises a spinal nerve, the spinal cord, or its roots. Pressure can be applied to the nerve root by a disc protrusion without direct compression of the nerve against the posterior elements of the foramen (Spencer et al., 1983).

Diagnosis

- Clinical examination can reliably identify HNP. Additional investigations, such as electromyography and radiology, are not essential for the purpose of diagnosis. Computed tomographic (CT) scans, magnetic resonance imaging (MRI), and myelography are of value in determining the exact site and extent of the prolapse (Morris et al., 1986).

Treatment

- The objectives of treatment are: (1) relieving pain; (2) increasing mobility; and (3) minimizing the impairment and disability associated with LBP.
- For a small percentage of patients with HNP, surgical intervention may be necessary. The surgical intervention can include chemonucleolysis, percutaneous discectomy, microdiscectomy, open laminotomy, and laminectomy.

Hypochondriasis — a mental disorder characterized by morbid anxiety about health, body function, and sensations.

Iatrogenic — describing a disease or dysfunction that has resulted from treatment. Iatrogenic discitis is an intensely painful condition, resulting from a disc infected by bacteria introduced by needles used for discography.

Idiopathic — of unknown origin.

Innervation — the nerve supply for a body part.

Intra-abdominal pressure (IAP) — pressure that develops within the abdominal cavity secondary to forces produced by the diaphragm, chest wall, and abdominal musculature.

Mechanism

- A great deal of controversy exists whether IAP directly or indirectly produces a reactive force counter to the compressive forces developed in the vertebral column during lifting (Gracovetsky et al., 1995). Some research supports a contrary opinion that IAP under some conditions and postures may actually increase compressive loading (Nachemson et al., 1986).
- IAP can be generated by other means as well, such as the Valsalva maneuver. This occurs when the breath is held while "bearing down" as if one were to move one's bowels.
- The use of back belts during lifting activities is often recommended as an ergonomic intervention. Wearing a back belt while lifting constrains expansion of the abdomen. In theory this should result in an increase in IAP and a subsequent decrease in the compressive load upon the lumbar disc. The efficacy of a back belt is quite controversial at present, and the exact mechanism of IAP development is not clear.

Ischemia — insufficient blood flow or inadequate blood supply.

-itis — suffix meaning "inflammation of."

Kyphosis — a convex or posterior curvature of the spine; the normal curvature of the thoracic spine.

Lamina — a thin bony plate; part of the dorsal region of a vertebra.

Laminectomy — surgical excision of the posterior arch (lamina) of the vertebrae. This procedure is performed to relieve spinal or nerve root pressure secondary to a space occupying lesion (e.g., herniated disc). Laminectomy is a futile intervention for disc disruption, because the lesion lies within an otherwise intact disc (Wiltse, 1977).

Laminotomy — division of lamina of a vertebra without removal or excision of the arch. The rationale for this procedure is the same as for laminectomy or for exploration of the intervertebral space.

Lancinating pain — sharp stabbing or cutting pain.

Lifting (manual lifting) and low back pain — The current theory of the relationship between LBP and lifting is based on three assumptions (Buseck et al., 1988; Jäger and Luttmann, 1989):

- Manual lifting poses a risk of LBP to many workers (Bigos et al., 1986a). The risks vary due to postural characteristics, repetition rates, and the load of the specific task (Nordin et al., 1984).
- LBP is more likely to occur when workers lift loads that exceed their physical capacities (Battié et al., 1989a; Poulsen, 1981; Herrin et al., 1986).
- The physical capacities of workers vary substantially (Kishino et al., 1985).

The implications of these three assumptions are:

- LBP is a potential outcome of various activities. Lifting may be one of them.
- Manual lifting can be evaluated as a risk factor for LBD. The measurement of the exposure includes type (e.g., lifting or other activities) and virulence (e.g., lifting over the physical capacity).
- Each individual has his or her own physical capacity, and such capacity can be approximated by statistics (percentile), such as,

  The maximal permissible compressive forces on the spine (Chaffin and Park, 1973),
  The maximal permissible energy expenditure,
  The maximal acceptable weight of load (Troup et al., 1987; Snook and Ciriello, 1991).
- The deviation of physical capacities among individuals can be identified and measured.

Lifting technique and low back pain — several variables have been reported to be important in the evaluation of the spinal loading of various lifting techniques (Hart et al., 1987)

- Initial position of the load (Marras et al., 1993).
- Initial posture (stoop vs. squat) (Omino and Hayashi, 1992).
- Foot position (straddle vs. parallel) (Anderson and Chaffin, 1986).
- Velocity and acceleration of lift (Marras et al., 1993).
- Muscular recruitment and fatigue (Trafimow et al., 1992; Lavender et al., 1992; Schultz et al., 1982b).
- Compound movements (e.g., flexion with torsion) (Kelsey et al., 1984b; Markolf, 1972; Marras et al., 1993; Omino and Hayashi, 1992).
- Lumbodorsal curvature (lordotic vs. kyphotic) (Panjabi and White, 1980) — Controversy exists as to the role of lumbar lordosis in lifting. One theory is that preservation of the lordosis in conjunction with intra-abdominal pressure (IAP) strengthens the spine and is critical to prevention of injury during lifting (Aspden, 1989). To the contrary, another theory states that decreased lordosis may increase the tension in the posterior spinal ligaments thus decreasing the muscular forces required and subsequently decreasing the resultant compressive forces on the spine (Gracovetsky et al., 1989, 1990). The risk of injury may be influenced more by the degree of flexion than the choice of lifting posture (stoop vs. squat) (Potvin et al., 1991). The mechanical advantage provided the trunk muscles is also dependent on the assumed posture (Schultz et al., 1984).

Ligament — a band of strong, fibrous connective tissue that attaches bone to bone and serves to provide stability, transmit loads, and restrict motion (Yamamoto et al., 1990; Panjabi and White, 1980). The posterior ligaments of the vertebral column act passively to assist in lifting (Gracovetsky et al., 1981). The spinal ligaments have different biochemical composition and fiber density and orientation based on their specific function within the spinal column (Hukins et al., 1990) (see spinal ligament).

Ligamentum flavum — connects the laminae of adjacent vertebrae (see ligament, spinal ligament).

Lordosis — the inward curvature of the spine; the normal curvature of the cervical and lumbar spine (Gracovetsky and Farfan, 1986, Gracovetsky et al., 1987).

- Different postures result in different loading patterns of the passive tissues, such as the ligaments and fascia. In some circumstances these structures can contribute significantly to the development of axial torque (McGill, 1992).
- Posture can directly influence the amount of curvature. The height of the heel has been shown to change the curvature of the lumbar spine (Bendix et al., 1984).
- The degree of lumbar lordosis has not been directly correlated to low back pain (Hansson et al., 1985).

Low back pain (LBP) — a symptom of any one of several pathologies of the lumbosacral spine, with varying terminology, etiology, evaluation and medical management techniques (Cavanaugh and Weinstein, 1994; Ghosh, 1988; Hadler, 1987; Loeser, 1980; Troup, 1981; Waddell et al., 1980; White and Gordon, 1982).

Low back pain research — There are many different hypotheses as to the origin of LBP. In general the potential causes of LBP may be categorized as mechanical, chemical, biological, infectious, genetic, and degenerative (Boden et al., 1991; Carr et al., 1985; Ghosh, 1988). Ergonomists and engineers have focused on the mechanical basis of LBP since the variables in that domain are more readily defined and manipulated. Good scientific method dictates that research must build upon the knowledge at each of the levels below it in the critical path (Pope, 1990; Weinstein, 1988a; Zohn, 1988):

Level I: The pain — Is the pain of spinal origin, and is it real? Usually based on medical history, self reporting, standardized pain questionnaires, and physical exam.

Level II: The anatomy — Can the pain signal be transmitted? Is the presence of pain receptors consistent with the present understanding of the anatomy?

Level III: The autopsy and mechanical testing — What are the internal and external forces needed to cause sufficient injury to stimulate the pain receptors? The techniques used in cadaver lumbar motion segment studies have not been standardized, making prediction based on this body of research difficult (Keller et al., 1990).

Level IV: Biomechanical modeling — Can the forces required to cause damage as determined from mechanical testing, be generated in a person during work activities?

Low back pain impairment (LBPI) — refers to the physical limitations in movement, strength and function of an individual due to LBP (Waddell et al., 1984; Mayer et al., 1988; Deyo, 1988).

Low back pain disability (LBPD) — refers to the loss of capacity or time for an individual to function at work or in activities of daily living. It is not synonymous with LBPI, since impaired individuals may be able to function normally, and conversely an individual with minimal measurable impairment may be significantly disabled with respect to work and/or activity of daily living. This dichotomy is secondary to psychosocial factors, socioeconomic factors and specific job requirements (Andersson et al., 1983; Bigos et al., 1992; Deyo and Diehl, 1983; Barnes et al., 1989; Polatin et al., 1989; Rowe, 1971; Spengler et al., 1986a, 1986b, 1986c; Waddell et al., 1984; Waddell and Main, 1984).

Lumbosacral sprain/strain — one of the more common diagnoses associated with an episode of low back pain often related to a memorable injury of moderate severity. An imprecise diagnosis, it is often used prior to ruling out other, more specific diagnoses.

Manipulation/Mobilization/Chiropractic practices for low back pain — manual techniques involving movement of spinal segments for the purpose of adjusting spinal alignment, improving ROM, or decreasing pain (Cassidy and Kirkaldy-Willis, 1988; Hadler et al., 1987; Haldeman, 1980; Khalil et al., 1992).

Mechanical traction — the application of a tensile load along the long axis of the spine with the goal of promoting relaxation of the back musculature and decompressing the spine.

Mechanoreceptor — a specialized nerve ending that is stimulated by mechanical pressure, pressure changes, oscillation or movement.

Medical management of low back pain — the goals of intervention are the reduction of symptoms and the restoration of functional levels. The outcome of treatment for LBP is very variable, with a portion of the population typically becoming chronic (Gatchel et al., 1992; Wiesel et al., 1984; Spitzer et al., 1987; Deyo, 1983).

- Activity modification: (see bed rest, back school, and pain clinics).
- Oral medication (see pharmacological agents).
- Physical medicine: (see acupuncture, biofeedback, corsets/orthotics, exercise, mobilization/manipulation/chiropractic, mechanical traction, physical modalities, work hardening).

Meninges — the three membranes that cover the spinal cord and brain. They are the dura mater, pia mater, and arachnoid membrane.

Meniscus — a dense fibrocartilage pad present in some synovial joints (e.g., knee, facet) which acts to protect the cartilage surfaces of the joint and to distribute the load between the articular surfaces. A meniscus is typically attached at one end to the joint capsule and/or ligaments and thus receives some degree of vascular supply and innervation.

Micro-discectomy — the use of microsurgical techniques and a small incision to perform a partial removal of a herniated disc. The smaller incision and decreased disruption to surrounding tissues allows for a more rapid recovery phase.

Morbidity — state of being diseased.

Morbidity rate — the number of instances of a disease over a specified period of time per unit of population. It is usually expressed as x cases per year per 1,000, 10,000, or 100,000 of population.

Morphology — the science of forms and structures of organisms.

Myelography — a radiographic imaging technique for visualization of the spinal cord, involving the injection of a radiopaque dye into the area around the meninges of the spinal cord. It is particularly useful in detecting impingement on the spinal canal and nerve roots.

MRI (magnetic resonance imaging) — a noninvasive imaging technique using electromagnetic radiation particularly useful for visualization of soft (non-radio-opaque) tissues. The MRI provides greater resolution of intrinsic disc abnormality than a CT scan due to its ability to detect water content in the disc (Bernhardt et al., 1992; Hickey et al., 1986; Panagiotacopulos et al. 1987a 1987b) (see CT scan).

NCV (nerve conduction velocity) — a direct measure of the conductivity of motor nerves in the region being tested. Often performed in conjunction with EMG testing.

Necrosis — death of one or more cells.

Neoplasm — new growth of cells or tissues, may be benign or malignant (cancerous).

Neurological deficit — a defect or dysfunction of the central or peripheral nervous system, such as diminished deep tendon reflexes or decreased sensation.

Nociceptor — a specialized peripheral nerve ending which senses painful stimuli.

Nucleus pulposus — the central, more viscous portion of the disc.

Anatomy

- A semifluid mass of proteoglycans in a collagen fiber matrix.
- The nucleus is intrinsically cohesive and resists herniation, even under experimental conditions when a posterior channel is cut into the annulus fibrosus.

Immunological basis of spinal pain syndromes

- The nucleus pulposis lacks a blood supply and is never exposed to the circulation. In a disc that suffers an endplate fracture, proteoglycans may be exposed to the body's immune system for the first time, triggering an inflammatory response (Pennington et al., 1988). The gelatinous matrix of the nucleus has not been studied according to contemporary standards of immunology, but available evidence suggests that the matrix does have antigenic properties.
- Prolapsed nuclear material elicits an inflammatory response if it enters the epidural space or the vertebra spongiosa in the case of traumatic Schmorl's nodes. Patients exhibit changes in lymphocyte migration and antibody profiles consistent with an antigenic response.

Mechanism

- The main function of the nucleus pulposis is to transmit loads from one vertebra to another.
- Functioning as a volume of fluid within an enclosed space, the nucleus is incompressible. The attempted expansion of the nucleus pulposus effectively braces the annulus fibrosus from within, preventing it from buckling.
- When exposed to a compression load, the nucleus deforms by attempting to spread in a radial direction, but this displacement is arrested by the annulus fibrosus. Tension develops in the annulus fibrosus to resist radial expansion of the nucleus pulposus.

Orthosis — orthopedic device for assisting the function of part of the body without replacing it (see corset).

-osis — suffix meaning "disease of."

Ossification — the formation of bone.

Osteoarthritis — commonly described as a "wear and tear" disease of the joints. Chronic, degenerative disease particularly of weight-bearing joints characterized by erosion of the cartilage and bony overgrowth.

Osteomalacia — a softening of the bone related to demineralization of the matrix, often associated with a deficiency of vitamin D.

Osteopenia — loss of bone mass.

Osteophyte — a bony or osseous outgrowth or excrescence (tumor).

Osteoporosis — a decrease in bone mass, often related to prolonged bed rest, post-menopausal endocrine changes, and prolonged periods of weightlessness among astronauts. It is often treated with hormone therapy and dietary supplements in post-menopausal women. Weight-bearing activities are recommended for cases resulting from prolonged bed rest, as bone remodels in response to the stresses imposed on it.

Pain clinics and low back pain — specialized facilities utilizing a multidisciplinary approach to the treatment of pain. Emphasis is placed on improving functional levels, using activity and behavior modification, and pain management techniques rather than treatment of symptoms (Aronoff, 1982, 1983).

Percutaneous — through the skin.

Pharmacological agents for low back pain — the medical intervention for low back pain may involve the use of pharmacological agents. These drugs, prescribed for several reasons, are classified into three major categories:

- Analgesic — a category of drugs whose main action is reduction of pain. They can be narcotic (morphine based, and by prescription only) or non-narcotic. They may be prescription (e.g., Darvon™ — propoxyphene hydrochloride) or over the counter (e.g., Tylenol™ — acetaminophen, aspirin).
- Anti-inflammatory — a category of drugs whose main action is reduction of the inflammatory response with a secondary effect of pain reduction. These drugs are further divided into two subcategories based on their pharmacology:

  Steroidal — these are typically prescription corticosteroids or their derivatives which are analogs of the hormones produced by the adrenal cortex (i.e., prednisone). Their action is to directly modulate the inflammatory response via the endocrine system. There use is usually limited to severe episodes due to the significant number and severity of side effects.

  Nonsteroidal anti-inflammatory drug (NSAID) — a category of prescription such as Feldene™ — piroxicam, Voltaren™ — diclofene, Naprosyn™ — naproxen and nonprescription such as Motrin™, Advil™ — ibuprofen. The action of this class of drugs is to inhibit the inflammatory response. The side effects of NSAIDs tend to be less severe than steroids.

- Muscle relaxants — a category of drugs whose action is to decrease muscle spasm, thereby reducing associated muscle pain (e.g., Flexeril™, Soma™, Robaxin™).

Physical exam and history of low back pain (Hadler, 1984; Haig, 1992; McCombe et al., 1989; Roland and Morris, 1983a 1983b) —

- Patient self report
  Medical History (Bigos et al., 1992).
  Pain diagram — patient drawing of pain type and distribution.
  Pain questionnaire — developed to assess the nature of the pain as well as its psychological and psychosocial impact (Herron and Pheasant, 1982; Waddell et al., 1980).
  Visual analog scale — subjective patient report of pain level using a calibrated scale.
- Referral pain/trigger point — palpation (probing by touch) of the muscles in or around the location of pain, for local tenderness.
- Range of motion (ROM) — a measure of the freedom of movement of a joint (Keeley et al., 1986; Pope et al., 1985). A relationship between decreased lumbar ROM and a previous history of LBP is reported to exist, but no significant predictive value for future LBP has been demonstrated (Battié et al., 1990).
- Straight leg raise test (see straight leg raise test, SLR).
- Neurologic/reflex evaluation — assessment of sensory and motor function, including deep tendon reflexes (i.e., knee jerk), muscle strength. Provides an indication of the function of the nerves controlling the tested area.

Physical modalities — the application of physical agents with the goal of decreasing pain, decreasing inflammation, increasing local circulation to promote healing, and/or increasing flexibility. These include: electrical stimulation, heat or ice applications, and ultrasound (Kane and Taub, 1975).

Pia mater — the innermost of the three meninges (membranes) covering the spinal cord and brain.

Plexus — network of nerves or blood vessels.

Posterior longitudinal ligament — the ligament extending along the posterior surface of the vertebral body (within the spinal canal) from the cervical spine to the sacrum (see ligament, spinal ligament).

Proteoglycans — substance found in the matrix of connective tissues and in synovial fluid, vitreous humor of the eye, and mucous secretions.

Radiculopathy — any disease of the motor or sensory root of a spinal nerve. Distribution of symptoms and findings is characteristic for each spinal segment. There are mono- and polyradiculopathies (Saal and Saal, 1989).

Radiopaque — impenetrable by X-ray.

Ramus (**rami** pl.) — branch.

Rheumatoid arthritis — chronic systemic disease characterized by inflammation of the joint linings, bony changes and eventually deformity usually involving multiple joints.

Sacrum — the five fused vertebrae between the lumbar vertebrae and the coccyx. They comprise the posterior aspect of the pelvis.

Sagittal plane — the plane through the body which bisects it into left and right parts.

Scarification — making a number of superficial incisions in the skin or other tissue.

Schmorl's nodes — disc protrusions through the endplate and into the vertebral body (Hansson and Ross, 1983). They are reported to be healed fractures of the vertebral endplates (Farfan et al., 1972).

Sciatic nerve — longest nerve of the body formed from lower lumbar and sacral nerve roots and running deep in the posterior aspect of the thigh. It is responsible for sensation of the posterolateral thigh, calf, and parts of the foot, as well as motor control of some of the muscles of the foot and calf.

Sciatica — common terminology for the pain and sensory changes along the distribution of the sciatic nerve, usually resulting from compression or trauma to the nerve or the nerve roots that form it (Heliovaara et al., 1986; Allan and Waddell, 1989).

Scoliosis — abnormal curvature of the spine. This can be an accentuation or decrease of the normal regional curvatures, or rotation or curvature of the spine in an abnormal plane or direction.

Sequela (**sequelae** pl.) — disorder due to a preceding disease or accident.

Shear loading of the disc — the component of the resultant force acting on the disc which is perpendicular to the long axis of the spinal segment. It is the complement of the compressive force. All the factors contributing to disc compression may also contribute to shear loading. The magnitude of shear loading depends on the orientation of the disc and body posture. Because there is little dynamic information about trunk muscles and their orientation to the disc, for biomechanical modeling straight line representation of the muscle force perpendicular to the disc is assumed. As a result, shear loading may be underestimated.

Somatotype — the classification of persons by body shape or type (morphology). The categories are: ectomorph (thin, wiry), mesomorph (well proportioned), endomorph (soft, rotund).

Somatosensory evoked potentials (SEP) — an electrodiagnostic procedure used to assess sensory neurons in peripheral and spinal cord pathways. This technique is sometimes used to evaluate stenosis.

Somatization — conversion of anxiety into physical symptoms.

Spasm — involuntary muscular contraction.

Spinal fusion — the surgical fixation of two or more adjacent vertebrae to provide structural stability or maintain normal disc space (Abumi et al., 1989a). Fusion is indicated in the presence of clinical instability of one or more lumbar motion segments (Farfan and Gracovetsky, 1984). Fusion provides no significant advantage over laminectomy alone in the case of the simple herniated disc (White et al., 1987). The mechanics of spinal fusion are complex, and the procedure can produce adverse effects, such as higher stresses at the adjacent vertebrae (Goel et al., 1988). Likewise, clinical research has not yet demonstrated what degree of stability is necessary for optimal healing and progression to solid fusion (Panjabi, 1988; Panjabi et al., 1988a).

- Anterior interbody fusion — a transabdominal spinal fusion interferes little with the integrity of the lumbar spine as compared to the posterior approach. There are some advantages of this procedure: back muscles are not disabled, the posterior elements remain intact, and the contents of the vertebral canal are not exposed. This latter advantage avoids the potential complication of arachnoiditis and preserves the sinuvertebral nerves, avoiding the risk of neuroma formation (tumor of the nerve fiber).
- Posterior interbody fusion- spinal fusion surgery involving access to the vertebrae through the back musculature and ligaments.

Spinal ligaments — see ligament, anterior longitudinal ligament, posterior longitudinal ligament, ligamentum flavum, and supraspinous ligament (Panjabi et al., 1982; Ratcliffe, 1980).

Supraspinous ligament — the strong, fibrous cord joining the spinous processes from the seventh cervical vertebrae to the sacrum (see ligament).

Spinal nerve roots — the nerve trunks that exit and enter the spinal cord at the intervertebral space (Bogduk, 1983).

Spinal surgery — surgical intervention for the treatment of severe pain, spinal instability, and/or neurologic involvement (Pope and Panjabi, 1985; Posner et al., 1982). Surgery is usually the treatment of last resort, when conservative therapies have failed.

Spondylitis — inflammation of one or more vertebrae.

Spondylolisthesis — a condition of forward slippage of a lumbar vertebrae which may or may not be associated with instability (Grobler et al., 1993a, 1993b). It can be categorized as follows (Pearcy and Shepherd, 1985):

- Isthmic spondylolisthesis — the slippage is due to loss of the constraint by the neural arch, secondary to a spondylolysis, most commonly occurring in the last lumbar vertebrae (Grobler et al., 1993a).
- Degenerative/pseudo spondylolisthesis — forward slippage, typically at L_4 and L_5 levels due to a degenerative process, without any defect in the neural arch. This may also be due to a developmental disposition related to the orientation of the facet joint (Grobler et al., 1993b).
- Retrospondylolisthesis — misalignment of one vertebra relative to the adjacent vertebra characterized by backward displacement.
- Displastic, traumatic, and pathologic spondylolisthesis are less common diagnoses.

Spondylolysis — a defect in the neural arch, typically at L_5 and commonly believed to be due to repetitive low-grade trauma over prolonged periods. Short duration exposures may be of little clinical significance (Semon and Spengler, 1981; Crock, 1981).

Spondylosis — ankylotic state or fusion of the vertebrae. Disease of the spine often associated with disc degeneration. There is a spectrum of changes effecting one or multiple levels, initially signified radiologically by loss of disc height, osteophytes arising from the margins of the vertebral bodies, and osteoarthritic changes in the apophyseal joints (Edwards and LaRocca, 1985).

Spongiosa — spongy. The sustantia spongiosa ossium is the thin bony matrix that makes up the center of the intervertebral body.

Stenosis, spinal — narrowing of the spinal canal, nerve root canals, or tunnels of the intervertebral foramina. The narrowing not only reduces the anteroposterior and lateral diameters, but also alters the cross-sectional configuration of the spinal canal (Schonstrom et al., 1985; Spengler 1983). Decreased cross-sectional area causes an associated increase in local pressure, with a resultant effect on local nerve function (Schonstrom and Hansson, 1988; Grabias, 1980; Crock, 1981).

- When the spinal canal and lateral nerve root canal are narrowed, a relatively small disc herniation can produce clinically significant symptoms, such as sciatica, which would not have occurred if the canal had been of adequate dimensions. A nerve root may be compromised in a stenotic canal even during normal movement patterns (Panjabi et al., 1984).
- Stenosis may result from occupational factors or trauma, or from non-occupational factors such as malnutrition or genetic predisposition (Clark et al., 1985).
- There are gender related differences in the dimensions of the canal (Vanharanta et al., 1985).

Steroid — a lipid-based organic compound. Steroidal hormones are synthesized by the sex glands (e.g., estrogen, testosterone) and the adrenal cortex (corticosteroids, e.g., cortisone). There are a number of synthetic analogs which also have physiologic activity. Corticosteroids are potent anti-inflammatory agents, and in some cases may be given orally or by direct injection to the target site.

Straight leg raise (SLR) test — a clinical evaluation technique where the thigh with the knee extended straight is gradually bent (flexed) at the hip. Onset or aggravation of low back pain or sciatica during the maneuver is considered a positive test result and may be indicative of lumbar disc disease (Troup, 1981).

Subluxation — partial dislocation of a joint; bone ends are misaligned but still in contact.

Taxonomy — the science dealing with the identification, naming, and classification of organisms.

Thermography — an imaging technique currently being investigated as an indirect measure of autonomic nervous system function. It is a measure of skin surface temperature using the infrared spectrum.

Torsional loading of the disc — torque resulting from rotation about or near the long axis of the spinal column (Bogduk, 1991; Farfan, 1984). Torsional loading occurs frequently in common activities such as the rotation of the pelvis during walking, and rotation of the trunk when throwing a ball (Farfan et al., 1970; Farfan and Gracovetsky, 1984; Gracovetsky and Farfan, 1986; Marras and Mirka, 1992; Miller et al., 1988).

Mechanism (Panjabi and White, 1980)

- The rotational range of motion of a lumbar segment is limited to less than three degrees by the blocking of the apophyseal joint (Adams and Hutton, 1981). Both the disc and the apophyseal joint provide resistance to rotation. (Gunzburg et al., 1991a).
- When the apophyseal joint fails, the range of rotational movement increases. This occurs at the expense of increased axial torque and lateral shear forces on the disc itself (Adams and Hutton, 1981). Fracture or dislocation of this joint is a rare occurrence (Kramer and Levine, 1990).
- An isolated disc can tolerate only three degrees of axial rotation before the collagen fibers of the annulus suffer injury. For a given direction of rotation, only half of the collagen fibers in the annulus fibrosus are appropriately oriented to resist the rotation.

Failure

- Torsion injury to the annulus fibrosus is increased if axial rotation occurs in combination with flexion. Flexion prestresses the posterior annulus fibrosus. Consequently, far less axial rotation is required to strain collagen fibers beyond their physical limit.
- In flexion, the articular processes of the apophyseal joints are subluxated, and the apophyseal joints afford less resistance to axial rotation. As a result, during flexion–rotation, the annulus fibrosus is maximally stressed while least protected by the posterior elements (Miller et al., 1983). Trunk rotation ROM is decreased while the trunk is flexed (Gunzburg et al., 1991a, 1991b).
- Torsion is more recurrent and more detrimental than a compressive force. After injury, if the joint never heals fully, it is more likely to fail with repeated torsion. When torsion of the disc is combined with compression, the probability of failure increases.
- Torsional injury may include a rupture of ligamentous tissue, the annulus, and associated damage to the facet.
- Under continued rotation beyond three degrees, disc failure occurs with avulsion of the laminae of the annulus from the endplate. The nucleus and endplate remain intact. Gradually, the annulus develops radial fissures and the lumbar segment may become unstable. Furthermore, the torsional deformity can cause the neural elements in the intervertebral canal to displace to one side, stretching the nerve roots.

Diagnosis/Imaging

- Neurologic examination, CT scans, myelography, and MRI are normal because the lesion is restricted to the annulus fibrosus and does not involve nerve root compression.
- During discography, contrast medium and local anesthetic are injected into the painful site. If pain is relieved by the local anesthetic, and if the contrast medium is confined to the annulus fibrosus, the source of pain is assumed to be the annulus fibrosus.

Symptoms

- The pain will be aggravated by any movements that stress the annulus fibrosus, particularly flexion and rotation in the same direction that produced the lesion. Healing may require several months of protection from reinjury coupled with gradual and gentle reintroduction and restoration of function.

Trabecula — any of the bands of tissue that pass from the outer part of an organ to its interior, dividing it into separate chambers.

Trabecular bone — spongy bone.

Transverse plane — the cross-sectional plane through the body perpendicular to the long axis.

Trigger point — small, focal region of a muscle which is tender to touch. This sensitivity is assumed to be related or in reaction to regional pain.

Ventral — pertaining to the anterior or frontal surface of the body.

Work hardening — a therapeutic approach to functional training for return to work. This involves reduction of the patient's work tasks to component parts, progressive training in these components, followed by recombination of the work task and progression to maximum functional work level.

X-ray (radiography) — an imaging technique in which a body part is placed between an X-ray source and a photographic film plate. The X-rays are absorbed by the body part in direct proportion to the density of the tissue in its path. The radiograph (a photographic negative) is therefore a "map" of tissue density. Bone and mineralized tissue are well defined, while less dense substances such as body fluids and muscle are poorly imaged. X-ray can be used to estimate the dimension of vertebrae with adequate accuracy (Malmivaara et al., 1986).

Zygapophyseal — pertaining to an articular process of a vertebra.

References

Abumi, K., Panjabi, M. M. and Duranceau, J., 1989, Biomechanical evaluation of spinal fixation devices. Part III Stability provided by six spinal fixation devices and interbody bone graft, *Spine*, **14**, 1249-1255.

Abumi, K., Panjabi, M. M., Kramer, K., Duranceau, J., Oxland, T. and Crisco, J. J., 1990, Biomechanical evaluation of lumbar spinal stability after graded facetectomies, *Spine*, **14**, 1142-1147.

Adams, M. A. and Dolan P., 1991, A technique for quantifying the bending moment acting on the lumbar spine in vivo, *Journal of Biomechanics*, **24**, 117-126.

Adams, M. A. and Hutton, W. C., 1981, The relevance of torsion to the mechanical derangement of the lumbar spine, *Spine*, **6**, 241-248.

Adams, M. A. and Hutton, W. C., 1981, Prolapsed intervertebral disc: A hyperflexion injury, *Spine*, **7**, 184-191.

Adams, M. A. and Hutton, W. C., 1983a, The effect of posture on the fluid content of lumbar intervertebral discs, *Spine*, **8**, 665-671.

Adams, M. A. and Hutton, W. C., 1983b, The mechanical function of the lumbar apophyseal joints, *Spine*, **8**, 327-330.

Adams, M. A. and Hutton, W. C., 1985, Gradual disc prolapse, *Spine*, **10**, 524.

Adams, M. A., Dolan, P. and Hutton, W. C., 1987, Diurnal variations in the stresses on the lumbar spine, *Spine*, **12**, 130-137.

Adams, M. A., Dolan, P. and Hutton, W. C., 1988, The lumbar spine in backward bending, *Spine*, **13**, 1019-1026.

Adams, M. A., Dolan, P., Hutton, W. C. and Porter, R. W., 1990, Diurnal changes in spinal mechanics and their clinical significance, *The Journal of Bone and Joint Surgery* (Br), **72**, 266-270.

Agre, K., Wilson, R. R., Brim, M. and McDermott, D. J., 1984, Chymodiactin postmarketing surveillance demographic and adverse experience data in 29,075 patients, *Spine*, **9**, 479-485.

Allan, D. B. and Waddell, G., 1989, Understanding and management of low back pain, *Acta Orthop Scand*, **60**, 1-23.

Althoff, I., Brinckmann, P., Frobin, W., Sandover, J. and Burton, K., 1992, An improved method of stature measurement for quantitative determination of spinal loading: Application to sitting postures and whole body vibration, *Spine*, **17**, 682-683.

Anderson, C. K. and Chaffin, D. B., 1986, A biomechanical evaluation of five lifting techniques, *Applied Ergonomics*, **17**, 2-8.

Andersson, G. B. J., Schultz, A., Nathan, A. and Irstam, L., 1981, Roentgenographic measurement of lumbar intervertebral disc height, *Spine*, **6**,154-158.

Andersson, G. B. J., Svensson, H. and Oden, A., 1983, The intensity of work recovery in low back pain, *Spine*, **8**, 880-884.

Antti-Poika, I., Soini. J., Talroth, K., Yrjonen, T. and Konttinen, Y., 1990, Clinical relevance of discography combined with CT scanning, *The Journal of Bone and Joint Surgery (Br)*, **72-B**, 480-485.

Aronoff, G. M., 1982, Pain clinic #2 pain units provide an effective alternative technique in the management of chronic pain, *Orthopaedic Review*, **11**, 95-100.

Aronoff, G. M., 1983, The role of the pain center in the treatment for intractable suffering and disability resulting from chronic pain, *Seminars in Neurology*, **3**, 377-381.

Asfour, S. S., Khalil, T. M., Waly, S. M., Goldberg, M. L., Rosomoff, R. S. and Rosomoff, H. L., 1990, Biofeedback in back muscle strengthening, *Spine*, **15**, 510-513.

Aspden, R. M., 1989, The spine as an arch: A new mathematical model, *Spine*, **14**, 266-274.

Barnes, D., Smith, D., Gatchel, R. J. and Mayer, T. G., 1989, Psychosocioeconomic predictors of treatment success/failure in chronic low-back pain patients, *Spine*, **14**, 427-430.

Battié, M. C., Bigos, S. J., Fisher, L. D., Hansson, T. H., Jones, M. E. and Wortley, M. D., 1989, Isometric lifting strength as a predictor of industrial back pain reports, *Spine*, 14, 851.

Battié, M. C., Bigos, S. J., Fisher, L. D., Hansson, T. H., Nachemson, A. L., Spengler, D. M., Wortley, M. D. and Zeh, J., 1989, A prospective study of the role of cardiovascular risk factors and fitness in industrial back pain complaints, *Spine*, **14**, 141-147.

Battié, M. C., Bigos, S. J., Fisher, L. D., Spengler, D. M., Hansson, T. H., Nachemson, A. L. and Wortley, M. D., 1990, The role of spinal flexibility in back pain complaints within industry: A prospective study, *Spine*, **15**, 768-773.

Bendix, T., Sorensen, S. S. and Klausen, K., 1984, Lumbar curve, trunk muscles, and line of gravity with different heel heights, *Spine*, **9**, 223-227.

Bernhardt, M., Gurganious, L.R., Bloom, D. L. and White, A. A. III, 1993, Magnetic resonance imaging analysis of percutaneous discectomy: A preliminary report, *Spine*, **18**, 211-217.

Bigos, S. J., Battié, M. C. and Fisher, L. D., 1991a, Methodology for evaluating predictive factors for the report of back injury, *Spine*, **16**, 669-670.

Bigos, S. J., Battié, M. C., Fisher, L. D., Hansson, T. H., Spengler, D.M. and Nachemson, A. L., 1992. A prospective evaluation of preemployment screening methods for acute industrial back pain, *Spine*, **17**, 922-926.

Bigos, S. J., Battié, M. C., Spengler, D. M., Fisher, L. D., Fordyce, W. E., Hansson, T. H., Nachemson, A. L. and Worley, M. D., 1991b, A prospective study of work perceptions and psychosocial factors affecting the report of back injury, *Spine*, **16**, 1-6.

Bigos, S. J., Battié, M. C., Spengler, D. M., Fisher, L. D., Fordyce, W. E., Hansson, T., Nachemson, A. L. and Zeh, J., 1992, A Longitudinal, Prospective Study of Industrial Back Injury Reporting, *Clin Orthop*, **279**, 21-34.

Bigos, S. J., Spengler, D. M., Martin, N. A., Zeh, J., Fisher, L., and Nachemson A., 1986b, Back injuries in industry: a retrospective study: III. employee-related factors, *Spine*, **11**, 252-256.

Bigos, S. J., Spengler, D. M., Martin, N. A., Zeh, J., Fisher, L., Nachemson, A. and Wang, M. H., 1986a, Back injuries in industry: a retrospective study: II injury factors, *Spine*, **11**, 246-251.

Boden, S. D., Wiesel, S. W., Laws, E. R. and Rothman, R. H., 1991, *The Aging Spine*, (Philadelphia: W. B. Saunders Co.), 21-22.

Bogduk, N. and Engle, R., 1984, The menisci of the lumbar zygapophyseal joints: A review of their anatomy and clinical significance, *Spine*, **9**, 454-460.

Bogduk, N. and Macintosh, J. E., 1984, The applied anatomy of the thoracolumbar fascia, *Spine*, **9**, 164-170.

Bogduk, N., 1983, The innervation of the lumbar spine, *Spine*, **8**, 286-293.

Bogduk, N., 1991, The lumbar disc and low back pain, *Neurosurgery Clinics of North America*, **2**, 791-803.

Bogduk, N., Macintosh, J. E. and Pearcy, M. J., 1992, A universal model of the lumbar back muscles in the upright position, *Spine*, **17**, 897-913.

Bogduk, N., Tynan, W. and Wilson, A. S., 1981, Innervation of lumbar intervertebral disks, *Anat. Soc. G.B. & I*, 39-56.

Brinckmann, P. and Grootenboer, H., 1991, Change of disc height, radial disc bulge, and intradiscal pressure from discectomy: An *in vitro* investigation on human lumbar discs, *Spine*, **16**, 641-646.

Brinckmann, P. and Horst, M., 1985, The influence of vertebral body fracture, intradiscal injection, and partial discectomy of the radial bulge and height of human lumbar discs, *Spine*, **10**, 138-145.

Brinckmann, P., 1986, Injury of the annulus fibrosus and disc protrusions: an *in vitro* investigation on human lumbar discs, *Spine*, **11**, 149-153.

Brinckmann, P., Biggemann, M. and Hilweg, D., 1989, Prediction of the compressive strength of human lumbar vertebrae, *Spine*, **14**, 606-610.

Brinckmann, P., Frobin, W., Hierholzer, E. and Horst, M., 1983, Deformation of the vertebral end-plate under axial loading of the spine, *Spine*, **8**, 851-856.

Bromley, J. W., Varma, A. O., Santoro, A. J., Cohen, P., Jacobs, R. and Berger, L., 1984, Double-blind evaluation of collagenase injections for herniated lumbar discs, *Spine*, **9**, 486-488.

Buseck, M., Schipplein, O. D., Andersson, G. B. J. and Andriacchi, T. P., 1988, Influence of dynamic factors and external loads on the moment at the lumbar spine in lifting, *Spine*, **13**, 918-921.

Butler, D., Trafimow, J. H., Andersson, G. B. J., McNeill, T. W. and Huckman, M. S., 1990, Discs degenerate before facets, *Spine*, **15**, 111-113.

Calhoun, E., McCall, I. W., Williams, L. and Pullicino, V. N., 1988, Provocation discography as a guide to planning operations on the spine, *The Journal of Bone and Joint Surgery* (*Br*), **70-B**, 267-271.

Carr, D., Gilbertson, L., Frymoyer, J., Krag, M. and Pope, M., 1985, Lumbar paraspinal compartment syndrome: A case report with physiologic and anatomic studies, *Spine*, **10**, 816-820.

Carrera, G. F., 1980a, Lumbar facet joint injection in low back pain and sciatica, *Radiology*, **137**, 661-664.

Carrera, G. F., 1980b, Lumbar facet joint injection in low back pain and sciatica, *Radiology*, **137**, 665-667.

Cartas, O., Nordin, M., Frankel, V. H., Malgady, R. and Sheikhzadeh, A., 1993, Quantification of trunk muscle performance in standing, semistanding, and sitting postures in healthy men, *Spine*, **18**, 603-609.

Cassidy, J. and Kirkaldy-Willis, W., 1988, Manipulation in *Managing Low Back Pain*, Ed. William H. Kirkaldy-Willis, New York: Churchill Livingstone. 287-291.

Cavanaugh, J. M. and Weinstein, J. N., 1994, Low back pain: epidemiology, anatomy and neurophysiology, Ed. Patrick D. Wall and Ronald Melzack. *Textbook of Pain*, 3rd edition, Churchill Livingstone, New York, NY. 441-454.

Chaffin, D. B. and Park, K. S., 1973, A longitudinal study of low-back pain as associated with occupational weight lifting factors, *American Industrial Hygiene Association Journal*, **34**, 513-525.

Clark, G. A., Panjabi, M. M. and Wetzel, F. T., 1985, Can infant malnutrition cause adult vertebral stenosis, *Spine*, **10**, 165-170.

Crock, H. V., 1981, Normal and pathological anatomy of the lumbar spinal nerve root canals, *The Journal of Bone and Joint Surgery*, **63-B**, 487-490.

Damkot, D. K., Pope, M. H., Lord, J. and Frymoyer, J. W., 1984, The relationship between work history, work environment and low-back pain in men, *Spine*, **9**, 395-399.

Davis, A. A. and Carragee, E. J., 1993, Bilateral facet dislocation at the lumbosacral joint: A report of a case and review of literature, *Spine*, **18**, 2540-2544.

Davis, P. R., 1981, The use of intra-abdominal pressure in evaluating stresses on the lumbar spine, *Spine*, **6**, 90-92.

Deyo, R. A. and Diehl, A. K., 1983, Measuring physical and psychosocial function in patients with low-back pain, *Spine*, **8**, 635-642.

Deyo, R. A., 1983, Conservative therapy for low back pain, *JAMA*, **250**, 1057-1062.

Deyo, R. A., 1986, Diehl AK; Rosenthal M. How many days of bed rest for acute low back pain? A randomized clinical trial. *NEJOM*, **315**, 1064-1070.

Deyo, R. A., 1988, Measuring the functional status of patients with low back pain, *Arch Phys Med Rehabil*, **69**, 1044-1053.

Dillard, J., Trafimow, J., Andersson, G. B. J. and Cronin, K., 1991, Motion of the lumbar spine: Reliability of tow measurement techniques, *Spine*, **16**, 321.

Dolan, P., Adams, M. A. and Hutton, W. C., 1988, Commonly adopted postures and their effect on the lumbar spine, *Spine*, **13**, 197-201.

Dong, G. X. and Porter RW., 1989, Walking and cycling tests in neurogenic and intermittent claudication, *Spine*, **14**, 965-969.

Dvorak, J., Panjabi, M., Gerber, M. and Wichmann, W., 1987, CT-functional diagnostics of the rotatory instability of upper cervical spine: 1. An experimental study on cadavers, *Spine*, **12**, 197-205.

Edwards, W. C. and LaRocca, S. H., 1985, The developmental segmental sagittal diameter in combined cervical and lumbar spondylosis, *Spine*, **10**, 42-49.

Eisen, A. and Hoirch, M., 1983, The electrodiagnostic evaluation of spinal root lesions, *Spine*, **8**, 98-106.

Eismont, F. J. and Currier, B., 1989, Surgical management of lumbar intervertebral-disc disease, *The Journal of Bone and Joint Surgery*, **71-A**, 1266-1271.

Elnaggar, I. M., Nordin, M., Sheikhzadeh, A., Parnianpour, M. and Kahanovitz, N., 1991, Effects of spinal flexion and extension exercises on low-back pain and spinal mobility in chronic mechanical low-back pain patients, *Spine*, **16**, 967-972.

Fager, C. A., 1984, The age-old back problem new fad, same fallacies, *Spine*, **9**, 326-328.

Fardon, D., Pinkerton, S., Balderstron, R., Garfin, S., Nasca, R. and Salib, R., 1993, Terms used for diagnosis by English speaking spine surgeon, *Spine*, **18**, 274-277.

Farfan, H. F. and Gracovetsky, S., 1984, The nature of instability, *Spine*, **9**, 714-719.

Farfan, H. F., 1984, The torsional injury of the lumbar spine, *Spine*, **9**, 53.

Farfan, H. F., 1985, The use of mechanical etiology to determine the efficacy of active intervention in single joint lumbar intervertebral joint problems: Surgery and chemonucleolysis compared: A prospective study, *Spine*, **10**, 350-358.

Farfan, H. F., Cossette, J.W., Robertson, G. H., Wells, R. V. and Kraus, H., 1970, The effects of torsion on the lumbar intervertebral joints: the role of torsion in the production of disc degeneration, *The Journal of Bone and Joint Surgery*, **52-A**, 468-497.

Farfan, H. F., Huberdeau, R. M. and Dubow, H. I., 1972, Lumbar intervertebral disc degeneration the influence of geometrical features on the pattern of disc degeneration a post mortem study, *The Journal of Bone and Joint Surgery*, **54-A**, 492-510.

Fraser, R. D., Osti, O. L. and Vernon-Roberts, B., 1987, Discitis after discography, *The Journal of Bone and Joint Surgery*, **69-B**, 26-35.

Friedman, J. and Goldner, M. Z., 1955, Discography in evaluation of lumbar disk lesions, *Radiology*, **65**, 653-661.

Frymoyer, J. W. and Gordon, S. L., 1989, Research perspectives in low-back pain, Report of a 1988 workshop, *Spine*, **14**, 1384-1390.

Frymoyer, J. W., 1981, The role of spine fusion, *spine*, **6**, 284-307.

Gatchel, R. J., Mayer, T. G., Hazard, R. G., Rainville, J. and Mooney, V., 1992, Functional restoration: pitfalls in evaluating efficacy, *Spine*, **17**, 988-995.

Ghosh, P., 1988, *The Biology of the Intervertebral Disc, I & II*, Boca Raton, FL: CRC Press, Inc.

Goel, V. K., Goyal, S., Clark, C., Nishiyama, K. and Nye, T., 1985, Kinematics of the whole lumbar spine: Effect of discectomy, *Spine*, **10**, 543-554.

Goel, V. K., Kim. Y. E., Lim, T-H. and Weinstein, J. N., 1988, An analytical investigation of the mechanics of spinal instrumentation, *Spine*, **13**, 1003-1010.

Goel, V. K., Kong, W., Han, J. S., Weinstein, J. N. and Gilbertson, L. G., 1993, A combine finite element and optimization investigation of lumbar spine mechanics with and without muscles, *Spine*, **18**, 1531-1541.

Goel, V. K., Nishiyama, K., Weinstein, J. N. and Liu, Y. K., 1986, Mechanical properties of lumbar spinal motion segments as affected by partial disc removal, *Spine*, **11**, 1008-1012.

Gordon, S. J., Yang, K. H., Mayer, P. J., Mace, A. H. Jr., Kish, V. L. and Radin, E. L., 1991, Mechanism of disc rupture: A preliminary report, *Spine*, **16**, 450-456.

Goren, G. J., Baljet, B. and Drukker, J., 1990, Nerves and nerve plexuses of the human vertebral column, *The American Journal of Anatomy*, **188**, 282-296.

Grabias, S., 1980, Current concepts review the treatment of spinal stenosis, *The Journal of Bone and Joint Surgery*, **62-A**, 308-313.

Gracovetsky, S. and Farfan, H., 1986, The optimum spine, *Spine*, **11**, 543-573.

Gracovetsky, S., Farfah, H. and Helleur, C., 1985, The abdominal mechanism, *Spine*, **10**, 317-324.

Gracovetsky, S., Farfan, H. F. and Lamy, C., 1977, A mathematical model of the lumbar spine using an optimization system to control muscles and ligaments. *Orthop Clin North Am*, **8**, 135-153.

Gracovetsky, S., Farfan, H. F. and Lamy, C., 1981, The mechanism of the lumbar spine, *Spine*, **6**, 249-262.

Gracovetsky, S., Kary, M., Levy, S., Said, R. B., Pitchen, I. and Helie, J., 1990, Analysis of spinal and muscular activity during flexion/extension and free lifts, *Spine*, **15**, 1333-1339.

Gracovetsky, S., Kary, M., Pitchen, I., Levy, S. and Said, R. B., 1989, The importance of pelvic tilt in reducing compressive stress in the spine during flexion-extension exercises, *Spine*, **14**, 412-416.

Gracovetsky, S., Zeman V., Carbone, A. 1987, Relationship between lordosis and the position of the center of reaction of the spinal disc. *J Biomed Eng*, **9**, 237-248.

Grobler, L., Robertson, P. A., Novotny, J. E. and Ahern, J. W., 1993a, Decompression for degenerative spondylolisthesis and spinal stenosis at $L_{4\text{-}5}$: The effects of facet joint morphology, *Spine*, **18**, 1475-1482.

Grobler, L., Robertson, P. A., Novotny, J. E. and Pope, M. H., 1993b, Etiology of spondylolisthesis: Assessment of the role played by lumbar facet morphology, *Spine*, **18**, 80- 91.

Gunzburg, R., Gunzburg, J., Wagner, J. and Fraser, R. D., 1991a, Radiologic interpretation of lumbar vertebral rotation, *Spine*, **16**, 660-664.

Gunzburg, R., Hutton, W. and Fraser, R., 1991b, Axial rotation of the lumbar spine and the effect of flexion: an in vitro and in vivo biomechanical study, *Spine*, **16**, 22-28.

Hadler, N. M., 1984, *Diagnosis and Treatment of Backache. Medical Management of the Regional Musculoskeletal Diseases*, New York, NY: Grune & Stratton Inc., 3-52.

Hadler, N. M., 1987, Regional musculoskeletal diseases of the low back. Cumulative trauma versus single incident, *Clinical Orthopaedics and Related Research*, **221**, 33-41.

Hadler, N. M., Curtis, P., Gillings, D. B. and Stinnett, S., 1987, A benefit of spinal manipulation as adjunctive therapy for acute low-back pain A stratified controlled trial, *Spine*, **12**, 703-706.

Haig, A. J., 1992, Diagnoses and treatment options in occupational low-back pain, *Occupational Medicine*, **7**, 641-653.

Haig, A. J., Weismann, G., Haugh, L. D., Pope, M. and Grobler, L. J., 1993, Prospective evidence for change in paraspinal muscle activity after herniated nucleus pulposus, *Spine*, **18**, 926-930.

Haldeman, S., 1980, *Spinal Manipulative Therapy in the Management of Low Back Pain, Low Back Pain*, 2nd ed. Philadelphia, PA: J.B. Lippincott Company, 245-275.

Haldeman, S., 1984, The electrodiagnostic evaluation of nerve root function, *Spine*, **9**, 42-48.

Hampton, D., Laros, G., McCarron, R. and Franks, D., 1989, Healing potential of the annulus fibrosus, *Spine*, **14**, 398.

Hansson, T. and Ross, B., 1983, The amount of bone mineral and Schmorl's nodes in lumbar vertebrae, *Spine*, **8**, 266.

Hansson, T. H., Keller, T. S., Panjabi, M. M., 1987, A study of the compressive properties of lumbar vertebral trabeculae: effects of tissue characteristics, *Spine*, **12**, 56-62.

Hansson, T. H., Bigos, S., Beecher, P. and Wortley, M., 1985, The lumbar lordosis in acute and chronic low-back pain, *Spine*, **10**, 154.

Hart, D. L., Stobbe, T. J. and Jaraiedi, M., 1987, Effect of lumbar posture on lifting, *Spine*, **12**, 138-145.

Heliovaara, M., Vanharanta, H., Korpi, J. and Troup, J. D. G., 1986 Herniated lumbar disc syndrome and vertebral canals, *Spine*, **11**, 433.

Herrin, G. D., Jaraiedi, M. and Anderson, C. K., 1986, Prediction of overexertion injuries using biomechanical and psychophysical models, *Am. Ind. Hyg. Assoc. J*, **47**, 322-330.

Herron, L. D. and Pheasant, H. C., 1982, Changes in MMPI profiles after low-back surgery, *Spine*, **7**, 591-597.

Hickey, D. S. and Hukins, D. W. L., 1982, Aging changes in the macromolecular organization of the intervertebral disc: An X-ray diffraction and electron microscopic study, *Spine*, **7**, 234-242.

Hickey, D. S. and Hukins, W. L., 1980, Structure of fetal annulus, *Anat. Soc. G.B. & I*,81-89.

Hickey, D. S., Aspden, R. M., Hukins, D. W. L., Jenkins, J. P. R. and Isherwood, I., 1986, Analysis of magnetic resonance images from normal and degenerate lumbar intervertebral discs, *Spine*, **11**, 702-708.

Holmes, A. D., Hukins, D. W. L., Freemont, A. J., 1993, End-plate displacement during compression of lumbar vertebra-disc-vertebra segments and the mechanism of failure, *Spine*, **18**, 128-135.

Holt, E. P., 1968, The question of lumbar discography, *The Journal of Bone and Joint Surgery*, **50-A**, 720-726.

Horst, M. and Brinckmann, P., 1981, Measurement of the distribution of axial stress on the end-plate of the vertebral body, *Spine*, **6**, 217-232.

Hsu, K. Y., Zucherman, J. F., Derby, R., White, A. H., Goldthwaite, N. and Wynne, G., 1988, Painful lumbar end-plate disruptions: A significant finding, *Spine*, **13**, 76-78.

Hsu, K. Y, Zucherman, J., Shea, W., Kaiser, J., White, A., Schofferman, J. and Amelon, C., 1990, High lumbar disc degeneration: Incidence and etiology, *Spine*, **15**, 679-682.

Hukins, D. W. L., Kirby, M. C., Sikoryn, T. A., Aspden, R. M. and Cox, A. J., 1990, Comparison of structure, mechanical properties, and functions of lumbar spinal ligaments, *Spine*, **15**, 787-795.

Hutton, W. C. and Adams, M. A., 1982, Can the lumbar spine be crushed in heavy lifting? *Spine*, **7**, 586-590.

Inoue, H., 1981, Three dimensional architecture of lumbar intervertebral discs, *Spine*, **6**, 139-146.

Jager, J. and Luttmann, A., 1989, Biomechanical analysis and assessment of lumbar stress during load lifting using a dynamic 19-segment human model, *Ergonomics*, **32**, 93-112.

Kane, K. and Taub, A., 1975, A history of local electrical analgesia, *Pain*, **1**, 125-138.

Keeley, J., Mayer, T. G., Cox, R., Gatchel, R. J., Smith, J. and Mooney, V., 1986, Quantification of lumbar function: part 5: reliability of range-of-motion measures in the sagittal plane and an *in vivo* torso rotation measurement technique, *Spine*, **11**, 31-35.

Keller, T. S., Hansson, T. H., Abram, A. C., Spengler, D. M. and Panjabi, M. M., 1989, Regional variations in the compressive properties of lumbar vertebral trabeculae. Effects of disc degeneration, *Spine*, **14**, 1012-1019.

Keller, T. S., Holm, S. H., Hansson, T. H. and Spengler, D. M., 1990, 1990 Volvo award in experimental studies: The dependence of intervertebral disc mechanical properties on physiologic conditions, *Spine*, **15**, 751-761.

Kelsey, J. L., 1975, An epidemiological study of the relationship between occupations and acute herniated lumbar intervertebral discs, *International Journal of Epidemiology*, **4**, 197-205.

Kelsey, J. L., Githens, P. B., O'Conner, T., Weil, U., Calogero, J. A., Holford, T. R., White, A. A. III., Walter, S. D., Ostfield, A. M. and Southwick, W. O., 1984, Acute prolapsed lumbar intervertebral disc: An epidemiologic study with special reference to driving automobiles and cigarette smoking, *Spine*, **9**, 608-613.

Kelsey, J. L., Githens, P. B., White, A. A., Holford, T. R., Walter, S. D., O'Connor, T., Ostfeld, A. M., Weil, U., Southwick, W. O. and Calogero, J. A., 1984, An epidemiologic study of lifting and twisting on the job and risk for acute prolapsed lumbar intervertebral disc, *Journal of Orthopaedic Research*, **2**, 61-66.

Khalil, T. M., Asfour, S. S., Martinez, L. M., Waly, S. M., Rosomoff, R. S. and Rosomoff, H. L., 1992, Stretching in the rehabilitation of low-back pain patients, *Spine*, **17**, 311-317.

Khatri, B. O., Baruah, J. and McQuillen, M. P., 1984, Correlation of electromyography with computed tomography in evaluation of lower back pain, *Arch Neurol*, **41**, 594-597.

Kim, Y. E., Goel, V. K., Weinstein, J. N. and Lim, T-H., 1991, Effect of disc degeneration at one level on the adjacent level in axial mode, *Spine*, **16**, 331-335.

Kishino, N. D., Mayer, T. G., Gatchel, R. J., Parrish, M. M., Anderson, C., Gustin, L. and Mooney, V., 1985, Quantification of lumbar function: Part 4: Isometric and isokinetic lifting simulation in normal subjects and low-back dysfunction patients, *Spine*, **10**, 922-927.

Krag, M. H., Cohen, M. C., Haugh, L. D. and Pope, M. H., 1990, Body height change during upright and recumbent posture, *Spine*, **15**, 202-207.

Kramer, K. M. and Levine, A. M., 1989, Unilateral facet dislocation of the lumbosacral junction. A case report and review of the literature, *The Journal of Bone and Joint Surgery*, **71-A**,1258-1261.

Ladin, Z., Murthy, K. R. and De Luca, C. J., 1989, 1989 Volvo award in biomechanics: Mechanical recruitment of low-back muscles: Theoretical predictions and experimental validation, *Spine*, **14**, 927-938.

Lantz, S. A. and Schultz, A. B., 1986a, Lumbar spine orthosis wearing I. Restriction of gross body motions, *Spine*, **11**, 834-837.

Lantz, S. A. and Schultz, A. B., 1986b, Lumbar spine orthosis wearing II. Effect on trunk muscle myoelectric activity, *Spine*, **11**, 838-842.

Lavender, S. A., Marras, W. S., Miller, R. A., 1993, The development of response strategies in preparation for sudden loading to the torso, *Spine*, **18**, 2097-2105.

Lavender, S. A., Tsuang, Y. H., Andersson, G. B. J., Hafezi, A. and Shin, C. C., 1992, Trunk muscle cocontraction: the effects of moment direction and moment magnitude, *Journal of Orthopaedic Research*, **10**, 691-700.

Lin, H. S., Liu, Y. K. and Adams, KH., 1978, Mechanical response of the lumbar intervertebral joint under physiological (complex) loading, *The Journal of Bone and Joint Surgery*, **60-A**, 41-55.

Macintosh, J. E. and Bogduk, N., 1991, The attachments of the lumbar erector spinae, *Spine*, **16**, 783-792.

Macintosh, J. E. and Bogduk, N.,1987, 1987 Volvo award in clinical sciences: The morphology of the lumbar erector spinae, *Spine*, **12**, 658-660.

Macintosh, J. E., Bogduk, N. and Pearcy, M. J., 1993, The effects of flexion on the geometry and actions of the lumbar erector spinae, *Spine*, **18**, 884-893.

Magnusson, M., Granqvist, M., Jonson, R., Lindell, V., Lundberg, U., Wallin, L. and Hansson, T., 1990, The loads on the lumbar spine during work at an assembly line: The risks for fatigue injuries of vertebral bodies, *Spine*, **15**, 774-779.

Malmivaara, A., Videman, T., Kuosma, E. and Troup, J. D. G., 1986, Radiographic vs. direct measurements of the spinal canal of the thoracolumbar junctional region (T10-L1) of the spine, *Spine*, **11**, 574-578.

Markolf, K. L. and Morris, J. M., 1974, The structural components of the intervertebral disc: A study of their contributions to the ability of the disc to withstand compressive forces. *The Journal of Bone and Joint Surgery*, **56-A**, 675-687.

Markolf, K. L., 1972, Deformation of the thoracolumbar intervertebral joints in response to external loads, *The Journal of Bone and Joint Surgery*, **54-A**, 511-533.

Marras, W. S. and Mirka, G. A., 1992, A comprehensive evaluation of trunk response to asymmetric trunk motion, *Spine*, **17**, 318-326.

Marras, W. S. and Reilly, C. H., 1988, Networks of internal trunk-loading activities under controlled trunk-motion conditions, *Spine*, **13**, 661-667.

Marras, W. S., King, A. I. and Joynt, R. L., 1984, Measurements of loads on the lumbar spine under isometric and isokinetic conditions, *Spine*, **9**, 176-187.

Marras, W. S., Lavender, S. A., Leurgans, S. E., Rajulu, S. L., Allread, W. G., Fathallah, F. A. and Ferguson, S. A., 1993, The role of dynamic three-dimensional trunk motion in occupationally-related low back disorders: The effects of workplace factors, trunk position, and trunk motion characteristics on risk of injury, *Spine*, **18**, 617-628.

Mayer, T. G., Barnes, D., Kishino, N. D., Nichols, G., Gatchel, R. J., Mayer, H. and Mooney, V, 1988, Progressive isoinertial lifting evaluation: I. A standardized protocol and normative database, *Spine*, **13**, 993-997.

McCarron, R. F., Wimpee, M. W., Hudkins, P. G. and Laros, G. S., 1987, The inflammatory effect of nucleus pulposus: A possible element in the pathogenesis of low-back pain, *Spine*, **12**, 760.

McCombe, P. F., Fairbank, J. C., Cockersole, B. C. and Pynsent, P. B., 1989, Reproducibility of physical signs in low-back pain, *Spine*, **14**, 908-918.

McGill, S. M., 1991, Kinetic potential of the lumbar trunk musculature about three orthogonal orthopaedic axes in extreme postures, *Spine*, **16**, 809-815.

McGill, S. M., 1992, The influence of lordosis on axial trunk torque and trunk muscle myoelectric activity, *Spine*, **17**, 1187-1193.

McNally, D. S. and Adams, M, A., 1992, Internal intervertebral disc mechanics as revealed by stress profilometry, *Spine*, **17**, 66-73.

McNally, D. S., Adams, M. A. and Goodship, A. E., 1993, Can intervertebral disc prolapse be predicted by disc mechanics? *Spine*, **18**, 1525-1530.

Melzack, R. and Wall, P. D., 1965, Pain mechanisms: a new theory, *Science*, 150, 971-979.

Miller, J. A. A., Haderspeck, K. A. and Schultz, A. B., 1983, Posterior element loads in lumbar motion segments, *Spine*, **8**, 331-337.

Miller, J. A. A., Schmatz, C. and Schultz, A. B., 1988, Lumbar disc degeneration: Correlation with age, sex, and spine level in 600 autopsy specimens, *Spine*, **13**, 173-178.

Mills, G. H., Davies, G. K., Getty, C. J. M. and Conway, J., 1986, The evaluation of liquid crystal thermography in the investigation of nerve root compression due to lumbosacral lateral spinal stenosis, *Spine*, **11**, 427-432.

Mirka, G. A. and Marras, W. S., 1993, A stochastic model of trunk muscle coactivation during trunk bending, *Spine*, **18**, 1396-1409.

Mooney, V., 1987, Presidential address, International Society for the Study of the Lumbar Spine, Dallas, 1986 — Where is the pain coming from? *Spine*, **12**, 754-759.

Mooney, V., Haldeman, S., Nasca, R. J., White, A. H., Nix, J. E., Wiltse, L. L., Selby, D. K., Kostuik, J. P., Krag, M. H., Ray, C. D., Simmons, J. W., Drabing, J., Yong-Hing, K., Russell, G. S., Cauthen, J. and Saal, J. A., 1988, Position statement on discography, *Spine*, **13**,1343.

Morris, E. W., DiPaolo, M., Vallance, R. and Waddell, G., 1986, Diagnosis and decision making in lumbar disc prolapse and nerve entrapment, *Spine*, **11**, 436-439.

Mundt, D. J., Kelsey, J. L., Golden, A. L., Pastides, H., Berg, A. T., Sklar, J., Hosea, T. and Panjabi, M. M., 1993, Northeast Collaborative Group on Low Back Back Pain. An epidemiologic study of non-occupational lifting as a risk factor for herniated lumbar intervertebral disc, *Spine*, **18**, 595-608.

Nachemson, A. L., 1985, Advances in low-back pain, *Clinical Orthopaedics and Related Research*, **200**, 266-278.

Nachemson, A. L., 1989, Editorial comment lumbar discography — where are we today? *Spine*, **14**, 556-557.

Nachemson, A. L., Andersson, G. B. J. and Schultz, A. B., 1986, Valsalva maneuver biomechanics; Effects on lumbar trunk loads of elevated intrabdominal pressures, *Spine*, **11**, 476-479.

Nachemson, A. L., Troup, J. D. G., Videman, T., Bigos, S. J., Battié, M. C., Fisher, L. D., Cats-Baril, W. L., Frymoyer, J. W., Pope, M. H., Nelson, R. M., Spangfort, E. and Waddell, G., 1991, Symposium: research methods in occupational low-back pain, *Spine*, **16**, 667-686.

Ninomiya, M. and Muro, T., 1992, Pathoanatomy of lumbar disc herniation as demonstrated by computed tomography/discography, *Spine*, **17**, 1316-1322.

Nordby, E. J., 1983, Chymopapain in intradiscal therapy, *The Journal of Bone and Joint Surgery*, **65-A**, 1350-1353.

Nordin, M., R. Ortengren, Andersson, G. B. J., 1984, Measurements of trunk movements during work, *Spine*, **9-5**, 465-469.

Noren, R., Trafimow, J., Andersson, G. B. J. and Huckman, M. S., 1991, The role of facet joint tropism and facet angle in disc degeneration, *Spine*, **16**, 530-532.

Omino, K. and Hayashi, Y., 1992, Preparation of dynamic posture and occurrence of low back pain, *Ergonomics*, **35**, 693-707.

Õrtengren, R., Andersson, G. B. J. and Nachemson, A. L., 1981, Studies of relationships between lumbar disc pressure, myoelectric back muscle activity, and intra-abdominal (intragastric) pressure, *Spine*, **6**, 98-103.

Ostgaard, H. C., Andersson, G. B. J. and Karlsson, K., 1991, Prevalence of back pain in pregnancy, *Spine*, **16**, 549.

Ostgaard, H. C., Andersson, G. B. J., Schultz, A. B. and Miller, J. A. A., 1993, Influence of some biomechanical factors on low-back pain in pregnancy, *Spine*, **18**, 61-65.

Oxland, T. R., Crisco, J. J. III., Panjabi, M. M. and Yamamoto, I., 1992, The effect of injury on rotational coupling at the lumbosacral joint: A biomechanical investigation, *Spine*, **17**, 74-80.

Panagiotacopulos, N. D., Pope, M. H., Bloch, R. and Krag, M. H., 1987, Water content in human intervertebral discs: Part II. Viscoelastic behavior, *Spine*, **12**, 918-924.

Panagiotacopulos, N. D., Pope, M. H., Krag, M. H. and Bloch, R., 1987, Water content in human intervertebral discs: Part I. Measurement by magnetic resonance imaging, *Spine*, **12**, 912-917.

Panjabi, M. M. and White, A. A., 1980, Basic Biomechanics of the Spine, *Neurosurgery*, **7**, 76-93.

Panjabi, M. M., 1988, Biomechanical evaluation of spinal fixation devices: I. A conceptual framework, *Spine,* **13**, 1129-1134.

Panjabi, M. M., Abumi, K., Duranceau, J. and Crisco, J. J., 1988a, Biomechanical evaluation of spinal fixation devices: II Stability provided by eight internal fixation devices, *Spine,* **13**, 1135-1140.

Panjabi, M. M., Geol, V., Oxland, T., Takata, K., Duranceau, J., Krag, M. and Price, M., 1992b, Human lumbar vertebrae: Quantitative three-dimensional anatomy, *Spine,* **17**, 299-306.

Panjabi, M. M., Krag, M. H. and Chung, T. Q., 1984, Effects of disc injury in mechanical behavior of the human spine, *Spine,* **9**, 707-713.

Panjabi, M. M., Oxland, T., Takata, K., Goel, V., Duranceau, J. and Krag, M., 1993, Articular facets of the human spine: Quantitative three-dimensional anatomy, *Spine,* **18**, 1298-1310.

Panjabi, M. M., Takata, K. and Goel, V. K., 1983, Kinematics of lumbar intervertebral foramen, *Spine,* **8**, 348-352.

Panjabi, M. M., Brown, M., Lindahl, S., Irstam, L. and Hermens, M., 1988b, Intrinsic disc pressure as a measure of integrity of the lumbar spine, *Spine,* **13,** 913-917.

Panjabi, M. M., Chang, D. and Dvorak, J., 1992a, An analysis of errors in kinematic parameters associated with *in vivo* functional radiographs, *Spine,* **17**, 200-205.

Panjabi, M. M., Goel, V. and Takata, K., 1982, Physiologic strains in the lumbar spinal ligaments: An *in vitro* biomechanical study, *Spine,* **7**, 192-203.

Pearcy, M. and Shepherd, J., 1985, Is there instability in spondylolisthesis, *Spine,* **10**, 175-177.

Pearcy, M., Portek, I. and Shepherd, J., 1985, The effect of low-back pain on lumbar spinal movements measured by three-dimensional X-ray analysis, *Spine,* **10**, 150-153.

Pedrini-Mille, A., Weinstein, J. N., Found, E. M., Chung, C. B. and Goel, V. K., 1990, Stimulation of dorsal root ganglia and degradation of rabbit annulus fibrosus, *Spine,* **15**, 1252-1256.

Pennington, J. B., McCarron, R. F. and Laros, G. S., 1988, Identification of IgG in the canine intervertebral disc, *Spine,* **13**, 909-912.

Polatin, P. B., Gatchel, R. J., Barnes, D., Mayer, H., Arens, C. and Mayer, T. G., 1989, A psychosociomedical prediction model of response to treatment by chronically disabled workers with low-back pain, *Spine,* **14**, 956-961.

Pope, M. H. and Panjabi, M., 1985, Biomechanical definitions of spinal instability, *Spine,* **10**, 255-256.

Pope, M. H., 1990, Bioengineering-The bond between basic scientists, clinicians, and engineers: The 1989 presidential address, *Spine,* **15**, 214-217.

Pope, M. H., Bevins, T., Wilder, D. G. and Frymoyer, J. W., 1985, The relationship between anthropometric, postural, muscular, and mobility characteristics of males ages 18-55, *Spine,* **10**, 644-648.

Pope, M. H., Svenson, M., Andersson, G. B. J., Broman, H. and Zetterberg, C., 1987, The role of prerotation of the trunk in axial twisting efforts, *Spine,* **12**, 1041-1045.

Porter, R. W., Adams, M. A. and Hutton, W. C., 1989, Physical activity and the strength of the lumbar spine, *Spine,* **14**, 201-203.

Posner, I., White, A. A. III., Edwards, W. T. and Hayes, W. C., 1982, A biomechanical analysis of the clinical stability of the lumbar and lumbosacral spine, *Spine,* **7**, 374-389.

Potvin, J. R., McGill, S. M. and Norman, R. W., 1991, Trunk muscle and lumbar ligament contributions to dynamic lifts with varying degrees of trunk flexion, *Spine,* **16**, 1099-1107.

Poulsen, E., 1981, Back muscles strength and weight limits in lifting burdens, *Spine,* **6**, 73-75.

Ratcliffe, J. F., 1980, The arterial anatomy of the adult human lumbar vertebral body: a microarteriographic study. *J. Anat.,* 131, 57-59.

Roland, M. and Morris, R., 1983a, A study of the natural history of low-back pain: part II: development of guidelines for trials of treatment in primary care, *Spine,* **8:2,** 145-150.

Roland, M. and Morris, R., 1983b, A study of the natural history of low-back pain: part I: development of a reliable and sensitive measure of disability in low-back pain, *Spine,* **8:2**, 141-144.

Rowe, M, L., 1971, Low back disability in industry: update position, *JOM,* 13, 476-478.

Saal, J., and Saal, J., 1989, Nonoperative treatment of herniated lumbar intervertebral disc with radiculopathy: an outcome study, *Spine,* **14**, 431-437.

Sachs, B. L., Vanharanta, H., Spivey, M. A., Guyer, R. D., Videman, T., Rashbaum, R. F., Johnson, R. G., Hochshuler, S. H. and Mooney V., 1987, Dallas discogram description: A new classification of CT/Discography in low-back disorders, *Spine*, **12**, 287-294.

Sairanen, E., Brushaber, L. and Kaskinen, M., 1981, Felling work, low-back pain and osteoarthritis, *Scand J Work Environ Health*, **7**, 18-30.

Sampson, H. W. and Davis, J. S., 1988, Histopathology of the intervertebral disc of progressive ankylosis mice, *Spine*, **13**, 650-654.

Sandover, J., 1988, Behaviour of the spine under shock and vibration: a review, *Clin Biomech*, **3**, 249-256.

Scavone, J. G., Latshaw, R. F. and Rohrer, G. V., 1981, Use of lumbar spine films, Statistical evaluation at a university teaching hospital, *JAMA*, **246**, 1105-1108.

Schonstrom, N. and Hansson, T., 1988, Pressure changes following constriction of the cauda equina: An experimental study in situ, *Spine*, **13**, 385-388.

Schonstrom, N. S. R., Bolender, N-F. and Spengler, D. M., 1985, The pathomorphology of spinal stenosis as seen on CT scans of the lumbar spine, *Spine*, **10**, 806-811.

Schultz, A. B. and Andersson, G. B. J., 1981, Analysis of loads on the lumbar spine, *Spine*, **6**, 76-82.

Schultz, A. B., 1986, Loads on the human lumbar spine, *Mechanical Engineering*, 36-41.

Schultz, A. B., Sorensen, S-E. and Andersson, G. B. J., 1984, Measurements of spine morphology in children, ages 10-16, *Spine*, **9**, 70-73.

Schultz, A. B., Andersson, G. B. J., Ortengren, R., Bjork, R. and Nordin, M., 1982a, Analysis and quantitative myoelectric measurements of loads on the lumbar spine when holding weights in standing postures, *Spine*, **7**, 390-397.

Schultz, A. B., Andersson, G., Ortengren, R., Haderspeck, K. and Nachemson, A., 1982b, Loads on the lumbar spine — Validation of a biomechanical analysis by measurements of intradiscal pressures and myoelectric signals, *The Journal of Bone and Joint Surgery*, **64-A**, 713-720.

Schultz, A. B., Haderspeck, K. and Takashima, S., 1981, Correction of scoliosis by muscle stimulation: biomechanical analyses, *Spine*, **6**, 468-476.

Semon, R. L. and Spengler, D., 1981, Significance of lumbar spondylolysis in college football players, *Spine*, **6**, 172.

Shirazi-Adl, S. A. and Drouin, G., 1987, Load-bearing role of facets in a lumbar segment under sagittal plane loadings, *Journal of Biomechanics*, **20**, 601-613.

Shirazi-Adl, S. A., Shrivastava, S. C. and Ahmed, A. M., 1984, 1983 Volvo award in biomechanics: Stress analysis of the lumbar disc-body unit in compression: A three-dimensional nonlinear finite element study, *Spine*, **9**, 120-134.

Skovron, M. L., 1992, Epidemiology of low back pain, *Baillierss Clinical Rheumatology*, **6**, 559-573.

Snook, S. H. and Ciriello, V. M., 1991, The design of manual handling tasks: revised tables of maximum acceptable weights and forces, *Ergonomics*, **34**, 1197-1213.

Spencer, D. L., Irwin, G. S. and Miller, J. A. A., 1983, Anatomy and significance of fixation of the lumbosacral nerve roots in sciatica, *Spine*, **8**, 672-679.

Spencer, D. L., Miller, J. A. A. and Bertolini, J. E., 1984, The effect of intervertebral disc space narrowing on the contact force between the nerve root and a simulated disc protrusion, *Spine*, **9**, 422-426.

Spencer, D. L., Miller, J. A. A., and Schultz, A. B., 1985, The effects of chemonucleolysis on the mechanical properties of the canine lumbar disc, *Spine*, **10**, 555-561.

Spengler, D. M. and Freeman, C. W., 1979, Patient selection for lumbar discectomy: An objective approach *Spine*, **4**, 129-134.

Spengler, D. M., 1982a, Lumbar discectomy: Results with limited disc excision and selective foraminotomy, *Spine*, **7**, 604-607.

Spengler, D. M., 1982b, *Low Back Pain, Assessment and Management*, Grune & Stratton, New York, N.Y.

Spengler, D. M., 1983, The clinical spectrum of lumbar spinal stenosis, *Orthopaedic Surgery*, **2**, 2-6.

Spengler, D. M., Bigos, S. J., Martin, N. A., Zeh, J., Fisher, L. and Nachemson, A., 1986a, Back injuries in industry: a retrospective study I. Overview and cost analysis, *Spine*, **11**, 241-245.

Spengler, D. M., Bigos, S. J., Martin, N. A., Zeh, J., Fisher, L. and Nachemson, A., 1986c, Back injuries in industry: a retrospective study III. Employee-related factors, *Spine*, **11**, 252-256.

Spengler, D. M., Bigos, S. J., Martin, N. A., Zeh, J., Fisher, L., Nachemson, A. and Wang, M. H., 1986b, Back injuries in industry: a retrospective study II. Injury factors, *Spine*, **11**, 246-251.

Spitzer, W. O., Leblanc, F. E., Dupuis, M., Abenhaim, L., Belanger, A. Y., Bloch, R., Bombarider, C., Cruess, R. L., Drouin, G., Duval-Hesler, N., Laflamme, J., Lamoureux, G., Nachemson, A., Page, J. J., Rossignol, M., Salmi, L. R., Salois-Arsenault, S., Suissa, S. and Wood-Dauphinnee, S., 1987, Scientific approach to the assessment and management of activity-related spinal disorders: A monograph for clinicians: Report of the Quebec task force on spinal disorders, *Spine*, **12**, 1S.

Stubbs, D. A., 1981, Trunk stresses in construction and other industrial workers, *Spine*, **6**, 83-89.

Svensson, H. O., Andersson, G. B. J., Hagstad, A. and Jansson, P-O., 1990, The relationship of low-back pain to pregnancy and gynecologic factors, *Spine*, **15**, 371-375.

Svensson, H. O. and Andersson, G. B. J., 1983, Low-back pain in 40 to 47 year old men: Work history and work environment factors, *Spine*, **8**, 272-276.

Tan, J. C., Parnianpour, M., Nordin, M., Hofer, H. and Willems, B., 1993, Isometric maximal and submaximal trunk extension and different flexed position in standing: Triaxial torque output and EMG, *Spine*, **18**, 2480-2490.

Tanaka, M., Nakahara, S. and Inoue, H., 1993, A pathologic study of discs in the elderly. Separation between the cartilaginous endplate and the vertebral body, *Spine*, **18**, 1456-1462.

Thompson, J. P., Pearce, R. H., Schechter, M. T., Adams, M. E., Tsang, I. K. Y. and Bishop, P. B., 1990, Preliminary evaluation of a scheme for grading the gross morphology of the human intervertebral disc, *Spine*, **15**, 411-415.

Tonzola, R. F. and Ackil, A. A., 1981, Usefulness of electrophysiological studies in the diagnosis of lumbosacral root disease, *Annals of Neurology*, **9**, 305-308.

Trafimow, J. H., Schipplein, O. D., Novak, G. J. and Andersson, G. B., 1992, The effects of quadriceps fatigue on the technique of lifting, *Spine*, **18**, 364-367.

Triano, J. J. and Schultz, A. B., 1987, Correlation of objective measure of trunk motion and muscle function and low-back disability ratings, *Spine*, **12**, 561-565.

Troup, J. D. G., 1981, Briefly noted: Straight-leg raising (SLR) and the qualifying tests for increased root tension: Their predictive value after back and sciatic pain, *Spine*, **6**, 526-527.

Troup, J. D. G., 1991a, Briefly noted: Definitions of occupational low-back pain in Great Britain, *Spine*, **16**, 667-668.

Troup, J. D. G., 1991b, Measurement of strength and endurance: The psychophysical lift test, *Spine*, **16**, 679.

Troup, J. D. G., Foreman, T. K., Baxter, C. E. and Brown, D., 1987, 1987 Volvo award in clinical sciences: the perception of back pain and the role of psychophysical tests of lifting capacity, *Spine*, **12**, 645-657.

Troup, J. D. G., Martin, J. W. and Lloyd, D. C. E. F., 1981, Back pain in industry: A prospective survey, *Spine*, **6**, 61-69.

Tyrrell, A. R., Reilly, T. and Troup, J. D. G., 1985, Circadian variation in stature and the effects of spinal loading, *Spine*, **10**, 161-164.

Van Schaik, J. P. J., Verbiest, H. and Van Schaik, F. D. J., 1985, The orientation of laminae and facet joints in the lower lumbar spine, *Spine*, **10**, 59-63.

Vanharanta, H., Guyer, R. D., Ohnmeiss, D. D., Stith, W. J., Sachs, B. L., Aprill, C., Spivey, M., Rashbaum, R. F., Hochschuler, S. H., Videman, T., Selby, D. K., Terry, A. and Mooney, V., 1988, Disc deterioration in low-back syndromes: A prospective, multi-center CT/discography study, *Spine*, **13**, 1349-1361.

Vanharanta, H., Korpi, J., Heliovaara, M. and Troup, J. D. G., 1985, Radiographic measurements of lumbar spinal cord size and their relation to back mobility, *Spine*, **10**, 461-466.

Vanharanta, H., Sachs, B. L., Ohnmeiss, D. D., Aprill, C., Spivey, M., Guyer, R. D., Rashbaum, R. F., Hochschuler, S. H., Terry, A., Selby, D., Stith, W. J. and Mooney, V., 1989, Pain provocation and disc deterioration by age: A CT/Discography study in a low-back pain population, *Spine*, **14**, 420-423.

Vanharanta, H., Sachs, B. L., Spivey, M. A., Guyer, R. D., Hochschuler, S. H., Rashbaum, R. F., Johnson, R. G., Ohnmeiss, D. and Mooney, V., 1987, The relationship of pain provocation to lumbar disc deterioration as seen by CT/discography, *Spine*, **12**, 295-298.

Vanharanta, H., Sachs, B. L., Spivey, M., Hochschuler, S. H., Guyer, R. D., Rashbaum, R. F., Ohnmeiss, D. D. and Mooney, V., 1988b, A comparison of CT/discography, pain response and radiographic disc height, *Spine*, **13**, 321-324.

Videman, T., Nurimen, M. and Troup, J. D. G., 1990, 1990 Volvo award in clinical sciences: Lumbar spinal pathology in cadaveric material in relation to history of back pain, occupation, and physical loading, *Spine*, **15**, 728-740.

Videman, T., Nuriminen, T., Tola, S., Kuorinka, I., Vanharanta, H. and Troup, J. D. G., 1984, Low-back pain in nurses and some loading factors of work, *Spine*, **9**, 400-464.

Waddell, G. and Main, C. J., 1984, Assessment of severity in low-back disorders, *Spine*, **9**, 204-208.

Waddell, G., Main, C. J., Morris, E. W., Di Paola, M. and Gray, I. C. M., 1984, Chronic low-back pain, psychologic distress, and illness behavior, *Spine*, **9**, 209-213.

Waddell, G., McCulloch, J. A., Kummel, E. and Venner, R. M., 1980, Nonorganic physical signs in low-back pain, *Spine*, **5** (2), 117-125.

Wall, P. D. and Melzack, R., 1994, *Text Book of Pain*, Churchill Livingston, Edinburgh, U.K.

Walsh, T. R., Weinstein, J. N., Spratt, K. F., Lehmann, T. R., Aprill, C. and Sayre, H., 1990, Lumbar discography in normal subjects, *The Journal of Bone and Joint Surgery*, **72-A**, 1081-1088.

Webster, B. and Snook, S., 1990, The cost of compensable low back pain, *JOM*, **32** (1), 13-15.

Wehling, P., Pak, M. A., Cleveland, S. J. and Schulitz, K. P., 1989, The influence on spinal cord evoked potentials of chymopapain applied to the rat lumbar spine canal, *Spine*, **14**, 65-67.

Weinstein, J. N., 1988, The perception of pain, *Managing Low Back Pain*, New York: Livingston Churchill, 83-90.

Weinstein, J. N., Claverie, W. and Gibson, S., 1988a, The pain of discography, *Spine*, **13**, 1344-1348.

Weinstein, J. N., Pope, M., Schmidt, R. and Seroussi, R., 1988b, Neuropharmacologic effects of vibration on the dorsal root ganglion: An animal model, *Spine*, **13**, 521-525.

White, A. A. III. and Gordon, S. L., 1982, Synopsis: Workshop on idiopathic low-back pain, *Spine*, **7**, 141-147.

White, A. H., Von Rogov, P., Zucherman, J. and Heiden, D., 1987, Lumbar laminectomy for herniated disc: A prospective controlled comparison with internal fixation fusion, *Spine*, **12**, 305-307.

Wiesel, S. W., Feffer, H. L. and Rothman, R. H., 1984, Industrial low-back pain: A prospective evaluation of a standardized diagnostic and treatment protocol, *Spine*, **9**, 199-203.

Wilder, D. G., Woodworth, B. B., Frymoyer, J. W. and Pope, M. H., 1982, Vibration and the human spine, *Spine*, **7**, 243-254.

Williams, M. M., Hawley, J. A., McKenzie, R. A. and Van Wijmen, P. M., 1991, A comparison of the effects of two sitting postures on back and referred pain, *Spine*, **16**, 1185-1191.

Wiltse, L. L., 1977, Surgery for intervertebral disk disease of the lumbar spine, *Clinical Orthopaedics and Related Research*, **129**, 22-45.

Yamamoto, I., Panjabi, M. M., Oxland, T. R. and Crisco, J. J., 1990, The role of the iliolumbar ligament in the lumbosacral junction, *Spine*, **15**, 1138-1141.

Yang, K. H. and King, A. I., 1984a, Mechanism of facet load transmission as a hypothesis for low-back pain, *Spine*, **9** (6), 557-565.

Yoshizawa, H., O'Brien, J. P., Smith, W. T. and Trumper, M., 1980, The neuropathology of intervertebral discs removed for low-back pain, *J. Pathology*, 132, 95-104.

Zohn, D. A., 1988, *Musculoskeletal Pain: Diagnosis and Physical Treatment*, Boston: Little Brown, 185-247.

Part IV
Administrative Controls

Section I
Ergonomics Surveillance

64

Fundamentals of Surveillance for Work-Related Musculoskeletal Disorders

Vern Putz-Anderson
U.S. Department of Health and Human Services

Katharyn A. Grant
U.S. Department of Health and Human Services

64.1 Introduction

Scope

This chapter will address occupational **surveillance*** as it relates to the practice of industrial ergonomics and the prevention of work-related musculoskeletal disorders. Surveillance is a "continuous analysis, interpretation, and feedback of systematically collected data, generally using methods distinguished by their practicality, uniformity, and frequently their rapidity, rather than by their accuracy or completeness" (Last, 1995). Occupational surveillance provides the data needed to identify, control, and prevent work-related injuries and illnesses. Epidemiologists also use surveillance data to study "the distribution and determinants of health-related states or events in defined populations." Methods for conducting occupational surveillance programs are well documented, and reviews of surveillance concepts and methods have been presented elsewhere (Baker et al., 1989; Baker and Matte, 1994; Greife et al., 1995). This chapter will limit the discussion of surveillance to those issues that affect the work of the industrial ergonomist and the prevention of musculoskeletal disorders.

*This chapter contains a glossary of the terms in bold type.

0-8493-2641-9/99/$0.00+$.50

Purpose

The purpose of occupational surveillance is to track patterns of health and disease in groups of workers and to identify risk factors that influence these trends. Ultimately, this information should be used to direct the implementation of measures to prevent and/or control work-related disorders. Surveillance can answer questions such as, "Is there a problem?" or "Is there *still* a problem?" Establishing surveillance procedures is generally the first step in establishing an industrial ergonomics program. Surveillance serves to stimulate and focus prevention efforts and provides a method to assess the impact of corrective action.

64.2 Collecting, Analyzing, Intervening

Effective surveillance systems include the following components: (1) data collection, (2) analysis of the data, i.e., a mechanism to evaluate the meaning of the health/injury or hazard data, and (3) some action or response to ensure that surveillance activities are translated into preventive action. The response action may be directed toward individuals (e.g., providing medical treatment for symptomatic workers), or toward groups of workers (e.g., eliminating hazardous workplace conditions).

Several factors govern the strategy for collecting, analyzing, and acting upon surveillance data. These factors include the scope and urgency of the perceived problem, the resources available to the investigators, and the types of information systems already set in place. Many surveillance programs begin with efforts to document the number of work-related musculoskeletal disorders that have occurred in recent history (i.e., establish a baseline). Data for this effort can be obtained from a review of **existing records.** If existing records are incomplete or are unreliable as sources of surveillance information, or if additional information about current (rather than historical) conditions is needed, **questionnaires** or **worker surveys** may be used to determine how many workers are experiencing symptoms that could be caused by work activities. Additionally, because prevention of work-related musculoskeletal disorders depends on the identification and elimination of hazardous working conditions, **hazard surveys** or simple checklists can be used to identify jobs where risk factors for musculoskeletal disorders may be present.

For example, during an annual review of the OSHA 200 logs, an occupational nurse notices there has been a sudden increase in the number of workers reporting severe back strains on a new job. The surveillance plan would likely call for an immediate inspection of the worksite, along with interviews of workers and the supervisors to assist in identifying hazards. Once hazard control techniques had been identified and implemented, worker and hazard surveys would be administered at periodic intervals to determine the long-term effectiveness of the intervention.

Additional information about each of these data sources (existing records, worker surveys, and hazard surveys) will be provided later in this chapter.

Establishing Definitions and Criteria

Before data can be collected and analyzed, it must be defined. An obvious (and unfortunately common) problem in the surveillance of work-related musculoskeletal disorders is the lack of consistency or standardization in the way these disorders are defined. The resulting confusion can lead to inaccuracies or inconsistencies in the data that make comparisons between locations and over time difficult if not impossible.

Defining what is meant by terms like "musculoskeletal disorder" or "ergonomic hazard" can be especially difficult. "Musculoskeletal disorders" encompass a broad spectrum of illnesses; in a recent review, NIOSH (1993) listed more than 160 different musculoskeletal conditions that can be caused or aggravated by various work activities. Diagnosis of these conditions is often complicated since many of these disorders have nonspecific symptom patterns, long latency periods and complex etiologies (Schierhout and Myers, 1996). For surveillance purposes, work-related musculoskeletal disorders have been defined from self-reported symptoms, clinical signs, specialized medical tests, impairment, or disability.

TABLE 64.1 Examples of Case Definitions Used in Studies of Work-Related Musculoskeletal Disorders

Disorder	Defined as:	Determined by:	Applied to:	Reference
Carpal tunnel syndrome (CTS)	Pain or numbness in the hands or wrists, pain in the hand or wrists that awakens at night, or tingling in the hands and fingers.	Questionnaire	Supermarket cashiers	Margolis and Kraus, 1987
Cumulative trauma disorders (CTDs)	Inflammation or irritation of joints, tendons, or muscles (excluding strains, sprains, or dislocations), resulting from overexertion or nonimpact repetitive motion, occurring over a protracted or unknown period of time.	Review of workers' compensation records	Industrial workers in Ohio	Tanaka et al., 1988
Work-related carpal tunnel syndrome	Combination of 1. symptoms affecting the median nerve distribution of the hand (e.g., pain, paresthesia, numbness), 2. objective physical or electro-diagnostic findings suggestive of CTS, and 3. history of employment in a job involving frequent or repetitive hand movements, forceful hand exertions, awkward hand postures, use of vibrating tools, or prolonged pressure over the wrist or palm of the hand.	Clinical history and physical examination	Patients referred for neurophysio-logic testing	Katz et al., 1991
Shoulder tendinitis	Pronounced palpable pain of the muscle attachment or pronounced pain reaction to isometric contraction in any of the rotator cuff or biceps muscles.	Clinical examination	Construction workers	Stenlund et al., 1993
Work-related musculoskeletal disorders	Symptoms of pain, aching, stiffness, numbness, tingling, or burning: 1. Not due to accident or sudden trauma; 2. Developed since working in current job; 3. Occurring within past year; 4. Lasting more than one week, or occurring at least once a month; 5. Reported as moderate or worse on a 5-point intensity scale.	Questionnaire	Newspaper workers	Bernard et al., 1994

Examples of **case definitions** used in various studies are found in Table 64.1. Two important considerations in the selection of a case definition are its **sensitivity** and **specificity**. The sensitivity of a case definition is the likelihood that the definition will identify diseased individuals. The specificity of a case definition is the probability that it will not label healthy individuals as diseased. In general, case definitions based solely on symptoms (e.g., pain, tingling in a particular body part) tend to be more sensitive but less specific than definitions that rely on physical findings or diagnostic tests; however, for surveillance purposes, it may be preferable to use definitions that are more inclusive or sensitive (Katz et al., 1991). In the face of uncertainty, the use of sensitive definitions will encourage investigators to examine suspect jobs and begin interventions at the earliest indication of a problem.

Implementing Standardized Reporting Procedures

Providing information in an accessible and usable format is critical to the success of a surveillance effort. Unfortunately, the data needed for surveillance are usually not centralized in a single location. Performing a comprehensive analysis of costs associated with absenteeism, turnover, injury morbidity, compensation, lost productivity, and poor quality can require a search of numerous record systems, many of which employ their own coding procedures. Although relational database management systems can greatly simplify this process, their success depends on the establishment of common linkages between record systems. One approach employed by some plant medical departments and occupational health clinics is to use the coding system provided in the International Classification of Diseases (ICD). Such a coding system should lead to greater consistency in collecting, analyzing, and reporting health data.

Calculating Rates and Percentages

Surveillance data are often reported as rates or percentages. Expressing the occurrence of injuries, illnesses, symptoms, complaints, etc. as a rate or percentage allows investigators to compare the occurrence across jobs/departments/plants that employ different numbers of workers and across periods of fluctuating employment. When the surveillance goal is to identify high-risk jobs that need attention, rates are most useful when they are computed for individual departments or even on similar jobs within a department.

The **incidence rate** is the rate of work-related musculoskeletal disorders appearing for the first time during a specific period (usually a year). The value commonly used in occupational health is the number of illnesses per 100 full-time workers per year and is calculated as follows:

$$\frac{\text{\# new cases during the past 12 months} \times 200{,}000 \text{ hours}}{\text{\# work hours during the past 12 months}}$$

The information needed to compute the denominators for the rates can usually be obtained from personnel or payroll records. The assumption is that each employee works 2000 hours per year (8 hours a day, 5 days a week, 50 weeks a year). If the number of hours worked in the past 12 months is not known, it can be estimated by multiplying the number of full-time equivalent workers employed in the job, department, or plant by 2000 hours.

The **prevalence rate** is the percentage of all cases of disease/symptoms/complaints at a specific instance in time, regardless of when they first appeared. It is calculated as follows:

$$\frac{\text{total \# cases at a given point in time}}{\text{\# workers at the same point in time}}$$

The **severity index** uses the number of lost or restricted workdays due to illness as a surrogate for the seriousness of the disorder. The severity index is calculated as follows:

$$\frac{\text{total \# lost or restricted workdays in the past 12 months} \times 200{,}000 \text{ hours}}{\text{\# work hours during the past 12 months}}$$

The magnitude of the severity index can be influenced by medical treatment practices, the health benefits available to employees, and the opportunity for transfer to jobs that are less stressful. The severity index can be skewed by unusually long illnesses suffered by a few employees.

Interpreting Incidence, Prevalence, and Severity

Although closely related, incidence and prevalence rates convey somewhat different information. Prevalence depends on the incidence and the duration of the disease from its onset to its resolution. For example, if the incidence of back pain is low, but recovery is slow, the prevalence will be high relative to the incidence. Conversely, if the incidence of musculoskeletal disorders among workers is high, the prevalence may be low relative to the incidence if workers recover quickly, or if they leave the workforce because of their condition (Hennekens and Buring, 1987).

From a surveillance perspective, incidence and prevalence rates serve as valuable *prevention tools* for guiding interventions to mitigate hazardous workplace conditions. Incidence and prevalence rates, however, are also being used as *management tools* to gauge the performance of supervisors and health and safety staff who are held responsible for workplace injuries and illnesses. Such measures, when used as an indicator of performance or compliance, can inadvertently penalize employers or safety and health personnel who have recently introduced an ergonomics program. Typically, the incidence and prevalence rates will initially increase in organizations in which ergonomics programs are implemented. This is often due to an increase in training-related awareness by the employees of the work-relatedness of their

conditions and represents better reporting rather than a true increase in the disorders. Moreover, the goal of an ergonomics program is to encourage employees to report problems early to allow immediate treatment and intervention, and thus reduce lost time and the risk of permanent injury. This goal may be undermined when incidence and prevalence rates are used as the sole measure to evaluate the effectiveness of an ergonomics program. For this purpose, the severity index may provide a more appropriate yardstick, since it reflects failures in early detection and prevention that represent real costs to the company. These failures are evident in the records of workers whose injuries have gone undetected, unrecognized, or unreported until the injuries reach a level of severity where restricted duty or lost time result.

Establishing Priorities for Intervention

Ultimately, surveillance should direct the allocation of resources toward groups at highest risk for musculoskeletal injury. A common question that arises in discussions of surveillance is "When does surveillance data become compelling?" i.e., when is there enough information to warrant intervention? Some suggest that even a single reported musculoskeletal disorder should be sufficient cause to trigger more focused evaluations of workplace conditions (ANSI, 1996). This recommendation is based on the recognition that formal surveillance activities usually detect only a small proportion of the musculoskeletal problems and that one reported case may lead to several times as many unreported cases.

If multiple problem areas are identified and resources are constrained, it is often necessary to rank order jobs for further analysis and intervention. Jobs can be ranked by incidence, prevalence, or hazard severity; many survey instruments are designed to provide guidance for directing intervention efforts. For example, a checklist developed for the automobile industry uses a series of questions and a three-level scale to rate the postural stress associated with a specific job (Keyserling et al., 1992; 1993). Jobs that receive one or more "stars" (indicating significant exposure to postural stress) are considered to have priority for additional investigation. Giving precedence to jobs that employ many people, or jobs where major changes are already planned, can also be a sensible and cost-effective approach.

64.3 Data Sources

Data sources for conducting occupational surveillance can be conveniently grouped into three classifications: (1) existing records, (2) worker surveys, and (3) hazard surveys.

Existing Records

Record-based surveillance involves reviewing and analyzing existing records or data systems for evidence of work-related musculoskeletal disorders. Because of its availability, these data can provide an initial gauge of the status of workers' health without a substantial investment of time or labor. Potential information sources include Occupational Safety and Health Administration (OSHA) 200 logs, on- and off-site medical records, workers' compensation records, and insurance claim data. Other records that can provide helpful information include absentee records, job transfer applications, employee grievances, or job satisfaction surveys. Although this information has been described as "passive surveillance data" (Fine et al., 1984), this description is not meant to minimize the importance of reviewing available records. Record reviewing is far from a passive endeavor, requiring great diligence in interpreting and coding the information in a consistent manner. Rather, the term serves to differentiate record reviewing from the "active" process of generating data of interest using targeted surveys.

OSHA Records. OSHA requires most employers to maintain records of work-related injuries and illnesses. Since 1978, the standard form for keeping these records has been the *Log and Summary of Occupational Injuries and Illnesses* (OSHA No. 200). In February 1996, OSHA proposed changes in record keeping requirements, including replacement of the OSHA 200 form with a new recording form, known as OSHA No. 300. These changes had not been adopted at the time of this writing. Whether or not new

requirements are enacted, OSHA-required records should be maintained at each facility or establishment where work activities are performed.

Under OSHA record keeping guidelines, musculoskeletal disorders must be recorded if (1) they were caused or exacerbated by work activities, and (2a) there is at least one physical finding (e.g., positive Tinel's or Phalen's test, swelling, loss of motion), or (2b) there is at least one subjective symptom (e.g., pain, numbness, burning) that resulted in medical treatment, lost or restricted workdays, or transfer/rotation to another job (OSHA, 1990). Determination of these conditions may be made by a physician, nurse, or other health care provider. Musculoskeletal conditions are generally recorded on the existing OSHA 200 form as an occupational illness under column "7f" ("disorders associated with repeated trauma"). These disorders are caused, aggravated, or precipitated by repeated motion, vibration, or pressure.

A review of OSHA logs will generally yield a count of the number of musculoskeletal disorders recorded within a given time frame. Although almost all worksites keep an OSHA log, its utility for musculoskeletal disorder surveillance can be limited. The form does not require a detailed description of the worker's job or the disorder, making it difficult to determine the exact cause and nature of the injury. Also, because most musculoskeletal problems tend to develop over time (i.e., do not result from a specific event), workers or medical personnel may not realize that these disorders are work-related, and hence, recordable.

The OSHA logs do provide a convenient basis for making internal comparisons within a company. The rates of injuries and illnesses can be compared over time and at different sites to assist in determining trends in injuries and illnesses. An industry can also use the data to compare itself against a national benchmark, such as the experience of other companies in the same Standard Industrial Classification (SIC).

For example, Company XYZ, a nursing home provider, employed 250 workers in 1991. This same year, workers reported 20 injuries and illnesses that resulted in a total of 100 lost workdays. Using payroll records, the company determined that employees had worked 450,000 hours during this period. They calculated an injury severity rate of 42.1 lost workdays per 100 full-time employees. Although the severity rate is high, this rate is less than the national severity rate of 61.2 for companies in SIC 805, individual and family services.

To find the national rate for the company's SIC code, refer to the Department of Labor's *Occupational Injuries and Illnesses in the United States by Industry,* available from the U.S. Government Printing Office each spring.

On-and Off-site Medical Records. Some companies maintain an onsite health clinic; others contract with external medical providers to provide care to workers injured at the job site. In either case, reports from these services, whether they are first-aid reports, dispensary logs, or employee medical records may provide useful information about potential work-related musculoskeletal disorders. At a minimum, these records may supplement information contained in the OSHA log.

For example, a review of the dispensary records at a poultry processing plant revealed no OSHA recordable injuries among workers on the trimming line in the previous year. However, the first aid medical reports showed that several employees who worked on that line on a daily basis requested anti-inflammatory medications and ice packs.

The utility of medical records for surveillance purposes can vary — they may or may not describe the reason for the visit or the condition underlying the prescribed treatment. Also, unless the information is routinely summarized, reviewing many medical records can be highly inefficient. Because medical records can contain sensitive information about employees, their contents must be treated confidentially. Routine access should be limited to health care personnel and public health agencies. Others should not have access without consent of the affected individuals.

Workers' Compensation Records. In many states, companies are required to obtain workers' compensation insurance to cover the medical and indemnity costs of employees who sustain injuries arising out of, and in the course and scope of their work. Where they exist, workers' compensation records can provide valuable information about the direct costs (medical and indemnity) associated with work-related musculoskeletal disorders. These costs include (1) payments made to outside hospitals, clinics, physicians, and other licensed medical personnel for the diagnosis, treatment (including surgery), and rehabilitation

of the injury, (2) payments made to the injured worker as compensation for lost work time, and (3) payments made as a lump sum settlement for permanent disability. The actual cost of musculoskeletal disorders is often higher than those covered by workers' compensation insurance. For example, the cost of medical treatments rendered directly by the employer or charged to the employee's health plan is not included. As a result, the total financial burden of these disorders to a company's balance sheet often goes unrecognized and unaddressed.

Workers' compensation claims are filed under specific rules and regulations that vary from state to state. Each workers' compensation law specifies what, how, and when work-related injuries and illnesses must be reported. To be covered, the worker must first choose to report the injury. Once the worker files a claim, there are issues of eligibility that govern whether the worker will receive benefits. Surveillance data based on the number of claims filed will differ from those based on the number of claims paid. As a result, workers' compensation data will often underestimate the true rates of work-related injuries and illnesses (Schwartz, 1987).

Misclassification and coding errors are also serious problems in using workers' compensation data. In an examination of workers' compensation data in Washington State, Franklin et al. (1991) found that only 72% of claims for carpal tunnel surgery were given an ICD code compatible with carpal tunnel syndrome.

Despite these limitations, workers' compensation data provide some significant advantages as a source of surveillance information. First, all records in the data set relate to conditions of suspected occupational origin. Also, workers' compensation records usually describe the circumstances of the disorder in a way that provides an understanding of the cause of the condition (Baker and Matte, 1994). If case identification leads to improvement of workplace conditions, the cost of compensation should be greatly reduced in future years.

Payroll Records. Payroll records are useful from two standpoints. First, payroll records can be used to determine the number of hours worked by employees on a particular job. This value is often used as a crude measure of employees' exposures to job hazards and is required for the incidence/prevalence/severity rate calculations described previously in this chapter. Second, payroll records can be used to identify jobs or departments where absenteeism or turnover is high. Although high turnover can result from several causes, physical stress is a common reason for leaving a job. If less stressful jobs are available, workers may choose to quit or bid out of their present, more stressful jobs. In a study by Lavender and Marras (1994), high rates of job turnover were identified as a useful indicator of jobs that posed a risk of overexertion injury leading to low back disorders. Although turnover can be affected by factors beyond the physical hazards posed by the job (e.g., the availability of higher paying positions, psychosocial factors), the results suggest that the sensitivity of surveillance programs based on existing records can be improved by using turnover rates to supplement injury rate data.

Summary. The quality and utility of existing record systems for surveillance purposes can vary. Musculoskeletal disorders often go unreported, and depending on the data source, striking differences in the incidence of work-related musculoskeletal disorders can be found (CDC, 1989). Even when records are complete, a lag will often exist between the appearance of a hazard and the onset of injury; therefore, records may not give an accurate picture of the current situation. Finally, linking injury/disease data with exposure to a hazard can be especially challenging. Job titles are often poor indicators of exposure to risk factors for musculoskeletal disorders. Additional data needed to link disorders to specific tasks or job processes may not exist.

Worker Surveys

Most work-related musculoskeletal disorders produce symptoms of pain or discomfort. Likewise, workers can provide valuable information about job attributes that cause fatigue or pain. Therefore, one of the most direct and effective methods for collecting surveillance data is to administer questionnaires or other surveys to workers. These techniques are sometimes called "active surveillance" measures because the investigator plays an active role in soliciting and collecting information specifically for surveillance

purposes. (In traditional public health surveillance, e.g., for infectious diseases such as tuberculosis, "active surveillance" means going out to hospitals and clinics to review patient records to detect cases. To avoid confusion, we do not use the term "active surveillance" here.)

Worker surveys can take several forms. They can be lengthy or quite short; they can be oral (i.e., administered by an interviewer) or written; and they may rely heavily on pictures or charts. Examples of worker surveys include the "body part discomfort scale" (Corlett and Bishop, 1976), the *Nordic Musculoskeletal Questionnaire* (Kuorinka et al., 1987), the *NIOSH Health Hazard Upper Extremity Questionnaire* (Hales et al., 1992), and the computerized "discomfort assessment system" (Saldana et al., 1994). Common features of surveys include the following:

- Use of body part diagrams, where workers can indicate the location of pain or other symptoms;
- Questions about the onset and duration of the symptoms;
- Questions about the nature of job activities;
- Use of numerical rating scales to indicate the severity of pain, fatigue, or discomfort.

Worker surveys are sometimes underrated as useful sources of surveillance data. A main reason stems from concerns that workers either may under- or over-report their symptoms. Inaccurate reporting may be more of an indication of a poorly worded survey, a lack of understanding, fear of job loss or recrimination, or simply a stoic attitude about discomfort. In situations in which worker surveys yield evidence of over-reporting of an occupational problem, investigators should not dismiss the findings until they are certain they understand the reason for their findings. In such cases organizational problems may be intertwined with safety and health problems (Schierhout and Myers, 1996).

To avoid concerns about reporting biases in worker surveys, some investigators have combined questionnaires with physical examinations of workers to identify musculoskeletal conditions (Bernard et al., 1994; Hales et al., 1994). These studies show that while symptoms of discomfort or pain may not always reflect an underlying pathology, the vast majority of employees with moderate to severe musculoskeletal symptoms have at least one positive physical finding on a concurrent physical examination (Baron et al., 1992).

Worker surveys should be conducted (1) when there is evidence from any data source of increased musculoskeletal injury or illness in the facility, (2) when new jobs or tasks are begun, or (3) when employees change jobs. The latter two surveys provide useful baseline data for determining the impact of the change on employee health. For example, one method to evaluate the effectiveness of a redesigned workstation would be to compare the pattern of shoulder, neck, and back discomfort recorded before the intervention with the pattern of discomfort for those same body segments after the improvements have been installed.

The advantages of using worker surveys as a source of information include the following: (1) The investigator can exert more control over the data collected. Once the investigator has decided what questions need to be answered, he or she can select the survey items that will elicit the most complete information that is needed. Because the investigator has control over the survey, there is also less chance that the information will change or become biased between administrations of the survey. (2) Surveys are easy to administer in the field — workers can complete the surveys at their convenience, and responses can be kept anonymous. These features can encourage high participation rates and candid responses, although the opportunities for follow-up surveillance are more limited. (3) Worker surveys can be readministered periodically to allow for early recognition of musculoskeletal disorders.

Worker surveys also have some inherent limitations. First, the time and resources required to develop and administer surveys may make them more costly than surveillance activities that rely on existing records. Second, the survey information may be unreliable if there is a lack of trust between management and workers. Third, the design of the questionnaire can have a dramatic effect on the quality of the responses and the number of workers who complete the survey. Key design issues include the length of the questionnaire, the phrasing of the questions, and method of administration. In a diverse or multilanguage

workforce, translating the questionnaires into several languages may be necessary. Finally, responses can be influenced by the workers' expectations about how much effort is required to perform a certain job. For example, a young warehouse worker may rate his job as only moderately strenuous, whereas an older worker using the same scale in a busy office environment may perceive his work to be very stressful in terms of work demands.

Hazard (Risk Factor) Surveys

Ideally, actions to prevent work-related musculoskeletal disorders should proceed before injuries and/or symptoms develop. In the last decade, a significant amount of research has been undertaken to improve our understanding of the risk factors that lead to the development of work-related musculoskeletal disorders (Gerr et al., 1991; Rempel et al., 1992). The process of examining jobs for these risk factors is known as hazard surveillance. Even without clear medical evidence that musculoskeletal problems exist, hazard surveillance activities can provide the data needed to begin an effective primary prevention program.

Because surveillance for ergonomic hazards is a relatively new notion, few industries maintain records that explicitly identify hazards for each job. However, many companies maintain job descriptions that identify the skills or abilities required to perform jobs in their facilities. In recent years, many industries have updated their job descriptions to include statements defining "Essential Job Functions" to comply with the Americans with Disabilities Act (ADA). According to the ADA, the descriptions should identify essential functions, or fundamental job duties, and the physical and mental abilities needed to perform these functions. If functional job descriptions are available, they can provide useful information for identifying potentially stressful jobs or jobs requiring unique skills or special endurance.

For example, a job description in a large assembly plant may list the ability to perform manual lifts, assemble parts, and use an impact wrench as requirements. Based on this description, an investigator could infer that the job might require high force exertions, awkward postures, repetitive hand movements, and exposure to segmental hand vibration.

Whether or not job descriptions exist, hazard surveillance efforts depend heavily on walk-through surveys. Investigators need to observe job activities, speak with workers and supervisors to obtain job information not apparent from observation, and use checklists to score job features against a listing of risk factors. The walk-through survey is differentiated from more formalized job analysis efforts by the amount of detailed information collected. The purpose of the walk- through is to identify risk factors that might otherwise go unnoticed and provide additional basis for prioritizing jobs for further evaluation.

Examples of checklists that might be used for hazard surveillance are provided in Tables 64.2, 64.3, and 64.4. Although most of the checklists are designed for use by non-experts, some minimal level of training is usually needed to use checklists properly. Hazard checklists can vary in length and in scope. At one extreme, there are "generic" checklists that are widely applicable, i.e., for use on nearly all jobs in all industries (OSHA, 1995). These checklists can be contrasted with more focused checklists, designed to evaluate a specific job or industry, e.g., carpenters or nurses' aides (Bhattacharya, 1992; Engkvist et al., 1995). While some checklists are intended to serve primarily as mnemonics (i.e., to remind users to evaluate a particular job characteristic), other checklists incorporate a scoring system for indicating the risk associated with a particular job or process. Attempts to validate various checklists as scoring instruments have yielded mixed results (Lifshitz and Armstrong, 1986; Keyserling et al., 1992; 1993; Shierhout et al., 1994; Kemmlert, 1995).

Hazard surveys should be administered (1) whenever a job, task, or process is changed substantially, (2) when new jobs are introduced, and (3) periodically (especially after new cases of musculoskeletal disorders are reported) to detect whether trends exist across jobs that use similar equipment, tools, or processes (ANSI, 1996). Hazard surveys can also be incorporated into regular safety and health inspections at the facility, expanding the scope of these inspections to include identification of musculoskeletal disorder risk factors.

TABLE 64.2 Ergonomic Hazard Identification Checklist*

[Answer Questions Based on the Primary Job Activities of Workers in Facility]

Never — Worker is never exposed to the condition

Sometimes — Worker is exposed to the condition less than 3 times daily.

Usually — Worker is exposed to the condition 3 times or more daily.

	Never	Some < 3	Usually > 3	If USUALLY, list jobs to which answer applies here
1. Do workers perform tasks that are externally paced?				
2. Are workers required to exert force with their hands (e.g., gripping, pulling, pinching)?				
3. Do workers use hand tools or handle parts or objects?				
4. Do workers stand continuously for periods of more than 20 mins?				
5. Do workers sit for periods of more than 30 mins without a chance to stand or move around freely?				
6. Do workers use keyboards, mice, joysticks, etc. for continuous periods of more than 30 mins?				
7. Do workers kneel for more than 5 min (one or both knees)?				
8. Do workers perform activities with hands raised above shoulder height?				
9. Do workers perform activities while bending or twisting at the waist?				
10. Are workers exposed to vibrations?				
11. Do workers lift or lower objects between floor and waist height, or above shoulders?				
12. Do workers lift or lower objects more than once/min. for continuous periods of more than 15 minutes?				
13. Do workers lift, lower, or carry objects weighing >12 lbs that are not held close to body?				
14. Do workers lift, lower or carry objects weighing more than 50 lbs.?				

*Developed by Grant, Habes, Fernandez, and Putz-Anderson, NIOSH, Cincinnati, Ohio, 1994.
Any response of: "*Usually*" to Questions 1-14 = Potential Risk Present: Follow-up

64.4 Conclusions

Surveillance is essential to the prevention/control of musculoskeletal disorders in the workplace. Unfortunately, surveillance efforts will ultimately be wasteful and ineffective if there are major breakdowns in any of the three major components: data, analysis, or intervention. In general, no single data source provides enough information to direct a program for preventing work-related musculoskeletal disorders. Therefore, effective surveillance programs make use of multiple data sources to identify problem areas and determine intervention priorities. Even in the absence of health data, hazard surveys conducted in workplaces where there are significant or well-defined hazards can provide the data needed to mount an effective primary prevention program for work-related musculoskeletal disorders.

Once established, surveillance should become an ongoing process. As corrective actions are taken, surveillance data can provide the information needed to show the beneficial effects of these efforts. By integrating surveillance efforts with existing quality assurance and cost containment programs, their utilization and success will be maximized.

TABLE 64.3 Manual Material Handling Checklist for Lifting, Carrying, Pushing, or Pulling*

Risk Factors		Yes	No
1. General			
1.1	Does the load handled exceed 50 lbs.?	[ ]	[ ]
1.2	Is the object difficult to bring close to the body because of its size, bulk, or shape?	[ ]	[ ]
1.3	Is the load hard to handle because it lacks handles or cutouts for handles, or does it have slippery surfaces or sharp edges?	[ ]	[ ]
1.4	Is the footing unsafe? For example, are the floors slippery, inclined, or uneven?	[ ]	[ ]
1.5	Does the task require fast movement, such as throwing, swinging, or rapid walking?	[ ]	[ ]
1.6	Does the task require stressful body postures, such as stooping to the floor, twisting, reaching overhead, or excessive lateral bending?	[ ]	[ ]
1.7	Is most of the load handled by only one hand, arm, or shoulder?	[ ]	[ ]
1.8	Does the task require working in environmental hazards, such as extreme temperatures, noise, vibration, lighting, or airborne contaminants?	[ ]	[ ]
1.9	Does the task require working in a confined area?	[ ]	[ ]
2. Specific			
2.1	Does lifting frequency exceed 5 lifts per minute?	[ ]	[ ]
2.2	Does the vertical lifting distance exceed 3 feet?	[ ]	[ ]
2.3	Do carries last longer than 1 minute?	[ ]	[ ]
2.4	Do tasks which require large sustained pushing or pulling forces exceed 30 seconds duration?	[ ]	[ ]
2.5	Do extended reach static holding tasks exceed 1 minute?	[ ]	[ ]

Comment: "Yes" responses are indicative of conditions that pose a risk of developing low back pain. The larger the percentage of "yes" responses, the greater the possible risk.

* Developed by Thomas Waters, Ph.D., CPE, NIOSH, Cincinnati, Ohio, 1994.

TABLE 64.4 Physical Exertion Questionnaire

The purpose of this questionnaire is to assess the amount of exertion you use in your job. Exertion is defined by (1) force, (2) repetitive motion, and (3) whole-body activity. The following pages contain three lists of activities, one for force, one for repetition, and one for whole-body activity. Use the activities in these lists to estimate the amount of exertion you use in your job.

To complete the questionnaire, read the activities on each page. Then compare the amount of force, whole-body activity, or repetitive motion (according to the particular list) the activities require, with the amount of force, whole-body activity, or repetition your work activities require. If you don't spend much work time doing **exactly** these activities, you may spend time doing activities that require **about the same amount** of exertion as the activities in the lists. Read the activities and think about an average workday. Then, in the boxes beside the activities, write in the number of hours and minutes during a typical workday, on the average, you spend doing activities that require about the same amount of exertion as those at each intensity level. **The total amount of time on each page should add up to about the length of your average workday.** Some of the intensity levels do not have activities to describe them. These levels stand for an amount of exertion that is halfway between the activities above and below. Write in the amount of time on those lines also. Think carefully about the amount of time you spend in low intensity activities, which may be more difficult to remember than high intensity activities.

As an example of completing the questionnaire, how much time during an average workday do you spend in activities that require about the same amount of whole-body activity as running up stairs? If you usually spend about 4 hours and 30 minutes during an average day, write this in the box beside the activity. Then estimate the amount of time you spend doing activities at the other intensity levels and write the time in the boxes. Please do not leave any of the boxes empty. Put a 0 in the box if you don't spend time doing activities at that intensity level.

***Remember*:** The activities are only **examples** of amounts of exertion. The question is "Do you use the same amount of exertion as required by the different activities and not whether you perform the activities listed." The activities are only a guide for you to help you estimate your level of exertion, relative to the listed activity.

Whole-Body Activity Scale

On this page, you will indicate how much whole-body activity your job requires. Whole-body activity is any activity that increases heart rate and breathing, and involves movement mostly of the legs, such as pedaling a bicycle, or movement of the entire body. The activities below require whole-body activity; those near the top of the list require more whole-body activity than those near the bottom.

TABLE 64.4 (continued) Physical Exertion Questionnaire

Compare your work activities to the activities below in terms of how much whole-body activity your job requires. Then decide how much time on the average, you spend doing activities that require about the same amount of whole-body activity as those at each intensity level. *To help you estimate the intensity of your whole-body activity, think about how your body feels when you perform the activities in the scale, and compare this to the way your body feels when you perform your work activities.*

Remember: The activities below are only examples of amounts of whole-body activity for you to compare with your work activities. We are NOT interested in whether you perform the specific activities listed in the scale. Also, think only about the amount of whole-body activity used to perform these activities, and not about the amount of force or repetitive motion the activities require.

Duration	Level	Equivalent Amount of Whole-Body Activity
	6	Running up stairs
	5	Between Level 6 and Level 4
	4	Climbing a vertical ladder
	3	Carrying boxes or packages from your house/apt to your car
	2	Making a bed (straightening the sheets and blankets)
	1	Rolling over from back to stomach while lying in bed
	0	No whole-body activity

The total number of hours and minutes on this page should equal about the length of your average workday.

Repetitive Motion Scale

On this page, we are interested in how much repetitive motion your job requires. Repetitive motion is movement that requires use of the same muscles and body parts to perform the same movements or sequence of movements over and over.

The activities below require repetitive motion. Those near the top of the list require more repetitive motion, meaning that the movements or sequence of movements take only a short time to complete before being repeated. Activities near the bottom of the list require less repetitive motion, and it takes longer before the same movement is repeated.

Please read the activities below and think about your **average** workday. Then write in the number of hours and minutes during a typical workday, **on the average**, you spend doing work activities that require about the same amount of repetitive motion as those at each intensity level. Please remember: The activities below are only examples of repetitive motion to be compared with your work activities. We are NOT interested in whether you perform the specific activities listed in the scale. Also, think only about the amount of repetitive motion used to perform these activities, and not about the amount of whole-body activity or force the activities require.

Duration	Level	Equivalent Amount of Repetitive Motion
	6	Stirring water with a spoon Manually grating cheddar cheese using a cheese grater
	5	Between Level 6 and Level 4
	4	Hammering nails into soft wood Cleaning windows/mirrors with a cloth or paper towel
	3	Between Level 4 and Level 2
	2	Taking notes using a pen/pencil and paper Removing groceries from a paper grocery bag Answering telephones and writing messages
	1	Between Level 2 and Level 0
	0	No repetitive motion

The total number of hours and minutes in the table should equal about the length of your average workday.

TABLE 64.4 (continued) Physical Exertion Questionnaire

Force Scale

On this page, you will indicate how much muscle force your job requires. In the boxes below enter the number of hours and minutes during a typical workday, on the average, you spend doing activities that require about the same amount of force as those at each intensity level.

The activities near the top of the list require more force, and activities near the bottom of the list require less force.

To help you estimate the amount of force you use in your job, think about how your body feels when you perform the activities in the scale, and compare this to the way your body feels when you perform your work activities. Remember: The activities are only examples of amounts of force. We are NOT interested in whether you perform the specific activities listed in the scale. Also, think only about the amount of force used to perform these activities, and not about the amount of whole-body activity or repetitive motion the activities require.

Duration	Level	Equivalent Amount of Force
		Pushing a refrigerator on a flat, smooth surface, such as tile, linoleum, or a wood floor (the refrigerator is not on wheels, a cart, or a dolly)
		Between Level 6 and Level 4
		Scooping hard, frozen ice cream out of a container Opening a new jar (a jar that has never been opened) of pickles or jelly
		Lifting a 12-pack of beer or pop using one hand Unscrewing a bottle cap on a new bottle or container or pop (a bottle or container of pop that has never been opened)
		Crushing an aluminum soda can or beer can with one hand (crushing it side-to-side, not top-to-bottom) Lifting a full gallon of milk Opening a car door with one hand
	1	Pulling the end of a cord out of an electrical wall outlet Lifting a telephone receiver Turning a doorknob
	0	Lifting a quarter Bending a piece of typing paper or notebook paper Lifting a cottonball

The total number of hours and minutes in this table should equal about the length of our average workday.

Cole, L.L. (1996). *Construction and Validation of a Musculoskeletal Risk Questionnaire.* Unpublished doctoral dissertation, University of Cincinnati, Cincinnati, Ohio.

Defining Terms

Case definition: Set of decision-making criteria, intended to assist health care providers in identifying work-related injuries and illnesses so that further investigation and preventive activities can be initiated.

Existing records: Records created for other purposes that may be useful in surveillance efforts. These records can include OSHA 200 logs, medical or health care records, workers' compensation claims, payroll records, sickness and accident reports, etc.

Hazard surveys: Assessments of jobs, workplaces, or processes for the purpose of identifying risk factors (e.g., biomechanical stress, vibration) that can lead to development of musculoskeletal disorders. Can be used to identify intervention targets before injuries or diseases have occurred.

Incidence rate: The rate of work-related musculoskeletal disorders appearing for the first time during a specific period.

Prevalence rate: The percentage of all cases of disease/symptoms/complaints at a given point in time.

Sensitivity: The probability that a case definition or screening procedure will identify individuals with the condition (disease) of interest.

Severity index: The rate of lost or restricted workdays due to musculoskeletal illness occurring within a specific period.

Specificity: The probability that a case definition or screening procedure will not label a healthy individual as diseased.

Surveillance: The continuous analysis, interpretation, and feedback of systematically collected data, essential to the planning, implementation and evaluation of occupational safety and health programs. Surveillance methods are often distinguished by their practicality, uniformity, and rapidity, rather than by their accuracy or completeness.

Worker surveys: Questionnaires and interview procedures developed to solicit information about signs, symptoms, and risk factors for musculoskeletal disorders from workers.

References

ANSI, 1996. *ANSI Z-365 Control of Work-Related Cumulative Trauma Disorders, Part 1: Upper Extremities* (Working Draft, January 1, 1996). National Safety Council, Itasca, Illinois.

Baker, E.L., Honchar, P.A., and Fine, L.J. 1989. Surveillance in occupational illness and injury: concepts and contents. *Am. J. Public Health* 79(suppl):9-11.

Baker, E.L. and Matte, T.P. 1994. Surveillance for occupational hazards and disease, in *Textbook of Clinical Occupational and Environmental Medicine*, Eds. L. Rosenstock and M.R. Cullen, p. 61-67. W.B. Saunders Company, Philadelphia.

Baron, S., Hales, T., and Fine, L. 1992. Evaluation of a questionnaire to assess the prevalence of work-related musculoskeletal disorders. *Arbete Och Halsa* 17:39-41.

Bhattacharya, A. 1992. *Walkthrough Ergonomics Checklist for Carpentry Tasks.* Greater Cincinnati Occupational Health Center, Cincinnati, Ohio.

Bernard, B., Sauter, S., Fine, L., Petersen, M., and Hales, T. 1994. Job task and psychosocial risk factors for work-related musculoskeletal disorders among newspaper reporters. *Scand. J. Work Environ. Health* 20:417-426.

Centers for Disease Control, 1989. Occupational disease surveillance — carpal tunnel syndrome. *MMWR* 38:485-489.

Corlett, E.N. and Bishop, R.P. 1976. A technique for assessing postural discomfort. *Ergonomics* 19(2):175-182.

Engkvist, I.L., Hagberg, M., Wigaeus-Hjelm, E., Menckel, E., and Ekenvall, L. 1995. Interview protocols and ergonomics checklist for analyzing overexertion back accidents among nursing personnel. *Appl. Ergonom.* 26(3):213-220.

Fine, L.J., Silverstein, B.A., Armstrong, T.J., Anderson, C.A., and Sugano, D.S. 1986. Detection of cumulative trauma disorders of the upper extremity in the workplace. *J. Occup. Med.* 28:674-678.

Franklin, G.M., Haug, J., Heyer, N., Checkoway, H., and Peck, N. 1991. Occupational carpal tunnel syndrome in Washington State, 1984-1988. *Am. J. Public Health* 81:741-746.

Gerr, F., Letz, R., and Landrigan, P.J. 1991. Upper-extremity musculoskeletal disorders of occupational origin. *Annu. Rev. Publ. Health* 12:543-566.

Greife, A., Halperin, W., Groce, D., O'Brien, D., Pedersen, D., Myers, J.R., and Jenkins, L. 1995. Hazard surveillance: its role in primary prevention of occupational disease and injury. *Appl. Occup. Environ. Hyg.* 10(9):737-742.

Hales, T., Sauter, S., Peterson, M., Fine, L., Putz-Anderson, V., Schleifer, L., Ochs, T., and Bernard, B. 1994. Musculoskeletal disorders among visual display terminal users in a telecommunications company. *Ergonomics* 37(10):1603-1621.

Hennekens, C.H. and Buring, J.E. 1987. *Epidemiology in Medicine.* Little, Brown and Company, Boston.

Katz, J.N., Larson, M.G., Fossel, A.H., and Liang, M.H. 1991. Validation of a surveillance case definition of carpal tunnel syndrome. *Am. J. Public Health* 81:189-193.

Kemmlert, K. 1995. A method assigned for the identification of ergonomic hazards — PLIBEL. *Appl. Ergonom.* 26(3):199-212.

Keyserling, W.M., Brouwer, M., and Silverstein, B.A. 1992. A checklist for evaluating risk factors resulting from awkward postures of the legs, trunk and neck. *International J. of Industrial Ergonomics* 9:282-301.

Keyserling, W.M., Stetson, D.S., Silverstein, B.A., and Brouwer, M.L. 1993. A checklist for evaluating ergonomic risk factors associated with upper extremity cumulative trauma disorders. *Ergonomics* 36(7):807-831.

Kuorinka, I., Jonsson, B., Kilbom, Å., Vinterberg, H., Biering-Sorensen, F., Andersson, G., and Jorgensen, K. 1987. Standardized Nordic questionnaire for the analysis of musculoskeletal symptoms. *Appl. Ergonom.* 18(3):233-237.

Last, J.M. (Ed.) 1995. *A Dictionary of Epidemiology,* 3rd ed. Oxford University Press, New York.

Lifshitz, Y. and Armstrong, T.J. 1986. A design checklist for control and prediction of cumulative trauma disorders in intensive manual jobs. *Proceedings of the Human Factors Society, 30th Annual Meeting.* Human Factors Society, Santa Monica, pp. 837-841.

Margolis, W. and Kraus, J.F. 1987. The prevalence of carpal tunnel syndrome symptoms in female supermarket cashiers. *J. Occup. Med.* 29(12):953-956.

NIOSH, 1993. Comments from the National Institute for Occupational Safety and Health on the Occupational Safety and Health Administration Proposed Rule on Ergonomic Safety and Health Management. 29 CFR Part 1910, Docket No. S-777.

OSHA, 1990. *Ergonomics Program Management Guidelines for Meatpacking Plants.* U.S. Department of Labor, Occupational Safety and Health Administration, Washington, D.C.

OSHA, 1995. OSHA Draft Proposed Ergonomic Protection Standard: Summaries, Explanations, Regulatory Text, Appendices A and B, March 20, 1995. *Occupational Safety and Health Reporter 24*(42):S3-S248.

Rempel, D.M., Harrison, R.J., and Barnhart, S. 1992. Work-related cumulative trauma disorders of the upper extremity. *JAMA* 267(6):838-842.

Saldana, N., Herrin, G.D., Armstrong, T.J., and Franzblau, A. 1994. A computerized method for assessment of musculoskeletal discomfort in the workforce: a tool for surveillance. *Ergonomics* 37(6):1097-1112.

Schierhout, G.H., Myers, J.E., and Bridger, R.S. 1993. Musculoskeletal pain and workplace ergonomic stressors in manufacturing industry in South Africa. *Int J. Ind. Ergon.* 12:3-11.

Schierhout, G.H. and Myers, J.E. 1996 Is self-reported pain an appropriate outcome measure in ergonomic-epidemiologic studies of work-related musculoskeletal disorders? *Am. J. Ind. Med.* 30:93-98.

Schwartz, E. 1987. Use of workers' compensation claims for surveillance of work-related illness — New Hampshire, January 1986-March 1987. *MMWR* 36(43):713-720.

Stenlund, B., Goldie, I., Hagberg, M., and Hogstedt, C. 1993. Shoulder tendinitis and its relation to heavy manual work and exposure to vibration. *Scand J. Work Environ. Health* 19:43-49.

Tanaka, S., Seligman, P., Halperin, W., Thun, M., Timbrook, C.L., and Wasil, J.J. 1988. Use of workers' compensation claims data for surveillance of cumulative trauma disorders. *J. Occup. Med.* 30(6):488-492.

65

Injury Surveillance Database Systems

Carol Stuart-Buttle

Stuart-Buttle Ergonomics

65.1 Introduction

A database is a collection of organized, related data typically in electronic form that can be accessed and manipulated by computer software. When the interface between the data and user is developed beyond just managing the data, but with a particular function in mind, a database system evolves. Features of the interface may include data dictionaries, data security, statistical and graphical capabilities, and report writing.

Injury surveillance database systems focus on the collection of data that are pertinent to injuries and illnesses, the management of that data, and data interpretation. The goals of interpretation may include determining trends, promoting prompt treatment and directing prevention strategies. The general use of the term injury refers to all types of injuries and illnesses. Conducting epidemiological studies or other research that attempts to find causal relationships of injuries may be practical with large organizations but not for most of industry; therefore research use of databases are not discussed in this chapter. Injury

0-8493-2641-9/99/$0.00+$.50

surveillance databases and the relationship of surveillance to ergonomics are addressed from the practicing industrial safety and health professional's perspective.

Surveillance of the workplace for injuries and potential injuries is an important aspect of the ergonomics process as injuries are an indicator of poor design. Often, the effectiveness of ergonomic processes are measured by the rates of injury and lost workdays. Most employers are required to record work-related injuries and illnesses in accordance with the Occupational Safety and Health (OSH) Act (OSHA, 1970). As personal computers (PCs) have become commonplace in industry, the most basic information pertaining to the OSH Act can be easily computerized; in other words, a company database can be established. Considerable time and effort can be saved in the update and analysis of the data with the right software interface.

The rate of change and growth in the computer software market is so great that for many industries the task of exploring, evaluating, and choosing a database system for the company is formidable. This chapter does not attempt to predict the future of information systems, nor does it try to summarize the existing safety and health software since such information rapidly changes. However, some safety and health software directories are provided. The main points and discussion of this chapter focus on the fundamental components of injury surveillance as they relate to ergonomics and tips on how to assess commercial injury surveillance database systems. A section touches on the development of an injury surveillance database only to provide an opportunity to understand injury surveillance in the context of system development.

65.2 Injury Surveillance

History

"Surveillance is the ongoing and systematic collection, analysis, and interpretation of data related to health." (Baker, 1989a). Surveillance historically began in the public health setting, as a method of watching out for certain serious diseases or illnesses such as syphilis and smallpox, so that isolation could be instituted at the first signs of symptoms. Since 1950 the term "surveillance" was formally applied to the systematic collection of relevant disease data with the purpose to improve the control of specific diseases. During the 1960s, surveillance in the U.S. expanded such that it applied not only to communicable diseases but also to noninfectious diseases including environmental and occupational hazards (Langmuir, 1976)

During the 1960s, some states developed reporting requirements for selected occupational diseases or used data from the workers' compensation system for surveillance. Not until the federal OSH Act of 1970 was there a standardized scheme to monitor injuries and illnesses (Wegman, 1985). It was not surprising that the chemical industry was a pioneering group in developing occupational surveillance systems, since there was much potential for disease. Hazard surveillance, as a means of predicting work-related health problems in the chemical industry, was developed to complement the basic surveillance of recording injuries and illnesses required by the OSH Act. Many companies then started hazard profiles that included information such as industry demographics (employment size in the geographic area), use of chemicals in the workplace, levels of the chemicals to which workers were exposed, and data on the dose–response relationship for each chemical (Froines, 1986).

The hazard surveillance model has been applied to ergonomics-related issues in the workplace. One problem in applying such a model to ergonomics is that suspected risk factors for musculoskeletal disorders are treated discretely without taking into account the interactions of the risk factors. The effects of the interactions and the dose–response relationships are still being discovered.

"Monitoring" is another term that refers to the methods to anticipate a disease or medical disorder before it occurs (Yodaiken, 1986). Monitoring complements medical surveillance in a way similar to that of hazard surveillance. The difference lies in semantics and the use of the word "hazard" which implies that the dose–response is known. As stated above, it is difficult to determine when one or more risk factors are hazardous for musculoskeletal disorders.

Monitoring may be considered as a more general term, one that refers to looking at the workplace in order to anticipate problems and prevent their occurrence. An illustration of the practical difference between medical surveillance and monitoring is the distinction between a symptom survey and discomfort survey (Stuart-Buttle, 1994). A symptom survey identifies the employees with physical problems through the collection of detailed information about symptoms. On the other hand, a discomfort survey attempts to find out the discomfort that employees are experiencing while performing their jobs. Discomfort indicates a physical stress at performing the job, not necessarily an active medical problem. The implication of a discomfort survey is that discomfort could be a precursor to injury, hence a discomfort survey is a monitoring method for determining where to focus prevention efforts.

Another common distinction of the term surveillance is whether it is "passive" or "active" (BNA, 1993a). Passive surveillance is the analysis of data from existing databases, such as the Occupational Safety and Health Administration (OSHA) 200 logs that record work-related injuries and illnesses, or workers' compensation records. Active surveillance is collecting data to "actively" seek information, such as using surveys to determine the physical conditions that individuals have not reported, or defining aspects of the jobs that may be risk factors for injury. The distinction between passive and active surveillance is almost the same as that between surveillance and monitoring, or medical surveillance and hazard surveillance.

Goals and Objectives

Injury surveillance should entail both active and passive surveillance. Existing injury data as well as additional information collected about the workers and their jobs must be analyzed. The goals of a surveillance system include "the prevention of occupational disease *(injuries)* through control of the causative agents" (Sundin, 1986). Surveillance helps to direct the prevention programs to control or eliminate preventable disorders (Baker, 1989b).

The objectives to fulfill the goals of surveillance are to locate and monitor groups of workers who are exposed to risk, to determine the risk factors and take corrective action, or reduce exposure to those factors. Discovering the relationships between exposure and the injury is a research domain which may be undertaken by some corporations with substantial databases or as state or university projects. The interaction and dosage of risk factors for musculoskeletal disorders are not yet known, but industry can take action based on the best knowledge to date and refine the decisions as research progresses. Gerard Scannell, president of the National Safety Council asks, "How do you direct your safety and health program if you don't know what is happening in your workplace? The proper analysis of recorded injuries and illnesses will drive the programs" (Smith, 1995). In addition, surveillance is the primary method of assessing the efficacy of prevention measures that are instituted to control the identified problems (Hanrahan, 1989). Therefore, a company benefits from collecting the data that are useful in reducing the risk of injuries and improving job performance.

65.3 Sources of Primary Injury Surveillance Data

Injury Data

There are several primary sources of injury data that companies either keep or to which they have access without having to survey the workforce.

- Injury logs that are mandated by OSHA
- Workers' compensation records
- Medical records
- Accident records

OSHA Recordkeeping

Since the OSH Act of 1970 companies have had to keep a record of their injuries. In 1981 OSHA began to look at the log data as indicators for an inspection. Industry responded to this new focus in one of two ways. They either improved the safety and health in the facility or became more selective in what was recorded on the logs. OSHA began to discover significant under-reporting on companies' logs, and stronger enforcement action was instituted in 1986 (Tyson, 1991). Total penalties have become so substantial that companies now pay much more attention to keeping accurate records (BNA, 1993b, 1994). However, some of the difficulties of accurate recordkeeping have come to light and influence injury surveillance.

The increase in cumulative trauma disorders in the workplace highlights the difficulty of keeping the records up to date. As a medical condition progresses, the diagnosis sometimes changes, but the OSHA records may not reflect the new diagnosis. For example, if an injury originally recorded as a strained wrist eventually led to carpal tunnel syndrome, the diagnosis is typically not changed in the logs. This may be a common oversight if the records are kept by nonmedical personnel or if there is poor communication with the health care professional (HCP). Computerized databases can help a company, even if there is no medical department, as the system could be designed to prompt for confirmation of diagnosis while a case is open.

Despite guidelines on recordkeeping (BLS, 1986), there remains considerable confusion among employers about when and how to record occupational injuries and illnesses. The ambiguity of some of the requirements has been a point of litigation, and one company issued a digest of official interpretations of the recordkeeping guidelines primarily based on legal decisions in enforcement cases or in the courts (Duvall, 1996). OSHA proposed revised recordkeeping requirements and a new guidebook on how to interpret the requirements (OSHA, 1996). The proposal suggested significant improvements that included new versions of the OSHA 200 and 101 forms and simpler recording requirements.

In addition to recording injuries, OSHA and other government and state agencies mandate that several types of records be kept by the person or people responsible for safety and health. Records pertaining to hearing conservation, tracking of and training in the use of personal protective equipment (PPE), and compliance with confined space are just a few examples of the requirements. These records may be easily tracked and integrated into the safety and health database system.

Workers' Compensation Data

Small industries commonly rely on workers' compensation data supplied by the company's insurance carrier as an injury surveillance system. The benefit of this practice is the convenience of the data, especially if the insurance carriers summarize the information. There are several downsides to using only these data. Not all injuries become workers' compensation cases, and therefore the injury profile is only partial. By the time a claim is processed and appears on a summary, a considerable length of time has passed during which preventative action could have been initiated. However, the claims on the workers' compensation insurance originate in the company, so that the company can have immediate access to the information if good records are kept. But, if the workers' compensation data can be tracked effectively by the company, then so can the accident data that might not become workers' compensation. The company would have richer data on which to base prevention and improvements.

The cost of injuries is one type of information that workers' compensation data do provide that is difficult to obtain any other way. Cost data are very useful for calculating cost benefit equations for improvements. Since workers' compensation databases are often large, they can be used effectively for major epidemiological studies or to complement large-scale surveillance (Seligman, 1986).

Medical Records

Medical records can be the most reliable source of injury data since there is no interpretation required for recording as there is for the OSHA logs and workers' compensation filing. As discussed later in this chapter, a difficulty that can arise is access to the medical records. Very few workplaces have onsite medical services, and medical records can reflect both work- and non-work- related injuries which can confuse data collection for injury surveillance.

First aid treatment is often recorded by a company but may not be entered in an employee's medical record. Although by definition the incidents are minor, there may be indication of a design problem. For example, it is not uncommon to find frequent treatment for grazed knuckles. If this is happening to the same employee or to a group of employees under the same circumstances, it could be a clearance or access issue.

Accident Records

Accident reports are filled out by many companies and are the earliest records of an injury within a company's safety and health process. Not all injuries from accidents are recorded on the OSHA forms or claimed on workers' compensation, so that accident reports potentially provide the most comprehensive record of injuries, although maybe without medical detail. If good root cause analyses are conducted and recorded on the accident reports there may be valuable information indicating job redesign.

Job Descriptions

Job information is useful from many standpoints. The causes of injuries need to be identified in order to prevent them. Although the causes for musculoskeletal injuries are not well defined, identifying risk factors of the job can be helpful. For example, risk factors for cumulative trauma disorders include extreme postures, high forces, high repetitions, inadequate recovery times and excessive duration. Some of the reasons to consider building a database of job descriptions are to:

- Have a list of risk factors present in each job
- Track employees' exposure
- Aid in the initial placement of employees
- Determine a job rotation system to create a balance of the job demands
- Provide information to the HCPs to help them with decisions regarding the work-relatedness of conditions
- Provide the context for functional capacity evaluations or to tailor workhardening or conditioning programs
- Guide appropriate placement of employees returning to work and to determine temporary alternative jobs or jobs that could be modified
- Help determine jobs that may be modified for accommodating a person under the Americans with Disabilities Act (ADA) (EEOC, 1990).

The downside of developing a job description database is the expense of establishing it and keeping it up to date. Such costs will vary considerably according to the type and size of industry. Some industries change their processes frequently, perhaps seasonally. To keep job descriptions current in a rapidly changing environment may not be practical.

Computerized injury surveillance database systems could provide a tremendous benefit for a small company that is communicating with offsite services. If an electronic network is established, the records can be efficiently updated. Information about the jobs could be on-line for the medical service to access, including formats such as computer-captured video.

Worker Tracking

Keeping up with job descriptions may be a challenge for some industries, as is tracking the worker. Many of the companies that developed computerized databases early on stated that tracking the workers was the most complex and difficult task (Sugano, 1982; Hagstrom, 1982; Smith, 1982; Kuritz, 1982). Effective tracking required frequent updating of employee assignments including the length of time at the job in the specific area. In addition, each job in every area required a profile so that there was a history of exposure. The job profiles or descriptions also needed to be kept up to date by periodic job analyses or audits.

Employee tracking remains worthwhile, except that the experience of the companies that were in the forefront of developing surveillance databases, found that not only was employee tracking difficult but

also very costly to implement. Therefore, each company must assess the importance of worker tracking, that is, to determine the type and extent of the hazards of the industry and to develop surveillance that is the most reasonable, practical, and cost effective. The trade-offs to consider with the four basic qualities of an employee tracking system follow (Sugano, 1982):

- *Uniformity within the exposure group.* Are the exposures the same across many jobs, and if so, may some of the descriptions be combined?
- *Accuracy of tracking the worker.* What are the means for communicating when an employee moves area or job? How does job rotation affect exposure? What differences are there between plants?
- *Adequacy of job and location description.* How often does the job change, and how frequently should the database be updated? Is environmental monitoring conducted regularly? How difficult is it to keep the physical, environmental, and material descriptions up to date?
- *Cost effectiveness.* How much money and manpower is needed to develop and maintain the database? Is too much data being collected and too often?

At present, worker tracking for exposure to risk factors for musculoskeletal injuries remains an area of research. There is no model by which to analyze the data and to provide an answer to the question "how much exposure to each risk factor or combination of factors is too much." Risk models that do exist are based on task analyses rather than monitoring the worker throughout a shift.

65.4 Benefits of an Injury Surveillance Database System

A complete and standardized reporting system for occupational injuries does not exist in the U.S. (Brewer, 1990). Almost all employers with more than 11 employees are mandated by OSHA to keep a current record of injuries and illnesses and details of how the injury occurred. The government forms may be substituted by ones that are equivalent. Exposure information, such as the demands of the job, a history of the different jobs the employee has held, or the job characteristics such as the state of the environment or the presence of chemicals, is typically not included. However, even without the enhanced job information, the two mandated OSHA forms contain considerable amounts of data that are not easily analyzed by hand. Computerization of only the OSHA forms can be helpful. What is important to remember is that the quality of the data depends on the information that is entered. The OSHA forms, filled thoroughly and thoughtfully, provide a better company database. A basic computerized injury surveillance database system can have the following benefits:

- Standardized and accurate recordkeeping, improving the quality of health records, and compliance with OSHA
- Efficient handling of data, reducing paper work and data re-entry
- Easy and effective reporting system
- Quick statistical analyses for the evaluation of potential problems and the measurement of effectiveness of interventions
- Enhanced tracking ability, particularly of employees' job exposures, and such tracking also facilitates epidemiological studies
- Quick verification in disability and compensation cases
- Potential for integration with other data, e.g., those used by the safety or industrial hygiene professional
- Ease of cost tracking for cost benefit analyses
- Efficient corporate response to litigation, community complaints, and public relations requirements.

There are many commercial software packages available for computerizing the OSHA forms combined with basic surveillance functions. The sophistication of the available programs vary considerably, from large database systems with an injury surveillance module combined with advanced statistical analyses,

to simple OSHA form recording and tracking. A few software directories that are regularly updated are listed in For Further Information.

Cost Benefit Analysis of Database Investment

A cost benefit analysis for investing in an injury surveillance database system is not easy to conduct (Wrench, 1990). Some of the benefits are listed above but assigning a dollar value can be difficult. At the planning stage the time saved can only be anticipated; however, software vendors can sometimes provide a percentage figure of time saved by their systems. Wrench (1990) listed some cost aspects to consider in the cost benefit equation.

- *Direct costs*
  - Initial costs that include the stages of development, such as needs and requirements analysis, systems analysis and design, programming and software development, hardware, software, installation, training, and system implementation.
  - Operating costs that include personnel to use and maintain the system, backup systems, training, and supplies. In addition there are costs related to updating the system as software changes, or when regulations alter, which affect decision algorithms, data entry, and form outputs of a database.
- *Indirect costs*
  - Space requirements may increase with purchase; air-conditioning may be needed for larger systems; security measures and additional insurance may be necessary.

Integration of Database Systems

The data collection mandated by OSHA may be useful, but, most industries would benefit from a broader collection of data and greater integration with other company functions (Finucane, 1982; Committee Report, 1992; Holzner, 1993; Dieterly, 1995). Early on, during the development of surveillance within the chemical industry it was realized that exposure measures were needed to interpret the medical findings, thus integrating medical surveillance programs with industrial hygiene surveillance (Parkinson, 1986). A survey conducted by OSHA also found medical surveillance programs were most effective when they were integrated into a comprehensive and systematic approach for identifying and addressing workplace hazards (BNA, 1991, 1993c). A comprehensive approach requires the involvement of different expertise within a company and hence involves more than one discipline.

Injury surveillance can be incorporated into comprehensive safety and health programs. The Occupational Medical Practice Committee (OMPC) of the American College of Occupational and Environmental Medicine (ACOEM) provided guidelines for the scope of occupational and environmental health programs. The committee identified 14 essential components of a program (Committee Report, 1992).

1. Health and evaluation of employees.
2. Diagnosis and treatment of occupational and environmental injuries or illnesses, including rehabilitation.
3. Emergency treatment of nonoccupational injury or illness.
4. Education of employees in jobs where potential occupational hazards exist, including job-specific instruction, instruction on methods of prevention, and recognition of possible adverse health effects.
5. Implementation of programs for the use of indicated personal protective devices — ear protection (plugs, muffs), safety spectacles, respirators, etc.
6. Evaluation, inspection, and abatement of workplace hazards.
7. Toxicological assessments, including advice on chemical substances that have not had adequate toxicological testing.
8. Biostatistics and epidemiology assessments.

9. Maintenance of occupational medical records.
10. Immunization against possible occupational infections.
11. Medical interpretation and participation in development of governmental health and safety regulations.
12. Periodic evaluation of the occupational or environmental health program.
13. Disaster preparedness planning for the workplace and the community.
14. Assistance in rehabilitation of alcohol and drug-dependent employees or those with emotional disorders.

Many of these items can be addressed by other professionals such as those in industrial hygiene, safety, or ergonomics. The OMPC stressed that the practitioner in occupational medicine should understand how to enlist and collaborate with the skills of colleagues from the many other professions involved with industry. The above list is a useful one for a company to consider. Some elements will be more important than others, depending on the industry. If many services are contracted offsite, the overall program may not be managed by an HCP but perhaps by another professional, for example in safety or industrial hygiene.

65.5 Ergonomics and Injury Surveillance

Injury surveillance is of interest to several professions in industry such as safety, industrial hygiene, medicine, and risk management, as well as ergonomics. The different disciplines may have a part in contributing to the database, or one department may be responsible for it overall, while others access it. Different professions may develop their own interface or module with the same database so as to manipulate and interpret the data as they wish. If a person wears the hats of several job functions, for example all aspects of safety and health, he or she they might work with a number of discrete modules, such as safety, industrial hygiene, and ergonomics. However, a small company may choose to integrate the modules to a basic safety and health one. The scale of module for each discipline depends on the level of detail desired or degree of expertise of the person using the system. If expertise is outsourced, then an in-house module in that field may not be needed.

Many injuries are related to poor design of the workplace or jobs, which is the domain of ergonomics. In particular, musculoskeletal injuries such as low back disorders and cumulative trauma disorders have become informally termed *ergonomics-related disorders*. This is a limited view as other injuries can occur due to poor design; for example, accidents can arise because of inadequate accessibility or because of an error provoked by poor layout and labeling of controls and displays. However, there is considerable focus on musculoskeletal injuries as they are one of the 10 leading work-related health problems affecting workers in the U.S. and such injuries are the leading cause of disability of people in their working years. The cost, based on lost earnings and workers' compensation, exceeds that of any single health disorder (NIOSH, 1986). The most common source of injury and illness for lost workday cases in manufacturing was reported to be the "worker motion or position." Overexertion was the primary event causing injury and illness across all the divisions of industry (Synergist, 1995). Injury surveillance, therefore, is an important method to determine indicators of poor job design and the need for ergonomics intervention. The data are also important as an effectiveness measure of interventions in an ergonomics process.

Ergonomics Process

An ergonomics process in a company commonly has components of medical management which includes injury surveillance; worksite analysis which is active surveillance for risk factors; prevention and control which also entails measures of effectiveness of the controls; and training and education (OSHA, 1990). Good injury surveillance promotes prompt treatment and directs the prevention programs. The greater the understanding of the causes of the injuries, the more proactive can be the prevention. If prevention measures are not taken, then the system consists only of treatment and remains reactive to the occurrences

of the workplace. Surveillance systems that are set up from only a medical standpoint can be weak on the prevention aspect, even if there are good data. The "who, what, when, and where" (Garrett, 1982) of injuries can often be captured on the accident reports and on the OSHA forms or equivalent if they are thoroughly completed. Typically, the "how and why" of an injury are not well reported and are dependent upon the quality of the accident investigation and whether it includes an ergonomics perspective. Some caution is needed in relying upon a medical focus for improving the workplace. The workplace system may suffer if workstation changes are in the form of accommodation or excessive tailoring on a case by case basis, as a response to medical injuries. A balance must be kept between appropriate design of workstations and the overall function of the system, including organizational design aspects. Involvement of an interdisciplinary team helps to ensure the work system is addressed. Disciplines such as professionals in safety, engineering, management, health care, industrial hygiene, and ergonomics, as well as the operators often participate in the teams.

Responsibility has to be assigned to an individual or a group within the company to apply the results of the surveillance data by looking at the workplace, identifying the problems and root causes, and developing potential solutions. The process does not stop there, as interventions have to be implemented and then followed to ensure effectiveness. It is more likely that prevention measures will not be taken when the medical facilities are offsite, which is the case for the majority of workplaces. It is the company's responsibility to bridge the communication gap between the company and offsite services. Assigning a company contact person often helps in promoting communication.

Ergonomics Module of a Database System

An ergonomics module that interfaces with an injury surveillance database system has its own list of needs to fulfill the ergonomics function. The main tasks are:

- Collecting data, apart from injury data, that indicate ergonomics intervention or help with the interpretation of injury surveillance data
- Interpretation of the data, for example, prioritization methods and graphical trends or statistical analysis of surveys
- Tracking of information such as, project status, timelines, workplace changes, effectiveness, and training

To accomplish data collection and interpretation in the context of task analyses, an ergonomics module may also interface with technical software such as the National Institute of Occupational Safety and Health (NIOSH) Lifting Index. Job profiles may be generated from job analyses or analyses results added to profiles that may have been developed by another department, such as human resources. Data that can be gathered apart from injuries include:

- Turnover rates
- Absenteeism rates
- Productivity rates and quotas
- Quality-related figures
- Workers' compensation costs
- Job profiles or descriptions
- Employees' jobs record or work history
- Discomfort survey results
- Ergonomics audit results
- Employee suggestions or reports of problems
- Ergonomic analyses results

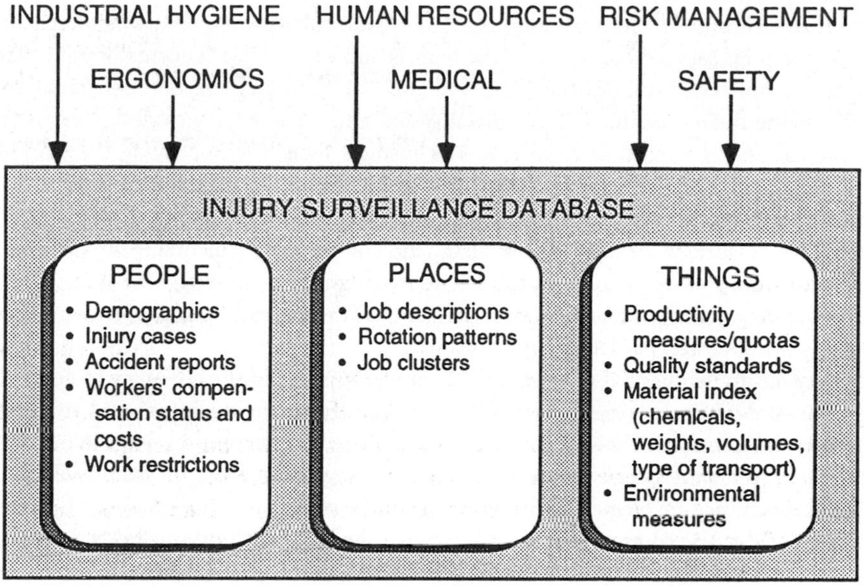

FIGURE 65.1 A fundamental model of an injury surveillance database system and possible user groups. Listings under People, Places, and Things are examples of database information of common interest to the users. (Adapted from Hillman, G. 1982. ECHOES: IBM's environmental, chemical and occupational evaluation system, in *Medical Information Systems Roundtable.* p. 827-835. *JOM* 24(10) Supplement.)

A simple way to categorize most of this information is by using three groups: People, Places and Things as illustrated in Figure 65.1. An ergonomics module would add ergonomics software tools of choice, statistical analysis programs, analyses and survey results, and tracking systems. Specific customization will be required to interface the desired software of an ergonomics module to an injury surveillance database, whether that database is company developed or a commercial package. The degree of software tailoring and the extent of the company's capabilities and resources will determine how much would be in an ergonomics module. It may be practical to keep some components independent such as the ergonomics tools.

65.6 Early Databases

The earliest computerized surveillance databases were purchased or developed during the 1970s primarily by the chemicals and petroleum industry. Computer-based information systems became a necessity to manage the data growth more effectively. Manual data systems in large corporations were hand-written and paper-based, lacked consistency and standardization and could not adequately capture the data on important monitoring aspects such as job history and workplace exposures (Finucane, 1982). The models that were conceived were comprehensive and sophisticated, and many currently continue their development (Holzner, 1993). The focus of the systems were occupational health and, at a minimum, incorporated (Finucane, 1982):

- Detailed worker and job histories and demographic data
- An inventory of potential exposures and associated adverse health effects related to specific workplace location
- Worksite exposure data
- Employee medical information collected throughout the worker's career

These extensive models depended on effective interaction between the departments and the readily available retrieval and analysis of the information. Wolkonsky (1982) described the following information as necessary for a health data system at Amoco:

- *Employee* — job history, location codes, sickness, and disability
- *Claims* — workers' compensation payments, death certificates, internal cost reports
- *Medical* — personal and family history, physical examination, laboratory results, immunizations
- *Industrial Hygiene* — personal samples, area samples, potential exposures by location, exposure levels
- *Toxicology* — animal research, materials safety data sheets
- *Safety* — accident data, physical conditions, supervisor data, DOT data

These early medical surveillance database systems remain but are continually up-graded and developed as computer hardware and software change, regulations alter, and knowledge about exposures and risk assessment models progress through research.

Challenges that Were Met

Descriptions of the early computerized systems highlighted many difficulties in establishing comprehensive database systems. Of the large and sophisticated models being developed by the major corporations in the country, none were the same. Those who attempted to purchase another's system ended up making significant modifications and customizations to fit the needs of the company (Joiner, 1982). However, despite the individuality of the systems there were some common factors that made them successful. The common factors of the systems were: flexibility, interaction, user friendliness, modular design, valid databases, economy, innovation, key staff, commitment, and phased approach.

One of the first difficult decisions for the developers was whether to develop a system in-house or buy a commercial system and modify it to the meet the company's needs. Building the system gradually by modules was a common approach, but the modules of the database systems varied considerably between companies. On a basic level, the modules can be summarized as People, Places, and Things (Hillman, 1982). Deciding how much information about each area was needed and was cost effective, and what hardware to accomplish the goal varied according to company philosophy and existing computer experience or investments.

Early developers reported some modules as challenging to design. In particular, tracking workers was the most complex and difficult task (Sugano, 1982; Hagstrom, 1982; Smith, 1982; Kuritz, 1982). Centralization of information, such as personnel data, was another issue for a few companies. Despite the large corporate size, not all systems, such as the personnel system, were consistent or centralized across facilities (Hagstrom, 1982). Privacy and confidentiality of medical information were of general concern and were resolved in different ways. Some companies encrypted information in case it was routed to the wrong terminal, while others limited access and used passwords. Many of the issues and difficulties that were met by those who developed the early medical surveillance systems are pertinent today. Some of these issues are further addressed in the context of developing or choosing an injury surveillance database system.

65.7 Defining the Database Model

For a company to define the model of an injury surveillance database system to be developed or purchased requires more than a clear understanding of the goals and objectives of the system. Development of a system must be a team effort (Wrench, 1990). The computer software professionals need to work closely with the safety and health professionals.

An assessment of the company itself is necessary to ensure that the goals of the surveillance system integrate with the company's needs, vision, and resources. There are common phases of development for a computerized health surveillance program. Designing the system and defining the components are

just one stage (Joiner, 1982). If commercial software packages are being deliberated, take particular care to consider them in context of the overall system because the result can be a different program for each departmental function. If later the decision is to combine the programs, it may be technically too difficult to do so and efficiency is lost in the long run. The following program development phases are also pertinent for defining programs to purchase:

- Feasibility study
- Planning and identifying potential users
- Designing the system and defining components (system development and modification)
- Implementing
- Operating
- Maintaining
- Upgrading the system

Designing the System and Defining Components

The step of designing the system and defining the components requires particular comment because it is important for determining the scope of an injury surveillance system. Deciding whether injury surveillance should be part of a larger system or if it should be a small discrete system may not be an easy task for a company. This step of program development is especially useful also for the decision process necessary prior to purchasing a program.

1. Define the present injury surveillance system in the company, however manual that may be, and critique for aspects that you wish to maintain and determine areas of improvements.
2. Define the basic components of the injury surveillance database system that you would like to have. Keep in mind the primary goal of an injury surveillance database system; that is, determining the who, what, where, when of employee exposure (Garrett, 1982) and using that information to guide prevention. As a database, this may be organized conceptually as three subsets: People, Places, Things (Hillman, 1982) (Figure 65.1).
3. Determine the functions or modules of the groups that could utilize the database and compare to the needs of the company; e.g., an industrial hygiene function is possible but may not be needed as a full module by the company.
4. Compare the present surveillance system in the company with the defined database model. Prioritize the development of the modules.
5. Assess the status of computerization in the company. Allocate which modules warrant computerization now vs. in the future. Not all aspects of a system need to be computerized. Some aspects may not be cost effective to develop, depending on the size of the company, the number of people using the information, and the criticality of the information to risk exposure for the workers in the industry. However, allow for growth in both software and hardware.

Company Size

The size of a company influences the models and decisions about injury surveillance database systems. In the past, database systems utilized mainframes and analyzed volumes of data, putting them on a scale and at a cost prohibitive to smaller industries. With the advent of powerful personal computers (PC) and easy networking, computerization is available to most companies. Despite the availability, many companies have not devoted resources to developing injury surveillance database systems despite the time-saving advantages. One possible reason is that the system is not developed in a manual form in those companies. Basic recordkeeping that entails filling out the OSHA forms does not constitute a simple injury surveillance system, unless there is some analysis of the data and preventive action is taken. Another

possible reason that resources have not been devoted to surveillance is that a company experiences so few recordable injuries that attention to injury surveillance is deemed unwarranted. A third reason may be that improving the efficiency of surveillance is not a top priority relative to other company improvements. It is important to remember, however, that computerization only recently has been so available and relatively affordable.

Large Companies

Large corporations were the first to develop injury surveillance database systems. This is not surprising for several reasons. There was a large volume of data due to the size of the company. The companies had a number of potential exposures from chemicals and the environment, and they had the resources for epidemiological studies.

Small and Medium Companies

Establishment sizes of 50 to 249 employees incur the highest injury incidence rates (per 100 full-time workers). Often, less than 100 employees is considered a small company and up to 1000 employees is medium-sized. The Small Business Administration (SBA), however, defines small as typically less than or equal to 500 employees (Stuart-Buttle, 1996). Regardless of definition, companies of about 50 to 500 employees in size should be particularly vigilant of their injury statistics. Despite national statistics, there will be some small companies with such low injury prevalence that a formal injury surveillance system is not necessary (other than mandated recordkeeping).

Some of the characteristics of a small industry indicate that conducting injury surveillance may be difficult (Stuart-Buttle, 1996). The main common characteristics are:

- Less formality
- Responsibility for several positions
- Greater responsiveness
- Less specific knowledge
- More management involvement
- Less data-oriented approach

However, it is precisely some of these characteristics, such as the responsibility for several positions, that would make an investment in an injury surveillance database system such an aid. The system could provide the benefits listed earlier and enhance an individual's and the company's efficiency.

Onsite or Offsite Medical and Surveillance Services

Only 6.3% of U.S. workplaces provide onsite medical surveillance that includes treatment and recording of occupational injuries (BNA, 1995). These facilities are large corporations and medium-sized companies with a sufficient number of employees or injuries or both for an onsite medical department to be cost-effective. The remainder of the workplaces use offsite treatment services that may or may not track data for the company. Usually a company employee, typically a person without medical knowledge, fulfills the basic clerical duty of recording the OSHA logs and filling out workers' compensation forms.

If the OSHA logs and accident records are the minimum data for surveillance tracking, the quality of the information is very important if it is to be useful. The questionable quality of offsite surveillance has been raised by NIOSH. Murthy of NIOSH suggested that physicians away from the worksite were less likely to probe a patient about work history and often had very little occupational medicine training to assess work-related conditions effectively (BNA, 1995). Although the company is dependent upon the HCP for accuracy and detail of diagnosis, the designated contact person for the company can exert some control by communicating the requirements of the company to the medical service (Stuart-Buttle, 1996). Close communication between the medical service and company is needed for all injury cases to keep the information accurate and up to date.

Brewer et al. (1990) described a clinic-based occupational injury surveillance system that served many industries. The system connected several clinics that treated employees for even minor injuries. All injuries, regardless of whether they were compensable or recordable on the OSHA log, were recorded. Statistical analyses were routinely conducted and results provided to the companies. The companies learned what type of injuries were occurring, the frequency, part of the body, severity, and how the injury was caused. Of all the cases analyzed within the system, 78.4% of the patients received care for minor traumas which were classified as cuts or lacerations, sprains, strains, or contusions. Such analyses showed that collecting data of first aid treatment is beneficial and that a company could reduce costs by preventing the occurrence of minor trauma. The study by Brewer et al. also demonstrated the feasibility of clinic-based occupational injury surveillance as a means of assisting employers with the control of work-related conditions.

Medical screening is often used to complement environmental control measures (Baker, 1990). The purpose of screening is to detect disorders early and implement appropriate intervention. Screening also serves as a monitor of the effectiveness of prevention measures already instituted. Many medical screening functions, such as auditory testing and pulmonary function screening, can be conducted and supported by an offsite contract service. Tracking and time tabling the routine screening of employees is easily performed by a database.

Types of Database Models

An injury surveillance database system can be developed anywhere along the continuum of simple to sophisticated. It may be a system that is customized, commercially purchased, or contracted as an offsite service. Whichever type of system is chosen, it is most likely to be modular. The most basic injury surveillance database model fulfills the recordkeeping function and provides some descriptive statistics of the data. Many PC packets provide the interface to enter the information required for the OSHA forms, analyze the data, and produce summary graphs by department, for the month, quarter, or year and sometimes across years. Printouts of the filled forms are also commonly available.

Beyond a recordkeeping level, the fundamental model can be represented by the three interacting modules of Hillman (1982): Places, People, Things (Figure 65.1). This still could be kept simple: Places includes a basic coding of the areas or jobs, possibly with a description; People includes demographics (gender, age, date of hire, job), and health data, both evaluations and incidents; and Things includes information files of chemicals or materials that are in the company. A small company may have very little information in the Places and Things files. A large corporation could expand this model with considerable databases of information under all three modules.

With this concept, potentially there may be many users of the database, depending what data is entered in the first place. Human resources, safety, medical, industrial hygiene, ergonomics, and risk management are some of the possible parties interested in an injury surveillance database system. Each group interfaces with the system from its own perspective, using its own programs to extract appropriate data, perform statistical functions and risk assessments as required, and present data in the format desired and appropriate for the user's job function. Injury surveillance is dependent upon someone being responsible for looking at the data of the injury reports that has been entered into the system. The responsibility may be almost any one of the users indicated, as the job function depends on the company size and structure. Each interface of a user group is a module that could be developed as the system evolves according to the company's needs.

Confidentiality

A database that is shared between many users raises concern about confidentiality and the need to limit access to certain information. Systems with distinct modules for each user group typically make each module a limited access to a defined group of people, usually by use of a password. A network can be designed to monitor passwords and terminal identification. Medical data are particularly sensitive information. Joyner

and Pack (1982) described the Shell Oil Company's system that encrypts the data and stores it centrally. Decryption programs were installed in the medical department to decode output to the terminal, therefore precluding disclosure of confidential information in the event the output inadvertently arrived at another terminal. Another technique that has been adopted to ensure confidentiality is the scrambling of social security numbers (Hillman, 1982; Wolkonsky, 1982).

Data Entry

Many of the early large and sophisticated systems were invested with efficient methods of data entry. For example, the Amoco occupational health system recorded a patient's medical history and examination on mark-sense forms that were computer input sheets that were entered by optical scanner. This eliminated any possible error from transcription (Wolkonsky, 1982). Questionnaires or surveys can be efficiently collected in a similar way. An overall benefit of a database, however, is the reduced re-entry of data. Personnel data, in particular, is typically entered many times if there is no network to a personnel database.

Computerization

There are many commercial software programs which can suit a variety of companies. (See For Further Information for software directories.) However, some industries choose to develop their own injury surveillance database system. The level of computerization that already exists in the company influences the choice of a customized program. Often there are many programs that already exist in a company, for example, production and quality control systems, payroll, accounting, and maintenance department systems. With the increase of computers in the workplace many companies have invested in management information system (MIS) departments to develop and maintain the hardware and software. Older systems were originally written for mainframes and minicomputers, and most newer systems are written for PCs. If an additional system is intended to be added to a network, it is sometimes easier to develop it rather than purchase a program that may be difficult to make compatible and to support in the future.

The compatibility of commercial software to an existing system is influenced by the age of the system and to what extent the software language has been updated. Therefore, companies with relatively new computer systems are more likely to be able to take advantage of the large commercial market, tailor the packages with fewer technical problems, and support the systems with in-house personnel. Considering the available choices of sophisticated programs, purchase of a system can save considerable design and development costs. The choice of system also depends upon the level of computer skill of the users or resources of the company. The Centers for Disease Control and Prevention (CDC) developed "Epi Info," a series of public-domain PC software programs in IBM-compatible format (DOS) for word processing, database, and statistics work in public health. (See For Further Information.) The programs are designed to be easily modified to meet company needs but still require some computer knowledge, especially since many computer users are only familiar with the Microsoft Windows operating system. In addition, before modifying the program, the user must clearly define the functions desired of Epi Info which requires knowledge of injury surveillance. Despite being in the public domain and having excellent technical support, the programs may not be compatible with some companies.

Not every company has extensive computerization or a network. Some may have computerization but choose to have an independent safety and health system. As stated above, the majority of workplaces do not have onsite medical services or extensive exposure issues, so the company's needs are primarily to track and analyze first aid logs, OSHA logs, and accident-related records. Job descriptions would be beneficial for injury analysis, but, as previously discussed, they can be difficult to maintain for some industries. Many commercial software programs are available that provide a recordkeeping function with OSHA log entry and analysis. (See For Further Information.) There are several safety or industrial hygiene functions that could be efficient to computerize depending on the company size, and for which there is available software, for example, maintaining Material Safety Data Sheets (MSDS), tracking PPE, and tracking employee training (Johnson, 1996).

An expanding area of computers is internet capability. The internet provides a source of information for data such as MSDSs, as well as updates on regulations. As more professional associations have representation on the world wide web, the internet may serve to help resolve technical questions.

65.8 Statistical Treatment of Data

A computer can perform sophisticated data analyses only if specifically instructed to do so. A good understanding of statistical methods is required to develop a data analysis program, and if necessary, a statistician should be consulted. There are commercial statistical packages that may be interfaced with a database system, and some of the packages guide a user in interpreting data. There is potential for a vast array of analyses of the data in an injury surveillance database, depending on the extent and sophistication of the overall system. For example, if there is a focus on industrial hygiene issues, numerous calculations may be required to interpret data from workplace monitoring. The safety and health professional needs some knowledge of statistics to interpret data beyond the OSHA logs. The commercial software market is rapidly responding to demand by producing programs specifically for industrial hygienists, such as those that assist with industrial hygiene formulae (Cameron, 1996). There are other programs on the market designed to meet the technical needs of the ergonomist and safety professional.

The majority of workplaces are small businesses in which those responsible for safety and health often do not have a background in statistics. In addition, they may not know what to look for in the basic data collected from OSHA logs and accident records. Some fundamental information is, what volume and type of injuries are occurring, to which parts of the body, while performing which jobs or tasks. Most of these questions can be answered by descriptive statistics. The commercial market offers many programs that focus on OSHA recordkeeping and provide basic data analysis with graphic output. After careful assessment of these programs there is likely to be a software package that meets the company's needs. However, even more useful information can be obtained if data collection and analysis extend beyond the minimum requirements for compliance. The following are a few points that will help to make the data more useful. The points may also help when assessing software programs.

- Convert the number of injuries into an incidence rate as described in OSHA's recordkeeping guideline (BLS, 1986). This provides a figure that has a common denominator with other companies and national statistics. The national injury incidence rate calculated by the Bureau of Labor Statistics (BLS) or the rate for the company's Standard Industrial Classification (SIC) code can serve as benchmarks for the company. Incidence rate can be calculated for groups of types of injuries and also for departments. Direct comparison between departments can be made when the numbers are expressed as incidence rates.
- The number of lost workdays is usually considered an indication of severity of the injury. An incident rate can also be calculated for the lost workdays. At times a given incidence may be so severe that it accounts for most of the recorded lost workdays. In such situations the incidence rate should be calculated with and without that particular case. The proposed recordkeeping revision does not require the number of restricted days to be recorded, however, a company may find it useful to record them (OSHA, 1996). When injuries begin to be detected early and undergo conservative treatment, then lost workdays are less likely. However, an increase in restricted workdays is more probable. Recording the restricted workdays provides some measure of progress, especially if the number of restricted workdays are then improved.
- Record the side of the body that was injured. This is not often recorded yet it can be useful to know. If the information is related to a record of the employee's dominant hand a useful picture of the stresses and injuries at the job begin to be built, which helps with job analyses and prevention strategies.
- Look at the incidence rate by age, gender, length of time on the job, and whether the employees are full-time or part-time. All of these analyses will help to determine if a certain group is

susceptible. Those who are very new on the job are typically suffering from lack of conditioning. If this is found to be the case, strategies can be implemented to gradually introduce the worker to the job. For example, orientation sessions could be scheduled throughout the first week rather than in the first couple of days.

- Look at the injury rate by supervisor or line leader if these are different groupings from the department or area. This can be very useful in determining if a certain team has a problem and suggests there may be some administrative issues to investigate. Similarly, data should be compared between shifts. Often any differences are related to management.
- Incorporate workers' compensation costs into the database and determine the mean for type of injury and cause and for each area or job. This helps with cost benefit analyses and prioritizing areas to be improved. Sometimes the incidence rate in a plant has increased, but the workers' compensation costs decreased. This may indicate that more people are reporting before the condition is severe and more expensive to treat.
- Track the work status of all those injured, whether they are off work on workers' compensation or on work restrictions. Develop a list on a regular basis, such as once a week, to ensure that each case is kept up to date and progressed toward full return to work, preferably at the previous job. It is not unusual for an employee to "fall between the cracks" and be on workers' compensation for a period of time without any functional change or decision about the case.
- Identify the causes and mechanisms of injuries based on the accident record, and in particular, relate the findings to the areas of high incidence rates.
- Look at the data to see if there are patterns according to time of day, days of the week, the month, quarter, or year. Compare across the years as well.
- If possible, collect first aid data as well as the data mandated by OSHA. Treating minor cuts and bruises takes time and there can also be an associated loss of productivity. If this information is collected and analyzed similar to the logs, there may be some useful emerging patterns that can help with identifying the causes and lead to prevention measures. Examples of common problems are blunt tools that cause the user to slip, or inadequate clearance around equipment that provokes bruising.

More sophisticated analyses can be conducted on data from OSHA forms. However, some understanding of statistics is needed to make use of such features in a software program, even if the interface is user friendly. A larger company may benefit from a data analysis that indicates when the occurrence of injuries is "out of control" relative to a determined baseline. Boyd (1988) presented one approach based on injury costs per exposure hour. Goldberg, (1993) describes an injury surveillance database with some useful, advanced analysis features. A description of some of these features follows.

- Actual exposure hours were collected from payroll to calculate the incidence rate. This allowed for a better picture of incidence rate related to overtime hours, especially since overtime was inconsistent and potentially a reason for increased injuries.
- Regression analysis helped determine the factors leading to lost and restricted workdays. The software interface displayed a list of independent variables and interactions from which to choose. For example, the number of restricted workdays could be looked at by gender, age, and length of employment. Regression analysis could also be used to predict the number of expected lost or restricted workdays for an injured employee, and this helped management plan for the situation.
- Time series modeling provided insight as to whether there were certain periods of higher incidence rates. Accidents are sometimes based on cyclical seasonal weather change, employment relationships, or perhaps rapid expansion.
- A forecasting feature smoothed the data to remove known cyclical fluctuations and provided indication of the data trend, whether the incidence rate was increasing or decreasing.
- Inclusion of cost and billing information was used to generate cost/benefit analyses.

Report Generation

Except for mandated government forms, reports are primarily for company use, (e.g., OSHA forms). The reports are not generally required for scientific presentations but rather analysis and communication. Two levels of reports are needed: one for the professional to explore and interpret the data, and another that summarizes the findings for presentation to management. The system needs to be flexible enough such that a summary report can be focused on specific findings.

The maxim "pictures speak a thousand words" is true when it comes to concise presentation. Good graphic presentation straight from the software program saves considerable time, especially when the alternative is transporting the data to another program. However, it is possible to set up standard links to a program that produces charts and graphs, so that graphs can be routinely produced.

65.9 Choosing a Commercial Injury Surveillance Database

One size does not fit all when it comes to setting up an injury surveillance database system (Wrench, 1990). If a commercial program is purchased, it will require professional tailoring to fit the needs of the company. For example, along with social security numbers, company identification numbers are commonly used for personnel tracking, yet every company has a different numbering system which requires modification of that data field. Likewise, the job coding system is different in every company.

Involve the employees as much as possible in decisions about a database. The employees know the information needed for their jobs and can contribute significantly to the needs assessment and system design and to evaluating existing programs. In addition, the success of a system is dependent upon their acceptance of it.

The following are some aspects to consider when choosing a commercial software program (Menzel, 1994).

Evaluating needs and resources

- Identify data management goals and objectives.
- Define requirements. (This may be accomplished by mapping out the existing manual system.) Include anticipated growth and potential new functions.
- Define company constraints, such as type of hardware system, technical equipment, and the characteristics of other company databases.
- Assess company resources (personnel, budget, hardware, and software).
- Develop a list of criteria, based on the needs and resources evaluation, by which to assess the software programs. There may not be a program that satisfies each criterion so developing a spreadsheet can be useful for deciding the trade-offs for a final selection. A scoring system could also be used to determine several final contenders.

Finding the software

- Use directories that may be available through professional safety and health associations (see For Further Information).
- Research articles on software in safety and health magazines.
- Respond to software advertisements in occupational safety and health magazines.
- Ask colleagues about the software they use and their experience with it.
- Request demonstration disks and onsite demonstrations.

Evaluating the software

- Assess the basic features of the programs against the criteria. This will narrow the selection for more critical view.
- Evaluate for qualities such as:
  - ease of use
  - provision of high quality, onsite training

  - variety and quality of reports
  - ability to transfer data from other programs
  - potential degree of customization
  - presence of security systems to safeguard confidentiality and data integrity
  - reasonable initial and continuing cost
- Select a few final contenders based on the extent to which criteria are satisfied.
- With program specifications attached to the proposal, request price proposals with details of features, service, and renewal fees.
- Request names of current users that can be contacted.
- Assess the responsiveness and quality of the support available.
- Review the length of time each company has been in operation.
- Assess their financial stability and long-term viability. (This is not easy but the number of clients being supported may give an indication.)

Selecting the vendor

- Negotiate the price
- Negotiate the extent of customization and support
- Negotiate for additional features such as training

Evaluating the effectiveness

- Ensure that the program is thoroughly reviewed before the first anniversary (when maintenance fees usually begin).
- If necessary, negotiate for modifications before paying maintenance fees.

When implementing computerization of a system ensure that parallel systems are run for at least a month after full installation, even if the computerized system is off-the-shelf (Bonnett, 1982). The accuracy of the system must be checked before abandoning the old methods. If there are software modifications or connections made to an existing network, it is better to assure that the system is running smoothly before relying upon it. New hardware should be tried also before being relied upon.

65.10 Summary

The cost benefits of injury surveillance database systems are becoming more apparent since there has been greater focus on accurate recordkeeping, and as computerization has become more affordable. There are several benefits including standardized and accurate recordkeeping, time saving and efficient data entry and analysis, enhanced tracking abilities, and effective guidance to prevention strategies.

Considerable and careful thought should be given by a company to the design of the injury surveillance database system. The design is dependent upon the company's needs, size, constraints, potential growth, and resources. An assessment of the company's needs prevents a quick purchase of a commercial system that might not have the growth potential, adequate support and customization services to make the program a successful addition to the company.

References

Baker, E. L., Honchar, P. A., and Fine, L. J. 1989(a). I. Surveillance in occupational illness and injury: concepts and content, in *Surveillance in Occupational Safety and Health*, Ed. E. L. Baker, p. 9-11. *AJPH* 79 Supplement.

Baker, E. L. 1989(b). IV. Sentinel Event Notification System for Occupational Risks (SENSOR): The concept, in *Surveillance in Occupational Safety and Health*, Ed. E. L. Baker, p. 18-20. *AJPH* 79 Supplement.

Baker, E. L. 1990. Role of medical screening in the prevention of occupational disease. *JOM.* 32(9):787-788.

Brewer, R. D., Oleske, D. M., Hahn, J., and Leibold, M. 1990. A model for occupational injury surveillance by occupational health centers. *JOM.* 32(8):698-702.

Bonnett, J. C. and Pell, S. 1982. Du Pont's health surveillance systems, in *Medical Information Systems Roundtable.* p. 819-823. JOM 24(10) Supplement.

Boyd, A. H. and Herrin, G. D. 1988. Monitoring industrial injuries: A case study. *JOM.* 30(1):43-48.

Bureau of Labor Statistics (BLS). 1986. *Recordkeeping Guidelines for Occupational Injuries and Illnesses.* OMB. No. 1220-0029, U.S. Government Printing Office, Washington, D.C.

Bureau of National Affairs (BNA). 1991. Preliminary results from OSHA survey said to show benefit of "integrated" plan. *Occupational Safety and Health Reporter.* 5-22-91:1712-1713.

Bureau of National Affairs (BNA). 1993a. All worksites need passive surveillance, job checklist for risk factors, ANSI group told. *Occupational Safety and Health Reporter.* 1-20-93:1444-1445.

Bureau of National Affairs (BNA). 1993b. OSHA's egregious penalty policy for recordkeeping violations upheld. *Occupational Safety and Health Reporter.* 2-10-93:1601-1602.

Bureau of National Affairs (BNA). 1993c. Comprehensive risk management programs effective at finding illness, draft report says. *Occupational Safety and Health Reporter.* 3-31-93:1899-1900.

Bureau of National Affairs (BNA). 1994. Recordkeeping violations affirmed against general dynamics as non-serious. *Occupational Safety and Health Reporter.* 3-16-94:1353-1354.

Bureau of National Affairs (BNA). 1995. NIOSH staffer says onsite surveillance plays important role in treating workers. *Occupational Safety and Health Reporter.* 11-8-95:817.

Cameron, M. 1996. IH calculator. *The Synergist.* April:17-18.

Committee Report. 1992. Scope of occupational and environmental health programs and practice. *JOM.* 34(4):436-440.

Dieterly, D. L. 1995. Industrial injury cost analysis by occupation in an electric utility. *Human Factors.* 37(3):591-595.

Duvall, M. N. 1996. Digest of official interpretations of the bureau of labor statistics recordkeeping guidelines for occupational injuries and illnesses (second revised edition). *Occupational Safety and Health Reporter.* 1-31-96:1144-1171.

EEOC, 1990. *Americans with Disabilities Act of 1990.* Public Law 101-336 101st Congress, July, 26. U.S. Government Printing Office, Washington, D.C.

Finucane, R. D. and McDonagh, T. J. 1982. Foreword, in *Medical Information Systems Roundtable.* p. 781-782. *JOM* 24(10) Supplement.

Froines, J. R., Dellenbaugh, C. A., and Wegman, D. H. 1986. Occupational health surveillance: A means to identify work-related risks. *AJPH.* 76(9):1089-1096.

Garrett, R. W. 1982. Environmental tracking at Eli Lilly and Company, in *Medical Information Systems Roundtable.* p. 836-839. *JOM* 24(10) Supplement.

Goldberg, J. H., Leader, B. K., and Stuart-Buttle, C. 1993. Medical logging and injury surveillance database system. *Int J of Indus Ergonomics.* 11:107-123.

Hagstrom, R. M., Dougherty, W. E., English, N. B., Lochhead, T. J., and Schriver, R. C. 1982. SmithKline Environmental Health Surveillance System, in *Medical Information Systems Roundtable.* p. 799-803. JOM 24(10) Supplement.

Hanrahan, L. P. and Moll, M. B. 1989. VIII. Injury surveillance, in *Surveillance in Occupational Safety and Health,* Ed. E. L. Baker, p. 38-45. *AJPH* 79 Supplement.

Hillman, G. 1982. ECHOES: IBM's environmental, chemical and occupational evaluation system, in *Medical Information Systems Roundtable.* p. 827-835. *JOM* 24(10) Supplement.

Holzner, C. L., Hirsh, R. B., and Perper, J. B. 1993. Managing workplace exposure information. *Am Ind Hyg Assoc J.* 54(1):15-21.

Johnson, L. F. 1996. A world of great choices. *Occupational Health and Safety.* October: 70 and 107.

Joiner, R. L. 1982. Occupational health and environmental information systems: basic considerations, in *Medical Information Systems Roundtable.* p. 863-866. *JOM* 24(10) Supplement.

Joyner, R. E. and Pack, P. H. 1982. The Shell Oil Company's computerized health surveillance system, in *Medical Information Systems Roundtable*. p. 812-814. JOM 24(10) Supplement.

Kuritz, S. J. 1982. The Ford Motor Company environmental health surveillance system, in *Medical Information Systems Roundtable*. p. 844-847. *JOM* 24(10) Supplement.

Langmuir, A. D. 1976. William Farr: founder of modern concepts of surveillance. *Int. J of Epidemiology*. 5(1):13-18.

Menzel, N. N. 1994. Occupational health software: Selecting the right program. *AAOHN Journal*. 42(2):76-81.

National Institute for Occupational Safety and Health (NIOSH). 1986. *Proposed National Strategy for the Prevention of Musculoskeletal Injuries*. DHHS (NIOSH) No. 89-129. Cincinnati, OH.

Occupational Safety and Health Administration (OSHA). 1970. *Occupational Safety and Health Act*. Public Law 91-596 91st Congress S. 2193 Dec. 29. U.S. Government Printing Office, Washington, D.C.

Occupational Safety and Health Administration (OSHA). 1996. OSHA notice of proposed rulemaking to revise occupational injury and illness recording and reporting requirements. (61 FR 4030, Feb. 2, 1996) *Occupational Safety and Health Reporter*. 2-7-96:1219-1270.

Occupational Safety and Health Administration (OSHA). 1990. *Ergonomics Program Management Guidelines for Meatpacking Plants*. OSHA 3123. Department of Labor/OSHA, Washington, D.C.

Parkinson, D. K. and Grennan, M. J. Jr. 1986. Establishment of medical surveillance in industry: Problems and procedures. *JOM*. 28(8):773-777.

Saldaña, N. 1990. DAS: A graphical computer system for the collection of musculoskeletal discomfort data. *Proceedings of the Human Factors Society 34th Annual Meeting*. p. 1097.

Seligman, P. J., Halperin, W. A. E., Mullan, R. J., and Frazier, T. M. 1986. Occupational lead poisoning in Ohio: Surveillance using workers' compensation data. *AJPH*. 76(11):1299-1302.

Smith, R. B. 1995. Recordkeeping rule aims for accuracy, wiser use of injury and illness data. *Occupational Health and Safety*. January:37-68.

Smith, F. R., Gutierrez, R. R., and McDonagh, T. J. 1982. Exxon's health information system, in *Medical Information Systems Roundtable*. p. 824-826. JOM 24(10) Supplement.

Stuart-Buttle, C. 1994. A discomfort survey in a poultry-processing plant. *Applied Ergonomics*. 25(1):47-52.

Stuart-Buttle, C. 1999. How to set up ergonomic processes: A small industry perspective, in *The Occupational Ergonomics Handbook*, Eds. W. Karwowski and W. S. Marras, CRC Press, Boca Raton, FL.

Sugano, D. S. 1982. Worker tracking — a complex but essential element in health surveillance systems, in *Medical Information Systems Roundtable*. p. 783-784. *JOM* 24(10) Supplement.

Sundin, D. S., Pedersen, D. H., and Frazier, T. M. 1986. Occupational hazard and health surveillance. Editorial. *AJPH*. 76(9):1083-1084.

The Synergist, 1995. Graphically speaking. *The Synergist*. October:7.

Tyson, P. R. 1991. Record-high OSHA penalties. *Safety and Health*. March:17-20.

Wegman, D. H. and Froines, J. R. 1985. Surveillance needs for occupational health. Editorial. *AJPH*. 75(11):1259-1261.

Wolkonsky, P. 1982. Computerized recordkeeping in an occupational health system: The Amoco system, in *Medical Information Systems Roundtable*. p. 791-793. *JOM* 24(10) Supplement.

Wrench, C. P. 1990. *Data Management for Occupational Health and Safety: A User's guide to Integrating Software*, Van Nostrand Reinhold, New York.

Yodaiken, R. E. 1986. Surveillance, monitoring, and regulatory concerns. *JOM*. 28(8):569-571.

For Further Information

ACOEM. 1996. *Directory of Occupational Health and Safety Software*. Version 9.0. American College of Occupational and Environmental Medicine, Arlington Heights, IL. Telephone: 847 228-6850.

Brauer, R. (Ed). 1996. *Directory of Safety Related Computer Resources*. American Society of Safety Engineers, Chicago, IL. Telephone: 847 699-2929.

Canadian Centre for Occupational Health and Safety (CCOHS). 1997. Directory of Health and Safety Software (Diskette). AIHA Press, Fairfax, VA. Telephone: 703 849-8888.
Centers for Disease Control and Prevention (CDC). 1996. *Epi Info, Version 6.04a.* Brixton Books, 740 Marigny Street, New Orleans, LA 70117. (504) 944-1074. Internet: http://www.cdc.gov

66

OSHA Recordkeeping

Stephen J. Morrissey

Oregon OSHA Consultative Services

66.1 Introduction

The history of occupational safety has been recorded as progress primarily initiated by injury, loss, and disaster. Despite many well-meaning efforts at controlling the human and social–economic losses in the workplace, it was not until the passage of the Occupational Safety and Health Act of 1970 that a comprehensive, integrated safety and health program at the national level was present in the United States. The OSHA Act, while safety and health oriented, has clear social and economic goals in its basic philosophy as reflected in the enabling legislation which in part states that the purpose of the Act was "... to assure so far as possible every working man and woman in the Nation safe and healthful working conditions and to preserve our human resources."

Over the years the regulatory aspects of OSHA have become more complicated and widespread with changing industries and job demands, public pressures for changes in safety law, and as the work environments have become better understood. This increased regulatory presence has not been without industry complaint. However, since OSHA has been in existence there have been clear drops in the numbers of workers injured, killed, and suffering adverse health effects, all of which translate into lower costs of doing business (Gray and Scholz, 1993; U.S. Department of Labor, 1995, 1997).

Fundamental to the effective control of job-related injuries and illnesses has been the development of a comprehensive safety and health recordkeeping standard. The Occupational Safety and Health Act of 1970 gave OSHA the authority to develop such a program, and in 1971 the occupational injury and illness recordkeeping regulations, 29 CFR Part 1904, were published. These regulations set comprehensive and mandatory requirements for reporting job-related injuries and illnesses. The basic goals of OSHA recordkeeping are to establish the risks present in the work environment; to direct employer attention to tasks and areas that need attention or which have unacceptable injury–illness levels; and to help guide research and enforcement activities to better serve industry and the employee.

0-8493-2641-9/99/$0.00+$.50

However, despite the breadth and detail of these recordkeeping goals, the overall recordkeeping requirements have been kept limited to reduce the burden on employers and to help safety and health professionals identify and then respond to high-risk industries and workplaces. Over the years, OSHA has set forth and enforced the recordkeeping regulations. The Bureau of Labor Statistics (BLS) has prepared survey and recordkeeping forms and information and conducted the Annual Survey of Occupational Injuries and Illnesses. Information from this survey is used by OSHA to identify industries with high injury rates. In 1990, the duties performed by the BLS were transferred to OSHA.

This chapter will discuss the basics of occupational safety and health recordkeeping as mandated by OSHA. It will not discuss the parallel systems used by workers' compensation systems, as these systems differ widely by state and many major changes in compensation systems are being undertaken or planned. There are also major changes proposed in the actual mechanics of OSHA recordkeeping and in definitions of critical factors that are used to record data. At the time of writing, only preliminary proposals have been presented for these changes, and these proposed changes in recordkeeping will be identified and discussed later in this paper. To obtain the most current information, the interested reader is encouraged to contact his or her local or regional OSHA office or access the OSHA world wide web site at: http://www.osha.gov or http://www.osha-slc.gov.

66.2 Who Must Keep Records

Compliance with OSHA recordkeeping is required of private sector employers in the 50 States, the District of Columbia, Puerto Rico, the Virgin Islands, American Samoa, Guam, and the Trust Territories of the Pacific Islands. In general, employers who employed at least 11 or more employees at any time in the previous calendar year must comply with OSHA recordkeeping regulations. If employers have more than one establishment or place of business, it is the combined total number of employees at all places of work that is considered in determining the number of employees in the previous calendar year. If this number is 11 or more at any time in the previous calendar year, then the company must comply with OSHA recordkeeping requirements, and records must be kept at *each* individual place of employment. Employers with fewer than 11 employees at any one time in the previous calendar year are generally exempt from recordkeeping, as are the self-employed, partners with no employees, those that employ domestic help in their private residences, and employers engaged in religious activities, services, or rites. However, on occasion, the Bureau of Labor Statistics (BLS) will require companies with fewer than 11 employees or in other generally noncovered occupations to keep records in order to develop more accurate hazard data for their Annual Survey of Occupational Injuries and Illnesses.

Employers who are required to keep records must do so at each of their establishments, regardless of the number of employees at the site, but they *do not have to keep records for employees of other firms, independent contractors, or companies temporarily present on the worksite.* If an injury or illness occurs to the employees of the other firms on the worksite, these injuries or illnesses must be recorded on the records of the company for which the injured employee works, or that of the company which has the most immediate control and supervision of work activities. Thus, if the employer is using temporary or leased workers, and has immediate control of their day-to-day work activities, then the employer must record any injuries or illness on the employer's own records. Employers must also present copies of the OSHA 200 and 101 upon request of OSHA inspectors and may, under the direction of other codes, have to provide these data to employees or their legal representatives.

66.3 Which Employers Do Not Have to Keep Records

Despite the apparent simplicity of the recordkeeping rule of 11 or more employees, not all industries or occupations are required to follow OSHA recordkeeping requirements. Most manufacturing, agricultural, construction, and service industries are required to keep records if they have the required number of employees. However, OSHA recordkeeping is generally not required for employers in what are considered

low-risk occupations. These low-risk industries or occupations include the retail trade, finance, insurance, real estate, and various service industries in the Standard Industrial Codes (SIC) 52-89. However, there are some employers within this group who must follow recordkeeping regulations. These covered industries are building materials and garden supplies, SIC 52; general merchandise and food stores, SIC 53 and 54; hotels and other lodging places, SIC 70; repair services, SIC 75 and 76; amusement and recreation services, SIC 79; and health services, SIC 80. Also exempt from recordkeeping requirements are those industries which have an average lost workdays cases injury rate over a three-year recording period that is at or below 75% of the U.S. private sector injury rate. Even though these industries are exempt from general federal recordkeeping requirements, they may be subject to state recordkeeping requirements, must comply with all other safety and health requirements set by OSHA, and as noted above, may be required by the BLS to participate in special studies to establish injury and illness rates.

Employers who do not have to keep formal OSHA records either because of the number of employees or by virtue of their SIC code still must meet reporting and recordkeeping requirements in the event of the death of an employee or if multiple employees are hospitalized as a result of a work-related incident or exposure. Basically these requirements are to notify the local OSHA office within eight hours in the event of an employee death or within eight hours in the event of an incident that results in the hospitalization of three or more individuals.

Employers who are subject to other federal government safety and health regulations must also comply with basic OSHA recordkeeping requirements. However, these employers can use modified reporting forms as long as they are equivalent to the required OSHA 200 and 101 forms discussed later in this paper.

Churches or other religious organizations or their affiliated operations do not have to comply with recordkeeping requirements for those individuals who perform religious rites or activities. However, churches or other religious organizations must comply with OSHA recordkeeping requirements for those individuals not engaged in performance of religious rites or activities. In some cases, volunteers who are not clergy or direct participants in religious services, may be covered by recordkeeping requirements.

Charitable and non-profit organizations, if they meet the number of employees criteria, must comply with recordkeeping requirements.

More detailed information on specific employers, industries, and SIC codes that are required to keep records, and to what extent, is available in the two primary documents on recordkeeping (U.S. Department of Labor, 1986 a,b).

It must be noted again that this discussion applies only to OSHA mandated recordkeeping. Recordkeeping requirements mandated by insurance or workers' compensation laws are often separate and different from those of OSHA, and employers should consult with their own state to determine applicable laws and codes.

66.4 What is An Employer? An Employee?

A critical aspect of recordkeeping is the determination of who is an employer, an employee, or what is a place of employment. In Section 3(5) of the Act, an employer is defined as "…a person or persons engaged in business that directly or indirectly affects commerce and has one or more employees and who is not the United States or any State or political subdivision of a State."

An employee is any individual who receives compensation of any form from the employer for services. This also applies to part-time, temporary workers or leased workers or employees. Volunteers are exempt from recordkeeping if they serve of their own free will and do not receive any compensation for their efforts. Temporary workers are a particular issue these days as more and more employers use them to fill staffing needs. Despite common views, work-related injuries and illness experienced by temporary employees are reported on the records of the employer who has the most immediate supervision, direction, or control of them. This is almost always the employer who is using the temporary worker. Eventual liability for any citable violations of safety and health codes may, however, be distributed between both the immediate employer and the temporary worker supplier.

There are many more highly specific questions that can be raised that relate to the role and definitions of an employer and employee, and many of these are discussed in the recordkeeping reference, *Recordkeeping Guidelines for Occupational Injuries and Illnesses* (U.S. Department of Labor, 1986b). These documents are available from the U.S. Government Printing Office, local and regional OSHA offices, and from OSHA's world wide web sites (http://www.osha.gov; http://www.osha-slc.gov).

A place of employment is any location at which services or industrial operations are performed or where employees report to work, work at, or where they are paid. Records have to be kept for each individual workplace.

66.5 What Records Have to be Kept, Should be Kept?

While the regulatory coverage under OSHA is extensive and sometimes complicated, the recordkeeping is relatively simple, with only two forms required. These two forms are:

OSHA 200: Log and Summary of Occupational Injuries and Illnesses: The OSHA 200 form is widely available and will not be reproduced here because of its large size (11 inches by 17 inches). The OSHA 200 is used to record and classify all recordable injuries and illness and contains three basic sections. Section I records information on the date of the injury, who was injured, how, and where (Columns A–F). Section II records the nature and severity of the injury/illness, how much time was lost, and whether the injury resulted in a fatality (Columns 1–6). The third section (Columns 7–13) contains information on occupational illnesses experienced by the employee. Section 7 has seven subsections (a to g) that further define the nature of the occupational illness.

Data for any injury or illness must be filled in within six days after the incident is discovered and, in the case of a fatality or multiple hospitalizations (three or more workers), the local OSHA office must be notified within eight hours. At the end of each calendar year, the data on this form are totaled at the bottom, signed, and dated. The summary sections describing overall lost workdays (Columns 1–6), must be posted in a conspicuous location at each work location for the entire month of February in the following calendar year. The completed forms must be kept at each work location for the next five calendar years, even if an establishment changes owners. If, in years subsequent to the one in which an injury or illness is recorded, there need to be changes in a recorded entry based on new information, then the original entry must be lined out (not erased) and a corrected entry made. If errors in recording are discovered, or if unreported cases are identified, these need to be entered.

Conversely, if it is later found that an entry is a nonreportable entry, it should be lined out but not erased. While only recordable injuries have to be entered on the OSHA 200, many companies enter all injuries and illnesses on the form to help identification, understanding, and control of hazards that may be present in the workplace.

OSHA 101, Supplemental Record of Occupational Injuries and Illnesses: For every recordable event entered on the OSHA 200 log, it is necessary to record additional information on the OSHA 101 form or its equivalent. This form gives additional information that describes how the accident or illness occurred, the work processes involved, substances or exposures present, and more detail on the worker and the exact nature of the injuries or illnesses experienced. This form is also commonly used by workers' compensation carriers and various state organizations for filing and evaluating claims. In general, entries on the OSHA 101 must be completed within six working days of the date the employer received information on the event. A major problem with the OSHA 101 and 200 forms is that the data from these forms are not always properly interrelated, or even related to information describing the actual accident. This lack of correspondence can complicate or slow down understanding and control or elimination of hazards.

66.6 Posting Requirements

Employers who are required to keep records must keep OSHA 101 and 200 forms for each physical location where work is performed. If an employer has multiple locations, then separate sets of records

must be kept for each distinct location. Further, if within a physical work location distinctly different or separate activities are performed, separate records must be kept for each area. As noted earlier, the OSHA 200 must be summarized and posted for the entire month of February in a place that is used by all employees.

Failure to keep proper records can lead to substantial fines from OSHA. Over the past few years, some recordkeeping penalties have approached a million dollars, so the consequences can be substantial from a regulatory aspect. The more real penalties that occur from improper recordkeeping are the loss of control of the economics of production. Studies have found that companies that keep good records have good safety and health programs, lower overall production costs, and higher profitability (Gray and Scholz, 1993). The changes in OSHA regulations proposed *at the time of the writing of this chapter* all support a reduction in penalties or enforcement schedules for companies that have a well-developed safety and health program. OSHA considers careful recordkeeping and tracking of progress made in addressing accidents, injuries, and illness as the cornerstone of any such program.

66.7 Essential Definitions and Concepts for Recordkeeping

Recordable Injury/Illness: By OSHA definition, a recordable event is any "... work-related death and illness, and those work-related injuries which result in: Loss of consciousness, restriction of work or motion, transfer to another job, or require medical treatment beyond first aid." (U.S. Department of Labor, 1986a). It is important to understand that a case being recordable does not necessarily mean that the employer is at fault, has violated any OSHA standards, that the worker is at fault, or that any injuries are of a compensable nature. Recording an event only means that the consequences to the worker meet the legal definitions for being recorded. Once a case is determined to be recordable, it must be recorded in the correct columns on the OSHA 200 and 101 forms or their equivalent.

Other tests to define the recordability of an injury or illness are that an injury or illness is recordable if it results in *any* of the following: (1) the inability of the worker to perform all parts of the normal work assignments with the same ability as before the incident; (2) transfer of the worker, even temporarily to another job; (3) physical damage to the structures of the body, such as fractures, cuts requiring stitches, burns and bruises requiring repeated visits to a health care provider, and continuing infections; (4) loss of consciousness; or (5) an illness or injury that requires treatment by licensed medical personnel or physicians.

Work Related: An injury or illness is considered to be work-related if it resulted from a single or cumulative exposure or event in the work environment. The work environment is the immediate workplace and any and all locations where employees are engaged in work-related activities, or have to be present at, as part of their normal work activities. When workers normally have to travel between different worksites as part of work or to perform a work duty, the transportation may also be considered as a work activity, and injuries that occur during this transportation may be recordable. With many types of injuries and illnesses, determination of work-relatedness can be complicated, and the interested reader is referred to either of the documents on recordkeeping (U.S. Department of Labor, 1986 a,b) for more detailed information. Determination of work-relatedness is usually easier for injuries than for illnesses, particularly for those illnesses whose symptoms only appear after long exposures or long latency periods after a critical exposure.

First Aid: With respect to OSHA, first aid is "...any one-time treatment, and any follow up visit for the purpose of observation of minor scratches, cuts, burns, splinters, and so forth, which do not ordinarily require medical care. Such one-time treatment and follow up visits are considered as first aid, even though provided by a physician or registered professional personnel."

The critical distinction between medical treatment (a recordable case) and first aid treatment (a nonrecordable case) depends not only on what treatment is provided but also on the severity of the injury involved. A brief listing of different types of injuries and whether they classify as being first aid or as medical treatment is presented as Table 66.1 and further discussion of the recordability of injuries and illnesses is given in the two basic documents on recordkeeping (U.S. Department of Labor, 1986 a,b).

TABLE 66.1 Determining an Injury's Recordability

Type of Injury	Type of Treatment Needed and Subsequent Recordability First Aid Only, Not Recordable	Requires Medical Treatment Recordable
Cuts, lacerations, abrasions, splinters, punctures	Bandaging on any visit to health care provider (HCP) Application of antiseptic on first visit to HCP Application or ointments on any visit to prevent drying or cracking of skin Removal of foreign bodies by tweezers or other simple techniques Removal of non-embedded foreign bodies in eye by irrigation	Stiches Butterfly sutures Medical treatment of infections or application of antiseptic on second or subsequent visits to HCP Removal of foreign bodies requiring the skilled services of physician
Fractures	X-ray is taken as a precaution and found to be negative.	X-ray shows fracture. Cast or other professional method of immobilization of limb is applied.
Strains, sprains, and dislocations	Use of an elastic bandage on strain on the first visit to HCP Use of hot or cold compresses on strain on the first visit to HCP	Application of casts or other professional means of immobilizing injured part, including rigid splints Use of hot or cold compresses for treatment of strains, sprains, or dislocation on second or subsequent visits to HCP Use of diathermy and whirlpool treatment Hot wax treatments
Thermal or chemical burns. Any burn is recordable if the worker cannot perform any of his or her normal duties even if medical treatment is not required.	Treatment by HCP for first degree burn	Treatment of all second- and third-degree burns.
Bruises, contusions. Any bruise or contusion is recordable if the worker's range of motion is affected in any manner which prevents him or her from performing any of the regularly assigned duties, regardless of whether medical treatment is rendered.	Soaking or application of cold compresses to a bruise on the first visit to HCP	Treatment of a bruise by draining of collected blood Soaking or application of cold compresses on second or subsequent visits to HCP
Miscellaneous procedures	Tetanus shots Observation of injury on subsequent visits	Any hospitalization, even if only for observation will usually result in lost workdays and is thus recordable Use of prescription drugs

Injury: Is any injury or disorder such as a cut, fracture, sprain/strain, etc., that results from a work accident or from a *single, instantaneous exposure* to the work environment. This includes exposures to all work processes, insects, animals, and chemical or toxic agents.

Illness: Is any abnormal condition or disorder other than an injury that is a consequence of exposure to environmental factors related to work and the work environment. This includes cumulative, acute and chronic illness or diseases resulting from any exposure to agents in the work environment by inhalation, absorption, ingestion, direct contact, or direct exposure. There are seven different categories of occupational illness which are recorded on the OSHA 200 form. These illness categories are: Occupational skin diseases or disorders; dust disease of the lungs; respiratory conditions due to toxic agents; poisoning (systematic effects of toxic materials); disorders due to physical agents other than toxic materials; disorders associated with repeated trauma (noise-induced hearing loss, vibration-related injuries such as Raynaud's

phenomenon, and other conditions arising from repetitive motion such as bursitis, synovitis, tenosynovitis); and, all other occupational illnesses.

Criteria to identify and evaluate the work-relatedness of injuries or illnesses related to repeated trauma (cumulative trauma), such as tendinitis, synovitis, bursitis, back injuries, and carpal tunnel syndrome, are more fully developed in the *Ergonomics Program Management Guidelines for Meatpacking Plants* (U.S. Department of Labor, 1993). This document, which is available on OSHA's world wide web servers or from OSHA centers, describes the stages in developing an ergonomics program to reduce the losses associated with ergonomics injuries.

Restricted Work or Motion: A work or motion restriction occurs when, as a consequence of a work-related injury or illness, the employee cannot perform any part of, or all of, the normal job assignments during any part of, or all of, a workday or workshift. The common practice of having someone move to another job for a while to rest up thus constitutes a work restriction and requires the accident or illness to be recorded. The key point to understand is that if, as a consequence of a work-related accident, the employee cannot do the same job or work assignments at the same rate as before the accident, then the accident is recordable.

66.8 Recording Data on OSHA Forms

Once an injury or illness is determined to be recordable it must be evaluated to determine its type and outcomes. There are three types of recordable cases: Fatalities; accidents resulting in lost workdays; and cases that are recordable but do not result in lost workdays. Each case must be entered into the correct columns on the OSHA 200 form.

For fatalities, the entry is obvious, but determination of what is to be entered for lost workdays can be more complicated. Lost workdays occur either because the worker could not be at work due to the injury or illness, was unable to perform the normal job duties for a period of time, or the worker was assigned to another job until he or she recovered (light duty, modified duty, restricted work). In all cases, the total number of days to be recorded is the total number of days that worker was not present due to the injury or illness not including the initial day of injury/illness, or any days that the worker would not have normally worked (holidays, weekends, special days off, vacations) as a result of the injury or illness. If the worker's injury extends into another year, the amount of time that will be lost in the next year must be estimated and entered for the year in which the injury occurred.

Once the total number of days missed is correctly entered, it becomes important to properly and fully detail the remainder of the OSHA 200 and 101 to facilitate tracking of problems in the workplace as well as to meet recordkeeping requirements. While the various columns on the OSHA 200 form are self-evident, a common error is to incorrectly or too briefly describe the events that led to the accident. As much as possible, the physical location where an accident occurred (loading dock station 3; not "receiving"), actual job title of the individual involved in the accident (assistant press operator, machine #3; not machine operator), and a clear description of the injury (laceration to top of left hand; not cut hand) should be included on the 200 form to facilitate tracking of accidents, injuries, illnesses, and losses.

Data Developed From OSHA 200 Form: There are several statistics that are developed from the OSHA 200 form to describe the relative and absolute injury and illness experience for a particular facility, company, or industry. These data are used by OSHA to direct inspections and research and by insurance providers in rate setting.

All OSHA statistics are based on an Incidence Rate statistic (Equation 1) that is the number of injuries, illness, or lost workdays normalized to reflect a worksite with 100 full-time employees. As all employers who must comply with OSHA must use the same basic criteria for recordkeeping, this allows direct comparison of individual company loss experience with respect to the respective industry or region, across time, or even within a company. The statistic that is most commonly used by OSHA to direct inspections, define research needs and also by insurers to set and adjust rates is the Lost Workdays Case Incident Rate (LWDCIR). The LWDCIR is calculated by entering the appropriate date into the following equation:

$$\text{Incidence Rate} = \frac{N \times 200{,}000}{HW \times 2{,}000} \quad (1)$$

where

N = total number of recordable injuries, illnesses, or lost workdays (sum of Columns 2 and 9)

HW = total number of hours worked by all full-time and part-time employees during the previous calendar year as derived from payroll data

200,000 = the normalizing factor to correct data to a 100 full-time employee workforce that works 40 hours a week, 50 weeks a year

66.9 Proposed Changes in OSHA Recordkeeping

Over the past 10 years, numerous criticisms have been leveled against OSHA's recordkeeping requirements by industry, government, and academic policy and study groups. Some of the most common criticisms included that the current system was too complicated, particularly with respect to definition of injury and illness classifications; that the current system encouraged under-reporting or misreporting of injuries and illnesses; tracking of changes in the system was hard to follow; difficulty of the forms to facilitate good accident investigation; and, the difficulty of establishing "work-relatedness" of injuries and illnesses.

In February of 1996, OSHA published its proposed modifications to recordkeeping, "29 CFR Parts 1904 and 1952: Occupational Injury and Illness Recording and Reporting Requirements; Proposed Rule" for comment in the February 2, 1996, *Federal Register.* This proposal, which will be debated and discussed for the next year or so, involves major changes in several aspects of recordkeeping. Important definitions and criteria that are being proposed for change are briefly described in the following paragraphs. In addition to changed definitions, the recording forms would be changed to the OSHA 300 (replacing the OSHA 200) and the OSHA 301 (replacing the OSHA 101). These proposed forms (from the *Federal Register* of February 2, 1996) are shown as Figures 66.1 and 66.2. Bearing in mind that this legislation is being discussed and changed *at the time of publication of this book,* the interested reader should consult the *Federal Register* or a local OSHA office for more information.

OSHA 300 and 301 Forms: The new OSHA 300 form (Figure 66.1) is designed to fit on standard 8½ inch by 11 inch paper and is significantly streamlined in definitions and instructions for use as well as providing a column for the employer's use. Some of these improvements are described in the following sections and are designed to make recordkeeping easier and more consistent. The use of computerized equivalents is specifically allowed in the proposed regulations to increase speed and accuracy of recordkeeping as well as to facilitate better database manipulation and study. Similarly, the OSHA 301 (Figure 2), has been redesigned to facilitate ease and consistency of data entry and linking to other reports.

66.10 Specific Changes in Recordkeeping Requirements

All employers must maintain an OSHA 300 log or its equivalent at each establishment. Employers with multiple establishments may maintain a consolidated log for establishments with no more than 20 employees. Recordable injuries or illness must be recorded within seven calendar days of receiving information that a recordable injury or illness has occurred. The employer will also maintain an Injury and Illness Incidence Record, OSHA 301 or its equivalent, and it must be updated within seven calendar days of receiving information that a recordable injury or illness has occurred.

Before the end of January of each calendar year, the employer will post a year-end summary of all occupational injuries and illness for each establishment for the previous year. This form will remain posted for the *entire year* and be updated as new information is determined. This form *must be signed by a responsible company official indicating that they have examined the summary log and that the year-end summary is true, accurate, and complete.* Records must be kept for three years following the calendar year they represent rather than the current five-year period. During this retention period, the OSHA 300 and 301 must be updated if new information is discovered.

This form is not an insurance form. Cases listed below are not necessarily eligible for Workers' Compensation or other insurance. Listing a case below does not necessarily mean that the employer or worker was at fault or that an OSHA Standard was violated.

OSHA Injury and Illness Log and Summary

Public Law 91-596 and 29 CFR 1904 require you to:

- Enter all recordable occupational injuries and illnesses. (See instructions on back.)
- Update and retain completed form for three years.

Failure to complete, update and post can result in the issuance of citations and penalties.

U.S. Department of Labor
Occupational Safety and Health Administration

Form approved O.M.B. No. 1218-0000
See O.M.B. disclosure statement on back.

For calendar year ____

Page ___ of ___

Establishment Name __________

Establishment Address __________

Mailing Address if different __________

Industry description and Standard Industrial Classification (SIC) if known (e.g. Manufacture of motor truck trailers, SIC 3715)

	CASE IDENTIFICATION				CASE DESCRIPTION	CASE CLASSIFICATION (Check only one)					OTHER
A. Employee's Name (e.g. Doe, Jane B.)	B. Case Number (e.g. 1, 2,3...)	C. Date of injury or illness (m/d)	D. Department and location where event occurred (e.g. loading dock north end)	E. Regular job title (e.g. Welder)	F. Description of injury or illness; part(s) of body affected, and object/substance which directly injured or made employee ill (e.g. Second degree burns on right forearm from acetylene torch)	G. Death (X)	H. Involving Days Away (X)	H. Involving Days Away (# Days)	I. Without Days Away: Restricted Work Activity (X)	I. Without Days Away: Other (X)	J. Employee Use
						☐	☐		☐	☐	
						☐	☐		☐	☐	
						☐	☐		☐	☐	
						☐	☐		☐	☐	
						☐	☐		☐	☐	
						☐	☐		☐	☐	
						☐	☐		☐	☐	
						☐	☐		☐	☐	

YEAR END SUMMARY
Complete the year end portion of this form, even if there were no cases during the year. Fold along line to the right and post this form from February 1 to January 31 where

Employees, former employees, and their representatives have the right to review all OSHA Injury and Illness Records, in their entirety, for this establishment.

Year end totals ___ ___ ___ ___ ___

Annual average number of employees ______

Total hours worked by all employees ______

I have examined this Log and Summary and certify its accuracy and completeness X________ (Responsible Company Official) Title ______ Phone ( ) ______ Date / /

Knowingly falsifying this document can result in fine, imprisonment, or both. Draft OSHA Form 300 (10/95)

FIGURE 66.1 Proposed OSHA 300 form that will replace OSHA 200. (From *Federal Register*, Friday, February 2, 1996; 29 CFR Parts 1904 and 1952: Occupational Injury and IllnessRecording and Reporting Requirements; Proposed Rule; pages 4030-4067.)

OSHA Injury and Illness Incident Record

Public Law 91-596 and 29 CFR 1904 require you to update and retain completed form for three years.

Failure to complete this form can result in the issuance of citations and penalties.

Employees, former employees, and their representatives have the right to review all OSHA Injury and Illness Records, in their entirety, for this establishment.

This form is not an insurance form. Cases listed below are not necessarily eligible for Workers' Compensation or other insurance. Listing a case below does not necessarily mean that the employer or worker was at fault or that an OSHA Standard was violated.

U.S. Department of Labor
Occupational Safety and Health Administration

Form approved O.M.B. No. 1218-0000
See O.M.B. disclosure statement on back.

_____ Case number from OSHA Form 30

Employee

1. Last name First name MI
2. Male ☐ Female ☐ 3. Date of birth / /
4. Home address
5. Date hired / /

Health Care Provider

6. Name of health care provider
7. If treatment off-site, facility name and address

8. Hospitalized overnight as in-patient? (If emergency room only, mark "no") yes ☐ no ☐

Employer Use (Optional)

Completed by

Name Title

Phone ( ) Date

Illness or Injury

9. Specific injury or illness
   (e.g. Second degree burn or Toxic hepatitis)
10. Body part(s) affected *(e.g. Lower right forearm)*
11. Date of injury or illness: / / 12. If employee died, date of death / /
13. If the case involved days away from work or restricted work activity, enter the date the employee returned to work at full capacity: / /
14. Time of event: (Specify a.m. or p.m.) 15. Time employee began work: (Specify a.m. or p.m.)
16. All equipment, materials, or chemicals employee was using when the event occurred.
    (e.g. Acetylene cutting torch, metal plate)
17. Specify activity the employee was engaged in when the event occurred
    (e.g. Cutting metal plate for flooring) Indicate if activity was part of normal job duties.
18. How injury or illness occurred. Describe the sequence of events and include any objects or substances that directly injured or made the employee ill.
    (e.g. Worker stepped back to inspect work and slipped on some scrap metal. As she fell, worker brushed against the hot metal)

Draft OSHA Form 301 (10/95)

FIGURE 66.2 Proposed OSHA 301 form that will replace OSHA 101. (From *Federal Register*, Friday, February 2, 1996; 29 CFR Parts 1904 and 1952: Occupational Injury and Illness Recording and Reporting Requirements; Proposed Rule; pages 4030-4067.)

If an establishment changes hands, the new owners are responsible for recording only those injuries and illnesses that occur after the purchase date. They still have to keep the previous owners' records for the required period of time.

Access to Records: The employer must, upon request by a government representative, provide copies of OSHA Forms 300 and 301 or their legal equivalents, year-end summaries for their own employees, and injury and illness records for "subcontractor employees." When the request for records is in person, the information must be provided in hard copy within four hours. If the request is in writing, the information must be made available within 21 days unless requested otherwise.

Upon request, employees, former employees, and/or their representatives may see the OSHA 300 and 301 forms or their legal equivalents. These forms must be made available for viewing by the close of business on the next scheduled workday. An individual or his or her representative may also request to see the OSHA 301 form for his or her own injury or illness.

Days Away from Work (Lost Workdays): is defined as the total number of days the employee could have worked but did not because of the injury/illness. These days do not include the day the worker was injured or days the employee would not have normally worked because of holidays or weekends. The maximum number of days away from work will be entered as 180 or "180+" in the days lost column. For injuries that last into the next calendar year, the number of days lost should be estimated and entered into the days lost column for the year in which the injury/illness occurred. The 180 or 180+ days lost limits still apply.

Employee: an employee is any full-time, part-time, temporary, leased, limited service, "independent contractor" subject to day-to-day supervision, and those corporate officials who receive compensation of any form. Day-to-day supervision is defined as specifying how all aspects of output, production, and work processes are done. This definition now formally includes temporary or leased workers, independent contractors, or migrant laborers.

First Aid: refers to any treatment in the following comprehensive list that is provided by any heath care provider for a work-related injury or illness:

1. Visit(s) to health care provider limited to observation.
2. Diagnostic procedures including use of prescription medicines solely for diagnostic purposes.
3. Use of nonprescription medications, including antiseptics.
4. Simple administration of oxygen.
5. Administration of tetanus or diphtheria shot(s) or booster(s).
6. Cleansing, flushing, or soaking wounds on skin surface.
7. Use of wound coverings such as bandages, gauze pads, etc.
8. Use of any hot/cold therapy for local relief, except for musculoskeletal disorders. (Musculoskeletal disorders are covered in a special section, Appendix B, in the Proposed Rules).
9. Use of any totally non-rigid, non-immobilizing means of support, such as elastic bandages.
10. Drilling of a nail to relieve pressure for subungual hematoma.
11. Use of eye patches.
12. Removal of foreign bodies not embedded in the eye, if only irrigation or removal with a cotton swab is needed.
13. Removal of splinters or foreign material from areas other than the eyes by irrigation, tweezers, cotton swabs, or other simple means.

Any procedure other than those listed above is considered as being medical treatment, and thus defines the injury or illness as recordable.

Injury or Illness: is any sign, symptom, or laboratory abnormality which indicates an adverse change in an employee's anatomical, biochemical, physiological, functional, or psychological condition.

Work-Related: is still a very complicated question, but under the proposed rules, an injury or illness is work-related if an event or exposure in the *work environment* either caused or contributed to the

resulting condition or aggravated a pre-existing condition. The events or exposures must occur at the employer's establishment or as a consequence of a work activity performed as a condition of employment. More detailed descriptions of the criteria for work-relatedness are given in Appendix A of the Proposed Rules.

Restricted Work Activity: means that the employee is not capable of performing his or her normal work activities or the activity he or she was performing at the time of the injury or illness, at full capacity, for a full workshift.

Recordable Injury or Illness: a recordable injury or illness is one that meets all of the following four criteria:

1. An injury or illness, as defined above, exists.
2. The injury or illness is work related.
3. The injury or illness is new. A new injury or illness does not result from the recurrence of a pre-existing condition if no new or additional workplace incident or exposure or occurs.
4. The injury or illness meets one or more of the following conditions:
   a. results in death or loss of consciousness,
   b. results in day(s) away from work, restricted work activity or job transfer,
   c. requires medical treatment beyond first aid,
   d. is a recordable condition as listed in Appendix B of the proposed rules.

Only if an injury or illness meets *all* of these tests is it recordable.

66.11 Summary and Conclusions

The recordkeeping requirements developed by OSHA set the basic criteria for controlling economic and human loses in the workplace. While recordkeeping may seem to be a nonproductive, time-consuming task, it is the law, and when properly used, will result in a more profitable organization.

References

Gray, W. and Scholz, J., Does regulatory enforcement work? *Law and Society Review*, July, 1993.

U.S. Department of Labor, Bureau of Labor Statistics (BLS), 1997. Workplace Injuries and Illnesses in 1995. Statistical Report from the Bureau of Labor Statistics, Safety and Health Statistics, March 12, 1997.

U.S. Department of Labor, Bureau of Labor Statistics. *A Brief Guide to Recordkeeping Requirements for Occupational Injuries and Illnesses.* June, 1986a. U.S. Government Printing Office.

U.S. Department of Labor, Bureau of Labor Statistics. *Recordkeeping Guidelines for Occupational Injuries and Illnesses.* September, 1986b. U.S. Government Printing Office.

U.S. Department of Labor, Occupational Safety and Health Administration, 1995. The Cumulative Impact of Current Congressional Reform on American Working Men and Women. Special Report prepared for presentation by Robert B. Reich, Secretary of Labor, August, 1995.

U.S. Department of Labor, Occupational Safety and Health Administration. *Ergonomics Program Management Guidelines for Meatpacking Plants (OSHA 3123).* 1993 (reprinted). U.S. Government Printing Office.

67

Body Discomfort Assessment Tools

Leon M. Straker
Curtin University of Technology

67.1 Overview

This chapter reviews various body discomfort assessment tools that can be used in industry for the purpose of preventing work-related musculoskeletal disorders. Prior to describing the various tools available the theoretical background to discomfort assessment is reviewed to assist the ergonomics practitioner with decisions about when to use discomfort as a measure and which tool to use.

The aim of ergonomics is to provide a match between the person and the environment. In order to evaluate the success of any ergonomics input or intervention, tools are required which provide information on the degree of match or mismatch.

One aspect of the match between people and their environment that has become increasingly important is the physical aspect. An important reason for this is that it has been claimed that the high rates of work-related musculoskeletal disorders (such as work-related back problems and work-related neck and upper limb disorders) are at least partially attributable to physical mismatches.

Unfortunately, the etiology of these work-related musculoskeletal disorders is complex and controversial. This lack of etiology clarity makes an assessment of risk difficult. However, it is widely believed that discomfort is a useful risk indicator.

Conceptually, discomfort is an attractive risk indicator as it uses the body's own feedback system to detect possible problems.

Possible sources of discomfort resulting from musculoskeletal stress include: tension in muscles, nerves, blood vessels, ligaments, and joint capsules; compression of the same body tissues; local chemical changes associated with muscle fatigue; local chemical changes related to restricted blood flow and partial ischemia; disruption of nerve conduction resulting from pressure; and secondary inflammation.

As mechanical stress on tissue and local chemical changes are both thought to be sources of tissue damage and pathology, the potential utility of discomfort as a risk indicator is clear.

0-8493-2641-9/99/$0.00+$.50

However, discomfort may also be influenced by psychological and social factors. This is seen by some as a major disadvantage, but perhaps it should not be seen as a disadvantage as the level of disability a certain pathology creates is also related to psychological and social factors.

Discomfort is thought to be especially useful for assessing situations where the impact of physical mismatch may be greatest on small muscles and where static muscle activity is required. This is beneficial, because small muscle problems are not detected well with other common risk assessment tools, such as biomechanical modeling, and gross physiological indicators, such as heart rate and body temperature.

The following section reviews some concepts fundamental to the use of discomfort as an indicator of risk of musculoskeletal disorder.

67.2 Fundamental Concepts

The Discomfort Phenomenon/Construct

There has been little ergonomics research and discussion reported about the phenomenon of discomfort. Therefore, many of the fundamental concepts of discomfort measurement discussed here are drawn from the extensive research on the measurement of pain.

One of the important issues from pain measurement research is the specificity of terms. Pain researchers have found that different people interpret terms differently. This has important implications for discomfort measurement in ergonomics as the ergonomics literature has not been very specific about the use of the term discomfort. For example, some authors have considered pain and discomfort as synonymous. However, the error of this assumption is well illustrated in a study by Bates et al. (1989).

The study by Bates et al. was mainly aimed at investigating the effect of arm posture on worker performance at a computer task. They therefore had their 32 subjects perform a computer-based choice reaction time task for 8 minutes while working in both a "good" arm posture (no shoulder flexion) and a "poor" arm posture (45° shoulder flexion). However, they were also interested in the possible interactions of discomfort (from poor posture) and performance. The ergonomics literature available at the time of their study used discomfort and pain interchangeably; however, their understanding of the pain literature suggested this was not valid. To investigate the interchangeability assumption, their subjects were asked to rate both their shoulder discomfort and their shoulder pain, with a balanced order of recording.

Figure 67.1 shows the relationship between the discomfort and pain intensities recorded by Bates et al. It is clear that discomfort intensity tended to increase before pain did, suggesting discomfort is more sensitive at lower noxious stimuli levels.

However, Figure 67.1 also shows that the relationship between discomfort and pain was a nonlinear relationship. This suggests that different subjects interpreted the two terms, pain and discomfort, differently. The implication of this is that scales which use multiple nouns may not be valid.

Figure 67.2 shows the visual analog scale intensity ratings of discomfort and pain at the end of working in the 45° shoulder posture. The significant difference (38mm, $t_{31} = -6.04$, $p = .0001$) confirms that discomfort is a more sensitive scale than pain, although they may both be accessing the same feelings. This study also confirmed the utility of discomfort in gathering early information about possible physical ergonomics mismatches.

Another example of possibly invalid discomfort measurement scales is where some authors have considered discomfort and comfort to be on the same continuum. However, there is little evidence to suggest that this is true and some ergonomists have suggested that perhaps comfort is a separate entity which is more (or less?) than just the absence of discomfort. Branton (1969) illustrates this with the analogy that the absence of pain does not necessitate the presence of pleasure.

A final point about the importance of terminology in scales is that the interpretation of terms is likely to not only vary between individuals, but also between cultural groups. Thus, multiple term scales which may be valid in one culture may not be valid in another culture. With many modern societies being comprised of individuals from varied ethnic and cultural backgrounds, the clear implication for ergonomics is that

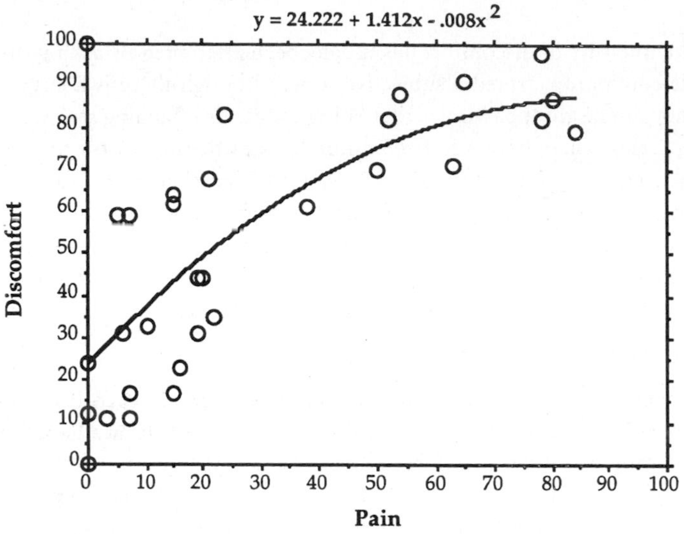

FIGURE 67.1 Relationship between discomfort and pain.

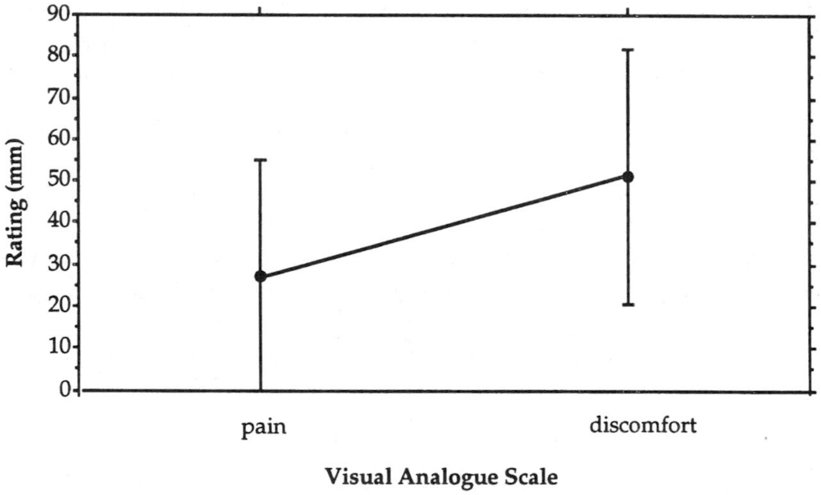

FIGURE 67.2 Pain and discomfort ratings (mean and standard deviation) after working in 45° shoulder flexion for 8 minutes.

widely applicable discomfort assessment tools should restrict the number of terms (nouns) used to ensure more consistent use of the tool across cultural groups.

The Description of Discomfort

To adequately describe discomfort four aspects need to be covered — intensity, quality, location, and temporal pattern. For example, sitting on a hard chair for several hours may result in discomfort which could be described as a numb, cold feeling in the areas extending approximately 15 cm out from each ischial tuberosity which is of low to moderate intensity and which began after about 15 minutes of sitting and increased to the end of the first hour then remained at a constant level until arising from the chair, when the discomfort subsided to minimal intensity after 5 minutes. The following sections describe how the different attributes of discomfort can be measured.

Intensity

Measurement of the intensity of discomfort has usually been attempted by asking the worker to rate the intensity on a scale commonly termed a subjective scale. Although there is a large number of possible subjective scales, they can be grouped into: verbal rating scales; visual analog scales; numeric rating scales; and graphic rating scales. There have also been a number of attempts to try to use a more "objective" measure of intensity. Thus, discomfort intensity is inferred from changes in behavior (using a behavior rating scale), or changes in correlated biomechanical and physiological entities. Examples of possibly suitable correlates are: estimates of static muscle tension (using biomechanical modeling); and estimates of muscle fatigue (using amplitude and frequency shifts in power spectrums of muscle electrical activity). The various types of scales and their relative advantages and disadvantages are described below.

Biomechanical and Physiological Correlates

If discomfort is thought to arise from mechanical loads around joints, then it is reasonable to estimate those loads using position data and biomechanical modeling. Some studies have demonstrated a good correlation between joint load and discomfort ratings (Boussenna, Corlett, and Pheasant 1982). Similarly, if discomfort is thought to be due to sustained or high-magnitude muscle activity, then electromyography can provide an objective measure. Other physiological correlates which could be used are heart rate, blood pressure, respiratory rate, skin conductance, sweat rate, and skin temperature.

The advantage of these measures is their lack of reliance on worker reports.

However, it should always be remembered that it is not discomfort which is being measured. Rather, a correlation is assumed between the measure taken and discomfort. Another disadvantage is the potentially culture-specific nature of any correlation. For example, people in some cultures understand comfort to equate to dynamic balance (and resultant moderate muscle activity) and not the lack of muscle activity commonly accepted as "comfort" by people in Western cultures.

Behavior Rating Scales

Some ergonomists have suggested measuring discomfort intensity by using observation of behavior thought to be indicative of discomfort reaching a certain intensity, such as fidgeting. For example, Branton (1969) suggested sitting was for a purpose and that discomfort can be seen as an interference which, when it reaches a sufficient intensity, results in changes to sitting posture. Thus, an increased number of postural changes would be considered to indicate an increase in discomfort intensity.

Shackel et al. (1969) considered the use of time-sampling of posture changes and duration as an objective measure of discomfort. However, the labor intensive nature of such observation was not thought to be feasible for their purposes. However, with newer technologies of electrogoniometry and digital motion analysis this may now be more feasible.

Corlett and Bishop (1976) recorded machine use and machine idle times for two weeks before and after ergonomic intervention. The increased overall percentage of work vs. idle time and the increased length of work periods were interpreted as suggesting a decrease in physical stress. That is, a reduction of discomfort allowed the workers to work productively for longer periods which increased overall production. It was also used to argue the cost-benefit of the changes as the cost of the machine changes were paid for in a few days of higher productivity.

One advantage of behavioral scale assessment is that it is independent of a worker's capacity and willingness to verbalize feelings. It also provides task interference information and thus can be more readily used in productivity-based justifications for ergonomic interventions. A major disadvantage is the assumption that the postural movements are due to discomfort reaching a certain intensity. For example, frequent movements could be the result of a good work habit of not maintaining static postures, and be a desirable characteristic to be encouraged in any ergonomic intervention.

Verbal Rating Scales

There are two types of verbal rating scale: one in which a single noun is used to describe the construct (in this case "discomfort") and multiple adjectives are used to indicate changes in intensity, and another in which different nouns are used.

no discomfort	minimal discomfort	moderate discomfort	severe discomfort	maximal discomfort

FIGURE 67.3 Single noun verbal rating scale for discomfort intensity.

relaxed	comfortable	neutral	uncomfortable	painful

FIGURE 67.4 Multiple noun verbal rating scale for discomfort intensity.

FIGURE 67.5 Visual analog discomfort scale.

Figure 67.3 shows a single noun, multiple adjective verbal rating scale for discomfort which uses "no, minimal, moderate, severe, and maximal" to indicate increasing intensities of discomfort. Commonly five or seven categories are used.

Figure 67.4 shows an example of a multiple noun scale where the terms "relaxed, comfortable, neutral, uncomfortable, and painful" are used to indicate increasing intensities of discomfort.

For both types of verbal rating scale a rating of discomfort intensity is collected from the worker either by their circling a descriptor or their verbally reporting a descriptor. Analysis of these data is by frequency distributions and rank order nonparametric statistics.

The advantages of verbal rating scales are that they are relatively straightforward and easy for workers to understand.

One disadvantage of verbal rating scales is the relatively small number of points on the scale, resulting in only gross changes in the intensity of discomfort being detected. Another disadvantage is the assumption that feelings like discomfort can be verbalized (which has led some ergonomists to trial cross modality matching and suggest behavioral scales). Multiple noun scales have the added disadvantage of introducing error from different interpretations of the different nouns. For example, one worker may consider "numb" to equate to a higher intensity of discomfort than "stiff," while another worker may interpret them in the opposite order. Such errors will hinder the evaluation of ergonomic interventions because an improvement post intervention may not be detected due to the different uses of the scale by workers.

Visual Analog Scales

Visual analog scales consist of a line, usually 100 mm in length, with a label at each end (often termed "anchors"). Figure 67.5 shows a Visual Analog Discomfort Scale which uses the anchors "no discomfort" and "extreme discomfort" to indicate the ends of the continuum. Another common anchor for the high intensity end is "discomfort as bad as it could be." Both vertical and horizontal lines can be used.

To indicate the level of discomfort, a worker places a mark on the line to indicate the intensity. The intensity rated is then measured as the distance from the left-hand end of the line to the mark placed by the worker. When the measuring is recorded in mm, the scale effectively has 101 levels of discomfort. Robust parametric statistics, such as analysis of variance, are often used for analysis, although the data are strictly not interval data and may be skewed (for example, if most workers record very low or no discomfort).

The advantages of visual analog discomfort scales include their ease of administration, sensitivity, and amenability to statistical analysis.

Pain measurement research has suggested that a possible disadvantage of visual analog scales is that some workers may have difficulty conceptualizing how to indicate perceptions of discomfort intensity

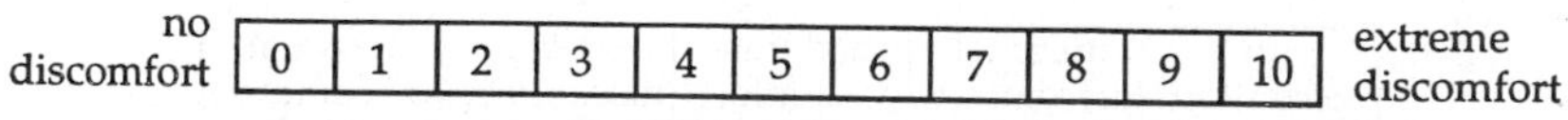

FIGURE 67.6 Visual numeric rating scale.

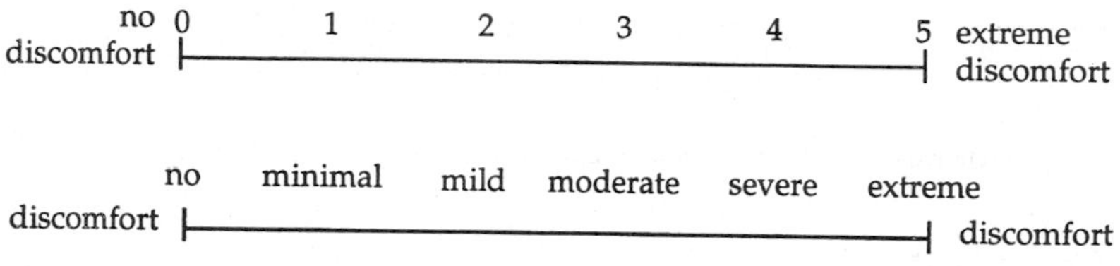

FIGURE 67.7 Graphic rating scales.

on a line. Groups thought to have greater difficulty are older people and those without formal education (those who may be less familiar with abstracting concepts).

Numeric Rating Scales

Numeric rating scales are similar to visual analog scales except they have a discrete number of categories and can be either visual or verbal. Common examples use 0 to 10 in one-unit intervals to give an 11-unit scale or 0 to 100 in one-unit intervals to give a 101 unit scale. The scale has anchors similar to analog scales. Figure 67.6 shows a visual numeric rating scale with 11 levels of intensity.

Workers rate their discomfort intensity either by marking a number or verbally reporting a number. Data collected are less parametric than visual analog scale ratings so nonparametric statistical analysis should be used.

Advantages of numeric rating scales include that they are simple to administer and the verbal scale can be used during a manual task without interference with posture.

Disadvantages include that the 0 to 10 point scale has limited sensitivity, and workers often tend to clump ratings on the 1 to 100 scale around deciles.

Graphic Rating Scales

Graphic rating scales are a mixture of a visual analog scale and either a numeric or verbal rating scale. The scale thus consists of a vertical or horizontal line with anchors (as for a visual analog scale) with the addition of either numbers or adjectives along the line. Figure 67.7 shows examples of both types.

Workers place a mark on the line to represent their rating of discomfort intensity. Nonparametric statistical tests should be used.

An advantage of graphic rating scales is that the extra labels may assist a worker having difficulty using a visual analog scale.

However, it has been demonstrated that there is a problem with clustering of results around the labels along the line.

Quality

The quality of discomfort can probably only be assessed by allowing different nouns to be used by the worker. Different qualities of discomfort may include: tingling; burning; searing; numbness; coldness; stiffness; heat; cramping; prickling; stabbing; and gnawing. Although quality of pain is widely used in health assessment, quality of discomfort has not been regularly used by ergonomists. Perhaps this is because the implications of the different possible qualities are unclear, whereas the implications of intensity, location, and temporal pattern are usually clear.

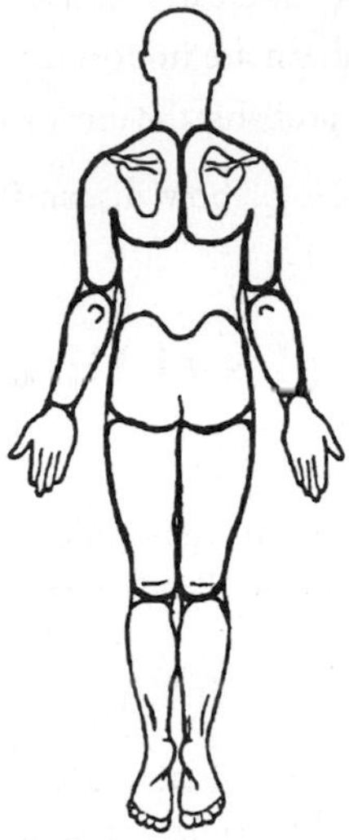

FIGURE 67.8 Body map for reporting discomfort location.

Location

The location of discomfort is commonly collected either through the use of a body map or by specific reference to a body part. Thus, where only the discomfort of a specific body part is of interest, it is usually made clear to the worker that the other information collected about discomfort (intensity, quality, and temporal pattern) is just related to that body part. However, when discomfort at a number or sites, or at any site, is of interest, the worker is usually asked to indicate where each and every discomfort is felt. Workers can indicate the body parts either by shading on a body map or by reporting the name or number of a body part. Figure 67.8 shows an example of a body map which separates the body into 13 parts.

Temporal Pattern

The temporal pattern of discomfort is often measured by collecting information about discomfort at different times. Depending on the reason for investigating discomfort, the time between collections may vary from a number of minutes (presumably for quite severe or intense tasks), to a number of hours (if a daily fluctuation is of interest), to a number of days or longer. Multiple recording of aspects of discomfort can be achieved by either using separate data collection sheets for each time (thus keeping the worker blind to previous recordings), or by recording on the same data collection sheet (enabling the worker to compare with previous recordings).

Another issue related to assessing the temporal pattern of discomfort is the importance of the period between worker experience and data collection. Branton (1969) suggested that because post-experience discomfort reporting relies on kinesthetic memory, discomfort information should be collected while the worker is experiencing the discomfort. Pain research has clearly demonstrated the importance of immediacy for best validity.

Summary of Important Fundamentals of Discomfort Measurement

Based on the discussion above, the following fundamentals of discomfort measurement can be distilled:

- Discomfort measurement is likely to be useful in the assessment of information about physical matches and mismatches.
- Consistent use of the sole noun "discomfort" will assist the validity of assessment.

- Discomfort is a subjective experience and can therefore only be measured by worker report.
- Intensity, location and temporal pattern are important attributes of discomfort.
- A Visual Analog Discomfort Scale is probably the most widely applicable discomfort intensity scale.

The following section provides examples of how discomfort tools have been used in ergonomics research and practice.

67.3 Application — Examples of Discomfort Tools Used in Ergonomics

Some examples of the use of discomfort tools in ergonomics are described below. A list of other reports of discomfort tool use is provided in the Further Reading section.

Whole Body Discomfort Scales

Shackel et al. (1969) used a multiple noun graphic rating scale, which they called a "General Comfort Scale," to measure the "discomfort" caused by different chair designs. The rating scale had 11 items arranged at 10 mm intervals on a vertical line as shown in Figure 67.9. Chair raters were asked to place a mark on the vertical line to express their rating. Scoring was achieved by rounding the mark to the nearest 5 mm, giving a 0 to 20 scale. Using nonparametric statistics, Shackel et al. were able to rank chairs for "discomfort" and compare user ratings with expert opinion.

Although the scale appeared to work reasonably well for the purposes of the study, the earlier discussion presented in this chapter suggests that a single noun scale would have been more valid, and may have allowed better differentiation of the chairs.

Corlett and Bishop (1976) evaluated ergonomic changes to pedestal spot welder machines by measuring "discomfort" before and after machine modifications using an "Overall Comfort Scale." Three operators were asked to rate the intensity of their overall comfort using a seven-point graphic rating scale with end labels "extremely comfortable" and "extremely uncomfortable." Coding of intensity was achieved by dividing each scale interval into 2 to give 14 measurement units. Prior to modification, one machine resulted in mean operator ratings of around 12, which was reduced to around 7.5 after modification. From the earlier discussion, use of a continuum from extreme comfort to extremely uncomfortable may not be valid because comfort may be a separate state.

Bhatnager et al. (1985) used a slightly different approach to assessing whole body discomfort. They used Drury and Coury's (1982) "Body Part Discomfort Frequency" (total number of body parts with non-zero discomfort) and "Body Part Discomfort Severity" (mean severity of body parts with non-zero discomfort) to gain an indicator of overall body discomfort. Their fascinating study showed a negative relationship between discomfort and productivity at an inspection task. There has been some discussion

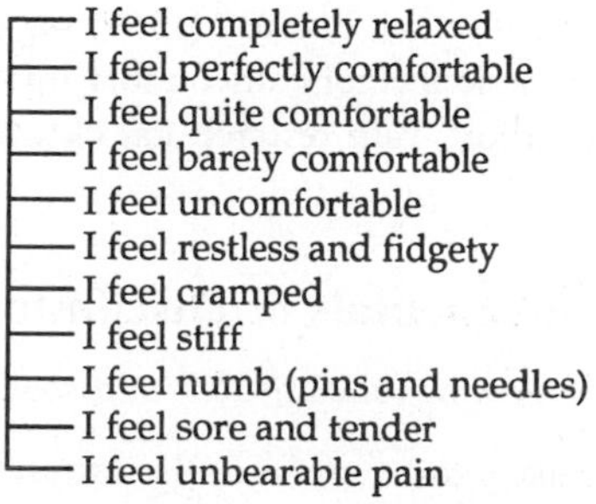
I feel completely relaxed
I feel perfectly comfortable
I feel quite comfortable
I feel barely comfortable
I feel uncomfortable
I feel restless and fidgety
I feel cramped
I feel stiff
I feel numb (pins and needles)
I feel sore and tender
I feel unbearable pain

FIGURE 67.9 General Comfort Scale. (From Shackel, B., Chidsey, K. D., and Shipley, P. (1969). The assessment of chair comfort. *Ergonomics*, 12 (2), 269-306. With permission.)

in the ergonomics literature about whether it is the number of body parts experiencing discomfort, or the average severity of discomfort in various body parts, or the peak severity of discomfort in one body part which determines overall ratings of discomfort. Until this issue is resolved, it is probably useful to use a tool which collects the specific information required so that no unproven assumptions are needed.

Visser and Straker (1994) collected general body discomfort intensity ratings using a "General Body Visual Analog Discomfort Scale." The written instructions to workers were to "Place a dash (/) at a point along the line that best corresponds to your *present* feeling of *overall* discomfort." The scale was the same as shown in Figure 67.5. The tool was easy to use as was demonstrated by the remarkably high compliance rate for the study. (Other details of this study are described below.)

Body Part Discomfort Location and Intensity Scales

Besides collecting whole body discomfort, Shackel et al. (1969) also collected body part discomfort data. Chair raters were shown a body map divided into 15 numbered body parts. Alongside the map were 5 boxes, the first of which was marked "3 most comfortable" and the last marked "3 most uncomfortable." Chair raters were asked to identify the 3 most comfortable body parts and note these in the first box and then cross them off on the body map. They were then to identify the next 3 most comfortable and record the numbers for those body parts in the next box and cross them off the body map. This was continued until all body parts were noted. Chair raters were also allowed to rate the 3 most comfortable then the 3 most uncomfortable at each successive round of rating, if that was found easier. Shackel et al. tried to rank chairs by various body part discomforts, but as the buttocks experienced the most discomfort for each chair the differences between chairs were not able to be detected. The authors also reported the technique to be extremely laborious. The difficulty of use, lengthy procedure, and lack of sensitivity were disadvantages of this assessment tool.

Corlett and Bishop (1976) also collected body part discomfort data. Their method was to ask operators to identify the body part with the most discomfort, then the body part with the next most discomfort, and so on until all body parts had been ranked. The data was used to compare the locations of discomfort experienced when using 2 different machines. A machine which required considerable right foot force showed asymmetry in discomfort and greater lower limb discomfort compared with a machine which required a much smaller foot force. The data were therefore useful in assisting to identify the features of the machinery which contributed to a physical mismatch with the worker. The assessment tool was easier to use and quicker than that of Shackel et al., though it still lacked some sensitivity.

Straker et al. (1997b) collected ratings of body part discomfort from manual handlers performing a range of tasks. The data collection form used a body part map for location information and multiple visual analog scales for intensity information. Figure 67.10 shows the form used.

Besides statistical analysis of the information, Straker et al. used graphic representation to facilitate comparison of the location and intensity of discomfort between different tasks. For example, the body diagram on the left side of Figure 67.11 shows the group average intensity for each of 13 body parts during a task requiring box carrying at shoulder height (high carry). The areas of highest discomfort are the shoulders and wrists/hands. This can be compared to the body on the right side in Figure 67.11 which summarizes the group discomfort data for a task requiring box carrying in a stooped posture (low carry). The change of highest discomfort areas to low back, pelvis, and thighs in the low carry task is consistent with biomechanical modeling of areas of greatest stress. This assessment tool was easy and quick to administer and provided a high level of sensitivity.

Temporal Pattern

Visser and Straker (1994) used a "General Body Visual Analog Discomfort Scale" to collect ratings from 56 dental workers at 6 different times over a working day (on arrival, morning break, prior to lunch break, upon returning from lunch break, mid afternoon, and on completion of work). Figure 67.12 shows some of the results of Visser and Straker's study and clearly indicates both the trend for increasing discomfort as the workday progressed and the ameliorative effect of the lunch break.

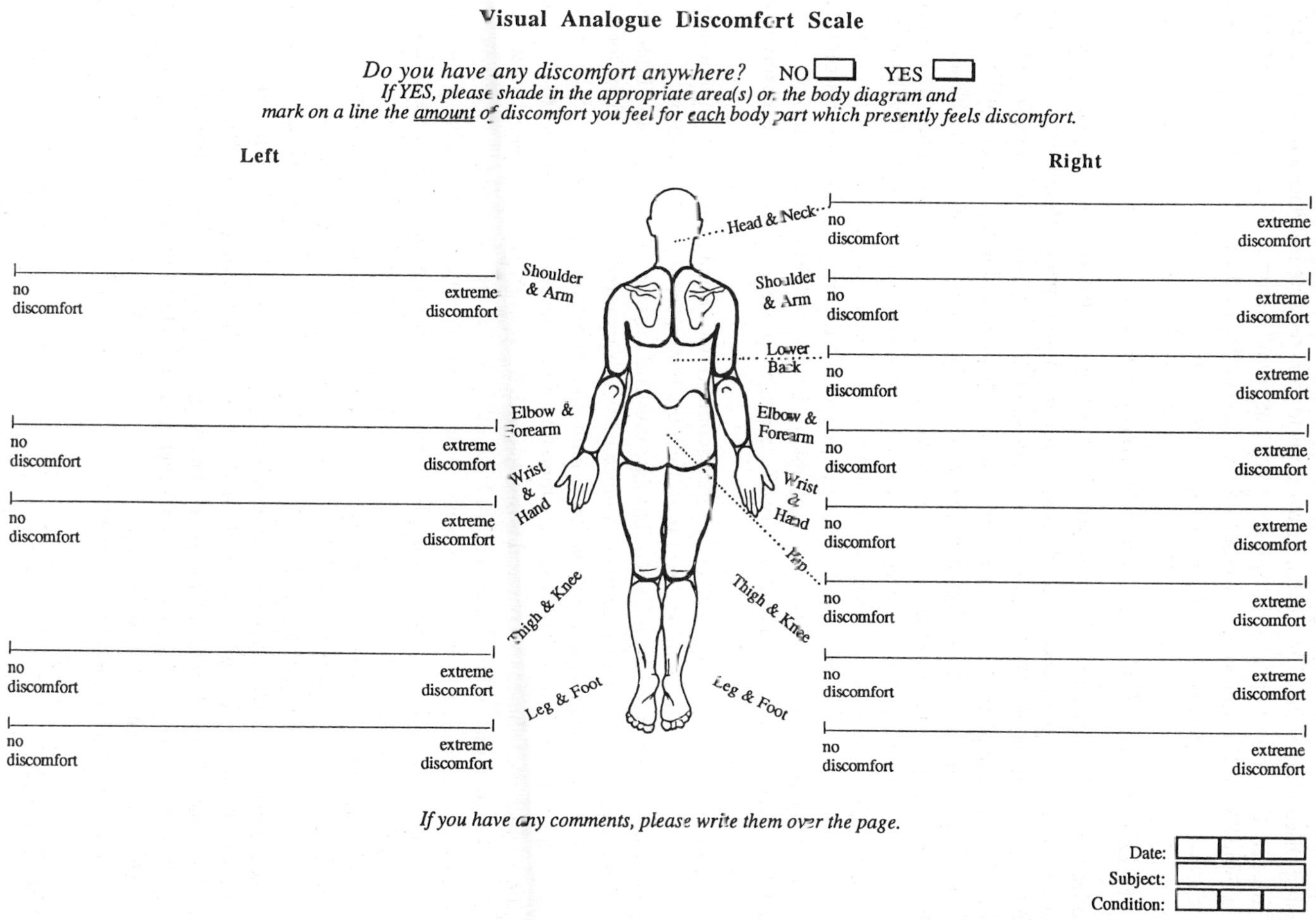

Visual Analogue Discomfort Scale

Do you have any discomfort anywhere? NO ☐ YES ☐

If YES, please shade in the appropriate area(s) on the body diagram and mark on a line the amount of discomfort you feel for each body part which presently feels discomfort.

Left

Shoulder & Arm: no discomfort — extreme discomfort

Elbow & Forearm: no discomfort — extreme discomfort

Wrist & Hand: no discomfort — extreme discomfort

Thigh & Knee: no discomfort — extreme discomfort

Leg & Foot: no discomfort — extreme discomfort

Right

Head & Neck: no discomfort — extreme discomfort

Shoulder & Arm: no discomfort — extreme discomfort

Lower Back: no discomfort — extreme discomfort

Elbow & Forearm: no discomfort — extreme discomfort

Wrist & Hand: no discomfort — extreme discomfort

Hip: no discomfort — extreme discomfort

Thigh & Knee: no discomfort — extreme discomfort

Leg & Foot: no discomfort — extreme discomfort

If you have any comments, please write them over the page.

Date:

Subject:

Condition:

FIGURE 67.10 Body part discomfort location and intensity form.

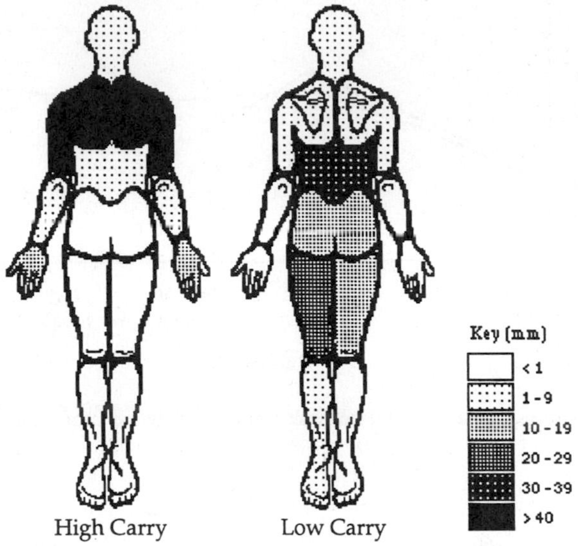

FIGURE 67.11 Body part discomfort for high and low carry tasks.

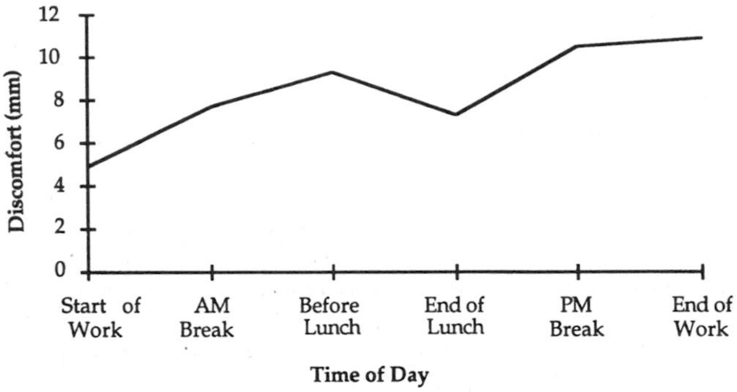

FIGURE 67.12 Mean general body discomfort of dental professionals for 6 different times of day.

Visser and Straker also compared the differences in the development of whole body discomfort when dental workers where provided with a specially designed chair which incorporated an arm rest. Figure 67.13 clearly shows the benefit of the introduction of an arm rest in reducing discomfort in the afternoon.

Straker et al. (1997a) conducted a study to further investigate the relationship between arm posture and discomfort and computer operator performance, as begun by Bates et al. They were interested in the changes in discomfort over a 20-minute task. A verbal "Numeric Rating Scale-101" was used which allowed the operators to verbally report the intensity of the discomfort they were experiencing without needing to change postures. Straker et al.'s scale was an adaptation of the scale recommended by Jensen et al. (1986) for pain. Operators were verbally instructed to "Indicate to me the number between '0' and '100' that best describes your right shoulder discomfort. A '0' would mean no discomfort and a '100' would mean discomfort as bad as it could be at your right shoulder." These ratings were collected every 5 minutes from 21 female operators working in either no shoulder flexion or 30° shoulder flexion. Figure 67.14 shows how the data collected was able to clearly show a more rapid rise in discomfort in the poorer posture.

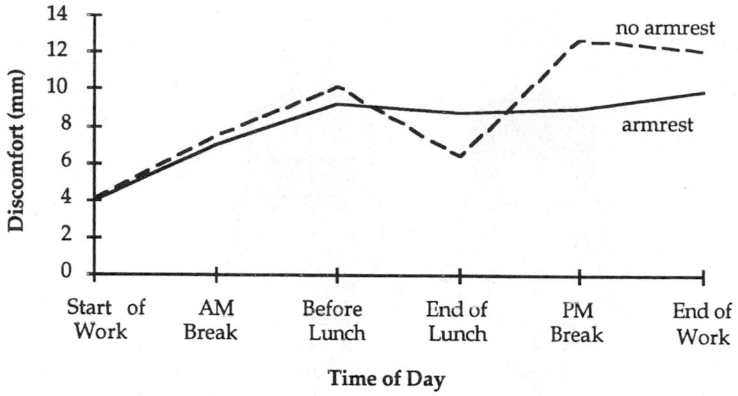

FIGURE 67.13 Comparison of mean general body discomfort of dental professionals across the day with and without an armrest.

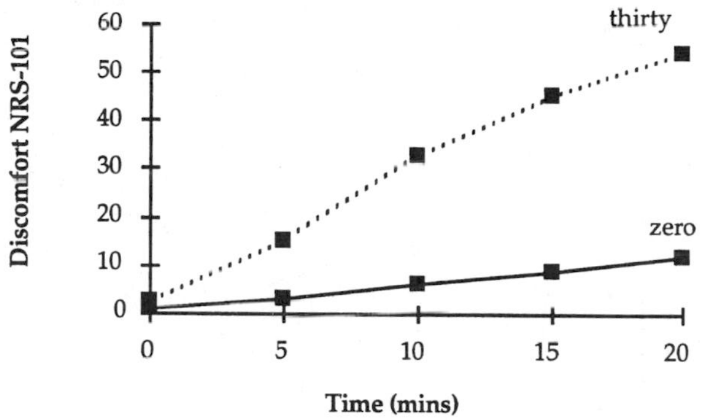

FIGURE 67.14 Changes in shoulder discomfort intensity while working in 0° and 30° shoulder flexion.

These examples have shown that a wide range of discomfort tools have been used in ergonomics research and practice to provide information about the physical match between workers and their environment.

To assist ergonomists in the decision of which discomfort assessment tool to use, a number of criteria are presented.

Criteria for Selection of an Appropriate Tool

A discomfort assessment tool should have high utility, validity, and sensitivity. Utility can be considered in two phases, data collection and data analysis. High utility in data collection requires the tool to be easy for workers to use correctly, quick to administer, and with minimal interference with workers' performance of their tasks. One aspect of widespread ease of use is that the tool should require minimal language skills. Thus, a tool used in an English-speaking country should require little English competence so that the tool can be used across workgroups without difficulty and without jeopardizing validity. Ease of use will improve the quality of collected data by minimizing errors.

High utility in data analysis requires the data to be readily amenable to statistical analysis and graphical representation. Numerical data with either nonparametric characteristics or close to parametric characteristics will facilitate easy statistical analysis. Currently, parametric analysis of complex statistical models tends to be easier, giving an advantage to a tool which collects parametric-like data. Graphical representation of data is important for the communication of ergonomics findings to managers and workers.

Validity is critical, as no validity means the data are of no value. As previously mentioned, the only valid measure of a person's experience is the report of that person. Although there is little discomfort-specific research on tool validity, pain measurement research has produced considerable evidence on the validity of certain tools, such as the visual analog scale for intensity assessment.

However, because discomfort is commonly used in ergonomics to imply a problem in the physical match between worker and work, the strong correlation of discomfort to biomechanical and physiological risk indicators such as joint torque and electromyograph power spectral shifts provides corroboratory evidence of the validity of discomfort.

Reliability is an important corequisite of validity. Van der Grinten (1991) provides evidence of reasonable reliability of a discomfort assessment tool, and this was the only study of discomfort tool reliability found.

Finally, a discomfort assessment tool needs to have sensitivity appropriate to the workers' capacities for discrimination and the assessment purpose. It is unlikely that workers can reliably differentiate 1000 levels of discomfort intensity, and even 100 levels may be ambitious. However, 10 levels are often not sensitive enough for comparisons between work situations unless radical ergonomic interventions have occurred.

67.4 Conclusion and Recommendations

To help ergonomists choose and use an appropriate assessment tool, this chapter has provided a concise review of the fundamental concepts, provided examples of how discomfort assessment tools have been used, and provided criteria to select an appropriate assessment tool.

Discomfort is a valuable variable for ergonomists to use to assess the physical match between workers and their work. Several decades of practical experience by ergonomists and research in the area of pain, has resulted in the development of easy-to-use, valid, and sensitive discomfort assessment tools.

This author recommends the use of a Visual Analog Discomfort Scale or Verbal Numerical Rating Scale for assessment of intensity; body map or specific instructions for assessment of discomfort location; and repeated measurements for the assessment of the temporal pattern of discomfort.

Acknowledgments

The author would like to thank past students, Grace Szeto, Michael Bates, Marshall Stockden, Mark Petrich, Jodie Visser, and Jean Mangharam for stimulating the conceptualization of this chapter.

References/Further Reading

Studies Cited in Chapter

Bates, M., Petrich, M., and Stockden, M. (1989) *Posture, pathology, pain and performance.* Bachelor of Applied Science Research Report, Curtin University of Technology, Perth Australia.

Bhatnager, V., Drury, C. G., and Schiro, S. G. (1985). Posture, postural discomfort and performance. *Human Factors,* 27 (2), 189-199.

Boussenna, M., Corlett, E. N., and Pheasant, S. T. (1982). The relation between discomfort and postural loading at the joints. *Ergonomics,* 25 (4), 315-322.

Branton, P. (1969). Behaviour, body mechanics and discomfort. *Ergonomics,* 12, 316-327.

Corlett, E. N., and Bishop, R. P. (1976). A technique for assessing postural discomfort. *Ergonomics,* 19 (2), 175-182.

Drury, C. G., and Coury, B. G. (1982). A methodology for chair evaluation. *Applied Ergonomics,* 13 (3), 195-202.

Jensen, M. P., Karoly, P., and Braver, S. (1986). The measurement of clinical pain intensity: a comparison of six methods. *Pain,* 27, 117-126.

Shackel, B., Chidsey, K. D., and Shipley, P. (1969). The assessment of chair comfort. *Ergonomics*, 12 (2), 269-306.

Straker, L. M., Pollock, C. M., and Mangharam, J. (1997a). The effect of shoulder posture on performance, discomfort and muscle fatigue while working on a Visual Display Unit. *International Journal of Industrial Ergonomics*, 20, 1-10.

Straker, L. M., Stevenson, M. G., and Twomey, L. T. (1997b). A comparison of single and combination manual handling tasks risk assessment: 2. discomfort, rating of perceived exertion and heart rate measures. *Ergonomics*, 40(6), 656-669.

van der Grinten, M. P. (1991). Test-retest reliability of a practical method for measuring body part discomfort, in Y. Quennec and R. Daniellou (Ed.), *Designing for Everyone. Proceedings of the 11th Congress of the International Ergonomics Association*, (pp. 54-56). Paris: Taylor and Francis.

Visser, J. L. and Straker, L. M. (1994). An investigation of discomfort experienced by dental therapists and assistants at work. *Australian Dental Journal*, 39 (1), 39-44.

Further Reading

Cameron, J. A. (1996). Assessing work-related body-part discomfort: Current strategies and a behaviorally oriented assessment tool. *International Journal of Industrial Ergonomics*, 18, 389-398.

Chapman, C. R., Casey, K. L., Dubner, R., Foley, K. M., Gracely, R. H., and Reading, A. E. (1985). Pain measurement: An overview. *Pain*, *22*, 1-31.

Hagberg, M., and Sundelin, G. (1986). Discomfort and load on the upper trapezius muscle when operating a wordprocessor. *Ergonomics*, 29 (12), 1637 1645.

Life, M. A. and Pheasant, S. T. (1984). An integrated approach to the study of posture in keyboard operation. *Applied Ergonomics*, 15 (2), 83-90.

Marley, R.J. and Kumar, N. (1994). An improved musculoskeletal discomfort assessment tool, in F. Aghazadeh (Ed.). Advances in Industrial Ergonomics and Safety VI (pp. 45-52). London: Taylor and Francis.

Scott, J., and Huskisson, E. C. (1976). Graphic representation of pain. *Pain*, 2, 175-184.

Van der Grinten, M. P. and Smitt, P. (1992). Development of a practical method for measuring body part discomfort, in F. Adhazadeh (Ed.), *Advances in Industrial Ergonomics and Safety IV* (pp. 311-318). London: Taylor and Francis.

Section II
Medical Management Prevention

68

Medical Management of Work-Related Musculoskeletal Disorders

Thomas Hales
National Institute for Occupational Safety and Health (NIOSH)

Patricia Bertsche
The Ohio State University

68.1 Introduction

The Bureau of Labor Statistics (BLS) reports that in 1994 nearly two thirds of the workplace illnesses were disorders associated with repeated trauma (one category of musculoskeletal disorders) (BLS 1995). These figures do not include low back disorders associated with overexertion, which accounted for 380,000 lost time cases in 1993. The number of repeated trauma cases reported in 1994 was 332,000, a 10% increase from the 1993 figure. In fact, since 1982, the number of reported disorders associated with repeated trauma has been increasing each year (BLS 1995). Not surprisingly, many health care providers (HCPs) find evaluating and treating these employees consumes an increasing proportion of their time and energy.

To prevent or reduce symptoms, signs, impairment, or disability associated with work-related musculoskeletal disorders (WRMSDs), employers, in collaboration with HCPs, should develop a medical management program which is outlined in Figure 68.1. This chapter provides assistance to employers setting up a medical management program and to HCPs managing these cases in two ways — first, by outlining the general principles and listing the components of a program needed to adequately evaluate

0-8493-2641-9/99/$0.00+$.50

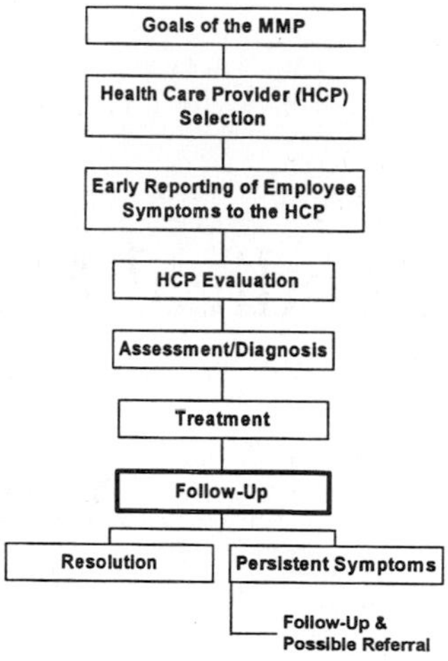

FIGURE 68.1 Overview of a medical management program (MMP).

and treat affected employees; second, by providing HCPs with practical guidance and forms to collect the appropriate information. These forms can then be incorporated into the employee's medical record.

68.2 Terminology

Before addressing the various components of a medical management program, the term musculoskeletal disorder must be defined. MSDs are disorders of the muscles, tendons, peripheral nerves, or vascular system not directly resulting from an acute or instantaneous event (e.g., slips or falls). These disorders are considered to be work-related when the work environment and the performance of work contribute significantly, but as one of a number of factors, to the causation of a multifactorial disease (WHO 1985). Physical risk factors that cause or aggravate MSDs and that may be present at the workplace include, but are not limited to: repetitive, forceful, or prolonged exertions; frequent or heavy lifting; pushing, pulling, or carrying of heavy objects; fixed or awkward work postures; contact stress; localized or whole-body vibration; cold temperatures; and poor lighting leading to awkward postures. These workplace risk factors can be intensified by work organization characteristics, such as inadequate work–rest cycles, excessive work pace and/or duration, unaccustomed work, lack of task variability, machine-paced work, and piece rate.

68.3 Selection of a Health Care Provider (HCP)

An HCP is a practitioner operating within the scope of his or her license, registration, certification, or legally authorized practice. The evaluation and treatment of employees with WRMSDs should be performed by an HCP with experience and/or training in managing these disorders. Many HCPs are capable of providing these services, including physicians, occupational health nurses, physical therapists, occupational therapists, and hand therapists. Employers and employees may be more familiar with the services of physicians, therefore Table 68.1 provides information regarding some of the other HCPs who might be directly providing the care or coordinating the care of employees with WRMSDs. Considerations for the employer to use in selecting an HCP include:

- Specialized training and experience in ergonomics and the treatment of work-related musculoskeletal disorders
- Current working knowledge of the worksite and the specific industry
- Willingness to periodically tour the worksite

TABLE 68.1 Non-Physician Health Care Providers Who Might Be Involved in the Medical Management of Work-Related Musculoskeletal Disorders (Not intended to be all-inclusive)

Profession	Scope of Practice	Training/Experience	Services They Provide
Occupational Health Nurse (OHN)	An OHN is a Registered Nurse (RN), independent licensure with scope defined by individual state boards of nursing; certification is voluntary (COHN); Advanced practice nurses (nurse practitioners) treat independently or provide medical treatment with protocol depending on requirements of state licensing board. RNs refer to physicians and other health care providers when treatment beyond their scope of practice is required.	Basic education includes complete assessment (history and physical examination) of all body systems; OHNs have academic and/or continuing education in assessment of the musculoskeletal and nervous systems and diagnosis, treatment, and rehabilitation of work-related musculoskeletal disorders.	Assessment, treatment of common work-related musculoskeletal disorders, particularly in early stages (under protocol when required by state statute), referral to other appropriate health care providers as needed, and rehabilitation including case management; Preventive services include trend analysis, education and training, and involvement in the job improvement process including job analysis.
Occupational Therapist (OT)	49 states, the District of Columbia, Guam, and Puerto Rico have laws regulating the profession; The American Occupational Therapy Certification Board's national certification exam is a basic requirement in the states/jurisdictions that license or certify OTs. Generally, an OT may independently provide services, however, in certain states, occupational therapy laws/regulations require physician referral for services for specific medical conditions.	OTs have either a bachelor's or master's degree and pursue continuing education and extensive on-the-job training to specialize in work-related musculoskeletal disabilities; OTs have a comprehensive background in the biological and behavioral sciences; knowledge and application of the components of human performance including psychosocial, neurological, cognitive, perceptual, and motor function.	OTs use standardized tests, observational skills, activities and tasks designed to evaluate specific work-related skills, functional abilities, physical abilities, and behaviors. Examples of assessments include: functional capacity evaluation, physical capacity testing, examination of essential functions of a job. Other services include work hardening and involvement in the job improvement process such as job analysis and workstation and tool modification.
Physical Therapist (PT)	PTs licensed in all states, the District of Columbia, Puerto Rico, and the U.S. Virgin Islands; Direct physician oversight is not required. Of the 53 jurisdictions, 44 permit physical therapy evaluation without physician referral.	PTs have either a bachelor's or graduate degree and pursue continuing education to specialize in prevention and rehabilitation of work-related musculoskeletal disorders. PTs' basic education includes courses in anthropometrics, biomechanics, ergonomic interventions, kinesiology, movement and posture analysis, the components of human psychophysical performance, orthotic prescription, fabrication, and application of supportive devices.	PTs evaluate a variety of conditions such as abnormalities of body alignment and movement patterns; impaired motor function and learning; impaired sensation limitations of joint motion; muscle weakness; and pain. PTs perform tests and measures such as batteries of work performance; assessment of work hardening or conditioning; determination of dynamic capabilities and limitations during specific work activities. Involvement in the job improvement process including analysis of jobs or activities, and workstation or tool modifications.
Hand Therapist (HT)	A Hand Therapist is either an OT or PT who voluntarily becomes certified by the Hand Therapy Certification Commission. Certified HTs specialize in upper extremity rehabilitation.	HTs have specialized training and experience in assessment and rehabilitation of work-related musculoskeletal disorders.	Services include diagnostic work up of quantitative sensory testing to determine peripheral neuropathy, grip strength, and motor testing to determine the localization of muscular tenderness areas of inflammation; physical or functional capacity evaluations. HTs apply treatments such as thermotherapy, ultrasound, and electric stimulation; re-education home exercise programs, splintage, pain management, soft tissue mobilization and myofascial release. HTs are skilled in work task analysis and therefore are well suited for involvement in the job improvement process.

- Willingness to communicate with the employer and employees (Louis, 1987; Haig et al., 1990)
- Experience in the case management of work-related musculoskeletal disorders
- Willingness to consider conservative therapy prior to surgery
- History of successful treatment of work-related musculoskeletal disorders

68.4 Early Reporting of Symptoms and Access to Health Care Providers

The case management process begins with an employee informing his or her employer of the presence of musculoskeletal symptoms or signs. Generally, the earlier that symptoms are identified, an evaluation completed, and treatment initiated, the likelihood of a significant disorder developing is reduced. Early treatment of many MSDs has been shown to reduce their severity, duration of treatment, and ultimate disability (Haig et al., 1990; Wood, 1987; Wiesel et al., 1984; Mayer et al., 1987). There can be various workplace situations influencing an employee's decision to report symptoms. These situations can result in employees over-reporting, or under-reporting, symptoms. In either case, to prevent severe disorders from occurring, employees must not be subject to reprisals or discrimination based on reporting symptoms to their supervisors.

Supervisors and foremen are not trained to evaluate and assess MSDs. To prevent supervisors or other plant personnel from performing triage, employees reporting persistent musculoskeletal symptoms (e.g., symptoms lasting seven days from onset, or symptoms that interfere with the employee's ability to perform the job) should have the opportunity for a prompt HCP evaluation. If an HCP is available at the workplace, this initial assessment should be offered when the employee reports symptoms or at least within two days. If the HCP is offsite, the employer should make available an assessment to the employee promptly, but no later than a week after the signs or symptoms are reported. This is not meant to imply that employers should wait seven days from onset of all employee's symptoms before referring the employee to an HCP. There are foreseeable circumstances where immediate evaluation by an HCP would be warranted. For example, an employee who reports to the supervisor that he/she is experiencing severe low back pain with numbness and tingling radiating down his/her leg, an inability to sleep due to the pain, and obvious difficulty walking should immediately be referred to the HCP.

68.5 Health Care Providers Being Familiar with the Employee's Job

HCPs who evaluate employees, determine an employee's functional capabilities, and prepare opinions regarding work-relatedness and work-readiness, must be familiar with employee jobs and job tasks. Being familiar with employee jobs not only assists HCPs in making informed case management decisions, but also demonstrates to employers and employees the importance HCPs place on making informed decisions, assists with the identification of workplace hazards that cause or aggravate MSDs, assists with the identification of alternate duty jobs, and can help establish the proper diagnosis for the employee's condition.

Critical to this process is open lines of communication with the employer, employee, and the HCP. The employer should appoint a contact person who is familiar with plant jobs and workplace risk factors to communicate and coordinate with the HCP. In addition, HCPs should perform a plant walk-through. Once familiar with plant operations and job tasks, the HCP can periodically revisit the facility to remain knowledgeable about working conditions. Other approaches to become familiar with jobs and job tasks include review of job analysis reports, job surveys or risk factor checklists, detailed job descriptions, job safety analyses, photographs and/or videotapes accompanied by narrative or written descriptions, and interviewing the employee.

68.6 Evaluation of the Employee

The HCP evaluation of the symptomatic employee should contain a relevant occupational and health history, a physical examination, laboratory tests appropriate to the reported signs or symptoms, and conclude with an initial assessment/diagnosis. If the HCP providing the initial evaluation does not have the training or experience to make a preliminary assessment or diagnosis, the employee should be referred to an HCP with such training and experience. The content of the evaluation is outlined below with a recording form available (see Form 1).

1. Characterize the symptoms and history
   - Onset (date; circumstance; abrupt vs. gradual, etc.)
   - Duration and frequency
   - Quality (pain; tingling; numbness; swelling; tenderness, etc.)
   - Intensity (mild; moderate; severe; other rating scales)
   - Location
   - Radiation
   - Exacerbating and/or relieving factors or activities (both on-the-job and off-the-job)
   - Prior treatments
2. Relevant considerations:
   - Demographics (e.g., age; gender; hand dominance)
   - Past medical history (e.g., prior injuries or disorders related to the affected body part)
   - Recreational activities, hobbies, household activities
   - Occupational history with emphasis on the (a) job the employee was performing when the symptoms were first noticed, (b) prior job if the employee recently changed jobs, (c) amount of time spent on that job, and (d) whether the employee was working any other "moonlighting" or part-time jobs.
3. Characterize the job:

   Becoming familiar with an employee's job is a critical component of the HCP evaluation and treatment process. In addition to collecting the information from the plant contact person and plant walk-through (described above), employees should be interviewed regarding their work activities. The employee should be asked to describe their required job tasks with respect to known workplace risk factors for MSDs and the duration of exposure such as hours per day, days per week and shift work. Workplace risk factors for MSDs include repetitive, forceful, or prolonged exertions; frequent or heavy lifting or lifting in awkward postures (e.g., twisting, trunk flexion, or lateral bending); pushing, pulling, or carrying of heavy objects; fixed or awkward work postures; contact stress; localized or whole-body vibration; cold temperatures; and others. The employee should also be asked if there has been any recent changes in their job, such as longer hours, increased pace, new tasks or equipment, or new work methods which may have caused or contributed to the current illness.
4. Physical examination:

   The physical examination should be targeted to the presenting symptoms and history. Components of the exam include inspection (redness, swelling, deformities, atrophy, etc.), range of motion, palpation, sensory and motor function (including functional assessment), and appropriate maneuvers (e.g., Finkelstein's). It is important to note that clinical examinations may not identify the specific structure affected, nor find classic signs of inflammation (e.g., redness, warmth, swelling). This should not be surprising since the role of inflammation in the pathophysiology of these disorders is unclear (Nirschl 1990). For further information on the content of an appropriate exam, or the technique to perform the exam, please consult the following references: AHCPR, 1994; ASSH, 1990; Hoppenfield, 1976; Tubiana et al., 1984.

5. Assessment and diagnosis:

   For each employee referred for an assessment, the HCP should make a specific diagnosis consistent with the current International Classification of Diseases, or the HCP should summarize the findings of his or her assessment. Terms such as repetitive motion disorders (RMDs), repetitive strain injury (RSI), overuse syndrome, cumulative trauma disorders (CTDs), and work-related musculoskeletal disorder (WRMSD) are not ICD diagnoses and, although useful as general terms, should not be used as medical diagnoses. Given the difficulty in establishing the specific structure affected, many diagnoses should describe the anatomic location of the symptoms without a specific structure diagnosis (e.g., unspecified neck symptoms or disorders should be listed as ICD-9 723.9; unspecified disorders of the soft tissues should be listed as ICD-9 729.9). When a specific anatomical structure can be ascertained, most of these conditions involve the muscles or tendons (unspecified disorders of muscle, ligament, and fascia should be listed at ICD-9 728.9; unspecified disorders of synovium, tendon, and bursa should be listed as ICD-8 727.9). Table 68.2 provides a listing of ICD-9 codes.

   The HCP should assist in determining whether occupational risk factors are suspected to have caused, contributed to, or exacerbated the condition. Factors helpful in making this determination are:

   - Is the medical condition known to be associated with work?
   - Does the job involve risk factors (based on job surveys or job analysis information) associated with the presenting symptoms?
   - Is the employee's degree of exposure consistent with those reported in the literature?
   - Are there other relevant considerations (e.g., unaccustomed work, overtime, etc.)?

68.7 Treatment of the Employee

Before initiating treatment, the HCP should document the specific treatment goals (e.g., symptom resolution or restoring of functional capacity), expected duration of treatment, dates for follow-up evaluations, and time frames for achieving the treatment goals. Resting the symptomatic area, and treatment of soft tissue and tendon disorders are the mainstays of conservative treatment. Despite the wide application of some therapeutic modalities, many are untested in controlled clinical trials.

Resting the Symptomatic Area

Reducing or eliminating employee exposure to musculoskeletal risk factors through engineering and administrative controls in the workplace is the most effective way to rest the symptomatic area while allowing employees to remain productive members of the workforce (Upfal 1994). Until effective controls are installed, employee exposure to workplace risk factors can be reduced through restricted duty and/or temporary job transfer. The specific amount of work reduction for employees on restricted duty must be individualized; however, the following principles apply: the degree of restriction should be proportional to the condition severity and to the frequency and duration of exposure to relevant risk factors involved in the original job. HCPs are responsible for determining the physical capabilities and work restrictions of the affected worker. The employer is responsible for finding a job consistent with these temporary restrictions. The employer's contact person (who is knowledgeable about the employee's job requirements and their associated risk factors) is critically important to this process. The contact person should communicate and collaborate with the HCP so that appropriate job placement of the employee occurs during the recovery period. Written return-to-work plans ensure that the HCP, the employee, and the employer all understand the steps recommended to promote recovery, and ensure that the employer understands what his or her responsibility is for returning the employee to work. A form is included to collect and distribute this written plan (Form 2). The HCP is also responsible for employee follow-up to document a reduction in symptoms during the recovery period.

FORM 1 — Occupational and Health History Recording Form for Musculoskeletal Disorders

Name: ________________ **Dept:** ________________ **Job Title:** ________________

Age: ___ **yrs** **Gender:** ___ **F** ___ **M** **Length of time at the plant:** _____ **mo/yrs**

Dominate Hand: ___ **R** ___ **L** ___ **Both** **Length of time on-the-job:** _____ **mo/yrs**

Symptom Characterization:

Onset: Date: ____________ Abrupt vs. Gradual: ____________

Quality: (let employee describe, check all that apply)

___ pain ___ tenderness ___ weakness ___ soreness ___ numbness

___ tingling ___ burning ___ swelling ___ cramping ___ throbbing

Duration: ____________ Frequency: ____________

Intensity: (mild, moderate, or severe) ____________

Location: (R = right, L = left) ___ neck ___ upper arm ___ lower arm ___ back ___ upper leg ___ foot

(Check all that apply) ___ shoulder ___ elbow ___ hand/wrist ___ hip ___ lower leg

Radiation: (R = right, L = left) ___ neck ___ upper arm ___ lower arm ___ back ___ upper leg ___ foot

(Check all that apply) ___ shoulder ___ elbow ___ hand/wrist ___ hip ___ lower leg

Exacerbating or relieving activities (both on-the-job and off-the-job):

Exacerbating: 1) ____________ 2) ____________ 3) ____________

Relieving: 1) ____________ 2) ____________ 3) ____________

Past Medical History (prior injuries or disorders):

1) ____________ 3) ____________

2) ____________ 4) ____________

Recreational Activities, Hobbies, Household Activities:

1) ____________ 3) ____________

2) ____________ 4) ____________

Occupational History:

1) ____________ 3) ____________

2) ____________ 4) ____________

Characterize the Job:

Forceful, repetitive or sustained **exertions** can be estimated from production standards, employee ratings of efforts required to complete job tasks, descriptions of work objects and tools, weights of work objects and tools, and length of the workday. Extreme, repetitive or sustained **postures** can be estimated from a description of work methods and equipment. Employees can demonstrate the posture required for each step of the job task, or simulate the workstation in the examining room. Insufficient rest, pauses, or **recovery time** and be estimated from a description of rest breaks, production standards, work flow, and work organization factors. Extreme levels, repeated or long exposure to **vibration** can be estimated from a description of hand tools, or equipment. **Cold temperatures**, repeated or long exposure to cold can be based on temperature measurements, estimated from a description of the work environment, and the duration of time spent in cold areas.

Physical Stress	Property		
	Magnitude	Repetition Rate	Duration
Force			
Joint Angle			
Recovery			
Vibration			
Temperature			

TABLE 68.2 Specific ICD-9 Diagnoses Referred to as Musculoskeletal Disorders by ICD-9 Numbers

Tendon, synovium, and bursa disorders	**727**
Trigger finger (acquired)	727.03
Radial styloid tenosynovitis (deQuervain's)	727.04
Other tenosynovitis of hand and wrist	727.05
Specific bursitides often of occupational origin	727.2
Unspecified disorder of synovium, tendon, and bursa	727.9
Peripheral enthesopathies	**726**
Rotator cuff syndrome, supraspinatus syndrome	726.10
Bicipital tenosynovitis	726.12
Medial epicondylitis	726.31
Lateral epicondylitis (tennis elbow)	726.32
Unspecified enthesopathy	726.9
Disorders of muscle, ligament, and fascia	**728**
Game-Keepers thumb	728.8
Muscle spasm	728.85
Unspecified disorder of muscle, ligament, and fascia	728.9
Other disorders of soft tissues	**729**
Myalgia, myositis, fibromyositis	729.1
Swelling of limb	729.81
Cramp	729.82
Unspecified disorders of soft tissue	729.9
Osteoarthritis	**715**
Mononeuritis of upper limb	**354**
Carpal tunnel syndrome (median nerve entrapment)	354.0
Cubital tunnel syndrome	354.2
Tardy ulnar nerve palsy	354.2
Lesions of the radial nerve	354.3
Unspecified mononeuritis of upper limb	354.9
Peripheral vascular disease	**443**
Raynaud's syndrome	443.0
Hand-Arm Vibration Syndrome	443.0
Vibration White Finger	443.0
Arterial embolism and thrombosis	**444**
Hypothenar hammer syndrome	444.2
Ulnar artery thrombosis	444.21
Nerve root and plexus disorders	**353**
Brachial plexus lesions:	353.0
Cervical rib syndrome	353.0
Costoclavicular syndrome	353.0
Scalenus anticus syndrome	353.0
Thoracic outlet syndrome	353.0
Unspecified nerve root and plexus disorder	353.9
Spondylosis (inflammation of the vertebrae)	**721**
Cervical without myelopathy	721.0
Cervical with myelopathy	721.1
Thoracic without myelopathy	721.2
Lumbarsacral without myelopathy	721.3
Thoracic or lumbar with myelopathy	721.4
Intervertebral disc disorders	**722**
Displacement of cervical disc	722.0
Displacement of thoracic or lumbar disc	722.1
Degeneration of the cervical disc	722.4
Degeneration of the thoracic or lumbar disc	722.5
Intervertebral disc disorder with myelopathy	722.17
Disorders of the cervical region	**723**
Cervicalgia (pain in neck)	723.1
Cervicobrachial syndrome (diffuse)	723.3
Unspecified neck symptoms or disorders	723.9
Unspecified Disorders of the Back	**724**
Low back pain	724.2

FORM 2 — Musculoskeletal Disorder Management Plan

Forward Only Work Related Medical Information to the Employer

Date of Assessment: ________________

Name: ______________ Date of Birth: ____________

Employer: ___________ Contact Person: ____________________ Phone: ___________ FAX: ___________

Diagnosis/Assessment: ____

____

Treatment Plan: (e.g., medications/dosage, splints, physical or occupational therapy including frequency and duration of treatment, etc.) ____

____

Next Appointment: ________________

Other Scheduled Appointments: ____

WORK STATUS

Is the Employee able to perform his/her regular work?

___ Yes, Full duty
___ No, Remove from Work Environment until ____________
___ No, Modified or Alternate Work until ____________

(Complete Activity Checklist below for Job Modifications)

Name: ________________________

Description of Restricted Work Activity

Activity	Duration	Frequency
a. Sitting	___ Hrs. Per Day	___ Hrs. at a Time
b. Standing	___ Hrs. Per Day	___ Hrs. at a Time
c. Walking	___ Hrs. Per Day	___ Hrs. at a Time
d. Lift/Carry:___ lbs.	___ Hrs. Per Day	___ Times Per Hr.
e. Climbing Stairs	___ Hrs. Per Day	___ Times Per Hr.
f. Climbing Ladders	___ Hrs. Per Day	___ Times Per Hr.
g. Kneeling	___ Hrs. Per Day	___ Times Per Hr.
h. Bending at Waist	___ Hrs. Per Day	___ Times Per Hr.
I. Squatting	___ Hrs. Per Day	___ Times Per Hr.
j. Twisting	___ Hrs. Per Day	___ Times Per Hr.
k. Pull/Push: ___ lbs.	___ Hrs. Per Day	___ Times Per Hr.
l. Reach Above Shoulder	___ Hrs. Per Day	___ Times Per Hr. ___ L ___ R
m. Extended Reaching	___ Hrs. Per Day	___ Times Per Hr. ___ L ___ R
n. Neck bend/twisting	___ Hrs. Per Day	___ Times Per Hr.
o. Elbow/Forearm Twist	___ Hrs. Per Day	___ Times Per Hr. ___ L ___ R
p. Hand/Wrist Bending	___ Hrs. Per Day	___ Times Per Hr. ___ L ___ R
q. Pinch Gripping	___ Hrs. Per Day	___ Times Per Hr. ___ L ___ R
r. Forceful Grasping	___ Hrs. Per Day	___ Times Per Hr. ___ L ___ R
s. Continuous Keyboard Use	___ Hrs. Per Day	___ Times Per Hr.
t. Vibrating Tool/Equip Use	___ Hrs. Per Day	___ Times Per Hr. ___ L ___ R
u. Ankle/Foot Bend/Twist	___ Hrs. Per Day	___ Times Per Hr. ___ L ___ R
v. Cold Temperature	___ Hrs. Per Day	

Other Restricted Job Tasks (including frequency and duration): ______________________

____

Other Specific Job Recommendations: ______________________________

____

Health Care Provider Name: ____________________
Address: ________________________________
City/State/Zip: ____________________________
Phone: (___)____________ FAX: (___)____________
Copy of Form Given to Employee: ____ Yes ____ No
Health Care Provider Signature: ______________ Date: ____________

Complete removal from the work environment should be avoided unless the employer is unable to accommodate the prescribed work restrictions. Research has documented that the longer the employee is off work, the less likely he/she will return to work (Vallfors 1985). In these cases, the employer's contact person and the employee should be in day-to-day contact, and the employee can be encouraged to participate in a fitness program that does not involve the injured anatomical area.

Wrist immobilization devices, such as wrist splints or supports, can help rest the symptomatic area in some cases. These devices are especially effective off the job, particularly during sleep. They should be dispensed to individuals with MSDs only by HCPs with the training and experience in the positive and potentially negative aspects of these devices. Wrist splints, typically worn by patients with possible carpal tunnel syndrome, should not be worn at work unless the HCP determines that the employee's job tasks do not require wrist deviation or bending. Struggling against a splint can exacerbate the medical condition due to the increased force needed to overcome the splint. Splinting may also cause other joint areas (elbows or shoulders) to become symptomatic as work technique is altered. Recommended periods of immobilization vary from several weeks to months depending on the nature and severity of the disorder. Immobilization should be prescribed judiciously and monitored carefully to prevent iatrogenic complications (e.g., disuse muscle atrophy).

The *prophylactic* use of immobilization devices worn on or attached to the wrist or back is not recommended. Research indicates wrist splints have not been found to prevent distal upper extremity musculoskeletal disorders (Rempel 1994). Likewise, there is no rigorous scientific evidence that back belts or back supports *prevent* injury, and their use is not recommended for prevention of low back problems (NIOSH 1994; Mitchell et al., 1994). Where the employee is allowed to use a device that is worn on or attached to the wrist or back, the employer, in conjunction with a HCP, should inform each employee of the risks and potential health effects associated with their use in the workplace, and train each employee in the appropriate use of these devices. (McGill 1993)

The HCP should advise affected employees about the potential risk of continuing non-modified work, or spending significant amounts of time on hobbies, recreational activities, and other personal habits that may adversely affect their condition (e.g., requires the use of the injured body part). However, as mentioned above, the employee should engage in a fitness program designed for exercise and aerobic conditioning that does not involve the injured anatomical area.

Thermal (more frequently cold) Therapy

Such treatment is generally considered useful in the acute phase of some MSDs. Cold therapy may be contraindicated for other conditions (e.g., neurovascular).

Oral Medications

Aspirin or other nonsteroidal anti-inflammatory agents (NSAIA) are useful in reducing the severity of symptoms either through their analgesic or anti-inflammatory properties. Their gastrointestinal and renal side effects, however, make their prophylactic use among asymptomatic employees inappropriate, and may limit their usefulness among employees with chronic symptoms. In short, NSAIAs should not be used prophylactically.

It must be noted that the effectiveness of Vitamin B-6 for treatment of musculoskeletal disorders has not been established (Amadio 1985; Stransky et al., 1989; Spooner et al., 1993). Additionally, at this time there is no scientifically valid research that establishes the effectiveness of Vitamin B-6 for *preventing* the occurrence of musculoskeletal disorders.

Stretching and Strengthening

A valuable adjunct in individual cases, this approach should be under the guidance of an appropriately trained HCP (e.g., physiatrists, physical and occupational therapists). Exercises that involve stressful motions or an extreme range of motions, or that reduce rest periods may be harmful.

Hot Wax

At this time there is no scientific evidence regarding the effectiveness of hot wax treatments as a preventative measure or as a therapeutic modality.

Steroid Injections

For some disorders resistant to conservative treatment, local injection of a corticosteroid by an experienced physician may be indicated. The addition of a local anesthetic agent to the injection can provide valuable diagnostic information.

Surgery

With an effective ergonomics and medical management program, surgery for work-related MSDs should be needed rarely. Surgical intervention should be used for objective medical conditions and should have proven effectiveness. While the indications for prompt or emergency surgical intervention may still be present (e.g., ulnar artery thrombosis), surgery should be reserved for severe cases (e.g., very high levels of pain resulting in significant functional limitations) not responding to an adequate trial of conservative therapy.

68.8 Follow-up and Return to Work

Follow-up

Many, if not most, WRMSDs improve with conservative measures. HCPs should follow up the symptomatic employee to document improvement, or to reevaluate employees who have not improved. The time frame for this follow-up depends on the symptom type, duration, and severity. A clinical exam or telephone contact with the employee should be made once a week, followed by a complete reevaluation within ten days from the last examination if the employee's symptoms are not improving. Where HCPs are available at the workplace, monitoring the symptomatic employee should occur every 3 to 5 working days depending on the clinical severity of the disorder (Wiesel et al., 1984; Wiesel et al., 1994).

In reassessing employees who have not improved, the following should be considered:

- Is the diagnosis correct?
- Are the treatment goals appropriate?
- Have the MSD risk factors on and off the job been addressed?
- Is referral appropriate?

If the job's relevant risk factors have been eliminated but the employee's symptoms persist, it is important for the HCP to realize that employee reactions to pain and functional limitations may prolong the recovery period. Strategies to help the employee cope with the pain and stress associated with these disorders should be incorporated into the employee's treatment plan. The time frames for considering referral depends on the primary HCP's training and expertise, in addition to the type, duration, and severity of the condition. In general, severe symptoms with objective physical examination findings interfering with an employee's ability to perform his/her job should be referred to an appropriate HCP specialist sooner than milder symptoms without objective findings.

Return to Work

If an employee's treatment plan required time away from work, the next step is to return the employee to work in a manner that will minimize the chance for re-injury. Employees returning to the same job without a modification of the work environment are at risk for a recurrence. Key to the return to work process is open communication among the employee, the HCPs, and management. This will allow:

(1) prompt treatment, (2) an expedient return to work consistent with the employee's health status and job requirements, and (3) regular follow-up to manage symptoms and modify work restrictions as appropriate. The principles guiding the return to work determination include the type of MSD condition, the severity of the MSD condition, and the MSD risk factors present on the job.

Employees with MSDs who have difficulty remaining at work or returning to work in the expected timeframes are candidates for rehabilitation therapy. Rehabilitation refers to the process in which an injured worker follows a specific program that promotes healing and helps him or her return to work. During the rehabilitation process, psychosocial factors (factors present both on the job, and off the job, that can compromise an individual's ability to cope with symptoms, physical disorders, and functional limitations) should be addressed.

68.9 Screening

Currently there is no scientific evidence that validates the use of preassignment medical examinations, job simulation tests, or other screening tests as a valid predictor of which employees are likely to develop MSDs (Frymoyer 1992; Werner et al., 1994; Cohen et al., 1994). Literature findings are mixed on the use of preplacement strength testing as a valid predictor of back injury.

68.10 Conclusion

The financial and human costs of work-related musculoskeletal disorders to our society are staggering. This chapter on the medical management of these disorders should help employers and HCPs wishing to prevent or reduce the severity of these disorders, resulting in a healthier, more productive workplace.

Acknowledgments/Disclaimer

We would like to thank the individuals and professional associations who contributed to the draft ANSI Z-365 standard, and the draft OSHA Ergonomic Protection Standard. Ms. Bertsche's contribution to this chapter occurred while a visiting scientist at the National Institute for Occupational Safety and Health (NIOSH), from the U.S. Department of Labor Occupational Safety and Health Administration (OSHA). This chapter represents the views of the authors and does not constitute official policy of NIOSH.

References

AHCPR (1994). Acute Low Back Problems in Adults: Assessment and Treatment. U.S. Dept of Health and Human Services, Public Health Service, Agency for Health Care Policy and Research. Rockville, MD. Publication No. 95-0643.

Amadio PC (1985). Pyridoxine as an adjunct in the treatment of carpal tunnel syndrome. *J Hand Surg;* 10A:237-241.

ANSI (1995). ANSI Z-365 Control of Work-Related Cumulative Trauma Disorders Part 1: Upper Extremities. American National Standards Institute. Chicago, IL: Working draft 4/17/95.

ASSH (1990). *The Hand: Examination and Diagnosis,* 3rd ed. American Society for Surgery of the Hand. New York, NY; Churchill Livingstone.

BLS (1995). Workplace injuries and illnesses in 1994. U.S. Department of Labor, Bureau of Labor Statistics. Washington, D.C.

Bongers PM, De Winter CR, Kompier MA, Hildebranndt VH (1993). Psychosocial factors at work and musculoskeletal disease. *Scand J Work Environ Health;* 19:297-312.

Cohen JE, Goel V, Frank JW, Gibson ES (1994). Predicting risk of back injuries, absenteeism, and chronic disability. *J Occup Med*; 36(10):1093-1099.

Day DE (1987). Prevention and return to work aspects of cumulative trauma disorders in the workplace. *Seminars in Occup Med*; 2(1):57-63.

Frymoyer JW (1992). Can low back pain disability be prevented? *Bailliere's Clinical Rheumatology*; 6(3):595-607.

Haig A, Linton P, McIntosh M, Moneta L, Mead P (1990). Aggressive early medical management by a specialist in physical medicine and rehabilitation: effect on lost time due to injuries in hospital employees. *J Occup Med*; 32(3):241-244.

Hoppenfield S (1976). *Physical Examination of the Spine and Extremities.* Norwalk, CT: Appleton Century Crofts.

Kiefhaber TR, Stern PJ (1992). Upper extremity tendinitis and overuse syndromes in the athlete. *Clinics in Sports Med*; 11(1):39-55.

Louis DS (1987). Cumulative trauma disorders. *J Hand Surg*; 12A(5):823-825.

Mayer TG, Gatchel RJ, Hayer H et al. (1987). A prospective two-year study of functional restoration in industrial low back injury: an objective assessment procedure. *JAMA*; 258:1763-1767.

McGill SM (1993). Abdominal belts in industry: A position paper on their assets, liabilities, and use. *Am Indust Hygiene Assoc*; 54(12):752-554.

Mitchell LV, Lawler FH, Bowen D, Mote W, Ajundi P, Purswell J (1994). Effectiveness and cost-effectiveness of employer issued back belts in areas of high risk for back injury. *J Occup Med*; 36(1):90-94.

NIOSH (1994). Workplace use of back belts. Review and recommendations. U.S. Department of Health and Human Services, Public Health Service, Centers for Disease Control and Prevention, National Institute for Occupational Safety and Health. Cincinnati, OH; Publication No. 94-122.

Nirschl RP (1990). Patterns of failed healing in tendon injury, in Leadbetter WB, Buckwalter JA, Gordon SL (eds). Sports-Induced Inflammation. Park Ridge, IL: American Orthopaedic Society, pp 577-585.

OSHA (1995). Draft Ergonomics Protection Standard. U.S. Department of Labor, Occupational Safety and Health Administration. Washington, D.C. March, 1995

Ranney D (1993). Work-related chronic injuries of the forearm and hand: Their specific diagnosis and management. *Ergonomics*; 36(8):871-880.

Rempel D, Manejlovic R, Levinsohn DG, Bloom T, Gordon L (1994). The effect of wearing a flexible wrist splint on carpal tunnel pressure during repetitive hand activity. *J Hand Surg*; 19(1):106-110.

Spooner GR, Desai HB, Angel JF, Reeder BA, Donat JR (1993). Using pyridoxine to treat carpal tunnel syndrome. Randomized control trial. *Canadian Family Practice*; 39:2122-2127.

Stransky M, Rabin A, Leva NS, Lazaro RP (1989). Treatment of carpal tunnel syndrome with vitamin B-6: A double blind study. *Southern Med J;* 82(7):841-842.

Tubiana R, Thomine JM, Mackin E (1984). *Examination of the Hand and Upper Limb.* Philadelphia, PA: W.B. Saunders Co.

Upfal M (1994). Understanding medical management for musculoskeletal injuries. *Occup Hazards*; Sept:43-47.

Vallfors B (1985). Subacute and chronic low back pain; Clinical symptoms, absenteeism, and work environment. *Scand J Rehab Med*; Suppl 11:1-99.

Werner RA, Franzblau A, Johnston E (1994). Quantitative vibrometry and electrophysiological assessment in screening for carpal tunnel syndrome among industrial workers. *Arch Phys Med Rehabil*; 75:1228-1232.

WHO (1985) Identification and control of work-related diseases. World Health Organization Geneva: WHO Technical report; 174:7-11.

Wiesel SW, Feffer HL, Rothman RH (1984). Industrial low back pain, a prospective evaluation of a standardized diagnostic and treatment protocol. *Spine*; 9:199-203.

Wiesel SW, Boden SD, Feffer HL (1994). A quality based protocol for management of musculoskeletal injuries. *Clinical Orthopaedics and Related Research*; 301:164-176.

Wood DJ (1987). Design and evaluation of a back injury prevention program within a geriatric hospital. *Spine*; 12:77-82.

[illegible]

69

Ergonomic Programs in Post Injury Management

Tarek M. Khalil
University of Miami

Elsayed Abdel-Moty
University of Miami

Renee Steele-Rosomoff
University of Miami

Hubert L. Rosomoff
University of Miami

0-8493-2641-9/99/$0.00+$.50

69.1 Introduction

The management of work-related injuries and illnesses continues to be a challenge to ergonomists and health and safety professionals. This chapter outlines a selected number of ergonomic contributions in rehabilitation and disability prevention. The rationale, concepts, and methods developed for patient evaluation, work conditioning and job simulation, workplace design, worksite analysis, neuromuscular conditioning, and other topics are presented. Many of these intervention techniques were developed and tested by the ergonomics team at the University of Miami Comprehensive Pain and Rehabilitation Center (CPRC), and have proven valuable to post-injury management and return to work. It is difficult to provide detailed descriptions of the various ergonomic interventions in post-injury management in a few pages. Therefore, the authors have included an expanded list of references that may help the reader get more detailed information about the approaches used.

69.2 Workplace Injuries

Injuries or pain happen either due to specific accidents (trauma or a single stress of high magnitude) or cumulative events (a series of repeated stressors or micro traumas). If the resulting effects of an injury extend beyond three months, an acute injury is classified as a chronic condition (Pain Commission, 1987).

Injuries in the workplace often occur due to inadequate design of work environments, tools, and tasks. Workplace inadequacy also means that environments, tools, and tasks are designed without consideration for the people who must use them. Injuries can also develop due to a variety of causes including unsafe acts or human errors. These can also be dealt with through ergonomic interventions. According to the Bureau of Labor Statistics, there were approximately 6,252,000 occupational injuries and 514,000 occupational illnesses in 1994 (Safety Management, 1996). These figures translate into a total rate of 8.4 occupational injuries or illnesses per 100 full-time workers. There are certainly many more injuries and ergonomic disorders in the nonoccupational environment.

69.3 Ergonomics Disorders

Poor ergonomic design of tasks, equipment, or workplaces have resulted in a newly recognized group of injuries recently called ergonomic disorders. These are disorders of the musculoskeletal and nervous systems occurring in upper or lower extremities. They may be caused by repetitive motions, forceful exertions, vibration, sustained or awkward postures, and/or mechanical compression. Among ergonomic disorders are: low back pain, soft tissue injuries, eye fatigue and strain, sensitivity to light, blurred vision, changes in color perception, numbness of fingertips, headache, neck strain, skin problems, stress, and cumulative trauma disorders (also known as repetitive motion injuries, overuse disorders, or motion strain.)

Musculoskeletal injuries are the most common workplace injuries with back pain leading the charts in terms of incidence, cost, and suffering. Backaches can strike almost anyone. Although the exact incidence of low back pain is unknown, it is obviously high. It has been estimated that 70 to 80% of the population will develop some form of back pain during their life time. Annual estimates of the new cases of low back pain (LBP) have ranged from 10 to 15% of the United States population. LBP accounts for 25% of all disabling injuries (Rosomoff and Rosomoff, 1991). It has been reported to consume about 80% of total health care expenditures associated with musculoskeletal injuries.

Impairments of the back are the most frequent cause of activity limitation in persons under the age of 45, and they rank as the third most common cause of disability after heart and arthritic conditions in people age 45 years and over. The National Institute for Occupational Safety and Health (NIOSH) cited that back injuries account for about 20% of all injuries and illnesses in the workplace and that these injuries cost the nation an estimated $20 to 50 billion per year (NIOSH, 1991).

Cumulative trauma disorders (CTD) have been identified as the "industrial disease of the information age" and the "work-related disease of the 1980s and 1990s." CTDs are the most frequently reported on-the-job ailment. Millions of Americans have been afflicted by this family of disorders. It has been estimated that by the year 2000, over half the workforce could be vulnerable.

The latest federal statistics put the number of work-related ergonomic CTD injury cases at 514,000, up from just 30,000 in the mid-1980s (Safety Management, 1996). Carpal tunnel syndrome (CTS) is a frequently reported work illness. CTS is a special case of CTDs. It refers to injury to the median nerve of the wrist. Associated symptoms include numbness, tingling, burning sensations in the fingers, pain and wasting of muscles at the base of the thumb, dry or shiny palms, and clumsiness. CTS develops when the effective cross section of the carpal tunnel in the wrist is decreased or there is an increase in the volume and pressure within the tunnel due to inflammation or edema. While there is an increasing number of CTS cases which have been associated with workplace factors, some believe that the majority of CTS cases are not occupationally related and that certainly not all wrist problems are necessarily CTS.

The staggering statistics make rehabilitation and effective management of low back pain, cumulative trauma disorders, and all types of musculoskeletal or soft tissue injury a challenge to health care professionals and ergonomists.

69.4 Causes of Workplace Injuries/Illnesses

Workplace injuries can be caused by a host of factors. Musculoskeletal work injuries afflict people who perform light tasks (sedentary) and those involved in heavy labor. The magnitude of the stressor is an important factor, however a common characteristic among injury profiles is that job tasks require excessive bending, twisting, reaching, and deviation from neutral postures.

Most ergonomic disorders have been associated with injuries to the soft tissues in the human body. Soft tissue injury (for example: tension, sprain, spasms, weakness, strain, contracture, and inflammation) is the most common cause of pain (Quebec Study, 1987).

Soft tissue injury (STI) can result from activities such as lifting heavy objects, slipping and falling, bending and twisting, driving, accidents, and the manner by which work activities are carried out (also known as body mechanics). Soft tissue injuries can also develop gradually due to unknown causes and due to disuse and muscle weakness (Abdel-Moty et al., 1992a). In one study, we have shown that 20% of a sample of 200 patients with LBP were not able to attribute their painful condition to a specific incident, indicating that "pain developed gradually" (Abdel-Moty et al., 1990a).

STI can also develop gradually as a result of the "wear and tear" when activities are done with forceful muscular exertion, awkward postures, fast actions, sustained activity, and/or with frequent repetitive motion (case of CTDs) (Abdel-Moty et al., 1991b). Several human activities and job tasks are, therefore, more likely to result in STIs. Examples are: checkout scanning in supermarket, keypunching, working at an assembly line, manufacturing, meat processing, butchers, hairdressers, sewing and knitting, packing, stapling, tight grasping at the steering wheel, polishing and buffing, surface grinding, painters, dentists, interpreters, postal workers, milking cows, use of a spray gun, and sports activities.

Work-related injuries can also develop due to medical conditions and dysfunction of body structures and systems (e.g., infections, congenital disorders, neurological factors) (Khalil et al., 1993a).

In particular, computers have been blamed for causing or contributing to workplace injuries. Computers are becoming a household item in millions of offices, business, and homes for the purpose of increasing productivity. However, the effects on health have been negative. Prolonged keypunching, constant switching of the eyes from the document to the computer screen, improper combination of computer tasks and other activities, and inadequate visual attributes of the operator are but some of the conditions that have resulted in disabling physical and psychological conditions of the workplace user.

Numerous guidelines have been recommended to deal with workplace and ergonomic injuries. These will be discussed throughout this chapter.

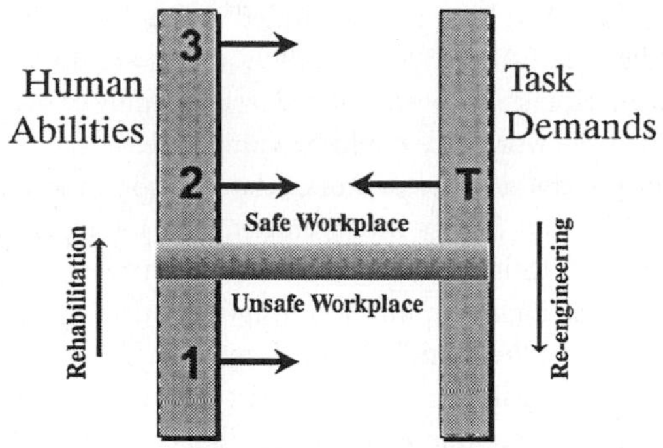

FIGURE 69.1 The premise of the ergonomic approach to injury/reinjury prevention is matching human abilities to task demands. When task demands exceed human abilities (T>1), human safety is at risk. There is a need, then, to improve human abilities through conditioning, exercise, and rehabilitation. When human abilities meet or exceed task demands (T = 2 or T<3), human safety is improved. Task demands can be decreased through ergonomic, engineering, and management controls.

69.5 Prevention of Workplace Injuries — The Ergonomic Approach

The thrust of the ergonomic approach to injury prevention is matching human abilities/limitation to the demands of the task (Figure 69.1). In doing so, ergonomists utilize knowledge of engineering, health, behavioral, physical, and management sciences in an integrated fashion for analyzing, designing, testing, and evaluating systems to improve human performance, reduce risk, improve productivity, and reduce workers' compensation costs.

Ergonomic contributions to injury prevention can be identified under three stages:

1. Primary prevention stage — where the prevention of the onset of injury becomes a priority and a desired goal for the avoidance of subsequent disability and lost workdays.
2. Secondary prevention stage — where the goal is the prevention of disability, quick functional restoration, and an expeditious return to work and productive life style of the injured worker.
3. Tertiary prevention stage — where prevention of re-injury becomes essential in order to ensure no recurrence of the injury-disability cycle.

Ergonomics contributions at each level of injury prevention are discussed below with emphasis on secondary and tertiary prevention approaches, herewith referred to as post-injury management.

69.6 Primary Prevention of Workplace Injuries

By far, this is considered to be the most effective approach to curing the problem of work injuries and illnesses. However, despite the increasing efforts to make workplaces safer, injuries still happen at an alarming rate. In response to the problem, several government, academic, commercial, and industrial entities have devoted tremendous resources toward injury prevention. The underlying approach has been early identification and recognition of the problem. This is important especially if done at a stage that allows prevention of injury through ergonomic work reorganization and work redesign (Kroemer, 1994).

To enhance efforts associated with primary prevention of injury, several bills, regulations, and standards have been issued by local, state, and federal agencies. In 1990, OSHA introduced its "OSHA Program Management Guidelines" which emphasize the need for:

1. Routine worksite analysis to recognize, identify, and correct ergonomic hazards including review and analysis of injury and illness records.
2. Hazard prevention and control through design measures to prevent or control hazards; engineering controls for workplace design; work practice controls such as work rest schedule; use of personal protective equipment; and administrative controls.
3. Medical management; injury/illness recordkeeping; early recognition and reporting; systematic evaluation and referral; treatment and return to work; systematic monitoring; and adequate staffing and facilities.
4. Training and education (both general and job specific) of the supervisors, managers, and engineers and maintenance personnel.

Also, NIOSH declared that the most effective way to prevent back injury is to implement an ergonomic program that focuses on redesigning the work environment and work tasks to reduce the hazards. One such effort was the introduction of the "lifting equation" and guidelines for manual material handling. Repetitive material handling has been considered to increase vulnerability to injuries/claims involving the back. NIOSH, therefore, developed its *Work Practice Guide* which has undergone two revisions since 1981. The guide provided direction as to the amount of weight to be handled under various conditions. Obviously, this biomechanical approach has its practical limitations. Employee capabilities, skills, past injuries, stress, and similar factors should be considered together with this mathematical determination of lifting limits. Additionally, guidelines alone are not an effective substitute for employee training in safety, a well-engineered workplace, and a "fit" worker.

The American National Standards Institute (ANSI) is another agency which advocates human factors in workplace design for health and safety. ANSI published technical standards for implementation of human factors engineering principles and practices in the design of visual display terminals (VDT), associated furniture, and the office environment in which they are placed (ANSI, 1988).

69.7 Post Injury Management — Prevention of Disability

The primary objective of ergonomic contributions in this stage is "to design effective intervention strategies for the restoration of functional abilities and immediate return to work and productive life style." Another objective is deterrence of further aggravation of an existing condition or injury.

This philosophy was adopted by the University of Miami Comprehensive Pain and Rehabilitation Center as early as 1982 when ergonomics was first introduced in a multidisciplinary pain management team. This marked the beginning of a new era for ergonomics research and application in health care systems and pain treatment in particular. The basis for introducing ergonomics in post-injury management is demonstrated in Figure 69.2. In this model, injury management according to the traditional medical approach alone may be effective in treating the symptoms of pain, but it may fail to address and correct the cause of the injury. Through the use of a "rehabilitation through technology" approach, ergonomists assist in effective treatment as well as returning the patient to a modified work environment with improved safety and better ergonomic design. Therefore, the potential for reinjury is greatly reduced or eliminated. Since its introduction in pain management, ergonomics has gained wide acceptance, and its contributions have proved vital in addressing issues of primary importance. Working with other health care professionals who are usually involved in musculoskeletal and low back pain rehabilitation, ergonomists have played equally important roles in the successful rehabilitation and management of injuries.

The Role of the Medical Community in Post-Injury Management

Several medical treatment approaches have been advocated by different care providers to deal with musculoskeletal injuries; especially low back pain problems. Successful rehabilitation and management of musculoskeletal injuries should integrate the different medical treatment disciplines in order to accomplish such goals as: restoration of function; pain reduction; and, consequently, increasing productivity

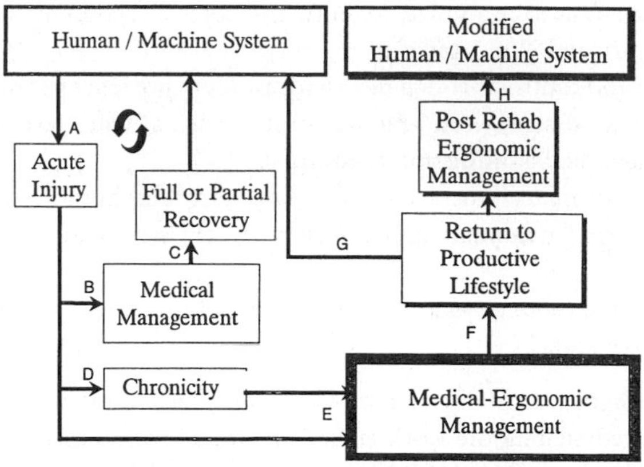

FIGURE 69.2 (A) Injury occurs as a result of a faulty human-machine system. (B&C) Medical management attempts to resolve acute injuries in order to return the injured worker to the *same* job. Since no measures are taken to change the cause of injury, the worker is at a risk of reinjury. (D) Acute injuries may develop into chronic conditions if left untreated or when mismanaged. (E&F) The medical–ergonomic management of both acute and chronic injuries or illnesses will assist the injured individual to return to a productive life style. (G) The inclusion of ergonomics in the post-rehabilitation stage will ensure the individual's return to an ergonomically correct human machine system, thus minimizing the potential of reinjury.

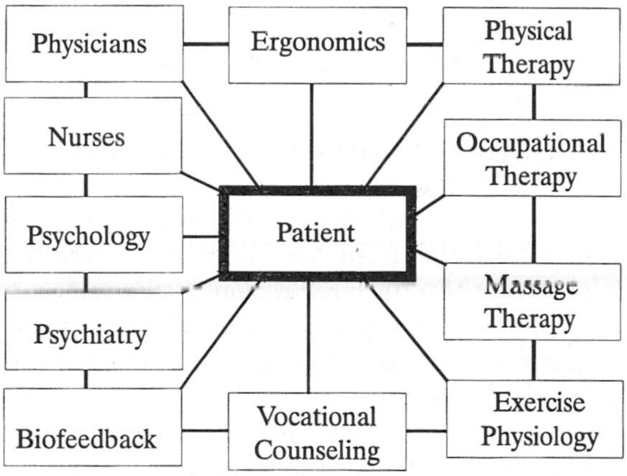

FIGURE 69.3 Comprehensive pain and rehabilitation programs employ a multidisciplinary approach. In this model, ergonomists work jointly with other health care providers to address patient issues and needs for full restoration.

and reducing rising disability costs. The health care professions that are usually involved in dealing with an important problem such as low back pain rehabilitation include: the physician(s), physical therapists, occupational therapists, nurses, vocational counselors, psychologists, and psychiatrists. Another profession which has shown to be equally important in a successful low back pain rehabilitation effort is ergonomics (Figure 69.3) (Khalil et al., 1993a).

The Role of the Clinical Ergonomist in Injury Management

The application of ergonomics principles and methods to the injured population is far from simple. A major difficulty often encountered in applying ergonomic methods in a clinical setting is that these methods are developed for the "healthy" population, representing military personnel, industrial workers, and volunteers. Therefore, it becomes important to bridge the gap between traditional ergonomic research

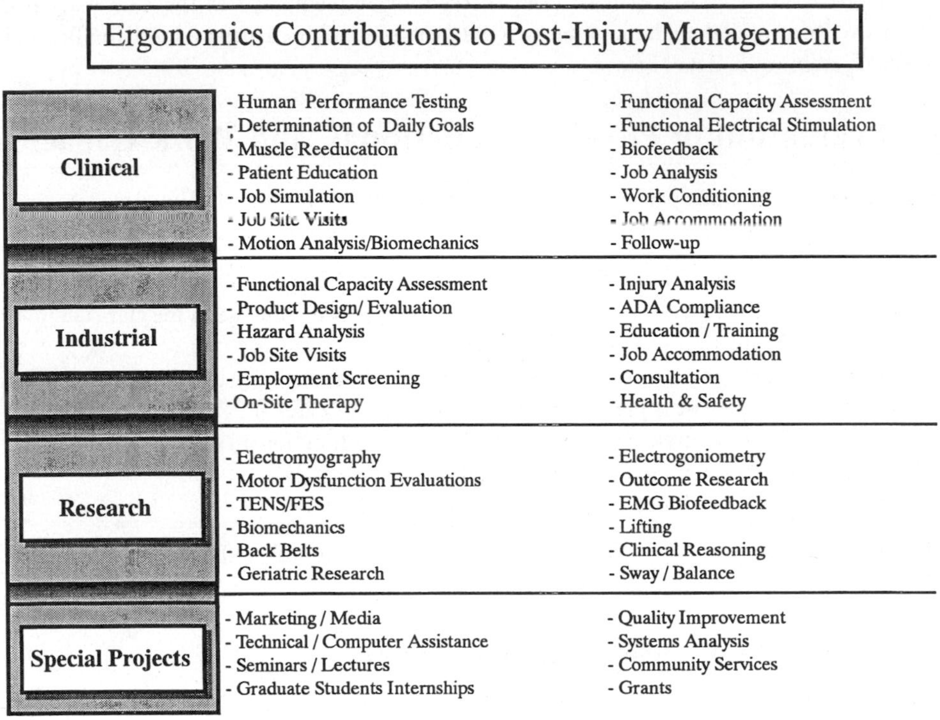

FIGURE 69.4 Listing of ergonomic contributions to post-injury management in the various areas.

and its application in health care systems. This requires a thorough understanding of the type of injury, the specifics of the population at hand, and of medical, social, and legal issues (Khalil et al., 1991b; Khalil et al., 1992b). For example, in ergonomic research, the evaluation of muscle strength has been based on the use of maximum voluntary exertion protocols where subjects are asked to increase their efforts to a maximum level. Applying this protocol to patients with back injury may be detrimental to their health and their safety, may be psychologically unacceptable to the patient, or may result in irreversible damage leading to medical as well as legal complications. In this case, it becomes necessary to develop methods of strength testing which are more appropriate and specific to the type of injury such as the acceptable maximum effort procedures developed by Khalil et al. (1987). Conversely, and as their awareness of the role of ergonomics is increased, the medical staff will begin to redefine their roles and contributions in order to make use of available ergonomic approaches not previously known to them.

69.8 Ergonomic Interventions in Post-Injury Management

Ergonomic contributions to injury management can be classified under the following categories in the order of their application (Abdel-Moty et al., 1990b; Khalil et al., 1993b) (Figure 69.4):

1. Physical, functional, and work capacity assessments.
2. Determination of patients' daily treatment goals.
3. Electrically assisted methods to improve neuromuscular functioning.
4. Work conditioning and job simulation.
5. Workplace design and analysis including job site visits.
6. Biomechanical analysis of human motion (gait, sway, kinetics, and kinematics).
7. Patient education (safety, body mechanics, ergonomics).
8. Applied clinical research to develop innovative solutions to clinical problems (electronic goniometry, stretching, traction, outcome).

The methods used in each area have been successfully implemented at the University of Miami Comprehensive Pain and Rehabilitation Center (CPRC) at South Shore Hospital in Miami Beach, Florida and are discussed in more detail below.

69.9 Evaluation of Physical, Functional, and Work Capacities

Objective evaluation of human performance is a step in an ergonomic program aimed at matching people to their environments, machines, and tasks. In a rehabilitation setting, evaluations are used in order to:

1. Assess the current level of function and the degree of the loss of function due to injury or illness. This helps the patient understand the degree of their functional loss and the clinician to establish a baseline for evaluation of treatment efficacy
2. Evaluate and monitor patients' progress during rehabilitation to ensure that treatment goals are met and that the patient is part of the therapeutic plan of care
3. Direct treatment to address problem areas and focus on functional restoration
4. Evaluate treatment outcome following rehabilitation
5. Examine the patient's ability to re-enter a productive life style

Also, objective assessment of functional capacities can be a useful tool in making administrative, clinical, and legal decisions regarding work readiness.

Components of the Human Performance Profile

The patient's performance profile may consist of one or more of the following measures (Abdel-Moty et al., 1991c; Abdel-Moty et al., 1992b; Fishbain et al., 1994, 1995):

a. Measures of *physical* capacities such as isometric and dynamic strength, flexibility, mobility, posture, sway and balance, psychomotor abilities, and gait.
b. Measures of *physiological* capacities such as muscular endurance and cardiovascular endurance.
c. Measures of *functional* capacities such as tolerance to sitting, standing, walking, climbing, lifting, carrying, pushing, pulling, driving, stooping, crouching, and squatting.
d. Measures of work capacities specific to the patient's ability to perform job demands.

Measurements of the human performance profile are reported relative to basic scores upon initiation of treatment, progress from beginning to final scores, and in comparison to "norms." Behaviors during evaluation are observed and recorded with respect to cooperation, consistency, effort, motivation, patient comments, and pain behavior. At the CPRC, this profile is established for each patient at three intervals: upon admission to the rehabilitation program (inpatient or outpatient); two weeks into the program; and at the end of the 4-week treatment program. The profile can also be compared to the physical demands dictated by the job thus assisting in job placement for the prevention of further injury.

The issues surrounding patient evaluation are numerous and can be found in Abdel-Moty et al. (1996). Among these are:

1. Operational definitions with respect to the types of evaluations (physical capacity evaluation vs. functional capacity assessment, work-related assessments)
2. Patient's issues (pain, perceptions, behaviors, motivation, effort, secondary gains, use of medications, prior exposure to similar testing, contraindications, use of assistive devices),
3. Relationship to the Americans with Disabilities Act (ADA) — the ADA requires that employment/post-hire evaluations using functional capacity assessment (FCA) to be job specific and task oriented (Abdel-Moty et al., 1993a).
4. Administrative issues of insurance, authorization, referrals, and scheduling
5. Evaluator's issues (bias, objectivity, training, experience, qualifications)
6. Methodological issues (patient safety, testing equipment, protocols, instructions, sequence of testing)

7. Statistical issues (validity, reliability, sensitivity)
8. Other issues (testing environment, presence of third-party, such as attorneys, during testing, reporting of results)

By accurately establishing performance parameters of the patients, it is possible to chart their course of treatment through the use of the concept of daily goals.

69.10 Scientific Determination of Patients' Daily Treatment Goals

Restoration of functional abilities of patients requires knowledge about patients' baseline level of performance as well as target goals. Baseline values represent the levels of functional abilities at which patients enter the rehabilitation process. Typically, baseline values are determined for categories of muscle strength, ranges of motion, tolerances, mobility, and soft-tissue factors, as well as a host of behavioral and vocational variables. Treatment goals are usually determined based on vocational and/or avocational objectives as well as on expected levels of functional abilities representing norms for healthy subjects (e.g., the expected angle for lumbar flexion is 90 degrees).

The objective of rehabilitation is to treat, condition, and build the patients' functional abilities in order for them to achieve the desired levels of performance constituting their treatment goals.

In most rehabilitation settings, the progression from the initial (admission) levels to the final goals (i.e., daily achievement during treatment) is determined through subjective clinical judgment of the patient's current status. Additionally, in some instances patients' progression may be seen as tied to the length of hospitalization or rehabilitation.

In today's atmosphere of health care management awareness, cost effectiveness, and health reforms, optimization of treatment approaches and maximization of service delivery becomes a priority. Using a daily goal model can help maximize treatment plans and determine optimal pathways patients should follow during rehabilitation. A prediction modeling approach provides the treating team, as well as the patient, with scientific tools for determining, realistically and objectively, patients' daily performance achievement during treatment. It also enables effective utilization of resources, and hence cost effectiveness, for both the patient and the health care delivery system.

At the CPRC, ergonomists developed a novel approach to maximizing patients' progress during rehabilitation through the use of computerized modeling and optimization approaches. Named the computerized daily goals (CDG) model, it was developed to determine optimal pathways a patient should follow during chronic low back pain rehabilitation. The model utilizes statistical projection methods which take into consideration the patient's initial performance level and the desired goals (Abdel-Moty et al. 1993b) (Figure 69.5).

This nonlinear model derives its coefficients from retrospective data collected on over 1000 patients with chronic pain who were admitted for treatment at the CPRC, successfully completed the 4-week rehabilitation program, and returned to a productive life style. All patients underwent evaluations of functional abilities at three points during treatment: upon admission, after 2 weeks, and at the end of the 4-week treatment program. Patients were evaluated and their performance was analyzed in the following categories:

1. Tolerances (sitting, standing, walking, squatting, kneeling, stair climbing).
2. Strength (lifting, carrying, reaching, pulling, pushing).
3. Flexibility (trunk, cervical).

The purpose of this type of classification was to determine quantitative trajectories of progress in the various categories as a result of the intervention protocols. Following necessary testing and validation, a model was constructed and implemented on a personal microcomputer. Currently, once initial levels of performance have been measured and treatment goals have been determined, the daily goals program is used to provide a personalized print-out of the expected daily performance in the different categories.

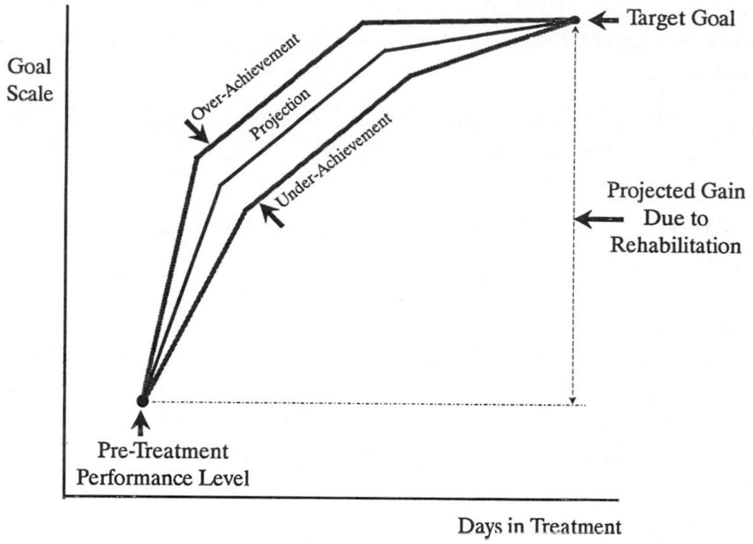

FIGURE 69.5 The scientific determination of patients' daily projected goals and clinical pathways requires accurate determination of a patient's initial performance levels as well as knowledge of a desired. Projections are then made over the length of the patient's treatment. The "range of projection" allows for flexibility in order to accommodate individual patients.

Final goals are reached just prior to patients' discharge from the program. The daily goals program predicts daily increments and corresponding dosage of patients' therapeutic activities. Optimal progression from initial tolerances to final goals is then printed and used by the patient and the medical staff to determine activities needed to achieve desired daily performance in the various categories throughout treatment.

In order to demonstrate its usefulness, a sample of 200 patients who completed the treatment program was analyzed. About 82% of patients who were studied followed the projected levels, with another 9% exceeding projections and showing improvements at rates higher than expected. In about 3% of the cases examined, the "printed" goals were lowered due to unexpected changes in the patients' medical condition. In another 5% of the cases the projected goals were modified (mostly decreased) due to changes in return-to-work plans.

69.11 Electrically Assisted Methods (EMG, Biofeedback, Muscle Reeducation, and Functional Electric Stimulation)

Soft tissue injury and the resulting pain often produce compensatory changes in posture and in the manner by which patients perform daily activities. The failure to treat chronic pain patients effectively is, in part, due to an inability to identify and pinpoint the source of these altered motor patterns affecting functional abilities. In order to assist the patient and the medical team with treatment goals, several technology-based methods and intervention protocols were developed by clinical ergonomists. Among these are protocols for biofeedback, neuromuscular reeducation, and functional electrical stimulation.

One method was developed to identify and effectively treat motor dysfunction in patients with chronic pain conditions. This innovative multidisciplinary evaluation/treatment approach is based on the use of on-line computerized electromyographic (EMG) methods to study multiple muscles involved in a chain of precise activities specific to recruiting these muscles. The EMG signals of the various muscles are examined for: baseline activity, symmetry, magnitude, frequency contents, synchrony, timing, and patterns. Patients' behaviors are also observed. EMG findings are then compared to relevant clinical findings. The team of medical and ergonomics professionals discusses the significance of the findings, composes

interpretations, and recommends treatment plans or addresses outcomes as applicable. Patient-specific, as well as condition-specific, multidisciplinary approaches are then generated to deal with the problems during daily treatment. The overall objective is to decrease pain and improve function. This is accomplished through using EMG and other electrically assisted methods to increase sensory perception (of muscles and joints); increase neuromuscular recruitment; increase strength and endurance; and reestablish synchrony, symmetry, pattern, and synergy of muscle activity to increase functional capacities. These protocols are incorporated into patients' daily treatment and clinical pathways. These treatment approaches fall under the categories of functional electric stimulation, biofeedback, and neuromuscular reeducation.

Functional Electric Stimulation

Functional electric stimulation (FES) is used when minimal muscle recruitment or reduced voluntary control is detected upon medical or ergonomic testing. FES is a good example of an approach for muscular conditioning and strengthening through which muscles can be strengthened "passively" without placing excessive demands on the patient especially in the presence of pain. FES is the process of applying an external electrical stimulus to a muscle or muscle groups in order to induce muscle contraction. FES found many applications in rehabilitation, physical therapy, sports medicine, and recently in physical fitness. Its usefulness ranged from increasing local blood flow to strengthening weak and healthy muscles and enabling paraplegics to walk. We have also used FES successfully to treat conversion-disorders-type paralysis, for electric testing of motor responses, and as a motor dysfunction treatment method. The strength of FES arises from its ability to induce maximal muscular contraction without any voluntary effort on the part of the treated individual. This later aspect is of value for patients with chronic pain whose pain is often aggravated through regular exercise or who are unable to initiate the voluntary effort necessary for muscular conditioning due to disuse.

The use of FES as a muscle strengthening alternative proved useful and shows great potential in the rehabilitation and physical restoration of muscles weakened by disuse or pain. The current technology has allowed for the development of electrical stimulation units that are safe for use with minimal adverse effects. Studies on FES indicate that this passive intervention strategy is quite effective in numerous cases (Khalil et al., 1991a; Abdel-Moty et al., 1994). The use of FES may be less effective in some cases due to such reasons as fear of electricity, impaired sensory perception, skin conditions, lesions, muscles of the chest wall, and cervical paraspinals.

It should be emphasized here that FES is not a substitute for regular exercise. FES is used to "jump start" the neuromuscular system function. Once the patient has gained sufficient power to initiate voluntary movement comfortably, he or she is to engage in active exercises of strengthening, flexibility, and endurance training. In this regard recent advertisements which market muscle stimulators so "you never have to do sit-ups again" and "the cellulite breakthroughs" must be evaluated objectively and scientifically, especially with respect to patients' perception of such products.

Electromyography and Biofeedback

Another electrically assisted tool for improving functional abilities of injured individuals is the use of electromyography (EMG). Recently there has been a renewed interest in the use of surface electromyographic (EMG) recording for the assessment and evaluation of patients with chronic pain. In general, a carefully processed EMG signal can be a useful tool in the quantitative measurement of muscular performance. In chronic low back pain rehabilitation, we found EMG to be valuable in the study of muscle dynamics, for purposes of muscle reeducation and biofeedback, and in the evaluation of patients' response to specific treatments (Khalil et al., 1991a). In earlier reports, it has been shown that the initial EMG levels of pain patients upon admission to the rehabilitation program were low and increased following muscle reeducation and physical restoration (Khalil et al., 1993a). The significant increase in EMG was associated with significant increase in strength and reduction of pain.

Biofeedback (biological feedback) is the process of using specialized instruments to give people information about their biological systems (temperature, heart rate, muscle activity, etc.). It is a set of treatment/training techniques used to increase awareness and voluntary control of biological conditions and relate them to human physical and emotional well being. Biofeedback (BF) is useful in the relief of stress, tension, headaches, muscular dysfunction, and for improvement of muscle strength (Khalil et al., 1987; Asfour et al., 1990).

Using EMG biofeedback, patients perform therapeutic maneuvers while affected muscles are monitored. Muscles are then compared bilaterally, and the information is used to facilitate the patient's awareness of muscular performance. Patients are also evaluated without the use of feedback to determine carry-over and awareness level. Ergonomists contribute in this area by making the technology and measurement methods available to clinicians. The developed methods can then be used to improve the patients' ability to coordinate muscle activity, re-establish proper reciprocal inhibition, re-establish functional synergy including appropriate force couples and sequential contractions, decrease the need for inappropriate muscular or postural compensations, and achieve recruitment levels and baseline activity.

Work Conditioning and Job Simulation

The concepts of job simulation and work conditioning are important to effective return-to-work strategies. The objective here is to improve the physical and functional abilities of post-injury individuals while practicing job tasks. It is important to raise the individual's tolerances to adequate levels in order to maximize his or her potential for return to work. This is accomplished through exercise, training, work conditioning, and job simulation. In clinical settings, the ergonomists assist the health care delivery team in developing realistic job simulations within the rehabilitation establishment to permit patients to perform critical job tasks, preferably under medical supervision. An individual or a patient is taught to perform his or her job task properly, which assures them that he or she is capable of carrying them out. This also allows the treating physician to certify that the patient has been physically rehabilitated to handle task demands.

During this activity, the ergonomist evaluates motions, time, and forces required to perform the various tasks and relates these demands to the functional abilities of the individual. For this purpose, it is not possible, nor necessary, to replicate jobs. It is also not necessary to simulate a full workday unless a work tolerance evaluation is requested.

Work conditioning programs are highly structured, task focused, goal oriented, individualized, and multidisciplinary/interdisciplinary in nature. The following are factors which affect the design of a work conditioning program:

1. Intensity (the level of exertion at which a patient is expected to perform). The amount of weight to be carried, distance to be covered, and other physical demands are a function of, not only the patient's status during treatment, but also his or her projected goal. The ergonomist seeks input from the rehabilitation team regarding the patient's medical status in order to determine work intensity in job simulation.
2. Environment (actual or simulated). In many instances the rehabilitation team may decide that it is necessary to observe the patient performing work tasks on the job. This gives an insight into work behaviors, ability to manage stress on the job, and helps develop realistic treatment plans and ergonomic recommendations.
3. Personnel involved (medical, ergonomic). In most cases, the ergonomist designs and conducts work conditioning activities with input from the multidisciplinary team. It is not unusual to find a team of the ergonomist, the occupational therapist, physical therapist, the vocational counselor, the biofeedback specialist, and a psychologist working simultaneously with a patient during job simulation. Each professional brings to this process a unique input and perspective and assists in immediate problem solving. The degree of involvement of each discipline depends on the activities required.

4. Approach (progressive nature of work simulation sessions). A patient who is deconditioned and who has a physically demanding job to return to (e.g., heavy work), will need a different approach to work conditioning than a patient who will return to a sedentary job. For example, at the CPRC, the objective of rehabilitation and ergonomic interventions is to raise the patient's abilities to maximum attainable physical and functional abilities. The inclusion of work demands in these abilities will modify the approach to development of the plan of care for the patient. The plan of care also includes essential goals (or prerequisites) to be achieved before work conditioning activities begin. These are:
   a. Increase patient's awareness of posture and body mechanics.
   b. Increase flexibility and mobility.
   c. Increase strength and endurance.
   d. Improve stress management skills.
   e. Improve pain control, safety, and preventive medicine techniques.
   f. Modify behaviors toward work, employment, and return to work.
   g. Involve patient in work conditioning activities early in the treatment program.

The ergonomist provides direct supervision during the work conditioning sessions as well as encouragement, support, and reinforcement to the patient while monitoring behaviors.

Workplace Design and Job Site Analysis

This type of ergonomic intervention aims at assessing the relationship between human characteristics (e.g., posture, body mechanics) and musculoskeletal stresses with emphasis on work issues. This is a rather complex process due to the large number of human and environmental factors involved (Khalil et al., 1990). The ergonomic premise here is that awkward postures and poor body mechanics can result in a multitude of health problems. With patients who already have an injury or illness, the aches and pains are magnified. The task of the ergonomist is to assist the patient or employer to design/modify the workstation to ensure:

1. Proper engineering design from an ergonomic point of view.
2. Proper posture of the workplace user.
3. Good body mechanics when performing job tasks.

These issues are also interrelated. For example, poor workplace design can result in poor work habits by not allowing good posture and proper body mechanics. Additionally, proper design does not guarantee stress-free environment if good posture and proper body mechanics are not practiced.

The process of workplace design within a clinical setting consists of the following components:

1. Preliminary data collection of patients' information, job-related information, and a self description of a typical workday.
2. Initial evaluation. This is a session of simulated activities during which essential job tasks are conducted, recorded, and analyzed. Emphasis is on posture, body mechanics, and muscular work during repetitive activities. This evaluation aims at identifying tasks as potentially stressful. During this session, no feedback or suggestions for modifications or adjustments are given. Observations and analysis of video graphic data are used to isolate the critical risk factors with respect to the worker, the tools, the furniture, and other workplace parameters.
3. Intervention phase. Once risk factors are identified, the ergonomist begins to develop the interventions specific to reducing/resolving the risk factors. Once more, the interventions usually address the engineering design of the workplace as well as the posture, body mechanics, and patients' ability to follow through (learning). Engineering measures aim at adjusting, modifying, changing, and/or replacing current heights, layout, equipment, tools, and design characteristics. It is important at this stage to explain to the patient the rationale behind the recommendations in order to increase his or her awareness and ability to generalize to situations outside work. It is

also not necessary to always make recommendations for new equipment or assistive technologies (e.g., cushions, ergonomic chairs, etc.). In most cases, workplaces can be reasonably accommodated to meet ergonomic needs. An essential ergonomic rationale is that people may not have the "ideal ergonomically correct" setup wherever they go. It is, therefore, the responsibility of the ergonomist to teach patients methods of improving safety and comfort without the need for expensive adjustments.

A tool we have used to aid in the process of analyzing and recommending workplace adjustments and modifications is SWAD (sitting workplace analysis and design) (Abdel-Moty and Khalil, 1991). SWAD is a computer program resembling artificial intelligence. The inputs to the program are workplace users' demographics, 16 anthropometric dimensions, workplace dimensions, work tools, and the priority and frequency of use of each work element. SWAD then combines this information with a knowledge base of ergonomic principles and guidelines and a set of inference procedures. The output of SWAD is the recommended workplace dimensions, heights, reaches, foot rest, chair parameters, VDT parameters, and optimal placement of all work tools. SWAD packages anthropometric data and ergonomic principles into a software available to the medical community and to workplace users. It enables customization of workstation adjustment without trial and error.

Also, in this intervention phase, patients' posture and body mechanics are corrected and the effect on reduced muscle activity is demonstrated through the use of EMG. Patients are shown how to correct movements, minimize awkward postures, and refine their body mechanics. In most instances, patients can now perform the same job tasks with minimum discomfort.

4. Follow-up evaluation. In order to document the effectiveness of the interventions, a follow-up evaluation is performed which consists of the patient performing the same simulated activities performed during the initial evaluation. Also, during this session, no feedback is given. Data are then analyzed and results are compared to baseline information. In all cases, significant resolution in the risk factors are observed, muscle activity is decreased reflecting improved economy of motions and less stress, and patients are able to perform essential work functions with no increase in pain or discomfort.

Job Site Analysis

The analysis of worksites can be extremely helpful and should follow a systematic method. The process starts with a data collection phase during which health records are examined in order to identify priority problem areas/tasks. Data collection also includes photographing work areas, recording postures and motions, measuring forces and vibration data, and documenting repetitiveness and frequencies of job cycles.

The data collected are then analyzed and examined to determine the risk factors specific to motion economy, postures, and body mechanics. Quantitative data are also analyzed to determine if, for example, forces and vibration data are within exposure ranges.

Recommendations can then be offered to improve working conditions and/or reduce repetitiveness, vibration, forces, etc. For example, in order to reduce vibration the ergonomists may recommend the use of dampen technology, reduce contact surface area, reduce exposure to driving force, isolate vibration or use material that absorbs vibration. Another common risk factor that usually requires ergonomic intervention is task repetitiveness. Common recommendations to reduce the exposure to high-frequency activities can be: reducing the number of cycles, augmenting human activities with machines through automation and mechanization (e.g., use of scanners), enlarging jobs (e.g., providing help to workers), and when possible alternating use of extremities and job rotation.

Recommendations often include workers' training in proper posture and body mechanics for performing job tasks. Workers are either taught to maintain the neutral postures of their body joints or are provided with technological aides to minimize awkward deviations of joints.

In order to reduce muscle force, the ergonomists may recommend using machines to replace human effort, improving quality of tools to reduce force, changing tool design to allow use of stronger muscles (e.g., the thumb is stronger than any single finger), modifying design to use multiple muscle groups (e.g., group fingers are stronger than the thumb), distributing force over a larger area (e.g., using trigger strips), reducing force by having a slip prevention design in handheld tools, improving tool balance, use of proper gloves (e.g., not too tight at the wrist), and utilizing spring action in tools such as scissors and clippers.

A useful tool in data collection is the use of ergonomic checklists. These are specially designed forms that facilitate problem identification and documentation. While it may be possible to prepare a generic checklist, job environments are not alike, thus necessitating customized checklists. An ergonomic checklist should be designed to focus on the human component of the workplace rather than the workplace itself. For example, it should be asked "Does work surface height permit satisfactory postures of the arms?" rather than "What is work surface height?"

Biomechanical Analysis, Motion Analysis

Biomechanics is the discipline dedicated to the study of the living body as a structure which can function properly only within the confines of both the laws of Newtonian mechanics and the biological laws of life. Occupational biomechanics deals with understanding the complex mechanisms of human interaction with the industrial environment. A number of biomechanical models have been developed which allow the prediction of tissue load indices. Some of these mathematical models deal with evaluating the muscle force and joint reactions for different static postures or during motion. Other models of the lumbosacral region are used to study the combined stresses and strains on the local ligaments, muscles, and disc tissue (Khalil and Ramadan, 1987; Andersson and Chaffin, 1986; Adams et al., 1994; Hindle et al., 1990; McGill and Norman, 1987).

Biomechanical models use different criteria for the evaluation of the muscle and joint forces necessary for maintaining the human body's equilibrium during various activities. Some researchers use such models for the design of manual handling tasks and for the development of a physical job evaluation methodology. Biomechanical models are usually validated mathematically, experimentally, or through the use of EMG. The use of biomechanical models in post-injury management is helpful for patient education and for worksite analysis.

Patient Education

Another task of the clinical ergonomist is to develop patient education programs and materials based on ergonomics concepts. Patient education is an essential component in the comprehensive rehabilitation process. This is especially important with chronic pain patients who have been treated and evaluated in many facilities and have failed various rehabilitation efforts. In addition to the education on pain, myofascial syndrome, health, diet, relaxation, dealing with flare-ups, activities of daily living, and stress management issues, patients are educated in workplace/home safety. Patient education takes many forms: lecture type, group discussions, and hands-on training.

Applied Clinical Research: Innovations in Problem Solving

In the area of applied clinical research, ergonomists evaluate the effectiveness of treatment regimens upon the restoration of functional abilities and upon the reduction of pain, thus providing objective rationalization and justification of treatment. Research is also performed to develop and evaluate devices useful in diagnosis and treatment. Quantitative methods based on recognized approaches are used to assist medical professionals in identifying the usefulness of tools and equipment often prescribed for use in pain management (Zaki et al., 1990).

Another type of research has been necessitated by today's health care environment. This is clinical outcome research. This involves the development of a computerized program evaluation system to

manage and analyze data regarding patient treatment, satisfaction, and the effectiveness of rehabilitation (Zaki et al., 1992). Data collection instruments and questionnaires are developed to assess patients' self-report of pain and functional status upon admission to the program, at the conclusion of treatment, and at regular intervals following discharge (3 months, 6 months, and 1 year). Data are used, not only to evaluate program effectiveness, but also to measure and improve the quality of care. Ergonomists provide the research methods, computer expertise, and statistical approach to manage this type of information.

69.12 Ergonomics and the ADA

The Americans with Disabilities Act (ADA, 1992) is a civil rights protection act that provides comprehensive protection to individuals with disabilities. The ADA addresses disability issues in relation to employment, public accommodation, transportation, and telecommunication. The application of ergonomics methods to individuals with disabilities necessitates complete knowledge of the ADA and its provisions. A review of the ADA is beyond the scope of this chapter; however, a selected number of relevant issues are discussed here.

According to the ADA, an individual with a disability is (ADA, 1992):

1. A person who has a physical or mental impairment that substantially limits one or more major life activities
2. Person who has a record of physical or mental impairment that substantially limits one or more major life activities
3. A person who is regarded as having such an impairment

The ADA prohibits discrimination in all employment practices, including job application procedures, hiring, firing, advancement, compensation, and training. It applies to recruitment, advertising, tenure, layoff, leave, fringe benefits, and all other employment-related activities. An employer may not make a pre-employment inquiry on an application form or in an interview as to whether, or to what extent, an individual is disabled. The employer may ask a job applicant whether he or she can perform particular job functions. If the applicant has a disability known to the employer, the employer may ask how he or she can perform job functions that the employer considers difficult or impossible to perform because of the disability, and whether an accommodation would be needed. A job offer may be conditioned on the results of a medical examination, provided that the examination is required for all entering employees in the same job category regardless of disability, and that information obtained is handled according to confidentiality requirements specified in the ADA. After an employee enters duty, all medical examinations and inquiries must be *job related* and necessary for the conduct of the employer's business. The employment provision of the ADA will, therefore, govern the ergonomist's approach to assisting an employer in designing job applications, job descriptions, and employment offers. It also affects the approach to functional capacity assessment since evaluations have to be job specific. This, in turn, means that the use of computerized testing equipment for the evaluation of "generic" functional abilities may no longer be valid. Evaluation setting should simulate the work situation. Testing protocols will need to be changed and modified to meet compliance. The ergonomist can also assist the employer in developing practical "post-offer" screening tools in order to test the applicant's ability to perform essential job functions (with or without reasonable accommodations).

The ADA expressly permits employers to establish qualification standards that will exclude individuals who pose a direct threat — i.e., a significant risk — to the health and safety of others, if that risk cannot be lowered to an acceptable level by reasonable accommodation. However, an employer may not simply assume that a threat exists; the employer must establish through objective, medically supportable methods that there is genuine risk that substantial harm could occur in the workplace. By requiring employers to make individualized judgments based on reliable medical evidence rather than on generalizations, ignorance, fear, patronizing attitudes, or stereotypes, the ADA recognizes the need to balance the interests of people with disabilities against the legitimate interests of employers in maintaining a safe workplace. The

ergonomist can assist in this area by establishing workplace safety criteria in relation to human factors while taking medical information into account.

Another core issue in the ADA is reasonable accommodation. Reasonable accommodation is any modification or adjustment to a job or the work environment which should be made to enable a qualified applicant or employee with a disability to perform essential job functions. This is another area of primary relevance to ergonomics. Examples of reasonable accommodation include making existing facilities used by employees readily accessible to and usable by an individual with a disability; restructuring a job; modifying work schedules; acquiring or modifying equipment; providing qualified readers or interpreters; or appropriately modifying examinations, training, or other programs. Reasonable accommodation also may include reassigning a current employee to a vacant position for which the individual is qualified, if the person becomes disabled and is unable to do the original job. An employer, however, is not required to make an accommodation if it would impose an "undue hardship" on the operation of the employer's business.

69.13 Post Injury Management of the Aging Population

Another population which requires special ergonomic attention is the injured elderly. One of the most pressing needs in the 1990s is to develop effective ways to serve an increasingly aging population.

In the coming decades the greatest growth is projected for the population aged 55 and over. In fact between the years 1990 and 2000 this group will increase by 11.5%, a gain of over 6 million persons. Within this group the greatest growth will be among persons 75 and over, an increase of 26.2% (U.S. Bureau of the Census, 1988). A question of major policy and fiscal importance is how well will the large population of elderly be able to live and function independently? It is well established that age-related changes in functional capacities affect the ability of older individuals to successfully complete daily living activities.

Many elderly persons reside in nursing homes and need some type of assistance to perform basic activities of daily living. Many others have difficulty functioning well at home and experience a high rate of accidents. The most frequent types of accidents are falls associated with stairs and steps, floors and floor coverings, bathtubs and showers, and ladders (Consumer Product Safety Commission, 1985). Many of the accidents could be prevented with the implementation of appropriate ergonomic intervention strategies. A wide variety of remedial techniques is already available for hazard reduction. The role of the ergonomist is to provide the information and products regarding older individuals to caregivers and medical professionals. Currently there are more than 15,000 assistive products on the market and a number of home safety catalogs available; however, most elderly home care providers and health care specialists are unaware of their existence. The ergonomist assists with the training and education. Not only must older people be aware of assistive tools and safety strategies, they must also understand how to use them. Elderly people must also be aware of safety issues in the living environment and must learn how to deal with them. Certainly, technological aids are not effective without appropriate training. Furthermore, this training needs to be directed at the health care specialist, designers of homes for the elderly, as well as potential users.

Understanding the human factors-related changes in performing daily activities for the aged population is one of the major aims of the ergonomist (Khalil et al., 1992b). Application of this knowledge to the design of tasks, equipment, and the home environment in which the elderly live contributes to the enhancement of their quality of life.

The ergonomic approach in this case will consist of:

1. Studying and understanding accident and hazardous environments of the elderly population (e.g., stairs and steps, bathtubs and showers, use of tools and appliances)
2. Identifying appropriate hazard intervention strategies
3. Developing and evaluating education and safety training material to ensure successful implementation of the hazard interventions
4. Developing techniques to educate and train relevant individuals caring for the elderly

69.14 Ergonomic Interventions Post-Rehabilitation (Prevention of Reinjury)

The continuous spectrum of including ergonomics in injury prevention goes beyond the clinical setting into post-rehabilitation. The thrust of the ergonomic approach continues to follow up on the worker once on the job or at home. Strategies used in the post-rehabilitation stage may include:

1. Immediate matching of the worker to the work.
2. Job modification based on the recommendations of the clinical ergonomist.
3. Ongoing screening of worker's capabilities.
4. Development of effective health maintenance programs including onsite stretching and home exercise programs.
5. Continuity of care through follow-up questionnaires at predetermined intervals. (This is a requirement of the Commission on the Accreditation of Rehabilitation Facilities, CARF.)
6. Potential involvement with the employer to extend the ergonomic safety interventions to other workers.

69.15 Who Should Do Ergonomics in Post-Injury Management Programs?

This is by no means a rhetoric question. It represents the reality of practicing ergonomics in clinical settings. The dilemma facing the ergonomics professions recently has been that anyone can say he or she is an ergonomist after attending brief training seminars. The field has become attractive to unqualified individuals who claim association to the profession regardless of education or experience. Recently, efforts have been made to address this issue. However, more needs to be done to combat the problem of unqualified individuals presenting themselves as ergonomists.

69.16 Who Pays for Ergonomics Service in a Clinical Setting?

Rehabilitation settings and hospital facilities follow a stringent system of reimbursement through different types of insurance (e.g., self-insurance, third-party insurance). In most states, workers' compensation provisions allow for reimbursement of necessary and justifiable ergonomic services. There must be a physician referral for the service, and authorization must be obtained in advance. Charges are then posted according to specific codes describing the procedure(s) performed, the length of the session, work performed, patients' interaction, and outcome data. In doing so, the ergonomist follows the hospital/rehabilitation facility's policies and procedures regarding documentation, reporting, and billing systems.

69.17 Summary

Ergonomics is the missing piece of the puzzle in post-injury management programs. Since 1982, the University of Miami Comprehensive Pain and Rehabilitation Center realized the importance of including ergonomics rationale, expertise, personnel, and resources in the day-to-day struggle to return injured individuals to a productive life style.

The clinical ergonomists have demonstrated that it is possible to return the disabled individual to full function with a properly designed multidisciplinary therapeutic rehabilitation program. The contribution of ergonomics to post-injury management has proved to be quite valuable. At the CPRC, ergonomists interact with patients as well as other members of the health care team to restore function to the patient. The involvement of ergonomists in activities of patient evaluation, job simulations, body mechanics teachings, patient education, and rehabilitation research should be an integral part of pain management settings.

The capacity to utilize ergonomics in a clinical setting extends beyond treatment to optimization of the treatment and prognosis of treatment outcome. It also extends to the post-rehabilitation stage where re-injury can be avoided without the need for expensive technology to solve workplace problems.

It remains the responsibility of the ergonomics profession, its organizations, and governing bodies to protect the field from less qualified individuals who encroach on the practice of ergonomics in general, and in clinical settings in particular.

In order to be involved in post-injury management, ergonomists will need to become familiar with health care trends, reforms, and language. They will require training in and exposure to a host of rehabilitation and medical issues ranging from hospital policies to documentation and health insurance. Above all, they will need to adapt to the clinical setting and develop interactive ties with other health care providers. The reward of helping patients overcome their disability and restore their identity not only justifies but compels the inclusion of the science and practice of ergonomics in rehabilitation and all phases of post-injury management.

For Further Information

The book *Ergonomics in Back Pain: A Guide to Prevention and Rehabilitation* (Van Nostrand Reinhold, NY, 1993) provides detailed descriptions of the methods and approaches presented in this chapter.

Questions

For specific questions, call Elsayed Abdel-Moty, Ph.D., at (305) 532-7246, or write to the Ergonomics Division, University of Miami Comprehensive Pain and Rehabilitation Center, 600 Alton Road, Miami Beach, Florida 33139 USA.

References

Abdel-Moty, E. and Khalil, T.M., 1991. Computer-aided design and analysis of the sitting workplace for the disabled. *International Disability Studies.* 13:121-124.

Abdel-Moty, E., Khalil, T., Goldberg, M., Rosomoff, R. and Rosomoff, H., 1990a. Posture and pain: health effects and ergonomics interventions, in *Advances in Industrial Ergonomics and Safety II*, Ed. B. Das, p. 117-124, Taylor & Francis, London.

Abdel-Moty, E., Khalil, T.M., Rosomoff, R.S. and Rosomoff, H.L., 1990b. Ergonomics considerations and interventions, in *Painful Cervical Trauma: Diagnosis and Rehabilitative Treatment of Neuromusculoskeletal Injuries*, Eds. C. D. Tollison and J.R. Satterthwaite, p. 214-229, Williams & Wilkins, Baltimore.

Abdel-Moty, E., Khalil, T.M., Asfour, S.S., Sadek, S., Rosomoff, R.S. and Rosomoff, H.L., 1991a. Workers compensation and non-worker's compensation chronic pain patients responsiveness to rehabilitation, in *Advances in Industrial Ergonomics & Safety III*, Eds. W. Karwowski and J.W. Yates, p. 467-474, Taylor & Francis, London.

Abdel-Moty, E., Khalil, T., Diaz, E., Sadek, S., Rosomoff, R. and Rosomoff, H., 1991b. Ergonomic job analysis for patients with chronic low back pain during rehabilitation, in *Designing for Everyone*, Eds. Y. Queinnec and F. Daniellou, p. 1638-1640, Taylor & Francis, London.

Abdel-Moty, E., Khalil, T.M., Fishbain, D., Rosomoff, R.S. and Rosomoff, H.L., 1991c. Functional capacity assessment of low back pain patients, in *Advances in Industrial Ergonomics & Safety III*, Eds. W. Karwowski and J.W. Yates, p. 475-482, Taylor & Francis, London.

Abdel-Moty, E., Diaz, E., Khalil, T.M., Abou Elseoud, M., Steele-Rosomoff, R. and Rosomoff, H.L., 1992a. Ergonomic job analysis for patients with cervical trauma during rehabilitation, in *Advances in Industrial Ergonomics and Safety IV*, Ed. S. Kumar, p. 1195-1200, Taylor & Francis, London.

Abdel-Moty, E., Khalil, T.M., Sadek, S., Dilsen, E.K., Fishbain, D., Steele-Rosomoff, R. and Rosomoff, H.L., 1992b. Functional capacity assessment: a test battery and its use in rehabilitation, in *Advances in Industrial Ergonomics and Safety IV*, Ed. S. Kumar, p. 1171-1178, Taylor & Francis, London.

Abdel-Moty, E., Fishbain, D., Khalil, T., Sadek, S., Cutler, R., Steele Rosomoff, R. and Rosomoff, H., 1993a. Functional capacity and residual functional capacity and their utility in measuring work capacity. *The Clinical Journal of Pain.* 9:168-173.

Abdel-Moty, E., Khalil, T., Steele-Rosomoff, R. and Rosomoff, H., 1993b. Maximizing progress during low back pain rehabilitation, in *Advances in Industrial Ergonomics and Safety V,* Eds. R. Nielsen and K. Jorgensen, p. 331-335, Taylor & Francis, London.

Abdel-Moty, E., Fishbain, D., Goldberg, M., Cutler, B., Zaki, A., Khalil, T., Peppard, T., Steele Rosomoff, R. and Rosomoff, H., 1994. Functional electrical stimulation treatment of postradiculopathy associated muscle weakness. *Arch Phys Med Rehabil.* 75: 680-686.

Abdel-Moty, E., Compton, R., Steele-Rosomoff, R., Rosomoff, H.L. and Khalil, T.M., 1996. Process analysis of functional capacity assessment. *J Back and Musculoskeletal Rehabil,* 9:223-236.

Adams, M.A., McNally, D.S. and Dolan, P., 1994. Posture and the compressive strength of the lumbar spine. *Clin Biomech,* 9:5-14.

Americans with Disabilities Act, 1992. *Technical Assistance Manual on the Employment Provision (Title) of the ADA.* U.S. Equal Employment Opportunities Commission.

Andersson, C.K. and Chaffin, D.B., 1986. A biomechanical evaluation of five lifting techniques. *Appl Ergonom,* 17:2-8.

Asfour, S.S., Khalil, T.M., Waly, Goldberg, M.L., Rosomoff, R.S. and Rosomoff, H.L., 1990. Biofeedback in back muscle strengthening. *Spine,* 15(6):510-513.

ANSI, 1988. *American National Standard for Human Factors Engineering of VDT Workstations.* Human Factors Society, Inc., Santa Monica.

Consumer Product Safety Commission January 1985. *Safety for Older Consumers: Home Safety Checklist.* Washington, D.C.

Department of Commerce, Bureau of the Census (1988). *Statistical Abstracts of the United States.* Washington, D.C.: U.S. Government Printing Office.

Department of Labor, Bureau of Labor Statistics, 1982. *Back Injuries Associated with Lifting.* Bulletin No. 2144, Washington, D.C.

Federal Register, 1992. *Ergonomics, Safety and Health Management. Advanced Notice of Proposed Rulemaking.* Vol 57, No. 149, August 3.

Fishbain, D.A., Abdel-Moty, E., Cutler, R., Khalil, T., Sadek, S., Steele Rosomoff, R. and Rosomoff, H., 1994. Measuring residual functional capacity in chronic low back pain patients based on the Dictionary of Occupational Titles. *Spine,* 19(8):872-880.

Fishbain, D.A., Khalil, T.M., Abdel-Moty, E., Cutler, R., Sadek, S., Steele Rosomoff, R. and Rosomoff, H.L., 1995. Physician limitations when assessing work capacity: A review. *J Back and Musculoskeletal Rehabil,* 5:107-113.

Hindle, R.J., Pearcy, M.J., Cross, A.T. and Miller, D.H.T., 1990. Three-dimensional kinematics of the human back. *Clin Biomech,* 5:218-228.

Khalil T.M. and Ramadan M.Z., 1987. Biomechanical evaluation of lifting tasks: A microcomputer-based model. *Comput Indust Eng,* 14:1.

Khalil, T.M., Goldberg, M.L., Asfour, S.S., Abdel-Moty, E., Steele, R. and Rosomoff, H.L., 1987. Acceptable maximum effort (AME): A psychophysical measure of strength in back pain patients. *Spine,* 12(4):372-376.

Khalil, T.M., Abdel-Moty, E. and Asfour, T.M., 1990. Ergonomics in the management of occupational injuries, in *Industrial Ergonomics: Case Studies.* Eds. B.M. Pulat and D.C. Alexander, p. 41-53, Industrial Engineering and Management Press, Norcross, GA.

Khalil, T.M., Abdel-Moty, E., Diaz, E., Rosomoff, R.S. and Rosomoff, H.L., 1991a. Electromyographic symmetry pattern in patients with chronic low back pain and comparison to controls, in *Advances in Industrial Ergonomics & Safety III,* Eds. W. Karwowski and J.W. Yates, p. 483-490, Taylor and Francis, London.

Khalil, T.M., Asfour, S.S. and Moty, E.A., 1991b. Clinical ergonomics: ergonomic practice in health care setting, in *Designing for Everyone*, Eds. Y. Queinneç and F. Daniellou, p. 314-316, Taylor & Francis, London.

Khalil, T.M., Abdel-Moty, E., Sadek, S., Dilsen, E.K., Steele-Rosomoff, R. and Rosomoff, H.L., 1992a. Postural sway and balance in healthy subjects and in patients with chronic pain, in *Advances in Industrial Ergonomics and Safety*, Ed. S. Kumar, p. 925-932, Taylor & Francis, London.

Khalil, T.M., Abdel-Moty, E., Zaki, A. M., Dilsen, E.K., DeVito, C., Steele-Rosomoff, R. and Rosomoff, H.L., 1992b. Reducing the potential for fall accidents among the elderly through physical restoration, in *Advances in Industrial Ergonomics and Safety IV*, Ed. S. Kumar, p. 1127-1134, Taylor & Francis, London.

Khalil, T.M., Abdel-Moty, E., Zaki, A. M., Velez, B., Dilsen, E.K., Diaz, E., Steele-Rosomoff, R. and Rosomoff, H.L., 1992c. Effect of secondary gain issues on performance and response to rehabilitation of workers compensation chronic low back pain patients, in *Advances in Industrial Ergonomics and Safety IV*, Ed. S. Kumar, p. 1187-1194, Taylor & Francis, London.

Khalil, T.M., Abdel-Moty, E., Rosomoff, R.S. and Rosomoff, H.L., 1993a. *Ergonomics in Back Pain: A Guide to Prevention and Rehabilitation.* Van Nostrand Reinhold, New York.

Khalil, T.M., Abdel-Moty, E., Steele-Rosomoff, R. and Rosomoff, H., 1993b. The role of ergonomics in the prevention and treatment of myofascial pain, in *Diagnosis and Comprehensive Treatment of Myofascial Pain: Handbook of Trigger Point Management.* Ed. E.S. Rachlin, p. 487-523, Mosby Year Book, St. Louis.

Khalil, T.M., Abdel-Moty, E., Diaz, E., Steele-Rosomoff, R. and Rosomoff, H., 1994. Efficacy of physical restoration in the elderly. *Experimental Aging Research*, 20(3):189-199.

Kroemer, K., Kroemer, H. and Kroemer-Elbert, K., 1994. *Ergonomics: How to Design for Ease and Efficiency.* Prentice Hall, Inc., Englewood Cliffs, N.J.

McGill, S.M. and Norman, R.W., 1987. Effects of anatomically detailed erector spinae model on L4/L5 disc compression and shear. *J Biomech*, 20(6):591-600.

NIOSH, 1981. *A Work Practice Guide for Manual Lifting, Technical Report #81-122.* National Institute for Occupational Safety and Health, U.S. Department of Health and Human Services, Cincinnati, Ohio.

NIOSH, 1991. *Work Practices Guide for Manual Lifting. NIOSH Technical Report Draft*, National Institute for Occupational Safety and Health, U.S. Department of Health and Human Services, Cincinnati, Ohio.

OSHA, 1990. *Ergonomics Program Management Guidelines.* U.S. Department of Labor, Occupational Safety and Health Administration.

Pain Commission, 1987. *Pain and Disability: Clinical, Behavioral, and Public Policy Perspectives.* Institute of Medicine, Committee on Pain Disability and Chronic Illness Behavior. ed. M. Osterweis, A. Kleinman and D. Mechan. National Academy Press, Washington, D.C.

Quebec Study, 1987. Scientific approach to the assessment and management of activity-related spinal disorders. A monograph for clinicians. Report of Quebec Task Force on Spinal Disorders. *Spine*, 12-7S.

Rosomoff H.L. and Rosomoff R.S., 1991, Comprehensive multidisciplinary pain center approach to the treatment of low back pain. *Neurosurgery Clinics of North America*, 2(4):877-890.

Safety Management, 1996. Bureau of Business Practice, Waterford, CT, pp. 8.

SSA, 1986. *Social Security Administration: Report of the Commission on Evaluation of Pain.* Department of Health and Human Services, Washington, D.C.

Zaki A.M., Goldberg M.L., Khalil, T.M. et al., 1990, Comparison between the one and two inclinometer techniques for measuring range of motion, in *Advances in Industrial Ergonomics and Safety II*, Ed. B. Das, p. 135-142, Taylor & Francis, New York.

Zaki, A.M., Khalil, T.M., Abdel-Moty, E., Steele-Rosomoff, R. and Rosomoff, H.L., 1992. Profile of chronic pain patients and their rehabilitation outcome, in *Advances in Industrial Ergonomics and Safety IV*, Ed. S. Kumar, p. 1179-1186, Taylor & Francis, New York.

70

Physical Ability Testing for Employment Decision Purposes

Charles K. Anderson
Advanced Ergonomics, Inc.

70.1 Introduction

Much of ergonomic activity is focused on designing or altering the demands of the job so that there is better match with the capabilities of the workforce. Sometimes, this approach reaches a point where further change in the job is either cost-prohibitive or technically infeasible at the moment. An alternative approach is to consider matching workers to the job demands on the basis of their physical abilities. For instance, if the job requires individuals lift cases weighing 60 pounds and there is no way to reduce the case weight, one approach would be to assess job candidates' ability to lift 60 pounds as part of the process of determining whether the individual is able to perform the job.

A number of studies have documented the effectiveness of physical ability testing as a means of identifying individuals who will be able to safely perform a given job (Anderson et al., 1980; Anderson et al., 1987; Anderson, 1989a,b; Anderson, 1992a,b; Arnold et al., 1982; Ayoub 1983; Cady et al., 1979; Chaffin et al., 1977; Herrin et al., 1982; Jackson et al., 1984; Jackson et al., 1991a,b; Jackson et al., 1992; Keyserling 1979; Keyserling et al., 1980a,b; Laughery et al., 1984; Laughery et al., 1988; Liles et al., 1984; Reilly et al., 1979). The experience of the author is that injury rates for new hires typically fall 20 to 40% with the implementation of testing (Anderson, 1996).

There are numerous issues that need to be considered when implementing a physical ability test battery to assure that it will be effective. One of the first concerns is to assure that the test battery will truly assess what is intended to be assessed — the individual's ability to perform the job. This means that a thorough job analysis must be performed, tests carefully chosen, and efforts taken to validate that the battery truly predicts job performance. Without this foundation, the employer may find that the battery provides no useful information while being a source of additional hiring cost and time delay.

In many countries, including the United States, there is a legal mandate that test batteries be valid predictors of job performance, particularly if it can be anticipated that protected groups, such as females and older individuals, will be less likely to pass the battery. In the United States, lack of compliance with the various pieces of legislation addressing employment testing can result in a company having to pay back wages to all individuals denied employment on the basis of the test, being required to offer those

0-8493-2641-9/99/$0.00+$.50

individuals employment, and perhaps in being liable for punitive damages of up to $300,000. Hence, there are moral, financial, and legal considerations when implementing a physical ability test battery.

The courts prefer that the emphasis be on the ability to perform the job rather than the risk of an injury or illness that may occur sometime in the future. The basis for this preference is the observation that employers have sometimes decided not to hire individuals with particular health conditions, such as a "bad back," because of the employer's perception that the individual is at increased risk of injury. If the basis for establishing the battery is risk of injury, the technical assistance manual for Title I of the Americans with Disabilities Act, which deals with employment discrimination, emphasizes that injury must be expected to occur in the near future, be severe, and have a significantly higher likelihood in the population to be denied employment than the risk in the general population (EEOC, 1992).

All of these considerations can be met with a carefully constructed implementation and ongoing management practice. The implementation consists of the job analysis, test battery design, and validation. Ongoing practices refer to the steps taken to assure that all individuals are given equal consideration and the manner in which other information, such as prior experience on a similar job, is integrated in the decision-making process.

70.2 Implementation

Job Analysis

One of the most critical aspects of the job analysis is to identify the essential functions of the job. These are the functions that define the purpose of the job and are typically the elements that have to be performed in order for an employee to be considered satisfactory. In most cases, the essential functions are the ones that are performed frequently, though there may be situations where a function that is performed infrequently is still considered essential. For example, the essential function of a warehouse selector is to lift cases from storage racks to a pallet that will be shipped to a client. A second essential function is to drive a motorized vehicle that transports the pallet throughout the warehouse. These tasks comprise the bulk of the selector's activity over the course of the shift. As a counter-example, one of the essential functions of a firefighter is to respond to emergencies. The amount of time actually spent in emergency response may be small, but it is essential that the firefighter be able to perform these duties. An example of a nonessential function for the warehouse selector would be the task of cleaning up broken cases since selectors rarely need to perform that task and it is an essential function of the janitor.

When developing a physical ability test battery, the degree to which it can be anticipated that there will be a substantial portion of individuals who will be unable to perform the task is also an important consideration. If the task can be performed by virtually everyone, then there is little economic justification for testing applicants since virtually everyone would pass the test.

The cost/benefit analysis can be quantified by calculating the cost of not being able to perform the task and the probability of an individual not being able to perform the task, and then balancing these against the cost of testing all applicants for the ability to perform the task. For instance, consider a situation where 100 employees are hired per year. If there is a 1% probability that an individual could not perform the task, and the inability to perform the task results in a $2,000 loss associated with hiring and training the individual, then there is an expected cost of $2,000 per year for not using the test (100 new-hires × 1% chance of not being able to perform the job × $2,000 cost). If the cost of the test is $30 per applicant and 101 applicants must be tested to find 100 who pass (99% pass rate), the expected cost of testing would be $3,030 per year. Hence, in this example, it is less costly to not test and accept the risk that 1% of all the workers will not be able to perform the job (expected cost of $2,000 per year) rather than test all candidates ($3,030 per year). In contrast, if it can be anticipated that only 80% of new-hires truly have the ability to perform the job, then the cost of not testing rises to $40,000 per year (100 new-hires × 20% chance of not being able to perform × $2,000 cost). The cost of testing would now be $3,750 ($30 per test × 125 applicants tested to find 100 who pass). Under this scenario there is better than a 10-to-1 benefit/cost ratio for implementation of testing ($40,000 without testing vs. $3,750 with testing).

A third consideration during the job analysis is whether the essential functions that are potentially difficult for a substantial portion of the population can be modified through some form of ergonomic intervention (i.e., reasonable accommodation). For instance, wholesale grocery delivery drivers have to lift cases weighing up to 60 pounds on a routine basis. In addition, to access the cases, they must raise the rear door on their trailer. When brand-new trailers arrive from the manufacturer, virtually no force is required to open the door. However, the door track can become bent from forklifts running into the track while loading pallets onto the trailer. This can lead to a situation where a force of up to 200 pounds has to be exerted to open the door. Obviously, an essential function of the driver's job is to open the trailer door, but is it reasonable to design a test that assesses the driver's ability to exert 200 pounds to open a jammed door? An alternative is to require preventative maintenance rather than require the driver to exert 200 pounds. This would reduce the demand for this essential function back into the region where it can be anticipated that virtually all candidates will be able to perform the task. Hence, in this example, the critical task for the purposes of test battery design for these drivers would be the ability to lift 60-pound cases.

Test Battery Design

It has already been mentioned that tests should be included only for those essential functions that will be difficult for a substantial portion (5 to 10% or more) of the applicant population. There are three additional considerations in designing the test battery.

Job-Relatedness

First, it is critical that the tests bear a high degree of job-relatedness to the essential functions of the job that are used as the basis for the test battery design. For instance, when testing the ability of an individual to lift a 60-pound case, the most job-related test would be to have the individual literally demonstrate the ability to lift a 60-pound case by having him or her move the case through the region (for example, floor to table level) that is found on the job. In contrast, having the applicant perform a series of exercises such as push-ups, chin-ups, and sit-ups would not be a job-related test of the ability to lift a 60-pound case.

Basis for Establishing the Cutoff Score

The second consideration in test selection is the ease with which the cutoff score can be determined. The cutoff score is the level of test performance required in order to be considered as having demonstrated the ability to perform the job. Ideally, there is a clear relationship between the cutoff score and the performance of the job. For instance, if the test for the ability to lift a 60-pound case consists of lifting a case from floor level to table level, the most job-related cutoff for weight lifted would be 60 pounds.

In contrast, an alternative method is to use normative data. For example, some companies strap the applicant into a fixture that isolates his or her movement to flexion and extension about the low back. The test consists of measuring the moment, or torque, that the applicant can then generate with these muscles. It is difficult to directly measure the moment about the low back required for a given individual to lift a 60-pound case, since the moment will be a function of the posture used, how rapidly the individual lifts the case, the anthropometry of the individual, and a host of other factors. The only alternative is to measure the moment created by a number of incumbents and develop a distribution of their scores. Conceivably, one could then choose the lowest score of the incumbents and require that the applicant at least demonstrate the ability to demonstrate that level of strength. A problem with this approach is that there is no guarantee that the scores demonstrated by the incumbents have a bearing on what the job requires. For instance, if all of the incumbents are regular participants in heavy weight training, their strength may be a reflection of the weight training rather than the job requirement.

A second example illustrating the problem of using cutoff scores based on normative data comes from test batteries used for evaluating police officer candidates. Some batteries for police officers require that applicants perform a run over a short distance to simulate the essential function of chasing suspects fleeing on foot. The score for the test is the time taken to cover the distance. The question becomes "How fast must an applicant complete the distance in order to be able to perform the job?" One approach that

has been used is to determine whether the applicant's time is above the median for individuals of their sex and age. The justification has been that one would want their police officers to be at least as fit as an average person of their sex and age. Unfortunately, this does not speak to whether the officer can perform the essential function of the job.

Most jobs lend themselves to determining the cutoff scores directly from the job requirements, as described in the first example given in this section — handling 60-pound cases. Likewise, there typically are testing formats available that will allow the test administrator to assess the ability at issue in a manner that can be directly compared to the job requirement. Hence, the problems that arise with utilizing the "normative data" approach can generally be avoided once the issue is recognized.

Accuracy, Reliability, and Safety

The third consideration is that the tests have a high degree of accuracy, reliability and safety. An accurate test is one that precisely measures the attribute it purports to measure. A reliable test is one that yields the same results when it is repeated over time and by different test administrators. Reliability is enhanced when objective test measures are used rather than subjective assessments or opinions of the test administrator.

Safety is an important consideration, particularly if maximal strength tests are being used. Typically, steps can be taken to reduce the risk of injury during the test. For instance, an alternative to having the participant demonstrate the maximum amount of weight they can lift is to ask them to only demonstrate the ability to lift the weight required by the job. A second safety modification would be to have the participant first demonstrate the lift with an empty case, and then add weight to the box in fixed increments so that there is the chance to detect inability to lift the required weight without actually having the individual attempt the lift with the full weight required on their first effort.

Validation

Statistical Validation Studies

The strongest form of validation is a prospective statistical analysis. The typical form of this type of analysis involves administering the test to a group of individuals who are about to be hired into the position in question, although the test scores are not used as part of the hiring decision. The performance of these individuals is then tracked over the course of their employment. Measures of performance might be productivity, retention, injury rate, or supervisor evaluation. Care must be taken when selecting a measure that it reflect the important aspects of job performance and be reliable. Performance for individuals who pass the test would then be compared to performance for individuals who fail. As an example, Figure 70.1 shows the worker compensation injury rate for new-hires who failed a physical test battery compared to the injury rate for their peers who passed the battery designed for grocery warehouse selectors (Anderson, 1989a). The injury rate for those who failed the test was about double the rate for those who passed.

The effectiveness of the battery can be projected by comparing the performance of the entire group of new-hires, which reflects the performance anticipated without implementation of the battery, to the performance of those individuals who passed the battery. This latter group reflects the performance that would be expected with implementation of the battery. Table 70.1 summarizes the results calculated in this manner for three prospective statistical studies performed by the author (Anderson, 1989b; Anderson, 1992a,b).

A less strong form of statistical validation is to compare performance for new-hires who began employment in time periods prior to and after implementation of the battery. The primary shortfall of this approach is that there is no control for factors that did not stay constant between the two time periods, such as work load, job structure, or management policies. The benefit of this type of analysis is that there is at least some indication of the effectiveness of implementing the battery when it is technically or economically infeasible to perform a prospective validation study. As noted earlier, the author has found that worker compensation incidence rates for new-hires typically fall 20 to 40% with implementation of a well-designed test battery (Anderson, 1996).

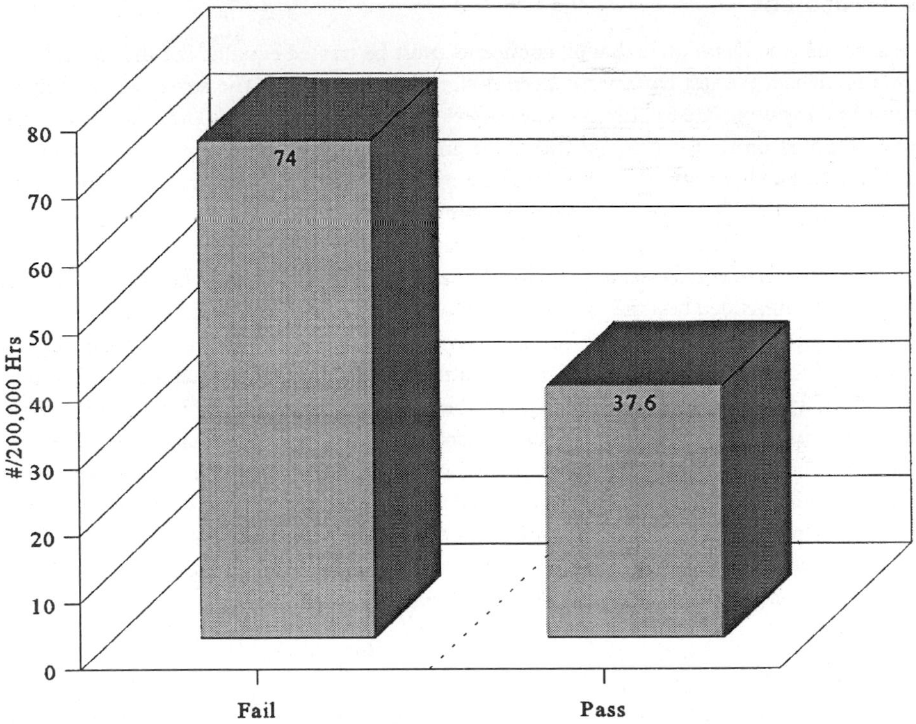

FIGURE 70.1 Worker compensation incidence rate by pass/fail status.

TABLE 70.1 Summary of Expected Impact of Test Battery Implementation

	Grocery Distribution	Soft Drink Distribution	Retail Distribution
Decrease in Incidence Rate	18%	26%	17%
Decrease in Lost Time Due to Injuries	n/a	11%	27%
Decrease in Workers' Compensation Cost	n/a	55%	n/a
Increase in Retention	10%	15%	7%
Increase in Productivity	n/a	n/a	7%

n/a — data not available for all locations in the study

Content Validation Studies

The alternative to a statistical validation is a content validation. A content validation study consists of demonstrating that the content of the job is reflected in the content of the test battery, and that the cutoffs on the tests reflect the job demands. It is valuable to be able to also make reference to any statistical validation studies in similar industries that further support the content validity of the battery at issue.

In the United States, the Equal Employment Opportunity Commission states that a content validation is acceptable when a statistical validation is technically infeasible (EEOC, 1978). For example, this could be the case when there is a small number of new-hires on which to base the study, or the job demands are not homogeneous.

70.3 Ongoing Practices

It is critical to not only carefully construct the test battery, but also the management policies that surround its use. Four aspects are particularly crucial to assure that the manner in which the test battery is used is fair to all applicants.

Equal Treatment

The fundamental consideration is that all applicants must be treated equally. For instance, all applicants for the job for which the test battery has been designed must be given the same opportunity to test. It is not considered appropriate to only give the battery to females, individuals over the age of 40, or those considered disabled under the premise that these are individuals who are most likely to be unable to perform the job. Furthermore, they should all be given the same information prior to the test. This information might include instructions to wear loose-fitting clothes, avoid a heavy meal prior to the test, and have adequate rest the night prior to their appointment.

Another dimension of equal treatment is that the same scoring process must be used for all individuals. It is generally recommended that the same cutoff be used for all individuals without exception. Problems arise when managers attempt to factor in their own judgment by allowing some individuals who fall below the cutoff to be employed, particularly when the decision as to who shall benefit is arbitrary.

Finally, all individuals who fail should be given the same opportunity to be reconsidered. This opportunity could be structured on the basis of their performance, but must be equally applied. For instance, all individuals who score within 95% of the cutoff might be offered the chance to retest immediately. Those who score between 80 and 95% might be allowed to retest after four weeks, whereas those who score less than 80% have to wait at least eight weeks to retest. There should be a valid basis for the time periods, such as the percent of improvement in performance that could be expected at particular points in time with a well-designed physical conditioning program.

Prior Experience

There is only one exception to the policy of rigid interpretation of the cutoff score which occurs when an individual who fails the test has demonstrated the ability to perform the job during prior experience in similar employment. It is important to remember that the best indicator of an individual's ability to perform a job is actual demonstration of that performance. Lacking direct evidence of ability to perform the job, one might consider performance on a test battery designed to provide information about that ability. The test battery is not necessarily a perfect indicator, so it is possible that an individual who can perform the job might fail the battery. Clearly, the battery should be designed so that this possibility is minimized. This is the reason for selecting tests that bear the closest resemblance to the way the job is actually performed, carefully selecting the cutoff score and taking all possible steps to assure that the test scores are valid predictors of job performance.

In the circumstance where there is additional information available about an individual's ability to perform the job, such as prior experience, this information can be used to override a failing test result. It is critical, though, that all individuals in similar circumstances be treated the same way. This can be promoted by establishing clear definitions of what would be considered adequate previous experience that can be consistently applied in each case that arises. For instance, if an individual applying for a warehouse selector position has previous experience as a selector yet fails the battery, the manager might ascertain if the previous experience involved handling cases of similar weight at a similar pace for a similar shift length using similar equipment (e.g., manual lifting vs. picking full pallets with a forklift). If all aspects of the previous job and the job at issue are similar, then the prior experience could be used as a legitimate basis for hiring the individual in spite of the failing test results.

70.4 Summary

Physical ability testing can be an effective way to improve the match between job demands and worker ability, thereby enhancing performance and reducing the risk of injury. Thorough job analysis is required to guarantee that the test battery selected is a valid predictor of the ability to perform the essential functions of the job. Furthermore, it is important to assure that the management policies related to the utilization of the battery guarantee equal application of its use and interpretation. This includes to whom

the battery is administered, the information that applicants are given prior to the test, the way their results are interpreted, and the opportunity for retesting that is extended to those who fail the test. When the test battery is designed with these factors in mind, it can meet the legal requirements and be an effective management tool. Results from a wide range of applications indicate that there typically is a 20 to 40% reduction in worker compensation injuries associated with implementation of a well-designed test battery.

References

Anderson, C.K. Strength and endurance testing for pre-employment placement. K. Kroemer, Ed. *Manual Material Handling: Understanding and Preventing Back Trauma*, American Industrial Hygiene Association, Akron, 1989a.

Anderson, C.K. Impact of physical ability screening for grocery warehousing. Technical Report, Advanced Ergonomics, Inc., 1989b.

Anderson, C.K. Impact of physical ability testing on injuries and retention for the Coca-Cola Bottlers Association Technical Report, Advanced Ergonomics, Inc., 1992a.

Anderson, C.K. Impact of physical ability testing on workers' compensation injury rate and severity: target distribution centers. Technical Report, Advanced Ergonomics, Inc., 1992b.

Anderson, C.K. The advanced ergonomics physical ability testing program: a seven year review. Technical Report, Advanced Ergonomics, Inc., 1996.

Anderson, C.K.; Catterall, M.J. The impact of physical ability testing on incidence rate, severity rate and productivity. S.S. Asfour, Ed. *Trends in Ergonomics/Human Factors IV*, Elsevier Science Publishers, North Holland, 1987.

Anderson, C.K.; Herrin, G.D. Validation study of pre-employment strength testing at Dayton Tire and Rubber Company. Technical Report, University of Michigan Center for Ergonomics, 1980.

Arnold, J.D.; Rauschenberger, J.M.; Soubel, W.G.; Guion, R.M. Validation and utility of a strength test for selecting steel workers. *Journal of Applied Psychology*, 67(5): 588-604, 1982.

Ayoub, M.A. Design of a pre-employment screening program. Kvalseth, T.O., Ed. *Ergonomics of Workstation Design*, Butterworths & Co. London, 1983.

Cady, L.D.; Bischoff, D.P.; O'Connell, E.R.; Thomas, P.C.; Allan, J.H. Strength and fitness and subsequent back injuries in fire fighters. *Journal of Occupational Medicine*, 21(4): 269-272, 1979.

Chaffin, D.B.; Herrin, G.D.; Keyserling, W.M.; Foulke, J.A. *Pre-Employment Strength Testing in Selecting Workers for Materials Handling Jobs.* Cincinnati, OH, NIOSH Physiology and Ergonomics Branch. Contract No. CDC 99-74-62, 1977.

Equal Employment Opportunity Commission (EEOC). *Title I of the Americans with Disabilities Act: EEOCs Technical Assistance Manual.* 1992.

Equal Employment Opportunity Commission (EEOC). *Uniform Guidelines on Employee Selection Procedures.* Title 29, CFR, Part 1607, 1978.

Herrin, G.D.; Kochkin, S.; Scott, V. Development of an employee strength assessment program for United Airlines. Technical Report, The University of Michigan Center for Ergonomics, 1982.

Jackson, A.S.; Osburn, H.G.; Laughery, K.R. Validity of isometric strength tests for predicting performance in physically demanding tasks. *Proceedings of the Human Factors Society 28th Annual Meeting*, 452-454, 1984.

Jackson, A.S.; Osburn, H.G.; Laughery, K.R. ;Validity of isometric strength tests for predicting endurance work tasks of coal miners. *Proceedings of the Human Factors Society 35th Annual Meeting*, 753-767, 1991a.

Jackson, A.S.; Osburn, H.G.; Laughery, K.R.; Vaubel, K.P. Strength demands of chemical plant work tasks. *Proceedings of the Human Factors Society 35th Annual Meeting*, 758-762, 1991b.

Jackson, A.S.; Osburn, H.G.; Laughery, K.R.; Vaubel, K.P. Validity of isometric strength tests for predicting the capacity to crack, open, and close industrial valves. *Proceedings of the Human Factors Society 36th Annual Meeting*, 688-691, 1992.

Keyserling, W.M. *Isometric Strength Testing in Selecting Workers for Strenuous Jobs.* Ph.D. Dissertation, University of Michigan, Center for Ergonomics, 1979.

Keyserling, W.M.; Herrin, G.D.; Chaffin, D.B. Isometric strength testing as a means of controlling medical incidents on strenuous jobs. *Journal of Occupational Medicine,* 22(5): 332-336, 1980a.

Keyserling, W.M.; Herrin, G.D.; Chaffin, D.B.; Armstrong, T.A.; Foss, M.L. Establishing an industrial strength testing program. *American Industrial Hygiene Association Journal,* 41: 730-736, 1980b.

Laughery, K.R.; Jackson, A.S. Pre-employment physical test development for roustabout jobs on offshore platforms. Technical Report, Kerr McGee Corporation, 1984.

Laughery, K.R.; Jackson, A.S.; Fontenelle, G.A. Isometric strength tests: Predicting performance in physically demanding transport tasks. *Proceedings of the Human Factors Society 32nd Annual Meeting,* pp. 695-699, 1988.

Liles, D.H.; Deivanayagam, S.; Ayoub, M.M.; Mahajin, P. "A job severity index for the evaluation and control of lifting injury. *Human Factors,* 26(6):683-693, 1984.

Reilly, R.R.; Zedeck, S.; Tenopyr, M.L. Validity and fairness of physical ability tests for predicting performance in craft jobs. *Journal of Applied Psychology,* 64(3): 262-274, 1979.

71

Preplacement Strength Screening

Valerie J. Rice
U.S. Army Research Institute of Environmental Medicine

According to federal employment laws, it is illegal to disqualify an employee for reasons of race, color, religion, gender, national origin, or disability status. Accordingly, employers are using preplacement screening tests to determine an individual's ability to safely perform job tasks. This chapter reviews the pertinent literature related to preplacement screening and derives guidelines for developing job-specific screening. Both management and workers benefit from the increased productivity and safety resulting from a well-constructed preplacement screening process.

71.1 Introduction

When hiring new personnel, management has a vested interest in having workers succeed in their jobs. Workers who will have an acceptable level of productivity and will perform the job safely are needed to run an efficient operation. In physically demanding jobs, there is a concern that smaller, weaker individuals may be at greater risk for injury and may be less productive. Job redesign, worker selection, and training are used in an attempt to ensure safe execution of job requirements. Of these approaches, it is preferred to design the equipment, job, or task so the majority of workers can perform the task safely and efficiently. In this way, the risks of injury are controlled by reducing the probability of occurrence. If the task or tools cannot be redesigned, then selection procedures can be used to identify personnel who meet the physical requirements of the job. For example, the tasks of fire fighters, police, rescue workers, and soldiers cannot always be anticipated or redesigned in order to permit persons with lower levels of physical fitness and strength to complete them. Instead, worker selection procedures are used to select personnel for these physically demanding jobs.

Pre-employment and preplacement screening refer to the process of administering a test or set of tests to job applicants in order to discern whether they can safely perform the job in question. Three assumptions must be accepted before a strength-based screening program is introduced. First, it is assumed the job and job tasks have been evaluated to determine whether ergonomic redesign was appropriate and any necessary redesign is complete. Second, the individuals designing the preplacement strength testing

0-8493-2641-9/99/$0.00+$.50

program must believe that they can accurately test a person's physical performance in some manner and use those test results to predict successful job performance. Third, it is assumed that an inverse relationship exists between injury risk and an individual's ability to meet the physical requirements of a job, and that this relationship can be demonstrated objectively.

The ergonomist must also be acutely aware of the legal issues involved in preplacement strength screening. The majority of screening programs have the appearance of inherent bias for age, gender, and disability classification. Therefore, legislation has been enacted in an attempt to eliminate these biases and provide guidance to base personnel selection criteria on sound business and ethical considerations.

71.2 Literature Review

A number of researchers have examined methods of determining and/or predicting a person's ability to perform a job or task. Static strength tests (Chaffin, Herrin, and Keyserling, 1978; McMahan, 1988) dynamic strength tests (Kroemer, 1983; Menon and Freivalds, 1985; Pytel and Kamon, 1981), work simulations (Arnold, Rauschenberger, Soubel, and Guion, 1982), calisthenics (Sharkey, 1981), aerobic capacity (Sharp, Wright, Vogel, Patton, Daniels, Knapik, and Kowal, 1980), body composition and anthropometrics (Misner, Plowman, and Boileau, 1987), and other physical performance measures (Reilly, Zedeck, and Tenopyr, 1979) have been suggested as evidence of ability to perform physically demanding work. The research is limited, however, as many of the predictions have centered on simple lifting capacity, rather than multistep complex tasks such as those involved in most jobs.

For example, Pytel and Kamon (1981) examined the relationship between several dynamic strength measures (lift, back and arm strength) and a one repetition maximum (1-RM) lift from the floor to a 113 cm height with a psychophysical assessment of a maximal acceptable lift to the same height at a rate of 6 lifts/min. A prediction equation was developed using dynamic lift strength and gender, accounting for 94% of the variance. The authors concluded the study demonstrated the usefulness of a simple, portable "isokinetic" dynamic strength measuring device; however, it is unknown whether the same device can be used to predict actual job performance. Teves et al. (1985) examined the use of body composition and various strength measures to predict a one repetition maximum lift to 132 cm for military soldiers. This is similar to lifting a box onto the bed of a 2½ ton truck. They found that lean body mass and a lift to a height of 183 cm, using an incremental dynamic lift device, were the best predictors ($R^2 = 0.47$). They predicted the ability to complete one lift, and no task analysis was mentioned; therefore, it is unknown whether this was an essential task for the military. Similar difficulties are seen in other studies predicting individual (Kumar, 1995; Mital and Ayoub, 1980) and team lifting capacities (Fox, 1982; Karwowski, 1988; Rice, Sharp, Nindl, and Bills, 1995).

At face value, perhaps the best measure of job performance is to have the applicant perform the job. However, having each applicant work at the job, even for a short period, is time consuming, costly, and may put the applicant at undue risk of injury. Instead, criterion tasks, in the form of work simulations or task components, are used to predict job performance. It is assumed that testing applicants on an aspect of job performance that is critical to job success is a valid approach, as it seems to satisfy content requirements (Wigdor et al., 1991). It also appears to ensure the constructs of job performance are represented and as long as characteristic samples of *each* critical element of a job are included, the validity "should be axiomatic" (Wigdor et al., 1991, p. 59). In addition to using work simulations to predict work performance, many screenings use work simulations as the performance to be predicted. In other words, instead of being on the predictor side of the equation, the work simulation is on the criterion side of the equation as the standard against which the preplacement screening measures will be tested. In these cases, various constructs such as strength or aerobic capacity are used to predict performance on simulated tasks.

The use of isometric strength tests to predict performance is appealing because they are relatively easy to administer, standardized procedures have been established, test–retest reliability has been established (Hershenson, 1979; Keyserling, Herrin, and Chaffin, 1980a), and they have been shown to be safe in a series of industrial studies (Chaffin, 1974; Hershenson, 1979; Kamon and Goldfuss, 1978). They have

also been found to be predictive of performance on dynamic tasks (Aghazadeh and Ayoub, 1985; Ayoub, Mital, Asfour, and Bethea, 1980; Kamon, Kiser, and Pytel, 1982; Kroemer, 1982; Kroemer, 1983; McDaniel, Kendis, and Madole, 1983). However, accurate predictions of dynamic tasks from isometric strength tests are not always found, as the differences in mechanical and physiological processes that occur during static and dynamic strength testing make prediction difficult (Kroemer, 1970). In addition, strength in one working posture is not necessarily correlated with strength in another posture (Hershenson, 1979), and maximal isometric lifting strength is not necessarily a predictor of psychophysically acceptable levels of lifting (Garg, Mital, and Aurelius, 1980).

Predicting Job Performance from Preplacement Screening

Simulated job performance. A review of over 21 research studies designed to predict work performance revealed the use of static (isometric), dynamic (isotonic), and combined strength tests (task simulations) to predict job performance, with varying degrees of success (see Table 71.1 at end of chapter). Predictions accounted for anywhere from 12 to 90% of the variance. The most successful predictions, those that accounted for 60% of the variance or more demonstrated several consistencies (Arnold et al., 1982; Celentano, Nottrodt, and Saunders, 1984; Davis, Dotson, and Santa Maria, 1982; Jackson, Osburn, and Laughery, 1984; Rice and Sharp, 1994; Robertson, 1982; Robertson and Trent, 1985; Schonfeld, Doerr, and Convertino, 1990; Sharkey, 1981; Stalhammar, Leskinen, and Nurmi, 1992; Wilmore and Davis, 1979). All performed a well-constructed job analysis, with a breakdown of job tasks into component parts, using multiple assessment techniques: interviews with ratings of key tasks for frequency, duration, and difficulty; direct observation, videotaping, measurements of masses moved and forces exerted; and identification of pace and frequency. As a result, all of the highly successful evaluations examined multiple constructs determined from the job analysis.

Sharkey (1981) (Table 71.1) found that unless the full complement of pertinent measures (with a combination of constructs) was used, screening did not guarantee positive results. The inclusion of multiple factors increases the predictive ability, due to correctly identifying essential components of the task. For example, Rice and Sharp (1994) evaluated seven strength measures, three Army Physical Fitness Test scores, and three physical descriptors for their ability to predict performance on two stretcher-carry tasks (Table 71.1). The best predictors for repeated short distance stretcher-carrying were 2-mile run time (aerobic factor) and peak handgrip (strength factor) (R^2 range from 0.74 to 0.78). For a continuous carry, a sustained mean grip strength was the best predictor of performance (R^2 range from 0.74 to 0.78). The results illustrate the necessity for tailoring preplacement tests to accurately reflect job demands, as different predictors were selected for each type of stretcher-carrying task.

It also appears that use of consolidated or summed scores either as the dependent or independent variable may strengthen the predictive ability of a model. This procedure has been used with an isometric test battery administered to coal miners (Laughery, Hayes, Jackson, Osburn, and Hogan, 1986), roughneck and roustabout positions in oil drilling and production operations (Jackson, 1984), and operator positions in refining and chemical manufacturing (Jackson, 1986 as cited in Jackson, 1994). In their study of oil drilling operations, Jackson, Osburn, and Laughery (1984) used three strength tests and one arm endurance test to predict performance in several work simulations (Table 71.1). A composite score for the strength battery was used, resulting in correlations between the isometric strength tests and work sample tests ranging from 0.67 to 0.93, with equal validity for men and women. In another study by the same authors (Jackson, Osburn, Laughery, and Vaubel, 1991) (Table 71.1), a composite of the isometric strength tests was found to be a valid predictor of endurance work tasks, again with the measure being valid for both men and women (accounting for up to 87% of the variance, Table 71.1). Other studies have also used this technique with positive results (Laughery, Jackson, and Fontenelle, 1988), with one study concluding that task performance was dependent on strength rather than gender, and the higher the sum of an individual's isometric strength, the greater the probability of success on work simulations (Jackson et al., 1991). The commonalities in the studies conducted by these authors are their meticulous

attention to task analysis, inclusion on all pertinent performance constructs, and the use of a composite score to predict performance on job simulations. However, since the use of a composite score can artificially raise the R^2 value, it is debatable whether its use adds or detracts from a prediction. Arnold, Rauschenberger, Soubel, and Guion (1982) used a composite score on the other side of the equation (Table 71.1). General physical abilities thought to be required for work simulations of moving/carrying, lifting, wheelbarrowing, and shoveling were used as a test battery to predict a composite score derived from several work simulations. Each of the simulations was evaluated by observers and objective measures were used for as many tasks as possible (strength, number of objects moved during a timed session). The best single predictor was the strength measure attained with the arm dynamometer, accounting for 67% to 72% of the variance of a composite score.

On-the-job performance. Table 71.1 also contains studies that examine the ability to predict performance and productivity while in training or on the job (Anderson, 1988; Anderson and Catterall, 1989; Arnold et al., 1982; Bernauer and Bonanno, 1975; Doolittle, Spurlin, Kaiyala, and Sovern, 1988; Hogan, 1985; Reilly et al., 1979). Certainly attrition is one aspect of productivity, since retraining is required following most job changes. In one study of transfers from jobs with heavy demands vs. jobs with light demands, it was found that men left the heavy demand jobs at a rate almost double to those with light demands (34% vs. 57%), and women's transfers from heavy jobs were more than triple those from light jobs (23% vs. 75%, Kowal, 1983). Robertson (1992) discussed the idea that personnel whose strength capacity marginally meets the strength requirements of a task may fatigue at a significantly faster rate. For jobs requiring repeated heavy strength exertions, it can be assumed individuals with lower strength will be put at a performance disadvantage and increased risk of injury.

However, results of other studies do not support the relationship between worker strength and productivity. The job demands and loads on the lumbosacral spine of National Guard mechanics were compared to static strength physical capacity testing of the same workers (arm lift, back lift, and floor lift) (Pedersen, Clark, Johns, White, and Hoffman, 1989). Results indicated up to a 38% job mismatch between requirements and capabilities; however, it was not clear how the workers were able to perform the job with such a mismatch. They were currently working (and presumably performing successfully) in those jobs, although they reported a high incidence of low back pain. It is possible they completed their work through development of alternate methods for task completion or through informal selection (letting the stronger individuals perform the heavier jobs), but the usefulness of the study results are questionable without this information. General physical fitness has not been found to be predictive of successful job performance. Wilmore and Davis (1979) found no difference in general physical fitness levels between California Highway Patrol officers who received high vs. low supervisory ratings. Nor did they find differences in the physical fitness of police officers compared with other researcher's results for the general public.

Researchers have also examined the ability to predict success during training. Bernauer and Bonanno (1975) investigated applicants for pole climbing jobs using body fat, arm strength, trunk strength, cardiovascular fitness, response time, and body balance. The mean scores for the step test and balance were found to differ for successful and unsuccessful applicants. Hogan (1985) also evaluated successful training completion (Table 71.1), using a test battery of 23 measures and five widely recognized physical fitness batteries. Performance results were compared with receiving either a pass or fail, and with nine performance rating scales (completed by three independent instructors). Multiple regression revealed that cardiovascular fitness, muscular endurance, and flexibility were highly related to program completion and overall potential ratings (Table 71.1). Results from the five physical fitness batteries yielded multiple R's ranging from 0.38 to 0.60 for prediction of final outcome (pass/fail). Although a comparison with actual job performance might be preferable, these results indicate that models can be developed for predicting training attrition.

Few studies have gone the extra distance to try and predict actual job performance. Instead many researchers assume that being able to complete a task simulation of the most demanding portion of the job is indicative of an individual's ability to perform the job. This may be accurate, but becomes questionable if a one-time administration of a task is used as the criterion, when repeated performances

are required on the job. Those research evaluations that have included this last step to ensure that their measure is predictive of successful job performance appear more persuasive for court challenges of preplacement testing. Anderson and Catterall (1989) studied job productivity and turnover of initial applicants (Table 71.1). They conducted a prospective validation involving grocery warehousemen and demonstrated that use of a screening battery decreased turnover and increased productivity. It was also demonstrated that the more stringent the cutoff criteria, the greater the productivity. The test battery consisted of two isometric strength tests and a step test, based on a thorough job analysis. Criteria were established based on the heaviest cases to be moved (60 to 100 lbs) and an estimated aerobic capacity three times the average metabolic requirement of the job (15 kcal/min).

Considerations in predicting performance. One difficulty in reviewing the literature is that researchers vary in the methods they use and how they report their data. For example, several authors did not report the r values or used correlations without any regression analysis. Although a correlation or prediction may be significant ($p < 0.05$), if it only accounts for 30% of the variance, the model may be questionable in court challenges of its ability to foretell job performance.

Another challenge is where to place the cutoff criterion. It would appear that determining the cutoff criterion by the exact job requirements would be the ideal. For example, if the job requires a 75-lb. lift, then have the job applicant perform a 75-lb. lift. The more challenging position comes with tasks requiring repetitive heavy physical demands, as it is unlikely that preplacement testing could incorporate four lifts per minute for a full day due to time, financial constraints, and safety concerns. Instead, if some examiners observe an applicant straining, out of breath, or demonstrating a high heart rate and slow recovery, they report that information to the physician for consideration in the final medical report. Should an applicant be denied a particular position based on this analysis and challenge the medical and preplacement screening, the problem will be in demonstrating that although the person could perform the basic task, these observations would indeed indicate an inability to do the job or increased risk to themselves or others. A better procedure would be to establish a rigorous statistical model that is predictive of performance. For example, Anderson and Catterall (1989) demonstrated the more stringent the pass cutoff score, the more likely personnel were to stay for eight weeks and the more productive they were.

Ayoub (1982) introduced a method to determine cut scores for exercise intensity relative to the individual. He suggested that cutoff scores for heart rate be based on a percentage of their heart rate maximum. He suggested that cutoff scores for heart rate as "those falling below 75% are acceptable, those between 75% and 80% are marginal, and those above 80% are unacceptable." For strength, he suggests a cutoff of 50% of a person's maximum strength; "applicants falling below 50% are acceptable, those between 50% and 60% are marginal, and those above 60% are unacceptable." Dr. Ayoub noted the rationale for pre-employment screening, and these cutoff scores are explainable, integrative (consider all aspects of the physical and physiological properties of human work capacity), and can be used for jobs characterized by brief exertions as well as sustained performance. However, his recommendations are conservative and appear more focused on sustained activity rather than brief periods of high activity.

Caution should be used if assessing personnel who are currently performing the job to establish cutoff criteria. This could lead to misleading conclusions, as these persons have developed the skills, strength, and knowledge regarding task requirements. Use of a cutoff score that allowed for improvement through a training program could be one solution. The military services accept the fact that physical strength will be necessary and include physical training as part of a soldier's entry level instruction, job requirements, and performance assessment. Civilian counterparts may be less likely to invest the time and funding in such an effort, instead recommending that persons who are unable to qualify for a given position may be able to train on their own and reapply for a position.

Using Preplacement Screening to Predict Injury Rates

There is some evidence for an association between physically demanding jobs and injury rates, which could lead to the assumption that workers who are better able to meet the physical demands are less

prone to injury. Physical fitness was found to be predictive of back injuries for fire fighters; however, the injuries sustained by the fit group were more severe than those of the "unfit" group (Cady, Bischoff, O'Connell, Thomas, and Allan, 1979). Load transportation in narrow spaces was found to be associated with musculoskeletal (especially spinal) injury (Stalhammar et al., 1992), and increased incidence and severity of low back pain has been found in people who work in jobs requiring lifting and moving of heavy loads (Wickstrom, 1978; Kosiak, Aurelius, and Harfiel, 1968; Rowe, 1969). Although most individuals with back problems cannot associate the onset of their symptoms with an accident or unusual activity, those who can cite a specific event, cite lifting (Cady et al., 1979; Snook et al., 1978). In a retrospective review of back injuries, it was demonstrated that the most common event associated with back injury claims was manual materials handling, and improper lifting was most frequently listed as the cause (Bigos et al., 1986).

A previous history of back pain (Biering-Sorensen, 1984; Bigos, Spengler, and Martin, 1986) and/or a change of occupation to heavier work (Nordgren, Schele, and Linroth, 1980) have been suggested as potential predictors of consequent back pain. Persons experiencing back pain within the last 6 months have been found to have a higher incidence of subsequent back problems (Battié et al., 1989), which is consistent with the idea that recurrences (of back injury) are more likely in the 2 to 3 years following an injury. However, Venning, Walter, and Stitt (1987) found work-related factors such as the job category and exposure levels have greater contributions to back injury than did a previous history of back pain.

Table 71.2 contains a synopsis of studies relating preplacement screenings with injuries. Strength has been identified as a potential risk factor for back injuries by a number of researchers (Biering-Sorensen, 1984; Cady et al., 1979; Chaffin et al., 1978; Doolittle and Kaiala, 1986; Keyserling et al., 1980a; Nachemson and Lindh, 1969). In fact, it has been over 20 years since researchers identified an association between an increased chance of injury with strength capabilities of workers (Chaffin, 1974; Chaffin, 1971; Rowe, 1971). However, Hagberg, Silverstein, Wells, Smith, Hendrick, Carayon, and Perusse (1995) reviewed 25 articles dealing with preplacement or pre-employment screening and concluded that "at present there is no scientific evidence to support the use of pre-employment or preplacement screening tests to predict the development of work-related musculoskeletal disorders, nor are they justified from an ethical and, in some countries, legal point of view" (p. 336).

Several articles provide direct support for this viewpoint. Battié, Bigos, Fisher, Hansson, Jones, and Wortley (1989) examined isometric lifting strength as a predictor of industrial back pain. During a 4-year follow up, those with *greater* isometric strength were found to be at greater risk for injury than were weaker workers. However, after controlling for age, no significant difference was seen. These researchers concluded that isometric lifting strength testing was ineffective in identifying individuals at risk for industrial back problems. Anecdotally, in military operations, it has been suggested that natural selection occurs with stronger workers executing the heavier work more often. Dueker, Ritchie, Knox, and Rose (1994) measured isokinetic trunk flexion and extension, as well as repetitive isotonic lifting (Table 71.2), and monitored the workers for 6 years. No difference was found between any isokinetic scores of injured vs. non-injured workers. Mostardi, Noe, Kovacik, and Porterfield (1992) used an isokinetic (Cybex) lifting device to measure lifting capacity of nurses. After following the workers for 2 years, the authors concluded that isokinetic lifting strength was not related to occupational back injury. Several elements were missing from these studies. The design of the strength measures was not based on a precise task analysis (at least none was reported), and no other constructs (aerobic capacity, movement quality) were included.

Several studies have demonstrated the positive effect of an individual's strength capacity meeting or exceeding job requirements (Chaffin, 1974; Chaffin et al., 1978). In Chaffin's seminal work, he examined subjects' maximal acceptable isometric lifting capacity and compared it with job requirements. The hypothesis was that those with higher job strength ratios (job lift requirements ÷ average acceptable lift) would be more likely to experience low back pain. The hypothesis was upheld, as a sharp increase in the mean low back pain incidence rates was found for those jobs populated by persons without strength equal to or exceeding that required by the job. The medical incidence rate of workers with insufficient strength to meet job demands was nearly three times the rate of workers who were matched to their jobs (Chaffin, 1974). Three subsequent prospective studies used isometric tests that were similar to job

requirements and compared those results with the job requirements (Chaffin et al., 1978; Keyserling et al., 1980a; Keyserling, Herrin, Chaffin, Armstrong, and Foss, 1980b). Results revealed greater injury rates when the job demands approached or exceeded the person's abilities. The authors concluded that workers whose strength abilities were less than job requirements suffered higher incidence rates in general than workers whose strength matched or exceeded job demands. However, excess strength was not found to be of protective value. Anderson (1988) found incidence rates were 34% greater for persons who did not meet job strength requirements (below criteria) compared with those who met or exceeded the requirements. He also noted that workers below criteria had more lost duty time and higher worker compensation costs (although the data were not statistically analyzed), and workers above criteria were more productive. The authors suggested the use of a screening battery could result in an increase in productivity and a reduction of back injuries, over-exertion injuries, incurred compensation costs, and lost time. Although the results demonstrated that workers who met the criteria were less likely to report back or over-exertion injuries, this was not true for all five jobs studied, even though they were considered physically demanding.

Other methods of decreasing injuries in the workplace have been suggested in early reports by Snook, Campanelli, and Hart (1978) and later by Doolittle and Daniel (1989). Their suggestions focused on limiting the requirements of the task. For Snook et al. (1978) the limitations were based on his research using maximal acceptable strength tasks, while Doolittle and Daniels' limitations used the NIOSH lifting formula and previous research by Ayoub et al. (1984). Snook et al. (1978) determined that a worker is three times more susceptible to low back injury if exposed to manual handling tasks that less than 75% of the population find acceptable and, therefore, that 67% of the injuries could be avoided through ergonomic design. Troup, Foreman, Baxter, and Brown (1987) also found that lifting capability, expressed as a rating of acceptable load, was related to the development of future low back pain of bakery workers. The physically demanding tasks were compared with the 1981 NIOSH lifting guide, and the authors' recommended preassignment screening to select workers who are physically capable of performing the job in order to prevent injuries. They further recommended that workers not exceed 75% of maximum capability based on Ayoub's recommendations (Ayoub, 1982). Rather than a pure cutoff score, they suggested ranking candidates into four performance levels: below minimum (applicant taxed at more than 40% maximum aerobic capacity and more than 95% strength capacity), marginal (40% aerobic capacity and 85% strength), acceptable (33% to 40% aerobic and 75% strength), and better qualified (<33% maximum aerobic capacity and <65% strength). Similar procedures were used for screening fire fighters (Doolittle and Kaiyala, 1987) and for identifying physically demanding jobs for line workers (Doolittle et al., 1988); however, neither article included research to demonstrate that their method actually reduced injuries of workers.

One of the more convincing reports was published in a recent study by Karwowski, Caldwell, and Gaddie (1994) (Table 71.2). They examined the relationship between the NIOSH 1991 Lifting Index, estimated compressive and shear forces on L_5/S_1, and back injury incidence rates. Strong, positive correlations were found between the maximum compressive forces and the maximum lifting index for the job ($r = 0.88$, $p = 0.05$), with similar findings between the average compressive forces and the average lifting index. Strong correlations were also found between the average recommended weight limit (RWL) and the maximum incidence rate of low back injury ($r = .97$, $p = 0.004$), and between the Lifting Index (ratio between the actual load lifted and the RWL) and incidence rate for 1992 ($r = .94$, $p = 0.1$) and 1993 ($r = .97$, $p = 0.007$). Although no prediction equations were used, Reimer, Halbrook, Dreyfuss, and Tibiletti (1994) found that use of a worker fitness evaluation, based on job-related lifting and trunk function, reduced injury rates over a 4-year follow-up period.

It appears that there is a relationship between preplacement strength measures and injuries, but the associations were not as strong or as consistent with those found between preplacement testing and performance. Once again, it appears that basing the criterion tasks on a well-designed task analysis is essential. Although not as obvious, it also appears beneficial for other constructs such as aerobic capacity to be included in the assessment. Certainly, additional rigorous evaluation of this assumption is warranted.

71.3 Establishing a Preplacement Screening Program

The goal of a preplacement program is to match the job requirements with a worker's capacity in order to ensure productivity and safety. As stated by Keyserling et al. (1980b), preplacement testing should be safe, reliable, quantitative, practical (time, equipment, financial expenses, supervision/manpower–ease of administration, availability), related to job requirements, and predictive of job performance and/or risk of future illness or injury. Preplacement screening may result in fewer older individuals, women, and persons with disabilities being hired for certain jobs. As long as the assessment batteries are carefully developed, validated, and alternate methods of completing a task are considered so that it is evident the preplacement screening is based on the job requirements, this should not be an issue of undue concern.

The steps in developing a preplacement screening program are in outline form in Appendix A (preceding the Reference list for this chapter). During the job analysis, it is important to consider the skill levels and degree of competency required in each job. In order to do this, ergonomists must be directly involved in the job analysis. They should interview the subject matter experts (i.e., the workers, supervisors, administrators, and safety personnel; consult training manuals), observe, and conduct careful assessments of the job to include frequency, duration, and difficulty, measurements of masses moved and forces exerted; and identification of pace, frequency, and postural constraints. The ergonomist should also be aware of confounding factors such as noise and calibration of equipment. It is important to reiterate that each preplacement test will most likely differ for distinct jobs, all of which may be physically demanding. No studies have shown that the same preplacement/prescreening evaluations have been predictive of job performance or injury rates in multiple physically demanding jobs. Instead, the more successful reports have come from authors who tailored their assessments to the jobs in question (Jackson, 1994).

From the job analysis, an accurate description of the task requirements (essential job functions) with "limiters" should be identified. Every essential function does not need to become part of the preplacement evaluation. For example, if one of the essential functions requires the worker to lift a 40-lb. box, carry it 15 ft, and place it on a 4-ft. shelf and a second essential function requires the worker to lift a 30-lb. box to 2 ft, it can be assumed that an applicant who can complete the former task, can also complete the latter task. The first task (lift and carry) is a "limiter" Once a task analysis is complete, the description of the tasks should be presented to the company subject matter experts and revised as necessary. The preplacement job simulations and test battery should then be derived from the task descriptions of the essential functions, and the model of evaluation and criterion measures should be identified.

Kroemer, Snook, Meadows, and Deutsch (1988) described various models to identify limiting factors, such as a linear anatomical model of the long bones, joints, and muscles to describe volume, mass, and muscle strength; physiological models using oxygen consumption, energy expenditure, and circulatory demands; orthopedic models using a prediction of musculoskeletal injuries; biomechanical models which may use anatomical, anthropometric, and orthopedic measures; and psychophysical models which evaluate synergistic mental and physical functions. One of these models, or a combination of them, must be selected for use in developing a preplacement screening evaluation.

The type of model selected will determine the test measures that will be used. For example, a medical examination may be used when employing an orthopedic model, and a lumbar motion monitor may be used to assess biomechanical limitations, such as the load-bearing capacities of the spine (Marras, Fathallah, Miller, Davis, and Mirka, 1992; Marras, Lavender, Leurgans, Rajulu, Allread, Fathallah, and Ferguston, 1995). Hogan (1991) conducted a review of literature and a series of evaluations on the physical requirements of tasks as reflected in job analysis. As a result of these evaluations, Hogan suggested that the structure of physical performance has three major components: strength, endurance, and movement quality. Hogan concluded that all occupational task performances fit within these three constructs. Although examination of specific subcomponents may be necessary, Hogan (1991) suggests using this taxonomy to organize worker selection procedures, and hence the test measures, for physically demanding jobs. Each model and test measure has its own strengths and weaknesses and its own proponents and adversaries.

Generally, it may be easier to develop a job simulation and consequently identify subelements that can then be used in a test battery. Many researchers prefer the use of a test battery, rather than task

simulations in their preplacement screening. Test batteries are repeatable, require less skill from the participants (and often from the evaluators), can be used in a variety of locations, and require less equipment compared with task simulations. Of course, task simulations have greater face validity. Once the job simulation and test batteries have been developed, they should be compared using statistical analysis to determine whether one is representative of the other. If they are closely related, then only one need be used to predict on-the-job performance. Although the ADA and EEOC requirements do not include validation studies *per se*, it behooves those conducting the preplacement screening to be aware of the predictive ability of their simulations and test batteries. If possible, it would be wise to conduct a validation study in which either the job simulation, the test battery, or both are statistically examined for their relationship (association and predictive ability) with on-the-job performance. Thus, the ergonomist can establish that the preplacement testing does what it purports to do, and potentially, the results will be more defensible.

Once a job simulation or test battery is selected and implemented, it is desirable to conduct a follow-up analysis on the effectiveness of the preplacement screening through the evaluation of both productivity and injury data. Although the investigators can select any amount of time for a follow-up period, it is recommended that it be a minimum of 18 months to attempt to eliminate a placebo effect.

As mentioned previously, the preplacement tests should be reliable (giving the same evaluation at different times should yield the same result), and the safety of the applicants should be of utmost importance. Isometric strength tests have exhibited test–retest coefficients-of-variation ranging between 5% and 15% (Hershenson, 1979, Keyserling et al., 1980b), and have been shown to be safe in a series of industrial studies (Hershenson, 1979; Chaffin, 1974; Kamon et al., 1978). For task simulations, test–retest reliability evaluations may become part of the validation process. Simulations must also be fair. If they are too complex, learning and skill development can contaminate the assessment capabilities.

It will not always be possible to include each of the steps in Appendix A, due to funding or contract restraints of the hiring company. It is imperative that the researcher/ergonomist developing the preplacement screening evaluation be comfortable with the level and stringency of the particular program. It is possible that both the hiring company and the involved contractor could be questioned, should an applicant decide to challenge the validity of the preplacement process. The "developer" of the preplacement screening evaluation is not relieved of the duty to meet test validation requirements, even if the test user (hiring company) did not request the test be validated or asked for a lesser standard.

An inordinate amount of testing is undesirable due to time and funding costs, and the use of validation testing should assist in honing the preplacement evaluation. A direct relationship between the preplacement evaluation and task demands should be unequivocal. However, the requirement for business necessity may be more lenient if there is a high degree of risk to the job or if all or substantially all, women would be unable to perform the job safely and efficiently because of the strenuous manual labor required. For this reason, a validation of injury potential should be just as rigorous as the validation of the prediction of work performance.

71.4 Summary

According to federal employment laws, it is illegal to disqualify an employee for reasons of race, color, religion, gender, national origin, or disability status. Accordingly, employers are using preplacement screening tests to determine an individual's ability to safely and effectively perform a job. To date, there have been successful challenges to preplacement screening which failed to test important job actions (Jackson, 1994). Using a well-defined, scientific approach, it is possible to design a preplacement screening battery that is directly related to (and predictive of) job performance. It is also possible to use the preplacement screening to prospectively determine its impact on injuries, workers' compensation, lost workdays, and turnover. Following the evaluation and completion of ergonomic design alterations, a preplacement screening program should be the next step in ensuring a match between work demands and worker abilities for physically demanding positions. Both management and workers will benefit from the resulting increases in productivity and safety.

Appendix A

I. Job Analysis
   A. Job descriptions
   B. Observation/Videotaping
   C. Interview/Survey (subject matter experts, management/workers)
      1. Difficulty
      2. Frequency
      3. Importance
      4. Other means to accomplish tasks
   D. Direct measurement
      1. Heights
      2. Weights
      3. Distances
      4. Frequency
      5. Pace
   E. Develop task list/description of essential functions

II. Presentation and Alteration
   A. Present to company
      1. Officials
      2. Supervisors
      3. Workers
   B. Modify task list, as needed
   C. Suggest uses of task list
      1. Preplacement screening
      2. Return to work of injured employees
      3. Graded introduction to task demands for new workers
      4. Americans with Disabilities Act compliant job advertisements

III. Preplacement Test Development
   A. Develop simulation (criterion tasks with standards of performance) and may include "task limiters"
      1. Lifting, carrying, pushing/pulling, digging, or climbing,
         a. Single
         b. Repetitive
         c. Location of beginning/ending object placement
         d. Object description (handles, size, mass, etc.)
         e. Posture requirements
         f. Distance
      2. Aerobic demands
      3. Other
   B. Identify task elements from task simulation
      1. Range of motion
      2. Strength
      3. Endurance
      4. Coordination
      5. Positioning/posture
   C. Development of preplacement test battery from task elements
   D. Development of necessary instrumentation

IV. Test new (avoid "cumulative" injury confounds) and current workers
   A. On job task simulations
   B. On preplacement test battery

V. Validation of task simulation, test battery, and on-the-job performance
   A. Compare scores on task simulation and test battery
      1. Multiple correlations
      2. Regression equations
      3. Other statistical methods as needed
   B. Validation of task simulation and test battery with on-the-job work performance
      1. Follow workers for specified period of time and record on-the-job data
         a. Descriptive data (age, weight, job, etc.)
         b. Work performance (productivity, supervisor ratings)
         c. Lost duty time
         d. Turnover/retention
      2. Record potential confounding factors, such as pertinent changes at company
         a. New safety programs
         b. Reorganization
         c. Other
      3. Compare scores on task simulations and test battery with on-the-job work performance
         a. Multiple correlations
         b. Regression equations
         c. Other methods as needed

VI. Validation studies of test battery and task simulation with injury data
   A. Follow-up study for specified period of time (suggested *minimum* of 18 months)
      1. Tabulate data of new and current workers
         a. Descriptive data (age, weight, job, etc.)
         b. Injury rates (categorized by type of injury/illness)
         c. Lost duty time due to injuries
         d. Turnover/retention
      2. Record potential confounding factors, such as pertinent changes at company
         a. New safety programs
         b. Reorganization
         c. Other
   B. Compare test battery and task simulation with injury data
      1. Multiple correlation
      2. Regression
      3. Other methods as needed

VII. Report findings to company management

VIII. Revise preplacement strength test battery and/or simulation in accordance with findings

IX. Repeat steps IV-X, as necessary

References

Aghazadeh, F. and Ayoub, M. M. (1985). A comparison of dynamic and static strength models for prediction of lifting capacity. *Ergonomics*, 28, 1409-1417.

Anderson, C. K. (1988). *Strength and endurance testing for pre-employment placement.* (Available from C.K. Anderson, Back Systems, Inc., 5520 LBJ Freeway #200, Dallas, TX 75240.)

Anderson, C. K. and Catterall, M. J. (1989). A prospective validation of pre-employment physical ability tests for grocery warehousing, in A. Mital (Ed.), *Advances in Industrial Ergonomics and Safety* (pp. 57-63). New York: Taylor and Francis.

Arnold, J. D., Rauschenberger, J. M., Soubel, W. G., and Guion, R. M. (1982). Validation and utility of a strength test for selecting steelworkers. *Journal of Applied Psychology*, 67(5), 588-604.

Ayoub, M. A. (1982). Control of manual lifting hazards: III. Preemployment screening. *Journal of Occupational Medicine*, 24(10), 751-761.

Ayoub, M. M., Mital, A., Asfour, S. S., and Bethea, J. J. (1980). Review, evaluation, and comparison of models for predicting lifting capacity. *Human Factors*, 22, 257-269.

Battié, M. C., Bigos, S. J., Fisher, L. D., Hansson, T. H., Jones, M. E., and Wortley, M. D. (1989). Isometric lifting strength as a predictor of industrial back pain reports. *Spine*, 14(8), 851-856.

Bernauer, E. M. and Bonanno, J. (1975). Development of physical profiles for specific jobs. *Journal of Occupational Medicine*, 17(1), 27-33.

Biering-Sorensen, F. (1984). Physical measurements as risk indicators for low back trouble over a one-year period. *Spine*, 9, 106-119.

Bigos, S. J., Spengler, D. M., and Martin, N. A. (1986). Back injuries in industry: A retrospective study II Injury factors. *Spine*, 11(3), 246-251.

Cady, L. D., Bischoff, D. P., O'Connell, E. R., Thomas, P. C., and Allan, J. H. (1979). Strength and fitness and subsequent back injuries in fire fighters. *Journal of Occupational Medicine*, 21, 269-272.

Celentano, E. J., Nottrodt, J. W., and Saunders, P. L. (1984). The relationship between size, strength, and task demands. *Ergonomics*, 27(5), 481-488.

Chaffin, D. B. (1971). Human strength capability and low back pain. *Journal of Occupational Medicine*, 16, 248-254.

Chaffin, D. B. (1974). Human strength capability and low back pain. *Journal of Occupational Medicine*, 16(1), 248-254.

Chaffin, D. B., Herrin, G. D., and Keyserling, W. M. (1978). Preemployment strength testing: An updated position. *Journal of Occupational Medicine*, 20(6), 403-408.

Davis, P. O., Dotson, C. O., and Santa Maria, D. L. (1982). Relationship between simulated fire fighting tasks and physical performance measures. *Medicine and Science in Sports and Exercise*, 14(1), 65-71.

Doolittle, T. L. and Kaiyala, K. (1986). Strength and musculo-skeletal injuries of fire fighters. *Proceedings of the 19th Annual Conference of the Human Factors Association of Canada* (pp. 49-52). Vancouver: British Columbia: Human Factors Association of Canada.

Doolittle, T. L. and Kaiyala, K. (1987). A generic performance test for screening fire fighters, in S. Kumar (Ed.), *Trends in Ergonomics and Human Factors IV.* New York: Elsevier.

Doolittle, T. L. and Daniel, B. (1989). Physical demands of bakery workers. *Proceedings of the Human Factors Society 33rd Annual Meeting* (pp. 682-686). Santa Monica, CA: Human Factors Society.

Doolittle, T. L., Spurlin, O., Kaiyala, K., and Sovern, D. (1988). Physical demands of lineworkers. *Proceedings of the Human Factors Society 32nd Annual Meeting* (pp. 632-636). Santa Monica, CA: Human Factors Society.

Dueker, J. A., Ritchie, S. M., Knox, T. J., and Rose, S. J. (1994). Isokinetic trunk testing and employment. *Journal of Occupational Medicine*, 36(1), 42-48.

Fox, R. R. (1982). *A psychophysical study of bimanual lifting.* Masters thesis in industrial engineering, Texas Tech University, Lubuck, TX.

Garg, A., Mital, A., and Aurelius, J. R. (1980). A comparison of isometric strength and dynamic lifting capability. *Ergonomics*, 23, 13-27.

Hagberg, M., Silverstein, B., Wells, R., Smith, M., Hendrick, H. W., Carayon, P., and Perusse, M. (1995). *Work Related Musculoskeletal Disorders (WMSDs): A Reference Book for Prevention.* London: Taylor and Francis.

Hershenson, J. D. (1979). Cumulative injury: A national problem. *Journal of Occupational Medicine,* 21(10), 674-676.

Hogan, J. (1985). Tests for success in diver training. *Journal of Applied Psychology,* 70(1), 219-224.

Hogan, J. (1991). Structure of physical performance in occupational tasks. *Journal of Applied Psychology,* 76(4), 495-507.

Jackson, A. S., Osburn, H. G., and Laughery, K. R. (1984). Validity of isometric strength tests for predicting performance in physically demanding tasks. *Proceedings of the Human Factors Society 28th Annual Meeting* (pp. 452-454). Santa Monica, CA: Human Factors Society.

Jackson, A. S., Osburn, H. G., Laughery, K. R., and Vaubel, K. P. (1991). Strength demands of chemical plant work tasks. *Proceedings of the 35th Annual Human Factors Society Meeting* (pp. 758-762). Santa Monica, CA: Human Factors Society.

Jackson, A. S. (1994). Preemployment physical evaluation, in J. O. Holloszy (Ed.), *Exercise and Sport Sciences Reviews* (pp. 53-90). Philadelphia: Williams and Wilkins.

Kamon, E. and Goldfuss, A. J. (1978). In-plant evaluation of the muscle strength of workers. *American Industrial Hygiene Association Journal,* 39(10), 801-807.

Kamon, E., Kiser, D., and Pytel, J. (1982). Dynamic and static lifting capacity and muscular strength of steelmill workers. *American Industrial Hygiene Association Journal,* 43, 853-857.

Karwowski, W. (1988). Maximum load lifting capacity of males and females in teamwork. *Proceedings of the Human Factors Society 32nd Annual Meeting* (pp. 680-682). Santa Monica, CA: Human Factors Society.

Karwowski, W., Caldwell, M., and Gaddie, P. (1994). Relationships between the NIOSH (1991) lifting index, compressive and shear forces on the lumbosacral joint, and low back injury incidence rate based on industrial field study. *Human Factors and Ergonomics Society 38th Annual Meeting* (pp. 654-657). Santa Monica, CA: Human Factors and Ergonomics Society.

Keyserling, W. M., Herrin, G. D., and Chaffin, D. B. (1980a). Isometric strength testing as a means of controlling medical incidents on strenuous jobs. *Journal of Occupational Medicine,* 22, 332-336.

Keyserling, W. M., Herrin, G. D., Chaffin, D. B., Armstrong, T. J., and Foss, M. L. (1980b). Establishing an industrial strength testing program. *American Industrial Hygiene Association Journal,* 41, 730-736.

Kosiak, M., Aurelius, J. R., and Harfiel, W. F. (1968). The low back problem: An evaluation. *Journal of Occupational Medicine,* 10, 588-593.

Kowal, D.M. (1983). Validation and utility of a work capacity test battery for selection and classification of military personnel. Washington, D.C.: DCP Office of Assistant Secretary of Defense (MRA & L).

Kroemer, K. H. E. (1970). Human strength: Terminology, measurement and interpretation of data. *Human Factors,* 12, 279-313.

Kroemer, K. H. E. (1982). *Development of LIFTTEST, A dynamic technique to assess the individual capability to lift material.* (Final Report, NIOSH Contract 210-79-0041). Blacksburg, VA: Ergonomics Laboratory, Industrial Engineering and Operations Research Department, Virginia Polytechnical Institute and State University.

Kroemer, K. H. E. (1983). An isoinertial technique to assess individual lifting capability. *Human Factors,* 25, 493-506.

Kroemer, K. H. E., Snook, S. H., Meadows, S. K., and Deutsch, S. (1988b). *Ergonomic Models of Anthropometry, Human Biomechanics, and Operator-Equipment Interfaces.* Washington, D.C.: National Academy Press.

Kumar, S. (1995). Development of predictive equations for lifting strengths. *Applied Ergonomics,* 26(5), 327-341.

Laughery, K. R., Hayes, T. L., Jackson, A. S., Osburn, H. G., and Hogan, J. C. (1986). Physical abilities and performance tests for coal miner jobs. *Proceedings of Human Factors 30th Annual Meeting* (pp. 377-391). Santa Monica, CA: Human Factors Society.

Laughery, K. R., Jackson, A. S., and Fontenelle, G. A. (1988). Isometric strength tests: Predicting performance in physically demanding transport tasks, in *Human Factors and Ergonomics Society 32nd Annual Meeting* (pp. 695-699). Santa Monica, CA: Human Factors Society.

Marras, W. S., Fathallah, F. A., Miller, R. J., Davis, S. W., and Mirka, G. A. (1992). Accuracy of a three-dimensional lumbar motion monitor for recording dynamic trunk motion characteristics. *International Journal of Industrial Ergonomics*, 9, 75-97.

Marras, W. S., Lavender, S. A., Leurgans, S. E., Rajulu, S. L., Allread, W. G., Fathallah, F. A., and Ferguston, S. A. (1995). Biomechanical risk factors and trunk motion. *Ergonomics*, 38(2), 377-410.

McDaniel, J. W., Kendis, R. J., and Madole, S. W. (1983). *Weight Lift Capabilities of Air Force Basic Trainees.* (TR No. 83-0001). Wright Patterson Air Force Base: Air Force Aerospace Medical Research Laboratory.

McMahan, P. B. (1988). Strength testing may be an effective placement tool for the railroad industry, in F. Aghazadeh (Ed.), *Trends in Ergonomics/Human Factors V* (pp. 787-794). North-Holland: Elsevier.

Menon, K. and Freivalds, A. (1985). Repeatability of dynamic strength tests. *Proceedings of the Human Factors Society 29th Annual Meeting* (pp. 517-520). Santa Monica, CA: Human Factors Society.

Misner, J. E., Plowman, S. A., and Boileau, R. A. (1987). Performance difference between males and females on simulated fire-fighting tasks. *Journal of Occupational Medicine*, 29, 801-805.

Mital, A. and Ayoub, M. M. (1980). Modeling of isometric strengths and lifting capacity. *Human Factors*, 22, 285-290.

Mostardi, R., Noe, D., Kovacik, M., and Porterfield, J. (1992). Isokinetic lifting strength and occupational injury. *Spine*, 17, 189-193.

Nachemson, A. L. and Lindh, M. (1969). Measurement and abdominal and back muscle strength with and without low back pain. *Scandinavian Journal of Rehabilitation Medicine*, 1, 60-65.

Nordgren, B., Schele, R., and Linroth, K. (1980). Evaluation and prediction of back pain during military service. *Scandinavian Journal of Rehabilitation Medicine*, 12, 1-8.

Pedersen, D. M., Clark, J. A., Johns, R. E., White, G. L., and Hoffman, S. (1989). Quantitative muscle strength testing: a comparison of job strength requirements and actual worker strength among military technicians. *Military Medicine*, 154, 14-18.

Pytel, T. L. and Kamon, E. (1981). Dynamic strength test as a predictor for maximal and acceptable lifting. *Ergonomics*, 24, 663-672.

Reilly, R. R., Zedeck, S., and Tenopyr, M. L. (1979). Validity and fairness of physical ability tests for predicting performance in craft jobs. *Journal of Applied Psychology*, 64(3), 262-274.

Reimer, D. S., Halbrook, B. D., Dreyfuss, P. H., and Tibiletti, C. (1994). A novel approach to preemployment worker fitness evaluations in a material-handling industry. *Spine*, 19(18), 2026-2032.

Rice, V. J. and Sharp, M. A. (1994). Prediction of performance on two stretcher-carry tasks. *Work: A Journal of Prevention Assessment and Rehabilitation*, 4(3), 201-210.

Rice, V. J., Sharp, M. A., Nindl, B., and Bills, R. (1995). Prediction of two-person team lifting capability. *Proceedings of the Human Factors and Ergonomics Society 39th Annual Meeting* (pp. 645-649). Santa Monica, CA: Human Factors and Ergonomics Society.

Robertson, D. W. (1982). *Development of an Occupational Strength Test Battery (STB).* (AD-A114 247). San Diego: CA, Navy Personnel Research and Development Center.

Robertson, D. W., and Trent, T. (1985). *Documentation of Muscularly Demanding Job Tasks and Validation of an Occupational Strength Test Battery (STB).* (MPL TN 86-1). San Diego: CA, Navy Personnel Research and Development Center.

Robertson, D. W. (1992). Development of job performance standards for muscularly demanding military tasks, in Kumar S. (Ed.), *Advances in Industrial Ergonomics and Safety IV* (pp. 1299-1304). Washington, D.C.: Taylor and Francis.

Rowe, L. M. (1969). Low back pain in industry. *Journal of Occupational Medicine*, 11, 161-169.

Rowe, M. L. (1971). Low back disability in industry: Updated position. *Journal of Occupational Medicine*, 13, 476-478.

Schonfeld, B. R., Doerr, D. F., and Convertino, V. A. (1990). An occupational performance test validation program for fire-fighters at the Kennedy Space Center. *Journal of Occupational Medicine*, 32(7), 638-643.

Sharkey, B. J. (1981). Fitness for wildland fire fighters. *The Physician and Sports Medicine*, 9(4), 93-102.

Sharp, D. S., Wright, J. E., Vogel, J. A., Patton, J. F., Daniels, W. L., Knapik, J., and Kowal, D. M. (1980). *Screening for Physical Capacity in the U.S. Army: An Analysis of Measures Predictive of Strength.* (T8/80). Natick, MA: Exercise Physiology Division, U.S. Army Research Institute of Environmental Medicine.

Snook, S. H., Campanelli, R. A., and Hart, J. W. (1978). A study of three preventive approaches to low back injury. *Journal of Occupational Medicine*, 20, 478-481.

Stalhammar, H. R., Leskinen, T. P. J., and Nurmi, P. A. (1992). Psychophysical tests, isometric and dynamic muscle force measurements as determinants of aircraft loaders' functional capacity, in S. Kumar (Ed.), *Advances in Industrial Ergonomics and safety IV* (pp. 683-691). Washington, D.C.: Taylor and Francis.

Teves, M. A., Wright, J. E., and Vogel, J. A. (1985). *Performance on Selected Candidate Screening Test Procedures Before and After Army Basic and Advanced Individual Training.* (T13/85). Natick, MA: Exercise Physiology Division, U.S. Army Research Institute of Environmental Medicine.

Troup, J., Foreman, T., Baxter, C., and Brown, D. (1987). The perception of back pain and the role of psychophysical tests of lifting capacity. *Spine*, 7, 645-657.

Venning, P. J., Walter, S. D., and Stitt, L. W. (1987). Personal and job-related factors as determinants of incidence of back injuries among nursing personnel. *Journal of Occupational Medicine*, 29(10), 820-825.

Wickstrom, G. (1978). Effect of work on degenerative back disease: A review. *Scandinavian Journal of Work Environment and Health*, 4(Suppl. 1), 1-12.

Wigdor, A. K. and Green, B. F. (1991). *Performance Assessment for the Workplace.* Washington, D.C.: National Academy Press.

Wilmore, J. H. and Davis, J. A. (1979). Validation of a physical abilities field test for the selection of state traffic officers. *Journal of Occupational Medicine*, 21(1), 33-40.

TABLE 71.1 Prediction of Performance

		Initial Assessment and Tests			Relationships/Predictions		
Study	Subjects	Job Analysis	Work Simulation	Tests	Work Simulation	On-the-Job Performance	Results
Anderson, 1988	613	not described		strength-arm pull, aerobic step test		productivity	Chi²: $p < 0.05$ for drivers, $p < 0.02$ for selectors
Anderson and Catterall, 1989	♂518	critical tasks: onsite analysis, measurements, posture analysis		pull-up, low bar pull, step test		probability of staying on job 8 wks, productivity levels	Chi²: $p < 0.05$. Lower pass rates more likely to quit by 8 wks, higher pass scores had higher avg. productivity at week 8
Arnold, Rauschenberger, Soubel, and Guion, 1982	♂168 ♀81	Interview, task quantification		arm, back, and leg strength; balance; leg lifts; push-ups; squat thrusts; pull-ups; flexibility; and a step test	composite score: move/carry, lift, wheelbarrow, and shovel	observer ratings	regression: arm dynamom. → 67 – 72% variance composite score
Bernauer and Bonanno, 1975	241 pole climber applicant	not mentioned		40 item test battery			
	300 applicant ♂♀			body fat, body weight, grip, sit-up, recovery HR reaction time, beam walking		successful completion of climbing school	difference in mean scores between success/no-success applicants w/step test & balance ($p < 0.05$)
Celentano, Nottrodt, and Saunders, 1984	♂23 ♀18	interviews, task quantification		static & dynamic strength, aerobic endurance, anthropometry	breech-block trade task		regression: body weight & lifting → 76% var. body weight, lifting & gender → 92% variance
Davis, Dotson, and Santa Maria, 1982	100 fire-fighters (♂?)	not mentioned		anthropometrics, grip strength, sit-ups, chin-ups, standing long jump, push-ups, hamstring flexibility, physiological measures	ladder extension & retraction, carry, hose pull, rescue, forcible entry		Laboratory strength/power → 90% variance Non-laboratory → 54% variance Laboratory fatigue resist. → 80% variance Non-laboratory → 60% variance

Doolittle, Spurlin, Kaiyala, and Sovern, 1988	48 line-workers (♂?)	interviews, observation, movement classification, forces & masses, pace & frequency, energy cost via heart rate	modified biceps curl, lateral rise, row, lat pull, chin-ups, submaximal aerobic step test		performance ratings on productivity, work w/others, supervision, safety, physical ability, technical	correlation: composite score → overall perform rating 35% variance
Hogan, 1985	♂46	not mentioned	23 fitness measures and 5 physical fitness batteries		Successful completion of diving training, supervisor rating scales	regression: cardiovascular fitness, muscular endurance, and flexibility → program completion 40% variance & → overall potential 41% variance
Jackson, Osburn, and Laughery, 1984	♂25 ♀25	conducted, process not described	isometric strength:grip strength, arm and back lift, arm endurance (arm VO^2 max)	roof bolting, bag lift & carry, block carry, shoveling		regression values not reportedmult. correlation: composite stren. → bag carry 41% variance, → block carry 69% variance, → shoveling 56% variance composite stren. + arm endur → shovel 64% variance
	♂25 ♀25	conducted, process not described	isometric strength:grip strength, arm and back lift, arm endurance (arm VO^2 max)	pipe transport		composite stren. → pipe transport ♂77-87% var. ♀46-54% var.
Jackson, Osburn, Laughery, and Vaubel 1991	♂118 ♀66	conducted, process not described	isometric grip, arm lift, and torso lift summed	tool lift, scaffold building, railroad tank car cap replacement, one and two arm tool hold. Composite strength score predictive of work simulations		ANOVA: $p < 0.01$ isometric str. higher for those passing work sample testregression: equations given, probability of passing work sample given for each strength category
Jackson, Osburn, and Laughery, 1991b (U. Houston/Rice)	♂28 ♀28	conducted, process not described	isometric grip, arm lift, torso lift, and arm endurance (arm VO^2 max), body weight (fat, lean)	shoveling, bag lift & carry		regression: R^2 not reportedmult. correlation:composite strength score → shovel 56% variance, comp str + arm endurance → shovel 56% variance

TABLE 71.1 (continued) Prediction of Performance

		Initial Assessment and Tests			Relationships/Predictions		
Study	Subjects	Job Analysis	Work Simulation	Tests	Work Simulation	On-the-Job Performance	Results
Laughery, Jackson, and Fontenelle, 1988	♂25 ♀25	interviews, observations, physical measurements		isometric strength tests: grip arm lift, back lift, arm press.	transporting a 15.9 kg box up and down a stair, also a 22.7 kg box		mult. correlation: composite score with work power (kg/m/min) 22.7kg/15.9kg both genders → 41/41% var. ♂= 18/12% var ♀= 49/42% var
Oseen, Singh, Chahal, Lee, Reid, Couture, and Wheeler, 1992	♀45			static and dynamic strength, anthropometrics, body fat, and body image	casualty evacuation; jerry can lift, carry, and empty; box lift, a trench dig, weighted load march		correlation(r = ?, $p < 0.05$)
Reilly, Zedeck, Tenopyr, 1979	♂83 ♀45	interviews, task quantification		static (grip, arm) and dynamic (trunk, arm) strength, reaction time, balance, aerobic step-test, flexibility, body density	overall total performance (OTP): pole testing, pole climbing (2), ladder placing (2), pole climb with drop wire removal	6 months job completion (retention), supervisor job ratings ($r = 0.36$)	regression: dynamic arm strength → OTP 12% variance; dynamic stren. → retention 13%; dynamic arm strength, reaction time, step test → OTP 17% variance
	♂132 ♀78	interviews, task quantification		static (grip, arm) and dynamic (trunk, arm) strength, reaction time, balance, aerobic step-test, flexibility, body density	time to complete training task (TTC), supervisor rating, training completion, tasks w/o errors, accident in 1st 6 months	6 months job completion (retention)	regression: body density, balance, static stren. (PAT)→ TTC 20% var.; → train. dropout 14% var, → tasks w/o errors 13%, → OTP 10%, → retention 11%, → accidents 2% var
Rice and Sharp, 1994	♂12 ♀11	process not described		7 strength tests, 3 fitness scores (sit-up, push-up, run), 3 physical descriptors	repeated stretcher carry — mass casualty, continuous stretcher carry — remote site		regression: grip + run → repeated carry 74% var for harness & 78% var for hand, grip → continuous hand carry 74% var, bench press → cont. harness carry 33% variance
Robertson, 1992	♂274 ♀259	criterion tasks, survey, interview, observation, measurement (force, grip points, distance, direction)		anthropometrics, static, dynamic, and power strength measures	carrying, lifting, and pulling		regression: separate gender groupings validity coefficient = 0.30's — 0.50's, combined gender 0.60's — 0.80's

Robertson and Trent, 1985	♂274 ♀259	survey, observation, measurement	anthropometrics, static, dynamic, and power strength measures	carrying, lifting, pushing and pulling	predictive validity for separate gender groupings ranged from 0.3's to 0.50s, combined ranged 0.60's — 0.80's. Greatest single predictor = arm pull
Schonfeld, Doerr, Convertino, 1990	♂20		treadmill VO² max, isokinetic knee flexion and extension, power arm and leg tests, grip strength, push-ups, sit-ups, sit-and reach, body composition	stairclimb, chopping simulation, and victim drag	regression: most powerful predictor = treadmill time. Various test combinations → simulations 2–91% var.
Sharkey, 1981	121	process not described	body fat, chin-ups, sit-ups in 60 sec, push-ups in 60 sec, and a 50 lb pack test	hiking, building a fireline, deploying & moving hoses, shoveling & throwing sand, carrying	Combined tests → 71% variance Muscular fitness + experience → 59% variance
Stalhammar, Leskinen, and Nurmi, 1992	♂103	not mentioned	anthropomentrics, skinfold body fat, hand grip (static, dynamic, endurance), max. voluntary contraction of trunk flexor/extensor, range of motion	- rating of acceptable lifting load for an 8 h day - rating of acceptable lifting load for simulated postal parcel sorting	correlation & regression - left lateral bend, dynamic grip endurance, and back stren/wt ratio → 63% variance - dynamic grip endurance, back stren/wt ratio, and trunk flexor stren/wt ratio → 44% variance
Wilmore And Davis, 1979	♂217 ♀13	critical job task analysis	laboratory: strength (bench press, leg press, two arm curl and grip), flexibility, body comp., treadmill max-aerobic.field: strength (bench press, grip), flexibility, power (vertical jump), run time, predicted body composition	barrier surmount; dummy drag from car	correlation & regression: laboratory test → barrier surmount & dummy drag (46 & 44% var).Field test → barrier surmount & dummy drag (38 & 32%).

→ = predicted

TABLE 71.2 Prediction of Injury

		Initial Assessment and Tests			Relationships/Predictions		
Study	Subjects	Job Analysis	Work Simulation	Tests	Work Simulation	Work-Related Injury	Results
Anderson, 1988	613 total	not described		strength-arm pull, aerobic step test		contact, back, and over exertion injury reports	Chi²: $p < 0.05$ for 1 job for back, over-exertion injuries, 2 jobs for all injuries reported
Battié et al., 1989	♂1726 ♀452	not mentioned		Isometric torso, arm, and leg lift strength		reported back problems (pain)	correlation: back injury not related to strength
Chaffin, Herrin, and Keyserling, 1978	900 jobs ♂446 ♀105	biomechanical evaluation, hand placement during heavy strength demands	hand placement during lifting (job position test)	Torso, arm, leg, and job position lifting tests		contact, back, musculoskeletal injuries	As strength requirement approaches/exceeds strength, incident and severity rates increase 3:1 Statistical sig. not reported
Chaffin and Park, 1973	411 total ♂& ♀ 103 jobs	most stressful lifting tasks (weight, frequency)		age, weight, stature, previous back injuries isometric lift	lifting strength rating (max load required ÷ strength of large/strong man	low back incidence rates	higher strength requirements, very low and very high frequency rates, previous low back pain, and low lift ratio related to higher incidence rates. statistical sig. not reported
Doolittle, Spurlin, Kaiyala, and Sovern, 1988	48line-workers (♂?)	interviews, observation, movement classification, forces & masses, pace & frequency, energy cost via heart rate		modified biceps curl, lateral rise, row, lat pull, chin-ups, submaximal aerobic step test		lost duty days	multiple correlation: composite score → lost days (21% of variance)
Griffin and Troup, 1984	♂234 ♀116			anthropometry, psychophysical lifting capacity, weight estimation, handling accuracy, isometric lifting strength, trunk strength			Diff. in mean scores between those with and those without back pain. Both genders: acceptable lift, accep lift/body wt. Chi² for situps for men only

Karwowski, Caldwell, and Gaddie, 1994	not reported			NIOSH lifting index, compressive and shear forces on lumbosacral joint		low back injury incidence rate	multiple correlation: (86% of variance)
Keyserling, Herrin, and Chaffin, 1980	♂54 ♀27	biomechanical job analysis, observation by subject experts, measurements		task based requirements for strength: arm, back, push, pull		control group did not have to pass cutoff score (actual job force requirements), experimental group did	Chi²: marginal significance ($p < 0.1$) with all medical visits, not for musculo-skeletal visits
Keyserling, Herrin, Chaffin, Armstrong, and Foss, 1980	♂309 ♀35	biomechanical job analysis, observation, measurements		9 strength tests: lift, push, pull, arm, back		all medical visits, musculoskeletal visits	Chi²: weaker group had between 1.25 and 2.71 times the incidence rate for all visits. Between 1.6 and 3.1 for musculoskeletal visits ($p < 0.05$)
Reilly, Zedeck, Tenopyr, 1979	♂83 ♀45 ♂132 ♀78	interviews, task quantification	six tests of pole testing, pole climbing, and ladder placing	static and dynamic strength, reaction time, balance, aerobic step-test, flexibility	task perform ($r = 0.45$), time-to-complete a training task ($r = 0.45$), training completion ($r = 0.38$)	6 months job completion ($r = 0.36$), supervisor job ratings ($r = 0.36$)	regression: number of accidents ($r = -0.15$)
Reimer, Halbrook, Dreyfuss, and Tibiletti, 1994	♂33 control ♀1001 workers	onsite job analysis (observation, measurement)		trunk flex/ext, body wt, isokinetic lifting, range of motion		pre- and post-implementation low back injury rates - total injury rates	yr 1 = 32% decreased injury rate yr 2 = 41% decr injury rate yr 1 = 13% yr 2 = 21% yr 3 = 23% yr 4 (projected) = 37%

72

Assessment of Worker Functional Capacities

Glenda L. Key
KEY Method

72.1 Introduction

There are nearly 500,000 U.S. workers each year who, as a result of injury, are unable to resume their jobs for long periods of time. Low back pain will afflict roughly 80% of the population at some time in their lives (Wheeler and Hanley, 1995). Workers' compensation costs have reached $70 billion per year, tripling since 1980. The direct cost of the situation is staggering and in each case the degree of loss is unpredictable.

The knowledge of what a worker's functional capabilities are, is one of the most valuable pieces of information that a professional can have for reducing workers' compensation costs and prevention of injury in the workplace today. There are two occasions in the employment of a worker when it is critical to know just what the physical work capability of that worker is. One occurs during the decision to hire, and the other is upon return to work of an injured employee. The focus of this chapter will be on these two occasions as highlighted in Figure 72.1, the Worker Care Spectrum representing assessment of worker functional capabilities.

72.2 Worker Assessments

Definitions

Functional Capacity Assessment

The functional capacity assessment (FCA) is a return-to-work testing process that determines an individual's physical functional work-related capabilities through measuring, recording, and analyzing data gathered during a standardized physical testing procedure. There are many different formats of functional

0-8493-2641-9/99/$0.00+$.50

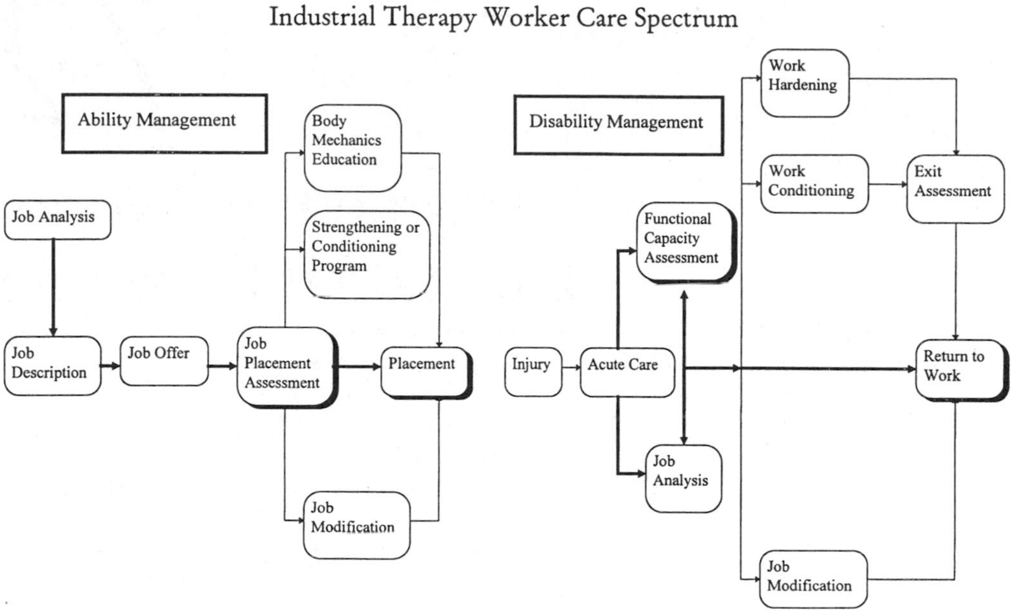

FIGURE 72.1

capacity assessments and almost as many names for them. The term used may be physical capacity assessment, functional capacity evaluation, work capacity evaluation, or work assessment. For the purposes of this chapter, the term functional capacity assessment is used.

The activities of an FCA include materials handling functions of lifting, carrying, pushing, and pulling. It also includes tolerances of posture such as standing, sitting, kneeling, crawling, and activities of walking and stair climbing (Key, 1995a).

Simulation of the movements and postures during the weighted activities and standardizing for consistency of instruction, weight loading protocols, and determining termination points are primary issues that need to be considered in a functional capacity assessment. An FCA allows for the testing of the individual to confirm meeting or not meeting the minimum physical requirements of the job.

As quickly as possible, the workers should be returned to their jobs and workplaces — even if they are unable to be fully productive. Athletes do not wait until they are fully recovered before they start back in practice or training. They work their bodies to the point where they still maintain safety and will not reinjure themselves. Their capabilities are built up gradually until they are back in their "game." It does not mean that they play without pain. It does not mean that they can do everything as well as or in the same manner as before. This approach applies to the injured worker as it does to the athlete.

Job Placement Assessment

The job placement assessment (JPA) provides data prior to hiring that assists managers in making decisions resulting in significantly decreased incident rates in the future. When armed with the information of the requirements of the job as provided through a job analysis and job description, and the physical abilities of the individual as provided through a job placement assessment, it is relatively easy to determine whether persons should be hired into a specific job or not. If they meet the requirements of the job, the answer is "Yes." If they do *not* meet the requirements of the job, the answer is "No."

"The job placement assessment is a series of specific, objective, and standardized protocols followed in a consistent progression to allow for objective, accurate, and repeatable results" (Frey, 1995). These results identify the prospective employee's work capabilities in the areas of lifting, carrying, pushing, pulling, and other job-specific activities.

The continuous and significant increase cost of health care and workers' compensation has added to the already inherent need to have healthy, productive employees. This process begins with hiring individuals who are able to physically perform the requirements of the job. How can employers tell if their next hire will be their best employee or become their worst workers' compensation and litigation nightmare?

The employee has the most to gain and lose if decisions are made based on anything other than accurate data. At the time of hire, the prospective employee has the right to be judged through a nondiscriminatory process. Generally, his or her personal need is to be employed and be a productive part of society. The expectation is that the hiring process will be fair and that the job activity will not be a precursor of an injury. Once injured, whether on the job or not, the worker relies on the medical system and those individuals carrying out company policies to establish the return-to-work process. The desire to be productive, the fear of reinjury, the unfamiliar role of patient, and the acceptability of not working can all be contributors to the complex course of action. It is of substantial importance that the match of the worker to the work be accurate to prevent any reinjury yet place the worker in the highest level of productivity possible — for economic and social reasons.

At these two points in the Worker Care Spectrum, the employee is especially vulnerable to the decisions of others. Employees often report that their experience in a worker assessment, JPA or FCA, is the first time that decisions are being made based on their actual participation, with themselves and their activities in control of the end points. To have their capabilities objectively tested in a participatory manner is important to them.

Components

Recommendations of an individual's physical work ability are made based on assessing components in the following three categories.

1. Weighted capabilities
2. Tolerance and endurance parameters
3. Validity of participation determinants

Weighted Capabilities

These tests include analyzing an individual's ability to perform specific work-related materials handling functions such as lifting at three standard heights, carrying, pushing, and pulling. A detailed list of components is found in Figure 72.2. Cardiac responses, kinesiological changes and repetition of posture and posture changes are documented and included in the formulas of decision of the individual's capabilities. This further confirms cardiovascular and body mechanics safety of the employee or prospective employee at that level of physical activity. Important questions are answered through these thorough testing techniques. For example, can the employee lift the 48 lbs. that are required for the job? How can one know unless the worker is tested?

Tolerances and Endurance

Testing includes and recommendations are given in the categories of standing tolerance, sitting tolerance, and workday tolerance. Kneeling, crawling, and activities such as walking and stair climbing are also covered. A detailed list of components is found in Figure 72.2. For example, can your employee return to an eight hour workday or is he or she limited to a lesser amount? And if so, what is the safe limitation of a workday?

Validity of Participation Determinants

It is important to objectively, scientifically, and statistically determine and demonstrate when individuals are not being forthright in the representation of their capabilities (Grossman, 1985; Osterweis et al., 1987). It is known that some individuals will demonstrate less than their full capability. They may be attempting to prolong their time off work or may be attempting to falsely maximize their case closure settlement.

Weighted Capabilities

Lifting desk to chair 30" to 18"	Lifting chair to floor 18" to 0"
Carrying	Lifting above shoulder 30" to 60"
Pulling	Pushing

Tolerances and Endurance

Balancing	Bending
Cervical mobility	Circuit board tolerance
Climbing	Grip strength
Crouching	Crawling
Fine manipulation	Fastener board tolerance
Keyboard tolerance	Firm grasping
Simple grasping	Kneeling
Repetitive foot motion	Reaching
Sitting	Simple grasping
Standing	Squatting
Tool station work tolerance	Stooping
Walking	Work day tolerances

FIGURE 72.2

The percentage of people participating in this kind of exaggeration and falsification is less than the public generally perceives it to be. Figure 72.3 presents the percentages in each category of participation determinant based on a sample of 43,000 injured clients (Worker Data Bank, 1994–1997).

It is important to the valid participators to have their honest attempts proven to be full effort demonstrations. The injustice is not only when the conscious and deceptive low participator is not revealed, but also when the honest, full effort, low capability level participator is misjudged.

Validity of participation determinants have been developed by one vendor of FCA equipment and protocols (Gilliand et al., 1986; Personnel Decisions, Inc., 1994). The individual's level of participation is determined through the use of algorithms and decision science objectively identifying the degree of an individual's consistency or inconsistency. The formula includes data from a database made up of over 100,000 cases of this standardized, protocolized, FCA (Worker Data Bank, 1994–1997). These data have provided valuable correlations, one of which is between the amount someone can lift at specific heights and how much they can push and pull. Scientific literature and the same database have provided statistics on certain vital signs, such as heart rate, which respond predictively as the individual approaches and reaches exertion level (Astrand and Rhyming, 1954; Gilliand et al., 1986; Worker Data Bank, 1994–1997).

The delineations of performance levels as introduced by KEY Method in Minneapolis, Minnesota are (Gilliand et al., 1986; Key, 1984; Worker Data Bank, 1994–1997):

1. Valid participation. This determinant means that the individual participated with full effort and the results and recommendations reflect safe capabilities.
2. Invalid participation. This as a determinant means that the participant consciously and intentionally provided less than full effort. The numbers produced by this individual reflect less than full capability. The results can be used in addressing the issue directly with the participant or, if necessary, in litigation.
3. Conditionally valid participation. An individual with this determinant has demonstrated less than full capability. The results reflect their *perception* of capability even though they can physically do more. The numbers and capabilities are, therefore, safe, but may also reflect an unconscious psychological barrier.
4. Conditionally invalid participation. The individual with this determinant has demonstrated work levels that are beyond what would be considered full, safe levels for extended work periods. When left to their own end point determinations of work activity, they exceed the safe levels.

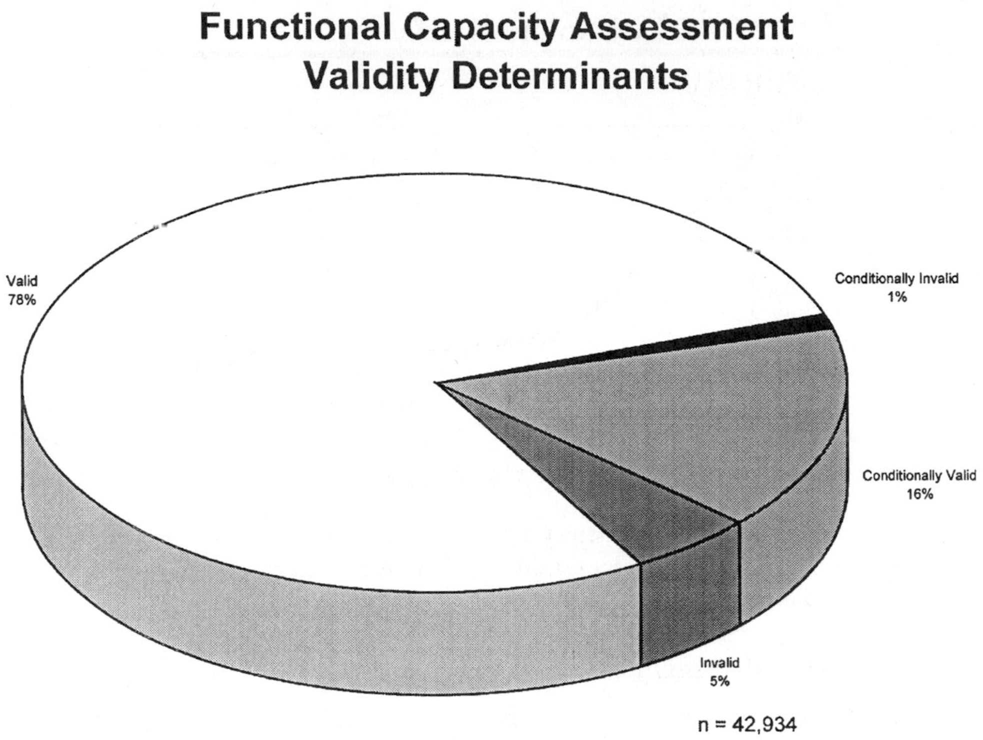

FIGURE 72.3

TABLE 72.1 Consistency of KEY Method Assessment Results

	Validity by Percent in Each United States Territory			
	West	Central	North East	South East
Valid	78	78	78.5	76
Invalid	5	3	5	5
Conditionally Valid	15	17	15	17
Conditionally Invalid	1	2	1.5	1

Freidman two way analysis of variance by ranks. Region by participation (%)
$k = 4$, $n = 4$; $X^2 = 7.5\ X_r^2 + .9$

The importance of standardization, statistical analysis of data, and carrying the level of data in a database to produce determinants of participation is paramount.

Standardization

When looking for an assessment, standardization and validity of participation determinants are of consummate importance. They are the primary elements of defense against the occurrence of litigation and the primary elements of defense should the case eventually go to court. One of the standardized systems maintains a complete data bank of all assessments performed (Worker Data Bank, 1994–1997).

To be assured of such level of standardization, studies should be provided to those using the assessment that demonstrate reliability and consistency. The consistency of the KEY Method assessment results was analyzed and is represented in Table 72.1. Analysis using 2-way analysis by ranks found "no differences" in providers' results across the United States. Statistically, any differences were found to be *extremely not significant* (Aitken, 1996; Portney and Watkins, 1993, Worker Data Bank, 1994–1997).

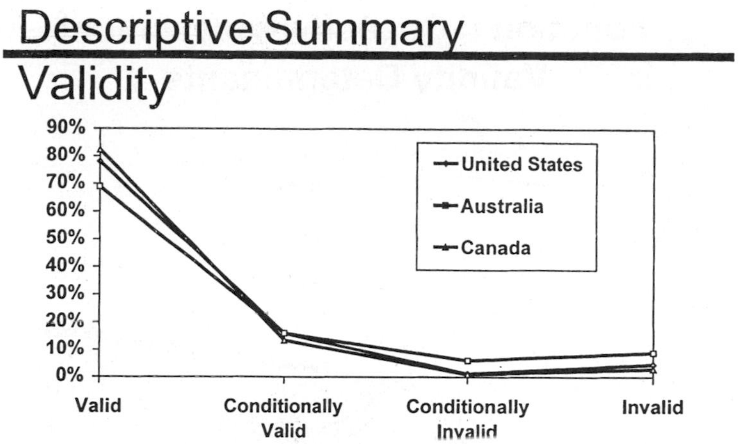

FIGURE 72.4

Another analysis demonstrating consistency in the assessment results, representing standardization, is the graph in Figure 72.4. This confirms the consistency of the results of the assessment when administered in three different countries: the United States, Canada, and Australia.

Worker Assessment Principles

Principles to look for in selecting a functional capacity assessment include:

1. The assessment must contain standards for identifying validity of participation of the individual being tested
2. The methodology administered must be consistent from tester to tester and test to test
3. Standardized equipment must be used and the same procedure followed with each assessment
4. The administrator of the assessment must be thoroughly trained and objective
5. The processing of the information gathered during testing must be standardized

The psychology of the individual needs to recognized in the assessment as well as the kinesiology of the activities (Schmidt et al., 1989). There is a growing body of literature supporting the theory that a statistical relationship does exist between low back pain and an individual's psychological factors as tested by the Minnesota Multiphasic Personality Inventory (MMPI) (Block et al., 1996). The MMPI has also been able to predict the occurrence of job-related low back pain or when poor response to surgery would be the outcome (Bigos et al., 1991; Blair et al., 1994; Block et al., 1996; Keller and Butcher, 1991; Schmidt et al., 1989).

Assessment Reports

A report should provide details of the results of the assessment and then compare those results with the physical demands of the job. Through this matching exercise, the decision maker can make an informed, unbiased, and defensible decision relating to the return to work of a previously injured employee.

It is also helpful to receive a visual display of the major highlights of the results, as in Figure 72.5. Graphic displays of these major areas can be most helpful in assisting the decision-making process for rehire or case management. The individual's assessment results are compared with the same profile of individual in the data bank. They also may be compared with the job requirements. Another option is to compare norms from the injured database with similar profiles of individuals in the uninjured database.

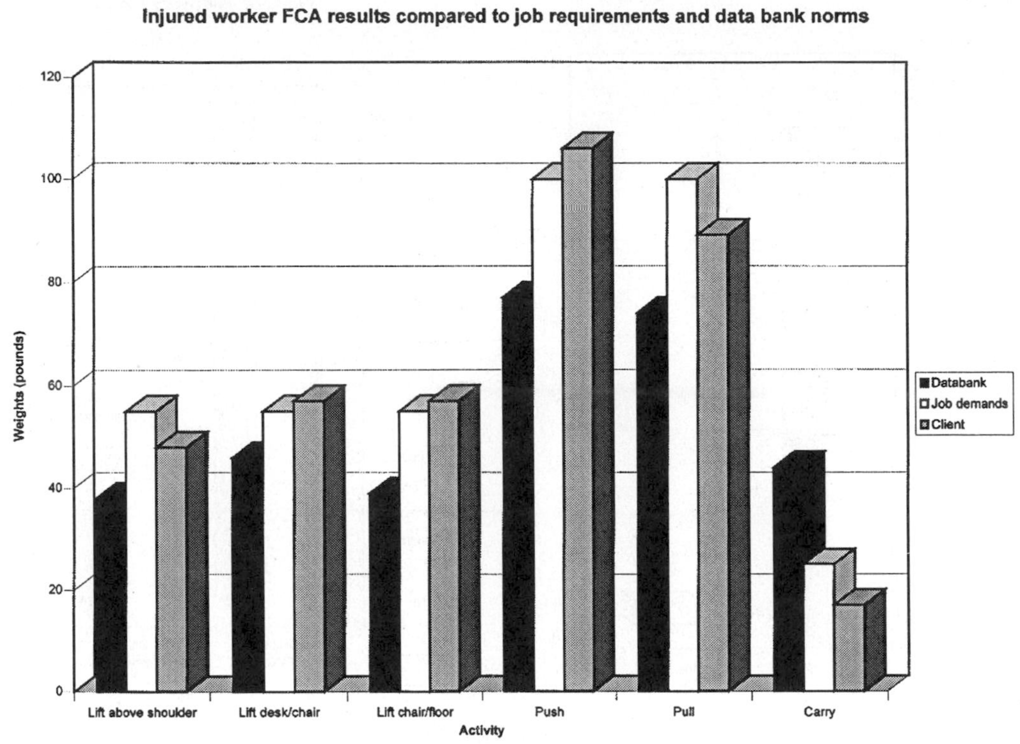

FIGURE 72.5

72.3 The Provider of Worker Assessments

The provider of these assessments can be instrumental in increasing return-to-work percentages, reducing reinjury rates, decreasing lost workdays, and yielding other short and long-term cost reductions.

Physical and occupational therapists are the primary providers of worker assessments, job placement assessments, and return-to-work functional capacity assessments. The motive for providing quality results in functional assessments lies primarily in the accuracy of the results and the defensibility in the courts. The therapist needs to provide the most objective, unbiased data. These data are then used as an assist in the hiring and returning-to-work decisions. As a result, therapists need to be able to support their method with objectivity, standardized equipment, and consistent protocols and procedures.

72.4 Outcomes

Outcome surveys or studies should be available for review for one to trust the results and the legitimacy of the recommendations of a worker assessment (Dobrzykowski, 1995; Key, 1995b). The predictive ability of an FCA needs to be demonstrated based on a track record of return-to-work without reinjury.

The primary outcomes that one should be looking for include:

1. Decreased reinjury rates
2. Decrease lapse of time from date of injury to date of return to work
3. Decrease incidents and cost of litigation

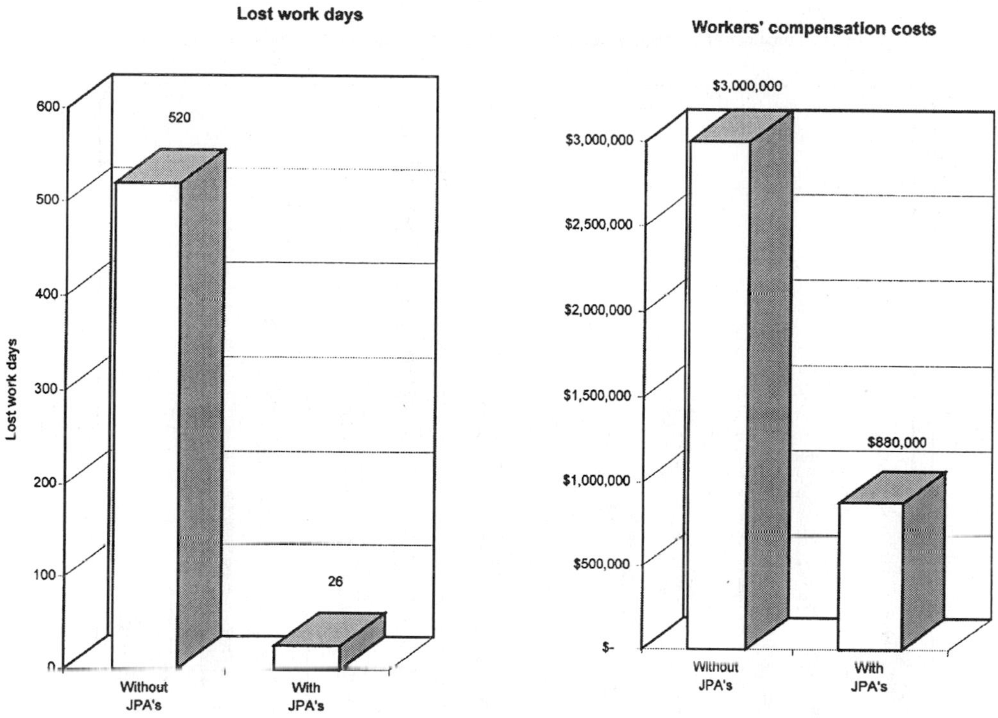

FIGURE 72.6

The State of Colorado studied the impact of FCAs on shortening the amount of time required for vocational evaluation of injured workers. H. D. Waite found that the inclusion of the KEY Method FCA resulted in a median of 18 fewer days of rehabilitation. This would save the State of Colorado over $200,000 annually (Waite, 1987).

JPA Outcomes

Having the data and using them to safely match workers' capability levels to the job demands, or being in a position to know the modifications necessary to facilitate a match results in fewer injuries and subsequent lower health care costs, lower workers' compensation complications, preserved productivity, worker morale, and job satisfaction. These issues along with that of the assessments' predictive capabilities regarding future injuries are the primary issues in outcome reports. Examples assist in demonstrating the effects (Karwowski and Salvendy, 1998).

Outcome — Case A

In 1988 a paper manufacturer in Minnesota instituted job placement assessments to help stem the costs related to workers' compensation claims and lost workdays. In analysis two years later, the experiences of 70 employees hired before the use of JPAs were compared to 70 hired after the initiation of administering JPAs. Use of the JPAs lowered both lost workdays and workers' compensation costs (see Figure 72.6) (Frey, 1995).

Outcome — Case B

A state transportation authority discovered that job placement assessments enable them to predict if an employee is at risk of injury. Of the 36 employees injured from 1985 to 1992, 75% had been categorized as "at risk" by JPAs performed when they were hired. Fourteen providers across the state administered the same system of JPA. Personnel Decisions, Inc. reports, "While the analysis is based on a relatively few

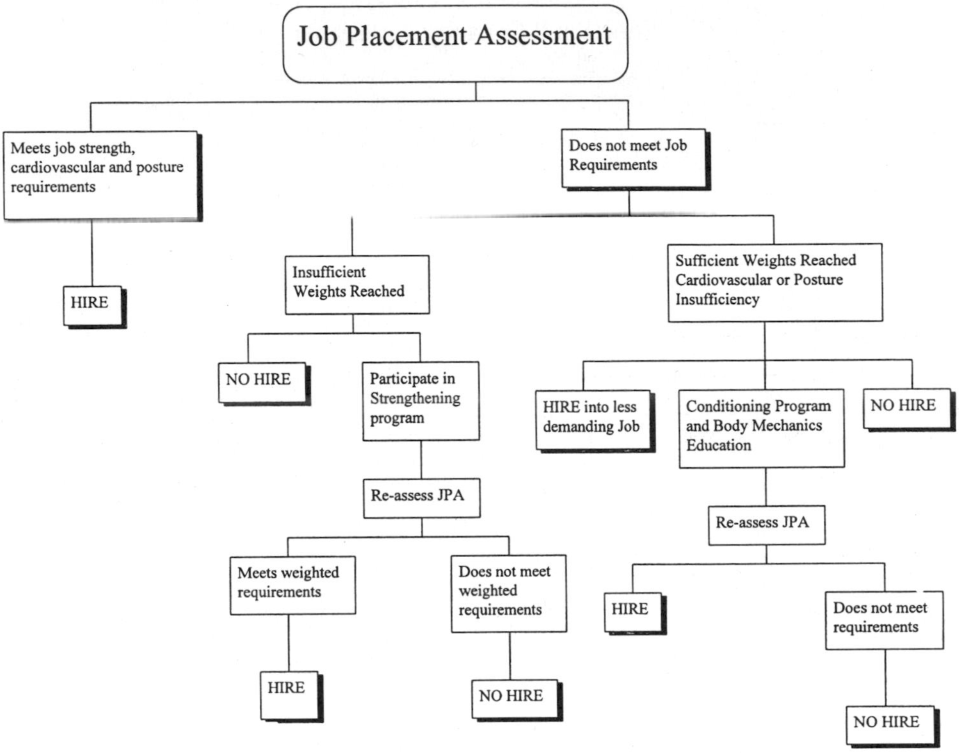

FIGURE 72.7

cases, the results are statistically significant. The chi square for this cross-tabulation is 15.4, which is significant at p = .00045" (Personnel Decisions, Inc., 1994).

Outcome — Case C

A major trucking firm used the JPA for two separate hiring locations over a continuous eighteen-month period. Since implementing the administration of the JPA on each candidate and hiring based on the results, there were no new injuries (Worker Data Bank, 1994–1997).

72.5 After the Assessment Results Are In

The results of job placement assessments and functional capacity assessments offer options for pathways to follow, depending on the data assembled. The decision tree layout of Figures 72.7 and 72.8 demonstrate optional pathways available.

72.6 Diversification Options

As change occurs in industry so will industry's approach to minimizing work-related injuries. To meet these changing needs medical providers are diversifying the services they now offer to industry and the mode of delivery of those services. Today's marketplace offers some examples of successful diversification.

Mobile Assessments

For those who offer functional capacity assessments or job placement assessments, a beneficial adaptation has been to perform these assessments at the job site rather than in a clinic. This is done by utilizing a

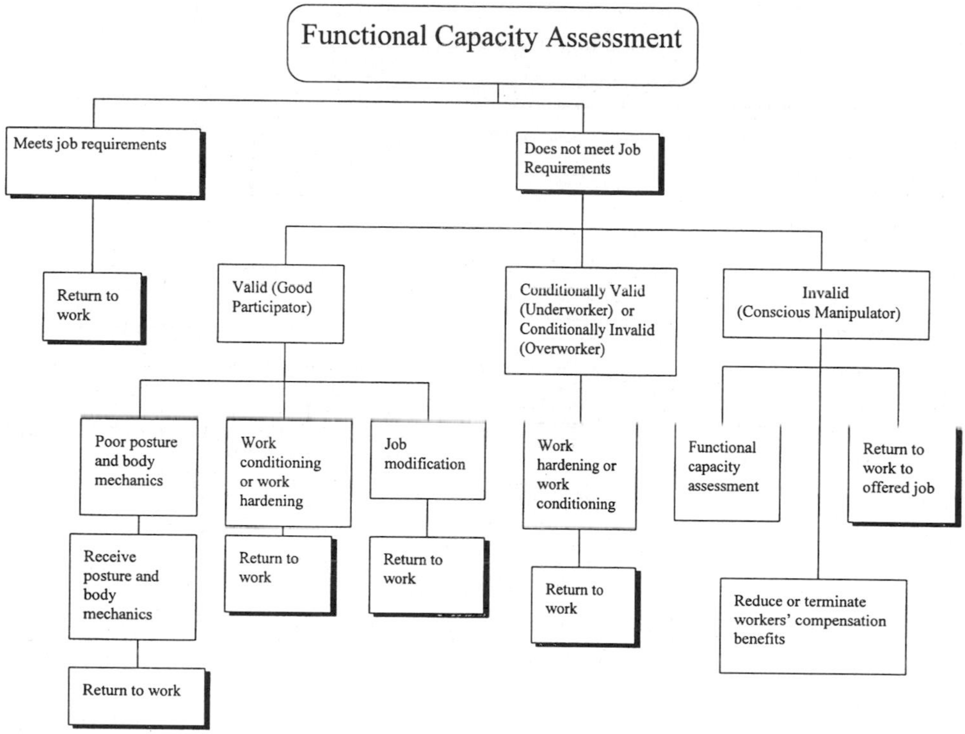

FIGURE 72.8

mobile testing unit that uses the same standardized test procedures as those which are practiced in the clinical setting. By having the JPAs performed onsite, the individual responsible for the "hire/no-hire" decision has immediate access to the assessment results. Because of this, no administrative or production time is lost waiting for reports. They can be printed up immediately upon completion of the JPA. With today's tight economic issues and tight worker market it is important that the hire be made quickly. Onsite JPAs provide a system by which many employees can test through and begin work the same day.

Having the return-to-work assessment, the FCAs, performed onsite promotes the bond between the injured worker and the workplace. It assists in the process of employee acceptance, co-worker to co-worker. Management and nonmanagement observe that the individual is indeed participating in the return-to-work program. Bringing the worker back to the site for testing also allows the employee to reunite with other colleagues and co-workers.

Additionally, therapists are able to develop a closer working relationship with management and thereby better understand and meet the company's needs.

Mobile Occupational Health Clinic

Taking this idea a step further, some have adapted large vans or buses to house a mobile occupational health clinic. This includes facilities for examinations and physicals, functional capacity assessments, job placement assessments, job analysis, ongoing treatment, and educational materials, as well as facilities for drug, hearing, and vision screenings. This mobility presents industry with a broad offering for delivery of services. One such mobile unit travels up to 200 miles to deliver such services to hundreds of industries in a rural region.

Onsite Services

Onsite industrial therapist services are becoming a very cost-effective way for industries to more thoroughly meet the needs of their company. When the provider of services shows up for work every day at

the plant site, there is a deepened knowledge level of the problems and a quicker pathway to the solutions. Services onsite may be delivered by contracting with a local clinic or therapist. It may be to the company's advantage to have the individual or individuals be employees of the company.

Onsite services can vary from full-time positions to a few hours per week depending on the size of the company and the workers' risk for injury. This approach requires that the company designate space for the medical provider to assess and treat workers. The most apparent benefits of onsite therapy services are the immediacy of services and the decreased worker down time. Musculoskeletal injuries (i.e., repetitive motion injuries, back injuries, sprain/strain injuries) receive treatment faster and therefore allow the patient to return to work faster.

Prompt treatment is the key to returning the worker to work as quickly as possible. In addition to helping the worker overcome injuries, prompt treatment can also impact the bottom line by cutting the costs of finding a replacement worker and reducing workers' compensation premiums. An intangible benefit of onsite services is the therapist's ability to effectively communicate the workers' situations to administration. This independent opinion can alleviate tensions which often develop between injured workers and their supervisors. It is these tensions that can lead to exaggerated symptoms and litigation.

The services that can be provided to industry and the worker through this approach are limited only by the willingness of the parties involved. Most services offered in a medical clinic can be offered at the plant site. The number of employees and the incident rate usually dictate the choice of services brought in-house.

The entire Worker Care Spectrum as represented in Figure 72.1 can be provided along with drug, hearing and vision screening, medical physicals, ergonomics consulting, fitness and wellness programs, inoculations, and traditional general medical treatment. The company also experiences economic benefits through the immediacy of evaluation and treatment of resultant earlier return-to-work, less time off work, and maintaining the bond between the worker and the worksite. Diversification can take many forms. These are but a few examples which are being offered to industry in today's market and demonstrate insight of how challenges will be met in the future.

72.7 Conclusion

Financial reward for losses as result of work activities is evidenced as far back as Greek warriors and pirates. More that 2000 years ago, the families of warriors who lost their lives in battle were compensated. Pirates were also provided a scheduled award for loss of a limb, but only after a successful attack and after the captain had taken his share of the recovered booty. Today's workers' compensation system is much more complicated. In 1986, Liberty Mutual reported $6807 as the mean cost per case for *low back pain.* In 1989, this cost had risen to $8321. This 1989 amount is more than twice the amount for the average workers' compensation claim which was $4075 (Webster and Snook, 1990). Liberty estimates the cost of time lost by injured workers to be $18.7 billion. Liberty also reports the cost of disability for corporations to be 6 to 12 percent of their corporate payroll. In 1993, 8.5 million people had disabling injuries (Liberty Mutual, 1998). It has been shown that the majority of costs incurred when an employee is injured comes from nonmedical costs (Quebec Task Force, 1987). A 1987 study revealed that only 14% of the total cost incurred was direct medical. The remaining 86% was attributed to indirect costs including the costs surrounding the replacement of an employee. Considering that industry directly pays for lost time, it is incumbent that industry know what an individual's capability is upon hiring and upon return-to-work from an injury.

Armed with accurate data, the human resource department can be sure that potential employees who do not have the capacity for the work will not be hired. Also, when an injured worker returns to work, it will be as soon as they are able to do so without risk of reinjury.

The employer needs increasingly more objective determination of job requirements and of worker capabilities to defensibly support their decisions of hire and of return-to-work once injured. By focusing on what the worker *can* do, job placement assessments and functional capacity assessments can trigger a more accurate and proactive response from the employer. This includes proper placement upon hiring and easier placement and accommodation, if necessary, once injured (Key, 1995a).

References

Aitken MJ, Creighton University School of Pharmacy and Allied Health, Department of Occupational Therapy, Omaha, Nebraska June, 1996.

Astrand PO, Ryhming I, A nomogram for calculation of aerobic capacity from pulse rate during submaximal work, *J of Applied Physiology* 7:218-221, 1954.

Bigos SJ, Battié MC, Spengler DM, et al. A prospective study of work perceptions and psychosocial factors affecting the report of back injury. *Spine* 16:1-6, 1991.

Blair JA, Blair RS, Rueckert P, Pre-injury emotional trauma and chronic back pain — an unexpected finding. *Spine,* 19: 10, May 15, 1994.

Block AR, Vanharanta H, et al. Discogenic pain report, influence of psychological factors, *Spine* 21(3):334-338, February 1996.

Dobrzykowski E, Data collection and use in industrial therapy, in *Industrial Therapy,* Key GL (Ed.) St. Louis, Mosby, 1995, p 42-60.

Frey DH, Job placement assessments and pre-employment screening, in *Industrial Therapy,* Key, GL (Ed.) St. Louis, Mosby, 1995, pp 110-122.

Gilliand RG, Sevy BA, Ahlgren A, *A Study of Statistical Relationships Among Physical Ability Measures of Injured Workers Undergoing KEY Functional Assessments,* Minneapolis, Minnesota 1986.

Grossman P, Respiration, stress, and cardiovascular function, *Psychophysiology* 20(33):284-300, 1983.

Karwowski E, Salvendy G, Ed. *Ergonomics and Manufacturing: Raising Productivity Through Work Place Improvement,* Society of Manufacturing Engineers, 1998.

Keller LS, Butcher JN, *Assessment of Chronic Pain Patients with the MMPI-2.* Minneapolis, MN: University of Minnesota Press, 1991.

Key GL, Functional capacity assessment, in *Industrial Therapy,* Key GL (Ed.) St. Louis, Mosby, 1995a.

Key GL, The impact and outcomes of industrial therapy, in *Industrial Therapy,* Key GL (Ed.) St. Louis, Mosby, 1995b, p 220-254.

Key GL, *Key Functional Assessment Policy and Procedures Manual and Training and Resource Manual,* Minneapolis, MN, 1984.

Liberty Mutual, For Your Employees: Workers' Compensation, available at: www.libertymutual.com/business/workcomp.html, accessed 10-6-98.

Osterweis M, Kleinman A, Mechanic D, *Pain and Disability, Clinical, Behavioral, and Public Policy Perspectives,* Washington, DC, National Academy Press, 1987.

Personnel Decisions Inc. *Key Functional Assessment Pre-employment Screening Battery as a Predictor of Job Related Injuries,* Minneapolis, MN, January, 1994.

Portney LG, Watkins MP, *Foundations of Clinical Research,* Norwalk, Conn., Appleton & Lange, 1993.

Quebec Task Force Study: Scientific approach to the assessment and management of activity-related spine disorders, *Spine,* 12:7S, 1987.

Schmidt AJM, Gierlings EH, Madelon LP, Environmental and interoceptive influences on chronic low back pain behavior. *Pain* 38:137-43, 1989.

Waite HD, *Use of a New Physical Capacities Assessment Method to Assist in Vocational Rehabilitation of Injured Workers,* University of Colorado, Thesis, 1987.

Webster BS, Snook, SH, The cost of compensable low back pain, *J Occup Med,* 32:13-15, 1990.

Wheeler AH, Hanley EN, Spine update: nonoperative treatment for low back pain — rest to restoration, *Spine,* 20(3): 375-378, February 1, 1995.

Worker Data Bank, KEY Method, Minneapolis, Minnesota. 1994, 1995, 1996, 1997.

73

Ergonomics and Rehabilitation

Ewa Nowak
Institute of Industrial Design

73.1 Introduction

Based on a criterion of body "efficiency," the human population can be divided into three main groups:

- People of the highest (extreme) efficiency
- People of an average efficiency (within normal range)
- People of efficiency below normal range — these are the *disabled*

Efficiency undergoes changes during ontogenesis. It is determined genetically, but can also change (within the limits of so-called reaction standard) depending on the environmental, social, material, cultural, and other conditions. Body efficiency can be improved through physical exercises, better nutrition, and a more hygienic way of life. In the case of the disabled, *rehabilitation* is in charge of this question.

Disability, according to International Classification posed by WHO (1980) is defined as follows: "A restriction or lack of ability (resulting from impairment) to perform an activity in the manner or within the range considered normal for a human being."

Rehabilitation (after Kumar, 1989) is defined as "a science of systematic multidimensional study of disordered human neuropsychosocial and/or musculoskeletal function(s) and its/their remediation by physicochemical and/or psychosocial means." Rehabilitation can be treated as a process that, by making use of the latest medical, technical, and social sciences achievements, restores, within the optimum limits, the efficiency of *impaired* or *handicapped* organs. The aim, and at the same time the result, of rehabilitation is enabling the man to come back to normal life in society. Two steps can be distinguished in this process. At the same time they constitute two basic aims accepted by the World Health Organization in the World Program of Action (1984) (Figure 73.1).

The first step is rehabilitation. As a result of rehabilitation, the disabled person regains the optimum level of mental, physical, and social functioning. Of course, the process, in addition to being enabling, also strives to reduce pain and suffering. The second step is equalization of opportunities for disabled people. This involves a process through which physical environment, housing and transportation, social and health services, educational and work opportunities, cultural and social life, including sports and recreational facilities, are made accessible to all (Kumar, 1992).

0-8493-2641-9/99/$0.00+$.50

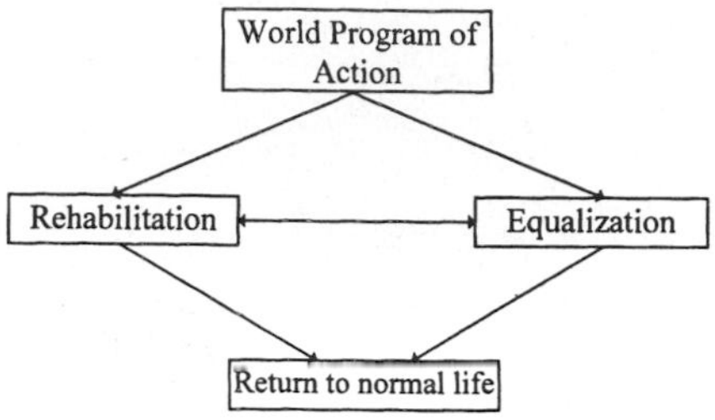

FIGURE 73.1 Activities of the World Program of Action.

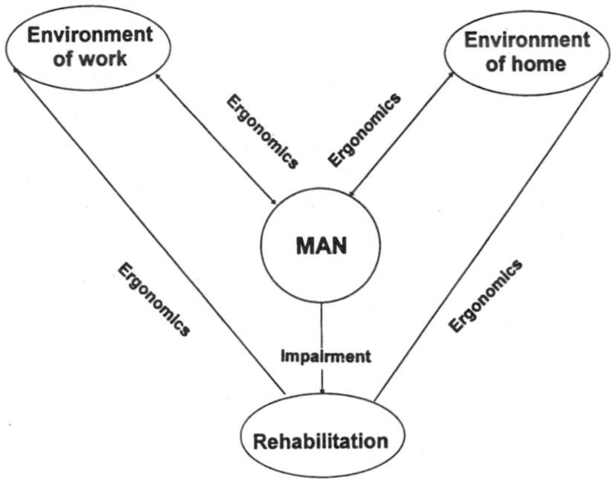

FIGURE 73.2 Correlation of ergonomics and rehabilitation in home and work environments.

Particularly helpful in this second step is ergonomics. While rehabilitation improves the efficiency of the body, ergonomics, through adjusting and optimizing external environmental factors to the capabilities of the person, enhances work comfort and increases its efficiency and effectiveness. Investigations of Canadian scientists (Kumar, 1992) as well as my own observations confirm the fact that rehabilitation and ergonomics applied together considerably intensify and advance both the progress of the rehabilitation process and the return to normal life.

Exercises, as well as medical treatment, supported by properly constructed workstands and objects of everyday use, intensify significantly the effect of rehabilitation. Under the influence of cumulated factors an impaired organ or function attain their initial capabilities to a higher degree. Ergonomics is of extreme importance in attaining "equal chances." This is of assistance in the process of integrating the disabled into society. It helps to achieve the goal of the World Health Organization, i.e., full participation of the disabled in social life and development. Figure 73.2 illustrates interrelation between ergonomics and rehabilitation in work and home environments.

It seems that ergonomics and its possibilities are not sufficiently used for the needs of the disabled. Several methods and measuring techniques applied in ergonomics can be used, e.g., in rehabilitation. Some of these should be completed, partially changed, and adapted. It is necessary to develop closer cooperation between designers and specialists from the fields of ergonomics and rehabilitation.

Even at the present moment there appears a new direction of ergonomics aimed at the needs of disabled people. This is *rehabilitation ergonomics* (Kumar, 1992; Nowak, 1993). This can be roughly defined as an interdisciplinary field of science that aims at adjusting tools, machines, equipment, and technologies as well as material work and life environments including objects of daily use and rehabilitation equipment to the psychophysical needs of the disabled. Rehabilitation ergonomics takes part both in the rehabilitation process and in equalizing chances. This chapter presents ergonomic possibilities and their achievements in these spheres.

73.2 Ergonomics and Equalization of Opportunities

Somatic Characteristics of the Disabled as a Determinant of Spatial Structures

Defining the possibilities and necessities of the disabled is a necessary condition for designing the material work and life environment for this population. Data that characterize somatic structure constitute basic information. The influence of disabilities on shaping the body structure was studied by numerous anthropologists (Floyd et al., 1966; Pheasant, 1986; Boussena and Davis, 1987; Goswami et al., 1987, 1994; Laubach et al., 1981; Molenbroek, 1987; Nowak, 1988, 1989, 1994; Samsonowska-Kreczmer, 1988; Jarosz, 1996; Mieçsowicz, 1990; Łuczak et al., 1993; Das and Kozey, 1994; Lebiedowska et al., 1994).

The largest disproportions between the healthy and the disabled population can be found in a group of people with motor dysfunction. This is understandable since the dysfunction results from the past or currently developing diseases that lead to joint disturbances of the osseous, ligament and joint, muscular, and nervous systems. These disturbances lead to deformities and somatic changes of particular parts of the body, and this affects the final shape and dimensions of the body and its motorics.

Other factors that restrain the development and growth of the body including restriction of motion activities, neglected nursing, improper or lack of rehabilitation, as well as stresses connected with pain, frequent stays in hospitals and rehabilitation centers etc., accompany the pathological process. Descriptions of investigations of people with lower extremity dysfunction are usually found in the literature, and they usually are wheelchair users. It should be realized, however, that this group embraces people with various degrees of motor efficiency limitations. This depends not only on a type and stage of a disease but also on the time of its appearance. Therefore, researchers dealing with this problem face great difficulties in selecting subjects and in describing results scientifically. This may be the cause of the small number of studies undertaken in this field. This particularly concerns studies, the results of which are to provide data for designing.

It turns out, that height measurements are measured in relation to various reference bases. This fact creates a difficulty not only for comparison purposes, but also creates a difficulty for designers who would like to utilize investigation results. A synthesis of existing data was done for design purposes.

Table 73.1 comprises height measurements measured in relation to the seat plane (Bs), and Table 73.2 to the floor level (B). *Anthropometric* data differ significantly regardless of the fact that these concern various populations. The influence of diseases resulting in the necessity of using the wheelchair affects shaping the body. Pheasant (1986) indicates that the body proportions of wheelchair users resemble those of elderly people above age 65. This was confirmed by the results of investigations carried out by Molenbroek (1987).

Not only anthropometric measurements of the disabled are important for the needs of design but also differences between the disabled and healthy population. The majority of authors indicate that the body structure of disabled men and women differs significantly from the able-bodied population (Pheasant, 1986; Samsonowska-Kreczmer, 1988; Nowak, 1988, 1989; Jarosz, 1996; Das and Kozey 1994). This problem was illustrated by the example of data on the Polish population (Jarosz, 1996). Using the threshold values of the 5th percentile, a linear anthropometric model of disabled women in the sitting position was drawn. This model was compared to the model of the 5th percentile of the able-bodied (Figure 73.3).

It appears that the above measurements show smaller values for the disabled than for healthy people. The differences are significant and amount to 110 mm for the seated stature (Bs-V), for eye level (Bs-en) to 113 mm, for shoulder height (Bs-a) to 126 mm, and for elbow height (Bs-r) to 57 mm. For arm reach measurements the differences amount for arm reach forward (Bs-phIII) to 204 mm, and to 90 mm for arm overhead reach (Bsd-phIII). Similar results were obtained by Nowak (1988, 1989) by comparing the population of disabled young people aged 15 to 18 with the lower extremity dysfunction, with young of the same age representing the Polish population (Nowak, 1988).

Floyd (1966) and Bouisset and Moynot (1985) indicate that the smaller seated stature (Bs-v) measurements of the disabled can result from the deformities of the osseous system as well as from the fact that,

TABLE 73.1 Structural Anthropometric Data for Males and Females with Respect to the Seat

		Author										
		Boussena and Davis		Das and Kozey		Goswami et al.	Jarosz		Molenbroek		Nowak	
Dimension per/mm/		Men	Women	Men	Women	Men	Men	Women	Men	Women	Men	Women
Seated Stature	5	824	794	734	647	—	769	668	761	702	744	708
	95	962	912	963	857		960	894	919	858	972	890
Eye Height	5	—	—	496	546	—	667	570	643	585	630	592
	95			717	744		857	789	810	763	857	783
Shoulder	5	—	—	468	423	330	495	433	520	479	474	461
Height	95			676	597	564	682	619	649	601	647	592
Elbow Height	5	177	176	108	105	136	144	133	168	156	158	139
	95	269	266	312	257	212	297	281	289	270	289	309
Knee Height	5	483	450	—	—	—	468	407	—	—	453	442
	95	586	539				605	530			572	532
Popliteal	5	381	364	—	—	343	383	315	401	361	386	371
Height	95	473	453			465	513	454	503	460	502	462
Trunk Depth	5	—	—	198	143	—	180	191	211	219	165	182
	95			281	182		340	315	344	368	270	286
Popliteal Depth	5	421	418	—	—	356	461	418	401	405	435	424
	95	522	516			447	636	571	525	524	555	545
Shoulder	5	383	368	354	291	—	353	310	—	—	337	316
Breadth	95	482	434	439	355		425	394			439	410
Overhead	5	—	—	1072	947	—	1028	882	828	733	1022	963
Reach	95			1415	1090		1324	1192	1214	1113	1320	1195
Reach Forward	5	568*	552ˣ	—	—	—	653	558	—	—	668	617
	95	677*	630ˣ				840	713			861	768

* measured from the acromiale point
ˣ measured from the acromiale point

TABLE 73.2 Reported Anthropometric Data of Disabled People with Respect to the Floor

		Floyd et al		Jarosz	
Dimensions per/mm/		Men	Women	Men	Women
Floor to vertex	5	1260	1180	1299	1198
	95	1410	1355	1490	1424
Floor to eye	5	1150	1080	1197	1100
	95	1290	1235	1387	1319
Floor to shoulder	5	965	910	1025	963
	95	1100	1065	1212	1149
Floor to elbow	5	625	610	674	663
	95	745	730	827	811
Floor to top of height	5	620	565	607	598
	95	680	635	672	667
Floor to top of foot	5	120	165	—	—
	95	180	215		
Floor to vertical grip reach	5	1550	1460	1558	1412
	95	1785	1680	1854	1722

as a result of back muscle paralysis, difficulties with maintaining the straight position of the body appear. This problem seems to be very serious for paraplegics. A similar interpretation can be accepted for reach measurements. According to Pheasant (1986) the same analogies of changes in body proportions occur in wheelchair users and in elderly people. In old age, similarly to people with motor organ dysfunction, deficient muscular tonicity of the chest and belly appears. This results in pectoral kyphosis increasing.

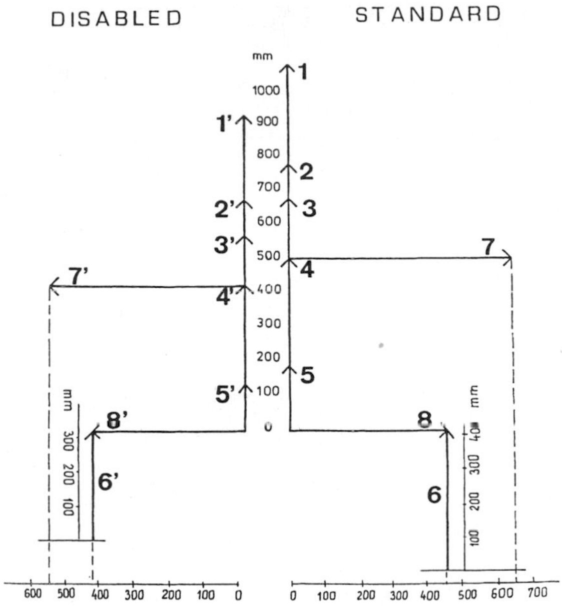

FIGURE 73.3 Anthropometric model 5th percentile of disabled and standard women (in the sitting position), where 1,1′ — Arm overhead reach; 2,2′ — Stature; 3,3′ — Eye height; 4,4′ — Shoulder height; 5,5′ — Elbow height; 6,6′ — Popliteal height; 7,7′ — Arm reach forward; 8,8′ — Popliteal depth. (From Dr. Emilia Jarosz. With permission.)

At the same time, the processes of intervertebral cartilage flattening and of the back shortening occur. This leads to the C shape of the spine (Nowak, 1980, 1988). Samsonowska-Kreczmer (1988) indicated that the stooping back and sunken chest occur in young people with motor organ dysfunction. Another reason for diminishing the seated stature measurements is that the buttock and thigh muscles atrophy, resulting from immobility of the back and the lower extremities. Studies conducted by Jarosz (1993) indicate that for 55% men and 65% women thigh thickness measurement is below the lower limit of the healthy population standard. Similar results concerning adults were obtained by Goswami et al. (1987), and concerning the young by Nowak (1988, 1989) and Miesowicz(1990). In comparison with the stature and reach characteristics, shoulder breadth characteristics (a-a) appear different. Most scientists confirm the fact that the value of this characteristic is higher for wheelchair users than for the healthy population (Goswami et al., 1987; Boussena and Davies, 1987; Nowak, 1989; Jarosz, 1996). Wężyk (1989) and Mięsowicz (1990) point out that children with cerebral palsy have larger shoulder breadth values in comparison to those of healthy children. Without exploring the reason for this phenomenon, this is, without doubt, important information for clothes designers. This question will be discussed in a later section.

Essential characteristics exerting an influence on workspace shaping are functional characteristics of the upper extremities, i.e., reaches. Values observed in these characteristics (Tables 73.1 and 73.2) are significantly lower in persons with lower extremity dysfunction, although their upper extremities are qualified as "efficient." Lower reach values result not only from lower values of the arm and forearm length, but also from limitations in shoulder and elbow joints. In connection with the above, the disabled have difficulty in performing the movements of abduction and extension (Nowak, 1988). Grabowska et al. (1986) found that in the case of persons suffering from rheumatoid arthritis, workspace of the upper extremity is 7 to 10% lower if a shoulder joint is constrained and 25 to 33% lower when there is a constraint of the elbow joint movements. It is obvious that disorders of these two joints increase the limitation of the upper extremity movements, and thus the efficient workspace is significantly reduced. This is confirmed by investigations conducted by Nowak (1988, 1989) and Jarosz (1996). It should be pointed out that particular values of reach of the disabled refer to the straight position of the body. Thus, they can be increased through trunk movements forward and lateral. Floyd's investigations (1966) proved

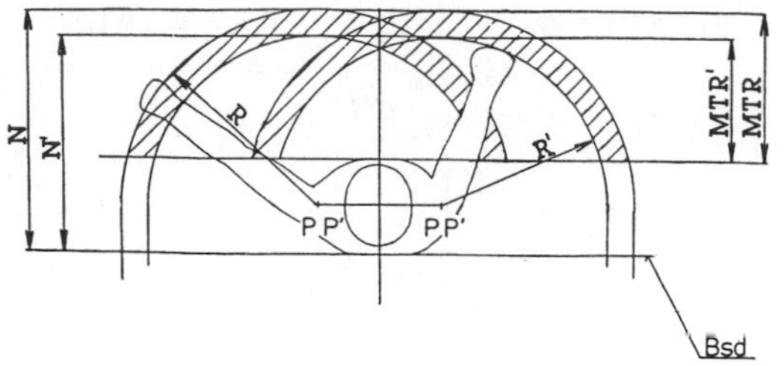

FIGURE 73.4 Maximum Transverse Reach for the all-Polish population (MTR) and the disabled population (MTR′), where R,R′ — reach radius; P,P′ — hypothetical axis of rotation; N,N′ — reach forward.

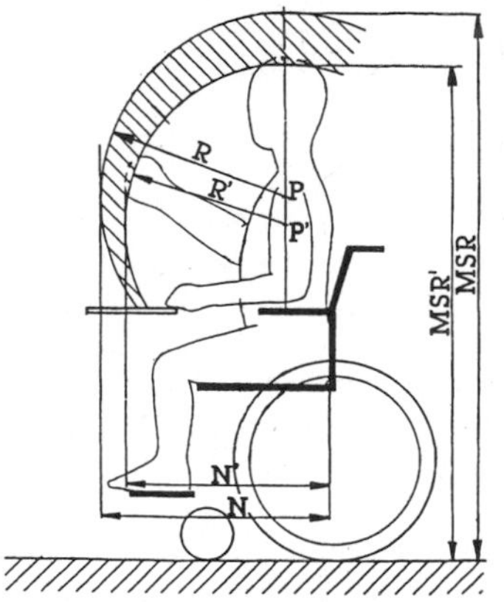

FIGURE 73.5 Maximum sagittal reach for the all-Polish population (MSR) and the disabled population (MSR′), where R,R′ — reach radius; P,P′ — hypothetical axis of rotation; N,N′ — reach forward.

that the difference between reach forward of disabled women in the straight position of the body and the same reach in the position of maximum bend forward amounted to 142 mm, and in case of the lateral reach with the trunk lateral bend — 135 mm. Reaches forward of the disabled are limited by the foot-rest of the wheelchair. The difference between the point of reach of disabled women investigated and the beginning of the wheelchair amounted to 242 mm. That is, it exceeded the difference between reaches in the straight position and reaches in the bend forward stated by Floyd. That means that the wheelchair user is not able to reach any devices in the frontal position if there is not enough space for the wheelchair under the working plane.

Nowak (1989), based on the Das and Grady's method (1983), developed a simple method of defining workspace for arms. This space was determined for disabled young people with dysfunction of the lower extremities, using the wheelchair. Figures 73.4 and 73.5 show the graphic way to determine this space in the sagittal plane (maximum sagittal reach — MSR) and in the transverse plane (maximum transverse reach — MTR). These figures show the difference for both reaches between disabled and healthy young people. Differences in maximum reach measurements were significant and amounted for the 5th percentile to 300 mm. Results of the study were used for ergonomic analysis of schools and for designing school workshops, laboratories, and rehabilitation centers. This method can be recommended for workspace design for the disabled. It is simple and easy to use. Using only five anthropometric characteristics, one can obtain the graphic representation of workspace of any population or an individual person.

While designing the functional space for the paraplegic, one should bear in mind that they have difficulty with the stability of position, and they can easily fall out of the wheelchair if they bend forward too much (Przybylski, 1979). Upper extremities moving forward disturb the balance of the body, which is more difficult to redress because of the paralysis of the muscles of the spine and pelvis, stabilizing the body in the sitting position (Bouisset and Moynot, 1985).

Molenbroek's investigations (1987) provide data determining the values of reach of the elderly. Limitation of the movement range of particular joints and decrease in the values of reach increase as age increases. Molenbroek (1987) proved that differences between the values of the arm reach overhead for 50-year-old and 95-year-old Dutchmen amount (for the value of the 5th percentile) to approximately 300 mm. These are the directions for designing living interiors meant for the elderly. It is important for determining the ability to reach the top shelf of a wardrobe, the highest buttons in a lift, or a window handle. These data prove that while designing for people aged 50 to 110, one should place all kinds of switches at the height not exceeding 1474 mm. Above this height, the elderly have difficulty operating them.

According to Pheasant (1986), a wheelchair user (whose upper extremities are unimpaired) can reach a zone from around 600 to 1500 mm in a sideways approach, but considerably less "head on." It may well be that the location of fittings within this limited zone will prove entirely acceptable for the ambulant users of the building, but in case of working surface heights no such easy compromise is possible.

Another task is determining appropriate steps and handrails adjusted to the dimensions of the disabled, who can move without aid. Petzal (1993) recommends installing a handrail at the height of 900 mm above the floor. Based on the investigations of different step variants, he suggests the height of a step within 150 to 200 mm and the width from 250 to 300 mm.

The above author, on the basis of anthropometric data, research tests, and observations, determined placement of handrails in a bus adjusted to the needs of ambulant disabled people, as well as the size of seats and the distance from one to another.

As the review of investigations has proved, the body structure of disabled men and women differs considerably from that of the healthy population. Ergonomics, providing data concerning the body structure of disabled people, makes it possible to adjust designs of products and spatial structures to the abilities and predispositions of this group of users.

Hand Characteristics as the Basis for Designing Objects of Everyday Use

Statistics of many countries report that a significant proportion of the population with disabilities requires help from family, friends, volunteers, or paid care-givers. Almost one half (45.6%) of all disabled people require assistance with heavy household work, and 22.4% with daily housework (Kumar, 1995). In terms of the cost, though the figures were not given, it is surmised that it is significant. Many able-bodied people have to spend their time to assist. The cost of this time along with the cost of the paid health care worker can add up to a significant magnitude. It may be useful to point out that the larger cost for these people is not medical but maintenance, attendant care, nursing home, and home care expenses on top of the loss of productivity due to inability to work. Environment modification and assistive devices may offer disabled people a chance to decrease these costs and be productive members of society.

Therefore, the problem of making daily life more efficient for the disabled is of extreme importance from the point of view of economics. The possibility of designing objects of every day use with the help of ergonomics can play a significant part in this field. One of the basic aims of ergonomics is to define the needs and capabilities of the user. These capabilities, in comparison with the able bodied, are usually limited and result from dysfunction of a particular organ. The dysfunction of performing manual functions is felt most acute during everyday activities. The hand is a unique organ, adjusted to performing manipulation activities. Efficiency of performing these activities depends on the grasping capabilities of the hand. The grasping function is one of the most important functions allowing a person to perform daily activities. Dysfunction in grasping makes it impossible for a person to perform the majority of activities, including such fundamental ones as eating and drinking. Figure 76.6 shows the types of hand grips used (Nowak, 1993). The possibility of performing particular grips depends on the type and degree of hand dysfunction. Thus, it is important to know handgrip capabilities of the disabled user, and these capabilities should be taken into consideration by a designer while designing an object. Stabilization of an object in the disabled hand should be secured by the grip, that at the same time allows an object to be manipulated. Normally this is attained by adjusting the shape and size of an object or its handle to

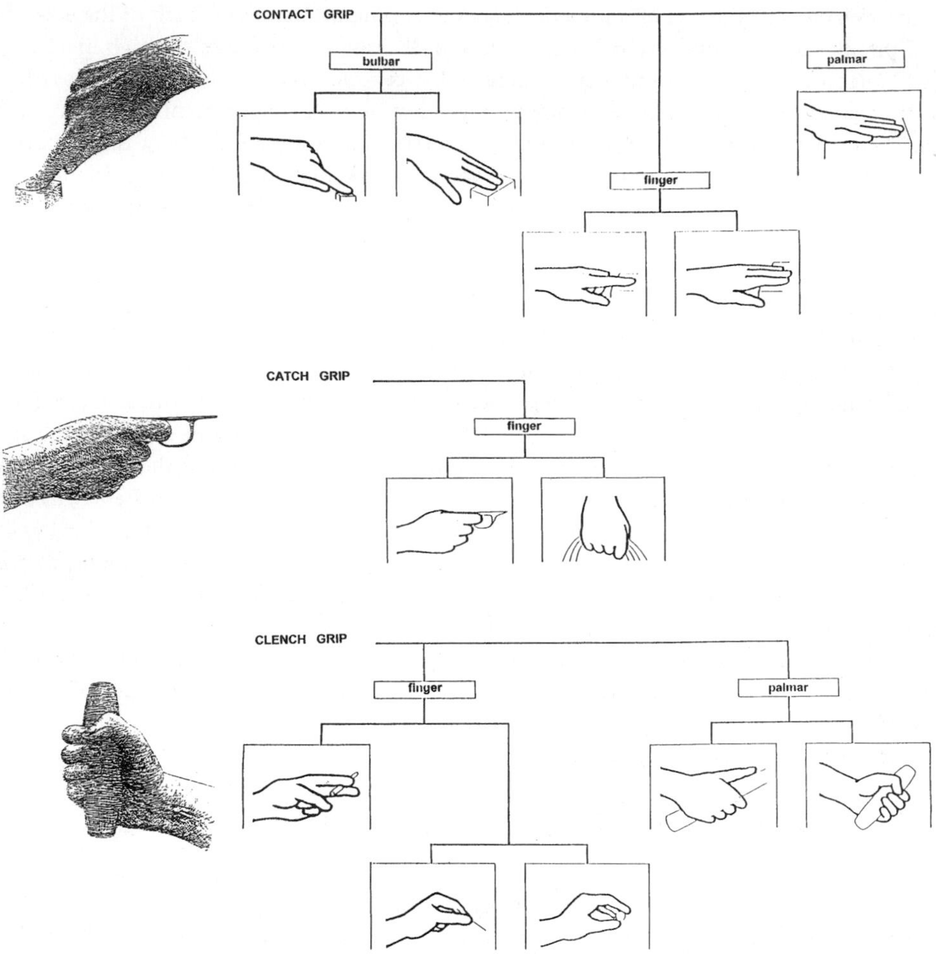

FIGURE 73.6 Hand grips classification. (From Nowak, E. 1993. Hand grip classification, in the catalogue *Human Dimension — Ergonomics and Design*, Institute of Industrial Design, Warsaw: 30 [in Polish]. With permission.)

the capabilities of the deformed hand. Investigations that define the contact area of the hand and the object or its handle, depending on its shape and size, are quite popular in this field. Figure 73.7 (Frejlich, 1993) shows pictures of the contact area of the hand and the handle with the shape of a circle cross section that depends on diameter measurements of the handle. The larger cross section, the larger the contact area. Frejlich proved (1983) that handles with square, rectangular, and triangle cross sections attain smaller contact areas in comparison to handles with circle cross sections. This is important information for designers, since the shape and size of handles are of great significance for an object to be held, stabilized, and manipulated. This was confirmed by Benktzon's investigations (1993). The most important for her was to define the appropriate contact between the hand and a handle. Benktzon, with the cooperation of numerous specialists from the Ergonomi Design Gruppen (Benktzon, 1993), analyzed objects of everyday use and offered new solutions for these objects. These can be used both by the elderly and the disabled. Several of these objects were presented by the Ergonomi Design Gruppen at the exhibition "Design and Ergonomics," organized in Warsaw in 1993. Figure 73.8 presents a new solution for a walking stick. The design was developed basing on a keen analysis of the manner of gripping by the disabled with hand dysfunction. The contact area between the hand and the new anatomical handle has increased considerably, thus providing much better support. In practice, seemingly small differences can be of vital importance for function, safety, and comfort.

The new design (Fenix) is characterized by:

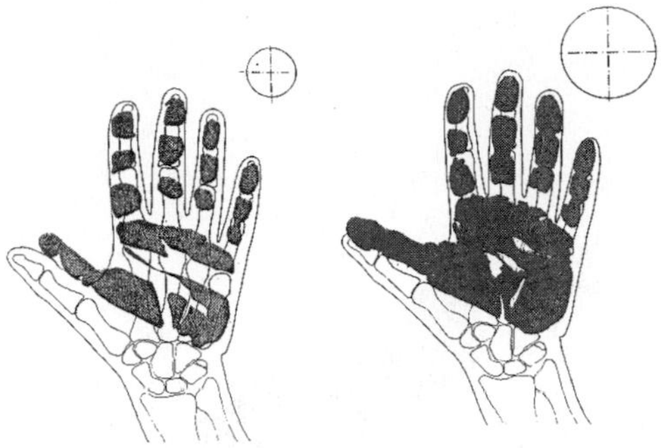

FIGURE 73.7 Pictures of contact area of the hand with the handle of a circular cross section. (From Frejlich, C. 1993. Analysis of the active contact surface of the hand with a handle depending on its cross-section shape and size, in the catalogue *Human Dimension — Ergonomics and Design*, Institute of Industrial Design, Warsaw: 100 [in Polish]. With permission.)

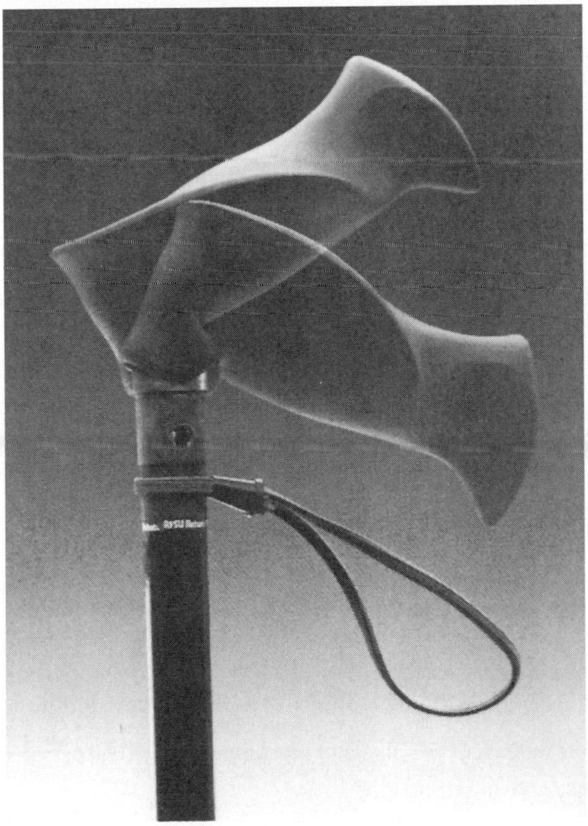

FIGURE 73.8 The walking stick handles; made by RFSU, Sweden, designed by Ergonomi Design Gruppen. (From *Catalogue of the exhibition "Human Dimension -Design and Ergonomics."* 1993. Institute of Industrial Design, Warsaw. With permission.)

- Larger load-bearing areas for the hand, without the usual consequence of the thumb being pushed out of place
- The stick being positioned at the back of the handle which allows the fingers to be flexed together
- Good friction and soft grip surface to enhance comfort
- The angle of the handle being adjustable to each individual by means of a ball-and-socket joint

FIGURE 73.9 Cutlery for the disabled using one hand; made by RFSU, Sweden, designed by Ergonomi Design Gruppen. (From *Catalogue of the exhibition "Human Dimension -Design and Ergonomics."* 1993. Institute of Industrial Design, Warsaw. With permission.)

Figure 73.9 presents a design developed by the same group. This is a set of cutlery for people using only one hand. The fork has small prongs, and the spoon has a serrated edge that allows cutting of food. There is also a range of cutlery designed to accommodate many "special needs," along with a goblet and a curved rimmed plate. All these designs demonstrate the goal of integrating style and "special needs" design; people should be not stigmatized because their needs can be classed as "special."

Elements of Ergonomics in Clothing Design

Clothes for the disabled constitute a social problem of particular importance. A person suffering from disability or permanent illness should get such clothing that makes his or her life easier, is functional, meets his or her emotional needs, and, at the same time, is adjusted to the limitations caused by a disability. People with motion dysfunction have the most problems with clothes. Clothing designed for this group of people should assure appropriate physiohygienic properties, heat comfort, as well as ease of manipulation. Concepts of model-constructions solutions should take into account not only the usefulness of products, but also the economic effect. Designs must be simple in construction and easy to manufacture. The form of clothing is to integrate the group of the disabled with society, to conceal a disability, and to give the feeling of satisfaction at possessing clothes appropriate to needs and expectations.

Basing on ergonomic criteria, we can determine two main types of clothing:

- Clothing for totally inefficient people, who are attended by other persons
- Clothing for people able to self-service

In both these cases ergonomic investigation are helpful.

Determination of the somatic characteristic of the disabled is the starting point for clothing design. Figure 73.10 (a,b) shows the measurements that determine the necessary dimensions of a working suit. As was proved earlier, disability and the motion limitations connected with it result in changes in the shape, dimensions, and proportions of the body.

Investigations of the young aged 15 to 18 (Nowak, 1988, 1989) with motor dysfunction of the lower extremities proved that the body dimensions of this group differ considerably from those of the healthy population of the same age. Their body weight and height are lower. This was confirmed by other scientists' investigations (Malinowski, 1975; Pheasant, 1986; Samsonowska-Kreczmer, 1990; Jarosz, 1993).

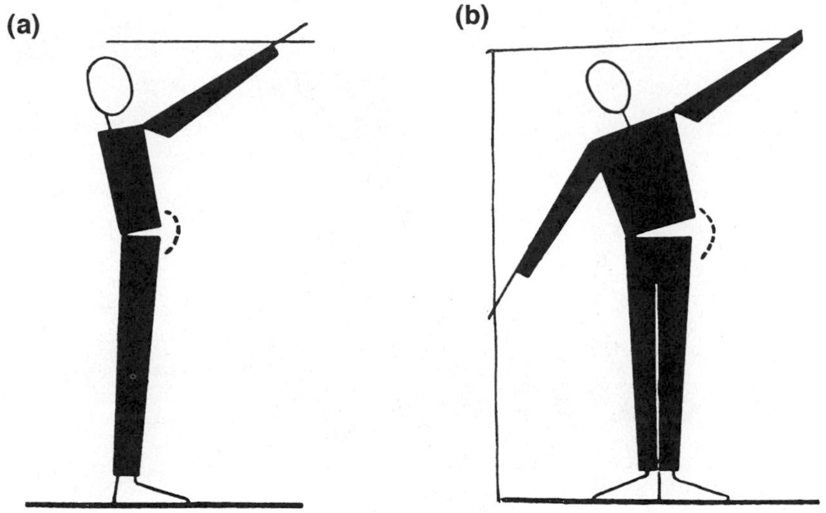

FIGURE 73.10 Examples of anthropometric characteristics defining clothes dimensions. (a) at maximum backward trunk bending; (b) at maximum lateral trunk bending. (From Batogowska, A. 1993. Selected maximum measurements for the needs of clothing design. In the catalog *Human Dimension — Ergonomics and Design*, Institute of Industrial Design, Warsaw: 64 (in Polish). With permission.)

On the other hand, it was proved that this group of people has considerably higher values of shoulder breadth characteristic. Walking with the help of crutches and walking sticks, or driving a wheelchair is performed thanks to the work of the upper extremities and the trunk muscles. It is also supposed that intensive physical training of these muscles during rehabilitation can stimulate the growth of the clavicles in length. This results in the considerable increase in the shoulder breadth characteristic values. In addition to the body dimensions, determination of motor abilities is of vital importance. Figure 73.11 shows the measurement of the upper extremity movement range (Nowak, 1978). Basing on ergonomic investigations, assumptions for clothing design are determined. For example, clothing for the wheelchair user must fulfill (among others) the following requirements: (1) appropriate clearance allowing the hands to move easily up and forward and appropriately larger shoulder breadth. This is ensured by the special construction of sleeve and armpit cut; (2) increased transverse dimensions related to the enlarged, muscular chest; (3) widened sleeve finish related to the considerably enlarged biceps and triceps; (4) adjustment of the bottom part of clothing length and cut (in case of a blouse, shirt, and vest) to the sitting position of its user. This is attained through the removal of excess fabric in front. Figures 73.12, 73.13, and 73.14 show example solutions of the upper and lower parts of clothing for a wheelchair user. The projects were designed in the clothing department of the Institute of Industrial Design in Warsaw (Dą browska-Kiełek; Szymańska-Petrykowska, 1991).

Interesting ergonomic investigations for the needs of clothing design for the disabled were conducted by Sperling and Karlsson (1989). They determined demands on clothing fasteners for long-term care patients:

Functional Demands

The clothing fastener should:

- Be located in the patient's optimum grip area (Figure 73.3)
- Be easy to understand and identify, visually as well as tactally
- Be possible to handle with one hand
- Be easy to grip and hold
- Not demand more grip strength or precision handling than the patient is able to produce
- Stand body movements without opening
- Have dimensions suitable for handling as well as clothing construction

FIGURE 73.11 Measurement of upper extremity movements range performed in working clothes.

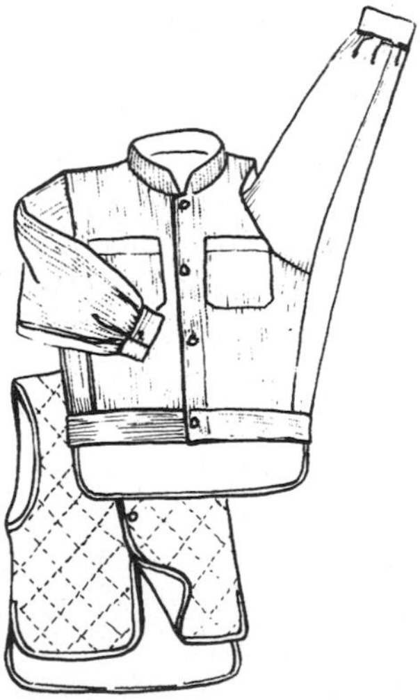

FIGURE 73.12 Exemplary solution of the upper part of clothes for men using a wheelchair; designed by Jolanta Szymańska-Petrykowska, made by the Institute of Industrial Design. (From Dąbrowska-Kiełek, J., Szymańska-Petrykowska, J. 1991. Clothes and shoes. *Designer's Vademecum — Problems of the Disabled*, Vol. 5 (in Polish). With permission.)

Comfort and Safety Demands

The closure should:

- Not scratch or rub
- Not cause such a pressure against the skin that might lead to pressure-sores

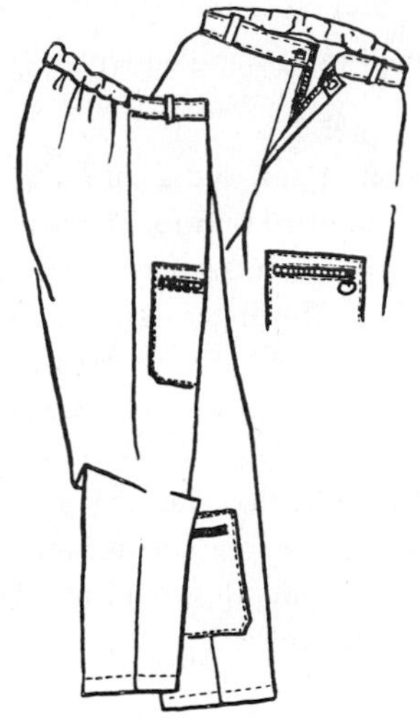

FIGURE 73.13 Design of trousers for men and women using a wheelchair; designed by Jolanta Szymańska-Petrykowska, made by the Institute of Industrial Design. (From Dąbrowska-Kiełek, J., Szymańska-Petrykowska, J. 1991. Clothes and shoes. *Designer's Vademecum — Problems of the Disabled,* Vol. 5 (in Polish). With permission.)

FIGURE 73.14 Design of a skirt for women using a wheelchair; designed by Jolanta Szymańska-Petrykowska, made by the Institute of Industrial Design. (From Dąbrowska-Kiełek, J., Szymańska-Petrykowska, J. 1991. Clothes and shoes. *Designer's Vademecum — Problems of the Disabled,* Vol. 5 (in Polish). With permission.)

Sewing and Maintenance Demands

The closure should:

- Be easy to fasten
- Be lasting and easy to replace
- Stand washing (chemicals, high temperature)
- Be inexpensive

The aim of this study was to facilitate a more independent daily life for long-term care patients. For this group of people it is of great importance to be able to dress and undress without assistance. The design and position of clothing fasteners often lead to a restriction of the functional capacity of the patient. Three different types of fasteners were evaluated:

- Oval asymmetric button for a horizontal buttonhole (25 × 19 mm)

- Oval symmetric button for a vertical buttonhole (28 × 19 mm)
- A finger strap, which was designed to be handled with one hand and nonprehensile movements (36 × 14 mm)

In a subsequent study, an adapted oval button and a "finger strap" alternative to hook-and-eye were designed and evaluated, together with a standard button. The oval button in combination with a vertical buttonhole improved the function for most patients, and the front position for the fasteners was superior to a diagonal or lateral position on the chest. The finger strap was of advantage to patients with hemiplegia and joint complaints but was difficult for many of the subjects to understand, being a technical innovation.

Many investigators and designers of clothing for the elderly and the disabled (Benktzon, 1980; Dąbrowska-Kiełek; Szymańska-Petrykowska, 1991; Nowak, 1996) underline the fact that in the process of clothing design they are governed by the idea that clothing, in addition to its functional characteristics, should give its user physical comfort. The disabled want to live and work among healthy people. Therefore, they have to accept the clothes they wear. Clothing should not deform the shape of the body, but just the opposite, it should cover anatomical defects.

73.3 Ergonomic Methods for the Needs of Rehabilitation

Rehabilitation is a complex and complicated process. It consists of restoring, to a maximum possible degree, motor efficiency of the human body. Rehabilitation employs many means and research techniques used in various fields of science, such as medicine, sociology, psychology, mechanics, and biology. It would seem that rehabilitation processes could also benefit from drawing more freely on achievements of anthropometry. An objective of this subsection is to present methods of dynamic anthropometry used in ergonomics, and to put forward proposals for suitable modification of those methods to satisfy the requirements of rehabilitation processes.

Anthropometry can be particularly useful in diagnosing and assessing the motor efficiency of the human body. Rheumatic diseases and mechanical injuries result in pathological changes in joints, ligaments, tendons, muscles, and lead to considerably restricted movement ranges. Thus, the assessment of motor efficiency of the affected joints is essential for monitoring the rehabilitation processes.

The simplest way of assessing motor efficiency is by comparing a restricted motion range of an affected joint with its initial range of motion, i.e., before injury or disease. Unfortunately, after injury or disease has already prevailed, it is impossible to determine what the initial motion range was in the healthy patient. Also, it may hardly be expected that a rehabilitated patient had his initial motion ranges measured just before the injury.

Thus, motor efficiency of a rehabilitated patient can only be assessed based upon the data of the healthy population. A method based upon this kind of data and allowing the quantitative assessment of rehabilitation progress was developed by Nowak (1992). The results of investigations conducted for the needs of ergonomics were used. The investigations comprised ranges of motion of arm, leg, hand, foot, and head.

The following maximal movements were measured:

- Flexion and extension of the arm and leg
- Flexion and extension, adduction and abduction, supination, and pronation of the extended hand
- Flexion and extension, supination, and pronation of the grasping hand
- Flexion and extension, adduction, and abduction of the foot
- Flexion and extension, bending to the right and left, right and left turning of the head

Motion ranges for these movements were measured by means of a set of measuring devices. The investigations embraced 355 men and 215 women aged 18 to 65. The results of the investigations were used to calculate standards, i.e., biological bases of reference for particular motion ranges. Standards were developed for three age categories and adequate subordinate classes of movements.

These are:

Age Groups	Motion Classes
1. 18 up to 30 years	1. Wide range of motion (W)
2. 31 up to 40 years	2. Average range of motion (A)
3. 41 up to 65 years	3. Small range of motion (S)

The following formulae were used to calculate the ranges of motion in three classes (Batogowska, 1977):

$$\begin{array}{lll} W & \text{— Wide range of motion} & W = (\beta_1, \beta_2) \\ A & \text{— Average range of motion} & A = (\beta_2, \beta_3) \\ S & \text{— Small range of motion} & S = (\beta_3, \beta_4) \end{array} \tag{1}$$

Values of β for i = 0, 1, 2, 3, 4 were found from equations:

$$\begin{aligned} \beta_1 &= P_{95} \\ \beta_2 &= x + 1/2\ s \\ \beta_3 &= x - 1/2\ s \\ \beta_4 &= P_5 \end{aligned} \tag{2}$$

where

P_5 = value of the 5th percentile
P_{95} = value of the 95th percentile
x = average value
s = standard deviation

Angular values of the hand grasping characteristic are shown in Table 73.3. It includes the values attained by a healthy population of adults aged 18 to 65. The values of movement ranges are grouped in three motion classes, i.e., from the maximum to the minimum value— classes W, A, and S. In each class, minimum and maximum extreme values are given, calculated according to formulae given earlier. Three age categories are subordinated to the classes of movement shown. The youngest persons, aged 18 to 30, belong to class W, characterized by the maximum values. Class A includes adults aged 31 to 40, and class S includes the eldest subjects, aged 41 to 65, whose movement range in particular joints has the smallest values.

TABLE 73.3 Movement Ranges of the Grasping Hand for Three Classes of Motion: Wide (W), Average, (A), and Small (S) for Polish Population 18–65 Years

		W 18–30 years		A 31–40 years		S 41–65 years	
		Men	Women	Men	Women	Men	Women
Flexion	min.	96°	99°	75°	75°	50°	48°
	max.	119°	126°	95°	98°	74°	74°
Extension	min.	65°	62°	46°	43°	31°	28°
	max.	88°	85°	64°	61°	45°	42°
Pronation	min.	100°	108°	81°	87°	64°	64°
	max.	122°	129°	99°	107°	80°	86°
Supination	min.	94°	100°	69°	82°	59°	65°
	max.	112°	118°	93°	99°	68°	81°

The values of movement ranges for particular classes shown in the table can be practically applied in the rehabilitation process. If the age of a patient is known, one can read from the table the value of a range of movement in a given joint that can be attained by the joint, i.e., the range of movement that it should have had prior to the injury. The maximum value is particularly significant in our case. It shows the angular value that should be aimed at in the rehabilitation process. The maximum value of range movement provides a basis for the evaluation of rehabilitation progress in a selected age group and is expressed by ϕ.

After assessing the value ϕ, the value of the range of movement of a subject investigated before the rehabilitation process should be defined. This state is denoted by the symbol ϕ' and is indicated by measurement, using a suitable protractor.

The final step is to define the "decrease in the movement range." Knowing the values ϕ and ϕ' one can calculate the absolute decrease in the movement range (Ar), and a percentage decrease in the movement range (Pr) using the following formula:

$$\mathrm{Ar} = (\phi - \phi') \tag{3}$$

$$\mathrm{Pr} = \phi - \phi'/\phi \times 100 \tag{4}$$

The value of a percentage decrease in movement range (Pr) is inversely proportional to the ability of movement (Ab) which can be defined by the equation:

$$\mathrm{Ab} = 100 - \mathrm{Pr}\% \tag{5}$$

The above values Ar, Pr, and Ab provide a basis for a direct evaluation of the progress of the rehabilitation process. Based upon the above, the motion ability can be assessed in many various stages of rehabilitation. This assessment may be employed in:

- Diagnosing a disease
- Evaluating progress and monitoring the rehabilitation process
- Determining a permanent decrease in movement ranges

The assessment may be either continuous or administered at random, depending upon the requirements.

The application of the method described above is presented in a graph (Figure 73.15) which shows the rehabilitation process of a wrist joint. A considerable movement restriction was caused by injury of the hand in the wrist joint and surgery that followed the injury. Results of systematically taken measurements of hand flexion (a) and extension (b) shown in Figure 73.15, illustrate a process of gradual improvement. Percentage of angular decrease in the range of hand flexion was 14.6% and hand extension 30.8%. Thus, the post-injury efficiency of the wrist joint can be determined as 85.4% during flexion and 69.2% during extension.

Motion efficiency of individual movements can be assessed using different scales, depending upon requirements, disease and general mobility. The scale is also dependent upon the individual preferences of the assessor. Scales of three, five, or ten grades may be used in assessing the motion efficiency. A five-grade scale may be considered optimal after Zeyland-Malawka et al. (1968):

- Very good: decrease up to 20% — efficiency from 80%
- Good: decrease up to 40% — efficiency from 60%
- Poor: decrease up to 60% — efficiency from 40%
- Very poor: decrease more than 60% — efficiency less than 40%

The above proposal does not preclude applications of other assessment scales, which may be found convenient.

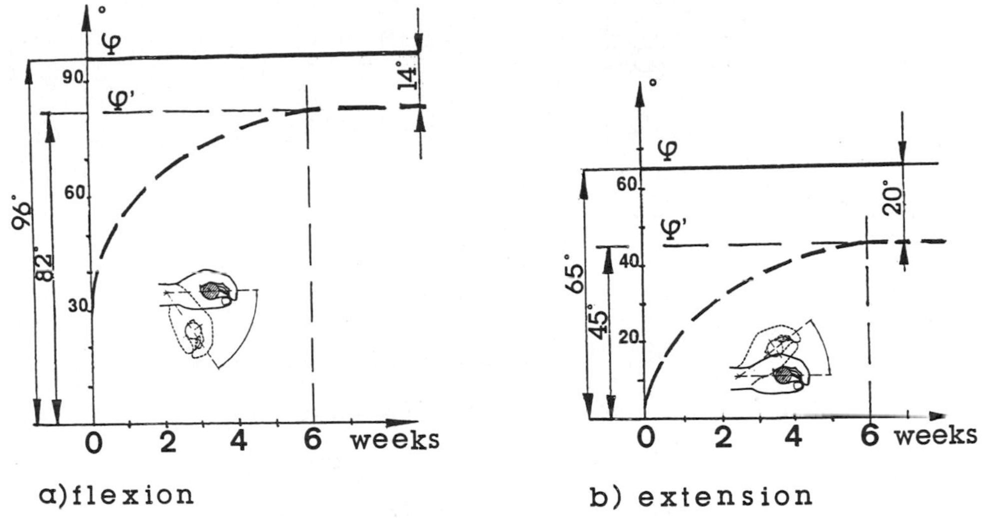

FIGURE 73.15 An example of the wrist joint rehabilitation at the movements of hand flexion (a) andextension (b).

The measurement of force is, besides motion ranges, one of the parameters determining the efficiency of the person. Ergonomics studies the amount of force exerted by the hand or the foot. The results of investigations are used in designing hand- or foot-operated control devices. Some of the results can be used in rehabilitation for the assessment of the efficiency of an impaired organ or organ with dysfunction resulting from a disease. The assessment is based, as in the case of motion ranges, on the standards of force range, developed with the use of data concerning the healthy population. These standards make biological bases of reference to be strived for by a person increasing the efficiency of an impaired organ. Investigations conducted by many scientists confirm the fact that there is a correlation between the exerted force and circumferential measurements as well as body weight (Laubach, 1966; Hunsicker, 1970; Malinowski, 1975; Jarosz, 1993; Łuczak et al., 1993; Batogowska, 1993).

Persons with motor dysfunction, as it was proved in an earlier section, have lower values of somatic characteristics — the same concerns children and young people. Rehabilitation makes it possible to increase the efficiency of particular motor organs and activate respective groups of muscles. Measurement of force can be an excellent indicator of the rehabilitation progress. Following this line of thought, the measuring stand to test the hand efficiency was developed in the Institute of Industrial Design in Warsaw (Nowak, 1995). This efficiency is determined, among others, by measurement of force exerted by the hand grasping a cylindrical grip. Figure 76.16 shows the schematic diagram of the measuring stand. The principle of measurement is based upon the hydraulic system. The measuring system cooperates with a computer unit. The appropriate program makes it possible to calculate and visualize tested parameters. The subject can watch the effect of measurement on the monitor. This can increase motivation to achieve better results when the stand is used as a rehabilitation device, creating natural biofeedback. In the case of children, rehabilitation progress can be stimulated by games. There is a special computer program developed for this purpose.

The increase of motor efficiency is an important factor in the rehabilitation of children with cerebral palsy. Myszkowski and Stężała (1989) put forward a proposal of an interesting testing stand. It consists of the set of control devices — manipulators using the handgrip function to perform repeated movements. The set of manipulators and the respective psychomotor computer tests allow the investigation and rehabilitation of children regardless of the degree of motor dysfunction of the hands.

The examples reported in this subsection prove that methods used in ergonomics are useful in the rehabilitation process. These methods, as well as the results of investigations, can be used for assessing an impaired or injured organ. Measuring stands can serve as rehabilitation devices.

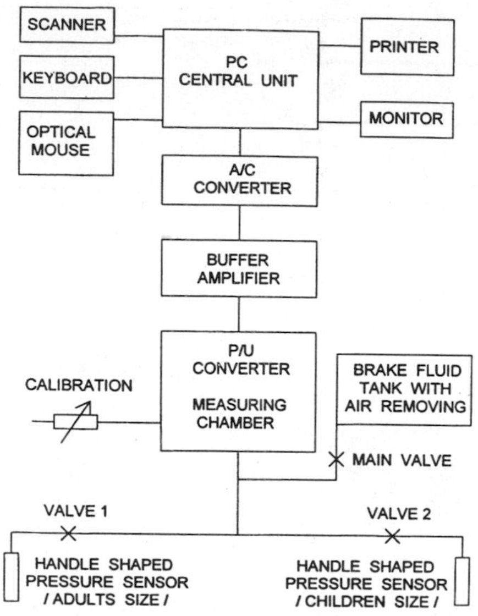

FIGURE 73.16 Schematic diagram of measuring stand for estimation of hand capabilities.

Defining Terms

Anthropometry: (anthropos — a man, metreo — I measure). A set of techniques used in anthropological investigations, it embraces methods of physical measurements of the man.

Rehabilitation: A science of systematic multidimensional study of disordered human neuropsychosocial and/or musculoskeletal function(s) and its (their) remediation by physicochemical and/or psychosocial means.

Impairment: Any loss or abnormality of psychological and anatomical structure or function (WHO, 1980).

Disability: A restriction or lack of ability (resulting from impairment) to perform an activity in the manner or within the range considered normal for a human being (WHO, 1980).

Handicap: A disadvantage for a given individual, resulting from an impairment or disability that limits or prevents the fulfillment of role that is normal (depending on age, sex, and social and cultural factors) for that individual.

Rehabilitation ergonomics: An interdisciplinary field of science aiming at adjusting material work and life environment, including articles of everyday use and rehabilitation equipment, to the psychophysical requirements of the disabled. Rehabilitation ergonomics assists the rehabilitation process, accelerating integration of the disabled into society.

References

Batogowska, A. 1993. Selected maximum measurements for the needs of clothing design. In the catalog *Human Dimension — Ergonomics and Design*, Institute of Industrial Design, Warsaw: 64 (in Polish).

Benktzon, M. 1993. Designing for our future selves: the Swedish experience. *Applied Ergonomics.* 24(1): 19-27.

Bouisset, S., Moynot, C. 1985. Are paraplegics handicapped in the execution of a manual task? *Ergonomics.* 28(7): 299-308.

Boussena, M., Davies, B.T. 1987. Engineering anthropometry of employment rehabilitation centre clients. *Applied Ergonomics.* 18(3): 223-228.

Catalogue of the exhibition "Human Dimension -Design and Ergonomics." 1993. Institute of Industrial Design, Warsaw.

Das, B., Grady, M. 1983. Industrial workplace layout design. An application of engineering anthropometry. *Ergonomics.* 26(5): 433-447.

Das, B., Kozey, J. 1994. Structural anthropometry for wheelchair mobile adults. *12th Triennial Congress of IEA, Toronto. vol.3 (Rehabilitation Ergonomics)*: 63-65.

Dąbrowska-Kiełek, J., Szymańska-Petrykowska, J. 1991. Clothes and shoes. *Designer's Vademecum — Problems of the Disabled,* Vol. 5 (in Polish).

Floyd, W.F. 1966. A study of the space requirements of wheelchair users. *Paraplegia.* 1(4): 24-37.

Frejlich, C. 1993. Analysis of the active contact surface of the hand with a handle depending on its cross-section shape and size, in the catalogue *Human Dimension — Ergonomics and Design,* Institute of Industrial Design, Warsaw: 100 (in Polish).

Goldsmith, S. 1967. *Designing for the Disabled.* London: RIBA Publications Ltd.

Goswami, A., Ganguli, S., Chatterjee, B.B. 1987. Anthropometric characteristics of disabled and normal Indian men. *Ergonomics.* 30(5): 817-823.

Grabowska, Z., Salwa, J., Seyfried, A. 1968. Studies on the work zone limitation in patients with rheumatoid arthritis. *Informative Newsletter, Research Department of ZSI,* No 5: 7-11.

Holden, J.M., Fernie, G. 1989. Specification for a mass producible static lounge chair for the elderly. *Applied Ergonomics.* 20(1): 187-199.

Hunsicker, P., Greey, G. 1970. Studies in human strength. *Research Quarterly,* No 2: 109-122.

Jarosz, E. 1996. Determination of the workspace of wheelchair users. *International Journal of Industrial Ergonomics,* 17: 123-133.

Kumar, S. 1989. Rehabilitation and ergonomics: Complimentary disciplines. *Canadian Journal of Rehabilitation,* 3: 99-111.

Kumar, S. 1992. Rehabilitation: An ergonomic dimension. *International Journal of Industrial Ergonomics.* Elsevier. 9(2):97-108.

Kumar, S. 1995. Rehabilitation Ergonomics: Rationale, Means and Justification. *IEA World Conference, Rio de Janeiro*: 84-89.

Laubach, L.L. 1981. Anthropometry of aged male wheel-dependent patients. *Annals of Human Biology,* 8(1): 25-29.

Lebiedowska, M.K., Graff, K., Syczewska, M., Kalinowska, M., Polisiakiewicz, A., Lebiedowski, M.J. 1994. Mathematical models as a method of child growth process assessment. *12th Triennial Congress of IEA, Toronto, Vol. 3 (Rehabilitation Ergonomics):* 52-54

Łuczak, E., Mięsowicz, I., Szczygieł, A. 1993. Somatic characteristics of children and the young with celebral palsy. AWF Cracov, *Scientific Annals,* vol. XXVI: 121-142.

Malinowski, A. 1975. Differentiation of muscular force of the right and left hand in adult men and women according to the age. Poznań, Academy of Physical Training. *Monographs,* No 8.

Mięsowicz, I. 1990. Somatic development of handicapped children. *Annals of Special Pedagogics,* vol. 1: 157-170.

Molenbroek, J.F.M. 1987. Anthropometry of elderly people in the Netherlands: research and applications. *Applied Ergonomics,* 18(3): 187-199.

Myszkowski, R., Stężała, D. 1989. Scientific system for measuring the level of psychomotoric efficiency of children with celebral palsy. *Technical Tasks of Medicine* XX(3): 166-176.

Nowak, E. 1976. Determination of the upper extremities workspace for the needs of workstands design. Institute of Industrial Design, *Works and Materials,* vol. 30. Warsaw.

Nowak, E. 1978. Determination of the spatial reach area of the arms for workspace design purposes. *Ergonomics,* No 7: 493-507.

Nowak, E. 1988. Physical development of children and the young aged 4-18. Data for design purposes. Institute of Industrial Design, *Works and Materials,* vol. 75. Warsaw.

Nowak, E. 1988. Method of workspace determination. *Institute of Industrial Design News-Design,* vol.1: 3-8.

Nowak, E. 1989. Workspace for disabled people. *Ergonomics,* No 9: 1077-1088.

Nowak, E. 1992. Practical application of anthropometric research in rehabilitation. *International Journal of Industrial Ergonomics*, No 9: 109-115.

Nowak, E. 1993. Ergonomics and design for the needs of the disabled. *Materials of the Conference "Equalizing opportunities of the disabled."* Warsaw (in Polish).

Nowak, E. 1993. Hand grip classification, in the catalogue *Human Dimension — Ergonomics and Design*, Institute of Industrial Design, Warsaw: 30 (in Polish).

Nowak, E. 1994. Anthropometric measurements of the young for the needs of clothing design. *12th Triennial Congress of IEA, Toronto, vol.3 (Rehabilitation Ergonomics)*: 58-59.

Nowak, E. 1995. Workstand for measuring hand efficiency. *Materials of the Conference of PTF Physiotherapeutists of the Hand*, Poznań, September 21-23.

Nowak, E. 1996. The role of anthropometry in design of work and life environments of the disabled population. *International Journal of Industrial Ergonomics*, 17: 113-121.

Petzall, J. 1993. Ambulant disabled persons using buses: experiments with entrances and seats. *Applied Ergonomics*, No 24: 313-326.

Pheasant, S. 1986. *Bodyspace. Anthropometry, Ergonomics and Design*. Taylor and Francis. London and Philadelphia.

Przybylski, B. 1979. *Flats for the Disabled*. Publishing House of CZSBM, Warsaw.

Samsonowska-Kreczmer, M. 1988. Measurements of disabled young people. Data for clothing design. *Institute of Industrial Design News — Design*, No 4: 5-8.

Samsonowska-Kreczmer, M. 1990. *Anthropometric Measurements of Children with Celebral Palsy*. Institute of Industrial Design, Warsaw.

Sperling, L., Karlsson, M. 1989. Clothing fasteners for long-term-care patients. Evaluation of standard closures and prototypes on test garments. *Applied Ergonomics*, 20(2): 97-104.

Wężyk, E. 1989. *Physical Development of Children and the Young with Celebral Palsy*. Wrocław: Department of Anthropology of Wrocław University. (Master's thesis).

World Program of Action for the Benefit of the Disabled. *Problems of Occupational Rehabilitation* CNB CZSI, No 4.

Thoren, M. 1994. Clothing made to fit the disabled users. *12th Triennial Congress of IEA, Toronto, vol.3 (Rehabilitation Ergonomics)*: 187-189.

Zeyland-Malawka, E., Gładkowska, E., Henicz, T., Domańska, B. 1968. Method of investigation and assessment of motion range and usefulness of the hand. *Physical Culture*, No 6: 246-256.

74

Update on the Use of Back Belts in Industry: More Data — Same Conclusion

Stuart M. McGill
University of Waterloo

The use of abdominal belts in industrial settings continues to be the topic of lively debate. The premier question still remains "Should abdominal belts be prescribed to workers in industry to perform manual materials handling tasks?" In a paper published a few years ago (McGill, 1993), I reviewed the available scientific literature pertaining to the use of back belts in industry with the objective of formulating a policy for belt prescription. The intent was not to take a position, either unconditionally for or against belt usage, but rather weigh the potential assets and liabilities for the development of belt prescription guidelines. The purpose of this chapter is to briefly review the literature previously reported, together with the most recent scientific data to see if our position on belt prescription has changed.

Abdominal belts and lumbar supports continue to be sold to industry in the absence of a regulatory requirement to conduct controlled clinical trials similar to that required of drugs and other medical devices. Many claims have been made as to how abdominal belts could reduce injury. For example, some have suggested that the belts remind people to lift properly. Some have suggested that belts may possibly support shear loading on the spine that results from the effect of gravity acting on the handheld load and mass of the upper body when the trunk is flexed. Compressive loading of the lumbar spine has been suggested to be reduced through the hydraulic action of increased intra-abdominal pressure associated with belt wearing. Belts have been suspected of acting as a splint, reducing the range of motion, and thereby decreasing the risk of injury. Still other hypotheses as to how belts may affect workers include (1) providing warmth to the lumbar region, (2) enhancing proprioception via pressure to increase the perception of stability, and (3) reducing muscular fatigue. These issues, together with others, will be addressed in this chapter.

A recent publication from NIOSH (1994) entitled *Workplace Use of Back Belts,* contained critical reviews of a substantial number of scientific reports evaluating back belts and concluded that back belts do not prevent injuries among uninjured workers nor do they consider back belts to be personal protective equipment. While this is generally consistent with our position stated in 1993, my personal position for belt prescription is somewhat more moderate.

0-8493-2641-9/99/$0.00+$.50

The following sections have subdivided the scientific studies into clinical trials and those that examined biomechanical, psychophysical, and physiological changes from belt wearing. Finally, based on the evidence, guidelines are recommended for the prescription and usage of belts in industry.

74.1 Field Trials

Many clinical trials that were reported in the literature were fraught with methodological problems and suffered from the absence of a matched control group, no post-trial follow-up, limited trial duration, and insufficient sample size. While the extreme difficulty in executing a clinical trial is acknowledged, only a few trials will be reviewed in this chapter.

The first trial reviewed here was reported by Walsh and Schwartz (1990), in which 81 male warehouse workers were divided into three groups: a control group (n = 27); a group that received a half-hour training session on lifting mechanics (n = 27); and a group that received the one-hour training session and wore low back orthoses while at work for the subsequent six months (n = 27). Instead of using more common types of abdominal belts, this research group used orthoses with hard plates that were heat molded to the low back region of each individual. Given the concern that belt wearing was hypothesized to cause the abdominals to weaken, the abdominal flexion strength of the workers was measured both before and after the clinical trial. The control group and the training-only group showed no changes in abdominal flexor strength nor any change in lost time from work. The third group, which received both training and wore the belts, showed no changes in abdominal flexor strength or accident rate, but did show a decrease in lost time. However, it appears that the increased benefit was only to those workers who had a previous low back injury.

In a larger clinical trial reported by Reddell (1992) and colleagues, 642 baggage handlers who worked for a major airline were divided into 4 treatment groups; a control group (n = 248); a group that received only a belt (n = 57); a group that received only a one-hour back education session (n = 122); and a group that received both a belt and a one-hour education session (n = 57). The trial lasted eight months and the belt used was a fabric weight lifting belt, 15 cm wide posteriorly and approximately 10 cm wide anteriorly. There were no significant differences between treatment groups for total lumbar injury incident rate, lost workdays, or workers' compensation rates. While the lack of compliance by a significant number of subjects in the experimental group was cause for consideration, those who began wearing belts but discontinued their use had a higher lost-day case injury incident rate. In fact, 58% of workers belonging to the belt-wearing groups discontinued wearing belts before the end of the eight-month trial. Further, there was an increase in the number and severity of lumbar injuries following the trial of belt wearing.

The clinical trial reported by Mitchell and colleagues (1994) was a retrospective study administered to 1,316 workers that performed lifting activities in the military. While this study relied on self-reported physical exposure and injury data over six years prior to the study, the authors did note that the costs of a back injury that occurred while wearing a belt were substantially higher than if injured otherwise.

A most recent study, reported by Kraus and colleagues (1996), surveilled the low back injury rates of nearly 36,000 employees of the Home Depot Stores in California from 1989 to 1994. As reported, the company implemented a mandatory back belt use policy. Although the authors claim that belt wearing reduced the incidence of low back injury, analysis of the data and methodology suggests a much more cautious interpretation may be warranted. The concern is based on two issues: the lack of a robust effect and co-interventions (in addition to belts). The data suggest that the beneficial effect was limited to men with 1 to 3 years of employment (but not for longer or shorter lengths of employment) and for women employed 1 to 2 years — this is a very narrow band of affected employees. However, of greatest concern is the lack of scientific control over co-interventions to ferret out the true belt wearing effect — there was no comparable non-belt wearing group, which is critical given that the belt wearing policy was not the sole intervention at Home Depot. For example, over the period of the study, the company increased the use of pallets and forklifts (changing physical demands), installed mats for cashiers, implemented post-accident drug testing, and enhanced worker training. In fact, the company made a conscious attempt to enhance safety in the corporate culture. This was a large study and the authors deserve credit for the massive data reduction and logistics. However, despite the title and claims that back belts reduce low back injury, this uncontrolled study cannot answer the question about the effectiveness of belts.

In summary, difficulties in executing a clinical trial are acknowledged: the placebo effect is a concern, as it is difficult to present a true double-blind paradigm to workers since those who receive belts certainly know so; and there are logistical constraints on duration, diversity in occupations, and sample size. However, the data reported in the better-executed clinical trials cannot support the notion of universal prescription of belts to all workers involved in manual handling of materials to reduce the risk of low back injury. There is weak evidence to suggest that those already injured may benefit from belts (or molded orthoses) with a reduced risk of injury recurrence. However, there does not appear to be support for uninjured workers wearing belts to reduce the risk of injury, and in fact, there appears to be an increased risk of injury during the period following a trial of belt wearing. Finally, there appears to be some evidence to suggest that cost per back injury may be higher if the worker was wearing a belt than if injured otherwise.

74.2 Biomechanical Studies

Biomechanical studies have examined changes in low back kinematics, posture, and issues of specific tissue loading. Two studies in particular have suggested that wearing an abdominal belt can increase the margin of safety during repetitive lifting: Lander et al. (1992) and Harman et al. (1989). Both of these papers reported ground reaction force and measured intra-abdominal pressure while subjects performed repeated lifting of barbells. Both reports observed an increase in intra-abdominal pressure when abdominal belts were worn. These researchers assumed that intra-abdominal pressure is a good indicator of spinal forces, which is highly contentious. Nonetheless, they assumed the higher recordings of intra-abdominal pressure indicated an increase in low back support which, in their view, justified wearing belts. Spinal loads were not directly measured or calculated in these studies.

Several studies have questioned the hypothesized link between elevated intra-abdominal pressure and reduction in low back load. For example, using an analytical model and data collected from three subjects lifting various magnitudes of loads, McGill and Norman (1987) noted that a build-up of intra-abdominal pressure required additional activation of the musculature in the abdominal wall, resulting in a net increase in low back compressive load and not a net reduction of load as had been previously thought. In addition, Nachemson et al. (1986) published some experimental results that directly measured intradiscal pressure during the performance of valsalva maneuvers documenting that an increase in intra-abdominal pressure increased, not decreased, the low back compressive load. Therefore, it would seem erroneous to conclude that an increase in intra-abdominal pressure due to belt wearing reduces compressive load on the spine. In fact, it may have no effect or may even increase the load on the spine.

In another study McGill, Norman, and Sharratt (1990) examined intra-abdominal pressure and myoelectric activity in the trunk musculature while six male subjects performed various types of lifts both wearing and not wearing an abdominal belt (a stretch belt with lumbar support stays, Velcro tabs for cinching, and suspenders for when subjects were not lifting). Wearing the belt increased intra-abdominal pressure by approximately 20%. Further, it was hypothesized that if belts were able to help support some of the low back extensor moment, one would expect to measure a reduction in extensor muscle activity. There was no change in activation levels of the low back extensors nor in any of the abdominal muscles (rectus abdominis or obliques).

In a recent study that examined the affect of belts on muscle function, Reyna and colleagues (1995) examined 22 subjects for isometric low back extensor strength and found belts provided no enhancement of function (although this study was only a four-day trial and did not examine the affects over a longer duration). Ciriello and Snook (1995) examined 13 men over a four-week period lifting 29 metric tonnes in four hours twice a week both with and without a belt. Median frequencies of the low back electromyographic signal (which is sensitive to local muscle fatigue) were not modified by the presence or absence of a back belt, strengthening the notion that belts do not significantly alleviate the loading of back extensor muscles. Once again this trial was not conducted over a very long period of time.

In 1986, Lantz and Schultz observed the kinematic range of gross body motions while subjects wore low back orthoses. While they studied corsets and braces rather than abdominal belts, they did report restrictions in the range of motion, although the restricted motion was minimal in the flexion plane. In a more recent study McGill, Seguin, and Bennett (1994) tested flexibility and stiffness of the lumbar torsos of 20 male and 15 female adult subjects, both while they wore and did not wear a 10 cm leather abdominal belt. The stiffness of the torso was significantly increased about the lateral bend and axial twist axes but not when subjects were rotated into full flexion. Thus, it would appear from these studies that abdominal belts assist to restrict the range of motion about the lateral bend and axial twist axes but do not have the same effect when the torso is forced in flexion, as in an industrial lifting situation. Posture of the lumbar spine is an important issue in injury prevention for several reasons, but in particular Adams and Hutton (1988) have shown that the compressive strength of the lumbar spine decreases when the end range of motion in flexion is approached. Therefore, if belts restrict the end range of motion one would expect the risk of injury to be correspondingly decreased. While, the splinting and stiffening action of belts occurs about the lateral bend and axial twist axes, stiffening about the flexion–extension axes appears to be less. A most recent data set presented by Granata, Marras, and Davis (1997) supports the notion that some belt styles are better in stiffening the torso in the manner described above, namely the taller elastic belts which span the pelvis to the rib cage. Furthermore, they also documented that a rigid orthopedic belt generally increased the lifting moment, while the elastic belt generally reduced spinal load, but a wide variety in subject response was noted (some subjects experienced increased spinal loading with the elastic belt). However, even in well-controlled studies, it appears that belts can modulate lifting mechanics in some positive ways in some people and in negative ways in others.

74.3 Physiological Studies

Blood pressure and heart rate were monitored by Hunter and colleagues (1989) while five males and one female subject performed dead lifts, bicycle riding, and one-armed bench presses, while wearing and not wearing a 10-cm weight belt. A load of 40% of each subject's maximum weight in the dead lift was held in a lifting posture for two minutes. The subjects were required to breathe throughout the duration so that no valsalva effect occurred. During the lifting exercise blood pressure was significantly higher (up to 15 mmHg), while heart rate also was higher when the belt was worn. Given the relationship between elevated systolic blood pressure and an increased risk of stroke, Hunter et al. (1989) concluded that individuals who may have cardiovascular system compromise are probably at greater risk when undertaking exercise while wearing back supports.

Recent work conducted in our own laboratory (Rafacz and McGill, 1996) investigated the blood pressure of 20 young men performing sedentary and very mild activities both with and without a belt (the belt was the elastic type with suspenders and Velcro tabs for cinching at the front). Wearing this type of industrial back belt significantly increased diastolic blood pressure for quiet sitting and standing both with and without a handheld weight, during a trunk rotation task, and during a squat lifting task. There is increasing evidence to suggest belts increase blood pressure!

Over the past three or four years I have been asked to deliver lectures, and participate in academic debate on the back belt issue. On several occasions, occupational medicine personnel have approached me after hearing the effects of belts on blood pressure and intra-abdominal pressure, and have expressed suspicions that long-term belt wearing at their particular workplace may possibly be linked with higher incidents of varicose veins in the testicles, hemorrhoids, and hernias. At this point in time, there has been no scientific and systematic investigation of the validity of these claims and concerns. Rather than wait for strong scientific data to either lend support to these conditions or dismiss them, it may be prudent to simply state concern. This will motivate studies in the future to track the incidents and prevalence of these pressure-related concerns to assess whether they are indeed linked to belt wearing.

74.4 Psychophysical Studies

Studies based on the psychophysical paradigm allow workers to select weights that they can lift repeatedly using their own subjective perceptions of physical exertion. McCoy et al. (1988) examined 12 male college students while they repetitively lifted loads from floor to knuckle height at the rate of 3 lifts per minute for a duration of 45 minutes. They repeated this lifting bout three times, once without a belt and once each with two different types of abdominal belts (a belt with a pump and air bladder posteriorly, and the elastic stretch belt previously described in the McGill, Norman, and Sharratt [1990] study). After examining the various magnitudes of loads that subjects had self-selected to lift in the three conditions, it was noted that wearing belts increased the load that subjects were willing to lift by approximately 19%. There has been some concern that wearing belts fosters an increased sense of security, which may or may not be warranted. This evidence may lend some support to this criticism of wearing abdominal belts.

74.5 Back Belt Prescription

My earlier report (McGill, 1993) presented data and evidence that neither completely supported nor condemned the wearing of abdominal belts for industrial workers. Definitive laboratory studies that describe how belts affect tissue loading and physiological and biomechanical function have yet to be performed. In addition, clinical trials of sufficient scientific rigor to comprehensively evaluate the epidemiological risks and benefits from exposure to belts must be done. The challenge remains to arrive at the best strategy for wearing belts. Therefore, the available literature will be interpreted and given placement, and also combined with "common sense" to derive the most sensible position on prescription.

Given the available literature, it would appear the universal prescription of belts (i.e., providing belts to all workers in a given industrial operation) is not in the best interest of globally reducing both the risk of injury and compensation costs. Uninjured workers do not appear to enjoy any additional benefit from belt wearing and in fact may be exposing themselves to the risk of a more severe injury if they were to become injured and may have to confront the problem of weaning themselves from the belt. However, if some *individual* workers perceive a benefit from belt wearing, then they may be allowed to conditionally wear a belt, but only on trial. The mandatory conditions for prescription (*for which there should be no exception*) are as follows:

1. Given the concerns regarding increased blood pressure and heart rate, and issues of liability, all those who are candidates for belt wearing should be screened for cardiovascular risk by medical personnel.
2. Given the concern that belt wearing may provide a false sense of security, belt wearers must receive education on lifting mechanics (back school). All too often belts are being promoted to industry as a quick fix to the injury problem. Promotion of belts conducted in this way is detrimental to the goal of reducing injury as it redirects the focus from the cause of the injury. Education programs should include information on how tissues become injured, techniques to minimize musculoskeletal loading, and what to do about feelings of discomfort to avoid disabling injury.
3. No belts will be prescribed until a full ergonomic assessment has been conducted of the individual's job. The ergonomic approach will examine, and attempt to correct, the cause of the musculoskeletal overload and will provide solutions to reduce the excessive loads. In this way, belts should only be used as a supplement for a few individuals while a greater plant-wide emphasis is placed on the development of a comprehensive ergonomics program.
4. Belts should not be considered for long-term use. The objective of any small-scale belt program should be to wean workers from the belts by insisting on mandatory participation in comprehensive fitness programs and education on lifting mechanics, combined with ergonomic assessment. Furthermore, it would appear wise to continue vigilance in monitoring former belt wearers for a period of time following belt wearing, given that this period appears to be characterized by elevated risk of injury.

References

Adams, M.A. and Hutton, W.C. Mechanics of the intervertebral disc, in *The Biology of the Intervertebral Disc*, Ed. P. Ghosh, Boca Raton, FL:CRC Press, (1988).

Ciriello, V.M. and Snook, S.H. The effect of back belts on lumbar muscle fatigue. *Spine* 20(11):1271-1278 (1995).

Granata, K.P., Marras, W.S., and Davis, K.G. Biomechanical assessment of lifting dynamics, muscle activity and spinal loads while using three different style lifting belts, *Clin. Biomech.* 12(2): 107-115 (1997).

Harman, E.A., Rosenstein, R.M., Frykman, P.N., and Nigro, G.A. Effects of a belt on intra-abdominal pressure during weight lifting. *Med. Sci. Sports Exercise* 2(12):186- 190 (1989).

Hunter, G.R., McGuirk, J., Mitrano, N., Pearman, P., Thomas, B., and Arrington, R. The effects of a weight training belt on blood pressure during exercise. *J. Appl. Sport Sci. Res.* 3(1):13-18 (1989).

Kraus, J.F., Brown, K.A., McArthur, D.L., Peek-Asa, C., Samaniego, L., and Kraus, C. Reduction of acute low back injuries by use of back supports. *Int. J. Occup. Environ. Health* 2:264-273 (1996).

Lander, J.E., Hundley, J.R., and Simonton, R.L. The effectiveness of weight belts during multiple repetitions of the squat exercise. *Med. Sci. Sports Exercise* 24(5):603-609 (1992).

Lantz, S.A. and Schultz, A.B. Lumbar spine orthosis wearing. I. Restriction of gross body motion. *Spine* 11(8):834-837 (1986).

McCoy, M.A., Congleton, J.J., Johnston, W.L., and Jiang, B.C. The role of lifting belts in manual lifting. *Int. J. Ind. Ergonomics* 2:259-266(1988).

McGill, S.M. Abdominal belts in industry: A position paper on their assets, liabilities and use. *Am. Ind. Hyg. Assoc. J.* 54(12):752-754 (1993).

McGill, S.M. and Norman, R.W. Reassessment of the role of intra-abdominal pressure in spinal compression. *Ergonomics* 30(11):1565-1588 (1987).

McGill, S., Norman, R.W., and Sharratt, M.T. The effect of an abdominal belt on trunk muscle activity and intra-abdominal pressure during squat lifts. *Ergonomics* 33(2):147-160 (1990).

McGill, S.M., Seguin, J.P., and Bennett, G. Passive stiffness of the lumbar torso in flexion, extension, lateral bend and axial twist: The effect of belt wearing and breath holding. *Spine* 19(6):696-704 (1994).

Mitchell, L.V., Lawler, F.H., Bowen, D., Mote, W., Asundi, P. and Purswell, J. Effectiveness and cost-effectiveness of employer-issued back belts in areas of high risk for back injury. *J. Occup. Med.* 36(1):90-94 (1994).

Nachemson, A.L., Andersson, G.B.J., and Schultz, A.B. Valsalva maneuver biomechanics. Effects on lumbar trunk loads of elevated intra-abdominal pressures. *Spine* 11(5):476-479 (1986).

NIOSH -Workplace use of back belts, U.S. Department of Health and Human Services, Centres for Disease Control and Prevention. National Institute for Occupational Safety and Health. July, 1994.

Rafacz, W. and McGill, S.M. Abdominal belts increase diastolic blood pressure. *J. Occup. Environ. Med.* 38(9):925-927 (1996).

Reddell, C.R., Congleton, J.J., Huchinson R.D., and Mongomery J.F. An evaluation of a weightlifting belt and back injury prevention training class for airline baggage handlers. *Appl. Ergonomics* 23(5):319-329 (1992).

Reyna, J.R., Leggett, S.H., Kenney, K., Holmes, B., and Mooney, V. The effect of lumbar belts on isolated lumbar muscle. *Spine* 20(1):68-73 (1995).

Walsh, N.E. and Schwartz, R.K. The influence of prophylactic orthoses on abdominal strength and low back injury in the workplace. *Am. J. Phys. Med. Rehab.* 69(5):245-250 (1990).

75

The Influence of Psychosocial Factors on Sickness Absence

Chris J. Main
Hope Hospital, Salford, U.K.

A. Kim Burton
University of Huddersfield

Michele C. Battié
University of Alberta

75.1 Introduction

Despite increasing advances in medical technology, the cost of musculoskeletal incapacity, particularly low back pain, in terms of sickness benefits, invalidity benefits, and associated allowances has led to a fundamental reconsideration of the nature of chronic incapacity. Recent reports from the United Kingdom (CSAG, 1994), the United States (Bigos et al., 1994), and New Zealand (ACC and the National Health Committee, 1997) in their recommendations for a comprehensive multidisciplinary assessment

0-8493-2641-9/99/$0.00+$.50

for patients still symptomatic at six weeks, are based on the clear assumption that a significant proportion of chronic incapacity is preventable. Such a proposition represents a fundamental challenge to much of current medical practice.

Incapacity, as defined clinically is seen as the result of a series of processes in which nature of injury, rate of recovery, and extent of recovery are significantly intermeshed with the nature of medical management and patients' beliefs about the nature of their injury, confusion about the nature of hurting and harming, and mistaken beliefs that physical activity will lead to further damage. Such beliefs may also be the product of the organizational culture in which they work and affected by workplace practices in which persistence of incapacity is viewed with suspicion and conditions of service dictate that "100%" recovery is required before the employer is willing to undertake the "risks" of permitting an injured employee to return to work.

There has been a tendency also to predicate beliefs about incapacity on an incorrect and outmoded view of the relationships among physical and psychological factors in the context of injury. In simple terms, the mind and the body have become divorced. Psychological influences are viewed with suspicion either as a sign of mental infirmity or as malingering. The adoption in particular of a blue-collar macho culture has made it difficult to admit that there are many factors affecting successful recovery, and these are not simply confined to the parameters of medical treatment and associated rehabilitative efforts.

The costs of incapacity have led to a considerable financial commitment to the investigation and amelioration of risk factors for chronicity. It has been argued in the previous chapter that the concept of risk has been confined almost entirely to biomechanical and ergonomic perspectives. Psychosocial factors, similarly, have been investigated principally as aspects of "stress," a multifaceted concept which has incorporated a wide range of perspectives including management style, cognitive demands of tasks, fatigue, the social climate of work (whether supportive or antagonistic), and the recognition of stress responses in individuals. A wide-ranging survey of the literature has revealed direct relevance to the *interaction* of occupational and clinical factors in recovery from injury and return to work.

It is the intention of this chapter to address such issues from the perspective specifically of *psychological obstacles to rehabilitation.* An attempt will be made to offer a new conceptual framework and suggest a set of specific procedures and guidelines which may be of assistance in considering afresh the problem of occupational incapacity, using pain-associated incapacity (and back pain in particular) as an example.

75.2 Workloss and Suboptimal Work Functioning: The Extent of the Problem

The extent of the problem has been addressed in our previous chapter, and the issue will be addressed here specifically only to the extent that a specific psychosocial perspective is relevant.

Influence of Psychological Factors on Accidents and Reinjury

Psychological factors have been implicated in the occurrence of accidents, both in terms of their frequency and in the actual nature of the accidents. "Accident-proneness" has been appraised from both the statistical and the psychological or physiological points of view. A number of specific psychological characteristics have been associated with the occurrence of accidents. These can be conceptualized first in terms of central nervous system dysfunction in terms of the processing of information, and second in terms of performance or execution of tasks.

Apparent accident-proneness may be indicative of or may result from some sort of impaired function in the interpretation or execution of tasks (Porter, 1988). Fatigue and mood disturbance can affect concentration. Difficulties in concentration in turn can affect the accurate coding and decoding of information., which may lead to misinterpretation of information with associated performance compromise. She concludes that individuals do appear to differ in their propensity toward minor accidents, and this is most likely the result of an attention deficit, which is itself a consequence of stress. Such an effect may be particularly noticeable in complex tasks or tasks requiring sustained vigilance. The complexity of the task also appears to be important rather than the intelligence of the subject.

Execution of tasks can also be affected by impaired neuromuscular coordination as a result of fatigue or disturbances to muscle control. Possibly mediated by sympathetic over-activity such as heightened muscle tension or tremor. The specific importance of such factors, over and above central processing effects, on accident-proneness does not seem specifically to have been evaluated.

Whatever the scientific basis for the term "accident-proneness," it would appear to be the case, however, that some individuals *believe* themselves to be more prone to accidents, and as such engage in a variety of superstitious behaviors designed to reduce the likelihood of accidents. Such behaviors may range from relatively benign rituals to marked avoidance of a range of situations. To the extent that normal functioning is not compromised, such behavior can be viewed as harmless eccentricity, but where it develops a compulsive quality and interferes significantly with effective functioning or psychological well-being then it may constitute a clinical disorder necessitating treatment. Attempts to return to work may be hindered by such beliefs.

More common, however, are specific fears of hurting and harming when engaging in specific activities rather than worries about unforeseen accidents. Such fears can either be about the activity being painful or about the dangers of further damage or reinjury. Such fears may lead to excessive caution in the execution of tasks, disturbances to normal biomechanical rhythm, and thereby increased the actual risks of accident or re-injury. (The evaluation of the nature and significance of such beliefs and fears is discussed further below.)

Costs and Effects of Long-Term Disability

The economic costs of pain-associated incapacity have been discussed in earlier chapters. In addition to the costs to industry of lost productivity and wage-replacement, there is of course a cost to the injured worker and his/her family. This "cost" has to be understood not only in financial terms but also in terms of its psychological effects. In addition to the pain and limitations on everyday living, absence from work can be demoralizing. Boredom, loss of identity, and diminished self worth can create stress and depression, which can have a major impact on the individual and the family. Major psychological effects may not be evident until the worker has been off work for some time and faces the prospect of losing his/her job. Persistence of pain, disturbance to normal activities (particularly sleep) and routines of daily living, and concerns about their recovery and satisfactory return to work can produce frustration and irritability even in the short term. In order to understand the complex interaction between medical, psychological, and occupational factors, it is necessary to reconsider the nature of incapacity and its relationship with sickness absence from work.

75.3 Models of Incapacity

The evidence for and against biomechanical and ergonomic models has been reviewed in our related chapter (Burton et al. 1997). Any workplace will have its own culture or *zeitgeist* concerning the nature of injury and the specific role of ergonomic and biomechanical factors. These maxims may be "enshrined" in specific documents concerning work practices or be the subject of health and safety training or procedures required by the organization to reduce the risk of injury. As mentioned, the actual *evidence* for the efficacy of primary prevention is in fact weak, and whether workers understand, adhere to, or even accept the assumptions behind such regulations is unknown. To the extent that workers do form a view about the nature of injury and the process of recovery, it seems likely that they would adopt a biomechanical or ergonomic view of causation rather than a more complex biopsychosocial formulation.

If a clearly identified injury has occurred, and this has resulted in a medical assessment, then the person may have been offered a diagnosis implying a structural basis to their complaint. It may be implied that some sort of permanent damage might have occurred and a series of impressive high-technology investigations may be recommended. Even if no permanent physical impairment in terms of structural damage to bone or nerves is identified, the notion of a significant weakness and therefore occupational vulnerability may be promulgated. Either of these views *may* of course have substance if a significant trauma has occurred, but usually have not. Even in the context of minor injury, it may be implied that the work accident (or incident) simply has been "the straw that broke the camel's back." It may be implied that there has been a cumulative impact of work leading finally to tissue breakdown. Osteoarthritic changes on X-ray may be adduced as evidence of this, despite the fact that only age-related changes may be apparent.

A worker may thus come to believe that his or her problem can only be understood in straightforward biomechanical, ergonomic, or medical terms. The worker may believe that the nature of the "injury" has specific implications for the nature of treatment and rehabilitation. His or her understanding of these issues concerning causality may make it difficult to accept the worth of and indeed necessity for, a speedy return to normal work. He may be resistant to rehabilitative efforts based on a self-help approach and may resist vigorously the suggestion that successful return to work may be hindered or indeed be prevented by psychosocial factors.

It is, in our view, necessary to consider disability from a wider perspective. Research into the nature of disability in clinical settings has suggested the need for a wider perspective on pain-associated incapacity than the traditional medical model. In formulating the biopsychosocial model of disability (Waddell, 1992), Waddell and his colleagues demonstrated that level of disability could by adequately understood only if consideration was given to psychological and social as well as physical factors. In an earlier publication (Waddell et al., 1984), they critically compared the influence of a large number of psychological factors on reported disability. They found that patients' current level of distress and behavioral responses to pain were far more important as predictors of disability than general personality traits, specific hypochondriachal fears, or generalized beliefs about patients' perceived degree of control over their lives. The biopsychosocial model, as it came to be known, was later further refined by the incorporation of beliefs about pain into a dynamic model addressing the transition from acute to chronic incapacity (Waddell et al., 1993). The relative importance of specific psychological factors is described in the next section. Recently it has been suggested (Main and Watson, 1996) that it may be helpful to consider incapacity as a developing process rather than as some sort of "end-state."

Most injuries seem to be over-reliant on concepts of tissue-healing time, where it is assumed that after about six weeks tissue should have "healed" and by implication the physical status should have returned to its preinjury status. This perception has the unfortunate concept that pain and pain-associated incapacity persisting beyond that time may be viewed with suspicion. Clearly a proportion of patients remain incapacitated after injury and theoretical models are needed which address the *process* of chronicity.

Injury and Incapacity: A Dynamic Model

Such perspectives of chronic incapacity as a *dynamic* process have not to date been integrated adequately into the psychology of return-to-work after injury. Before addressing occupational factors specifically, it is necessary from the psychological point of view to consider further the nature and influence of psychological factors in relation to pain and incapacity.

75.4 Psychological Factors and Pain

Nature of Psychological Factors

During the last 15 years, the role of psychological factors in the genesis and maintenance of pain problems has been increasingly recognized (Main and Spanswick, 1991). Early research into personality traits and identifiable psychiatric illness has been followed by investigation of more specific psychological characteristics such as psychological distress, pain behavior, beliefs about pain, and pain coping strategies. A range of psychometric tests, behavioral measures, and more general tools assessing function or impact of pain have been developed both for the assessment of pain and the investigation of specific psychological dimensions and mechanisms.

Clinical Perspectives on Pain and Disability

There is a wide range of psychological tests available. For the purpose of this review, psychological aspects of pain will be considered under a number of headings. The focus of this review will be primarily on clinical and occupational applications. A comprehensive analysis of the relative merits of objective and subjective measures in chronic pain from a methodological point of view is presented elsewhere (Dworkin and Whitney, 1992).

Evaluation of Pain

There is no simple relationship between the amount of tissue damage and the sensation of pain since the experience of pain is affected by a number of physiological and psychological factors. It cannot therefore be assessed as simply, for example, as temperature since it has both sensory and emotional components. Pain has been defined as: *An unpleasant sensory and emotional experience associated with actual or potential damage, or described in terms of such damage...pain is always subjective* (International Association for the Study of Pain, 1979). The emotional content can be appraised by investigation of the vocabulary patients use to describe their pain (Melzack, 1987; Melzack and Katz, 1992) or the way they complete pain drawings (Ransford et al., 1975; Parker et al., 1995). Thus, while the assessment of pain may appear to be fairly straightforward, an understanding of its significance requires a more comprehensive assessment of the pain *patient.*

Influence of General Psychological Characteristics

In general clinical practice there is a range of goals for psychological assessment including influences on pain perception, adjustment to pain, selection for particular types of treatment, and the design of treatment interventions. Methods of assessment include administration of questionnaires, psychodiagnostic clinical interviews, videotape analysis, and behavioral observation. No single assessment tool or indeed method of assessment is suitable for all purposes. The use of interviews and self-report inventories in the evaluation of pain patients is reviewed comprehensively elsewhere (Bradley et al., 1992). The focus of many of these studies has been on the nature of *pain perception.* For the purpose of this review, specific attention will be directed toward influences on incapacity, recovery from injury, and *obstacles to rehabilitation.*

It is important to consider how patients react to the whole treatment process, and the establishment of a successful "therapeutic alliance" between doctor and patient is extremely important (Hazard et al., 1994). With regard to pain patients in particular they considered that this was best achieved by setting mutually agreed goals concerning pain relief, improved functional capacity, and occupational reengagement. Mayer and his colleagues (Mayer and Gatchel, 1988) have consistently advocated functional restoration as the *primary* goal for rehabilitation for pain-associated incapacity. While acknowledging the tremendous value and importance of functional restoration in treatment of the chronically incapacitated patient, a degree of preselection is involved in all treatment programs and the noncommitted might feel that it cannot be assumed that all psychological obstacles can necessarily be dissolved or rendered inert by a robust functional restoration approach. Nonetheless, the functional restoration approach has been hugely important in shifting emphasis from "psychopathology" to function.

Although the more psychiatrically derived and complex psychological dimensions have been as yet of only limited value in the prediction of outcome of treatment, it would appear that simpler measures of the individuals' response to pain and incapacity are helpful.

In back-associated disability, patients' level of distress, in the form of heightened somatic awareness and depressive symptomatology, has been shown to be more closely associated with reports of disability than general personality characteristics or hypochondriacal fears (Main and Waddell, 1987). The two distress questionnaires later were combined into the Distress Risk Assessment Method or DRAM (Main et al., 1992), which yields a fourfold classification of patients in terms of their level of distress. Although the test does not offer an evaluation specifically of the relationship between distress and incapacity, scores on this test have been shown to be predictive of the development and continuation of chronic incapacity in patients with back pain (Burton et al., 1995). The relationship between generalized distress and more specific fears of hurting and harming (see below) may merit further investigation.

Pain Behavior

Although it is not possible to experience another's pain, people can communicate their pain either verbally or nonverbally. During the last 30 years, the importance of pain behavior has been increasingly recognized (Fordyce, 1976). It is important to recognize that the explanation for a pain behavior is not necessarily immediately apparent. Indeed a clear formulation is usually considered to require a detailed functional behavior analysis to identify the *precise* circumstances in which the behavior is occurring. The type of analysis will depend on the behavior in question. Typical pain behaviors might include verbalized pain responses, nonverbalized pain behaviors, patterns of activity, or health care usage (e.g., medication use). The behavioral view has been hugely influential as an alternative to the "disease-dominated" medical model and has led to significant changes in the manner and content of health care delivery (as evidenced by the development of pain management programs and functional restoration programs).

Pain behavior may be assessed in a number of ways, and a variety of methods have been devised for use in different settings Clinical approaches have included assessment of pain behavior during clinical examination (Waddell et al., 1980; Waddell et al., 1984) and use of videotaped ratings of pain behavior during a standardized sequence of movements (Keefe and Hill, 1985), but simpler rating scales have also been devised (Richards et al., 1982).

Although the interpretation of pain behavior is not without controversy, it should form an important aspect of assessment following injury and may contain important features indicative of obstacles to successful rehabilitation. Only the most committed of behavioral theorists would probably now argue that pain behavior can be adequately understood without an evaluation of specific beliefs, fears, and coping strategies. Nonetheless, in many contexts, self-report may be insufficiently informative, and an adequate crystallization of the problem may require a systematic behavioral analysis.

Cognitive Factors in Pain

The study of cognition has become increasingly complex, but it is possible in general to subdivide the field into three distinct fields of enquiry: beliefs or appraisal; cognitive processes; and coping strategies. Such perspectives have led to the investigation of cognitive factors in pain, adjustment to pain, and in the development of chronic incapacity. The complexity of these interrelationships is indicated in the following quotation:

"Patients who believe they can control their pain, who avoid catastrophizing about their condition; and who believe they are not severely disabled appear to function better than those who do not. Such beliefs may mediate some of the relationships between pain severity and adjustment" (Jensen et al., 1991, p. 249).

Specific Beliefs About the Nature of Pain

There are a number of different types of belief or appraisal about the nature of pain, and the development of new tests assessing different aspects of beliefs has become something of a "growth industry." A comprehensive review of the field is presented elsewhere (DeGood and Schutty, 1992). While many of these tests seem to be of value, there is clearly much conceptual overlap among them and further research is needed to determine the precise value of each of these tests for particular purposes and contexts. For this review, a number of tests will be chosen to illustrate the types of belief which have been found in the literature to be of relevance in the assessment of the psychological impact of pain.

In the psychological literature there has long been interest in the extent to which individuals believe they can control or gain control over aspects of their life. Researchers have assessed specific beliefs about health (Wallston et al., 1978). The same scale was then adapted for the study of pain by simple substitution of "pain" for "health" throughout the questionnaire (Toomey et al., 1991; Crisson and Keefe, 1988). Other "pain locus of control" scales have been developed (Main and Waddell, 1991) and used specifically in occupational settings. In one study, negative pain locus of control beliefs were significantly associated with absence from work (Symonds et al., 1996), but in a related study, the introduction of educational leaflets produced increased optimism about the control of pain (Symonds et al., 1995).

Locus of control beliefs would seem to have some influence on response to treatment and return to work, but the relationship of such beliefs to other beliefs merits further examination, and a critical comparison of these questionnaires in a wider range of occupational contexts is needed. Their specific value in occupational rehabilitation has yet to be demonstrated.

Specific Beliefs About Outcome of Treatment

Specific beliefs about treatment outcome can also be investigated. Such beliefs are often referred to as "self-efficacy beliefs." Self-efficacy has been defined in terms of the belief that one was capable of producing a behavior which was necessary to achieve a certain outcome (Bandura, 1977) This general idea has been investigated in terms of the relationship between such beliefs and resultant behavior in relation to treatment outcome.

According to DeGood and Shutty (1992), "These concepts suggest that pain patients who perceive themselves lacking the capacity to acquire self-management skills might be less persistent, more prone to frustration, and more apt to be noncompliant with treatment recommendations. Hence, some patients might demonstrate adequate understanding of particular treatment rationale, yet be noncompliant due to their perceived inability to produce the behavior necessary to follow treatment recommendations" (DeGood and Shutty, 1992, p. 221).

The Self-Efficacy Questionnaire or SEQ (Nicholas et al., 1992) is a 10-item scale yielding a single score giving a measure of the patient's confidence in the outcome of pain management (or rehabilitation in which they have to play an active part). This questionnaire seems promising but requires further scientific evaluation. Such beliefs would also seem to be important in the context of occupational rehabilitation. It has been shown, for example, that patients' beliefs about whether or not they would be able to return to work were the best predictor of return to work after rehabilitation (Sandstrom and Esbjornsson, 1986).

There are as yet few questionnaires specifically validated for the assessment of self-efficacy beliefs in occupational settings. Recently one such questionnaire has been developed specifically to investigate beliefs about back pain. Unlike the previous measures it has been used primarily in occupational settings. The Back Beliefs Questionnaire or BBQ (Symonds et al., 1996) includes scales specifically concerning the perceived inevitability of back pain and its future course. They found that workers with a previous history of back pain were more likely to believe their backs would be problematic in the future. More pessimistic beliefs about the controllability of pain and less willingness to take responsibility for it were associated with a higher number of spells of back pain and more back-associated work loss in the past.

Specific Fears of Hurting, Harming, and Further Injury

It is crucial to recognize the role of fear and avoidance as obstacles to rehabilitation following injury. Behavioral theorists such as Fordyce (1976) explained the development of "avoidance learning" where successful avoidance of pain established a behavioral pattern which was successful in reducing pain but with the cost of maintaining the "disability." Letham et al. (1983) and Slade et al. (1983) incorporated these ideas into a "fear avoidance model of exaggerated pain perception" in chronic low back pain. In their view, such beliefs were central in the development of pain avoidance behavior. Letham et al. (1983) described patients as *confronters* or *avoiders*, and it has been observed (Waddell et al., 1993) that fear and avoidance of pain can become more disabling than pain itself, since although avoidance at early stages may reduce nociception, the avoidance behaviors may persist in anticipation of pain rather than simply as a response to it.

Fear, specifically of movement and reinjury, may also become established. If an injured worker mistakenly believes that resumption of a particular activity will actually lead to an increase in pain or produce further damage, this obstacle to rehabilitation will have to be removed if the worker is to be rehabilitated. The genesis of such fears has also been investigated. Vlaeyen et al. (1995a) found that fear of movement and reinjury was more related to depressive symptoms and catastrophizing than to pain itself. Finally, in a study of 300 patients attending their family doctor with acute back pain, fear–avoidance beliefs predicted outcome at two and twelve months (Klenerman et al. 1995). The importance of addressing such beliefs and associated psychological mechanisms within a reactivation framework would seem to be compelling, but development of accurate assessment instruments is still at an early stage.

The Fear–Avoidance Beliefs Questionnaire or F.A.B.Q (Waddell et al., 1993) has two scales. The first (physical activity scale) assesses beliefs about the influence of physical activity on pain, while the second (work scale) examines specifically beliefs about the influence of work on pain. The Tampa Scale of Kinesophobia or T.S.K. (Miller et al., 1991) has also been recently developed. A recent study (Vlaeyen et al., 1995b) has shown that fear of movement may have a major impact on behavioral performance. This topic would seem to merit further investigation.

Coping Strategies

Coping can be considered in terms of what people do to try to diminish the occurrence or impact of unpleasant events of various sorts. Patients may employ a wide range of behavioral and coping strategies in order to limit the effects of pain. Choice of strategies will depend on patient's beliefs about pain, on their confidence in being able to influence events, and of course on their repertoire of coping behaviors. In addition to differentiating behavioral from cognitive coping strategies, researchers have also described coping behaviors in terms of their appropriateness and effectiveness, specifically in relation to pain and adjustment. Thus, Brown and Nicassio (1987) made the important distinction between active and passive coping strategies. Perhaps the most widely used questionnaire is the Coping Strategies Questionnaire or CSQ (Rosenstiel and Keefe, 1983). The 50 items give scores of six cognitive scales (diverting attention, reinterpreting pain sensation, coping self-statements, ignoring pain sensations, passive praying, or hoping and catastrophizing), two behavioral coping scales (increasing activity level and increasing pain behaviors), and two single items concerning perceived control over pain.

The questionnaire offers a general evaluation of the relative use of effective (or appropriate) coping strategies and ineffective (or inappropriate) coping strategies. Several studies have shown that negative or ineffective coping strategies such as catastrophizing ("fearing the worst") are associated with higher levels of self-reported disability or adjustment (Jensen et al., 1991; Keefe et al., 1989; Main and Waddell, 1991) The catastrophizing scale has also shown impressive predictive value in outcome of treatment, particularly in patients with acute pain (Burton et al., 1995). Brandtstadter (1992) distinguished two fundamentally different ways of coping with chronic pain: *assimilative coping,* involving active attempts to alter circumstances in line with personal preferences and *accommodative coping* or "downgrading" of goals or expectations when goals are seen to be unattainable through active coping efforts. Two scales: Tenacious Goal Pursuit (TGF) and Flexible Goal Adjustment (or FGA) have been developed to measure these aspects of coping (Brandstadter and Renner, 1990). In a recent study of coping style and pain-associated distress, it was concluded "accommodative coping functions as a protective resource by preventing global losses in the psychological functioning of chronic pain patients and maintaining a positive life perspective. Most important, the ability to flexibly adjust personal goals attenuated the negative impact of the pain experience (pain intensity, pain-related disability) on psychological well-being (depression). Furthermore, pain-related coping strategies led to a reduction of disability only when accompanied by a high degree of flexible goal adjustment" (Schmitz et al., 1996; p. 41).

These research findings are consistent with early work by demonstrating that depression is primarily a function of pain-associated incapacity (or interference with life) rather than a direct result of pain itself (Rudy et al., 1988)). These findings suggest that, in the design of a return-to-work rehabilitation program for the injured worker, a degree of flexibility should be seriously considered.

Conclusion

There is now wide evidence that psychological factors influence pain perception, level of incapacity, adjustment to disability, and response to treatment in clinical settings. A number of key psychological dimensions have been identified. Such factors are important not only for understanding the impact of pain on the individual but also for understanding the nature of pain-associated incapacity and recovery from injury. A number of recent studies have suggested that a number of these psychological factors are also predictive of the *development* of chronic incapacity and as such can be considered as risk factors for chronicity.

Determinants of Chronicity in Clinical Settings

Psychological factors, however, need to be understood not only in terms of their influence on pain perception and on response to treatment, but also on the process of chronicity (or conversely, recovery from incapacity).

Several studies have compared the *relative* importance of physical and psychological factors on the prediction of outcome of treatment. As has been stated elsewhere, most treatment and intervention is based on concepts of tissue damage and amelioration of biomechanical dysfunction following injury. Many studies have tried to identify the importance of all sorts of anthropomorphic, clinical, and biomechanical factors in terms of risk of injury, or risk of continued incapacity (Halpern, 1992). Multivariable analyses of such data suggest that the relative contribution of the different classes of variable to prediction of chronicity is particularly complex (Burton and Tillotson, 1991). In general, however, it may be the (clinical) historical and psychosocial information rather than the socio-occupational or biomechanical variables that are associated with lack of recovery (Burton A.K. et al., 1999).

In chronic back pain patients, levels of distress at time of initial assessment were found to be as important as known risk factors (such as previous lumbar surgery) in outcome from conservative treatment 2 to 4 years later (Main et al., 1992). The particular importance of psychological factors in *chronic* pain patients may not be particularly surprising, but recent studies on patients with acute pain have come up with similar findings. In a one-year follow-up cohort study of patients with back pain attending for osteopathic treatment, cognitive factors (specifically negative or inappropriate coping strategies) were found to be by far the most powerful predictors of outcome; particularly in the subgroup of patients with acute back pain in which it was possible to explain 47% of variance in outcome from negative cognitive coping strategies alone (Burton et al., 1995). In another study (Main et al., *submitted*) psychological factors had a powerful influence on outcome of treatment in patients with acute back pain, whether treated by McKenzie physiotherapy or by a nonsteroidal anti-inflammatory drug, both immediately following treatment and at one-year follow-up.

Such factors have not been investigated as yet with sufficient precision in the context of occupational rehabilitation, but seem to merit investigation. It would appear, nonetheless, that levels of distress or dysfunctional thoughts may be important factors in the *development* of chronicity and thus represent psychological obstacles to recovery which need to be addressed as part of the rehabilitative process.

Mechanisms of Chronic Incapacity

The importance of psychological factors has been clearly shown, but most of the above studies have been either cross sectional in nature or involved analysis of outcome of treatment and have not addressed, therefore, the possible *mechanisms* involved. The original Gate Control Theory or GCT (Melzack and Wall, 1965) postulated an interaction between nociception and central psychological factors, but most research studies have treated physical and psychological factors as if they were quite independent. The relationship between emotion and physiological activity has been shown however in a series of investigations of peripheral and central physiological responses to experimental stimuli. (Flor et al., 1989). In an early experiment (Flor et al., 1985) they had shown heightened levels of muscle activity in the paraspinal muscles to personally relevant stressors in patients with low back pain.

Although there is some evidence for differences in resting levels of SEMG between back pain patients (Ahern et al., 1988), guarded movements of various sorts, while protective during the acute stages of recovery from injury, are clearly evident in many chronic pain patients. Guarded movements can be investigated using surface electromyography (SEMG) from the lumbar muscles, and recently a specific measure of the guarded movement during forward flexion the Flexion Relaxation Ratio (FRR) has been developed and validated for use with low back pain patients (Watson et al., 1997a). In a further study using the FRR, the relationship between muscle activity and response to treatment was investigated (Watson et al., 1997b).

A significant correlation was found initially between fear avoidance beliefs and abnormalities of muscle action prior to treatment (a three-week pain management program). Following the pain management program there were significant correlations between changes in the muscular abnormalities and changes in fear avoidance beliefs as well as between changes in self efficacy beliefs and changes in muscular abnormalities. The results raise the possibility that following injury, or a significantly painful episode, some people develop persistent abnormalities in muscle electrical activity suggestive of a persistent physiological abnormality around the painful site.

They conclude:

"The extent to which these abnormalities are mediated by fears of hurting and harming is not yet entirely clear, but changes in SEMG patterns towards the normal following pain management are associated with reductions in fear of hurting and harming and increase in self-confidence as measured by self-efficacy beliefs."

Conclusions from Clinical Studies

It would seem, therefore, that several conclusions can be reached from the above research findings.

1. Psychological factors are implicated in the perception of pain.
2. Reactions to pain, in terms of distress, specific beliefs, coping strategies and the development of pain behavior, seem likely to be more influential than more widespread psychological constructs or perspectives.
3. It has been possible to identify specific psychological factors which place individuals at risk of the development of both the persistence of and the development of chronic incapacity.
4. It would appear that these psychological factors are more important than clinical history or physical examination factors in the prediction of outcome.
5. Psychologically oriented pain management program can reduce these risk factors.
6. Laboratory investigations have shown than reduction in specific fears of hurting and harming are associated with physiological changes in electrical activity in muscle.
7. It would appear that future research might usefully be directed at secondary prevention in patients in whom chronic incapacity has not yet become established.

75.5 Psychological Aspects of Work

Difficulties in the Identification of Work-Related Stress

The complex sociopsychological characteristics of organizations has made the investigation of work-related stress extremely difficult. Most of the studies have tended to focus on specific features of the organization or particular job characteristics. Kasl (1992) identifies, for example: organizational size and structure, cumbersome or arbitrary procedures and practices, and role-related issues. Cox (1993) discusses the significance of role ambiguity, role conflict, and role insufficiency as aspects of the latter. Responsibility for people (Leiter, 1991), lack of job security, and failure to realize job expectations are among other factors considered to be of importance. A number of recent studies have examined the nature of work stress (Schaefer and Moos, 1993; Crum et al., 1995; Alliger and Williams, 1993). Other studies have examined more specific aspects, such as job satisfaction (Medcof and Hausdorf, 1995), job opportunity/frustration (Noack, 1994), and job predictability and turnover (Pearson, 1995).

Organizational stressors therefore can be identified, and certain characteristics of the working environment constitute risk factors in their own right. The extent to which an individual copes with such risk factors will influence job satisfaction, morale, and psychological well-being. Extended absence from work, however, can have a marked effect on the perception of the work environment. Interpersonal relationships at work for example may be an important determinant of successful return to work. Good relationships with colleagues, a sense of being valued, and worry about letting one's colleagues down may act as important incentives to return to work after injury. An unpleasant or difficult interpersonal environment may represent a significant obstacle to return to work. Appraisal of such factors, combined with a loss of self-confidence and anxiety about not being able to perform satisfactorily on return to work may also represent a specific hindrance. According to Bigos et al. (1990) "once an individual is off work, perceptions about symptoms, about the *safety* of return to work, and about the impact of return to work on one's personal life can affect recovery even in the most well-meaning worker," (p. 854; Bigos' italics).

Approaches to the Definition and Study of Work Stress

For the purpose of this chapter an attempt will be made to examine the nature of work stress from a general perspective, derived from Cox (1993). Finally, it has to be recognized that although it is possible to identify work stress, there may be an interdependence with other life stresses. Long working hours, for example, can significantly interfere with family life.

There have been three different but overlapping approaches to the definition and study of stress (Cox, 1993). According to the "engineering model," stress is conceptualized as an aversive or noxious characteristic of the work environment and is construed essentially as an environmental *cause* of ill health. The model has been criticized on the grounds of oversimplicity, since specific stimuli such as noise may be stressful or beneficial depending on a number of other factors. Furthermore, according to Douglas (1992), perception of risk and risk-related behavior are not adequately explained by the natural science of objective risk and are strongly determined by group and cultural biases.

In the "physiological model," stress is defined in terms of the common physiological effects of a wide range of aversive and noxious stimuli; i.e., stress is the physiological *response* to a threatening or dangerous environment. The model has been criticized because differences in physiological responses to the same stressor, and difficulties in identifying unambiguously physiological responses which are specifically stress responses. According to Cox (1993), there are subtle but important differences so that noradrenaline activation may be more related to the physical activity inherent in various tasks, while adrenaline activation may be more related to feelings of effort and stress.

More fundamentally, both models have been criticized for their reliance on an over simplistic stimulus-response paradigm, while ignoring individual differences in the cognitive and perceptual processing of information; ignoring the interactions between the individual and his/her environment; and in their lack of a "systems-based" approach, ignore the psychosocial and organizational contexts of work stress.

The third model, the "psychological" model, conceptualizes stress in terms of the *dynamic interaction* between the person and the work environment. Interactional models focus on the structural features of the person's interaction with the work environment, while transactional models are concerned principally with psychological mechanisms such as cognitive appraisal and coping which underpin the interactions.

The Person–Environment Fit or P–E fit model (French et al., 1974) identifies two aspects of *fit*:

1. The degree to which an employer's attitudes and abilities meet the demands of the job, and
2. The extent to which the job environment meets the workers' needs.

Lack of fit, whether defined objectively or subjectively (in terms of the individual's perceptions) has been recognized as a potential stress. Job demands and job decision latitude have also been identified as powerful factors in psychological reaction to work (Karasek, 1979).

According to transactional models, stress involves elements of both cognition, such as appraisal, and emotion. According to Lazarus and Folkman (1984), *primary appraisal* involves monitoring the situation and may lead to identification of a problem. such problem recognition may be accompanied by unpleasant emotions or general discomfort. *Secondary appraisal*, following after recognition of the problem, involves attempts to generate a set of coping strategies. According to Cox (1993): "The experience of stress is therefore defined, first, by the realization that they are having difficulties in coping with demands and threats to their well being and, second, that coping is important and the difficulty in coping depresses or worries them" (p. 17).

Finally, the appraisal process may be thought of in five stages, beginning with no more than a hazy recognition of the existence of a problem, and leading via an *on going* process of interaction between coping strategies, reappraisal, and redefinition of the problem.

Coping with Occupational Stress

A wide range of coping strategies has been identified. They have been differentiated, for example, into task-focused or emotion-focused (Dewe, 1987) and into strategies changing meaning or managing stress symptoms (Pearlin and Schooler, 1978). Individual characteristics such as informational styles (*blunters* and *monitors*) have been identified (Miller et al., 1988), and concepts of vulnerability or job hardiness (Kobasa, 1979; Kobasa et al., 1981) have also been considered. Finally, individual differences have been investigated both as *components* of the appraisal process, or as *moderators* of the stress–health relationship (Payne, 1988).

Behavioral characteristics such as "Type A behavior" (Friedman and Rosenman, 1974) were identified originally as risk factors for coronary heart disease. Since then concepts such as work commitment and time-urgency have become an integral part of the psychology or work. Gaining control is an important coping strategy in dealing with work difficulties. If as a coping strategy it is ineffective and the problem persists, frustration or anger can result in significant work stress, deterioration in health, and absence from work.

Predictors of Chronic Incapacity in Occupational Settings

Cats-Baril and Frymoyer (1991), in a study of employees with between 2 and 6 weeks' incapacity, identified four main risk factors for long-term disability. These included job characteristics (work status at time of the survey, work history, and type of occupation); job satisfaction factors (including preretirement policies and benefits; perception about whether the injury was compensable; who was at fault and whether a lawyer had been contacted); past hospitalizations; and educational level. The utility of their predictive model, however, was criticized on methodological grounds specifically because of their failure to control for duration of disability, and because of the likely influence of psychological factors in terms of expectation.

"The population of patients who have already incurred two weeks of off-work time secondary to low back trouble probably are anticipating long-term disability, and, perhaps more importantly, so are their health care providers, employers, and insurers. These expectations may trigger illness behaviors that help establish long-term disability" (Lehmann et al., 1993, p. 1110).

In a recent review of risk factors in industrial low back pain, four major types of risk factors have been implicated in industrial low back pain (Bigos et al., 1990). These were individual factors (mainly demographic and anthropomorphic), physical findings (mainly radiographic), workplace factors (including various work characteristics), and psychological factors (determined from psychological tests or evidence of substance abuse). In their view, however, methodological problems significantly compromised accurate evaluation of these factors.

According to a recent review:

"Monotonous work, high perceived work load, and time pressure are related to musculoskeletal symptoms. The data also suggest that low control on the job and lack of support by colleagues are positively associated with musculoskeletal disease. Perceived stress may be an intermediary in the process" (Bongers et al., 1993, p. 297). Specific psychological features have been found in a number of prospective studies. Carosella et al. (1994) found that patients with low return-to-work expectations, heightened perceived disability, pain, and somatic focus had problems complying with an intensive work rehabilitation program. Haazen et al. (1994) have shown that change in distorted pain cognitions, workers' compensation status, and use of medication were the most important predictors in behavioral rehabilitation of low back pain. (They were, however, pessimistic about the overall level of prediction achieved.) Bigos and colleagues (1991) found that psychological distress, low job satisfaction, and a history of back trouble were the strongest predictors of the report of back pain in the future. The above studies, although tantalizing, do not identify the putative psychological factors with sufficient degree of accuracy to evaluate their specific importance. There is a need for the development of psychometrically sound occupationally focused instruments to assess aspects of job satisfaction; perception of safety; and the perceived consequences of sickness absence for job security and financial stability. In addition to investigation of the employee, it may be necessary furthermore to consider characteristics of the organization.

In the clinical assessment of low back pain, the lack of a simple relationship between physical signs and resulting incapacity is well recognized and different sorts of factors have been adduced to explain the apparent mismatch. Recent clinical studies into the outcome of treatment for low back pain have however offered a more specific evaluation of the role of different sorts of psychological variables in the prediction of chronic incapacity (as determined by self-reported disability) than is currently available in most occupational studies. The findings of such studies for our understanding of the nature and development of chronic incapacity may have relevance also to secondary prevention in occupational settings and so some of these studies will now be reviewed.

In a study of attitudes, beliefs, and absenteeism among workers in a biscuit manufacturing factory, Symonds et al. (1996) showed that workers who had taken in excess of one week's absence due to low back trouble had significantly more negative attitudes and beliefs (when compared with workers who had taken shorter absences, or with those who reported no history of back trouble). Beliefs about the inevitability of back pain, fears of hurting or harming, and perceived disability were significantly associated with absenteeism. In an associated study (Symonds et al., 1995), introduction of a psychosocial pamphlet, designed to correct mistaken beliefs about back pain (e.g., confusing hurting with harming) and reduce avoidance behavior, successfully reduced extended sickness absence resulting from low back trouble.

75.6 The Socioeconomic and Medicolegal Dimensions

Since the turn of the century, the nature of chronic incapacity has been considered in the medicolegal literature, but little systematic research into the role of socioeconomic factors has emerged. The dichotomous view of pain as *either* physical *or* psychological has prevailed. The persistence of pain without a clear and unambiguous physical lesion has been considered either as a manifestation of psychiatric disorder or as evidence of simulation (malingering). The conceptual confusion evident in the investigation of pain and incapacity in the legal system is reviewed elsewhere (Main and Spanswick, 1995). For the purpose of this chapter, it can be stated that in occupational rehabilitation, the specific influence of socioeconomic factors in general, and of medicolegal factors in particular awaits further research; both in terms of their specific influence and in terms of their possible interactions with physical and psychological factors.

75.7 Return to Work: Prevention of Chronic Incapacity

The Role of Education

Results of back schools based mainly on medical and ergonomic principles have been somewhat disappointing. As far as *primary prevention* is concerned, according to Nordin et al. (1992) there is little evidence for sole efficacy of back schools, but they may be beneficial within the wider context of a company-wide effort (Wood et al., 1987) where management and unions participate and the program is geared to the needs of the employee (Versloot et al., 1992). Yet somewhat surprisingly, in a recent U.K. study carried out in a light industrial environment, mere distribution of a simple educational pamphlet designed specifically to reduce fear–avoidance beliefs, produced a substantial reduction in extended absence associated with LBT (Symonds et al., 1995). As far as *secondary prevention* is concerned, the picture is somewhat confused, perhaps because of the variety of approaches adopted. Bergquist-Ullman and Larsson (1977), in a randomized controlled trial, showed that back school was superior to physical therapy treatment or placebo; but recent studies (Berwick et al., 1989; Stankovic and Johnell, 1990) have failed to replicate the findings. Methodological difficulties in the evaluation of their efficacy are presented elsewhere (Schlapbach and Gerber, 1991). According to Scheer et al. (1995): "Although an ergonomic education for recently injured workers makes inherent sense, there is *not* incontrovertible published evidence that back school is more efficacious than placebo for acute LBP" (p. 970).

Physical Therapy

Physical therapists employ a wide range of techniques ranging from passive modalities such as heat treatment, ice packs, interferential stimulation, and a range of manipulative techniques. The approaches are detailed elsewhere (Tan et al., 1992; Ottenbacher and Fabio, 1985). As far as return to work is concerned, the evidence for efficacy is unproven, perhaps because there is no requirement for the individual to be *actively* involved in the treatment process.

Functional Restoration (Sports Medicine Approach)

Functional restoration programs (FRP) place a particular emphasis on the restoration of strength and the development of fitness. As such, they are more appropriate for the chronically incapacitated worker in whom a "disuse syndrome" may have developed, than for a secondary prevention approach in the more recently incapacitated worker. They do have an advantage over the individualized physical therapy approach, however, in that they require a specific Functional Capacity Evaluation or FCE (Mayer and Gatchel, 1988) which can be more closely matched with specific occupational requirements than a more generalized and less focused clinical approach.

Specific Psychological Approaches

A number of individual approaches to the management of pain are available (Gatchel and Turk, 1996) and may be offered as part of routine pain management following injury. As far as return to work is concerned, there is no more evidence for individualized psychological therapy such as psychotherapy or stress reduction as a *single* treatment approach than for individualized physical therapies. The recent development of graded activity programs with a behavioral therapy approach, however, has shown much more promise (Lindstrom et al., 1992). In the latter study, patients were randomized to either traditional physiotherapeutic care or to a graded activity program with a behavioral therapy approach. The graded activity program proved to be a successful method of restoring occupational function and facilitating return to work in subacute low back patients.

Conclusion

It would seem that the traditional content of workplace education concerning injury, pain, and rehabilitation needs radical revision. Ergonomic interventions alone may be insufficient to prevent the development and/or persistence of musculoskeletal problems (Burton et al., 1997b). It is clear that the problem of return to work is multifaceted. It seems that to enable successful return to work, occupational rehabilitation requires a multidisciplinary perspective with an integrated model of work disability such as the "Rochester model" (Feuerstein, 1991). This "highlights the potential interaction of a patient's medical condition and ability to meet physical and psychological demands at work. It also emphasizes the importance of psychological factors (such as fear of reinjury, expectations of return to work, perception of disability, illness behavior, and pain and stress coping skills) in rehabilitation efforts directed at RTW (Feuerstein and Zastowny, 1996, p. 463).

75.8 Implications

Reappraisal of the Nature of Work-Related Incapacity

In this chapter we have attempted specifically to consider the psychological influences of "injury" on absenteeism and return to work. They have been considered from a biopsychosocial framework, recognizing the interactions among physical, psychological, and socio-occupational factors. It is considered that models of incapacity constructed solely from biomechanical and ergonomic principles are inherently limited since they are based on a partial understanding of the nature of absenteeism (Burton et al., 1997b). It follows from the above that the concept of risk analysis merits fundamental reconsideration.

If an individual has no interest, for whatever reason, in successful and satisfactory return to work, then consideration of psychological obstacles to return to work is an irrelevance. It is important to distinguish generalized occupational stress from specific injury-associated distress. While both may present obstacles to return to work, they require different solutions. Furthermore, it is important to recognize that in the well-intentioned worker psychologically mediated work incapacity should not be confused with irredeemable work incapacity. It has been shown that lengths of absence from work are associated both with characteristics of the individual and features of the working situation. It is necessary therefore to consider absenteeism both from a system and an individual perspective (Versloot et al., 1992).

Solutions for the Individual Worker

In general it is not in any worker's interests to become chronically incapacitated. Although most workers recover from injury, unnecessary delay in return to work is costly and potentially psychologically damaging. Once an individual has become injured, the assessment of the risk of chronic incapacity needs to considered. Such risks need to be understood not only from the biomechanical and ergonomic perspectives, but also from the psychological perspective. Clinical studies have identified early psychosocial risk factors for chronicity. Consideration should be given to a system of routine evaluation of such risks if the worker does not make a speedy recovery.

Assessment of the individual worker cannot be undertaken without consideration of the context of assessment. Whether or not an individual becomes chronically incapacitated depends not only on his/her actual and perceived clinical status, but also on the socioeconomic system. Differences in social policy, in remuneration for sickness, and in industrial agreements regarding the protection of the injured worker will all have a profound effect on sickness absence and chronicity (I.A.S.P., 1995). The importance of the socioeconomic implications of continued incapacity should not be underestimated.

Features of the occupational environment can be more or less conducive to return to work. The issue of whether it is right to expect someone to be *fully fit* merits some discussion. From an employer's perspective, in terms of job performance this is clearly desirable, since it obviates the need for special consideration, discussion of temporary restricted work, with the associated problem of preventing such arrangements being viewed adversely by workmates. The increase in all sorts of litigation in industrial society is a matter of fact, irrespective of one's political stance on it. The issue is increasingly seen as a risk management issue. The easiest way to eliminate all risk simply would be to refuse to employ or re-employ anyone who has had any sort of injury. Such a solution however, in a skilled workforce with a relatively high rate of minor job-associated injury, could be extremely expensive as well as socially unacceptable. On the other hand, identification of *specific role conflicts* likely to follow injury and subsequent absence from work could, at relatively low cost, be specifically addressed in a time-limited workplace rehabilitation program for patients with specific musculoskeletal rehabilitation needs. Success of such an endeavor would of necessity have to take into consideration reaction of managers, their management style, and the reaction of workmates as well as specific task functioning and general job performance.

Given the costs of extended absenteeism, specific attention might be given to a number of the other risk factors for occupational stress (discussed above). While accepting that a fully comprehensive psychological evaluation is not a practical proposition, specific attention might be directed at the assessment of *changes* in the perception of the work environment by the injured worker in terms of:

a. Job satisfaction
b. Job stress (including relationships with colleagues)
c. Specific requirements regarding job performance
d. Perceived role conflict

In fact a new instrument has been recently developed specifically for use in occupational settings. The Psychosocial Aspects of Work (PAW) questionnaire measures three different types of attitude about work. It contains a 7-item job satisfaction scale, a 4-item social support scale, and a 4-item mental stress scale (Symonds et al., 1996). Although at an early stage of development, this questionnaire would seem to have some potential as a general screening instrument.

The identification of specific psychological obstacles to work requires specific evaluation. Since such obstacles are probably evident soon after injury, they may be perhaps best tackled close to the work environment immediately after injury during the phase of active treatment. Since most musculoskeletal injury, however, if managed properly, should resolve successfully, provision of "front-line" treatment within occupational health settings has an obvious appeal, and many employers have employed physical therapists from various professional disciplines for this purpose. Such an approach, provided it is not seen as a thinly disguised management ploy to reduce costs or identify malingerers, has an intuitive appeal in that it is preventing the "distancing" between the injured worker and his/her worksite. Unfortunately, since such workplace-based treatment has been offered primarily with an ergonomic or biomechanical emphasis, specific psychological obstacles to return to work have not usually been incorporated into the format of the rehabilitation program.

Although a number of clinical measurement tools (as identified above) are now available, further research is needed into specific occupational applications. The following clinical dimensions would seem to be of particular importance and perhaps should be incorporated routinely when a worker has been off for more than a few days, or where occupational records indicate a pattern of increasing spells of work absence with nonspecific musculoskeletal complaints.

1. General level of distress.
2. Specific beliefs and fears about pain and rehabilitation
3. Pain coping strategies
4. Fear of physical activity

Solutions for a "Sick Workplace"

There is an urgent need to develop sound measurement instruments for the perception of work. In development of new tests, however, it should be remembered that jobs differ significantly in their psychological characteristics and so occupationally sensitive instruments with adequate normative data will be required.

Assessment of Psychological Risk Factors

Traditionally, assessment of risk in occupational settings has tended to focus on environmental characteristics, such as safety, or on ergonomic dangers, such as poor lifting and handling techniques. In this chapter it has been argued that the concept of psychological risk must be recognized. The reduction of unnecessary absenteeism needs to be considered both from an occupational and a clinical perspective.

Reduction of Occupational Stress

It would seem likely on a simple intuitive level that perceived occupational stress will delay return to work after injury. While the widespread introduction of mental health promotion (Kasl, 1992) in industry as an answer to occupational stress in industry: is perhaps more of a pious hope than a realistic possibility, specific attention to the company "ethos" regarding injury and pain-associated incapacity might merit consideration. It has recently been recommended that stress management programs which encourage individuals to change workplace factors through innovation would be a useful addition to interventions that emphasize individual adaptation to stressful work environments (Bunce and West, 1994). According to Cox and Griffiths (1995): "The challenge for the late 1990s is a practical one: to develop effective systems for assessing and managing the problem in organizations, and then to educate those organizations in using those systems" (p. 3). It is being suggested that workplace rehabilitation after injury *also* should incorporate such perspectives.

Establishment of a Work-Based Rehabilitation Policy

Identification of possible changes to working practices necessary prior to full recovery and development of a clear work-based rehabilitation policy will reduce anxiety about return to work. Working practices clearly are job specific, but it is known that certain occupations carry higher risks for back injury, even though the physical demands may differ (Halpern, 1992). Functional Capacity Evaluation is an integral part of the "sports medicine approach" (Mayer and Gatchell, 1988), but the emphasis is very much on the capacity of the individual. Traditionally, graded work practices have taken place in rehabilitation settings, but it may be possible to establish closer integration between rehabilitation and occupational health in the design of *work-based* graded return to work, as has been evident in Scandinavia. While a clear understanding of the biomechanical demands and ergonomic features of the jobs concerned is essential, such considerations have to be located within an appropriate model of rehabilitation (Feuerstein, 1991; Feuerstein and Zastowny, 1996).

Management Style and Conditions of Employment

Return to work is affected first and foremost by conditions of employment which may have a specific influence on rate of return to work irrespective of severity of injury. Management style is a key ingredient of both job satisfaction and job stress (Cox, 1993). Specific attention to the management of return to work may pay dividends. In one North American study (Wood, 1987), for example, a simple telephone call from a line manager offering good wishes and expressing concern for the absentee's welfare significantly increased rate of return-to-work.

The Design of Context-Sensitive Interventions

A recent review of RCTs of early intervention in industrial low back pain was pessimistic (Scheer et al., 1995), but frequently the design of such interventions has been inappropriate in that they have been over-reliant on biomechanical and ergonomic principles (Burton et al., 1997). A number of recent studies have suggested that *specifically designed* and *appropriately targeted* interventions may be effective. Such interventions have ranged in complexity from the introduction of educational pamphlets designed to change beliefs (Symonds et al., 1995) and telephone calls (Wood, 1987); via graded exercise programs (Lindstrom et al., 1992) and occupationally specific secondary prevention programs (Linton et al.), to major tertiary rehabilitation programs (Mayer and Gatchell, 1988).

75.9 Conclusion

The rising costs of back-associated incapacity has led to a number of reports which have offered solutions to the problem of long-term incapacity (CSAG, 1994; Bigos et al., 1994; ACC and the National Health Committee, 1997). These reports essentially have addressed the problems of secondary incapacity rather than primary prevention, and made a number of clinical recommendations involving assessment and management. A recent report by the International Association for the Study of Pain (IASP, 1995) has offered a radical "solution" to work absence from nonspecific back pain. It sees long-term absenteeism primarily as a consequence of the "over-medicalization" of pain. It recommends that after six weeks' work absence, such patients be redefined as "activity-intolerant." The incapacity is then no longer defined as a medical problem. Whatever economic advantages such a "solution" may offer in answer to the increasing costs of back-pain associated incapacity (which has to be addressed), the report has not met with universal support. It has been criticized mainly on the grounds that it seems to rest on old fashioned concepts of tissue-healing time and does not offer either an adequate clinical or an adequate occupational answer to the problem of persistent incapacity (Turk and Main, 1997).

Traditionally, clinical treatment has in general been considered in complete isolation of the occupational context. Conversely, the psychological aspects of return to work, if considered at all, have tended to omit consideration of such clinical factors. In terms of psychological mechanisms, however, it is clear that an effective rehabilitation program has to address both perspectives. In this chapter an attempt has been made to identify the sorts of psychological factors which influence return to work following injury. It would appear that the traditional separation into clinical and occupational factors needs to be reconsidered. The cost of unnecessary absenteeism is considerable. Adequate attention to the psychological risk factors highlighted in this chapter might be of considerable economic benefit in the reduction of unnecessary absence from work. The development of context-sensitive interventions, while recognizing the powerful influence of psychosocial risk factors within an overall disability management framework, may offer a better way forward than the narrow biomechanical and ergonomic-based approaches, or the fairly drastic "social policy" solutions offered by *Pain in the Workplace* (IASP, 1995).

Acknowledgments

Sincere thanks to Dr. Ana-Paola Viera for assistance with the background literature.

References

ACC and the National Health Committee. 1997. *New Zealand Acute Low Back Pain Guide.* Wellington, New Zealand.

Ahern D.K., Follick M.J., Council J.R., Laser-Wolston N., and Litchman H. 1988. Comparison of lumbar paravertebral EMG patterns in chronic low back pain patients and controls. *Pain* 34: 153-160.

Alliger G.M. and Williams K.J. 1993. Using signal-contingent experience sampling methodology to study work in the field: A discussion and illustration examining task perceptions and mood. *Personnel Psychol.* 46: 525-549.

Bandura A. 1977. Self-efficacy: Toward a unifying theory of behavioral change. *Psych. Rev.* 84: 191-215.

Bergquist-Ullman H. and Larsson U. 1987. Acute low back pain in industry. *Acta Orthop. Scand.* 170 (Suppl). 1-117.

Berwick D.M., Budman S., and Feeldstein M.1989;. No clinical effect of back schools in an HMO. A randomised prospective trial. *Spine* 14: 338-349.

Bigos S, Bowyer O, and Braen G. 1994. Acute Low Back Problems in Adults. *Clinical Practice Guidelines No. 14* AHCPR Publications No. 95-0645 Agency for Health Care Policy and Research. U.S Department of Human Health and Human Services.

Bigos S.J., Battié M.C., Nordin M., Spengler D.M., and Guy D.P. 1990. Industrial low back pain, in Weinstein J.N. and Weisel S.W. (Eds.) *The Lumbar Spine.* p 846-859. Philadelphia: WB Saunders, and Co.

Bigos S.J., Battié M.C., Spengler D.M., Fisher L.D., Fordyce W.E., Hansson T.H., Nachemson A.L., and Wortley M.D. 1991. A prospective study of work perceptions and psychological factors affecting the report of back injury *Spine* 11: 252-255.

Bongers P.M., de Winter C.R., Kompier M.A.J., and Hildebrandt V.H. 1993. Psychosocial factors at work and musculoskeletal disease. *Scand. Journ. Work Environ. Health* 19: 297-312.

Bradley L.A., Haile J.McD., and Jaworski T.M. 1992. Assessment of psychological status using interviews and self-report instruments, in Turk D.C. and Melzack R. (Eds.) *Handbook of Pain Assessment* pp 193-213. New York: The Guilford Press.

Brandstadter J. and Renner G. 1990. Tenacious goal pursuit and flexible goal adjustment: explication and age-related analysis of assimilative and accommodative strategies of coping. *Psychol. Aging.* 5: 58-67.

Brown G.K. and Nicassio P.M. 1987. The development of a questionnaire for the assessment of active and passive coping strategies in chronic pain patients. *Pain* 31:53-65.

Bunce D. and West M. 1994. Changing work environments: innovative coping responses to occupational stress. *Work and Stress* 8: 319-331.

Burton A.K., Battié M.C., and Main C.J. (1999) The relative importance of biomechanical and psychosocial factors in low back injuries, in Karwowski W. and Marras W.S. (Eds.) *The Occupational Ergonomics Handbook.* Boca Raton, Florida: CRC Press Inc.

Burton A.K., Symonds T.L., Zinzen E., Tillotson K.M., Caboor D., Van Roy P., and Clarys J.P. 1997b. Is ergonomic intervention alone sufficient to limit musculoskeletal problems in nurses? *Occup. Med.* 47: 25-32.

Burton A.K., Tillotson M.K., Main C.J., and Hollis S. 1995. Psychosocial predictors of outcome in acute and sub-chronic low back trouble. *Spine* 20 722-728.

Burton A.K. and Tillotson K.M. 1991. Prediction of the clinical course of low back trouble using multivariable models. *Spine* 16: 7-14.

Carosella AM, Lackner J.M., and Fuerstein M. 1994. Factors associated with early discharge from a multidisciplinary work rehabilitation program for chronic low back pain *Pain* 57: 69-76.

Cats-Baril W.L. and Frymoyer J.W. 1991. Identifying patients at risk of becoming disabled because of low-back pain: the Vermont Rehabilitation Engineering Center predictive model. *Spine* 16: 605-607.

Clinical Standards Advisory Group. 1994. *Back Pain: Report of a CSAG Committee on Back Pain,* London: HMSO.

Cox T. 1993. Stress research and stress management: putting theory to work. *HSE Contract Research Report no 61.* Sudbury, Suffolk: HSE Books.

Cox T. and Griffiths A. 1995. "Editorial: Guidance for U.K. employers on managing work-related stress. *Work and Stress.* 9: 1-3.

Crisson J.E. and Keefe F.J. 1988. The relationship of locus of control to pain coping strategies and psychological distress in chronic pain patients. *Pain* 35: 147-154.

Crum R.M., Muntaner C., Eaton W.W., and Anthony J.C., 1995; Occupational stress and the risk of alcohol abuse and dependence. *Alcoholism Clinical and Experimental Research* 19: 647-655.

DeGood D.E. and Shutty M.S. Jr. 1992. Assessment of pain beliefs, coping and self-efficacy, in Turk D.C. and Melzack R. (Eds.) *Handbook of Pain Assessment* pp 214-234. New York: The Guilford Press.

Dewe P.1987. New Zealand ministers of religion: identifying sources of stress and coping strategies. *Work and Stress* 1: 351-363.

Douglas M. 1992. *Risk and Blame.* London: Routledge.

Dworkin S.F. and Whitney C.W. 1992. Relying on objective and subjective measures of chronic pain, in Turk D.C. and Melzack R. (Eds.) *Handbook of Pain Assessment,* pp 429-446. New York: The Guilford Press.

Feuerstein M. 1991. A multidisciplinary approach to the prevention, evaluation and management of work disability. *Journ. of Occup. Rehab.* 1: 5-12.

Feuerstein M. and Zastowny T.R. 1996. Occupational rehabilitation, in Gatchel R.J. and Turk D.C. (Eds) *Psychological Approaches to Pain Management: A Practitioner's Handbook.* New York: The Guilford Press.

Flor H. and Turk D.C.1989. Psychophysiology of chronic pain: do chronic pain patients exhibit symptom specific psychophysiological responses? *Psychol. Bull.* 105: 215-259.

Flor H., Turk D.C., and Birbaumer N. 1985. Assessment of stress-related psychophysiological stress reactions in chronic back pain patients. *Journ. of Consult. & Clin. Psychol.* 53: 354-364.

Fordyce W.E. 1976. *Behavioural Methods for Chronic Pain and Illness.* St. Louis: CV Mosby.

French J.P.R., Rogers W., and Cobb S. 1974. A model of person-environment fit, in Coehlo G.W., Hamburg D.A., and Adams J.E. (Eds.) *Coping and Adaptation.* New York: Basic Books.

Friedman M. and Rosenman R.H. 1974. *Type A: Your Behaviour and Your Heart.* New York: Knoft.

Gatchel R.J. and Turk D.C. (Eds). 1996. *Psychological Approaches to Pain Management: A Practitioner's Handbook.* New York: The Guilford Press.

Haazen I.W.C.J., Vlaeyen J.W.S., Kole-Snijders A.M.K., van Eek F.D., and van Es F.D. 1994. Behavioural rehabilitation of chronic low back pain: searching for predictors of treatment outcome *Journ. of Rehab. Sci.* 7: 34-43.

Halpern M. Prevention of low back pain: basic ergonomics in the workplace. 1992, in Nordin M. and Vischer T.L. *Clinical Rheumatology: International Practice and Research. Common Low Back Pain: Prevention of Chronicity* 6: 705- 730. London: Bailliere Tindall.

Hazard R.G., Haugh L., Green P., and Jones P. 1994. Chronic low back pain: the relationship between patient satisfaction and pain, disability and impairment outcomes. *Spine*: 19: 881-887.

I.A.S.P., 1995. *Back Pain in the Workplace: Management of Disability in Nonspecific Conditions.* Seattle: IASP Press.

International Association for the Study of Pain. 1979. Pain terms: a list with definitions and notes on usage. *Pain* 6: 249

Jensen M.P., Turner J.A., Romano J.M., and Karoldy P. Coping with chronic pain: a critical review of the literature. *Pain* 1991: 249-283.

Kobasa S. 1979. Stressful life events, personality and health: an enquiry into hardiness. *Journ. of Pers. and Soc. Psychol.* 37: 1-13.

Karasek R.A. 1979. Job demands, job decision latitude and mental strain: implications for job redesign. *Admin. Sci. Quart.* 24: 205-308.

Kasl S.V. 1992. Surveillance of psychological disorders in the workplace, in Keita G.P. and Sauter S.L. (Eds.) *Work and Wellbeing. An Agenda for the 1990s.* Washington, D.C.: American Psychological Association.

Keefe F.J. and Hill R.W. 1985. An objective approach to quantifying pain behaviour and gait patterns in low back pain patients. *Pain* 21: 153-161.

Keefe F.J., Brown G.K Wallaston K.A., and Caldwell D.S. 1989. Coping with rheumatoid arthritis pain: catastrophising as a maladaptive strategy *Pain* 37: 51-56.

Klenerman L., Slade P.D., Stanley I.M., Pennie B., Riley J.P., Atchison L.E., Troup J.D.G., and Rose M.J. 1995. The prediction of chronicity in patients with an acute attack of low back pain in a General Practice setting. *Spine* 20: 478-484.

Kobasa S., Maddi S., and Courington S.1981. Personality and constitution as mediators in the stress-illness relationship. *Journ. of Health and Soc. Behav.* 22:368-378.

Lazarus R.S. and Folkman S.1984. *Stress, Appraisal and Coping.* New York, Springer Publications.

Lehmann T.R., Spratt K.F., and Lehmann K.K. 1993. Predicting long term disability in low back injured workers presenting to a spine consultant. *Spine* 18: 1103-1112.

Leiter M. 1991. The dream denied: professional burnout and the constraints of human service organisation. *Canad. Psychol.* 32: 547-558.

Letham J., Slade P.D., Troup J.D.G., and Bentley G. 1983. Outline of a fear-avoidance model of exaggerated pain perception. Part 1. *Behav. Res. and Ther.* 21: 401-408.

Lindstrom I., Ohlund C., and Eek C.E. 1992. Graded activity of subacute low back pain patients. A randomised prospective clinical study with an operant conditioning behavioural approach. *Physical Therapy* 72: 279-293.

Linton S.J., Bradley L.A., Jensen I., Spangfort E. and Sundell L. 1989. The secondary prevention of low back pain: a controlled study with follow-up. *Pain* 54: 353-359.

Main C.J., Hollis S. and Roberts. A. (submitted). Psychological predictors of outcome of acute low back pain. *Journal of Psychosomatic Research.*

Main C.J. and Waddell G. 1987. Personality assessment in the management of low back pain. *Clin. Rehab.* 1 139-142.

Main C.J., Wood P.L.R., Hollis S, Spanswick C.C., and Waddell G. 1992. The distress assessment method: A simple patient classification to identify distress and evaluate risk of poor outcome. *Spine* 17: 42-50.

Main C.J. and Spanswick C.C. 1991. Pain: psychological and psychiatric factors. *Br. Med. Bull.* 47:732-742.

Main C.J. and Spanswick C.C. 1995. Functional overlay and illness behaviour in chronic pain: distress or malingering? Conceptual difficulties in medico-legal assessment of personal injury claims. *Journ. of Psychosom. Res.* 39: 737-753.

Main C.J. and Waddell G. Cognitive measures in pain. 1991. *Pain* 46: 287-298.

Main C.J. and Watson P.J. 1996. Guarded movements and the development of chronicity. *Journ. of Musculoskeletal Pain* 4:163-170.

Mayer T. and Gatchel R.J. 1988. *Functional Restoration for Spinal Disorders: the Sports Medicine Approach.* Philadelphia: Lea and Febiger.

Medcof, J.W. and Hausdorf P.A.1995. Instruments to measure opportunities to satisfy needs, and degree of satisfaction in the workplace. *Journ. Occup. and Organis. Psychol.* 68:193-208.

Melzack R. 1987. The short form McGill Pain Questionnaire. *Pain* 30: 191-197.

Melzack R and Katz J. 1992. The McGill Pain Questionnaire: Appraisal and current status, in Turk D.C. and Melzack R. (Eds.) *Handbook of Pain Assessment* pp. 152-168. New York: The Guilford Press.

Melzack R. and Wall P.D. 1965. Pain mechanisms: a new theory. *Science* 150: 971-979.

Miller R.P. Kori S.H., and Todd. D.D 1991. The Tampa Scale. Unpublished report. Tampa Florida (reported in Vlaeyen et al., 1995b).

Miller S., Brody D., and Summerton J. 1988. Styles of coping with threat: implications for health. *Journ. of Pers. and Soc. Psychol.* 54: 142-148.

Nicholas M.K., Wilson P.H., and Goyen J. 1992. Comparison of cognitive-behavioural group treatment and an alternative non-psychological treatment for chronic low back pain. *Pain* 48: 339-347.

Nordin M., Cedraschi C., Balangue F., and Roux E.B. 1992. Back schools in the prevention of chronicity, in Nordin M. and Vischer T.L. *Clinical Rheumatology: International Practice and Research. Common Low Back Pain: Prevention of Chronicity* 6: 685-703. London: Bailliere Tindall.

Ottenbacher K. and DiFabio R.P. 1985. Efficacy of spinal manipulation/mobilization therapy — a meta-analysis. *Spine* 10:833-837.

Parker H, Wood P.L.R., and Main C.J. 1995. The use of the pain drawing as a screening measure to predict psychological distress in chronic low back pain. *Spine* 20: 236-243.

Payne R.1988. Individual differences in the study of occupational stress, in Cooper C.L. and Payne R. (Eds.) *Causes, Coping and Consequences of Stress at Work.* Chichester: Wiley and Sons.

Pearlin L. and Schooler C. 1978. The structure of coping. *Journ. of Health and Soc. Behav.* 19: 2-21.

Pearson C.A.L. 1995. The turnover process in organisations: An exploration of the role of met-unmet expectations. *Human Relations* 48: 405-420.

Porter C.S. 1988. Accident Proneness: a Review of the Concept. *Int. Revs. of Ergonomics* 2:177-206

Ransford A.O., Cairns D., and Mooney V. 1976. The pain drawing as an aid to the psychological evaluation of patients with low back pain. *Spine* 1: 127-134.

Richards J.S., Nepomuceno J.A., Riles M., and Suer Z. 1982. Assessing pain behaviour: the UAB pain behaviour scale *Pain* 14: 393-398.

Rosenstiel A.K. and Keefe F.J. 1883. The use of coping strategies in chronic low back pain patients: relationship to patient charactereistics and current adjustments. *Pain* 17: 33-34.

Rudy T.E., Kerns R. D., and Turk D.C. 1988. Chronic pain and depression: towards a cognitive mediation model. *Pain* 35: 129-140.

Sandstrom J. and Esbjornsson E. 1986. Return to work after rehabilitation: the significance of the patient's own prediction. *Scand. Journ. Rehab. Med.* 18:29-33.

Schaefer J.A. and Moos R.H. 1993. Relationship, task and system stressors in the health care workplace. *Journ. of Community and Applied Soc. Psychol.* 3: 285-298.

Scheer S.J., Radack K.L., and O'Brien D.R. 1995. Review article: Randomised controlled trials in industrial low back pain relating to return to work. Part 1. Acute interventions. *Arch. Phys. Med. Rehab.* 76:966-973.

Schlapbach P. and Gerber N.J. 1991. Back school. *Rheumatology* 14: 25-33.

Schmitz U., Saile H., and Nilges P.1996. Coping with chronic pain: flexible goal adjustment as an interactive buffer against pain-related distress. *Pain* 67: 41-51.

Slade P.D., Troup J.D.G., Letham J., and Bentley G. 1983. The fear-avoidance model of exaggerated pain perception. Part 2. *Behav. Res. and Ther.* 21: 409-416.

Stankovic R. and Johnell O. 1990. Conservative treatment of acute low back pain: a prospective randomised trial. Mckenzie method of treatment vs. patient education in "Mini Backschool." *Spine* 15:120-123.

Symonds T.L., Burton A.K., Tillotson K.M., and Main C.J. 1995. Absence resulting from low back pain can be reduced by psychosocial intervention at the workplace. *Spine* 20: 2738-2745.

Symonds T.L., Burton A.K., Tillotson K.M. and Main C.J. 1996. Do attitudes and beliefs influence work loss due to low back trouble? *Occup. Med.* 46: 25-32.

Tan J.C., Roux E.B., Dunand J., and Vischer T.L. 1992. Role of physical therapy in the management of common low back pain, in Nordin M. and Vischer T.L. *Clinical Rheumatology: International Practice and Research. Common Low Back Pain: Prevention of Chronicity* 6: 629-655. London: Bailliere Tindall.

Toomey T.C., Mann J.D., Abashian S., and Thompson-Pope S. 1991. Relationship between perceived self-control of pain, pain description and functioning. *Pain* 45: 129-133.

Turk D.C. and Main C.J. 1997. Resource review: back pain in the workplace. *American Pain Society Bulletin.* 7(3) 11-13.

Versloot J.M., Rozeman A., vanSon A.M., and van Akkerveeken P.F. 1992. The cost-effectiveness of a back school program in industry. 1992. *Spine* 17: 22-27.

Vlaeyen J.W.S., Kole-Snijders A.M.K., Rotteviel A.M., Ruesink R., and Heuts P.H.T.G. 1995b. The role of fear of movement/(re)injury in pain disability. *Journ. Occup. Rehab.* 5: 235-252.

Vlaeyen J.W.S., Kole-Snijders A.M.J., Boeren R.G.B., and van Eek H. 1995a. Fear of movement/(re)injury in chronic low back pain and its relation to behavioural performance, *Pain* 62: 363-372.

Waddell G.1992. Biopsychosocial analysis of back pain, in Nordin M. and Vischer T.L. (Eds.) *Common Low Back Pain: Prevention of Chronicity,* pp 523-557. London: Bailliere Tindall.

Waddell G., McCulloch J.A., Kummel E., and Main C.J. 1984. Symptoms and signs: physical disease or illness behaviour. *British Medical Journal* 289: 739-741.

Waddell G., Somerville D., Henderson I., Newton M., and Main C.J. 1993. A fear avoidance beliefs questionnaire (FABQ) and the role of fear avoidance beliefs in chronic low back pain and disability. *Pain* 52: 157-168.

Waddell G., McCulloch J.A., Kummell E., and Venner R.M. 1980. Nonorganic physical signs in low back pain. *Spine* 5:117-125.

Wallston K.A., Wallston B.S., and DeVellis R. 1978. Development of the multidimensional health locus of control scale (MHLC). *Health Educ. Monogs.* 6: 160-170.

Watson P.J., Booker C.K., Main C.J., and Chen A.C.N. 1997a. Surface electromyography in the identification of chronic low back pain patients: the development of the flexion relaxation ratio. *Clin. Biomechan.* 12: 165-171.

Watson P.J., Booker C.K., and Main C.J. 1997b. Evidence for the role of psychological factors in abnormal paraspinal activity in patients with chronic low back pain (CLBP). *Journ. of Musculoskeletal Pain.* 5: 41-56.

Wood D.J. (1987) Design and evaluation of a back injury prevention program within a geriatric hospital. *Spine* 12: 77-82.

76

Back Pain in the Workplace: Implications of Injury and Biopsychosocial Models

Michele C. Battié
University of Alberta

Chris J. Main
Hope Hospital, Salford, U.K.

A. Kim Burton
University of Huddersfield

The growth of back pain complaints and associated long-term disability in the industrial insurance system defies simple explanations based on concomitant changes in physical environments, work demands, or increases in degenerative or pathological changes to the spine. As has been eloquently chronicled in an historical overview by Allen and Waddell (1989), back symptoms have been present throughout the ages. What appears to have changed dramatically in many societies is public perception of back pain as an "injury" or medical problem, and its effect on work and life in terms of disability.

Long-term disability from low back pain is a relatively new phenomenon in the history of Western civilization, and its dramatic growth in the last half century suggests that factors other than physical pathology may be influencing its development. This is not to say that back pain is not genuinely experienced by large numbers of people, or that severe symptoms do not occasionally cause physical limitations of some duration. On the contrary, numerous health surveys indicate that back symptoms are extremely common in both developed and third-world countries (Waddell, 1987; Anderson, 1984; Honeyman and Jacobs, 1996; Näyha et al., 1991; Bierine-Sørensen, 1984; Troup, 1987; Videman, 1989), and that they begin at an earlier age than once thought (Balague et al., 1988; Burton et al., 1996; Salminen et al., 1995). Yet, only a small percentage of persons experiencing back symptoms each year file industrial injury claims or seek medical care.

Back injury reporting and disability are clearly influenced by other factors than purely the presence of pain. For example, the severity of symptoms and the physical demands of the workplace are likely to interact to affect reporting and absenteeism. While an individual with light physical job demands may be able to continue work through an episode of back pain, the same condition may be sufficiently exacerbated by heavy loading or work in awkward postures, such that the individual is unable to continue with normal work tasks during recovery. In addition to the effects of symptom severity and job demands, other factors of likely importance to the onset and persistence of disability are cultural norms (Anderson,

0-8493-2641-9/99/$0.00+$.50

1984; Waddell, 1987; Honeyman and Jacobs, 1996), socioeconomic conditions (Volinn et al., 1988; Volinn, 1991) and opportunities for compensation (Loeser et al., in press; Leavitt, 1992), attitudes and beliefs about back problems (Symonds et al., 1995, 1996), emotional distress (Bigos et al., 1992; Klitzman et al., 1990), and job satisfaction and the social work environment (Bigos et al., 1992; Linton et al., 1989; Linton and Kamwendo, 1989; Linton, 1990). Fears that continuing work activities will cause further harm also may influence the decision to file a complaint or miss work. Innumerable factors could affect the response to back symptoms, many relating to perceptions of the problem and the perceived benefits and costs of various options available in the social systems and cultural norms in which the person lives and works.

Considering the factors noted above, a biomedical model, focusing purely on physical pathology, has long been recognized as inadequate to explain back pain disability or formulate a sufficient management strategy to combat it. Focusing simply on physical pathology and biomechanical causes and solutions to the problem would appear to be a gross, and occasionally misleading, oversimplification of the back pain experience in the developed countries of the world. While such factors should not be ignored, a broader perspective is needed if growth in work-related back pain disability is to be curbed. Some suggest that biopsychosocial (Waddell, 1987) or biocultural models (Honeyman and Jacobs, 1996) may provide better frameworks for, and insights into, the development and management of this growing problem. In this chapter we will examine the industrial injury model and explore the shift in paradigms over the past decade toward biopsychosocial and sociocultural models of back pain disability and their implications.

76.1 Back Pain in the Workplace: An Injury Model

As more is learned about the epidemiology and natural history of back symptoms, that they are endemic and episodic in nature, the more difficult it becomes to apply a simple injury model. The challenges of applying such a model in the working environment were demonstrated by Robertson and Keeve (1983), who reported on the results of efforts to prevent industrial back injury claims by minimizing physical hazards at the workplace through establishing and enforcing safety and health regulations. Although these efforts led to a decrease in objectively verifiable injuries, such as fractures and lacerations, more subjective injury reports, including back strains, were unaffected.

Most back symptoms are of gradual onset, and are not associated with a traumatic accident or unusual activity (Videman et al., 1989; Rowe, 1969; Välfors, 1985). Therefore, it may be inappropriate to refer to all back symptoms occurring in the workplace as injuries, a common practice to meet the demands of the industrial insurance system. Among first-year nurses, who often have relatively high and unpredictable loading conditions in patient handling, symptoms were found to be of sudden onset with or without a specific inciting event in only 28% of cases cited (Videman et al., 1989). The great majority of back pain episodes were of gradual onset. Thus, it may be more appropriate to reserve the term “injury,” and the images that the term elicits, for back symptoms of sudden onset following a specific inciting event, such as a fall, when direct trauma is more likely. As Allen and Waddell outline in their historical review of back pain, this condition has not always been viewed in an injury context (Allen and Waddell, 1989).

Hadler noted that once an inguinal herniation was termed a “rupture” and then considered an injury, and a ruptured disc was accepted as a source of back symptoms in the 1930s, back pain also received injury status in the workers’ compensation system (Hadler, 1995). He goes on to note that since that time much of what is costly and contentious in workers’ compensation systems relates to back injuries. Furthermore, Americans, in general, have come to perceive musculoskeletal discomfort in an injury context. As will be further discussed in later sections, the injury model has other consequences.

The Influence of Physical Loading

If we followed the paradigm that back pain and resultant disability are consequences of injury caused by mechanical overload, to be prevented by lessening loads on the spine, then it would follow that populations with heavier loads would have a higher incidence of back pain and subsequent disability. This clearly is

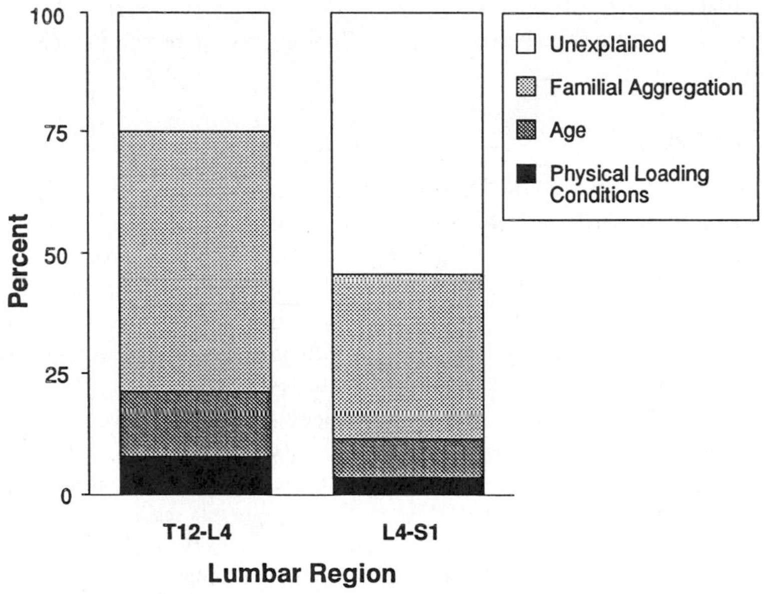

FIGURE 76.1 The graph displays the percentage of variability in disc degeneration in the T_{12}-L_4 and L_4-S_1 spinal regions explained by specific occupational and leisure physical loading variables relative to that explained by familial aggregation which represents a combination of genetic and shared childhood environmental influences (beyond that explained by age). (Adapted from Battié MC, Videman T, Gibbons LE et al.: Determinants of lumbar disc degeneration. *Spine* 1995; 20:2601-2612.)

not the case, however, as has been discussed previously in Chapter 61 (Burton et al.). It is puzzling that instead, back pain complaints and disability are highest in countries with generally less extreme occupational physical loading conditions and high utilization of medical services.

Volinn (1997) studied the point prevalence of low back pain in low and middle income countries, where a far greater portion of the population is engaged in heavy labor, and often into old age, as compared to relatively high income countries. He hypothesized that if back pain were caused by heavy physical loading, that low income countries with substantially more heavy labor would have a higher point prevalence. Instead, a considerably lower back pain prevalence was found among populations of low income farmers as compared to wealthy populations of Northern European countries that had two to three times higher prevalences. He concluded that "hard physical labor itself is not necessarily related to LBP." He did, however, find that workers from low income countries working in urban "enclosed workshops," such as sewing machine operators, had the highest prevalence of back pain.

As we discussed in Chapter 61 (Burton et al.), even when physical loading is investigated in the etiology of degenerative changes in the lumbar spine, its effects are seldom clear. In a recent study of determinants of disc degeneration (Battié et al., 1995), occupational and leisure time physical loading exposures in adulthood were found to have a minor influence on degree of degenerative findings, particularly when compared to familial aggregation, which represented a combination of genetic and childhood environmental influences (Figure 76.1). It was suggested that this relatively modest influence of physical loading may explain the mixed results of earlier studies of the effects of physical loading on degenerative findings. The investigators concluded, that disc degeneration may result primarily from a combination of genetic influences, childhood exposures, and unidentified factors, which may include complex, unpredictable interactions.

While the role of occupational physical loading may be modest in terms of accelerating disc degeneration, certainly some physical activities may aggravate back symptoms, and ergonomic interventions may make it easier for someone experiencing back problems to continue work or return sooner following an acute or recurrent episode. Appropriate interventions may include accommodations to the work environment or task that minimize extreme loads or awkward postures. It is important, however, to keep the role of ergonomic interventions in perspective when developing more comprehensive management strategies.

Perceptions of Fault

The common view of back pain as an industrial injury may lead to the perception that fault lies with the work tasks or environment, particularly if symptoms occur during the course of normal work activities. Furthermore, affected workers may feel as though their "injuries" are the result of negligence on the part of their supervisors or employers for placing them in hazardous situations. It has been hypothesized that such perceptions of fault may influence the attitudes and expectations of workers experiencing back symptoms. Placing blame on external sources also may influence the affected worker's sense of responsibility in promoting the recovery process (Bigos and Battié, 1987).

DeGood and Kiernan (1996) were interested in investigating the influence of perceptions of fault on expectations of recovery, psychological distress, and behavioral disturbances. When they grouped patients under their care for chronic pain according to whether they believed that their employer, another (primarily doctors or another driver from a MVA), or that no one was really responsible for their pain, they found the groups were similar with respect to current pain intensity and level of activity limitations. However, those who blamed their employer for their current pain problems were significantly more likely to be unemployed, to report negative responses from past treatments and expect the same from future therapy, to perceive that they had been treated unfairly by their employer before and after the onset of their pain, and to exhibit greater psychological distress than those who felt that no one was to blame. The retrospective nature of this study limits the conclusions, but the findings do support the hypothesis that perceptions of fault may influence pain behavior, psychological disturbances, and poor expectations of recovery.

It also has been suggested that individuals who report back injuries may be more inclined to forget having had minor back symptoms related to performing common activities prior to their "injury," which may reinforce perceptions that the injury is responsible for any subsequent back symptoms (Fagan, 1995).

Back Pain in the Workers' Compensation System

When a person experiences what is believed to be work-related back symptoms and chooses to file an injury claim he or she enters the workers' compensation paradigm. This paradigm defines work tasks as causal, infers structural damage, and requires medical testimony of injury. Unfortunately, conditions that are identified primarily through symptom complaints, where medicine has yet to clearly identify the underlying pathology, do not fit well into this paradigm. Affected individuals often find themselves in the position of having to demonstrate to health care providers and insurance adjusters that they are indeed injured and worthy of receiving the benefits awarded to such a designation. This situation can place the worker in an adversarial position where his or her integrity is questioned, such that an appropriate response is to demonstrate that the health problem is serious enough for acceptance by the system. Such patients are sometimes viewed with suspicion, and as Leavitt (1992) once stated, "The unfortunate problem is that stereotypes have consequences. Doubts raised by labels often shape evaluation and treatment of industrial workers in problematic ways to the extent that their integrity and status as patients are challenged." This is counter-productive to recovery.

Several studies have suggested poorer outcomes in patients receiving workers' compensation. Studies have found patients on compensation were less likely to benefit from lumbar interbody fusion for "degenerative disc disease" as compared to patients not receiving compensation (Flynn and Hoque, 1979). Compensation also has been associated with residual back pain following surgery (Hanley and Shapiro, 1989) and delayed recovery from conservative treatment (Greenough and Frazer, 1989). Leavitt (1992) pointed out, however, that researchers typically do not account for the influence of physical labor on the outcome of return to work, and that compensation and noncompensation groups usually differ on occupational physical job demands. In a study designed to determine the effects of compensation on work loss while controlling for differences in occupational physical demands, he determined that irrespective of the job performed work-related back pain was associated with prolonged disability duration (Leavitt, 1992).

A later study of 1,366 self-referred patients with back pain of less than eight weeks' duration, was conducted to investigate whether workers with acute low back pain were better served if they were insured through a workers' compensation insurance policy (Hadler et al., 1995). Those with and without workers' compensation insurance were deemed to be similar in clinical status and demographic variables at entry. Yet, while most in both groups returned to full function and work activities rapidly, those with workers' compensation insurance returned to work less convinced that they had recovered to the level of health experienced before the onset of their back pain episode. This perception of compromised health remained even after perceived job characteristics were controlled, diminishing the chances that this finding could be due to differences in work-related physical demands.

Loeser et al. (1995) note that discussions around the future of workers' compensation insurance programs center around understanding: (1) personal and social factors that lead an individual to injury or illness reporting and the adoption of a sick role, with its associated disability and health care utilization, and (2) factors that influence the behavior of workers in a compensation system. They proceeded to conduct a comprehensive review looking at the effects of wage replacement levels on claims behavior. As part of the review, they conducted some reanalyses of previously reported data in an effort to control for possible confounding factors that may have influenced the original interpretation of results. They included 24 studies that directly addressed the relationship between wage replacement level and workers' compensation claims incidence and duration.

In their synthesis of the literature Loeser at al. (1995) conclude that, when other factors are held constant, both the incidence and duration of industrial injury claims in the workers' compensation system are associated with the level of wage replacement benefits. The effect on duration appears to be somewhat larger than that for incidence. The methodologically stronger studies provided estimates of a 1.83 to 11.28% increase in claim duration as a result of a 10% increase in workers' compensation benefits. Evidence was mixed with respect to a relationship to underlying injury rates. The challenge would appear to be finding the appropriate balance between financially supporting an injured worker to prevent major financial hardship during recovery, while providing financial incentive for timely return to work and discouraging unnecessarily extended disability.

Hadler et al. (1995) urge an examination of whether or not regional backache belongs in the workers' compensation paradigm, under the "injury rubric." They suggest that the industrial injury distinction received by back symptoms has not led to a safer, healthier workplace, nor has it forwarded the understanding or management of this problem. Over past decades the frequency of workplace injuries has been relatively constant, but their consequences for workers have grown dramatically. As evidence, Hadler et al. note that disability from work-related injuries has steadily climbed, as measured through workdays lost. In addition, the cost of workers' compensation in the United States in the two decades following 1971 increased 12-fold, and went from 0.5% to 2.5% of the national payroll. Such increases are not exclusive to the United States; similar growth in back-related disability has been noted in other Western countries as well.

A controversial discussion document, *Back Pain in the Workplace: Management of Disability in Nonspecific Conditions* (1995), reached similar conclusions to those discussed above. Yet, reconstructing a more successful and humane system of dealing with disability or work incapacity associated with nonspecific symptom complaints faces many challenges. Hopefully, such documents as *Back Pain in the Workplace*, while stirring controversies, will also advance the discussion of constructive reforms in the medical management and social systems that have been dismally inadequate in dealing with this problem.

76.2 Back Pain in the Workplace: A Biopsychosocial Model

A decade ago, in a Volvo Award winning paper, Waddell (1987) presented a strong argument for the adoption of a biopsychosocial model of illness for the management of low back problems. At the core of this model is the acknowledgment and integration of physical, psychological, and social aspects of illness. Such a model demands a broad approach to considering factors influencing illness and recovery. Since the time Waddell encouraged the use of a biopsychosocial model, evidence supporting the importance of psychosocial factors in back pain reporting and disability has steadily grown. Much of this evidence has been presented in the previous chapters.

Most studies of back problems and suspected risk factors have been retrospective or cross-sectional in design and have been quite limited with respect to inferences regarding risk. Currently, a number of prospective cohort studies specifically investigating predictors of work-related back pain complaints have been reported (Battié et al., 1989a; Battié et al., 1989b; Battié et al., 1990; Bigos et al., 1992; Cady et al., 1979; Chaffin and Park, 1973; Dueker et al., 1994; Mostardi et al., 1992; Ready et al., 1993; Rossignol et al., 1993; Venning, 1987). Most of these studies have focused on the effects of individual physical factors and medical history. One of the most comprehensive prospective studies of predictors of work-related back pain complaints and disability was conducted among Boeing employees (Bigos, 1992). This study demonstrated that the role of psychosocial factors in risk of future injury complaints and disability was greater than that of individual physical factors, such as lifting strength, aerobic capacity, and range of motion which proved to be very poor predictors (Battié et al., 1989a, 1989b, 1990), findings that have been supported by subsequent prospective cohort studies (Mostardi et al., 1992; Ready et al., 1993; Dueker et al., 1994). Specifically, emotional distress, as determined from the Minnesota Multiphasic Personality Inventory, and poorer job task enjoyment and co-worker relationships were implicated in risk of reporting work-related back problems. A later analysis looking at perceptions of the work environment in a broader context of work-related injury and illness reporting revealed that lower job task enjoyment and co-worker relations had a similar influence on the reporting of other types of injuries, as well (Battié et al., 1993). Thus, there is no indication from the Boeing study that workers who file back injury claims differ significantly from workers claiming other types of injuries with respect to these work perceptions. It should be emphasized, however, that such workplace perceptions, while associated with a higher risk of reporting, are poor predictors of precisely who will and will not report a future industrial injury. While most studies that have included factors related to job satisfaction have found associations with back pain reporting and work loss (Bigos et al., 1992; Ready et al., 1993; Magnusson et al., 1996), all have not (Symonds et al., 1996). The inconsistency may be partly due to differences in question presentation and partly to differing philosophies among occupational groups; these issues require further study.

As was discussed in the previous chapter, other perceptions or beliefs have been linked to the development of long-term disability. The fear–avoidance model was first proposed by Lethem et al. in 1983 to explain exaggerated pain perception and behavior resulting from relatively minor nociception and underlying pathology. Fear of pain is the central concept of the model, with the two extreme responses being confrontation or avoidance. Confrontation results in resumed social and physical activities which regularly tests the pain relative to activities and allows the individual to calibrate a response, such that activity and pain level remain in synchrony. On the other hand, avoidance leads to withdrawal of normal activities and an increased likelihood of encountering and adapting to various positive and negative reinforcers of chronic pain. They hypothesized and provided data to support that confrontation or avoidance responses depend on the psychosocial context (stressful life events, personal pain history, personal coping strategies, and personal characteristics) in which the symptoms originate and management occurs (Lethem et al., 1983; Slade et al., 1983). Philips (1987) later hypothesized that "avoiding stimulation plays an active part in reducing the sufferer's sense of control over pain, and in increasing his or her expectation that exposure will increase pain. These cognitive changes encourage further withdrawal from normal activities and a growing intolerance of stimulation." Over the past decade the fear–avoidance model has been studied and supported as having an important role in the development of chronic pain (Klenerman et al., 1995).

Szpalski et al. (1995) conducted a population-based survey in Belgium that revealed that health beliefs related to whether back problems experienced would be lifelong or not was significantly associated with current health behaviors and history of health care utilization. Another study examining beliefs about back problems among British office and manufacturing workers found those who had missed more than one week of work due to low back problems, as compared to those with less time loss or no reported back problems, were significantly more likely to have negative attitudes and beliefs (Symonds et al., 1996). Specifically, they were more likely to believe that negative consequences of back pain were inevitable and to feel less control and responsibility over pain. It is not clear, however, from such a retrospective study design whether such beliefs influenced the time loss or whether the episodes that led to greater time loss led to more negative beliefs. However, a prospective study designed to change negative beliefs about back problems suggests that such a change may positively influence subsequent back-related work loss, strengthening the hypothesis that beliefs about back problems may directly influence illness behavior (Symonds et al., 1995).

Sociocultural Influences

It is generally accepted that back pain is common in populations throughout the world, yet the interpretation of the experience and associated behaviors differ significantly depending upon cultural norms. When studying health problems in rural Nepal, Anderson (1984) reported that although back symptoms were very common, virtually no one sought care for the symptoms when medical services were made available. It appeared that back pain problems were not perceived as a medical problem, but rather a common consequence of normal living and aging. Later, Waddell (1987) reported that before the introduction of Western medicine in Oman, extended disability for back pain problems was virtually nonexistent. Similar to the experience in Nepal, while persons were crippled by such diseases as polio or tuberculosis and injuries involving spinal fractures, they did not stop daily life or become permanently disabled from idiopathic back pain problems. Back pain reportedly was, however, a very common condition there, as is the case in Western countries.

In a study of back pain in an Australian Aboriginal community (Honeyman and Jacobs, 1996), nearly one third of men and half of women reported problems with long-term low back pain when privately questioned. However, the Aboriginals did not perceive back pain as a health issue, and consequently did not report such symptoms openly, display pain-related behaviors, or seek medical care. The result of this study led Honeyman and Jacobs to suggest that "cultural beliefs and practices influence how people respond to back pain in themselves and in others, including how and whether they present to health professionals or seek involvement of others."

Sanders et al. (1992) investigated cross-cultural similarities and differences in Sickness Impact Profile scores of patients seeking help for back pain problems at pain clinics in the United States, Japan, Mexico, Columbia, Italy, and New Zealand. They found levels of psychosocial, vocational, and avocational impairment to be greatest in the Americans, followed by the New Zealanders and Italians. The authors note that many sociocultural factors could be responsible for these differences, such as work ethic, social support and acceptance, economic stability and entitlement issues, beliefs and coping strategies, to name a few. The numbers of representative patients included in the study were small, however, and further work will be needed to more clearly establish differences in chronic low back pain patients between these cultures and to explore the causes of such differences.

Similarly, an earlier study found more interference with social, sexual, recreational, and vocational activities in Americans attending a pain clinic for chronic low back pain than their counterparts in New Zealand (Carron et al., 1985). The study investigators hypothesized that substantial differences in the disability systems, rather than inherent cultural differences, may have accounted for the differences between patient groups. The study's limitations, however, make such inference speculative. Interestingly, while a similar percentage of patients from each group attributed the onset of their pain to a specific incident (74% vs. 71%), patients in the United States were three times as likely to cite a work-related accident, whereas New Zealanders, with a no fault insurance system were just as likely to cite a non-work-related incident.

Work-Related Disability Management Interventions

There have been several interventions designed to minimize disability from back problems that appear promising and worthy of further examination. They are based on a biopsychosocial model and the belief that psychosocial factors, as well as physical factors, influence response to injury and illness, particularly return to work. The interventions have several commonalities; they are employer-based, involve changing attitudes and responses to back symptom complaints, and focus on decreasing the negative consequences of back problems once reported. They concentrate on improving the quality and quantity of communication with the "injured" worker and other parties involved, and support temporary work modifications, if necessary.

Two reports in the mid 1980s described a program designed to mitigate the negative consequences of back problems among employees of a shoe-parts manufacturer (Fitzler and Berger, 1982, 1983). The employer previously had been active in ongoing safety training and enforcing adherence to regulations and had redesigned work tasks to eliminate unnecessary or frequent heavy exertions, but continued to have significant problems with back injury complaints and subsequent disability. Therefore, the company began a new strategy that emphasized changing attitudes and responses of management personnel toward back pain reporting and encouraging prompt, appropriate medical attention. Management training emphasized that back pain symptoms are very common in the adult population and that malingering is rare. Recognizing employee needs also was emphasized, along with responding empathetically. To meet the goal of prompt medical attention, symptom relief agents were provided onsite through the employer's medical service. The affected employee also was provided with education in back care, including treatment and recovery expectations and a discussion of any activity restrictions. If in-house care was deemed insufficient, referral was made to an outside physician. Attempts were made to modify work activities, if needed, to allow the employee to continue working or return to work more quickly during recovery.

There was a dramatic and sustained decrease in workers' compensation costs for the three years following the implementation of the program, as compared to the three prior years. Workers' compensation costs decreased during the first year of the program to approximately 20% of what they had been in each of the prior three years, and further decreased to 10% during the second year. This decrease was sustained during the third year of the program. Because the efforts of the program were determined simply by examining the experiences with back problems within the company before and after the program was implemented, it is possible that other changes occurring at the same time may have been partly responsible for the apparent effects of the program. There is also the Hawthorne effect to consider, employee behavior may have been affected temporarily simply in response to a change in the system. Yet, there are several points in support of the effect being due to the intervention. First, there was a substantial decrease in related work loss and medical and indemnity costs, but not in the rate of back symptom reporting. This result supports the specific intent of the program to minimize the negative consequences of back problems once reported, and not to influence reporting. Second, the apparent effects of the program endured over the three years of observation following implementation.

Subsequently, Battié reported on an intervention with several similar traits conducted in a long-term health care facility (Battié, 1994). The goal was to investigate the effects of an intervention designed to create a more supportive and accommodating work environment following injury reporting. The intervention focused on improved communication and provisions for temporary accommodations in work activities, if needed, to allow an earlier return to the workplace. A system was put in place to ensure regular contact was maintained with employees throughout recovery. The contacts were made by either the employee's supervisor or the safety and health representative, and expressed concern for the employee's well-being, offered assistance, and let the employee know that he or she was missed. A special training session was held for all supervisors to improve their understanding of back problems and encourage empathetic attitudes and responses to back symptom complaints. There was a 77% reduction in the mean back injury claim cost during the intervention year, as compared to the claims filed in the prior three years. Furthermore, none of the back injury claims filed were over $5,000, as compared to 14% of claims filed in the prior three years at the intervention site or among other long-term health care facilities in the region during the year of intervention. This intervention was of a short duration (only one year) and results may be due, in part, to the normal fluctuations commonly experienced in injury rates and costs. It is interesting to note, however, that similar to the previously reported experience in the shoe-parts manufacturing company, the incidence of back injury claims appeared unaffected during the intervention. It was primarily work loss and associated costs that decreased, which were the outcomes specifically targeted by the intervention.

In contrast, Greenwood et al. (1990) reported on an early intervention program among coal miners, initiated by the West Virginia Workers' Compensation Fund that involved early case management by an independent rehabilitation firm that acted as an advocate for the patient. Unlike employer-based disability management approaches, the intervention failed to lower medical costs or prevent extended disability. It would appear that the support of the workplace may be of critical importance in the success of disability management approaches that rely on such elements as improved communication and the availability of modified work duties, when needed.

Wood (1987) also reported on the effects of an intervention in a group of long-term health care facilities designed to decrease the duration of wage loss claims, similar to the subsequent study by Battié (1994), which was described previously. The "personnel program," as it was called, focused on improving communications between all the parties involved in the recovery of the affected employee, including: the employee, his or her doctors, the Workers' Compensation Board, and the hospital personnel. A hospital representative immediately contacted the employee and the workers' compensation representative after the report of back symptoms. The employee then was contacted regularly. The tone of the communications was empathetic and helpful, and included the message, "You are a vital part of the hospital team. Your work is important and your job is waiting." The hospital representative served as a liaison between the various parties involved. After the adoption of the "personnel program," the proportion of high time-loss claims significantly decreased, as compared with the prior period (7.1 vs. 1.7%). There also was a reversal in the trend of increasing claims rates over the prior years.

Wood also examined the effects of a training program to improve work practices. Although the training program was popular among the staff, no added benefit beyond that received from the "personnel program" could be determined based on quantitative data. In general, despite its popularity, there has been little evidence to suggest that educational approaches presenting biomedical and ergonomic principles alone are effective in primary prevention of back symptom reports and disability (Scheer et al., 1995). There have been reports of other approaches to education, however, that appear more promising. They are based on a biopsychosocial model of the problem.

Symonds et al. (1995) implemented an education program among workers in a British biscuit manufacturing factory involving the distribution of written information aimed at changing common misconceptions about back problems. The information was based on the fear–avoidance model and aimed at enforcing active coping strategies, rather than passive, avoidance behavior. When comparing the employee group receiving the educational information to a control group, a decrease in incidence of back pain reports requiring brief absences was observed in both groups during the follow-up period, but the intervention group had significantly less extended sickness absence resulting from low back trouble.

Another educational program with encouraging results was reported by Versloot et al. (1992). The findings of the longitudinal field study in a Dutch bus company provided some evidence to support a decrease in overall length of absenteeism related to the program, although the incidence of time loss episodes was unaffected. The program emphasized psychosocial factors, including responsibility of one's own health, interactions between mind and body in relation to illness, managing stress and coping strategies, as well as ergonomic advice specifically tailored to work and leisure activities.

Recognizing that back problems are most often recurrent in nature, several worksite programs have aimed to decrease subsequent problems in employees with back problem histories by encouraging regular exercise. Although the benefits of exercise in the prevention and treatment of back problems can be debated, and appear to vary depending upon the stage of symptoms (Fass, 1996), several workplace exercise interventions targeting employees with previous back problems report positive effects (Donchin et al., 1990; Kellett et al., 1991; Gundewall et al., 1993). The various exercise programs were offered through sessions at the workplace during regularly scheduled work time. In two of the programs, exercise compliance and greater gains in "fitness" outcomes were found along with fewer symptoms and work loss (Donchin et al., 1990; Gundewall et al., 1993). However, another study reporting significantly less work loss attributable to program participation did not find a concomitant change in "fitness" measures (Kellett et al., 1991). This raises questions about whether physical or psychosocial aspects of the program were most responsible for the observed decrease in subsequent work loss. These interventions suggest that employer-supported activity programs specifically for employees with a history of back problems may decrease subsequent problems and be cost-effective. Whether apparent benefits are due to the physiological effects of exercise or the many possible psychosocial effects of the programs, including a more positive attitude toward the employer, is unclear.

Other interventions to control disability from work-related back problems have aimed at improving the quality of medical care through the use of quality-based, standardized diagnostic and treatment protocols. Among the motivations for developing such protocols was to minimize apparently haphazard variations in care. Wiesel and his coworkers (1984) were among the first to develop and investigate the effects of using such a system for the medical management of industrial back incident reports. They demonstrated a significant and sustained decrease in lost workdays and medical and compensation costs following the implementation of the program. Interestingly, fewer back injury incident reports also were reported, an unexpected finding given that the intervention was designed to improve care once back problems were reported. Since Wiesel et al. presented their approach, other similar models have been developed in an effort to improve worker/patient outcomes (Wiesel et al., 1985, 1994; Lonstein and Wiesel, 1988; Tufo et al., 1991), but not all have achieved similar findings (Battié et al., 1995). One aspect of the intervention reported by Wiesel et al. that may have significantly influenced the success of their program was their strong working relationship with the employers of the intervention sites. They worked closely with the employers and had support and cooperation when, for example, temporary work accommodations were needed to encourage earlier return to work.

Practical Considerations for Workplace-Based Management of Low Back Problems

Judging the effectiveness of workplace interventions is challenging. There can be difficulties in selecting adequate control or comparison groups and good results often observed during short-term follow-up may be simply a brief manifestation of the "Hawthorne effect." The many unpredictable changes that can occur in the work environment related to labor–management relations, administrative processes in the industrial insurance system, as well as the imposition of other safety and health initiatives, to name just a few, further complicate the evaluation of specific interventions. Yet the examples mentioned above do warrant further attention and point to several practical considerations for safety and health professionals involved in workplace-based interventions to minimize back-related disability.

First they suggest the need to adopt the broader perspective of a biopsychosocial model in developing injury management strategies. Such a model recognizes that a variety of psychological and sociocultural factors, as well as physical factors, can influence back problems and related work loss. Several specific considerations follow.

1. Improve the quality and quantity of communication between the workplace and the employee following an injury report. Early contact should express genuine concern and offer assistance, if possible. Contact should be maintained with employees throughout recovery to let them know they are missed and will be welcomed back when well enough to return (Wood, 1987; Battié, 1994). Management training may be a necessary step toward changing negative attitudes and beliefs about back injury claimants to create a more responsive environment of trust and employee advocacy (Fitzler and Berger, 1982, 1983; Battié, 1994).
2. The importance of maintaining as normal a life style as possible during recovery should be stressed. Recent randomized clinical trials have demonstrated that even in the acute stages of back problems, advising the patient not to fear activity, but rather to maintain as much normal activity as tolerated

is more effective in promoting recovery than prescribing specific exercises or other common treatment approaches (Fass, 1996; Malmivaara, 1995). Along the same line, temporary work modifications should be made when possible to allow an employee to return to activities in the workplace as soon as medically advisable (Fitzler and Berger, 1982, 1983; Wood, 1987; Battié, 1994).

3. Educational approaches should include messages that promote realistic, positive attitudes and beliefs about back problems and their consequences. In particular, consider promoting positive coping strategies and an active approach to confronting and managing symptoms to discourage fear–avoidance reactions and maladaptive responses (Versloot et al., 1992; Symonds et al., 1995). Information on biomechanics and ergonomic advice should be practical and tailored specifically to the employees' work and leisure activities.
4. If health promotion programs involving exercise are being considered, it may be most cost-effective to target individuals with a history of recent or recurrent back problems (Keller et al., 1991; Gundewall et al., 1993). While the employee may be expected to do some exercise on his or her own time, having some sessions during work hours, or offering other types of employer support, may send a powerful message about the employer's commitment to employee well-being.
5. Consider ways of bolstering employees' perceptions of the workplace and their personal resources to discourage maladaptive approaches to handling problems. Management style and labor–management relations would appear to be important aspects of such perceptions, influencing both job satisfaction and stress (Cox, 1993).
6. Implement any new intervention in such a way that its effects can be evaluated. The interventions discussed earlier, while promising, are in need of further study to verify their effects in other environments and to better understand the specific mechanisms through which they influence back-related disability and work loss. There are many examples of approaches that have reported significant positive results in some environments, but have not been successful when duplicated in others.

76.3 Summary

The aim of this chapter was to highlight some of the apparent strengths and shortcomings of two competing models of back problems in the workplace, rather than offering an extensive literature review. We have had greater than a half-century of experience using the injury model, stressing the role of biomechanical factors in the etiology and management of back problems, and it has been an unequivocal failure. This is not to say that biomechanical considerations and ergonomic interventions have no value in the management of back problems in the workplace, but rather that they are insufficient in preventing the development and persistence of long-term work incapacity. It is time for a new paradigm. Many psychosocial and cultural factors influence the response to back symptoms, in terms of reporting and disability. Many of the factors appear to relate to attitudes and beliefs about back problems and the perceived costs and benefits of the various options available in the sociocultural systems in which we live and work. In this broader perspective of the problem lie opportunities to develop more effective approaches to curbing the impact of back problems in the workplace.

References

Allen BA, Waddell G: An historical perspective on low back pain and disability. *Acta Orthop Scand Suppl* 1989; 60:1-23.

Anderson RT: An orthopaedic ethnography in rural Nepal. *Med Anthropol* 1984; 8:46-59.

Balague F, Dutoit G, Waldburger M: Low back pain in school children. *Scand J Rehab Med* 1988; 20:175-179.

Battié MC, Bigos SJ, Fisher LD, Hansson TH, Nachemson AL, Spengler DM, Wortley MD, and Zeh J: A prospective study of the role of cardiovascular risk factors and fitness in industrial back pain complaints. *Spine* 1989a; 14(2):141-147.

Battié MC, Bigos SJ, Fisher LD, Hansson TH, Jones ME, and Wortley M: Isometric lifting strength as a predictor of industrial back pain reports. *Spine* 1989b; 14(8):851-856.

Battié MC, Bigos SJ, Fisher LD, Spengler DM, Hansson TH, Nachemson AL, and Wortley MD: The role of spinal flexibility in back pain complaints within industry: A prospective study. *Spine* 1990; 15:(8):768-773.

Battié MC, Bigos SJ, Fisher LD, Fordyce WE, and Gibbons LE: *The Effect of Psychosocial and Workplace Factors on Back Related and Other Industrial Injury Claims.* The International Society for the Study of the Lumbar Spine, Marseilles, France, June 15-19, 1993.

Battié MC: *The Effect of Improved Communication and Work Accommodations on Back Injury Claims.* The International Society for the Study of the Lumbar Spine, Seattle, WA, June 21-25, 1994.

Battié MC, Videman T, Gibbons L, Fisher L, Manninen H, and Gill K: Volvo Award in Clinical Sciences 1995. Determinants of lumbar disc degeneration: A study relating lifetime exposures and MRI findings in identical twins. *Spine* 1995; 20:2601-2612.

Bergquist-Ullman M, Larsson U: Acute low back pain in industry. A controlled prospective study with special reference to therapy and confounding factors. *Acta Orthop Scand* 1977; 170:1-117.

Berwick DM, Budman S, Feldstein M: No clinical effect of back schools in an HMO. A randomized prospective trial. *Spine* 1989; 14(3):338-344.

Biering-Sørensen F: Physical measurements as indicators for low back trouble over a one-year period. *Spine* 1984; 9:106-119.

Bigos SJ, Battié MC: Acute care to prevent back pain disability: Ten years of progress. *Clin Orthop* 1987; 221:121-130.

Bigos SJ, Battié MC, Spengler DM, Fisher LD, Fordyce WF, Hansson T, Nachemson AL, Zeh J: A longitudinal, prospective study of industrial back pain reporting. *Clin Orthop Relat Res* 1992; 279:21-34.

Burton AK, Clarke RD, McClune TD, Tillotson KM: The natural history of low back pain in adolescents. *Spine* 1996; 21(20):2323-2328.

Cady LD, Bischoff DP, O'Connell ER, et al: Strength and fitness and subsequent back injuries in firefighters. *J Occup Med* 1979; 21:269-272.

Carron H, DeGood DE, Tait R: A comparison of low back pain patients in the United States and New Zealand: Psychosocial and economic factors affecting severity of disability. *Pain* 1985; 21:77-89.

Chaffin DB, Park KS: A longitudinal study of low back pain as associated with occupational weight lifting factors. *Am Ind Hyg Assoc J* 1973; 34:513-525.

Cox T: Stress research and stress management: putting theory to work. HSE Contract Research Report No. 61. Sudbury, Suffolk: HSE Books, 1993.

DeGood DE, Kiernan B: Perception of fault in patients with chronic pain. *Pain* 1996; 64:153-159.

Donchin M, Woolf O, Kaplan L, Floman Y: Secondary prevention of low back pain. A clinical trial. *Spine* 1990; 15(12):1317-1320.

Dueker JA, Ritchie SM, Knox TJ, et al: Isokinetic trunk testing and employment. *J Occup Med* 1994; 36:42-48.

Fagan JM, Rehm A, Ryan WG, Hodgkinson JP: The perceived relationship between back symptoms and preceding injury. *Injury* 1995; 26:335-336.

Fass A: Exercises: which ones are worth trying, for which patients, and when? *Spine* 1996; 21(24):2874-2879.

Fitzler SL, Berger RA: Attitudinal change: The Chelsea back program. *Occup Health Saf* 1982; 51:24-26.

Fitzler SL, Berger RA: Chelsea back program: One year later. *Occup Health Saf* 1983; 52:52-54.

Flynn JC, Hoque MA: Anterior fusion on the lumbar spine. End-result study with long-term follow-up. *J Bone Joint Surg* (Am) 1979; 61(8):1143-1150.

Fordyce WE (ed): *Back Pain in the Workplace: Management of Disability in Nonspecific Conditions.* Seattle, International Association for the Study of Pain Press, 1995.

Greenough CG, Fraser RD: The effects of compensation on recovery from low back injury. *Spine* 1989; 14:947-955.

Greenwood JG, Wolf HJ, Pearson JC, et al: Early intervention in low back disability among coal miners in West Virginia: Negative findings. *J Occup Med* 1990; 32:1047-1052.

Gundewall B, Liljeqvist M, Hansson T: Primary prevention of back symptoms and absence from work. A prospective randomized study among hospital employees. *Spine* 1993; 18(5):587-594.

Hadler NM, Carey TS, Garrett J: North Carolina Back Pain Project: The influence of indemnification by workers' compensation insurance on recovery from acute backache. *Spine* 1995; 20:2710-2715.

Hanley EN, Shapiro JA: The development of low back after excision of a lumbar disc. *J Bone Joint Surg* (Am) 1989; 71:719-721.

Honeyman PT, Jacobs EA: Effects of culture on back pain in Australian Aboriginals. *Spine* 1996; 21:841-843.

Kellett KM, Kellett DA, Nordholm LA: Effects of an exercise program on sick leave due to back pain. *Phys Ther* 1991; 71(4):283-291.

Klenerman L, Slade PD, Stanley IM, Pennie B, Reilly JP, Atchison LE, Troup JD, Rose MJ: The prediction of chronicity in patients with acute attack of low back pain in a general practice setting. *Spine* 1995; 20(4):478-484.

Klitzman S, House JS, Israel BA, et al: Work stress, non-work stress and health. *J Behav Med* 1990; 13:221-243.

Leavitt F: The physical exertion factor in compensable work injuries. A hidden flaw in previous research. *Spine* 1992; 17:307-310.

Lethem J, Slade PD, Troup JD, Bentley G: Outline of a fear-avoidance model of exaggerated pain perception. *Behav Res Ther* 1983; 21(4):401-408.

Linton AJ, Bradley LA, Jensen I, et al: The secondary prevention of low back pain: A controlled study with follow-up. *Pain* 1989; 36:197-207.

Linton SJ, Kamwendo K: Risk factors in the psychosocial work environment for neck and shoulder pain in secretaries. *J Occup Med* 1989; 26:23-28.

Linton SJ: Risk factors for neck and back pain in a working population in Sweden. *Work and Stress* 1990; 4:41-49.

Loeser J, Henderlite S, Conrad D: Incentive effects of workers' compensation benefits: A literature synthesis. *J Health Politics and Law* 1995; 52: 34-59.

Lonstein MB, Wiesel SW: Standardized approaches to the evaluation and treatment of industrial low back pain (review). *Occup Med* 1988; 3(1):147-156.

Magnusson ML, Pope MH, Wilder DG, Areskoug B: Are occupational drivers at an increased risk for development musculoskeletal disorders. *Spine* 1996; 21(6):710-717.

Main CJ, Spanswick CC: "Functional overlay" and illness behaviour in chronic pain: distress or malingering? Conceptual difficulties in medico-legal assessment of personal injury claims. *J Psychosomatic Res* 1995; 39(6):737-753.

Malmivaara A, Hakkinen U, Aro T, et al: The treatment of acute low back pain: Bedrest, exercises, or ordinary activity? *N Engl J Med* 1995; 332:351-355.

Mostardi RA, Noe DA, Kovacik MW, et al: Isokinetic lifting strength and occupational injury. A prospective study. *Spine* 1992; 17:189-193.

Näyha S, Videman T, Laasko M, Hassi J: Prevalence of low back pain and other musculoskeletal symptoms and their association with work in Finnish reindeer herders. *Scand J Rheumatol* 1991; 20(6):406-413.

Philips HC: Avoidance behaviour and its role in sustaining chronic pain. *Behav Res Ther* 1987; 25:273-279.

Ready AE, Boreskie SL, Law SA, et al: Fitness and life style parameters fail to predict back injuries in nurses. *Can J Appl Physiol* 1993; 18:80-90.

Robertson LS, Keeve JP: Worker injuries: the effects of Workers' Compensation and OSHA inspections. *J Health Politics, Policy and Law* 1983; 8(3):581-597.

Rossignol M, Lortie M, Ledoux E: Comparison of spinal health indicators in predicting spinal status in a 1-year longitudinal study. *Spine* 1993; 18:54-60.

Rowe ML: Low back pain in industry: Updated position. *J Occup Med* 1969; 11:161-169.

Salminen JJ, Erkintalo M, Laine M, Pentti J: Low back pain in the young. A prospective three-year follow-up study of subjects with and without low back pain. *Spine* 1995; 20(19):2101-2107.

Sanders SH, Brena SF, Spier CJ, Beltrutti D, McConnell H, Quintero O: Chronic low back pain patients around the world: Cross-cultural similarities and differences. *Clin J Pain* 1992; 8:317-323.

Scheer SJ, Radack KL, O'Brien DR Jr: Randomized controlled trials in industrial low back pain relating to return to work. Part 1. Acute interventions. *Arch Phys Med Rehabil* 1995; 76(10):966-973.

Schlapbach P, Gerber NJ: Back school. *Rheumatology* 1991; 14:25-33.

Slade PD, Troup JDG, Lethem J, Bentley G: The fear-avoidance model of exaggerated pain perception - II. Preliminary studies of coping strategies for pain. *Behav Res Ther* 1983; 21:409-416.

Stankovic R, Johnell O: Conservative treatment of acute low back pain. A prospective randomized trial: McKenzie method of treatment versus patient education in "mini back school" *Spine* 1990; 15(2):120-123.

Symonds TL, Burton AK, Tillotson KM, Main CJ: Absence resulting from low back trouble can be reduced by psychosocial intervention at the workplace. *Spine* 1995; 20(24):2738-2745.

Symonds TL, Burton AK, Tillotson KM, Main CJ: Do attitudes and beliefs influence work loss due to low back trouble? *Occup Med* 1996; 46:25-32.

Szpalski M, Nordin M, Skovron ML, Melot C, Cukier D: Health care utilization for low back pain in Belgium. *Spine* 1995; 20:431-442.

Troup JD, Foreman TK, Baxter CE, et al: 1987 Volvo award in clinical sciences: The perception of back pain and the role of psychophysical tests of lifting capacity. *Spine* 1987; 12:645-657.

Tufo HM, Rothwell MG, Frymoyer JW: Managing the quality of care for low back pain, in Frymoyer JW (Ed.): *The Adult Spine: Principles and Practice.* New York, Raven Press Ltd., 1991, pp 61-75.

Välfors B: Acute, subacute and chronic low back pain: Clinical symptoms, absenteeism and working environment. *Scand J Rehabil Med* 1985; Suppl 11:1-98.

Venning PJ, Walter SD, Stitt LW: Personal and job-related factors as determinants of incidence of back injuries among nursing personnel. *J Occup Med* 1987; 29:820-825.

Versloot JM, Rozeman A, van Son AM, van Akkerveeken PF: The cost-effectiveness of a back school program in industry. A longitudinal controlled field study. *Spine* 1992; 17(1):22-27.

Videman T, Rauhala H, Asp S: Patient-handling skill, back injuries and back pain. An intervention study in nursing. *Spine* 1989; 14:148-156.

Volinn E, Lai D, McKinney S, et al: When back pain becomes disabling: A regional analysis. *Pain* 1988; 33:33-39.

Volinn E: Back sprain in industry: The role of socioeconomic factors in chronicity. *Spine* 1991; 16:542-548.

Volinn E: The epidemiology of low back pain in the rest of the world: A review of surveys in low and middle income countries. *Spine* 1997; 22(15): 1798.

Waddell G: A new clinical model for the treatment of low back pain. *Spine* 1987; 12:632-644.

Wiesel SW, Feffer HL, Rothman RH: Industrial low back pain: a prospective evaluation of a standardized diagnostic and treatment protocol. *Spine* 1984; 9(2):199-203.

Wiesel SW, Feffer HL, Rothman RH: The development of a cervical spine algorithm and its prospective application to industrial patients. *J Occup Med* 1985; 27(4):272-276.

Wiesel SW, Boden SD, Feffer HL: A quality-based protocol for management of musculoskeletal injuries: a ten-year prospective outcome study. *Clin Orthop Relat Res* 1994; 301:164-176.

Wood DJ: Design and evaluation of a back injury prevention program within a geriatric hospital. *Spine* 1987; 12:77-82.

77

Upper Extremity Support

Carolyn M. Sommerich
North Carolina State University

77.1 Introduction

A concise review of the literature on upper extremity supports is presented in this chapter. Broadly speaking, there are two categories of upper extremity supports. One category contains devices that are used as *medical treatment* to relieve symptoms or improve functionality. For example, wrist splints and elbow braces are commonly prescribed in the conservative treatment of carpal tunnel syndrome and lateral epicondylitis (tennis elbow), respectively. The other category contains items that are used as *workstation modification* in order to minimize joint and soft tissue loading or to reduce effects of physiologic tremor on precision tasks. Arm rests on chairs and palm rests on newer computer keyboards theoretically provide users with opportunities to reduce loads on postural muscles in the shoulder and forearm. Both aspects of upper extremity support, as medical treatment and as workstation modification, are discussed in this chapter. Situations wherein usage overlaps, such as wearing wrist splints during work, are discussed as well. Additionally, the use of arm supports for some specific work activities, including keyboarding and microsurgery, are discussed in some detail. Where sufficient evidence exists, recommendations are provided for circumstance-specific use of upper extremity supports.

77.2 Fundamentals

In this section, basic theories of upper extremity support are explored, including the ways in which various designs provide support, and under what circumstances they may be employed.

Splinting and Bracing as Medical Treatment

Wrist Splints. In the treatment of fractures, splinting had long been known to provide rest and relief from pain and disability. By analogy, splinting has been applied to arthritis (Ehrlich, 1968), where the treatment is typically prescribed to relieve pain and inflammation, but may also be prescribed to prevent deformities (contracture), or increase function (Spoorenberg, Boers, and van der Linden, 1994). Wrist splinting is also used in conservative treatment of carpal tunnel syndrome (CTS). When CTS is attributed to pregnancy or

0-8493-2641-9/99/$0.00+$.50

to work which is highly repetitive or requires sustained non-neutral wrist postures, splinting may reduce pain and paraesthesia (Ditmars, 1993). Symptoms would not be expected to respond to splinting when CTS is due to vibration exposure, lumbrical muscles entering the distal carpal tunnel during pinch grips, pathologic encroachment (carpal bone fracture or dislocation), or external pressure applied to the palm (from hand tools with sharp edges or short handles, or from repetitive activation of palm buttons).

Splinting the wrist in extension (using a cock-up splint) is a common practice. This practice is based on studies of wrist position during common activities, which have shown the centroid of motion for most common tasks to be an extended position (Palmer, Werner, Murphy, and Glisson, 1985). However, minimum pressure in the carpal canal, and therefore on the median nerve, has been shown to occur when the wrist is in a neutral orientation (in both flexion–extension and radial–ulnar axes) (Weiss, Gordon, Bloom, So, and Rempel, 1995). Splinting the wrist in extension may be effective for treating early stage (fully reversible inflammatory) lateral epicondylitis (Nirschl, 1985). Passive extension of the wrist relieves tensile stress at the common origin of the wrist and finger extensor muscles, the loci of pain in epicondylitis.

Splints are categorized as either rigid or flexible. Rigid splints do not permit any wrist motion, while flexible splints permit a limited range of motion, thereby preserving more hand function. Rigid splints may also be referred to as immobilizing, resting, or static splints. Flexible splints are also referred to as activity or working splints, or wrist supports. Splints usually have a plate or bar in or across the palmar region which may interfere with grasping activities. Both prefabricated and custom-made splints are used in treatment protocols, depending on the preference of the medical provider. Wearing recommendations may be for night use, usage during painful activity, or continuous usage, depending on the protocol recommended by the medical provider.

Elbow Bracing. A variety of occupations and tasks have been associated with tennis elbow, including meat processing, carpentry, plumbing, short cycle repetitive tasks, typing, and writing (Nirschl, 1985). Pain associated with tennis elbow (medial or lateral epicondylitis) may be relieved by wearing a support band just distal to the elbow, that covers the upper part of the forearm. Typically, about 5 cm in width, the band inhibits maximum contraction of the extrinsic wrist and finger muscles, which reduces tensile stress at the origins of those muscles (the tendons, aponeuroses, and epicondyles thought to be the sites of involvement). A band may be prescribed for use only during strenuous activity or throughout the day (Froimson, 1971; Valle-Jones and Hopkin-Richards, 1990).

Arm Supports as Ergonomic Modifications to a Workstation

Work in constrained postures is an acknowledged risk factor for neck and shoulder disorders (Hagberg, 1984), particularly when joint angles deviate significantly from neutral (resting) orientations. In reviewing numerous studies of the role of posture in neck and upper limb disorders, Wallace and Buckle (1987) concluded that posture may not always be a sufficient cause, but may interact with or add to other physical or psychosocial risk factors. Constrained postures impose continuous loads (stress) on muscles, tendons, ligaments, and joints. When they are of sufficient duration, even low levels of stress can lead to adverse health outcomes. Aarås (1994) reported a significant reduction in lost time associated with musculoskeletal illness at an electronics assembly facility when static loads on workers' trapezius muscles were reduced from levels of 4 to 6% of maximum voluntary contraction (% MVC) to below levels of 2% MVC, through workstation modification. For reference, simply positioning the arm for typing (upper arm vertical, elbow bent 90 degrees) has been estimated to require trapezius muscle activation of 3% MVC (Hagberg and Sundelin, 1986).

Potential pathologic reactions from sustained work-related stress on various musculoskeletal elements in the neck and shoulder include degenerative joint disease; tendon disorders, including degeneration and tendinitis; and problems with muscles, including myofascial syndrome (Hagberg, 1984). Development of localized muscular fatigue may be an immediate consequence of sustained muscle loading. Symptoms of fatigue may include discomfort or pain, reduced strength capacity, increased time for hand-eye coordination, increased hand tremor, and, when severe, difficulties in hand positioning (Chaffin,

1973). Low level muscle contractions (5% MVC) sustained for one hour have been shown to result in a 12% reduction in MVC, and shifts in the spectral frequency of the electromyographic (EMG) signal that are associated with localized muscle fatigue (Jørgensen, Fallentin, Krogh-Lund, and Jensen, 1988). Considering typical work break schedules (morning and afternoon work blocks, each four hr in length and each interrupted by a single 15 min break), maintaining a work posture for an hour would not be extraordinary, especially for individuals performing precise, hand-intensive work, or work requiring intense concentration.

Working with the upper arm unsupported and out of the vertical plane has been associated with symptoms of musculoskeletal disorders in the neck and shoulder region, with duration of exposure a key element (Melin, 1987; Jonsson, Persson, and Kilbom, 1988). Ergonomic hazard surveillance tools assign penalties for arms positioned out of the vertical or not supported (McAtamney and Corlett, 1993). Working with arms unsupported, significant levels of static activity have been recorded from muscles in the shoulder and forearm of typists performing typical keyboard operations (Onishi, Sakai, and Kogi, 1982; Sommerich, Marras, and Parnianpour, 1995). Positioning an unsupported arm or arm segment in any posture away from vertical requires activation of shoulder and arm muscles in order to oppose gravitational forces. Concomitant tensile loads are imposed on ligaments and tendons in the upper extremity, including the shoulder. These active and passive internal loads contribute, along with any external loads at the hand, to compressive and shear loading of the joints of the upper extremity. However, it is important to recognize that joint posture is not the sole factor affecting muscle activity requirements. Joint posture interacts with arm segment weight and the weight of any handheld objects. Specifically, it is the moments that result from these interactions that directly determine muscle activity requirements. These levels of muscle contraction, and associated soft tissue and joint loads, are necessary in order for an individual to accommodate the physical conditions of his or her work, including work space layout and tools. However, levels of muscle activity that exceed physical work requirements are not uncommon, and may occur due to operator training or technique, environmental conditions, task complexity, or personality (Lundervold, 1958; Westgaard and Bjørklund, 1987; Goldstein, 1964).

Joint and Soft Tissue Stress Reduction. The use of arm supports may be an effective method for reducing muscle activity requirements imposed by physical work conditions. Effectively, arm supports reduce arm segment weight. This reduces joint moments due to external forces, which, in turn, reduces moments required from muscles and may reduce stress on other soft tissues. Frequently supporting hands and arms on the work surface was found to be associated with lower incidence of pain in the neck, shoulder, and arms in groups of professional keyboard operators performing different types of keyboarding tasks (Hünting, Läubli, and Grandjean, 1981). In a laboratory-based study where subjects performed a simulated soldering task, Schüldt, Ekholm, Harms-Ringdahl, Németh, and Arborelius (1987) demonstrated that supporting the upper extremity through suspending the elbow or resting the elbow on a support resulted in marked reductions in shoulder muscle activity. Muscle activity was reduced whether subjects sat upright or sat with trunks flexed over the work surface. Sitting in a slightly reclined position reduced muscle activity so significantly that arm support did not provide further reductions in activity in that condition.

Based on force plate measurements, Occhipinti, Colombini, Frigo, Pedotti, and Grieco (1985) found that 4 to 7% of body weight was supported through resting wrists on the keyboard support surface, while 7 to 14% was supported when resting forearms on the same surface. Ranges reflect differences due to changes in unsupported trunk postures. Based on biomechanical modeling, they estimated reductions in spinal loading at the level of the third lumbar vertebra (L3) ranging from approximately 25 to 100 kg, depending upon type of arm support and trunk posture. In a related study, Colombini, Occhipinti, Frigo, Pedotti, and Grieco (1986) reported 6.6% of body weight supported through a forearm support when the back was also supported.

Methods of Supports. There are many ways to provide direct support for the upper extremity, and many points at which a support can contact the upper extremity including the elbow, along the forearm, at the wrist, or at the base of the palm. Objective effectiveness and user acceptance are both important issues in determining which methods may be appropriate for a particular situation. For example, as mentioned above, researchers have demonstrated reductions in shoulder muscle activity when the elbow was supported via suspension, or through fixed support located directly under the elbow during a soldering task (Schüldt, Ekholm et al., 1987). Chaffin (1973) demonstrated that time to fatigue was extended by 2 to 4 times with the use of an elbow support. Alternative keyboards are now available with built-in wrist rests. In an early study of alternative keyboards, subjects tended to prefer a split keyboard with a large built-in forearm–wrist support over both a split keyboard with a small built-in support and a traditional keyboard with a large support (Nakaseko, Grandjean, Hünting, and Gierer, 1985). Subjects were found to exert about twice as much force on the large support (about 35 N), compared to the small one (about 17 N), meaning that more upper body weight was supported by the larger support. Efficacy and preference for various types of supports will be examined in more detail in the Applications section of this chapter.

Influence of Support on Posture. Supports may be useful in helping muscles to maintain or improve the position of the supported limb. However, other body parts may be impacted by support, as well. For example, during a series of six 10-minute long typing periods, Weber, Sancin, and Grandjean (1983) found arm abduction was greater when trained typists keyed with a forearm–wrist support, when compared to unsupported keying. Subjects also sat nearer to the keyboard support surface when arms were supported.

In theory, one effect of wrist pad usage should be a reduction of wrist extension (if the pad is positioned appropriately). However, in a study of 12 office employees, Paul and Menon (1994) reported wrist extension ranging from 29 to 41 degrees when subjects used five different wrist pads that varied in shape, width, and compressiveness. However, no unsupported condition was provided to show whether extension with any of the pads was different from extension when no pad was used.

77.3 Applications

The theoretical benefits of various upper extremity supports were discussed in the previous section. Results from numerous studies are summarized in this section in order to demonstrate how and under what circumstances various types of upper extremity supports have been shown to be effective.

Splinting and Bracing

Effectiveness as Medical Treatment.

Wrist splints for disorders affecting the wrist. There is a consensus that usage of a splint should be based on a medical opinion, and that the usage should be supervised by a medical provider (Falkenburg, 1987). However, usage protocols and treatment effectiveness seem to vary widely across providers. Results of several studies of various protocols are provided in the following paragraphs.

Use of a working splint for one week, at night and during stressful activities, was found to reduce, at the end of the week, subjective assessments of pain, numbness, and tingling in a group of CTS patients with abnormal nerve conduction study results, when compared with symptoms in a similar group of patients who did not receive splints (Dolhanty, 1986). Wrists were splinted at 0 to 5 degrees of flexion.

The efficacy of conservative treatment of CTS based on steroid injection followed by continuous use of a neutral position, rigid splint for four weeks was tested prospectively by Weiss, Sachar, and Gendreau (1994). No patients with advanced CTS or associated medical conditions (such as diabetes, pregnancy, or arthritis) were admitted to the study. The authors found that only 13% of the treated hands were cured (defined as symptom free) at final follow-up, which occurred between 6 and 18 months after the injection.

A similar success rate (17%) was achieved by Banta (1994), with a regime of ibuprofen and three weeks of continuous, neutral positioned, wrist immobilization for 23 hands with early-mild CTS. A second stage treatment consisting of one week of iontophoresis and splinting followed by two more weeks of continuous splinting resulted in an improvement in 58% of the remaining hands. The author's overall success rate was 65%, based upon absence of symptoms at six months follow-up.

From a retrospective review of 105 CTS patients treated with a neutral angle rigid splint designed to permit hand function, Kruger, Kraft, Deitz, Ameis, and Polissar (1991) determined that 67% of the patients received symptom relief attributable to the splint. There was a significant decrease in the median sensory latency group average between pre- and post-treatment measurements. Splinting seemed to be most effective for patients who had experienced symptoms for a short period of time (1 to 3 months). Patients were told to wear the splints at night and during the day as much as possible, however, the authors were not able to confirm the wearing patterns of the patients. Though no lower limit was provided for post-treatment measurements, some occurred as long as 17 months after receiving the splint.

Based on a review of 363 hands with CTS, Kaplan, Glickel, and Eaton (1990) identified five factors which were related to the successful medical management of CTS. The five factors were age over 50 years, symptom duration exceeding 10 months, constant paresthesia, stenosing flexor tenosynovitis, and a positive Phalen's test in less than 30 s. Treatment success was defined as a patient remaining symptom free for six months following conservative treatment with a rigid wrist splint (neutral wrist position; worn at night and during the day when symptomatic) and anti-inflammatory medication. Follow-up averaged 15.4 months, but ranged from 6 to 48 months. Although their overall success rate was only 18.4%, there was a 66.7% success rate in the subgroup of patients who did not have any of the five factors. In the subgroup of patients who had one of the five factors, the success rate dropped to 40.4%; in the group with two, the rate was 16.7%, and was only 6.8% in the group with three factors. No patients with more than three factors were successfully treated.

Splinting may be specifically prescribed to relieve nocturnal CTS symptoms. This is based on the theory that patients sleep with wrists curled (flexed), which results in elevated pressure on the median nerve. Luchetti, Schoenhuber, Alfarano, DeLuca, De Cicco, and Landi (1994) found no statistically significant differences in carpal tunnel pressure (CTP) due to splinting, either during the day or throughout the night, between two groups of CTS patients (one splinted in 20 deg of extension, and the other not splinted). It is not known whether splinting would have had an effect if splints had maintained the wrists in neutral rather than extended postures.

The question of appropriate splint angle was addressed in a clinical trial in which patients' subjective responses indicated that a two-week regime of splinting in a neutral position was more effective in relieving CTS symptoms than was splinting in 20 degrees of extension (Burke, Burke, Stewart, and Cambré, 1994). Both splint designs appeared to be more effective for relief of nocturnal symptoms than daytime symptoms. Also of interest was the finding that only 8% of patients who continued to use their splints for two months experienced further symptom relief.

Compliance is an important part of any medical treatment. In their study of splint usage in the treatment of rheumatoid arthritis, Spoorenberg, Boers et al. (1994) found that patients were less likely to adhere to wearing advice for an immobilization splint than for an activity splint, although none of the patients who were told to wear their splints at night did so. Only 17% of the patients with immobilization splints who rested often (and therefore had the opportunity to wear the splint often), did so. In contrast, 57% of the patients with activity splints wore them often. This is, undoubtedly, at least partly due to patients' perceptions of the splints. While 96% of the rheumatologist responding to the survey prescribed both immobilization and activity splints for reduction of pain, only 44% of patients with immobilization splints found they relieved pain, and only 26% found they improved hand function. Seventy-eight percent of those patients found the rigid splints both unwieldy and ugly. In contrast, 75% of the patients with activity splints found they relieved pain and improved hand function. Only 29% thought they were ugly, but 63% found even the activity splints to be unwieldy.

Wrist splints for tennis elbow. As in the case with most musculoskeletal disorders and their associated risk factors, unless exposure to the risk factors associated with tennis elbow are reduced or eliminated, after immobilization treatment pain returns once a patient resumes the activities that lead to the development of tennis elbow (Nirschl, 1985). Therefore, Nirschl (1985) advocated for alteration or elimination of abusive activity rather than for immobilization of either the elbow or wrist joint. He cautioned that rigid immobilization could result in muscle atrophy. He also recommended changes in training techniques and equipment, along with counterforce bracing once activities were resumed. Little (1984) did not find splints to be effective in the treatment of tennis elbow. In two groups of subjects who both received cortisone and ultrasound treatment for tennis elbow, he found essentially no difference in recovery time (time when subjects became symptom-free) though the patients in one group were given splints and told to rest, while members of the other group were told to continue their normal activities.

Elbow bracing. Chen (1977) found reduced pain in patients with tennis elbow who used an elbow brace. Patients were reported to be able to perform many normal activities without experiencing symptoms, and were even able to participate in racquet sports. Apparently, however, an elbow brace may not provide instant relief from tennis elbow pain. Wadsworth, Nielsen, Burns, Krull, and Thompson (1989) were not able to demonstrate a statistically significant reduction in pain in a group of patients with tennis elbow who performed grip and extension strength tests, with and without an armband. Although patients reported feeling more comfortable with the brace. For the arm affected by tennis elbow, patients demonstrated a significant increase in wrist extension strength, and a lesser increase in grip strength when wearing the armband. No such increases occurred in the unaffected (control) arm.

Use on the Job. The science of ergonomics seeks to match workplace requirements with human capabilities and limitations. The preferred method of matching is through engineering controls, by which the work or work environment is altered to fit the worker. Use of wrist splints on the job is, instead, a method of altering the worker. Falkenburg (1987) described wrist splinting as "an industrial treatment to keep the employee on a job." She specifically mentioned, however, that splinting interferes with hand grasp, thereby reducing hand strength and interfering with holding or grasping tasks. Such functionality problems are documented in some of the splint research studies that are summarized in the following paragraphs. Most of the research on splints has been conducted in laboratory settings using healthy subjects, so results are only suggestive of what may occur in actual employment situations.

In an early field study of splinting as an intervention for CTS, results from a battery of objective and performance-based tests demonstrated that CTS was actually aggravated by the use of a rigid splint during work (Armstrong, 1981). In contrast, time off and light duty were found to be effective intervention methods.

One reason splints might be thought to be useful on the job for patients with CTS would be to reduce pressure on the median nerve by restricting patients from deviating wrists too far from a neutral posture. However, in a laboratory-based study using only healthy subjects, Rempel, Manojlovic, Levinsohn, Bloom, and Gordon (1994) were not able to demonstrate any difference in carpal tunnel pressure during a hand-intensive task between conditions when subjects wore a flexible wrist splint and when they did not, even though wrist motion, in both flexion–extension and radial–ulnar directions, was restricted by the splint. Rempel, Manojlovic et al. (1994) did find, however, that simply putting on the splint significantly raised CTP in each subject.

Perez-Balke and Buchholz (1995) also observed reduced wrist motion in subjects performing a repetitive pick and place task while wearing a rigid splint (with a malleable metal bar across the forearm, wrist, and palm), compared to performing the same task without the splint. They found reductions in peak grip force when subjects were splinted. Female subjects were particularly handicapped by the splint. When wearing the splint, their grip strength was reduced by a greater percentage than the males in the study (Perez-Balke and Buchholz, 1994; 1995). Their results might be explained, at least in part, by the work of Fransson-Hall and Kilbom (1993), who recorded significantly lower hand pain threshold tolerances and shorter times to experience pain in female subjects who experienced pressure applied to various points over the palm, fingers, and thumb, compared with results from male subjects who experienced the same test protocol. Another important finding from Perez-Balke and Buchholz (1995) was the increase in activity in the deltoid muscle and forearm flexor muscle group during the pick and place task when the splint was worn. In other words, subjects were less efficient when wearing the splint (more input for the same output).

In another task-based study using healthy subjects, Carlson and Trombly (1983) found that wearing a rigid splint slowed the performance of several manual tasks. Increased performance times were recorded for each of seven different tasks in the Jebsen Hand Function Test (which includes writing, manipulating small and medium-sized objects, picking up heavy cans, and feeding) when subjects wore a rigid splint, compared to when no splint was worn. The authors suggested that increased task performance times would be particularly problematic for an employee wearing a splint on a job with speed requirements (such as a highly repetitive assembly task), or for an employee who fatigued easily. They also made note of remarks of fatigue in the shoulder and upper trunk from several subjects, following task performance with the splint. Obviously, whether or not a person is wearing a wrist splint, the hand still needs to be in the same spatial location to perform a given task. If the wrist is not able to position the hand, the positioning task falls to the more proximal joints (elbow, shoulder, spine), and the muscles that control those joints. This is another example of the worker adapting to the workplace.

Stern, Sines, and Teague (1994) also employed the Jebsen Hand Function Test to evaluate the impact on performance due to splint usage in a group of healthy subjects. In contrast to the previous study, these authors tested several styles of commercial, nonrigid splints, which they referred to as "static wrist extensor orthoses," devices designed to support rather than immobilize the wrist. Only in manipulating small objects was there a significant difference in performance between the splinted and free hand conditions. In the other six tests, performance when wearing at least one of the five test splints did not differ from performance in the free hand condition, though it was not the same commercial splint which matched barehanded performance in each test. Subjects also wore each splint for a day, and reported differences in subjective experiences among the different splints. Splints differed in terms of the types of daily tasks with which they interfered (such as hygiene, housekeeping, or driving). Splints also differed in terms of comfort, temperature, and pain experienced by the subjects.

Possible Adverse Results. No reports were found in the literature which associated wrist splint usage with accidental injury. However, one report described a fracture of the distal radius in a gymnast wearing a standard gymnast's wrist support, which caught while he worked on the high bar. This experience should cause workers, employers, and medical providers to consider whether a wrist support could pose a danger in some situations. Wearing a splint might increase the chances of an employee becoming caught in rotating machinery, either directly by catching the splint, or indirectly by restricting the employee's movements and thereby restricting his or her ability to avoid contact with the equipment.

Arm Supports as Ergonomic Modifications to a Workstation

Arm support is typically considered for work that requires steady positioning of the hands for some period of time, such as electronics assembly, keyboarding, or surgery. One brief, but interesting mention was made by Jex and Magdaleno (1978) regarding their efforts to model the potential vibration damping effects of elbow rests for pilots flying high-performance aircraft. As in the case of wrist splints, few intervention studies have been performed to examine the effects of arm support on the job. Most of the research has been conducted in laboratories, on healthy subjects with limited exposure to each test condition, making it difficult to predict either short- or long-term effects from use of the various types of arm support. Results of many of those studies are summarized below.

Support During Specific Work Tasks

Assembly work. Eighteen months after installing arm suspension devices at an electronics plant, Harms-Ringdahl and Arborelius (1987) found that workers continued to choose to use the devices, which may have been due to the reductions in shoulder and neck discomfort experienced by the workers. Compared to before the suspension was introduced, average discomfort intensity was reduced by one-half to two-thirds. Before the suspension was introduced, about 40% of the workers rated their afternoon shoulder and neck discomfort greater than 50 on a scale of 0 to 100, whereas after 18 months of suspension use, only about 14% rated their discomfort that high.

In making ergonomic modifications to several workstations in an electronics assembly plant, Aarås (1994) recognized the importance of reducing shoulder load moments in workers. They achieved this through a number of engineering controls, including the provision of arm rests, either positioned on the chair or work surface, for any tasks carried out above elbow height.

Keyboarding. When given an opportunity to work at an adjustable workstation, either with or without a forearm–wrist support, two-thirds of a sample of 67 keyboard operators preferred the support (Grandjean, Hünting, and Pidermann, 1983). Seventy-eight percent of that sample did not find the support hindered their work on the keyboard. Based on observations of the operators, when the forearm–wrist support was present 80% of the subjects used the support. When no formal support was present, 50% of the subjects still rested their forearms or wrists on the keyboard support surface. A recent cross-sectional epidemiological study of VDT work seemed to objectively confirm the importance of arm supports for keyboard operators. Bergqvist, Wolgast, Nilsson, and Voss (1995) found neck/shoulder discomfort in a group of VDT operators was associated with a combination of three factors: more than 20 hr per week of VDT work, limited rest break opportunity, and working with the lower arm unsupported.

In a study of the effects of palm rest height and profile, 40 typists were each allowed to work with nine different palm rests during a one-week period, and were asked to use the one they preferred for at least half a day (Parsons, 1991). Though it is unclear how the supports compared physically, unlike the outcome of the study by Grandjean, Hünting et al. (1983), in this study only four of the 40 typists found the rests to be useful, while seven commented about increased discomfort. Several operators suggested that arm rests on their chairs might have been more comfortable because they would have provided support while enabling freedom of wrist and hand movement.

Relying on objective electromyographic data, two studies have demonstrated reduced activity in forearm extensor muscles with use of one particular alternative keyboard with a built-in wrist support (Smith and Cronin, 1993; Gerard, Jones, Smith, Thomas, and Wang, 1994), although a portion of that reduction was apparently due to reductions in typing speed which occurred with the alternative keyboard. However, palm rests were not shown to reduce activity in any of the forearm, shoulder, or back muscles studied by Fernström, Ericson, and Malker (1994). The authors discussed a couple of important limitations of their study which may have impacted their results: none of their subjects had prior experience with palm rests; and both practice (10 to 20 min) and testing (5 min) times in the study were limited. The authors suggested that people unfamiliar with palm rests tend to avoid using them (which may be important to remember when evaluating other support studies, or when introducing supports into the workplace).

Bendix and Jessen (1986) also found that forearm muscle activity (extensor carpi radialis) was not affected by the use of a wrist support, in a situation wherein subjects typed on electric typewriters, although trapezius muscle activity actually increased somewhat with use of the support. Both performance and subjective assessments of conditions with and without the support were similar. All subjects in the study were professional secretaries who had experienced discomfort in the neck and shoulder or elbow region for a substantial portion of the twelve-month period preceding the study. Prior to data collection, subjects had been allotted two weeks to adapt to the wrist support. In spite of the lack of objective or subjective beneficial effects of the support, nine of the twelve subjects wanted to keep the support once the study was completed.

Unlike wrist pads which are fixed relative to the keyboard, full motion forearm supports are fixed relative to the user's forearms. They are designed specifically to facilitate unrestricted motion of the arm and hand in a fixed horizontal plane, while maintaining support of the forearm and hand. Nonetheless, Powers, Hedge, and Martin (1992) found that subjects, described as office workers with substantial computer experience, using full motion forearm supports thought that the supports slowed their typing and increased their errors, although objective measurements showed this not to be the case. The authors did not find any postural benefits to using this type of support, based on a limited assessment of wrist and elbow postures.

Erdelyi, Sihvonen, Helin, and Hänninen (1988) studied the effects of fixed arm supports and arm suspension on experienced keyboard operators, some of whom were experiencing shoulder and neck pain. Both support methods were effective in reducing trapezius activity in the group with pain. In contrast, the group of healthy subjects experienced either no reduction in muscle activity, or, in some conditions, an increase in activity with the supports. In each condition, pain sufferers tended to display higher levels of muscle activity than did healthy subjects. Neither group preferred working with the supports. Compared to these findings with experienced keyboard operators, Sihovonen, Baskin, and Hänninen (1989) found a significant reduction in trapezius activity with the use of moveable arm supports in a group of healthy, novice word processors.

Medical procedures. Several different wrist and forearm supports have been designed for use during microsurgical, plastic, and ophthalmic surgical procedures (Halliday, 1988; Bustillo, 1968). The common objectives of each are to provide fatigue relief for shoulder and back muscles during procedures that require surgeons to remain almost immobile for long periods of time, as well as to provide a steady base of support for control of postural tremor. Unfortunately, there are other restrictions in an operating room which may diminish the impact of arm supports, including the adjustability and height requirements of the surgical microscopes and the height and thickness of the patient support surface.

An electromyographic study of dentists performing a variety of procedures on patients revealed fairly high levels of activity in the trapezius and extensor carpi radialis muscles (Milerad, Ericson, Nisell, and Kilbom, 1991). A second study was performed in order to determine the effect of several factors on muscle activity during dental procedures, including different types of upper extremity support (Milerad and Ericson, 1994). Compared to a no-support condition, arm support provided at the elbow significantly decreased muscle activity in the deltoid and trapezius muscles. Support at the elbow appeared to be more effective than support applied at the hand in reducing shoulder muscle activity. Whether such a support would be acceptable in practice is not clear from this study, since dentists were not used as subjects.

Possible Adverse Results. Most concerns are for the effects of localized pressure at contact sites between the support and the user's body. Direct pressure applied to the base of the palm was shown to raise the pressure in the carpal tunnel in cadaver specimens (Cobb, An, and Cooney, 1995), with point of application appearing to be the crucial factor. A 1 kg force applied just proximal to the distal wrist crease resulted in a mean increase in carpal tunnel pressure of 9 mmHg (median value of 1 mmHg), whereas that same force applied just distal to the distal wrist crease resulted in a mean pressure of 77 mmHg (median value of 29 mmHg) (Cobb, An et al., 1995). Pressure in the range of 20 to 30 mmHg begins to impact nervous and circulatory function and performance. *In vivo*, CTP has also been shown to increase when typing while resting the wrist, either on a wrist rest or on the keyboard support surface, in comparison to unsupported typing (Horie, Hargens, and Rempel, 1994). However, those authors did not report wrist posture in the three test conditions. If changes in wrist extension occurred between those three conditions, it would be difficult to determine whether the pressure change was due to external pressure on the wrist or change in wrist extension.

Pressure is also a concern at the elbow. Working with continuously flexed elbows, especially if supported at the elbow such that the ulnar nerve is compressed, may result in the development of cubital tunnel syndrome (compression of the ulnar nerve at the cubital tunnel) (McPherson and Meals, 1992).

77.4 Summary

In reviewing the literature there seems to be little consensus on the effectiveness or acceptance of upper extremity supports. However, some summary statements and recommendations can still be made.

Wrist splints and elbow braces are forms of medical treatment. As such, usage is an issue which should be decided between a medical provider and patient. Individuals should not self-prescribe these devices. If an individual's symptoms are the result of physical exposures, either in the workplace or outside of work, there is no reason to believe the symptoms will not return if the individual continues to be exposed to the same risk factors. Wrist splints may, in fact, cause a disorder to migrate to a more proximal joint that is forced to compensate for the reduced functionality imposed on the injured wrist by the splint. As such, the use of wrist splints at work should not be encouraged.

Wrist splints used at work would, at best, be considered an administrative control measure. However, nowhere are splints categorized as personal protective equipment (PPE), and they may actually make hand-intensive tasks more difficult to perform. The decision to wear a splint on the job should be made by an occupational health professional who has firsthand knowledge of the patient's workstation, tools, and tasks. Wrist splints may treat symptoms, but they do not address the root cause of an individual's discomfort. Thoughtful engineering controls are the most effective method for reducing an individual's exposure to physical risk factors associated with work-related upper extremity symptoms and disorders, such as carpal tunnel syndrome.

Wrist pads, forearm supports, and arm rests can all be categorized as engineering controls. Yet, based on the literature, their presence does not guarantee benefits or even usage. First, a worker's need for support must be balanced with the need for freedom of movement of the upper extremity. Continuous support methods would be appropriate if arms and hands are maintained in the same spatial location for extended periods of time. However, if movement is required, then the need is for a support that provides a resting point during mini-breaks (a few seconds in length) and longer pauses in activity. This resting support should not hinder the individual during work periods. Additionally, the worker's preference must be considered. For example, in modifying VDT workstations, once chair, monitor, and keyboard heights have all been established, operators should be provided with a variety of support options from which to choose. They should receive proper instructions in the ways the various supports should be used. Operators should also be encouraged to try out the supports at their own workstations.

There are a myriad of commercially available ergonomic support devices for keyboard workstations. The key to evaluating the efficacy of these products is to determine how and where the support is delivered, what postural changes might occur when using the device, and whether operators will find the support interferes with task performance. Based on the information in this chapter and some trial-and-error user-testing effective support devices may be found for those operators interested in utilizing them. For industrial settings, appropriate commercial supports may not exist. However, with some basic ergonomics training engineers, maintenance personnel, and operators may be able to devise unique support systems tailored to particular workstations or operators.

The goal of any upper extremity support is to aid an individual in achieving or maintaining a desired posture with less effort or less discomfort. Given that joints are designed to move, and most tasks require motion, there is often a conflict between the desire for support and the need for motion. Identifying a support which will provide support without interfering with motion requirements is important for the successful employment of upper extremity supports.

References

Aarås, A. 1994. The impact of ergonomic intervention on individual health and corporate prosperity in a telecommunications environment. *Ergonomics.* 37(10):1679-1696.

Armstrong, T. 1981. *Investigation of Occupational Wrist Injuries in Women.* 2 R01 OH 00679, U.S. Department of Health and Human Services, Centers for Disease Control, National Institute for Occupational Safety and Health.

Banta, C.A. 1994. A prospective, nonrandomized study of iontophoresis, wrist splinting, and anti-inflammatory medication in the treatment of early-mild carpal tunnel syndrome. *JOM.* 36(2):166-168.

Bendix, T. and Jessen, F. 1986. Wrist support during typing — a controlled, electromyographic study. *Appl Ergo.* 17(3):162-168.

Bergqvist, U., Wolgast, E., Nilsson, B., and Voss, M. 1995. The influence of VDT work on musculoskeletal disorders. *Ergonomics.* 38(4):754-762.

Burke, D.T., Burke, M.M., Stewart, G.W., and Cambré, A. 1994. Splinting for carpal tunnel syndrome: in search of the optimal angle. *Arch Phys Med Rehabil.* 75(Nov):1241-1244.

Bustillo, J.L. 1968. Hand and arm support in ophthalmic surgery. *Am J Ophthal.* 66(2):345-346.

Carlson, J.D. and Trombly, C.A. 1983. The effect of wrist immobilization on performance of the Jebsen Hand Function Test. *Am J Occ Therapy.* 37(3):167-175.

Chaffin, D.B. 1973. Localized muscle fatigue — definition and measurement. *J Occup Med.* 15(4):346-354.

Chen, S. 1977. A tennis elbow support (Letter to the editor). *Br Med J.* (1 Oct):894.

Cobb, T.K., An, K.-N., and Cooney, W.P. 1995. Externally applied forces to the palm increase carpal tunnel pressure. *J Hand Surg.* 20A(2):181-185.

Colombini, D., Occhipinti, E., Frigo, C., Pedotti, A., and Grieco, A. 1986. Biomechanical, electromyographical, and radiological study of seated postures, in *The Ergonomics of Working Postures*, Ed. N. Corlett, J. Wilson, and I. Manenica, p. 331-344. Taylor & Francis, London.

Ditmars, D.M. 1993. Patterns of carpal tunnel syndrome. *Hand Clinics.* 9(2):241-252.

Dolhanty, D. 1986. Effectiveness of splinting for carpal tunnel syndrome. *CJOT.* 53(5):275-280.

Ehrlich, G., E. 1968. Splinting for arthritis. *Medical Times.* 96(5):485-489.

Erdelyi, A., Sihvonen, T., Helin, P., and Hänninen, O. 1988. Shoulder strain in keyboard workers and its alleviation by arm supports. *Int Arch Occup Environ Health.* 60:119-124.

Falkenburg, S.A. 1987. Choosing hand splints to aid carpal tunnel syndrome recovery. *Occupational Health & Safety.* (May):60, 63-64.

Fernström, E., Ericson, M.O., and Malker, H. 1994. Electromyographic activity during typewriter and keyboard use. *Ergonomics.* 37(3):477-484.

Fransson-Hall, C. and Kilbom, Å. 1993. Sensitivity of the hand to surface pressure. *Appl Ergo.* 24(3):181-189.

Froimson, A.I. 1971. Treatment of tennis elbow with forearm support band. *J Bone Joint Surg.* 53-A(1):183-184.

Gerard, M.J., Jones, S.K., Smith, L.A., Thomas, R.E., and Wang, T. 1994. An ergonomic evaluation of the Kinesis Ergonomic Computer Keyboard. *Ergonomics.* 37(10):1661-1668.

Goldstein, I.B. 1964. Role of muscle tension in personality theory. *Psych Bull.* 61(6):413-425.

Grandjean, E., Hünting, W., and Pidermann, M. 1983. VDT workstation design: preferred settings and their effects. *Human Factors.* 25(2):161-175.

Hagberg, M. 1984. Occupational musculoskeletal stress and disorders of the neck and shoulder: a review of possible pathophysiology. *Int Arch Occup Environ Health.* 53:269-278.

Hagberg, M. and Sundelin, G. 1986. Discomfort and load on the upper trapezius muscle when operating a wordprocessor. *Ergonomics.* 29(12):1637-1645.

Halliday, B.L. 1988. A new surgical head rest. *Br J Ophthal.* 72:284-285.

Harms-Ringdahl, K. and Arborelius, U.P. 1987. One-year follow-up after introduction of arm suspension at an electronics plant, in *Proceedings of the Tenth International Congress, World Confederation for Physical Therapy*, p. 69-73, Sydney.

Horie, S., Hargens, A., and Rempel, D. 1994. Effect of keyboard wrist rest in preventing carpal tunnel syndrome, in *Proceedings of the Marconi Keyboard Research Conference*, Marshall, CA.

Hünting, W., Läubli, T., and Grandjean, E. 1981. Postural and visual loads at VDT workplaces. 1. Constrained postures. *Ergonomics.* 24(12):917-931.

Jex, H.R. and Magdaleno, R.E. 1978. Biomechanical models for vibration feedthrough to hands and head for a semisupine pilot. *Aviat Space Environ Med.* (Jan):304-316.

Jonsson, B.G., Persson, J., and Kilbom, Å. 1988. Disorders of the cervicobrachial region among female workers in the electronics industry. A two-year follow up. *Intl J Indust Ergo.* 3:1-12.

Jørgensen, K., Fallentin, N., Krogh-Lund, C., and Jensen, B. 1988. Electromyography and fatigue during prolonged, low-level static contractions. *Eur J Appl Physiol.* 57:316-321.

Kaplan, S.J., Glickel, S.Z., and Eaton, R.G. 1990. Predictive factors in the non-surgical treatment of carpal tunnel syndrome. *J Hand Surg.* 15B(1):106-108.

Kruger, V.L., Kraft, G.H., Deitz, J.C., Ameis, A., and Polissar, L. 1991. Carpal tunnel syndrome: objective measures and splint use. *Arch Phys Med Rehabil.* 72(June):517-520.

Little, T. 1984. Tennis elbow — to rest or not to rest? (Letter to the editor). *The Practitioner.* 228:457.

Luchetti, R., Schoenhuber, R., Alfarano, M., DeLuca, S., De Cicco, G., and Landi, A. 1994. Serial overnight recordings of intracarpal canal pressure in carpal tunnel syndrome patients with and without wrist splinting. *J Hand Surg.* 19B(1):35-37.

Lundervold, A. 1958. Electromyographic investigations during typewriting. *Ergonomics.* 1(3):226-233.

McAtamney, L. and Corlett, E.N. 1993. RULA: a survey method for the investigation of work-related upper limb disorders. *Appl Ergo.* 24(2):91-99.

McPherson, S.A. and Meals, R.A. 1992. Cubital tunnel syndrome. *Orthop Clin of NA.* 23(1):111-123.

Melin, E. 1987. Neck-shoulder loading characteristics and work technique. *Ergonomics.* 30(2):281-285.

Milerad, E. and Ericson, M.O. 1994. Effects of precision and force demands, grip diameter, and arm support during manual work: an electromyographic study. *Ergonomics.* 37(2):255-264.

Milerad, E., Ericson, M.O., Nisell, R., and Kilbom, Å. 1991. An electromyographic study of dental work. *Ergonomics.* 34(7):953-962.

Nakaseko, M., Grandjean, E., Hünting, W., and Gierer, R. 1985. Studies on ergonomically designed alphanumeric keyboards. *Human Factors.* 27(2):175-187.

Nirschl, R.P. 1985. Muscle and tendon trauma: tennis elbow, in *The Elbow and Its Disorders*, Ed. B.F. Morrey, p. 537-552. W.B. Saunders, Philadelphia.

Occhipinti, E., Colombini, D., Frigo, C., Pedotti, A., and Grieco, A. 1985. Sitting posture: analysis of lumbar stresses with upper limbs supported. *Ergonomics.* 28(9):1333-1346.

Onishi, N., Sakai, K., and Kogi, K. 1982. Arm and shoulder muscle load in various keyboard operating jobs of women. *J Human Ergol.* 11:89-97.

Palmer, A.K., Werner, F.W., Murphy, D., and Glisson, R. 1985. Functional wrist motion: a biomechanical study. *J Hand Surg.* 10A(1):39-46.

Parsons, C.A. 1991. Use of wrist rests by data input VDU operators, in *Contemporary Ergonomics*, Ed. E.J. Lovesay, p. 319-321. Taylor & Francis, London.

Paul, R. and Menon, K.K. 1994. Ergonomic evaluation of keyboard wrist pads, in *Proceedings of the 12th Triennial Congress of the International Ergonomics Association*, p. 204-207. Toronto.

Perez-Balke, G. and Buchholz, B. 1995. A study of the effect of a wrist splint on extrinsic flexor and anterior deltoid electromyography during a pick and place task, in *Proceedings of the Human Factors and Ergonomics Society 39th Annual Meeting*, p. 958. San Diego.

Perez-Balke, G. and Buchholz, B.O. 1994. A study of the effect of a "resting splint" on peak grip strength, in *Proceedings of the Human Factors and Ergonomics Society 38th Annual Meeting*, p. 544-548. Nashville.

Powers, J.R., Hedge, A., and Martin, M.G. 1992. Effects of full motion forearm supports and a negative slope keyboard support system on hand-wrist posture while keyboarding, in *Proceedings of the Human Factors Society 36th Annual Meeting*, p. 796-800. Atlanta.

Rempel, D., Manojlovic, R., Levinsohn, D.G., Bloom, T., and Gordon, L. 1994. The effect of wearing a flexible wrist splint on carpal tunnel pressure during repetitive hand activity. *J Hand Surg.* 19A(1):106-110.

Schüldt, K., Ekholm, J., Harms-Ringdahl, K., Németh, G., and Arborelius, U.P. 1987. Effects of arm support or suspension on neck and shoulder muscle activity during sedentary work. *Scand J Rehab Med.* 19:77-84.

Sihovonen, T., Baskin, K., and Hänninen, O. 1989. Neck-shoulder loading in wordprocessor use: effect of learning, gymnastics and arm supports. *Arch Occup Environ Health.* 61:229-233.

Smith, W.J. and Cronin, D.T. 1993. Ergonomic test of the Kinesis keyboard, in *Proceedings of the Human Factors and Ergonomics Society 37th Annual Meeting*, p. 318-322. Seattle.

Sommerich, C.M., Marras, W.S., and Parnianpour, M. 1995. Activity of index finger muscles during typing, in *Proceedings of the Human Factors and Ergonomics Society 39th Annual Meeting*, p. 620-624. San Diego.

Spoorenberg, A., Boers, M., and van der Linden, S. 1994. Wrist splints in rheumatoid arthritis: a question of belief? *Clin Rheum.* 13(4):559-563.

Stern, E.B., Sines, B., and Teague, T.R. 1994. Commercial wrist extensor orthoses: hand function, comfort, and interference across five styles. *J Hand Ther.* 7(Oct-Dec):237-244.

Valle-Jones, J.C. and Hopkin-Richards, H. 1990. Controlled trial of an elbow support ('Epitrain') in patients with acute painful conditions of the elbow: a pilot study. *Current Med Res Op.* 12(4):224-233.

Wadsworth, C.T., Nielsen, D.H., Burns, L.T., Krull, J.D., and Thompson, C.G. 1989. Effect of the counterforce armband on wrist extension and grip strength and pain in subjects with tennis elbow. *JOSPT.* 11(5):192-197.

Wallace, M. and Buckle, P. 1987. Ergonomic aspects of neck and upper limb disorders, in *International Reviews of Ergonomics*, Ed. D.J. Oborne, p. 173-200. Taylor & Francis, London.

Weber, A., Sancin, E., and Grandjean, E. 1983. The effects of various keyboard heights on EMG and physical discomfort, in *Ergonomics and Health in Modern Offices*, Ed. E. Grandjean, p. 477-483. Taylor & Francis, London.

Weiss, A.-P.C., Sachar, K., and Gendreau, M. 1994. Conservative management of carpal tunnel syndrome: a reexamination of steroid injection and splinting. *J Hand Surg.* 19A(3):410-415.

Weiss, N.D., Gordon, L., Bloom, T., So, Y., and Rempel, D.M. 1995. Position of the wrist associated with the lowest carpal-tunnel pressure: implications for splint design. *J Bone Joint Surg.* 77-A(11):1695-1699.

Westgaard, R.H. and Bjørklund, R. 1987. Generation of muscle tension additional to postural muscle load. *Ergonomics.* 30(6):911-923.

Part V
Organizational Design

Section I
Ergonomics Quality and Cost-Benefit Issues

78

Human Factors and TQM

Colin G. Drury
State University of New York at Buffalo

78.1 Introduction: TQM and Human Factors Programs in Industry

Over the past decade the pace of change in industry has been remarkable. Whether in manufacturing or service, industry has moved on an unprecedented scale to new technologies, new forms of organization, and new programs (Mize, 1992). It has not done this from an innate love of change, but because of strategic imperatives. The removal of tariff barriers and creation of trading blocs (EU, NAFTA, etc.) in the 1980s and 1990s has exposed even the smaller companies to unprecedented competition. One response in industrial countries has been to join the competition rather than fight it, for example by using manufacturing (and service) facilities in areas of relatively low labor costs. Thus, European and Japanese automobile plants have appeared in the U.S., while American apparel plants have been built in Central and South America. Even service operations, such as data entry and computer programming have been moved "offshore" using modern communication links.

However, many companies have chosen to remain in their traditional locations and compete by application of more advanced knowledge to their business. For example, Kleiner and Drury (1996) show how a number of companies in a rust-belt region chose to remain and expand by exploiting regional knowledge and skills.

One area in which global competition has benefited companies has been the free flow of ideas, matching the freer flow of goods and services. Thus, developments in microprocessor based technology, productivity software, organizational change, cellular manufacturing, and quality solutions arising in one county have been rapidly emulated throughout the developed (and now the developing) world. A major movement within this has been the quality imperative — the realization that without high quality, products will not sell, and the simultaneous realization that organizing for quality will produce benefits in productivity, efficiency, and safety (Crosby, 1979; Dobyns and Crawford-Mason, 1991; Krause, 1993; Deming, 1986).

Through the quality imperative in particular, companies have realized the importance of process control, i.e., ensuring that the process produces its intended output in a highly reliable manner. As more is learned about the process through quality methodologies (e.g., Statistical Process Control: Grant and

0-8493-2641-9/99/$0.00+$.50

Leavenworth, 1995; Designed Experiments: Taguchi, 1986), so the process can change from closed loop control using performance feedback, to open loop control using valid prediction models (Drury and Prabhu, 1994). Predictive control allows the process to operate with minimum setup time after a product change, thus facilitating moves toward just-in-time manufacturing.

The other company response to the quality imperative is the active management of quality. This has comprised both the realization that managerial leadership is important (Witcher, 1995) and specific policies for managing quality (Deming, 1986). An obvious policy is the use of teams as both a change agent (Blest, Hunt, and Shadle, 1992) and as a natural group for controlling a process (Brennan, 1990).

At the same time that these strategic-driven changes in quality have been taking place, there have been simultaneous programs at other levels. Thus, new technology has been introduced to reduce labor costs and/or improve process capability. In response to both rising costs and public/government pressure, there has been a movement toward managing the costs associated with human errors and injuries. Company responses here have been injury reduction/safety programs (Rahimi, 1995), ergonomics programs (Liker, Joseph and Armstrong, 1984) focusing on workforce injury reduction, and similar programs for reduction of the consequences of human error (e.g., Taylor, 1990). These latter are usually termed "human factors" programs but have many characteristics in common with "industrial ergonomics" programs. Indeed, programs incorporating both injury and error reduction are possible (Drury, 1995). In this chapter the terms "ergonomics" and "human factors" will be used interchangeably.

With few exceptions, programs arising from the quality movement and the human factors movement have been simultaneous but unrelated in industry. They have many similarities and some obvious differences, but there is no *a priori* reason for them to be separate. The remainder of this chapter takes up the managerial challenge of integrating these largely parallel programs so as to gain additional benefits. For a more detailed comparison and discussion of their linkages, see Drury (1996). In particular, that paper looks at many facets of the quality movement, such as TQM, Quality awards, the ISO-9000 series, and just-in-time manufacturing, while in the current chapter we concentrate on just one (TQM) for simplicity.

78.2 Fundamentals: The Basic Tenets of TQM and Human Factors

Before we can discuss interactions between the quality movement (e.g., TQM) and human factors/ergonomics, we must at least review their basic beliefs. Both programs "work" in the sense of improving industrial performance. For example, see Larson and Sinha (1995) for TQM and Drury (1992) for ergonomics.

TQM is not a monolithic philosophical structure, but rather a set of beliefs built upon the largely parallel efforts of a number of early practitioners. Rather than debate the merits of including each tenet, a recent review paper will be used to provide a convenient synopsis. Hackman and Wageman (1995) provide a thoughtful review of TQM so that we will use their structure of TQM. Table 78.1 provides this in outline form and does not appear to contradict the writings of most TQM practitioners, e.g., Deming's fourteen points (Deming, 1986). Hackman and Wageman (1995) also note two enhancements routinely used by (at least) U.S. practitioners of TQM: competitive benchmarking and employee involvement (EI). Benchmarking is the measurement of the level of performance of equivalent parts of other organizations to provide goals for those processes in your own organization. Goals are typically seen as being the best available rather than the average. Employee involvement is a generic title for a movement common in at least the larger companies to extend the employee voice in organizational affairs beyond the traditional union/management bargaining and beyond the roles specified in TQM (Russell, 1991). In fact, many companies see benchmarking and EI as integral parts of the TQM process.

Finding an equivalent set of basic beliefs or tenets of ergonomics has proven rather more difficult. As with TQM, human factors has been a largely empirical discipline, which defines what it *does* rather than its basic beliefs. Thus, societies within the International Ergonomics Association have their own definitions of ergonomics, as do textbooks and journals. Although some authors have begun to consider the

TABLE 78.1 Tenets of TQM

Assumptions	1. Good quality is less costly to an organization than is poor workmanship. 2. Employees naturally care about quality and will take initiatives to improve it. 3. Organizations are systems of highly interdependent parts: problems cross functional lines. 4. Quality is viewed as ultimately the responsibility of top management.
Change Principles	1. Focus on the work processes. 2. Uncontrolled variability is the primary cause of quality problems: it must be analyzed and controlled. 3. Management by fact: use systematically collected data throughout the problem-solving cycle. 4. The long-term health of the organization depends upon learning and continuous improvement.
Interventions	1. Explicit identification and measurement of customer requirements. 2. Creation of supplier partnerships. 3. Use of cross-functional teams to identify and solve quality problems. 4. Use scientific methods to monitor performance and identify points for process improvement. 5. Use process-management heuristics to enhance team effectiveness.

Adapted from Hackman, J. R. and Wageman, R. 1995. Total quality management: empirical, conceptual, and practical issues, *Administrative Science Quarterly,* 40: 308-342.

TABLE 78.2 Tenets of Ergonomics/Human Factors

Assumptions	1. Errors and stress arise when task demands are mismatched. 2. In any complex system, start with human needs and system needs, and allocate functions to meet these needs. 3. Honor thy user: use measurements and models to provide the detailed technical understanding of how people interact with systems. 4. Changing the system to fit the operator is usually preferable to changing the operator to fit the system. At least develop personnel criteria and training systems in parallel with equipment, environment, and interface. 5. Design for a range of operators rather than an average; accommodate those beyond the design range by custom modifications to equipment. 6. Operators are typically trying to do a good job within the limitations of their equipment, environment, instructions, and interfaces. When errors occur, look beyond the operator for root causes.
Change Principles	1. Begin design with an analysis of system and human needs using function and task analysis. 2. Use the task analyses to discover potential as well as existing human/system mismatches. 3. Operators have an essential role in designing their own jobs and equipment, and are capable of contributing to the design process on equal terms with professional designers. 4. Optimize the job via equipment, environment, and procedures design before optimizing the operator through selection, placement, motivation, and training. 5. Use valid ergonomic techniques to measure human performance and well-being before and after the job change process.
Interventions	1. Prepare well for any technical change, especially at the organizational level. 2. Involve operators throughout the change process, even those in identical jobs and on other shifts. 3. Use teams comprising operators, managers, and ergonomists (at least) to implement the change process.

underpinnings of the discipline (e.g., Karwowski, Marek, and Noworol, 1988; Meister, 1996), there is no simple list of tenets similar to Table 78.1. As a working list, Table 78.2 is proposed, keeping the structure of the equivalent TQM list to facilitate comparison. Note that this listing is biased toward design ergonomics, rather than more overtly sociotechnical systems approaches (e.g., Taylor and Felten, 1993).

As is obvious from Table 78.2, ergonomics is a human-oriented process, using detailed knowledge of human functioning as a basis for designing high-performance, safe systems. Indeed, the current book gives many examples of both the detailed human knowledge, and its use in design. We now need to consider the linkages between TQM and ergonomics explicitly, to find how to manage both programs together.

78.3 Applications of TQM and Ergonomics to Each Other

As a point of departure, a comparison of Tables 78.1 and 78.2 is useful in considering the matches between TQM and ergonomics. It is immediately obvious that there are both similarities and differences between the two lists. Under Assumptions, both consider the complexity of the system (#3 for TQM, #2 for ergonomics) as an explicit element in design and analysis. Both also have a belief in the integrity of the human operator in the system (#2 in TQM, #6 in ergonomics).

At the level of change principles, TQM starts with a focus on the work process, typically the *existing* work process (#1 in TQM). Ergonomics similarly starts from the work process (#1 and #2 in ergonomics), but this is typically advocated at a function level in the sense of all possible processes which *could* perform the task, rather than in the sense of the current process.

Neither ergonomics nor TQM advocates starting from the existing *jobs*, i.e., what individuals currently do in the process. This does not stop much of current ergonomics practice being oriented toward small changes in existing jobs. Also within the Change Principles, both TQM and ergonomics advocate a measurement-based approach (#3 in TQM, #5 in ergonomics).

For Interventions, the main point of similarity is in the use of small teams to control the change process (#3 in TQM, #3 in ergonomics). In a similar vein, specific measurements are stressed as #4 in TQM corresponding to #5 in ergonomics, listed, however, under Change Principles.

In those tenets where TQM and ergonomics differ, it is primarily due to differences in level of application. TQM is concerned with the company as a whole, its customers, and suppliers. It advocates a managerial approach, emphasizing responsibility, overall costs, continuous improvement, and managerial heuristics. Human factors/ergonomics in contrast deals with the system defined in narrower mission-oriented technical terms. It advocates particular solutions (hardware before training), detailed task analysis, user involvement and use of specific data on human functioning. Ergonomics is still a technical discipline, perhaps more at level equivalent to statistical quality/process control in TQM, than at the level of management intervention. As a single glaring example, Drury (1996) notes the almost complete absence of leadership considerations in the ergonomics/human factors literature.

With these similarities and differences in mind, we can explore some of the reported interactions between ergonomics and TQM. Practitioners of TQM have been noticeably silent on ergonomics/human factors. In the management literature many papers examine the impact of TQM on their discipline, for example Waldman (1994) on a theory of work performance, Grant, Shari, and Krishnan (1994) on management theory, or Costigen (1995) on human resource management, but nothing on ergonomics. However, the human factors literature has reported on how the quality movement affects the human factors profession, for example Zink, Hauer, and Schmidt (1994) on quality awards or Wilson, Neely, and Chew (1993) on the effects of modern manufacturing on worker well-being. This latter is taken further by Bjorkman's (1996) analysis of the similarities and differences between various quality movements and the tenets of modern work organization design. However, there are two relatively well-developed sets of ergonomics/TQM studies. The first uses TQM ideas in safety, while the second provides evaluations of joint TQM/ergonomics programs.

In the traditional safety area, a number of authors have pointed out the similarities between safety and quality (Rahimi, 1995; Roughton, 1993; Smith and Larson, 1991; Krause, 1993). One similarity is that both have departments (safety department, quality department) which may no longer be needed with a TQM approach (Rahimi, 1995). Krause (1993) shows their similarity in the measures they use. Traditionally, both have focused on downstream measures (defects reaching the customer, injury-producing accidents), but both should move toward process measures (SPC charts, behavioral measures) to help control the downstream events nearer to their source.

Two recent papers show how these ideas have progressed in Finland. Vainio and Mattila (1996) integrated safety concerns within the TQM system for an electrical utility. They made safety and health an integral part of TQM largely by addressing safety and health issues within the total quality handbook. More evaluation data were provided by Saari and Laitinen (1996) in a manufacturing setting. They set up continuous improvement teams for safety, defining best work practices in each area. Then, using the measurement-based TUTTAVA system, the teams set goals, made continuous improvements, and validated the results. A posture survey across the whole plant showed considerable improvement over the course of the project. In addition, injury and illness days lost were reduced by about 90% over three years.

Beyond safety is the safety role of ergonomics, typically designing to avoid injury. Here also a considerable literature is developing. Stuebbe and Houshmand (1995) characterize the production system as an inadvertent injury-producing system and advocate applying quality control approaches such as control charting, Pareto analysis, etc., to an "integrated ergonomic-quality system." This consists of analysis of the task, worker, and environment using these quality control techniques. Getty, Abbott, and Getty (1995) link quality initiatives to ergonomic projects, showing how an intervention to control cumulative trauma disorder in a panel drilling task also had a substantial effect on quality and productivity.

A major program in Sweden, the Quality, Working Environment and Productivity (QPEP) project (Axelsson, 1994; Eklund, 1995) examined specifically the linkages between quality and ergonomics in a car assembly plant. In eight departments, they produced an inventory of ergonomically demanding jobs, both those which were physically demanding and those causing production problems. Two different measures of quality showed significant differences between ergonomically good and ergonomically poor tasks, indicating the close link between ergonomics and quality.

One of the most integrated quality ergonomics efforts so far appears to be the implementation of ergonomic change within the TQM philosophy at the mail order clothing manufacturer and distributor, L. L. Bean (Rooney and Morency, 1992; Rooney, Morency, and Herrick, 1993). Their ergonomic objective was initially to eliminate the cumulative trauma disorder exposures of repetitive sewing production in a 400-person manufacturing plant. TQM was seen as defining the mission, objectives, and responsibility for safety with line management. Ergonomics moved over a six-year period from reacting to employee injuries, through proactive job design using teams, to now become part of the management and employee performance expectations and rewards.

In a follow-on paper, Rooney et al. (1993) were able to tackle some of the more deep-seated problems of repetitive work. They redesigned payment systems (with active operator involvement), replacing direct piece-rates with an annual appraisal system in which units produced were only 35 to 33% of the weighting. More complexity was built into jobs, by using cross-training and team work. Management and supervisor commitment for the ergonomics program was shown by their active support. Rooney et al. (1993) see these changes as a way of incorporating the musculoskeletal injury reduction aspects of ergonomics into a wider framework based upon macroergonomics (Hendrick, 1992) and TQM principles.

We can, however, go beyond these examples to provide managerial advice on TQM/human factors interactions, making use of similarities where they exist and exploiting differences to enhance each program. Starting with the similarities between the tenets of TQM and ergonomics (Tables 78.1 and 78.2) we can suggest, with some combining of categories:

1. *Study and Measure the Process.* Start from a systems focus rather than the current process (also advocated in Business Process Reengineering, Hammer and Champy, 1990). Use this as the basis for a detailed quantitative understanding of the process. Standard quality techniques should be used to measure process parameters, and models of human performance and well-being to measure and understand the role of the operator in the system. Use these measurements as the basis for directing and quantifying continuous improvement.
2. *Honor Thy User.* (To quote Kantowitz and Sorkin, 1987). Respect the operators in the system as people trying to do their best, and having an inherent stake in performing well. Do not necessarily blame the operator alone for poor quality/productivity/safety. Tap the potential for operator-empowered improvement by giving real power to small teams which include operators. The rewards will be improvements in performance, safety, and job satisfaction.

From differences between TQM and ergonomics we can show first how ergonomics can learn from TQM practice. These represent largely a shift from a technical process level of intervention to a more strategic, managerial level. (Longer discussions of each issue are presented in Drury, 1996.)

3. *Consider the Strategic Level.* Understand the forces beyond the process within the factory, such as requirements of the ultimate customer, and active management of the supply chain. Ensure that ergonomic interventions are truly customer-driven by explicitly measuring customer needs.
4. *Understand Leadership.* Any change activity needs responsibility of managers, up to the highest level. Do not take the mechanistic view of an organization which defines each manager by function. Understand the principles of leadership, recognize leaders, and practice leadership. All change projects need a powerful champion.
5. *Use Well-Developed Team Skills.* TQM, and many other change disciplines have standard methods of starting, organizing, and running successful teams. Use these methods where they are appropriate. At least understand these methods so that you can build on the teamwork training existing within the organization from TQM programs.

Where TQM can learn from ergonomics is in the area of technical knowledge of the human operator, and how to incorporate this in process design.

6. *Use Allocation of Function Techniques.* A basic building block of human factors is the concept of function allocation, i.e., permanent or flexible assignment of logical functions between human and machine. This has been used by ergonomists at levels ranging from the whole complex system (Older, Clegg, and Waterson, 1996) to a single human–machine system (Drury, 1994). Without an explicit treatment of function allocation, technology can easily fail. For example, consider the baggage handling system at the Denver International Airport.
7. *Error-Free Manufacturing/Service.* While TQM is calling for drastic reductions in error rates, human factors is coming to grips with the causes of human error (e.g., Reason, 1990). In airline flight operations (Wiener and Nagel, 1988) and maintenance (Wenner and Drury, 1997) we have classified errors and derived logical interventions, moving from a consideration only of the accident-precipitating event to a study of root causes and latent pathogens.
8. *Interface Design.* From physical workplace layout to reduce injuries (e.g., Kroemer, Kroemer, and Kroemer-Elbert, 1994) to the interface between software and the user (Helander, 1988), human factors engineers have been designing less error-prone interfaces between people and systems. This set of techniques is largely ignored in the TQM literature, despite the latter's emphasis on error reduction, parts per million, and six-sigma processes.

78.4 Summary

In this chapter we have examined the relationship between quality programs, specifically TQM, on the one hand and ergonomics/human factors programs on the other. Simple listing of their tenets, although these may still be arguable, led to recognition of the similarities and differences between the programs. Examples of use of ergonomics within a TQM context showed that sensible linkages had already been reported.

The aim of the chapter was to find prescriptions which would help the manager exploit the similarities and differences, so as to find new linkages between human factors and TQM. Seven prescriptions are given which can lead to greater integration between the two programs in the future. Readers who *do* use these for successful integration of the human factors and the quality imperative are urged to continue to report their work in the open literature and continue the integration process for the benefit of all.

References

Axelsson, J. R. C. 1994. Ergonomic aspects on design and quality, *IEA'94*, Vol. 4: 18-21.

Bjorkman, T. 1996. The rationalisation movement in perspective and some ergonomic implications, *Applied Ergonomics*, 27.2: 71-77.

Blest, J. P., Hunt, R. G. and Shadle, C. C. 1992. Action teams in the total quality process: experience in a job shop, *National Productivity Review/Spring 1992*, 195-202.

Brennan, L. 1990. The human dimension to statistical process control within advanced manufacturing systems, in *Ergonomics of Hybrid Automated Systems II*, Eds. W. Karwowski and M. Rahimi, p. 527-534, Elsevier Science Publishers, London.

Costigan, R. D. 1995. Adaptation of traditional human resources processes for total quality environments, *Quality Management Journal 95 Spring*, 7-23.

Crosby, 1979. *Quality is Free, The Art of Making Quality Certain*, McGraw-Hill, New York.

Deming, W. E. 1986. *Out of the Crisis*, Massachusetts Institute of Technology, Cambridge, Mass.

Dobyns, L. and Crawford-Mason, C. 1991. *Quality or Else. The Revolution in World Business*, Houghton Mifflin Company, Boston.

Drury, C. G. 1992. Ergonomics of job and equipment design, *Impact of Science on Society*, 165: 41-52.

Drury, C. G. 1994. Function allocation in manufacturing, in *Proceedings of the Ergonomics Society 1994 Annual Conference*, Ed. S. A. Robertston, University of Warwick, 19-22 April 1994, p. 2-16.

Drury, C. G. 1995. Work design, in *Human Factors Guide for Aviation Maintenance*, Ed. M. Maddox, Chapter 6, Federal Aviation Administration/DOT, Washington, D.C.

Drury, C. G. 1996. Ergonomics and the quality movement. The 1996 Ergonomics Society Lecture, Leicester, U.K., 10-12 April 1996.

Drury, C. G. and Prabhu, P. V. 1994, Human factors in test and inspection, in *Design of Work and Development of Personnel in Advanced Manufacturing*, Eds. G. Salvendy and W. Karwowski, Chapter 13, p. 355-492, John Wiley & Sons, New York.

Eklund, J. A. E. 1995. Relationships between ergonomics and quality in assembly work. *Applied Ergonomics*, 26(1): 15-20.

Evans, J. R. and Lindsay, W. M. (1995). *The Management and Control of Quality*, Third Edition, West Publishing Company, Minneapolis/St. Paul, MN, 143.

Getty, R. L., Abbott, W. L. and Getty, J. M. 1995. ISO 9000 methodology enhances ergonomics effort: ergonomics becomes a tool for continuous improvement, *ASQC 49th Annual Quality Congress Proceedings*, 904-913.

Grant, E. L. and Leavenworth, R. S. 1995. *Statistical Quality Control*, McGraw-Hill, 1984, New York.

Grant, R. M., Shari, R. and Krishnan, R. 1994. TQM's challenge to management theory and practice, *Sloan Management Review/Winter*, 25-35.

Hackman, J. R. and Wageman, R. 1995. Total quality management: empirical, conceptual, and practical issues, *Administrative Science Quarterly*, 40: 308-342.

Hammer, M. and Champy, J. 1993. *Reengineering the Corporation*, Harper Business, New York.

Helander, M. Ed. 1988. *Handbook of Human-Computer Interaction*, Elsevier Science Publishers B.V., Amsterdam, The Netherlands.

Hendrick, H. W. 1992. A macroergonomic approach to work organization for improved safety and productivity, in *Advances in Industrial Ergonomics and Safety IV*, Ed. S. Kumar, p. 3-10, Taylor & Francis, Hampshire.

Kantowitz, B. H., and Sorkin, R. D. 1987. Allocation of function, in *Handbook of Human Factors*, Ed. G. Salvendy, p. 355-369, John Wiley & Sons, New York.

Karwowski, W., Marek, T. and Noworol, C. 1988. Theoretical basis of the science of ergonomics. *Ergonomics International 88, Proceedings of the 10th Congress of the International Ergonomics Association*, Sydney, Australia, 1-5 August 1988, Eds. A. S. Adams, R. R. Hall, B. J. McPhee and M. S. Oxenburgh, p. 756-758, Taylor & Francis, London.

Kleiner, B. M. and Drury, C. G. 1996. Macroergonomics in regional planning and economic development in O. Brown, Jr. and H.W. Hendrick (Eds.), *Human Factors in Organizational Design and Management*, 523-528.

Krause, T. R. 1993. Safety and quality: two sides of the same coin, *Occupational Hazards*, April 1993, 47-50.

Kroemer, K.H.E., Kroemer, H.J. and Kroemer-Elbert, K.E. 1994. *Ergonomics: How to Design for Ease and Efficiency*, Prentice Hall, Englewood Cliffs, NJ, 430-441.

Larson, P. D. and Sinha, A. 1995. The TQM impact: a study of quality managers' perceptions, *Quality Management Journal 95 Spring*, 53-66.

Liker, J. K., Joseph, B. S. and Armstrong, T. J. 1984. From ergonomic theory to practice: Organizational factors affecting the utilization of ergonomic knowledge, in *Human Factors in Organizational Design and Management*, Eds. H. W. Hendrick and O. Brown, Proc. 1st Symp., Honolulu, HI, August 1984, North-Holland, Amsterdam.

Meister, D. 1996. A new theoretical structure for developmental ergonomics. Paper to *4th Pan-Pacific Conference on Occupational Ergonomics*, Taipei 1996.

Mize. J. H. 1992. Constant change, constant challenge, in *Manufacturing Systems, Foundations of World-Class Practice*, Eds. J. A. Helm and W. D. Compton, p. 196-203, National Academy Press, Washington, D.C.

Older, M., Clegg, C. W. and Waterson, P. E. 1996. Task allocation in complex systems, in *Advances in Applied Ergonomics*, Eds. A. F. Ozok and G. Salvendy, p. 471-474, U.S.A. Publishing Corp, IN.

Rahimi, M. 1995. Merging strategic safety, health and environment into total quality management, *International Journal of Industrial Ergonomics*, 16: 83-94, Elsevier, London.

Reason, J. 1990. *Human* Error, Cambridge University Press, Cambridge, U.K.

Rooney, E. F. and Morency, R. R. 1992, A practical evaluation method for quantifying ergonomic changes at L. L. Bean, in *Advances in Industrial Ergonomics and Safety IV*, Ed. S. Kumar, p. 475-481, Taylor & Francis, New York.

Rooney, E. F., Morency, R. R. and Herrick, D. R. 1993. Macroergonomics and total quality management at L. L. Bean: a case study, in *Advances in Industrial Ergonomics and Safety V*, Eds. R. Nielsen and K. Jorgensen, p. 493-498, Taylor & Francis, New York.

Roughton, J. 1993. TQM Integrating a total quality management system into safety and health programs. *American Society of Safety Engineers*, June 1993, 32-37.

Russell, S. 1991. Employee involvement aspects of total quality management, *P+ European Participation Monitor*, (2): 29-32.

Saari, J. and Laitinen, H. 1996. Towards continuous improvement of workplace, in *Advances in Applied Ergonomics*, Eds. A. F. Ozok and G. Salvendy, p. 82-87, U.S.A. Publishing Corp, IN.

Smith, T. J. and Larson, T. L. 1991. Integrating quality management and hazard management: a behavioral cybernetic perspective, *Proceedings of the Human Factors Society 35th Annual Meeting — 1991*, 903-907.

Stuebbe, P. A. and Houshmand, A. A. 1995. Quality and ergonomics, *Quality Management Journal 95 Winter*, 52-64.

Taguchi, G. 1986. *Introduction to Quality Engineering*, UNIPUB, White Plains, NY.

Taylor, J. C. and Felten, D. F., 1993. *Performance by Design*, Prentice Hall, NJ.

Taylor, J.C. 1990. Organizational context for aircraft maintenance and inspection, in *Proceedings of the Human Factors Society 34th Annual Meeting, Volume 2*, 1176-1180.

Vainio, P. and Mattila, M. 1996. Development of a safety and ergonomics oriented total quality system for an electricity company, in *Advances in Applied Ergonomics*, Eds. A. F. Ozok and G. Salvendy, p. 43-46, U.S.A. Publishing Corp, IN.

Waldman, D. A. 1994. The contributions of total quality management to a theory of work performance, *Academy of Management Review 1994*, 19(3): 510-536.

Wenner, C. and Drury, C. G. 1997. Deriving targeted interventions for ground damage, *Proceedings of the 1997 SAE Airframe Engine Maintenance & Repair Conference* (AEMC '97) August 1997, SAE Technical Paper Series 972591, Warrendale, PA.

Wiener, E. L. and Nagel, D. C. Eds. 1988. *Human Factors in Aviation*. Academic Press, Inc., San Diego.

Wilson, J. R., Neely, A. D. and Chew, T. 1993. Human and production requirements in modern manufacturing: complementary or contradictory? *Journal of Design and Manufacturing (1993)*, 3: 167-175.

Witcher, B. 1995, The changing scale of total quality management, *Quality Management Journal 95 Summer*, 9-29.

Zink, K. J., Hauer, R. and Schmidt, A. 1994. Quality assessment: instruments for the analysis of quality concepts based on EN 29000, the Malcolm Baldridge Award and the European Quality Award, in *Total Quality Management*, Ed. G. K. Kanji, 5(5): 329-343, Carfax Publishing Company, UK.

For Further Information

The ideas in this chapter were based on the rather fuller treatment in Drury (1996), and the concept of tenets of the discipline reported by Hackman and Wageman (1995). The latter is a good and thoughtful review of TQM from a management viewpoint.

Standard works on TQM are Deming (1986), Evans and Lindsay (1995), and Taguchi (1986).

Excellent evaluations of the social role of TQM can be found in Wilson, Neely, and Chew (1993) and Bjorkman (1996). Comments on TQM from a sociotechnical systems viewpoint are given by Taylor and Felten (1993).

79

Participatory Ergonomics

Kageyu Noro
Waseda University

79.1 Introduction: The Concept of Participatory Ergonomics and its Expanded Interpretation

This chapter begins with an explanation of the conventional participatory ergonomics as applied to workers and users of a product. Then an expanded interpretation of participatory ergonomics as applied to the general public is discussed. Participatory ergonomics enables development of an information loop that facilitates ergonomic activities. Figure 79.1 shows the difference between conventional ergonomics and participatory ergonomics, using a workplace as an example. The worker becomes an actor in the process. A manager can be either an agent or an actor.

The two-way information flow shown in Figure 79.1 is achieved through the workers' active involvement at the workplace. Participatory ergonomics is the workers' active involvement in implementing ergonomic knowledge and procedures in their own workplace.

0-8493-2641-9/99/$0.00+$.50

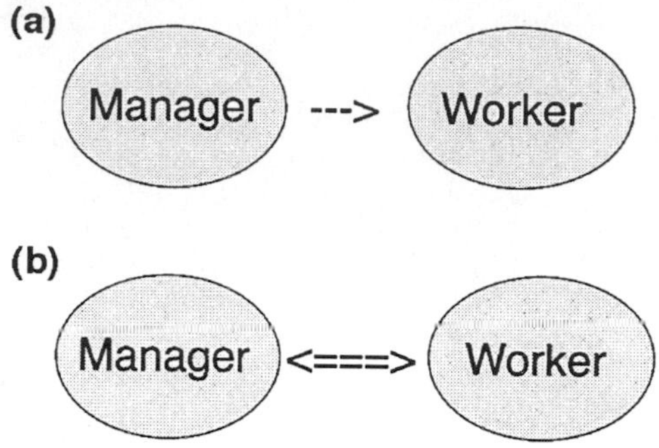

FIGURE 79.1 Two-way information flow in participatory ergonomics.

The concept of participatory ergonomics has its origins in discussions between Noro and Kazutaka Kogi in Singapore in 1983. The term *participatory ergonomics* was proposed by Kogi in a subsequent discussion (Noro and Kogi, 1985). The concept was further solidified in a workshop held by Noro (Noro, 1984) in Toronto. Noro and Imada's joint work began with informal discussions in Sacramento, California, in 1984. This was followed by a workshop which they jointly held at the first ODAM meeting in Hawaii later that year. Noro proposed the term *participatory approach* or *participatory ergonomics* for the Hawaii workshop (Noro and Imada, 1984). The participatory approach has been a major topic at all the IEA Congresses, Human Factors and Ergonomics Society Annual Meetings, and ODAM Symposia from the IEA Congress in Bournemouth in 1985, through the present day.

The original context of participatory ergonomics was workers at workplaces. This means that participatory ergonomics was interpreted as a company-wide activity. Recently, specialists have started to extend the range of application of participatory ergonomics from factory workers to product designers and users. In this latter case, participatory ergonomics aims to enhance the mutual participation of product designers and manufacturers together with the users of the product in each other's "scene." The term *scene* is used to mean the otherwise separate activities of specialists working on designing and manufacturing a product and users using the product.

The range of persons involved is expanded to include the users of a product, and the degree of reliance of the system on organizational design and management is decreased drastically. Instead, in order to function actively, the system relies heavily upon a network with an unspecified number of users and upon a common description method that the nonspecialist users can easily understand.

It is clear that the spread of the participatory approach is closely related to the interest in strengthening management of factories and enterprises since the early 1980s. While participatory ergonomics was first applied in the closed loops and controlled environments of such enterprises, it has recently been applied in the more open environment of the wider society. This change has been made possible largely by having the technological means — for example, email, inter/intranet and virtual reality — to communicate the message of participatory ergonomics. Participatory ergonomics has expanded from a typically company-wide activity to one involving the more diverse activities of members of society as a whole. Since ergonomics plays an increasing role in all aspects of our daily life, this sort of expansion of its approach and application is natural and is set to become a dominant development in ergonomics from now on.

The two types of participatory ergonomics summarized above are further described — first in the context of companies and enterprises and then in the wider context.

79.2 Participatory Ergonomics as a Company-Wide Activity

Noro stated (1991) that participatory ergonomics across the company was rooted in the influence of "small group activity" which started to be employed as part of the product quality control system (kaizen) that had been one of the basic technologies of Japan's rapid economic growth from the 1970s to the 1980's. Noro (1988) carefully examined 313 kaizen-related reports at the Nippon Steel Corporation's

TABLE 79.1 Recommended Steps for Participatory Ergonomics

Step 1: Selecting a theme
Step 2: Establishing a goal
Step 3: Understanding the situation and analyzing the factors
Step 4: Identifying the problems
Step 5: Developing and improving measures to solve the problems
Step 6: Confirming the effects of the measures taken

Source: Noro, K. 1991. Concepts, methods and people, in *Participatory Ergonomics,* Eds. K. Noro and A. Imada. p. 13 Taylor & Francis, London.

Yawata Works in the early 1980s. He found that one third of all the reports were about ergonomics. These reports written by factory workers were included in the first chapter of the resulting book (Noro, 1991). The ergonomic tools described there are a variety of creative and well-devised ideas that actually came from factory workers at their own workplaces and not from specialists or managers. The workers are, in this case, the actors typically working in small groups. Their ideas range from a campaign called "0.1-s Operation" to avoid wasting even 0.1 seconds during work to a fault tree drawn by the workers themselves.

79.3 Steps for Participatory Ergonomics

Noro (1991) recommended the steps listed in Table I for participatory ergonomics. Table 79.1. As already mentioned, the participatory approach has become one of the most discussed topics in ergonomics, and a number of reports on the subject have been published since the late 1980s. A selection of these publications is reported here. These cases are significantly different from the reports included in Noro's study (1991) mentioned above. Noro included in his report information conveyed from the workers to the management that would be too inconspicuous for a specialist to take notice of but is actually very important. In the following cases specialists, namely "agents," have carried out participatory ergonomics.

In addition to those listed in Table 79.1, there are other suggestions for the general process of participatory ergonomics. Kuorinka (1995) suggested the following outline for the general process:

1. Clarify the essence of the problem and establish a goal
2. Generalize and prioritize the measures
3. Implement the measures
4. Follow up

A step-by-step approach according to Vink (1995) would be to

1. Prepare (decide the objective and the framework of the project)
2. Analyze work and health
3. Select measures
4. Implement the measures
5. Evaluate

79.4 Education

It is important that not only actors but also agents participate in education in order to carry out participatory ergonomics smoothly. Garmer et al. (1995) report on a coeducational program between workers, technicians, and managers. The program was implemented using a participatory ergonomics approach to promote participation of workers at an automobile assembly factory. In this factory, a tilting car body mounting system was being introduced that rotates and allows workers to adjust its height so that the workers can do 80% of the assembly work while standing upright. Alongside the introduction of this mounting device, a coeducational program between workers, technicians, and managers was developed using a dialogue-style model. In the model, personnel and labor management took the role of the agent and the operator, and the production engineer took the role of the actor.

It is important to give information about better working postures and warnings about postures that are harmful when using the lifting devices in the factory. A real-time posture input system (Kawano, 1996) was developed to instruct workers and to help them to easily and quickly understand the study results while participating for a short period of time. The device allows a worker to check his or her posture with a puppet-like device; the torques acting at body joints are displayed on a monitor. This acted as a support system for consultation regarding working posture.

79.5 Team Building

Building a team generally plays a part in participatory ergonomics as typified in this example from a red meat packing company. The team analyzed musculoskeletal injuries and hazards (Moore and Garg, 1996). The objective was to address previously unsolved problems and a high job turnover rate. A problem-solving method was identified using principles related to the product quality improvement process. The workers participated in the problem solution process. The largest obstacle for this team was in scheduling the meetings.

79.6 Measures for Occupational Safety and Health Problems and Human Errors

Participatory ergonomics is suitable for solving occupational safety, health problems, and human error. Kuorinka (1995) reports an example of a team activity to reduce human errors in handling radioactive material. The essential factors for this team activity were a formal procedure to serve as a guideline for participants, a process analysis methodology creating a customer-oriented culture among the engineering staff and team building techniques (Caccamise, 1995).

Participatory ergonomics is also effective for office work. Participatory ergonomics employed in a salary distribution department in a Dutch government ministry is reported by Vink et al. (1995). The purpose was to reduce mental and physical workload. A strong commitment from management was obtained in this example which used a step-by-step approach. The merit of this step-by-step approach in participatory ergonomics was that it was possible to systemize the process and make all the participants aware of the necessity for improvement. The drawback was that the step-by-step approach took too much time.

79.7 Participatory Ergonomics for Product Development and its Users

Although VDTs and chairs are used largely in the workplace, computers have evolved to be used not only in workplaces but in various other situations. Guidelines on how to use VDTs have been published in many countries by various organizations since the 1960s. Those, however, are for specialists. Guidelines have been developed for use by nonspecialists at their own workplace. These guidelines are introduced below.

79.8 Dialogue-style VDT Guidelines

A great deal of research has been done on chairs and VDTs, and many geometrical models have been developed of the relationships between workers and their computers in office workplaces. These models attempt, for each office worker, to specify the ideal values for predefined parameters of a workstation by processing selected physical measurements of the worker. These models are thus called "parametric models" (Noro, 1994). The physical dimensions of each worker and the dimensions of chairs, desks and appliances are all easily measured by a specialist. However, it is often "too much trouble" for the user of a chair to measure its dimensions. Therefore, it will be too difficult for a user to use these models to select the right settings of an adjustable chair to suit his/her own physical dimensions.

Noro et al. (1995) proposed a nonparametric method which does not rely on dimensions. The method is based on a database of an abundant record of measurements compiled using parametric models. It does not, however, simply present these measurements to a user so that he/she can adjust the dimensions to fit his/her own dimensions by himself/herself. After the final set of optimum measurements is derived, the user expresses his/her sensation into words. The aim is to make adjustments to the chair based on the verbal values without specific reference to the resulting metric values. This sort of approach will become increasingly important as participatory ergonomics deals more and more with the general public, as will be discussed later. *A Guideline for VDT Operators* (Waseda University, 1995) published by Waseda University in April 1995 is a VDT guideline based on this method.

The university established a VDT committee which is a team composed of eight workers, selected by the labor union as representatives of VDT users at work, and eight managers. The members of the committee were all nonspecialists. As an advisory body to this committee, a group of specialists was appointed. The guideline leaflet employs a dialogue-style approach with text such as "Which posture is the closest to your present posture?" and a procedure similar to that of a board game; the dialogue proceeds in the same manner in which a game piece moves toward its goal. Figure 79.2 shows the opening of this guideline leaflet.

Each factor from the sitting posture through the keyboard to the display adjustment is checked in a progressive order. It is a pictorial VDT guideline and when the user reaches the goal he/she is led to the desired posture. Figure 79.3 shows an information loop for communicating research results to individuals who are nonspecialists by converting experimental values into words.

79.9 Product Test Charter

It is difficult to measure and analyze personal information such as a person's opinion, sensations, and feelings. In participatory ergonomics more effort should go into developing methods to incorporate peoples' opinions, sensations, and feelings. The product test laboratory in Sony Corp. has a "Product test charter" (Noro, 1992). The following is an extract from the charter. "We shall not use any measuring instruments when we communicate with a user. We shall value the user's feelings about the product and his/her hearing and visual senses. We shall value the user's memory and emotional impressions. We shall test a product in an environment where the product will actually be used." This charter is an example of the effort to enlarge the pipeline of information from the user to an enterprise and encourage the participation of the user.

Participatory ergonomics is directed toward the general public. The concept of networking is central to activities in expanded participatory ergonomics. The development of interactive technology and media that support the concept is, therefore, important. Electronic networks — namely, e-mail, the internet and intranet — and virtual reality technology provide participatory ergonomics with powerful supportive means.

79.10 Taking Advantage of the Internet

The internet is an extremely effective tool for participatory ergonomics with the general public. Miyamoto created a home page on the internet as a means to offer and collect information more easily and efficiently. The home page offers ergonomic knowledge regarding chairs to the internet public and collects information to survey the situation regarding chairs by having the internet user answer a questionnaire at the same time. (Hata and Miyamoto, 1996).

79.11 Common Knowledge

There is ideally a common knowledge or tacit understanding involved in communication between workers in related fields whether the field is product development, office work, or factory work. For example, with the latest optical character recognition systems (OCRs) the tacit understanding is to use material with current commercial laser printer print quality. OCRs' ability to read documents printed by a

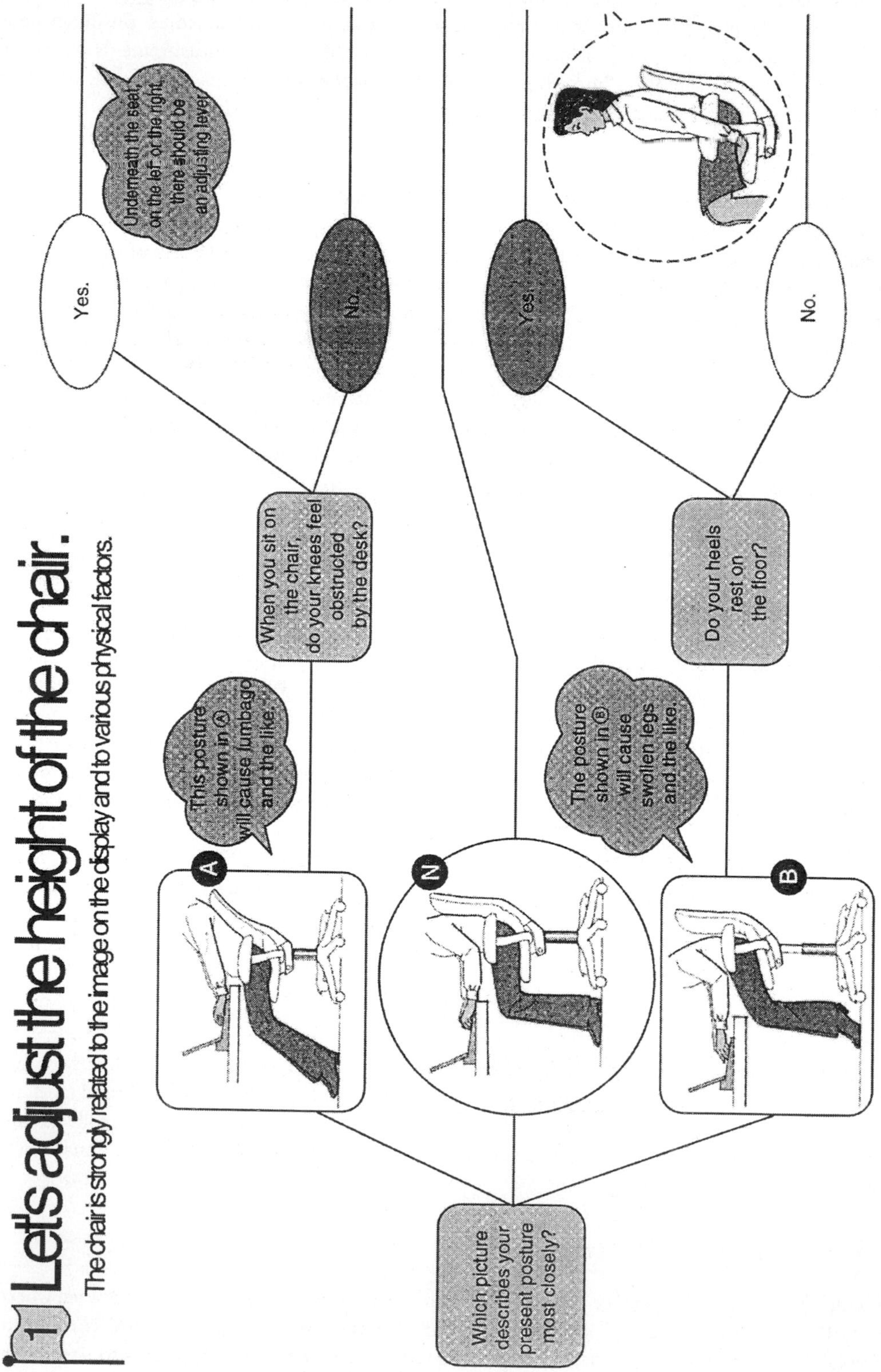

FIGURE 79.2 The opening part of the non-parametric VDT guideline.

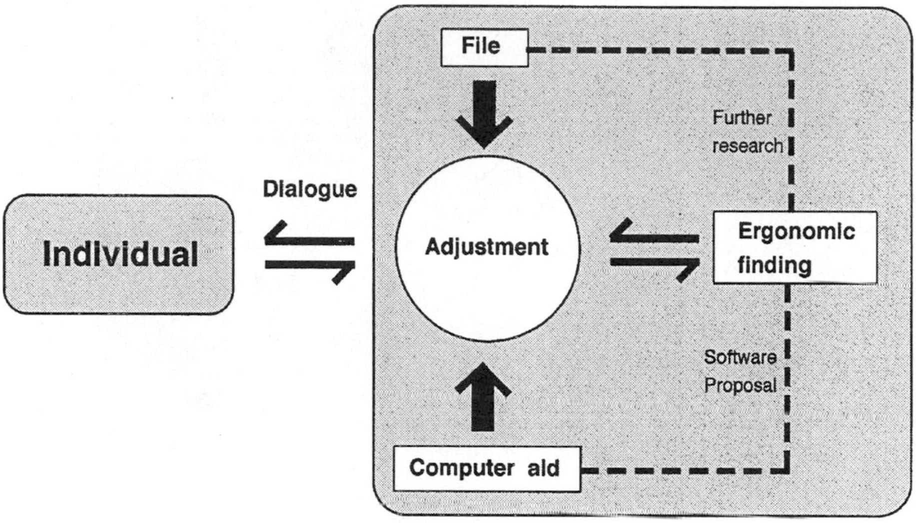

FIGURE 79.3 Information loop for making adjustments.

10-year-old dot matrix printer is drastically lower. The extent of such common knowledge is an aspect which ergonomics must clarify in the 1990s, an era of internationalization, in which countries with differing cultures communicate with each other at an unprecedented level of intensity.

Gill stated (1990) that there is, in a knowledge-based system, a limit to the art of acquisition in trying to understand active knowledge and skills. Based on the experience of conducting case studies with insurance agents and consultants, Gill points out the importance of knowledge of the tacit dimension such as a metaphor, a speaker's emotion, and cultural references that appear in conversation. This concept is called a "human mergence" by technicians who are not familiar with ergonomics (Ebukuro, 1994).

79.12 Participatory Ergonomics in Virtual Environment

Presence Communication and VR

Since the start of the 1990s the capability of computers has increased and a diverse range of peripheral equipment is employed for the user interface. The content transmitted by the interface is also changing, from a dialogue between a user and a computer to communication between many users via computers. A representative form of communication between users is the audiovisual teleconference. The conversation between users in a teleconference is, however, somewhat awkward for several reasons and the will to participate is dampened. A system in which this awkwardness is reduced and a conversation much like face-to-face conversation is possible and is called "presence communication" or "telepresence" (Noro, 1996) (Figure 79.4).

Advent of a Virtual Factory

Kao Corporation is planning a "virtual factory" in which nine factories scattered throughout Japan are efficiently managed simultaneously as a single factory. Experiments have been carried out in factories in Wakayama and Kyushu and the experiments on remote control have been successful.

The larger plan is likely to be realized in the future when communication costs go down. The scope of subjects of participatory ergonomics is expanding from specified participants in a single factory or workplace to a wide variety of people across a vast area such as a whole nation, or even beyond national boundaries. In a virtual factory like this one, teamwork will be carried out as if two factories 600 km apart were a single workplace. Presence communication will be required in order for people to participate in the teamwork.

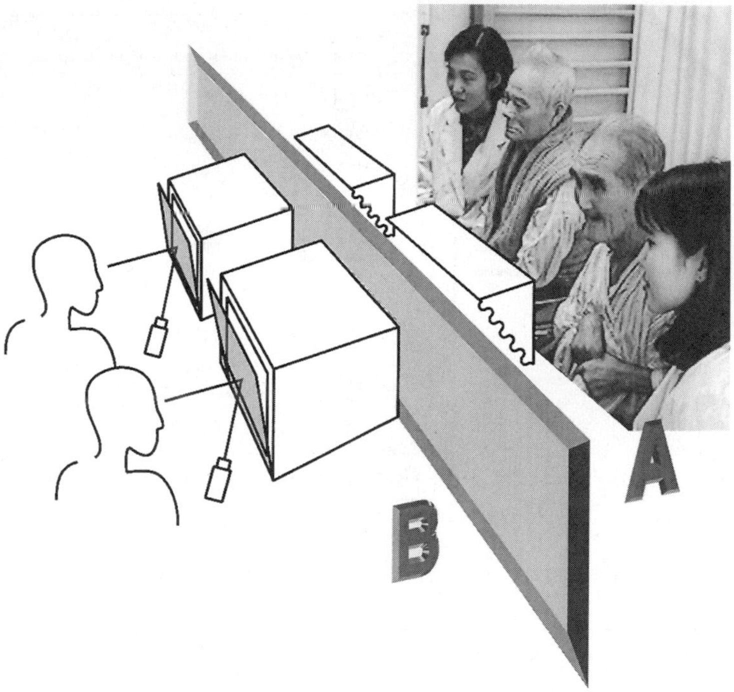

FIGURE 79.4 Presence communication network where participants exchange views. System A and system B shown here together are actually located far apart but connected.

79.13 Conclusion: Participatory Ergonomics as Future Ergonomics

Worker-oriented Participatory Ergonomics

In participatory ergonomics, as a part of the activities of macroergonomics and organizational design and management (ODAM), the worker should be the central figure. Past reports have tended to be procedural theories or success stories from the point of view of a researcher.

The Difference Between Participatory Ergonomics and Macroergonomics or ODAM

Macroergonomics and ODAM have a different concept from participatory ergonomics. It is not enough to merely treat participatory ergonomics as a part of the activities in macroergonomics or ODAM. Doing so neglects the extensive applicability of participatory ergonomics and may hinder its future development.

More Effective Methods

One of the tasks for researchers is to develop a more effective method to implement participatory ergonomics.

Activity Linking the Community, Home, and Hospitals

Participatory ergonomics is becoming an activity that extends beyond the boundary of a workplace. Home care is an example of an activity that links the community, home, and hospitals. Ergonomics should increasingly become a field where specialists and nonspecialists have an equal share.

Participatory ergonomics is evolutional in the sense that it designs the diverse flows of information that relate to ergonomics. Electronic networks that can help to promote these flows are now becoming widespread. Participatory ergonomics should not be confined in a bulky archive of files with records of improvements made at workplaces.

Participatory ergonomics will be acknowledged by people in ergonomics-related fields as the main concept of ergonomics in the networking era.

References

Caccamise, D. J. 1995. Implementation of a team approach to nuclear criticality and safety: The use of participatory methods in macroergonomics. *International Journal of Industrial Ergonomics* 15, 5, 397-409.

Ebukuro, R. 1994. Macro-logical management approaches, selected paper of ICC&IE '94, *Computers Industrial Engineering* 27, 1-4, 321-326.

Garmer, K., Dahlman, S. and Sperling, L. 1995. Ergonomic development work: Co-education as a support for user participation at a car assembly plant. A case study. *Applied Ergonomics* 26, 6, 417-423.

Gill, S. P. 1990. A dialogical framework for participatory KBS design. Cybernetics and Systems 90. 1039-1046. European Meeting on Cybernetics and Systems Research (10th) Vienna.

Hata, M. and Miyamoto, H. 1996. a home page offering ergonomic knowledge regarding chairs. Annual Kanto chapter meeting, Japan Ergonomics Society, Chiba.

Kawano, T., Nishida, S., Hashimoto, M., Yamamoto, C., Iwata, K. and Onosato, M. 1996. Study of Computer Aided Consultation System for Working Posture, Using Posture Input Puppet-Like Device and Visualization of Joint Torque. *Journal of Precision Engineering Society* 62, 5, 1996.

Kuorinka, I. and Patry, L. 1995. Participation as a means of promoting occupational health *International Journal of Industrial Ergonomics* 15: 365-370.

Moore, J. S., Garg, A. 1996. Use of participatory ergonomics teams to address musculoskeletal hazards in the red meat packing industry. *American Journal of Industrial Medicine* 29 (4): 402-408.

Noro, K. 1984. A workshop, International Conference on Occupational Ergonomics. May. Toronto.

Noro, K. and Imada, A. 1984. A workshop on Participatory Approach of Ergonomics, Hawaii. 1st ODAM meeting.

Noro, K. and Kogi, K. 1985. Invitation to participatory ergonomics, *Anzen* 36 53-60.

Noro, K. 1988. Participatory ergonomics, *Japanese Journal of Ergonomics,* 24, 5-10.

Noro, K. 1991. Concepts, methods and people, in *Participatory Ergonomics,* Eds. Noro, K. and Imada, A. Taylor & Francis, London.

Noro, K. 1992. Ergonomics for daily living, University of the Air.

Noro, K. 1994. Sashaku: A user-oriented approach for sealing, in *Hard Facts About Soft Machines* Eds. R. Lueder and K. Noro. Taylor & Francis, London.

Noro, K., Kawai, T. and Takao, H. 1996. The development of a dummy head for 3-D audiovisual recording for transmitting telepresence. *Ergonomics,* 39, II, 1381-1389.

Vink, P., Peeters, M., Grundemann, R. W. M., Smuldrers, P. G. W., Kompier, M. A. J., and Dul, J. 1995. A participatory ergonomics approach to reduce mental and physical workload. *International Journal of Industrial Ergonomics* 15, 389-396.

Waseda University. 1995. *A Guideline for VDT Operators,* Waseda University.

80

Tuttava: A Participatory Method to Improve Ergonomics and Safety Through Better Housekeeping

Jorma Saari
University of Waterloo

80.1 Strategies for Improving Workplaces

The purpose of this chapter is to show that housekeeping, when properly understood, can be a very useful vehicle for improving workplace ergonomics and safety. This chapter will describe a process that has been used successfully in hundreds of companies and in several countries. Its primary purpose is to involve employees in analyzing and improving their jobs and workstations, and to initiate a process toward continuous improvement.

First of all, it is important to understand the nature of housekeeping properly. Materials and tools are the core of housekeeping. They also are the core of any manufacturing process. Workers handle tools and materials every day. This gives them the expertise which makes it easy for them to talk about materials and tools. Therefore, tools and materials are a very good topic for a participatory workplace improvement process.

Three groups of people may initiate workplace improvements: management, experts, and employees. In recent years, it has become more widely understood that change processes initiated by employees offer many advantages. Employees know the exact problems of work processes. They accept changes more easily if they have participated in the planning.

0-8493-2641-9/99/$0.00+$.50

Several mechanisms can trigger a process toward workplace improvements. (1) Laws and regulations establish limits, such as exposure values, after which law enforcing authorities force management to take action. Standards, codes of practice, and professional rules may have similar effects. Internal inspections, audits, and monitoring systems ensure the compliance with these external norms. The problem is that, for many exposures, we do not know the limit values (Westgaard and Winkel, 1996). (2) Epidemiological studies or routine monitoring of injuries can make ergonomists, occupational health care professionals, or safety experts aware of a problem, and they then convince management to take action. This is an expert strategy. (3) Management and other professionals in the workplace initiate changes continually. The problem, however, is that both management and experts have different perceptions of problems and needs than the employees. Even when recognizing a problem, management may estimate its magnitude quite differently. In a top-down driven environment, employees initiate changes by presenting complaints, or refusing to perform a dangerous job. Complaints create an immediate negative tone for the situation, and positive solutions become less likely. (4) As employees spend all their working time at the actual place of work, they possess a lot of information on problems at work. Those problems need not be severe enough for workers to present complaints. They may, however, be sufficient to prompt ideas for better ways of getting the work done.

Employee involvement and empowerment have become more topical recently (Björkman, 1996). More people understand that workers have more education and training and a heightened need for the possibility of having an influence on their own working situation. Employee participation has positive effects in productivity (Doucouligiaos, 1995) and satisfaction (Wagner, 1994).

Management and experts do not have enough capacity to become aware of all opportunities for improvements. Workers' observations and ideas are valuable assets in finding better ways of doing the job. The question is *how can an organization create conditions where workers share their ideas and where these ideas lead to action?* The technique I am going to describe provides an answer.

80.2 What Housekeeping Really Can Offer for Workplace Improvement

Housekeeping is a word one can find in almost every textbook on safety. It is one of those words which has been around for so long that people do not realize its full potential. "A place for everything and everything in its place" is a maxim that most managers think they have accomplished in their plant. Housekeeping is, according to textbooks, one of the cornerstones of good safety. However, the concept of housekeeping is in many texts quite narrow and relates primarily to the physical appearance of a workstation, not to the underlying processes. A guide defines housekeeping as (Stewart, 1990):

- Day-to-day clean-up
- Waste disposal
- Removal of unused materials
- Inspections to ensure that clean-up is complete

In this case, housekeeping means housecleaning. One also finds general checklists, such as the one in Table 80.1. These are often prepared for inspectors for auditing purposes. Again, the focus is often on cleanliness. The list in Table 80.1 is broader, but the view is an external auditor's view.

The traditional view emphasizes the maintenance aspect of housekeeping. Is the workplace kept in good condition? There is no doubt that this is an important aspect and that it quickly gives an auditor an overall impression of the quality of maintenance operations.

The other aspect of housekeeping, the most important one, is that *housekeeping actually reflects the quality of management and production processes* at the workplace. The flow of materials and the use of tools are in close relation to these processes. This is an aspect that is often ignored. For example, the checklist in Table 80.1 addresses this aspect with one question only: "Has excess material in-process collected around machines?"

TABLE 80.1 A Housekeeping Checklist (Ontario Meat Packers Safety Council)

- Are aisles clearly marked and free from stored materials, idle hand trucks, projecting piles of stock, etc.?
- Are stairs and ramps free from wall obstructions and stored items? Are hand rails and stair treads in good repair?
- Are suitable cleaning materials on hand to clean up spills?
- Has excess material in-process collected around any machines?
- Are there means for disposing of soft-drink bottles, milk cartons, lunch wrapping, etc.?
- Has scrap been allowed to accumulate in work areas?
- Does final disposal of scrap produce fire or other hazards?
- Are hand and power tools properly stored when not in use?
- Are there any tools left on machines, overhead beams, or in other locations where they could be jarred off?
- Is there good personal housekeeping at individual workstations?
- Are all goggles, face shields, aprons, or other personal protection equipment clean and in acceptable condition for use?
- Are all cables, ropes, chains, slings, or other gear properly stored when not in use and regularly inspected and repaired?
- Is all flammable material stored in fireproof receptacles with fusible links or other devices for emergency enclosure?
- Are all warning and direction signs on machinery and elsewhere clean and legible?
- Are the bulletin boards clean and attractive?
- If first aid materials are kept in the department, are they sanitary, of amply supply and fresh?
- Are there leakages, either from overhead shafting, machine packings, or elsewhere, which are causing hazards?
- Does the department need a paint job to enhance both appearance and lighting?
- Is oil and grease regularly removed from all machinery?
- In wet processes, is the drainage good around machines? Are slat platforms in good condition?
- Are windows and skylights clean and in good repair?
- Is the exhaust equipment reasonably free from accumulated dust, lint, or soil?
- Are lighting fixtures cleaned and up to maximum efficiency? Including emergency lighting installations?

The amount, the placement, and the condition of materials in a workstation are some of the most essential aspects of housekeeping, and at the same time, a visible reflection of the quality of the production process. They even reflect the overall organizational performance. Too many components in a workstation may tell us that the purchasing process does not function on time.

Materials and tools provide a window into the "heart" of the whole production system. At the same time, they are in close relation to the ergonomic demands of the workers. Therefore, focusing on materials handling offers a possibility of improving the ergonomic quality of a job and the efficiency of production simultaneously (Salminen and Saari, 1995).

Materials and tools have another nice characteristic. Both *are visible and touchable*, and therefore they are easy *starting points for discussions.* Operators do not usually know enough, for example, about a company's purchasing process. They do see the timing when the materials are received, and this makes it possible for the operators to contribute to the analysis of purchasing processes.

Housekeeping is a visible indicator for obscured organizational conditions. In the tool department of a plant, I heard continuous complaints about not having enough room for everything; machines, work benches, parts, components, etc. The cupboards were full, neat and in good order though. The top of every cupboard was also full of materials. I told the workers, that the top of a cupboard is not a good storage place for two reasons: the objects on the top will often be forgotten, and it is difficult to take anything down safely because of poor visibility.

The workers accepted these arguments but continued to complain about the lack of space. However, this apparently innocent observation and discussion revealed a major flaw in the management culture — conflicting goals. Management expected two incompatible accomplishments. They expected the tool department to remove any problems immediately and not to store so many spare parts, components, and materials. However, the delivery of items which were not in stock could take hours. During this time, production might be down which is absolutely unacceptable for management. A little visual signal led us to the some of the root problems of this organization.

The observation of parts and components in inappropriate places actually unveiled organizational values more broadly than a discussion or an interview could do. Therefore, the conflicts between production and safety goals could be discussed more broadly. Materials and tools have tight connections to the management system and to the production system. Therefore, they provide an excellent starting point

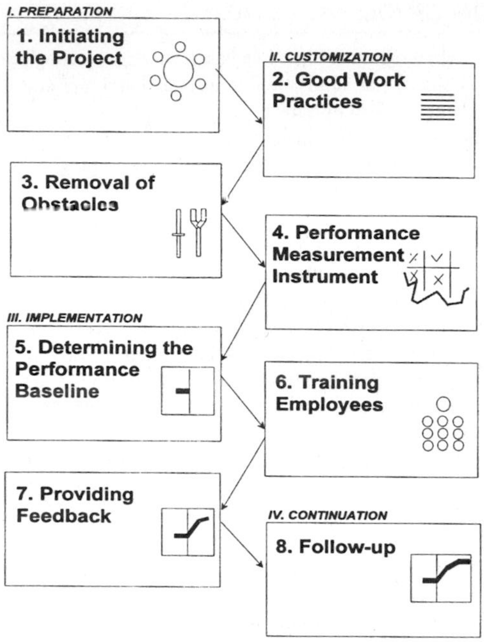

FIGURE 80.1 The steps of Tuttava.

for workplace improvement. They lead straight to systemic core issues and provide a window into the management processes.

This window is important in several ways. (1) In spite of being directly linked to the most important aspects of any production system, *tools and materials are a relatively neutral topic.* (2) It offers a manageable theme for discussions with the workers about workplace problems.

80.3 Tuttava — A Strategy to Initiate Workplace Improvements

The best strategy to initiate positive changes at a workplace is *induction by success.* It means going through several small steps that give *positive experiences* and *reinforce the desire to go further.* The experiences of success lead to more sustainable results than other strategies do. Stepwise progress leads to continuous improvements. Positive results enhance the development of management–labor relations, especially when the improvements result from joint efforts.

To this end, we developed a method and a process which focuses on tools and materials, the most important aspect of housekeeping. It is called *Tuttava,* which is an acronym coming from Finnish words which mean "*safely productive work habits.*" Tuttava is a method for initiating the change toward a better workplace (Figure 80.1). It was developed about ten years ago and it has become widely implemented in Finland (Saari and Näsänen, 1989; Saari 1996). There are also several successful applications internationally. Tuttava has the following distinctive characteristics:

- It affects all kinds of health and safety factors, including ergonomics
- It also helps improve quality and production
- It is a model for participatory improvement
- Its main principle is to keep the process positive
- Using a positive approach, it initiates small changes which then reinforce further improvements

Tuttava is a project-like method. It takes from four to twelve months to complete. The exact time depends on the participants' previous experience with participatory methods and on the organization's ability to adopt new directions.

Tuttava is still considered a project even if some companies have also used it as a tool for continuous performance measurement. The purpose is to introduce a permanent change which lasts without further efforts. A clear end signals the attainment of a reasonable new performance level, and thereby provides satisfaction. After the end, other techniques, such as detailed ergonomic analyses, etc., can utilize the new foundation for further improvements. Tuttava has formed a good basis for the implementation of new technologies, new production methods, etc.

Organizing Tuttava

Tuttava starts as any other workplace improvement project does. *The decision to start* is the first step. Experience shows that a company has to "mature" for the process (Harper et al., 1996). Workplace improvement must be perceived as a desirable goal. The company has to be reasonably healthy. If it has major problems, it may not be able to focus on this process, or any other process, intensively enough. Problems may be related to job security, to market, to the reengineering of the company, etc. Workplace issues seem not to be an overriding priority under such circumstances.

Tuttava is well suited to companies with poor management–labor relations. Tuttava is designed to ease problems between management and labor. Actually the best results we have obtained are in companies with rather tense relations. Otherwise the company has to be fairly healthy. Each company has a different culture which determines the roles of management and labor. In a top-down type of culture, consistent commitment and continuous support from the management are essential. The project may fail if the management gives negative reinforcement. In the case of Tuttava, middle managers and supervisors may be afraid of loosing part of their power. If so, they have to be put at ease.

I have also seen situations in which employees have initiated the process without input from upper management. In a big company, the workers' health and safety representative started the process, and the top management came onboard only when positive results were quite clear. The program became a big success in this company, and they use it permanently in their different locations. In this case, the workers' OHS representative took the leading role. It seems to be important that there is a dedicated person who maintains the momentum. The company is multinational. It has two major locations in Finland. Tuttava has been used with great success in one location where the OHS representative took the leading role. The top management, after becoming convinced, put pressure on the second location to implement the process. In the absence of a leading person, the implementation remained superficial. The good results, however, encouraged other locations in other countries to adopt the process.

It is important to *find someone who will become the Tuttava leader.* Who this person is depends on the local culture. However, if this person's "ownership" in the process becomes too strong, it may drive users away.

An external expert may be necessary as a neutralizer in those companies where relations are tense or where no previous experience of a similar method exists. Tense relations do not apply just to situations between management and labor. On the contrary, it seems that more often the biggest problems exist between organizational units on the same level, such as production and maintenance. Also it is often found that some individuals cannot "come along."

Companies have used different models to organize project teams. Normally, *the project team consists of representatives from the project department, both workers and supervisors, as well as representatives from management.* OHS representatives, representatives from maintenance, occupational health services, or the personnel department sometimes supplement the core team. In other cases, all the workers belong to the implementation team. However, a smaller core team makes the preparations easier even if this is the case.

It is important that the group has access to funds or services which make technical and other corrective actions possible. The purpose of the process is to initiate all kinds of changes, new work habits, and technical/organizational improvements. When the workers improve their performance, it is only fair that the management makes funds available for technical improvements. The purpose is to initiate a comprehensive process of improvements. This is important for making the improvements visible, which then reinforces other improvements.

One team covers a work area of 5 to 30 people. We have used Tuttava in larger areas too. However, the effectiveness tends to deteriorate since individuals tend to lose sight of their meaning for the whole. It is easy for them to think that their contribution does not matter. The lower limit is to keep the focus on groups instead of individuals. Because the process should be positive, it should not point a finger at any individual. In a group that is too small, it is difficult to meet this requirement.

To make sure that a participatory Tuttava succeeds, some conditions have to be fulfilled. (1) There must be a mechanism to elect worker representatives so that the other workers accept them as their representatives (Saari, 1996). (2) Management representatives in team must have a clear mandate from their superiors.

Especially in hierarchical organizations, a participatory program can induce unnecessary fears. If management thinks a participatory approach takes away their power, an expert-driven implementation might be more justified. The same applies in those cases when there is no way of ensuring that workers accept the worker representative(s) as their representatives. For example, other workers may see volunteers representing management views more than their views.

An expert-driven implementation may be more justified if the team members cannot obtain a clear mandate from their constituents. Even in this case, a team can and should be formed to advise the expert. The biggest risk in an expert approach is that the expert does not understand what is important at work and what obscured obstacles may exist to deter changes.

A participatory approach gives the best results. However, the expert approach is more advisable, if a participatory implementation team cannot make decisions as a team, and if the other workers or managers do not accept those decisions freely. An expert may have a role in the participatory approach too. In this case, the expert acts as a coach and provides the team(s) with sufficient training. In the following section, I assume that a participatory team implements the program.

Job Analysis and the Identification of Improvement Goals

The first task of the implementation team is *a job and workstation analysis.* Tools and materials are the keywords. The team makes an *inventory of good work practices for tools and materials.*

Hand tools and various work equipment are obvious tools. However, there also are tools that are used infrequently or never. Ladders, work platforms, carts, lifting appliances, etc., are examples of tools that may mostly stay in storage. Fire extinguishers and hoses, emergency exits, eyewash stations, other emergency equipment, including emergency lights and exit signs, are tools for those undesirable extreme situations easily forgotten in the daily routine. Access to this special equipment should be free, and it should be properly maintained.

In one case, an implementation team had a fire extinguisher replaced only one day before a painting box broke into flames. This was lucky timing, since the previous fire extinguisher was of the wrong type and would have only made the fire worse. The team noticed the incompatibility of the extinguisher which had been hanging on the wall of the painting area for a long time.

Typical good work practices for the use of tools are, for example, "put tools back in their designated places after use," "clean and store tools properly," "coil hoses and cables if not in use," "keep access clear to fire extinguishers and other similar emergency equipment."

Some of the following tasks are done in every workstation: receiving, storing, moving, handling, disposing, and shipping of materials. Storing is often the most essential work practice. Because of traffic requirements some areas need to stay open all the time. Many times there are no clear rules for appropriate storage places. Typical outcomes of this discussion are: keep aisles open and other areas meant for traffic, keep the access to shelves and cupboards free, etc.

Usually, the identification of good work practices is an easy task for the implementation team. Most often, the result is nothing dramatic. Table 80.2 gives an example from a printing ink factory. The list of good work practices may not include a single new practice. However, these good work practices may not be fully in use. It is common that everybody can accept the list without hesitation.

TABLE 80.2 An Example of Good Work Practices in a Department at a Printing Ink Factory

1.	Keep aisles open
2.	Always when possible put covers on containers to prevent solvents from evaporating into air
3.	Close bottles after use
4.	Clean and return tools after use
5.	Ground containers when moving flammable substances
6.	Use personal protector specified in the recipe
7.	Use local exhaust
8.	Store in working areas only materials and substances needed immediately
9.	Use only the designated forklift truck in the department making flexographic printing inks
10.	Label all containers

There may be several reasons why good work practices are not in use. Some of them may not be technically feasible. Once we studied the equipping phase of a ship. A problem was that workers brought too many materials to the ship. Many jobs were slowed down as materials blocked aisles, or materials had to be removed first. This caused extra work. It also caused unnecessary stress on the musculoskeletal system. To reduce the amount of supplies in the ship, the work practices specified "Store in the ship only materials needed for one day's work."

Why this ideal was not met depended on several factors. Bringing more materials to the ship was faster for the workers. The management was happy, because the workers did not go onshore so frequently. On the other hand, they were unhappy about the extra work. There were conflicting motivations, because management had never really specified the practices. In this case, the primary obstacle for good work practices was the lack of set procedures, in other words an organizational deficiency.

In an other case, welding light was a problem. The company had movable curtains for preventing the light from reaching the eyes of workers nearby. It was never really thought out who should put up the curtain, the welder or the person hit by the light. Industrial engineers did not give a respective time allowance to either of the workers when setting up time standards which were the basis of wages. Another factor was that the curtains were not technically most suitable for those conditions, and they often fell down even if time was initially spent to put them up. In this case, the obstacles were both organizational and technical.

A very common example of an obstacle is the use of lifting devices. The workers often do not use lifting devices or help from another person even if technically possible. For example, nurses in hospitals lift patients alone or without a lifting device. The background factor often is that they expect their supervisors will praise them for working fast.

These kinds of problems deter the good work practices from becoming routine. The problem sometimes is technical, sometimes just the lack of agreement, or another type of organizational deficiency. Not knowing the best work practice usually is not the obstacle. These obstacles need to be identified and fixed before a permanent improvement in working habits is possible.

Good work practices should be (a) specific, (b) positive and make work easier, (c) generally acceptable, (d) simple and short statements, (e) started with action verbs, and (f) easy to observe and measure. They should give specific instructions; they should not be general warnings or bans.

Good work practices may require modifications later when the team devises a measurement system. The good work practices should be partially in use. In some parts of the application area they may be in use all the time; in some other parts partially or not at all. The purpose is to give a positive outset for the process of change. If all good work practices were totally new, or never in use, the process would not appreciate current achievements.

The Removal of Technical or Organizational Obstacles

Simultaneously with the drafting of good work practices, a discussion starts about possible obstacles deterring their use. It would be best, if the analysis were done workstation by workstation. This way it

TABLE 80.3 Examples from Two Departments of a Shipyard. The Percentage of Respondents Accepting After the Implementation the Statement: "Housekeeping Has Improved"

Department	Relation to the Intervention Process	Agree	Undecided	Disagree
A	Group member	100	0	0
A	*Not member*	*57*	*43*	*0*
J	Group member	72	14	14
J	*Not member*	*35*	*47*	*25*

From Saarela, K.L. 1990. An intervention program utilizing small groups. *Journal of Safety Research* 21:149-156. With permission.

is possible to integrate everyone into the process. This may be essential for the acceptance of the change. Table 80.3 shows a result of an interview study after an intervention at a shipyard (Saarela, 1990). Saarela formed representative improvement groups for each department. Afterwards, the group members and other workers responded to a questionnaire about the results of the process. The group members clearly perceived more improvements than the other employees. Participation strengthens a change process.

If every workstation undergoes an analysis first, then the final list of good work practices will be combined from the individual lists. In similar workstations, often the same obstacles are discussed and more ideas for solutions may surface. The workers often have ideas they have not presented. The analysis may provide a channel for those ideas to surface.

Tuttava has been especially successful in opening development gridlock. Management is often unwilling to invest money on technical improvements, since they believe the workers will not change their work habits accordingly. Workers are unwilling to change their work habits, as management does not invest in permanent improvements. This vicious circle may continue for years.

Because Tuttava emphasizes a positive approach and stepwise progress, it has been a way to break the circle and to learn mutual trust. Therefore, it may not be essential to remove all the obstacles immediately. Sometimes, those obstacles represent major investments or changes in thinking. Then it may be wiser to do small things first and let the bigger ones follow later.

This approach proved to be highly efficient in an engineering workshop employing about 250 people (Laitinen et al., 1995). A Tuttava program leading to some permanent improvements proved to be a positive experience and provided more mutual trust. In the beginning, the management was quite reluctant to invest in technical improvements. In the late phase of the first set of departmental Tuttava projects, the management attitude changed. Especially, after the good results became visible, it was very easy to get money for technical improvements, and an intensive innovation cycle started. The whole process took two to three years.

Behavioral Changes

Theoretical Background

Tuttava does three main things. Those are: (a) the identification of goals and objectives for workplace improvements, both work habits and structural changes, (b) the identification of opportunities for improvements and using the opportunities, and (c) the change of work habits as indicated by the goals and objectives. To make the change of work habits happen, Tuttava utilizes a version of behavior modification.

Bird and Schlesinger (1970) outlined the application of behavior modification for safety. Komaki et al. (1978) and Sulzer-Azaroff (1978) published independently the first empirical applications. The idea of behavior modification is to provide more positive consequences for safe behavior. Too often unsafe behavior leads to positive consequences and safe behavior to negative ones. Thus, people more likely choose the unsafe way of working.

The great benefit of this technique is the use of positive consequences only. Therefore, the process gets a positive flavor and the consequent changes get associated with positive factors. Another great benefit is that very small consequences change behavior. Information about the current performance alone is enough to prompt an increase in safe behaviors.

A behavior modification program requires that (1) safe behaviors are identified and specified (good work practices in Tuttava), and (2) a measurement technique for measuring the prevalence of safe behaviors is available. The measurement technique is similar to safety sampling which was developed in the sixties (Pollina, 1962; Rees, 1967).

In behavior modification, we assume that *peoples' behavior is determined more by the consequences of the behavior than the antecedents of the behavior* (Komaki, 1986). People adopt unsafe behaviors because they expect more positive than negative consequences from those. For example, it is common that people do not use existing lifting equipment. Some of the reasons may be: (1) it is faster to lift manually, (2) the local industrial climate favors faster lifting, and rewards come from the person's supervisor or co-workers' positive comments, (3) the person's friends value strength, and manual lifting provides "free exercise," etc. Several positive consequences from unsafe lifting override the only negative consequence, low back pain. This consequence does not even materialize after each lift.

The theory of behavior modification tells us to change the balance of consequences to remove unsafe behaviors. The most desirable strategy is to introduce new positive consequences for safe behavior. More negative consequences for unsafe behavior would be another strategy. It is less desirable because the positive strategy leaves better feelings behind and these feelings may support later actions.

A wide variety of positive consequences is possible. The possibilities include various privileges (time off, extra break, etc.), tokens, promotional items, chance to win a contest (lottery, bingo, etc.), social attention, etc. Some of these are straightforward but difficult to administer. Tokens and promotional items are an example. Some are controversial — for example, social attention.

As the first empirical researchers, Komaki et al. (1978) and Sulzer-Azaroff (1978) showed that information on current performance is sufficient to alter behavior. Performance feedback has several advantages. (1) It is easy to administer. (2) It can be made objective. (3) The costs are low. (4) It is effective. Many researchers have shown that performance feedback works (for reviews see: McAfee and Winn, 1989; Sulzer-Azaroff et al., 1994). For these reasons, we decided to use performance knowledge as the new consequence for safe behavior in Tuttava. The consequences can be given to individuals or to groups. In our case, we wanted to provide the consequences to groups for enhanced group performance, in the attempt to promote team building and the cohesion of the organizational unit.

Tuttava deviates considerably from a behavior modification program. Behavior modification for safety obtains the feedback from the observation of people's behavior (Krause, 1990, 1995; McSween, 1995). In Tuttava, observations do not focus on behavior but on the traces of behavior. In other words, conditions are being observed. However, those conditions are brought about by behavior. Using conditions instead of behaviors is to depersonalize the problems in order to make the request for new behaviors easier to accept. As the purpose is to help improve management–labor relations, this supports team building and promotes cohesion. Also it is assumed that conditions lead to lasting effects (Ray et al. 1993).

Performance Measurement

To provide feedback, the implementation team writes a checklist to measure performance at any given time for providing feedback. The checklist is based on good work practices. It should consist of approximately 100 items to make the measurement accurate enough. Even with this number of observation items, the measurement is still reasonably fast. One observation round does not usually take more than 30 minutes.

Each item of the checklist has only two possible answers "correct" or "incorrect." After an observation round in the area, the observers can calculate a simple performance index, which is the percentage of "correct" items. This gives a comprehensible indicator which tells to what extent the good work practices are in use.

To devise the checklist, the team divides the implementation area into sections which form logical units. The sections can be single workstations or larger areas. The purpose of the sections (called observation areas later) is twofold. (1) They make it easy to develop the checklist. (2) They allow the use of a weighing procedure to prioritize the importance of work practices.

When the team has defined the observation areas, it writes a checklist for each area. A good work practice may generate from zero to several items into the checklist. For example, "keep the access to fire extinguishers free" generates two items in the checklist, if there are two fire extinguishers in the observation area. These items are: "Is access to fire extinguisher #1 free," and "Is access to fire extinguisher #2 free."

In principle, *these questions have only two possible answers: correct or incorrect.* If a fire extinguisher has been removed from the area for maintenance, then there is one more alternative, "not observable." To limit the possible answers to two, it may be necessary to define some words. In this example, "free" may require a definition. It may be necessary to specify the area to be kept free in units of length. The need for specifications depends on the homogeneity of peoples' interpretations. If everybody gives similar answers during a test, there is no need for specifications.

Some work practices yield several items for the checklist. Some work practices may not apply to all observation areas. For example, the implementation teams usually write some kind of good work practice related to aisles. Aisles may not run through every observation area. Other work practices may lead to different items. "Store tools on the tool shelves" would lead to such items as, "are there tools only on the shelves," "are the tools in the marked places," "are there any other objects put on the tool shelves," etc.

Some practices are more important than others. Therefore, the team may want to generate more items relating to those. One way of achieving this is to *change the size of observation areas.* If, for example, there is a work practice related to aisles, the team can define observation areas so the number of items relating to aisles is at a maximum.

Another technique is to *use several items relating to an important practice.* A company had lots of problems because of wrong labels put on products. The first cure was to mark the place of each roll of labels on the shelves. Because this was a problem, they wanted to have an observation item for each roll of labels checking that it was in the correct place. If there were 20 rolls of labels, there would be twenty observation items, too. If the correct placement of labels had not been a big problem, they would have had one item in the checklist, checking that all rolls not in use are put back on shelves.

When we have several observation areas, it is easy to develop the full checklist. When the list for one area is ready, much of it can usually be copied to the following area. The team can produce the checklist of the next area just by removing the irrelevant items and by adding any items which did not apply to the previous area. Therefore, the number of items varies from area to area. At the same time, the items are similar in different areas, and learning to use the checklist goes fast.

Measure Baseline

When the team has the first draft of the checklist, they should test it. By comparing results and notes after the first observation round, it is easy to identify the weaknesses of the checklist. Usually, it is necessary to make several changes to the checklist. However, it is also customary that the checklist is ready for use after a couple of test rounds.

When the checklist works well enough, the team should carry out measurements for a few weeks to establish a baseline. The baseline is important, since it allows the comparison of present performance with previous performance. This makes any improvements more visible and gives more satisfaction as the difference is known.

When there is a sufficient number of baseline measurements, it is time to arrange a meeting for everyone working in the area. The team explains the best work practices, the principle of measurement, and the baseline results. During a discussion period, the attendants may express their views and opinions about needs, problems, and potential solutions in housekeeping.

Feedback

After the meeting, the team puts a feedback chart in a visible spot in the area. When observation rounds continue, the observers post the result immediately in the chart. They post only the performance index which is the percentage of items in the checklist marked correct of all items observed. This performance index applies to the whole group of people working in the feedback area. As there should be at least five people working, nobody should be able to recognize one person's contribution to the performance index.

The team is strongly encouraged not to release any negative information, such as detailed observation results, etc. Being positive is important in this method. Therefore, the baseline should be somewhere between 50 and 60% to give a positive starting point for further improvements. To get a suitable baseline, the teams sometimes have to modify the list of good work practices or the observation checklist during the first baseline measurements.

Any other feedback, except the chart, is usually not used. Primarily, the motivation to change behavior comes from the knowledge of positive development. It would be possible to use other rewards, but we have preferred not to.

Usually, a sufficient feedback period is from six to eight weeks. Employees adopt the new behaviors well enough during this time, and they learn to read the index from their surroundings directly. The index seems to get better either quite quickly, even in a couple of weeks (Figure 80.2), and then stabilize, or it may take a period of several weeks (Figure 80.3). This depends on the number and type of obstacles deterring the use of specified good work practices.

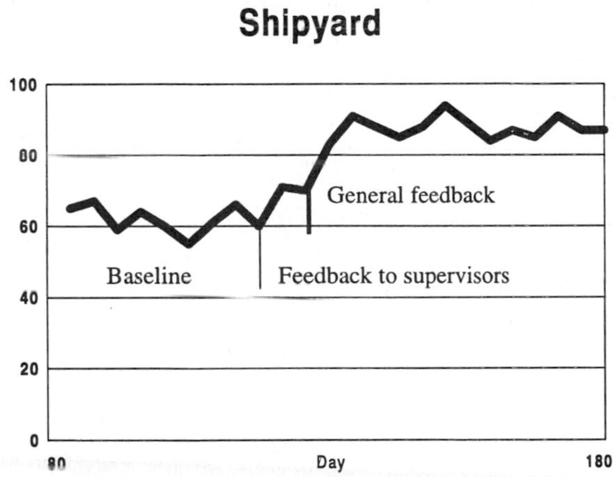

FIGURE 80.2 Performance index at a shipyard.

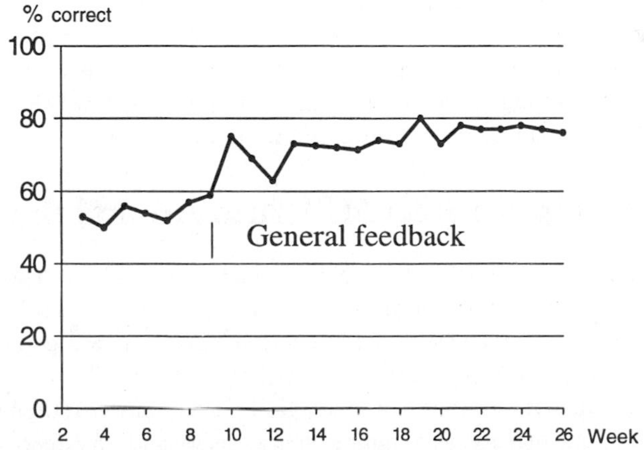

FIGURE 80.3 Performance index in a plant where improvements happened over many weeks.

Follow-up

When the index reaches a stable level and stays there for a few weeks, the team can terminate the feedback. In most cases, the index fluctuates a few percentage points, reflecting production conditions. It will not be a straight line. The right time to end the feedback period is when the average of the index does not seem to change any more.

The team should conduct follow-up measurements for a few months, once a month or every second month. If a relapse happens, the team can provide more feedback or take other actions. Usually, there is no relapse. The new level has lasted as long as two years (Saari and Näsänen, 1989). Various production factors seem to change within a couple of years, making some good work practices less relevant and making the performance index obsolete. The team can start a similar project again or use other techniques.

80.4 Why It Is Important to Observe Conditions Instead of Behaviors

An important difference between Tuttava and behavioral safety programs utilizing feedback (e.g., Krause, 1990; McSween, 1995) is that behaviors are not observed directly in Tuttava. The items in the observation checklist deal only with physical and material conditions. This offers several advantages. It depersonalizes the feedback and helps avoid blaming individuals. An important advantage is that the checklist measures what anyone can see at any moment. Behaviors often last just seconds, and afterwards the connection between the performance index and the real performance is not verifiable. This way, people learn to read the feedback directly. When the team takes the feedback chart down, everybody has learned to read their performance index from the visible conditions directly.

A very important factor is that there are visible improvements if the curve goes up. Both management and labor can see that things happen. One of the common frustrations in many attempts to improve is that advancements come slowly or they cannot be seen. For example, if a company implemented a better procedure to investigate injuries and accidents, most employees might not become aware of the improvement at all.

The visibility of improvements, and the possibility to read the feedback directly without a chart are factors which make Tuttava results sustainable. If a safety program focuses on behaviors, it is more likely that the results will not stop after the cessation of the program (Ray et al. 1993) unless the program led into a cultural change (Geller 1990).

Visible improvements encourage both management and labor to make new efforts. Employees become convinced that something really happens. Frustration "because nothing happens" is common in any company and in any country. Managers, on the other hand, may not believe that employees will change their behavior if they introduce technical improvements. Because they do not spend much time in the production area, they may not see the change in the actual behaviors. When the focus is in the physical conditions, they can see the change without seeing the corresponding new behavior. This helps develop mutual trust and willingness to invest in technical improvements.

80.5 Tuttava Helped a Steel Mill Turn Around the Safety Culture

A large steel mill in Finland with about 4000 employees, had a few fatal accidents in the late 1970s. The president of the company decided that safety must improve. He ordered his subordinates to take better care of safety. This sent a strong message to the line and the response was good, as Figure 80.4 shows. The injury rate declined quite nicely in the following years.

A problem arose when the president retired in 1984. The safety officer retired the same year, and the company lost both the strong management commitment to safety and experience in safety management. The injury rate started to rise again. The situation arose largely because, after the retirements, line managers focused on production and gave a lower priority to safety.

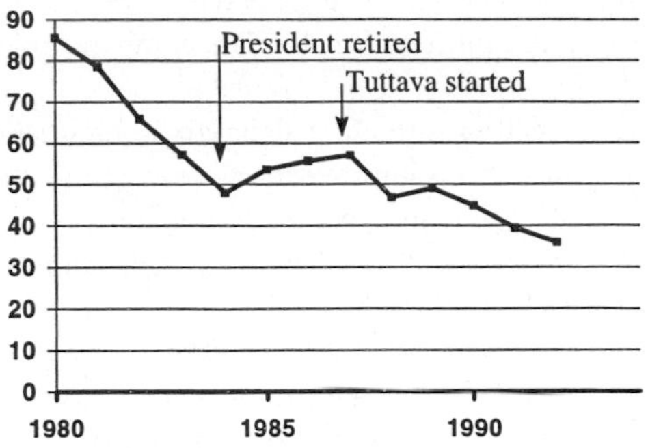

FIGURE 80.4 Injury rate at a steel mill.

The president's order had come with an unexpressed threat, "If safety in your area is not good, you as the responsible manager will face negative consequences." Safety culture had a negative tone. The organization was forced to safety. Safety did not offer positive sentiments. When the new management took office, they did not realize how strongly they should have expressed their commitment to safety. The new management did not have the same political power and charisma as the retired president had. As a result, line managers reduced their work load by giving less attention to safety, which was very human and natural.

Then the new safety officer heard about Tuttava and wanted to try it. Because the steel mill was large, the implementation had to be done department by department over several years. The first year, they organized Tuttava in 10 departments, the next year in seven departments, etc. Each department formed an implementation team. The management formed a steering committee consisting of management representatives, workers' safety delegates, and representatives from the safety office. The steering committee had the political role, and it existed over the whole five years' implementation period. The departmental implementation teams had an operational role, and they existed only for the implementation period which usually lasted about six months. During the implementation period, the teams were called together four times for meetings which lasted a couple of hours. They had to report on what they had done after the previous meeting.

The results were quite remarkable. The injury rate declined by almost 50%. It is quite obvious, that Tuttava did not cause the decline directly. It served as a vehicle for a cultural change. Because Tuttava applies a positive approach and because it produces different types of benefits, it offers positive experiences for everyone. The workers benefit from better safety and easier working. The supervisors appreciate productivity and quality improvements. Other managers may see other benefits, such as the better physical appearance of the plant, etc.

Tuttava helped the different levels of management to associate a safety program with several positive improvements. Safety was not a threat but an opportunity. The implementation of Tuttava turned the safety culture around, making it possible to implement other safety processes. For example, a new safety information system requiring thorough incident investigations, safety in job descriptions, risk analyses, safety instructions for new employees, etc., were renewed or adopted. Some of these required considerable effort. The good response was possible only because the line management had more positive expectations of safety programs.

80.6 Other Proven Benefits

Tuttava is a vehicle for a cultural change, as it results in the different types of benefits. Improved safety and ergonomics are only two of those. In a company, the maintenance service department employing 300 people reported a reduction of lost time occupational injuries from 600 days to 200 days per year. In addition, the absenteeism rate fell by one percentage point which was caused by the positive psychosocial effects of Tuttava.

Another company employing 1500 people reported the release of 15,000 m^2 of production area, since materials, equipment, etc., were stored more rationally. The company paid 1.5 million U.S. dollars per year less in rent.

The appearance of production facilities is the first signal to a potential customer or employee about a company's functional quality. Tuttava helps improve the appearance. At the steel mill, a supervisor gave an example. His department manufactures steel plates. Sometimes the operators walked on the plates and their shoes left clear marks on plate surfaces. The users will sandblast the plates before painting them. The marks did no real harm. However, when a customer came for a visit and saw marks left by operators' shoes "on his plates" he was dissatisfied. As a result of Tuttava, the operators stopped walking on the plates, which was also good for safety, as the risk of tripping was lower on aisles than on plates.

An engineering workshop had a high absenteeism rate. It belonged to a large corporation which had transferred some older employees there from other locations. These employees had many musculoskeletal injuries and a lowered work motivation for several reasons. The jobs of the engineering workshop were physically heavy and put employees often into awkward postures. Low motivation, previous injuries, and bad jobs were a combination which made the decision to stay at home easy.

The workshop started a several years' long project to improve ergonomics, the risk of musculoskeletal injuries, and working conditions. The relations between management and union were not very good. Management wanted to form a better partnership with the union and to encourage a participatory approach in solving problems. The core of their strategy was to improve relations first by implementing Tuttava department by department. This was aimed to produce some visible changes and to reduce peoples' frustrations from some previous projects. Tuttava achieved this goal. Previous common mistrust started to ease. Visible improvements encouraged more demanding ergonomic innovations. Finally, all the departments invented and implemented a large number of ergonomic improvements, psychosocial working conditions became much better, and absenteeism started declining (Laitinen et al.).

The ultimate factor often seems to be *the need for a cultural change which makes ergonomics and safety positive and more desirable.* Change strategies emphasizing legislation and regulations easily lead to negative sentiments about safety and ergonomics. "You have to" mentality is not appealing to many managers. Therefore, Tuttava has been a successful tool, as it is a way to initiate a change process toward a more positive safety culture. Housekeeping, especially materials and tool, are a good theme for the process, since they are in direct relation to the core factors of any production system.

References

Bird, Jr., F.E. and Schlesinger, L.E. 1970. Safe-behavior reinforcement. *ASSE Journal* 15(June): 16-24.

Björkman, T. 1996. The rationalization movement in perspective and some ergonomics implications. *Applied Ergonomics* 27: 111-117.

Doucouliagos, C. 1995. Worker participation and productivity in labor-managed and participatory capitalist firms: A meta analysis. *Industrial & Labor Relations* 49:52.

Geller, E.S. 1990. Performance management and occupational safety: start with a safety belt program. *Journal of Organizational Behavior Management* 11(1):149-174.

Harper, A., deKler, N., Osborne, D., Robinson, L., Cordery, J., Sevastos, P., Gunson, C., Sutherland, M. and Colquhun, J. 1996. Effectiveness of behavior based safety — the Curtin industrial injury trial, in *The Proceedings of 1996 Occupational Injury Symposium*, February 24-27, 1996, Sydney, Australia, 73.

Hog dressing safety and health guide. *Ontario Meat Packers Safety Council/IAPA*, 29-30. Toronto, Ontario TS200, 781.

Janssens, M., Brett, J.M. and Smith, F.J. 1995. Confirmatory cross-cultural research: Testing the viability of corporation-wide safety policy. *Academy of Management Journal* 32: 364-382.

Komaki, J., Barwick, K.D. and Scott, L.R. 1978. A behavioral approach to occupational safety: pinpointing and reinforcing safe performance in a food manufacturing plant. *Journal of Applied Psychology* 63(4):434-445.

Komaki, J. 1986. Promoting job safety and accident prevention, in *Health and Industry, A Behavioral Medicine Perspective,* Eds. M.F. Cataldo and T.J Coates, 301-319. John Wiley & Sons, New York.

Krause, T.R., Hidley, J.H. and Hodson, S.J. 1990. *The Behavior-Based Safety Process.* Van Nostrand Reinhold, New York.

Krause, T.R. 1995. *Employee-Driven Systems for Safety.* Van Nostrand Reinhold, New York.

Laitinen, H., Kivistö, M., Kuusela, J., Rasa, P.L. and Saari, J. 1995. Removing the cultural obstacles to better ergonomics with behavior modification programme, in *Second International Conference on Prevention of Work-Related Musculoskeletal Disorders.* IRRST, Montréal, 495-494.

Laitinen, H., Saari, J. and Kuusela J. 1997. Initiating effective change process for improved working conditions and ergonomics with participation and performance feedback: A case study in an engineering workshop. *International Journal of Industrial Ergonomics* 19(4):299-305.

McAfee, R.B. and Winn, A.R. 1989. The use of incentives/feedback to enhance workplace safety: A critique of the literature. *Journal of Safety Research* 20(1):7-19.

McSween, T. 1995. *The Values-Based Safety Process.* Van Nostrand Reinhold, New York

Pollina, V. 1962. Safety sampling. *Journal of the American Society of Safety Engineers* 7(8):19-22.

Ray, P.S., Purswell, J.L. and Bowen, D. 1993. Behavioral safety program: Creating a new corporate culture International. *International Journal of Industrial Ergonomics* 12:193-198.

Rees, A.G. 1967. Safety sampling — a technique for measuring accident potential *The British Journal of Occupational Safety* 7(79):190-195.

Saarela, K.L. 1990. An intervention program utilizing small groups. *Journal of Safety Research* 21:149-156.

Saari, J. and Näsänen, M. 1989. The effect of positive feedback on industrial housekeeping and accidents; a long-term study at a shipyard. *International Journal of Industrial Ergonomics* 4:201-211.

Saari, J. 1996. Participatory workplace improvement process, in Stellman, J.M. (Ed.) *Encyclopedia of Occupational Health and Safety,* 4th edition, International Labour Office, Geneva 1997. 59.11-59.15.

Saari, J. 1996. Use of a safety program in two different organizational cultures. CybErg 96, http://www.curtin.edu.au/conference/cyberg/.

Salminen, S. and Saari, J. 1995. Measures to improve safety and productivity simultaneously. *International Journal of Industrial Ergonomics* 15:261-269.

Sulzer-Azaroff, B. 1978. Behavioral ecology and accident prevention. *Journal of Organizational Behavior Management* 2:11-44.

Sulzer-Azaroff, B., Harris, T.C. and McCann, K.B. 1994. Beyond training: organizational performance management techniques. *Occupational Medicine: State of the Art Reviews* 9(2): 321-339.

Stewart, K. 1990. *Guide to Workplace Housekeeping.* Canadian Centre for occupational Health and Safety, Hamilton.

Westgaard, R.H. and Winkel, J. 1996. Guidelines for occupational musculoskeletal load as a basis for intervention: a critical review. *Applied Ergonomics* 79-88.

Wagner III, J.A. 1994. Participation's effects on performance and satisfaction: a reconsideration of research evidence. *Academy of Management Review* 19(2):312-330.

81

Quality and Ergonomics: Application of Ergonomics to Continuous Improvement Is Integral to the Goals of Business

Robert L. Getty
Lockheed Martin Tactical Aircraft Systems

81.1 Introduction

A holistic view is essential for quality initiatives such as Total Quality Management (TQM), ISO 9000, Concurrent Engineering, Business Reengineering, and Business Process Improvement. The challenge is knowing how to transition from this theoretical concept to implementation. The relationship between quality interest and an ergonomics program will be the focus. An ergonomics-oriented safety improvement program includes (1) ergonomics or fitting the job to the person, (2) integration of operations management, safety engineering, medical management, and employees as co-owners of the process,

0-8493-2641-9/99/$0.00+$.50

(3) the emphasis of ergonomic precepts in the engineering of new processes and improvement of current processes, and (4) the emphasis of employees taking responsibility for their own well-being and the improvement of their work environment. The parallel between the continuous improvement process delineated by the quality-system requirements in ANSI/ASQC Q9001-1994 and the improvement contributions of ergonomics are very revealing (Getty, Abbott, and Getty, 1995). It is the contention of this approach that if the precepts of ergonomics were applied to the work environment, it would support the objective of world-class quality and productivity, resulting in improved global competitiveness of businesses.

When improving processes for TQM or continuous improvement, the focus is on how to improve the process in order to achieve a quality product or service. In contrast, the focus in ergonomics is on how to improve the process to make it more compatible with the person performing the process or task by fitting the job to the individual. From the ergonomics approach, the question of the worker is how we can improve the process to make this task easier, more comfortable, and more satisfying. This approach, because of its intuitive nature and easy identification, achieves the very objective that the conceptual approach of TQM or continuous process improvement may not. The focus on people involved with work processes, as well as the administration of meeting their needs, clearly indicates the requirement for integration of all facets of these processes. The physical demands of work and the identification of ergonomic hazards or physical stressors (1) define the treatment that medical professionals prescribe for the injured employee, (2) describe those areas that are incompatible with the worker's physical or cognitive capability, and thus (3) reveal areas needing improvement, (4) provide useful tools for management in assignment of work that will reduce the risk potential and gain efficiency, and finally (5) portray, to designers of new processes, techniques to gain quality, productivity, and safety objectives (Getty, 1994).

The approach of ISO 9000/ANSI/ASQC Q9000 emphasizes the mechanisms needed to create a continuous improvement environment and then an audit process to verify the effectiveness of such an environment. This model could be very useful for an ergonomics process that must fit the requirements of both large and small companies. It may encourage large companies to aid the ergonomic development of tools, packaging, and processes that are purchased from suppliers.

81.2 Human Element Links Clearly to Continuous Improvement

Sparks and Dorris (1990) are scholars and practitioners in the fields of organizational behavior/development and industrial psychology and have written much about organizational change. Their work focuses on individuals, groups, the macro organization — or alternatively, systems — and on the management of change in people in organizations. A model is necessary that incorporates the issues of human behavior, the concerns for improved functional performances within organizations, the need to improve quality and productivity in the workplace, and the need to manage change within institutions and organizations. TEAMS (Training for Excellence in American Manufacturing and Services, Inc.) has developed a model which provides a conceptual basis for transforming an organization from one that manages for short-term profits into a productive, forward-looking, competitive business. The goal of continuous improvement of productivity through quality is at the heart of the model which is derived by blending several theories and methods. Continuous quality improvement has as its underlying philosophy the works of Deming, Juran, Taguchi, Tribus, and Crosby. Leadership, people, method, and strategy are essential for realizing the focus, achieving the goal, and putting the philosophy into action. Of the three current philosophies — defect detection, defect prevention, and continuous improvement — continuous improvement is the most advanced relative to seeking to control products or services that are defective or of lower quality than desired. Continuous improvement suggests that improving the production system is a never-ending process.

Hatch (1993) developed a cultural dynamics model showing how the processes of manifestation, realization, symbolization, and interpretation provide a framework for discussing the dynamism of organizational cultures. Van Donk and Sanders (1993) found it is possible to improve quality management and its implementation through the study of organizational culture. Goldberg (1992) suggested that classic change management techniques are no longer adequate because change is occurring at a much faster pace. According to classic theory, change management requires several steps, specifically, unfreezing the organization's existing culture, creating cognitive recognition to open the workforce to what is new, and refreezing the culture once the change has been accepted. Today's new framework for change, the Static stage, also calls for the unfreezing of the current organizational culture. In the Fluid stage, employees begin to understand the changes that will benefit them as well as the organization. In the final stage — the Dynamic stage — people work with the new processes and await the next change that will be made. The implementation involved opening channels of communication, creating visionaries and change agents, developing a learning environment, providing training, and establishing a team approach.

Ergonomics provides the method to maintain quality systems that produce quality products. Designing quality into production or service processes can best be achieved by considering the capabilities and capacities of human performance. All processes in both service and manufacturing industries are completely dependent on effectively meeting the needs of those performing the tasks. Assessing those needs and designing processes that satisfy them enhance the talents of the worker. Manufacturing processes often are more clearly descriptive of the process improvement methods, since the idea of value added to the product at each process is more intuitive. However, from an ergonomic perspective, the processes within the service industries more clearly depict the human input to the development and delivery of the service. When the human impact of developing a service is neglected, then the individual's needs are not met and the input to the development of the service suffers. The end result is a dissatisfied or excessively fatigued worker who is incapable of adding sufficiently to the service for it to meet the needs of the customer (Getty and Getty, 1994).

The goals for improving processes to achieve better quality can best be realized by integrating and applying ergonomic precepts so that those carrying out the process are accommodated. The definitions of the inputs for early involvement in process design can be best formulated through ergonomics. When the total organizational culture and structural elements are considered, then individual roles will have stronger and more meaningful input. Those who perform, manage, and interact with the operational processes are the experts in process improvement. The application of the ergonomics principles provides the means to design quality processes (Getty and Getty, 1994).

The human element and the culture of the organization are a common thread in continuous improvement initiatives (Getty, 1996). Considering these human elements is the foundation for human factors engineering of the workplace, which is also known as ergonomics. There are numerous resources available to organizations to initiate an improvement process that addresses quality, productivity, and safety. The essential resource for bringing it all together resides within the organization (Getty, 1992b).

The improvement change process occurs when management commitment and action coincide, as well as when employee self-interest and quality goals are in agreement. When the management and employee roles are addressed together through principles of ergonomics, clearer understanding of needs and steps to solve them are realized. The process of developing a continuous improvement program must be an integrated approach. Many resources are available from various experts residing in multiple areas of the company as well as in many professional societies. Both research and the experience of ergonomics practice clearly show the necessity for holistic thinking and willingness to look beyond one's immediate capability and resources to achieve continuous improvement.

81.3 Organizational Process Orientation

Organizations are made up of processes and streams of processes (Conti, 1993). Processes are entities in an organization over which a manager has complete visibility and effective control from inputs to outputs. When organizational boundaries occur and more processes are present, then there are multiple processes or streams of processes. This concept taken together with the structure and culture of an organization can provide focus for those in the structure to be instrumental in the change process. Each process owner

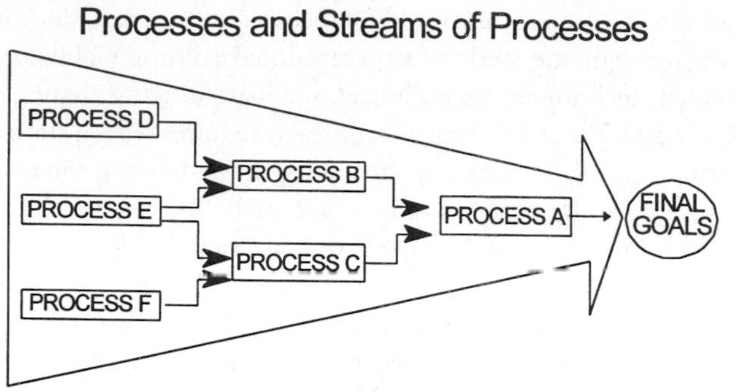

FIGURE 81.1 Processes are not clearly defined. Process owners must be properly reoriented. (From Getty, R. L., Cost justification of ergonomics improvements. In O. Brown and H. W. Hendrick (Eds.) *Human Factors in Organizational Design and Management - V,* (1996), pages 417 to 420, with kind permission from Elsevier Science - NL, Sara Burgerhartstraat 25, 10055 KV Amsterdam, The Netherlands.)

with the proper orientation and knowledge of available tools can become the mechanism for change for that process (Getty, 1992a).

In the industrial setting with so many demands for meeting individual and organizational objectives, there are not enough resources to successfully pursue all the improvement changes plus an ergonomics program at the same time. Frequently, ergonomists or others promoting improvement to company processes focus only on their program. In the enthusiasm of making improvements, the tools such as ergonomics seem to be more important than the processes of the company. Furthermore, when work demands are high, there is little time to support the extra effort of improvement programs. Even though these programs have goals that will improve company processes in the long run, the short-term urgency of getting a product out the door takes precedence. The key to achieving improvement is to blend together work processes with the improvement effort. Improvement cannot be an add-on to normal process work, but must be one and the same. Tools for improvement, including ergonomics do not produce the products, the processes do. In the area of ergonomics cost justification, there must be clear improvement to processes since this is the foundation for the central purpose of the company.

Organizational Processes

An organization consists of processes and streams of processes (Figure 81.1). In order for the organization to accomplish its purposes, various processes are present. Frequently, these processes are not clearly defined, and an initial step to properly orient process owners for change is for them to become aware of the organization from this perspective. The processes are the focus of the behavioral groups and provide cohesiveness.

Process Focus

Processes provide the focus for establishing the individual process goals that are derived from the overall organizational goal (Figure 81.2). In order for change to occur, there must be clear relevance to the goals of the processes. An important step, once the behavioral and process aspects are recognized and understood, is to delineate how individual goals contribute to the organizational goals. Once this occurs, there is purpose provided to initiate change. The strength of the desire to change and improve is directly correlated to how well the processes are meeting their goals. It may well be that there must be an evaluation of how well goals are being met before any suggestion for change can occur.

Organizational Structure and Processes

The organizational structure relates to the organizational processes (Figure 81.3). Often in the change endeavor the first attention is given to a criticism of the organizational structure. That structure has been

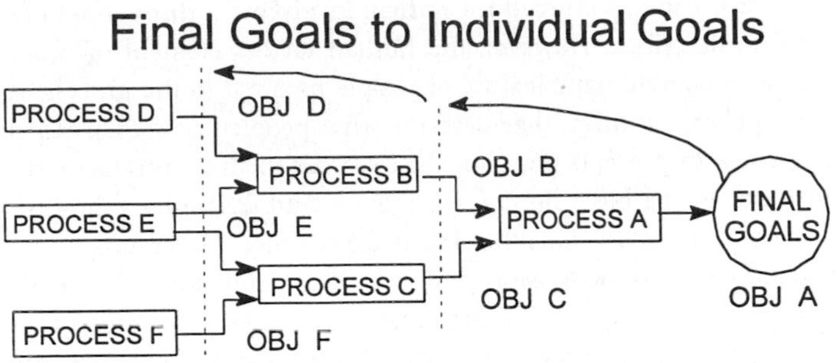

FIGURE 81.2 Processes provide the focus for establishing the individual process goals that are derived from the overall organizational goal. (From Getty, R. L., Cost justification of ergonomics improvements. In O. Brown and H. W. Hendrick (Eds.) *Human Factors in Organizational Design and Management - V,* (1996), pages 417 to 420, with kind permission from Elsevier Science - NL, Sara Burgerhartstraat 25, 10055 KV Amsterdam, The Netherlands.)

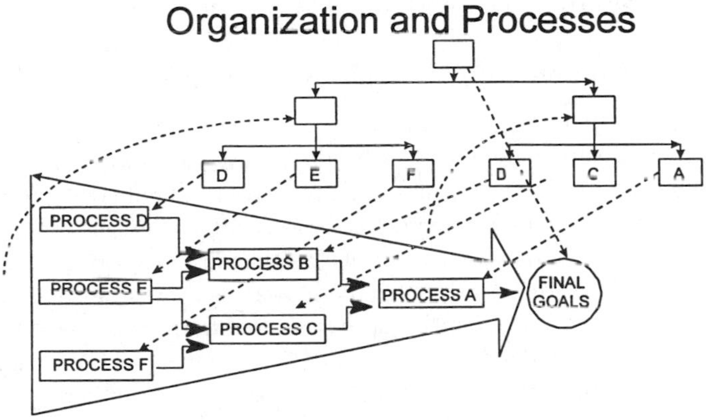

FIGURE 81.3 The structure of an organization seems to depict that the goals of lower levels are to satisfy those assigned to upper levels. The view of processes and streams of processes, changes the focus to the demands of the next process. (From Getty, R. L., Cost justification of ergonomics improvements. In O. Brown and H. W. Hendrick (Eds.) *Human Factors in Organizational Design and Management - V,* (1996), pages 417 to 420, with kind permission from Elsevier Science - NL, Sara Burgerhartstraat 25, 10055 KV Amsterdam, The Netherlands.)

derived over time and may be a major part of the organization's culture. The structure initially must be identified to the processes that accomplish the organization's purpose. Due to the way the structure of an organization is depicted, it appears that the goals of lower levels are to satisfy the requirements of those assigned to upper levels. Once one considers processes and streams of processes, the focus is to meet the demands of the next process. If there are difficulties in meeting the next process demands, then the management level responsible for the stream of processes becomes involved.

81.4 Utilizing Ergonomic Precepts to Design Processes

A Major Element for Achieving Quality

Ergonomic precepts address human physical and cognitive skills. When applied to process development, quality is significantly improved. These essentials are presented for service and manufacturing industries. A model is presented for the initial design of processes or for review of existing processes.

The goal of quality is to meet requirements 100% — zero defects (Cosby, 1979). To prevent quality problems, human skills need to be matched to processes. In addition, according to the principles of concurrent engineering, all previously downstream activities should be part of the design process. For this activity to be effective, the total picture of the human involvement throughout production or service processes must be clearly described. However, the human factors element of process design is often considered a mere given or an automatic feature of people involved in the process. It is true that many of the ergonomic concepts are intuitive; they describe what people know they can and cannot do. Yet, there are specific laws, rules, or precepts that must be followed to attain productivity, quality, and safety goals. These precepts refer to the basic human capabilities and capacities exhibited in the workplace. These goals can only be achieved by soliciting input from process operators. Otherwise, the process methods become a management-only decision, and individual productivity is stifled.

This discussion will present the application requirements of ergonomic precepts to process design. A review of human capabilities will illustrate all aspects of the skills that must be considered. Both production and service processes will be discussed to show the relationships to the human element. Finally, a model will be presented to be followed for the initial design of processes or review of existing processes. The best aspect of this approach is that any individual can be an ergonomic practitioner, with a thorough orientation to ergonomic precepts coupled with experience, to provide the necessary input to realize the benefits of applying these principles. At the same time, these precepts must be followed or all the technical process development and analysis will be deficient and flawed when they are executed.

The Goals of Early Involvement to Achieve Quality

Research and development (R&D) must be aware of the manufacturing commitment of uninterrupted output and the anxiety caused by retraining of skills, new status, and communication patterns (Steele, 1989). Consequently, manufacturing people must be involved in technology development. In order to achieve producibility and productivity, designers must be aware of all the elements of the manufacturing processes and effectively utilize all resources and talents (Stephanou and Spiegl, 1992). In addition, future factories will require that all the human elements (both social and physical) as well as economic and technological aspects be addressed. The way the Japanese moved from mass production to their current manufacturing approach is discussed in *The Machine that Changed the World* (Womack, Jones, and Roos, 1990). The authors indicated that the Japanese could not afford to invest in mass-producing machinery to the same extent that the United States did. Consequently, *people* became more important. They added that when the companies agreed to provide life-time employment, the employee was expected to perform whatever tasks were necessary and contribute expertise to improvement. "So it made sense to continuously enhance the worker's skills and to gain the benefit of their knowledge and experience as well as their brawn" (Womack, Jones, and Roos, p. 55). This last statement is probably the key to the Japanese success story. Although life-time employment cannot always be provided in the present Japanese economy, this role of the individual worker continues to be sustained, and the Japanese are able to maintain their competitive position. If other world economies genuinely accepted input from the worker with equal importance as any management activity or technical innovation, the goals of quality, productivity and safety would be realized. Ergonomic precepts provide the tools for such involvement.

Basic Human Capabilities and Capacities

It is obvious that the human element must be considered when designs are developed. Without awareness of human capabilities, the impact of the human role is not sufficiently evaluated. Designers underestimate variability and its importance in industry (Garrigou, 1991). Ergonomists should, in addition to contributing their knowledge, "create design situations which enable the use of operators' and designers' knowledge and their confrontation in order to establish forecasts of future work situations concerning health and efficiency criteria in order to transform them" (Garrigou, 1991, p. 1666). The focus of this "confrontation" should be work activity and not technology. Some work activity topics include: succession of operations, processing of information by the operator, physical strain (efforts, posture), and exposures to environmental factors. These discussions would lead to the evaluation of processes in terms of health and efficiency. This, in turn, would lead to modification and improvement.

Much of this discussion may appear to be simply "doing the right thing," but failing to account for ergonomics is antagonistic to the business goals of productivity, quality, and safety. Any change that takes ergonomics into account will cost money but should lead to reduced injury and cost, as well as improve efficiency (Alexander, 1986). Ideally, there should be ways of predicting benefits and developing and implementing workable recommendations. It is important to note that there are both tangible, business reasons for changes, and intangible benefits to the worker. Alexander states, "One must learn to distinguish between value and the ability to measure. In some cases, the value is there but it is just not worth much to develop the quantitative measures necessary to prove it" (1986, p. 4). Those involved in reviewing causes for quality defects, missed production rates, or dissatisfied customers can relate to a need to fix the obvious cause rather than spending extensive amount of time pondering all the possible causes.

Many, when determining the area that ergonomics or human engineering covers, may focus only on a single element such as physical aspects. Ergonomic principles cover *all* aspects of human skills. Alexander (1986) explains that industrial ergonomics problems should be characterized according to the type of body system that is affected. These body systems include: (1) Physical Size: Anthropometric, (2) Endurance: Cardiovascular, (3) Strength: Biomechanical, (4) Manipulative: Kinesiology, (5) Environmental: External, and (6) Cognitive: Thought. These systems should be considered as engineering principles to be followed, not soft, optional choices that are weighed against other factors. Once ergonomic precepts are made part of the design process, a new awareness of human capabilities and limitations emerges and the desired outcome of the processes is achieved.

Application of Ergonomic Precepts to Processes

Service Processes

There are a number of essential aspects to be considered in the design of service operations (Armistead, 1985). Elements of service processes include: (1) The interaction of the customer, either as the only receiver of the service or as integrally involved with the service delivery, (2) the behind-the-scene (back room) activities to produce the service, (3) the visible interface with the customer (front office) activities to deliver the service, (4) the different focus of the customer who may desire only parts of the same service, and (5) the simultaneous production and consumption of services to be delivered.

The need for input of the human service worker becomes apparent throughout these elements. Service delivery is very labor intensive. People skills are of greater importance than technological capability. Understanding the intangibility of the service processes can be assisted by evaluating the role of human delivery and by role-playing the customer's participation. Although the skill levels of the service provider vary enormously from a doctor to a cashier, the human capability is the major factor in the delivery of services. Some service processes that develop in the back room actually resemble much of the processes found in production industries. However, task analysis of service processes have an additional dimension of customer interface. The quality of this encounter is unpredictable. To improve the process, the expertise of service delivery personnel and customers should be included in the design of service processes. This utilization of ergonomic precepts enhances the service delivery personnel/customer interface and produces processes that result in satisfied customers.

Production/Manufacturing Processes

Edosomwan (1989), in his discussion of rules to follow for emerging technology suggests, "The designer must combine and use interdisciplinary knowledge to understand all issues involved in the interface between man and technology" (p. 43). He sees four phases: (1) information gathering, (2) planning, assessment, and measurement, (3) selection, and (4) testing and evaluation. He feels that designers, manufacturers and potential users of technology should be involved throughout the design processes.

As production processes are designed and developed, the impact of workers and their supervisors should be evaluated by actually running a simulation and having the affected workers assess the effects. The cost associated with "make-it-work" changes have traditionally been a major part of manufacturing process development. These changes have often been considered part of the business cost of developing new technology. However, hidden in these costs are the quality and productivity costs that are not always apparent during the design phases. Further removed from the design arena are costs that are impacted by repeated exposure to manufacturing processes that fail to take into account human capability and capacity. These processes, over a period of time, cause cumulative trauma injuries. Costs associated with injuries are high and so are the costs due to poor quality and productivity. Human errors caused by process designs that exceed human cognitive skills lead workers into a vicious cycle of repeated injury, inefficiency, and quality problems. To break the cycle, involve the worker in evaluating the processes, and many of these effects will be avoided. The production worker may have a simple input such as, "the task is uncomfortable," or, "is hard to understand." These inputs boldly signal the need for change.

81.5 Method of Developing Ergonomic-Oriented Processes

The precepts of ergonomics are frequently taken for granted and not given adequate attention in the design process. This may be due to a large extent to their intuitive nature. However, this characteristic of ergonomic principles makes them easy to apply. Imada (1991) discusses that the application of ergonomics precepts should be participatory, and noted three reasons for involving people in the design process: (1) Ergonomics is intuitive from workers' experience; (2) ownership enhances implementation; and (3) end-user participation causes flexible problem-solving.

Resources for change exist *within* the organization and any attempt to hand this activity over to others greatly dilutes the effort (Getty and Getty, 1993). Liker and Joseph (1986) suggest most companies do well to use their in-house personnel as the source for analyzing and redesigning jobs. This capability to be "ergonomic practitioners" comes from their experience and the knowledge that was gained in brief ergonomic training sessions. It has been shown that those with greater industrial experience within their own company are better able to identify ergonomic stresses than outsiders with academic background oriented to factory operations. Traditionally, companies call in consultants. However, this practice causes dependency and does not recognize the worker as a valuable resource (Imada, 1991).

Successful adherence to ergonomic precepts involves utilizing the expertise that is most closely familiar with the operational processes. A model for applying ergonomic precepts is portrayed in steps one through eight below. Tables 81.1 through 81.8 provide check sheets for each respective step. The elements in this model are similar to any evaluation of processes for improvement. The emphasis in this model is the human element. It takes those involved in the process design through a series of steps that begins with training and carries them through the development of an ergonomically sound process. Ergonomics then becomes a methodology for learning required skills, utilizing human capability in process design, and performing root cause analysis in order to take corrective action and improve processes. The ergonomic-oriented individual who is familiar with the manufacturing and service processes is the most capable one to apply this model, since this person views work situations from the "eye" of the worker.

Step One — Training

Training of those involved in company processes must occur first. Training does not take place just because classes are held. Before any training can take place, the need to learn must be established. This need can take place by developing an awareness of what can be better than the existing condition. This can take the form of learning what other similar groups have attained or that the current safety, productivity, quality, or service delivery is deficient. The next aspect of awareness is that those involved in the processes are the best resource for identifying the solutions. In the safety arena, describing the potential for working without pain develops motivation for improvement. When considering productivity, the awareness of how the individual shares in the success of the company creates motivation. In order to improve service delivery, the awareness of how the customer and the service provider have common interests, reveals the need for improvement. According to Getty (1993), the closer the service provider is to the customer, the higher the perception of service quality. Training topics must provide the participants with the tools, the awareness of the work environment, and a clear understanding of their individual roles of improving the processes through the use of ergonomic precepts.

TABLE 81.1 Train

TRAINEES
Those Familiar with Process — Operators.
Engineers — Process/Product Designers.
Management — Process Supervisors.
SUBJECTS
Ergonomics Awareness.
Human Systems.
Methods of Collecting Data.
Specific Organization's Plan for Ergonomics Improvement.
Identify Tools and Resources such as guidebooks and checklists.

From Getty, R. L and Getty, J. M., Significance of approaching participatory ergonomics from the macroergonomics perspective: A continuous improvement process. In F. Aghazadeh (Ed.) *Advances in Industrial Ergonomics and Safety VI,* (1994) pages 182 to 186 and Figure 1, with kind permission from Taylor & Francis, 1 Gunpowder Square, London.

TABLE 81.2 Task Analysis

Review Areas with Highest Productivity, Quality or Injury Problems.
Incorporate in Development of New Process.
Review Flow of Processes.
Define Inputs and Outputs of Process Increments.
Determine Interfaces: Man–Machine–Customer

From Getty, R. L and Getty, J. M., Significance of approaching participatory ergonomics from the macroergonomics perspective: A continuous improvement process. In F. Aghazadeh (Ed.) *Advances in Industrial Ergonomics and Safety VI,* (1994) pages 182 to 186 and Figure 1, with kind permission from Taylor & Francis, 1 Gunpowder Square, London.

TABLE 81.3 Human Capability

Compare Posture, Force, Repetition Requirements for Each Task.
Identify Specific Special Skills for Each Tasks.
Delineate Human System Requirements:
1. Physical
2. Endurance
3. Strength
4. Manipulative
5. Environmental
6. Cognitive
Emphasize Observation from Experience Rather Than from Analysis.

From Getty, R. L and Getty, J. M., Significance of approaching participatory ergonomics from the macroergonomics perspective: A continuous improvement process. In F. Aghazadeh (Ed.) *Advances in Industrial Ergonomics and Safety VI,* (1994) pages 182 to 186 and Figure 1, with kind permission from Taylor & Francis, 1 Gunpowder Square, London.

TABLE 81.4 Perceptual Queues

Define Resources that Assist Understanding.
Delineate Input Information that Triggers Response.
Evaluate Sufficiency of Available Queues.
Determine Flexibility to Change Physical Activity.
Assess Empowerment, Level of Decision Making and Self-Determination.

From Getty, R. L and Getty, J. M., Significance of approaching participatory ergonomics from the macroergonomics perspective: A continuous improvement process. In F. Aghazadeh (Ed.) *Advances in Industrial Ergonomics and Safety VI*, (1994) pages 182 to 186 and Figure 1, with kind permission from Taylor & Francis, 1 Gunpowder Square, London.

TABLE 81.5 Potential Assistance

Work Aids to Provide Better Understanding.
Mechanical Assistance for Physical Demands.
Work Rate Flexibility.
Methods to Eliminate Potential Error, Such as:
1. Better Flow
2. Only One Way to Install
Improvement of Work Environment.

From Getty, R. L and Getty, J. M., Significance of approaching participatory ergonomics from the macroergonomics perspective: A continuous improvement process. In F. Aghazadeh (Ed.) *Advances in Industrial Ergonomics and Safety VI*, (1994) pages 182 to 186 and Figure 1, with kind permission from Taylor & Francis, 1 Gunpowder Square, London.

TABLE 81.6 Task Assignment

Assign Task that Conform to Match of Operator, Machine or Customer Skills.
Design Flow that Maximizes the Success of Assigned Task.
Perform Task Analysis on New or Changed Process.
Review Task by Process Designers, Operators and Management.
Perform Trade-off Assessment for Business and Human Benefit.

From Getty, R. L and Getty, J. M., Significance of approaching participatory ergonomics from the macroergonomics perspective: A continuous improvement process. In F. Aghazadeh (Ed.) *Advances in Industrial Ergonomics and Safety VI*, (1994) pages 182 to 186 and Figure 1, with kind permission from Taylor & Francis, 1 Gunpowder Square, London.

TABLE 81.7 Simulation

Test and Evaluate Each Stage of Product/Process Development.
Involve Operator, End-user, Customer.
Provide Sufficient Realism to Represent Processes.

From Getty, R. L and Getty, J. M., Significance of approaching participatory ergonomics from the macroergonomics perspective: A continuous improvement process. In F. Aghazadeh (Ed.) *Advances in Industrial Ergonomics and Safety VI*, (1994) pages 182 to 186 and Figure 1, with kind permission from Taylor & Francis, 1 Gunpowder Square, London.

TABLE 81.8 Modification

Modify As Soon as Correction is Identified.
Constantly Review Processes Based on Performance.
Solicit Modification suggestions from All Sources:
1. Management
2. Designers
3. Process Operators
4. Customer
Overcome Status Quo Attitude.

From Getty, R. L and Getty, J. M., Significance of approaching participatory ergonomics from the macroergonomics perspective: A continuous improvement process. In F. Aghazadeh (Ed.) *Advances in Industrial Ergonomics and Safety VI,* (1994) pages 182 to 186 and Figure 1, with kind permission from Taylor & Francis, 1 Gunpowder Square, London.

Step Two — Task Analysis

The process of task analysis performed by those most familiar to the processes takes on a new meaning than when an engineer unfamiliar with the process environment performs this analysis. The research performed to develop the needs for improvement and the elements of the process increases one's awareness of activity that in the past was taken for granted. By reviewing elements of the processes, the required interfaces, and coordination process, participants develop an important capability to suggest or initiate realistic changes.

Step Three — Human Capability.

Considering the capability of those performing a process by individuals familiar with the processes is the most logical and efficient approach. It is in this area that the intuitive nature of ergonomics is most clearly illustrated. There are areas of biomechanics and anthropometry that approach the technical and must be applied by qualified individuals. However, process-operators with proper guidance can more clearly identify the details of the tasks. In addition any peculiarities of the tasks that are unique from the textbook approach will be missed by those not familiar with the process.

Step Four — Perceptual Queues

Clearly, those familiar with processes can identify the reality and sufficiency of perceptual queues required for their accomplishment. Others not familiar with specific tasks may well see what should be available to the worker rather than what is actually present. The preferred approach would be an individual performing the process to team with another who is an observer and can inspire insight into the elements that constitute the job. Individuals in the same workplace or service process, following training observing others in different processes than their own, can detect details not seen by those performing that process. Observers can watch another process and then have others watch their process.

Step Five — Potential Assistance

Again process-operators follow through on their experience after an orientation into ergonomic precepts. They identify improvements that they may have observed in the past. Deficiencies of processes may have been considered numerous times by those performing the process, but they have not been provided the opportunity, nor the skills, to articulate viable solutions.

Step Six — Task Assignment

A team provided by process designers and process-operators can best complete task assignments. Awareness of the work environment will provide the realism to detailed task assignments. Process-operators who have been involved with the application of ergonomics precepts from the start develop qualifications that are not easily obtainable in the average continuous improvement effort. The process itself elevates the capability of workers, service providers, and management in the performance of continuous improvement efforts.

Step Seven — Simulation

Simulation designed and accomplished by process-operators solves the majority of the deficiencies found in unrepresentative simulations. For improvement of existing processes and for the design of new processes workers become active participants in improvement.

Step Eight — Modification

When participatory ergonomics becomes the methodology of continuous improvement, then modification becomes an ongoing activity. Becoming totally involved in the application of ergonomic precepts by process-operators, managers, and customers creates an awareness of the human element that is vital to the success of all the processes that produce an organization's product or delivers its services.

The application of ergonomic precepts becomes a clear method of identifying with many of the goals of continuous improvement that are often missed by those closest to the processes slated for improvement. First the application of this model must have an atmosphere that is developed by approaching change from the macroergonomic perspective. Then the participation of all elements throughout the organizational structure will follow.

81.6 Measuring Results

Unique Contribution of Ergonomics to Design Activity and to Company Processes

As industry moves into a world-class competitive environment, continuous improvement efforts must consider the human element in order to be successful. Even in the most automated factory or the most labor-intensive service processes, the common denominator is the capability of the worker. With clear descriptions of the physical demands and identification of ergonomic hazards or physical stressors in the workplace, better processes can be developed to achieve the organization's purpose. By continually improving the fit between the job and the worker, processes are continuously being improved.

Essential elements of ergonomics include: (1) Ergonomics design function throughout product or service life cycle from conception, design, development/production, delivery, support, continuing through obsolescence; (2) concentration on the human–machine system; (3) machine aspects cover display, information, and control processes; (4) human aspects entail sensory mechanisms, information/decision processing, and alternatives/effectors for control movement and command processes; (5) environmental aspects address the work space, the environment, and the work organization; (6) application of the ergonomic tools of task analysis and task allocation; (7) extensive training of process owners of the application of ergonomics to meet the objectives of the organization, processes, and individual workers.

Ergonomics principles must be integrated with: (a) all functions of the manufacturing process, (b) the design of new manufacturing processes during product design, (c) the process analysis techniques to improve manufacturing processes, (d) the root-cause analysis of workmanship defects, (e) the purchase of tools and equipment for facilities, plant maintenance, and factory workers, and (f) the improvement of the office environment throughout the company. The primary goal will be to enhance the workplace environment and improve productivity, quality and minimize potential for injury.

Value Added to Design Activity and to Company Processes

Performance metrics show the value added as one progresses through the phases of system design, manufacturing development, and implementation. The following delineate the benefits to industry from the consideration of human–system interaction in manufacturing processes with metrics that can be used to verify the success of a design.

Include ergonomic principles in the development of new processes before they reach the factory floor. Metric: reduced make-it-work engineering changes.

Lower overall costs of tooling, fixtures, and processes due to reduced rework. Metric: lower tool change orders.

Reduced schedules due to better match with human skills by integrating ergonomics methods with management of cost, quality, and schedule. Metric: on-schedule and in-station work flow.

A continually evolving simplification of physical demands that improves productivity and quality is consistent with continuous improvement. Metric: improving trends for productivity and quality.

Provide ergonomic inputs to design of processes to attain increased productivity and quality at less cost. Metric: improved productivity.

Quality improvements through reduction of rework due to human error. Metric: decreasing quality defects requiring rework.

Physical demands data become available for the supervision of exposure to the work hazards. Metric: reduced lost time due to fatigue and injury.

Return-to-Work (RTW) gains by an ergonomic focus by pulling together the elements that must be integrated. Metric: reduced lost workdays.

Individuals are able to understand the various factors that expose them to cumulative trauma and have a role in reducing their exposure. Metric: fewer cumulative trauma incidences.

Physical description of the tasks can be used to improve medical treatment. Metric: lower workers' compensation costs.

81.7 Conclusion

Designing quality into production or service processes can best be achieved by considering the capabilities and capacities of human performance. All processes in both service and manufacturing industries are completely dependent on effectively meeting the needs of those performing the tasks. Assessing those needs and designing processes that satisfy them enhances the talents of the worker. The goals for improving processes to achieve better quality can best be realized by integrating and applying ergonomic precepts so that those carrying out the process are accommodated. The definitions of the inputs for early involvement in process design will be best formulated through ergonomics. Those who perform, manage, and interact with the operational processes are the experts in process improvement. The application of the principles of ergonomics provide the means to design quality processes.

This chapter has focused on the techniques of enhancing the quality effort with the principles of ergonomics. Indeed both initiatives improve total business processes. Additional approaches for blending these efforts can be seen from the quality orientation, (Stuebbe and Houshmand, 1995) and from the view of comparing major trends that have occurred in both ergonomics and quality (Drury, 1997). Both of these approaches agree with the premise of this discussion, namely, that human-oriented ergonomics is closely related and an integral part of continuous improvement processes.

References

Alexander, D. C., *The Practice and Management of Industrial Ergonomics*, Prentice-Hall, Englewood Cliffs, 1986.

Armistead, C. G., Design of service operations, in *Operations Management in Service Industries and the Public Sector*, Voss, C., Armistead, C., Johnston, R. and Morris, B., Eds., Wiley, New York, 1985.

Conti, T., *Building Total Quality: A Guide for Management*, Chapman & Hall, New York, 1993.

Crosby, P. B., *Quality is Free*, McGraw-Hill, New York, 1979.

Drury, C. G., Ergonomics and the quality movement, *Ergonomics*, 40-3, March 1997.

Edosomwan, J. A., *Integrating Innovation and Technology Management*, Wiley, New York, 1989.

Garrigou, A., The role of the ergonomist in the case of workers' participation in the design of complex industrial installations, in *International Ergonomic Association Congress Proceedings*, IEA, Paris, 1991.

Getty, J. M., *An Investigation of the Perception of Delivered Quality at Different Levels of Organizational Hierarchy in Services*, Dissertation, University of North Texas, Denton, Texas, 1993.

Getty, R. L., Cost justification of ergonomics improvements, in *Human Factors in Organizational Design and Management — V*, Brown, O. and Hendrick, H. W., Eds., Elsevier Science B. V., Amsterdam, 1996.

Getty, R. L., Physical demands of work are the common reference for an integrated ergonomics program, in *Proceedings of the Human Factors and Ergonomics Society 38th Annual Meeting*, HF&ES, Santa Monica, 1994.

Getty, R. L., Continuous improvement efforts often dictate organizational change: What are the mechanisms for change? in *Proceedings of the Human Factors Society 36th Annual Meeting*, HF&ES, Santa Monica, 1992a.

Getty, R. L., Clearing the fog of continuous improvement, *Manage* 44, 1992b.

Getty, R. L., Abbott, W. L. and Getty, J. M., ISO 9000 methodology enhances ergonomics effort: Ergonomics becomes a tool for continuous improvement, in *The 49th Annual Quality Congress Transactions*, ASQC, Milwaukee, 1995.

Getty, R. L. And Getty, J. M., Significance of approaching participatory ergonomics from the macroergonomics perspective: A continuous improvement process, in *Advances in Industrial Ergonomics and Safety VI.*, Agahazadeh, F., Ed., Taylor & Francis, London, 1994.

Getty, R. L. and Getty, J. M., Organizations hold the initiative for improvement: The need for self-assessment by process owners, in *The 47th Annual Quality Congress Transactions*, ASQC, Milwaukee, 1993.

Goldberg, B., Manage change — Not the chaos caused by change, *Management Review* 81, 1992.

Imada, A. S., The rationale and tools of participatory ergonomics, in *Participatory Ergonomics*, Noro, K. and Imada, A., Eds., Taylor and Francis, London, 1991.

Hatch, M. J., The dynamics of organizations culture, *Academy of Management Review* 18, 1993.

Liker, J. K. And Joseph, B. S., Ergonomic knowledge in a U.S. automotive plant: Correlates of practical and theoretical job analysis capabilities, in *Human Factors in Organizational Design and Management — II*, Brown, O., Jr. and Hendrick, H., Eds., Elsevier Science Publishers B. V., North Holland, 1986.

Sparks, R. And Dorris, J. M., Organizational transformation: Continuous improvement of productivity through quality, *Advanced Management Journal* 55, 1990.

Steele, L. W., *Managing Technology*, McGraw-Hill, New York, 1989.

Stephanou, S. E. And Spiegl, F., *The Manufacturing Challenge: From Concepts to Production*, Van Nostrand Reinhold, New York, 1992.

Stuebbe, P. A. and Houshmand, A. A. (1995). Quality and ergonomics, *Quality Management Journal*, Winter, 1995.

Van Donk, D. P. and Sanders, G., Organizational culture as a missing link in quality management, *International Journal of Quality and Reliability Management* 10 (5), 1993.

Womack, J. P., Jones, D. T. and Roos, D., *The Machine that Changed the World*, Rawson Associates, New York, 1990.

82

Corporate Cost Avoidance Using Sound Ergonomics Technology and Quality-Based Customer Services

Dwight P. Miller
Sandia National Laboratories

82.1 Introduction

Although Sandia National Laboratories (SNL) has been performing ergonomic assessments and studies since the late 1970s, it was not until the spring of 1992 that SNL initiated a concerted effort to develop a corporate ergonomics program. The program has been developed with great success. Data from the 1993–96 period indicate that the once increasing rate of musculoskeletal injuries and illnesses has been turned around into a decreasing rate. Associated costs have been lowered, while employee satisfaction with services and worker productivity has remained high. This chapter describes the plan, quality approach, processes, and tools developed, and reviews performance statistics that qualify Sandia's corporate ergonomics program as a success.

82.2 The Project Plan

The program began with assigning a project manager and forming a process management team to identify needs and plan the program's development. We felt it was important to include representatives from all departments affected by or contributing to the activity. By the fall of 1992, a project plan was written

0-8493-2641-9/99/$0.00+$.50

and approved by management. Phase one was dedicated to the development of program infrastructure. That is, staff had to be trained to provide services and standard processes had to be developed and documented. Care was taken not duplicate a mistake made by others, in overselling the services before capacity had been grown, thereby creating a backlog that might undermine credibility.

Phase two was to identify areas at SNL that needed ergonomics help the most and to train management on the benefits of ergonomics and the services made available. Because 90% of Sandians work in offices with computers, emphasis was put on creating helpful literature for them and developing a standardized, detailed checklist for performing office worksite evaluations.

Phase three embodied moving from a predominantly reactive to a more preventive approach. This meant getting the workers involved and teaching them how to avoid poor work design. Special project teams were formed with the line organizations to work together toward risk identification and reduction. Employee and contractor training was instituted en masse, but on a voluntary basis, so that ergonomics would avoid the stigma of being just another enforced safety and health requirement.

82.3 A Quality Approach

We did not set out to embody the "Q word" when we began to pull together a program; we just wanted to do what was right. We wanted to know how others had done it, so we called 56 companies and visited eight sites with established programs (McKeen and Miller, 1993). We found out later this was called "benchmarking." We also wanted to know what was expected of the team as we formed a charter and outlined tasks; this was called "interviewing stakeholders." We systematically evaluated ergonomic task chairs and VDT workstations in two independent public events that involved general laboratory staff, facilities representatives, and folks from purchasing. We learned that this was called "establishing requirements with your suppliers" (Leger et al, 1989).

After a while, we started consciously utilizing the quality approach and began to design services with future metrics in mind. We established an ergonomics coordinator and designed a triage procedure to prioritize customer requests on the basis of medical need. Existing services were documented via flowcharts as were newly developed alternative procedures. A customer survey was designed and implemented to quantify customer satisfaction with various aspects of the program and collect input on what needed improvement (see Section 82.7). Team members who could not live up to the performance standards established by the group were ask to find other work that had lower standards. We tracked accident and illness data to monitor what impact we had on work at SNL. Eventually, after several years of process refinement, the program earned a reputation for quality and won some prestigious awards. Because quality is an approach that applies to all aspects of the program, additional details can be found in the following sections.

82.4 The Corporate Ergonomics Group

As an outgrowth of the process management team, a Corporate Ergonomics Group (CEG) was formed with similar professional diversity. Members came from existing corporate resources in safety, medical, education, facilities, industrial hygiene, and human factors disciplines. Each recruit/volunteer either had a background directly relevant to ergonomics or one conducive to working in ergonomics (e.g., the training specialist was previously a Registered Nurse). Members from our California laboratory were involved from the earliest activities through 1994, when our budget was separated and they became their own separate entity. Many of our processes, course work, and written documents were based on the California group's precedents and contributions.

A consistent philosophical and methodological approach was fostered by sending members to three short courses taught by internationally renowned university professors: one in advanced VDT workstation design, and two covering general occupational ergonomics. All CEG members were committed to acquiring the required training, performing customer services, and attending CEG meetings. Teamwork was nurtured by holding annual off site strategic planning meetings, weekly CEG meetings, practicing group problem-solving, and by developing a group mission statement, values list, logo, and tag line (Figure 82.1).

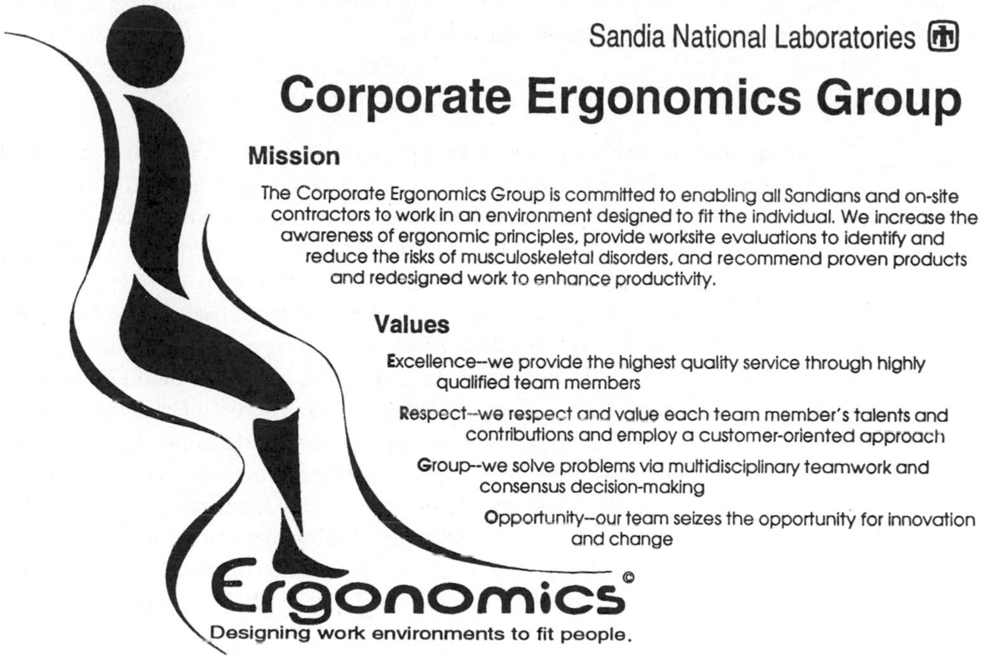

FIGURE 82.1 Corporate Ergonomics Group logo, mission statement, and values.

Consistent with a quality approach, when CEG meeting productivity began to wane, sub-teams were formed on a voluntary basis with specific responsibilities. The sub-teams elected their leaders, prioritized tasks, met whenever necessary, and reported progress to the rest of the group. The CEG meetings were consequently held only monthly and followed a strict, timed agenda to ensure high productivity. Guest speakers and suppliers were periodically invited to address topics of interest or demonstrate new products.

82.5 Services Developed

During phase one of the project plan, several customer services were developed, based on customer needs. Throughout the project plan, service processes were modified based on experience and customer input.

Ergonomics Coordinator

Interdisciplinary teams are fine, but when the members are not collocated, communication can suffer. We established an ergonomics coordinator (EC) to handle all incoming calls from customers and match their needs to our resources. This required some basic ergonomics training so that a triage procedure could be used by the EC to determine the level of priority and urgency of response. Upon receiving a call, the EC filled out a form with all of the relevant customer and problem information. She then assigned a priority category, based on symptoms, and scheduled one of the CEG members to respond, based on pre-established time blocks of member availability. All information relating to symptoms was handled as private. A memorandum was then faxed to the caller's supervisor stating that one of the services had been arranged, what it consisted of, and how much it could potentially cost. The service was free, but any purchases were paid for by the customer. The EC also used judgment in matching CEG members' skills with the nature of the customers' needs.

The EC coordinated paperwork resulting from worksite evaluations (WSEs), kept files, and maintained a database on service calls. Periodically the EC would inform members which of their customers were returning required paperwork and which were not. A second, back-up coordinator was added to cover for the EC when she was not at the office. This kept returned-call and scheduling delays to a minimum and helped maintain high customer satisfaction. The EC also helped the project manager communicate with the CEG, schedule meetings, and distribute the minutes.

Chair Fittings

In addition to the standard office seating provided by the facilities people, the program offered 10 alternative models through a chair-fitting process. Customers set up times with the EC to visit either our ergonomics resources room or the medical center. Each had a full complement of sample chairs, so location was determined by proximity to the customer's office and medical requirements. A CEG member would meet the customer, proceed through a series of questions concerning the job requirements, functional loss assessment, height and weight, and look up the ideal subset of chairs on a chart and then invite the customer to try out the models in that subset. After some sitting and some adjusting, the fitter and the sitter would mutually agree on the best chair. In many cases, models would be ordered with larger or smaller seat pans, different foam, adjustable lumbar supports, or other options based on customer needs and preferences. Affordability and appropriateness were determined by the customers' management. Using this process, the CEG reduced the delay in acquiring custom ergonomic task chairs from 2 to 9 months to 4 to 6 weeks, with special medically related rush orders taking only two weeks. Company policy allowed customers to keep their individual chairs regardless of subsequent moves within the company. In the rare event that a customer was dissatisfied after chair delivery, we determined if the CEG had made an error in prescribing the chair, and if so, the chair was reapplied and another chair was provided at no charge.

Worksite Evaluations

As in any active ergonomics program, CEG members went to the customers' worksites to assess the materials, tools, processes, and environment for musculoskeletal stressors or other factors that could lead to work-related musculoskeletal disorders or inefficient operations. This service was considered to be the backbone of the program. The flow of the process, from initial call, to final paperwork is shown in Figure 82.2.

Having not found a suitable checklist off-the-shelf, the CEG developed its own four-page, graphically based, field checklist, partially displayed in Figure 82.3. It served as a guidance and data-collection tool that was refined iteratively, using field experience. In addition to customer information; existing ergonomic equipment and furniture, tasks and durations, postures, office layout, and existing workstation measurements were recorded. The last section provided space for recommendations. A similar checklist was developed for non-office work environments.

Prior to performing WSEs, new evaluators attended professional training classes, read a guidelines document, and accompanied senior ergonomists on WSE calls. Eventually, the new evaluators performed the WSE under the guidance of the senior member, and when both were confident, the fledgling was allowed to fly solo. The author felt this was an important safeguard to reduce the liability of giving a customer bad advice on how to work. Typically, a CEG member performed about three WSEs a week, which would take about a half-day. This relatively light level of effort allowed the program to tap into existing laboratory personnel resources and avoid the need to hire new staff.

Occasionally, an entire work area or department was identified for evaluation. In these cases, several team members met with department representatives to form a joint project team to investigate ergonomic issues. We found that without involvement in developing solutions from the host organization, the best ideas had almost no chance of acceptance.

Resources Room

A resources room was set up to house the physical assets of the program (pamphlets, alternative keyboards, training materials, etc.) and provide space for chair fittings and analysis of videotaped jobs. The ergonomics video and text library was also housed in the resources room. A company vehicle was provided for transportation to customer appointments throughout the technical areas of the laboratories, which cover 2830 acres.

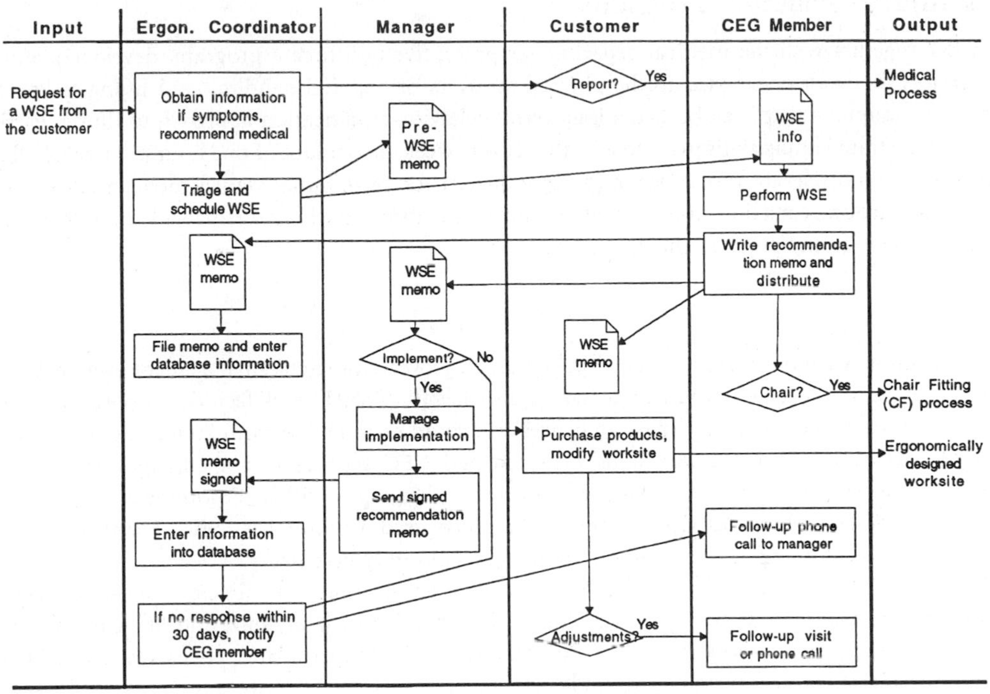

FIGURE 82.2 Worksite evaluation process flow chart.

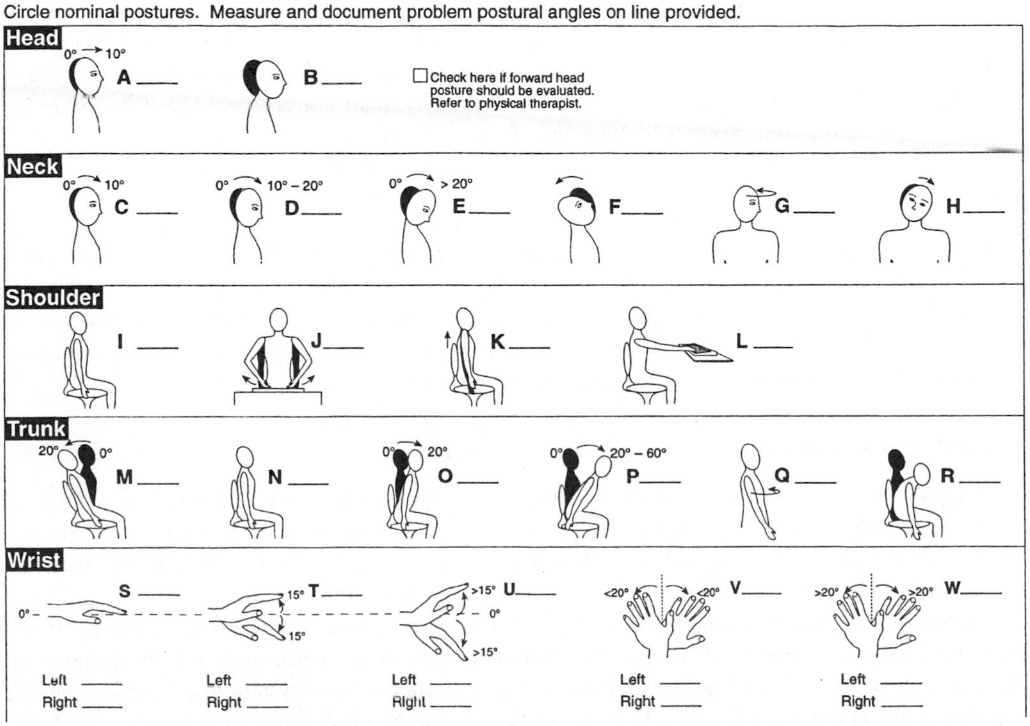

FIGURE 82.3 Section of worksite evaluation checklist. (From Nina Stewart-Poppelsdorf, CIH, CPE. Copyright Sandia National Laboratories, 1994. With permission.)

Back Injury Reduction Program

The CEG, together with the medical department's preventive health care program, developed courses and services for workers exposed to the physical stressors of material-handling and lifting tasks. One course was unique in that it embodied a long-term, behavior-modification approach to lifting. Participants demonstrated lifting skills prior to and after a four-week curriculum of one hour per week. Follow-up sessions on job-relevant topics occurred every quarter for 18 months. Management incentives were designed to encourage maximum participation and onsite lifting coaching was provided on demand by exercise-physiology and physical-therapy interns.

Training

Three ergonomics courses were developed in-house: employee awareness, management awareness, and advanced ergonomics for environmental, safety, and health (ES&H) and facilities professionals. The employee course was designed for mass audiences and could be modified from 20 minutes to one hour in length. Musculoskeletal stressors were identified and CEG services described. Approximately ten minutes of the class were devoted to playing an in-house videotape entitled "Ergonomics Detective C.T. Dodd" (Corporate Ergonomics Group, 1995). The video used a humorous "Mike Hammer" style to convey the concepts of ergonomics without inducing involuntary narcolepsy.

The managers' course was designed to be given in small groups, as in director's staff meetings. In addition to the basic concepts, injury and illness statistics for each organization are emphasized to cultivate sensitivity to lost-workday costs. Additionally, simplified surveillance checklists were added to the training materials so that managers could walk their spaces and identify musculoskeletal stressors before they became physical symptoms. The advanced course, which was four hours long, covered all the essentials and provided hands-on workstation evaluation and redesign techniques for staff who could use the skills in their regular work.

The CEG also instituted the Ergonomics Colloquium. About once a year, a well-known ergonomist would visit the CEG to consult on its development and present a one-hour colloquium to the laboratory staff on a technical topic of interest. Speakers were videotaped for people who could not attend the live colloquium. Dr. Roger Stephens from OSHA and Professor William Marras from The Ohio State University were the colloquium speakers in 1994 and 1995, respectively.

Consulting

In addition to the bread-and-butter services outlined above, occasionally the CEG was called upon to provide consulting in unique situations. Several times we were asked to join teams tasked with redesigning work rooms or entire work areas, such as the classified document vault and document delivery service center in the technical library. Other projects included the mail room, shipping/receiving, the corporate computing center, and a hazardous waste management facility.

Medical Management

Communication and coordination on medical issues were facilitated by the fact that four members of the CEG were employed by the medical department: one doctor, two nurses, and a physical therapist. Screening and diagnosis procedures for work-related musculoskeletal disorders were developed, documented, and used by the primary care physicians. Onsite physical therapy was available, as well as an onsite optician, who filled prescriptions for VDT glasses. Due to lack of empirical evidence justifying lumbar belt usage, the company denied employees lumbar belts except when healing from an injury, and only if prescribed and fit by a health-care professional. A mutual referral system was instituted whereby if a symptomatic employee called for a WSE, he was referred to Medical, and if he showed up in Medical, he was referred to the EC for a WSE appointment.

Written Materials

In addition to the many off-the-shelf educational pamphlets and booklets used by the program to educate Sandians, the CEG developed several of its own. The most popular was a tri-fold that discussed ergonomics for computer users. Based on a similar pamphlet developed at the Lawrence Berkeley Laboratory, it used graphics from Apple Computer Inc. to demonstrate proper workplace biomechanics and text to explain use of accessories and who to call for assistance (Mulligan et al., 1994). In addition, two booklets were developed, complete with photographs, detailing safe exercises to use to stretch and strengthen musculoskeletal subsystems undergoing stress associated with work (Suzuki et al., 1995a,b). They were written by two physical therapists who researched the literature on exercises and found that many being prescribed actually exacerbate work-related strain by exercising the affected tissues in ways similar to the work itself. One booklet was targeted directly at office-related tasks, while the other addressed stressors encountered at home. Finally, as is true of most large institutions, Sandia has a comprehensive ES&H Manual. The CEG, with help from its California colleagues, developed a manual chapter containing a complete set of ergonomic guidelines for office work (McMahon and Miller, 1995). The manual is now issued electronically on the Sandia intranet. Although none of these products was particularly innovative, excepting perhaps the exercise booklets, the concise format, familiar style, and repackaging of information made them extremely useful to Sandia employees.

Software

Throughout the program's life various software packages were evaluated for both CEG and customers' use. We always looked at price and ease of use as foremost considerations in addition to functionality and time savings. Some of the packages we evaluated were more vaporware than actual product, and many of the legitimate, finished, commercial products were extremely difficult to use. Several were so costly that we decided we could do the work manually. We chose LifeGuard® for reminding serious computer users to take alternative-work breaks occasionally, and ErgoSmart® for general information in a question/answer format. Site licensing made these products affordable. In an unrelated feasibility project, the WSE checklist was programmed into a pen-based portable computer, designed for field data collection and automated database entry. Despite showing initial promise, problems with battery life and handwriting recognition software in the feasibility testing precluded implementation.

82.6 Accomplishments and Performance Metrics

By any standard, the program was immensely successful. The teamwork demonstrated within the CEG set a new benchmark for interdisciplinary matrixed projects within the ES&H organization. Customers loved the individual attention and expert advice they received. The injury/illness statistics dropped consistently for four years. The money spent on program development and operation was recovered easily in reduced lost-time illnesses. Effectiveness was maintained even during budget reductions after the program was developed. The following sections discuss specific accomplishments and various performance metrics in more detail.

Creating and Responding to Demand

As plotted in Figure 82.4, the number of WSEs increased initially as services were introduced and then reached asymptote in 1995. When benchmarked against other similar institutions, SNL had the highest asymptotic level of WSEs provided. Chair fittings were introduced in 1993 as a standard service and have increased steadily since. Worksite evaluation performance not only increased in numbers (up by a factor of three), but also responsiveness. According to the 1994 survey, 57% of the customers received WSE services within 2 weeks, and 82% within one month of the initial call. The new triage priority system

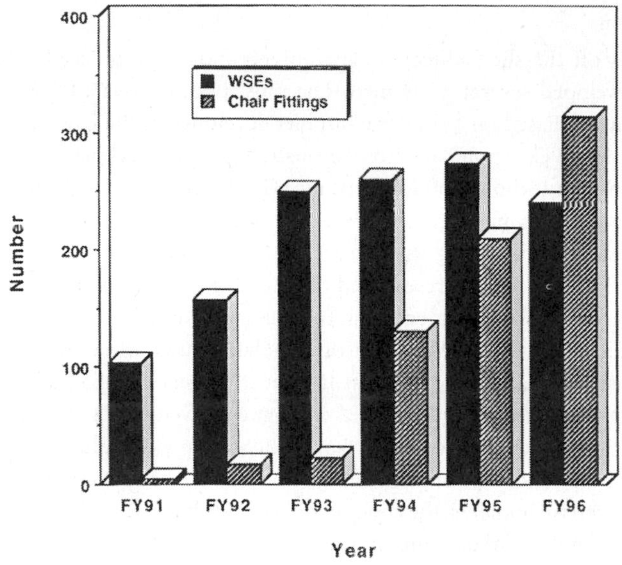

FIGURE 82.4 Number of worksite evaluations and chair fittings performed.

ensure a five-day response to symptomatic callers. Customers previously received documentation of recommendations in 1 to 3 months due to dictation delays. Feedback time was reduced to less than one week by using a computer template distributed to all CEG members.

In the last two years, training has reached approximately half of the Sandia population. This is not remarkable unless you consider that it was not required training, except for incoming secretaries. Feedback on the instructors and content have been very favorable, and most attendees loved the aforementioned video. A few managers have adopted the surveillance checklists in their yearly walkthroughs, generating additional requests for WSEs.

Reductions in Pain and Suffering

Four out of five (79%) symptomatic customers, who made the recommended ergonomic improvements experienced symptom relief in the 1994 quality survey. This figure was up from the 1992 figure of 70% (see Figure 82.5). From a medical case standpoint, before we could calculate our gains on work-related musculoskeletal disorders, we had to establish criteria for case inclusion. We defined them as cumulative trauma illnesses or musculoskeletal soft-tissue injury (such as a strain) obtained while performing a repeated, regular job task in a less than ideal manner (a manner that could be improved using common ergonomic practices). We also put on the stipulation that the injury or illness occurred while performing assigned work, leaving out the weight-room lifting injuries of the guard force, and the occasional luggage-toting strain experienced during business travel. These restrictions tended to make our performance metrics conservative.

Using the data-inclusion criteria, we calculated numbers of cases, and associated costs for four years of the program (Figure 82.6). As can be easily observed, both the number of cases and the associated costs have consistently dropped over the four-year period. The costs were calculated using a conservative formula developed for Department of Energy (DOE) contractor facilities in 1988, accounting for lost work time, lost productivity, and medical expenses. In the years 1993 to 1995, the costs associated with cumulative trauma were 30% higher than other cases, and required almost 16 days away from work, compared to 9 days away for other types of work-related injuries and illnesses (Figure 82.7).

As might be expected in a research and development laboratory, upper extremity cases consistently outnumbered spinal problems in the three years audited (Figure 82.8). When analyzed for job location and type of work being performed, the pattern shown in Figure 82.9 obtained. Office workers suffered

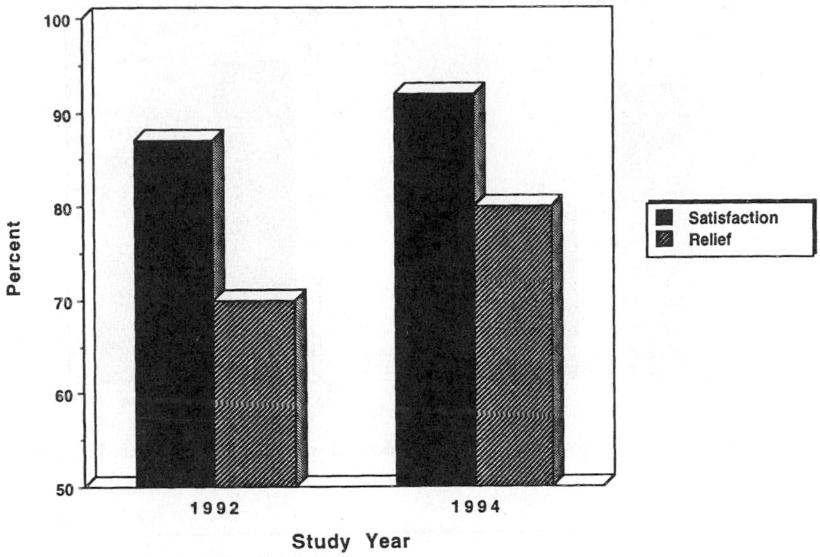

FIGURE 82.5 Relief of symptoms and overall customer satisfaction.

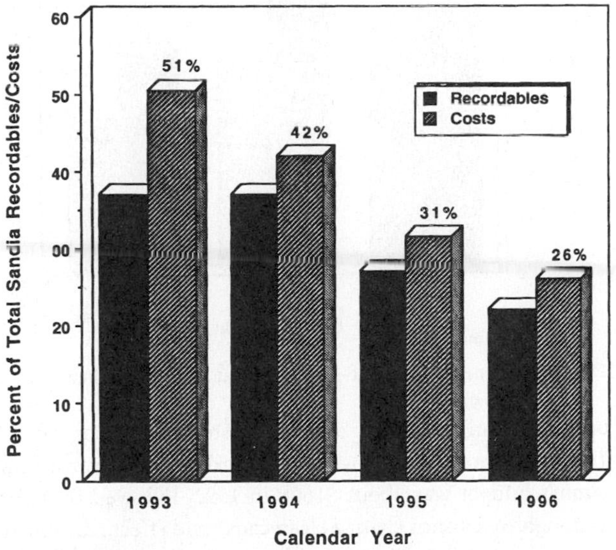

FIGURE 82.6 Ergonomics-related cases as percent of total recordable cases and total costs.

the most, followed by material handlers and laborers, with laboratory technicians and crafts/trades people showing the fewest WRMSDs.

Was There Return on Investment?

Perhaps the ultimate metric for program success is a benefit-cost analysis, taking into account how much money was expended and what was gained as a result. Figure 82.10 compares program budget figures and costs associated with work-related musculoskeletal disorders for the years 1993 through 1996. (The program costs do not include equipment purchases made by employees, or costs involved with remodeling

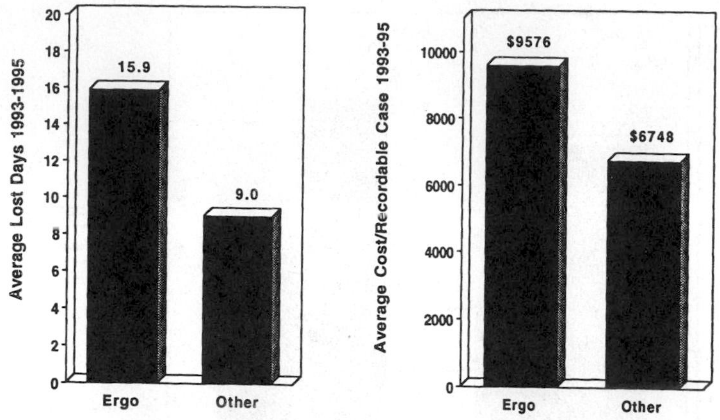

FIGURE 82.7 Average lost days and costs associated with ergonomics-related illnesses.

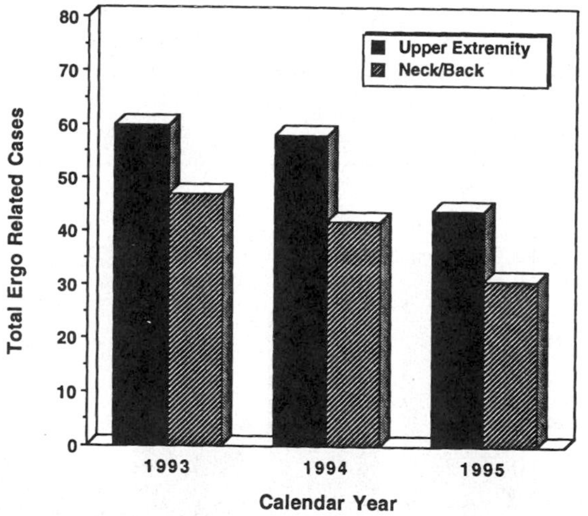

FIGURE 82.8 Number of upper-extremity and spine-related cases by year.

work areas, only the money spent on developing and administering the program.) As the figure suggests, the costs associated with work-related musculoskeletal disorders were over a million dollars each year prior to 1995. The program's budget was about $160k in 1992, followed by $280k and $585k in 1993 and 1994 respectively. Although we cannot claim a clean cause and effect relationship (we had no control over other factors), the precipitous fall in illness costs in 1995 and 1996 suggests that the program had some effect in turning around a dangerous trend of increasing costs in the early 1990s. By 1996, the costs associated with work-related musculoskeletal disorders had been reduced to approximately one sixth of what they were in 1994. The program's budget was trimmed back systematically in 1995, 1996, and 1997, taking advantage of the early development work and stressing implementation of the processes created in prior years. Budget projections for 1997 are just above the $100k figure, signaling an 80% decrease since 1994.

Was the money well spent? I am sure the employees and contractors who received the advice and work-design benefits would reply with an emphatic "yes." Approximately 80% of the recommendations made to employees were implemented, despite customers having to purchase their own furniture and accessories. Statistically, a rapidly increasing trend in work-related musculoskeletal disorders was turned

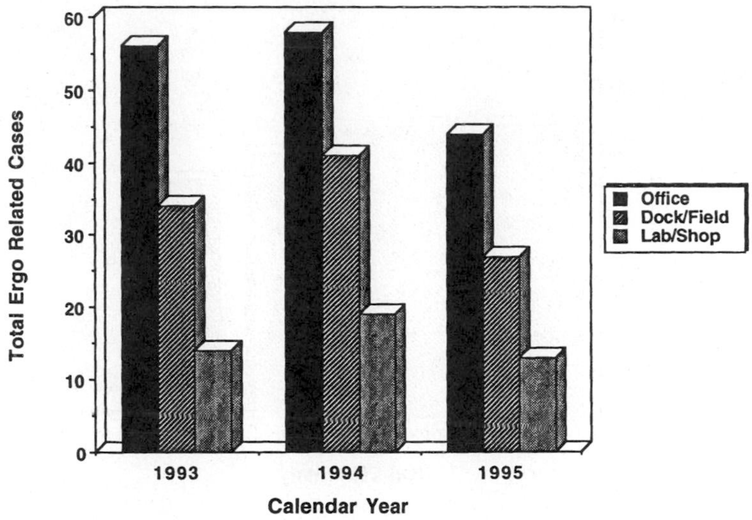

FIGURE 82.9 Number of ergonomics-related cases by job location.

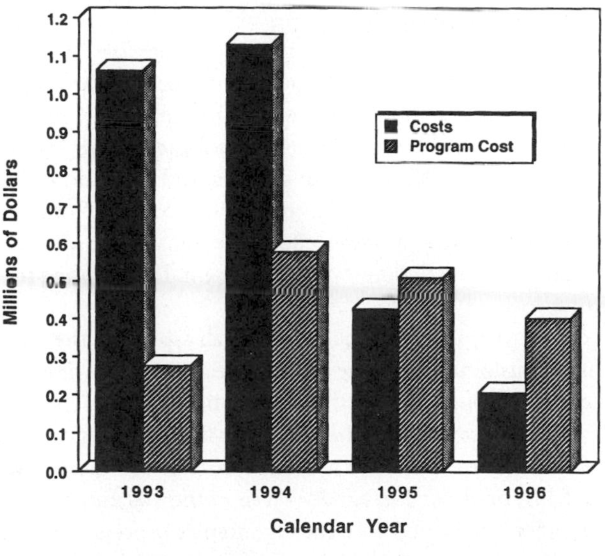

FIGURE 82.10 Comparison of illness costs and program costs from 1993 to 1996.

around to an even more rapidly decreasing trend. Over the four years analyzed, the money saved was equivalent to the money spent. However, if the costs associated with work-related musculoskeletal disorders remained at the 1996 level, cost savings would amount to about $800k per year, using 1993 as the reference point. At the current funding level of $120k, this would amount to over a 6.7:1 return on investment. We also discovered in our quality surveys that half of our customers perceived their productivity to increase after having implemented ergonomic improvements. Subjective estimates ranged from 5 to 15%. Assuming 135 people, paid $40,000 in salary increased their productivity 7%, the company would have gained another $378k per year in increased or improved work. With estimates of increased productivity figured in, the return on investment increases to just under 10:1.

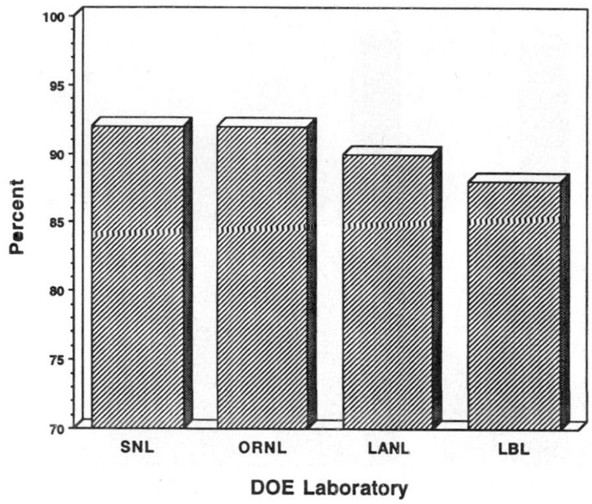

FIGURE 82.11 Percent overall customer satisfaction with ergonomics services at DOE National Laboratories benchmarked in 1995.

82.7 Quality Metrics and Awards

In addition to objective data on services performed and case reduction, the program initiated telephone surveys to evaluate the customers' perceptions of the quality of our work. Our first survey, done in 1992, consisted of calling 30 people who had received benefits from the old system in 1991 and 1992. Seven questions surveyed symptom relief, satisfaction with services, and duration of waiting period for ergonomic task chairs. We discovered that on the whole, customers recovered from symptoms well, and were satisfied with the individualized services; however, they were dissatisfied with the long waits associated with chair selection and delivery. These findings were instrumental in planning the program and creating an efficient chair-fitting process.

In 1994, we redesigned the survey to include questions on all aspects of the program and administered it via telephone to 100 of our customers from the previous year. The following results are extracted from the resulting Sandia Technical Report (Longbotham and Miller, 1995). The 1994 survey respondents recorded 81% "pleased" or "very pleased" with the new chair-fitting process. Most of the remaining 19% were not happy with the WSE requirement for a new chair, so the requirement was dropped. The return rate on chairs was 6 out of 331, or about 2%. Ninety-seven of the 100 customers thought the WSE went smoothly, and overall customer satisfaction with the ergonomics processes increased from 87% in 1992 to 92% in the 1994 survey (Figure 82.5). Corroborating evidence from the 1994 Medical Center Quality Survey indicated that 156 respondents rated the program 3.97 out of 5 for responsiveness and 4.13 out of 5 for ergonomics professionalism (Sanderville, 1994). Comparable satisfaction metrics from sister DOE laboratories, benchmarked in 1995, are shown in Figure 82.11.

Metrics on training effectiveness and satisfaction were collected immediately after training sessions. Manager satisfaction with ergonomics training was high, with 100% reporting that notebook materials would be useful back on the job and that the instructor communicated well. Nine out of ten thought the instructor was prepared, knowledgeable, enthusiastic, encouraged interaction, and treated students with respect.

In 1995, the program won the Sandia President's Quality Award. The award was based on a subset of criteria taken from the Malcolm Baldridge Award application process. Also in 1995, the videotape entitled

"Ergonomics Detective C.T. Dodd," won The Communicator Crystal Award for its effectiveness in communication and education. Lockheed-Martin identified the program as one of only three Industrial Hygiene/Safety programs worthy of receiving an "excellent" rating in its 1995 audit. Last, in an external appraisal conducted by the DOE in 1994, the program received "noteworthy recognition." It was very satisfying for the entire CEG to experience good work being recognized, but the rewards were in knowing that pain was reduced, comfort enhanced, and productivity increased during this period.

82.8 Community Outreach

Recent reductions in budget necessitated fewer formally trained people working on the program and more leveraging the knowledge and techniques to other ES&H staff. The author has moved on to his next challenge in the company and is no longer managing the program. However, he remains involved in transferring the technologies to other companies, small and large. He is currently the Energy and Environment Sector Representative in Lockheed-Martin Corporation's Ergonomics Task Force and is pursuing the diffusion of ergonomics technologies to other companies within the corporation. He is also pioneering the application of ergonomics principles to architecture. The checklists, course materials, and training video described above are available for licensing by contacting Sandia's Licensing and Partnership Department. Please feel free to contact the author at (505) 845-9803 or dpmille@sandia.gov with questions regarding the program or the techniques developed.

Acknowledgments

The author would like to acknowledge and express gratitude to the CEG members who unselfishly gave of themselves to make the program a success. Some of them put helping Sandians find relief above doing what was politically correct for their careers. Special thanks also go to the Sandians at the California Laboratory who showed us the way by spearheading several major efforts. Our project would never have gotten off the ground without management support from Judith Mead, Larry Clevenger, Allan Fine, Joe Stiegler, Bill Burnett, and Lynn Jones. Thanks also go to those who helped us start up by letting us benchmark their operations back in 1992, or attended Energy Facility Contractors' Group meetings from 1994–1996 to share their knowledge and materials. In many cases it was their innovative ideas that helped to make our program a success. Those who directly contributed to the program's development are listed below:

Eric Grose
Deborah Mulligan
Hugh Whitehurst
Margarita Ferguson
Joey Boyce
Nancy Spear
Larry Powell
David O'Brien
Patrick Girault
Patricia DeVivi
Charmaine DeWerff
Don Bridgers
Carol Meincke
Mary Gould
Sal Sabasco
Terry McMahon
Nita Archambault
Linda Edlund
Sam Walters
Kevin Babb
Sandra Hansen
David Kessel
Tom Faturos
Patsy John
Claudia Hawkes
William Ormond
Deana Butler
Gloria Holtzclaw
Larry Suzuki
Nina Stewart-Poppelsdorf
Gordon McKeen
Lori Longbotham

The author dedicates this chapter to Allan Fine, who will be remembered for his trusting support, sincere compassion, and timely humor.

References

Corporate Ergonomics Group, C. T. *Dodd, Ergonomics Detective,* video, Sandia National Laboratories, Albuquerque NM, 1995.

Leger, E., Ackerman, R., Coleman, R., and MacDorman, J., *Process Quality Management and Improvement Guidelines,* AT&T Bell Laboratories, Indianapolis, 1989.

Longbotham, L. and D. P. Miller, *1994 Ergonomics Program Quality Evaluation,* SAND95-0937, Sandia National Laboratories, Albuquerque NM, June 1995.

McKeen, R. G. and D. P. Miller, *Benchmarking Corporate Ergonomics Programs for Sandia National Laboratories,* SAND92-2667, Sandia National Laboratories, Albuquerque NM, April 1993.

McMahon, T. and Miller, D. P., *Office Ergonomics,* Chap. 3A, Sandia ES&H Manual, Rept. MN471001, Sandia National Laboratories, Albuquerque, NM, 1995.

Mulligan, D. R., Stewart-Poppelsdorf, N., and Suzuki, L., *Ergonomics for Computer Users,* pamphlet, Sandia National Laboratories, Albuquerque, NM, 1994.

Sanderville, K., *Quality Survey on Medical Services,* Sandia National Laboratories, Albuquerque NM, 1994.

Stewart-Poppelsdorf, N., *Ergonomic Worksite Evaluation,* Office Checklist, Sandia National Laboratories, December, 1994.

Suzuki, L., Mulligan, D. R., and Hansen, S., *Ergonomics Exercises at Work,* booklet, Sandia National Laboratories, Albuquerque, NM, 1995.

Suzuki, L., Mulligan, D. R., and Hansen, S., *Ergonomics Exercises at Home,* booklet, Sandia National Laboratories, Albuquerque, NM, 1995.

83

Economic Analysis for Ergonomics Programs

Carolyn M. Sommerich
North Carolina State University

83.1 Introduction

Good ergonomics often has a positive economic effect. Hendrick (1996) recounted several examples of ergonomic projects that each resulted in significant economic benefits that well outweighed initial project costs. However, unless an assessment method is in place to document both the costs and savings (or income) associated with ergonomics-related activities, only the costs may be readily apparent. Ergonomics then appears as just one more expense burden. The benefits of ergonomics, such as increased productivity and reduced expenses, must be objectively documented, just as costs are, in order to change this view of ergonomics. This chapter presents a method for continuous economic assessment of an ergonomics program.

The method presented in this chapter facilitates a comprehensive assessment of a program over time, and assessment of individual constituent projects, from proposal, to implementation, and throughout each project's life. A two-tier recording system, for tracking costs and benefits for individual projects and the program as a whole, forms the basis of the method. The chapter is divided into three main sections. The first section contains explanations of program-level cost and benefit line items and provides equations for calculations. Methods for evaluating individual projects, including capital expenditures, small projects, and light duty assignments, are also presented in that section. The next section of the chapter

0-8493-2641-9/99/$0.00+$.50

contains sample calculations that demonstrate application of the material in the first section. Alternative evaluation methods are briefly discussed in the third section of the chapter (such as absenteeism rates and machine availability).

83.2 Fundamentals: Identifying Costs and Benefits of Ergonomics Activities

There are various recommended metrics by which ergonomic programs or individual projects can be evaluated. Though there is a natural aversion to reducing health and safety issues to financial terms, money is the metric of business. Increasingly, the impetus is toward evaluation of health and safety issues, including ergonomics-related activities, in monetary terms. To date, relatively little has been written on the subject of economic analysis of ergonomics-related activities. The accounts that do appear in the literature, however, describe positive economic effects from ergonomics-related activities (Helander and Burri, 1995; Hendrick, 1996; DeKraker, Lindstrom-Hazel, Cooper, and Ambrosius, 1995).

Andersson (1992a) suggested focusing on savings associated with increased productivity and decreases in absenteeism, rework and spoilage, and turnover. However, for a fair and complete analysis, that can address the question, "At what cost did we achieve _______________ (outcome), and was that the best investment of company resources?" savings must be assessed relative to associated costs. Alexander (1996) recommended changing metrics as a program matures, from activity-oriented metrics (for example: program implementation milestone completions, number of projects completed successfully) at inception, to results-oriented metrics (for example: changes in health and safety statistics), and finally to systems-oriented metrics that correspond with institutionalizing the program's practices. In accordance with these recommendations, this chapter presents a systems-oriented method for predicting, tracking, and assessing costs and savings associated with an ergonomics program and its various constituent projects.

The method presented is reasonably complete, in that it accounts for a wide range of costs and benefits associated with a program (refer to Figure 83.1). A Program Record (PR, Table 83.1) is used for program-level analysis, and an Ergonomic Project Worksheet (EPW, Table 83.2) is used to assess individual proposals and projects. Figure 83.2 depicts the flow of information within and between the forms. The forms can easily be adapted to a computer spreadsheet or database. However, the validity of the output is ultimately dependent upon the quality of the monetary estimates and other values supplied by the user. Potential sources have been provided to help the reader identify various costs or savings, or constituent values. Certain values may already be standardized within some companies. Where values are not available, they should be developed in order to perform fair and comprehensive analyses. Depending on the scope or scale of a particular program or project, some line items that appear on the forms may not require entries. Only those items that are affected by and change as a result of the ergonomics program need to be tracked. In some parts of the predictive evaluation, providing estimates of expected changes from the current situation may suffice (Konz, 1995).

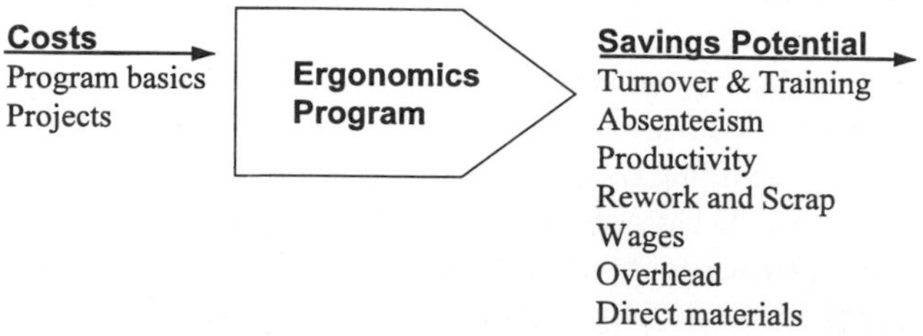

FIGURE 83.1 Categorical model of costs, savings, and income-generating areas to consider in assessing the economic effects of an ergonomics program.

TABLE 83.1 Program Record

	Costs	Benefits
A. Ergonomics Program — regular activities		
1. Meetings		
a. Lost production	______	______
b. Other ______________	______	______
2. Training		
a. Lost production	______	______
b. Trainer/consultant costs	______	______
c. Materials	______	______
d. Other ______________	______	______
3. Medical Management		
a. PT/OT, nurse (specific to program)	______	______
b. Lost production		
i. Consultation	______	______
ii. Treatment	______	______
c. Back belts, wrist supports	______	______
d. Other ______________	______	______
B. Turnover and Training/Replacement		
1. Acquisition		
a. Recruitment	______	______
b. Selection	______	______
c. Hiring		______
2. Development		
a. Orientation	______	______
b. On-the-job training		
i. Lost production due to lack of proficiency	______	______
ii. Overtime to compensate for lost production	______	______
c. Off-the-job training	______	______
3. Separation		
a. Severance	______	______
b. Productivity decrement	______	______
c. Open position	______	______
C. Absenteeism		
1. Compensation costs		
a. Wages (including taxes, fringe)	______	______
b. Insurance charges or fees	______	______
2. Medical costs		
a. Payments to providers	______	______
b. Insurance charges or fees	______	______
3. Replacement costs (refer to Turnover and Training)	______	______
D. Productivity		
1. Permanent changes	______	______
2. Temporary changes	______	______
E. Rework and Scrapped Product		
1. Rework		
a. Direct labor	______	______
b. Lost production or Overtime	______	______
c. Direct materials	______	______
d. Other ______________	______	______
2. Scrapped product		
a. Direct labor lost	______	______
b. Overtime	______	______
c. Direct material loss, less salvage value	______	______
d. Other ______________	______	______

TABLE 83.1 (continued) Program Record

	Costs	Benefits
F. Wages		
1. Direct labor, including benefits, taxes, insurance	______	______
2. Supervisory and administration charges	______	______
G Overhead		
Items: ____________	______	______
H. Direct materials (not listed elsewhere)	______	______
I. One-time costs (such as initial project costs)		
1. Equipment	______	______
2. Jigs and fixtures	______	______
3. Installation	______	______
4. Engineering time	______	______
5. Operator training	______	______
6. Other	______	______

TABLE 83.2 Ergonomic Project Worksheet

Step 1. Perform economic analysis on a proposal using payback period or one of the four methods that account for the time value of money (NPW, NAW, B/C, or ROI). Use the first set of columns on the worksheet to identify and document anticipated costs and benefits.

Step 2. Track all project costs, savings, and income throughout the life of the project. Use the second set of columns, which will facilitate comparison between projected and actual values. This will prove useful for future evaluations.

Step 3. Transfer all actual costs and benefits onto the Program Record, in order to provide information that will facilitate the evaluation of the program as a whole.

Item*	Expected Costs	Expected Benefits	Actual Costs	Actual Benefits
B. Turnover and Training				
C. Absenteeism.				
....				
I. Project costs				

Perform economic decision analysis for proposal in the area below:

State all assumptions:

Calculations:

Outcome:

Decision:

* This table is presented in an abbreviated format. It should contain items B-I that appear on the Program Record (Table 83.1).

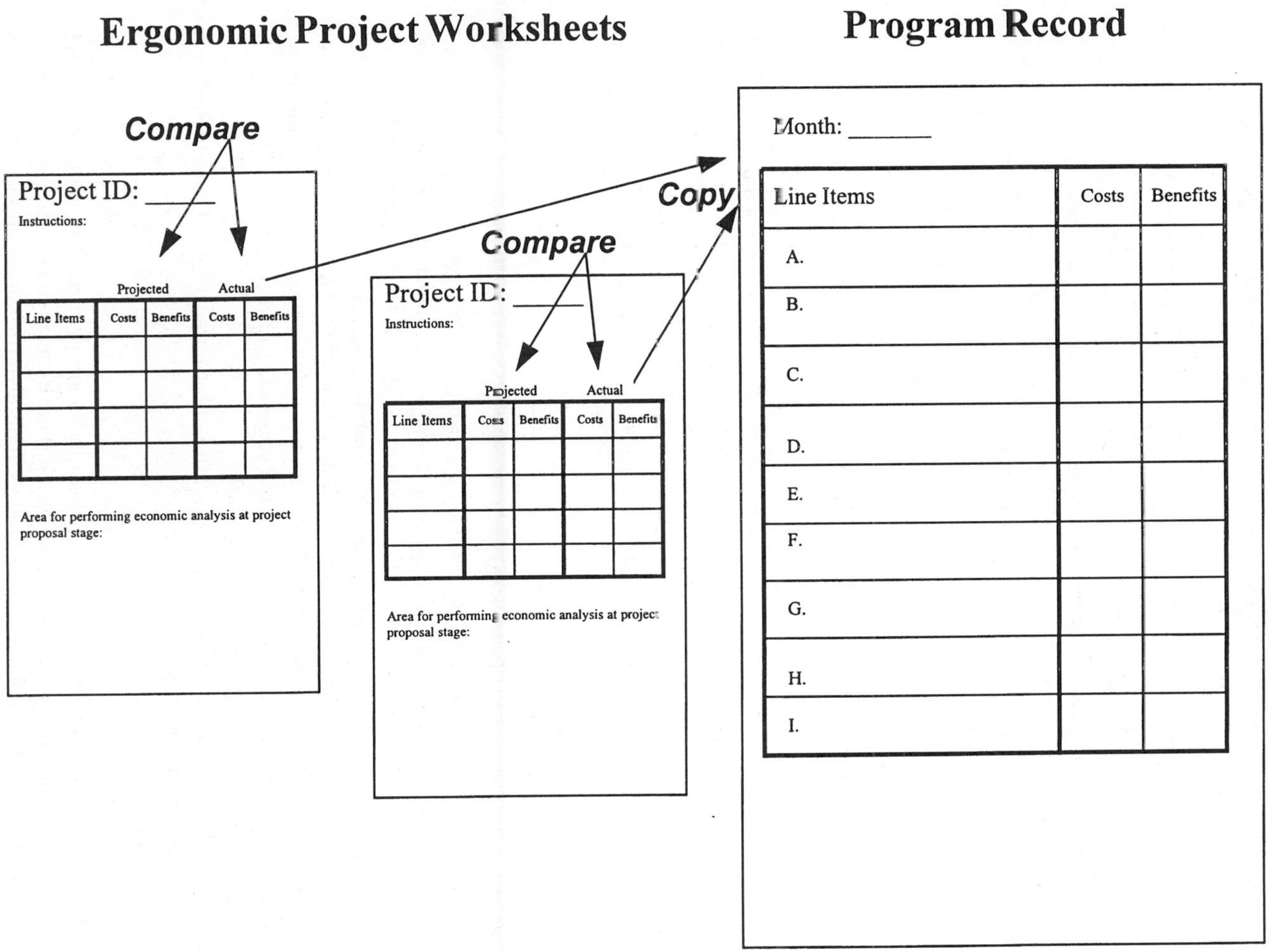

FIGURE 83.2 Information flows from individual Ergonomic Project Worksheets to the Program Record.

Program Record

The costs and savings included in this analysis method are primarily those discussed in Konz (1995), Oxenburgh (1997), Andersson (1992a), and Andersson (1992b), although modifications have been made to some of their methods and cost definitions. Additionally, items specifically associated with the ergonomics program and team have been added. Refer to Table 83.1 when proceeding through the discussion that follows.

A. Ergonomics program — regular charges. These are costs associated with the maintenance and day-to-day functioning of the program.

1. *Meetings.* The ergonomics team meets on a regular basis. While in these meetings, members are not working on production, therefore, lost production costs are associated with the meetings. Overhead charges may be assessed as well, for indirect labor for nonproduction employee members. Line items and suggested calculation method:
   a. Lost production

$$C_{LP} = W_h \times T_h \tag{1}$$

   where
   C_{LP} = Cost of lost production, \$
   W_h = Hourly wage rate*, \$/hr
   T_h = Time (length of meeting), hr

   b. Other

2. *Training.* Ergonomics training, specific to an employee's responsibilities, is a fundamental element of an ergonomics program. Lost production, consultant fees, and materials are all costs associated with training. The costs of training for indirect labor (such as engineers and managers) should also be accounted for. Line items and suggested calculation method:
   a. Lost production: calculated per Equation 1, above, substituting length of training time for meeting time.
   b. Trainer/consultant costs
   c. Materials: books, pamphlets, etc.
   d. Other

3. *Medical management.* Medical management is a key element to an ergonomics program. Any costs that can be directly attributed to this element, as well as any savings, should be accounted for. One example would be contracting with a physical therapist to work onsite for a half day each week to answer employee questions or assist with treatment regimens. Lost production associated with these visits should also be accounted for. If as a result of the ergonomics program, a facility eliminated the practice of providing back belts on demand to workers, the cost savings from that decision should be accounted for. Line items and suggested calculations:
   a. PT/OT, nurse (for effort specific to program)
   b. Lost production: calculated per Equation 1, above, substituting length of consultation or treatment session for meeting time.
      i. Consultation
      ii. Treatment
   c. Back belts, wrist supports purchases: If these items are purchased in conjunction with the program, the line item would appear as a cost. If these items are no longer purchased as a result of the program, the elimination of the average monthly expenditure on these items should be recorded as a savings.
   d. Other

*In all calculations where wages appear, include pay and fringe benefits. However, do not include overhead (burden). That is treated in item G. of the EPW.

B. Turnover and Training/Replacement. These are the costs associated with replacing workers due to turnover (Andersson, 1992a). Companies may have standardized these costs for various classifications of workers. Replacement costs, per Andersson (1992a), include costs of acquiring and developing new employees, as well as costs associated with separation of departing employees.

1. *Acquisition costs.* These are the costs associated with acquisition of new employees, and may include costs associated with personnel recruitment, selection, and/or hiring. Recruitment costs include costs such as advertising, agency fees, brochures, travel, meals, and administrative costs. Selection costs include interviewing, pre-employment testing, and administrative costs. Hiring costs might include costs such as medical screening and administrative costs. Line items:
   a. Recruitment
   b. Selection
   c. Hiring
2. *Development costs.* These are the costs associated with bringing an employee "up to speed" on a job which an employee has left. The most common costs associated with development are associated with orientation, on-the-job-training, and off-the-job training. Orientation costs might include costs for materials and trainers. On-the-job training would result in a loss of productivity due to lack of expertise of the new employee and possibly the need for overtime in order to compensate for lost production. Off-the-job training costs might include materials, costs for trainers, and loss of productivity while in training sessions. Line items and suggested calculations:
   a. Orientation
   b. On-the-job training
      i. Lost production due to lack of proficiency

$$C_{LC} = W_d \times (1 - \text{eff}) \times T_d \tag{2}$$

where
C_{LC} = Lost production due to learning curve effects, \$
W_d = Daily wage rate (includes benefits), \$/day
eff = New employee performance, relative to seasoned (range 0–1)
T_d = Time (learning period in days), days

      ii. Overtime to compensate for lost production

$$C_{OT} = W_{Eh} \times T_h \tag{3}$$

where
C_{OT} = Cost of overtime, \$
W_{Eh} = Hourly excess wage rate for overtime, \$/hr
T_h = Time (number of overtime hours), hr

   c. Off-the-job training (formal instruction away from the workstation)
      i. Trainer and material costs
      ii. Lost production: Calculate per Equation 1
      iii. Overtime to compensate for lost production: Calculate per Equation 3
3. *Separation costs.* Typical costs associated with the departure of an employee include severance costs, reduced productivity, and open position costs. Severance costs include compensation, settlement, or benefits paid to employee upon departure from company. Productivity may be reduced prior to employee's departure (from employee and from co-workers). Open position costs might include overtime for other employees to make up for the vacant position, or lost productivity if lost productivity results in income loss. Line items and suggested calculations:

a. Severance
b. Productivity decrement (similar to calculation in Equation 2)
c. Open position
 i. Lost production

$$C_{LPO} = W_d \times T_h \tag{4}$$

where
C_{LPO} = Cost of lost production due to open position, $
W_d = Daily wage rate*, $/day
T_d = Time (number of day position remains open), day

 ii. Overtime to compensate for lost production: Calculate per Equation 3.

C. Absenteeism. Costs associated with work absences that are both work-related and associated with ergonomic hazards of the job are referred to as absenteeism costs (Andersson, 1992a). They include workers' compensation costs, medical expenses, and replacement costs (Andersson, 1992a). Note that sick days (nonrecordable lost time days) may also be reflective of poor ergonomic conditions, and, therefore, might be expected to respond to ergonomic modifications. However, rather than making this assumption, sick absence data should be tracked before and following modifications, in order to determine: (1) if this is the case, and (2) what standardized value might be used in economic analyses of proposed capital expenditures in the future.

1. *Workers' compensation costs*: Monies paid to ill or injured employees during their absence from work are accounted for here. These may be represented by or substituted with insurance fees. Also include payroll taxes, benefits, etc. Line items:
   a. Wages (including taxes, fringe)
   b. Insurance charges or fees

2. *Medical expenses*: These are expenses associated with treatment for illness or injury, and may also be substituted with insurance fees. Line items:
   a. Payments to providers
   b. Insurance charges or fees

3. *Replacement costs*: Review costs associated with turnover (Category B, described above) for costs associated with loss of production, overtime for other employees, hiring of replacement or temporary employee, etc. Line items: Refer to Turnover and Training (Category B, above)

D. Productivity. Productivity may be adversely impacted by poor work conditions, or positively impacted by ergonomic controls. Equipment that requires frequent maintenance or adjustment may reduce productivity when it is running (suboptimally) and when down for maintenance or adjustment. Productivity changes can be estimated from predetermined time systems, historical records, experience, or from pilot time studies. Actual changes can be determined from time study, work sampling methods, or production records. In calculating productivity savings, actual production time requirements should be used in assessing productivity changes, rather than standard times unless production rate is equivalent to the standard. Time standard allowances are one place to look for improvements in productivity. Often allowances are built in for less than optimal working conditions (such as poor lighting or strenuous, fatiguing work). Modifications that target allowances are likely to improve productivity. Line items and suggested calculation method:

*In all calculations where wages appear, include pay and fringe benefits. However, do not include overhead (burden). That is treated in item G. of the EPW.

1. *Permanent changes in productivity*

$$S_U = W_h \times \Delta_{eff} \times T_{RU} \tag{5}$$

where

S_U = Savings, \$/production unit

W_h = Hourly wage rate, \$/hr

Δ_{eff} = Improvement in production time (percentage time reduction from reference time)

T_{RU} = Reference time/unit

2. *Temporary changes in productivity:*
   a. Lost production time for maintenance, repair, or adjustment. Refer to Equation 1.
   b. Overtime to compensate for down time. Refer to Equation 3.

E. Rework and Scrapped Product. These are costs associated with products that must be reworked, reduced to salvage, or written off (Andersson, 1992a). Per piece costs and rates are required to perform the calculations.

1. *Rework.* Direct labor and material costs, and costs associated with making up for lost productivity, such as overtime charges; extra costs associated with warranty claims may contribute to rework costs. An inventory fee may be charged if product is stored for a significant period of time before being reworked. Line items and suggested calculations:
   a. Direct labor

$$CR_{dl} = W_h \times R_{pu} \times T_{ru} \tag{6}$$

where

CR_{dl} = Cost of direct labor for rework, \$/production unit

W_h = Hourly wage rate, \$/hr

R_{pu} = Rework rate, units to rework/units of production

T_{ru} = Time for rework, hr/unit

   b. Lost production or overtime: Refer to Equations 1 or 3.
   c. Direct materials
   d. Other

2. *Scrapped goods.* Direct labor and material costs, less salvage value of materials; costs associated with making up for lost productivity, such as overtime charges, contribute to the costs of scrapped goods. Overhead charges may also be included. Line items:
   a. Direct labor cost
   b. Overtime
   c. Direct material loss less salvage value
   d. Other

F. Wages. Wage costs to the company include wages for productive work hours (no absences or vacation hours included), and additional charges associated with wages (taxes, benefits). Personnel and administrative charges are included in the methodology of Oxenburgh (1997). They are included as a separate line item in the current method, per recommendation of Konz (1995), for the following reasons: (1) it is unlikely that these charges will change as a result of ergonomic changes, and (2) savings may be unfairly inflated by working with a larger labor cost. Line items:

1. Direct labor, including benefits, taxes, insurance, etc.
2. Supervisory and administration charges

G. Overhead. Included in overhead that might be likely to be affected by ergonomics-related changes would be items such as utility costs, insurance fees, and OSHA fines. Activity-based costing is promoted as a method for allocating overhead costs (Riel and Imbeau, 1995).

H. Direct materials. Any changes in direct material costs not included elsewhere would appear here.

I. One-Time Costs. These costs are typically incurred in conjunction with capital expenditures (investments), and include costs for equipment, jigs and fixtures, installation, engineering time, and operator training costs (refer to items B.2.b and c for costs associated with training). Capital expenditures are discussed in more detail in the section on project calculations. Line items:

1. Equipment
2. Jigs and fixtures
3. Installation
4. Engineering time
5. Operator training
6. Other

Supporting Data

In order to estimate or calculate costs and savings for projects, information may be required on several of the following items (Andersson, 1992b):

- Total hours, and hours categorized as productive vs. nonproductive (absences, vacation, training, illness, injury): for machine(s), process(es), individual operator(s)
- Wage rates
- Direct and indirect labor hours for a product or operation
- Labor hours and costs associated with specific absences, by illness or injury, by department, by job, by time period
- Turnover by department or job, by time period
- Accounting of rework and scrapped material, at particular stages of production
- Time standards

Data Sources. Simpson and Mason (1995) provided suggestions for locating these supporting data. When values are not available in house, they either need to be calculated from available data, or, if applicable, industry standard values might be used.

- *Personnel department*: Costs and/or rates of absenteeism, turnover, training, workers' compensation.
- *Industrial or process engineering*: Allowances, time standards (and assistance with time standard estimates for proposals), production variation, rejection rates, inspection costs, excessive downtime, equipment maintenance costs and time requirements, scrap and rework rates and costs. Simpson and Mason (1995) provided a 30% of ownership cost rule-of-thumb estimate for machine maintenance.
- *Medical department*: Nature and severity of illnesses and injuries, nature and length of health-related absences.

Additionally, Simpson and Mason (1995) identified other resources within a facility upon which the ergonomics team may draw information or support:

- *Safety department*: To help with locating areas within the plant with ergonomic hazards.
- *Medical department*: Type and frequency of minor injuries, type and frequency of reported musculoskeletal-related symptoms.

Simpson and Mason (1995) also suggested execution of an ergonomic survey, to obtain information of job satisfaction, risk-taking behaviors, and near misses, any or all of which may point to areas in the facility in need of the attention of the ergonomics team.

Sources of data and calculations used in the analysis should be clearly documented. This facilitates future analyses, keeps analyses on common ground, and facilitates retracing steps if questions arise, especially if proposal estimates and actual project costs or savings do not match. Riel and Imbeau (1995) recommended development of a three-component support system that would provide decision support to personnel involved in ergonomic investment decision-making. The system included a model of costs (which costs to include; how are they analyzed; how they behave), a safety information system (a database of injury and risk factor information), and an appropriate user interface (once the system is established, computerize it in order to facilitate its use).

Project Evaluations

Ergonomics team activity is often project based. Each proposed project should be evaluated before being undertaken, not only in terms of ergonomic benefits, but in economic terms as well. Methods for performing economic analyses on small projects, large projects, and light duty assignments are examined in this section. Simpson and Mason (1995) cautioned analysts to be conservative in their estimations of benefits from proposed projects. They found it preferable to underestimate, rather than overestimate, in order to reduce the likelihood of implementing projects that turn out to be poor investments of company resources.

Small Projects. Small projects may be evaluated by using payback period analysis as an accept/reject criteria, recognizing that this does not account for the time value of money. Payback period is simply the length of time required for benefits for equal initial project costs, without consideration of interest. If savings are equivalent in each time period, then Equation 7 can be used to calculate payback period.

$$PB = C/S \tag{7}$$

where
PB = payback period, in months, years, etc., corresponding to S
C = one-time project cost, $
S = periodic savings (or income), per month, year, etc.

Capital Expenditures. Projects that require significant monies to be spent at the beginning, with the expectation of benefits (savings, income, or both) in the future are referred to as capital expenditures. There are several ways to evaluate whether or not a proposed project would be a good investment of a company's funds. The first assessment is simply a go–no go decision, in which estimated costs are compared with estimated future benefits (income or savings). If costs exceed benefits, then from an economic standpoint, the project would not be a sound investment of the firm's money. If the benefits at least equal the costs and the company has limited funds, then the project should also be compared with other projects in which the company could invest its resources. If the company cannot undertake all the acceptable projects, then, according to economic theory, the costliest project that will provide the acceptable *incremental* return, among all acceptable projects, should be chosen.

In performing capital expenditure evaluations, the *time value of money* (interest rates and acceptable rate of return for the company) should be considered for projects that have extended lives. That is where benefits may not be realized immediately, where benefits may be realized for an extended period of time, or where there are recurring costs associated with a project (for example: periodic replacement of parts, or maintenance costs).

There are four popular methods for performing economic analysis of capital projects: Net Present Worth determination (NPW), Net Annual Worth determination (NAW), Benefit/Cost Ratio (B/C), and

TABLE 83.3 Economic Analysis Methods

Method	Calculation	Condition	Decision Criteria
NPW	$NPW^* = \frac{A[(1+i)^n - 1]}{i(1+i)^n} - C$	Independent alternatives	Select those with NPW > $ 0
		Dependent alternatives	Select alternative with maximum NPW
NAW	$NAW = A - \frac{Ci(1+i)^n}{[(1+i)^n - 1]}$	Independent alternatives	Select those with NAW > $ 0
		Dependent alternatives	Select alternative with maximum NAW
B/C	$\frac{NPWB}{NPWC}$ or $\frac{NAWB}{NAWC}$	Independent alternatives**	Select those with B/C ≥ 1
ROI	***	Independent alternatives**	Select those with ROI ≥ MARR

where

A = savings or income that remains the same each year
i = interest rate (company's MARR), annual, in decimal format (ex for 20% use 0.20)
n = project life, years
C = project cost, assumed to be single cash outlay at beginning of project
NPWB = Net Present Worth of Benefits (savings or income) = NPW + C
NPWC = Net Present Worth of Costs = C
NAWB = Net Annual Worth of Benefits = A
NAWC = Net Annual Worth of Costs = A — NAW

* If savings or income is not projected to be the same in each year calculate NPW as follows:

$NPW^* = \Sigma (S_t/(1 + i)^t) - C$, where the summation extends over the life of the project. S_t is savings or income for any particular year t of the project.

** For the case of dependent alternatives, refer to one of the books mentioned in the For Further Information section of this chapter.

*** ROI is determined through the use of any of the following equations. ROI is that i that satisfies the equations, and is determined through trial and error.

NPWB – NPWC = 0
NAWB – NAWC = 0
NAWB/NAWC = 1
NPW = 0

Return-on-Investment (ROI). NPW and NAW are generally easier methods to apply if a set of mutually exclusive alternatives is assessed, because pairwise comparisons among the alternatives occurs automatically. Both methods combine all costs, income, and savings for a proposal into a single monetary value that is dependent upon the company's (or analyst's) minimum acceptable rate of return (MARR). The largest positive NPW or NAW value identifies the best alternative. B/C and ROI methods are easily applied in situations where independent alternatives are assessed (where selecting one proposal does not influence the opportunity to select any other), but involve multistep analysis processes when mutually exclusive proposals are evaluated. B/C ratio and ROI provide intuitively appealing outputs and criteria. For analysis of independent projects, any project with a B/C that meets or exceeds 1 is accepted. An interest or investment rate is the outcome of ROI analysis. If the rate is acceptable, the proposal should be undertaken. Equations for performing each of these analyses are presented in Table 83.3, along with decision criteria and assumptions regarding application.

Andersson (1992a) recommended checking the stability of the outcome of any economic analysis by performing sensitivity analyses. That is, determining how the outcome of the analysis is affected by changes in the estimated values used in the original calculation. This will provide a view of the stability of the projected outcome of the project.

Light Duty Evaluation. The following calculation can assist in the economic assessment of a light duty program, or in the decision to bring an individual employee back to work on light duty. This assessment is based, in part, on calculations in Ritzel and Allen (1988).

$$C_{LD} = \{[P_w \times V - (P_w - P_a)] \times D\} + C_m \quad (8)$$

where

C_{LD} = Cost of light duty assignment, $

P_w = Regular wage rate for employee, including benefits and taxes, $/day

V = Valuc to light duty work, relative to regular work, percentage

P_a = Payments to worker when on sick absence*, $/day

D = Number of days on light duty

C_m = Costs of modifications to workplace, $

Medical costs should also be accounted for, including rehabilitation costs. A review of Table 83.1 and company policies regarding early return to work may identify other line items that should be included in this evaluation.

83.3 Method Application: Sample Calculations for Evaluating and Tracking Costs and Benefits of Proposals, Projects, and Programs

A series of examples is presented to illustrate the application of the equations presented in the previous section. The reader is reminded that the primary purpose for the chapter is to present a model of cost categories, and that specific equations are only suggested methods for obtaining line item values for the Program Record and Ergonomic Project Worksheets. Companies may have established methods for estimating many of these line items. Additionally, the references that are cited throughout the chapter may provide some alternative calculation methods that are better suited for a particular company.

Example 1: A. Ergonomics Program — Ergonomics Team Meetings

Company XYZ's ergonomics team meets once a week for 1 hour. The team consists of three hourly operators (at $10/hr), one engineer, the plant nurse, the plant manager, the human resources manager, and one first line supervisor. Accounting, the plant manager, and the HR manager have decided that a charge of $25/hr per nondirect team member is appropriate. The cost of each meeting would then be calculated as follows, using Equation 1:

$$C_{LP} = W_h \times T_h$$

$$\begin{aligned} \text{Meeting cost} &= C_{\text{LP-direct labor}} + C_{\text{LP-indirect labor}} \\ &= (\$10/\text{hr} \times 1\ \text{hr/meeting} \times 4\ \text{meetings/month} \times 3\ \text{operators}) + \\ &\quad (\$25/\text{hr} \times 1\ \text{hr/meeting} \times 4\ \text{meetings/month} \times 5\ \text{members}) \\ &= \$620/\text{month} \end{aligned}$$

*Note that this cost could be workers' compensation, benefits, and taxes paid directly by the company, or this cost could be a portion of the insurance premium the company pays out that can be fairly attributed to the employee's absence.

Example 2: B. Turnover and Training — On-the-job training

Operation E has an annual turnover rate of three employees. The hourly wage cost to the company is $16/hr (includes benefits, taxes); employees work 8 h per day. It takes about 90 days for an employee to become fully productive on this job. Over the 90 days, a new employee typically performs at about 70% of the efficiency of a seasoned employee. The annual cost of this turnover would be calculated as follows per Equation 2:

$$C_{LC} = W_d \times (1 - \text{eff}) \times T_d$$

$$\text{Annual cost/replaced employee} = (\$16/\text{hr} \times 8\,\text{hr/day}) \times (1 - .7) \times 90\ \text{days}$$

$$= \$3456/\text{employee/year}$$

Therefore, the annual cost to replace three employees would be $10,368 (= $3456 × 3).

If turnover were reduced to one employee per year, the savings attributed to the ergonomics program would be calculated as follows:

$$\text{Savings} = \text{Cost}_{\text{after change}} - \text{Cost}_{\text{before change}}$$

$$= -\$3456 - -\$10,368 = \$6912/\text{year}$$

Example 3. B. Turnover and Training — Overtime to compensate for lost production

Management estimates that Operation K runs ten Saturdays each year to compensate for reduced production associated with training new employees. Hourly wage is $18/hr; overtime is time and a half. Two operators are required. Use Equation 3 to calculate overtime costs.

$$C_{OT} = W_{Eh} \times T_h$$

$$\text{Cost} = \$9/\text{hr} \times 8\ \text{hr/day} \times 10\ \text{days} \times 2\ \text{employees}$$

$$= \$1440/\text{year}$$

Assume employees were able to learn the job more quickly, as a result of changes in training methods or production methods, for example. If new employees were able to learn the job more quickly, and which resulted in a reduction of overtime to only four Saturdays per year, the following recurring savings could be attributed to the ergonomics program:

$$\text{Savings} = \text{Cost}_{\text{after change}} - \text{Cost}_{\text{before change}}$$

$$= -\$576 - -\$1440 = \$864/\text{year}$$

Example 4: B. Turnover and Training — Costs due to an open position

Part a: Lost production due to an open position. Operation G loses 3 days worth of production for each turnover due to a temporary vacancy. This production is not made-up through overtime. The turnover rate is 4 employees/year. The hourly wage cost to the company is $11/hr (includes benefits, taxes); employees work 8 h per day. Lost production could be calculated using Equation 4.

$$C_{LPO} = W_d \times T_h$$

$$\text{Cost of lost production/turnover} = (\$11/\text{hr} \times 8\,\text{hr/day}) \times 3\text{ days}$$
$$= \$264/\text{turnover}$$

Therefore, the total cost of lost production for turnover rate of four per year would be \$1056 (= \$264 × 4). If turnover were reduced to 2 employees per year, the savings attributed to the program would be calculated as follows:

$$\text{Savings} = \text{Cost}_{\text{after change}} - \text{Cost}_{\text{before change}}$$
$$= -\$528 - -\$1056 = \$528/\text{year}$$

Part b. Overtime to compensate for lost production. If, instead, management opts to utilize overtime to make up for the lost production, costs and savings would be calculated as follows, based on the same turnover rate reduction from 4 per year to 2, an hourly wage cost of \$11/hr, and 3 days of overtime per turnover. Using Equation 3,

$$C_{OT} = W_{Eh} \times T_h$$

$$\text{Savings} = \text{Cost}_{\text{after change}} - \text{Cost}_{\text{before change}}$$
$$= (\text{Hourly excess overtime rate} \times \text{hours} \times \text{days} \times \text{number of turnovers})_{\text{after change}} -$$
$$(\text{Hourly excess overtime rate} \times \text{hours} \times \text{days} \times \text{number of turnovers})_{\text{before change}}$$
$$= (\$5.50/\text{hr} \times 8\text{ hr} \times 3\text{ days} \times 2\text{ turnovers}) -$$
$$(\$5.50/\text{hr} \times 8\text{ hr} \times 3\text{ days} \times 4\text{ turnovers})$$
$$= -\$264 - -\$528 = \$264/\text{year}$$

Example 5: C. Absenteeism

Department S has an average of 6 lost time back injuries per year. The company is self-insured, and so pays all medical and compensation to sick or injured employees. Average lost time for a back injury in this department is 10 days; average medical costs are \$5000. Absent workers are not replaced. Compensation costs are ⅔ the usual \$9/hr labor cost.

$$\text{Annual costs/back injury} = \text{Cost}_{\text{compensation}} + \text{Cost}_{\text{medical}} + \text{Cost}_{\text{lost production}}$$
$$= \$6/\text{hr} \times 8\text{ hr/day} \times 10\text{ days} +$$
$$\$5000 +$$
$$\$9/\text{hr} \times 8\text{ hr/day} \times 10\text{ days}$$
$$= \$480 + \$5000 + \$720 = \$6200/\text{back injury}$$

Therefore, total cost for six back injuries in a year would be \$37,200 (= \$6200 × 6). A proposed change in methods is expected to reduce the number of back injuries to 3 per year. The expected savings would be:

$$\text{Savings} = \text{Cost}_{\text{after change}} - \text{Cost}_{\text{before change}}$$

$$= -\$18,600 - -\$37,200 = \$18,600$$

Example 6: D. Productivity

Production was 80 units per 8 hours for Operation F (10 units/hour, or 0.1 hour/unit). A 13% reduction in per unit production time was realized following methods improvements. Direct labor costs are \$12/hr. Using Equation 5,

$$S_U = W_h \times \Delta_{eff} \times T_{RU}$$

$$\text{Savings per unit} = \$12/\text{hr} \times 13\% \times 0.1\,\text{hr/unit}$$

$$= \$0.156/\text{unit}$$

Annual savings would be based on the number of units produced.

Example 7. E. Rework and Scrapped Product — Rework — Direct labor costs/savings

Records show that on average 1 unit in 50 requires rework in Department C. The production rate is 800 units/day; labor costs are \$10/hr. On average rework time is 1 min/unit. Using Equation 6,

$$CR_{dl} = W_h \times R_{pu} \times T_{ru}$$

$$\text{Rework direct labor cost} = \text{labor cost} \times \text{rework rate} \times \text{time requirement}$$

$$= \$10/\text{hr} \times 1\ \text{unit}/50\ \text{production units} \times 1\ \text{min/unit} \times 1\ \text{hr}/60\ \text{min}$$

$$= \$0.0033/\text{production unit}$$

The annual rework cost for a production rate of 800/day, running 250 days per year would be \$667 (= \$0.0033 × 800 × 250). Note that if a production worker is pulled off production work to perform rework tasks, lost production should also be accounted for.

Example 8: E. Rework and Scrapped Product — Scrapped Product

Records show that on average 1 unit in 400 is scrapped in Department C. The production rate is 800 units/day; labor costs are \$10/hr; direct material costs are \$1.20/unit. Salvageable materials are valued at \$.30/unit

$$\text{Scrapped product cost/unit} = (\text{direct labor} + \text{direct materials} - \text{salvage value})$$

$$= \left[(\$10/\text{hr} \times 8\ \text{hr/day})/(800\ \text{units}) + \$1.20/\text{unit}\right] - \$0.30/\text{unit}$$

$$= \$1.00/\text{unit}$$

The annual scrapped cost for a production rate of 800 production units/day, running 250 days per year would be \$500 (=\$1.00 × 2 scrapped units/day × 250 days).

Example 9: Small Project

The purchase of tool belts for 10 furniture company employees is expected to reduce the need for employees to repeatedly walk between the line and their work benches for various hand tools they use in the clean-up and repair department. Tool belts cost \$12 each, and are expected to last one year. Productivity is expected to increase by 1%. Wage costs are \$15/hr to the company. Average time per unit is currently 20 min. Using Equation 7,

$$\text{PB} = \text{C/S}$$

$$\text{Project cost, C} = \$12/\text{belt} \times 10 \text{ belts} = \$120$$

$$\text{Savings/unit} = \$15/\text{hr/employee} \times 10 \text{ employees} \times 1\% \times 0.33 \text{ hr/unit} = \$0.495/\text{unit}$$

$$\text{Annual savings, S} = \$0.495/\text{unit} \times 24 \text{ units/day} \times 250 \text{ days/yr} = \$2970$$

$$\text{Payback period, PB} = \$120/\$2970 = 0.04 \text{ years, or about } 1/2 \text{ month.}$$

Example 10: Capital Expenditure

Scenario. BVB Company is evaluating the purchase of a lift table for a particular location in its plant. The rate of occurrence of back injury at that location averages one in three years. The average cost of a back injury is \$5200, per the company's insurance carrier and the human resources director (Category C. Absenteeism). The lift table will cost \$3200 (Category I. One-Time Costs), and is expected to last for 5 years, with proper maintenance (expected maintenance costs of \$500/year, charged under Category G. Overhead). The lift table is expected to improve productivity by 5% (current production rate is 40 units per hour, 7 hours per day, 250 days per year; wages for employees at that location are calculated at \$15/hr; appears under Category D. Productivity). To be conservative, these are the only benefits that are used to evaluate the purchase. If the company has established an MARR of 10%, should the lift table be purchased, based on economic projections?

Solution. Economic decision analysis always involves comparison of alternatives, even if there is apparently only one proposal, because Do Nothing, is always considered an alternative.

First consider the Do Nothing option. If back injury rate is one in three years, with an average cost of \$5200, the annual cost of a back injury can be assessed as \$5200/3, or \$1733/year. Use a cash flow diagram to depict this alternative (refer to Figure 83.3a).

Next, evaluate each of the applicable line item categories for the lift table alternative. The initial cost is \$3200, but maintenance costs of \$500/year are also anticipated. Productivity is anticipated to improve by 5%. Using Equation 5,

$$S_U = W_h \times \Delta_{eff} \times T_{RU}$$

$$\text{Savings per unit} = \$15/\text{hr} \times 5\% \times (1 \text{ hr}/40 \text{ units}) = \$0.0188/\text{unit}$$

$$\text{Annual savings} = \$0.0188/\text{unit} \times 280 \text{ units/day} \times 250 \text{ days/yr} = \$1316/\text{yr}$$

The costs and benefits associated with the lift table alternative are presented in Figure 83.3b. Figure 83.3c combines the two, and is the cash flow diagram that will be analyzed. Each of the four methods of analysis will be demonstrated. A project life of 5 years will be used, since that is the life of the table.

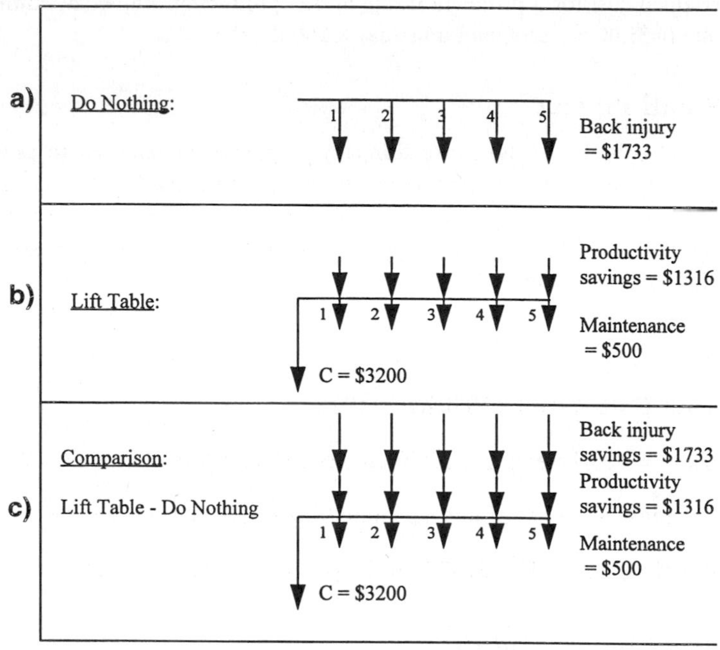

FIGURE 83.3 Cash flow diagrams for economic assessment of a) Do Nothing alternative, b) lift table purchase, and c) comparison of lift table purchase to Do Nothing alternative.

$$NPW = PW_{\text{savings in back injury costs}} + PW_{\text{productivity savings}} - PW_{\text{lift table cost}} - PW_{\text{maintenance costs}}$$

$$= 1733\frac{\left[(1+i)^n - 1\right]}{i(1+i)^n} + 1316\frac{\left[(1+i)^n - 1\right]}{i(1+i)^n} - 3200 - 500\frac{\left[(1+i)^n - 1\right]}{i(1+i)^n}$$

Setting i = 10%, n = 5 years, results in a NPW = $6463. Since NPW is greater than $0, the project should be undertaken.

$$NAW = AW_{\text{savings in back injury costs}} + AW_{\text{productivity savings}} - AW_{\text{lift table cost}} - AW_{\text{maintenance costs}}$$

$$= 1733 + 1316 - 3200\frac{i(1+i)^n}{\left[(1+i)^n - 1\right]} - 500$$

Setting i = 10%, n = 5 years, results in a NAW = $1705. Since the NAW is greater than $0, the project should be undertaken.

B/C analysis can be performed with PW or AW calculations. Using AW, i = 10% and n = 5 years:

$$AWB = AW_{\text{savings in back injury costs}} + AW_{\text{productivity savings}} = 1733 + 1316 = 3049$$

$$AWC = AW_{\text{lift table cost}} + AW_{\text{maintenance costs}} = 844 + 500 = 1344$$

Therefore, B/C = 3049/1344 = 2.3. Since this exceeds 1, the project should be undertaken.

ROI analysis can be performed using any of several equations presented in Table 83.3. Using NAWB — NAWC = 0, determine by trial and error the i that satisfies the equation,

$$AW_{\text{savings in back injury costs}} + AW_{\text{productivity savings}} = AW_{\text{lift table cost}} + AW_{\text{maintenance costs}}$$

$$1733 \qquad +1316 \qquad = 3200\frac{i(1+i)^n}{\left[(1+i)^n - 1\right]} + 500$$

i turns out to exceed 60%. Since i ≥ MARR, the project should be undertaken.

83.4 Additional Metrics: Results-Oriented Statistics

A number of other metrics can be used to assess the effectiveness of an ergonomics program, including injury and illness rates, severity rates (referring to lost workdays or restricted workdays), absenteeism, productivity (described in units of production), and machine availability (which signals reduced downtime). Illness, injury, and severity rates are usually normalized to a standard, so that rates can be compared across departments or plants, even if numbers of employees or work hours vary across departments or plants. Equations for illness and injury rate calculation and for severity rate calculation appear in Equations 9 and 10, respectively.

$$IR = \frac{I \times 200{,}000 \text{ hrs}}{N_e \times H_a} \tag{9}$$

where
IR = Injury or illness rate normalized to 100 people working 2000 h per year
N_e = Number of employees in department or plant
H_a = Average number of hours worked per year per employee

$$SR = \frac{D \times 200{,}000 \text{ hrs}}{N_e \times H_a} \tag{10}$$

where
SR = Rate of days lost or restricted normalized to 100 people working 2000 h per year
D = Days lost or restricted
N_e = Number of employees in department or plant
H_a = Average number of hours worked per year per employee

Increased machine availability may be a goal of an ergonomics program. Improvement in equipment access can encourage regular maintenance activities and reduce repair time. Simpson and Mason (1995) presented an equation for calculating machine availability:

$$MA = \frac{MTBF}{MTBF + MTTR + MTPM} \tag{11}$$

where
MA = machine availability
MTBF = mean time between failure
MTTR = mean time to repair
MTPM = mean time for preventive maintenance

83.5 Concluding Remarks

In order to support the process of economic justification of health and safety (H&S) projects, including those related to ergonomics, Riel and Imbeau (1995) recommended developing a "comprehensive support system..." that included "a model of H&S costs, a safety information system, and a proper user interface, all of which make the analysis of H&S investments efficient and effective." One element of that support system, a categorical model of cost and saving sources, was presented in this chapter. Suggestions for calculating those costs and savings were also provided. However, performing specific cost and savings calculations, using those equations or methods specific to a given facility, will depend upon the quality of information available within the facility. The effort to develop and implement a systems approach to ergonomics that includes provisions for supporting economic analysis will provide a sound foundation for an ergonomics program.

References

Alexander, D.C. 1996. The ergonomics program report card. *Workplace Ergo.* (Jan/Feb):26.

Andersson, E.R. 1992a. Economic evaluation of ergonomic solutions: Part I — Guidelines for the practitioner. *IJIE.* 10:161-171.

Andersson, E.R. 1992b. Economic evaluation of ergonomic solutions: Part II — The scientific basis. *IJIE.* 10:173-178.

DeKraker, M., Lindstrom-Hazel, D., Cooper, R., and Ambrosius, F. 1995. Ergonomic modifications: examining cost and effectiveness. *Work.* 5:123-131.

Helander, M.G. and Burri, G., J. 1995. Cost effectiveness of ergonomics and quality improvements in electronic manufacturing. *IJIE.* 15:137-151.

Hendrick, H.W. 1996. *Good Ergonomics is Good Economics.* 1996 HFES Presidential Address.

Konz, S. 1995. *Work Design: Industrial Ergonomics,* 4th. Publishing Horizons, Inc., Scottsdale.

Oxenburgh, M.S. 1997. Cost-benefit analysis of ergonomics programs. *AIHAJ.* 58(2):150-156.

Riel, P.F. and Imbeau, D. 1995. Making the best ergonomics investment. *IIE Solutions.* (June):30-33.

Ritzel, D.O. and Allen, R.G. 1988. Value of work — a way of evaluating a light duty program in a work setting. *Prof. Safety.* (Nov):23-25.

Simpson, G. and Mason, S. 1995. Economic analysis in ergonomics, in *Evaluation of Human Work,* Ed. J.R. Wilson and E.N. Corlett, p. 1017-1037. Taylor & Francis, London.

For Further Information

A number of papers on economics and health and safety or ergonomics have appeared in the *International Journal of Industrial Ergonomics.*

What Every Engineer Should Know about Economic Decision Analysis by D.S. Shupe, and *Engineering Economic Analysis* by D.G. Newman are useful references books on economic analysis methods.

84

Economic Justification of the Ergonomics Process

David C. Alexander
Auburn Engineers, Inc.

Thomas J. Albin
3M Center

84.1 Background

This chapter will present some basic thoughts and information on how to justify ergonomics activity from a corporate perspective. Ergonomics, as a term, is frequently perceived as solely concerned with efforts to control work-related musculoskeletal disorders; yet in its true sense, it is the technology of enhancing all facets of human performance in the workplace, and is not limited to just safety, health, and comfort. An ergonomist would be equally at home with issues such as improving product quality or decreasing the likelihood of errors through the application of ergonomic techniques.

Many individuals charged with implementing ergonomics programs in corporations are unfamiliar with the culture of business organizations in general and may often be unfamiliar with the culture within their own corporation in particular. Accordingly, a brief introduction to some concepts within corporate culture may be beneficial.

A business corporation is a legal entity that resembles a person. Its reason for being is to generate profits for its shareholders. It may enter into contracts or litigation. It is comprised of a number of individuals who work together toward a common goal.

When you propose an action, you have done so because you perceive a need to change something, generally in order to improve it. You justify doing this because there is some added value in achieving the desired goal. Because there is often great difference in what is valued among individuals, you must persuade all those who control the resources you wish to employ so that they will also value the end

0-8493-2641-9/99/$0.00+$.50

results sufficiently to devote resources to them. It must be understood that management is not the only resource controller within a corporation.

Since resources are limited, you will find yourself in competition with other proposals which seek to employ the same resources that you are requesting. You must make the case that your proposal outweighs the others, or find a way to combine your proposal with others to make joint use of the resources in order to achieve mutual goals. As an example, of how such goals may make use of the resources in order to achieve mutual goals consider the following example.

Production equipment was installed in a factory without sufficient attention paid during design and construction to health and safety issues. A plant health and safety team refuses to allow the equipment to be run until several deficiencies are corrected. As a result, the equipment sits on the plant floor for two weeks until the corrections are made. The manufacturing manager must absorb the cost of this equipment during this time without the offset of income from product manufactured, has to rearrange product delivery schedules, pay for raw materials held in stock, etc.

If this situation occurs frequently within a corporation, it is to the mutual advantage of the health and safety function and the manufacturing function to ensure that health and safety issues are addressed before the production equipment reaches the production facility. In fact, economics will suggest that these issues are most efficiently addressed when the production equipment is in the concept phase, rather than when it has already been built. While the benefit of this combined effort is seemingly obvious, the benefit may not always be so clear. How can you build a case to convince management that a course of action offers benefits?

84.2 Management Duties in a Corporation

Managers of a corporation have a number of duties to the corporation, which have been classified under the headings of loyalty, obedience, and due care.[1] They incur personal liability for failure to perform those duties, as well as potentially being subjected to fines or other penalties set by law if they break a law. Of these responsibilities, obedience and due care have direct bearing on the justification of ergonomics programs.

Obedience. Corporate managers have a responsibility to ensure that the corporation obeys the law. This is sometimes referred to as managing for compliance. Consequently, a manager would be expected to implement a recommendation for action which would enable a corporation to comply with a law or regulation in the workplace.

Due Care. A manager is expected to use good judgment. This judgment is to be exercised in determining actions that affect the well-being of the corporation. Courts have generally held the standard of behavior expected as that of a normally prudent person. Since the business of business is business, a reasonably prudent person would be expected to act in a financially responsible manner. A corporate manager is expected and required to act in a financially responsible manner. There are other things which, although less tangible than finances, materially affect the well-being of a corporation. A corporation, is for example, generally seeking to both act as, and be seen as, a good citizen. It will place great value on its reputation.

Responsibilities of an Ergonomist. In addition to the responsibility to point out regulatory requirements, an ergonomist must be able to make a case for the action which he/she proposes. This justification may be based on tangible benefits such as cost savings, intangibles such as corporate reputation, or some combination of both. The justification may be for a specific project, or for a corporation-wide effort.

84.3 Intangible Benefits

While corporations may not totally agree with Shakespeare that, "He who steals my purse steals trash," they are certainly in agreement with him regarding the loss of their good name. There is often a sense of shared purpose or *esprit de corps* among all employees of a company. Certainly no manager worthy of the responsibility, would intentionally seek to injure or place their employees at risk. Consequently, it may be possible to justify some programs because they are "the right thing to do." On the flip side, if such programs are regarded as nice to do, then they may be limited and not realize their full potential for mutual benefit to both employees and the company.

84.4 Tangible Information

An ergonomist will typically need to utilize tangible information regarding the way the corporation is conducting its business in order to build a case. A common starting point is to examine the potential benefit of reducing the cost of compensating injured workers. Since there are legal requirements regarding the documentation of these cases, it is relatively easy to identify at least a significant portion of the case costs.

Research conducted by the Liberty Mutual Research Center has identified the average and median cost for two categories of cases commonly associated with ergonomic risk factors, upper extremity cumulative trauma disorders (UECTDs) and low back cases. UECTD[2] and low back cases[3] were found to average $8,070 and $6,807, respectively. Because the costs are skewed to the high side, the median costs are much lower, $824 for UECTD and $391 for low back cases.

Case cost information of this sort, paired with information regarding the number of cases occurring in the corporation, offers a good first approximation of the cost of ergonomic cases to the corporation. This might be used to develop justification for funding of pilot projects to demonstrate the effectiveness of ergonomic analysis and modification of high-risk jobs.

84.5 Cost Justification Methods

There are a number of ways to justify ergonomic spending. Some of these methods are discussed below.

Benefit/Cost Ratio

This simple method calculates the ratio between the benefits of an ergonomic project and the cost to implement it. The idea is that an ergonomics program that pays for itself is a good investment.

How do you calculate the economic benefit of an ergonomic solution? One way of calculating the economic benefit of an ergonomic program is to look at the cost of injuries associated with ergonomic problems. You make the assumption that implementing the ergonomic program will generate solutions which will prevent future injury. The calculation of the benefits to the cost ratio is shown below:

$$\text{Benefit to cost ratio} = \frac{\text{Value of benefits}}{\text{Cost of changes}}$$

Example:

An ergonomic injury has occurred once a year on average for a particular material handling job. The injuries have occurred when employees move bundles to and from sewing machine workstations. There are four material handlers in the area, and one injury has occurred in each of the last three years. The average cost of the injuries is $20,000.

The cost of the solution is $5,000. This money will purchase material handling equipment, raise platforms beside the sewing machines, and large work tables to store materials near waist height before they are distributed to the sewing operators.

$$\text{Benefit to cost ratio:} \frac{\$20{,}000}{\$5{,}000} = 4.0$$

In other words, the benefits of the ergonomics project are worth four times its cost. This is a sound investment.

Payback Period

You can also look at the payback period for an ergonomic improvement. Payback period refers to the length of time it will take to recover the costs of improvements.

First you must determine the costs and benefits associated with the ergonomics improvement. Then you can calculate the time it will take to offset the cost to implement. To calculate the payback period, use the following equation:

$$\text{Payback period (in years)} = \frac{\text{Costs per year}}{\text{Benefits per year}}$$

Example:

In the previous example, the costs of the injuries that the solution will prevent average $20,000. The improvements cost $5,000. How long does it take to recover the cost of the ergonomics solution?

$$\text{Payback period in years} = \frac{\text{Costs per yr.} = \$5{,}000}{\text{Benefits per yr.} = \$20{,}000} = 0.25 \text{ years (3 months)}$$

Another straightforward way of evaluating the benefit of an ergonomics program is to use return-on-investment analysis. Its calculation is shown below:

$$\text{Return on investment} = \frac{\text{Return to company}}{\text{Investment}} \times 100\%$$

Example:

$$\text{Return on investment} = \frac{\$20{,}000}{\$5{,}000} \times 100\% \text{ for a ROI of } 400\%$$

Using the data from the cost-to-benefit example, the return is the benefit, or $20,000 for the avoided injury. The investment is the cost of the ergonomics solution or $5, 000 for the material handling changes. The return on investment is 400%.

As was shown earlier, this is a sound investment.

Losses vs. Goods Sold

You can also evaluate an ergonomic solution by determining the sales volume required to offset an injury. While the calculations for this method of analysis are relatively easy, you must know the profit margin for the business. The calculation is shown below:

$$\text{Volume of sales required to offset loss} = \frac{\text{Cost of losses}}{\text{Profit margin}}$$

Once the sales volume needed to offset the cost of the ergonomic injuries is determined, then the same effort used to generate this sales volume should be used to correct the ergonomics program.

Example:

A carpal tunnel surgery case has cost your company $18,000 for medical and workers' compensation costs. The profit margin for your company has been 5% for the past three years.

$$\text{Volume of sales to offset loss} = \frac{\text{Cost of losses}}{\text{Profit margin}} = \frac{\$18{,}000}{.05} = \$360{,}000$$

This tells us that the effort to prevent an ergonomic injury for this company should be equivalent to the effort used to generate $360,000 in sales.

Present Value (Time Value of Money)

You can enhance the previous cost justification techniques by including the time value of money in your calculations. With the technique, both the value of the savings and the value of the costs are reviewed over the economic life of the project, which may encompass several years. The technique takes interest effects into account. A dollar benefit realized at a future date is equivalent to a smaller amount of money invested now at interest. This equivalent value now of a future benefit is commonly known as the present value of a future benefit. This provides a common reference point to compare both the future benefits and the present costs.

The time value of money assumes that the value of a dollar today will be different a few years from now. For example, $100 today, when invested at an interest rate of 8% per year, will have a value of $108 in one year. This calculate is known as the *future value of present sum*. This is sometimes abbreviated as F/P.

Likewise, $100 one year from now is only worth $92.59 invested at 8% interest per year today. This calculation is known as the *present value of a future sum* (abbreviated as P/F). As you may suspect, they have a reciprocal relationship. The present value of a future sum may be used to express varying future benefits in terms of a present sum. This might be encountered when some years will produce more savings than others.

Often there will be a series of equal benefits over a number of equivalent periods which comprise the economic life of the project. Because these time periods are often, although not necessarily, years, the benefits are described as annuities. In the case of equivalent benefits occurring at equal time periods, two calculations are of use, the *present value of an annuity* (abbreviated P/A) and the *future value of an annuity* (abbreviated F/A).

The tables for determining the time value of money can be found in computer spreadsheet programs or in financial or engineering economy books. In order to select the appropriate table, you must know both the interest rate and the economic life of the project. Most companies have an interest rate that is used to evaluate projects in this manner. The tables list multipliers for the appropriate combination of interest rate and year. The cost or benefit is multiplied by the multiplier value given in the table in order to determine the present value.

Once the present values are determined, then the benefit to cost ratio or the return on investment calculation can be made.

Example:

A certain family of jobs has a recurring injury pattern. This is a sewing job, and there are 40 workers on the day and evening shifts. Approximately three lost time injuries have occurred each year. The injuries have required medical treatment and have sometimes resulted in lost workdays. The average cost of the lost time and medical treatment is $1,000 per injury or $3,000 per year.

Minor changes are needed to correct the ergonomics problem. The sewing machine head will be modified with a different guide, and the bundle table will be altered to make room for more asiding. These changes will cost $500 per sewing machine workstation. The cost of the changes is $500 per workstation times 20 workstations, for a total cost of $10,000. This cost will be incurred immediately, so the present value of the cost is also $10,000.

The value of the benefits is an annual savings of $3,000. Based on outstanding contracts there is reason to expect that the benefits will continue to accrue for a period of at least five years. Using a computer spreadsheet, we find the appropriate multiplier for an annuity at 8% annual interest for 5 years is approximately 3.993. We multiply this times the average savings of $3,000 per year to obtain the present value of the savings over the five-year economic life of the project and find that it is $11,979.

Determining the benefits to cost ratio is now simple:

$$\text{Benefit to cost ratio} = \frac{\text{Benefits}}{\text{Costs}} = \frac{\$11,979}{\$10,000} = 1.2$$

This is a sound investment as the benefits exceed the costs by 20%.

The return on investment can also be calculated using the present values of the benefits and costs. The calculations are shown below:

$$\frac{\text{Return on investment} = \$11{,}979 \times 100\% = 120\%}{\$10{,}000}$$

assuming that a family of jobs has experienced only one injury recently and that methods training has been suggested to avoid a similar injury in the future. The injury was a carpal tunnel surgery, and it was a serious compensation claim. The full cost of the injury was $20,000. It is estimated that another injury would occur approximately every five years.

The methods training that has been recommended is relatively short and will only require each worker to spend one hour per year for the training. Since the training is done on the job a trainer will also be required for each hour of training. The full cost of the training then will be 40 hours total for the 40 workers plus an additional 40 hours for a trainer. The average cost of a worker's time is $6.50; and there is an additional cost for employee benefits of 25%. The cost of the trainer is $8.00 per hour plus the employee benefit cost of 25%.

The annual cost of the training is:

Item	Costs
Workers: 40 workers × 1 h./worker × $6.50/hour × 1.25	$325.00
Trainer: 40 hours × 8.00/hr. × 1.25	$400.00
Annual Training Costs	**$725.00**

The present value of the recurring costs for 5 years at 8% interest is the same as that calculated above. For a five-year period, it is 3.993 times the annual cost for a total of $725 × 3.993 = $2,894.93.

$$\text{F/A}\,(8\%,\ 5\ \text{years})\ \$725.00 = 3.993 \times \$725.00 = \$2{,}894.93$$

We assume, for simplicity, that the future benefit of avoiding a carpal tunnel case is a sum that will be realized five years from now. Looking in our reference, we find that the appropriate multiplier for the present value of a future sum (P/F) at 8% interest for five years is 0.6805. Multiplying the future value of $20,000 by 0.6805 we obtain the present value of the benefit.

$$\text{P/F}\,(8\%,\ 5\ \text{years})\ \$20{,}000 = 0.6805 \times \$20{,}000 = \$13{,}611$$

The cost to benefit ratio is:

$$\frac{\text{Cost to benefit ratio} = \$2{,}895 = 4.7}{\$13{,}611}$$

This is a sound investment.

84.6 How Do You Determine the Full Cost of an Injury?

A common problem in the justification of ergonomic improvements is that the benefits appear not to offset the costs. Typically, this is because the full dollar amount of the benefits has not been determined. You must capture all of the costs associated with the ergonomic injury. Some of the additional costs that are often omitted include cost of replacement workers, lowered productivity, lowered quality and increased supervisory costs.

Cost of Replacement Workers

If a worker is injured and unable to work, then someone has to assume his/her function. This may be covered by offering overtime to other employees or by hiring more workers, perhaps on a temporary basis. In either case, there are extra costs involved which are directly linked with the injury case. Because the injured worker receives compensation to make up his/her lost wage, an employer, in effect pays double for an off-the-job worker.

Lowered Productivity and Quality

Productivity may also be affected by an ergonomic case. If the worker is not replaced, fewer products are made; yet the overhead cost remains the same. This overhead may consist of the cost of production equipment, supplies purchased and on hand, heat and light, etc., as well as continuing employee benefits. The result is less product available for sale to defray overhead costs.

It is very likely that the worker covering for the injured individual is less familiar with the job and is more likely to make mistakes. This increased likelihood of error has associated costs. At best, it will result in an increase in the amount of rework or salvage operations necessary in order to produce a salable product. The extra labor and material cost of the rework are a result of the ergonomic injury. At worst, the defective product is beyond salvage. All the value of the previous work done to the product to convert it from raw material to goods in process is then a lost cost attributable to the ergonomic injury.

Increased Supervisory Requirements

A worker unfamiliar with a job will require more supervision or assistance from co-workers in order to perform a job. This may be as simple as not knowing where tools are stored or general unfamiliarity with the production process which leads to more required support time.

In addition, there are supervisory costs associated with the time required in preparing the paperwork necessary to process an injury case such as injury and illness reports, developing staffing schedules to cover for injured workers, etc.

Training

A worker unfamiliar with a job will need to be trained in how to perform it. During this time their production may be lowered, or nonexistent. There is also a cost associated with the time of the person who conducts the training.

Example:

An injury requires employers to work overtime to meet product demands. The ergonomic injury was not a major one, but it did result in lost time for six workdays. What are the costs in addition to the medical costs and lost time costs for the injured worker?

Employees earn an extra 50% for overtime pay. Since the replacement worker is from another department, he/she produces up to 12 fewer units per day for the six-day period. In addition, there were some quality problems–one reject at a cost of $36 and five pieces which required rework at a cost of $10 each. A summary of the costs is shown below:

Item	Costs
Overtime (0.5 × $8.00/hr. for 6 days)	$192.00
Lower productivity (12 units/day × $4.00 profit/unit for 6 days)	$288.00
Quality (1 reject)	$ 36.00
Rework (5 units at $10 each)	$ 50.00
Total Additional Costs	**$566.00**

Example:

An assembly task requires a worker to mount a component to a chassis with a single screw using a power driver. The two workers assigned to this area report two issues which have made this assembly operation a problem.

First, the screws frequently stop turning before being fully seated in the chassis. Because the screws are difficult to drive, operators have adjusted the clutches on their power drivers to the maximum torque limit. The power screw driver "kicks" or exerts a powerful torque on operators' arms when the screw stops turning. This has caused some discomfort and minor injuries, but no lost time. To correct this, the operator reverses the screw out, then redrives it. This happens on every other unit. The second concern is that when the screw stops turning the power driver often jumps out of the drive slots on the screw head. This happens about once for every five units. Occasionally, it strips out the screw hole in the chassis. This happens about once every 20 units. The standard rate is to expect the worker to finish three complete assemblies each hour during a ten-hour shift. Production is expected to continue indefinitely.

The first case requires the removal of the damaged screw and replacement. The second event requires the operator to redrill the hole and replace the screw with an oversized one. Customers complain of "rattling" noises during operation if the assembly is not fastened correctly.

It takes approximately 30 seconds to successfully drive a screw, five minutes to remove a damaged screw, and one-half minute to redrive a new screw. Redrilling the screw hole takes an additional five minutes.

Examination of the screws used reveal that they are not self-threading as the specification calls for. Utilization of the appropriate type of screw takes approximately 20 seconds, 10 seconds less than previously. The problems of improper seating, head stripping, and hole stripping are resolved by utilization of the appropriate screw. Adjusting the clutches on the power drivers eliminates the kickback torque to the workers' arms. A summary of the costs and benefits is shown below.

Item	Cost
1000 screws discarded	$18.51
1000 screw purchase	$18.51
Total Costs	**$37.02**

Item			Cost Savings
	Rework time per day		
Remove and replace			
(30 units/day × 0.2 units reworked × 0.092 h × $10.00 h		=	$5.50/day)
Redrill			
(30 units/day × 0.05 units reworked × 0.092 h. × $10.00 h		=	$1.38/day)
	Time saving per day		
Improved drill time			
(30 units/day × 0.0027 h saving × $10.00 h		=	$0.83/day)
Production time gained			
(30 units day × 0.25 units × 0.092 h × $10.00 h		=	$6.88/day)
Total Cost Savings	**$14.59/day per worker × 2 workers**	=	**$29.18 day**

In these examples, by carefully measuring overtime, productivity, and quality, you can capture additional costs, even in the absence of an injury. This means that additional funds are available to offset the injury, regardless of the cost justification technique used.

84.7 How Do You Predict Future Injuries?

One of the most common questions asked during the cost justification of ergonomic changes is, "How do you know that another injury will occur?" One way is to simply see if injuries are, in fact, occurring annually. If they are, then demonstrating this pattern may be the most convincing method.

If the following is the recent pattern of compensable injuries, then you will likely see another injury unless the process is changed.

Year	1992	1993	1994	1995	1996
Compensable injuries	1	2	1	1	1

Often the trend is not as obvious, and predicting future injuries becomes more difficult. To respond to this, one can look at the pattern of less serious injuries, such as first aid visits; if that is stable, then the overall injury pattern is probably stable, as well.

In the data below, the pattern of compensable injuries appears to be inconsistent; the pattern of first aid visits and OSHA 200 log entries, however, is very consistent. Using the stability of these injuries permits a prediction of the more serious injuries.

Year	1992	1993	1994	1995	1996
Compensable injuries	1	0		0	1
OSHA 200 Log entries	4	5	4	3	5
First Aid Visits (nonrecordable)	9	10	10	8	10

In general, there is a consistent trend in less severe injuries, with OSHA 200 log entries occurring at a rate of roughly 4 to 5 per year, and nonrecordable first aid visits occurring at a rate of roughly 9 to 10 visits per year. This clarifies the injury pattern, which is not evident by simply maintaining a count of the compensable injuries. It also provides an excellent illustration of the ratio of minor to major injuries.

In this case, there are approximately three compensable injuries for every 21 OSHA 200 log entries and for every 47 first aid visits. Since both OSHA 200 log entries and nonrecordable first aid visits are relatively stable, we would expect a recurring compensable injury once every two years.

84.8 What Other Costs Should You Consider?

To justify ergonomics changes, look for productivity benefits which may occur. Many ergonomic changes will increase output, lower cycle time, and improve quality. For example, the use of guides and fixtures can control the pinch forces and hand/finger movements of sewing machine operators. Fixtures can also reduce the time to position and guide a product through the sewing head, improving the quality of the finished product. More comfortable seating can increase the amount of time spent in the sewing workstation, and this can be measured in daily output. Also, modifying inspection to improve the employee's comfort and lower the risk of injury can enhance inspection quality, resulting in fewer customer complaints.

Turnover can be a major cost, and is often overlooked. Turnover cost can be measured, and it is frequently a major cost. To determine the cost of the turnover, look at hiring costs, training costs, costs of rejects and lower output, and the opportunity cost of lost profits for products that are not shipped.

Example:

An ergonomic review was done on a key sewing job, and the costs of retrofit improvements were about $50,000 for 50 workstations. The changes recommended included:

- Enhance the layout
- Position the person appropriately in front of the machine
- Modify lifting aids for large bundles
- Ensure appropriate material flow
- Alter the sewing machine orientation
- Tilt the sewing machine surface to better accommodate vision and reach
- Improve the quality of seating

Initially, it did not appear that costs associated with injury reductions, productivity improvements, or quality enhancements would be great enough to justify the ergonomic improvements. However, turnover was high on this job.

A task called final sewing was a key operation for this job. It required great skill and was a bottleneck for product output. Workers are sensitive to their working conditions: Since there are a number of apparel manufacturers in the area it is easy for people to shift to jobs where conditions are better. Final sewing for this employer is a high turnover job. While there are approximately 80 people on two shifts assigned to this job, turnover occurs at a rate of 10 to 20 people per year.

The training time to become fully proficient is approximately one full year. The worker is paid a training rate until incentive pay can be achieved. Also, because of the high skills required, substantial trainer time is needed to aid in skill development. Finally, many products are rejected or sent back for rework during the initial training period. Later in the training year, the major cost becomes the lost sales due to lower output from the new workers. The approximate costs to develop an experienced sewing machine operator who can perform the final sewing operation for this job is shown below:

Item	Costs
Training pay (wages not offset by production prior to full rate production)	$1,500.00
Training time (80 h at 6.00/hr)	$480.00
Trainer time (training and auditing time is 210 hr/trainee at $9.00/hr)	$1,080.00
Lower productivity	$10,000.00
Rejects and rework	$3,000.00
Lost profit	$2,500.00
Turnover Costs	**$18,650.00**

At an annual turnover rate of 10 to 20 workers per year, the full cost of the turnover is $185,600 to $371,200 per year. Reduction of turnover by only three individuals per year would pay back the cost of the modifications to the workstations.

84.9 Conclusions

There are a number of ways to justify expenditures for ergonomics improvements. Many methods are straightforward and simple to use as the preceding examples have shown.

When justifying ergonomic improvement, remember to:

- Make sure all the benefits are identified and fully measured.
- Determine the value of recurring injury costs. Do not short change yourself by counting only the first year.
- Implement the lowest cost solution which corrects the ergonomic problems, though it may not be the most sophisticated or elegant solution.
- Avoid high cost solutions such as automation.
- Ask if another plant has already solved the problem. Reinventing the solution is expensive and time consuming.

- Share information with other plants about the problems you have solved.
- Use cost measurements and justification on every project. The information will be valuable in justifying other ergonomics projects and in justifying your ergonomics program.

References

1. *Business Law: Text and Cases*, Howell, Allison, and Prentice. The Dryden Press, Hinsdale, IL, 1988, pp. 870-879.
2. "The Cost of Compensable Upper Extremity Cumulative Trauma Disorders," Webster B.S., Snook S.H., *Journal of Occupational Medicine*, 36, 1994, pp. 713-717.
3. "The Cost of Compensable Low Back Pain," Webster B.S., Snook S.H., *Journal of Occupational Medicine*, 32, 1990, pp. 13-15.

85

Economic Aiding and Economic–Ergonomic Interactions in Design and Management

James R. Buck
The University of Iowa

85.1 Introduction

Studies of Economics preceded Ergonomics by a couple of centuries. That fact is one of the reasons that some economic notions are better understood than those in ergonomics. It also shows that the general public and management are influenced far more by economics. As a result, ergonomists must be prepared to meet the tests of economics if ergonomic recommendations are to find funding. As Rose et al. (1992) stated, "To succeed in introducing a new, ergonomically better working method, that method must also have economic advantages." Accordingly, part of this chapter addresses the question of how ergonomic

0-8493-2641-9/99/$0.00+$.50

proposals can be analyzed economically so that ergonomists can gain management's attention by showing appropriate economic evidence. Part of the analysis lies in estimating costs and benefits due to ergonomic interventions. Some new and improved methods of economic analysis are shown below that help in the estimation of appropriate costs and benefits as well as their analysis.

Numerous economic principles are very useful in ergonomic design, problem solving, and management. Some such principles are particularly useful during the establishment of a new ergonomics program in a company, while others tend to find greater merit later in the life of such programs. Numerous other interactions between ergonomics and economics are discussed in this chapter as well. In fact, some theories in ergonomics and engineering psychology stem directly from economics. For example, the ergonomic notion of human resource expenditures assumes that people behave as if they were expending their internal resources in an economic manner. Another example of the interrelationship between economics and ergonomics is the notion of concurrent tasking (i.e., attending to two tasks during the same time frame such as driving and talking). The Navon–Gopher model was heavily influenced by economic theory.[†] The following discussion explores these and other such interrelationships between economics and ergonomics.

85.2 Economic Analysis Begins with the Identification of Costs, Benefits, and Required Investments

Once an ergonomic project proposal is begun, one must start to mentally track the costs and gains for the proposal. There are a number of typical cost and benefit categories. A common example is labor savings as a result of new tools, a machine, or a new procedure of operation. The investment is the amount of money the company must pay for the tools, machine, or other facilities which generally last over a year. Items that the company buys that last less than a year are usually classified as expenditures, rather than investments, and are treated differently because of taxation laws and because short-term purchase prices are usually much less. Some companies, however, treat all expenditures as investments and require a more formal quantitative analysis. Accordingly, the following discussion makes no distinction between short-term and long-term projects.

At the outset of a project, one must identify each individual benefit and cost, estimate the cash magnitudes and occurrence times. For example, suppose an assembly operation is using a manually operated screwdriver to insert 6 screws in each assembly of a particular job. The total assemble takes 6 minutes for the entire assembly using the manual screwdriver. An ergonomist proposes substituting a power screwdriver for the manual one, a change that affects only the installation of the screws. The ergonomist measures the time the operator spends putting in the 6 screws manually, including the time it takes to pick up the tool and lay it back down. Suppose that the time required for this manual operation is about 5 seconds per screw for a total of 30 seconds plus about 10 seconds for pickup and drop off of the screwdriver. Then the ergonomist tries a power screwdriver and finds that it takes only 4 seconds per screw with the power screwdriver and the pickup and drop-off time was about the same. The apparent savings derived from the power tool is only 6 seconds per assembly, but that measurement fails to consider the fatigue effect of manual screwdrivers. The inclusion of this cost reveals that the power tool yields an average saving in direct time assembly of 3 seconds per screw or 18 seconds per assembly. Since the screw insertions are only part of the total assembly time of 6 minutes per assembly, the total time benefit per assembly is 0.3 minutes, and at $12.00 per hour, the savings appear to be only $0.06 per assembly. This calculation is not quite accurate, however, because it fails to consider the assembly in the context of the typical workshift. If the expected working time per shift is 408 minutes (480 minutes per shift excluding 15% for allowances) the production rate of a manual process is 408 min./6.0 min. per assembly or 68 assemblies per shift. With the new power screwdriver, the rate would be 408 min./(6.0 – 0.3 min.) or

[†]See Wickens (1992) or Navon and Gopher (1979) for the concurrent tasking model and the resulting effect on performance.

71.58 assemblies per shift. This calculation shows that 3.58 more assemblies would occur during the shift if power tools are used. Now by examining the production run size of 10,000 assemblies, it is clear that the manual method would take 147.06 shifts @ 68 assemblies per shift but with the power screwdriver, it would only take 139.70 shifts. The savings over the production run are as follows:

$$147.06 \text{ shifts} \times 8 \text{ hours/shift} \times \$12.00/\text{hr} = \$14{,}117.65$$

$$139.70 \text{ shifts} \times 8 \text{ hours/shift} \times \$12.00/\text{hr} = \$13{,}411.57$$

According to these calculations, the total cost savings during the production run is $706.08, not counting costs. An often overlooked point is that direct labor cost savings can occur only about 85% of the working time because of delay, fatigue, and personal allowances. That benefit stream would appear as a uniform cash flow totaling $706.08:

$$139.70 \text{ shifts} \times 8 \text{ hours/shift}/2000 \text{ hours/year} = 0.5588 \text{ years}$$

or $706.08/0.5588 years or $1, 263.56 per year (but lasting a bit over half a year). It is assumed here that there is a single shift operation for this assembly over the year.

There is a required investment of about $150.00 for a heavy-duty industrial-grade power screwdriver occurring at the beginning of the project, and the cost of the electricity over the life of the project must also be calculated. And finally, the expected life of the screwdriver is several years at the anticipated usage level. Using a $0.05 per minute cost for power, the total cost over the production run is:

$$\frac{24 \text{ sec}}{\text{assembly}} \times \frac{71.58 \text{ assemblies}}{\text{shift}} \times \frac{139.7 \text{ shifts}}{\text{run}} \times \frac{1 \text{ min.}}{60 \text{ sec.}} \times \frac{\$0.05}{\text{min.}} = \frac{\$199.99}{\text{run}}$$

This calculation yields $357.90 per year uniform cash flow over the 0.5588 years. Overlooking the time differences between these different cash flows and tallying them algebraically with outflows represented as negative outflows and inflows represented as positive, the net worth of this project based on one assembly station is as follows:

$$-\$150.00 \text{ investment} + \$705.23 \text{ benefits} - \$199.99 \text{ power} = +\$355.24$$

Most companies would be convinced at this point that the proposed project was economically sound, providing the estimates are accurate. However, as the following discussion reveals, the calculations are flawed in a number of ways.

85.3 Interest Calculations and Discounted Cash Flows

When an amount of capital is invested for a specified period of time, it normally earns interest. Traditionally, interest earned over a one-year time interval is indicated as i %; this is known as the *nominal* interest rate. The simplest interest calculations assume a single compound interest period each year. An investment of **P** principal earns interest at the end of the year as **Pi**, and so the value of the account after one year is P (1+i), often noted as $\mathbf{F_1}$ denoting a future cash amount in 1 year. Both **P** and **F** are quantities of monetary value (e.g., dollars). When F_k is a future sum of money k years later and k > 1, then F_k is even larger than F_1 because of the longer interest-earning time. The relationship between P and F_1 or F_k turns out to be as follows:

$$F_1 = P\,(1 + i) \quad \text{or} \quad F_k = P\,(1 + i)^k \tag{1}$$

When interest is compounded k times each year, the relationship between F and P is calculated with the following equation:

$$F = P\left[1 + \frac{i}{k}\right]^k \tag{2}$$

According to equation (2), when compounding is semiannual, k is two and F after one year is $(1 + i/2)^2 P$. Note that the bracketed part of equation (2) is greater than the $(1 + i)$ value shown in equation (1) for any value of $i > 0$. In fact, as k gets larger and larger, F increases but with a decreasing rate. For example, if the nominal rate i is 12% and P is $1, then:

time period =	annual	semiannual	quarterly	—	bimonthly	monthly
k =	1	2	3	4	6	12
F =	1.1200	1.1236	1.1249	1.1255	1.1262	1.1268

The *upper limit* of **F** is 1.1275, and that limit occurs when compounding is continuous; that is, compounding periods are instantaneous. Notice that the bracketed amount on the right-hand side of equation (2) approaches $e^{0.12}$ or 1.1275 in this example as k approaches positive infinity. Clearly, then, more frequent compounding creates more interest, but an upper limit exists with continuous compounding.

It also follows that continuous compounding can occur over a year's time, which is equal to any discrete interest rate and compounding period. This notion of equivalency is becoming better known as new laws require banks, credit card companies, and other financial institutions to show equivalent annual percentage rates. The method of calculating this equivalency can be demonstrated with the following example, where continuous interest at rate j yields exactly 12% over one year with a single compounding period as shown below:

$$e^j = (1 + i) = 1.12 \tag{3}$$

By taking the natural logarithm (Ln) of each side of equation (3), it follows that:

$$j = \text{Ln}\left[\left(1 + \frac{i}{k}\right)^k\right] \tag{4}$$

When k is one, the *equivalent continuous interest rate* for 12% annually is 11.3333%. For the nominal rate of 12% and various values of k, equivalent continuous interest rates are:

k =	1	2	3	4	6	12
j % =	11.333%	11.654%	11.769%	11.823%	11.885%	11.934%

That observation can be verified by replacing **j** in equation (4) with the tabled values above. It follows that equivalent continuous interest rates for nominal rates stated by banks are as follows:

Nominal Interest Rate	6.0%	8.0%	10.0%	12.0%	14.0%	16.0%	18.0%	20.0%
Equivalent Continuous Interest Rate j =	5.83%	7.70%	9.53%	11.3%	13.1%	14.8%	16.6%	18.2%

Although one does not often think of continuous compounding, as opposed to annual, semiannual, or quarterly periodic compounding, the option to use it for computational convenience does not affect the computed interest when one uses an equivalent continuous interest rate. Also daily or weekly interest periods advertised by some banks are vastly closer to continuous interest than annual interest rates which many engineering economic textbooks allude to as the norm. Also, when interest is compounded periodically, none of the cash flows within a compounding period should have interest accredited until the very end of the time period. That is, if compounding is annual, events which occur during the year are treated as if they occurred only at the end of the year. Banks maintain savings accounts for people with the promise that they will periodically find the amount of money in the account and credit the account with interest earned over that time period. The account holder controls the cash flow process by putting money into the account and taking it out, but interest earning is a separate process controlled by the bank. Theoretically, monies deposited into the account or taken out during the same interest-generating time period merely accumulate algebraically (inflows add and outflows subtract). However, this calculation does not hold when the money flows in during one interest-earning time period and out during another. Hence, dollars are said to have *time value.* Because of the time value of money, dollars at one point in time cannot be simply added or subtracted to dollars at another point in time. The principal implications here are that dollar values at an equivalent point in time can be added or subtracted and that cash flows at other points in time can be converted to equivalence through equation (1) for 1 or k years and then added or subtracted. Without the conversion to equivalent values, cash flow cannot be added or subtracted.

85.4 Present Worth of Cash Flow Series and Functions

Present worth is a monetary value that occurs at the beginning of a project or a value that is equivalent to the algebraic sum of all the cash flows (inflows are + and outflows are –) in a project including accrued interest. Since invested capital earns money at the rate of $\mathbf{i}$ or an equivalent $\mathbf{j}$ percent or more, a cash flow of $\mathbf{F_k}$ dollars after $\mathbf{k}$ time periods have elapsed will have a present value that is calculated as follows:

$$P = F_k\, e^{-jk} = F_k\, (1 + i)^{-k} \tag{5}$$

It may also explain that when $\mathbf{i}$ or $\mathbf{j}$ is the *minimally acceptable (or attractive) rate of return* (MARR), the resulting present value is the least current value acceptable for any future value. Thus, if all future values were rescaled at equivalent current value by multiplying the actual cash flow that occurs at $\mathbf{k}$ by the factor $\mathbf{e^{-jk}}$ or $\mathbf{(1 + i)^{-k}}$, the *algebraic* sum of these equivalent current values represents the total present worth. The typical convention is to sign all outflows as negative and inflows as positive. Projects with greater positive present worth are more preferred than those with smaller positive present worths, since all cash inflows add to wealth and all outflows detract from wealth, and the accumulation of present worth of inflows minus the worth of outflows represents the surplus current value. If the net present worth is zero, the investment is paid off exactly at the minimum attractive rate of return (MARR). The surplus in present worth tells one how much more the future returns on the investment accumulate beyond the minimum investment coverage.

Discrete Serial Models for Discounted Cash Flows

The future cash flows shown above consist of a single cash flow at the beginning of a project which is equivalent to several future cash flows. In most cases the equivalent present worth was found by finding the present worth equivalent to each separate future cash flow and then accumulating these equivalents algebraically using the sign convention of positive inflows and negative outflows. While that computational practice is correct, it is inefficient. An alternative procedure that is computationally efficient is the use of present worth factors for future time series. There are factors to several cash flow series which convert each member of the series into an equivalent present worth accumulation over the series. These factors provide computational convenience for analysts because the computation of a single formula handles the entire series rather than treating each member of the series individually.

These series also come with different functional forms over time, as shown in Table 85.1. The most elementary is the **step** function series which consists of a uniform series of **C** dollars at the end of each time period and going on for **k** serial flows. The present worth of the step series over k time periods is:

TABLE 85.1 Five Functional Series, Serial Magnitudes, Cumulative Quantities Over k Periods, and Present Worths.

Functional Series	Magnitude at Time k	Cumulative Cash Flow from Time 0 to k	Present Worth at the Minimum Attractive Rate of Return i or P(i) =
STEP	$f(k) = C$ typical of operating cost	$\sum_{x=1}^{k} f(x) = Ck$	$\frac{C}{i}\left[1-(1+i)^{-k}\right]$
UP-RAMP	$f(k) = Ck$ typical of maintenance and deterioration costs	$\sum_{x=1}^{k} f(x) = C\frac{k(k+1)}{2}$	$\frac{C}{i^2}\left[1-(1+i)^{-k}\right]+\frac{C}{i}\left[1-(k+1)(1+i)^{-k}\right]$
DOWN-RAMP	$f(k) = R - Ck$ typical of lost sales due to fewer customers ($R > Ck$)	$\sum_{x=1}^{k} f(x) = Rk - \frac{Ck(k+1)}{2}$	$\frac{Ri-C}{i^2}\left[1-(1+i)^{-k}\right]+\frac{C}{i}\left[1-(k+1)(1+i)^{-k}\right]$
DECAY	$f(k) = Ce^{-rk}$ startup costs	$\sum_{x=1}^{k} f(x) = C e^{-r}\left[\frac{e^{-rk}-1}{e^{-r}-1}\right]$	$\frac{C}{1+i-e^{-r}}\left[e^{-r}-\frac{e^{-(k+1)r}}{(1+i)^k}\right]$
GROWTH	$f(k) = R - Ce^{-rk}$ maintenance costs of aging equipment	$Rk - Ce^{-r}\left[\frac{e^{-rk}-1}{e^{-r}-1}\right]$	$\frac{R}{i}\left[1-(1+i)^{k}\right]-\frac{C}{1+i-e^{-r}}\left[e^{-r}-\frac{e^{-(k+1)r}}{(1+i)^k}\right]$

$$P(i) = C\sum_{k=1}^{K}(1+i)^{-k} = C\frac{1-(1+i)^{-k}}{i}$$

where the final term in the equation above is the sum of the series $(1+i)^{-1}+(1+i)^{-2}+\ldots+(1+i)^{-K}$. Table 85.1 shows the step and four additional functional cash flow series where each successive time in the series is an interest compounding time in which cash flow events occur. The final column in Table 85.1 describes the present worth formula for the step and other series. Actual time series of values and typical kinds of cash flows where the series is often appropriate are shown in the second column. The fourth column in Table 85.1 is simply the cumulative sum of undiscounted cash flows. These cumulative sums are useful in estimating cash flow constants **R** and **C** in these series from accounting or other data sources.

In the case of a step series of $100 each time period for 3 years, with a minimum attractive rate of return of 10%, the formula and computation of present worth is as follows:

$$P(i) = \frac{C}{i}\left[1-(1+i)^{-k}\right] = \frac{\$100}{0.1}\left[1-(1+0.1)^{-3}\right] = \$248.69$$

In effect, this is equivalent to 3 individual cash flows of $100 each at the end of years 1, 2, and 3. Individual computations over the three cash flows are more cumbersome, as the following equation demonstrates:

$$P(0.1) = 100\,(1.1)^{-1}+100\,(1.1)^{-2} + 100\,(1.1)^{-3} = 90.91 + 82.64 + 75.13 = \$248.69$$

Obviously, the computational savings of a series improves with the length of the series, and that is one important advantage of series formulae.

While the process of cash flow generation is really a separate process from interest generation, calculations describing these processes in Table 85.1 are *not* totally separable. The reason is because the fundamental time period between successive time points in a series must correspond to the interest-generating process, as this correspondence is the basis of compounding. When that fact is clear and unambiguous, there is no difficulty in using these serial descriptions of cash flow.

Other functional series are shown in Table 85.1 and some comments on them are appropriate. There are two varieties of **ramp** series. The first is the up-ramp, which starts flowing in the first time period at C and then continues at 2C, 3C, 4C, ... dollars in time periods 2, 3, 4, and so forth respectively. As Table 85.1 notes, up-ramp series are often used to describe increasing costs over time due to deterioration and required maintenance of equipment. Many textbooks on engineering economy or financial calculations† present a functional series known as a *gradient* that also increases uniformly the flow amount of money over time However, there is no cash flow for the gradient at the first time period and the flows at time periods 2, 3, and 4 are respectively C, 2C, and 3C dollars. Accordingly, the up-ramp series differs from the gradient, although it bears similarities. The **down-ramp** series is nothing but a step function of **R** dollars per period minus an up-ramp of **$C**, so long as Rk > Ck. A common example of a down-ramp series is reduced revenues expected as the result of deteriorating machinery or seasonal decreases in revenues from the sales of off-season products. Two other serial functions, the **decay** and **growth** series, are shown in Table 85.1. Startup or learning costs can often be captured by decay series, and maintenance costs frequently follow a growth series. Here again, present worth formulae are presented in Table 85.1 for these series as well.

In addition to the present worth formulae, Table 85.1 describes the magnitude of the series at specific points in time (in column 2) and the cumulative sums over time from t = 1 to t = k (in column 3). As noted above, before the economics of the situation can be analyzed, the benefits and costs must be identified and estimated. These other formulae in Table 85.1 are useful in this estimation once the different benefit and cost streams are identified. In fact, part of the identification involves a recognition of the general nature of the costs and benefits over time. For example, a maintenance cost would be expected to increase more with time and so an up-ramp series or a growth series would be likely candidates for describing this situation. Then it is merely a case of finding which functional series best describes these costs. If past records indicate increasing values at a diminishing rate over time, the growth series would be a better choice than the up-ramp. By fitting the function f(nt) for the magnitude of that series to previously recorded data on costs, the following series of equations will result:

$$f(1t) = R - Ce^{-r} \text{ and } f(2t) = R - Ce^{-2r} \text{ and } f(3t) = R - Ce^{-3r}$$

The values of **R**, **C**, and **r** that best satisfy those estimates are the most reasonable values. More data facilitates a statistically better fit. Sometimes available data (e.g., accounting records) show cumulative expenditures, rather than serial expenditures. In such a case, the cumulative values can be used as follows:

$$\sum_{n=1}^{K} f(t) = R - Ce^{-r}, \sum_{n=1}^{2} f(t) = R2 - Ce^{-r}\left[\frac{e^{-2r}-1}{e^{-r}-1}\right], \sum_{n=1}^{3} f(t) = R3 - Ce^{-r}\left[\frac{e^{-3r}-1}{e^{-r}-1}\right]$$

Any combination of the series magnitudes and cumulative serial amounts can be used.

As stated above, the use of a series carries the assumption that the time period between successive cash flows is constant and can be designated as the interest-generating time period. If one chooses to use a quarterly time period, the interest rate charged per time period should be the nominal MARR divided by four. It also follows that the series magnitudes are the costs or benefits over that quarter-of-the-year time interval or whatever period the analyst chooses. Although interest and cash-flow generating processes are theoretically separate, the algebra describing a series makes it difficult to separate them. One could use monthly series with an interest rate per month as one-third of a quarterly interest rate. That computation would contain but a very slight error and one far less than the expected errors of estimation, but highly frequent series computations become cumbersome.

†See Au & Au (1992), Newnan (1996), Fabrycky, Thuesen, and Verma (1998).

Continuous Models of Discounted Cash Flow

An alternative to modeling cash flows as series is to model cash flows as continuous flow functions over time. For example, a continuous step function accumulates \$C over a year and the accumulation each month is 1/12th of C. One of the distinct differences between series modeling and continuous cash flow modeling is that series parameters correspond to the individual serial flows, but continuous model parameters pertain to the flow *over a year.* For example, a continuous up-ramp that has a parameter of \$C per year means that the flow magnitude each year increases the flow density by \$C more each year. At the beginning of the first year the flow was \$0 per year and by the end of the year cash is flowing at \$C per year. During that first year the accumulated cash flow is (\$0 + \$C)/2 or \$C/2 because the flow goes from \$0 to \$C can be represented as a triangle with a base of 1 year and an area that contains \$ C/2. That same up-ramp reaches a magnitude of \$2C over two years and the accumulation is \$4 C/2 or \$2C. It is often helpful to think of continuous cash flows over time as analogous to putting a pail below a water faucet. The magnitude of the cash flow rate at any point in time is analogous to the amount the faucet value is open, but the amount of flow into the pail depends upon how long the faucet is set at a given opening. Step flows describe the case in which the faucet is set at a constant opening over time. An up-ramp flow is analogous to putting the pail below the faucet and then uniformly increasing the faucet opening. The reverse is true for a down-ramp flow. When that down-ramp represents benefits, the stream of benefits gets thinner and thinner, supporting less and less business operations. Continuous cash flows accord with the notion of stream flow and this form of modeling has some mathematical and conceptual advantages. For one thing, a continuous flow can be stopped at any point without concern about the next serial value, and whether or not it should be included in the calculation. Continuous flows carry continuous compounding so that compounding time periods are not a problem; the continuous interest rate must be set to be equivalent with the nominal rate or with a natural monthly, quarterly, or semiannual rate as shown above. Moreover, some of the formulae for continuous cash flows are a bit simpler than the discrete series, yet they yield the advantage of describing flows over various time periods.

Table 85.2 shows formulae for computing the present worth of continuous cash flow functions that start immediately, accumulate up to time **k**, and then stop. In all of those formulae, it is assumed that both the pattern of changing cash flow and the start of cash flow occurs at time zero relative to the project start. As with the Table 85.1 formulae, Table 85.2 shows cash flow densities at any future point in time as well as cumulative functions of cash flow. Those density magnitude and cumulative flow formulae are particularly useful in setting the C and r parameters in these continuous flow models.

Note that when **k** is a very long time, the value of $\mathbf{e^{-jk}}$ in the formula in Table 85.2 approaches zero. Similar to the series formulae, this situation greatly simplifies the formula in Table 85.2. In the case of a step function, the present worth calculated for a very long time period approaches **C/j**. With an up-ramp function, the present worth approaches $\mathbf{C/j^2}$.

The formulae for continuous cash flows in Table 85.2 correspond closely to those for discrete cash flow series in Table 85.1. For example, in the step function with a constant flow per year over k years at **j** percent interest. If j is set at 9.53%, the equivalent continuous interest rate for 10%, the present worth over three years is calculated as follows:

$$P(j) = C\frac{1-e^{-jk}}{j} = \$100/\text{year}\frac{1-e^{-0.0953(3)}}{0.0953} = \$260.93$$

The reader may recall that a similar calculation using the Table 85.1 formula and 10% annual discrete interest yielded only \$248.69 rather than \$260.93. Since the exponential term in the fraction part of the equation above is equal to the numerator in the discrete series step function, the only cause of a difference is the j in the denominator, in contrast to the i in the denominator for a series step formula. In other words, \$260.93 times 0.0953/0.1 is \$248.67. This example demonstrates that present worths of continuous flows and discrete serial flows are different, even when the interest rates are termed "equivalent." The amount of this difference, however, is typically small.

TABLE 85.2 Five Continuous Functions of Cash Flow, Magnitudes at Specific Points in Time, Cumulative Quantities Over k Time Periods, and Present Worths.

Continuous Function	Magnitude at Time k	Cumulative Cash Flow from Time 0 to k	Present Worth at the Minimum Attractive Rate of Return j (Continuous Compounding) or P(j) =
STEP	f(k) = C typical of operating cost	$R\int_0^k dx = R\,k$	$\frac{R}{j}\left[1-e^{-jk}\right]$
UP-RAMP	f(k) = C k typical of maintenance and deterioration costs	$C\int_0^k x\,dx = C\frac{k^2}{2}$	$C\left(\frac{1-e^{-jk}}{j^2}-\frac{ke^{-jk}}{j}\right)$
DOWN-RAMP	f(k) = R – Ck typical of lost sales due to fewer customers (R > Ck)	$R\int_0^k dx - C\int_0^k x\,dx = \left(R-\frac{Ck}{2}\right)k$	step less up-ramp R > Ck
DECAY	$f(k) = Ce^{-rk}$ startup costs	$C\int_0^k e^{-rx}dx = C\left[\frac{1-e^{-rk}}{r}\right]$	$\frac{C}{j+r}\left[1-e^{-(j+r)k}\right]$
GROWTH	$f(k) = R - Ce^{-rk}$ maintenance costs of aging equipment	$C\int_0^k \left(1-e^{-rx}\right)dx = C\left[\frac{rk-1+e^{-rk}}{r}\right]$	step less decay

Estimating Parameter r in Serial and Continuous Models

Sometimes analysts have difficulty estimating the exponential parameter r in decay and growth models. One aid to making these parameter estimates is to find the expected cumulative cash flows over one and two time periods. Theoretically it does not make any difference what time period is selected, but in practice a period long enough to reveal changes in the flow is most desired. Thus, if the changes were relatively rapid, monthly time periods would be adequate. However, quarterly, semiannual, or annual periods are preferable for flows that change less frequently over time. Note in Tables 85.1 and 85.2 that the cumulative cash flow of a *decay* cash flow over k time periods as follows:

$$S(k) = \frac{C}{-r}\left[e^{-kr} - 1\right]$$

Now consider the ratio S(1)/S(2) describing the occurrence of a two-period flow during the first period:

$$\frac{S(1)}{S(2)} = \frac{\left[e^{-r} - 1\right]}{\left[e^{-2r} - 1\right]} \tag{6}$$

In a similar fashion, the growth model has the following ratio:

$$\frac{S(1)}{S(2)} = \frac{e^{r} - 1}{e^{2r} - 1} \tag{7}$$

TABLE 85.3 Ratios of S(1)/S(2) for the Growth and Decay Continuous Cash Flow Functions Corresponding to Selected Values of Parameter r.

parameter r =	0.01	0.03	0.05	0.07	0.09	0.10	0.20	0.30
growth S(1)/S(2)	.2508	.2525	.2542	.2558	.2574	.2567	.2660	.2742
decay S(1)/S(2) =	.5050	.5086	.5026	.5176	.5228	.5251	.5499	.5745
parameter r =	0.40	0.50	0.60	0.70	0.80	0.90	1.00	1.10
growth S(1)/S(2) =	.2820	.2895	.2969	.3041	.3109	.3176	.3141	.3303
decay S(1)/S(2) =	.5987	.6225	.6156	.6682	.6900	.7109	.7310	.7502
parameter r =	1.20	1.30	1.40	1.50	1.60	1.70	1.80	1.90
growth S(1)/S(2) =	.3362	.3419	.3475	.3528	.3579	.3623	.3674	.3719
decay S(1)/S(2) =	.7685	.7859	.8022	.8176	.8320	.8455	.8581	.8699
parameter r =	2.00	2.10	2.20	2.30	2.40	2.50	2.60	2.70
growth S(1)/S(2) =	.3761	.3802	.3841	.3879	.3914	.3949	.3981	.4012
decay S(1)/S(2) =	.8808	.8909	.9003	.9089	.9168	.9241	.9309	.9370

If follows with the step function that the S(1)/S(2) ratio is 0.5, and with an up-ramp function it is 0.25. A growth function with a very small **r** value increases in time almost uniformly, and so that ratio should be similar to the up-ramp when **r** is small. The **r** values of growth functions thus lie between 0.25 (with a small r) and 0.5 (with a large r). In a similar manner, decay functions with a small r behave similarly to step functions. Hence, the S(1)/S(2) ratios of decay functions vary from about 0.50 with a very small r and increase up to 1.00 as r increases. Table 85.3 verifies these observations by describing associated S(1)/S(2) ratios for growth and decay functions at various r values from 0.01 to 2.70.

Delayed Cash Flow Streams

All of the present worth functions in Tables 85.1 and 85.2 are assumed to start immediately and continue for **k** years. If the stated pattern of cash flow is delayed **b** time units before the pattern starts but the pattern is otherwise exactly the same *after starting* i, then the present worth without delay (P′) can be computed exactly as if it had started immediately. To find the correct present worth for the pattern delay of **b** time units, the following formula may be used:

$$P = e^{-jb}P' \tag{8}$$

For instance, suppose that a 3-year-long step function of $100 each year were to experience a 2-year delay but remain otherwise unchanged. The present worth could then be figured out as $260.93, as shown earlier, and the delay of 2 years is computed as follows:

$$P = e^{-0.11333(2)}\, 260.93 = \$\, 208.01$$

Note that once a cash flow pattern is recognized and fitted with parameter values to reflect actual cash flows, those patterns and parameters can be directly used along with the equivalent continuous interest rate to compute present worths. In this use, continuous interest rates are considered to be the minimum acceptable rate of return on company-invested capital.

Repeated Cash Flows Over Time

A typical situation in modeling the cash flows of a project finds that costs or returns associated with maintenance, production, and such other processes repeat themselves over time. Such situations are similar to that of a homeowner who must perform maintenance on the heating and air-conditioning system twice a year, year in and year out. In that situation the present worth of the first event in the series can be multiplied by factors which change future cash flows at each future occurrence into equivalent present worth amounts as follows:

$$PW\left[1+e^{-jk_1}+e^{-jk_2}+e^{-jk_3}+e^{-jk_4}+\ldots+e^{-jk_{(n-1)}}\right]$$

This formula describes the **n** recurrences of that cash flow at future times 0, k_1, k_2, k_3,..., k_{n-1}; each member of the series is e^{-jk_1} times the previous serial value. Whenever there are very many of these future times, it is inconvenient to compute each of them individually and then sum up the bracketed expression. Since this expression is a geometric progression (series), the sum over the first n term is as follows:

$$PW\left[1+\sum_{i=1}^{n-1} e^{-jk_i}\right]=PW\left[1+e^{-jk_1}+\ldots+e^{-jk_{(n-1)}}\right]=PW\left(\frac{1-e^{-jn}}{1-e^{-j}}\right) \tag{9}$$

Accordingly, it is useful to recognize repeated cash flow patterns over time and use the relationship above to simplify calculations.

85.5 Time Savings of Human Operators

A typical human operator in industry works a 480-minute shift each of 5 days per week for about 50 weeks per year. This working arrangement amounts to about 2000 paid hours per human operator per year. When the allowances total about 15%, the remaining 85% of effective work hours equals 1700 effective hours per year. If that operator is paid at a stated rate of $K per hour, the actual cost per hour over the year is:

$$K' = \frac{2000\ \text{hrs/yr}}{1700\ \text{hrs/yr}}\left[\$\frac{K}{\text{hr}}+\frac{\text{Other Benefits}}{\text{yr}}\right]$$

Note that if other benefits per year are zero, the cost per actual worked hours (**K′**) is about 1.1765 $K. Savings in performance time of a human operator can only be calculated on the actual time worked, not the total time.

The traditional manner for analyzing the economics of new tools and devices to aid production is to cost out the method with and without the new tool or device. The cost of the new tool or device is the principal P, which is subtracted from the present worth of cost difference or cost savings. For example, consider a particular item that requires 2 minutes/piece (0.0333 hrs. per piece) when 100,000 pieces are made per year in a single production run. This product requires 3,333 active assembly hours, 3,333/0.85 = 3,921 actual production hours; which requires 3,921/20,000 per year or 0.19606 years. At a stated rate of $12/hour, the cost is $47,052 over the course of the year. If the new tool or device saved 10% of the actual production time, the active time hours are 90% of the 3,333 h or 3,000.0 active hrs, which amounts to 30000/0.85 or 3,529. actual production hours (i.e., 0.17645 years). The cost of production with the tool or device over the course of the year is $12 × 3,529 h or $42,353 per year, yielding a cost difference of $4,699 on the investment principal of $200 in less than a year. The economic advantage of this tool or device is so obvious, that further analysis seems unnecessary. Yet some problems remain. One is that the actual time of production differs, so it is difficult to use discrete interest rates to evaluate present worth. Some analysts ignore time differences within a year and assume all within-year costs as occurring at the end of the year. These analyses might calculate the present worth of the savings at a 12% MARR as follows:

$$P = \$4,699\,(1.12)^{-1} - \$200 = \$3,996$$

In other words, this investment repaid the principal and 12% interest on the principal, plus an additional $3,996.

Some Alternative Forms of Analysis with This Example

An alternative but analogous form of analysis is to use continuously compounded interest at the equivalent MARR or 11.333% in this example. The present worth of existing operations without the tool or device can thus be computed as follows:

$$P = \$24,000 \frac{\left[1 - e^{-0.11333(0.19606)}\right]}{0.11333} = \$4,654.$$

With the tool or device, the present worth calculation is:

$$P = \$24,000 \frac{\left[1 - e^{-0.11333(0.17645)}\right]}{0.11333} + 200. = \$4,393.$$

The resulting difference of \$261 is the net present worth of savings over the principal. Although, this second form of analysis gives the same result, accounting for the actual time differences within the year shows that ignoring within-year time effects can sometimes mislead the decision maker.

An associated form of analysis is annual worth analysis. This annual worth amount can be computed directly from the present worth using the annual-worth-to-present-worth factor as follows:

$$P = C \frac{\left[1 - e^{-jk}\right]}{j} = C \frac{\left[1 - e^{-.11333(1)}\right]}{0.11333} = 261.$$

The value of C that satisfies the above equation is the annual worth, or \$276. One could compute the equivalent annual worth without and with the tool or device as follows:

$$C_{w/o\ device} \frac{\left[1 - e^{-.11333(1)}\right]}{0.11333} = 4,654.$$

$$C_{w\ device} \frac{\left[1 - e^{-.11333(1)}\right]}{0.11333} = 4,392$$

where the two respective C values are $C_{w/o\ device}$ = \$4,922. and $C_{w\ device}$ = \$4,646. The difference between these two annual worths is the net annual worth of the cost savings which is \$262 per year.

Another form of analysis which is quite popular but sometimes very misleading is the internal rate of return basis for comparing alternative projects. In such cases the internal rate of return over an entire year is found by setting the present worth of the savings, without specifying the interest rate, against the investment principal. For the present example, the calculations could be:

$$200\ (1 + i) = \$4,699 \quad \text{or} \quad i = 22.495$$

This result shows that this investment earned 2,250% of the supplied principal. Here again this analysis ignores within-year time differences. While the internal-rate-of-return method of analysis has difficulties, it does point out those actions that should be avoided. If the internal rate of return falls below MARR, that action should *not* be undertaken. Also, with two alternative actions that essentially do the same thing but one costs more and provides a better return, one can test the more expensive alternative to see if the *added* return earns at least the MARR on the added investment. Only when it does can the more expensive alternative be considered economical.

85.6 Economics of Learning, Forgetting, and Training

One of the better known topics of ergonomics is learning. People perform tasks better with practice and the effect is known as *learning*. When practice on a task is stopped for a protracted time, performance on that task deteriorates, and the effect is known as *forgetting*. Accordingly, learning is performance improvement with practice, and forgetting is performance deterioration due to lack of practice. In industry, learning is described by shorter performance times or greater production rates as more and more production occurs. One model of learning† is based on performance time on the nth sequential unit of production:

$$y_{n+1} = \alpha\, y_n + \beta \tag{10}$$

In equation (10), α and β are constant parameters of the model. In effect, this model predicts performance on the next product unit as the current performance time multiplied by the constant α plus a constant β. Since learning is improvement, α is defined as a fraction ($0 < \alpha < 1$). The values of $\mathbf{y_{n+1}}$ decrease progressively with each new unit produced but at a decreasing rate which approaches the asymptote **y***:

$$y^* = \frac{\beta}{1-\alpha} \tag{11}$$

The asymptote associated with this learning curve provides an advantage in using this model of learning over many others.‡ A more effective prediction equation for this discrete model is as follows:

$$y_n = \alpha^n \left[y_1 - y^* \right] + y^* \tag{12}$$

Equation (12) describes the performance time on the *n*th production unit based on performance on the first production unit and the asymptote y*. Another useful feature of this learning curve model is the differences between sequential performance times, calculated as follows:

$$\Delta y_n = y_{n+1} - y_n = \alpha\, y_n + \beta - y_n = (\alpha - 1)\, y_n + \beta \tag{13}$$

If performance times are generally decreasing, the average Δy_n is negative. Also, as equation (13) denotes, Δy_n is a linear function of y_n. With a short sequence of performance times, then one can compute the forward differences $\mathbf{\Delta y_n}$ as a function of $\mathbf{y_n}$. For instance, suppose that the following performance times were recorded on the first five production units:

	n				
	1	2	3	4	5
y_n time (hours) =	10.000	9.920	9.842	9.765	9.689
Δy_n =	–0.0800	–0.078	–0.077	–0.0755	—

†The discrete exponential model is shown by Pegels (1969) and by Buck, Tanchoco, and Sweet (1976) which is based on the first-order difference equation with constant coefficients which Goldberg (1961) describes in detail. Hutchings and Towill (1975) describe a similar but not identical model.

‡The powerform learning curve, noted by Snoddy (1926) and later by Wright (1936), which is much better known in the U.S., does not have a natural asymptote and so an artificial asymptote at n = 1000 is often used. Hax and Majluf (1982) used this same model for describing improvements over time by whole industries.

TABLE 85.4 Some Basic Data on the Learning Curve in the Numerical Example.

	n						
	10	20	30	40	50	60	70
Total Hours Yn =	96.59	186.48	270.90	350.86	427.16	500.49	571.38
Trans.. Hours	36.59	66.48	90.90	110.86	127.17	140.49	151.38
% Trans.	37.9	35.9	33.2	31.6	29.8	28.1	26.5

When Δy_n is plotted as a linear function of y_n, the following equation results:

$$\Delta y_n = 0.1055 - 0.01853\ y_n$$

This result is obtained by using a simple linear regression routine on a calculator where the coefficient of determination is r = 0.99. Based on equation (13), $\alpha = 0.98$ and $\beta = 0.12$. By substituting these values in equation (11), one finds the asymptote y^* is 6.0 hours per unit and the production time on each production unit can be described with the following equation:

$$y_n = 0.98^n[10.00 - 6.00] + 6.00$$

With this discrete exponential model of learning, the total production time required to produce a total of n units is as follows:

$$Y_n = \frac{1-\alpha^n}{1-\alpha}[y_1 - y^*] + y^* n \tag{14}$$

In the example above, the total production time can be described with the following equation:

$$Y_n = \frac{1-0.98^n}{1-0.98}[10.0 - 6.0] + 6.0\ n$$

In equation (14) the last term on the right-hand-side is the asymptotic values times the number of items produced. In essence, that value is the steady-state value and the first term in that equation shows the transition time required by the learning process. For the example above, the transition and total production hours for selected values of n are shown in Table 85.4. The last row of this table reports the transition production time (added time for learning) as a function of the total time. As the production run get longer, a smaller and smaller fraction of production time is required for learning.

An economic question with production is the economic production run length. Stated otherwise, "How many items should the company make each time they start up production of this item?" Typically, a company builds up a component that is used in several products over the course of a year. If it makes the entire yearly production in one single run, the company takes advantage of the learning process and minimizes the number of production setups needed per year. But with a single annual production run, many items must be placed in inventory until needed and the inventory costs are often substantial. Consider as an example the product unit made by the company which needs 80 lots per year uniformly over the year. This usage rate is about 1.6 units per week or 6.67 per month, averaging 0.04 per production hour. Now let us consider several production strategies in separate sections below.

Produce Lots As Needed

At one extreme, the company could make a unit of the product during the first 10 hours and then cease production. After every 15 production hours, the company could start producing another unit and then after 10 production hours another lot is ready. In this case the learning is minimal and production is spread out to the maximum so that inventory is least. Crews who set up the production runs are repeating their jobs frequently. To put this situation into a specific numerical example, suppose that production work per operator cost $12/hour. Over the course of a year (or 2000 production hours) the uniform annual cost is $24,000. Building one lot at a time is expected to require 10 hours each time because there is no experience time for learning. Each production run of 10 hours has a present worth of:

$$\text{PW(production)} = \frac{\$24{,}000}{0.11333}\left(1 - e^{-0.11333\frac{10}{2000}}\right) = \$119.97$$

if it started immediately. However, only the initial run of one lot starts immediately, the remaining 79 lots occur every 15 production hours. Accordingly, the total PW of production is:

$$\text{Total PW} = \$119.97\left[1 + e^{-.11333\frac{25}{2000}} + e^{-.11333\frac{50}{2000}} + \ldots + e^{-.11333\frac{1975}{2000}}\right]$$

$$\text{Total PW} = \$119.97\left[\frac{e^{-.11333\frac{25}{2000}80} - 1}{e^{-.11333\frac{25}{2000}} - 1}\right] = \$9{,}073.71$$

However, before each production run can occur, a setup crew comes in and gets everything ready. Setup costs are estimated at 15 operator hours times $12/hr or $180 each, and each is done off shift just prior to production. So the setup cost component present worth is:

$$\text{Total PW} = \$180.00\left[\frac{e^{-.11333\frac{25}{2000}80} - 1}{e^{-.11333\frac{25}{2000}} - 1}\right] = \$13{,}613.97$$

The only other cost component to be considered is the inventory cost (also known as storage or holding cost). In the particular case where only one lot is produced at a time and production is timed to finish just as the lot is issued, there is no inventory and zero inventory costs. The other two components sum to $22,687.68 each year.

Make the Entire Year's Production at One Time

At the other extreme, the company could make the entire year's requirements in a single production run. If the learning curve continued as described, the required production time yielded by equation (14) would be computed as follows:

$$Y_{80} = \frac{1 - 0.98^{80}}{1 - 0.98}[10.0 - 6.0] + 6.0(80) = 640.27 \text{ hours}$$

For a company which works single shifts all year, this production strategy requires 16.65 weeks, 83.24 shifts, or 0.32 years. Actual production costs are $12/hour or $24,000 per year at a uniform rate so that present worth of the production operation is a step function of $24,000/year over the 0.32 years as:

$$PW_{\text{production}} = \$24000\left(\frac{1 - e^{-.11333(.32)}}{.11333}\right) = \$7{,}542.41$$

During these 640.27 production hours, the company issues some components from production, and inventories the production surplus. The number of issues during production can be described as follows:

$$640.27 \text{ production hours} \times 0.04 \text{ lots per hour} = 25.6 \text{ lots}$$

If 80 lots are produced during production but during the same duration 25.6 are dispensed, the net inventory at the end of production is 54.4 lots. Inventory increases from the start of production. Although the rate of lot production is not constant, due to the learning curve, an average rate would be 80/640.27 or about 0.125 per production hour. Lots are issued out of inventory at the rate of 0.04 per hour so that inventory builds during production at approximately 0.085 per hour. Accordingly, the inventory cost component during the production of 80 lots is approximated by a ramp which increases at the rate of 0.085 per production hour or 170 lots per year times $1000 per lot per year. The present worth of this inventory buildup is:

$$PW_{inv} = \$170,000\left[\left(\frac{1-e^{-.11333(.32)}}{.113333^2}\right)-\left(\frac{0.32\ e^{-.113333(.32)}}{.11333}\right)\right] = \$8,496.4$$

There is also the single setup cost of $180 at time zero. The inventory cost from the end of production over the rest of the year decreases uniformly, and so the inventory cost can be described as a down-ramp (see Table 85.2). The step portion of this down-ramp has an initial inventory level of 54.4 lots at a $1000 per year per lot is:

TABLE 85.5 The Effect of the Production Run Size on Cumulative Storage (Inventory) Requirements.

	no. made at-a-time							
	10	20	30	40	50	60	70	80
Manuf. hours	96.59	186.48	270.90	350.86	427.16	500.49	571.38	640.27
Issued during manuf.	3.86	3.60	3.38	3.56	3.05	2.93	2.84	2.76
Stored during manuf.	6.14	6.40	6.62	6.44	6.95	7.07	7.16	7.24
Cumul. storage	6.14	12.54	19.16	25.60	32.55	39.62	46.78	54.02

At the end of the production (0.32 years) inventory is at the level of 54.02 lots. The present worth for the remaining inventory can be found by computing the present worth of the maximum inventory continuing over the remainder of the year minus the present worth of a ramp that describes the issuing rate from the inventory. The first part of that computation is a step of 54.4 times $1000, or:

$$PW_{step} = 54400\left[\frac{1-e^{-.11333(.68)}}{.11333}\right] = \$\,35,603.$$

The other part is the ramp or:

$$PW_{ramp} = \$\,80,000\left[\left(\frac{1-e^{-.11333(.68)}}{.11333^2}\right)-\left(\frac{0.68\ e^{-.11333(.68)}}{.11333}\right)\right] = \$17,572.$$

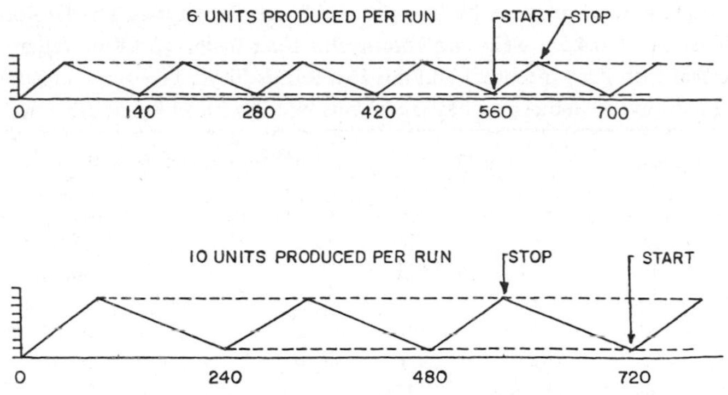

FIGURE 85.1

where the difference is $18,031. However, that final inventory cost starts at 0.32 years so that the present worth is:

$$PW_{downramp} = \$18,031\ e^{.11333(.32)} = \$17,389.$$

The total inventory present worth using a single production run is calculated as the sum of the first phase during production and the second phase after production or:

$$PW_{inventory} = \$8,496. + \$17,389 = \$25,885.$$

Two other cost components are the production work present worth of $7,742 and the single and setup of $180 for a grand total of:

$$PW_{total} = \$7,742 + 25,885 + 180 = \$33,807$$

Alternative Strategies Between These Extremes

There are many alternative production strategies, starting with making 2 items in each run and restarting production every 40 hours of operation time. Based on the learning curve above, the initial production would require 19.92 hours. During the 20.08 hours before restarting production (i.e., 40-hours less 19.92 hours), forgetting would be expected to be very small and so the next learning curve is estimated to start at 9.92 hours for the first unit. Continuing this strategy of 2 lots-at-a-time would result in the annual 2000 h divided by 40-hour cycles between successive production starts for a total of 50 cycles a year. Other strategies of 3, 4, 5, ..., 10 lots of production in each cycle are shown in Figure 85.1. Longer production runs are illustrated along with longer gaps between subsequential runs. Table 85.6 provides additional data on these alternative strategies. It is assumed in this example that the first cycle is the initial production run, so the second cycle production is affected by forgetting after the first cycle, but all subsequent cycles have production times similar to those shown in the second cycle. These cycle times were based on the first cycle's initial time of 10 hours and on the fact that the second and subsequent cycles started production 10 hours before the next lot was needed for issue.

In this general situation, an initial cycle follows the learning curve stated above. Following that, a forgetting curve occurs that shows the amount of learning lost before production restarts. This forgetting curve is based on the calculation 150 operating hours (3 months) as the longest time knowledge is retained. It is assumed here that the time gap between ending production on the nth run and restarting production on the next run will increase forgetting (or decrease learned skill). For the time gap g, the

TABLE 85.6 Production Run-Size Problem with Learning and Forgetting Curves. The 1st Run is 10 hours Short to Produce a Minimum Inventory of 0.4 Lots. The 2nd Through the Next-to-the-Last Run, Where the Forgetting Curve Occurs, Completes the Year with Zero Inventory and this Last Run is Either Like the 2nd But with 10 Hours More Time or the Last Run Produces the Remainder of the 80 Units Not Produced Previously with Fixed Run Sizes.

	No.	Cycle	1st Run		2nd Run to N-1th. Run			Last Run	
Run Length	Runs	Time	Duration	Gap	Duration	1st Unit	Gap	Duration	Gap
1	80.0	25	10.0	5.00	9.925	9.925	15.07	9.925	25.075
2	50.0	50	19.92	20.08	19.65	9.863	30.35	19.65	40.35
3	26.67	75	29.76	35.24	29.23	9.820	45.77	19.64	40.36
4	20.0	100	39.53	50.47	38.72	9.793	61.28	38.72	71.28
5	16.0	125	49.22	65.78	48.18	9.785	76.82	48.18	86.82
6	13.33	150	58.83	81.48	57.63	9.789	92.37	19.58	40.42
7	11.42	175	68.31	96.69	67.38	9.84	107.62	29.46	55.55
8	10.0	200	77.78	112.22	74.52	9.84	125.48	74.52	135.48
9	8.89	225	87.18	127.82	86.44	9.90	138.56	77.77	132.23
10	8.0	250	96.59	143.41	96.27	9.96	153.73	96.27	163.73

fraction (1- g/150 h) is multiplied by ($y_1 - y_m$) and that value is subtracted from y_1. That computed value is repeated in the next learning curve until the actual curve is lower. This computational procedure of forgetting brings the process back to an earlier value on the original learning curve. For example, suppose the company produced 7 lots at a time. The initial production follows the learning curve from 10.0 down to 9.543 hours. The amount learned is 10.0 – 9.473 h or a learning effect of 0.5275 hours. The time gap between two successive production runs is 107 hours. Thus, the fraction computed is (1 – 107/150) or 28.67% of the 0.5275 hours or 0.1512 hours retained. This is 10.0 – 0.5275 = 9.4725 hours for the restart initial time. Inasmuch as Arzi and Shtub (1996) and Dar-El and Vollichman (1996) found that interrupted tasks which are restarted come back to the original learning curves after a short transition period. Their transition (forgetting) period was not precisely described and the one shown above is my conjecture. Table 85.6 shows production durations on the 1st run where the initial learning occurred. On the second and subsequent runs there is some forgetting as reflected by the time required to make the first unit, and the combination of learning and forgetting is reflected by the duration of the 2nd run compared to the 1st run. Note in the 2nd run how the time required to make the first unit, decreases down to the case of 6 lots production and thereafter increases. Longer run lengths have greater learning but also longer time gaps between production runs. Between runs of length 6 and 7, the gap effect overcame the longer learning effect.

Specific production and inventory levels are given in Figure 85.2 for the case of producing 7 lots in a production run. Note that this figure shows the 1st and 2nd production runs at the top. At the bottom of this figure are part of the 10th run, the 11th run and part of the last (12th) run. Runs 2 through 11 are identical in character. Since 11 runs each produce 7 lots, the combined production of 77 lots is 3 short of the required 80 lots per year and the final production run completes those 3 lots and inventory goes to zero on New Year's eve. As there are only 3 different types of runs, economic calculations for each of the 3 types are computed separately in Table 85.7. Each separate calculation starts with the temporary assumption that the production started at time zero. Within each production run the initial phase of manufacturing and dispensing inventory, while the final phase is only dispensing inventory. These two phases are computed separately in Table 85.7. Note that all of the coefficients are based on an annual basis and so the time units in the calculations are yearly fractions. These units must agree. Production operations cost of $24,000 per year uniformly over the actual length of the production phase and the present worth calculations are for a step function. Inventory costs during the production and dispensing phase is an up-ramp whereas during the dispensing phase there is a down-ramp, where present worth is computed as a step at the maximum inventory in that run minus an up-ramp at a cost of $80,000 per year. Each type is similarly calculated. However, the 2nd run has 10 replications and setup costs, $180 each, for runs 2 through 12 have 11 replications. The present worth of the final production run is shown

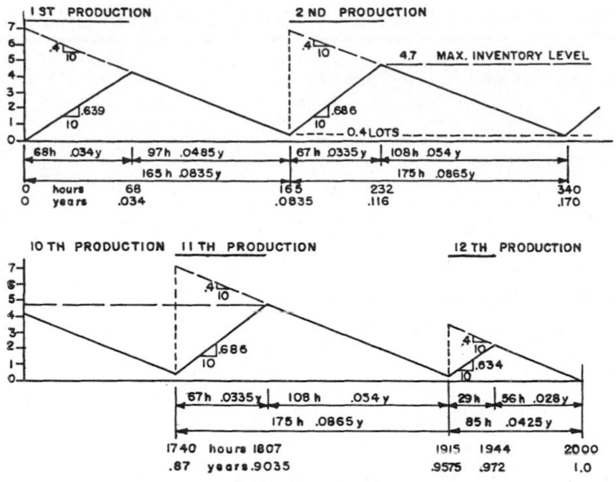

FIGURE 85.2

TABLE 85.7A Present Worth Computations of the First (1st) Production Run.

GENERAL COMPUTATIONS OF THE PRODUCTION-RUN-SIZE PROBLEM

1st Production Run —

Production Labor Cost Present Worth

\$12.00/hr × 2000 hrs/year = \$24,000/year

$$\text{PW} = \$24{,}000/\text{year}\left(\frac{1-e^{-0.11333(.0342)}}{0.11333}\right) = \$818.25$$

Inventory Cost During Production — Up Ramp 0 to 0.0342 y-

@ .06247 lots/hr × 2000hrs/year × \$1000/lot = \$124,940/year

$$\text{PW} = \$124{,}940\left(\frac{1-e^{-.1133(.0342)}}{.1133^2} - \frac{.0342e^{-.1133(.0342)}}{.1133}\right) = \$72.63$$

Inventory Cost During Dispensing Down Ramp from .0342 to .04834 y

@ 0.04 lots/hr × 2000hrs/year × \$1000/lot = \$80,000/year reaching 4.27 lots inventory

$$\text{PW}_{\text{step}} = \$4{,}270/\text{year}\left(\frac{1-e^{-0.11333(.04834)}}{0.11333}\right)e^{-.1133(.03416)} = \$204.99$$

$$\text{PW}_{\text{ramp}} = \$80{,}000/\text{yr}\left(\frac{1-e^{-.1133(.0483)}}{.1133^2} - \frac{.0483e^{-.1133(.0483)}}{.1133}\right)e^{-.1133(.0342)} = \$74.82$$

The Total for the 1st. Run excluding Setup Cost is:

\$818.25 + 72.82 +204.99 – 74.82 = \$1,021.05

under the temporary assumption of a zero time start and then it is corrected for starting 0.9575 years instead. Finally, present worth computations of all 3 types of runs, including the initial setup cost, are combined to show the overall present worth for a year.

TABLE 85.7B Present Worth Calculations for the 2nd Production Run of the Learning and Forgetting Example.

2 nd Production Run (Tentatively Consider Start at Time 0)

Production Labor Cost Present Worth

\$12.00/hr × 2000 hrs/year = \$24,000/year

$$PW = \$24,000/\text{year}\left(\frac{1-e^{-0.11333(.0337)}}{0.11333}\right) = \$806.99$$

Inventory Cost During Production- Up Ramp 0 to 0.0337 years

@ .06389 lots/hr × 2000hrs/year × \$1000/lot = \$127,780/year

$$PW_{ramp} = \$127,780\left(\frac{1-e^{-.1133(.0337)}}{.1133^2} - \frac{.0337e^{-.1133(.0337)}}{.1133}\right) = \$72.33$$

Inventory Cost During Dispensing Down Ramp from .0337 y to .0875 y

@ 0.04 lots/hr × 2000hrs/year × \$1000/lot = \$80,000/year reaching 4.27 lots inventory

$$PW_{ramp} = \$80,000\left(\frac{1-e^{-.113(.0538)}}{.1133^2} - \frac{.0538e^{-.113(.0538)}}{.1133}\right)e^{-.1133(.0337)} = \$114.92$$

This gives a total cost for the 2nd run, excluding setup costs, consists of:

806.99 + 72.33 + 170.72 – 114.92 = \$935.12

Combined Costs of Production Runs 2 through 11, which are each identical in cost but are 175 hours [a]or 0.0865 years) apart temporarily let the start of the 2nd run as time zero with 10 repetitions:

$$PW = \$935.12\left\{\frac{1-e^{-.1133(0.0865)10}}{1-e^{-.1133(0.0865)}}\right\} = \$8,951.20$$

Setup Costs of \$180.00 Occur at the Beginning of Each of these Runs, plus at the end of the 11th. run, and at the beginning of the first run. Excluding the initial setup, the total amount of PW at the beginning of the 2nd run is

$$PW = \$180\left\{\frac{1-e^{-.1133(0.0865)11}}{1-e^{-.1133(0.0865)}}\right\} = \$1,886.20$$

These two cost components sum to \$8,951.20 + \$1,886.20 = 10,837.40

Table 85.8 summarizes the production, setup, and inventory costs present worth for each of the alternative production run lengths from 1 at-a-time to 10 at-a-time production. All of the calculations follow similarly to those in Table 85.7 covering a single year. These data show the minimum cost strategy of those considered is making run length of 7 lots per run.

There is an economic principle which reappears in the literature of practitioners, but it is not frequently stated. The principle is, "It is rarely very important to be precise in finding the optimal economic alternative, but it is imperative to be near." The total cost differences between the least cost alternative strategies at producing 7 lots during a production run and those quite similar are quite small. In fact, those differences are likely much smaller that the precision associated with the cost estimates. But differences between the least cost and those alternatives much farther from the minimum cost are very striking. Hence, it is very important to identify the approximate optimum economic choices. There is an ergonomic principle authored by Helson (1949) which is very similar. It is known as his U-Hypothesis and it is: "For most variables of concern in ergonomics, performance, as an inverted function of that variable, is U-shaped. The bottom of the U is nearly flat but the extremes rise almost vertically." Here again, the point is the same. One location at the bottom of the U is much the same as another, but being

TABLE 85.7C General Computations of Present Worth for the Final Production Run at the Minimum Attractive Rate of Return of 11.333% (Continuously Compounded).

The LAST Production Run — temporarily assume 0.9575 years is time 0

Production Labor Cost Present Worth

\$12.00/hr × 2000 hrs/year = \$24,000/year

$$PW = \$24{,}000/\text{year}\left(\frac{1-e^{-0.11333(.0145)}}{0.11333}\right) = \$347.71$$

Inventory Cost During Production- Up Ramp 0 to 0.0342 y-

@ .0634 lots/hr × 2000hrs/year × \$1000/lot = \$126,800/year

$$PW = \$126{,}800\left(\frac{1-e^{-.1133(.0145)}}{.1133^2} - \frac{.0145e^{-.1133(.0145)}}{.1133}\right) = \$13.31$$

Inventory Cost During Dispensing Down Ramp from .972 y to .1.0 y where the maximum inventory there is 2.23 lots, decreasing at 0.04 lots per hour.

@ 0.04 lots/hr × 2000hrs/year × \$1000/lot = \$80,000/year reaching 2.23 lots inventory

$$PW_{step} = \$2{,}238.60/\text{year}\left(\frac{1-e^{0.11333(.028)}}{0.11333}\right)e^{-.1133(.0145)} = \$62.58$$

$$PW_{ramp} = \$80{,}000/\text{yr}\left(\frac{1-e^{-.1133(.028)}}{.1133^2} - \frac{.028e^{-.1133(.028)}}{.1133}\right)e^{-.1133(.0145)} = \$31.24$$

The Total for the 1st. Run excluding Setup Cost is:

\$347.71+ 13.31 + 62.58 – 31.24 = \$392.35

Now correcting for the temporary assumption and finding the present worth of the last production run in terms of the actual time zero is:

\$392.35 $e^{-.11333(0.9575)}$ = \$350.31

Completing the Calculations

Adding the Initial Setup Cost of \$180.00 plus the 1 st Run Cost Plus the Last Run Cost Plus the Intermediate Runs

\$180.00 + 1,021.05 + 10,837.40 $e^{-.1133(0.0835)}$ + 350.31 = \$12, 106.68

TABLE 85.8 Summary of Production and Inventory Costs.

Number Made per Run	Production Work PW	Setup Cost PW	Inventory PW	Total PW
1	9,074	13,614	10.00	22,688
2	8,858	6,726	666	16,251
3	10,321	5,246	742	16,309
4	9,825	3,789	840	14,454
5	9,790	2,970	908	13,668
6	9,547	2,441	1,033	13,020
7	8,853	2,069	1,469	12,396
8	9,397	1,802	1,459	12,658
9	9,932	1,593	1,660	13,186
10	8,845	1,428	3,134	13,310
...				
80	7,742	180	25,885	33,807

TABLE 85.9 Evaluating Alternative Learning Curves from Component Speeds and Asymptotes.

Asymptote	$\alpha = 0.98$	$\alpha = 0.95$	Speed Savings
$y^* = 6$ hrs	0.320 yrs $7546	0.279 yrs $6599	.041 yrs $947
$y^* = 5$ hrs	0.300 yrs $7083	0.249 yrs $5897	.051 yrs $1186
Asymptote Savings	0.0200 yrs $463	0.030 yrs $702	.071 yrs $1649

outside of the U-bottom can be a design tragedy. This observation describes a strong ergonomic–economic interaction. Nadler (1970) strongly advocated searches for optimum designs.

Evaluating Training Programs

One feature of training programs that have economic consideration is the time change that a training program can have on production work following training. It should be stated that other features have economic significance as well. When the training program reduces errors in the operations, that error reduction is a valuable asset as well, but this discussion focuses only on the performance time. A rather elementary approach to evaluating training programs is to find Present Worth changes in the learning curve with and without the training program. In the situation above, for example, the full learning curve over 80 lots was evaluated at:

$$PW = \frac{\$24,000}{.11333}\left[1 - e^{.11333(0.32014)}\right] = \$7,546.$$

just for direct labor costs during a single annual production run.

In considering a training program, the concentration could be on faster learning and a reduction in the asymptote. A lower asymptote usually involves finding a change in the method of assembly and perhaps new jigs and fixtures, but quicker learning can sometimes be achieved by pointing out the things to be aware of and some of the things assemblers do wrong. Let us say that the ergonomic manager suggested that they should seek both a quicker rate of learning and a lower final asymptote. For an improvement rate the ergonomic team targeted the learning rate improvement in terms of lowering α from 0.98 to 0.95. They also targeted an asymptote reduction from 6 hours per lot to 5 hours. If the team is able to achieve both targeted features in their training program, then the total learning curve at $\alpha = 0.95$ and $y^* = 5$ hours is:

$$Y_{80} = \frac{1 - 0.95^{80}}{1 - 0.95}[10 - 5] + 5(80) = 498.\text{ hours}$$

which is equivalent to 0.2492 years based on 2000 production hours per year. The present worth of just the direct labor cost is:

$$PW = \frac{\$24,000}{0.11333}\left[1 - e^{-0.11333(0.24917)}\right] = \$5,897.$$

As a first-pass economic test, this training program initially passes this test if the cost of it was less than the cost savings of $7,546 – $5,897 or $1,649 per presentation. It is interesting and often useful to investigate saving due to the learning-speed improvement and the asymptote improvement as shown in

Table 85.9. The two different alpha values describe speed differences, with a smaller alpha denoting faster learning and the lower y^* describing a lower asymptote. It is clear from a review of the results in Table 85.9 that savings are not linear with the causes of savings. But it is equally clear from a review of those tabled data that a smaller α and a smaller y^* have synergistic effects. The same reduction in α causes a greater effect on performance times with a smaller y^* and conversely.

85.7 Effects of the Lack of Manufacturing Specification on Repair

One of the cost components that is gaining larger importance because of rapid increases over the past several decades is the cost of repair. Repair times are becoming particularly large due to the lack of ergonomic attention associated with repair and maintenance activities during new product design. A case in point is when companies in charge of the final component assemblies fail to recognize what they are doing in design is insufficient. Many companies who are trying to sell a product competitively against most aggressive competitors often tell the vendors that as long as they match interconnecting features and keep the price least, that they will be favored. In this situation things look good until there is a repair to be made. It is then that the repair persons suddenly find that component piece has many different bolt or screw sizes. As a result the repair person needs additional wrenches in the repair. As the repair person is disassembling the component to get to the failed component part, he or she tries to use the wrench in hand to remove a nut but finds that it is the wrong size. This situation requires added time from the operator doing the repair work to discard the wrong tool and to identify the correct tool size before starting to remove that component. For every change in bolt size, there is at least a change in tool and often several before the correct one is found. In addition, many repair crews have assistance made up by the support people who put together the kits those people use. When the repair people do not have the correct size or type of wrench in their kit, they either must work with an inferior wrench (screwdriver or other tool) while the activity is being performed or they must stop that repair activity and come to the tool dispensary and retrieve the correct tool before returning to the repair activity and correctly performing the repair activities as it should be done. That result is often required because the company simply ignores what vendors do on product designs that do not affect the interface connections of the prime contractor.

There are numerous repair principles for greater economic effectiveness. One is simply, "*Arrange activities in a sequence or create sequential activities that require the same tools.*" Note that the reason for this principle is that every time this rule is violated, the repair person must put down one tool and select another for use. That simple tool put down and subsequent pickup simply takes unnecessary time to perform. When someone asks me if they need to think about time-and-motion principles all the time, I merely mention that the answer is "yes" if they wish to be considerate and more so if they wish to be successful. Now one almost always faces the tradeoff of an existing choice of some design, bolt size, or screw type, with regard to additional connector decisions. That leads to another principle which is, "*Keep the number of different types of screws, nails, and other connectors as small as effectively possible so that the tool kits can also be minimized.*" Tool kit sizes are particularly important when the item to be repaired is large enough that the repair person must be transported to it rather than the other way around. In that case a missing tool in the tool kit necessitates another trip from the object or thing under repair to the tool repository and back again.

Another ergonomic feature of maintenance and repair is that those maintenance and repair activities are so infrequent that memory often fails the repair person. This is a good situation to use a **job aid**, that is, a device which enables one to perform a task which one could not do before and to do it in the first try (Wulff and Berry, 1962) Job aids can be as simple as a chart that shows garage mechanics where to lubricate the suspension of an automobile of a particular make, year, and model. More extensive job aids may consist of elaborate productions which tell an operator how to disassemble, diagnose, fix, and reassemble a complex machine.

85.8 Parallel Processing Economics

Ergonomists frequently face the task of assigning multiple assemblers, pairs of machines and their operators, and multiple processors to a job because no single assembler, machine, or processor could handle the requirements alone. The principal ergonomic–economic question is which pairs should be assigned to the job? An answer lies with an old economic principle which states, "When multiple devices must be used concurrently, then the optimum economic relationship is for the multiple devices to operate such that marginal costs of each are equal." This principle doesn't specify the number of devices except being more than one (see Buck and Askin (1983) for a more complete development of this principle.) Practical experience suggests that the number should be minimum to handle the job.

In the ergonomics of "crews" or "teams," it should be recognized that crew activities in most instances involve parallel operations. Boat crews have crew members, rowing and sailing crews have seamen attending to the sails, rudder, and boat trim. Similarly, in industry, order-picking crews take customers' orders and go out into the inventories and fill them. Loading and unloading crews take objects from one or more locations and put them into a vessel for shipment or vice versa. In any case, one must select N crew personnel and assign them to the activities to be performed. The question addressed here is the selection of personnel or personnel and machines where it is assumed that all crew personnel essentially perform the same functions. That is, specialties among crew members (e.g., a punter in football or a center in basketball) is not addressed.

In the discussion to follow, the subscript symbol **i** is an identifier of a processor consisting of a particular operator and a particular machine-tool. While a different human operator with the same machine-tool will have a different identifier, a distinct pair of processors are two different persons and two different machine tools. To illustrate this situation, let us assume that the cost of each parallel processor followed the function†:

$$y_i(x_i) = \alpha_i^{x_i} y_i(0) + \beta_i \frac{1 - a_i^{x_i}}{1 - a_i} \tag{15}$$

where

i = an index denoting specific processors,
x_i = the workload assigned to the *ith* processor,
α_i = the rate-of-change parameter as a function of the exponential workload x_i ($\alpha_i > 1$),
β = is the scale parameter, and
$y_i(0)$ = the setup cost.

Although this cost function appears to differ from the one used to describe the learning curve, it is in the same family and the right-hand-side equates to:

$$y_i(x_i) = \alpha_i^{x_i} y_i(0) - \frac{\alpha_i^{x_i}(\beta_i)}{1 - \alpha_i} + \frac{\beta_i}{1 - \alpha_i}$$

where the last term is equivalent to y* and the middle term is essentially α^{xy^*}. Accordingly, equation (15) is the same as equation (12), verifying that the function is of the same family. Marginal costs of this function are shown in equation (13) or:

$$\Delta y_i(x_i) = \alpha_i^{x_i}(\alpha_i - 1)[y_i(0) - y^*_i] \tag{16}$$

†The fundamental function here is the same as the one used as a learning curve above. Goldberg (1961) provides details. A principal difference here is that the alpha (α) parameter here is a value greater than unity so that costs increase with the amount of work assigned x in an exponential manner away from y*. Accordingly, y* is not an asymptote in this model except in a beginning sense but it is an important parameter.

where y^*_i is defined as:

$$y^*_i = \frac{\beta_i}{1-\alpha_i} \tag{17}$$

Equation (16) describes the marginal costs of different combinations of human operators with differences in skills and different machine-tools which also have different capabilities. If two distinct pairs of processors are under consideration (say i = 1 and i = 2) and their marginal costs are equal, then it follows that:

$$\alpha_i^{x_1}(\alpha_1 - 1)[y_1(0) - y_1^*] = \alpha_2^{x_2}(\alpha_2 - 1)[y_2(0) - y_2^*]$$

Bring the two alpha factors with exponents to the left-hand-side and the rest to the right as:

$$\frac{\alpha_i^{x_1}}{\alpha_2^{x_2}} = \frac{(\alpha_2 - 1)\,[y_2(0) - y_2^*]}{(\alpha_1 - 1)\,[y_1(0) - y_1^*]}$$

Everything on the right-hand-side of the equation above is known so that it results in the constant which can be called K. The result, with a little simplification, is:

$$\alpha_1^{x_1} = \alpha_2^{x_2}\ K$$

When the above equation is converted to logarithmic form, x_1 and be written in terms of x_2 as:

$$x_1 = x_2 \frac{\log \alpha_2}{\log \alpha_1} + \frac{\log K}{\log \alpha_1} \tag{18}$$

Equation (18) shows the workload to processor 1 (or x_1) as a linear function of the workload to processor 2 (or x_2) because the two fractions in equation (18) are constants. So long as the workloads to the two processors maintains the relationship in (18), no better combination of those two processors can be obtained. Figure 85.3 describes the workload assigned to processor 1 as a function of that assigned to processor 2. At x_2 equals 0, equation (18) has the constant ratio of log K/log α_1 as the intercept. Then for each unit increase in the workload to processor two ($x_{2)}$, the workload to processor one ($x_{1)}$ changes at the rate of log α_2/log α_1. That optimum x^*_1 to x^*_2 relationship is:

$$x_1^* = \frac{\text{Log}\ \alpha_2}{\text{Log}\ \alpha_1} x_2^* + \frac{\text{Log}\ K}{\text{Log}\ \alpha_1} = 0.6949\ x_2^* - 3.0908$$

where the intercept of the $x^*_1 - x^*_2$ relationship is negative but the slope is positive. The equation above also denotes that the minimum amount of workload to x_2 is 4.45 units when x_1 and x_2 are used together (otherwise x_1 or x_2 are used separately). Those downward sloping lines in Figure 85.3 describe contours with equal sums of x_1 and x_2. When the sum of x_1 and x_2 adds up to the needed amount Q, then those values of workload describe the workloads which will complete the job, and the sum of the two cost functions describe the cost. However, the cost sum of both processors, as Figure 85.3 shows, is only optimum economically for that combination of humans and machines. If those two person–machine–tool combinations are the only ones available, then no other combinations of two person–machine–tools are less costly, but clearly the use of a single person and machine tool can be. In fact, costs of processors 1 and 2 are described individually below as a function of the assigned workload. Results above show that processor 1 starts off less costly but between 6 and 7 units of workload individually, processor 2 becomes

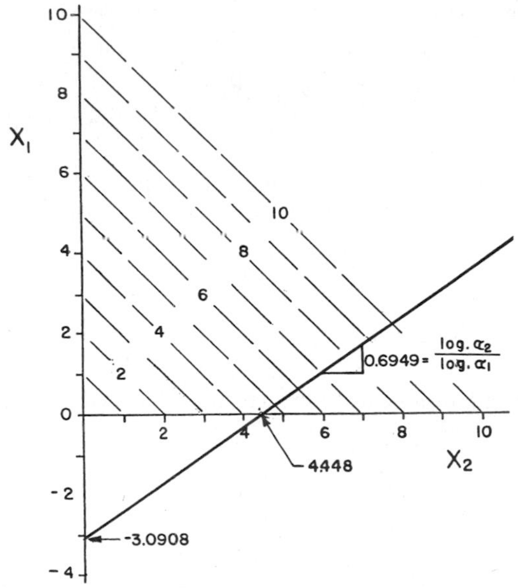

FIGURE 85.3

less costly. In fact, at about 6.32 units of workload operating costs of the two individual processors are approximately equal. Accordingly, it is more economical to use processor 1 alone so long as the workload is below 6.32 units. From that level upward, first use processor 2 alone. At about 10.7 units of workload, it becomes more economical to use both processors 1 and 2 together at the optimal combination. At that point forward, both processors should be used together as stated above. While it is purely conjecture, one may wonder if people similarly use a form of concurrent processing of attention by allocating attention between pairs of tasks on an approximate procedure resembling the principle above.

x_i =	1	2	3	4	5	6	7	8
$y_1(x_i)$ =	33.0	36.9	42.0	48.6	57.1	68.3	82.8	101.6
$y_2(x_i)$ =	43.0	46.6	50.9	56.1	63.3	69.8	78.8	89.5

85.9 Using Economic Principles to Guide Ergonomic Studies of Automated Highway Design

On occasion, ergonomists are called upon to help develop new public systems or to assist in creating new laws regulating industrial operations, usually in health and safety. Those situations usually do not have easy precedence to follow. The only other recourse is for ergonomists to follow reasonableness and evolution. There is perhaps no better guide than selecting the less costly alternative as a starting point. The case below involves the design of a new form of highway system where the conceptual design is uncertain. Using the reasonableness and evolution principle, it was assumed that the new system would be as close to our existing systems as was possible without jeopardizing the fundamental nature of the concept.

For a couple of decades, highway people have dreamed of having an automated highway system (AHS) where a computer could control the vehicles. Part of this dream was a fantastic reduction in automobile accidents. Another part was an equally fantastic increase in highway capacity; particularly around our large cities where huge numbers of drivers enter the city in the mornings and leave each evening. It is even more obvious that the urban AHS also has the problem of very expensive land so that conventional expressways cannot be economically used to solve the problem of excessive traffic volumes. However, it is even less clear what alternative design of an AHS is even reasonably feasible. The story to follow is a

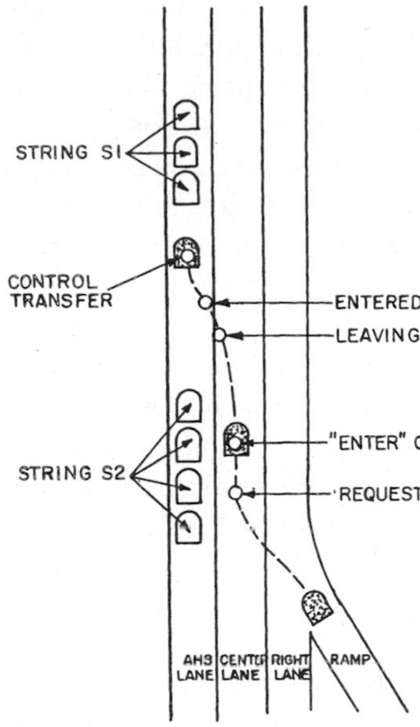

FIGURE 85.4

result of an initial effort by the Federal Highway Administration (FHA) of the U.S. government in an ergonomic research effort to explore this question. (See Buck and Yenamendra, 1996.)

In an initial examination of the highway economics by the AHS research group, it became evident that it might be possible for a relatively low-cost version of the AHS concept to work and so this group of researchers first focused on that concept. If the initial experimentation proved otherwise, then an alternative AHS with greater investments and expenses could be examined. Although we are referring to the initial version as low-cost, that is only relative to the other versions. Costs of this version will be substantial because of extensive computers and sensor equipment, as well as at least one added traffic lane on the existing expressways and bridges. The envisioned concept was that the traditional two-lane (each way) expressway would be expanded by a single third lane in each direction which would be used exclusively by AHS-outfitted vehicles. This added lane is expected to cost at least \$3 to $\$4 \times 10^6$ per mile in each direction for just the concrete. While this configuration is not inexpensive, it is vastly less expensive than duplicating the existing interstate roadways as an alternative plan proposed.

Figure 85.4 illustrates this AHS concept with strings of AHS vehicles traveling in the left-most lane with gaps between successive strings. The right two lanes contain non-AHS vehicles and AHS vehicles either waiting to join the other AHS vehicles, departing from the AHS lane for an exit ramp, or AHS vehicles who simply choose to remain in manual mode for one reason or another. That schematic in Figure 85.4 illustrates the entering AHS vehicle coming from the on-ramp to the right lane of the expressway, crossing over to the center lane before attempting entry into the AHS lane. It was assumed that the entry procedure required the driver to first request entry, then the central computer would check the vehicle for adequacy to drive in AHS mode, and if the checkout was adequate, the computer would direct the vehicle to enter immediately behind a passing string of AHS vehicles. The driver of the entering vehicle was instructed to accelerate as rapidly as possible after entry until the computer took over. One of the modes of transferring control from the driver to the central computer was automatic as soon as the first wheel crossed into the AHS lane. The other mode was where the driver manually shifted control by pushing a button on the steering wheel. It was felt that there may be legal reasons for a manual mode of control transfer even though that mode may be slower.

It was assumed that AHS cars left the AHS lane by requesting to exit, splitting the string of vehicles so that the exiting vehicle was last, and then manually driving into the center lane. Experiments in an automobile simulator verified that both older and younger drivers could perform the maneuvers of both entering and exiting AHS. Almost subjects experienced difficulties during exiting. The difficulties encountered during exiting were almost always due to difficulties in seeing slower manually controlled vehicles in the middle lane due to the vehicles immediately ahead of the exiting car. As the driver of the exiting car was seated on the left-side of the car and exiting maneuvers necessitated rightward movements of the vehicle, another car could be a short distance ahead and not be very visible due to intervening cars. This difficulty was compounded with greater speed differences between the AHS design speeds and those allowed for manually controlled traffic because of the short time between entry and encountering a vehicle in the middle lane. Those difficulties appeared to be correctable with some form of vision aiding.

Entering maneuvers posed a different approach. The maneuvers necessitated the gaps between successive strings of vehicles driving along the AHS lane in order to get the entering vehicle safely into that lane. If the gap was too small, then the entering vehicle could not accelerate rapidly enough to the AHS design speed before the following vehicles would catch up to the entering vehicle. Clearly, the greater the differential in speed between the center lane and the AHS design speed, the longer it takes a vehicle to accelerate to the AHS design speed and the greater the gap needed between successive AHS strings of vehicles. In fact, the entering car was traveling at 55 miles per hour in the center lane and then acceleration to any of three different AHS design speeds of 65, 80, and 95 miles per hour requires respective minimum string-to-string distances of 32.5, 121.5, and 335.0 meters. That is, the gap at the AHS design speed of 80 miles per hour is almost 3.75 times that at 65 miles per hour and the one at 95 miles per hour is 2.76 times the length of the gap at 80 miles per hour or 10.3 times the one at 65 miles per hour. It was envisioned that the controlling computer could create an opening for a new vehicle by slowing down vehicles behind the spot targeted for the entering vehicle to a sufficiently long gap. As the vehicle entered at a lower speed and started to accelerate, the faster vehicle behind would catch up and close the gap. Our experiments on people accommodating with this entry play proved that they could perform the maneuvers extremely well. In fact the degree to which they could perform the entries came as a surprise.

After the vehicles entered the AHS lane, they were placed in collections of up to four-car strings which would move at the AHS design velocity with this minimum gap. Minimum gaps were about 1/16 seconds and that time interval translated into about 3 to 5 meters separation distance. When those four-car-strings have minimum gaps within strings and inter-string distances as shown above for these three design speeds, the number of vehicles per hour in the AHS lane at this maximum theoretical case would be 7,551, 3,531, and 1,825 vehicles per hour for the AHS design speeds of 65, 80, and 95 miles per hour, respectively. As a result, the lowest of these three AHS design speeds was recommended with this particular design for greater highway capacities. Note that the capacities described above are theoretical upper limits rather than reasonable capacity estimates. But at the recommended AHS design speed, that theoretical capacity is 441% greater than the upper limit of conventional traffic, which is about 1712 vehicles per hour. If the practical upper limit of AHS capacity was only 3/4 of the theoretical limit and there were ample AHS outfitted vehicles to use the single AHS lane, the increase in capacity due to AHS capability would be expected to be over 175% of that with an added third lane of conventional traffic. This economic information would have been impossible without the ergonomic information that people could perform the entry and exit tasks and the delay in AHS traffic they cause during the entry.

These data demonstrate that the AHS configuration is feasible and that this lowest-cost plan can be economically viable, but it does not show that another configuration may be better. For example, another configuration could consist of a partial fourth lane which is added at locations where entry and exit maneuvers are to be performed and that lane would be long enough to allow vehicles to accelerate to the AHS design velocity and decelerate from it to 55 miles per hour. This new configuration would increase AHS capacity more at the higher speeds, but that added capacity is conjecture without confirming ergonomic data. If the ergonomic testing had been carried on assuming a more expensive version of AHS, the public would have insisted on running further ergonomic tests at lower-cost versions of AHS before starting the long political journey toward the AHS. As it is, future ergonomic studies can proceed

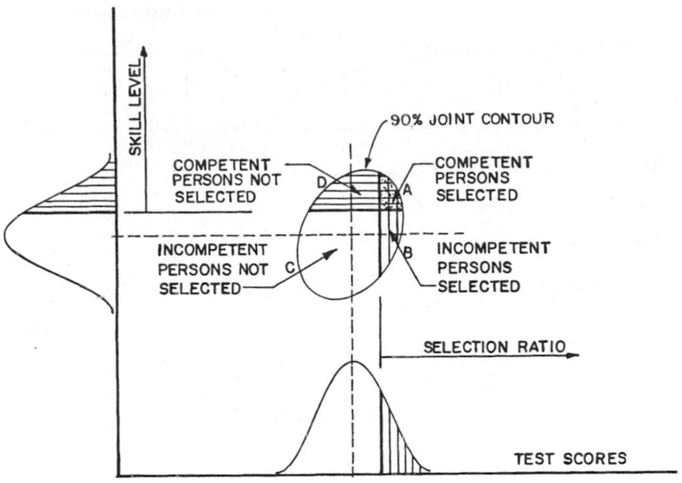

FIGURE 85.5

to fine-tune the eventual AHS configuration, (e.g., the acceleration lane addition) now that initial feasibility appears evident.

85.10 Personnel Selection Economics

Personnel selection involves a selection basis that may consist of a test, questionnaire, and/or interview evaluations. It is assumed here that applicants for the jobs have been administrated one or more tests, questionnaires, and/or an interview and their composite scores are shown as the horizontal axis in Figure 85.5. The vertical axis of this figure denotes job performance scores, and in the middle, between these axes, are the relative frequencies of persons with competent and subcompetent performance such as the upper and lower regions around the dashed line. The composite scores from the interview–testing instruments separate the persons selected on the basis of those scores on the right of the dashed line who were selected from those on the left region where people were not. A vertical line separating left and right regions shows the percentage of persons selected on the basis of these scores (i.e., selection ratio).

The composite of the selection ratio and the minimum competency level yields four regions which are marked:

A. Competent persons selected
B. Incompetent persons selected
C. Incompetent person rejected by the test/interview
D. Competent persons rejected

The ellipse shown in this figure is a 95% contour line of a bivariate normal distribution. The narrowness of the shorter ellipse axes relative to the longer axes depends upon the correlation coefficient **r** between test scores and competency level. That coefficient is often called the *validity coefficient.* As r approaches unity, a far greater proportion of the selected persons fall in the A and C regions rather than the B and D regions. In fact a special statistic **Z** is the fraction of persons selected who are expected to be competent. Of the persons in regions A or B, Z is the expected fraction who are in region A. This Z statistic changes with the three variables:

1. Selection ratio
2. Validity coefficient
3. The proportion of the population who can perform satisfactorily on the job

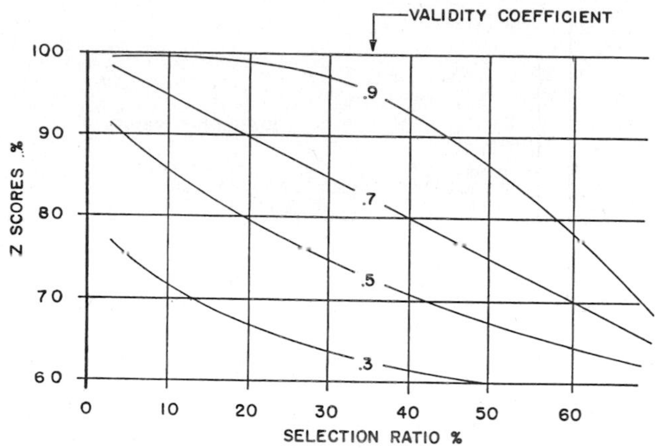

FIGURE 85.6

Figure 85.6 shows **Z** statistics as a function of the first two variables but with the third set with the competency level 50% of the population. That figure shows slopes decrease with greater selection ratios (i.e., lower percentages with greater fractions selected). Figure 85.6 also shows that Z-scores drop more gradually with small selection ratios when validity coefficients are large, but when r is small, there is greater immediate decrease in Z-scores with low selection ratios. For example, if there were 50 applicants and a selection ratio of 40%, 20 persons with the highest test–interview scores would be selected. If that test–interview rating has a validity coefficient r = 0.7 for a job where 50% of the people could perform it satisfactorily, Figure 85.6 shows that one should expect that 80% of those selected would perform in a satisfactory manner, or 16 of the 20 persons. Without the test and interview rating, the best one could expect is that 10 of the 20 persons could perform the job. In that and other cases of 50 applicants, all with jobs which 50% of the population could perform, the expected numbers of those selected who are expected to be satisfactory (S) and unsatisfactory (U) are indicated as:

	Selection Ratios							
Validity Coeff. r	40%		30%		20%		10%	
	S	U	S	U	S	U	S	U
0.9	18.4	1.6	14.4	0.6	9.8	0.2	4.95	0.05
0.7	16.0	4.0	12.6	2.4	9.0	1.0	4.75	0.25
0.5	14.4	5.6	11.1	4.9	7.8	2.2	4.25	0.75
0.3	12.4	7.6	9.6	5.4	6.7	3.3	3.60	1.40
0.1	11.0	9.0	8.4	6.6	5.6	4.4	2.85	2.15
Total Selected	20.0		15.0		10.0		5.0	

If a company observed that new hires who did not seem to perform well at the start would simply quit during the first two weeks, they would recognize that the cost associated with hiring people who could not perform well was nearly a constant equal to about two weeks pay and benefits, plus the test cost. Thus, the expected cost per selected applicant is approximately **C (1 – Z)**. At approximately $12 per hour; two weeks pay amounts to around $960. Now as management selects a better test–interview procedure, **Z** increases and this cost decreases. While a better test–interview procedure decreases the cost associated with hiring the wrong people, it also increases the cost of testing. If the cost of testing was estimated to be a function of the validity coefficient or **25** e^{3r}, the question remains as to the most

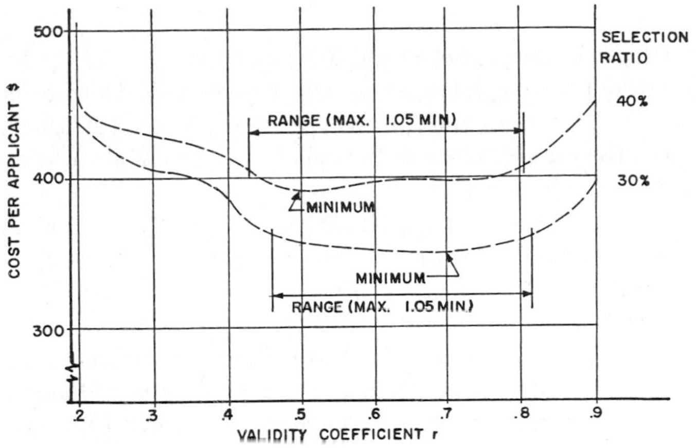

FIGURE 85.7

economical cost of the coefficient **r**. Figure 85.7 shows cost as a function of the validity coefficient **r** for the sum of the selection error cost and the test cost separately at 30% and 40% selection ratios. This illustration describes a decreasing total cost as long as the decreasing error cost is greater in magnitude than the added test cost and that total cost decreases to a minimum at r = 0.7 (for 30% selection ratio) and at **r** = 0.5 (at a 40% selection ratio). That figure also shows a range of r values where the total cost is within 5% of the minimum cost. This range of validity coefficient values is from a mid 0.4 value to slightly more than 0.8. Since it is difficult to estimate costs closer than 5%, total costs are frequently uncertain within 5%, so it is unimportant to find the optimum **r** value precisely, but it is important to find the approximate range of optimum r values. Here again, economics is like Harry Helson's (1949) U-Hypothesis in ergonomics in which he states that "Human performance is an inverted U-shape as a function of most variables of concern; the bottom of the U is nearly flat but the extremes fall strikingly." In both disciplines one seeks to be at the nearly flat U-bottom but the precise optimum is unimportant.

85.11 Final Remarks

Ergonomists compete with many parts of the company for limited amounts of capital available to improve the company. In order to compete effectively, ergonomists must identify and quantify the economic advantages and show them to be more important than other contenders if they expect to receive the investment funds. Accordingly, ergonomists must learn to identify important economic opportunities and they must learn to seize those with potential and then to get the greatest potential from them. While that requires art and ergonomical technology to find appropriate solutions, it also requires knowledge, prediction capability, and the forms of analysis shown above.

One of the pragmatic principles of economics is to look for some short-range optimum economic solutions to be sure the optimum range is identified, but one does not have to be dogmatic about specifying the very best. Solutions in the optimum ballpark are often indistinguishable from one another, and usually they are quite stable under expected variations. Another economic principle is to use single resources if possible, but if not, use multiple resources with balanced marginal costs.

Economics was formed from the study of people and logic. Things which are economically sound are usually humanistically balanced if realistically represented. For these very reasons, some of the ergonomic theories of how people behave comes from economic theory and related principles. That is why there are economic–ergonomic interactions and why both specialists of economics and ergonomics should be looking for them.

References

Anderson, D. (1990) *Design for Manufacturability*, CIM Press, USA.

Arzi, Y. and Shtub, A. (1996) Learning and forgetting in high cognitive and low cognitive tasks, *Advances in Applied Ergonomics*, A. F. Ozok and G. Salvendy (editors), U.S.A. Publishing, 672-675.

Au, T. and Au, T.P. (1992) *Engineering Economics for Capital Investment Analysis*, Second edition, Prentice Hall, Englewood Cliffs, New Jersey.

Bevis, F. W., Finniear, C., and Towill, D. (1970) "Prediction of Operator Performance During Learning of Repetitive Tasks," *International Journal of Production Research*, 8, 293-305.

Buck, J. R. and Hill, T. W., Jr. (1975) Alpha/beta difference equations and their zeta transforms in economic analysis, *American Institute of Industrial Engineers Transactions*, 7, 3, 330-358.

Buck, J. R., Tanchoco, J. M. A., and Sweet, A. L. (1976) Parameter estimation methods for discrete exponential learning curves, *American Institute of Industrial Engineers Transactions*, 8, 2, 184-194.

Buck, J. R. and Askin, R. G. (1983) On the economic loading of parallel processes, *The Engineering Economist*, 29, 1, 13-32.

Buck, J. R. and Cheng, S. W. J. (1993) Instructions and feedback effects on speed and accuracy with different learning curve functions, *Institute of Industrial Engineers Transactions*, 25, 6, 34-47.

Buck, J. R. and Yenamedra, A, (1996) Ergonomic issues on entering the automated highway system, Chapter 18 in *Ergonomics and Safety of Intelligent Driver Interfaces*, I. Noy, editor, Lawrence Erlbaum Associates, Mahwah, NJ, 309-328.

Dar-El, E. and Vollichman, R. (1996) speed v/s accuracy under continuous and intermittent learning, *Advances in Applied Ergonomics*, A. F. Ozok and G. Salvendy (editors), U.S.A. Publishing, 676-682.

DeGarmo, E. P., Sullivan, W. G., and Canada, J. R. (1984) *Engineering Economy*, 7th Edition, Macmillan, New York.

de Jong, J. R. (1957) The effect of increasing skill on cycle time and its consequences for time standards, *Ergonomics*, 1, 51-60.

Drury, C. G. (1988) The human as optimizer, *Contemporary Ergonomics*, E. D. Megaw, editor, Taylor & Francis, London.

Fabrycky, W. J., Thuesen, G. J., and Verma, D. (1998) *Economic Decision Analysis*, Third Edition, Prentice Hall, Upper Saddle River, New Jersey.

Fabrycky, W. J., Thuesen, G. J., and Verma, D. (1998) *Economic Decision Analysis*, Third Edition, Prentice Hall, Upper Saddle River, New Jersey.

Glover, J. (1966) Manufacturing process functions ii: selections of trainees and control of their progress, *International Journal of Production Research*, 8, 293-305.

Goldberg, S. (1961) *Introduction to Difference Equations*, John Wiley & Sons, Inc., New York.

Hax, A. C. and Majluf, N. W. (1982) Competitive cost dynamics: the experience curve, *INTERFACES*, 12, 5.

Helson, H. (1949) Design of equipment and optimal human operation, *American Journal of Psychology*, LXII, 4, 473-497.

Hill, T. W., Jr. and Buck, J. R. (1974) Zeta transforms, present value, and economic analysis, *American Institute of Industrial Engineers Transactions*, 16, 2, 120-124.

Hutchings, B. and Towill, D. R. (1975) An error analysis of the time constraint learning curve model, *International Journal of Production Research*, 13, 2, 105–135.

Nadler, G. (1970) *Work Design: A Systems Concept*, Revised Edition, R. D. Irwin, Inc., Homewood, IL.

Nanda, R. and Adler, G. editors, (1977) *Learning Curves Theory and Applications*, Monograph 6, American Institute of Industrial Engineers, Norcross, GA.

Navon, D. and Gopher, D. (1979) On the economy of the human processing systems, *Psychological Review*, 86, 254-255.

Newnan, D. G. (1991) *Engineering Economic Analysis*, 4th edition, Engineering Press, San Jose, CA.

Newnan, D. G. (1996) *Engineering Economic Analysis*, Engineering Press, San Jose, CA.

Park, C. S. and Sharpe-Bette, G. P. (1990) *Advanced Engineering Economics*, John Wiley & Sons, New York.

Pew, R. W. (1969) The speed-accuracy operating characteristic, *Acta Psychologia*, 30, 16-26.

Pegels, C. C. (1969) On startup learning curves: an expanded view, *American Institute of Industrial Engineers Transactions*, 1, 3, 216-222.

Remer, W. S., Tu, J. C., Carson, D. E., and Ganiy, S. A. (1984) The state of art of present worth analysis of cash flow distribution, *Engineering Cost and Production Economics*, 7, 4, 257-278.

Rose, L., Ericson, M., Glimskar, B., Nordgen, B., and Ortengren, R. (1992) Ergo-index. development of a model to determine pause needs after fatigue and pain reactions during work, *Computer Applications in Ergonomics, Occupational Safety and Health*, M. Mattila and W. Karwowski, editors, North Holland, Amsterdam, The Netherlands.

Snoddy, G. S. (1926) Learning and stability, *Journal of Applied Psychology*, 10, 1-36.

Tanchoco, J. M. A., Buck, J. R., and Leung, L. C. (1981) Modelling and discounting of continuous cash flows under risk, *Engineering Costs and Production Economics*, 5, 205-216.

Theusen, G. J. and Fabrycky, W. J. (1989) *Engineering Economy*, 7th edition, Prentice-Hall, Englewood Cliffs, NJ.

White, J. A., Agee, M. H., and Case, K. E. (1989) *Principles of Engineering Economic Analysis*, 3rd edition, John Wiley & Sons, New York.

Wickens, C. D. (1992) *Engineering Psychology and Human Performance*, Second Edition, HarperCollins Publishers, New York.

Wright, T. P. (1936) *Factors Affecting the Cost of Airplanes, Journal of Aeronautical Science*, 3, 122-128.

Wulff, J. J. and Berry P. C. (1962) Aids to job performance, *Psychological Principles in System Development*, R. M. Gagne and A. W. Melton, editors, Holt, Rinehart and Winston, New York.

86

The Cost Benefit of Ergonomics: A Corporate Perspective

William E. Lischeid
Travelers Insurance Company

David J. Roy
Travelers Insurance Company

"I have not been able to discover that repetitive labor injures a man in any way. I have been told by parlor experts that repetitive labor is soul, as well as body-destroying, but that has not been the result of our investigations." — Henry Ford[1]

The great automotive pioneer Henry Ford expressed the above opinion in 1922. What we know today about the soul and body-destroying nature of "repetitive labor" and its associated science, ergonomics, is much more definitive than Henry Ford could have ever imagined. While recent ergonomic investigations, discoveries, and understandings of the adverse affects of repetitive tasks conducted by Ford Motor Company and researchers worldwide have moved us far beyond the opinions of a true industrialist, there

0-8493-2641-9/99/$0.00+$.50

are still too many senior executives and managers in business and industry today echoing these past opinions. They have subscribed to the Fordian school of thought of yesteryear without appreciating the evolution of the science of ergonomics and the significant cost that repetitive labor can have on the risk of injury and their own bottom line.

Without a carefully conceived method of justifying the cost of ergonomic improvements, ergonomists will continue to find resistance to the science and the benefits that it can bring to the modern workplace. The ability to quantify as well as qualify the cost benefit of ergonomic improvement initiatives is a critical skill that will differentiate those ergonomists who are viewed as being a source of value to the bottom line from those who are viewed as a corporate irritant.

In the past, many safety and health professionals were hesitant to use cost benefit analysis because it was ethically and morally challenging to place a value on the life and health of an individual. Ergonomics is a science that is based on both documented scientific research and sound engineering principles. Ergonomics is far more than a specialty within the safety and health field. It is a set of design principles that, when applied to the workplace, can create jobs that are within human capabilities and limitations.

The practical application of the science of ergonomics in the vast majority of situations has the potential to maximize the contribution of the worker to the work process, without compromising safety and health. We cannot forget that despite the advances in technology and automation, the single greatest resource available to most organizations is the human resource. Ergonomics can increase productivity and human performance by making jobs more physically and psychologically efficient as well as reducing the risk factors that can and often do result in lost work time and turnover. It is therefore very appropriate to use cost benefit analysis techniques to document the costs of ergonomic initiatives and quantify the benefits to safety and health as well as company productivity and profitability.

Many of us in the ergonomics community intuitively know that the benefits of most ergonomic improvements far out weigh the human and capital costs. Why, then, is it so difficult to sell ergonomics to management?

86.1 Why Do Organizations Feel They Need Ergonomics?

Why do companies today get interested in ergonomics? Typically, we have found from our consulting work with many clients that the driving factors are (in order of perceived importance):

- Money — specifically the money associated with ergonomically related workers' compensation losses.
- Employee complaints
- Regulatory concerns — primarily OSHA citation potential under the General Duty clause of 29CFR 1910 or in anticipation of an OSHA ergonomics standard
- Productivity/Quality concerns

Money

Escalating workers' compensation insurance costs, lost workdays, absenteeism, and turnover all represent direct costs to the organization's bottom line. Reducing these costs is typically a top concern among CFOs, human resource managers, and risk managers.

Employee Complaints

Based on feedback from our customers, many indicated that they had started ergonomic initiatives because of employee complaints. The human body is a wonderful barometer of workplace physical and mental stress. Employees have become increasingly educated through the media, unions, and corporate communication policies to know that if their bodies are experiencing pain during the course of work, that there may be a compensable causal relationship present. For management to ignore these complaints in the hopes that they will go away is simply naive in today's world. Addressing the complaints in a proactive, compassionate, and constructive manner has a definite, positive impact on employee relations.

Regulatory Concerns

A proposed OSHA ergonomics standard has been talked about for the past two administrations. Regulatory pressure has sent many employers the message, "Begin addressing ergonomic issues before the compliance officer shows up at the front door with the new standard in hand." Although a common consulting practice is to use the threat of regulatory penalties to drive behavior, we have found that such an approach typically also fosters a minimalist attitude. Management often loses sight of the fact that any standard is a "minimum standard" and although it may produce some improved results it generally will not achieve maximum results. Using the regulatory stick is not the best way to convince management of the cost benefit of ergonomics. Driving management to care about ergonomics through fear often results in programs that are not truly functional or integrated into the business operations.

Production/Quality Concerns

Some companies initiate ergonomics programs for productivity and quality reasons. These organizations typically end up with an ergonomics process that is truly integrated into the day-to-day business operations. Organizations that realize the most benefits from ergonomics will be those who drive the responsibility for successful program implementation to line management, where it truly belongs.

86.2 What Are The Costs Associated with Ergonomic Mismatches?

Workers' Compensation Claim Costs

Job related injuries and illnesses can result in either the filing of workers' compensation (WC) claims or employees and employers seeking redress through the group health insurance program if there is not a clear causal link to the workplace. When there is work relatedness, the claim is typically filed through the workers' compensation program. For any claim there are associated direct and indirect costs, often referred to as "insured" and "non-insured" costs respectively.

Direct Claim Costs

Workers' compensation claim expenses may include immediate and long-term medical expenses resulting from the injury or illness, as well as payment of loss of income benefits if an employee is not able to work. In more serious cases, there may be payments for permanent or partial, full or temporary, disability. These costs may increase the (WC) premiums by increasing the experience modification factor. The experience modification factor is calculated and used by insurance carriers to determine a company's premium based on its own prior 3 years of loss experience. The rising costs of workers' compensation insurance is often one of the driving forces behind an employer's desire to implement safety and health program improvements.

Indirect Claim Costs

Indirect costs are not covered by (WC) insurance. Instead, they come directly off the bottom line profit of the organization. Although real costs that have a significant impact on profits and expenses, they are hard to systematically capture and quantify and are therefore frequently ignored. Examples of indirect costs include:

- Time lost by personnel who were not injured but may be called in to assist the injured employee seek medical treatment or investigate the occurrence
- Cost of damaged material that resulted from the accident
- Cost of lower productivity while replacement workers learn new job skills
- Costs to make up lost productivity through increased overtime

Although direct and indirect costs may be used by safety and health professionals to justify an ergonomic initiative these variables are widely misused and misunderstood. A better understanding of these expense variables coupled with the expertise to include them in the preparation of a business case, can help the ergonomist build personal credibility in the financial ranks of the organization.

Measuring Claim Cost

A dollar spent today on a workers' compensation claim will buy more treatment than a dollar spent two years in the future on that same claim. For claims with minor medical expenses and only a few days of lost time, the claim expenses are largely known. Generally once a claim is closed, no additional expenses can be added to the file, assuming the file was closed properly. In more difficult cases such as those associated with cumulative trauma disorders, the claim file is usually open longer, due to the length of disability and the more complex nature of the medical treatment. In these cases, the longer the case is open the more likely significant expenses will be incurred. In the insurance industry it is said that the longer the claim is active, the longer the "tail" it has, "tail" referring to the billing tail.

To accurately assign value to the "ultimate" expense associated with a workers' compensation claim, we need to take expense growth into account. Some claims may close quickly with all "incurred" expenses paid out. For others, expenses may continue to be billed for many years. In a recent consulting project, a review of a client's open claims found one "open" broken toe case from 1957 which had resulted in multiple failed surgical attempts at correction. Needless to say, the 40-year "tail" on this claim was far too long.

The Concept of Fully Developed Losses

Because ergonomically related injuries and illnesses tend to have long claim tails, the ergonomist needs to carefully compute the costs associated with these claims. For instance, suppose we analyze WC expenses for 1991, 1992, 1993, 1994, and 1995, as of some date in 1996 to make a business case for an ergonomic improvement program. It would be erroneous to assume that the total expenses that were incurred (as of our 1996 analysis date) were the ultimate costs of ergonomic losses, particularly if there were any cases still open in those time periods. Claims filed in any of those years may continue to incur expenses for years to come. Therefore, we must review claims from a specific policy year at a common point in time when performing any trending or development analysis. For example, to better estimate what the ultimate expenses might be for these claims we would want to review them as of one year from the end of the policy period; 1991 claims as of 12/31/1992, 1992 claims as of 12/31/93, 1993 claims as of 12/31/94, and so on. Using this valuation method we would have a more meaningful estimate of what an appropriate trending factor for more recent claims with only a 12 month period of development.

Projecting what the ultimate loss cost could be for the open claims when they are fully developed in the future is what insurance actuaries do. They determine what the appropriate loss development factor is for a given situation. Actual development factors are very complex to compute and are dependent on many variables such as the mix of accident types, the state workers' compensation laws where the losses occurred, and the cost of medical treatment in that region. It is best to have the workers' compensation carrier provide this information to the corporate ergonomist for his or her use in performing a cost benefit analysis.

TABLE 86.1 Estimated Development Factors Example — ABC Company

Number of Months From Injury Period	Estimated Development Factor
12	3
24	2.4
36	2
48	1.7
60	1.5
72	1.4

TABLE 86.2 Total Incurred Loss Development Triangle Data as of 12/31/96 — ABC

	Incurred Costs Months After End of Injury Period					Projected* Developed
Injury Period	12 months	24 months	36 months	48 months	60 months	Loss Costs
01/01/91to12/31/91	$208,479	$491,458	$486,729	$610,929	**$644,170**	$625,437
01/01/92to12/31/92	$372,350	$696,184	$771,243	**$898,899**		$1,117,050
01/01/93to12/31/93	$341,006	$495,115	**$611,806**			$1,023,018
01/01/94to12/31/94	$212,734	**$331,337**				$638,202
01/01/95to12/31/95	**$255,169**					$765,507

* Note: the above projected developed loss costs were calculated using the costs incurred 12 months after the injury period (column 2 of Table 86.2) multiplied by the corresponding 12 month estimated development factor (column 2 from Table 86.1).

Table 86.1 is an example of estimated development factors that were used to project what the fully developed losses for a given injury period would be when all the claims are closed. These factors are gross estimates.

While this approach will work on workers' compensation claims, the ergonomist should not forget that some ergonomically related claims may be reported through the group health system. Actuarial development of loss factors may not be applied on the group benefit side of the house or if it is, it may be harder to obtain those development factors from the group insurance carrier, particularly where employers offer flexible options to a host of competing health care providers and health maintenance organizations.

Table 86.2 illustrates the importance of using fully developed losses to assist in financial justifications. These data represent actual incurred dollars of a manufacturing organization with approximately 1000 employees.

Using the data in Table 86.2, one can clearly see that if the ergonomics consultant does not use estimated, fully developed loss costs he/she can greatly underestimate the actual anticipated costs of the losses to the organization. If the ergonomist were to simply query the WC data for all incurred expenses for 1991–1995 as of 12/31/96, he/she would get the sum of the diagonal values in Table 86.2 to get a value of $2,741,381. The earlier years have matured and have values closer to what their expected ultimate should be. The younger accident periods are considered "green" or not near ultimate. When development factors are applied to the data, the expected cost of the losses at ultimate is estimated at $4,169,214. By not considering the development of the more recent injury periods the total expected cost was underestimated by $1,427,833, or nearly 35%!

Salary Continuation Expenses (LTD/STD)

For those injuries and illnesses where workers' compensation income benefits (in most states this is generally a statutory maximum benefit of 2/3 of the average weekly wage up to a predetermined ceiling) have been paid, it is important to remember to take into account the statutory waiting period and the retroactive period for income benefit payment. These periods, which vary by state, can affect salary continuation payments through employer long-term disability (LTD) or short-term disability (STD) programs under group insurance. A few examples may be helpful to illustrate this consideration.

Connecticut has a 3 day waiting period before an injured employee is eligible to receive income benefit payments and a retroactive period of 7 days. The first 3 days of the absence which is work related are not paid under workers' compensation but may be covered by other means such as sick time or short-term disability (STD). On the 4th through 6th days the employee would be eligible for WC benefit payments for those 3 days only. If the employee is out 7 days, then all lost time WC income benefits going back to the original date of injury are covered under WC.

Florida has a waiting period of 7 days and a retroactive period of 21 days. For all WC claims with a duration of less than the retroactive period (21 days), the first 7 days of the absence are not eligible for income benefits under WC although they may be compensable under some other plan such as sick time or STD. Again, on the 8th day through the 20th day, the employee would be paid WC wage loss benefits only for those 13 days. Once the claim has a duration greater than 21 days all lost time after the original date of injury is covered under WC.

Using WC indemnity or income benefit payments as the sole means to quantify expenses associated with a WC case can result in under estimating the total cost to the organization. In the Florida example, for every WC claim with a duration of less than 21 days, WC income benefit is paid for a maximum of 14 days. The remaining 7 days are not covered by the WC insurance but may be paid by other group or benefit funds of the employer. This "group benefit" can become substantial when considering injuries involving permanent partial disability and if the employer does not return the employee to work.

Some companies pay their employees full salary while they are out of work recuperating. In such a case WC payment from the WC insurer may be used to pay back the group policy. For example, if the worker is paid in full out of the group STD plan, the carrier will pay back the group plan the statutory amount (66% in most states) of lost wages. The remaining (34%) will be incurred by the group plan. Depending on the employer's STD plan, this could continue up to 26 weeks.

Insurance industry studies have consistently shown a direct correlation on claim costs between the date a claim occurs and the date it is actually reported to the carrier. Industry studies have shown that the longer the lag time in claim reporting, the higher the costs of settling those claims, due to the lapse in time in establishing and monitoring appropriate medical treatment. Most ergonomists know that the sooner a worker reports job discomfort from repetitive motion tasks, the greater the chance of treating the job symptoms through early ergonomic intervention rather than waiting until the repeated trauma manifests itself into a more chronic case of RSI.

With a thorough understanding of the impact of ergonomically related injuries and illnesses on group disability and workers' compensation programs, the ergonomist can be proactive and work with the risk manager to establish policies which minimize the financial impact of the state WC statute on the company. These revised policies can be a significant benefit to the ergonomics effort in the areas of injury management and return to work.

Lost Workdays/Restricted Workdays

Production suffers when employees are injured and either away from the job or in a restricted duty capacity. For example, an employee working 40 hours/week with two weeks vacation has a maximum "full potential" for productive work of around 2000 hours/year. Subtracting from this full productive potential time items such as vacation, paid holidays and other paid leave days, non-work-related illnesses, short- and long-term disability (either group or workers' compensation), continuing education days, and other planned absences, we find a realistic available productive time of closer to 1500 to 1700 hours productive time/employee/year.

To determine the actual amount of time it takes to make up for 100 hours of lost productive time the following formula can be used (100 h ÷ realistic productive hrs/maximum potential hours). Using the example of 1700 realistic productive hours, and 2000 maximum potential hours a loss of 100 hours would require the company to recoup nearly 118 hours to get back the productivity lost due to the injury. Many loss analyses and cost benefit analyses do not take this fact into consideration. Therefore, they underestimate the actual costs of an injury or illness. The burden rate for the employees in the form of benefits, paid leave and vacation, corporate overhead, taxes, and pension contributions must be considered as well.

The Cost of Turnover

Ergonomic interventions can impact employee turnover rates. Little, if any, empirical research data are available on average rates of turnover by industry types due solely to ergonomic issues. However, we've found that human resources managers can frequently relate higher than normal turnover rates to those "tough jobs" which are physically demanding and/or involve a considerable amount of repetitive, labor-intensive work. We have worked with clients who have experienced high turnover rates in particular departments because of recurring back injury and repetitive strain problems. Injured employees may not "turnover" in the sense that they leave the company, but they may have moved on to other jobs within the same company that are less physically demanding or that could accommodate the physical restrictions resulting from their injuries.

It is important that a company capture the relationship between ergonomic issues and employee turnover. Where a positive correlation is identifiable, ergonomic cost/benefit analysis must consider the cost impact of ergonomic-related turnover. Typically, those costs include higher recruitment costs (advertising, interviewing, processing of new employees), higher training costs (basic orientation and job-specific skills training), lower productivity costs (as the new worker progresses up the learning curve), and higher costs due to quality problems, rework, scrap, and equipment misuse. Costs may be less if the turnover can be filled with internal candidates. Quantification of turnover costs is possible over time provided that there are tracking mechanisms in place to measure these costs. Employers may be able to identify ergonomic-related turnover during exit interviews.

Estimating the Probability and Value of Loss

A continual struggle for ergonomists and safety professionals alike is estimating the value of losses that might have occurred had effective loss prevention efforts not been in place. This intangible benefit — less risk of injury — can however be estimated when credible loss data exist from recent history for the organization. With good loss data the ergonomist should be able to determine the frequency of different types of ergonomic losses on a calendar or policy year basis for at least the last 5 years. The values of those fully developed losses (direct claim costs) should be known. If ergonomic-related claims are still open they should be projected to ultimate value. An illustration is helpful. One client put together a table (see Table 86.3). Using historical loss data for a prior 10-year period it was possible to determine the frequency and probability of a $60,000 back injury occurring (1 in 10 chance per year). By stratifying back injury statistics by severity (i.e., $0–$10,000, $10,001–$20,000, $20,001–$40,000, and over $40,000) and frequency in prior years the probability factors were determined. For example, if you were looking at the 1994 calendar year, the data would appear as in Table 86.3.

TABLE 86.3 Estimated Probability of Occurrence of Different Valued Claims

Type of Injury	Claim Value	Est. Indirect Cost	Total Claim Value	Claim Severity Range	Probability of Occurrence
Back injury — 1/5/94	$45,675	$15,000	$60,675	over $40,000	1:10
Back injury — 6/3/94	$15,270	$4,000	$19,270	$10,001–$20,000	1:5
Back injury — 7/20/94	$6,508	$2,200	$8,708	$0–$10,000	1:2
Back injury — 9/15/94	$27,300	$10,400	$37,700	$20,001–$40,000	1:7
Back injury — 11/12/94	$19,450	$8,000	$27,450	$20,001–$40,000	1:7
Total	$114,203	$41,600	$155,803		
Average Value	$22,840	$8,320	$31,160		

TABLE 86.4 Extrapolation to Predict Future Loss Information

Year	Clinic Visits	Doctor Cases	LWD Cases
1990	5	2	1
1991	6	2	2
1992	8	1	2
1993	6	3	1
1994	7	2	1
Total	32	10	7
Expected	**6**	**2**	**1**

From this type of information you could then make a more reasonable estimate of the benefit of preventing "X" number of back injury claims as a result of a specific ergonomic intervention.

In 1995, David Alexander[2] proposed and additional method for predicting future injuries based on historical claim data. He suggests that simple extrapolation of the frequencies of injuries can help establish a likely level of injuries for the upcoming year. Table 86.4 illustrates his concept.

86.3 Methods for Justifying Ergonomic Interventions

In addition to loss analysis skills and having an understanding of true loss costs, ergonomists still need to have a basic understanding of traditional accounting methods for determining cost benefit. While a detailed understanding of these accounting methods would certainly be beneficial, generally that type of expertise already resides in the accounting departments in many organizations. The corporate ergonomist should be interested in forming relationships with the true accounting professionals in their organizations, to assist him or her in developing a fiscally sound approach to cost benefit analysis. Ergonomists will still need to address the nonfiscal, less quantifiable, value-based benefits in most cases, but at least they will be starting from a sound accounting base.

What is a Cost Benefit Analysis?

In his text book, *Essentials of Engineering Economics*, James L. Riggs[3] identifies the origin of benefit analysis back to 1844. He also outlines the use of this economics tool in the U.S. and refers to the Rivers and Harbors Act of 1902, and the Flood Control Act of 1936. This benefit analysis technique required the Army Corps of Engineers to compare the expected results of alternative solutions. From these early uses of economic benefit analysis two traditional types of benefit analysis are illustrated by Professor Riggs. The first, cost benefit analysis, evaluates all of the costs of the alternatives for the same project against the expected benefits, and the second, cost effectiveness analysis, compares the costs of various methods to reach a common business goal where the return may or may not have been determined.

Ossler,[4] in 1984 outlined the eight steps performed to conduct a cost benefit analysis. These steps can be a framework to approach ergonomic justification and cost benefit analysis (see Table 86.5).

TABLE 86.5 Ossler's Eight Steps of Cost Benefit Analysis

	Steps Performed in Cost Benefit and Cost Effectiveness Analysis
Step I	Identify the stimulus or problem for conducting the economic analysis.
Step II	Specify the objectives or goals to be achieved by the program under study.
Step III	Determine the alternative means of attaining those objectives or goals.
Step IV	Enumerate the costs and benefits or effects of each alternative.
Step V	Assign monetary values to the costs and benefits or assign units of effectiveness.
Step VI	Perform discounting.
Step VII	Calculate the cost benefit or cost effectiveness ratio.
Step VIII	Compare these ratios by use of a decision matrix and feed this information into the decision-making process.

In addition to the traditional cost benefit analysis there are some additional economic methods that may be used to justify ergonomic initiates. They include: the payback method, present-value method, and the internal rate of return method. Details on these methods can be found in many accounting and management texts. A brief review of each method follows.

Payback Method

In the payback method the number of years required to recover the cash outlay of an ergonomic improvement, such as purchasing adjustable workstations for data entry operators, would be weighed against some predetermined cutoff point or date that management has considered to be acceptable.[5] For example, in one organization a payback period of 5 years may be the norm against which decisions on capital and labor-intensive projects are based. In another organization that is more focused on short-term results, the acceptable payback period may be considered 3 years. Determining what are acceptable payback periods may represent one of the "values" of the organization, i.e., "If we are going to invest in a project we want a quick payback."

Internal Rate of Return

One of the most commonly used methods for rating and evaluating potential improvement projects is the internal rate-of-return. This method uses a discounted cash flow that ranks competing projects against each other. The internal rate-of-return is defined as the discount rate that renders the net present value of a stream of cash flows equal to zero.

Present-Value Method

The present-value method is a discounted cash flow method that considers the decreasing time value of any monetary investment. It is based on the assumption that a dollar today has greater value than a dollar at some point in the future. The net present value of a proposed ergonomic improvement would be determined by discounting future cash flows by appropriate factors and then algebraically adding all the discounted flow. The result is that the cost benefit ratio equals the net present value of the inflows divided by the net present value of the outflows. If this ratio is greater than one, generally the project would be considered acceptable.

Ergonomics and Productivity

Many ergonomic practitioners attempt to sell ergonomics using primarily injury loss data. This is an economically and socially *post facto* "reactive" approach which seems to overlook the fact that ergonomics and human factors have their roots in human performance. Enhanced performance improves productivity, and ultimately has an economic impact on operating costs and financial earnings. As an industry and a science, we must stress the simple truth that ergonomics assists the human machine to perform work more efficiently, within human capabilities and limitations. It is critical that we not lose our ability to justify ergonomic interventions on the basis of performance and productivity. Otherwise, how will we assist employers who have few or no workers' compensation losses to build a preventive, proactive, ergonomics management system? Waiting until employees are hurt or injured is like waiting to perform maintenance on a valuable piece of equipment until it breaks. To reduce "human scrap" and maximize results we must intervene earlier in the process to maximize results.

Oxenburgh,[6] in his book uses a model to quantify those productivity costs related to safety and health. His model is a very thorough process by which the financial impact of safety-related losses (direct and indirect) can be shown to affect the overall productivity of an organization. Framing your ergonomic argument in terms of actual productivity data will open many doors and may help you convince skeptical managers about the true value of ergonomics and safety and health strategies. The following is a broad overview of the stages of Oxenburgh's model:

1. Calculate the productive hours worked and paid for by the employer
2. Calculate the wage or salary costs
3. Calculate employee turnover and training costs
4. Calculate productivity shortfall due to absences
5. Determine total costs for employment and productivity short-fall
6. Estimate health, safety, and productivity benefits
7. Calculate cost of improvements
8. Determine payback period

As we will mention in the next section of this chapter, decisions made in any organization are strongly influenced by the company values and culture. If you find yourself working in an organization where the culture is production oriented then Oxenburgh's model can prove very beneficial. Many of the variables used to calculate the effective wage costs, as compared to the nominal wage of the employee, may not appear to be readily available. Additional digging is often required. If actual figures are not available the model is very flexible if assumptions or estimates have to be used. Dr. Oxenburgh goes a long way to prove his thesis that, "Good health and safety practices are good for business."

Depending on the audience to whom the cost benefit analysis is being presented, each of the above models will have more or less value to them. For example, a department manager will generally place more value on the payback period, breakeven analysis, or return on investment. They are normally not accountable for the time value of money, but they are accountable for annual budgets and short-term results. At the corporate treasury and finance level, there would be more interest in the present value or internal rate of return methods because of their concern for cash flows and the cost of borrowing money.

86.4 The Influence of Corporate Culture on Cost Benefit Analysis

The influence of an organization's inherent culture or personality will greatly affect how the cost benefit analysis financial tool can or should be used. "Costs and benefits can only be defined in the context of each organization's unique culture." Can this statement be true? Clearly one could successfully argue that the simple direct costs of repetitive strain injuries, back injuries, and other "ergonomically related" injuries are not culturally dependent. While this is most likely true, the culture comes into the cost benefit analysis consideration when other soft and hidden costs and benefits are likely to be perceived by management as credible and value adding. Because of this it is important to view the total set of costs and benefits as closely related to what the organizational value system is. The quantification of costs and benefits is especially sensitive to those organizational values. It becomes especially important when the costs and benefits are less tangible and often indirect.

The concept of "value-focused thinking"[7] has been presented in the literature as a different way to approach decision making. When faced with making a decision, such as "Should we conduct this ergonomic intervention in the workplace to reduce cumulative trauma injuries?," the traditional approach is to consider the problem from the perspective of the alternatives and then considering objectives or criteria to evaluate each alternative (cost/benefit ratio). With value-focused thinking, the values that are important to the organization become the principal measure against which alternatives and their consequences are evaluated. Ralph L. Keeney, professor of systems management at the University of Southern California, states in his article, "It is these values that are fundamentally important in any decision situation, more fundamental than alternatives, and they should be the driving force for our decision making. Alternatives are relevant only because they are means to achieve values."

Organizational Culture and Values

Culture is defined as the set of key values, guiding beliefs and understandings that are shared by most members of an organization. The culture defines the basic organizational values. It helps to communicate to new members the "correct" way to think and act, and how to get things done — "the unwritten rulebook." Culture represents the feeling part of the organization and provides members with a sense of identity. When the culture is shared and accepted by the members of the organization it generates a commitment to support the beliefs and values that are essential to sustain the performance of the group. It should be noted that "performance" may range from poor to outstanding.

Organizational values are the underlying beliefs or philosophies of the organization. The values of the organization are the core of the culture and influence decision making. Some authors describe values as:

- "The shared ideals, either explicit or implicit, that guide the organizational choice and behavior."[8]
- "The basic concepts and beliefs of an organization, such that they form the heart of the corporate culture. Values define "success" in terms for employees — 'If you do this you too will be a success' — and establish standards of achievement within the organization."[9]
- "Shared values. What the company stands for — stated and implied, good and bad. What a company is proud of or would like to be proud of."[10]

Tom Peters and Bob Waterman stated, that their "one all purpose bit of advice for management was to "figure out your value system" as they pursued excellence.[11] Those companies with excellent management have one thing in common: a shared understanding of what their organizational value system is and what their companies stand for. Values influence individual behavior and decision making. They can shape how management will define those variables that they are willing to use for an ergonomic cost benefit analysis. To understand how to quantify costs and benefits, the ergonomist must first determine what variables and values that the organization considers credible and will allow in the calculation. The so-called soft costs or savings, although recognized as real, may not be allowed by some organizations in the cost benefit analysis.

Performance-related values define the orientation of the organization toward issues involving finance, productivity, quality, and health and safety. Organizations that do not have a balanced set of values that place the well-being of their employees on the same level as the fiscal values for increasing stockholder wealth and productivity, will create a value of conflict when considering ergonomic benefits to the organizations. It is crucial that the corporate ergonomist or consultant determine the value set of the organization prior to making his or her business case for the recommended improvements. One way to determine that value set is to look at examples of other business cases that were successfully presented to and accepted by management. A business case that does not mention safety and health or quality may be an indication that you are dealing with a very fiscally based value system. Knowing this you would want to build your business case primarily on those hard, direct tangible benefits that might be expected from the ergonomics intervention.

Demonstrating and Communicating Safety and Health Values

The existence of shared safety and health values and norms that clearly indicate the importance of safety as a function of good business is essential. These values must be clear so employees will understand what choices to make if conflicts between safety and other business priorities arise. The following are benchmark values that are common among companies with excellent safety records.

- Nearly all work-related injuries and illnesses are preventable.
- Management is responsible for creating a safe work environment.
- All employees are accountable for following established safety procedures.
- All workplace safety and health exposures can be controlled.
- Safety training is essential
- Safety is an integral part of the business plan

Many U.S. corporations do not have documented statements that communicate their corporate values, let alone their safety and health values. Lacking these statements people create their own perceptions or interpretations of what they think the safety values of the company are. For example, faced with making a decision to get help, use a mechanical lifting aid, just say that the object is too heavy to lift, or take a chance that they can handle the lift alone, a worker may choose the latter based on the perception that speed takes precedence over safety. He or she may suffer a serious back injury as a result of this decision. When questioned, management would likely state that they do not tell people to place themselves at risk to save a few minutes. However, the reality is that risk-taking behavior has been condoned (although maybe indirectly by the lack of verbal or visible management action when risk-taking behavior has been observed in the past) while getting the job done quickly has been rewarded. Therefore, speed over safety has become a value that is supported in the organization and in the minds of the employees.

Advances in Economic Analysis

Professors John K. Shank and Vijay Govindarajan[12] present a case for a new method of looking at the benefit of technological investments, which they call "Strategic Cost Analysis." "A project-level net present value (NPV) framework, it is argued, places such a premium on short-term financial results, and so little emphasis on difficult-to-quantify issues, such as quality enhancement or manufacturing flexibility, that major manufacturing breakthroughs do not pass the NPV test."[13] The Strategic Cost Analysis (SCM) perspective may have more widespread applicability to the issues facing corporate ergonomists in trying to quantify the benefits of ergonomic interventions and improvements. SCM is a blend of three approaches to cost analysis:

- Value chain analysis
- Cost driver analysis
- Competitive advantage analysis

Value Chain Analysis

Value chain analysis is a more holistic approach that would take into consideration the impact of the ergonomic improvement throughout the value chain of its goods and services. Rather than asking what value would be added in this department or this job by the intervention, value chain analysis would ask what impact will this change have on the consumers of the goods and services, on the company's suppliers, on other employees in different departments, and on the rest of the organization.

Cost Driver Analysis

Cost driver analysis looks at what Shank and Govindarajan refer to as "structural drivers" and "executional drivers." Structural drivers are a function of the:

- Scope of the ergonomic improvement (one workstation or one thousand workstation upgrades)
- Complexity of the new technology that may be employed (redesign of an automated material handling system)
- Other factors such as experience of the designers and installers

On the other hand, executional drivers would force organizations to look at issues such as how will this strategy be implemented in a TQM or traditional business environment, how can employee and management participation and acceptance be enhanced, and are the necessary resources to execute the strategy effectively available and committed?

Competitive Advantage Analysis

Finally, the competitive advantage analysis would address the issues of whether the ergonomic improvement will enhance the ability of the organization to compete based on cost or some other means of market differentiation. The retention of skilled employees is a clear example of how an ergonomic intervention can give the organization a competitive advantage over another employer that experiences a high rate of turnover and lost time within its workforce of knowledge workers due to ergonomically related injuries. The employer that reduces ergonomic risk will experience less new hire training expense and will have a larger percentage of skilled, experienced employees in the workforce.

Summary of Costs and Benefits

Regardless of whether a simple cost benefit method or combined value/cost based approach is used to sell the need for an ergonomics intervention, the ergonomist must be capable of communicating objectively and clearly to management about the anticipated costs and benefits of their proposal. The following table is a summary matrix of some of the cost drivers and benefits. These are easily quantifiable and typically associated with an ergonomics intervention, and therefore, should be considered in the cost benefit discussion.

TABLE 86.6 Summary of Potential Cost Benefit Variables

Potential Cost Variables	Actual Costs	Potential Benefit Variable	Actual Benefits
Employee training		Reduced injury costs (medical, claim, expense)	
Outside consultant fees		Increased potential for reduced WC experience modifier	
Capital equipment purchases		Reduction in non-value added material handling tasks	
Facilities design changes		Improved job cycle time	
Procedural changes cost		Improved productivity by hourly workers	
New equipment installation costs		Improved productivity by administrative workers	
Future equipment maint.costs		Reduced scrap rates	
Other costs:		Improved product and service quality	
		Reduced employee turnover	
		Reduced training costs resulting from accidents	
		Reduction in overtime charges to replace injured worker production	
		Reduced potential for regulatory fines and related productivity loss	
		Improved employee morale	
		Enhanced customer service capability	
		Enhanced product and service differentiation potential	

Summary and Conclusion

In today's globally and locally competitive marketplace it is natural and logical that senior management will ask the question "What's in it for us?" when faced with a request to commit resources to ergonomic improvements. Traditional methods of defining cost benefit in terms of expected return on investment in the shortest possible time period are still important principles that the ergonomist should know about.

The ability to understand and effectively communicate the financial ramifications of ergonomically related workers' compensation losses or losses that are put in against a company's group insurance program are critical skills that the corporate ergonomist must have. As insurance programs and risk financing methods become more complex, being able to talk one on one with risk managers and financial managers about direct and indirect costs that they understand will go a long way toward securing buy-in for ergonomic interventions.

A new twist to the cost benefit puzzle that needs to be explored and incorporated in more benefit presentations is the "values based" approach. Whether we like to admit it or not, corporate values have a very powerful influence on what gets done or doesn't get done in organizations. Without a clear understanding of what those internal value systems are, both organizationally as well as individually, the ergonomist may find himself or herself with what appears to be a fiscally sound intervention proposal that goes nowhere. The ability to appeal to values when the more traditional approaches to salesmanship have failed is what will differentiate the ergonomist of the future.

86.5 Justifying Ergonomics Initiatives — Case Studies

Travelers Tennessee Nursing Home Study[14]

This project was undertaken with six nursing homes in Tennessee. They were purchasing their workers' compensation insurance through the assigned risk plan because of previous high injury rates and elevated workers' compensation experience modification factors. Back injuries among nurses and nurses aides were the major cost drivers behind the experience rating. The project focused on three areas of concern:

- Resident patient transfer methods
- Resident transfer lifting and patient handling equipment issues
- Medical management of injured workers

In their study the authors reported that "One of the major stumbling blocks to implementation of mechanical resident transfer equipment was the perception that equipment could not be justified. An example cost justification for mechanical resident transfer equipment is shown here. The justification is based on a five-year lease, assuming a yearly interest rate of 9%."

Based on an initial estimate of four lifts in one of the nursing homes in the study, a five-year lease will cost approximately $650 per month, based on equipment cost estimates and interest on the lease. The lease analysis assumes an average of the five-year injury cost history ($5454/month) and a 15% minimum return. For the five-year lease period, assuming no escalation in workers' compensation costs, the net present value of all costs and benefits are:

BENEFITS = $154,616 (workers' compensation back injury reduction)
COSTS = $25,430

To be conservative, 50% of the potential workers' compensation back injury reduction benefit was used, 50% × $154,616, or $77,308. The cost benefit ratio (CBR) for all the participants in this study are shown in Table 86.7.

Table 86.7 Cost Benefit Ratios for Project Participants

Facility Number	Cost Benefit Ratio
Facility 1	3.04
Facility 2	3.47
Facility 3	3.25
Facility 4	Data not available
Facility 5	2.10
Facility 6	0.5 *

* The losses for facility six did not justify equipment purchase based on the study's assumptions

Other benefits cited were:

- Certified Nurse's Aides (CNA)s should be able to safely transfer residents alone, where now two, three, or four are required. (**Benefits:** A reduction in manpower requirements. Better utilization of resources. Potential for improving quality of care to other residents with more availability of nursing staff. The time required to get help was also drastically reduced.)
- The concept of using three or four CNAs to move a resident could be eliminated. Even though using additional help may help in lessening the load, generally one or more of the CNAs get in each other's way. The biomechanical load is imposed unequally, usually on one or two of the CNAs. (**Benefits:** Reduced potential for back injury due to unequal biomechanical loading plus the other benefits cited above.
- All of the nursing homes implemented all or part of the improvement strategy. The total injury incidence rates were reduced compared to the previous year's rates. Two of the six facilities were taken out of the Tennessee assigned risk pool and were able to return to the competitive insurance market for pricing quotes. (**Benefit:** Generally the competitive market results in lower premiums, thereby reducing operating expenses and improving cash flow.)
- Two other facilities have current experience modifiers of 0.85. This means that because they are approximately 15% better than the industry average their workers' compensation insurance premium earns a 15% credit. (**Benefit:** Reduced operating expenses. Possibly translating into better allocation of funds to facilities and staff improvements that benefit the nursing home customers, i.e., more value-added services or reduction in charges resulting in more competitive market position.)
- Subsequent interviews with CNAs have suggested greater job satisfaction, as a great deal of the difficult portion of their job has been reduced due to implementation of mechanical lift equipment. (**Benefit:** Improved employee morale.)

Vehicle Subassembly Plant Ergonomics Success Story

This case study took place in a vehicle subassembly plant that essentially attached vehicle bodies to a preassembled chassis. Operation began early in 1987 and by November that year they hit the production level of 95 units per day. As the operation began to grow through 1989 so did the incidence of back injuries and cumulative trauma disorders.

Late that year the corporate parent company began a TQM initiative and started to identify work processes that had opportunities for improvement. In the 1st quarter of 1990 the lost time frequency rate at this location approached 19.4. Local management saw ergonomics as an area of importance and decided to address these issues using the corporate TQM approach. At that time they developed an ergonomics problem-solving team and allocated $100,000 for ergonomic improvements.

Soon after team formation each member received 2.5 days of ergonomics training. The teams initially were focused on reactive problem solving using loss analysis as a guide to direct them to problem areas. Specific solutions at this stage included:

- Work tables
- Carts
- Platforms
- Floor mats

By June of 1990 the team began to use proactive techniques to address ergonomic issues by identifying problem tasks through observation of risk factors, conducting task analyses and requesting employee feedback. Through this process the team determined that supervisory training was needed. A training plan to train all supervisors was created and implemented. At the end of 1990, another $100,000 was allocated for ergonomic improvements.

TABLE 86.8 Cost Benefit Analysis (1989 to 1992 as of 1992 year end)

Loss Area	Ergonomic Expenditures	Total Savings	Return on Investment
All Injuries	$300,000	$657,618	119%
Back Injuries	$160,000	$349,315	118%

Early in 1991, ergonomics was considered to be a critical work process consideration and became part of the 1995 vision statement. After the supervisory training was conducted, supervisors were integrated into the teams and employee feedback was incorporated into the problem-solving efforts. At this time the ergonomics team had implemented many solutions including:

- Rivets squeezers
- Improved tooling and fixturing
- Ergonomic seating
- Powered pallet movers
- Hydraulic lift tables
- Overhead cranes and hoists
- Vacuum manipulators

The ergonomics process was now integrated into the business operation and the team began to measure and monitor results of the interventions. Administrative controls were considered in those tough areas were the engineering controls were not completely addressing the issues. In the paint department, for instance, a job rotation process was piloted and later implemented. In addition, a formal medical case management process was implemented which facilitated employee return to work and conservative medical treatment. Preliminary results indicated that the frequency rate had decreased to 9.57. A presentation was made to senior management with these results and another $100,000 was allocated for the 1992 budget.

Through 1992 employee and supervisory training continued and more strides to involve employees in the process were made. Job rotation was embraced by management, and initial rotation strategies were expanded to include 85% of the production area. By the end of 1992 over 90% of all repetitive bending, twisting, and heavy lifting was eliminated. Lost time frequency continued to fall to a level of 5.89 and for the first quarter of 1993 reached the lowest level since the inception of the ergonomic efforts. Management was very pleased with these results and allocated an additional $85,000 for the 1993 budget.

Measured Results (loss data collected on 12/31/92)

- Back injuries per year were reduced from 85 to 11
- Upper extremity CTDs per year were reduced from 105 to 54
- Lost workdays were reduced from 1,402 to 476
- Lost time frequency reduced from 19.39 to 4.4 (data as of 3/93; see Table 86.8)

Other Tangible and Intangible Benefits which Were Cited but not Quantified

- Increased quality
- Increased productivity
- Decreased inventory levels
- Decreased scrap
- Decreased turnover/absenteeism
- Increased morale

Keys to Success

- Extremely strong senior management support
- Identified as a critical work process

- Senior management focused on performance of the ergonomics team, implementation was stressed
- Strong TQM culture facilitated problem solving
- Ergonomics team was multidisciplinary
- Ergonomics team was empowered to solve problems and given the resources to execute
- All employees and management received appropriate ergonomics training

Post-Injury Ergonomics Intervention For A Chronic Carpal Tunnel Case

Ergonomic interventions can be useful to facilitate the return to work of employees once an injury has occurred. This approach can be valuable for repetitive strain injuries to the upper extremities or to the back. Ergonomic interventions may also offer value to many of the traumatic injuries reported on the OSHA 200 log. Estimates indicate that up to 25% of the OSHA recordable injuries have root causes that are ergonomic related.

This case study illustrates the importance of return to work processes as part of an overall ergonomics program. Return to work efforts become more complex when the employee injury is cumulative in nature. In these cases it is essential that the employer not only provide the employee with exemplary medical care but identify and control those workplace factors that were likely causes of the injury. In return-to-work scenarios such as this, controlling the injured employee's exposure to those job-related risk factors can be either engineering or administratively based.

This specific case involved an employee who was diagnosed with chronic carpal tunnel syndrome. The employee was a computer operator who performed keying up to 6 hours a day, 5 days a week. After a trial of conservative treatment, the employee's symptoms persisted. Electrodiagnostic evaluation indicated that there was significant median nerve pathology, and surgical intervention was recommended. Post-surgical response was poor and the patient was later diagnosed with reflex sympathy dystrophy (RSD).

The employer was very cooperative with the employee up to the diagnosis of RSD. At that time the treating physician gave the employee permanent work restrictions that included a maximum of 1 hour of keying a day. It was felt that this restriction would continue as long as the employee had pain. At this time the employer made the decision to terminate the employee because they did not have any jobs that this employee was qualified to do with this keyboarding restriction. After the employee's 26 weeks of short term disability was completed the termination process was initiated.

The employer's insurance provider was informed of this decision and contacted management of the employer to discuss this case. The carrier felt this course of action would put the employer at legal risk and increase the eventual costs of this claim. The carrier carefully evaluated the financial impact of this decision and prepared the two scenarios that are diagrammed in Table 86.9 for the employer.

As can be seen from the data the employer would save up to $255,294 if they accommodate the employee. In this case these figures helped convince the employer to consider providing the employee with alternative productive duty. In fact the employer did create a job for the employee that did not require extensive keying and that the employee was qualified to perform. At the time of this report, the

TABLE 86.9 Comparison of Two Approaches to a Chronic Workers' Compensation Case

Type of Cost	Scenario 1 — Terminate Employee after 26 Weeks of STD	Scenario 2 — Accommodate the Employee	Saving Realized by Accommodating the Employee
Workers' Compensation Expenses (Projected Medical and Indemnity	$86,125	$48,711	$37,414
Short-Term Disability Minus Any WC Offsets	$6,932	$5,532	$1,400
Projected Long Term Disability	$116,480	0	$116,480
Potential Legal Exposure (ADA, Wrongful Termination, Retirement Settlements)	$100,000	0	$100,000
Total	**$309,537**	**$54,243**	**$255,294**

employee was still experiencing pain from the CTS and RSD but remained gainfully employed and most importantly happy to contribute to the employer where he/she had provided many years of loyal service prior to the injury.

References

1. Ford, Henry, Crowther, Samuel, 1922, *My Life and Work,* Garden City Publishing Co., Inc., Garden City, New York.
2. Alexander, David, C., 1995, *The Economics of Ergonomics: Part II,* Proceedings from the Human Factors and Ergonomics Society 39th Annual Meeting, Santa Monica, CA.
3. Riggs, J. L., 1982, *Essentials of Engineering Economics,* McGraw-Hill, New York,.
4. Ossler, C. C, 1984, *Cost Benefit and Cost Effectiveness Analysis In Occupational Health,* Occupational Health Nursing, January.
5. AMA Management Handbook, 2nd Edition, copyright 1983, AMACOM, New York, NY
6. Oxenburgh, M, 1991, *Increasing Productivity and Profit through Health and Safety,* CCH International, Chicago, IL.
7. Keeney, Ralph L, *Creativity in Decision Making with Value-Focused Thinking,* Sloan Management Review/Summer 1994.
8. Connor, Patrick E. and Lake, Linda K., 1988, *Managing Organizational Change,* Praeger Publishers, New York.
9. Deal, Terrance E., Kennedy, Allen A., 1982, *Corporate Cultures — The Rites and Rituals of Corporate Life,* Addison — Wesley Publishing Co., Reading, MA.
10. Waterman, Robert H. Jr., 1987, *The Renewal Factor,* Bantam Publications
11. Peters, T. and Waterman, R., 1988, in *Search of Excellence,* Warner Books.
12. Shank, John, Govindarajan, Vijay, *Strategic Cost Analysis of Technological Investments,* Sloan Management Review/Fall 1992, pp. 39 — 51.
13. Hayes and W. Abernathy, *Managing Our Way to Economic Decline,* Harvard Business Review, July-August 1980, pp. 66-77.
14. Wyatt, R.S. and Booth, C., *Reducing Employee Back Injuries in Skilled Nursing Facilities,* 1993 Internal Thesis, Travelers Indemnity Company.

Section II
Ergonomics Processes

87

Success Factors for Industrial Ergonomics Programs

David C. Alexander

Auburn Engineers, Inc.

Gary B. Orr

USDOL/OSHA

87.1 Introduction

Many ergonomics programs are not successful. A survey performed by Auburn Engineers (Auburn, Alabama) found that only 25% of the ergonomics programs they surveyed were successful. The data, shown in Figure 87.1, separates the organizations into small, medium, and large sizes. Four different outcomes are possible:

- Successful
- Too new to call
- Floundering due to management issues
- Floundering due to technical issues

The term floundering was chosen to indicate that more effort is going into the ergonomics program than is appropriate for the results being achieved. While this is not technically a failure, it is clearly headed in that direction. Floundering due to management issues was the result of lack of vision or program direction, inadequate resources, lack of coordination, and other management issues. Floundering due to technical issues included programs where there was a fundamental lack of technical skills such as job analysis or ability to generate appropriate solutions.

0-8493-2641-9/99/$0.00+$.50

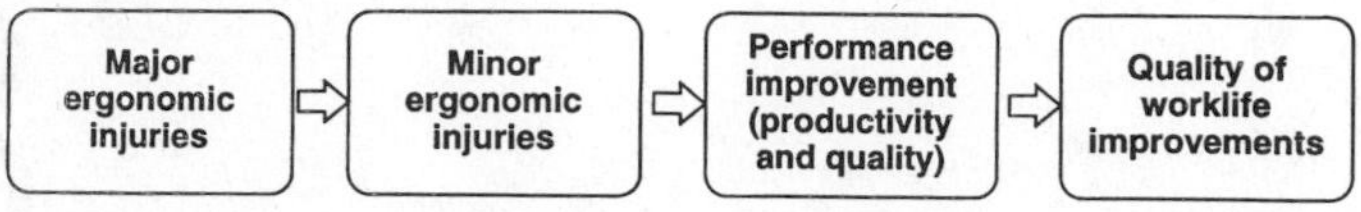

FIGURE 87.1 Progression of the results of ergonomics projects.

This chapter lists a number of success factors which will make an ergonomics program effective. It is not easy to develop a list such as this. After reviewing hundreds of programs, success factors and common flaws begin to appear. Many of these are found in this chapter. At the same time, a successful ergonomics program does not have to include all of these factors. Many, but not all, of these factors are found in successful programs.

The success factors can be roughly divided into four groups:

- Meet business needs
  1. Emphasize business objectives
  2. Avoid too many low-value/high-cost solutions
  3. Ensure that ergonomics projects are evaluated quantitatively
  4. Maintain a tabulation of the cost of projects
  5. Use resources efficiently (the self-help/skilled-help/expert-help strategy)
- Avoid common traps
  6. Identify and overcome barriers
  7. Training should be supported by suitable infrastructure
  8. Avoid using "ergo-babble"
- Create a strong purpose
  9. Clearly define the purpose of your ergonomics program
  10. Plan the stages of the ergonomics culture change
  11. Create a strategic plan
- Maintain the program
  12. Understand the difference between an ergonomics program and the practice of ergonomics
  13. Create a tactical plan
  14. Ensure that there are regular quantitative evaluations of the overall ergonomics program
  15. Do not wait for top management to push the program down
  16. Maintain political support

87.2 Discussion of Success Factors

Emphasize Business Objectives

For an ergonomics program to sustain itself over the long term, it must be anchored to business objectives. The best way to do this is to ensure that its results improve the business objectives of the organization. There is a progression in the types of results achieved, as shown in Figure 87.1. Most ergonomics applications are initially targeted toward the elimination of major injuries, and then minor injuries become important. Once injuries are under control, the emphasis should shift to improving performance in the areas of productivity and quality, and eventually, improvements in the quality of worklife should occur. A successful ergonomics program will, as soon as practical, ensure that business objectives are improved, documented, and shared with management.

This may be difficult to do because many ergonomists only see ergonomics as a technology and not as business enhancement tools. In business, however, technology is a tool to achieve business objectives. Unfortunately, when ergonomics is applied primarily as a technology without firm business goals, the results rarely amount to more than training and scattered job analyses.

FIGURE 87.2 The money pit.

Fortunately, though, when ergonomics is viewed as a tool which helps drive important business measures, it continues to be important over an extended time period, and will be retained. However, when ergonomics is seen either as an "add-on" with little business value or as a "must do" from a compliance standpoint, an ergonomics program will only be mildly supported and eventually discontinued.

Avoid Too Many Low-Value/High-Cost Solutions

Cost is or soon becomes an issue for most ergonomics programs. When costs are too high relative to the value received, the ergonomics program is regarded as a money pit and is stopped or slowed down (see Figure 87.2).

Expensive solutions usually result from a misunderstanding of the role of people and equipment such as an "automation mentality" which requires that automation be used to remove the person totally from the job. In the office area, many office ergonomists go on a "chair buying binge" and spend too much on new seating with correcting the hand/wrist problems common in office areas.

Typically, these problems result from less experienced ergonomists who have difficulty creating low-cost solutions that address the root cause. Many ergonomists have developed skills to identify and analyze ergonomics problems. Unfortunately, if they have not developed skills for the efficient resolution of those problems, the result can easily be overly expensive solutions.

When the common solution for ergonomics problems is to "automate the job," a single solution can be very costly, often in the range of $100,000 to $1,000,000. It is difficult to offset this high cost with the benefits gained, and this gives management the perception that ergonomics is (and always will be) prohibitively expensive. It does not take management long to tire of these types of "low-value/high-cost solutions." And once that occurs, the ergonomics program usually has a short future.

An example can illustrate this concept. Suppose that a material handling problem has been uncovered, and the question is how to resolve it. A less experienced ergonomist might use automation, when other alternatives such as scissors-lifts, spring-loaded levelers, and turntables are available. Typically, an administrative control is also considered, and one might even add the consideration of back belts to the decision. For many lifting situations, simply getting the lift to correct height and close to the body will resolve the problem. The two-dimension matrix in Figure 87.3 easily shows where the value/cost benefit lies. Figure 87.4 provides a generic view of the value cost matrix and how to assess different solutions for their own value/cost relationship.

Ensure That Ergonomics Projects Are Evaluated Quantitatively

Only the most successful ergonomics programs have instituted a systematic method for quantitatively evaluating individual ergonomic projects. Ergonomic projects should be evaluated for both ergonomic improvement and for cost/benefits. The degree of ergonomic improvement can be measured by changes in such lagging indicators as incident rate, severity rate, or losses for workers' compensation. It is also

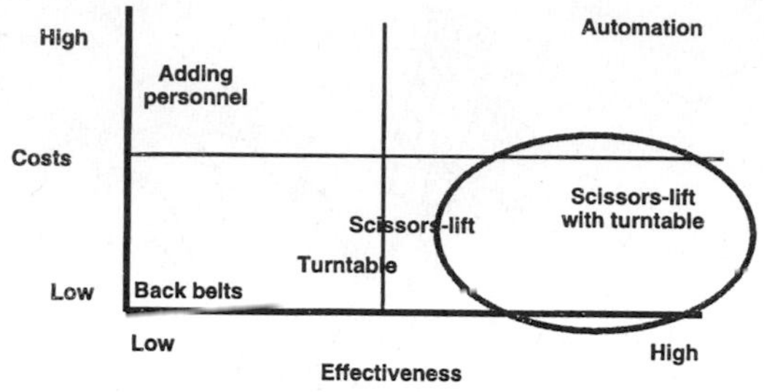

FIGURE 87.3 Possible solutions for a material handling problem.

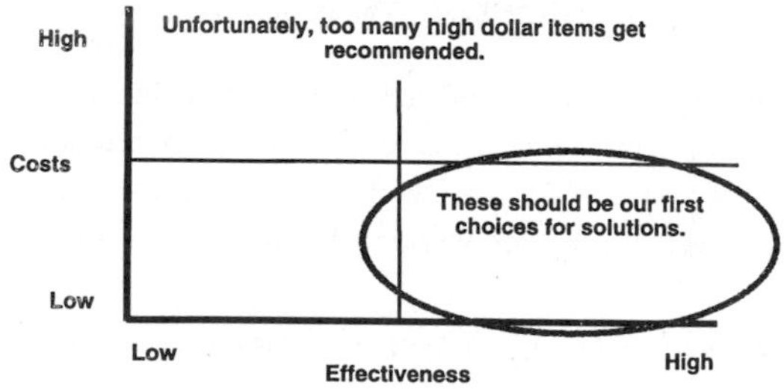

FIGURE 87.4 Avoid too many low-value/high-cost solutions.

possible to use leading indicators such as a symptoms survey or pain/discomfort body parts survey. It is best to use these surveys more than once, and a suitable timetable for surveys is based on "Dave's Rule of Twos"— perform a survey after the changes have been implemented for two days, then again after two weeks, and finally, after two months.

In addition, each ergonomics project should also be evaluated financially. The costs and benefits can be measured for each project. These dollar figures for costs and benefits can be translated into net overall improvement for all the ergonomics projects, or a cost/benefit ratio for each project can be maintained. Each project does not have to pay off, but, overall, the program should be able to pay for itself.

These basic items can become part of composite measures such as those listed below:

- Cumulative stress reduction (CSR) index — Index of stress reduction × number of people affected
- Cumulative stress reduction per $1,000 — Ratio of CSR index to costs

One additional project evaluation measure which has created a lot of interest is the time to complete each project. Usually this is just the elapsed time from the day the ergonomic project is initiated until the recommended solution is implemented. If the ergonomic problem-solving skills are increasing, then the time to complete individual projects should be decreasing.

Maintain a Tabulation of the Cost of Projects

Cost is an important issue for many ergonomics programs and one additional method of cost measurement is valuable. Since so many projects are incorrectly assumed to be overly expensive, a simple tabulation of the costs of a number of projects can help dispel the notion that all projects must be costly. Table 87.1 shows the costs of 29 ergonomics projects completed by Auburn Engineers in late 1994 and early 1995. The interested reader will note that about half of these project solutions cost less than $500 per project. Most of these projects (98%) cost less than $5,000.

Tracking costs like this places an emphasis on the low-cost (yet effective) solutions, and it sends a clear and simple message to ergonomic problem solvers. The message is "Solve the problem, but spend as little money as possible. If we save money on one project, then we have more left for additional projects."

TABLE 87.1 Cost of Ergonomics Solutions

Cost per Project	Number of Projects	Cumulative % of All Projects
Less than $100	11	22%
$100 to $500	12	47%
$500 to $1,000	11	69%
$1,000 to $2,000	7	84%
$2,000 to $5,000	7	98%
Over $5,000	1	100%

TABLE 87.2 Efficiently Using Problem-Solving Skills

Skill Level	Self-Help Problems	Skilled-Help Problems	Expert-Help Problems
Occurrence	50–70% of problems	20–40% of problems	5–15% of problems
Typical problem	Simple and "Quick Fix" problems	Multiple workplace changes or complex diagnosis	Most complex problems, unique problem, complex multi-part solution, expensive solutions
Typical solution	Adjust workplace, proper pace, use proper tools, awareness of early warning symptoms	Workstation redesign, modification of production process, new assembly tools	Unique tool, design of entire factory, redesign of production process
Expertise	With minor training, can be resolved by many people	Requires training, practice and possible guidance	Require special expertise or professional judgment
Typical training	2–4 hours of ergonomics awareness training	2–3 days of team-based ergonomics problem solving	Graduate degree plus professional experience

Use Resources Efficiently (the Self-Help/Skilled-Help/Expert-Help Strategy)

A successful ergonomics program will ensure that it uses resources as efficiently as possible. The major costs are personnel and hardware. Hardware costs were outlined above. Personnel costs can be controlled by delegating the ergonomics problems to the correct skill level. Too many ergonomists get involved in projects that do not require their level of skill, and which they should delegate to others.

Efficient problem solving uses a stratification based on the difficulty of problems using three levels called self-help problems, skilled-help problems, and expert-help problems. For most organizations, the number of self-help problems is the largest. Following is a list of the three types of problem-solving groups, a brief description of each type, and the training required:

- *Self Help.* Self-help is the lowest cost method to resolve an ergonomic problem. Self help also requires the lowest level of problem-solving skill. Self help solutions are usually generated by a worker along with his or her supervisor. Awareness training provides the workers with the necessary skills to determine the self help solutions by familiarizing them with symptoms of musculoskeletal disorders, workplace risk factors, and ways to reduce risk factors.
- *Skilled Help.* This is the second lowest cost method of problem-solving help. Skilled help typically involves a problem-solving team comprised of workers who have had ergonomics problem-solving training. Awareness training is inadequate to perform at this level.
- *Expert Help.* Expert help is provided by an expert ergonomist, typically a corporate ergonomist or an outside ergonomics consultant. This is the most expensive level of help and should be utilized for problems that are too difficult to solve or go beyond the knowledge and skill of the other two help levels.

A summary is shown in Table 87.2.

TABLE 87.3 Four Typical Barriers

Barriers	Methods to Overcome Barrier
Not enough time	• Determine "Top Five" or "Dirty Dozen" problem areas • Avoid "paralysis by analysis" • Enable others and get them involved • Buy additional time
Too little money	• Use low-cost/high-value solutions • Use nickel and dime solutions • Avoid cost/benefit justification • Cluster projects • Get "refillable pot" of funds • Use two-step solutions
Gaps in skills	• Provide specific training (and only as needed) • Look for existing solutions (remember worker modified solutions) • Use teams for simple problems and experts for difficult problems
Management concerns	• Propose a specific plan • Answer the 5 questions managers ask • Develop an ergonomics culture • Understand change management

Identify and Overcome Barriers

Successful ergonomics programs identify and overcome barriers to their success. There are many barriers which occur with any new initiative, and to be successful, the barriers must be identified, resolved, and eliminated. Unsuccessful ergonomics programs will either be stopped by these barriers, or will be overcome by them. There are four typical barriers within an ergonomics program as shown in Table 87.3. In addition, some typical methods to overcome these barriers are also listed.

It is important to address barriers to the success of the ergonomics program early and often. The most successful programs will address barriers during the initial strategic planning, and will discuss barriers during each planning session. Once the barriers are identified, corrective actions are planned and implemented, and follow-up is done to ensure that the barriers are not impeding progress with installation.

Training Should Be Supported by Suitable Infrastructure

Training is a valuable part of an ergonomics program, but it can be done before the organization is ready for it. One important objective of an ergonomics program is to ensure that people are aware of the signs and symptoms of musculoskeletal disorders. General awareness training of these disorders is an appropriate way to meet that need. But what happens too often is that this awareness training occurs much too early in the program. Once this training takes place, many ergonomic concerns quickly come to the surface — some very important, but many of less significance. The dilemma is that each situation should be evaluated reasonably soon or the ergonomics program loses credibility. When an operator recognizes an ergonomics problem and asks for help, there is a limited amount of time before the worker suspects that ergonomics is just another management fad.

When numerous situations surface at once, particularly if the ergonomics program is new, there are insufficient resources to deal with everything in a timely manner. This creates a big problem for the ergonomics program — too many requests, not enough time, and lots of frustration, discontent, and loss of credibility.

This is clearly a situation which a successful ergonomics program should avoid, and the recommendation to wait for the ergonomics program to mature a bit and develop a suitable infrastructure of people with some basic skills prior to conducting widespread training. An organization which has performed passive or active surveillance is well aware of the more serious problems anyway, and is probably already dealing with them. Therefore, little is lost by waiting to perform training, and the ergonomics program will avoid generating substantial negative publicity and discontent.

Avoid Using "Ergo-Babble"

One common contributor to the lack of success of an ergonomics program success is the extensive use of technical jargon. This "ergo-babble" is rarely understood and results in a loss of support by managers and the line workforce. Often, after a long (and difficult-to-understand) medical term has been used to describe a workplace illness, the operator is left in the dark and is only too willing to leave the situation to the "doctors and engineers" to fix. Unfortunately, many ergonomic problems which could be corrected by operators are left untouched once this scenario begins to play out.

It often takes more effort to describe a musculoskeletal illness in terms a lay person can understand, but the effort is worthwhile in gaining support for the ergonomics program.

Clearly Define the Purpose of Your Ergonomics Program

The successful ergonomics program has a clearly defined objective. Programs which flounder and fail have unclear objectives. Programs which have been put in place to "do ergonomics" will likely fail relatively quickly. Defining the objectives may take the form of a mission statement or vision statement. Typically, vision statements are longer than mission statements, and "paint a picture of what life will be like at some point in the future." Some examples are provided in Tables 87.4 and 87.5.

TABLE 87.4 Examples of Mission Statements

Mission Statement	Comments/Observations
Example #1 The Ergonomics Committee will develop systems for the multi-disciplinary study of the problems that exist between people, the tools and machinery they use and their work environment. These systems will initially focus on reducing injuries/illnesses related to cumulative trauma disorders of the upper extremities and backs by hazard prevention and control, medical management, and training/education. This will lead to an increased level of comfort at work, improved quality of product, and greater productivity. This mission will be obtained with management support and associate involvement.	Positive Points 1. Takes ergonomics beyond injury/illness into quality and productivity 2. Will not do all the work itself, but will develop systems for ... Areas of Concern 1. Over emphasis on CTDs 2. Somewhat long 3. Mentions how this will be achieved which may be unnecessary
Example #2 The Ergonomics Committee will develop and manage systems for the improvement of the conditions between people, tools, machinery, and their work environment.	Positive Points 1. Short 2. Will develop systems for... Areas of Concern 1. When is the mission completed?
Example #3 The ergonomics program provides education, analysis, and guidance to prevent and alleviate ergonomically related stress and illness in order to protect the health, and further, the productivity of the plant.	Positive Points 1. Mentions prevention 2. Relatively short Areas of Concern 1. The program should set up systems to provide... because when the program ends ergonomics will be part of the culture 2. Ergonomics can go beyond stress and illness (it's not just CTDs)
Example #4 To put into place the will and skill to eliminate and/or prevent ergonomics problems (pain, illness, injuries), and to capture quality and efficiency benefits so that ergonomics becomes institutionalized.	Positive Points 1. Will put into place ... 2. Relatively short 3. ... becomes institutionalized. Areas of Concern 1. A little wordy 2. Stilted language

TABLE 87.5 Examples of Vision Statements

Example	Vision Statement
Vision Statement #1	The vision was addressed from two different time frames — long term and over the next 12 months. Both are important because they provide the framework necessary to build the appropriate program. Long Term • Ergonomics is part of our culture. We don't think about it separately any more. • Ergonomics improves safety and health, improves plant performance (worker productivity, product quality, cost control) and improves the quality of worklife (QWL) of workers. • Ergonomics is used before as well as after injuries/illnesses occur. Prevention is common. • All aspects of human performance are considered part of ergonomics, including such issues as heat stress and human error. Within One Year • Ergonomics will be more commonplace with a great deal more awareness. There will be successful projects completed, ergonomics reviews of new designs, supervisors evaluation of jobs, and illness investigation procedures. • Ergonomics will still be primarily a safety and health issue. • The plant will be addressing all known problems. • The major areas of emphasis will be cumulative trauma disorders along with manual material handling type injuries.
Vision Statement #2	The program will undergo a change in focus and activities. The focus will go from pain reduction to maximizing effectiveness on the job. Maximizing effectiveness on the job includes all aspects of performance, such as a safe and healthy workplace, the ability to produce high quality goods, highly productive workplaces that don't waste time and energy of the workers, and high quality of worklife. Ergonomics is one of several tools used to maximize effectiveness on the job. During and after the transition, there will be technical ergonomic resources available to the plants for projects and for auditing assistance. Auditing for ergonomic concerns will become part of normal auditing procedures used for other safety and health audits.
Vision Statement #3	The guiding principles involved with this organization are: • to push problem solving down to the working level in the organization, • to spread ergonomics throughout the organization by heavily involving others, • to avoid making ergonomics an "overlay" that just adds work, and • to ensure that ergonomics solutions dovetail with other changes being made. These principles dictate that the ergonomics task force be more of a facilitator, technical resource and trainer than a problem solving group. Where appropriate, the Ergonomics Committee will either work with a group who is already studying a job or task, or will request a specific problem solving team to consider the ergonomics issues. If a team is not available, then the Ergonomics Committee may work on the project itself.

Plan the Stages of the Ergonomics Culture Change

Successful ergonomics programs are guided by a knowledge of organizational culture change models. There is a body of knowledge generally titled "change management" which deals with change in organizations. One especially helpful part of this technology is that it outlines the steps which must be followed before commitment to successful organizational change occurs. As the ergonomics program matures, it goes through six distinct stages, each with separate concerns and issues. The six stages, using layman's language, are:

1. *Awareness* that a change is necessary (e.g., injuries are excessive)
2. *Acceptance* of ergonomics as a tool that can help
3. *Trial* using ergonomics to see if it works
4. *Regular use* of ergonomics because it does work
5. *Procedures* written to include ergonomics
6. And finally, a *culture* that is totally supportive of the use of ergonomics

These stages are outlined in Table 87.6, along with brief comments about some key issues.

TABLE 87.6 Stages of an Ergonomics Program

Area:	Stage 1 Awareness	Stage 2 Acceptance	Stage 3 Trial	Stage 4 Regular Use	Stage 5 Procedure	Stage 6 Culture
Brief description of stage	Learning about ergonomics	Positive image of ergonomics	Willingness to give it a try	Multiple ergonomics projects	Ergonomics in operating and design	Inconceivable not to use ergonomics
Ergonomics	"Ergo-What?"	Oh, yes, ergonomics. That sounds interesting	Ergonomics — it should reduce injuries	Ergonomics is more than injury prevention	Ergonomics is human performance	Ergonomics helps with every aspect of our business.
Results	None	None (But wants to hear about success stories from others in this industry)	Very limited (Results only with specific projects — still going on faith)	Paying off (Still used mainly for injury prevention)	Solid benefits (Results in safety and health, performance, cost reduction)	Solid benefits (But little need to measure benefits any longer)
Management feelings	Skeptical	Acceptance (grudging to willing)	Prove it to me! (On our site.)	Yes, it works, but can you do it again? And lets show some payoff.	This stuff really works. I'll have everyone use it.	"And why didn't you think of ergonomics? We always use it."
Ergonomics Committee feelings	Why us?	Learning	OK I hope it works.	I hope these other people understand it like we do.	All we do is training. Will we ever get done?	That was a great committee. I'm glad I was on the team!
Role of Ergonomist	Advocate	Assurance of others	Leading the effort	Facilitator and training	Builder of others; ensures systems in place	Maintenance

Unfortunately, few people are familiar with the stages of change or with the pitfalls that occur when the stages are not followed. The most common problem results from the tendency to skip the first two stages of creating awareness and acceptance. The next most common problem is to attempt to jump over stages, for example by going from "Stage 1 — Awareness" to "Stage 5 — Procedure." While jumping any stage causes resistance to the change, jumping multiple stages creates even more resistance.

Another way to look at the program stages is to think in terms of child development, from infant, toddler, child, adolescent, young adult, and finally mature adult. As the child grows, its needs change and it no longer responds to things the way it once did. An ergonomics program is very similar.

Create a Strategic Plan

Successful ergonomics programs have a strategic plan. A strategic plan is necessary to guide the ergonomics program. The strategic plan defines what the ergonomics program intends to accomplish over the long term. Some organizations use vision statements or mission statements to describe program objectives, and these are very helpful concepts for the program.

However, a mission or vision statement, by itself, is not sufficient to fully describe the strategic plan. To develop the strategic plan, the following questions need to be discussed and answered:

A. *What do we want the ergonomics program to do?* This usually becomes the mission, vision and scope of the ergonomics program.
B. *How do we monitor results?* What data do we measure to demonstrate progress with ergonomics?
C. *What are the barriers?* And how can they be overcome?
D. *What policy issues* are likely to be affected?
E. *Who is/should be involved, and what are their roles?* This includes both the ergonomics committee and ergonomics problem-solving groups.
F. *What is the priority?* How long should it take to finish the job? What resources are available to us?
G. *When and how should we review our plans* and progress with management? Are there others who need to hear our story?

Understand the Difference Between an Ergonomics Program and the Practice of Ergonomics

Many ergonomists have had training in the identification, analysis, and resolution of ergonomics problems. They have not typically had training in the management of complex programs such as the ergonomics program. Without this background, they have difficulty visualizing the roll-out of the program, planning for resources, estimating degrees of success, and conceptualizing how the ergonomics program will mature and what it will be like once it is completed.

It is important to clearly distinguish the two areas and the skills necessary to be successful at each. If it is necessary, provide both program management support and ergonomics technical support for the ergonomics committee. Ergonomics technical support may be provided by the ergonomist while program management support is provided by a mid-level manager. Figure 87.5 provides a distinction between the two areas.

Create A Tactical Plan

The strategic plan is necessary but not sufficient for ergonomics program success. There are two types of plans which are necessary for a successful ergonomics program:

- Strategic plans determine what are we trying to accomplish
- Tactical plans determine specifically how we accomplish those goals

A tactical plan is the month-to-month and week-to-week plan which outlines the jobs to be analyzed, the procedures to be written or reviewed, the training to be accomplished, and the solutions to be implemented. Many ergonomics programs fail to develop tactical plans. Some people develop initial plans but then fail to maintain them.

FIGURE 87.5 Ergonomics practice and ergonomics programs.

High-quality tactical plans should be developed initially, then monitored to ensure that the planned activities are accomplished. Experience with the most successful ergonomics programs indicates that tactical plans must be reviewed monthly, revised quarterly, and fully reviewed and updated semiannually.

Sound tactical plans are the key to obtaining funding and personnel to accomplish the ergonomics program goals. Once the tactical plans are developed, management can easily see the costs (financial resources and personnel) as well as the benefits (expected improvements from proposed projects). This makes approval of the ergonomics program plans easier, and allows management to clearly understand what to expect from the ergonomics program.

In developing the tactical plan, the following questions must be answered many times:

- What activities should be done?
- When should they be done?
- Who should do them?
- What are the quality standards?

An example of some activities found in a tactical plan are shown in Table 87.7.

Ensure that There Is Regular Quantitative Evaluations of the Overall Ergonomics Program

Successful ergonomics programs measure themselves. And they do it regularly and often. The adage "You can't manage what you don't measure" is certainly true when it comes to ergonomics programs. Yet, few ergonomics programs are quantitatively measured on any regular schedule. With beginning programs, rigorous evaluations are seldom required. However, as an ergonomics program begins to mature and the "honeymoon period" ends, management often asks about progress and results. Without ongoing evaluations, it is difficult to respond with any specificity to these questions. There are a number of audit and assessment tools available to evaluate ergonomics programs, and an example is provided using the assessment tool *Assessing Your ERGONOMICS Program* developed by Auburn Engineers. This audit tool is widely used, in part because of its simplicity and ease of use. It has 50 multiple choice statements, and requires less than one hour to complete.

An example of the scoring provided by *Assessing Your ERGONOMICS Program* is shown in Figure 87.6.

The interpretation of this assessment is equally easy. From the bars shown in Figure 87.6, these conclusions can be drawn:

1. The areas of organization, medical management, and correction are adequate.
2. There is a need for more balance in this program because there is a large gap between the best and worst areas assessed.
3. There is a need for more work on prevention and on demonstrated results.
4. This is average progress for an ergonomics program which is 6 to 12 months old.

TABLE 87.7 Abbreviated Example of a Tactical Plan

WHAT?	WILL BE DONE BY WHOM?	BY WHEN?	Check When Complete
1. Plant strategy developed	Committee	May 5	
2. Ergonomics policy written	Committee	May 5	
3. Roles/responsibilities developed	Committee	May 5	
4. First problem solving training	Don	June 3	
5. Initial 2 projects started	Todd, Mary and teams	June 6	
6. Baseline data collected for measurement systems	Alice	June 12	
7. Initial project recommendations implemented	Todd, Mary	July 15	
8. Next 2 projects initiated	Maria, Jude and teams	July 6	
9. Illness investigation procedure tested	Bruce	July 20	
10. Project recommendations implemented	Maria, Jude	August 20	
11. Next 2 projects initiated	Cindy, Ron and teams	August 4	
12. Detailed surveillance completed	Bruce	August 16	
13. Policy issues identified	Mike	August 24	
14. Project recommendations implemented	Cindy, Ron	September 24	
15. Next 2 projects initiated	Todd, Maria and teams	September 7	
16. Plans for proactive ergonomic program developed	Ivan	September 9	
17. Project recommendations implemented	Todd, Maria	October 12	
18. Next 2 projects initiated	Mike, Randy and teams	October 5	
19. Engineering training started	Ivan	October 14	
Continued			

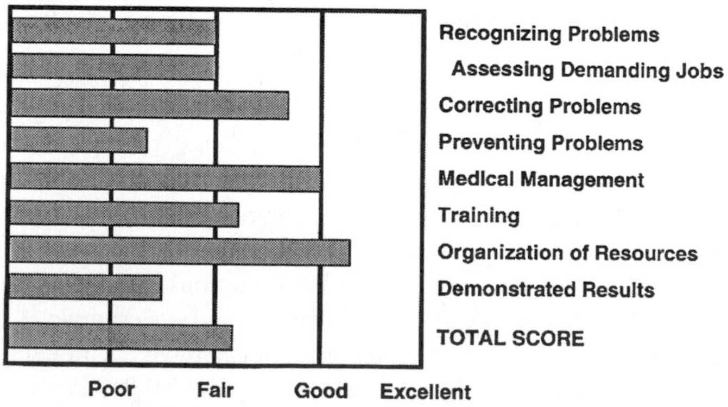

FIGURE 87.6 Scoring the success of an ergonomics program.

More information on ergonomics program evaluation can be found in the chapter titled "Evaluation of Ergonomics Programs."

Do Not Wait for Top Management to Push the Program Down

Unfortunately, for many programs, a litmus test of acceptance is the degree to which top management supports the program. Top management support is too often seen as a requirement for ergonomics program success, and without an endorsement, nothing happens.

Successful ergonomics programs do not require top management support. A caveat is important: this is not to suggest that one simply take on ergonomics projects with no management support or against management's directions. There are usually several layers of management within an organization, and one does not need to wait for top management endorsement to begin. Even with the support of lower level management, much can be accomplished, thus laying the foundation for a larger program later on. This initial success with positive activities and projects now permits the easy endorsement of the program by management.

There are several distinct advantages to getting things going, at least on a small scale, with some important projects. Clearly, this will begin the process of eliminating some difficult tasks, thus making the workplace a little safer. By starting with some projects, one can determine how to go about the process of solving ergonomic problems, thus gaining valuable experience. Many of these projects become examples of the types of changes which can take place with ergonomics, and therefore serve as good illustrations to use once the program begins to grow.

Finally, it is important to note that good ideas "catch on" on their own, while poorly conceived ideas are stopped or die from lack of interest. If ergonomics is seen to work well, it will take hold and people will soon be asking for help with more ergonomic projects. However, if the ergonomics program seems to go nowhere without the push from top management, then perhaps it is a poorly planned and ill-managed program.

Maintain Political Support

Successful ergonomics programs have internal political support when they need it. It is necessary to develop and maintain strong political support with the organization's safety and health committee, with senior management, and with key staff groups like engineering and health services.

This is done with frequent contact, seeking and using input, and by openly sharing what is happening. Even small successes should be shared with others, and credit should passed liberally around the organization. Publicity plans should be developed which permit everyone who contributes to share in the limelight. If these things are done as the program grows, then the political support will be there when it is needed.

87.3 Conclusion

There is enough information regarding ergonomics programs to know what factors contribute to their success. An ergonomics program manager should review the sixteen factors, assess their presence in the program, and if missing, seek to implement them as soon as practical.

For Further Information

The Practice and Management of Industrial Ergonomics, David C. Alexander, Prentice-Hall, 1986.
The Top Ten Reasons Ergonomics Programs Fail, Auburn Engineers, Inc., Auburn, AL 1994.
Assessing Your ERGONOMICS Program, Auburn Engineers, Inc., Auburn, AL.
The Economics of Ergonomics, Auburn Engineers, Inc., Auburn, AL 1994.
The Economics of Ergonomics, Part II, Auburn Engineers, Inc., Auburn, AL 1995.
Advanced Techniques for Managing Your Ergonomics Program, a short course sponsored by Auburn Engineers, Inc. Some specific items used for this paper which are included in that course are:

- Selling Ergonomics To Management
- A Model Ergonomics Program
- Stages Of An Ergonomics Program
- Defining The Ergonomics Culture
- Building Commitment To Ergonomics

88

Elements of the Ergonomic Process

Åsa Kilbom
National Institute for Working Life, Solna, Sweden

Nils F. Petersson
National Institute for Working Life, Solna, Sweden

88.1 Introduction

During the ergonomic process, problems, potential or already apparent, are gradually worked out and brought to a solution. Problems may be similar but the contexts in which they appear are almost unique. Thus, the ergonomic process will hardly ever be the same, and experiences gained in one case cannot be applied mechanically to another place.[15] Furthermore, there is very seldom only one possible solution, but many and probably quite different ones, depending on the culture and awareness at the workplace, its size and level of technology, and the human and financial resources available. Moreover, the chosen solution must comply with the aims of the organization which again influences the choice of solution. The ergonomic process in practice will consequently take different ways and differ considerably from time to time. However, some important steps or phases in the process can be traced in most cases[14,19] and need to be handled for a successful outcome. These phases are: organization of the process, identifying the problem, analyzing the problem, developing a solution, implementing the solution, and evaluating the result (Figure 88.1).

All of these phases normally have to be present to achieve a good result. In most cases it is important to treat the different phases separately from each other and not to begin one phase until the preceding is completely finished. Too often solutions are presented before a thorough analysis is carried out, maybe resulting in a suboptimization.

Schneider[24] suggests three basic elements to implement an effective ergonomics program:

1. Establish ergonomics as a business function
2. Establish a predefined return on investment profile for workplace improvements
3. Establish goals and measure performance

He also argues for not using the word *ergonomics* as it often seems to be synonymous with costs; instead it should be emphasized that healthy people perform and work better. Improvements based on ergonomic investments as well as other investments should be evaluated on the basis of value. Other,

0-8493-2641-9/99/$0.00+$.50

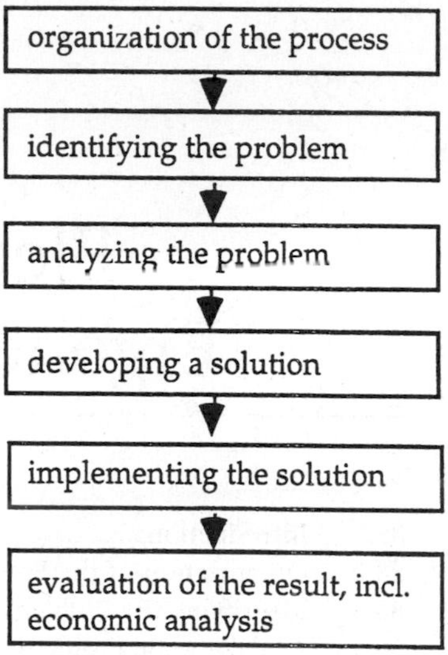

FIGURE 88.1 Important steps in the ergonomic process.

more complex elements of the ergonomic process have been reviewed by Wilson, e.g., the ergonomic design process.[29]

Apart from the phases included in the process (Figure 88.1), a routine to take advantage of the experiences for the next project is also desired.

Ergonomic programs need time, and an underestimation may lead to failure to fulfil the program. One problem is modern management's focus on short-term goals and profit which can be hard to handle.[3]

Finally, an ergonomic process is characterized by its comprehensive view and multidisciplinary approach, taking both productivity and human aspects into consideration. The multidisciplinary approach must be considered in forming the team of the process, taking the participants' background into account.[23] Actually the next most important requirement, after involvement of the employees and management commitment, for a successful implementation of ergonomics, is the multidisciplinary approach.[2]

88.2 Organization of the Process

Although the organization of the project or the process* many times does not start until the problem is identified or during this step, it is advisable to treat it separately from the problem analysis and problem-solving phases. It is common to have some sort of a fixed organization taking care of ergonomics. Big companies often have their ergonomics department or a health care service with ergonomic expertise. For example, Faville recently presented one approach for a large manufacturing company faced with a high rate of musculoskeletal problems.[8] Another example is the nationwide occupational health service for the construction industry "Bygghälsan" which has been in operation in Sweden since the late 1960s.[7] Through medical check-ups every 2 to 3 years a high prevalence of musculoskeletal disorders were identified. An ergonomic program was linked to the health surveys and gradually introduced improved work practices and technical improvements at the building sites.

Obviously, the preexistance of an organization dedicated to ergonomics will facilitate and speed up the identification, analysis, and implementation phases. Ideally, such an organization will already have

*Throughout this text we use the term *project* for time-limited activities and *process* for those activities going on without any predetermined termination.

developed methods for surveillance of risk factors and health, a system for collaboration with production engineering and personnel departments, and pathways for feedback to management and employees.

In many countries, e.g., the Scandinavian nations, formal joint union–management committees, taking care of ergonomics, are prescribed and in other countries voluntary union–management groups appear in companies.[18,22]

The process of continuous improvement, kaizen, has been introduced in many western companies and ergonomic projects may be integrated in these processes. When a preexisting organization is not in place, the emergence of an ergonomic problem can be an important factor triggering its development.

Lacking a preexisting organization, many different ways to organize an ergonomics project or process exist. One extreme is to hire an external consultant expert of ergonomics, and the other is to involve all concerned employees in the process without an expert. Although both systems appear and may be adequate in some cases the most effective organization, in general, lies somewhere in between, taking the advantages of both. It is hardly possible for employees to master all different fields of ergonomics, and therefore not using an expert may result in nonoptimal solutions. Especially for small companies with limited technical expertise, the implementation of high-tech solutions without external know-how may result in costly mistakes. On the other hand, not using the knowledge of the employees can imply that important basic factors are not taken into account. A process where employees take an active part in all phases of the problem is usually called "participatory ergonomics" and is used successfully all over the world.[13,20] Involving the employees most often facilitates the implementation through a greater acceptance of change, and adds problem-solving capabilities as the person doing the job often has the best insights in how to improve the work.[3,11] By teaching workers fundamental ergonomic principles they can become responsible partners in the ergonomic process.[3] The involvement of workers in the process leads to empowered workers,[3,26] who can interpret accurately ergonomic needs and who become increasingly responsible for pushing the processes. The ergonomic training has to be "just-in-time-training" adapted to the real acute situation. Starting the training from one's own situation, e.g., an evaluation of the worker's own workstation, seems to be an appropriate method.[9] Apart from training in ergonomics, training in team work and communication is also important.[11]

Another prerequisite for success is management commitment.[3] An ergonomic program is doomed to failure without visible support of the management and financial backing.[23] Schneider[24] too puts a heavy emphasis on management commitment to ergonomic improvements and argues that the onus for managing the ergonomic agenda is with line management. Active participation of all involved including management and supervisors has been found to be the most important factor in implementing ergonomics.[2] Chavalitsakulchai et al.[2] also argue for regarding government officers as vital for ergonomics intervention programs.

An organization pleaded for by some authors is the "system group" approach.[1] Representatives of all those concerned with the problem, i.e., ergonomics or health department, management, employees, production engineering, personnel, sales representatives, and customers (of the product or the service) form a group that meets to analyze the problem and develop a solution.

Important issues for team building and participative approaches have recently been summarized in a report from NIOSH[11] and are:

- *Management commitment,* from top management to supervisors
- *Training* in ergonomics, communication, feedback skills, technology, all tailored to the participants' skills
- *Composition* of teams, tailored to the problem at hand
- *Information sharing,* within teams, to and from management and employees
- *Activities and motivation,* includes meetings, data gathering and analysis, and planning of remedies. Motivation is achieved by goal setting and feedback, commitment from management and rewards
- *Evaluation,* e.g., of team efforts and outcome of activities

In most cases a change agent, e.g., a production engineer or a safety professional, is recommended. Unfortunately, ergonomists often occupy low organizational positions which makes their negotiating position weak.[14] Apart from the change agent, dedicated persons are of great value and an early identification of them — if they are present — is recommended.

88.3 Identifying the Problem

The title of this phase is not entirely appropriate as many ergonomic processes do not start with a real "problem" but are part of a development strategy. In fact the reasons to start a process or project can vary widely and be divided either regarding their abode — a productivity or a health issue — or in how they occur. For example, the problem can appear as an acute incidence, e.g., an accident which has to be remedied. It can also emerge after a survey to identify critical factors, e.g., a bottleneck in the production or a work task creating musculoskeletal disorders. Third, the problem can be identified as the result of a conscious and continuous improvement procedure, e.g., activities in quality circles. Depending on what type of problems and by whom they are identified, the ensuing process will differ regarding both organization and solutions identified.

Both for a temporary survey and a long-lasting improvement procedure some supporting devices exist to facilitate the problem identification. Group discussions based upon photos or videos are often efficient. Brainstorming is another method both for the problem identification and also later in the solution phase. Short courses in ergonomics, including a workplace-based project, can make problems evident. An ergonomic training program including an evaluation of workstations can be effective. The evaluation of the worksite should include not only physical aspects of the workplace but also analysis of work methods, product flow, and maintenance of tools.[23]

Other sources for problem identification may be internal statistics, e.g., sick-leave or numbers of health care visits. Benchmarking and other quality methods[5] are still other ways to be aware of problems.

It is important to have systems which encourage workers who have early symptoms to report these.[3]

The problem phase may end up with frame settings regarding the costs and time accepted for the project.

88.4 Analyzing the Problem

The analysis phase includes, apart from a thorough analysis of all the components of the problem, also the analysis of the consequences if the problem remains unsolved and the obstacles remain for a solution.

Both in the step of identifying the problems as well as in the analyzing phase one way to start is to ask (1) What is the purpose of the work performed? (2) How are the functions in the work process allocated between humans and technology?[26] For work-related problems a task analysis is a good base[27] and in product development a function analysis is recommended.[29]

When analyzing the components of the problem, it is important not to limit the scope only to the imminent problem. Sometimes the optimal solution is not confined to the work process where the problem was identified, but to work processes or technology used in a preceding process.

The analysis should also contain the goals and the criteria for the solution. Goals are preferably expressed in measurable quantitative terms, e.g., a certain increase in productivity or reduction of sick leave. Fuzzy goals have to be operationalized, i.e., translated and expressed into concrete ones.[17] Skill in defining goals is essential for measuring ergonomic progress.

88.5 Developing a Solution

As stated in the introduction there is seldom only one solution. If the analysis is carried out thoroughly, the solutions are normally easy to find.

Solutions are traditionally subdivided according to their approach into engineering, administrative, and behavioral.[12] Engineering approaches can be redesign of a machine, a workplace or a tool; administrative approaches are changes of work processes, e. g. job rotation, job enlargement, or reallocation of tasks between machines and humans; and behavioral approaches attempt to influence attitudes or behaviors toward risks and changes at work. Training is sometimes classified as an administrative, and sometimes as a behavioral approach depending on the content.

The problem analysis phase should ideally identify the most promising and feasible approach in the individual case. A cost benefit analysis of the chosen solution should be undertaken, and in the case of several feasible approaches such an analysis can assist in choosing the best. Although one approach usually dominates, the solution commonly contains (and should contain) elements of all three approaches. Most administrative approaches focus on work organization which is one part of the macroergonomic concept. Commonly engineering and behavioral solutions are the first ones considered, especially in organizations with limited experience in ergonomic problem solving. However, in a process of continuous changes it will soon be evident that administrative approaches are also necessary. These changes sometimes meet more resistance in an organization and are more far-reaching, and they often require more experience.

The analysis phase also includes a thorough time planning and an allocation of tasks to those concerned. Information about the stages of the analysis, from concepts to detailed plans, is important for acceptance by all those directly or indirectly involved.

Many of the methods used in the problem identification phase are also appropriate in the solving phase, e.g., group discussions and brainstorming. Other means are sketches, models, full-scale mock ups. Literature review, net search, and visits to other sites should not be neglected.

88.6 Implementing the Solution

In many cases the implementation phase seems to be the most critical one, calling for special care and time. Many projects have turned out unsuccessful due to an underestimation of the problems in the implementation phase. This is common in projects carried out by an external expert with no or little involvement by the employees/users during the preceding phases.

A change process, generative or innovative,[6] is very seldom a straightforward action which can be planned in detail. Instead it is a movement with many loops and to's and fro's. Those responsible must be adaptable and ready to change the plans, keeping the main purpose in mind.

As mentioned above, all projects include organizational changes, even those regarded as purely technical projects. Neglecting this fact may be disastrous. Organizational changes are a threat to most people. According to Gardell, resistance to change originates in a threat to the following circumstances: job security, material standard, social status, social relations and freedom of movement.[10] Szilagyi and Wallace, using a somewhat different classification, distinguish the following reasons for resistance: fear of economic loss, potential social disruption, inconvenience, fear of uncertainty, and resistance from groups.[28]

The resistance to change is a serious threat to the implementation of a program, and the best way to handle it is by continuous information. Access to necessary information is one of the most important factors promoting a program.[2] Misunderstandings are often a source of resistance. The main goals must be stated clearly and very early to all concerned.

A participative approach or a system group approach with representatives among all those involved in the change is very useful to forestall the resistance. The following approaches will facilitate the implementation:

- Provide possibilities to influence the solutions
- Offer longer time to comprehend
- Reconcile differences between different personnel categories
- Set up effective channels for information and communication

Resistance may also occur due to a fear of new technical equipment if those involved are not given enough and appropriate training. Training should be planned for and incorporated in all projects of change. The training has to be adjusted to the special situation and occur at the proper occasion, and should preferably be conducted at the worksite rather than in classrooms.[16]

Another source of resistance may be former changes which have not been accomplished in a proper way or have resulted in deteriorations.

88.7 Evaluating the Result

Commonly, the evaluation of an ergonomic change process is based only on perceptions and random observations, without quantitative data support. Such an evaluation implies a risk that unspecific, short-lasting effects are recorded as specific consequences of the change (Hawthorne effects). As a consequence, the effectiveness of the change can be seriously misjudged.

Although ergonomic changes at the workplace are usually not research projects, some lessons can be learned from intervention research.[25] One important experience is to plan for the evaluation as soon as after the analysis stage, when the goals of the process have been identified. Moreover, data describing the situation before the change process are needed for the evaluation. Since the implementation of changes is a dynamic process, the evaluation should ideally be a continuous process where short- as well as long-lasting effects are monitored. Efforts to push the change process may result in a temporary dip of productivity during the changes. Such dips must not be interpreted as a failure. In line with the multi-disciplinary character of ergonomics the evaluation should include productivity and economic, as well as health aspects. Obviously, this evaluation is vastly facilitated if a change agent, or an ergonomics department, is in charge of the program and if a system for monitoring is already in place.

The outcome of the ergonomic process should also be evaluated in economic terms.[23,24] The easiest way for this analysis is to balance the costs of implementing the changes, including the investments, against the likely savings, i.e., reduced incidence of injuries, increases in productivity and quality, reduced staff turnover (including training new staff), etc. In a similar way, alternative approaches to improvements can be compared to select the most cost-effective solution. As demonstrated in many case studies, the payback period for ergonomic improvements is frequently only a few months.[21]

88.8 Using the Results and Experiences for the Next Process

The ergonomic process creates a vast amount of experience and knowledge among all concerned. This experience must not be discarded but should be used for future processes. The evaluation of the program should focus not only on the outcomes in quantitative terms. The process of development and implementation of the program can also be assessed and expressed in qualitative rather than quantitative terms.[4] A protocol where the process is described and where the experience obtained is documented is therefore a valuable tool for future work. It enables those responsible to analyze the reasons for successes and failures, and in this way the ergonomic process can achieve results far beyond those of a single project.

References

1. Andersson E. A systems approach to product design and development; an ergonomic perspective. *Int J Indust Ergon* 1990; 6:1-8.
2. Chavalitsakulchai P, Okubo T, Shahnavaz H. A model of ergonomics intervention in industry: case study in Japan. *J Human Ergol* 1994; 23: 7-26.
3. Dawkins S. Does ergonomics work? *Manag Office Tech* 1995; 40(3): 12-14.
4. Dehar M-A, Casswell S, Duignan P. Formative and process evaluation of health promotion and disease prevention programs. *Evaluation Review* 1993; 17(2): 204-220.
5. Drury C. Ergonomics and the quality movement. *Ergonomics* 1997; 40(3): 249-264.
6. Ellegård K. Förändringsarbete och förändring — perspektiv på Volvo Uddevallaverken i efterhand. (The process of changes and its result — perspectives of the Volvo Uddevalla plant. In Swedish). Nordiska Ergonomisällskapets Årskonferens NES '94, 1994: 170-173.

7. Engholm G, Englund A. Morbidity and mortality patterns in Sweden. *Occ Med State of the Art reviews.* Construction safety and health. 1995; 10: 261-268.
8. Faville B. One approach for an ergonomics program in a large manufacturing environment. *Int J Indust Ergon* 1996; 18: 373-380.
9. Fragala G. Get more from your ergonomics training. *Manag Office Tech* 1995; 40(11): 45-46.
10. Gardell B. Arbetsanpassning och teknologisk miljö (Work adaptation and technological environment. In Swedish). In Luthman G, Åberg U, Lundgren N, Eds. *Handbok i ergonomi.* 2nd ed. Stockholm: Almqwist och Wiksell, 1969: 546-573.
11. Gjessing C, Schoenborn T, Cohen A. *Participatory Ergonomics Interventions in Meatpacking Plants.* NIOSH, Cincinnati, 1994:
12. Goldenhar L, Schulte P. Methodological issues for intervention research in occupational health and safety. *Am J Indust Med* 1996; 29: 289-294.
13. Imada A. The rationale and tools of participatory ergonomics, in Noro K, Imada A, Eds. *Participatory Ergonomics.* London: Taylor & Francis, 1991: 30-49.
14. Kuorinka I. Tools and means of implementing participatory ergonomics. *Int J Indust Ergon* 1997; 19: 267-270.
15. Kvarnström S. Organizational approaches to reducing stress and health problems in an industrial setting in Sweden. *Conditions of Work Digest* 1992; 11(2): 227-232.
16. Luopajärvi T. Workers' education. *Ergonomics* 1987; 30: 305-311.
17. McCoy T. Getting to the seat of the matter: Ergonomics can improve worker safety and productivity. *Rehab Management* 1994; 2: 116-119.
18. Moir S, Buchholz B. Emerging participatory approaches to ergonomic interventions in the construction industry. *Am J Indust Med* 1996; 29: 425-430.
19. Moore J, Garg A. Use of participatory ergonomics teams to address musculoskeletal hazards in the red meat packing industry. *Am J Indust Med* 1996; 29: 402-408.
20. Nagamachi M. Requisites and practices of participatory ergonomics. *Int J Indust Ergon* 1995; 15: 371-377.
21. Oxenburgh M. *Increasing Productivity and Profit Through Health and Safety.* Sydney: CCH International, 1991.
22. Reid P. *Well Made in America: Lessons from Harley-Davidson on Being the Best.* New York: McGraw-Hill, 1990.
23. Ross P. Ergonomic hazards in the workplace; assessment and prevention. *AAOHN Journal* 1994; 42(4): 171-176.
24. Schneider F. Targeting ergonomics in your business plan. *Manag Office Tech* 1995; 40(9): 28-30.
25. Skov T, Kristensen T. Etiologic and prevention effectiveness intervention studies in occupational health. *Am J Indust Med* 1996; 29: 378-381.
26. Springer T. Managing effective ergonomics. *Manag Office Tech* 1994; 39(3): 19-24.
27. Stammers R, Shepherd A. Task analysis, in Wilson J, Corlett E, Eds., *Evaluation of Human Work.* London: Taylor & Francis, 1995: 144-168.
28. Szilagyi A, Wallace M. *Organizational Behavior and Performance.* (5th ed.) Glenview: Foresman, 1990.
29. Wilson J. A framework and a context for ergonomics methodology, in Wilson J, Corlett E, Eds. *Evaluation of Human Work.* London: Taylor & Francis, 1995: 1-29.

89

An Ergonomics Process: A Large Industry Perspective

Bradley S. Joseph
Ford Motor Company

Susan Evans
Sue Evans & Associates, Inc.

89.1 Introduction

Background

Ergonomics in the manufacturing arena is a fairly new science and not well known. Even though it has been taught in various forms through traditional academic institutions, it has just recently been viewed as a useful approach to "fitting jobs to people" in order to reduce work-related injuries and illnesses, and to produce a host of other operational benefits. Although almost every facet of a manufacturing company stands to benefit from improved job design through ergonomics, certain organizational and knowledge-based barriers to change make the adoption and wide application of ergonomic principles a difficult process.

Many recommendations in this chapter are based on research into traditional and more modern theories of organizational politics and of organizational change. In addition, these recommendations are based on a set of basic assumptions and beliefs that are reinforced by experiences in the auto industry.

0-8493-2641-9/99/$0.00+$.50

- The practice of ergonomics works most effectively through the participative, rather than the expert, approach
- Adequate training of participants at all levels is essential. General introductory training should be provided for all participants, while special topic training should be available for designated individuals
- The structuring of ergonomics committees, teams, task forces or other groups must suit the organization within which they function
- The selection of committee members is critical. Representatives from workers, supervision, management, and union should be included; all areas affected by ergonomics should be represented
- Securing top management commitment is essential for success

Chapter Organization

This chapter is divided into five main sections:

- The introduction is devoted to background information on ergonomics.
- Section two discusses the theory and practice of participative management in large organizations.
- The third section, Recommended Methods for Implementing a Plant Ergonomics Program, outlines a protocol for the establishment and maintenance of an effective participative ergonomics program.
- The fourth section, Recommended Methods of Data Collection, details existing and proposed data collection methods, for use in identifying ergonomics problems, justifying their correction, and evaluating their effectiveness.
- The fifth and final section discusses computer, internet, and communication issues in an ergonomics program.

Statement of Need

Why is a chapter like this needed in an ergonomics handbook for practitioners? Neglecting ergonomic considerations in the workplace contributes heavily to the incidence of musculoskeletal disorders, e.g., low back pain and upper extremity cumulative trauma disorders. In addition, poor job design has been blamed for reduced productivity, poor quality, and increased absenteeism. In some industries, these disorders can occur in epidemic proportions, costing billions of dollars in health costs annually. Research findings suggest that changes in work practices and equipment design can substantially reduce the number of musculoskeletal injuries and substantially increase productivity and improve quality.

This concept has led to great advances in ergonomics research and knowledge. The output of this research is often in the form of models and guidelines to aid engineers in designing machines. However, despite increasing knowledge about ergonomics, changes to the workplace, whether to new or existing machines, still largely fail to incorporate ergonomic principles.

Therefore, even as more research is completed and more is known about the limits of the human body in the workplace, without an understanding of the organizational process that is now widely used, ergonomic factors will continue to be inadequately used in the workplace.

Definition of Ergonomics

Ergonomics is defined as the study of work. Chaffin and Andersson (1984) further define ergonomics as "fitting the work to the person." Ergonomics is concerned with the problems and processes involved in designing things for effective human use, and in creating environments that are suitable for human living and work. It recognizes that work methods, equipment, facilities, and tool design all influence the worker's motivation, fatigue, likelihood of sustaining an occupational injury or illness, and productivity.

Properly designed workplaces, equipment, facilities, and tools can:

1. Reduce occupational injury and illness.
2. Reduce workers' compensation and sickness and accident costs.
3. Reduce medical visits.
4. Reduce absenteeism.
5. Improve productivity.
6. Improve quality and reduce scrap.
7. Improve worker comfort on the job.

The primary goal of ergonomics is "improving worker performance and safety through the study and development of general principles that govern the interaction of humans and their working environment" (Chaffin and Andersson, 1984). Rohmert (1985) states that ergonomics "deals with the analysis of problems of people in their real-life situations." Further, he urges that ergonomists "design these relations, conditions, and real-life situations with the aim of harmonizing people's demands and capacities, claims and actualities, longings and constraints."

Ergonomics should not be associated with the measurement of work in the traditional sense. The ergonomist does not measure work only to set standards of time and productivity. That is a task best left to the industrial engineer, who performs motion and time studies using scientific measuring systems. Instead, the ergonomist identifies elements of the job that reduce the quality of the interface between the human operator and the workstation. A poor interface can cause unnecessary stress to the operator, leading to an increased risk of injuries and errors (which in turn may lead to an accident, poor quality, or a loss in productivity).

89.2 Organizational Issues in Developing a Comprehensive Ergonomics Process

Why has the workplace been deprived of something so useful as ergonomics? One explanation is that there is something internal to industrial organizations, such as communication breakdowns between designers and operators of workstations, or that there is something lacking in the knowledge base of the key actors, such as engineers who do not understand the principles of anthropometry, that erects barriers to proper ergonomic design. The operators, who often know the problems associated with workstations, do not communicate with the designers of the workstations early enough, and the designers of the workstation may not have the ergonomic expertise to design the workplace properly. This leads to design flaws that can contribute to ergonomic stress.

It is important to understand the complexity of industrial organizations when attempting to implement a health and safety (or any) program in the workplace. These organizations are usually very large and resist change. In order to be effective, the ergonomist or the organizer of the ergonomics program must understand how the organizations work and what can be done to overcome the barriers to change.

Below is a summary of basic organizational theory. This summary will help to explain and justify the need for developing specific types of ergonomics programs in plants.*

Organizational Models — Traditional and New

Traditional organizational theory is largely based on the assumption that organizations are rational entities (Shafritz and Whitbeck, 1978). Allison (1971) calls this the Rational Actor Model. That is, an organization is a "system within which individuals and groups will act in internally consistent ways to

*Note that parts of this section can be found in "A Participative Ergonomic Control Program in a U.S. Automotive Plant: *Evaluation and Implications*" by Bradley S. Joseph.

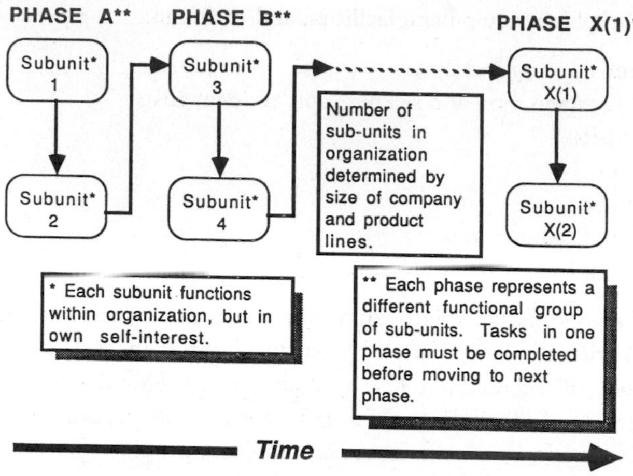

FIGURE 89.1 Simplified organizational politics model.

reach explicit objectives" (Tushman and Nadler, 1980). Therefore, organizations and their structures are "planned and coordinated for the most efficient realization of explicit objectives" (Tushman and Nadler, 1980). The theory blames behavior that violates these assumptions on ignorance, miscalculation, or managerial error within the organization, that is, error independent of either the organizational structure or the management approach. The theory assumes that organizational directions are explicitly and rationally planned. That is, organizations rationally choose goals that optimize an objective function. Managers' roles, as defined by traditional organizational theory, are rationally to plan, organize, coordinate, and control the organization's objectives (Koontz, 1964).

Traditionally, organizations are hierarchical systems involving top-down decision making. Top management decides on an objective for the organization (i.e., what it should be pursuing over the next several months) and instructs subordinates to follow a particular plan designed to reach that objective. These organizations are characterized by an extensive division of labor, including detailed job descriptions, tightly controlled departmental budgets, and narrow spans of supervisory control. This pattern effectively limits the opportunity of people to interact with one another either vertically or horizontally within the organization.

This model has done a satisfactory job in explaining why and how an organization, as a whole, reacts to a particular stimulus (e.g., a change in raw material prices). However, it often fails to explain or predict important aspects of organizational life (Tushman and Nadler, 1980). For example, while the organization as a whole may appear to be reacting consistently to crises or other stimulation from the environment, components within the organization often do not react in consistent ways. Too often, these inconsistencies cannot be explained solely by ignorance, miscalculations, or management error. Instead, the inconsistencies happen so often and are so overwhelming that they cause changes in the organization. Therefore, a new organizational model must be developed to explain these disparities.

One model that explains many of the disparities in industrial settings is called the Organizational Politics Model (see Figure 89.1). Allison (1971) describes this model (based on the work of Cyert and March, 1963, and March and Simon, 1958) in a book which examines the decisions leading up to and during the 1962 Cuban Missile Crisis. He argues that in this case study the rational actor model does not adequately explain many of the critical events or answer critical questions.

The basic principle of the organizational politics model is the concept of organizational units. Within the bureaucracy, e.g., a corporation, these organizational units will act in their own self-interest to achieve their desired goals. This behavior will lend to conflicts between units with different goals, differing perceptions of how to reach a common goal, or joint dependence on scarce resources (March and Simon, 1958; Pfeffer, 1977; Schmidt and Kochan, 1972).

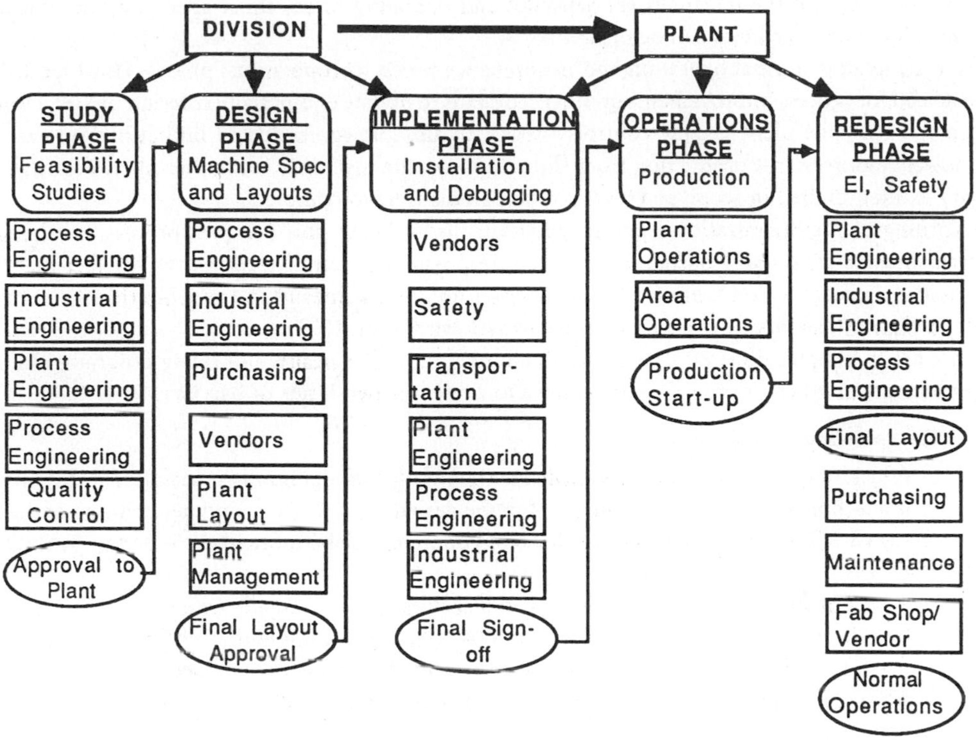

FIGURE 89.2 Organizational process for the installation of new and the maintenance of existing equipment.

The organizational politics model fits well in current U.S. industrial settings, particularly now that budget cutting and increased workloads are placing increasing pressures on individual subunits. All these pressures have occurred without subsequent change in the basic organizational structure. Units are still expected to operate with the same or even greater productivity as before the budget cuts. Because they fear for their very survival, subunits consult their self-interest to a point where it is detrimental to the effective operation of the whole organization. Often these subunits may be organized into a matrix that attempts to break down individual unit self-interest. However, this only works if subunits work with each other as a team.

An example of the organizational politics model is shown in Figure 89.2. The organizational process of a division and plant that produces chassis and suspension components and rear axles was analyzed. Figure 89.2 shows the complexity of the organizational process required for the installation of new and the maintenance of existing equipment within this division.

In this example, a series of units are grouped together into phases. The phases represent periods of time during which groups of units have to complete a task before passing it on to the next phase. First, new processes are studied (study phase) and designed (design phase) at division engineering. Little plant input is solicited in this phase.

Next, division and plant engineering together install and debug machinery (implementation phase). This procedure involves a complex series of actions whereby process plans are sent to selected vendors, interpreted and built to specifications, delivered, and installed in the plant, using resources from the plant, the vendor, and the division. Unless the plant is willing to bear costly delays and excessive expenditures, few changes can be made on machinery, for ergonomic or any other reason, between the time the vendor builds and the time he delivers it, because the vendor has signed off on machine specifications and is under contract to build it to standards agreed upon by the plant and division. Consequently, the plant must wait until the machines are delivered and operating under those specifications (known as final sign-off) before changes can be made.

After debugging, normal operation and maintenance proceeds (operations phase). However, there is often a need for process improvements or other redesigns to update equipment (redesign phase). Depending on the cost, the plant usually controls these activities. However, due to limitations on cash and manpower resources, lost production from shutting down the machines, and other plant priorities, this activity is often limited in scope and takes a considerable time to complete.

According to the organizational politics model (Allison, 1971), this complex process, involving the interdependence of so many parties, offers a certain prospect of conflict. Moreover, the fact that those who make key decisions on manufacturing processes are geographically and organizationally separated creates a high probability of communications breakdown (Allen, 1977).

What are the implications of all this for ergonomics and other health and safety programs? In order to implement sound ergonomic design, we have to overcome two kinds of barriers:

- *The knowledge-based barriers.*
  1. A lack of general ergonomics knowledge (knowledge of ergonomic principles) — Ergonomics is a technical science. Persons involved in the designing, operating, and maintaining of machinery who lack the technical knowledge of ergonomics will be more likely to design workstations poorly.
  2. A lack of specific job knowledge by workplace designers — People who operate jobs are most familiar with them. People who design jobs often do not know the specific information that pertains to daily operation (e.g., the process sheets and industrial engineering studies may vary from the designer to actual operation). This information is important when trying to determine job stresses, etc.
- *The organizational-based barriers.*
  1. A lack of communication between personnel involved in workplace designs — Players in each organizational unit must be able to interact with adjacent units to ensure that ergonomics is properly transferred along the organizational pathway. If for some reason this does not occur, then ergonomics, and many other considerations, may not be incorporated into the new job design.
  2. A conflict between subunit interests — Each subunit has its unique set of goals. Therefore, things like budget, manpower, and time can all be important aspects of subunit performance and may reduce the cooperation between competing and/or successive units along the job installation pathway.

In order to correct this situation and eliminate the barriers, ergonomic or other changes to the workplace will best be accomplished by organizational interventions in combination with technical ergonomics training. Several mechanisms for managing technical change have been researched. They can be grouped into two categories: expert methods and participative methods. These will be discussed below.

Ways to Effect Change in Industry

Traditional Ways to Effect Change — The Expert Approach

Traditionally, major operational changes in large organizations are effected with the help of professionally trained experts. In the case of ergonomics, most manufacturing plants have to import this expertise from the outside. The experts bring their special knowledge to the plant, collect data, return to their labs to analyze it, and make recommendations for change based on their investigations. Once the experts have done their work, there is likely to be no one in the plant with sufficient initiative and interest to follow through with improvements.

Lack of involvement is more often responsible than lack of knowledge for this failure to follow up. In fact, many people in the plant possess potentially beneficial knowledge and skills, but they are rarely asked to play a part in implementing ergonomic changes. For example, workers who do a job every day know it better than anyone else yet they are usually excluded from the job design process. As a result, workers may resist job changes and workplace designs that make sense technically, but in which they have had no stake.

New Ways to Effect Change — The Participative Approach

Worker–management participation itself is not new. However, industry in the United States has only recently started using worker–management participation on a large scale. In fact, there are ongoing debates if participation should be used in all industries. These debates are discussing a variety of issues including — who owns the problem (labor or management) and do these committees undermine the collective bargaining process. This trend has primarily resulted in growing realization that American productivity, labor–management communication, and the overall competitiveness of American goods worldwide have not been keeping up with the world pace (Peters and Waterman Jr., 1982). By looking at companies in other countries, notably Japan, American managers have learned to make increasing use of their employees as a source of information for all areas of plant operations. Truly enlightened managers perceive people as their most important resource.

Experience suggests that the participative approach can ensure the effective continuation of a program long after the expert or consultant is gone. However, when contemplating the use of participation, management and labor must consider how effective it will be in accomplishing *specific* goals of the workplace.

Participation in Health and Safety Programs

Recently, health and safety issues have become of greater concern in industry, partly because of the creation of the Occupational Safety and Health Administration in 1970 and partly because of the realization that health and safety is a core process that can affect the bottom line as much or more than other traditional programs. OSHA has increased both management's and labor's awareness of employee health and safety rights under the law, and legislation has inspired workers to take a more aggressive stance against observed violations. In fact, once a labor contract has been negotiated and ratified, most United Auto Worker shops throughout the country can strike for only one of two reasons: health and safety issues and productivity issues. This emphasis on health and safety has given managers who were reluctant to act an incentive to solve health and safety problems.

Several studies on the effectiveness of worker participation have been conducted. The W. E. Upjohn Institute funded a study (Kochan, Dyer, and Lipshy, 1977) to survey plants with union/management health and safety committees. Its intent was to determine how these committees function and to make a preliminary judgment regarding their effectiveness. General findings indicate that the committees with a high degree of continuity or high levels of interaction exist where OSHA pressure is strong, the local union itself is strong, rank and file involvement in health and safety is substantial, or management approaches health and safety in a problem-solving manner. This indicates that the most important attribute predicting a successful program was management and union commitment to solving health and safety problems rather than objective attributes such as frequency and length of meeting, number of members, existence of an agenda, and whether the committee was mandated by the collective-bargaining agreement.

Obstacles to Effective Use of Participative Problem Solving

Participative management and participative problem solving are not universal answers. Participation should be used for the kinds of programs in which it is known to be effective and for problems which are best addressed by a group process.

The quality circle is a good example of how participation can both succeed and fail (Lawler and Mohrman, 1985). In a manufacturing plant, a quality circle is a group which concentrates on solving workplace problems, usually those affecting the quality of the product, the quality of worklife, and working conditions. The quality circle works well with such problems, especially in early stages, and especially with easy problems. However, when the problems become more difficult or if the quality circle program is expanded too rapidly, the confidence and the effectiveness developed early on quickly becomes eroded. Frustration and the increasing cost of the expanded program usually spell disaster.

An additional threat to quality circles, and one which is relevant to the participative approach in ergonomics, comes in the form of supervisor resistance. Supervisors, like many middle managers, often feel that quality circles (and other participative programs) undermine their authority and control. Their unwillingness to support the quality circle program greatly diminishes its possible effectiveness.

Another barrier to success in participative programs arises in the differing perceptions of what constitutes participation, and how it should be administered. Workers often have unrealistic expectations about what they can do. Managers often treat participation as a special program or campaign rather than a viable technique; or else they abruptly embrace participation, something which can throw the workforce into confusion. Finally, managers may not have the patience to wait for the long-term benefits of participation to appear before they scrap a program when they fail to secure early success.

In her book *The Change Masters: Dilemmas of Managing Participation*, Rosabeth Moss Kanter provides guidelines for appropriate and inappropriate uses of participation. These include:

Appropriate Use of Participation:

1. To assemble sources of expertise and experience among the workforce.
2. To tackle a problem that no one "owns" by organizational assignment.
3. To address conflicting approaches or views.
4. To develop and educate people through their participation (i.e., to develop new skills, acquire new information, and make new contacts).

Inappropriate Use of Participation:

1. When there is a "hip pocket solution" (i.e., the manager already knows the solution).
2. When nobody really cares much about the issues.
3. When there is insufficient time for discussion and the group process.

Kanter also draws attention to five critical challenges, or what she calls dilemmas, which any participatory group must face and which must be overcome if the participative program is to succeed. These are the situations which have no easy resolution. They are as follows:

1. The beginning or setting up the program.
2. The organization of the program, in terms of structure and management.
3. The prioritizing of issues to be addressed.
4. The linking of teams with their environment so as to make them compatible with the existing organization.
5. The evaluation process, that is, determining whether the program is working.

Concluding Statement on Needs of a Participative Program

Role of Training in Participation

One of the most important requirements for success of any participative program is adequate training. In general, effective workplace education should consist of a process of instruction, reinforcement, and establishment of norms of behavior for workers (Vojtecky, 1985; Klein, 1984). It should provide guidelines on problem-solving skills and techniques in running a meeting. In essence, it gives participants the tools to perform their required functions in such programs. Many types of participative training programs are available. For example, Ford Motor Company trains all its employees before they are involved in participative problem-solving groups, emphasizing the basic skills outlined above.

For effective application of ergonomics, Shackel (1980) suggested that six factors must be addressed:

1. Ergonomics should be considered a science and a technology

2. Ergonomists should be researchers and practitioners
3. Ergonomics training and its content need constant updates and review
4. Presentation of data must be in a usable form for engineers, designers, and producers
5. The status of ergonomics must be high enough in the organization to make an impact
6. The ergonomists must have the necessary social skills to use ergonomics in the organization

Training addresses several of these factors. For example, training should teach theory and practice in order to give participants the skills to conduct research and to implement practical changes.

Even though ergonomics is a complex science, often requiring specialization in one area for the development of expertise, the complexity should not discourage the training of persons with average to lower levels of education. Knowledge is not necessarily a function of education level — a lot of knowledge comes from practical experience. The worker does not need to understand ergonomic models that explain the biomechanical cause of injury. Rather, the training needs to emphasize workplace configurations that lead to health problems and to provide understanding of how to reduce the risk associated with poor ergonomic design. Technical experts should be available to aid participants if they request more information or need more knowledge (Allen, 1977); they should play a resource, not an expert, role.

89.3 Recommended Methods for Implementing a Plant Ergonomics Program

This section outlines a recommended methodology for implementing a participative in-plant ergonomics program. The description of the three major steps is followed by a discussion of training needs. Note that although the steps are presented sequentially, some activities are best carried out concurrently in order to increase efficiency and reduce project overhead.

It is the opinion of the author that a participative approach to ergonomics is most effective. If your plant decides against the participative approach, only Step One (Securing Top Management Support) and Part Three of Step Two (Training) are relevant.

Setting Up and Implementing a Plant Ergonomics Program

All plants possess an operating organization which directs daily procedures. This organization requires that the proper authority be secured to begin implementing a program. In particular, health and safety programs require top management and labor support. Figure 89.3 outlines a process by which a program should develop at a plant.

Step 1: Secure top management and labor support — In order for a health and safety, in particular an ergonomics, program to be successful, management and labor must commit to the following:

1. Both management and labor must agree that the problem exists.
2. Management and labor must agree that the problems can be corrected.
3. Management and labor must agree that they will work together on solving the problems.
4. Top management must commit to the program by giving a high priority to the implementation of job changes recommended by the program. This includes, but is not limited to, a commitment from the maintenance department to fabricate and install the changes, a commitment of plant funds to pay for the changes, and a commitment from management to install the changes in a timely manner.
5. Top labor officials must commit and give a high priority to implementing and to using the changes.

Therefore, top plant management and labor must be in agreement that this program is important to the overall operation of the plant. In order to get this buy-in, it is often necessary to educate them. This education usually involves a presentation that defines the program, describes how it fits into existing plant programs and plans, and outlines its benefits and risks. Often this educational process can be supplemented by showing the audience case studies from their own and other facilities. These case studies

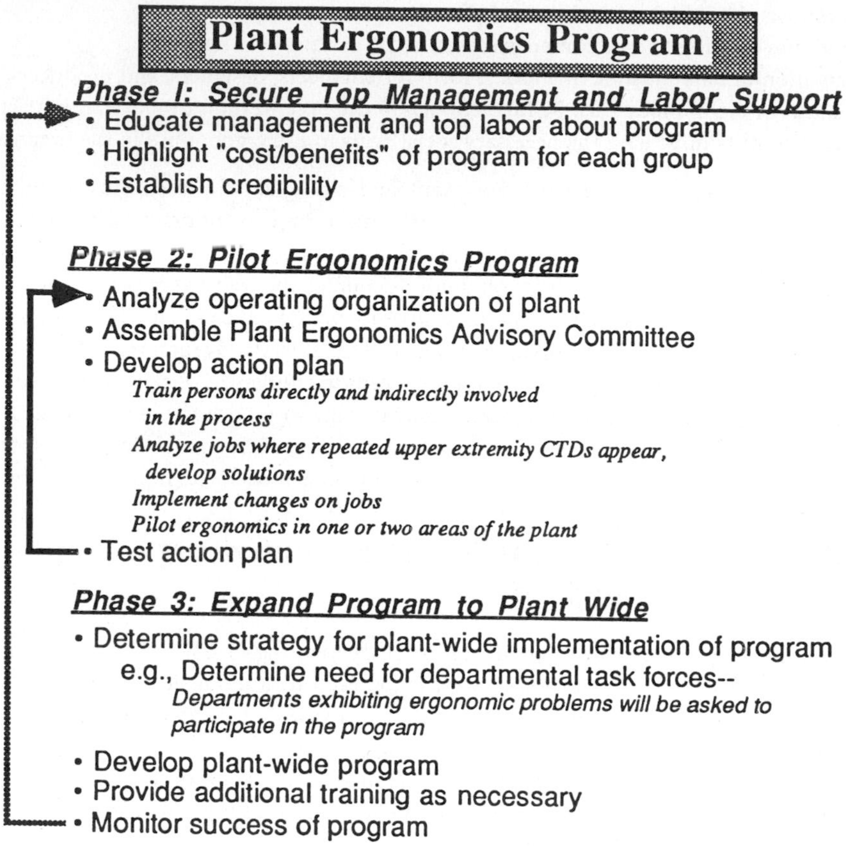
Plant Ergonomics Program
Phase I: Secure Top Management and Labor Support
• Educate management and top labor about program
• Highlight "cost/benefits" of program for each group
• Establish credibility
Phase 2: Pilot Ergonomics Program
• Analyze operating organization of plant
• Assemble Plant Ergonomics Advisory Committee
• Develop action plan
Train persons directly and indirectly involved in the process
Analyze jobs where repeated upper extremity CTDs appear, develop solutions
Implement changes on jobs
Pilot ergonomics in one or two areas of the plant
• Test action plan
Phase 3: Expand Program to Plant Wide
• Determine strategy for plant-wide implementation of program
e.g., Determine need for departmental task forces--
Departments exhibiting ergonomic problems will be asked to participate in the program
• Develop plant-wide program
• Provide additional training as necessary
• Monitor success of program

FIGURE 89.3 Ergonomics process flow chart.

demonstrate that the problems are real and widespread. The use of an in-house expert or outside consultant is helpful in conducting these case studies.

The presentation to plant management and labor representatives must be designed to demonstrate the "costs and benefits" of the programs and to highlight areas that positively affect each group's self-interest. This task is relatively easy for an ergonomics program because all parties stand to benefit.

Dollar costs are those that affect the bottom line of the plant. They can be assessed through the use of traditional accounting techniques and cost/benefit analysis. For example, successful implementation of proper ergonomic design may reduce the numbers of injuries and costs associated with them, while increasing quality and productivity. Simple rate of return charts can be used to demonstrate these costs. An example of a rate of return chart is shown in Figure 89.4. With this chart, an experienced person, knowing the cost of poor job design and the associated plant profitability margin, can show the sales necessary to offset the costs.

Emphasis must also be placed on "people benefits." These include reductions in injury prevalence and increases in employee job satisfaction. Accurate records for employee injuries and worker satisfaction are often difficult to obtain, but are important assets to the success of the program.

Step 2: Pilot ergonomics program — Only after top management and labor support have been granted should you proceed to the Step 2. Often this support is conditional — they will want to review the results of the program after a trial period. In order to perform this step, several activities must be initiated.

First, a thorough study of the plant's organizational structure must be done. Typically, a manufacturing plant has a bipartite structure: production and support. Production's purpose is to produce whatever the plant sells. Support's purpose is to provide the expertise and facilities to ensure a smooth production

Rate of Return Calculation for Health and Safety

Accident Costs (Dollars)	Company Profit Margin 2%	4%	6%	8%	10%
$50,000	2,500,000	1,250,000	833,000	625,000	500,000
$100,000	5,000,000	2,500,000	1,667,000	1,250,000	1,000,000
$250,000	12,500,000	6,250,000	4,167,000	3,125,000	2,500,000
$500,000	25,000,000	12,500,000	8,333,000	6,250,000	5,000,000
$1 Million	50,000,000	25,000,000	16,667,000	12,500,000	10,000,000
$10 Million	500,000,000	250,000,000	166,667,000	125,000,000	100,000,000
$20 Million	1,000,000,000	500,000,000	333,333,000	250,000,000	200,000,000

Sales Necessary to Offset the Cost of Injuries at Different Profit Margins

FIGURE 89.4 Rate of return table.

process. Some plants are highly automated, and have few production personnel but a large support function. Figure 89.5 shows the operational organization of a large automotive assembly plant.

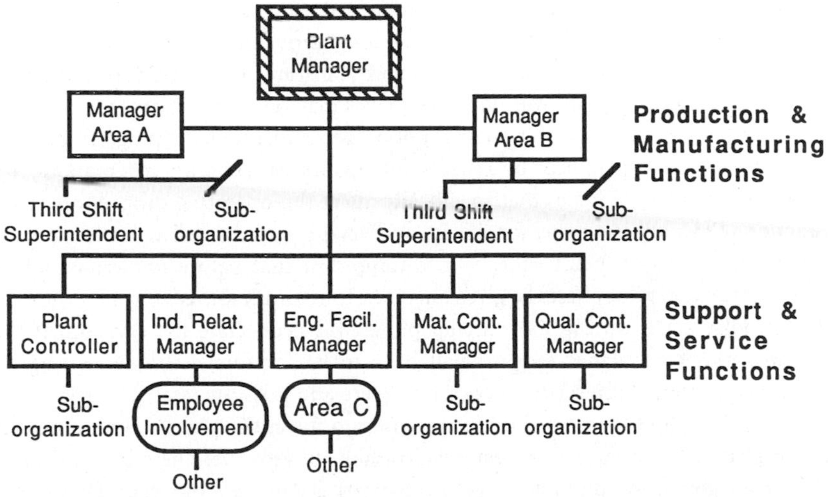

FIGURE 89.5 Functional operating organization of a large industrial facility.

Second, a team must be assembled to form the plant ergonomics committee. The members should be selected by high-level management and union representatives. These leaders should also be represented on the team or be closely associated with it. This support helps to ensure the timely implementation of ergonomic improvements. A letter, signed by these leaders and sent to the appointed individuals and their supervisors, helps to reinforce support for the program.

One criterion for membership should be a familiarity with the plant's culture, since this will ease the team's way in getting things done. Because of the interdisciplinary nature of ergonomics, the committee should consist of people responsible for:

- Identifying problem jobs (a representative from the medical department)

- Determining job stresses (labor representative from the jobs, engineering representative, health and safety engineer, and union health and safety representative)
- Developing solutions (process engineers, manufacturing engineers, maintenance)
- Implementing change (maintenance, industrial engineering, labor representative)
- Follow-up (medical department, etc.)
- Facilitating meetings (a facilitator or group leader to organize meetings and document changes)

A possible list of candidates for a plant ergonomics committee are:

- Industrial Relations Manager or Plant Manager
- Union chairperson
- Plant safety engineer
- Union health and safety representative
- Process engineering supervisor, industrial engineering supervisor, or manufacturing engineering supervisor
- Maintenance supervisor
- Union committee representative from each of the plant area units or departments
- Hospital representative

Third, participants should be trained in the basics of ergonomics. Different levels of training will be required by different people associated with the program. See section on *Training Needs*, below.

Fourth, the program concept must be introduced and piloted in one or two representative areas of the plant. This start-up phase is extremely important for the long-term success of the program. In this phase, management and union support for the program is confirmed. Although the initial step was to gain verbal support for the program, this phase actually secures their long-term commitment, which is necessary for the expansion of the program throughout the plant. Action plans are developed stipulating the "rules and regulations" under which the program will run. These rules may be changed several times before they are agreed upon by all interested parties. If this phase is not properly nurtured, the program may never mature.

Figure 89.6 shows an action plan used by a successful ergonomics program in a manufacturing facility. This action plan includes global issues that specifically affect the operating procedures of the program. Note that the action plan addresses issues for *existing* and *new* projects, as well as for the *people* of the plant.

Step 3: Expand the program plant-wide — It is important that before implementing the program plant-wide, the start-up and pilot phases be given adequate time and resources for the various components to mature. Often, false starts during the start-up phase will necessitate the revision of the rules and regulations. Management and labor support will have to be reinforced. Implementing a plant-wide program during one of these false starts can cause a severe and possibly fatal set-back.

The right time and method for expanding the program to the entire plant will vary, depending on the plant organization, culture, union/management relationship, etc. However, the expansion may involve the development of area ergonomics committees assigned to particular areas of the plant. These area committees may be organized around departments that manufacture and assemble unique products (e.g., building a line of gauges for a particular vehicle) or around areas in the plant that perform a specific set of operations as a subset of the entire assembly and manufacturing process (e.g., the body shop, paint shop, and the trim department of a large auto assembly plant). Because each plant area involves different ergonomic stresses, these teams can concentrate their efforts on identifying their unique problems and developing solutions.

It is often necessary to develop separate guidelines under which each team operates. These guidelines help define the scope of the each team's responsibilities as well as outlining the way the program operates. These guidelines, together with the plant action plan, will help resolve questionable issues. The guidelines must outline when an ergonomics problem will be reviewed by the committee, methods used to evaluate the job, and actions taken (including time limits) for problem resolution.

Action Plan

Organization

The Ergonomics Organization will consist of an Area Ergonomics Committee.

Training

Provide education and training to increase ergonomics awareness.

New Projects

All new projects will utilize ergonomic principles and considerations in the design of products, manufacturing and assembly processes, and equipment.

Existing Projects

Based on medical data, safety considerations, or employe responses, existing processes identified as ergonomically stressful will be reviewed for ergonomic problems and improved

People

Employes exhibiting the effects of ergonomic problems in their work will be encouraged to identify those problems to the ergonomics task force and participate in their solution.

FIGURE 89.6 Ergonomics action plan from manufacturing facility.

Figure 89.7 and 89.8 display two ergonomics programs in large industrial plants. Note that both plants decided to expand their programs to the area levels. In the assembly plant, the expansion involved forming departmental ergonomics committees; in the manufacturing facility, where different processes are needed to make a single product, ergonomics committees were formed in each of the seven manufacturing areas.

The composition of area committees depends on the size of the respective areas. Typically, a member of supervision, the union committee person, a maintenance person associated with the job, and several hourly employees are associated with such committees.

Training Needs for a Plant Ergonomics Program

One of the significant barriers to the successful implementation of plant ergonomics programs is a lack of ergonomic knowledge. Studies (Joseph, 1986) indicate that training can increase the ergonomics knowledge of all participants, regardless of education level or job status. The nature of the training program depends on the needs of the plant. In general, it is suggested that at least two levels of training be used — awareness training and expert training.

Awareness Training

The basic purpose of the awareness training is to make plant persons familiar with the program and to get buy-in from key members of the plant staff. This buy-in is essential in launching and sustaining ergonomics programs. Training should be short, less than two hours, and should be designed to give participants a basic understanding of the principles of ergonomics, an outline of methodologies for identifying job stresses, and an overview of alternative solutions. In addition, this training should address the procedure by which the ergonomics program and its organization will function within the plant

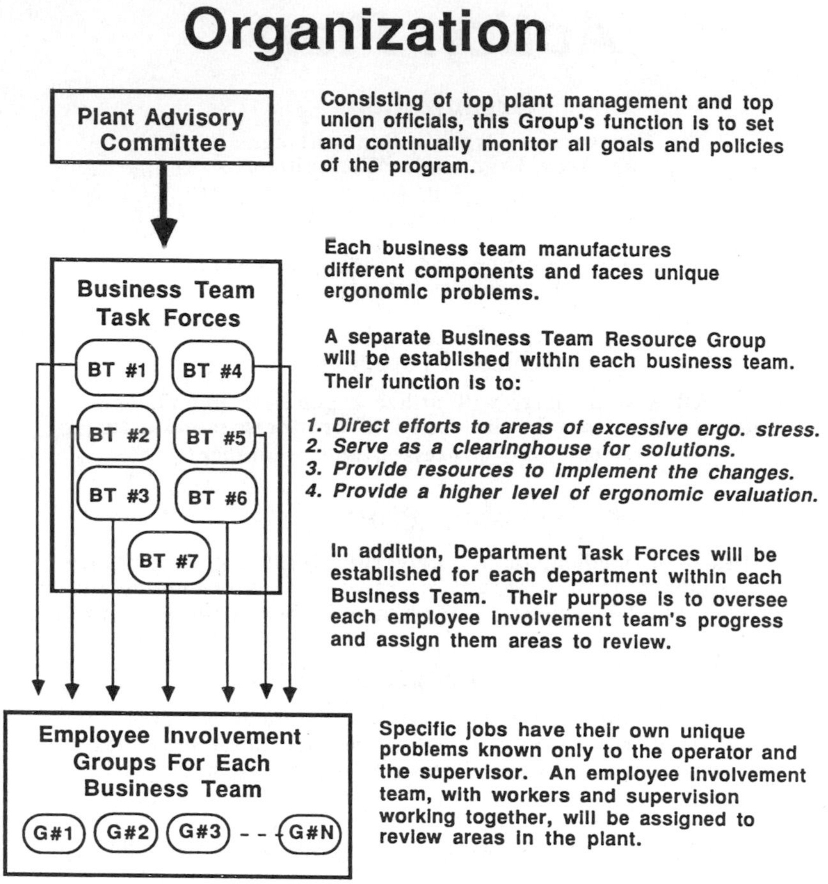
Organization
Plant Advisory Committee
Consisting of top plant management and top union officials, this Group's function is to set and continually monitor all goals and policies of the program.
Business Team Task Forces
BT #1
BT #4
BT #2
BT #5
BT #3
BT #6
BT #7
Each business team manufactures different components and faces unique ergonomic problems.
A separate Business Team Resource Group will be established within each business team. Their function is to:
1. Direct efforts to areas of excessive ergo. stress.
2. Serve as a clearinghouse for solutions.
3. Provide resources to implement the changes.
4. Provide a higher level of ergonomic evaluation.
In addition, Department Task Forces will be established for each department within each Business Team. Their purpose is to oversee each employee involvement team's progress and assign them areas to review.
Employee Involvement Groups For Each Business Team
G#1
G#2
G#3
G#N
Specific jobs have their own unique problems known only to the operator and the supervisor. An employee involvement team, with workers and supervision working together, will be assigned to review areas in the plant.

FIGURE 89.7 Ergonomics organization for a manufacturing plant.

organization. Training should be made available to everyone in the plant to make them aware of the efforts being undertaken. In particular, management personnel responsible for implementing and paying for projects, and union representatives responsible for securing worker buy-in to the program should receive awareness training.

Expert Training

The expert training should be designed to give participants the necessary skills to understand the theory behind an ergonomic problem, the procedures for job analysis, the problem-solving skills to correct job stresses, and methodologies to evaluate the solution. All members of the ergonomics committee should receive this training.

This training may be further divided into two categories:

- *General introductory training* — which covers a broad range of topics
- *Special topic training* — which focuses on one or two special topics in ergonomics that are necessary for risk factor analysis and correction of job stresses.

The length of the *general introductory* training can vary between a day and a week, depending on the number of topics covered. Regardless of length, the training should combine both lecture-based and hands-on sessions. The contents should include at least the following components:

1. Introduction and General Principles — A short session on general ergonomic principles, including an explanation of the theoretical man–machine interface. Often these principles are demonstrated

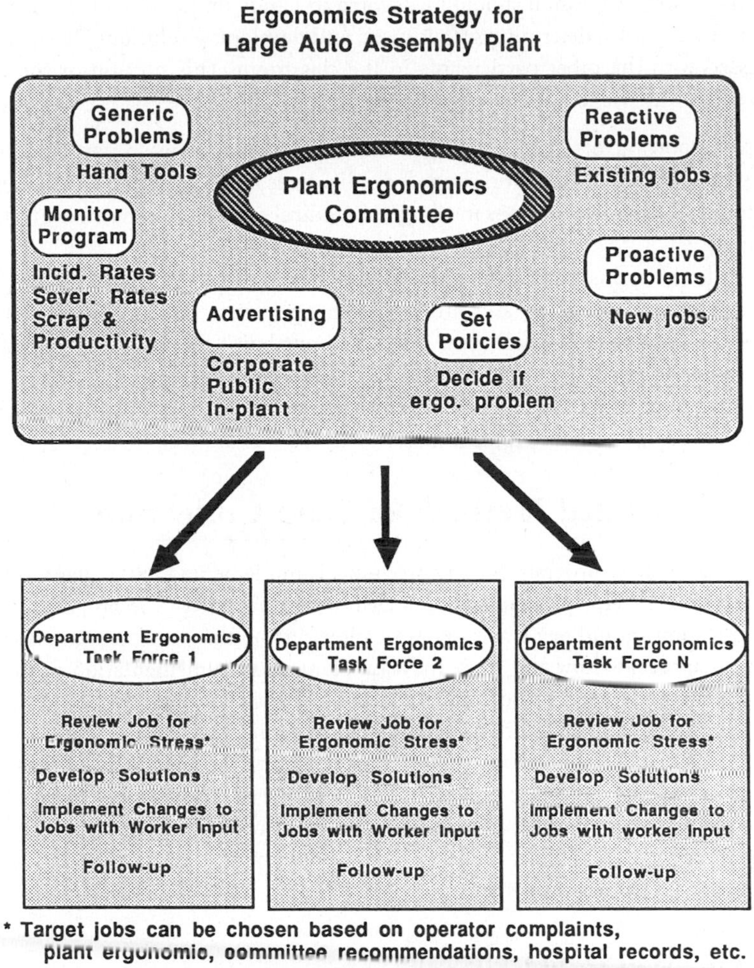

FIGURE 89.8 Ergonomics organization for a large auto assembly plant.

by defining the man–machine interface, the components that make up the interface, and the consequences of poor design.

2. Risk Factors in Poor Job Design — A session on specific problems associated with poor ergonomic design. Generally, four types of problems are highlighted, along with their associated risk factors:
   1. Upper extremity cumulative trauma disorders, including localized fatigue, shoulder, hand, and wrist disorders, etc.
   2. Manual material handling, including problems with lifting, whole-body fatigue, heat stress, etc.
   3. Controls, displays and information processing problems, including problems with machine repair, quality control, etc.
   4. General considerations in workplace design and layout, including problems with lighting, seating, etc.
3. Job Analysis and Solutions — A session on methods to use in assessing ergonomic stresses, and on developing solutions. Obviously, the introductory level of training will not teach sophisticated job analysis techniques. Instead, it should equip committee members to analyze jobs for basic ergonomic stress. The results of this analysis can often be used to reduce the exposures. However, the training should also emphasize that in some cases, more sophisticated job analysis techniques will be necessary, requiring the assistance of in-house or outside experts.

4. Hands-on — The final session should be devoted to hands-on exercises. Participants should have practice analyzing jobs, determining exposures, and developing solutions. These solutions should be discussed with the other participants in the classroom. This portion of the training can be supplemented with videotapes or job-site visits.

The *special topics* portion of expert training should be designed to teach participants advanced job analysis techniques. It should include workshops for individuals in the program who have very specialized tasks — for example, the medical persons in charge of medical surveillance may need training in record-keeping techniques.

It is recommended that this training be given to committee members who require special skills and who are in charge of implementing and fitting the changes onto the shop floor. This includes the practitioners in the program who will have direct responsibility for analyzing jobs and developing solutions (e.g., engineers, etc.).

In all levels of training, it should be emphasized that ergonomics is a broad-based science requiring a team approach to identify and fix problem jobs.

89.4 Recommended Methods of Data Collection

Whenever a health and safety program (or any program) is implemented in the plant, constant feedback is essential. This feedback helps participants to determine the success of the program in accomplishing its goals. Data collection should begin as early as possible during the start-up phase because the results can help management decide how and when to begin plant-wide implementation of the program.

Data collection methods need to be compatible with the plant's operating procedures in order to minimize the resistance to change so often encountered. In some cases, existing data collection methods can be adapted for the requirements of the project, thus obviating the need to invent new ones. However, any method used in the data collection process has to be reviewed, and possibly changed, to ensure that it suits both the plant's system and the needs of ergonomic recordkeeping.

There are two important areas of data collection that should be studied to accurately assess a program. They are Process and Outcomes measures.

Proposed Methods of Data Collection

Accurate and complete data collection is essential to the success of an ergonomics program, both to ensure that efforts to reduce injuries and illnesses are working, and to demonstrate this success to a facility's decision makers. This is especially true in large organizations where "corporate management" is often removed from the daily operations.

Data should be collected in two areas: Process measures and outcome measures. Process measures determine if the system is performing properly. It evaluates systems and organizations to determine if they are doing what they are suppose to be doing. Outcome measures determine if the process is delivering the correct product to the customer. In the case of ergonomics programs, at least two different measures should be used — medical and intervention information.

Medical data: Medical reporting and recording systems in most industrial plants are inadequate for the purposes of effective ergonomics data collection. For the purpose of identifying problem jobs, the chief weakness of the system is the difficulty of tracing a particular injury to a particular job. Another problem is that the nature of ergonomics-related injuries and illnesses makes "early detection" difficult. It is almost impossible, in the present state of affairs, to assess the real numbers of work-related injuries and illnesses.

It is recommended that existing medical data collection be supplemented by active surveillance (diagnostic examinations and self-administered questionnaires are discussed) and by the use of other data sources, such as job process sheets. It is further recommended that medical costs be more accurately tracked through the use of a relational database system for the recording of medical visits.

The report reviews the sources currently available which reflect operational changes due to ergonomic improvements. Suggestions are made for recommendations in procedures for collecting the following data:

- Absenteeism
- Scrap/quality measurables
- Productivity
- Worker satisfaction
- Reductions in known risk factors for ergonomics-related injuries

Intervention costs: The report discusses the need for and the difficulties in securing accurate cost/benefit information for ergonomic changes. Recommendations are made for the development of a form which details the various costs involved in implementing job changes. Such costs should be weighed against the documented reductions in medical and other costs associated with work-related injuries in particular jobs, departments, or areas of the plant.

Data Collection for Medical

One of the primary benefits of health and safety programs is the reduction in job-related injuries. Therefore, the proof of any program's effectiveness depends on a demonstration that injuries and the associated costs have declined. Evaluating the effectiveness of an ergonomics program means showing that injuries and medical costs have gone down for jobs or areas where ergonomic improvements have been made. The present medical recording system makes this task very difficult. In many plants, after an employee develops a medical condition, the procedure works as follows:

Step 1. Employee reports a medical problem in four different ways:
   a. Reports to foreman and gets permission to go to medical.
   b. Goes straight to Emergency Room of affiliated hospital.
   c. Reports to main hospital next day before shift.
   d. Report problem to family doctor — If the doctor judges it to be work-related, then the employee will usually report the problem to plant medical within a few working days.

Step 2. Plant hospital case history or equivalent — Once the employee reports the problem to the medical department, a case history report or equivalent is filled out. Typically, this form has two sides or has multiple pages. On one of the additional pages or the back side of the form is information about the injury itself and any follow-up notes concerning the progress of the injury and the employee's recovery. On the front side of the form is information that identifies the employee by his/her name, social security/employee number, home address, telephone number, sex, age, shift, plant department number, date/time of injury, date/time injury reported, cause of accident/injury, the job classification at time of injury, and injury/illness code. In some cases, these reports will include a statement from the employee as to the cause of the incident, including a description of how it occurred.

Step 3. Depending on the severity of the case an accident investigation may be done. This investigation often has two parts — one done by the supervisor and another by the safety engineer. The purpose of this report is to determine exactly how the incident occurred and what corrective actions are being taken to prevent further occurrences. If the case results in lost or restricted workdays, then the number of days will be recorded.

Step 4. Application for compensation and lost time — If the injury is sufficiently severe to require medical attention beyond initial treatment at the medical office, or if the injury prevents the employee from performing normal work duties, then the employee can apply for workers' compensation. Depending on the state and the rules, compensation for specific types of injuries may be available. Usually every claim must be reviewed by the plant compensation office before being accepted or rejected by workers' compensation. Often workers' compensation records only the most severe cases and does not provide an accurate picture of the true incidence of illness or injury in a plant.

This medical recording system is often called passive surveillance. It has a number of weaknesses for use as a basis for evaluating ergonomics programs.

First, the system is not designed to associate a specific injury with a specific job. Instead, it provides a picture of "global" trends of injury for large areas of the plant. This picture makes it possible to evaluate the overall effectiveness of health and safety efforts, and to pinpoint large "hot spots" where further attention is needed. But because one cannot use the data to relate the incidence of injuries to specific jobs, its value for use in cost analysis of ergonomic improvements is limited.

Second, passive surveillance systems alone, or analysis of existing medical records, may not indicate the extent of injury. Cumulative trauma disorders and related musculoskeletal disorders often have nonspecific symptoms that occur after hours and on weekends. Employees often do not relate these symptoms to the job and do not seek medical attention until after the symptoms have progressed enough to hamper their work efforts. Often these late-stage cases indicate only the tip of the iceberg of job effects.

For every compensable case, there are many more cases of employees with subclinical complaints. Consequently, OSHA logs and workers' compensation data often identify late-stage (tip of iceberg) disorders and complaints, reflecting only a subset of the population afflicted by these disorders. However, the plant medical case reports may be useful as an early indicator of medical incidents. Because these reports are filled out for all cases other than simple first aid, they can give a more accurate indication of the number of work-related injuries and illnesses in the plant. Still, the accuracy of these records depends on many uncontrollable factors.

Third, the period elapsing from the time when the operator notices the initial symptoms of the injury and when he seeks medical attention often is lengthy, and the employee may have moved to a new department or job. Consequently, it is hard to pinpoint the job that caused the injury or to correlate the job type with specific injury type.

Fourth, passive surveillance systems depend on the employee to report the injury to the plant medical department. If employees are not knowledgeable about the symptoms and do not associate them with their work activity, or if employees do not have a good relationship with the plant medical department, they may neglect to report them. This can result in an under-estimation of the numbers of work-related injuries in the plant.

Fifth, the use of medical records in passive surveillance systems may hamper the participative approach since only certain personnel have access to them.

An alternative method for determining the extent of work-related injuries is active surveillance. Active surveillance can be done in a number of ways, of which two are discussed here.

One method involves noninvasive diagnostic examinations by trained medical professionals. The examinations are designed to detect symptoms of cumulative trauma disorders and other ergonomics-related injuries. Although accurate, this method is expensive and inconvenient. The employee must be off the job for the period of the examination. Furthermore, it requires the services of trained medical specialists.

The other method uses self-administered questionnaires. The main advantage of this method is that it is inexpensive. The employee can fill out the questionnaire on his or her own time. No special personnel are required to administer or interpret the results. However, the questionnaires are probably not as reliable as the examinations in producing data about subclinical cases. Furthermore, the success of the program depends on the timeliness and accuracy with which the questionnaires are completed and returned.

The best strategy is to combine these two methods. For example, one approach is to administer the questionnaire to all employees, and then, based on their responses, to select a subset of employees who have a high probability of disease. These employees are then examined by medical specialists in order to confirm the diagnosis.

As one might suspect, even though active surveillance systems are more accurate than passive surveillance, they can be costly to administer. In addition, they are time consuming and may disrupt normal plant operations. Therefore, even if the resources are available to conduct active surveillance properly (e.g., a well-trained staff, arrangements for workers to be off the job for up to an hour), it should be used with discretion.

Because of the complications involved in establishing an active surveillance method for ergonomics, it is often desirable, at least initially, to adapt the plant's existing system. Typically, the information in the existing plant medical system is sufficient for the current needs of the medical department and any reports they have to generate. However, for evaluation purposes, there needs to be a way to identify the individual job. One cannot simply develop a new medical recordkeeping system without adequate support from the medical department, plant management, and the corporation. This support can take months or years to gain, and once there, the system itself can take longer to implement. In the meantime, it is proposed that modifications should be made to the current system.

Because of the availability of the plant case history report and the frequency with which it is used, most modifications should be tied to this report. These modifications should help link reported injuries to specific jobs, and they should include at least the following information. Below is a summary of proposed modifications.

1. Job identification system — Most plants have a structure for dividing up responsibilities on the production floor. Typically, this structure has at least three levels: superintendent, general supervisor, and supervisor. The superintendent has responsibility for all the general supervisors in a particular area, which is defined by various factors such as similarity of products or operations. A general supervisor controls several supervisors, each of whom oversees from 20 to 30 operators. Therefore, a method should be developed to match a particular employee to a specific supervisor so that a medical incident can be tied to 1 of 20 jobs.

It should be noted that a mechanism already exists in most plants to help identify specific jobs — namely, the job process sheets completed by the plant engineers. These sheets are typically developed and maintained by industrial and process engineering. They outline the specific tasks the operator must go through to complete the job. All jobs have a unique sheet.

2. Medical cost system — There are several costs associated with a medical incident, whether it involves a single medical visit or lost-time compensation. The cost of a medical visit includes the time away from the job, the time for the doctor or nurse to make an examination, treatment time, and materials associated with treatment. (In fact, some studies estimate that medical visit costs may average $50.) These costs should be recorded, on existing medical record forms, along with the major costs associated with workers' compensation, days restricted, or days lost from work.

Because of the nature of these injuries, it is important to use "relational" database systems. Many of the larger database systems used in corporations (payroll, workers' compensation) are relational. A relational database is very powerful and allows one to record multiple injuries for a single individual. These injuries may be differentiated by time of occurrence or by body location. Whichever is the case, accurate records of the data can only be done efficiently if the data collection system relates each incident of an indicator to a common denominator. These denominators can either be the person or the job the person is working on. (In many cases, it is both.) A relational database system can record such data properly.

Figure 89.9 depicts the logic of a simple relational database that can be used for recording medical records. There are two components — employee information and injury data. For each employee, demographic information identifying the employee and worker number are recorded to identify exactly who the injured is and where he/she was working at the time of injury. For each injury (there may be more than one injury for each employee), the type of injury, the location of the job at the time of injury, the date of the injury, and the cost of immediate medical attention can be recorded. If the employee receives lost time or restriction, then the employee number can be used to access an existing medical records management system.

Success in using the modified medical records system depends on certain characteristics of the plant. First, it is necessary that the project be implemented in a plant where the medical department and the employees have a good relationship, and where they are aware of and sensitive to the symptoms and causes of CTDs and low back pain.

Second, only plants located where the workers' compensation system recognizes that CTDs and low back pain can be work-related can employ this modified system. For example, in the state of Ohio,

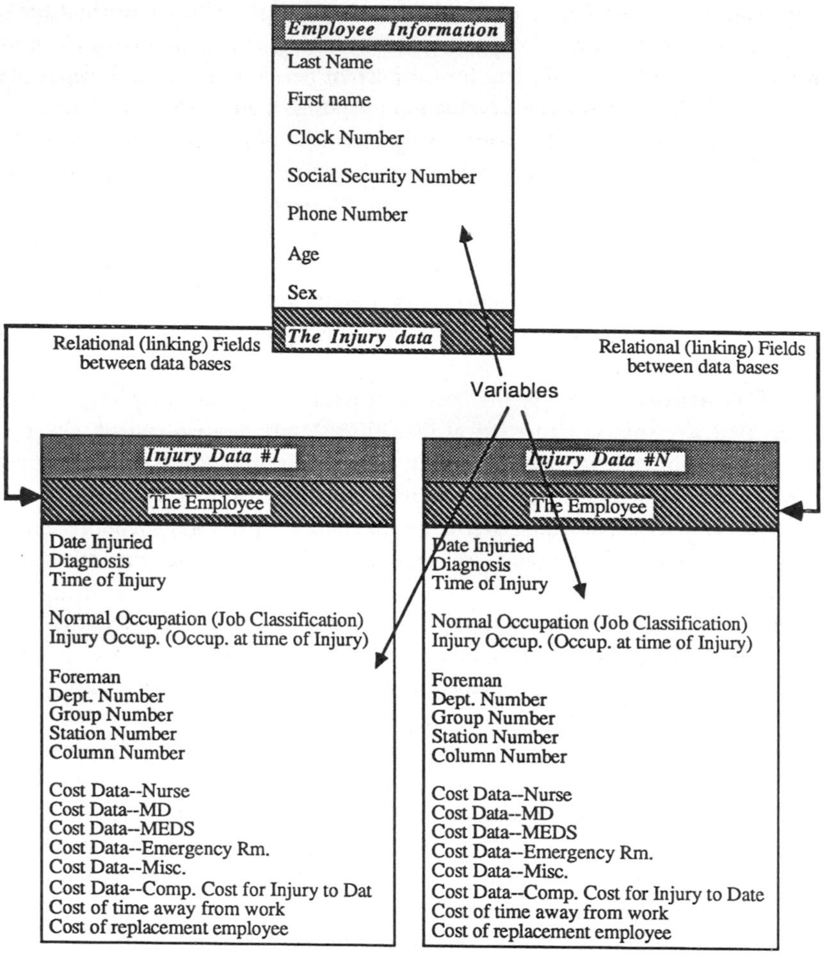

FIGURE 89.9 Database logic for personal computer-based medical records analysis system.

because of a recent state Supreme Court ruling in favor of allowing workers' compensation for CTDs, employers and employees alike are keenly aware of the problem. Consequently, they are reporting the problems earlier to the medical department.

Data Collection for Outcomes of Ergonomics Projects that Benefit the Plant

In order to determine whether ergonomic improvements have the desired effect, it is essential to record before and after measurements of certain operational variables of the job. Desired effects include:

- Lower absenteeism rates
- Reduction in errors and scrap (quality measurables)
- Improved productivity
- Greater job satisfaction
- Reduction in known risk factors leading to musculoskeletal disorders

Analysis of all these variables should be done as a part of the job analysis, and again later as a part of project evaluation.

The current sources for this information in most manufacturing facilities are as follows:

1. *Absenteeism.* Absenteeism records are typically kept for each department by the foreman. These records are sent to the personnel office and recorded.
2. *Scrap/Quality.* Scrap rates are kept for each operation. A scrap budget is set, by operation, for each department. This scrap rate is standardized by comparing it to the number of vehicles assembled.
3. *Productivity.* Productivity rates are determined by traditional time study procedures. The plant operates at a specific rate, expressed in the number of units per hour. These rates are changed only after engineering has made a process change that warrants a reevaluation of the time standard.
4. *Worker Satisfaction.* In many plants, no data are *routinely* collected.
5. Reduction in the Risk Factors Contributing to CTDs and Low-Back Pain. No data are routinely collected.

Data for scrap rates, quality rates, and productivity are currently collected in most industrial plants. However, the data are usually collected by department and sent to a central processing area to be summarized. This makes it difficult to associate the data with a particular job change. In addition, data on worker satisfaction and ergonomic risk factors are not currently collected in most plants. These data need to be collected before and after job changes to determine if the projects are reducing exposures to known risk factors.

In all cases, current operating procedures should be maintained. Any new collection systems should be developed to work within the existing systems. With this in mind, it is proposed that the following changes be made to the existing databases or other data collecting mechanism in order to facilitate data collection. A form should be developed to record and compare the data before and after projects have been implemented on the floor. Every entry onto the form should record the department and specific job location.

1. *Absenteeism.* Current operating procedures should be maintained. A copy of the data should be intercepted and entered into the project database for analysis by department and, if possible, by job.
2. *Scrap/Quality.* The ergonomics task force should devise a new form to keep track of scrap and quality measurements, by department and by job, and to determine improvements on jobs changed by the ergonomics committee. Since plants measure quality in different ways, a QC inspector should be assigned to the team to help with the data collection.
3. *Productivity.* Current operating procedures should be maintained. The industrial engineering process studies and process sheets should be used. A copy of this data should be intercepted and entered into the project database, by department and, if possible, by job before and after job changes.
4. *Worker Satisfaction.* Worker satisfaction may give an indication of the success of the job change since ergonomics is a tool to improve the quality of worklife. The ergonomics task force should devise a form that assesses participant satisfaction in the program. This form may be modeled after one used by Bradley Joseph in his study of an ergonomics program in an automotive plant.
5. *Reduction in the Risk Factors Contributing to CTDs and Low-Back Pain.* This information will be reflected in the data collection for medical because job improvements should show a related reduction in the medical incidence rates. However, because of the nature of these injuries and the time it takes for them to develop, this information may take months or years to collect. Therefore, in the meantime, it is important to keep accurate records on changes, by department and by job.

   A standardized job analysis system should be developed using existing technology. For example, the NIOSH Work Practices Guide and the University of Michigan Static Strength Model can assess many of the risk factors associated with low back pain. These instruments can be used together to analyze stresses to the low back and determine the static strength requirements of the job.

   Analysis of the risk factors for upper extremity cumulative trauma disorders can be done in a variety of ways. However, an economical and convenient way uses a checklist approach that assesses the repetitiveness and postural requirements of the job. The forcefulness of the job is more difficult to assess without the aid of electronic equipment. In addition, ongoing research is being conducted that may lend insight into estimating force through the measurement of several simple variables.

If this research is not available at the time of the study, an estimation of force can still be made by comparing the EMG values to several known forces and stating if it is above or below the known level.

Finally, it is proposed that several characteristics of the job before and after the changes are made will be recorded on paper and videotaped for archival purposes, for further analysis when new risk information becomes available, and for use in the estimation of other risk factors, including force.

Data Collection for Intervention Costs

Interventions or job changes represent a major cost of any ergonomics program. Often, before management will allow the installation of any new project, it has to be cost-justified. Therefore, since cost justification of projects can often decide their fate, it is important that these costs be monitored during the program. In most plants, the current procedure for implementing a project is as follows:

Step 1. Write up proposal— At this stage, the project specifications are developed and summarized through engineering on a project form. A cost justification (cost analysis) statement is attached that allows the decision-makers to determine if the costs of the project will justify the benefits— projects often have to be justified on the basis of traditional cost/benefit analysis and computed in terms of productivity and completed pieces per hour rather than in terms of health and safety costs (see discussion below).

Step 2. Approval of the project— There are two stages in this step.

Stage 1: Corporate Approval— Depending on the estimated cost of the project, engineering and other approval, have to be obtained either at the corporate level or at the plant level. For example, some plants specify a dollar value cutoff above which the project must be approved by the corporation, and below which the project must be approved through normal plant channels. If a project requires corporate approval, then the corporation pays for the project. However, if a project does not require corporate approval, then the project must be paid through existing plant resources. Projects requiring corporate approval must also go through plant approval.

Stage 2: Plant Approval— If necessary, and once approval has been obtained from the corporate engineering functions, the project must also be approved through various plant functions. The director of manufacturing at the plant, industrial engineering or equivalent, and the controller's office must all approve the project.

Regardless of the stage, the project form is sent through proper channels for approval. If approved, the responsible supervisor signs the form and retains a copy.

Step 3. Purchasing— The purchasing department sends the completed project out for bids.

Step 4. Vendors bid on project— Usually the lowest bid with the highest quality wins the bid and builds the project. It should be noted that this step may be omitted if the work is done in-house.

Step 5. Installation of project— The project may be installed with existing plant resources or, in special cases, with contracted professional services.

Step 6. Payment for project— Depending on the costs, either the corporation or the plant pays for the completed project.

Intervention costs can be assessed through normal operating procedures. If the project is approved, the project form should be copied by a member of the plant ergonomics team (and the departmental task force, if there is one). This information will be useful in determining the direct costs (materials, design time, etc.) to implement the job changes.

Problems with Cost/Benefit Analysis for Ergonomics Projects

Currently, as in most manufacturing facilities and depending on the costs, all business projects must go through normal purchasing channels to be approved for funding (see above discussion). Unless costs are nominal, these projects must be reviewed for cost/benefits. Funding is awarded based on traditional cost/benefit analysis calculations and expected savings due to work standards, work practices, or quality.

Below is a list of some of the costs involved in installing new equipment. All these costs should be considered in order to determine accurately the costs of implementing ergonomics projects and changes on the plant floor. It is recommended that a form be developed that records these costs for later analysis.

1. *Design time*— The time and resources involved in designing projects.
2. *Engineering time*— The time and resources involved in engineering the project.
3. *Tool change*— The fabrication costs and time necessary to fabricate a set of tools for the project.
4. *Skilled trades time*— Manpower needs for installing, testing, and maintaining the projects.
5. *Materials*— Cost of materials for the new project.
6. *Machine down time*— If the project is going to directly affect an existing line, that line may have to schedule down time to properly install the project. Therefore, down time and lost production must be budgeted into the installation costs.
7. *Training*— When new equipment and/or processes are implemented on the plant floor, operators responsible for running and maintaining the equipment must receive training.

It may be difficult to use traditional cost systems to justify an ergonomics project. This is because ergonomics projects often do not show significant savings, in the traditional sense, immediately after installation. Instead, the type of savings often seen in ergonomic projects are reductions in health care costs. These are often difficult to justify when the relationship between injuries and the responsible jobs is not well established (see above section).

This lack of an obvious link between an injury and a job yields two results: First, medical costs associated with worker accidents and chronic musculoskeletal disorders are usually not charged directly to the production department responsible for causing the injury. Instead, they are charged to a separate central account in the plant's Industrial Relations Department (or equivalent), thereby partitioning the true costs over the entire plant. This makes it difficult to justify a job change because the benefits are hidden. Consequently, projects often have to be justified on the basis of traditional cost/benefit analysis and computed in terms of plant-wide and area productivity (e.g., or completed pieces per hour). Projects that cannot show a cost/benefit advantage based on these measures often have little chance of implementation.

The second result of the absence of a known relationship between injuries and the production department deals with the poor recording of data. Often, CTDs are recorded only on sickness and absence reports with little or no follow-up. Consequently, employers have little data to go on in establishing a relationship between a job and injury.

Figure 89.10 depicts the relationship between the cost and benefits of ergonomics. Because of the problems of using traditional cost/benefit analysis, it becomes more important to document all the costs associated with poor job design and all the benefits after ergonomic intervention. Therefore, it is often best to make simple, inexpensive changes first. As poorly designed jobs are identified, the data (as outlined above) should be collected and analyzed before and after the proposed job changes. As more data are collected and the cost/benefit equation becomes better defined, it should become less difficult to justify job changes.

89.5 Computers, Internets, and Ergonomic Communication

Introduction

Communication breakdowns between workstation designers and operators has been stated as a barrier to proper ergonomic design. Effective ergonomic programs rely on communication to train the operators to recognize ergonomics problems, provide appropriate measurement systems for detection and reme-

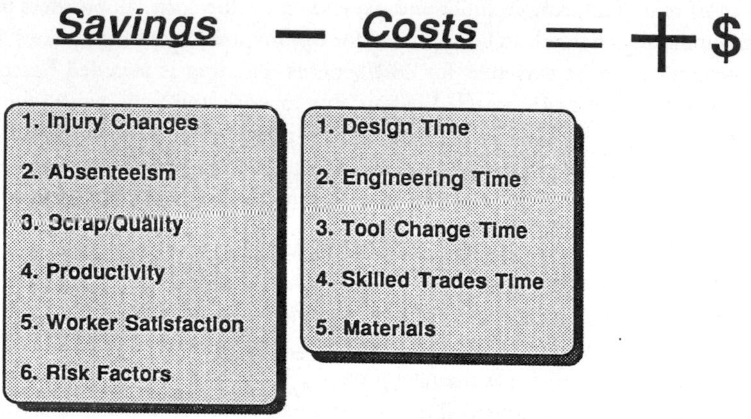

FIGURE 89.10 Cost/benefit summary.

diation, and ensure timely feedback to engineers and designers to ensure that existing problems are not replicated and new designs fully consider ergonomic factors.

This section explores the use of computer technology to overcome communications breakdowns within a plant ergonomics committee and across the entire corporation. Characteristics of participative ergonomics programs which impact communication and the flow of information are presented. Challenges to effectively manage communication throughout corporate ergonomics programs are presented, followed by an approach which uses computer technology to overcome some of these challenges.

Communication and Information Flows

The characteristics of the in-plant ergonomics program discussed in this chapter have several direct implications on the role of communication to ensure its effectiveness. These characteristics include:

- *Monitoring and feedback systems* to track and record injuries, costs, and benefits of job changes
- *Participative teams*, involving representatives from all sectors of the manufacturing and support areas, which are tailored to the existing organizational structures
- *Available training* to attain general and specialized levels of ergonomics expertise among team members

The implications deal with what, who, or how ergonomics program information is being communicated. A considerable amount of data must be collected, analyzed, and reviewed to ensure the program is working. Having consistent procedures in place across the organization helps to ensure that plants can talk to other plants and share information, lessons learned, and discuss problems from a common base of understanding. Table 89.1 suggests several implications for effective communication in a corporation-wide ergonomic program.

Gathering and tracking the myriad data needed to support an effective ergonomics program are no small feat. Examples of the types of information flowing through these programs are presented in Table 89.2. All the information is useful within the plant committee to manage the program. Much of these data are also of interest to a broader corporation-wide ergonomics community. This utility is indicated in the table.

The benefits for sharing information across the corporation are significant. Developing effective mechanisms for achieving this is but one challenge to those developing corporate ergonomics programs.

TABLE 89.1 Ergonomics Program Communication Characteristics and Implications

Characteristic	Implication
Monitoring and feedback systems	Process and procedures must be in place to document incidents, causes, solutions, and effectiveness, and report to appropriate communities of interest.
	Data collection methods must be identified, tested, and implemented within the plant to gather medical, cost intervention, and outcome data.
	Paper and pencil or automated tracking systems are needed to log incidents, assign responsibility, and track progress.
	Mechanisms for tracing design issues back to the designer, along with any in-plant solutions and operator feedback must be established.
Participative teams	Mechanisms must be established for selecting team members with an appropriate cross-functional background.
	Team-building exercises and management commitment documents are encouraged to ensure common goals and objectives are adopted among members from divergent experience and background.
	Team member rosters must be kept current and include active and adjunct members, and their position on the team as well as contact data.
Available training	Training classes must be made available to all team members to ensure informed participation and common bases for problem solving.
	Training must be consistent across teams to achieve common understanding.

TABLE 89.2 Ergonomic Process Information and the User Community

Information used by →	In-plant committee	Corporate community
Best Practices	1	1
Case Studies	1	1
Committee Meeting Minutes	1	
Cost data (injury, design solutions)	1	1
Engineering Drawings	1	1
Ergonomic Incidence Logs	1	
Ergonomic Indicators	1	1
Ergonomic Risk factors	1	1
Ergonomic Standards	1	1
Incidence Action Tracking Log	1	
Job/operation codes	1	1
Lessons Learned	1	1
Management commitment documents	1	1
Medical data (visits, case reports)	1	
OSHA Logs	1	1
Part code, characteristics	1	1
Team roster/phone directory	1	1
Training classes, schedules	1	1
Vendor/Supplier data	1	1
Workforce population data	1	1
Workers Compensation data	1	1
Workstation layouts	1	1
Workstation/Ergonomic checklists	1	1

Challenges to Effective Communication

The task of keeping ergonomics programs running smoothly are full of communications challenges. Challenges relevant to the plant-level programs, and those which occur as a result of adopting the program corporation-wide are discussed.

Setting up in-plant programs clearly must accommodate plant-specific cultures, management styles, and work patterns. Maintaining effective plant programs faces the challenges of:

1. Applying appropriate ergonomics expertise to a broader community of cross-functional team members, with safeguards to prevent misapplication (through training).
2. Effectively supporting an administrative documentation process without burdening committee members.
3. Allowing plant-control of work in progress but provide sufficient corporate oversight to ensure accurate data gathering and consistent process documentation.

All of these challenges pose a manpower resource dilemma for ergonomics programs. Unless the information is captured and documented, the process cannot document its effectiveness. Nor can it provide insights and lessons to use later. The documentation becomes the database. Supporting the documentation process through labor-saving means should become a number-one priority.

As the process is adopted for corporation-wide use, an additional set of challenges surfaces. These include:

1. Providing access to corporate repositories for common data (e.g., manufacturing part information, sanitized workers' compensation costs, approved ergonomic risk factors, etc.).
2. Applying consistency to the ergonomics process to ensure an acceptable level of corporation-wide legal compliance.
3. Providing for informal, risk-free but timely means of communication across committees.
4. Providing means and media to share information on ergonomic best practices, lessons learned, costs, and recurring problems.

The common threads among these challenges seem to be

- Access to data
- Access to people
- Minimize the impact on resources

An effective solution to these challenges must weave these issues together.

Using Computers to Overcome Challenges

The traditional mode of communication employed in participative teams has been face-to-face. The time constraints and geographic dispersion of today's corporate work environment (even for in-plant programs) demand alternative modes of collaboration to achieve effective communication. Just as counterparts in the manufacturing, production, and back-office components of the corporation have adopted computer technology as the means to do more with less, so must the ergonomist employ the computer as a tool to leverage resources effectively.

Experience in developing computer tools for ergonomics processes has taught us several things:

1. The applications must effectively reduce the administrative burden at the plant level first (i.e., the tools must reduce the documentation workload and free the committee for solving ergonomic problems).
2. Users from several local committees must be involved in the design, pilot testing, and fielding to ensure broadest acceptance.
3. The life-cycle costs of fielding, training, distributing, and supporting traditional software applications can far exceed the cost of initial development.
4. Internet technology presents significant advantages for addressing communication challenges and managing application life cycle costs.

Figure 89.11 presents the components for a proven computer-based solution. The approach is based around a centralized ergonomics evidence database application and uses corporate internet technology to support plant access and data exchange.

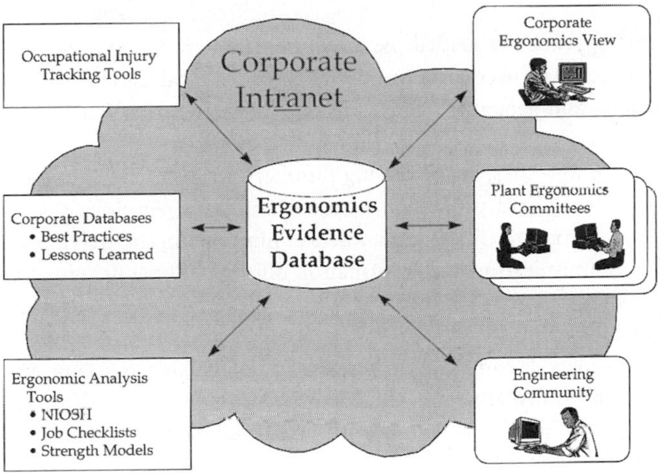

FIGURE 89.11 Components of a computer-assisted solution for ergonomic processes.

The central database allows all plants to read and write to the database as they access and manage their ergonomic documentation data. Data sharing across plants is permitted only at the discretion of the source plant. Once the documentation process is completed for an ergonomics incident, the plant can release the data for corporation-wide viewing.

The Automated Ergonomics Evidence Book (AEEB) uses an advanced internet browser front end to guide the user through the features and functions of the application. It links all plants to the central database, but maintains plant-level control over documentation in process. Builds an electronic ergonomics incident evidence book and easily supports the administrative tasks of maintaining the book, documenting the progress, and generating required and *ad hoc* reports. It also reduces the time spent documenting by sharing common data across reports.

By using a centralized database, data generated by or of use to a corporate audience can be maintained centrally but accessed globally. Corporate commitment documents are updated centrally. Schedules for ergonomic training classes, as well as corporation- or division-wide ergonomics news releases are maintained in one place, but are accessible for all local committees.

The database application also lets the plants track and maintain their roster of plant committee members. These rosters are viewable by other plants and provide an effective means of connecting one with another. Other electronic communication aids provided by the system are bulletin boards for posting news releases, and discussion databases to share ergonomics problems and solutions over an informal but timely communication channel.

The centralized database also permits corporate oversight through read-only access to plant data. Corporate and division ergonomics personnel can conduct audits of each plant's electronic evidence book remotely. For global corporations, this can reduce travel costs and improve process effectiveness significantly. Spot-audits conducted remotely can permit remedial coaching and process improvement on a more timely basis.

This approach has provided a flexible, accessible, and maintainable solution with key features. These include:

1. Centralized data and software location reduces maintenance and distribution costs as well as simplifying corporate oversight and process audits

2. Leverage of existing corporate intranet technology for data access, distribution, and sharing
3. Enhanced support for communication among diverse corporate groups, including the ergonomics, engineering, and health and safety communities
4. Customized computer support for all the documentation requirements of the established ergonomics process.

As the AEEB internet approach is fielded, new requirements are being added to expand access to a broader range of data sources and user communities. Web-based databases for engineering data, material safety, process descriptions, and personnel rosters are all potential sources which can be accessed through corporate internets.

In addition to corporate web sites, an increasing number of ergonomic sites exist on the world wide web. Commercial, government, and professional organizations are employing the WWW to promote data sharing and communication to the widest audience. Jumping on the web has appeal as a near-term solution to solving the communication and information dilemma. Some degree of corporate ergonomic oversight and control is still advisable to ensure that the information derived from public sites is consistent with the corporate vision or intent.

In order for the ergonomics documentation process to be effective, the data must be accessible and usable by the broadest community. However, the documentation process itself must not be so burdened by recordkeeping that it fails under its own weight. Well-designed computer applications have been proven to be an effective means to the end.

While they may sound obvious, the following lessons bear repeating here:

1. Develop solutions for nontechnical users.
2. Understand the user's requirements thoroughly — this is an ergonomics solution.
3. Realize that local plant committees will not have the high-tech hardware of engineers and designers at their disposal.
4. Develop a single solution with sufficient flexibility to be applicable across the organization.
5. Try to share data from existing corporate databases wherever possible.
6. In the days of global corporations, consider foreign language issues in the interface.
7. Factor the distribution, training, and ongoing support costs into the budget.
8. Dealing with corporate gear-heads presents new challenges!

By employing the power of the web and corporate internet resources in developing computer solutions for ergonomics, the challenges of data access, communication, and cost constraints can be effectively managed.

89.6 Lessons Learned

First, it takes a considerable amount of time to start up, pilot, and expand an ergonomics (or any health and safety) program in a large industrial facility. All parties must exercise patience.

Second, the start-up and pilot phase are important steps. Do not attempt to expand the program before these steps have matured.

Third, developing plant monitoring systems is crucial to the cost justification process. Because traditional methods of cost justification do not readily fit ergonomics projects, "cost justification" must be based on other indicators. Proper monitoring can help show how these indicators help in the overall operating efficiency of the plant — people benefits and cost benefits.

Fourth, a plant action plan is very important to develop. Without it, the program will lose focus and direction. This should be jointly developed with management and labor.

Fifth, and most important, top management and top labor support are essential to launch the program. Without this support, the program will have too low a priority to justify participants' time and resources.

Because of competitive pressures and quality specification and because employees are no longer willing to accept the pain and suffering resulting from poorly designed jobs, these and other safety programs will become the norm rather than the exception. It is hoped that with this report, other plants will develop successful ergonomics programs.

Acknowledgments

The authors would like to thank the American Automobile Manufacturers Association (AAMA) for their support in this effort. The research behind this chapter would have not been possible without the commitment from the plant where the research was done and the support of the AAMA. A list of those who contributed to this effort would be too long to include here, but a special thanks must go to the Plant Safety Supervisor, the Engineering Staff, the workers, and all those individuals and groups that are part of a joint effort. The views, opinions, and conclusions expressed herein are, of course, those of the author and are not necessarily those of the supporting institutions.

References

Allen, T. *Managing the Flow of Technology.* Cambridge, MA: The MIT Press, 1977.

Allison, G. *Essence of Decision.* Boston: Little, Brown and Co., 1971.

Cyert, R. and March, J. *The Behavioral Theory of the Firm.* Englewood Cliffs, NJ: Prentice-Hall, 1963.

Chaffin, D., and Andersson, G. *Occupational Biomechanics.* New York: John Wiley & Sons, 1984.

Kanter, R.M. *The Change Masters.* New York: Simon and Schuster, 1983.

Klein, J.A. Why supervisors resist employee involvement. *Harvard Business Review,* Sept.-Oct. 1984.

Kochan, T., Dyer, L., and Lipski, D. *The Effectiveness of Union Management Safety and Health Committees.* Michigan: W.E. Upjohn Institute for Employment Research, 1977.

Koontz, H. and O'Donnell, C. *Principles of Management* (3rd Ed.). New York: McGraw-Hill Book Co., 1964.

Lawler, E., III, and Mohrman, S.A. Quality circles after the fad. *Harvard Business Review,* Jan.-Feb., 65-71, 1985.

March, J. and Simon, H. *Organizations.* New York: John Wiley & Sons, 1958.

Peters, T. and Waterman, R, in *Search of Excellence: Lessons from America's Best-Run Companies.* New York: Harper and Row, 1982.

Pfeffer, J Power and resource allocation in organizations, in *New Directions in Organizational Behavior* (M.B. Straw and G. Salancik, Eds.). Chicago: St. Clair Press, 1977.

Rohmert, W. Ergonomics and manufacturing industry. *Ergonomics,* 28(8), 1115-1134, 1985.

Schmidt, S. and Kochan, T. Conflict: toward conceptual clarity. *Administrative Science Quarterly,* 17, 359-370, 1972.

Shafritz, J.M. and Whitbeck, P.H. *Classics of Organization Theory.* Oak Park, Ill.: Moore Publishing Co., 1-67, 1978.

Shackel, B. Factors influencing the application of ergonomics in practice. *Ergonomics,* 23(8), 817-820, 1980.

Tushman, M. and Nadler, D. Implications of political models of organization. Reprinted from *Resource Book in Macro-Organizational Behavior* (R.H. Miles, Ed.). Santa Monica, CA: Goodyear Publishing, 170-190, 1980.

Vojtecky, M.A. Workplace health education: principles in practice. *Journal of Occupational Medicine,* 27(1), 1985.

90

How to Set Up Ergonomic Processes: A Small-Industry Perspective

Carol Stuart-Buttle
Stuart-Buttle Ergonomics

90.1 Introduction

The ergonomics process for a small business differs from that of a large corporation. The main components of the ergonomics process are similar, but the characteristics and limitations related to the size of a business or industry need to be taken into account. Each of the primary elements of the ergonomics process are discussed in this chapter in terms of the associated difficulties and benefits related to industry size.

Many of the issues raised may pertain to a small plant that is part of a large corporation. The main difference between a small industry and a small facility of a larger entity is the latter usually has a safety and health process defined or mandated at the corporate level. The small facility may turn to the corporation for guidance and information and perhaps assistance and expertise. This chapter focuses on the issues faced by small businesses but does not attempt to contrast these issues with those of large industry.

90.2 Definition of Small Business

Small businesses account for 99.7% of the workplaces in the U.S. and employ 54% of the workforce. About 6.5% of all businesses (20% of the workforce) are in manufacturing, 98.5% of which have less than or equal to 500 employees (7.4% of the workforce) (Armstrong, 1995). The U.S. Small Business Administration (SBA) based the definition of small business on the Standard Industrial Classification (SIC) codes (OMB, 1987), assigning each code in the industry division a maximum number of employees.

0-8493-2641-9/99/$0.00+$.50

Typically, a small business is defined as having less than or equal to 500 employees (SBA, 1993), although a maximum is 1000 employees for a few manufacturing SIC codes (Rose, 1993).

The contribution of small businesses to the overall economy is important. In the U.S. more than half the workforce is employed by small businesses, and since 1989 nearly 100% of new jobs have been created by businesses of less than or equal to five employees (Armstrong, 1995). Small to medium companies are viewed as vitally important to the economic health of other industrialized nations (Wortham, 1994).

90.3 Characteristics of Small Industry

Implementing an ergonomics program and evolving a process are difficult for any company. Large companies meet real-world constraints when establishing a program (Elson, 1994), and small industries have unique characteristics which carry associated constraints. These constraints impact the decisions that are required to develop the best ergonomics program model to fit the culture and business needs of the company. Small industries often have the following characteristics (Stuart-Buttle, 1993):

- *Less formality.* Procedures and communication methods are informal and often have minimal documentation. Teams or task forces are often used but are loosely structured and formed *ad hoc.*
- *Responsibility for several positions.* Personnel perform several job functions, and therefore ergonomics is often included at the same time as addressing other issues. There is reduced team input because there are fewer people in different positions, although the range of perspectives remains.
- *Greater responsiveness.* The company tends to be project based. Therefore, it gives focus to issues and brings them to conclusion.
- *Less specific knowledge.* Having personnel with several job positions means that less time can be spent focused on one specific area such as ergonomics. Therefore, the degree of in-house expertise is usually limited.
- *More management involvement.* Management tends to be involved with the details of plant activities and shows their responsiveness to projects by making decisions and coordinating project efforts. However, the culture of the company also affects the extent of management commitment and employee involvement.
- *Less data-oriented approach.* Once the company has decided on a course of action and is convinced that the approach is a good one, quantification is typically down-played. This reduces the data available for prioritizing problem areas, cost benefit decisions, and determining project effectiveness.

90.4 Reasons to Implement an Ergonomics Process

A productive, competitive business is important for all sizes of business (SBA, 1993) Workers' compensation costs can be a primary drain on a company's productivity. High workplace injuries and illnesses were the reason for the passing of the Occupational Safety and Health Act of 1970 in the U.S. (OSHA, 1970). The Occupational Safety and Health Administration (OSHA) recognized that there were several elements in management practice of safety and health that affected prevention of injuries and illnesses. As a result, OSHA issued management guidelines (OSHA, 1989) and continues to research an ergonomics standard. Some industries, especially small ones, may have lower costs associated with workers' compensation cases, yet their competitiveness may remain suboptimal because the production process is inefficient and potentially unsafe due to poor design. Ergonomics offers more to industry than reduced injuries and illnesses. It is a science that applies knowledge about people to the design of the workplace. When the workplace is designed so that people can perform at their best and machines at their most effective, then productivity is optimized.

Competitiveness

The greatest return from and the prime reason for addressing ergonomics is the increase in the bottom line through production and quality. This may be partly realized by the reduction in workers' compensation and related costs or by the increase in work by those no longer in discomfort. However, making the job easier for the worker by improving the work layout or reducing rehandling also improves production. According to the government report on the state of small business: "The drive for manufacturing competitiveness in both domestic and overseas markets means that there must be fewer person hours per unit of production" (SBA, 1993, p. 45). Too often operating with fewer workers and automating production are seen as the primary means to increase competitiveness. The effects of automation may not always be positive. Remaining jobs may be more monotonous. Greater skills may be required to interface with the automation and new types of injuries could occur (Järvinen, 1991). This efficiency is best achieved by improving the existing workplace which can generate large production savings (Oxenburgh, 1991).

Quality is part of the productivity equation. If products are reworked or wasted, then the unit cost is increased. Quality is influenced by human performance, and that performance is directly impacted by the design of the job and workplace. Insufficient time to perform the task, unsuitable lighting and environment, and ineffective training are examples of aspects that can affect quality. Excessive reaching and awkward postures increase the time it takes to perform the task, and quality standards may not be met. If a worker begins to fatigue at the task, then he or she will slow down or may be likely to make mistakes, hence affecting quality and production.

"Money signs and hassles are what most small business owners see when you talk about health and safety" (Synergist, 1994). This is a common view, especially when the full cost benefits are not understood or considered when assessing a safety and health problem and the potential solutions. When the only measure of benefit is the return from reduced workers' compensation costs, the solution may sometimes appear expensive. This may be particularly noticed in a small company with just one or two low-cost medical cases. It is important to consider the full cost benefits of workplace improvement. If the solution or improvement is not cost-effective with a return on investment in a reasonable time frame, then the approach may not be the right one, and an alternative needs to be sought.

Workers' Compensation Costs

For a number of years there has been a trend across most industry divisions that shows the highest injury incidence rates (per 100 full-time workers) are incurred by establishment sizes of 50 to 249 employees (BLS, 1992, BLS, 1995). Figure 90.1 illustrates the injury incidence rates for manufacturing industries by employment size.

Days away from work are particularly expensive to a company. The most common source for injury and illness for lost workday cases in manufacturing was reported to be the "worker motion or position." Overexertion was the primary event causing injury and illness across all the divisions of industry (Synergist, 1995). Such data strongly suggest the presence of design issues and the job not matching worker capabilities, i.e., poor ergonomics.

Although the national statistics show higher incidence rates for small and medium-sized companies (Figure 90.1), small companies often have only a few cases. This precludes identifying trends and makes it difficult to decide if there is a larger problem. A proactive approach, that is not dependent upon the medical or workers' compensation data, helps to prevent the company from becoming entangled in such trend debates. Workers' compensation and the safety and health of the workforce are important, and the jobs associated with the problems should be addressed first.

Lost workdays are expensive for any company, but the burden may be greater on a small industry because there are fewer employees. If an employee serves several job functions, the loss may be even greater. The hidden or indirect costs that go hand in hand with a workplace injury should also be remembered in the cost benefit equation. Examples of cost include training a temporary replacement, the associated lost production, productive time lost while attending to medical needs, and the time for accident investigation.

Compliance

Compliance to government regulations should not be the sole reason for implementing an ergonomics program. If the workplace is designed well for most people to perform their jobs effectively on a long-term

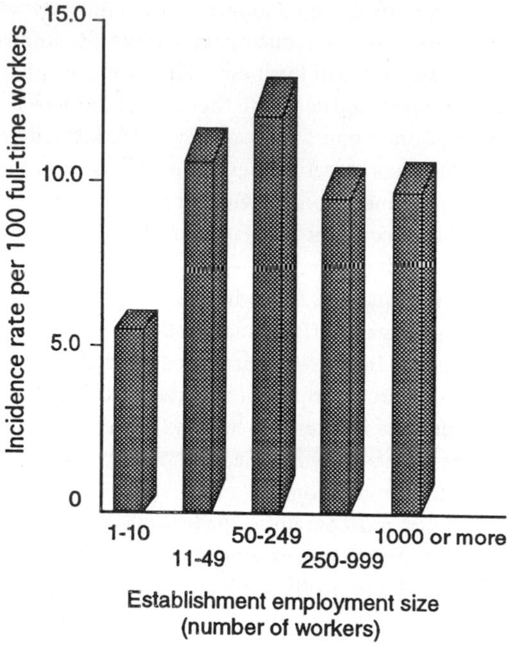

FIGURE 90.1 1994 injury incidence rates by industry employment size for the manufacturing industry division. (Data from Bureau of Labor Statistics (BLS). 1995. Excerpts from annual BLS survey; occupational injuries and illnesses in 1994. *Occupational Safety and Health Reporter.* 12-20-95:1002.).

basis, then the workplace will be safe and healthy. However, there are laws that oblige employers to provide a safe and healthy workplace. In the U.S., the OSH Act of 1970 has a general duty clause (Section 5 (a) (1)) that requires employers to furnish each employee a place of "employment which is free from recognized hazards that are causing or are likely to cause death or serious physical harm" (OSHA, 1970). There are also ergonomics standards that are being developed at federal and state levels that will require compliance. Some guidelines, although not mandated standards, may be used in the future for compliance if they reflect a consensus of best practice. If an ergonomics program is implemented with compliance as the only goal, then redesigns are likely to fall short of the best solutions to the problems and limit the efficiency improvements.

At present, in the U.S. there are no national ergonomics or safety and health program standards, however there are some guidelines. Many companies have incorporated ergonomics into general safety and health programs following the *Safety and Health Management Guidelines* issued by OSHA in 1989 (OSHA, 1989). The following year OSHA published similar ergonomics program guidelines for meat-packing plants which have since been adopted by other industries (OSHA, 1990). The same basic program components appear in the *OSHA Handbook for Small Businesses* (OSHA, 1992). The Voluntary Protection Program (VPP) is also based on the *Safety and Health Management Guidelines.* A recent survey of 650 companies indicated a 26% decline in workers' compensation costs, due in part to safety and health programs (BNA, 1996). Companies that have participated in VPP have had on average a 50% drop in workers' compensation premiums since 1989 (Esposito, 1996). Total quality improvement or management programs incorporate the primary elements encouraged by OSHA in their guidelines (Robinson, 1991). However, the effectiveness of an ergonomics program is dependent on how well it is implemented.

There are some additional guidelines that are being written in the U.S. that may affect how a company addresses ergonomics. The American National Standards Institute (ANSI) is developing a guideline for the control of work-related cumulative trauma disorders (ANSI Z365) (ANSI, 1996). ANSI and The Human Factors and Ergonomics Society are updating the 1988 standard for Human Factors Engineering of Visual Display Terminals (ANSI, 1988).

Other government regulations which require compliance and are related to ergonomics are the laws of Equal Employment Opportunity (EEO) (EEOC, 1964) and the Americans with Disabilities Act (ADA) which falls under the EEO Commission (EEOC, 1990). Many manufacturing tasks are performed by men and are difficult for most women to undertake. However, if a job is improved so that more of the population, including women, can perform it, then costs associated with finding the right people are lowered and compliance with the EEO is not in question. Likewise, if the workplace is designed to accommodate as much of the general population as possible, then specific accommodation for the disabled is easier because the workplace is more flexible. The ADA requires employers to make "reasonable accommodation" for a worker with a disability. During the first 18 months in which the ADA has been in effect, about 12,000 charges were filed. Nearly 20% dealt with back ailments (BNA, 1993). When a company develops a culture that attempts to accommodate those with disabilities, it also makes it easier for people to return to work. The company becomes more creative and flexible, so that jobs can be modified allowing injured workers to be reintroduced gradually to a fully productive job (Olsheski, 1996). Therefore, there are many benefits to designing well for a wide sector of the population, so compliance in itself need not be a burden to industry.

90.5 Elements of the Ergonomics Process

To establish an ergonomics process a program has to be initiated. The process sustains the program using methods such as systematic evaluation and revision based on effectiveness. In the long-term, a program without a process is not successful. The model of the program and the degree of integration into other company processes is dependent upon several factors, such as company size, responsibilities of the personnel, available resources and the company culture. For success, it is important for a company to find an approach that is the most suitable and effective, rather than to apply a theoretical model that may conflict with the organization's needs and culture. Whatever the model, the ergonomics program goal remains the same: Design the workplace so that it is healthy, safe, and optimally productive.

Choosing the Ergonomics Approach

The primary four elements in the *Safety and Health Management Guidelines* are: management commitment and employee involvement; worksite analysis; hazard prevention and control; and safety and health training (OSHA, 1989). These components are generally considered essential in any approach to ergonomics in industry. Medical management and a written program are elements in the meatpacking guidelines and are often included in companies' programs (OSHA, 1990). The success of ergonomics in a company remains dependent on how well the program is structured and carried out. A "one size fits all" method does not work when implementing an ergonomics program. A program must be tailored to fit the company culture. The following factors influence the development of a program and should be taken into consideration:

- Size
- Culture
- Resources
- Type of industry
- Types of ergonomics controls implemented
- Compatibility with other programs and processes

A program does not have to reach full integration with other company processes to be successful. The above factors and particularly the characteristics of small industry determine the appropriate extent of integration. If a program is minimally integrated into the company, it is more likely that the program remains driven by reacting to problems and precludes the development of a proactive approach. In other words, job improvements tend to be made in response to problems and not before problems occur. Incorporating ergonomics proactively requires shared responsibility, involvement, and commitment

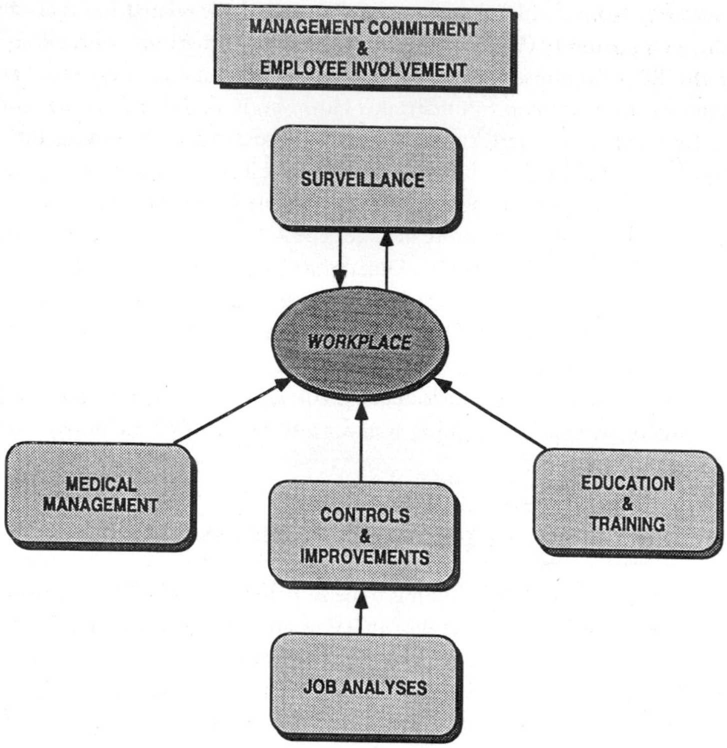

FIGURE 90.2 Diagram of the elements of an ergonomics process.

throughout departments. The more ergonomics becomes part of the company culture the greater its effectiveness. But it takes time to evolve and integrate a process into a system. An ergonomics process is dynamic, and regular evaluation and revision is necessary to maintain or improve effectiveness.

Process Elements

The elements of an ergonomics process are discussed from the perspective of small business, highlighting the problems and advantages due to size. The order in which the elements are discussed is the recommended order that small industry should follow when establishing their ergonomics programs. To begin with management commitment and employee involvement is essential for any degree of success with an ergonomics program. If there are many work-related medical cases, establish or enhance medical management because addressing this component early in program development can have a dramatic affect on employees' well-being and workers' compensation costs. Next, acquire some education about ergonomics and initiate a plan for training, then having learned more about what to monitor in the workplace, establish a surveillance system. Conduct job analyses as prioritized by the surveillance system and proactively assess new designs. Finally, implement improvements for a safe, healthy, and productive business. Figure 90.2 illustrates the process components of which discussion follows.

1. Management Commitment and Employee Involvement

A. Management commitment

Management commitment is fundamental to the success of an ergonomics program. Several aspects of commitment are discussed.

- *Is essential.* Lack of management commitment undermines any ergonomics program, however well that program may have been conceived and initiated. Small companies have an advantage, as top management is more involved in the details of the production or service and closer alliances with the line workers are forged. Demonstrate commitment to the ergonomics process by presenting a positive attitude. Management attitude indicates the true priorities of the company to the employees. Commitment also affects whether a program matures to become a way of doing business. If a culture change is required for the success of the program, then the vision, expertise, and example of management is essential.

- *Must be communicated.* To enhance communication consider the following points.
  - Verbally communicate the importance of ergonomics as the means to a safe, healthy and efficient workplace.
  - Express interest to give a message that the program is important and to instill energy, pride, and quality work in employees. Reinforce the commitment by an ergonomics policy or including the statement in the overall company policy and mission statement.
  - State the program objectives clearly to ensure that employees understand the reasons for changes.
- *Is set by example.* The behavior of an owner and the managers will convey more than a written statement or formally expressed policy. Set an example by displaying an active interest in how the work is performed by the employee and whether the job demands are reasonable. Closely monitor the injuries and illnesses records and follow safety requirements when in the work areas, such as donning personal protective equipment. Convey the importance of ergonomics in productivity by showing concern for the best environment and methods when looking to improve the process and production.
- *Entails employee involvement.* Employees are a valuable source of information and ideas. Involve employees in the process to accomplish effective ergonomics. This also demonstrates the respect and expectations management has for employees.
- *Requires clear responsibilities.* Involve the employees in program development to initiate active participation. As the program develops, clearly delineate the roles, assignments, and responsibilities of those involved. A common downfall of an ergonomics process is the lack of project follow-through for implementation and confirmation of effectiveness. Establish responsibility for project follow-through.
- *Entails the provision of resources.* Provide employees with the training and the time to contribute to the program. Provide authority for decisions to be made and money allocated to implement the projects to meet the program objectives.

B. Employee involvement

Management should support and facilitate employee involvement. The employees know the jobs best. They are an excellent source of ideas for improvements and are the ones required to work with any implemented changes.

- *Set up a core task force.* Certain knowledge and perspectives are needed to contribute to an ergonomics program. How the individuals are included depends on the structure of the program and the formality of a task force or team. To establish a core task force, designate a person to monitor the program and individuals to be responsible for specific aspects of the program. Clearly define the roles and responsibilities. The identification of the primary problems, their root causes and potential solutions come from the contribution of many perspectives during problem solving. The following are the departments or areas that have personnel who provide a particular perspective and should be considered on an ergonomics team. In a small business, some of these departments are combined.
  - *Operators.* Those who perform the job usually know the most about it. How the job is done is essential information and may be contrary to the way it was theoretically designed. The operators know the difficulties in performing a job and often have ideas for improvement. The operators will continue doing their jobs after improvements are made so their acceptance of any changes is important.

    - *Human Resources.* In small companies information about turnover, absenteeism, accident records, OSHA logs, and workers' compensation records are often kept in the human resources office. Turnover is typically low in small companies and less likely to be an indicator of a problem job. Absenteeism and sick days, however, should be monitored, as they may reflect an employee's attempt to counteract the high demands of a job.
    - *Safety and Health/Medical.* Accident information and related records may reside in the safety and health department rather than in human resources. Apart from medical treatment, medical personnel generally handle return-to-work issues and job modifications to accommodate medical restrictions.
    - *Engineering/Maintenance/Facilities.* Knowledge about the mechanical aspects of the production system and the environment helps when determining the workplace redesign potentials and the related costs. The involvement of engineering personnel is also necessary to ensure that ergonomics is incorporated.
    - *Production Engineers and Quality Control.* Personnel overseeing production and quality contribute information that helps to identify problem areas. Their perspective is useful in problem solving and assessing potential solutions.
    - *Management/Supervisors.* In small companies management is likely to be involved, therefore providing decision-making authority. Management support is necessary for effective implementation of administrative controls.
    - *Purchasing.* Involvement of the purchasing department ensures that the best products are purchased based on their function and quality as well as cost. For example, an inexpensive drill bit may seem a good bargain at first, but uneven wear of the bit may cause the worker to use excess force and may increase the risk of injury from the drill slipping.
- *Set clear responsibilities.* An ergonomics process that is fully integrated into the company processes is likely to be the responsibility of everyone. However, one person or group usually becomes the resident "expert" or ergonomics resource. That person or group may be responsible for establishing and keeping ergonomics "alive" within the company.
- *Use existing task forces or teams if possible.* Combining committees or teams helps to prevent redundancy of efforts and reduces the time employees are away from production. However, this lessens the opportunity for employee involvement. A rotation onto a task force and *ad hoc* involvement helps to include more people.

C. Written program

A written program is often emphasized as essential for a successful program. However, the typical reaction of a small business to this suggestion is "more paper work." There are many advantages to putting down in writing the basics, such as the objectives and who is involved. Writing the program basics helps to:

- Get the program started more efficiently
- Organize thoughts and the best plan of action
- Clearly communicate the process
- Make it easier to introduce the process to a newcomer
- Establish the goals and achievements by which the program can be assessed for success and improvement

The written program need not be a lengthy document but rather a clear statement conveying the objectives, goals and processes to establish and continue ergonomics within the company. A schematic in Figure 90.2 helps provide an overall picture. Consider writing the document so that it includes other functions such as training. For example, a section that outlines the medical management procedure for a work injury could be used during orientation training of new employees. A written program need not be an independent document but may be incorporated into other processes within the company, such as the safety and health program or quality management process. Consider including the following in the document:

- An overall schematic of the components and process
- Program objectives
- A list of those involved (or of job positions) and their responsibilities
- The program process (e.g., who gets training, to what extent, and by whom)
- A section for the project action list, analyses, and record of changes and their effectiveness (these could be the minutes from project meetings)

In addition, it is important to record the goals and action items to establish the program, so that there is accountability.

2. Medical Management

A primary responsibility of a company is to respond to work-related medical problems reported by employees. The medical conditions require treatment and correction of the cause to prevent recurrence. Medical management *per se* is not in the realm of ergonomics, however, early detection, prompt treatment and quick return to work have direct bearing on the recorded injuries and illnesses. If poor design at work caused or contributed to the medical condition, then improvement of the workplace is essential to prevent recurrence. Good communication between the medical community and the company is important to successfully improve the incidence of work-related injuries.

An industry with a large number of medical cases should start its ergonomics process by addressing medical management. Many companies have dramatically reduced their workers' compensation costs in as short a time as one year by improving the medical management of existing and emerging cases (Stuart-Buttle, 1993). Small industries do not usually have in-house medical departments, thus it can be difficult to control medical management. For a company with 600 employees or more, the employment of an in-house health care professional (HCP) may be beneficial, particularly if there are a large number of employees requiring treatment. Contracting with a medical group to provide onsite services using an occupational medicine group in the community are possible alternatives. Industrial or occupational medicine centers are becoming more common. A small company may have difficulty in finding the services, or having access to the service, especially if the company is located in a rural area. However, it may be possible to develop a relationship with a local medical group that is prepared to acquire the expertise the company needs. Trade groups, professional groups, or the company's workers' compensation carrier may be able to provide guidance and referral to occupational medical groups.

The company has some control over the quality of the services rendered. The following suggestions may help to establish a good working relationship:

- *Establish close communication.* Assign a primary contact person in the company and have him or her establish a rapport with the medical group or HCP.
- *Communicate the expertise expected of the medical group.* The company should make clear the type of medical expertise that is required. For example, if vibration exposure is inherent in the job, then the medical practice needs to have some experience or willingness to acquire training in screening for vibration white finger.
- *Introduce the medical group to the plant's culture and processes.* When the HCP understands the jobs, he or she may be able to give specific job restrictions and helpful guidance on job modifications or alternative jobs. This promotes return to work more quickly and successfully. The HCP may also assist in progressing the worker through various modified jobs until the worker is able to return to work at the previous job or an equivalent.
- *Consider the medical community as a training resource.* The local medical or rehabilitation group may also be a source for education of employees in health-related topics, such as back care and exercise programs.

3. Education and Training

First, some general education in ergonomics is needed by someone in the company so as to plan an ergonomics program. In-depth training is recommended for a specified team, and general awareness training is recommended for managers, supervisors, and line workers. The extent of investment in training may depend upon the financial resources and the turnover rate. A smaller team may be trained to save costs, but everyone should receive some awareness training. If a company suspects there are many problems requiring improvement, it may be helpful to train gradually. The team should be trained to respond effectively to the increase in reports that may occur after general awareness training of the employees. Lack of a response of the team may invite cynicism and reduce participation of the line workers.

A. In-depth training

At the very least in-depth training is recommended for the person or team primarily responsible and involved with ergonomics. This does not mean that the recipients have to become ergonomists but rather the training objectives are to:

- Understand the overall program objectives, goals, and process
- Understand the injury and illness system for treatment, return to work, and job modifications
- Be able to correctly record and interpret medical records and OSHA logs for surveillance purposes
- Know how to conduct basic problem-solving job analysis for ergonomics issues
- Recognize risk factors for injury and illness in workplace design
- Be able to develop, implement, and affirm effectiveness of solutions to basic problems
- Understand basic ergonomics principles to apply to solutions and new designs
- Be familiar with outside resources and methods for finding resources

B. Awareness training

An awareness level of ergonomics training prepares employees to participate in the ergonomics process. A program can become less reactive and more prevention-oriented when there is greater understanding and contribution of all the employees. The objectives of an awareness level of training are for the employees to:

- Generally understand the ergonomics program
- Appreciate their role and responsibilities in the program
- Recognize the early indicators of physical problems
- Understand the company medical management system
- Understand basic risk factors for injuries and illnesses
- Know basic ergonomics principles
- Understand their participation in job analyses

To train those employees who work in jobs with identified physical risk factors may cause the company to lose the opportunity to improve areas from a production or quality standpoint. An awareness level of training is recommended for all employees. New employees should be trained to maintain the knowledge base in the workforce. There are many resources for basic training. Commercial videos and training programs are available, as are companies that conduct training programs. Those who received in-depth training may also be able to provide the awareness training. It is not unusual to give separate programs for supervisors, management, and production workers as there may be differences in educational levels and perspectives of each group.

C. Refresher training

Refresher sessions maintain interest in ergonomics. As with the ergonomics program or process itself, the refresher sessions can be part of other processes in the company, such as continuous improvement or safety and health.

Additional training may be beneficial for either the task force or the in-house "expert." The extent of extra training can vary according to the degree of investment the company wishes to make. Some companies have found that when employees receive training and are empowered to contribute, many issues are addressed by the employees themselves (Burson, 1993).

4. Surveillance

An initial step to improving the workplace is to identify the areas with problems or the potential for improvement. This entails looking at data that have already been collected by the company. Such data include information on injuries and accidents, production and quality measures, and personnel records.

A. Methods

- *Medical information.* The first priority are the jobs at which injuries or illnesses have been reported. The OSHA 200 logs, workers' compensation records, first aid logs, and accident records are all sources of data from which to determine problem areas. For small companies, the numbers can be small and hard to interpret. For example, only one person in a department may have a physical problem, which may or may not mean other employees will also develop a problem. Statistical analyses on small samples are less feasible, making trends difficult to determine. Although it may not be statistically feasible to detect trends, it may be useful to informally determine if there are patterns in data or performance.

  Government statistics for given SIC codes can be used as a benchmark against which company performance can be compared. If the number of cases on the OSHA 200 logs is at least five for the year, and this figure is converted to an incidence rate, it can be compared to the annually published figures of the Bureau of Labor Statistics (BLS). The BLS can also be telephoned to find the rate for a given SIC code if it has not been published. Calculation of incidence rates is described in the recordkeeping guidelines for the OSHA 200 logs (BLS, 1986).
- *Discomfort surveys.* Occupational medical cases often go unrecorded, because employees tend to report to their personal physicians rather than to the company. This can occur especially in smaller companies in which there is no medical department. Furthermore, the collaborative nature of a small company often deters absence from the job and hence the reporting of a problem. Finding out who is working with discomfort attributable to the workplace and treating those conditions promptly are important for two main reasons:

  - The medical outcome is more successful and occurs more quickly, and it prevents the person from experiencing unnecessary hardship.
  - The treatment and lost time costs are lower when the condition is less severe.

  The main way to encourage early reporting is through education of the employees so that they understand the benefit in the long term. In addition, supervisors and managers need to understand the benefit of early treatment as well as encourage the employees to report. Detailed symptom surveys can be used as a formal mechanism for finding those with work-related medical problems. However, such surveys may be of limited value to the small company due to statistical constraints and the need for medical knowledge for interpretation.

  Improving the jobs that provoke discomfort and fatigue can reduce the likelihood of an illness or injury occurrence. These jobs provoke a performance decrement, for example, a reduction in productivity, a high number of rejects, or an increase in absenteeism or sick days. Data demonstrating reduced performance and subjective discomfort can be collected either informally through interviews and focus groups or formally through questionnaires and discomfort surveys (Stuart-Buttle, 1994). A discomfort survey typically contains a simple body diagram that is shaded by the respondent to indicate areas of discomfort. Ratings of the intensity of the discomfort is also marked for each shaded area. A valuable addition to a survey is questions about the cause of the problem and requests for suggestions for improvement (Burson, 1993; Stuart-Buttle, 1994).

To attempt to uncover physical issues before they become medical problems does not necessarily mean "opening a can of worms" necessitating responses to a flood of reports. If there is likelihood of too many reports to handle at one time, the departments or areas can be surveyed sequentially. Soliciting information consecutively by area, or specifically by task, helps to gather information pertinent to improvements. Surveying the whole plant provides a useful baseline and helps to prioritize the departments or jobs to be assessed. If the number of jobs is large, the circumstances of the job may change before the area is addressed, thereby making the survey less useful. Job rotation can make the interpretation of surveys difficult as the connection of discomfort to a particular task is less distinct. There are also inherent problems with discomfort surveys, for example, discomfort would not be reported by those who are "survivors" at the job.

- *Absenteeism and turnover.* Both of these can be indicators of difficult or stressful jobs that may warrant redesign. There may be insufficient data due to company size upon which to draw formal conclusions, but exit interviews and informal investigation may indicate the roots of a problem.
- *Production and quality data.* Data related to worker performance can be indicators of a mismatch of the workplace to the employee. Production may be lower than necessary if the worker takes longer to perform the job because of awkward postures or an inefficient layout. A change in production rate can also occur if the worker slows due to fatigue or discomfort. Rework, rejects, and mistakes are measurable aspects of performance that job redesign can improve. Errors may be due to a variety of causes including physical fatigue, cognitive or perceptual limitations, environmental distractions, or psychological and social stresses. So finding the primary cause is important to improve the job.
- *Accident investigation.* Traditionally, accident investigations tend to guide a company to accept the workplace design and look for behavioral shortcomings or system and equipment failures. The existing design falls short of being questioned during the investigations. Accidents and near misses should be investigated with ergonomics in mind because an inadequate design may be a contributing factor.
- *Audits.* Conducting an audit to determine problem areas or areas of potential improvement is proactive. The scale of a small business lends itself to auditing ergonomics at the same time as other system checks. Consider including ergonomics audits with safety and production improvement reviews. A checklist approach is common, and there are many checklists available on the market. However, consider tailoring one to capture the industry-specific issues. Checklists should be used only as a reminder of what to look for since they do not adequately address the interactions of risk factors or of the worker with the workplace.
- *Interview and employee reports.* Open-ended interviews, focus groups, and employee suggestions are other ways to identify problem areas.

B. Prioritizing

Problem areas need to be prioritized to develop an effective action plan. A scoring system, which assigns a number for the presence of an indicator of a problem, is a questionable method by which to prioritize, because very different data are gathered as indicators of a problem area. Another approach that can be useful is to develop a spreadsheet with types of indicators or data by departments, areas, or jobs. The amount and extent of the indicators can be used to qualitatively rank the problem areas in conference with other team members or employees. Consider factors such as the anticipated scope and difficulty of the project or whether the area or job is about to be changed for production reasons, because these factors also influence the priority. At the beginning of a program, undertake projects that appear to be relatively easy and inexpensive so there are some early successes.

5. Job Analyses

Job analyses can be conducted from a reactive or proactive stance. A reactive analysis is evaluating a job known to have problems. A program usually starts from a reactive standpoint in response to injuries and illnesses that have occurred. A proactive analysis is looking at a new design, recent installation, or redesign to anticipate problems and ensure that the workstation incorporates ergonomic principles. A proactive approach should be implemented as early as possible in an ergonomics program not only to prevent new problems arising, but also to avoid remaining in a reactive position which is more expensive in the long run compared to designing well initially.

Start the program with small and simple projects and complete many of them. This is an effective way to involve employees from many areas and to build interest and full support for ergonomics. Learn by doing, getting more training and taking on more challenging projects as experience builds. However, remember that using an expert may be necessary for complex problems and may be a time- and cost-saving measure.

A. Responsibility for conducting analyses

One of the decisions in designing the ergonomics process centers on who will conduct the actual ergonomics analyses. A team approach can be used in which the group collects the data, looks at the problem, brainstorms for the root causes, and generates potential solutions. Alternatively, one person may actually go and collect the data using interviews or videotaping for example, and present to the team for a group brainstorming. It is not uncommon, particularly in a small company, for a process to develop in which one person collects data, analyses it, and generates the solutions with informal input from others. Caution should be used when adopting an isolationist approach, as over time, any joint responsibility for ergonomics might lessen because of a lack of involvement, ergonomics principles will perhaps no longer be applied by everyone, and the program might be perceived as one person's responsibility. When this occurs, ergonomics has not been sufficiently integrated into the company processes.

There are many sources of data that indicate ergonomics issues. Who collects the data depends on company resources and the complexity of the problems. In small companies an individual may be responsible, for example, for human resource issues, safety and health, as well as fulfill the function of sales manager. Owing to the multiple positions typically held, the ergonomics team would be small with less input, although with multiple perspectives.

Some companies choose not to invest in in-depth education of an employee but rather work with an external expert who gets to know the company process and culture. The decision about whether to have an internal or external "expert" depends upon the size and resources of the company and the other responsibilities of the "expert" within the company. The approach chosen also depends on the amount and complexity of the issues to be addressed. There may be benefit in having more in-house knowledge if the production process changes frequently. The company needs sufficient understanding of ergonomics to know when assistance is needed, how to find it, and where to get further information to address the issues.

B. Problem solving

The root causes of poor design are identified through careful analyses. A good problem list helps generate the best possible solutions. For example, after detailed analyses it may be determined that a material is rehandled unnecessarily and that the handling can be eliminated, whereas a more casual look at the job may have focused on improving the handling, incurring higher costs to make improvements and less productivity savings by not eliminating the step. Ineffective or expensive improvements may also be decided upon unnecessarily if a casual approach to analyses is adopted. The traditionally informal structure of small industry does not preclude good problem solving, nor does careful, detailed problem solving always require extensive quantification. A key to ergonomic problem solving is to repeatedly question if a job can be performed a better way. If the issue does not appear straightforward, assistance should be sought.

C. Quantification

Quantification assures that what is perceived as a problem is in fact a problem, and to what extent. Quantification also helps to assess improvements to determine the best alternative, and it provides a measure of effectiveness and cost benefit. However, quantification is unusual in small businesses and is discouraged, especially when the decision makers are already convinced of the benefit. Measurement should always be selective, that is, collecting necessary data rather than what is possible to collect. Unnecessary measurements increase the cost of analysis. A simple problem with straightforward improvements may be more informally approached although thorough problem solving is always essential. When the situation is more complex, selective quantification assures appropriate design decisions are made and helps prevent the generation of new problems.

As Brough (1996) pointed out, the overall program objectives need to be kept in mind during analyses, that is, to design a safer, more productive workplace. The budget can easily be consumed by using elaborate materials and computer programs that produce more data than needed and that require more time for analysis than correcting the workplace problems.

D. Analysis

Discussion of the many analysis methods and tools is outside the scope of this chapter. However, some discussion about checklists is appropriate since they are widely used. An overview of the basic steps in an applied analysis is provided as a guide.

- *Checklist.* A checklist is a popular applied method for identifying and prioritizing the problem areas. However, caution should be taken when using a checklist as an analysis tool, because it does not address the interactions in a job. Checklists may serve as a reminder of the issues that indicate a poor design. A checklist can be useful for assessing a new design.
- *Workstation analysis.* The following steps provide a general overview of an approach in the analysis of a workstation.
  - Clearly define the job function and the tasks, so that they are understood in context with the overall system.
  - Collect pertinent job information such as performance rates and quality expectations.
  - Interview employees.
  - Describe the component actions of each task (possibly videotape them if the task is complex or fast).
  - Identify the risk factors and job components that place excessive demand on the worker or that make the job awkward or inefficient.
  - Assess the risk factors and job demands quantitatively and qualitatively.
  - Determine the root causes of the risk factors, job demands, and awkward methods.
  - Develop a primary problem list with possible causes.
  - Brainstorm for several short- and long-term solutions to the problems, especially ones that are inexpensive.
  - Assess the cost benefits of alternative solutions.
  - Develop an implementation plan including a trial stage if necessary.
  - Reassess the solution after implementation to determine its effectiveness.
  - Record the project.
- *New design.* Incorporate ergonomics into the process of purchasing new equipment, designing new layouts, or redesigning an existing area or workplace in the plant. If ergonomics is addressed at this stage there will be minimal subsequent concerns and better productivity from the beginning of equipment startup. A checklist may help to consider all the ergonomic aspects, but the interaction of the operators with the equipment or layout should be especially anticipated and critiqued. Company engineers should have at least basic knowledge of ergonomics and work closely with other personnel to ensure that all safety and quality standards are met.

  Maintenance requirements are commonly forgotten. Evaluate new designs for access and ease of maintenance. The time lost from awkward access, for example, the need to fetch and use a ladder to reach a control, can be considerable. The physical toll on the employee from difficult maintenance is also a cost to the company. Set up a preventative maintenance schedule for the equipment and facility. Equipment that is not properly maintained can increase the forces or control required of the operator, making the job harder to perform.

At times corporate memory can be very short. Design mistakes can be repeated, often when there are new people who are unaware of the history of changes at a workplace. A small industry does not experience such forgetfulness as much as a larger company because turnover or movement within the company is less. However, a project record may still be helpful to prevent repeating earlier trials and errors.

6. Controls and Improvements

Effective problem solving usually generates more than one solution, typically long-term and short-term ones of various expense. The solutions may incorporate administrative and engineering approaches. Although engineering solutions are perceived as more permanent and less dependent upon human behavior, it is not uncommon to need an administrative measure to accompany an engineering change. An advantage in a smaller company is that employees are more likely to be cross-trained so that administrative controls, such as job rotation, are easier to implement.

A. Engineering changes
Engineering solutions vary considerably by type of industry. The company has knowledge of equipment that is specific to the industry but often finds it difficult to locate any other equipment that might be a design solution. A small company has less time to search the market due to the multiple job functions held by personnel. Therefore, a company may find it worthwhile to develop outside resources that can reduce search time.

B. Administrative changes
Administrative changes can entail organizational and policy changes. Quota systems, break patterns, rotation methods, and overtime are common aspects that are addressed to improve the balance of job demands. Sometimes changes in purchasing policies may be indicated. The common practice of buying the cheapest is not always cost effective, particularly if the item is of poor quality and adds stress and time to do the job.

C. Follow-through
A small company has the advantage of closer communication and more management involvement so there is focus when the decision is made to address a project. Therefore, follow-through to ensure implementation and effectiveness is easier than in a large corporation, where often many jobs are analyzed but few changes are made. Even so, roles must be defined clearly for accountability, particularly if there is a team approach. After ergonomics awareness training many small improvements can be made from the plant floor by employees and supervisors.

D. Cost benefits
In a small industry the cost benefit of an improvement cannot be based on injury and illness alone because there are fewer incidents and they may be scattered throughout the facility. Therefore, gains in productivity and quality become important in the cost benefit equation. Of course productivity and quality are important components for all businesses but companies with more employees often have more incidents (but not necessarily higher incidence rates) and potentially more people that may be affected, so that injury and illness costs may be the basis for justification.

90.6 Summary

When a small industry sets up an ergonomics process there are many factors that influence the development of the process. The main factors relate to the unique characteristics associated with small company size. Characteristics commonly found are less formality, responsibility for several positions, greater responsiveness, a less specific knowledge, more management involvement and a less data-oriented approach. By being aware of these influences a small business may successfully incorporate ergonomics, especially if it focuses on the productivity and quality benefits.

References

American National Standards Institute (ANSI). 1988. *American National Standard for Human Factors Engineering of Visual Display Terminal Workstations.* ANSI/HFS 100-1988. The Human Factors Society, Inc., Santa Monica, CA.

American National Standards Institute (ANSI). 1996. *Control of Work-Related Cumulative Trauma Disorders. Part 1: Upper Extremities.* Draft, August 29. ANSI Z365. National Safety Council, Itasca, IL.

Armstrong, J. 1995. Personal communication. U.S. Small Business Administration.

Brough, W. R. 1996. Make your ergonomics program a powerful tool. *Safety and Health.* February:109.

Bureau of Labor Statistics (BLS). 1986. *Recordkeeping Guidelines for Occupational Injuries and Illnesses.* OMB. No. 1220-0029, U.S. Government Printing Office, Washington, D.C.

Bureau of Labor Statistics (BLS). 1992. Occupational Injuries and Illnesses in the United States by Industry, 1990. Bureau of Labor Statistics, April, Bulletin 2399:37.

Bureau of Labor Statistics (BLS). 1995. Excerpts from annual BLS survey; occupational injuries and illnesses in 1994. *Occupational Safety and Health Reporter.* 12-20-95:1002.

Bureau of National Affairs (BNA). 1993. Effective programs reduce liability from OSHA- and ADA-related charges, group told. *Occupational Safety and Health Reporter.* 10-13-93:527.

Bureau of National Affairs (BNA). 1996. Safety programs cited as factor in 18 percent drop in employer costs. *Occupational Safety and Health Reporter.* 1-3-96:1053.

Burson, R. 1993. Find it and fix it. *Ohio Monitor.* November/December:8-11.

EEOC, 1990. *Americans with Disabilities Act of 1990.* Public Law 101-336 101st Congress, July, 26. U.S. Government Printing Office, Washington, D.C.

EEOC, 1964. *Title VII, Civil Rights Act of 1964. U.S.* Government Printing Office, Washington, D.C.

Elson, I.J. 1994. Implementing an ergonomics program under real world constraints, in *Advances in Industrial Ergonomics and Safety VI,* Ed. F. Aghazadeh, p. 177-180. Taylor and Francis, Philadelphia, PA.

Esposito, P. 1996. Program management guidelines: A safety and health management system. *The Synergist.* August:27-29.

Järvinen, J., Karwowski, W., and Lepistö, J. 1991. The overexertion injury due to manual lifting: another irony of automated manufacturing, in *Advances in Industrial Ergonomics and Safety III,* Eds. W. Karwowski and J.W. Yates, p. 201-207. Taylor and Francis, Philadelphia, PA.

Occupational Safety and Health Administration (OSHA). 1970. *Occupational Safety and Health Act.* Public Law 91-596 91st Congress S. 2193 Dec. 29. U.S. Government Printing Office, Washington, D.C.

Occupational Safety and Health Administration (OSHA). 1989. Safety and health program management guidelines. *Federal Register,* 54(16):3904-3916.

Occupational Safety and Health Administration (OSHA). 1990. *Ergonomics Program Management Guidelines for Meatpacking Plants.* OSHA 3123. Department of Labor/OSHA, Washington, D.C.

Occupational Safety and Health Administration (OSHA). 1992. *OSHA Handbook for Small Businesses.* OSHA 2209. Department of Labor/OSHA, Washington, D.C.

Office of Management and Budget (OMB). 1987. *Standard Industrial Classification Manual,* PB 87-100012. National Technical Information Service, Springfield, VA.

Olsheski, J.A. and Breslin, R.E. 1996. The Americans with Disabilities Act: Implications for the use of ergonomics in rehabilitation, in *Occupational Ergonomics: Theory and Applications,* Eds. A. Bhattacharya and J.D. McGlothlin, p. 669-683. Marcel Dekker, Inc., New York, NY.

Oxenburgh, M. 1991. *Increasing Productivity and Profit Through Health and Safety,* CCH International, Australia.

Robinson, A. (ed.) 1991. *Continuous Improvement in Operations: A Systematic Approach to Waste Reduction,* Productivity Press. Cambridge, MA.

Rose, S. 1993. Personal communication. Regional Size Program Manager, U.S. Small Business Administration.

Small Business Administration (SBA). 1993. *The State of Small Business: A Report of the President,* U.S. Government Printing Office, Washington, D.C.

Stuart-Buttle, C. 1993. *Small Industry Programs.* Paper presented at "Low Back Injury Symposium" at Ohio State University, Columbus, Ohio.

Stuart-Buttle, C. 1994. A discomfort survey in a poultry-processing plant. *Applied Ergonomics.* 25(1):47-52.

The Synergist, 1994. Reach out to small business owners to improve health and safety, IH professionals told. *The Synergist.* June/July:12.

The Synergist, 1995. Graphically speaking. *The Synergist.* October:7.

Wortham, S. 1994. Britain's small businesses face big safety issues. *Safety and Health.* February:66-67.

91

The Role of Ergonomics Training in Industry

Marilyn Joyce
The Joyce Institute
Arthur D. Little, Inc.
Cambridge, MA

91.1 Introduction

In recent years, ergonomics training has become recognized as a critical element in implementing an ergonomics program within an organization. The type of training, the relative importance placed on it, the expectations for outcomes and the audience vary widely among organizations.

Ergonomics training, when it is part of a comprehensive, systematic approach to integrating ergonomics into an organization, can play a key role in enabling an organization, regardless of its size, to gain "ergonomic self-reliance." Often, when people first think of ergonomics training, they think only of "worker" training, when, in reality, training workers without first training others responsible for ergonomics can be counterproductive. As we will detail later in the chapter, training needs to be tailored to each of the

0-8493-2641-9/99/$0.00+$.50

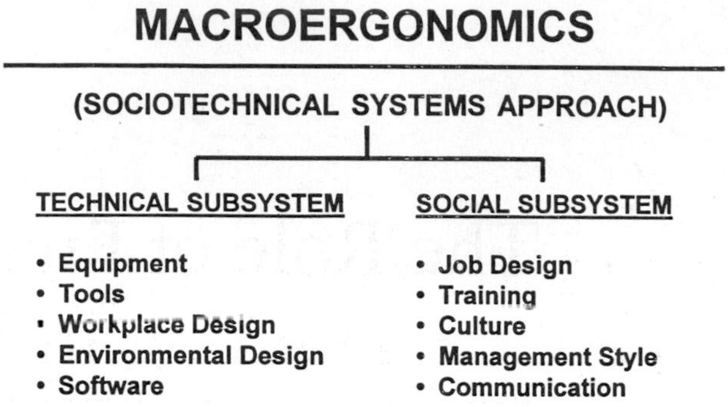

FIGURE 91.1

following audiences within the organization: senior management; design engineers; manufacturing, process, industrial, and facilities engineers; health and safety professionals; human resources and risk management; operations managers and supervisors; and buyers.

Ergonomics training involves developing the attitudes and competencies at all levels within the organization so that people have the knowledge, skills, and motivation to fulfill their role in ergonomics implementation. Of course, training is a means to an end, not the end itself. In other words, "providing training" does not constitute implementing ergonomics; the only reason to train is to enable the participants to make changes consistent with their role in the organization. The purpose of this chapter on training is to provide the rationale for training and the practical guidelines for designing ergonomics training in such a way that participants make changes that can be measured in terms of their impact on the workers, the jobs, and the organization.

91.2 Training and the Macroergonomics Approach

For an ergonomics program to sustain itself over a long period, it must be perceived as a value-added element in the organization and must be integrated in the nature of the organization. Training is the catalyst for that process. The purpose of training is to bring about change in the workplace that can be measured in terms of safety, productivity, and quality. In order to achieve this goal, ergonomics training needs to be part of a comprehensive ergonomics plan, based on a macroergonomics approach.

Macroergonomics is an organizational design approach that includes both the technical subsystems (design of environments, equipment, workstation, etc.) and social subsystems (job design, training, management style, relationships between management and workers, communication, etc.) In other words, a macroergonomics approach deals with the impact of work on both the physical and the cognitive aspects of employees. (See Figure 91.1. It includes what is referred to as both ergonomics and human factors.)

Training provides the means by which people's performance and well-being are enhanced to maximize an organization's investment in people and technology. If a manufacturer has its design engineers designing consumer products, but does not train them in ergonomics, then the company is inhibiting the designer from producing the best product available because the consumer's needs may not be met; if an employer purchases ergonomically sound power tools and provides reasonable workstations, but does not train the managers and supervisors in the basics of job design, then injuries could still occur because the needs of workers may not be met.

The key questions to be asked then, are these:

- How does ergonomics relate to and contribute to our core business functions?
- Whom, when, and what do we train?
- How do we set up the metrics to document the results of training?

91.3 Justifying the Need for Ergonomics Training

Even though by definition "ergonomics addresses human performance and well-being in relation to the job, the equipment, the tools and environment," many organizations only associate ergonomics with safety and injury prevention. The focus is sometimes so strong, that it obscures the impact that ergonomics training can have on the primary business of the organization. Ergonomics training can contribute to the core business functions by addressing the work process, the workplace, and the work product.

An organization can determine if it needs ergonomics training on the basis of contributing to its core business functions by asking the questions listed next.

Is the Organization in a Highly Competitive Market for its Products or Services?

If so, ergonomics training that improves the product design, decreases turnaround times, saves materials costs, or minimizes employee turnover can help the organization be more competitive.

Can Human Error Cause Severe Consequences?

If there are a lot of near misses occurring or if human error can result in severe injuries to employees, damage to products, or damage to the plant, then ergonomics training may need to focus on those involved in changing the design of the workplace or process or on those actually doing the job. For example, in chemical plants or petrochemical sites where alertness and attentiveness to visual displays in a control room can mean the difference in reaction time and appropriate decisions, the design of the control room itself, the nature and position of the controls and displays, and the work organization issues, all need to be addressed through training in ergonomics/human factors.

Do the Manufacturing Engineers and Workforce Struggle to Manufacturer What the Designers Design?

Ergonomics training can be helpful in facilitating the manufacturing process if there is enhanced cooperation between the designers and those responsible for producing the product. Examples of issues include: determining whether assembly processes should be done in a series of subassemblies to minimize awkward postures for extended periods of time; examining whether the number of fasteners used is really needed or whether the number or type can be reduced to minimize the use of power tools.

Are Changing Technologies and Customer Demand Posing Major Challenges?

The rapidly changing technologies in all types of work environments need to be considered in terms of the human interaction with those technologies if the capital investment is to have the desired results. In offices, for example, the increasing sophistication of computer systems and uses can overwhelm many workers. Ergonomics training can address the interface by addressing job design issues and workplace design issues.

For example, a Fortune 100 company reported that they minimized lost time and workers' compensation costs by implementing ergonomics in two of its units that have high production computer areas. They did so by training managers and supervisors in ergonomics so that they were aware of the importance of work organization issues, rest pauses, and attitudes toward workers; they also trained maintenance personnel so that they know how to respond to employees' needs for adjustments; and they trained in-house trainers to conduct ergonomics awareness training for the employees.

Are the Demographics of the Current Workforce Changing?

One of the triggers for initiating ergonomics training is a change in the demographics of a workforce. For example, a reorganization often results in having a high percentage of aging workers performing a wide range of tasks because younger workers have been laid off. Engineers and human resource professionals need ergonomics training so that they can design jobs and workplaces to accommodate the limitations of those workers, while capitalizing on their capabilities that experience brings. Also, the workers need training in ergonomics and in the job functions.

Other changes in the makeup of a workforce can also be triggers. If women or men of smaller stature are now expected to do jobs that had been designed for larger, stronger workers, injuries can skyrocket. It is important to provide ergonomics training to the engineers and managers so they can make changes to the workplace, the tools, the work organization, and the processes to enable these workers to be productive and yet safe; and it is important to train the workers in ergonomics.

Is There a High Demand for Productivity and Quality?

If the demands are increasing, and yet down time for product changeover, routine maintenance, etc., are hurting productivity, ergonomics training may be one way of addressing the issues.

Do High Workers' Compensation and Medical Costs as well as High Absenteeism Disrupt the Business on a Daily Basis?

Also, is animosity developing among the injured employees and their co-workers and management? Do co-workers resent those who are on "light duty" jobs? If so, ergonomics training may be one of the solutions to the problem. Training needs to be provided to the health and safety professionals, and case managers to address the issues of the currently injured workers and, at the same time, training needs to be provided to engineers, managers, and workers to make changes that will prevent future injuries.

If ergonomics training is justified in terms of the overall contribution that it can make to the business, it will often gain the support that it needs to succeed.

91.4 Identifying the Audience and Type of Training

As we have discussed, training applies to various audiences within the organization. Regardless of the size of the organization, it is important to design ergonomics training so that the organization, after a period of time, can handle most of its ergonomics problems internally, without relying on an ergonomics consultant for all types of implementation. One approach that is effective is to look at training in three phases: first, use a qualified consultant to provide all the initial senior management, management, ergonomics task force/teams/committees, engineering training, and to customize training for supervisors and workers; the second phase consists of training internal trainers to conduct training for workers; the third phase consists of developing a relationship with the consultant so that when problems arise that are beyond the level of knowledge of those being trained (for example, designing a new plant, or making major capital purchases of equipment or furniture), the consultant can work with the organization to assist in the problem-solving effort. The consultant can also be brought in from time to time to train additional engineers, health and safety professionals, and others requiring technical training. The goal of the training remains "ergonomics self-reliance" so that ergonomics becomes an integral part of how the organization functions (see Figure 91.2).

The training needs to be designed so that the information provided is consistent with the roles and responsibilities of each individual. The actions that will occur, or be expected as a result of the training, should also be consistent within the scope of each individual's job or position. As part of the planning process, timelines need to be established to indicate at what point in the entire process each group needs to be trained. In fact, the timing of the various levels of training is critical to the success of the program. For example, if the ergonomics steering team thinks that senior management may be unsupportive of a major ergonomics effort, often it works well to begin with a pilot project, that is within the budget constraints of a sponsoring group, such as the Director of Health and Safety, in conjunction with an

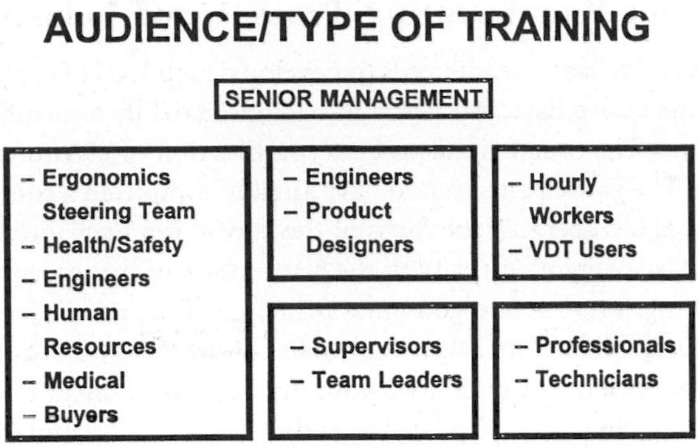

FIGURE 91.2

operations unit. If some successes can be demonstrated on a small scale, then it may be easier to make a business case for ergonomics training and implementation on a grander scale.

Briefing for Senior Management

The briefing for senior management needs to occur early in the process, after a pilot in some cases, to secure commitment of resources in personnel and capital expenditures necessary for the success of the program. The session generally needs to be one to two hours long and can be given by an ergonomics consultant or by a senior person within the organization. Since it needs to focus on the strategic importance of ergonomics within the organization, the presentation needs to include the following topics: the trends in ergonomics nationally/internationally and within the industry; the regulatory climate; the need for ergonomics in terms of the impact on this organization (hard data on workers' compensation or medical costs, down time, near misses, turnover, excessive scrap or rework, success of any pilot projects in ergonomics in the plant or company); proposed action plan; estimated costs; anticipated return on investment. If an overall ergonomics plan has not yet been developed, it is important to do so before making the presentation to management so that the training effort can be seen in context.

Training for Health and Safety Professionals, Engineers, Ergonomics Steering Team, and Facilities Managers, Buyers

The training for this group generally needs to be the first step because it provides the expertise necessary to initiate an ergonomics program and develop changes. Usually this type of training needs to be two to three days long conducted by a qualified ergonomist who is knowledgeable not only about technical issues, but also about organizational issues. If there is a corporate ergonomics policy, then the course needs to begin with a review of the policy. If there is no policy, then the course needs to include information on structuring an ergonomics plan. The course content needs to include the following: background on the underlying sciences of ergonomics, namely, anthropometry, biomechanics, physiology, and others; an overview of macroergonomics so that participants know the importance of addressing both physical and psychosocial issues; types, symptoms, and causes of musculoskeletal and stress disorders; information on how to review workers' compensation and medical logs; ergonomics principles and criteria; basic worksite analysis procedures that enable participants to identify and quantify the risk factors of the jobs so that corrective actions can be prioritized; prevention and control strategies including application of engineering, work practice, and administrative controls; documentation procedures for tracking and reporting on measurable results. The training should be as specific as possible to the plant, with videos and slides of the facility used to illustrate ergonomics principles and serve as the basis of discussion for solutions. Most important, some type of prioritized action plan should emerge from the training.

Advanced Training for Manufacturing, Process, and Industrial Engineers

If an organization is to be self-reliant, then it needs to develop a high level of expertise. Advanced training usually requires about four to five days and needs to be conducted by a qualified ergonomist who has an engineering background. The course needs to be a combination of classroom training and practical application in the facility. The participants should have already completed a course in ergonomics before proceeding to this advanced level because the focus of the course is on advanced problem solving rather than just the basic knowledge of ergonomics. With the assumption of the prerequisite, the course content should include the following: a review of ergonomics principles and criteria; advanced worksite analysis methods, including postural, biomechanical, energy expenditure, and others; engineering and process solutions for problem jobs, including workplace layout, tool and equipment design and selection; techniques for estimating the cost and benefits of implementing the solutions; and use of software to analyze and solve problems. One of the focuses of the training can also be on designing for new facilities, new-product production processes, and new equipment so that problems are prevented.

Training for Supervisors, Ergonomics Team Members, and Labor Representatives

Supervisors, ergonomics team members and labor representatives need to participate in one-day training sessions conducted by an ergonomics consultant. The training should enable participants to: identify the causes of musculoskeletal injuries/illnesses; assist the health and safety professionals in identifying "problem" jobs, perhaps by using a checklist; make simple modifications and adjustments to the workplace; apply the principles of ergonomics as they are setting up jobs or developing work procedures; communicate with workers to support design changes and to report symptoms of discomfort early; provide input to the engineers. Supervisors can play an important role in the ergonomics process, if they are trained appropriately. They need to learn to work constructively with workers. In office environments, for example, supervisors can be trained to assist computer users to adjust their chairs and to place their hands in correct postures. They can encourage workers to take short rest pauses to give the body a chance to recover; they can keep the workload balanced so that workers can work together to get tasks completed; and they can encourage employees to use their tools and equipment properly and to follow procedures, such as buddy lifts, that are intended to keep workers from injury.

Training for Hourly Workers

Employee involvement is a critical part of any successful ergonomics implementation. The more involved workers are in the identification and correction of hazards, the more likely that the solutions will be appropriate and will be accepted by the workers. Therefore, effective worker training is essential. Initial training for hourly employees generally needs to be about two hours and should occur after training for others in the organization who will be responsible for making changes has been completed. (It can be counterproductive to begin an ergonomics training process by training hourly workers, especially in industrial environments because if participants request any changes based on what they learn in the course, their requests will need to be supported by knowledgeable supervisors, safety personnel, or engineers.) There is also a tendency in some organizations to train workers *instead of* implementing engineering or work practice controls. In other words, the entire responsibility for working safely is placed on the worker. Although the underlying assumption for worker training is that employees need to take responsibility for their own well-being, it needs to be recognized that they may need resources to assist in that process. For example, if a job requires that an employee lift 50-lb. rolls of cable from the floor to a shipping container, no matter how well the employee follows safe lifting guidelines, he may still develop problems. The problem can only be solved by someone who questions why the cable is on the floor, whether there is an alternate way of getting the cable into the container, etc. In other words, the worker will need support in changing the job so that it has minimal risks.

Ergonomics training should include: reminders that the employees are responsible for their own safety and well-being on the job; information about the body's capabilities and limitations; appropriate positions and movements; appropriate use of tools, equipment, etc.; stretches and massages which can be used to prevent the onset of muscular fatigue and discomfort; information on ergonomics principles so that they can provide valuable input to the health and safety committees and engineers during the solution design process; information on the importance of early reporting of symptoms of discomfort so the problem can be solved before serious problems develop; and tips for their overall health and well-being on and off the job, such as regular exercise, good eating habits, proper rest, attention to hobbies or second jobs that may be contributing to discomfort, etc.

Workers who are trained can really be helpful to the worksite analysis process since the process involves the use of worker surveys and/or concern logs to identify potential hazards. Workers should be consulted during the detailed task analysis for additional means of identifying potential solutions.

Training for Design Engineers

For companies that design and manufacturer products for use by other industrial companies or by consumers, ergonomics training for design engineers can offer a significant competitive edge. The training needs to be about two to three days long, taught by an ergonomics consultant experienced in product design. It needs to address the entire product life cycle, beginning with the manufacturing and assembly processes, since design decisions greatly impact the workers who are making the product. Other considerations such as distribution, installation, and disassembly can be important, depending on the nature of the product.

Training for Risk Managers, Department Managers, Human Resource Managers, Occupational Physicians

Depending on the size of the company and the division of responsibilities, department managers, technical managers, human resource managers, and buyers can attend one of the two-day sessions with the ergonomics steering team and the others. Generally a one- or two-day course can provide the background that they need to support the ergonomics efforts and to incorporate the principles into their roles. Those involved in setting policies regarding organizational structures, return to work, interface with medical management, job design and work organization, staffing, and training need to have a knowledge of: human capabilities and limitations; ergonomics principles; types and causes of musculoskeletal and stress disorders; preventive strategies; means of documenting and tracking measurable results of ergonomics implementation.

Comprehensive training to a wide range of employees, combined with a systematic approach to implementing changes can be effective in generating a significant impact on the organization.

91.5 Measuring the Impact of Ergonomics Training

The benefits of training can be measured in several ways. These can be summarized as the Impact on the Person, Impact on the Job, and Impact on the Organization. The most significant in the Impact on the Organization since it is the culmination of the all the changes (see Figure 91.3).

Impact on the Person

In the world of training, there are several approaches to measuring the effectiveness of the training on the person who participated in the training. In ergonomics training, those that apply are measures that are observable in demonstrations in the classroom or on the plant floor during the training; results of questionnaires completed by participants; and changes that occur on the job after the training is completed. For example, engineers who are being trained can complete a job analysis of a videotaped job

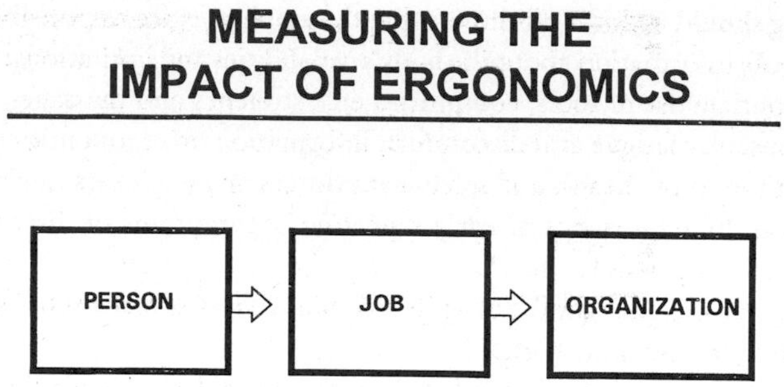

FIGURE 91.3

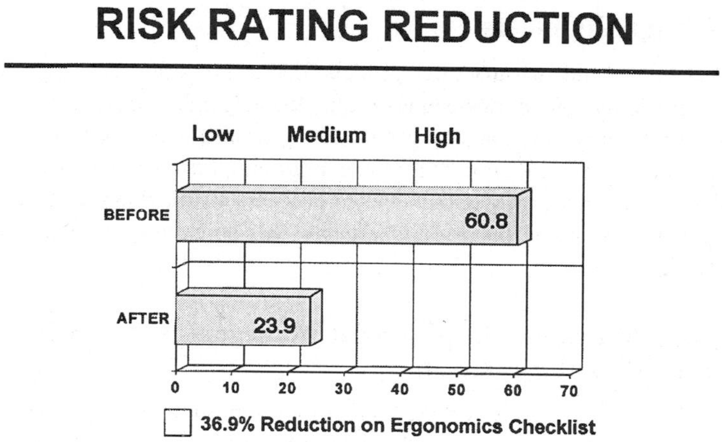

FIGURE 91.4

during a workshop segment of the training and make practical recommendations for solving the problem; supervisors can complete a checklist while they are observing a task that an office worker is doing; a safety person can assist a worker in adjusting a workstation; a worker can work with his hands in a neutral position while using a screwdriver; a worker can adjust the height of the chair to improve his or her posture; a product design engineer can eliminate the need for 30% of the fasteners required for an assembly task. In many cases, the impact on the worker can be measured by the decrease in risk factors, using a checklist, and in other cases the impact of the change is directly observable (see Figure 91.4). For example, in a large reel handling job in a cable manufacturing plant, a small ramp was installed at each stand for a cost of $30 each. The numbers of lost time injuries was reduced to zero during the first year after the installation.

Impact on the Job

The impact of training on the job can be measured by establishing a baseline and tracking individual project results. These results can be particularly effective when they are measured in terms of core business functions. For example, after training a group of engineers, managers, supervisors, and workers in one plant, a problem that was identified was that the locks that were used to keep the machine guards closed were substandard. This cause increased physical effort on the part of the workers to open and close the locks and also caused scrap. After spending $700 to change the locks, the department saved $6,090 in scrap and increased productivity by 20 million sheets per year, which was worth $263,640.

A games manufacturer had been experiencing an unacceptably high number of games being returned because of damaged packaging. They also found that on the packaging line they had a high number of

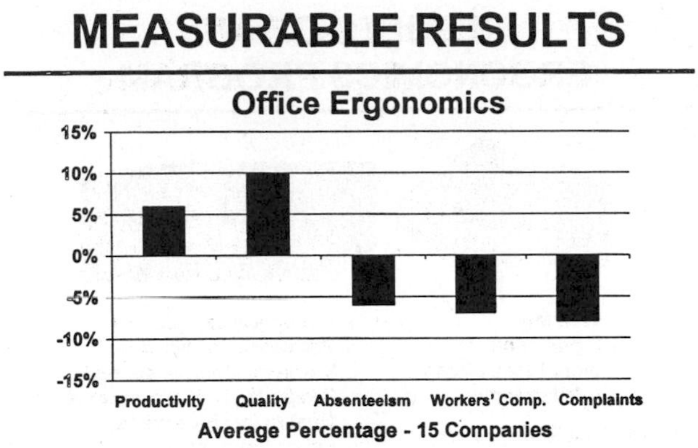

FIGURE 91.5

lost time injuries involving the shoulders and wrists. After receiving ergonomics training, the engineers decided to reconfigure the packing lines, slightly modify the cartons, and train the workers in package insertion. As a result, they improved their quality by 90%, as measured by the decrease in the returns of damaged packages, and virtually eliminated injuries to workers on that job. In office environments many types of results are documented, including increases in productivity (see Figure 91.5).

The key to these and many other similar types of measurable results is providing training that not only provides the technical expertise to make the change, but to train the people to track the changes by involving the engineers responsible for production and quality performance and looking beyond only the safety issue.

Impact on the Organization

The most meaningful results are those that are measured by their *impact on the organization* because these are the culmination of the others and are the most likely to sustain long-term interest in and support for an ergonomics program. These can be measured by relying on company data, such as reduction in workers' compensation costs or lost days; increase in the capacity of the facility or reduced delivery time to client.

For example, an automotive parts manufacturer actively trained the key players in ergonomics, including advanced training for engineers and worker training. Over a three-year period, about 150 changes have been made to jobs. Consequently, there has been a 50% reduction in repetitive strain injuries with a 75% reduction in workers' compensation costs. The plant received an award from its corporate headquarters for Most Improved Safety Record. Another plant, a manufacturer of office products reported a savings of $1.7 million in two and a half years as the result of implementing the ergonomics strategies that the engineers, managers, supervisors, and workers learned through ergonomics training (see Figure 91.6).

In each of these cases, the lessons learned in the training were implemented and the changes tracked and documented according to the process taught in the courses. For training to be effective, the expectation of accountability must be set from the beginning.

91.6 Characteristics of Good Training

Good training is based on adult learning theory. That means that adults learn best when the knowledge is related to their needs and draws upon their experience and expertise. The design needs to be highly interactive, including not only lectures, but hands-on activities, and slides and videos of the particular work environment. The training materials need to be totally integrated as a manual or package, not just a collection of random articles or papers. Finally, the structure needs to develop around specific objectives so that participants leave with an action plan in place.

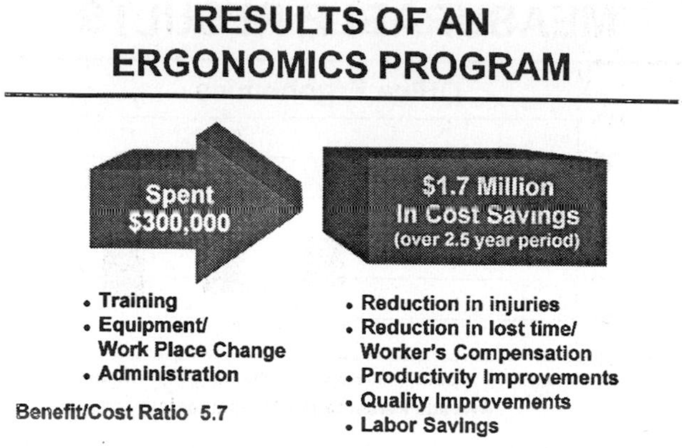
RESULTS OF AN
ERGONOMICS PROGRAM
Spent
$300,000
$1.7 Million
In Cost Savings
(over 2.5 year period)
• Training
• Equipment/
Work Place Change
• Administration
Benefit/Cost Ratio 5.7
• Reduction in injuries
• Reduction in lost time/
Worker's Compensation
• Productivity Improvements
• Quality Improvements
• Labor Savings

FIGURE 91.6

91.7 Conclusion

With training as the cornerstone of a comprehensive ergonomics program, organizations can experience significant decreases in injuries, improvements in productivity and quality, and a positive impact on the core business functions.

References

Bradley, Gunilla. 1989. *Computers and the Psychosocial Work Environment,* Taylor & Francis, Bristol.

Colombini, D., Occhipinta, E., and Menoni, O. 1993. Health education and training for workers involved in manual handling tasks, in *The Ergonomics of Manual Work,* Eds. W.S. Marras, W. Karwowski, J.L. Smith and L. Pacholski, pp. 503-506. Taylor & Francis Ltd., Bristol, PA.

Cook, R.E. and Marcotte, A. 1990. Ergonomic improvement in games manufacturing: a case study, in *Proceedings of the Human Factors Society 34th Annual Meeting,* pp. 707-709. Santa Monica, CA: Human Factors Society.

Hendrick, Hal W. 1986. Macroergonomics: a conceptual model for integrating human factors with organizational design, in *Human Factors in Organizational Design and Management — II,* Eds. O. Brown, Jr. and H.W. Hendrick North-Holland, Amsterdam, pp. 467-477.

Imada, A.S. 1991. The rationale and tools of participatory ergonomics, in *Participatory Ergonomics,* Eds. Kageyu Noro and Andrew Imada, Taylor & Francis, London, pp. 30-49.

Joyce, M. and Wallerstriner, U. 1989. *Ergonomics: Humanizing the Automated Office.* South-Western Publishing Co., Cincinnati, OH.

Joyce, M. 1994. Ergonomics training: the cornerstone of a successful ergonomics program." In *BNAC Communicator.*

Joyce, M. and Marcotte, A. 1996. The Business Benefits of Ergonomics, in *PRISM*; 63-71, First Quarter issue.

Joyce, M. and Marcotte, A. 1994. *Writing and Implementing and Ergonomics Plan,* The Joyce Institute/A Unit of Arthur D. Little, Seattle, WA.

Knowles, Malcolm 1980. *The Adult Learner: A Neglected Species,* Houston: Gulf Publishing Co.

Robinson, D. and Robinson, J. 1989. *Training for Impact.* Jossey-Bass Inc., San Francisco, CA.

Wilson, John R. 1990. A framework and a context for ergonomics methodology, in *Evaluation of Human Work: A Practical Ergonomics Methodology,* Eds. John R. Wilson and E. Nigel Corlett, Taylor & Francis, London, pp. 1-29.

92

Training Issues in Industrial Ergonomics

Richard J. Koubek
Wright State University

Sheau-Farn Liang
Purdue University

92.1 Introduction

Engineering controls and administrative controls are two broad categories for the industrial ergonomics. While engineering controls such as job design and workplace layout are frequently suggested, training, the major approach of administrative controls, should accompany engineering controls so that the optimal result can be achieved. It is true that a training program should always be conducted after the design has been examined since training cannot overcome a bad design. Yet, a good design also needs an effective training program to accomplish its objectives. For example, the DVORAK keyboard is a better design than the popular QWERTY keyboard since it reduces finger movements by the different arrangement of keys. However, it has not been widely accepted because most users are too resistant to change their typing habits. That is, from another perspective, no effective training program made this change happen. In summary, training issues must be considered together with engineering controls for the improvement of the quality of work.

The goal of training programs in ergonomics is to promote productivity and occupational health through the prevention of work-related disorders. Generally, work-related disorders include the work-related musculoskeletal disorders (WMSDs), low back pain (LBP), and other health disorders due to exposures to the hazardous conditions of the workplace. WMSDs, such as tendinitis and carpal tunnel syndrome (CTS), are defined as the disorders and diseases involving the musculoskeletal system of the neck, shoulder, and upper limb, whereas LBP deals with the lumbar spine (Kuorinka and Forcier, 1995). Compared to other occupational injuries and illnesses, work-related disorders rank first in frequency of occurrence in the United States (Chenoweth, 1995). As a result, they have a great influence on the quality of life as well as on the operation of health care resources (AAOS, 1992).

The causality between occupational risk factors and work-related disorders is complicated due to the multifaceted nature of risk factors. Risk factors may have direct effects to work-related disorders but they may also have indirect effects. Further, risk factors are not independent of one another. The symptom of disorders may come from the interactions of risk factors (Kuorinka and Forcier, 1995). Although the exposure–symptom association is not fully clarified, an empirical category of risk factors associated with

0-8493-2641-9/99/$0.00+$.50

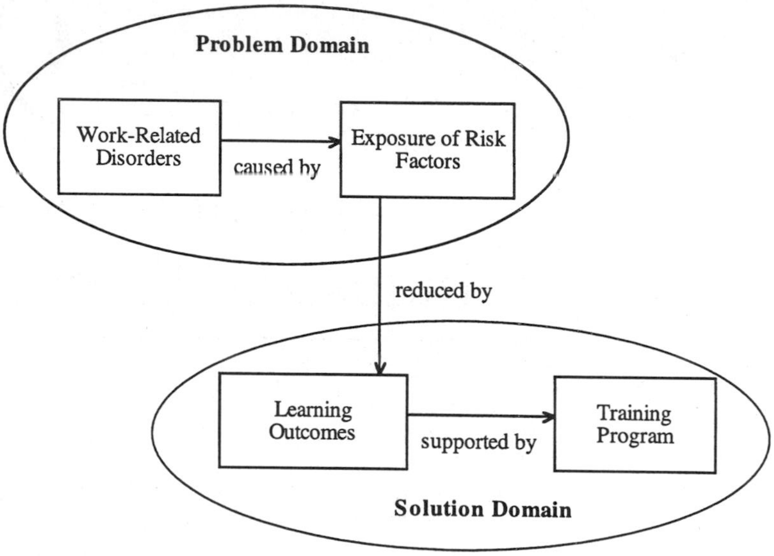

FIGURE 92.1 The conceptual model for the function of training programs in ergonomics.

the occurrence of work-related disorders is useful for the design of training programs in ergonomics. Basically, this category can be classified by two dimensions: types of risk factors (physical vs. psychological) and location of risk factors (person vs. task). Note that this category is not complete.

- Physical–person: physical fitness, capability, and limitation.
- Physical–task: working postures; task load, movement, repetition, and duration; environmental factors, such as temperature, vibration, and light.
- Psychological–person: cognitive ability; attitudes.
- Psychological–task: cognitive demands; organizational climate; psychosocial characteristics.

To design an effective ergonomic training program, the first step is to know the work-related disorders and possible risk factors. From this, the desired learning outcomes can be identified to reduce the exposure of risk factors. Finally, a training program can be developed based on these desired learning outcomes. Figure 92.1 presents this conceptual model for the function of training programs in ergonomics. There are two domains in this model: the problem domain and the solution domain. In the solution domain, the purpose of a training program is to support the learning process and facilitate the learner to achieve the desired learning outcomes. Between these two domains, it is assumed that the exposure of risk factors will be reduced if trainees acquire these learning outcomes. As a result, the corresponding work-related disorders can be reduced or avoided in the problem domain.

In this chapter, a systems approach for the design of training programs is addressed. Based on this scientific background, literature on training programs in ergonomics is reviewed and interpreted. The details of work-related disorders and risk factors are not the focus in this document.

92.2 Systematic Design of Training Programs

In order to achieve an effective training program, the first objective is to determine the knowledge, attitudes, and skills desired in the trainee. With this as the target, the second objective is to understand the current state of the trainee on these features. Upon completion of these two activities, the current status and desired goals are explicitly detailed, and a training system can then be developed to efficiently and effectively bring the trainee from the current to desired state. The following discussion first provides

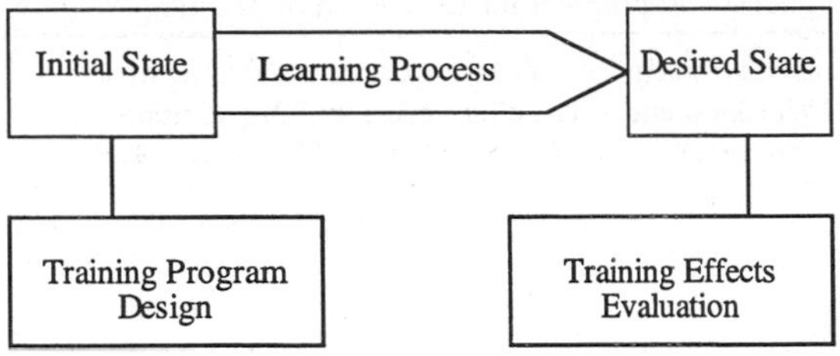

FIGURE 92.2 Training framework of instructional psychology.

an overview of the systems approach to training, followed by a more detailed description of relevant literature in each step of the training system process.

Current progress in training research has provided more perspectives. Tannenbaum and Yukl (1992) presented pretraining environment and post-training environment as important factors in training effectiveness. Organizational and social factors were also considered to impact training effectiveness (Latham and Crandall, 1991). Research on transfer of training emphasized the transfer of trainees' knowledge, skills, abilities, and other characteristics (KSAOs) from the training stage to the job (Baldwin and Ford, 1988; Ford and Sego, 1990; Ford and Wroten, 1984; Holding, 1991). Research on individual differences suggested that trainees' characteristics are important criteria for evaluating training effectiveness (Ackerman and Kyllonen, 1991). Further, research on instructional systems development (ISD) provide several training program design models based on learning theories and task analysis (Gagné, 1985; Gagné and Briggs, 1974; Gagné and Glaser, 1987; Reigeluth, 1992; Reigeluth and Curtis, 1987; Ryder and Redding, 1993; Wilson and Cole, 1992). In summary, a vast literature can be found in training research, and many training system models have been built with different emphases.

While many systematic approaches for the design of training programs have been proposed, three essential components of instructional theory from instructional psychology research appears and can be considered a simplistic framework of the training system. According to Glaser's (1976) view, these three essential components were defined as follows:

1. Initial state: the trainees' knowledge, skills, abilities, and other characteristics (KSAOs) prior to training.
2. Desired state: the competent performance that we want trainees to acquire after training.
3. The process of learning: the transition from initial state to desired state that can be achieved by training.

Snow and Swanson (1992), based on Glaser's views, added two additional components to make this framework more comprehensive. These two components are training program design and training effects evaluation. This framework is presented in Figure 92.2.

Mager (1988) classified the instructional process into four phases: analysis, development, implementation, and improvement. In the analysis phase, training objectives are identified by the analyses of training performance, task, and goal. The training prerequisites and target population are also defined in this phase. In the development phase, training evaluation criteria, relevant practice, and training content are derived. Furthermore, the instruction modules, sequence, and the delivery techniques are decided. Following these two phases, the training program is ready to be implemented to trainees. The characteristics of the instructor are important factors in this phase. Finally, evaluation of the training program is emphasized in the improvement phase to make sure that the program works and is up to date.

For training in the organization, Goldstein (1993) states that a systematic approach for the design of a training program should include three phases. The first is the assessment phase. The components of this phase are the assessment of training needs and the derivation of training objectives. The next is the development phase. Training conditions and learning principles are considered in this phase. The last is the evaluation phase. Training effectiveness and validity are evaluated in this phase.

TABLE 92.1 Summary of Systematic Approaches for the Design of Training Programs

Stage	Instructional Psychology (Glaser, 1976; Snow and Swanson, 1992)	Behavioral Learning Theory (Mager, 1988)	Training in Organization (Goldstein, 1993)	Safety Training (OSHA, 1995)
Pretraining stage	Initial state	Analysis	Assessment	Determining if training is needed Identifying training needs Identifying goals and objectives
Training development stage	Training program design Learning process	Development Implementation	Development	Developing learning activities Conducting the training
Post-training stage	Desired state Training effects evaluation	Improvement	Evaluation	Evaluating program effectiveness Improving the program

For safety training, a model with seven guidelines has been suggested by OSHA (1995). These seven training guidelines are: (1) determining if training is needed, (2) identifying training needs, (3) identifying goals and objectives, (4) developing learning activities, (5) conducting the training, (6) evaluating program effectiveness, and (7) improving the program.

Based on time spent in the training process, the training system can be divided into three stages: pretraining stage, training stage, and post-training stage. Each stage has different types of information to convey and different techniques to analyze information. Based on these stages, a summary of systematic training approaches is provided in Table 92.1. Following sections use this classification to present the details in each stage.

Desired Learning Outcomes

In this chapter, the term *learning outcomes* is synonymous with the *learned capabilities* or *trainee's knowledge, skills, and abilities after training.* Generally, learning outcomes include three domains: cognitive, affective, and psychomotor domains (Bloom, Madaus, and Hastings, 1981; Reigeluth and Curtis, 1987; Singer, 1972). Research in the cognitive domain concentrates on intellectual knowledge and skills, studies in the affective domain deal with the trainee's feelings and choices of actions, whereas physical movements and positions are the focus in the psychomotor domain.

Based on different purposes and perspectives, particular taxonomies of learning outcomes have been developed in each domain. However, the first integrated taxonomy which included all three domains was proposed by Gagné (Driscoll, 1994). Gagné (1985) classified all of learning outcomes into five main capabilities: intellectual skills, cognitive strategies, verbal information, motor skills, and attitudes. Any learned capability can be categorized to one or another of these varieties. Since the training strategy employed is a function of the desired learning outcome, a brief discussion is provided to outline the three major categories of learning outcomes. This categorization will be implemented in the review of the literature of training programs in ergonomics. The associated example and domain of learning outcomes for each capability are presented in Table 92.2.

TABLE 92.2 Gagné's Taxonomy of Learning Outcomes

Capability	Example	Domain
Intellectual Skills	Demonstrate how to plan a production schedule.	Cognitive
Verbal Information	State major parts of a product.	Cognitive
Cognitive Strategies	Originate a new production method by invention.	Cognitive
Attitudes	Choose products from certain companies as the benchmarking competitors.	Affective
Motor Skills	Execute an assembly procedure in a production line.	Psychomotor

Adapted from Gagné, R. M. 1985. *The Conditions of Learning and Theory of Instruction,* 4th ed. Holt, Rinehart and Winston, Orlando, FL.

TABLE 92.3 Bloom's Taxonomy of Learning Outcomes in Cognitive Domain

Knowledge	The recall of specific or general information, ways and means of dealing with this information
Comprehension	The translation, interpretation, and extrapolation of information
Application	The use of information in particular and concrete situations
Analysis	The clarification the elements, relationships, and organizational principles of information
Synthesis	The development of a plan, a set of operations, or a set of abstract relations
Evaluation	The Judgment in terms of internal and external criteria

Adapted from Bloom, B. S. 1956. *Taxonomy of Educational Objectives: The Classification of Educational Goals*, Handbook 1, *Cognitive Domain*, McKay, New York.

TABLE 92.4 The Summary of the Types of Learning Outcomes in Cognitive Domain

Learning Outcome	Knowledge	Gagné's Taxonomy	Reigeluth's Taxonomy	Bloom's Taxonomy	Merrill's Taxonomy
Know-What	Declarative	Verbal Information	Conceptual	Knowledge	Remember
Know-How	Procedural	Intellectual Skills	Procedural	Comprehension	Use
		Discrimination		Application	
		Concrete concept		Analysis	
		Defined concept		Synthesis	
		Rule		Evaluation	
		Higher-order rule			
Know-Why		Cognitive Strategies	Theoretical		Find

Cognitive Domain

In the cognitive domain, three types of learning outcomes can be found in many studies. That is, the learning outcomes of "know-how," "know-what," and "know-why." For example, the corresponding capabilities in Gagné's taxonomy are intellectual skills, verbal information, and cognitive strategies. The classification of conceptual, procedural, and theoretical goals in Reigeluth's Elaboration Theory (ET) also follows this category (Reigeluth and Curtis, 1987). The distinction between procedural and declarative knowledge is associated with the learning outcomes of "know-how" and "know-what" (Driscoll, 1994). Furthermore, the hierarchical categories of learning outcomes can be found in the literature. Bloom (1956) classified the learning outcomes of know-what and know-how into six levels: knowledge, comprehension, application, analysis, synthesis, and evaluation. The description of each level is shown in Table 92.3.

Also, the intellectual skills in Gagné's taxonomy are divided into five hierarchical levels: discriminations, concrete concepts, defined concepts, rules, and higher-order rules (Gagné, 1985). Moreover, the performance-content matrix in Merrill's (1983) Component Display Theory provided ten hierarchical combinations of learning outcomes. A summary of these approaches is presented in Table 92.4.

For training programs in ergonomics, the desired learning outcomes of cognitive domain often include the understanding of the musculoskeletal system, the recognition of work-related disorders and risk factors, and the comprehension of ergonomics principles. An introduction of the anatomy and physiology of the musculoskeletal system is frequently suggested and used to enhance trainees' knowledge of the operation of their bodies (Genaidy, Karwowski, and Mousavinezhad, 1990; Luopajärvi, 1987; Tomer, Olson, and Lepore, 1984; Van Akkerveeken, 1985). The statistics of disorders and the description of possible risk factors are provided for the discussion in the training (Genaidy, Karwowski, and Mousavinezhad, 1990; Linton, 1991). Based on the literature in biomechanics, psychophysics, and physiology, ergonomics principles are derived. Principles on the appropriate body mechanics and the adjustment of workload are taught to trainees (Chaffin, Gallay, Woolley, and Kuciemba, 1986; Colombini, Occhipinti, and Menoni, 1993; Davies, 1978; Genaidy, Karwowski, and Mousavinezhad, 1990; Luopajärvi, 1987; Parenmark, Engvall, and Malmkvist, 1988; Snook, 1978; St-Vincent and Tellier, 1989; Tomer, Olson, and Lepore, 1984; Van Akkerveeken, 1985).

TABLE 92.5 Taxonomy of Learning Outcomes in Affective Domain

Receiving (Attending)	The awareness, willingness, and controlled or selected attention of receiving certain information
Responding	The acquiescence, willingness, and satisfaction in responding
Valuing	The acceptance, preference, and commitment for a value
Organization	Conceptualization of a value and organization of a value system
Characterization by Value	Generalization and Characterization of value sets

Krathwohl, D. R., Bloom, B. S. and Masia, B. B. 1964. *Taxonomy of Educational Objectives: The Classification of Educational Goals*, Handbook 2, *Affective Domain*, McKay, New York.

Affective Domain

For the affective domain, Krathwohl, Bloom, and Masia (1964) classified learning outcomes into five levels: receiving (attending), responding, valuing, organization, and characterization by a value or value complex. The definition for each level is presented in Table 92.5.

Another approach was proposed by integrating the affective domain into the instructional design process (Main, 1992). Keller's ARCS model of motivation for learning was applied in this approach. The ARCS model includes four components: attention, relevance, confidence, and satisfaction, whereas the five instructional design phases are analysis, design, development, implementation, and evaluation. Based on four components of motivation for learning, the affective considerations in each of instructional design phases can be determined.

This dimension addresses the willingness of the learner to accept and adopt the training methods. This important factor cannot be overlooked in a successful training program. The motivation of the trainees is considered an important criterion in the ergonomic training programs (Luopajärvi, 1987; St-Vincent and Tellier, 1989; Van Akkerveeken, 1985). The characteristics of the instructor and the reaction of the trainees are the key issues in this domain (Linton, 1991; Luopajärvi, 1987).

Psychomotor Domain

From different perspectives, variant dimensions of learning outcomes have been proposed in the psychomotor domain. Task characteristic is one factor to identify the dimensions of learning outcomes in psychomotor domain. Schmidt (1991) classified motor skills according to three dimensions: open–closed, discrete–serial–continuous, and motor–cognitive dimension. An open skill and closed skill are distinguished by their environment. When the environment is variable and unpredictable, the skill is an open skill, such as playing football. On the other hand, when the environment is relatively stable and predictable, the skill is called a closed skill, such as typing. The second dimension is based on the distinction of tasks. A discrete skill has a distinct beginning and end, while there is no particular beginning and end for a continuous skill. A serial skill with discrete actions linked together falls between these two ends of dimension. A similar dimension can also be found in Fitts's taxonomy (Singer, 1972). Finally, the third dimension is defined by involvement of motor skill and cognitive skill. Examples of motor skills are running or jumping, whereas playing chess or controlling a robot are cognitive skills. Other dimensions, such as fine–gross, force–accuracy can be found in Cratty's taxonomy (Singer, 1972). The summary of each dimension and associated example is presented in Table 92.6.

Other factors for the identification of dimensions of learning outcomes are the part of body involved and the type of movement in the execution of actions. Haslegrave (1994) categorized task postures into five parts: head and neck, hand and arm, leg, trunk, and whole body. The three types of movement defined by Haslegrave are duration, repetition, and movement. Similar taxonomies can also be found in the models of Cratty, Guilford, and Fitts (Singer, 1972).

The skills of using proper working postures and the maintenance of physical fitness by means of exercising are the common desired outcomes in this domain for the training program in ergonomics. To gain these working skills, the "learning by doing" approach is the general method (Chaffin, Gallay, Woolley, and Kuciemba, 1986; Genaidy, Karwowski, and Mousavinezhad, 1990; Luopajärvi, 1987; Stubbs, Buckle, Hudson, and Rivers, 1983; Tomer, Olson, and Lepore, 1984).

TABLE 92.6 Summary of Dimension and Example in Psychomotor Domain

Dimension	Example
Open	Playing tennis
Closed	Playing golf
Discrete	Kicking a ball
Serial	Assembly in a production line
Continuous	Driving a car
Motor	Running
Cognitive	Playing chess
Fine	Typing
Gross	Jogging
Accuracy	Darts
Force	Javelin

The following sections of this document will discuss relevant literature which provides the background for assessing the training needs and designing the training system.

Pretraining Stage

During the assessment phase of the training process, three questions must be answered sequentially (OSHA, 1995):

1. Whether training is needed?
2. What training, if any, is needed?
3. What are training goals and objectives?

The preliminary assumption for the first question is that there is a problem to be solved or a desired goal to be achieved. Also, the first question implies that training is not the only means to achieve desired goals. Other methods, such as job redesign or personnel selection, could be considered at the same time.

If the answer to the first question is "yes," then the following step is to assess training needs and decide training evaluation criteria.

Assessment of Training Needs

McGehee and Thayer (1961) provided a framework with three components of training needs assessment: organization analysis, task analysis, and person analysis. The definitions are as follows:

- Organization analysis: Determining where in the organization training emphasis can and should be placed.
- Task analysis: Determining what should be the training contents in terms of what an employee must do to perform a task, job, or assignment effectively.
- Person analysis: Determining what knowledge, skills, or attitudes an employee must develop for performing his/her job.

Through organization analysis, organizational goals, resources, and the allocation of these resources should be identified. Psychosocial characteristics, such as organization climate, the support from manager, supervisors and peers, need also be assessed (Latham and Crandall, 1991). The support from higher levels of an organization is suggested as an important factor for the effective training program in ergonomics (Kroemer, 1992; Linton, 1991; Tomer, Olson, and Lepore, 1984). Another influential component for the effectiveness of training is the communication between the training designers and trainees (Van Akkerveeken, 1985). The involvement of trainees in the early training design stage usually leads to a successful training program.

In the task analysis, training goals and objectives are specified and the training content is determined. Two essential factors need to be examined in task analysis. One is the complexity of the to-be-learned task, the other is the similarity between the to-be-learned task and previous task. It seems that a better training result is obtained on a simple task rather than a difficult one (Chaffin, Gallay, Woolley, and Kuciemba, 1986). With proper analysis, the task demands on knowledge, attitudes, and skills can be determined and the irrelevant training content or the content that trainees have already known can be avoided. Current progress in cognitive engineering provides useful models for the analysis of tasks (Morrison, 1991; Ryder and Redding, 1993). However, task analysis in most ergonomic training programs focuses on the physical aspect such as the working movement patterns and associated musculoskeletal group (Doolittle and Kaiyala, 1991; Parenmark, Engvall, and Malmkvist, 1988).

For the person analysis, the current and desired capabilities in the three learning outcome domains are identified. Research on trainee characteristics has emphasized the trainees' capabilities prior to training as important variables for designing an effective training program (Ackerman and Kyllonen, 1991; Snow and Swanson, 1992). The homogeneity of trainees also needs to be considered. For instance, the training results may be different between experienced workers and newcomers (Parenmark, Engvall, and Malmkvist, 1988). Trainees with various backgrounds of ability and experience may need different training programs. An adaptive training program is suggested to fulfill the different requirements of trainees (Kroemer, 1992; Van Akkerveeken, 1985).

Through the assessment of training needs, the training goals and objectives are characterized, and desired learning outcomes are defined. Objectives are defined as measurable statement of intention, whereas goals are relatively general statements (Mager, 1988). Usually, goals are more abstract and qualitative. To proceed with the training evaluation phase, these goals are transferred to more concrete and quantitative objectives for the selection of training effectiveness criteria.

Criteria of Training Effectiveness Evaluation

According to the training objectives, training effectiveness criteria and training contents are determined in the development phase. The prevalent framework of training effectiveness criteria is Kirkpatrick's (1959) typology with four criteria: reaction, learning, behavior, and results. The definitions of these four training effectiveness criteria are as follows (Kirkpatrick, 1987):

- Reaction: How well did the trainees like the training program?
- Learning: What principles, facts, and techniques were learned? What attitudes were changed?
- Behavior: What changes in job behavior resulted from the training program?
- Results: What were the tangible results of the training program in terms of reduced cost, improved quality, improved quantity, etc.?

For training programs in ergonomics, the reaction of trainees is used as an evaluation criterion (Linton, 1991). The comprehension of the knowledge, attitudes, and skills is the popular area for the evaluation of training effectiveness. Different criteria can be found, such as the knowledge of safety information (Saari, Hryniewiecki, Bédard, Dufort, and Thériault, 1994), the changes of attitudes and behavior (Linton, 1991); and the change of working postures (Genaidy, Karwowski, and Mousavinezhad, 1990; Goggins and Robertson, 1994; Luopajärvi, 1987; St-Vincent and Tellier, 1989). Furthermore, the physiological measurements are used as the criteria of evaluation, for example, the electromyography (EMG) (Parenmark, Engvall, and Malmkvist, 1988) and the intra-abdominal pressure (IAP) (Stubbs, Buckle, Hudson, and Rivers, 1983). Finally, the productivity is considered as the factor. The reduction in cost, the occurrence of work-related disorders, and the absenteeism are applied to be the criteria (Goggins and Robertson, 1994; Snook, 1978; Tomer, Olson, and Lepore, 1984). These criteria are provided for the evaluation of training effectiveness in the post-training stage. The components in the pretraining stage are summarized in Figure 92.3.

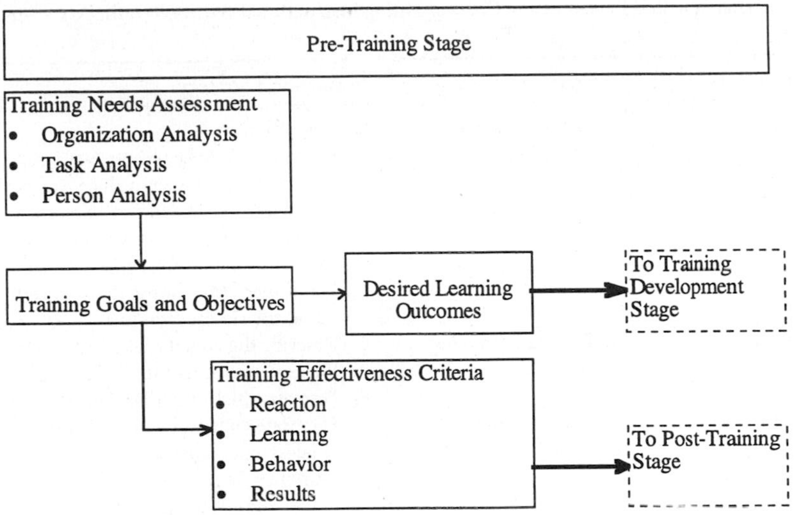

FIGURE 92.3 Components in pretraining stage.

Training Development Stage

The training program is developed and implemented in this stage. Based on the results of the pretraining stage, instructional strategies are applied sequentially to each training activity. Then, the proper training delivery techniques are selected. With the completion of the previous steps, the training program is ready for implementation.

Instructional Strategies

Based on the internal learning processes of trainees, nine corresponding external instructional events are defined (Gagné, 1985). Furthermore, different types of learning outcomes need different instructional strategies to optimize the learning process. The eight instructional steps (Gagné, 1985) are the fundamental elements and are listed below. Note that these are not the steps in developing a training program, but rather, the elements that should exist in a training program.

1. Gaining the attention of the trainee.
2. Preparing the trainee for new information (expectancy).
3. Using recall or transferring from existing experience of the trainee.
4. Presenting the training content.
5. Providing learning guidance or elaboration.
6. Allowing performance and providing feedback.
7. Evaluating performance.
8. Facilitating transfer of the training to actual task performance.

Table 92.7 presents the internal learning processes, the corresponding instructional steps, and the associated instructional strategies.

Training Delivery Techniques

After the instructional steps have been employed, the next step is to select the training delivery techniques. A delivery technique is the combination of media and other resources that will facilitate the delivery of the instruction. Hinrichs (cited by Borman, Peterson, and Russell, 1991) classified delivery techniques into three categories:

TABLE 92.7 Internal Learning Processes, the Corresponding Instructional Events, and the Associated Instructional Strategies

Internal Learning Process	Instructional Event	Instructional Strategy
Reception	1. Gaining attention	Change stimulus
Expectancy	2. Preparing the new information	C: Demonstrate the concepts, rules, procedures, or strategies that trainees will be expected to learn. A: (Trainees are to be informed later). P: Demonstrate the expected motor skills.
Retrieval to Working Memory	3. Recall or transfer of experience	C: Recall the prior knowledge. A: Recall the action choice made in certain situations. P: Recall the skill subroutines.
Selective Perception	4. Presenting training content	C: Describe the content with distinctive features. A: Present general nature of the action choice. P: Display initial situations and executive subroutines.
Semantic Encoding	5. Learning guidance or elaboration	C: Elaborate content by providing examples and using images, mnemonics. A: Demonstrate the action choice made by human model. P: Practice continually.
Responding and Reinforcement	6. Performance and feedback	Ask trainees to perform and provide effective feedback.
Retrieval and Reinforcement	7. Evaluation	Test trainees' performance with feedback.
Retrieval and Generalization	8. Transfer of Training	Enhance practice variety and the number of cues.

Note: **C**: Cognitive Domain; **A**: Affective Domain; and **P**: Psychomotor Domain.
Adapted from Gagné, R. M. 1985. *The Conditions of Learning and Theory of Instruction,* 4th ed. Holt, Rinehart and Winston, Orlando, FL.

TABLE 92.8 Summary of Training Delivery Techniques in Each Learning Domain

Cognitive Domain	Affective Domain	Psychomotor Domain
• Manuals or printed instructions • Audiovisual instruction • Lecture • Discussion or conference • Tutoring • On-the-job training • Computer-assisted instruction (CAI) • Case study • Simulations or games	• Discussion or conference • Tutoring • On-the-job training • Case study • Simulations or games • Role playing • Behavior Modeling	• On-the-job training • Practice • Coaching • Simulators

1. Content oriented: techniques designed to transmit knowledge on a cognitive level.
2. Process oriented: techniques designed to change attitudes, develop awareness, and enhance interpersonal skills of trainees.
3. Mixed: techniques that combine content-oriented techniques and process-oriented techniques.

In this categorization, content-oriented techniques are appropriate for enhancing cognitive learning outcomes, while affective learning outcomes benefit from process-oriented techniques. For the learning outcomes in the psychomotor domain, some "learning by doing" techniques, such as practice and simulator, are useful. The common training delivery techniques based on the three learning outcomes domains are presented in Table 92.8.

For training programs in ergonomics, printed instructions such as handouts and posters are used (Davies, 1978; Colombini, Occhipinti, and Menoni, 1993). Videotape is another tool to facilitate the accomplishment of cognitive learning outcomes (Chaffin, Gallay, Woolley, and Kuciemba, 1986). However, the most common delivery method for cognitive domain is the lecture. The lecture can be in a classroom as well as in the actual workplace. For the desired learning outcomes in affective domain, the

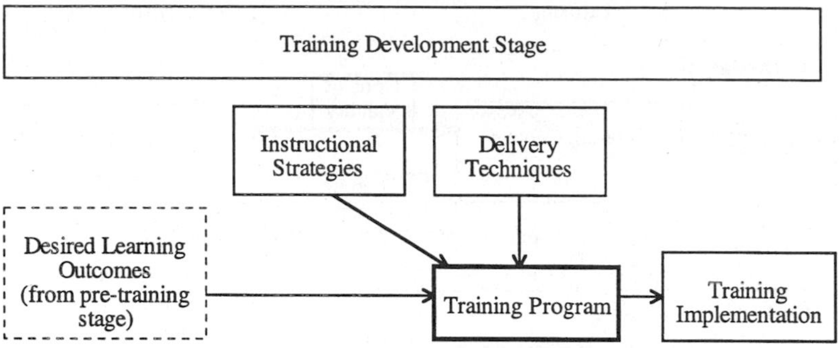

FIGURE 92.4 Components in training development stage.

discussion between the instructor and trainees is the common method for improving the motivation of trainees (Linton, 1991). The stress management course is also used in some training programs (Tomer, Olson, and Lepore, 1984; Van Akkerveeken, 1985). Finally, the achievement of desired psychomotor learning outcomes is usually supported by on-the-job training and practice (Genaidy, Karwowski, and Mousavinezhad, 1990; Luopajärvi, 1987; Tomer, Olson, and Lepore, 1984). Each delivery technique has its strengths and limitations. The choice of delivery techniques depends on the desired training objectives and current state of the trainee's knowledge, attitudes, and skills. For example, lecture is suitable for large audiences, but the audiences are relatively passive and the trainee's attention is difficult to organize (Gordon, 1994).

Training Implementation

There are two major formats of training implementation. One is instructor-controlled, the other is performance-controlled (Mager, 1988). In instructor-controlled format, content-oriented delivery techniques are usually used, and the instructor is the primary resource of information. On the other hand, performance-controlled training is conducted in a self-paced style. According to their capabilities, trainees can decide their learning pace. The instructor does not lead trainees to learn but acts as a coach to support the trainee to learn. Process-oriented delivery techniques are often associated with this format. The instructor-controlled format is the major format for most training programs in ergonomics.

Another important component in this stage is the characteristics of the instructor, especially the characteristics of presentation. In general, the effective instructor should speak clearly and understandably, and provide a positive learning environment (Mager, 1988). For ergonomic training, expertise in the work-related disorders (Tomer, Olson, and Lepore, 1984) and familiarity with the task (Luopajärvi, 1987) are also important for a successful training program. The components of this stage are presented in Figure 92.4. Following this stage, the evaluation of training effectiveness is conducted in the post-training stage.

Post-Training Stage

Training effectiveness evaluation and transfer of training are major issues in this stage. The following sections first provide the theoretical background for effectiveness evaluation and the transfer of training, then the common methods of the evaluation of ergonomic training effectiveness are presented.

Evaluation of Training Effectiveness

According to Salas, Burgess, and Cannon-Bowers (1995), training effectiveness can be defined as: the extent to which training brings desired or appropriate outcomes. Research on instructional psychology defined these "desired or appropriate outcomes" as the "desired state" (Glaser, 1976; Glaser and Bassok, 1989). These desired learning outcomes can be seen as training goals and objectives. When training results match training goals and objectives, the training program can be concluded to be effective. As shown in the pretraining stage, the most popular framework for evaluating training effectiveness is Kirkpatrick's (1959) four criteria.

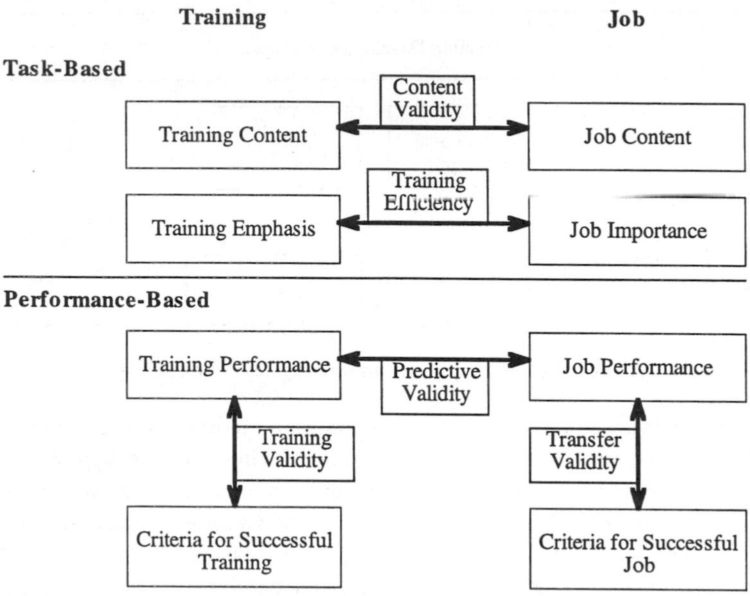

FIGURE 92.5 Relation of purposes of evaluation and information required. (Adapted from Ford, J. K. and Sego, D. 1990. *Linking Training Evaluation to Training Needs Assessment: A Conceptual Model,* Technical Paper, 90-69, Air Force Human Resources Laboratory, Texas.)

When we use criteria to perform the evaluation, criterion reliability, criterion relevancy, criterion deficiency, and criterion contamination must be considered (Goldstein, 1993). Criterion reliability is the consistency of the criteria measures. Criterion relevancy is the degree to which components are identified in training effectiveness and are represented by the criteria. Criterion deficiency is the degree to which components are identified in training effectiveness but are not represented by the criteria. Finally, criterion contamination is the degree to which components are not identified in training effectiveness but are represented by the criteria.

Besides Kirkpatrick's (1959) framework, Ford and Sego (1990) provide a conceptual model for evaluating training effectiveness by linking the purposes of training effectiveness evaluation and the type of information that is required by the evaluation. These five major purposes for conducting training evaluation are: content validity, training efficiency, training validity, transfer validity, and predictive validity. The definitions of these purposes are as follows:

1. Content validity: the relevancy between training content and job.
2. Training efficiency: over- or under-training by the training program.
3. Training validity: the extent of trainees' learning of training material.
4. Transfer validity: trainees' job performance after training.
5. Predictive validity: the prediction of job performance by training performance.

To fulfill these purposes, different types of information are required for different purposes. Ford and Sego (1990) provide two dimensions of information. One is the source of information with two domains: training domain and job domain. The other is the type of information included: task-based (what) and performance-based (how well). The relationship between purposes for conducting training evaluation and their required types of information is presented in Figure 92.5.

Compared to other components in the training system, content validity can be linked to task analysis for training needs assessment in the pretraining stage. Also, training efficiency relates to organization analysis and person analysis for training needs assessment in the pretraining stage. Training validity is similar to the learning component in Kirkpatrick's (1959) category. Finally, transfer validity and predictive validity have derived another important issue: transfer of training.

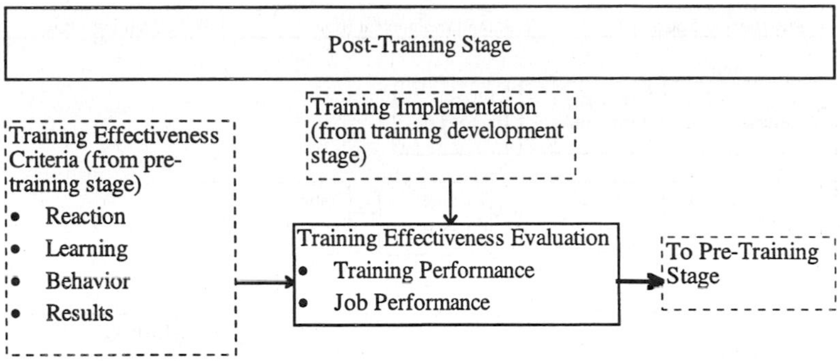

FIGURE 92.6 Components in post-training stage.

Transfer of Training

Transfer of training can be defined as the degree to which trainees effectively apply the knowledge and skills acquired from training to the job (Salas, Burgess, and Cannon-Bowers, 1995). Since training activities are not totally equal to a job, a success in training does not guarantee success in performing the job. Baldwin and Ford (1988) provided two criteria to examine the transfer of training: generalization and maintenance. The definitions of these two criteria are provided as follows:

- Generalization: the degree to which learned knowledge, attitudes, and skills are transferred to the actual job.
- Maintenance: the length of time that learned knowledge, attitudes, and skills continue to be used on the job.

To enhance the transfer of training, trainee characteristics, the training program, and organizational environment must be considered. More details of training evaluation are provided in the next section.

Evaluation Methods for Training Effectiveness

The common evaluation tools are questionnaires, tests, interviews, observation, and performance records (Phillips, 1991). Different methods are suitable for different evaluation purposes. For the criteria related to the desired learning outcomes, tests, interviews, and performance records are adequate for cognitive learning outcomes evaluation. Questionnaires and interviews are the common methods for the evaluation of affective learning outcomes. In psychomotor domain, tests, observation, and performance records are used frequently. In ergonomic training programs, learning outcomes from cognitive and affective domains are usually measured by questionnaires (Linton, 1991; Saari, Hryniewiecki, Bédard, Dufort, and Thériault, 1994), whereas psychomotor learning outcomes are evaluated by using the observation method from videotaping (Genaidy, Karwowski, and Mousavinezhad, 1990) or with a specific observational grid to measure the working postures of trainees (Luopajärvi, 1987; St-Vincent and Tellier, 1989).

For the measurement of physiological factors, special equipment is necessary (Parenmark, Engvall, and Malmkvist, 1988; Stubbs, Buckle, Hudson, and Rivers, 1983). The performance recordings are useful for the evaluation of long-term training effectiveness (Goggins and Robertson, 1994; Snook, 1978; Tomer, Olson, and Lepore, 1984). Figure 92.6 presents components concerned in the post-training stage. The summary of relationships among pretraining stage, training development stage, and post-training stage is presented in Figure 92.7.

92.3 Overview of Literature on Training in Ergonomics

Literature is categorized and reviewed according to the task domain and the evaluation criteria of training effectiveness.

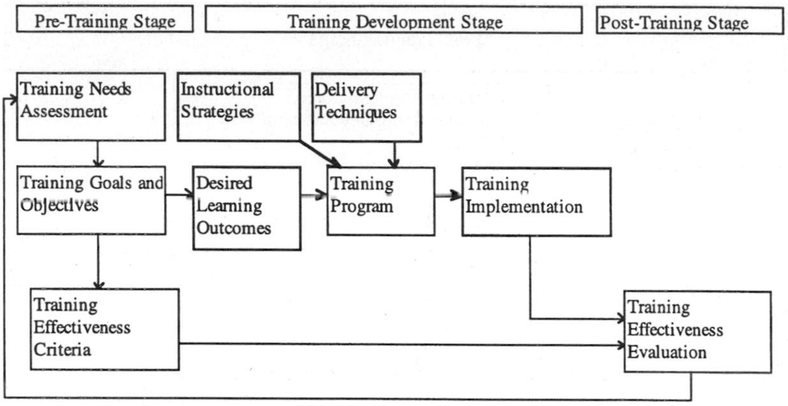

FIGURE 92.7 Components in training system.

Task Domain

For training programs in ergonomics, two task areas have received most attention: the manual handling and VDT-related tasks. In the manual handling task domain, material handling is the most popular area (Chaffin, Gallay, Woolley, and Kuciemba, 1986; Colombini, Occhipinti, and Menoni, 1993; Davies, 1978; Kroemer, 1992; NIOSH, 1981; Snook, 1978), whereas nursing training and patient handling is another focused area (Colombini, Occhipinti, and Menoni, 1993; St-Vincent, Tellier, and Lortie, 1989; Stubbs, Buckle, Hudson, and Rivers, 1983). Regulations and standards for the measurement of capabilities and limitations of workers in manual handling to prevent work-related disorders have been provided over the past two decades. "Training of workers" has become one aspect in *Work Practices Guide for Manual Handling* (NIOSH, 1981).

Since the advance of technology and the expansion of the computer, ergonomic training on VDT-related tasks has been emphasized in recent years (Goggins and Robertson, 1994; Luopajärvi, 1987; Mill, 1994; Robertson, 1994; Pecina and Bojanic, 1993). Standards such as the *American National Standard for Human Factors Engineering of Visual Display Terminal Workstations* (HFS, 1988) has been proposed to provide the standards for a working environment, visual display, keyboard, and furniture. However, training is not included.

Evaluation of Training Effectiveness in Ergonomics

Training effectiveness is the greatest concern in ergonomic training programs. Various criteria are proposed to evaluate the effectiveness of training programs. Based on the measurements of these criteria, a positive result concludes that the training program is effective, but the answer remains unknown for an insignificant result. However, the common conclusion is an unsuccessful training program. Without a systematic framework to guide the selection of evaluation criteria, both claims are questionable and may find alternative explanation for the result. In this section, the evaluation criteria of previous studies are reviewed based on the conceptual model (Figure 92.1) provided at the beginning of this chapter. This conceptual model was developed specifically for the training in ergonomics to provide a framework for the selection of training evaluation criteria.

As discussed before, the engineering controls always have the priority compared to the administrative controls. Snook's (1978) study is just another confirmation of this statement. With the full examination of approaches of engineering controls, training programs can be developed. In the conceptual model, the causality among work-related disorders, exposure to risk factors, learning outcomes, and the training program is defined. The criteria of evaluation can be the measurements related to the work-related disorders, the exposure to risk factors or the learning outcomes. The reaction and the comprehension of the knowledge, attitudes, and skills are the criteria for measuring the learning outcomes (Genaidy,

Karwowski, and Mousavinezhad, 1990; Goggins and Robertson, 1994; Linton, 1991; Luopajärvi, 1987; Saari, Hryniewiecki, Bédard, Dufort, and Thériault, 1994; St-Vincent and Tellier, 1989). The measurements of physiological factors belong to the exposure to risk factors (Parenmark, Engvall, and Malmkvist, 1988; Stubbs, Buckle, Hudson, and Rivers, 1983). Finally, work records can be the criteria related to the disorders (Goggins and Robertson, 1994; Snook, 1978; Tomer, Olson, and Lepore, 1984).

Usually, it is easier to explain the results when the criteria are related to the learning outcomes. The reason is that there is a closer linkage between the training program and the learning outcomes. However, the disadvantage is the lack of information for the effects on the exposure of risk factors and work-related disorders. On the other hand, criteria for the work-related disorders are most powerful to evaluate the training effectiveness. However, the uncertainty is increased and a long-term evaluation period may be necessary.

92.4 Summary

Although some training approaches have shown positive results, effects of most studies are uncertain and inconsistent (Kroemer, 1992; Kuorinka and Forcier, 1995). To overcome this deficiency, current research suggests that a holistic and systematic approach is essential for effective training (Kroemer, 1992; Kuorinka and Forcier, 1995; Robertson, 1994). In this chapter, a conceptual model for the function of training programs in ergonomics and a systems approach for the design of training programs are proposed to build up a framework for designing an effective training program. In the conceptual model, the work-related disorders can be reduced only because the exposure to associated risk factors has been reduced. That is, work-related disorders are the results of the exposure to their risk factors. From the other side, the training program is designed to support trainees to acquire the desired learning outcomes. In other words, the learning outcomes are the result of ergonomic training programs and are used to reduce the exposure of risk factors. Consequently, the requirement of a successful ergonomic training program is to make sure that there are well-assessed linkages between the training program and learning outcomes, between learning outcomes and risk factors, and between risk factors and work-related disorders. In contrast, the reasons for an unsuccessful ergonomic training program can be the failures of each linkage or of the combination of these linkages. For instance, the trainees may not obtain the desired learning outcomes after the training, the learning outcomes cannot reduce the exposure of risk factors, or the focused risk factors have little contribution to the occurrence of corresponding work-related disorders.

In conclusion, due to the prevalence of work-related disorders and the complexity of problem domain, the design of ergonomic training programs for the prevention of work-related disorders is a significant and rich research area. It is important not only because of the economic aspect, but also the influence on quality of life. Training is not a panacea but just a part of a more global solution for the prevention of work-related disorders (Kuorinka and Forcier, 1995). Other methods such as engineering controls should be considered simultaneously. Moreover, the multidisciplinary approach is needed with the integration of human factors, psychology, education, epidemiology, and pathology to understand the detail structure of the conceptual model provided in this chapter.

References

Ackerman, P. L. and Kyllonen, P. C. 1991. Trainee characteristics, in *Training for Performance: Principles of Applied Human Learning,* Ed. J. E. Morrison, p. 193-229. John Wiley & Sons Ltd., New York.

American Academy of Orthopaedic Surgeons. 1992. *Musculoskeletal Conditions in the United States,* AAOS, Park Ridge, IL.

Baldwin, T. T. and Ford, J. K. 1988. Transfer of training: A review and directions for future research. *Personnel Psychology.* 41: 63-105.

Bloom, B. S. 1956. *Taxonomy of Educational Objectives: The Classification of Educational Goals,* Handbook 1, *Cognitive Domain,* McKay, New York.

Bloom, B. S., Madaus, G. F. and Hastings, J. T. 1981. *Evaluation to Improve Learning,* McGraw-Hill, New York.

Borman, W. C., Peterson, N. G. and Russell, T. L. 1991. Selection, training, and development of personnel, in *Handbook of Industrial Engineering,* 2nd ed. G. Salvendy, p. 882-915. John Wiley & Sons Ltd., New York.

Chaffin, D. B., Gallay, L. S., Woolley, C. B. and Kuciemba, S. R. 1986. An evaluation of the effect of a training program on worker lifting postures. *International Journal of Industrial* Ergonomics. 1: 127-136.

Chenoweth, D. 1995. Worksite health promotion and injury, in *Worksite Health Promotion Economics: Consensus and Analysis,* Ed. R. L. Kaman, p. 117-130. Human Kinetics, Champaign, IL.

Colombini, D., Occhipinti, E. and Menoni, O. 1993. Health education and training for workers involved in manual handling tasks, in *The Ergonomics of Manual Work,* Eds. W. S. Marras, W. Karwowski, J. L. Smith and L. Pacholski, p. 503-506. Taylor & Francis, London.

Davies, B. T. 1978. Training in manual handling and lifting, in *Safety in Manual Materials Handling,* Ed. C. G. Drury, p. 175-178. NIOSH Publication, No. 78-185, Cincinnati, OH.

Doolittle, T. L. and Kaiyala, K. 1991. Screening and training as administrative controls, in *Industrial Ergonomics: Case Studies,* Eds. B. M. Pulat and D. C. Alexander, p. 61-73. McGraw-Hill, New York.

Driscoll, M. P. 1994. *Psychology of Learning for Instruction,* Allyn and Bacon, Boston.

Ford, J. K. and Sego, D. 1990. *Linking Training Evaluation to Training Needs Assessment: A Conceptual Model,* Technical Paper, 90-69, Air Force Human Resources Laboratory, Texas.

Ford, J. K. and Wroten, S. P. 1984. Introducing new methods for conducting training evaluation and for linking training evaluation to program redesign. *Personnel* Psychology. 37: 651-665.

Gagné, R. M. 1985. *The Conditions of Learning and Theory of Instruction,* 4th ed. Holt, Rinehart and Winston, Orlando, FL.

Gagné, R. M. and Briggs, L. J. 1974. *Principles of Instructional Design,* Holt, Rinehart and Winston, New York.

Gagné, R. M. and Glaser, R. 1987. Foundations in learning research, in *Instructional Technology: Foundations,* Ed. R. M. Gagné, p. 49-83. Lawrence Erlbaum Associates., Mahwah, NJ.

Genaidy, A. M., Karwowski, W. and Mousavinezhad, S. H. 1990. Computer-aided ergonomics: A tool for control of musculoskeletal injuries, in *Computer-Aided Ergonomics: A Researcher's Guide,* Eds. W. Karwowski, A. M. Genaidy and S. S. Asfour, p. 18-27. Taylor & Francis, London.

Glaser, R. 1976. Components of a psychology of instruction: Toward a science of design. *Review of Educational Research.* 46 (1): 1-24.

Glaser, R. and Bassok, M. 1989. Learning theory and the study of instruction. *Annual Review of Psychology.* 40: 631-666.

Goggins, R. W. and Robertson, M. M. 1994. The use of instructional systems design and performance analysis to design and evaluate office ergonomics training. *Proceedings of the Human Factors and Ergonomics Society.* 38: 952.

Goldstein, I. L. 1980. Training in work organizations. *Annual Review of Psychology.* 31: 229-272.

Goldstein, I. L. 1993. *Training in Organizations,* 3rd ed. Brooks/Cole Publishing, Pacific Grove, CA.

Gordon, S. E. 1994. *Systematic Training Program Design,* Prentice Hall, Englewood Cliffs, NJ.

Haslegrave, C. M. 1994. What do we mean by a working posture? *Ergonomics.* 37 (4): 781-799.

Hedge, J. W., Borman, W. C. and Carter, G. W. 1994. Personnel selection and training, in *Design of Work and Development of Personnel in Advanced Manufacturing,* Eds. G. Salvendy and W. Karwowski, p. 187-218. John Wiley & Sons Ltd., New York.

HFS 1988. *American National Standard for Human Factors Engineering of Visual Display Terminal Workstations,* ANSI/HFS 100-1988, Human Factors Society, Inc., Santa Monica, CA.

Holding, D. H. 1991. Transfer of training, in *Training for Performance: Principles of Applied Human Learning,* Ed. J. E. Morrison, p. 93-125. John Wiley & Sons Ltd., New York.

Kirkpatrick, D. L. 1959. Techniques for evaluating training programs. *Journal of the American Society of Training Directors.* 3: 21-26.

Kirkpatrick, D. L. 1987. Evaluation, in *Training and Development Handbook,* 3rd ed. R. L. Craig, p. 301-319. McGraw-Hill, New York.

Krathwohl, D. R., Bloom, B. S. and Masia, B. B. 1964. *Taxonomy of Educational Objectives: The Classification of Educational Goals*, Handbook 2, *Affective Domain*, McKay, New York.

Kroemer, K. H. E. 1992. Personnel training for safer material handling. *Ergonomics*. 35 (9): 1119-1134.

Kuorinka, I. and Forcier, L. 1995. *Work Related Musculoskeletal Disorders (WMSDs): A Reference Book for Prevention*, Taylor & Francis, London.

Latham, G. P. and Crandall, S. R. 1991. Organizational and social factors, in *Training for Performance: Principles of Applied Human Learning*, Ed. J. E. Morrison, p. 259-285. John Wiley & Sons Ltd., New York.

Linton, S. J. 1991. A behavioral workshop for training immediate supervisors: The key to neck and back injuries? *Perceptual and Motor Skills*. 73: 1159-1170.

Luopajärvi, T. 1987. Workers' education. *Ergonomics*. 30 (2): 305-311.

Mager, R. F. 1988. *Making Instruction Work or Skillbloomers*, Lake Publishers, Belmont, CA.

Main, R. G. 1992. *Integrating the Affective Domain into the Instructional Design Process*, Technical Paper, AL-TP-1992-0004, Technical Training Research Division, Brooks Air Force Base, Texas.

McGehee, W. and Thayer, P. W. 1961. *Training in Business and Industry*, John Wiley & Sons Ltd., New York.

Morrison, J. E. 1991. Introduction, in *Training for Performance: Principles of Applied Human Learning*, Ed. J. E. Morrison, p. 1-12. John Wiley & Sons Ltd., New York.

NIOSH 1981. *A Work Practices Guide for Manual Lifting*, National Institute for Occupational Safety and Health Publication, No. 81-122, Cincinnati, OH.

OSHA 1995. *Training Requirements in OSHA Standards and Training Guidelines*, Occupational Safety and Health Administration, Washington, D.C.

Parenmark, G., Engvall, B. and Malmkvist, A-K. 1988. Ergonomic on-the-job training of assembly workers. *Applied Ergonomics*, 19 (2): 143-146.

Patrick, J. 1991. Types of analysis for training, in *Training for Performance: Principles of Applied Human Learning*, Ed. J. E. Morrison, p. 127-166. John Wiley & Sons Ltd., New York.

Pecina, M. M. and Bojanic, I. 1993. *Overuse Injuries of the Musculoskeletal System*, Ed. CRC Press, Boca Raton, FL.

Pheasant, S. 1991. *Ergonomics, Work and Health*, Aspen Publishers, Gaithersburg, MD.

Phillips, J. J. 1991. *Handbook of Training Evaluation and Measurement Methods*, 2nd. ed. Gulf Publishing, Houston.

Reigeluth, C. M. 1992. Elaborating the elaboration theory. *Educational Technology, Research and Development*. 40 (3): 80-86.

Reigeluth, C. M. and Curtis, R. V. 1987. Learning situations and instructional models, in *Instructional Technology: Foundations*, Ed. R. M. Gagné, p. 175-206. Lawrence Erlbaum Associates, Mahwah, NJ.

Robertson, M. M. 1994. Designing VDT operator training programs for preventing work-related musculoskeletal disorders. *Proceedings of the Human Factors and Ergonomics Society*. 38: 429-433.

Ryder, J. M. and Redding, R. E. 1993. Integrating cognitive task analysis into instructional systems development. *Educational Technology, Research and Development*. 41 (2): 75-96.

Saari, J, Hryniewiecki, J, Bédard, S, Dufort, V. and Thériault, G. 1994. Efficacy of training procedures in implementing a legislated safety program. *Applied Ergonomics*, 25 (2): 116-118.

Salas, E., Burgess, K. A. and Cannon-Bowers, J. A. 1995. Training effectiveness techniques, in *Research Techniques in Human Engineering*, Ed. J. Weimer, p. 439-475. Prentice-Hall, Englewood Cliffs, NJ.

Schmidt, R. A. 1991. *Motor Learning and Performance: From Principles to Practice*, Human Kinetics Books, Champaign, IL.

Singer, R. N. 1972. *The Psychomotor Domain: Movement Behaviors*, Lea and Febiger, Philadelphia.

Snook, S. H. 1978. The design of manual handling tasks. *Ergonomics*. 21 (12): 963-985.

Snow, R. E. and Swanson, J. 1992. Instructional psychology: Aptitude, adaptation, and assessment. *Annual Review of Psychology*. 43: 583-626.

St-Vincent, M. and Tellier, C. 1989. Training in handling: an evaluative study. *Ergonomics*. 32 (2): 191-210.

Stubbs, D. A., Buckle, P. W., Hudson, M. P. and Rivers, P. M. 1983. Back pain in the nursing profession. II. The effectiveness of training. *Ergonomics*. 26 (8): 767-779.

Tannenbaum, S. I. and Yukl, G. 1992. Training and development in work organizations. *Annual Review of Psychology.* 43: 399-441.

Tomer, G. M., Olson, C. N. and Lepore, B. 1984. Back injury prevention training makes dollars and sense. *National Safety News.* Jan.: 36-39.

Van Akkerveeken, P. F. 1985. Teaching aspects. *Ergonomics.* 28 (1): 371-377.

Wilson, B. and Cole, P. 1992. A critical review of elaboration theory. *Educational Technology, Research and Development.* 40 (3): 63-79

Part VI
Environmental Issues in Ergonomics

93

Noise in Industry: Auditory Effects, Measurement, Regulations, and Management

John G. Casali
Virginia Polytechnic Institute and State University

Gary S. Robinson
Virginia Polytechnic Institute and State University

93.1 Introduction

The din of noise emanating from industrial processes pervades many occupational settings, and its effects on workers range from minor annoyance to major risk of hearing damage. Unfortunately, at least within the current limits of technology, noise is a by-product of many industries, such as manufacturing, especially those which use high-energy or impact processes such as metal cutting and mineral refinement, and service-related industries, such as air transport, construction, and farming. Workers complain about the negative effects of noise on their abilities to communicate, hear warning and other signals, and concentrate on tasks at hand. However, the effect which has been of most concern to industry has been permanent *noise-induced hearing loss*, or NIHL.

The primary intent of this chapter is to provide an introductory overview of the basic properties, measurement, effects on hearing, government regulations, and abatement of industrial noise, with a particular focus on reducing the physiological damage potential of noise as it impacts the human hearing organ. While the effects of noise exposure are serious and must be reckoned with by the hearing

0-8493-2641-9/99/$0.00+$.50

conservationist or safety professional, one fact is encouraging: process/machine-produced noise is a physical stimulus that can be avoided, reduced, or eliminated; therefore, occupationally related NIHL in workers is completely preventable with effective abatement and protection strategies. Total elimination of NIHL should thus be the only acceptable goal.

93.2 Sound and Noise

Because almost all aspects of hearing conservation and noise abatement in industry rely upon accurate quantification and evaluation of the noise itself, a basic understanding of sound parameters and sound measurement is needed before delving into other noise issues.

Basic Parameters

Sound is a disturbance in a medium (in industry, most commonly air or a conductive structure such as a plant floor) that has mass and elasticity. For example, an industrial metal-forming process wherein a hydraulic ram impacts a plate of sheet metal with great force causes the plate to oscillate or vibrate. Because the plate is coupled to the air medium, it produces a pressure wave that consists of alternating compressions (above ambient air pressure) and rarefactions (below ambient pressure) of air molecules, the *frequency* (f) of which is the number of above/below ambient pressure cycles per second, or *hertz* (Hz). The reciprocal of frequency, *1/f*, is the *period* of the waveform. The waveform propagates outward from the plate as long as it continues to vibrate, and the disturbance in air pressure that occurs in relation to ambient air pressure is heard as sound. The linear distance traversed by the sound wave in one complete cycle of vibration is the *wavelength.* As shown in the following equation, wavelength (λ in meters or feet) depends on the sound frequency (f in Hz) and velocity (c in m/sec or ft/sec; in air at 68°F and pressure of 1 atm, 344 m/sec or 1127 ft/sec) in the medium. The speed of sound increases about 1.1 ft/sec for each increase of 1°F.

$$\lambda = c/f \tag{1}$$

Noise can be loosely defined as a subset of sound; that is, noise is sound that is undesirable or offensive in some aspect. However, the distinction is largely situation- and listener-specific, as perhaps best stated in the old adage "one person's music is another's noise."

Unlike some common ergonomics-related stressors such as repetitive motions or awkward lifting maneuvers, noise is a physical stimulus that is readily measurable and quantifiable using transducers (microphones) and instrumentation (sound level meters) that are commonly available. Aural exposure to noise, and the damage potential therefrom, is a function of the *total energy* transmitted to the ear. In other words, the energy is equivalent to the product of the noise intensity and duration of the exposure. Several metrics which relate to the energy of the noise exposure have been developed, most with an eye toward expressing the exposures that occur in industrial or community settings. These metrics are covered later in this chapter. But first, the most basic unit of measurement must be understood, namely, the *decibel.*

Physical Quantification: Sound Levels and the Decibel Scale

The unit of *decibel,* or 1/10 of a *bel,* is the most common metric applied to the quantification of noise amplitude. The decibel, hereafter abbreviated as dB, is a measure of *level,* defined as the logarithm of the ratio of a quantity to a reference quantity of the same type. In acoustics, it is applied to sound level, of which there are three types.

Sound power level is the most basic quantity, is typically expressed in dB, and is defined as:

$$\text{Sound Power Level in dB} = 10 \log_{10} Pw_1/Pw_r \tag{2}$$

where Pw_1 is the acoustic power of the sound in Watts, or other power unit; Pw_r is the acoustic power of a reference sound in Watts, usually taken to be the acoustic power at hearing threshold for a young, healthy ear at the frequency of maximum sensitivity, or the quantity 10^{-12} Watts.

Sound intensity level, following from power level, is typically expressed in dB, and is defined as:

$$\text{Sound Intensity Level in dB} = 10 \log_{10} I_1/I_r \tag{3}$$

where I_1 is the acoustic intensity of the sound in Watts/m^2, or other intensity unit; I_r is the acoustic intensity of a reference sound in Watts/m^2, usually taken to be the acoustic intensity at hearing threshold, or the quantity 10^{-12} Watts/m^2.

Within the last decade, sound measurement instruments to measure sound intensity level have become commonplace, albeit expensive and relatively complex. Sound power level, on the other hand, is not directly measurable but can be computed from empirical measures of sound intensity level or sound pressure level.

Sound pressure level (SPL) is also typically expressed in dB. Since power is directly proportional to the square of the pressure, SPL is defined as:

$$\text{SPL in dB} = 10 \log_{10} P_1^2/P_r^2 = 20 \log_{10} P_1/P_r \tag{4}$$

where P_1 is the pressure level of the sound in microPascals (μPa), or other pressure unit; P_r is the pressure level of a reference sound in μPa, usually taken to be the pressure at hearing threshold, or the quantity 20 μPa, or 0.00002 Pa. Other equivalent reference quantities are: 0.0002 dynes/cm^2, 20 μNewtons/m^2, and 20 μbars.

The application of the decibel scale to acoustical measurements yields a convenient means of collapsing the vast range of sound pressures which would be required to accommodate sounds that can be encountered into a more manageable, compact range. As shown in Figure 93.1, using the logarithmic compression produced by the decibel scale, the range of typical sounds is 120 decibels, while the same range measured in pressure units (Pa) would be 1,000,000. Of course, sounds do occur that are higher than 120 dB (for instance, artillery fire) or lower than 0 dB (below normal threshold on an audiometer). A comparison of decibel values of example sounds to their pressure values (in Pa) is also depicted in Figure 93.1.

In considering changes in sound level measured in decibels, a few numerical relationships emanating from the above decibel formulae are often helpful in practice. An increase (decrease) in SPL by 6 dB is equivalent to a doubling (halving) of the sound pressure. Similarly, on the power or intensity scales, an increase (decrease) of 3 dB is equivalent to a doubling (halving) of the sound power or intensity. This latter relationship gives rise to what is known as the "equal energy rule or trading relationship." Because sound represents energy which is itself a product of intensity and duration, an original sound which increases (decreases) by 3 dB is equivalent in total energy to the same original sound which does not change in decibels but decreases (increases) in its duration by half (twice).

Psychophysical Quantification: Loudness Scales

While the decibel is useful for quantifying the amplitude of a sound on a physical scale, it does not yield an absolute or relative basis for quantifying the human perception of sound amplitude, commonly called loudness. However, there are several psychophysical scales which are useful for measuring loudness, the two most prominent being *phons* and *sones.*

Phons

The decibel level of a 1,000 Hz tone which is judged by human listeners to be equally loud to a sound in question is the phon level of the sound. The phon levels of sounds of different intensities are shown in the top panel of Figure 93.2; this family of curves is referred to as the *equal loudness contours.* On any given curve, the combinations of sound level and frequency along the curve produce sound experiences

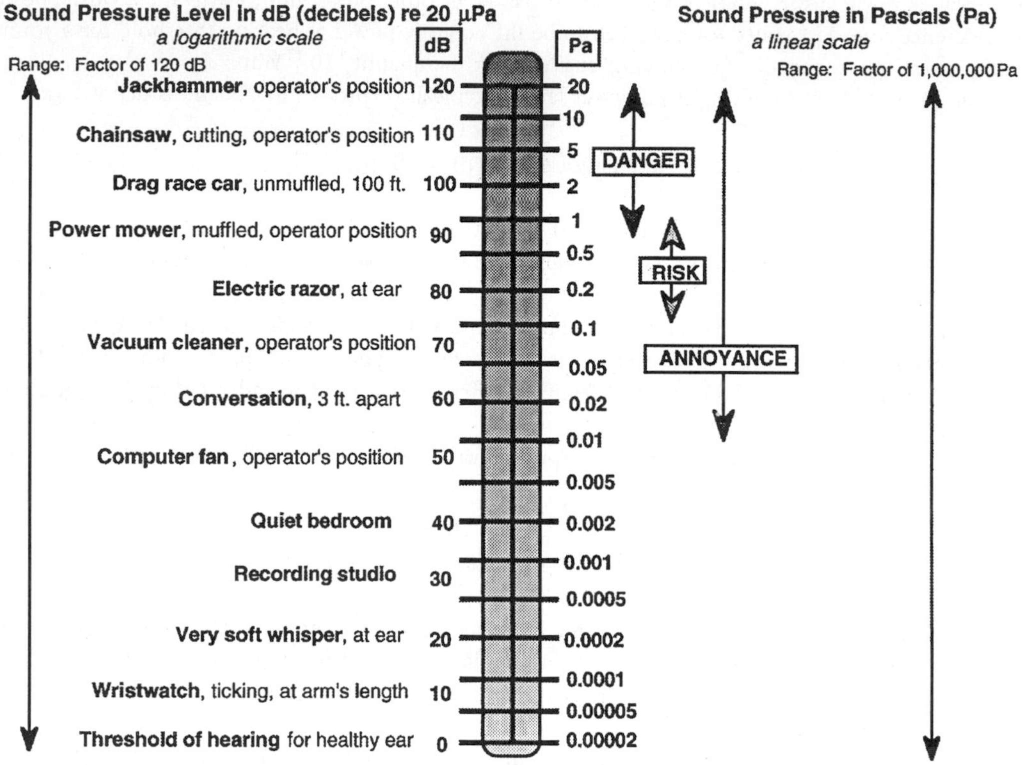

FIGURE 93.1 For typical sounds, sound pressure level values in dB and sound pressure values in Pascals.

of equal loudness to the normal-hearing listener. Note that at 1,000 Hz on each curve the phon level is equal to the dB level. The threshold of hearing for a young, healthy ear is represented by the 0 phon level curve. The young, healthy ear is sensitive to sounds between about 20 Hz and 20,000 Hz, although, as shown by the curve, it is not equally sensitive to all frequencies. At low- and mid-level sound intensities, low frequency and to a lesser extent high frequency sounds are perceived as less intense than sounds in the 1,000 to 4,000 Hz range, where the undamaged ear is most sensitive. But as phon levels move to higher values, the ear becomes more linear in its loudness perception for sounds of different frequencies.

Because the ear exhibits this nonlinear behavior, several frequency weighting functions have been standardized for use with sound level meters. The most common curves are the A, B, and C curves, with the corresponding dB measurement denoted as dBA, dBB, and dBC, respectively. If no weighting function is selected on the meter, the notation dB or dB(linear) is used, and all frequencies are processed without weighting factors. The actual weighting functions for the three suffix notations A, B, and C are superimposed on the phon contours of the top panel of Figure 93.2, and also depicted as actual frequency weighting functions in the bottom panel. In nearly all U.S. measurements of industrial noise made for assessment of exposure risk to workers, the dBA scale is used, and the meter is set on the "slow" dynamic response setting, which produces slow exponential averaging of a one-second window. For determination of the adequacy of hearing protection for a particular noise, and for application of noise control measures, the C-weighted level is often taken in addition to the A-weighted level.

Sones

While the phon scale provides the ability to equate the loudness of sounds of different frequencies, it does not afford an ability to describe how much louder one sound is compared with another. For this, the *sone* scale is needed (Stevens, 1936).[1] One sone is defined as the loudness of a 1,000 Hz tone of 40 dB SPL. In relation to one sone, two sones are twice as loud, three sones are three times as loud, one-half

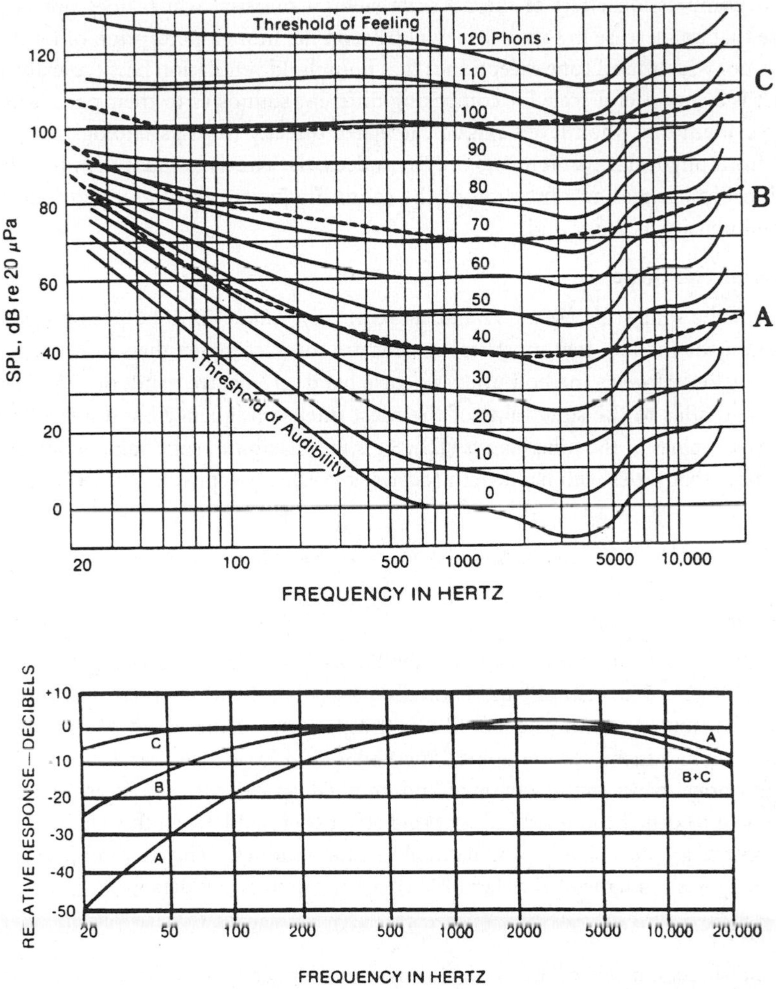

FIGURE 93.2 Top: Equal loudness contours based on the psychophysical phon scale, with sound level meter frequency weighting curves superimposed. Bottom: dB vs. frequency values of A, B, and C sound level meter weighting curves. (Adapted from Earshen, J. J. [1986]. Sound measurement: Instrumentation and noise descriptors, in E. H. Berger, W. D. Ward, J. C. Morrill, and L. H. Royster [Eds.], *Noise and Hearing Conservation Manual* [pp. 38-95]. Akron, OH: American Industrial Hygiene Association. With permission.)

sone is half as loud, and so on. Phon level (L_P) and sones are related by the following formula for sounds at or above a 40 phon level:

$$\text{Loudness in Sones} = 2^{\frac{(L_p - 40)}{10}} \tag{5}$$

According to formula 5, 1 sone equals 40 phons and the number of sones doubles with each 10-phon increase; therefore, it is straightforward to conduct a comparative estimate of loudness levels of sounds with different decibel levels. The "rule-of-thumb" is that each 10 dB increase in a sound (that is, one which is above 40 dB to begin with) will result in a doubling of its loudness. For instance, a conference room which is at 45 dBA may currently be comfortable for communication. However, if a new ventilation system increases the noise level in the room by 10 dBA, the occupants will experience a doubling of loudness and will likely complain about the effects of the background noise on conversation in the room.

Once again, the compression effect of the dB scale yields a measure which does not reflect the much larger influence that an increase in sound level will have on the human perception of loudness. Although the sone scale is not widely used (one exception is that household ventilation fans typically have voluntary sone ratings), it is a very useful scale for comparing different sounds as to their perceived loudness.

It should be evident that phon levels can be calculated directly from psychological measurements in sones, but not from physical measurements of SPL in decibels. This is because the phon-based loudness and SPL relationship changes as a function of the sound frequency and the magnitude of this change depends on the intensity of the sound.

Modifications of the Sone

A modification of the sone scale (*Mark VI* and subsequently, *Mark VII sones*) was proposed by Stevens (1972)[2] to account for the fact that most real sounds are more complex than pure tones. Utilizing the general formula below, Steven's method utilizes octave band, 1/2 octave band, or 1/3 octave band noise measurements, and adds to the sone value of the most intense frequency band a fractional portion of the sum of the sone values of the other bands (ΣS). S_m is the maximum sone value in any band, and k is a fractional multiplier that varies with bandwidth (octave, $k = 0.3$; 1/2 octave, $k = 0.2$; 1/3 octave, $k = 0.15$).

$$\text{Loudness in Sones} = S_m + k(\Sigma S - S_m) \tag{6}$$

Zwicker's Method

The concept of the *critical band for loudness* formed the basis for Zwicker's method of loudness quantification (Zwicker, 1960).[3] The critical band is the frequency band within which the loudness of a band of continuously distributed sound of equal SPL is independent of the width of the band. The critical bands widen as frequency increases. A graphical method is used for computing the loudness of a complex sound based on critical band results obtained and graphed by Zwicker. The noise spectrum is plotted and lines are drawn to depict the spread of masking effect (defined later in this chapter). The result is a bounded area on the graph which is proportional to total loudness. The method is relatively complex and the reader may wish to consult Zwicker (1960)[3] for computational detail.

Noisiness Units

Loudness and noisiness are related but not synonymous. Noisiness can be defined as the "subjective unwantedness" of a sound. Perceived noisiness may be influenced by a sound's loudness, tonality, duration, impulsiveness, and variability (Kryter, 1994)[4]. Whereas a low level of loudness might be perceived as enjoyable or pleasing, even a low level of unwantedness, that is, noisiness, is by definition undesirable. Equal noisiness contours, analogous to equal loudness contours, have been developed based on a unit (analogous to the phon) called the *perceived noise level* (PN_{dB}), which is the SPL in dB of a 1/3 octave band of random noise centered at 1,000 Hz which sounds equally noisy to the sound in question. Also, an *N*- (later D-) sound level meter weighting curve was developed for measuring the perceived noise level of a sound. A subjective noisiness unit analogous to the sone, the noy, is used for comparing sounds as to their relative noisiness. One noy is equal to 40 PN_{dB}, and two noys are twice as noisy as one, five noys are five times as noisy, and so on. Similar to the behavior of sone values for loudness, an increase of about 10 PN_{dB} is equivalent to a doubling of the perceived noisiness of a sound.

93.3 Effects of Noise in Occupational Settings

Nonauditory Effects

Noise exposure in industry has been linked to several deleterious effects, some of which are nonauditory and thus beyond the scope of this chapter. However, it is at least important to recognize that noise can degrade operator task performance. Research studies concerning the effects of noise on performance are primarily laboratory-based and task/noise specific; therefore, extrapolation of the results to actual industrial settings is somewhat risky (Sanders and McCormick, 1993).[5] Nonetheless, on the negative side, noise is known to mask task-related acoustic cues, as well as cause distraction and disruption of "inner speech," while on the positive, noise may at least initially heighten operator arousal and thereby improve performance on tasks which do not require substantial cognitive processing (Poulton, 1978).[6] To obtain reliable effects of noise on performance, except on tasks which rely heavily on short-term memory, the level of noise must be fairly high, usually 95 dBA or greater. Tasks which are simple and repetitive often show no deleterious performance effects (and sometimes improvements) in the presence of noise, while difficult tasks that rely on perception and information processing on the part of the operator will often exhibit performance degradation (Sanders and McCormick, 1993).[5] It is generally accepted that unexpected or aperiodic noise causes greater degradation than predictable, periodic, or continuous noise, and that the startle response created by sudden noise can be disruptive.

Furthermore, noise has been linked to physiological problems other than those of the hearing organ, including hypertension, heart irregularities, extreme fatigue, and digestive disorders. Most physiological responses of this nature are symptomatic of stress-related disorders. Because the presence of high noise levels often induces other stressful feelings (such as sleep disturbance and interference with conversing in the home, and fear of missing oncoming vehicles or warning signals on the job), there are second-order effects of noise on physiological functioning that are difficult to predict.

Signal Detection and Communications Effects

Interference and the Signal-to-Noise Ratio

One of the most noticeable effects of noise is its interference with speech communications and the hearing of nonverbal signals. Workers often complain that they must shout to be heard and that they cannot hear others trying to communicate with them. Likewise, noise interferes with the detection of workplace signals such as alarms for general area evacuation and warnings, annunciators, on-equipment alarms, and machine-related sounds which are relied upon for feedback. The ratio (actually the algebraic difference) of the speech or signal level to the noise level, termed the *signal-to-noise ratio* (S/N) is the most critical parameter in determining whether speech or signals will be heard in noise. An S/N of 5 dB means that the signal is 5 dB greater than the noise, while an S/N of –5 dB means that the signal is 5 dB lower than the noise. Hearing protection is often blamed for exacerbating the effects of noise on the audibility of speech and signals, although, at least for individuals with normal hearing, protectors may actually facilitate hearing in some noisy situations, particularly those above about 90 dBA.

Masking

Masking is technically defined as the tendency for the threshold of a desired signal or speech (*the masked sound*) to be raised in the presence of an interfering sound (*the masker*). As an example, in the presence of a noisy airport waiting area, a pay telephone's earphone volume must often be increased to enable the listener to hear the party on the line, whereas a lower volume will be more comfortable while affording audibility when there is no crowd or public address system noise present. The *masked threshold* is defined as the SPL required for 75% correct detection of a signal when that signal is presented in a two-interval task wherein, on a random basis, one of the two intervals of each task trial contains the signal and the noise and the other contains only noise. In a controlled laboratory test scenario, a signal that is about 6 dB above the masked threshold will result in near perfect detection performance (Sorkin, 1987).[7] Analytical prediction (as opposed to actual experimentation with human subjects) of the interfering effects of noise on speech communications may be conducted using the Articulation Index (AI) technique defined in ANSI S3.5-1969 (R1986).[8] Essentially, this relatively complex technique utilizes a weighted sum of the speech-to-noise ratios in specified frequency bands to compute an AI score ranging between 0.0 and 1.0, with higher scores indicative of greater predicted speech intelligibility. Nonverbal signal detectability predictions can also be made analytically, with the most comprehensive computational technique, based on a spectral analysis of the noise, appearing in ISO 7731-1986.[9] While a full discussion of these analytical procedures is beyond the scope of this chapter, the reader is referred to the individual ANSI and ISO standards for detail. The AI and masked threshold computational techniques provide better resolution and accuracy for speech intelligibility and signal detectability predictions than a simple evaluation of broadband S/N ratios because the techniques incorporate the frequency-specific information that simple S/N ratios do not reflect. However, the following general principles regarding masking effects on nonverbal signals and speech can be used for general guidance.

1. The greatest increase in masked threshold occurs for nonverbal signal frequencies which are equal or near to the predominant frequencies of the masking noise; this is called *direct masking*. Therefore, warning signals should not utilize tonal frequencies equivalent to those of the masker. Preferably, the signal should be in the most sensitive range of human hearing, approximately 1,000 to 4,000 Hz, unless the noise energy is intense at these frequencies.
2. If the signal and masker are tonal in nature, the primary masking effect is at the fundamental frequency of the masker and at its harmonics. For instance, if a masking noise has primary frequency content at 1,000 Hz, this frequency and its harmonics (2,000, 3,000, 4,000, etc.) should be avoided as signal frequencies.
3. The greater the SPL of the masker, the more the increase in masked threshold of the signal. A general rule-of-thumb is that the S/N ratio at the listener's ear should at a minimum be about 15 dB above masked threshold for reliable signal detection. However, in noise levels above about 80 dBA, the signal levels required to maintain an S/N ratio of 15 dB above masked threshold may increase the hearing exposure risk, especially if signal presentation occurs frequently. Therefore, if lower S/Ns become necessary, it is best to construct signals which are unlike the masker in frequency and which have modulated or alternating frequencies to grab attention.
4. Warning signals should not exceed the masked threshold by more than 30 dB to avoid verbal communications interference and operator annoyance (Sorkin, 1987).[7]
5. As the SPL of the masker increases, the primary change in the masking effect is that it spreads upward in frequency, often causing signal frequencies which are higher than the masker to be missed. This is termed *upward masking*. Since most warning signal guidelines recommend that the midrange and high-frequency signals (about 1,000 to 4,000 Hz) be used for detectability, it is important to consider that the masking effects of industrial noise of lower frequencies can spread upward and cause interference in this range. Therefore, if the noise has its most significant energy in this range, a lower frequency signal, say 500 Hz, may be necessary. However, it must be kept in mind that the ear is not as sensitive to low frequencies, so the signal level must be carefully set to ensure reliable audibility.
6. Masking effects can also spread downward in frequency, causing signal frequencies below those of the masker to be raised in threshold. This is called *remote masking* and the effect is most prominent at signal frequencies which are subharmonics of the masker. With typical industrial noise sources, remote masking is generally less of a problem than direct or upward masking.
7. In extremely loud environments of about 110 dB and above, nonauditory signal channels such as visual and vibro-tactile should be considered as alternatives to auditory displays.
8. Speech intelligibility in noise depends on a combination of complex factors and, as such, predictions based on simple S/N ratios should not be relied upon. However, in very general terms, S/N ratios of 15 dB or higher should result in intelligibility performance above about 80% words correct for normal-hearing individuals in broadband noise (Acton, 1970).[10] Above speech levels of about 85 dBA, there is some decline in intelligibility even if S/N ratio is held constant (Pollack, 1958).[11] In very high noise levels, it is impractical and may pose additional hearing hazard risk to amplify the voice to maintain the high S/N ratios necessary for good intelligibility performance. The S/N ratio required for reliable intelligibility may be reduced via the use of certain techniques such as reduction of speaker-to-listener distances, use of smaller vocabularies, provision of contextual cues in the message, use of the phonetic alphabet, and use of noise-attenuating headphones and noise-canceling microphones in electronic systems.

9. Electronic speech communications systems should reproduce speech frequencies in the range of 500 to 5,000 Hz, which encompasses the most sensitive range of hearing and includes the speech sounds important for message comprehension. More specifically, because much of the information required for word discrimination lies in the consonants, which are in the higher end of the frequency range and of low power (while the power of the vowels is in the peaks of the speech waveform), the use of electronic peak-clipping and reamplification of the waveform may improve intelligibility because the power of the consonants is thereby boosted relative to the vowels. Furthermore, it is critical that frequencies in the region of 1,000 to 4,000 Hz be faithfully reproduced in electronic communication systems to maintain intelligibility. Filtering out of frequencies outside this range will not appreciably affect word intelligibility, but will influence the quality of the speech.
10. Actual human speech results in higher intelligibility in noise than computer-generated speech; therefore, especially for critical message displays and annunciators, live, recorded, or digitized human speech is preferable over synthesized speech (Morrison and Casali, 1994).[12]

Noise-Induced Hearing Loss

Scope of Hearing Loss in the U.S.

Noise-induced hearing loss (NIHL) is one of the most widespread occupational maladies in the U.S., if not the world. In the early 1980s, it was estimated that over 9 million workers are exposed to noise levels averaging over 85 dBA for an 8-hour workday (EPA, 1981).[13] Today, this number is likely to be higher because the control of noise sources, both in type and number, has not kept pace with the proliferation of industrial and service sector development. Due in part to the fact that before 1971 there were no U.S. federal regulations governing noise exposure in general industry, many workers over 50 years of age now exhibit hearing loss that results from the effects of occupational noise. Of course, the total noise exposure from both occupational and nonoccupational sources determines the NIHL that a victim experiences. Of the estimated 28 million Americans who exhibit significant hearing loss due to a variety of etiologies, such as pathology of the ear, ototoxic drugs, and hereditary tendencies, over 10 million have losses which are directly attributable to noise exposure (NIH, 1990).[14] Therefore, the noise-related losses are preventable in nearly all cases. The majority of losses are due to on-the-job exposures, but leisure noise sources do contribute a significant amount of energy to the total noise exposure of some individuals.

Types and Etiologies of Noise-Induced Hearing Loss

Although the major concern of the industrial hearing conservationist is to prevent employee hearing loss that stems from occupational noise exposure, it is important to recognize that hearing loss may also emanate from a number of sources other than noise, including: infections and diseases specific to the ear, most frequently originating in the middle or conductive portion; other bodily diseases, such as multiple sclerosis which injures the neural part of the ear; ototoxic drugs, of which the mycin family is a prominent member; exposure to certain chemicals and industrial solvents; hereditary factors; head trauma; sudden hyperbaric- or altitude-induced pressure changes; and aging of the ear (presbycusis). Furthermore, not all noise exposure occurs on the job. Many workers are exposed to hazardous levels during leisure activities, from such sources as automobile/motorcycle racing, personal stereo headsets and car stereos, firearms, and power tools. The effects of noise on hearing are generally subdivided into the following three categories (Melnick, 1991).[15]

Acoustic Trauma

Immediate organic damage to the ear from an extremely intense acoustic event such as an explosion is known as *acoustic trauma.* The victim will notice the loss immediately and it often constitutes a permanent injury. The damage may be to the conductive chain of the ear, including rupture of the eardrum or dislodging of the ossicles (small bones) of the middle ear. Conductive losses can, in many cases, be compensated for with a hearing aid and/or surgically corrected. Neural damage may also occur, involving a dislodging of the hair cells and/or breakdown of the neural organ (Organ of Corti) itself. Unfortunately, neural loss is irrecoverable and not typically compensable with a hearing aid. Acoustic trauma represents a severe injury, but fortunately its occurrence is uncommon, including in the industrial setting.

Noise-Induced Threshold Shift

A *threshold shift* is defined as an elevation of hearing level from the individual's baseline hearing level and it constitutes a loss of hearing sensitivity. *Noise-induced temporary threshold shift* (NITTS), sometimes referred to as "auditory fatigue," is by definition recoverable with time away from the noise. The elevation of threshold is temporary, and usually can be traced to an overstimulation of the neural hair cells (actually, the stereocilia) in the Organ of Corti. Although the individual may not notice the temporary loss of sensitivity, NITTS is a cardinal sign of overexposure to noise. It may occur over the course of a full workday in noise or even after a few minutes of exposure to very intense noise. Although the relationships are somewhat complex and individual differences are rather large, NITTS does depend on the level, duration, and spectrum of the noise, as well as the audiometric test frequency in question (Melnick, 1991).[15]

Prevention of *noise-induced permanent threshold shift* (NIPTS), for which there is no possibility of recovery, is the primary target of the industrial hearing conservationist. NIPTS can manifest suddenly as a result of acoustic trauma; however, industrial noise problems that cause NIPTS most typically constitute exposures that are repeated over a long period of time and have a cumulative effect on hearing sensitivity. In fact, the losses are often quite insidious in that they occur in small steps over a number of years of overexposure and the worker is not aware of the problem until it is too late. This type of exposure produces permanent neural damage, and although there are some individual differences as to magnitude of loss and audiometric frequencies affected, the typical pattern for NIPTS is a prominent elevation of threshold at the 4,000 Hz audiometric frequency (sometimes called the 4 kHz notch), followed by a spreading of loss to adjacent frequencies of 3,000 and 6,000 Hz. From a classic study on workers in the jute weaver industry, Figure 93.3 depicts the temporal profile of NIPTS as the family of audiometric threshold shift curves, with each curve representing a different number of years of exposure (Taylor, Pearson, Mair, and Burns, 1964).[16] As noise exposure continues over time, the hearing loss will spread over a wider frequency bandwidth inclusive of midrange and high frequencies, and encompassing the range of most auditory warning signals. In some cases, the hearing loss renders it unsafe or unproductive for the victim to work in certain occupational settings where the hearing of certain signals are requisite to the job. Unfortunately, the power of the consonants of speech sounds, which heavily influence the intelligibility of human speech, also lie in the frequency range that is typically affected by NIPTS, compromising the victim's ability to understanding speech. This is the tragedy of NIPTS in that the worker's ability to communicate is hampered, often severely and always irrecoverably. Furthermore, unlike blindness or many physical disabilities, hearing loss is not overt and therefore often goes unrecognized by others. Thus, it is a particularly isolating disability because the victim is unintentionally excluded from conversations and may miss important auditory signals because others are either unaware of the loss or simply forget about the need to compensate for it.

Concomitant Auditory Maladies

Following exposure to high-intensity noise, some individuals will notice that ordinary sounds are perceived as "muffled," and in some cases, they may experience a ringing or whistling sound in the ears, known as *tinnitus.* These manifestations should be taken as serious indications that overexposure has occurred, and that protective action should be taken if similar exposures are encountered in the future. Tinnitus may also occur by itself or in conjunction with NIPTS, but in any case it is thought to be the result of *otoacoustic emissions,* which are essentially acoustic outputs from the inner ear that are audible to the victim, apparently resulting from mechanical activity or *microphonics* of the neural cells. Some individuals report that tinnitus is always present, pervading their lives. It thus has the potential to be quite disruptive and in severe cases, debilitating.

More rare than tinnitus, but typically quite debilitating is the malady known as *hyperacusis,* which refers to hearing that is extremely sensitive to sound. Hyperacusis can manifest in many ways, but a number of victims report that their hearing became painfully sensitive to sounds of even normal levels

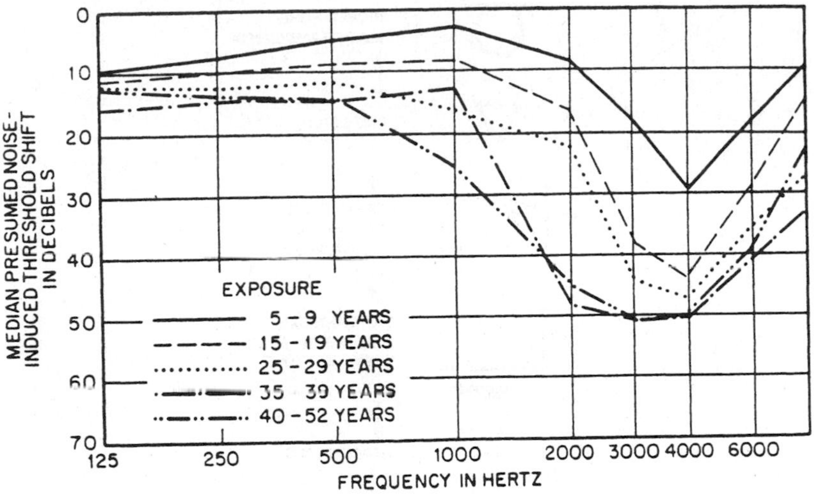

FIGURE 93.3 Cumulative auditory effects of years of noise exposure in a jute weaving industry. (Adapted from Taylor, W., Pearson, J., Mair, A. and Burns, W. [1964]. Study of noise and hearing in jute weavers. *Journal of the Acoustical Society of America*, 38, 113-120. With permission.)

after exposure to a particular noise event. Therefore, at least for some, hyperacusis can be directly traced to noise exposure. Sufferers typically must use hearing protectors when performing normal activities, such as walking on city streets, visiting movie theaters, or washing dishes in a sink, because such activities produce sounds which are painfully loud to them. It should be noted that hyperacusis sufferers often exhibit normal audiograms; that is, their thresholds are not typically better than those of "normal hearers," even though their reaction to sound is one of hypersensitivity.

It is important that the industrial hearing conservationist be aware of these hearing-related maladies that may or may not arise as a result of on-the-job noise exposure, but which may influence the worker's ability to perform certain jobs or work in certain environments.

93.4 Measurement and Quantification of Noise Exposures

Basic Instrumentation

Measurement and quantification of sound exposure levels provide the fundamental data for assessing hearing exposure risk, speech and signal masking effects, hearing conservation program needs, and engineering noise control strategies. A vast array of instrumentation is available for sound measurement; however, for monitoring and assessment of most noise exposure situations, a basic understanding of three primary instruments (sound level meters, dosimeters, and real-time spectrum analyzers) and their data output will suffice. In instances where noise is highly impulsive in nature and/or selection and development of situation-specific engineering noise control solutions is anticipated, more specialized instruments may be necessary.

Because sound is propagated as pressure waves which vary over space and in time, a complete quantification would require simultaneous measurements over the continuous time periods (representing complete operator exposure durations) at all points of an occupational sound field to exhaustively document the noise level in the space. Clearly, this is typically cost- and time-prohibitive, so one must resort to sampling strategies for establishing the observation points and intervals. The hearing conservationist must also decide whether detailed, discrete time histories are needed (such as with a noise-logging *dosimeter*, discussed later), if averaging over time and space with long data records is required (with an averaging/integrating dosimeter), whether discrete samples taken with a short-duration moving

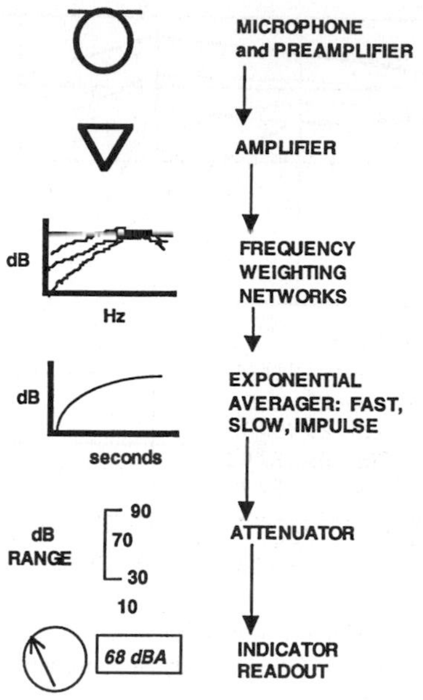

FIGURE 93.4 Block diagram of the basic functional components of a sound level meter.

time average (with a basic sound level meter) will suffice, or if frequency-band-specific SPLs are needed for selecting noise abatement materials (with a spectrum analyzer). Following is a brief discussion of the three primary types of sound measurement instruments and the noise descriptors that can be obtained therefrom.

Sound Level Meter

Most sound measurement instruments derive from the basic sound level meter (SLM), a device for which four grades (and associated performance tolerances that become less stringent as the grade number increases) are described in ANSI S1.4-1983 (R1994).[17] Type 0 instruments have the most stringent tolerances and are for laboratory use only. Other grades include Type 1, intended for precision measurement in the field or laboratory, Type 2, intended for general field use, especially where frequencies above 10,000 Hz are not prevalent, and Type S, a special purpose meter which may perform at grades 0 through 2, but may not include all of the operational functions of the grade. A grade of Type 2 or better is needed for occupational exposure measurements.

Components of a Sound Level Meter

A block diagram of the functional components of a generic SLM appears in Figure 93.4. At the top, a microphone/preamplifier senses the pressure changes caused by an airborne sound wave and converts the pressure signal into a voltage signal. Because the pressure fluctuations of a sound wave are small in magnitude, the corresponding voltage signal must be preamplified and then input to an amplifier which boosts the signal before it is processed further. The passband, or range of frequencies which are passed through and processed, of a high-quality SLM contains frequencies from about 10 Hz to 20,000 Hz, but depending on the frequency weighting used, not all frequencies are treated the same. A selectable frequency weighting network, or filter, is then applied to the signal. These networks most commonly include the A-, B-, and C-weighting functions shown in the bottom panel of Figure 93.2. For OSHA noise monitoring measurements, the A-scale, which de-emphasizes the low frequencies and to a smaller extent the high frequencies, is used. In addition to the common A scale (which approximates the 40 phon level of hearing) and C scale (100 phon level), other scales, including dB(linear) may be included in the meter.

Next, (not shown) the signal is squared to reflect the fact that sound pressure level in decibels is a function of the square of the sound pressure. The signal is then applied to an exponential averaging network, which defines the meter's dynamic response characteristics. In effect, this response creates a moving-window, short-time average display of the sound waveform. The two most common settings are defined as FAST, which has a time constant of 0.125 second (s), and SLOW, which has a time constant of 1.0 s. These time constants were established decades ago to give analog needle indicators a rather sluggish response so that they could be read by the human eye even when highly fluctuating sound pressures were measured. Under the FAST or SLOW dynamics, the meter indicator rises exponentially toward the decibel value of an applied constant SPL. In theory, when driven by an exponential process, the indicator would reach the actual value at infinite time; however, the time constant defines the time period within which the indicator reaches 63% of the maximum value in response to a constant input. For OSHA measurements, the SLOW setting is used, and this setting is best when the average value or average as it is changing over time is desired. The FAST setting is more appropriate when the variability or range of fluctuations of a time-varying sound is desired. On certain SLMs, a third time constant, IMPULSE, may also be included for measurement of sounds which have sharp transient characteristics over time and are generally less than one second in duration, exemplified by gun shots or impact machinery such as drop forges and embossing processes. The IMPULSE setting has an exponential rise time constant of 35 ms and a decay time of 1.5 s. It is useful to afford the observer the time to view the maximum value of a burst of sound before it decays, and is more commonly applied in community and business machine noise measurements than in industrial settings.

Microphone Considerations

Most SLMs have interchangeable microphones which offer varying frequency response, sensitivity, and directivity characteristics (Peterson, 1979).[18] The *response* of the microphone is the ratio of electrical output (in volts, V) to the sound pressure at the diaphragm of the microphone. Sound pressure is commonly expressed in Pascals (Pa) for free-field conditions (where there are no sound reflections resulting in reverberation), and the free-field voltage response of the microphone is given as mV/Pa. When specifications for *sensitivity* or *output level* are given, the response is based on a pure tone sound wave input. Typically, the output level is provided in dB re 1 V at the microphone electrical terminals and the reference sensitivity is 1 V/Pa.

Most microphones that are intended for industrial noise measurements are essentially *omnidirectional* (that is, nondirectional) in their response for frequencies below about 1,000 Hz. When the physical diameter of the microphone is comparable in length to the wavelength of the sound frequency (as occurs at higher frequencies), the microphone, even an omnidirectional one, will exhibit some directionality. This means that depending on the angle of the microphone's diaphragm in relation to the noise, the measurement readout can be less than or even greater than the true value. The 360-degree response pattern of a microphone is called its *polar response*, and the pattern is generally symmetrical about the axis perpendicular to the diaphragm. Some microphones are designed to be highly directional; one example is the cardioid design which has a heart-shaped polar response wherein the maximum sensitivity is for sounds whose direction of travel causes them to enter the microphone at 0 degrees (or the *perpendicular incidence response*), and minimum sensitivity is for sounds entering at 180 degrees behind the microphone. The response at 90 degrees, where sound waves travel and enter parallel to the diaphragm, is known as the *grazing incidence response.* Another response pattern, called the *random incidence response*, represents the mean response of the microphone for sound waves that strike the diaphragm from all angles with equal probability. This response characteristic is the most versatile, and thus it is the response pattern most often applied in the U.S. Hypothetical response characteristics for different sound wave incidences are shown in Figure 93.5.

Because most U.S. SLM microphones are omnidirectional and utilize the random-incidence response, it is best for an observer to point the microphone at the primary noise source and hold it at an angle of incidence of approximately 70 degrees. This will produce a measurement most closely corresponding to the random-incidence response. Care must be taken to avoid shielding the microphone with the body

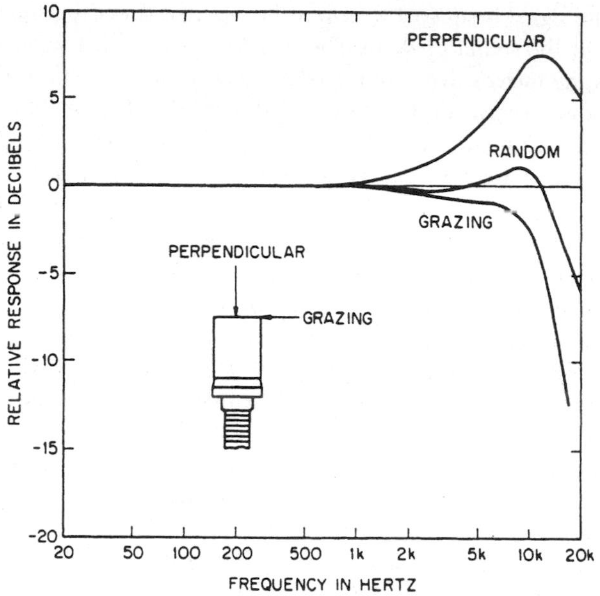

FIGURE 93.5 Frequency response of a hypothetical microphone for three angles of incidence. (Adapted from Peterson, A. P. G. [1979]. Noise measurements: Instruments, in C. M. Harris [Ed.] *Handbook of Noise Control* [pp. 5-1–5-19]. New York: McGraw-Hill. With permission.)

or other structures. The response of microphones can also vary with temperature, atmospheric pressure, and humidity, with temperature being the most critical factor. Correction factors for variations in decibel readout due to temperature effects are supplied by most microphone manufacturers. Atmospheric effects are generally only significant when measurements are made in aircraft or at very high altitudes, and humidity has a negligible effect except at very high levels. In any case, microphones must not be exposed to moisture or large magnetic fields, such as those produced by transformers. When used in windy conditions, a foam windscreen should be placed over the microphone. This will reduce the contaminating effects of wind noise, while only slightly influencing the frequency response of the microphone at primarily high frequencies. In an industrial setting, the windscreen offers the additional benefit of protection of the microphone from damage due to striking and/or airborne foreign matter.

Root Mean Square

Because sound consists of pressure fluctuations above and below ambient air pressure for which the arithmetic average is zero, a *root mean square* (rms) averaging procedure is applied within the SLM when FAST, SLOW, or IMPULSE measurements are taken. In effect, each pressure (or converted voltage) value is squared, the arithmetic sum of all squared values is then obtained, and finally the square root of the sum is computed to provide the rms value. The rms value is what appears on the meter's display.

True Peak SLM

Some SLMs include an unweighted *TRUE PEAK* setting which does not utilize the rms measurement averaging technique, but instead provides an indication of the actual peak SPL reached during a pressure impulse. This measurement mode is necessary for determining if the OSHA limit of 140 dB for impulsive exposure is exceeded. A Type 1 or 2 meter must be capable of measuring a 50-μs pulse. It is important to note that the rms-based IMPULSE dynamics setting is unsuitable for measurement of TRUE PEAK SPLs.

Analog vs. Digital Readouts

In regard to the final component of an SLM shown in Figure 93.4, the indicator display or readout, much debate has existed over whether an analog (needle pointer or bar "thermometer-type" linear display) or digital (numeric) display is best. Ergonomics research indicates that while the digital readout affords higher precision of information to be presented in a smaller space, its Achilles heel is that the digits (particularly the least significant position) become impossible to read when the sound level is fluctuating rapidly. Also, it is more difficult for the observer to capture the maximum and minimum values of a sound, as is often desirable using the FAST response, or the maximum impulse peak attained, with a digital readout. On the other hand, if very precise measurements down to a fraction of a dB are needed, the digital indicator is preferable as long as the meter incorporates an appropriate time integrating or averaging feature or "hold" setting so that the data values can be captured by the human eye. Because of the advantages and disadvantages of each type of display, some contemporary SLMs include both analog and digital readouts.

Sound Level Meter Applications

It is important to note that the standard SLM is intended to measure sound levels at a given moment in time, although certain specialized devices can perform integration or averaging of levels over an extended period of time to provide a long-term descriptor of the noise. When the nonintegrating/averaging SLM is used for noise exposure measurements in the workplace, it is necessary to sample and make multiple manual data entries on a record to characterize the exposure. This technique is usually best limited to area sampling, not individual employee measurements, because it is difficult for the observer to hold the microphone near the employee's ear and to closely shadow the employee as he/she moves about the workplace. Furthermore, the sampling process becomes more difficult as the fluctuations in a noise become more rapid and/or random in nature.

Dosimeter

The *"audio-dosimeter"* or more simply, *"dosimeter,"* is a battery-powered, highly portable device which is derived directly from an SLM but also features the ability to obtain special measures of noise exposure (discussed later) which relate to regulatory compliance and hearing hazard risk. Dosimeters are very compact and are generally worn on the belt or in the pocket of an employee, with the microphone generally clipped to the lapel or shoulder of a shirt or blouse. The intent is to obtain a noise exposure log or record over the course of a full or partial workshift, and to obtain, at a minimum, a readout of the TWA exposure and noise dose for the period measured. Depending upon the features, the dosimeter can log the time history of exposure, providing a running histogram of noise levels on a short time interval (such as one-minute) basis, compute statistical distributions of the noise exposures for the period, flag and record exposures which exceed OSHA maxima of 115 dBA continuous or 140 dB TRUE PEAK, and compute average metrics using 3 dB, 5 dB, or even other time-versus-level exchange rates. The dosimeter eliminates the need for the observer to set up a discrete sampling scheme or follow the worker, both of which are necessary with a conventional SLM. However, it is important that the observer establish rapport and gain the confidence of the worker wearing the dosimeter, and convey at least the following information: (1) to behave normally as to the work activity, (2) to not tamper with the dosimeter or microphone, (3) to return the device when visiting restrooms or entering damp areas, (4) to return the device if there is a need to approach large transformers or other magnetic fields, and (5) to understand the purpose of the dosimetry. Since they are designed to be worn on the noise-exposed employee, dosimeters are typically thought of as devices for personal measurements, but they may also be tripod-mounted or held by an observer for area or survey measurements and are very useful for obtaining community noise measurements as well.

Spectrum Analyzer

The *spectrum analyzer* is an advanced SLM which incorporates selective frequency-filtering capabilities to provide an analysis of the noise level as a function of frequency. In other words, the noise is broken down into its frequency components and a distribution of the noise energy in all measured frequency bands is available. Bands are delineated by upper and lower edge or cutoff frequencies and a center frequency. Different widths and types of filters are available, with the most common width being the *octave filter,* wherein the center frequencies of the filters are related by multiples of two (that is, 31.5, 63,

125, 250, ..., 4,000, 8,000, and 16,000 Hz), and the most common type being the center frequency proportional, wherein the width of the filter depends on the center frequency (as in an octave filter set, in which the passband width equals the center frequency divided by $2^{1/2}$). The *octave band,* commonly called 1/1 octave filter, has a center frequency, cf, which is equal to the geometric mean of the upper (f_u) and lower (f_l) cutoff frequencies. The formulae to compute the center frequency for the octave filter, as well as the band edge frequencies, are:

$$\text{center frequency, cf} = \left(f_u * f_l\right)^{1/2} \quad \text{upper cutoff, } f_u = (\text{cf})2^{1/2} \quad \text{lower cutoff, } f_l = (\text{cf})/2^{1/2} \tag{7}$$

More precise spectral resolution can be obtained with other center frequency proportional filter sets with narrower bandwidths, the most common being the 1/3-octave, and with constant percentage bandwidth filter sets, such as 1 or 2% filters. Note that in both types, the filter bandwidth increases as the center frequency increases. Still other analyzers have constant bandwidth filters, such as 20 Hz-wide bandwidths which are of constant width regardless of center frequency. While in the past most spectrum analyzer filters have been analog devices with "skirts" or overshoots extending slightly beyond the cutoff frequencies, digital computer-based analyzers are now very common. These "computational" filters use fast Fourier transform (FFT) algorithms to compute sound level in a prespecified band of fixed resolution. FFT devices can be used to obtain very high resolutions of noise spectral characteristics using bandwidths as low as one Hz. However, in most industrial noise applications, a 1/1 or 1/3-octave analyzer will suffice unless the noise has considerable power in near-tonal components which must be isolated. One caution is in order: if a noise fluctuates in time and/or frequency, an integrating/averaging analyzer should be used to achieve good accuracy of measurements. It is important that the averaging period be long in comparison to the variability of the noise being sampled.

Inexpensive spectrum analyzers sometimes have filter sets which must be addressed individually in obtaining a measurement. Such devices are called *sequential analyzers* and the operator must manually (or via computer control) step through each filter separately and then read the result. Obviously, sequential filters are problematic when applied to the measurement of a fluctuating noise. On the other hand, *real-time analyzers* incorporate parallel banks of filters which can process all frequency bands simultaneously, and the signal output may be controlled by a SLOW, FAST, or other time constant setting, or it may be integrated or averaged over a fixed time period to provide L_{OSHA}, L_{eq}, or other average-type data.

While occupational noise is monitored with a dosimeter or SLM for the purpose of noise exposure compliance (using A-weighted broadband measurement), or assessment of hearing protection adequacy (using C-weighted broadband measurement), both of these applications can also be addressed (in some cases more accurately) with the use of spectral measurements of the noise level. For instance, the OSHA occupational noise exposure standard (OSHA, 1983)[19] allows the use of octave band measurements reduced to broadband dBA values to determine if noise exposures exceed dBA limits defined in Table G-9 of the standard. Furthermore, Appendix B of the standard concerns hearing protector adequacy and allows the use of an octave band method for determining, on a spectral rather than a broadband basis, whether a hearing protector is adequate for a particular noise spectrum. It is also noteworthy that spectral analysis can help the hearing conservationist discriminate noises as to their hazard potential even though they may have similar A-weighted SPLs. This is illustrated in Figure 93.6, where both noises would be considered to be of equal hazard by the OSHA-required dBA measurements (since they both are 90 dBA), but the 1/3-octave analysis demonstrates that the lowermost noise is more hazardous as evidenced by the heavy concentration of energy in the midrange and high frequencies.

Perhaps the most important application of the spectrum analyzer is to obtain data that will provide the basis for engineering noise control solutions. For instance, in order to select an absorption material for lining interior surfaces of a workplace, the spectral content of the noise must be known so that the appropriate density and thickness of material may be identified. If the noise is found to be primarily of low frequency, the absorption techniques may not provide adequate reduction because low frequencies are more difficult to absorb than high frequencies.

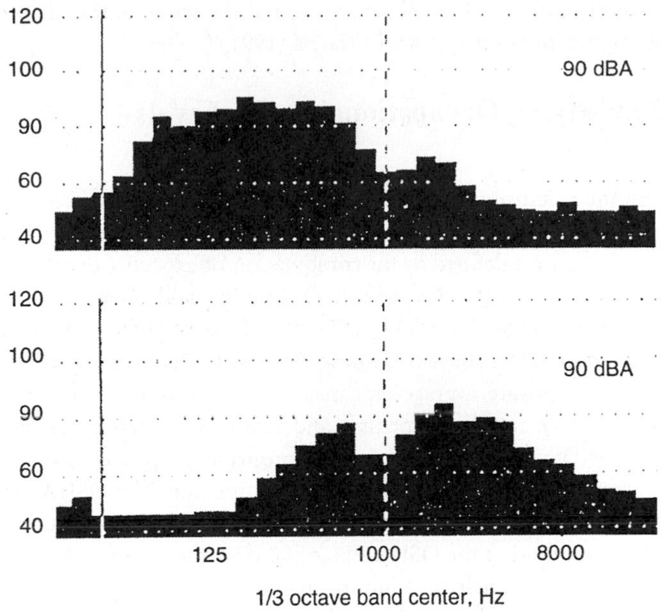

FIGURE 93.6 Spectral differences for two different noises which have the same dBA value.

Lacking a spectrum analyzer, the hearing conservationist can obtain a very rough indication of the dominant spectral content of a noise by using an SLM and taking measurements in both dBA and dBC for the same noise. If the (dBC – dBA) value is large, that is, about 5 dB or more, then it can be concluded that the noise has considerable low frequency content. If, on the other hand, the (dBC – dBA) value is negative, then the noise clearly has strong midrange components, since the A-weighting curve exhibits slight amplification in the 2,000 to 4,000 Hz range. Such rules-of-thumb rely on the differences in the C- and A- weighting curves shown in Figure 93.2. However, they should not be relied upon in lieu of a spectrum analysis if the noise is believed to have high frequency or narrow band components that need noise control attention.

Acoustical Calibrator and Microphone Calibration

Each of the instruments described above contains a microphone which transduces the changes in pressure and inputs this signal into the electronics. While modern sound measurement equipment is generally stable and reliable, calibration is necessary to match the microphone to the instrument so that the accuracy of the measurement is assured. Because of its susceptibility to varying environmental conditions and damage due to rough handling, moisture, and magnetic fields, the microphone is the weakest link in the measurement equipment chain. Therefore, an acoustical calibrator should be applied before and after each measurement with an SLM. The pre-test calibration ensures that the instrument is indicating the correct SPL for a standard reference calibrator output at a specified SPL and frequency (most often 94 dB at 1,000 Hz). The post-test calibration is done to determine if the instrumentation, including the microphone, has drifted during the measurement and if so, if the drift is large enough to invalidate the data obtained. Calibrators may be electronic transducer-type devices with loudspeaker outputs from an internal oscillator, or "pistonphones" which use a reciprocating piston in a closed cavity to produce sinusoidal pressure variations as the cylinder volume changes. Both types include adapters which allow the device to be mated to microphones of different diameters. Calibrators should be sent to the factory for annual calibration. SLMs and dosimeters used for occupational noise measurements should also be factory-calibrated on an annual basis.

There are many other issues which bear on the proper application of sound level measurement equipment, such as microphone selection and placement, averaging time and sampling schemes, and statistical data reduction techniques, all of which are beyond the scope of this chapter. Further coverage of measurement and instrumentation appears in Harris (1991).[20]

Measures for Quantifying Occupational Noise Levels

Exchange Rates

As alluded to earlier in the discussion on the OSHA occupational noise exposure standard, most noise regulations stipulate that a worker's exposure may not exceed a maximum daily accumulation of noise energy, and that the total energy is defined by the combination of exposure duration and intensity of the noise. In other words, in OSHA terms the product of duration and intensity must remain under the regulatory cap or *permissible exposure limit* (PEL) of 90 dBA time-weighted average (TWA) for an 8-hour work period, which is equivalent to a 100% noise dose. Because both noise amplitude and noise duration determine the energy in the exposure, average-type measures of exposure are based on simple algorithms or "exchange rates" which trade amplitude for time and vice versa. Much debate has occurred over the past several decades as to which exchange rate is most appropriate for prediction of hearing damage risk, and most countries currently use either a 3-dB or 5-dB relationship. The OSHA exchange rate is 5 dB, which means that an increase (decrease) in decibel exposure by 5 dB is equivalent to a doubling (halving) of exposure time. For instance, using the OSHA PEL of 90 dBA for 8 hours, if a noise is at 95 dBA, the allowable exposure per workday is half of 8 hours, or 4 hours. If a noise is at 85 dBA, the allowable exposure time is twice 8 hours, or 16 hours. These allowable reference exposure durations (T values) are provided in Table A-1 of the OSHA (1983)[19] regulation, or they may be computed using the formula for T which appears below as Equation 14. The 5-dB exchange rate is predicated on the theory that intermittent noise is less damaging than continuous noise because some recovery from temporary hearing loss occurs during quiet periods. Arguments against it include the fact that an exchange of 5 dB for a factor of two in time duration has no real physical basis in terms of energy equivalence. Furthermore, there is some evidence that the quiet periods of intermittent noise exposures are insufficient in length to allow for recovery to occur. The 5-dB exchange rate is used for all measures associated with OSHA regulations, including the most general average measure of L_{OSHA}, the TWA referenced to an 8-hour duration, and noise dose in percent.

Most European countries use a 3-dB exchange rate, also known as the "equal energy rule." In this instance, a doubling (halving) of sound intensity, which corresponds to a 3 dB increase (decrease), equates to a doubling (halving) of exposure duration. The equal energy concept stems from the fact that if noise energy is doubled or halved, the equivalent decibel change is 3 dB. An exposure to 90 dBA for 8 hours using a 3 dB exchange rate is equivalent to a 120 dBA exposure of only 0.48 minute. Because each increase in decibels by 10 corresponds to a ten fold increase in intensity, the 30 dB increase from 90 to 120 dBA represents a 1,000-fold (10^3) increase in sound intensity, from 0.001 to 1 W/m^2. The 90 dBA exposure period is 8 hours or 480 minutes, and this must be reduced by the same factor as the SPL increase, so 480/1,000 equals 0.48 minute or 29 seconds. The 3-dB exchange rate is used for all measures associated with the equivalent continuous sound level, or L_{eq}.

Average and Integrated SPLs

As discussed earlier, conventional SLMs provide "momentary" dB measurements that are based on very short moving-window exponential averages using FAST, SLOW, or IMPULSE time constants. However, since the majority of noises fluctuate over time, one of several types of average measurements, discussed below, is usually most appropriate as a descriptor of the central tendency of the noise. Averages may be obtained in one of two ways: (1) by observing and recording conventional SLM readouts using a short time interval sampling scheme, and then manually computing the average value from the discrete values, or (2) by using an SLM or dosimeter which automatically calculates a running average value using microprocessor circuitry which provides either a true continuous integration of the area under the sound pressure curve or which obtains discrete samples of the sound at a very fast rate and computes the average. Generally, average measures obtained by method 2 yield more representative values because they are based on continuous or near-continuous sampling of the waveform, which the human observer cannot perform. For sounds which are constant or slowly fluctuating in level, either method should provide representative values, although method 1 necessitates continuous vigilance by an observer.

The average metrics discussed below are generally considered as the most useful for evaluating noise hazards and annoyance potential. In most cases for industrial hearing conservation as well as community noise annoyance purposes, the metrics utilize the A-weighting scale. The equations are all in a form where the data values are considered to be discrete sound levels. Thus, they can be applied to data from conventional SLMs or dosimeters. For continuous sound levels (or when the equations are used to describe true integrating meter functioning), the Σ sign in the equations would be replaced by the integral sign, $\int_0^T$ and the t_i replaced by dt.

Variables used in the equations are as follows:

L_i = dB level in measurement interval i
i_i = length of measurement interval i
Q = exchange rate in dB
N = number of intervals
T = total measurement time period
t_i = time period of interval i

$$q = Q/\log_{10}(2)\begin{cases}\text{for 3 dB exchange, } q = 10.0\\ \text{for 4 dB exchange, } q = 13.3\\ \text{for 5 dB exchange, } q = 16.6\end{cases}$$

The general form equation for *average SPL*, or $L_{average}$, L_{av}, is:

$$L_{av}(Q) = q\,\log_{10}\left[\frac{1}{T}\sum_{i=1}^{N}\left(10^{\left(\frac{L_i}{q}\right)} * t_i\right)\right] \tag{8}$$

The *equivalent continuous sound level*, or, L_{eq}, equals the continuous sound level which, when integrated or averaged over a specific time, would result in the same energy as a variable sound level over the same time period. The equation for L_{eq}, which uses a 3-dB exchange rate, is:

$$L_{eq} = L_{av}(3) = 10\,\log_{10}\left[\frac{1}{T}\sum_{i=1}^{N}\left(10^{\left(\frac{L_i}{10}\right)} * t_i\right)\right] \tag{9}$$

In applying the L_{eq}, usually the individual L_i values are in dBA. Equation 9 may also be used to compute the overall equivalent continuous sound level (for a single site or worker) from individual L_{eq}'s that are obtained over contiguous time intervals by substituting the L_{eq} values in the L_i variable. L_{eq} values are often expressed with the time period over which the average is obtained; for instance, L_{eq} (24) is an equivalent continuous level measured over a 24-hour period. Another average measure which is derived from the L_{eq} and often used for community noise quantification is the L_{dn}. The L_{dn} is simply a 24-hour L_{eq} measurement with a 10-dB penalty added to all nighttime noise levels from 10 P.M. to 7 A.M. The rationale for the penalty is that humans are more disturbed by noise, especially due to sleep arousal, during nighttime periods.

The equation for the *OSHA average noise level*, or L_{OSHA}, which uses a 5-dB exchange rate is:

$$L_{OSHA} = L_{av}(5) = 16.61\,\log_{10}\left[\frac{1}{T}\sum_{i=1}^{N}\left(10^{\left(\frac{L_{iA}}{16.61}\right)} * t_i\right)\right] \tag{10}$$

where L_{iA} is in dBA, slow response.

OSHA's *time-weighted average* (TWA) is a special case of L_{OSHA} which requires that the total time period always be 8 hours, that time is expressed in hours, and that sound levels below 80 dBA, termed the *threshold level,* are not included in the measurement:

$$TWA = 16.61 \log_{10}\left[\frac{1}{8}\sum_{i=1}^{N}\left(10^{\left(\frac{L_{iA}}{16.61}\right)} * t_i\right)\right] \tag{11}$$

where L_{iA} is in dBA, slow response; T is always 8 hours; only $L_{iA} \geq 80$ dBA are included.

OSHA's *noise dose* is a percentage representation of the noise exposure, where 100% is the maximum allowable dose, corresponding to a 90 dBA TWA referenced to 8 hours. Dose utilizes a *criterion sound level,* which is presently 90 dBA, and a *criterion exposure period,* which is presently 8 hours. A noise dose of 50% corresponds to a TWA of 85 dBA, and this is known as the OSHA *action level.* Calculation of dose, D, is as follows:

$$D = \frac{100}{T_c}\sum_{i=1}^{N}\left(10^{\left(\frac{L_{iA}-L_c}{q}\right)} * t_i\right) \tag{12}$$

where L_{iA} is in dBA, slow response; L_c is the criterion sound level; is the criterion exposure duration; only $L_{iA} \geq 80$ dBA are included.

Noise dose, D, can also be expressed as follows, for a constant sound level over the workday:

$$D = 100 * \left(\frac{C_1}{T_1} + \frac{C_2}{T_2} + \ldots + \frac{C_n}{T_n}\right) \tag{13}$$

where C_i is the total time (hours) of actual exposure at L_i; T_i is total time (hours) of reference allowed exposure at L_i, from Table G-16a of OSHA, 1983;[19] $\frac{C_i}{T_i}$ represents a partial dose at sound level i.)

T, the *reference allowable* exposure for a given sound level, can also, in lieu of consulting Table G-16a in OSHA (1983),[19] be computed as:

$$T = \frac{8}{2^{(L-90)/5}} \tag{14}$$

where L is the measured dBA level.

Two other useful equations to compute dose, D, from TWA and vice versa are:

$$D = 100 * 10^{\left(\frac{TWA-90}{16.61}\right)} \tag{15}$$

$$TWA = \left[16.61 \log_{10}\left(\frac{D}{100}\right)\right] + 90 \tag{16}$$

TWA can also be found for each value of dose, D, in Table A-1 of OSHA (1983).[19]

A final measure that is particularly useful for quantifying the exposure due to single or multiple occurrences of an acoustical event (such as a complete operating cycle of a machine, a vehicle drive-by, or aircraft flyover), is the *sound exposure level,* or SEL. It has also been suggested for use in exposure regulations for industry, but to date has not been incorporated into OSHA requirements. The SEL represents a sound of one second length that imparts the same acoustical energy as a varying or constant sound that is integrated over a specified time interval, t_i, in seconds. Over t_i, an L_{eq} is obtained, which indicates that SEL is used only with a 3-dB exchange rate. A reference duration of one second is applied for t_0 in the following equation for SEL:

$$SEL = L_{eq} + 10 \log_{10}\left(\frac{t_i}{t_0}\right) \tag{17}$$

where L_{eq} is the equivalent sound pressure level measured over time period t_i.

Example Computational Problems

Because the majority of industrial noise exposure problems in the U.S. involve measurements to determine OSHA compliance and hearing conservation program needs, the most common measurements from those discussed above entail calculation of the OSHA dose and TWA. Therefore, example computational problems using these measures follow.

Example 1. Workshift less than 8-hours, reading from SLM
Exposures comprising a 7-hour workday consist of 1 hour at 95 dBA, 2 hours at 90 dBA, and 4 hours at 85 dBA, with measurements taken from an SLM. What is the dose and TWA?

Use Equation 14 (or OSHA, 1983,[19] Table G-16a) to determine that 95 dBA is allowed for 4 hours, 90 dBA for 8 hours, and 85 dBA for 16 hours. Then use Equation 13 to determine the partial doses associated with each exposure and the total dose for the workday:

$$D = 100 * \left(\frac{1}{4} + \frac{2}{8} + \frac{4}{16}\right) = 75\%$$

Since 50% action level is exceeded, a hearing conservation program (HCP) is needed.

Equation 16 (or OSHA, 1983,[19] Table A-1) is then used to compute the TWA:

$$TWA = \left[16.61 \log_{10}\left(\frac{75}{100}\right)\right] + 90 = 87.9 \text{ dBA per 8-hour day}$$

Note: As shown in this example, regardless of the total workday, the OSHA method references everything to an 8-hour criterion, with PEL of 90 dBA TWA. This problem could also have been solved by application of Equations 11 and 12.

Example 2. Workshift greater than 8-hours, reading directly from dosimeter
A dosimeter is set up to run for a 12-hour shift, and the readout at the end of the period is D = 300%. If the dosimeter is programmed for an 8-hour criterion exposure duration, a 90 dBA criterion sound level, an 80 dBA threshold sound level, and a 5-dB exchange rate, then the OSHA dose may be read directly from the meter regardless of the fact that the total measurement period is 12 hours. The TWA can then be computed using Equation 16 (or OSHA, 1983,[19] Table A-1):

$$TWA = \left[16.61 \log_{10}\left(\frac{300}{100}\right)\right] + 90 = 97.9 \text{ dBA}$$

Example 3. Workshift greater than 8-hours, reading from SLM
A sound level meter is used to measure exposures in a 12-hour workshift and the average levels obtained over the four time periods sampled are 3 hours at 92 dBA, 2 hours at 98 dBA, 6 hours at 96 dBA, and 1 hour meal time at 75 dBA.

First, the 75 dBA period is deleted in the TWA computation (but not in L_{OSHA} if it is to be calculated) since it is less than the OSHA 80 dBA threshold. Then, Equation 11 is used to compute the TWA, which is based on an 8-hour criterion (therefore, T = 8):

$$TWA = 16.61 \log_{10}\left[\frac{1}{8}\left(3*10^{92/16.61} + 2*10^{98/16.61} + 6*10^{96/16.61}\right)\right]$$

$$= 16.61 \log_{10}\left[\frac{1}{8}(1037416.8 + 1588876.7 + 3612451.8)\right]$$

$$= 16.61 \log_{10} 779843.2$$

$$= 97.9 \text{ dBA}$$

Next, Equation 15 is used to compute the dose from the TWA:

$$D = 100*10^{\left(\frac{97.9-90}{16.61}\right)}$$

$$D = 299\%$$

Example 4. Workshift greater than 8-hours, dosimeter measurement for only partial workshift
A dosimeter is worn by an employee for 7 hours of a 12-hour workshift. It was not possible to apply the dosimeter for the full shift, but it has been determined, based on discussion with employees and direct observation, that the entire workshift is consistent in regard to work activity. The dose measured for the 7 hour period is 115%. Note that this dose is based on only 7 hours of data and that the OSHA criterion exposure period of 8 hours is reflected in the dose calculation from the meter. Since only 7 hours of data are included, the dose is lower than that which would occur during a full 12-hour shift.

Because the entire workshift is consistent with respect to noise-producing work activity, it is reasonable to assume that the same rate of dose per hour would continue through the complete shift.

The 7-hour sampling period included: (1) one 15-minute rest break, and (2) one 30-minute meal break. The remaining 5-hour period that was not sampled does include one 15-minute break.

7 hours sampled, less the total of meal/breaks of 45 minutes = 375 minutes in noise

Total 12-hour shift = (12 * 60) – 60 minutes of meal/breaks = 660 minutes in noise

The 12-hour shift dose can be computed via either of the following methods:

1. Set up a proportional relationship as follows:
   $$\frac{115\% \text{ dose}}{375 \text{ minutes}} = \frac{D\% \text{ dose}}{660 \text{ minutes}}$$
   375D = 75900
   D = 202.4%
   Applying Equation 16 (or OSHA, 1983,[19] Table A-1):
   TWA = 95.1 dBA
2. Calculate a rate of dose per minute:
   $$\frac{115\% \text{ dose}}{375 \text{ minutes}} = 0.3067\% \text{ dose per minute}$$
   D = 660 minutes * 0.3067% dose/minute = 202.4%
   Applying equation 16 (or OSHA, 1983,[19] Table A-1):
   TWA = 95.1 dBA

93.5 Industrial Noise Regulation and Abatement

Indicators of the Need for Attention to Noise

The need for management, or perhaps more appropriately, abatement of industrial noise is indicated when: (1) noise creates sufficient intrusion and operator distraction such that job performance (and even job satisfaction) are compromised; (2) noise creates interference with important communications and signals, such as inter-operator communications, machine- or process-related aural cues, and/or alerting/emergency signals; and (3) noise exposures constitute a hazard for noise-induced hearing loss in workers. While this chapter primarily targets problem 3, which is governed by OSHA federal regulations in general industry (OSHA, 1983)[19] and MSHA (Mine and Safety Health Administration) regulations in mining, the principles of noise measurement, management, and abatement discussed herein may also be applied in mitigating problems 1 and 2.

OSHA Noise Exposure Limits

In regard to combating the hearing loss problem, in OSHA terms if the noise dose exceeds the OSHA action level of 50%, which corresponds to an 85 dBA TWA, the employer must institute a hearing conservation program (HCP) which consists of several facets, to be discussed later (OSHA, 1983).[19] (It is noteworthy that the OSHA regulation specifically exempts employers in oil and gas well drilling and servicing from the HCP requirements, although they are subject to the 100% dose criterion.) If the criterion level of 100% dose is exceeded (which corresponds to the Permissible Exposure Level of 90 dBA TWA for an 8-hour day), the regulations specifically state that steps must be taken to reduce the employee's exposure to the PEL or below via administrative work scheduling and/or the use of engineering controls. It is specifically stated that hearing protection devices (HPDs) shall be provided if administrative and/or engineering controls fail to reduce the noise to the PEL. Therefore, in applying the letter of the law, HPDs are only intended to be relied upon when administrative or engineering controls are infeasible or ineffective. The final OSHA noise level requirement pertains to impulsive or impact noise, which is not to exceed a TRUE PEAK SPL limit of 140 dB.

Hearing Conservation Programs and the Systems Approach

Shared Responsibility between Management, Workers, and Government

A successful HCP, which includes many facets relating to the measurement, management, and control of noise, depends upon the shared commitment of management and labor, as well as the quality of services and products provided by external noise control consultants, audiology or medical personnel who conduct the hearing measurement program, and vendors (for example, hearing protection suppliers). Furthermore, government regulatory agencies, such as OSHA and MSHA, have a responsibility to maintain and disseminate up-to-date noise exposure regulations and HCP guidance, to conduct regular in-plant monitoring of noise exposure and quality of HCPs, and to provide strict enforcement where inadequate noise control and hearing protection exists. And finally, the "end-user" of the HCP, that is, the worker him/herself, must be an informed and motivated participant. For instance, if a fundamental component of the HCP is the personal use of hearing protection devices (HPDs), the effectiveness of the program in preventing NIHL will depend most heavily on the worker's commitment to properly and consistently wear the HPD. Failure by any of these groups to carry out their responsibilities can result in HCP failure and worker hearing loss.

Hearing Conservation Program Structure and Components

Hearing conservation in industry should be thought of as a strategic, programmatic effort that is initiated, organized, implemented, and maintained by the employer, with cooperation from other parties as indicated above. A well-accepted approach is to address the noise exposure problem from a *systems* perspective, wherein empirical noise measurements provide data input which drives the implementation

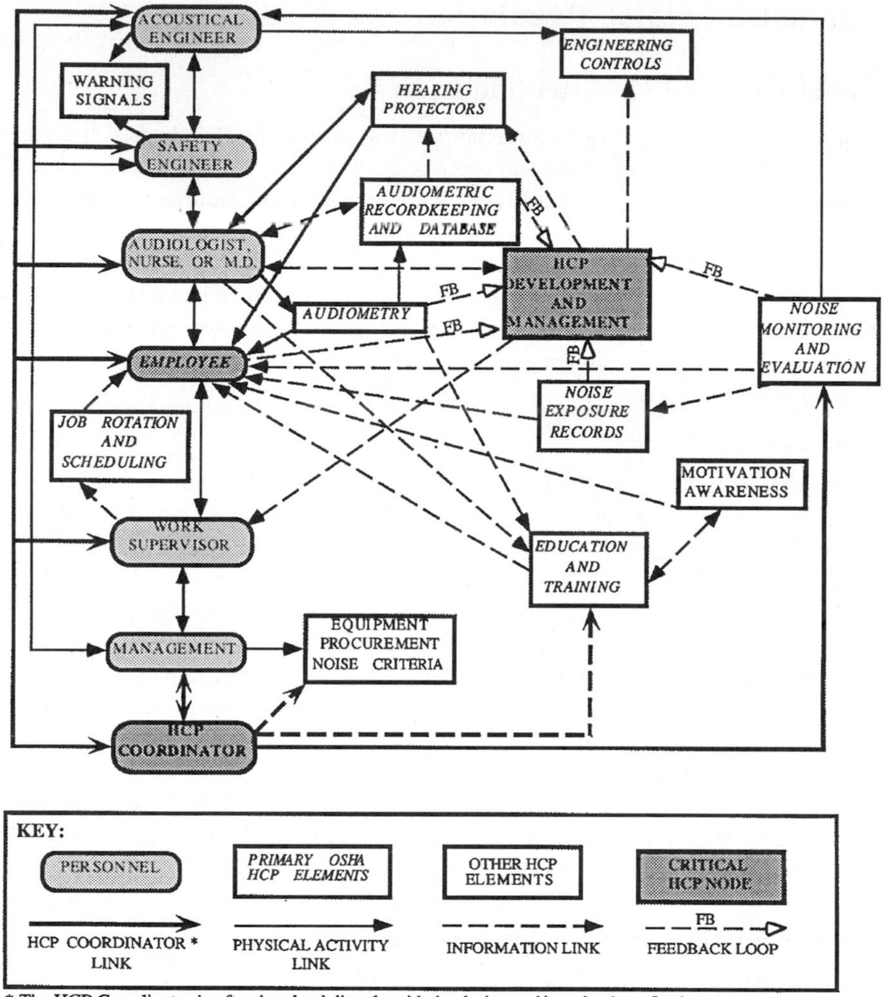

* The HCP Coordinator is often involved directly with the design and/or selection of noise countermeasures.

Note: All elements of the HCP program may act as throughput nodes, in that incoming information may travel through the intermediate node and out to another adjacent node (node-to-node links are not necessary, as long as the nodes are joined through an intermediate node).

FIGURE 93.7 System and components of an industrial hearing conservation program.

of countermeasures against the noise (including engineering controls, administrative strategies, and personal hearing protection). Subsequently, noise and audiometric data, which reflect the effectiveness of those countermeasures, serves as feedback for program adjustments and improvements. Figure 93.7 illustrates the human and other system components that are typically included in an HCP, along with the links between components. Not all programs will include all of these components; for instance, personal hearing protection may be unnecessary if engineering controls provide sufficient noise reduction. A brief discussion of the major elements of an HCP, as dictated by OSHA (1983),[19] follows.

Monitoring

Noise exposure monitoring is intended to identify employees for inclusion in the HCP and to provide data for the selection of HPDs. The data are also useful for identifying areas where engineering noise control solutions and/or administrative work scheduling may be necessary. All OSHA-related measurements, with the exception of the TRUE PEAK SPL limit, are to be made using an SLM or dosimeter set on the dBA scale, SLOW response, using a 5-dB exchange rate, and incorporating all sounds whose levels are from 80 to 130 dBA. It is unspecified, but must be assumed that sounds above 130 dBA should also be monitored. (Of course, such noise levels represent OSHA noncompliance since the maximum allowable continuous sound level is 115 dBA.) The measurement instrument should be ANSI Type 2[17] or better and calibrated to a known standard level before and after noise measurement. Monitoring strategies must take into account the effects of worker movement and noise level variation over time. Although no specific time interval between consecutive monitoring samples is specified, new samples should be taken whenever alterations in equipment or production produce changes in noise exposure. Appendix G of the OSHA regulation suggests that monitoring be conducted at least once every one or two years.

Relating to the noise monitoring requirement is that of notification. Employees must be given the opportunity to observe the noise monitoring process, and they must be notified when their exposures exceed the 50% dose (85 dBA TWA) level.

Audiometric Testing Program

All employees whose noise exposures are at the 50% dose level or above must be included in a pure-tone audiometric testing program wherein a baseline audiogram is completed within six months of the first exposure, and subsequent tests are done on an annual basis. Prior to the baseline audiogram, the worker must avoid workplace noise exposure for 14 hours, or alternatively, use HPDs. Annual audiograms are compared against the baseline to determine if the worker has experienced a *standard threshold shift* (STS), which is defined as an increase in hearing threshold level relative to the baseline of an average of 10 dB at 2,000, 3,000, and 4,000 Hz in either ear. The annual audiogram may be adjusted for age-induced hearing loss (presbycusis) using gender-specific correction data found in Appendix F of the regulation. All OSHA-related audiograms must include 500, 1,000, 2,000, 3,000, 4,000, and 6,000 Hz, in comparison to most clinical audiograms which extend from 125 to 8,000 Hz. If an STS is revealed, a licensed physician or audiologist must review the audiogram and determine the need for further audiological or otological evaluation, the employee must be notified of the STS, and the selection and proper use of HPDs must be revisited. An annual audiogram is substituted for the original baseline when the STS is determined to be persistent or when the annual audiogram indicates significant improvement over the baseline.

Hearing Protection Devices

A selection of HPDs that are suitable for the noise and work situation must be made available to all employees whose TWA exposures meet or exceed 85 dBA. *Earplugs* consist of vinyl, silicone, spun fiberglass, cotton/wax combinations, and closed-cell foam products that are inserted into the ear canal to form a noise-blocking seal. Proper fit to the user's ears and training in insertion procedures are critical to the success of earplugs. A related device is the *semi-insert* or *ear canal cap* which consists of earplug-like pods that are positioned at the rim of the ear canal and held in place by a lightweight headband. The headband is useful for storing the device around the neck when the user moves out of the noise. *Earmuffs* consist of earcups, usually of a rigid plastic material with an absorptive liner, that completely enclose the outer ear and seal around it with foam- or fluid-filled cushions. A headband connects the earcups, and on some models this band is adjustable so that it can be worn over-the-head, behind-the-neck, or under-the-chin, depending upon the presence of other headgear, such as a welder's mask. In general terms, as a group, earplugs provide better attenuation than earmuffs below about 500 Hz and equivalent or greater protection above 2,000 Hz. At intermediate frequencies, earmuffs typically have the advantage in attenuation. Earmuffs are generally more easily fit by the user than earplugs or canal caps, and depending on the temperature and humidity of the environment, the earmuff can be uncomfortable (in hot or high-humidity environments) or a welcome ear insulator (in a cold environment). Semi-inserts generally offer less attenuation and comfort than earplugs or earmuffs, but because they are readily storable around the neck, they are convenient for those workers who frequently move in and out of noise. A thorough review of HPDs and their application may be found in Berger and Casali (1997).[21] Recent new technologies in hearing protection have emerged, including electronic devices offering active noise cancellation, communications capabilities, and noise-level-dependent attenuation, as well as passive, mechanical HPDs which offer level-dependent attenuation and near flat or uniform attenuation spectra; these devices are reviewed in Casali and Berger (1996).[22]

Regardless of its general type, HPD effectiveness depends heavily on the proper fitting and use of the devices (Park and Casali, 1991).[23] Therefore, the employer is required to provide training in the fitting, care, and use of HPDs to all affected employees (OSHA, 1983).[19] Hearing protector use becomes mandatory when the worker has not undergone the baseline audiogram, has experienced an STS, or has a TWA exposure which meets or exceeds 90 dBA. In the case of the worker with an STS, the HPD must attenuate the noise to 85 dBA TWA or below. Otherwise, the HPD must reduce the noise to at least 90 dBA TWA.

The protective effectiveness or adequacy of an HPD for a given noise exposure must be determined by applying the attenuation data required by the EPA (1979)[24] to be included on protector packaging. These data are obtained from psychophysical threshold tests at nine 1/3 octave bands with centers from 125 to 8,000 Hz that are performed on human subjects, and the difference between the thresholds with and without the HPD on constitutes the attenuation at a given frequency. Spectral attenuation statistics (means and standard deviations) and the single number noise reduction rating (NRR), which is computed therefrom, are provided. The ratings are the primary means by which end-users compare different HPDs on a common basis and make determinations of whether adequate protection and OSHA compliance will be attained for a given noise environment.

The most accurate method of determining HPD adequacy is to use octave band measurements of the noise and the spectral mean and standard deviation attenuation data to determine the *protected exposure level* under the HPD. This is called the *NIOSH long method* or the *octave band* method. Computational procedures appear in NIOSH (1975).[25] Because this method requires octave band measurements of the noise, preferably with each noise band's data in TWA form, the data collection requirements are large and the method is not widely applied in industry. However, because the noise spectrum is compared against the attenuation spectrum of the HPD, a "matching" of exposure to protector can be obtained; therefore, the method is considered to be the most accurate available.

The NRR represents a means of collapsing the spectral attenuation data into one broadband attenuation estimate that can easily be applied against broadband dBC or dBA TWA noise exposure measurements. In the calculation of the NRR, the mean attenuation is reduced by two standard deviations; this translates into an estimate of protection theoretically achievable by 98% of the population (EPA, 1979).[24] The NRR is primarily intended to be subtracted from the dBC exposure TWA to estimate the protected exposure level in dBA, as via the following equation:

$$\text{Workplace TWA in dBC} - \text{NRR} = \text{Protected TWA in dBA} \tag{18}$$

Unfortunately, because OSHA regulations require that noise exposure monitoring be performed in dBA, the dBC values may not be readily available to the hearing conservationist. In the case where the TWA values are in dBA, the NRR can still be applied, albeit with some loss of accuracy. With dBA data, a 7 dB "safety" correction is applied to the NRR to account for the largest typical differences between C- and A-weighted measurements of industrial noise, and the equation is as follows:

$$\text{Workplace TWA in dBA} - (\text{NRR} - 7) = \text{Protected TWA in dBA} \tag{19}$$

While the above methods are promulgated by OSHA (1983)[19] for determining HPD adequacy for a given noise situation, a word of caution is needed. The data appearing on HPD packaging are obtained under optimal laboratory conditions with properly fitted protectors and trained human subjects. In no way does the "experimenter-fit" protocol and other aspects of the current test procedure (ANSI S3.19-1974)[26] represent the conditions under which HPDs are selected, fit, and used in the workplace (Park and Casali, 1991).[23] Therefore, the attenuation data used in the octave band or NRR formulae shown above are highly inflated and cannot be assumed as representative of the protection that will be achieved in the field. The results of a review of research studies in which manufacturers' on-package NRRs were compared against NRRs computed from actual subjects taken with their HPDs from field

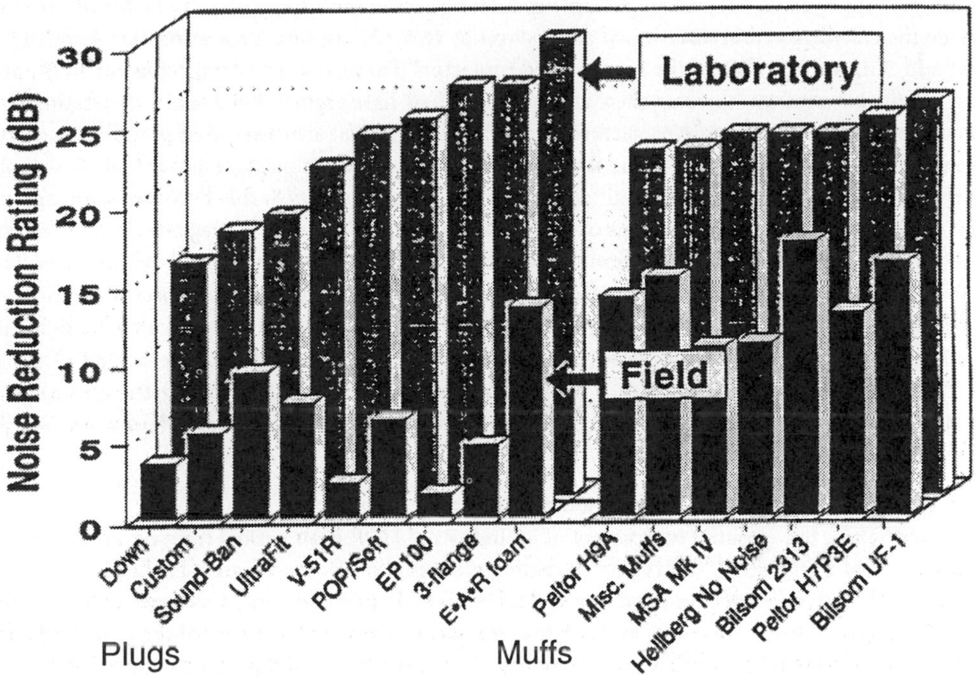

FIGURE 93.8 Comparison of hearing protection device NRRs by device type: manufacturers' laboratory data vs. real-world "field" data. (Adapted from Berger, E. H., Franks, J. R., and Lindgren, F. [1996]. International review of field studies of hearing protector attenuation, in A. Axelsson, H. Borchgrevink, R. P. Hamernik, P. Hellstrom, D. Henderson, and R. J. Salvi [Eds.], *Scientific Basis of Noise-Induced Hearing Loss.* [pp. 361-377], New York: Thieme Medical Publishers, Inc. With permission.)

settings is shown in Figure 93.8 (Berger, Franks, and Lindgren, 1996).[27] Clearly, the differences between laboratory and field estimates of HPD attenuation are large, and the hearing conservationist must take this into account when selecting protectors. Recent efforts by ANSI Working Group S12/WG11 have focused on the development of a testing standard which utilizes subject (not experimenter) fitting of the HPD and relatively naive (not trained) subjects to yield attenuation data that are more representative of those achievable under workplace conditions wherein an HCP is operated (described in Royster et al., 1996).[28] However, at press time this draft standard had not been adopted into law promulgating its use in producing the data to be utilized in labeling HPD performance.

If the currently available HPD attenuation data are inaccurate, what steps should be taken to gain a more accurate estimate of the NRR for use in determining protected exposure levels? The OSHA (1989)[29] Field Operations Manual of the Office of General Industry Compliance Assistance indicates: "Citations for violations of 29CFR 1910.95(b)(1) shall be issued when engineering and/or administrative controls are feasible, both technically and economically; and (1) Employee exposure levels are so high that hearing protectors alone may not reliably reduce noise levels received by the employee's ear to the levels specified in Tables G-16 or G-16a of the Standard. Given the present state of the art, hearing protectors which offer the greatest attenuation may not reliably be used when employees' exposure levels border on 100 dBA." This guideline alludes to the importance of engineering controls as a primary countermeasure against high noise levels. The OSHA (1990)[30] Technical Manual of the Directorate of Technical Support states: "OSHA experience and the published scientific literature indicate that laboratory-obtained real-ear attenuation data for hearing protectors are seldom achieved in the workplace." ... Under "Field Attenuation of Hearing Protection": "When analyzing the attenuation a personal hearing protector may afford a noise-exposed employee in an actual work environment, the hearing protector shall be evaluated as follows: ... (2) To adjust for the lack of attainment of the laboratory-based noise reduction calculated according to Appendix B (laboratory ratings) estimating techniques, apply a safety factor of 50%; that is, divide the calculated laboratory-based attenuation by two. (3) For dual protection (i.e., earplugs and muffs) add 5 dB to the NRR of the higher-rated protector." For case 2, the derating factor may appear to be a reasonable strategy; however, these authors and others have argued that a constant derating factor is not appropriate because certain protectors (for example, earmuffs) are easier to fit properly than others (for example, user-formed earplugs), and thus the differences between laboratory and actual in-workplace performance will not be the same for all devices. In perusing Figure 93.8, this becomes quite apparent in that the laboratory NRRs for earplugs overestimate the field NRRs by an average of about 75%, while the laboratory NRRs for earmuffs overestimate the field NRRs by an average of only about 40%. These data would argue for the use of derating factors that differ by device type, not a constant derating such as the 50% OSHA recommendation. But in any case, the use of derating factors or other modifications of the NRR to adjust it for field applications is tenuous at best and should not be expected of the end user. The best solution is to establish a testing standard (and attenuation rating therefrom) which accurately predicts workplace protection achieved by HPDs, and this is the ANSI standard work described in Royster et al. (1996).[28]

Training Program and Access to Information and Materials

An oft-overlooked, but essential component of an industrial HCP is an annual training program for all workers included in the HCP. The required training elements to be covered are: (1) the effects of noise on hearing; (2) purpose, selection, and use of HPDs; and (3) purpose and procedures of audiometric testing. It is essential to the success of an HCP that workers become acutely aware of the need for hearing conservation, understand and believe in the merits of the program, and develop a commitment to and the motivation for protecting their hearing. Employers must make the OSHA regulations available to affected employees and, upon request, make all training materials available to OSHA representatives.

Recordkeeping and Intra-Program Feedback

Accurate records must be kept of all noise exposure measurements, at least from the last two years, and audiometric test results for the duration of the worker's employment. It is important, but not required by OSHA, that noise and audiometric data be used as feedback for improving the program as shown in the feedback loops of Figure 93.7. Because the primary goal of the HCP is to prevent NIHL for employees, the program's effectiveness can be evaluated via *audiometric database analysis* (ADBA) for employees as a group, as opposed to individuals. By using population statistics from and inferential analysis of the database for exposed employees, problems can be identified early and corrective actions taken before significant threshold shifts appear in a number of individuals (Royster and Royster, 1986).[31] ADBA, however, is not a substitute for annual individual audiogram review and comparison against baseline. As discussed previously, this type of intra-worker analysis is essential for identifying threshold shifts and implementing preventative measures that are specific to the individual worker and job environment. Also, discussion of individual audiogram data with employees can aid in motivating them to exercise care in their daily hearing conservation practices, and audiometric feedback, sometimes posted anonymously by code number but including each individual's HPD use information, has been experimentally demonstrated to be an effective means of establishing higher HPD usage rates (Zohar, Cohen, and Azar, 1980).[32]

Noise exposure records may be used as feedback to identify machines that need maintenance attention, to assist in the relocation of noisy equipment during plant layout efforts, to provide information for future equipment procurement decisions, and to target plant areas that are in need of noise control intervention. Some employers plot noise levels on a "contour map," delineating floor areas by their dB levels. When monitoring indicates that the noise level in a particular contour has changed, it is taken as a sign that the machinery and/or work process has changed in the area and that further evaluation may be needed.

Engineering Noise Control

While OSHA does not stipulate the level of effort to be devoted to engineering noise controls or the types of controls which should be applied, the physical reduction of the noise energy, either at its source, in its path, or at the worker, should be a major focus of noise management programs. Hearing protection and/or administrative controls should not supplant noise control engineering; the best solution, because it does not rely on employee behavior, is to reduce the noise itself, preferably at the emission source. However, in many cases where noise control is ineffective, infeasible (as on an airport taxi area), or prohibitively expensive, HPDs become the primary countermeasure.

There are many techniques used in noise control, and the specific approach must be tailored to the noise problem at hand. A noise control engineer is typically consulted to assist in the measurements, usually taken from spectrum analyzers, and in the selection of control strategies. Example noise control strategies include: (1) *isolation of the source* via relocation, enclosure, or vibration-damping using metal or air springs (below about 30 Hz) or elastomer (above 30 Hz) supports; (2) *reduction at the source or in the path* using mufflers or silencers on exhausts, reducing cutting, fan, or impact speeds, dynamically balancing rotating components, reducing fluid flow speeds and turbulence, absorptive foam or fiberglass on reflective surfaces to reduce reverberation, shields to reflect and redirect noise (especially high frequencies), and lining or wrapping of pipes and ducts; (3) *replacement or alteration of machinery*, examples include belt drives as opposed to noisier gears, electrical rather than pneumatic tools, and shifting frequency outputs such as by using centrifugal fans (low frequencies) rather than propeller or axial fans (high frequencies), keeping in mind that low frequencies propagate further than high frequencies, but high frequencies are more hazardous to hearing; and (4) *application of quieter materials*, such as rubber liners in parts bins, conveyors, and vibrators, resilient hammer faces and bumpers on materials handling equipment, nylon slides or rubber tires rather than metal rollers, and fiber rather than metal gears. Further discussion of these techniques may be found in Bruce and Toothman (1986),[33] and an illustration of implementation possibilities in an industrial plant appears in Figure 93.9. A final approach which has just recently become available to industry is *active noise reduction* (ANR) in which an electronic system is used to transduce an offensive noise in a sound field and then process and reintroduce the noise into the same sound field such that it is exactly 180 degrees out-of-phase with, but of equal amplitude to the original noise (Casali and Berger, 1996).[22] The superposition of the out-of-phase "anti-noise" with the original noise causes physical cancellation of the noise in a target zone of the workplace. For highly repetitive, predictable noises, synthesis of the anti-noise, as opposed to transduction and reintroduction, may also be used. At frequencies below about 1000 Hz, the ANR technique is most effective, which is fortuitous since the passive noise control materials to combat low frequency noise, such as absorptive liners and barriers, are typically heavy, bulky, and expensive. At higher frequencies and their corresponding shorter wavelengths, the processing and phase relationships become more difficult and cancellation is less successful, although the technology is rapidly improving.

In designing and implementing noise control hardware, it is important that ergonomics be taken into account. For instance, in a sound-treated booth to house an operator, the ventilation system, lighting, visibility outward to the surrounding work area, and other considerations relating to operator comfort and performance must be considered. With regard to noise-isolating machine enclosures, access provisions should be designed so as to not compromise the operator/machine interface. In this regard, it is important that production and maintenance needs be met. If noise control hardware creates difficulties for the operators in carrying out their jobs, they may tend to modify or remove it, rendering it ineffective.

Personnel

As shown in Figure 93.7, multiple individuals play important roles in an industrial HCP, and the program should filter down from management personnel who must demonstrably support it. The key individual is the HCP coordinator (at the lower left in Figure 93.7), typically a permanent employee of the company but sometimes an outside consultant, who serves as the responsible individual and overseer for the program as well as its internal "champion." This individual, if properly qualified, may also be responsible for implementation of certain aspects of the program, including noise monitoring, audiometry on employees, selection and purchase of hearing protection devices (HPDs), and other functions. The HCP coordinator often heads a hearing conservation committee with representatives from labor, management, plant engineering, and safety. The coordinator also serves as a link between management and the

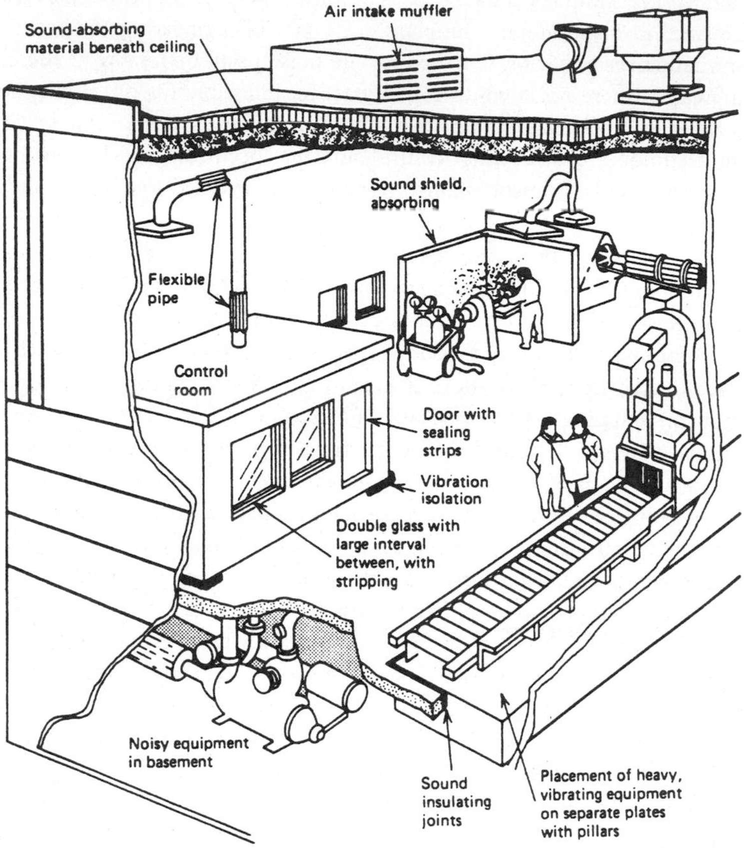

FIGURE 93.9 Examples of noise control implementation in an industrial plant. (Adapted from OSHA (1980). *Noise Control, A Guide for Workers and Employers.* Occupational Safety and Health Administration Report No. 3048. Washington, D.C.: U.S. Department of Labor.)

workforce, and generally participates in management decisions which impact the noise environment or the HCP itself. For instance, one means of noise control is to establish a procurement policy which limits the decibel output of new equipment to a prespecified level; the HCP coordinator should be involved in such purchase decisions and in ensuring that criteria for noise emissions are met.

An audiologist, nurse, otolaryngologist, or other physician may conduct audiometric tests on employees and maintain a database for the test records. Industrial audiometry for OSHA purposes may also be conducted by a technician who is certified by the Council of Accreditation in Occupational Hearing Conservation (CAOHC), but this individual must ultimately be responsible to a professional audiologist or physician. The person who performs the audiometric test function may also be involved in helping the worker select an appropriate HPD (with input from the noise exposure records) and in educating and training the worker about the hazards of noise and the proper use of protection.

The work supervisor or foreman may also provide input to the HCP. For instance, in cases where workers are rotated on and off noisy machines to limit their exposures (a type of administrative countermeasure), the supervisor should be consulted to determine feasible rotation schemes. Furthermore, it is imperative that the supervisor exhibit good hearing conservation practice him/herself and provide specific feedback to the HCP coordinator about occurrences which impact the success of the HCP, such as a machine which has become noisy due to lack of maintenance or a worker who is uncomfortable with his/her assigned HPD and therefore repeatedly takes it on and off. Because of his/her close relationship and proximity to production employees, the foreman or supervisor can serve as a key individual in helping to motivate the workers to exercise good hearing conservation practice, both by serving as a role model and an information resource.

Some large companies have an acoustical engineer on staff while others may need to hire such an individual when engineering noise control becomes necessary. The acoustical engineer can perform in-depth spectral analyses of specific noise sources and design noise control solutions. Furthermore, acoustical engineers can be helpful in the overall design of the HCP, in that the specialized knowledge they possess will be useful in considering tradeoffs in dollar cost-to-dB reduction benefits when comparing various countermeasure strategies.

If the company has a safety engineer on staff, this individual should serve on the hearing conservation committee and participate in noise-related decisions that impact safety in other ways. For instance, if noise levels increase in an area where acoustic alarms signal the approach of an automated material transport vehicle, the safety engineer will need to work to increase the alarm's output to maintain detectability and/or use an alternate warning system, such as a flashing strobe, to maintain vehicle conspicuity. The safety engineer may also work with the HCP coordinator in selecting appropriate hearing protection for employees who must maintain communications in hazardous areas. In some small companies, the safety engineer may, in fact, have responsibility for the HCP itself.

Involvement and commitment of the proper hearing conservation and safety personnel, support of company management, and a trained and motivated workforce are all important to the success of a properly designed and implemented industrial hearing conservation program. Such a program can markedly reduce noise-induced distractions and interference on the job, and above all, prevent the tragic and irrecoverable occurrence of occupational hearing loss in workers.

References

1. Stevens, S. S. (1936). A scale for the measurement of a psychological magnitude: Loudness. *Psychological Review,* 43, 405-416.
2. Stevens, S. S. (1972). Perceived level of noise by Mark VII and decibels (E). *Journal of the Acoustical Society of America,* 51(2, pt. 2), 575-601.
3. Zwicker, E. (1960). En verfahren zur berechnung der lautstarke. *Acustica,* 10, 304-308.
4. Kryter, K. D. (1994). *The Handbook of Hearing and the Effects of Noise.* New York: Academic Press.
5. Sanders, M. S. and McCormick, E. J. (1993). *Human Factors in Engineering and Design,* 7th Edition, New York: McGraw-Hill.
6. Poulton, E. (1978). A new look at the effects of noise: A rejoinder. *Psychological Bulletin,* 85, 1068-1079.
7. Sorkin, R. D. (1987). Design of auditory and tactile displays, in Salvendy, G. (Ed.), *Handbook of Human Factors,* (pp. 549-576). New York: McGraw-Hill.
8. ANSI S3.5-1969 (R1986). *Methods for the Calculation of the Articulation Index.* New York: American National Standards Institute, Inc.
9. ISO 7731-1986 (E) (1986). *Danger Signals for Workplaces- Auditory Danger Signals.* Geneva, Switzerland: International Organization for Standardization.
10. Acton, W. I. (1970). Speech intelligibility in a background noise and noise-induced hearing loss. *Ergonomics,* 13(5), 546-554.
11. Pollack, I. (1958). Speech intelligibility at high noise levels: Effects of short-term exposure. *Journal of the Acoustical Society of America,* 30, 282-285.
12. Morrison, H. B. and Casali, J. G. (1994). Intelligibility of synthesized voice messages in commercial truck cab noise for normal-hearing and hearing-impaired listeners. *Proceedings of the 1994 Human Factors and Ergonomics Society 38th Annual Meeting,* Nashville, Tennessee, October 24-28, 801-805.
13. EPA (1981). *Noise in America: The Extent of the Noise Problem.* Environmental Protection Agency Report No. 550/9-81-101. Washington, D.C.: EPA.
14. National Institutes of Health (NIH) Consensus Development Panel. (1990). Noise and hearing loss. *Journal of the American Medical Association,* 263(23), 3185-3190.

15. Melnick, W. (1991). Hearing loss from noise exposure, in C. M. Harris (Ed.), *Handbook of Acoustical Measurements and Noise Control.* (pp. 18.1-18.19). New York: McGraw-Hill.
16. Taylor, W., Pearson, J., Mair, A., and Burns, W. (1964). Study of noise and hearing in jute weavers. *Journal of the Acoustical Society of America,* 38, 113-120.
17. ANSI S1.4-1983 (R1994). *Specification for Sound Level Meters.* New York: American National Standards Institute, Inc.
18. Peterson, A. P. G. (1979). Noise measurements: Instruments, in C. M. Harris (Ed.) *Handbook of Noise Control* (pp. 5-1–5-19). New York: McGraw-Hill.
19. OSHA (1983). 29CFR1910.95. *Occupational Noise Exposure; Hearing Conservation Amendment; Final Rule.* Occupational Safety and Health Administration. *Code of Federal Regulations,* Title 29, Chapter XVII, Part 1910, Subpart G, 48 FR 9776-9785. Washington, D.C.: Federal Register.
20. Harris, C. M. (1991). *Handbook of Acoustical Measurements and Noise Control.* New York: McGraw-Hill.
21. Berger, E. H. and Casali, J. G. (1997). Hearing protection devices, in M. J. Crocker (Ed.) *Encyclopedia of Acoustics.* New York: Wiley.
22. Casali, J. G. and Berger, E. H. (1996). Technology advancements in hearing protection: Active noise reduction, frequency/amplitude-sensitivity, and uniform attenuation. *American Industrial Hygiene Association Journal,* 57, 175-185.
23. Park, M. Y. and Casali, J. G. (1991). A controlled investigation of in-field attenuation performance of selected insert, earmuff, and canal cap hearing protectors. *Human Factors,* 33(6), 693-714.
24. EPA (1979). 40CFR211, Noise labeling requirements for hearing protectors. Environmental Protection Agency, *Federal Register,* 44(190), 56130-56147.
25. NIOSH (1975). *List of Personal Hearing Protectors and Attenuation Data.* National Institute for Occupational Safety and Health-HEW Publication No. 76-120, 21-37. Washington, D.C.
26. ANSI S3.19-1974 (1974). *Method for the Measurement of Real-Ear Protection of Hearing Protectors and Physical Attenuation of Earmuffs.* New York: American National Standards Institute, Inc.
27. Berger, E. H., Franks, J. R., and Lindgren, F. (1996). International review of field studies of hearing protector attenuation, in A. Axelsson, H. Borchgrevink, R. P. Hamernik, P. Hellstrom, D. Henderson, and R. J. Salvi (Eds.), *Scientific Basis of Noise-Induced Hearing Loss.* (pp. 361-377), New York: Thieme Medical Publishers, Inc.
28. Royster, J. D., Berger, E. H., Merry, C. J., Nixon, C. W, Franks, J. R., Behar, A., Casali, J. G., Dixon-Ernst, C., Kieper, R. W., Mozo, B. T., Ohlin, D., and Royster, L. H. (1996). Development of a new standard laboratory protocol for estimating the field attenuation of hearing protection devices. Part I. Research of Working Group 11, Accredited Standards Committee S12, Noise. *Journal of the Acoustical Society of America,* 99(3), 1506-1526.
29. OSHA (1989). OSHA Instruction CPL 2.45B, June 15. *Field Operations Manual.* (pp. IV-33 — IV-35), Rockville, MD: Government Institutes, Inc.
30. OSHA (1990). OSHA Instruction CPL 2-2.20B, February 5. *Field Technical Manual.* (pp. 4-1 — 4-15), Rockville, MD: Government Institutes, Inc.
31. Royster, J. D. and Royster, L. H. (1986). Audiometric database analysis, in E. H. Berger, W. D. Ward, J. C. Morrill, and L. H. Royster (Eds.), *Noise and Hearing Conservation Manual* (pp. 293-317). Akron, OH: American Industrial Hygiene Association.
32. Zohar, D., Cohen, A., and Azar, N. (1980). Promoting increased use of ear protectors in noise through information feedback. *Human Factors,* 22(1), 69-79.
33. Bruce, R. D. and Toothman, E. H. (1986). Engineering controls, in E. H. Berger, W. D. Ward, J. C. Morrill, and L. H. Royster (Eds.), *Noise and Hearing Conservation Manual* (pp. 417-521). Akron, OH: American Industrial Hygiene Association.
34. Earshen, J. J. (1986). Sound measurement: Instrumentation and noise descriptors, in E. H. Berger, W. D. Ward, J. C. Morrill, and L. H. Royster (Eds.), *Noise and Hearing Conservation Manual* (pp. 38-95). Akron, OH: American Industrial Hygiene Association.
35. OSHA (1980). *Noise Control, A Guide for Workers and Employers.* Occupational Safety and Health Administration Report No. 3048. Washington, D.C.: U.S. Department of Labor.

94

Vibrometry

Donald E. Wasserman
D.E. Wasserman, Inc.

David G. Wilder
University of Iowa

94.1 Introduction

There are some 8 million workers[1] in the U.S. exposed to occupational whole-body vibration (WBV) or hand-arm vibration (HAV) with resulting severe medical consequences of WBV or HAV exposures (see text Chapter: Occupational Vibration and Cumulative Trauma Disorders). The ability to measure, quantify, and evaluate the vibration impinging on the human body and relating these results to the disease processes it produces is essential to understanding both dose–response relationships and methods for controlling human vibration exposure. The purpose of this chapter is thus threefold: (1) To provide an introduction to the occupational vibration measurement process; (2) to provide a basic understanding of the occupational WBV and HAV health and safety standards/guides currently in use in the U.S.; and (3) to demonstrate the interrelationships between these measurements and their respective WBV and HAV standards/guides.

94.2 Vibration Basics[2,3]

Vibration is a description of motion. As such, this motion is characterized by its direction *and* a corresponding magnitude; thus, by definition vibration is a *vector quantity.* A total of six vectors are needed to describe vibrating motion measured at any one point; three of these vectors portray "linear motion" and are situated mutually perpendicular to each other; the remaining three vectors portray the rotational motion around each of these linear vectors and are called pitch, yaw, and roll. Currently pitch, yaw, and roll are not measured; only the three perpendicular linear vectors are measured and evaluated in human vibration work. Figure 94.1 shows the mutually perpendicular measure coordinate system used for WBV measurements. Similarly, Figure 94.2 shows the two coordinate systems used to measure HAV. We define the directions of motion as follows: The "Z axis" motion is in the long (head-to-toe) WBV direction, and for HAV measurements the motion is a direction parallel to the hand/arm long bones. Similarly, the "Y axis" motion is in the direction across the shoulders for WBV measurements, and for HAV measurements the motion is across the knuckles of the hand. Finally, the "X axis" motion is in the front-to-back direction (through the sternum) for WBV, and for HAV measurements the motion is through the palm of the hand.

Having defined the directions of motion, the vibration magnitude or intensity parameter(s) must be specified. We can choose between three mathematically interrelated quantities: displacement, velocity, or acceleration. Displacement is merely the distance moved away from some reference position. Velocity (or speed) is the time-rate-of-change of displacement. *Acceleration* is the time-rate-of-change of velocity.

0-8493-2641-9/99/$0.00+$.50

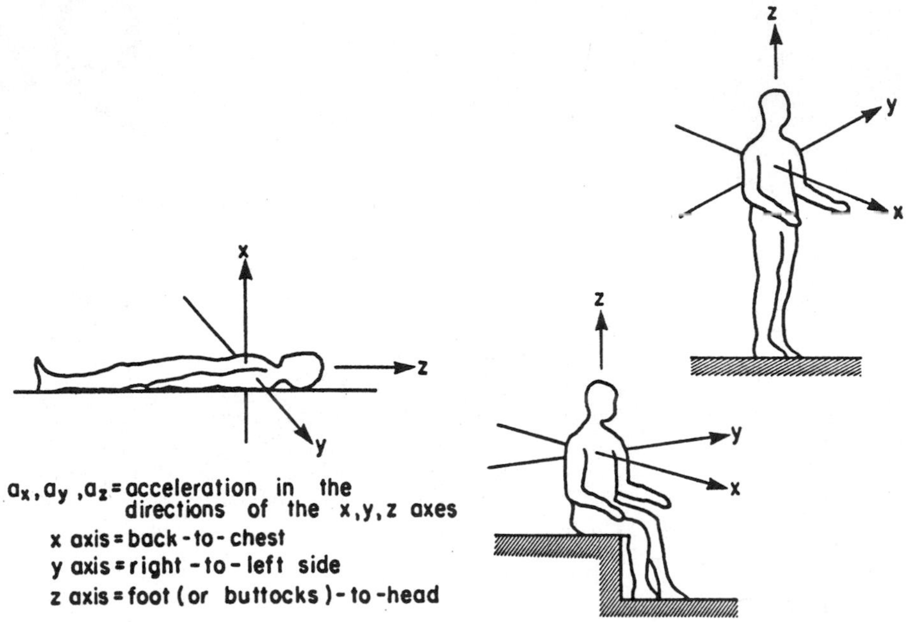

FIGURE 94.1 Whole-body vibration measurements coordinate system. (ANSI S3.18, ISO 2631, ACGIH-TLV (WBV), EU)

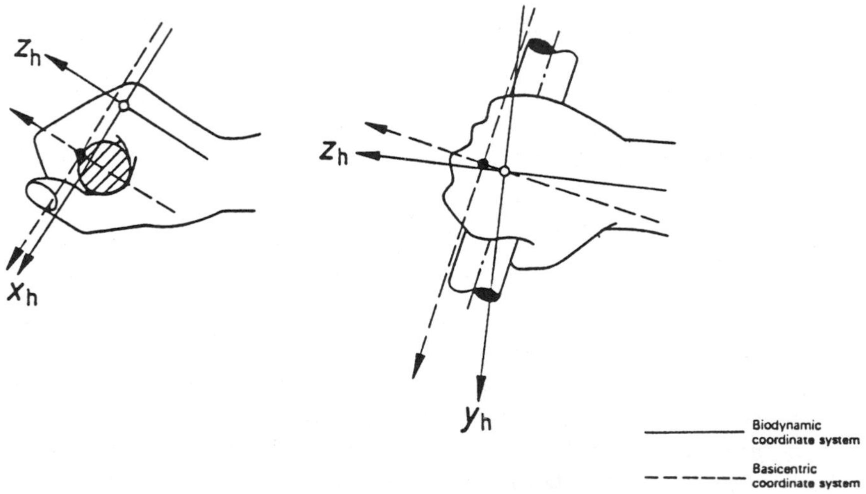

FIGURE 94.2 Hand-arm vibration measurements coordinate systems. (ANSI S3.34, ACGIH-TLV (HAV), EU, NIOSH #89-106)

Acceleration is usually the magnitude/intensity parameter of choice for several reasons which include ease of measurement and the belief that acceleration is both a hard and soft tissue stressor. Acceleration is expressed in units of meters/sec./sec. or in terms of gravitational *g* units, where $1g = 9.81$ meters/sec./sec. The "peak" acceleration or maximum values are not usually evaluated, rather an average acceleration parameter called *root-mean-squared* or *rms* acceleration is measured and evaluated and is relatable directly to the human vibration standards. In the rms process the measured values of acceleration are squared and subsequently averaged to get its mean value. Finally, the square root is determined resulting in an acceleration value proportional to the vibration signal's energy content (see equation 1).

$$a_{rms} = \sqrt{\frac{1}{T}\int_0^T a^2(t)dt} \tag{1}$$

where a = acceleration (in *g*'s or m/sec²); t = time (in seconds).

Vibration motion can repeat itself. This is called periodic motion. Motion need not repeat itself; it can be random or nonperiodic. The most basic form of motion is sinusoidal, or a pure tone if it were audible, which usually repeats itself; the rms value of a sinusoid is about 70% of its peak or maximum value. If a sinusoid completes its cycle in one second, before it begins to repeat itself, its vibration *frequency* is 1 Hertz (Hz); 10 Hertz simply means that 10 complete cycles have occurred in one second; two kilohertz (2kHz) means that two thousand complete cycles have occurred in one second, etc. For WBV we are interested in a vibration frequency range (or bandwidth) of 1 Hz to 80 Hz; for HAV the bandwidth of interest is from about 6 Hz to 1400 Hz or 1.4kHz; sometimes even higher extending up to 5000 Hz or 5kHz. Since most motion appearing in the industrial environment is a compound mixture of many vibration frequencies at various acceleration levels, it is necessary to mathematically sort these frequencies into their individual sinusoidal frequencies and their corresponding magnitudes (see equation 2). This computer process is called *Fourier Spectrum Analysis* and is required by most human vibration standards before they can be used.

$$\begin{aligned} F(t) = a_0 + a_1 \sin wt + a_2 \sin 2\,wt + a_3 \sin 3\,wt + \ldots + a_n \sin(n)\,wt + \\ b_1 \cos wt + b_2 \cos 2\,wt + b_3 \cos 3\,wt + \ldots + b_n \sin(n)\,wt \end{aligned} \tag{2}$$

where a and b = amplitude values of each sinusoid at specific frequencies composing the spectrum; a_0 = dc term or zero Hertz value.

Specialized computers called Fast Fourier Transform (FFT) analyzers or real time analyzers (RTA) are used to transform the vibration mix into its discrete frequencies.[4] Each such frequency is graphically displayed as a series of vertical lines or spectra; the position of each line identifies its vibration frequency in Hz and thus its place in the spectrum; the height of each line is a measure of its individual vibration acceleration intensity in *g* units or meters/sec./sec. The entire spectrum is the sum of all these lines. Vibrating tool spectra are quite unique, depending on the tool. Vehicle spectra from trucks, buses, trains, etc., are also unique.

The final concept to be discussed is called *resonance* (or natural frequency) which is an unwelcome situation where the conditions for transferring vibration from its source (i.e., tools, vehicles, etc.) to the human receiver are optimal. Thus, a very small magnitude of vibration impinging on a human or a structure (such as a bridge) causes an uncontrollably amplified response by the human or structure. This is the reason why bridges collapse if soldiers march in cadence across them. Unfortunately, we humans have resonances too, namely, in WBV 4 to 8 Hz for Z axis vertical vibration, 1 to 2 Hz in both the X and Y axes. Spinal resonance is 4.5 to 5.5 Hz.[5] The hand-arm system seems to resonate in the 150 to 250 Hz range.[3] In general the larger the mass or weight of a structure, the lower the resonant frequency. Equation II is the resonance equation for a simple single-degree of freedom system consisting of motion in one direction consisting of a mass, spring, and damping element.

$$W = \sqrt{\frac{k}{m}} \tag{3}$$

where W = $2\pi f$; k = spring constant; m = mass.

Resonance thus represents the Achilles heel of human response to vibration. As is the auditory system to sound, human response to vibration is therefore frequency dependent and nonlinear because at resonance the impinging vibration finds its easiest pathway to the person; at other vibration frequencies, the vibration pathway is not as easy and thus it requires more acceleration at a nonresonant frequency to produce the same level of human response.

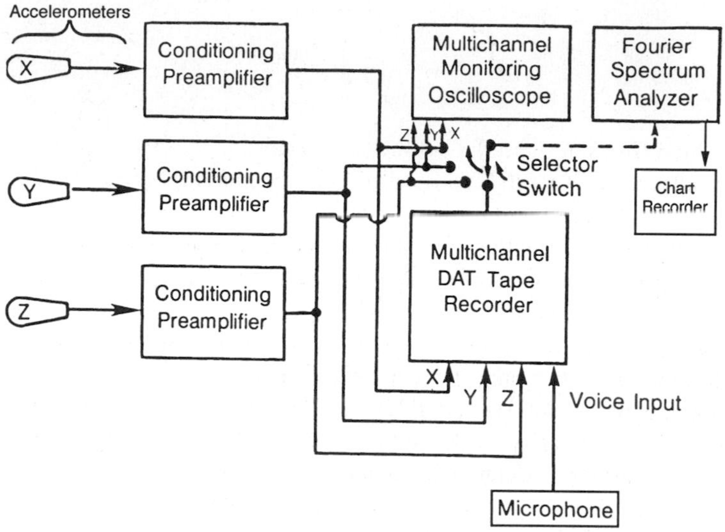

FIGURE 94.3 Basic triaxial vibration acceleration measuring system.

94.3 Vibration Measurements Basics[2-4]

The major reason for performing occupational vibration measurements is to evaluate the vibration impinging on persons. Evaluations are performed using the various WBV and HAV standards/guides. These standards/guides are the critical link between the various health and safety effects of WBV or HAV and the vibration hazard levels experienced by workers. It is important to note that there are many esoteric types of vibration measurements (i.e., mechanical: impedance, mobility, stiffness, compliance, etc.) and other methods of data analysis, such as modal analysis, but intentionally in this chapter we briefly describe only performing acceleration measurements as required by the applicable health and safety standards/guides in order to use them. Finally, note that since displacement, velocity, and acceleration are all mathematically linked, then from a measurement of acceleration, the velocity function can be derived by electronic integration; repeating the integration next on the velocity function yields the displacement function. Thus, if desired, an acceleration measurement can yield additional data.

Figure 94.3 shows a basic vibration acceleration measurement setup. Since we must *simultaneously* but separately measure in all three X,Y,Z axes acceleration data from a vibrating tool, for example, or from a driver's truck seat, three separate data channels are needed. Three perpendicularly mounted lightweight accelerometers are used to measure each axis acceleration, followed by three appropriate preamplifiers to amplify and electronically condition the tiny millivolt signals coming from each accelerometer. The outputs of each of these three X,Y,Z preamplifiers are then individually recorded and stored on a multitrack tape system known as a *digital audio tape* (DAT) for later Fourier spectrum analysis. It is also desirable to have: (1) a microphone/voice track on the DAT to note the chronology of events being recorded, and (2) an oscilloscope or similar device monitoring the X,Y,Z acceleration axes for possible signal overload conditions leading to distortion of the recorded signal(s) and resulting in erroneous data processing results. After the three channel or "triaxial" data have been measured, stored, and recorded, then a Fourier spectrum analysis can be performed separately on each data channel. Each spectrum is next separately evaluated using the appropriate standard(s). A key element in these measurements is the triaxial accelerometer.

Accelerometers are devices which convert mechanical motion into a corresponding electrical signal. Two distinctly different devices are used for occupational vibration measurements. For WBV measurements piezoresistive accelerometers are used; for HAV measurements piezoelectric or crystal accelerometers are used.

The former device works on the principle of an electronic four arm electronic balanced bridge; P or N semiconductors form each of the bridge arms. All of these arms are bonded to one end of a tiny metal beam. The other end of this beam is bonded to a tiny metal mass or weight to which the force of vibration is applied. With no vibration present, the bridge is "balanced," yielding zero output voltage. When vibration is applied to this tiny mass/beam combination, we obtain the acceleration of this mass against the beam because of the bending motion of the beam compressing some of the beam arms, thereby unbalancing the bridge, resulting in an electrical voltage proportional to acceleration; in effect we are calculating Newton's second law, F = ma, with a vibration force and a known mass bonded to the metal beam. The acceleration signal is very small (millivolts) and needs to be amplified using a so-called difference or differential amplifier which measures and amplifies the voltage or potential difference across the arms of the Wheatstone bridge. The amplifier's output voltage is next recorded by the DAT.

HAV measurements require a different type of accelerometer called a crystal, which is commonly found in nature. The crystal has a phenomenon called the piezoelectric effect; if a moving force is applied to the crystal, the crystal responds by generating a small electrical voltage across its face. The more intense the force, the larger the voltage generated. A crystal is a force-measuring device found in nature, not an accelerometer *per se*. If, however, a small weight or mass is bonded to the motion-sensitive surface of the crystal and the force of vibration is applied to this tiny mass, we have once again created an accelerometer, since, as before, it is the vibration force accelerating this mass against the crystal face which results in a corresponding voltage and charge proportional to acceleration. The acceleration signal is very small and needs to be amplified by a special charge amplifier whose output can then be recorded on a DAT recorder.

In either WBV or HAV measurements the accelerometers must be of very light weight (less than 15 grams) and small; if not, measurement errors called mass loading result, the rule is that the total weight of the triaxial accelerometers, mounting fixture, cables, etc., must collectively be less than 10% of the weight of the object whose vibration is to be measured — a tool handle, for example. In all cases great care must be taken to avoid cable entanglement and/or breakage; the shortest possible cable length and integrity must be maintained especially from accelerometers to preamplifiers to ensure high quality and low electrical noise signals. In the case of HAV measurements, some crystal accelerometers can be purchased with built-in charge amplifiers to avoid some of these problems.

WBV measurements are usually made using an instrumentated hard rubber disc about the size of a pie plate (see Figure 94.4). The disc is placed between the top of the driver's seat cushion and the buttocks. The center of the disc is hollow and contains three tiny accelerometers, mounted mutually perpendicular to each other to a small metal cube. The three piezoresistive accelerometer cables lead from the disc to preamplifiers and on to a DAT.

HAV measurements use three small lightweight crystal accelerometers mounted to a small metal cube, which in turn is usually welded to an inexpensive automotive hose clamp as shown in Figure 94.5. This hose clamp/accelerometer assembly is next clamped around the vibrating tool handle with the accelerometers placed very close to where the operator grasps the tool. Once again great care must be taken in arranging the three accelerometer cables such that the tool operator is free to perform the job safely during the vibration measurements.

To summarize, piezoresistive accelerometers are best suited to performing WBV measurements which are inherently very low frequency, low acceleration level measurements. Piezoelectric or crystal accelerometers are best suited to performing HAV measurements which require a wide bandwidth from a low of about 6 Hz to as high as 5,000 Hz; tool acceleration levels can be very high (several hundred *g*'s or more) and these devices must be able to measure these high *g* levels; these devices must be rugged too. In all cases care with the accelerometer cabling must be taken. It is certainly not advisable to drop any of these devices on the ground or else they can be severely damaged and/or lose their calibration. Generally the manufacturer will supply a calibration sheet with a newly purchased accelerometer. It is advisable to use a portable calibrator as an added calibration safety measure just in case the accelerometer has been unknowingly damaged.

Obtaining vibration measurements requires careful planning and first performing a walk-through tour of the worksite to be measured, or the course a vehicle takes if its vibration is to be measured. Many

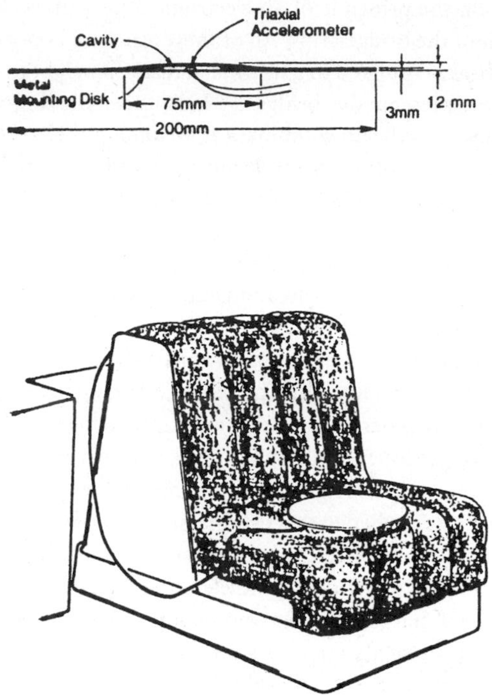

FIGURE 94.4 Whole-body vibration vehicle measurements using an instrumentated seat disc (Adapted from Wasserman, D. 1987. *Human Aspects of Occupational Vibration,* Elsevier Publishers, Amsterdam, The Netherlands.)

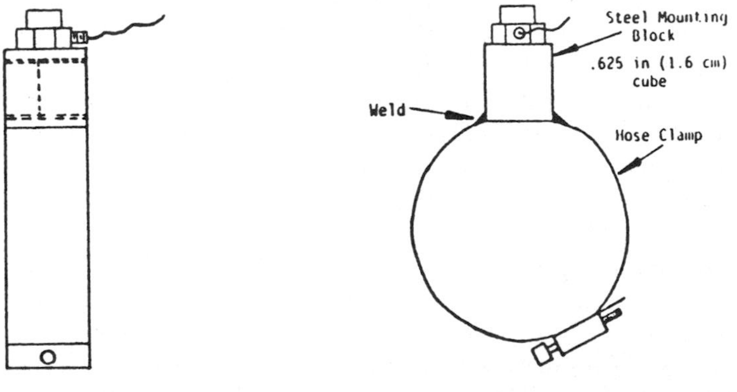

FIGURE 94.5 Hand-arm vibration tool measurements using an instrumentated automotive hose clamp (Adapted from Wasserman, D. 1987. *Human Aspects of Occupational Vibration,* Elsevier Publishers, Amsterdam, The Netherlands.)

times, first using a video camcorder is a good way to record the details of how workers function on the job. Using the camcorder in real time while vibration measurements are obtained is also very useful for recalling the chronology of events of the test day. Finally, the minimum test time that vibration data are gathered and recorded is usually specified by the standard(s) which will be used. For example, most HAV standards require that a minimum time of one minute of continuous triaxial vibration acceleration data be collected and tape recorded per tool tested. The differences in WBV work situations and the so-called duty cycle to a large extent determine the minimum vibration measurement time. For example, the length of a complete work cycle for a delivery truck, or the duty cycle of a large vibrating metal stamping machine in a plant are quite different and should be considered individually.

Finally, a word about handheld portable human vibration meters. We have briefly described measurement methods which will yield maximum usable information for the time and expense spent in gathering, recording, and analyzing vibration data and then applying it to the human vibration standards (to be discussed next). These methods provide: (1) a permanent tape recording of the vibration data; (2) a computer spectrum analysis which provides a graphical picture of the vibration frequencies which comprise the spectrum; (3) the interaction and comparison of these spectra with these standards; and (4) numerical results indicating the total rms acceleration of the spectra. However, if only a single number total rms acceleration value for each axis is required, then there are handheld instruments available from two commercial manufacturers at this writing. The problem is that some of these instruments measure only one acceleration axis and the testing is stopped. The one accelerometer is reoriented in another axis and the testing is resumed. This is repeated until all three axes are recorded. This is *not* desirable since vibration virtually always moves *simultaneously* in all three axes; thus data can be lost as the one accelerometer is reoriented over and over again. One of the available commercial instruments has in a single handheld meter the desirable three accelerometers for simultaneous measurements of either WBV or HAV and also has triple output jacks for DAT recording and later spectrum analysis of the data. Thus, the reader should be very careful in the selection of a handheld vibration meter. Further, be aware that with the advent of miniaturized, high-density/high-speed, surface-mounted electronics technology, many of the above-mentioned functions (i.e., triaxial accelerometer and signal conditioning, data collection, analog-to-digital conversion, data storage, initial unweighted spectrum analysis/display, radio frequency remote control of functions) can all be performed onsite using rugged, battery operated/stackable, miniature solid-state modules.

94.4 Occupational Vibration Standards/Guides*

There are four whole-body vibration standards/guides and four hand-arm vibration standards/guides now in use in the U.S.:

1. International Standards Organization (Geneva, Switzerland), ISO 2631: *Guide to the Evaluation of Human Exposure to Whole-Body Vibration,* 1972-85.
2. American National Standards Institute (New York, NY), ANSI S3.18: *Guide to the Evaluation of Human Exposure to Whole-Body Vibration,* 1979.
3. American Conference of Government Industrial Hygienists (Cincinnati, Ohio), *ACGIH-Threshold Limit Values for: Whole-Body Vibration,* 1995-96.
4. European Union (EU, Luxembourg, Belgium), #89/392, 91/368EEC: (Whole-Body) *Vibration Standard,* 1989-91.
5. American National Standards Institute (New York, NY), ANSI S3.34: *Guide for the Measurement and Evaluation of Human Exposure to Vibration Transmitted to the Hand,* 1986.
6. American Conference of Government Industrial Hygienists (Cincinnati, Ohio), *ACGIH-Threshold Limit Values for: Hand-Arm Vibration,* 1984.
7. National Institute for Occupational Safety and Health (Cincinnati, Ohio), NIOSH Document #89-106: *Criteria for a Recommended Standard for Hand-Arm Vibration,* 1989.
8. European Union (EU, Luxembourg, Belgium), #89/392, 91/368EEC: (Hand-Arm) *Vibration Standard,* 1989-91.

Whole-Body Vibration Standards/Guides Used in the U.S.

ISO 2631 is the oldest standard, initially introduced in 1972. There have been several revisions over the years, but the basic evaluation criteria remain the same. In 1979, ANSI S3.18 was introduced; this document is virtually identical to the revised 1978 version of ISO 2631. In 1989–91 the EU essentially

*Because of the differences and complexity of each occupational vibration standard/guide, the reader is encouraged to obtain, read, and understand the standard(s) which are to be used *before* collecting vibration data. Herein we can only discuss some of the major elements contained in these standards.

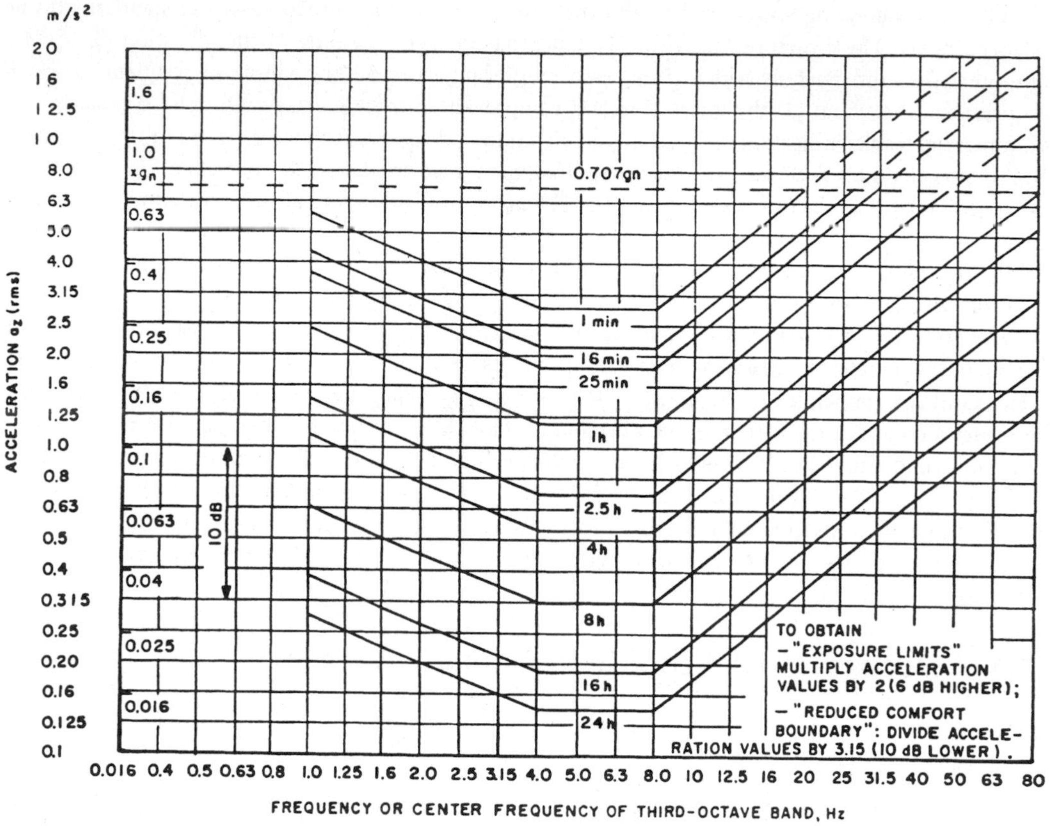

FIGURE 94.6 (FDP) Whole-body vibration curves for Z axis rms acceleration evaluation. (ANSI S3.18, ISO 2631, EU)

agreed with ISO 2631 and adopted a (weighted) vector sum triaxial acceleration level of 0.5 meters/sec./sec. as an action level for an 8 hr/day workplace WBV exposure level. In 1995–96, the ACGIH–TLV for WBV was introduced; it too uses the basic ISO 2631 curves, but the focus is mainly on occupation health and safety criteria and calculations, while ignoring the so-called comfort criteria. Since all of these standards use the same shape weighting curves, we begin there (refer to Figures 94.1, 94.6, and 94.7. All WBV acceleration measurements used in all these standards use the biodynamic coordinate system defined in Figure 94.1. Figure 94.6 is used to evaluate the Z axis (vertical) rms acceleration data. Figure 94.7 is used to separately evaluate: (1) the X axis rms acceleration data, and then (2) the Y axis rms acceleration data. In order to use these WBV standards, each axis of vibration acceleration data must be converted from the time domain in which it was collected to the frequency domain, which says a Fourier spectrum analysis must be performed separately for the X,Y,Z axes. Each spectrum is then formatted into 1/3 octave bands before it can be applied to the WBV standards. Once the foregoing has taken place the data are then compared to (i.e., overlayed) the "weighted family of curves" shown in Figures 94.6 and 94.7. The abscissa in each of these figures is 1/3 octave band "vibration frequency" from 1 to 80 Hz. The ordinate in each of these figures is "vibration intensity" in rms acceleration in both meters/sec./sec. or *g*'s, where 1 $g = 9.81$ m/sec./sec. Within each graph in Figures 94.6 and 94.7 are families of "weighted" parallel-time-dependent daily exposure curves which are called FDP or fatigue decreased proficiency curves. There are three levels of comparing these spectra to the ISO and ANSI standards, the acceleration data can be separately compared to: (1) the FDP curves as shown; (2) if we divide each of the FDP acceleration values by 10dB (3.15) we generate another family of similar curves called RC or reduced comfort; (3) if we double each value of the original FDP curves, we generate another family of similar curves called EL or exposure limits. These standards tell us that the RC curves should be used

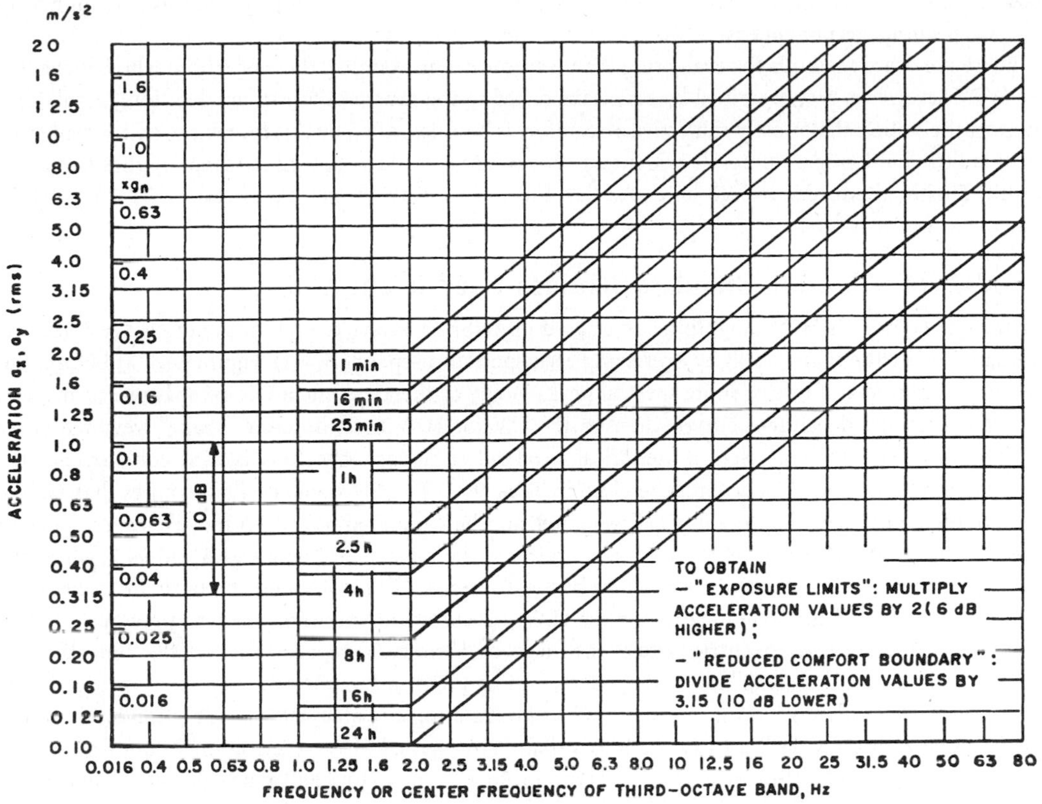

FIGURE 94.7 (FDP) Whole-body vibration curves separately used for X, Y axes rms acceleration evaluations. (ANSI S3.18, ISO 2631, EU)

for WBV *comfort* criteria, such as in a vehicle ride situation; the FDP curves are operator fatigue level curves where *safety* may well be the issue; and finally the EL curves are concerned with WBV *health effects*. The U-shape of the family of the Z axis curves in Figure 94.6 emphasizes that resonance occurs in the 4 to 8 Hz band as shown by the trough of the curves for each of the daily exposure times. Higher daily rms acceleration levels are allowed at frequencies lower than 4 Hz and greater than 8 Hz since at resonance smaller input acceleration levels produce larger responses than would occur at other frequencies. Similarly, the format for Figure 94.7 shows elbow-shaped curves where the resonant frequency range in either the X or Y axes occurs at 1 to 2 Hz; higher acceleration levels are allowed for frequencies greater than 2 Hz. In actual use, vibration spectra are overlayed on, say, the FDP curves separately for the Z axis, Y axis, and X axis. If, for example, all of the Z spectra fall below the Figure 94.6 FDP curves, then the standard has not been exceeded for that axis; if one or more spectral peaks touches and/or exceeds an FDP weighted curve, then the standard has been exceeded for that daily exposure time. The most severe axis is defined by the highest spectral peak(s) which intersect the FDP curves. As a matter of practice, FDP curves are mostly used for *both* health and safety, and the EL curves are *not* used because researchers believe these curves are not protective enough.[2,6] Thus, the ACGIH-TLV for WBV uses only the FDP curves for health and safety, and they totally eliminate both the EL and RC curves. Further, the ACGIH-TLV for WBV then recommends using a weighted vector-sum calculation for all three axes to obtain a single number which is then compared to the 0.5 m/sec./sec. action level established by the EU. In all of the cited WBV standards if in any of the three axes, the vibration crest factor (defined as the peak acceleration divided by the rms acceleration in the same direction) is less than or equal to six, the standard can be used; values greater than six cause the standard to *underestimate* the true severity of the vibration hazard. This is particularly troublesome when a vehicle, for example, goes off road and traverses numerous very steep bumps at fast speeds.

There are other methods for evaluating WBV exposure. For example, there are those who believe that the WBV severity is best described by equations raised to the fourth power of acceleration;[7] actual data support the notion that mostly subjective discomfort is best described by this fourth power concept since there is little hard epidemiological evidence at this writing to show that this concept applies to worker health. Finally, the reader should be aware that there are various proposals to revise ISO 2631, which may occur in the future.

Hand-Arm Vibration Standards/Guides Used in the U.S.

Figure 94.8 shows the HAV weighting curves used in ANSI S3.34 where each of the three X,Y,Z axes are evaluated using the same graph by overlaying each spectrum separately over Figure 94.4; as before, this standard is exceeded if one or more spectral peaks in any of the axes touches or exceeds one or more of the exposure time dependent curves; the ACGIH-TLV for HAV uses the same "shape" weighted curve but requires that each axis yield a numerical weighted sum, each of which is next compared to the acceptable values of HAV daily exposure given in Table 94.1. The EU standard also requires that each of these numerical weighted values or their weighted sum be compared to the 2.5 meters/sec./sec. "action level." Notice that the format of Figure 94.8 is similar to the formats of Figures 94.6 and 94.7, where the abscissa is vibration frequency, in 1/3 octave bands, from 5.6 to 1250 Hz and the ordinate is vibration intensity in acceleration. All standards use the HAV measurement coordinate system previously shown in Figure 94.2 with the "basicentric system" the method of choice. Except for NIOSH #89-106, all of the above standards use this same "elbow shaped" weighting given in Figure 94.8.

The NIOSH standard is an interim standard without stating any acceptable acceleration level limit at any frequency; this standard asks for each axis that: (1) weighted HAV acceleration values from 5.6 to 1250 Hz be calculated, (2) unweighted acceleration values from 5.6 to 5000Hz be calculated, and (3) the weighted and unweighted be compared in view of the severity of the prevalence of the hand-arm vibration syndrome (HAVS) determined by using the tool(s) from whence these acceleration measurements were made. NIOSH has chosen to issue this interim standard because there is an anomaly in the other HAV standards, namely that the HAV weighting network shown in Figure 94.4 was originally developed using older vibrating tool types commonly found in the workplace. Over the last few years, some very high speed vibrating hand tools have been introduced, some of which have spectral peaks extending to 5000 Hz and above. Current standards end at 1250 Hz, and hence in a few instances the current standards would rule these very high-speed tools as acceptable when that may not be the case. NIOSH has chosen to keep their interim standard, until this anomaly is resolved.

The architects of the other HAV standards are carefully making adjustments to their standards with regard to the special case of these very high-speed tools as the vibration data and corresponding HAVS health data become available. We recommend the use of ANSI S3.34, ACGIH-TLV for HAV and the EU criteria since all provide good overall guidance, and in the special case where very high-speed tools are to be tested *caution* is advised when evaluating the triaxial vibration test data.

94.5 Summary

In this chapter the basic concepts of displacement, velocity, acceleration, resonance, coordinate systems for measurements, and spectrum analysis are presented and integrated for an understanding of their application to whole-body and hand-arm occupational vibration. Generic acceleration measurement systems and methods are discussed, which included piezoelectric and piezoresistive accelerometers, conditioning preamplifiers, data recording systems, and Fourier spectrum computers. The chapter concludes with a discussion of the various occupational whole-body and hand-arm vibration standards/guides currently used in the U.S. and their application to the evaluation of triaxial acceleration data from the workplace; because of the complexity of each standard, users are encouraged to obtain copies of standards which are to be used before obtaining vibration data.

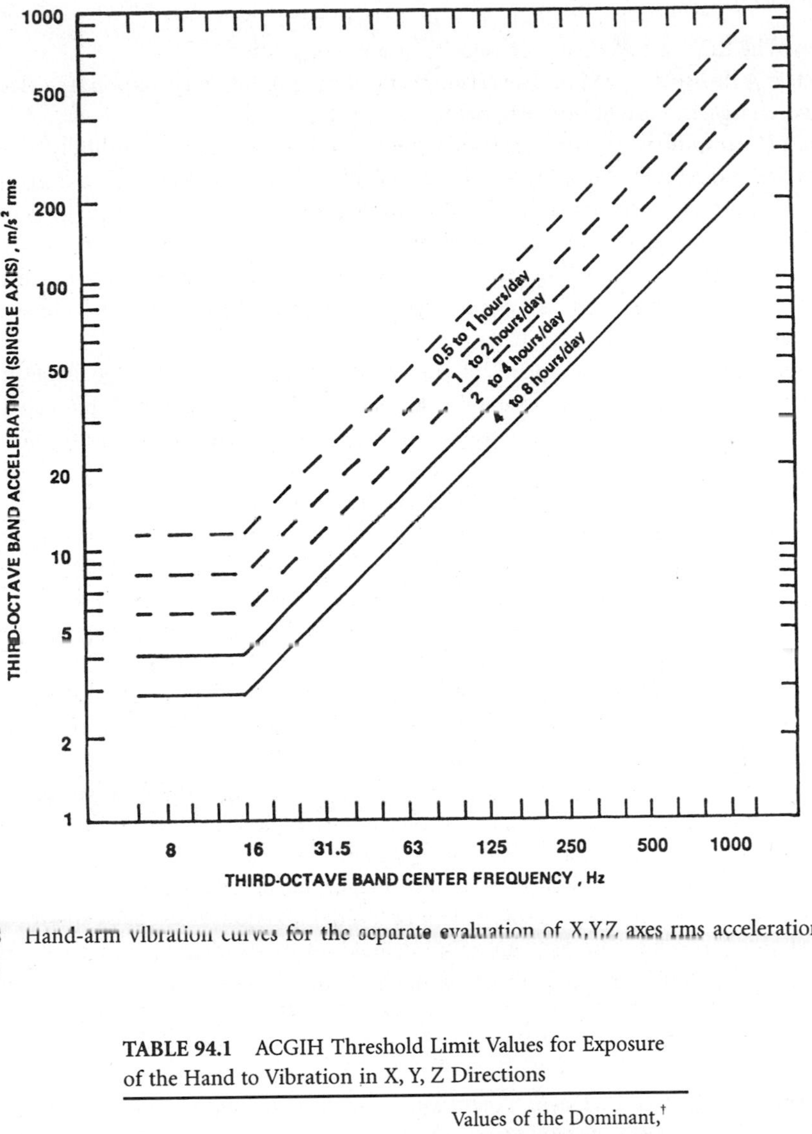

FIGURE 94.8 Hand-arm vibration curves for the separate evaluation of X,Y,Z axes rms accelerations (see text). (ANSI S3.34)

TABLE 94.1 ACGIH Threshold Limit Values for Exposure of the Hand to Vibration in X, Y, Z Directions

	Values of the Dominant,† Frequency-Weighted, rms, Component Acceleration Which Shall Not Be Exceeded a_K, ($a_{K_{eq}}$)	
Total Daily Exposure Duration*	m/s²	g
4 hours and less than 8	4	0.40
2 hours and less than 4	6	0.61
1 hour and less than 2	8	0.81
Less than 1 hour	12	1.22

Note: 1 g = 9.81 m/sec²

* The total time vibration enters the hand per day, whether continuously or intermittently.

† Usually one axis of vibration is dominant over the remaining two axis.

If one or more vibration axis exceeds the Total Daily Exposure then the TLV has been exceeded.

Courtesy of ACGIH

Defining Terms

Acceleration: The time rate of change of velocity of a moving object.

Accelerometer: A device designed to convert mechanical motion into a corresponding electrical analog voltage, charge, or current proportional to acceleration.

Conditioning Preamplifier: An electronic solid state amplifier designed to faithfully amplify, both in amplitude and frequency bandwidth, the minute electrical signals emanating from an accelerometer. Some preamplifiers are called "charge amplifiers" and convert the voltage generated across the face of a crystal (piezoelectric) accelerometer into corresponding charge, thereby allowing long cables to be used for measurements without loss of signal. Other preamplifiers are called "differential amplifiers," which act as an amplifying voltmeter when used with "piezoresistive" type accelerometers.

DAT or **Digital Audio Tape:** A new type of instrumentation tape system with large dynamic input range and wide frequency bandwidth, whereby an analog input signal is converted and stored on a cassette tape in digital format. The original signal so stored can be retrieved either in digital format or reconverted again into its original analog version.

Displacement: Movement traversed away from a reference position.

Fourier Spectrum Analysis: The analysis of vibration data by mathematically converting time domain information into its corresponding frequency domain; the underlying assumptions are that the data are linear and that time domain information can be dissected and represented as a mathematical series of elemental sines and cosines. Computers which perform this function are called Fast Fourier Transform (FFT) analyzers or Real Time Analyzers (RTA).

Resonance: The tendency of an object to (1) move in concert with an external vibrating source and (2) to internally amplify the impinging vibration from that source; resonance is the optimum energy transfer condition between the source and the receiver.

Vector Coordinate System: A mutually perpendicular set of vectors, originating at the same motion point, which define the vector motion of that point. Typically, there are three linear and three rotational vectors which comprise motion at a point.

Vector: A mathematical quantity defined by both its magnitude and direction.

Velocity: The time rate of change of displacement of a moving object. Also called speed.

Vibration: At any one point, vibration is motion defined by six vectors, three mutually perpendicular linear vectors and three rotational vectors moving around these linear vectors (pitch, yaw, roll).

References

1. Wasserman, D., Badger, D., Doyle, T., and Margolies, L. 1974. Industrial vibration — An overview. *J. Am. Soc. Safety Engrs.* 19(6):38-43.
2. Wasserman, D. 1987. *Human Aspects of Occupational Vibration,* Elsevier Publishers, Amsterdam, The Netherlands.
3. Pelmear, P. and Wasserman, D. 1998. *Hand-Arm Vibration: A Comprehensive Guide for Occupational Health Professionals,* Second edition, OEM Press, Beverly Farms, MA.
4. Thalheimer, E. 1996. *Practical Approach To Measurement and Evaluation of Exposure to Whole-Body Vibration in the Workplace.* Conference Proceedings: Seminars in Perinatology, International Conference on Pregnant Women in the Workplace Sound and Vibration Exposure, Univ. of Florida, Gainesville, FL.
5. Wilder, D., Wasserman, D., Pope, M., Pelmear, P., and Taylor, W. 1994. Chapter 4: Vibration, in *Physical and Biological Hazards of the Workplace,* Eds. P. Wald and G. Stave, p. 64-83. Van Nostrand Reinhold, New York, NY.
6. Bovenzi, M. and Betta, A. 1994. Low-back disorders in agricultural tractor drivers exposed to whole-body vibration and postural stress. *Appl. Ergonomics* 25(4): 231-241.
7. Griffin, M. 1990. *Handbook of Human Vibration.* Academic Press, London, U.K.

For Further Information

Some of the cited references are comprehensive and excellent sources of information. Reference 2 is principally a WBV and HAV measurements, evaluation, and control textbook. Reference 3 is principally a medical textbook on HAV with some measurements and control information; reference 5 is a basic and comprehensive book chapter for both WBV and HAV; reference 7 is a very complete tome on WBV and HAV, but definitely not for the beginner.

The WBV or HAV standards cited in this chapter can be obtained from the following:

TLVs from: American Conference of Government Industrial Hygienists, 1330 Kemper Meadow Drive, Cincinnati, Ohio 45240 (Telephone: 513-742-2020).

ISO and ANSI Standards from: Standards Secretariat, Acoustical Society of America, 120 Wall St., 32nd. Floor, New York, NY 10005

(Telephone 212-248-0373).

NIOSH HAV Standard #89-106 write to: NIOSH Publications Dept. Taft Labs. 4676 Columbia Parkway, Cincinnati, Ohio 45226.

European Union Standards write to: Commission des Communautes Europeennes, Direction generale emploi, relations industrielles et affaires sociates: Batiment Jean Monnet L-2920 Luxembourg, Belgium.

95

On the Behavioral Basis for Stress Exposure Limits: The Foundational Case of Thermal Stress

Peter A. Hancock
University of Minnesota

Ioannis Vasmatzidis
iNautix Technologies, Inc.

Abstract

In this work, we seek to expand the basis for understanding the limits for worker stress exposure. We wish to challenge the fundamental basis upon which many occupational exposures are founded. It is our contention that performance level should be considered as the principal criterion for exposure. Change in behavioral performance efficiency is the most sensitive reflection of human response to stress. Understanding these forms of response is a critical component in the development of indices of incipient stress effects and should expand and augment our traditional measurements of physiological functioning. Efficient and error-free performance is the principal criterion of contemporary work, especially in high-technology systems. Therefore, continuing exposure after work performance efficiency begins to fail, but before current physiological limits are reached, is inappropriate for both the safety and the productivity of the individual worker, their colleagues, and the systems within which they operate. Behavioral performance assessment should therefore be integrated with current physiological assessments to provide the primary exposure criterion. An examination of these performance thresholds for heat stress is presented together with

0-8493-2641-9/99/$0.00+$.50

its fundamental theoretical foundation. This foundation is applicable to all forms of stress. These performance limits are of growing and now primary importance for prescriptions to all forms of occupational exposure and are critical necessities for future statements concerning comprehensive protective safety standards.

95.1 Introduction

The origins of contemporary occupational stress exposure standards are founded upon a knowledge of the characteristics and limitations of different facets of the human physiological system. The study of such processes and responses has traditionally fallen within the realm of occupational health and safety. Together, the ergonomist, the safety specialist and those in occupational medicine have sought to enact standards which protect the individual worker against physiological damage. This perspective directly accords with the early nature of industrial work, where the principal demand was for physical effort. In essence, the currency of heavy industrial work was physiological effort. However, contemporary commercial work demands have changed from largely physical to largely cognitive requirements. Despite this fundamental change, current stress standards are still based upon assumptions principally concerning physiological detriment. Unfortunately, this is now insufficient as a basis for a full exposure standard for the larger spectrum of contemporary work.

We propose that health and safety exposure standards should be based primarily upon behavioral response, that being measures of task performance itself. Such measures provide a more sensitive metric of performer condition and are directly relevant to all aspects of human work, be they predominantly physical or cognitive in nature. Since performance productivity is the central measure of output across a wide spectrum of industrial work, the transfer of focus from physiological protection to performance assessment is liable to experience ready acceptance. In addition, setting standards with respect to performance begins to deal with the highly problematic issue of interactions between stresses. These are typically referred to in many standards documents; however, the lack of data and readily observable and consistent patterns of interactive effects means that, in reality, most efforts are directed to understanding one single source of occupational stress at a time. Obviously, this uni-dimensional approach does not include the combination of stresses actually experienced at work, and the use of behavioral outcomes can consequently begin to present a more complete picture of protection against the spectrum of stresses that do occur. In this chapter, we present a detailed articulation of this perspective in respect of one form of occupational exposure namely, heat stress. Our purpose is to show how this specific stress can represent a prototype for safety and health criteria for a wide range of occupational conditions. The general approach upon which this position is based is discussed, as is the global application of this concept to widespread forms of occupational stress. However, since the purpose of the present text is to provide researchers and practitioners alike with a complete picture of heat stress effects, we begin with an overview of the traditional physiological approaches to understanding thermal problems and protection against their adverse effects in occupational environments.

95.2 Physiological Measures and Standards

Despite the improvements in environmental controls, heat stress problems are not uncommon in contemporary workplaces. Thermal stress represents a threat to the human physiological system, and prolonged exposure can cause permanent physical damage and even death. As we show, experimental studies have shown that exposure to heat stress below the level of physiological threat have detrimental effects on cognitive performance and may result in operator error which leads to accidents that can affect more than the exposed worker alone. We (Hancock and Vasmatzidis, 1998) have proposed, and continue to argue here, that performance-based criteria should be developed to at least augment the existing physiology-based criteria. These additional criteria are presented later in this work. However, to understand the physiological effects of heat stress, the ergonomics practitioner has to have a basic knowledge of the thermoregulatory mechanisms of the human body, and the avenues by which humans exchange heat with the environment. Several engineering and administrative control methods can then be implemented to optimize this heat exchange and thus reduce or eliminate the adverse effects of heat stress.

Heat Exchange with the Environment

The avenues by which the human body exchanges heat with the environment are:

Conduction. Heat transmission by conduction occurs during direct contact between the skin and various objects (floor, stone, hot kitchen appliances, etc.). Heat flows from the skin to colder objects (sensation of cold), or vice versa (sensation of warmth). Heat exchange by conduction is in general of low magnitude due to the insulating properties of clothing. In cold workplaces, however, the hands, which are frequently the main source of a worker's interaction with the environment, can come into contact with tools and cold surfaces. In extreme conditions, conductive heat loss through the hands should be prevented or reduced by using gloves.

Convection. Is essentially similar to conduction, however, the medium of the exchange of heat is through the air. The magnitude of convection effects depends upon the temperature difference between the skin and the surrounding air, as well as the air velocity.

Evaporation. When conduction and especially convective heat loss are insufficient to remove body heat, a further avenue of heat loss, evaporation, begins to come into operation. Sweat evaporation is the diffusion of water vapor through the epidermis (the outer level of skin) and water vapor exhaled from the lungs during respiration. Sweat evaporation is the predominant heat loss mechanism. As sweat evaporates, it absorbs heat from the skin which generates a cooling sensation. One effect of continued sweat evaporation is fluid loss. If this loss continues without replacement for an extended time, some symptoms of heat illness begin to appear. The ergonomist should be careful to provide fluid replacement, and despite many claims to the contrary, the most easily obtainable and useful replacement is water. The ergonomist should be aware that voluntary drinking does not necessarily always replace all the fluid lost. Indeed, in very hot environments such replacement is physically difficult to do, requiring almost constant drinking. Therefore, much care should be exercised in supervising workers who are exposed on a daily basis to very high temperatures since chronic fluid loss over a prolonged period can result in some of the heat illnesses we discuss below.

Radiation. Radiation represents transmission of heat from hot to cold objects by means of electromagnetic waves of relatively long wavelength. Colder objects absorb the wavelength transmitted by hotter objects, and convert them into thermal energy. Radiation depends mainly on the temperature difference between the skin surface and the surrounding surfaces. In general terms, the most potent source of radiant heat exposure for most workers is the sun. Prolonged exposure to its harmful level of various forms of radiation will result in adverse effects beyond heat problems alone.

In calculating the effects of different forms of thermal stress, it is possible to use the simple heat balance equation developed by Burton (1934), which defines the heat balance state of the human body as:

$$S = M - E \pm R \pm C - W$$

where

S = is bodily heat storage expressed either as the rate of heating (+) or cooling (–) in the body,
M = is the rate of bodily metabolic heat production,
E = is the rate of evaporative heat loss,
R = is the rate of heat gained (+) or lost (–) by radiation,
C = is the rate of heat gained (+) or lost (–) by convection, and
W = is the rate of work accomplished.

When $S = 0$ the body is said to be in the state of balance or thermal equilibrium. In conditions of heat stress, heat storage takes place, and S is positive. In conditions of cold stress, the body looses heat to the environment, and S is negative.

Human Thermoregulation

Human beings are *homeothermic.* They maintain a constant internal body temperature, also known as the *core* temperature of approximately 98.6°F or 37°C. Human body temperature is a distributed function and varies dynamically across the surface of the body and within the body as immediate and long-term adjustments are made to the conditions the individual faces. Deep body temperature, or core temperature, is the value most individuals refer to when they consider human thermal state. Such core temperature is usually measured experimentally by means of a probe which assesses rectal or ear (tympanic membrane) status. These provide a measure as close as possible of the temperature of the main organs of the body. However, these are often either uncomfortable or unacceptable as measures for prolonged working conditions and more recent sophisticated techniques are now available to provide remote monitoring of worker physiological status. In normal (i.e., thermoneutral) conditions, the skin temperature of a clothed person ranges between 31°C and 33°C. The necessary heat to maintain this level of temperature is generated as a by-product of metabolic processes which convert the chemical energy of foods into mechanical energy (work). Under fluctuations of the environmental thermal conditions, body temperature is regulated through change of the blood flow to the periphery (skin); sweating and shivering.

The hypothalamus of the brain is considered to be the coordinating center for thermoregulation. Thermoregulatory mechanisms are activated in two ways. First, by peripheral input provided by heat sensitive nerves (thermoreceptors) in the skin. Second, by direct stimulation of the hypothalamus which contains cells that are extremely sensitive to the blood temperature perfusing the hypothalamus. In either case, the hypothalamus responds by invoking those mechanisms that will maintain the core temperature at a dynamically constant level. In thermally neutral environments (air temperature about 20°C to 23°C for a resting and normally clothed person), the heat balance in the body is maintained by regulation of the blood flow. In warm environments, the body's first line of defense is increased blood circulation toward the periphery. Increased blood flow in the skin dissipates the excessive body heat into the environment, and maintains the heat balance of the body. At higher temperatures (or during intense exercise), sweating is invoked. Evaporation of sweat is more efficient under conditions of low relative humidity. Under excessive heat stress, the sweating mechanism fails and heat accumulates in the body over time. This condition is known as *hyperthermia* and is characterized by a life-threatening rise in the core temperature.

In slightly cold environments the body responds by restricting heat flow toward the skin therefore preventing bodily heat from dissipating into the environment. At lower temperatures, the body generates additional heat through rapid muscular contractions known as shivering. Shivering is an action designed specifically to raise the temperature of the body and is one of the few cases of reflexive movement with no goal of affecting the external environment. Under extreme cold stress, thermoregulation fails, and the body is continuously losing heat to the environment. This condition, known as *hypothermia,* eventually causes death. The mechanisms by which this happens are complex, and some experiments to determine such tolerance have an unusual and reprehensive history in the story of scientific research (see Burton and Edholm, 1955).

The range of environmental conditions invoking the various types of thermoregulatory responses can be described by three zones (Grandjean, 1988; see also Figure 95.1):

1. *Zone of thermal comfort or vasomotor regulation.* Within this zone the body maintains thermal equilibrium by regulating the flow of blood to different parts of the body. For a clothed and resting person in winter this zone lies between 20°C and 23°C.
2. *Zone of heat regulation.* Within this zone, increasing ambient temperature invokes the thermoregulatory mechanisms against heat stress: (a) increased blood circulation in the skin, and (b) sweating. If heat stress continues to increase and exceeds the level of heat tolerance, the core temperature rises steeply and death by heat stroke is possible.
3. *Zone of bodily cooling.* Within this zone, decreasing ambient temperature invokes the thermoregulatory mechanisms against cold stress: (a) restricted blood circulation in the skin, and (b) shivering. As cold stress progresses, body temperature becomes lower than the body can tolerate, and death due to freezing is possible.

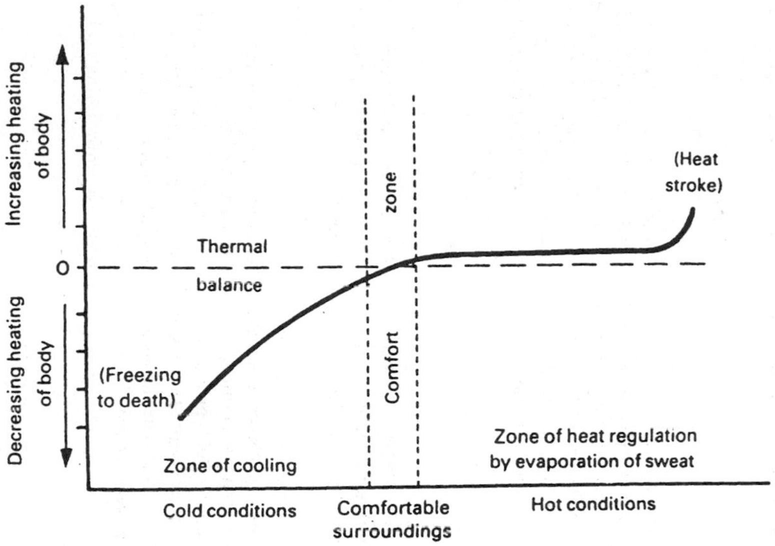

FIGURE 95.1 Zones of thermoregulation. (From Grandjean, E. 1988. *Fitting The Task to the Man.* 4th Edition. Taylor and Francis: New York. With permission.)

Effects of Heat Stress on Health

Exposure to heat stress can cause the following heat disorders and illnesses:

1. *Behavioral disorders.* These include transient heat fatigue that impairs skill and sensorimotor performance, and chronic heat fatigue resulting in permanent reduction of performance capacity.
2. *Skin eruptions.* They can take the form of heat rashes (tiny raised red vesicles) or anhidrotic heat exhaustion (gooseflesh areas of skin that do not sweat on heat exposure).
3. *Heat cramps.* Painful spasms of muscles used during work caused by depletion of salt.
4. *Heat exhaustion.* Caused by dehydration, depletion of circulating blood volume, and/or competing demands for blood flow by the skin and active muscles (includes fatigue, nausea, headache, and giddiness).
5. *Heat syncope.* Caused by pooling of blood in dilated vessels of skin and lower parts of the body. It may cause fainting while standing erect and immobile in the heat.
6. *Heat stroke.* This condition represents failure of the sweating mechanism and is accompanied by a rectal temperature of 40.5°C or higher. Symptoms include confusion, loss of consciousness, convulsions, and continuous rise of core temperature. Heat stroke can cause death if treatment is not applied immediately.

Table 95.1 summarizes the range of heat-related disorders and illnesses, their predisposing factors, symptoms, treatments, and means of prevention.

Heat Acclimatization

Exposure of the human body to a hot environment over a long period of time brings about a physiological adaptation known as heat acclimatization. A heat-acclimatized body can better cope with the adverse effects of heat stress than a non-acclimatized body. Acclimatization can take place naturally; and we all develop some degree of heat acclimatization over a hot summer. In hot workplaces, however, acclimatization of the worker can be attained by a gradually increasing exposure to the heat over a period of weeks. In this case, the body develops the following adaptations:

TABLE 95.1 Heat Stress Illnesses and Disorders

Category and Clinical Features	Predisposing Factors	Underlying Physiological Disturbance	Treatment	Prevention
1. Temperature Regulation Heatstroke				
Heatstroke: (1) Hot dry skin usually red, mottled or cyanotic; (2) Rectal temperature 40.5 C (104 F) and over; (3) Confusion, loss of consciousness, convulsions, rectal temperature continues to rise, fatal if treatment delayed	(1) Sustained exertion in heat by unacclimatized workers; (2) Lack of physical fitness and obesity; (3) Recent alcohol intake; (4) Dehydration; (5) Individual susceptibility; and (6) Chronic cardiovascular disease	Failure of the central drive for sweating (cause unknown) leading to loss of evaporative cooling and an uncontrolled accelerating rise in t_{re}, there may be partial rather than complete failure of sweating	Immediate and rapid cooling by immersion in chilled water with massage or by wrapping in wet sheet with vigorous fanning with cool dry air, avoid overcooling, treat shock if present	Medical screening of workers, selection based on health and physical fitness, acclimatization for 5–7 days by graded work and heat exposure, monitoring workers during sustained work in severe heat
2. Circulatory Hypostasis				
Heat Syncope				
Fainting while standing erect and immobile in heat	Lack of acclimatization	Pooling of blood in dilated vessels of skin and lower parts of body	Remove to cooler area, rest recumbent position, recovery prompt and complete	Acclimatization, intermittent activity to assist venous return to heart
3. Water and/or Depletion				
(a) Heat Exhaustion				
(1) Fatigue, nausea, headache, giddiness; (2) Skin clammy and moist; complexion pale, muddy, or hectic flush; (3) May faint on standing with rapid thready pulse and low blood pressure; (4) Oral temperature normal or low but rectal temperature, usually elevated (37.5–38.5 C) (99.5–101.3 F); water restriction type: urine volume small, highly concentrated; salt restriction type: urine less concentrated, chlorides less than 3g/L	(1) Sustained exertion in heat; (2) Lack of acclimatization; and (3) Failure to replace water lost in sweat	(1) Dehydration from deficiency of water; (2) Depletion of circulating blood volume; (3) Circulatory strain from competing demands for blood flow to skin and to active muscles	Remove to cooler environment, rest recumbent position, administer fluids by mouth, keep at rest until urine volume indicates that water balances have been restored	Acclimatize workers using a breaking-in schedule for 5–7 days, supplement dietary salt only during acclimatization, ample drinking water to be available at all times and to be taken frequently during work day
(b) Heat Cramps				
Painful spasms of muscles used during work (arms, legs, or abdominal); onset during or after work hours	(1) Heavy sweating during hot work; (2) Drinking large volumes of water without replacing salt loss	Loss of body salt in sweat, water intake dilutes electrolytes, water enters muscles, causing spasm	Salted liquids by mouth, or more prompt relief by I-V infusion	Adequate salt intake with meals; in unacclimatized workers supplement salt intake at meals

4. Skin Eruptions				
(a) Heat Rash (miliaria rubra; "prickly heat")				
Profuse tiny raised red vesicles (blister-like) on affected areas prickling sensations during heat exposure	Unrelieved exposure to humid heat with skin continuously wet with unevaporated sweat	Plugging of sweat gland ducts with retention of sweat and inflammatory reaction	Mild drying lotions, skin cleanliness to prevent infection	Cool sleeping quarters to allow skin to dry between heat exposures
(b) Anhidrotic Heat Exhaustion (miliaria profunda)				
Extensive areas of skin which do not sweat on heat exposure, but present gooseflesh appearance, which subsides with cool environments; associated with incapacitation in heat	Weeks or months of constant exposure to climatic heat with previous history of extensive heat rash and sunburn	Skin trauma (heat rash; sunburn) causes sweat retention deep in skin, reduced evaporative cooling causes heat intolerance	No effective treatment available for anhidrotic areas of skin, recovery of sweating occurs gradually on return to cooler climate	Treat heat rash and avoid further skin trauma by sunburn, periodic relief from sustained heat
5. Behavioral Disorders				
(a) Head Fatigue — Transient				
Impaired performance of skilled sensorimotor, mental, or vigilance tasks, in heat	Performance decrement greater in unacclimatized and unskilled worker	Discomfort and physiologic strain	Not indicated unless accompanied by other heat illness	Acclimatization and training for work in the heat
(b) Heat Fatigue — Chronic				
Reduced performance capacity, lowering of self-imposed standards of social behavior (e.g., alcoholic over-indulgence), inability to concentrate, etc.	Workers at risk come from temperate climates, for long residence in tropical latitudes	Psychosocial stresses probably as important as heat stress, may involve hormonal imbalance but no positive evidence	Medical treatment for serious cases, speedy relief of symptoms on returning home	Orientation on life in hot regions (customs, climate, living condition, etc.)

From *Criteria for a Recommended Standard: Occupational Exposure to Hot Environments*, revised criteria, NIOSH publication No. 86-113, 1986.

a. Gradual increase of sweating. A heat acclimatized worker can loose up to 2 liters of sweat per hour, and up to 6 liters of sweat per workday.
b. The sweat becomes more dilute, i.e., presents a lower salt concentration. This change helps to prevent heat disorders such as heat cramps.
c. Gradual reduction of body weight. This reduction allows a more efficient heat exchange with the environment due to reduced fat, and reduces energy consumption.
d. Gradual reduction of heart rate and body temperature.

Both exposure to the heat and physical activity are necessary for acclimatization to take place. In general, exposure to the hot job for approximately two hours per day will result in complete acclimatization in a period of two to three weeks. Exposure to the job for more than two hours a day will speed up acclimatization somewhat. Termination of exposure to the heat stress, will result in lost acclimatization in a period of 2 to 3 months. However, a significant loss of acclimatization takes place in the first week of this period. Thus, workers who have been removed from the hot job for a period of one week or more (say for a vacation), and return to that job, will need to get acclimatized again. Much of the work on heat acclimatization has come from the need to rapidly acclimatize a changing work force in deep mining. Hence, some account must be taken of the work which the individual is expected to perform in the acclimatized state.

There is one final stage to this process of adaptive change which occurs at a genetic level whereby those best able to combat the adverse effects of heat exposure are subject to forms of selection which promote certainly bodily traits. These can be seen in the peoples of the world as they have adapted to differing geographical and climatological conditions. However, since this is not a part of worker acclimatization in the manipulable sense, we do not dwell on this aspect of adaptation.

Measurement of Heat Stress

The four environmental variables that affect the sensation of comfort and the heat exchange process with the environment are: air temperature, relative humidity, air velocity, and temperature of the surrounding surfaces. Those variables, combined with the amount of physical work performed by an individual, and the amount of clothing worn, largely define the thermal environment. In the past 50 years, various indices have been developed to assess or predict the level of environmental heat stress. Some of these indices take into consideration only a single environmental factor, whereas others combine the effects of all four environmental variables directly or indirectly. Finally, the level of metabolic heat production is not considered, with the exception of heat stress index (HSI). In considering a number of these heat stress indices, we indicate how measures of the four components of the thermal environment are typically measured.

1. *Dry bulb temperature.* The dry bulb temperature is the simplest practical index of cold and warmth under usual room conditions. It is measured with a dry bulb thermometer, such as an ordinary mercury-in-glass thermometer and is significant in judging comfort under cold conditions.

2. *Wet bulb temperature.* The psychometric wet bulb temperature is a useful index of severe heat stress, especially when the body is near its upper limits of temperature regulation by sweating. It is obtained by drawing air at a velocity of at least 1000 feet per minute over a wetted wick covering a mercury-in-glass thermometer effectively shielded from radiation. The natural wet bulb temperature is obtained when the wetted wick covering a mercury-in-glass thermometer is exposed to natural air movement unshielded from radiation.

3. *Black globe temperature.* The equilibrium temperature of a black globe has been used as a single temperature index describing the combined physical effect of the dry bulb temperature, air movement, and radiant heat received from surrounding conditions. The black globe temperature is measured with the globe thermometer. It consists of a 6-in. diameter thin copper sphere, the outside of which is painted matte black. A mercury-in-glass thermometer, having a range of 30°F to 220°F with 1°F graduation and accurate to 1°F is inserted through a rubber stopper in a hole in the top of the shell, and the thermometer bulb is located at the center of the globe.

These first three indices represent values derived directly from physical instruments used to assess particular parts of the thermal environment. However, there are several indices which derive their values from various combinations of these measures. One of the first of these is the effective temperature (ET) index derived by Houghten and Yagloglou (1923).

4. *Effective Temperature* (*ET*). ET has been one of the best known and most widely used of all thermal indices. It combines into a single value the effects of air temperature, humidity, and air movement on the body. According to this index, different combinations of these values which provide the same perceived level of warmth are given the same value of effective temperature. In cases where radiant heat is present and the person is fully clad, the corrected effective temperature (CET) may be used instead. CET substitutes black globe temperature for the dry bulb temperature of the original effective temperature scale to correct for effects of any intense radiant heat source in the surrounding environment, such as the opening to a blast furnace.

5. *Wet Globe Temperature* (*WGT*). The wet globe temperature index was developed by Botsford (1971). Only one temperature is required to determine this index. A thermometer is placed in the center of a hollow 2.5-in. diameter black globe covered with a wetted black cloth. When placed in the hot environment, an equilibrium is obtained between evaporative cooling, convective heating and radiant heating. The equilibrium temperature measured after approximately 15 minutes is the wet globe temperature.

6. *Heat Stress Index* (*HSI*). The rationale behind the heat stress index proposed by Belding and Hatch (1955) is the concept that in order to maintain body temperature within the safe range, the body's heat loss must equal or exceed heat gain. Their heat stress index is the ratio of the body's heat load from metabolism, convection, and radiation E_{req} to the evaporative cooling capacity of the environment E_{max}. By multiplying this ratio by 100, the HSI index is obtained. This can be expressed in equation form as follows:

$$HIS = E_{req}/E_{max} = 100\ (M + C + R)E_{max}$$

where M = is the heat from metabolism, C = is the heat gain or loss from convection, and R = is the heat gain or loss from radiation.

7. *Wet Bulb Globe Temperature* (*WBGT*). Yagloglou and Minard (1957) developed the wet bulb globe temperature (*WBGT*) index in order to take into consideration radiant heat and air velocity. Originally, it was based on the psychometric wet bulb temperature t_{wb} and the black globe temperature t_g as follows:

$$WBGT = 0.7\,t_{wb} + 0.3\,t_g$$

Subsequently, the t_{wb} was replaced by the natural wet bulb temperature t_{nwb}, so today *WBGT* is a function of t_{nwb}, t_g, and for outdoors the dry bulb temperature t_{db}. The weighting of these temperatures is expressed according to the following formulas:
For indoor conditions:

$$WBGT = 0.7\,t_{nwb} + 0.3\,t_g$$

and for outdoor conditions:

$$WBGT = 0.7\,t_{nwb} + 0.2\,t_g + 0.1\,t_{db}$$

The natural wet bulb temperature is measured by a mercury-in-glass thermometer having a range of 30°F to 120°F with 0.5°F graduations and accurate to 0.5°F. An absorbent cotton wick is used to cover the thermometer bulb at least 1.25 inch above the bulb. The lower end of the wick is immersed in a reservoir of distilled water at the temperature of the work area, and there is about 1 inch of wetted wick exposed to the air between the top of the reservoir and the bottom of the thermometer bulb. The wick

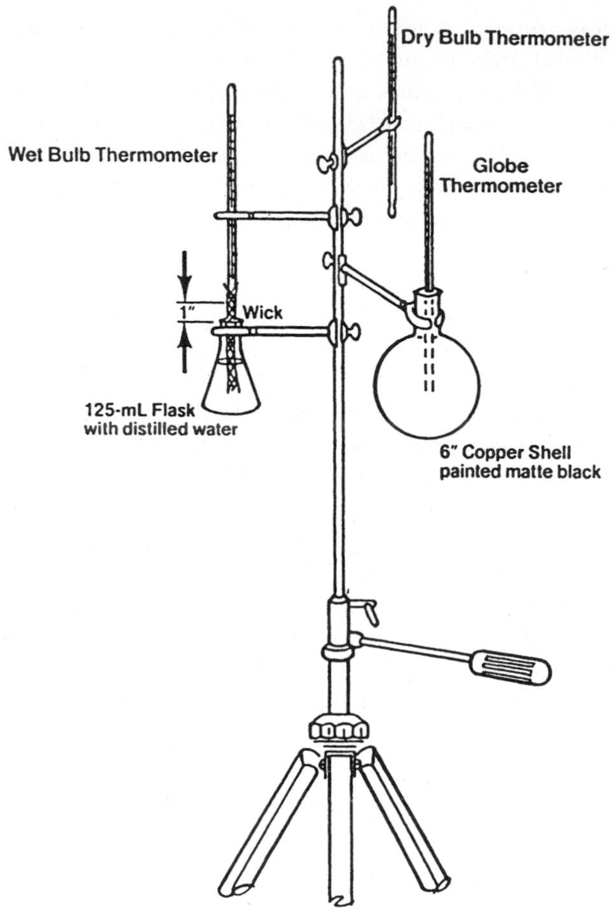

FIGURE 95.2 Instrument arrangement for outdoors WBGT measurements. (From *Fundamentals of Industrial Hygiene,* National Safety Council, 3rd edition, 1988, p. 269.)

must be wet at all times in order to obtain accurate readings. Figure 95.2 shows an instrument arrangement that can be used to obtain measurements of the various outdoor temperatures. Today, several portable instruments are available to measure the WBGT index.

Metabolic Heat Measurement

Metabolic heat production is combined with the effect of the environmental factors, and adds to the level of heat stress imposed on the worker. Thus, it is necessary to know the metabolic energy expended during routine physical activities. Table 95.2 provides estimates of energy metabolism of various types of activities for light, moderate, and heavy work. In cases where the worker is engaged in various types of activities over the 8-hour shift, a time-weighted average (TWA) energy expenditure for these activities should be calculated.

Physiological Exposure Limits to Heat Stress

From the physiological standpoint, it is generally agreed that exposure to occupational heat stress should be terminated when the rectal temperature reaches the upper limit of 38°C (Dukes-Dobos, 1976; Fanger, 1972). Grandjean (1988) proposed upper temperature limits for daytime work, which are reproduced in

TABLE 95.2 Energy Metabolism Values for Various Activities

		Metabolic Rate, M			
	Activity	Btu/hr	Watts (W)	kcal/hr	kcal/min
	Sleeping	250	73	63	1.05
	Sitting quietly	400	117	100	1.75
Light Work	Sitting, moderate arm and trunk movements (e.g., desk work, typing)	450–550	130–160	113–140	1.8–2.3
	Sitting, moderate arm and leg movements (e.g., playing organ, driving car in traffic)	550–650	160–190	140–160	2.3–2.7
	Standing, light work at machine or bench, mostly arms	550–650	160–190	140–160	2.3–2.7
Moderate Work	Sitting, heavy arm and leg movement	650–800	190–235	165–200	2.8–3.3
	Standing, light work at machine or bench, some walking about	650–750	190–220	165–190	2.8–3.2
	Standing, moderate work at machine or bench, some walking about	750–1,000	220–290	190–250	3.2–4.2
	Walking about, with moderate lifting or pushing	1,000–1,400	290–410	250–350	4.2–5.8
Heavy Work	Intermittent heavy lifting, pushing or pulling (e.g., pick and shovel work)	1,500–2,000	440–590	380–500	6.3–8.3
	Hardest sustained work	2,000–2,400	590–700	500–600	8.3–10.0

Note: Values apply for a 70-kg (154 lb) man, and do not include rest pauses.
From *Fundamentals of Industrial Hygiene*, National Safety Council, 3rd edition, 1988, p. 272.

TABLE 95.3 Temperature Limits for Daytime Work in Hot Environments

Overall Consumption of Energy (KJ/h)	Examples	Upper limit of temperature (°C) Effective Temperature	Temp. with 50% RH
1600	Heavy work; walking, with 30 kg load	26–28	30.5–33
1000	Moderately heavy work; walking at 4 km/h	29–31	34–37
400	Light sedentary work	33–35	40–41

From Grandjean, E. 1988. *Fitting The Task to the Man.* 4th Edition. Taylor and Francis: New York.

TABLE 95.4 Permissible Working Times for Heavy Work

Wet-Bulb Temperature (°C)[a]	Permissible Working Time (min)
30	140
32	90
34	65
36	50
38	39
40	30
42	22

From Grandjean, E. 1988. *Fitting The Task to the Man.* 4th Edition. Taylor and Francis: New York.

Table 95.3. If the heat load, i.e., consumption greater than 1900 kJ/hr, exceeds the values of Table 95.3, and cannot be reduced by technical means, Grandjean suggests shortening the working time according to Table 95.4.

In general, the upper physiological limits that should not be exceeded in hot industrial places can be expressed in terms of the physiological parameters heart rate, rectal temperature, and sweat evaporation as follows (Grandjean, 1988):

Heart rate:	100–110 beats/min (daily average)
Rectal temperature:	38°C
Evaporation of sweat:	0.6 l/h.

NIOSH (1986) has recommended alert limits (RALs) for the average, unacclimatized individual and exposure limits (RELs) for the acclimatized individual in terms of WBGT as a function of work activity and hourly exposure to the heat. These limits apply to a standard worker of 70 kg (154 lb) body weight and 1.8 m^2 (19.4 ft^2) body surface area. These curves are illustrated in a subsequent section in which we address the rationale for a greater use of behavioral indicators. In hot workplaces where meeting this limits is a problem, a variety of engineering and administrative controls can be implemented. These controls are listed in Table 95.5.

TABLE 95.5 Engineering and Administrative Controls Recommended for Control of Heat Stress

Engineering Controls

- Install air condition in the workplace.
- If air temperature is below 35°C, increase cooling by enhanced general or local ventilation by means of fans and blowers.
- Install line-of-sight radiant reflective shields around radiant heat sources.
- Where possible, apply coating to the radiant heat source to reduce its emissivity.
- Where possible, eliminate water-vapor sources in the workplace.
- Automate or mechanize physical components of work to reduce physical activity demands.
- Provide heat-protective clothing (goggles, heat reflective garments, etc.) where necessary.
- Train workers on how to wear such clothing.

Administrative Controls

- Schedule hot activities during the cooler parts of the day.
- Schedule routine maintenance and repair activities in hot spots in the cooler seasons of the year.
- Increase personnel to reduce heat stress exposure time per worker.
- Make water or other fluids readily available to hot workplaces.
- Encourage frequent water intake in hot industrial workplaces. Water should be taken at least once every hour.
- Implement a heat-acclimatization program by gradually exposing the worker to the hot job. Heat acclimatization can be induced in 5-7 days of exposure to the hot work environment. A few variations of heat acclimatization programs have been proposed. According to the National Safety Council, for workers with previous experience with the job, acclimatization is induced using the following schedule of exposure: 50% on day one, 60% on day two, 80% on day three, and 100% on day four. For workers without previous experience with the job the exposure should be 20% on day one with 20% increase on each additional day.
- Encourage workers to participate in physical fitness programs.
- For heat unacclimatized workers on a restricted salt diet, additional salt intake during the first 2 days of exposure to the hot job may be required; the worker's physician however, should be consulted.
- Implement a heat stress training program in the workplace. Such a program should teach supervisors and workers on recognizing early signs and symptoms of heat-induced illnesses, and administration of first-aid procedures.
- Implement a buddy system; Under such a system supervisors and workers trained in recognizing heat-induced signs and in providing first-aid procedures, are assigned the task of periodically observing fellow workers for early signs of heat illnesses. When such signs are present, the worker should not be allowed to continue work, and should be sent to the first-aid station for a more thorough evaluation.
- Screen worker medical records for heat intolerance. Workers who have experienced heat illnesses in the past are likely to be less tolerant to heat stress.

(For Comparisons see: Millican, Baker, and Cook, 1981).

In this section, we have briefly examined the basis for the physiological limits for heat stress exposure. It is not our main purpose here to provide an elaborate evaluation of these concerns, especially since there are many sources of information which provide further detailed coverage. (See for example, ACGIH, 1986; Hardy, Gagge, and Stolwijk, 1970; ISO, 1989; Kantowitz and Sorkin, 1983; Kerslake, 1972; Konz, 1983; McIntyre, 1980; NIOSH, 1972; 1986; Parsons, 1993; Ramsey and Kwon, 1992; Rohles and Konz, 1987; Sanders and McCormick, 1987.) What we seek to establish is the importance of behavioral indices as indicators of the stress effects of heat and potentially the whole spectrum of occupational sources of stress. In claiming this ubiquity and the central importance of this approach, it is important to provide a full foundation for this claim, and this we do in the following sections.

95.3 The Prevalence of Heat Stress

In their publication, the National Institute for Occupational Safety and Health (NIOSH) estimated that, as a conservative figure, some five to ten million workers in United States may be exposed to heat stress as a potential safety and health hazard (NIOSH 1986). From geographical considerations alone, this figure is liable to be by population proportion, higher for workers on a worldwide scale, particularly given the predominance of agrarian economies in equatorial and tropical regions. In order to alleviate potential harm from heat stress and to protect exposed workers, as we have discussed above, a number of exposure criteria have been promulgated (see Parsons, 1995). In general, their principal aim is to provide guidelines for acceptable exposure to heat through designation of environmental conditions, exposure duration, work composition, and some rudimentary information concerning the status of the individual worker. NIOSH (1986) has more recently issued a revision of these criteria. Examination of this latter document highlights the careful and thorough evaluation of the physiological consequences of exposure to heat, and the derivation of criteria designed to protect the worker engaged in heavy physical labor from heat stress related illnesses such as heat stroke, heat syncope, heat exhaustion, and other disabling conditions as we have noted. In this respect, the revised criteria admirably attain this aim. Such protection guards workers engaged in traditional industrial activities which are composed primarily of differing degrees of physical activity.

However, the revised criteria neglect a large and growing segment of the industrial population whose job is to perform more cognitively demanding operations frequently within the confines of complex operational systems. It is also the case that numerous workers have to combine different duties, where for one period they may be actively engaged in exacting physical labor, and the next moment, monitoring automated systems for critical failures. The deletion of cognitive performance limits under heat stress in the NIOSH (1986) document was intentional (Ramsey, 1995). It was suggested (NIOSH, 1980) that an insufficiently clear picture had been developed concerning such a relationship. We suggest a resolution is now possible. Consequently, one particular purpose of the present work is to explore heat stress exposure limits not founded upon the concept of physiological injury or medical illness, but predicated upon performance change. As the growth of cognitive, information-processing work characteristics is ubiquitous throughout contemporary industry, the necessity to consider behavior-related exposure limits applies across the wide spectrum of occupational stress sources and is not restricted to heat stress alone. The generality of this important observation is explored in the final section of the chapter.

95.4 The Foundation for Exposure Criteria to Occupational Stress

As indicated by Millar (1986), the revised heat stress criteria (NIOSH, 1986) were generated as part of programmatic efforts initiated by the Occupational Safety and Health Act (U.S. Public Law 91-596, 1970). Documents were designed to:

> ...provide *medical* criteria which will assure insofar as practicable that no worker will suffer diminished *health, functional capacity, or life expectancy* as a result of his (or her) work experience. (italics and parenthetical inclusion added.)

Following the wording of the above designation, current exposure limits for occupational stress are founded upon medical evaluations of the impact of components of the ambient surroundings upon physiological functioning. This natural line of development emanates from the fact that traditional concern for worker safety and health is founded in, and focused through the medical profession, whose knowledge of physiological systems and the disturbances to which they are vulnerable is preeminent. Consequently, evolving criteria limits have been based upon the premise of avoidance of acute or chronic physiological insult to the whole organism or component sensitive organs. Such sentiments are to be applauded and our purpose here is not to disparage this approach in any way. Rather, it is to extend this protective rationale by providing a more sensitive tool for the use of practitioners. That is, the assessment of behavioral efficiency.

Development of performance-based criteria is neurobiologically justified since the central nervous system displays particular sensitivity to disturbance due to stress, and its behavioral manifestation in performance capability represents an avenue through which to evaluate early and less obvious effects due to occupational stress compared with traditional gross physiological manifestations. From the managerial and safety standpoint such an approach has at least two useful facets. First, productivity, as generated by performance, is a prime concern of management and such data are constantly under scrutiny. Second, with the changing nature of work, the failure of cognitive decision-making activities and operator error are becoming of greater concern as the concentration upon heavy physical work wanes.

As medical criteria, the NIOSH (1986) recommendations are conservative and justified. However, the criteria also seek to:

> Protect against the risk of heat-induced illness and unsafe acts. ...prevent possible harmful effects from interactions between heat and toxic, chemical, and physical agents.

However, it is commonly the case that efficient performance on a task fails before physiological limitations are reached or physiological systems are perturbed to the boundaries of their region of steady-state operation. Consequently, if the workers are no longer able to adequately discharge their duties by performing the task in question in the stressful condition, why is continued exposure permitted?

This latter statement explicitly recognizes limitations to safe functioning as given in the criteria aims cited above. To elaborate with an example, if a worker is required to monitor a display for a critical signal to initiate an important response (cf., Hancock, 1984), is it advisable to continue exposure (within physiological limits) if that worker is missing a significant and potentially catastrophic number of signals for response? The submission here is that, although insufficient to cause physiological distress, such conditions cause unacceptable hazard to the worker, his or her colleagues, and the system within which they are operating. We suggest that, as performance is commonly the most sensitive systemic response to imposed environmental stress, and is the key reason for exposing the worker to many occupational sources of stress in the first place, exposure guidelines should be formulated to deal with such restrictions in appropriate work conditions. It is the lack of information upon those forms of performance variation that represents one major limit of the current heat stress criteria document revision (see NIOSH 1986, p. 32).

95.5 Current Criteria: Derivations and Limitations

The first criteria developed by NIOSH (1972) did give recommended limits for a threshold of unimpaired mental performance. This illustration is produced in Figure 95.3. This curve was a simple transcription of that described by Wing (1965) from his review of then existing studies. Having examined the extrapolations and inferences Wing made from a survey of existing studies, Hancock (1981) re-evaluated this threshold and provided a revision of these tolerance limits based on correction of factual errors and suspect interpretations. This comparison was incorporated into a number of texts. For example, Kantowitz and Sorkin (1983) observed that:

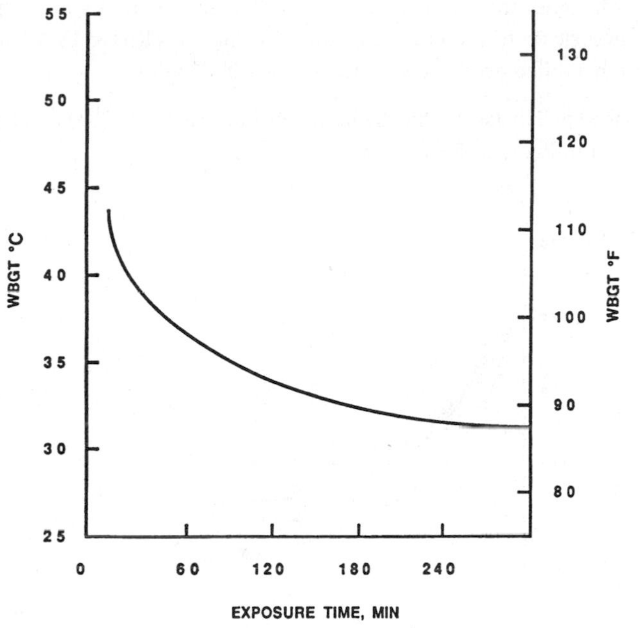

FIGURE 95.3 Upper limits of exposure for unimpaired mental performance. (From National Institute for Occupational Safety and Health [NIOSH] 1972, *Criteria For a Recommended Standard-Occupational Exposure to Hot Environments* [NIOSH Publication No. 72-10269]).

There is currently strong disagreement about the effects of heat stress on the mental efficiency of sedentary workers. Wing (1965) performed the classic study of this problem. Wing found that mental tasks such as arithmetic and memory tasks were not affected for very short exposures (6 minutes) but that exposures of about 43 minutes to effective temperatures over 100°F did cause impairment. Wing proposed an exponential function relating exposure time, and an effective temperature (see Figure 19-6) later adopted by the National Institute for Occupational Safety and Health (NIOSH, 1972) as the recommended standard for the lower limit of heat-impaired mental performance. These limits are reached before the physiological tolerance limit of the human. However, this standard has been strongly criticized as being far too conservative (Hancock, 1980). A detailed reanalysis of Wing's data has led to the line in Figure 19-6 labeled Hancock. It is close to the physiological limits, shown as the line labeled Taylor in Figure 19-6. Until this debate is resolved, cautious Human Factors Specialists will base decisions on a curve falling between the Hancock and the Wing stress tolerance functions; extremely cautious designers may prefer to use the more conservative Wing tolerance curve.

In his work on a wide variety of stress effects, Hockey (1986) commented:

Wing (1965) summarized the results of 15 studies of performance on sedentary tasks. These data have become the basis of a well-known guideline for heat limits in Industry (NIOSH, 1972), illustrated in Figure 44.7. The limit of unimpaired performance at any combination of temperature and duration is well below that for physiological tolerance. A recent reevaluation of this evidence (Hancock, 1981) suggests that the upper limit is very close to that for physiological collapse. This conclusion is based on a detailed analysis of only one study (Blockley and Lyman, 1951), however, and may not be representative of the effects of thermal stress on performance.

Finally, Sanders and McCormick (1987) observed:

Note that Wing (1965) had developed a curve of tolerance limits for performing mental activities similar to the curve for mental and cognitive tasks shown in Figure 15-9. However, Wing's curve was somewhat lower than Hancock's — actually a bit below the curve for tracking tasks. In discussing this difference Hancock argues that, on the basis of his synthesis, mental and cognitive abilities can be performed at a level closer to the physiological limits than is reflected by Wing's presentation. The clarification of such a difference, however, probably is still dangling.

The graphs described by Hockey (1986), by Kantowitz and Sorkin (1983) and Sanders and McCormick (1987) are reproduced in Figures 95.4 and 95.5.

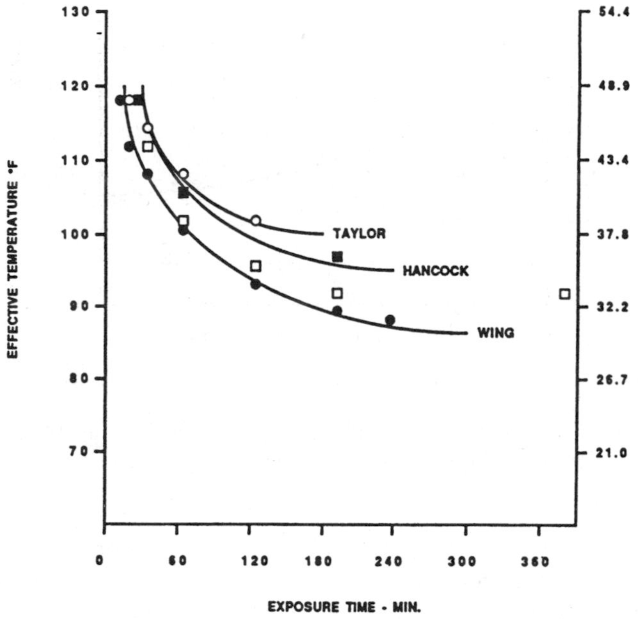

FIGURE 95.4 Revised heat stress limits of exposure for unimpaired mental performance (Hancock curve). The Taylor curve indicates physiological tolerance to heat stress (Taylor, 1948), and the Wing curve represents the exposure limits adopted by NIOSH, 1972 (see Figure 95.3).

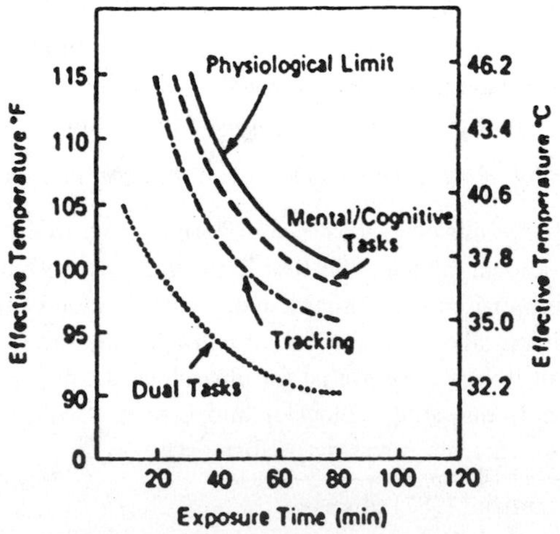

FIGURE 95.5 Revised heat stress limits of exposure for unimpaired mental performance. (From Sanders, M.S. and McCormick, E.J. 1987. *Human Factors in Engineering and Design*, 6th Edn, McGraw-Hill, New York. With permission.)

It is important to address the issues raised by these commentators, as their summaries are often the first, and on occasion the only source of information available to individuals who are concerned with broad interdisciplinary inquiries as typified in ergonomics and human factors. The observations made by Kantowitz and Sorkin (1983) are factually correct, although one inference they make might be regarded as slightly misleading. In the original work, Hancock (1980) criticized the factual basis for the foundation of Wing's curve. However, the use of the two respective curves and their functions as protective standards was not considered exhaustively in the latter work since the focus was on tolerance itself, not on recommended limits. One obvious drawback in the use of Wing's curve is that NIOSH (1972) simply transcribed Wing's original curve for their use, and in so doing apparently equated effective temperature and WBGT, which is a highly questionable procedure. However, this does not mean that either Wing (1965) or Hancock (1980) expressly advocated their respective curves as tolerance criteria. Rather, they are the limits at which significant impairment in mental performance can be expected. As pointed out above, operating at ceiling limits as a protective standard is a highly inadvisable strategy. Clearly, it is appropriate in setting tolerance criteria to adopt conservative standards. As Konz (1983) and others have pointed out, information from studies which underlie such criteria are often taken from well motivated, young college subjects, who are exposed on only one or two occasions to the debilitating effects of the stress. However, the subsequent protection is often directed toward older individuals who experience the potential for chronic as well as acute exposures and whose motivation to perform cannot be sustained at the same level for extended periods of operation. Compounded with these are additional concerns such as gender, fitness, task performance capability and numerous other individual differences that have to be understood before a fully comprehensive standard can be finalized (see also Enander and Hygge, 1990). As pointed out by Hancock (1980) then, the standard was rightly conservative, but for essentially the wrong reason, being founded on fallacious information. This is more than mere polemics in that subsequent differentiation of tasks (Hancock 1982, 1984) indicated that some forms of performance could be expected to suffer impairment at time/intensity combinations well below those designated by Wing's curve.

Hockey's (1986) comments indicated misapprehension by suggesting that the re-evaluation by Hancock (1981) was based on only a single study, that being the report by Blockley and Lyman (1951). As perusal of the original work shows, the re-evaluation was actually based upon the analysis of all the studies cited by Wing (1965), and a number of reports that appeared since that time. It is the case that one clear factual mistake made by Wing (1965) was in his interpretation of the data given by Blockley and Lyman (1950) (not Blockley and Lyman 1951 as indicated by Hockey 1986), an observation that certainly was given prominence by Hancock (1981). However, Wing's interpretation of Blockley and Lyman's (1950) data was an important point in that it established a critical exposure at the one-hour duration. It is not correct to assert that Hancock's (1981) whole argument was based on the results from this single study.

Sanders and McCormick (1987) pointed to the problem of task differentiation. They noted that the curves subsequently developed by Hancock (1982), and discussed in more detail below, crossed the single function given by Wing. It is important to note that in his work, Wing did not make any explicit differentiation for task performance category. In so doing he implied the equivalence between tasks that require simple mental operations (e.g., mental arithmetic as reported by Blockley and Lyman, 1950) and those involving some more complex motor responses (e.g., tracking as reported by Pepler, 1958). Hancock (1982) argued that such tasks could not be considered of homogeneous cognitive demand and had to be differentiated. This tactic had also previously been advocated by other researchers (e.g., Grether, 1973; Ramsey and Morrissey, 1978). When such a differentiation was made, the tolerance limit curves for each category of performance tasks showed common shapes, although occurring at different absolute tolerance levels, see Figure 95.5. This important observation becomes the subject of more critical evaluation later in the present work.

Each of the above reviews rightly indicated the danger of extrapolating thresholds from a sparse database, and in considering the different interpretations that can be made from each supportive study.

However, some of their respective equivocation does not sufficiently reflect the fact that Hancock (1980, 1981) pointed to a number of simple *factual* mistakes made by Wing (1965) in his original interpretation. Given some of these problems it is perhaps not surprising that the limits illustrated in Figure 95.4, as included in the first NIOSH criteria document, were not reproduced in the subsequent revised criteria (NIOSH 1986).

95.6 Single Versus Multiple Thresholds for Criteria

As is readily apparent from the above argument, one of the limitations to Wing's curve was the way in which all forms of performance tasks were included in the derivation of a single threshold, regardless of their actual composition. A different perspective on performance limits was given by Ramsey and Morrissey (1978). Following the notion developed by Grether (1973), they developed a description based on different task categories. Essentially, they distinguished two groups of tasks, one consisting principally of mental performance, and a second consisting of psychomotor performance. They rightly pointed out that for each respective group, a single delimiting curve could represent only the dichotomous differentiation into decrement and no decrement. Consequently, they developed isodecrement curves based upon the probability of performance failure at particular time–temperature conditions. Examples of such curves are given in Figures 95.6 and 95.7. As can be seen by comparing the respective figures, there are radically different limits depending on the nature of the task. In subsequent work, Hancock (1982) sought commonalities across these different limits as they applied to the respective tasks, particularly at the upper extremes of exposure. The illustration in Figure 95.8 shows the detailed foundation of this subsequent synthesis. As can be seen, each of the curves presents a similar shape, including the limits for physiological tolerance, and these accord with the limits presented by Meister (1976).

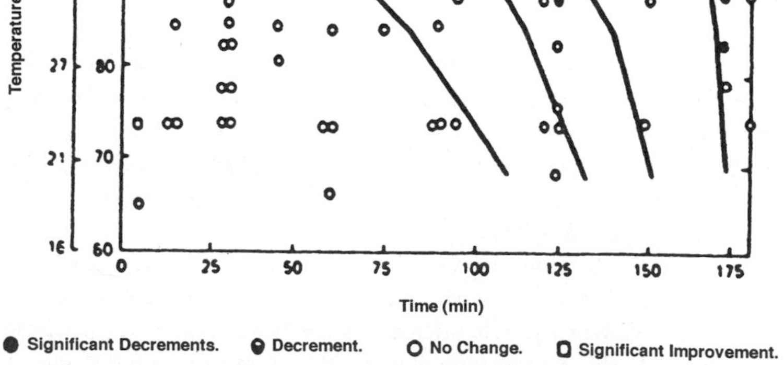

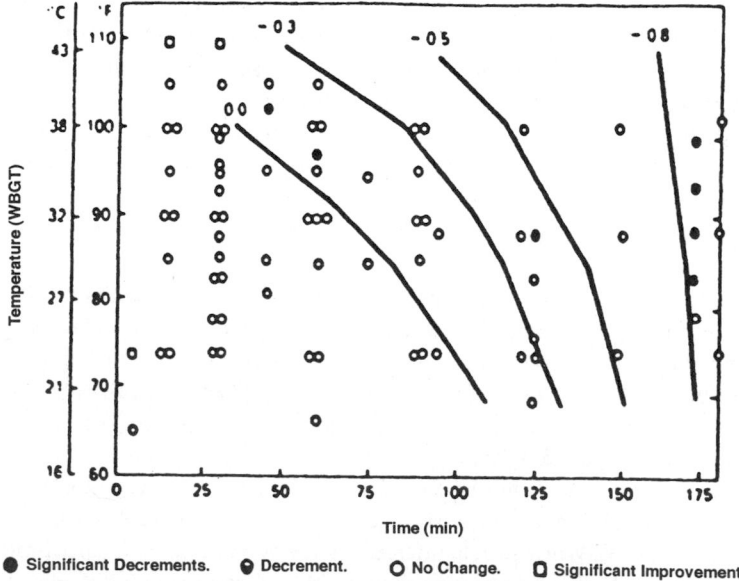

FIGURE 95.6 Isodecrement curves for mental performance tasks. The numbers represent level of performance decrement in the range from 0.0 (no change) to -1.0 (definite significant decrement in task performance). (From Ramsey, J.D. and Morrissey, S.J. 1978, Isodecrement curves for task performance in hot environments, *Applied Ergonomics*, 9, 66-72. With permission.)

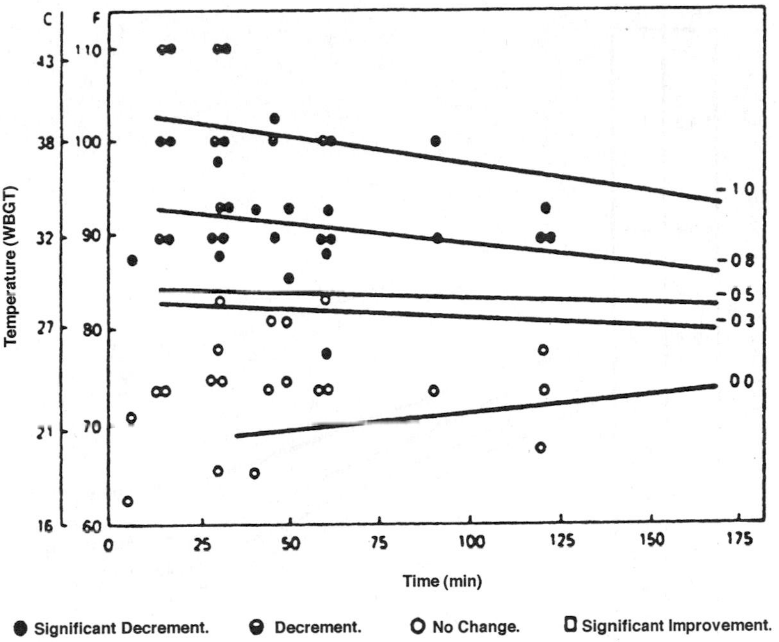

FIGURE 95.7 Isodecrement curves for tracking tasks. The numbers represent level of performance decrement in the range from 0.0 (no change) to −1.0 (definite significant decrement in task performance). (From Ramsey, J.D. and Morrissey, S.J. 1978, Isodecrement curves for task performance in hot environments, *Applied Ergonomics*, 9, 66-72. With permission.)

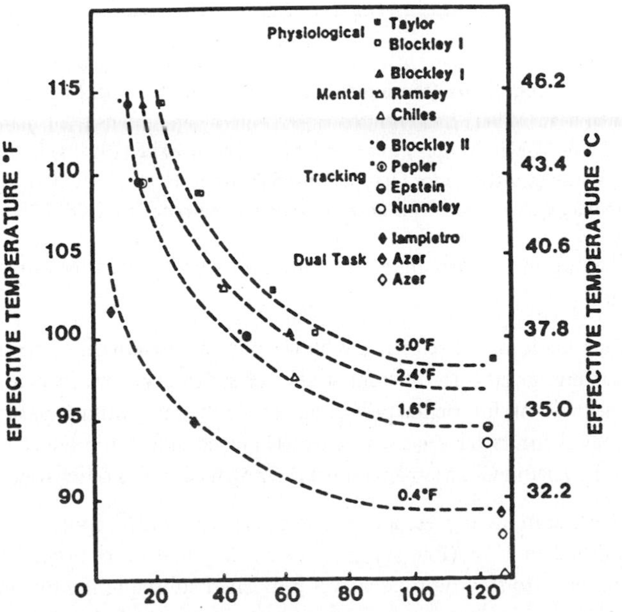

FIGURE 95.8 Heat stress limits for unimpaired performance for mental and cognitive tasks (triangular symbols), tracking tasks (circular symbols), and dual-tasks (diamond symbols). Square symbols represent physiological tolerance. Superimposed are dashed lines representative of prescribed rises in deep body temperature which accrue from time, ET intensity specifications outlined. These absolute values for the rise of body temperature are given on each curve. Names are for the first author for each study. (From Hancock, P.A. 1982, Task categorization and the limits of human performance in extreme heat, *Aviation, Space and Environmental Medicine*, 53, 778-784. With permission.)

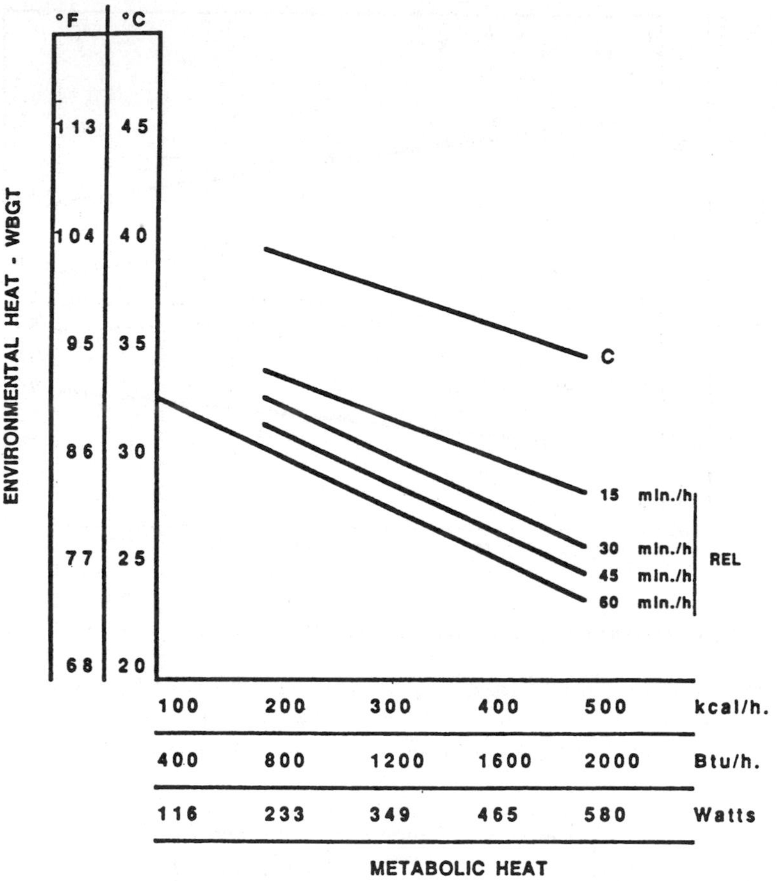

FIGURE 95.9 Revised heat stress limits (recommended exposure limits or REL) for acclimatized workers as a function of environmental heat (WBGT), intermitted work (min./h) and intensity of manual work as reflected by metabolic energy expenditure. The limits apply to a standard worker of 70 kg (154 lbs) body weight and 1.8 m^2 (19.4 ft^2) body surface. (From National Institute for Occupational Safety and Health [NIOSH] 1986, *Criteria for a Recommended Standard: Occupational Exposure to Hot Environments* revised criteria [NIOSH Publication No. 86-113]).

Different limits based upon performance differentiation were quoted by Konz (1983) in his text on work design. He stated:

> Wing and Touchstone made a 162-reference bibliography on the effects of temperature on human performance. Wing, summarizing 15 different studies of sedentary work in heat, gave Figure 20.4 as the temperature-time trade-off for mental performance; he noted that human performance deteriorates well before physiological limits have been reached. Hancock shows the effect on different tasks in Figure 20.5. Ramsey, Dayal and Ghahramani report their own data plus other support for Wing's curve.

Konz is correct in his statements, except that because the study reported by Ramsey, Dayal and Ghahramani was published in 1975 (Ramsey et al. 1975) they had no opportunity to comment on the relationship of their findings to the multiple curves published subsequently by Hancock (1982), but see Ramsey (1995).

The revised criteria (NIOSH 1986) give recommended alert limits and exposure limits from knowledge concerning the human physiological tolerance to heated conditions. Figures 95.9 and 95.10 present these thresholds for heat acclimatized and heat unacclimatized workers, respectively. They represent functions of an environmental heat load, expressed in WBGT units, and worker-generated heat load. Limits are given in terms of continuous and intermittent work schedules and between illustrations for differing

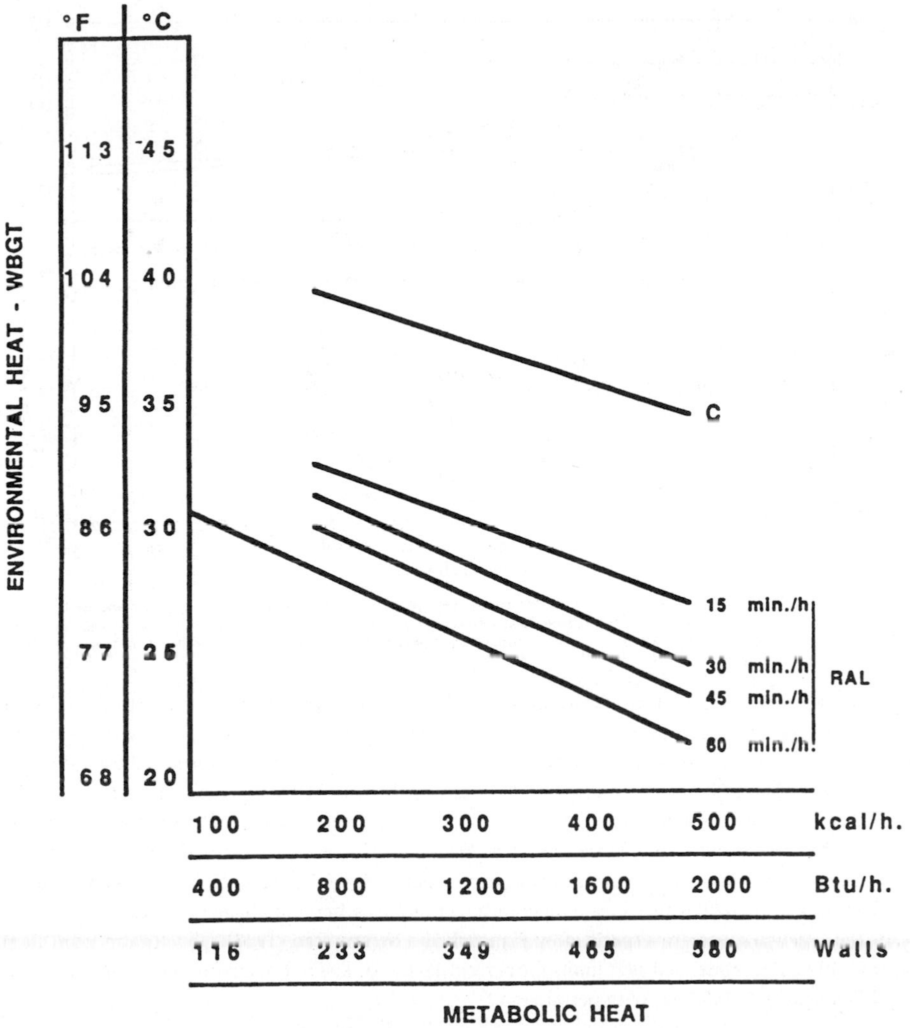

FIGURE 95.10 Revised heat stress limits (Recommended Alert Limits or RAL) for unacclimatized workers as a function of environmental heat (WBGT), intermitted work (min./h) and intensity of manual work as reflected by metabolic energy expenditure. The limits apply to a standard worker of 70 kg (154 lbs) body weight and 1.8 m^2 (19.4 ft^2) body surface. (From National Institute for Occupational Safety and Health [NIOSH] 1986, *Criteria for a Recommended Standard: Occupational Exposure to Hot Environments* revised criteria [NIOSH Publication No. 86-113]).

states of worker acclimatization. However, as can be seen, these limiting functions make no reference to cognitively demanding work as did its predecessor.

In their article, Ramsey and Kwon (1992) examined results from more than 150 studies in which the impact of differing heat intensities and exposure times had been evaluated on differing forms of performance task. In keeping with their previous observations (Ramsey and Morrissey, 1978), they divided these tasks into two major categories, those requiring simple mental performance, and those requiring psychomotor response. Within these categories, they established whether the examined studies showed obvious and statistically significant decrement, marginal decrement, no evidence of performance change, or performance enhancement. No comparable marginal category was given for the case of enhanced efficiency. Of critical importance for these cross study comparisons was the establishment of a common heat stress index which combines the characteristics of the thermal surround. As Ramsey and Kwon clearly pointed out, much of the experimental work in this area is two to three decades old, since contemporary human

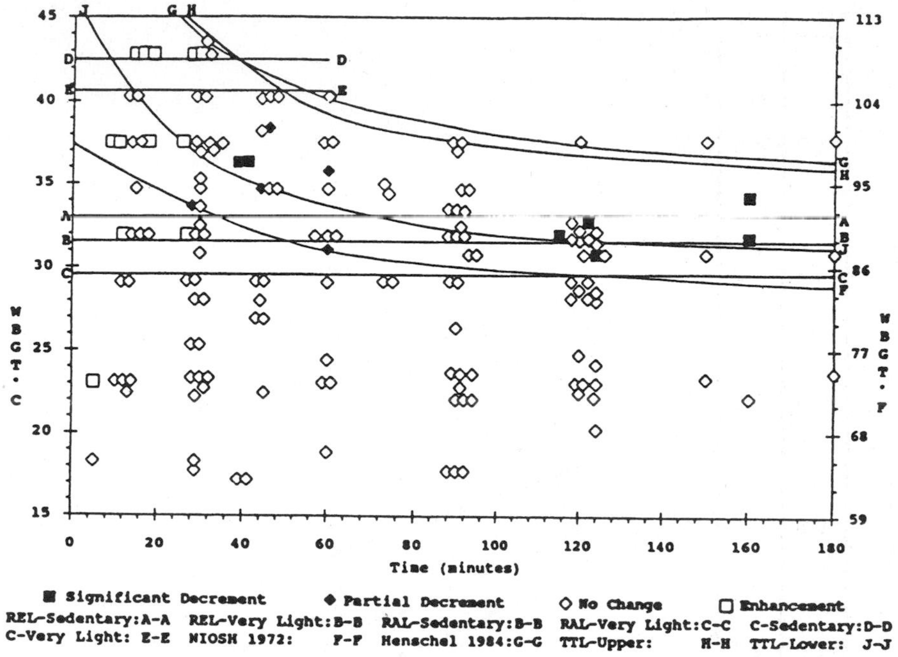

FIGURE 95.11 Mental or simple task performance in the heat and proposed temperature- time limits for human responses. REL, or recommended exposure limit, applies to heat acclimatized workers, and RAL, or recommended alert limit, applies to heat unacclimatized workers. Curves A-A, B-B, and C-C are exposure limits based on one-hour time-weighted averages for the case of unsteady thermal conditions. Curves D-D and E-E are the NIOSH 1986 recommended ceiling limits for sedentary and very light work, respectively. Curve G-G was derived based on personal communication between Ramsey and Kwon 1992 and Henschel. Curve H-H defines the upper thermal tolerance limits (TTL-Upper) for unimpaired neuromuscular performance as specified by Hancock and Vercruyssen 1988. Curve J-J (TTL-Lower) describes the time–temperature conditions where no change in deep body temperature is expected for the sedentary worker as specified by Hancock and Vercruyssen (1988). (Illustration from Ramsey, J.D. and Kwon, Y.G. 1992, Recommended alert limits for perceptual motor loss in hot environments, *International Journal of Industrial Ergonomics,* 9, 245-257. With permission.)

subject restrictions do not allow severe experimental heat exposures. As many early studies used the effective temperature (ET) scale, a perceptually oriented rather than an environmentally oriented scale, some assumptions and translations were needed to establish intensity levels on a common scale. Quite properly, Ramsey and Kwon chose the WBGT scale and used the NIOSH 1973 nomogram to convert ET to CET (or corrected effective temperature) units, from which WBGT values may be derived. This use of WBGT allows comparison with many current criteria documents (e.g., ISO 1982, NIOSH 1986, ACGIH 1988, see Parsons, 1995). Their findings are summarized in Figures 95.11 and 95.12, which are reproduced below.

Their findings indicate that for the category of very simple mental performance, there is little evidence of decrement across the range of intensities and exposure times surveyed and that on many occasions such capabilities are enhanced during brief exposures. These results confirm a previous assertion that there is only a slight mental performance decrement before impending physiological collapse (Hancock 1981, p. 180). However, as with Ramsey and Kwon's (1992) additional observations, it has been noted that most tasks which require constituents of motor performance are more susceptible to heat (Hancock 1981, p. 180). Hancock (1982) divided psychomotor performance on the basis of single and dual-task demands and found a number of consistencies in the data which are illustrated in Figure 95.8.

Following the careful division of the performance tasks and the establishment of curves whose function were supported by experimental observations at specific conditions, Hancock (1982) used the function

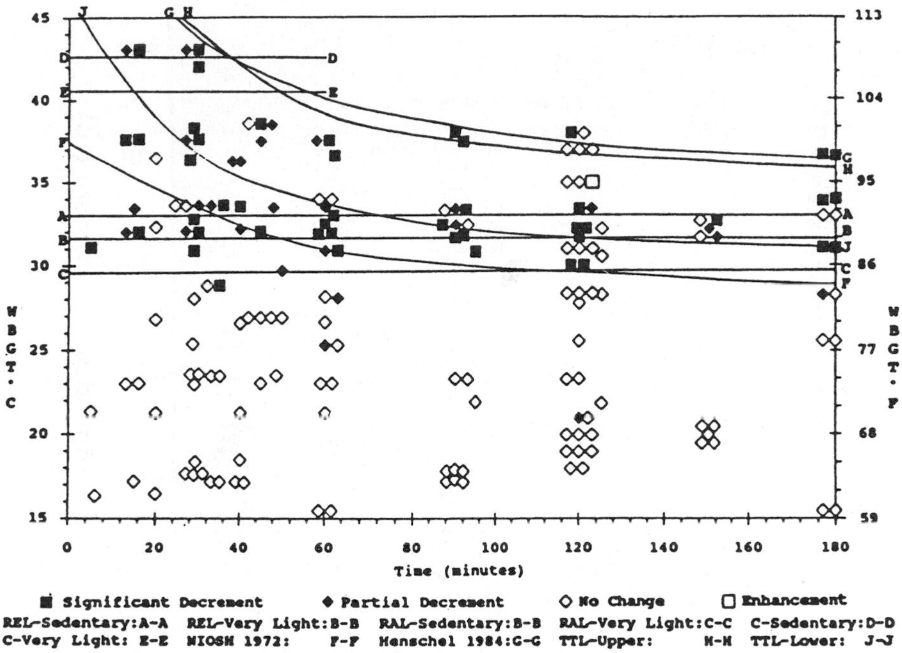

FIGURE 95.12 Perceptual motor task performance in the heat and proposed temperature–time limits for human responses. REL, or recommended exposure limit, applies to heat acclimatized workers, and RAL, or recommended alert limit, applies to heat unacclimatized workers. Curves A-A, B-B, and C-C are exposure limits based on one-hour time-weighted averages for the case of unsteady thermal conditions. Curves D-D and E-E are the NIOSH 1986 recommended ceiling limits for sedentary and very light work respectively. Curve G-G was derived based on personal communication between Ramsey and Kwon 1992 and Henschel. Curve H-H defines the upper thermal tolerance limits (TTL-Upper) for unimpaired neuromuscular performance as specified by Hancock and Vercruyssen (1988). Curve J-J (TTL-Lower) describes the time–temperature conditions where no change in deep body temperature is expected for the sedentary worker as specified by Hancock and Vercruyssen (1988). (Illustration from Ramsey, J.D. and Kwon, Y.G. 1992, Recommended alert limits for perceptual motor loss in hot environments, *International Journal of Industrial Ergonomics,* 9, 245-257. With permission.)

described by Houghten and Yagloglou (1923) to establish that the boundary functions of each performance category are each represented by different dynamic changes to operator deep body temperature.[1] So, different levels of dynamic change to deep body temperature could be substituted for the distinct time/ET functions, illustrated as the limits for each task, and these values for each respective category were added to each boundary in Figure 95.8. A number of studies have used impermeable garments which manipulated dynamic change to deep body temperature, but without variation in environmental temperature, provided data that supported the observed limits of different deep body temperature values (Allan, Gibson, and Green, 1979). This convergent evidence established that it is *change* in the dynamic thermal state of the operator which influences performance, not manipulation of the physical environment *per se.* This is important when considering performance variations due to individual differences in capacities such as acclimatization and task skill (Enander and Hygge, 1990; Hancock 1986a).

It is important to provide the rationale under which previous syntheses of empirical data relating heat stress to performance variation has been developed. In analogy with physical effort, Hancock (1982)

[1]It should be noted that elevation to a higher steady state value, as occurs when the individual can use physiological mechanisms to partially neutralize the impact of the increased ambient thermal load, does not represent dynamic, or uncompensable change. So, it is the degree of change and its derivative, rate of change, that dictate dynamic oscillations in deep body temperature.

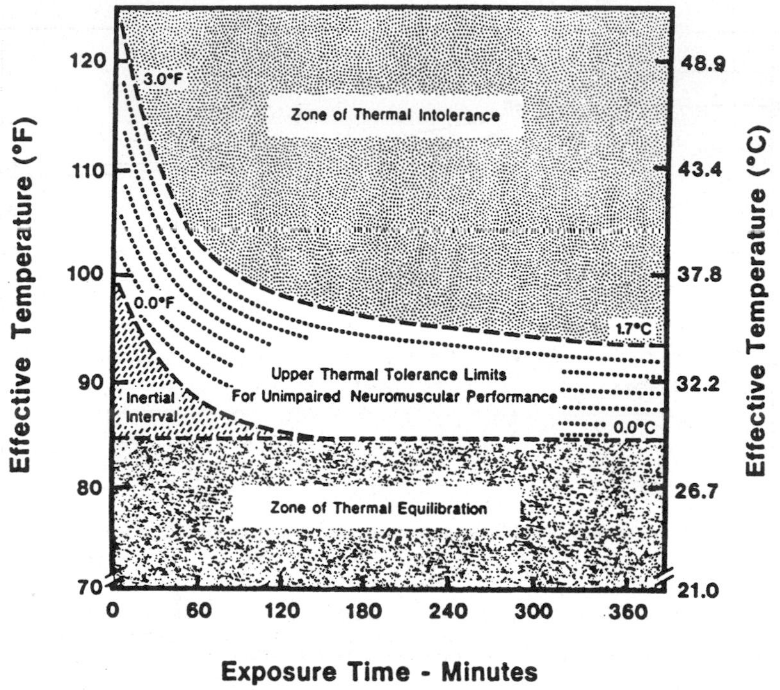

FIGURE 95.13 Differing performance zones identified by Hancock and Vercruyssen (1988). Numbers within the figure represent dynamic rise in deep body temperature.

suggested that the more demanding an information-processing task, the less heat strain could be sustained before performance interruption would occur. Therefore, as expressed in Figure 95.8, a task requiring considerable cognitive effort is constrained by a more conservative limit than one requiring comparatively little cognitive effort. Hancock (1982) indicated overall performance limits in a similar manner where simple mental tasks could be performed under conditions impermissible for a complex or dual task situation which required the simultaneous performance of two unrelated tasks. In this way, the attention demands of an information-processing task were equated with the physical demands of a material handling task. This is *not* to suggest that each may be directly related to metabolic demand as this implies an unfounded argument for an isomorphism between biochemical energy and information processing capacity (Hancock, 1986b). Further, in some tasks such as monitoring, very little effort is apparent but the hidden demands of such vigilance tasks make them particularly vulnerable to heat effects (Hancock, 1984, 1986b; Ramsey and Morrissey, 1978). As a result, Hancock (1982) plotted isodecrement curves for performance failure in heat which could be interpreted as limits, expressed also as dynamic change in deep body temperature. These were elaborated into descriptive zones of heat stress designation for the worker engaged in low metabolic demand activity (Hancock and Vercruyssen, 1988) (see Figure 95.13).

95.7 A New Descriptive Framework

There is, however, a more parsimonious way to describe human performance limits under heat stress. The approach is illustrated in Figures 95.14 and 95.15 and involves an alternative representation of known performance limit curves in different task categories. As shown in Figure 95.15, the characteristic of the primary (horizontal) axis is exposure time. The secondary (vertical) axis is thermal intensity of the environment expressed first in terms of the traditional effective temperature (ET). The temporal axis extends from the brief pulse-like exposures (approximately 3 minutes) to a time approximating a common shift, excluding meal breaks and start-up and shut-down time (approximately 7 hours). The vertical axis

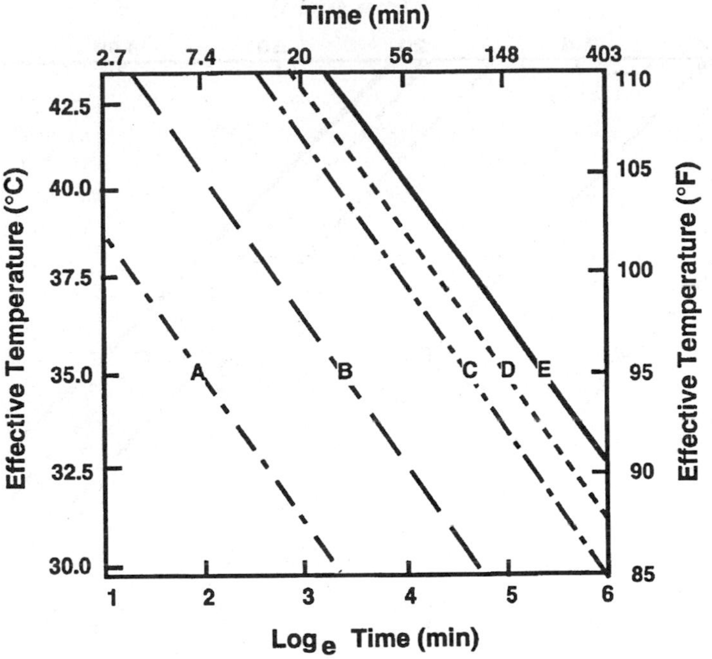

FIGURE 95.14 Human performance limits in (ET)/log$_e$(Time) Cartesian space (A: vigilance performance, B: dual-task performance, C: tracking performance, D: simple mental performance, E: physiological tolerance). (From Hancock, P.A. and Vasmatzidis, I. 1998. Human occupational and performance limits under stress: The thermal environment as a prototypical example, *Ergonomics,* 44, 1169–1191. With permission of Taylor & Francis.)

extends from marginally tolerable conditions for any exposure period (i.e., 114°F ET) to the limits of Lind's (1963) "prescriptive zone" (85°F ET). The prescriptive zone has more recently been termed the zone of thermal equilibration, to denote the absolute boundary of steady-state, thermoregulatory capacity (Hancock and Vercruyssen, 1988).

Within this framework, performance limit curves are plotted as parallel lines. The general form of the equation describing these performance thresholds is:

$$ET = a - b \log_e T \tag{1}$$

where ET is effective temperature, T is exposure time, and a and b are empirically determined constants. In the above relationship, parameter b, the slope of the equation, is equal to 4.094 and remains constant for each task category curve. Parameter a, the intercept of the lines with the thermal intensity axis, reflects the attentional involvement required by each task category plotted. The higher the value of parameter a, the higher the respective performance limit and the lesser the attentional demand placed on an individual by the task.

In Figure 95.14, performance limit curves are drawn from right to left in an increasing attentional demand order. Initially, to the right is line E indicating the physiological tolerance ceiling. Immediately below this absolute ceiling is the performance limit for simple mental tasks represented by line D (Hancock, 1981; Ramsey and Kwon, 1992), followed by tasks requiring neuromuscular coordination (line C). Next is line B, the threshold for dual-tasks combining each of the requirements in the two latter categories (Hancock, 1982). A final line (line A) formed from empirical data and using the summary as presented in Hancock (1986b), describes the tolerance of sustained attention, also known as vigilance, or more commonly in industry as monitoring and inspection (see also Hancock, 1984). Note that vigilance is particularly vulnerable to heat effects. This failure of monitoring-type behavior is particularly pertinent

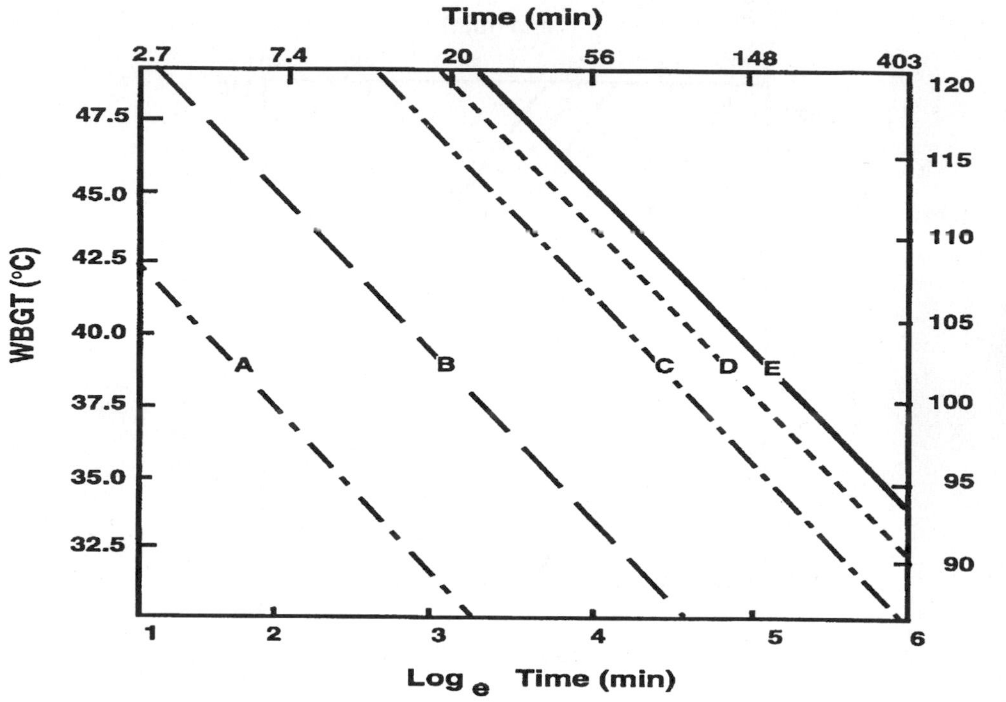

FIGURE 95.15 Human performance limits in the WBGT–log$_e$(Time) Cartesian space (A: vigilance performance, B: dual-task performance, C: tracking performance, D: simple mental performance, E: physiological tolerance). (From Hancock, P.A. and Vasmatzidis, I. 1998. Human occupational and performance limits under stress: The thermal environment as a prototypical example, *Ergonomics,* 44, 1169–1191. With permission of Taylor & Francis.)

TABLE 95.6 With respect to Figure 95.14, since the slope is common at –4.094, designation of each performance limit curve, in °C, can be specified by a single intercept value. There are two crucial issues to note. As the zero point on the logarithmic base goes to infinity, the intercepts shown are purely pragmatic and are used to plot the lines within the time/intensity boundaries shown. Thus the tolerances should not be extended beyond the time/intensity limits illustrated without further experimental validation. Second, two intercepts have been presented. The first is derived solely from the empirical data, the second contains a conservative adjustment so that the designations can be used for acceptable tolerance standards.

Curve	Task type	Empirical intercept	Tolerance adjusted intercept
A	Vigilance performance	42.82	41.0
B	Dual-task performance	48.59	47.0
C	Tracking performance	53.96	53.0
D	Simple mental performance	55.81	54.0
E	Physiological tolerance	57.06	55.0

From Hancock, P.A. and Vasmatzidis, I. 1998. Human occupational and performance limits under stress: The thermal environment as a prototypical example, *Ergonomics,* 44, 1169–1191. With permission of Taylor & Francis.

to the design of operator tasks in future systems. Table 95.6 provides empirical and tolerance standard adjusted intercept values for each performance curve of Figure 95.14 when ET is expressed in terms of °C.

Although the linearity across the present plot allows for a simple mathematical description of tolerance, where task category performance threshold is defined by the intercept value, this linearity is not the only significance of the illustration. Each threshold also describes a particular dynamic rise in deep body temperature which corresponds to the limit of efficient performance on that task. This property permits

TABLE 95.7 With respect to Figure 95.15, the slope is common at –5.435. Again, since the zero point on the logarithmic base goes to infinity, the intercepts shown are purely pragmatic and are used only to plot the lines within the time/intensity boundaries shown. The tolerances therefore, may not be extended beyond the time/intensity limits illustrated without further experimental validation. Again, the empirical and tolerance adjusted intercepts are specified.

Curve	Task type	Empirical intercept	Tolerance adjusted intercept
A	Vigilance performance	48.02	46.0
B	Dual-task performance	55.68	54.0
C	Tracking performance	63.11	62.5
D	Simple mental performance	65.33	64.0
E	Physiological tolerance	66.56	65.0

From Hancock, P.A. and Vasmatzidis, I. 1998. Human occupational and performance limits under stress: The thermal environment as a prototypical example, *Ergonomics*, 44, 1169–1191. With permission of Taylor & Francis.

a transcription of performance limits from the effective temperature–time domain to the wet bulb globe temperature (WBGT)-time domain. Since WBGT has replaced ET as the principal measure of the thermal load in most of the experimental studies conducted during the last decades, this transcription is illustrated next. Consequently, WBGT was the index incorporated in the two most recent NIOSH heat stress recommended standards (NIOSH, 1972, 1986) and several other international standards (Parsons, 1995).

A direct translation between ET and WBGT cannot be accomplished through the physical properties of the atmosphere without knowledge of the radiant heat value in the respective ET environment. In the paper by Ramsey and Kwon (1992) this translation was accomplished by estimation of conditions in the absence of reported data. However, a knowledge of the rate of rise of body temperature against WBGT can act as an alternate source of translation. This translation is accomplished by a link derived from the work of Jensen and Heims (1976). The procedure presented here, and the estimates of Ramsey and Kwon (1992) are both superior to the unfounded equivalence between ET and WBGT assumed in the NIOSH (1972) document for drawing the curve concerning the heat stress-related performance limits as given by Wing (1965). This whole question of index selection and transcription process is worthy of further study (Parsons, 1993). For the purpose of the present work, the transcription of performance limits was accomplished by employing the information in Jensen and Heims (1976). The new threshold curves are presented in Figure 95.16, where the respective thresholds for vigilance, dual-tasks, neuromuscular coordination tasks, simple mental performance and physiological tolerance are 0.055°C, 0.22°C, 0.88°C, 1.33°C, and 1.67°C dynamic increase in body temperature, respectively. The negative slope b in Equation 2 is 5.435. The empirical and tolerance adjusted intercept values for each performance category in terms of WBGT as °C values are provided in Table 95.7. The equation describing the lines in Figure 95.15 takes the general form of:

$$\mathrm{WBGT} = a - b \log_e T \tag{2}$$

with the empirical constants as noted in the text and indicated in the appropriate table.

Within the scale ranges described earlier, it is suggested that the performance limits in Figures 95.14 and 95.15 provide the upper tolerance levels of performance in each of the task categories. These limits represent the points of statistical degradation when compared to performance in a thermoneutral condition. We do not enter here into arguments that have surrounded the methodological limitations inherent in designs which use such pairwise within-subject comparisons, nor in the argument which contrasts statistical against substantial real-world performance degradation. Suffice it to note that these are not necessarily coincident. Rather, the boundaries should be thought of as critical failure points along an exponential curve relating performance to stress intensity, the latter being the product of exposure time and exposure temperature. Figure 95.15 indicates that, as with the observations of Ramsey and Morrissey (1978), there is a series of contours which describe states of performance degradation for which the threshold of significant decrement can be regarded as one major feature. However, unlike

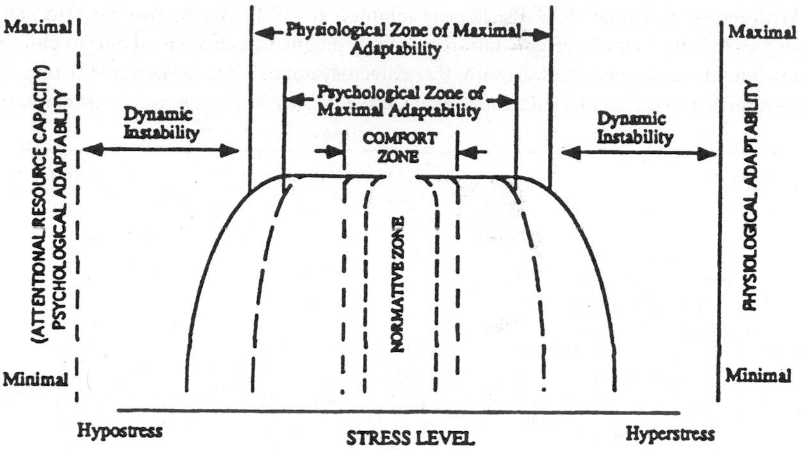

FIGURE 95.16 The maximal adaptability model. (From Hancock, P.A. and Warm, J.S. 1989, A dynamic model of stress and sustained attention, *Human Factors*, 31, 519-537. With permission.)

Ramsey and Morrissey (1978), we observe that the pattern in which these contours occur represents geometric degradation rather than the linear relationships shown in their work. The tasks referred to in the latter part of this chapter are all performed without substantive levels of muscular work. However, boundary conditions to performance decrement in terms of dynamic increases in deep body temperature may serve as useful limits even if one source of heat stress, e.g., the environment, is augmented with another source, e.g., physical activity. In essence, the present limits can accommodate situations in which the worker has to be both physically and mentally active at the same time.

95.8 Theoretical Foundation of the Limits Derived

The physiological limits of tolerance to heat and associated criteria as developed in the most recent NIOSH document are founded upon a solid body of knowledge concerning human physiological response. Thus, it is incumbent to demonstrate comparable theoretical foundations for performance-based criteria. This has been presented in recent work (Hancock and Warm, 1989) and is considered here. The theory is founded upon a direct link to physiological degradation. The lack of a single coherent theoretical framework that accounts for experimental findings is the weakest point in heat stress literature as related to mental performance and for that matter in the stress literature in general. By far, the most popular theory has been the behavioral arousal account (Duffy, 1962; Provins, 1966; Poulton, 1977) which postulates an inverted-U relationship between the arousal level of the individual and the level of environmental stress. However, this position is unfalsifiable and can account for almost any pattern of data in a *post hoc* manner. Thus, it has often been invoked by researchers unable to find any other explanation for seemingly contradictory sets of data (see Hancock, 1987 for a critique of behavioral arousal theory). Without reference to a theoretical structure, there is no rationale, other than empirical separation, for dividing results into different performance categories.

In this work we have divided tasks on the basis of attentional demands. Therefore, we have used attentional characteristics and their variation under stress as the theoretical basis for our defined limits. This variation of attentional characteristics in the presence of environmental stress is accounted for in detail by the maximal adaptability model (Hancock and Warm, 1989) which is reproduced in Figure 95.16. This model assumes that heat exerts its detrimental effects on performance by competing for and eventually draining attentional resources (Kahneman, 1973).

Briefly, in Figure 95.16, the base axis is similar to Selye's (1956) conception of stress ranging from extremes of underload (hypostress) to extremes of overload (hyperstress). In the middle of this range is an area of minimal stress (normative zone) which requires no active compensatory effort on the part of the individual. Surrounding the normative zone is the comfort zone in which cognitive adjustments to task demands are easily obtained, and therefore performance remains near to optimal. As the level of stress increases away from this zone, attentional resources are progressively drained. Initially, the remaining resources are efficiently used by the individual, with the net result being no performance decrement, and, on occasion, some performance enhancement. This behavioral pattern is a reflection of psychological adaptability and is observed inside the zone of maximal psychological adaptability. At higher stress levels, depletion of attentional resources causes progressive failure of task efficiency (see dashed line in Figure 95.17 comprising the boundary of the psychological zone of maximal adaptability; see also Hancock, 1986b for a more detailed discussion). Finally, extreme levels of stress move the individual outside the zone of homeostasis (physiological zone of maximal adaptability), toward the region of dynamic instability, a life-threatening condition, such as is found in heat stroke for example.

As illustrated in Figure 95.16, rather than an inverted U-shaped function, the maximal adaptability model proposed an extended U-shaped function in which three modes of operation are represented. The first, the flat ceiling of the extended-U, represents a mode of operation in which dynamic stability predominates. Second, at the shoulders of the extended-U, there are regions of transitions which, in the generic case of multiple stress exposure, might be represented as discontinuities using the morphology of catastrophic failure (Zeeman, 1977). Finally, the arms of the extended-U are dynamically unstable and in the present circumstances represent incipient failure. Another characteristic of the model is its symmetrical nature. Using engineering terms, this feature implies that strain increases symmetrically with progressive deviation from the central, normative zone. The function of this increase in strain is given by the solid line in Figure 95.17, and replicates the geometric functions noted above for heat stress. The strain function is the same for both physiological and behavioral degradation, but the parameters of the curves differ, signifying that behavior is affected before physiological effects are observed. The maximal adaptability model assumes initially that heat drains attentional resources from a single, undifferentiated resource pool. A methodology for expanding the model to incorporate multiple resource pool theories was provided by Vasmatzidis and Schlegel (1994). Recently, the maximal adaptability model has served as the basis for an experimental investigation of multiple task performance under heat stress (Vasmatzidis et al., 1995) and extensive work on the effects of stress and fatigue on a variety of real-world tasks (Desmond and Matthews, 1996; Matthews and Desmond, 1995; Matthews, Sparkes, and Bygrave, 1996). For full details of the model, see Hancock and Warm (1989).

95.9 Implications for All Stress Criteria

The present work has implications for developing exposure criteria for all types of occupational stress. Previous derivations of exposure criteria have been founded upon medical science, and the developed criteria are intended to ensure the healthy functioning of the human physiological system. Ergonomists are comfortable with this approach since physiology is the basis of historical ergonomics standards. However, the change in the nature of contemporary work has broad and far-reaching effects. With respect to the present work, the principal effect is the transformation of the currency of work from physical energy to information. This is not to say that there are not many situations in which physical work is not of continued importance; there are. However, this is to recognize that the predominant currency of the work environment is now information in nature. Consequently, cognitive performance and associated mental error are major concerns for today's human factors and systems specialist, as physical injury has been for the traditional ergonomist.

In the past, injuries were physical, and criteria were derived to protect against such physical sources of occupational threat. Today, in large part the emphasis has shifted from physical over-exertion to cognitive over-exertion or more appropriately, maladaptation (Hancock and Warm, 1989; see also Miller, 1960). In this chapter we have argued that heat stress exposure criteria which are designed to protect against physical harm do not always suffice for cognitive work. Our assertion is that the active central nervous system and its performance output is the most vulnerable element of the worker. Thus, it is performance that needs to be protected and consequently should comprise the focus of future heat stress exposure criteria.

In addition, we wish to propose that the present log-linear space description is not constrained to heat stress alone. Rather, we suggest that many sources of stress such as noise and vibration can be captured using the same form of description and the same underlying model, although the parameters of the space are expected to change depending on the specific nature of stress involved. Thus, the present framework can serve as a basis for understanding the multiple interaction of stresses which has always been a problematic issue (Hancock and Pierce, 1985). The great advantage of the present construct is the employment of attention as the basis for our formulation. Consequently, the different cognitive demands of the tasks, which themselves are frequently the main source of occupational stress, can be easily incorporated into our framework. No previous proposals have allowed the inclusion of such a critical contemporary work factor into exposure limitations.

95.10 Summary and Conclusions

The 19th and the early 20th centuries have been characterized as the industrial age, whereas the late 20th century and the coming millennium are better described as the information age. The associated transformation of human work has had its effects on all segments of society but on none so much as on those who seek to understand, describe, and use the laws of work to our collective benefit. In its foundation and growth, ergonomics has sought to protect the worker from sources of threat so that work can be carried on in a safe and productive manner. However, many of the stress exposure criteria that seek to attain this goal need to be supplemented in the light of the changing nature of work. We argue, and have elaborated on the specific example of exposure to heat stress, that the basis of worker protection must shift solely from physiological concerns to embrace psychological issues. Thus, cognitive science with its study of neuropsychological capabilities must play a much greater, if not primary, role in the development of the exposure criteria than they have done to date.

Fortunately, physiological and psychological response to stress can be promoted within a common framework which emphasizes their similarities, not their disparities. Thus, exposure criteria can be developed which integrate an understanding of behavioral response to work demands with physiological adjustment to environmental conditions. Such a unified framework promises to provide a new foundation for establishing worker protection from all sources of threat to health, safety, and performance.

Acknowledgment

We have, in this work, purposefully, relied heavily upon our previously reported work concerning the issue of behavioral limits, since we view this development as crucial for practicing occupational ergonomists. We would, therefore, very much like to thank the editors and publishers involved for their kind permission to reproduce such work. In particular we thank Taylor & Francis as the publisher and the journal editor for this valued agreement.

References

American Conference of Governmental Industrial Hygienists, ACGIH 1986, *Threshold Limit Values and Biological Exposure Indices for 1986-87* (Cincinnati, Ohio).

Allan, J.R., Gibson, T.M., and Green, R.G. 1979, Effects of induced cyclic changes of deep body temperature on task performance, *Aviation, Space and Environmental Medicine,* 50, 585-589.

Azer, N.Z., McNall, P.E., and Leung, H.C. 1972, Effects of heat stress on performance, *Ergonomics,* 15, 681-691.

Belding, H.S. and Hatch, T.F. 1955. Index for evaluating heat stress in terms of resulting physiological strains, *Heating, Piping and Air Conditioning,* 129-136.

Blockley, W.V. and Lyman, J.H. 1950, *Studies of Human Tolerance for Extreme Heat: III. Mental Performance Under Heat Stress as Indicated by Addition and Number Checking Tests* (USAF technical report 6022, Wright-Patterson Air Force base, Ohio).

Blockley, W.V. and Lyman, J.H. 1951, *Studies of Human Tolerance for Extreme Heat: IV. Psychomotor Performance of Pilots as Indicated by a Task Simulating Aircraft Instrument Flight,* (USAF technical report 6521, Wright-Patterson Air Force base, Ohio).

Botsford, J.H. 1971. A wet globe thermometer for heat stress evaluation. *American Industrial Hygiene Association Journal,* 32, 17-25.

Burton, A.C. 1934, The application of the theory of heat flow to the study of energy metabolism, *Journal of Nutrition,* 7, 497.

Burton, A.C. and Edholm, O.G. 1955, *Man in a Cold Environment.* Arnold: London.

Chiles, W.D. 1958, Effects of elevated temperatures on performance of a complex mental task, *Ergonomics,* 2, 89-96.

Desmond, P.A. and Matthews, G. 1996, Task-induced fatigue effects on simulated driving performance, in A.G. Gale (Ed.). *Vision in Vehicles VI,* (Amsterdam: North-Holland).

Duffy, E. 1962, *Activation and Behavior.* Wiley, New York.

Dukes-Dobos, F.N. 1976. Rational and provisions of the work practices standard for work in hot environments as recommended by NIOSH, in *Standards for Occupational Exposures to Hot Environments,* (S.M. Horvath and R.C. Jensen, Eds.) NIOSH publication No. 76-100, Cincinnati, OH.

Enander, A.E., and Hygge, S. 1990, Thermal stress and human performance, *Scandinavian Journal of Work and Environmental Health,* 16, 44-50.

Epstein, Y., Keren, G., Moisseiev, J., Gasko, O., and Yachin, S. 1980, Psychomotor deterioration during exposure to heat, *Aviation, Space and Environmental Medicine,* 51, 607-610.

Fanger, P.O. 1972. *Thermal Comfort.* McGraw-Hill, NY.

Grandjean, E. 1988. *Fitting The Task to the Man.* 4th Edition. Taylor and Francis: New York.

Grether, W.F. 1973, Human performance at elevated environmental temperatures, *Aerospace Medicine,* 44, 747-755.

Hancock, P.A. 1980, Mental performance impairment in heat stress, *Proceedings of the Human Factors Society,* 24, 363-366.

Hancock, P.A. 1981, Heat stress impairment of mental performance: A revision of tolerance limits, *Aviation, Space and Environmental Medicine,* 52, 177-180.

Hancock, P.A. 1982, Task categorization and the limits of human performance in extreme heat, *Aviation, Space and Environmental Medicine,* 53, 778-784.

Hancock, P.A. 1984, Effects of environmental temperature on display monitoring performance: An overview with practical implications, *American Industrial Hygiene Association Journal,* 45, 122-126.

Hancock, P.A. 1986a, The effect of skill on performance under an environmental stressor, *Aviation, Space and Environmental Medicine,* 57, 59-64.

Hancock, P.A. 1986b, Sustained attention under thermal stress, *Psychological Bulletin,* 99, 263-281.

Hancock, P.A. 1986c, Stress and adaptability, in G.R.J. Hockey, A.W.K. Gaillard and M.G.H. Coles (Eds.), *Energetics and Human Information Processing* (Martinus Nijhoff, Dordrecht, The Netherlands), 243-251.

Hancock, P.A. 1987, Arousal theory, stress and performance: Problems of incorporating energetic aspects of behavior onto human–machine systems function, in L.S. Mark, J.S. Warm and R.L. Huston (Eds.), *Ergonomics and Human Factors: Recent Research* (Springer-Verlag, New York), 170-179.

Hancock, P.A., and Pierce, J.O. 1985. Combined effects of heat and noise on human performance: A review, *American Industrial Hygiene Association Journal,* 46, 555-566.

Hancock, P.A. and Warm, J.S. 1989, A dynamic model of stress and sustained attention, *Human Factors,* 31, 519-537.

Hancock, P.A. and Vasmatzidis, I. 1998, Human occupational and performance limits under stress: The thermal environment as a prototypical example. *Ergonomics,* 41, 1169-1191.

Hancock, P.A. and Vercruyssen, M. 1988, Limits of behavioral efficiency for workers in heat stress, *International Journal of Industrial Ergonomics,* 3, 149-158.

Hardy, J.D., Gagge, A.P., and Stolwijk, J.A.J. 1970. *Physiological and Behavioral Temperature Regulation.* Thomas: Springfield, IL.

Hockey, G.R.J. 1986, Changes in operator efficiency as a function of environmental stress, fatigue and circadian rhythms, in K.R. Boff, L. Kauffman and J.P. Thomas (Eds.), *Handbook of Perception and Human Performance* (Wiley, New York), Chapter 44.

Houghten, F.C. and Yagloglou, C.P. 1923, Determining lines of equal comfort, *Transactions of the American Society of Heating and Ventilating Engineers,* 29, 163-176.

Houghten, F.C. and Yagloglou, C.P. 1923. *Determination of the comfort zone. American Society of Heating and Ventilating Engineers,* 29, 515-536.

Iampietro, P.F., Chiles, W.D., Higgins, E.A., and Gibbons, H.L. 1969, Complex performance during exposure to high temperatures, *Aerospace Medicine,* 40, 1331-1335.

International Organization for Standardization, ISO (7243) 1989, *Hot Environments-Estimation of the Heat Stress on Working Man Based on WBGT Index,* ISO 7243 (Geneva, Switzerland).

Jensen, R.C. and Heims, D.A. 1976, *Relationships Between Several Prominent Indices* (NIOSH-Pub No. 77-109).

Kahneman, D. 1973, *Attention and Effort.* Englewood Cliffs, Prentice Hall, New Jersey.

Kantowitz, B.H. and Sorkin, R.D. 1983, *Human Factors: Understanding People-System Relationships.* John Wiley & Sons, New York.

Kerslake, D. McK. 1972. *The Stress of Hot Environments.* Cambridge University Press: Cambridge.

Konz, S. 1983, *Work Design: Industrial Ergonomics,* 2nd Ed., John Wiley & Sons, New York.

Lind, A.R. 1963, Physiological effects of continuous or intermittent work in the heat, *Journal of Applied Physiology,* 18, 57-60.

Matthews, G. and Desmond, P.A., 1995, Stress as a factor in the design of in-car driving enhancement systems, *Le Travail Humain,* 58, 109-129.

Matthews, G., Sparkes, T.J., and Bygrave, H.M., Stress, attentional overload, and simulated driving performance, *Human Performance,* 9, 77-101.

McConnell, W.J. and Yagloglou, C.P. 1925. Work tests conducted in atmospheres of high temperatures and various humidities in still and moving air, *Journal of the American Society of Heating and Ventilation Engineers,* 31, 217-221.

McIntyre, D.A. 1980. *Indoor Climate.* Applied Sciences Publishers: Essex, England.

Meister, A. 1976, Heat exposure, performance reliability and efficiency, *Zeitschrift fur Psychologie Mit Zeitschrift fur Angewandte Psychologie,* 184, 63-72.

Millar, J.D. 1986, Foreword in: *Criteria for a Recommended Standard: Occupational Exposure to Hot Environments,* revised criteria (NIOSH pub No. 86-113).

Miller, J.G. 1960, Information input overload and psychopathology, *American Journal of Psychiatry,* 116, 695-704.

Millican, R., Baker, R., and Cook, G. 1981. Controlling heat stress: Administrative vs. physical control, *American Industrial Hygiene Association Journal,* 42, 411-415.

National Institute for Occupational Safety and Health (NIOSH) 1972, *Criteria For a Recommended Standard-Occupational Exposure to Hot Environments* (NIOSH Publication No. 72-10269).

National Institute for Occupational Safety and Health (NIOSH) 1973, *The Industrial Environment, Its Evaluation and Control.* U.S. Government Printing Office, Washington, D.C. (p. 419).

National Institute for Occupational Safety and Health (NIOSH), 1981, *Proceedings of a Workshop on Recommended Heat Stress Standards.* F.N. Dukes-Dobos and A. Henschel (Eds.). National Institute for Occupational Safety and Health, Cincinnati, OH 81-08.

National Institute for Occupational Safety and Health (NIOSH) 1986, *Criteria for a Recommended Standard: Occupational Exposure to Hot Environments* revised criteria (NIOSH Publication No. 86-113).

National Safety Council. 1988. *Fundamentals of Industrial Hygiene.* 3rd Edition, Washington, D.C.

Nunneley, S.A., Dowd, P.J., Myhre, L.G., Stribley, R.F., and McNee, R.C. 1979, Tracking-task performance during heat stress simulating cockpit conditions in high-performance aircraft, *Ergonomics*, 22, 549-555.

Parsons, K.C. 1995, International heat stress standards: A review, *Ergonomics*, 38, 6-22.

Parsons, K.C. 1993, *Human Thermal Environments*. Taylor & Francis, London.

Pepler, R.D. 1958, Extreme warmth and sensorimotor coordination, *Journal of Comparative and Physiological Psychology*, 14, 383-486.

Poulton, E.C. 1977, Arousing stresses increase vigilance, in R.R. Mackie (Ed.) *Vigilance: Theory, Operational Performance and Physiological Correlates* (Plenum Press, New York), 423-459.

Provins, K.A. 1966, Environmental heat, body temperature and behavior: An hypothesis, *Australian Journal of Psychology*, 18, 118-129.

Ramsey, J.D. 1995, Task performance in heat: A review, *Ergonomics*, 38, 154-165.

Ramsey, J.D., Dayal, D., and Ghahramani, B. 1975, Heat stress limits for the sedentary worker, *American Industrial Hygiene Association Journal*, 36, 259-265.

Ramsey, J.D. and Kwon, Y.G. 1992, Recommended alert limits for perceptual motor loss in hot environments, *International Journal of Industrial Ergonomics*, 9, 245-257.

Ramsey, J.D. and Morrissey, S.J. 1978, Isodecrement curves for task performance in hot environments, *Applied Ergonomics*, 9, 66-72.

Rohles, F.H., and Konz, S.A. 1987. Climate, in G. Salvendy. *Handbook of Human Factors*, Wiley, New York.

Sanders, M.S. and McCormick, E.J. 1987. *Human Factors in Engineering and Design*, 6th Edn, McGraw-Hill, New York.

Selye, H.A. 1956, *The Stress of Life*. McGraw-Hill, New York.

Taylor, C.L. 1948, *Interim Report to the Committee on Aviation Medicine*, National Research Council (Washington, D.C.).

Vasmatzidis, I. and Schlegel, R.E. 1994, A methodology for investigating heat stress selectivity effects on mental performance, *Proceedings of the Human Factors and Ergonomics Society 38th Annual Meeting* (Human Factors and Ergonomics Society, Santa Monica, CA), 510-514.

Vasmatzidis, I., Schlegel, R.E., Purswell, J.L., and Hancock, P.A. 1995. Cognitive resource depletion under heat stress, *Proceedings of the Konz/Purswell Occupational Ergonomics Symposium*, 91-98.

Wing, J.F. 1965, Upper thermal tolerance limits for unimpaired mental performance, *Aerospace Medicine*, 36, 960-961.

Yaglou, C.P. and Minard, D. 1957. Control of the heat casualties at military centers *AMA Archives of Industrial Health*, 16, 302-310.

Zeeman, E.C. 1977, *Catastrophe Theory: Selected Papers 1972-1977* (Addison-Wesley, Reading, Massachusetts).

96

Work Shift Usability Testing

Donald I. Tepas
University of Connecticut

96.1 Work Shift Usability

Introduction

The term *shift* is used in a variety of ways. In the most general usage, shift refers to the hours of a given day that an individual or a group of individuals is scheduled to be in the workplace. Often, this term is used in a more restricted way, where shift refers only to a specific category of work hours, defined by practices at a specified location. In many cases these hours are viewed as undesirable, nontraditional, exceptional, or unusual. The more general usage will be used here, since process and practice developments in recent years make it very difficult to define what is exceptional or unusual. Presser (1995) has clearly demonstrated that work hours for the majority of Americans are not restricted to the traditional image of regular daytime weekday work of 35 to 40 hours per week.

The industrial revolution and the factory system began with long workdays and gradually moved to a workday of eight or fewer hours. It was the hope of many workers that continued innovation would lead to further reductions in work hours and an increase in worker participation in work hour selection (Tepas, 1985). For the most part, these expectations have *not* been realized in the United States. Instead, we are now in the early days of a new technological revolution whose key components often include automation, flexible manufacturing systems, and around-the-clock operations. In many of these cases, the resulting work schedules are driven by computer systems that do not consider human limitations or preferences. All too often, this failure to address the limits of the human components results in workers with less work hour flexibility, around-the-clock work hours, extended workdays, and/or irregular work hours. These failures suggest the "misapplication hypothesis." This hypothesis proposes that the inappropriate application of advances in technology often decreases workplace performance and increases worker stress, fatigue, illness, and injury (Tepas, 1994).

0-8493-2641-9/99/$0.00+$.50

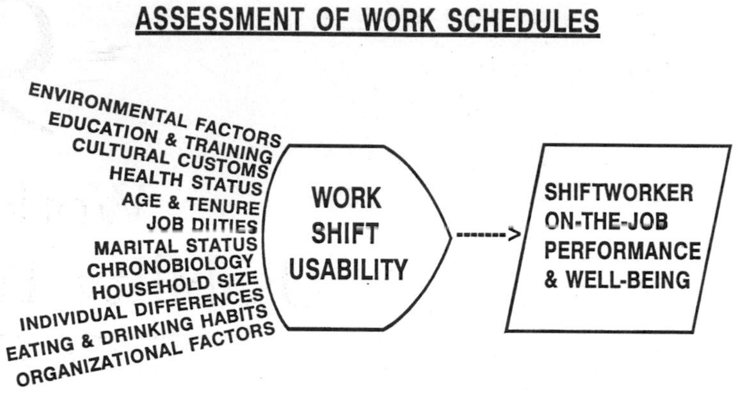

FIGURE 96.1 The major variables which have been related to shift worker performance and well-being.

History

Although night work has existed for centuries, the laws governing night and shift work in the United States cover only a very small percentage of the total workforce (Steinberg, 1982). These laws originated mainly prior to World War II, and they are based mostly on political and economic issues present at the time they were enacted. Research interest in shift work issues evolved after World War II, as the scientific and business community became more sensitive to the problems of night work. On an international level, work shift specialists have been organized since 1957, meeting on a regular basis as the Scientific Committee on Night and Shiftwork of the International Commission on Occupational Health. These meetings have resulted in a regular stream of dedicated book and journal publications. The result is an international and interdisciplinary literature which has been instrumental in increasing our awareness of the cultural, biological and psychosocial issues which have a demonstrated impact on the health, safety and/or performance of shift workers.

It is now quite clear that impact of night and shift work is always determined by multiple variables. Thus, a systems approach to work schedule assessment, design, and selection is in order. Figure 96.1 provides a listing of some of the major variable categories which have been related to shift worker on-the-job performance and well-being. The combined effect of these variables determines work shift usability. *Work shift usability testing* is aimed at the assessment and installation of work schedules whereby there is a good match between work shift usability and worker performance and well-being. This is the primary topic of this chapter. It is beyond the scope of this chapter to review all of the evidence confirming each of the categories listed in Figure 96.1, their operation, or all of the variable interactions which occur. Instead, some basic shift work concepts will be presented and an approach to the assessment of work shift usability testing will be outlined.

In the past, work shift assessment and the design of work shift systems have been purveyed by conventional managers or subordinates directed by these managers. This do-it-yourself approach is no longer appropriate or adequate. The work shift literature is much too vast, diverse, and difficult to locate and comprehend. Work shift options are expanding and fraught with hazards to be avoided. Contemporary work shift assessment and/or design projects require the skills of a trained work shift usability specialist. This chapter provides an introduction to the methods, terminology, and concepts of a work shift usability specialist. It should be viewed as a guide and introduction to a complex problem area, not an exhaustive review of the literature.

Terminology

The impact of cultural and social variables on the evolution of contemporary shift work systems is evident in the diversity of the terminology and practices which exist in the workplace. Table 96.1 provides some operational definitions for the shift work terms to be used in this chapter. Shift work as a term applies

TABLE 96.1 Definitions for Shift Work Terms, as Used in this Chapter

Term	Definition
First shift	Often called the day or morning shift, this is a work time period of about 8 h in duration which mainly falls between the hours of 0600 and 1700.
Second shift	Often called the afternoon-evening shift or swing shift, this is a work time period of about 8 h in duration which mainly falls between the hours of 1500 and 0100.
Third shift	Often called the night or graveyard shift, this is a work time period of about 8 h in duration which mainly falls between the hours of 2200 and 0700.
Shift break	A time period within a shift when work is not scheduled, under usual conditions. Usually less than 1 h long. When longer, it is often termed a split-shift.
Permanent hours	Fixed work hours whereby the shift work time period of an individual does not change under usual conditions.
Rotating hours	Changing work hours period whereby an individual works more than one shift time period, with the sequence of consecutive work time period types follows a set and regular schedule under usual conditions.
Irregular hours	Changing work hours whereby the hours an individual works vary, or are perceived to vary, in a very irregular or unpredictable way as a usual condition.
Compressed hours	Any shift type which on a regular basis includes one or more work time periods longer than 8 h which results in a workweek of less than 5 full days.
Continuous operations	A workplace or position which employs shifts around-the-clock and on every day of the week.
Discontinuous operations	A workplace or position which does *not* employ shifts around-the-clock and on every day of the week
Non-workday	Also called and off-day or day-off, this is a calendar day in which no work is scheduled. For some schedules this is a time period less than 24 h in duration.

when more than one of these terms applies to a given workplace. Definitions like these are needed if one is to begin to comprehend the international literature on shift work and attempt to apply it in an intelligent manner.

The definition for the *first shift* (day shift) provides a good example of the diversity of practice which exists in the literature, and why definitions are needed. Contrary to the definition in Table 96.1, a few workplaces in the U.S. refer to the night shift as the first shift, starting around midnight. Consistent with our definition, however, most United States workplaces start the first shift around 0700 to 0800. In Scandinavian countries a 0500 starting time is common. Spain provides a further contrast, since there a starting time of 0900 or later does not appear to be unusual.

Another example of differences in practice relates to how the hours of work are usually scheduled. *Permanent* hours are common in the United States, but approached with caution and special laws in Austria. The rate at which *rotating hours* are changed in the United States is much slower than that practiced by many European workplaces. In the United States, changing shifts once per week is usually considered to be a rapid rotation rate. European investigators term this a slow rotation rate, since some change shifts several times each week. Some specific examples of these, as well as a more detailed notation system, are described in Tepas, Paley, and Popkin (1997).

Differences in practice and/or preference like these and others make it quite difficult for the inexperienced reader to accurately compare shift work findings. One cannot simply accept the labels attached by users or experts to the work schedules they note in their publications and reports. In using work hour labels and definitions, one should recognize that they frequently use historical and colloquial terms which may not match your own usage. As the diversity and variety of work shift systems increase, the importance of using a detailed notation system increases. Work shift presentations without operational definitions and work hour details can be misleading and perhaps dangerous.

Basic Concepts

Chronobiology is one of the shift work variables listed in Figure 96.1. Although it should never be considered the main or only variable to be considered in the design of a work shift system, it is the basis of some special problems which all night workers must face. Just as many animals are nocturnal beings, research in the last 20 years has clearly demonstrated that people are *diurnal.* That is, each of us has an internal biological clock. This endogenous clock has a natural cycle of about 25 h, but on earth it adjusts to around 24 h as it is entrained by the daily changes in light and dark. Concomitant with this cycle are daily variations in most, if not all, biological systems. These variations are termed *circadian* (from the Latin *circa,* about, and *dies,* a day) variations, since they cycle daily. For the human, this usually translates into an animal who is most likely to be working during the daylight and sleeping during the night. Consistent with this, human societies and cultures are, for the most part, circadian and diurnal.

Changing the shift one works, for example changing from the first shift to the third shift, may require many changes. For this example, after the change one would work during the night and sleep after work during the day (Tepas and Carvalhais, 1990). This is a change in both the order and time-of-day of sleep. Although the human biological clock does make small adjustments every day to keep in phase with the seasonal changes in daylight, research on humans in controlled environments suggests that large changes like this are very difficult and take many days. Thus, it may be that in a real workplace *total* adjustment of the biological clock to a new work shift is limited and may not be desirable.

Since we most often live in a society and culture which is mainly day-orientated, total adjustment of social circadian variations may be even more limited. Thus, the demands of a change in work shift are at their least difficult for some, and at their worst may be impossible for others. For the human factors specialist, two work shift design strategies are suggested by these biological and social limitations. On the one hand, one might design and implement schedules which *minimize* the need for large adjustments in biological and social variables, thereby trying to maintain existing circadian variations. At the other extreme, one might implement work shift designs and conditions which require large adjustments *and* at the same time *facilitate changes* in the circadian variations. Both of these alternative strategies have been used, and each has advantages and disadvantages. Work shift usability testing should keep these two strategies in mind, since they make clear that there are often several alternative solutions to a given work shift design problem.

Sleep as a Benchmark

The concept of a *benchmark* comes to us from the work of land surveyors. A benchmark is a discretionary mark made by a surveyor on a permanent landmark that has a known position and altitude. The benchmark is then used as a reference point in determining other locations. When the appropriate reference marks are needed, the sleep length of workers has proven to be a fairly robust variable for surveying the impact of work shifts.

In general, there are large individual differences in the usual sleep length of humans. These differences make it very difficult to predict, in advance, all work shift problems within an individual. On the other hand, sleep length has a demonstrated value as a good tool for the *actuarial* analysis (group prediction) of work shift impact (Dekker, Tepas, and Colligan, 1996). For discussion purposes, sleep length will be used as a marker as this chapter provides the reader with an introduction to shift work issues.

Figure 96.2 shows the major sleep length periods of a diverse sample of experienced shift workers on discontinuous operations with *permanent* hours (Tepas and Carvalhais, 1990). Mean sleep length is presented for each of the three shift types on workdays and non-workdays. The differences between shift types on workdays are significant and similar to those reported in many studies: workers on the third shift sleep least; those on the second shift sleep the most; and the first shift is somewhere in between. For all three shift types, non-workday sleep is significantly longer than workday sleep, but the differences between shift types on non-workdays are *not* significant. Workday sleep length has a small but significant *positive* correlation with non-workday sleep length. Thus, workday sleep reductions are not fully replaced by non-workday sleep.

A number of additional characteristics of shift work are demonstrated in Figure 96.3. These data are from fire fighters on *rotating* hours (Paley and Tepas, 1994). For this analysis, the fire fighters were divided into senior and junior groups. Junior fire fighters had from 1.2 to 3.7 years of shift work exposure, and

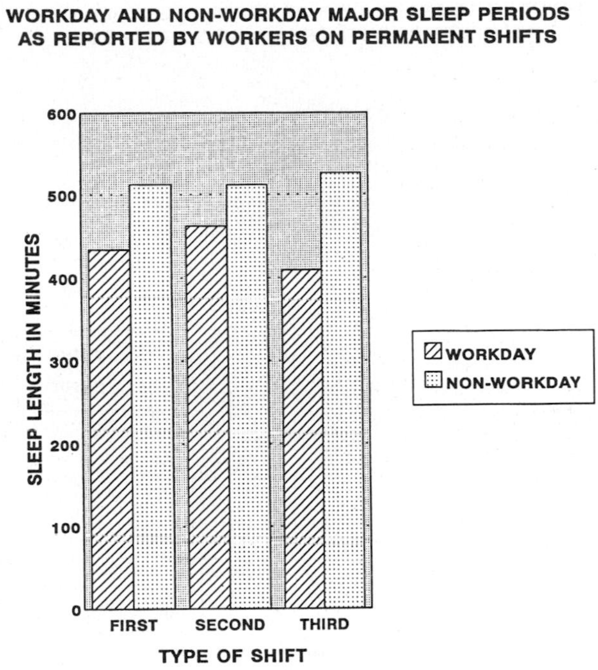

FIGURE 96.2 Mean major sleep length on workdays and non-workdays. Data graphed from a study of permanent discontinuous hourly shift workers. (From Tepas, D.I. and Carvalhais, A.B. (1990) Sleep patterns of shift workers, in A.J. Scott (Ed.) *Shiftwork.* Philadelphia: Hanley and Belfus. With permission).

ranged in age from 22 to 43 years. Senior fire fighters had from 10.8 to 19.6 years of shift work exposure, and ranged in age from 30 to 59 years. The sleep length data for junior fire fighters are *not* significantly different from those of senior fire fighters. However, the type of shift sleep differences are significant and parallel those of permanent shift workers as demonstrated by Figure 96.2. It should also be noted that the third shift sleep of rotating workers is shorter than that of the permanent third shift workers.

Longitudinal studies of shift worker sleep have also confirmed that the reduction in sleep length associated with third shift work does *not*, as a general rule, disappear with repeated exposure. Radosevic-Vidacek, Vidacek, Kaliterna, and Prizmic (1995) have demonstrated that this reduction in sleep length associated with night work remains significant after five years of exposure to rotating hours. Gersten (1987), using age-matched samples of permanent hour first shift workers for additional control for age changes, confirms that the reduction in sleep length associated with third shift work remains significant after over six years of exposure to permanent hours. Interestingly, age and gender (Tepas, Duchon, and Gersten, 1993; Dekker and Tepas, 1990) do significantly interact with the third shift tenure effect.

In sum, sleep length is a significant benchmark for the impact of shift work on real shift work users. The data reviewed clearly indicate that work has an impact on worker sleep, that shift type is a significant factor, and that the daily impact of a shift schedule does not appear to fully disappear with exposure. This suggests that sleep length differences can be used to assess or monitor the impact of a work schedule on groups of workers for actuarial purposes, using appropriate norms.

Work Schedule Impact Model

Shift workers, as a rule, do *not* suffer from a sleep disturbance or disorder. They do frequently complain about their sleep and have sleep problems, as most people do (Howarth, Pratt, and Tepas, 1997). There is little evidence, however, to support the notion that most healthy workers when placed on shift work immediately manifest a disorder or disturbance which requires medical treatment or a special training program (Mahan, Carvalhais, and Queen, 1990; Cole, Loving, and Kripke, 1990; Tepas, 1993). In addition,

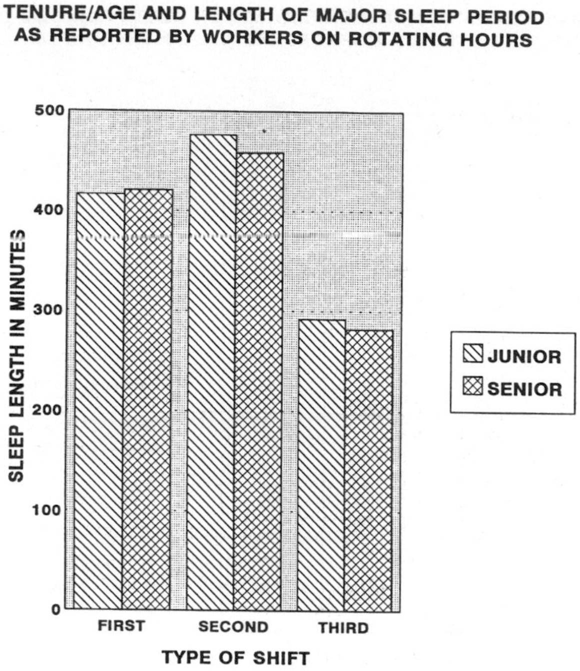

FIGURE 96.3 Mean major sleep length for fire fighters on rotating continuous shift work. The senior and junior groups differ in both their tenure on this schedule and their age. (Data graphed from Paley, M.J. and Tepas, D.I. (1994) The effect of tenure on fire fighter adjustment to shift work. Paper presented at the 23rd International Congress of Applied Psychology, Madrid, Spain.).

it is important to note that short- and long-term shift tolerance, like sleep deprivation, can be influenced by the impact of social factors (Monk, 1989). Evidence does suggest that *acute* (short-term) exposure to some work shifts and/or normal circadian variations in alertness does *increase* the probability of accidents and injury (Mitler, et al., 1988).

In the long run, however, *chronic* exposure to some work schedules most probably does lead to illness *and* injury (Knutsson, 1989; Kawachi, et al., 1995; Harma, Seitsamo, and Ilmarinen, 1997). Thus, in practice, the impact of chronic exposure to a given work schedule may be quite different from that of an acute exposure (Tepas, 1982). The distinction between chronic and acute work shift impact is diagrammed in Figure 96.4. Bias from years of quality laboratory research has led many to overlook this distinction and assume that the impact of shift work can be easily predicted from existing laboratory research. This is not true. Field research results sometimes differ from laboratory results. A conservative user must assume that the impact of a work schedule varies with duration of exposure to shift work.

Using the sleep length data presented earlier (Tepas and Carvalhais, 1990) as a benchmark, Figure 96.5 provides an estimate of the *total accumulated sleep loss* (TASL) one might expect during 30 days on permanent discontinuous shift work. This is a frequently used shift work schedule in the United States. The reconstruction shown assumes that non-workday sleep length is a reasonable estimate of the amount of sleep needed, and it is based upon group means for these large samples. Since the three shift types do not vary in the length of their non-workday sleep length but do vary significantly in their workday sleep length, sleep loss carry-over or debt (TASL) increases as consecutive days on this schedule increase.

Given the increased health and safety risk exhibited in experienced long-term shift workers, it seems reasonable to assume that the chronic carry-over demonstrated by sleep length data may also hold for many of the variables listed in Figure 96.1. Figure 96.6 provides a graphical view of this general work shift carry-over model. The model assumes that health and safety risk increases whenever there is a negative carry-over of exposure impact from one day to the next day. It also assumes that multiple

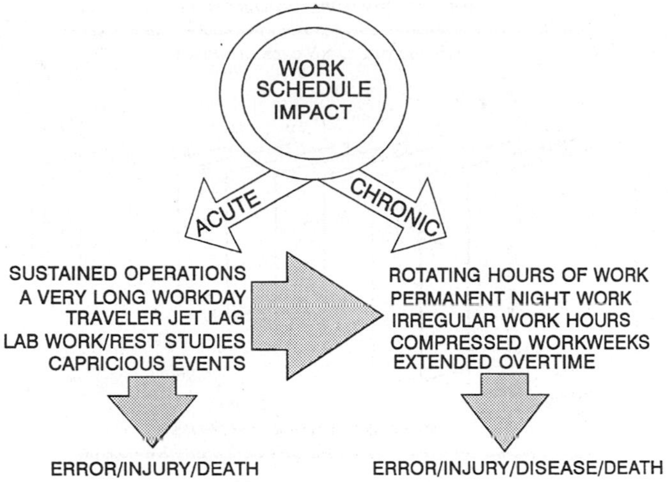

FIGURE 96.4 The distinction between chronic and acute work shift impact. The impact of a work shift may vary with duration of exposure.

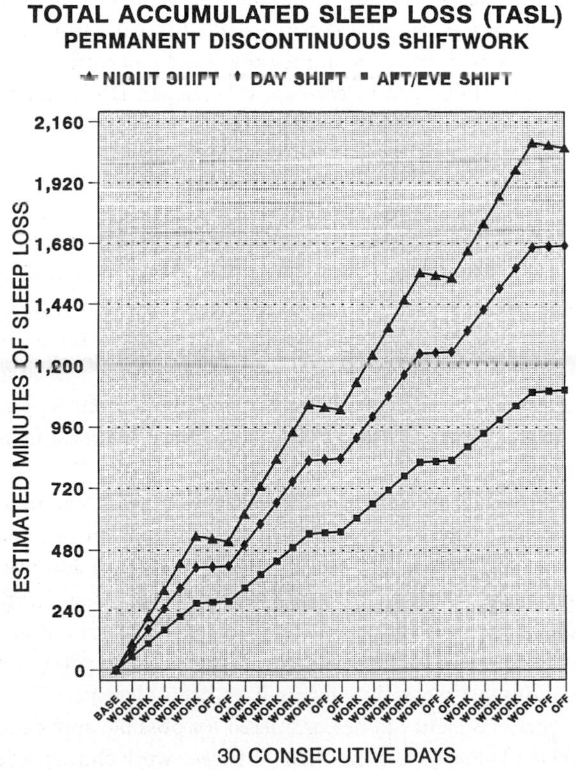

FIGURE 96.5 Estimated total accumulated sleep loss for workers on permanent discontinuous shift work over 30 consecutive days on this schedule. (Estimates based on data from Tepas, D.I. and Carvalhais, A.B. (1990) Sleep patterns of shift workers, in A.J. Scott (Ed.) *Shiftwork*. Philadelphia: Hanley and Belfus. With permission).

variables, both social and biological, may simultaneously exhibit carry-over, sum, and thereby increase or decrease overall cumulative health and safety risk.

In summary, it is reasonable to expect that the acute and chronic impact of work shifts may differ and must be evaluated. Obviously, it is also true that very long continuous operations and/or work shifts

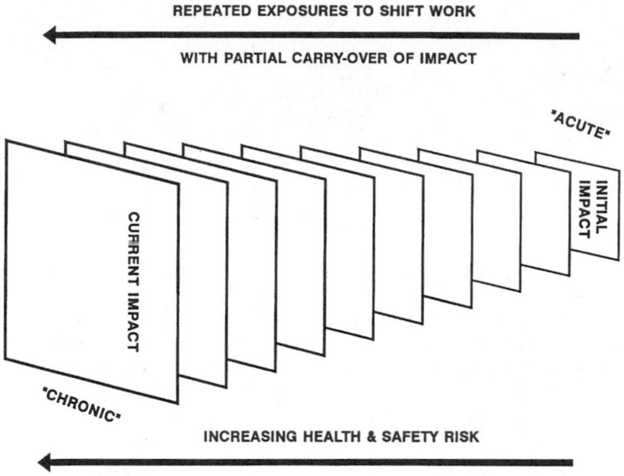

FIGURE 96.6 The work shift carry-over model. Repeated exposure to a shift with partial carry-over increases health and safety risk.

have an acute impact on performance which can terminate or flaw behavior (Kreuger, 1989). In the field, the chronic impact of work shifts may be more difficult to identify, since those who cannot tolerate night work may elect to leave shift work. It has been argued that experienced shift workers should be viewed as a survivor population (Koller, Kundi, and Cervinka, 1978; Tepas, Duchon, and Gersten, 1993).

Work Shift Usability — Summary

Advances in our knowledge of work shift variables during the last 20 years have slowly led to changes in traditional expectations. Early shift work investigators expected that their efforts would identify universal work shift *hazards*, and they assumed that once these hazards were identified they would work on minimizing their use. Most managers, on the other hand, expected that research would provide them with universal *best ways* to achieve maximum shift worker productivity. Both perspectives assumed that in the future they would be able to apply universal limitations or principles in a fairly mechanical way. It is now quite clear that both of these viewpoints were quite naive and incorporated expectations which are neither appropriate nor practical. Unfortunately, many managers and some investigators continue to hold these naive perspectives.

Why are the more traditional approaches wrong? Four developments lead to a change in outlook and the contemporary approach taken in this chapter. *First* is the expansion in our knowledge with regard to the range of complex variables impacting shift worker performance and well-being. This multiple-variable context makes precise prediction and application difficult. *Second*, interaction between these complex variables sometimes reverses the valence sign of a given work shift. That is, a number of evaluation studies have demonstrated that the same work shift may be good in one workplace and bad in another. *Third*, many additional work shift forms have been identified. These are alternative work shifts which have already been used and can be considered for possible application at new sites. Knauth (1997) estimates that worldwide there are over 10,000 different work shift schedules being used. This is probably a conservative underestimation. *Fourth*, the evolution of new manufacturing and scheduling technologies makes the use of some innovative shifts practical or required. In some cases, these are shifts which were previously judged to be impractical or unlikely.

The contemporary approach to work shifts recognizes the developments outlined above. Informed work shift investigators admit that the shift design task may sometimes be more art than science. The search for the ideal work schedule, a suit that fits without tailoring, has ended. Informed managers recognize the complexity of the work shift problem, the importance of the task, and the need for expert help. When informed, both shift experts and managers are working toward shift solutions with objectives which include improved health, safety, and productivity. Most important, there is a recognition that apparent solutions require objective assessment and evaluation. Work shift usability testing is one approach to achieving these goals.

96.2 Work Shift Assessment

Introduction

Usability testing is not a new approach for the ergonomics professional. It has been used for years with many problems including the design and evaluation of tools, devices, displays, and software. This usability testing refers to the methods used to ascertain the value of a design, not simply an effort to assess the quality or ability of an individual. The user aids in the identification of desirable and undesirable designs, but in most cases does *not* select the final design. Alternative designs are often considered. A fundamental feature of this approach to design is the assumption that *use testing with real users is a required and a fundamental approach* in most cases.

Work shift usability testing retains this fundamental assumption that real users should participate in assessing the value of a design. In addition, it assumes that a macroergonomic approach is needed. That is, one must recognize that work shifts, as well as changing work shifts, do impact and interact with many of the basic ways in which an organization operates. Within this approach, work shift assessment is the first phase of work shift usability testing, and it can be practiced as a three-step process: preliminary assessment, initial assessment, and formal assessment.

Preliminary Assessment

As an introduction to a given work shift usability testing effort, the work shift specialist completes an informal guided walk-through of the complete workplace. Following this first step, the work shift usability testing scenario begins with a *preliminary assessment* of the work shift *perceptions* present in the workplace. This preliminary assessment is often coupled with a general review for users of what experts now know about work shift impact, and a more specific discussion of the methods which might be used by the visiting specialist. When practical, this should be done by presentations and discussions within three meetings: *first*, a meeting with the highest level resident manager; *second*, a meeting with a representative group of middle managers; and *third*, a meeting with local union representatives. If the workers are not represented by organized labor, the *middle* level managers should be asked to supply a worker group which they feel represents all facets of worker opinion and job duties.

For the work shift specialist, the primary purpose of these meetings is to assess whether the culture and customs of the workplace are receptive to the methods required for work shift usability testing. A prudent approach recommends that one *not* commit to doing work shift usability testing until after completing and evaluating the results of the preliminary assessment. Since the users are to be participants in the usability testing, one must assess whether the required methods are feasible and appropriate for the specific workplace under consideration. A secondary goal is to gather some initial preliminary impressions as to what and how these individuals currently perceive their own work shift issues and problems. Finally, these meetings provide the specialist with an opportunity to learn more about the technology, tasks, practices, and limitations of this specific workplace.

Figure 96.7 lists some of the important features, common to quality work shift usability testing, which should be communicated in each of these meetings. These points should be communicated to all parties as the essential features required to assure a state-of-the-art work shift usability testing effort. Obviously, these presentations provide the specialist with a key opportunity to begin to earn the trust of the organization at all levels. Each presentation should clearly communicate to those present what will be required from all parties if work shift usability testing is to be done. It is appropriate to present these features to everyone as requirements, rather than as options which they might select from. Obviously, additional requirements should be added to this list when warranted by local conditions.

REQUIREMENTS FOR WORK SHIFT USABILITY TESTING

- √ **Time: A quality assessment takes time.**
- √ **Participation: Universal participation is possible and needed.**
- √ **Trust: The Work Shift specialist must be trustworthy and objective.**
- √ **Voluntary & Anonymous: Only group data will be revealed.**
- √ **Alternatives: All suggestions will be evaluated and more than one alternative solution will probably be made.**
- √ **Selection: The Final solution selection is not made by the Work Shift Specialist.**
- √ **Solution: The Work Shift Specialist helps install a changes, if it is needed.**
- √ **Evaluation: If any changes are installed, they should be evaluated by the Specialist..**

FIGURE 96.7 Some of the important features, common to most work shift usability testing, which should be communicated to potential users in advance.

For those attending these meetings, the work shift specialist also provides a general presentation of the potential benefits of a good work shift design, as well as a general discussion of how bad work shift design can lead to problems. This is followed by the detailed discussion of the *methods* which are used by this work shift usability specialist. Both the presentation and the discussion should be conducted in a manner which invites questions and provides honest answers. Each of the three preliminary assessment meetings should include an overt attempt by the work shift usability testing specialist to determine what those present feel are important issues, likely outcomes, and interest in further participation.

Initial Assessment

Following completion of the preliminary assessment, the *initial assessment* begins. At this point, the work shift usability specialist has not as yet agreed to do work shift usability testing. Using the data gathered during the preliminary assessment, the specialist must now determine if usability testing is practical in this situation and how it should be done. The features listed in Figure 96.7 provide a good outline for this initial assessment by the specialist. User interest, participation, and rights must be assured at all levels within the organization before the specialist proposes to do a complete and professional usability testing.

In some cases, one or more of the features in Figure 96.7 may not be met. Should this be true, the specialist should consider this as a possible and adequate reason for project termination. Alternatively, one or more of these features may be the basis of additional discussions and lead to subsequent change. For example, a manager may indicate that s/he is simply interested in obtaining a stamp of approval for a new work shift s/he has already selected on some basis other than discussion with the specialist. In this case, the specialist may be able to broaden the manager's perspective by providing several additional examples of possible alternative work shifts that should be considered.

On ethical grounds, the right to voluntary and anonymous participation should never be negotiated. Given the need for additional data and the potential for error, *evaluation* is another feature which has major merit and should be retained whenever possible. In general, the features listed in Figure 96.7 should be viewed as *requirements* needed to do quality work shift usability testing. Explanation and discussion is appropriate, when possible, but these points should not be presented as special options which are not needed. When the specialist concludes that work system usability testing is not recommended, a brief written report explaining why these services are not recommended is appropriate.

When work system usability testing appears to be practical, the specialist should attempt to categorize the work shift problems of interest. Since many in the workplace have a rather limited knowledge of work shifts and what we know about them, their initial identification of what they want may or may not be in error. The task of placing the work shift problems into a category often helps clarify *to the users* what they should realistically expect from work shift usability testing. In addition, categorization often promotes user participation, since it may be the first time specific goals have been stated. Given our current knowledge of work shifts, four categories are suggested. These *categories of evaluation* are: acute exposure hazard analysis, chronic exposure hazard analysis, acute exposure schedule change, and chronic exposure schedule change.

Acute exposure hazard analysis. The work shift to be assessed is *not* experienced by an individual on a daily or frequent basis, but many individuals may be exposed to this work shift over time on an actuarial basis. This category includes work shifts which may or may not constitute a threat to the health, safety, and/or performance of workers. The work shift usability specialist is trying to find out if any of these threats exist or is likely. Knowing what the threat or potential threat is may suggest a prophylactic measure or therapeutic method. In some cases it might be a proscriptive work shift which is to be avoided whenever possible. Some likely examples: travelers with many rapid time zone changes (jet lag), sustained emergency medical operations, search and rescue efforts, laboratory shift work studies, and, sleep deprivation studies.

Chronic exposure hazard analysis. The work shift is experienced by individuals on a daily or frequent basis. This category includes work shifts which may or may not constitute a threat to the health, safety, and/or performance of the workers. The work shift usability specialist is trying to find out if any threats exist or should be anticipated as a future development. The employer does not have an interest in change, but wants to know the current status of the work shift system. Finding that a threat exists may, of course, suggest that a new work shift is needed. An educational program and/or other benefits may also provide solutions. Some likely examples: any continuous operation, permanent night shift work, workers employed at remote locations, work in a harsh environment, permanent part-time workers, and on-call-when-needed workers.

Acute exposure schedule change. The work shift is *not* experienced by an individual on a daily or frequent basis, but over time many individuals may be exposed to this work shift on an actuarial basis. This category includes work shifts in which the employer wishes to propose a new shift remedy. The need to change can be related to health, safety, performance, legal, and/or economic reasons. The work shift usability specialist is to assess the current status of things and then suggest a remedy. Among the remedies are increasing the number of workers to decrease frequency of exposure, changing the manner in which workers are assigned to these shifts, installation of rules governing frequency of exposure, and an improved work shift design. Examples of this are the same as for those given for acute exposure hazard analysis. The difference is that for one reason or another the employer elects to make a work shift system change.

Chronic exposure schedule change. The work shift is experienced by individuals on a daily or frequent basis. This category includes work shifts which the employer thinks need changing due to health, safety, performance, legal, and/or economic reasons. The work shift usability specialist is to assess the current state of things and then suggest work shift changes. In addition to a new work shift design, an educational program and other benefits may also be suggested. Examples of this are the same as those given for chronic exposure hazard analysis. Again, the difference is that for one reason or another the employer elects to explore the possibility of making a work shift system change.

Proposal and schedule. The initial assessment, when additional work is appropriate, concludes with a proposal and schedule for a *formal assessment* and subsequent activities. The content and schedule of this proposal will vary as a function of categorization (above). Obviously, a formal assessment may or may not fully confirm the impressions of the preliminary or the informal assessments. In practice, much of the formal assessment method does *not* vary with categorical assignment. What does vary with categorical assignment is the information and recommendations generated from the subsequent analysis of the formal assessment data. In a very real way, the proposal should make the objectives and requirements of the process very clear to both employer and user.

When only a *hazard analysis* is requested, a formal assessment is proposed, but no new work shift suggestions are included in the venue. For an *acute* hazard analysis report where a hazard is detected, the employer and users should expect recommendations for a prophylactic preparation or therapeutic treatment via a training program or support services. Managers and users should also be warned in advance that these educational programs and support services, if implemented, will need evaluation since they are largely unproved (Tepas, 1993). For a *chronic* hazard analysis report which identifies the presence of a hazard, neither employer nor participants should be led to expect that a treatment or training program might overcome the problems associated with a bad work shift schedule. An improved work schedule should always to be given priority over the installation of a treatment or training program.

When a *schedule change* is requested, a formal assessment is also proposed. The employer and users should be led to expect that more than one new work shift design suggestion will be included within the results of the formal assessment report. They should also be informed that the report will include a plan to evaluate the impact of the new schedule. For an *acute* exposure schedule change, the strategy is to eliminate or minimize the occurrence of a hazard by eliminating or minimizing the use of the hazard-producing shift. Many *chronic* exposure effects (for example, working at night) cannot be eliminated. For these chronic impact cases, the employer and users should expect to be supplied with alternative work shift design recommendations. Each of these alternative work system designs will have both advantages and disadvantages which they will evaluate.

Formal Assessment

A variety of methodological approaches are available for the formal assessment data collection effort. These include the study of personnel and safety records (Colligan, Smith, Hurrell, and Tasto, 1979), laboratory testing and field interviews (Walsh, Gordon, Maltese, McGill, and Tepas, 1979), and questionnaire surveys (Smith, Colligan and Tasto, 1979; Gordon, Tepas, Stock, and Walsh, 1979). With each of these approaches, the objective is to gather representative data about users and their duties in a valid and reliable manner. Each of these methodological approaches has advantages and limitations which are reviewed in these publications. The questionnaire survey approach appears to be the method of choice, since practice shows that nearly all workers respond to these instruments when the requirements listed in Figure 96.7 are met (Tepas, Armstrong, Carlson, Duchon, Gersten, and Lezotte (1985).

Figure 96.8 provides a listing of some of the more important dimensions which should be included within this formal assessment. The order in which these dimensions are listed is not significant, and this is not an exhaustive listing. It is recommended that the actual variables used with any of these approaches be developed in an *iterative* process within small group meetings of the work shift specialist with managers and users. This iterative process promotes progress toward three goals. First, it helps confirm the specialist's status as an independent third-party *expert* who values participation and confidentiality. Second, it is one way of making sure that the assessment examines all variables relevant to the workplace being assessed. Third, the review of items with users assures that the vocabulary level and technical terminology used in the survey matches the ability of the user and local workplace term usage.

Although the selection of items for this formal analysis is participatory and iterative, it must be tempered by using standard and proven items whenever possible. There are two major reasons for this. First, it often allows one to assume item validity generalization on the basis of previous findings. Second, collecting quantitative data in a standard way allows one to directly compare the data from one workplace to those gathered in another. Using standard items on the survey provides the specialist with access to the benchmarks needed to evaluate the data collected.

Experience with the sleep length variable provides a good example of why validity, reliability, and benchmarks are needed. All of the sleep length data presented earlier in this chapter are subjective data gathered by asking the same standard questions every time. Sleep length was calculated from times produced by workers when they were asked when they usually go to bed and get up. These times are

DIMENSIONS FOR WORKER ASSESSMENT

- **Demographic reports of age and gender**
- **Usual work, work hours and overtime**
- **Sleep, fatigue and stress reports**
- **Eating behavior and appetite**
- **Drug, alcohol and tobacco use**
- **Medical problems and treatment**
- **Marital, family and household status**
- **Work schedule history and preferences**
- **Non-workplace activities and preferences**

FIGURE 96.8 Some of the more important dimensions which should be included within the formal user assessment.

significantly and highly correlated with polysomnographic, log, and interview estimates of sleep length (Tepas, Walsh, and Armstrong (1981). However, simply asking workers how many hours they usually sleep often fails to yield valid estimates. It is not surprising to learn that some investigators (for example, Frese and Harwich, 1984) have had trouble relating their measure of sleep length to shift schedule differences.

As noted earlier, the major goal of formal assessment is the production of representative quantitative data. These data aid the work shift usability specialist in designing appropriate alternative work schedules, and supporting their use. Sometimes managers or users attribute additional goals to this effort. For example, the term "testing" leads users to conclude that this assessment is a competitive evaluation or examination which they can pass or fail. Other users conclude that the assessment is an election wherein they are voting for or against some work shift schedule. Neither of these attributions is correct or helpful, and overt efforts to minimize their face validity are warranted.

Whether the task at hand is limited to the evaluation of a practiced work shift design or it also includes the design of a new work shift system, Figure 96.9 provides a listing of some of the basic options which a work shift usability specialist should consider when designing a new work shift system. This listing is not exhaustive and the order of the variable listing is not significant. Given the fact that these options can combine to form thousands of different work shift systems, the suggestion that the design process be performed by a specialist is appropriate. Similar to many ergonomic design problems, several work shift designs will appear to be reasonable proposals.

When the category of evaluation is schedule change, the work shift usability specialist will usually produce a written *work shift assessment report* which includes several design alternatives which merit workplace assessment. This written report to managers and users should make an overt effort to link the designs recommended to the objective data collected in the workplace under study. It should also relate the data collected in this workplace to other similar databases. The advantages and disadvantages of each work shift design proposal should be stated in a manner which will ensure that all of the suggested design alternatives are given a full workplace assessment.

96.3 Workplace Assessment

Introduction

When testing software for usability, the human factors specialist knows that some changes s/he might recommend will *not* be accepted because they are too costly. The work shift usability specialist faces

OPTIONS FOR WORK SHIFT DESIGN

- **Permanent or rotating hours**
- **If rotating, direction of rotation**
- **If rotating, rate of rotation**
- **Length of workday**
- **Number of consecutive workdays**
- **Number of consecutive non-workdays**
- **Time-of-day for shift changes**
- **Rules and laws for work shift use**
- **Special provisions for shiftworkers**

FIGURE 96.9 Some of the more important options which should be considered in designing work shift systems.

similar and even more complex problems when testing the usability of a work shift. Every work shift has both a long-term and an immediate monetary cost associated with implementation and use. Some work shift designs are simply not as cost-effective as other designs. Historically, there is reason to suggest that work shift changes were often introduced without taking the time to do a comprehensive workplace assessment. Improved work shift systems can significantly reduce costs, but there are many hazards which must be avoided by good planning, if these changes are to be profitable.

For example, the introduction of a continuous work shift may require an inventory increase due to a failure to consider receiving and shipping limitations. Also, the work shift system may require equipment maintenance down-time that makes it too expensive since it would require additional capital equipment investment. User work shift preferences may evoke many complex long-term and immediate costs. For example, overtime payments required by law might make some workdays or workweek features too expensive. Also, the chronic toxic substance exposure levels of an extended workday or workweek might increase sickness or increase physical agent carry-over to harmful levels. Many additional examples are possible.

Work Shift System Selection

Just as work shift assessment and recommendations require the services of a work shift usability specialist, the final selection of a work shift for use in a specific workplace requires special skills. A diverse management team is required, and this methodology may be appropriate even when the category of evaluation is one of hazard analysis rather than schedule change. When possible, this team should include the work shift usability specialist, appropriate managers, and representative users. This is the *workplace assessment team.* The initial task of this team is the completion of a *workplace assessment report* to complement the work shift assessment report prepared (alone) by the work shift usability specialist.

Using a systems approach, the workplace assessment team evaluates the work shift designs recommended (or the existing work shift system) by the work shift usability specialist. Figure 96.10 provides a listing of system dimensions which should be reviewed by the team as it completes the workplace assessment. This listing should not be regarded as either an exclusive or an all-inclusive list. An early task of the team should be a review of these dimensions to decide if any dimensions should be removed or added to their assessment task. A review of the list should make it clear that many of these items require special expertise outside the purview of the work shift usability specialist.

DIMENSIONS FOR WORKPLACE ASSESSMENT

- **Community and organization shiftwork history**
- **Health, safety and legal issues and records**
- **Present and future demand for product/services**
- **Present and future job market status**
- **Maintenance and support service requirements**
- **Utility, shipping and receiving costs/problems**
- **Supervision and manpower requirements**
- **Personnel data collection and analysis methods**
- **Measurement and use of productivity indicants**

FIGURE 96.10 Some dimensions which should be reviewed by the workplace assessment team when evaluating alternative work shift designs.

The methods and product of the workplace assessment will vary from workplace to workplace, molded to a considerable degree by the characteristics of the organization and the category of evaluation. Four alternative goals should be considered in each case: (1) a recommendation that the present work shift system continue to be used; (2) a change in work shift is in order, and the work shift specialist should offer some designs; (3) a recommendation that one of the alternative work shift designs offered be implemented; and (4) a recommendation that none of the alternatives presented is acceptable, but change is needed and additional alternatives should be examined.

96.4 Work Shift Implementation

When workplace assessment leads to the installation of a new work shift system, an *implementation plan* should be developed and put into practice. In many cases, this is done by the workplace assessment team which now becomes the *implementation team.* In any case, the work shift usability specialist should be an active participant in the development and practice of the implementation plan. Often, eagerness to try a new work shift system results in implementation without planning. This is not recommended.

The implementation team has four primary goals. First, a communication to all users which clearly states the rationale for selecting their work shift system, the rules which will govern work shift operation, and how the system will be evaluated. Second, the development of a training program for users which will instruct them about their individual work schedule and what they might do to improve their ability to cope with it. It is often beneficial to include spouses in this training program, and to involve users in an evaluation of the training program. Third, the development of a plan to install any administrative or physical changes in the workplace is required for the new work shift system to become fully operational. Fourth, the design of a plan to evaluate the impact of the new work shift system.

Designing a research plan for the evaluation of a change, prior to the actual implementation of the new work shift system, serves several purposes. As a workplace announcement, the evaluation plan makes it clear to users that evaluation of a new system will take time, requires a significant test period, and assures that faults will draw attention. In addition, designing evaluation in advance makes it more likely that appropriate data are collected and methodological pitfalls are avoided. The research plan should be designed to evaluate both the acute and chronic effects of the work shift system installed on worker productivity, safety, and health.

96.5 Summary

Work shift usability testing is a macroergonomic process aimed at the design, installation, and evaluation of work shifts and shift systems. Today, multiple variables affect people and are associated with the use of thousands of different work shifts. Work shift usability testing involves work shift assessment, workplace assessment, and the assessment of work shift interventions when they are made. Recognizing the complexity of the ergonomic problem, the approach recommends the employment of a third-party independent specialist. The methods outlined apply to a wide range of work shift applications using actuarial prediction.

Whenever possible, work shift usability testing should be a participatory process involving real users. Research during the last decade or so makes it quite clear that work hours have both acute *and* chronic impacts on the safety, health, happiness, and productivity of work shift users. Since the paradigm addresses both acute and chronic changes, it can also be used to address unique problems such as those traditionally associated with hours of service issues or emergency service. Given the fact that humans are diurnal animals with biological clocks which are difficult to reset, the mastery of time is a difficult task. The realities of an international market and global organizations suggests that the mastery of around-the-clock operations may be the ultimate challenge of the new technological workplace.

The work shift usability testing model outlined has been applied to a wide range of situations and industries. For example, the same general approach can be applied to commercial transportation workers (Tepas, Popkin, and Dekker, 1990), traditional factory workers (Tepas, 1993), and workers on sustained operations (Carvalhais, Tepas, and Paley, 1994). In each case, the impact of work schedules is complex and the variety of work hour solutions is immense. The merits and limitations of this approach must be tested, and they will be most evident when the work shift usability testing includes a good evaluation research plan.

References

Carvalhais, A.B., Tepas, D.I., and Paley, M.J. (1994) An evaluation of a coast guard "live-aboard" concept: Can crews adapt to a restricted living and operational environment? *Proceedings of the Human Factors and Ergonomics Society 38th Annual Meeting,* 873-877.

Cole, R.J., Loving, R.T., and Kripke, D.F. (1990) Psychiatric aspects of shift work, in A.J. Scott (Ed.) *Shiftwork.* Philadelphia: Hanley and Belfus.

Colligan, M.J., Smith, M.J., Hurrell, J.J., and Tasto, D.L. (1979) Shift work: a record study approach. *Behavior Research Methods and Instrumentation,* 11, 5-8.

Dekker, D.K. and Tepas, D.I. (1990) Gender differences in permanent shift work sleep behavior, in G. Costa, G. Cesanna, K. Kogi, and A. Wedderburn (Eds.) *Shiftwork: Health, Sleep and Performance.* Frankfurt-am-Main: Peter Lang.

Dekker, D.K., Tepas, D.I., and Colligan, M.J. (1996) The human factors aspects of shift work, in A. Bhattacharya and J.D. McGlothlin (Eds.) *Occupational Ergonomics: Theory and Applications.* New York: Marcel Dekker.

Frese, M. and Harwich, C. (1984) Shift work and the length and quality of sleep. *Journal of Occupational Medicine,* 26, 561-566.

Gersten, A.H. (1987) *Adaptation in Rotating Shift Workers: A Six Year Follow-Up Study.* Unpublished doctoral dissertation. Illinois Institute of Technology, Chicago.

Gordon, G.C., Tepas, D.I., Stock, C.G., and Walsh, J.R. (1979) Gaining access to shift workers through labor unions. *Behavior Research Methods and Instrumentation,* 11, 14-17.

Harma, M., Seitsamo, J., and Ilmarinen, J. (1997) Shift work, aging and health: an 11-year follow-up study of nurses. *Shiftwork International Newsletter,* 14 (1), 127.

Howarth, H., Pratt, J., and Tepas D. (1997) Maritime search and rescue crew member reports of sleep duration and related problems. *Shiftwork International Newsletter,* 14 (1), 82.

Kawachi, I., Colditz, G.A., Stampfer, M.J., Willett, W.C., Manson, J.E., Speizer, F.E., and Hennekens, C.H. (1995) Prospective study of shift work and risk of coronary heart disease in women. *Circulation*, 92, 3178-3182.

Knauth, P. (1997) Innovative working times. *Shiftwork International Newsletter*, 14 (1), 2.

Knutsson, A. (1989) Shift work and coronary heart disease. *Scandinavian Journal of Social Medicine*, Supplementum 44.

Koller, M., Kundi, M., and Cervinka, R. (1978) Field studies of shift work in an Austrian oil refinery I: Health and psychosocial wellbeing of workers who drop out of shift work. *Ergonomics*, 21, 835-847.

Krueger, G.P. (1989) Sustained work, fatigue, sleep loss and performance: A review of the issues. *Work and Stress*, 3, 129-141.

Mahan, R.P., Carvalhais, A.B., and Queen, S.E. (1990) Sleep reduction in night-shift workers: Is it sleep deprivation or a sleep disturbance disorder? *Perceptual and Motor Skills*, 70, 723-730.

Mahan, R.P., Tepas, D.I., and Carvalhais, A.B. (1987) Morningness-eveningness norms for industrial shift workers, in A. Oginski, J. Pokorski, and J. Rutenfranz (Eds.). *Contemporary Advances in Shiftwork Research: Theoretical and Practical Aspects in the Late Eighties.* Krakow: Medical Academy.

Mitler, M.M., Carskadon, M.S., Czeisler, C.A., Dement, W., Dinges, D.F., and Graeber, R.C. (1988) Catastrophes, sleep, and public policy: consensus report. *Sleep*, 11, 100-109.

Monk, T.H. (1989) Social factors can outweigh biological ones in determining night shift safety. *Human Factors*, 31, 721-724.

Paley, M.J. and Tepas, D.I. (1994) The effect of tenure on fire fighter adjustment to shift work. Paper presented at the 23rd International Congress of Applied Psychology, Madrid, Spain.

Presser, J.B. (1995) Job, family and gender: Determinants of nonstandard work schedules among employed Americans in 1991. *Demography*, 32, 577-598.

Radosevic-Vidacek, B., Vidacek, S., Kaliterna, L., and Prizmic, Z. (1995) Sleep and napping in young shift workers: A 5-year follow-up. *Work and Stress*, 9, 272-280.

Smith, M.J., Colligan, M.J., and Tasto, D.L. (1979) A questionnaire survey approach to the study of the psychosocial consequences of shift work. *Behavioral Research Methods and Instrumentation*, 11, 9-13.

Steinberg, R. (1982) *Wages and Hours: Labor and Reform in Twentieth-Century America.* New Brunswick, New Jersey: Rutgers University Press.

Tepas, D.I. (1982) Work/sleep time schedules and performance, in W.B. Webb (Ed.) *Biological Rhythms, Sleep, and Performance.* New York: John Wiley & Sons.

Tepas, D.I. (1985) Flextime, compressed workweeks and other alternative work schedules, in S. Folkard and T.H. Monk (Eds.) *Hours of Work.* New York: John Wiley & Sons. pp. 147-164.

Tepas, D.I. (1993) Educational programs for shift workers, their families, and prospective shift workers. *Ergonomics*, 36, 199-209.

Tepas, D.I. (1994) Technological innovation and the management of alertness and fatigue in the workplace. *Human Performance*, 7, 165-180.

Tepas, D.I. and Carvalhais, A.B. (1990) Sleep patterns of shift workers, in A.J. Scott (Ed.) *Shiftwork.* Philadelphia: Hanley and Belfus.

Tepas, D.I. and Mahan, R.P. (1989) The many meanings of sleep. *Work and Stress*, 3, 93-102.

Tepas, D.I., Duchon, J.C., and Gersten, A.H. (1993) Shift work and the older worker. *Experimental Aging Research*, 19, 295-320.

Tepas, D.I., Paley, M.J., and Popkin, S.M. (1997) Work schedules and sustained performance, in G. Salvendy (Ed.) *Handbook of Human Factors and Ergonomics, Second Edition.* New York: John Wiley & Sons.

Tepas, D.I., Popkin, S., and Dekker, D. (1990) A survey of locomotive engineers on irregular schedules and their spouses: A preliminary report, in G. Costa, G.C. Cesanna, K. Kogi, and A. Wedderburn (Eds.) *Shiftwork: Health, Sleep and Performance.* Frankfurt-am-Main: Verlag Peter Lang.

Tepas, D.I., Walsh, J.K., and Armstrong, D.R. (1981) Comprehensive study of the sleep of shift workers, in Johnson, L.C., Tepas, D.I., Colquhoun, W.P., and Colligan, M.J. (Eds.) *Biological Rhythms, Sleep and Shift Work.* New York: SP Medical and Scientific Books.

Tepas, D.I., Armstrong, D.R., Carlson, M.L., Duchon, J.C., Gersten, A., and Lezotte, D.V. (1985) Changing industry to continuous operations: different strokes for different plants. *Behavior Research Methods, Instruments, and Computers,* 17, 670-676.

Walsh, J.R., Gordon, G.C., Maltese, J.W., McGill, W.L., and Tepas, D.I. (1979) Laboratory and field interview methods for the study of shift workers. *Behavior Research Methods and Instrumentation,* 11, 18-23.

Part VII
Ergonomics and the Working Environment

Section I
The Office Environment

97

Ergonomics of Seating and Chairs

Marvin J. Dainoff
Miami University

97.1 Introduction

In seated posture, the chair is, by definition, critical. For any work situation in which sitting is involved, the chair represents the primary support system which puts the user in contact with the workstation. This support function is even more important for those tasks, increasingly characteristic of modern workplaces, which require precise coupling of hands and tools and high degrees of visual attention for prolonged periods of time. A large proportion of such tasks, of course, include those involving video display terminals.

Designing effective seating is not a trivial problem. Eminent designers have attested to the problems of chair design. Frank Lloyd Wright complained that: "…all my life my legs have been banged up somewhere by the chairs I designed." Mies van der Rohe asserted that: "…a skyscraper seemed almost easier to design than a chair" (quoted in Stewart, 1987). The issues in chair design have often reflected competing interests and agendas, in which aesthetic and status considerations have often overpowered functionality. As a case in point, the famous Barcelona chair — designed for a 20-second seated appearance by the King of Spain–has had a major impact on the design community, although it is functionally better fitted to look at rather than sit in (Stewart, 1987; Pheasant, 1986). There is no inherent reason why functionality, aesthetics, and cost must inevitably conflict, but such conflicts have, unfortunately, been common. Perhaps part of the problem has been that there has been no coherent or integrated statement of the functional and physiological requirements for seated posture arising from an applied

0-8493-2641-9/99/$0.00+$.50

science (ergonomic) framework. Accordingly, individual chair designers have been required to create their own theoretical model of the seated user, incorporating intuitive concepts of both anatomical structure and user behavior into their designs. The goal of this chapter is to provide an organizing framework within which to discuss critical ergonomic issues applicable to both designers and customers.

Since the end of the 19th century, sitting in erect postures has, at least in Western societies, been considered both socially acceptable and biologically healthful (Zacharkow, 1988; Kroemer et al., 1994). *Erect posture* refers to a situation in which the head, trunk, and lower legs are vertical, and the upper legs and arms horizontal. This posture has been nicknamed the "cubist" work posture, since it is rather easy to describe with a series of right angles (Mark and Dainoff, 1988). Prior to the early 1980s, most ergonomic recommendations for "task" chairs (appropriate to office or industrial work) utilized a cubist model. A state of the art ergonomic task chair was one that had height adjustability of the seatpan, a rounded forward edge of the seatpan, and possibly some lumbar support.

However, with the advent of large-scale office automation, and the associated rise of musculoskeletal complaints among office workers, a greater degree of attention began to be focused on the functional requirements of seating (Dainoff and Dainoff, 1986). In particular, the cubist assumption was challenged. Grandjean et al. (1983) called the upright posture "wishful thinking," and indicated that the natural working posture would be one in which an adjustable backrest allowed the trunk to be inclined rearward. Mandal (1981, 1986), on the other hand, argued that efficient working posture required the trunk to be upright, but this could be achieved efficiently if the seatpan were sloped forward. Dainoff and Mark (1986) argued that both postures had their advantages, depending on the demands of the task. In fact, chairs which could allow both the Mandal and the Grandjean postures started to become available in the marketplace during the middle and late 1980s, and are now commonplace.

As additional research continues to emerge, however, it is clear that there is much left to understand about seated posture and chair design. Grieco's (1986) review paper indicated that static models of seated posture — involving specification of optimal postures for given tasks–were inadequate, and that movement itself was a primary requirement in seated work. Consequently, there continues to be active investigation involving issues relating to the dynamics of seated posture. These include examination of the relationships between seatpan and backrest angles adjustability, and the nature and extent of user adjustability.

The materials which follow cannot, at this stage, reflect a fully developed set of principles and recommendations. Rather, they will complement the chapter on "Office Ergonomics" by reviewing the underlying ergonomic issues and concerns which must be addressed by designers of chairs, as well as designers and users of the work environments within which the chairs are incorporated. While much of the focus will be on office workplaces, many of the same issues apply to industrial workplaces, and these will be addressed where appropriate.

97.2 Fundamentals

This chapter parallels and expands on both the content and the conceptual framework of *ecological ergonomics* laid down in "Office Ergonomics." It would be helpful to the reader to have read that chapter first, since issues of chair design and seated posture cannot be understood without considering the overall workplace context. Some of that framework will be briefly reviewed here.

Ergonomics can be described as the fit between people and the elements of the physical environment with which they interact (Dainoff and Dainoff, 1986). As such, ergonomics is inherently relationship-oriented in that absolute dimensions and physical characteristics of objects in the work environment must be defined relative to the relevant characteristics of the user. The ecological perspective of J.J. Gibson (1979) provides a principled approach to conceptualizing such relationships.

A key element in this approach involves the terms *affordance* and *effectivity.* Affordances refer to the characteristics of the physical environment measured with respect to the individual or "actor." Affordances thus represent, in physical terms, the potential for action afforded by the environment for a particular individual. Effectivities are complementary in that they refer to action capabilities of individuals measured with respect to the physical environment. In the case of seating, for example, an ordinary office chair (physical object) affords sitting (action) for an adult but not a typical two-year-old child. The adult and child differ in their effectivities (action capabilities) because of their different sizes.

In order to carry out goal-directed actions in the world, the individual actor must be able to perceive the affordances of that world. (A 250-lb. adult will most likely not choose to sit in a chair made for a two year old!) However, most real-world work environments consist of collections of affordances, some of which may be nested within others. To meet the challenge of characterizing such environments — to allow perception of complex sets of affordances by the user/actor — Vicente and Rasmussen (1990) have utilized a conceptual tool called the *means–end abstraction hierarchy*. Central to this concept is the notion that groups of affordances can be hierarchically organized by function. At any point in the hierarchy, the affordances at one level act as the ends or goals with respect to the affordances (means) at the next level down.

The rest of the material in this chapter will be, to some extent, organized as a means–end abstraction hierarchy. The first three sections under Fundamentals will include the three highest levels of the hierarchy. The remaining sections, under Applications, will consist of the lower levels. The relationships between levels will be addressed as they occur.

System Goals in Seated Posture (Functional Purpose)

The highest level in the hierarchy reflects the overall goals or functional purpose of the system. Branton's seminal 1976 paper conceptualized seated posture as homeostatic in nature involving two complementary demands: the need for structural support in order to provide stability for the body and the need to dissipate fatigue resulting from the effects of prolonged compression of body tissues against the chair surfaces. Thus, *stability* and *movement* can be considered overall goals of seated posture.

This assumes that seated posture is, in fact, appropriate for a given work task. This may be obvious in office work, but is more problematic in industrial settings. (See Kroemer et al. ([1994, p. 365] and Helander [1995, Ch. 6] for decision criteria for choosing situations in which sitting rather than standing is appropriate.)

Priority-Defining Constraints and Mechanisms (Abstract Function)

The abstract function level of the hierarchy describes the underlying conceptual and scientific framework within which the overall system goals can be achieved. This framework includes proposed postural mechanisms which are subject to certain constraints which, in turn, provide a method for prioritizing chair design decisions.

Statement of the Problem

The human body can be considered a biomechanical system with multiple degrees of freedom. That is, in principle, each muscle–joint combination can be independently controlled by the brain–nervous system, resulting in a virtually infinite number of body postures. Each such posture represents an equilibrium state (balance of internal, applied, and reactive forces) acting against the force of gravity. In practice, this large number of degrees of freedom is greatly reduced by a series of constraints. These constraints can be categorized as follows:

Task Constraints

The demands of each individual task will naturally specify certain postural requirements. For example, a data entry task requires that the fingers be in operational contact with a keyboard and that the copy and screen should be within the central field of vision. These task constraints significantly reduce the number of possible postural orientations and are of great importance in determining functional requirements for chairs.

Environmental Constraints

The number of possible postural orientations is further reduced by the design limitations of the objects, tools, or furniture within the environment with which the user must interact. For an office environment, for example, the linear and angular dimensions, geometrical configuration, and degree of adjustability of chair, worksurfaces, keyboards, etc., constrain operating postures. In the case of data entry, for example, a standing workstation (e.g., airline reservation desk) allows a rather limited range of postural variations (slumping, bending, moving the legs). On the other hand, nonadjustable chairs and workstations will allow a different but also limited range of postural shifts. Reach envelopes, defining the three-dimensional space which can be easily reached by the seated worker without bending, are a key component.

Human Constraints

Each of the above sets of constraints can be considered affordances. However, the individual user's effectivities — those personal characteristics, preferences, and abilities of each individual — will finally determine the postures which can be seen in the workplace. These characteristics can be divided into two categories:

Physical: includes anthropometric dimensions, strength, and range of limb motion (relevant to operating controls), and somatic/physiological status (e.g., physical disability).

Psychological: includes awareness of body posture, state of discomfort, willingness to shift body positions, and, in particular, knowledge of how adjustability controls of ergonomic furniture work and knowledge of why they should be used. Finally, the question of operator intentionality is crucial; knowledge of how and why must be combined with motivation to act if the controls are to be used effectively.

Functional Requirements in Chair Design (General Function)

Moving down the hierarchy, the conceptual framework described above, becomes, in turn, a set of goals or ends for which the means are described at the general function level. Within this framework, the problem of ergonomic design of workspaces can, broadly stated, be conceptualized as the design/organization/arrangement of environmental constraints in such a way as to ensure that the fit between person and components of work environment is optimal. What follows represents a conceptual basis for attaining person–environment fit.

97.3 Envelope of Postural Orientations

Once the set of tasks is defined, we can define an envelope of permissible postural orientations. The basic criterion implicit in specifying a given postural orientation as permissible is that it minimizes biomechanical load on the musculoskeletal system. At the same time, it must be emphasized that static biomechanical models are only a first approximation. A more realistic analysis would view the human operator from the perspective of a complex biological adaptive control system. The act of changing postural orientation should be viewed as an optimization strategy involving the search for zones of biomechanical equilibrium, in which biomechanical load is temporarily minimized (cf. Nubar and Contini, 1961). From the operator's perspective, these may be considered zones of perceived comfort.

Practically, this translates to a concern for shifting among multiple working postures, as well as consideration of rest breaks, task duration, and other temporal issues. It is recognized that in some cases there may be conflicting constraints. For example, data entry work requires tight coupling between fingers and keyboard and eye and copy. Consequently, requirements for postural stability — demanded by the task — may preclude frequent postural shifts (cf. Branton, 1976; Mark et al., 1991).

97.4 Biomechanical Efficiency Assumption

It is assumed that reducing load will result in increased biomechanical efficiency. Biomechanical efficiency will, in turn, lead to: (a) reduced muscular fatigue; (b) decreased perception of discomfort/pain;

(c) improved work performance. A further assumption is that symptoms of perceived discomfort/pain can be precursors of musculoskeletal disorders. Finally, comfort is conceptualized as the relative absence of discomfort/pain.

There is a scientific literature related to biomechanical factors which impact musculoskeletal load during sitting. This material is reviewed in Kroemer et al. (1994, pp. 429-498) and Chaffin and Andersson (1991, Ch. 9). See also Zacharkow (1988). Grieco's (1986) review of the epidemiological literature on musculoskeletal disorders comments that:

> ...nearly all authors have shown a relationship between prolonged sitting posture and disturbances of the osteo-articular apparatus that was statistically significant compared with control groups....

It must be emphasized, however, that there is no agreed-upon unified definition of biomechanical efficiency. Such a definition would imply that not only are the individual underlying mechanisms well understood, but that the ways in which they interact are equally understood, allowing an integrated approach to biomechanical efficiency through optimization. We are far from this understanding (Kroemer et al., 1994, pp. 346) Nevertheless, the concept of biomechanical efficiency can serve as a useful theoretical concept in attempting to understand the literature. (Lederman, 1993, has made a similar argument in the realm of particle physics.)

What follows will be a brief overview of these factors. This overview will be based on Kroemer et al. (1994, pp. 429-498), Chaffin and Andersson (1991, Ch. 9), and Grieco (1986). Citations of individual studies will not be attempted.

Unsupported Seated Posture

Imagine sitting in a doctor's office on the very edge of an examining table. Your feet are hanging freely; there is no back support. The entire body is supported by the ischial tuberosities (which can be thought of as a pair of inverted pyramids) plus a small amount of thigh surface. This posture is, in fact, unstable, and one tends to rock either forward or backward. Stability can be achieved by holding on to the side of the table, or by shifting one's weight backward to increase the surface area. A natural tendency is to slump forward a bit, which has the effect of lowering the center of gravity.

Spine and Discs

When the body is standing erect, the spine — which is composed of bony vertebrae and flexible discs — takes on a characteristic S curve. The lower (lumbar) region of the spine is concave inward and the pelvis is angled forward. This inward posture is called *lordosis.* Upon sitting on a flat seat, the pelvis rotates backward on the ischial tuberosities, and the lordotic curve tends to flatten or reverse into a *kyphosis* (outward concavity.) This kyphotic tendency is enhanced when the trunk slumps forward and the arms are unsupported. Large increases in disc pressures have been observed in such postures. On the other hand, if an inclined backrest is provided and the trunk can be angled backward opening up the thigh–trunk angle, disc pressure is greatly decreased. Figure 97.1 illustrates examples of lordotic and kyphotic postures. Please note that the chair characteristics illustrated are not meant to characterize recommended or nonrecommended chair features.

Additionally, nutrients are provided to the disc via osmotic pressure which can only occur when the discs are alternately loaded and unloaded. Hence, there can be deleterious effects on disc metabolism regardless of postural load, if the sitter does not move periodically.

Muscles and Internal Organs

Prolonged muscle contraction along with maintaining an awkward unsupported sitting posture may result in reduced blood flow and consequent local degeneration of muscle tissue. This may be a particular problem in neck and shoulder muscles during certain seated tasks. On the other hand, electrical activity in muscles has been shown to decrease when the trunk— thigh angle is opened up, as when leaning

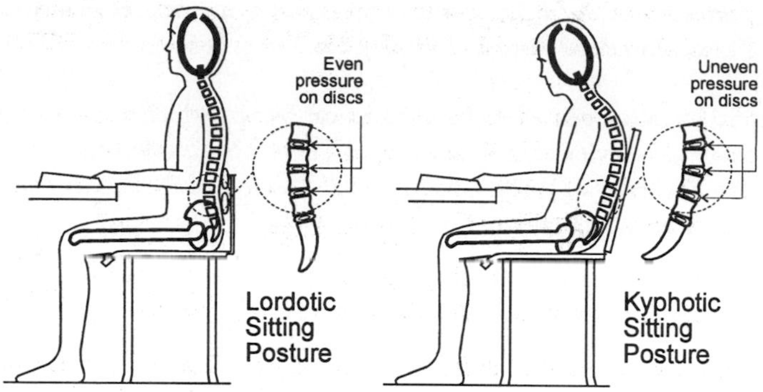

FIGURE 97.1 Illustrations of lordotic and kyphotic seating postures. Chair features are illustrative only, and do not reflect necessary associations between features and postures.

against a backrest. Results from a variety of sources including studies of astronauts in weightless conditions seem to indicate that optimum (or neutral) posture appears to occur at 120 degrees.

When the trunk — thigh angle is less than 90 degrees, pressure on internal organs of the body is increased.

Pressure on Thighs and Buttocks and Lower Legs

Local pressure gradients on thighs and buttocks which are too steep can result in impaired blood flow to these regions and to the lower legs as well. This can occur from the edge of a seatpan which is not rounded. The effect is increased if the seatpan is too high (in which case the weight of the leg pulls the underside of the thigh against the edge of the seat) or too low (in which case the region of the buttocks in contact with the seatpan is reduced — thereby increasing the pressure gradient).

97.5 Design Specifications

Specific design specifications of workstation components and arrangements can be attained by combining the envelope of permissible postures, and set of task constraints (affordances) determined above, with a set of specific individual physical constraints (effectivities) while attempting to optimize biomechanical efficiency. Design constraints are specified by a given percentile range (e.g., 5th to 95th) of appropriate anthropometric dimensions for an appropriate user population. If the user population includes individuals with disabilities, additional design constraints are required. The outcome will be a set of workplace dimensions (constraints) which allow individuals within this population range to attain the specified permissible postures.

Chair specifications form a critical component of the workplace. In line with the overall system goals, the chair will afford both stability and the opportunity for movement. These goals appear to be most easily met when the trunk–thigh angle approaches 120 degrees — the so-called neutral posture. Opening this angle can be achieved by either inclining the trunk backward or by keeping the trunk erect and sloping the seatpan forward.

97.6 Multiple-Degrees-of-Freedom Problem Spaces

The art of designing/integrating workspaces, and of designing the seating and other furniture which populate such workplaces, requires that the multiple sets of constraints must be satisfied. However, the operator must also deal with a multidimensional problem space — albeit on an individual scale. To the extent that required adaptability and flexibility is attained by adjustable workplace dimensions, the operator must cope with a multiple-degrees-of-freedom control. A fully adjustable workstation and chair may have upwards of 10 different controls or adjustments which must be coordinated (see Dainoff and Mark, 1989). Hence, issues of control mechanisms and relationships (e.g., articulation between seatpan and backrest) ought to have the same consideration as biomechanical concerns in the design solution. At the same time, training requirements must go beyond mere lists of simple adjustment procedures to provide the operator with a more conceptual and functional understanding of the overall problem space within which she/he is operating (Rasmussen et al., 1995; Vicente and Rasmussen, 1990).

97.7 Applications and Examples

The remaining sections of this chapter will address the means by which the general functional ends are achieved. The functional requirements described above must now be translated into specific hardware requirements for office furniture. As above, this discussion will focus on issues rather than prescriptions. Where appropriate, reference to ANSI HFES 100/1988, The American National Standard for Human Factors Engineering of Video Display Workstations will be made. However, this standard is currently in the process of revision. In this revision, a more recent anthropometric survey of U.S. Army personnel (Gordon et al., 1988) was utilized; thus, numerical values for some of the design specifications are likely to differ from the 1988. Moreover, the methods of application of anthropometric data to design specifications are themselves under review. In particular, when two or more anthropometric dimensions must be combined to provide such specifications (as is the case for seatpan height), it is inappropriate to simply add 5th or 95th percentile values of such dimensions (Robinette and McConville, 1981; Zehner, Meindl, and Hudson, 1993; Nemeth and Dainoff, 1997). Thus, specifications based on such combinations should be considered tentative. This is an area of currently active investigation.

Individual Chair Components

Backrest and Lumbar Support

Traditional task chairs for secretaries and clerks have tended to be little more than upright support pads for the lumbar region of the back. These designs seems to have been predicated on the assumption of a 90 degree (cubist) work posture. A more functional backrest will support a greater area of the trunk while inclined backward.

The role of the lumbar support is more problematic. It has typically been assumed that providing a padded surface in the lumbar region will function to restore the lumbar lordosis while the trunk is erect. However, this assumption requires that the trunk remain in close contact with the lumbar pad, and that the pad is properly adjusted so it is, in fact, adjacent to the user's lumbar spine (see Corlett and Eklund, 1984). Hence, the lumbar pad must be height adjustable— either independently, or as part of an adjustable backrest to accommodate anthropometric variability in lumbar height. The only data currently available are those of Branton (1984), from a sample of British railway workers. Computing the lumbar heights of 5th percentile females and 95th percentile males, from means and standard deviations yields a range of 2.2 to 37.3 cm above seatpan height.

However, there is a variety of approaches to providing padding (or the lack of padding) in the lumbar region of the chair, including a concavity along the vertical axis of the backrest. There is no evidence for the relative effectiveness of these different approaches. In fact, experimental investigations by Bendix et al. (1996) question the basic assumption that such backrests restore lumbar lordosis.

A wide backrest may interfere with tasks which require lateral movement of arms and shoulder. Thus, the possibility of a narrower backrest which still provides vertical support to the trunk should be considered.

Seatpan

ANSI-HFES 100/1988 specifies that the depth of the seatpan be between 38 and 43 cm. This is based on the anthropometric dimension buttock–popliteal length. It is recognized that a seatpan which is too long for a small person (e.g., 5th percentile female) would interfere with seated posture, whereas a seatpan which is too short for a large person (e.g., 95th percentile male) would not provide adequate support. From the Gordon et al. (1988) survey, the 5th to 95th percentile range for buttock-popliteal length is 44.0 to 54.5 cm. In addition, it is recommended that the front of the seatpan be rounded (the "waterfall" front) in order to avoid pressure gradients on the underside of the thigh.

ANSI-HFES 100/1988 specifies a seatpan width of 45 cm. This is based on the hip breadth of the 95th percentile female. From the Gordon et al. (1988) survey, this value is 41.2 cm.

In traditional task chairs, the seatpan was sloped backward 3 to 5 degrees in order to allow the trunk to rest against the backrest. More recently, it has been recognized that tilting the seatpan forward will allow the trunk–thigh angle to open up (Mandal, 1986; Bridger, 1988). However, this effect may be countered by a tendency to slide out of the chair. The sliding tendency can be countered by use of kneepads, by training the user to exert counter-pressure by the legs (Dainoff and Mark, 1987), by building in a saddle-like contour in the front end of the seatpan (Corlett and Gregg, 1994), and by keeping the rear part of the seatpan level and inclining the forward part (Graf et al., 1993).

ANSI-HFES 100/1988 specifies that the backward inclination of the seatpan must be within 0 to 10 degrees. Forward inclination is permitted, but no ranges are specified.

Armrests

Whether or not armrests are provided, and if so, what their configuration should be, is an important issue in chair design. If fixed height armrests are provided, they cannot be higher than the 5th percentile female's seated elbow height without interfering with seated posture of smaller users. This value, from Gordon et al. (1988), is 17.6 cm. The case against fixed armrests is that larger individuals using them for support will be inadvertently forced into awkward forearm posture. Also, unless carefully designed, armrests may keep the user from getting close to the worksurface, or catch on clothing. On the other hand, if users are properly instructed regarding working posture, fixed height armrests may be very helpful in supporting the arms during pauses between periods of keying or other work.

Adjustable height armrests are now widely available. They seem to be particularly useful with the advent of large-scale use of mice, trackballs, and other non-keyboard input devices. However, here again, proper instruction against misuse is important.

Adjustability Ranges

Achieving the functional goals of designing seating affordances within task and environmental constraints while accommodating individual variability (effectivities) is difficult to accomplish without some degree of adjustability of chair components. However, because of the complexity of the design problem space (the intersection of task, environmental and personal constraints), there is still much room for debate and disagreement as to which components should be adjustable and how much adjustment is necessary and desirable. This section will discuss the issue from an anthropometric and biomechanical perspective; the subsequent section will focus on more behavioral concerns.

Seatpan Height

The height of the seatpan would seem to be a straightforward question of locating a seating surface at the level of the underside of the thigh. This would allow the user's thighs and buttocks to be firmly supported while their feet could rest firmly on the floor. The corresponding anthropometric dimension — popliteal height — has a 5th female–95th male percentile range of 35.1 to 47.6 cm according to Gordon et al. (1988), to which a correction for shoes (typically 2.5 cm) is added. However, several other constraints must be considered. Most important, the chair can be considered alone, but must be viewed within the context of the overall workplace. Thus, if fixed height worksurfaces are present, a small person's elbow height would be well below effective working height. Published research summarized by Cornell and Kokot (1994) indicated that users tend to adjust their chairs above (i.e., 5 cm) their measured popliteal heights.

Furthermore, appropriate working heights for elbows and hands will be affected by whichever solution is employed to open up the thigh–trunk angle (see above). A forward-sloping seatpan will tend to raise the elbow height, while keeping the trunk erect; this may necessitate an increase in worksurface height. On the other hand, a backward inclination of the trunk may also require a corresponding backward inclination and lowering of the seatpan. This would tend to lower the effective working height of the elbows.

In addition, seatpan height adjustment strategies must take into account clearance considerations under the worksurface. A forward-sloping seatpan may result in interference of the top of the thigh with the underside of the worksurface. With a backward inclined seat, the problem might be with the height of the knees.

The solution proposed by ANSI-HFES 100/1988 is to require seatpan height adjustability within a range of 40.6 to 52 cm. This is designed to accommodate both small females and large males at a fixed height (76.2 cm) worksurface in terms of working elbow height. It is recognized that a footrest would be essential for smaller persons under these conditions.

Backrest Angle

ANSI-HFES 100/1988 specifies that, if adjustable, backrest angles should incorporate some part of the range between vertical and 15 degrees backward from vertical. Larger inclinations of up to 30 degrees behind vertical can be beneficial in reducing disc pressure and muscle activity (Andersson et al., 1979).

Armrest Height

In considering height adjustment ranges for armrests, the applicable anthropometric dimension is seated elbow height (above seatpan surface). Applicable 5th female–95th male values from Gordon et al. (1988) are 17.6 to 27.4 cm.

Control Mechanisms

Providing user adjustability necessitates a consideration of control mechanisms. The following section provides a brief description of some issues relating to the functional characteristics of chair control mechanisms. This discussion will be expanded in a following section on system integration.

Helander, Zhang, and Michel (1995) listed 10 separate chair controls which could be considered characteristic of the best chairs available in the market at that time. These include: seat height, seatpan tilt, backrest height, backrest angle, back tension, armrest height, lumbar height, lumbar depth, seatpan tilt tension, and seat depth. This list will obviously need to be modified as new chair designs reach the marketplace, but it represents a good functional overview of the potential opportunities for flexibility and control available to the user.

In general, providing height adjustability has tended to be straightforward. Either a compressed gas mechanism controlled by a lever or button allows the height to be adjusted by the seated operator, or a mechanical mechanism is employed which requires the operator to leave the chair.

A larger variety of design solutions has been seen in the interaction between seatpan and backrest. Either one or both of these components can be made adjustable. There is a variety of adjustment possibilities. *Active* mechanisms require that the control mechanism must be activated while the surface (seatpan or backrest) is moved by applying force from the appropriate body surface. *Passive* mechanisms have no control mechanisms but adjust solely by force from body surfaces against some resistance. Mixed controls allow a combination of both. More complex mechanisms provide for the possibility of locking out the forward slope function.

In addition, where both seatpan and backrest are adjustable, the two components can either be independently controlled, or are mechanically linked. In the case of the latter, there is typically a ratio of 2 or 3 to 1 degrees of movement between backrest angle and seatpan angle.

Tension adjustments are sometimes provided to take into account the variation of user weight and strength relative to the force required to move seatpan and backrest surfaces.

System Integration: The Ecology of Seated Posture

The previous discussion has focused primarily on mechanical attributes of the chair, and biomechanical/anthropometric characteristics of the user. At this point, the focus of discussion will be expanded to conceptualize the seated worker within an overall workplace system.

Intentionality of the User: Task Demands and User Performance

Referring to the means–end abstraction hierarchy discussed in an earlier section, the overall system goal for the chair is to provide stability while the user is carrying out one or more sets of tasks.

At the same time, some degree of movement must be afforded, both to alleviate fatigue, and to respond (in most cases) to task demands which require changing postures. However, in framing this discussion, it is important to emphasize that, from the user's perspective, sitting is rarely a goal in itself. The chair, instead, must be regarded as a tool which allows accomplishment of a given task. This, of course, can include the task of resting, or even taking a nap! Accordingly, in considering task demands, the user's *intention* becomes of prime importance.

The incorporation of intentionality into the question of the ergonomics of sitting poses some interesting problems. For most users, most of their life experience has been with "ordinary" (i.e., non-adjustable) chairs. Hence, postural adjustments tend to driven strictly by task constraints; for example, leaning forward to read a document lying flat on a table. The chair is a passive support structure which allows (affords) certain working postures; in another case (e.g., the document is too far to reach), the user simply leaves the chair. These postural adjustments operate primarily at an automatic level in the sense that the adjustments are below the level of conscious attention. However, the situation becomes more complicated once adjustable chairs are provided. Now, user intentionality must, at least initially, involve deliberate postural regulation through active chair adjustment. In a sense, operation of the chair becomes incorporated as part of the overall task demands. Accordingly, a human performance perspective must be added to the biomechanical focus previously adopted.

What Is the Operator's Task in Using an Adjustable Chair?

Helander et al. (1995) have characterized the operator's task in terms of the following sequential stages of problem-solving:

a. Search: Can the chair control be located?
b. Feedback: Can the mode of operation of the control be understood?
c. Compatibility: Does the control allow adjustment to the desired position?

The introduction of an adjustable chair into a workplace necessitates that these additional task demands be incorporated in work design and training so that the added value — in terms of improved performance and reduced fatigue — which accompanies the introduction of this new productivity tool can actually be attained.

The landmark work of Rasmussen et al., (1995) on levels of performance can be helpful in this discussion. Rasmussen identified three distinct levels at which human performance can occur. The first or *skill* level, reflects largely automatic behavior governed by preprogrammed instructions. *Rule-based* performance applies to those kinds of familiar situations where behavior is controlled by stored sets of rules of the form "If A then B." The third level is *knowledge-based* performance involving novel situations in which conscious analysis and planning is required.

The postural adjustments which occur in non-adjustable chairs typically exist at the mostly automatic skill-based level (e.g., bending forward to see better). When an adjustable chair is first encountered, a set of rules is developed which encompasses the first two stages of Helander et al. That is, the user learns *where* the controls are (search task) and *how* (feedback task) they function. An example of these stages in combination is the following: *IF* I press the button on the left, *THEN* the seatpan will lower.

This is the kind of information which typically is found on the hang tags or other instructional materials which accompany most ergonomic chairs, but which, unfortunately, are often removed before they reach the end user. For this reason, chair designers often have as a goal, the design of controls which are "intuitive" in the sense that the "where" and the "how" tasks could be accomplished without written instructions.

To some extent, this is a reasonable requirement. Helander et al., (1995–Experiment 1) conducted an experiment in which 5 chairs, representing a range of control configurations, were presented to a group of 20 naive users. The users, who had no previous exposure to the chairs, were asked to identify and operate the controls. The chairs were independently rated by the authors as to how well they afforded accomplishment of each of the three tasks. The results indicated that the users were better able to identify the controls on the chairs which had the highest expert ratings on the first two tasks ("where" and "how"). Thus, chair controls which three professional ergonomists agreed were easy to find, and easy to operate were in fact, easy to find and operate by a group of naive users.

However, the third stage of Helander et al. (1995) is more problematic. The question of compatibility assumes that the user's intention to place the chair in a given orientation can, in fact, be achieved by the chair characteristics. This is a more subtle issue, presupposing that the user understands, for each combination of chair control and task characteristics, what the desired postures should be. Coordinating the operation of multiple chair controls so as to optimize posture for a given set of task constraints can be considered knowledge-based performance, at least initially.

The Problem of Coordinating Multiple Controls

The extent of this problem can be seen in the case of those chairs which include all 10 of the chair adjustment controls described by Helander et al. (1995). While the tension, lumbar depth, and seat depth controls might be considered one-time only settings, there remain six independent adjustments which must be operated in coordinated fashion. This presents the operator with what has been discussed earlier as the *degrees-of-freedom problem* (Mark, Dainoff, Moritz, and Vogele, 1991). Each separate control can be considered as an independent degree of freedom. It is possible that the user will understand the need to systematically explore the possible combinations of working postures which can result and intuitively find those which are optimum. A more likely possibility is that the operator will simply become overwhelmed and either ignore, or worse, misadjust control mechanisms. Karwowski (1991) has provided a more general description as to how this might occur.

A more reasonable approach is to try to understand the relationships between working posture and task constraints, and to then communicate these relationships as part of a training program for operators. Looking just at the relationship among three controls — seatpan angle, backrest angle, and seatpan height — Dainoff and Mark (1987) designed two simulated VDT tasks which differed in task constraints. High-speed data entry work required the fingers to be in close contact with the keyboard, the eyes in close contact with the paper source document. Editing was completely done on the display screen; the keying rate was less. In the analysis of the authors, the constraints of the editing task dictated an upright posture with the backrest vertical, the seatpan angled forward (for lumbar support), and the seatpan raised to correct for the angled seatpan. In this case, the visual demands were critical. The source document characters were approximately half the size of the screen characters. Consequently, the editing task, which was completely screen-based, allowed a longer viewing distance. Consequently, a more "relaxed" working posture in which both the backrest and seatpan angle was reclined rearward could be employed.

Utilizing this analysis, subjects were trained to use an ergonomic chair which had all three of the above controls. The alternative postures described above were suggested as hypotheses. Following training, subjects worked three hours per day for four days; alternating between entry and editing tasks every 30 minutes. An incentive pay system was utilized, and fatigue was assessed throughout the course of each day's session. Results indicated that the subject consistently used a forward-sloping upright backrest position for the entry task and a rearward inclined position for the editing task. In addition, although fatigue ratings increased over the course of each daily session, the average daily fatigue ratings *decreased* over the four days. An alternative explanation for the postural data might be that the subjects were simply following instructions blindly (the so-called demand characteristics explanation). However, this would argue that subjects would deliberately work for 12 hours in suboptimal postures at the cost of higher incentive pay.

The results of Dainoff and Mark (1987) were followed up by two separate investigations. Helander et al. (1995–Experiment 3) had users evaluate four chairs in which the number of degrees of freedom of adjustability varied systematically. Chair #1 had height adjustment only with some flex in the seatpan and backrest. Chair #2 had a height adjustment, and a single adjustment in which seatpan and backrest were linked. Chair #3 had independent adjustments for height, seatpan tilt, and backrest tilt, but a single lock for seatpan and backrest. Chair #4 had independent adjustments for all three functions, and independent locking mechanisms for seatpan and backrest. Subjects were given essentially the same kinds of instructions as in Dainoff and Mark (1985). They were then asked to adjust each of the five chairs to both backward and forward postures and then rate the overall comfort of the chair. Videotape analysis was used to measure total adjustment time and number of adjustments made.

The results indicated a linear relation between chair complexity and rated comfort, time to adjust, and number of adjustments. Thus, these subjects seemed to be willing to take the additional time, and make the additional adjustments required in order to achieve a more comfortable posture.

The Question of Active or Passive Movement of Chair Surfaces

The second follow-up investigation involved a simulated work situation in which 12 operators worked for 180-minute sessions on two separate days alternating between 90-minute sessions of entry and editing (Mark, Dainoff, Moritz, and Vogele, 1991). The ergonomic chair used in the previous study (Dainoff and Mark, 1987) was modified with potentiometers attached to seatpan and backrest controls, so that the angular position of seatpan and backrest surface could be continuously tracked throughout the study. In addition, the basic design of this chair included a capability to put the seatpan and backrest control into passive or "floating" mode. The previous studies had utilized only the active control model. However, in this study operators were assigned to both active and passive chairs on alternative days. As before, a day of training preceded the experiment in which the relationships between working posture (forward leaning for entry, backward leaning for editing) were proposed as recommendations which could be either accepted or not by the operator.

When the operators were using the active controls, the observed working postures were quite similar to the earlier results of Dainoff and Mark (1987). The operators seemed to follow the training recommendations in that the forward-leaning posture was used in the entry task, and the backward-leaning was used in the editing task. However, this did not occur when the same operators used the passive controls. The primary difference was found in the entry condition; the forward seatpan angle was not used, but positions were much more variable. The data appear to reflect attempts by the operators to search for a stable position. This was confirmed by comments of the operators in this condition in which terms like "rocking horse" and "seasick" were used.

The results from Mark et al. (1991) were subjected to additional analyses by Gardner, Mark, Dainoff, and Xu (1995). In these analyses, the relationship between seatpan and backrest angles was plotted individually for each operator. Of particular interest were the data from the data entry tasks when the chair was in passive mode. The seatpan–backrest angle relationships could be classified into two basic categories. Figures 97.2 and 97.3 illustrate typical results for both types. In each case, a characteristic pattern of activity is the linear relationship between both angles centered around the (zero degree) "cubist" position in which seatpan is horizontal and backrest vertical. These data are consistent with the "hunting" interpretation characteristic of an unstable posture. The operator attempts to achieve the forward-leaning posture by moving the seatpan and backrest forward, but this posture is perceived as unstable, so the seatpan and backrest are moved in the opposite direction, but not too far. These linear data are descriptive of a rocking motion around the cubist position. For comparison, a typical data pattern from the Entry Active condition is shown. Here, the seatpan remains clearly inclined downward, while some experimentation with the backrest is carried out. Finally, the vertical "tail" in Figure 97.3 reflects a second group of operators who appeared to "give up" the search for a stable equilibrium position close to the vertical.

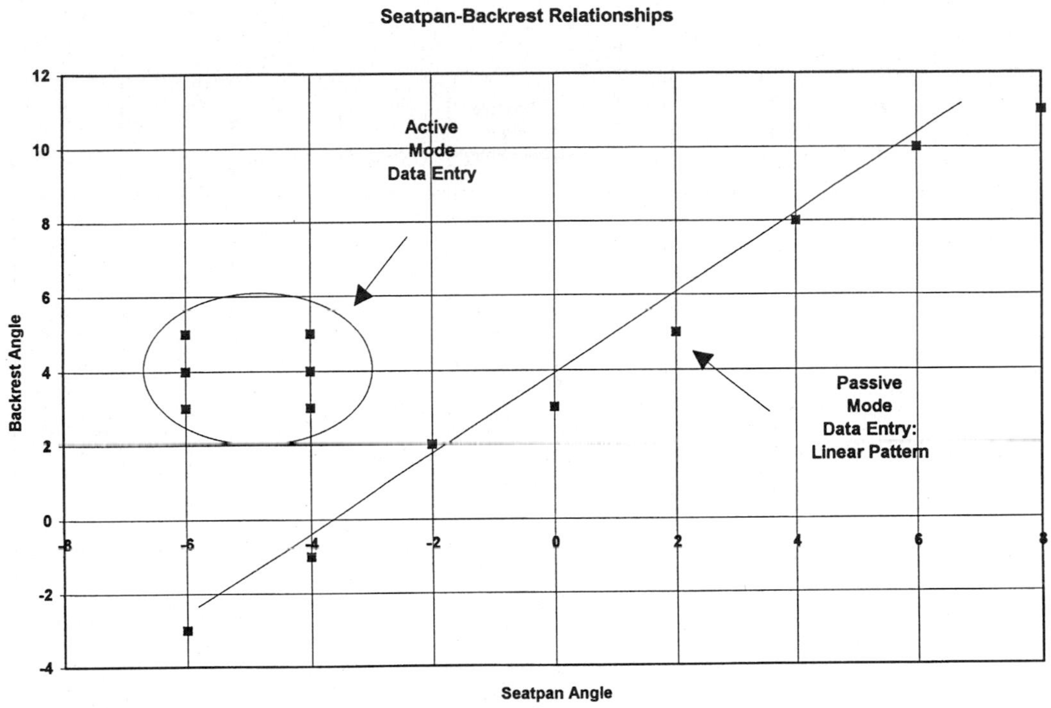

FIGURE 97.2 Typical patterns of results from experiment of Gardner et al. (1995). Illustrated are relationships between seatpan and backrest angles during data entry work during which chair was in active and passive modes of adjustment.

These people moved the seatpan to its maximum (but stable) backward inclination, and varied their posture only by moving the backrest.

In the editing tasks, operators in both active and passive modes tended to look like the vertical portion of Figure 97.3.

A second goal of this study was to utilize these results as a means of assessing the design feature of linked seatpan and backrest angles. For such chairs, the industry standard appears to require ratios of backrest to seatpan angles of either 2:1 or 3:1. That is, for every 1 degree of rotation of the seatpan, the backrest will move either 2 or 3 degrees. Note that linkage can be found in either active or passive modes of adjustment. In the present study, if the chair mechanism had been linked, the resulting postures would have fallen along a straight line with a slope of either 2 or 3. (See Figures 97.2 and 97.3). To assess how many of the operators would have normally adjusted their chairs to approximate these ratios, the slopes of the linear portions of the curves were computed for each of the individual plots. The results indicated that most (73%) of the resulting slopes were less than 2. This outcome, together with the appearance of a nonlinear component in many of the plots, indicates that these users adjusted their chairs to postural orientations which would have been impossible had they been using a chair with a linked mechanism.

While these studies utilized somewhat artificial laboratory tasks, the characteristics of the study was carefully chosen to represent two extremes of keyboard-intensive work. The entry task involves close visual attention to paper copy and rapid keying. The editing task requires close visual attention to the display screen, but much less typing. It is assumed that typical office work will represent a mix between these two extremes, and would, consequently, involve an alternation of working postures. The outcome of these studies would seem to argue that, for such computer-intensive tasks, an ergonomic chair with active controls which independently operate seatpan and backrest angles will, with proper training, allow the user to effectively attain a desired work posture. The key element, however, is proper instruction.

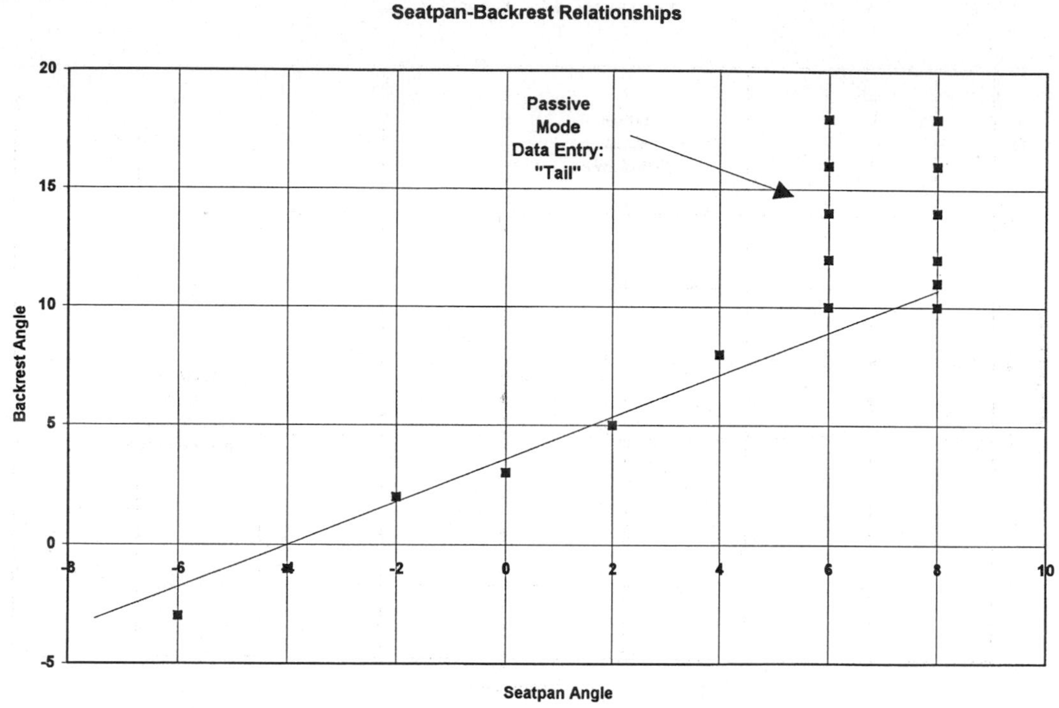

FIGURE 97.3 Typical pattern of results from experiment of Gardner et al. (1995). Illustrated is a second example of the relationship between seatpan and backrest angles during data entry work during which chair was in the passive mode of adjustment.

Thus, the advantage of the linked mechanism in terms of simplifying the user's control problem, must be balanced against the demonstration that such mechanisms exclude postural orientations which might otherwise be desired. Likewise, passive controls eliminate levers and buttons, but at the potential cost of postural instability.

Relationship of Seated Posture to User's Reach Capabilities

A final set of research results is relevant. The concept of *reach envelopes* is fundamental to workplace design and layouts. Reach envelopes refer to regions in three-dimensional space where task-related devices, such as controls, are reachable without bending or moving. Reach envelopes have been derived from anthropometric models, empirically derived reach contours (in which a representative range — typically 5th percentile female to 95th percentile male — of users attempts a set of standardized reaches), or a combination of the above. (See, for example, Kroemer et al. [1994; pp. 29-35] or Sanders and McCormack [1993, Chapter 13].) However, these approaches have been based strictly on static postures and have failed to take seating variables into account.

A recent set of studies by Mark and his colleagues (1997) has brought a new, more dynamic, perspective to this question. Consider a person who initially starts walking slowly, and then gradually begins to increase speed. At some point, the pattern of movement we call "walking" will become inefficient, and the person will switch to a new *action mode* which might be called "jogging." As speed increases even further, jogging will give way to "running." It should be noted that fast walking and jogging can coexist at the same speed, but jogging will be more physiologically efficient. (These arguments are based on the classic work of Hoyt and Taylor (1981) who measured energetic changes as horses moved from walk to trot to canter to gallop.) Mark et al. (1997) applied the same analysis to seated reach. They defined the *absolute critical boundary* as the maximum distance, for any given person, which can be reached without moving the shoulder away from the backrest (action mode 1). Beyond this point, a second action mode comes into play — namely, reaching which includes shoulder rotation and/or bending of the trunk (action mode 2).

However, in parallel to the work of Hoyt and Taylor (1981), Mark et al. (1997–Experiment 1) discovered the presence of a *perceived critical boundary* — where the individual started to switch action modes from the first to the second action modes. This transition occurred, on the average, at 85% of each person's absolute critical boundary. In the same experiment, the actors were asked to maintain the arm-only action mode for reach distances out to the absolute critical boundary while making comfort ratings of each reach. The rate of decrease in mode 1 comfort ratings with reach distance corresponded exactly with the (previously established) rate of decrease in mode 1 reaches. These results support the interpretations that even though the actor is physically capable of reaching objects within the last 15% of the distance to his/her maximum capability using action mode 1, it is more efficient/comfortable to switch to action mode 2.

The practical consequences of this finding relate to workstation layout. The actual distance between perceived and absolute critical boundaries is on the order of 8 cm. This distance can be of practical significance in a crowded work area. A target located in the zone between the perceived and actual critical boundaries might result in arm muscle strain, if the act is repeated over prolonged periods with insufficient rest breaks.

Furthermore, and more to the point of this chapter, the location of the perceived critical boundary is effected by the characteristics of the chair. Mark et al. (1997–Experiment 2) repeated the reach experiments using three different chair configurations. An ergonomic chair was used to position the actors in the following positions: forward leaning (backrest vertical and seatpan inclined 8 degrees forward); upright (backrest vertical and seatpan horizontal) and extreme backward leaning (backrest at 28 degrees from vertical; seatpan and 24 degrees from horizontal). The results indicated that the preferred critical boundary appeared earliest for the forward leaning position, next for the upright leaning position, and essentially corresponded with absolute critical boundary for the backward leaning position. The importance of this finding is the demonstration that not only does the reach envelope (location of the absolute critical boundary) decrease with backward inclination, but that there is a corresponding decrease in the difference between perceived and absolute critical boundaries.

Moreover, the next experiment in this series (Mark et al., 1997–Experiment 3), found the relationship between perceived and absolute critical boundaries was subject to biodynamic manipulation. Experiment 3 was a replication of Experiment 2 except that the "chair" was simply two flat but padded boards, in which the backrest could be inclined from 30 degrees backward to vertical. The result of this manipulation was that virtually all actors maintained their mode 1 reaches up to their absolute critical boundary regardless of the backrest inclination. It appears that the lack of padding and contoured surfaces had the effect of reducing the stable support structure from which the more complex but efficient mode 2 reach act could be organized. This finding is of some theoretical importance, but also presents a practical opportunity. There is much debate among chair designers as to how much contouring and padding is appropriate. This experimental procedure has great promise as an assessment tool for comparing alternative chair designs.

97.8 Discussion

This chapter has attempted to present a systematic, integrated overview of issues related to the seated workplace. Working within the structured framework of the means–end abstraction hierarchy laid out in the chapter *Office Ergonomics*, it was argued that in order to achieve the overall system goals of stability and movement for the seated operator, it is necessary to consider interactions among task demands, user characteristics, and chair features. When the task requires close linkages between eye and display screen/copyholder, and fingers and keyboard/mouse, these demands tend to drive postural orientations. To the extent that the chair can provide support for these postural orientations through adjustability, it can be considered an effective ergonomic chair for that situation. However, increased adjustability is achieved at the cost of complexity. If the adjustment possibilities are designed into the chair itself, through linked controls and/or passive movement, the operator has less cognitive demand, but also has less control in the sense that certain postures which he or she may desire are likely to not be available. Putting the operator more in control, with active controls, also requires a higher degree of training and education. However, such training would seem to be a reasonable investment in achieving the real benefits of an adjustable chair as an essential component of the modern workplace.

Defining Terms

Active Control Mechanism: Adjustment mechanism of chairs which require the operator to perform a positive action on the control mechanism to move the chair. See **Passive Control Mechanism.**

Affordance: Attributes of the surrounding environment which allow a given individual to perform specific actions. Physical properties measured with respect to an individual.

Degrees of Freedom Problem: Generalized description of the requirement for the operator to coordinate multiple adjustment mechanisms simultaneously.

Effectivity: Characteristics of the individual which allow or constrain his/her capabilities for actions within a given environment.

Kyphosis: Flattening of the curvature of lower (lumbar) region of the spine. May occur during unsupported sitting. See **Lordosis.**

Lordosis: Inward curvature of the lower (lumbar) region of the spine which occurs naturally when standing.

Means-End Abstraction Hierarchy: Conceptual tool for analyzing complex systems. Consists of a hierarchy of affordances in which lower levels are the means for allowing desired actions (ends) to be realized.

Passive Control Mechanism: Adjustment mechanism of a chair in which the chair surface may move as the operator moves his/her body. See Active Control Mechanism.

Reach Envelope: A three-dimensional region in front of an operator within which objects can be manipulated without bending or stretching.

References

Andersson, G.B.J. and Ortengren, R. 1974. Lumbar disc pressure and myoelectric back muscle activity during sitting. *Scandinavian Journal of Rehabilitation Medicine*, 6, 115-121.

Andersson, G.B.J., Murphy, R.W., Ortengren, R., and Nachemson, A.L. 1979. The influence of backrest inclination and lumbar support on lumbar lordosis. *Spine*, 4, 52-58.

Bendix, T., Poulsen, V., Klausen, K., and Jensen, C. 1996. What does a backrest actually do to the lumbar spine? *Ergonomics*, 39, 533-542.

Branton, P. 1976. Behaviour, body mechanics and discomfort, in E. Grandjean (Ed.), *Sitting Posture.* London: Taylor and Francis, Ltd.

Chaffin, D.B. and Andersson, G. 1991. *Occupational Biomechanics.* (2nd Ed.) Wiley-Interscience. New York.

Corlett, E.N. and Eklund, J.A.E. 1984. How does a backrest work? *Applied Ergonomics*, 15, 111-114.

Corlett, E.N. and Gregg, H. (1994) Seating and access to work, in R. Lueder and K. Noro (Eds.) *Hard Facts About Soft Machines: The Ergonomics of Seating.* London: Taylor and Francis.

Cornell, P. and Kokot, D. 1994. Use of adjustable VDT stands in an office setting, in R. Lueder and K. Noro (Eds.) *Hard Facts About Soft Machines: The Ergonomics of Seating.* London: Taylor and Francis.

Dainoff, M.J. and Dainoff, M.H. 1986. *People and Productivity: A Manager's Guide to Ergonomics in the Electronic Office.* MDA Books, Cincinnati, Ohio.

Dainoff, M.J. and Mark, L.S. 1987. Task and the adjustment of ergonomic chairs, in B. Knave and P.-G. Wideback (Eds.) *Work with Display Units 86.* Amsterdam: Elsevier (North-Holland).

Dainoff, M.J. and Mark, L.S. 1989. Instabilities in seated posture and the ergonomics of perception and action. Fifth International Conference on Event Perception and Action. Miami University, Oxford, Ohio

Gardner, D.L., Mark, L.S., Dainoff, M.J., and Xu, W. 1995. Considerations for linking seatpan and backrest angles. *International Journal of Human Computer Interaction*, 7, 153-165.

Gibson, J.J. 1979. *The Ecological Approach to Visual Perception*. Boston: Houghton Mifflin.

Gordon, C.C., Churchill, T., Clauser, T. E., Bradtmiller, B., McConville, J.T., Tebbetts, I., and Walker, R.A. 1989. 1988 Anthropometric Survey of U.S. Army Personnel: Summary Statistics Interim Report (Technical Report NATICK/TR-89-027). Natick, MA: United States Army Natick Research, Development and Engineering Center.

Graf, M., Guggenbühl, U., and Krueger, H. 1993. Investigations on the effects of seat shape and slope on posture, comfort, and back muscle activity. *International Journal of Industrial Ergonomics*, 1993, 12, 91-103.

Grandjean, E., Hunting, W., and Pidermann, M. 1983. VDT workstation design: preferred settings and their effects. *Human Factors*, 25, 161-175.

Grieco, A. 1986. Sitting Posture: An old problem and a new one. *Ergonomics*, 29, 345-362.

Helander, M. 1995. *A Guide to the Ergonomics of Manufacturing*. London: Taylor & Francis.

Helander, M.C., Zhang, L., and Michel, D. 1995. Ergonomics of ergonomic chairs: a study of adjustability features. *Ergonomics*, 38, 2007-2029.

Hoyt, D.F. and Taylor, R. 1981. Gait and the energetics of locomotion in horses. *Nature*, 292, 239-240.

Karwowski, W. 1991. Complexity, fuzziness, and ergonomic incompatibility issues in the control of dynamic work environments. *Ergonomics*, 34, 671-686.

Kroemer, K.H.E., Kroemer, H.B., and Kroemer-Ebert, K.E. 1994. *Ergonomics*, Englewood Cliffs, N.J.: Prentice-Hall.

Lederman, L. 1993. *The God Particle*. New York: Dell.

Lueder, R. and Noro, K. *Hard Factors about Soft Machines: The Ergonomics of Seating*. London: Taylor and Francis.

Mandal, A.C. 1981. The seated man (homo sedens). The seated work position. Theory and practice. *Applied Ergonomics*, 12.1, pp. 19-26.

Mandal, A.C. 1986, the influence of furniture height on back pain, in B. Knave and P.-G. Wideback (Eds.) *Work with Display Units 86*. Amsterdam: Elsevier (North-Holland).

Mark, L.S. and Dainoff, M.J. 1988. An ecological framework for ergonomic research. Innovation, *The Journal of the Industrial Designers Society of America*, 7.2, 8-11.

Mark, L.S., Dainoff, M.J., Moritz, R., and Vogele, D. 1991. An ecological framework for ergonomic research and design, in R.R. Hoffman, and D.A. Palermo (Eds.). *Cognition and the Symbolic Processes: Volume III*. Hillsdale, NJ: Erlbaum Associates.

Mark, L.S., Nemeth, K., Gardner, D., Dainoff, M.J., Paasche, J., Duffy, M. and Grandt, K. 1997. Postural dynamics and the preferred critical boundary for visually-guided reaching. *Journal of Experimental Psychology: Human Perception and Performance*, 25, 1365-1379.

Nemeth, K.J. and Dainoff, M.J. 1997. Inappropriately applying anthropometric methods for ergonomic design testing. Proceedings of the International Ergonomics Association 13th Triennial Congress. Tampere, Finland.

Nubar, Y. and Contini, R. 1961. A minimal principle in biomechanics. *Bulletin of Mathematical Biophysics*, 23, 377-391.

Pheasant, S. 1986. *Bodyspace: Anthropometry, Ergonomics and Design*. London: Taylor and Francis.

Rasmussen, J, Pejtersen, M., and Goodstein, L. 1995. *Cognitive Systems Analysis*. New York: Wiley.

Robinette, K.M. and McConville, J.T. 1981. An Alternative to Percentile Models. Society of Automotive Engineers Technical Paper Series, International Congress and Exposition. Detroit, Michigan.

Sanders, M.S. and McCormick, E.J. 1993. *Human Factors in Engineering and Design*. (7th ed.) New York: McGraw-Hill.

Stewart, D. 1987. Modern designers still can't make the perfect chair. *Smithsonian*, 1, 97-105.

Vicente, K.J. and Rasmussen, J. 1990. The Ecology of Human–Machine Systems II: Mediating "Direct Perception" in Complex Work Domains. *Ecological Psychology*, 2, 207-249.

Zacharkow, D. 1988. *Posture: Sitting, Standing, Chair Design and Exercise.* Springfield, Il.: Charles C Thomas.

Zehner, G.F., Meindl, R.S., and Hudson, J.A. 1993. *A Multivariate Anthropometric Method for Crew Station Design: Abridged.* (Tech. Report AL-TR-1992-0164). Wright-Patterson Air Force Base, Ohio: Air Force Material Command, Armstrong Laboratory.

For Further Information

In addition, Zacharkow (1988) and Lueder and Noro (1994) provide extensive discussion of seating issues.

98

Seating and Posture in VDT Work

Antonio Grieco
University of Milan

Giovanni Molteni
University of Milan

98.1 Introduction

Proper workplace design is a necessary prerequisite for persons whose jobs require them to spend considerable periods of time seated in front of a video display terminal (VDT). To be considered proper, a workplace should at least be adaptable to meet the anthropometric characteristics of 90% of the potential users, and should be physiologically comfortable. The application of correct ergonomic criteria has led to considerable workplace improvements for VDT users, decreasing the musculoskeletal disorders due to unhealthy posture.

However, it has been shown, investigating the effect on posture of changing from an old to a new ergonomically designed workstation in a switchboard control room, that ergonomically designed workplaces increase postural fixity. A prolonged fixed posture can in itself be considered a risk factor, particularly for the lumbar spine, where correct intervertebral disc nutrition depends mainly on alternating hydrostatic pressures, above and below a critical hydrostatic pressure value (Cantoni, 1984; Kraemer, 1977).

The changes in work processes are rapidly leading us to spend more and more of our time in positions that tend to be fixed, both at work and at leisure, and in this respect, postural fixity may be a very important adverse factor for our osteomuscular apparatus.

These two brief considerations move our attention to the fact that ergonomically designed workplaces are useful for avoiding unhealthy posture, but do not solve all the problems related to posture in VDT work.

0-8493-2641-9/99/$0.00+$.50

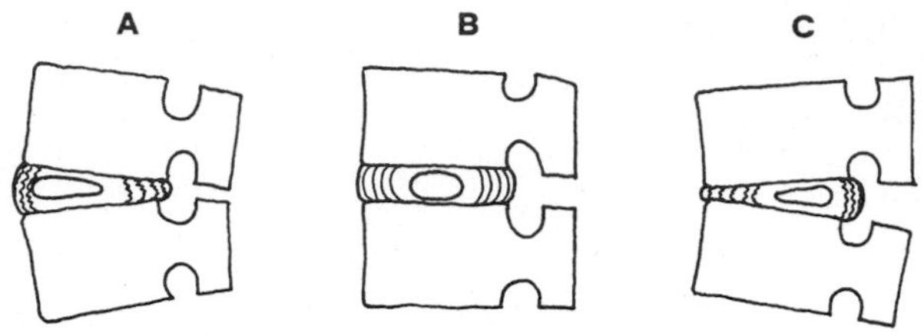

FIGURE 98.1 Disc displacement in extended (A), neutral (B), flexed (C) trunk position.

98.2 Elements of Spine Biomechanics

In order to stand erect, the spine had to be repositioned from horizontal, in the primatial posture, to vertical position in erect posture. This could not be achieved by simply rotating the pelvis on the vertical femur by 90 degrees until the trunk is erect, because the extension of the femora, required for walking, is then impeded by the ischium. Further, the sacrum cannot be rotated backward with respect to the ileum and, in this condition, the birth canal would be obstructed by the coccyx. The evolutionary solution seems to have been rotating the sacrum forward on the ischium and increasing the extension of the lumbar spine. This is the origin of the lumbar lordosis (Bridge, 1991).

In the standing adult, the lumbar lordosis supports the upper body so as to minimize the bending moment of the spine, and the body line of gravity passes through the facet joints of the fourth and the fifth lumbar vertebra (Klausen, 1968).

The spine itself lacks intrinsic stability and short and long back muscles together with anterior abdominal muscles are responsible for the stabilization of the spine as a whole, when the line of gravity of the upper body parts passes ventrally or dorsally to the axis of movement of the lumbosacral joint. This means that posture should be analyzed not only in terms of joint angles and depth of spinal curves but also in terms of the strain imposed on the stabilizing muscles (Nachemson, 1976).

Sitting position ensures less total body energy expenditure and major body stability. When the person is seated, the thigh is horizontal, the hip joint is flexed, and the pelvis has a sloping axis. In this condition, the lumbar spine straightens from its normal lordotic curve and exhibits a kyphosis curve when bending forward. In both cases, the disc protrudes posteriorly from its normal position between the vertebral bodies and impinges on the spinal cord. Bending forward in the seated position, moves the anterior part of the vertebral bodies closer together, increasing the posterior force on the disc, which may increase the stress on spinal tissue (Figure 98.1.).

98.3 Nutrition Mechanism of the Intervertebral Disc and Paravertebral Soft Tissue

Examining the genesis of spine pain ascribed to soft tissue showed how some of the pain was essentially due to a process of protracted isometric contraction by the paravertebral muscles (Caillet, 1973). In some cases, the pain is directly generated by the overuse of the myofascial formations where the paravertebral muscles intervene at the periosteal level. In the majority of cases, however, prolonged isometric contraction causes a constant increase in endomuscular pressure, with relative constriction of the blood vessels and consequent ischemia. The resulting pain is due to a relative oxygen deficiency, to the action of irritating metabolites, to the accumulation of lactic acid, and to the reduced intracellular concentration of potassium. The condition of localized hypoxemia may be the cause for muscular degeneration, that may lead to a fibrotic reaction of both the muscle and the surrounding tissue (Figure 98.2) In light of these comments, it should be kept in mind that while isometric contractions up to 20% of the maximum voluntary contraction are accompanied by an increase in blood supply to the muscle, for isometric contractions above this level, a decrease in the blood circulation and thus, a condition of relative

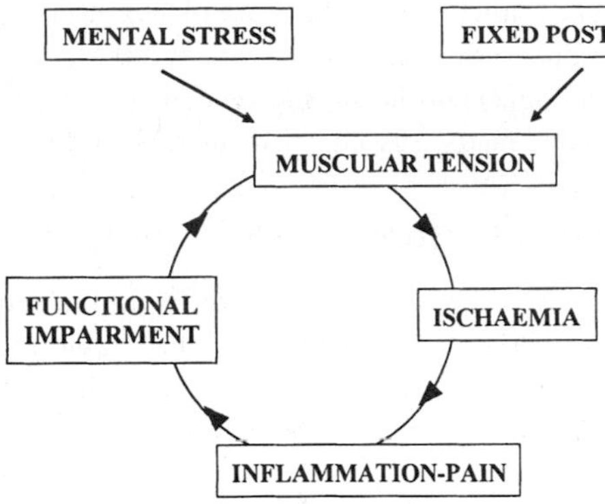

FIGURE 98.2 Role of fixed posture and mental stress in the pathogenesis of functional muscular impairment and pain.

hypoxemia, begins to occur (Barnes, 1980; Edwards, 1972). The level of isometric contraction near or above 20% of the MVC is easily reached, for example during digital keyboard operations, in the trapezius and elevator muscles of the arm, particularly when the upper limbs are unsupported. The result is that, especially in the paravertebral muscles and the shoulder girdle muscles, nonmaximal protracted isometric contractions, which are typical of fixed postures, may lead not only to sensations of discomfort and pain in the short term, but may eventually lead to the onset of a real disease due to alterations of the soft tissue in the long term.

The second structure worth considering in this context is the intervertebral disc. This structure is avascular in the adult, and therefore its nutrition depends on the process of diffusion of substances from adjacent tissues.

Disc nutrition through diffusion depends on a complex relationship between hydrostatic pressure and osmotic pressure inside and outside the disc itself; a decrease in hydrostatic pressure promotes the input of nutritional substances in the disc, and slows down the expulsion of catabolites, while an increase causes an inverse process (Kraemer, 1977; Nachemson, 1968; Ogata, 1981). In the first condition, the input of nutritional substances is accompanied by an increase in disc volume; in the second case, however, the disc tends to lose water and decreases in volume. This phenomenon is so well known that some authors (Eklund, 1984), have decided to use it as a measurement of spinal loading effects.

On the basis of this knowledge, it becomes clear that the optimum in the disc nutritional process is determined also by the constant alternation between conditions of loading and unloading of the disc itself. On the contrary, prolonged conditions of overloading or underloading of the disc, as can occur in prolonged fixed posture, obstruct nutritional exchange and can, in the long term, promote degeneration processes of the disc. There is a threshold value of intradiscal lumbar pressure (80 kg) which acts as a discriminating factor between conditions of overload and underload. When the lumbar spine and upper limbs are completely supported, the lumbar load is below the threshold value, while with the spine nonsupported and upper limbs raised, the lumbar load is above the threshold value (Colombini, 1985).

It can, therefore, be concluded that ergonomically designed chairs with supports for the trunk and upper limbs improve general comfort, but do not essentially alter the problems of lumbar disc metabolism.

98.4 Musculoskeletal Disorders in VDT Work

Musculoskeletal disorders among VDT operators are very diffused, even if the actual prevalence is not well known. In fact, according to various studies, the prevalence rates vary from 1% to more than 50% (Silverstein, 1986). The symptoms are reported in the form of pain, soreness, aching, stiffness, numbness, tremors, fatigue, and cramps that occur in specific parts of the body, but more frequently in the neck, back, and upper limbs. Also the occurrence varies daily, weekly, monthly, or occasionally.

The variability in form, occurrence, and anatomic location of symptoms indicates that the etiology of musculoskeletal disorders may be due to different risk factors and that, at the present, it is not yet well understood. However, three factors appear to be mostly related to musculoskeletal disorders in VDT operators: task- and workplace-related factors, psychosocial factors, and non-task-related factors.

Task- and Workplace-Related Factors

It has been shown that the various office jobs can give rise to different incidences and anatomic locations of musculoskeletal disorders (Hunting, 1981). The highest incidence was found in jobs involving repetitive keyboard work (data entry operators) with anatomic location mainly in neck and upper limbs, while in less repetitive work (conversional operators), the incidence was lower and equally located in different parts of the body.

The daily hours and years of computer use showed a statistically significant positive relationship with musculoskeletal disorders. It is worth pointing out that as the average daily hours increase, the percentage of operators with complaints increases at a greater rate than for years of computer use (Hochanadel, 1995).

As far as the types of workstation furniture and the perception of low back support are concerned, pneumatic adjustable chairs were judged to ensure good back support by 80% of users, manual adjustable chairs by 51%, fixed height chairs by only 41%, and the occurrence of symptoms in back and neck came out to be inversely correlated. On the other hand, no significant difference was found when comparing symptom responses in VDT operators using adjustable keyboard tables with those using fixed-height tables (Hochanadel, 1995).

Improper keyboard height relative to the operator's stature seems to be responsible for a high percentage of symptoms, particularly in neck and upper limbs, and for reduced work efficiency.

Of the mechanical problems that can be found in workstations, the more frequent are inadequate legroom, drawer under the keyboard, and tables with sharp edges. When these three mechanical problems are present at the same workstation, the percentage of symptoms rises to 75% (Hochanadel, 1995).

The use of non-keyboard computer input devices, like the computer mouse, has been associated with symptoms in the shoulder, wrist, and fingers.

The use of the mouse requires exertion of force with the finger and palm to overcome the button, gravity, inertia, and friction forces. The force may be transferred through the surface of the skin and causes stress on the contact area. In addition, the side-to-side movement of the mouse requires radial and ulnar deviation of the wrist, together with flexion and extreme extension and static load on the shoulder region. These two risk factors, force and *non-neutral* postures of the wrist, if frequency and duration of exposure are elevated, are associated with upper limb musculoskeletal disorders, in particular with carpal tunnel syndrome (Armstrong, 1995). In fact, the static load imposed on the upper part of the trapezium muscle fibers, with induced ischemia and release of metabolites such as bradykinin, may induce persistent pain; extreme wrist positions may stimulate pain receptors in the joint capsules or ligaments, rapidly causing pain (Hagberg, 1995).

Psychosocial Factors

In addition to biomechanical and ergonomic risk factors, psychosocial work factors and psychological stress have been postulated to be related to musculoskeletal disorders in VDT operators. Recently, it has been recognized that musculoskeletal and stress problems may be interrelated: they may influence each other or may have some common causes (Carayon, 1995). Psychosocial work factors such as workload, work pressure, fast work pace, time pressure (high demand), low control over work, poor intellectual discretion, and technological obsolescence (poor resources) are related to job stress, which in turn is related to musculoskeletal problems. Psychological and physiological pathways have been put forward to explain that job stress may change the perception of symptoms or induce psychological changes that may result in musculoskeletal problems. One of the hypotheses is that job stress may increase static muscle activity. Prolonged exposure to static muscle activity may lead to increased lactic acid and lack of nutrients and, in the long term, provoke pain due to the inability of, or insufficient time for, the muscles to recover from fatigue (Bongers, 1995; Lim, 1995). On the other hand, job stress could have another type of indirect effect on musculoskeletal disorders via ergonomic risk factors. VDT operators under stress may adopt awkward posture or may use higher force to press on the key of the keyboard or perform their activities at a higher level of repetitiveness (Carayon, 1995).

Non-Task-Related Factors

All the symptoms and medical conditions which have been attributed to VDT use occur within the population with a certain baseline frequency, and it would be surprising if VDT users did not experience everyday ailments, symptoms, and adverse health outcomes with at least the same frequency as any other comparable group of workers (Pearce, 1995). Biodemographic factors, such as age and sex, previous musculoskeletal injuries, emotional stress, family burden, or environmental risk factors can play an important role.

The question is whether VDT operators experience significantly higher levels of musculoskeletal symptoms and adverse health outcomes than any other comparable group of workers. In this sense, it is important to emphasize that in epidemiological studies among VDT operators, there is a need for appropriate reference data, stratified for age and sex, collected among a working population not exposed to occupational risk factors, such as heavy manual handling, awkward and prolonged fixed postures, and whole body vibration. Comparisons are thus possible between frequency of disorders in the exposed and nonexposed subjects; in cases when a significant excess of disorders is demonstrated in the exposed as compared with the nonexposed subjects, ground could be laid for speculation concerning the etiology of the link between occupational risk and disorders. However, with cross-sectional studies, which are the most frequent in the literature, it is not possible, except for particular cases, to reach definite conclusions on such links, which should normally be obtained from longitudinal studies.

98.5 Seating and Posture in VDT Work

Three different and alternative sitting postures have been proposed for a VDT operator: the upright posture (ANSI, 1988), with the joints of the hip, knee, and ankle at right angles, the backward leaning (Grandjean, 1983), and the forward tilted (Mandal, 1991).

The first one seems not to be supported by any physiological or orthopedic reason and, in real life conditions, is quite seldom spontaneously chosen by VDT operators.

The second one has been experimentally demonstrated to reduce the pressure on the intervertebral disc and the stress on back muscle (Grandjean, 1983). However, this posture increases the viewing distance and forces VDT operators to flex the neck, increasing the risk of pain in this part of the spine.

The forward-tilted posture is suggested because in this position the pelvis rotates forward and the lower vertebral bodies are kept vertical, reducing in this way the intervertebral pressure and the lumbar share forces (Mandal, 1991). This posture is unusual and requires supporting the body weight with the feet placed on the floor or the knee on a special pad, and thus it may not be suitable for all VDT operators.

Advantages and disadvantages of the three sitting postures might be increased or reduced in relation to the tasks performed and the working conditions, respectively (Dainoff, 1987).

The upright posture has been suggested as the best for typing tasks, the forward-tilted posture when the viewing distance is the most important factor and the need to write on paper is frequent, while the backward-leaning posture, which decreases the pressure on the intervertebral discs but increases the viewing distance, is considered to be good for tasks mostly requiring screen work and for which the viewing distance is not a critical factor (Mandal, 1991).

Specific solutions should obviously be tailored to a specific task, but general guidelines are necessary, taking into account that: *the VDT workplace configuration is to be designed in such a way as to avoid forcing the operator into awkward or fixed postures; there is not a single working posture that is optimal for all VDT*

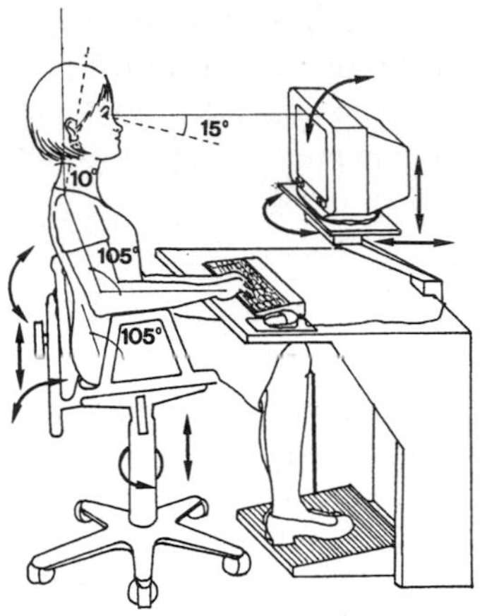

FIGURE 98.3 Ideal static reference posture for VDT operator.

tasks, and finally even the most correct working posture becomes uncomfortable if maintained for prolonged periods of time.

An ideal static reference posture, based on ergonomic criteria, should be defined, not intended as fixed but to provide a generally corrected basis and a *starting posture* able to allow VDT operators to easily change the working position. From this reference posture, requirements for furniture and workstation configuration should be derived. Finally, linear dimensions, and profile adjustability features of the components of the workplace should be established on the basis of general population anthropometric parameters to satisfy the needs of 90% of the potential users and to facilitate the movement of the body.

According to biomechanical and psychological criteria and to a recent study (Hochanadel, 1995) the ideal static posture should be characterized by *neutral wrist posture* to reduce pressure in the carpal tunnel, *shoulders relaxed* to prevent static muscle activity in the neck and shoulder, *trunk slightly reclined and relaxed,* to minimize intradiscal pressure at lumbar spine level, limited *neck bending,* to avoid increase in tone of the anterior neck musculature, and *knees at the level of or slightly higher than hips and feet, supported* to decrease pressure on the thigh.

The resulting position of the VDT operator can be described as: elbows at keyboard height with forearms parallel to floor, upper limbs in line with trunk, trunk reclined between 100° and 110°, neck flexion not exceeding 15°, eyes in line and at VDT screen level, knees at the level of or slightly higher than the hips, and feet on the floor or footrest (Figure 98.3).

To ensure this ideal posture, the essential required characteristics of workplace components are: proper adjustable chair with back and arm–wrist support, table with adequate height, depth, and legroom, and a footrest, if needed.

98.6 Requirements and Characteristics of a VDT Workstation

A basic VDT workstation is composed of a chair, desk, video display, and keyboard.

Chair

The chair is the most important component, since it interacts with other components and significantly influences operator comfort. Recent changes in the manufacturing and marketing of VDT chairs have led to an increasingly widespread exploitation of the ergonomic quality of chairs; in other words, the ergonomic characteristics have become a determining factor in marketing. But what are the characteristics and requirements on the basis of which a work chair can be defined as ergonomic? In the literature, ergonomic rules relative to the characteristics of the chair have so far been expressed too briefly and in a disorganized way, mainly referring to dimensions and comfort rather than to safety or performance.

To overcome this fragmented approach, all aspects should be properly considered when evaluating work chair ergonomics.

The basic requirements that the VDT chair should meet in order to be defined as ergonomic are (Drury, 1982; Drury, 1985; Occhipinti, 1993):

Safety: the chair must not be a source or cause of accidents.

Adaptability: the chair and its components should have the correct dimensions and be easily adjustable to meet the anthropometric needs of a wide range of users (normally at least 90% of the potential users).

Comfort: the chair and its components should be upholstered, contoured, and reciprocally adjustable so as to meet the physiological needs and characteristics of many different "body curves and shapes."

Practicality: the chair and its components must be easily adjustable by the user; the covering materials should be hygienic.

Solidity: the chair, its components, and relative adjustment controls should be reliable, maintaining the same performance over time.

Safety

- In plan view, the area of the supporting base should contain the surface area of the seat to ensure chair stability.
- Pressurized gas springs for adjusting the seat and the inclination of the backrest should be approved and tested by qualified standards authorities.
- It should not be possible to activate the chair adjustment controls unintentionally, especially if they are of the mechanical type.
- The components should be made of nonflammable materials.
- A range of casters should be available with braking and anti-skid features for different type of floors.
- There should be no sharp edges.
- Armrests should not be open in front and in the back.

Adaptability

Adaptability is ensured when the range of variables concerning the measurements and adjustment positions of the chair components meet the variability of potential users. In Table 98.1 recommendations for chair adjustability are given.

The anthropometric distribution considered is that of an adult western population (Pheasant, 1986). A fixed working surface height of 72 to 75 cm has been considered, since it represents a physiologically and economically satisfactory solution and doesn't allow only a minimum percentage of taller men to cross their legs.

- Seatpan height range has been defined considering the trend of growth of the popliteal height parameter in the population and the thickness of footwear (2 to 5 cm).
- Seatpan width corresponds to the 95th percentile of anthropometric parameter "hip width" in women.
- Seat depth in chairs without an adjustable depth is particularly important. This parameter should not exceed 41 cm to meet dimensions of 5th percentile subjects.
- Generally, backrests that only support the lower back are defined as "low." Backrests supporting the trunk up to the maximum thoracic kyphosis are considered "medium," while those which exceed this height are considered "high."

  When the backrest is height-adjustable, it should be at least 32 cm high, and its upper edge should be capable of reaching 48 cm from the seat plane.
- Backrest width: to properly satisfy the anthropometric parameters (considering the values for the 95th percentile male), the width should be at least 33 cm in the lower back segment, and 38 cm in the thoracic segment.

TABLE 98.1 Recommendations for the Adjustability of Chair Components and Corresponding Anthropometric Parameters

	Recommended Dimensional Range (cm)	Corresponding Anthropometric Parameters
Seat height	39–52	Popliteal height
Seat depth	39–55	Buttock popliteal length
Seat width without armrests	≥47	Hip width
Seat width + armrests	>49	Hip width
Backrest height	32–50	Max kyphosis/seat plane height
Backrest width		
Lumbar region	≥33	Minimum width lumbar lordosis
Thoracic region	38	Max width at thoracic kyphosis
Armrest		
Height	16–23	Elbow height to seat plane
Depth	20–25	
Width	≥5	

TABLE 98.2 Main Characteristics for Chair Comfort

	Recommended Dimensional Range
Seatpan max concavity (from backrest)	≤10 cm
Curvature of front edge (radius)	4–12 cm
Seat plane angle (degrees)	3–10°
Lumbar support height from seat plane depth	17–28 cm
	3–5 cm
Backrest angle net angle	90–110°
Horizontal profile (radius)	40–80 cm
Padding	Semi-rigid
Covering materials	Porous to prevent sliding

Armrests are useful for supporting upper limbs while work is not being done, but it is not clear whether or not it is desirable to provide support for the arms. It might be appropriate for tasks in which typing is intermittent, otherwise it should be an optional feature. The length of the armrest should be at least 20 cm and no more than 25 cm in order to position the chair close to the desk.

Comfort

The chair should have such a shape and profile as to meet "body curves and sizes" of different users. In Table 98.2. main characteristics for chair comfort are given.

- The backrest adjustment control should allow the backrest to be set at any desired inclination, or at a wide range of preset inclinations (e.g., at 5 degree intervals).
- Backrests that incline only by putting pressure on them should be avoided (even those with adjustable resistance): blocking devices to fix desired inclinations are preferred.
- The front edge of the seat should be rounded. The curvature should have a radius between 4 and 12 cm and height of approximately 4 cm.
- The center of the seatpan concave section should be placed within 10 cm of the most protruding point of the lumbar support when the backrest is in its normal position.

  Lumbar support height should be adjustable between 17 and 28 cm. The maximum protrusion for nonadjustable backrests should be fixed somewhere between 20 and 24 cm from the seat plane. The length of lumbar support should range between 20 and 30 cm for medium-sized backrests, and the shape should be vertically convex and horizontally concave.

TABLE 98.3 Recommendations for Desk, VDT Screen, and Keyboard

	Recommended Dimensional Range
Desk	
Surface height	72–75 cm
Surface width	150 cm
Surface depth	90 cm
Legroom	80 cm
Width	60 cm
Depth at knee	80 cm
Depth at feet	
Keyboard	
Slope	5°–30°
Thickness	3–5 cm
Screen height	90–110 cm

Practicality and Solidity

- Adjusting the chair's various components should be made as easy as possible for the user. If the controls are difficult to maneuver because they are hard to reach, nonresponsive, or require too much strength, they will never be used. The adjustment controls should be responsive, precise, and easy to reach from a seated or semi-seated position and should not require strength. Any knobs or handles should not come off easily.
- The solidity of the chair and its components is an essential requisite, not only from the marketing point of view, but also for ensuring continuous ergonomic performance over time.

Desk

The second important component of the VDT workstation is the desk. The height or work surface is the most discussed parameter since improper height leads to development of musculoskeletal problems and hinders a correct allocation of legs (ANSI, 1988). Also the use of an adjustable desk as opposed to a fixed one gives rise to controversial opinions. As previously mentioned, a fixed working surface height is a physiologically and economically satisfactory solution. In fact, with a fixed desk height of 72 to 75 cm, if a footrest is available for the lower stature operators, almost any postural combination which can be obtained with an adjustable desk, is possible. It has also been observed that most adjustable desks are quite often unstable, that operators very rarely adjust them, and that no significant difference in musculoskeletal problems has been found between adjustable and fixed-desk users (Hochanadel, 1995).

In Table 98.3 recommendations for desk, VDT screen, and keyboard are given.

Desk width and depth are very important factors not sufficiently investigated in the literature. A width of 150 cm minimizes the proper placement of all the working instruments, while a depth of 90 cm is essential for positioning the VDT in front of the operators; commercially available desks have a depth of 75 cm, which does not allow, considering the dimensions of VDT and keyboard, enough space in front of the desk for supporting at least half of the forearm and/or for writing tasks. In this condition, the operators usually move the VDT to the left or right side of the desk, and consequently adopt unnatural and unfavorable neck posture.

98.7 Organizational Factors

As underlined in the introduction, ergonomically designed workplaces avoid awkward and unhealthy postures but do not solve all postural problems. Some other aspects, such as psychosocial and biodemographic factors and individual characteristics, have an impact on VDT operators' comfort.

In addition, organizational factors like information and training, work duration, pauses, and job rotation, taken into appropriate consideration, can reduce fatigue and musculoskeletal disorders.

Information should be given to VDT operators on physiopathology of musculoskeletal disorders, how to sit correctly, how to properly adjust the workplace components, and how to reduce exposure to VDT work risk factors. However, information by itself gives poor results if operators are not well-trained and their learning is not verified. In other words, one must avoid giving oral information or informative booklets without checking that operators have well understood the contents and are able to put them into practice.

Job variety, task rotation, and breaks, as means to reduce the duration of fixed posture, are useful in interrupting the isometric tension of muscles required to maintain a mostly fixed posture. To prevent the possible consequences of postural fixity, VDT work must be designed in such a way that postural pauses are planned and used, in particular when work organization and times are for the most part fixed. In other situations, where the operators have higher levels of discretion in performing tasks, adequate information should be given on how to use the equipment, how to alternate work periods with pauses, and what to do during the pauses (Grieco, 1986).

The need for more dynamic activity to reduce the stress of sedentary work has given rise to a proliferation of exercise programs designed to reduce musculoskeletal discomfort from VDT work, both at work or before and after the workshift.

A review (Lee, 1992) of 14 exercise programs for VDT users with regard to the usability and physiotherapeutic/safety criteria showed that less than 50% of exercises were satisfactory, and could be easily performed at the workstation, without being embarrassing or significantly disrupting the work routine.

In addition, a number of exercises exacerbated biomechanical stresses common to VDT work or were even contraindicated for people with health problems. These conclusions indicate the need for great caution in suggesting physical exercises to VDT operators.

98.8 Conclusions

Physical attributes of VDT workplaces, organizational and psychosocial factors together with individual characteristics are recognized as potential risk factors for the musculoskeletal apparatus responsible for the diffusion of musculoskeletal disorders among VDT operators.

To be successful, a preventive approach should take into consideration all the risk factors in an integrated manner. The following 10 rules will help in minimizing the risks associated with VDT work.

1. Workplaces should be dimensionally designed to allow not only physiologically comfortable posture but also body movement for a wide variety of body types and dimensions (from 5th percentile female to 95th percentile male).
2. Adjustable chair and desk, with a proper footrest provide the best solution. Also an adjustable chair and a footrest with fixed desk height ensure numerous postural combinations to meet the variability of potential users, being, in addition, an economically satisfactory solution.
3. The chair is the most important VDT workplace component: for this reason it must have multiple adjustments, shapes, and profiles, to ensure adaptability and comfort. In addition, an adaptable and comfortable chair should be practical, solid, and safe in order to be easily adjustable from a seated position, capable of ensuring ergonomic performance over time and not being a cause of accidents.
4. Keyboard and non-keyboard input devices should facilitate neutral postures and avoid excess of force application.
5. Desk depth is a very important factor quite often overlooked; desk depth less than 90 cm hinders operators to position the VDT in front of them. Moving the VDT to the left or right side of the desk can frustrate all the efforts made for providing ergonomic workplaces.
6. Job demand, in terms of workload, work pressure, and workplace should conform to the individual resources in order to avoid psychological stress, which can increase the risk of musculoskeletal disorders.

7. Frequent breaks can reduce fatigue and musculoskeletal disorders. However, operators should know what to do during the pauses so that they do not remain seated, maybe reading a paper, having a work pause but not a postural (and visual) pause.
8. When the operator has little discretion in the operation performance, the job should be designed so that alternative postures and body mobilization are required.
9. Minimizing direct and indirect glare helps avoiding awkward postures.
10. Information and training are essential for the successful implementation of ergonomic intervention. Information and training are aimed at introducing or modifying behavior in situations made up of real people. For this purpose, information and training should be integrated, and operators should be tested on their understanding of the content and their ability to put it into practice.

98.9 Summary

Musculoskeletal disorders among VDT operators are very diffused, even if the actual prevalence is not well known. The etiology is not well understood, but three factors appear to be mostly related to musculoskeletal disorders in VDT operators: task- and workplace-related factors, psychosocial factors, and non-task-related factors.

The risk factors and their effect on musculoskeletal apparatus are discussed, and recommendations on workplace and dimension characteristics are given. It is underlined that to be successful a preventive approach should take into account all the risk factors in an integrated manner.

This means that psychological and organizational factors, like workload, work pressure, work pauses, information, training, and individual characteristics should be taken into account together with workplace physical characteristics.

Acknowledgments

Thanks are due to Arch. P. Lavano for graphical assistance and computer drawings and to R. Mancra for their helpful support.

References

American National Standards Institute, 1988. American National Standard for Human Factors Engineering of Visual Display Terminal Workstations. *The Human Factors Society,* Santa Monica.

Armstrong, T. J., Martin, B. I., Franzblau, A., Rempel, D. M., Johnson, P. W. 1995. Mouse input and work-related upper limb disorders. *Work With Display Units 94.* p. 375-385. Elsevier Science B. V., Amsterdam.

Barnes, W. S. 1980. The relationship between maximum isometric strength and intramuscular circulatory occlusion. *Ergonomics.* 23: 351-357.

Bongers, P. M., Houtman, I. L. D. 1995. Psychosocial aspects of musculoskeletal disorders. *Premus95.* p. 172-174. IRSST. Montreal.

Bridge, R. S. 1991. Some fundamental aspects of posture related to ergonomics. *International Journal of Industrial Ergonomics.* 8:3-15.

Caillet, R. 1973. *Il dolore cervico-brachiale.* Leonardo Ed. Scientifiche. Roma.

Cantoni, S., Colombini, D., Occhipinti, E., Frigo, E., Pedotti, A. 1984. Posture analysis and evaluation at the old and new workplace of a telephone company. in *Ergonomics and Health in Modern Offices.* p. 456-464. Edited by Grandjean, E. Taylor & Francis. London.

Carayon, P. 1995. Work pressure as a determinant of job stress and cumulative trauma disorders in automated offices. *Premus95.* p. 172-174. IRSST. Montreal.

Colombini, D., Occhipinti, E., Frigo, F. Pedotti, C., Grieco A. 1985. Biomechanical, electromyographical and radiological study of seated postures. *International Occupational Ergonomics Symposium.* Zara. 14-16 April.

Dainoff, M. J., Mark, L. S. 1987. Task and adjustment of ergonomic chair. *Work with Display Units 86.* p. 294-302. Knave, B., Wideback, P. G. Eds., North-Holland. Amsterdam.

Drury C. G., Coury, B. G. 1982. A methodology for chair evaluation. *Applied Ergonomics.* 13: 195-202.

Drury C. G., Frencher, M. 1985. Evaluation of a forward sloping chair. *Applied Ergonomics.* 16:41-47.

Edwards, R. H. T., Hill, D. K., McDonnell, M. 1972. Myothermal and intramuscular pressure measurements during isometric contractions of the human quadriceps muscle. *Journal of Physiology.* 224:58-59.

Eklund, J. A. E., Corlett, E. N. 1984. Shrinkage as a measure of the effect of load on the spine. *Spine.* 9:189-194.

Grandjean, E., Hunting, W., Piderman, M. 1983. VDT workstation design: preferred settings and their effects. *Human Factors.* 25: 161-175.

Grieco, A. 1986. Sitting and posture: an old problem and a new one. *Ergonomics.* 29:345-362.

Hagberg, M. 1995. The "mouse-arm syndrome" — concurrence of musculoskeletal symptoms and possible pathogenesis among VDU operators. *Work With Display Units 94.* p. 381-385. Elsevier Science B. V. Amsterdam.

Hochanadel, C. D. 1995. Computer workstation adjustment: a novel process and large sample study. *Applied Ergonomics.* 26(5):315-326.

Hunting, W., Laubli, T., Grandjean, E. 1981. Postural and visual loads at VDT workplaces I. Constrained postures. *Ergonomics.* 24:917-931.

Klausen, K., Rasmussen, B. 1968. On the location of the line of gravity in relation to L5 in standing. *Acta Physiological Scandinavica.* 72:45-52.

Kraemer, J. 1977. Pressure dependent fluid shifts in the intervertebral disc. *Orthopaedic Clinics of North America.* 8:211-216.

Lee, K., Swanson, N., Sauter, S., Wickstrom, R., Waikar, A., Mangum, M. 1992. A review of physical exercises recommended for VDT operators. *Applied Ergonomics.* 23(6):387-408.

Lim, S. Y. 1995. Psychosocial and work stress perspectives on musculoskeletal discomfort. p. 175-177. *Premus95.* IRSST. Montreal.

Mandal, A. C. 1991. Investigation of the lumbar flexion of the seated man. *International Journal of Industrial Ergonomics.* 8:75-87.

Nachemson, A. 1976. The lumbar spine: an orthopaedic challenge. *Spine.* 6:93-302.

Nachemson, A. 1968. The possible importance of the psoas muscle for stabilisation of the lumbar spine. *Acta Orthopaedica Scandinavica.* 39:47-57.

Occhipinti, D., Colombini, D., Molteni, G., Grieco, A. 1993. Criteria for the ergonomic evaluation of work chairs. *La medicina del lavoro.* 84:274-285.

Ogata, K., Whiteside, L. A. 1981. Nutritional pathways of the intervertebral disc. *Spine.* 6:211-216.

Pearce, B. 1995. An alternative perspective on the VDT health hazards debate: folk hazards and health panics. *Work With Display Units 94.* p. 15-20. Elsevier Science B. V., Amsterdam.

Pheasant, S. T. 1986. *Body Space.* Taylor & Francis Ltd. Philadelphia.

Silverstein B. 1986. Carpal tunnel syndrome: Causes and a preventive strategies. *Seminars in Occ. Med.* 1:213-219.

For Further Information

Ergonomic Aspects of Visual Display Terminals. E. Grandjean and E. Vigliani. Eds., Taylor & Francis. London 1980. Proceedings of the First International Workshop on "Ergonomic Aspects of Visual Display Terminals," Milan, Italy. March 17-19, 1980.

Ergonomics and Health in Modern Offices by E. Grandjean. Ed., Taylor & Francis. London and Philadelphia, 1984. Proceedings of the International Scientific Conference on Ergonomic and Health Aspects in Modern Offices, held in Turin, Italy. November 7-9, 1983, under the auspices of the International Ergonomics Association, the Permanent Commission and International Association of Occupational

Health, the University of Turin, the Società Italiana di Medicina del Lavoro e di Igiene Industriale and the Regione Piemonte.

Ergonomics in Computerized Offices by E. Grandjean. Taylor & Francis. London and Philadelphia, 1987. This book is a milestone by an early pioneer in researching human capabilities and limitations within industry, and applying that knowledge to the design of industrial work situations.

Work With Display Units 86 by B. Knave and P.G. Wideback. Eds., Elsevier Science B.V. North-Holland, 1987.

Work With Display Units 89 by L. Berlinguet and D. Berthelette. Eds., Elsevier Science B.V. North-Holland, 1990.

Work With Display Units 92 by H. Luczak, A. Cakir and G. Cakir. Eds., Elsevier Science B.V. North-Holland, 1993.

Work With Display Units 94 by A. Grieco, G. Molteni, E. Occhipinti, and B. Piccoli. Eds., Elsevier Science B.V. North-Holland, 1995. Selected papers of WWDU Conferences are published by Elsevier Science B.V. These selected papers document the latest developments in the field of work with display units.

99

Human–Computer Interface Requirements

Ahmet Çakir
Ergonomics Institute
Berlin

99.1 Introduction

General

Human–computer interaction is a frequently used but not adequately defined term. The main reason for this may be found in the rapid changes of forms of human–computer interaction over time. While some users still remember pushing buttons on the housing of a computer to enter hexadecimal codes, other users expect computers to respond simply to their gestures. Even barring these extremes, a great variety in the way of interacting with computers remains: while some users work with computers using a simple keyboard and a single screen, others utilize highly sophisticated input/output media and multiple screens. Moreover, the workstation, formerly being the "terminal" (or "end station") of a single computer, may today be used for simultaneous access to networks that may be distributed over the entire world. Thus, while formulating requirements in the human–computer interface a great diversity of tasks needs to be considered as well as user populations with considerably different characteristics.

For defining and confining the term "requirement," different approaches can be taken, e.g., relying on one's own expertise, deriving them from consistent guidelines, for example Smith and Mosier (Smith and Mosier 1986), or drawing conclusions from relevant literature. The latter approach was taken by expert groups who prepared the standard ISO 9241 after almost a decade of discussions. Our recent experience with the application of this standard for testing workplaces in German industries suggests that most user requirements are covered by the provisions of ISO 9241. Thus, this section will concentrate on this standard with its 17 parts. However, the part 1 will not be addressed since it gives introduction to other parts of the standard. Part 2 will also not be addressed due to the objective of this part which is giving guidance on task design.

0-8493-2641-9/99/$0.00+$.50

The entirety of "normative documents" applicable to human–computer interface design goes well beyond ISO 9241. Various organizations throughout the world are involved in creating new normative documents (see below) of different nature and legal relevance, an ongoing process which is described in Smith (1996) with focus on international standardization and U.S. involvement. For project managers who want to integrate ergonomics within the activities for designing or redesigning interactive computer systems, the future standard ISO 13 407 will provide guidance for human-centered design of such systems.

Should System Designers Adhere to Human–Computer Interface Standards?

A "standard" in general is a set of normative rules applicable to single products, systems, services, or even behavior of organizations, to name the most important. They represent a consensus met within a trade or business area, in a country or in an international context. Although there is no obligation for written standards, in today's world, standards are documents which result from an agreement by a group of sanctioned persons. In most countries, standards are established by private organizations, and their use is voluntary unless they are legislated.

Both the growth of world trade and communication amplified the need for international agreements in the form of standards. However, such agreements may be of different quality and, most important, flexibility. While some standards specify all relevant features of the object in question, e.g., the transmission quality of a telephone line, others may give guidance similar to a code-of-practice. Some standards even simply serve as documents of "understanding": they describe their objective, but they do not contain any requirement or recommendation concerning this objective. Since this understanding of standardization is not common in the international community, e.g., the word "standard" may be perceived differently in the U.S. and Europe, the role and nature of standards may be misunderstood also. For example, Smith and Mosier (1986) make a distinction between standards, guidelines, design rules, and algorithms as means for design guidance, with the standards being the most rigid. For example, with regard to "aids for tailoring," the provisions of standards are declared as "none," while algorithms "might have optional parameters." The conclusion of the authors is: "Our present knowledge supports development of flexible design guidelines for user interface software, but does not justify imposition of standards."

Today, more than 10 years later, our widely improved knowledge still does not support the imposition of standards in the rigid understanding of the word "standard," however, the lesson learned from the past is that any rigid standard based on design parameters or technical features will always fail since it will be subject to aging (see "The Approach of ISO 9241"). The standards in question are even more flexible than guidelines from the past. The surprising effect of the new approach in standardization is that system developers are often disappointed by the lack of mandatory requirements concerning their specific object of interest. There are still various good reasons for observing the rules set by the standards without adhering to their perceived nature as "mandatory and rigid" rules. The main reasons for this are:

- The standards described here represent the result of an agreement within the international community.
- They focus on user comfort and performance rather than on product features (human-centered approach instead of product-centered).
- They allow developing different methods of evaluating usability of systems on the basis of a consistent method.
- They do not prefer any technical solution, and, thus, are friendly toward diversity.
- To demonstrate compliance with legislation, now existing in Europe, thinkable for many countries outside this region, will only be possible on the basis of standards.

ISO 9241 replaced the first national standard on human–computer interface, DIN 66 234, in Germany. The European standardization will accept it as unique standard for this area. For the U.S., a future standard (ANSI/HFS 200) comparable with ISO 9241 is in preparation. It must be noted, however, that the focus of ISO 9241 is usability, and, usability is only one of many quality criteria for software and system design.

The Systematic Approach: Focusing on Tasks Rather than on Products

From the perspective of work system design, human–computer interaction does not constitute the primary design goal. In the development of work systems, the first phase is their specification regarding the system task, i.e., the entirety of all activities required to achieve the intended outcome of the work system. In the second phase, the necessary system functions are allocated to machines and humans with some of these functions to be performed with human–machine interaction. Those functions allocated to humans and those to be performed with human–machine interaction constitute the work task of the user. The starting point for the human–computer interface requirements is the work task. This means that such requirements can only be formulated or justified on the basis of task requirements. Following this undisputed rationale, the starting point for the most elaborate standard on human–computer interface, ISO 9241, is the task itself (see ISO 9241-2, Guidance on Task Requirements).

Although the approach seems obvious, its application to system or software development is not easy since many tasks "emerge" after the software is introduced and users of a certain system may perform different tasks with it (e.g., intended usage, customizing, or maintenance). Companies developing their own software or software for the market are interested in product features rather than in analyzing task requirements. Thus, their interest focuses on usage-independent criteria. Following this rationale, some existing sets of requirements, e.g., "style guides," checklists etc., consist of descriptions of features without a specific consideration of the user's task. To some extent, such requirements can be justified if their objectives are user and usage independent, e.g., "Error messages shall give information on how to recover from the error state." However, limiting the term "information" in this requirement is task and user dependent, thus, making it difficult to specify a requirement on the extent of the error message.

For the reason given above, the requirements formulated by different groups of people display a great diversity, e.g., a list of requirements set up by the manufacturer of certain products can be very specific, while requirements of international standards need to be generic and technology independent. To satisfy this need, the requirements of the International Standard 9241 are based on the "performance" approach, i.e., they specify the goals to be achieved by the use of a product rather than the features to be realized to reach that goal. In practice this means, however, an operationalization of the standard is needed. Most of the standards of the series ISO 9241, and none of the standards on software contain requirements. Instead, they give recommendations. For one of them (ISO 9241-14 on menus), the procedure of operationalization is described as follows (Harker, 1995): "If a product is claimed to have met the applicable recommendations in ISO 9241-14, the procedure used in establishing requirements for, devel oping, and/or evaluating the menus shall be specified. The level of specification of the procedure is a matter of negotiation between the involved parties." The same applies to other parts of the standard that contain no requirements.

In general, the recommendations in ISO 9241 are formulated as guidelines, e.g., "dialogue principles" (see ISO 9241-10). However, those recommendations are turned into legal requirements for the so called "operator/computer interface" by the EU-Directive 90/270/EWG (so called "VDT-Directive"), valid for all countries of the European Union. This Directive requires that "the principles of software ergonomics must be applied, in particular to human data processing." The requirements of the Directive themselves are even less specific than those of the standards; nevertheless, they are valid legal requirements for all workplaces with computers with a small number of exceptions.

Some important human requirements cannot be found in standards or other regulations on computer applications but elsewhere since they are generic for human work, and not for computerized work only. In some respects, it is claimed, some relevant criteria for human work have been "forgotten" while formulating software-related ergonomic standards or guidelines. One of the sources for relevant information on designing systems is ISO 6385, which constitutes the general framework for all ergonomic standards. From this standard, a number of directly applicable requirements for the human–computer interface can be derived although it does not refer to computerized work.

99.2 Requirements Based on ISO 6385

Unlike most sources of requirements which focus on the "user" and "usability," ISO 6385 states "The observance of ergonomics principles applies not only to the intended use of equipment, but also to its installation, adjustment, maintenance, cleaning, repair, removal, and transport." This is clearly a systems-oriented approach that considers all those involved with the utilization of a product and not the intended users only. The importance of this approach can easily be understood by people responsible for PC-applications, for whom the effort and costs for installation, customizing, updating, deinstalling, etc., go well-beyond the price of the application, and where the workload of persons responsible for accomplishing such tasks is much higher than that of the intended users.

The requirements concerning the human–computer interface derived from this statement of ISO 6385 are as follows:

- The human–computer interface of a system shall be evaluated for all persons involved in the utilization of a computer system, and not for the intended users only.
- While designing the human–computer interface of a system all tasks associated with different phases of the life cycle shall be considered, and not the phase of the usage by the intended users only.

Prior to the design of the human–computer interface, the work task should be considered, e.g.,:

1. Recognize the experience and capabilities of the working population.
2. Provide for the application of an appropriate variety of skills, capabilities, and activities.
3. Provide people with an appropriate degree of autonomy in deciding priority, pace, and procedure.
4. Provide sufficient feedback in meaningful terms to those performing the task.
5. Provide opportunities for the development of existing skills and the acquisition of new skills with respect to the tasks concerned.

One important consequence of giving priority to task design over interface-related issues is that even highly complex interfaces can be considered ergonomic if the task at hand itself is complex. In other words, task adequateness takes precedence over ease of use. Following this rationale, various requirements can be derived from ISO 6385, e.g.,:

re 1. The design and use of relevant elements of the interface (e.g., screen layout, input/output language, graphical representations of objects, form and contents of help information, response times, etc.) shall take the existing knowledge of the intended users and their capabilities into account.
Note: Taking into account does not necessarily mean the designers should stick to what is already known to the intended users. Carroll and Olson (1991) describe two experiments in which the prospects of such an approach were tested. The successful strategy was that designers evaluated the input from the potential users and made their own decisions.

re 2. Applications shall be sufficiently flexible to accommodate a variety of user needs depending on their skills, capabilities, and activities, e.g., users can select different styles of interaction.

re 3. The interface shall permit the users to control their application efficiently.

re 4. Each user action shall be followed by an adequate feedback.

re 5. The interface shall be designed to promote learning; e.g., by making user actions reversible, and thus encouraging them to learn by trial-and-error.

As can be seen from these examples, the general requirements of ISO 6385 can be broken down to more specific requirements which go well beyond concepts that consider the intended users only. By an adequate interpretation of the entirety of ISO 6385, systematic sets of rules for both the software and its interface can be derived.

TABLE 99.1 Three Types of Performance

Meaning	Example
Behavior of an object	The selected menu option is echoed at the display as highlighted.
Process conducted by the user	The user discriminates available and unavailable menu options.
Success a user should achieve	A user is able to select the intended menu option within a menu panel.

99.3 Requirements and Recommendations of ISO 9241

The Approach of ISO 9241

The requirements of ISO 9241 are mostly defined in terms of "performance" instead of attributes (Çakir and Dzida, 1997). Performance has a threefold meaning (see Table 99.1).

The rationale behind the performance-based approach can be explained using the second example: The performance requirement stipulates that the user be able to discriminate available and unavailable menu options. A technical standard or a style-guide would require "graying out the unavailable options." This requirement would, however, become obsolete when a designer finds a better way of discriminating the state of the menu options. Standards based on attributes rely on technical properties and are therefore subject to aging since aging is the most prominent feature of technology. Moreover, such standards are likely to impede the creative potential of designers. Mainly for these reasons, ISO 9241 describes goals rather than means.

Coverage of ISO 9241

ISO 9241 covers all topics related to the human–computer interface including task design, hardware (screens, input devices), software, and work environment. Some parts provide general guidance to be considered in the design of equipment, software, and tasks, while other parts include more specific design guidance and requirements. Since covering a wide range of information within a single standard was considered impractical, the topics were broken down to 17 parts, out of which 7 deal with software issues (see Table 99.2), some others have some relationship to software, e.g., part 8 on displayed colors. Each part of ISO 9241 is formally a separate standard. The numbers of the different parts of ISO 9241 do not reflect its structure since this was changed occasionally. The original objective of the standard was to provide requirements for the displays, the keyboard and the workplace. Later, the objective was expanded to cover task requirements, work environment and software-related issues. The need to cover usability as a general aspect lead to the introduction of a specific standard (part 11).

The parts of ISO 9241 can be assigned to three groups dealing with (classification after Çakir and Dzida 1997):

- Work, organization and their role in usability specification (Parts 2 and 11)
- Workplace and work environment (Parts 5 and 6)
- Interactive equipment/tools

  Parts 3 and 7 (on visual displays)
  Parts 4 and 9 (on input devices, keyboard, and non-keyboard input devices)
  Parts 10 and 12 up to 17 (on software interfaces)
  Part 8 (on color, both for visual displays and software)

This classification is based on the overall nature of each part from a systemic point of view that considers hardware, software, and their interaction as a whole. For example, part 3 dealing with visual display requirements requires primarily the consideration of hardware features (monitor); however, fulfilling visual display requirements as described in this standard requires the consideration of software,

TABLE 99.2 Parts of ISO 9241 and Their Status as of September 1998

Part	Title	Status[1]
1	General introduction	IS, FDIS, second revision
2	Guidance on task requirements	IS
3	Visual display requirements	IS
4	Keyboard requirements	IS
5	Workstation layout and postural requirements	IS
6	Guidance on the work environment	DIS
7	Display requirements with reflections	IS
8	Requirements for displayed colors	IS
9	Requirements for non-keyboard input devices	DIS
10	Dialogue principles	IS
11	Guidance on usability	IS
12	Presentation of information	FDIS
13	User guidance	IS
14	Menu dialogues	IS
15	Command dialogues	IS
16	Direct manipulation dialogues	DIS
17	Form filling dialogues	IS

[1] IS = Standard; DIS = Draft International Standard; FDIS = Final Draft International Standard

also. The same applies to keyboards and to non-keyboard input devices if the focus of interest lies on performance and comfort.

If the main focus of each part is of interest, the structure of the standard may be seen as follows:

- Parts dealing *mainly* with *hardware concerns* include: visual display requirements (3), issues related to reflections (7), keyboard (4), workstation layout and posture (5), and non-keyboard input devices (9)
- Parts dealing *mainly* with *software concerns* include: general dialogue principles (10), presentation of information (12), user guidance (13), menu dialogues (14), command dialogues (15), direct manipulation dialogues (16), and form-filling dialogues (17)
- Part dealing with the *use of color* (8), related both to hardware and software
- Parts dealing with *tasks and usability*: guidance on task requirements (2), usability (11)

The standards related to visual displays and input devices, to software, and to usability will now be discussed in greater detail.

Requirements and Recommendations for Displays and Input Devices

Requirements for Visual Displays

The requirements for visual displays are distributed over three parts of the standard, parts 3, 7, and 8. In addition, ergonomic requirements for the use of flat panel displays are regulated in a separate standard (ISO 13406). Although the title of part 3 is "Visual display requirements," displays conforming to this standard do not necessarily meet all needs of users of visual displays. For example, part 3 does not deal with an important issue, the anti-glare treatment of the screen surface. In addition, some relevant characteristics of displayed images, e.g., the character size, have been standardized on the basis of research on monochrome character images. A number of experiments have shown, however, that legibility is lower for color displays (see Widdel and Post, 1992). Thus, character sizes for such displays must be larger than on monochrome displays. This is considered in part 8 of the standard. (See "Requirements for displayed colors.")

The performance requirement for Part 3 is "that VDTs shall be legible, readable, and comfortable in use." Conformance with this standard can be demonstrated either by meeting the design requirements given in the standard or by passing the comparative user performance test which forms an informative annex. Part 3 contains design requirements related to the following topics:

- Flicker (as temporal instability and jitter as spatial instability)
- Legibility (design viewing distance and character height, as well as character design (stroke width, width-to-height ratio, between-character, -word and -line spacing, linearity, and orthogonality))
- Luminance (display luminance, luminance contrast and balance, image polarity, luminance uniformity, and coding)

The comparative user performance test reflects the approach of ISO 9241 following which any device equal to or better than another device complying with the standard is also in compliance with it. The compliance is demonstrated by passing the test which compares the device under consideration with a reference by means of user performance and comfort, the performance requirements of ISO 9241-3. Currently, the test method is informative. However, it will become normative after revision. Although the test method was introduced for formal testing, it can be extremely useful for other purposes, e.g., for testing different fonts on the same screen or for comparing the legibility of the same font on different displays, etc.

The measurement techniques needed for most of these design parameters are so complex that only few laboratories are able to measure them. Measurements with less qualified equipment are likely to yield wrong results. Consequently, it is not advisable to check the reliability of the measurements performed by laboratories. It is more promising to check whether the performance requirement is met under the conditions of the specific workplace(s) because most items addressed in this standard can be affected either by the operating system, the application, the font used, or the graphic board, or they can be easily changed unfavorably by the user. In addition, the standard does not consider the optimum visual distance for office work which is approximately 600 mm (see ISO 9241-5). Instead, the design viewing distance of a display may be selected to 400 mm by the manufacturer. This means that the characters may not be legible at the actual reading distance.

The following design parameters of a visual display are influenced by the selected settings of the graphic board and should therefore be checked at the workplace:

Flicker: Most monitors allow the user to select the resolution of the image (e.g., 800 × 600 or 1600x1200 pixels) according to the requirements of the application at hand, but only if the display refresh rate is changed. In many cases, the refresh rate will be too low for a flicker-free image. ISO 9241 contains a formula for calculating the minimum refresh rate to meet the requirement for a flicker free image for 90% of the intended user population. According to this formula, a refresh rate of approximately 72 Hz is needed for displays with a dark background ("negative" screens). If the background is light ("positive" screens) the refresh rate should be at least 80 Hz to achieve a flicker-free display. This means, for example, that many VGA-displays with 60 Hz do not comply with the standard.

Legibility related parameters: Many parameters that influence the legibility of alphabetic characters but also those with an effect on the visibility of graphical objects in general, (e.g., visibility of icons) depend on the actual settings of the graphic board. For a given font, even the contrast of the characters may be changed if some settings of the graphic board are changed. Under some circumstances this is even true for settings not related to character formation. The graphical representations that may be influenced or affected in any case are those generated by the operating system, e.g., menus, directory listings, icons, etc.

Checking a display at a certain workplace can be accomplished either visually, or by performing a user test applying the method described in the standard. For checking the most important visual parameter for legibility, the character size, for compliance with the standard under the conditions of a specific workplace, the actual character size (in mm) is to be multiplied with 170. The result should be bigger than the actual visual distance, e.g., a character height of 3 mm is sufficient for visual distances up to 510 mm. For usual office workstations, the characters should have a height of 3 mm to 4 mm. The luminance contrast of character details (contrast ratio) shall not be less than 3:1 (part 3); however, it is not possible to measure this at the workplace with simple equipment.

Luminance: Issues related to the luminance of displays are addressed in parts 3 and 7 of ISO 9241. Part 3 requires the display be capable of a display luminance of at least 35 cd/m^2. If luminance coding is used, 35 cd/m^2 specifies the minimum for the lower luminance. It is noted in the standard that operators often prefer substantially higher display luminance levels (e.g., 100 cd/m^2), particularly in conditions of high ambient illuminance.

The approximate luminance contrast can be calculated by measuring the display luminance when the screen is active and the background luminance caused by ambient light. The correct methods for measuring this feature are given in part 3 and part 7 The latter specifics the measurement including glare guarding (filters, treatment of the screen surface, etc.)

Part 7 of ISO 9241 deals solely with glare-guarding of the display. It proposes three classes of displays. Its performance criterion is that VDTs shall be legible and comfortable to use. The basic requirement is that the luminance ratio of the image, including superimposed specular and diffuse luminances shall be equal to or greater than 3, i.e., the minimum contrast ratio according to ISO 9241-3 shall be maintained under the lighting conditions at the workplace. Class I displays which comply with the requirement under the conditions specified in the standard (luminance of extended glare sources ≤200 cd/m^2, that of small sources ≤2000 cd/m^2) are considered suitable for general office use; classes II and III require a specially controlled luminous environment for use. Since the measurements required in part 7 cannot be performed at the workplace possible actions of user organizations are limited either to selecting the appropriate class for the existing visual environment or to creating the environmental conditions for utilizing class II and class III displays (see ISO 9241-6).

No provisions are made in any standard for one of the most important design parameters of visual displays, their physical size. The same is true for the resolution of the monitor. This does not mean that there are no user requirements concerning these parameters. It must be noted, however, that it is extremely difficult to specify the needs of different user groups with different tasks. For the same reason, there is also no specification for the "optimum" paper size needed for office work. Instead, an "unofficial" standard exists stating that the size "A4" is to be used unless some other considerations suggest any other format. Even "A4" is not the same in all countries.

For computer applications with a character-based interface, a similarly informal standard for the display was found (25 lines with 80 characters each). The physical size of such displays is approximately 14 inch to 15 inches, ensuring a fair legibility under office conditions with character sizes between 3 and 4.5 mm depending on the actual size of the display area. The users do not have to bother about the screen resolution since the character generator is integrated in the monitor hardware.

For modern applications with graphical interfaces, however, the aspects to be considered are very different. Regarding the question of resolution, one can require the same resolution as on print. However, the current display technology for electronic displays is not suitable to meet such a requirement. While a satisfactory print medium (e.g., a laser printer) offers a resolution of 600 dpi, the usual screen resolution lies somewhere between 50 dpi and 80 dpi. Extremely good displays may offer 150 dpi or even 200 dpi. However, the bulk of existing computer displays rarely exceeds a resolution of 60 dpi. This means that a laser print contains one hundred times more picture elements than the average screen on an area of the same size. One square inch of laser print has more pixels than an electronic visual display with 640 × 480 pixels. Poor resolution constitutes one of the most important reasons for the poor readability of computer screens which was demonstrated by various authors (e.g., Wright and Lickorish, 1983; Gould et al., 1987). An overview of studies related to this issue was published by Dillon et al. (1988). Gould et al. demonstrated that the performance decrement vanishes if the overall quality of the electronic display including its resolution is improved. In general, research findings suggest that a higher screen resolution will improve both performance and comfort of the users.

Surprisingly, the vast literature on VDT use makes no mention of the required physical size of the display area although complex programs like Excel™ or Freehand™ may clutter up to one third of the available display with their rulers, palettes, etc. leaving a small proportion of the screen for the task. Although there is no requirement based on standards and also no evidence based on research, it is advisable to consider this issue while designing systems and workplaces. According to our experience, applications running under Windows™ need a 17" display while applications that include document imaging require a monitor size of at least 21". For complex programs with various palettes, it is even advisable to use two screens.

Requirements for Displayed Colors

Part 8 of ISO 9241 deals with characteristics of displayed colors. However, the performance objective of this standard is not color use in general, but representation of data coded with color. The objective is obtained if colors can easily be detected, identified, and discriminated and if the assignment of meaning to color is appropriate to the task. The requirements of this standard are related to hardware as well as to software.

The standard requires a default set of colors for all applications that require the user to discriminate or identify colors. If the colors can be altered by the user, the default set of colors shall be retrievable and restorable. In general, the standard requires the number of colors presented simultaneously to be minimized. For accurate identification, the standard recommends that the default color set should consist of no more than eleven colors for each set. However, when a rapid visual search is required or the meaning of a set of colors is to be recalled from memory, no more than six colors should be used. Software applications that require the meaning of each color of a set of more than six colors to be recalled shall make the associated meaning of each color accessible.

For the size of images respecting height of characters, the standard requires different sizes depending on the relevance of the visual objects: For alphanumeric character strings, the character height shall subtend at least 20 minutes of arc, while for accurate color identification of a single object (a character or a symbol) 30 minutes of arc is required as a minimum; 45 minutes of arc is recommended. The use of extreme blue should be avoided for small images subtending less than 2°.

The luminance contrast of multicolor displays shall conform to ISO 9241-3, i.e., they must have a contrast ratio of 3:1. Spectrally, extreme blue shall not be used on a dark background.

Requirements for Input Devices

The requirements for input devices are divided into two standards with ISO 9241 part 4 dealing with keyboards, and part 9 with other input devices (non-keyboard).

The provisions of ISO 9241-4 are for conventional (linear) keyboards, but they do not rule out that any other design can comply with the standard if it meets the performance requirement of this standard: To be usable for the designated purpose, i.e., if users can achieve a satisfactory level of keying performance on a given task and maintain a satisfactory level of effort and comfort. The objective is obtained either by meeting the design requirements of the standard or passing the usability test for data or text input or both. The layout of the keyboard is not regulated in this standard but in ISO 9995. However, one important issue for user organizations throughout the world, the layout of the numeric keypad vs. the layout of touch telephones, has been solved: ISO 9241-4 recommends the telephone layout also for computer keyboards.

Most provisions of ISO 9241-4 are addressed to the manufacturer, e.g., the use of graphical symbols, durability of legends throughout the intended life of a product or a matte finish of the keys for better legibility. Given the fact that the matchless price war in the computer business forces the manufacturers to reduce costs wherever possible, user organizations should demand that the keyboards offered them comply with this standard.

For user organizations the most important provisions of the standard are:

- Minimum footprint: The overall dimension of the keyboard should not exceed the minimum space determined by the number of keys needed for its designated purpose and by the requirements for grouping the keys.
- Low profile: The height of the keyboard should be 30 mm or less and shall not exceed 35 mm.
- Ease of placement: The design of the keyboard shall permit it to be easily repositioned on the work surface. The keyboard shall be detachable except for special applications with clearly defined tasks.

- Adjustability: The keyboard should be adjustable in slope.
- Legibility of legends: All legends on keys shall be legible from the design reference position, i.e., users should not be forced to change their posture to view the key legends.
- Cursor keys: Keys for the control of cursor movement shall be provided.

ISO 9241-4 together with ISO 9995 aims to ensure that keyboards used in a given country have a uniform layout, a certain quality for keying performance and standardized use of symbols.

For other input devices, relevant requirements will be formulated in ISO 9241-9. The purpose of this standard is to formulate generic requirements for any input device including future devices, and to formulate specific requirements for existing devices.

Requirements and Recommendations for Software

Introductory Remarks

Seven parts of ISO 9241 (10, 12 to 17) are dedicated to software issues. As explained above, they will not contain requirements. However, the recommendations given in each part can be converted to requirements by applying the method described in part 11. All parts from part 12 contain a checklist including all recommendations of each part and guidance on applying the method of part 11 to the specific topics of the part under consideration. In addition, for the countries of the EU, the VDT-Directive may establish a legal status for the provisions of the standard (see "Relationship Between Dialogue Principles and Legal Requirements").

One of the main reasons for not including requirements in some parts of ISO 9241 lies in the ISO Regulations which demand that any requirement shall be accompanied by sufficient information on how to comply with it. In general, this is a method for measurement and a statement on conformance. For example, if a standard requires a certain length of an object, a measurement method (e.g., ruler, microscope) shall be specified. The statement of conformance could be "Objects of a length of 20" ± 0.05" comply with the requirements of this standard." Given the fact that today such demands cannot be met even for technical issues related to software, it is impossible to formulate mandatory requirements for ergonomics of software and applications within the ISO framework.

In other regulations, it is permissible that requirements can be formulated without specifying the means for demonstrating compliance with them. For example, the IEC-standard on the application of visual display units in the main control rooms of nuclear power plants requires: "Information shown on VDU shall be clearly understood in any operating conditions" (IEC 1772, 1995) The standard does not contain exhaustive and precise methods for measuring "information" and "understanding." Instead, it refers loosely to an annex of the main standard on the design for control rooms of nuclear power plants (IEC 964, 1989).

In such cases, however, a high conflict potential exists since involved parties of an agreement may interpret relevant items of a standard differently. For example, a former national standard required that the screen size be sufficient to display the information to be viewed simultaneously without providing further specification of the object to be viewed, however. The intention of the requirement was to prevent that screens designed to be viewed without scrolling or selecting pages would be broken down into smaller units because of insufficient display size. However, it was not intended to keep designers from placing different pieces of information in consecutive screens if this makes more sense than stuffing them into a single screen. The message of the requirement was not understood. Thus, it created various conflicts. Moreover, imprecisely defined and confined requirements are likely to cause even more trouble in an international standard than in a national regulation. Thus, the ISO Regulations are bureaucratic but also very helpful in avoiding conflicts in general. In the case of software, however, the conflicts may have been postponed or shifted from the standardization to users and software vendors. In the case of IEC 1772, the parties involved in utilizing VDUs in nuclear power plants need an agreement on the conditions under which the specific requirement is considered properly addressed by the system.

TABLE 99.3 Dialogue Principles of ISO 9241-10 and Their Definitions

Dialogue Principle	Definition
Suitability for the task	A dialogue is suitable for a task when it supports the user in the effective and efficient completion of the task.
Self-descriptiveness	A dialogue is self-descriptive when each dialogue step is immediately comprehensible through feedback from the system or is explained to the user on request.
Controllability	A dialogue is controllable when the user is able to initiate and control the direction and pace of interaction until the point at which the goal has been met.
Conformity with user expectations	A dialogue conforms with user expectations when it is consistent and corresponds to the user characteristics, such as task knowledge, education, experience, and to commonly accepted conventions.
Error tolerance	A dialogue is error tolerant if, despite evident errors in input, the intended result may be achieved with either no or minimal corrective action by the user.
Suitability for individualization	A dialogue is capable of individualization when the interface software can be modified to suit the task needs, individual preferences, and skills of the user.
Suitability for learning	A dialogue is suitable for learning when it supports and guides the user in learning to use the system.

A conflict potential exists at least for the European Union and its fifteen member states due to legal requirements on the human–computer interface. The EU-Directive 90/270/EEC ("VDT-Directive") contains legal requirements concerning the visual quality of displays including the readability of characters on the screens and the stability of the image as well as the ergonomic quality of the "operator/computer interface" in general. In practice, the required quality can be achieved as a result of hardware features (e.g., quality of the monitor and graphic card, computing power of the system and transmission capabilities of networks) and software features (e.g., features of the application and communication software, network software). In most cases, the overall quality will also depend on the interaction of soft- and hardware. Thus, meeting the legal requirements is a question of system design. The lack of agreed methods for demonstrating compliance with the legal requirements is likely to create conflicts.

ISO 9241 part 11, which gives guidance on usability, may help overcome possible problems. However, the guidance given in this standard is not exhaustive, the parties involved in utilizing computer systems also need an agreement on the basis of this standard.

Provisions of ISO 9241-10, "Dialogue Principles"

Part 10 of the standard provides ergonomic principles for dialogue design in general terms, that is, the principles are presented without considering situations of use, specific applications, and environments. These principles are intended to be used in specifications, design, and evaluation of dialogues. The standard provides guidance on principles but not on their application to product attributes. This task is taken care of in parts 12 to 17. The idea of breaking down the term "user friendliness" into principles stems from an early research work by Dzida et al. (1977) which formed the basis of the German standard DIN 66234-8. This standard introduced five of the seven principles of ISO 9241-10 under the same name.

The term "dialogue" instead of "software" was chosen to reflect the knowledge that the relevant characteristics of the human–computer interface are a result of software and hardware features and their interaction. "Dialogue" is defined as an interaction between a user and a system to achieve a particular goal.

The standard defines seven principles which are named and defined as displayed in Table 99.3.

For further guidance, each principle is explained by a number of examples, which are not meant as recommendations for real applications but as illustrations for possible implementations. The way in which each dialogue principle can be applied will depend on the characteristics of the intended users of a system, the tasks, the environments, and the specific dialogue technique used. Guidance on identifying relevant aspects of the users tasks and the environment of use is given in ISO 9241-11. Specific guidance on the application of these principles to techniques such as menus, command languages, direct manipulation, and form-based entry is given in ISO 9241 parts 14 to 17. However, the standard does not specify whether any of these techniques is preferable over others in general terms. Instead, each standard specifies the conditions under which the particular dialogue technique or style is considered appropriate.

Table 99.4 gives some examples for each principle to illustrate how they can be applied. The examples are not exhaustive. The same is true for the examples in the standard. Their sole purpose is to show what is meant by the specific principle.

TABLE 99.4 Applications of Dialogue Principles and Examples

Application of the Principle	Examples
	Suitability for the task
The format of input and output is appropriate to the given task and user requirements.	• The precision of input is equal to the precision required by the task; e.g., if a line with an approximate length is to be drawn, the input can be accomplished by dragging the mouse. If the line must have a precise length and position, the relevant data can be entered by a suitable device (e.g., tablet or keyboard).
The dialogue supports the user when performing recurrent tasks.	• The system allows sequences of activities to be saved and allows the user to reuse them (e.g., usage of self-defined macros or scripts).
Actions that can appropriately be allocated to the system for automatic execution are carried out by the system without user involvement.	• System start-up procedures are automatically processed. • If installing an application does not require the selection of options, all procedures related to this action are executed by the system.
	Self-Descriptiveness
Feedback or explanations are presented in a consistent terminology which is derived from the task environment rather than from system technology.	• The technical terms used in the dialogue are those actually used in the specific field of application. • An object has only one name; different objects are named differently. • Abbreviations used in the dialogue stem from the specific field of application.
Feedback or explanations are related to the situation for which they are needed.	• The dialogue system offers context-sensitive help.
The dialogue provides adequate feedback for all user actions.	• Keying activity is echoed within the time period required for efficient eye–hand coordination. • If an object is deleted, it disappears from the screen. • If a requested action cannot be performed, the error message explains why and describes the actions needed for recovery.
	Controllability
It is possible to undo at least the last dialogue step for any reversible action, if the task permits.	• The dialogue system offers the possibility of accessing deleted objects. • The system offers the possibility of selecting the adequate number of steps for undoing.
The user can determine the point of restart if the dialogue has been interrupted.	• After a system crash, the user can decide on certain conditions for restarting the dialogue (e.g., "revert to saved," use "save file" or just continue).
The level and methods of interaction can be selected to meet user needs and characteristics.	• The system offers menus and accelerators for novice and experienced users. • The system offers user selectable levels of detail in help to correspond to different levels of expertise.
	Conformity with User Expectations
Dialogue behavior and appearance within the system are consistent.	• System status messages always appear on the same line. • The same key or command is always used to terminate the dialogue.
Dialogues used for similar tasks should be similar so that the user can develop common task-solving strategies.	• To activate an application in a system with different applications the user is always required to double click on the icon of the specific application. • The function of the backstep key is the same in all system modes.

TABLE 99.4 (continued) Applications of Dialogue Principles and Examples

Application of the Principle	Examples
	Error Tolerance
Error correction is possible without switching the dialogue system states, where the task permits.	• During an entry into a form the user can type over incorrect characters without going to an editing mode. • During the use of a help facility explaining the error condition, the system state permits entries.
Error messages are explained to help the user in recovering from the error state.	• The error messages contain information on error occurrence, type of error, and possible methods of correction.
	Suitability for Individualization
The dialogue system allows the user to adapt it to her/his language and culture, individual knowledge and experience of task domain, perceptual, sensory–motor, and cognitive abilities.	• The mouse can be adapted for left- or right-hand usage. • The keyboard can be adapted for one-handed use. • The formats for time, date, and the count of weeks of the year are adaptable to the needs of users in different countries.
The system allows the users to set up operational time parameters to match their individual needs.	• The speed of scrolling of windows can be selected by the user.
The system provides different dialogue techniques for selected tasks.	• The system allows the user to start a dialogue function either by entering a command or selecting a menu option.
	Suitability for Learning
Relevant learning strategies are supported by the system.	• Interactive online tutorials support learning by doing. • Dialogue steps can be reversed ("undo") unless the user is warned (support for learning by trial-and-error).
The use of the system takes advantage of what the users already know.	• "Population stereotypes" are considered in the design of screens, control buttons, and other visual elements. • Metaphors from everyday life are used to convey the meaning of new functions to the user (e.g., envelope for e-mail).

Relationship Between Dialogue Principles and Legal Requirements

Although ISO 9241-10 provides no requirements, it has generated some important legal implications. These implications are not caused by the legal status of the standard itself or by its contents but by the lack of further normative documents with requirements on the same subject. First, in many countries of the world legal requirements exist forcing product designers to consider the "state-of-technology" or "state-of-the-art." In these countries, software manufacturers may be held responsible for not considering the provisions of ISO 9241-10 since this standard may be considered "state-of-the-art." Second, in the member states of the EU, the "VDT Directive" requires taking into account principles closely related to the dialogue principles of ISO 9241-10 in designing, selecting, commissioning, and modifying software, and in designing tasks. Since there is no other normative agreement on the corresponding requirements of the VDT-Directive it is likely that ISO 9241-10 may be used instead. These requirements are directed to the employers, who try, however, to pass the unpleasant task to the manufacturer asking for a statement that the specific product is in compliance with the requirements of the Directive. Even if the employer is reluctant to do this, she or he may be forced by the employees who have the right to reject new applications that do not conform with legal requirements. Moreover, employers who are reluctant to observe the mandatory rules, may be fined or even sentenced to prison. While such an event is extremely unlikely, the most probable effect may be delays in the introduction of new systems and more bureaucracy. The best solution would be an agreement between the involved parties on how to take into account the relevant items.

For the time being, there is no formal relationship between the dialogue principles of ISO 9241-10 and the principles of the VDT-Directive. The following listing shows the correspondence of the principles:

Principle of the Directive 90/270/EEC	Dialogue principle of ISO 9241-10
Software must be suitable for the task.	Suitability for the task
Software must be easy to use and, where appropriate, adaptable to the operator's level of knowledge or experience;	Suitability for individualization Conformity with user expectations
no quantitative or qualitative checking facility may be used without the knowledge of the workers.	No correspondence
Systems must provide feedback to workers on their performance.	Suitability for learning Self-descriptiveness Controllability
Systems must display information in a format and at a pace which are adapted to operators.	Suitability for individualization Suitability for the task
The principles of software ergonomics must be applied, in particular to human data processing.	All principles

As can be seen from the correspondence list, all principles of ISO 9241-10 are subject to legal requirements within the EU, though, with a reference to tasks. In addition, no method for demonstrating conformance is specified. This gives each country the freedom to define under which circumstances a system can be assumed to comply with the Directive. For Germany, a set of rules is in preparation which is formulated in consideration of ISO 9241-10 (VBG, 1995). For formal reasons, these rules are titled as "Accident Prevention Rules" which is the official name for (mandatory) regulations for health and safety. The most important function of these rules is limiting the application of ergonomic principles to certain aspects specified within the framework of the regulation, thus enabling vendors and user organizations to set up procedures for checking their software and applications.

In areas where such rules do not exist, rules for testing software can be formulated after the guidance given in ISO 9241-11 and in consideration of parts 12 to 17 of ISO 9241. It should be noted, however, that a product has no intrinsic usability, only a capability to be used in a particular context (ISO 9241-11). Thus, usability cannot be assessed by studying a product in isolation. In this respect, there is no difference between usability and other quality criteria.

In addition, it should be noted that usability is only one criterion for software quality and other quality criteria defined in ISO/IEC 9126, such as functionality, reliability, and computer efficiency all contribute to the quality of the work system in use.

Recommendations for Presentation of Information (ISO 9241-12)

This standard provides information relevant for all dialogue techniques in three general categories: organization of information, design of graphical objects, and coding techniques. Even though the standard is for software, the information it provides may be useful to a great extent for anyone presenting information or preparing documents. For the evaluation of products, the users of the document can utilize the procedures provided as an annex.

An overview of the recommendations of the standard and their relationship to attributes of presented information is given in Table 99.5. For each item, recommendations are given. The method given for the use of the standard specifies recommended methods for assessing the applicability of each item in consideration of the task and a corresponding method for the evaluation of compliance. For the evaluation of a product, the compliance of the applicable items is to be evaluated, e.g., for applications without windows, the items on windows do not need to be evaluated.

Since presenting information constitutes one of the basic "technologies" for all human civilizations, the standard is not exhaustive. However, an annex provides lists of references sorted for each topic for further information. For designers who need specific information on the interface design for graphical user interfaces, useful information is given by Galitz (1994). Useful information on how to present information on paper and online is given by Horton (1991).

TABLE 99.5 Relevant Attributes of Presented Information and the Specific Ways to Present Information

Attributes of Presented Information	Provisions of ISO 9241-12, Specific Ways to Present Information
Clarity (the information content is conveyed quickly and accurately)	Windows, Location of information
Discriminability (the displayed information can be distinguished accurately)	Areas
Conciseness (users are not overloaded with extraneous information)	Input/output area
Consistency (the same information is presented in the same way throughout the application, according to the user's expectation)	Groups
Detectability (user's attention is directed towards information required)	Lists
Legibility (information is easy to read)	Tables
Comprehensibility (meaning is clearly understandable, unambiguous, interpretable, and recognizable)	Fields, Labels
	General recommendations for graphical objects
	Cursors and pointers
	Syntactic aspects of alphanumeric coding
	Semantic aspects of alphanumeric coding
	Abbreviations
	Graphical coding
	Color coding, Markers
	Coding with other visual techniques

Recommendations for User Guidance (ISO 9241-13)

Part 13 of ISO 9241 provides recommendations for user guidance attributes of software interfaces and their evaluation. It also is applicable to interaction components that aid users in recovering from error conditions. The provisions of this standard may be applied to all dialogue techniques. Most of the recommendations of this standard could also be requirements because they are technology and user independent, and express the presumably best way of guiding the user in the specific context.

Recommendations are given for

- Common guidance (e.g., phrasing of user guidance)
- Prompts (e.g., generic vs. specific prompts, location of prompts, cues, etc.)
- Feedback (e.g., appropriate system response time, type of feedback)
- Status (e.g., continuous presentation of status information, conditions under which automatic presentation is needed)
- Error management (e.g., error prevention, error correction by the system, error management by the user, error messages)
- Online help (e.g., system-initiated help, user-initiated help, context sensitive help, help navigation)

For the evaluation procedure no specific guidance for the assessment of the applicability of the items is given since most recommendations can be evaluated without using a specific method. In addition, the majority of the recommendations is likely to be found "applicable." Assessment of compliance can be accomplished either by observing a representative user or by reviewing user guidance during self-use (walk-through).

Recommendations on Specific Dialogue Techniques (ISO 9241-14 to 17)

The last four standards in the series deal with particular dialogue styles, namely menus, commands, direct manipulation and form-filling. Each standard provides information on the conditions under which the particular style is considered appropriate (user and organizational characteristics, task requirements, and system capabilities) and recommendations for this style. Actual design processes may be different, though. Complex applications regularly employ two or more of these styles. Thus, testing them for compliance may form a task for which all four standards need to be considered.

Part 14 (menu dialogues) provides information covering various features of menus including windowing, panels, buttons, and fields. The recommendations relate to interaction design or input/output design, e.g., placement of menu structure, levels, grouping options, syntax rules for textual and graphic options. In some respects, the provisions of part 14 overlap with those of part 13 which is not surprising since interaction and input/output design under consideration of graphical rules and means always aim at guiding the user visually.

Recommendations are given for

- Menu structure (e.g., structuring into levels, grouping options within a menu, sequencing options within groups)
- Menu navigation (e.g., navigational cues, rapid navigation)
- Option selection and execution (e.g., selection methods, minimizing keystrokes, using function keys)
- Menu presentation (e.g., option accessibility and discrimination, placement, consistency of layouts, text option syntax, graphic option structure and syntax, auditory option structure and syntax).

The application of the standard by designers, procurers, and evaluators is described in an annex.

Part 15 (Command dialogues) provides information covering ergonomic aspects of command dialogues and gives guidance on when to use this dialogue technique (user characteristics, task requirements). The consideration of system capabilities is absent from the condition list since this dialogue style is usable even with "dumb terminals." It must be noted, however, that this style is recommended if extendibility of a system is required for which it is the only available dialogue form.

Recommendations are given for

- Structure and syntax (e.g., internal consistency, command macros, argument structures)
- Command representation (e.g., command names, abbreviations, function and hot keys)
- Input and output considerations (e.g., command reuse, command queuing, defaults, customization)
- Feedback and help (e.g., command acceptance, error feedback, error highlighting, long parameter lists).

This part of ISO 9241 provides extensive information on compliance methods which is also applicable to other parts. The annex with the applicability and compliance checklist also gives guidance for applicable methods for assessment.

Part 16 (Direct manipulation dialogues) provides information covering relevant issues for direct manipulation, the most recent style in interaction with computers. The list of conditions describing when this style is considered appropriate is the longest of all standards; in particular the system requirements are the highest.

Recommendations are given for

- General considerations (e.g., metaphors, appearance of objects, feedback, input devices)
- Manipulation of objects (e.g., pointing and selecting, dragging, sizing of objects)
- Manipulation of text objects (e.g., pointing and selecting, sizing of objects)
- Manipulation of windows (e.g., moving windows, sizing, scaling)
- Manipulation of control icons (e.g., indicating manipulation types, indicating user tasks)

Part 17 (Form-filling dialogues) provides information for one of the oldest styles of interaction. This style is considered appropriate for data entry tasks requiring input or modification of multiple data items.

Recommendations are given for

- Form-filling structure (e.g., visual coding, layout, fields, and labels)
- Input considerations (e.g., cursor movement, default values, alphanumeric entry, choice entries, control, validation criteria)
- Feedback (e.g., echoing, cursor position, field errors, transmission acknowledgment, feedback)
- Navigation (e.g., initial cursor position, movement between fields, tabbing, scrolling, form selection)

This part of ISO 9241 provides extensive information on compliance methods which is also applicable to other parts. The annex with the applicability and compliance checklist also gives guidance for applicable methods for assessment.

Guidance on Usability

In ISO 9241-11, the objective of designing and evaluating for usability is to enable users to achieve goals and meet needs in a particular context of use. Usability is to be measured in terms of user performance and satisfaction. Guidance is given on how to describe the context of use of the product (hardware, software, or service) and the required measures of usability in an explicit way as a part of a quality system. However, the standard does not detail all activities to be taken.

The components of usability are effectiveness, efficiency, and satisfaction. Effectiveness is defined as the accuracy and completeness with which users achieve specified goals. Efficiency represents the resources expended in relation to the accuracy and completeness with which users achieve goals. Both efficiency and effectiveness, are components of user performance, whereas satisfaction is defined as the freedom from discomfort and positive attitudes to the use. The context of use consists of descriptions of the users, goals, tasks, equipment (hardware, software, and materials), and the physical and social environments in which a product is used.

Guidance is given for

- Benefits and rationale of the usability assessment procedure
- Framework for specifying usability
- Specifying the context of use
- Usability measures (effectiveness, efficiency, satisfaction)
- Interpretation of measures
- Specification and evaluation of usability during design
- Specifying and measuring the quality of a work system in use
- Relationship to other standards (ISO 9126, ISO 9241 parts 10, 12 to 17)

ISO 9241-11 does not contain any provisions on the human–computer interface; it describes the methodological framework for applying the rest of the standard on work systems.

99.4 Future Requirements for the U.S.

Currently, standardized requirements for the user interface in the U.S. are provided by ANSI/HFS 100 standard from the year 1988 which covers hardware (displays, keyboards, furniture) and environmental issues only. This standard is under revision and will cover similar topics like ISO 9241 except for software interface issues. These will be covered by a new standard ANSI/HFES 200 which is intended to be a "Human Computer Interaction" (HCI) standard. The standard is likely to include (Smith, 1996):

- Software terminology
- Software usability
- Menu layouts
- Effective use of color
- Command syntax
- Graphical user interfaces
- Command line interfaces
- Voice input/output
- Special considerations for people with disabilities. (Smith, 1996)

The last two items have not been covered in other standards. While the voice input/output may be considered not a principal issue, the absence of special considerations for people with disabilities is a major flaw of existing standards in general.

Both U.S. standards are likely to be comparable with ISO 9241, however, the differences in the understanding of the term "standard" between different countries and the specific legal implications of standards in the U.S. may force the working committees responsible for the revision of ANSI/HFES 100 and for processing ANSI/HFES 200 to introduce some substantial differences.

In the U.S., the requirements of military applications formed the most important driving force for the development of ergonomics in general. In this connection, it should be mentioned that the type of computers we use today throughout the world, the universal programmable engine, was once created following ideas of the military. U.S. military standards have a long tradition and have influenced even civilian standards outside this country. One of the most important standards from this point of view is MIL STD 1472 "Human Engineering Design Criteria for Military Systems, Equipment and Facilities," which formed a basis for many civilian standards including ANSI/HFS 100 (Smith, 1996). This and other military standards will play a role both by their existence for their area of application and by their direct and indirect contribution to civilian standards.

99.5 Conclusion

ISO 9241 provides the most extensive information on human–computer interaction or interface to be found in standardization. The entire standard represents the result of discussions on relevant interface issues since 1983 in the context of an international ergonomic standard. Unlike technical standards which are subject to change when their object, the technology under consideration, changes, ergonomic standards should remain constant. For this reason, the performance approach formed the main basis for this standard. Designers expecting statements on product attributes comparable to those in technical standards may be disappointed when reading a statement such as "A keyboard shall be usable for its designated purpose." However, just this form of expressing the objectives of a standard gives the freedom to design completely new products with features unknown to anyone when a relevant standard was formulated, and still be in compliance with this standard. The price to be paid lies in the effort to understand the objectives of the standard instead of applying what it requires explicitly.

Future regulations in single states of the European Union will adopt the parts of this standard. In addition, future U.S. standards for human–computer interface will at least be compatible with the performance approach of ISO 9241. Since the processing of these standards is synchronized, it is likely that their provisions will be very similar.

With respect to software, various parts of the standard 9241 recommend certain product attributes. The question whether a product shall comply with the recommended form of a particular attribute can only be answered in consideration of the task characteristics and the particular interaction technique employed for the application. Although this approach seems new, in principle, it dates back to Aristotle the philosopher who coined the term *qualitas* for quality. According to the understanding of quality, a product has no intrinsic quality but only with regard to specified requirements it shall comply with. In the case of software, the requirements have to be formulated considering user and task characteristics. For this reason, it cannot be claimed for a software or user interface to be "ergonomic" without specifying for whom and for what.

References

Çakir, A. and Dzida, W. 1997 in Helander, M.G., Landauer, T.K., Prabhu, P.V., *Handbook of Human-Computer Interaction*. North-Holland. International Ergonomic HCI Standards.

Carroll, J.M. and Olson, J.R. 1991. Applying what we know of the user's knowledge to practical problems, in Helander, M. (Ed.): *Handbook of Human-Computer Interaction*. North-Holland.

Dillon, A, McKnight, C. and Richardson, J. *Reading From Paper vs. Reading From Screen.* The Computer Journal, 31, 457-464.

DIN 66234-8 Bildschirmarbeitsplätze — Grundsätze der Dialoggestaltung.

Dzida, W., Herda, S., Itzfeldt, W.D., Schuberth, H. 1977. Zur Benutzerfreundlichkeit von Dialogsystemen, GMD, St. Augustin.

EU-Directive 90/270/EEC — Council Directive of 29 May 1990 on the minimum safety and health requirements for work with display screen equipment.

Galitz, W. O. 1994. It's time to clean your windows — designing GUIs that work, John Wiley & Sons, New York, Chichester, Brisbane, Toronto, Singapore.

Gould, J.D., Alfaro, L., Barnes, V., Finn, R., Grischkowski, N. and Minuto, A. 1987. Reading is slower from CRT displays than from paper: attempts to isolate a single variable explanation. *Human factors,* 29 (3), 269-299.

Harker, S. The development of ergonomics standards for software, *Applied Ergonomics,* 26, 4, 275-279.

Horton, W.1991. *Illustrating Computer Documentation,* John Wiley & Sons, Inc., New York.

IEC 964 Design for control rooms of nuclear power plants, 03.1993.

IEC 1772 Nuclear power plants — Main control room — Application of visual display units, 08.1995.

ISO 6385 Ergonomic principles in the design of work systems, draft, 10.1995.

ISO 9241-2 Ergonomic requirements for office work with visual display terminals (VDTs) — Guidance on task requirements.

ISO 9241-3 Ergonomic requirements for office work with visual display terminals (VDTs) — Visual display requirements.

ISO 9241-4 Ergonomic requirements for office work with visual display terminals (VDTs) — Keyboard requirements.

ISO 9241-5 Ergonomic requirements for office work with visual display terminals (VDTs) — Workstation layout and postural requirements.

ISO 9241-7 Ergonomic requirements for office work with visual display terminals (VDTs) — Display requirements with reflections.

ISO 9241-8 Ergonomic requirements for office work with visual display terminals (VDTs) — Requirements for displayed colours.

ISO/DIS 9241-9 Ergonomic requirements for office work with visual display terminals (VDTs) — Requirements for non-keyboard input devices.

ISO 9241-10 Ergonomic requirements for office work with visual display terminals (VDTs) — Dialogue principles.

ISO 9241-11 Ergonomic requirements for office work with visual display terminals (VDTs) — Guidance on usability.

ISO/FDIS 9241-12 Ergonomic requirements for office work with visual display terminals (VDTs) — Presentation of information.

ISO/FDIS 9241-13 Ergonomic requirements for office work with visual display terminals (VDTs) — User guidance.

ISO/FDIS 9241-14 Ergonomic requirements for office work with visual display terminals — Menu dialogues.

ISO/FDIS 9241-15 Ergonomic requirements for office work with visual display terminals — Command dialogues.

ISO/DIS 9241-16 Ergonomic requirements for office work with visual display terminals — Direct manipulation dialogues.

ISO/DIS 9241-17 Ergonomic requirements for office work with visual display terminals — Form-filling dialogues.

ISO 9995 Information technology — Keyboard layouts for text and office systems. Parts 1 to 8.

ISO/DIS 13407 Ergonomics of human system interaction — Human centred design processes for interactive systems.

ISO/IEC 9126 Information technology — Software product evaluation — Quality characteristics and guidelines for their use.

ISO 13406 Ergonomic requirements for the use of flat panel displays.

MIL STD 1472 Human engineering design criteria for military systems, equipment and facilities.

Smith, W. J. 1996. *ISO and ANSI Ergonomic Standards for Computer Products — A Guide to Implementation and Compliance*, Prentice Hall International, Upper Saddle River.

Smith, S.L. and Mosier, J.N. 1986. Application of guidelines for designing user interface software. *Behaviour and Information Technology*, Vol. 5 No. 1.

VBG 1995. Unfallverhütungsvorschrift Arbeit an Bildschirmgeräten (VBG 104) (Accident Prevention Rule "Working with Visual Displays), final draft.

Widdel, H.; Post, D. 1992. *Colour in Electronic Displays*. Plenum Press, New York.

Wright, P. and Lickorish, A. Proof-reading texts on screen and paper. *Behaviour and Information Technology*, 2 (3), 227-235.

100

Psychosocial Factors and Musculoskeletal Disorders in Computer Work

Naomi G. Swanson
National Institute for Occupational Safety and Health

Steven L. Sauter
National Institute for Occupational Safety and Health

100.1 Introduction

Physical factors and their influence on health among computer or video display terminal (VDT) workers have been studied since the 1970s, and a range of workstation and physical environmental factors have been linked with musculoskeletal problems in VDT work (Cakir et al., 1978; Hunting et al., 1981; Maeda et al., 1980; Ong et al., 1981; Onishi et al., 1973; Sauter et al., 1991a). There have been significant improvements in the ergonomic aspects of office environments since personal computers were first introduced. However, musculoskeletal problems among computer users are still commonplace. In recent years, increasing attention has been given to psychosocial stressors for their possible etiological role in musculoskeletal disorders among computer users. However, this work is still formative, lacking a widely accepted model of how psychosocial stressors act to influence musculoskeletal outcomes. For that matter, even the concept of psychosocial factors seems elusive to many ergonomists and occupational health professionals. The present chapter provides a definition of workplace psychosocial factors, suggests pathways by which psychosocial factors can influence musculoskeletal disorders, and provides research evidence for these pathways.

What are psychosocial factors? We define psychosocial factors as attributes of the job and the individual that influence psychological demands and thus contribute to job stress (Sauter and Swanson, 1996). These factors include aspects of job content, such as workload, skill usage, clarity of demands, control over tasks; organizational aspects of the job, such as participation in decision making and career issues; interpersonal relationships, such as co-worker and supervisor support, and availability of feedback; and temporal aspects

0-8493-2641-9/99/$0.00+$.50

of the job, such as pacing and hours of work. Also included are individual factors, such as age, marital status, prior learning and experience, coping strategies, and personality factors, although we treat these factors as intervening variables that can moderate or modify the demands imposed by the job characteristics or stressors mentioned above. In other words, the primary emphasis is on the influence of job or work environment stressors, not individual factors, as risk factors for job stress and associated health disorders.

Changes in the office psychosocial work environment related to computerization. The introduction of computers into the office environment has greatly changed the way in which work is accomplished. While office automation holds the promise of task and skill enlargement, and more worker control over tasks, there are indications that this promise is not met with some types of VDT work. Early studies by NIOSH of clerical workers who used computers and their counterparts who did not found that the computer users reported significantly greater work pressure and supervisory control and less autonomy, role clarity, and support from co-workers (Smith et al., 1981; Sauter et al., 1983a, 1983b). More recent confirming evidence was reported by Bammer (1987) and Asakura and Fujigaki (1993) who found that the introduction of computers into office work was related to increased time/job pressures, lessened job discretion, and a reduction in task diversity. Korunka et al. (1995) reported differential effects of the introduction of new computer technologies on jobs, depending on the initial characteristics of the jobs. For those jobs which were monotonous and relatively menial (e.g., cashier work, telephone information work), introduction of computerized technologies led to a deterioration of working conditions, while for jobs that were more challenging (e.g., computer-aided drawing), the introduction of new technology led to an improvement in working conditions (e.g., greater participation, control, skill usage, etc.).

100.2 How Psychosocial Factors May Influence Musculoskeletal Disorders

Although the relationship between psychosocial stressors and musculoskeletal problems in computer users has gained increasing attention in recent years, there is uncertainty regarding the pathways which relate the two. Sauter and Swanson (1996) and others (Bongers et al., 1993; Sauter, 1991b; Sauter et al., 1983a; Smith and Carayon, 1996) have proposed two major pathways by which psychosocial factors may influence musculoskeletal disorders. In the first pathway, psychosocial factors themselves create physiological strain via a generalized stress response. This stress-related physiological strain can then exacerbate task-related biomechanical strains. For example, stress resulting from work-related psychosocial demands may produce increments in muscle tension that add to the muscle loads and symptoms related to physical task demands. In the second pathway, psychosocial factors may influence physical work demands directly. For example, increased fractionation of tasks can result in increased repetitiveness.

Although the evidence to date cannot fully verify these pathways (i.e., due primarily to the cross-sectional design of most of these studies, which limits causal determination), there is still a wide range of studies that provide support for these pathways. This research is summarized below.

Pathway 1: Psychosocial Demands Creating Physiological Strain

Studies relating physical and emotional stressors to physiological strain date from the early part of this century (Cannon, 1929, 1935; Selye, 1936, 1946). These early studies have demonstrated that exposure to stress results in increases in blood pressure, corticosteroids, peripheral neurotransmitters, and muscle tension. All of these physiological changes ready the organism to respond to threatening situations. More recent work by Johansson and Aronsson (1984), Frankenhaeuser and Johansson (1986), and Lundberg et al. (1989, 1993) have indicated that work-related psychosocial stressors such as low decision latitude, and boring and repetitive tasks, can result in similar physiological strain as evidenced by outcomes such as increases in blood pressure, heart rate, and corticosteroid levels. Smith and Carayon (1996) hypothesize that these physiological reactions to psychosocial stressors can increase the susceptibility of nerves and muscles to damage. For example, increases in fluid retention in peripheral body tissues due to increased levels of corticosteroids might exacerbate nerve compression in structures such as the carpal tunnel. Smith and Carayon suggest that if the exposure to the psychosocial stressor(s) is chronic or prolonged, permanent tissue damage may result. To date, however, this hypothesis has not been tested empirically.

Increased muscle tension has been the physiological response that has received the most attention as a possible mechanism connecting stress to musculoskeletal disorders. Certain psychological states, such as anxiety, have long been associated with muscle over-activity (Jacobsen, 1931; Sainsbury and Gibson, 1954). Recent studies of office workers have demonstrated that stressful task demands result in increased muscle tension that is unrelated to the physical demands of the job. For example, Westgaard and his colleagues (Waersted et al., 1987, 1991; Westgaard and Bjorkland, 1987) have reported sustained attention-related muscle loads of 0.5 to 3% maximum voluntary contraction (MVC) during the performance of VDT-based psychophysical tasks. These results are consistent with an earlier study by Weber et al. (1980) who reported increases in neck muscle activity and perceived tension among subjects completing cognitively complex tasks. Ekberg et al. (1995) and Lundberg and Melin (1995) have also reported higher static muscle loads among subjects exposed to psychologically stressful tasks.

Both Waersted et al. (1991) and Lundberg and Melin (1995) hypothesize that jobs which are psychologically stressful or demanding, even though they may not be demanding physically, may carry a risk for musculoskeletal disorders due to the sustained elevations in muscle tension induced by the psychological demands. Several investigators have found a relationship between sustained low-level static muscle activity and discomfort or sick leave in the workplace (Aaras, 1994; Veierstad, 1994), although not all investigations have been able to replicate this finding (e.g., Westgaard and Vasseljen, 1995).

Pathway 2: Direct Effects of Psychosocial Factors on Physical Demands

Studies that have examined the relationship between musculoskeletal disorders and both physical and psychosocial factors in the office environment have generally found that both sets of factors are related to musculoskeletal problems (Bernard et al., 1992; Bergqvist et al., 1995; Faucett and Rempel, 1994; Hoekstra et al., 1995; Kamwendo et al., 1991; Pot et al., 1987; Ryan and Bampton, 1988). A number of these studies report interactive effects between the psychosocial and physical factors. In other words, the psychosocial stressors appear to change the physical demands of the job. Several studies offer support for this premise.

Smith et al. (1981) queried office workers about psychosocial stressors, ergonomic factors, and health outcomes. Three groups of workers were examined — clericals who did not use VDTs, and clericals and professionals who did use VDTs. The clericals who used VDTs reported the highest level of psychosocial stressors and the most musculoskeletal symptoms, while the professionals reported the lowest level of psychosocial stressors and the least musculoskeletal symptoms. An examination of the jobs revealed important differences in job content that may have accounted for these differences. The clericals were subjected to rigid work procedures, high production standards, high pressure to complete their work, and had little control over their work. The professionals, on the other hand, had a great deal of flexibility and control over their jobs, and made use of their training and skills in their jobs. It was conjectured that these differences in job content influenced the physical demands of the job, translating into differences in workload, work pace, repetitiveness, and time on the VDT, and thus probably influenced the experience of musculoskeletal symptoms among the workers.

A study by Lim and Carayon (1993; 1995) provides additional evidence that psychosocial stressors may change or exacerbate ergonomic stressors among office workers. In a field study of office/VDT workers, they found that psychosocial factors influenced musculoskeletal symptoms via effects on ergonomic factors. In other words, psychosocial factors, such as work pressure, production standards, and task control, directly influenced ergonomic aspects of the job such as repetitiveness and work postures. The study found, for example, that individuals with more control over their work performed less repetitive work, and those who worked under production standards sat in more fixed, static work postures. Awkward work postures and repetitiveness, in turn, predicted musculoskeletal symptoms.

Does the Experience of Musculoskeletal Problems Increase the Reporting of Psychosocial Stressors?

One issue that researchers have struggled with is the direction of the relationship between musculoskeletal disorders and psychosocial factors. The studies reported above are all cross-sectional in design and do not answer whether the experience of symptoms results in a perception of a less favorable psychosocial work environment, or if the poorer psychosocial work environment precedes the development of musculoskeletal symptoms. While the answer to this question can ultimately only be answered definitively with long-term studies simultaneously examining the development of musculoskeletal symptoms and perceptions of the psychosocial work environment, evidence from two studies suggests that the latter relationship holds. These studies examined only asymptomatic worker's assessments of the psychosocial environment in office settings with high or low rates of musculoskeletal problems. (Thus, study results could not be influenced by the experience of musculoskeletal problems in the workplace.) Both Hopkins (1990) and Stephens and Smith (1996) found that worksites with high rates of musculoskeletal symptoms had lower levels of co-worker support, less control and autonomy, less clarity about their jobs, more work pressure and stress, and less job satisfaction.

100.3 Measuring and Controlling Psychosocial Work Factors

Various instruments and methods for assessing the psychosocial work environment, as well as methods for managing and controlling psychosocial stressors, are provided in Chapter 15 by Carayon and Lim.

100.4 Conclusions

Although not conclusive, current evidence suggests that workplace psychosocial stressors influence musculoskeletal disorders in computer workers via the two pathways discussed in the present chapter. Other pathways, such as the influence of the psychosocial work environment on the perception and reporting of symptoms have been proposed (see Sauter and Swanson, 1996), although the evidence in support of these pathways is more limited. It is clear that more work is needed to better elucidate mechanisms linking psychosocial factors and musculoskeletal symptoms. Longitudinal studies are also needed, as well as better exposure assessment methods. However, the evidence to date points to the need for a holistic assessment of the computerized workplace in order to determine which aspects of the workplace need to be modified. While the importance of physical or ergonomic factors is not in question, there are indications that changes in the physical work environment without attention to the psychosocial work environment may not be sufficient to prevent or reduce musculoskeletal disorders in the computerized workplace (Spillane and Deves, 1988).

References

Aaras, A. 1994. Relationship between trapezius load and the incidence of musculoskeletal illness in the neck and shoulder. *International Journal of Industrial Ergonomics*, 14, 341-348.

Asakura, T. and Fujigaki, Y. 1993. The impact of computer technology on job characteristics and workers' health, in *Human–Computer Interaction: Applications and Case Studies*, Eds. M.J. Smith and G. Salvendy, p. 982-987. Elsevier, New York.

Bammer, G. 1987. VDUs and musculo-skeletal problems at the Australian National University — a case study, in *Work with Display Units 86*, Eds. B. Knave and P.-G. Wideback, p. 279-287. Elsevier Science Publishers B.V. (North Holland), Amsterdam.

Bergqvist, U., Wolgast, E., Nilsson, B. and Voss, M. 1995. Musculoskeletal disorders among visual display terminal workers: individual, ergonomic, and work organizational factors. *Ergonomics*, 38(4), 763-776.

Bernard, B., Sauter, S.L., Fine, L.J., Petersen, M.R. and Hales, T.R. 1992. Psychosocial and work organization risk factors for cumulative trauma disorders in the hands and wrists of newspaper employees. *Scandinavian Journal of Work, Environment and Health*, 18 (Suppl. 2), 119-120.

Bongers, P.M., De Winter, C.R., Kompier, M.J. and Hildebrandt, V.H. 1993. Psychosocial factors and work and musculoskeletal disease. *Scandinavian Journal of Work, Environment and Health*, 19(5), 297-312.

Cakir, A., Hart, D.J. and Stewart, D.F.M. 1978. *Research into the Effects of Video Display Workplaces on the Physical and Psychological Function of Persons.* Federal Ministry for Work and Social Order, Bonn, West Germany.

Cannon, W.B. 1929. *Bodily Changes in Pain, Hunger, Fear and Rage.* C.T. Branford, Boston.

Cannon, W.B. 1935. Stresses and strains of homeostasis. *American Journal of Medical Science*, 189, 1-14.

Ekberg, K., Eklund, J., Tuvesson, M., Ortengren, R., Odenrick, P. and Ericson, M. 1995. Psychological stress and muscle activity during data entry at visual display units. *Work and Stress*, 9(4), 475-490.

Faucett, J. and Rempel, D. 1994. VDT-related musculoskeletal symptoms: Interactions between work posture and psychosocial work factors. *American Journal of Industrial Medicine*, 26, 597-612.

Frankenhaeuser, M. and Johansson, G. 1986. Stress at work: psychobiological and psychosocial aspects. *International Review of Applied Psychology*, 35, 287-299.

Hoekstra, E., Hurrell, J. and Swanson, N. 1995. Evaluation of work-related musculoskeletal disorders and job stress among teleservice center representatives. *Applied Occupational and Environmental Hygiene*, 10(10), 812-817.

Hopkins, A. 1990. Stress, the quality of work, and repetition strain injury in Australia. *Work and Stress*, 4(2), 129-138.

Hunting, W., Laubli, T. and Grandjean, E. 1981. Postural and visual loads at VDT workplaces: I. Constrained postures. *Ergonomics*, 24, 917-931.

Jacobsen, E. 1931. Electrical measurements of neuromuscular states during mental activities. *American Journal of Physiology*, 97, 200-209.

Johansson, G. and Aronsson, G. 1984. Stress reactions in computerized administrative work. *Journal of Occupational Behavior*, 5, 159-181.

Kamwendo, K., Linton, S.J. and Moritz, U. 1991. Neck and shoulder disorders in medical secretaries. Part I. Pain prevalence and risk factors. *Scandinavian Journal of Rehabilitative Medicine*, 23, 127-133.

Korunka, C., Weiss, A., Huemer, K.H. and Karetta, B. 1995. The effect of new technologies on job satisfaction and psychosomatic complaints. *Applied Psychology: An International Review*, 44(2), 123-142.

Lim, S.Y. and Carayon, P. 1993. An integrated approach to cumulative trauma disorders in computerized offices: The role of psychosocial work factors, psychological stress and ergonomic risk factors, in *Human-Computer Interaction: Applications and Case Studies*, Eds. M.J. Smith and G. Salvendy, p. 880-885. Elsevier, New York.

Lim, S.Y. and Carayon, P. 1995. Psychosocial work factors and upper extremity musculoskeletal discomfort among office workers, in *Work with Display Units 94*, Eds. A. Grieco, G. Molteni, B. Piccoli and E. Occhipinti, p. 57-62. Elsevier Science B.V., Amsterdam.

Lundberg, U. and Melin, B. 1995. Stress, muscular tension and musculoskeletal disorders, in *Premus 95: Second International Conference on Prevention of Work-Related Musculoskeletal Disorders*, Montreal, Quebec, September 24-28, p. 167-168.

Lundberg, U., Granqvist, M., Hansson, T., Magnusson, M. and Wallin, L. 1989. Psychological and physiological stress responses during repetitive work at an assembly line. *Work and Stress*, 3, 143-153.

Lundberg, U., Melin, B., Evans, G.W. and Holmberg, L. 1993. Physiological deactivation after two contrasting tasks at a video display terminal: learning vs. repetitive data entry. *Ergonomics*, 36(6) 601-611.

Maeda, K., Hunting, W. and Grandjean, E. 1980. Localized fatigue in accounting-machine operators. *Journal of Occupational Medicine*, 22, 810-816.

Ong, C.N., Hoong, B.T. and Phoon, W.O. 1981. Visual and muscular fatigue in operators using visual display terminals. *Journal of Human Ergology*, 10, 161-171.

Onishi, N., Normura, H. and Sakai, K. 1973. Fatigue and strength of upper limb muscles of flight reservation system operators. *Journal of Human Ergology*, 2, 133-141.

Pot, F., Padmos, P. and Brouwers, A. 1987. Determinants of the VDU operator's well-being, in *Work with Display Units 86*, Eds. B. Knave and P.G. Wideback, p. 16-25. North-Holland, Amsterdam.

Ryan, G.A. and Bampton, M. 1988. Comparison of data process operators with and without upper limb symptoms. *Community Health Studies*, 12(1), 63-68.

Sainsbury, P. and Gibson, J.G. 1954. Symptoms of anxiety and tension and the accompanying physiological changes in the muscular system. *Journal of Neurology, Neurosurgery and Psychiatry*, 17, 216-224.

Sauter, S.L., Schleifer, L.M. and Knutson, S.J. 1991a. Work posture, workstation design, and musculoskeletal discomfort in a VDT data entry task. *Human Factors*, 33(2), 151-167.

Sauter, S.L. 1991b. Ergonomics, stress and the redesign of office work. Presentation at *The National Occupational Musculoskeletal Injury Conference*, Ann Arbor, MI.

Sauter, S.L. and Swanson, N.G. 1996. An ecological model of musculoskeletal disorders in office work, in *Beyond Biomechanics: Psychosocial Aspects of Musculoskeletal Disorders in Office Work*, Eds. S. Moon and S.L. Sauter, p. 3-21. Taylor & Francis, London.

Sauter, S.L., Gottlieb, M.S., Jones, K.C., Dodson, V.N. and Rohrer, K. 1983a. Job and health implications of VDT use: Initial results of the Wisconsin-NIOSH study. *Communications of the Association for Computing Machinery*, 26(4), 284-294.

Sauter, Gottlieb, M.S., Rohrer, K.M. and Dodson, V.N. 1983b. *The Well-Being of Video Display Terminal Users: An Exploratory Study*, University of Wisconsin Department of Preventive Medicine, Madison, WI.

Selye, H. 1936. A syndrome produced by diverse noxious agents. *Nature*, 138, 32.

Selye, H. 1946. The general adaptation syndrome and diseases of adaptation. *Journal of Clinical Endrocrinology*, 6, 217-230.

Smith, M.J. and Carayon, P. 1996. Work organization, stress and cumulative trauma disorders, in *Beyond Biomechanics: Psychosocial Aspects of Musculoskeletal Disorders in Office Work*, Eds., S.D. Moon and S.L. Sauter, p. 23-42. Taylor & Francis, London.

Smith, M.J., Cohen, B.G.F., Stammerjohn, L.W. and Happ, A. 1981. An investigation of health complaints and job stress in video display operations. *Human Factors*, 23(4), 387-400.

Spillane, R.M. and Deves, L.A. 1988. Psychosocial correlates of RSI reporting. *Journal of Occupational Safety*, 4(1), 21-27.

Stephens, C. and Smith, M. 1996. Occupational overuse syndrome and the effects of psychosocial stressors on keyboard users in the newspaper industry. *Work and Stress*, 10(2), 141-153.

Veierstad, K.B. 1994. Sustained muscle tension as a risk factor for trapezius myalgia. *International Journal of Industrial Ergonomics*, 14, 333-339.

Waersted, M., Bjorklund, R.A. and Westgaard, R.H. 1987. Generation of muscle tension related to a demand of continuing attention, in *Work with Display Units 86*, Eds. B. Knave and P.-G. Wideback, p. 288-293. North-Holland, Amsterdam.

Waersted, M. Bjorkland, R.A. and Westgaard, R.H. 1991. Shoulder muscle tension induced by two VDU-based tasks of different complexity. *Ergonomics*, 34(2), 137-150.

Weber, A., Fussler, C., O'Hanlon, J.F., Gierer, R. and Grandjean, E. 1980. Psychophysiological effects of repetitive tasks. *Ergonomics*, 23(11), 1033-1046.

Westgaard, R.H. and Bjorkland, R. 1987. Generation of muscle tension additional to postural muscle load. *Ergonomics*, 30(6), 911-923.

Westgaard, R.H. and Vasseljen, O. 1995. The relationship between perceived tension, observed muscle tension in the trapezius muscle and psychosocial factors, in *Premus 95: Second International Conference on Prevention of Work-Related Musculoskeletal Disorders*, Montreal, Quebec, September 24-28, p. 181-183.

For Further Information

An excellent multidisciplinary overview of the influence of psychosocial factors on musculoskeletal disorders in office work is provided by *Beyond Biomechanics: Psychosocial Aspects of Musculoskeletal Disorders in Office Work*, Eds. S.D. Moon and S.L. Sauter. Taylor & Francis Ltd., New York, 1996.

The proceedings of several human factors/ergonomics conferences routinely report the results of studies examining the effects of psychosocial factors on musculoskeletal disorders in office environments. These include *The Human Factors and Ergonomics Society* and *The International Ergonomics Association*, which meet annually, and *Work with Display Units* and *Human–Computer Interaction*, which meet biannually or triennially.

Scientific studies on workplace psychosocial factors are published in a wide array of human factors/ergonomics, stress, and public health journals, including *Human Factors, Ergonomics, International Journal of Human-Computer Interaction, Journal of Occupational and Environmental Medicine, Scandinavian Journal of Work, Environment and Health,* and *Work and Stress.*

101

User-Centered Software Development: Methodology and Usability Issues

Udo Konradt
Ruhr-University of Bochum

Bernhard Zimolong
Ruhr-University of Bochum

Barbara Majonica
Ruhr-University of Bochum

101.1 Introduction

Many authors have argued that the small to medium-sized batch manufacturing sector will be characterized by increased competition, smaller batch sizes, and increased demand for customization and product variety (Brödner, 1990; Clegg and Symon, 1989). Market requirements will be accompanied by more nonroutine variances, higher capital cost of production systems, and relatively lower labor costs. As a response to this challenge, the production island concept is increasingly gaining in significance as an efficient production concept. Production islands, with their quasihomogeneous qualification structure, can — from a manufacturing perspective — react more flexibly to the varying requirements of the market than centralized and hierarchical organizations. Most organizations adopt a traditional technology-centered approach to the design and implementation of new manufacturing systems. The technical aspects command most resources and are considered first in the planning and design stage. Human and organizational factors are only considered relatively late in the process, and sometimes only after the system is operational (Corbett, Rasmussen, and Rauner, 1991). Countervailing approaches, e.g., the sociotechnical and human-centered systems design share the idea that people and organizational structures must be designed parallel to the technical structures. Concepts such as "Kaizen" (Imai, 1994), "Business Reengineering" (Hammer and Champy, 1993), or "Agile Manufacturing" (Kidd, 1994) emphasize the role of organizational and human factors in the production process.

At the University of Bochum, an interdisciplinary research group, including engineers, social scientists, and economists developed and evaluated flexible manufacturing structures on the basis of semiautonomous

0-8493-2641-9/99/$0.00+$.50

production islands (Zimolong, 1996). This approach reflects, unlike the centralized system architecture, commonly associated with the technology approach, organizational structures and processes that are integrated in terms of production islands. It emphasizes the reintegration of design, planning, and manufacturing. In particular, tasks of quality assurance, maintenance, and repair are integrated into shop floor activities. Although the computer controls routine operations, the planning and decision making is left to the personnel. Consequently, human–machine interface has to be designed to allow the worker to understand the workings of the machine and to develop his or her experience and skill in the machining process.

The increasing complexity of machine tools requires an information database and decision support system (DSS) to aid workers in quality assurance, diagnosis, and repair. This paper describes the incremental development of a decision support system for maintenance tasks in semiautonomous production islands that fosters the acquisition of abilities and the human learning process and that is adaptive to the individually preferred methods of working. At first, we explicate the design philosophy, thereafter the main design steps, and finally the usability results gained so far.

101.2 Software Design is Work Design

The use of information technology and the design of software should be part of an in-plant process of innovation and organizational growth that follow strategic decisions made by the management. Strategic decisions suggest the portfolio of important future markets the firm wants to secure or develop. At the tactical level, decisions are concerned with structural aspects of work organization, technical ways to support them, and personal development. The use of personnel, technology, and organization must be subordinate to the strategies of the business. The succeeding tactical plans and operative conversions must be judged with reference to these plans and policies.

In Figure 101.1 a scheme is presented reflecting major concepts and their relationships in system development. The "Process-Oriented Task Analysis" (PTA, Konradt, 1996) stresses practical application dealing with characteristics of users, tasks, and organizational forms as the fundamental aspects of system development. In PTA, management decisions that define business processes are embedded in a top-down perspective. Task structures of users are embedded through action strategies in a bottom-up perspective. The design of user-centered software requires taking into account the user's professional competence in addition to the degree of experience and types of cognitive preferences. Starting from the goals of the organization, objects of the working and recreational world of the user are brought forth and implemented in an evolutionary and participatory system design process.

101.3 Some Stages of the Software Life Cycle: A Case Study

The development of software systems is composed of different phases that create the software life cycle. Although the models differ in types of classification, procedures, and details, they usually contain requirements analysis, design, implementation, and evaluation.

To our view, requirements analysis is a central part in the software development process, because it determines the decomposition of problems, the hierarchical structure of the system, the specification of main features, and the design of user interfaces. Usability problems often originate from flaws in the requirements analysis, which may be properly conducted with respect to the functional needs, but not with respect to the organizational and psychological requirements of the user (see Table 101.1). Object-oriented analysis and design (OOAD) is a methodology to support the translation of user demands more smoothly in prototypes (Monarchi and Puhr, 1992) and an attempt to bridge the gap between analysis and design. In OOAD, data structures are defined in user language, and actions are subsequently matched. Domains of users are represented through specification of a set of semantic objects that are revealed in cognitive task analysis. In this case, objects were derived from strategies of operators. Figure 101.2 presents major steps in the project model.

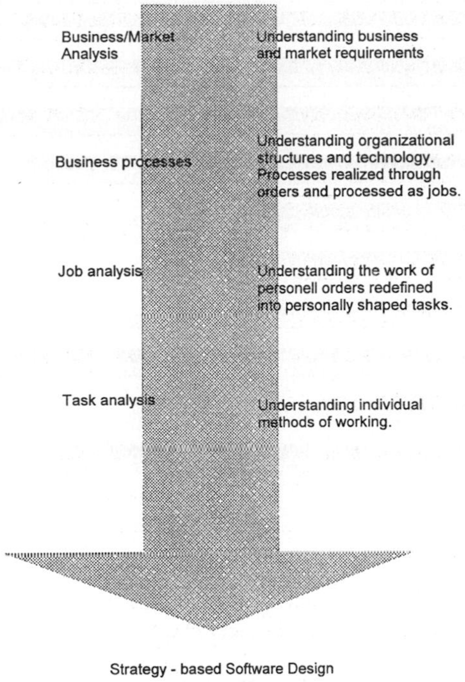

FIGURE 101.1 Simplified model of the process-oriented task analysis

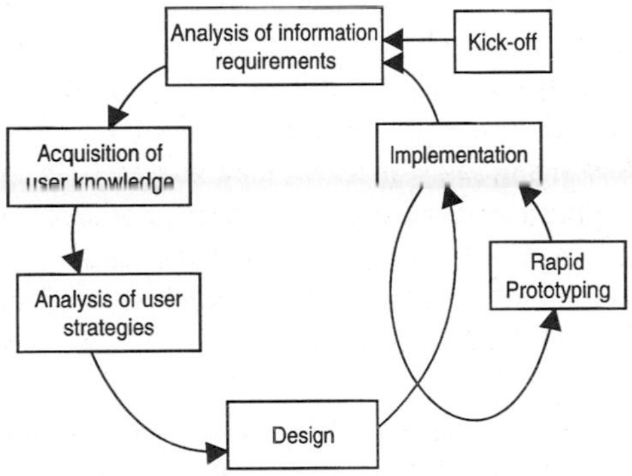

FIGURE 101.2 Major steps in strategy-based software development

TABLE 101.1 Organizational, Functional, and Psychological Aspects of Task Analysis

Job and Task Analysis Maintenance, Diagnosis, Repair	
Organizational analysis	Type of work organization; collaboration central maintenance/machine operators; information requirements, allocation of responsibilities.
Functional analysis	Preventive maintenance tasks; failure analysis; deviations, failures, breakdowns; fixing, replacement of parts.
Psychological analysis	Cognitive and mental strategies, rules and procedures; information needs; user qualification; user acceptance.

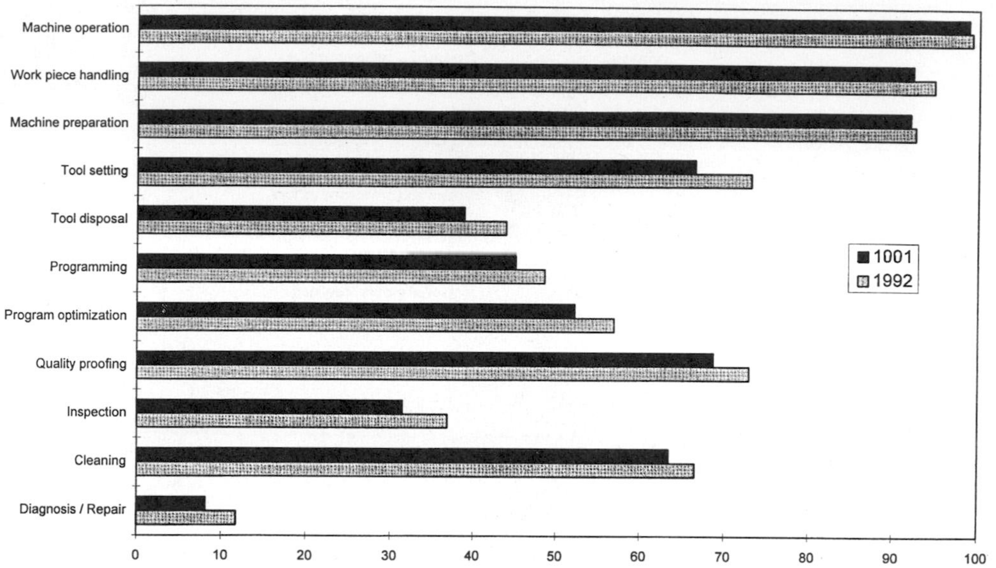

FIGURE 101.3 Functional task analysis of operators at numerical controlled machines in 1991 and 1992. Results of a representative survey in Germany in the machining industry.

Understanding Business and Market Requirements: Supporting Tasks by Software

Following the concept of semiautonomous work groups, a variety of tasks which are commonly allocated to different departments are reintegrated into manufacturing units, such as planning, scheduling, preventive maintenance, troubleshooting, and even repair. Figure 101.3 shows the results of a representative survey of the Sonderforschungsberüch (SFB) 187, which encompasses about 2.200 firms of the metal-manufacturing industry in Germany (Ostendorf and Seitz, 1992; Zimolong and Saurwein, 1995). In 1991, 8.2% of the companies reported that operators at the shop floor were completely or to a greater part engaged in maintenance and servicing tasks. In 1992 there was even an increase of 3.7 to 11.9%. The results stress the increasing significance of the integration of maintenance and repair tasks into the work of operators. In small and medium-sized companies, maintenance and repair tasks are usually performed by operators and not by specialists. It remains unaddressed whether operators usually are qualified to handle preventive maintenance, troubleshooting, and in some cases even repair (see the in-depth work analysis in the following chapter).

Maintenance tasks usually pose high information and decision demands on the technicians. The number of alternatives and the interconnectivity of components, symptoms, and possible operations show that maintenance tasks usually are complex problem-solving tasks. Moreover, business processes in maintenance are closely interconnected with other different processes usually carried out by different departments, like quality assurance and operation scheduling. Assuring operativeness, as a process in maintenance, for example, has the preconditions of keeping the flow time in operations scheduling and keeping the date of completion in quality assurance, but is often impeded by functions and divisions. The organizational concept of semiautonomous work groups, responsible for production planning, time-scheduling of material, manufacturing, quality control, maintenance, and repair has the potential to overcome some of these problems. On the other hand, in work groups the need for information and decision support on the shop floor increases, leading to the usage of decentralized information systems at the shop floor. In that sense, information systems incorporating decision support facilities are not only potentive tools but also a precondition for the decentralization of maintenance tasks. Support systems specifically promote a lot of aspects of semiautonomous working groups, including:

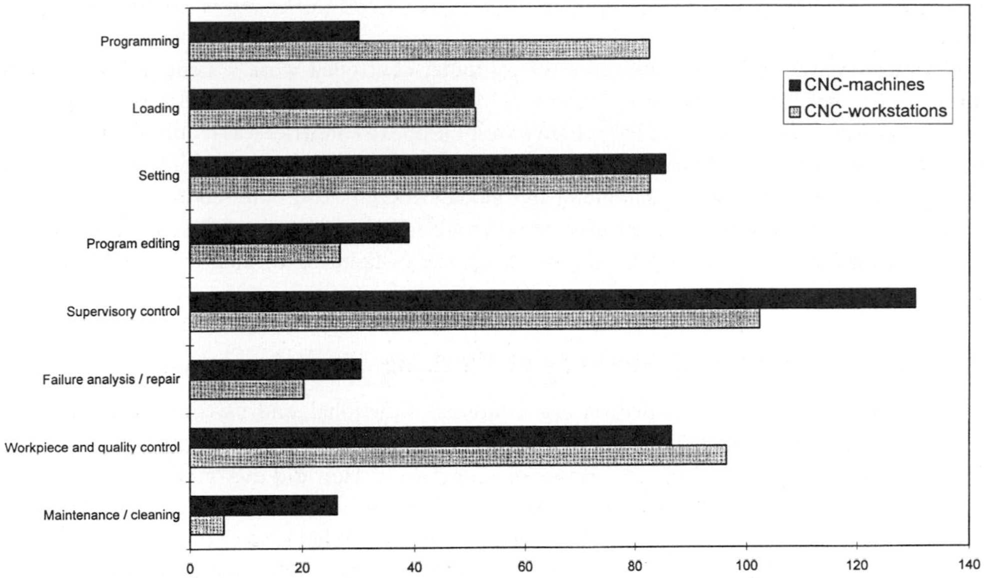

FIGURE 101.4 Task and time study of operators in computer-controlled work systems (in minutes).

- The allowance to integrate workers with very different qualifications, e.g., apprenticeships and highly qualified maintenance technicians
- The reduction in diagnosis and repair time
- The increase of qualification of the working group members through the accesses of the jointly acquired knowledge base

Understanding the Work: Job Analysis

The analysis of market and business requirements leads to the conclusion that decentralized maintenance tasks accomplished by machine operators in work groups may well be supported by an information and decision support system. Looking at the job in more detail should reveal what kind of typical tasks are fulfilled by operators, what amount of time they need, and what kind of qualifications they have.

Consequently, a job analysis was carried out to analyze typical functions of operators and qualification requirements in small and medium-sized metal-manufacturing companies. Seven small and medium-sized companies of the metal-manufacturing industry were examined. A sample of 21 computer-controlled work systems out of 105 systems was chosen for the work analysis. The analysis included eleven CNC-machines and eight CNC-workstations. The results of the 19 work systems are representative for the 86 computer-controlled work systems. A more detailed description of the sample studied can be found in Konradt and Zimolong (1993).

Main tasks and average working time for each task is indicated in Figure 101.4. Eight main tasks of operators were classified. As the results show, operators mainly perform supervisory tasks (CNC-machines: 24.9%, CNC-workstation: 17.5%), followed by workpiece and quality control tasks (15.7%; 19.4%). Results correspond closely to those reported in Fix-Sterz, Lay and Schultz-Wild (1986).

A detailed functional analysis of the task "failure analysis and repair" was carried out. These tasks consume 30.6 min (7.6%) of the job time at CNC-machines and 20.3 min (4.4%) at CNC-workstations. Diagnosis and repair activities fill out nearly half of the time (57.5%; 43.3%). Note the considerable amount of time (30%) operators spent to support diagnosis activities of maintenance personnel. Work time of operators required for failure diagnosis and repair differs considerably between both work systems.

Operators are less involved in troubleshooting activities at CNC-workstations due to their complex technology.

In general, the availability and utilization of computer-controlled work systems in industry differ significantly. For flexible manufacturing systems (FMS) a technical availability between 70 and 98% is reported. A survey among 17 users of FMSs in five West European countries was reported by Shah (1987), indicating a range of actual utilization between 67 and 95%. An overview on availability, actual utilization and stoppage causes is provided in Zimolong and Duda (1992). In computer-controlled work systems, operators spent a considerable amount of time on troubleshooting activities. The particular activities which were identified by our in-depth study are diagnoses of failures, working on repair, and support for external repair personnel.

Understanding Individual Methods of Working: Task Analysis

A comprehensive task analysis is based on organizational, functional, and psychological requirements (refer to Table 101.1 for details). The functional analysis is concerned with the questions: What needs to be done? The psychological analysis asks: How can people do it? How did they accomplish similar tasks in the past? How will they do it in the future? The organizational analysis asks questions such as: What kind of responsibilities should be allocated to the job incumbents? What kind of information is required? Who else should be involved in the accomplishment of the task? In the following section, we concentrate on the psychological requirements analysis.

With respect to their qualification levels, kind of experience and situational constraints, maintenance personnel develop specific strategies to locate failures of stoppages or breakdowns. They hesitate to engage immediately in systematic reasoning and failure testing. Instead, they use a variety of quick opening strategies before entering into more systematic checks and tests. In an empirical study the application of strategies in failure diagnosis in computer-controlled manufacturing systems was analyzed with dependence on the user and task variables (Konradt, 1995).

The analysis included 22 electricians and mechanics from the central maintenance and repair departments of five companies in the metal-manufacturing industry. Age varied between 20 and 57 years, on average 33.7 years. Professional experience varied between one and 40 years, on average 15.5 years. Beginners had up to 5 years of experience, advanced 6 to 20 years, and experts more than 20 years of work experience. Their rules of troubleshooting were recorded by means of verbal interviews using a scenario technique. Interviews were conducted for three types of failure cases at CNC-machines: rare, less frequent and frequently emerging failures and breakdowns. In total, 69 interviews were conducted and 182 rules were identified and classified into 15 diagnosis strategies.

Results show that the most frequent troubleshooting strategies were "Historical information," "Least effort," "Reconstruction," and "Sensory check" (Figure 101.5). Historical information about types and frequencies of breakdowns of the machine tools is used through maintenance records, failure statistics, and contact with colleagues. Least effort means that simple checks are used first. Reconstruction requires that the operator or a colleague is asked about the failure course. By sensory check, symptoms such as loose connections, odors, or sounds are located. It is not affordable to use measuring instruments. In contrast, strategies such as "Information uncertainty" and "Split half," which lead to a binary reduction of the problem space only play a minor role in troubleshooting machine tools.

In diagnostic search two basically different maintenance and repair strategies were proposed (Rasmussen, 1981). "Symptomatic search" comprises a set of symptoms collected by the operator, and this set is matched to a set of symptoms under normal system state. If the search is performed in the actual system or physical domain, it is called "topographic search." Our results suggest a further class of generalized strategies called "case-based." In case-based strategies, symptoms available in the diagnostic situation are collected and compared to those in similar cases. In novel failure cases topographic search dominates. In routine failures case-based strategies are most frequently used and the importance of topographical strategies is diminished. Finally, topographical and case-based strategies are dominant in familiar failure cases.

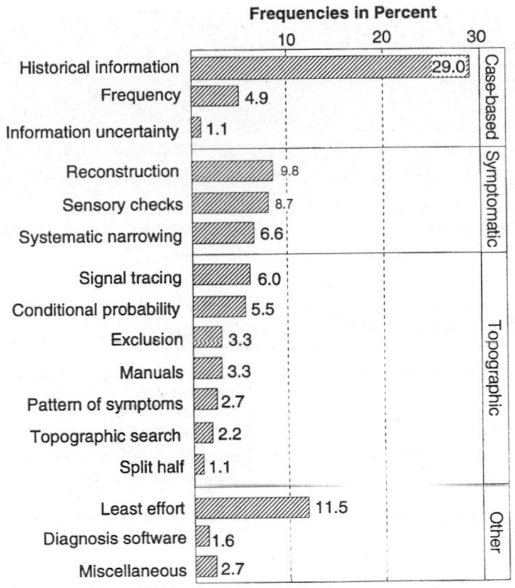

FIGURE 101.5 Frequencies of failure diagnosis strategies and their relation to generalized strategy classes.

Results of task analysis of maintenance personnel mirrors a distinctive and complex picture of troubleshooting procedures. Maintenance experts do not engage immediately in systematic reasoning and failure testing due to time constraints and the heavy burden they put on short term memory and decision making. In contrast, they use a variety of quick-opening strategies before entering into more systematic checks and tests. These strategies require specific kinds of information. For example, the strategy "historical information" demands a list of failures and technical changes made in the life cycle of the machine. The particular information and documentation demands of novices and experts, including explanations and support in planning, can be inferred from strategies, too.

Bridging the Gap Between Requirements and Design

The three main troubleshooting strategies — symptomatic, topographic, and case-based search — led to the result that it is essential to support all strategies and their various rules and procedures. All of them were translated into semantic objects, which were then transformed into the objects of the software design. Objects were grouped into four modules: The *system module* gives an overview of the system and its components; the *information module* contains the basic structures of the machine tool, circuit diagrams, technical data, and descriptions of single elements; the *diagnosis module* comprises all data on symptoms and failures; the *logbook* stores the breakdowns, failure types, their related symptoms and statistics.

In detail, data structures were derived through specification of a set of semantic objects that are necessary to carry out a strategy revealed in cognitive task analysis. The methodology is called "Strategy-Based Software Design" (SSD, Konradt, 1995). It is assumed that software should be compatible to the operator and an appropriate tool in working tasks. Moreover, a continuous transition between analysis results and design should be guaranteed.

In Figure 101.6, the three stages of SSD are represented. The support of the strategy "Sensory checks," for example, requires the implementation of objects generally used by the operator pursuing this strategy. Typical objects are components of the machine, symptoms, lists of identified components and symptoms, failures, and causes of failures. Operations such as selection, search, input, and relation are defined according to the objects. This procedure is put forth with every one of the other 14 strategies. It is evident that several strategies require the same objects. It is therefore logical to take advantage of common elements from different strategies rather than handling each strategy separately. The elements of the strategies provide the bonds in a network

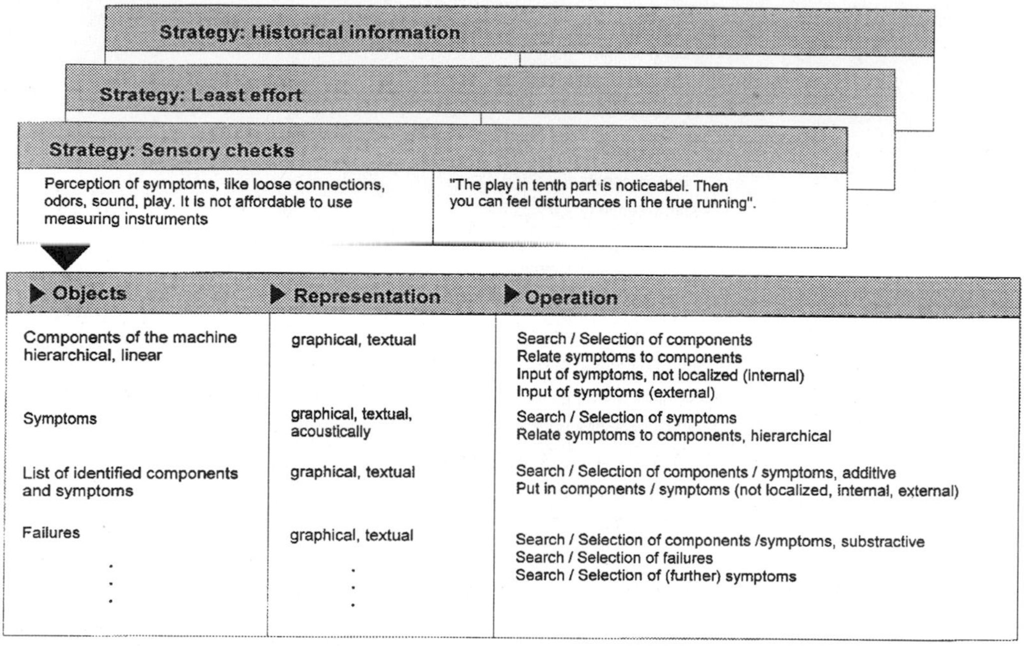

FIGURE 101.6 Three stages of strategy-based software design. The strategy: Sensory checks require the implementation of objects, type of representation and of operation (From Konradt, U. (1995). Strategies of failure diagnosis in computer-controlled manufacturing systems: Empirical analysis and implications for the design of adaptive decision support systems. *International Journal of Human-Computer Studies*, 43, pp. 503-521. With permission.)

whose connections arise out of the individual strategies. The resulting network offers the user the chance to switch between different strategies or combine them with each other.

A final usability study concerning the proposed efficiency and the use of different types of information related to different strategies are presented in the next section.

Understanding the Usefulness of the Application: Usability Study

During the design cycle, the Diagnosis Information System (DIS) was continuously refined by user participation. At the stage of a smoothly running prototype, an intensive usability study has been conducted with respect to efficiency, enhancement of users knowledge, flexibility of the system, and user acceptance. In this paper, results on efficiency and flexibility in comparison with a manual will be presented. Additional results are outlined in Majonica and Zimolong (1994) and Majonica (1996).

An overview on the study is given in Figure 101.7. It was carried out with 60 subjects in cooperation with work-training centers, and small metal-manufacturing and automobile manufacturing companies. Age of participants was between 17 and 62 years; average was 29 years. Mean professional experience was 7 years; 35% of the subjects had experience with CNC-machine tools, 65% had none. Ten subjects (17%) conducted failure diagnosis tasks at CNC-machines in their own company. No one was familiar with the specific type of CNC-machine used in this study.

The study consisted of five parts: (1) introduction of DIS or Manual and exercises; (2) first knowledge test; (3) individual solution of five diagnostic tasks; (4) second knowledge test, and finally (5) overall assessment of the system. The total efficiency of the system was tested with a control-group design. Forty subjects used DIS as a tool to solve five diagnostic tasks, and 20 subjects used the manual, which contained the equivalent information. Between three and ten solutions had to be found for each of the five different tasks. Subjects spent more than three hours on the study. Knowledge tests and assessment of the system were presented as paper and pencil tests. Log-file data were automatically recorded. Analysis of data

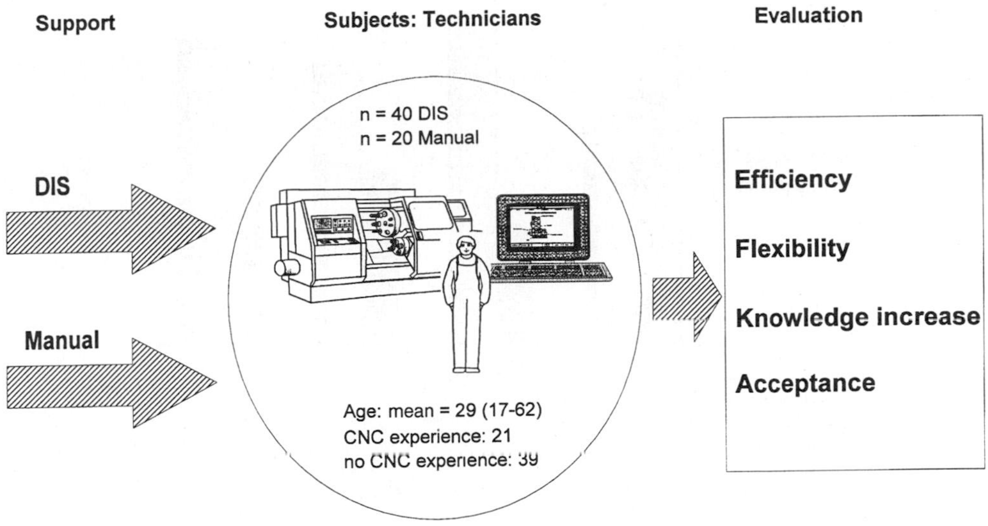

FIGURE 101.7 Design of the usability study. DIS = Diagnosis Information System; Manual of the system.

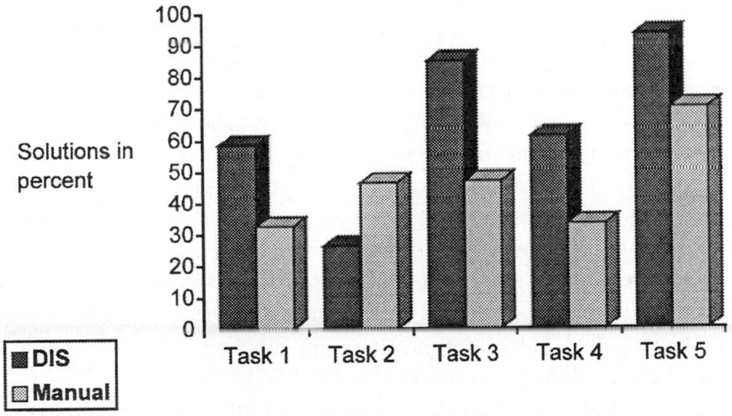

FIGURE 101.8 Frequency of solutions obtained with the support of DIS or manual. Except for task 2, subjects performed better with DIS.

allowed investigators to identify the frequency of use of the system pages and to compute the amount of time spent per page within the four modules of DIS. The use of the manual was recorded with a videorecorder and after that analyzed with respect to frequency and time spent with different pages and chapters.

Total efficiency was measured as frequency of solutions obtained with DIS and manual (Figure 101.8). Except for task 2, subjects gained better results with DIS. They solved 65% of the tasks as compared to 46% of tasks in the manual condition. Results of the efficiency analysis show that DIS significantly improves the correct solutions of the diagnostic tasks. Compared with the use of a manual, 19% more correct solutions were found on average. Only in task 2 the manual was superior. Without task 2 there was a gain of 29% in performance against the manual.

No difference could be found between the two samples with respect to experience with troubleshooting or professional experience. As an important result, DIS offers equally good support to both users with domain-specific knowledge and beginners: there was no difference in the number of solutions found by the two groups.

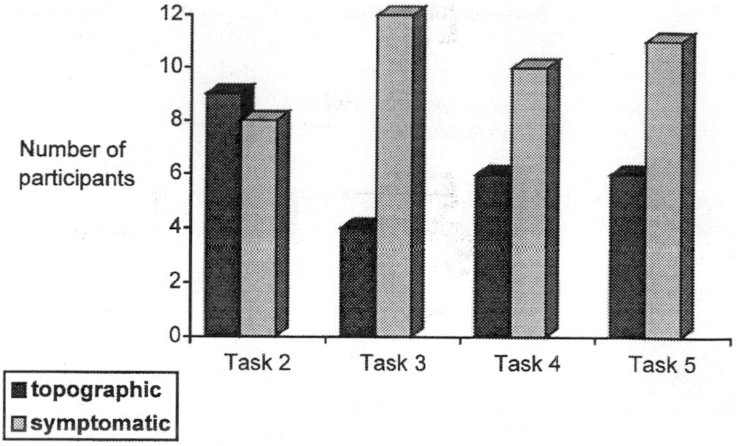

FIGURE 101.9 Strategies of participants.

Psychological flexibility of a system means the degree of support a system offers to foster the application of individual strategies and cognitive preferences of users. The flexible use of the system was measured as frequencies and amount of time spent with the different modules of DIS or with different chapters of the manual to solve the diagnostic tasks. Significantly more time was spent with the four modules of DIS as compared to the equivalent chapters of the manual. Users more often switched back and forth from one module to the next, performing significantly more transitions between modules ($x = 3.44$) as compared to the transitions between chapters of the manual ($x = 1.17$).

Finally, the flexibility of the system was evaluated. Log-file data of 19 subjects were analyzed with respect to the transitions from one page of the system to the next one. The various steps of the individual solution paths of four tasks were identified separately for each subject. The steps served as entries of a cluster-analysis which was computed with respect to an analysis of the main strategies "topographic," "symptomatic," and "case-based" search. Results showed that seven of the participants pursued a topographic strategy, and twelve a symptomatic strategy. No case-based strategy was found. Five users switched from topographic to symptomatic search and vice versa while performing their analysis. One of the strategies could not be identified. Thus, the system not only supports different strategies of the users but also allows switching between different strategies (Figure 101.9).

Results of the usability study demonstrated that users with different backgrounds in professional as well as in diagnostic experience were equally well supported by DIS. The system allowed them to choose their own way of solving the tasks. Data from questionnaires show that they also improved their knowledge about the machine structures, components, and failure causes. No case-based strategy was found. This is probably due to the characteristics of the sample. Subjects were not familiar with this specific type of CNC-machine and therefore could not use their experience with previous failures and breakdowns. The usability study showed very clearly the advantages of the psychological requirement analysis and the strategy-based software design. SSD strongly improved the efficiency of the system, supported the personal strategies of the participants, and led to an overwhelming acceptance of the system by the users.

101.4 Conclusions

This chapter has presented an overview of an integrated approach to software design that considers relevant organizational, human, and technical issues in software development. The software development process starts with an analysis and understanding of business and market requirements. Process-oriented task analysis considers at an early point in the software design cycle the organizational, functional and psychological aspects of work. Three field studies were reported, covering these aspects. First, organizational aspects of decentralized manufacturing units and semiautonomous work groups were studied. In particular, the allocation of functions to the units with respect to planning, scheduling, maintenance, and repair were discussed. From this analysis, requirements of information flow and decision support for work groups were derived. Second, a task analysis with special emphasis on cognitive preferred working methods of maintenance personnel was performed. The study revealed 14 strategies, rules, and procedures on how maintenance personnel performed troubleshooting activities. The message from the findings was that in order to support these strategies efficiently, all information required has to be stored in the DSS. This procedure is called "strategy-based software design." Data structures are derived through specification of a set of semantic objects that are necessary to carry out one of the strategies. Finally, a usability study revealed the usefulness and efficiency of the approach.

An empirically oriented and practically applicable theory of design that covers business processes, principles of sociotechnical design, such as cooperative work and user participation, and information technology which is sufficiently detailed, simple, and structured is still outstanding. In our approach an integrated attempt has been made to identify, connect and handle some important structural variables that may improve a unified understanding toward design and application of computer-based working tools. In this way it can be treated as a reference model for a future generation of flexible software support systems.

References

Brödner, P. (1990). *The Shape of Future Technology. The Anthropocentric Alternative.* Springer, New York.

Clegg, C. and Symon, G. (1989). A review on human-centred manufacturing technology and a framework for its design and evaluation. *International Review of Ergonomics,* 2, pp. 15-47.

Corbett, J.M., Rasmussen, L.B. and Rauner, F. (1991). *Crossing the Border. The Social Engineering Design of Computer Integrated Manufacturing Systems.* Springer, London.

Fix-Sterz, J., Lay, G. and Schultz-Wild, R. (1986). *Flexible Fertigungssysteme und Fertigungszellen.* VDI-Z, 128, pp. 369-379.

Hammer, M. and Champy, J. (1993). *Reengineering the Corporation.* New York, Harper Collins Publ.

Imai, K. (1994). *Kaizen: Der Schlüssel zum Erfolg der Japaner im Wettbewerb.* Berlin/Frankfurt, M. Ullstein.

Kidd, P.T. (1994). *Agile Manufacturing: Foregoing New Frontiers.* Addison-Wesley, Reading.

Konradt, U. (1995). Strategies of failure diagnosis in computer-controlled manufacturing systems: Empirical analysis and implications for the design of adaptive decision support systems. *International Journal of Human-Computer Studies,* 43, pp. 503-521.

Konradt, U. (1996). *Gestaltung gebrauchstauglicher Anwendungssysteme.* Wiesbaden, Deutscher Universitäts-Verlag.

Konradt, U. and Zimolong, B. (1993). Arbeitstätigkeiten, Qualifikationsanforderungen und Organisationsformen an CNC-gesteuerten Werkzeugmaschinen und Bearbeitungszentren. *Zeitschrift für Arbeitswissenschaft,* 47, pp. 71-78.

Majonica, B. (1996). Evaluation eines Informations-Systems für die Unterstützung von Instandhaltungsaufgaben. Münster, Waxmann.

Majonica, B. and Zimolong, B., (1994), Evaluation of a diagnosis information system for semi-autonomous production islands, in *Proceedings of IEEE International Conference on Systems, Man, and Cybernetics,* San Antonio, Texas, pp. 741-746.

Monarchi, D.E. and Puhr, G.I. (1992). A research typology for object-oriented analysis and design. *CACM,* 35(9), pp. 35-47.

Ostendorf, B. and Seitz, B. (1992). Alte und neue Formen der Arbeitsorganisation und Qualifikation — Ein Überblick, in J. Schmid and U. Widmaier (Eds.). *Flexible Arbeitssysteme im Maschinenbau.* Opladen, Leske + Budrich, 1992, pp. 75-89.

Rasmussen, J. (1981). Models of mental strategies in process plant diagnosis, in J. Rasmussen and W. B. Rouse (Eds.). *Human Detection and Diagnosis of System Failures.* Plenum Press, New York, pp. 241-258.

Shah, R. (1987). Flexible Fertigungssysteme in Europa. *Verfahren der Anwender.* VDI-Z, 129, pp. 13-21.

Zimolong, B. (Ed.) (1996). *Kooperationsnetze, Teilautonome Flexible Fertigungsstrukturen und Gruppenarbeit.* Opladen, Leske + Budrich.

Zimolong, B. and Duda, L. (1992). Human error reduction strategies in advanced manufacturing systems, in M. Rahimi and W. Karwowski (Eds.), *Human-Robot Interaction* (pp. 242-265). London: Taylor & Francis.

Zimolong, B. and Saurwein, R.G. (1995). Maschinenbau zwischen CIM und Gruppenarbeit. *Zeitschrift für Arbeitswissenschaft*, 21 (4), pp. 226-232.

Zimolong, B. and R. Trimpop (1994). Managing human reliability in advanced manufacturing systems, in G. Salvendy and W. Karwowski (Eds). *Design of Work and Development of Personnel in Advanced Manufacturing.* Wiley, New York, pp. 431-461.

Section II
Manufacturing Systems

102

Kansei Engineering: A New Consumer-Oriented Technology for Product Development

Mitsuo Nagamachi
Hiroshima University

Abstract

Kansei Engineering was developed as a consumer-oriented technology for new product development. It is defined as "ergonomic translation technology of the consumer's feeling and image of a product into physical design elements including mechanical function." Kansei Engineering (KE) technology is classified into five types, KE Types I through V. KE Type I is a category classification of the new product toward the design elements. Type II utilizes the current computer technologies such as Expert System, Neural Network Model, and Genetic Algorithm. Type III is concerned with a mathematical modeling of Kansei Engineering. Type IV consists of two sorts of Kansei Engineering, forward and backward reasoning systems; and finally Type V utilizes a combination of virtual reality technology and Kansei Engineering.

Kansei Engineering has permeated Japanese industries, including automotive, home electric appliance, construction, clothing, and so forth. The successful companies using Kansei Engineering benefited from good sales due to the new consumer-oriented products.

102.1 Introduction

Currently, the industrial age is changing from "a product-out concept" to "a market-in concept" regarding new product development. In the 1970s, manufacturers produced a volume of products and people bought them. At that time, the manufacturers designed the products according to their own concept and strategy. However, the consumers had to buy them regardless of their personal preferences. At present, they have a lot of things at home they do not want anymore. Sophisticated consumers desire products that match their own feelings of design and functional requirements as well as price.

0-8493-2641-9/99/$0.00+$.50

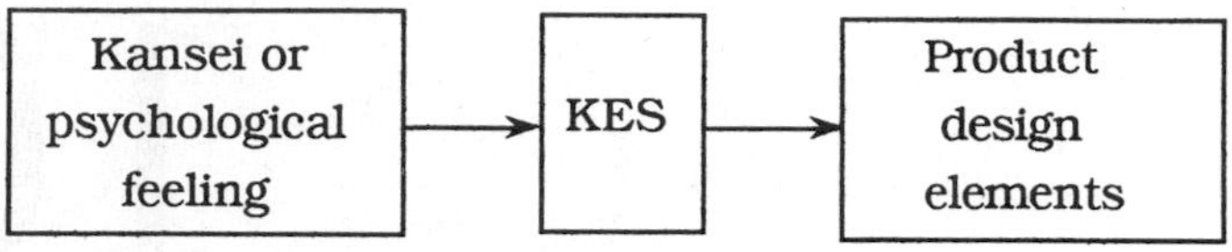

FIGURE 102.1 A diagram of a process of the Kansei Engineering System.

The product-out strategy means production by a manufacturer based on its own design strategy regardless of the consumer's demand and preference. On the other hand, the market-in strategy implies production based on the current consumer's desire and preference. Nowadays, consumers are stringent in choosing products in terms of their demand and preference. The manufacturers should change their production strategy and attitude to a more consumer-oriented one.

Nagamachi developed Kansei Engineering as an ergonomic consumer-oriented technology for new product development (Nagamachi, 1986; 1988a, b, c; 1989a, b; 1990a, b; 1991a, b: 1993a, b, c; 1994a, b, c, d; 1995a, b; Nagamachi et al., 1974; 1977; 1985; 1986; 1987; 1988; 1989, Enomoto et al., 1993, 1995; Fukushima et al., 1995: Imamura et al., 1994; Ishihara et al., 1993a, b; 1994; 1995; Jindo et al., 1991: 1995; Matsubara et al., 1994a, b; 1995; Tsuchiya et al., 1995). Kansei is a Japanese term which means a consumer's psychological feeling and image regarding a new product. When a consumer wants to buy something, he or she has an image of the product, such as "luxurious, gorgeous, and stable." Kansei Engineering technology enables his or her image and feeling to be used in the new product and to produce a good product that fits the customer's image. Kansei Engineering is defined as "translating technology of a consumer's feeling (Kansei) of the product into the physical design elements," as shown in Figure 102.1.

Kansei Engineering aims to produce a new product based on the consumer's feeling and demand. There are four points concerning this technology; (1) how to grasp the consumer's feeling (kansei) about the product in terms of ergonomic and psychological estimation, (2) how to identify the design characteristics of the product from the consumer's kansei, (3) how to build a Kansei Engineering system as an ergonomic technology, and (4) how to adjust the product design to the current societal change in people's preferences.

Concerning the first point, we use the semantic differentials (SD) developed by Osgood and his colleagues (Osgood et al., 1957) as a main technique to grasp the consumer's kansei. In Kansei Engineering, we collect the kansei, or the consumer's words, from shops and from industrial magazines. At first, we collect 600 through 800 kansei words and then select approximately 100 of the most relevant words.

In regard to the second point, we conduct a survey or an ergonomic experiment to look at the relation between the kansei words and the design elements. Regarding the third point, we utilize advanced computer technologies to build a systematic framework of the Kansei Engineering technology. Artificial intelligence, neural network model, and genetic algorithm as well as fuzzy set theory are utilized in the Kansei Engineering computerized system to construct the concerned databases and the inference system. Finally, we attempt to maintain the fit of the databases of the Kansei Engineering system and the consumer's current kansei trend by inputting new kansei data from the consumer every three or four years.

102.2 Types of Kansei Engineering

There are five technical styles of Kansei Engineering; Type I through Type V. Type I Kansei Engineering means Category Classification from zero- to *n*th- category. Type II uses a computer-aided system. Type III utilizes a mathematical framework to reason the appropriate ergonomic design. Type IV refers to Kansei Engineering system constructed by forward and backward reasoning, and Type V combines Kansei Engineering technology with virtual reality.

FIGURE 102.2 Miata made by Mazda.

Type I: A Category Classification

The category classification is a method by which a kansei category of a planned target is broken down in a tree structure to determine the physical design details. The following is a good example of the application of Kansei Engineering Type I. A Japanese car maker, Mazda, has developed a new sports car named "Miata" (Eunos Roadster in Japanese) (Figure 102.2) which was derived from Kansei Engineering. Kenichi Yamamoto, the chairman of Mazda was much interested in this new ergonomic technology (Yamamoto, 1986). As a result, Nagamachi was asked to teach Kansei Engineering to Mazda's development engineers and often visited Mazda. Mazda's people came to Hiroshima University to discuss Kansei Engineering. At present, Kansei Engineering has become the fundamental technology for new product development at Mazda.

Mr. Hirai, a manager of a new brand car, and his project team decided to implement Kansei Engineering in the development of Miata and based the zero-level category of the new model "Human–Machine Unity," after a vast of market survey and analysis of young drivers maneuvering. This concept, HMU, implies that a young driver feels a unity between himself or herself and the car when driving. The driver feels that his or her body might be unified with the car and controls the machine as if controlling his or her own body freely. Human–Machine Unity at the start of Kansei Engineering is just the concept of a new brand car. It tells nothing about the car design details such as engine type, car size, exterior design, interior design, and so forth. In Kansei Engineering Type I, the zero-level concept should be broken down into clearly meaningful subconcepts to determine the real design details.

The project team members started to classify the zero-level concept into the subconcepts, that is, 1st, 2nd, …, and *n*th subconcept until they obtained the car design specifications. The procedure used in Type I category is shown in Figure 102.3. In regard to Miata, the zero-level kansei was classified into four subconcepts in the first level; "Tight-feeling," "Direct-feeling," "Speedy-feeling" and "Communication." Tight-feeling implies the feeling of "fitting closely to the machine and "not large nor small." With this subconcept, the team decided that the car length should be around 4 meters, based on experience in car development. As a result, it was set at as 3.98 m due to the chassis length. If four seats would be mounted in this car, the driver would feel "narrow" and this feeling would not fit in with the subconcept. So it was decided that Miata would be a two-seat sports car. The above-mentioned story is an example of how the concepts were broken down into more detail until the team reached the final physical design elements.

Figures 102.3 and 102.4 explain how the Type I procedure is transferred into the subconcepts to the design detail. Figure 102.3 shows the break-down procedure of "Tight-feeling" into physical traits. Figure 102.4 illustrates the translation of kansei concepts into physical traits, ergonomic specifications,

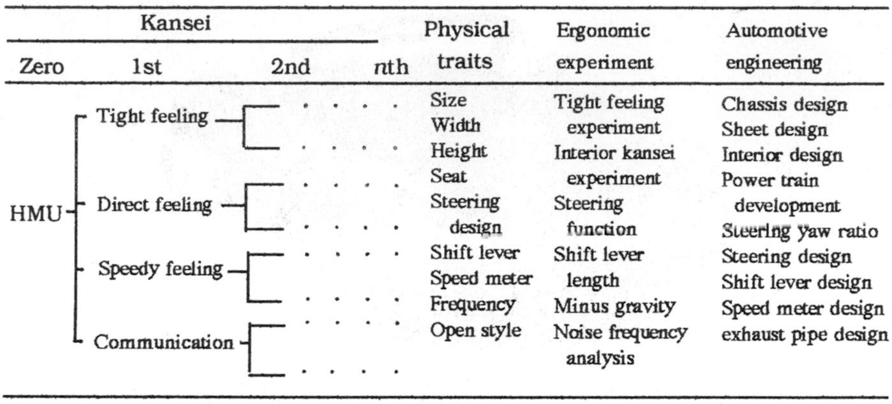

FIGURE 102.3 The translation of kansei into car physical traits in the case of the Miata.

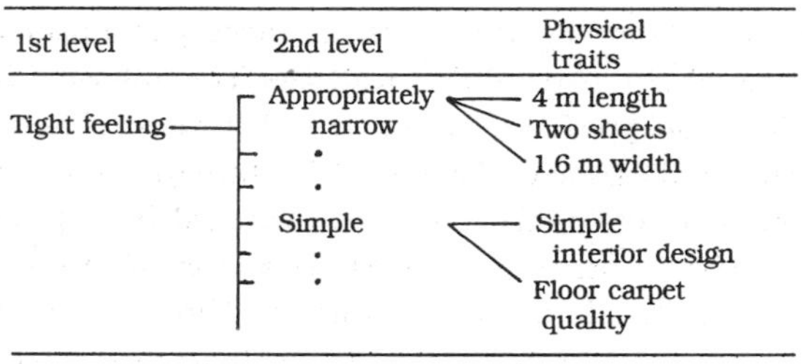

FIGURE 102.4 An example of refining the physical traits from kansei subconcepts.

and automotive engineering designing. The physical traits decided upon according to this breakdown procedure are tested ergonomically to find ergonomic specifications, which will be explained later. Then, the automotive engineering development and design attempt to realize the ergonomic characteristics. In this stage, the team creates a great number of new patents.

Mazda's project team succeeded in developing the new sports car, Miata, and it became a best-seller in the U.S. as well as in Japan. Miata is a typical example of the application of Kansei Engineering Type I.

Type II: Kansei Engineering Computer System

Kansei Engineering Type II is a computer-aided system concerning Kansei Engineering. Kansei Engineering System (KES) is a computerized system with an Expert System that supports the transfer of the consumer's feeling and image into physical design elements. The KES basically has four databases and an inference engine in the structure, shown in Figure 102.5.

Kansei Databases

First of all, kansei words used in a new product domain are collected from shops or from the concerned industrial magazines. These are mostly adjectives and sometimes nouns. If it is concerned with a passenger car, customers use words such as e.g., "speedy," "easy control," "gorgeous," "highly qualified," etc. These kansei words are analyzed by multivariate analysis like factor analysis, cluster analysis, and others, after an ergonomic evaluation has been conducted. The Kansei database consists of these statistically analyzed data. Kansei Engineering evaluation is mostly conducted through the Semantic Differential (SD) Method with five-point scaling.

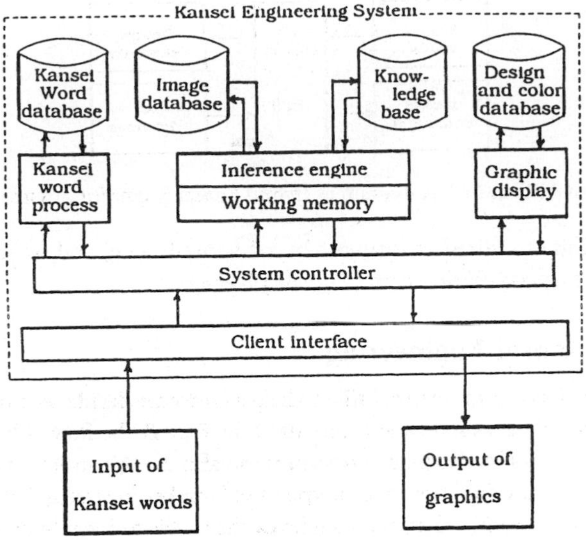

FIGURE 102.5 A system structure of the Kansei Engineering System.

Image Database

The data evaluated by the SD scale are analyzed by Hayashi's Quantification Theory Type II (Hayashi, 1976), which is a type of multiple regression analysis for qualitative data. After this analysis, we are able to construct the statistical relations between the kansei words and the design elements. This is the image database. Hence, we can identify the contributory items in the design elements to specific kansei words and vice versa. For instance, if a consumer or a designer wants to decide the design fit to "speedy feeling," he or she checks the image database inputting these words into the system and finds the design element most contributory to this feeling.

Knowledge-Base

The knowledge-base consists of a group of rules in the form of "if, then" and the rules for controlling the image database. It also includes the color conditioning principle, the round design guideline and so forth.

Shape and Color Databases

The design details are implemented in the shape design database and in the color painting database, separately. All design details consist of the aspects of design that are correlated as a total shape with each kansei word. The color database consists of full colors that are also correlated to the kansei words. The combined design with shape and colors is extracted by the specific inference system based on the rule-base and displayed in graphics on the screen.

Type III: Kansei Engineering Modeling

Kansei Engineering Type III utilizes a mathematical modeling constructed in the computerized system in spite of the rule-base. In this technique, the mathematical model works as if it is a kind of logic like the rule-base.

Sanyo Electric Co. attempted to implement kansei fuzzy logic as machine intelligence in a color printer (Fukushima et al., 1995). The intelligent color printer consisted of a camera, a computer, and a color printer system with fuzzy logic and was able to diagnose the color of the original picture to print out the more beautiful color picture in terms of fuzzy logic. Nagamachi has developed a computerized diagnosis system about Japanese language feeling in terms of fuzzy integral and fuzzy measure logic

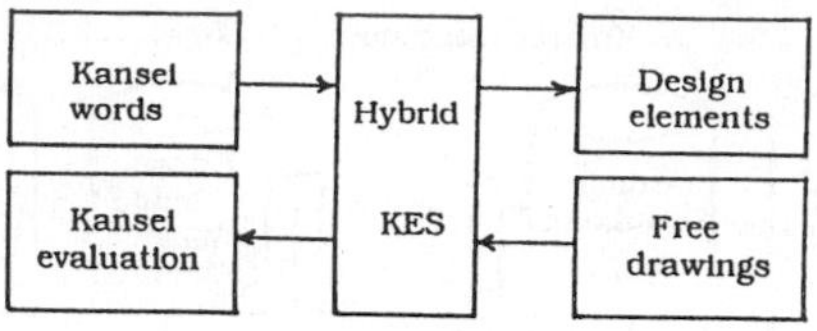

FIGURE 102.6 A diagram of Hybrid Kansei Engineering System.

(Nagamachi, 1993c). This is utilized to diagnose brand name feeling, and several Japanese companies use this system to select a good-feeling name for new products.

Type IV: Hybrid Kansei Engineering

Kansei Engineering Type II is a computerized KES to help a customer decide on a product that has the best fit for his or her product feeling, as shown in Figure 102.1. In Type II, the flow of kansei translation is from the customer's kansei to the design details. The designers can also use this style to decide the design. If he or she wants to design a kansei product from the viewpoint of "market-in strategy," the designer settles his or her design image of the new product first, and then he or she is able to identify the design specifications of the product with support of the KES. This direction is named "forward Kansei Engineering."

However, the designer sometimes wants to know how well his or her image and design fit the kansei. This computerized suggestion gives the designers good suggestions on how to create his or her design much more than the Type II can do. This direction is called "backward Kansei Engineering." Figure 102.6 shows both kansei translation directions in kansei engineering. The combination is called "Hybrid Kansei Engineering."

Using the Hybrid Kansei Engineering System, a designer is able to get the design specifications from the kansei words through the Forward Kansei Engineering. From the output displayed by the system, the designer designs more creatively based on his or her own idea and refers to the suggestion given by the system. Then, the designer inputs his or her new sketch of the product into the system, and the Backward Kansei Engineering recognizes the designer's sketch by means of an image recognition system. Finally, the system diagnoses the inputted sketch with reference to the kansei database, and the designer can evaluate his or her own sketch that has been formed by his or her creative sense.

Type V: Virtual Kansei Engineering

Kansei Engineering Type V is a new technique combining kansei engineering and virtual reality technology. Virtual reality is now a very popular technology in which people can virtually walk through the design created by the computer and displayed by means of HMD (Head Mounted Display) and data gloves.

In this technique, the kansei environment including the new product is constructed in the system, and the customer is invited to the virtual space with the new product. The virtual space and the product is decided by kansei engineering. Therefore, the customer can check the new product inside the virtual space. For instance, Nagamachi constructed a virtual kansei engineering kitchen system with Matsushita Works Co. (Nagamachi et al., 1996). In this system, the kitchen is designed by the kansei engineering system to fit the customer's kitchen image, and then the customer walks through the computer graphics and evaluates the virtual kitchen with reference to his or her image.

102.3 Applications of Kansei Engineering

Applications of Type I

Applications to the Automotive Industry

We have some applications of Kansei Engineering Type I in the automotive industry. The first application of kansei engineering in Japan has been conducted by Mitsubishi Motors Co. Mitsubishi applied kansei engineering technology to the dashboard design of a passenger car. Mazda implemented kansei engineering in the interior design of the Persona, in which case Mazda has introduced the concept of "Interiorism." This means to implement the atmosphere of a house reception room with sofas inside a vehicle. Mazda next tried to implement the complete technique and methodology of kansei engineering in the development of Miata. Therefore, as mentioned above, Miata exhibits many applications of kansei engineering.

When deciding the shift-lever length of Miata, for instance, the concern has been mainly the driver's kansei or feeling of "self-controlling the car." Then Mazda conducted an ergonomic experiment in which the subjects consisting of the designers and researchers had to estimate a variety of shift-lever lengths on the SD scale of the "self-controlling" feeling. After the calculation of experimental results, they noticed that the shift-lever length should be 9.5 cm to satisfy the drivers' "self-controlling" feeling. This is just an example of the application of kansei engineering. After Miata's success, the kansei analysis became important at Mazda. Now, Mazda generally starts from the kansei concept construction for every new product.

Wacol Co. which is a well-known maker of ladies' lingerie, tried to introduce kansei engineering in developing a new brassiere. As in the procedure for Miata, it tried to settle the first concept for the new brassiere from a survey of 2,000 women. From the questionnaire, Wacol found that their aim was to be "beautiful" and "elegant." However, the findings did not apply to design details. Then, Wacol invited all subjects to its ergonomic laboratory and asked them to wear a variety of test brassieres and to estimate them on the "beautiful" and "elegant" SD scale. The experimental results were analyzed through Quantification Theory Type I, and Wacol succeeded in finding the physical traits regarding the new brassiere; that is, (1) the breasts when wearing the brassiere should be kept within both body lines, and (2) two nipples should be in parallel and a little upward. Wacol endeavored to realize these two principles using the new materials and design. Wacol found after many trials that amorphous fiber would be best for women's very soft breasts and designed the brassiere to pull the breasts toward a central position.

Figure 102.7 shows the estimated results analyzed by the moiré-topography of a popular and finally designed brassiere. The lower silhouette in Figure 102.7 illustrates the popular brassiere moiré and the upper one represents the new brand moiré.

The upper figure satisfies two principles of a kansei brassiere, that is, the two breasts are kept within both body lines and the nipples are directed toward the reader's face. This was named "Good-up Bra" and has succeeded in achieving the highest sales record.

Applications of Type II

Kansei Engineering Type II is a computer supporting system for a designer designing kansei products. The KES utilizes expert system and neural networks, as well as genetic algorithm as the knowledge engineering technologies.

Figure 102.5 illustrates an example of KES, which has the kansei word, image, form, and color databases, and knowledge-base as well. When a designer inputs his or her kansei words in the system, the KES calculates them through the inference engine using the databases, and outputs a graphic as the outcome of system calculation. Figure 102.8 shows an example of (painted) kitchen design produced by the KES, which is called HULIS, HUman LIving System (Nagamachi, 1991a). The KES has been applied to the design of a house, costume, aluminum entrance door, passenger car interior, construction machine interior, and so forth.

Applications of Type III

Kansei Engineering Type III implies a mathematical modeling of KES. A good example of this was developed by Sanyo Co. Sanyo attempted to develop a new style of an intelligent color printer which enables production of a more beautiful picture than the original one. In general, the color copy machine faithfully prints the original picture. However, this new type of color printer can change, for example,

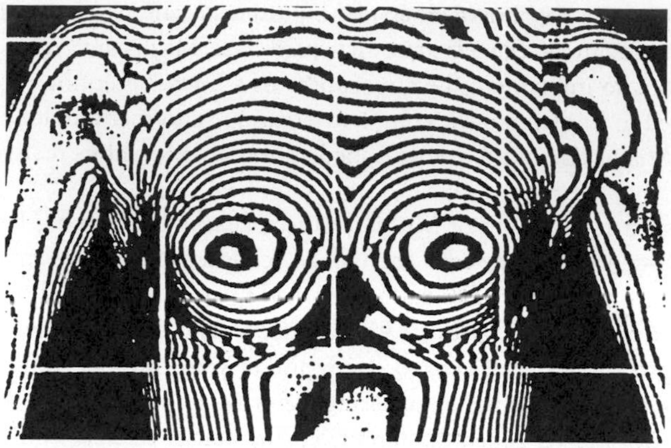

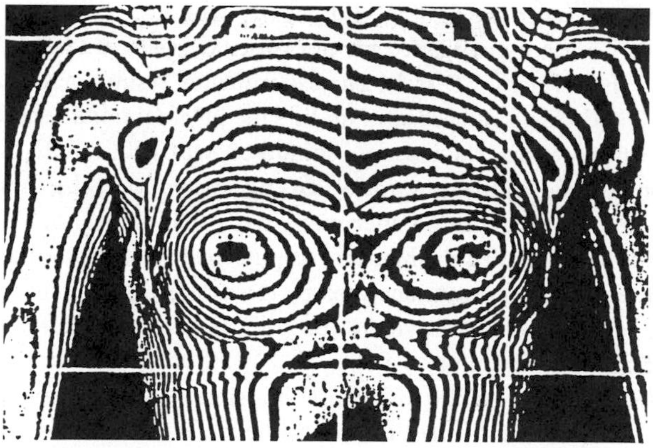

FIGURE 102.7 The lower figure is the moire-topography for ordinary brassiere and the upper for the new brand one.

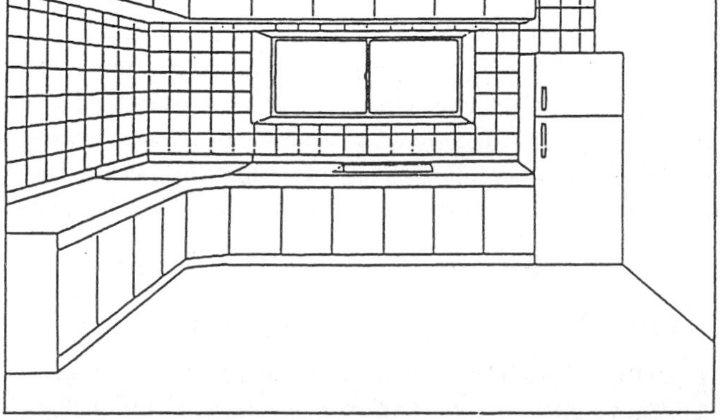

FIGURE 102.8 The displayed "kitchen" (painted) outputted by HULIS.

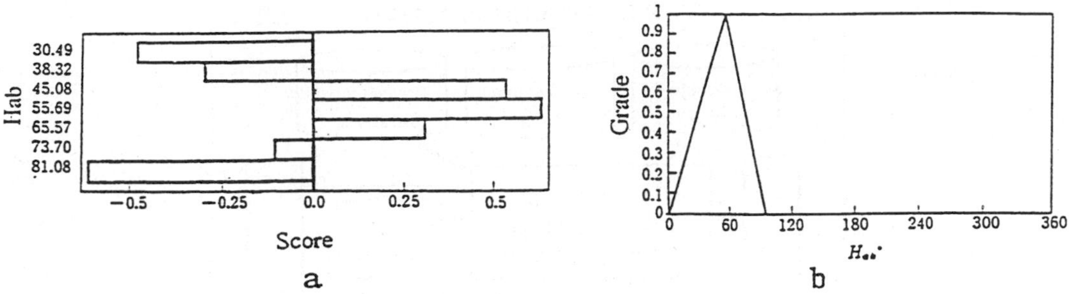

FIGURE 102.9 The experimental result of hue of face color (a) and the triangle Fuzzy expression of Fuzzy Membership Function of it (b).

face color to be more beautiful, because it has an intelligent system to recognize the face color and changes it in terms of a mathematical structure.

The KES color printer of Sanyo consists of a camera, image scanner, CPU recognition system, and color print system. We conducted ergonomic research on face color analysis using the SD scale. Sanyo researchers as the subjects evaluated a variety of face colors on the 5-point scale of "healthy and beautiful," and we found kansei dimensions of hue, brightness, and saturation concerning face color could be represented by a triangle fuzzy membership function of fuzzy set theory, as shown in Figure 102.9.

The tuning system of face color was implemented in the CPU of the new color copy machine, and we succeeded in producing healthier looking and more beautiful color copy (see Fukushima et al., 1995).

Nagamachi also used fuzzy logic structure as a model to diagnose Japanese word feeling. The new computerized word diagnosis system, WIDIAS, Word Intelligent DIAgnosis System, was developed at Hiroshima University using kansei engineering technology, and it has been used in Japanese companies to select appropriate brand names for newly developed products (Nagamachi, 1993c).

Applications of Type IV

The KES, in general, implies a decision procedure, kansei engineering, from the input of kansei words to the output of design details. However, a designer sometimes desires to have an assessment of his or her creation with reference to kansei engineering output.

The KES Type IV, Hybrid Kansei Engineering has two directions, Forward Kansei Engineering which means the flow of KES from kansei words to design details, and Backward Kansei Engineering which means the procedure flows from the designer's creation with reference to the output due to Forward KES to the assessment of feeling about his or her design creation. Figure 102.6 illustrates the Hybrid Kansei Engineering System. The upper flow of Hybrid KES is the procedure used to find kansei design details by inputting kansei words. The lower represents an opposite direction, from the designer's sketch toward the assessment or diagnosis of the designer's sketch. The designer can get a final decision about the new product design after cyclic usage of the Forward KES and the Backward KES.

Figure 102.10 shows the Backward KES. It has a computerized image recognition system which recognizes the inputted sketch created by the designer. It also has an assessment system which is able to evaluate the inputted sketch with reference to kansei databases. A Hybrid KES computer system is shown in Figure 102.11 which is a combination of KES (the Forward KES) shown in Figure 102.5 and the Backward KES shown in Figure 102.10 (Matsubara and Nagamachi 1995). Hybrid KES is utilized at Nissan for designing steering wheels and at Komatsu for designing a construction machine interior. For instance, Nissan uses Forward KES to get design details for a steering wheel, and the designer creates a more up-to-date design with reference to the KES conclusion. Then the designer inputs his or her sketch of the new steering wheel in to Hybrid KES. It evaluates the designer's sketch based on Hybrid KES rulebase and presents the kansei evaluation to the designer according to Backward inference.

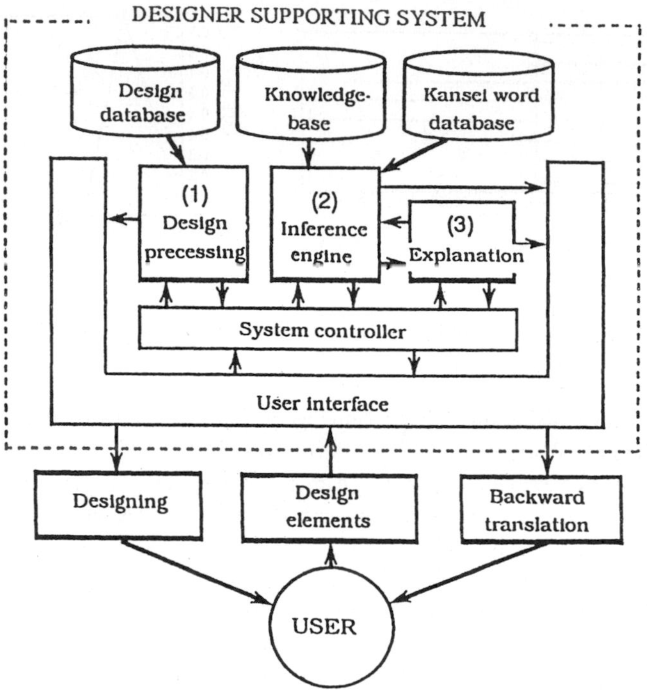

FIGURE 102.10 A diagram of the Backward Kansei Engineering System.

Figure 102.12 illustrates the use of the Backward KES for a steering wheel conducted at Nissan. The sketch created by a designer was inputted into the Hybrid KES by an automatic image recognition system (the right side in the figure) or a menu procedure (the left side of the figure). The system constructed a list of sketched design characteristics and diagnosed the kansei level of the sketch using the kansei diagnosis indices, with 0.0 to 1.0 applied to each kansei word, shown in the lower figures in Japanese in Figure 102.12.

Application of Type V

Virtual reality technology is well known about realization of virtual experience. We attempted an integration of virtual reality and kansei engineering regarding a house kitchen design, which is called Virtual Kansei Engineering. The Virtual KES is able to select a kitchen style in terms of a computer system with kitchen databases, when a customer addresses his or her height data and life style concerning kitchens. After that, the customer can walk through the computer graphics of the virtual kitchen to check whether it fits his or her image and feeling about the kitchen (Enomoto et al., 1993).

The Virtual KES is utilized in Matsushita Electric Works sales shops, in Tokyo, Osaka, and Hiroshima, Japan, and it has good reputation among customers.

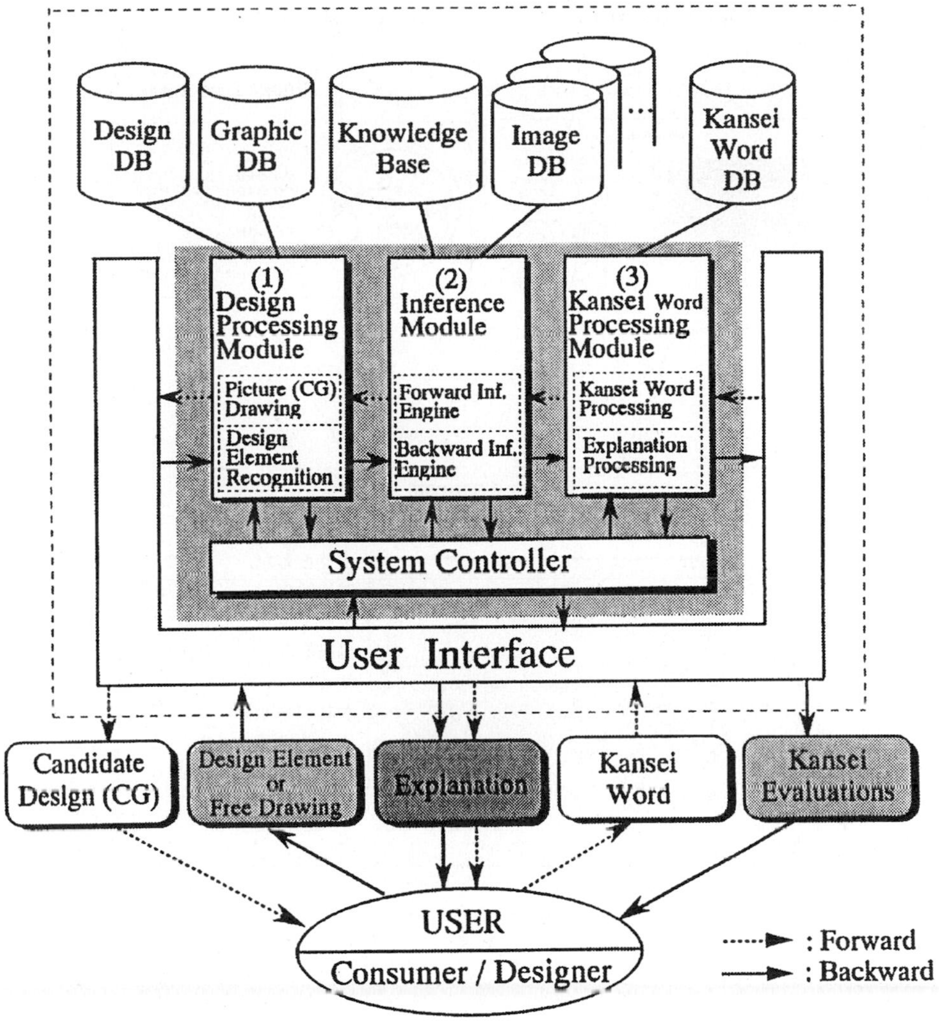

FIGURE 102.11 A diagram of the Hybrid Kansei Engineering System.

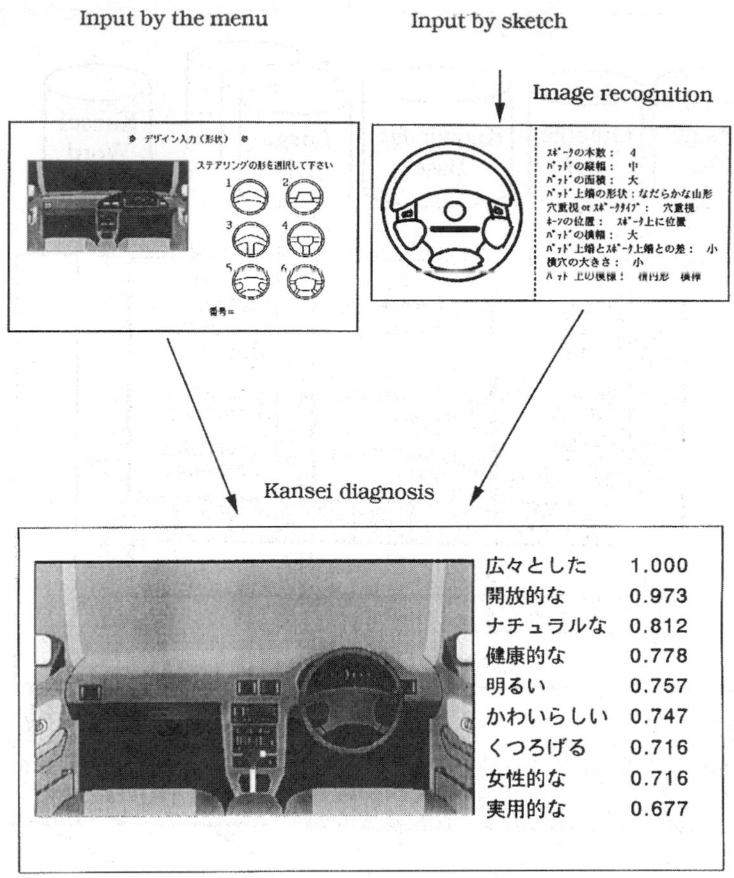

FIGURE 102.12 An example of Hybrid KES applied to steering wheel design. The inputted sketch is diagnosed to the kansei implication in terms of Hybrid KES.

References

Enomoto, N., Nagamachi, M., Nomura, J. and Sawada, K., 1993, Virtual kitchen system using Kansei Engineering, in *Human Computer Interaction: Software and Hardware Interface,* G. Salvendy and M.J. Smith (Eds.) (Elsevier, Amsterdam), pp. 657-662.

Enomoto, N., Nomura, J., Sawada, K., Imamura, K. and Nagamachi, M., 1995, Kitchen planning system using Kansei Virtual Reality. *Symbiosis of Human Artifact,* 179-184.

Fukushima, K., Kawata, H., Fujiwara, Y. and Genno, H., 1995, Human sensory perception oriented image processing in color copy system. *International Journal of Industrial Ergonomics,* 15(1), 63-74.

Hayashi, C., 1976, *Method of Quantification,* Toyokeizai, Tokyo.

Imamura, K., Nomura, J., Enomoto, N. and Nagamachi, M., 1994, Virtual space support system using kansei engineering, in *Japan-USA Symposium on Flexible Automation,* pp. 549-555.

Ishihara, S., Hatamoto, K., Nagamachi, M. and Matsubara, Y., 1993a, ART1.5SSS for Kansei Engineering Expert System, in *Proceedings of International Joint Conference on Neural Networks,* pp. 2512-2515.

Ishihara, S., Ishihara, K. and Nagamachi, M., 1993b, AKSYONN: Automated kansei engineering expert system builder by self-organizing neural network, in *Human Factors in Organizational Design and Management-IV,* pp. 485-490.

Ishihara, S., Ishihara, K., Nagamachi, M. and Matsubara, Y., 1994, Automatic analysis on Kansei Engineering experiment by self-organizing neural network, in *Japan-USA Symposium on Flexible Automation,* pp. 541-544.

Ishihara, S., Ishihara, K., Nagamachi, M. and Matsubara, Y., 1995, An automatic builder for a kansei engineering expert system using self-organizing neural networks. *International Journal of Industrial Ergonomics,* 15(1), 13-24.

Jindo, T. and Nagamachi, M., 1991, The development of a car interior image system incorporating knowledge engineering and computer graphics, in *Proceedings of the 11th Congress of the International Ergonomics Association,* pp. 625-627.

Jindo, T., Hirasago, K. and Nagamachi, M., 1995, Development of a design-support system for office chairs using 3-D graphics. *International Journal of Industrial Ergonomics,* 15(1), 49-62.

Matsubara, Y. and Nagamachi, M., 1994a, An application of image processing technology in kansei engineering, in *Proceedings of 12th Triennial Congress of the International Ergonomics Association,* pp. (4) 127-131.

Matsubara, Y. and Nagamachi, M., 1995, Hybrid kansei engineering system and design support. *Symbiosis of Human and Artifact,* 161-166.

Matsubara, Y., Nagamachi, M. and Jindo, T., 1994b, Kansei engineering as an artificial intelligent system, in *Human Factors in Organizational Design and Management-IV,* pp. 473-478.

Nagamachi, M., 1986, Image technology and its application. *Japanese Journal of Ergonomics,* 22(6), 316-324.

Nagamachi, M., 1988a, Image technology based on knowledge engineering and its application to design consultation, in *Proceedings of the 10th Congress of International Ergonomics Association,* pp. 72-74.

Nagamachi, M., 1988b, Image technology. *Journal of Electronics, Information and Communication Engineers,* 71(3), 245-247.

Nagamachi, M., 1988c, Assessment of human feeling and image technology. *Journal of Japanese Society of Mechanical Engineers,* 91(838), 955-961.

Nagamachi, M., 1989a, Kansei engineering approach to automobile. *Journal of the Society of Automotive Engineers of Japan,* 43(1), 94-100.

Nagamachi, M., 1989b, *Kansei Engineering.* (Kaibundo, Tokyo).

Nagamachi, M., 1990a, Kansei engineering and development of new product. *Journal of Japanese Industrial Engineering Association,* 41(4B), 66-71.

Nagamachi, M., 1990b, Procedure of new product development in terms of Kansei Engineering. *Engineers,* 504-1-7.

Nagamachi, M., 1991a, An image technology expert system and its application to design consultation. *International Journal of Human-Computer Interaction,* 3(3), 267-279.

Nagamachi, M., (Ed.), 1991b, *Comfort Science.* (Kaibundo, Tokyo).

Nagamachi, M., 1993a, *A Study of Kansei Merchandise.* (Kaibundo, Tokyo).

Nagamachi, M., 1993b, Kansei Engineering and its intelligent implication in product development, in *Proceedings of the Second China-Japan International Symposium on Industrial Management,* pp. 539-544.

Nagamachi, M., 1993, Kansei engineering study on word sound. *Journal of the Acoustic Society of Japan,* 49(9), 638-644.

Nagamachi, M., 1994a, Kansei engineering: A consumer-oriented technology, in *Human Factors in Organizational Design and Management-IV,* pp. 467-472.

Nagamachi, M., 1994b, Kansei engineering: An ergonomic technology for a product development, in *Proceedings of 12th Triennial Congress of the International Ergonomics Association,* pp. 120-122.

Nagamachi, M., 1994c, Kansei engineering: An intelligent system in product development, in *Japan-USA Symposium on Flexible Automation,* pp. 537-540.

Nagamachi, M., 1994d, Implication of kansei engineering and its application to automotive design consultation, in *Proceedings of the 3rd Pan-Pacific Conference on Occupational Ergonomics,* pp. 171-175.

Nagamachi, M., 1995a, Kansei Engineering: A new ergonomic consumer-oriented technology for product development. *International Journal of Industrial Ergonomics,* 15(1), 3-12.

Nagamachi, M., 1995b, *Introduction of Kansei Engineering.* (Japan Standard Association, Tokyo).

Nagamachi, M., Senuma, I. and Iwashige, R., 1974, A study of image technology. *Japanese Journal of Ergonomics,* 10(2), 121-130.

Nagamachi, M., Senuma, I. and Iwashige, R., 1977, An analysis of feeling on room atmosphere. *Japanese Journal of Ergonomics,* 13(1), 7-14.

Nagamachi, M., Ito, K., Fukuba, Y., Tsuji, T. and Kawamoto, H., 1986, Development of interior consultation system in terms of knowledge engineering. *Japanese Journal of Ergonomics,* 22(1), 1-7.

Nagamachi, M., Ito, K., Fukuba, Y., Tsuji, T., Tabuchi, Y. and Irieda, T., 1985, Image technological study on room lighting. *Japanese Journal of Ergonomics,* 21(5), 265-270.

Nagamachi, M., Ito, K., Tsuji, T. and Chino, T., 1988, A study of costume design consultation system based on knowledge engineering. *Japanese Journal of Ergonomics,* 24(5), 281-290.

Nagamachi, M., Kaneda, Y. and Matsushima, K., Automobile and kansei engineering. *Automobile Research,* 11(1), 2-6.

Osgood, C. E., Suci, G. J. and Tannenbaum, P. H., 1957, *The Measurement of Meaning* (University of Illinois Press, Urbana).

Tsuchiya, T., Matsubara, Y. and Nagamachi, M., 1995, A study of kansei rule generation using genetic algorithm. *Symbiosis of Human and Artifact,* 191-196.

Yamamoto, K. 1986, *Kansei Engineering — The Art of Automotive Development at Mazda.* Special Lecture at The University of Michigan, Ann Arbor.

103

Design for Human Assembly (DHA)

Martin G. Helander
Linköping Institute of Technology

Bengt Å. Willén
Linköping Institute of Technology

103.1 Introduction

Ergonomics professionals working for manufacturing companies have in the past specialized in two areas: design of industrial workstations and design of products to improve functionality and usability. In this chapter we are proposing a new field of activity: the study of effects of product design on the type of jobs created in assembly of the product. The basic concept is that through product design, jobs are created in manufacturing. It is then important to design products so they are easy to assemble. In most manufacturing companies there is a choice of using manual labor and automated processes. One must then distribute the manufacturing tasks in an optimal way. This task allocation must be productive for the company, and it must create satisfying jobs for the employees.

Design for Automation (DFA) became of great interest in the early 1980s when robots were first being used for assembly (Boothroyd and Dewhurst, 1982). There was a realization that robots can handle only fairly simple assembly tasks; those that do not require great precision in movement and are also easy to program. In order to enhance the utility of robots, it became necessary to redesign products so they were

0-8493-2641-9/99/$0.00+$.50

easy to assemble. More complicated assembly tasks — tasks that were left over — had to be done manually, which often created monotonous or machine-paced tasks for the employees.

It turned out that DFA guidelines, which were then intended to simplify automatic assembly, also simplified manual assembly. In one study a product was redesigned for automation, but it became so easy to assemble the product manually, that automation did not pay off (Helander and Domas, 1986). It may seem ironic that only the introduction of robots and consideration of their requirements in manufacturing made us reflect on human requirements. One main structural problem in manufacturing is that the primary responsibility of engineers and designers is technology, and there is not an equally detailed analysis of human labor. In the future, however, as the critical trade-offs between machine labor and manual labor are recognized, engineers must take a broader view and consider task allocation between humans and machines — who shall do what.

A narrow focus on automation and technology does not bring about either productivity or job satisfaction. Unfortunately, automation is what engineers take an interest in — modeling and planning of human work is not a priority. To optimize systems performance, engineers need to understand and consider all subsystems including human work capabilities and principles for task allocation.

This chapter will provide an overview of guidelines that may be used in product design to simplify automated assembly as well as manual assembly. The information is presented under three headings: (1) Boothroyd's method for redesign of products, (2) use of predetermined time systems to diagnose product design, and (3) human factors design principles applied to product design.

103.2 Product Design for Automation Using Boothroyd's Principles

The design principles formulated by Boothroyd and Dewhurst (Boothroyd and Dewhurst, 1982, Boothroyd, Dewhurst, and Knight, 1994) have been extremely influential in industry. Several companies, including Hitachi, Black & Decker, General Electric, General Motors, IBM, and Xerox have used these principles to develop corporate guidelines (Gager, 1986; Holbrook and Sackett, 1988).

In Boothroyd's technique an existing product is disassembled (Boothroyd, Dewhurst, and Knight, 1994). The necessity of each part is then analyzed. First one must decide if a part is necessary for assembly or disassembly. A part may be eliminated or integrated with another part if:

1. There is no relative motion between the two parts.
2. The materials of the two mating parts do not have to be different.

For each part the assembly time is measured. Boothroyd made the assumption that an "ideal" assembly time for a part is three seconds. This is reasonable for a part that is easy to handle and insert. A measure of the manual assembly design efficiency E_m is then obtained from the following equation:

$$E_m = \frac{3N_m}{T_m}, \qquad (1)$$

where N_m = minimum number of parts, and T_m = total assembly time.

If $E_m < 1$ then the design is inefficient, and if $E_m > 1$ the design is efficient. An example of this methodology is given in Figures 103.1a and 103.1b, and Tables 103.1a and 103.1b.

The value of E_m is not always conclusive. Complex electromechanical products that require extensive wiring tend to have low design efficiencies, even when well designed. On the other hand, simple products with few parts can have a high design efficiency, due to their simplicity rather than their DFA virtues. In their handbook, Boothroyd et al. (1994) provided many examples of successful redesigns where productivity gains of 200 to 300% were obtained.

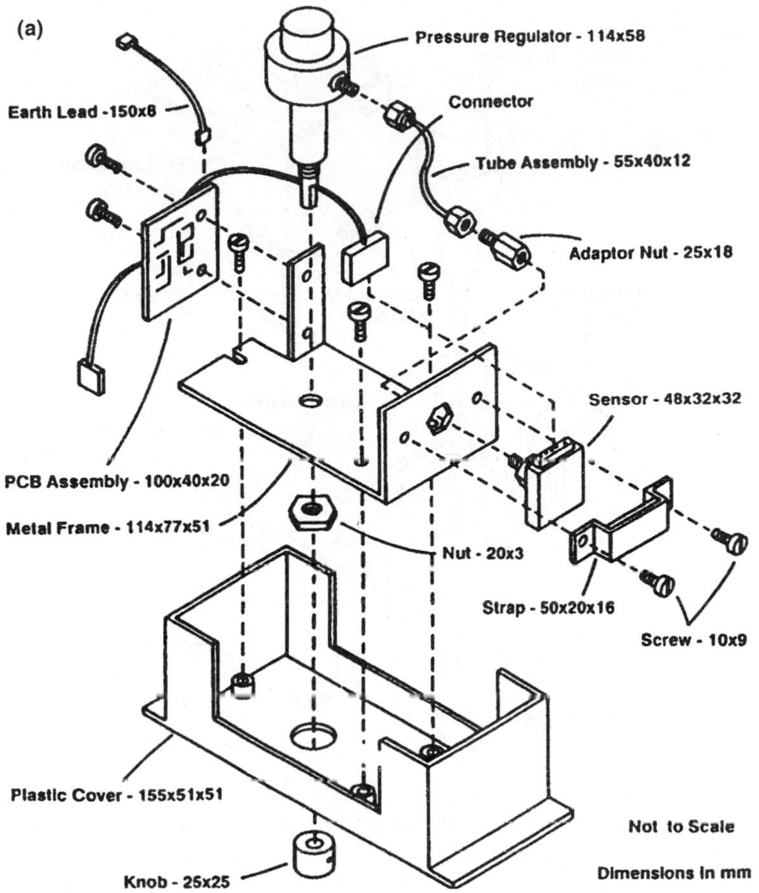

FIGURE 103.1a Example of product redesign for ease of assembly. The old design of the controller had 19 parts. The calculated assembly times are given in Table 103.1a. (From Boothroyd, G., Dewhurst, P. and Knight, W. 1994. *Product Design for Manufacture and Assembly*, Marcel Dekker, New York, NY. With permission.)

103.3 MTM Analysis of an Assembly Process

Boothroyd's technique is useful for redesign of existing products, but is difficult to use for conceptual design of new products. Predetermined time and motion studies (PTMS) can be used to suggest what types of parts should not be used and how parts as well as the assembly process can be redesigned to reduce assembly time. As a basis for our analysis below we use Motion Time Measurement (MTM) (e.g., Konz, 1990). In MTM and other PTMS methods the time for manual assembly is from tables which list the manual assembly for a variety of different cases. These times are based on extensive studies of operators performing assembly in the factory environment.

Assembly time depends not only on how the parts of a product are designed, but also on how they are stored and presented to the operator, and how they are moved to the point of assembly. In MTM, an assembly is broken down into several elemental tasks including: reach, grasp (pickup and select), move, position part, and insert. MTM specifies the amount of time it takes for a trained worker to do each of these elemental tasks. However, the assembly time depends on how the product is designed. Table 103.2 illustrates time savings for a "best design case" as compared to a less efficient design. For example, reaching to a fixed location is the "best case" and takes about 30% less time than reaching to

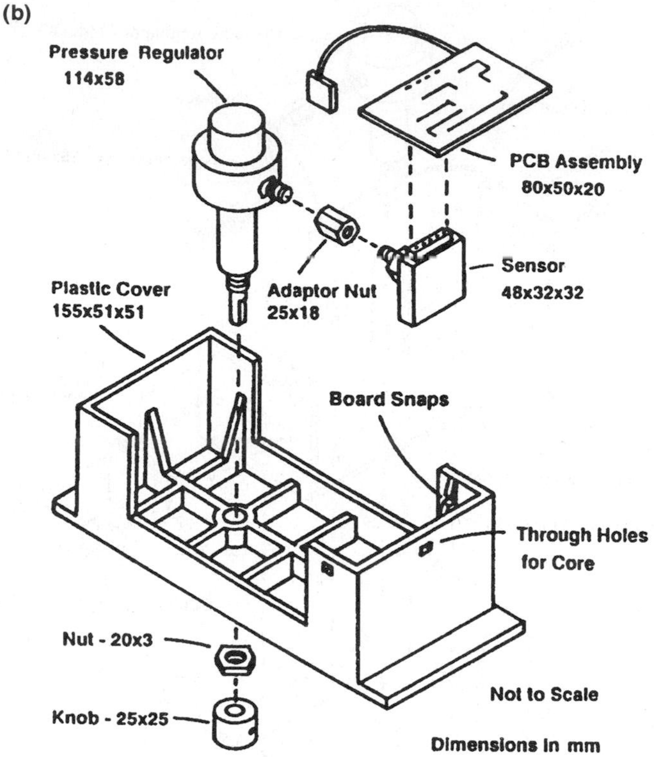

FIGURE 103.1b The new design of the controller had only 7 parts. The calculated assembly times are given in Table 103.1b. (From Boothroyd, G., Dewhurst, P. and Knight, W. 1994. *Product Design for Manufacture and Assembly,* Marcel Dekker, New York, NY. With permission.)

a variable location or to small and jumbled parts. "Grasping" of easily picked up parts is 75% faster than parts that are not easily grasped. Hence, the design engineer should design parts that are easily reached and easily grasped. This can be done, for example, by using large parts rather than small parts.

The parts should be presented at a fixed location. This can be accomplished for example, by using part feeders. Much research has been performed to develop part feeders for robots and automation (Boothroyd, 1982). They can also be used for manual assembly. A cost-benefit calculation can easily determine whether a parts feeder for manual assembly is cost-efficient. One can simply calculate the savings in time and money for the assembly and compare to the cost for a parts feeder. The information in Table 103.2 can be used for this calculation.

Following the "pick-up" the part has to be transported and positioned for the final insertion step. Table 103.2 illustrates that moving a part against a stop (Case A) requires about 15% less time, than when a part is moved to a location without a stop (Case B). In the latter case the absence of tactile feedback requires greater manual control. Ironically, most products are assembled as in Case B. One objective of good design must therefore be to incorporate stops which provide tactile feedback (Furtado, 1990).

In MTM the parts insertion or mating is described using a position element composed of three complex motions; align, orient, and engage. "Align" is the time required to line up the insertion axes of the two parts, like a pen into a cap. "Orient" describes the basic motions required to geometrically match the cross sections of the two parts, like a key into a lock. "Engage" consists of motions required to insert a part. Alignment is effected by asymmetry of the part, and Table 103.2 exemplifies a 20% time savings for symmetrical parts.

TABLE 103.1a Worksheet for Design for Manual Assembly; the Old Design Had 19 Parts and Required 227.4 s for Assembly

Part. Sub. or PCB assembly or Operation No.	No. of items RP	Handling time per item (s) TH	Insertion time per item (s) TI	Total oper'n cost — cents TA×CP TA	Total oper'n time RP×(TH+TI) CA	Minimal No. parts NM	Operator rate OP: 0.83 cents/s
1. Pressure regulator	1	1.95	1.5	3.5	2.9	1	place in fixture
2. Metal frame	1	1.95	5.5	7.4	6.2	1	add
3. Nut	1	1.13	8.0	9.1	7.6	0	add & screw fasten
4. Reorientation	1	—	9.0	9.0	7.5	—	reorient & adjust
5. Sensor	1	1.95	6.5	8.4	7.0	1	add
6. Strap	1	1.80	6.5	8.3	6.9	0	add & hold down
7. Screw	2	1.80	8.0	19.6	16.3	0	add & screw fasten
8. Apply tape	1	—	12.0	12.0	10.0	—	special operation
9. Adaptor nut	1	1.50	10.5	12.0	10.0	0	add & screw fasten
10. Tube assembly	1	3.00	4.0	7.0	5.8	0	add & screw fasten
11. Screw fastening	1	—	5.0	5.0	4.2	—	standard operation
12. PCB assembly	1	5.60	6.5	12.1	10.0	1	add & hold down
13. Screw	2	1.80	8.0	19.6	16.3	0	add & screw fasten
14. Connector	1	1.95	5.0	6.9	5.8	0	add & snap fit
15. Earth lead	1	5.60	5.0	10.6	8.8	0	add & snap fit
16. Reorientation	1	—	9.0	9.0	7.5	—	reorient & adjust
17. Knob assembly	1	1.95	6.5	8.4	7.0	1	add & screw fasten
18. Screw fastening	1	—	5.0	5.0	4.2	—	standard operation
19. Plastic cover	1	1.95	6.5	8.4	7.0	0	add & hold down
20. Reorientation	1	—	9.0	9.0	7.5	—	reorient & adjust
21. Screw fastening	3	1.80	10.5	36.9	h30.8	0	add & screw fasten
				T_m:	C_m:	N_m:	Design efficiency:
				227.4s	\$1.90	5	$\frac{3N_m}{T_m} = 0.07$

Adapted from Boothroyd, G., Dewhurst, P. and Knight, W. 1994. *Product Design for Manufacture and Assembly*, Marcel Dekker, New York, NY.

TABLE 103.1b The Improved Designed Had Only 7 Parts with an Estimated Assembly Time of 83.8 s

Part. Sub. or PCB assembly or Operation No.	No. of items RP	Handling time per item (s) TH	Insertion time per item (s) TI	Total oper'n cost — cents TA×OP TA	Total oper'n time RP×(TH+TI) CA	Minimal No. parts NM	Operator rate OP: 0.83 cents/s
1 Pressure regulator	1	1.95	1.5	3.5	2.9	1	place in fixture
2 Plastic cover	1	1.95	5.5	7.4	6.2	1	add & hold down
3 Nut	1	1.13	8.0	9.1	7.6	0	add & screw fasten
4 Knob assembly	1	1.95	6.5	8.4	7.0	1	add & screw fasten
5 Screw fastening	1	—	5.0	5.0	4.2	—	standard operation
6 Reorientation	1	—	9.0	9.0	7.5	—	reorient & adjust
7 Apply tape	1	—	12.0	12.0	10.0	—	special operation
8 Adaptor nut	1	1.50	10.5	12.0	10.0	0	add & screw fasten
9 Sensor	1	1.95	8.0	9.9	8.3	1	add & screw fasten
10 PCB assembly	1	5.60	2.0	7.6	6.3	1	add & snap fit
				T_m:	C_m:	N_m:	Design efficiency:
				83.8s	$0.70	5	$\frac{3N_m}{T_m} = 0.18$

Adapted from Boothroyd, G., Dewhurst, P. and Knight, W. 1994. *Product Design for Manufacture and Assembly*, Marcel Dekker, New York, NY.

TABLE 103.2 Examples of Time Savings Obtained with the "Best Case" as Compared to Less Efficient Alternatives. The Cases and Codes Used in MTM for Classification are also Mentioned

	Best Case	Comparison	Time Increase from Best Case to Comparison
Reach	To fixed location (case A)	To variable location (case B)	30%
		To small or jumbled objects (case C)	40%
Grasp			
1. Pickup	Easily grasped (case 1A)	Object on flat surface (case 1B)	75%
		Small object, 1/2 inc. (case 1C2)	400%
2. Select (For jumbled objects only)	Large jumbled objects (case 4A)	Object smaller than 1x1x1 in. (case 4B, 4C)	50%
Move	Against a stop or to other hand (case A)	To exact location without a stop or physical barrier (case C)	15%
Position Part			
1. Symmetrical part	Symmetrical, e.g., round peg in round hole (code S)	Semi-symmetrical, 45 degree turn typical (code SS)	20%
		Non-symmetrical, 75 degree turn typical (code NS)	30%
2. Depth of insertion	No depth	4" insertion	100%
3. Pressure to fit	Gravity, no pressure (code 1)	Light Pressure (code 2)	210%
		Heavy Pressure (code 3)	500%
Disengage (two parts)			
1. Class of fit	Loose	Tight	500%
2. Ease of handling	Easy	Difficult	40%

103.4 Human Factors Principles in Design for Assembly

PTMS does not consider the time required for human information processing and decision making. Yet, even in simple, repetitive manual tasks operators make decisions, and this takes time (Fish, 1993). Since the overall goal in DFA is to reduce the time for the assembly, one has to consider both mental and manual aspects of the assembly task. For the human operator, there are three sequential time components in information processing and task execution: time required for (1) perception, (2) decision making, and (3) manipulation. Each of these can be minimized by thoughtful product design, and for each of the three time components there are different design methods to consider in product design. Table 103.3 lists the three time components in the "Why" column, their implications in terms of behavioral principles in the "What" column, and the types of measures that may be implemented in the "How" column. Table 103.3 hence establishes 3 levels of abstraction going from the overall purpose to the implementation of human factors principles applicable to Design for Human Assembly (DHA).

103.5 Minimize Perception Time

Visible Parts

Visibility is important in assembly. Everything that is used in a manufacturing task should be fully visible. Hidden or invisible parts or tools are difficult to refer to; they cannot be pointed at. They become abstract and difficult to think of. The design of the product should be such that the assembly can be performed with full visibility of parts and tools, see Figure 103.2. Hidden or invisible parts are also difficult to operate on and they are sometimes forgotten.

TABLE 103.3 Human Factors Principles in DHA

Why	What	How
Minimize perception time	Visible parts Visual discrimination Tactile discrimination	Nothing hidden Size, color Texture, size
Minimize decision time	Ease the formation of a mental mode Reduce choice reaction time Spatial compatibility Visual, auditory and tactile feedback	Visible parts Minimize number of parts Collocation of associated items Assembly looks different, auditory and tactile snaps
Minimize manipulation time	Ease of manipulation Physical affordances and constraints Design for transfer of training	Fixture to hold parts, parts that are easy to grip and don't tangle, fasteners that are easy to use Self-locating parts, increase tolerances New product similar to old

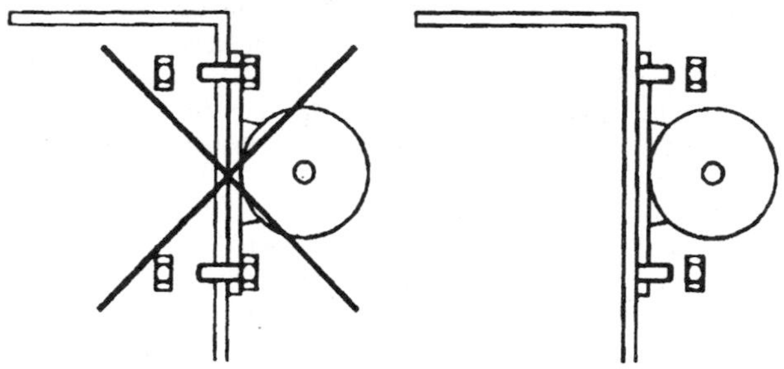

FIGURE 103.2 Avoid designs that lead to hidden assembly. From Willkrans, R. 1995. *Design for Easy Assembly,* IVF, publ. 95825, Gothenburg, Sweden. With permission.)

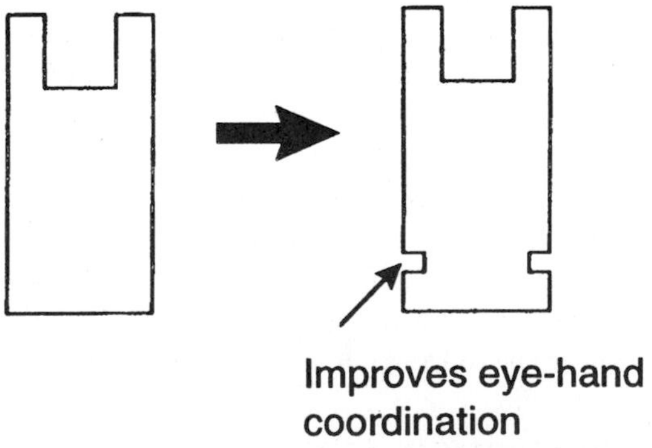
Improves eye-hand coordination

FIGURE 103.3 Exaggerated asymmetry may enhance stimulus–response compatibility.

Visual Discrimination

Parts may be perceptually organized by using different shapes, sizes, and colors. For example, color coding of parts may be used to form families of parts, that is, parts which belong together in a subassembly. Color coding will also enhance stimulus–response compatibility in assembly and results in reduced reaction time and better eye–hand coordination.

If asymmetric parts are used (see Figure 103.3), it may be advantageous to exaggerate the asymmetry to improve visual cues (Chhabra and Ahluwalia, 1990).

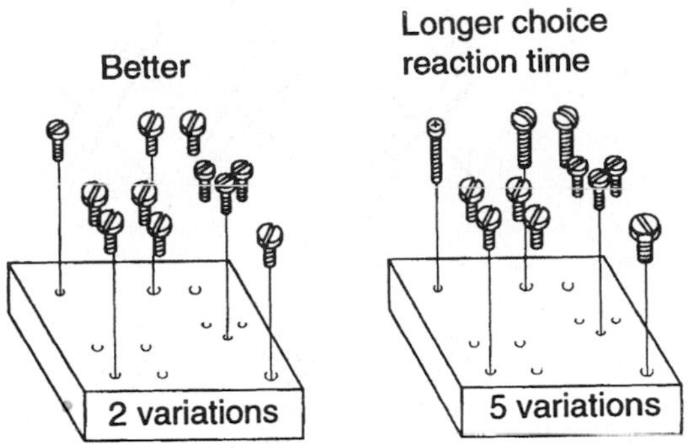

FIGURE 103.4 Minimize the various types and sizes of screws.

Tactile Discrimination

Parts shall be designed so that they can be discriminated by touch alone. This can be accomplished by using parts of different size, texture, and so forth.

103.6 Minimize Decision Time

Ease of Formation of a Mental Model

Workers develop mental models of the task they are performing; they think of an assembly in a certain way. Mental models are important since they help the worker to organize the execution of a task. However, things that are difficult to see become abstract and it is then difficult to form a mental model. Task visibility is important and helps the creation of a mental model (Prabhu, Helander, and Shalin, 1995). The concept of mental models has been used extensively in human–computer interaction. Software programmers have a different mental model than users of the same software. Therefore, programmers fail to consider the needs of the user. Similarly, in manufacturing the product designer may fail to consider the operators' mental model. In fact, there are many different tasks in manufacturing with different mental models (Baggett and Ehrenfeucht, 1991). A person assembling a product would have a different mental model than a person responsible for quality control of the same product. They look for different things and they do different things, and the priorities are different. This observation is contrary to the notion that assembly operators should exercise their own quality control; it may be difficult to change mindset (Shalin, Prabhu and Helander, 1996).

Reduce Choice Reaction Time

Minimize the Number of Components and Parts

An example of how to reduce the number of components and parts is to minimize the various types and sizes of screws. Fewer parts reduces the number of parts bins, which in turn reduces the operators' choice reaction time (see Figure 103.4). It will also save space. In addition, fewer parts will reduce the number of hand tools, which in turn decreases choice reaction time and space requirements.

Use Symmetrical Parts Because They Are Easy to Orient

The use of symmetrical parts reduces information processing time, since the operator does not have to decide whether to turn the part around. It also reduces manual handling time (see Figure 103.5).

Integrate or Combine Parts

Integrate or combine parts, since they simplify the operators' mental model, reduce choice reaction time, and often take less time to assemble. In some cases an entire subassembly can be replaced by a single

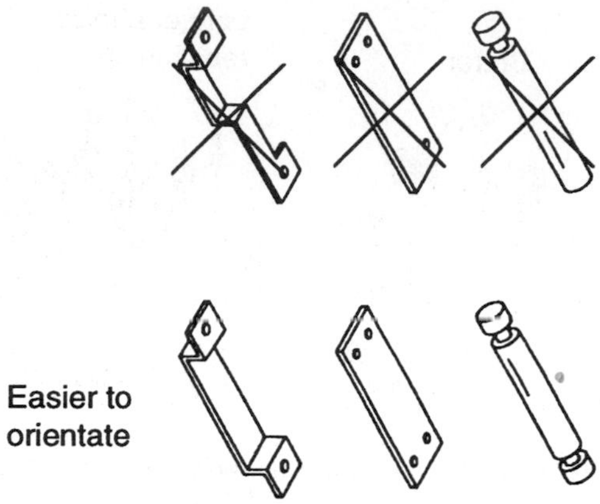

FIGURE 103.5 Use parts that are easy to orient, such as symmetrical parts.

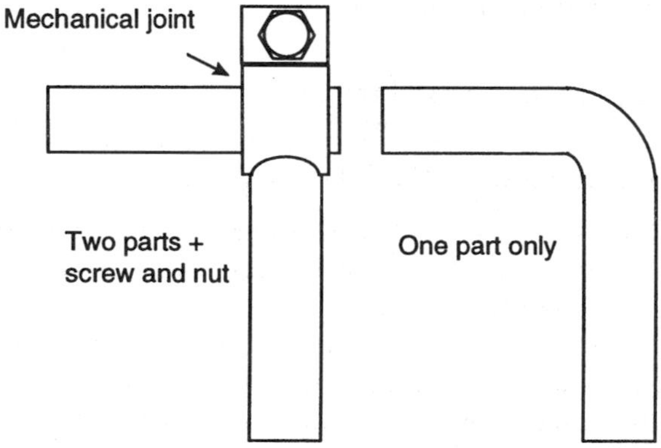

FIGURE 103.6 Integrate or combine parts.

part (compare modular design in electronics). Integrated parts may be difficult to handle manually, but overall they are efficient, since they reduce the number of operations (Figure 103.6).

Holbrook and Sackett (1988) noted that it is difficult to combine parts if:

- Parts move relative to each other
- Parts are required to be of different materials
- Parts must be separate for maintenance and service reasons
- Parts are necessary to enable the assembly of remaining parts

Combined parts can often be fabricated using plastic injection moulding. Another advantage with plastic parts is that they can easily be provided with chamfers, notches, and guides which are helpful in assembly. Metal parts can also be moulded or mounted into plastic parts. The elastic property of thermoplastics, (e.g., nylon) can be utilized to form snap joints, integral springs, and integral hinges. Thermoplastics can also be utilized to straighten other parts and to eliminate clearances.

Spatial Compatibility

Spatial compatibility has to do with the spatial layout of a workstation. Part bins can be located in sequential order so the operator can pick parts from left to right in the same order as used in the assembly.

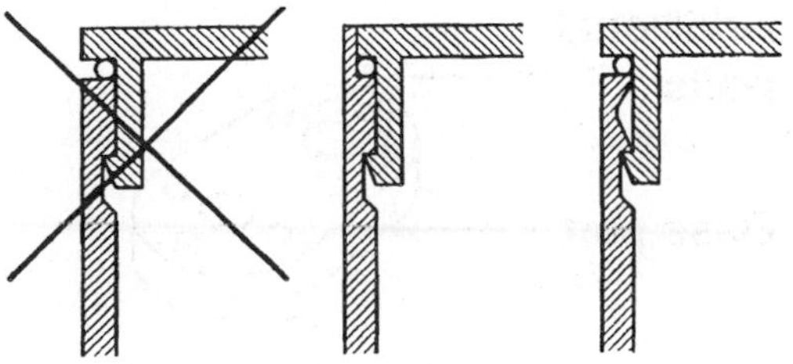

FIGURE 103.7 Integral fasteners and clips should be designed so that the operator can see, feel, or hear that the assembly has succeeded. (From Willkrans, R. 1995. *Design for Easy Assembly*, IVF, publ. 95825, Gothenburg, Sweden. With permission.)

Part bins can also be arranged so that their location mimics the product design. This could, for example, be used with components that are inserted in an electronic board. The best arrangement depends on the product design and the number of parts used. Obviously, product design should consider spatial com patibility (Helander and Waris, 1993).

One should also consider the locations of hand tools and controls. Typically, items that belong together in task execution should be physically close.

Visual, Auditory and Tactile Feedback

When a task has been completed, there should be visual feedback — in other words, something should look different. Sometimes in automobile assembly a piece of tape is put on top of an assembly to indicate it is finished.

An example of tactile feedback is the use of physical stop barriers. When a part is moved against a stop there is a sensation in the fingers — tactile feedback which indicates that the task was completed. Auditory feedback is helpful not only with parts but also for hand tools and controls and for hand tools operating on parts. In this case a sound is produced that indicates task completion. For example, the clicking sound of a switch or the ricketing noise of a hydraulic screwdriver indicates that the task has been completed (see also Figure 103.7).

103.7 Minimize Manipulation Time

In this section we provide examples of product design features that simplify assembly by ease of manipulation. Many of them are used for automation and have been published in DFA guidelines for DFA. They, however, apply equally well to manual assembly (Helander and Nagamachi, 1992).

Ease of Manipulation

Use a base part as product foundation and fixture. Design the product with a base part as foundation and fixture for other parts. It should be possible to assemble the other parts from one direction, preferably from above (see Figure 103.8). It is also advantageous to use fasteners which are inserted from one direction, either from the front or from above. The base part should also serve as a fixture. If this arrangement is not feasible, pins can be used so the base part can be easily positioned on a fixture as in Figure 103.8. If this is not possible, a specially designed fixture is used. To make the product easy to transport, it should have a flat bottom and a simple shape.

Use Parts that Are Easy to Grip and Don't Tangle

Improve parts handling by using parts that are easy to grip (see Figure 103.9).

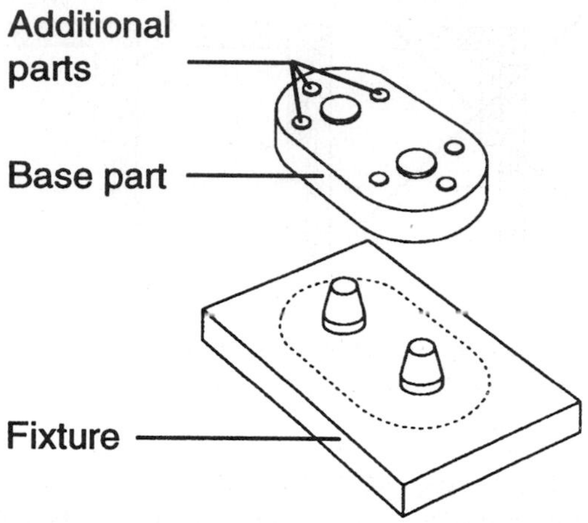

FIGURE 103.8 Provide a simple and reliable fixture for the base part. If possible, the base part should also serve as a fixture.

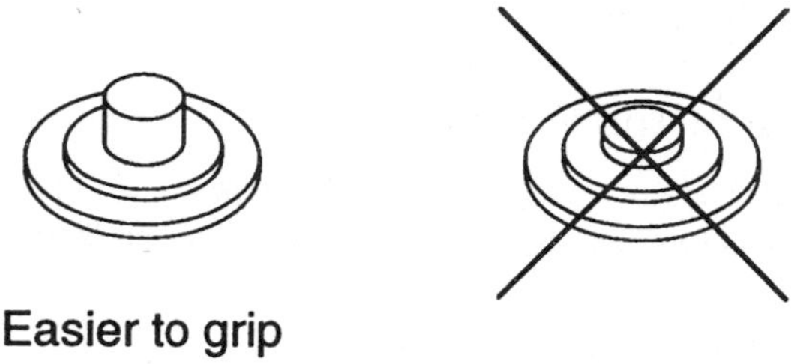

FIGURE 103.9 Improve parts handling by making parts easy to grip.

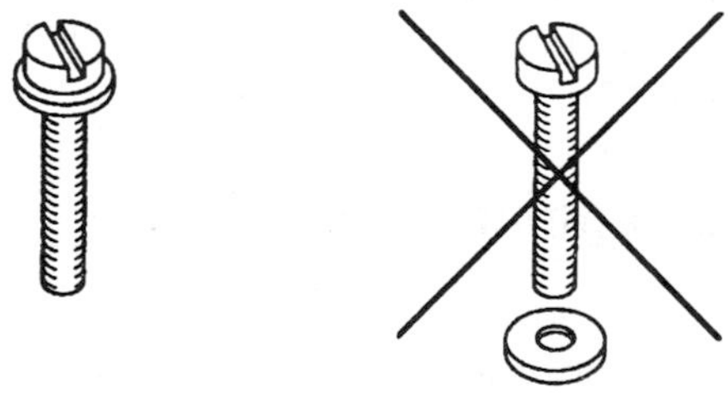

FIGURE 103.10 Do not use small parts that are difficult to handle, such as washers.

Don't Use Small Parts, such as Washers

Avoid using separate washers which increases manual handling time. This requirement, which is a necessity in robotics assembly, simplifies manual assembly as well (see Figure 103.10). The use of washers increases manual handling time. It may also make it necessary for the operator to use pinch grips, which may increase the risk for cumulative trauma disorder.

Design the Process to Enhance Grasping

Luczak (1993) presented several methods which simplify the grasping of a part to be assembled (see Figure 103.11).

This is a complementary approach. The parts are not redesigned to be easier to grasp, rather the process (of grasping) is redesigned. Process design is typically more abstract than product design and

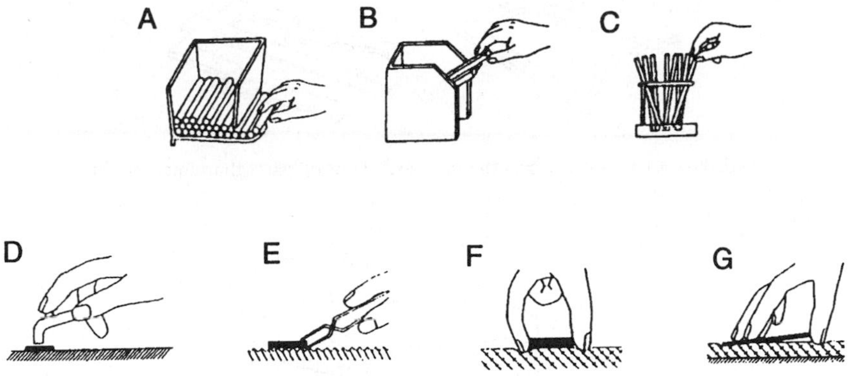

FIGURE 103.11 Gripping aids for an assembly workstation. (A) Self-feeding container. (B) Container with inclined opening. (C) A ring holder with smaller bottom diameter. (D) Use of vacuum gripper. (E) Tweezers or tongs used against a rippled table surface. (F and G) Gripping against a soft surface.

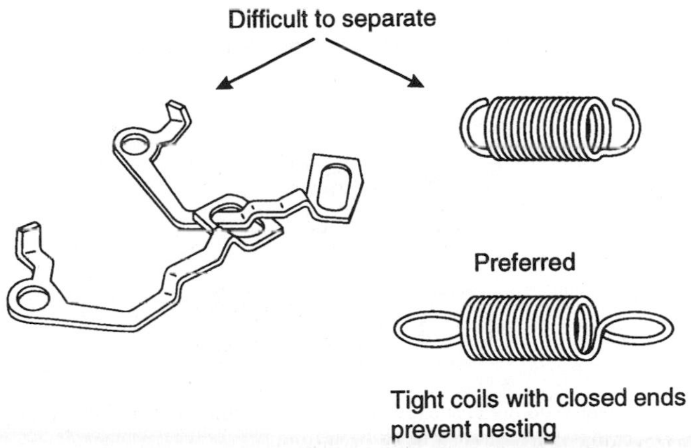

FIGURE 103.12 Avoid parts that nest or tangle.

therefore more difficult to think of and to suggest innovative solutions. In Figure 103.11 the process is redesigned for the purpose of improving the interactivity of products and processes so that manual assembly is made easier.

Avoid using flexible parts, such as wires, cables, and belts, because they are difficult to handle. Sometimes components can be plugged together in order to eliminate the use of connecting wires.

Avoid Parts that Nest or Tangle

Close open ends and make part dimensions large enough to prevent tangling. For example, use springs with closed ends rather than open ends (see Figure 103.12).

Consider the Physical Integrity of the Parts

Parts that are weak or easily bent are difficult to assemble (see Figure 103.13). These parts often cause extra work in quality control, visual inspection, and replacement. Grossmith (1992) noted that many microscope inspection tasks can be avoided if product designers choose materials that are less likely to chip and crack.

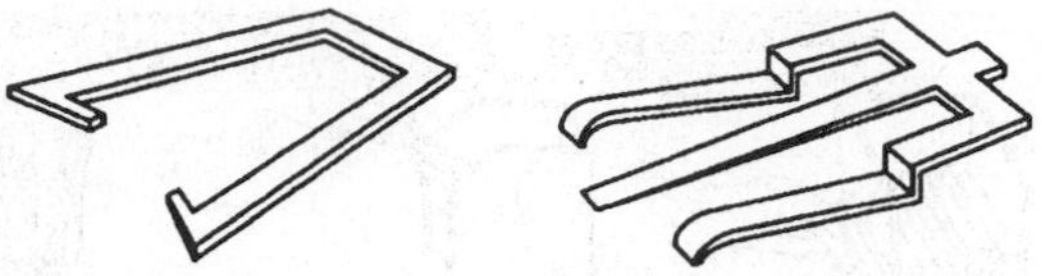

FIGURE 103.13 Avoid parts that are easily bent or parts that crack or chip.

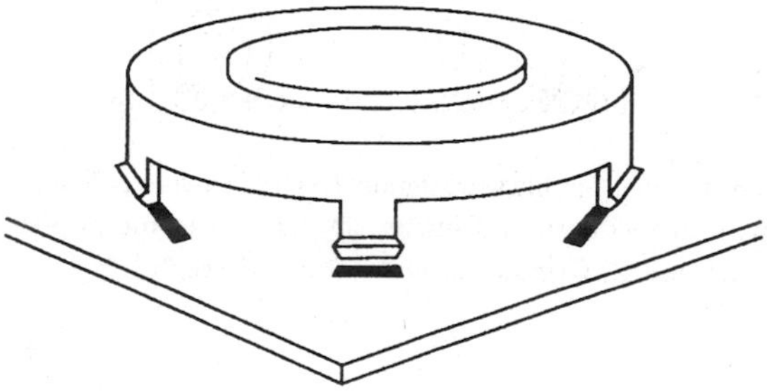

FIGURE 103.14 Use snap and insert assembly.

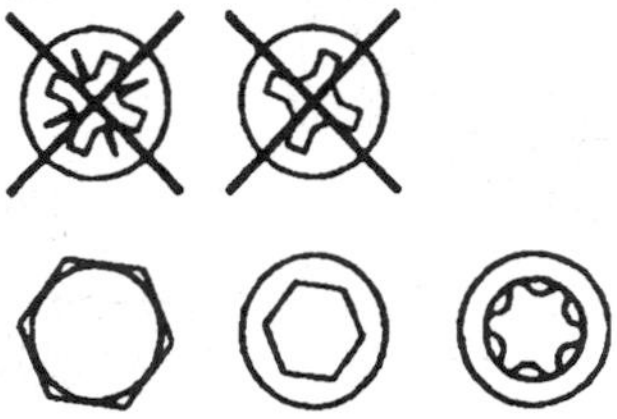

FIGURE 103.15 Use screws which do not create lifting forces on the tool. (From Willkrans, R. 1995. *Design for Easy Assembly*, IVF, publ. 95825, Gothenburg, Sweden. With permission.)

Use Fasteners that Are Easy to Assemble

If practical avoid fasteners by using snap-and-insert assembly. Design integral fasteners and clips into parts so that no screws are required as in Figure 103.14.

If screws are used, it is better to use screws that do not create lifting forces on the tool as in Figure 103.15. Thereby the tool will not so easily disengage from the screw. Musculo-skeletal discomfort is also reduced (Cederqvist and Lindberg, 1993).

Use Physical Constraints and Affordances to Simplify Assembly.

Use self-locating parts, such as parts with chamfers, notches, and guides for self-location to simplify assembly (see Figure 103.16). The use of chamfers, for example, reduces the amount of manual precision necessary to insert the part. This is obvious from Fitts's law, since a larger target area (chamfer) reduces the time for manual handling (Fitts and Posner, 1973).

Reduce Tolerances in Part Mating

Part mating becomes easier if tolerances are reduced. For example, Figure 103.17 illustrates how a slotted hole may be used to simplify positioning and relax accuracy requirements.

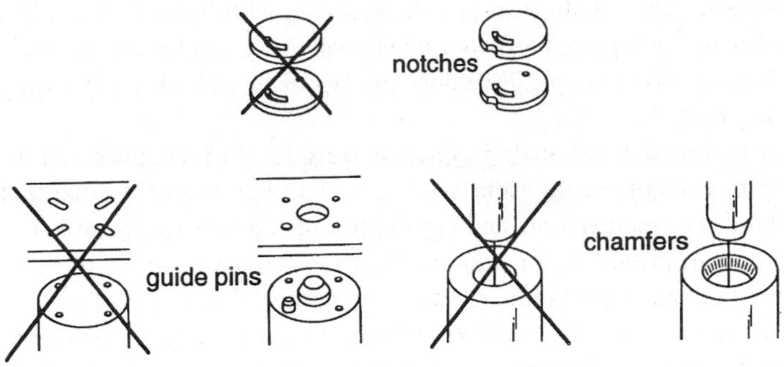

FIGURE 103.16 Facilitate assembly by using self-locating parts.

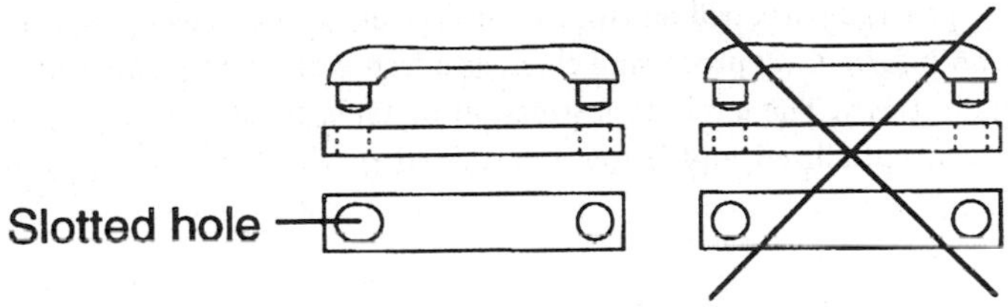

FIGURE 103.17 Reduce tolerances in part mating.

Transfer of Training

Transfer of training applies when a new product has only small modifications compared to the old product. A worker can then apply his previously acquired skills to the new product. Differences in product design and workstation layout, on the other hand, may create confusion, and assembly times can increase drastically. Product designers have a responsibility here, so that they try to make the assembly of new products similar to the assembly of previous products (Shalin, Prabhu and Helander, 1996).

103.8 Discussion

This paper has presented three different principles, which can be used for design for human assembly: Boothroyd's principles, PTMS, and DHA. Boothroyd's work, the principles for design for automation (DFA), also have implications for design for human assembly (DHA). We noted in the introduction that principles for human assembly were inspired by the introduction of robots, when it became obvious that product design can be changed to fit both the requirements of automation and human assembly. The present interest in concurrent engineering makes it necessary to consider many criteria in design, such as the trade-offs between automation and human labor. In small batch manufacturing, automation will be less important and much assembly will be performed by manual labor. Yet there are manufacturing processes where automation is necessary, and for other tasks human labor is essential. These are important arguments for design engineers to comprehend and utilize principles for DHA. Yet engineers may not take an interest. The education of engineers emphasizes hardware and software design of systems, which are considered high priorities. To optimize systems performance it will be necessary for engineers to include principles of human factors engineering.

One traditional engineering tool, which is informative for product design is the use of PTMS to model the time elements in assembly for alternative product design. We are not interested in the assessment of the performance of individual workers, which is what PTMS was created for, but rather as a modeling tool in engineering design. PTMS does not need an existing design for analysis, whereas DFA analyzes an existing design as a point of departure for a new design. With DFA principles one can make more concrete suggestions than with PTMS. This is to be expected with iterative design. PTMS broadens the scope compared to Boothroyd's principles, since PTMS will also consider workstation design parameters which affect the assembly. For example, how does the layout of part bins affect the time for reaching, grasping, and moving parts.

DHA offers a much more detailed analysis, since we consider the three phases in the human information processing chain: perception, decision making, and motor response (manipulation). It is then possible to understand the requirements and capabilities of the human operator for each of the three steps, and propose human principles for product design to reduce assembly time. Visibility and visual and tactile discrimination are important to enhance perception. Visibility is of course a requirement for perception, but it is also important for decision making, since it is easier to create a mental model of an assembly and support memory. Likewise, several of the other principles in Table 103.3 have dual use; a reduction in the number of parts will reduce the time for decision making as well as assembly time. Feedback of any kind is essential to decide if the task has been finished, but feedback must of course first be perceived. The use of large parts will enhance visibility and facilitate assembly, and so forth.

By using the human information processing chain as a frame of reference we can readily interpret the human requirements in DHA. The levels of abstraction in Table 103.3 can be used to derive principles, which can then be conceptualized and better interpreted in the light of our knowledge of human information processing.

Acknowledgment

Both authors assume equal responsibility for this text.

References

Baggett, P. and Ehrenfeucht, A. 1991. Building physical and mental models in assembly tasks, *International Journal of Industrial Ergonomics.* 7:217-228.

Boothroyd, G. 1982. *Design for Assembly Handbook*, Department of Mechanical Engineering, University of Massachusetts, Amherst, MA.

Boothroyd, G., Dewhurst, P. and Knight, W. 1994. *Product Design for Manufacture and Assembly*, Marcel Dekker, New York, NY.

Cederqvist, T. and Lindberg, M. 1993. Screwdrivers and their use from a Swedish construction industry perspective. *Applied Ergonomics.* 24(3):148-157.

Chhabra, S.L. and Ahluwalia, R.S. 1990. Rules and guidelines for ease of assembly. *Proceedings of the International Ergonomics Association Conference on Human Factors in Design for Manufacturability and Process Planning*, p. 93-99. Human Factors and Ergonomics Society, Santa Monica, CA.

Fish, L. 1993. *A Cognitive Model for Assembly.* Dissertation. State University of New York at Buffalo, Department of Industrial Engineering, Buffalo, NY.

Fitts, P.M. and Posner, M. 1973. *Human Performance*, Prentice-Hall, Englewood Cliffs, NJ.

Furtado, D. 1990. Principles of design for assembly. *Proceedings of the International Ergonomics Association Conference on Human Factors in Design for Manufacturability and Process Planning*, p. 147-152. Human Factors and Ergonomics Society, Santa Monica, CA.

Gager, R. 1986 Design for productivity saves millions, *Appliance Manufacturer*, Jan. 46-51.

Genaidy, A.M., Duggai, J.S. and Mital, A. 1990. A comparison of robot and human performances for simple assembly tasks. *International Journal of Industrial Ergonomics*, 5:73-81

Grossmith, E.J. 1992. Product design considerations for the reduction of ergonomically related manufacturing costs, in *Design for Manufacturability. A Systems Approach to Concurrent Engineering and Ergonomics*, Eds. M.G. Helander and M. Nagamachi, Taylor & Francis, London.

Helander, M.G. and Domas, K. 1986. Task allocation between humans and robots in manufacturing. *Material Flow*, 3:175-185.

Helander, M.G. and Nagamachi, M. 1992. *Design for Manufacturability. A Systems Approach to Concurrent Engineering and Ergonomics*, Taylor & Francis, London.

Helander, M.G. and Waris, J. 1993. Effect of spatial compatibility in manual assembly performance, in *Ergonomics of Manual Work*, Eds. W.S. Marras, W. Karwowski, J.L. Smith and L. Pacholski, p. 421-424, Taylor & Francis, London

Holbrook, A.E.H. and Sackett, P.J. 1988. Design for assembly guidelines for product design. *Assembly Automation*. 8:202-211.

Konz, S. 1990. Work Design: *Industrial Ergonomics*, Publishing Horizons, Worthington, OH.

Luczak, H. 1993, *Arbeitswissenschaft*, Springer-Verlag, Berlin.

Mital, A. 1991 Economics of flexible assembly automation: Influence of production and market factors, in *Economic Aspects of Advanced Production and Manufacturing Systems*, Eds. H.R. Parsei and A. Mital, Chapman & Hall, London.

Prabhu, G.V., Helander, M.G. and Shalin, V. 1995. Cognitive implications of product structure on manual assembly performance. *International Journal of Human Factors in Manufacturing*. 5:149-161.

Shalin, V., Prabhu, G.V. and Helander, M.G. 1996. A cognitive perspective on manual assembly. *Ergonomics*. 39:107-126.

104

Human Factors in Agile Manufacturing

Chris Forsythe
Sandia National Laboratories

Eric Grose
Sandia National Laboratories

104.1 Introduction

In *21st Century Manufacturing Enterprise Strategy*, Nagel and Dove[1] introduced "Agile Manufacturing" as a production paradigm that would replace mass production as the dominant industry paradigm around the globe. Since this original postulation, government and industry have championed agility, as agile manufacturing has come to be known, as the solution for reinvigorating U.S. industrial competitiveness, and substantial efforts have been devoted to developing the requisite technologies and experimenting with various implementations of agility concepts.[2-5] Although it may have different manifestations across various industries, agility is commonly characterized by each of the following:

- Cycle time reduction for new product development
- Flexible and continually evolving manufacturing base
- Focus on product customization
- Reliance on information-driven manufacturing[6]
- Virtual co-location achieved through communications technologies
- Opportunistic partnering to fill market demands

In the migration from existing product development and production processes, and accompanying business practices, the greatest changes will occur with the human interfaces that exist throughout an enterprise. These interfaces will often describe the points of interaction between people and technology, but equally, if not more often, the human interfaces will be between different people, within and across enterprises, and between people and engineering, production, and businesses practices. The following sections identify human interfaces important to the realization of agility and the application of human factors to these interfaces.

0-8493-2641-9/99/$0.00+$.50

104.2 Product Specification and Automated Design

Central to the concept of agility is the notion of "mass customization," or providing products that are finely tuned to the individualized requirements of a specific customer.[1,7] Mass customization, in a greatly simplified substantiation, is not a new concept. Pizza makers have long provided the opportunity for customers to select from alternative sizes and combinations of toppings. In contrast, agility goes well beyond providing a fixed set of alternatives from which the customer may choose, to custom fitting the product to the unique requirements of a given customer. This level of customization is demonstrated by National Bicycle and their capacity to rapidly produce a bicycle custom fit to the body dimensions of a specific customer.[8] For many products, such as bicycles, specification of a custom product may be a simple process of matching dimensions of the product to those required by the customer. For other products, the relationship between design variations and product interfaces or product performance are considerably more complex. For example, it may be necessary for a product to function as a component of an overall system, interfacing with other components for which chemical, thermal, and other considerations must be accommodated. Similarly, complex interactions may exist between certain design attributes of a given product (e.g., weight reduction to enhance mechanical speed of operation may reduce product robustness). Where product specification goes beyond allowing the customer to choose from a fixed set of alternatives or simple matching of one or more design attributes, such as physical dimensions, designers need an interface that meaningfully represents the range of product variations, and makes the process of specifying new products from the range of variations intuitive and readily accomplished.

The set of design attributes that are varied to meet different customer requirements may be thought of as design parameters.[5,7] The design parameters that are varied for a given product define a design parameter space within which alternative designs may be represented as points defined by their values on each of the design parameters. Thus, it may be asserted that the process of specifying a given design is really one of selecting a point (design) from within the parameter space. The first step in developing an interface by which a given product (or point design) may be specified is to define the parameter space for that product. For all but the simplest products, definition of a parameter space will require preliminary research and development to determine the range of anticipated customer applications, and identify the relationships between design parameters and performance variables, and the bounds placed on design parameters by the limitations of manufacturing processes.[7] This preliminary work may also establish relevant cost and schedule relationships.[5,7] Once defined, the parameter space provides the basis for the product specification interface.

A decision tree offers a simple interface by which designers work their way through a series of decisions, until the complete design specification has been attained. As certain decisions constrain the array of available design options, these constraints may be represented by the number or range of options available at specific points in the tree. A significant drawback of the decision tree concept is that it offers little to aid the recognition and appreciation of design trade-offs. Tweedie et al.[9] have demonstrated the "Influence Explorer," an interface that very effectively depicts design trade-offs by maintaining an ongoing graphical representation of design alternatives available, following each design decision. As design trade-offs are encountered, partially or nonoverlapping ranges of available design alternatives are shown. Furthermore, a graphical interface such as the Influence Explorer may be combined with knowledge elicitation techniques (e.g., ELK[10]) to assure the proper breadth and depth of information is obtained during product specification.

Once requirements are known, values must be determined for design attributes to define a concept design satisfying customer requirements. For many simple, well-understood products, the process of generating design attributes from customer requirements may be automated. Taking as an example a simple mounting plate, if interface constraints and design attributes such as weight, physical dimensions, and center of gravity are known, and environmental factors such as maximum shock, vibration, and acceleration are specified, an automated routine may be developed whereby design attributes of the mounting plate such as size and thickness, and number, size and position of connectors are automatically generated. A spreadsheet offers an effective interface in that the designer need only enter necessary design attributes, and based on these entries, design dimensions may be automatically generated.[11] Such routines may be interfaced with Computer-Aided Design (CAD) software to create electronic representations of the part, from which 3-D models may be easily derived.[5,11]

While automated design routines may greatly streamline the design process, extensive automation of design is rarely practical or desirable.[12] In selecting facets of design for automation, design tasks that are highly detailed and time consuming, yet mundane, are prime candidates for automation.[5,13,14] Within an agile enterprise, it is preferred that human knowledge and skills be applied to challenging aspects of product development and not wasted on routine, monotonous tasks that often require considerable skill, due to their dependence on proficiency with Computer Aided Design (CAD) software.

104.3 Product Data Management

An agile enterprise must have mechanisms in place to capture corporate knowledge and then, with each new product development, deploy that corporate knowledge to its maximum benefit. This reliance on corporate knowledge makes it essential that there be an effective interface for locating and retrieving relevant product data. Forsythe and Ashby have demonstrated an interface whereby product archives may be searched on the basis of design attributes.[15] The designer may enter specific values for attributes such as voltage and use relational operators (e.g., $>$, $<$, $\leq$) for other attributes (e.g., minimum operating temperature, maximum sustained vibration). Given the desired combination of design attributes, the product data manager (PDM) searches through archived design files to return all designs with the specified design attributes. Where there are no known designs that meet the search criteria, using engineering judgment, the designer may relax one or more attributes to find those designs that approximate the desired characteristics. Similarly, where there is a failure to comply with precise physical dimensions, copies of design files may be obtained and appropriate modifications made. Where there is a failure to comply with performance attributes, such as environmental requirements, analysis and testing files associated with archived designs may be located and often, information contained therein may provide the necessary data to determine how to modify the previous design to meet current requirements. For example, the results of thermal stress analysis may be reviewed to identify weak points in the mechanical assembly that may be redesigned to provide the necessary level of robustness.

Along with design files, and associated analysis and testing data, the PDM also provides associative links to manufacturing data, machine code, and robot assembly code.[15] Manufacturing data may be used to evaluate design tolerances, and machine code and robot assembly code may be reused where no or minimal design changes have been made. Additionally, the PDM allows documentation to be associatively linked to design files, preserving a design history of each new product development effort. Although there exists nearly limitless potential to manage product data as it is generated and reuse that product data during subsequent development, significant trade-offs must be made. In particular, the reusability of product data is proportional to the extent to which meta-data or data attributes are specified. To be maximally reusable, it is necessary to provide rich descriptions of product data using numerous attributes. However, this introduces considerable overhead to the process of entering product data into the PDM, automated capture of attributes considered. Consequently, in designing the PDM interface, it is essential to weigh the value of reusing product data against the overhead incurred in entering product data.

The design of the PDM interface must also accommodate the range of disciplines that must employ the system, and the unique interests, terminology, and work processes of each of these disciplines.[15] While much of the benefit derived from the PDM is attributable to having a single data repository, the need to support a diverse group of users may necessitate the development of customized interfaces for each discipline. For example, whereas the design engineer may need to search product data on the basis of specific design attributes, the test engineer may have little use for this utility, but instead, may need to search product data on the basis of attributes describing alternative test configurations.

Beyond capturing and enabling the reuse of corporate knowledge, the PDM also accomplishes two additional objectives essential to agility. First, within a fast-paced product development environment, much of the work of designers, manufacturing and assembly engineers, quality assurance, and others occurs in parallel.[5] Under these conditions, numerous modes of failure exist, whereby multiple versions of a design can enter into circulation. Through strict control of read/write permissions and check-in/check-out protocols, the PDM introduces a structured process that assures all participants in the product development effort work with the latest version of design files. Second, within an agile production environment, every hour can be precious. Concurrent engineering must permeate all facets of production, since late design changes can have devastating effects on the ability to meet tight schedules. Consequently, it is important that all participants in the production process be kept aware of ongoing changes as they are made, so that there is an opportunity to affect those changes at the earliest possible point. Through automatic notification, all participants in the product development effort are alerted to design changes and provided regular status updates. After making changes to a design file, on returning the file to the PDM, e-mail is automatically sent to each participant, alerting them to the changes. Both version control and automatic notification serve valuable functions enabling the tight coordination of parallel efforts by representatives from multiple disciplines; but most important, information currency is assured with minimum overhead and distractions from basic engineering activities.

104.4 Virtual Protoype Testing

During the course of design development, many questions may arise regarding the feasibility or performance of design alternatives. Traditionally, such questions would be addressed by either adopting a conservative, low-risk, but limited design strategy, or conducting time-intensive, often costly physical testing. As a result, design innovation has often been sacrificed for the sake of cost and budget concerns.

Agility incorporates virtual prototype testing, when posed with design decisions for which there may be considerable risks. Virtual prototype testing may take various forms. One example would be engineering analysis codes that using three-dimensional design models, assess characteristics such as static and shock loading, vibration, operational environment, and thermal environment. Dynamic simulations that allow functionality to be evaluated on the basis of 3-D models represent another example. In a further example, stereolithography allows design alternatives to be assessed on the basis of physical representations.

Virtual prototype testing contributes to decision processes by providing a mechanism that enables hypothesis testing. Whereas traditionally, design has involved an inductive process, whereby design is largely an act of creation and inductive reasoning,[16] and evolves through small gradations, virtual prototyping makes it possible for designers to readily apply deductive processes, generating and testing hypotheses as they seek innovative design solutions. To be effective, certain conditions must be met. First, there needs to be a rapid turnaround. From the time that a design question is posed, an answer must be available within days, if not hours. Given that the primary bottleneck encountered will likely result from the large number of CPU cycles required by some analysis routines and the availability of requisite supercomputing resources, other aspects of the virtual prototyping process must be streamlined. This includes: definition of the analysis, preparation of the analysis, and communication of results. In defining the analysis, concurrent engineering processes should emphasize a close working relationship between designers and engineering analysts, from the earliest design stages. This allows the designer to take advantage of the analyst's expertise to narrow analysis options to those most likely to yield success. Second, since the basic design representation used by the designer will be a CAD model, processes should be in place to minimize or essentially eliminate the often tiresome steps necessary to convert CAD representations into a format suitable for analysis. This may include routines for the translation of CAD representations into 3-D solid models, and from there, the generation of finite element models.[5] Furthermore, mechanisms may need to be implemented to validate analyses prior to execution so as to avoid lost time and computing resources that may result if parameters are incorrectly defined. Finally, visualization techniques may be employed to aid in the interpretation and communication of virtual prototype testing results.

TABLE 104.1 Differences Between Communications in an Agile and a Traditional Enterprise

Traditional Enterprise	Agile Enterprise
• Geographical Co-Location	• Geographical Separation
• Solitary Work	• Collaborative Work
• Sequential Flow of Information	• Parallel Flow of Information
• Time is Negotiable	• Time is Critical
• Standardization of Technology	• Opportunistic Use of Technology
• Static Product Artifacts	• Dynamic Product Artifacts
• Info Flow w/Org Structure	• Info Flow w/Project Structure
• Extensive Use of Hard Media	• Nearly Exclusive Use of Electronic Media
• Constant, Known, Internal info Sources of Info	• Diverse, often Unknown, External Info Sources

104.5 Visualization

As virtual prototype testing is employed to quickly and efficiently test design hypotheses, various visualization routines may be utilized to test and verify manufacturability and assemblability of design, prior to production. Since agility requires that products be transitioned from design to production at the earliest practical point, it is critical that steps be taken to prevent delays incurred during production. Visualization accomplishes this objective in two ways. First, it may be employed as an interface to aid in the communication of manufacturing and assemblability concerns, and second, it may be used to verify manufacturability and assemblability.

During the design process, once CAD models of part designs have been developed, Computer-Aided Manufacturing (CAM) applications may be employed to quickly determine machine processes that will be required for parts fabrication. Starting with models of material stock, animated sequences of machine tool paths may be generated to illustrate the fabrication and problems that may be introduced by different facets of the design. Such illustrations provide an interface with which manufacturing may visualize the fabrication process in performing their own assessments, but equally important, it offers an interface that allows manufacturing to effectively communicate their concerns to design engineers.

Similarly, once models of a design assembly have been developed, automated routines may be used to generate assembly plans, and using these assembly plans, provide animated illustrations of the assembly process.[17] Furthermore, where robot assembly will be employed, automatically generated assembly plans may be utilized to produce animated illustrations of robot assembly operations, for evaluation of the impact of a design on robot assembly workstations.[17]

104.6 Communications

A substantial portion of the reductions in product development time and improvements in quality realized with agility may be attributed to the application of communications and information technologies. The previous sections have discussed several such innovations: e.g., PDM and automatic notification of design changes.

104.7 Differences Between Agile and Traditional Enterprises

Table 104.1 summarizes several of the differences between communications within an agile and a traditional enterprise. The following sections discuss the implications of each of these differences.

Geographical Separation

Within traditional product development paradigms, often a new product was developed by a single company, with team members located in a single building or complex of buildings, utilizing local suppliers for a customer within close proximity. All of these conditions minimized the burdens faced by geographically dispersed team members, organizations, and companies. In recognition of the inefficiencies introduced by geographical separation, "co-location" has often been recommended for teams as a means of enhancing concurrent engineering.

While co-location may facilitate communication between team members, enhance awareness of ongoing developments, and reduce bottlenecks to information flow, it assumes a business model that in many ways, contradicts key premises of agility. In particular, agility recognizes that it is not practical or cost-effective for a given company to maintain all of the capabilities necessary to bring many products to market. Instead, to attain a competitive advantage, companies must focus on their core competencies. As market opportunities develop, these companies may opportunistically combine their talents with other companies who have complementary capabilities, to form a Virtual Corporation.[1] For the life-span of the product, this Virtual Corporation would remain intact, and once this life-span comes to a close, the Virtual Corporation would dissolve, with each of its constituents going its own path.

Prohibitive costs and delays would be incurred were participants within a Virtual Corporation to attempt co-location. However, co-location does offer many benefits and with the capabilities introduced by communication technologies (e-mail, voice-mail, file transfer, desktop videoconferencing, collaborative work tools, internet and web-based technologies) and information technologies (PDMs, electronic product data, translators, automated tracking), Virtual Co-location provides a mechanism to allow distant and diverse companies to collaborate in new product development.[18]

The technology for Virtual Co-location is available, however the technology alone is not sufficient, as the need to work with geographically distant co-workers introduces many human factors and social obstacles.[19] With co-location, communication between co-workers may occur in real-time, with immediate feedback. These conditions promote a very natural and efficient exchange of information. Electronic communications are not nearly as efficient in that communication often occurs asynchronously (e.g., a sequence of voice-mail exchanges), anything more than one-to-one communications is impractical (e.g., conference calls may be burdensome due to the need for only one person to speak at a time), and compared to face-to-face interchanges, electronic communications have impoverished information content (e.g., facial and gestural expressions are either not transmitted or are poorly communicated). Furthermore, it has been reported that users are generally less satisfied collaborating using videoconferencing and related technologies, as compared to face-to-face collaboration.[20]

An additional challenge posed by Virtual Co-location relates to the absence of familiarity between collaborators. Workers who see each other daily and interact with one another on a regular basis develop a familiarity that may greatly enhance the efficiency with which they communicate.[21]

Various techniques may be employed to overcome the challenges posed by Virtual Co-location. Desktop videoconferencing removes much of the overhead typically associated with systems that require collaborators to leave their personal work spaces for specialized facilities. By combining telephone for voice and videoconferencing for visual, the drawbacks of digital voice communications may be avoided. Systems, such as Cruiser[22] and Telepresence,[23] incorporate a variety of mechanisms to aid in the day-to-day, mutual awareness of co-workers. Collaborative work tools, such as Interactive Collaborative Environments (ICE), allow sharing of work tools (e.g., CAD software) between two or more geographically separated co-workers.[24] An additional advantage of collaborative work tools is that whereas specialists from various disciplines have traditionally been restricted to the role of reviewers, they may readily be drawn into the design process as co-designers by simply opening a shared session in which the designer and one or more others jointly work on a CAD representation of the design. Finally, specific measures (e.g., team building exercises) may be taken to help create a sense of familiarity between co-workers who will need to work together collaboratively using electronic media.[25]

Collaborative Work

Traditional work processes have been dominated by a pattern in which team members meet in one place to review progress, discuss current issues, and assign work to various individuals. Afterwards, the group disperses and the overwhelming majority of work is accomplished by individuals, working alone, with little or no contact between team members. This approach emphasizes solitary work, a mode of operation often greatly favored by engineers due to the lack of distractions to the sustained concentration required to perform highly analytical tasks, and the opportunity to focus on tasks, without the cognitive overhead associated with social interactions.

While solitary work is still a common mode of operation within agile enterprises, there is a substantially greater emphasis on collaborative work. This emphasis results from the need to condense the design phase of product development into a very short period of time. To shorten the design phase, two things have to happen. First, delays associated with obtaining needed information or input from specialists must be minimized. Second, if a design path is not feasible, this needs to be determined at the earliest possible point, before precious time and resources have been committed. Both of these factors point to the need for collaborative work. Through collaborative work, input from specialists may be obtained almost immediately; time is not lost providing specialists with the background and design rationale they require to contribute to design; and potential problems and concerns may be identified and resolved before investing considerably in a particular design approach.

Reluctance, sometimes bordering on aversion, to collaborative work poses a significant problem. Within this reluctance, there may be an unwillingness by designers to relinquish the power to control design through their design decisions, and inclusion or exclusion of other disciplines, and an unwillingness among representatives of specialty disciplines to accept the responsibility for the success or failure of design that goes with being a collaborator in the design process.[19] A culture of solitary work may be deeply entrenched, and the skills required for productive collaborative work may need to be trained and nurtured. Consequently, large-scale measures may need to be taken to bring about the changes in corporate culture necessary to thoroughly integrate collaborative work into the product development process.

Parallel Flow of Information

A key characteristic of an agile enterprise is that information is where it is needed, when it is needed.[26] Therefore, an important step in the migration to agility is the elimination of information bottlenecks within the product development process. Within traditional enterprises, information often flows sequentially. Information passes from person A to person B to person C to person D, instead of flowing to all four at once. Consequently, the transmission of information is slowed, with it often being irrelevant by the time it reaches individuals late in the routing scheme. Likewise, as information is re-transmitted by different individuals, it may often be corrupted, either intentionally or nonintentionally. Furthermore, power is often derived from withholding information, resulting in information being withheld for the benefit of a few, but to the detriment of many.

Agility introduces a transition from sequential to parallel information flow. One example, previously discussed, is the use of automatic notification of design changes. E-mail and voice-mail have both provided mechanisms that enable information to be easily broadcast throughout an organization. With the parallel flow of information and the accompanying ease with which information may be broadcast, concern arises that information overflow may become an unwelcome by-product of these technical innovations. Currently, available mechanisms for filtering incoming information are at best, partially effective. This results from their reliance on either keyword-based sorting routines or sender-assigned prioritization. With either approach, their effectiveness is directly proportional to the effectiveness with which users can either select an effective set of keywords or deploy a prioritization scheme. In neither case, is there evidence that users are particularly effective.

All of the approaches discussed in the previous paragraph rely on a push-model of information distribution, information is pushed from a single point out along defined distributions. An alternative that alleviates some of the problems of information overload is to apply a pull-model, in which information is made available at a single or multiple source points (multiple source points may be desirable when delays are likely due to heavy network traffic), and individual users download the information at their own convenience. Corporate, web-based intranets provide an effective interface for implementing such a pull-model for information distribution within an enterprise. The primary drawback to the pull-model is that users must know the information is available in order for it to be requested. Consequently, it becomes necessary to develop user interface concepts that allow users to quickly and efficiently review available information and of equal, if not greater importance, provide sufficient cues for the users to make their decision regarding whether the information will be of sufficient value to justify its download.

Time Criticality

The relationship between cost, schedule, and performance whereby a reduction in cost leads to an increase in schedule and/or decrease in performance is an often-cited axiom of engineering management. As evidence of this three-way relationship, schedule is often negotiated when stringent cost and, especially, performance requirements are introduced by the customer. Agility seeks to violate this axiom by applying technical, organizational, and process innovation to drastically reduce schedule, without affecting cost and if anything, improving quality (however, it may be argued that the axiom still applies since investments are required up front to attain agile capabilities). To negotiate schedule is contrary to the spirit of agility, since the general aim is invariably to produce quality, custom product as fast as possible. Therefore, in transitioning to agility, every effort should be made to identify bottlenecks in the product development process that delay new product development, and either eliminate or substantially reduce those delays.

The identification of bottlenecks in the product development process requires a systematic analysis of the sequence of events necessary to bring new product to market.[18] Task analysis offers an effective analytic approach for obtaining the information needed to streamline product development. Through interviews, and review of corporate guidelines and process specifications, a detailed task analysis may be developed that illustrates the sequence of events by which product development occurs. Typically, simple qualitative assessments will yield a number of process improvements. For further process optimization, the task model of the product development process may be incorporated into a network simulation to quantitatively evaluate the elapsed time and utilization of resources throughout the process. Similarly, methods from Human Reliability Analysis[27] may be applied to estimate the likelihood of human error at various points throughout the process and the overall ramifications of specific human errors.

Opportunistic Use of Technology

Technology standardization is a common approach applied in the attempt to realize enterprise-wide efficiency and reductions in cost. Encouragement to pursue this approach comes from vendors, who often offer substantial price reductions for large purchases. Although, on the surface standardization seems like a straightforward solution, it is often not conducive to agility. Within computer user communities with no or minimal corporate pressure for standardization, heterogeneity is typically the rule as different users gravitate to platforms and software applications that best meet the requirements of their jobs, skills, experience, and knowledge, and with which they are generally satisfied. Where corporations apply pressure for standardization, it has been observed that users will invariably find ways to violate the standard, and introduce nonstandard hardware and software. Likewise, users express strong, sometimes nearly religious, loyalties to their preferred computer platform and software applications, resulting in considerable discontent when efforts are undertaken to deny users their preferences for the sake of standardization. Additionally, standardization hampers the efficiency with which a company may interface with other companies, excepting the improbable case where two companies have adopted comparable standards.

For the above reasons, the agile enterprise tends not to invest monetary and human resources in standardization, but instead pursues an opportunistic approach to technology utilization. Thus, there is an acknowledgment of heterogeneity and investment in solutions that allow people to work together, despite their using different computer platforms and software applications.

Dynamic Product Artifacts

Moving at a much slower pace, the artifacts generated following traditional design approaches are generally rather static. Great emphasis is placed on developing requirements that are firm, with considerable negotiation surrounding any change, once requirements have been fixed. During early conceptualization, design concepts may evolve fairly rapidly, but generally, acceptance of a design concept occurs fairly early, after which, there may be an extended progression from conceptual to detailed design. During this progression, there is a pervasive tendency to commit to various facets of design, and often, dissatisfaction is expressed when design team members attempt to revisit design decisions. In general, economy is sought by making design decisions early, and sticking to those design decisions, so that there is minimal effort expended on design paths that are ultimately rejected.

With agility, two factors lead to a more dynamic design process. First, design automation and reuse of product data reduce much of the pressure to push forward the beginning of detailed design development, allowing greater freedom to explore alternative design concepts. Second, virtual prototype testing makes it practical to explore innovative design concepts that otherwise would have been too risky to invest the time and resources. Consequently, in the conceptual design phase, a given product design may undergo numerous permutations, but once a design has been accepted, fabrication may begin almost immediately.

However, with an agile enterprise, there is no less concern for time and resources being wasted on unproductive design paths. Thus, during conceptual development, when design is in rapid flux, the design team must function as a coordinated unit, moving to apply the appropriate resources, but carefully steering and gauging those resources to maximize the return on investment. To function in such a manner requires that members of the design team have an ongoing situation awareness of the design progression, the efforts of their colleagues, and as needs arise that they must fill, an anticipation of those needs, so that they may move quickly and effectively to provide the needed inputs to the process.

Information Flows in Accordance with Project Structure

As stated previously, information is often synonymous with power and may be used as currency within the social dynamics of an organization. To the detriment of the organization, information may not flow along paths that are most expedient and productive, but follow paths defined by organizational or other political structures. For example, it is not uncommon for Engineer A to inform his superior that assistance is needed from Engineer B, who then contacts Engineer B's superior about securing the services of Engineer B. An agile enterprise cannot afford the inefficiency of such an exchange, but instead, would encourage direct communication between Engineer A and B. To supersede politically defined paths of communication, an enterprise must instill a culture wherein there is sufficient empowerment of workers to enable direct lines of communication. Likewise, embodied within this culture is an openness and willingness to share information that far surpasses that found within the vast majority of traditional enterprises.[26]

Electronic Media

While it may not be practical for an enterprise to eliminate all paper transactions and become truly paperless, there are countless efficiencies that may be gained from eliminating paper exchanges. All facets of the enterprise may be affected. Administrative functions may be streamlined through providing open access to corporate data (e.g., financial records, procurement processes, property management). Human Resource functions may be brought on-line through open access to employee benefits, vacation balances, and to the extent allowed by confidentiality, personnel records. Policies, Handbooks, Guides and similar corporate documents may be made widely accessible through electronic media, achieving the added benefit of assuring that all employees have the most current versions.

Whereas traditional enterprises see extensive exchange of paper between workers in technical positions (e.g., memos, reports, schedules, etc.), most of these exchanges could be easily accomplished through electronic media. For example, e-mail may be employed for most correspondence. Similarly, web-based intranets enable the posting of most documents, so that those interested may readily view the contents, and if sufficiently interested, download or print the document.

Applied to product development, electronic media enable a near-seamless progression from concept to finished deliverable. Hand-drawn sketches may be transitioned into CAD models. These CAD models may be converted to true 3-D representations, suitable for generation of finite element models on which virtual prototype testing may be conducted.[5] Similarly, CAD models may serve as a basis for developing machine tool paths, from which machine code may be derived. Furthermore, 3-D representations may be used to generate assembly plans and from there, robot code for the final assembly. In each of these cases, through the use of electronic media combined with various translation, conversion, and automation routines, design progresses through the various stages of product development, seamlessly, without the laborious activities associated with developing hard media and converting from hard media to either another form of hard media or electronic media.

Diverse, Often Unknown, External Information Sources

With traditional enterprises, information sources remained relatively constant from project to project. The same engineering groups, with mostly the same personnel, relied on the same support groups, similarly, with mostly the same personnel. Although obtainable information may have been limited to the capabilities and resources of the people providing information, there was constancy and familiarity. Within Virtual Corporations, personnel will vary from project to project, as will the information sources that are utilized. Familiarity of project personnel was discussed in an earlier section. With regard to variability of information sources, several concerns arise.

With the agile enterprise, there are more plentiful and richer sources of relevant information than has ever been the case. This information may be internally derived, such as corporate archives, or externally derived, such as public or commercial databases. The first problem concerns the filtering of available information to locate relevant data. Numerous utilities have been offered for searching large information spaces. However, most of these employ keyword-type searches, and although Boolean operators may be used to define relatively sophisticated search conditions, skills are required that exceed those of the vast majority of users. Furthermore, the effectiveness of such interfaces hinges on the user's ability to define successful search parameters, a task that is often difficult for even skilled information retrieval specialists. The second problem relates to the presentation of search results, a problem that is exacerbated by the inadequacy of interfaces for search definition. With most currently available utilities, search results are presented as a list of hits that may be ordered with regard to the incidence and placement of keywords, and may provide a brief excerpt. Given the investment of time associated with retrieving and reviewing each returned item, such an interface is highly inefficient for extracting answers to specific questions. However, as illustrated by the TileBars[28] interface, noteworthy progress has been made with regard to interface concepts for display of search results and similar progress may be anticipated for interfaces to aid in extracting information, once source documents have been located and retrieved.

104.8 Ergonomics of the Agile Shop Floor

Most discussions of agile manufacturing focus on its impact on design and business operations. The view of agile manufacturing from the shop floor is quite different than that of a product designer, manager, or administrator.

Human errors during production can be tremendously costly and will be more so as production runs grow smaller and more customized. The actual cost of human error would be useful to know, unfortunately it is difficult to gather this information. There are several reasons for this. First, there is a stigma associated with error, such that neither technicians nor supervisors are eager to admit or discuss their mistakes. Second, at some level in a company, there may be an urgent desire to understand and contain costs, but there is little reward for making such information public. Third, even industries that are well suited to capture the cost of human error on the factory floor are not very good at doing this. In the semiconductor industry, for example, the lack of standardized human–computer interfaces in semiconductor factories costs millions in "mis-processing" errors. Even though there exist formal organizations in this industry whose sole purpose is to produce and promote industry standards, to date this problem has largely gone unaddressed.[29]

The causes of human error on the production floor are numerous. First, the agile manufacturing paradigm, by its nature, assumes rapid response to customer demands. Agility is characterized by a remarkable confluence of people, skills, materials, facilities, and equipment to produce a product as quickly as possible. The dominant reality on the factory floor, therefore, is often that the production schedule drives everything. This is an environment highly prone to produce numerous human errors. The frenetic atmosphere that accompanies agile manufacturing is an over-arching factor that constrains every decision, pushing everyone at all levels away from optimal solutions to merely pragmatic ones. Some of the most potent sources of error in an agile manufacturing environment are discussed below. They are mentioned briefly here, although much has been written about each.

Operator Selection and Training

Successful manufacturing requires operators who bring basic skills to the job. These people are also specially trained to operate particular pieces of equipment. The operator selection process ranges from informal, where an experienced operator might take on an apprentice who seems to "have his wits about him," to formal, where minimum scores are necessary on performance tests to be considered for a job. In any case, by some process, people are chosen to work on the factory floor.

Regarding training, one of the most important determinants of success is that the training accurately represents the actual performance situation. Therefore, hands-on training to a criterion level of performance is essential for classroom training. On-the-job training is often effective (with quality feedback), particularly in conjunction with in-class training, provided close supervision is maintained for long enough (exactly how long is specific to the task being trained). It is known, for example, that around 100 hours of flight time is a troublesome period for new pilots as they transition to a more sophisticated level of performance beyond rote performance. A similar phenomenon should be anticipated for shop floor tasks, as well.

Procedures

One of the most potent factors shaping human performance on the factory floor is written work procedures. Work is often performed, unfortunately, without written work procedures of any kind. At the least, such situations run the risk of producing substandard products. More significantly, workers may begin to disregard safety and health risks, as such risks become routine through day-to-day exposure.

Shift Work

Taking a broad definition of shift work as anything outside of 7am to 6pm, approximately 20 to 30% of the workforce participates in shift work.[30] In fact, this percentage should be higher in agile manufacturing environments because of the time pressures discussed earlier. Shift work has an inherently negative affect on human performance, because it disrupts the body's normal biological rhythm. All things being equal, performance should be expected to suffer somewhat among shift workers, particularly on tasks that require decision-making.[31]

Human-Computer Interfaces

Software that is difficult to use promotes human errors. Most modern process machines now use computer screens as their dominant user interface. Also, procurement and planning software tools are an integral part of the modern factory. Travelers, work instructions, etc., are displayed from shop floor computers. For the most part, these interfaces have not met their potential to help operators control processes or make better decisions, though this is slowly changing. The fact is, most of these large tools are purchased on the basis of factors other than their usability, and most buyers have no way to determine the ergonomics of a user interface. If the vendor says it's user friendly, that's usually as far as it goes, if the question is even asked.

Safety Procedures

Accidents happen, and sometimes when they do, they cause damaged product, thus affecting the success of agile manufacturing. Many accidents happen because of nonconformance with published safety policies, and often management must bear some blame for this fact. It is often the case that, in the drive to meet schedule, plant management pushes priorities that send the message to the floor that safety can be compromised.

Shop Floor Culture

In any complex manufacturing enterprise, social groupings form based on educational background, status, and function. Usually, an informal caste system evolves with production technicians at the lowest level, engineers and other specialists in the middle, and management at the top. Production technicians, because they work together and often share the belief that they are under-appreciated by members of the other social groupings, tend to develop a unique shop floor culture. This is important to understand, because it can hurt or help an agile enterprise. Shop floor technicians are often very aware of ergonomically poor situations. However, as a group, typically, they are proud of their abilities and may take it as a suggestion of their own shortcomings that a situation is pronounced "error-likely." A specialist, intent on making ergonomic improvements, must gain the trust of the technicians, being careful to give them ownership in the result.

104.9 Conclusion

It has been often reported that major technological innovations have met with unsatisfactory results due to inattention to human factors.[31] The previous sections have summarized many of the human factors faced as enterprises migrate from traditional to agile modes of operation. Particular emphasis has been placed on the product development process, communication, and information utilization, all areas that have proven significant during agile manufacturing efforts undertaken at Sandia National Laboratories.[5] Admittedly, some significant issues have been omitted, most notably those associated with developing an "agile workforce",[32] and sociotechnical issues[33] related to the introduction of new technologies and business practices. Furthermore, with the spectacular rate of innovative growth in information technologies, the factors identified here are likely to be amplified, while many new considerations are introduced.

Acknowledgment

This work was supported by the United States Department of Energy under Contract DE AC04-95AL85000.

References

1. Nagel, R.N. and Dove, R., *21st Century Manufacturing Enterprise Strategy*, Iacocca Institute, Lehigh University, Bethlehem, PA, 1992.
2. Shunk, D. L., Patterson, W. and Ames, J. G., A report on the results of an agile pilot project: The Nationwide Electronics Industry Sector Pilot (NEISP), in *Proc. Fifth National Agility Conference*, Boston, MA, 1996, 9.
3. Hoy, T., VanderBok, R. and Margolin, D., Manufacturing Assembly Pilot (MAP) improving supply chain performance, in *Proc. Fifth National Agility Conference*, Boston, MA, 1996, 14.
4. Kinsella, E. M., Kinsella, M. D., Swanson, E. T. and Etue, J., Size doesn't matter: How a very small (30 employees) businesses used agile practices to help Whirlpool Corporation launch their state-of-the-art dryer in record time, in *Proc. Fifth National Agility Conference*, Boston, MA, 1996, 68.
5. Forsythe, C, Ashby, M. R., Diegert, K. V., Parratt, S. W., Benavides, G. L., Jones, R. E. and Longcope, D. L., *A Process for the Agile Product Realization of Electro-Mechanical Devices*, SAND Report Number 95-0536, Sandia National Laboratories, Albuquerque, NM, 1995.

6. Brost, R.C., Strip, D.R. and Eicker, P.J., The technology base for agile manufacturing, in *Proc. of the NASA Workshop on Space Operations*, Houston, TX, 1992, 522.
7. Diegert, K. V., Easterling, R. G., Ashby, M. R., Benavides, G. L., Forsythe, C., Jones, R. E., Longcope, D. L. and Parratt, S. W., Achieving agility through parameter space qualification, in *Proc. Fourth National Agility Conference*, Atlanta, GA, 1995, 331.
8. Comerford, R., The flexible factory: case studies, *IEEE Spectrum*, 30 (9), 28, 1993.
9. Tweedie, L., Spence, B., Dawkes, H. and Su, H., The Influence Explorer (video) — a tool for design, in *Proc. Human Factors in Computing Systems*, Vancouver, BC, 1996, 390.
10. Hauge, P. L. and Stauffer, L. A., ELK: A method for eliciting knowledge from customers, in *Proc. Design Theory and Methodology*, Albuquerque, NM, 1993, 73.
11. Ramaswamy, R. and Ulrich, K., A designer's spreadsheet, in *Proc. Design Theory and Methodology*, Albuquerque, NM, 1993, 105.
12. Helander, M., Models of design for concurrent engineering, in *Advances in Agile Manufacturing: Integrating Technology, Organization and People*, Kidd, P. T. and Karwowski, W., Eds., IOS Press, Washington, D.C., 1994, chap 4.
13. Nakazawa, H., Human oriented manufacturing systems, in *Advances in Agile Manufacturing: Integrating Technology, Organization and People*, Kidd, P. T. and Karwowski, W., Eds., IOS Press, Washington, D.C., 1994, chap 2.
14. Forsythe, C. and Ashby, M.R, Human factors in agile manufacturing, *Ergonomics in Design*, 4(1), 15, 1996.
15. Forsythe, C. and Ashby, M.R., User-driven product data manager system design, in *Proc. Human Factors and Ergonomics Society*, San Diego, CA, 1995, 1170.
16. Goel, V., *Sketches of Thought: A Study of the Role of Sketching in Design Problem Solving and Its Implications for the Computational Theory of Mind*, Dissertation, University of California, Berkeley, CA, 1991.
17. Jones, R.E., Ames, A.L., Calton, T. L., Kaufman, S.G., Laguna, C.A., Rebeil, J.P., Saavedra, M., Wilson, R.H. and Woods, R.O., *The Automated Assembly Team Contributions to the {A-PRIMED) Agile Manufacturing Project*, SAND Report Number 95-1340, Sandia National Laboratories, 1995.
18. Forsythe, C. and Ashby, M.R, *Developing Communications Requirements for Agile Product Realization*, SAND Report Number 94-048, Sandia National Laboratories, Albuquerque, NM, 1994.
19. Forsythe, C., Human factors in agile manufacturing: a brief overview with emphasis on communications and information infrastructure, *International Journal of Human Factors in Manufacturing*, 1997, 3-10.
20. Olson, J.S., Olson, G.M. and Meader, D.K., What mix of video and audio is useful for small groups doing remote real-time design work? in *Proc. CHI 96 Human Factors in Computing Systems*, Vancouver, BC, 1996, 362.
21. Krause, R. M. and Fussell, S. R., Mutual knowledge and communication effectiveness, in *Intellectual Teamwork*, Erlbaum, Hillsdale NJ, 1990, 111.
22. Rosenberg, J., Kraut, R.E., Gomez, L. and Buzzard, C.A., Multimedia communications for users, *IEEE Communications*, 1992, 20.
23. Buxton, W.A.S., Telepresence: integrating shared task and person spaces, in *Proc Graphic Interface 92*, Vancouver, BC, 1992, 11.
24. Ashby, M.R. and Lin, H.W., Interactive Collaborative Environments (ICE): platform independent X application sharing and multi-media over wide area networks at Sandia National Laboratories. in *Proc. Fifth International Symposium on Robotics and Manufacturing*. Maui, HI, 1994.
25. Forsythe, C. and Ashby, M.R., Ramping up for agility: development of a concurrent engineering communications infrastructure, in *Proc. Fifth National Agility Conference*, Boston MA, 1996, 1062.
26. Goldman, S. and Preiss, K., *21st Century Manufacturing Enterprise Strategy. Volume 2. Infrastructure*, Lehigh University, Bethlehem, PA, 1992.
27. Gertman, D.I. and Blackman, H.S. *Human Reliability and Safety Analysis Data Handbook*, John Wiley, New York, 1994.

28. Hearst, M.A., TileBars: visualization of term distribution information in full text information access, in *Proc. CHI 96 Human Factors in Computing Systems*, Vancouver, BC, 1996, 59.
29. 1. Miller, D.P. and Whitehurst, H.O., *Preventing User-Hostile Interfaces in IC-Fab Equipment: Ergonomic Approaches for Preventing Ten Frequent User-Interface Problems*, Sematech Technology Transfer Report #92091299A-ENG, Sandia National Laboratories, Albuquerque, NM, 1992.
30. Monk, T.M. and Folkard, S., *Making Shift Work Tolerable*, Taylor & Francis, London, 1992.
31. Mitton, R. and McLoughlin, I., "A lack of fit?" the adoption of Japanese style manufacturing techniques in Britain, in *Advances in Agile Manufacturing: Integrating Technology, Organization and People*, Kidd, P. T. and Karwowski, W., Eds., IOS Press, Washington, D.C., 1994, chap 11.
32. Plonka, F.E., Developing a lean and agile workforce, *International Journal of Human Factors in Manufacturing*, 1997, 11-20.
33. Baba, M.L. and Mejabi, O., Advances in Socio-technical systems integration: object-oriented simulation modeling for joint optimization of social and technical subsystems, *International Journal of Human Factors in Manufacturing*, 1997, 37-61.

105

Occupational Safety in Robotics

Mansour Rahimi
University of Southern California

105.1 Introduction

The robotics industry is expecting a doubling in total revenues between 1994 and 2001 (*Robotics World*, 1995). Even though the annual growth rate is expected to average 9.6% for all types of robots, the percent of growth will be substantial in material handling (28.6%), welding (25.7%), and material application (17.2%) by the year 2001. It is expected that the new business is likely to come from the implementation of long-term, customer-based strategies rather than from the aggressive marketing that took place in the market's infancy during the 1980s. According to McAlinden (1995), the world now uses more than 0.5 million robots. The robotics market analysts believe that such an increase in robotic applications will only be sustained by the ability of robot designers and producers to diversify outside the traditional automotive industry into nonautomotive areas such as: food/beverage and consumer products, pharmaceuticals, computer and consumer electronics, and telecommunications. Also, a paradigm shift has also been noticed in the use of robots from an individual robot cell application to a broader integration into the overall manufacturing system including product and process flow integration, generating 20% of revenues in 1994.

This sharp increase in the use of robotic systems in the past three years has renewed interest in the occupational safety aspects of these sophisticated machines. The following two incidents highlight the potential *hazards* that a computer-controlled robot may pose to people who work with (and around) these machines. These incidents occurred in a robotics laboratory and are directly quoted from a news bulletin at RISKS@SRI-CSL.ARPA. "In the first incident, a 68000 board in our system failed and caused the processor to jump to (of all places) a robot move routine. We were all standing around the *emergency stop* button looking at a terminal, and Jeff and Talia got to the button within a few milliseconds of hearing the crunching noise which marked the premature demise of a small jack belonging to the lab. With our sensor mounted on the *end-effector* as it was, it could have been a lot worse if we had been further from a kill button." The second incident was described as "Some drive motor cards in the Cincinnati-Milacron box failed and joints 5 and 6 began jerking around randomly. Again, the kill button was nearby … It could have been any other joint — including the base or the shoulder. And someone could have been standing next to it. We do that all the time. It can and does happen. Watch yourselves around robots."

0-8493-2641-9/99/$0.00+$.50

TABLE 105.1 Some Causes of Robot-Related Accidents

Design and Engineering Factors
Robot system failures (mechanical, electrical, etc.)
Peripheral machine malfunctions
Inadvertent robot movement
Software design and control malfunctions
Failure to halt all robot movements
Unexpected starts and halts
Gripper design problems
Improper design of control panel and its functions
Inadvertent contact with start buttons and other switches
Unreliability of the existing robot safety devices
Excessive robot arm speed
Existence of other sources of hazards (e.g., chemical, noise)
Behavioral and Organizational Factors
Worker entering danger zone of a halted robot for:
trouble-shooting
repair
testing
maintenance
teaching
Unauthorized entry into general robot work area
Inappropriate robot start-up procedure
Inappropriate work-piece adjustment and repositioning
Inappropriate robot teaching mode and maintenance procedures
Inappropriate change of tooling
Worker unaware of robot's programmed (unpredictable) movement
Intentional disabling of safety devices
Improper lockout/tagout procedure
Lack of targeted safety training for managers, engineers and operators

105.2 Robot Accidents and Safety Analysis

Statistics on robotic accidents and injuries are spurious and somewhat unreliable. Robot manufacturers and *users* seldom publish any relevant robot safety and health related information. Similarly, mandated federal and state accident reporting systems do not contain a significant number of accidents labeled as robot injuries. However, due to the nature of robots and the unpredictability of their motion sequences, some robot injuries may be severe with major disruptions in the company's operation. For example, a death caused by a robot resulted in a $10 million law suit and the closure of an automotive plant. The only significant robot injury data sources and analyses have been from Backstrom and Harms-Ringdahl (1983), NIOSH (1984), and Sugimoto (1987).

The type of injuries from robots may be classified as direct impact (e.g., cuts, lacerations, crushing), electrocutions, airborne contaminants (e.g., welding gases, acid fumes, irritants, systemic poisons), hazardous material (e.g., process chemicals), nonionizing and ionizing radiation, high temperature and fires, noise, and biohazards (for a description of each hazard see Brauer, 1990). This chapter deals with hazards from stationary (nonmobile) production robots only.

Since some robot injuries may be serious, a number of hazard analysis techniques have been suggested to assess the hazardous conditions of a robotic system. A robot hazard analysis technique was introduced by Rahimi (1986) using the concept of energy *barrier* analysis. This study introduced the sources of robot hazards into two categories listed in Table 105.1. The study suggested a five-step approach to minimize the robotic system hazards:

a. Improving mechanical reliability and hardware component design of robotic systems.
b. Incorporating software safety controls for all phases of robot operations.
c. Developing a robot *presence-sensing* capability to detect human proximity.
d. Incorporating ergonomic design considerations for human–robot workstation layout and operations.
e. Using effective safety training systems for individuals associated with robot operation, testing, and maintenance.

In another study, Seward, Bradley, and Margrave (1994) present a safety life cycle model which involves a two-step analysis process. Step 1 uses the data from five input variables: robot physical characteristics, robot goal specification, environmental considerations, safety criteria, and safety regulations and constraints. The risk analysis in Step 1 produces the functional control system specifications and safe and unsafe states of the robotic system. Then the outcomes from Step 1 are inputted into Step 2: analysis of control system specification which produces any revisions to the current functional control system specification, specific safety requirements, and information requirements and robot control interfaces with other machines and tools. In order to perform specific safety analyses in Step 1, Seward et al. (1994) suggest a number of techniques regularly used in system safety analysis (see also Rahimi, 1995). The following is a short description of some of these techniques.

Preliminary Hazard Analysis. This simple technique is usually used as the first step to prepare for more detailed analysis. It is an organized way of collecting crude data on system components and their hazardous characteristics. The purpose is to generate a list of hazards inherent in the system. A good starting point is to generate the list based on the sources of potentially hazardous *energy source* in the system.

Failure Mode, Effects, and Criticality Analysis. Unlike PHA, which lists hazards in the overall system, FMECA analyzes the components of the system and all of the possible failures that can occur in each component. In this analysis each item's function must be determined. Once this is done, the causes and effects of the failure of the components are indicated. Then, the criticality factor for each failure is determined, and a quantified severity rating is given to the factor. Since the frequency of each potential occurrence is also an important factor, a risk assessment matrix can be used to codify the risk assignment (see Rahimi, 1995). A team of experts can then use this information to redesign robot work cell components or configurations to reduce the criticality ratings that are high or unacceptable.

Hazard and Operability Study. HAZOP is a systematic study of the hazards in a complex system. The system operation is broken down into a set of nodes. Each node contains information on the procedure or specific machines and is then interconnected logically with other nodes in the system. A team of specialists is formed who systematically question every aspect of every node of a system and its operation using a set of key "guide words" (e.g., More, Less, As well as). Using these guide words on every node will determine the possible deviations from the planned operation which may cause hazardous conditions. Each deviation is then considered in turn to establish how it is caused and what the consequences are. Unrealistic causes and trivial consequences are ignored. Important causes and consequences are examined at a later stage to establish how they can be eliminated or reduced to an acceptable level.

Fault Tree Analysis. FTA is a top-down or deductive reasoning approach to the elements of a chain of events which leads to an undesired event (e.g., accident, injury). These events are linked by an event tree structure (pyramid style) to the top undesired event using logic connections such as AND or OR functional interpretations. There are four main types of events in any fault tree: (a) a basic event which is the final event in the fault tree, initiating the failure process as a root cause, (b) a fault event which is considered to be the in-between event (not an end event), (c) a normal event which is an existing event which may or may not contribute to the fault propagation, and (d) an undeveloped event which requires more investigation because if its complexity or lack of analytical data. For example, in its simplest form, a robotic injury may be linked to a software control failure and a mechanical stop switch failure. A fault tree can be constructed in which an AND gate connects these two failures, leading to the possibility of such injury. Probabilities of robot injuries can also be studied using quantitative FTA for specific robot installations (Rahimi and Roland, 1989).

105.3 Risk Zones for Stationary Robots

Robot workstations are temporally and spatially dynamic. Work spaces within a robot workstation can be defined and modified based on the speed of the robot arm (e.g., see Karwowski and Rahimi, 1991), number and size of the robot tool exchange mechanisms, variabilities in the part and workpiece feeders and extractors, design of the conveyor systems, cooperation with the adjacent robots, and the most variable of all these elements, the operator's behavior. Therefore, allocating risk (danger) zones to a robot work zone becomes a major determinant of the overall safety of a robot workstation. One can take a sophisticated approach of mathematically formulating and predicting the probability of operator collision in the vicinity of a stationary robot workstation (Giusti and Tucci, 1994). Since robot workstations are highly flexible, the use of such time-consuming mathematical computations on actual factory floors may be impractical. A more practical approach is to follow a simple heuristic to determine robot danger zones and possible means of protecting workers, machines, and workpieces around robots. Each zone and a generic approach to its *safeguarding* is presented here.

1. *Adjacent to the Workstation.* Workers may walk around the general vicinity of a robot for a variety of reasons. A hazard prevention approach is to attract the worker's attention by appropriately designed robot warning signs and robot arm color coding.

2. *Approaching the Workstation.* The worker is still at a distance from the robot actual work zone. Therefore, there is no need for a robot controller action, except a flashing yellow warning light which can be seen form all different locations surrounding the robot.

3. *Zone 3* (*envelope*). This working zone is inside the robot work envelope plus any hazardous area within the zone caused by operation of peripheral machines and work-piece handling devices. This zone is usually marked by perimeter fencing, fixed and removable *barriers* and gates, modular hand rails, woven wire partitions, in addition to safety sensors (e.g., pressure sensitive mats, photoelectrics). Entering this zone should also trigger warning buzzers, yellow flashing lights, and/or red lights. The robot speed of motion should also be adjusted to reduce potential impact hazards.

4. *Zone 2* (*restricted envelope*). This zone encompasses all points where the robot arm and its associated equipment can travel in a three-dimensional space. (**Note:** majority of everyday robot operations do not extend the robot arm at its maximum distance.) In addition to red flashing lights and a warning buzzer, this zone should contain both pressure mats and light curtains. The robot control routines should be designed to freeze the arm. However, such an action should not create other dangers to the intruding worker, such as release of workpiece, or release of other environmental hazards. The control reactivation should allow normal production once the worker leaves this zone.

5. *Zone 1* (*operating envelope*). This zone contains the immediate proximity to the robot arm, gripper mechanisms, actual piece being handled, and other sources of hazardous energy associated with the robot work cell. Intrusion into this zone should place the robot into an emergency stop and activate all warning lights and buzzers at their maximum capacity. To prevent severe impact injuries the robot arm should have proximity and/or touch sensors with a *fail-safe* release mechanism. A formal restart procedure is necessary to resume the operation. Safeguarding this zone becomes a formidable task if a robot end-effector is cooperating with a human worker to perform a specific task. In cases where multiple human–robot work tables are being used for production (e.g., a shift workstation), robot controlled light curtains may be recommended. The light curtain limiting switch is turned on when the robot approaches the work table to perform its task routine, leaving the other work table free for the human worker to operate.

A word of caution is that each robot work-cell is different in terms of its design and the layout of its associated machinery and equipment. Each workstation design must be carefully studied for a comprehensive robot safety analysis. The next section briefly discusses the minimum requirements (standards) for robot safety.

105.4 Robotic Safety Standards

Countries with large populations of robots (i.e., United States, Japan, United Kingdom, Germany, France, Russia) have developed regulations, codes, and standards to govern their robot safety practices. In many respects, these documents differ in their formats and contents. In the United States, a voluntary standard has been produced by the American National Standards Institute with cooperation from Robotics Industries Association (ANSI/RIA, 1992). As of now, a mandatory U.S. federal standard specifically designed for robotic systems has not been developed. However, the general machine guarding section of the Occupational Safety and Health Administration, General Industry document (OSHA 29CFR1910-212) can be applied to robotic safeguarding. This clause states that: "... One or more methods of machine guarding shall be provided to protect the operator and other employees in the machine area from hazards such as those created by point of operation, in-going nip points, rotating parts, flying chips, and sparks. Examples of guarding methods are: barrier guards, two-hand tripping devices, electronic safety devices, etc." Therefore, it is the duty of every robot user to protect the workers from potential robot hazards. The means by which such protection can be assessed should be accomplished by careful analysis of the robotic workstation and application of relevant concepts such as the ones stated in the ANSI/RIA standard. Table 105.2 is a selected (partial) list of items in this standard. The original section numbers have been kept for ease of reference. For a discussion of the standards from other countries, the reader is referred to a publication by the International Labor Office (ILO, 1989). The ILO document also contains a comparison table as a cross reference checklist of standards from the six countries mentioned above. As mentioned earlier, following these standards appears to constitute only minimum requirements for the safety of workers around industrial robots. Also, these standards are not designed for a myriad of other robot applications such as construction, health care, mobile robots, etc. Figure 105.1 is presented to demonstrate the robot danger zones and the various safeguarding devices used in a typical robot installation. Figure 105.2 is presented to illustrate the concept of a human–robot cooperative work-cell and a solution to protect workers from possible injuries.

Defining Terms

Awareness device or signal: A device that by means of audible sound or visible light warns a person of an approaching or present hazard.

Barrier: A physical means of separating persons from the restricted envelope (space) where a danger may be present.

Emergency stop: The operation of a circuit using hardware-based components that overrides all other robot controls, removes drive power from the robot actuators, and causes all moving parts to stop.

End-effector: An accessory device or tool specifically designed for attachment to the robot wrist or tool mounting plate to enable the robot to perform its intended task. Examples may include gripper, spot weld gun, arc weld gun, spray paint gun, or any other application tools.

Energy source: Any electrical, mechanical, hydraulic, pneumatic, chemical, thermal, potential, kinetic, or other source of energy.

Envelope (space), maximum: The volume of space encompassing the maximum designed movements of all robot parts including the end-effector, workpiece, and attachments.

Fail-safe: A mechanism by which any failure of the robot components will place the robot into a safe or zero energy state.

Hazard: A condition that is likely to cause personal harm or damage to equipment.

Interlock: An arrangement whereby the operation of one control or mechanism allows or prevents the operation of another.

TABLE 105.2 A Partial List of Items Found in the U.S. ANSI/RIA R15.06 Voluntary Standard

ANSI/RIA Section Numbers and Section Title	
4.2	Hazard to personnel
4.2.1	Moving parts
4.2.2	Component malfunction
4.2.3	Sources of energy
4.2.4	Stored energy
4.2.5	Electromagnetic interference, radio frequency interference, and electrostatic discharge
4.3	Movement without drive power
4.4	Actuating controls
4.4.1	Protection from unintended operation
4.4.2	Status indication
4.4.3	Labeling
4.4.4	Remotely located controls
4.5	Emergency stop
4.6	Pendant
4.6.1	Automatic
4.6.2	Motion control
4.6.3	Pendant emergency stop
4.6.4	Single point of Control
4.6.5	Initiated motion
4.7	Attended continuous operation
4.8	Slow speed control
4.9	Mechanical stops
4.10	Provisions for lifting
4.11	Electrical connectors
4.12	Hoses
4.13	Power loss or change
4.14	Failures
4.15	Required Information
5	Installation of robots and robot systems
5.1	Installation specification
5.2	Grounding requirement
5.3	Power requirements
5.4	Control location
5.5	Robot system clearance
5.6	Power disconnect
5.7	Limiting devices
5.8	Environmental conditions
5.9	Associated equipment shutdown
5.10	Precautionary labels
5.11	Restricted envelope (space) identification
5.12	Dynamic restricted envelope
5.13	Robot system emergency stop
5.14	End-effector power loss/change
6	Safeguarding personnel
6.1	Responsibility
6.1.1	Risk assessment
6.1.2	Stages of development
6.2	Clearance
6.3	Safeguarding devices
6.3.1	Limiting devices
6.3.1.1	Mechanical limiting devices
6.3.1.2	Non mechanical limiting devices
6.3.2	Presence-sensing safeguarding devices
6.3.2.1	Component failure
6.3.2.2	Ambient factors
6.3.2.3	Resumption of robot motion
6.3.2.4	Muting

TABLE 105.2 (continued) A Partial List of Items Found in the U.S. ANSI/RIA R15.06 Voluntary Standard

		6.3.3	Barrier
	6.4	Awareness devices	
		6.4.1	Perimeter guarding
		6.4.2	Awareness barrier
		6.4.3	Awareness signal
			6.4.3.1 Failure to reach intended location signal
	6.5	Safeguarding the teacher	
	6.6	Safeguarding the operator	
	6.7	Safeguarding during attended continuous operation	
	6.8	Safeguarding maintenance and repair personnel	
7	Maintenance of robots and robot systems		
8	Testing and start-up of robots and root systems		
	8.1	Interim safeguarding	
	8.2	Restricted envelope (space)	
	8.3	Manufacturers' instructions	
	8.4	Initial start-up procedure	
9	Safety training of personnel		
	9.1	Training requirements	
	9.2	Training program content	
	9.3	Retraining requirements	

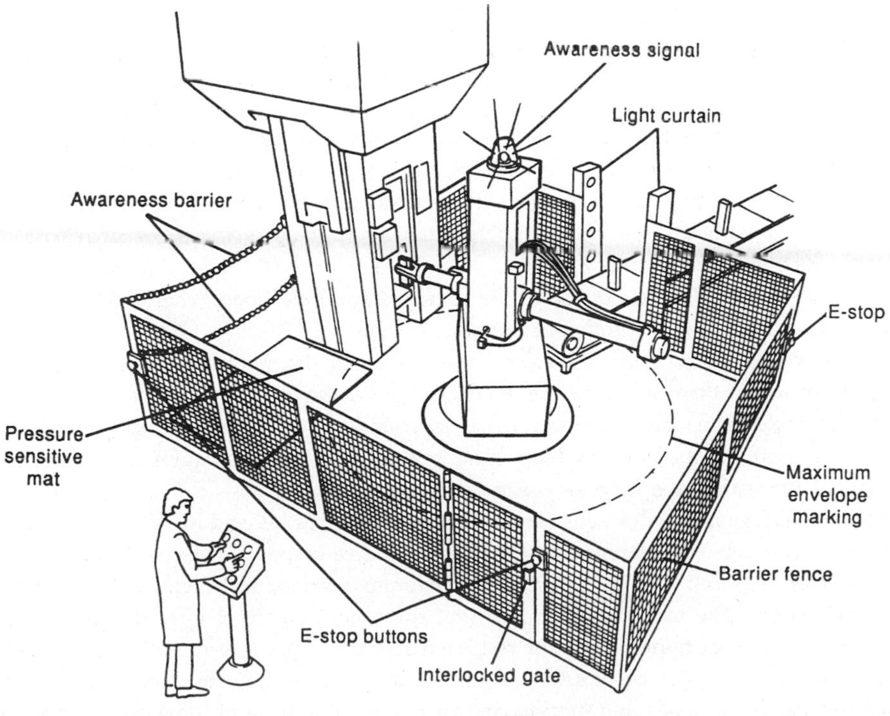

FIGURE 105.1 A schematic of a robotic workstation. This illustration includes a cylindrical robot, a conveyor belt feeding parts to the robot, an associated machine working on the workpiece, an operator with a computer control panel, and a typical array of safeguarding devices for worker protection in all zones of the workstation. (From *American National Standard for Industrial Robots and Robot Systems — Safety Requirements.* ANSI/RIA R15.06-1992, New York. With permission.)

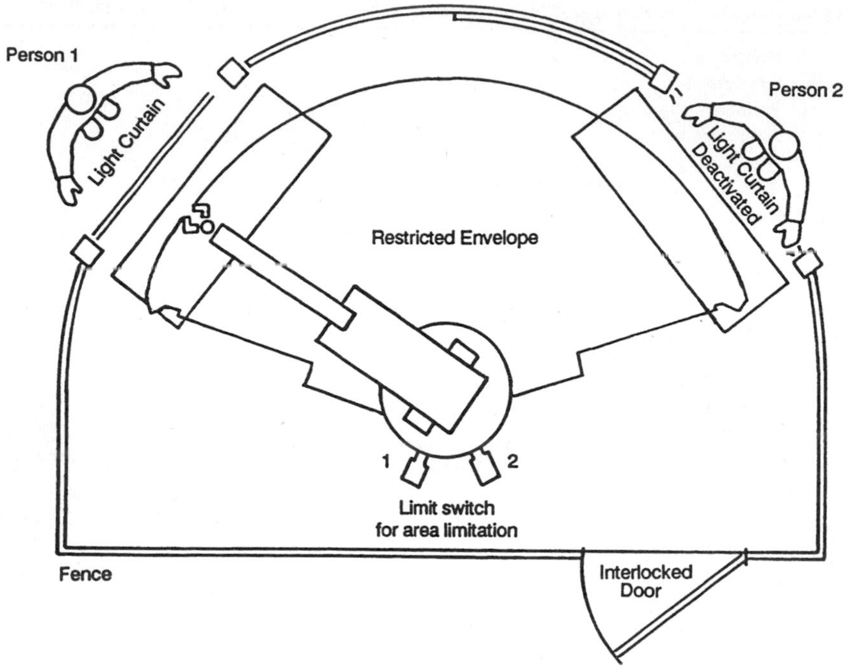

FIGURE 105.2 A schematic of a dynamic restricted human-robot cooperative workstation. The hazards for this cooperative workstation can be reduced by providing a robot controlled two-limit switch assembly. Each limit switch activates a light curtain where the robot arm is moving into for its routine task, while the other limit switch is deactivated allowing the second worker to perform the necessary tasks. (From *American National Standard for Industrial Robots and Robot Systems — Safety Requirements.* ANSI/RIA R15.06-1992, New York. With permission.)

Lockout/tagout: A procedure to ensure that the robot and its associated equipment is stopped, isolated from all potentially hazardous energy sources, and locked out or tagged out before any employees can enter the robot danger zones to perform any servicing or maintenance activities. This procedure is needed to prevent any unexpected energization or start-up of the robot or release of stored energy which could cause injury.

Operating envelope (space): That portion of the restricted envelope (space) that is actually used by the robot while performing its programd motions.

Pendant: Any portable control device, including teach pendants, that permits an operator to control the robot from within the restricted envelope (space) of the robot.

Perimeter guarding: A rigid fence-like structure that surrounds the restricted envelope (space) of a system of one or more robots and may have entry openings for process equipment, material, and/or personnel authorized to operate or maintain the robot system.

Presence-sensing safeguarding device: A device designed, constructed, and installed to create a sensing field or area to detect an intrusion into such field or area by personnel, robots, or other objects.

Restricted envelope (space): That portion of the maximum envelope to which a robot is restricted by limiting devices. The maximum distance that the robot can travel after the limiting device is actuated defines the boundaries of the restricted envelope (space) of the robot.

Safeguard: A barrier guard, device, or safety procedure designed for the protection of personnel.

Slow speed control: A mode of robot motion control where the velocity of the robot is limited to allow persons sufficient time to either withdraw from hazardous motion or stop the robot.

Teach mode: The control state that allows the generation and storage of positional data points effected by moving the robot arm through a path of intended motions.

User: A company, business, or person who uses robots, or who contracts, hires, or is responsible for the personnel associated with the robot operation.

Other Defining Terms

The following terms are important concepts that are frequently used in other robot safety literature. Due to space limitations, they were not used in this chapter.

Attended continuous operation: The time when robots are performing production tasks at a speed no greater than slow speed through attended program execution.

Automatic mode: The robot state in which automatic operation can be initiated.

Awareness barrier: Physical and visual means that warn a person of an approaching or present hazard.

Control device: Any piece of control hardware providing a means for human intervention in the control of a robot system, such as an emergency stop or a selector switch.

Control program: The inherent set of control instructions that define the capabilities, actions, and responses of the robot system. This program is usually not intended to be modified by the end user.

Limiting switch: A device that restricts the maximum envelope (space) by stopping or causing to stop all robot motion and which is independent of the control program and the application programs.

Muting: The deactivation of a presence-sensing safeguarding device by design during a portion of the robot cycle or during noncyclic activities such as teaching mode.

References

American National Standards for Industrial Robots and Robot Systems — Safety Requirements. ANSI/RIA R15.06-1992, American National Standards Institute, New York, N.Y.

Backstrom, T. and Harms-Ringdahl, L. 1983. A statistical study on control systems and accidents at work. *Proceedings of the International Seminar on Occupational Accidents,* Stockholm, Sweden, 312-328.

Brauer, R.L. 1990. *Safety and Health for Engineers.* Van Nostrand Reinhold, New York, N.Y.

Giusti, R. and Tucci, M. 1994. Improving active accident prevention in robotized manufacturing plants. *The International Journal of Robotics Research,* 13(4): 334-342.

International Labor Office, 1989. *Safety in the Use of Industrial Robots.* Occupational Safety and Health Series No. 60, Geneva, Switzerland.

Karwowski, W. and Rahimi, M. 1991. Worker perception of safe speed and idle condition in simulated monitoring of two industrial robots. *Ergonomics,* 34(5), 531-546.

McAlinden, J.J. 1995. Using robotics as an occupational health and safety control strategy. Industrial Robot, 22(1): 14-17.

National Institute for Occupational Safety and Health, 1984. *Request for Assistance in Preventing the Injury of Workers by Robots.* Publication No. 85-103, NIOSH Publications Dissemination, DSDTT, Cincinnati, Ohio: U.S. Government Printing Office.

Rahimi, M. 1986. System Safety for Robots: An energy barrier analysis. *Journal of Safety Science.* 8: 127-138.

Rahimi, M. 1995. Hazard identification and control, in *The Engineering Handbook,* Ed. R.C. Dorf, P. 1897-1903. CRC Press, Boca Raton, FL.

Rahimi, M. and Roland, H.E. 1989. A quantitative fault tree analysis for safety of human-robot interactions. *Proceedings of the 9th International System Safety Conference,* Long Beach, California, 236-240.

Robotics World, 1995. News of the Industry, 13(4): 8.

Seward, D.W., Bradley, D.A., and Margrave, F.W. 1994. Hazard analysis techniques for mobile construction robots, in *Automation and Robotics in Construction XI,* Ed. D.A. Chamberlain, p. 35-42. Elsevier Science B.V., Amsterdam, Netherlands.

Sugimoto, N. 1987. Subjects and problems of robot safety technology, in *Occupational Health and Safety in Automation and Robotics,* Ed. K. Noro, p. 175-183.

For Further Information

Publications

Proceedings of the Robotic Industries Association: Robot Safety Conference
Proceedings of the British Robot Association Annual Conferences
Proceedings of the International Seminar on Safety in Advanced Manufacturing
Proceedings of the Robots (x) Conference
Proceedings of the System Safety Conference
Proceedings of the Annual Reliability and Maintainability Symposia
Proceedings of the International Conference on Robotics and Factories of the Future
Proceedings of the IEEE International Conference on Robotics and Automation
Proceedings of the Annual Meeting of the Human Factors and Ergonomics Society
Proceedings of the Robot (x) Conference of the Society of Manufacturing Engineers
Proceedings of the International Conference on Human Factors in Manufacturing
Proceedings of the International Conference on Human Aspects of Advanced Manufacturing and Hybrid Automation
Safety, Reliability, and Human Factors in Robotic Systems (Van Nostrand Reinhold)
Human-Robot Interaction (Taylor & Francis)
Robot Safety (Springer-Verlag)
Occupational Health and Safety in Automation and Robotics (Taylor & Francis)
International Encyclopedia of Robotics (Wiley Interscience)
Handbook of Industrial Robotics (John Wiley)
Safety in the Use of Industrial Robots (International Labor Office)
National Safety News
Robotics World
The Industrial Robot
Robotics Today
Professional Safety
Journal of Safety Science
IEEE Transactions on Industry Applications
IEEE Transactions on Automatic Control
The International Journal of Robotics Research
Robotics Engineering
Robotics Age
Plant Maintenance
Plant Engineering
Ergonomics
International Journal of Industrial Ergonomics
Behavior and Information Technology

Organizations

Robotic Industries Association
Subcommittee R15.06 on Safety
900 Victors Way/P.O. Box 3724
Ann Arbor, Michigan 48106

Japan Industrial Robot Association (JIRA)
Kikai Shinko Kaikan Building
3-5-8 Shiba-kown
Minato-ku
Tokyo 105, JAPAN

British Robot Association
35-39 High Street
Kempston Bedford MK42 7BT
United Kingdom

Fraunhofer Institute of Manufacturing
Engineering and Automation
Nobelstrasse 12
7000 Stuttgart 80
Germany

106

Ergonomics Issues in Mining

Sean Gallagher
National Institute for Occupational Safety and Health

106.1 Introduction

It was not too long ago that coal and ore from underground mines were shoveled or manually loaded onto carts drawn by horse or mule (Sanders and Peay, 1988). As recently as the mid-1950s, almost a third of all coal produced in the U.S. was still hand-loaded. In the years prior to the introduction of mechanization, mining was truly backbreaking work. The principal tools of the miner were the pick and shovel, powered solely by raw muscle. Figure 106.1 illustrates a common task of the underground miner of this period: undercutting a coal face in preparation for blasting. As shown in Figure 106.1, this task was performed while the miner was lying on his side, using a pick to hew the coal and a shovel to support the body. Miners could spend three to six hours undercutting a coal face, using their picks to make a one meter deep horizontal incision at the base of the mineral seam.

0-8493-2641-9/99/$0.00+$.50

FIGURE 106.1 Miners undercutting a coal face prior to mechanization. (From National Archives. With permission)

Advances in mechanization in the second half of the 20th century have greatly reduced physical demands on the mine worker. Even so, mining remains among the most physically demanding occupations. While the overall magnitude of physical work performed by the miner has been reduced, many unique physical and environmental demands remain. For example, miners may have to deal with restricted workspace, less than desirable illumination, muddy or wet floor conditions, high levels of whole-body vibration, and considerable heavy lifting. Of the stressors listed above, the most demanding environmental characteristic of underground mines is undoubtedly the limited vertical workspace in which many miners must function. The impact of this single factor on human-centered design is extraordinary. The significant injury experience in the mining industry is undoubtedly the consequence of the multiplicity of risk factors present in this environment.

106.2 Epidemiology of Work-Related Musculoskeletal Disorders (WMSDs) in Mining

There is compelling evidence that work-related musculoskeletal disorders (WMSDs) affect mineworkers to a greater degree than workers in other industries (Lockshin et al., 1969). For example, studies have shown that miners experience more disability from knee and back pain (Lawrence, 1955; Lawrence and Aitken-Swan, 1952), more absenteeism (Duthie and Anderson, 1962), more osteoarthritis (Kellgren and Lawrence, 1952; Schlomka et al., 1955), and more disk degeneration (Kellgren and Lawrence, 1952) than comparison industrial populations. Back injuries emerge as a particularly serious problem in mining. A study by Klein et al. (1984) reported that the mining industry had the second highest incidence ratio for back injuries (1.5 claims/100 workers), trailing only the construction industry (1.6 claims/100 workers). Such injuries are consistently the single leading cause of lost-time injuries in U.S. coal mines (Peay, 1983), an experience shared by their international counterparts (Leigh et al., 1991). These injuries typically result from overexertion during the performance of manual materials handling tasks (Peters, 1983).

Manual handling of heavy materials is a pervasive activity in mining, and has been identified as a major contributing factor to sprain and strain injuries (Peay, 1983). The combination of heavy lifting and punishing environmental constraints has been linked to spinal changes in some studies. Lawrence (1955) examined British coal miners to identify factors related to degenerative disk changes, and found that injury, duration of heavy lifting, duration of *stooping*, and exposure to wet mine conditions were the factors most associated with spinal changes. Another study investigating spinal changes in miners was reported by MacDonald et al. (1984). These investigators used ultrasound to measure the spinal canal diameter of 204 coal miners and found that those with the greatest morbidity had significantly narrower spinal canals. The study by Lawrence (1955) and other evidence suggests that the seam height of the mine has a marked influence on the incidence of low back disorders. In general, compensation claims appear to be highest in seam heights of 0.9 to 1.8 meters (where stooping is prevalent). Claims are slightly lower in seams less than 0.9 meters (where kneeling and crawling predominate), and are substantially reduced when the seam height is greater than 1.8 meters. The finding of increased low back claims in conditions where stooping predominates is in concert with other evidence relating non-neutral trunk postures to low back disorders (Punnett et al., 1991). It is not surprising, given the physical demands and environmental constraints, that a field survey performed by the National Institute for Occupational Safety and Health (NIOSH) found that exposure to ergonomic hazards for miners was high compared to nonmining industries (Winn and Biersner, 1992).

While back injuries are more frequent in mining than in other industries, upper extremity CTDs do not appear as severe. This should not be too surprising since mining does not typically require highly repetitive or forceful exertions by the hands. As a result, the incidence rate of carpal tunnel syndrome is relatively small in mining (Hudock and Keran, 1992); however, it must be noted that a steady increase in the number of reported cases has been observed in recent years (MSHA, 1991). It is not clear whether this increase is due to changes in workplace factors (i.e., increasing use of remote controls or repetitive exertions in r*oof bolter's* tasks), or is a reporting artifact resulting from increased media attention given the disorder during this period.

106.3 Characteristics of the Underground Mining Workforce

Demographics

In 1986, the U.S. Bureau of Mines conducted a probability sample survey, the Mining Industry Population Survey (MIPS), to assess demographic characteristics of the United States mining workforces (Butani and Bartholomew, 1988a, 1988b). While the MIPS is by now somewhat dated, it remains our most recent look at the U.S. mining population, and contains several notable demographic findings. One finding that stands out is the almost exclusive dominance of male workers in the coal mining workforce. Ninety-eight percent of the coal workforce is male, and of the 2% that are female, only half work underground. Almost as dramatic is the dominance of Caucasians, comprising 94% of the total coal workforce. The cultural makeup of metal and nonmetal mines was a bit more varied, with whites, blacks, and Hispanics representing 82, 7, and 8% of the workforce, respectively. The difference in cultural makeup between coal and metal/nonmetal mines may be largely the result of the geographic location of the various types of mines. At the time of the survey, the average age of the coal workforce was 39; however, recent anecdotal evidence suggests that the mean age of the coal miners is now well into the 40s. Thus, ergonomics researchers and industry committees must consider the physical and cognitive effects of an aging workforce when carrying out ergonomics interventions.

Anthropometry

Physical characteristics of the underground mining population have been reported by several authors (Moss, 1934; Gary et al., 1955; Humphreys and Lind 1962, Ayoub et al., 1981b; Ayoub et al. 1984; Gallagher and Hamrick, 1992). By far the most in-depth studies (in terms of measures taken and sample size) are those by Ayoub and colleagues (Ayoub et al., 1981; Ayoub et al. 1984). These authors collected a battery of 42 anthropometric measurements for two major segments of the mining population: (1) low coal miners, and (2) non-low coal miners. Data for each of these populations were compared with those of other occupational groups to detect whether significant differences were evident. Comparison of the anthropometry of low coal miners vs. comparison industrial groups showed that miners were heavier,

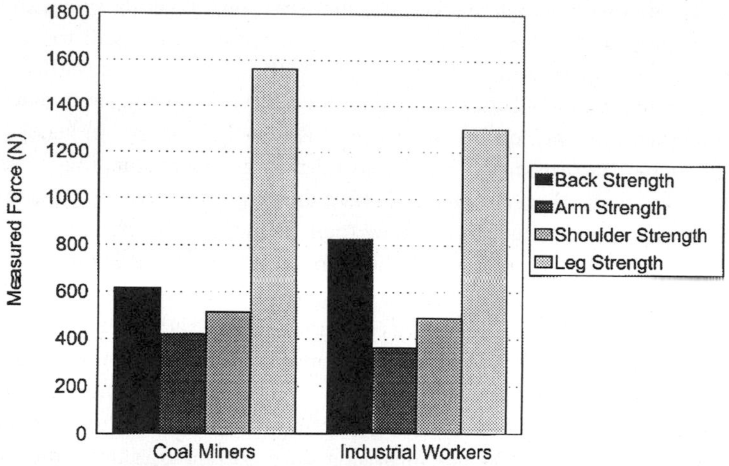

FIGURE 106.2 Comparison of strength measurements between low-seam coal miners and a traditional industrial population. (From Ayoub, M. M., Bethea, N. J., Bobo, M., Burford, C. L., Caddel, K., Intaranont, K., Morrissey, S., and Selan, J. 1981a. *Mining in Low Coal. Volume 1: Biomechanics and Work Physiology.* Final Report — U.S. Bureau of Mines Contract No. HO3087022. Texas Tech University, Lubbock, TX.)

and exhibited a related increase in the circumferences of the torso, arms, and legs (Ayoub et al., 1981b). Male low coal miners were also somewhat heavier than miners not working in low coal; however, the opposite trend was observed with female miners. Females working in non-low coal were also larger in all measures of circumference. This difference may be attributed to tasks performed, and varied geographical and ethnic makeups of the two groups.

Strength Characteristics of Miners

Various measures of isometric strength of underground miners were also obtained by Ayoub et al. (1981a) and Ayoub et al. (1984). These included back strength, shoulder strength, arm strength, sitting leg strength and standing leg strength. When compared with a sample of industrial workers (Ayoub et al., 1978), low-seam coal miners were found to have significantly lower back strength, but much higher leg strength (Figure 106.2). The authors ascribed the decrease in back strength to unspecified factors related to the postures imposed by the low-seam environment. Indeed, there is evidence to support this position. Low coal miners may be obliged to work in a stooping posture for extended periods. In this posture, the spine is largely supported by ligaments and other passive tissues, "sparing" the use of the back muscles. Studies of lifting in the stooping posture suggest that the gluteal muscles and hamstrings provide a large share of the forces in this position (Gallagher et al., 1988). The results of Ayoub et al. (1981a) may be due to a relative deconditioning of back muscles when stooping (due to relative inaction), and an increased reliance on the leg and hip musculature to perform underground work tasks (producing an increase in leg strength).

Aerobic Capacity

Several studies have investigated the maximal aerobic power of underground miners, using estimation techniques (Ayoub et al., 1981; Ayoub et al., 1984) or direct measurement (Kamon and Bernard, 1975; Kamon, Doyle, and Kovac, 1983). Most studies appear to agree that underground miners are inclined to have lower than average aerobic capacity compared with population norms and to comparison groups. The trend is evident for both genders (Ayoub et al., 1981a; Ayoub et al., 1984), and might be related to the finding, reported above, that underground miners exhibit increased body weight when compared with other groups. Kamon and Bernard (1975) found a steeper drop in maximal oxygen uptake and heart rate with age in miners than in other published data. However, it should be noted that other data has not shown as steep a decline (Ayoub et al., 1984).

106.4 Demands of Physical Work in Underground Mining

Imagine arriving at work one day to find that the ceiling of your workplace had been inexplicably lowered to 120 cm (approximately 4 ft.) above the floor. The impact of this restriction in workspace on the ability to perform normal work functions becomes immediately apparent. What once were routine tasks (for example, simply walking down the hall) suddenly become enormously demanding. Instead of walking erect, one is forced to walk fully bent over at the waist. Imagine further that part of your job for the day required considerable manual handling of heavy materials, for example, lifting or carrying 23-kg bags from one end of the hall to the other. As this scenario is contemplated, one can begin to get a picture of the unique physical demands that are present in the coal mining environment. As difficult as it may be to believe, the environmental restrictions described above might seem luxurious to some miners. Occasionally, miners perform physical work in vertical space restrictions so severe that crawling is not even possible. While this represents an extreme case, it is not at all uncommon for the mine to be any higher than 1.2 meters. In fact, about half of all coal mines in the U.S. fit this category. As will be discussed in this section, the physiological and biomechanical demands of doing manual work in such an environment are much greater than if this constraint were not present.

Daily Energy Expenditure of Miners

Before mechanization, the energy expenditure of underground coal miners remained relatively high throughout the workday. A study by Moss (1934) showed that the average daily energy expenditure for a coal miner before mechanization was approximately 4500 kilocalories per day. Rest periods were not of sufficient length to bring the oxygen consumption back down to a normal resting level. Modern mining, on the other hand, is characterized by short bursts of high energy expenditure tasks, interspersed with periods of rest or lower energy tasks. Figure 106.3 illustrates the ventilation volume and oxygen uptake for a roof bolter helper in a low-seam environment (Ayoub et al. 1981a). As can be seen in this figure, the roof bolting cycle contains periods where the energy expenditure is greater than 2 liters/min, and other periods where recovery is possible. The introduction of more frequent rest breaks from increased mining mechanization appear to be reflected in the reduction in shift energy expenditure of the coal miner. Depending on the specific job title of the miner, shift energy expenditures for miners in the late 1970s were found to range between 2100 and 2800 kcals (Ayoub et al., 1981a). However, it should be noted that these values are still at or above proposed maximum permissible limits for daily energy output for men (Banister and Brown, 1968; NIOSH, 1981).

Energy Expenditure for Specific Mining Tasks

Several studies have examined the energy expenditure of performing specific underground mining tasks (Ayoub et al., 1981a; Ayoub et al., 1984; Durnin and Passmore, 1967; Moss, 1934; Gary et al., 1955). Table 106.1 provides a summary of energy expenditure data for mining tasks from these sources. As can be seen from this table, many mining tasks fit into the category of heavy work (5.0 to 7.5 kcals/min), or very heavy work (7.5 to 10.0 kcals/min), based on the classification suggested by Astrand and Rodahl (1977). The table presented here is not exhaustive. Additional data on energy expenditure requirements for mining tasks are available (Durnin and Passmore, 1967; Ayoub et al., 1984).

Effects of Posture on Metabolic Cost

The posture adopted in the performance of a work task has a decided influence on the metabolic demands incurred by an individual. Nowhere is this more evident than in the evaluation of metabolic demands of working in constricted mining workspace (Moss, 1934; Bedford and Warner, 1955; Humphreys and

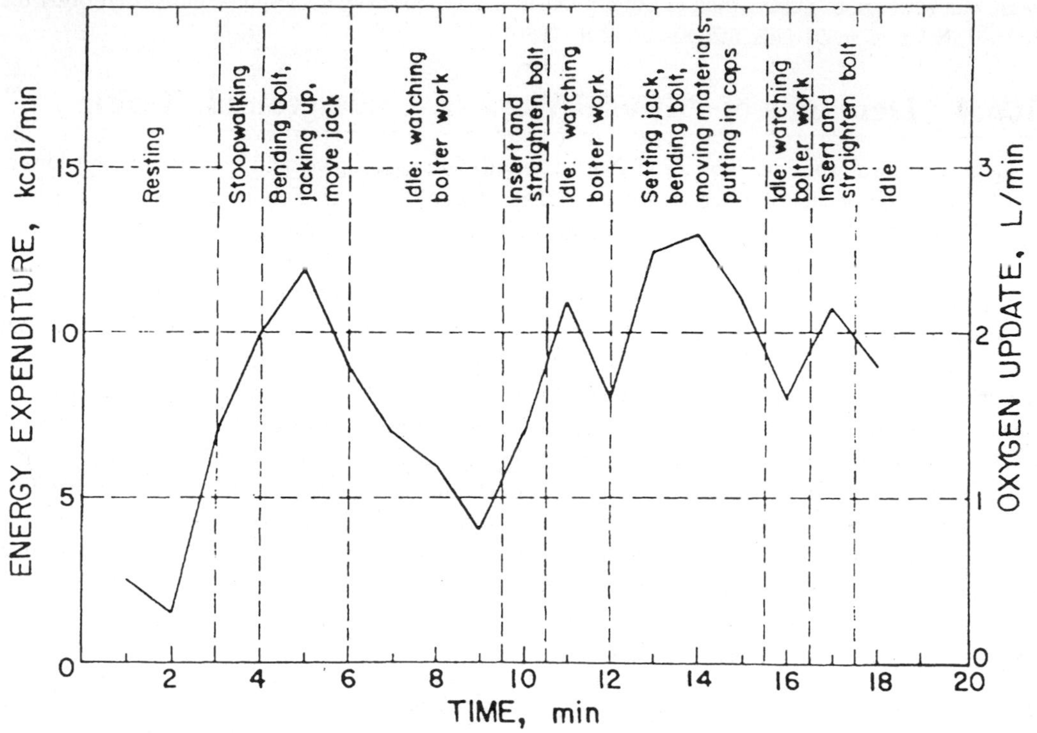

FIGURE 106.3 Energy expenditure and activity profile for a roof bolter helper. (From Ayoub, M. M., Bethea, N. J., Bobo, M., Burford, C. L., Caddel, K., Intaranont, K., Morrissey, S., and Selan, J. 1981a. *Mining in Low Coal. Volume 1: Biomechanics and Work Physiology.* Final Report — U.S. Bureau of Mines Contract No. HO3087022. Texas Tech University, Lubbock, TX.)

TABLE 106.1 Energy Expenditure for Selected Mining Tasks

Activity	Mean Energy Expenditure (kcal/min)	Standard Deviation	Range
Shoveling Coal			
Durnin and Passmore (1967)	7.0	—	5.1–9.4
Ayoub et al. (1981)	9.3	3.0	—
Garry et al. (1955)	6.9	0.9	—
Erecting Roof Supports	5.7	—	4.2–10.1
Helping			
Low Coal (Ayoub et al., 1981)	7.2	1.7	—
Non-Low Coal (Ayoub et al., 1984)	5.7	2.1	—
Roof Bolting			
Low Coal (Ayoub et al., 1981)	4.9	1.5	—
Non-Low Coal (Ayoub et al., 1984)	5.6	1.5	—
Timbering			
Ayoub et al. (1981)	6.0	2.4	—
Garry et al. (1955)	5.7	1.5	—
Humphreys and Lind (1952)	6.0	—	—
Scaling Roof (Ayoub et al., 1984)	8.5	—	—
Rock Dusting (Ayoub et al., 1984)	7.6	—	—
Moving Cables (Ayoub et al., 1984)	7.1	—	—
Jackleg Drilling (Ayoub et al., 1984)	7.1	—	—
Machine Maintenance (Ayoub et al., 1984)	6.2	—	—
Track Maintenance (Ayoub et al., 1984)	7.2	—	—

TABLE 106.2 Physiological Cost of Erect Walking, Stoopwalking, and Crawling

Task	Sex	Heart rate (beats/min)	Ventilation volume (L/min)	Percent Work Capacity	Oxygen uptake ($ml * kg^{-1} * min^{-1}$)
Normal Walk	Male	89.2 (5.4)	10.6 (0.4)	10.9 (0.9)	5.0 (0.9)
	Female	89.7 (3.6)	9.6 (0.7)	11.6 (2.2)	4.4 (0.6)
90% Stoopwalk	Male	96.0 (9.3)	12.8 (0.9)	12.5 (2.0)	5.7 (1.4)
	Female	107.5 (6.8)	12.4 (1.8)	15.3 (2.9)	5.8 (0.4)
80% Stoopwalk	Male	86.8 (15.8)	13.9 (1.8)	14.7 (2.3)	6.8 (1.5)
	Female	92.0 (12.7)	12.0 (0.6)	15.2 (2.2)	5.8 (0.2)
70% Stoopwalk	Male	82.2 (7.2)	13.2 (1.7)	15.1 (4.1)	6.8 (1.5)
	Female	89.9 (11.1)	11.0 (1.2)	15.7 (3.5)	6.0 (1.0)
60% Stoopwalk	Male	88.5 (7.2)	17.0 (2.3)	18.1 (1.4)	8.3 (1.0)
	Female	100.5 (21.6)	16.2 (5.3)	21.3 (5.0)	8.1 (1.8)
Crawling	Male	81.3 (11.3)	12.5 (1.3)	15.5 (2.3)	7.0 (0.5)
	Female	87.4 (7.8)	10.3 (1.0)	14.8 (2.7)	5.7 (1.8)

Numbers in parentheses represent the standard deviation.

Source: Ayoub, M. M., Bethea, N. J., Bobo, M., Burford, C. L., Caddel, K., Intaranont, K., Morrissey, S., and Selan, J. 1981a. *Mining in Low Coal. Volume 1: Biomechanics and Work Physiology.* Final Report — U.S. Bureau of Mines Contract No. HO3087022. Texas Tech University, Lubbock, TX.

Lind, 1962; Ayoub et al., 1981; Morrissey et al., 1985). Moss (1934) examined the physiological cost associated with normal walking, a "half-stoop" (80% of full stature), a "full-stoop" (60% of full stature), and walking on "all-fours" (50% of full stature) for eight experienced mining subjects. His finding showed that the half-stoop, full-stoop, and all-fours conditions increased the metabolic demands of walking 3.5 mph by 21%, 65%, and 73%, respectively. Similar trends were shown by Humphreys and Lind (1962), while the data from Bedford and Warner (1955) showed much higher increases in metabolic cost with *stoopwalking*. The most thorough experiment of the effects of stoopwalking and crawling was done by Ayoub et al. (1981a) and also reported by Morrissey et al. (1985). This study illustrated a progressive trend toward increasing metabolic cost as stooping becomes more severe (Table 106.2). Not only is the metabolic cost increased as stooping becomes more severe, the maximum speed attainable by subjects is reduced, particularly in stoopwalking at 60% stature and in crawling tests.

The metabolic cost of manual materials handling in restricted postures (stooping and kneeling) has also been studied (Gallagher and Bobick, 1988; Gallagher et al., 1988; Gallagher and Unger, 1990; Gallagher, 1991; Freivalds and Bise, 1991; Gallagher and Hamrick, 1992). These studies suggest that the metabolic cost of manual materials handling is not predominantly influenced by posture, but by an interaction between the posture adopted and the task being performed. For example, the kneeling posture is more costly than stooping when a lateral transfer of materials is done (Gallagher and Bobick, 1988; Gallagher et al., 1988; Gallagher and Unger, 1990). However, other studies have illustrated that kneeling can be more economical when the task requires increased vertical load displacement (Gallagher, 1991; Freivalds and Bise, 1991; Gallagher and Hamrick, 1992). A study of shoveling tasks in different postures by Morrissey et al. (1983) found no difference in energy expenditure in standing, stooping, and kneeling postures; however, only five subjects participated in this study and it may suffer from a lack of sufficient power to detect differences.

Manual Materials Handling in Restricted Postures

Mining is essentially an exercise in materials handing, some of which has been automated (especially the revenue-producing mineral extraction and transport segment), but much of which has not (movement of mining supplies, maintenance work, etc.). The amount of manual work that must be done in underground mines would be demanding enough without imposing restrictions in vertical workspace. As discussed below, such restrictions influence human strength capabilities, psychophysically acceptable workloads, and lifting biomechanics.

Effects of Restricted Postures on Strength

Studies examining static or dynamic strength capabilities in unusual or restricted postures are relatively rare. Haselgrave et al. (1987) reported results of isometric strength tests in kneeling vs. standing postures. These authors reported that lateral exertions were weaker when kneeling; however, pushing forces were equivalent in both postures. Pulling and lifting forces in the kneeling posture exceeded those in the standing position, by 25% and 44%, respectively. Results obtained by Gallagher (1989) were similar; however, this author reported higher pushing forces when kneeling on two knees than when standing. This study also studied a maximum upward push (with the force exerted upon a lifting handle at eye height), for which strength was not dependent on posture.

A study by Gallagher (1997) investigated trunk extension strength and electromyography of eight trunk muscles in standing and kneeling postures. Findings of this study showed that trunk extension strength is reduced by 16% in the kneeling posture in comparison with standing. However, normalized trunk muscle EMG was not significantly different between the two postures. Gallagher (1997) speculated that the reduction in trunk extension strength in the kneeling posture may be the result of a reduced capability to perform a strong rotation of the pelvis when the kneeling posture is adopted.

Psychophysically Acceptable Loads in Restricted Postures

Several studies of psychophysical lifting capacity in restricted postures have been done. Ayoub et al. (1987) reported psychophysical limits for a variety of unusual lifting postures including kneeling, sitting, lying down, and others. These data were collected with the assumption that atypical postures would be used only infrequently, for example, a one-time lift. Unfortunately, such postures may be used by miners for more prolonged bouts of manual lifting. As a result, the U.S. Bureau of Mines has examined the lifting capacity of underground miners in a variety of postures and vertical space constraints for tasks of longer duration (Gallagher et al., 1988; Gallagher and Unger, 1990; Gallagher, 1991; Gallagher and Hamrick, 1992). These investigators particularly wanted to quantify lifting capacity in the two most common postures used for lifting in *low-seam mines*— stooping and kneeling on both knees (Bobick, 1987). Results of these studies showed a significant decrease in lifting capacity in the kneeling posture, ranging from 8 to 18% (Gallagher et al., 1988; Gallagher and Unger, 1990; Gallagher, 1991; Gallagher and Hamrick, 1992). This deficit may be due to restrictions in the forces provided by the powerful hip and leg musculature in the kneeling posture, as compared with standing or stooping. In the stooping posture, the hamstrings and gluteal muscles appeared to provide a great deal of the force for the lifting tasks. Subjects consistently identified these muscle groups as those most sore after periods of lifting in the stooping posture (Gallagher et al., 1988). However, the effects of posture can be overridden by other MMH variables. For example, lack of a good handhold may reduce the acceptable load to a point that the differences related to posture are no longer evident (Gallagher and Hamrick, 1992).

A surprising, and somewhat disturbing, result of these studies is that psychophysically acceptable loads in prolonged stooping were generally on par with that achieved in the unencumbered standing position (Gallagher and Hamrick, 1992). Given the association of trunk flexion and incidence of low back disorders (Punnett et al., 1991), one would envision reduced load acceptability in this posture. However, as Snook (1985) has noted previously, the psychophysical approach does not seem sensitive to bending and twisting motions often associated with low back pain. Results of psychophysical studies in restricted postures seem to confirm this assessment. Further, these studies raise the issue of what drives subjective assessments of acceptable loads. Such assessments might be largely based on strength capabilities and physiological workload, rather than responses to the strain experienced by the low back. It may be advisable to base lifting limits for the stooping posture on biomechanical parameters, rather than relying on psychophysical estimates.

Biomechanics of Restricted Postures

Several methodologies have been employed to evaluate the biomechanical strain experienced during manual work in restricted postures. These have included use of intra-abdominal pressure (IAP),

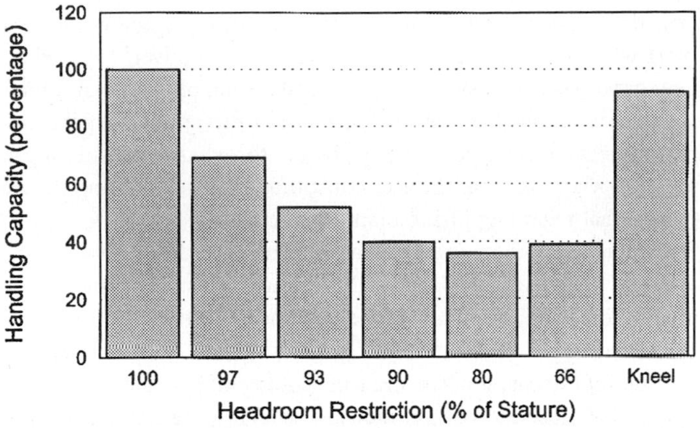

FIGURE 106.4 Handling capacity for stooping (percentages of full stature) and kneeling postures according to the IAP criterion. (From Ridd, J.E. 1985. Spatial restraints and intra-abdominal pressure. *Ergonomics*, 28(1): 149-166. With permission.)

electromyography (EMG), and estimations of L_5-S_1 moments. Because of their unique features, the restricted postures employed in underground mining have been found to present challenges to many traditional biomechanical models. As discussed in the following sections, there are often serious concerns whether certain ergonomic models are valid tools for the analysis of atypical postures.

Intra-abdominal Pressure (IAP)

Evaluation of spinal loading using the IAP criterion has been described by several authors (Davis and Troup, 1966; Davis and Ridd, 1981; Ridd, 1985; Sims and Graveling, 1988). Ridd (1985) found an almost linear decrement in lifting capacity with progressively lower vertical workspace up to 90% of stature, after which the decrement began to level off (Figure 106.4). At standing positions ranging from 66 to 90% of stature, the decrease in lifting capacity was 60%, according to the IAP criterion. The kneeling posture was found to incur only an 8% decrease in lifting capacity where the space restriction was equivalent to 75% of stature, according to Davis and Ridd (1981). Ridd (1985) also described the effects of asymmetric lifting activities on IAP. There was some indication that lifting asymmetrically is less stressful than sagittal plane activities in restricted postures. This result is in accord with psychophysical data showing that subjects were willing to accept greater loads asymmetrically in restricted postures (Gallagher, 1991). The reason may be that sagittal plane motions are precisely those most inhibited by vertical space constraints. Asymmetric motions are less affected by this restriction, leading to increased lifting capacity and decreased IAP responses.

There is currently much controversy regarding the role of IAP in spinal biomechanics. The original belief that IAP reduces the compressive loading on the lumbar spine has been disputed recently by many authors (Grillner et al., 1978; McGill and Norman, 1985; Nachemson et al., 1986). A particular concern is that the increase in IAP requires higher abdominal muscle activity, resulting in an additional compression penalty on the lumbar spine. Furthermore, the IAP does not always appear to respond to situations where spinal loading is known to be high, for example, when the spine is loaded asymmetrically (Andersson, 1982). In fact, IAP does not always produce consistent results with flexed postures (Sims and Graveling, 1988). At any rate, our understanding of the role of IAP in spinal biomechanics seems far from complete. As a result, some have recommended caution in using this mechanism to establish safe handling limits (Andersson, 1982; Ayoub and Mital, 1989).

Biomechanical Modeling

As mentioned previously, the robustness of many traditional biomechanical models may be put to the test in the analysis of atypical postures. As an example, while the spinal muscles provide the majority of lumbar support when the trunk is erect or moderately flexed, the spine employs a distinct "passive" mechanism of support when the trunk is fully flexed (Floyd and Silver, 1951; Floyd and Silver, 1955; Silver, 1954). While progress has been made in our understanding of this passive loading in recent years (Dolan et al., 1994), there remains a great deal to learn about this spinal support mechanism. Similarly, the kneeling posture is atypical of the type of lifting posture for which many biomechanical models have been formulated, most of which assume the feet constitute the base of support. Adopting a kneeling position changes a great deal regarding lifting biomechanics; for example, reducing leg muscle contributions to the lift. This may result in an increased agonist role for the erectores spinae.

An analysis reported by Gallagher and Unger (1988) illustrates some typical problems encountered with use of traditional biomechanical models in the analysis of restricted postures. These authors used the optimization model developed by Schultz and Andersson (1981) to estimate the compressive and shear loading in stooping and kneeling positions. The validity of this model in restricted postures was suspect, as model estimates of muscle forces contradicted EMG responses in the two postures. The analysis done in this study was admittedly beyond the scope of the validation performed by Schultz et al. (1982), where the most severe trunk flexion angle was 30 degrees. However, this example points out the problems run into repeatedly by the author. Many models available simply have not been validated for the analysis of unusual or atypical lifting postures.

Similar problems exist with the use of EMG-assisted biomechanical models with restricted postures (Gallagher et al., 1994). EMG-assisted models are greatly advancing our knowledge of trunk muscle responses and loading on the lumbar spine to manual materials handling activities (Marras and Sommerich, 1991a; Marras and Sommerich, 1991b). However, such models are generally valid under conditions where trunk muscles are the structures called upon to provide a restorative moment. However, the reliance on passive mechanisms (as opposed to muscles) in the stooping posture make it less amenable to such an approach. Use of an EMG-assisted model to evaluate restricted postures has uncovered complex muscle recruitment patterns (Gallagher et al., 1994); however, the issue of the passive component, estimated to account for 16 to 31% of the extensor moment in full flexion (Dolan et al., 1994) still needs to be resolved when using EMG-assisted models in the stooping posture.

A recent study of the mining task of handling heavy electrical cables examined the estimated peak L5-S1 moments resulting from hanging the cable under a variety of postures and vertical space limitations (Gallagher et al., 1995). The major finding of this study was that the peak moment when hanging cable was more highly related to the restriction in vertical space than to the posture employed. As can be seen in Figure 106.5, the greater the restriction in vertical space, the higher the peak moment experienced by the subject. It should be noted that this trend is evident even in the face of a contrary trend. That is to say, the higher the cable is lifted off the ground, the more weight is handled. Thus, lower moments were experienced with greater vertical space, though more weight was lifted under these conditions. Contrasts examining differences in peak moment in stooping and kneeling postures showed no difference in peak loading between these postures.

106.5 Summary of Physical Work in Restricted Postures

From the literature reviewed above, we can begin to develop an understanding of the stresses associated with working in restricted postures. Clearly, working in such postures exacts many tolls, including increased metabolic demands, decreased lifting capacity, and increased loading on the lumbar spine. What should also be clear, however, are some limitations of the parameters that we rely upon to provide insight regarding physical capabilities of workers. As discussed above, some of our parameters may disagree about which posture has the greatest limitation in handling capacity. If one were to look only at IAP data, we would conclude that handling capacity is only slightly reduced when kneeling. A different conclusion is reached when the psychophysical approach is used. Similarly, examination of the stooping posture using psychophysics may not suggest a deficit in acceptable workloads compared with standing, whereas biomechanical measures display increased strain on the low back. Clearly, reliance on only one measure or technique may not be sufficient to develop a full appreciation of the capabilities and/or

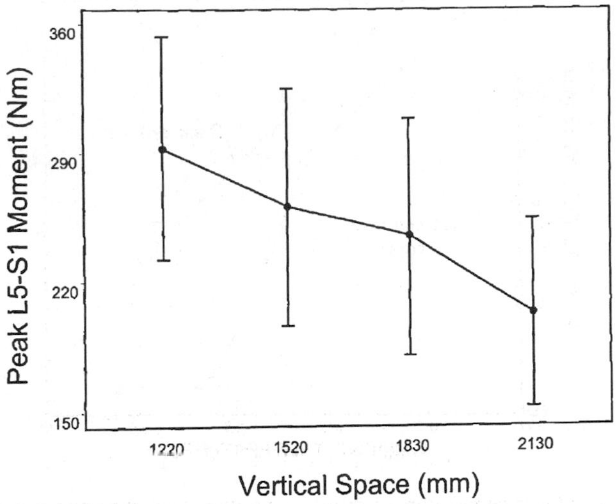

FIGURE 106.5 Relationship between vertical space restrictions and peak moment experienced during a cable hanging task. (From Gallagher, S., Hamrick, C.A., Cornelius, K., and Redfern, M. 1995. "Peak L5-S1 Moments Associated with a Cable Hanging Task," Poster Presentation at the 39th Annual Meeting of the Human Factors and Ergonomics Society Conference, San Diego, CA, October 9-13 1995.)

limitations associated with a particular working position. Examining a variety of measures, on the other hand, appears to help develop a better understanding of the multivariate stressors that workers experience, and may also point out the strengths and limitations of our own analysis techniques.

106.6 Heat Stress

Mining has long figured prominently among occupations routinely exposed to high heat stress (Martinson, 1977). This is particularly true for *deep mines* (for example, South African gold mines), mines sunk in hot countries, or in mines situated along zones where high heat flow from the earth occurs (Misaqi, 1991). Medical experts recognize that exposure to hot, humid conditions is both unhealthy and unproductive. Figure 106.6 illustrates the decreasing productivity of mineworkers loading mine cars and drilling rock as the ambient temperature is increased. Of course, many serious health problems are associated with heat exposure, including heat cramps, heat exhaustion, and heat stroke. The latter condition can often lead to death. However, heat stress can also have a significant impact on safety even below levels that may cause actual physical harm (Hancock and Vercruyssen, 1988). With even moderate heat exposure, workers may ignore unsafe working conditions, have decreased dexterity, coordination and cognitive ability, and are more apt to act emotionally. This may lead to rash acts by people performing hazardous jobs.

A great deal of our knowledge of the effects of heat stress, and on methods to control these effects, may be credited to the Human Sciences Laboratory of the Chamber of Mines of South Africa (Wyndham et al., 1973). In the 1920s, as the gold mines in the Witwatersrand were sunk to depths greater that 1800 meters, virgin rock temperatures continued to climb, and the ambient wet bulb temperatures in the mine began to exceed 30 degrees C. Heat stroke became an alarming problem; in 1930 alone, 27 deaths from heat stroke occurred. This rash of deaths prompted an intensive period of research, resulting in several control measures to protect mineworkers against the adverse effects of physical work in high heat and humidity conditions. These included better acclimatization to heat, recognition of heat intolerant individuals, definition of safe heat stress limits, and adaptation of microclimate cooling systems for use in the gold mining industry. Of these, the most important advances were in the field of heat acclimatization (Strydom, 1966). Initially, the acclimatization regimen was accomplished over a 12 to 14 day period by exposing miners to cooler production areas of the mine and progressing them to the hotter production

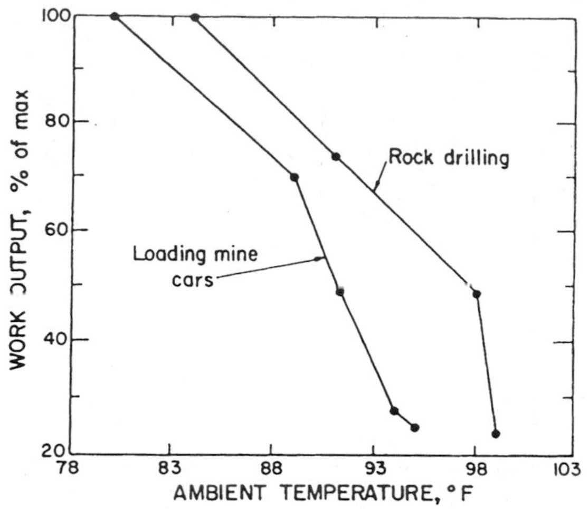

FIGURE 106.6 Effects of ambient temperature on miners' capacity to load mine cars and drill rock. (From Misaqi, F. L. 1984. *Heat Stress in Mining.* Safety Pamphlet No. 6, Mine Safety and Health Administration, Beckley, WV.)

areas. However, the laboratory continued to refine the procedure and ultimately initiated an 8-day regimen where workers were acclimatized by bench-stepping at a workload of 1 liter O_2/min in environmental chambers set at a wet bulb temperature of 32 degrees C. This procedure has been very successful in controlling the incidence of heat-related illnesses.

106.7 Equipment and Tool Design

Environmental conditions in underground mines not only affect the physical capabilities of underground workers, they also have a profound impact on the design of underground equipment and tools. For example, the restricted spaces in coal mines present a huge challenge for developing appropriately designed operator compartments and/or workstations. Underground mobile equipment may be so low profile that operators must lie completely down on their sides or back to operate the machine. This presents extreme problems for visibility and operator fatigue. Furthermore, mobile underground equipment may not have sufficient vertical space to provide systems for shock absorption, leading to serious whole-body vibration exposure. Add to this the problems of seating in mobile underground equipment. Often, seating consists of a steel seat welded to the frame of the vehicle. Illumination is also a critical design issue in underground mines, which are wholly dependent on artificial lighting systems. Other issues of concern include the design of hand tools and personal protective equipment used by miners.

Mobile Underground Equipment Design

Development of ergonomically designed operator compartments and workstations for underground mining equipment is an imposing task. The interaction of the confined space of the mine and the massive equipment required to mine the coal often results in operator compartments cramped and poorly designed. It is common to find operator compartments less than 76 cm in height and less than 61 cm wide. Visibility is almost inevitably influenced by the cramped conditions, often requiring the operator to lean out of the cab, which may expose him to hazards. Illumination systems in these confined spaces often cause the operator to be exposed to disability glare, further restricting visibility. In many cases, controls are designed and/or placed awkwardly. This increases the chance that controls may be improperly activated (or not activated at all) in emergencies. This is clearly a recipe for disaster (Conway and Unger, 1988). Fortunately, a great deal of international attention has been focused on the ergonomics of operator compartment and workstation design during the last 25 years. Major findings of this research are reported below.

Visibility

In the early 1980s, the U.S. Bureau of Mines sponsored research to learn critical visibility requirements for three common varieties of mobile underground equipment: shuttle cars, continuous miners, and scoops (Sanders and Kelley, 1981). Structured interviews and task analytic procedures were employed to evaluate the required visual information to perform tasks such as loading, hauling, or dumping. From this analysis, specific points were identified that must be visible to satisfy visibility requirements. These points, called visual attention locations (or VALs), were defined with reference to generic machine locations. For example, machine operators must be able to view a point on the ground sufficiently far away to stop their vehicle to avoid collision. The results of this VAL research have recently been incorporated into a computerized analysis package. Visibility analysis is automated to determine, for example, the relative visibility rating for a 5th percentile female or a 95th percentile male operator. This allows the designer easily to manipulate and optimize the visibility design of the operator's compartment, without the need to build expensive and time-consuming mockups.

Whole-Body Vibration

A study of mobile underground equipment operators in the early 1980s indicated that between 33 and 39% of operators are exposed to levels of whole-body vibration exceeding the ISO fatigue-decreased proficiency level (Remington et al., 1984). Between 7 and 14% were exposed to levels exceeding the ISO exposure limit. Other tests have been reported that miners may experience nearly 35% of the ISO 8-hour exposure limit simply riding for the 30 minutes it takes to get to the working face of the mine (Love et al., 1992). The effects of whole-body vibration have been studied in several Bureau of Mines experiments. A study by Bobick et al. (1988) examined the effects of 30 minutes of random whole-body vibration typical of a mine haulage vehicle on several measures, including back strength, dexterity, stature, heart rate, blood pressure, and subjective discomfort. Vibration was found to increase HR and BP, and discomfort ratings, but had no influence on back strength, stature, or dexterity. A subsequent study (Bobick et al., 1989) showed no compromise of back muscle strength nor endurance related to short-term vibration exposure. These authors suggest that the lack of effect on back strength suggests that the low back pain associated with whole-body vibration may depend on postural and mechanical effects rather than any change in the function of the back musculature. The effects of WBV exposure on postural stability have also been investigated (Cornelius et al., 1994). Balance seemed unaffected by short-term exposure to whole-body vibration.

Seating Design

There is an increasing awareness in the mining industry of the importance of providing appropriate seating for operators of underground mobile equipment. This is no doubt due to the sequelae of exposure to whole-body vibration. While this awareness is slowly infiltrating the industry, many seating designs for underground equipment currently remain relatively primitive. In fact, even some latest mining equipment provides only a bent steel plate bolted to the machine frame as the only means of operator support (Love et al., 1992). When this situation is combined with the inability to provide shock absorption systems (again, due to lack of sufficient clearance to dissipate the energy), the effects should become apparent. The restricted headroom in many underground mines provides additional challenges for the equipment designer. In very low mines, equipment operators may have to lie down to run mobile equipment. When conditions are not quite as severe, the operator may have to assume a reclining position that affects visibility. Given the problems enumerated above, it is not surprising that past attempts at improving seating by equipment manufacturers have not met with a great deal of success.

The goals for seating design in underground mobile equipment are similar to those in other applications: provide a stable position from which to control the machine, provide some isolation from vibration and jolting, and reduce the risk of postural fatigue (McPhee, 1993; Collier et al., 1986). There have been guidelines put forth for the design of seating in underground equipment for various seam heights (Collier

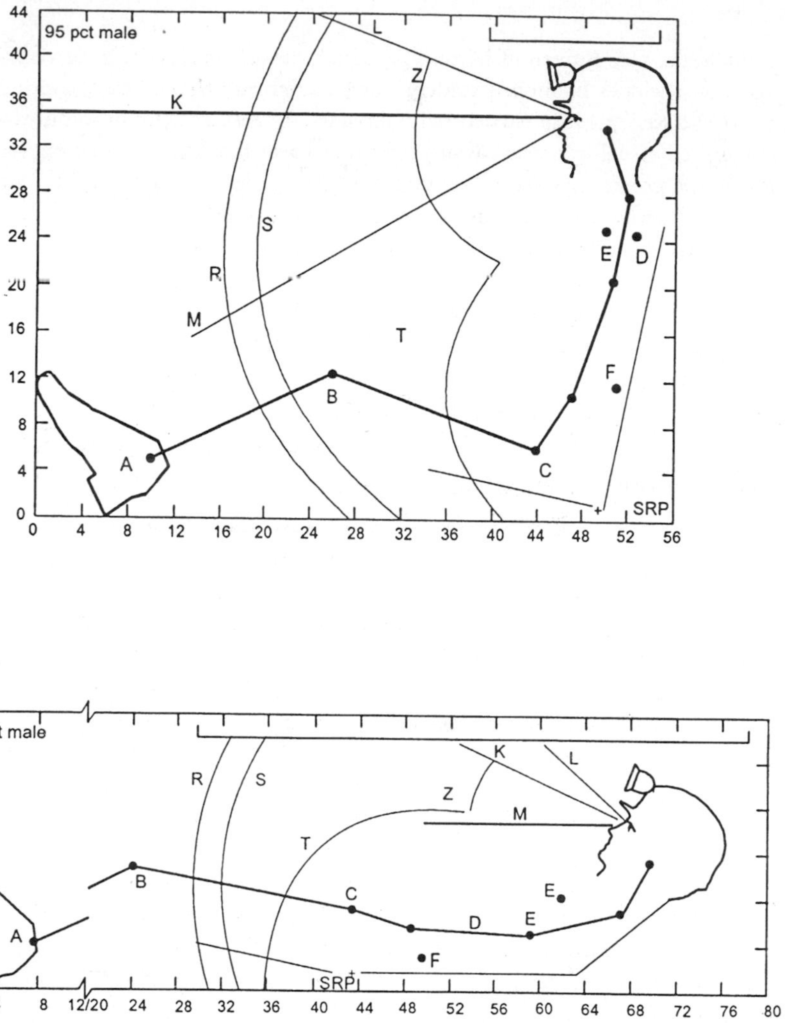

FIGURE 106.7 Seating space envelopes for 95th percentile miners in 107 cm (top) and 56 cm (bottom) cab heights. (From Canyon Research Group Inc. 1982. *Human Factors Design Guidelines for Personnel Carriers,* 37 pp., Canyon Research Group, Westlake Village, CA.)

et al., 1986; Mason, 1992; Canyon Research Group, Inc., 1982). Figure 106.7 illustrates the effects of vertical space restrictions on seating envelopes for the 95th percentile miner for two cabs: 107 and 56 cm in height (Canyon Research Group, Inc., 1982). Comparison of the two cabs clearly illustrates the increased cab length, reduced reach envelopes, and restricted field of vision associated with a reclined seating posture. Collier et al. (1986) present three design options for underground equipment seating. These included "normal" seating (for canopies > 1460 mm, of roof clearances > 1610 mm), a version with an increased backrest angle for canopies between 1440 and 1460 mm, and a "constant eye-height" option for canopies between 1135 and 1285 mm. A special consideration for mining equipment seating is the provision of sufficient space so that the operator's *caplamp battery* and *self-contained self-rescuer* (*SCSR*) can be worn on the belt. An example of such a design is provided in Figure 106.8 (McPhee, 1993).

Illumination

Proper design of illumination systems is a critical issue in the mining environment. In fact, underground mines (and surface nighttime operations) are often completely dependent upon artificial lighting systems.

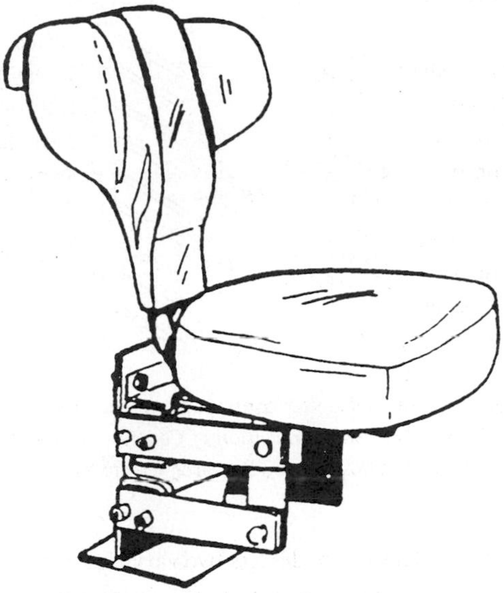

FIGURE 106.8 A prototype mining operator's seat developed using ergonomics principles and operator feedback. Note cutouts in the seatback, accommodating belt-worn personal protective equipment. (From McPhee, B. 1993. Ergonomics for the Control of Sprains and Strains in Mining. 36 pp., Handbook published by the National Occupational Health and Safety Commission (Worksafe Australia), Sydney. With permission.)

As humans receive the bulk of their information visually, the quantity and quality of illumination provided by lighting systems are extremely important for safety, productivity, and morale of the mining workforce (Sanders and Peay, 1988). Part 75.1719 of the Code of Federal Regulations discusses illumination requirements for different areas of underground mines, along with electrical standards, and other requirements to increase visibility in the mine environment (Lewis, 1986). The primary illumination standard states that underground equipment at the working face must provide *luminance* of at least 0.06 foot-Lamberts on the coal face. The illumination required to provide 0.06 foot-Lamberts (two footcandles) is adequate for most mining tasks, but is not so bright that severe adaptation problems will occur when the miner must go into darker areas of the mine (Lewis, 1986).

Presently, laboratory mockups are a major part of the process lighting equipment manufacturers (LEMs) must go through to design an approved underground machine-mounted illumination system. Moreover, when a lighting system is modified by using different *luminaires* within an existing configuration or by changing a luminaire's location or orientation, additional laboratory measurements may be necessary to approve the system. The U.S. Bureau of Mines, aware of the difficulties described above, has developed a PC-based computer model that will enable users to easily design, alter, and evaluate underground machine-mounted illumination systems without having to build a mockup or prototypes. A new method of modeling illumination provided by underground mobile equipment luminaires has been developed as part of this model (Gallagher et al., 1996). The Mine Safety and Health Administration has recently authorized the use of this computer model in the certification of mobile underground equipment illumination systems.

Control Design

An analysis of underground mining fatalities for the years 1972 and 1979 indicated that more than 7% of mining fatalities in underground coal mining during that period were associated with improper design of controls. The number of nonfatal injuries caused by poor control design remains unknown; however, it is safe to assume that this number is also substantial (Sanders and Peay, 1988). More recently, a spate of injuries among roof bolter operators also implicated control design as a contributing factor. A typical problem observed in mining equipment is lack of control standardization (Helander et al., 1980). An

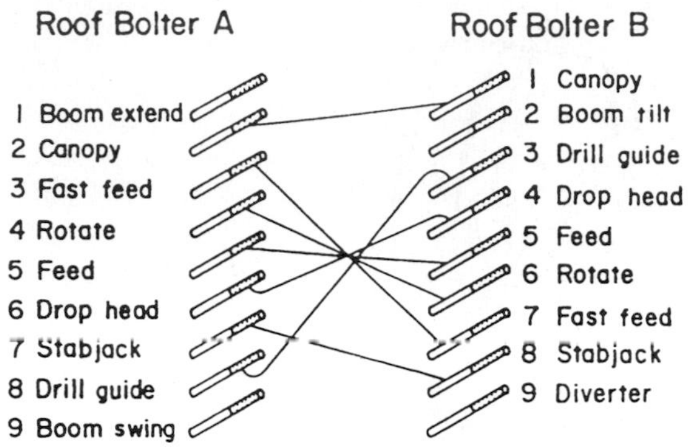

FIGURE 106.9 Different arrangements of roof bolter controls from the same manufacturer. (From Helander, M., Conway, E.J., Elliott, W., and Curtin, R. 1980. *Standardization of Controls for Roof Bolter Machines. Phase I. Human Factors Engineering Analysis* (contract HO292007, Canyon Research Group Inc.). Bureau of Mines PFR 170-82, NTIS PB 83-119149, 192 pp.)

example is provided in Figure 106.9. This figure depicts two varying arrangements of roof bolter controls, both coming from the same manufacturer. It is not an uncommon occurrence to have machines with different control arrangements working in the same mine, but perhaps in different *working sections.* If a worker who usually works on roof bolter A is suddenly called upon to fill in on roof bolter B, the risk of improper control activation is greatly increased. Many similar situations may be cited. Unfortunately, this is not an area where a great deal of progress has been made in mining. It is important that human factors design principles be considered in the design of new equipment, and, what is more important, that mines insist on good human factors design in their procurement process.

Maintainability Design

In the early days of mining automation, equipment consisted of relatively simple machines that could be easily maintained using simple hand tools. These have since been replaced by increasingly powerful and complex mining systems (Conway and Unger, 1991). The demands on the maintenance function have increased concomitantly. Unfortunately, little regard is given in the design of this equipment for ease of maintenance and serviceability. The following problems are most frequently observed in mining equipment (Long, 1983): (1) poor access to machine parts or areas for routine maintenance tasks, (2) inadequate access openings to reach parts needing repair or replacement, (3) need to remove or dismantle ancillary components to gain access to the failed unit, (4) inadequate provisions for safe handling of heavy or large parts, and (5) inadequate tools to perform required maintenance tasks. As a result of such design deficiencies, relatively simple maintenance tasks are turned into complex, time-consuming procedures. Recently, some recommendations for improving the maintainability design of mining equipment have been published (Conway and Unger, 1991). These recommendations contain both maintainability engineering information for equipment manufacturers and a buyer's guide for the evaluation of the maintainability design of mining equipment.

Ergonomics Design Guidelines

Some mining operations now require ergonomics evaluations to be performed on major items of underground equipment prior to their purchase (Mason, 1992). This is a positive development in terms of implementing ergonomics in mining, but also demands that appropriate design information be provided to engineers, designers, and purchasers of this equipment. In the late 1980s and early 1990s, a number of ergonomics design handbooks were developed for specific pieces of underground machinery (Canyon

Research Group, 1982; Collier et al., 1986; Mason, 1992). An effort to develop a generalized ergonomics measurement tool resulted in the Bretby Operability Index (Mason, 1992). This index provides a means through which an initial ergonomics screening of new equipment can be performed. Areas requiring a more detailed assessment can be quickly identified, so that ergonomists can focus developing recommendations in these areas.

106.8 Hand Tool Design

Most analyses indicate that hand tools are involved with between 7 and 10% of all nonfatal lost-days accidents in the mining industry (Marras et al., 1988a, 1988b; Sanders and Peay, 1988). Most of these injuries (approximately 75% or more) are associated with nonpowered hand tools. Many traditional hand tools are used in the mining industry. Of these, hammers, wrenches, and knives are most commonly implicated in accidents (Sanders and Peay, 1988). However, there are many specialty tools used for specific mining tasks, for example, the *scaling bar* or the *jack leg drill.* Many of these specialty tools also appear to have very high frequency and severity rates associated with them (Marras et al., 1988a, 1988b). The types of injuries associated with mining hand tools are most often struck-by and overexertion. There is some indication that the awkward postures observed in low-seam coal mines may contribute to both types of injuries. Research in this area has resulted in recommended design changes that may be useful in improving handtool design (Marras and Lavender, 1988). As an example, a counterbalanced scaling bar was developed which was found to significantly reduce compressive forces on the spine compared with the conventional scaling bar (Marras and Lavender, 1991).

106.9 Personal Protective Equipment

In response to the multiplicity of mining hazards, miners are equipped with an extensive array of personal protective equipment (PPE) (Sanders et al., 1981). This equipment includes (at a minimum) ear protection, safety glasses, respirator, hard hat, cap lamp and battery, mining belt, overalls, gloves, safety boots, self-rescue device, and in low-seam mines, knee pads. Unfortunately, much of this equipment has not received proper consideration with regard to ergonomics nor the unique environmental conditions present underground. However, in the past 15 years, a number of studies have been performed to address issues pertaining to miners' PPE. This research has resulted in the development of an improved slip-resistant tread design for mining boots, improved overall designs with retro reflective materials for increased detectability, and improved knee pad designs (Sanders et al., 1981).

Another item of PPE that is currently generating some controversy is the introduction of belt-worn self-contained self rescuers (SCSRs). These devices are heavier than the *filter self rescuers* (*FSRs*) they replaced, but have the benefit of immediately providing the miner with an hour's worth of oxygen to escape a mine fire. The FSRs worn previously were lighter; however, they were not protective from all noxious gases produced by a mine fire and still required the user to don another device before escape was possible. Some miners feel the new devices are too heavy and cumbersome, and are resisting the change. Research is currently under way to examine methods of improving the ergonomics design of these devices.

106.10 Status of Knowledge and Unresolved Issues

Our knowledge of ergonomics issues in mining has increased a tremendous amount over the last two decades. However, we would be sadly mistaken if we were to imply that all (or even a sizable portion) of the relevant issues were resolved. There remains much to be learned and surely some to be discarded from what we believe we know presently. The following list discusses some of the most important unresolved issues, in the author's mind, that need to be addressed:

1. Ergonomics research in mining has far outstripped its implementation by the industry. While there have been significant ergonomics success stories in mining, by and large effective mining ergonomics committees are relatively rare. How can we improve the dissemination of ergonomics information and facilitate the development of committees in this industry, particularly among smaller mines with limited resources?
2. The unique stresses of working in restricted postures do not appear to be well-addressed by many of our current ergonomics models. As we continue to develop our knowledge of the ergonomics and biomechanics of traditional industrial tasks, it is important not to limit our research too narrowly. We should continue to broaden the applicability of our models to atypical postures and unusual situations. This process would undoubtedly provide us with better insight into the adaptive mechanisms used by the body, and may well spawn new ergonomics and biomechanics paradigms.
3. Recent years have seen a rapid growth in the development of new mining technologies. However, human-centered design principles have often been neglected in the design and implementation of new equipment and new technologies (Randolph and Love, 1991). Increased technology transfer efforts must be focused on the manufacturing sector of the mining industry, so that equipment can be designed to facilitate increased productivity, decreased risk of accidents, and improved worker satisfaction and comfort.

Defining Terms

Caplamp: The lamp worn by a miner on his safety hat or cap for illumination purposes. The caplamp is powered by a rechargeable battery worn on the miner's belt.

Deep mine: A mine where the coal or mineral deposit is at a depth exceeding 915 m (3000 ft). Some gold mines have been sunk to depths exceeding 3050 m (10,000 ft).

Filter self-rescuer (FSR): A protective device, worn on the miner's belt, to be worn in the event of a mine fire or explosion. This unit protects the miner from the potentially lethal effects of carbon monoxide inhalation.

Jack leg drill: A percussive type of automatically rotated rock drill that is worked by compressed air. This drill has a supporting bar (leg) that allows the drill to be used to drill into vertical mineral faces.

Low-seam mine: In general, a mine where the mineral seam is less than 1.2 m (4 feet) in thickness.

Luminaire: A complete lighting unit. These are mounted on underground mobile equipment to ensure compliance with mine lighting regulations.

Luminance: The luminous intensity of a surface in a given direction per unit of surface area.

Roof bolter: In bituminous coal mining, a machine used to drill holes and install bolts into the roof of the mine to prevent rock and slate falls. The term may also refer to the operator of this machine.

Scaling bar: A barlike implement used to remove loose rock from the roof of the mine in order to prevent this rock from falling unexpectedly and injuring a worker.

Self-contained self-rescuer (SCSR): A self-sufficient breathing apparatus which provides respiratory protection in oxygen-deficient or highly toxic atmospheres. In contrast to the filter self-rescuer (which only provides protection from carbon monoxide), this unit isolates the wearer's lungs completely from the toxic atmosphere.

Stoop: A working posture involving bending the trunk forward and down, sometimes simultaneously bending the knees, commonly used in the cramped spaces in underground coal mines.

Stoopwalking: Walking in a stoop posture.

Working section: The area of the mine where the coal, ore, or mineral is being mined.

References

Andersson, G.B.J. 1982. Measurement of loads on the lumbar spine, in *Symposium on Idiopathic Low Back Pain.* p. 220-251, Mosby, St. Louis.

Astrand, P.O. and Rodahl, K. 1977. Textbook of Work Physiology, McGraw-Hill Book
Company, New York, NY.

Ayoub, M.M. and Mital, A. 1989. *Manual Materials Handling.* Taylor and Francis, London.

Ayoub, M.M., Bethea, N.J., Deivanayagam, S., Asfour, S.S., Bakken, G.M., Liles, P., Mital, A., and Sherif, M. 1978. Determination and modeling of lifting capacity. Final report, Grant #5R010H-0054502, HEW, NIOSH. Texas Tech University, Lubbock, TX.

Ayoub, M. M., Bethea, N. J., Bobo, M., Burford, C. L., Caddel, K., Intaranont, K., Morrissey, S., and Selan, J. 1981a. *Mining in Low Coal. Volume 1: Biomechanics and Work Physiology.* Final Report — U.S. Bureau of Mines Contract No. HO3087022. Texas Tech University, Lubbock, TX.

Ayoub, M. M., Bethea, N. J., Bobo, M., Burford, C. L., Caddel, K., Intaranont, K., Morrissey, S., and Selan, J. 1981b. *Mining in Low Coal. Volume II: Anthropometry.* Final Report — U.S. Bureau of Mines Contract No. HO3087022. Texas Tech University, Lubbock, TX.

Ayoub, M. M., Selan, J.L., Burford, C. L., Intaranont, K., Rao, H.P.R., Smith, J.L., Caddel, D.K., Bobo, W.M., and Bethea, N.J. 1984. *Biomechanical and Work Physiology Study in Underground Mining Excluding Low Coal.* Final Report — U.S. Bureau of Mines Contract No. JO308058. Texas Tech University, Lubbock, TX.

Ayoub, M.M., Smith, J.L., Selan, J.L., Chen, H.C., Lee, Y.H., Kim, H.K., and Fernandez, J.E. 1987. Manual materials handling in unusual postures. Technical Report, Department of Industrial Engineering, Texas Tech University, Lubbock, TX.

Banister, E.W. and Brown, S.R. 1968. The relative energy requirements of physical activity. Chapter 10 in *Exercise Physiology* Academic Press, Inc., New York, NY.

Bedford, T. and Warner, C.G. 1955. The energy expended while working in stooping postures. *Br. J. Ind. Med.* 12: 290-295.

Bobick, T. G. 1987. Analyses of materials-handling systems in underground low-coal mines, in *Proceedings of Bureau of Mines Technology Transfer Seminar, Bureau of Mines Information Circular 9145*, p. 13-20, U.S. Bureau of Mines, Pittsburgh, PA.

Bobick, T. G., Gallagher, S., and Unger, R.L. 1988. Pilot subject evaluation of whole-body vibration from an underground mine haulage vehicle, in *Trends in Ergonomics/Human Factors V*, Ed. F. Aghazadeh, p. 521-528. Elsevier Science Publishing Co., Amsterdam.

Bobick, T.G., Gallagher, S., and Unger, R.L. 1989. Effects of random whole-body vibration on back strength and back endurance, in *Advances in Industrial Ergonomics and Safety I*, Ed. A. Mital, p. 537-544. Taylor and Francis, London.

Butani, S.J. and Bartholomew, A.M. 1988a. Characterization of the 1986 coal mining workforce. *Bureau of Mines Information Circular 9192*, 57 pp., U.S. Bureau of Mines, Minneapolis, MN.

Butani, S.J. and Bartholomew, A.M. 1988b. Characterization of the 1986 metal and nonmetal mining workforce. *Bureau of Mines Information Circular 9193*, 49 pp. U.S. Bureau of Mines, Minneapolis, MN.

Canyon Research Group Inc. 1982. *Human Factors Design Guidelines for Personnel Carriers*, 37 pp., Canyon Research Group, Westlake Village, CA.

Collier, S.G., Chan, W.L., Mason, S., and Pethick, A.J. 1986. *Ergonomic Design Handbook for Continuous Miners.* 128 pp., Institute of Occupational Medicine, Edinburgh, Scotland.

Conway, K. and Unger, R.L. 1991. Ergonomic guidelines for designing and maintaining underground coal mining equipment, in *Workspace Equipment and Tool Design*, Ed. A. Mital and W. Karwowski, p. 279-302. Elsevier Sciences Publishing Co., Amsterdam.

Cornelius, K.M., Redfern, M.S., and Steiner, L.J. 1994. Postural stability after whole-body vibration exposure. *Int. J. Ind. Ergon.* 13(4): 343-352.

Davis, P.R. and Troup, J.D.G., 1966. Effects on the trunk of erecting pit props at different working heights. *Ergonomics.* 9(6): 475-484.

Davis, P.R. and Ridd, J.E. 1981. The effect of restricted headroom on acceptable lifting capacity. *Ergonomics.* 24: 239.

Dolan, P., Mannon, A.F., and Adams, M.A. 1994. Passive tissues help the back muscles to generate extensor moments during lifting. *J. Biomech.* 27(8): 1077-1085.

Durnin, J.V.G.A. and Passmore, R. 1967. *Energy, Work, and Leisure.* London: Heinemann Educational Books, Ltd., 166 pp.

Duthie, J.J.R. and Anderson, M.A.D. 1962. Social and economic effects of rheumatic disease. *Arch. Environ. Health,* 4: 511-518.

Floyd, W.F. and Silver, P.H.S. 1951. Function of the erectores spinae in flexion of the trunk. *Lancet,* Jan 20: 133-138.

Floyd, W.F. and Silver, P.H.S. 1955. The function of the erectores spinae muscles in certain movements and postures in man. *J. Physiol.,* 129: 184-203.

Freivalds, A. and Bise, C.J. 1991. Metabolic analysis of support personnel in low-seam coal mines. *Int. J. Ind. Ergon.,* 8(2): 147-155.

Gallagher, S. 1989. Isometric pushing, pulling, and lifting strengths in three postures. *Proceedings of the Human Factors Society 33rd Annual Meeting,* pp. 637-640, Human Factors Society, Santa Monica, CA.

Gallagher, S. 1991. Acceptable weights and physiological costs of performing combined manual handling tasks in restricted postures. *Ergonomics.* 34(7):939-952.

Gallagher, S. 1997. Trunk extension strength and trunk muscle activity in standing and kneeling postures, *Spine.* 22(16): 1864-1872.

Gallagher, S. and Bobick, T.G. 1988. Effects of Posture on the Metabolic Expenditure Required to Lift a 50-Pound Box, in *Trends in Ergonomics/Human Factors V,* Ed. F. Aghazadeh, p. 927-934, Elsevier Science Publishing Co., Amsterdam.

Gallagher, S. and Unger, R.L. 1988. A biomechanical analysis of manual materials handling tasks in restricted working postures. *Proceedings of the Human Factors Society 32nd Annual Meeting, p.* 670-674, Human Factors Society: Santa Monica, CA.

Gallagher, S. and Unger, R.L. 1990. Lifting in four restricted lifting conditions. *Appl. Ergon.* 21(3):237-245.

Gallagher, S. and Hamrick, C.A. 1991. The kyphotic lumbar spine: Issues in the analysis of the stresses in stooped lifting. *Int. J. Ind. Ergon.* 8:33-47.

Gallagher, S. and Hamrick, C.A. 1992. Acceptable workloads for three common mining materials. *Ergonomics.* 35(9):1013-1031.

Gallagher, S., Marras, W.S., and Bobick, T.G. 1988. Lifting in stooped and kneeling postures: Effects on lifting capacity, metabolic costs, and electromyography at eight trunk muscles. *Int. J. Ind. Ergon.* 3(1):65-76.

Gallagher, S., Hamrick, C.A., Love, A.C., and Marras, W.S. 1994. Dynamic biomechanical modeling of symmetric and asymmetric lifting tasks in restricted postures, *Ergonomics.* 37(8): 1289-1310.

Gallagher, S., Hamrick, C.A., Cornelius, K., and Redfern, M. 1995. "Peak L5-S1 Moments Associated with a Cable Hanging Task," Poster Presentation at the 39th Annual Meeting of the Human Factors and Ergonomics Society Conference, San Diego, CA, October 9-13 1995.

Gallagher, S., Mayton, A.G., Unger, R.L., Hamrick, C.A., and Sonier, P. 1996. Computer Design/Evaluation Tool for Illuminating Underground Coal Mining Equipment. *J. Illum. Eng. Soc.* 25(1): 3-12.

Gary, R.C., Passmore, R., Warnock, G.M., and Durnin, J.V.G.A. 1955. Studies on expenditure of energy and consumption of food by miners and clerks, Fife, Scotland 1952. Medical Research Council Special Report Series.

Grillner, S.J., Nilsson, J., and Thorstensson, A. 1978. Intra-abdominal pressure changes during natural movements in man. *Acta Physiol. Scand.* 103: 275-283.

Hamrick, C.A, Cornelius, K.M., Rossi, E.W., and Unger, R.L. 1993. Appropriate ingress/egress dimensions for mobile underground mining equipment. *Int. J. Ind. Ergon.* 11:13-18.

Hancock, P.A., and Vercruyssen, M. 1988. Limits of behavioral efficiency for workers in heat stress. *Int. J. Ind. Ergon.* 3(2): 149-158.

Haselgrave, C.M., Tracy, M., and Corlett, E.N. 1987. Biomechanical effects of force exertions while kneeling. Paper in *Proceedings of the 31st Annual Meeting of the Human Factors Society*, p. 318-322, Human Factors Society, Santa Monica, CA.

Helander, M., Conway, E.J., Elliott, W., and Curtin, R. 1980. *Standardization of Controls for Roof Bolter Machines. Phase I. Human Factors Engineering Analysis* (contract HO292007, Canyon Research Group Inc.). Bureau of Mines PFR 170-82, NTIS PB 83-119149, 192 pp.

Hudock, S.D. and Keran, C.M. 1992. Risk profile of cumulative trauma disorders of the arm and hand in the U.S. mining industry. *Information Circular 9319*, p. 1-5, U.S. Bureau of Mines, Minneapolis, MN.

Humphreys, P.W. and Lind, A.R. 1962. The energy expenditure of coal-miners at work. *Br. J. Ind. Med.*, 19: 264-275.

Kamon, E. and Bernard, T. 1975. *Physiological Responses of Coal Miners to Emergency.* Annual Technical Report, Grant No. GO-115006, The Noll Laboratory for Human Performance Research, The Pennsylvania State University, University Park, PA.

Kamon, E. Doyle, D., and Kovac, J. 1983. The oxygen cost of an escape from an underground coal mine. *Am. Ind. Hyg Assoc. J.* 44(7): 552-555.

Kellgren, J.H. and Lawrence, J.S. 1952. Rheumatism in coal miners. Part II. X-ray study. *Br. J. Ind. Med.* 9: 197-207.

Klein, B.P., Jensen, R.C., and Sanderson, L.M. 1984. Assessment of workers' compensation claims for back strains/sprains. *J. Occ. Med.* 26(6):443-448.

Lawrence, J.S. 1955. Rheumatism in coal miners. Part III. Occupational factors. *Br. J. Ind. Med.* 12: 249-261.

Lawrence, J.S. and Aitken-Swan, J. 1952. Rheumatism in miners. I. Rheumatic complaints. *Br. J. Ind. Med.* 9: 1-18.

Leigh, J., Mulder, H.B., Want, G.V., Farnsworth, N.P., and Morgan, G.G. 1991. Sprain/strain back injuries in New South Wales underground coal mining. *Safety Sci.* 14:35-42.

Lewis, W.H. 1986. *Underground Coal Mine Lighting Handbook* (In two parts: 1. Background and 2. Application) USBM, IC 9074, 1986, 42 and 89 pp., U.S. Bureau of Mines, Pittsburgh, PA.

Lockshin, M.D., Higgins, I.T.T., Higgins, M.W., Dodge, H.J., and Canale N. 1969. Rheumatism in mining communities in Marion County, West Virginia. *Am. J. Epidemiology* 90(1):17-29.

Long, D.A. 1983. Solving the problem of getting on and off large surface mining equipment. Paper in *Bureau of Mines Information Circular 8947*, pp. 3-16, U.S. Bureau of Mines, Minneapolis, MN.

Love, A.C., Unger, R.L., Bobick, T.G., and Fowkes, R.S. 1992. *A Summary of Current Bureau Research into the Effects of Whole-Body Vibration and Shock on Operators of Underground Mobile Equipment*, U.S. Bureau of Mines Report of Investigations 9439, p. 1-15, Pittsburgh, PA.

MacDonald, E.B., Porter, R., Hibbert, C., and Hart, J. 1984. The Relationship Between Spinal Canal Diameter and Back Pain in Coal Miners. *J. Occ. Med.* 26(1): 23-28.

Marras, W.S. and Lavender, S. 1988. *An Analysis of Hand Tool Injuries in the Underground Mining Industries.* U.S. Bureau of Mines Final Report JO348043, 234 pp. The Ohio State University, Columbus, OH.

Marras, W.S., Lavender, S., Bobick, T., Rockwell, T., and Lundquist, R. 1988. Risks of hand tool injury in U.S. underground mining from 1978 through 1983, Part II: Metal-nonmetal mining. *J. Safety Res.*, 19(3):115-124.

Marras, W.S., Bobick, T.G., Lavender, S.A., Rockwell, T.H., and Lundquist, R.L. 1988. Risks of hand tool injury in U.S. underground mining from 1978 through 1983: Part I: Coal mining. *J. Safety Res.*, 19(2):71-85.

Marras, W.S. and Lavender S. 1991. The effects of method of use, tool design, and roof height on trunk muscle activities during underground scaling bar use. *Ergonomics.* 34(2):221-232.

Marras, W.S. and Sommerich, C.M. 1991a. A three dimensional motion model of loads on the lumbar spine: I. model structure, *Human Factors.* 33, 123-138.

Marras, W.S. and Sommerich, C.M. 1991b. A three dimensional motion model of loads on the lumbar spine: II. model validation. *Human Factors.* 33, 139-150.

Martinson, M.J. 1977. Heat stress in witswatersrand gold mines. *J. Occ. Acc.* 1:171-193.

Mason, S. 1992. Improving the ergonomics of British Coal's mining machinery. *Appl. Ergon.* 23(4):233-242.

McGill, S.M. and Norman, R.W. 1987. Reassessment of the role of intraabdominal pressure in spinal compression. *Ergonomics.* 30(11): 1565-1588.

McPhee, B. 1993. Ergonomics for the control of sprains and strains in mining. Handbook published by the National Occupational Health and Safety Commission (Worksafe Australia), 36 pp.

Misaqi, F.L. 1984. *Heat Stress in Mining.* Safety Pamphlet No. 6, Mine Safety and Health Administration, Beckley, WV.

Morrissey, S., Bethea, N.J., and Ayoub, M.M. 1983. Task demands for shoveling in non-erect postures. *Ergonomics.* 27: 847-853.

Morrissey, S.J., George, C.E, and Ayoub, M.M. 1985. Metabolic costs of stoopwalking and crawling. *Appl. Ergon.*, 16(2): 99-102.

Moss, K.N. 1934. The energy output of coal miners during work. *Trans. Inst. Min. Eng.* 189: 132-149.

MSHA 1991. *Carpal Tunnel Syndrome.* Safety Manual No. 23, Mine Safety and Health Administration, Beckley, WV.

Nachemson, A.L., Andersson, G.B.J., and Schultz, A.B. 1986. Valsalva maneuver biomechanics: effects on lumbar trunk loads of elevated intraabdominal pressure. *Spine.* 11(5): 476-479.

NIOSH 1981. *Work Practices Guide for Manual Lifting.* NIOSH publication 81-122, NTIS PB 82-178948, 183 pp.

Peay, J.M. 1983. Introduction. Paper in *Back Injuries. Bureau of Mines Information Circular 8948*, p. 2, U.S. Bureau of Mines, Pittsburgh, PA.

Peters, R.H. 1983. Activities and objects most commonly associated with underground coal miners' back injuries. Paper in *Back Injuries. Bureau of Mines Information Circular 8948*, p. 22-31, U.S. Bureau of Mines, Pittsburgh, PA.

Punnett, L., Fine, L.J., Keyserling, W.M., Herrin, G.D., and Chaffin, D.B. 1991. Back disorders and nonneutral trunk postures of automobile assembly workers. *Scand. J. Work Environ. Health.* 17: 337-346.

Randolph, R.F. and Love A.C. 1991. Ergonomics in mining: Human factors and organizational research on high-technology mining. *Appl. Occ. Environ. Hyg.* 6(7):577-580.

Remington, P.J., Andersen, D.A., and Alakel, M.N. 1984. *Assessment of Whole-Body Vibration Levels of Coal Miners. Volume II: Whole-Body Vibration Exposure of Underground Coal Mining Machine Operators* (contract JO308045, Bolt, Beranek, and Newman, Inc.). Bureau of Mines OFR 1B-87, NTIS PB 87-144119, 114 pp.

Ridd, J.E. 1985. Spatial restraints and intra-abdominal pressure. *Ergonomics.* 28(1): 149-166.

Sanders, M.S. and Kelley, G.R. 1981. *Visual Attention Locations for Operating Continuous Miners, Shuttle Cars, and Scoops. Volume I* (contract JO387213) Bureau of Mines OFR 29(1)-82, NTIS PB 82-187964, 142 pp.

Sanders, M.S. and Peay, J.M. 1988. *Human Factors in Mining.* U.S. Bureau of Mines Information Circular 9182, U.S. Bureau of Mines, Pittsburgh, PA.

Sanders, M.S., Peay, J., and Bobick, T.G. 1981. Research on the development of personal protective equipment for underground mines. Paper in *Bureau of Mines Information Circular 8866*, p. 70-83, U.S. Bureau of Mines, Pittsburgh, PA.

Schlomka, G., Schroter, G., and Ocherwal, A. 1955. Uber derbedeutung der berufkischer Belastung fur die entsehung der degenerativen Gelenkleiden. *Z. Gesamte Inn Med.* 10:993-999.

Schultz, A.B. and Andersson, G.B.J. 1981. Analysis of loads on the lumbar spine. *Spine.* 6(1): 76-82.

Schultz, A., Andersson, G., Ortengren, R., Haderspeck, K., and Nachemson, A. 1982. Loads on the lumbar spine: Validation of a biomechanical analysis by measurements of intradiscal pressures and myoelectric signals. *J. Bone Joint Surg.* 64A(5): 713-720.

Silver, P.H. 1954. Direct observations of changes in tension in the supraspinous and interspinous ligaments during flexion and extension of the vertebral column in man. *J. Anat.* 88: 550-551.

Sims, M.T. and Graveling, R.A. 1988. Manual handling of supplies in free and restricted headroom. *Appl Ergon.* 19(4): 289-292.

Snook, S.H. 1985. Psychophysical considerations in permissible loads. *Ergonomics.* 28(1): 327-330.

Strydom, N.B. 1966. The need for acclimatising labourers for underground work. *J. Mine Vent. Soc. S. Afr.* 19: 124-125.

Winn, F.J. and Biersner, R.J. 1992. "Exposure probabilities to ergonomic hazards among miners." Poster presented at 36th Annual Meeting of the Human Factors Society, Atlanta, GA, October 12-16, 1992.

Wyndham, C.H., Strydom, N.B. Benade, A.J.S., and Rensburg, A.J. 1973. Limiting rates of work for acclimatization at high wet bulb temperatures. *J. Appl. Physiol.* 35: 454.

For Further Information

An excellent source of information pertaining to ergonomics in mining is the text *Human Factors in Mining* by Sanders and Peay. This text is becoming a bit dated (it was published in 1988); however, it remains among the most comprehensive treatments available.

107

Risk Assessment and Safety Management in Industry

Juraj Sinay
Technical University of Košice

All human activities carry a risk. The risk can apply to health and safety, e.g., immediate health damage or a gradual negative impact to health due to harmful substances or it can cause economical losses, e.g., as a result of machine failure or destruction due to fire or explosion. The aim of the activities in the field of risk management is to control, eliminate, or minimalize the possibility of the risk of death, injury, illness, or damage to technical equipment or environment. At the same time the improvement of safety conditions and health protection at work in the man–machine–environment system is also observed with

0-8493-2641-9/99/$0.00+$.50

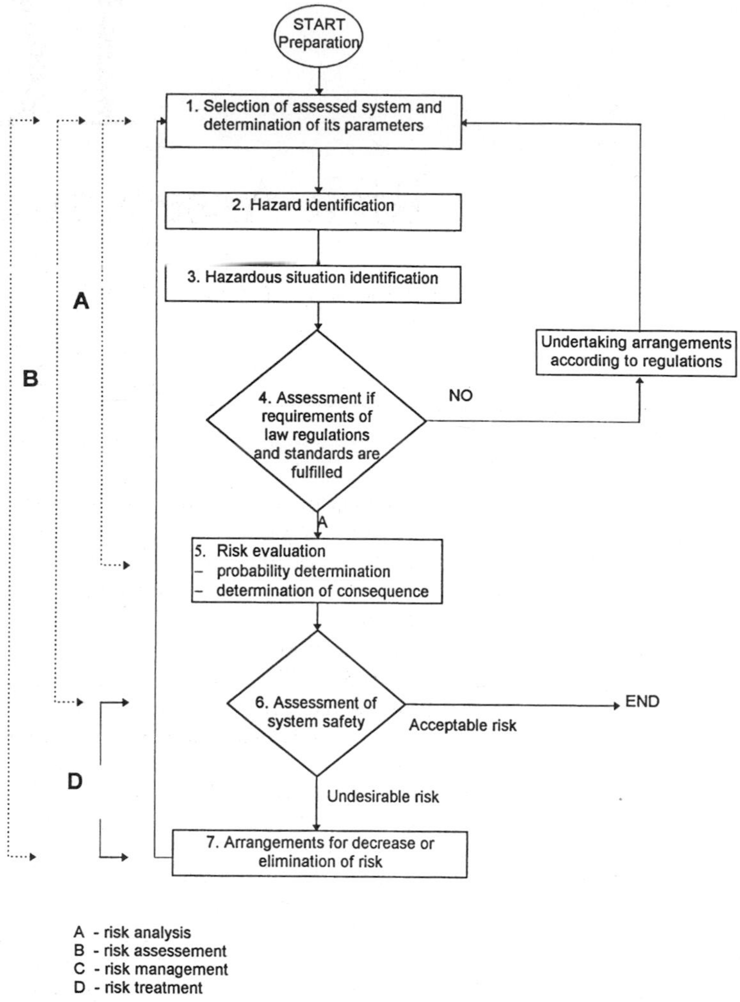

FIGURE 107.1 Risk Management scheme.

the goal to prevent economical losses due to an ineffective system of managing the work safety. There is also a legal requirement to present to inspection organs a systematic fulfillment of safety requirements according to the established legal regulations.

However, it is the task of an employer, or all the units of the man–machine–environment system, to keep the risk under control, i.e., to create conditions so that it achieves the value of an acceptable risk at most. This requires the possibility to analyze and classify risks in such a way that it is possible to carry out arrangements for their control with the aim to eliminate or minimalize them, i.e., to use the activities complex from the risk management system[1,2] (Figure 107.1). It is also possible to use simpler approaches to risk management in smaller and middle-sized companies which are shown in Reference 1.

The activities in the framework of risk management are performed in the man–machine–environment system which involves people, technological procedures, used materials, tools, machines and devices, software, fauna, and flora. Risk management mainly includes the following activities:

- Definition of the aim the analysis is performed for
- To set up the time plan and strategy of the risk assessment
- Establishing a working group and other persons who will participate in the process of risk assessment and management as well as providing their training

- To provide the essential information for risk assessment
- Description of the method used for the risk assessment
- To involve the executives in the accomplishment of the results of risk assessment and monitoring
- Installation of the risk assessment results into work organization and staff training
- To provide systematic repeated risk assessment

Risk assessments can be of different scope, content, form, method, and various procedures may be used. It depends on:

- *The purpose the analysis is to be done for.* There is one procedure when a designer evaluates the machine construction, and a different one when an operator detects that the machine in operation endangers somebody. There is a different procedure to find out the reasons for injuries, and a different one to demonstrate the accordance with safety regulations.
- *The kind of hazard and hazardous situation.* Different methods are used for chemical operations, different ones for machinery operations, different ones for special activities, e.g., chimney destruction.
- *The size of the operation, the branch of economical activity.* The larger the company, the more complex the work organization and the higher risk possibility, the more detailed will be the assessment.

The risk assessment strategy is based on the purpose of the analysis. In the framework of the strategy there is a determined proceeding for the risk assessment, definition of the assessed systems, danger detection, especially with regard to the operation site area, definition of all the operation procedures and operation activities. It is necessary to involve all the staff and other persons who might be exposed to hazard.

The time plan will put the single stages of the risk assessment into accordance as well as monitor the assigned tasks accomplishment.

107.1 Who Performs the Activities Within the Scope of Risk Assessment and Management?

The performer of the activities within the scope of risk management must be a competent specialist with professional education and rich experience in the job as well as in the field of work safety. The employer might manage the risk assessment, especially if the operation is small and simple. In more complex operations, where there are more technical installations and technologies, and various activities are performed, it is necessary to assess the risks in a complex and systematic way. A group of experts from the following areas should deal with risk assessment:

a. System analysis
b. Theory of probability and mathematical statistics
c. Technical areas, e.g.,: mechanics, material engineering, elasticity and strength, electrotechnics, nuclear technic, machine building
d. Natural-science branches, e.g., chemistry, physics, biology
e. Medical field, toxicology and epidemiology included
f. Sociology, national economy branches, psychology, ergonomy
g. Management theory

The risk assessment consists of a set of logical steps, however, its efficiency depends greatly on experience as well as decision-making skills of the assessors. Hence, the selection of these people and fulfilling the following principles are important:

- A working group is authorized to assess the risks, not only one person — thus the subjectivity of a human factor is eliminated;
- There should be an expert in the risk assessment theory as well as those who have knowledge of the operation, technology, or installation being assessed;

- It is necessary to involve the staff in risk assessment — technologists, technicians, executives, as well as employees working directly in the operation;
- It is an advantage to increase the number of external specialists since it provides an independent, professional view, free from "operation blindness";
- From two to five coordinators should work in a working group, and other specialists will be called in according to the kind of hazard;
- It is necessary for the assessors to be qualified in their field.

They should have the ability to identify the hazard, to analyze how this hazard could endanger a man and to be capable of assessing the probability and consequence of the negative event:

- The employer should enable these people to take part in professional training not only in technical areas, but also in the methods of logical analysis, modeling, and evaluation
- Regardless of who does the risk assessment, the employer is responsible for the level of the risk assessment and assumed arrangements in every case

107.2 Provision of Information for Assessors

The persons who are to assess the risk should have the knowledge of a wide range of information, such as:

- Work organization and operation procedures;
- Machines, installations, technologies, and materials being utilized;
- Statistics, analysis, and injury accident rate;
- Sources of risk, risks already known;
- Relationship between the source of risk and its impact;
- Probability and consequence of hazards;
- Number of endangered persons, the extent of expected damages;
- Regulations, standards, and demands on safety.

This information may be obtained from the following sources:

- Technical and operation documentation of machines and technologies, organizational and technical regulations of the company, written directions and instructions, and operation procedures
- Data on injury accident and disease rate, undesirable events, failures, information on "almost accidents"
- Records from internal and external audits
- Consultations with specialists and staff
- Legislative regulations, standards, including the European ones, technical and scientific literature, instructions

Efficient risk control requires an analysis of the risk as the first step. When planning the goals that are to be achieved by a risk analysis, it is necessary to take the following activities into consideration:

- Selection of the assessed system and determination of its parameters
- Identification of hazard
- Identification of hazardous situation
- Assessment of the extent of the compliance with the law regulations
- Risk evaluation

All the stages of the technical life of the machine have to be the object of the analysis of the risk of technical systems (Figure 107.2).

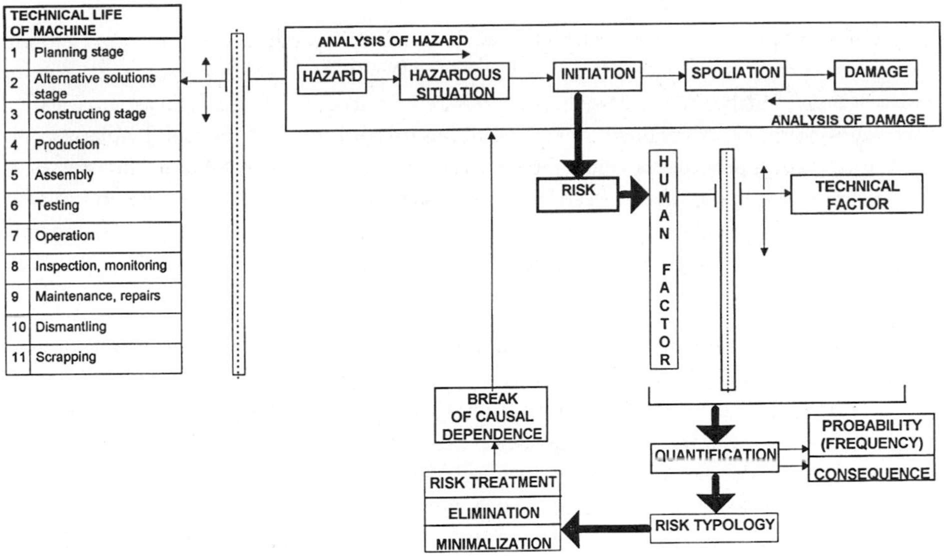

FIGURE 107.2 Scheme of analysis of technical systems risk.

a. Risks at the stage of projecting and constructing
b. Risks at the production stage
c. Risks at the stage of assembly and dismantling
d. Risks during operation
e. Risk during maintenance and repairs
f. Risks at re-valorization of a part of the system or at its depreciation (scrapping, destruction)

Risk evaluation is the final stage of the risk analysis and it requires an analysis of probability of a negative event arising, the analysis of frequency, analysis of possible consequences, and their mutual relationship.

The procedure creates conditions for defining risk factors, i.e., the parameters that influence the probability and consequence of a negative event during the whole technical life of the installation.

It is possible to use the results of the risk analysis for the following:

- Determination of the conditions for risk acceptability
- Comparison of the risk with risks in the framework of other technologies, or activities
- Working out the proceedings for introduction of the arrangements for decrease or elimination of the risk at technical risks, e.g., by way of efficient maintenance activities
- Assessment of the efficiency of chosen arrangements for minimalization or elimination of the risk
- Preventing the rise of economic losses

The quality of the undertaken analysis requires perfect knowledge of the examined system, e.g., the machinery installations complex, or complex technology. The knowledge of the system altogether with verification of the suitability of the used kind of risk analysis enable the utilization of the results obtained from the analysis of one object with a certain level of adjustment also for similar installations or objects.

107.3 Selection of Assessed System and its Definition

The first step of the systematic risk assessment is a selection of the assessed system. The assessed system might be a machine, installation, technology, working space, working activity, used material, etc., which is to be examined. A precise definition of the assessed system will show where a hazard occurs.

There are actually two possible ways of selection:

a. To list all the operation spaces, machines, installations, technological knots, working activities and materials where it is possible to suppose hazard for the life and health of people. Each item of the list will be an independent assessed system where it is necessary to perform an analysis.
b. According to the recommended general list of hazards, it is possible to find the places at the sites and in operation procedures where these hazards occur. These places will be an assessed system. For example, for the hazard of electric current it is necessary to find out where and in what kind of activities it is possible to suppose the hazard for life and health of people. The advantage of this is that the next steps of the analysis will be the same for more assigned places — assessed systems.

In larger companies these lists could be endless, therefore it depends on the assessors to decide where there is the highest presupposition of damage. The employees could be particularly helpful at this selection. It is necessary to take into regard not only common working process, but also exceptional activities, (e.g., if a heavy burden is being carried up the stairs, special repairs, etc.). It is necessary to pay particular attention to dangerous substances whose undesirable impact can cause relevant losses. (A special regime is established for the prevention of major industrial breakdowns.)

Defining the assessed system, i.e., definition of its parameters, should also be a part of this step (for example, tension, speed of lifting, concentration, temperature, etc.). It is important due to the fact that in a situation when it would be necessary to take arrangements, one of the possible ways could also be the change of parameters of the assessed system that may be considered to be the risk factors.

The level of operators or users of the installation is also a part of the system characteristics. The essential thing is if the operators or users are trained, experienced or there might appear third persons, i.e., nonprofessional people.

107.4 Identification of Hazard

If a chosen system is being assessed (e.g., machine, activity, working space), it is essential to identify the properties in this system that might cause a negative event in the form of injury, health endangering, or machine failure.

The proceeding is as follows: the assessors discuss with designers and schemers, technicians, maintenance men, executives, and workmen at the site about how they perceive the real condition of a machine during its operation, as well as what they know of particular dangers at the site and of their negative impact.

Another possibility is a systematic investigation of all the aspects of the assessed system according to the documentation, injury accident rate statistics, and other data, and to search for the dangers directly at the site.

The records on injury accidents that have already happened, as well as experience obtained from the results of the risk analyses performed in the past, can provide usable information for identification of hazard. It is necessary to realize that assessing these figures has a high level of subjectivity and therefore the identified hazards might not be the only ones that occur in the system.

The lists of kinds hazard and hazardous situation are included in standards and manuals that might be a good tool for orientation in this field.

As for practical use, some examples of hazard can be used that apply to some working activities and situations:

a. Working installations
   - Insufficient protection of rotating and movable parts
   - Free motion of parts or material (falling, rolling, sliding, tilting over, flying away, swinging, warping) that might strike a person
   - Motion of machines and means of transport

- Danger of fire or explosion (friction, pressure tanks)
- Capture, cutting, pulling-in, stabbing, stroke, bruising, amputation (mechanical hazardous situations)

b. Working habits and arrangement of sites
- Dangerous surfaces (sharp edges, corners, points, rough surfaces, slippy surfaces of outstanding parts)
- Work in altitudes
- Work in unsuitable position (one-sided load)
- Limited space (work among fixed parts)
- Stumble and slip (damp and slippy surfaces)
- Stability of worker
- Impact of use of OOPP
- Working techniques and methods
- Entrances and work in closed space

c. Use of electricity
- Electric switches of machines
- Electric installation
- Electric equipment, operating devices, isolation
- Mobile electric equipment
- Electric energy that might cause fire or explosion
- Overhead electric lines

d. Exposure to substances dangerous health
- Breathing-in, consummation, or absorption of dangerous substances through skin, aerosols and fine parts included
- Use of inflammable and explosive materials
- Oxygen shortage
- Presence of agents
- Reactive substances
- Irritant substances

e. Exposure to physical factors
- Electromagnetic radiation (heat, X-ray, ionizing)
- Lasers
- Noise and ultrasound
- Hot substances and environment
- Cold substances and environment
- Media under pressure (pressed air, steam, liquids)

f. Exposure to physical factors
- Risk of infection by micro-organisms (exo- and endotoxines)
- Presence of allergens

g. Factors of environment and working climate conditions
- Unsuitable lighting
- Unsuitable temperature, humidity, ventilation
- Pollution, mess

h. Relation between working place and human factor
- Safety system depends on obtaining and processing precise information
- Dependence on staff knowledge and skills
- Dependence on good communication and right instructions for change of conditions
- Consequence of supposed not complying with the safe operation procedures
- Suitability of personal working preventatives
- Weak motivation to work safely
- Ergonomic factors

i. Psychological factors
   - Working loading (intensity, monotony)
   - Site dimensions, e.g., claustrophobia, loneliness at the site
   - Impact of conflicts
   - Impact of decision-making in a stress (loading) situation
   - Low level of work management
   - Reactions in case of an emergency situation

j. Work organization
   - Factors of working process (night work, rest, ...)
   - Management of safety and health protection at work
   - Strategy and management of maintenance, especially safety equipment
   - Providing the investigation of injury accidents and extraordinary situations

k. Other factors
   - Dangerous acting of other persons
   - Work with animals
   - Unfavorable weather conditions
   - Changing the sites
   - Work under water, etc.

107.5 Identification and Analyses of Hazardous Situations

Once the hazards have been identified it is necessary to determine how they can cause an injury, damage, failure — negative event. One or more hazardous situations can be derived from one hazard. When identifying hazardous situations, the assessors should consider the following aspects:

- Who might be exposed to the effect of the hazard?
  Not only production staff is taken into regard, but also auxiliary and service activities — maintenance men, cleaners, workers of other firms and operations, visitors, emergency service and rescuers, excursions, etc.
- What is the reach of the hazard impact?
  It is necessary to know about hazardous situation zones, as well as the hazard impact; for example, the zone of hazard in case of escape of a dangerous substance, dangerous reach space of a crane.
- Characteristics of hazard and the way of initiation, creating dangerous situations and level of protection.
  Hazardous situation also depends on the parameters of the assessed system and hazard. For example, other possibilities of injury arise at higher speed. The extent of hazardous situation also depends on the level of possible protection. For example, if the hazard is electric current that is objectively present in the investigated system, it is not necessary to take the hazardous situation into consideration if the isolation of conductors is sufficient and circuit protected.

The methods for hazardous situations identification can be divided into two groups:

a. Comparative methods — the method of questionnaires and method of catalogue pages for hazardous situations, or risk and methods based on the use of data from the past
b. Basic methods based on the answers to "What happens if...". It is possible to use HAZOP (Hazard and Operability Studies) and FMEA (Fault Modes and Effect Analysis).

Inductive methods use assumed course of the system behavior. These methods use combinations of possible negative events in the system and basing on them the result at the end of the whole process is assessed — prospective methods. The methods ETA (Event Tree Analysis) and PHA (Preliminary Hazard Analysis).

107.6 When Are the Hazardous Situation Analyses Performed?

a. If it is necessary to obtain data for decision-making in the field of system safety
b. If the existence of hazard in individual working systems was confirmed on the basis of various data
c. If an obvious increase of work injuries, accidents or failures, or occupational diseases occurs at particular work places
d. If the risk management system is to be applied

A Particular Example of Identification and Analysis of Hazardous Situations for a Steel Plant[3]

A great significance is attached to the determination of a hazardous situation at a site as a basis for risk evaluation. There is a duty for the employer to find the hazardous situation at the site that emerges from the European Union law No. 9, paragraph 1a and Guidelines 89/391/EU. To comply with this requirement it is necessary to determine the kinds of hazardous situation that are characterized by:

- Dependence on injury accidents — only hazardous situations that have already led to damage are registered
- Indirectness — coming out of the damages that have occurred, the energy that caused them is assessed
- Limitation — only the hazardous situation of direct injury and not the hazardous situation of health are taken into consideration

It is possible to determine a hazardous situation by way of analysis that is:

- Independent of the kind of negative event — the hazardous situations that have not caused any injury or material damage are traced
- Direct — coming out of the known energy, the possibilities of a negative event are assessed
- Extensive — direct injuries as well as indirect injuries (almost injuries) are taken into account

The basis for carrying out the hazardous situation analysis is as follows:

- Information on the consequences of negative events that have already occurred, e.g., injury accidents
- Information on diseases that arose during work
- Information from the staff of the site where the analysis is being done as well as information obtained by systematic assessing of the safety of machines and installations

To ensure that the activities in the framework of safety assessment and health protection are efficient, it is essential that hazardous situation analyses be oriented toward the already arisen damages, and mainly toward the operations where there is the highest level of hazard, i.e., risk. In the area of metallurgical technologies it is the operation "Steel plant," where hazardous situation analysis has been carried out.[3] Its accomplishment comes out of the following presumptions:

- Performance of a preliminary analysis
- Determination of hazardous situation on the basis of information from four companies of metallurgical industry
- Detection of hazardous situation will only be limited to normal operation; the conditions of failure will not be considered in the given example
- Information on sources of negative events will be used in hazardous situation determination

TABLE 107.1 Phases of Production Technology in the Operation of Converter Steel Plant

Main system	Converter steel plant
Subsystem	1. Filling pit/pig iron mixer
	2. Pig iron processing
	3. Manipulation with slag
	4. Steel alloying
	5. Melting process — convertors
	6. Steel treatment — secondary metallurgy

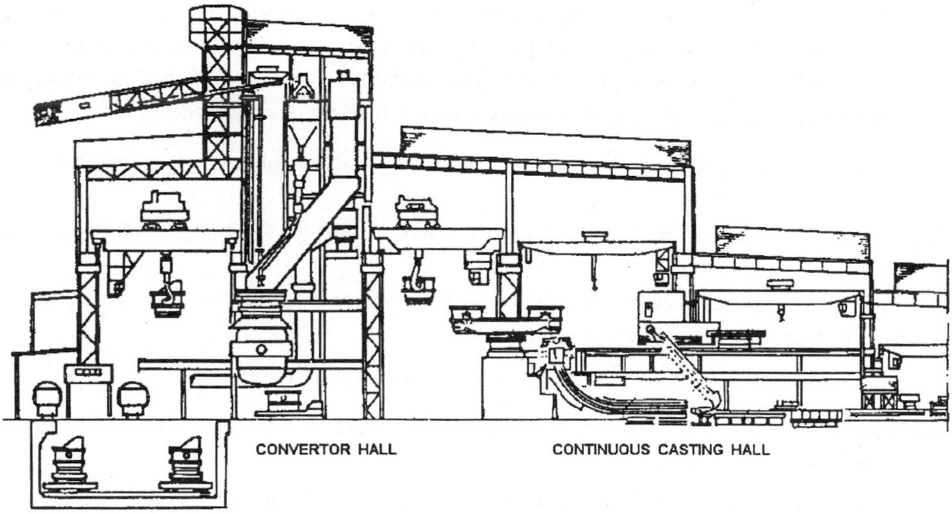

FIGURE 107.3 Converter steelplant scheme.

Description of Installations Used in the Analyzed Object

Technological installations of steel plants can be divided into the following main units:

1. Converter steel plants
2. Electric steel plants
3. Casting
4. Lining unit
5. Manipulation with slag

and at the same time the hazardous situation analysis is performed for the converter steel plant. The operation of converter steel plant can be divided into the following phases of production process (Table 107.1).

Converter Steel Plants

Using the oxidation process, steel is produced from pig iron in converter steel plants. The process of pig iron processing in a converter is preceded by its treatment. Blowing in the converter is followed by final steel treatment by means of so-called secondary metallurgy. The scheme of the production process is shown in Figure 107.3, and the course of technological process is apparent from the Figure 107.4.

Stages within the scope of carrying out the hazardous situation analysis
Hazardous situation analysis includes the following stages:

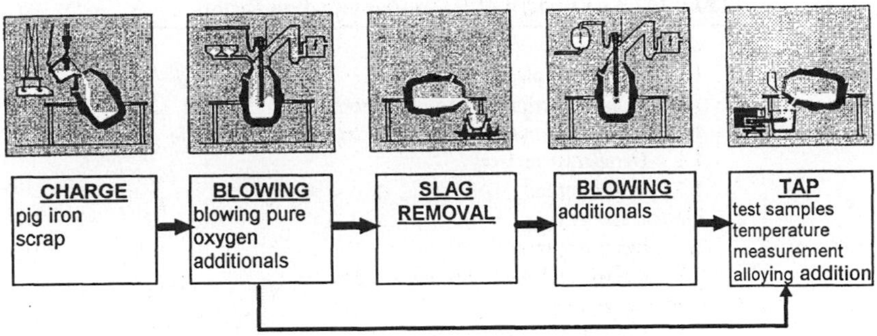

FIGURE 107.4 Scheme of blowing process.

TABLE 107.2 Form for Performing the Hazardous Situation Analysis

FORM: Hazardous situations analysis

Production phases
Main system:
Subsystem:

Page:
Date:
Emploee:

	Production stage	Hazardous situations complex	Goal of protection	Regulations	Arrangements

1st stage Determination of the production process phases — (Table 107.1.)

2nd stage Attaching the information on hazardous situation to individual phases of production process. The information on occurred injury accidents and diseases will be attached to individual phases of production process, defined in the 1st stage, and the damages will be described.

3rd stage Detection of a hazardous situation — real state. The phases of production process according to the 2nd stage are investigated and they are attached direct hazardous situations. Besides the injury accidents, the information on almost injuries is obtained, i.e., on hazardous situations that have not led to a negative event.

4th stage Determination of the goals of protection. The required state is derived and determined on the basis of information about the real state. The determination of the goals of protection assigns the future state of the installation safety.

5th stage Determination and performance of safety arrangements. Determination of arrangements is actually a decision about the kind and way of eliminating the difference between required and real state.

6th stage Monitoring of the impact of undertaken arrangements. The monitoring will be done by means of monitoring steps if the undertaken arrangements led to the achievement of the state required.

The form "Hazardous situation analysis in a steel plant"
Hazardous situations, goals of protection and arrangements are efficiently recorded in forms, e.g., according to Table 107.2. With regard to the fact that it is a rough analysis, it is possible to use the term *hazardous situation complex* instead of the term *hazardous situation*.

TABLE 107.3 Checklist of Hazardous Situation Factors

1	Mechanic energy	
	1.1	Dangerous places
	1.2	Rolling, frictional and tilting motions
	1.3	Motions of machines and vehicles
	1.4	Dangerous surfaces
	1.5	Hazard of fall
2	Electric energy	
	2.1	Electric current
	2.2	Electric and electromagnetic fields
3	Chemical energy	
	3.1	Hazard of fire and explosion
	3.2	Dangerous substances
4	Thermal energy	
	4.1	Hot substances/environment
	4.2	Cold substances/environment
5	Other energies/factors	
6	Factors of working environment	
	6.1	Lighting
	6.2	Climate
	6.3	Noise
	6.4	Mechanic vibration
	6.5	Radiation
7	Physiological factors	
	7.1	Muscular work
	7.2	Antropometrical parameters
8	Psychological factors	
	8.1	Receiving the information, processing, communication
	8.2	Motoric habits
	8.3	Knowledge and skills
	8.4	Acceptation of safe behavior
	8.5	Assessment of hazardous situation
9	Organizational factors	
	9.1	Working process factors
	9.2	Team/individual work
	9.3	Management/Organization of work safety
10	Combined factors	
	e.g., control/regulation systems	

For application of the analysis results, e.g., in a group of the steel plant executives or for the representatives of trade unions, it is an advantage to present in the form also the information on the determined goals of the protection, e.g., safety regulations.

An aid is necessary for systematic hazardous situations detection that will help the elaborator distinguish the hazardous situations. A checklist from Nohl and Thiemeck[4] should be used for this purpose (Table 107.3).

With this control sheet it is possible to examine systematically the existence of the hazardous situation factors in every phase of the production process. Using this procedure the probability that some of the factors will not be taken into consideration is very low.

Determination of the complex of hazardous situation and defining the goals of protection

Table 107.4 shows the example of a preliminary analysis, then hazardous situation complexes are defined and goals of protection determined, and arrangements for their achievement are presented.

To determine the goals of protection, the following procedures can be used according to Meisenbach[5] and Schneider[6]:

1. Working out the review of possible goals of protection based on:
   - Elimination of hazardous situation
   - Technical prevention of hazardous situation impacts

TABLE 107.4 Example of a Preliminary Analysis

FORM SHEET: Hazardous situation analysis

Production phases

Main system:
Converter steel plant
Subsystem:
FILLING PIT

Page: 1
Date: 04.03.1997
Worker:

Production stage	Complex of hazardous situation	Goal of protection	Regulations	Arrangements
Delivery of liquid pig iron in a tandish by railway	Railway entering steel plant hall and in the reach of it occurring: • persons • vehicles • material	Rails • keep free • enter for specified persons exclusively		Safety signs
		Test of vehicles		
		To assure • advising people • assurance of stopping before obstacles in time	Appropriate intraplant regulations and articles containing definitions of individual goals of protection	Executive instructions for railway To keep the rule "step speed" in hall
	Derailment of carriage at the end of rail	Safety stops at the end of rail		Buffer gear
	Movement of laid-by carriages when loading and unloading tandish	Fixing the carriages		Executive regulation for attendance
	Hazard of burning due to pouring out liquid pig iron from surcharged tandish	Keeping the appointed level Assurance of transport on rails in case of surcharged tandish Assurance of safe distance when transferring tandish by crane	Appropriate intraplant regulations and articles containing definition of individual goals of protection	Regulation of - BF operation - railway transport Regulation in steel plant

- Prevention of or influencing the hazardous situation due to use of personal working preventatives
- Controlling the hazardous situation by safe behavior

2. Determination of expected level of safety taking into regard the kind, consequence, and probability of possible injury
3. Consideration of hazardous situation by safe behavior

Determination and performance of arrangements
The determination and performance of arrangements is the task of the steel plant executives taking into regard the rights of employees. When formulating decisions, the management is left to the professional standpoint of experts in the area of safety and protection of health at work. The workers of the operation being analyzed also have to be members of the teams.

The results of the preliminary "rough" analysis
When taking the phases of the production process in a converter steel plant into consideration, the hazardous situations shown in Table 107.5 are for injuries and in Table 107.6 for occupational diseases. The results do not offer the total review of all the hazardous situations in the scope of technological process, neither do they determine their extent, i.e., risks. Experts use them for directing the resulting activities in the framework of risk management.

107.7 Assessment if the Requirements of Obligatory Regulations and Standards Are Fulfilled

This step is not often included in the algorithm of risk management. It is often an obvious assumption that the assessed system complies with the safety regulations given by laws, public notices, guidelines, technical standards, etc. From experience it appears to be favorable to include this step even before the risk evaluation. Respecting the law enactments it is possible to influence the risk parameters to a great extent. If the state is put into accordance with safety regulations in this step, it will not be necessary to take into regard the risk that these regulations deal with.

Thus, in this step the assessors compare if the given regulation, technology, space, etc., comply with the requirements of actual safety regulations and standards, but also technical documentation and instructions of the producer.

TABLE 107.5 Injuries as a Result of Hazardous Situation

	HAZARDOUS SITUATIONS — INJURIES
TECHNOLOGICAL OBJECT: **STEELPLANT**	
Main system	**Converter steel plant**
Subsystems	***Complexes of hazardous situations***
1. Filling pit/pig iron mixer	A. Transport of liquid metal by:
	crane
	special vehicles
	railway
	charging equipment in:
	+ tandishes
	+ buckets
	hazardous situation due to
	RE sprayer
	hot liquid metal spilled in the hall — pouring out of tandish
	slag falling out of tandish
	effect of heat to clamping devices when transporting by crane
	surcharged tandishes

TABLE 107.5 (continued) Injuries as a Result of Hazardous Situation

	hazard of fall in area of filling pits
	collision between attendance and vehicles, as well as between vehicles
	power failure at mobile mixer during rotation
2. Pig iron treatment	places of pressing and cutting at output
	hazard of burning at sample withdrawal
	RE sprayer
	unsuitable floors for transport of additionals by means of transport
	concentration of humidity in buckets for slag
	dismounting of rinsing jet
	filling and storing of CaC_2 _ hazard of creation of acetylene
	removal of sediments by hand tools and crane
	transport of liquid metal by crane
	collision between people and spacial transporters
3. Scrap unit	falling pieces of scrap
	explosives in scrap
4. Melting process Converter	falling pieces of scrap
	ignition in converter due to existence of explosives or hollow bodies
	ignition in converter due to damp scrap or alloying addition
	insufficient co-ordination between a craner and vectorer of converter
	boling in converter
	transport of liquid metal by crane
	falling sediments
	motion of converter and protective gates
	delay of reaction in converter
	oxygen
	jet cooling
	jet mouth
	suspension devices
	hazard of ignition and poisoning conditioned by creation of CO in combustions of converter
	boiler house — cooling water
	collision of persons and vehicles
	steel sprayer
	ignition at deslagging with accumulated humidity
5. Alloying	alloying additions falling down from lorry
	fall into deep storage tank
	automatic drive in of transport means
	collision of persons and vehicles
	manipulation with barrels
	automatic motion of conveyors
	objects falling down from conveyors
	transport of fine grain alloying additions — explosion due to dust
	emptying the storage vehicles
6. Additional steel treatment — desulfurization	steel sprayer
	flames — desulfurization
	hazard due to dust explosion
	hazard of ignition at uncontrollable output of desulfurization aid
	break of tandish in pit
	falling sediments
	dangerous anchorage of tandish at scavenging by argon
	transport of tandish by crane
	CO gas
	leaking spots on systems of inert gas

TABLE 107.6 Occupational Diseases as a Consequence of Hazardous Situation

HAZARDOUS SITUATION — OCCUPATIONAL DISEASES	
***TECHNOLOGICAL OBJECT:* STEEL PLANT**	
Main system	**Converter steel plant**
Subsystems	***Complexes of hazardous situations***
1. Filling pit/mixer of pig iron	high temperature
	infrared radiation
	dust
	noise
2. Pig iron treatment	high temperature
	infrared radiation
	gases, smoke and dust
	noise
	dangerous substances — additionals
	lifting and transfer of burdens
3. Scrap unit	dust
	noise
	radioactively contaminated scrap
4. Melting Converter	high temperature
	gases and smoke
	CO — gas
	radioactively contaminated scrap
5. Alloying	dust
	noise
	lifting and transfer of burdens
6. Additional steel treatment	high temperature
	dust and smoke
	creation of acetylene CaC_2
	creation of CO gas

This decision block appoints another procedure in the framework of the risk analysis as follows:

- If the legislative requirements are not fulfilled, it is necessary to accomplish the arrangements according to regulations, and to check again if the parameters of the assessed system have not changed, as well as what hazard and hazardous situation it implies
- If the legislative requirements are fulfilled, further step is taken

107.8 Risk Evaluation

The risk (R) is expressed by probability of arise (P) and at the same time consequence of a possible undesirable event (D):

$$\text{Mathematically expressed: } \mathbf{R} = \mathbf{P} \times \mathbf{D}$$

The sign × expresses the function according to the kind of evaluation. It can be a matrix or product.

The risk evaluation can be performed in two forms and it depends on the obtained information, potentiality of assessors, but also on the purpose of the risk assessment, kind of hazardous situations, etc.

The risk evaluation can be as follows:

- **Qualitative** evaluation uses verbal expression to describe different levels of probability and consequences. It is mostly used for obtaining the general view of risks in case of a simple operation or in case of lack of figures for quantitative evaluation.
- **Semiquantitative** evaluation is a procedure in which the qualitatively described scales are attached numerical values by combination of which the level of hazardous situation is determined, the

value of risk is determined. It is an ideal method for verification of risks at the site serving as bases for safety arrangements in operation

- **Quantitative** evaluation uses numerical values of probability (1× per 100,000 cycles, 1 injury per 100,000 employees, etc.) and of consequence of an undesirable event (value in currency, degree of health damage, political damage, ecological, etc.). It is used for precise and consistent risk evaluation, especially for machine construction, use of dangerous substances, etc.

Probability of a Negative Event

The assessors have to do an expert estimate, i.e., what is the probability of accident? It can be expressed either by proportional values: frequent, occasional, rare, or by a number expressing that the accident happens once per a certain number of events or per a time unit. In a practical life as well as in professional literature[7,8] the term frequency of hazardous situations, or a negative event occurrence is also used. Basically there is no difference between probability and frequency. The frequency is expressed in the form of discrete values, and it is a presupposition for determining the probability.

Probability is expressed in a more general way, e.g., in percents, and it can be functionally dependent. Frequency expresses the intensity of the hazardous situations that have been determined on the basis of the analysis of evaluated or supposed negative events. It is possible to express it by an integer: e.g., 10^{-5} means that a negative event has occurred 1× from the total number of 100,000 events.

It is favorable to determine (estimate) the probability (frequency) even in the stage of identification of hazard or hazardous situation. This procedure can to be performed by:

A. Using the information about events that have already happened. Using the statistical data processing it is possible to determine the probability (frequency) of the event occurrence in the past, and then to use this information for risk assessments (evaluation, estimate) also in future analyses. This can be used for the risk evaluation of similar objects if it is possible to determine the level of similarity. This method uses the *post factum* procedure.
B. The prediction of the probability value of the negative event based on some suitable methods, e.g., tree analysis. The frequency values can be determined on the basis of the figures that are defined for single components, subsystems, or whole constructions and at the present time they are also worked up in a form of catalogues, e.g., NASA, DAIMLER, or for common components.[8] In some concrete cases retrospective data are also available that can be used with the greatest degree of approaching the real values. The method is based on the *ante factum* principle.
C. Use of the knowledge and experience of experts who obtain reliable data on the probability of a negative event on the basis of questions formulated in appropriate way.

It is possible to use particular procedures either individually or in mutual combinations. The first two procedures complete each other in a suitable way, and utilizing them in the framework of one analysis increases the statement reliability of the obtained results. In case neither of these two methods can be used, it is necessary to rely on the estimates of experts.

As for practical expression of frequency, it is possible to use, e.g., the procedure that uses "frequency,"[20] while the basis is the recognition that the absolute data only have a very limited statement value. It is therefore purposive to choose a base value, and to compare it with absolutely obtained values according to the relation:

$$\text{Frequency "P"} = \frac{\text{number of negative events } 10^6}{\text{total number of hours worked}} \tag{1}$$

The base value has to be linked with the number of hours worked (denominator in the formula (1)), e.g., for work at a lifting machine it is not the time from the beginning of shift that is calculated, but the truly worked hours on the lifting machine that are possible to be defined by computer of work cycles.

The factors that influence the probability of an undesirable event can be summed up into the following items:

A. Measurable factors:
   - Length of hazard acting, time of exposure
   - System parameters (speed of machine, etc.)
   - Rapidity of the event arise
B. Nonmeasurable factors:
   - Human factor — qualifications, attention, stress, etc.
   - Level of maintenance activities
   - Quality of control, revision and testing activities
   - Non-failure and observance of safety arrangements
   - Ability to identify existence of hazard

The determination of the impact of particular factors on the frequency of a concrete negative event is the contents of professional discussions of assessors. At the same time the assessors consider the necessity of regarding other factors when merging the frequency into groups. The factors can depend on the kind of activity or type of technology.

Consequence of a Negative Event

An estimate of impact of a negative event on people, technical objects, and surroundings will be performed within the framework of the consequence of the negative event analysis in case it really occurs. Then the attention has to be paid to the kind of hazardous situation, as well as to the kind of human damage that might happen — light, severe, mortal injury — or to what the value of technical (material) damage would be expressed in financial units. To assess the consequences of a negative event, e.g., in the chemical industry, it is possible to use in a suitable way the methods of mathematical simulation that enable us to establish how many persons will be affected or what area will be contaminated. Similar models can also be used for an estimate of damage consequences in substantially simpler cases, e.g., for swinging a burden on a suspension device where using mathematical models it is possible to determine the area to which people should be denied entry.

For the analysis of possible consequence due to damage, it is necessary to consider the following circumstances:

a. Undesirable negative events causing damage are the base for analysis
b. All the consequences of a negative event have to be identifiable and describable
c. All the protective arrangements enabling reduction (minimalization) of the negative event consequences have to be taken into regard
d. It is necessary to assign criteria that are the base for assessing the negative event consequences
e. The existence of not only the immediate direct consequences of an event, but also of the ones that might arise as their impact later, after a certain time
f. The possibility of secondary consequences, e.g., hazardous situation of other elements in the framework of the investigated man–machine–environment system

At the same time the factors that influence the probability of an accident should be taken into consideration:

A. Measurable factors
   - Kind of injury — light, severe, mortal
   - Number of endangered people
   - Financial loss also involving all the costs of the operation revival; parameters of the system (height of the site, weight of the manipulated burden, speed of motion)

B. Nonmeasurable factors
- Relation between hazard and its effect
- Emergency arrangements, crisis plans
- Complexity of technology or machines

Some of the common forms of expressing the negative event consequence are as follows:

a. Kind of injury — light, severe, mortal
b. Financial loss expressed by a complete calculation taking into regard all the necessary costs of the revival of — defined by technical conditions — operation state
c. Value according to the formula[9]

$$\text{Consequence "D"} = \frac{\text{Number of days of stoppage}}{\text{Total number of negative events}} \tag{2}$$

or

$$\text{Consequence "D"} = \frac{\text{Number of days of stoppage } 10^3}{\text{Real number of operation hours}} \tag{3}$$

To compare individual values it is possible to use:

$$\text{Consequence "D"} = \frac{\Sigma \text{ real operation hours}}{\Sigma \text{ stoppage hours}} \tag{4}$$

In the scope of the assessment the negative event consequences, i.e., assessment of arisen damage, these types of risk have to be taken into regard:

- Individual risk — impact on one person from a group
- Risks at the site — risk of health damage at work
- Risk at general public activities — impact on all persons
- Risks linked with technical objects — losses from downtime, penalties, etc.
- Risks of impact on environment — water, soil, air, fauna and flora

Extended Risk Definition

Overall risk evaluation, however, also requires consideration of other parameters characterizing the impact on the negative event. This includes:

a. Continuation of conditions for a negative event — time exposure "E" (the longer the duration of the conditions for the event, the higher probability that the event arises, and vice versa)
b. The possibility to use protective arrangements "O" in the stage of a hazardous situation

Then according to Reference 10, it is possible to use a so-called extended definition for the risk evaluation in the form of:

$$R = P \times D \times E \times O$$

which is a base for the use of some methods for risk evaluation, e.g., methods of net graphs.

Information in the Framework of Risk Assessment

The possibility to use an optional method of analysis and accordingly risk assessment requires data about the negative event in a form so that they are the most possibly precise and well-turned, but at the same time they have to be reproducible.

When evaluating the information on negative events it can be stated that it is not possible to carry out the comparison of obtained information among subjects in the framework of one country as well as among states mutually. This claim can be made clear by the fact that in Germany a negative event in the form of an injury is only registered when disability is longer than 3 days. There are countries where every negative event connected to injury during working process is registered to be an injury accident. Also the definition of a severe injury is not established explicitly, and owing to this it is assessed in different ways in practical life. This fact may be made even more clear by the fact that while it is not possible to appoint particular parameters sharply at classification of probability and consequences, there is a space for subjective qualitative evaluation.

Therefore, it is recommended that the analyses in the framework of one subject (firm, company, group of citizens) are carried out by one specialist or one team of specialists. In case there is an attempt to compare several subjects, it is possible to have the analysis carried out by one independent institution where it is guaranteed that the qualitative assessment of the risk parameters will comply with particular criteria.

Risk assessment has to be performed in a comprehensible form. The statement power and boundary values of parameters have to be clearly explained and the scope of possible inaccuracies has to be defined, and all the information has to be published in a comprehensible language.

Activities risk assessment should be documented on some of the available media, the most easily by way of questionnaires or protocols that have to include the following entries:

- Title page
- Summary of the problem
- Contents
- Goal and significance of the analysis
- Limitations, presuppositions and reasoning of limitations
- Description of the system and its functional structures
- Description of the used method of risk analysis
- Results of the hazards identification
- Definition of hazardous situations
- Description of the used models (in case they have been used)
- Summary of essential information and its sources
- Risk evaluation
- Proposal of methods for risk management

In case the risk assessment is a part of a continuous process of work safety management, it has to be performed in a way that during all the technical life of a technological system, installation, or during whole period of operating the most modern procedures will be used and current information will be at disposal, and at the same time it will be carried out after every technical or personnel change undertaken in the man–machine–environment system.

Procedures for Performance of the Risk Analysis

The following conditions are a base for selection of a method for risk analysis and assessment:

- The method has to be scientific and it has to comply with the analyzed system
- The results of the assessment have to be in a form that enables to describe the risk in a comprehensible way and to propose arrangements for its minimalization
- It has to be repeatable, comprehensible, and controllable

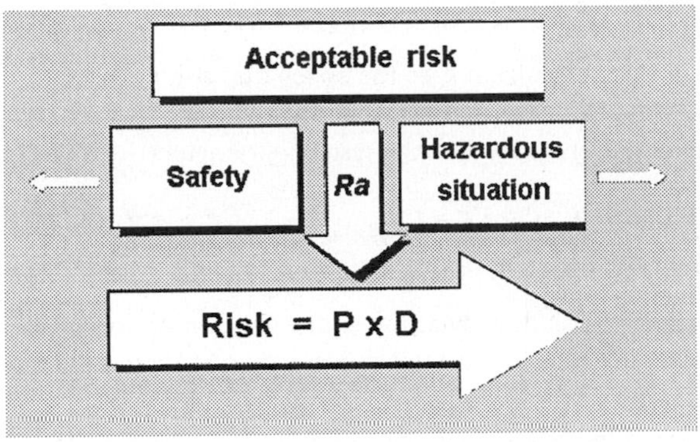

FIGURE 107.5 Significance of acceptable risk.

107.9 Assessment of System Safety

The risk assessment of a chosen system and its ranking into the scale of risks also appoints the criteria of safety of the system being assessed. It is a decision-making step (Figure 107.1, Block 6).

The criteria of the system safety evaluation and qualification of risks comes out of the risk boundary value that is considered acceptable. This value is not defined precisely and it is determined by the level of science, culture of work relationships, legislative requirements, operation intensity, etc. (Figure 107.5).

Acceptable risk is understood to be a risk that the involved persons are willing to bear taking into regard all the operation as well as human conditions. The science of technical systems safety and work safety[11] proved there is no 100% safety in a functional system that means there is no zero risk — it only can approach zero.

107.10 Arrangements for Decrease or Elimination of Risk

If the system safety assessment showed that the risk is high than acceptable, or assessors came to this conclusion by a qualified estimate, it is necessary to suggest arrangements for total elimination or decrease of the risk.

The risk can to be eliminated totally if the hazard was eliminated (e.g., dangerous chemical glue would be replaced by another, harmless one), or if the hazardous situation was eliminated (people and/or technical objects would be excluded from the dangerous space). If there is a hazard in the system being assessed that induces hazardous situation with higher risk than acceptable, it is necessary to suppose that it will cause an injury or damage if no safety arrangements are assumed.

To decrease the risk it is necessary to assign safety arrangements in a careful and professional way. It depends on the invention of people who are to propose them. Two framework procedures can be presented for application in real conditions:

Method of Safety Arrangements Priority

The principle of priority is held when assuming safety arrangements. The arrangements of collective protection are assumed preferentially, and in case it is not possible to achieve it by available means, individual protection succeeds.

Another principle is that risks are decreased preferentially by design and project solutions. If the required level of safety has not been achieved the use of safety arrangements will be suggested. If the protection is not perfect, another step is individual protection of workers and organizational arrangements. Residual risks are solved by safety instructions, operation procedures, and systems of workers training.

Method of Adaptation of Risk Parameters and Risk Factors

The use of this method is advantageous in case the systematic risk assessment has been performed and doing so it is possible to assign from the analysis what mostly influences the risk parameters (probability and consequence of a negative event) as well as the system parameters or risk factors that mostly influence the high value of risk.

The main principle of this method is the change or adjustment of risk factors so the risk is decreased. The priorities of activities system can be summed up into the following procedures:

1. Preferential limitation of risk directly at the source — elimination of danger. (If electricity is the danger in the system, the voltage 220 can be changed to 24 Volts — danger is eliminated. Accordingly, if harmful chemical substance is replaced by harmless.)
2. Possibility of change of the assessed system. (Deceleration, decrease of potential energy, introduction of protective installations, etc.)
3. Minimalization of probability of an undesirable event arise. (Reduction of exposure, attendance training, improvement of maintenance and checks, etc.)
4. Minimalization of consequences of a possible undesirable negative event. (Excluding the endangered persons, introduction of emergency arrangements, etc.)

107.11 Establishment of Priorities

The complex risk assessment will enable ranking the priorities according to importance. The systems with high risk are to be solved immediately. Possible solutions could be to stop the operation, put the machine out of operation, or temporary arrangements.

Acceptable risks should also be included in the framework of performing the arrangements since they can be even more minimalized, working conditions as well as ambiance can be improved.

In the next step it has to be checked whether the risk would cease at the proposed arrangements. Therefore, it is necessary to carry out the procedure according to the presented algorithm, and to verify whether the residual risk is acceptable.

107.12 Example of Risk Assessment Methods

Combined Procedures

Combined methods for risk assessment consist of two independent procedures that complete each other. The first are the methods for risk identification, and the second one, methods for risk evaluation.

The record of causal dependence for the observed event in optional form has to include the hazard and hazardous situation description, as well as their resulting appearance — damage. Catalogue pages that are of different form are a typical representative of these procedures.

The evaluation of risk is an independent stage in the process of its assessment. There are various procedures for this purpose from the most simple methods up to complex analytic methods including detailed economical analyses. For the needs of operative risk assessment a simple point method of risk assessment was chosen — MIL- STD 882C — System safety program requirements.[12]

When selecting the combination of identification method and risk assessment, it is necessary to consider the requirements put on the risk assessment process and further utilization of results.

Catalogue Pages

The term catalogue pages is a form of record of causal dependence for the observed group of risks. In technical practice it is also possible to meet a term that expresses the description of a certain group of risks in a more appropriate way — cadastre of risks. Catalogue pages have to contain minimum information on the type of hazard, kind of hazardous situation, and its consequence. This information does not have a form laid down by a standard, and it is often completed with other information, e.g., what standards and regulations concern the observed risk, in which stage of technical life of the installation the observed risk occurs. The most simple way of recording with regards to further utilization is the database record with a precisely selected structure.

Utilization and Goal

The catalogue page can be used from a construction design of the system up to liquidation of the observed system. It can be used easily by projectors, inspection workers, as well as normal operation workers. The goal of the catalogue page is to quickly distinguish possible hazards and hazardous situations in a concrete operation as well as their displays. The catalogue page is a living material, and it is necessary to complete it gradually, which is also shown in.[13]

The set of catalogue pages forming a compact cadastre of risks for a specific group of machines or complex technology is extensive material.

Procedure of Catalogue Page Creation

Particular steps for creating a catalogue page consist of:

- Definition of assessed system
- Definition of catalogue page structure (kind and quality of information on causal dependence)
- Definition of structure of the catalogue page record with regard to user's requirements
- Way of completing the catalogue page
- Possibility of gathering catalogue pages into groups in the form of catalogues

Figure 107.6 shows a catalogue page for the group of lifting machines — cranes where there is a type of risk identified (break of suspension device).

Primary information essential for making the catalogue pages contains inspection records, injury records as well as recommendations of producer. The problem of creating the catalogue pages is demanding in time, extent, and quality of information. It is necessary to realize that there is typization, and thus also the possibility to gather single hazards and hazardous situations into groups which reduces the number of catalogue pages.

Summary

The advantage of the catalogue page is its simple usage not demanding special knowledge, usable in all stages of technical life of the observed system. Introduction of identification enables operative interventions from the lowest control levels of the system up to the level of medium risk management. The catalogue page serves as a suitable base for risk evaluation.

Risk Assessment

For an efficient risk assessment process it is necessary to attach a numeric value to the concrete risk. In general risk is a function of at least two basic parameters: frequency or probability, and consequence. Other measurable factors like exposure time and possibility of prevention are functions of probability and consequence. Financial expression of consequence appears to be problematic in some systems. It mostly concerns systems with human factors. In systems with insignificant influence of human factors, the consequence can be evaluated in financial units (technical risks).

A special group of risk managers deals with the process of risk evaluation. Basic system steps to creating the system of risk assessment and next management are in standards.[14,15] One of the partial outputs of these procedures is formation of a risk matrix. The risk matrix expresses double-parametric record of the probability and consequence category. The resulting product of combinations of probability and consequence category expressed by point value of a concrete risk is its ranking into criteria groups. The criteria group assigns the priority of intervention with the goal to minimalize the existing risk.

Utilization

Point evaluation of risk can be used in the stage of primary analyses as well as in the stage of detailed analyses. The point risk evaluation in the process of operative risk management in normal operation has

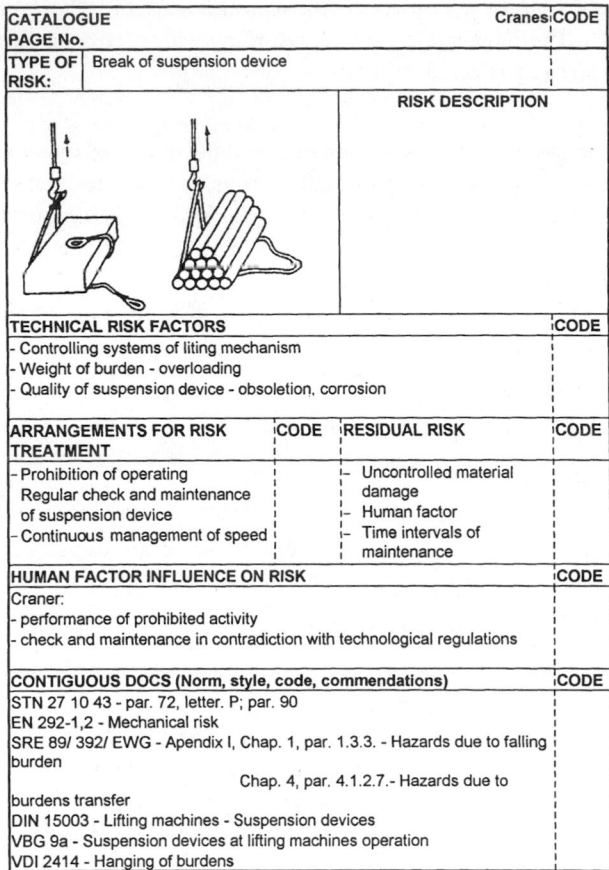

CATALOGUE PAGE No.	Cranes	CODE
TYPE OF RISK:	Break of suspension device	

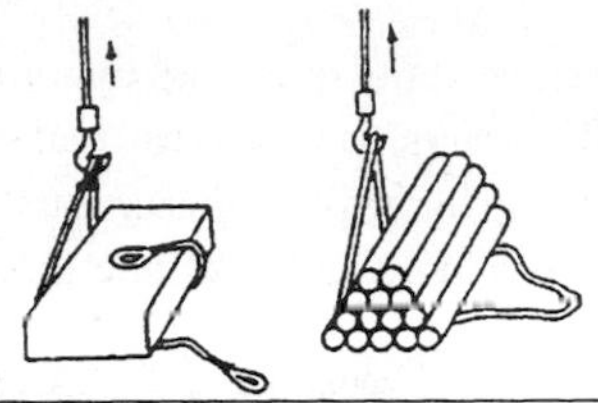

RISK DESCRIPTION

TECHNICAL RISK FACTORS	CODE
- Controlling systems of liting mechanism - Weight of burden - overloading - Quality of suspension device - obsoletion, corrosion	

ARRANGEMENTS FOR RISK TREATMENT	CODE	RESIDUAL RISK	CODE
- Prohibition of operating Regular check and maintenance of suspension device - Continuous management of speed		– Uncontrolled material damage – Human factor – Time intervals of maintenance	

HUMAN FACTOR INFLUENCE ON RISK	CODE
Craner: - performance of prohibited activity - check and maintenance in contradiction with technological regulations	

CONTIGUOUS DOCS (Norm, style, code, commendations)	CODE
STN 27 10 43 - par. 72, letter. P; par. 90 EN 292-1,2 - Mechanical risk SRE 89/ 392/ EWG - Apendix I, Chap. 1, par. 1.3.3. - Hazards due to falling burden Chap. 4, par. 4.1.2.7.- Hazards due to burdens transfer DIN 15003 - Lifting machines - Suspension devices VBG 9a - Suspension devices at lifting machines operation VDI 2414 - Hanging of burdens	

FIGURE 107.6 Catalogue page.

minor demands on theoretical knowledge, but on the other hand it requires perfect knowledge of the assessed technology since the assessor has to rank the probability and consequence into particular categories

Risk Matrix

The risk matrix is formed by the combination of probability categories (Table 107.7) and consequence categories (Table 107.8). The risk matrix is shown in Table 107.9.

Risk matrices based on a simple definition of risk fulfill the following general principles:

- Risk matrix is of a line form where single lines express the level of risk
- Individual linguistic variables describing the category of defined level of statement value
- The consequence value should have the dimension of standard hour, cycles, financial dimension; by checkback it is possible to define in a more accurate way particular linguistic variables or number sets related to them

The above-described point method recognizes four levels of risk in the range of 20 points. The risk level enables us to accept concrete arrangements leading to risk minimalization.

The example of risk matrix use in the risk management process is e.g., formation of responsibility hierarchy for the arrangements by appropriate organ and responsible worker for the risk assessment (Table 107.10).

TABLE 107.7 Probability Chart

Type	Level	Description for Event	Description Generally
frequent	A	it will probably arise often	continuously expected
probable	B	it will arise several times during life-time	frequent
occasional	C	it will arise occasionally during life-time	several times
rare	D	little probable, but possible	is expected seldom
improbable	E	almost excluded	possible very seldom

TABLE 107.8 Chart of Consequences

Type	Category	Description
catastrophic	I	death or loss of system
critical	II	severe injury, illness or extensive system spoliation
marginal	III	lighter injury, illness or smaller spoliation of system
negligible	IV	less than lighter injury, negligible system failure

TABLE 107.9 Resulting Matrix of Numeric Risk Assessment

	Risk Matrix			
	I	II	III	IV
Probability/consequence	Catastrophic	Critical	Marginal	Negligible
A frequent	1,00	3,00	7,00	13,00
B probable	2,00	5,00	9,00	16,00
C occasional	4,00	6,00	11,00	18,00
D rare	8,00	10,00	14,00	19,00
E improbable	12,00	15,00	17,00	20,00

Point Value	Risk Level
1–5	unacceptable
6–9	undesirable
10–17	acceptable with examinations
18–20	acceptable without examinations

TABLE 107.10 Ranking the Risk into Groups

	Risk Matrix			
	I	II	III	IV
Probability/Consequence	Catastrophic	Critical	Marginal	Negligible
frequent	high risk	high risk	high risk	medium risk
probable	high risk	high risk	medium risk	low risk
occasional	high risk	high risk	medium risk	low risk
rare	high risk	medium risk	low risk	low risk
improbable	medium risk	low risk	low risk	low risk

Risk Level	Responsible Organ
high risk	main authorized person
medium risk	program authorized person
low risk	program engineer

The risk matrix can also be used for the creation of the risk rate analysis program. It is a matter of general principle that the lower the point risk value, the more necessary is detailed risk analysis.

The risk matrix can be used in the process of the proposal of acceptable risk borders. From the point of view of system safety it is necessary to define the conditions regarded to be unacceptable, and when they occur, an intervention is necessary with the aim of minimalization of the risk to an acceptable level. In general, unacceptable conditions include:

- Failure of a elementary component, human fault, or construction characteristics that might cause an accident with critical or catastrophic relevance
- Dual failure of independent components, dual human fault, or their combination also including an incorrect command or check function that might cause a negative event with critical or catastrophic relevance
- Generation of dangerous radiations, e.g., ionizing radiation, if arrangements for protection of people or sensitive installation are not carried out
- Manipulation processes that might cause an accident to unprotected persons and installations
- Risk categories that are specified in contract as unacceptable. The other conditions are regarded to be acceptable and they do not require further analysis once they were checked and verified
- System structure that requires two or more independent human faults or failures or their combinations
- System structures that positively create fault prevention
- System structures that positively prevent spoliation of one part by another
- Limitations of system structure in operation, interaction or linkage that precedes the failure arise
- System structures that provide improved safety factor, or they keep it at acceptable level
- System structures where energetic flows might cause a failure, e.g., valves
- System structures where a failure in one part can be temporarily tolerated since residual strength or function complies with safety conditions
- System structures where the attendance can manage a risk situation according to the assumption
- System structures bordering or checking the use of dangerous materials

The next example shows the procedure of a human risk evaluation by way of application of the point method in the man–machine–environment system represented by a lifting machine (bridge crane). A similar analysis of injuries regarding the classification of injuries can be found in Figure 107.7.

With regard to the types of risks defined in reference 16, hazards and mechanic hazardous situations creating the risk have been analyzed in a more detailed way. Figure 107.8 presents the percentage representation of individual kinds of injuries caused by mechanic parts of the crane for particular parts of a human body.

The injury percentage does not provide a sufficient answer to the question about the height of particular risks. Only hazard and hazardous situations are sufficiently defined. The definition of consequences and their evaluation enables to assess the height of risk.

Table 107.11 presents the categories of probability and consequence. Graphic display of particular risks is presented in Figure 107.9.

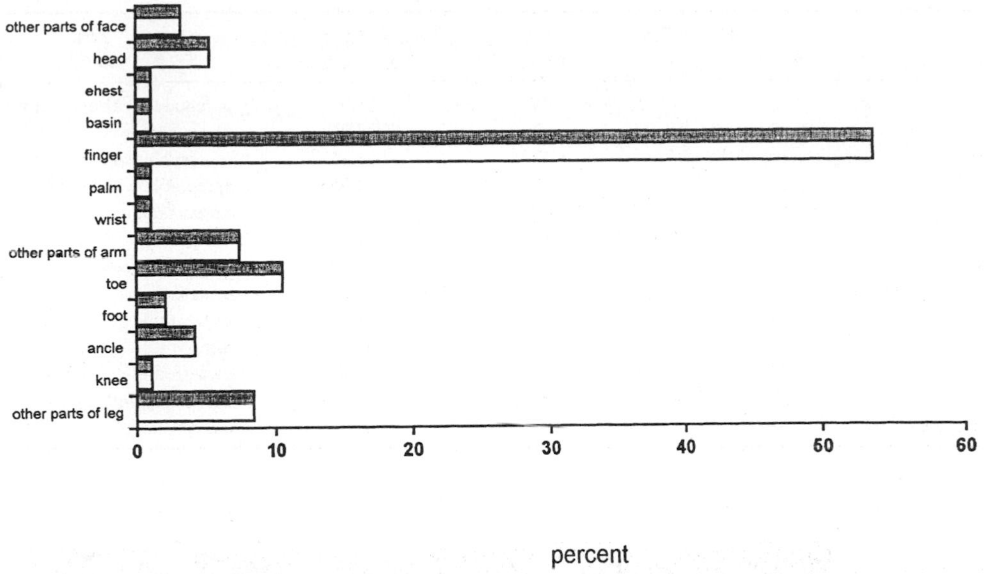

FIGURE 107.7 Analysis of risks in bridge crane operation.

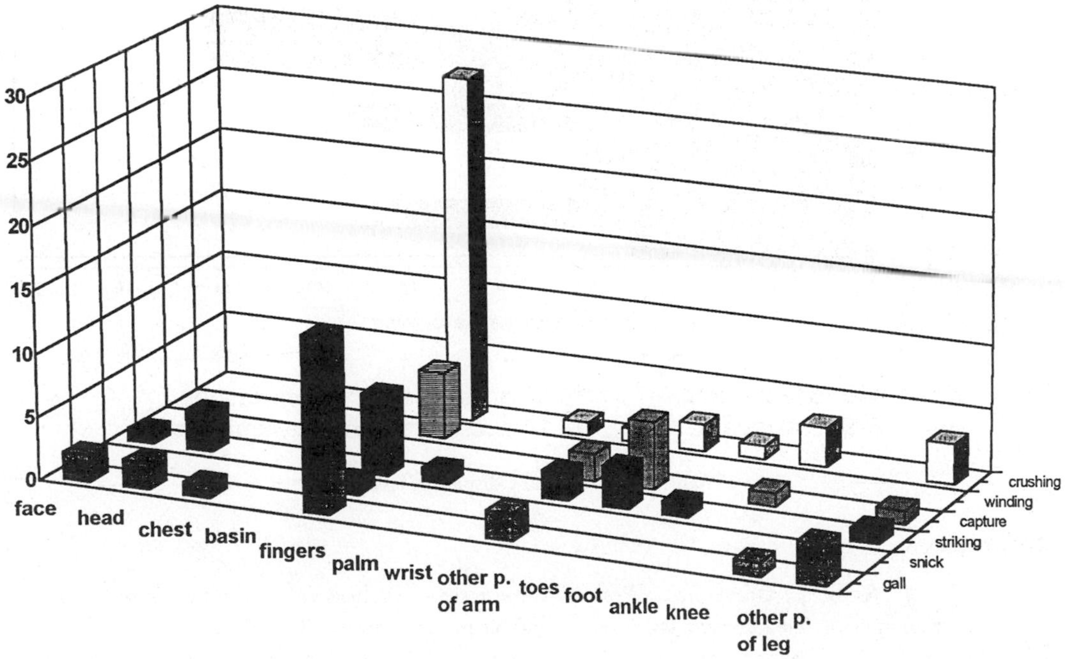

FIGURE 107.8 Injuries caused by mechanical parts of bridge crane.

TABLE 107.11 Risk Assessment

Part of Body	%	Probability Categ.	Conseq. Categ.	Point Method Point Value of Risk	Risk Level
other parts of face	3.20	C	IV	18,00	acceptable without examinations
head	5.30	B	II	5,00	unacceptable
chest	1.10	D	II	10,00	acceptable with exam.
basin	1.10	D	II	10,00	acceptable with exam.
fingers	53.70	A	III	7,00	undesirable
palm	1.10	D	III	14,00	acceptable with exam.
wrist	1.10	D	III	14,00	acceptable with exam.
other parts of arm	7.40	B	III	9,00	undesirable
toes	10.50	B	III	9,00	undesirable
foot	2.10	C	II	6,00	undesirable
ankle	4.20	C	II	6,00	undesirable
knee	1.10	D	II	10,00	acceptable with exam.
other parts of leg	8.40	B	II	5,00	unacceptable

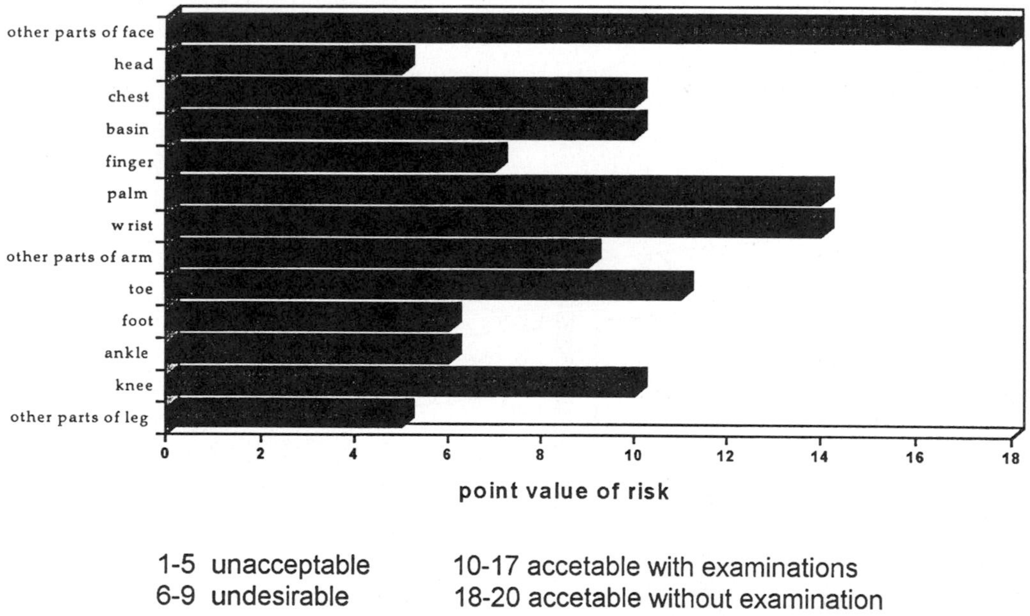

FIGURE 107.9 Graphic display of risk levels.

References

1. Majer, I., Sinay, J., Oravec, M., Posudzovanie rizík — východisko k účinnym bezpečnostným opatreniam/slov. Výskumný a vzdelávací ústav bezpečnosti práce SR, 1997.
2. Defren, W., Qualitätsmanagement — Strategie bei der Risiko-Analyse im Maschinenbau/ger. Prednáška na konferencii A+A, 1995, Düsseldorf.
3. Hoor, M., Umsetzung einer Gefährdungsanalyse am Beispiel "Stahlwerke."/ger. *Moderne Unfallverhüttung,* Heft 36, str.: 33-42 ISSN 0544-7119.
4. Nohl, J., Thiemeck, H., Durchführung von Gefährdungsanalysen -Teil1/ger. Schriftenreihe der BAA, Dortmund, 1988.
5. Meisebach, J., Strategie und Praxis wirksamer Sicherheitsarbeit/ger. Papiermacher BG, Mainz, 1988.

6. Schneider, B., Motivation der Arbeitsnehmer und Führungskräfte durch Ausbildung während des Übergangs der Belegschaft von Kokilen — zum Stranggiesverfahren/ger. Düsseldorf/Duisburg, 1986.
7. Merz, A.-H., Schneider, T., Bohnenblust, H., Bewertung von technischen Risiken/ger. ETH Zürich — vdf Hochschulverlag AG., 1995 ISBN 3-7281-2178-9.
8. Haibach, E., Betriebsfestigkeit/ger. VDI-Verlag Düsseldorf, 1989, ISBN 3-18-400828-2.
9. Strnad, H., Vorath, B.-J., Sicherheitsgerechtes Konstruieren/ger. Verlag TÜV Köln, 1992.
10. Kreutzkampf, F. Meffert, K., Risiken in Steuerungen — Grundlagen der Klassifizierung/ger. Moderne Unfallverhüttung, 34, str.: 16-20, ISSN 0544-7119.
11. Kuhlmann, A., Einführung in die Sicherheitswissenschaft/ger. Verlag TÜV Rheinland, Köln, 1981, ISBN 3-528-08495-2.
12. Kuhlmann, A., MIL-STD 882C — System safety program requirements APR 1993, USA.
13. Kuhlmann, A., STN-EN 292.1 — Bezpečnosť strojných zariadení, základné pojmy, všeobecné zásady navrhovania/slov ÚNMS SR Bratislava, Slovakia 11/96.
14. Kuhlmann, A., pr. EN 954-1 — Sicherheit von Maschinen — Sicherehits- bezogene Teile von Steuerungen, Teil 1 — Allgemeine Gestaltungsleitsätze/ger. Beuth-Verlag, Berlin, 1993.
15. Kuhlmann, A., DIN IEC 65A 123 — Funktionelle Sicherheit von elektrischen Systemen — Allgemeine Aspekte/ger. Beuth-Verlag, Berlin, 1992.

Section III
Service Systems

108

Ergonomics in Health Care Organizations

Roger C. Jensen
UES, Inc.

108.1 Introduction

The methods and tools of ergonomics apply to the health care industry. This chapter provides background information on the health care industry, discusses injury and illness concerns of different industry sectors, and summarizes examples of ergonomics programs in health care. The chapter concludes with a discussion of elements of ergonomics programs in health care organizations.

Health Care is a Large and Diverse Industry

National health care programs in many countries provide most health care through public institutions. In the United States, both public and private institutions provide health care. These health care providers include hospitals, nursing homes, home health care organizations, medical offices, and dental offices.

Health care providers are classified by the type of services provided. In the United States the Standard Industrial Classifications (SIC) for health care consists of the sectors identified in Table 108.1. Variations on the categories may be found in other countries.

Employment in the private sector health care industry makes up 10.4% of all employment in the United States.* Of the sectors listed in Table 108.1, hospitals employ the most people, followed by nursing and personal care facilities.

The hospital sector includes the subsectors: general medical and surgical hospitals, psychiatric hospitals, and specialty hospitals. The nursing and personal care sector includes the subsectors: skilled nursing care facilities, intermediate care facilities, and nursing and personal care facilities not elsewhere classified.

*Based on the 1990 census of the population, U.S. Bureau of the Census. Persons employed in the health services: 9,682,684. Persons employed in all private sector establishments: 95,449,300.

0-8493-2641-9/99/$0.00+$.50

TABLE 108.1 Health Care Industry Sectors in the United States

SIC Number	Description of Industry Sector
801	Offices and clinics of Doctors of Medicine
802	Offices and clinics of Dentists
803	Offices and clinics of Doctors of Osteopathy
804	Offices and clinics of Chiropractors, Optometrists, Podiatrists, and other health practitioners
805	Nursing and personal care facilities
806	Hospitals
807	Medical and dental laboratories
808	Home health care services
809	Other health and allied services

From U.S. Office of Management and Budget, Standard Industrial Classification Manual 1987. U.S. Government Printing Office, Washington, D.C.

Within the nursing and personal care sector, approximately 70% of employment is in skilled nursing care facilities. As a result, the nursing and personal care sector is often referred to as the nursing home sector.

Ergonomics Creeps Into Health Care

The health care industry has been a follower, not a leader, in applying ergonomics. Applications of ergonomics to aircraft design took hold during the 1940s, and expanded significantly during the 1950s and 1960s (Grether, 1986). These applications emphasized design to enhance crew performance. Applications of ergonomics in the automobile industry initially focused on product design to improve comfort and enhance driver performance. During the 1970s, a few automobile manufacturers started applying ergonomics to production operations. Joint labor–management ergonomics programs spread throughout the automobile manufacturing companies during the 1980s (Joseph and Jimmerson, 1996). Companies that supply parts to automobile manufacturers implemented similar programs during the early 1990s. These applications addressed employee safety, health, and comfort. In contrast, the health care industry showed limited interest in ergonomics prior to the 1990s.

For many years the health care industry has been aware of a back injury problem among nurses. Throughout the 1960s and 1970s, numerous articles in nursing magazines showed widespread concern for the high incidence of sprains and strains among nurses (Jensen, Nestor, Myers, and Rattiner, 1988). The articles also reflected an apparent acceptance of the belief that low back sprains and strains occur most often while performing patient-handling tasks (Jensen, 1985). Most authors advised nurses to avoid low back injury by maintaining personal fitness and always using proper body mechanics when transferring patients. Prior to 1980, authors of articles about back pain among nurses failed to mentioned the ergonomics approach of changing task demands to fit the capabilities and limitations of the nurses (Jensen et al., 1988).

During the 1980s, concern about back injuries among nurses continued. Other health care occupations also started receiving attention – particularly nursing assistants who work in nursing homes (Gagnon and Lortie, 1987; Jensen, 1987) and geriatric hospitals (Lortie, 1985, 1986). Several studies were published reporting increased back injury risk with increased exposure to patient handling (Jensen, 1990a).

Common Injuries and Illnesses

Injury and illness rates provided in Table 108.2 show that the extension of concerns beyond hospitals was fully justified. The rates are for injuries and illness that resulted in an employee being unable to work for a day or more. Each rate uses the same denominator, 10,000 full-time equivalent employees working for one year. The name for this is "days-away-from-work case rate." Rates in Table 108.2 apply to the year 1994 in the United States. Rates are not published for the sectors consisting primarily of small offices and clinics of dentists, osteopathic physicians, chiropractors, optometrists, and podiatrists – SIC 802, 803, 804.

TABLE 108.2 Rates for Days-Away-From-Work Cases in 1994 for All Industries and for Sectors of the Health Care Industry

Industry Sector	Overall Rate	Sprain & Strains	Carpal Tunnel Syndrome	Tendonitis
All industries combined	277.0	119.3	4.8	3.1
Offices and clinics of medical doctors	53.5	22.6	2.7	1.2
Nursing and personal care facilities	633.5	366.7	1.9	3.3
Hospitals	326.4	119.1	3.0	2.1
Home health care services	473.6	279.8	2.7	2.2
Other health and allied services	217.9	107.4	5.2	2.4

From U.S. Bureau of Labor Statistics. Annual Report on Workplace Injuries and Illnesses, 1994. Table 6.

The first row of data in Table 108.2 shows the 1994 rates for all industries combined. Rates for all types of injuries and illnesses, in the first column of data, indicate that the overall rate for all industries combined was 277. This rate was exceeded by nursing and personal care facilities, by hospitals, and by home health care services. For sprain and strain type injuries, the rate in nursing and personal care facilities was three times greater than for all industries combined. The rate of sprain and strain injuries in hospitals was no different than the rate for all industries combined.

For the carpal tunnel syndrome data shown in Table 108.2, the rate for all industries combined (4.8) was greater than that for each health care sector, except "Other health and allied services." The elevated rate for "Other health and allied services" was probably due to employment of large numbers of people who perform highly repetitive manual work such as medical transcribers and clerical personnel using computers to process insurance, billing, and medical records. For tendonitis, the rate for all industries combined (3.1) exceeded the rate for each of the health care sectors, except nursing and personal care facilities (3.3).

108.2 Fundamental Concerns of Health Care Sectors

The data in Table 108.2 show the large differences in days-away-from-work case rates among sectors of the health care industry. Each sector has some unique safety and health concerns. The following sections summarize some issues and literature unique to specific sectors of the health care industry.

Hospitals

Analyses of hospital injury records have identified the nursing department as a major concern. Hoover (1973) reported that of 85 lifting injuries, the nursing service contributed 43%, followed by housekeeping with 13%. Hubley-Kozey, Westers, Stanis, and Wall (1985) reported that of 171 back injuries, nursing personnel incurred 60%, and 70% of these occurred on the nursing wards. Lewy (1981) reported that nurses accounted for 59% of injuries even though they represented only one-third of the hospital workforce.

Back injuries have plagued hospital nurses for years. Nurses who suffer back injuries generally associate the injury with a patient-handling task (Jensen, 1985).

Nurses are not the only hospital employees with injuries. An analysis of injury and illness records of a large hospital in Israel found that most cases were among the housekeeping, maintenance, laundry, and catering workers (Pines, Cleghorn de Rohrmoser, and Pollak, 1985).

Nursing Homes

Sprains and strains accounted for 58% of the days-away-from-work cases in nursing and personal care sector, according to the 1994 data in Table 108.2. Sprains and strains accounted for 68% of the workers' compensation claims in nursing and personal care facilities, according to data for the year 1986 (Jensen, 1990b). The rate for sprains and strains in nursing and personal care facilities was three times that of all industries combined (Table 108.2).

Of the sprain and strain compensation claims reviewed by Jensen (1990b), 56% were for back injuries. Data in Table 108.2 indicate that the nursing and personal care sector had a lower rate of carpal tunnel syndrome than the other health care sectors, and lower than the rate for all industries. From all these data it is concluded that nursing and personal care facilities have a major problem with back sprains and strains, and comparatively little concern about carpal tunnel syndrome.

Home Nursing

Injury data from two home health care agencies were examined by Myers, Jensen, Nestor, and Rattiner (1994). They collected injury records for home health aides and a comparison group – nursing assistants and orderlies working in a large hospital. Back injury rates of the home health aides were almost three times that of the comparison group. For the home health aides, 64% of the activities associated with low back injuries involved patient handling. The specific patient handling activities that comprised the 64% were:

- Moving patient up in bed (21%)
- Helping patient in or out of bed (11%)
- Helping patient in or out of chair (9%)
- Catching patient starting to fall (9%)
- Helping patient in or out of tub (5%)
- Turning patient in bed (4%)
- Helping patient from bed (2%), and
- Helping patient on or off toilet (2%).

Ergonomics has not been applied, to a significant extent, to the work of home health aides. Their needs are different from corresponding workers in institutional settings (Owen, 1996). One difference is that most beds in health care facilities are adjustable in height. Most home patients have beds that are not adjustable in height. A low bed height means the health care provider must bend more to care for the patient, and the bending means more stress on the lower back.

Patient lifting equipment for the home needs to be usable by one care provider, whereas in nursing homes, nursing assistants generally work in pairs. Some home patients have beds with insufficient space underneath for the legs of a portable hoist. Therefore, patient transfers must be performed without a hoist.

Another difference is the matter of responsibility for providing patient-lifting equipment and other patient handling devices. The visiting care provider can bring some patient-handling devices into the home during each visit. Examples are walking belts and sliding boards. Portable patient hoists are too bulky and heavy to carry around from home to home on a daily basis. It is far more practical to have the patient lease or buy a hoist that can be kept in the home. Some home nursing organizations lease hoists to patients. By leasing, the home nursing organization controls the brands and models of hoists. In contrast, hoist selection by the patient or a family member may not consider characteristics such as difficulty of use, suitability for the patient's needs, or safety characteristics like stability.

Owen (1996) discusses a variety of devices that can be helpful for patient handling in the home. Devices discussed include: portable patient hoists, portable stand-assist lifts, transfer boards, portable commodes, chairs and cushions that lift, lift poles, riser attachment for toilet bowls, hospital type beds, and ceiling mounted hoists. Before selecting devices, the home care organization conducts a patient assessment. Part of the assessment addresses patient-handling devices that match the patient's needs. The assessment findings need to be clearly communicated to the home health aide, and the aide must be trained to properly use the patient-handling devices selected for the patient.

Dental Offices

Literature on the occupational injury and illness problems in dental offices reflects little concern about sprains and strains. There have, however, been some reports of disorders associated with maintaining awkward postures for extended periods of time, e.g., Adelman and Elsner (1982).

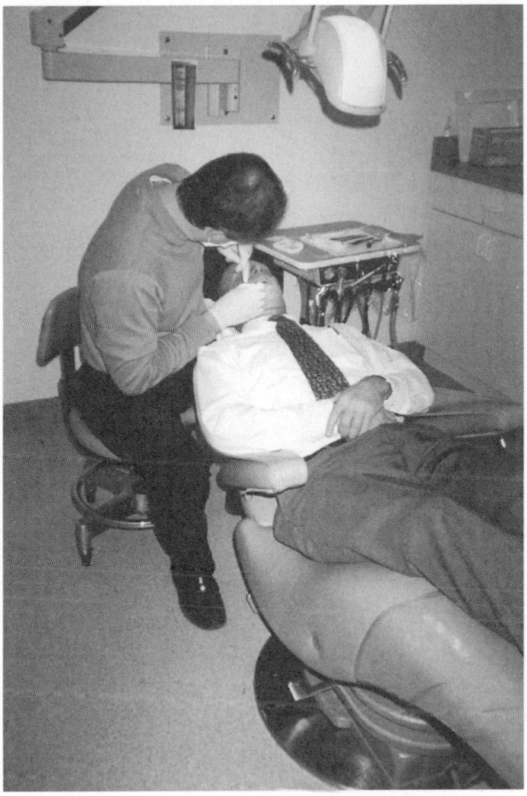

FIGURE 108.1

Dentists tend to work in static postures with low intensity muscle contraction. The photograph of a dentist in Figure 108.1 illustrates a working posture requiring constant tension of neck and shoulder muscles. His spine is rotated and his back muscles are tensed to maintain his head position.

In a study of postures during common dental procedures, Finsen and Christensen (1994) found that the eight dentists studied spent 82% of their time with their neck flexed more than 30 degrees, and they spent about 30% of their time working with their arms abducted more than 30 degrees. These times, however, include only direct patient care time. The periods between direct patient care allow some time for fatigued muscles to recover.

Data from several surveys indicate that dentists tend to suffer from musculoskeletal complaints (Shugars, Miller, Williams, Fishburne, and Strickland 1987; Murtomaa, 1982; Lehto, Helenius, and Alaranta, 1991). Representative findings from a survey of dentists in Denmark (Christensen and Finsen, 1994), were:

- 54% reported trouble with their neck muscles during the past year (This 54% prevalence may be compared to prevalence rates for neck troubles in the general working population in Denmark of 49% for females and 28% for males.)
- 59% reported low back trouble
- 39% reported shoulder trouble
- 20% reported trouble with their hands

Dental hygienists have also been studied. Their work tends to involve low load, static contraction of the trapezius muscles and other muscles that hold joints in fixed position. During patient care these periods of static contraction are interrupted occasionally with rest pauses of about one-second duration (Öberg, Karsznia, Kadefors, and Sandsjö, 1994). Figure 108.2 shows a dental hygienist working in a

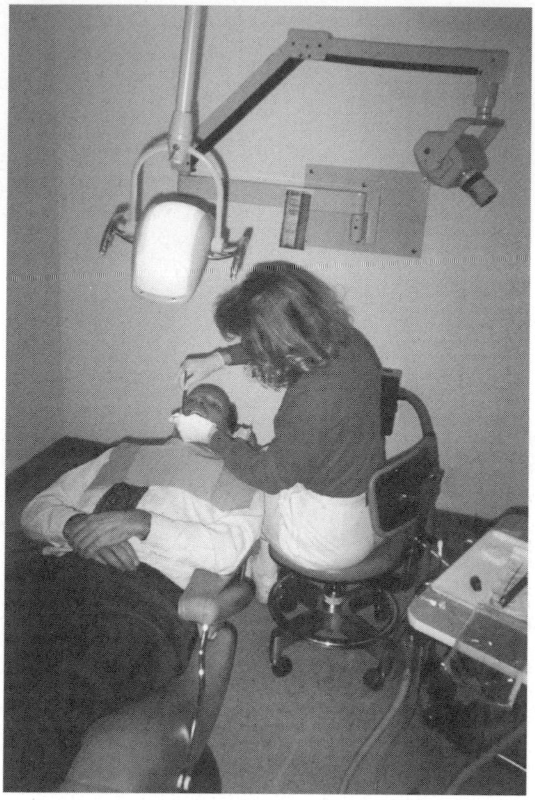

FIGURE 108.2

posture that requires sustained tension of the right shoulder muscles to support her abducted arm, and neck muscles to support the weight of her head. Her left wrist is extended and in ulnar deviation while using the instrument to scale the patient's teeth.

In a survey of Danish dental hygienists, 62% reported having neck complaints during the past year and 81% reported shoulder complaints (Öberg and Öberg, 1993). Both prevalence rates were considerably greater than those of the general Danish population of working women.

Some dental hygienists develop carpal tunnel syndrome. While scaling teeth hygienists exert force with their dominant hand, often with the wrist deviated considerably from the neutral posture. A survey of Canadian dental hygienists found that 7% had been told by a physician since starting work that they had carpal tunnel syndrome (Liss et al., 1995).

Medical Devices and Equipment

While the most widely recognized application of ergonomics in health care is for employee injury prevention, a less recognized area is for the design of medical devices. A full review of these applications is beyond the scope of this handbook on occupational ergonomics, but to illustrate the range of applications, some recent ergonomics contributions are listed below. These were published in an issue of *Human Factors* – the journal of the Human Factors and Ergonomics Society.

- Design of instructions for medication (Morrow et al., 1996)
- How users cope with poorly designed computer interfaces (Obradovich and Woods, 1996)
- Adapting to new technology in the operating room (Cook and Woods, 1996)
- Accuracy of pharmacists' prescriptions as affected by ambient noise (Flynn et al., 1996)

- Evaluation of self-reporting of deficiencies in airway management (MacKensie et al., 1996)
- Task complexity in emergency medical care (Xiao et al., 1996)

108.3 Ergonomics Programs in Health Care

Application of ergonomics in health care has evolved somewhat differently from the growth patterns in other industries. Back injury programs in hospitals have traditionally been initiated by nursing departments, sometimes in conjunction with employee health and physical therapy departments. Some of these programs only applied to nursing departments.

Historically, back injury prevention programs placed considerable emphasis on: instruction of nursing staff on personal physical fitness, proper techniques for manually transferring patients, and case management. Case management in these programs was a process of monitoring rehabilitation and facilitating the earliest practicable return to work of employees who suffer from back pain (Wood, 1987). Early return to work has traditionally meant accommodating a worker with lifting restrictions by temporarily omitting job duties that require lifting more than the restricted weight. These methods for controlling back injuries are not mainstream ergonomics. The process becomes ergonomics when the method for performing the lifting tasks is changed so the individual can perform it.

The introduction of a greater emphasis on changing the task demands for nurses appears to have evolved in the United Kingdom during the 1980s. Various types of patient-handling equipment were evaluated (Bell, 1984). Techniques for patient transfers were compared (Stubbs, Buckle, Hudson, and Rivers, 1983). Various surveys were conducted, e.g., Stubbs, Buckle, Hudson, Rivers, and Worringham (1983). These findings were incorporated into the second edition of a book *The Handling of Patients: A Guide for Nurses, 2nd ed.* (Lloyd, Tarling, Troup, and Wright, 1987). The book included traditional methods of using proper body mechanics, but it emphasized the use of patient handling equipment. This was a major departure from the first edition and from traditional nursing instruction that only addressed proper body mechanics.

Two examples of ergonomics programs in health care are described below. They are presented in order of time.

Example 1

A paper by Aird, Nyran, and Roberts (1988) describes an ergonomics program implemented by a medium-size nursing home in Ontario, Canada. Included in the intervention program were the following:

- An assessment of each lift required for each resident, taking into account caregiver capabilities, resident characteristics, environmental factors, and the availability of patient-handling devices
- Communicating through a set of logos the proper procedure for performing specific patient transfers
- Obtaining additional lifting devices

Program evaluation was based on the number of reported back injuries associated with patient contact. In the year preceding the intervention there were three. During the year of intervention there were four. In the two years after intervention there were two back injuries reported each year. This illustrates one of the difficulties encountered when trying to evaluate effects of an injury prevention program in a workplace. Injury reports provide an obvious measure, but their occurrence is affected by factors unaccounted for by the intervention. In this instance, Aird et al. (1988) noted that the four post-intervention back injury reports consisted of two from employees regarded as discipline problems, and two reports from an employee with an underlying non-work-related injury. An important lesson learned from this intervention project was that reported back injuries are fine as one measure of intervention effectiveness, but other measures to supplement injury reports are advisable.

Example 2

Garg and Owen (1992, 1994) reported results of an ergonomic intervention program in a nursing home. The National Institute for Occupational Safety and Health sponsored the intervention to demonstrate how ergonomics could change back-stressing tasks performed by nursing assistants. As a first step Garg and Owen had the nursing assistants rate the stressfulness of their various tasks to their lower back, upper back, shoulders, and whole body. These ratings were used to identify the tasks considered most stressful. The most stressful tasks for the lower back were transfers of the nursing home residents, in both directions, between wheelchair and toilet, between wheelchair and bed, and between wheelchair and bathtub.

Before starting the intervention, an ergonomic evaluation of the work performed by the nursing assistants was completed (Garg, Owen, and Carlson, 1992). Alternative approaches for making the transfers were investigated. Different brands of equipment and techniques were compared in a laboratory setting (Garg, Owen, Beller, and Banang, 1991a, 1991b).

Implementation began after the best alternatives were known. The alternatives were based on the capabilities of each resident. For example, residents who could not bear their own weight were transferred using a portable hoist. Many of the stressful transfers were eliminated. One example was the use of a shower instead of a bathtub for certain residents. They were transported to the shower on a portable shower chair, and washed with a flexible shower hose while remaining in the chair. Stressful tasks that could not be eliminated were made less stressful by introducing ergonomic belts that were put around the resident's waist before a transfer. These belts had loops on the sides and back for the nursing assistants to hold during transfers. Training was provided for the nursing assistants who were to use new equipment and techniques.

Evaluation was based on multiple measures. One measure was acceptability. Monitoring the nursing assistants revealed acceptability rates ranged from 87 to 100%. For those transfers not eliminated, a comparison of exposures before and after the intervention revealed a reduction in the perceived exertion of the low back muscles and a reduction in the biomechanical stresses in the lower backs of the nursing assistants. The back injury rate dropped from 83 to 47 per 200,000 hours.

108.4 Elements of Ergonomics Programs in Health Care

Different authors have different ideas about the elements of effective ergonomics programs in health care settings. This section summarizes the ideas of three authors and concludes with a tabular listing of the elements of an ergonomics program for health care facilities.

A back injury prevention program with an ergonomics emphasis was implemented at the University of Massachusetts Hospital. It focused on back injuries associated with patient handling tasks because such injuries constituted a large portion of the hospital's total injuries. The hospital had 32% of all injuries coming from the nursing department, and many of these were sprains/strains associated with patient handling. From the experiences with this program, Fragala (1994) suggests that hospitals include the following five phases in a back injury program with an ergonomics emphasis.

1. Risk identification and assessment.
2. Risk analysis.
3. Formulation of recommendations.
4. Implementation of recommendations.
5. Monitoring and evaluation.

Notice that these phases do not expressly mention the medical processes of health surveillance and medical-case management. Also training is not identified as a separate phase. However, health surveillance is the core of the second phase, and Fragala sees medical management and training as being incorporated into Phase 4 — implementation of recommendations. Thus, the Fragala model stresses the parts of an ergonomics program aimed at changing the high-risk tasks by making appropriate equipment available.

TABLE 108.3 Elements of Ergonomics Programs in Health Care Establishments

Ergonomics Committee Effectiveness	Processes to Affect Working Conditions	Processes to Reduce Lost Time
Management commitment	Hazard identification	Case management
Supervisory support and participation	Evaluation of risk	Rehabilitation program
Worker support and participation	Hazard control identification	Job descriptions with physical demands
Employee Health Dept. and/or Occ. Med. participation and support	Implementation approval and support	Policy for job modifications to facilitate early return to work
Committee leadership	Funding for changes	Monitoring
Training for committee members	Implementation	Evaluation
Written plan with realistic goals and delegation of work	Monitoring	
Performance evaluation	Evaluation	

He sees medical management as an already established program within the hospital that complements the ergonomics program.

MacLeod (1994) presented the following list as the elements of an ergonomics program for a hospital:

1. Organization
2. Training
3. Communication
4. Job Analysis
5. Job Improvements
6. Medical Management
7. Monitoring Progress

Garg and Owen (1994) presented a list of elements they found effective for their ergonomics program in a nursing home.

1. Secure management commitment
2. Enlist worker participation
3. Determine patient characteristics
4. Select and order equipment
5. Train nursing personnel
6. Gather feedback
7. Determine effectiveness

The three lists of program elements all differ. Much of the difference is due to failure to distinguish among elements of effective committee performance, processes for affecting improvements in working conditions, and processes for reducing lost time. Table 108.3 lists key elements for each of these areas.

108.5 Summary

Data cited in this chapter show how large and diverse the health care industry is. Of the different sectors within the health care industry, nursing homes have the greatest problem with back injuries. Hospitals have problems with back injuries as well, primarily among nursing staff. The home nursing sector has problems with back injuries among home health aides. Compared to back injuries, carpal tunnel syndrome is not a very significant problem in these health care sectors. However, carpal tunnel syndrome is a concern for dental hygienists.

Patient handling is the activity most closely associated with back injuries among nursing personnel. Traditionally, hospitals and nursing homes have attempted to prevent back injuries by emphasizing personal fitness and the use of proper body mechanics when performing patient-handling tasks.

The point is made that ergonomics has not been used in health care as extensively as in some other industries. During the 1980s ergonomics began getting recognition as a useful means of reducing the stresses imposed on the backs of nursing personnel when performing patient-handling tasks. A demonstration project in one nursing home showed how stressful patient-handling tasks can be eliminated or made less stressful.

Elements of ergonomics programs for health care facilities are discussed. A summary of elements is provided in a table arranged into three groups:

1. Elements of effective ergonomics committees
2. Processes to bring about improvements in working conditions
3. Processes to reduce lost time

References

Adelman, S. and Elsner, K. 1982. Arm pain in a dentist; pronator syndrome, *J. Am. Dental Assoc.* 105: 61-62.

Aird, J.W., Nyran, P., and Roberts, G. 1988. Comprehensive back injury program: an ergonomics approach for controlling back injuries in health care facilities, *Trends in Ergonomics/Human Factors V,* Ed. F. Aghazadeh, 705-712. Elsevier (North-Holland Div.), Amsterdam.

Christensen, H. and Finsen, L. 1994. Musculoskeletal disorders among dentists and lack of variation in the dental work, *Proceedings of the* Triennial *Congress of the International Ergonomics Association, Volume 2,* Ed. R. Norman, p. 105-107. Human Factors Association of Canada, Mississauga, Ontario, Canada.

Cook, R.I. and Woods, D.D. 1996. Adapting to new technology in the operating room, *Human Factors* 38(4): 593-613.

Finsen, L. and Christensen, H. 1994. Postural loads in the neck and shoulder region during dental work, in *Proceedings of the 12th Triennial Congress of the International Ergonomics Association, Volume 2,* Ed. R. Norman, p. 102-104. Human Factors Association of Canada, Mississauga, Ontario, Canada.

Flynn, E.A. et al. 1996. Relationships between ambient sounds and the accuracy of pharmacists' prescription-filling performance, *Human Factors* 38(4): 614-622.

Fragala, G. 1994. Applying ergonomics to back injury prevention programs in health care, in *Ergonomics in Health Care Facilities,* Ed. K.M. Tomasik, p. 7-14. Joint Commission on Accreditation of Healthcare Organizations, Oakbrook Terrace, IL.

Gagnon, M. and Lortie, M. 1987. A biomechanical approach to low-back problems in nursing homes, in *Trends in Ergonomics/Human Factors IV,* Ed. S.S. Asfour, p. 795-802.

Garg, A. and Owen, B.D. 1992. Reducing back stress to nursing personnel: an ergonomic intervention in a nursing home, *Ergonomics* 35: 1353-1375.

Garg, A. and Owen, B.D. 1994. An ergonomic approach to preventing back injuries from patient lifting, in *Ergonomics in Health Care Facilities,* Ed. K.M. Tomasik, p. 15-22. Joint Commission on Accreditation of Healthcare Organizations, Oakbrook Terrace, IL.

Garg, A., Owen, B.D., and Carlson, B. 1992. An ergonomic evaluation of nursing assistants' job in a nursing home, *Ergonomics* 35: 979-995.

Garg, A., Owen, B.D., Beller, D., and Banaag, J. 1991a. A biomechanical and ergonomic evaluation of patient transferring tasks: bed to wheelchair and wheelchair to bed, *Ergonomics* 34: 289-321.

Garg, A., B.D. Owen, B.D., Beller, D., and Banaag, J. 1991b. A biomechanical and ergonomic evaluation of patient transferring tasks: wheelchair to shower chair and shower chair to wheelchair, *Ergonomics* 34: 407-419.

Grether, W.F. 1986. The genesis of human engineering, *Aviation, Space, and Environmental Medicine,* 57(10, Suppl.): A37-43.

Hoover, S.A. 1973. Job-related back injuries in a hospital, *American J. of Nursing* 73: 2078-2079.

Hubley-Kozey, C.L., Westers, B.M., Stanis, W.D., and Wall, J.C. 1985. An investigation into the incidence of low back pain in hospital workers, *The Nova Scotia Medical Bulletin* 64(1): 8-10.

Jensen R.C. 1990a. Back injuries among nursing personnel related to exposure, *Appl. Occup. and Envir. Hyg.* 5(1): 38-45.

Jensen R.C. 1990b. The increasing occupational injury rate in nursing homes, in *Advances in Industrial Ergonomics and Safety Research II*, Ed. B. Das, p. 569-576. Taylor and Francis, London.

Jensen R., Nestor D., Myers A., and Rattiner J. 1988. *Low Back Injuries Among Nursing Personnel, An Annotated Bibliography.* Injury Prevention Center, School of Hygiene and Public Health, The Johns Hopkins University, Baltimore, Maryland.

Jensen R.C. 1987. Disabling back injuries among nursing personnel: research needs and justification, *Research in Nursing and Health* 10(1): 29-38.

Jensen R.C. 1985. Events that trigger disabling back pain among nurses, in *Proceedings of the Human Factors Society 29th Annual Meeting*, p. 799-801. Human Factors Society, Santa Monica, CA.

Joseph, B.S. and Jimmerson, G. 1996. Development and Implementation of an Ergonomics Process in the Automotive Industry, in *Handbook of Occupational Ergonomics*, Ed. A. Bhattacharya and J. McGlothlin, p. 501-517. Marcel Dekker, New York.

Lehto, T.U., Helenius, H.Y.M., and Alaranta, H.T. 1991. Musculoskeletal symptoms of dentists assessed by a multidisciplinary approach, *Community Dental Oral Epidemiology* 19: 38-44.

Lewy, R. 1981. Prevention strategies in hospital occupational medicine, *J. of Occupational Medicine* 23: 109-111.

Liss, G.M. et al. 1995. Musculoskeletal problems among Ontario dental hygienists, *Am. J. of Independent Med.* 28(4): 521-540.

Lloyd, P., Tarling, C., Troup, J.D.G., and Wright, B. 1987. *The Handling of Patients: A Guide for Nurses, 2nd ed.* Back Pain Association, Middlesex, England.

Lortie, M. 1985. L'identification des activities a risque en milieu hospitalier, in *Proceedings of the Human Factors Association of Canada, 18th Annual Meeting*, p. 119-122. Human Factors Association of Canada, Mississauga, Ontario.

Lortie, M. 1986. Analyse du travail de manutention de patients des aides-soignants dans un hopital pour soins prolongs, *Le Travail Humain* 49: 315-332.

MacKensie, C.F. et al. 1996. Comparison of self-reporting of deficiencies in airway management with video analyses of actual performance, *Human Factors* 38(4): 623-635.

MacLeod, D. 1994. Setting up ergonomics programs in health care organizations, in *Ergonomics in Health Care Facilities*, Ed. K.M. Tomasik, p. 33-38. Joint Commission on Accreditation of Healthcare Organizations, Oakbrook Terrace, IL.

Morrow, D.G., Leirer, V.O., Andrassy, J.M., Tanke, E.D., and Stine-Morrow, A.L. 1996. Medication instruction design: younger and older adult schemas for taking medication, *Human Factors* 38(4): 556-573.

Murtomaa, H. 1982. Work-related complaints of dentists and dental assistants, *Int. Arch. Occup. Environ. Health* 50: 231-236.

Myers, A., Jensen, R., Nestor, D., and Rattiner, J. 1994. Low back injuries among home health aides compared with hospital nursing aides, *Home Health Care Services Quarterly*, 14(2/3): 153-160.

Öberg, T., Karsznia, A., Kadefors, R., and Sandsjö, L. 1994. Muscular load, fatigue, and pause distribution patterns in dental hygiene: a study with a portable recording and analysis equipment, in *Proceedings of the 12th Triennial Congress of the International Ergonomics Association, Volume 2*, Ed. R. Norman, p. 99-101. Human Factors Association of Canada, Mississauga, Ontario, Canada.

Öberg, T. and Öberg, U. 1993. Musculoskeletal complaints in dental hygiene. *J. Dental Hygiene* 67, 257-261.

Obradovich, J.H. and Woods, D.D. 1996. Users as designers: how people cope with poor hci design in computer-based medical devices. *Human Factors* 38(4): 574-592.

Owen, B.D. 1996. Back injuries in the home health care setting. *Home Health Care Consultant* 3(3): 25-39.

Pines, A., Cleghorn de Rohrmoser, D.C., and Pollak, E. 1985. Occupational accidents in a hospital setting: an epidemiological analysis, *J. of Occupational Accidents*, 7: 195-215.

Shugars, D., Miller, D., Williams, D., Fishburne, C. and Strickland, D. 1987. Musculoskeletal pain among general dentists, *General Dentistry* 4:272-276.

Stubbs, D.A., Buckle, P.W., Hudson, M.P., Rivers, P.M., and Worringham, C.J. 1983. Back pain in the nursing profession: I. Epidemiology and pilot methodology, *Ergonomics* 26: 755-765.

Stubbs, D.A., Buckle, P.W., Hudson, M.P., and Rivers, P.M. 1983. Back pain in the nursing profession: II. The effectiveness of training, *Ergonomics*, 26: 767-779.

Wood, D.J. 1987. Design and evaluation of a back injury prevention program within a geriatric hospital. *Spine* 12(2): 77-82.

Xiao, Y-H. et al. 1996. Task complexity in emergency medical care and its implications for team coordination. *Human Factors* 38(4): 636-645.

109

Ergonomics Efforts in the Package Delivery Industry

Steven W. Thompson
United Parcel Service

109.1 Ergonomics in the Package Delivery Industry

Ergonomics in the service industry, particularly those that are heavily intensive in manual material handling, present a unique set of challenges in attempting to meet the needs of the office worker as well as the needs of the manual material handlers. This chapter will attempt to answer some of these challenges by discussing:

A. Technology-driven ergonomic change
B. Workplace statistics and job-requirement-driven ergonomic change

109.2 Technology-Driven Ergonomic Change

Change, especially technological change, presents an opportunity to enhance a work environment with the proper attention to ergonomics and the impact that it can have on the changing job. The following case illustrates this point.

Current Job and Work Methods

The current method requires the workers to locate individual shipments by thumbing through paper records manually to locate the delivery information. The worker sits at a standard metal office desk surrounded by cabinets containing the shipment records. These cabinets are generally six feet wide and eight feet tall. The top of the desk would commonly contain a telephone, computer, monitor, adding machine, and writing instruments. This job was scheduled to be combined with other functions as technology advances allowed the tracing of these shipments to be completed with a computer rather than manually.

0-8493-2641-9/99/$0.00+$.50

Methodology

A cross-functional team consisting of ergonomics, health and safety, industrial engineering, plant engineering, and the involved workers was formed. The first part of the plan was to determine the exact job requirements of each worker. These included written job procedures of current and future jobs as they continue to evolve into the future.

The next step was to gather information on the frequency and importance of each item located on the desk. These were placed in order of times used and the importance of being located within a functional reach. The workspace needed for each function was analyzed as well as close and remote storage capabilities. A benchmark of all activities was established and a monthly monitor system was established to track the progress of the group (Congleton, 1993).

A survey of body part discomfort was distributed to all employees. This survey along with written comments allowed all of the employees to rate their level of discomfort and to explain why they thought this level of discomfort existed. These surveys were collected and summarized by function to allow the team to concentrate on the body parts of concern.

All of the information above was used to help develop an ergonomic level of risk of discomfort. The three levels chosen were light, medium, and heavy.

This information was needed to help design new workstations and ergonomic accessories. Each work group was classified into a level of risk of discomfort, and a workstation was developed for that work group.

Training documents from the MTM Association for Research and Standards (1985) was used to develop the following variables (MTM, 1985):

Frequency of use of computer
- Infrequent — Less than ½ hour per day
- Frequent — ½ hour to 4 hours per day
- Constant — Greater than 4 hours per day

Mode of use of computer
- Entry — Continuous input of information; little analysis of display information
- Inquiry — Analysis of displayed information; entries to access files through menus
- Combination — Time is almost equally divided between entry and inquiry modes

Assessment

A final assessment was needed to determine how often an employee was required by their job to leave their chair. (Kroemer et al., 1994) This was categorized into less than once an hour, once an hour, and more than once an hour. All of these elements were combined to determine the ergonomic risk of discomfort.

Ergonomic Risk of Discomfort

Light	Medium	Heavy
Infrequent Use of Computer	Frequent Use of Computer	Constant Use of Computer
Inquiry Mode of Use of Computer	Inquiry Mode of Use of Computer	Entry or Combination Mode of Use of Computer
Leave Seat More Than Once an Hour	Leave Seat About Once an Hour	Leave Seat Less than Once an Hour

The furniture levels were determined for each level of ergonomic risk:

Light	Medium	Heavy
Adjustable chair with armrests (if in seat greater than 2 hours)		
	Adjustable chair with armrests	Adjustable chair with armrests
Work surface at a comfortable height for paperwork		
	Rear work surface for computer monitor	Rear work surface for computer monitor (with a range to accommodate sitting and standing postures for 5th% female to standing 95th% male (Eastman Kodak, 1983)
	Maintenance adjustable	Independently adjustable front surface for keyboard
Modular Panel Walls (height dependent on communication needs)		
	Modular Panel Walls (height dependent on communication needs)	Modular Panel Walls (height dependent on communication needs)
		Wrist rest
	Other accessories: footrest, wrist rest, glare screen	Other accessories: footrest, glare screen
	Document holder, headset (as needed)	Document holder, headset (as needed)

The workstation developed was a counterbalance mechanism that allows the worker to adjust the station with fingertip control. This will give the worker total control on the decision to sit or stand as changes in the job occur (Kroemer, 1983). The height of the back work surface and the front work surface are independently adjustable. They may be raised or lowered individually or in tandem to comfortably fit the sitting or standing posture of the worker. The modular walls were designed to reduce the noise level and create a sense of privacy. The height of the panel (42") allowed most workers to see above the panels when standing to refocus their eyes. This reduced the feeling of confinement that a taller panel produced. The lighting was designed for an office interior and supplemented with task lighting where needed for paperwork (Sanders and McCormick, 1987).

Training

A training program was developed to help train the workforce on all of the new ergonomic features. This program taught the basic principles of human factors and ergonomics and stressed how comfort could be attained through good work methods, micro stretches, and posture. A safe work methods evaluation was then developed to help the supervisors follow up on this training. The safe work methods evaluation stressed the need to use proper methods and to utilize the new workstations, chairs, and accessories properly.

Surveys

A survey was conducted to determine how much time employees spent standing at their workstation. The survey included full time as well as part-time employees. The average amount of time spent standing was 23% and they adjusted their workstations to a standing position an average of 3.6 times per day. According to the survey, almost every employee (91%) adjusted their workstations some time during their workday (Nerhood and Thompson, 1994). To determine the effectiveness of the new workstations and ergonomic accessories, the following items were tracked prior to, during, and following the introduction and training of the new equipment:

- Body part discomfort surveys
- Injury/illness rates
- Productivity
- Absenteeism

The last results were taken one year following the introduction of the new equipment. Body part discomfort ratings improved by 26% following the office improvements. The number of work-related injuries or illnesses decreased by 28%, and lost time injuries decreased by 82%. The cost associated with these occurrences decreased by 95%. The productivity increase was conservatively estimated at 17% when compared to a similar site that had not received the new equipment. Absenteeism and tardiness did not change significantly. A follow-up review of the use of the new furniture in three locations showed that the lost time injury frequency was 69% less than at similar sites that did not receive the new equipment. The non lost time injury frequency was 89% less in these sites versus similar sites that did not receive the new furniture.

Commitments

Based on the results of these ergonomic interventions, resources have been committed to provide the following:

- Adjustable chairs for employees seated for longer than 2 hours
- Adjustable workstations that allow employees to sit or stand while working and provide keyboard adjustments
- Training videos for chairs and adjustable workstations
- Office accessories catalog including choices for many items including chair, footrest, headset, monitor arm, task light, lumbar pillow, adjustable keyboard surface, etc.
- Classroom training in ergonomics for management and non-management employees
- Safety training program for the entire company including tips for office work area set-up and micro stretches

109.3 Workplace Statistics and Job Requirements

Workplace statistics and job requirements sometimes trigger review of a particular job. The trailer unload job is an example. The following factors led to review of this particular job:

1. Unacceptable injury frequency
2. High percentage of new employees starting in this job
3. Difficult to automate
4. Need for high production rates
5. Wide range of motions required

Overview

A cross-functional team comprised of health and safety, industrial engineering, operations, plant engineering, and our outside insurance carrier was formed. This team was tasked with reviewing the unload job and determining short-term and long-term changes to improve this working environment.

Over a period of six months, this team visited several different locations in various geographical regions. Prior to their arrival in a location, the management team would address all the work groups and explain that the team was coming. They would inform the workers that the team was specifically interested in their opinion of how the workplace could be improved.

Surveys

Several types of surveys and interviews were collected. All surveys were completed with a team member available to clarify any questions and collected immediately after completion. All of the surveys and interviews were completed with total confidentiality for the respondent. The first two surveys were for the part-time and full-time management. These surveys asked safety questions pertaining to:

- Training
- Performance
- Methods
- Job set-up

All available management people were interviewed.

The hourly employee surveys were divided into seniority and new employee surveys. These surveys included safety questions pertaining to:

- Methods
- Training
- Support
- Job set-up

The surveys also addressed the occurrences of injuries.

The health and safety groups for each work area participated in a focus group meeting to discuss their ideas. This was generally comprised of both management and hourly employees.

The use of videotape proved to be advantageous in this project. Two types of videotaping were utilized. The first was to fix several multiplex video cameras in place and film an unload area over several days. These tapes were then analyzed for any trends that developed over time. These tapes were also utilized for any unsafe working conditions or unsafe work practices. The second method was to utilize handheld video cameras to film specific actions and flow rates. Several buildings were selected from various parts of the country. All of the injury reports from their unload operations were analyzed. A statistical analysis of various "contributing factors" was developed. The final tool employed was an ergonomic model that calculated the kilocalorie burned for each portion of the unload job. All of the data were analyzed for trends, and solutions were sought for all of the areas that impacted the injury frequency.

Potential Solutions

A list of potential solutions was developed for each function. This list included development of new equipment, equipment modification, new method development, and new training programs. A significant contribution of this team was the development and deployment of a variable height extendor. This device extends into the trailer to carry the packages out of the trailer to the sorter. The variable height extendor adjusts from the floor to over the person's head and from side to side. A new unload stand was also developed to utilize with the variable height extendor. This new stand has a larger standing surface area and was more stable than the old unload stand. This helps the unloader to always handle packages in the power zone. The power zone is from the person's knuckle height to shoulder height. The following is a short description of the old equipment and method versus the new equipment and method.

Old Equipment and Method

- Equipment
  - Fixed height rollers
  - Conveyor with minimum mobility
  - Load stand

- Method
  - Overhead reaching
  - Low lifting
  - Twisting
  - Carrying

New Equipment and Method

- Equipment
  - Adjustable height, extending conveyor with side to side adjustment
  - New unload stand
- Method
  - Transferring — less bending
  - Leveraging
  - Pivoting

Results

This new extendor reduced the kilocalorie requirements and significantly reduced the twisting, turning, and bending requirements. The new extendor allowed greater quality, improved productivity, and required less effort.

The success of both of these projects was due to the involvement of the workers and the utilization of the cross-functional teams. Ergonomics is designing the workplace to fit the worker. It is most appropriate to involve workers in what fits best in their workplace. The cross-functional teams provide the expertise to make the necessary equipment changes, and provide the training and methods needed.

References

Congleton, J.J. (1993) *Awareness of Ergonomics in the Workplace.* College Station, TX: Author.

Eastman Kodak Company. (Vol.1, 1983; Vol. 2, 1986). *Ergonomic Design for People at Work.* New York: Van Nostrand Reinhold.

Kroemer, K.H.E., Kroemer, H.B., and Kroemer-Elbert, K.E. (1994). *Ergonomics: How to Design for Ease and Efficiency.* Englewood Cliffs, NJ: Prentice Hall.

Kroemer, K.H.E. (1983, March). Fitting the workplace to the human and not vice versa. *Industrial Engineering,* 55-61.

Nerhood, H.L. and Thompson, S.W. (1994). Adjustable sit-stand workstations in the office. *Proceedings of the Human Factors Society 38th Annual Meeting.* 668-672.

Sanders, M.S. and McCormick, E.J. (1987). *Human Factors in Engineering Design.* New York; McGraw Hill.

The MTM Association for Standards and Research. (1985). *Ergonomics and Human Factors Training Student Training Manual.* Fairlawn, NJ: Author.

110

Ergonomics in the Construction Industry

Scott P. Schneider*

The Center to Protect Workers' Rights

110.1 Introduction — The Size and Scope of the Problem

There is substantial evidence that musculoskeletal injuries are a major problem in the construction industry (Schneider, 1997a). The lost-time injury rate for "sprain and strain" injuries in construction is about 50% higher than that in manufacturing and second only to the rate for the transportation industry. In 1995, the construction rate was 158.7 lost workday cases per 10,000 full-time workers, or about 1.6 cases per 100 workers (BLS, 1997). The rate in private industry overall was 107.5 cases per 10,000 workers. While this rate dropped in 1995 by almost 12% from 1994, the rate is still very higher and much higher than other industries. The rates for cumulative trauma disorders, like carpal tunnel syndrome and tendonitis, on the other hand, tend to be much lower in construction than in manufacturing industries, most likely due to increased awareness in manufacturing and, perhaps, to the less repetitive nature of construction work, in general. In 1995, the rate of lost workday injuries for carpal tunnel was only 2.8 cases per 10,000 full-time workers, compared with 8.0 in manufacturing and 3.9 in all private industry. The 1995 rate for tendonitis was 2.1 for all construction, 5.5 for manufacturing, and 2.7 for all private industry.

110.2 Using Surveillance Data to Identify High-Risk Trades

While construction work, in general, is risky for musculoskeletal injuries, the industry is diverse and the risks vary depending on the trade and the type of work done by that trade. BLS Annual Survey data show the roofing, siding, and sheet metal industry to have the highest risk of sprains and strain lost workday injuries (234.2 per 10,000 in 1995), followed by masonry contractors (202.4) and plumbing and heating

* Currently with the Laborers' Health and Safety Fund of North America

0-8493-2641-9/99/$0.00+$.50

(190.8). Painting and electrical contractors had the lowest lost workday injury rates in construction for sprains and strains in 1995 (128.0 and 125.2, respectively). Rates for carpal tunnel syndrome and tendonitis also varied dramatically, with carpenters having the highest rate of carpal tunnel lost workday injuries (12.1 per 10,000) and masonry contractors having the highest rate for tendonitis (7.8 per 10,000).

Hsaio and Stanevich (1996) analyzed workers' compensation data from 21 states in 1987 (the BLS Supplementary Data System) to rank construction occupations in terms of injury risks to set priorities for future research and interventions. They identified construction laborers, carpenters, roofers and drywall installers as the four highest risk construction occupations. Overexertion injuries were the most common injury type for construction laborers, drywall installers, plumbers, electricians, and structural metal workers, (representing about 22 to 29% of injuries) and the second most common injury for roofers and carpenters.

Analyses of workers' compensation data from Washington state also show construction occupations to be risky in terms of musculoskeletal disorders (WA DL&I, 1996), showing wallboard installation, roofing and concrete construction to be three of the top 10 occupations in terms of incidence rates (first, third, and tenth, respectively).

Surveys of musculoskeletal symptoms among construction workers also show different prevalence patterns among different construction trades (Cook et al., 1996a; Engholm et al., 1997; Holmström et al., 1995, Rosecrance et al., 1977; Zimmermann et al., 1997a). Knee injuries are highest among plumbers, roofers, floorlayers, and sheet metal workers whose jobs require a lot of kneeling. Shoulder problems are most common among scaffold erectors, insulators, and painters who have to work overhead a great deal. Low-back problems are common among most trades, but highest among roofers, floor layers, and scaffold erectors who have to do a lot of heavy-materials handling and stooping. Bricklayers have high rates of elbow and shoulder symptoms apparently because of the awkward postures during work (Cook et al., 1996b). Operating engineers, who operate heavy equipment, had the lowest rates of musculoskeletal symptoms because of the nature of their work, primarily sedentary (Zimmermann et al., 1997b). These trade-specific profiles should be helpful in identifying high-priority trades and areas for intervention.

110.3 The Risk Factors Associated with Different Trades

Certain construction trades have been extensively studied while others have not. Some trades are easier to study in that they perform a more limited number of tasks, like bricklayers and concrete reinforcement workers. Other trades, like carpenters and construction laborers, perform a wide variety of work and only certain subtrades, like drywall installers, have been well studied. Table 110.1 summarizes the state of research on ergonomic problems in the construction trades (CPWR, 1996).

For most trades, there is substantial information of the types of musculoskeletal injuries occurring in that trade. For many of the trades there has been a delineation of the tasks they perform, and high-risk tasks have been identified. Interventions have been identified for some trades, but few have been evaluated. The largest problem remaining for research is identifying the barriers to adoption of these interventions and how these barriers could be overcome.

Trades that have been thoroughly studied include: concrete reinforcement workers (rodmen) (Saari et al., 1978; Wakula et al., 1997), bricklayers (Cook et al., 1996b; Schierhorn, 1996), carpet layers (Thun et al., 1987; Tanaka et al., 1989), operating engineers (Zimmermann, 1997b).

110.4 Exposure Assessment for Ergonomic Risk Factors in Construction

Most ergonomic exposure assessment for exposure to risk factors has taken place in industries where workers have relatively short cycle jobs, like assembly line work. Recently though there has been a growing interest in looking at long cycle jobs, like those in construction. The approach has generally been to look at the risk factors for individual tasks and either sum up the stresses based on estimates of time spent in the task or use the information to prioritize tasks for intervention. Systems that have been adapted or

TABLE 110.1 Status of Research on Ergonomic Problems in Construction by Trade

Trade	WMD	Tasks	Hi Risk	Intervention	Barriers
Bricklayers	√	√	√	√	*
Carpenters					
drywall	√	√	√	√/*	*
concrete form	√	√	√	*	*
frame	√	*	*	*	*
finish	√	*	*	*	*
pile driver	*	*	*	*	*
ceiling	√	√	√	*	*
tile, terrazzo	*	*	*	*	*
lather	*	*	*	*	*
cabinetmaker	*	*	*	*	*
scaffolder	√	√	√	*/√	*
carpet/floorlayer	√	√	√	*/√	*
diver	*	*	*	*	*
exhibit	*	*	*	*	*
millwright	*	*	*	*	*
maintenance	√	*	*	*	*
Cement Masons	√	√	*	*	*
Electricians	√	*	*	*	*
Elevator Constructors	*	*	*	*	*
Insulators	*	*	*	*	*
Ironworkers					
structural	√	√	*	*	*
rodmen	√	√	√	*/√	*
ornamental	*	*	*	*	*
shop	*	*	*	*	*
Laborers	√	*	*	*	*
Operating Engineers					
operators	√	√	√	√	*
stationary	*	*	*	*	*
Painters	√	*	*	*	*
carpet/floorlayer	√	√	√	*	*
glaziers	*	*	*	*	*
Plasterers	√	√	*	*	*
Plumbers	√	√	*	*	*
Pipefitters	√	√	*	*	*
Roofers					
residential	√	*/√	*/√	*	*
commercial	√	*/√	*/√	*	*
Sprinklerfitters	*	*/√	*/√	*	*
Sheet Metal					
shop	*	√	*	*	*
field	√	√	*	*	*
Teamsters	*	*	*	*	*

* = needs more research
√ = unknown

created for ergonomic exposure assessment in construction include the OWAS method (Kivi and Mattila, 1991; Mattila et al., 1993), ARBAN method (Wagenheim et al., 1986), PATH method (Buchholz et al., 1996), and MAS method (Schildge et al., 1997; Wakula et al., 1997). These methods tend to estimate the percentage of time spent in awkward postures during given tasks. Time spent in a task is estimated by diaries, self reports, or expert observers. Studies are now under way to validate these methods against more quantitative methods, like dosimeters. Checklists for exposure assessment in construction are also being studied (Everett, 1997; Buchholz et al., 1996). Exposure assessment in construction is still at an early stage. Its importance will grow, particularly as a tool to measure the efficacy of interventions.

110.5 Types of Interventions in Construction

Ergonomics has often been associated with manufacturing and office work environments. Consequently, consideration of ergonomics in construction appears disconcerting. Contractors and workers seem to believe that construction work is hard manual labor and there's nothing that can be done to change it. This is despite the evidence of how the work has actually changed over the past 10 to 15 years. Construction work, while it hasn't become automated like assembly line work, has become more mechanized. More and different kinds of equipment are now used on job sites to move materials and to do some of the heavy work. Hoists and cranes are commonly used for materials handling. Boom trucks are used to lift materials to roof level. Scissors lifts are commonly used to move workers up to work at heights. Motorized buggies are used to move materials around on sites. Powered equipment like roof cutters, powered brooms and gravel removal equipment, and powered roof tear-off equipment all have reduced the risk of strains and sprains in roofing tear-off work. Asphalt is now pumped to the roof rather than being mixed in small batches in a kettle. Circular saws, powered screw guns (battery-operated), and pneumatic nail guns have made carpenters' work easier. Robots even exist now for doing demolition work. Yet even with this new equipment there is still a high risk of injury and a significant amount of manual work and work in awkward postures. In some cases the contractors are too small to afford or use such equipment. In other cases, the demands of a particular job may not allow their use. In addition, many jobs cannot be mechanized.

Ergonomic interventions in construction don't all revolve around mechanization or automation. In fact, most "ergonomic" changes in construction involve little more than proper planning of the job.

Interventions to reduce the risk of musculoskeletal disorders in construction can be classified as follows: (A) new materials, (B) new tools and equipment, (C) improved work practices, (D) improved work organization and planning, (E) education and exercise, and (F) personal protective equipment.

A) New materials. Construction materials have changed over the past few decades. Drywall has essentially replaced plaster walls. Poured concrete has replaced a lot of brick walls. Many sections of houses now come prefabricated. Sometimes the changes are beneficial from an ergonomic point of view. Other times they trade one hazard for another. Drywall work, as mentioned earlier, is one of the most hazardous trades in construction for musculoskeletal injuries. The trend in newer materials can be useful when lighter-weight materials are designed. For example, the Army Corps of Engineers worked with the University of Nebraska to design a new masonry block, the Nebraska A block, which is half the weight of a traditional block but just as strong (Hooker, 1996). Another solution instituted in Germany was the design of a new masonry block that has hand holds to make it easier to handle (Kaiser and Linke-Kaiser, 1992). The weight and the diameter (normally 4 ft) of drywall makes its use hazardous. In Sweden, the industry has been promoting the use of 3 ft wide (90 cm) drywall boards, which are easier to install than the larger size, but consequently increase the amount of drywall taping and screwing required. (Isakson et al., 1992; Björklund et al., 1991). The use of fiberglass ladders reduces the weight of handling compared with wood ladders. Plastic pipe has also reduced the weight of materials for plumbers.

B) New tools and equipment are constantly being invented to make construction work easier. Tool catalogues from tool suppliers and trade publications are a good source for keeping up with such tools. These tools are designed to reduce the need for bending, e.g., allow for work from a standing height, like guns for fastening roofing insulation and automatic feeding screwguns for fastening flooring. Hand tools can reduce the stress on the hand by having softer, easier-to-hold surfaces. Handles are available to make carrying materials easier. Carts and dollies can be used to help move materials around a site and reduce the need for manual handling. Pulleys and hoists make it easier to lift materials. Stands, like pipe stands, can be used to bring work to waist height. Sit-stand stools or matting can reduce the risk of back injury from standing on concrete all day (Redfern, 1995). Racks can be used for storing materials at waist height.

Power tools can be purchased with vibration dampening to reduce the amount of vibration transmitted to the hands. Construction vehicles now have better-designed cabs available which are more comfortable and reduce the transmission of whole-body vibration through the seat. Many of the best tools have been designed or invented by tradespeople who felt they needed something new to do the job right or easier

(Wigmore and Moir, 1997). The proper design of tools requires usability testing with tradespeople to understand the way tools are used in the field and the demands placed on them by workers (Bobjer and Jansson, 1997). The design of any new tool or equipment should include an evaluation to demonstrate reduced risk of injury or, at least, a reduction in risk-factor exposure.

C) Improved work practices involve changing how the work is done. By substituting a scissors lift for a ladder, workers can get to overhead work more easily and position themselves closer to the work, requiring less work with arms above shoulder level. Getting two workers to carry drywall (e.g., using drywall handles on the front and back) or relying on carts and dollies can reduce the risk of injury from materials handling. For those tasks where manual handling is unavoidable, teaching better work technique is important. For example, lifting heavy bags from ground level should be done from a kneeling position by sliding the bag onto the knee and then standing. Drywall boards should first be tipped on end before being picked up (CSAO, 1991). Studies have shown that work technique of older, experienced workers may be more efficient and ergonomically preferable than those of apprentices or novices (Authier et al., 1996). Training in these techniques could help reduce the risk of musculoskeletal injuries although it is unclear at this point how effective such training is.

D) Improved work organization means changing the way the work is organized to reduce the risk of injury. The foreman and superintendent on a job have a major role to play in the proper planning of the job to make sure it gets done on time and gets done safely. Through proper planning these two goals can complement each other. They need to make sure that materials are delivered on time and as close to the work area where they will be used as possible. They also want to avoid ordering or delivering too much material at once which leads to problems with storage and a cluttered worksite. They need to make sure work is done by the various subcontractors on time in order to not delay the subsequent contractors and place them under heavy production pressures to catch up. A Swedish project called "Building for the 21st Century" has called for bringing workers into the production planning process as the best method to improve planning and reduce production pressures (Kortabyggtider).

Sufficient helpers, apprentices or materials handlers should be available to make sure workers are supplied with the materials and equipment they need when they need it. They need to ensure the availability and usability of materials handling devices (carts, dollies, hoists, cranes). Crane time can sometimes be at a premium and proper scheduling of crane time can significantly reduce manual materials handling. A proper break schedule is also critical. Insufficient rest breaks lead to fatigue and reduce productivity as well as increasing risk of injury. In addition to scheduled rest breaks, short mini-breaks (e.g., 30 second "micropauses") have been shown to reduce fatigue and increase productivity in drywall installers (Anderson, 1991). Piece rate work has been shown to be a risk factor for musculoskeletal injuries (Brisson et al., 1989). In construction, drywall installation is one of the few jobs which is commonly paid on a piece rate. This may be one reason it has one of the highest risks for injury. The distribution of workload is another important work organization issue that needs to be addressed.

Improved housekeeping has also been shown to be related to reduced risk of injury (Oxenburgh, 1991). This is an important work organization issue, because it requires the cooperation and coordination of all subcontractors on the site. Superintendents must make clear that each subcontractor is responsible for their own housekeeping to avoid creating a hazard for others. Sometimes they take on more responsibility and develop joint clean up crews to organize housekeeping on a site-wide basis.

Job rotation, e.g., rotating workers between physically demanding and less physically demanding jobs, can sometimes be done in construction. In addition teaming of workers who can often rotate tasks is possible. Another concern is that often young workers carry a disproportionate amount of the heavy work on a site. This increases their risk of injury, particularly the chronic injuries that will accumulate later in life. While the heavy work cannot be redistributed to the older or less physically capable workers, they need to be sensitive to not straining or pushing the younger workers too hard. By reducing the workload for all workers, the work can be distributed more evenly and more fairly.

Architects and engineers also have a major role to play upstream, while the project is being designed. By specifying lighter-weight materials, e.g., 3-foot-wide drywall, they can reduce the load on the construction workers. In Sweden, designers placed pipes on the wall of a utility tunnel instead of overhead, reducing the amount of overhead work (Björk, 1984). The more ergonomics can be considered in the design of buildings, the less we have to rely on *post hoc* changes as the building is being built. Superintendents and foremen also need to structure the job so that issues of ergonomics can be incorporated into their regular safety program and walk arounds. A construction ergonomics checklist has been designed specifically to get safety personnel thinking about ergonomic issues on their worksite (CPWR, 1997).

E) Education and exercise have also been suggested as effective interventions for ergonomic risk factors in construction. Ergonomics training is becoming more common in the construction trades. The Carpenters Union has developed a half-day training program on ergonomics for carpenters which is given through their apprenticeship schools (UBC Safety & Health Fund, 1996). The Building Trades Department of the AFL-CIO is developing a training module on ergonomics to be given to all apprentices as part of their safety training. Contractors who do ergonomics training tend to focus on proper lifting technique and stretching exercises, while these new training programs focus on identification of risk factors and problem solving to identify potential solutions. They also tend to include more participatory training techniques, where workers are active participants in the learning process, which appear to be more effective and allow workers to share their knowledge of conditions and experiences in crafting solutions (Shurman et al., 1994).

Labeling of the weights of materials to be manually handled may also help reduce the risk of musculoskeletal injuries (Butler et al., 1993). It has been suggested that construction materials be labeled where possible with the weight and color coded labels to indicate if it was safe to lift manually or alone (Schneider, 1994a).

Exercise or stretching programs have become popular among construction contractors over the past few years. They operate on the assumption that by stretching the muscles and tendons prior to work, they can better adjust or acclimate to the stresses placed on them later in the day. There have been two evaluations done on stretching programs in the construction industry, one in Sweden and one in the U.S. (Cederqvist 1994; Hecker and Gibbons, 1997). Both were surveys of workers involved in a pre-job stretching program. Both found positive results. Workers who did the stretching exercises liked them and felt that they helped in reducing fatigue and increasing awareness of ergonomic risk factors. In general they felt better at the end of the day. A large percentage also continued doing the exercises on weekends and said they would continue doing them on their next job. While this doesn't necessarily translate into lower injury rates, these positive indicators are some support for continuing these programs and further evaluation of their impact.

F) Personal protective equipment (PPE) is normally the last resort in terms of intervention strategies. It allows the continued presence of a risk factor and the worker must rely on the proper use of a device to intervene and modify the effect of the exposure. This is a less reliable strategy. Yet in construction, there are several instances where PPE can be necessary. While some work can be modified to allow work to be done from a standing height, there will still be some work required at floor level. Kneeling will have to be done at some point. Large amounts of time spent kneeling has been correlated with knee disorders. So clearly knee pads can play an important role in prevention of knee disorders in construction. The problem is that workers don't like to wear knee pads. The straps used to keep them on bind against the back of the legs and make them uncomfortable to wear. One possible solution is pants with knee pad pockets in front of the legs. Shoulder pads are also available for construction workers who have to carry materials on their shoulder, where contact stresses can pinch nerves and tendons and contribute to shoulder disorders. There are also shoe insole pads designed to reduce stress on the back for those workers who have to walk around all day on concrete.

The most controversial protective equipment issue in ergonomics for construction workers is the use of back belts. During the 1990s they have become increasingly popular among contractors. Yet there is little evidence to support their use in preventing back injuries in the first instance (NIOSH, Back Belt Working Group, 1994), although a recent study indicated they may have some efficacy (Kraus et al., 1996). If they are used, use should be voluntary and should be accompanied by an education and training program on proper use. They should also be used under the supervision of a physician who can certify that workers are fit to wear them. It should be emphasized that use of the belt does not allow a worker to lift more then they normally would. They should only be considered a supplement to a complete ergonomics program and not a substitute for one. There should also be a program of quality assurance to make sure people are using the belts properly.

110.6 Introduction and Implementation of Interventions in Construction

While there are many potential ergonomic interventions in construction, few have been adopted. Making ergonomic changes in construction is difficult, but there are many possible interventions (Schneider, 1995; Schneider et al., 1995). The construction industry is inherently conservative. Workers tend to be individualistic and safety culture is difficult to change. This is in part due to the nature of the work, where workers are changing jobs frequently and may also change employers. Ergonomic challenges also change as the job site changes. There are four different levels of ergonomic changes available in construction: industry-wide changes, company-wide changes, site-specific changes, and changes on the individual level (Schneider, 1996b). Industry-wide changes include: development of new ergonomically designed tools, changes in materials used (e.g., switching to 3-ft-wide (90 cm) drywall), and availability of new equipment (e.g., adjustable height scaffolding). These changes are the most difficult to accomplish, because of the investment required for the development of new tools and equipment and the level of proof required before contractors will adopt new methods. Company-wide changes depend primarily on individuals in authority within those companies who have the vision or commitment to safety and are willing to try out changes. Site-specific changes depend on the job superintendent, if he or she is willing to try out new methods and chance their potential impact on the production schedule. Sometimes change on an individual site depends as well on the owners of the project. If they are open to new methods and willing to pay for interventions (which hopefully will result in some payback in less lost work time and injuries), then superintendents are willing to go along with such programs. Changes on the individual level are the most difficult to sustain in that each individual's behavior must be changed and monitored. Workers who have been doing a job one way for many years are often reluctant to change. Such changes are most effective when introduced early in their careers when they are learning as apprentices. But teaching a person how to lift heavy materials properly doesn't solve the problem the way a foreman or superintendent can, by ensuring carts or dollies are available and in good working order or by ensuring that materials are delivered as close as possible to where they will be used. Also the fundamental problems posed by poor planning can only be solved by superintendents, working in conjunction with the workers, on a site-wide level. Changes on the site-wide or company-wide level have much more potential for reducing the risk of injury than individual changes.

There are numerous problems in making changes in construction. If changes are primarily motivated by the need to save money and the promise of increased productivity, workers may be resistant to changes, since higher productivity means doing more work with fewer workers. This is also true if changes, in making the jobs easier, also deskill the work. Lower skill required often means less pay for workers, which means workers can be expected to resist these changes as well. Recently, several participatory ergonomic projects have been tried in construction with great success (Bronkhorst et al., 1997; van der Molen et al., 1997b; Moir, et al., 1996). Workers have been included in the process of identification of high-risk tasks and potential solutions. Projects have led to the development of successful interventions for scaffold erectors, glaziers, and other trades. By including workers in each phase of the process, acceptance of interventions and changes is much easier.

110.7 Regulatory Standards

Another way to effect change in an entire industry is through regulation. For the past several years the federal government has been working on the development of an OSHA standard for the prevention of musculoskeletal injuries. A draft of this proposed standard was circulated in March 1995. The OSHA Advisory Committee for Construction Safety and Health (ACCSH) set up a work group which reviewed the draft and proposed changes to make it more useful in construction (Schneider, 1996a). Because of Congressional restrictions, a proposed standard cannot be issued before October 1998, and it is unlikely that the initial proposal will cover construction.

In 1997, California OSHA promulgated a standard for the prevention of cumulative trauma disorders. While this standard would have exempted much of the construction industry, because it exempted employers with nine or fewer employees, the California Supreme Court struck down that provision in September 1997.

In the meantime, the U.S. Army Corps of Engineers, one of the largest construction employers in the U.S., issued a "cumulative trauma prevention "standard in September 1996 which applies to all contractors doing work for the Corps (Schneider, 1997b). It is a programmatic standard requiring job hazard analyses before each job to identify potential ergonomic risk factors and potential solutions to be instituted. Workers must get ergonomic training. Contractors must also reduce vibration exposures to below the ACGIH TLV.

The ASC Z 365 committee is finalizing its draft standard for the prevention of cumulative trauma disorders of the upper extremity, which would also apply in the construction industry. It was approved in Spring 1998. While this is a voluntary standard, these standards do set some minimum expectations with regard to safety programs which often become industry-wide standard practice. This draft standard is also a programmatic one which requires employers to identify high-risk jobs or tasks, develop potential solutions, implement and evaluate those solutions, and develop a medical management program for injured workers to help them return to work (Armstrong, 1997).

Other countries have developed or implemented standards to prevent musculoskeletal injuries in construction. Germany and Sweden have issue rules limiting the weight of masonry blocks that can be lifted manually (Swedish Standards, Kuger et al., 1992). (Masonry union contracts in the U.S. also contain weight limits for blocks that can be lifted by one person.) In The Netherlands, the Stichtung Arbouw (a joint labor–management group) has developed "Guidelines for Physical Workload in the Construction Industry" which contain limits on lifting, carrying, pushing and pulling, static postures, and repetitive work. The guidelines prescribe red (interventions required), yellow (interventions should be planned), and green (permissible) levels of effort (Stichtung Arbouw, 1997, van der Molen, 1997a).

110.8 Conclusion

Musculoskeletal injuries are a major problem in the construction industry. They constitute a large percentage of the lost workday injuries and workers' compensation cases and costs. They also result in shortened careers for many construction workers. Ergonomics is the process by which these injuries can be addressed and many of them prevented. High-risk trades can be identified through injury and symptom surveillance and associated with specific types of injuries. High-risk tasks can be identified though observations and focus groups of workers and through quantitative measurements of exposures to well-known risk factors, like awkward postures. A wide range of interventions are available, although few have been tested for efficacy. Interventions include: materials handling equipment, improved tools, new work methods, better work organization and planning, worker training, exercise programs, and personal protective equipment. The most effective strategies are those which effect change in the design of the work and those which include workers in each and every step of the process (a participatory ergonomics program). There are several recent examples of such successful projects in the construction industry. But change in the construction industry is difficult. There has to be a an acknowledgment of the problem and a willingness to try new ideas and techniques, even though they may not prove effective. But those firms which are committed to attacking this problem properly will find a great potential for success.

Acknowledgments

This research was funded through a cooperative agreement (# CCU/02-312014) from the National Institute for Occupational Safety and Health (NIOSH) to the Center to Protect Workers' Rights (CPWR), the research arm of the Building and Construction Trades Department of the AFL-CIO.

References

Anderson, P: Manual screw tightening with and without micro pauses, *Bygghälsan Bulletine,* 91-09-16, 39-40, 1991.

Armstrong, T: Development of a voluntary standard for the control of work-related cumulative trauma disorders in the United States: ANSI Z365, *Proceedings of the 13th Triennial Congress of the International Ergonomics Association,* June 29–July 4, Tampere, Finland, 3: 483-485, 1997.

Authier, M; Lortie, M; Gagnon, M: Manual handling techniques: comparing novices and experts, *International Journal of Industrial Ergonomics,* 17: 419-429, 1996.

Björk, L: Survey of working environment in the plumbing sector, *Bygghälsan Bulletine,* 1984-09-01, pgs. 4-5, 1984.

Björklund, M; Helmerskog, P.; Nordberg/Bohlin, M. et al.: 90-sheets for the 90s, *Bygghälsan Bulletine* 91-09-16, pgs. 35-36, 1991.

Bobjer, O; Jansson, C: A research approach to the design of ergonomic hand tools. The 11-point approach, *Proceedings of 13th Triennial Congress of the International Ergonomics Association,* June 29–July 4, Tampere, Finland, 2: 193-195, 1997.

Brisson, C., A. Vinet, M. Vézina, S. Gingras. Effect of duration of employment in piecework on severe disability among female garment workers. *Scandinavian Journal of Work, Environment and Health,* 15:329-334, 1989.

Bronkhorst, RE; Vink, P; Koningsveld, EAP: Ergonomic improvements for the glazier — new working methods and tools do improve the working conditions, *Presentation to the 13th Triennial Congress of the International Ergonomics Association,* June 29–July 4, Tampere, Finland, 1997.

Buchholz, B; Paquet, V.; Punnett, L; Lee, D; and Moir, S; PATH: A work sampling-based approach to ergonomics job analysis for construction and other non-repetitive work, *Applied Ergonomics,* 27(3): 177-187, 1996.

Bureau of Labor Statistics (BLS), Data from the 1995 Annual Survey of Occupational Injuries and Illnesses, Data obtained from BLS Website (http://www.bls.gov).

Butler, D.; Andersson, GBJ; Trafimow, J; Schipplein, OD; Andriacchi, TP: The influence of load knowledge on lifting technique, *Ergonomics* 36 (12): 1489-1493, 1993.

Cederqvist, T: Prevention of musculo-skeletal injuries in the swedish construction industry- experience from five years of a pre-work warm-up program, *Proceedings of the 12th Triennial Congress of the International Ergonomics Association,* August 15-19, Toronto, Canada, 2: 60-62, 1994.

Center to Protect Workers' Rights, (CPWR). *Construction safety and health research: five years of progress and proposals for the future* — Report of the CPWR-NIOSH Program Planning Conference *Joint Strategies to Advance Research in the Construction Industry,* Cincinnati, OH April 1–2, 1996.

Center to Protect Workers' Rights, (CPWR). "Construction Ergonomics Checklist," 6 pages, 1997.

Construction Safety Association of Ontario (CSAO), *Stand, Lift, Carry,* Revised, 1993 edition, 41 pages, Toronto, Ontario.

Cook, TM; Rosecrance, JC; Zimmermann, CL: *The University of Iowa Construction Survey,* The University of Iowa and the Center to Protect Workers' Rights, Washington, D.C. April 1996a.

Cook, TM; Rosecrance, JC, Zimmermann, CL: Work-related musculoskeletal disorders in bricklaying: a symptom and job factors survey and guidelines for improvement, *Applied Occupational and Environmental Hygiene,* 11 (11): 1335-1339, November 1996b.

Engholm, G; Holmström, H: Physical exposures, psycho-social factors and patterns of prevalence of musculoskeletal disorders in various groups of swedish construction workers, Presentation to the 13th Triennial Congress of the International Ergonomics Association, June 29–July 4, Tampere, Finland, 1997.

Everett, JG: Application of U.S.A. OSHA Draft Ergonomic Protection Standard to Construction Tasks, *Proceedings of the 13th Triennial Congress of the International Ergonomics Association,* June 29–July 4, Tampere, Finland, 6: 93-95, 1997.

Hecker, S; Gibbons, B: Evaluation of a prework stretching program in the construction industry, *Proceedings of the 13th Triennial Congress of the International Ergonomics Association,* June 29–July 4, Tampere, Finland, 6: 115-117, 1997.

Holmström, E; Ulrich Moritz; Engholm, E: Musculoskeletal disorders in construction workers, *Occupational Medicine: State of the Art Reviews,* 10(2), April 1995, pgs. 295-312, *Construction Safety and Health,* Eds. K. Ringen, A. Englund, L. Welch, J. Weeks and J. Seegal, Hanley & Belfus, Philadelphia, 1995.

Hooker, KA: Specialty block for special uses, *Masonry Construction* 9(5): 198-201, May 1996.

Hsaio, H; Stanevich, RL: Injuries and ergonomic applications in construction, Chapter 27, Pgs. 545-568 in *Occupational Ergonomics: Theory and Applications,* Eds. A. Bhattacharya and J.D. McGlothlin, Marcel Dekker, New York, 1996.

Isakson, H; Kling, J.: Fitting 900mm and 1200 mm plasterboard sheets, *Bygghälsan Bulletine,* 91-09-16, pgs. 37-38, 1991.

Kaiser, R. and G. Linke-Kaiser: Verbesserung der Arbeitsbedingungen im Mauerwersbau *ErgoMed,* 16: 14-25, 1992.

Kivi, P; Mattila, M: Analysis and improvement of work postures in the building industry: application of the computerised OWAS method, *Applied Ergonomics,* 22(1): 43-48, 1991.

Kortabyggtider, Building for the 21st Century: Results from the Development Programme Tight Construction Schedules, a publication from a joint employer-employee working group in Sweden.

Kraus, JF; Brown, KA; McArthur, DL; Peek-Asa, C; Samaniego, L; Kraus, C: Reduction of acute low back injuries by use of back supports, *Int J Occup Environ Health,* 2: 264-273, 1996.

Mattila, M; Karwowski, W; Vilkki, M: Analysis of working postures in hammering tasks on building construction sites using the computerized OWAS method, *Applied Ergonomics,* 24(6): 405-412, 1993.

Moir, S; Buchholz, B: Emerging participatory approaches to ergonomic interventions in the construction industry, *Am. J. Industrial Medicine,* 29, 425-430, 1996.

NIOSH Back Belt Working Group, *Workplace Use of Back Belts: Review and Recommendations,* National Institute for Occupational Safety and Health, Centers for Disease Control, U.S. Department of Health and Human Services, July 1994.

Oxenburgh, M: *Increasing Productivity and Profit through Health and Safety,* CCH International, Chicago, 309 pgs., 1991.

Redfern MS: Influence of flooring on standing fatigue, *Human Factors,* 37(3): 570-581, 1995.

Rosecrance, JC; Cook, TM; Zimmermann, CL: Work-related musculoskeletal symptoms among construction workers in the pipe trades, *Work,* 7: 13-20, 1996.

Saari, J. and G. Wickström: Load on back in concrete reinforcement work, *Scan. J. Work Environ. Health,* 4 suppl. 1: 13-19 (1978).

Schierhorn, C: Jobsite ergonomics, *Masonry Construction,* 9(5): 202-207, May 1996.

Schildge, B; Wakula, J; Rohmert, W.: Ergonomic Analysis of load on the knees at the tile-setter's work, *Proceedings of the 13th Triennial Congress of the International Ergonomics Association,* June 29–July 4, Tampere, Finland, Volume 6: pgs. 163-165, 1997.

Schneider, S., Musculoskeletal injuries in construction: are they a problem?, *Proceedings of the 13th Triennial Congress of the International Ergonomics Association,* June 29–July 4, Tampere, Finland, Volume 6: pgs. 169-171, 1997a.

Schneider, S, The U.S. Army Corps of Engineers ergonomics standard, *Applied Occupational and Environmental Hygiene,* 12(7): 460-461, July, Ergonomics Column, 1997b.

Schneider, S., OSHA's draft standard for the prevention of musculoskeletal disorders: the construction revision, pp. 953-955, *AIHA Journal,* 57 (10), October 1996a.

Schneider, S., CYBERG 96 paper, September, *Making Ergonomic Changes in Construction,* 1996b.

Schneider, S., "Implement ergonomic interventions in construction, *Applied Occupational and Environmental Hygiene,* 10(11): 822-824, October, Ergonomics Column, 1995.

Schneider S., Punnett L., and Cook T, Ergonomics: applying what we know, in *Construction Safety and Health,* in the *Occupational Medicine: State of the Art Reviews* series, Vol. 10 (2):385-394, Eds. K. Ringen, A. Englund, L. Welch, J. Weeks and J. Seegal, Hanley & Belfus, Philadelphia, 1995.

Schneider, S., Ergonomic safety data sheets, *Applied Occupational and Environmental Hygiene,* 9(6): 402-403, June, Ergonomics Column, 1994a.

Schneider, S. and Susi, P., Ergonomics and construction: a review of potential hazards in new construction, July 1994, *American Industrial Hygiene Association Journal,* 55 (7): 635-649, 1994b.

Shurman, S; Silverstein, BA; Richards, SE: Designing a curriculum for healthy work: reflections on the United Automobile, Aerospace and Agricultural Implement Workers — General Motors Ergonomics Pilot Project, *Occupational medicine: State of the Art Reviews,* 9(2): 283-304, 1994.

Stichtung Arbouw, *Arbouw Foundation Guidelines on Physical Workload for the Construction Industry,* Amsterdam, May 1997, 19 pgs.

Swedish Standards for Masonry Block Work (Svenska Standard SS 22 72 30), Swedish Ministry of Labour.

Tanaka, S., S.T. Lee, W.E. Halperin, M. Thun, and A.B. Smith: Reducing knee morbidity among carpetlayers, *Am. J. Public Health,* 79(3): 334-335, 1989.

Thun, M., S. Tanaka, A.B. Smith, W.E. Halperin, S.T. Lee, M.E. Luggen, and E.V. Hess: Morbidity from repetitive knee trauma in carpet and floor layers. *Br. J. Indus. Med.,* 44: 611-620, 1987.

UBC Safety and Health Fund of North America, *Ergonomics for Carpenters,* Washington, D.C., 88 pgs., 1996.

van der Molen, HF, and Delleman, NJ, Arbouw guidelines on physical workload for the construction industry, *Proceedings of the 13th Triennial Congress of the International Ergonomics Association,* June 29–July 4, Tampere, Finland, Volume 6: pgs. 197-199, 1997a.

van der Molen, HF; Vink, P.; and Urlings, IJM; A participatory ergonomic approach to the redesign of scaffolders work, *Proceedings of the 13th Triennial Congress of the International Ergonomics Association,* June 29–July 4, Tampere, Finland, Volume 1: pgs. 150-152, 1997b.

Wagenheim, M.; Samuelson, Björn; and Wos, H.; ARBAN-A force ergonomic analysis method, Chapter 21 in *The Ergonomics of Working Postures,* International Occupational Ergonomics Symposium, April 15-17, 1985, Zadar, Croatia (Yugoslavia), Eds. Nigel Corlett, John Wilson, and Ilija Manenica, London, Taylor and Francis, pages 241-255, 1986.

Wakula, J.; Wimmel, F.; Linke-Kaiser, G,; Hoffman, G.; and Kaiser, R.: Ergonomic Analysis of load on the back in concrete work, *Proceedings of the 13th Triennial Congress of the International Ergonomics Association,* June 29–July 4, 1997, Tampere, Finland, Volume 6: pgs. 191-193, 1997.

Washington State Department of Labor and Industries (WA DL & I), *Work-Related Musculoskeletal Disorders: Washington State Summary 1992-1994,* October, 84 pages, 1996.

Wigmore, D; Moir, S; Bright ideas: construction worker's innovations on the job, Presentation to the American Industrial Hygiene Conference and Exposition, Dallas, TX, May 19, 1997.

Zimmermann, CL; Cook, TM; Rosecrance, JC: Trade-specific trends in self-reported musculoskeletal symptoms and job factor perceptions among unionized construction workers, *Proceedings of the 13th Triennial Congress of the International Ergonomics Association,* June 29–July 4, Tampere, Finland, Volume 6: pgs. 214-216, 1997a.

Zimmermann, CL; Cook, TM; Rosecrance, JC: Work-related musculoskeletal symptoms and injuries among operating engineers: a review and guidelines for improvement, *Applied Occupational and Environmental Hygiene,* 12 (7): 480-484, July 1997b.

111

Ergonomics Issues in Air Traffic Management

Karol Kerns
The MITRE Corporation

Philip J. Smith
The Ohio State University

C. Elaine McCoy
Ohio University

Judith Orasanu
NASA Ames Research Center

111.1 Introduction

The air traffic management (*ATM*) system poses a broad set of human factors challenges. These range from the design of tools to support work by individual controllers, to the development of procedures and tools to support cooperative work by flight crews, controllers, traffic managers, and dispatchers, to the allocation of tasks within an overall system architecture such that the cognitive demands of the work performed by individuals are at acceptable levels while still achieving a high level of system performance (Wickens, Mavor, and McGee, 1997).

This chapter describes a system-wide view of the human factors in the current and future ATM system. The term, air traffic management, is a relatively new one, that denotes both of the system's primary functions, air traffic control (*ATC*) and traffic flow management (*TFM*). This term also recognizes the ongoing changes in the organizational culture and operating philosophy of the system, with greater emphasis on service to and collaboration with system users, less on control. With this in mind, this chapter provides an overview of the main human factors topics in ATM and an explanation of the operational context in which they occur. A major focus of the chapter is on the current and future operating philosophies, processes, and technologies and their implications for human performance and human factors research and practice.

This chapter first briefly outlines the components of the system as it exists in the United States (which is the context that will be used for discussing human factors issues). Then, the air traffic control system is discussed, giving a broad survey of the current and emerging issues and an assessment of the role and status of human factors. Following that, the traffic flow management system is discussed, emphasizing the importance of the impact of the overall system architecture on the demands placed on individuals within the system. Finally, there is a discussion of the human factors implications of proposed future designs for the air traffic management system in the United States.

0-8493-2641-9/99/$0.00+$.50

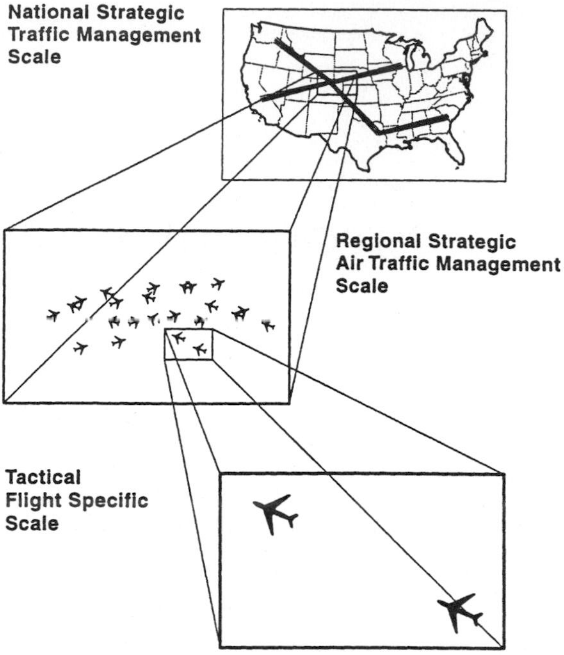

FIGURE 111.1 The hierarchical nature of air traffic management (From Klein, G. 1992. *The Human Air Traffic Management Role in a Highly Automated Air Traffic System.* MITRE Technical Report # ARDATMR-MTR-92W0000075, McLean, VA. With permission.)

111.2 System Overview

The design of the ATM system reflects important concepts about information management and social organization. Figure 111.1 illustrates the basic hierarchical nature of this design and of the management domains that make up the air traffic environment. Information regarding air traffic can be partitioned into a hierarchy of domains based on its quality and granularity. As is shown in the figure, long-run strategic planning based on aggregate traffic demand data must be done to make decisions about resource allocations and about rules and procedures, along with daily strategic traffic flow management decisions, while tactical activities, based on flight-specific data, must be done to assure separation of individual aircraft. The goal of these strategic planning activities is to ensure that controllers, pilots, and dispatchers can safely coordinate the activities of specific aircraft in order to assure safe, effective utilization of the airspace and airport facilities (Klein, 1992).

The ATM system is also organized hierarchically to take advantage of the informational structure (see Figure 111.2). At an organizational level, the ATM system has a hierarchical structure in which the tactical and strategic air traffic management functions are allocated among various ATC and TFM organizational units. Although it is useful to discuss these functions as if they were associated with discrete organizational units, in practice the elements of the functions are intertwined with overlapping and redundant responsibilities to ensure system reliability.

At the lowest level of organization, the ATM system is organized to assure aircraft separation throughout all phases of flight from takeoff to landing. To accomplish this, the ATC component of the system is composed of controllers and specialists working in several different types of facilities. These include:

1. Towers at airports, with air traffic controllers responsible for directing arrivals and departures
2. Terminal radar-approach control facilities (*TRACONs*), with controllers guiding flights for roughly the first and last 40 miles of flight
3. Air route traffic control centers (*ARTCCs*), with controllers in charge of flights while en route
4. Flight service stations (*FSSs*) with flight service specialists providing services such as flight plan filing preflight and inflight weather briefings to general aviation pilots

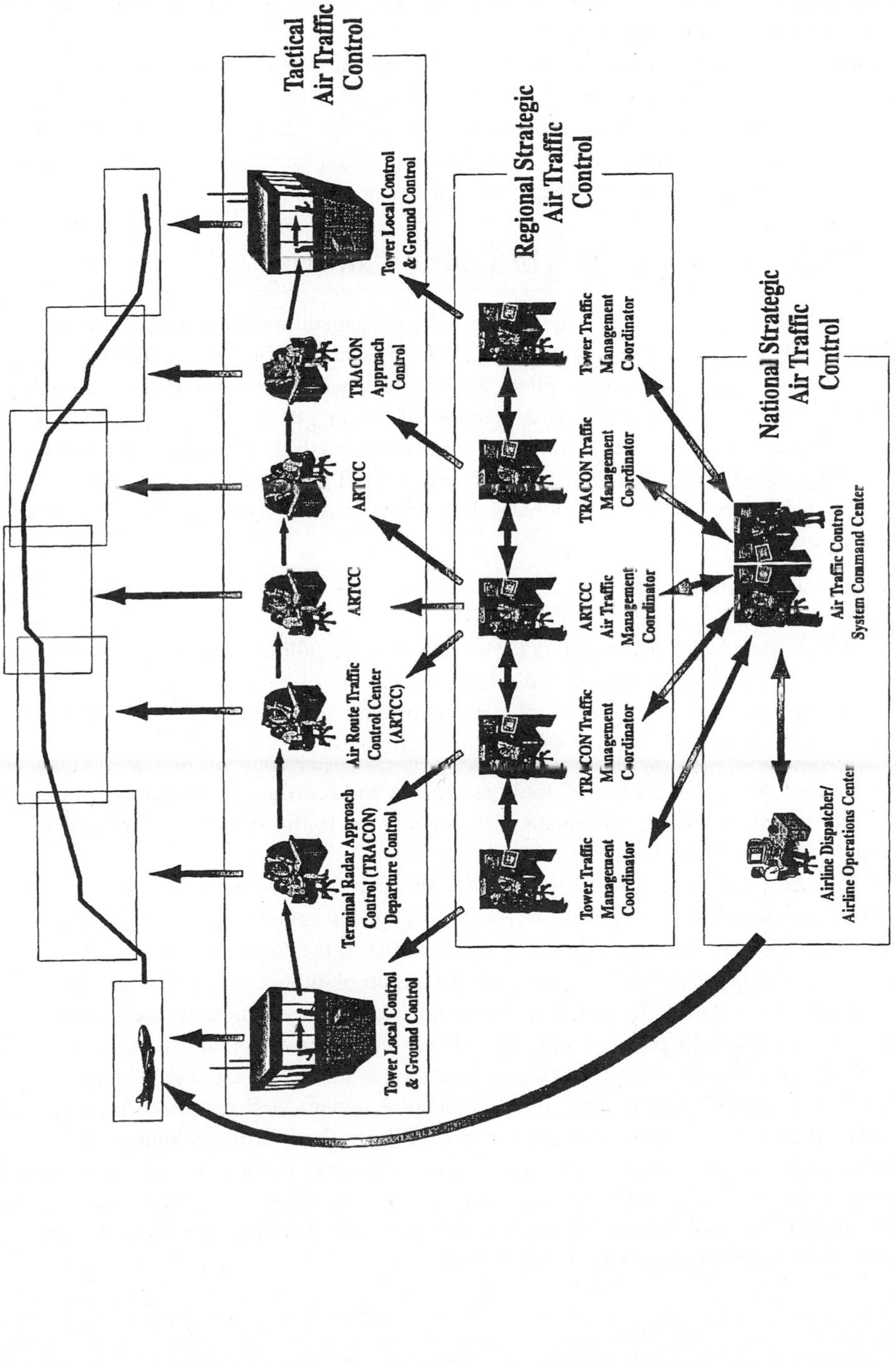

FIGURE 111.2 Representation of air traffic management organizational structure.

Pilots file a flight plan and obtain a clearance in order to fly through airspace controlled by the system. Once the initial clearance has been obtained, the pilot maintains radio communication with the controllers at these facilities to receive ATC services. The controllers monitor specific flights within the airspace (sectors) that they are in charge of and issue instructions to pilots in order to ensure safe separation and efficient use of the airspace.

At an intermediate level of organization, the ATM system includes local TFM units. Each ARTCC and TRACON also has a traffic management unit, and there are a few towers where traffic management coordinator positions have been established. These organizations are responsible for helping to plan and adjust the flow of traffic within their airspace.

At the national level, a centralized facility, the Air Traffic Control Systems Command Center (*ATCSCC*) coordinates the activities of the local units (Garland, Hopkin, and Muller, 1996). Finally, the dispatchers within airline operations centers (*AOCs*) have an increasingly significant impact on traffic flows, and thus must be considered in any discussion of the system. Details on the tactical and strategic operations within this structure are discussed in the following sections.

111.3 Human Factors in Air Traffic Control

The function of the ATC system is to provide safe, orderly, and expeditious flow of air traffic to its users. In accomplishing this function, the ATC system operates on two fundamental principles. First, the system exists to serve the users. The users are the pilots, aircrews, and passengers on board the aircraft being controlled by the system, plus the owners and operators of the aircraft. Control of aircraft by the ATC system also benefits operations of aircraft not being controlled directly, because ATC operations provide a systematic environment in which such "non-controlled" aircraft may be flown safely and efficiently.

Second, the system's direct service is provided by the controllers. This service consists of outputs which include the following:

1. Clearances and control instructions
2. Advisories on traffic or other flight conditions
3. Manual flight data handling, including keyboard entries, handling of entries on Flight Progress Strips or notepads or other flight data displays
4. Controller-to-controller voice coordination and controller–pilot voice communication
5. Plans for and selection of control techniques and actions

In addition, first-line supervisors in ATC facilities support and coordinate the operational outputs of the controller positions, including coordination of requests for traffic volume control and metering services (Kinney, Bell, and Ditmore, 1982).

As discussed earlier, the ATM system has evolved a specialized, hierarchical structure in which the traffic flow management (TFM) component works to simplify and expedite traffic flows, ensuring that the operational environment is compatible with the capabilities of the controllers and pilots responsible for conducting tactical flight operations. Safety or separation assurance is the first aim of the ATC component of the ATM system. Efficiency is second, implying orderliness more than expedience.

Controllers collaborate with pilots to provide ATC services to participating (commercial, general aviation, and military) aircraft during takeoff, while en route, and upon landing. The nature of the controller's mission, the ATC organization, the technological systems, and the operating environment vary across the different ATC facilities that support aircraft during each of these flight phases.

The following section describes the organizational, technological, and operational environments in which controllers are embedded. It also discusses factors such as mission requirements and operational constraints that affect the performance and workload of controllers. For each environment, examples of human factors and ergonomic issues are also discussed.

Type of Facility	Terminal Area Facilities		En Route Facilities	
Controlling Facility	Airport Traffic Control Towers	Approach Control Facilities	Air Route Traffic Control Centers (ARTCCs)	
Type of Control	Ground Traffic Control Takeoff and Landing Control	Approach and Departure Control	ATC during Transition and Cruise	
Airspace	Airport Traffic Area Typically 5 nmi and 3,000 ft AGL	Approach Control (Tracon Area) Typically Extending Up to 40 nmi and 10,000 ft msl from the Airport	En Route Airspace Transitional Phase Typically 50-150 nmi from Airport	Cruise Phase Up to 60,000 ft
Typical Flight Time	Typical Ground Time 5-10 min	Typical FlightTime 10-20 min	Typical FlightTime 10-20 min	Typical FlightTime 20 min to Several Hours
Flight Profile				

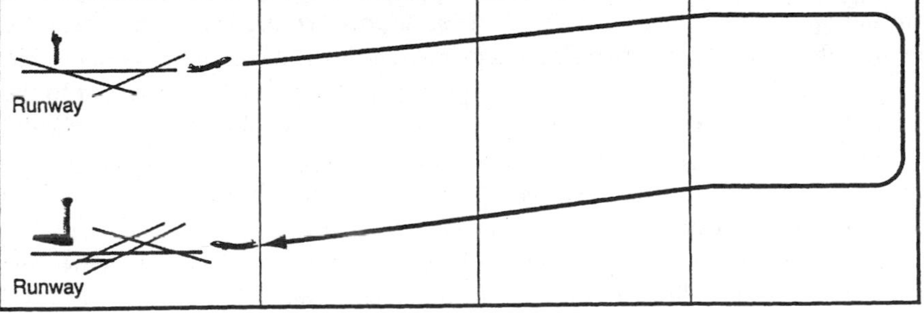

FIGURE 111.3 The role of terminal and en route facilities in ATC. (From Mundra, A.D. 1989. A Description of Air Traffic Control in the Current Terminal Environment. MTR88W167, The MITRE Corporation, McLean, VA. With permission.)

Missions and Tasks Within the ATC System

Air traffic control specialists or controllers may work in one of three areas: terminal, en route, or flight service. The terminal area includes controllers working in the airport traffic control towers (henceforth called simply towers) and the approach control environment. The en route area includes controllers working in ARTCCs or en route centers that manage the domestic and the oceanic airspace environments. The flight service area includes specialists working in FSSs which provide a variety of information and support services to general aviation pilots throughout the system. Because the flight service specialists do not provide services to active flights operating under the control of the ATC system, they are not discussed further in this section.

Figure 111.3 shows the role of the terminal and en route areas by depicting the ATC phases through which an aircraft passes from its point of origin to its destination (Mundra, 1989). Towers control the surface movement of aircraft and their landings and takeoffs. Once an aircraft takes off and before it lands the aircraft is under the control of the terminal radar approach control (TRACON). TRACON airspace generally extends about 40 nautical miles from the airport. Outside of this region an en route center has responsibility for control of the flight. The cruise phase of an aircraft may be conducted within one or more centers. For international flights, the aircraft may pass into oceanic airspace and be handed over to an international control facility. The oceanic airspace is divided into flight information regions operated by non-U.S. civil aviation authorities.

Terminal Environment — Tower Operations

The FAA operates over 400 towers to control traffic on airport runways and taxiways and in the immediate vicinity of the airport. ATC in the tower is based on the visual confirmation and pilot reports of aircraft locations and face-to face-interaction between controllers. Tower equipment types vary depending on each facility's operational needs, and the equipment configuration is often customized to a given tower's structure. As in all ATC environments, flight strips are used to record, maintain, and coordinate flight plan and clearance data. Many towers also have radar displays to aid visual acquisition of arriving aircraft, surface surveillance equipment for displaying aircraft locations on the airport surface, and a data link communications system for transmitting selected messages to pilots.

The tower may be staffed with a local, ground, clearance delivery/flight data, and supervisor position. Each position has different responsibilities depending on an aircraft's phase of flight. Only the local and ground controllers have traffic movement responsibility.

For departing aircraft, the flow of information and responsibility goes from flight data/clearance delivery to ground control and then to local control. The flight data/clearance delivery position processes flight plan data and manages the predeparture clearance (*PDC*) process. This process ensures that the aircraft has received current airport information (the Automatic Terminal Information Service (*ATIS*)) and an approved flight plan. At most major airports, controllers prepare digital messages and pilots request the ATIS and PDC by using a data link. The ground controller is responsible for issuing the taxi clearance and directing the aircraft through the system of taxiways that lead to the runway. The local controller has responsibility for the active runway and clears the aircraft for takeoff.

For arriving aircraft, the flow goes from local to ground control. The local controller clears the aircraft to land and issues instructions regarding where the aircraft should exit the runway. The local controller then directs the movement of the aircraft toward a gate.

Research on the tower operations highlights some characteristics of the environment that have important effects on controller performance and workload. Tower controllers spend a considerable amount of their task time, approximately one-third, visually tracking traffic movements outside the tower (Bruce, 1996). The task of visually tracking aircraft and knowing how to communicate with them is not a trivial one (Wickens, Mavor, and McGee, 1997). It entails extensive cross referencing of display aids, controller movement around the tower to view aircraft movements, and face-to-face interactions with the other controllers. Although tower controllers accomplish many tasks concurrently such as moving flight strips and cueing the radio microphone while looking out the tower cab window, they do not conduct keystroke entries or read display information.

For several years, the FAA and the National Transportation Safety Board have been concerned with reducing runway incursions and related surface incidents. These incidents may result from a variety of causes, including errors made by controllers as a result of reduced visibility. Display-based aids for detection of unauthorized movement of aircraft on runways are now being deployed by the FAA. However, because tower controllers have a major responsibility to continuously scan the terminal airspace, such aids must be carefully integrated with the controller's visual and auditory scanning tasks (Bales, Gilligan, and King, 1989).

Another important characteristic of the tower environment is that it is communications-intensive. Tower communications are generally time sensitive and often time critical. During busy periods, a tower controller accomplishes time-critical communications at a rate only marginally ahead of air traffic movements around the airport. Moreover, the only way tower controllers can manage their workload in heavy traffic situations is to delay aircraft movements (Bruce, 1996). Analyses of tower communications indicate that a variety of factors are adversely affecting the pilot's ability to correctly understand and remember controller instructions (Adam, Kelley, and Steinbacher, 1996; Burki-Cohen, 1995). In order to minimize time on the radio frequency at busy times, there is a tendency for controllers to speak more rapidly, making it difficult for pilots to understand messages. Furthermore, many of the messages transmitted by tower controllers, such as taxi instructions, are complex and lengthy, making them difficult for pilots to remember. Finally, frequency congestion, a pervasive problem at busy towers, often prevents pilots from reading back their instructions to ATC, eliminating the verification that the pilot has heard and understood the instructions.

Terminal Environment — Approach Control

Approach control facilities provide ATC services to arrival and departure aircraft transitioning between the tower and the en route airspace. In most busy terminals, the TRACON room is housed in the same building as the tower. ATC in the TRACON, as in the en route domestic airspace, is primarily radar based. However, the equipment, traffic environment, and control procedures used in the TRACON are quite different from those in the en route environment. The automated radar terminal system equipment which provides a radar display of air traffic is fairly standard across facilities, but system capabilities may vary somewhat depending on the level of traffic handled by the facility. Flight strips are also used in the TRACON.

A typical TRACON control room includes two arrival feeder, two final, and two departure positions. Each position is assigned responsibility for a sector of airspace with arrival and departure traffic segregated into dedicated airspace corridors. The arriving traffic in the terminal area funnels in from higher altitudes and speeds for landing. The arrival controllers sequence and space the aircraft. A pair of arrival or feeder controllers each work the arriving traffic for one side of the airspace, e.g., north or south arrivals, establishing the aircraft on their initial approach to the airport. Two final controllers, one for each side, are responsible for the final approach phase.

The departing traffic fans out to higher altitudes and speeds for the cruise phase. Departure traffic is also divided between two departure controllers who each work one side of the airspace, establishing the aircraft on headings to the planned route of flight.

The physical layout of the TRACON reflects the degree of coordination required between operational positions. The four arrival positions work as a team and are situated next to each other. The final controllers are next to each other, as they typically need to coordinate frequently. The arrival feeder controllers are each situated next to the final controller they feed. Departure positions are separated from the arrival positions, mirroring the segregation of arrival and departure traffic.

Mundra (1989) identified several characteristics of the TRACON operating environment that affect the performance and workload of the controllers. There is high frequency of aircraft maneuvering in altitude, speed, and heading, and a rapid convergence of many traffic streams. Once within the TRACON airspace aircraft are neither expected to nor allowed to follow a predefined route, instead control is exercised through headings (i.e., vectoring), altitude, and speed instructions. As a result, system performance is sensitive to controller skill level in planning and achieving throughput and aircraft efficiency (fuel-efficient speeds, altitudes, or paths).

The TRACON lacks planning data and a route structure on which to base flight planning for individual aircraft. Instead, it uses a vectoring plan that only the controller knows. Today, the path flown is a specific response to a tactical situation, and must be visualized in the controller's mind. The controller must also keep track of the situation and required actions under high workload conditions.

Because of the high frequency of controller-initiated maneuvering, voice communication is a significant portion of the controller's workload. This implies many of the same communication problems discussed in the preceding section.

The uncertainty and rapid pacing of activity in the TRACON environment requires flexibility and teamwork. The limited advance information on controlled traffic and the lack of any advance information on uncontrolled traffic means that controllers must respond quickly to adapt their plans and fit in new traffic and share or redistribute tasks.

En Route Environment — En Route Domestic

The FAA operates 22 en route centers to separate aircraft traveling between airports. The host computer system, display equipment, and facility designs in the en route centers are standardized. ATC in the en route centers is radar based and each operational position is equipped with a radar display of traffic and flight strips. However, this common equipment configuration accommodates a wider range of staffing plans, missions, and tasks than is found in other ATC environments. En route centers provide an ATC link with other centers, with TRACONs, and with towers. As in the TRACON, the physical layout of the sectors in the en route facility reflects the degree of coordination required between controller positions.

En route ATC operations are divided into various types of airspace sectors. Each operational position is assigned responsibility for a sector and each sector may be staffed with one to three controllers, depending on the volume of traffic. When multiple controllers work a sector, task assignments vary among teams. Typically, a radar (R) controller is in charge of the sector operation. This controller uses the radar situation display and voice communications to apply radar separation procedures and separate the aircraft from all others within the sector. A radar associate or data (D) controller is responsible for separation planning activities. The D controller uses flight strips which provide advance information on the aircraft's planned route and interphone communications with other controllers to identify potential problems and coordinate preventive control actions. When the sector is too busy for the R and D controllers to handle, a radar coordinator (tracker) or handoff (H) controller may share the R controller's load, serving as a redundant "set of eyes and ears" to support situation monitoring and as a second pair of hands to perform data entry and intersector coordination tasks.

En route sectors can be classified into types according to various sector and duty characteristics. High altitude sectors, generally above 24,000 feet, handle departure traffic climbing to cruise altitude, overflying en route traffic flying level, and traffic descending to a lower level in preparation for arrival. Low altitude sectors normally have a mixture of aircraft which fly at lower altitudes, slower speed arrival and departure traffic, and higher performance aircraft transitioning to the high altitude sectors. Within these two broad divisions, finer breakdowns can be made based on the homogeneity of the traffic flows and the sector's mission. For example, en route arrival/departure sectors coordinate traffic flow in and out of approach or airport control; transition high and transition low sectors have a majority of their traffic climbing or descending to reach cruise altitude; and en route high and en route low sectors have most of their traffic flying level.

Controller tasking and taskload in different sectors results from a combination of static and dynamic factors. Static factors, such as the type of traffic flows and the service that must be provided each aircraft, imply routine tasks that are known in advance and performed for every aircraft or for a specific subset of the traffic such as a particular flow. For example, en route controllers currently have a major responsibility to implement routine ATC procedures, such as routing and altitude constraints specified in interfacility directives, and flow management procedures to establish an orderly flow of traffic within the airspace. Dynamic factors are associated with the particular airspace delegated to the sector (e.g., size of the airspace for radar vectoring, time an aircraft remains in the sector), the number and characteristics of aircraft within the airspace, and the nature and frequency of requests by users to alter their flight plans. One dynamic factor thought to be an important determinant of complexity involves the number and pattern of potential conflicts that can occur within the sector airspace for a given period of time. The decision-making process associated with the detection and resolution of conflicts has a great impact on determining controller workload and sector capacity.

In the current environment, a mix of procedural solutions and automated capabilities are used to anticipate and manage complexity and controller workload. On a scheduled basis, controller positions are closed and opened and responsibility for sector airspace is combined and decombined under a position. Staffing arrangements also vary with sector load. During busy rush periods throughout the day, en route sectors are routinely staffed with two and sometimes three controllers. When traffic volume is expected to exceed the sector's capacity for an extended period of time, traffic flow management is alerted by an automated monitor alert function and procedures are activated to limit the volume of traffic handled by the sector.

The controller's ability to manage traffic in the current environment is also affected by equipment. The three-person sector team poses problems in performance because the sector equipment is set up and laid out for a two-person operation and is awkward for three-controller staffing (Kinney, 1977). For example, when the H controller sits down, the R controller moves farther away from the flight strips and has difficulty reading them. The three-controller operation also requires coordination between the R and

H controllers, a task that does not exist with the two-person sector. In general, the fact that equivalent tools and technological capabilities are not available on each side at the en route position limits the ability of the assistant controllers to contribute substantially to team performance (Shingledecker and Darby, 1995). Finally, the outdated message composition, editing, and list management capabilities available on the workstation may be contributing to deficiencies in controller performance with manual inputs. Early research on manual data entry tasks indicated that these tasks which make up a large part of the controller's workload are demanding and time consuming and that data entry errors appeared as contributing causes in many system error case histories (Kinney, 1977).

En Route Environment — En Route Oceanic

The FAA also provides ATC service within a large area of international airspace, including the western half of the Atlantic Ocean, the Gulf of Mexico, and a significant portion of the Pacific Ocean (Nolan, 1990). In contrast to the domestic en route airspace, oceanic airspace has no radar coverage over most of the area. For this reason, monitoring and control of the oceanic traffic is based on flight plan data, position estimates, and relayed voice position reports from the pilots. Route structures are used to impose order and separate the traffic flows. In addition, controllers apply a diverse set of non-radar separation criteria to provide ATC services over the ocean. Paper flight strips constitute the primary information display for aircraft separation; these are updated by controllers with the current flight plans and latest position reports for each aircraft. The oceanic computer system also provides a plan view display of traffic positions based on the pilot's reports. Unlike the other ATC environments, air-ground communications over the ocean are indirect. Controllers use a telecommunications processor to send messages to a service provider who relays them to pilots. The telecommunications processor is also capable of sending data link messages directly to aircraft via a satellite. In part of the airspace, controllers have begun to use the data link to communicate directly with pilots.

An oceanic controller is assigned responsibility for a sector of airspace. However, oceanic sectors are significantly larger than those in the domestic airspace environment (e.g., an oceanic sector may be several thousand miles in length) and a single controller may have responsibility for many more aircraft than in the en route environment. Procedural ATC requires large lateral and longitudinal separation between aircraft that takes into account the lack of timely, independent surveillance and the uncertainties and delays associated with the communications. Oceanic controllers plan for separation by mentally calculating fix crossing time differences between aircraft based on flight strip data. If aircraft are not traveling on structured routes, controllers may evaluate separation visually on the plan view display or estimate distance spacing by plotting position data on charts (Coulouris, 1985). To ensure that the planned separations are maintained, controllers monitor flight progress based on pilot reports submitted as they cross compulsory reporting fixes (hourly or every 10 degrees longitude/latitude). With each progress report, the controller reexamines the traffic situation and searches for potential conflicts.

Hamrick and Reierson (1993) discuss some characteristics of the oceanic environment that affect the controller's performance and workload. Separation planning is complex and must allow for uncertainty and delay. It often requires that controllers plan a sequence of clearances involving several aircraft in order to enable aircraft at different altitudes on the same route to climb or to merge traffic onto a single route.

Separation assessment entails significant mental calculations of fix crossing time differences and often requires cross referencing and integrating information from multiple sources to form a single "picture" of the situation. A variety of techniques and separation standards are currently applied in the oceanic airspace (FAA, 1992). Determining the applicable separation criterion is not a trivial task. It requires consideration of multiple factors, such as aircraft type, method of navigation, altitude, geographic region, speed, route of flight, flight origin, and destination. In some regions, many standards apply and the controller will select the standard that provides the best operational advantage in a given situation.

Today's oceanic controllers have few tools, and automation is limited to use for traffic visualization not separation. Use of flight strips as the primary means of separation means that controllers have a considerable taskload in bookkeeping pilot progress reports and monitoring for overdue reports. At times, oceanic sectors have so many flight strips that the controllers must stand and walk back and forth to examine and update the flight data. At the same time, the large sector sizes that must be accommodated on the plan view display cause aircraft positions to appear to be in close proximity, making it difficult to determine if aircraft are separated just by monitoring a situation display. Because the communications system is slow and cumbersome, controllers must take extra precautions in planning separation strategies that minimize the need for communication.

Missions and Tasks Within the ATC System — Summary

In all areas of ATC, controllers operate as part of a system of interacting components. These components include: the airspace and airport surface layouts, communications links, information display and management tools, and procedures, as well as pilots, other controllers, and traffic managers. Human factors and ergonomics issues in ATC arise as a result of controller interaction with each of these components. The present system is characterized by standard operating procedures that establish an orderly flow of traffic within the airspace, and controllers rely on this traffic organization to anticipate problems and manage separation. Yet, against this basic traffic organization, the process of separation planning and selection of control techniques is highly dependent on the controller and varies across environments with respect to the temporal demands and information processing and response resources involved. While the basic information required by controllers is fairly constant across environments, the quality and format of information available varies across the domains. Overall, the best information is available in the en route environment, but there is also a controller cost associated with maintaining the data. Throughout the system, air–ground communications is a demanding controller task and capacity limitations in the current communications links contribute to information transfer problems. For the most part, controllers are working with aging equipment. Equipment-related issues vary across environments, but they include lack of information integration and inefficient display formats, workstations, and equipment layouts that impede team coordination, and outdated data handling capabilities that induce errors and reentry of data. Controller and pilot workload tends to be concentrated in busy environments such as the tower and the approach control. In these busy environments, where there is a concomitant need for controllers and pilots to coordinate their activities, it is particularly difficult to time the information exchanges so that they do not interfere with higher-priority duties.

Ergonomic Issues in ATC

There is a sizable body of research literature on ergonomic issues in ATC (see, for example, Wickens, Mavor and McGee, 1997; Hopkin and Wise, 1996; Hopkin, 1988). This section focuses on the relevance and application of this research in the context of new and emerging ATC capabilities.

Some of the main ergonomic issues in ATC relate to the mental demands and uncertainties imposed by the ATC process and the environments in which the controller is embedded. Complexity management and control strategies are key issues in this area. Another source of issues relates to the cooperative nature of the activity (Leroux, 1995). Information transfer and communications are key issues in this area.

Complexity Management

Current air traffic control systems are functioning near, at, or even beyond their planned maximum traffic handling capacity, and all current projections foresee continuing and cumulative increases in air traffic for a long time (McAlindon and Gupta, 1993). Consequently, there has been a longstanding interest in understanding and quantifying the controller's traffic handling capacity and in predicting when this capacity breaks down. However, despite a lengthy history of research on controller workload and sector complexity, the operational relationship between the two concepts remains elusive, and practical application of complexity measures in the current system is limited. Today, the most widely used complexity measure is based exclusively on aircraft counts. In the field, operational decisions on complexity management continue to be made by instinct and local judgment.

Several factors make the analysis of controller workload in the current system a complex matter (Wickens, Mavor and McGee, 1997). Controller taskload has been analyzed extensively to derive traffic and airspace characteristics that predict workload; however, this approach seems to be most applicable to measurements of the observable, motor, and manual components of workload. Research on workload has also shown that successful controllers use various adaptive strategies to manage their performance in the face of increasing complexity (Sperandio, 1971). These adaptive strategies allow the controller to handle more aircraft without error or excessive workload. Furthermore, research relating complexity and workload to operational errors tends to confirm that this relationship is probably mediated by other factors, such as control strategies. Studies of operational error reports show that errors are not uniquely associated with high complexity and workload but also occur under low to moderate complexity and workload (Canadian Aviation Safety Board, 1990; Rodgers, 1993; Kinney, 1977). However, one of the chief difficulties encountered in interpreting the error reports is that there are no normative data. The relative frequencies of the levels of traffic volume and complexity are unknown, which precludes determining whether or not the reported frequencies of errors in these levels are disproportionately related to chance.

On one hand, these results emphasize the great need for better baseline and normative data on human and system performance as standards for evaluating causal relationships (FAA, 1995; Benel, 1995; Galushka et al., 1995). On the other hand, they may suggest a practical problem with the research questions (Sarter, 1996). From the standpoint of complexity management, it may be more useful to focus on prediction of complexity with respect to a specific operational context and decision. At a tactical level, controllers need detailed information on current and predicted sector complexity to plan control actions. Supervisors need information on predicted complexity for multiple sectors in order to plan for staffing. At a more strategic level, traffic managers need summary information predicted over a longer period of time and for a larger area to plan for flow management procedures and programs (Klein, 1992).

Controller needs for complexity information are being addressed by development programs in the U.S. and in Europe (Schultheis and Tucker, 1996; Makins and Drew, 1995). Both programs are developing controller tools for conflict prediction and resolution. With these tools, an indicator of sector complexity is provided in the form of display information on the number of predicted conflicts and their temporal distribution. Currently, both the U.S. and Europe rely on a second controller to plan traffic and ensure that the primary controller is not overloaded. Advance information on conflict resolution workload should allow the second controller to better schedule and manage conflict resolution tasks. Such tasks account for a significant and increasing component of the primary controller's work. Particularly in the U.S., where the ATC environment is gradually evolving away from structured routings toward user-preferred routings, the relative contribution of the conflict resolution to overall sector workload is likely to increase, making it a better predictor of complexity (Carlson, Rhodes, and Cullen, 1996). The ultimate goal for future ATC is a control-by-exception paradigm in which controller interventions are limited in extent and duration to correct identified problems (RTCA, 1995).

Control Strategies and Efficiency

The growing demand for aviation services and the constraints on budget that both the industry and the FAA face has stimulated interest in improving the efficiency of ATC. As mentioned in the preceding section, controllers cope with increasing traffic demand and complexity by employing adaptive strategies. Research on controller strategies has identified how specific adaptations tend to lower the efficiency of individual flights and the overall traffic flow when demand is high. New ATC capabilities are helping reduce the need for these specific adaptations and preserve flight efficiency.

In general, strategies that are economical for the controller are those that preserve the primary objective of safety but take less account of secondary objectives such as flight efficiency, user-preferred paths, and fuel economy path (Bruce, 1996; Bellorini and Decortis, 1995). Early on, Sperandio (1971) showed that controllers handled an unexpected increase in traffic load by adaptively decreasing the amount of time they spend on each aircraft. For example, the controller may structure the traffic so that all of the flights are in-trail and traveling at a uniform speed, thereby reducing the difficulty of monitoring for conflicts.

Another way controllers adapt their strategies is to focus on the immediate tactical situation and abandon planned strategies. To plan and execute control strategies that are more flight efficient, controllers must coordinate with each other. Under high workload, a shift from cooperative toward individual work has been observed in both the TRACON and en route environments. Bellorini and Decortis (1995) found that in the TRACON environment controllers shed coordination tasks under high demand, resulting in less efficient sequencing of arrival aircraft. Sperandio (1978) observed that in the en route environment, both sector controllers focus attention on the current tactical situation, reacting quickly and employing less efficient tactical control techniques. He further notes that an increase in the workload in the sector tends to make the support tasks more and more dependent on the central task and tends to overload the principal operator even more, so that the assistant becomes less and less efficient at a time when he is more and more necessary.

Controller–pilot coordination may also increase as controllers abandon planned strategies and react to the tactical situation. In many traffic situations, reactive strategies are communications intensive, with the controller assuming greater responsibility for flight paths and making continuous tactical adjustments. Increased communications tax the controller's perceptual, cognitive, and speech motor capacities and the pilot's ability to understand and respond to instructions, resulting in more frequent requests for repetition and clarification (Shingledecker and Darby, 1995; Adam and Kelley, 1996).

To preserve flight efficiency, a more proactive approach to ATC is desirable. Decision support tools for tactical planning and selection of control strategies are being tested in the en route and TRACON environments. As mentioned in the preceding section, the FAA is currently testing a conflict probe capability to provide early detection of conflicts and tools for conflict resolution in the domestic en route environment (Schultheis and Tucker, 1996). A similar capability is also under development for the oceanic environment (Hamrick and Reierson, 1993). This capability will reduce the mental calculations and extrapolations involved in separation monitoring and afford a longer lead time to plan resolution maneuvers that are less disruptive to the user's flight intent. Team performance will be aided by providing the assistant controller with more powerful tools for visualizing the future traffic situation and evaluating proposed maneuvers. In addition, an automated coordination aid will also allow controllers to share and approve plans.

In the TRACON, controller aids for merging flows and sequencing aircraft for approach are already in use or under test at field facilities (Mundra and Levin, 1990; Lee and Davis, 1995). These tools provide advance information on the predicted sequence of arrival aircraft. One tool, the converging runway display aid (*CRDA*), has been deployed to assist the controller in conducting staggered approaches. Staggered approaches require specific separations between aircraft landing on adjacent runways as well as between in-trail aircraft. Staggered approaches have been characterized as more complicated than simultaneous approaches which have only in-trail spacing requirements (FAA, 1991). The CRDA reduces the complex mental calculations and extrapolations involved in staggered approaches by projecting false targets or ghosts for aircraft arriving in one of two converging streams onto the other stream, thus allowing the controller to visualize and manage simple in-trail spacing on a single approach.

Another tool, the final approach spacing tool (*FAST*) is one element of a Center TRACON Automation System (*CTAS*). CTAS comprises a set of tools to assist controllers handling aircraft arrivals starting at about 200 n.mi. from the airport and continuing to the final approach fix. FAST provides the controller with landing sequence numbers and runway assignments to achieve an accurately spaced flow of traffic onto the final approach course (Davis et al., 1991). Based on the displayed sequence, the controller formulates appropriate instructions for merging and spacing the arrivals. Research indicates that FAST advisories improve the runway delivery precision and reduce controller workload by reducing the number of vectors issued to each aircraft and reducing the need for (verbal) coordination between controllers (Credeur et al., 1993; Lee et al., 1995).

Communications

There is ample evidence in the research literature that controller–pilot communications are a common and persistent problem in today's operations (Kerns, 1994). Analyses of incident and error reports and of recordings of routine controller–pilot communications offer a cogent explanation of how often and why problems occur. Field experience and simulation studies on data link offer a useful perspective on how this technology can be used in the operational environment to improve communications.

Some of the primary factors which contribute to communications problems arise from the use of spoken language to transfer information. ATC communications have been designed to ensure that spoken dialogues can be conducted efficiently and with minimum possibility of error or misunderstanding (Hopkin, 1988). Controllers and pilots have adopted a standardized phraseology, language conventions, and procedures which define the process for conducting the dialogue, including the cues that tell a listener when a transaction has been completed and whether a readback is required. However, despite years of refining the language and procedures, research confirms the intractable nature of many of the problems inherent in the exclusive use of spoken language for controller–pilot communications. Grayson and Billings (1981) analyzed aspects of human speech processing and conversational behavior that mediate communication performance. Their analysis found that a tendency to fill-in information, the expectancy factor, and timing problems were implicated in many types of controller–pilot communication problems. The expectation factor contributes to misinterpretations and inaccuracies because controllers and pilots sometimes hear what they expect to hear. This generates what have been called "readback and hearback" errors in which, respectively, a pilot perceives what he expected to hear in the instruction transmitted by the controller and a controller perceives what he expected to hear in the readback transmitted by the pilot.

Congestion on the voice radio frequency has also been implicated in communications problems. During busy periods, controllers issue longer, more complex messages in an attempt to minimize use of the radio frequency. However, as more transmissions are crowded onto the frequencies, the procedural steps (callsign identifications, readbacks) that assure communication are being dropped (Adam, Kelley, and Steinbacher, 1994; Cardosi, 1993). Analyses of routine communications (Morrow et al., 1993; Cardosi, 1993; Cardosi, Brett and Han, 1996) have highlighted the contribution of message complexity and the resulting memory burden to miscommunications. In both the terminal and en route environments, errors and procedural deviations increased as clearances increased in complexity.

The frequency congestion problem is most pronounced in the tower environment where controller transmission rates have been observed at 3.9 and 8 transmissions per minute for the local and ground controllers, respectively, as compared to 1.8 transmissions per minute in the en route environment (Burki-Cohen, 1995). A concentration of controller and pilot workload in the tower environment also accounts for failures to transmit information and untimely transmissions in busy environments. Because of high workload, controllers may fail to initiate lower priority traffic advisory messages, precisely when the pilot's need for this information is greatest. Conversely, pilots may be preoccupied with external vigilance and flight control tasks. They may not wish to receive messages and may fail to respond.

One of the ways the FAA is responding to communications problems is by developing alternative means of information transfer. Data link communications have already been introduced in environments where communications problems and limitations are the most severe: the tower and the oceanic en route. The PDC and the ATIS are being transferred via data link at many major airports. Direct controller–pilot communications via a satellite data link are being conducted in the oceanic airspace. The selection and design of these services have benefited directly from research on human interaction with voice and data link communication systems (Kerns, 1991;1994).

Research indicates that data link offers several advantages for transfer of lengthy, repetitive messages such as PDC and ATIS. Data link capabilities for message storage and retrieval reduce the controller's burden when preparing messages and the pilot's memory burden when receiving them. Data link also allows controllers and pilots to pace the transmission and processing of the information, thus avoiding conflicts with higher-priority tasks.

In the oceanic environment, the tempo and highly structured format of the controller–pilot communication is well suited to data link transmission. The design of the controller and pilot data link communication protocols and their message handling capabilities have also been guided by data link simulation research (Kerns, 1994). Based on the study results, operational communication protocols that minimize switching between voice and data link media have been implemented. Moreover, the procedural steps required to conduct communications within each medium are consistent. In terms of the level of automation, message handling capabilities have been designed so that controllers and pilots have the final authority to approve the transfer of information to each other and to their automation systems. At the same time, computer aiding of message composition has been implemented using menu style user interfaces to minimize input errors and relieve the human operators of these functions.

Although the experience and results to date indicate that these initial applications of data link offer important operational benefits at little or no cost to the human operators, future applications of data link must be selected carefully, addressing key human engineering challenges. Visual display and manual control of transmitted information may not be appropriate in environments where the controller or pilot visual and manual resources are already reaching an overload state. Delay factors associated with message composition may limit the utility of data link in rapidly changing conditions while transmission delays may limit its utility for time-critical transmissions. Finally, new procedures will be needed to maintain team performance when the communication medium is silent and may be less readily observable by multiple operators.

111.4 Human Factors in Traffic Flow Management

The function of the traffic flow management (TFM) system (Nolan, 1990; Odoni, 1987) is to provide strategic planning and control when necessary to try to avoid situations where potentially unsafe or inefficient operations are likely to arise. As discussed earlier, within the United States (which is the ATM system that will be used to provide examples in this chapter), the FAA has organized this system into a hierarchical structure including ATCSCC (which supervises and coordinates planning at a national level), traffic management units at en route centers and TRACONs, as well as FAA staff at towers and airport facilities. In addition, although not a formal part of the FAA's TFM system, AOCs (Airline Dispatchers Federation, 1995) play an increasingly important role in influencing traffic patterns.

Sample Control Methods Within the TFM System

To make the concept of TFM clearer, several examples are provided below.

Predeparture Interventions by ATCSCC

There is a variety of situations where the normal flow of traffic into an airport or some portion of the airspace needs to be restricted, and where information is available early enough to change plans before the affected flights depart. Such situations include forecasts of bad weather (thunderstorms, low visibility at an airport, etc.), runway restrictions or closures, and forecasts of heavy traffic congestion.

Under such circumstances, traffic managers at ATCSCC (in consultation with the affected traffic management units) may choose to employ a variety of procedural tools to reduce traffic. They may, for example, initiate a ground delay or ground stop for flights departing from airports within one or more centers. Alternatively, they may prevent AOCs from filing flights along a particular jet route or may reduce the flow of traffic along that route by requiring additional spacing between aircraft (for instance, requiring aircraft to fly 25 miles-in-trail). They may also give an airline several options to choose from in filing a particular flight.

Predeparture Interventions by AOCs

When capacity-limiting situations arise, the airlines may also choose to change their plans without any intervention by ATCSCC. They may, for example, cancel a number of their flights to a particular airport that is expected to have a reduced arrival rate due to bad weather, because a failure to do so could result in expensive diversions of some of their flights. Similarly, they may choose to file their flights along an alternative route because they are forecasting poor weather along the normally preferred routing.

At the other extreme, they may choose to file a few additional flights over and above the forecast arrival rate for an airport expected to have bad weather in order to provide a "reservoir" of flights to take advantage of the situation if the forecasted bad weather doesn't develop. (In this latter situation, they run the risk of having to divert these additional flights if the forecast weather does develop, and must fuel the aircraft to handle the resultant diversion to an alternate airport.)

Interventions While En Route

Situations also arise that require interventions while flights are airborne. As examples, traffic managers at en route centers may ask controllers to reroute certain flights to avoid predicted traffic congestion somewhere further along their routes, or may impose metering over an arrival fix, limiting the number of flights arriving at that fix within some period of time. Airline dispatchers may similarly request their flight crews to request rerouting if they foresee some problem further along the route.

Sample Actions — Summary

The point of these examples is first, that strategic planning decisions are made to try to avoid situations that could affect safety or efficiency. A second point is that there is a variety of alternative tools to influence traffic flow. A third is that, in the current TFM system, some of the decisions are made by FAA traffic managers, while others are made by individual AOCs (and that these decisions clearly interact with each other to determine the ultimate impact on traffic flow).

Traffic Flow Management — Conceptual Framework

To better understand the interactions of these organizational units that make up the TFM system, and to understand the performances of the individuals within them, it is useful to describe it as a distributed cognitive system, where the primary task is planning in the face of uncertain events. It is distributed in many senses, both within these organizational units and across them (Layton, Smith, and McCoy, 1994).

Distributed Problem-Solving

There are many senses in which TFM can be viewed as a distributed problem-solving task (Davis and Smith, 1983). First, the organizations (ATCSCC, ARTCCs, TRACONs, Towers, and AOCs) are geographically separated, so that direct face-to-face communication cannot occur. Second, in many cases information access is distributed among these different organizations, so that a traffic manager at an en route center may not have the same information as a dispatcher at an AOC or a specialist at ATCSCC. Third, different types of knowledge or expertise are distributed among these organizations. Fourth, different types of decision-support tools are available at the different organizations.

Tasks and responsibilities are not only distributed across organizations, they are also distributed within organizations. There are specialists at the command center to deal with weather forecasting, to design severe weather avoidance programs, etc. Similarly, dispatchers at AOCs have responsibility for developing flight plans for flights in different parts of the world, and interact with airline meteorologists and specialists with expertise in such things as aircraft maintenance, crew scheduling, and aircraft scheduling.

Competing and Complementary Goals

Another important characteristic of this cognitive system is that different organizations and individuals have different goals and priorities. FAA traffic managers have as their primary responsibility ensuring traffic flows that allow the safe separation of traffic from all sources (commercial airlines, general aviation, and Department of Defense flights). They are also concerned with the efficient use of airspace capacity. Some of the decisions they make further require consideration of equity among the different airlines.

AOC staff likewise have the safety of their flights as the most important consideration. However, within that constraint, their goal is the efficient and effective operation of their own airline's schedule, as they are in competition with the other airlines.

Thus, even from a broad systems perspective, disagreements between FAA traffic managers and AOC staff must be expected. When constraints such as workload and information access further intervene, or when there are differences of opinion about weather or traffic forecasts, such disagreements are even more likely to occur.

Decision-Making Under Uncertainty

A third major factor that must be considered to understand human performance within the TFM system is the high degree of uncertainty associated with decisions. Weather is a major issue (Andrews, 1993), whether it concerns a forecast regarding the development of a line of thunderstorms, a snowstorm closing east coast airports, strong crosswinds impacting a particular runway at an airport, or low visibility due to fog. There is also considerable uncertainty about traffic patterns. Flight departures are frequently delayed, for instance, which can influence traffic congestion at some further point along their routes, and there are numerous reasons why runway use at an airport must be restricted. Finally, controller staffing and workload limitations can introduce unexpected capacity constraints.

Conceptual Framework — Summary

In short, to cope with the complexities of managing traffic in the face of numerous sources of uncertainty, the TFM system has evolved into a very complex network of organizations and individuals where subtasks are allocated to specific individuals or organizations in order to reduce the cognitive demands on any one individual. This task decomposition includes considerable redundancy to catch errors and inefficiencies. This task decomposition also introduces all of the classic concerns associated with finding localized (suboptimal) solutions to problems, as well as the classic concerns regarding the strengths and weaknesses of group decision-making processes.

Ergonomics in TFM

The conceptual framework provided above begins to suggest some of the main ergonomic issues that arise in TFM (Wiener and Nagel, 1988). These concerns stem from the fact that this is a very complex group decision-making process under conditions with a high degree of uncertainty, and are highlighted in the subsections below. It is also worth noting that these same characteristics (the complexity resulting from a large number of factors interacting to influence performance, and the fact that it is the interaction of a network of organizations and individuals) have had a significant influence on the type of research conducted thus far. With a few exceptions (Layton, Smith, and McCoy, 1994), the research has been limited to descriptive observational studies of performance in field settings.

Goal Allocation as a Strategy to Reduce Complexity

To deal with complexity, the TFM system has been designed to distribute responsibility between FAA TFM facilities and the airlines by distributing responsibility for different goals. Traffic managers within the FAA have as their primary responsibility assuring safe and efficient use of overall system capacity. Thus, if an airport is predicted to have a reduced arrival rate due to a forecast of low visibility, ATCSCC, in consultation with the affected ATC facilities, will select a control strategy, such as an "all centers" ground stop for all flights to that airport for a certain period of time. This strategy is then simply imposed on the airlines, as ATCSCC has the ultimate authority in making such a decision. Under such a "control by directive" paradigm, the airlines' business concerns are considered only in the limited sense that safe, efficient overall use of the system's capacity is desirable from an airline perspective as well.

Such a paradigm reduces the complexity of the decision-making task faced by traffic managers, as they don't have to consider as many factors in making decisions. However, it also clearly can lead to suboptimal decisions from an industry perspective (as there can be several alternatives that are equivalent from a TFM perspective in terms of safety and overall use of system capacity, but which are quite different in terms of their economic impacts on individual airlines).

As a result, the system has been evolving toward formal procedures for incorporating airline preferences in TFM decisions (Lacher and Klein, 1993; Scardina, Simpson, and Ball, 1996; Wambsganss, 1995). Initial efforts in this regard focused on the process for approving the flight plans filed by airlines prior to departure. This evolution began in the early 1990s with the implementation of the National Route Program (Federal Aviation Administration, 1992). This gave the airlines a mechanism for requesting permission to file flights on routes other than FAA preferred routes. Such requests for "nonpreferred routes" were submitted to ATCSCC on a daily basis, and were evaluated for approval by command center specialists in consultation with traffic managers in the affected centers.

This "control by permission" paradigm maintained the distribution of goals, as FAA traffic managers did not have to directly consider airline business concerns in approving requests for nonpreferred routes. They just had to evaluate the flights plans submitted by the airlines in terms of their impact on safety and system utilization. Similarly, airline dispatchers didn't have to consider the impact of alternative routes for a flight on overall system utilization or on air traffic congestion, they just had to evaluate the routes in terms of safety and cost-effectiveness for their company. Hence, it provided a means for limiting the complexity of the decision-making task for any one individual (a traffic manager or a dispatcher), but provided a means for arriving at closer to optimal solutions.

The available data suggested that such a paradigm shift did indeed produce more cost-effective performance without any indication that safety was being compromised (McCoy, Smith, and Orasanu et al., 1995; Smith, McCoy, and Orasanu et al., 1996a). There were, however, two concerns expressed regarding the ultimate effectiveness of this approach:

1. Such interactions to get permission to fly non-preferred routes were costly in terms of the staffing requirements for both AOCs and the FAA
2. Continuing to give FAA traffic managers responsibility for evaluating airline requests provided insufficient impetus to explore the feasibility of new routings (as the traffic managers and controllers involved could achieve the goals for which they were primarily responsible — safe and efficient overall use of system capacity — without significantly changing the status quo).

As a result, a new order was implemented, referred to as the expanded NRP (Federal Aviation Administration, 1995), which gave the airlines greater autonomy (Denning et al., 1996). Under the expanded NRP, subject to certain constraints, the airlines were allowed to file flight plans (preflight) without seeking permission from ATCSCC (Carlson, Rhodes, and Cullen, 1996). This new paradigm was an example of "control by exception" (Sheridan, 1987, 1992), in the sense that, under its rules, the ATC system was supposed to intervene only after a flight had been launched, taking corrective actions to deal with traffic bottlenecks tactically, rather than preflight. (Under some circumstances, such as the development of a broad area of bad weather, traffic managers could still revert back to "control by directive" by canceling the NRP.) The available data indicate that this further paradigm shift has resulted in additional efficiencies for the airlines (Corlouris and Dorsky, 1995; Smith, McCoy, Orasanu et al., 1996b).

Distributing Information and Knowledge to Support Decision-Making

Although such paradigm shifts in the locus of control appear to have produced efficiencies, the available data indicate that, to maximize gains from such paradigm shifts, there is a need for improvements in access to information and knowledge, and for tools to help decision-makers cope with the added complexities that such shifts in control introduce. Indeed, after experience with this "control by exception" paradigm, dispatchers made such comments as "under the expanded NRP, it's like shooting ducks in the dark," and "it used to be the weather that was the biggest source of uncertainty. Now it's the air traffic system" (Smith, McCoy, Orasanu et al., 1996b).

Empirical studies indicate that such a lack of information about air traffic bottlenecks can result in fuel losses instead of fuel gains when flights filed under the expanded NRP along fuel efficient routes are rerouted by the ATC system to avoid excessive traffic congestion. As an example of a worst case scenario, Smith, McCoy, and Orasanu et al. (1996b) found that flights scheduled to fly NRP routes from Los Angeles to Dallas, and to arrive during the noon rush, on average burned 1.9% more fuel than they would have if filed on the FAA preferred route, instead of achieving the 4.5% fuel savings expected from the NRP route.

The implication of such studies is that, if the locus of control is shifted without a concomitant shift in access to information and knowledge (in this case dealing with ATC responses to air traffic bottlenecks), benefits may be less than expected.

Developing Alternative Models of Cooperative Problem-Solving to Cope with Complexity

The "control by permission" model described earlier represented a model for cooperative problem-solving that reduced complexity for any one individual (traffic manager or dispatcher) by continuing to distribute responsibility for certain goals between AOCs and traffic managers. The "control by exception" paradigm potentially reintroduces this complexity for the dispatcher, as, potentially, he or she is now expected to integrate a much broader set of factors in order to develop "optimal" plans to meet the needs of an airline's schedule. (As discussed earlier, traffic managers still serve as a safety net to identify and prevent situations where traffic flows could impact safety or overall efficiency in the use of system capacity, and do this by reverting back to a "control by directive" paradigm.)

A number of approaches are being explored to deal with this (Adams et al., 1996; Billings, 1997; McCoy et al., 1991; Pujet and Feron, 1996; RTCA, 1995; Smith, McCoy, Orasanu et al., 1996c). Some of them focus on enhancing data exchange and collaboration between AOCs and traffic managers, including the use of "white-boards" to enhance interactions and provide a framework for developing shared mental models and for sharing knowledge (McCoy, Orasanu et al., 1995; Orasanu, 1991). Others focus on distributing responsibilities within AOCs, essentially changing the organizational structures to reduce the complexity for any one individual. Still others are exploring the use of decision-support systems to reduce the workload and complexity for individuals. Finally, some proposals focus on reformulating the problem by changing the parameters of control. Under these proposals, instead of having the FAA impose detailed solutions, such as implementing an "all Centers" ground stop, the FAA would instead place a constraint on airline generated solutions, such as "each airline must reduce its arrival rate at a particular airport by 50%."

Ergonomic Issues In TFM — Summary

As outlined above, at a macroscopic level, the critical question regarding the design of the TFM system has been identifying the truly important determinants of performance. In recent years, a number of approaches for revising the design of the system have been discussed. Some of these, such as shifts in the underlying paradigm of control from "control by directive" to "control by permission" to "control by exception," have actually been implemented for some aspects of the TFM system. Others, such as changing the parameters for controlling airline performance, have thus far only been proposed. To understand the impact of such factors, a number of observational studies have been conducted. These studies indicate that such changes can in fact have a significant impact on performance, but because of the inherent confoundings in such studies, they do not definitively establish the relative contributions of the different changes that have been made to the system.

Thus, the fundamental human factors question remains: In an environment where capacity is limited and where there are competing goals (among the airlines), how do we design a system where the cognitive demands on individuals are reasonable, and yet still achieve high levels of performance?

111.5 Issues in the Design of the Future ATM System

Recently, the FAA and the aviation industry reached general consensus on a strategic direction for the ATM system in the United States called "free flight" (RTCA, 1995). The goal of free flight is an ATM system based on two fundamental premises:

1. The FAA retains and strengthens its safety mandate for separation of aircraft
2. The users of the system are given the flexibility to make decisions that allow them to extract the maximum economic benefit from the ATM system

Among other characteristics, free flight assumes a shift way from the current ground-based, tactical ATC operations toward a more cooperative arrangement in which users have greater flexibility to select and manage their flight paths and to participate routinely in airspace management decisions. As the ATM system evolves toward free flight, consideration of the human element will be critical to the realization of new operating philosophies, design of new functional architectures for integrating ATM system components, and application of advanced technologies to support ATM operations.

New Philosophies and Procedures

TFM

Under free flight, the fundamental goal for future TFM is to identify new operating philosophies and procedures that remove restrictions, allowing users to make choices based on business considerations subject to the constraints necessary to ensure safety.

Historically, TFM decision-making has been characterized by gaming, in which the users, armed with limited information about the state of the system and the rationale for denying or approving requests, attempt to achieve an advantage over their competitors and "the system" by hiding their true intentions and preferences. This gaming has led to mistrust and suspicion among the participants. In the future, a team perspective is likely to provide important insight into many of the issues in TFM. The team literature addresses the functional requirements for interpersonal collaboration as well as the social dynamics (Fleishman and Zaccaro, 1992; Scerbo, 1996). With regard to team functions, human factors knowledge and research can contribute to the definition of processes which meet system needs for acquiring and distributing information, coordinating the sequence and timing of participants' responses, evaluating team performance, establishing team objectives and means to resolve disputes, and monitoring for compliance with policies. With regard to social dynamics, a progressive research approach is needed to accumulate experience in real-world settings that will allow participants to practice team performance and assess the effectiveness and reliability of the collaborative process. Real-world experience will also help refine the decision-making process and reduce decision-maker biases and mistrust among the participants.

ATC

Currently in ATC, the basic rule of separation is that every controller is responsible for separation of participating aircraft for the duration of time the aircraft is within the controller's sector of responsibility (Nolan, 1990). In future ATC, the free flight goal is to evolve toward a more strategic operation that better accommodates user preferences. Under this philosophy, controllers may assume more responsibility for solving each other's problems as strategic predictions allow for early ATC interventions. In addition, as new technologies provide pilots with information and capabilities commensurate with or surpassing those of the controller, they may also be called upon to participate in solving ATC problems.

As in TFM, a team perspective is likely to be a major focus in addressing human factors issues in future ATC. Considerable research and analysis has addressed the controller's role (often as a single operator) within an ATC domain. There is much less work that addresses the multi-operator perspective (Benel, 1995). Wickens, Mavor, and McGee (1997) also note that team training is less formal in ATC than on the flight deck, but teamwork is likely to be a critical component of ATC for the foreseeable future. Drawing on the literature in team performance will help identify functional requirements for collaboration in ATC. In the initial stages of orientation and trust building, laboratory and field evaluations can be used to work out task redesigns, clarify goals and roles, and establish commitment. To sustain team performance, the collaborative philosophy of ATC will need to be clearly communicated to the controller workforce and reinforced through training, experience, and feedback programs that address concerns.

Functional Architecture

Already, new ATM technologies such as CTAS cut across traditional divisions of responsibility between en route and TRACON controllers, while strategic planning tools such as conflict probe tend to blur the distinction between ATC and traffic management responsibilities. It is also likely that workload concentrations in the tower environment will motivate further reallocation of tasks across ATC environments and additional capabilities for smoothing workload through greater human control over the tasks, their timescales, their scheduling, and their sequencing (Laios and Giannacourou, 1995). At the same time, imbalances in air and ground system capabilities in the oceanic environment will motivate exploration of alternative allocations of functions between controllers and pilots which increase the pilot's role in ATC. These forces, taken together with the new operating philosophies envisioned under free flight, will drive a progression toward new relationships among system participants.

Analyses of human factors issues in free flight identified the design of a new functional architecture as a key issue that cuts across all environments (Federal Aviation Administration, 1995). In TFM, human factors can help analyze the roles and interactions of TFM players, including central and local traffic managers and the AOC personnel. For example, laboratory simulations can resolve issues regarding human cognitive capacities and performance parameters, such as how much and how fast information can be assimilated. These results on the timing and quality of feedback needed by decision makers will help discriminate among alternative allocations of functions and define the interrelationships among groups of decision makers. In ATC, human factors can help evaluate the allocation of responsibility between controllers and pilots in all of the operational environments. And within the ATM organization, human factors can help analyze and evaluate the allocation of responsibility between controller team members and between controllers and traffic managers. For example, laboratory simulations can be used to explore the feasibility and effectiveness of new roles, such as a planner/coordinator controller role that uses automated tools to support multiple sector controllers. Such a role might combine functions currently performed by a D controller, supervisor, and traffic manager.

Advanced Technologies

At present, the technologies needed to support collaborative TFM are still in conceptual stages. Commercial-off-the-shelf technologies and prototypes exist for facilitating meetings between participants at remote locations, but their application to real-time operational environments such as TFM will take some exploration (Roberts, Zobell, and Blanchard, 1993). At this stage, human factors can assist in development of advanced technology that is designed with the express purpose of supporting teams. For example, research has shown that process behaviors tend to degrade under conditions of high workload. Automating team processes themselves may allow teams to perform more effectively (Bowers et al., 1996). In addition, distributed decision-making may require multiple modes of information transfer (e.g., text, graphics, and voice) to ensure successful communication (Scerbo, 1996).

In ATC, U.S. research and development on the next step in the evolution of ATC automation is aimed at automated advisories for final approach spacing and conflict resolution. This level of automation will have a profound impact on the way ATC is performed. Not surprisingly, there is considerable concern in the human factors community over the impact of such automation on the controller's situation awareness and ability to provide back-up for the automation system when necessary. An alternative approach is being pursued in Europe which focuses more on highly interactive problem-solving tools, including display formats that enable the controller to plan more efficient strategies (Makins and Drew, 1995). In the current environments, the en route and TRACON controllers' displays are optimized for vector solutions to separation and spacing problems (Kerns, 1994). Alternative formats, such as graphical displays of the temporal distribution of problems and predictive displays depicting a vertical view of the traffic situation, may facilitate situation awareness and strategy planning.

It is likely that both the U.S. and European approaches have application in the operational environment. Before questions on human roles can be addressed, more fundamental questions must be answered regarding the efficiency of the alternatives in different task environments and the nature of the back-up mechanisms. Laboratory simulations can help answer questions regarding the efficiency of controller tools under various traffic scenarios. Requirements for maintaining situation awareness and controller skills will vary according to the need for manual back-up in the future environment (Wickens, Mavor, and McGee, 1997). Human factors analyses can help specify these requirements under various back-up mechanisms.

111.6 Conclusion

There are major challenges ahead in analyzing functional requirements and the allocation of functions between human and automated elements of the future ATM system. At this point, there is an opportunity for proactive research in human factors to gain insight and perspective on issues surrounding collaborative ATM. Multiple research approaches, including the review and application of existing data and knowledge, will make it possible to prove the value of new concepts regarding human roles and interrelationships and their impact on the quality and efficiency of the ATM system. As the discussion above indicates, however, this is a system that has historically attempted to match the capabilities of the people with the demands of the system by distributing roles and responsibilities, and by imposing structure. Efforts to enhance efficiency through increased flexibility and shifts in the locus of control need to be based on an understanding of the complexity inherent in this system, and on a realistic assessment of the strengths and limitations of the technologies being considered to support controllers, pilots, traffic managers, and dispatchers in their new roles.

Defining Terms

AOC: Airline Operations Center
ARTCC: Air Route Traffic Control Center
ATC: Air Traffic Control
ATCSCC: Air Traffic Control Systems Command Center
ATIS: Automated Terminal Information Service
ATM: Air Traffic Management
CRDA: Converging Runway Display Aid
CTAS: Center TRACON Automation System
FAA: Federal Aviation Administration
FAST: Final Approach Spacing Tool
PDC: Predeparture Clearance
TFM: Traffic Flow Management
TRACON: Terminal Radar-Approach Control Facility

Acknowledgment

Portions of this chapter are based on research supported by the FAA Office of the Chief Scientist and Technical Advisor for Human Factors, the MITRE Corporation Center for Advanced Aviation System Development, and by the NASA Ames Research Center.

References

Adam, G.A., Kelley, D.R., and Steinbacher, J.G. 1994. *Reports by Pilots on Airport Surface Operations: Part 1. Identified Problems and Proposed Solutions for Surface Navigation and Communications.* The MITRE Corporation MTR94W60.V1.

Adam, G.A. and Kelley, D.R. 1996. *Reports by Pilots on Airport Surface Operations: Part 2. Identified Problems and Proposed Solutions for Surface Operational Procedures and Factors Affecting Pilot Performance.* The MITRE Corporation MTR94W60.V2.

Adams. M., Kolitz, S., and Odoni, A. 1996. Evolutionary concepts for decentralized air traffic flow management. *Proceedings of the 1996 AIAA Guidance Navigation and Control Conference.* San Diego, CA, 1-11.

Airline Dispatchers Federation and Seagull Technology, Inc. 1995. *Airline Operational Control Overview.* Cooperative Research and Development Agreement 93-CRDA-0034, Washington, D.C.

Andrews, J. 1993. Impact of weather event uncertainty upon an optimum ground-holding strategy. *ATC Quarterly,* 1: 59-84.

Bales, R.A., Gilligan, M.R., and King, S.G., 1989. An Analysis of ATC-Related Runway Incursions with Some Potential Technological Solutions, MTR-89W21, The MITRE Corporation, McLean, VA.

Benel, R. 1995. The human role in aircraft-air traffic automation integration: what we don't know can't help us, in *Human Factors in Aviation Operations: Proceedings of the 21st Conference of the European Association for Aviation Psychology,* Eds. Fuller, R., Johnston, N., and McDonald, N. p. 102-108.

Bellorini, A. and Decortis, F. 1995. A stress-based analysis in Air Traffic Control, in *Human Factors in Aviation Operations: Proceedings of the 21st Conference of the European Association for Aviation Psychology,* Eds. Fuller, R., Johnston, N., and McDonald, N. p. 95-102.

Billings, C. 1997. *Aviation Automation: The Search for a Human-Centered Approach.* Lawrence Erlbaum Associates: Mahwah, NJ.

Bowers, C., Oser, R.L., Salas, E., and Cannon-Bowers, J.A. 1996. Team performance in automated systems, in *Automation and Human Performance: Theory and Applications,* Eds. Parasuraman, R. and Mouloua, M. p. 243-266. Lawrence Erlbaum Associates: Mahwah, NJ.

Bruce, D.S. 1996. *Physical Performance Criteria for Air Traffic Control Tower Specialists.* Washington, D.C.: U.S. Department of Transportation, Federal Aviation Administration.

Burki-Cohen, J. 1995. *An Analysis of Tower Controller-Pilot Voice Communications* (DOT/FAA/AR-96/19) Washington, D.C.: U.S. Department of Transportation, Federal Aviation Administration.

Canadian Aviation Safety Board, 1990, Report on a Special Investigation into Air Traffic Control Services in Canada, No. 90-SO001. *Supply and Services Canada Catalog No. TU4-5/1990E.*

Cardosi, K. 1993. *An Analysis of En Route Controller-Pilot Voice Communications* (DOT/FAA/RD-93/11) Washington, D.C.: U.S. Department of Transportation, Federal Aviation Administration.

Cardosi, K., Brett, B., and Han, S. 1996. *An Analysis of TRACON Controller-Pilot Voice Communications* (DOT/FAA/AR-96/66) Washington, D.C.: U.S. Department of Transportation, Federal Aviation Administration.

Carlson, L., Rhodes, L., and Cullen, M. 1996. *Effects of Unstructured Routes on En Route Controllers' Work Activities and Operational Environment,* MITRE Technical Report MTR 96W0000019, Mclean VA.

Coulouris, G. 1985, *Study of U.S. Oceanic Airtraffic Control Consolidation and Automation* DOT/FAA/ES-85-5, U.S. Department of Transportation, Federal Aviation Administration, Washington, D.C.

Coulouris, G. and Dorsky, S. 1995. *Advanced Air Transportation Technologies (AATT) Potential Benefits Analysis.* Cupertino CA: Seagull Technology.

Credeur, L., Capron, W.R., Lohr, G.W., Crawford, D.J., Tang, D.A., and Rodgers, W.G. 1993. A comparison of final approach spacing aids for terminal atc automation, *Air Traffic Control Quarterly,* 1(2):135-177.

Davis, R. and Smith, R. 1983. Negotiation as a metaphor for distributed problem-solving. *Artificial Intelligence,* 20: 63-109.

Davis, T.J., Erzberger, H., Green, S.M., and Nedell, W. 1991. Design and Evaluation of an air traffic control final approach spacing tool. *Journal of Guidance Control and Dynamics.* 14: 848-854.

Denning, R., Smith, P.J., McCoy, E., Orasanu, J., Billings, C., Van Horn, A. and Rodvold, M. 1996. Initial experiences with the expanded national route program. *Proceedings of the 1996 Annual Conference of the Human Factors and Ergonomics Society.* Philadelphia: Sept. 2-6, 98-101.

Federal Aviation Administration 1992. *Concept of Operations: A Description of the Baseline Oceanic ATC System.* Air Traffic Plans and Requirements Service Department of Transportation Federal Aviation Administration Washington, D.C. DOT/FAA/AT-93-2

Federal Aviation Administration 1991. *Precision Runway Monitor Demonstration Report* (DOT/FAA/RD-91/5) Washington, D.C.: U.S. Department of Transportation, Federal Aviation Administration.

Federal Aviation Administration 1992. *National Route Program,* Advisory Circular 90-91, ATM 100, April 24, 1992.

Federal Aviation Administration 1995. *Advancing Free Flight Through Human Factors: Workshop Report.* U.S. Department of Transportation, Federal Aviation Administration: Washington, D.C.

Federal Aviation Administration 1995. *National Route Program (NRP)*, FAA Order 7110.128, Free Flight, ATM 100, effective Jan. 9, 1995.

Fleishman, E.A. and Zaccaro, S.J. 1992. Toward a taxonomy of team performance functions, in *Teams: Their Training and Performance,* Eds. Swezey, R.W. and Salas, E. p. 31-56: Ablex: Norwood, NJ.

Galushka, J., Frederick, J., Mogford, R., and Crois, P. 1995. *Plan View Display Baseline Research Report.* (DOT/FAA/CT-TN95/45) Washington, D.C.: U.S. Department of Transportation, Federal Aviation Administration.

Garland, D., Hopkin, V.D., and Muller, J. 1996. *The Air Traffic Control System Command Center's Operational Workplace: Volume II — Human Factors Recommendations.* Daytona Beach, FL: Embry-Riddle Aeronautical University.

Grayson, R.L. and Billings, C.E. 1981. Information transfer between air traffic control and aircraft: Communication problems in flight operations, in *Information Transfer Problems in the Aviation System* (NASA Technical Paper 1875), Eds. Billings, C.E. and Cheaney, E.S. Moffett Field, CA: NASA Ames Research Center.

Hamrick, L.Y. and Reierson, J.D., 1993, Operational Concept for the Enhanced Oceanic Conflict Probe, MTR93W67, The MITRE Corporation, McLean, VA.

Hopkin, V.D. 1988. Air traffic control, in *Human Factors in Aviation,* Eds. Weiner, E.L. and Nagel, D.C. p. 639-662. Academic Press, San Diego.

Hopkin, V.D. and Wise, J.A. 1996. Human factors in air traffic system automation, in *Automation and Human Performance: Theory and Applications,* Eds. Parasuraman, R. and Mouloua, M. p. 319-336. Lawrence Erlbaum Associates: Mahwah, NJ.

Kerns, K. 1991. Data link communication between controllers and pilots: a review of the simulation literature. *The International Journal of Aviation Psychology,* 1(3): 181-204.

Kerns, K. 1994. *Human Factors in ATC/Flight Deck Integration: Implications of Data Link Simulation Research* MP94W98 The MITRE Corporation, McLean, VA.

Kinney, G.C. 1977. *The Human Element in Air Traffic Control: Observations and Analyses of the Performance of Controllers and Supervisors in Providing ATC Separation Services,* MITRE Technical Report MTR 7655. The MITRE Corporation: McLean, VA.

Kinney, G.C., Bell, G.L., and Ditmore, M.A., 1983, *The Human Element in Air Traffic Control: Factors in System Recovery and Revitalization,* MTR-82W151, The MITRE Corporation, McLean, VA. (not in Public Domain).

Klein, G. 1992. *The Human Air Traffic Management Role in a Highly Automated Air Traffic System.* MITRE Technical Report # ARDATMR-MTR-92W0000075, McLean, VA.

Lacher, A. and Klein, G. 1993. *Air Carrier Operations and Collaborative Decision-Making Study,* MTR 93W0000244, The MITRE Corporation, Mclean VA.

Laios, L. and Giannacourou, M. 1995. Human factors issues of advanced ATC systems, in *Human Factors in Aviation Operations: Proceedings of the 21st Conference of the European Association for Aviation Psychology,* Eds. Fuller, R., Johnston, N., and McDonald, N., p. 108-113. Ashgate Publishing Limited: Cambridge, England.

Layton, C., Smith, P. J., and McCoy, E. 1994. Design of a cooperative problem-solving system for en route flight planning: An empirical evaluation. *Human Factors,* 36: 94-119.

Lee, K. K. and Davis, T.J. 1995. The development of the final approach spacing tool (FAST): A cooperative controller-engineer design approach. *Proceedings of the International Federation of Automatic Control Conference:* Automated Systems Based on Human Skill, Berlin, Germany. 219-226.

Lee, K.K., Pawlak, W.S., Sanford, B.D., and Slattery, R.A. 1995. Improved navigation technology and air traffic control: A description of controller coordination and workload, in *Proceedings of the Eighth International Symposium on Aviation Psychology,* Columbus, OH. 444-449.

Leroux, M. 1995. ERATO: cognitive engineering applied to ATC, in *Human Factors in Aviation Operations Proceedings of the 21st Conference of the European Association for Aviation Psychology,* Eds. Fuller, R., Johnston, N., and McDonald, N. 89-94.

Makins, N. and Drew, A. 1995. *Integrating FMS Data Into Air Traffic Control: A Prototyping Exercise Exploring the Operational Consequences and Possible Effects on Controller Roles.* EC Note No. 28/95 EUROCONTROL Experimental Center, Bretigny-sur-Orge, France.

McAlindon, P.J. and Gupta, U.G. 1993. A structured approach to the verification and validation of expert systems, in *Verification and Validation of Complex Systems: Additional Human Factors Issues,* Eds. Wise, J.A., Hopkin, V.D., and Stager, P., 15-23. Embry-Riddle Aeronautical University Press: Daytona Beach FL.

McCoy, E., Orasanu, J., Smith, P. J., VanHorn, A., Billings, C., Denning, R., Rodvold, M., and Gee, T. 1995. Situational awareness at different levels of abstraction: Distributed cooperative problem-solving in ATCSCC-airline interactions. *Proceedings of the International Conference on Experimental Analysis and Measurement of Situation Awareness.* Daytona Beach, FL, Nov. 1-3.

McCoy, C. E., Smith, P. J., and Layton, C. 1991. Research in cooperative problem-solving systems for aviation. *Proceedings of the 1991 Aviation Psychology Symposium,* Columbus, OH, 894-903.

McCoy, C. E., Smith, P. J., Orasanu, J., Billings, C., VanHorn, A., Denning, R., Rodvold, M., and Gee, T. 1995. Airline dispatch and ATCSCC: A cooperative problem-solving success story with a future. *Proceedings of the 1995 Aviation Psychology Conference,* Columbus, OH, 456-460.

Morrow, D., Lee, A., and Rodvold, M. 1993. Analysis of problems in routine controller-pilot communications. *The International Journal of Aviation Psychology,* 3(4): 285-302.

Mundra, A.D., 1989, A Description of Air Traffic Control in the Current Terminal Airspace Environment, MTR88W167, The MITRE Corporation, McLean, VA.

Mundra, A. D. and Levin, K. M. 1990. *Developing Automation for Terminal Air Traffic Control: Case Study of the Imaging Aid.* MP90W29, The MITRE Corporation, McLean, VA.

Nolan, M. 1990. *Fundamentals of Air Traffic Control.* Belmont, CA: Wadsworth.

Norman, D.A. and Bobrow, D.G. 1975. On data limited and resource limited processes. *Cognitive Psychology,* 7: 44-64.

Odoni, A.R. 1987. The flow management problem in air traffic control, in *Flow Control of Congested Networks,* Eds. Odoni, A.R., Bianco, L., and Szego, G. Springer-Verlag: Berlin.

Orasanu, J. 1991. Shared problem models and flight crew performance, in Aviation Psychology in Practice, Eds. Johnston, N., McDonald, N., and Fuller, R. Avebury: Brookfield, VT:, 255-285.

Pujet, N. and Feron, E. 1996. Flight plan optimization in flexible air traffic environments. *Proceedings of the 1996 AIAA Guidance Navigation and Control Conference.* San Diego, CA, 1-10.

Roberts, G. 1993. *Visual Collaboration in Air Traffic Control.* (MTR-W0000247). The MITRE Corporation: McLean, VA.

Roberts, G., Zobell, S. and Blanchard, J., 1993, *Visual Collaboration in Air Traffic Control,* MITRE Technical Report MTR-W000247, The MITRE Corporation, McLean, VA. (Not in Public Domain).

Rodgers, M.D., 1993, An Analysis of the Operational Error Database for Air Route Traffic Control Centers, DOT/FAA/AM, Federal Aviation Administration, Washington, D.C.

RTCA 1995. *Final Report of RTCA Task Force 3: Free Flight Implementation.* RTCA Incorporated, Washington, D.C.

RTCA 1995. *Report of the RTCA Board of Directors' Select Committee on Free Flight.* RTCA Incorporated, Washington, D.C.

Sarter, N.B. 1996. Cockpit automation: from quantity and quality, from individual pilot to multiple agents, in *Automation and Human Performance: Theory and Applications,* Eds. Parasuraman, R. and Mouloua, M. p. 267-280. Lawrence Erlbaum Associates: Mahwah, NJ.

Scardina, J., Simpson, T., and Ball, M. 1996. ATM: The only constant is change. *Aerospace America,* March, 20-40.

Scerbo, M.W. 1996. Theoretical perspective on adaptive automation, in *Automation and Human Performance: Theory and Applications,* Eds. Parasuraman, R. and Mouloua, M. p. 37-64. Lawrence Erlbaum Associates: Mahwah, NJ.

Schultheis, S. and Tucker, M. 1996. *User Request Evaluation Tool (URET) Evaluation Report,* MTR96W64, The MITRE Corporation, McLean, VA.

Sheridan, T. 1987. Supervisory control, in *Handbook of Human Factors,* Ed. G. Salvendy. Wiley: New York.

Sheridan, T. 1992. *Telerobotics, Automation and Human Supervisory Control.* Cambridge, MA: MIT Press.

Shingledecker, C.A. and Darby, E.R. 1995. Effects of data link ATC communications on controller teamwork and sector productivity. *The Air Traffic Control Quarterly,* 3(2), 65-94.

Smith, P.J., McCoy, C.E., Orasanu, J., Billings, C., Denning, R., Rodvold, M., Gee, T., and Van Horn, A. 1996a. Control by permission: a case study of cooperative problem-solving in the interactions of airline dispatchers with ATCSCC, *Air Traffic Control Quarterly,* 4(4): 229-247.

Smith, P.J., McCoy, E., Orasanu, J., Denning, R., Billings, C., Owlsey, T., Boeve, E., Bullington, R., and France, E. 1996b. *An Empirical Study of the Expanded National Route Program on Flight Planning and Performance.* Cognitive Systems Engineering Laboratory, Technical Report # 1996-18, The Ohio State University, Columbus OH and Ohio University, Athens OH.

Smith, P. J., McCoy, E., Orasanu, J., Billings, C., Denning, R., Rodvold, M., Van Horn, A., and Gee, T. 1996c. Cooperative problem-solving in the interactions of airline dispatchers with the Air Traffic Control Systems Command Center, in *Human Interaction with Complex Systems: Conceptual Principles and Design Practice,* Eds. Ntuen, C. and Park, E., p. 185-195. Kluwer Academic Publishers: Boston, MA.

Sperandio, J.C. 1971. Variation of Operator's strategies and regulating effects on workload. *Ergonomics,* 14: 571-577.

Sperandio, J.C. 1978. The regulation of working methods as a function of workload among air traffic controllers. *Ergonomics,* 21:195-202.

Wambsganss, M. 1995. *Collaborative Traffic Flow Management.* Metron: Washington, D.C.

Whitfield, D. 1979. A preliminary study of the air traffic controller's "Picture." *CATCA Journal,* Spring, 19-28.

Wickens, C., Mavor, A., and McGee, J. 1997. *Flight to the Future: Human Factors in Air Traffic Control.* National Academy Press: Washington, D.C.

Wiener, E. and Nagel, D., 1988. *Human Factors in Aviation.* Academic Press: San Diego, CA.

Indexes

Author Index

A

B

C

D

E

F

G

H

I

J

K

L

M

N

O

P

Q

R

S

T

U

V

W

X

Y

Z

Subject Index

Figures are noted by page references in **bold type**, and tables are noted by page references in *italic type*

A

D

E

I

M

N

O

P

R

S

T

U

V

W

Z